LIFE: THE SCIENCE OF BIOLOGY

FOURTH EDITION
LIFE
The Science of Biology

William K. Purves
Harvey Mudd College
Claremont, California

Gordon H. Orians
The University of Washington
Seattle, Washington

H. Craig Heller
Stanford University
Stanford, California

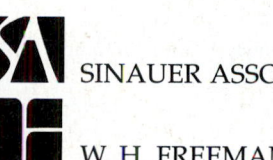

SINAUER ASSOCIATES, INC.

W. H. FREEMAN AND COMPANY

THE COVER

Elephants at a water hole in northern Botswana, Africa.
Photograph by Frans Lanting/Minden Pictures.

THE FRONTISPIECE

Scarlet ibis and cattle egrets at Hato el Frio, Venezuela.
Photograph by Art Wolfe.

LIFE: THE SCIENCE OF BIOLOGY, Fourth Edition
Copyright © 1995 by Sinauer Associates, Inc.
All rights reserved. This book may not be reproduced
in whole or in part without permission.

Address editorial correspondence to Sinauer Associates, Inc.,
Sunderland, Massachusetts 01375 U.S.A.

Address orders to W. H. Freeman and Co. Distribution Center,
4419 West 1980 South, Salt Lake City, Utah 84104 U.S.A.

Library of Congress Cataloging-in-Publication Data

Purves, William. K. (William Kirkwood), 1934-
    Life, the science of biology / William K. Purves,
  Gordon H. Orians, H. Craig Heller. -- 4th ed.
        p.   cm.
    Includes bibliographical references and index.
    ISBN 0-7167-2629-7
    1. Biology. I. Orians, Gordon H. II. Heller, H. Craig.
III. Title.
QH305.2.P87   1995
574--dc20   94-24802
                                  CIP

ABOUT THE BOOK

Editor: Andrew D. Sinauer

Project Editor: Carol J. Wigg

Developmental Editor: Elmarie Hutchinson

Copy Editor: Stephanie Hiebert

Production Manager: Christopher Small

Book Layout and Production: Janice Holabird

Art Editing and Illustration Program: J/B Woolsey Associates

Photo Research: Jane Potter

Book and Cover Design: Rodelinde Graphic Design

Composition: DEKR Corporation

Color Separations: Vision Graphics, Inc.

Prepress: Lanman Lithotech

Cover Manufacture: John P. Pow Company

Book Manufacture: R. R. Donnelley & Sons Company, Willard, OH

Printed in U.S.A.

4 3 2 1

To Jean, Betty, and Renu

ABOUT THE AUTHORS

William K. Purves

Bill Purves is Stuart Mudd Professor of Biology as well as founder and chair of the Department of Biology at Harvey Mudd College in Claremont, California. He received his Ph.D. from Yale University in 1959 under Arthur Galston. A Fellow of the American Association for the Advancement of Science, Professor Purves has served as head of the Life Sciences Group at the University of Connecticut, Storrs, and as chair of the Department of Biological Sciences, University of California, Santa Barbara, where he won the Harold J. Plous Award for teaching excellence. His research interests focus on the chemical and physical regulation of plant growth and flowering.

Professor Purves has taught introductory biology each year for over thirty years and considers teaching the course the most interesting and important of his professional activities. "I can't imagine a year without teaching it," he says. In describing his teaching philosophy, Purves states, "Students learn biological concepts much more rapidly and effectively if they understand where the concepts come from— what the experimental and conceptual background is. 'Facts' by themselves can be boring or incomprehensible, but they become exciting if given a context."

Gordon H. Orians

Gordon Orians is Professor of Zoology at the University of Washington. He received his Ph.D. from the University of California, Berkeley, in 1960 under Frank Pitelka. Professor Orians has been elected to the National Academy of Sciences and the American Academy of Arts and Sciences. He was President of the Organization for Tropical Studies from 1982 to 1994, and is currently President-elect of the Ecological Society of America. He is a recipient of the Brewster Medal from the American Ornithologists' Union, and in 1994 he received the Distinguished Service Award of the American Institute of Biological Sciences.

Professor Orians is a leading authority in ecology and evolution, with research interests in behavioral ecology, plant–herbivore interactions, community structure, and the biology of rare species. Like the other authors, he draws from his research to bring an added dimension to his teaching and writing. "Teachers who understand research because they are engaged in it can more easily communicate the excitement researchers feel as they discover new things," Orians says. "All three authors of *Life* have spent considerable time doing research and we have tried throughout the book to show the sources of our current understanding of biology."

H. Craig Heller

Craig Heller is Lorry Lokey/ Business Wire Professor of Biological Sciences and Human Biology and Associate Dean of Research at Stanford University, and is a popular lecturer on animal and human physiology. He received his Ph.D. from Yale University in 1970 and did postdoctoral research at the Scripps Institution of Oceanography on brain regulation of body temperature in mammals. He has continued this research since coming to Stanford in 1972, studying a variety of phenomena ranging from hibernating squirrels to sleeping college students to diving seals to meditating yogis. Professor Heller is a Fellow of the American Association for the Advancement of Science and a recipient of the Walter J. Gores Award for Excellence in Teaching.

"A first course in biology requires the student to learn more new words than a first course in a foreign language," Heller says. "The secret to teaching and learning biology is to focus on central and overarching concepts. Once you grasp the concept of how something works, you have a framework on which the facts and vocabulary fall into place. Conceptual understanding also helps you relate what you learn to the real world."

PREFACE

In revising this book, we have once again examined our goals and our hopes for it. Above all, we want to help students understand biological concepts and see where the concepts originate. For this reason we display biology as an experimental and observational science. Frequently we offer the student a chance to think—to figure out the next step rather than wait passively to learn it from us. For example, we ask the student to interpret experimental data used in elucidating the genetic code, in understanding the role of homeotic genes in flower development, and in working out the Calvin–Benson cycle. In Chapter 21, students are presented with real species with identified character traits and work through how to construct a phylogeny using cladistic methods. We use the cladistic material again in Chapter 27, where we explore human relationships with chimpanzees and gorillas. As yet another example, we lead the student to discover the physiological differences between reptiles and mammals by taking the reader through a series of experiments on the thermal biology of a mouse and a lizard.

Even when we present topics directly, we prefer to explain new material rather than serving it up as a cut-and-dried collection of "facts." Still, the study of biology requires exposure to a stunning number of new facts, and students are easily deterred by a bewildering excess of information. How can we deal with this problem? Our approach is to emphasize fascinating examples wherever possible, using these to engage the student's interest so that she or he wants to learn other related material.

The creation of a new edition provides opportunities to rethink how best to present existing material, as well as what from the exploding array of new information to include. Current advances in biomedical sciences present special challenges to the organizers of a course or a textbook. Diseases such as cancer and AIDS are of deep interest in relation to a wide variety of biological topics, including immunology, genetics, evolution, membrane biology, and virology; instructors often want to use the diseases as examples in discussing these topics. Yet students and instructors also want to see a disease and its biology considered in a single place in the book. For this edition, we have adopted an approach that we hope is effective. Recent advances in gene therapy and in the cloning of genes for particular diseases lend themselves to consideration in an all-new Chapter 15, "Genetic Disease and Modern Medicine." We consider some general principles and some widely applicable techniques in this chapter, which also includes much of our coverage of cancer. We build from the close of this chapter to begin the next, "Defenses against Disease," with an overview of AIDS as a worldwide problem. Before the end of Chapter 16, the student knows enough about the immune system to understand AIDS in greater biological detail.

Developmental biology continues to be one of the fastest-moving areas of biology; the newer work is reflected in Chapter 17. We pay particular attention to recent developments in the *Drosophila* larva and *Caenorhabditis elegans* systems. After giving genetic "instructions" for building a fly, we conclude the molecular section of the book with a transition from *Drosophila* larval genetics to evolutionary biology.

In the section on evolution we have expanded the coverage of methods of reconstructing phylogenies. We provide new treatments of cladistic methods and show how phylogenies are used to shed light on a wide variety of evolutionary questions. We use phylogenetic trees a number of times in subsequent chapters in the book to show how traits of organisms evolve under the influence of evolutionary agents.

Our restructuring of the chapters on plant anatomy and physiology resulted in the creation of a new Chapter 31, "Environmental Challenges to Plants," which deals with some of the ways plants cope with harsh envi-

UP-TO-DATE COVERAGE REFLECTS NEW DEVELOPMENTS IN BIOLOGY

ronments, predators, and pathogens. We have updated the plant chapters to include material on patch clamping, homeotic mutations, and other important phenomena and techniques.

Three aspects of the animal biology chapters that were significant in the third edition have been strengthened. First, we frequently use a comparative approach to help students understand basic principles and mechanisms as well as their evolutionary variations. Second, we emphasize experimental approaches so that the student learns *how* we know as well as *what* we know. Third, a capstone to the treatment of each physiological system is a discussion of how its contributions to homeostasis are controlled and regulated. Chapter 45, "Animal Behavior," has been revised to focus more strongly on the physiological mechanisms underlying behavior.

Because ecology is an increasingly experimental science, we have added descriptions of well-designed experiments that have been performed to demonstrate the causes of the evolution of traits used in courtship among animals and to assess the importance of predation and competition in structuring ecological communities. To keep pace with the increasing importance of environmental problems, we have expanded our treatment of lake eutrophication and of overexploitation of commercially important species.

PEDAGOGICAL INNOVATIONS ENRICH THE LEARNING PROCESS

Both as textbook authors and as teachers we want to help students in every way possible. In the next section ("To the Student") we offer some helpful advice we'd like students to consider as they begin their study of biology. This advice has helped many of our own students. Following "To the Student" is "*Life* at a Glance," in which we illustrate some of the pedagogic improvements that we discuss in the next few paragraphs.

A major source of the success of the third edition of this textbook was the exciting new art program developed by J/B Woolsey Associates. We have upgraded that already fine art for this edition. We have added many entirely new drawings and graphs, and virtually all the drawings in the book have been improved in one way or another by the development of a new artistic vocabulary for this edition.

Because learning is a visual as well as a verbal process, we have given much attention to creating illustrations that explain biological concepts clearly. To facilitate learning we use color consistently from illustration to illustration; for example, the outside of the cell is always represented by light red and the cytoplasm by pale blue. We have developed a set of icons (see "*Life* at a Glance") to represent biologically important molecules (such as water or ATP), active forms of enzymes, or activation and inhibition of pathways. We have used blocks of color to distinguish major pieces of information from details or to separate an illustration into parts representing discrete ideas; we think of these as "visual paragraphs." Finally, we have flagged via marginal arrowheads (also shown in "*Life* at a Glance") illustrations that are particularly significant. These figures illustrate and synthesize important concepts—protein synthesis, for example, or the life cycle of a flowering plant.

We have developed a new type of chapter-end summary, one that should help students in at least three ways. First, the student can skim the summary for orientation before reading the chapter. Second, after reading the chapter, the student can use the summary to review the material. Finally, each summary identifies the illustrations that give the best overview of the chapter and its most important concepts.

Each chapter starts with a brief introduction intended to catch the reader's interest by discussing a fascinating bit of biology. Each of these new introductions is supported by a striking photograph; an example is shown in "*Life* at a Glance."

Before beginning this edition, we asked 36 biologists, most of whom were using the third edition, to maintain "diaries" and record their ideas for improving the book. These diarists were enormously helpful in getting us started on the right foot for the new edition. Their guidance influenced the decisions to create the two new chapters mentioned earlier and to give priority to simplifying our prose. We are indebted to them.

As with the first three editions, many of our colleagues reviewed chapters or entire sections of this edition in manuscript. They and the diarists are listed below. The reviewers were helpful, thoughtful, and clearly dedicated to the success of this book, and we thank them all. We particularly thank Bob Cleland, Richard Cyr, Pat DeCoursey, Rob Dorrit, Art Dunham, Margaret Fusari, Harry Green, Ray Huey, Bob Jeffries, Jim Manser, William Milsom, Ron O'Dor, Dianna Padilla, Ronald Patterson, Zoe Roizen, Seri Rudolph, Michael Ryan, Iain Taylor, and David Woodruff. They gave us explicit recommendations for extensive improvements, helped simplify our writing style, and did so in ways that encouraged us to do our very best to live up to their expectations.

The third edition profited greatly from the prodigious efforts of its outstanding developmental editor, Elmarie Hutchinson. We were delighted when Elmarie agreed to take an even stronger role in the development of the fourth edition. Among her innovations are the new type of chapter summary and its suggested use as a chapter preview. Elmarie also helped us respond to diarists' requests for simpler language, better topic sentences, and more restrained use of boldface terms. Her suggestions and guidelines were implemented by our copyeditor, Stephanie Hiebert, whose sharp and prescriptive line editing has helped streamline the book's prose.

In a book like this the illustrations are as important as the prose, and here again Elmarie gave us outstanding input, scrutinizing every figure with an eye for its internal consistency and its agreement with the text. Her suggestions were incorporated by artists John Woolsey and Patrick Lane as they met with the authors to reconceptualize artwork. The task of coordinating and checking the changes made by editors, artists, and authors fell to Carol Wigg, who got the job once again of putting all the pieces together. In addition, previous users of the book will see a significant improvement in the photography program. Jane Potter has tapped important new sources, and we have gone to great lengths to seek out new photographs to illustrate important concepts and enliven the book's appearance.

We wish to thank W. H. Freeman's entire marketing and sales group. Their enthusiasm for *Life* helped bring the book to a wider audience and the efforts of several of the sales representatives put us in touch with a number of our colleagues who had specific questions or criticisms of the book. This contact has been fruitful, and we look forward to more of this "firing line" interaction with the fourth edition.

Finally, the opportunity to work with a publishing company whose president provides frequent personal contacts and feedback that is scientifically useful as well as production-wise, is a great privilege for us. Andy Sinauer is the ideal person for authors to deal with—firm but kind, involved but not overbearing, and friendly—hence, motivating. He also has a superb eye for good associates—the Sinauer team is first-rate!

William K. Purves   Gordon H. Orians   H. Craig Heller
September 1994

CAREFULLY REVIEWED WITH STUDENTS' NEEDS AND TEACHERS' CONCERNS IN MIND

THE EFFORTS OF MANY PEOPLE HELPED THE REVISION

Andrew R. Blaustein, Oregon State University

Daniel R. Brooks, University of Toronto

Andrew G. Clark, Pennsylvania State University

Rolf E. Christoffersen, University of California/Santa Barbara

Esther Chu, Occidental College

Patricia J. DeCoursey, University of South Carolina

T. A. Dick, University of Manitoba

Robert L. Dorit, Yale University

Arthur E. Dunham, University of Pennsylvania

Gisela Erf, Smith College

Russell G. Foster, University of Virginia

Michael Feldgarden, New Haven, Connecticut

William D. Fixsen, Harvard University

Gwen Freyd, Harvard University

William Friedman, University of Georgia

Margaret H. Fusari, University of California/Santa Cruz

Stephen A. George, Amherst College

Elizabeth A. Godrick, Boston University

Linda J. Goff, University of California/Santa Cruz

Harry W. Greene, University of California/Berkeley

Richard K. Grosberg, University of California/Davis

Marty Hanczyc, New Haven, Connecticut

Albert A. Herrera, University of Southern California

Raymond B. Huey, University of Washington

Robert L. Jeffries, University of Toronto

Cynthia Jones, University of Connecticut

David Kirk, Washington University/ St. Louis

Will Kopachik, Michigan State University

Arthur P. Mange, University of Massachusetts/Amherst

Jim Manser, Harvey Mudd College

Charles W. Mims, University of Georgia

Ron O'Dor, Dalhousie University

Judith A. Owen, Haverford College

Dianna K. Padilla, University of Wisconsin/Madison

Daniel Papaj, University of Arizona

Murali Pillai, Bodega Marine Laboratory

Martin Poenie, University of Texas

Roberta Pollack, Occidental College

Seri Rudolph, Bates College

Michael J. Ryan, University of Texas

Joan Sharp, Simon Fraser University

Dwayne D. Simmons, University of California/Los Angeles

Iain E. P. Taylor, University of British Columbia

Mark Wheelis, University of California/ Davis

Gene R. Williams, Indiana University

David S. Woodruff, University of California/San Diego

Ron Ydenberg, Simon Fraser University

Charles Yocum, University of Michigan/ Ann Arbor

Alice Alldredge, University of California/Santa Barbara

Daniel R. Brooks, University of Toronto

Margaret Burton, Memorial University of Newfoundland

Iain Campbell, University of Pittsburgh

Mark A. Chappell, University of California/Riverside

Robert E. Cleland, University of Washington

Richard J. Cyr, Pennsylvania State University

Wayne Daugherty, San Diego State University

William Davis, Rutgers University

Emma Ehrdahl, Northern Virginia Community College

Stuart C. Feinstein, University of California/Santa Barbara

Rachel Fink, Mt. Holyoke College

William Fixsen, Harvard University

R. M. Grainger, University of Virginia

Daniel Klionsky, University of California/Davis

James L. Koevenig, University of Central Florida

William Z. Lidicker, Jr., University of California/Berkeley

J. Richard McIntosh, University of Colorado/Boulder

William K. Milsom, University of British Columbia

Deborah Mowshowitz, Columbia University

Todd Newberry, University of California/Santa Cruz

Peter H. Niewiarowski, University of Pennsylvania

Larry D. Noodén, University of Michigan/Ann Arbor

Scott Orcutt, University of Akron

Ronald J. Patterson, Michigan State University

Zoe Roizen, Oakland, California

Steve Rothstein, University of California/Santa Barbara

Mark Shannon, University of Tennessee

Joan Sharp, Simon Fraser University

Steve Strand, University of California/ Los Angeles

Brian Taylor, Texas A & M University

Peter Webster, University of Massachusetts/Amherst

Nathaniel T. Wheelwright, Bowdoin College

Brian White, Massachusetts Institute of Technology

Thomas Wilson, University of Vermont

Ronald C. Ydenberg, Simon Fraser University

TO THE STUDENT

Welcome to the study of life! In our student days—and ever since—we have enjoyed studying the fascinating and fast-changing field of biology, and we hope that you will, too.

There are a few things you can do to help you get the most from this book and from your course. For openers, read the book actively—don't just read passively, but do things that force you to think as you read. If we pose questions, stop and think about them. If a passage reminds you of something that has gone before, think about that, or even check back to refresh your memory. Ask questions of the text as you go. Do you understand what is being said? Does it relate to something you already know? Is it supported by experimental or other evidence? Does that evidence convince you? How does this passage fit into the chapter as a whole? Annotate the book—write down comments in the margins about things you don't understand, or about how one part relates to another, or even when you find an idea particularly interesting. The point of doing these things is that they will help you learn. People remember things they think about much better than they remember things they have read passively. Highlighting is passive; copying is drudge work; questioning and commenting are active and well worthwhile.

For this edition we have developed new ways to help you read the book actively. The chapter-end summaries have been redesigned so that they may be used as both summaries and previews. To find out what a chapter covers, try reading the summary at the end of the chapter before you begin reading the chapter itself. Don't worry about unfamiliar terms in the summary, but notice them as terms you will need to learn. Just read all the statements as an overview and preview without studying the cited illustrations. Then, after reading the chapter, use the summary as a framework for your review. It is essential that you do study the cited illustrations and their captions as you review because important information that is covered in illustrations has been left out of the summary statements. Add concepts and details to the framework by reviewing the text.

Take advantage of our use of color and symbols in the illustrations. We generally use colors to mean the same thing from illustration to illustration, and we have developed a set of symbols (see the section "*Life* at a Glance") to represent biologically important molecules and phenomena. In many of the illustrations you will see blocks of color used to help you separate the illustration into parts representing discrete ideas. Also, some figures are identified by an arrow in the margin. These figures are particular significant; they illustrate and synthesize important concepts and retell visually the story you read in the text. Studying them will help you learn important biological concepts and systems. Going back to review them will help you to remember these concepts.

The chapter summaries will help you quickly review the high points of what you have read. A summary identifies particular illustrations that you should study to help organize the material in your mind. A way to review the material in slightly more detail after reading the chapter is to go back and look at the boldfaced terms. You can use the boldfaced terms to pose questions—and see if you can answer those questions. The boldfacing will probably be more useful on a second reading than on the first.

Use the self-quizzes and study questions at the end of each chapter. The self-quizzes are meant to help you remember some of the more detailed material and to help you sort out the information we have laid before you. Answers to all self-quizzes are in the Appendix. The study questions, on the other hand, are often fairly open-ended and are intended to cause you to reflect on the material.

Two parts of a textbook that are, unfortunately, often underused or even ignored are the glossary and the index. Both can help you a great deal. When you are uncertain of the meaning of a term, check the glossary first— there are more than 1,500 definitions in it. If you don't find a term in the glossary, or if you want a more thorough discussion of the term, use the index to find where it's discussed.

What if you'd like to pursue some of the topics in greater detail? At the end of each chapter there is a short, annotated list of supplemental readings. We have tried to choose readings from books and magazines, especially *Scientific American*, that should be available in your college library.

Most students occasionally have difficulty in courses, including biology courses. If you find that you are slipping behind in the course, or if a particular topic is giving you an unreasonable amount of trouble, here are some useful steps you might take. First, the basics: attend class, take careful lecture notes, and read the textbook assignments. Second, note that one of the most important roles of studying is to discover what you don't know, so that you can do something about it. Use the index, the glossary, the chapter summaries, and the text itself to try to answer any questions you have and to help you organize the material. Make a habit of looking over your lecture notes within 24 hours of when you take them—find out right away what points are unclear, and get them straightened out in your mind. We also call your attention to the Study Guide that accompanies *Life*. It is by Jon Glase at Cornell and Jerry Waldvogel at Clemson. It parallels this textbook and each chapter contains learning objectives, key concepts, activities, and questions with full answers and explanations.

If none of these self-help remedies does the trick, get help! Other students are often a good source of help, because they are dealing with the material at the same level as you are. Study groups can be very useful, as long as the participants are all committed to learning the material. Tutors are almost always helpful and useful, as are faculty members. The main thing is to get help when you need it. It is not a good idea to be strong and silent and drift into a low grade.

But don't make the grade the point of this or any other course. You are in college to learn, to pursue interesting subjects, and to enjoy the subjects you are pursuing. We hope you'll enjoy the pursuit of biology.

Bill Purves   **Gordon Orians**   **Craig Heller**

Life, Fourth Edition, is accompanied by a comprehensive set of supplements:

Study Guide . . . reviewed by students and extensively revised by Jon Glase of Cornell University and Jerry Waldvogel of Clemson University to help students master the textbook material.

Instructor's Manual . . . by Roberta Meehan of the University of Northern Colorado, featuring chapter objectives, chapter outlines, teaching hints and strategies, references, resources, and key terms.

Overhead transparencies and **slides** . . . a package of 300 full-color images from the book.

Transparency masters . . . of all text art figures not included in the overhead transparency/slide set.

Test bank . . . revised and updated, with at least 10 new questions per chapter and over 4,000 questions in total. Available in printed form, and in IBM and Macintosh formats.

Videodisc . . . for the first time the magnificent art program in *Life* comes to your classroom via laserdisc technology. The disc includes all the line art from the new edition, over 1,500 carefully selected still images, and more than 25 outstanding motion and animation sequences ranging from traffic through the membrane and electrophoresis to ecological succession.

Laboratory options . . . chosen from the following:

• The complete Abramoff and Thomson's *Laboratory Outlines in Biology VI*. The popular, critically acclaimed lab manual from W. H. Freeman, now in its new (1995) sixth edition.

• *Laboratory separates*. Select only those experiments you need from Abramoff and Thomson.

• *Customized laboratory package*. The separates of your choice, combined with your own laboratory exercises, notes, and other materials.

For information regarding policy on the educational use of these supplements, please contact your local W. H. Freeman representative.

SUPPLEMENTS

Videodisc Focus Group Participants
Tad Day, University of West Virginia
Guy Cameron, University of Houston
Valerie Flechtner, John Carroll University
Arnold Karpoff, University of Louisville
William Eickmeir, Vanderbilt University
Paul Ramp, University of Tennessee

Videodisc Reviewers
Stephen C. Adolph, Harvey Mudd College
Sally S. De Groot, St. Petersburg Junior College (retired)
Rachel Fink, Mt. Holyoke College
Nancy V. Hamlett, Harvey Mudd College
Brian A. Hazlett, University of Michigan
Martinez J. Hewlett, University of Arizona
Dan Lajoie, University of Western Ontario
Alfred R. Loeblich III, University of Houston
James R. Manser, Harvey Mudd College
Catherine S. McFadden, Harvey Mudd College
T. J. Mueller, Harvey Mudd College

Study Guide Reviewer
Wayne Hughes, University of Georgia

Instructor's Manual Reviewers
Erica Bergquist, Holyoke Community College
Nels H. Granholm, South Dakota State University

Transparency Reviewers
William S. Cohen, University of Kentucky
Anne M. Cusic, University of Alabama
Bruce Felgenhauer, University of Southwestern Louisiana
Alice Jacklett, State University of New York at Albany
Susan Koptur, Florida International University
Charles H. Mallery, University of Miami
Stephen P. Vives, Georgia Southern University

In addition, Tad Day, William Eickmeier, and Paul Ramp conducted student reviews of the videodisc, and Jon Glase had his students review the Study Guide. We greatly appreciate their efforts.

LIFE

They Are Not All the Same
When observed closely, the individuals in a population of red and green macaws vary a great deal.

19

The Mechanisms of Evolution

426

We are aware that no two people (unless they are identical twins) look exactly alike. We also recognize our pets as distinct individuals. But we have great difficulty in seeing differences among individuals of most other species of organisms. The brilliant red and green macaws feeding on a clay cliff in the Peruvian jungle may all appear identical to the untrained eye; scientists who study them closely, however, realize that each is unique. The colored feathers display many slight variations in pattern, and the black-and-white feathers that surround the birds' eyes form patterns that, like human fingerprints, are unique to the individual. Members of many groups, particularly among behaviorally sophisticated animals such as vertebrates, readily recognize one another and adjust their behavior accordingly.

Differences among individuals in local populations, even if they are subtle, are the raw material upon which evolutionary mechanisms act to produce the striking variability revealed by the multitude of organisms living on Earth today. A good fossil record can reveal much about when and how the forms of organisms changed. Fossils may also provide clues about the reasons for those changes, but they provide only indirect evidence of the causes of evolutionary change. To obtain direct evidence we must study evolutionary changes happening today. The study of variability is at the heart of investigations into the mechanisms of evolution.

In this chapter we discuss the agents of evolution and the short-term studies designed to investigate them. By testing hypotheses observationally and experimentally we can answer key questions about the processes guiding evolutionary changes. In later chapters we consider how we use this information to explain longer-term features of the evolutionary record.

Although ideas about evolution have been put forth for centuries, until the last one hundred years none of the hypotheses about the causes of evolutionary change ... tionary change ... his hypotheses ... 1), but he did ... basis of evolut ... vided by Greg ... and Darwin's ... twentieth cent ... evolutionary h ... test them.

WHAT IS EVOL...

The fossil reco... over time. The... anisms shared...

Intriguing chapter openers combine arresting photos and text to spark interest at the outset, setting the stage for the material that follows.

Consistent icons and color-coding clarify illustrations. Icons are used to represent biologically important molecules, active forms of enzymes, and the activation and inhibition of pathways. Color is used consistently throughout the text.

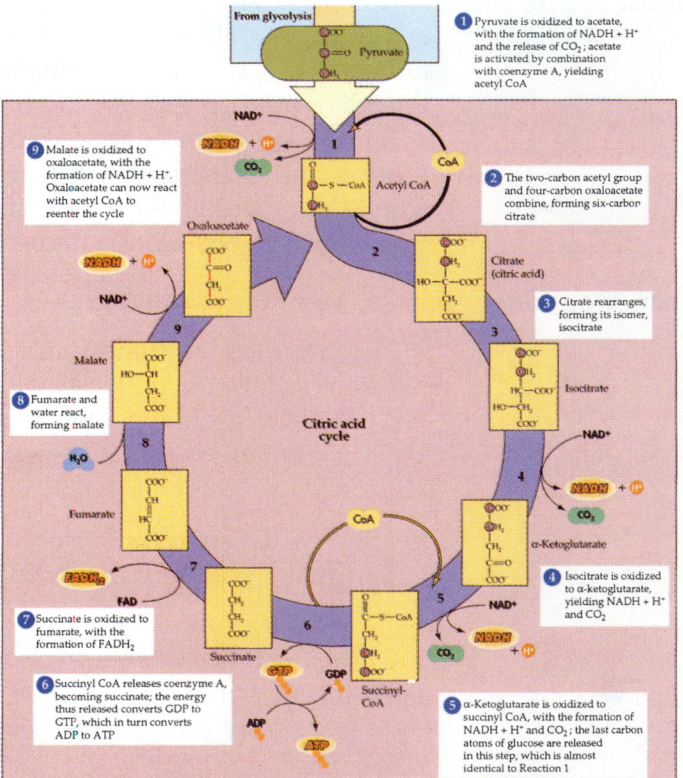

148 CHAPTER SEVEN

From glycolysis

Pyruvate

NAD+ NADH + H+ CO₂

Acetyl CoA

Citric acid cycle

Oxaloacetate

Citrate (citric acid)

Isocitrate

α-Ketoglutarate

Malate

Fumarate

Succinate

Succinyl-CoA

1. Pyruvate is oxidized to acetate, with the formation of NADH + H⁺ and the release of CO_2; acetate is activated by combination with coenzyme A, yielding acetyl CoA

2. The two-carbon acetyl group and four-carbon oxaloacetate combine, forming six-carbon citrate

3. Citrate rearranges, forming its isomer, isocitrate

4. Isocitrate is oxidized to α-ketoglutarate, yielding NADH + H⁺ and CO_2

5. α-Ketoglutarate is oxidized to succinyl CoA, with the formation of NADH + H⁺ and CO_2; the last carbon atoms of glucose are released in this step, which is almost identical to Reaction 1

6. Succinyl CoA releases coenzyme A, becoming succinate; the energy thus released converts GDP to GTP, which in turn converts ADP to ATP

7. Succinate is oxidized to fumarate, with the formation of $FADH_2$

8. Fumarate and water react, forming malate

9. Malate is oxidized to oxaloacetate, with the formation of NADH + H⁺. Oxaloacetate can now react with acetyl CoA to reenter the cycle

7.13 The Citric Acid Cycle
The first reaction produces the two-carbon acetyl CoA. Notice that the two carbons from acetyl CoA are traced with color through reaction 5, after which they may be at either end of the molecule (note the symmetry of succinate and fumarate). Reactions 1, 4, 5, 7, and 9 accomplish the major overall effect of the cycle—the storing of energy—by passing electrons to the carrier molecule NAD. Reaction 6 also stores energy.

AT A GLANCE

Dynamic visual paragraphs dramatize important concepts in a way that the text alone cannot. These are flagged with marginal arrows so the student is able to refer to them readily.

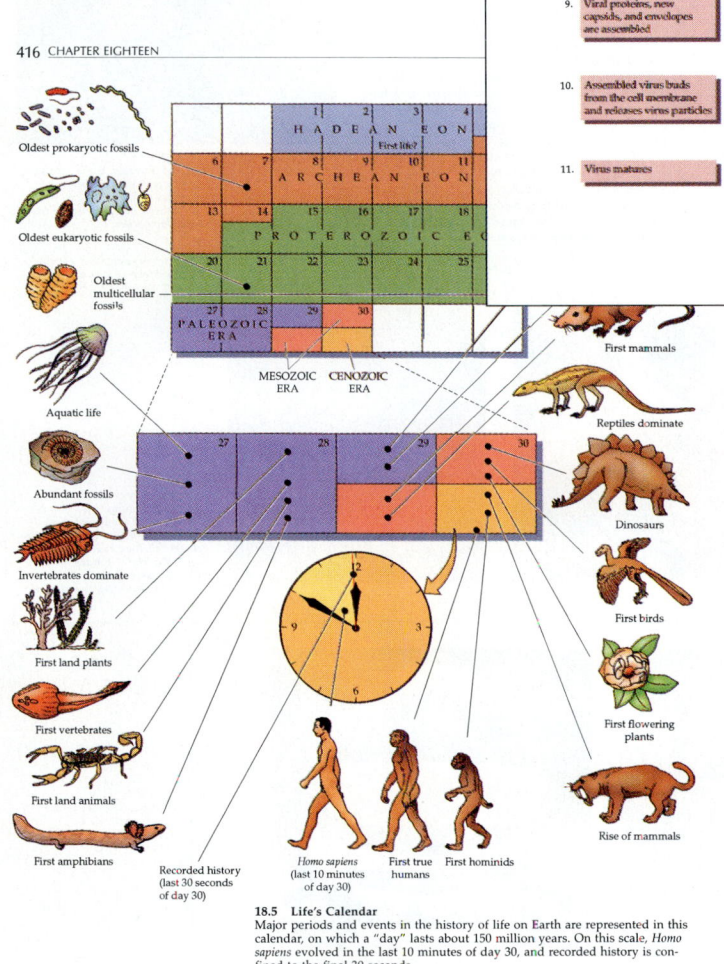

1. Retrovirus attaches to host cell at membrane protein CD4

2. The viral core is uncoated as it enters the host cell by endocytosis

3. Viral RNA uses reverse transcriptase to make complementary DNA

4. Viral RNA degrades

5. Single-stranded reverse transcript synthesizes second complementary DNA strand

6. DNA enters cell nucleus and is integrated into host chromosome, forming a provirus

7. Proviral DNA transcribes viral RNA, which is exported to cytoplasm

8. Viral RNA is translated

9. Viral proteins, new capsids, and envelopes are assembled

10. Assembled virus buds from the cell membrane and releases virus particles

11. Virus matures

16.27 The Life Cycle of HIV
The ultimate problem facing AIDS researchers is that of preventing the spread of HIV in the infected body. Almost every step in this complex life cycle can, in principle, be attacked by one or more potentially therapeutic agents. Which step will we manage to interrupt?

416 CHAPTER EIGHTEEN

Oldest prokaryotic fossils

Oldest eukaryotic fossils

Oldest multicellular fossils

HADEAN EON

First life?

ARCHEAN EON

PROTEROZOIC E

PALEOZOIC ERA

MESOZOIC ERA

CENOZOIC ERA

Aquatic life

Abundant fossils

Invertebrates dominate

First land plants

First vertebrates

First land animals

First amphibians

Recorded history (last 30 seconds of day 30)

Homo sapiens (last 10 minutes of day 30)

First true humans

First hominids

First mammals

Reptiles dominate

Dinosaurs

First birds

First flowering plants

Rise of mammals

18.5 Life's Calendar
Major periods and events in the history of life on Earth are represented in this calendar, on which a "day" lasts about 150 million years. On this scale, *Homo sapiens* evolved in the last 10 minutes of day 30, and recorded history is confined to the final 30 seconds.

(a)

(b)

(c)

25.31 Dicots
(a) The cactus family is a large grou... ...
about 1,500 species in the Americas...
takes its name from its scarlet flowe...
Anterrhinum majus. (c) These wood r...
stone National Park are members of...
as are the familiar roses from your l...
25.25 and 25.26a show other dicots.

Origin and Evolution of the Angic...

How did the angiosperms aris...
analyses (see Chapter 21) have s...
ing question. It is widely agreed t...
and two groups of gymnosperm...
and the long-extinct cycadeoids...
cycads, arose from a single ances...
rise to no other groups. A close r...
the angiosperms and the Gneto...

pected, primarily...
phyta have vesse...
angiosperms. In ...
by the light-micr...
tilization in *Ephe...*
The cycadeoids, ...
same time as dic...
portant character...
angiosperms. Th...
cycadeoids, altho...
with naked seed...
flower of *Magnol...*
   The next grea...
the question, Wh...

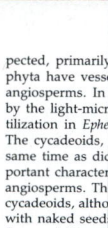
(a)

Hydrogen bond
between water
molecules

(c)

Hydrog...
in a prot...

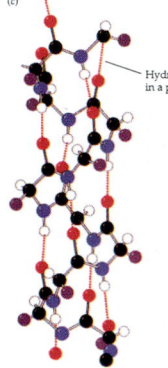

(b)

(c)

(a)

634

(b)

lineage gave rise to the prosimians—lemurs, tarsiers, pottos, and lorises (Figure 27.29). Prosimians were formerly found on all continents, but today they are restricted to Africa, tropical Asia, and Madagascar. All mainland species are arboreal and nocturnal, but on Madagascar, where there has been a remarkable prosimian radiation, there are also diurnal and terrestrial species. Until the recent arrival of humans, there were no other primates on Madagascar.

The anthropoids—monkeys, apes, and humans—evolved from an early primate stock about 55 million years ago in Africa or Asia. New World monkeys have been evolving separately from Old World monkeys long enough that they could have reached South America from Africa when those two continents were still connected. Perhaps because tropical America has

27.29 Prosimians
(a) The sifaka lemur, *Propithecus verreauxi,* is one of many lemur species of Madagascar, where they are part of a unique assemblage of plants and animals. (b) *Loris tardigradis,* the slender loris, of southern India. (b) In the rainforests of Borneo, this tarsier (*Tarsius bancanus*) seems otherworldly to our eyes.

27.30 Monkeys
(a) Golden lion tamarins (*Leontopithecus rosalia*) are New World monkeys, living in the trees of the coastal Brazilian rainforest. (b) Some Old World species, such as these Japanese macaques (*Macaca fuscata*) live and travel in groups.

(a)

(b)

2.18 Surface Tension
(a) A water strider "skates" along, supported by the surface tension of the water that is its home. (b) Surface tension demonstrated by a soap bubble on a teacup.

Boxes describe fascinating biological phenomena, highlighting special interest topics and expanding the discussion of many issues.

BOX 27.A

The Four-Minute Mile

Many mammals can run much faster, yet we humans are proud to have achieved a four-minute mile. Terrestrial vertebrates did not achieve such speeds easily. Amphibians and reptiles fill and empty their lungs using some of the same muscles they use for walking. In addition, because the limbs protrude laterally, their movement generates a strong lateral force that bends the body from side to side. Recent studies have shown that these animals cannot breathe while they walk or run. Therefore, they can operate aerobically only briefly. Because they depend upon anaerobic glycolysis while running, they tire rapidly.

In the lineage leading to dinosaurs and birds and in the lineage leading to mammals, the legs assumed more vertical positions, which reduced the lateral forces on the body during locomotion. Special ventilatory muscles that can operate independently of locomotory muscles also evolved. These muscles are visible in living birds and mammals. We can infer their existence in dinosaurs from the structure of the vertebral column and the capability of many dinosaurs for bounding, bipedal (using two legs) locomotion. The ability to breathe and run simultaneously, a capability we take for granted, was a major innovation in the evolution of terrestrial vertebrates.

A future four-minute miler?

Figure 28.12). The ability to move actively on land was not achieved easily. The first terrestrial vertebrates probably moved only very slowly, much more slowly than their aquatic relatives. The reason is that they apparently could not walk and breathe at the same time. Not until evolution of the lineages leading to the mammals, dinosaurs, and birds did special muscles evolve enabling the lungs to be filled and emptied while the limbs moved (Box 27.A). This ability enabled its bearers to maintain steady, high levels of activity, which generated enough heat to result in

subclass Aves) e embodies an scendants of a evolved in the (Figure 27.22), s and modern *teryx* was coveloped wings, much reduced may have been

ws features rehe modern

TABLE 49.4
Areas, Biomass of Plants, and Net Primary Production of Earth's Major Vegetation Zones

VEGETATION ZONE	AREA		MASS OF PLANTS		NET PRIMARY PRODUCTION	
	10^6 Km^2	PERCENT	10^9 TONS	PERCENT	10^9 TONS	PERCENT
Polar	8.05	1.6	13.77	0.6	1.33	0.6
Conifer forest	23.20	4.5	439.06	18.3	15.17	6.5
Temperate	22.53	4.5	278.67	11.5	17.97	7.7
Subtropical	24.26	4.8	323.90	13.5	34.55	14.8
Tropical	55.85	10.8	1,347.10	56.1	102.53	44.2
Total land	133.89	26.2	2,402.5	100	171.55	73.8
Glaciers	13.9	2.7	0	0	0	0
Lakes and rivers	2.0	0.4	0.04	<0.01	1.0	0.4
All continents	149.79	29.3	2402.54	100	172.55	74.2
Ocean	361.0	70.7	0.17	<0.001	60.0	25.8
Earth total	510.79	100	2,402.71	100	232.55	100

from the vents. Most of the other organisms of these ecosystems live directly or indirectly on the sulfur-oxidizing bacteria (see Figure 22.2).

This overview of the global pattern of biological production on Earth is sufficient to identify which processes limit primary production and nutrient cycling in different climatic zones and how they operate, but it does not give you a picture of what these ecosystems look like and how they function. Describing ecosystems is one of the goals of the next chapter.

SUMMARY of Main Ideas about Ecosystems

Ecosystems are powered by solar energy that first enters living organisms via photosynthesis at rates controlled by temperature and precipitation.
Review Figure 49.1

Food webs summarize who eats whom in ecological communities.
Review Figures 49.2 and 49.3

Because much of the energy taken in by an organism is used for maintenance and is eventually dissipated as heat, the efficiency of energy transfer to higher trophic levels is usually very low.
Review Figures 49.4 and 49.5

The main elements of living organisms—carbon, nitrogen, phosphorus, sulfur, hydrogen, and oxygen—cycle between organisms and other compartments of the global ecosystem.
Review Figures 49.10, 49.11, 49.12, 49.13, and Table 49.3

Human activity greatly modifies cycles of basic minerals on local, regional, and global scales.
Review Figures 49.14, 49.15, and 49.16

Earth's climate is determined primarily by the pattern of solar energy input at different latitudes and by Earth's rotation on its axis.

The directions of prevailing winds differ over the surface of Earth.
Review Figure 49.19

Surface winds drive global oceanic circulation.
Review Figure 49.20

The distribution of primary production on Earth is determined primarily by Earth's climate.
Review Figure 49.21

New chapter summaries synthesize information in the text and the art. Important concepts are encapsulated in clearly written summary sentences, with references back to key illustrations.

CONTENTS

IN BRIEF

CONTENTS

PART ONE
The Cell

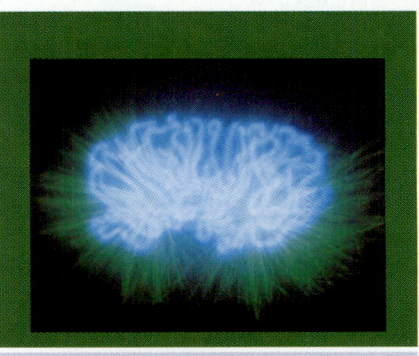

**PART TWO
Information
and Heredity**

PART THREE
Evolutionary
Processes

PART FOUR
The Evolution
of Diversity

PART FIVE
The Biology of Vascular Plants

PART SIX
The Biology
of Animals

PART SEVEN
Ecology and Biogeography

46. BEHAVIORAL ECOLOGY 1057

47. STRUCTURE AND DYNAMICS OF POPULATIONS 1080

48. INTERACTIONS WITHIN ECOLOGICAL COMMUNITIES 1100

49. ECOSYSTEMS 1127

A monarch butterfly feeds on nectar from flowers. She uses some of the energy from the nectar to power her flight and some of it to produce eggs. She lays her eggs on a milkweed plant, the prime food source for the caterpillars that will hatch from the eggs. After feeding, growing, and shedding its skin (under which a new, larger skin has developed), each caterpillar eventually changes into a pupa. The pupa is a nonfeeding stage encased within a protective cocoon held fast to a plant. Within the cocoon drastic reorganization changes the animal, and it emerges as an adult butterfly. At the end of the northern hemisphere's summer, surviving adult butterflies migrate south to traditional wintering areas in California and the mountains of Mexico. A suitable wintering site must have cool temperatures because the butterflies, which do not feed during the winter, must survive on stored food reserves that are used up more slowly at cool temperatures. If temperatures are too cold, however, the insects freeze. Very few places provide winter temperatures that are within the necessary narrow range most of the time. Butterflies that survive the winter migrate north in spring to initiate another annual reproductive cycle. We have no difficulty recognizing these colorful butterflies, and their pupae and caterpillars, as organisms—that is, as things that are alive. But how do we make this judgment?

The traits that lead us to say that certain things are alive are the processes that these things carry out. Monarch butterflies illustrate these processes, which we will give closer attention in order to understand more completely what it means to be alive.

CHARACTERISTICS OF LIVING ORGANISMS

Three processes characterize all organisms: metabolism, regulated growth, and reproduction. **Metabolism** is the sum of the chemical reactions taking place in an organism. All organisms depend on external sources of energy to fuel their chemical reactions. Some organisms—called **autotrophs**, which means "self-feeders"—synthesize their own organic molecules from simple raw materials. They do not need to eat other organisms to sustain themselves. Autotrophs obtain the energy to synthesize complex molecules from sunlight or, in a few cases, from the conversion of some very simple mineral substances (Figure 1.1). The remaining organisms, called **heterotrophs** ("other-feeders"), obtain energy from foods: complex chemical substances that were synthesized by other organisms (Figure 1.2). Heterotrophs break down these substances to release energy and to make the chemical building blocks for synthesizing other substances.

Monarch Butterflies Winter in the Highlands of Mexico

1

The Science of Biology

1.1   Exposed to the Sun
Autotrophs, such as these ferns, horsetails, and grasses growing in Alaska, usually present large surfaces to the sun, the source of their energy.

The metabolism of organisms proceeds well only within narrow ranges of internal physical and chemical conditions. Many of the mechanisms that organisms possess serve to maintain this relative internal constancy even when there are large changes in the surrounding environment. This maintenance of conditions at a steady state is called **homeostasis**. Homeostasis depends on an organism's ability to respond to the environment by changing the rates of its internal reactions or processes—in short, to regulate its metabolism. As we saw at the beginning of this chapter, monarch butterflies regulate their metabolism in winter by seeking places where temperatures remain within a relatively narrow range.

As a result of their metabolic activities, organisms may increase the number of molecules of which they are composed—that is, they grow. This is the second vital characteristic of living organisms: **regulated growth**. Organisms cannot achieve their adult shapes or function effectively unless their growth is carefully regulated. Uncontrolled growth, one example of which is cancer, ultimately destroys life.

Reproduction is the third major process characteristic of organisms. All organisms replicate themselves, but in many cases with variation—there are differences between parents and offspring (Figure 1.3). There are many modes of reproduction. Some organisms reproduce simply by dividing into two daughter cells. Other organisms reproduce by budding off small portions of their bodies to form new individuals. Most large organisms reproduce by means of special cells produced specifically for that purpose. All the information necessary to form a new individual is transmitted via these cells. Although it is a key characteristic of life, you should remember that reproduction is not essential for the survival of

(a)

(b)

1.2   Food from Many Sources
Heterotrophs feed on food substances synthesized by other organisms. (a) The African lion eats other animals. (b) The black-tailed deer feeds directly on plants.

1.3 Offspring Differ from Their Parents
These nursing puppies are members of a single litter, produced by the white female and fathered by a single male. Genetic variability among the offspring of two parents is the norm.

an organism. In fact, reproduction usually reduces survival rates.

Whatever its effects on the survival of individuals, reproduction with change makes possible the evolution of life and has produced one of the most distinctive features of life: **adaptation**. When we say that an organism is adapted to its environment, we mean that it possesses characteristics that enhance its survival and reproductive success in that particular environment. The caterpillars of monarch butterflies are adapted for feeding on milkweed leaves; the adults are adapted to extract nectar from flowers, to find mates, to find milkweed plants upon which to lay their eggs, and to migrate to suitable wintering areas.

Many adaptations fit organisms to their environments. The wings of birds, for example, are adapted for various types of flight, depending on the specific needs of the bird (Figure 1.4). Camouflaged animals blend into their backgrounds and are difficult for visually hunting predators to locate. Early efforts to explain adaptation implied that some purpose or foresight was involved, and these explanations did not provide testable hypotheses. Today biology proceeds from hypotheses that can be tested by scientific methods. Nearly a century and a half ago, Charles Darwin and Alfred Russel Wallace proposed the first scientifically testable theory about adaptation: evolution by natural selection (see Chapter 19).

THE HIERARCHICAL ORGANIZATION OF BIOLOGY

Biologists study processes ranging from the structure and function of simple molecules to the interactions among the hundreds or thousands of different types of organisms that live together in a particular region, as well as how these organisms evolve. Biology can be visualized as ordered into a hierarchy in which the units, from smallest to largest, are molecules, cells, tissues, organs, organisms, populations, communities, and biomes. These units interact with and adjust to one another as they seek and exchange matter, energy, and information. The organism is the central unit in biological investigations, even when a

(a)

1.4 Wings Adapted for Flight
Most birds use their wings for flight, but different birds fly in different ways. *(a)* The long, broad wings of the red-tailed hawk allow it to sustain a gliding flight above open country while it searches for prey with its keen eyes. *(b)* The action of hummingbird wings allows them to hover in front of flowers while they extract nectar.

(b)

study focuses on a unit distant from the organism in the hierarchy. Today many biologists investigate interactions among molecules and are concerned with processes that are closely related to physics and chemistry. However, when biologists study chemical structures and reactions, they ask different types of questions than chemists do. Biologists study chemical structures and reactions to discover the mechanisms upon which the life of an organism depends. At the higher levels in the hierarchy, biologists study how organisms interact with and adjust to one another to form social systems, populations, ecological communities, and biomes.

From Molecules to Tissues

The smallest entities biologists study are **molecules**, the basic structural units of chemical compounds (see Chapter 2). The principal chemical compounds constituting all living organisms are proteins, nucleic acids, lipids, and carbohydrates (see Chapter 3). Within living organisms many molecules join to form complex aggregates, such as membranes (see Chapter 5). These aggregates have properties that are essential to their functioning within the organism. These properties appear only when the aggregates form; that is, the isolated molecules do not exhibit them.

The **cell** is the fundamental unit of life because it is the simplest unit capable of independent existence and reproduction and because all organisms are composed of cells. Some organisms are single cells. Cells have many features in common (see Chapter 4), but they come in a wide variety of types, many of which are adapted for specific functions, such as secreting substances, storing and transmitting information, or capturing the energy of sunlight. In organisms consisting of many cells, cells of specific types are organized to form **tissues**. Familiar tissues are the muscles of animals and the wood of trees. Tissues may be organized into **organs,** such as skin, hearts, livers, roots, and leaves—structures composed of more than one tissue type.

From Organisms to Biomes

Each creature—each tree, each bacterium, each frog, each person—is an **organism**. Organisms are highly variable and they come in an enormous diversity of forms, to which we may apply names such as wine yeast, robins, garter snakes, howler monkeys, and sugar maple trees. Biologists call the different forms of organisms **species**; but, as we will see in Chapter 20, species are not easy to define. For our present purposes, a species can be defined as a total group of organisms, possibly living in many separate **pop-**

1.5 From Molecules to the Biosphere

The fish—an organism—belongs to a species that is a member of a coral reef community. The fish's molecules are organized into organelles (structures found in or on cells), cells, tissues, and organs, which work together in such a way that the fish can extract food from its environment, avoid its predators, and reproduce. The coral reef community exchanges energy and materials with other communities. Such exchanges unite communities in our biosphere.

ulations, capable of breeding with each other but not with other organisms. Individuals of many different species typically live together and interact to form **ecological communities** (all the different species living in a particular area) and **biomes** (the major types of ecological communities) (Figure 1.5). Together, Earth's biomes form the **biosphere**.

Life's Emergent Properties

Each level of biological organization has properties, called **emergent properties**, that are not found at lower levels. For example, cells and individual organisms exhibit properties that regulate their functioning within strict limits, but the molecules that make up the cells and organisms lack these properties. Individuals are born and they die, but an individual does not have a birth rate and a death rate. A population does. Other emergent properties of populations include age distributions, densities, and distribution patterns. Ecological communities may be described in terms of the number of species in them and the relative abundances of those species.

Emergent properties such as birth rates and species richness that appear at higher levels of organization do not violate principles that operate at lower levels of organization. Usually, however, emergent properties cannot be detected or even suspected from a study at lower levels. Biologists could never discover the existence of such emotions as hate, fear, and love by studying single nerve cells, even though they may eventually be able to explain those emotions in terms of interactions among nerve cells. The properties that emerge at the level of organization at which a biologist works often determine the types of research questions that are most appropriate.

Suppose, for example, a biologist walks along the edge of a pond and startles a frog that jumps into the water (Figure 1.6). Two obvious questions to ask are, *Why* did the frog jump into the water? and *How* did the frog jump? A biochemist might focus on the second question and answer it in terms of the molecular mechanisms underlying muscular contraction. A physiologist might also choose the second question, answering that the muscles in the frog's

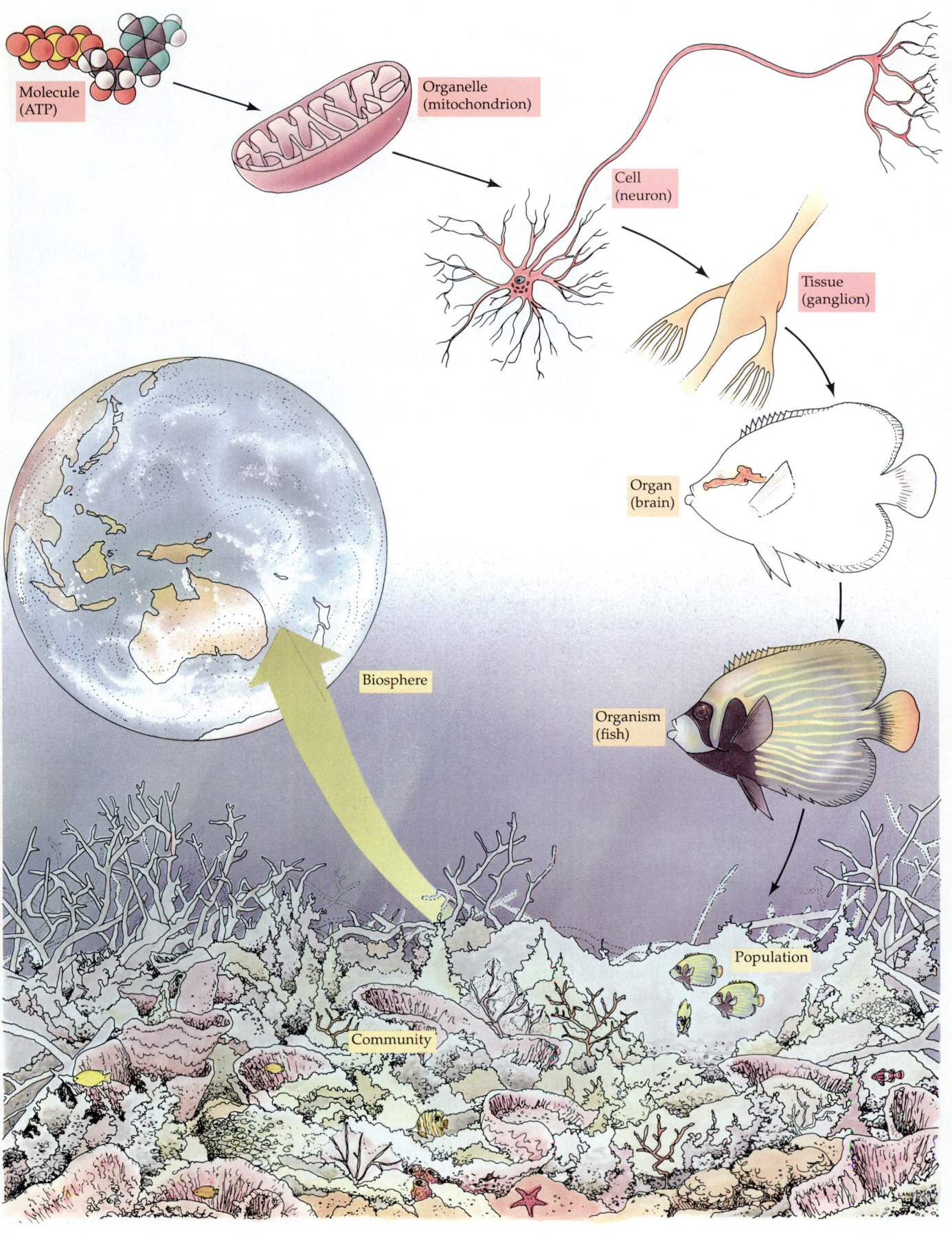

Molecule (ATP)

Organelle (mitochondrion)

Cell (neuron)

Tissue (ganglion)

Organ (brain)

Biosphere

Organism (fish)

Population

Community

1.6 Why and How Did the Frog Jump? Scientists from different disciplines are likely to answer only one of these questions and their answers are likely to be very different.

legs contracted because they were stimulated by motor nerves that synapse with the muscles, and that those nerves, in turn, fired as a result of stimuli initially received by the frog's eyes. A developmental biologist might answer the same question in terms of the embryological development of the neuromuscular wiring that underlies the physiological responses. An ecologist would be more likely to answer the first question and suggest that the frog jumped to escape from a potential predator. Finally, an evolutionist, also answering the first question, might offer some suggestions as to why frogs hop instead of walking and why they hide in water rather than in some protected site on land.

You might think that an act as familiar as a frog's jump would be completely understood, but you would be wrong. The answers biologists can give to either of these two questions at any level in the hierarchy are still incomplete. There is still much to learn about the living world.

Is either of the two questions about the frog's jumping more basic or important than the other? Is any one of the answers more fundamental or more important than the others? Not really. Both questions and all the different answers are essential parts of the full explanation of the jumping of frogs. This richness of answers to apparently simple questions makes biology a complex science, but also an exciting field. In this book, we begin with events and processes at molecular and cellular levels and end with processes at the levels of ecological communities and the biosphere. At all levels we pose both *how* and *why* questions.

THE METHODS OF SCIENCE

Science is a uniquely human activity (on this planet at least). As far as we are aware, no other animal practices science. Science, contrary to much popular opinion, is not merely a collection of facts about the world. Science is a process, a set of ways of discovering things about the universe.

Scientists employ a variety of methods in attempting to understand the structure and functioning of the universe and its components. Diverse methods are necessary because different scientific disciplines study such different subjects. Studying molecules, glaciers, climate, trees, or human social systems—to name only a few objects of scientific study—requires methods appropriate to each subject. Also, within disciplines many different questions can be asked about the same subject, as we saw with the jumping frog. And, not surprisingly, because they are people, scientists differ. They differ in the types of questions they are comfortable asking, in their skills with particular methods, and in how they interpret their results. Scientists agree, however, that the methods they use must be honest. Therefore, scientists worldwide will often accept each other's measurements and data, even though they may disagree on the conclusions to be drawn from those facts. Nonetheless, experiments yielding surprising or especially important data are often repeated by other scientists; the results may become widely accepted only after they have been obtained repeatedly.

Although science employs many methods, one general approach underlies most scientific work. This approach developed slowly over the centuries as scientists realized how it helped them make discoveries. One of its essential elements is that the methods used allow us to modify and correct our beliefs as new evidence becomes available. What is this approach?

The Hypothetico-Deductive Approach

The basic approach underlying most of science, despite the great variety of subjects studied and the methods by which they are investigated, is the **hypothetico-deductive approach**. This procedure has four stages: (1) making observations, (2) forming **hypotheses**, which are tentative answers to questions, (3) making predictions from the hypotheses, and (4) testing the predictions. Testing predictions, in turn, generates new observations, so the cycle continues indefinitely.

The following example shows how biologists apply the hypothetico-deductive approach. Biologists have long known that some caterpillars, such as those of monarch butterflies, are conspicuously colored. Others, such as the caterpillars of peppered moths, are cryptically colored—that is, they blend in with their backgrounds (Figure 1.7). Observations also revealed that conspicuously colored caterpillars often live in groups but are seldom attacked by birds. These initial observations were used to develop hypotheses, make predictions, and devise tests of those predictions. Let's examine how this was done.

GENERATING A HYPOTHESIS. The conspicuous appearance of some caterpillars, together with the observation that potential predators usually avoid them, suggested to some biologists that the bright color patterns of these caterpillars signal to potential predators that the caterpillars are distasteful or toxic. Another hypothesis is that inconspicuous caterpillars are

good to eat (palatable) and their coloration reduces the chance that predators will discover and eat them. For each hypothesis of an effect there is a corresponding **null hypothesis** that asserts that the proposed effect is absent. The null hypothesis for the hypotheses we have just stated is that there is no difference in palatability between colorful and cryptic caterpillars.

Notice that these hypotheses depend on certain assumptions or on previous knowledge. We assume, for example, that birds have color vision and can learn about the qualities of their prey by encountering and tasting them. If such assumptions are uncertain, they should be tested before predictions are made from the hypotheses.

MAKING AND TESTING PREDICTIONS. The hypotheses about colorful and cryptic caterpillars led to obvious predictions. Try to formulate them for yourself before you read further. The predictions were tested by presenting captive blue jays with both brightly colored monarch butterfly caterpillars and cryptically colored caterpillars. The blue jays were first deprived of food long enough to make them hungry, so they readily attacked the caterpillars. Ingesting even part of one monarch caterpillar caused a blue jay to vomit. Because the birds were housed individually, the experimenters knew which ones had previously tasted monarchs and which ones had not. They found that a single experience with a monarch caterpillar was enough to cause a blue jay to reject all other monarch caterpillars presented to it. In nature, monarch caterpillars live in groups, so a predator readily learns to avoid all group members after having tasted one. Cryptically colored caterpillars, on the other hand, were readily attacked and eaten, and the jays continued to eat these caterpillars without showing any

(a)

1.7 Caterpillars Can Be Easy or Hard to See
(a) Many caterpillars—like this peppered moth larva, which resembles a small green twig—blend into their surroundings because they are cryptically colored. *(b)* This larva will become a large tropical moth. Its defensive spines make it particularly conspicuous.

(b)

(a)

1.8   Research Is Essential in Biology
Research and experimentation in biology are carried out in the field and in laboratories. *(a)* Biologists who study the canopies of rainforest trees use special climbing equipment that allows them to collect data and carry out vital studies in the field. *(b)* Some scientists conduct laboratory studies of the properties of potentially dangerous chemicals or other substances that must be kept isolated and protected from contamination before using the substances in field experiments.

signs of sickness or discontent. These results supported the palatability hypothesis. The null hypothesis was thus rejected.

Hypothetico-deductive science uses a variety of methods to test predictions. Among these are laboratory and field experiments and carefully focused observations. Each method has its strengths and weaknesses. The key feature of **experimentation** is the control of most factors that might affect a result so that the influence of the factors that do vary can be seen more clearly. The advantage of working in a laboratory is that control of the environment is easier. Field experiments are more difficult because it is usually impossible to control more than a small part of the total environment. The conditions under which the laboratory experiments with blue jays were run allowed the investigators to reject alternative explanations. Their results, for example, could not have been due to lack of hunger on the part of the birds or their failure to see the caterpillars.

Nonetheless, field experiments have one important advantage. Their results are more readily applicable to what happens where the organisms actually live and evolve. Just because an organism does something in the laboratory does not mean that it behaves the same way in nature. A laboratory experiment demonstrates the *potential* for the organism to act in a certain way, but it does not demonstrate that it will or will not act that way. Because we usually wish to explain nature, not the behavior of organisms in the laboratory, combinations of laboratory and field experiments are needed to test most hypotheses about organismic and higher-level attributes of organisms (Figure 1.8).

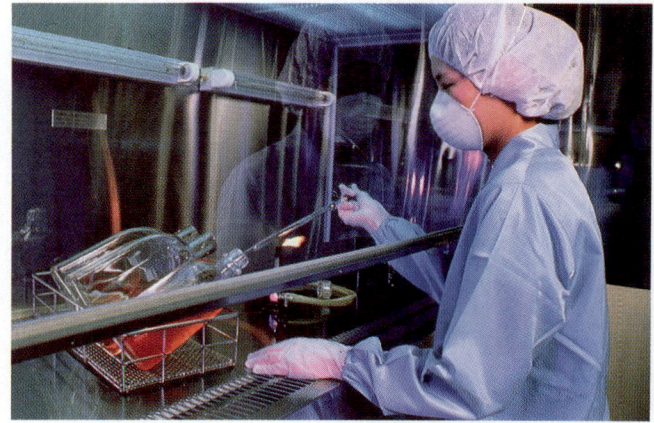

(b)

ACCEPTING A HYPOTHESIS. Scientists may disagree about the adequacy of the evidence in support of a particular hypothesis. Also, different scientists may interpret the same set of observations in different ways. Whereas most new factual discoveries in science are accepted quickly, scientists are slow to accept new *theories*, particularly those that are based on concepts that differ strikingly from ones that are commonly accepted. For example, Isaac Newton's theory that the motions of planets are determined by gravitational attraction was not universally accepted until about 80 years after he first proposed it. About 50 years elapsed between Alfred Wegener's first proposal of the theory of continental drift in 1912 and its general acceptance (see Chapter 28).

At any given moment, there are many hypotheses that are generally accepted as true or false by most scientists in the field. Others are regarded as not yet convincingly confirmed or rejected. The history of science shows us that generally accepted hypotheses are often overturned by newer discoveries. Sometimes hypotheses that have been "convincingly" rejected are resurrected by new discoveries. We can neither prove nor reject any hypothesis with absolute certainty. Nonetheless, the features of the hypothetico-deductive method allow us to develop a high degree of confidence in our interpretations of how the world works.

A single piece of evidence supporting a hypothesis rarely leads to its widespread acceptance, just as a single contrary result rarely leads to the abandonment of a hypothesis. Negative results can be obtained for many reasons, only one of which is that the hypothesis is wrong. For example, the error may be that incorrect predictions were made from a correct hypothesis. A negative result can also occur because of poor experimental design or because an inappropriate organism—one that does not fit the assumptions of the hypothesis—is chosen for the test. For example, a predator lacking color vision, or one that uses primarily its sense of smell, would not be appropriate for testing hypotheses about the colors of caterpillars.

A general textbook like this one is based on hypotheses that have been extensively tested and that are generally accepted. An extensively tested hypothesis is sometimes called a theory, but there is no rule that states how much evidence must be gathered before a hypothesis earns the status of a theory. When possible in this text, we illustrate hypotheses and theories with observations and experiments that support them, but we cannot, because of space constraints, detail all the evidence. Remember, as you read, that statements of biological "fact" are mixtures of observations, predictions, and interpretations.

Experimentation and Animals

Obtaining answers to many of the questions posed by biologists requires experimenting with plants, animals, fungi, and microorganisms. To study the antipredator adaptations of caterpillars, investigators had to keep blue jays in cages, make them hungry by depriving them of food, and feed them caterpillars. This procedure resulted in the deaths of some caterpillars and temporary stress for the jays. Determining which chemicals make the caterpillars toxic required the deaths of more caterpillars.

No amount of observation without intervention could possibly substitute for experimental manipulation. This does not mean, however, that scientists are insensitive to the welfare of the organisms with which they work. Most scientists who work with animals are continually alert to finding ways of getting answers that use the smallest number of experimental subjects and that cause the subjects the least pain and suffering.

WHY PEOPLE DO SCIENCE

Describing *how* people do science does not explain *why* they do it. The most important motivator of most biologists is curiosity. People are fascinated by the richness and diversity of life and want to learn more about organisms and how they function. Curiosity is an adaptive trait. Humans who were not curious about their surroundings and, therefore, were not motivated to learn about them, probably survived and reproduced less well, on average, than their more curious relatives. We hope this book will help you share the excitement biologists feel as they learn more about living organisms, develop hypotheses, and test them. Who knows, perhaps your curiosity will lead to an important new idea.

SIZE SCALES

Multicellular organisms are composed of many types of cells that perform different functions. Among the benefits of multicellularity are improved protection, the possibility of evolving into any of a wide variety of shapes, and the possibility of evolving to be large. Why can't evolution produce large unicellular (one-celled) organisms? A major reason relates to the changes in the surface-to-volume ratio of an object as it increases in size. A cubic cell measuring 100 micrometers (μm) along each edge has a volume (and mass) 1,000 times that of a cell measuring 10 μm along each edge, but its surface area is only 100 times greater than that of the smaller cell. Thus the surface-to-volume ratio of the smaller cell is greater than that of the larger cell (Figure 1.9).

How does the surface-to-volume ratio of a cell influence its functioning? As a cell metabolizes, it must exchange materials and heat with its environment. Although the number of chemical reactions a cell can carry out per unit time is proportional to its volume, its rate of exchange of nutrients and waste products with its environment is limited by its *surface area*. As a cell grows larger, its rate of production of wastes and its need for resources increase faster than the surface through which the cell must obtain resources and void its wastes. Substantial increases in overall sizes of organisms require the development of structures that transport food, oxygen, and waste materials to and from parts that are distant from their external surfaces. Multicellularity, with specialized cells to provide these functions, is what enabled large organisms to evolve.

Because the heat generated by the chemical reactions within an organism must be dissipated for the organism to survive, large animals have lower metabolic rates than small animals have. In total quantity, an elephant requires more food per day than a mouse

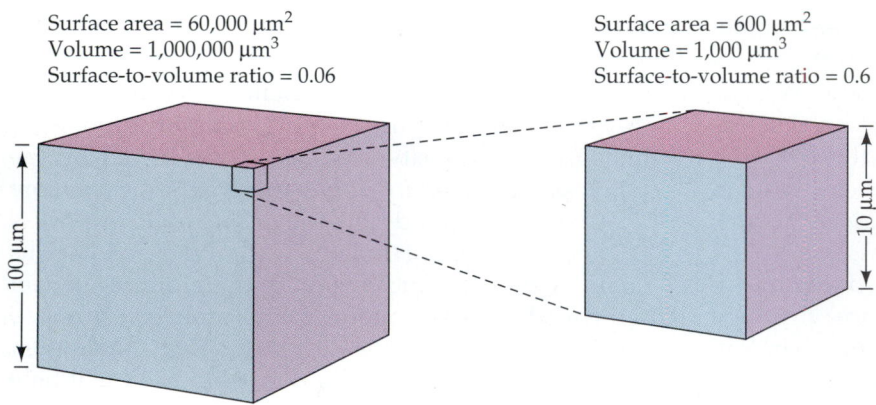

Surface area = 60,000 μm²
Volume = 1,000,000 μm³
Surface-to-volume ratio = 0.06

Surface area = 600 μm²
Volume = 1,000 μm³
Surface-to-volume ratio = 0.6

1.9 Small Objects Have Large Surface Areas
A cube 10 μm along each side has a much greater surface area in relation to its internal volume than a cube 100 μm along a side. Although the single large cube has the same volume as 1,000 of the small cubes, 1,000 small cubes have 10 times more surface area.

does; but gram for gram, the mouse requires more. Thus the metabolic rate of the mouse must be faster than that of the elephant. If the elephant had the same metabolic rate as the mouse, it would not be able to dissipate the heat produced by metabolism fast enough to avoid cooking itself! Conversely, a large animal does not cool off as rapidly as a small one. Thus a large lizard can heat its body in a warm area and then forage for a longer time in a cool place than a small lizard can.

Other properties of organisms also change with size. For example, the weight of an animal is related to its volume and is proportional to the cube of its linear dimensions. The legs of an organism, however, must be able to support its weight. The strength of bones is proportional to their cross-sectional area, which in turn is proportional to the square of their lengths. Therefore, an increase in size must be accompanied by a proportionally greater increase in leg diameter. Delicate, slender legs are characteristic of

lightweight impalas, not of heavy elephants (Figure 1.10).

Many ecological interactions among organisms are strongly affected by size. Size determines the food an organism can use, where it can hide, what it can mimic, the area it requires to obtain its food, and its abundance. Not surprisingly, then, biologists pay a great deal of attention to the distributions of sizes among organisms.

TIME SCALES

When Galileo studied the motion of a ball rolling down an inclined plane 375 years ago, he used his pulse to mark off equal intervals of time because no better device existed for measuring time. The rise of modern science has depended upon the invention of instruments capable of measuring time accurately. Scientists need to measure time spans both longer

1.10 Proportions Change with Size
Elephants have proportionally much more massive legs than do the slender impalas. An animal the size of an elephant with legs the shape of an impala's would collapse under its own weight.

and shorter than those we can perceive accurately with our unaided senses. In biology, the longer time spans have caused the greatest conceptual difficulties. Earth is a very old planet; the difficulty in perceiving its age delayed the recognition of evolutionary change for a long time.

Today, scientists measure long time spans by using naturally radioactive materials (see Chapter 18). Certain radioactive materials are incorporated into rocks and other materials when they are formed. As time passes, the radioactivity in these materials decreases at a regular rate. Scientists compare the proportions of radioactive and corresponding nonradioactive materials in particular rocks. From the observed ratio, they can estimate how long ago the rocks formed. This method gives us *absolute* ages. Indirect observations, such as the vertical positions of rock layers in relation to one another, help us assess the *relative* ages of materials. Young rocks lie on top of older ones (unless the rocks have been subjected to dramatic deformations, which are usually evident). By studying the remains of living things found in different layers of rock and correlating their distributions across many sites, we can determine the relative ages of different deposits even if absolute ages are not known.

Biologists working at different levels of organization may think about and study problems in different time scales. Biochemists and physiologists are concerned primarily with the time span relevant to the intervals required for chemical reactions and physiological changes within an organism. These intervals range from fractions of seconds to a day or a year. Studies of physiological processes such as aging may require observations extending over the lifetime of organisms, which may range up to centuries for some long-lived plants.

Studies of populations may extend over many generations of the organisms. The time span of generations ranges from less than an hour for some bacteria to centuries for some plants, but for any particular type of organism, the time required for changes in the sizes and distributions of populations is much longer than the time required for physiological changes in single individuals.

The study of changes in the genetic constitution of populations requires us to think in terms of microevolutionary time. Significant genetic changes usually take many generations, but they can happen quite rapidly when environmental conditions change abruptly. Organisms often change more substantially over spans of macroevolutionary time, covering thousands of generations or more. We can often observe microevolutionary changes directly, but we must measure macroevolutionary changes indirectly.

MAJOR ORGANIZING CONCEPTS IN BIOLOGY

Knowledge about organisms can be organized in many ways, and we use several methods in this book. In some chapters we look at the molecules of which organisms are composed. In other chapters we look at different groups of organisms and see how their structures, activities, and adaptations are related to their particular lifestyles. Other chapters focus on major processes carried out by organisms, such as digestion, movement, and reproduction. In still other chapters, we focus on the mechanisms of evolutionary change. Through all of the discussions, however, several major organizing concepts guide us.

The first of these concepts is that *all properties of organisms can be explained in physical and chemical terms.* Organisms appear to be triumphs of organic chemistry. The most complex biological activities, including the mysterious richness of our human emotions, are probably manifestations of underlying physicochemical systems. Because we use this concept, however, does not imply that we can understand all complex biological phenomena simply through the study of chemistry. Rather, it implies that biological phenomena must conform to the laws of physics and chemistry. Not too many years ago the idea that the properties of organisms resulted from physical and chemical interactions was vigorously debated, and lengthy books were written attacking and defending it. Today, however, most biologists accept the chemical basis of life and are attempting to identify the physical and chemical processes underlying specific biological phenomena. Some basic chemistry and biochemistry is presented in Chapters 2 and 3.

The second key organizing concept in biology came from physics and is a refinement of the understanding of energy. *Organisms can be viewed as systems that take in energy from their environments and convert it to biologically useful forms* (see Chapters 7 and 8). The notion of energy is highly abstract. Energy is weightless and occupies no space, yet it exists in many forms. Using experimentally derived formulas, we can calculate the equivalence of these forms. Energy can never be created or destroyed, but it can be converted from one form to another. If energy is weightless and occupies no space, how can we measure it and use it in meaningful ways? We measure energy by its *effects upon matter*, which can be weighed and measured.

The third organizing concept in biology is that *genetic information encodes and transmits information between generations.* You are probably familiar with this concept, having already heard many times that char-

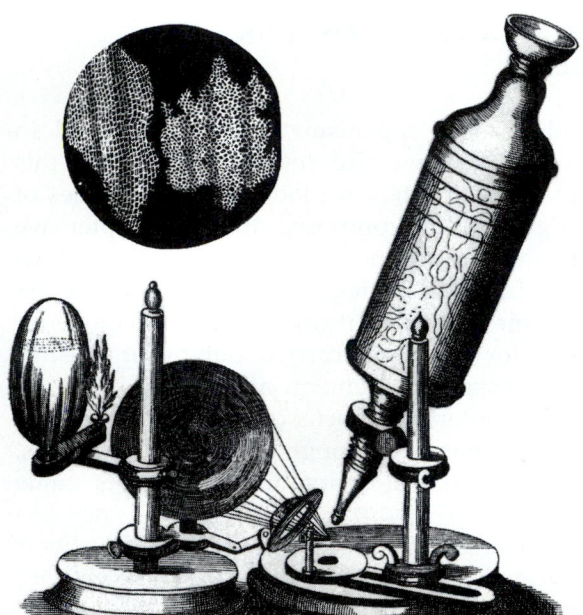

1.11 Hooke's Microscope and What He Saw
The microscope used by Robert Hooke. In the circle are Hooke's drawings of the empty plant cells found in the bark of a cork oak.

acteristics are transmitted from parents to their offspring by a genetic code, but widespread acceptance of the concept came only in this century. The discovery of the molecules that contain the hereditary information is even more recent (see Chapter 11).

The fourth organizing concept in biology, called the **cell theory**, states that *organisms are composed of cells* and that *the cell is therefore the basic building block of life*. Cells differ greatly in size and in the complexity of their internal structures, but all cells have certain characteristics that enable us to discuss many of the properties of organisms in terms of the way cells perform work and organize themselves. Viruses, which may or may not be considered alive, are not cells; they are too simple to survive without using the machinery of cells.

Magnifying devices existed for hundreds of years before anyone thought to look carefully at organisms with them. It was not until 1665 that the Englishman Robert Hooke noticed that cork, wood, and other plant tissues are made up of small, regularly shaped cavities surrounded by walls (Figure 1.11). Hooke called these cavities cells. Living, single-celled organisms were first observed a few years later by the Dutch naturalist Anton van Leeuwenhoek, who used a simple microscope of his own design. What these two men had observed was not fully appreciated for a long time. The first strong statement that *all organisms consist of cells* was made by the German physiologist Theodor Schwann in 1839. In 1858 the German physician Rudolf Virchow suggested that *all cells*

come from preexisting cells. Experiments by the French chemist and microbiologist Louis Pasteur between 1859 and 1861 provided generally accepted proof of this assertion. Since then the cell theory, as summarized by the two statements in the previous paragraph, has been a basic principle of biology.

The fifth organizing concept in biology is that *evolution by natural selection results in adaptation*. We will consider the mechanisms of evolution in detail in Part Three, but you need some knowledge of the mechanisms to understand material in chapters that precede Part Three. Fortunately, even though the details are complex, the basic principles are simple.

EVOLUTIONARY CONCEPTS

For more than 200 years biologists have suspected that the organisms they observed evolved from types unlike those currently living. In the 1760s, the French naturalist Count George-Louis Leclerc de Buffon wrote his *Natural History of Animals*, which contained a clear statement of the possibility of evolution. Buffon originally believed that all organisms had been specially created for different ways of life, but as he studied animals he observed that the limb bones of all mammals, no matter what their way of life, are remarkably similar in many details (Figure 1.12). If these limbs were specifically created for different ways of locomotion, Buffon reasoned, they should have been built upon different plans rather than being modifications of a single plan. He also noticed that the legs of certain animals, such as pigs, have

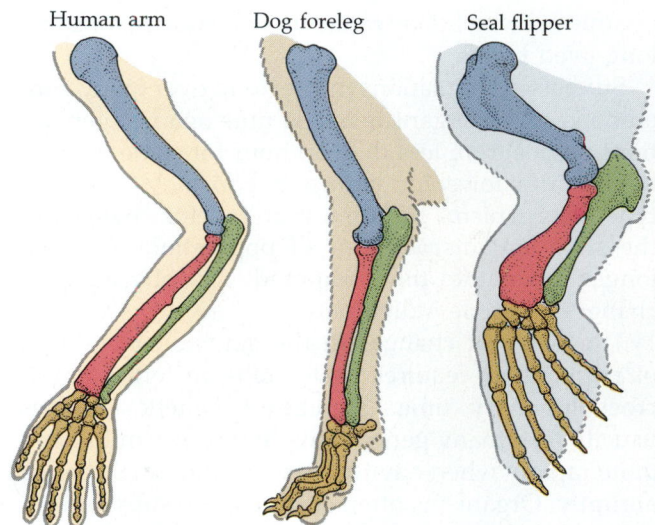

Human arm   Dog foreleg   Seal flipper

1.12 All Mammals Have Similar Limb Bones
Mammalian forelimbs have different purposes—humans use theirs for manipulating objects, dog forelimbs are used for walking on, and seals for swimming—but the number and type of their bones are similar. This shows bones of the same type in the same color.

THE SCIENCE OF BIOLOGY

toes that never touch the ground and appear to be of no use. Buffon found it difficult to explain the presence of these seemingly useless small toes by special creation. Both of these troubling facts could be explained if mammals had not been specially created in their present forms but had been modified from a common ancestor. Buffon therefore suggested that pigs have two functionless toes because they inherited them from ancestors in which the toes were fully formed and functional.

Buffon's student Jean Baptiste de Lamarck wrote extensively about evolution. Lamarck was the first person to support the idea of evolution with logical arguments. He was also the first person to propose a hypothesis concerning the mechanisms of evolutionary change. He suggested that lineages of organisms may change gradually over many generations as offspring inherit structures that have become larger and more highly developed as a result of continued use or, conversely, have become smaller and less developed as a result of disuse. For example, Lamarck suggested that aquatic birds extend their toes while swimming, stretching the skin between them. This stretched condition, he thought, could be inherited by the offspring, who would further stretch their skin during their lifetimes and would also pass this condition along to their offspring. According to Lamarck, birds with webbed feet would thereby evolve over a number of generations (Figure 1.13). He explained many other examples of adaptations in a similar way and showed how many domestic plants and animals have departed from the forms of their wild ancestors. We do not now believe that evolutionary changes are produced by the mechanisms proposed by Lamarck. However, Lamarck's ideas deserved more attention than they received from his contemporaries, most of whom believed in a young and unchanging universe.

1.14   Charles Darwin as a Young Man

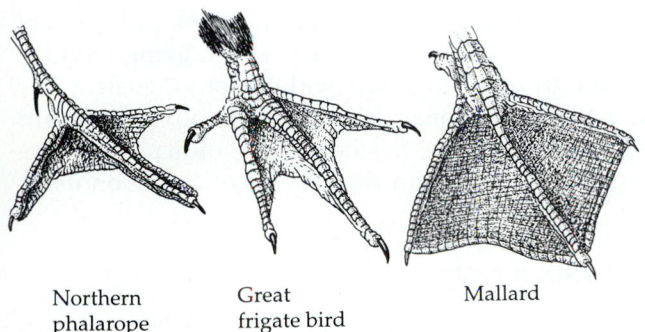

Northern
phalarope

Great
frigate bird

Mallard

1.13   Partially to Fully Webbed
Lamarck believed that the offspring of these birds, all of which stretch their feet while swimming, would inherit this stretched condition and eventually have webbed feet like the mallard's. Scientists today generally agree that the mechanism proposed by Lamarck does not explain evolutionary changes.

By 1858 the climate of opinion, among biologists at least, was receptive to the theory of evolutionary processes proposed independently by Charles Darwin (Figure 1.14) and Alfred Russel Wallace. By then geologists had shown that Earth had changed over millions of years, and many people were willing to think in terms of longer time spans. Thus, the presentation in the latter half of the nineteenth century of a well-documented and thoroughly scientific argument for evolution triggered a transformation of biology.

Darwin's Scientific Study of Evolution

Charles Darwin's approach to evolution incorporated the following hypotheses: (1) Earth is very old, and organisms have been changing steadily throughout the history of life. (2) All organisms are descendants of a common ancestor—that is, life arose only once on Earth. (3) Species multiply by splitting into daughter species, and such speciation has resulted in the great diversity of life found on Earth. (4) Evolution proceeds via gradual changes in populations, not by the sudden production of individuals of dramatically different types. (5) The major agent of evolutionary change is natural selection. Remarkably, these five hypotheses have all been supported by the mass of research that has been conducted since Darwin published his book *The Origin of Species* in 1859.

Darwin used some facts that were familiar to most of his fellow biologists. He knew that populations of all species, even those with very low reproductive rates, have the potential for exponential increases in numbers. Yet such rates of increase are rarely achieved in nature; populations of most species are relatively stable through time. Therefore, death rates in nature must be high (Figure 1.15). Without high death rates, even the most slowly reproducing forms would quickly reach enormous population sizes.

Darwin also noted that, although offspring tend to resemble their parents, the individuals of most types of organisms are not identical. He suggested that slight variations among individuals significantly affect the chance that a given individual will survive and reproduce. He called this differential reproductive success of individuals **natural selection**, probably because he was deeply interested in the artificial selection practices of animal and plant breeders. Indeed, many of Darwin's observations on the nature of variation came from domesticated plants and animals. Because Darwin himself was a pigeon breeder, he recognized close parallels between artificial selection by breeders and selection in nature.

Darwin argued his case for natural selection as follows:

> How can it be doubted, from the struggle each individual has to obtain subsistence, that any minute variation in structure, habits or instincts, adapting that individual better to the new conditions, would tell upon its vigour and health? In the struggle it would have a better chance of surviving; and those of its offspring which inherited the variation, be it ever so slight, would have a better chance. Yearly more are bred than can survive; the smallest gain in the balance, in the long run, must tell on which death shall fall, and which shall survive. Let this work of selection on the one hand, and death on the other, go on for a thousand generations, who will pretend to affirm that it would produce no effect, when we remember what, in a few years animal breeders effected in cattle, and . . . in sheep, by the identical principle of selection?

That statement, written more than 100 years ago, still stands as a good expression of the idea of evolution by natural selection.

The Importance of a World View

Biologists develop hypotheses, devise tests, and modify their ideas in the light of observational and experimental results, but they carry out these activities within a broader framework. All of us, whether scientists or not, operate within the framework of a general world view, which is sometimes called a **paradigm**. The paradigm determines which problems are interesting and which are not, and it strongly influences the responses we make to information that seems to oppose it.

Biology began a major paradigm shift a little over a century ago with the general acceptance of long-

1.15 Many Organisms Have High Reproductive Rates
A frog surrounded by the eggs it has laid. If all the offspring of frogs such as this one grew to adulthood, the world's frog population would be overwhelming. Many of these eggs will not survive, however. A rate of reproduction as high as this frog's is usually accompanied by a high mortality rate of eggs and young.

term evolutionary change and the recognition that natural selection is the primary agent producing adaptation. The shift took a long time because it required abandoning many components of a different world view. In the pre-Darwinian view, the world was thought to be a young one in which organisms had been created in their current forms. In the Darwinian view, the world is an ancient one in which both Earth and its inhabitants have been evolving from forms very different from the ones they now have. It is a world in which we would not recognize many organisms if we were transported far back into the past or far forward into the future. Accepting this paradigm means accepting not only the processes of evolution, but also the view that the living world is constantly evolving, but without any "goals." The idea that evolutionary change is not directed toward a final goal or state has been more difficult for some people to accept than the process of evolution itself.

LIFE'S SIX KINGDOMS

Perhaps as many as 30 million species of organisms inhabit Earth today. Many times that number lived in the past but are now extinct. To classify this past and present diversity of organisms, biologists have devised systems that reflect the evolutionary history of life. The details of the criteria biologists use today to classify organisms are given in Chapter 21, but because some key terms are used in the intervening chapters, we introduce the broad categories of the system here.

In the classification system used in this text, organisms are grouped into six large categories called **kingdoms** (Figure 1.16). Organisms belonging to the kingdoms **Archaebacteria** and **Bacteria** have distinctive cells referred to as **prokaryotic** ("prenuclear") because they lack a nucleus as well as some of the other internal structures found in the cells of members of other kingdoms. Prokaryotes are exceedingly abundant and are found everywhere on Earth where life can exist. Some are critical links in the biogeochemical cycles essential to all life (see Chapter 49).

More than a billion years ago, some prokaryotes invaded the cells of others. Over time, this relationship between hosts and parasites gave rise to organisms with structurally more complex **eukaryotic** cells. Eukaryotes are divided into four kingdoms, the first of which, the kingdom **Protista** (protists), includes many single-celled organisms. The remaining three kingdoms, whose members are all multicellular organisms, are believed to have arisen from ancestral protists. The kingdom **Fungi** includes molds, mushrooms, yeasts, and other similar organisms. Fungi absorb food substances from their surroundings and digest them within their cells. Many are important as decomposers of the dead bodies of other organisms. Most members of the kingdom **Plantae** (plants) convert light energy to the energy of chemical bonds by photosynthesis (see Chapter 8). The biological molecules they synthesize are the primary food for nearly all other living organisms. The kingdom **Animalia** (animals) consists of organisms that digest food outside their cells, and then absorb the products. Animals depend on other forms of life for most of their materials and energy.

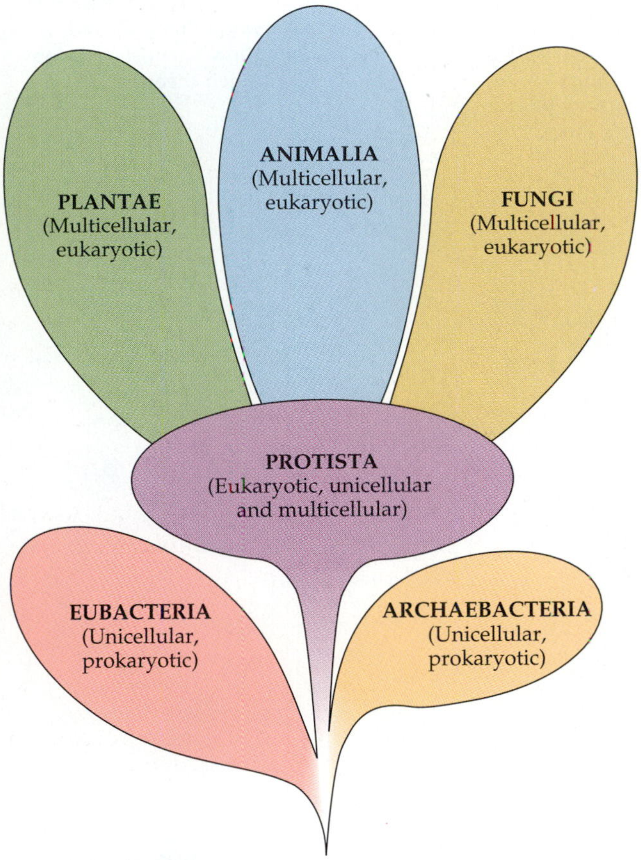

Sometimes organisms are referred to as "primitive" or "advanced." These terms can be useful when we are comparing ancestral and derived traits of organisms in the same evolutionary lineage. But they, and similar terms such as "lower" and "higher," are inappropriate in contexts where they imply that some organisms "work better" than others. The abundance of prokaryotes—all of which are relatively simple—readily demonstrates that they are highly functional, despite their simplicity.

SCIENCE AND RELIGION

Understanding the methods of science enables us to distinguish science from that which is not science. Recently some people have claimed that "creation science," sometimes called "scientific creationism," is a legitimate science that deserves to be taught in schools together with the evolutionary view of the world presented in this book. In spite of these claims, creation science is not science. Science begins with observations and the formulation of testable hypotheses. Creation science begins with the unsubstantiated assertion that Earth is only about 4,000 years old and that all species of organisms were created in approximately their present forms. This assertion is not presented as a hypothesis from which testable predictions are derived. Advocates of creation science do not believe that tests are needed because they assume the assertion to be true.

In this book we present evidence supporting the hypothesis that Earth is several billion years old, that today's living organisms have evolved from single-celled ancestors, and that many organisms dramatically different from those we see today lived on Earth in the remote past. All of this extensive scientific evidence is rejected by proponents of creation science in favor of a religious belief held by a very small minority of the world's human population. Evidence gathered by scientific procedures does not diminish the value of the biblical account of creation. Religious beliefs are not based on falsifiable hypotheses, as science is; they serve different purposes, giving meaning and guidance to human lives. The legitimacy of both religion and science is undermined when a religious belief is called science.

1.16 Six Kingdoms
In the classification system used in this book, Earth's organisms are first divided into these six groups, called kingdoms.

PLANTAE
(Multicellular,
eukaryotic)

ANIMALIA
(Multicellular,
eukaryotic)

FUNGI
(Multicellular,
eukaryotic)

PROTISTA
(Eukaryotic, unicellular
and multicellular)

EUBACTERIA
(Unicellular,
prokaryotic)

ARCHAEBACTERIA
(Unicellular,
prokaryotic)

FOR STUDY

1. Why is it so important in science that we design and perform tests capable of rejecting a hypothesis?

2. Some philosophers and practitioners of science believe that it is impossible to prove any scientific hypothesis —that instead we can only fail to find a cause to reject it. Evaluate this view. Can you think of a reason that we can be more certain about rejecting a hypothesis than we can about accepting it?

3. One hypothesis about the conspicuous coloration of caterpillars was described in this chapter, and some tests were mentioned. Suggest some other plausible hypotheses for conspicuous coloration in these animals. Develop some critical tests for one of these alternatives. What are the appropriate associated null hypotheses?

4. According to the theory of adaptation, an organism evolves certain features because they improve the chances that the organism will survive and reproduce. There is no evidence, however, that evolutionary mechanisms have foresight or that organisms can anticipate future conditions. What, then, do biologists mean when they say, for example, that wings are "for flying"?

5. Consider a single-celled organism. Explain why it is not feasible for this organism to grow to a size of 10 cm in diameter. Cover the following topics: (1) surface-to-volume ratio; (2) transport of nutrients; (3) gas exchange; (4) excretion; and (5) support.

READINGS

Darwin, C. 1859. *The Origin of Species by Means of Natural Selection.* John Murray, London. The book that set the world to thinking about evolution; still well worth reading. Many reprinted versions are available.

Irvine, W. 1955. *Apes, Angels, and Victorians.* McGraw-Hill, New York. A delightful account of the reactions of English society to the theory of evolution by means of natural selection.

Kuhn, T. S. 1970. *The Structure of Scientific Revolutions*, 2nd Edition. University of Chicago Press, Chicago. A widely discussed book that developed a view of science as a succession of paradigms.

Margulis, L. and K. V. Schwartz. 1988. *Five Kingdoms: An Illustrated Guide to the Phyla of Life on Earth*, 2nd Edition. W. H. Freeman, New York. A good introduction to the kingdoms of organisms, in which the two kingdoms of prokaryotes are united into one. Excellent examples and illustrations.

Mayr, E. 1991. *One Long Argument: Charles Darwin and the Genesis of Modern Evolutionary Thought*. Harvard University Press, Cambridge, MA. An excellent, concise account of the history of evolutionary thinking during the past century.

National Academy of Sciences. 1984. *Science and Creationism. A View from the National Academy of Sciences.* National Academy Press, Washington, D.C. A good summary of the evidence demonstrating the great age of Earth and the evolutionary processes that generated the modern Earth.

Young, J. Z. 1951. *Doubt and Certainty in Science.* Clarendon Press, Oxford. A discussion of how scientists develop confidence in their theories.

PART ONE
The Cell

What do you have in common with a rock? Like a rock, you have more oxygen in your molecules than anything else (you are about 65 percent oxygen, the rock about 47 percent). Inside both you and the rock, electrons spinning in orbits bind these oxygen atoms to other atoms. Electrons and other particles in the atoms determine their behavior and thus, ultimately, your behavior and that of the rock.

How do you differ from a rock (chemically, at least)? The rock may be about 28 percent silicon, but you have none, unless you have received a breast implant. Your body is 18 percent carbon—the second most common element in all living things—but the rock may contain no carbon. Probably the biggest difference between you and a rock is the number and speed of the chemical reactions going on inside.

Still, perhaps you have more in common with a rock than you think. What is Earth's matter—animal, vegetable, and mineral—composed of? What holds the units of matter together, what causes them to come apart, and how and why do they exchange parts?

In Part One of this book we begin with atoms (the lowest level of the hierarchical organization of biology) and travel up to the level of the cell. In this chapter we study the small molecules formed by the combination of atoms. Chapter 3 continues with the formation of large molecules from small molecules.

ATOMS

All matter, living and nonliving, is composed of **atoms**. More than a million million atoms could fit in a single layer over the period at the end of this sentence. Each atom consists of a dense, positively charged nucleus, around which one or more negatively charged electrons move. The nucleus contains one or more protons and may contain one or more neutrons. Electrons, protons, and neutrons are not indivisible particles. Each has a substructure, but that level of organization (the world of quarks) has no known consequence for biology.

The weight, or mass, of a proton serves as a standard unit: the atomic mass unit (amu) or dalton (after the English scientist John Dalton, who studied atoms two centuries ago). A single proton or neutron weighs one dalton, which is 1.7×10^{-24} gram (0.0000000000000000000000017 g), whereas the mass of an electron is 9×10^{-28} g (0.0005 dalton). Because they weigh so much less than protons and neutrons, electrons contribute only negligibly to the mass of an atom.

The positive electric charge on a proton is defined as a unit of charge. An electron has a charge equal and opposite to that of a proton. Thus the charge of

Two Accumulations of Matter Held Together by the Same Chemical Forces

2

Small Molecules

a proton is +1 unit, that of an electron is −1. The neutron, as its name suggests, is electrically neutral, so its charge is 0. The number of protons in an atom equals the number of electrons, so the atom itself is electrically neutral.

ELEMENTS

An **element** is a substance made up of only one kind of atom. The element hydrogen consists only of hydrogen atoms; the element iron consists only of iron atoms. An element cannot be broken down into simpler substances. Although there are 92 or more different chemical elements in the world, more than 99 percent of the living matter of all organisms (including humans) is composed of just six elements—carbon, hydrogen, nitrogen, oxygen, phosphorus, and sulfur.

In addition to the 92 elements that occur naturally on Earth, physicists have produced other elements by using cyclotrons and other particle accelerators. Of the natural elements, some, such as silver, gold, and thulium, are extremely rare; others, such as hydrogen, nitrogen, and oxygen, are abundant on this planet. Table 2.1 compares the proportions of some representative elements in the human body, in Earth's crust, and in the universe as a whole. Some elements, such as silicon—a major component of rocks and soils—are abundant in Earth's crust but are not present in significant concentration in living things.

How Elements Differ

An atom is identified by the number of its protons, which is unchanging. This number is called the **atomic number**. An atom of hydrogen contains 1 proton; an atom with 2 protons would be helium; carbon has 6 protons and plutonium has 94. Their atomic numbers are thus 1, 2, 6, and 94, respectively.

Every atom except hydrogen has one or more neu-

TABLE 2.1
Abundance of Some Chemical Elements

ATOMIC NUMBER	SYMBOL	ELEMENT	ATOMIC WEIGHT	UNIVERSE	EARTH'S CRUST	HUMAN BODY	ATOMIC NUMBER	SYMBOL	ELEMENT	ATOMIC WEIGHT	ABUNDANCE, AS % OF EARTH'S CRUST
1	H	Hydrogen	1.01	87	0.14	9.5	28	Ni	Nickel	58.70	0.01
2	He	Helium	4.00	12			29	Cu	Copper	63.55	0.01
3	Li	Lithium	6.94		0.01		30	Zn	Zinc	65.38	0.01
4	Be	Beryllium	9.01				31	Ga	Gallium	69.72	
5	B	Boron	10.81				32	Ge	Germanium	72.59	
6	C	Carbon	12.01	0.03	0.03	18.5	33	As	Arsenic	74.92	
7	N	Nitrogen	14.01	0.01		3.3	34	Se	Selenium	78.96	
8	O	Oxygen	16.00	0.06	46.6	65.0	35	Br	Bromine	79.90	
9	F	Fluorine	19.00		0.03		36	Kr	Krypton	83.80	
10	Ne	Neon	20.18	0.02			37	Rb	Rubidium	85.47	0.03
11	Na	Sodium	22.99		2.83	0.2	38	Sr	Strontium	87.62	0.03
12	Mg	Magnesium	24.31		2.09	0.1	39	Y	Yttrium	88.91	
13	Al	Aluminum	26.98		8.13		40	Zr	Zirconium	91.22	0.02
14	Si	Silicon	28.09		27.7		41	Nb	Niobium	92.91	
15	P	Phosphorus	30.97		0.12	1.0	42	Mo	Molybdenum	95.94	
16	S	Sulfur	32.06		0.05	0.3	43	Tc	Technetium	98.91	
17	Cl	Chlorine	35.45		0.03	0.2	44	Ru	Ruthenium	101.07	
18	A	Argon	39.95				45	Rh	Rhodium	102.91	
19	K	Potassium	39.10		2.59	0.4	46	Pd	Palladium	106.40	
20	Ca	Calcium	40.08		3.63	1.5	47	Ag	Silver	107.87	
21	Sc	Scandium	44.96				48	Cd	Cadmium	112.41	
22	Ti	Titanium	47.90		0.44		49	In	Indium	114.82	
23	V	Vanadium	50.94		0.02		50	Sn	Tin	118.69	
24	Cr	Chromium	52.00		0.02		51	Sb	Antimony	121.75	
25	Mn	Manganese	54.94			0.1	52	Te	Tellurium	127.60	
26	Fe	Iron	55.85		5.0		53	I	Iodine	126.90	
27	Co	Cobalt	58.93				54	Xe	Xenon	131.30	

This list contains only the first 54 of the 92 natural elements. Elements found in the human body are divided into three categories on the basis of abundance: ■, most abundant; ■, 0.1–3.3 percent; □, trace (less than 0.01 percent).

trons in its nucleus. The **mass number** of an atom equals the sum of the number of protons and neutrons in its nucleus. Because the mass of an electron is infinitesimal compared with that of a neutron or proton, electrons are ignored in calculating the mass number. The nucleus of a helium atom contains 2 protons and 2 neutrons; oxygen has 8 protons and 8 neutrons. Helium, therefore, has a mass number of 4 and oxygen a mass number of 16. The mass number may be thought of as the weight of the atom, in daltons.

Each element has its own one- or two-letter symbol. Thus H = hydrogen, He = helium, and O = oxygen. Some symbols come from other languages: Fe (from Latin *ferrum*) = iron, Na (Latin *natrium*) = sodium, and W (German *Wolfram*) = tungsten. Table 2.1 lists the symbols for 54 of the 92 natural elements. Sometimes the atomic and mass numbers of an element are written with the element's symbol, the atomic number at its lower left, and the mass number at its upper left. Thus hydrogen, carbon, and oxygen are written as $_1^1H$, $_6^{12}C$, and $_8^{16}O$.

Isotopes

We have been speaking of hydrogen and oxygen as if each had only one form. To be precise, we should have said "the common form of" oxygen or hydrogen because not all atoms of the same element have the same mass number. The common form of hydrogen is ^1H, but about one out of every 6,500 hydrogen atoms on Earth has a neutron as well as a proton in its nucleus and is thus ^2H, called deuterium. Furthermore, it is possible to create ^3H, tritium, which has *two* neutrons and a proton in its nucleus. Because all three types of hydrogen atoms have just one proton, however, they all have the atomic number 1. Deuterium, tritium, and common hydrogen have virtually identical properties, although ^2H is twice and ^3H three times as heavy as ^1H. Such multiple forms of a single element are called **isotopes** of the element (Figure 2.1). (The prefix *iso-*, encountered in many technical terms, means "same.")

Many elements exist in several isotopic forms in nature. For example, the natural isotopes of carbon are ^{12}C, ^{13}C, and ^{14}C. Unlike the hydrogen isotopes, the isotopes of most elements do not have distinct names but rather are written in the form shown here and are spoken of as "carbon-12," "carbon-13," and "carbon-14," respectively. Most carbon atoms are ^{12}C, but about 1.1 percent are ^{13}C, and a tiny fraction are ^{14}C. An element's **atomic weight** is the average of the mass numbers of a representative sample of atoms of the element, with all isotopes in their normal proportions. For example, the atomic weight of carbon is 12.011.

Some isotopes are radioactive: they spontaneously

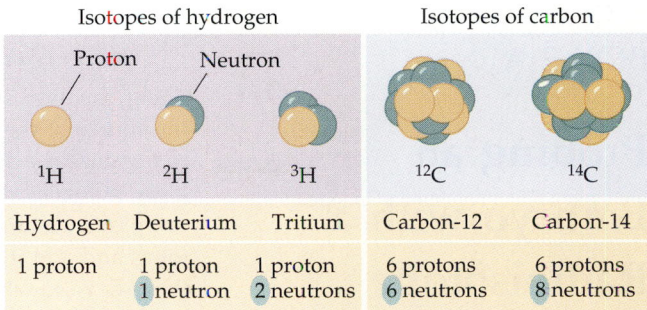

Isotopes of hydrogen			Isotopes of carbon	
Hydrogen	Deuterium	Tritium	Carbon-12	Carbon-14
1 proton	1 proton 1 neutron	1 proton 2 neutrons	6 protons 6 neutrons	6 protons 8 neutrons
^1H	^2H	^3H	^{12}C	^{14}C

2.1 Isotopes Have Different Numbers of Neutrons

give off energy or subatomic particles. Such isotopes are called **radioisotopes**. Radioactive decay transforms the original atom into another type, usually of another element. The radioisotope $_6^{14}C$, for example, is converted to $_7^{14}N$: Carbon becomes nitrogen by the emission of an electron as a neutron becomes a proton. In biological research, radioisotopes are employed as tracers of biochemical reactions. Some commonly used radioisotopes are tritium (^3H), ^{14}C, and ^{32}P (phosphorus-32).

Liquid scintillation counting is the most common method of measuring the amount of radioactivity in a sample. The sample is added to a solution containing a substance that emits light (scintillates) when it absorbs the products of radioactive decay. Like a light meter, the liquid scintillation counter measures the amount of light emitted (Figure 2.2). The amount of light detected corresponds directly to the amount of radioactive decay in the sample. Emitted particles can also be detected by imaging techniques such as **autoradiography**, which is described in Box 2.A.

Rare, nonradioactive isotopes, such as deuterium, do not decay and can be detected only by their dif-

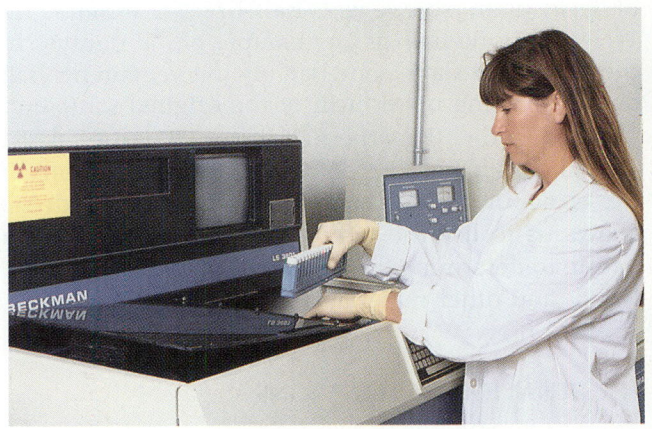

2.2 Measuring Radioactivity by Liquid Scintillation Counting
The machine shown here accepts hundreds of samples, each in its radiation-sensitive "cocktail," and lowers each sample in turn into a darkened well, where its emitted light is measured.

BOX 2.A

Probing an Embryo with a Radioisotope

One of the most exciting and puzzling of biology's phenomena is development—the complex series of progressive changes in an organism from its beginning as a single cell to its adult form. Radioisotopes are useful tools with which to tackle questions about development. They have been extensively used in experiments on the embryos of fruit flies.

An adult fruit fly's body is made up of 13 segments, each with a specific function. How do the different segments form and take on their dis-

tinct functions? We suspect that different chemical substances act during the course of the fly's development to organize the segments. To test this idea, biologists can label a particular chemical substance with radioisotopes and add it to the fly embryo; then they see where in the embryo the substance accumulates. They wash the embryo, press it against X-ray film, and leave the embryro and film in the dark for several days. Wherever a radioactive atom decays, the emitted particle exposes the film. When the film is developed, silver grains—seen in the resulting **autoradiograph** as intense black dots—appear over the parts of the embryo that have the radioactive substance.

The striking bands of silver grains in this autoradiograph of a fly embryo show that the particular substance being tested accumulated in seven segments of the developing fly and not in the others. This knowl-

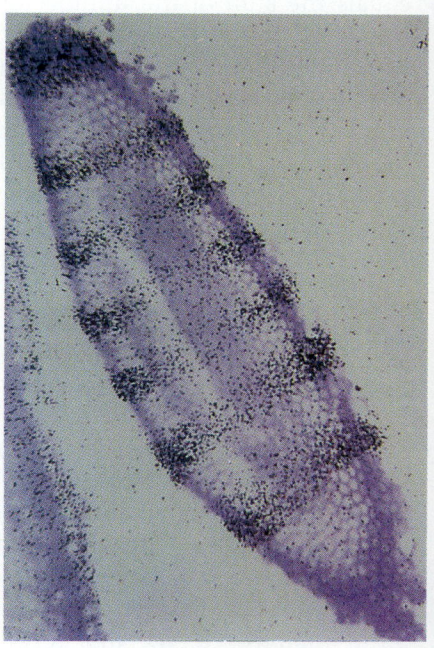

edge, when added to other similar discoveries, advances our still incomplete knowledge of what goes on in the developing insect body.

---

ferent mass. Measuring mass is more difficult than measuring radioactivity and usually requires the use of an expensive instrument called a mass spectrometer. Nevertheless, heavy water, which contains deuterium, is useful in studying biochemical reactions.

The decay of any radioisotope is regular. In successive, equal periods of time, the same fraction of the remaining radioactive material decays. The rate of decay of an isotope is its **half-life**. For example, in 14.3 days, one-half of any sample of ^{32}P decays. In the next 14.3 days, one-half of the remaining half decays, leaving one-fourth of the original sample of ^{32}P, and so on. Thus the half-life of ^{32}P is 14.3 days. Tritium has a half-life of 12.3 years; the half-life of ^{14}C is about 5,700 years. This regularity of decay allows us to use the radioactive isotopes present in nature to determine the ages of ancient bones, rocks, wood, and other materials (see Chapter 18).

THE BEHAVIOR OF ELECTRONS

The part of the atom of greatest biological interest is the electron. What happens in cells happens because of the way electrons behave. The characteristic number of electrons in each atom of an element determines how the atom reacts with other atoms. All **chemical reactions**, in cells or anywhere else, are

exchanges of electrons, or changes in the sharing of electrons between atoms.

Where a given electron in an atom is at any given time is impossible to say. We can only describe a volume of space within the atom where the electron is likely to be. The region of space within which the electron is found at least 90 percent of the time is the electron's **orbital** (Figure 2.3). An electron spins like a top—or like Earth on its axis—and, like a top, may spin in one of two directions: clockwise or counterclockwise. In an atom, a given orbital can be occupied by at most two electrons, which must spin in opposite directions. Thus any atom larger than helium (atomic number = 2) must have electrons in two or more orbitals. As shown in Figure 2.3, the different orbitals have characteristic forms. (Some atoms have more orbitals than are shown in the figure.)

The orbitals constitute a series of **shells** around the nucleus. The innermost shell, called the K shell, consists of only one orbital, an s orbital. The s orbital fills first, and its electrons have the lowest energy. Hydrogen ($_1$H) has one K-shell electron; helium ($_2$He) has two; all other atoms have two K-shell electrons and electrons in other shells as well. The L shell is made up of four orbitals (an s orbital and three p orbitals) and hence can hold up to eight electrons. The M, N, O, P, and Q shells have different numbers of orbitals.

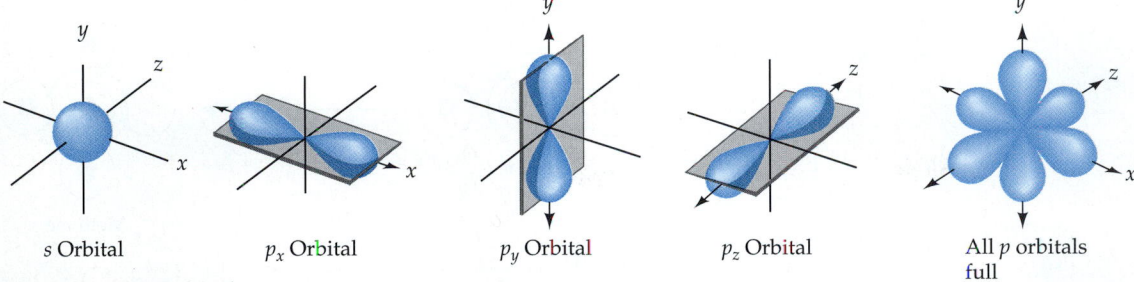

s Orbital   *p_x* Orbital   *p_y* Orbital   *p_z* Orbital   All *p* orbitals full

2.3 Electron Orbitals

Orbitals are the regions around an atom's nucleus where electrons are most likely to be found. The movements of the two electrons closest to the nucleus form a spherical *s* orbital. The next two electrons form a larger, spherical *s* orbital (not shown). The next six electrons fill three dumbbell-shaped *p* orbitals, one pair of electrons per orbital, oriented on the *x*, *y*, and *z* axes through a point in the center of the atom.

In any atom, it is the outermost shell of electrons that determines what the atom can do chemically. When the outer shell is full, the atom will not react with other atoms. Examples of some chemically inert elements (elements with full outer shells) are helium, neon, and argon. Other elements are reactive in various degrees, depending on the number of electrons in the outermost shell. They are, in a sense, seeking ways to fill their outer shells with electrons by combining with other atoms (Figure 2.4).

CHEMICAL BONDS

Two atoms can cooperate to give each atom a full outer shell of electrons. In the process the two atoms become joined. **Molecules** consist of two or more joined atoms.

Covalent Bonds

A hydrogen atom has one electron in its only shell. Picture two hydrogen atoms, initially far apart but coming closer and closer, until they begin to interact. The negatively charged electron of atom A is attracted by the positively charged proton in nucleus B, as well as by its own nucleus; electron B experiences similar attractions. So the two electrons spend time between the two nuclei. The atoms do not get *too* close to-

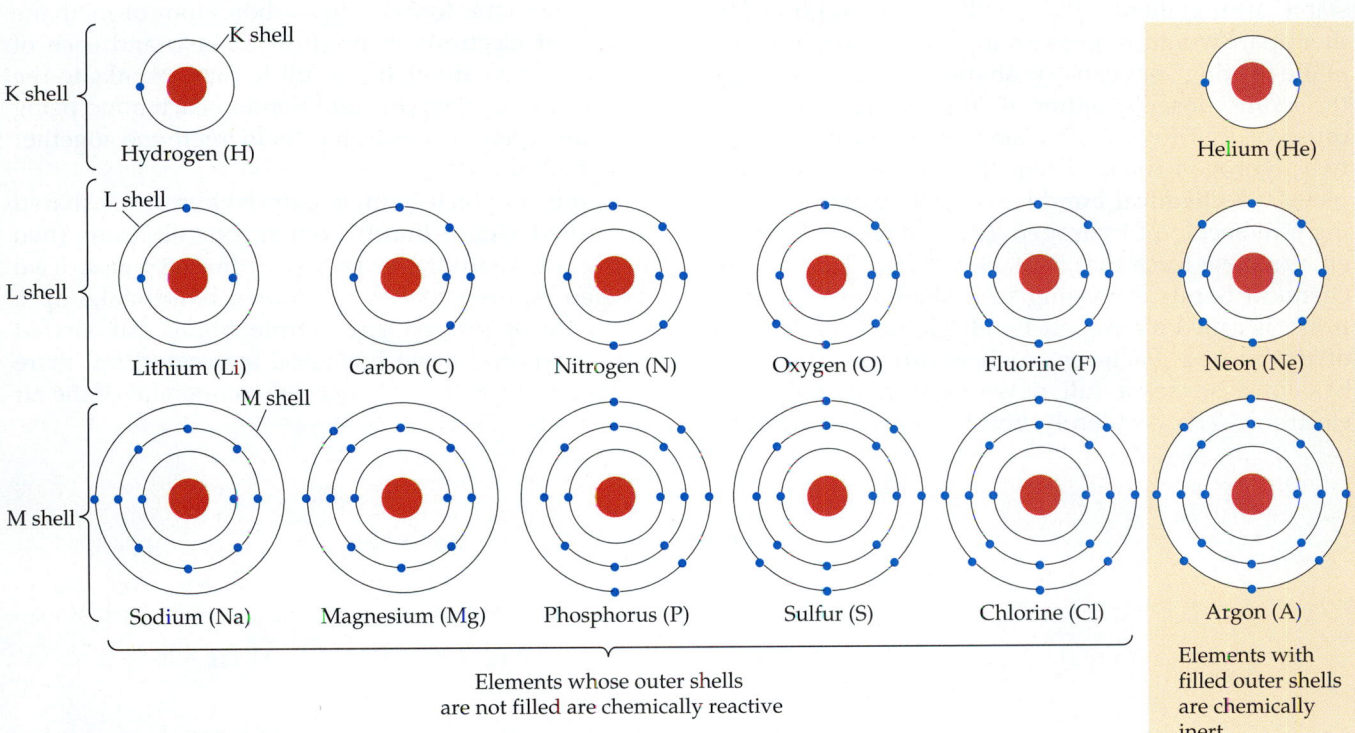

Elements whose outer shells are not filled are chemically reactive

Elements with filled outer shells are chemically inert

2.4 Electron Shells Determine the Reactivity of Atoms

Each shell can hold a specific maximum number of electrons: the K shell holds two; the L and M shells each hold eight. An atom with room for more electrons in its outermost shell may react with other atoms.

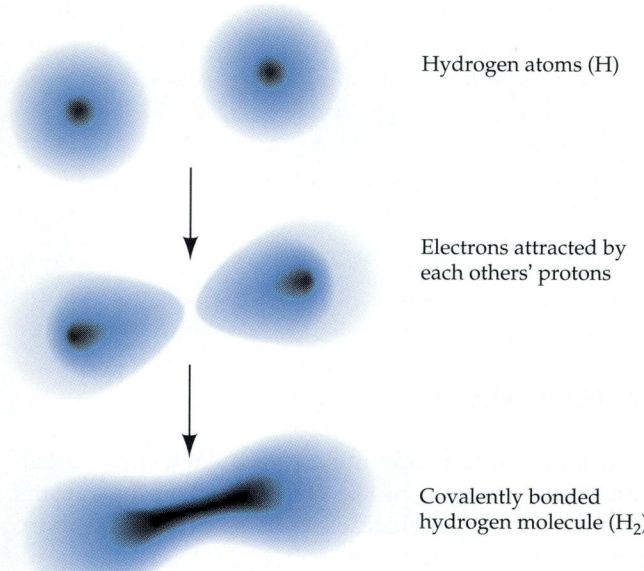

Hydrogen atoms (H)

Electrons attracted by
each others' protons

Covalently bonded
hydrogen molecule (H₂)

2.5 Electrons Are Shared in Covalent Bonds
Two hydrogen atoms combine to form a hydrogen molecule. Each electron is attracted to both protons, but the
two protons cannot come so close together that they repel
each other. A covalent bond forms when the electron orbitals in the K shells of the two atoms overlap.

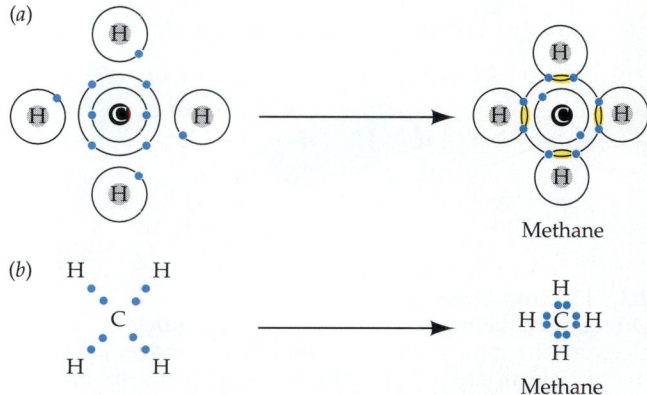

(a)

Methane

(b)

Methane

2.6 Covalent Bonding with Carbon
Carbon can complete its outer shell by sharing the electrons of four hydrogen atoms, forming methane. The
drawings show two methods of representing the formation of the covalent bonds. Note that (b) depicts only the
electrons from the initially unfilled shells (the two electrons in carbon's K shell are not shown).

gether, because their positively charged nuclei would
then repel each other strongly. A certain distance
between the coupled atoms, however, gives them a
minimum amount of energy, and this is the most
stable arrangement. (Pulling the atoms slightly farther apart would require an input of energy because
of the "gluing" effect of the shared electrons; pushing
the atoms closer together would require energy because of the mutual repulsion of the protons.) The
two hydrogen nuclei share the two electrons completely. A **chemical bond** joins the two atoms, forming a molecule of hydrogen gas. A chemical bond is
an attractive force that links two atoms. This type of
chemical bond, consisting of a shared pair of electrons, is called a **covalent bond** (Figure 2.5). Because
of the shared electrons, each hydrogen atom now
has, in a sense, a full outer shell containing two
electrons. The covalently bonded pair of hydrogen

atoms, with completely filled outer shells, is less reactive than the individual atoms, which have incomplete K shells.

A carbon atom has a total of six electrons: two in
its (full) K shell and four in its (outer) L shell. Because
the L shell can hold eight electrons, this atom can
share electrons with up to four other atoms. Thus it
can form four covalent bonds. When an atom of carbon reacts with four hydrogen atoms, a substance
called methane forms. The carbon atom of methane
has eight electrons in its (full) L shell, and each of
the hydrogen atoms has a full K shell, thanks to the
sharing. Thus four covalent bonds—each bond being
a shared pair of electrons—hold methane together
(Figure 2.6).

Bonds in which a single pair of electrons is shared
are called **single bonds**. When four electrons (two
pairs) are shared, the link is a **double bond**. Two
oxygen atoms joined by a double bond make up a
molecule of oxygen gas. **Triple bonds** (six shared
electrons) are rare in biological molecules, but there
is one in nitrogen gas, the chief component of the air
we breathe (Figure 2.7).

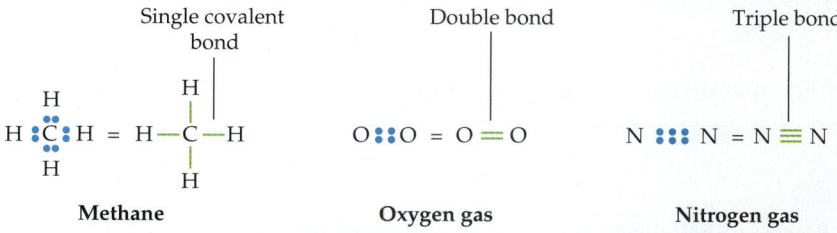

Single covalent bond   Double bond   Triple bond

Methane   Oxygen gas   Nitrogen gas

2.7 Single, Double, and Triple Bonds
Bonds may be shown by dots representing electron pairs (on the left in each
pair) or by solid lines (on the right).

The Covalent Bonds of Different Elements

The atoms of a given element tend to form a specific number of covalent bonds with other atoms. The number of electrons in the outer shell determines this number. For example, carbon tends to form four covalent bonds, nitrogen three, oxygen two, and hydrogen one. The Harvard biologist George Wald once suggested that one reason these four elements are so important to living things is that their atoms are the smallest ones that can fill their outer electron shells by gaining one, two, three, or four electrons. In larger atoms that form covalent bonds with their M-shell electrons, the shared electrons are "screened" from the positively charged nucleus by the K-shell and L-shell electrons. Hence their shared electrons are held less tightly by the nucleus and form less stable covalent bonds. Hydrogen, carbon, oxygen, and nitrogen form the most stable covalent bonds. Wald also pointed out that carbon, oxygen, and nitrogen are biologically significant partly because they are among the few elements that can form double or triple bonds (two others are sulfur and phosphorus, which are also essential to living organisms). Because of double bonding, for example, carbon and oxygen can combine to form carbon dioxide ($O{=}C{=}O$), a water-soluble gas readily taken up by plants.

Ions and Ionic Bonds

When dissolved in water, many substances ionize: Their component atoms break apart, gaining or losing electrons in the process. Because the number of protons and electrons is no longer balanced, these atoms or groups of atoms become electrically charged and are known as **ions**. If an ion carries one or more positive electric charges, it is a cation; if it is negatively charged, it is an anion. Ions play major roles in many biological events.

Hydrochloric acid (HCl) offers a good example of ionization in action. Composed of hydrogen and chlorine atoms, HCl is a gas at room temperature. The hydrogen and the chlorine atoms share a pair of electrons (one from each atom) that form a single covalent bond. When HCl is dissolved in water, the atoms separate, but the chlorine atom keeps *both* the originally shared electrons (Figure 2.8). The chloride ion (Cl^-) thus has one more electron than elemental chlorine (Cl), giving its outer, L shell a full, stable load of eight electrons. The hydrogen ion (H^+) has a single positive charge because it has lost an electron—actually, H^+ is just a lonely proton. It is stable because it has no incomplete electron shells.

Some elements form ions with multiple charges by losing or gaining more than one electron to fill or completely empty out shells. Examples are Ca^{2+} (the calcium ion; a calcium atom has lost two electrons), Mg^{2+} (magnesium ion), and Al^{3+} (aluminum ion). Groups of atoms can also form ions: NH_4^+ (ammonium ion), SO_4^{2-} (sulfate ion), PO_4^{3-} (phosphate ion). Two biologically important elements each yield more than one stable ion: iron yields Fe^{2+} (ferrous ion) and Fe^{3+} (ferric ion), and copper yields Cu^+ (cuprous ion) and Cu^{2+} (cupric ion).

Oppositely charged ions attract one another. If ions are densely concentrated, as when water or another solvent evaporates, crystals form. Solid table salt (sodium chloride, or NaCl) consists of sodium ions (Na^+) and chloride ions (Cl^-) in a highly ordered crystalline array (Figure 2.9). The array is held together by **ionic bonds**, chemical bonds in which the attractive force is the electrical attraction between cations and anions. Like the covalently bonded HCl, NaCl dissolves readily in water. Actually, there is no sharp dividing line between covalent bonds and ionic bonds. In some covalent bonds, the bond partners share the electrons equally. Other pairs of atoms share electrons unequally. An ionic bond is simply a case in which one of the partners has the "shared" electron pair *all* the time. In solution, an ionic bond is less than one-tenth as strong as a covalent bond that shares electrons equally, so the ionic bond can be broken much more readily.

Covalent and ionic bonds are the strongest forces that hold atoms together in molecules. However, other, weaker interactions are also important in mol-

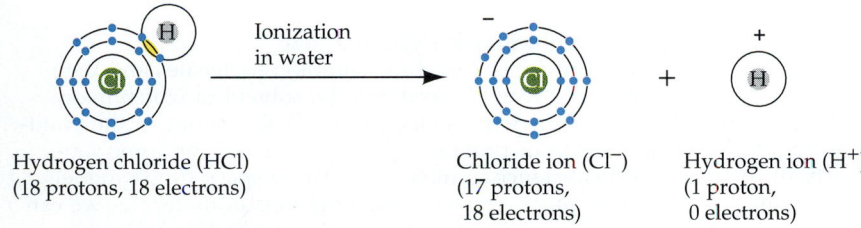

Hydrogen chloride (HCl)
(18 protons, 18 electrons)

Chloride ion (Cl^-)
(17 protons,
18 electrons)

Hydrogen ion (H^+)
(1 proton,
0 electrons)

Ionization
in water

2.8 Ionization of Hydrogen Chloride
When HCl—which has no electric charge—is dissolved in water, the chlorine atom retains *both* the shared electrons from the covalent bond and becomes a chloride ion (Cl^-). This ion is negatively charged because it contains one more electron than it does protons. The proton of the hydrogen atom, no longer balanced electrically by an electron, becomes a positively charged hydrogen ion (H^+).

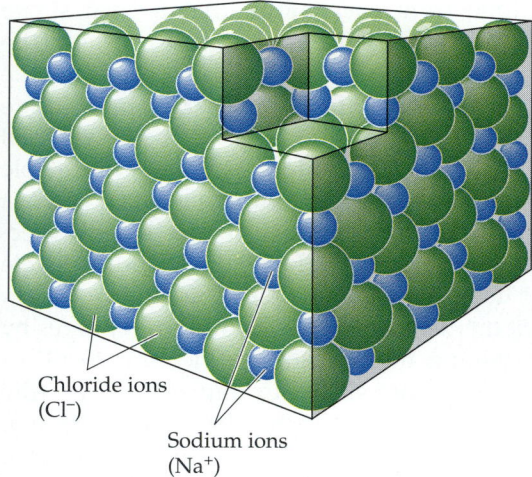

Chloride ions
(Cl⁻)

Sodium ions
(Na⁺)

2.9   Ionic Bonding
A crystal of sodium chloride (NaCl) is held together by ionic bonds between the sodium cations (Na^+) and the chloride anions (Cl^-).

ecules and cellular structures. The most important of these are hydrogen bonds and interactions between nonpolar molecules. Before describing them, we must introduce some chemical principles upon which they depend.

MOLECULES

A molecule consists of two or more atoms linked by chemical bonds. A substance whose molecules contain more than one kind of atom is a **compound**. Most biological substances are compounds. A substance (such as oxygen gas) that contains only one kind of atom is an **elemental substance**.

The **molecular formula** of a compound or an elemental substance shows how many atoms of each element are present in the molecule. This number is written to the lower right of the symbol. For example, the molecular formula for methane is CH_4 (each molecule contains one carbon atom and four hydrogen atoms), that for oxygen gas is O_2, that for nitric oxide is NO (Box 2.B), and that for sucrose (table sugar) is $C_{12}H_{22}O_{11}$. The hormone insulin is represented by the molecular formula $C_{254}H_{377}N_{65}O_{76}S_6$! Molecular formulas do not tell us anything about which atoms are linked to which. **Structural formulas**, as in Figures 2.6 and 2.7, give us this information.

Each compound has a **molecular weight**, which is simply the sum of the atomic weights of the atoms in the molecule. The atomic weights of hydrogen, carbon, and oxygen are, respectively, 1.008, 12.011, and 16.000. Thus the molecular weight of water (H_2O) is $(2 \times 1.008) + 16.000 = 18.016$, or about 18. What is the molecular weight of sucrose ($C_{12}H_{22}O_{11}$)? You can calculate this and find that the answer is

approximately 342. If you remember the molecular weights of a few representative biological compounds, you will be able to picture the relative sizes of molecules that interact with one another (Figure 2.10).

Suppose we want to compare how sodium chloride (NaCl), potassium chloride (KCl), and lithium chloride (LiCl) affect a biological process. You might at first think that we could simply give, say, 2 grams (g) of NaCl to one set of subjects, 2 g of KCl to another, and 2 g of LiCl to the third. But because the molecular weights of NaCl, KCl, and LiCl are different, 2-g samples of each of these substances contain different numbers of molecules. The comparison would thus not be legitimate. Instead, we want to give *equal numbers of molecules* of each substance so that we can compare the activity of one molecule of one substance with that of one molecule of another. But the weight of a single molecule of sodium chloride is 10^{-22} g—hardly a workable quantity.

Since we can neither weigh nor count individual molecules, we work with **moles** (also known as gram molecular weights). *One mole of a substance is an amount whose weight in grams is numerically equal to the molecular weight of the substance.* Potassium chloride (KCl) has a molecular weight of 74.55, so a mole of

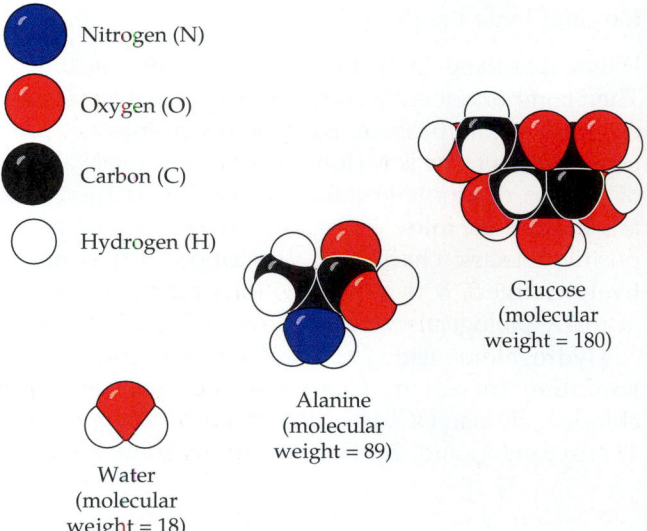

Nitrogen (N)

Oxygen (O)

Carbon (C)

Hydrogen (H)

Glucose
(molecular
weight = 180)

Alanine
(molecular
weight = 89)

Water
(molecular
weight = 18)

2.10   Molecular Weight and Size
The relative sizes of three common molecules and their molecular weights. Water is the solvent in which many biological reactions take place; alanine is one of the building blocks of proteins; glucose is a sugar, an important food substance in most cells. These space-filling models are the most realistic representations of molecules we can create; you will continue to see them in this and subsequent chapters. Space-filling models are particularly valuable in helping us understand how molecules interact. These color conventions for the atoms (along with yellow, which is used for sulfur and phosphorus atoms when they appear) are standard.

BOX 2.B

The Gas That Says NO

Nitrogen can react with oxygen to form any of three nitrogen oxides, depending on the reaction conditions. These oxides are nitric oxide (NO), nitrous oxide (N_2O, "laughing gas"), and nitrogen dioxide (NO_2). Until recently, we thought of NO primarily as a dangerous pollutant of the environment. It is a significant component of smog and of cigarette smoke. It contributes to the destruction of the ozone layer and to the formation of acid rain. We suspect that it can cause cancer.

What we did *not* suspect was that our bodies produce NO naturally and that NO plays a number of key roles in normal body functioning. In spite of its chemical simplicity, we began to recognize the biological functions of NO only in the late 1980s. Hundreds of technical papers on NO appeared in 1992 alone. What are the biological functions of NO? They are exceedingly diverse. NO helps defend the body against invading microorganisms: White blood cells called macrophages produce and release tiny quantities of NO, which destroys cells infected with bacteria. By releasing NO, macrophages also destroy tumor cells, thus perhaps contributing to the body's defenses against cancer.

In a dramatically different role, NO appears to participate in the formation of memories within the nervous system, as we will explain in Chapter 38. NO produced by cells lining blood vessels relaxes muscles in the vessel walls, helping regulate blood pressure. NO performs many other functions, among which is the translation of sexual excitement in male mammals into erection of the penis. In response to sexual stimulation, nerves in the pelvis release NO, which dilates certain blood vessels, allowing blood to rush in and produce the erection.

Natural regulators can wreak havoc if their production goes out of control. This is as true of NO as of other key regulators. Septic shock, a usually deadly condition that affects more than 300,000 people in the United States each year, results in part from a local excess of NO. NO is also implicated in some cases of diabetes and other diseases.

KCl weighs 74.55 g; a mole of NaCl weighs 58.45 g, and a mole of LiCl, 42.40 g. *A mole of one substance contains the same number of molecules as does a mole of any other substance.* This number, known as **Avogadro's number**, is 6.023×10^{23} molecules per mole. The concept of the mole is important for biology because it enables us to work easily with known numbers of molecules.

A solution containing one mole of solute per liter is a **molar** solution, 1 M. A solution containing one-half mole per liter is referred to as 0.5 M, or half-molar. How would you make 100 milliliters (ml) of a 0.02 M sucrose solution? The molecular weight of sucrose is 342, so one liter (1,000 ml) of a 1 M sucrose solution contains one mole, or 342 g, of sucrose. You were asked to make just 100 ml of 0.02 M solution. Since 34.2 g of sucrose would make 100 ml of 1 M sucrose, to make 100 ml of a 0.02 M sucrose solution you would use 0.02×34.2 g = 0.684 g of sucrose.

CHEMICAL REACTIONS

When atoms combine or change bonding partners, a chemical reaction is occurring. Consider the flame of a propane kitchen stove or water heater. When propane (C_3H_8) reacts with oxygen gas (O_2), the carbon atoms become bonded to oxygen atoms instead of to hydrogen atoms, and the hydrogen atoms become bonded to oxygen instead of carbon. This process is shown in Figure 2.11. As the covalently bonded atoms change bonding partners, the composition of the matter changes, and propane and oxygen gas become carbon dioxide and water. Chemical reactions in which covalently bonded atoms change partners are common and important in organisms.

The heat of the stove's flame and its blue light reveal that the reaction of propane and oxygen releases a great deal of energy. Changes in energy usually accompany chemical reactions: Energy may be given off to the environment, as in the reaction of propane with oxygen, or energy may be taken up from the environment (some substances will react only after being heated, for example).

We can measure the energy associated with chemical bonds. Work must be done to break a bond, and that work, or energy, can be expressed in calories. In most chemical reactions in which bond partners change, the total bond energies of the products differ from those of the reactants; these energies can also be expressed in calories. A calorie is the amount of heat energy needed to raise the temperature of 1 g of pure water (which contains no other substance) from 14.5°C to 15.5°C. (The nutritionist's Calorie, which biologists call a kilocalorie, is equal to 1,000 heat-energy calories.) Although defined in terms of heat, the calorie is a measure of any form of energy—mechanical, electric, or chemical.

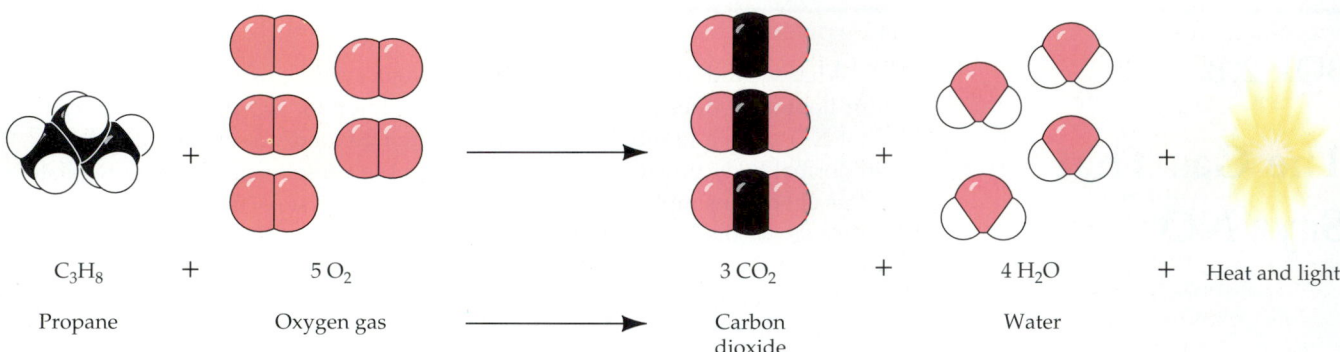

| C₃H₈ | + | 5 O₂ | | 3 CO₂ | + | 4 H₂O | + | Heat and light |

C_3H_8 + $5 O_2$ → $3 CO_2$ + $4 H_2O$ + Heat and light

Propane   Oxygen gas   Carbon   Water
                            dioxide

2.11 Bonding Partners and Energy May Change in a Chemical Reaction
One molecule of propane reacts with five molecules of oxygen gas to give three
molecules of carbon dioxide and four molecules of water. This particular reaction releases energy, in the form of heat and light.

WATER

One of the simplest compounds, yet one of the most
biologically important and chemically interesting, is
water (H_2O). Life originated in water; water covers
three-fourths of present-day Earth, and somewhere
between 45 and 95 percent of any active living organism's weight consists of water. We have all experienced the biological imperative of a raging thirst,
and some organisms must live out their lives in
water. No organism can remain biologically active
without water.

More kinds of substances dissolve in water than
in any other liquid, making it the most effective solvent known. The chemical reactions of interest to
biologists take place in solution, many of them in
watery or aqueous solutions (although many other
reactions occur in solutions in which fatty substances
are the solvent). Water itself takes part in a number
of important reactions.

How is water different from molecularly similar
substances (such as hydrogen sulfide, H_2S, a foul-
smelling gas that is poisonous to humans)? Water
has three different, temperature-dependent, physical
states—solid, liquid, and gas. Its solid state, ice, is
less dense than its liquid form, which is why ice floats
in water (Figure 2.12). If ice sank in water, as almost
all other solids do in their corresponding liquids,
ponds and lakes would freeze from the bottom up,
becoming solid blocks of ice in winter and killing
most of the organisms living in them. Once the whole
pond had frozen, its temperature could drop well
below the freezing point of water. In fact, however,

2.12 Water: Solid and Liquid
Solid water from a glacier floats in its liquid form. The clouds are also water,
but not in its gaseous phase: They are composed of fine drops of liquid water.

ice floats, forming a protective insulating layer at the top of a pond and reducing heat flow to the cold air above. Thus fish, plants, and other organisms in the pond can survive the winter without having to endure subfreezing temperatures. Unless the entire pond freezes, there will be a liquid portion no colder than 0°C, the freezing point of pure water.

As water changes from its liquid to its gaseous state, vapor, it uses heat. This is why sweating is a useful cooling device for humans—your body loses heat as the water in sweat evaporates. It also takes a relatively large amount of heat to raise the temperature of water. The temperature of a given quantity of water is raised only 1°C by an amount of heat that would increase the temperature of the same quantity of ethyl alcohol by 2°C, or of chloroform by 4°C. This important phenomenon contributes to the surprising constancy of the temperature of the oceans and other large bodies of water through the seasons of the year. This constancy is useful to the organisms living in lakes and oceans, for it means that they need not adapt to great variations in temperature. In addition, the relative constancy of water temperature helps to minimize variations in atmospheric temperature throughout the planet.

Water ionizes, but only to a limited extent. About one water molecule in 500 million is ionized at any one time. In a somewhat simplified form, the ionization of water may be represented as

$$H_2O \rightarrow H^+ + OH^-$$

H^+ is, of course, a hydrogen ion; the OH^- is known as a **hydroxide ion**. H^+ and OH^- ions participate in many important chemical reactions.

ACIDS, BASES, AND pH

In pure water, the concentration of hydrogen ions exactly equals that of OH^- ions, and this "solution" is said to be **neutral**. Now suppose we add some HCl (hydrochloric acid). As it dissolves, the HCl ionizes, releasing H^+ ions, so now there are more H^+ than OH^- ions. Such a solution is acidic. A basic, or alkaline, solution is one in which there are more OH^- than H^+ ions. A basic solution can be made from water by adding, for example, sodium hydroxide (NaOH), which ionizes to yield OH^- and Na^+ ions, thus making the concentration of OH^- ions greater than that of H^+ ions.

A compound that can *release* H^+ ions in solution is an **acid**. HCl is an acid, as is H_2SO_4 (sulfuric acid), one molecule of which may ionize to yield two H^+ ions and one SO_4^{2-} ion. Biological compounds such as acetic acid and pyruvic acid, which contain —COOH (the **carboxyl group**; see Figure 2.20) are also acids, because —COOH $\rightarrow$ —COO$^-$ + H^+.

Compounds that can *accept* H^+ ions are called **bases**. These include the bicarbonate ion (HCO_3^-), which can accept an H^+ ion and become carbonic acid (H_2CO_3); ammonia (NH_3), which can accept an H^+ ion and become an ammonium ion (NH_4^+); and many others.

Note that, although —COOH is an acid, —COO$^-$ is a base, because —COO$^-$ + H^+ $\rightarrow$ —COOH. Acids and bases exist as pairs, such as —COOH and —COO$^-$ because any acid becomes a base when it releases a proton, and any base becomes an acid when it gains a proton.

You may have noticed that the two reactions just discussed are the opposites of each other. The reaction that yields —COO$^-$ and H^+ is reversible and may be expressed as —COOH $\rightleftharpoons$ —COO$^-$ + H^+. A **reversible reaction** is one that can proceed in either direction—left to right or right to left—depending on the relative starting concentrations of reacting substances and products. In principle, *all* chemical reactions are reversible. Some consequences of this reversibility will be discussed in Chapter 6.

The terms acid*ic* and bas*ic* refer only to *solutions*. How acidic or basic a solution is depends on the relative concentrations of H^+ and OH^- ions in it. *Acid* and *base* refer to *compounds* and *ions*. A compound or ion that is an acid can donate H^+ ions; one that is a base can accept H^+ ions.

How do we specify how acidic or basic a solution is? First, let's look at the H^+ ion concentrations of a few contrasting solutions. In pure water the H^+ concentration is 10^{-7} M. In 1 M hydrochloric acid the H^+ concentration is 1 M; and in 1 M sodium hydroxide the H^+ concentration is 10^{-14} M. With its values ranging so widely—from more than 1.0 M to less than 10^{-14} M—the H^+ concentration itself is an inconvenient quantity. It is easier to work with the logarithm of the concentration, because logarithms compress this range.

How acidic or basic a solution is is indicated by its **pH** (a term derived from *p*otential of *H*ydrogen). The pH value is defined as the negative logarithm of the hydrogen ion concentration in moles per liter (molar concentration). In chemical notation, molar concentration is often indicated by putting brackets around the symbol for a substance: $[H^+]$ = the molar concentration of H^+. We can now write the equation

$$pH = -\log_{10}[H^+]$$

Since the H^+ concentration of pure water is 10^{-7} M, its pH is $-\log(10^{-7}) = -(-7)$, or 7. A smaller negative logarithm means a larger number; in practical terms, a lower pH means a higher H^+ concentration, or greater acidity. In 1 M HCl, the H^+ concentration is 1 M, so the pH is the negative logarithm of 1 ($-\log 10^0$), or 0. The pH of 1 M NaOH is the negative logarithm of 10^{-14}, or 14. A solution with a pH of

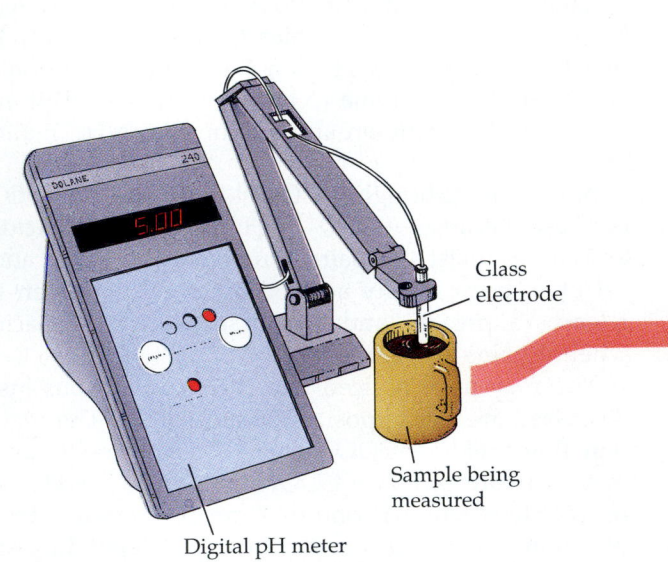

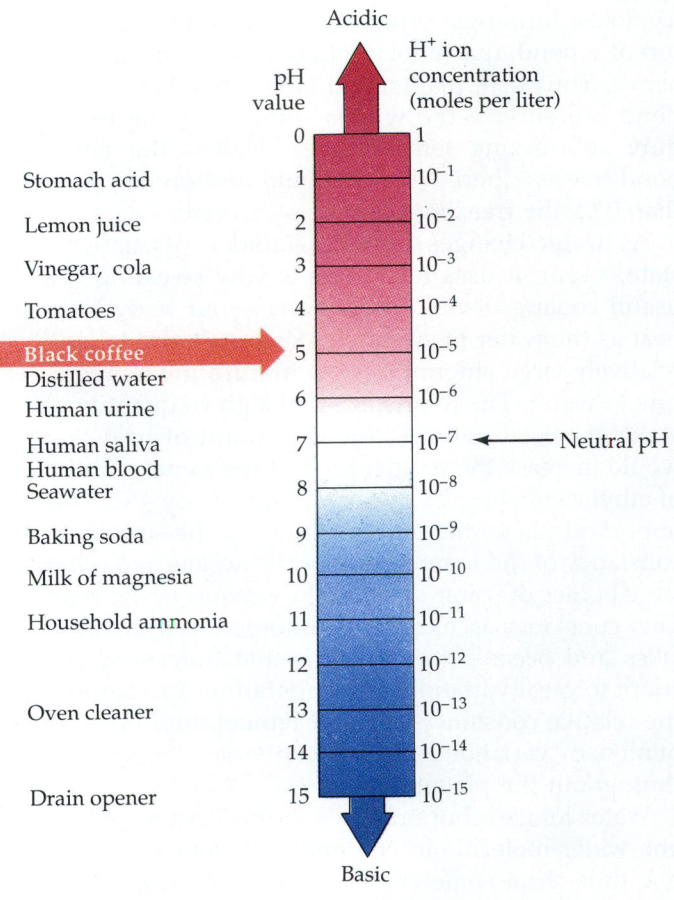

Acidic

pH value | H⁺ ion concentration (moles per liter)

	pH value	H⁺ ion concentration
	0	1
Stomach acid	1	10^{-1}
Lemon juice	2	10^{-2}
Vinegar, cola	3	10^{-3}
Tomatoes	4	10^{-4}
Black coffee	5	10^{-5}
Distilled water		
Human urine	6	10^{-6}
Human saliva	7	10^{-7} ← Neutral pH
Human blood		
Seawater	8	10^{-8}
Baking soda	9	10^{-9}
Milk of magnesia	10	10^{-10}
Household ammonia	11	10^{-11}
	12	10^{-12}
Oven cleaner	13	10^{-13}
	14	10^{-14}
Drain opener	15	10^{-15}

Basic

Glass electrode

Sample being measured

Digital pH meter

2.13 pH Values of Some Familiar Substances
A pH meter such as the one on the left tells us the pH of a solution. This scale reads from low (acidic) pH values at the top to high (basic) pH values at the bottom.

less than 7 is acidic: It contains more H⁺ than OH⁻ ions. A solution with a pH of 7 is neutral, and a solution with a pH value greater than 7 is basic. Because the pH scale is logarithmic, the values are exponential: A solution with a pH of 5 is 10 times more acidic than one with a pH of 6 (it has 10 times as great a concentration of H⁺); a solution with a pH of 4 is 100 times more acidic than one with a pH of 6. The pH values of some common substances are shown in Figure 2.13.

BUFFERS

An organism must control the chemistry of its cells—in particular, the pH of the separate compartments within cells. Animals must also control the pH of their blood. The normal pH of human blood is 7.4, and deviations of even a few tenths of a pH unit can be fatal. The control of pH is made possible in part by **buffers**, which are systems that maintain a relatively constant pH even when substantial amounts of acid or base are added. A buffer is a mixture of an acid that does not ionize completely in water and its corresponding base—for example, carbonic acid (H_2CO_3) and bicarbonate ions (HCO_3^-). If acid is added to this buffer, not all the H⁺ ions from the acid stay in solution. Instead, many of the added H⁺ ions combine with bicarbonate ions to produce more carbonic acid, thus using up some of the H⁺ ions in the solution and decreasing the acidifying effect of the added acid:

$$HCO_3^- + H^+ \rightleftharpoons H_2CO_3$$

If base is added, the reaction reverses. Some of the carbonic acid ionizes to produce bicarbonate ions and more H⁺, which counteracts some of the added base. In this way, the buffer minimizes the effects of added acid or base on pH. A given amount of acid or base causes a smaller change in pH in a buffered solution than in an unbuffered one (Figure 2.14). Buffers illustrate the reversibility of chemical reactions: The addition of acid drives the reaction in one direction, whereas addition of base drives it in the other direction.

POLARITY

In some molecules, called **polar molecules**, the electric charge is not distributed evenly in the covalent bonds. Water is an important polar molecule. In the O—H covalent bonds, the shared electrons are drawn more strongly to the oxygen nucleus, which has eight protons, than to the hydrogen nuclei, which have only one positive charge each. Because of this tendency of the electrons, the hydrogen atoms represent slightly positive regions of the water molecule. In addition, the two hydrogen atoms of water do not lie on directly opposite sides of the oxygen atom; rather, they are separated by an angle of only 104.5

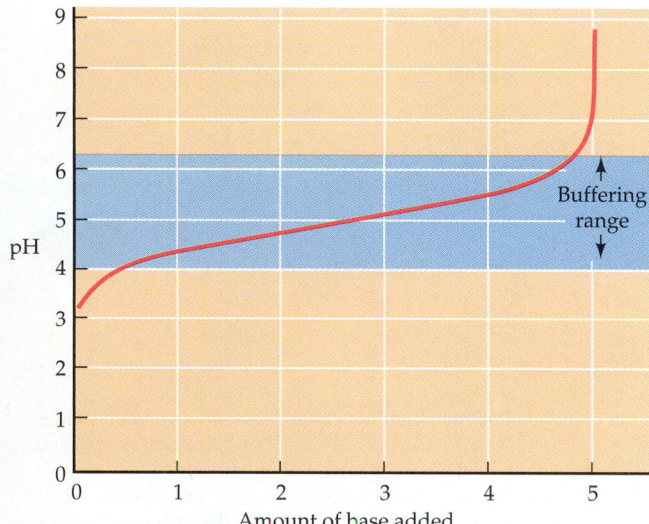

2.14 Buffers Minimize Changes in pH
Adding a base increases the pH of a solution. With increasing amounts of added base, the overall slope of a graph of pH is upward. In the buffering range, however, the slope is shallow. At high and low values of pH, where the buffer is ineffective, the slopes are much steeper.

this type of attraction—that between a slight positive charge on a hydrogen atom and a slight negative charge on a nearby atom—is only about one-tenth (or less) as strong as a covalent bond, it is strong enough to deserve a name: the **hydrogen bond**. It is also strong enough to be very important in biology (Figure 2.17). Hydrogen bonding plays major roles in determining the shapes of the giant molecules (for example, proteins and DNA; see Chapter 3) and in conserving and decoding genetic information (see Chapter 11). Compounds such as sugars dissolve readily in water because hydrogen bonds form between hydroxyl (—OH) groups on the sugars and the oxygen atoms of water. Hydrogen bonding accounts, too, for most of the unusual properties of water mentioned earlier.

Hydrogen bonding between the molecules of water in its liquid state give it a high **surface tension**. Surface tension creates an invisible "skin" that is so strong it permits some insects literally to walk on water (Figure 2.18). Hydrogen bonding causes **capillary action**—the rising of water and watery solu-

degrees. Because the electrons are drawn away from the hydrogen nuclei, the electron cloud is most dense in the opposite region, which therefore has a slightly negative charge. In contrast, the electric charge in **nonpolar molecules** is evenly balanced from one end of the molecule to the other. Ethane is an example of a nonpolar molecule (Figure 2.15).

Much of the excellence of water as a solvent is due to its polarity. Substances such as sodium chloride dissolve easily in water because the Na^+ and Cl^- ions become hydrated—that is, surrounded by water molecules (Figure 2.16). Because the water molecules shield the ions from interacting with one another, the ions are prevented from dropping back out of solution as solid particles of NaCl.

Hydrogen Bonds

Because of their polarity, many water molecules may become loosely attracted to one another. Although

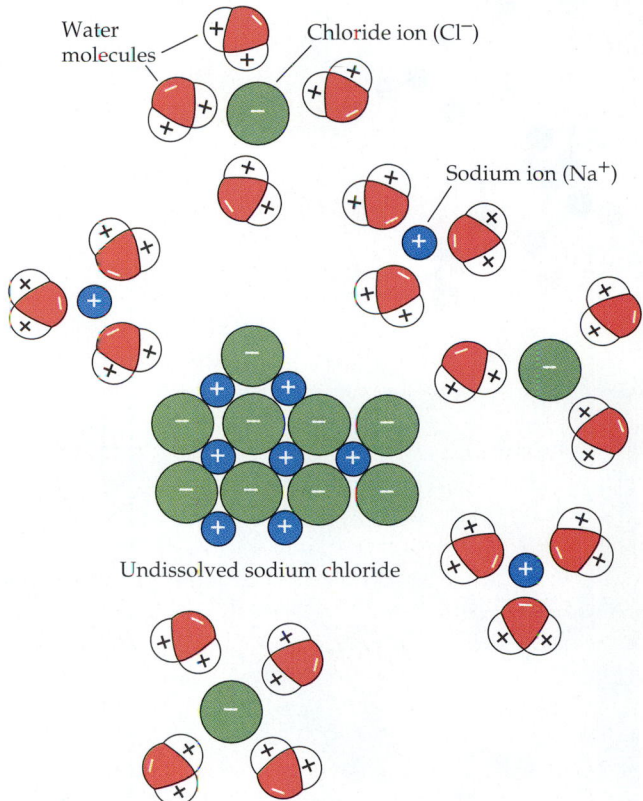

2.16 Water Hydrates Ions
Because water molecules are polar, they cluster around either cations or anions in solutions, blocking the reassociation of the dissolved ions. In this schematic representation of a solution, the negative ends of the water molecules are attracted to the sodium cations, whereas the positive hydrogen atoms in water are attracted to the chloride anions.

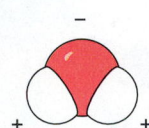 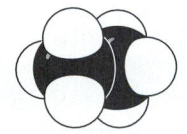

Water, a polar molecule   Ethane, a nonpolar molecule

2.15 Polarity of Molecules
The polar water molecule is slightly more positive near the hydrogen atoms and slightly more negative on the other side. The electrons are distributed evenly over the symmetrical surface of the nonpolar ethane molecule; there is no excess positive charge at any point.

(a)

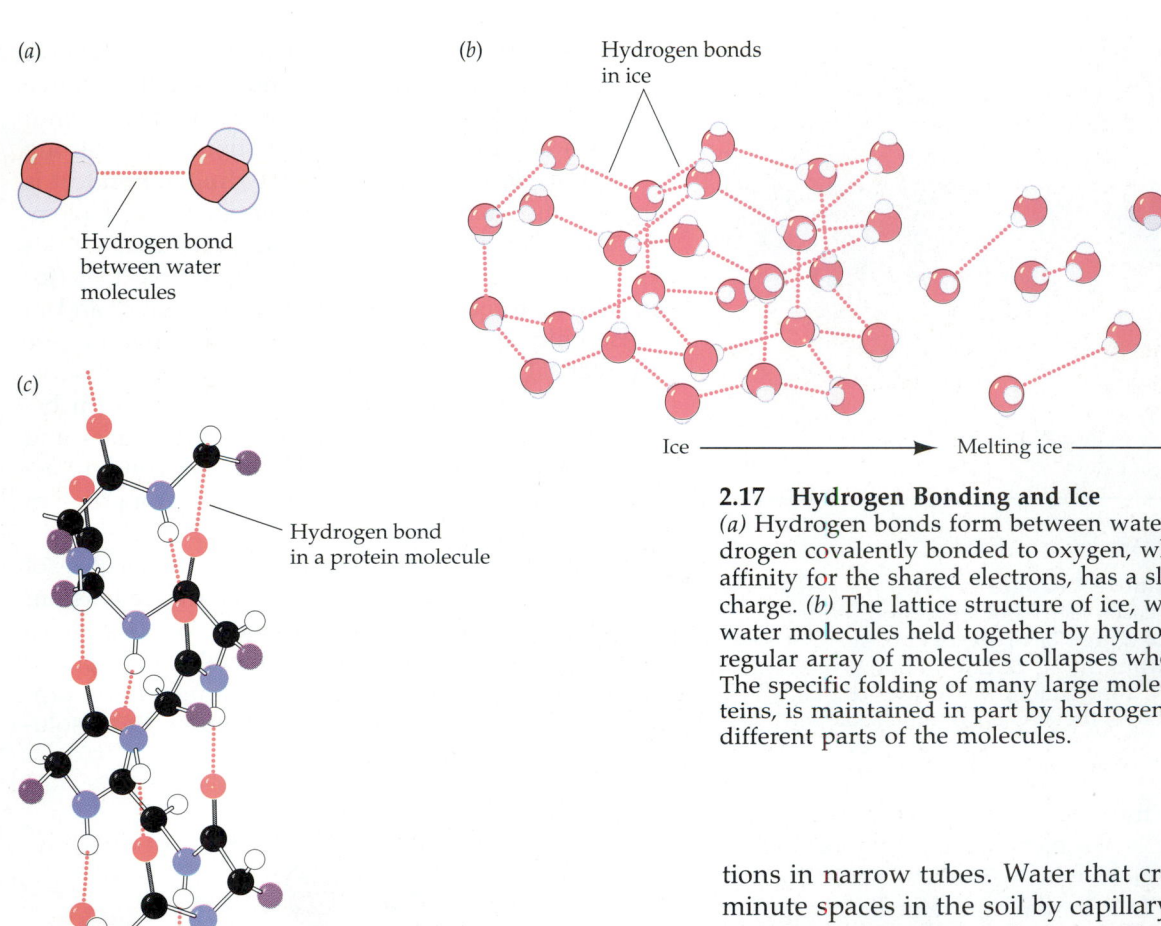

Hydrogen bond
between water
molecules

(b)

Hydrogen bonds
in ice

Ice ──────────→ Melting ice ──────────→ Water

(c)

Hydrogen bond
in a protein molecule

2.17   Hydrogen Bonding and Ice
(a) Hydrogen bonds form between water molecules. Hydrogen covalently bonded to oxygen, which has a greater affinity for the shared electrons, has a slight positive charge. *(b)* The lattice structure of ice, which consists of water molecules held together by hydrogen bonds. This regular array of molecules collapses when ice melts. *(c)* The specific folding of many large molecules, such as proteins, is maintained in part by hydrogen bonds between different parts of the molecules.

tions in narrow tubes. Water that creeps up through minute spaces in the soil by capillary action becomes available to the roots of plants. The ability of water to be pulled up through conducting tissues up to the tops of trees as tall as 100 meters (see Chapter 30) is also a result of the interaction between polar molecules known as hydrogen bonding.

(a)

(b)

2.18   Surface Tension
(a) A water strider "skates" along, supported by the surface tension of the water that is its home. *(b)* Surface tension demonstrated by a soap bubble on a teacup.

TABLE 2.2
Chemical Bonds

TYPE OF BOND	BASIS OF BONDING	ENERGY	BOND LENGTH
Covalent bond	Sharing of electron pairs	50–110 kcal/mol[a]	<0.2 nm
Ionic bond	Attraction of opposite charges	3–7 kcal/mol	0.28 nm (optimal)
Hydrogen bond	Sharing of H atom	3–7 kcal/mol	0.26–0.31 nm (between atoms that share H)
van der Waals interaction	Interaction of electron clouds	~1 kcal/mol	0.24–0.4 nm

[a] kcal/mol = kilocalories per mole; for other abbreviations of units of measurement see the inside front cover.

Interactions between Nonpolar Molecules

Nonpolar molecules also interact to form weak chemical bonds. When uncharged molecules, or parts of molecules, come so close together that their electron clouds touch, the electrons of one molecule are weakly attracted by the nuclei of the atoms in the other molecule. The force of this attraction, which exceeds the repulsive force between the electron clouds, is called a **van der Waals interaction**. Table 2.2 compares these interactions with the other chemical bonds we have discussed. Although only one-fourth to one-third as strong as hydrogen bonds, van der Waals interactions do contribute to the maintenance of the specific structures of large molecules.

Another type of weak attraction, comparable in strength to van der Waals interactions, is the **hydrophobic interaction**, which occurs when nonpolar molecules, or parts of molecules, that come together in the presence of water associate with one another in such a way as to minimize their exposure to the water. For example, molecules of oil in water minimize the area of oil–water contact by aggregating into droplets. This configuration requires the least energy to maintain; work must be done for the area of contact between the oil and the water to increase.

SOME SIMPLE ORGANIC COMPOUNDS AND FUNCTIONAL GROUPS

Organic compounds are made of molecules that contain the element carbon. Organisms produce several classes of organic compounds. The simplest class is the **hydrocarbons**, compounds composed of only hydrogen and carbon atoms. The hydrocarbons include methane (CH_4), ethane (CH_3—CH_3), propane (CH_3—CH_2—CH_3), and ethylene (CH_2=CH_2) (Figure 2.19). Methane, ethane, and propane are called **saturated hydrocarbons** because they contain no carbon–carbon double bonds and are thus saturated with hydrogens. Ethylene, in contrast, is **unsaturated**

because it can add more hydrogen to the carbon atoms connected by the double bond—thus becoming the saturated hydrocarbon ethane:

$$H_2C{=}CH_2 + H_2 \rightarrow H_3C{-}CH_3$$

We may write the formula for ethylene as $CH_2{=}CH_2$, which makes it easy to recognize that two identical parts are covalently bonded together, or, alternatively, as $H_2C{=}CH_2$, which reminds us it is the two carbon atoms that share the double bond.

Gasolines are hydrocarbons with six to ten carbon atoms arranged in a chain; a typical gasoline is octane, with eight carbon atoms. Motor oils have 12 to 20 carbon atoms, and waxy semisolids called paraffin waxes are longer-chain hydrocarbons. Polyethylene plastic is a large hydrocarbon, with chains thousands of carbon atoms long. Animal and plant fats also have long hydrocarbon chains (Chapter 3). Hydrocarbons are flammable, oily, and immiscible (they do not mix) with water. Things that dissolve in hydrocarbons ordinarily do not dissolve in water, and vice versa.

Several classes of biologically important compounds are formed by the replacement of a hydrogen atom on a hydrocarbon with any of several different functional groups (Figure 2.20). When the functional group is a **hydroxyl group** (—OH), for example, the product is an **alcohol**. Perhaps the most familiar alcohol is **ethanol** (ethyl alcohol). Small alcohols like ethanol are soluble in water, but larger alcohols are not soluble in water because of their long hydrocarbon chains.

Sugars contain both hydroxyl and carbonyl groups. The carbonyl group has a central carbon atom with a double bond to an oxygen atom. If one of the other two bonds on the carbon atom in a carbonyl group is to a hydrogen atom, the compound is an **aldehyde**; otherwise it is a **ketone**.

Molecules containing one or more carboxyl groups (—COOH) are acids because of the tendency of the carboxyl group to ionize. **Amines**, on the other hand, are organic bases. These compounds possess an **amino group** (—NH_2), which has a tendency to react

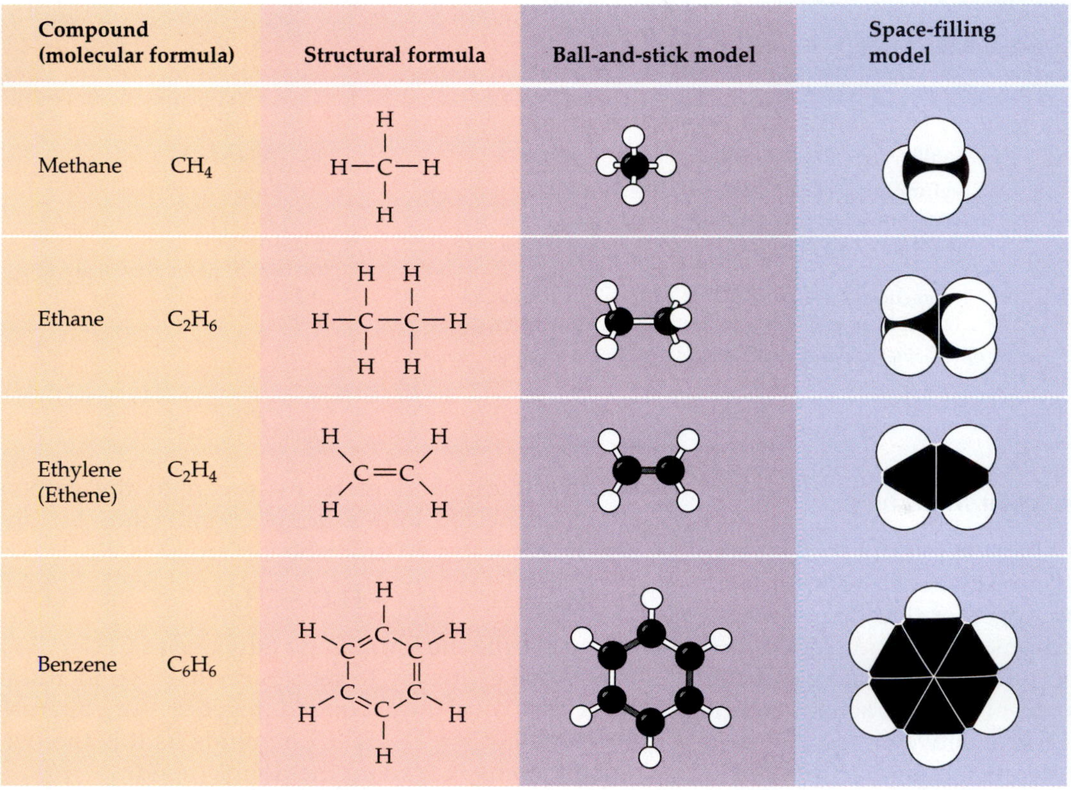

Compound (molecular formula)	Structural formula	Ball-and-stick model	Space-filling model
Methane CH$_4$			
Ethane C$_2$H$_6$			
Ethylene (Ethene) C$_2$H$_4$			
Benzene C$_6$H$_6$			

2.19 Some Small Hydrocarbons
Compare the structures of these hydrocarbons, noting which are saturated and which unsaturated. The molecules in the figure are represented in four different ways. The representations in the two right-hand columns emphasize the three-dimensional structure of the molecules; ball-and-stick models focus on bond angles, and space-filling models focus on the molecule's overall shape.

with H$^+$ to give the positively charged —NH$_3^+$ group. This H$^+$-accepting characteristic accounts for the classification of amines as bases.

Amino acids are important compounds that possess a carboxyl group *and* an amino group, both of which are attached to the same carbon atom, the α (alpha) carbon. Also attached to the α carbon atom are a hydrogen atom and a side chain (Figure 2.21). Twenty different amino acids constitute the building blocks of the giant protein molecules of living things. Each amino acid has a different side chain that gives it its distinctive chemical properties (Chapter 3). Because they possess both carboxyl and amino groups, amino acids are simultaneously acids and bases. At the pH values commonly found in cells, both the carboxyl and the amino groups are ionized: The carboxyl group has lost a proton, and the amino group has gained one.

Two other functional groups should be introduced here. The **sulfhydryl group** (—SH; Figure 2.20) is important in protein structure (Chapter 3) and in biochemical reactions (Chapter 7). The **phosphate group** (—OPO$_3^{2-}$) participates in many crucial reactions in which energy is transferred (Chapters 7, 8, 11, and 40). Phosphate groups are exchanged between sugars and many other compounds.

Isomers are compounds with the same chemical formula but different arrangements of the atoms. Whenever a carbon atom has four *different* atoms or groups attached to it, there are two different ways of making the attachments, each the mirror image of the other. Such a carbon atom is an asymmetric carbon, and the pair of compounds are optical isomers of each other (Figure 2.22). Your right and left hands are optical isomers. Just as a glove is specific for a particular hand, so some biochemical molecules can interact with a specific optical isomer of a compound but are unable to "fit" the other. The α carbon in an amino acid is an asymmetric carbon; hence, amino acids exist in two isomeric forms, called D- and L-amino acids (Figure 2.23). D- and L- are abbreviations for *dextro-* and *levorotatory*, referring to the directions (right or left, respectively) in which solutions of these compounds rotate the plane of polarized light; they

Functional group	Class of compounds	Structural formula	Example	Ball-and-stick model
Hydroxyl —OH	Alcohols	R—OH	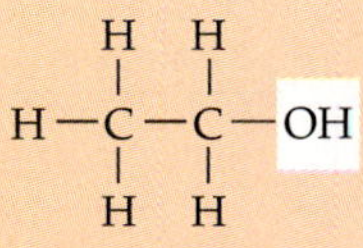 Ethanol	
Carbonyl —CHO	Aldehydes	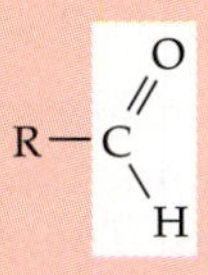 R—C(=O)H	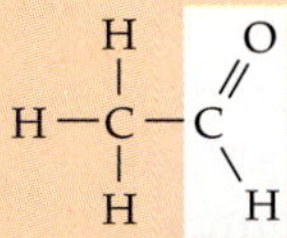 Acetaldehyde	
Carbonyl \CO/	Ketones	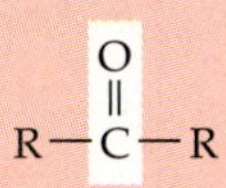 R—C(=O)—R	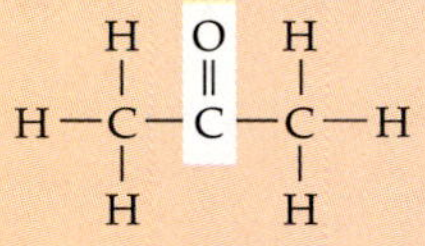 Acetone	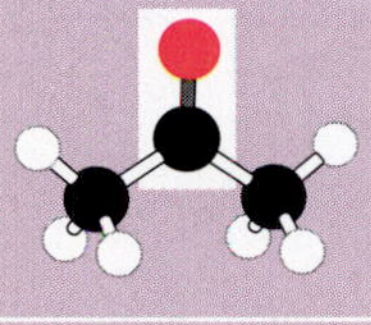
Carboxyl —COOH	Carboxylic acids	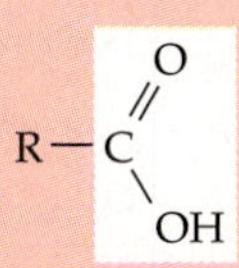 R—C(=O)OH	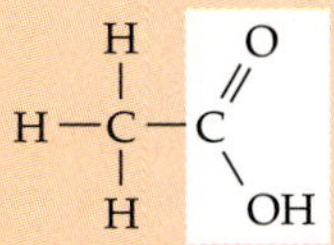 Acetic acid	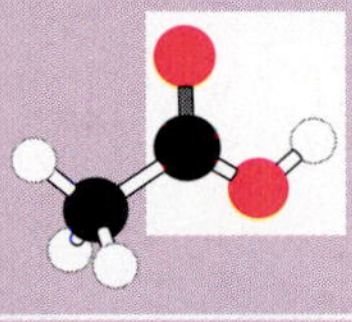
Amino —NH₂	Amines	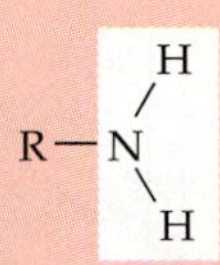 R—NH₂	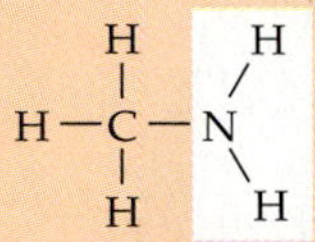 Methylamine	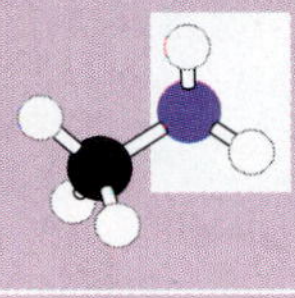
Phosphate —OPO₃²⁻	Organic phosphates	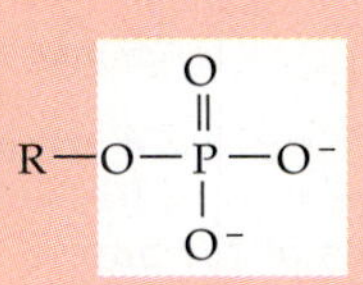 R—O—P(=O)(O⁻)—O⁻	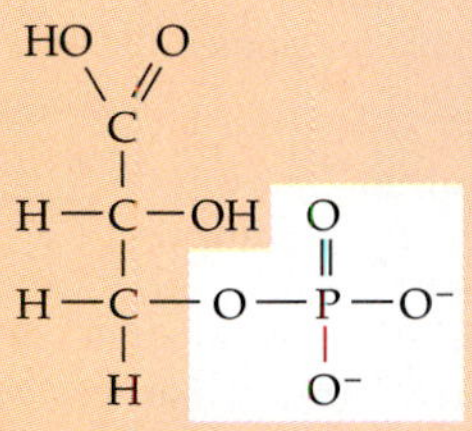 3-Phosphoglyceric acid	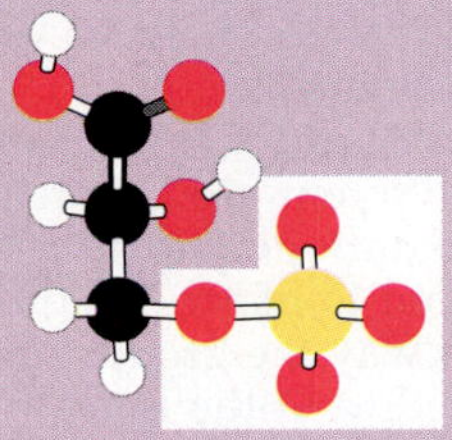
Sulfhydryl —SH	Thiols	R—SH	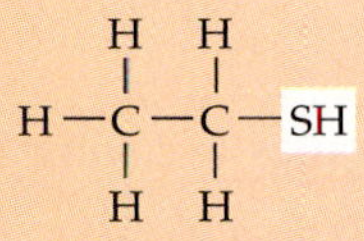 Mercaptoethanol	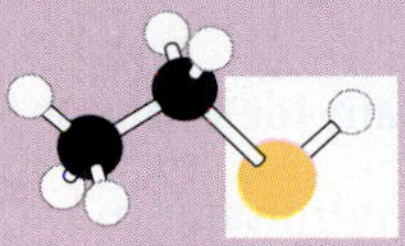

2.20 Simple Organic Compounds and Functional Groups
Compounds of the types shown here will appear throughout this book. The functional groups (highlighted) are the most common ones found in biologically important molecules. The term "R" represents the remainder of the molecule; it may be any of a large number of carbon skeletons or other chemical groupings.

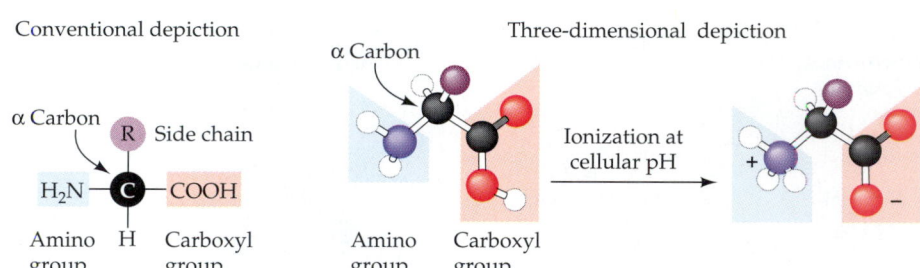

Conventional depiction   Three-dimensional depiction

2.21   Amino Acid Structure
Two depictions of the general structure of an amino acid. The side chain attached to the α carbon differs from one amino acid to another. At pH values found in living cells, both the carboxyl group and the amino group of an amino acid are ionized, as shown in the right-hand model.

refer to the "handedness" of a molecule with one or more asymmetric carbons. Only L-amino acids are commonly found in most proteins of living things.

The compounds discussed in this chapter include some of the more common ones found in organisms.

Between these small molecules and the world of the living stands another level, that of the giant macromolecules. These huge molecules—the proteins, lipids, carbohydrates, and nucleic acids—are the subject of the next chapter.

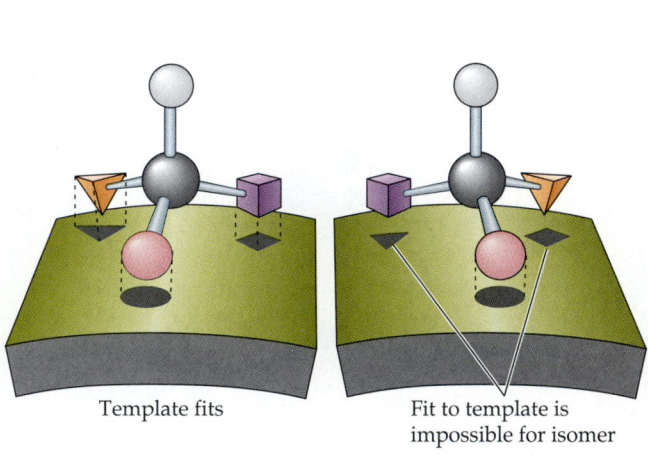

2.22   Optical Isomers
Optical isomers are mirror images of each other. They result when four different groups are attached to a single carbon atom (the dark gray sphere in the center). If a template is laid out to match the groups on one carbon atom, there is no way the groups on the mirror-image isomer can be rotated to fit the same template.

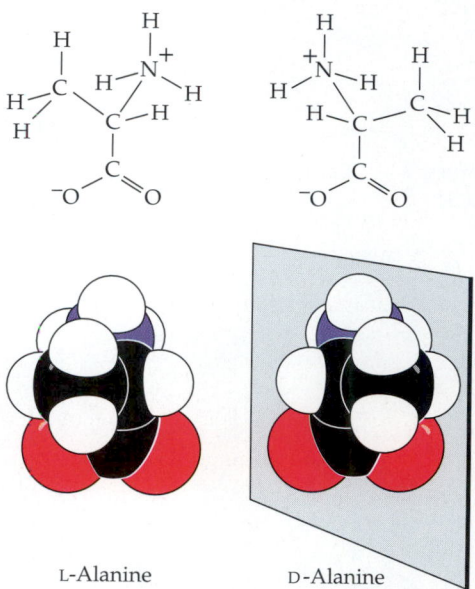

L-Alanine   D-Alanine

2.23   Isomers of Alanine
Structural formulas and space-filling models of the D- and L- forms of the amino acid alanine. Only L-alanine (on the left) is commonly found in living things.

SUMMARY of Main Ideas about Small Molecules in Living Things

Organisms, like nonliving matter, are composed of atoms bonded together as molecules.

Living things consist primarily of atoms of carbon, hydrogen, oxygen, nitrogen, sulfur, and phosphorus.

The numbers of electrons and protons they possess make each element unique.
   Review Figures 2.1, 2.3, and 2.4

A chemical reaction forms bonds between atoms, breaks such bonds, or does both.

The strongest chemical bonds are covalent bonds, in which two atoms share electrons.
   Review Figure 2.5

Carbon atoms form four covalent bonds, nitrogen three, oxygen two, and hydrogen one.
   Review Figures 2.6 and 2.7

Most biological compounds form by covalent bonding.

In a type of chemical reaction important in organisms, covalently bonded atoms change bonding

partners and the bond energies of the products differ from the bond energies of the reactants.
Review Figure 2.11

The molecules of many substances break apart in water into ions, which carry positive or negative electric charges.
Review Figures 2.8 and 2.16

Ions of opposite charge attract each other and may form ionic bonds.
Review Figure 2.9

Weaker bonds (hydrogen bonds, van der Waals interactions, and hydrophobic interactions) also help form large molecules and help molecules aggregate into larger structures.
Review Figure 2.17

Most of the chemistry of interest to biologists takes place in water.

The polarity of water makes it an exceptionally effective solvent.
Review Figures 2.15 and 2.16

Acids release hydrogen ions (protons), bases accept them.

The pH indicates how acidic or basic a solution is; buffers resist pH changes.
Review Figures 2.13 and 2.14

Various functional groups, such as hydroxyl, carbonyl, carboxyl, amino, sulfhydryl, and phosphate groups, give molecules some of their chemical properties.
Review Figure 2.20

SELF-QUIZ

1. The atomic number of an element
   a. equals the number of neutrons in an atom.
   b. equals the number of protons in an atom.
   c. equals the number of protons minus the number of electrons.
   d. equals the number of neutrons plus the number of protons.
   e. depends on the isotope.

2. The atomic weight of an element
   a. equals the number of neutrons in an atom.
   b. equals the number of protons in an atom.
   c. equals the number of electrons in an atom.
   d. equals the number of neutrons plus the number of protons.
   e. depends on the relative abundances of its isotopes.

3. Which of the following statements about all the isotopes of an element is *not* true?
   a. They have the same atomic number.
   b. They have the same number of protons.
   c. They have the same number of neutrons.
   d. They have the same number of electrons.
   e. They have identical chemical properties.

4. Which of the following statements about a covalent bond is *not* true?
   a. It is stronger than a hydrogen bond.
   b. One can form between atoms of the same element.
   c. Only a single covalent bond can form between two atoms.
   d. It results from the sharing of two electrons by two atoms.
   e. One can form between atoms of different elements.

5. Hydrophobic interactions
   a. Are stronger than hydrogen bonds.
   b. Are stronger than covalent bonds.
   c. Can hold two ions together.
   d. Can hold two nonpolar molecules together.
   e. Are responsible for the surface tension of water.

6. Which of the following statements about water is *not* true?
   a. It releases a large amount of heat when changing from liquid into vapor.
   b. Its solid form is less dense than its liquid form.
   c. It is the most effective solvent known.
   d. It is typically the most abundant substance in an active organism.
   e. It takes part in some important chemical reactions.

7. A solution with a pH of 9
   a. is acidic.
   b. is more basic than a solution with a pH of 10.
   c. has 10 times the hydrogen ion concentration of a solution with pH 10.
   d. has a hydrogen ion concentration of 9 molar.
   e. has a hydroxide ion concentration of 9 molar.

8. Which of the following compounds is an alcohol?
   a. O_2
   b. $CH_3CH_2CH_2OH$
   c. CH_3COOH
   d. C_3H_8
   e. CH_3COCH_3

9. Which of the following statements about the carboxyl group is *not* true?
   a. It has the chemical formula —COOH.
   b. It is an acidic group.
   c. It can ionize.
   d. It is found in amino acids.
   e. It has an atomic weight of 45.

10. Which of the following statements about amino acids is *not* true?
    a. They are the building blocks of proteins.
    b. They contain carboxyl groups.
    c. They contain amino groups.
    d. They do not ionize.
    e. They have both L- and D-isomers.

FOR STUDY

1. The elemental compositions of the universe, Earth's crust, and the human body differ sharply (Table 2.1). What factors might contribute to these differences?

2. Lithium (Li) is the element with atomic number 3. Draw the structures of the Li atom and of the Li^+ ion.

3. Draw the structure of a pair of water molecules held together by a hydrogen bond. Your drawing should indicate the covalent bonds.

4. The molecular weight of sodium chloride (NaCl) is 58.45. How many grams of NaCl are there in 1 l of a 0.1 M NaCl solution? How many in 0.5 l of a 0.5 M NaCl solution?

5. The side chain of the amino acid alanine is —CH_3 (see Figure 2.21). Draw the structures of the two optical isomers of alanine. The side chain of the amino acid glycine is simply a hydrogen atom (—H). Are there two optical isomers of glycine? Explain.

READINGS

Atkins, P. W. and J. A. Beran. 1992. *General Chemistry*, 2nd Edition. W. H. Freeman, New York. A first-rate textbook, beautifully illustrated.

Breed, A., T. Rodella and R. Basmajian. 1982. *Through the Molecular Maze*. William Kaufmann, Los Altos, CA. A short, inexpensive guide to the rudiments of chemical concepts and terminology needed by students in introductory courses on the life sciences.

Henderson, L. J. 1958. *The Fitness of the Environment*. Beacon Press, Boston. An essay written in 1912 about physical properties of water and carbon dioxide in relation to life. With a thought-provoking introduction.

Kotz, J. C. and K. F. Purcell. 1991. *Chemistry and Chemical Reactivity*, 2nd Edition. Saunders, Philadelphia. A well-illustrated modern textbook of general chemistry.

Lancaster, J. R. 1992. "Nitric Oxide in Cells." *American Scientist*, vol. 80, pages 248–259. A readable account of the multiple biological effects of this very small molecule.

Mertz, W. 1981. "The Essential Trace Elements." *Science*, vol. 213, pages 1332–1338. A review of the roles of more than a dozen elements needed in small amounts by animals if they are to function normally.

Zumdahl, S. 1992. *Chemical Principles*. D. C. Heath, Lexington, MA. A fine higher-level text for students with strong math backgrounds.

When we eat proteins, we are taking in molecules that were built up within the body of a plant or animal from smaller building blocks. In Chapter 2 we considered the structures of the smallest building blocks of organic molecules. In this chapter we see what organisms do with organic building blocks.

In consuming proteins, we first break them up into their constituent pieces and later reassemble the pieces into the chemically different proteins of our own bodies. This is not unlike picking up Lego toys that somebody else has made, taking them apart, and reassembling the parts to make the toys *we* want. One large molecule that our bodies must mass-produce (in kilogram amounts) is the hemoglobin that carries oxygen from lungs to working tissues via the bloodstream. Meat is mostly muscle, and muscle is mostly protein. We take apart the proteins in meat or vegetables and reassemble the pieces into our hemoglobin and other proteins.

In this chapter we take a brief look at the major classes of large molecules: lipids, carbohydrates, proteins, and nucleic acids. We begin to see that molecular structure (the way one piece fits with another) governs the way particular molecules function in the activities of living things.

LIPIDS

Our first class of large molecules is best understood by thinking about two behaviors that define these molecules: **Lipids** are insoluble in water but are readily soluble in organic (carbon-based) solvents such as ether, and they release large amounts of energy when they break down. Each of these properties is significant in the biology of these compounds. Because lipids do not dissolve in water and water does not dissolve in lipids, a mixture of water and lipids forms two distinct layers. Also, many biological materials that are soluble in water are much less soluble in lipids. Such materials include ions, sugars, and amino acids.

Suppose that you must design water-filled compartments, separated from each other and from their environment by barriers that limit the passage of materials. Based on the properties of lipids, a seemingly effective way to accomplish this is to use membranes containing lipids (Figure 3.1). This is, in fact, the system that has evolved in nature. Molecular traffic within an organism or into and out of its compartments is constrained by the properties of the lipid portion of the surrounding membrane. Compounds that dissolve readily in lipids can move rapidly through biological membranes, but compounds that are insoluble in lipids are prevented from passing, or

Different Plans for the Same Building Blocks

3

Large Molecules

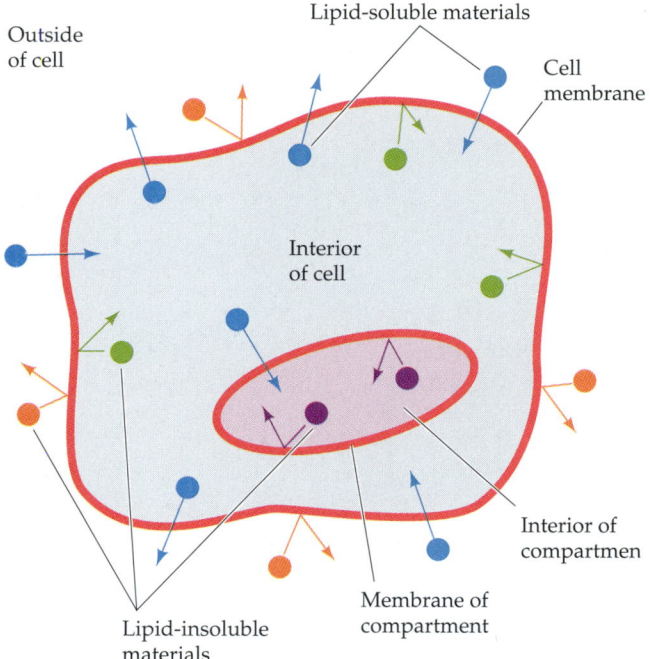

Outside of cell

Lipid-soluble materials

Cell membrane

Interior of cell

Interior of compartmen

Membrane of compartment

Lipid-insoluble materials

3.1 Lipids Assemble as Membranes Enclose Cells
Membranes made of lipids separate the cell from its environment; they also separate the contents of internal compartments from the rest of the cell. Materials that do not dissolve in lipids usually cannot pass through membranes, but lipid-soluble materials move through with relative ease.

must be transported across the membrane by specific proteins (see Chapter 5).

Lipids are marvelous storehouses for energy. By taking in excess food, many animal species deposit fat (lipid) droplets in their cells as a means for storing energy (Figure 3.2). Some plants, such as olives, avocados, sesame, castor beans, and all nuts, have substantial amounts of lipids in their seeds or fruits that serve as energy reserves for the next generation.

Triglycerides

One important group of lipids is the **triglycerides**, also known as *simple lipids*. Triglycerides that are solid at room temperature (20°C) are called **fats**; those that are liquid at room temperature are called **oils**. Triglycerides are composed of two types of building blocks: **fatty acids** and **glycerol**. Glycerol is a small alcohol with three hydroxyl (—OH) groups. Fatty acids are carboxylic acids with long hydrocarbon tails. Four typical fatty acids are shown in Figure 3.3. Palmitic acid is found in animal fats. Like palmitic acid, stearic acid is a **saturated** fatty acid because its hydrocarbon tail contains no double bonds. Oleic acid is **unsaturated**. Its double bond, which is near the middle of the hydrocarbon chain, causes a kink in the molecule. Fatty acids, such as linoleic acid, that

have more than one double bond are **polyunsaturated**. These molecules have multiple kinks. Unsaturated and polyunsaturated fatty acids can accept hydrogen atoms—that is, they can become hydrogenated. The addition of two hydrogen atoms across the double bond of oleic acid, for example, would produce stearic acid.

Three fatty acid molecules combine with a molecule of glycerol to form a molecule of a triglyceride (Figure 3.4). The three fatty acids in one triglyceride molecule are not always the same length, nor do they all have to be either saturated or unsaturated. The kinks associated with double bonds are important in determining the fluidity and melting point of a lipid. Triglycerides with short or unsaturated chains are usually oily liquids; those with long and saturated chains are waxy solids. Animal fats such as lard and tallow are usually solids with long-chain saturated or singly unsaturated fatty acids. In these fats, hydrocarbon chain lengths range from 10 to 20 carbon

3.2 Energy to Fight the Weather
These Alaskan walrus spend much of their time in frigid water. Lipids, deposited as layers of body fat, insulate their bodies against the cold and also store energy efficiently.

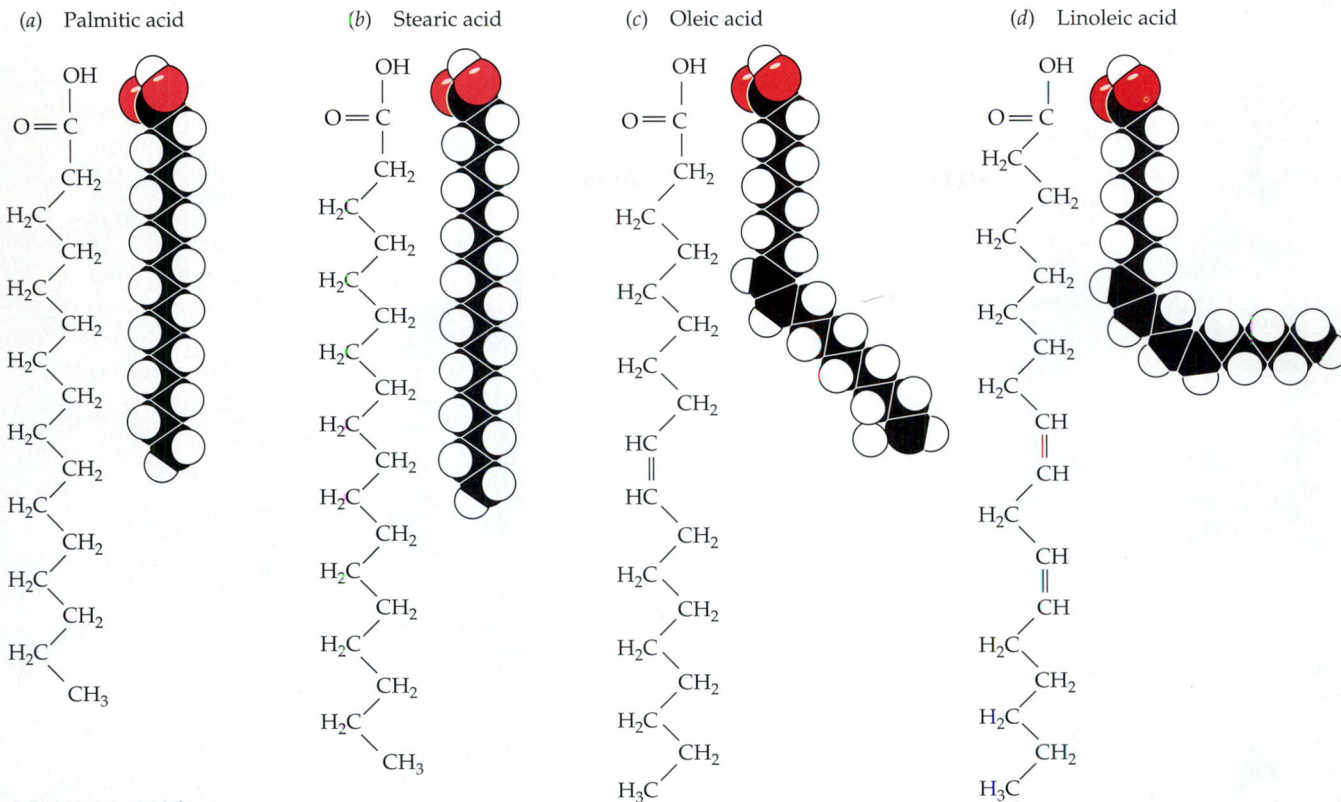

(a) Palmitic acid   *(b)* Stearic acid   *(c)* Oleic acid   *(d)* Linoleic acid

3.3 Fatty Acids

(a) The absence of double bonds between carbon atoms in the chain means that palmitic acid is a saturated fatty acid. The straight-chain configuration seen in the model of the molecule is characteristic of saturated fatty acids. *(b)* Stearic acid has two more carbons and four more hydrogens than palmitic acid and is also saturated. *(c)* Oleic acid has a double bond between two carbons in the chain and is therefore unsaturated. The double bond causes the molecule to bend. *(d)* With two double bonds in its chain, linoleic acid is polyunsaturated.

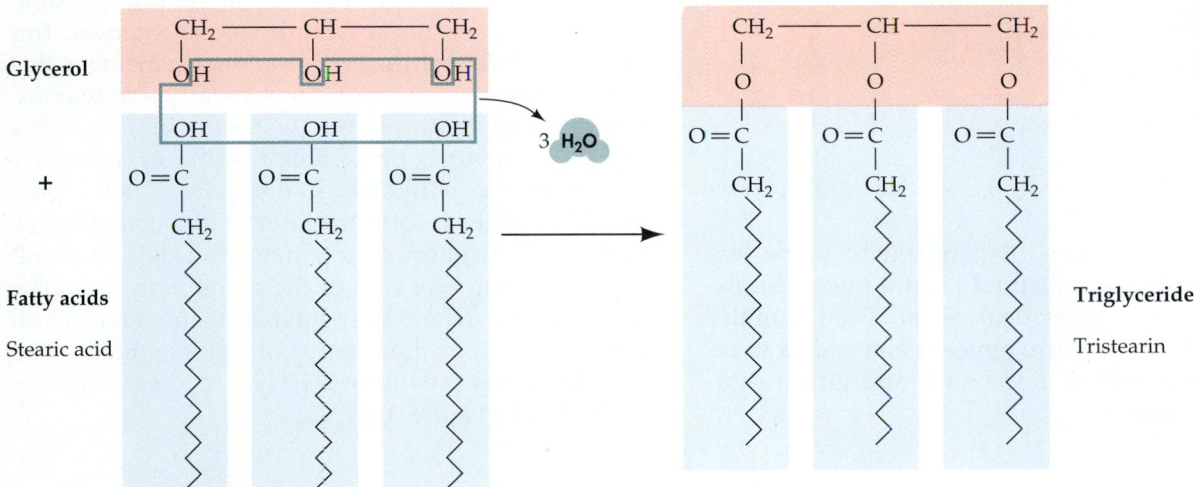

3.4 A Triglyceride and Its Components

In an example of triglyceride formation, tristearin forms from glycerol and three molecules of stearic acid by condensation. Condensations release water molecules. In living things the reaction is more complex, but the end result is as shown here. The jagged lines represent hydrocarbon chains.

Phosphatidate

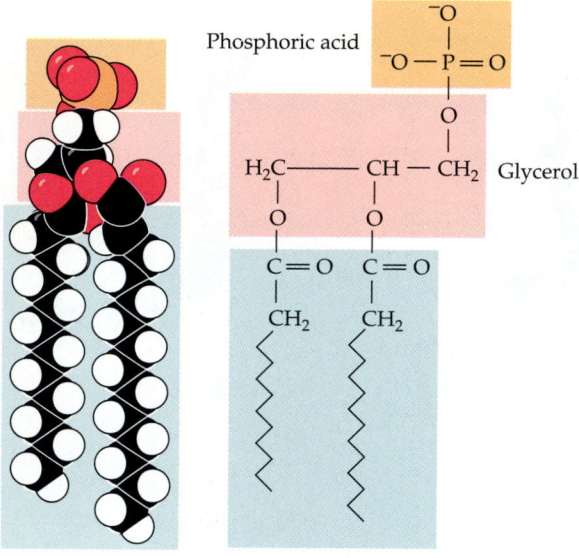

Phosphatidyl choline

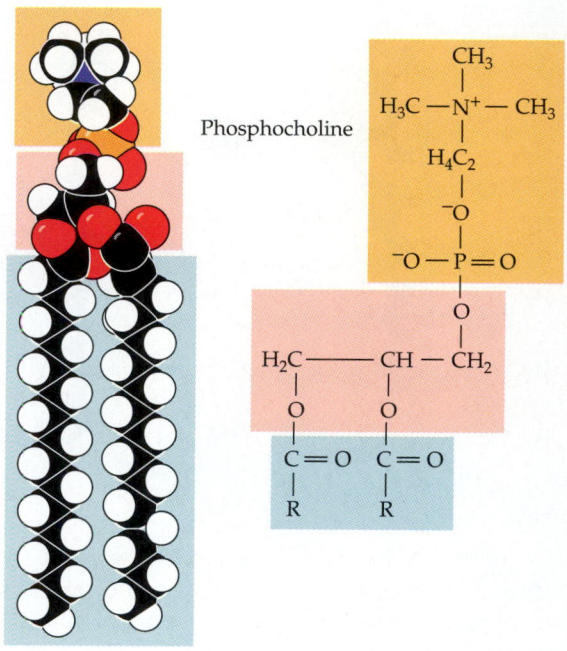

Phosphatidyl ethanolamine

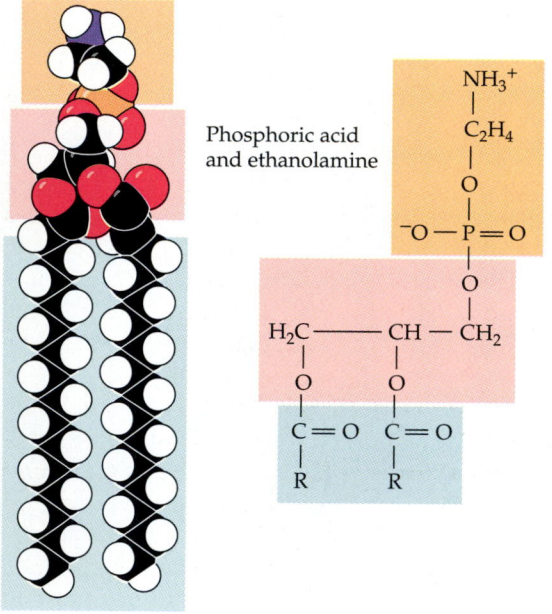

3.5 Some Phospholipids

Each phospholipid consists of a phosphorus-containing molecule (yellow shading), glycerol (red shading), and two molecules of fatty acid (blue shading). The fatty acid hydrocarbon chains may be shown as jagged lines or abbreviated to R (for "residue"). The hydrocarbon chains are nonpolar. The phosphorus-containing portions are electrically charged.

pholipids are important constituents of biological membranes. If you examine the structure of phospholipids closely, you will find it easy to understand how they are oriented in membranes. Because the phosphorus-containing portion of the phospholipid molecule carries one or more electric charges, this portion is **hydrophilic** (water-loving; remember that water is a polar molecule). The two fatty acid regions, however, are **hydrophobic** (water-fearing). In a biological membrane, phospholipids line up in such a way that the nonpolar, hydrophobic "tails" pack tightly together to form the interior of the membrane, and the phosphorus-containing "heads" face outward (some to one side of the membrane and some to the other), where they interact with water, which is excluded from the interior of the membrane. The phospholipids thus form a bilayer, a sheet two molecules thick (Figure 3.6).

atoms. The triglycerides of plants tend to be less saturated, oily liquids. Natural peanut butter, for example, contains a great deal of oil. Peanut butter manufacturers often hydrogenate their product to reduce the number of double bonds and give a saturated, solid product.

Phospholipids

Like triglycerides, **phospholipids** have fatty acids bound to glycerol. In phospholipids, however, any of certain phosphorus-containing compounds replaces one of the fatty acids (Figure 3.5). Many phos-

Other Lipids

The term *lipid* defines compounds not on the basis of structural similarity but in terms of their solubility. Remember that lipids are insoluble in water but are readily soluble in organic solvents such as ether, chlo-

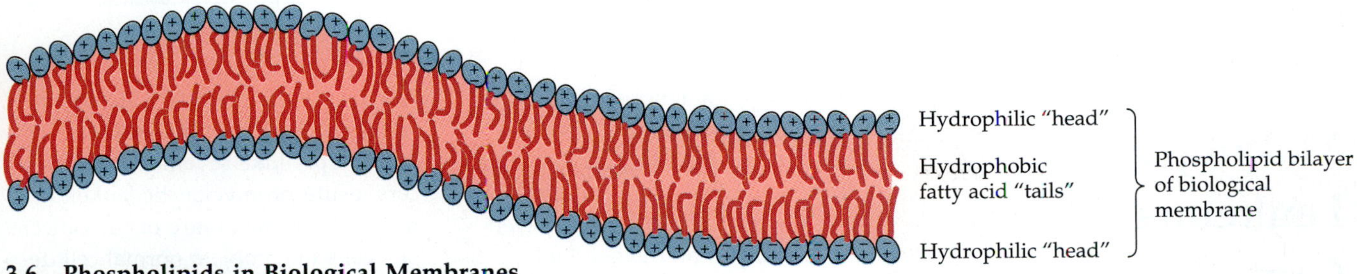

Hydrophilic "head"

Hydrophobic
fatty acid "tails"

Hydrophilic "head"

Phospholipid bilayer
of biological
membrane

3.6 Phospholipids in Biological Membranes
Hydrophobic interactions bring the "tails" together in the interior of a phospholipid bilayer. Hydrophilic "heads" face outward on both sides of the membrane. The details of this important structure are the subject of Chapter 5.

roform, or benzene. Two more groups of compounds with these properties (and hence classifiable as lipids) are the carotenoids and the steroids.

The **carotenoids** are a family of light-absorbing pigments found in plants and animals (Figure 3.7). Beta-carotene (β-carotene) is one of the pigments that traps light energy in leaves as part of the process of photosynthesis (see Chapter 8). It is the β-carotene in plants that senses light and causes their parts to grow toward or away from the light (a behavior called phototropism, discussed in Chapter 34). In humans, a molecule of β-carotene can be broken down into two vitamin A molecules, from which we make the pigment rhodopsin that is required for vision (Chapter 39). Another derivative of vitamin A is used in treating a form of cancer (Box 3.A). Carotenoids are responsible for the color of carrots, tomatoes, pumpkins, egg yolks, and butter.

The **steroids** are a family of organic compounds whose multiple rings share carbons (Figure 3.8). Some steroids are important constituents of membranes. Others are hormones, chemical signals that carry messages from one part of the body to another (Chapter 34). Testosterone is a steroid hormone that regulates sexual development in male vertebrates (animals with backbones); the chemically related estrogens play a similar role in females. Cortisone and related hormones play a wide variety of regulatory

roles in the digestion of carbohydrates and proteins, salt and water balance, and sexual development. Vitamin D is a steroid that regulates the absorption of calcium from the intestines. It is necessary for the proper deposition of calcium in bones; a deficiency of vitamin D can lead to rickets, a bone-softening disease. Irradiation of certain other steroids with sunlight or ultraviolet light converts them to vitamin D.

Cholesterol is synthesized in the liver. It is the starting material for making testosterone and several other steroid hormones and for the bile salts that help to get fats into solution so they can be digested. We absorb cholesterol from foods such as milk, butter, and animal fats. When we have too much cholesterol in our blood, it is deposited in our arteries (along with other substances), a condition that may lead to arteriosclerosis and heart attack.

The chemical structures of lipids are quite varied (see Figures 3.3 through 3.8). Their diversity matches the variety of their functions in living things: energy storage, digestion, membrane structure, bone formation, vision, and chemical signaling. Most lipids can be synthesized in the bodies of animals; the synthesis and storage of fats is an important means of locking energy away until it is needed. The few lipids that cannot be synthesized must be obtained in small amounts from the diet. For humans, the diet must include three particular unsaturated fatty acids and the fat-soluble vitamins: A, D, E, and K.

Although lipids are structurally and functionally diverse, there is one thing they share that sets them apart from the large molecules we will consider

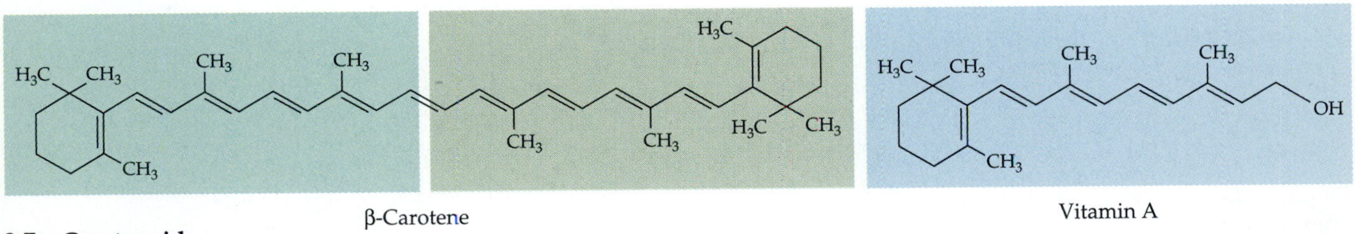

β-Carotene

Vitamin A

3.7 Carotenoids
These carotenoids have carbon atoms at each angle of the rings and chains. (Omitting the Cs that represent carbon atoms is standard chemical shorthand.) Methyl groups are indicated. Notice that β-carotene is symmetrical around the central double bond. Splitting β-carotene in the middle produces two vitamin A molecules.

BOX 3.A

Making a Leukemia Grow Up

Among the forms of cancer, leukemia has been one of the hardest to treat. In general, we treat cancers by radiation therapy or chemotherapy. Radiation therapy is most effective against localized cancers because only the particular part of the body affected needs to be subjected to radiation. Leukemia, however, is a cancer of the white blood cells, which travel throughout the body.

Recently, biomedical scientists have achieved great success with a novel treatment for a particular type of leukemia called acute promyelocytic leukemia. A substance derived from vitamin A, all-*trans*-retinoic acid, eliminates all symptoms of this leukemia in a high percentage of patients who take the acid orally. Medical researchers have learned that many victims of acute promyelocytic leukemia have a genetic defect. People with or without the defect produce retinoic acid naturally in their bodies, but people with the defect cannot process the acid normally. By ingesting additional retinoic acid as a supplement, patients with the defect can combat their cancer.

Why should this lipid, retinoic acid, act against certain cancer cells?

Does it kill them? In fact, it does *not* kill them. Rather, it makes them "grow up." Like several other cancers, acute promyelocytic leukemia results from the failure of the cancerous cells to complete normal cell development. Biologists have speculated for some time that retinoic acid promotes the normal development of cells in bird wings. Now, medical researchers propose that when it is given to patients with this type of leukemia, retinoic acid causes the cancerous cells to complete normal development and cease dividing.

Like many biologically active compounds, retinoic acid has multiple effects. For example, it is also a treatment for acne, in the skin cream Retin-A.

---

throughout the rest of the chapter: Lipid molecules do not form polymers in reactions of the type discussed in the next section.

FROM MONOMERS TO POLYMERS

The largest molecules covered in this chapter are called **macromolecules**—giant molecules, or aggregates of molecules, with molecular weights in excess of 1,000 daltons. All macromolecules are **polymers**: molecules made by the combination of many smaller molecules, called **monomers**. An **oligomer** contains only a few monomers. A cell can combine a limited variety of monomers into a near-infinite variety of polymers.

Macromolecules perform many essential functions in organisms. These functions arise directly from the structures of the molecules. Some macromolecules fold into globular forms with surface features that enable them to recognize and interact with certain other molecules. Others form long, fibrous systems that provide strength and rigidity to parts of an organism. Still others contract and allow the organism to generate the force to move itself. Some macromolecules aggregate to form structures that determine what materials enter or leave the compartments within an organism; others accelerate chemical reactions in cells. There is a flow of *information* among all the classes of macromolecules, but one class (the nucleic acids, discussed later in this chapter) specializes in information.

(a) Testosterone   (b) Cortisone   (c) Vitamin D   (d) Cholesterol

3.8 Steroids Share a Common Ring Structure
Among the important steroids in vertebrates are (a) the male sex hormone testosterone, (b) the hormone cortisone, (c) vitamin D, and (d) cholesterol.

Macromolecules in living things are polymers built from simpler monomers through a series of reactions called **condensations** or **dehydrations**. These reactions are of the general type

$$A—H + B—OH \rightarrow A—B + H_2O$$

(A—H is a molecule consisting of a hydrogen atom attached to another part, A; B—OH is a molecule consisting of an —OH group attached to another part, B.) The product A—B is formed, along with a molecule of water; the atoms of water are derived from the reactants—one hydrogen atom from one reactant, and an oxygen atom and the other hydrogen atom from the other reactant. **Reactants** are the molecules undergoing a chemical reaction.

The polymerization reactions that produce the different kinds of macromolecules differ in detail. In all cases, energy must be added to the system for polymers to form. Other kinds of specific molecules participate; their function is to activate the reactants—to provide the necessary energy for the reactions to be carried out. Large molecules are assembled through the repeated condensations of activated monomers.

CARBOHYDRATES

A **carbohydrate** is a compound based on the general formula CH_2O. Carbohydrates are a diverse group of compounds with molecular weights ranging from less than 100 to hundreds of thousands. They fall into three categories: the **monosaccharides**, or *simple*

sugars, which are monomers; the **oligosaccharides**, made up of a few monosaccharides linked together; and the **polysaccharides**, polymeric carbohydrates that include starches, glycogen, cellulose, and many other important biological materials. (*Mono-* means "single," *oligo-* means "few," and *poly-* means "many"; *saccharide* means "sugar.") There is no clear dividing line between a large oligosaccharide and a small polysaccharide, for these are simply terms of convenience used to separate "classes" within what is really a continuum of compounds of various sizes.

The general formula for carbohydrates, CH_2O, is true for monosaccharides, all of which have the same number of carbon and oxygen atoms but twice as many hydrogen atoms as oxygen atoms. The formulas of oligosaccharides and polysaccharides, however, differ slightly from this general formula.

Monosaccharides

All living cells contain the monosaccharide **glucose**, $C_6H_{12}O_6$. Green plants produce it by photosynthesis (see Chapter 8). Cells metabolize it to yield energy during cellular respiration (see Chapter 7). Glucose exists in both straight-chain and ring forms, in equilibrium with each other (Figure 3.9). The two distinct ring forms (α- and β-glucose) differ only in the placement of the —H and —OH attached to carbon 1. (The convention for numbering carbons shown in Figure 3.9 is used throughout this book). Although chemically and physically distinct substances, α- and β-glucose interconvert constantly in aqueous solution.

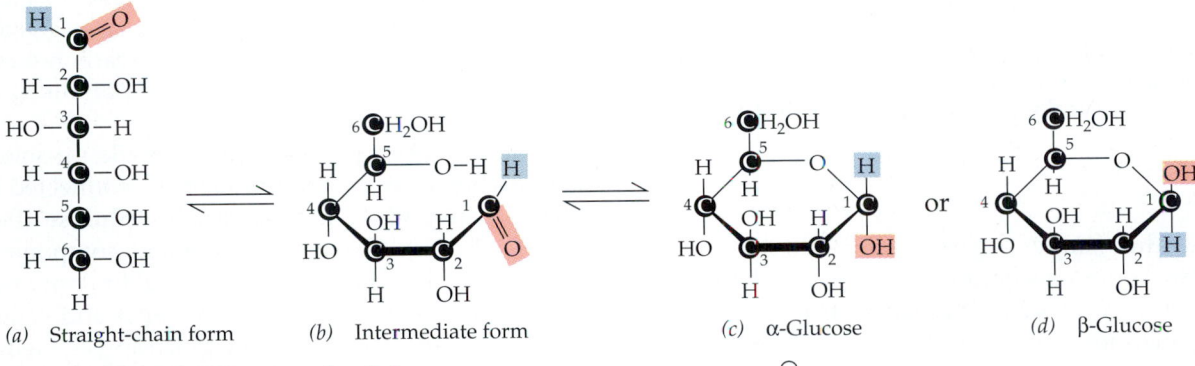

(a) Straight-chain form   (b) Intermediate form   (c) α-Glucose   (d) β-Glucose

3.9 Glucose: From One Form to the Other
Forms of glucose interconvert when dissolved in water. *(a)* The straight-chain form has an aldehyde group (shaded) at carbon 1. *(b)* A reaction between this aldehyde group and the hydroxyl group at carbon 5 gives rise to a ring form. (The darker line at the bottom of each ring indicates that that edge of the molecule extends toward you, while the upper edge extends back into the page.) *(c,d)* Depending on the orientation of the aldehyde group when the ring closes, either of two rapidly and spontaneously interconverting molecules—α-glucose or β-glucose—forms. The ball-and-stick model depicts α-glucose; it should be easy to visualize a comparable model of β-glucose.

Three-carbon sugar

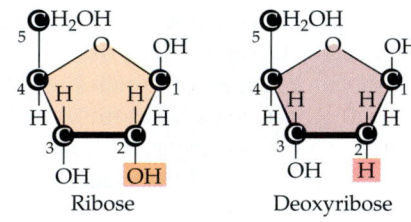

Glyceraldehyde

Five-carbon sugars

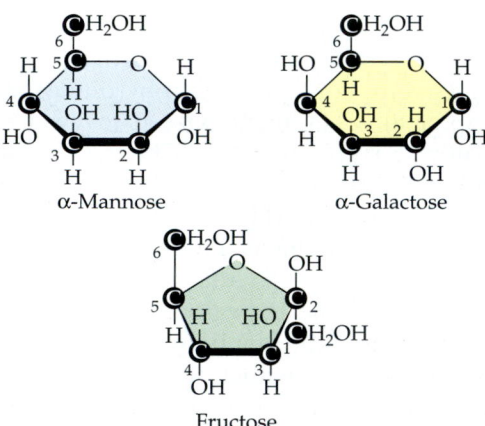

Ribose   Deoxyribose

Six-carbon sugars

α-Mannose   α-Galactose

Fructose

3.10   Monosaccharides

The three-carbon sugar glyceraldehyde has the formula $C_3H_6O_3$; its common form is the straight chain shown. Ribose and deoxyribose each have five carbons; their formulas are $C_5H_{10}O_5$ and $C_5H_{10}O_4$, respectively. All of the six-carbon sugars shown here have the formula $C_6H_{12}O_6$, but they are chemically and biologically distinct from one another.

Some other simple sugars are illustrated in Figure 3.10. Many monosaccharides have the same formula as glucose, $C_6H_{12}O_6$, including fructose ("fruit sugar"), mannose, and galactose. These compounds are isomers of each other: They are composed of the same kinds and numbers of atoms, but the atoms combined differently in each to yield different arrangements. Five-carbon sugars are referred to collectively as **pentoses**, and six-carbon sugars are called **hexoses**. Some pentoses are found primarily in the cell walls of plants, as are several of the hexoses. Two pentoses are of particular importance: **ribose** and **deoxyribose**, which form part of the backbones of RNA and of DNA, respectively. Ribose and deoxyribose differ by only one oxygen atom associated with one of the carbon atoms, carbon 2.

Disaccharides

Larger carbohydrates form by the bonding together of two or more monosaccharide molecules. The monosaccharides become covalently coupled by condensation reactions to form specific oligosaccharides and polysaccharides. The linkages between the monosaccharides are **glycosidic linkages**. The smallest oligosaccharides are the disaccharides and the trisaccharides, made up of two and three simple sugars, respectively. If one glucose molecule combines with another, as shown in Figure 3.11, the disaccharide product must be one of two types: α-linked or β-linked, depending on whether the molecule that bonds at the carbon 1 position is α-glucose or β-glucose. An α linkage with carbon 4 of a second glucose molecule gives maltose, whereas a β linkage gives cellobiose. Maltose and cellobiose are disaccharides, both with the formula $C_{12}H_{22}O_{11}$. Both are composed of two glucose molecules (minus one molecule of water), but they are different compounds; they are recognized by different enzymes and undergo different chemical reactions. Two other common disaccharides are sucrose and lactose. Sucrose (common table sugar; also $C_{12}H_{22}O_{11}$) is made from one molecule of glucose and one of fructose. Lactose (milk sugar) consists of glucose and galactose.

Polysaccharides

As we saw in Figure 3.11, maltose consists of two glucose units connected by an α linkage. Imagine a trisaccharide (three glucose units), a tetrasaccharide (four glucose units), and finally a giant polysaccharide consisting of hundreds or thousands of glucose units, each connected to the next by an α glycosidic linkage from carbon 1 of one unit to carbon 4 of the next. This polymer is **starch**, an important storage compound.

A similar giant polysaccharide, made up solely of glucose but with the individual units connected by β linkages instead of α linkages, is **cellulose** (Figure 3.12a). Cellulose is the predominant component of plant cell walls and by far the most abundant organic compound on this planet. Both starch and cellulose are composed of nothing but glucose; yet their biological functions and chemical and physical properties are entirely different. Enzymes that digest one do not affect the other at all.

Starch is not a single chemical substance; rather, the term denotes a large family of giant molecules of broadly similar structure. All starches are polymers of glucose with α linkages. All are large, but some are enormous, containing tens of thousands of glucose units. An important distinguishing characteristic of starches is the degree of branching. The starches that store glucose in plants, called **amylose**, are not highly branched (Figure 3.12b). The highly branched

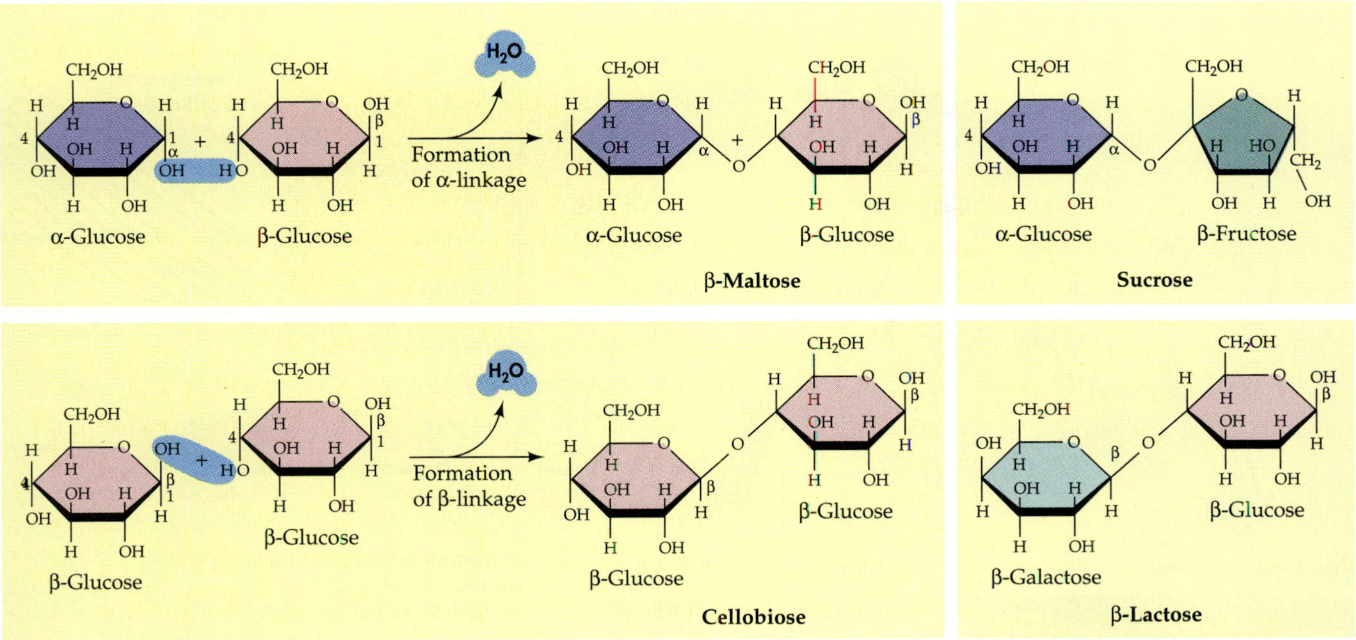

3.11 Disaccharides Are Composed of Two Monosaccharides

In the reaction shown at the top left (a simplified version of the reaction in nature), maltose is produced when an α-1,4 linkage forms between two glucose molecules, with the hydroxyl group on carbon 1 of one glucose in the α (down) position as it reacts with the 4-hydroxyl group of the other. In cellobiose (bottom left), the two glucoses are linked by a β-1,4 linkage. Lactose (bottom right) is made by a β linkage between carbon 1 of galactose and carbon 4 of glucose. In sucrose (top right), carbon 1 of glucose is joined by an α-1,2 linkage to carbon 2 of fructose.

polysaccharide that stores glucose in animals is **glycogen** (Figure 3.12c). Animals use glycogen to store energy in liver and muscle.

What do we mean when we say that starch and glycogen are storage compounds for energy? Very simply, these compounds can readily be depolymerized to yield glucose monomers. Glucose, in turn, can be further digested, or metabolized—that is, it can undergo chemical reactions—to yield energy for cellular work. Alternatively, glucose can be metabolized so that its carbon atoms are rearranged to form the skeletons of other compounds. Glycogen and starch are thus storage depots for carbon atoms as well as for energy. Each is chemically stable but is readily mobilized by digestion and further metabolism.

Derivative Carbohydrates

Derivative carbohydrates deviate from the general formula for carbohydrates by containing elements other than C, H, and O. Examples include sugar phosphates, amino sugars, and chitin (Figure 3.13). A number of sugar phosphates, such as fructose 1,6-bisphosphate, are important intermediates in cellular respiration (Chapter 7) and photosynthesis (Chapter 8). Sugar phosphates have phosphate groups attached to one or more —OH groups of the parent sugar. The two **amino sugars** shown in the figure, glucosamine and galactosamine, have an amino group in place of an —OH group. Galactosamine is a major component of cartilage, the material that forms caps on the ends of bones and stiffens the protruding parts of the ears and nose. The polymer chitin is made from a derivative of glucosamine. Chitin is the principal structural polysaccharide in the skeletons of insects and their relatives such as crabs and lobsters, as well as in the cell walls of fungi. Fungi and insects (and their relatives) constitute more than 80 percent of the species ever described, and chitin is another of the most abundant substances on Earth.

PROTEINS

In Chapter 2 we considered the amino acids (small molecules containing both carboxyl and amino groups). These are the monomers from which a fascinating set of polymers are formed—the **proteins**. Proteins account for many of the mechanical elements of living things, from parts of subcellular membranes to skin, bones, and tendons. In vertebrates, proteins called immunoglobulins form a major line of defense against foreign organisms. The specialized molecules needed to bring about all biochemical reactions make up a major class of proteins called enzymes. Our every movement results from the contraction and relaxation of muscles, resulting in turn from the delicately regulated sliding of particular pro-

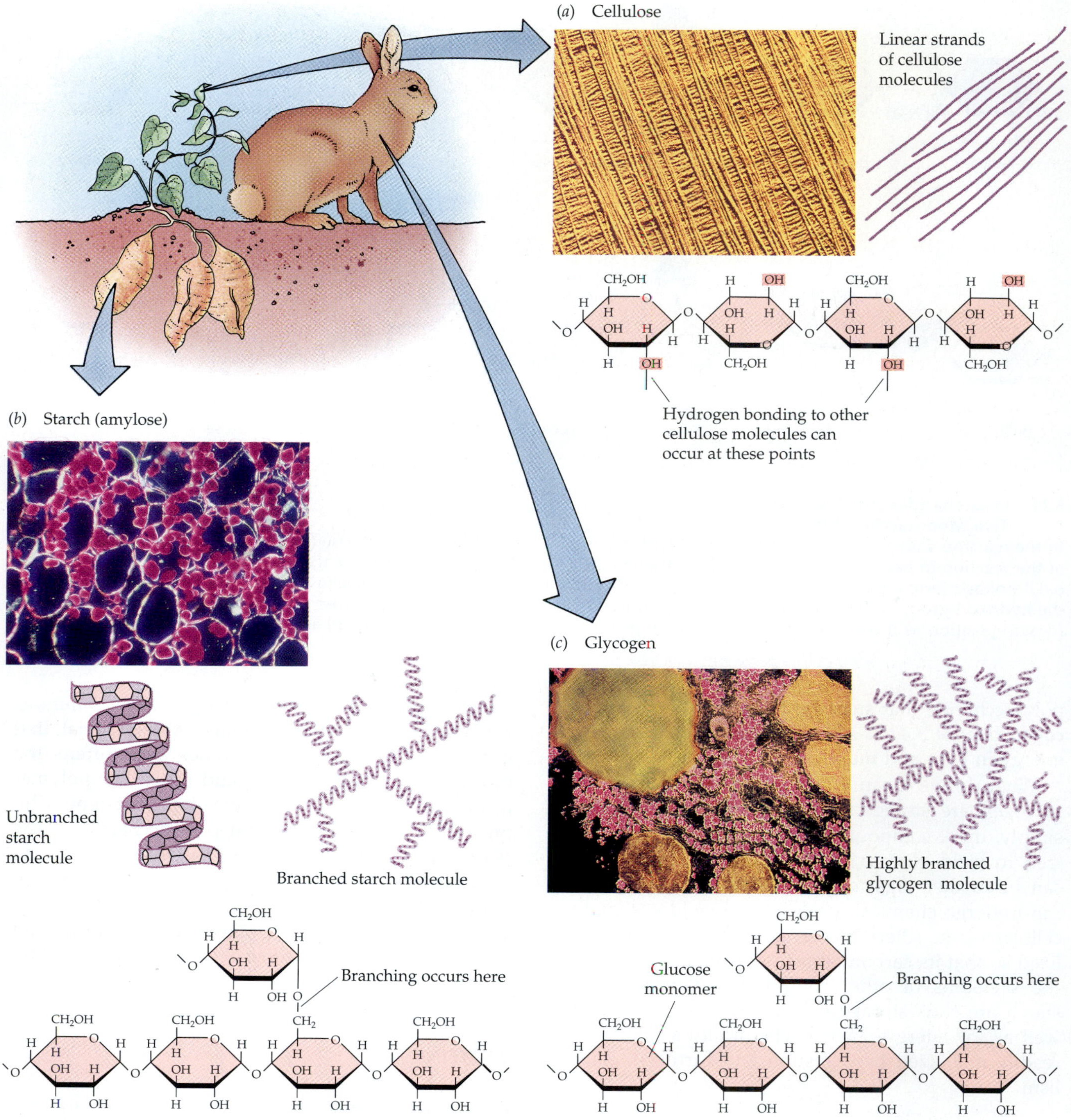

(a) Cellulose

Linear strands
of cellulose
molecules

Hydrogen bonding to other
cellulose molecules can
occur at these points

(b) Starch (amylose)

Unbranched
starch
molecule

Branched starch molecule

Branching occurs here

(c) Glycogen

Highly branched
glycogen molecule

Glucose
monomer

Branching occurs here

3.12 Representative Polysaccharides
(a) Cellulose is an unbranched polymer of glucose. Many adjacent cellulose
molecules form the cellulose fibrils in photosynthetic cells. (b) In starches such
as amylose, branching may occur. In the micrograph, a red dye stains the amylose grains in sweet potato cells. (c) Glycogen from animal cells differs from
plant starch only in being more extensively branched. The tinted electron micrograph shows part of a human liver cell; the pink-tinted bodies are glycogen
granules.

teins in muscle cells past one another. Still other proteins act as adjustable channels through which sodium ions (Na^+), potassium ions (K^+), and other ions are passed from one side of a nerve-cell membrane to the other, resulting in phenomena such as the transmission of electric signals along a nerve. To understand this stunning variety of functions, we must first explore the structure of these molecules.

Amino Acids

Twenty different amino acids are found in proteins. The **side chains** of amino acids show a wide variety of chemical properties. Side chains control the function of a protein—they are the reactive groups in proteins. The order of a protein's side chains determines how it folds into a three-dimensional configuration (discussed later in this chapter). Despite their importance, side chains are commonly left out of structural formulas, where they are represented simply by "R" (for "residue"); thus they are sometimes called R groups. Side chains are highlighted in the amino acid structural formulas in Table 3.1.

One useful classification of amino acids is based on whether their side chains are electrically charged, polar but uncharged, or hydrophobic. There are two groups of amino acids with electrically charged side chains: those with positive charges and those with negative charges. All five charged amino acids are very hydrophilic. The four amino acids with polar but uncharged side chains tend to form hydrogen bonds readily, both with water and with other molecules. They, too, are hydrophilic. The side chains of eight other amino acids either are hydrocarbon or are very slightly modified from hydrocarbons; hence they are hydrophobic.

Three amino acids—cysteine, glycine, and proline—are special cases, although their side chains are generally hydrophobic. Two cysteine side chains can lose hydrogen atoms so that their sulfur atoms are joined by a covalent bond in a **disulfide bridge** (Figure 3.14). Hydrogen bonds and disulfide bridges help determine how a protein chain folds. When cysteine is not part of a disulfide bridge, its side chain is very hydrophobic. The glycine side chain is just a hydrogen atom; thus glycines may fit into tight corners in the interior of a protein molecule, where a larger side chain could not fit. Proline differs from other amino acids because it possesses a modified amino group (see Table 3.1).

Peptide Linkages

When amino acids polymerize, the carboxyl group of one amino acid reacts with the amino group of another, undergoing a condensation reaction and forming a **peptide linkage**. Figure 3.15 gives a simplified description of the reaction (actually, other molecules must activate the reactants, and there are intermediate steps). A linear polymer of amino acids connected by peptide linkages is a polypeptide. A protein is made up of one or more polypeptides. At one end of the polypeptide molecules is a free amino group. This end is the N-terminus, named for the nitrogen atom in the amino group. At the other end of the polypeptide—the C-terminus—there is a free carboxyl group. The other amino and carboxyl groups are bound in peptide linkages. Thus a protein has direction. For one example, the dipeptide glycine–alanine, in which glycine has the free amino group, differs from alanine–glycine, in which alanine has the free amino group.

Fructose 1,6-bisphosphate

β-Glucosamine

β-Galactosamine

Chitin

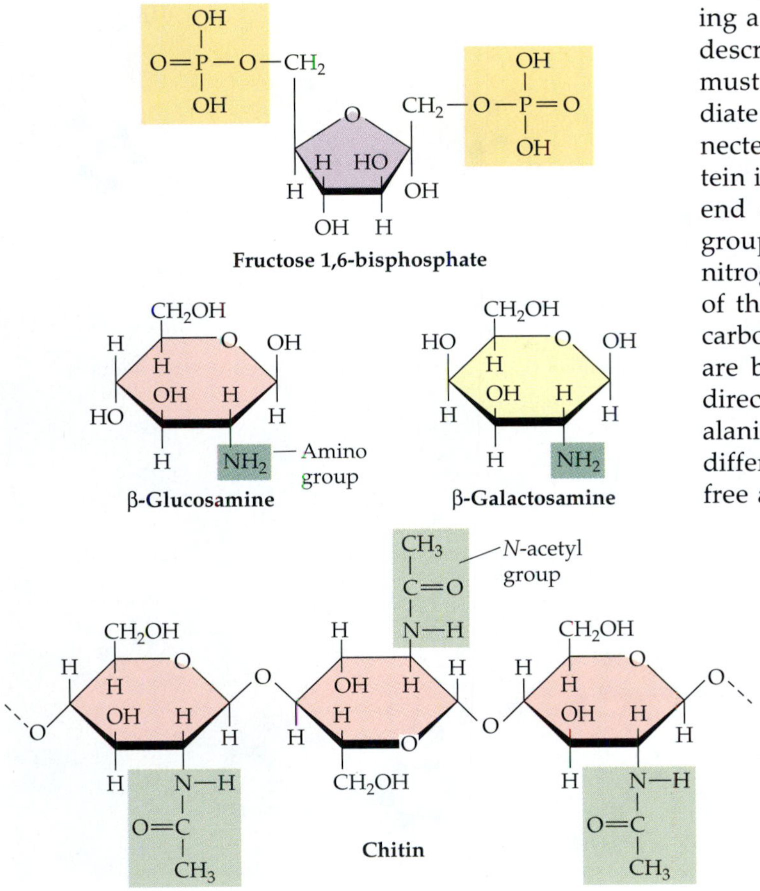

3.13 Derivative Carbohydrates
(a) Fructose 1,6-bisphosphate is a sugar phosphate; the numbers in its name refer to the bonding of the phosphate groups (shaded yellow) to carbons 1 and 6 of the sugar. (b) The amino groups on the amino sugars β-glucosamine and β-galactosamine are shown in green; recall that the β refers to the position of the –OH group on carbon 1. Chitin is a polymer of N-acetylglucosamine; N-acetyl groups are shown in pale green.

**TABLE 3.1
Twenty amino acids found in proteins**

A. *Amino acids with electrically charged side chains*

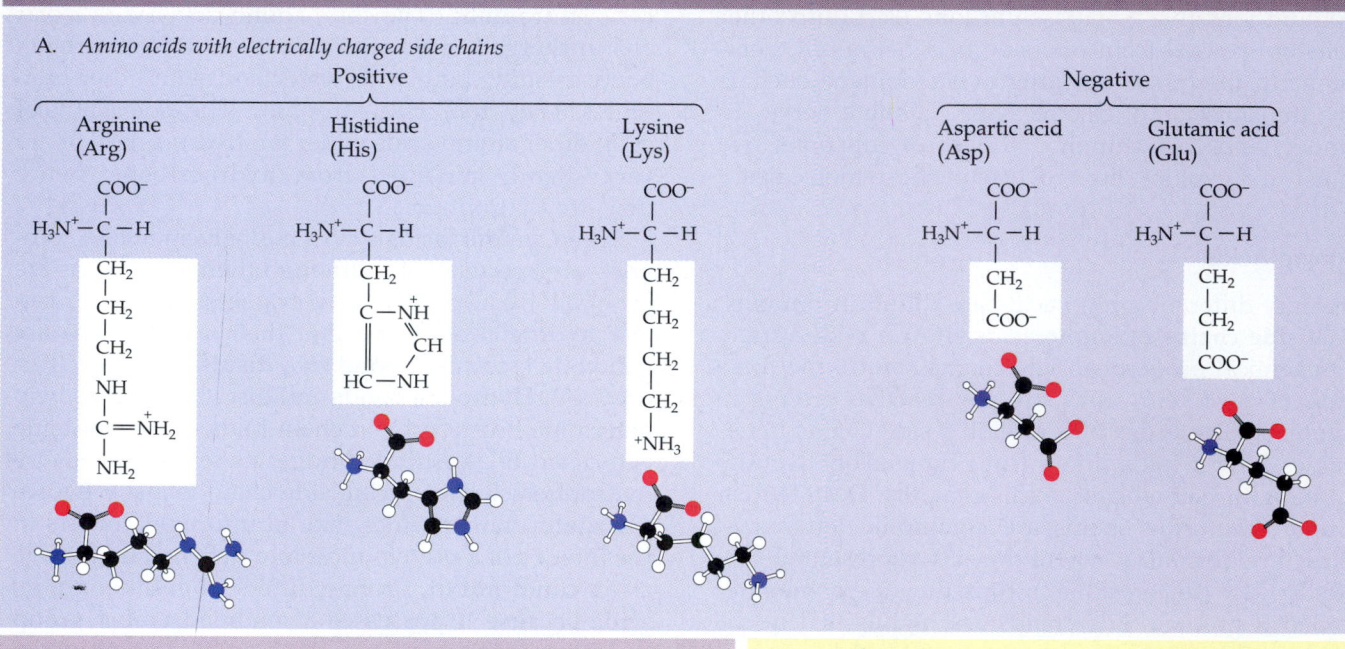

Positive

Arginine (Arg) — Histidine (His) — Lysine (Lys)

Negative

Aspartic acid (Asp) — Glutamic acid (Glu)

B. *Amino acids with polar but uncharged side chains*

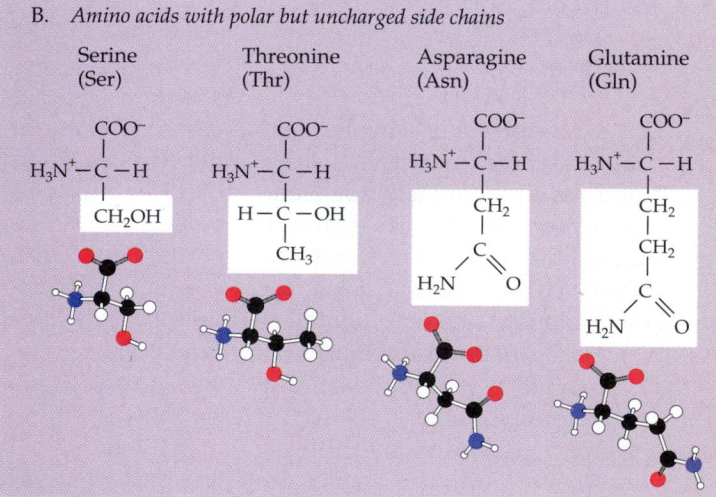

Serine (Ser) — Threonine (Thr) — Asparagine (Asn) — Glutamine (Gln)

C. *Special cases*

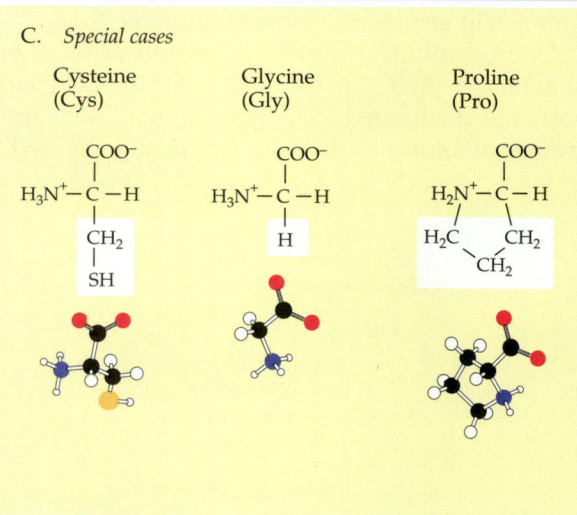

Cysteine (Cys) — Glycine (Gly) — Proline (Pro)

D. *Amino acids with hydrophobic side chains*

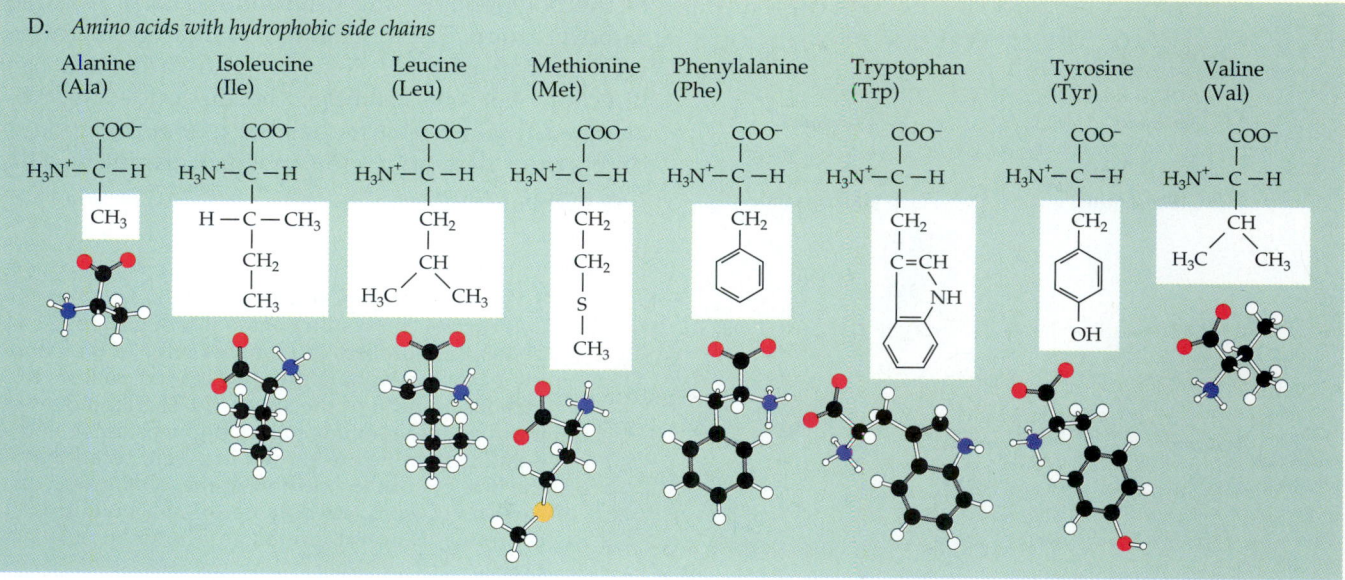

Alanine (Ala) — Isoleucine (Ile) — Leucine (Leu) — Methionine (Met) — Phenylalanine (Phe) — Tryptophan (Trp) — Tyrosine (Tyr) — Valine (Val)

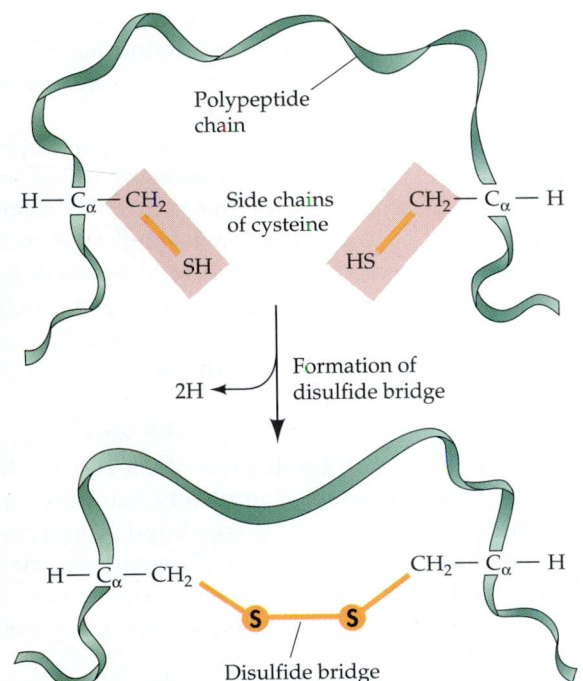

3.14 Formation of a Disulfide Bridge
The –SH groups on two cysteine side chains in a chain of amino acids can react to form a linkage between the two sulfur atoms. Such disulfide bridges are important in maintaining the proper three-dimensional shapes of protein molecules.

In the peptide linkage, the C═O oxygen carries a slight negative charge, whereas the N—H hydrogen is slightly positive. This asymmetry of charge favors hydrogen bonding (see Chapter 2) within the protein molecule itself and with other molecules, contributing to both the structure and the function of many proteins.

LEVELS OF PROTEIN STRUCTURE

Primary Structure

Protein structure is elegant and complex—so complex that it is described as consisting of several levels. The precise sequence of amino acids in the unbranched chain of a polypeptide (which is a linear polymer) constitutes the **primary structure** of a protein (Figure 3.16). This sequence is dictated by the precise sequence of monomers (which are called nucleotides) in a linear segment of a DNA molecule. The elucidation of this relationship between DNA primary structure and protein primary structure was one of the triumphs of molecular biology (it will be described in Chapter 11).

The theoretical number of different proteins is enormous. There are 20 different amino acids; this means there are $20 \times 20 = 400$ distinct dipeptides, and $20 \times 20 \times 20 = 8,000$ different tripeptides. Imagine this process of multiplying by 20 extended to a protein made up of 100 monomers (considered a small protein): There could be 20^{100} of these small proteins, each with its own distinctive primary structure. The higher levels of protein structure—from

4 Amino acids

Polypeptide

N-terminus

C-terminus

3.15 Formation of Peptide Linkages
In this depiction, four amino acids combine across their amino and carboxyl groups, as indicated in yellow; the resulting peptide linkages create a polypeptide. A molecule of water (blue) is lost as each peptide linkage forms. (In living things the reaction is substantially more complex, but the end result is as shown here.)

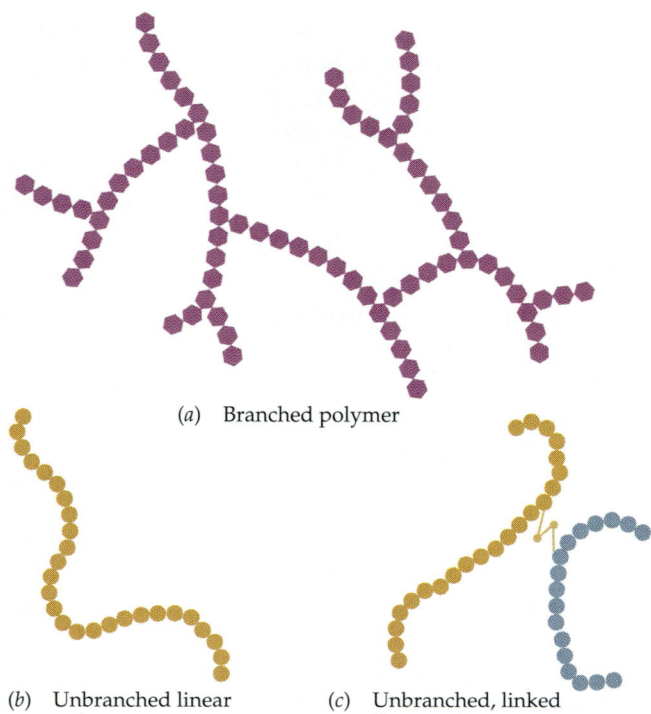

(a) Branched polymer

(b) Unbranched linear
polymer

(c) Unbranched, linked
linear polymers

3.16   Branched versus Linear Polymers
(a) Some biological polymers, such as the carbohydrate glycogen (see Figure 3.12c) are highly branched, as depicted in the generalized molecule here. Proteins, however, are unbranched (b), although the chains of amino acids may be linked together, as shown in (c).

local coiling and folding to the overall shape of the molecule—all derive from the primary structure. By presenting side chains of differing character (hydrophilic or hydrophobic, for example) in a specific and unique order, the precise sequence of amino acids in a given protein determines the ways in which the polypeptide chain can twist and fold. By twisting and folding, each protein adopts a specific structure that distinguishes it from every other protein.

Secondary Structure

Although the primary structure of each type of protein is unique, the secondary structure of many different proteins may be the same. A protein's **secondary structure** consists of regular, repeated patterns of orientation of parts of a polypeptide chain. One type of secondary structure, the **α helix** (alpha helix), is a right-handed coil "threaded" in the same direction as a standard wood screw. The twisting of a single polypeptide chain about its axis often allows hydrogen bonds to form between amino acids that are four monomers apart along the chain (Figure 3.17, top left). When this pattern of hydrogen bonding is established repeatedly over a segment of the protein, it stabilizes the twisted form, resulting in an α helix.

The ability of a protein to form an α helix depends on its primary structure: Certain amino acids have side chains that distort the coil or otherwise prevent the formation of hydrogen bonds.

Alpha helical secondary structure is particularly evident in the fibrous structural proteins called keratins. These include most of the protective tissues found in animals, such as fingernails and claws, skin, hair, and wool. Hair can be stretched because this requires breaking only hydrogen bonds in an α helix, rather than breaking covalent bonds; when the tension on the hair is released, both the helix and the hydrogen bonds re-form.

Silk is an example of a protein with another type of secondary structure, the **β-pleated sheet**. Here the protein chains are almost completely extended and are bound into sheets by hydrogen bonds connecting one chain to another (Figure 3.17, right). In many proteins, regions of β-pleated sheet are formed by bonding between different parts of the same polypeptide chain.

A third type of secondary structure, the triple helix, is found in collagen (Figure 3.17, bottom left). Collagen, an important protein found in cartilage, tendons, the underlayers of skin, and the cornea of the eye, consists of three polypeptide chains twisted around each other like the strands of a cable. Hydrogen bonds connect the chains, resulting in a structure that is strong, rigid, and unstretchable). The tail of a rat, under the skin, is almost pure collagen.

Tertiary Structure

The overall shape of a whole polypeptide molecule is its **tertiary structure**. The α helices and β-pleated sheets sometimes predominate throughout a protein molecule, determining the tertiary structure. More frequently, however, only limited portions of the molecule have these secondary structures, and they thus make only minor contributions to overall shape. A complete description of the tertiary structure specifies the location of every atom in the molecule in three-dimensional space, in relation to all the other atoms. The tertiary structure of the protein lysozyme is represented in Figure 3.18. Bear in mind that both this tertiary structure and the secondary structure emphasized in Figure 3.18c derive entirely from the protein's primary structure. If lysozyme is heated carefully, causing the tertiary structure to break down, the protein will return to its normal tertiary structure when it cools. The only information needed to specify the unique shape of the lysozyme molecule is the information contained in its primary structure.

Myoglobin is an important protein (Figure 3.19). Its function—to store oxygen in certain animal tissues—is discussed in Chapter 41. Myoglobin has 153 amino acids in its single polypeptide chain; there are

The α helix

The α helix is a secondary structure found in many proteins. The atoms in the relatively rigid plane of the peptide linkages are in color, and the hydrogen bonds that stabilize the helix are shown as red dotted lines.

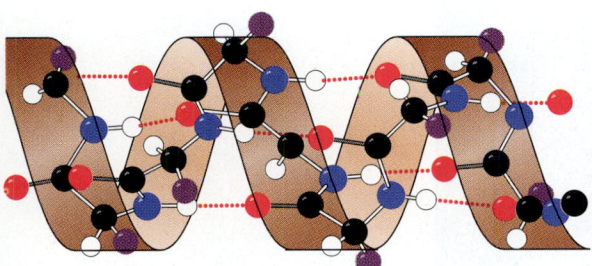

This computer drawing gives a three-dimensional sense of the relative positions of the atoms. Carbons are black, oxygens red, hydrogens white, and nitrogens blue. R groups are shown in purple.

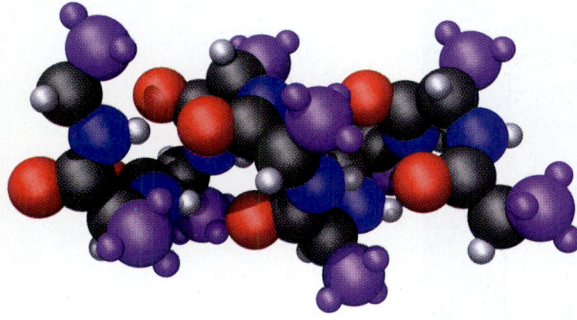

The triple helix

The secondary structure of the protein collagen is a triple helix of helical polypeptide chains (left). Such a triple helix is called tropocollagen. On the right, a number of triple helices of tropocollagen join in parallel fashion to create a strong, flexible collagen fibril; several collagen fibrils are shown here, magnified about 20,000 times. The spacing between black bands corresponds to the length of a single tropocollagen molecule.

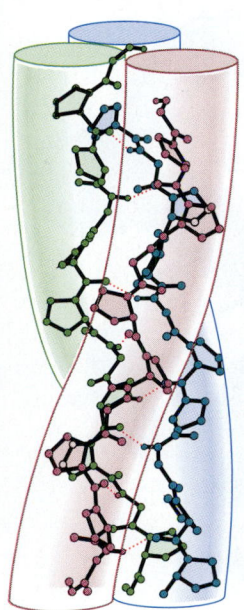

The β–pleated sheet

In the β-pleated sheet, polypeptide chains run side by side, linked by hydrogen bonds. The polypeptides are extended rather than coiling into a helix.

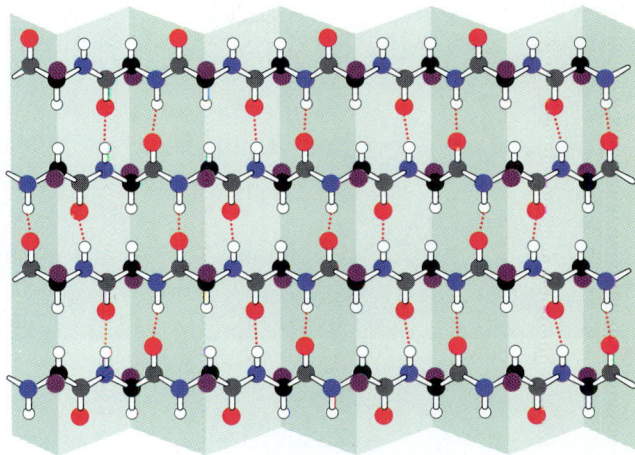

This computer drawing views a β-pleated sheet from the top. Four parallel strands of the polypeptide are joined by hydrogen bonds to form the sheet.

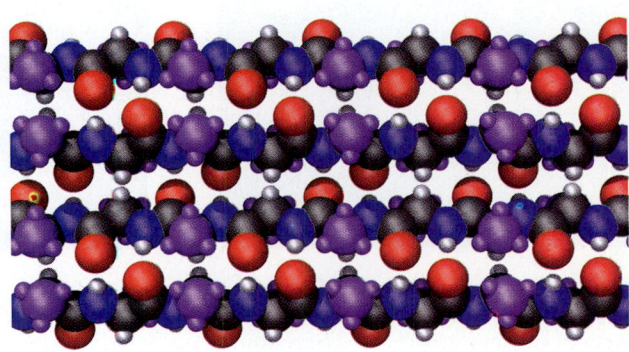

The computer drawing below shows the same material as the above drawing but viewed from the bottom edge to emphasize the "pleats" in the sheet.

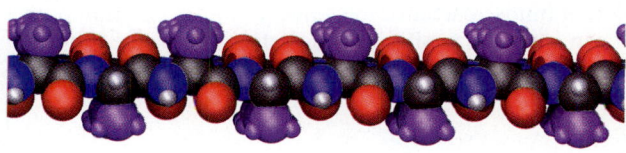

3.17 Forms of Protein Secondary Structure
Many proteins share local regions of similarity. These recognizable motifs are called secondary structure.

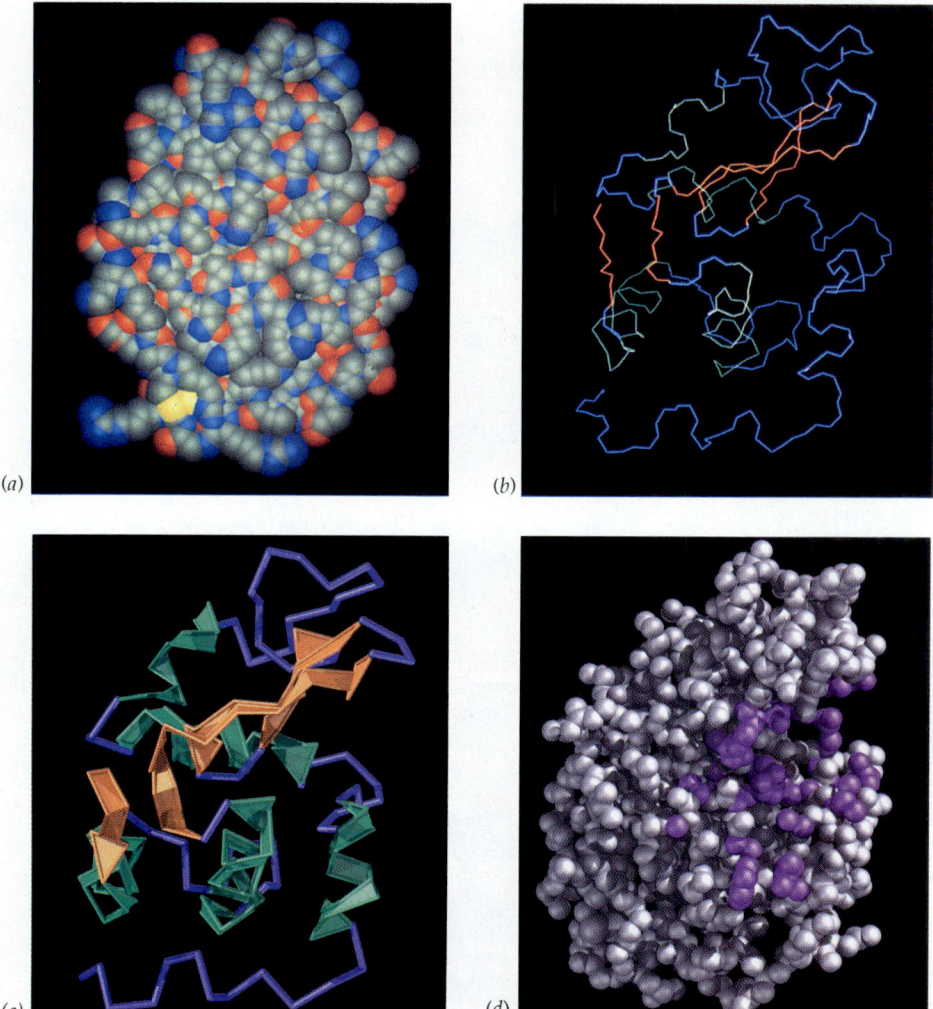

(a)

(b)

(c)

(d)

3.18 Four Representations of Lysozyme
Different molecular representations emphasize different aspects of tertiary structure. These four representations of lysozyme are similarly oriented. (a) This computer drawing gives the most realistic impression of lysozyme's tertiary structure, which is densely packed. (b) Another computer drawing emphasizes the backbone of the folded polypeptide. Regions in green have α-helical secondary structure; those in red constitute a β-pleated sheet. (c) The green coils here represent the α helices, and orange arrows represent the β-pleated sheet. (d) Another space-filling model, this one emphasizing the position (shown in purple) of the active site—the part of the enzyme molecule that binds reactant molecules. From its position here, you can infer the position of the active site in the other three representations.

3.19 Tertiary Structure of a Protein
Tertiary structure (the exact three-dimensional folding of a protein molecule) is illustrated here for the protein myoglobin. The individual atoms are not shown, nor are the individual amino acids, which form the coiled polypeptide chain. The chain is helical throughout most of its length. The blue shading shows the overall tertiary configuration of the molecule. The red structure in the upper part of the drawing is an iron-containing heme group.

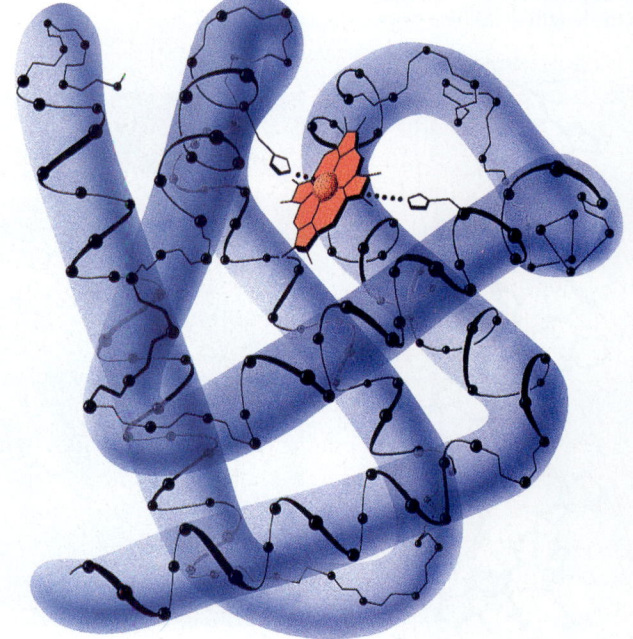

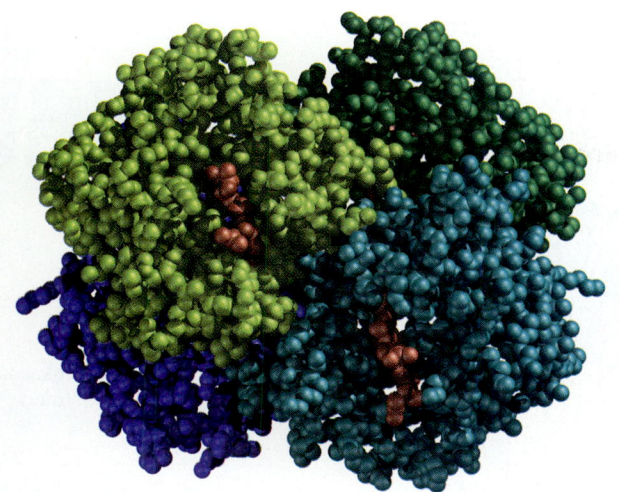

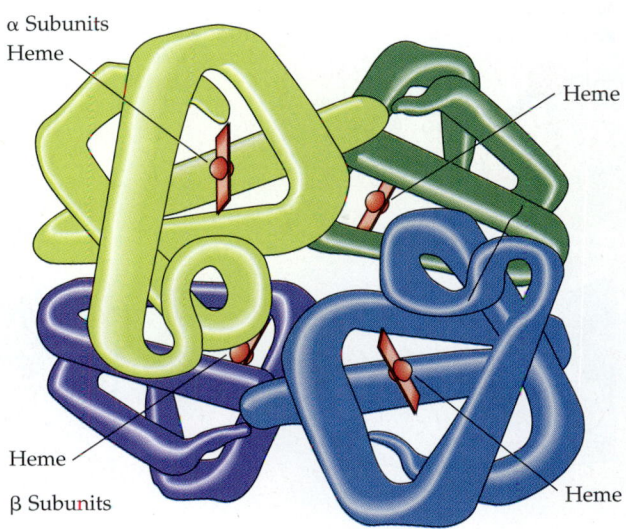

3.20 Quaternary Structure of a Protein
Hemoglobin consists of four folded polypeptide subunits that assemble themselves into the quaternary structure shown here. In these two analogous representations, each subunit is a different color. Note the heme groups, which are red.

no disulfide bridges, and the molecule is unusual in that it consists almost entirely of helices. The eight helical segments bend relative to each other and form a pocket that encloses a **heme group**: an iron-containing ring structure that binds O_2 (oxygen gas). Hydrophobic side chains on the inner sides of the helices help to ensure that the helices fold against one another correctly as the molecule is formed. Myoglobin and most other proteins are made from a single polypeptide chain.

Quaternary Structure

Some, although not most, proteins are made from two or more polypeptide chains. The overall shape of such a molecule, its **quaternary structure**, results both from how its subunits fit together and from how each subunit folds. Hemoglobin, a protein that brings oxygen from the lungs to the tissues and delivers it to myoglobin for storage, illustrates quaternary structure (Figure 3.20). Hydrophobic interactions, hydrogen bonds, and ionic bonds hold together four polypeptide chains (two each of two types). As the hemoglobin molecule takes up or releases oxygen, its four subunits shift their relative positions slightly, changing the quaternary structure. Ionic bonds are broken, exposing buried side chains that enhance the binding of molecular oxygen.

Each subunit of hemoglobin is folded like a myoglobin molecule, suggesting that both hemoglobin and myoglobin are evolutionary descendants of the same oxygen-binding ancestral protein. But on the surfaces where its subunits come in contact with each other—regions that on myoglobin are exposed to aqueous surroundings—hemoglobin has hydrophobic side chains where myoglobin has hydrophilic ones. Again, the chemical nature of side chains on individual amino acids determines how the molecule folds and packs in three dimensions.

The four levels of protein structure are summarized in Figure 3.21.

NUCLEIC ACIDS

The proteins of today exist because of the structures and activities of various **nucleic acids**. One group of these, the **DNAs**, or deoxyribonucleic acids, are giant polymers that carry the instructions for making proteins; another group, the **RNAs**, or ribonucleic acids, interpret and carry out the instructions coded in the DNAs.

Nucleic acids form from monomers called **nucleotides**, each of which consists of a pentose sugar, a phosphate group, and a nitrogenous (nitrogen-containing) base (Figure 3.22*a*,*b*). Molecules consisting of a pentose sugar and a nitrogenous base, but no phosphate group, are called nucleo*sides* (Figure 3.22*c*,*d*). In DNA, the pentose sugar is deoxyribose, which differs from the ribose found in RNA by one oxygen atom (see Figure 3.10).

The "backbones" of both RNA and DNA consist of alternating sugars and phosphates; the bases, which are attached to the sugars, project from the chain (Figure 3.23). The nucleotides are joined by **phosphodiester linkages** between the sugar of one nucleotide and the phosphate of the next. (The name *phosphodiester* comes from the fact that each phosphate is connected to two sugars.) Most RNA mole-

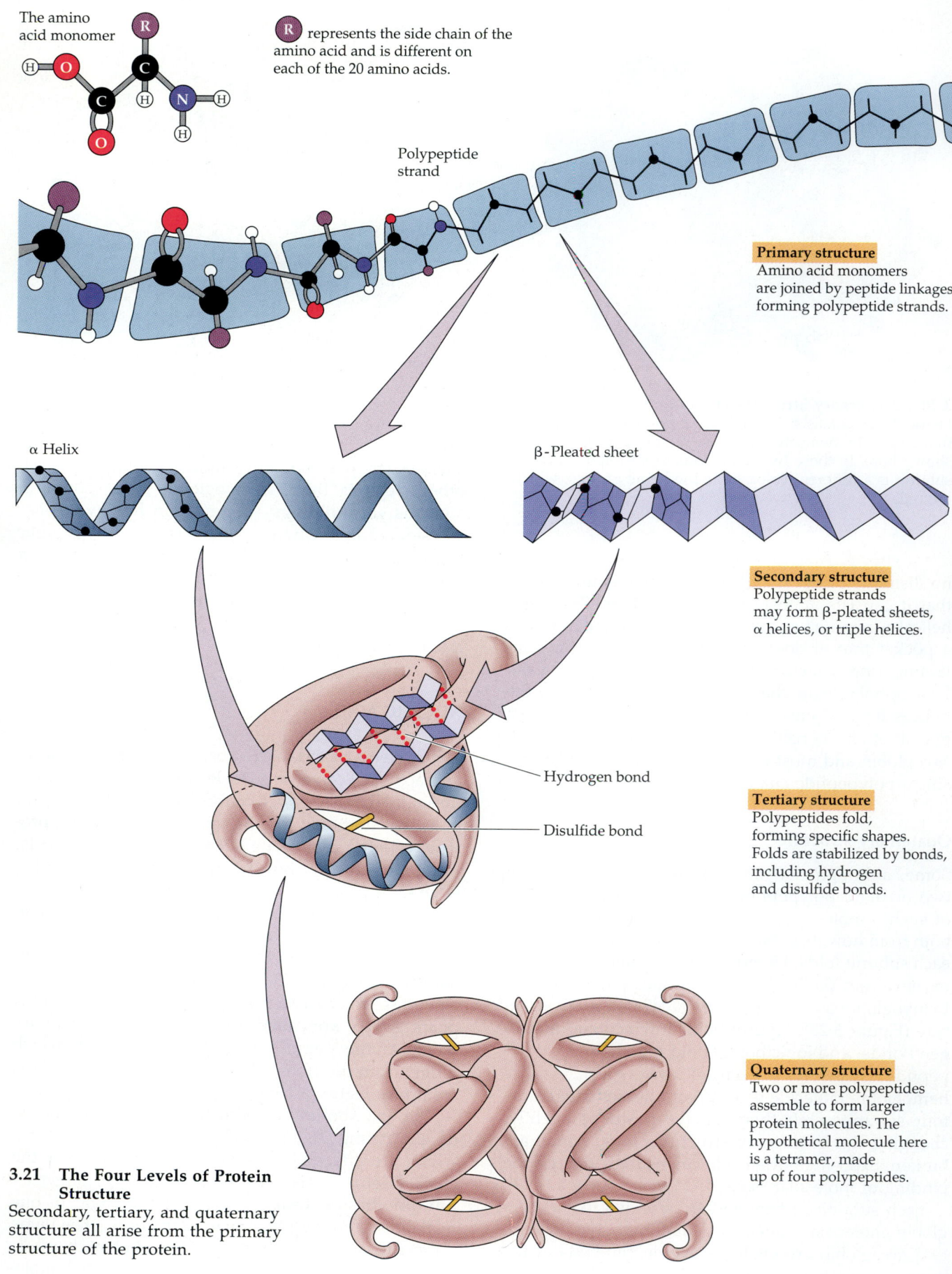

The amino acid monomer

R represents the side chain of the amino acid and is different on each of the 20 amino acids.

Polypeptide strand

Primary structure
Amino acid monomers are joined by peptide linkages, forming polypeptide strands.

α Helix

β-Pleated sheet

Secondary structure
Polypeptide strands may form β-pleated sheets, α helices, or triple helices.

Hydrogen bond

Disulfide bond

Tertiary structure
Polypeptides fold, forming specific shapes. Folds are stabilized by bonds, including hydrogen and disulfide bonds.

Quaternary structure
Two or more polypeptides assemble to form larger protein molecules. The hypothetical molecule here is a tetramer, made up of four polypeptides.

3.21 The Four Levels of Protein Structure
Secondary, tertiary, and quaternary structure all arise from the primary structure of the protein.

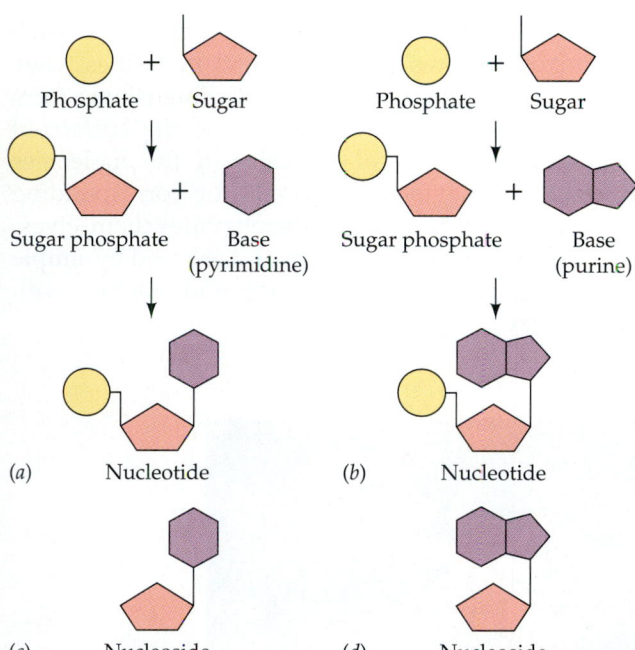

3.22 Components of a Nucleotide

(a,b) A sugar and a phosphate form a sugar–phosphate. In nucleotide synthesis, a nitrogen-containing base is then built on the sugar–phosphate in several steps not depicted here, forming the complete nucleotide monomer. The nitrogenous bases fall into two categories, as indicated here by their shapes. *(c,d)* A nucleo*side* consists of a sugar (*not* a sugar–phosphate) and a base.

cules are single-stranded: Each molecule consists of one polynucleotide chain. DNA, however, is usually double-stranded, with two polynucleotide chains held together by hydrogen bonding between their nitrogenous bases. The two strands are antiparallel—that is, they run in opposite directions.

Only four nitrogenous bases—and thus only four nucleotides—are found in DNA. The DNA bases are adenine, cytosine, guanine, and thymine. A key to understanding the structures and functions of nucleic acids is the principle of **complementary base pairing**: Particular bases pair only with certain other bases. In DNA, wherever one strand carries adenine, the other must carry thymine at the corresponding point. Wherever one chain has cytosine, the other must have guanine. The base-pairing rules for DNA and

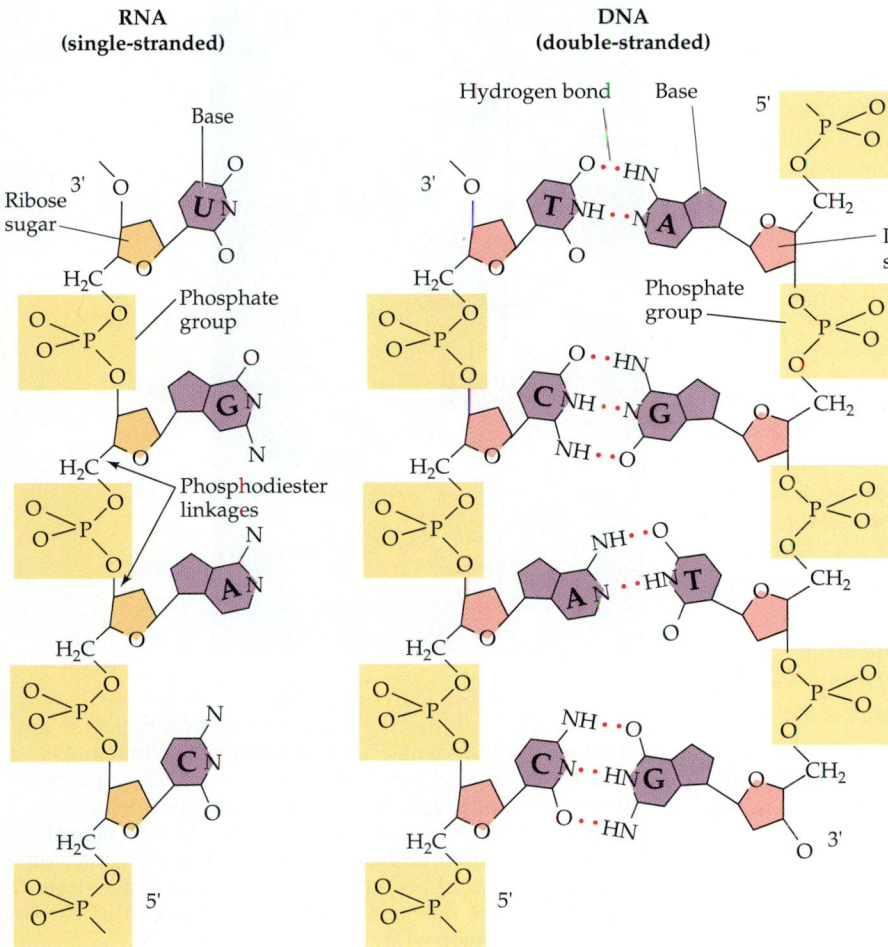

3.23 RNA versus DNA

A molecule of ribonucleic acid (RNA) is made up of a ribose sugar–phosphate backbone with a nitrogenous base attached to each sugar, as shown on the left. On the right is a portion of a double-stranded molecule of deoxyribonucleic acid (DNA); it has two deoxyribose sugar–phosphate backbones, running in opposite directions, with the bases between the strands. Red dots represent the hydrogen bonds that hold the strands together. For the bases, A = adenine, T = thymine, G = guanine, C = cytosine, and U = uracil.

TABLE 3.2
Base-Pairing Rules for the Nucleic Acids

IN DNA	WHEN RNA AND DNA INTERACT		WHEN RNA PAIRS WITH RNA
	RNA	DNA	
A pairs with T	A pairs with T		A pairs with U
T pairs with A	U pairs with A		U pairs with A
G pairs with C	G pairs with C		G pairs with C
C pairs with G	C pairs with G		C pairs with G

RNA are shown in Table 3.2. The pairing scheme maximizes hydrogen bonding between the two strands of DNA. Because one of the **purine** bases (adenine or guanine), which are large, always pairs with one of the **pyrimidine** bases (thymine or cytosine), which are small, all base pairs are the same size. Complementary base pairing between the two strands of the DNA molecule makes it possible to copy DNA molecules very faithfully (see Chapter 11).

Ribonucleic acids also have four different nucleotides, but the nucleotides differ from those of DNA. The **ribonucleotides** contain ribose rather than deoxyribose, and one of their four bases is different from that in DNA. The four principal bases in RNA are adenine, cytosine, guanine, and uracil (instead of thymine). Although RNA is generally single-stranded, complementary associations between nucleotides are important in the formation of new RNA strands, in determining the shapes of some RNA molecules, and in associations between RNA molecules during protein synthesis. Guanine and cytosine pair as in DNA, and uracil pairs with adenine. Adenine in an RNA strand can pair with either uracil (in an RNA strand) or with thymine (in a DNA strand).

The three-dimensional appearance of DNA is strikingly regular. The segment shown in Figure 3.24 could be from any DNA molecule. Through hydrogen bonding, the two complementary polynucleotide strands pair and twist to form a **double helix**. How regular this formation seems in comparison with the complex and varied structures of proteins! But this structural difference makes sense in terms of the functions of these two classes of compounds. DNA is a purely informational molecule. *The information carried by DNA resides simply in the sequence of bases carried in its chains.* This is, in a sense, like the tape of a tape recorder. The message must be read easily and reliably. A uniform molecule like DNA can be interpreted by standard molecular machinery, and the machinery can read any molecule of DNA—just as a tape player can play any tape of the right size. Proteins, on the other hand, have good reason to

differ so greatly. In particular, enzymes must each recognize their own specific "target" molecules. They do this by having a unique three-dimensional form that can match at least a portion of the surface of their targets; structural diversity in the molecules with which enzymes react calls for corresponding diversity in the structure of the enzymes themselves. DNAs are similar and uniform and are read by simple machinery; proteins are diverse and interact with many other compounds.

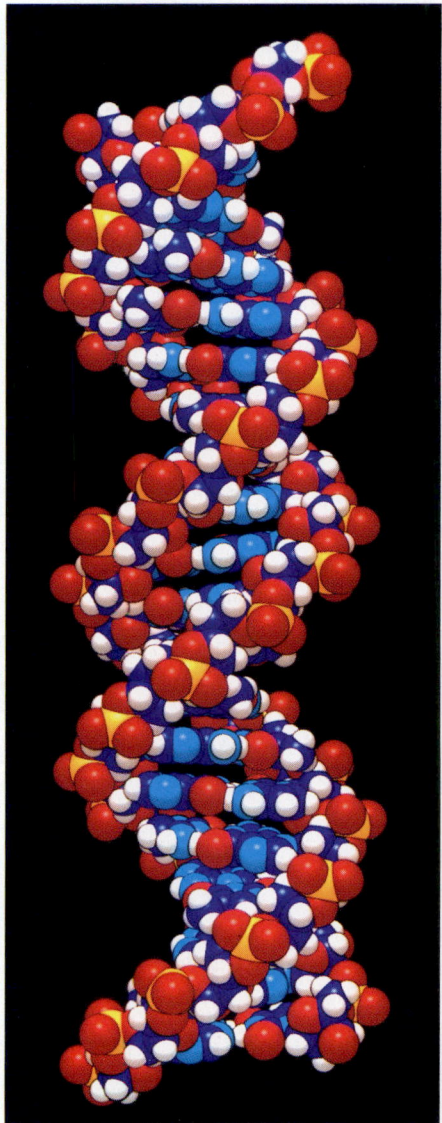

3.24 The Double Helix of DNA
The backbones of the two strands in a DNA molecule are coiled in a double helix. This computer drawing shows the atoms in a length of DNA containing 20 base pairs. Allow your eye to follow the yellow phosphorus atoms and their attached red oxygen atoms in the two helical backbones. The paired bases are stacked in the center of the coil and should become apparent if you concentrate on the light blue nitrogen atoms and the dark blue carbon atoms. The smaller white atoms are hydrogens.

GLYCOLIPIDS, GLYCOPROTEINS, LIPOPROTEINS, AND NUCLEOPROTEINS

We have been treating the classes of macromolecules as if each were completely separate from the others. In fact, certain macromolecules of different classes attach to one another to form covalently bonded products of great importance to cells. Many proteins have attached carbohydrate chains; this combination is called a glycoprotein. The carbohydrate chain often determines the placement of the glycoprotein within the cell. Other carbohydrate chains bind to lipids, resulting in glycolipids. Glycolipids usually reside in the membrane surrounding a cell, with the carbo-hydrate chain extending out into the cell's environment. The human blood group antigens, which determine the ABO blood types, are carbohydrates that combine with either proteins or lipids. When a cell becomes cancerous, both glycolipids and glycoproteins are modified.

Proteins bind DNA, forming nucleoproteins that regulate the activities of DNA. Still other proteins, in combination with cholesterol and other lipids, form lipoproteins. The lipoproteins make it possible to move very hydrophobic lipids through the predominantly hydrophilic environment of the human body, and they deliver cholesterol and certain other lipids to the appropriate cells and tissues within the body.

SUMMARY of Main Ideas about Large Molecules in Living Things

Organisms take in large molecules, break them down into their constituent parts, and from these parts assemble other necessary large molecules.

Proteins, nucleic acids, and large carbohydrates are polymers formed from monomers by condensation reactions, which release water.
Review Figures 3.4 and 3.15

Lipids are compounds that are insoluble in water but readily soluble in organic solvents.

Lipids differ from other molecules discussed in this chapter in that they do not form true polymers.

Triglycerides are composed of glycerol and fatty acids and serve as stored fuel.
Review Figures 3.3 and 3.4

Lipid molecules have large hydrophobic regions, so lipids present in an aqueous system tend to aggregate like oil in water.

Phospholipids tend to aggregate to form a continuous bilayer, as in biological membranes, and they control molecular traffic through the membranes.
Review Figures 3.1, 3.5, and 3.6

Carbohydrates (monosaccharides, oligosaccharides, and polysacchrides) have formulas based on CH_2O.
Review Figures 3.10, 3.11, and 3.12

Monosaccharides serve as fuel and as building blocks for polysaccharides.

The polysaccharides starch and glycogen are storage compounds.

Cellulose imparts strength to plant cell walls.

A protein is composed of one or more polypeptide chains. Amino acids are the monomers of polypeptides.
Review Figure 3.15 and Table 3.1

A protein's primary structure is the sequence of amino acids in its polypeptide chains.
Review Figure 3.21

Regular, repeated patterns of orientation such as α helices and β pleated sheets are a protein's secondary structure.
Review Figures 3.17 and 3.21

How a polypeptide is folded, its overall shape, is its tertiary structure.
Review Figures 3.18, 3.19, and 3.21

The tertiary structure of a polypeptide arises spontaneously from its primary structure.

For a protein consisting of more than one polypeptide chain, the spatial arrangement of the polypeptides relative to each other is the quaternary structure of the protein.
Review Figures 3.20 and 3.21

DNA molecules are composed of only four different monomers joined by complementary base pairing.
Review Figures 3.22, 3.24, and Table 3.2

Although the structure of DNA is very uniform, the sequence of monomers within the polymer contains an enormous amount of information.

RNA differs from DNA in the sugar component, in one of the bases, and sometimes in the number of polynucleotide stands.
Review Figure 3.23

The different classes of large molecules aggregate with one another to form important compounds such as glycolipids, glycoproteins, lipoproteins, and nucleoproteins.

The structures of large molecules are the keys to how they perform their functions in organisms.

SELF-QUIZ

1. All lipids
   a. are triglycerides.
   b. are polar.
   c. are hydrophilic.
   d. are polymers.
   e. are more soluble in nonpolar solvents than in water.

2. Which of the following is *not* a lipid?
   a. A steroid
   b. A fat
   c. A triglyceride
   d. A biological membrane
   e. A carotenoid

3. All carbohydrates
   a. are polymers.
   b. are simple sugars.
   c. consist of one or more simple sugars.
   d. are found in biological membranes.
   e. are more soluble in nonpolar solvents than in water.

4. Which of the following is *not* a carbohydrate?
   a. Glucose
   b. Starch
   c. Cellulose
   d. Hemoglobin
   e. Deoxyribose

5. All proteins
   a. are enzymes.
   b. consist of one or more polypeptides.
   c. are amino acids.
   d. have quaternary structures.
   e. have prosthetic groups.

6. Which statement is *not* true of the primary structure of a protein?
   a. It may be branched.
   b. It is determined by the structure of the corresponding DNA.
   c. It is unique to that protein.
   d. It determines the tertiary structure of the protein.
   e. It is the sequence of amino acids in the protein.

7. The amino acid leucine (Table 3.1)
   a. is found in all proteins.
   b. cannot form peptide linkages.
   c. is likely to appear in the part of a membrane protein that lies within the phospholipid bilayer.
   d. is likely to appear in the part of a membrane protein that lies outside the phospholipid bilayer.
   e. is identical to the amino acid lysine.

8. The quaternary structure of a protein
   a. consists of four subunits—hence the name *quaternary*.
   b. is unrelated to the function of the protein.
   c. may be α, β, or γ.
   d. depends on covalent bonding among the subunits.
   e. depends on the primary structures of the subunits.

9. All nucleic acids
   a. are polymers of nucleotides.
   b. are polymers of amino acids.
   c. are double-stranded.
   d. are double-helical.
   e. contain deoxyribose.

10. Which statement is *not* true of condensation reactions?
    a. Protein synthesis results from them.
    b. Polysaccharide synthesis results from them.
    c. Nucleic acid synthesis results from them.
    d. They consume water as a reactant.
    e. Different ones produce different kinds of macromolecules.

FOR STUDY

1. Phospholipids make up a major part of every biological membrane; cellulose is the major constituent of the cell walls of plants. How do the chemical structures and physical properties of phospholipids and cellulose relate to their functions in cells?

2. Suppose that, in a given protein, one lysine is replaced by aspartic acid (Table 3.1). Is this a change in primary or secondary structure? How might it result in a change in tertiary structure? In quaternary structure.?

3. If there are 20 different amino acids commonly found in proteins, how many different dipeptides are there? How many different tripeptides? How many different polypeptides composed of 200 amino acid subunits? If there are four different nitrogenous bases commonly found in RNA, how many different dinucleotides are there? How many different trinucleotides? How many different single-stranded RNAs composed of 200 nucleotides?

4. Contrast the structures of hemoglobin, a DNA molecule, and a protein that spans a biological membrane.

READINGS

Branden, C. and J. Tooze. 1991. *Introduction to Protein Structure*. Garland, New York. A well-illustrated book suitable for undergraduates.

Doolittle, R. F. 1985. "Proteins." *Scientific American*, October. A strikingly illustrated treatment of protein structure and evolution.

Stryer, L. 1995. *Biochemistry*, 5th Edition. W. H. Freeman, New York. A relatively advanced but beautiful reference on the subjects of this chapter; outstanding illustrations, concise descriptions, clear prose.

Voet, D. and J. G. Voet. 1990. *Biochemistry*. John Wiley & Sons, New York. A fine advanced textbook with outstanding illustrations.

Some free-living organisms are single cells. Even for multicellular organisms, the single cell, not the organism itself, is the basic unit of life. With a geranium from your garden you can demonstrate for yourself that the unit of life is something smaller than the whole organism. You can take a cutting consisting of a bit of stem and a leaf or two, put it in soil and care for it, and end up with an entire plant. Going a step further, scientists can isolate single cells from many plant species and induce them, by treatment with natural substances that control plant growth, to develop into whole plants.

If we try to go to a level below an entire cell, however, we come to the end of the line. Subcellular structures—organelles—such as nuclei and chloroplasts may be isolated from cells and induced to carry out their normal functions, but they can never be induced to regenerate whole cells, let alone an entire organism. The inability of even the nucleus to produce any sort of life form verifies that it is the whole cell that is the basic unit of life.

CELLS AND THE CELL THEORY

The basic unit of organization in living things is the cell. All organisms are composed of cells, and all cells come from preexisting cells—these two statements constitute the **cell theory** (Chapter 1). Even viruses, which are not cells themselves, are entirely dependent on the presence and chemical machinery of cells for their reproduction. Cells from each of the six kingdoms are shown in Figure 4.1.

Most cells are tiny. They have volumes from 1 to 1,000 μm^3. The eggs of some birds are enormous exceptions, to be sure, and individual cells of several types of algae are large enough to be viewed with the unaided eye. Neurons (nerve cells) have volumes that fit within the "normal" range, but they often have fine projections that may extend for meters, carrying signals from one part of a large animal to another. In spite of these special cases, we may generalize and say that cells are very small objects (Box 4.A). What else do cells have in common?

COMMON CHARACTERISTICS OF CELLS

Cells must do many things in order to survive. They must obtain and process energy, they must convert the genetic information of DNA into protein, and they must keep certain biochemical reactions separate from other, incompatible reactions that must occur simultaneously. Structures within cells carry out these functions, as we will examine in detail in the remaining chapters of Parts One and Two. We begin

Some Cells Can Give Rise to Whole Organisms
Each of these young carrot plants began as one or a few cells in culture.

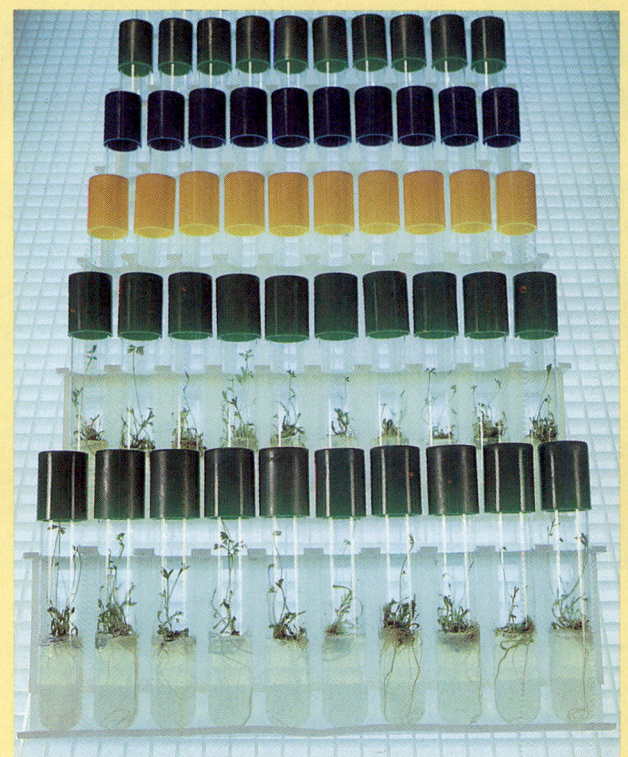

4

Organization of the Cell

4.1 Cells Come in Many Shapes

In these micrographs we see (a) a filamentous bacterium in the kingdom Archaebacteria; (b) two species of filamentous cyanobacteria (kingdom Eubacteria); (c) *Euglena*, a plantlike protist; (d) *Paramecium*, an animallike protist; (e) brewer's yeast, a fungus; (f) "leaf" cells of a moss, packed with green, photosynthetic chloroplasts; (g) blood cells of a frog; and (h) mammalian cells grown in culture in the laboratory.

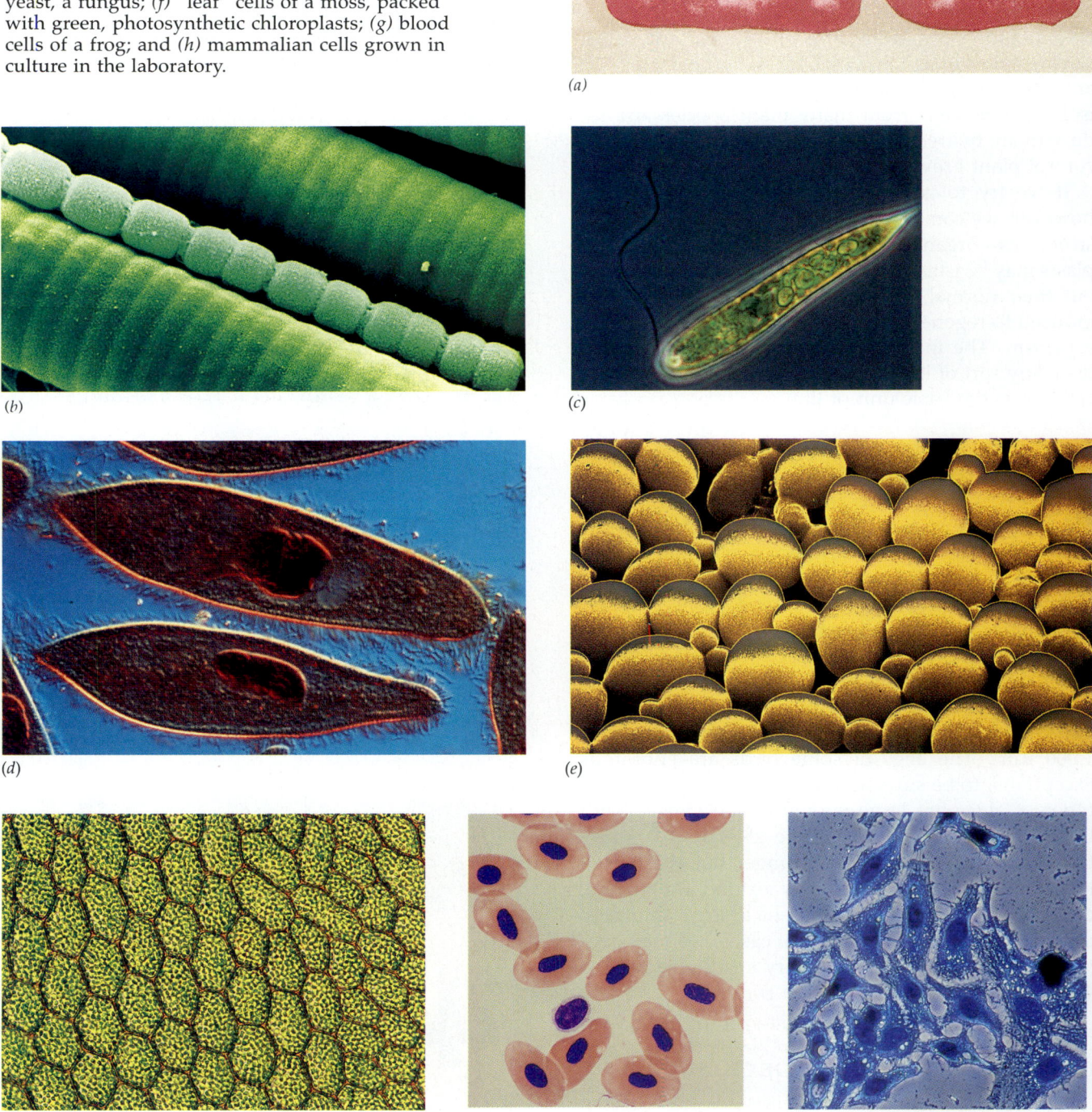

(a)

(b)

(c)

(d)

(e)

(f)

(g)

(h)

in this chapter by considering the component parts of cells.

Among all the kinds of cells there are two distinct general arrangements, with only a few intermediate forms in evidence. One general arrangement, the simpler, is the **prokaryotic cell,** characteristic of the kingdoms Eubacteria and Archaebacteria. Organisms in these kingdoms are called prokaryotes. Their single cells lack nuclear compartments and membrane-bounded internal compartments. The rest of the living world has **eukaryotic cells**—cells that contain membrane-bounded nuclei. Eukaryotic cells have

BOX 4.A

The Sizes of Things

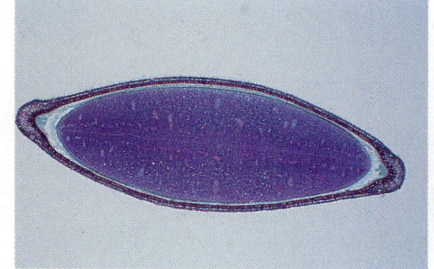 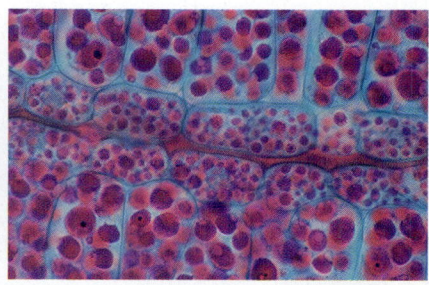

On the left, a pumpkin seed is shown magnified 10 times. When the same seed is magnified 1,000 times, we see the starch grains in the seed's cells as red balls.

Biologists study objects of very different sizes, ranging from molecules with diameters of less than one nanometer to organisms that are many meters long. You need a sense of the sizes of things to appreciate how they function and how their parts interact. The sizes of some objects are compared in the diagram, which also indicates the methods by which they are usually viewed. To help you develop a sense of sizes we will sometimes identify the magnifications of figures, as in the pictures above, taken with a light microscope. However, you will need to remember the approximate sizes of cells and their parts.

Inside the front cover of this book, there is a table of measurement units and their symbols. You may want to refer to it several times as you read this chapter. What is the size range 1 to 1,000 μm^3? If we assume that many cells are almost spherical, this range of volumes corresponds with a range of diameters of about 1.2 to 12 μm.

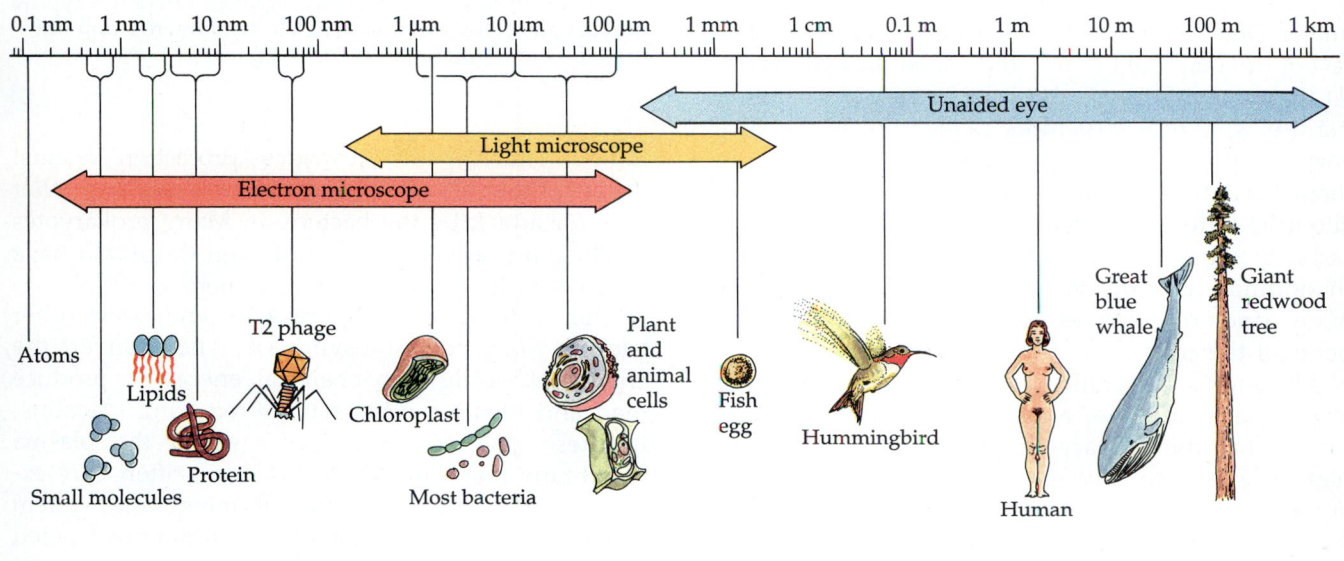

other internal compartments (called organelles) that, like the nucleus, are surrounded by membranes. Organisms with this type of cell are known as eukaryotes. Both prokaryotes and eukaryotes have prospered through many hundreds of millions of years of evolution, and both are great success stories.

PROKARYOTIC CELLS

We will first consider the characteristics that cells throughout the kingdoms Eubacteria and Archaebacteria have in common. Then, in the next section, we will discuss structural features that typify many, but not all, prokaryotes.

Features Shared by All Prokaryotic Cells

All prokaryotic cells have, without exception, a plasma membrane, a nucleoid, and cytoplasm filled with ribosomes. The **plasma membrane** separates the cell from its environment and regulates the traffic of materials in and out of the cell. The **nucleoid** is a relatively clear region (as seen under the electron microscope) that contains the hereditary material (DNA) of the cell. Each prokaryotic cell has at least

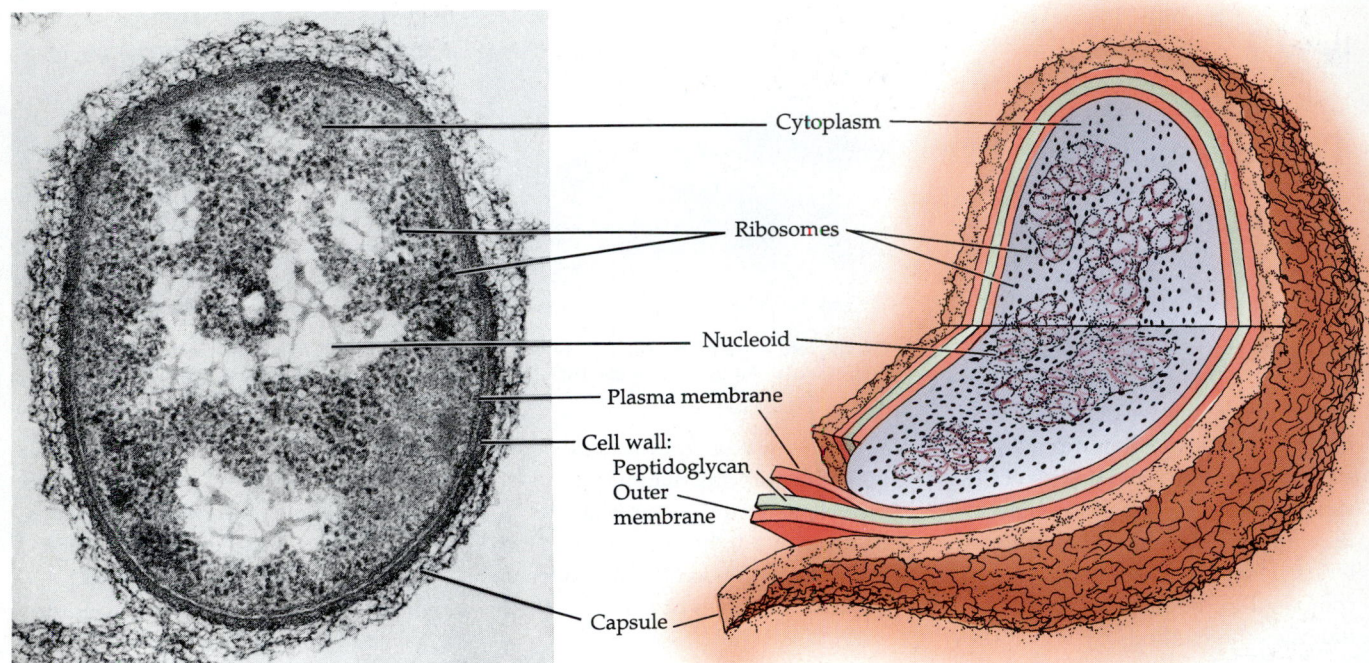

4.2 A Bacterial Cell
The eubacterium *Pseudomonas aeruginosa* illustrates typical prokaryotic cell structures. The electron micrograph on the left is magnified about 80,000 times.

one nucleoid; some cells have more than one. The rest of the material in the cell is called the **cytoplasm.** At high magnification, the cytoplasm appears full of minute, spherical structures called **ribosomes.** Prokaryotic ribosomes are approximately 15 to 20 nm in diameter and consist of three molecules of RNA and about 50 different protein molecules. Ribosomes coordinate the synthesis of proteins (see Chapter 11). In addition to ribosomes, the cytoplasm contains many kinds of enzymes and other chemical constituents of the cell.

Although structurally simple (consisting primarily of plasma membrane, nucleoid, and ribosome-filled cytoplasm), the prokaryotic cell is functionally complex. Enzymes in prokaryotic cells direct thousands of chemical reactions, with the cell's DNA serving as the molecular memory that allows successive generations of a given cell to be very much alike.

Other Features of Prokaryotic Cells

All but the simplest prokaryotic cells have at least a few more structural complexities. Most, for example, have a **cell wall** outside the plasma membrane (Figure 4.2). The rigidity of the wall lends support to the cell and determines its shape. The cell walls of most eubacteria consist primarily of **peptidoglycan,** a polymer of amino sugars, cross-linked to form a single molecule around the entire cell! Outside the cell wall there is often a layer of slime (composed mostly of polysaccharide), referred to as a **capsule.** The capsules of some bacteria may protect them from attack by white blood cells within the bodies of animals they

infect. The capsule provides protection against drying of the cell, and in some cases it may trap other cells for attack by the bacterium. Many prokaryotes produce no capsule at all, and even those that have capsules will not die if they lose them.

The cyanobacteria (Figure 4.1*b*) and some other bacteria carry on photosynthesis. They convert the energy of sunlight to chemical energy to produce food and to drive other energy-requiring reactions. In these photosynthetic prokaryotes, the plasma membrane folds into the cytoplasm—often very extensively—to form an internal membrane system containing chlorophyll and other compounds needed for photosynthesis (Figure 4.3). Other bacteria possess different sorts of membranous structures called **mesosomes,** which may function in cell division or in various energy-releasing reactions. Like the photosynthetic membrane systems, mesosomes are formed by infolding of the plasma membrane. They remain attached to the plasma membrane and never form the free-floating, isolated, membranous organelles that are characteristic of eukaryotic cells (Figure 4.4). The fluid portion of a cell's cytoplasm, in which ribosomes and membranous structures are found, is called the **cytosol.**

Some prokaryotes swim by using appendages called **flagella** (Figure 4.5*a*). A single flagellum, made of a protein called flagellin, looks something like a tiny corkscrew. It spins about its axis like a propeller,

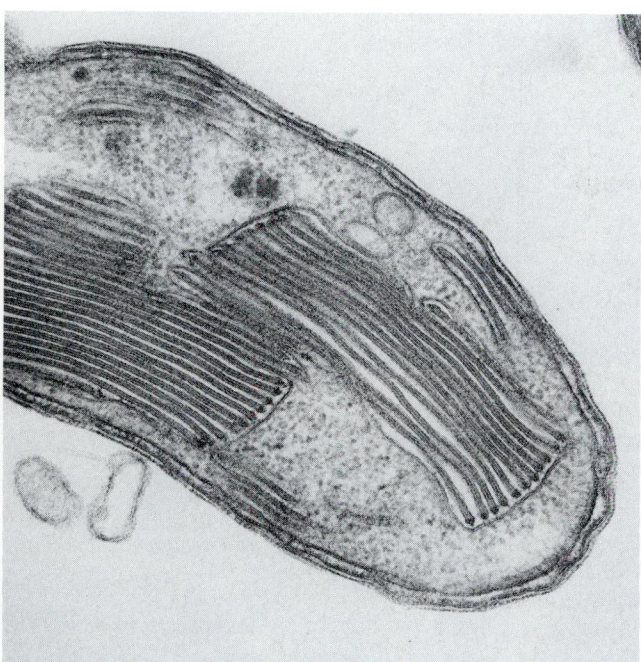

4.3   Photosynthetic Prokaryotes Have Extensive Internal Membranes
Photosynthetic membranes fold into "stacks" inside a bacterial cell; such organized collections of internal membranes contradict the mistaken notion that bacteria are nothing more than tiny bags of molecules.

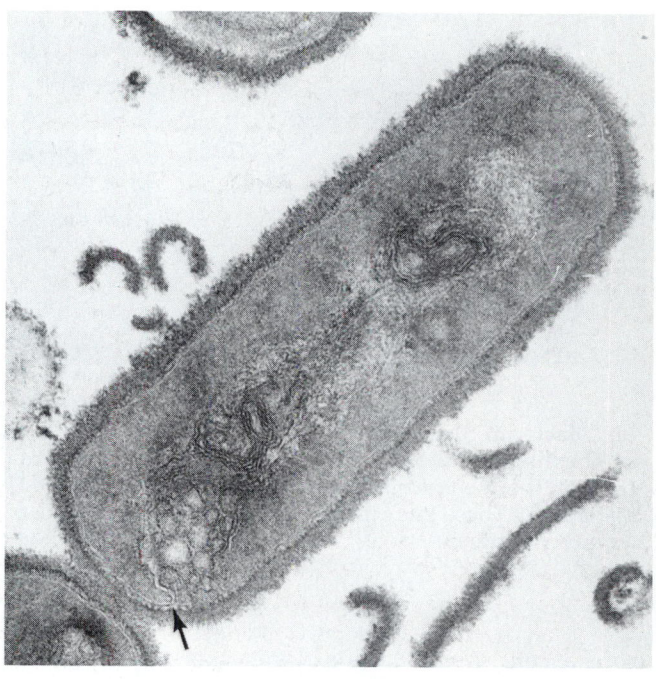

4.4   A Mesosome Remains Attached to the Plasma Membrane
Convoluted membranes of a large mesosome extend throughout this bacterial cell. At the lower left (arrow) you can see that the membrane of the mesosome is continuous with the plasma membrane.

driving the bacterium along. Ring structures anchor the flagellum to the plasma membrane and, in some bacteria, to the outer membrane of the cell wall (Figure 4.5*b*). The fact that flagella actually cause the motion of the cell can be shown by removing these

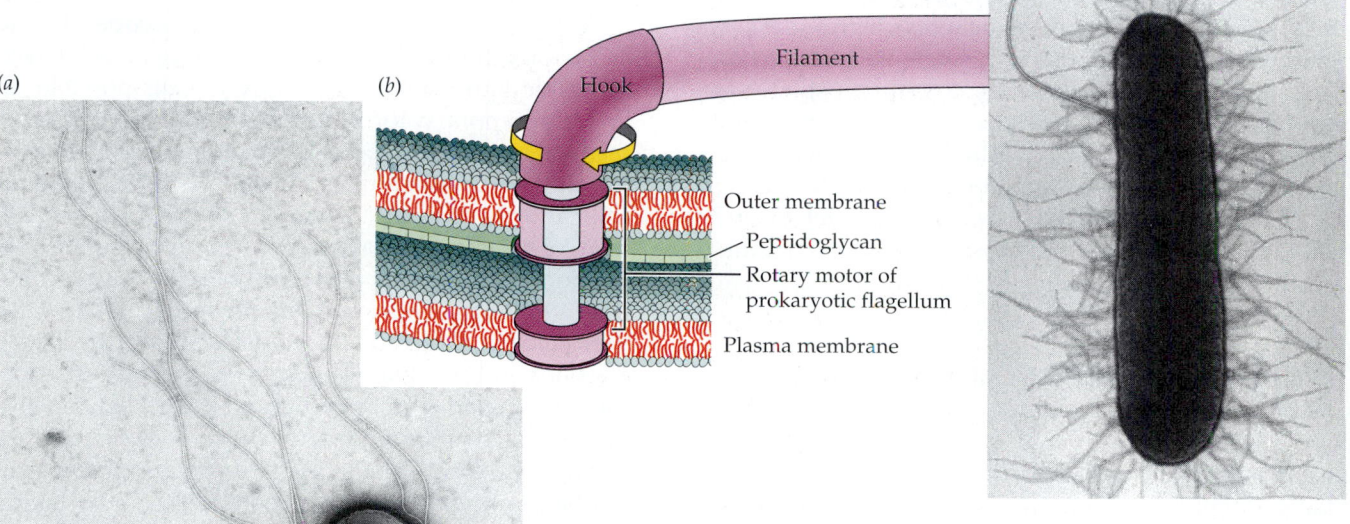

(a)   (b)   Filament   Hook

Outer membrane
Peptidoglycan
Rotary motor of prokaryotic flagellum
Plasma membrane

4.5   Prokaryotic Projections
(*a*) Wavy, whiplike flagella used in locomotion extend from this bacterium. (*b*) The basal ends of these tiny structures are complex; the mechanism by which bacterial flagella rotate is still under investigation. (*c*) The small, hairlike pili bristling from the surface of this cell of *Escherichia coli* help it adhere to other cells.

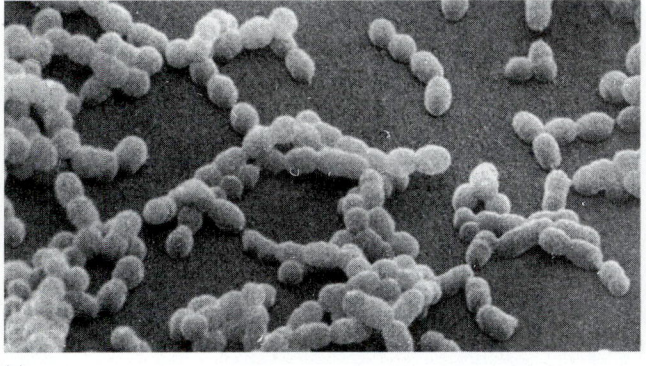

(a)

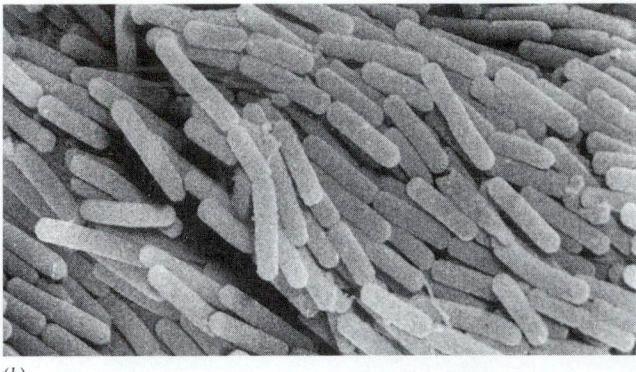

(b)

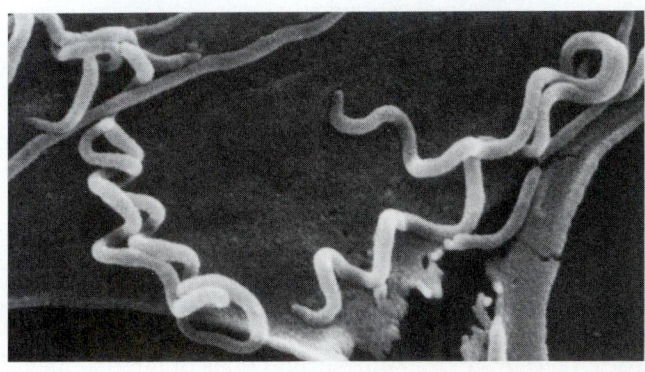

(c)

4.6 Bacterial Shapes and Aggregations
(a) These spherical cocci of an acid-producing bacterium grow on tooth enamel and cause decay. (b) Rod-shaped bacteria. (c) These spirochetes (*Treponema pallidum*) cause syphilis in humans.

from a cell. When this is done, the cell no longer moves. In addition, if the tip of a flagellum is attached to an immovable cell, the spinning of the flagellum causes the entire cell to rotate.

Pili project from the surface of some groups of bacteria (Figure 4.5c). Shorter than flagella, these threadlike structures seem to help bacteria adhere to other cells—to one another during mating, and also to animal cells.

Kinds of Prokaryotes

Bacteria are often categorized by the shapes of their cells. Spherical cells are **cocci;** rod-shaped cells are **bacilli;** and helical cells, sometimes coiled like corkscrews, are **spirilla** or **spirochetes** (Figure 4.6). Although each prokaryote is a single cell, some types are seldom seen singly. Many types are usually seen in chains, small clusters, or even colonies with hundreds of individuals. The diversity within the bacterial kingdoms is the subject of Chapter 22.

Many prokaryotes have been used extensively in biological research. The most familiar bacterium is *Escherichia coli* (Box 4.B).

PROBING THE SUBCELLULAR WORLD: MICROSCOPY

Many significant advances in our knowledge of cells have depended upon the **resolution,** or **resolving power,** of the instruments we use for magnifying tiny objects. We define the resolving power of a lens or microscope as the smallest distance separating two objects that allows them to be seen as two distinct things rather than as a single entity. For example, most humans can see two fine parallel lines as dis-

tinct markings if they are separated by at least 0.1 millimeter (0.1 mm); if they are any closer together, we see them as a single line. Thus, the resolving power of the human eye is about 0.1 mm, which is the approximate diameter of the human egg. To see anything smaller, we must use some form of microscope.

The invention of the **light microscope** (Figure 4.7a)—so called because it allows objects to be viewed in visible light—made the study of cells possible. In its contemporary form, the light microscope has a resolving power of about 200 nanometers (200 nm = 0.2 μm = 0.0002 mm), so it gives a useful view of cells and can reveal features of some of the subcellular organelles. Today, half a century after the invention of the more powerful electron microscope, the light microscope remains an important tool for the biologist. Many of the illustrations in this book are photographs taken through the light microscope; they are called photomicrographs. The light microscope has its limitations, however—principally its 200-nm resolving power. This resolving power cannot be improved by adding more lenses or by taking photomicrographs and then enlarging them. Such enlargements can be made, but they do not increase the resolution; as the images become larger, they simply become fuzzier.

Figure 4.2 and many other figures throughout this book show cellular structures that are far too small

BOX 4.B

The Best-Known Prokaryote: *Escherichia coli*

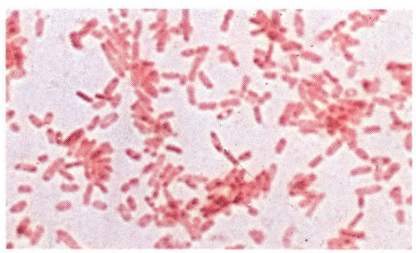

Cells of *E. coli* appear as red rods in this light micrograph of a stained preparation.

Without a doubt, the best-understood living creature is a humble bacterium living in our intestinal tracts: *Escherichia coli*—or, as it is commonly known, *E. coli*. This rod-shaped bacterium (shown in the figure) is about 2 µm in length and 0.8 µm in diameter, giving it a volume of about 1 µm³ and a weight of approximately 10^{-12} g (one-millionth of one-millionth of a gram). *E. coli* are about 100 times larger than the smallest living cells, the mycoplasmas (see Chapter 22). Immediately outside the *E. coli* plasma membrane is a cell wall about 10 nm thick, and projecting from the cell are flagella and pili. The flagella gather into a bundle and push the bacterium at a speed such that if the bacterium were magnified to human dimensions it would move at 30 miles per hour! Every second or so, the bundle of flagella separates and re-forms, causing the cell to change its direction.

An *E. coli* cell is approximately 70 percent water, 15 percent protein, 1 percent DNA, 6 percent RNA, 3 percent carbohydrate, 2 percent lipid, and 1 percent simple ions such as K^+ (potassium ions), as well as small amounts of other substances. It has 15,000 to 30,000 ribosomes and contains from one to four identical molecules of DNA. This genetic material is only about 1/500 as much DNA as is contained in a single cell of a human being. Nonetheless, as simple as it is, each prokaryotic cell of *E. coli* makes thousands of specific proteins.

E. coli is favorable for biological experimentation for several reasons. As noted, it is very small. Under the best conditions, it can divide once every 20 minutes, whereas most animal cells require about a day to go through a division cycle. Because of this rapid division, immense populations of *E. coli* can be grown very quickly. One cell can become 8 in an hour, 512 in 3 hours, over a billion in 10 hours, and more than 10^{21} in a day (in principle, that is, and with unlimited food and space). Its nutritional requirements are simple: water, some mineral ions, and an energy source such as glucose. Unlike some bacteria, most varieties of *E. coli* do not present a great health hazard, so they can be grown without extensive precautions. Many genetic strains with different, known characteristics are readily available. As a result of these and other advantages, *E. coli* has been used in countless investigations of genetics, biochemistry, and other areas of biology. It is extensively used in research on recombinant DNA ("genetic engineering"), a topic considered in detail in Chapter 14.

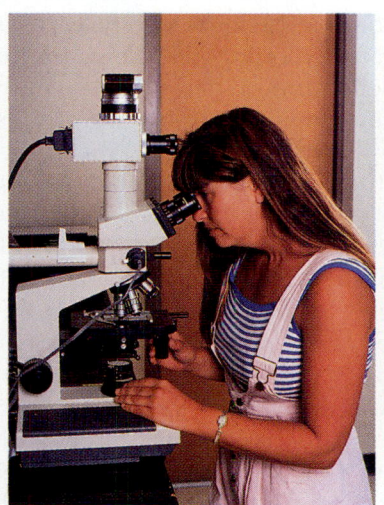

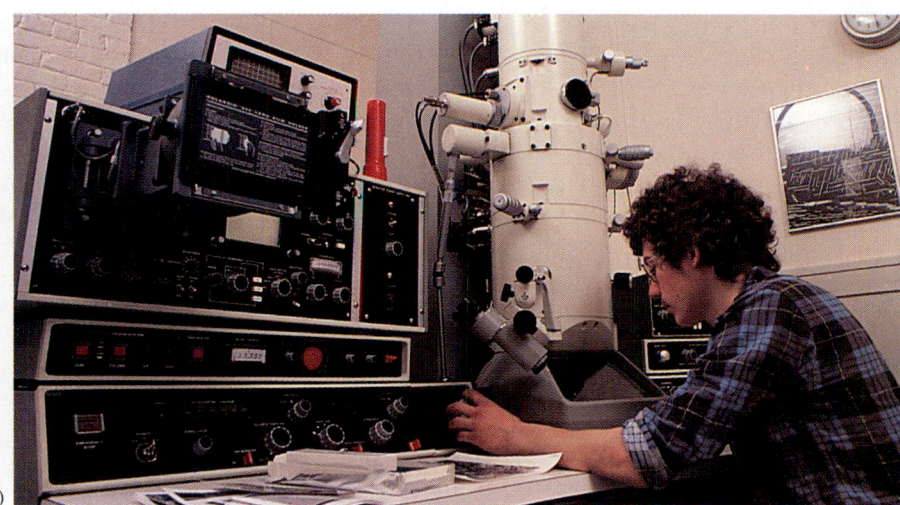

(a) (b)

4.7 Exploring Cells with Microscopes
(a) A research-quality light microscope. A camera is mounted at the top of the instrument for making photomicrographs. (b) A transmission electron microscope. The magnets that focus the electron beam are in the tall cylinder.

to be resolved with the light microscope. Ribosomes, for example, being 20 nm or less in diameter, cannot be resolved as individual objects under the light microscope. On the other hand, ribosomes are readily resolved with the **electron microscope** (Figure 4.7b). An electron microscope uses powerful magnets as lenses to focus an electron beam, much as the light microscope employs glass lenses to focus a beam of light. We cannot see electrons, however, so they are directed at a fluorescent screen or a photographic film to create an image we can see. The resulting images are called electron micrographs, many of which appear in this and later chapters. The resolving power of modern electron microscopes is about 0.2 nm, but no biological specimen has yet been seen in such detail with an electron microscope. One reason is that the energy of the electron beam at that power is so great that it destroys biological molecules before they can be seen. Because of this and other technical limitations, most electron micrographs resolve detail no finer than 2 nm, and even the best micrographs rarely resolve detail as fine as 1 nm. This corresponds to a resolving power about 100,000 times finer than that of the human eye.

There are two types of electron microscopy. In **transmission electron microscopy,** which produced Figures 4.3 and 4.4, electrons pass *through* a sample. Transmission electron microscopy is used to examine thin slices of objects—similar to the sections shaved off a material and placed on the slides used with a light microscope, but much thinner.

In **scanning electron microscopy,** electrons are directed at the surface of the sample, where they cause other electrons to be emitted; the scanning electron microscope focuses these secondary electrons on a viewing screen. Scanning electron microscopy reveals the *surface* structures of three-dimensional objects, such as the bacteria shown in Figure 4.6. A scanning electron microscope has a resolving power no better than 10 nm, so scanning electron micrographs are usually at a somewhat lower magnification than that of transmission electron micrographs.

You might think that with such a resolving power the electron microscope would be used for all microscopic studies, but this is not so. For some applications it would be sheer overkill, like using a magnifying glass to get an overall view of an elephant. A more important limitation is that biological samples have to be killed and dehydrated before they can be examined with an electron microscope. Light microscopy, on the other hand, allows us to observe *living* cells.

Samples for transmission electron microscopy have to be thinly sliced. Samples are also often sliced before examination under a light microscope. To get a reasonable three-dimensional view of large cells or tissues with a microscope, therefore, one looks at many successive slices, rather like examining successive slices of Swiss cheese to "see" one of the holes.

THE EUKARYOTIC CELL

The vast majority of living species, including all animals, plants, fungi, and protists, have cells that are structurally more complex than those of the prokaryotes. Compare Figures 4.8 and 4.9 with Figure 4.2 for a quick sense of the prominent differences. Eukaryotic cells are full of membranous structures of wondrous diversity. One or two membranes enclose each of many of these structures, which carry on particular biochemical functions. These structures are neatly packaged subsystems, with membranes to control their functions and to regulate what gets in and out. Some of the subsystems are like little factories that make specific products. Others are like power plants that take energy in one form and convert it to a more useful form (see Chapter 6). These membranous subsystems, as well as other structures (such as ribosomes) that lack membranes but possess distinctive shapes and functions, are called **organelles.** Like prokaryotic cells, eukaryotic cells have a plasma membrane, cytoplasm, and ribosomes.

Roles of Membranes in Eukaryotic Cells

In 1952 the first people to look at reasonably clear electron micrographs of eukaryotic cells were stunned by the complexity of what they saw. Based on chemical and biological observations, scientists had expected to find cells surrounded by plasma membranes, even though these structures were not resolved with the light microscope. It was also known that cells teemed with organelles, and it was suspected that at least some of these organelles were bounded by membranes. It is doubtful, however, that anyone expected membranes to be as profuse in the eukaryotic cell as they actually are. What are all those membranes for? How do they function? What is their structure—or structures, if all membranes are not alike? We will deal with these questions in Chapter 5, but will note a few of the most basic ideas here.

Biological membranes regulate molecular traffic from one side of the membrane to the other. The hydrophobic interior of the membrane is a barrier to the passage of many materials, especially polar materials that are readily soluble in water (see Figure 3.6). Many materials are transported through the membrane with the help of highly specific protein molecules. As discussed later in this chapter, the plasma membrane of some cells can fold inward and form compartments called vesicles in the cell to trap a portion of the cell's environment, as if taking a bite out of it.

Membranes participate in many activities besides transport. They are staging areas for interactions between cells. For example, immunologically active white blood cells recognize and interact with their targets by means of specific protein molecules built into their plasma membranes (see Chapter 16). The proper development and organization of multicellular animals depends upon recognition reactions between cells, which are mediated by the plasma membrane (see Chapter 17). Many intracellular membranes carry the components responsible for energy transformations in cells. Chlorophyll and other substances necessary for energy-capturing photosynthesis are bound in a specific way to membranes in chloroplasts, one type of organelle. The electron carriers that help transform food energy into a form the cell can use are organized as parts of the inner membrane of mitochondria, another organelle type. In many respects, a discussion of eukaryotic cells is a discussion of membranes that are specialized for various cellular activities.

INFORMATION-PROCESSING ORGANELLES

Living things depend on accurate, appropriate information. Information is *stored* as the sequence of bases in DNA molecules. The bulk of the DNA in eukaryotic cells resides in the nucleus. Information is *translated,* from the language of DNA into the language of proteins, on the surfaces of the ribosomes.

The Nucleus

Typically, the **nucleus** is the largest organelle in the eukaryotic cell (Figures 4.8 and 4.9). Most animal cells have a nucleus that is approximately 5 μm in diameter. The possession of a membrane-bounded nucleus is the defining property of the eukaryotic cell. (Remember that in prokaryotes there is no membrane separating the nucleoid from the surrounding cytoplasm.) As viewed under the electron microscope, a nucleus is surrounded by *two* membranes separated by a few tens of nanometers. The **nuclear envelope,** as this pair of membranes is called, is perforated by **nuclear pores** approximately 9 nm in diameter (Figure 4.10). Each pore is surrounded by eight large protein granules arranged in an octagon where the inner and outer membranes merge. RNA and water-soluble molecules pass through the pores to enter or leave the nucleus.

The outer membrane of the nuclear envelope sometimes folds outward into the cytoplasm and is continuous with the network called the endoplasmic reticulum (discussed later in this chapter). The endoplasmic reticulum and, to a lesser extent, the outer surface of the outer membrane of the nuclear enve-

lope often carry great numbers of ribosomes. There are no ribosomes on the inner surface of the outer membrane, or on either surface of the nuclear envelope's inner membrane.

In the nucleus, DNA combines with proteins in a fibrous complex called **chromatin.** We will consider the structure of chromatin in Chapter 9. Throughout most of the life cycle of the cell, the chromatin exists as exceedingly long, fine threads that are so tangled that they cannot be seen clearly with any microscope. When the nucleus is about to divide (that is, to undergo mitosis or meiosis; see Chapter 9), the chromatin condenses and coils tightly to form a precise number of readily visible objects called **chromosomes** (Figure 4.11). Each chromosome contains one long molecule of DNA. The chromosomes are the bearers of hereditary instructions; their DNA carries the information required to carry out the synthetic functions of the cell and to endow the cell's descendants with the same instructions. Between nuclear divisions, the chromatin attaches to a protein meshwork, the nuclear lamina, on the inside of the nuclear envelope. The chromatin detaches when nuclear division commences, and the envelope breaks up into vesicles.

During most of the nuclear cycle, dense, roughly spherical bodies called **nucleoli** are visible in the nucleus (see Figure 4.10). Taken together, the nucleoli contain from 10 to 20 percent of a cell's RNA. Ribosomes are assembled in the nucleolus. Protein molecules move into the nucleus and then into the nucleoli, where they combine with RNA molecules to form the cell's ribosomes. The ribosomes then move out of the nucleus. Each nucleus must have at least one nucleolus, and those of some species have several. The exact number of nucleoli in its cells is characteristic of a species.

Chromosomes and nucleoli float in a fluid called **nucleoplasm.** This fluid is a suspension of various particles, fibers, proteins, and other compounds. (Whereas in a solution solid particles dissolve, in a suspension they disperse but remain solid.) Recall that the fluid portion of the cytoplasm, in which the various organelles, particles, and fibers are suspended, is called the cytosol.

Nucleus and Cytoplasm

The nucleus contains most of the cell's DNA. The DNA encodes the information needed to make the cell's macromolecules, which carry out the activities of the nucleus and of the cytoplasm. Beyond this, what can we say about the relationship between the nucleus and the cytoplasm?

To answer this question, consider experiments performed with a giant single-celled alga of the genus *Acetabularia.* Cells of *Acetabularia* reach lengths of a

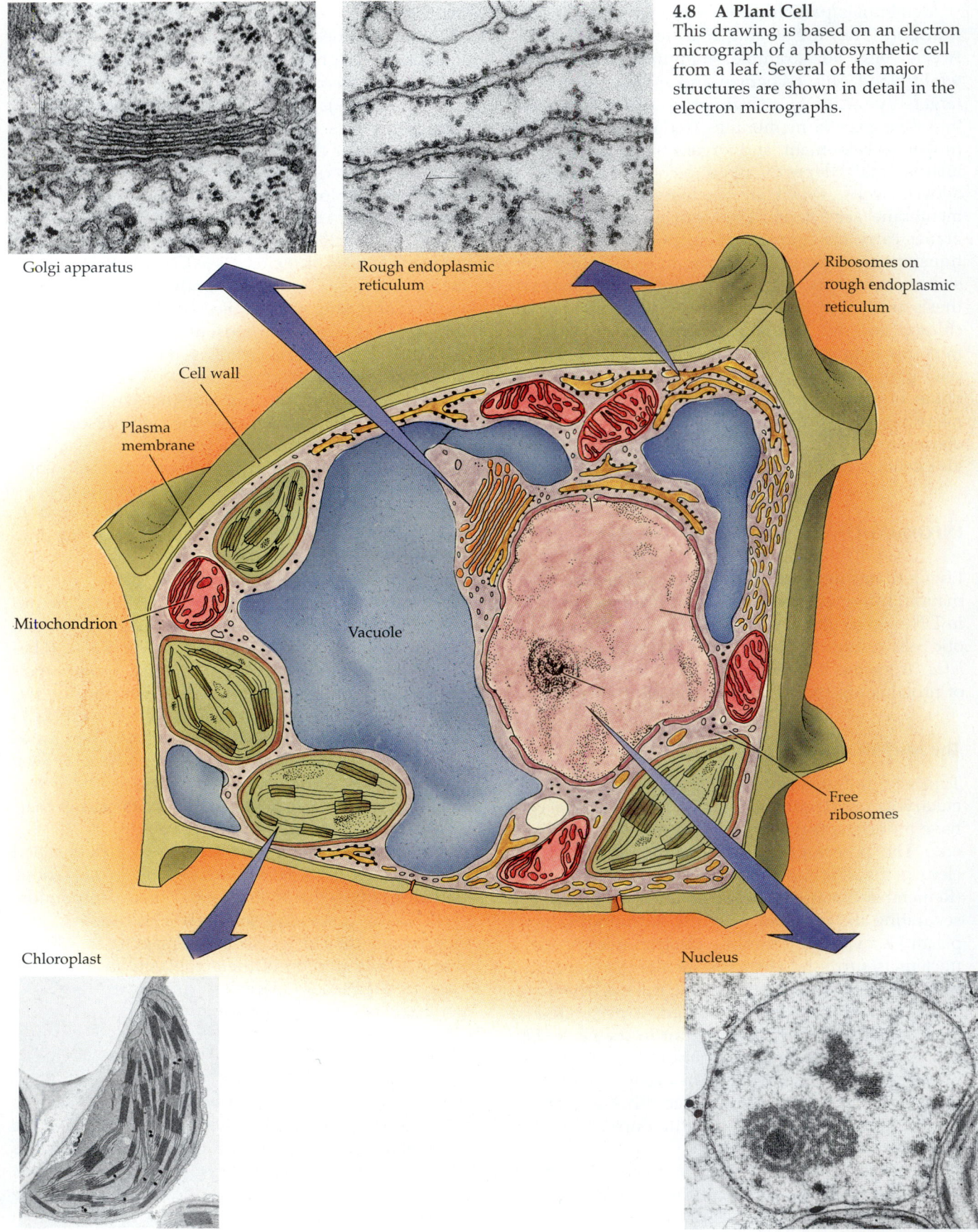

Golgi apparatus

Rough endoplasmic
reticulum

4.8 A Plant Cell
This drawing is based on an electron
micrograph of a photosynthetic cell
from a leaf. Several of the major
structures are shown in detail in the
electron micrographs.

Ribosomes on
rough endoplasmic
reticulum

Cell wall

Plasma
membrane

Mitochondrion

Vacuole

Free
ribosomes

Chloroplast

Nucleus

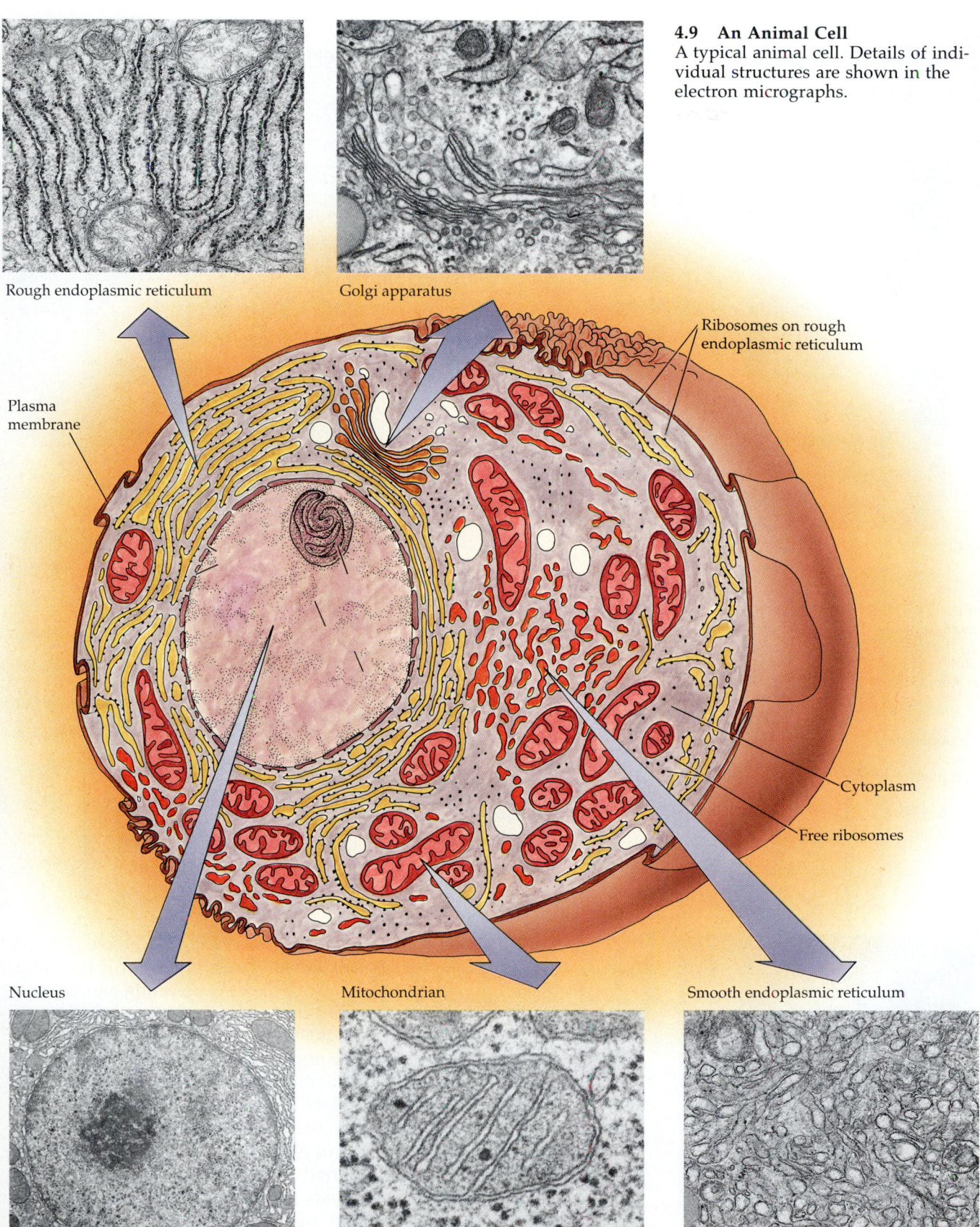

Rough endoplasmic reticulum

Golgi apparatus

4.9   An Animal Cell
A typical animal cell. Details of individual structures are shown in the electron micrographs.

Ribosomes on rough endoplasmic reticulum

Plasma membrane

Cytoplasm

Free ribosomes

Nucleus

Mitochondrian

Smooth endoplasmic reticulum

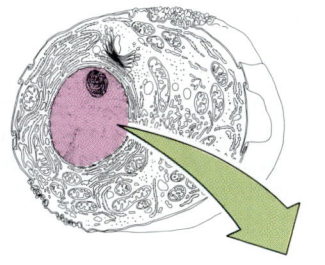

4.10 The Eukaryotic Nucleus Is Bounded by a Double Membrane

The electron micrograph shows the nucleus of an animal cell. Notice the double-membraned nuclear envelope, the nucleolus, and other common features of animal cell nuclei. The bottom two drawings illustrate pores in the nuclear envelope. Each nuclear pore complex comprises eight protein granules surrounding a pore.

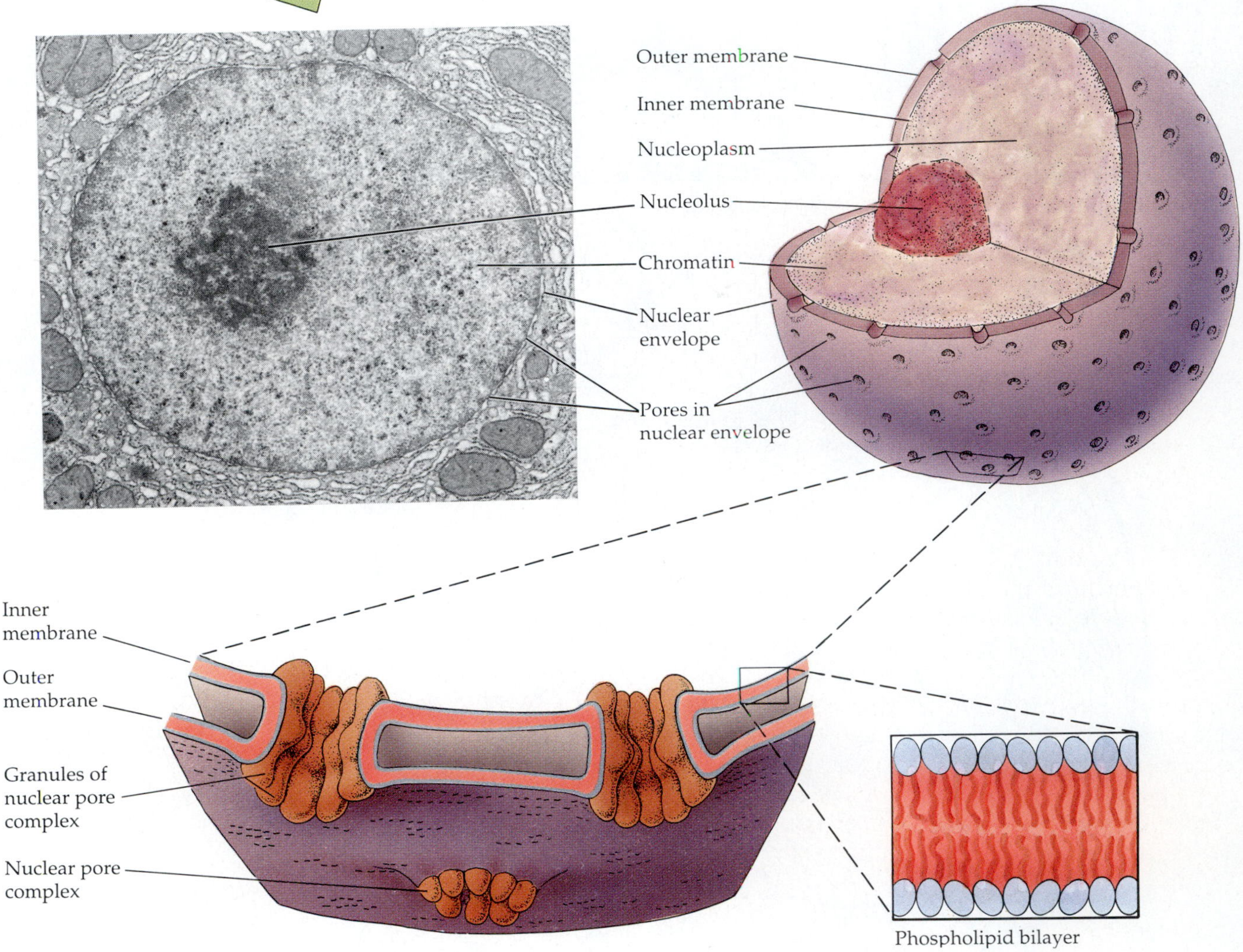

Outer membrane
Inner membrane
Nucleoplasm
Nucleolus
Chromatin
Nuclear envelope
Pores in nuclear envelope

Inner membrane
Outer membrane
Granules of nuclear pore complex
Nuclear pore complex

Phospholipid bilayer

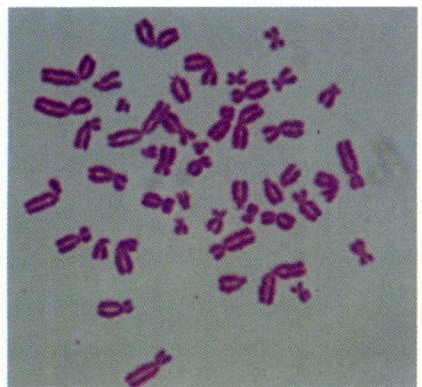

4.11 Humans Have 46 Chromosomes

The chromosome complement in a normal human cell. If a nucleus about to divide is ruptured and treated with certain stains, its chromosomes are readily visible under the light microscope, as shown here.

few centimeters and can readily be picked up and handled. A single cell of this organism is just large enough to be easy to use for dissecting and grafting experiments. The cells consist of three principal regions: the cap, the stalk, and the rhizoids. The rhizoids anchor the organism in its watery environment. The nucleus is within the rhizoid region throughout most of the life of the cell. Rhizoids have no ribosomes, and most of the cell's cytoplasm is in the stalk.

If the cap is cut off a cell of *Acetabularia*, a new cap forms over a period of several days, regenerated with proteins and lipids synthesized by the cytoplasm in the stalk. If this new cap is removed, still another forms, and so forth (Figure 4.12a,b).

Acetabularia cells of two different species can be grafted together to show the origin of the information for the cap structures. An *Acetabularia mediterranea* cap

4.12 Domination by the Nucleus

Grafting experiments with the giant protist *Acetabularia* point to the regulatory activity of the nucleus. The nucleus-containing rhizoids determine the type of cap produced by regenerating *Acetabularia*.

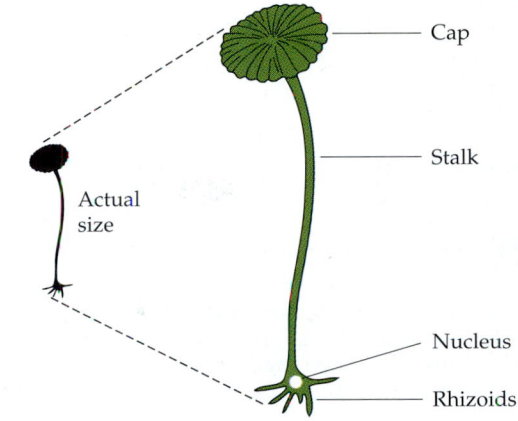

Cap

Stalk

Actual size

Nucleus

Rhizoids

looks like an umbrella, whereas an *Acetabularia crenulata* cap looks more like a bunch of bananas. If we cut up two *Acetabularia* and graft together the rhizoids from *A. mediterranea* and the stalk from *A. crenulata*, a new cap will form from the cytoplasm of the *A. crenulata* stalk. Which type will it be? The first new cap to form is of intermediate appearance. If we cut off this cap and wait for another to appear, the next cap looks like a typical *A. mediterranea* cap, even though it is made from the cytoplasm of the *A. crenulata* stalk (Figure 4.12c). With more refined experiments, we can show that it is the nucleus—which happens to lie in the rhizoids in both species—that provides the instructions for making a new cap. So the nucleus is the storehouse of information for the cell, even when the cell has been put together by grafting.

Can we be sure that we have interpreted the grafting experiment correctly? Perhaps *A. mediterranea* just happens to be somehow "dominant" over *A. crenulata*. We can test this possibility by turning the experiment around—by doing a reciprocal experiment. To do this, we combine *A. crenulata* rhizoids with an *A. mediterranea* stalk. Again, a cap of intermediate form is made first and is cut away; all subsequent caps regenerate as the *A. crenulata* type (Figure 4.12d). Now we can be more confident of our conclusion: The nucleus controls what the cytoplasm builds.

Ribosomes

In both eukaryotic and prokaryotic cells, ribosomes fill the need for a site where a crucial cellular activity—protein synthesis—can take place. Ribosomes reside in three places in eukaryotic cells: free in the cytoplasm; attached to the surface of endoplasmic reticulum, as will be described later in this chapter; and in the energy-processing organelles discussed in the next section. In each of these places the ribosomes provide the site where proteins are synthesized under the direction of nucleic acids (see Chapter 11).

The ribosomes of prokaryotes and of eukaryotes are similar in that they each consist of two different-sized subunits (Figure 4.13). Eukaryotic ribosomes are somewhat larger, but the structure of prokaryotic ribosomes is better understood. Chemically, ribosomes consist of a type of RNA to and around which are bound more than 50 different protein molecules. The ribosome temporarily binds two other types of RNA molecules (one large and one small) as it trans-

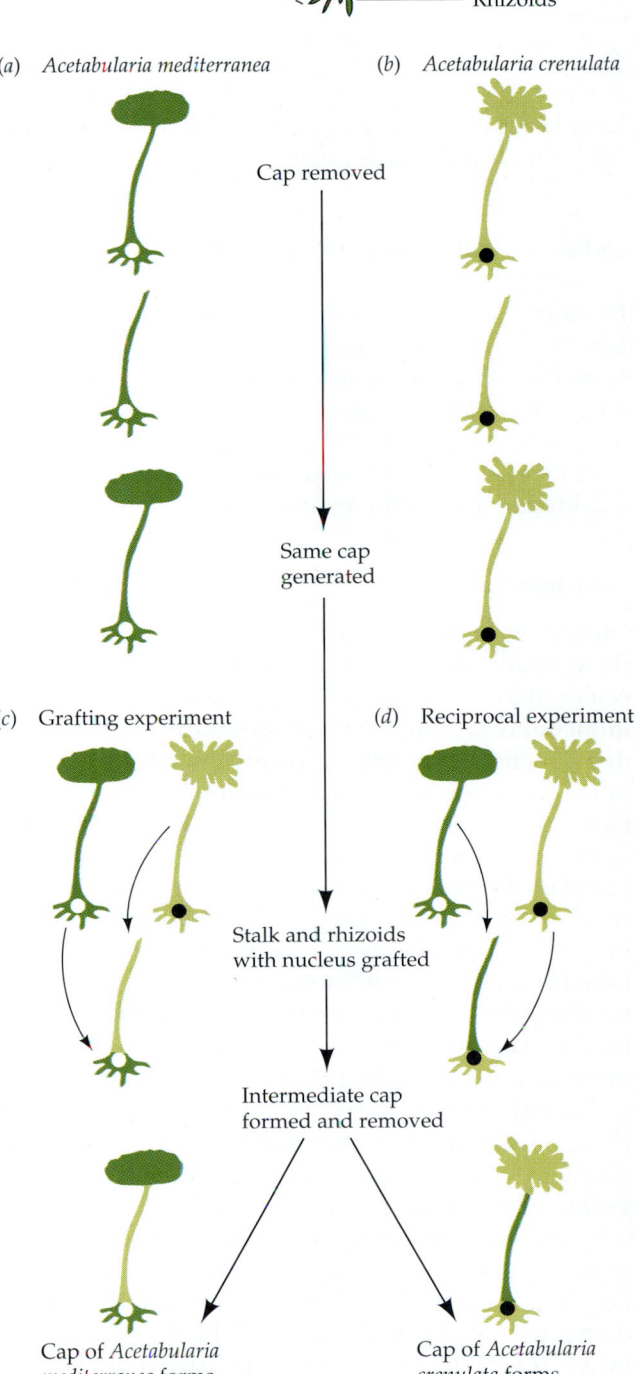

(a) *Acetabularia mediterranea*

(b) *Acetabularia crenulata*

Cap removed

Same cap generated

(c) Grafting experiment

(d) Reciprocal experiment

Stalk and rhizoids with nucleus grafted

Intermediate cap formed and removed

Cap of *Acetabularia mediterranea* forms

Cap of *Acetabularia crenulata* forms

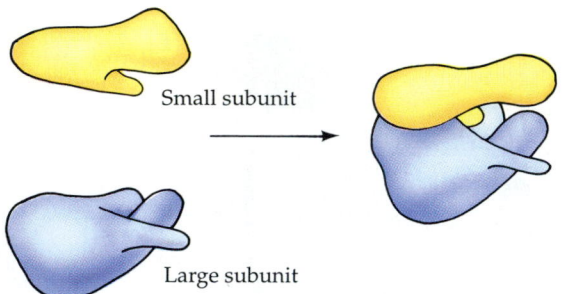

4.13 Tiny Ribosomes Have a Complex Structure
A ribosome consists of a large subunit and a small subunit; these subunits come together only when participating in protein synthesis. Ribosomes do not have membranes.

lates hereditary information into the structures of the cell's structural and regulatory proteins.

ENERGY-PROCESSING ORGANELLES

In addition to information, cells require energy. Eukaryotic cells have organelles for obtaining energy from food molecules, and some cells of plants and of some protists have organelles in which the energy of sunlight is captured. In prokaryotic cells energy is transformed on the plasma membrane, on membrane infoldings, and in the cytosol.

Mitochondria

Utilization of "fuels" for the eukaryotic cell begins in the cytosol, where chemicals from food become molecules that are then taken up by organelles called **mitochondria** (singular: mitochondrion). Mitochondria function primarily to convert energy from food into a form the cell can use. In mitochondria, energy-rich substances from the cytosol are oxidized—that is, electrons are removed from them (see Chapter 7). Some of the energy available from these electrons is used to make a substance—ATP—that stores energy in two special chemical bonds. The stored energy may be used either immediately or later to perform various kinds of work for the cell. The utilization of food in the mitochondria, with the associated formation of ATP, is called cellular respiration.

Typical mitochondria are small, somewhat less than 1.5 μm in diameter and 2 to 8 μm in length, about the size of many bacteria. Mitochondria are visible with a light microscope, but virtually nothing was known of their structure until they were examined with the electron microscope. Electron micrographs show that mitochondria have an outer membrane that is smooth and unfolded. Immediately inside this is an inner membrane that folds inward at many points, giving it a much greater surface area

than that of the outer membrane (Figure 4.14). In animal cells these folds tend to be quite regular, giving rise to shelflike structures called **cristae**. The mitochondria of plants also have cristae, but plant cristae tend to be much less regular in size and structure, and their inner membranes form both shelves and tubes. Special techniques and very high magnification have been used to show that the inner mitochondrial membrane contains large protein structures now known to participate in cellular respiration (see Chapter 7). The region enclosed by the inner membrane is referred to as the **mitochondrial matrix**. Within the matrix are ribosomes and DNA that make some of the proteins needed for the synthesis of mitochondria.

Almost all eukaryotes have mitochondria. The few exceptions are microscopic organisms that live in environments without oxygen, and parasites that exploit the energy resources of their hosts. The number

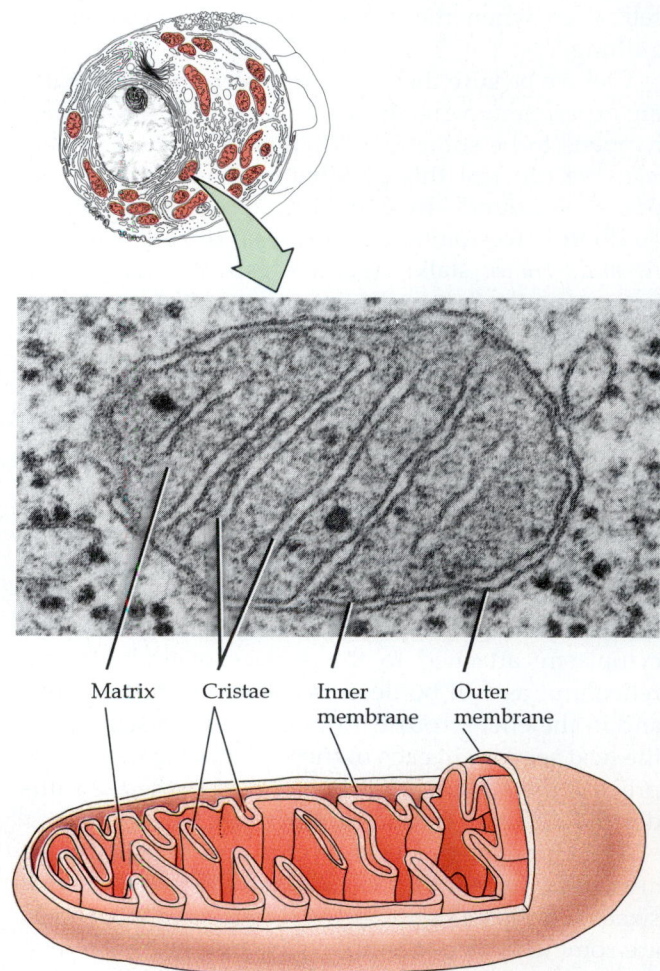

| Matrix | Cristae | Inner membrane | Outer membrane |

4.14 A Mitochondrion: Where Foods Yield Up Their Energy
A mitochondrion as seen by electron microscopy (top). The drawing shows the mitochondrion's surface cut away to expose internal structures.

of mitochondria per cell ranges from one contorted giant in some unicellular protists to a few hundred thousand in large egg cells. An average human liver cell contains more than a thousand mitochondria. The primary function of all mitochondria is cellular respiration, by which usable energy is derived from food materials and stored in ATP. The ATP is exported into the cytosol, where most of it is used; some also goes into the nucleus. The cells that require the most chemical energy tend to have more mitochondria per unit volume. In Chapter 7 we will see how different parts of the mitochondrion work together in the respiratory process.

Plastids

One class of organelles—the **plastids**—is produced only in plants and certain protists. The most familiar of the plastids is the **chloroplast,** which is the site of photosynthesis and contains all the chlorophyll in the plant or photosynthetic protist cell (Figure 4.15). Photosynthesis (see Chapter 8) is the process by which light energy is converted into the energy of chemical bonds. The molecules formed in photosynthesis provide food for the plant itself and for other organisms; directly or indirectly, photosynthesis is the energy source for much of the living world. The chloroplast has a number of other metabolic functions

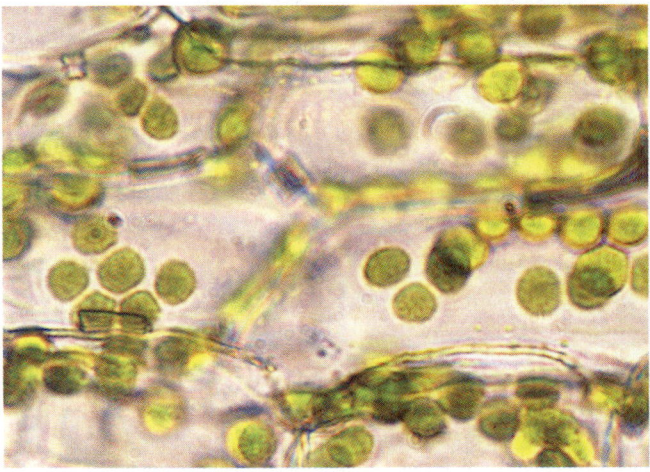

4.15 Chloroplasts in a Cell
Plant cells contains dozens of chloroplasts, the green organelles that carry on photosynthesis.

besides photosynthesis. For example, it helps make nitrogen available to the rest of the plant.

Like the mitochondrion, the chloroplast is surrounded by two membranes. However, both membranes are unfolded and surround the organelle as a smooth, closely fitting, double layer. Arising from the inner membrane is a series of discrete internal membranes, whose structure and arrangement vary from one group of photosynthetic organisms to another. As an introduction, we concentrate on the chloroplasts of the flowering plants. Even these show some variation, but the pattern in Figure 4.16 is reasonably typical.

4.16 The Organelle That Feeds the World
An electron micrograph of a section of a chloroplast from a leaf of corn. Note the stacks of thylakoids, called grana, and the membranous connections between the grana, making up an extensive network of photosynthetic membranes in the organelle. Only a thin layer of cytoplasm surrounds the chloroplast in this mature cell.

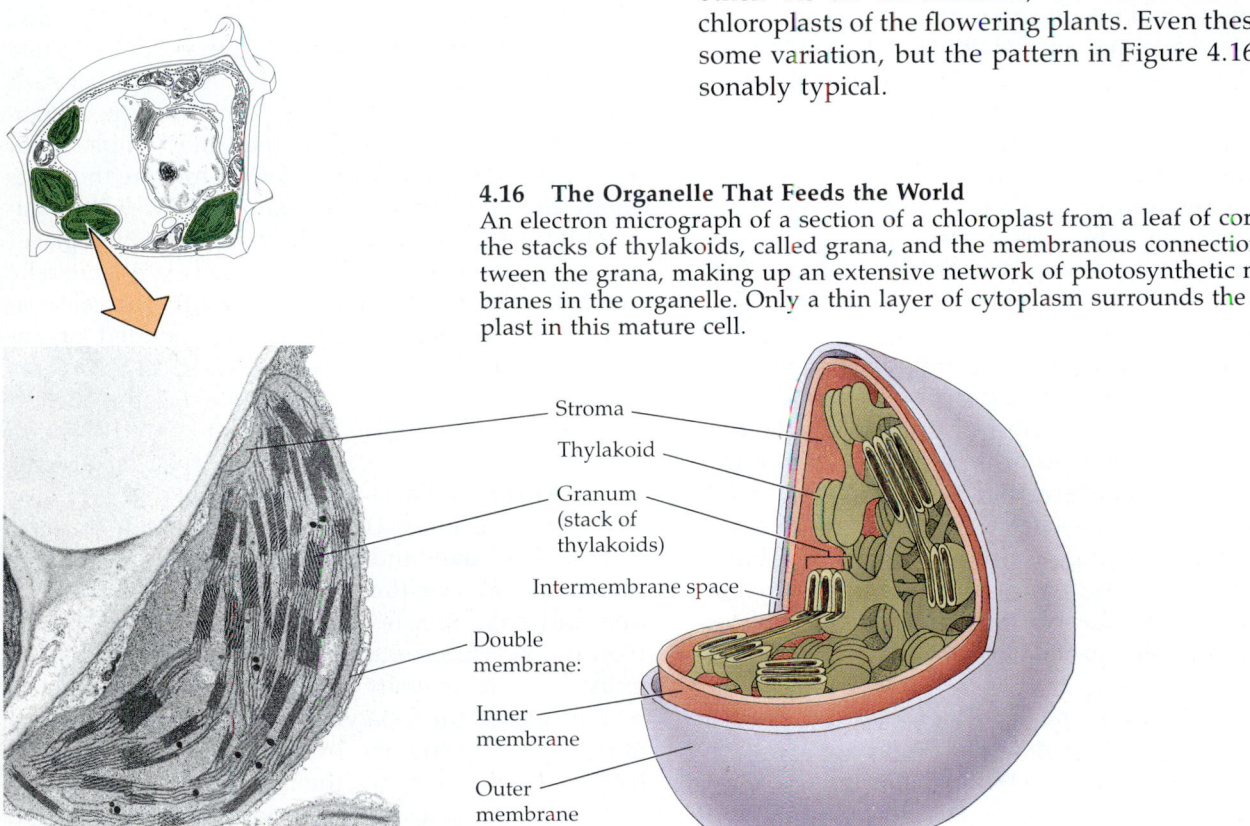

Stroma

Thylakoid

Granum
(stack of
thylakoids)

Intermembrane space

Double
membrane:

Inner
membrane

Outer
membrane

4.17 Living Together: Anemone–Alga Symbiosis
This giant sea anemone owes the intensity of its green
color to the chloroplasts in a unicellular alga that lives
and carries on photosynthesis within the tissues of the
anemone.

As seen in electron micrographs, chloroplasts contain structures that look like stacks of pancakes.
These stacks, called **grana** (singular: granum), consist
of a series of flat, closely packed, circular sacs called
thylakoids. Each thylakoid is a single membrane
composed of the usual membrane components (lipids
and proteins) to which have been added chlorophyll
and other substances needed for trapping photosynthetic energy and producing food. All the cell's chlorophyll is contained in the thylakoid membranes.
Thylakoids of one granum may be connected to those
of other grana (see Figure 4.16), making the interior
of the chloroplast a highly developed network of
membranes. The fluid in which the grana are suspended is referred to as **stroma.** Like the mitochondrial matrix, the chloroplast stroma contains ribosomes and DNA. These ribosomes and this DNA
provide some—but only some—of the proteins that
make up the chloroplast.

Chloroplasts are what give plant leaves their green
color. Looking at a thin slice of a leaf under the
microscope reveals that most of the leaf is quite colorless—the only green to be seen is contained in the
numerous chloroplasts within its cells. Not all plant
cells contain chloroplasts, however. Most roots, for
example, are colorless (or at least not green). This is
just as well, because it would be a waste of energy
and materials for the plant to provide chloroplasts
for cells that reside in the dark, since photosynthesis
requires light.

Animal cells do not *produce* chloroplasts, but some
contain functional chloroplasts. These organelles are
taken up either as free chloroplasts derived from
plants eaten as food, or as bound chloroplasts contained within unicellular algae living within the ani-

mal tissues. The green color of some corals and sea
anemones results from chloroplasts in algae that live
within the animals (Figure 4.17). The animals derive
some of their nutrition from photosynthesis carried
out by their algal "guests."

Chloroplasts are not the only plastids found in
plants. The red color of a ripe tomato results from
the presence of legions of plastids called **chromoplasts.** Just as chloroplasts derive their color from
chlorophyll, chromoplasts are red, orange, or yellow
because of the pigments (called carotenoids; see
Chapter 3) that they contain. Chromoplasts have no
known chemical function in the cell, but the colors
they give to some petals and fruits probably help
attract animals that assist in pollination or seed dispersal. On the other hand, there is no apparent advantage in a carrot root being orange. Other types of
plastids, called **leucoplasts,** serve as storage depots
for starch and fats. All plastid types are related to
one another. Chromoplasts, for example, are formed
from chloroplasts by a loss of chlorophyll and some
change in internal structure. All plastids develop
from **proplastids,** which are very simple in structure.

The Origins of Plastids, Mitochondria, and Eukaryotes

In the past, biologists tried to grow chloroplasts or
mitochondria in culture, outside the cells that they
normally inhabit. These organelles are about the size
of bacteria, they contain ribosomes and DNA, and
they divide within the cell—might they not be treated
like little cells in their own right? Although all such
efforts at organelle culture failed because the organelles depend on the cell's nucleus and cytoplasm for
some parts, the experiments helped nurture thoughts
about another important question: How did the eukaryotic cell with its organelles arise in the first place?
As we have seen, prokaryotic cells are generally
much simpler in structure than eukaryotic cells because prokaryotes lack membrane-bounded organelles. Prokaryotic fossils can be found in sediments
well over 3 billion years old, whereas the earliest
known eukaryotic fossils date back to only 1.4 billion
years ago. It is generally agreed that eukaryotes
evolved from prokaryotes. But how?

One suggestion that has been alternately popular
and scorned, over and over again for many years, is
the **endosymbiotic theory** of the origin of mitochondria and chloroplasts. An important current champion of and contributor to this theory is Lynn Margulis of the University of Massachusetts, Amherst,
who proposed the following idea: Picture a time, well
over a billion years ago, when only prokaryotes inhabited Earth. Some of them got their food by absorbing it directly from the environment, others were
photosynthetic, and still others fed by eating their

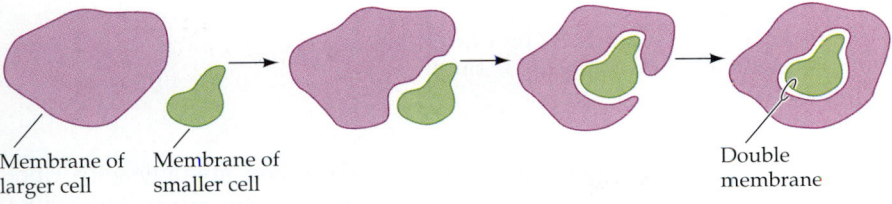

Membrane of larger cell

Membrane of smaller cell

Double membrane

4.18 Creation of a Double Membrane
Double membranes, such as those surrounding chloroplasts and mitochondria, might have been created when a larger cell (purple) engulfed a smaller cell (green) but did not digest the smaller cell. Thus the double membrane would consist of the portion of the larger cell's plasma membrane that enclosed the smaller cell and the plasma membrane of the smaller cell.

prokaryotic neighbors. Suppose that an occasional small, photosynthetic prokaryote was ingested by a larger one but did not get digested, so it sat trapped within the larger cell. Suppose further that the smaller prokaryote survived there and that it divided at about the same rate as the larger one, so successive generations of the larger prokaryote continued to be inhabited (or infected) by the offspring of the smaller one. We would call this endosymbiosis: "living within" another cell or organism, as, for example, certain algae live within sea anemones (see Figure 4.17).

Could the little green prokaryote "eaten" by the larger prokaryote have been the first chloroplast? A present-day chloroplast is surrounded by a double membrane. Such a structure might have arisen when, in the process of engulfing the photosynthetic cell, the membrane of the larger cell stretched around the plasma membrane of the smaller cell (Figure 4.18). The fact that chloroplasts contain ribosomes and DNA also fits the endosymbiotic theory because the chloroplast is proposed to have arisen from an engulfed prokaryote. In addition, whereas ribosomes in the eukaryotic cytosol are larger than those of prokaryotes, ribosomes in the chloroplast are similar in size to those of prokaryotes, further supporting the theory.

Arguments like these can be made for the proposition that mitochondria represent the descendants of respiring prokaryotes engulfed by, and ultimately endosymbiotic with, larger prokaryotes. Also, there are striking similarities between some functions of bacterial plasma membranes and mitochondrial inner membranes, and between the primary structures of certain bacterial and mitochondrial enzymes (see Chapter 3). Finally, a few modern cells do contain other, smaller cells as endosymbionts, suggesting that the endosymbiotic theory is plausible.

A spectacular example of endosymbiosis is found in the guts of certain Australian termites. We think of a termite as being able to digest wood; yet, strictly speaking, it cannot. Much of the digestive chemistry is accomplished by an endosymbiotic protist, *Mixo-tricha paradoxa*, that lives in the termite's gut. But that is far from the whole story. *Mixotricha* itself harbors an amazing colony of endosymbionts. *Mixotricha* swims around within the termite gut, apparently propelled by multitudes of flagella. Closer examination, however, shows that although there are a few true flagella in a tuft at one end of the organism, the hundreds of other flagellumlike propellers are not flagella at all—they are long, motile bacteria (spirochetes) that are attached at regular intervals to the plasma membrane of the *Mixotricha* cell and that beat just like real flagella. Also covering the surface of the protist, organized in a precise pattern, are other, smaller bacteria. *And* inside the *Mixotricha* are numerous bacteria of a third species, which are thought to help with the digestion of the tiny wood particles ingested by the protist—which, in turn, obtains them from the gut of the termite that is carrying this strange menagerie around inside itself. Perhaps termites are not all that special. If we carried colonies of *Mixotricha* in our digestive tracts, we could eat wood too. We do carry colonies of *Escherichia coli* (see Box 4.B) that aid in the digestion of our food, as discussed in Chapter 43.

We should stress that mitochondria and chloroplasts are not enough to make a prokaryote into a eukaryote. The origin of the nuclear envelope, as well as that of other important structures, including those responsible for nuclear division, still needs to be illuminated. Thus far, the endosymbiotic theory is incomplete, although suggestions have been made for its extension to deal with the origin of other eukaryotic organelles. Is the endosymbiotic theory true? Certainly it has not yet been proved. A number of compelling objections to the theory have been raised, among them the fact that the DNA responsible for the synthesis of most of the enzymes in chloroplasts and mitochondria resides in the nucleus. However the matter ultimately is resolved, the endosymbiotic theory is a good example of creative biological thinking; it gives us a useful perspective on the structures, functions, and origins of mitochondria and chloroplasts.

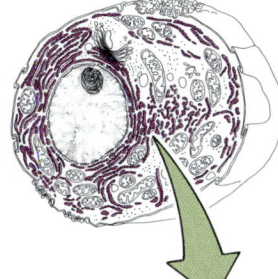

4.19 Endoplasmic Reticulum
The micrograph and drawing show rough ER on the left and smooth ER on the right. The "smoothness" is simply the absence of ribosomes.

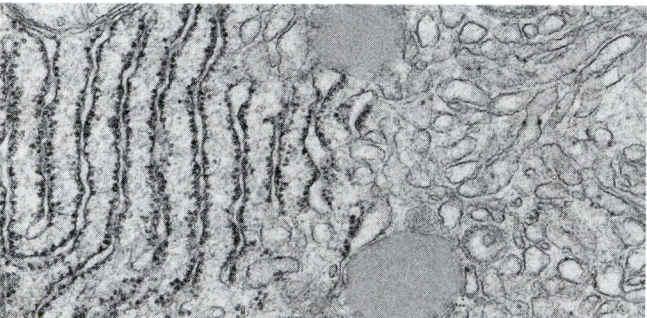

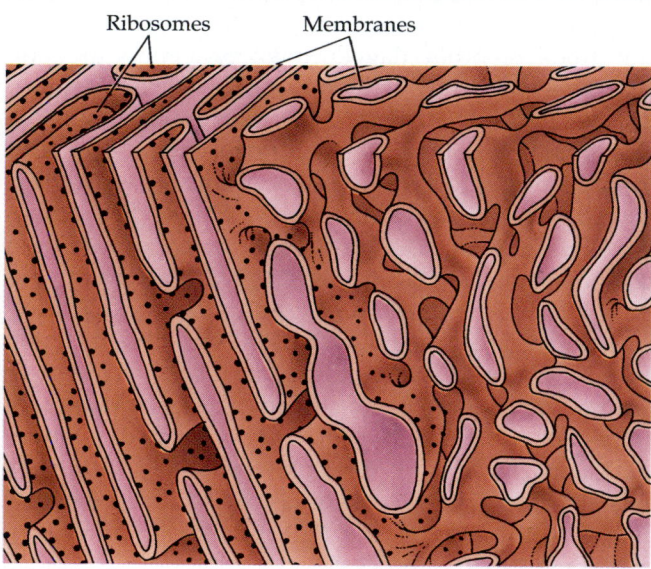

Ribosomes   Membranes

THE ENDOMEMBRANE SYSTEM

Much of the volume of a eukaryotic cell is taken up by extensive membrane systems that play numerous roles in the life of the cell. These membrane systems are closely interrelated and arise from one another. They are referred to collectively as the **endomembrane system.**

Endoplasmic Reticulum

Running here and there throughout the cytoplasm, branching and rejoining, is a network of tubes and flattened sacs called the **endoplasmic reticulum,** or **ER.** As we have noted, electron micrographs often show this membrane system to be continuous with the outer membrane of the nuclear envelope.

Parts of the ER are sprinkled liberally with ribosomes, which are attached to the outer faces of the flattened sacs. Because of their appearance in the electron microscope (Figure 4.19), these regions are called rough ER. The attached ribosomes are sites for the synthesis of proteins that function outside the cytosol—that is, proteins that are to be exported from the cell, incorporated into membranes, or moved into organelles of the endomembrane system. These proteins enter the lumen (interior) of the ER as they are synthesized, directed by a special sequence of amino acids known as the signal sequence (see Chapter 11). In the ER these proteins undergo a number of changes, including the formation of disulfide bridges (see Figure 3.14) and folding into their final, tertiary structures; some aggregate, forming quaternary structures (see Chapter 3). Some proteins gain carbohydrate groups in rough ER and thus become glycoproteins. The carbohydrate groups are part of an addressing system that ensures that the right proteins are directed to the right parts of the cell. Proteins that are to remain within the cytosol or to move into mitochondria and chloroplasts are synthesized on "free" ribosomes, that is, ribosomes that are not attached to the ER. These proteins lack the signal sequence that would otherwise direct them into the ER.

Other parts of the endoplasmic reticulum lack ribosomes and are referred to as smooth ER (see Figure 4.19). Smooth ER modifies proteins synthesized by rough ER. As you might guess from its function, the amount of ER in a cell is related to how busy the cell is in making proteins for export. Cells that are synthesizing a lot of proteins—glandular cells (see Chapter 36) or the immune system's plasma cells (see Chapter 16), for example—may be heavily packed with ER, whereas others with less work to do (such as food-storing cells) contain very little ER.

The Golgi Apparatus

In 1898 the Italian microscopist Camillo Golgi discovered a delicate structure in nerve cells. He reported that he always observed it near the nucleus. Unfortunately, his technique for staining this structure was tricky and often failed to work, so the **Golgi apparatus** was regarded by most biologists as a figment of Golgi's imagination. Work with the electron microscope in the late 1950s, however, showed clearly that these structures do exist—and not just in nerve cells, but in most eukaryotic cells.

The appearance of the Golgi apparatus varies from

(a)

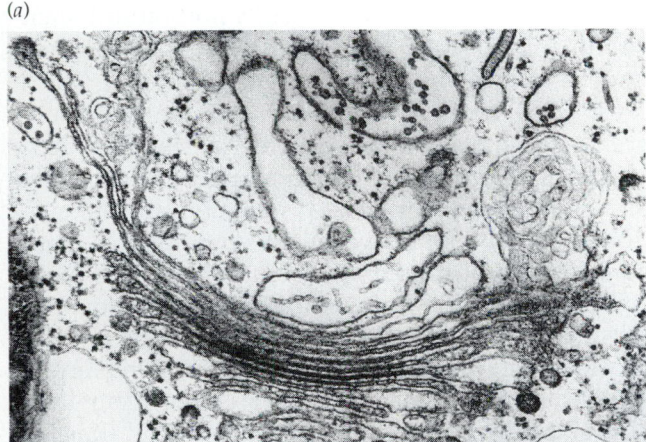

4.20 The Golgi Apparatus
(a) A Golgi apparatus in an alga. *(b)* Vesicles from the endoplasmic reticulum transfer substances to the apparatus by fusing with the *cis* region at the bottom of the stack, and vesicles with substances for export or transport to elsewhere in the cell leave the apparatus by budding off the *trans* region. The Golgi apparatus modifies incoming proteins and "targets" them to the correct addresses.

species to species, but a number of flattened sacs are always seen lying together like a stack of saucers (Figure 4.20*a*). In the cells of plants, protists, fungi, and many invertebrate animals these stacks are individual units scattered throughout the cytoplasm. In vertebrate cells, a few such stacks may form a more complex Golgi apparatus. The bottom saucers, constituting the *cis* region of the Golgi apparatus, lie nearest the nucleus or a patch of rough ER; the top saucers, constituting the *trans* region, lie closest to the surface of the cell. The saucers in the middle make up the *medial* region of the complex. These three parts of the Golgi apparatus contain different enzymes and perform different functions.

What are the functions of this organelle that Golgi discovered? The first clue comes from observing the relationships between the Golgi apparatus and other parts of the cell. Vesicles from the rough ER travel to and merge with the *cis* region of the Golgi apparatus. Other small vesicles move between the flattened sacs of the Golgi apparatus, always in the direction *cis* to *trans*, transporting proteins. Associated with the sacs, particularly those toward the *trans* region, are numerous tiny vesicles that pinch off from the sacs and then move away (Figure 4.20*b*). The vesicles sometimes merge with each other and finally merge with other organelles or with the plasma membrane, where they release their contents in a process called exocytosis (see next section). This behavior is the key to the important cellular functions of the Golgi apparatus. The Golgi apparatus serves as a sort of postal service depot in which some of the proteins synthesized on ribosomes on the rough ER are stored,

(b)

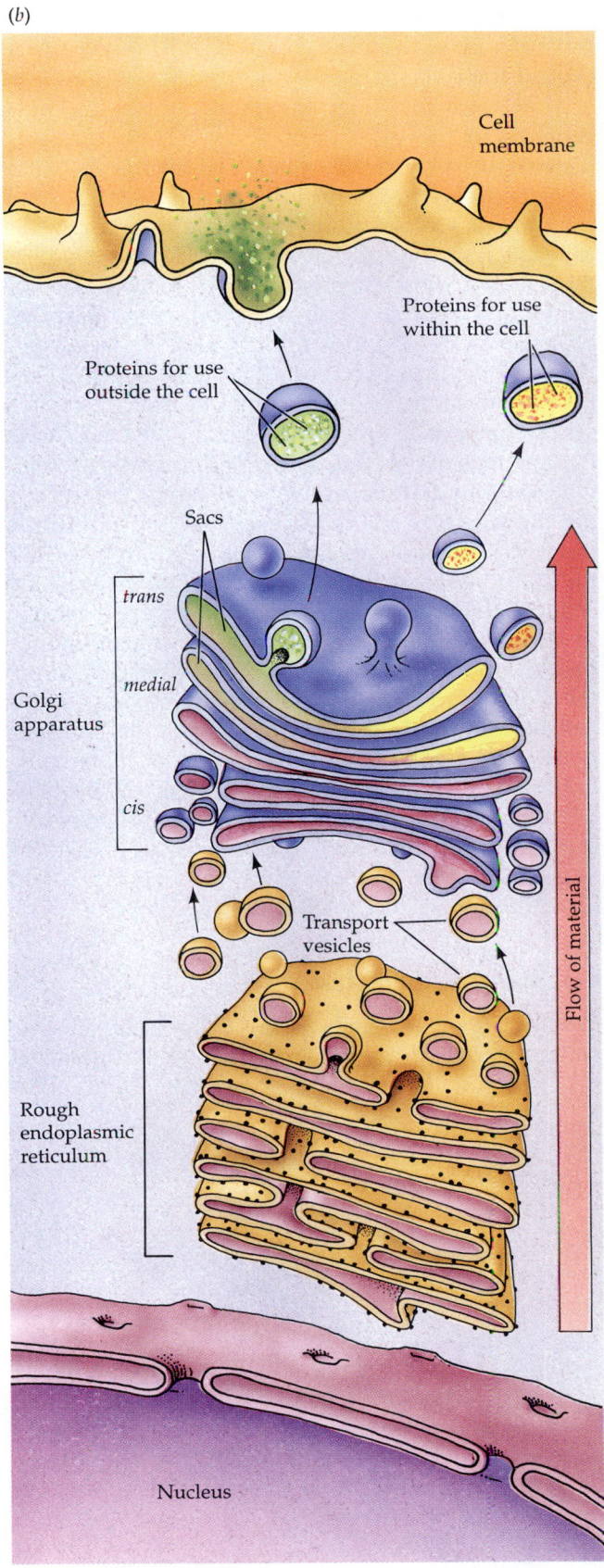

Cell membrane

Proteins for use within the cell

Proteins for use outside the cell

Sacs

trans

medial

Golgi apparatus

cis

Transport vesicles

Flow of material

Rough endoplasmic reticulum

Nucleus

chemically modified, and packaged for delivery to the environment outside the cell or to other organelles of the cell.

How does the Golgi apparatus address the right proteins to the right destinations? The addressing consists of a series of chemical reactions in which proteins gain specific chemical "address tags." For example, a protein destined for use in an organelle called a lysosome (discussed later in this chapter) possesses a signal sequence that directs it into the lumen of the rough ER as the protein is synthesized. Enzymes in the ER add an initial carbohydrate tag—the same tag that directs other proteins to the plasma membrane to be secreted from the cell. However, other enzymes in the Golgi sacs modify the carbohydrate tags of the lysosome glycoproteins—proteins with carbohydrate residues (see Chapter 3)—changing the structure of the tag and adding a phosphate group in a process called phosphorylation. When the phosphorylated glycoprotein reaches the *trans* region of the Golgi apparatus, it binds to a receptor protein in the Golgi membrane. A vesicle containing the bound, phosphorylated glycoprotein separates from the Golgi apparatus and delivers its contents to the developing lysosome. Comparable mechanisms, most using addressing systems other than carbohydrates, deliver other proteins to other parts of the cell. The Golgi apparatus exports some proteins constantly but retains others, releasing them only at the appropriate time. In these ways, the Golgi apparatus directs the molecular mail of the cell.

Exocytosis and Endocytosis

Certain organelles such as the Golgi apparatus secrete materials to the environment; other organelles acquire materials by enveloping a portion of the environment. In eukaryotic cells there is an ongoing traffic—in and out—of membrane-bounded "packages." The plasma membrane and the endomembrane system participate actively in shipping these packages. Macromolecules for export are contained in membranous vesicles that move toward the exterior of the cell and ultimately meet the plasma membrane. The phospholipid regions of the two membranes merge, and an opening to the outside of the cell develops. The contents of the vesicle are released to the environment, and the vesicle membrane is smoothly incorporated into the plasma membrane. This entire process—export of material and transformation of the membrane—is called **exocytosis** (Figure 4.21a).

Materials may be brought into the eukaryotic cell by a related process known as **endocytosis** (Figure 4.21b). The cell surface folds to make a small pocket that is lined by the plasma membrane. The folding increases until the pocket seals off, forming a vesicle whose contents are materials from the environment. This vesicle, enclosed in membrane taken from the plasma membrane, separates from the cell surface and migrates to the interior. Usually the vesicle fuses

4.21 Exocytosis and Endocytosis
(a) In exocytosis, a membrane-enclosed secretory vesicle containing substances for export fuses with the plasma membrane. First one layer of the membrane phospholipids of the two membranes fuses, then the other. The contents of the vesicle scatter, and the vesicle membrane becomes part of the plasma membrane. (b) In endocytosis, the plasma membrane surrounds a part of the exterior environment and the whole unit buds off to the interior as a membrane-surrounded vesicle. (c) In phagocytosis, a form of endocytosis, whole particles are engulfed. Here, an amoeba is engulfing another protist.

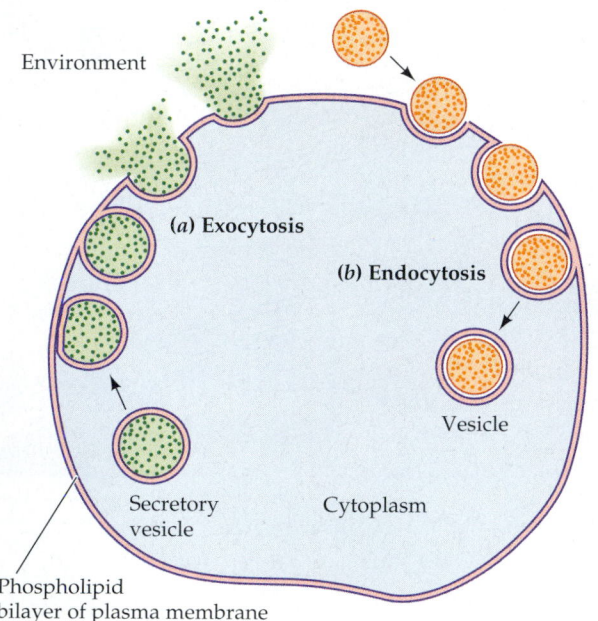

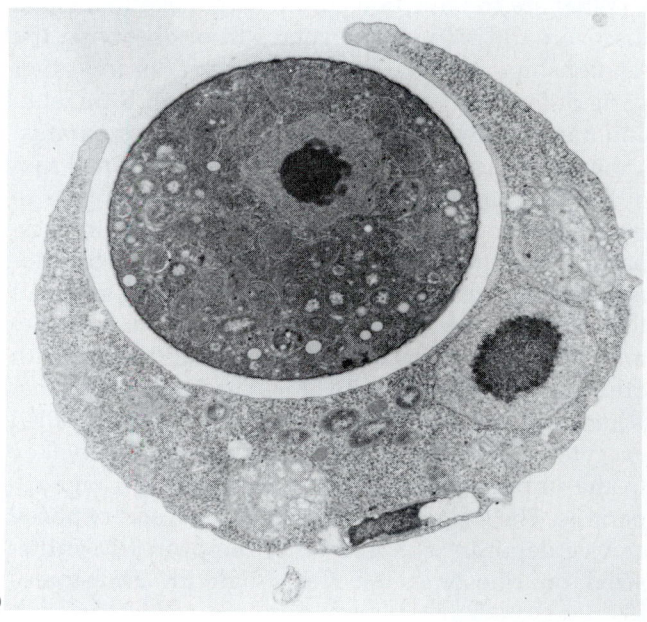

(c)

with a lysosome (see next section) and its contents are digested.

Endocytosis is a blanket term for several processes that have more specialized names. A distinction is often made between **pinocytosis** (cell drinking), in which tiny, liquid-containing vesicles are formed, and **phagocytosis** (cell eating), in which particles or even entire cells may be trapped in large vesicles (Figure 4.21c). Phagocytosis constitutes an important part of our immune system for defense against foreign cells (see Chapter 16). Many kinds of cells are able to engulf food materials from their environment, forming vesicles surrounded by pieces of the plasma membrane.

Lysosomes

Originating from the Golgi apparatus, organelles called **lysosomes** keep a cell from self-destructing by containing and transporting digestive enzymes that accelerate the breakdown of proteins, polysaccharides, nucleic acids, and lipids. Lysosomes are surrounded by a single membrane and have a densely staining, featureless interior (Figure 4.22a). The cells of animals, many protists, fungi, and a few plants have lysosomes.

Lysosomes are sites for the breakdown of food and foreign objects taken up by endocytosis. As Figure 4.22b shows, some of the vesicles that pinch off from the Golgi apparatus become primary lysosomes. After a primary lysosome fuses with a food-containing vesicle, the merged compartment is called a secondary lysosome. The effect of this fusion is rather like releasing hungry foxes into a chicken coop. The food particles are quickly digested by the enzymes in the secondary lysosome. The activity of the enzymes is enhanced by the mild acidity of the lysosome's interior, where the pH is lower than in the surrounding cytoplasm. The products of digestion exit through the membrane of the lysosome to be used by the rest of the cell. The "used" secondary lysosome then moves to the plasma membrane, fuses with it, and releases the remaining undigested contents to the environment by exocytosis.

This lysosomal compartmentalization is a very good arrangement. The digestive enzymes of the lysosome would be highly destructive if they were released into the cell, where they would attack the contents of the cytosol and the other organelles. Instead, enzymes are sealed in the lysosome, from which they cannot escape. The raw materials for enzymic breakdown are delivered tightly packaged in a vesicle that fuses with the lysosome; the useful products of digestion leak out; finally the enzymes, along with the unusable products, are thrown out of the cell.

(a)

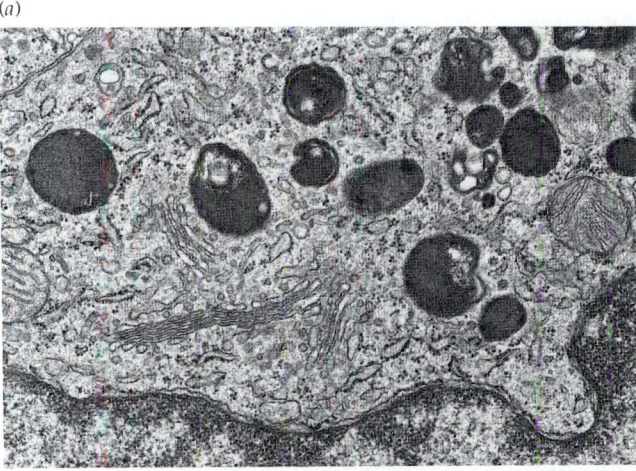

(b)

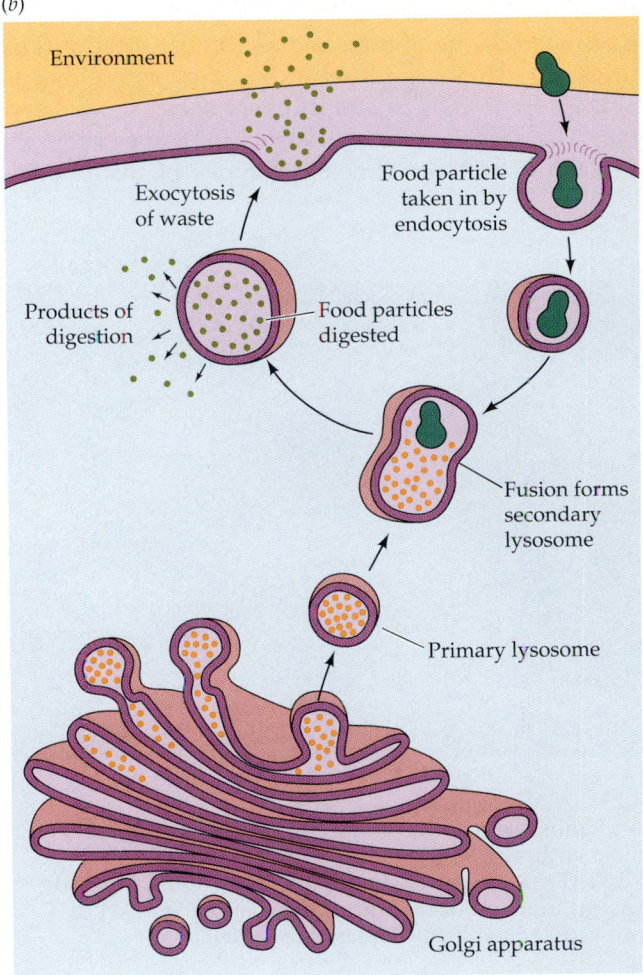

4.22 Lysosomes Keep Digestive Enzymes Where They Belong
(a) The darkly stained organelles in the upper half of this electron micrograph of a rat cell are secondary lysosomes with material being digested inside them. At the bottom is a portion of the cell's nucleus, above which are some of the flattened membrane sacs of the Golgi apparatus. (b) Food-containing vesicles fuse with enzyme-containing primary lysosomes. Digestive enzymes do not contact cell parts other than lysosome interiors.

Clearly, the consequences of digestive enzymes escaping from the lysosomes can be severe. At times, though, such digestive activity is appropriate, as during the development of a frog from a tadpole. The tadpole has a fleshy tail, whereas the mature frog has none. How does the tail disappear? Part of the job is accomplished by the breakdown of lysosomes within the tail cells of the tadpole, releasing enzymes that digest the cells themselves.

Why are the lysosomes not destroyed by these enzymes? It seems as if lysosomes themselves also should be subject to attack by the digestive enzymes they house, yet they survive and function. Despite much interest in this problem, no generally acceptable solution has yet been proposed.

OTHER ORGANELLES

In addition to the information-processing organelles (nucleus and ribosomes), the energy-processing organelles (mitochondria and plastids), and the organelles that form the endomembrane system of the cell (endoplasmic reticulum, Golgi apparatus, and lysosomes), there are several other organelles. **Microbodies** form by pinching off from rough endoplasmic reticulum. As seen with the electron microscope, these small organelles, 0.2 to 1.7 μm in diameter, have a single membrane and a granular interior (Figure 4.23). They are found at one time or another in at least some cells of almost every species of eukaryote. The same microbody structure serves different functions, and the functional types of microbodies

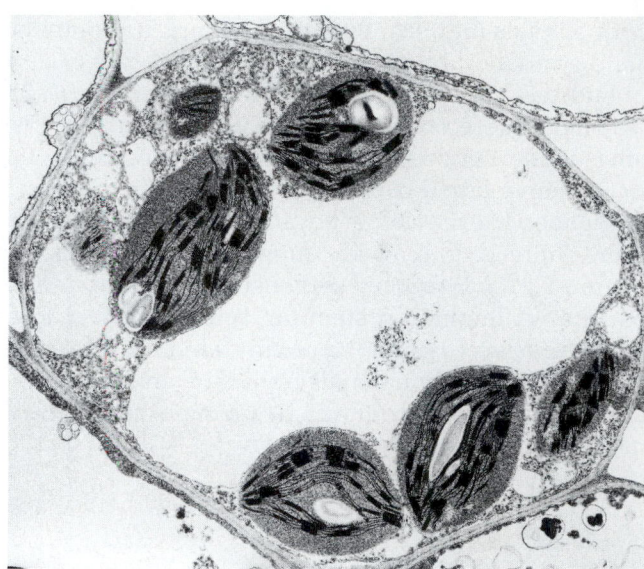

4.24 Vacuoles in Plant Cells Are Big
The large central vacuole is typical of mature plant cells. Smaller vacuoles are visible toward each end of the cell.

have their own names. Microbodies called **peroxisomes** house reactions in which toxic peroxides (such as hydrogen peroxide, H_2O_2) are formed as unavoidable side products of chemical reactions. Subsequently, the peroxides are broken down safely within the peroxisomes without mixing with other parts of the cell. Both plant and animal cells have peroxisomes, but another type of microbody, the **glyoxysome,** is found only in plants. Glyoxysomes, which are most prominent in young plants, are the sites at which stored lipids are converted into carbohydrates.

We think of organelles as self-contained compartments or "factories" for specific processes, but remember that materials are constantly being shuttled from one type of organelle to another. For example, there is considerable traffic of compounds among the organelles of plant cells in a process called photorespiration (see Chapter 8). The process of photorespiration begins in the chloroplasts. One intermediate product is transported from the chloroplasts to the peroxisomes for further chemical changes. Some of the products of peroxisome action are then passed to the mitochondria, whereas others are returned to the chloroplast for more changes.

In many eukaryotic cells, but particularly in those of plants and protists, are structures that look empty under the electron microscope. They are called **vacuoles** (Figure 4.24). They are not actually empty but are filled with aqueous solutions containing many dissolved substances. Each vacuole is surrounded by a single membrane.

For all their structural simplicity, vacuoles play a variety of crucial roles in the lives of cells. For example, consider one of the problems a plant faces.

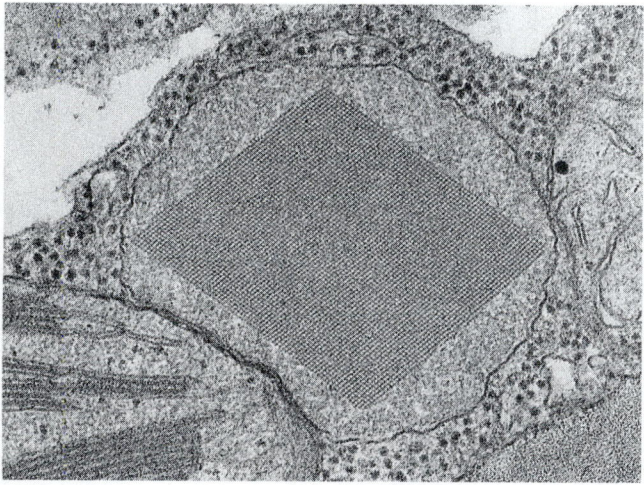

4.23 A Microbody
A diamond-shaped crystal, composed of an enzyme, almost entirely fills this rounded microbody in a leaf cell. The enzyme catalyzes one of the reactions fulfilling the special function of the microbody. The microbody is pressed against a chloroplast (lower left).

Like animals and other organisms, plants produce a number of by-products that would be toxic to the organism if not set aside. Animals have excretory mechanisms for getting rid of such wastes, but plants are not equipped in the same way. Although plants manage to secrete some wastes to their environment, many compounds must be stored within the cells. The solution to this storage problem is the vacuole. The vacuolar membrane can keep wastes away from the rest of the cell, preventing toxic reactions. The vacuoles of many plants store large amounts of chemicals that are poisonous or distasteful to herbivores (plant-eating animals), thus deterring animals from eating the plants.

In many plant cells, enormous vacuoles take up more than 90 percent of the total volume and grow as the cell grows. Vacuoles are by no means a waste of space, however, for the dissolved substances in the vacuole, working together with the vacuolar membrane, provide turgor, or stiffness, to the cell, which in turn supports the structure of nonwoody plants.

Some unicellular protists, sponges, and some of the more ancient invertebrate animals obtain nutrients by endocytosis. Particles trapped from the environment in this way end up in vesicles called **food vacuoles.** Many freshwater protists have a highly specialized **contractile vacuole** (see Chapter 23). Its function is to rid the cell of excess water that rushes in because of the imbalance in salt concentration between the relatively salty interior of the cell and its freshwater environment. Vacuoles even play a role in the sex life of plants. Some pigments (especially blue and pink ones) in petals and fruits (and sometimes in leaves) are contained in vacuoles. These pigments—the anthocyanins—serve as cues that encourage animals to visit flowers and thus aid in pollination, or to eat fruits and thus aid in seed dispersal.

THE CYTOSKELETON

Membranes divide the cytoplasm of eukaryotic cells into numerous compartments, as already described. But even the cytosol (the space and material inside the plasma membrane but outside the membrane-bounded organelles) is not a simple aqueous solution—far from it! In this compartment of the cell is a dynamic set of fibers—the **cytoskeleton**—that contributes to the cell's shape and physical texture. Some fibers also act as tracks for "motors" that help a cell to move. At least three components of the cytoskeleton are visible in electron micrographs: microtubules, microfilaments, and intermediate filaments.

Microfilaments are a common type of fiber in the cytosol. They drive many types of cellular movement.

Microfilaments are assembled from a protein called actin, often in combination with other proteins (Figure 4.25a). Each microfilament is 7 nm in diameter and several micrometers long. Microfilaments may be single or in bundles and networks. Often they are attached to the plasma membrane. In muscle cells the microfilaments are stable for days at a time, but in other cells they are more dynamic, forming and breaking down in minutes. Microfilaments help the cell to contract; many types of motion within the cell require their participation. Microfilaments take part in changes in cell shape (including cell length, as in the contraction of muscles), in the streaming of cytoplasm (a flowing movement observed in some cells), in movements of organelles and particles, and in pinching movements such as those that separate the daughter cells after an animal cell has undergone nuclear division. In muscles, filaments of a protein called myosin have "motor" projections that "row" the actin and myosin filaments past each other, pulling the ends of the cell together; the action of microfilaments in muscles will be discussed in detail in Chapter 40.

Filaments of another type, the **intermediate filaments,** play more static roles: They stabilize cell structure and resist tension. They are found only in multicellular organisms. Different kinds of cells contain different kinds of intermediate filaments. Although there are at least five distinct types of intermediate filaments, all share the same general structure (Figure 4.25b). Intermediate filaments are composed of fibrous proteins similar to those that make up hair and skin. In cells, these proteins are organized into tough, ropelike assemblages about 8 to 12 nm in diameter. In some cells, intermediate filaments end at the nuclear envelope and may maintain the positions of the nucleus and other organelles in the cell. Other types of intermediate filaments help hold a complex apparatus of microfilaments in place in muscle cells (see Chapter 40). Still other types help to stabilize and maintain rigidity in surface tissues by connecting "spot welds" called desmosomes (see Chapter 5). Rapidly growing or newly formed cells do not contain intermediate filaments.

Microtubules are long, hollow, unbranched cylinders about 25 nm in diameter and up to several micrometers long. Many of the microtubules in a cell radiate from a region called the microtubule organizing center. Microtubules are made up of many molecules of a protein called **tubulin.** Tubulin itself is a dimer consisting of two polypeptide subunits called α-tubulin and β-tubulin. Thirteen rows, or protofilaments, of tubulin dimers surround the central cavity of the microtubule (Figure 4.25c). Microtubules have a polarity—one end is called the + end and the other the − end. Tubulin dimers can be added to or subtracted from the + end, lengthening or shortening

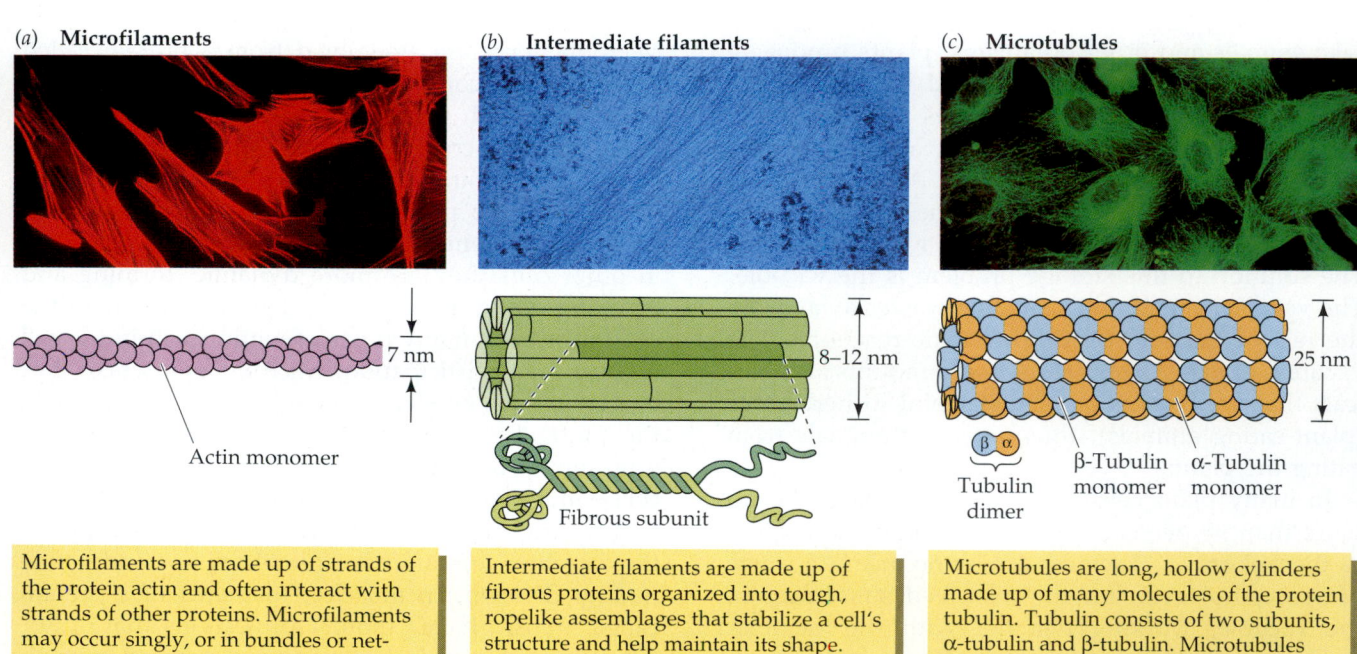

(a) **Microfilaments**

7 nm

Actin monomer

Microfilaments are made up of strands of the protein actin and often interact with strands of other proteins. Microfilaments may occur singly, or in bundles or networks. They change cell shape and drive cellular motion, including contraction, cytoplasmic streaming, and the "pinched" shape changes that occur during cell division. Microfilaments and myosin strands together drive muscle action.

(b) **Intermediate filaments**

8–12 nm

Fibrous subunit

Intermediate filaments are made up of fibrous proteins organized into tough, ropelike assemblages that stabilize a cell's structure and help maintain its shape. Some intermediate filaments hold neighboring cells together.

(c) **Microtubules**

25 nm

β α
Tubulin dimer
β-Tubulin monomer   α-Tubulin monomer

Microtubules are long, hollow cylinders made up of many molecules of the protein tubulin. Tubulin consists of two subunits, α-tubulin and β-tubulin. Microtubules lengthen or shorten by adding or subtracting tubulin dimers. Microtubule shortening moves chromosomes. Interactions between microtubules drive the movement of cells. Microtubules serve as "tracks" for the movement of vesicles.

4.25   Components of the Cytoskeleton

the microtubule and thus affecting the cell in various ways. The capacity to change length rapidly makes microtubules dynamic structures.

In plants, microtubules help control the arrangement of the fibrous components of the cell wall. Electron micrographs of plants frequently show microtubules lying next to the cell wall, and disruption of the cell's microtubules leads to a disordered arrangement of newly synthesized fibers in the cell wall. In animal cells, microtubules are often found in the parts of the cell that are changing shape, but the mechanisms by which the microtubules function are not yet known. In some specialized cells, such as nerve cells, the cytoplasm contains microtubules running parallel to the length of long cellular projections. Here the microtubules contribute to the mechanical stability of the projections. In these and some other cells, microtubules serve as tracks along which tiny "motors" carry protein-laden vesicles from one part of the cell to another (Box 4.C).

Many eukaryotic cells possess whiplike appendages, the flagella and cilia that are built from specialized microtubules. These organelles push or pull the cell through its aqueous environment or promote movement of the surrounding liquid over the surface of the cell (Figure 4.26a). Cilia and eukaryotic flagella are identical in internal structure but differ in relative lengths and patterns of beating. Longer appendages, usually single or in pairs, are called flagella; these

propagate waves of bending from one end to the other in snakelike undulation. The shorter appendages, usually present in great numbers, are called cilia; they beat stiffly in one direction and recover flexibly in the other direction (like a swimmer's arm), so the recovery stroke does not undo the work of the power stroke (see Chapter 40).

In cross section, a typical cilium or eukaryotic flagellum is seen to be covered by the plasma membrane and to contain what is usually called a 9 + 2 array of microtubules (Figure 4.26b). As you can see from the figure, this name is somewhat misleading: There are actually nine fused pairs of microtubules, called **doublets,** forming a cylinder, and one pair of unfused microtubules running up the center. The motion of cilia and eukaryotic flagella results from these microtubules sliding past one another, as described in Chapter 40. But what is the "motor" that drives this sliding? It is a protein called **dynein,** which can undergo changes in tertiary structure driven by en-

4.26   Cilia Move Cells   ▶
(a) Cilia covering the surface of this protist propel the unicellular organism through its watery environment. (b) Longitudinal section through three cilia on a protist cell. If one of these cilia and its basal body were viewed in cross section, the structures diagrammed would be seen. Movement is powered by "arms" of the protein dynein sliding the microtubules along.

BOX 4.C

Molecular "Motors" Carry Vesicles along Microtubules

A remarkable group of protein molecules serve as "motors" on the cytoskeleton. One such protein, kinesin, carries vesicles from the − ends to the + ends of microtubules. Other proteins carry other kinds of loads, such as mitochondria. Kinesin consists of four polypeptide chains: two identical heavy chains and two identical light chains. The "heads" of the heavy chains bind to microtubules and undergo changes in shape (tertiary structure). The shape changes cause the kinesin molecule to "walk" along the tubulin subunits of the microtubules. The light chains bind vesicles, which thus go along for the ride.

How were the functions of kinesin and the other motors discovered? Kinesin was isolated and purified by techniques that took advantage of its

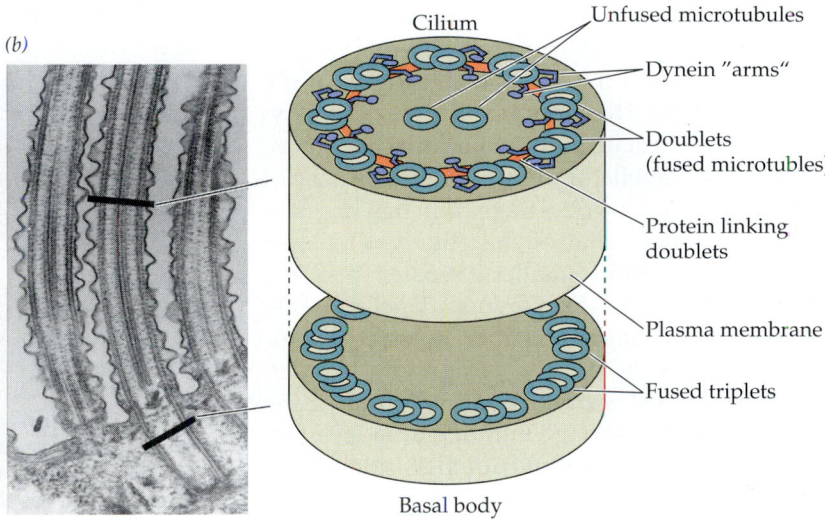

(a)

Kinesin receptor

Kinesin

Vesicle containing proteins

Vesicle movement

Microtubule does not move

(b)

Microtubule movement

Kinesin bound to glass slide

(a) Kinesin molecules "walk" along a microtubule, moving a vesicle along with them. (b) Diagram of the experiment that established the mechanisms of kinesin and other "motor molecules."

affinity for microtubules. The light chains of purified kinesin bind to glass surfaces, enabling the following elegant experiment: After binding kinesin to a glass slide and adding a solution containing microtubules and ATP, scientists discovered that the microtubules slithered along the slide, moving in the direction of their

− ends. Because the kinesin molecules were bound tightly to the glass, as their heads walked toward the + ends of the microtubules, the microtubules were pushed in the direction of their − ends. In other studies, scientists showed that kinesin can also carry plastic beads toward the + ends of isolated microtubules.

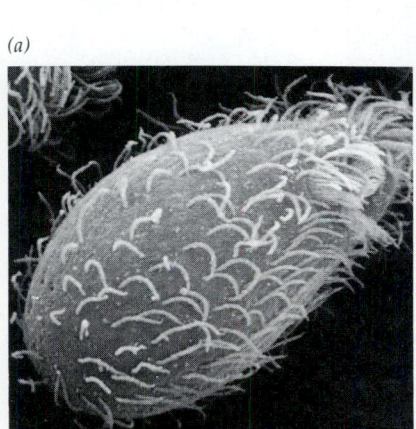

(a)

(b)

Cilium

Unfused microtubules

Dynein "arms"

Doublets (fused microtubules)

Protein linking doublets

Plasma membrane

Fused triplets

Basal body

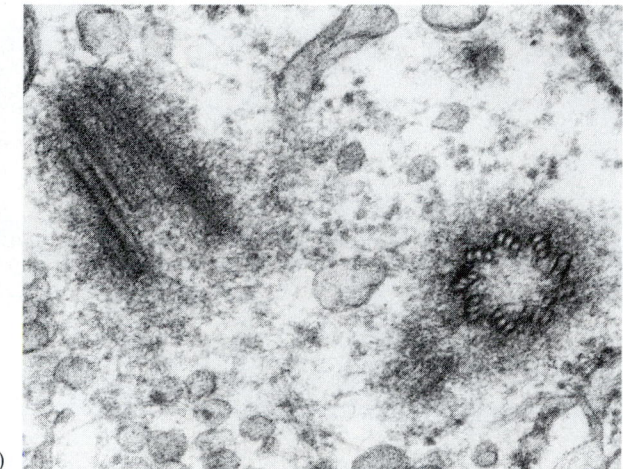

4.27 Centrioles Contain Triplets of Microtubules
Centrioles are found in the microtubule organizing center, a region near the nucleus. *(a)* This thin-section micrograph shows a pair of centrioles at right angles to each other. Nine sets of three microtubules are evident in the centriole on the right, which is seen in cross section. *(b)* The diagram emphasizes the three-dimensional structure of a centriole; compare with the diagram of the basal body in Figure 4.26.

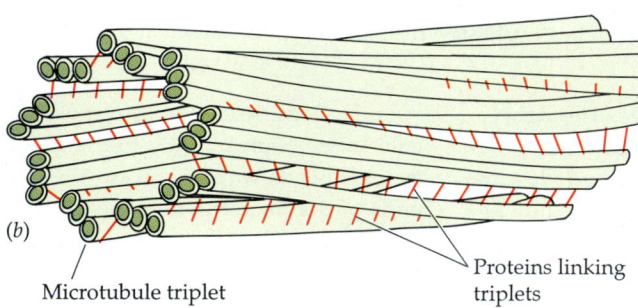

(b)
Microtubule triplet

Proteins linking
triplets

accompanied by another microtubule, making nine sets of *three* microtubules. The central, unfused microtubules of the cilium or flagellum do not extend into the basal body.

Centrioles are organelles that are virtually identical in structure to basal bodies. Centrioles are found in all eukaryotes except those that never produce cells with cilia or flagella (that is, the flowering plants, the pines and their relatives, and some protists). Under the light microscope, a centriole looks like a small, featureless particle, but the electron microscope reveals that it is made up of a precise bundle of microtubules, arranged as nine sets of three fused microtubules each (Figure 4.27). When present, two centrioles lie at right angles to each other in the microtubule organizing center in cells about to undergo division (see Chapter 9).

THE OUTER "SKELETON"—THE CELL WALL

The eukaryotic **cell wall** is a semirigid structure outside the plasma membrane of cells of plants (Figure 4.28*a*), fungi, and some protists. (There is no com-

ergy from ATP. Dynein molecules on one microtubule bind to a neighboring microtubule; as the dynein molecules change shape, they "row" one microtubule past its neighbor.

Although some prokaryotes also have flagella, as described earlier in this chapter, prokaryotic flagella are very different from the 9 + 2 arrangement of the eukaryotic flagellum: They lack microtubules and dynein. There is no structural or evolutionary relationship between the flagella of prokaryotes and those of eukaryotes. The prokaryotic flagellum has a much simpler structure than that of the eukaryotic flagellum (see Figure 4.5) and a smaller diameter than that of a single microtubule. Prokaryotic flagella rotate, whereas eukaryotic flagella beat in a wavelike (or undulating) motion.

An organelle called a **basal body** is found at the base of every eukaryotic flagellum or cilium (see Figure 4.26*b*). The nine microtubule doublets extend into the basal body. In the basal body, each doublet is

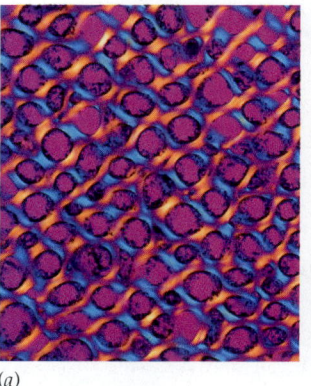
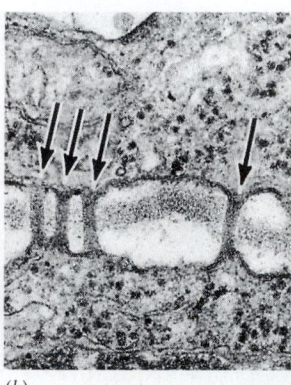

(a)   *(b)*

4.28 The Plant Cell Wall
(a) The brilliant gold and blue structures in this polarizing micrograph of seed tissue are the cell walls. Cell walls make up a substantial fraction of the tissue. (Compare with Hooke's drawing of cork in Figure 1.11.) *(b)* In this electron micrograph of a grass root, plasmodesmata are the gray channels (arrows) crossing the cell walls between the cytoplasm of the cell above and the cell below.

parable structure in animal cells.) The cell wall is made up primarily of polysaccharides. It provides support for the cell, limits the cell's volume, and in some instances restricts the flow of water into and out of the cell. In plants, modifications of the cell wall are important in determining the specific function of the cell.

Although it might seem that plant cells are completely isolated from one another by their cell walls, this is not the case. Plant cells are connected by channels, or **plasmodesmata,** each 20 to 40 nm in diameter, that extend through the walls of adjoining cells (Figure 4.28b). A strand of cytoplasm about 4 nm in diameter runs through most plasmodesmata. A plasmodesma allows relatively free passage of many molecules—not only is there no cell wall across the hole, but there is also no plasma membrane.

EUKARYOTES, PROKARYOTES, AND VIRUSES

Let's review the major differences between the cells of eukaryotes and those of prokaryotes. In contrast to prokaryotes, which have nucleoids, eukaryotes have a true, membrane-enveloped nucleus, and unlike the prokaryotes, eukaryotes have other membrane-bounded organelles that allow various cellular activities to be concentrated in specialized compartments. Eukaryotic cells, but not prokaryotic ones, can perform endocytosis and exocytosis. Eukaryotes also have many specialized molecules, such as tubulin and actin, that are used for movement and are not found in prokaryotes. The cell walls of prokaryotes differ structurally and chemically from those of eukaryotes.

Do such differences mean that eukaryotes are more advanced, higher, or more successful than prokaryotes? Not at all! Every surviving species is the product of eons of natural selection and is superbly adapted to its environmental niche. Each species has characteristics that enable it to live where and how it does and to compete successfully against other species. Both eukaryotic and prokaryotic cells are marvels of systematic organization to achieve complex function.

Where do the viruses fit into this picture? They are simpler in structure than the bacteria—so much simpler that they cannot be called cells. Viruses lack ribosomes and must use the ribosomes of a host cell (prokaryotic or eukaryotic) to synthesize the proteins they need. The hereditary material of a virus enters the host cell and there subverts the host's metabolic machinery to make new viruses. Except for a few complex forms, viruses are simply packets of hereditary material wrapped in coats of protein; in no way do they demonstrate the life processes of independent cells. Viruses are treated in detail in Chapter 22.

FRACTIONATING THE EUKARYOTIC CELL: ISOLATING ORGANELLES

Since organelles are so small (just 1 g of chloroplasts contains more than 10 billion individual chloroplasts of average size), isolating particular organelles for study is difficult. During the early days of cell biology in the nineteenth century, all that one could do with organelles—and only the largest ones at that—was look at them with a microscope as they sat within the cell. Later, with the electron microscope, it became possible to view cells at a higher magnification and to isolate and purify substantial quantities of specific organelles, enabling researchers to study their biochemical activities and physiological functions. Today, scientists who want to work with isolated organelles must accomplish two tasks: removing the organelles from cells, and separating the various types of organelles from one another. This process is called cell fractionation.

Rupturing the Cell

The first step in cell fractionation is opening up the cell; during this process, one must be careful not to burst the organelles. Various methods, most of which employ strong shearing forces, can be used to break cells open. The simplest methods use an old-fashioned mortar and pestle or a hand-operated glass homogenizer (which squeezes and shears cells between two tightly fitting, counter-rotating, ground-glass surfaces). Cells of some tissues can be opened by rapid chopping with a razor blade against a glass plate. Motor-driven homogenizers or blenders are also commonly used.

These methods rupture the cells by tearing their plasma membranes and (if present) cell walls. The rupturing of cells is conducted in a solution with a high concentration of solutes, which prevents the organelles from bursting. In a more dilute solution or pure water, the organelles would take up water and explode, as described in Chapter 5.

Separating the Organelles

Using techniques like those described in the previous section, one may reduce biological tissue to a crude suspension of mixed organelles, unbroken cells, and debris. But how can the components of such a suspension be separated from one another? The methods of choice are two types of centrifugation. The **centrifuge** is a laboratory instrument that can spin materials extremely rapidly about a fixed axis, which causes the particles in suspension to fall or rise more rapidly than they would under the force of gravity alone (Figure 4.29). Different classes of organelles sediment (settle out of a suspension) at different rates

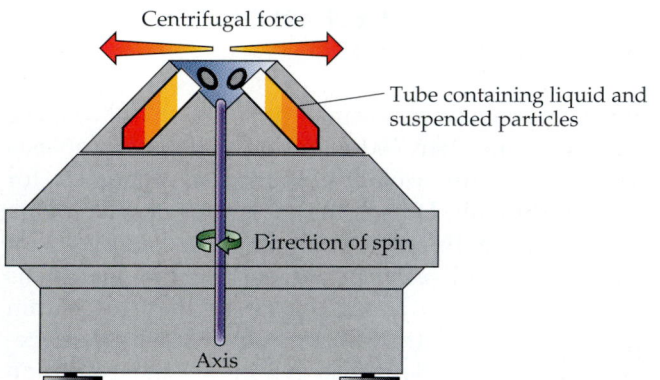

4.29 Separation by Centrifugation
Extremely rapid rotation of liquids in centrifuge tubes around an axis produces a force in the tubes analogous to that of gravity, but much more powerful. Suspended particles denser than the liquid in the tubes sink rapidly. Following centrifugation, the fluid can be poured off, leaving a dense pellet of particles in the bottom of the tube.

in a centrifuge. Factors determining the rate of sedimentation include the organelle's size (radius) and density (weight per unit volume). For example, an organelle that is more dense than the liquid in which it is suspended settles toward the bottom of the centrifuge tube. An organelle less dense than the liquid floats toward the surface. This is the same principle that explains why oil rises to the top of a mixture of oil and water that is shaken and then allowed to separate.

Equilibrium centrifugation can be used to separate two or more types of organelles that differ in density (Figure 4.30). First a centrifuge tube is filled with liquid whose density varies from the top to the bottom of the tube. This could be done, for example, by first putting in a small amount of a highly concentrated (and hence very dense) sugar solution—say, 60 percent sucrose. Next a small amount of a 50 percent sucrose solution is added, then 40 percent, and so on. Such a **density gradient** is usually con-

structed by an automatic device so that the gradient is smooth rather than erratic and jumpy. After the gradient is established, some of the mixture of, say, two organelles to be separated is carefully added to the top and the tube is centrifuged. Both populations of organelles sediment into the gradient as long as they are more dense than the surrounding liquid. When an organelle reaches the part of the gradient where its density matches that of the liquid, it stops sedimenting—it has reached buoyant equilibrium. If the organelle were pushed farther down, it would float back up to this point. If the two types of organelles have different average densities, they form separate bands in the liquid. After the tube is removed from the centrifuge, the organelles can be collected separately by the careful use of a pipette or by punching a hole in the bottom of the tube and collecting samples as the liquid slowly drips out.

A second approach, called **differential centrifugation,** depends on differences in either the radius or the density of the particles being centrifuged. No density gradient is used, and the liquid in the tube is usually of low density. A mixture containing various organelles is centrifuged briefly at low speed (and hence low relative centrifugal force). The largest and densest particles sediment, forming a pellet in the bottom of the tube; this pellet is left behind when the remaining liquid and its suspended contents (together called the supernatant) are poured off. The liquid and its remaining contents are then spun at a higher speed and for a longer time, causing other organelles to sediment. By repeating this procedure with ever-increasing centrifugal force, one can separate many organelles. Nuclei sediment into a pellet even when the centrifuge is spun slowly. The tiny ribosomes, on the other hand, require considerably longer centrifugation at extremely high speeds before they will sediment—in most rotors, the spinning rate must be about 40,000 revolutions per minute, giving forces as great as 100,000 times the force of gravity.

Partially purified organelles are obtained by equilibrium or differential centrifugation. Further purification is achieved by repeating the centrifugation

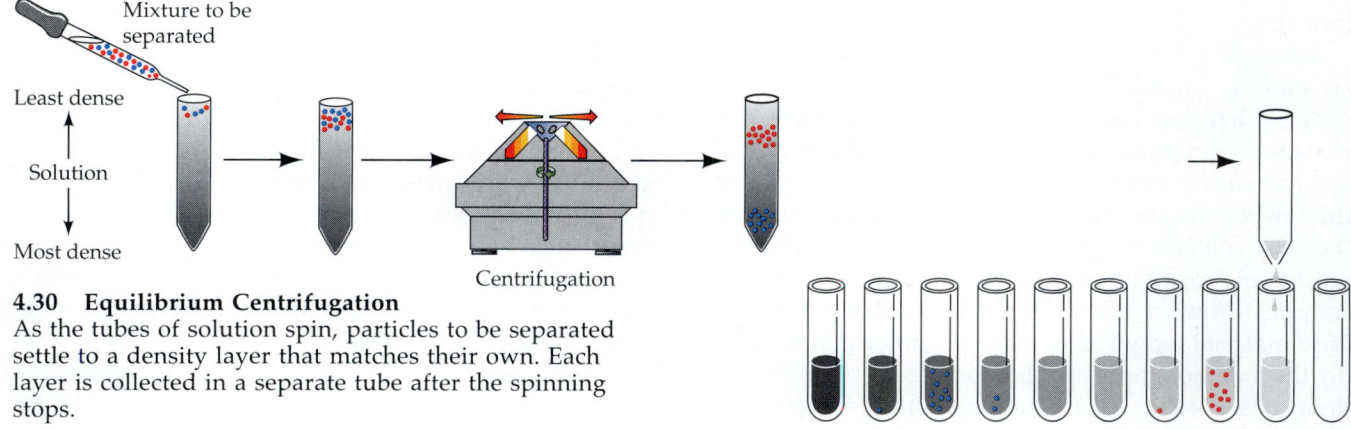

4.30 Equilibrium Centrifugation
As the tubes of solution spin, particles to be separated settle to a density layer that matches their own. Each layer is collected in a separate tube after the spinning stops.

routine. The purity may be determined by examining the final sample under the microscope—light or electron—or by testing the chemical activities of the sample in comparison with the known behavior of various organelles. We actually *can* get a 1-g sample of chloroplasts (or other organelles) in this way and study them to our heart's content.

Early work with isolated organelles focused on their chemical composition and then on the biochemical reactions that take place within them. Gradually we have come to have an extensive, but still partial, knowledge of the activities of the different organelles (as discussed in this and the next few chapters) and hence of the cell itself.

SUMMARY of Main Ideas about Cells

All organisms are composed of cells, and all cells come from preexisting cells.

In the kingdoms Eubacteria and Archaebacteria, each organism is a single, uncompartmented prokaryotic cell.

In the other kingdoms, each organism consists of one or more nucleated, compartmented eukaryotic cells.

All prokaryotic cells have a plasma membrane, ribosomes, and DNA in a nucleoid, and many have flagella, cell walls, capsules, and internal membranes formed from the plasma membrane.
 Review Figures 4.2, 4.3, 4.4, and 4.5

The nucleus of a eukaryotic cell contains chromosomes, nucleoli, and nucleoplasm, all surrounded by a nuclear envelope.
 Review Figures 4.8, 4.9, and 4.10

A cell's genetic information is stored in its nucleoid (prokaryotes) or nucleus (eukaryotes).

All cells synthesize proteins on ribosomes consisting of two subunits.
 Review Figure 4.13

Mitochondria convert energy into a usable form in eukaryotic cells.
 Review Figures 4.8, 4.9, and 4.14

Chloroplasts in photosynthetic eukaryotic cells capture light energy.
 Review Figures 4.8, 4.15, and 4.16

If the endosymbiotic theory is correct, mitochondria and chloroplasts evolved by the incorporation of smaller prokaryotic cells into larger ones.
 Review Figure 4.18

Endocytosis brings substances into eukaryotic cells, and exocytosis ejects substances from them.
 Review Figures 4.21 and 4.22

The endoplasmic reticulum and the Golgi apparatus sort, store, modify, and move proteins in eukaryotic cells.
 Review Figures 4.8, 4.9, 4.19, and 4.20

Reactions that either form toxic products or convert lipids to carbohydrates are housed, respectively, in peroxisomes or glyoxysomes; these are examples of microbodies.
 Review Figure 4.23

Vacuoles in eukaryotic cells store wastes and provide turgor.
 Review Figures 4.8 and 4.24

The cytoskeleton gives a eukaryotic cell shape and may help it move.
 Review Figure 4.25

Cilia and flagella propel eukaryotic cells or move substances past them.
 Review Figure 4.26

A cell wall surrounds the cells of plants, fungi, and some protists.

SELF-QUIZ

1. Which statement is true of both prokaryotic and eukaryotic cells?
   a. They contain ribosomes.
   b. They have peptidoglycan cell walls.
   c. They contain membrane-bounded organelles.
   d. They contain true nuclei.
   e. Their flagella have the 9 + 2 structure.

2. Which statement is *not* true of the nuclear envelope?
   a. It is continuous with the endoplasmic reticulum.
   b. It has pores.
   c. It consists of two membranes.
   d. RNA and some proteins pass through it to move in and out of the nucleus.
   e. Its inner membrane bears ribosomes.

3. Which statement is *not* true of mitochondria?
   a. Their inner membrane folds to form cristae.
   b. They are usually one micrometer or less in diameter.
   c. They are green because of the chlorophyll they contain.
   d. Energy-rich substances from the cytosol are oxidized in them.
   e. Much ATP is synthesized in them.

4. Which statement is true of plastids?
   a. They are found in prokaryotes.
   b. They are surrounded by a single membrane.
   c. They are the sites of cellular respiration.
   d. They are found in fungi.
   e. They are of several types, with different functions.

5. Which statement is *not* true of the endoplasmic reticulum?
   a. It is of two types: rough and smooth.
   b. It is a network of tubes and flattened sacs.
   c. It is found in all living cells.
   d. Some of it is sprinkled with ribosomes.
   e. Parts of it modify proteins.

6. The Golgi apparatus
   a. is found only in animals.
   b. is found in prokaryotes.
   c. is the appendage that moves a cell around in its environment.
   d. is a site of rapid ATP production.
   e. packages and modifies proteins.

7. Which of the following organelles is *not* surrounded by one or more membranes?
   a. Ribosome
   b. Chloroplast
   c. Mitochondrion
   d. Microbody
   e. Vacuole

8. Eukaryotic flagella
   a. are composed of a protein called flagellin.
   b. rotate like propellers.
   c. cause the cell to contract.
   d. have the same internal structure as cilia.
   e. cause the movement of chromosomes.

9. Microfilaments
   a. are composed of polysaccharides.
   b. are composed of actin.
   c. provide the motive force for cilia and flagella.
   d. make up the spindle that aids movement of chromosomes.
   e. maintain the position of the nucleus in the cell.

10. Which statement is *not* true of the plant cell wall?
    a. Its principal chemical components are polysaccharides.
    b. It lies outside the plasma membrane.
    c. It provides support for the cell.
    d. It completely isolates adjacent cells from one another.
    e. It is semirigid.

FOR STUDY

1. Which organelles and other structures are found in both plant and animal cells? Which are found in plant but not animal cells? Which in animal but not plant cells? Discuss, in relation to the activities of plants and animals.

2. Through how many membranes would a molecule have to pass in going from the interior of a chloroplast to the interior of a mitochondrion? from the interior of a lysosome to the outside of a cell? from one ribosome to another?

3. How does the possession of double membranes by chloroplasts and mitochondria relate to the endosymbiotic theory of the origins of these organelles? What other evidence supports the theory?

4. What sorts of cells and subcellular structures would you choose to examine by transmission electron microscopy? by scanning electron microscopy? by light microscopy? What are the advantages and disadvantages of each of these modes of microscopy?

5. Some organelles that cannot be separated from one another by equilibrium centrifugation can be separated by differential centrifugation. Other organelles cannot be separated from one another by differential centrifugation but can be separated by equilibrium centrifugation. Explain these observations.

READINGS

Alberts, B., D. Bray, J. Lewis, M. Raff, K. Roberts and J. D. Watson. 1994. *Molecular Biology of the Cell*, 3rd Edition. Garland Publishing, New York. An outstanding book in which to pursue the topics of this chapter in greater detail; authoritative treatment of modern cell biology and its experimental basis.

Allen, R. D. 1987. "The Microtubule as an Intracellular Engine." *Scientific American*, February. How microtubules cause two-way transport of materials in cells.

Brandt, W. H. 1975. *The Student's Guide to Optical Microscopes*. William Kaufmann, Los Altos, CA. A short, programmed guide for those interested in learning how to use a light microscope.

De Duve, C. 1975. "Exploring Cells with a Centrifuge." *Science*, vol. 189, pages 186–194. A discussion by a Nobel laureate of the uses of centrifugation in studies of cells.

Fawcett, D. W. 1981. *The Cell*, 2nd Edition. Saunders, Philadelphia. Beautiful electron micrographs of subcellular structures in animal cells.

Glover, D. M., C. Gonzalez and J. W. Raff. 1993. "The Centrosome." *Scientific American*, June. New findings about the structure and function of the organelle that directs the assembly of the cytoskeleton and controls cell division—when it is present.

Howells, M. R., J. Kirz and D. Sayre. 1991. "X-Ray Microscopes." *Scientific American*, February. A description of novel methods of microscopy afford striking improvements in resolution.

Lodish, H., D. Baltimore, A. Berk, L. Zipursky, P. Matsudaira and J. Darnell. 1995. *Molecular Cell Biology*, 3rd Edition. Scientific American Books, New York. Another excellent middle-level book; fine illustrations.

Margulis, L. 1993. *Symbiosis in Cell Evolution*, 2nd Edition. W. H. Freeman, New York. An authoritative and thought-provoking reference on the origin and evolution of eukaryotic cells by a leading student of the problem.

Rothman, J. E. 1985. "The Compartmental Organization of the Golgi Apparatus." *Scientific American*, September. Discusses the structure and function of the Golgi apparatus.

Weber, K. and M. Osborn. 1985. "The Molecules of the Cell Matrix." *Scientific American*, October. A clear treatment of microfilaments, intermediate filaments, tubulin, and the ways in which they are studied.

The Plasma Membrane
The edge of a red blood cell, magnified about 300,000 times by a transmission electron microscope, shows the bilayered phospholipid membrane as two dark lines reminiscent of railroad tracks.

5

Membranes

Poised between every cell and its environment is a filmy sheet so thin that it can be seen only with the aid of an electron microscope. Similar sheets within the cytoplasm divide many cells into compartments. The sheets are dynamic—forming, changing, merging, moving. Filmy and thin as they are, these sheets perform a sweeping array of vital functions. They recognize specific chemical substances as well as other cells, hold groups of molecules in place, and transmit signals along nerves. They let some substances pass through and keep other substances out.

How are such diverse tasks performed? How can some things pass through while others are stopped? These are the subjects of this chapter. The story begins with a close look at the filmy sheets and with consideration of the fundamental process of diffusion, followed by an explanation of other ways that particular substances move through and a description of more of the diverse tasks.

MEMBRANE STRUCTURE AND COMPOSITION

The filmy sheets described in our opening paragraphs are, of course, the plasma membrane and other **membranes,** all of which are thin, pliable bilayers of phospholipids with embedded proteins. (The phospholipid bilayer was introduced in Chapter 3, and membranous organelles are described in Chapter 4.) As we consider membranes, the relationship between the function and the physical structure of the membrane is particularly obvious; this relationship is evident at all levels, from the overall shape of the membrane down to its individual chemical components.

The chemical makeup, physical organization, and functioning of a biological membrane depend upon three classes of biochemical compounds: lipids, proteins, and carbohydrates. The lipids are an effective barrier to the passage of many materials between the inside and the outside of a cell or organelle. Many functions of membranes result from this **selective permeability**—some materials move through them more readily than others. The lipids also account for much of the physical integrity of the membrane. The lipids are present as a double layer that constitutes the continuous portion of the membrane. In this lipid "lake" floats a variety of proteins.

Some proteins reach from one side of a membrane to the other; others reside primarily on one side. The proteins are responsible for many of the specific tasks performed by membranes. Certain membrane proteins allow materials to pass through the membrane that cannot pass through the pure lipid bilayer. Other proteins receive chemical signals from the cell's ex-

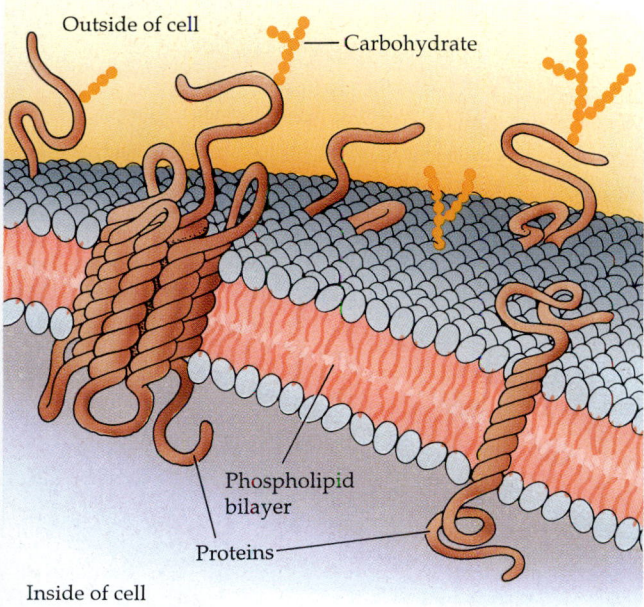

5.1 A Model of a Biological Membrane
The general molecular structure of biological membranes, sometimes called the fluid mosaic model, is a continuous phospholipid bilayer. Proteins are embedded in the bilayer. Carbohydrates may be attached to the proteins or to the phospholipids.

ternal environment and respond by regulating certain processes within the cell. Still other proteins accelerate chemical reactions on the membrane surface.

Like some proteins, the carbohydrates, the third class of compounds important in membranes, are crucial in recognizing specific molecules. The carbohydrates are attached to lipid or protein molecules, mostly on the outside of the plasma membrane, where they protrude into the environment, away from the cell.

Generalized membrane architecture is shown in Figure 5.1. The two sides of the membrane are not identical—in fact, the membrane is decidedly asymmetric. Not even all the lipids are the same: Those on the inward-facing half are different from those on the outward-facing half.

Membrane Lipids

Nearly all lipids in biological membranes are **phospholipids.** Recall that some compounds are hydrophilic ("water-loving") and others are hydrophobic ("water-fearing"). Phospholipids are both: They have hydrophilic regions and hydrophobic regions. The large, nonpolar fatty acid parts of phospholipid molecules associate easily with other fatty materials but do not dissolve in water. The phosphorus-containing region of the phospholipid is electrically charged and hence very hydrophilic. As a consequence, one way for phospholipids and water to coexist is for the

phospholipids to form a double layer with the fatty acids of the two layers pointing toward each other and the polar regions facing the outside (Figure 5.2). Artificial membranes with the same two-layered arrangement are made easily in the laboratory.

Both artificial and natural membranes form continuous sheets. Because of the tendency of the fatty acids to associate with one another and exclude water, small holes or rips in a membrane seal themselves spontaneously. This property helps membranes fuse during endocytosis, exocytosis, and cell fusion.

The phospholipid bilayer stabilizes the entire structure. At the same time, it makes the membrane fluid—about as fluid as lightweight machine oil—so materials can move laterally within the membrane. As we will see, some membrane proteins are relatively free to migrate about, and individual phospholipid molecules may also move. A phospholipid molecule in the plasma membrane of a bacterium may travel from one end of the bacterium to the other in a little over a second. On the other hand, it is *not* common for a phospholipid molecule in one half of the bilayer to flop over to the other side and trade places with another phospholipid molecule. For this to happen, the polar part of each molecule would

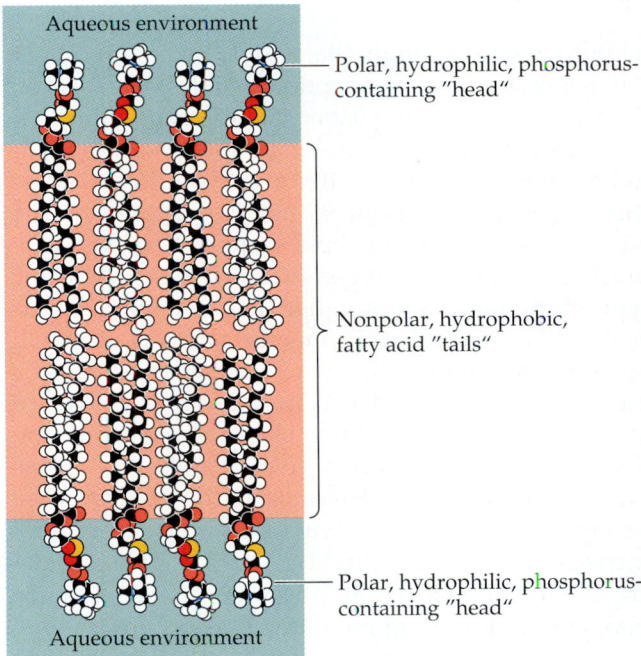

5.2 A Phospholipid Bilayer Separates Two Aqueous Regions
The six phospholipid molecules shown here represent a small section of a membrane bilayer. The charged, hydrophilic heads of the molecules orient toward the surfaces of the membrane. The hydrophobic fatty-acid tails mingle with one another in the interior.

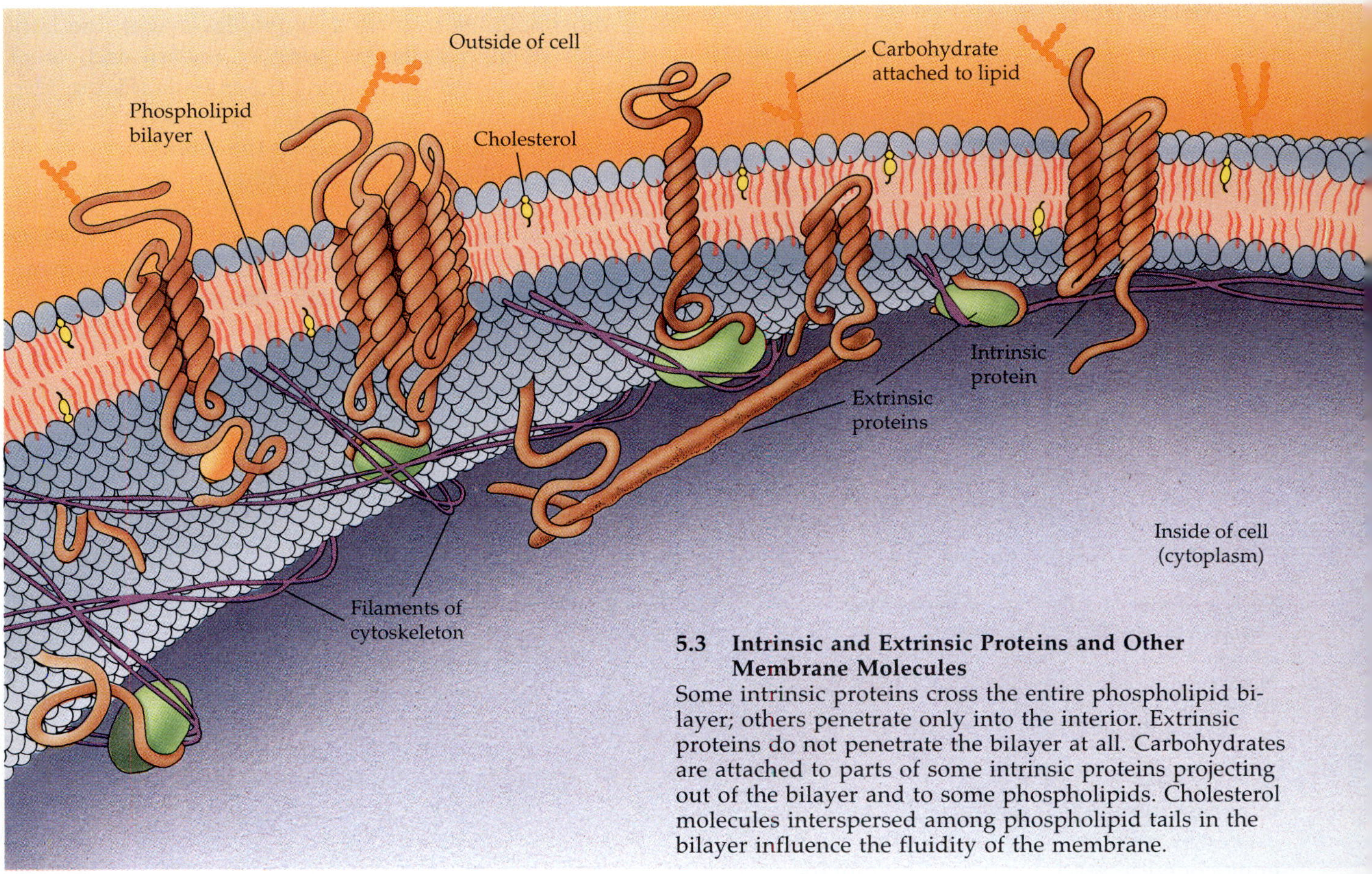

Outside of cell

Phospholipid bilayer

Cholesterol

Carbohydrate attached to lipid

Intrinsic protein

Extrinsic proteins

Inside of cell (cytoplasm)

Filaments of cytoskeleton

5.3　Intrinsic and Extrinsic Proteins and Other Membrane Molecules
Some intrinsic proteins cross the entire phospholipid bilayer; others penetrate only into the interior. Extrinsic proteins do not penetrate the bilayer at all. Carbohydrates are attached to parts of some intrinsic proteins projecting out of the bilayer and to some phospholipids. Cholesterol molecules interspersed among phospholipid tails in the bilayer influence the fluidity of the membrane.

have to move through the hydrophobic interior of the membrane. Since phospholipid flip-flops are rare, the two halves of the bilayer may be quite different.

All biological membranes have similarities, but different membranes may differ greatly in the details of their composition—even within the same cell. One big area of difference is lipid composition. The proportions of different types of lipids vary from one type of membrane to another. For example, 25 percent of the lipid in many membranes is cholesterol, but some membranes have no cholesterol at all. In a membrane, cholesterol is commonly next to an unsaturated fatty acid, and its polar region extends into the surrounding aqueous layer. Cholesterol plays more than one role in determining the fluidity of the membrane. Under some circumstances, cholesterol increases membrane fluidity; under others, it decreases membrane fluidity. Shorter fatty acid chains make for a more fluid membrane, as do unsaturated fatty acids. Organisms can modify their membrane lipid composition, thus changing membrane fluidity, to compensate for changes in temperature that are not too sudden. House plants adapted to indoor temperatures may die when accidentally left outdoors overnight. The sudden change in temperature does not give them time to adapt by adjusting their membrane lipid composition. Lipids constitute a major

fraction of all membranes, and they always form the continuous matrix into which the other chemical components become inserted.

Membrane Proteins

Each protein in a biological membrane is either an **intrinsic protein** or an **extrinsic protein** (Figure 5.3). Intrinsic proteins penetrate the phospholipid bilayer; many extend from one side of the membrane to the other. Extrinsic proteins are entirely outside the bilayer; they are attached to the surface of the membrane by weak (noncovalent) bonds with the exposed parts of the intrinsic proteins or with the hydrophilic parts of phospholipid molecules. These two types of proteins play different roles in membrane function.

The membranes of the various organelles differ sharply in protein composition. Different biochemical reactions (many of them requiring membrane-bound enzymes) occur in different organelles. In many of the important reactions of cellular respiration and photosynthesis, membrane-bound enzymes carry electrons from a donor to an acceptor molecule. Accordingly, both mitochondria and chloroplasts have highly specialized internal membranes, and these differ markedly (Figure 5.4).

Many membrane proteins move relatively freely

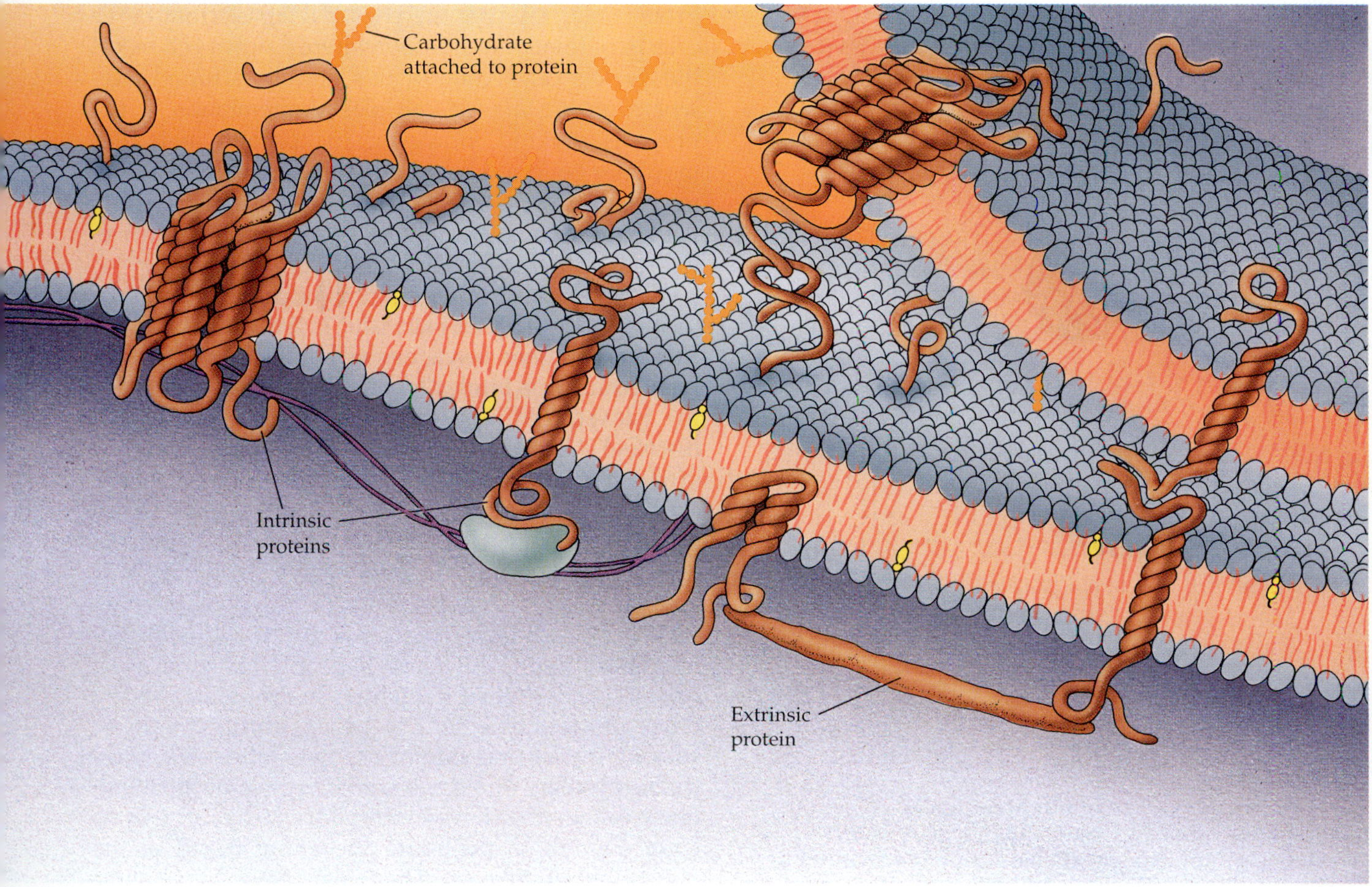

Carbohydrate
attached to protein

Intrinsic
proteins

Extrinsic
protein

within the phospholipid bilayer. Evidence of this migration is dramatically illustrated by experiments using the technique of cell fusion. Specially treated cells of two different species, such as human and mouse, can be fused so that one continuous membrane surrounds the combined cytoplasm and both nuclei. Ini-

(a)

(b)

5.4 Proteins in Specialized Membranes
(a) The outer membrane of this mitochondrion has been fractured away, exposing the inner membrane. The image has been magnified about 65,000 times. The particles giving the inner membrane a grainy appearance are proteins necessary for cellular respiration. (b) The distinct proteins embedded in these thylakoids from a spinach chloroplast, magnified about 70,000 times, are necessary for photosynthesis.

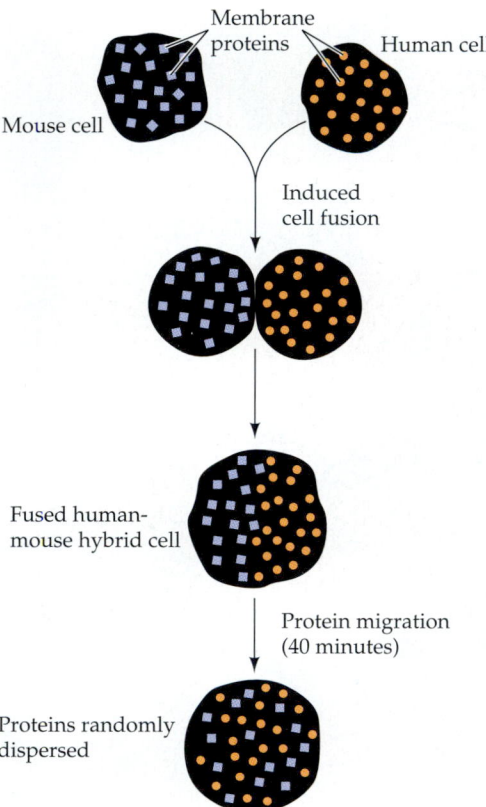

5.5 Proteins Move Around in Membranes
When specially treated mouse and human cells are joined, one continuous membrane surrounds the fused hybrid cell. Distinguished by different fluorescent dyes, many of the mouse and human membrane proteins mingle over time, demonstrating that some proteins move in the fluid phospholipid bilayer.

tially the experimenter can tell by the protein content which part of the plasma membrane came from which species, but the membrane proteins of the two cells migrate in the joint membrane until, after about 40 minutes, they are uniformly dispersed (Figure 5.5). On the other hand, there is also good evidence that other membrane proteins are *not* free to migrate at will—that they are, to an extent, held in place. These proteins are anchored by microfilaments, as described later in this chapter. Microtubules may also play a role.

Proteins are asymmetrically distributed in membranes. Many intrinsic proteins extend completely through a membrane and have specific parts of their primary structures on one side of the membrane, specific parts within the membrane, and specific parts on the other side of the membrane. All the other membrane proteins (intrinsic and extrinsic) are localized on one side of the membrane or the other, but not both.

The fact that molecules of one type of protein may be confined to one part of the cell surface, rather than scattered evenly about also contributes to asymmetry. In certain muscle cells, for example, the membrane

protein that receives the chemical signal from nerve cells is normally found only where a nerve cell meets the muscle cell. None of this protein is found elsewhere on the muscle cell plasma membrane. If the nerve is severed, however, the protein molecules may later be found evenly distributed over the entire membrane. If the nerve regenerates its attachment to the muscle, then the protein is once again limited to the junction area.

What determines whether a particular membrane protein is intrinsic or extrinsic? If it is intrinsic, what controls whether it reaches all the way through the membrane or is limited to one side? What keeps it in the bilayer, and what determines just how far in it reaches? All these questions are answered in terms of the tertiary structure of the protein (see Chapter 3). Recall that the side chains of the various amino acids in a protein differ chemically. What matters here is that some of the side chains are hydrophilic and others hydrophobic. After a polypeptide chain folds into the final tertiary structure, the protein may have both hydrophilic and hydrophobic surfaces. If one end of a folded protein is hydrophilic and the other hydrophobic, it will be an intrinsic protein, sticking out of one side of the membrane. Many intrinsic proteins that reach from one side of the membrane to the other have hydrophobic α-helical regions large enough to penetrate the entire depth of the phospholipid bilayer. They also have hydrophilic ends that protrude into the aqueous environments on either side of the membrane (Figure 5.6). Proteins like this resist being removed from the membrane. If the hydrophobic surface of such a protein is pulled partway out of the phospholipid bilayer, it is repelled by the aqueous environment. If an intrinsic protein is pushed farther into the membrane, its hydrophilic end is pushed back by the hydrophobic fatty acid region of the lipids. Thus such a protein may migrate

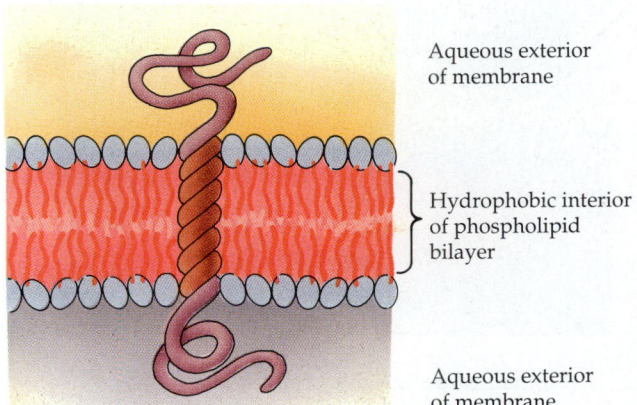

5.6 Controlling Surfaces of Intrinsic Proteins
This intrinsic membrane protein has hydrophilic surfaces (purple) made up of the hydrophilic side chains of some of its amino acids. A region consisting of hydrophobic side chains (brown) buries itself among the fatty acid tails in the bilayer's interior.

laterally in the membrane sheet, but it may not push through the membrane or pop out of it.

How does an intrinsic protein penetrate the membrane in the first place? Clearly, it cannot fold into its final shape and then be pushed into the membrane—for the same reasons that it cannot leave the membrane once in place. In fact, its insertion into the membrane usually proceeds while the protein is being synthesized. Targeting of the sort discussed in Chapter 4 plays a crucial role—specific sequences of amino acids in the growing protein interact with receptors on the membrane, causing the protein to enter the membrane (or not) and to exit on the other side (or not). Final folding is achieved after the different regions of the protein have been appropriately situated.

Membrane Carbohydrates

All plasma membranes and some other membranes contain substantial amounts of carbohydrate along with the lipids and proteins. For example, the plasma membrane of a red blood cell consists of approximately 40 percent lipid, 52 percent protein, and 8 percent carbohydrate by weight. The carbohydrates are recognition sites. Some of the membrane carbohydrates are bound to lipids, forming glycolipids. Glycolipids enable a cell to recognize other cells. Although there is still much to learn about glycolipids, we know that they are important—the structures of some glycolipids change when a cell becomes cancerous, for example.

Most of the carbohydrates in membranes are bound to proteins, forming glycoproteins. These bound carbohydrates are oligosaccharide chains, usually not exceeding 15 monosaccharide units in length. The glycoproteins enable a cell to recognize foreign substances. The oligosaccharide chains are added to the membrane proteins inside the endoplasmic reticulum and are modified in the Golgi apparatus.

The carbohydrates linked to membrane proteins and lipids are relatively small and made from only nine building blocks—the monosaccharides. Nevertheless, these carbohydrates are exceedingly diverse. To understand how this can be, recall that monosaccharides may join to form branched oligomers (see Figure 3.16). The possibility of different branching patterns greatly increases the diversity that can be achieved by the monosaccharide sequence of a carbohydrate alone. Also, monosaccharides may link together at any of several different carbons. All in all, membrane carbohydrates have great specificity and diversity. This structural diversity is important in all sorts of reactions at the cell surface in which different membrane carbohydrates recognize and react with specific foreign substances. All plasma membrane carbohydrates are on the *outside* of the plasma membrane, as befits their role as recognition sites for foreign substances and cells; none face into the cell (see Figure 5.3).

MICROSCOPIC VIEWS OF BIOLOGICAL MEMBRANES

Light microscopes cannot resolve anything as thin as the plasma membrane and the membranes within cells. Electron microscopes, however, offer various ways to examine membranes. The first visualizations were produced when very thin slices of tissue were made with diamond knives and the resulting sections were examined by transmission electron microscopy. The cuts were very clean, and membranes were often seen in cross section in electron micrographs of the slices. (Look back at the micrograph on page 92, noticing that the plasma membrane's phospholipid bilayer appears as two dark lines separated by a light region.) A more detailed understanding of membrane architecture, however, had to await methods for seeing surface views and for revealing the hydrophobic interior of the membrane.

The first successes were achieved by a technique known as **freeze-fracture.** In this procedure the tissue to be examined is frozen solid and then broken rather than cut. To visualize the consequences of this fracture, picture a chocolate bar with almonds. If the bar is cut carefully with a very sharp knife, the cut will pass cleanly through the nuts as well as the chocolate. If, on the other hand, the bar is broken, the break will pass around any almonds in its path and reveal them where they protrude from the chocolate. Similarly, in the freeze-fracture technique, the break tends to pass around membrane-encased organelles and to pass between the two halves of the phospholipid bilayer but around the proteins within it. When the exposed surfaces of the fractured bilayer are examined with the electron microscope, they appear bumpy—the intrinsic proteins are revealed (Figure 5.7a).

Further clarification of membrane structure is obtained by **freeze-etch.** In this method, a freeze-fractured sample is kept cold and put under a high vacuum for a minute or so, allowing some of the frozen water to evaporate, wherever it is exposed to the vacuum. The water evaporation reveals more of the texture of the membrane by uncovering surfaces that were covered by ice (Figure 5.7b).

Contrast in both freeze-fractured and freeze-etched preparations is enhanced by **shadowcasting** with platinum. The metal is sprayed on from an angle so that bumps and dips on the surface give shadow patterns that make them easier to see and interpret. Figure 5.7c is an example of a platinum-shadowed micrograph of a freeze-etched sample.

Freeze-fracture and freeze-etch were instrumental

(*a*) Freeze-fracture. Frozen cells are broken open, exposing membrane faces and interiors

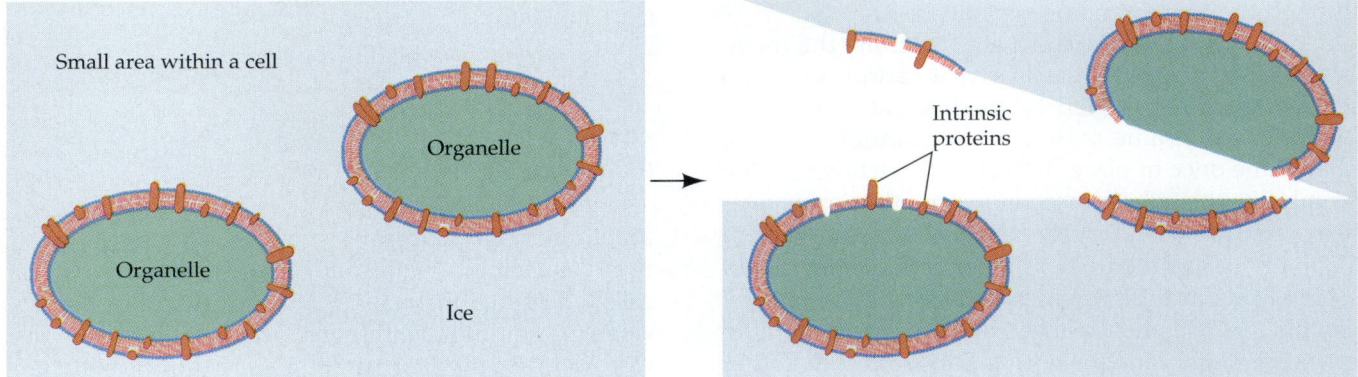

Small area within a cell

Organelle

Organelle

Ice

Intrinsic proteins

(*b*) Freeze-etch. Water is evaporated from surface of a freeze-fractured sample, exposing more structures

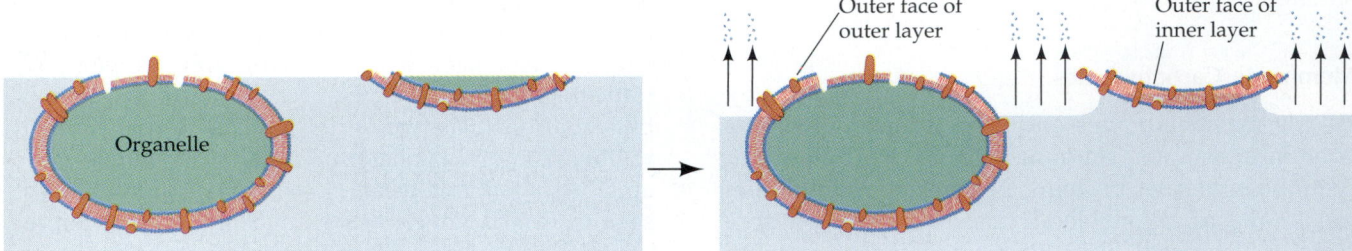

Organelle

Outer face of outer layer

Outer face of inner layer

5.7 Freeze-Fracture and Freeze-Etch

(*a*) Freeze-fracture of a cell often splits the lipid bilayer of a membrane, exposing membrane faces and intrinsic proteins. (*b*) Freeze-etch technique enhances the sample, revealing even more detail. (*c*) A freeze-etched sample of part of a photosynthetic protist cell, shadowed with platinum. We can see the Golgi apparatus (lower left), a chloroplast (upper right), and an inside view of the nuclear envelope, dotted with pores (upper left). Notice the abundant vesicles between the nucleus and the *cis* region of the Golgi apparatus. More examples of electron micrographs of freeze fractured preparations are found in Figures 5.4 and 5.8 (top photo).

(*c*) Freeze-etched sample shadowed with platinum reveals cellular structures in electron micrograph

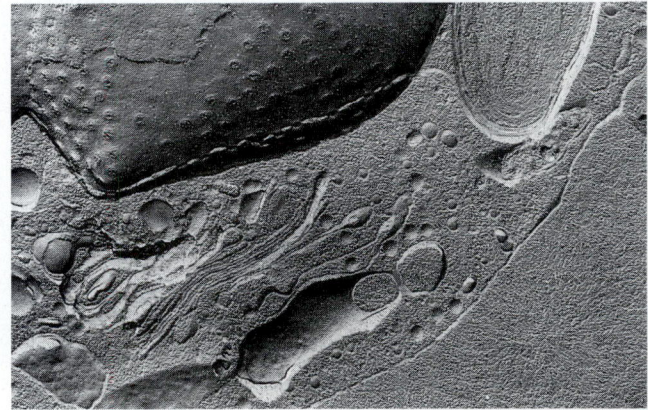

in formulating our current picture of membrane structure. In particular, they made it clear that the membrane is not just a continuous bilayer of phospholipids with the proteins spread on its surface, but that many of the proteins are embedded in the bilayer to various depths.

You now know quite a bit about the chemistry and appearance of plasma membranes. What happens when the plasma membranes of two cells meet? What can microscopic views tell us about this?

WHERE ANIMAL CELLS MEET

Plasma membranes of adjacent animal cells sometimes touch, but usually there is a space between the membranes. What occupies that space? Electron microscopic studies reveal that animal cells are surrounded by an **extracellular matrix** consisting of a fine meshwork of polysaccharides permeated by fibrous proteins. The most abundant protein in the

extracellular matrix is collagen, whose triple-helical structure was described in Chapter 3. One major function of the matrix is to hold the cells together as tissues. In addition, the matrix helps to direct the ways in which cells interact and to orient individual cells. This relationship is reciprocal because the cells themselves secrete the extracellular matrix and establish its orientation. Thus the organizational pattern of a tissue, once established, tends to be maintained as the tissue grows—the first cells orient the matrix, which in turn orients the new cells, and so forth.

Some epithelial cells (cells that line a body cavity or an exterior body surface) come in direct physical contact with one another and form special links, called **junctions.** By their structure and function, junctions fall into three categories: desmosomes,

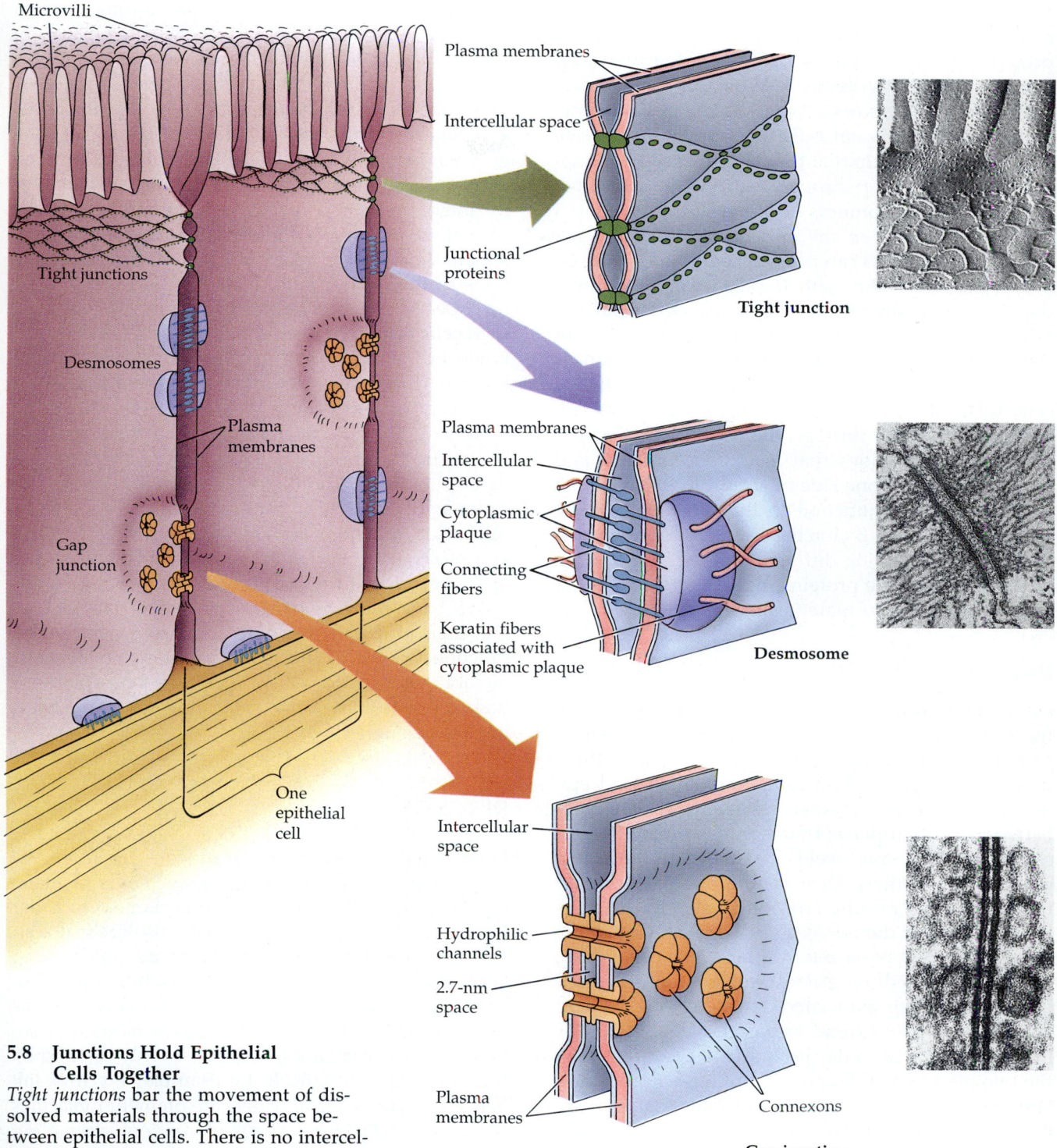

Microvilli

Plasma membranes

Intercellular space

Junctional proteins

Tight junction

Tight junctions

Desmosomes

Plasma membranes

Plasma membranes

Intercellular space

Cytoplasmic plaque

Connecting fibers

Keratin fibers associated with cytoplasmic plaque

Desmosome

Gap junction

One epithelial cell

Intercellular space

Hydrophilic channels

2.7-nm space

Plasma membranes

Connexons

Gap junction

5.8 Junctions Hold Epithelial Cells Together

Tight junctions bar the movement of dissolved materials through the space between epithelial cells. There is no intercellular space where there is a tight junction. Long rows of the tight-junction proteins form a complex meshwork, seen at the bottom of the freeze-etch micrograph. *Desmosomes* link adjacent cells but permit materials to move between them. Connecting fibers bind the neighboring cells to each other; cytoplasmic plaques anchor these fibers in both cells. *Gap junctions* let adjacent cells communicate. Dissolved molecules and electrical signals may pass from one cell to the other through the channels. Each channel is made of two connexons. A connexon reaches through the phospholipid bilayer of the membrane and extends into the cytoplasm on one side and into the external environment on the other side, where it abuts a connexon on an adjacent cell.

tight junctions, and gap junctions. As you read about each type of junction in the sections that follow, refer to the illustrations in Figure 5.8.

Tight Junctions

Tight junctions bind cells so closely that materials do not move between the joined cells; there is no space at all between cells at a tight junction. Tight junctions result from the mutual binding of strands of specific proteins in the plasma membranes of the two cells, forming belts that virtually fuse the two membranes. Epithelial cells are so extensively linked by tight junctions between adjacent cells that substances on one side of an epithelium (flat tissue composed of epithelial cells) cannot seep through to the other side. Thus, for example, the contents of the gut (digestive tract) cannot seep between the epithelial cells of the gut lining—the contents can pass through the epithelium only if they enter the epithelial cells from one side of the tight junction and are then released from the other ends of the cells. The selective permeability of the plasma membranes ensures that food substances from the digestive tract pass through the epithelial cells to the bloodstream and unwanted substances are unable to pass through the membrane. So tight are the tight junctions that membrane proteins and phospholipids on one side of a tight junction cannot flow through the junction to the other part of the plasma membrane. By forcing materials to enter some cells, and by allowing different ends of cells to have different membrane proteins, tight junctions help direct the transport of materials in the body.

Desmosomes

Unlike tight junctions, which cement cells so closely that the cells are sealed off from each other, **desmosomes** simply cause neighboring cells to adhere tightly. Although a desmosome holds adjacent epithelial cells firmly together, there is a 24-nm space between the two plasma membranes. Some desmosomes act like spot welds or rivets at individual points, while others form continuous belts around the outer ends of adjacent epithelial cells. At each weld, in each of the two cells, is a protein-containing mass called a cytoplasmic plaque. Desmosomes are easily recognized in electron micrographs by the plaques and their associated dense networks of keratin fibers, which extend into each of the cells. The fibers connect the cells through the space between the two plasma membranes. (Keratin is a fibrous protein that is classified as an intermediate filament; see Chapters 3 and 4. It makes up the bulk of our fingernails and hair.) Epithelial tissue is strengthened by the many desmosomes holding its cells together.

Gap Junctions

Adjacent cells in some animal tissues communicate through the third type of junction, called a **gap junction** because of the gap of 2.7 nm between the plasma membranes of two cells spanned by its many pipelike channels. The channels, called connexons, are made up of a specific protein. Connexons provide a cytoplasmic connection between cells, through which chemical substances or electric signals may pass.

In Chapters 38 and 40 we will see that the muscle cells of the vertebrate heart, many smooth muscles, and some nerve cells are connected by gap junctions, allowing the direct passage of an electric signal—the nerve impulse—from one cell to the next. Gap junctions are also important in embryonic development, for they appear at a specific developmental stage. If animal tissues are experimentally disrupted to dissociate the cells, the cells quickly form new gap junctions as they reassociate. In fact, cells isolated from one species of vertebrate readily form gap junctions with cells from other vertebrate species. Cancer cells, however, never develop gap junctions; presumably this means that they do not communicate with other cells as normal cells do.

Review the differences among the three types of junctions. Desmosomes allow cells to *adhere* strongly to one another. Tight junctions are *barriers* to the passage of molecules through the space between cells. Gap junctions provide channels for chemical and electric *communication* between the cells on the opposite sides of the junctions.

We have examined the structures of membranes and their junctions. Now it is time to consider their functions. Some of the most important functions of membranes depend on their selective permeability—their ability to allow some substances, but not others, to pass through.

DIFFUSION

How do molecules move in an aqueous environment? Before we discuss further the movements of molecules across membranes, it is important to consider this: Nothing in this world is ever absolutely at rest. Everything is in motion, though the motions may be very small. Molecules and ions in solution are constantly jiggling. An immediate consequence of this random jiggling is that all the components of a solution tend eventually to become evenly distributed throughout the system. If, for example, a drop of ink falls into a container of water, the pigment molecules of the ink will move about at random, spreading through the system until the concentration of pigment—and thus the intensity of color—is exactly the same in every drop of liquid in the container. A solution in which the particles are uniformly distributed is said to be at equilibrium, and the process of random movement toward a state of equilibrium is called **diffusion**.

In diffusion, the motion of each individual particle is absolutely random, even though the net movement of particles is directional until equilibrium is reached. Diffusion is this net movement—always in the direc-

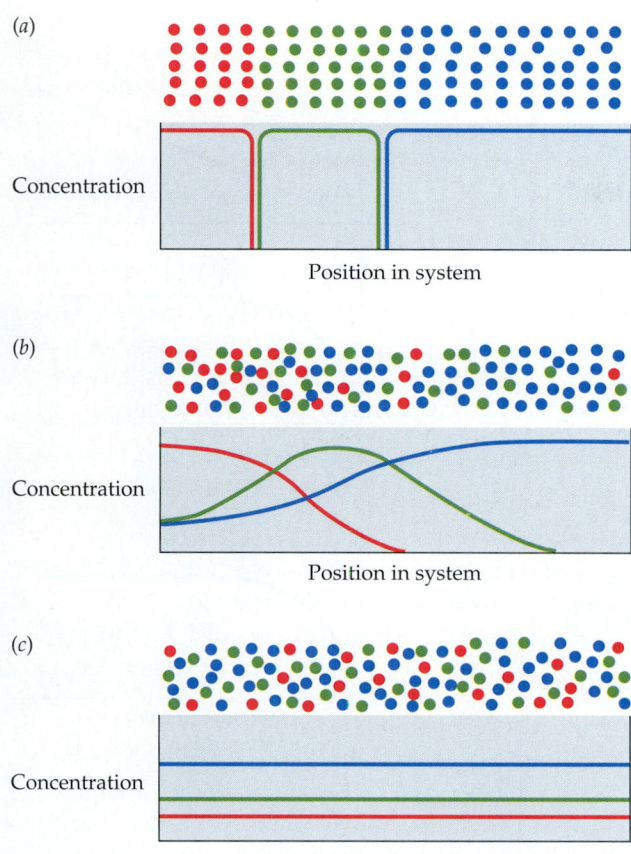

(a)

Concentration

Position in system

(b)

Concentration

Position in system

(c)

Concentration

Position in system

5.9 Diffusion Leads to Uniform Distribution
(a) Initially, each of three dissolved substances is highly concentrated in part of the solution in which they have been combined, and absent from the rest of it. (b) As the substances mix, each substance is at a higher concentration—its peak on the graph—in part of the solution but is also present elsewhere. (c) At equilibrium, all three substances are uniformly distributed throughout the solution, as shown by the lack of peaks on the graph.

tion *from greater* concentration *to lesser* concentration (Figure 5.9). In addition, in a complex solution the diffusion of each substance is independent of that of the other substances.

How fast substances diffuse depends on three physical properties—the diameter of the molecules or ions, the temperature, and the electric charge, if any, of the diffusing material—as well as on the **concentration gradient** in the system. The concentration gradient is the change in concentration with distance. The greater the concentration gradient, the more rapidly substances diffuse.

Within cells, where distances are very short, solutes distribute rapidly by diffusion. Small molecules and ions may move from one end of an organelle to another in a fraction of a millisecond, or from the center of a cell to its surface almost as fast. Diffusion of a chemical signal from one nerve cell to another takes less than a millionth of a second. On the other

hand, the usefulness of diffusion as a transport mechanism is diminished drastically as distances become greater. Diffusion over a centimeter may take an hour or more; over meters it takes years (it is assumed throughout this discussion that the fluid is not stirred or moved in any other way). Diffusion is not enough to distribute materials over the length of the human body, but within our cells or across layers of one or two cells it is rapid enough to distribute small molecules and ions almost instantaneously.

CROSSING THE MEMBRANE BARRIER

In principle, *all* substances diffuse, although the rates of diffusion vary. In a solution without barriers, all the solutes diffuse at rates determined by their physical properties and in directions determined by the concentration gradient of each solute. If a biological membrane (which is selectively permeable) is introduced as a barrier, the movement of the different solutes can be affected. Some solutes move fairly readily through the membrane, whereas others are prevented from crossing it. Molecules that can move through the phospholipid barrier diffuse from one compartment to the other until their concentrations are equal on both sides of the membrane. Molecules that cannot cross the membrane diffuse only within their own compartments, so their concentrations remain different on the two sides of the membrane.

Substances move through biological membranes in three ways: simple diffusion, facilitated diffusion, and active transport (Figure 5.10). In **simple diffusion,** small, nonpolar molecules pass through the lipid bilayer of the membrane. Equilibrium is reached when the concentrations of the diffusing substance are identical on both sides of the membrane. At equilibrium individual molecules are still passing through the membrane, but equal numbers of molecules are moving in each direction, so there is no change in concentration.

Facilitated diffusion (sometimes called carrier-mediated diffusion) also involves movement down a concentration gradient to produce equal concentrations of solute on the two sides of a membrane, but in contrast to simple diffusion, the solute molecules do not diffuse through the membrane on their own. Rather, they combine with **carrier molecules** in the membrane. By a mechanism that is still not understood, these carriers enable the solute molecules to pass through the membrane.

Both simple diffusion and facilitated diffusion permit the passage of a solute across a membrane down a concentration gradient—that is, from the side of higher concentration to the side of lower concentration. Neither of these mechanisms, however, allows for the transport of a solute *against* a concentration

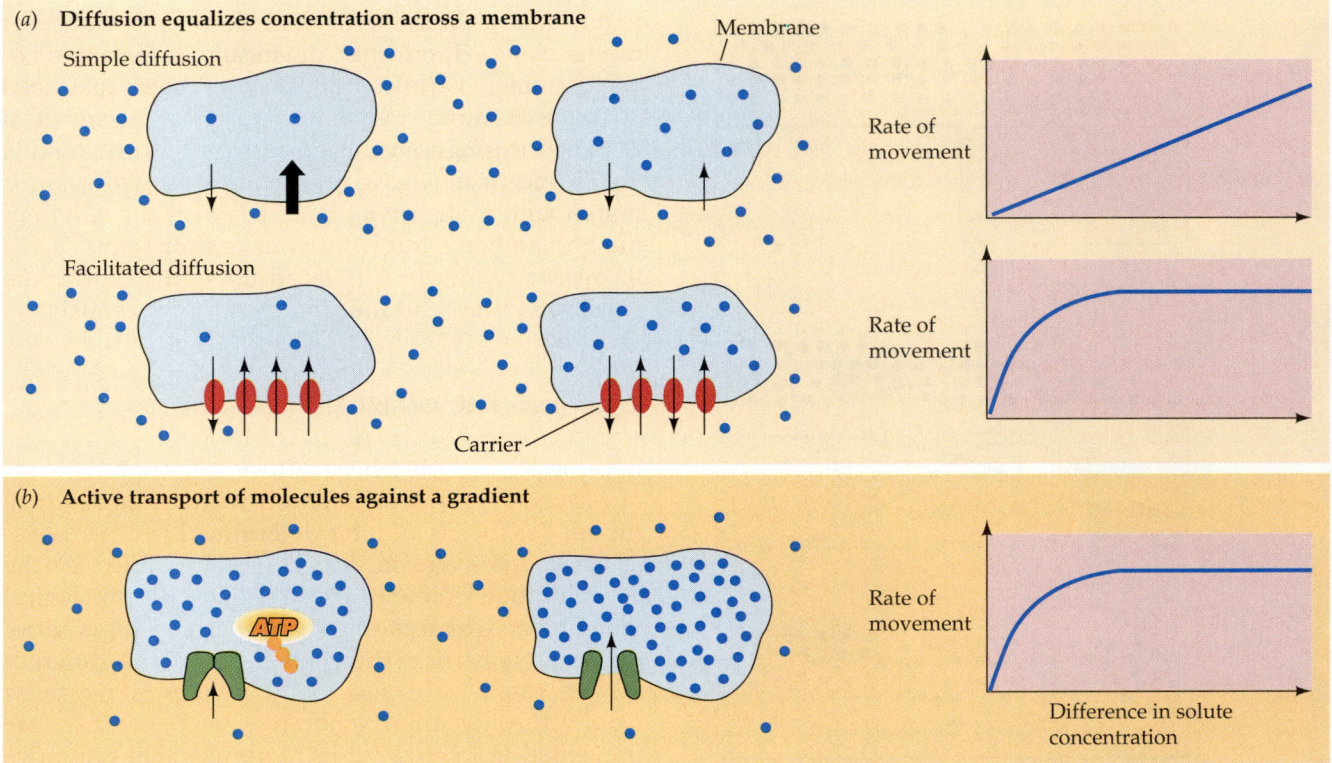

(a) **Diffusion equalizes concentration across a membrane**

Simple diffusion

Membrane

Rate of movement

Facilitated diffusion

Carrier

Rate of movement

(b) **Active transport of molecules against a gradient**

ATP

Rate of movement

Difference in solute concentration

5.10 Crossing Biological Membranes

(a) Both simple and facilitated diffusion equalize concentrations of a solute across a membrane. The rate of simple diffusion is directly proportional to the concentration difference, as the top graph shows. At equilibrium the concentration of solute inside the membrane equals that outside. With facilitated diffusion, equal concentrations are also reached, but a protein carrier in the membrane allows the rate of solute crossing to be greater. This rate reaches a maximum when all carriers are saturated with solute. (b) Active transport employs energy (ATP) and can move solutes against a concentration gradient. The rate of movement is similar to that of facilitated diffusion and reaches a maximum when membrane carriers are saturated, but the final concentrations of solute on either side of the membrane can be quite different due to the expenditure of energy.

movement also increases with the difference in solute concentration across the membrane, but a point is reached at which further increases in concentration difference are not accompanied by an increased rate (see Figure 5.10a). The facilitated diffusion system is said to be saturated at this high concentration. If there are only so many carrier molecules per unit area of membrane, then the rate of movement reaches a maximum when all the carrier molecules are fully engaged in moving solute molecules. In other words, at high solute concentration differences across the membrane, there are not enough carrier molecules free at a given moment to handle all the solute molecules. Like facilitated diffusion, the rate of active transport stops increasing at high solute concentrations (see Figure 5.10b).

Simple Diffusion

Whether a substance can pass through biological membranes depends on how soluble it is in lipids. The more lipid-soluble the compound, the more rapidly it diffuses (Figure 5.11). This statement holds true over a wide range of molecular weights. Only certain ions and the smallest molecules seem to deviate from this rule: Materials such as water, potassium ions (K^+), and chloride ions (Cl^-) pass through membranes much more rapidly than their solubilities in lipid would predict. Let's consider, then, how a membrane's chemical structure affects diffusion.

The key feature of membrane architecture, as we

gradient—that is, from the side of the membrane where the concentration is lower to the side where it is higher. Exactly this phenomenon, called **active transport,** is of extreme importance to living things. Like facilitated diffusion, active transport relies on carrier molecules. Unlike facilitated diffusion, this active process by which ions or molecules are moved against their own concentration difference requires a great deal of energy, which is provided by ATP obtained through cellular respiration (see Chapter 7).

In all three mechanisms of movement through membranes, the *rate* of movement depends on the concentration difference across the membrane. In simple diffusion, the net rate of movement is directly proportional to the concentration difference across the membrane. In facilitated diffusion, the rate of

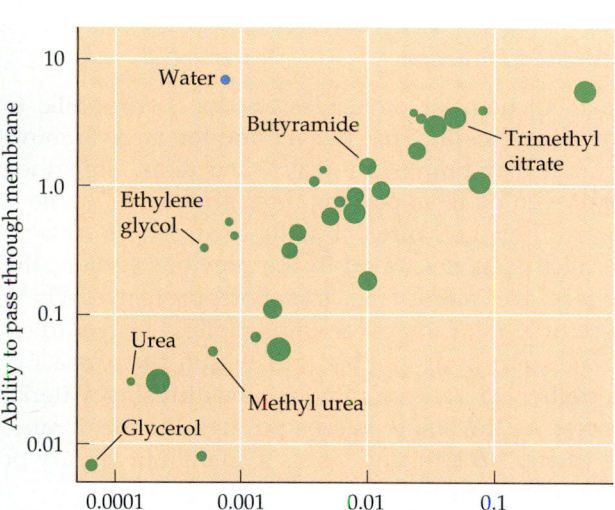

5.11 Membrane Permeability
Most substances cross membranes at rates relative to their solubility in lipids, as shown by the data points on this graph: The more lipid-soluble the molecule, the greater its ability to pass through the membrane. The sizes of the points correspond roughly to the sizes of the molecules studied; the assortment of molecules of all sizes along the curve indicates that size alone is not an important factor for permeability. The rate of diffusion of a few molecules, such as water, does not seem to be related to their solubility in lipids.

have seen, is the phospholipid bilayer that forms its framework (see Figure 5.1). The inner portion of the bilayer consists of the fatty acid chains of the phospholipids, along with cholesterol and other highly hydrophobic, nonpolar materials. When a hydrophilic molecule or ion moves into such a hydrophobic

region, it is rejected by the lipid layer and forced out. Such a molecule enters the hydrophobic region only when energy is available to push it in. On the other hand, a molecule that is itself hydrophobic, and hence soluble in lipids, enters the membrane readily and is thus able to pass through it.

This explanation accounts for most of the information in Figure 5.11, but it does not explain how water itself can move so rapidly through biological membranes—water is not hydrophobic enough to account for this flow. The diffusion of water into and out of cells is still under debate; many workers feel that the rapid movement of water through membranes can be explained in several ways.

We must also account for the rapid movement of certain ions through biological membranes. Ions pass through water-filled pores, or **channels,** in the membranes of all eukaryotic cells. There are specific channels for potassium, sodium, calcium, and chloride ions, allowing these ions to diffuse through membranes in spite of their hydrophilic character.

Membrane Transport Proteins

Both the channels through which certain ions diffuse across membranes and the carriers for facilitated diffusion and active transport are **membrane transport proteins:** intrinsic proteins that reach from one side of the membrane to the other. Different membrane transport proteins allow only specific substances to pass through, employing one of various methods of transport (Figure 5.12). The channels, for example, have an aqueous region through which ions can diffuse.

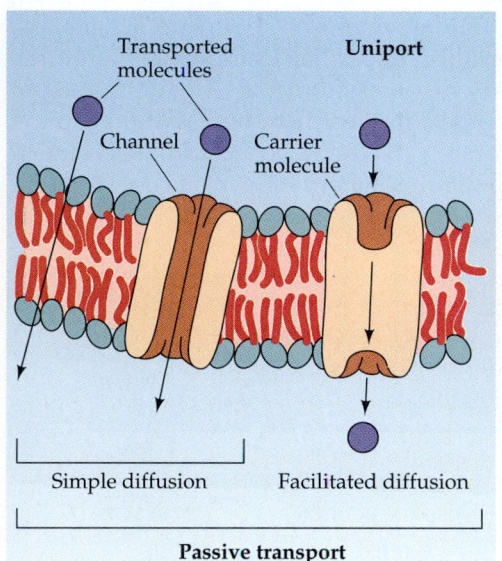

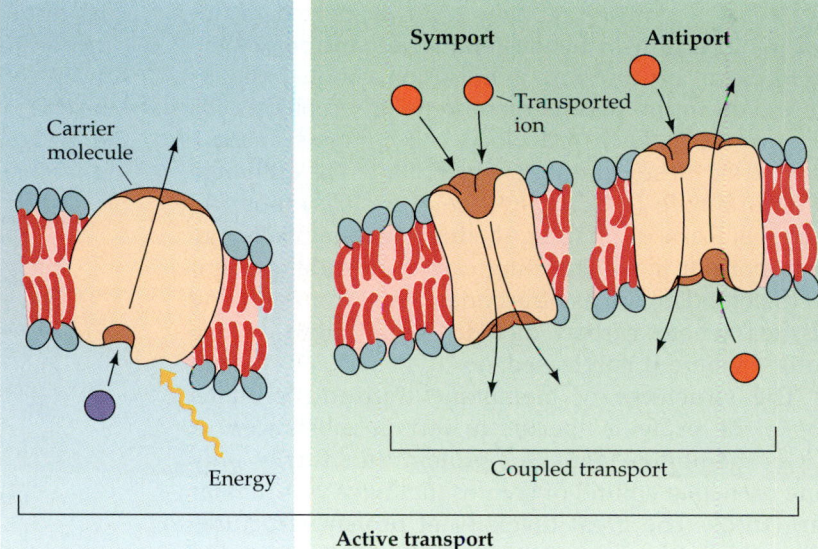

5.12 Getting Across the Membrane
Substances pass through biological membranes by many mechanisms, as illustrated here. Most of the mechanisms—all except for simple diffusion through the phospholipid bilayer, shown at the far left—make use of proteins that act as either channels or carriers.

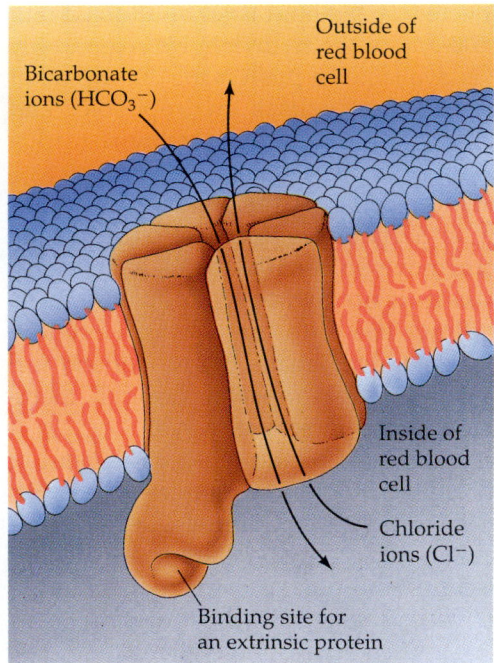

Outside of red blood cell

Bicarbonate ions (HCO₃⁻)

Inside of red blood cell

Chloride ions (Cl⁻)

Binding site for an extrinsic protein

5.13 An Anion Channel
Red blood cell membranes contain anion channels through which chloride and bicarbonate ions are exchanged. One end of the channel protein binds an extrinsic protein inside the red blood cell, anchoring the channel to a network of other proteins that stabilizes the positions of intrinsic proteins.

There are two classes of membrane transport proteins: uniports and coupled transport systems. **Uniports** transport a single type of solute. **Coupled transport systems** transport two or more different solutes, but neither solute can be moved by the system unless the other solute is also present. If the two coupled solutes are transported in the same direction, the system is a **symport.** If they are transported in opposite directions, the system is an **antiport.** One example of an antiport is an anion channel that is abundant in the plasma membrane of red blood cells (Figure 5.13). Each red blood cell contains about 1 million of these channels, which allow the exchange of bicarbonate (HCO_3^-) and chloride (Cl^-) ions. As you will learn in Chapter 41, this particular exchange is important in transporting carbon dioxide (present in the bloodstream as bicarbonate ions) from working tissues where carbon dioxide is produced to the lungs, where it is released.

The structures of membrane transport proteins make the proteins specific to certain substances. A given membrane transport protein thus carries only one particular solute, or solutes that have very similar structures. The great diversity of protein structures allows this high specificity for different transported solutes, just as it allows an enzyme to be highly specific for accelerating a particular chemical reaction.

Facilitated Diffusion

Most biochemical molecules are too hydrophilic to enter the phospholipid bilayer and too large to move through membranes the way water does. Thus they are prevented from passing through the membrane—unless they can interact with a carrier of suitable specificity. As discussed in the previous section, the carriers are membrane transport proteins. Where these proteins contact the phospholipid bilayer, their surfaces are hydrophobic, but within them is a hydrophilic region through which the diffusing material passes. As the solute passes through the membrane, the carrier protein undergoes a change in tertiary or quaternary structure.

In facilitated diffusion, the carrier proteins enable the solutes to pass in *both* directions. The net movement is toward the side where the solute concentration is lowest simply because on the side where the concentration is greater, the carriers encounter more solute molecules to transport.

Active Transport

Things are different in active transport. Active-transport carriers operate in one direction only. Typically, this direction is the one in which the transported substance is moved against a concentration difference—that is, from a region of low concentration to one of higher concentration. This "uphill" process requires an input of energy (see Chapter 6). There are two basic types of active transport: primary and secondary.

Primary active transport requires the direct participation of adenosine triphosphate (ATP), an energy-storing compound found in all cells (see Chapter 7). In primary active transport, energy released from ATP drives the movement of specific ions against a concentration difference. For example, compare the concentrations of potassium ions (K^+) and sodium ions (Na^+) inside a nerve cell and in the fluid bathing the nerve (Table 5.1). The K^+ concentration is much higher inside the cell, whereas the Na^+ concentration

TABLE 5.1
Concentration of Major Ions Inside and Outside the Nerve Cell of a Squid

ION	CONCENTRATION (MOLAR)	
	IN NEURON	IN BLOOD
K^+	0.400	0.020
Na^+	0.050	0.440
Cl^-	0.120	0.560

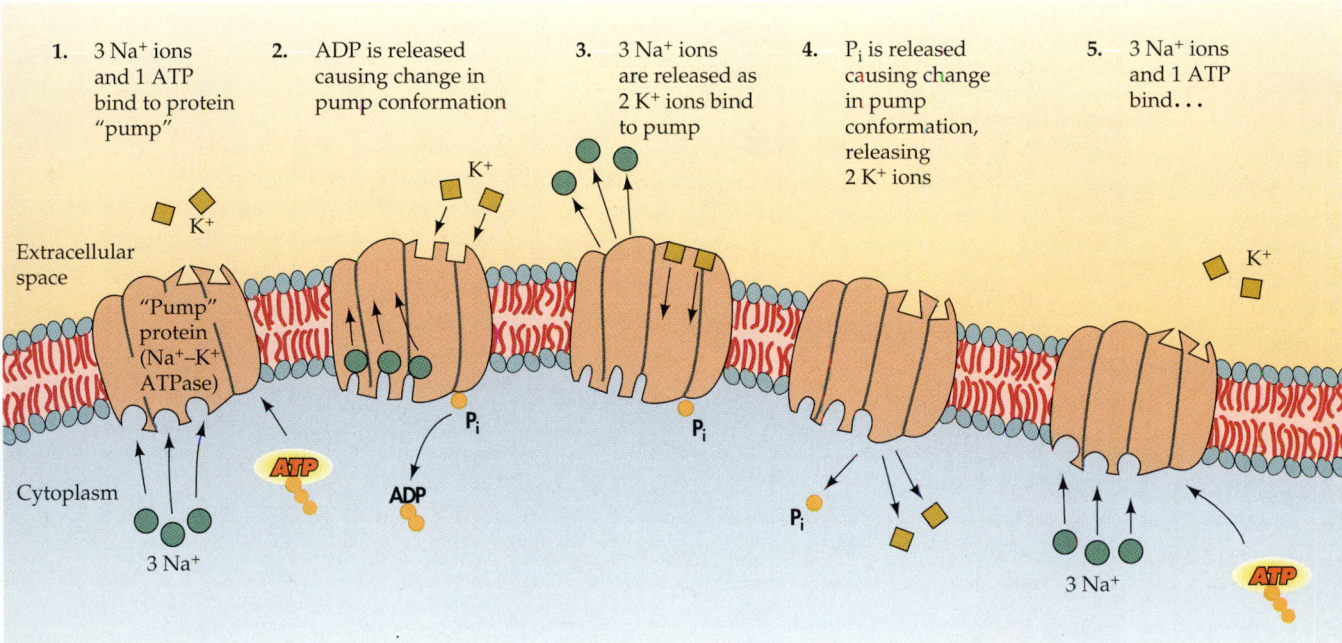

1. 3 Na⁺ ions and 1 ATP bind to protein "pump"

2. ADP is released causing change in pump conformation

3. 3 Na⁺ ions are released as 2 K⁺ ions bind to pump

4. P_i is released causing change in pump conformation, releasing 2 K⁺ ions

5. 3 Na⁺ ions and 1 ATP bind...

5.14 The Work of the Sodium–Potassium Pump Is Primary Active Transport
For each molecule of ATP used, 2 K⁺ are pumped into the cell and 3 Na⁺ are pumped out of the cell. The transport protein molecule—the pump—extends all the way through the phospholipid bilayer of the membrane; thus the highly hydrophilic Na⁺ and K⁺ ions need not interact with the hydrophobic center of the membrane.

is much higher outside. In spite of this, the nerve cells continue to pump Na⁺ out and K⁺ in, against these concentration differences, ensuring that the differences are maintained. This **sodium–potassium pump** is found in all animal cells and is an intrinsic membrane glycoprotein. It repeatedly breaks down one molecule of ATP to ADP (adenosine diphosphate), brings two K⁺ ions into the cell, and exports three Na⁺ ions (Figure 5.14). The sodium–potassium pump is thus an antiport. There are pumps for the transport of several other ions, but only cations are transported directly by pumps in primary active transport. The transport of other solutes is achieved by secondary active transport.

Unlike primary active transport, **secondary active transport** does not use ATP directly; rather, transport of the solute is tightly coupled to the difference in ion concentration established by primary active transport. The movement of particular solutes, such as sugars and amino acids, is regulated by coupled transport systems (some symports, some antiports) that move these specific solutes against their concentration difference, using energy "regained" by letting Na⁺ or other ions move *with* their concentration difference. Putting the two forms of active transport together, we see that energy from ATP is used in one example of primary active transport to establish concentration differences of potassium and sodium ions; the movement of some sodium ions in the opposite

direction provides energy for the secondary active transport of the sugar glucose (Figure 5.15). Other secondary active transporters are used for the uptake of amino acids and other solutes.

Osmosis

For years scientists disagreed about whether water only diffuses through biological membranes or is sometimes actively transported as well. It is now clear that water moves through membranes only by **osmosis,** the movement of a solvent through a membrane in accordance with the laws of diffusion. This process, in which no metabolic energy is expended, can be understood in terms of a very few principles, which we will develop here using two simple examples.

Red blood cells are suspended in a fluid called plasma that contains salts, proteins, and other solutes. If a drop of blood is examined under the light microscope, the characteristic shape of the red cells is evident. If pure water is added to the drop of blood, however, the cells quickly swell and burst (Figure 5.16). Similarly, if slightly wilted lettuce is put in pure water, it soon becomes crisp; by weighing it before and after, we can show that it has taken up water (Figure 5.17). If, on the other hand, the red blood cells and crisp lettuce leaves are placed in a relatively concentrated solution of salt or sugar, the

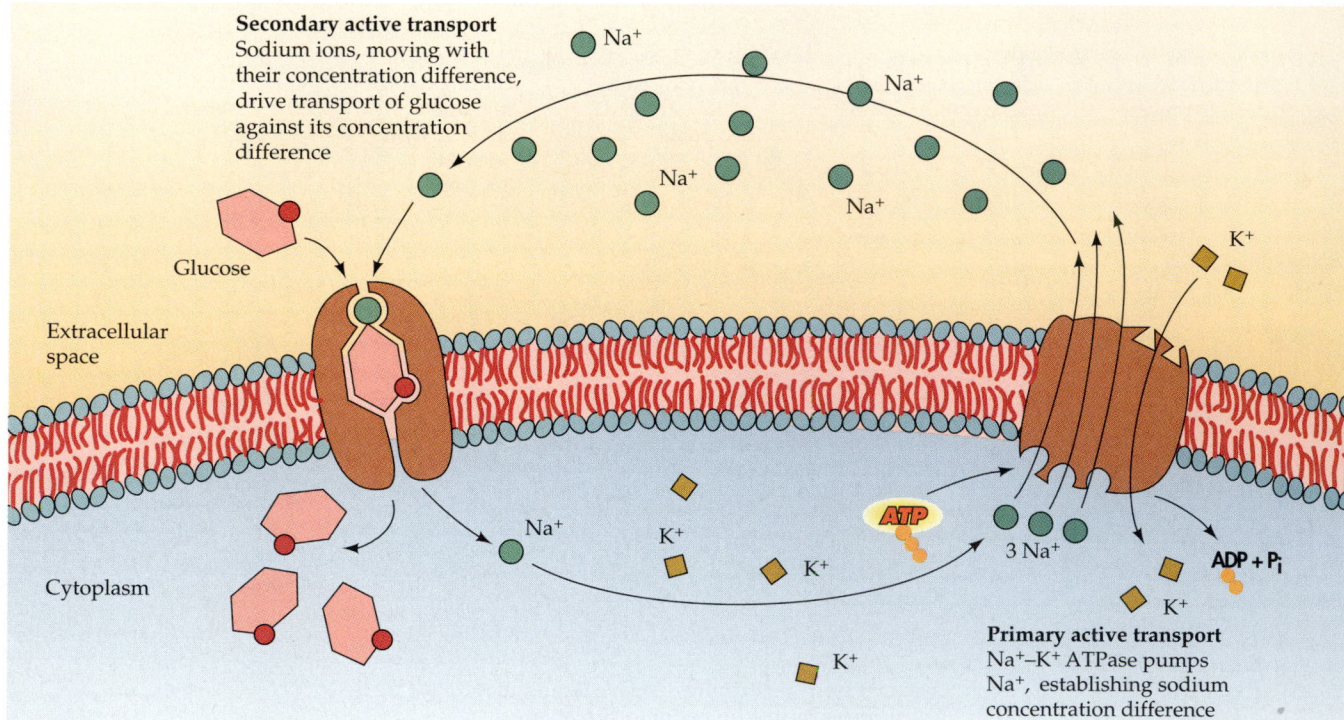

5.15 Secondary Active Transport
The Na^+ concentration difference established by primary active transport (right) powers the secondary active transport of glucose and some other substances. Glucose moves through the membrane against its concentration difference, accompanied by Na^+ ions that are moving with their concentration difference (left).

leaves become limp and the red blood cells pucker and shrink.

From these and other observations, we know that solute concentration is the principal factor in what is called the **osmotic potential** of a solution. The greater the solute concentration, the more negative the osmotic potential of the solution. Since pure water has nothing dissolved in it, its osmotic potential equals zero. Other things being equal, if two unlike solutions are separated by a differentially permeable membrane (one that allows water to pass through but not solutes), osmosis—movement of the water—proceeds toward the solution with the more negative osmotic potential.

If two solutions have identical osmotic potentials, they are **isotonic** to one another. (This is true even if their chemical compositions are very different). If they are not isotonic, then solution A with a more negative osmotic potential (that is, with a higher concentration of solutes) is said to be **hypertonic** to solution B; solution B is **hypotonic** to solution A. All three of these terms are strictly relative. They can be used only in comparing the osmotic potentials of two solutions; no solution can be called hypertonic, for example, except in comparison with another solution that has a less negative osmotic potential.

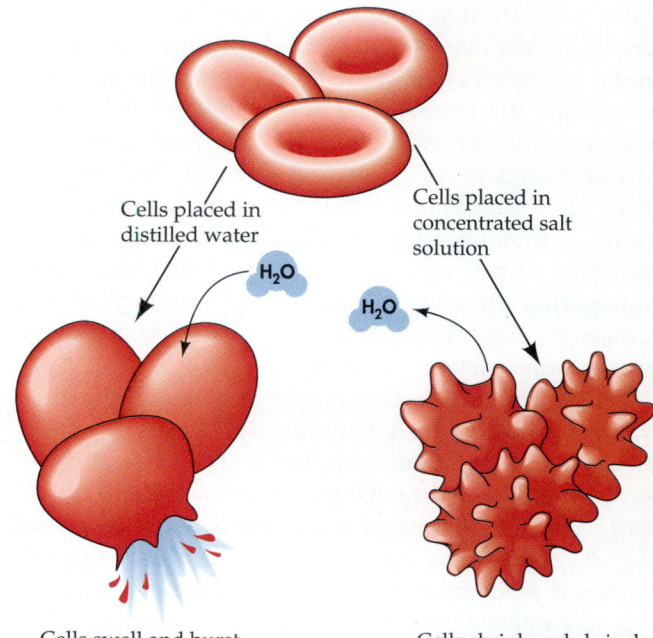

5.16 Osmosis Modifies Cell Shape
A mammalian red blood cell suspended in plasma has a biconcave shape (indented on both sides). If the cell is placed in distilled water, water enters by osmosis and the cell swells and bursts. If the cell is placed in a solution in which salts are more concentrated than they are in plasma, water leaves the cell by osmosis, causing it to shrivel.

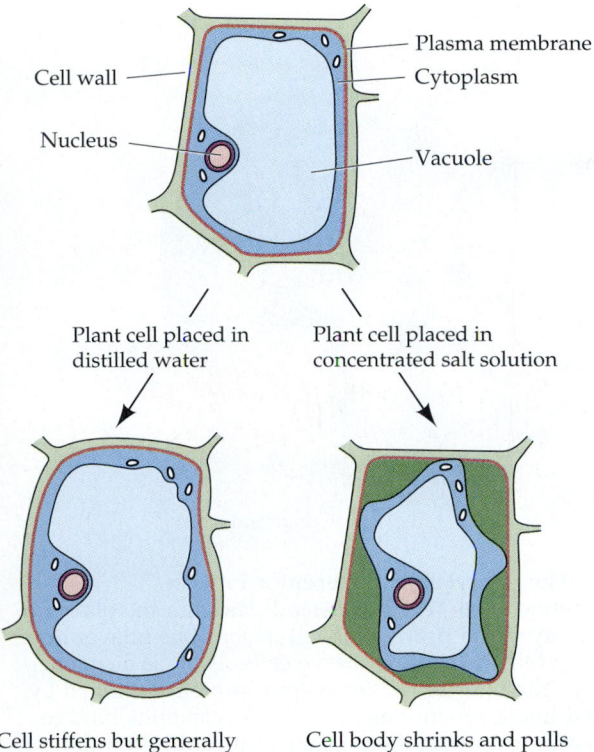

5.17 Osmosis and Cells with Walls
Water enters the vacuole of a plant cell placed in distilled water, but the cell retains its shape due to the rigid cell wall. When a plant cell is placed in a concentrated salt solution, the cell loses water to its surroundings; the cell wall retains its shape, but the plasma membrane shrinks away from the inside of the wall as the cell's volume is reduced.

Osmotic potentials determine the direction of osmosis in animal cells. A red blood cell takes up water from a solution that is hypotonic to the cell's contents. The cell bursts because its delicate plasma membrane cannot resist the swelling of the cell. The integrity of red blood cells and other blood cells is absolutely dependent upon the maintenance of a constant osmotic potential in the plasma in which they are suspended; the plasma must be isotonic with the cells if the cells are not to burst or shrink.

By contrast, the cells of plants, monerans, fungi, and some protists have cell walls that limit the volume of the cells and keep them from bursting. Cells with sturdy cell walls take up a limited amount of water and, in so doing, build up a pressure against the cell wall that prevents further water from entering. This pressure is the driving force for growth in plant cells—it is a normal and essential component of plant development. In cells with walls, then, osmosis is regulated not only by osmotic potentials but also by this opposed **pressure potential.** Osmotic phenomena in plants are discussed in Chapter 30.

MORE ACTIVITIES OF MEMBRANES

We have discussed some functions of membranes: the compartmentalization of cells, the regulation of traffic between compartments, and the active pumping of solutes. Membranes have many more functions. As discussed in Chapter 4, the membrane of rough endoplasmic reticulum is a site for ribosome attachment. Newly formed proteins are passed from the ribosomes through the membrane and into the interior of the endoplasmic reticulum for delivery to other parts of the cell. In nerve cells, the plasma membrane conducts the nerve impulse from one end of the cell to the other. The membranes of muscle cells, some eggs, and other cells are also electrically excitable. Many other biological activities and properties discussed in the chapters to follow are integrally associated with membranes.

To show the versatility of biological membranes, we will discuss here three more of their functions: mediating energy-trapping and energy-releasing reactions, recognizing materials at the cell surface, and helping cells associate as tissues.

Energy Transformations

Certain biological membranes are specialized to process energy—to convert light energy to the energy of chemical bonds, to trap energy released in oxidation–reduction reactions, and other vital activities. Why should membranes be involved in these activities? There are two reasons: structural organization and the separation of electric charges.

STRUCTURAL ORGANIZATION. Many processes in cells require various substances and take place step-by-step, with the products of one step being the reactants for the next step. If the necessary chemical substances for these reactions are all moving about at random, only chance collisions will bring them together and the processes will go forward slowly, if at all. If, on the other hand, the different substances are bound to a membrane (and especially if they are arranged in an orderly fashion), the product of one reaction can be released in close proximity to where it is needed for the next step in the pathway, and so forth—a virtual assembly line is established. In this sense, the membrane is a pegboard for the orderly attachment of specific proteins and other molecules.

SEPARATION OF CHARGES. A biological membrane can act like an electric battery. Work can be obtained from a battery by letting electrons flow from one of its terminals to the other by way of some device, like a motor or a light bulb, that uses the electric current. A similar process takes place in both photosynthesis and cellular respiration (see Chapters 7 and 8). Briefly, because of the limited permeability of the membranes in mitochondria and chloroplasts and because of the activities of certain electron carriers in those membranes, a substantial gradient of both electric charge and pH can be established across them. When these gradients are discharged by letting electric charge flow back through the membrane, the flow of charge can be used to do work or to form the energy-rich bonds of ATP. Without a membrane to enable the separation of charge, these reactions could not proceed.

Both the structure of membranes and their ability to separate electric charges relate to the properties of the two bulk components of membranes: lipids and proteins. The pegboard effect of the membrane's structure comes from the ability of the phospholipid bilayer to hold certain proteins in a defined plane so that they do not diffuse freely throughout the cell. The separation of charges is accomplished by certain membrane proteins and maintained by the insulating effect of the phospholipid bilayer.

Recognition and Binding

Membrane proteins and carbohydrates recognize and bind a variety of things to the outer surface of the plasma membrane. Antibodies (see Chapter 16) recognize target cells by virtue of specific proteins or carbohydrates on the surfaces of those cells. Viruses may begin their attack on intended host cells by attaching to carbohydrates on the host surface. Many hormones, including insulin, are recognized by membrane proteins that serve as receptors (see Chapter 36). Many nerve cells pass information to other nerve cells or to muscle fibers by means of a substance called acetylcholine (see Chapter 38). Acetylcholine activates these cells by attaching to a membrane protein—the acetylcholine receptor (Figure 5.18)—and changing the permeability of that membrane to ions. One of the most important classes of receptors binds substances, called growth factors, that regulate cell reproduction and differentiation (see Chapter 17).

Most animal cells have a mechanism, known as **receptor-mediated endocytosis,** that captures specific macromolecules from the cell's environment. The uptake is similar to endocytosis as described in Chapter 4, except that in receptor-mediated endocytosis parts of the plasma membrane contain specific receptor protein molecules. The parts of the membrane that contain receptor molecules are called coated pits because the inner surfaces of the membrane at these points are coated with several other, fibrous proteins, the best known of which is clathrin. When the receptor proteins bind the appropriate macromolecules

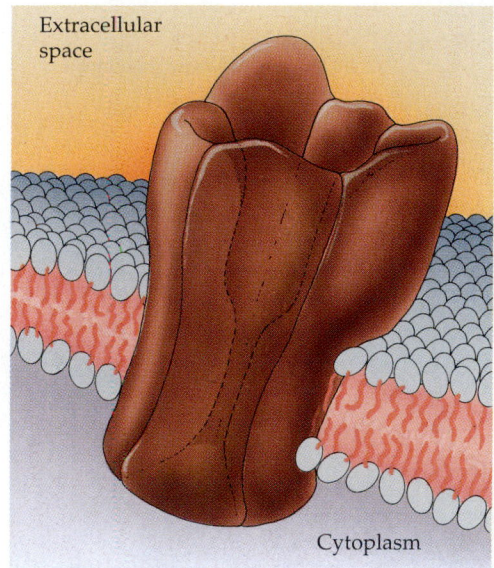

5.18   The Acetylcholine Receptor Protein
The acetylcholine receptor extends through the phospholipid bilayer and protrudes well beyond the bilayer into the extracellular space of nerve cells. There is a channel through the interior of the receptor protein (outlined by dashed lines); when molecules of acetylcholine bind to the receptor protein, the passage of ions through the channel is affected.

from the environment, the associated coated pit invaginates and forms a **coated vesicle** around the bound macromolecules. Strengthened and stabilized by clathrin molecules, this vesicle carries the macromolecules into the cell (Figure 5.19). Once inside, the vesicle loses its coat, making the macromolecules available to the cell. Because it is specific for particular macromolecules and can transport a group of macromolecules in one vesicle, receptor-mediated endocytosis is more rapid and efficient for taking up specific molecules than simple endocytosis is.

Receptor proteins must be intrinsic membrane proteins that span the entire thickness of the plasma membrane. Receptor carbohydrates are bound to such proteins or to phospholipid molecules. When a substance (hormone, virus, or other "visitor"), called a **ligand,** binds to its specific receptor, changes occur within the cell, on the cytoplasmic side of the membrane. Generally the visitor does not even cross the membrane to enter the cell. How can such a visitor produce an effect inside the cell if it does not enter it? Apparently the receptor protein undergoes structural changes. The portion of the protein inside the cell is altered, as is the way the protein functions in the cytoplasm.

The specificity of receptors (both proteins and carbohydrates) resides in their particular tertiary structures—that is, in their three-dimensional shapes. Some portion of the receptor protein or carbohydrate

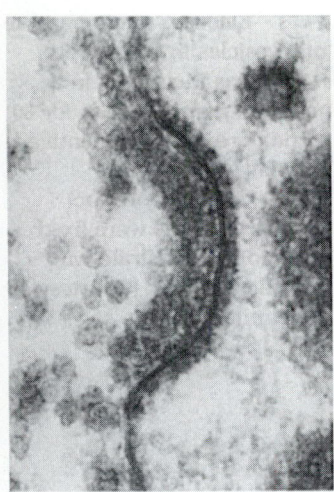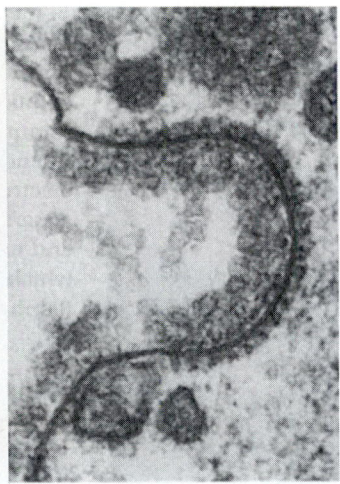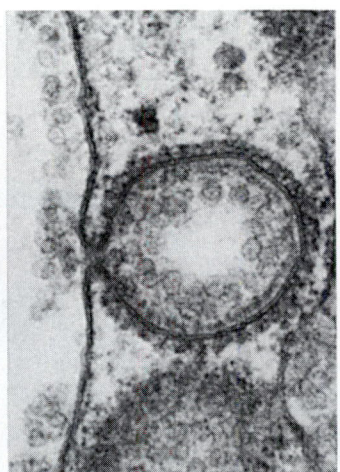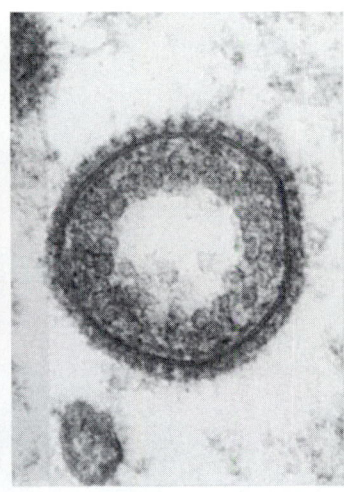

5.19 Coated Vesicle Formation
The micrograph at the far left shows a coated pit, with clathrin and other proteins concentrated against its cytoplasmic surface (arrow). The remainder of the sequence illustrates the development of a coated vesicle (far right) from the coated pit.

fits—hand in glove, as it were—the hormone, growth factor, part of a virus, or other ligand it is meant to recognize. We will discuss this sort of specificity in detail when we talk about enzymes in Chapter 6 and again in Chapter 16 when we talk about immune systems.

Cell Adhesion

During the growth of an animal embryo, specific membrane proteins known as **cell adhesion molecules** are crucial to the formation of the adult organism. As the embryo develops, its cells move about and associate with other specific cell types. This behavior is mediated by cell adhesion molecules in the membranes, which play roles in organizing the cells into tissues. One type of cell adhesion molecule, for example, organizes individual nerve cells into nerve cell bundles. Groups of cells that are about to migrate within the embryo lose their specific cell adhesion molecules; when they reach their new location, these cells regain their cell adhesion molecules and reorganize into tissue.

MEMBRANE INTEGRITY UNDER STRESS

Red blood cells appear fragile, yet they survive repeated compression and deformation as they squeeze through the finest of capillaries. This surprising resilience comes from certain intrinsic and extrinsic proteins in the plasma membrane of the cell (Figure 5.20). The extrinsic protein spectrin forms a meshwork of microfibrils on the cytoplasmic surface of the

plasma membrane. This meshwork provides structural support. Another extrinsic protein, ankyrin, anchors the spectrin to the membrane at many points by binding both the spectrin and an anion transporter that is an intrinsic membrane protein. (This anion transporter exchanges Cl^- for HCO_3^- as described earlier in this chapter.)

Genetic defects in spectrin and others of these proteins result in abnormal red blood cells and thus in various diseases. Mice with hemolytic anemia have spherical, fragile red blood cells. Their red blood cells have very little spectrin, but the cells take on a normal shape if spectrin is provided.

MEMBRANE FORMATION AND CONTINUITY

As we have seen, membranes are dynamic in that they participate in numerous physiological and biochemical processes. Membranes are dynamic in another sense as well: They are constantly being formed, transformed from one type to another, and broken down. In eukaryotes, phospholipids are synthesized within the sacs of the rough endoplasmic reticulum and are rapidly distributed to membranes throughout the cell. Membrane proteins are inserted into the sacs of the rough endoplasmic reticulum as they are formed on ribosomes. Sugars may be added to the proteins while they are in the endoplasmic reticulum. Next the proteins go to the Golgi apparatus, where some have carbohydrates added to them. The proteins then travel in Golgi-derived vesicles to the plasma membrane and are incorporated into it (see Chapter 4).

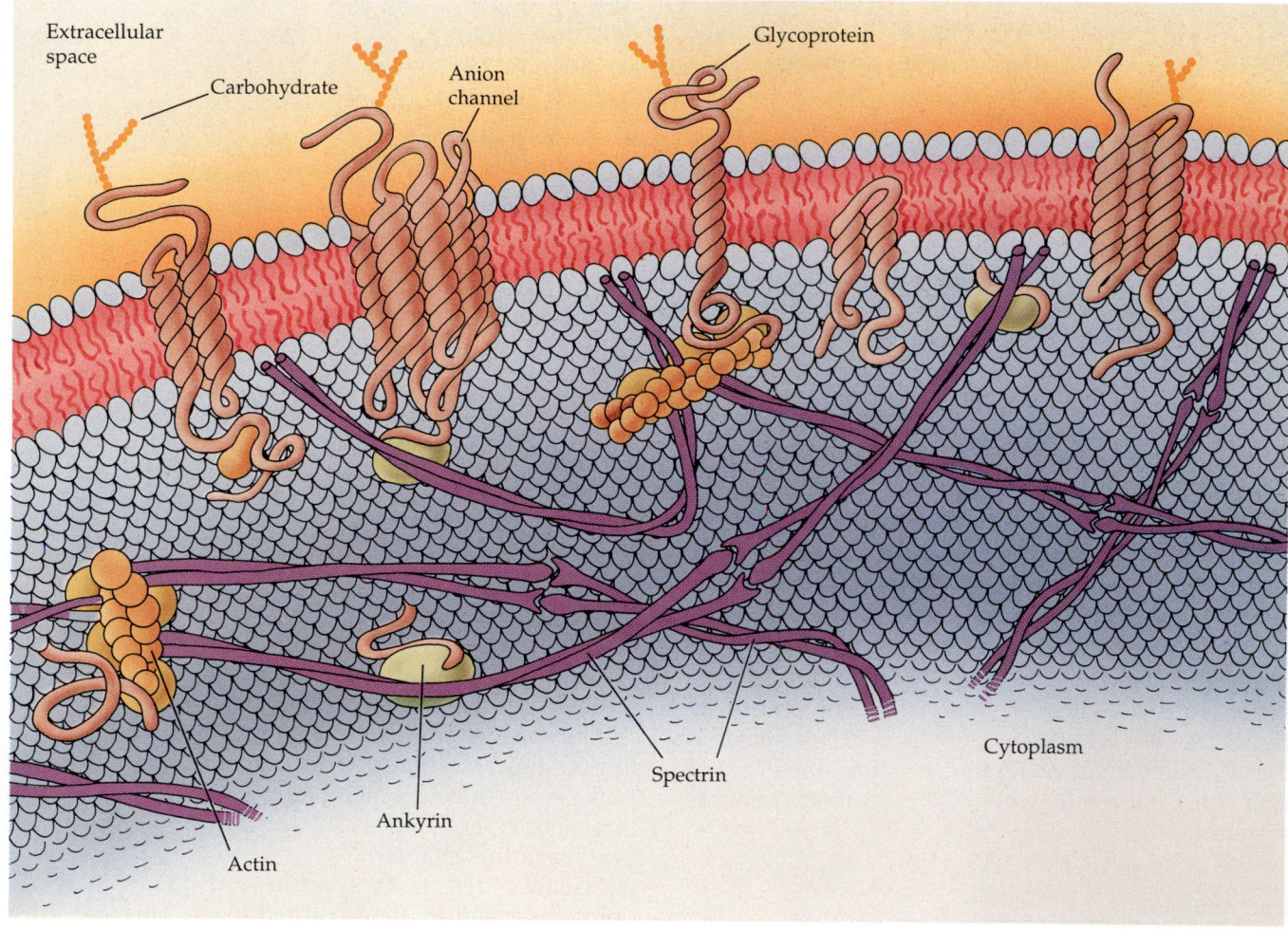

Extracellular space

Carbohydrate

Anion channel

Glycoprotein

Spectrin

Cytoplasm

Ankyrin

Actin

5.20 Some Proteins of the Red Blood Cell Membrane
Many intrinsic and extrinsic proteins contribute to this structure. Spectrin (purple) acts as a skeleton, binding the other proteins and thus strengthening the membrane that encloses the highly flexible red blood cell. Spectrin does not bind the membrane directly but connects by way of linker proteins such as ankyrin (green), which in turn binds the anion channel. Actin (yellow) also is a linker protein.

Functioning membranes move about within eukaryotic cells. For example, portions of the rough endoplasmic reticulum bud away from the endoplasmic reticulum and join the *cis* region of the Golgi apparatus (see Chapter 4). Rapidly—often in less than an hour—these segments of membrane find themselves in the *trans* regions of the apparatus, from which they bud away to join the plasma membrane (Figure 5.21). Bits of membrane are constantly merging with the plasma membrane in the process of exocytosis, but this is largely balanced by the removal of membrane in endocytosis. This removal by endocytosis affords a recovery path by which internal membranes are replenished. In sum, there is a steady flux of membranes as well as membrane components in cells.

Because we know about the constant interconversion of membranes, we might expect all subcellular membranes to be chemically identical. As you already know, this is not the case. There are major chemical differences among the membranes of even a single cell. Apparently membranes are changed chemically when they form parts of certain organelles. In the Golgi apparatus, for example, the membranes of the *cis* region are very similar to those of the endoplasmic reticulum, but the membranes of the *trans* region are more similar in composition to the plasma membrane. Ceaselessly moving, constantly carrying out functions vital to the life of the cell, biological membranes certainly are not the static, stodgy structures they once were thought to be.

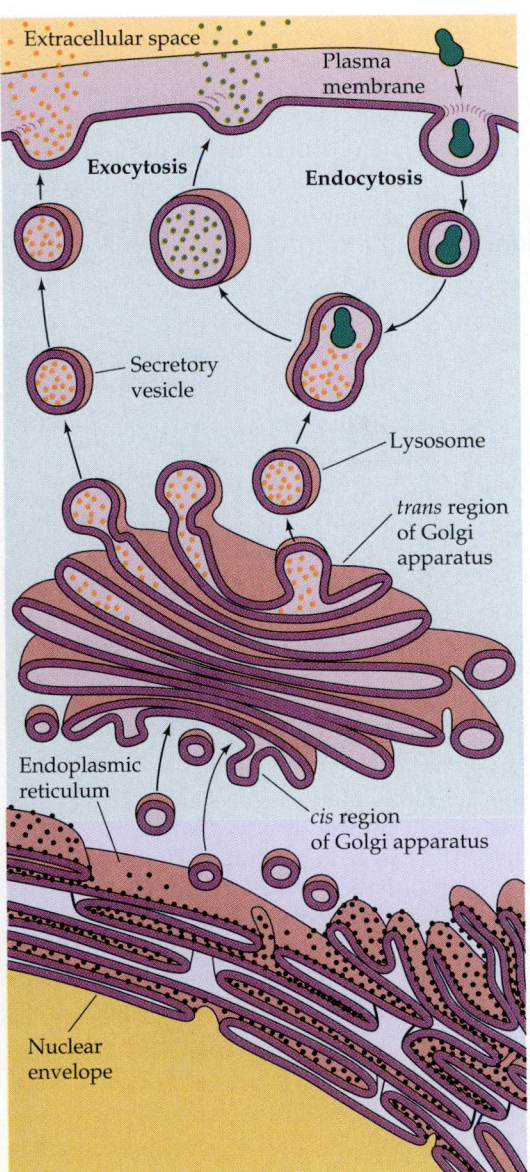

Extracellular space

Plasma membrane

Exocytosis

Endocytosis

Secretory vesicle

Lysosome

trans region of Golgi apparatus

Endoplasmic reticulum

cis region of Golgi apparatus

Nuclear envelope

5.21 Membrane Continuity in Cells

Arrows trace how membranes form, move, and fuse in cells. New stretches of membranes may be generated at certain locations, such as the outer membrane of the nuclear envelope. Vesicles budding from the *trans* region of the Golgi apparatus are also membrane-bounded. The vesicles may remain inside the cell as organelles, such as lysosomes, or they may fuse with the plasma membrane, delivering their contents to the exterior of the cell (exocytosis) and adding their membranes to the plasma membrane. Membrane is subtracted from the plasma membrane in the process of endocytosis.

SUMMARY of Main Ideas about Membranes

The membranes around and within cells are thin, pliable bilayers of phospholipids with embedded proteins and attached carbohydrates.
   Review Figures 5.1, 5.2, 5.3, and 5.6

Membrane lipids restrict the rates at which many solutes pass through membranes.

In a solution without barriers, solutes diffuse at rates determined by their physical properties and in directions determined by their concentration gradients.
   Review Figure 5.9

Solutes that can pass through a membrane do so by simple diffusion, facilitated diffusion, or active transport.
   Review Figure 5.10

Different membrane transport proteins allow specific substances to pass through in various ways.
   Review Figures 5.12 and 5.13

Only active transport can move solutes against a concentration difference.
   Review Figures 5.14 and 5.15

Water moves through membranes only by osmosis.

Water moves across the plasma membrane of an animal cell toward the side where the osmotic potential is more negative.
   Review Figure 5.16

Pressure potential as well as osmotic potential regulates osmosis in cells with walls.
   Review Figure 5.17

Carbohydrates attached to lipids or proteins on the outside surface of a plasma membrane are sites where the cell recognizes other cells or foreign substances.

Particular membrane proteins recognize other cells or molecules, regulate specific chemical reactions, or transmit information across the membrane.

Membranes within cells continually form, move, and fuse.
Review Figure 5.21

Where animal cells meet, junctions form, linking the plasma membranes: desmosomes for adherence, tight junctions that limit flow between cells, and gap junctions for communication.
Review Figure 5.8

SELF-QUIZ

1. Which statement is *not* true of membrane phospholipids?
   a. They associate to form bilayers.
   b. They have hydrophobic "tails."
   c. They have hydrophilic "heads."
   d. They give the membrane fluidity.
   e. They flop readily from one side of the membrane to the other.

2. The phospholipid bilayer
   a. is readily permeable to large, polar molecules.
   b. is entirely hydrophobic.
   c. is entirely hydrophilic.
   d. has different lipids in the two layers.
   e. is made up of polymerized amino acids.

3. Which statement is *not* true of membrane proteins?
   a. They all extend from one side of the membrane to the other.
   b. Some serve as channels for ions to cross the membrane.
   c. Many are free to migrate laterally within the membrane.
   d. Their position in the membrane is determined by their tertiary structure.
   e. Some play roles in photosynthesis.

4. Which statement is *not* true of membrane carbohydrates?
   a. Most are bound to proteins.
   b. Some are bound to lipids.
   c. Carbohydrates are added to proteins in the Golgi apparatus.
   d. They show little diversity.
   e. They are important in recognition reactions at the cell surface.

5. Which statement about animal membrane junctions is *not* true?
   a. Tight junctions are barriers to the passage of molecules between cells.
   b. Desmosomes allow cells to adhere strongly to one another.
   c. Gap junctions block communication between adjacent cells.
   d. Connexons are made of protein.
   e. The fibers associated with desmosomes are made of protein.

6. Which statement is *not* true of diffusion?
   a. It is the movement of molecules or ions to a state of even distribution.
   b. At the subcellular level it is a slow process.
   c. The motion of each molecule or ion is random.
   d. The diffusion of each substance is independent of that of other substances.
   e. Diffusion over meters takes years.

7. Which statement is *not* true of channels in membranes?
   a. They are pores in the membrane.
   b. They are proteins.
   c. All ions pass through the same type of channel.
   d. Some channels are gated.
   e. Movement through channels is by simple diffusion.

8. Facilitated diffusion and active transport
   a. both require ATP.
   b. both require the use of proteins as carriers.
   c. both carry solutes in only one direction.
   d. both increase without limit as the solute concentration increases.
   e. both depend on the solubility of the solute in lipid.

9. Primary and secondary active transport
   a. both generate ATP.
   b. both are based on passive movement of sodium ions.
   c. both include the passive movement of glucose molecules.
   d. both use ATP directly.
   e. both can move solutes against their concentration gradients.

10. Which statement is *not* true of osmosis?
    a. It obeys the laws of diffusion.
    b. In animal tissues, water moves to the cell with the most negative osmotic potential.
    c. Red blood cells must be kept in a plasma that is hypotonic to the cells.
    d. Two cells with identical osmotic potentials are isotonic to each other.
    e. Solute concentration is the principal factor in the osmotic potential.

FOR STUDY

1. How do freeze-fracture and freeze-etch techniques reveal aspects of membrane structure not revealed by transmission electron microscopy of thin sections of biological material?

2. In Chapter 40 we will see that the functioning of muscles requires calcium ions to be pumped into a subcellular compartment against a calcium concentration gradient. What types of chemical substances are required for this to happen?

3. Some algae have complex glassy structures in their cell walls. The structures form within the Golgi apparatus. How do these structures get to the cell wall without having to pass through a membrane?

4. Organisms that live in fresh water are almost always hypertonic to their environment. In what way is this a serious problem? How do some organisms cope with this problem?

5. Contrast simple endocytosis (see Chapter 4) and receptor-mediated endocytosis (this chapter) with respect to mechanism and to performance.

READINGS

Alberts, B., D. Bray, J. Lewis, M. Raff, K. Roberts and J. D. Watson. 1994. *Molecular Biology of the Cell*, 3rd Edition. Garland Publishing, New York. An outstanding general text in modern cell and molecular biology; Chapters 10 and 12 are particularly suitable for further study of biological membranes, and Chapters 17 and 19 are also useful.

Bretscher, M. S. 1985. "The Molecules of the Cell Membrane." *Scientific American*, October. A fine treatment of membrane chemistry, cell junctions, endocytosis, and other topics.

Lodish, H., D. Baltimore, A. Berk, L. Zipursky, P. Matsudaira and J. Darnell. 1995. *Molecular Cell Biology*, 3rd Edition. Scientific American Books, New York. Another fine general text; see Chapters 14, 15, and 16.

Lodish, H. F. and J. E. Rothman. 1979. "The Assembly of Cell Membranes." Scientific American, January. Good information on how membranes grow and why the two sides of a membrane differ.

Stryer, L. 1995. *Biochemistry*, 4th Edition. W. H. Freeman, New York. A reference for the major types of molecules found in membranes, with outstanding illustrations of phospholipids.

Unwin, N. and R. Henderson. 1984. "The Structure of Proteins in Biological Membranes." *Scientific American*, February. A clear presentation of how the proteins in membranes transport molecules.

In a single second, a typical cell carries out thousands of biochemical reactions. Why must it do this? Perhaps the main answer is that life depends on energy. Energy is available in the environment, and, as we will see in this chapter, energy can be released and converted among many forms by biochemical reactions. In certain reactions, cells break molecules apart, releasing energy that then fuels other reactions.

Via biochemical reactions, cells convert raw materials obtained from the environment into the building blocks of proteins and other compounds unique to living things. To maintain themselves, organisms must replace lost materials with new ones; they also grow and reproduce. All this is accomplished because cells carry out thousands of reactions per second.

Cells have a big advantage over nonliving matter when it comes to carrying out reactions. They have evolved the means to produce special substances, called enzymes, that enhance how reactions proceed. Enzymes and how they speed up energy conversions within cells are the main topic of this chapter.

ENERGY AND THE LAWS OF THERMODYNAMICS

The sum total of all uses of energy by a cell or by an organism is its **metabolism.** Metabolism consists of thousands of individual biochemical reactions that take place every second in most cells. Some of these reactions form more complex molecules from simpler ones (anabolism), and others break down complex molecules into simpler ones (catabolism). Some reactions release energy that is used to do physical work, such as the contraction of a muscle against a resisting load. Many other biochemical reactions, including those that synthesize macromolecules, proceed very slowly unless a source of energy is provided.

Energy has been defined both as the "capacity to do work" and as "heat or anything that can be transformed into heat." Energy comes in many forms: heat, light, electric, mechanical, chemical, nuclear, and others. Matter itself is a form of energy. An atomic explosion or a reaction in a nuclear power plant converts a small amount of matter into enormous amounts of energy. When all forms of energy are accounted for, the total amount of energy in the universe is unchanging: Energy can neither be created nor destroyed. This is the **first law of thermodynamics.** The first law holds for the universe as a whole or for any closed system within the universe. A **closed system** is one that is not exchanging energy with its surroundings.

Energy in one form can be converted into energy in another form (Figure 6.1). In solar batteries, light

Biochemical Self-Renewal
From biochemical reactions carried out in their cells, living things like this gecko replace worn out or lost materials with new ones.

6

Energy, Enzymes, and Catalysis

6.1 Biological Energy Transformations
The leaf traps light energy and produces food by photosynthesis. Sawfly larvae obtain energy by eating the leaf. As the larvae crawl and chew, they expend mechanical energy obtained from the chemical energy of food.

energy is converted into electric energy. Electric energy can be converted into light, heat, motion, and other forms of energy. Green plants convert light into chemical energy; in muscles, chemical energy is transformed into the energy of motion. None of these energy conversions is 100 percent efficient—some energy is always lost as heat. Heat, too, can be used to do work (think of a steam engine, for example), but here we run into limitations. The conversion of any other form of energy into heat is not fully reversible; that is, not all the heat can be converted back into the other forms of energy. Biological, chemical, and physical processes are often accompanied by the production of heat, not all of which can be made to do work.

Unusable heat is associated with an increase in disorder. Chemical changes, physical changes, biological processes, and anything else you can think of tend toward disorder, or randomness. A crystal of sodium chloride, which is highly ordered, will dissolve spontaneously in water to form a more random solution of sodium chloride. A sodium chloride solution, however, will not spontaneously reorder itself into a crystal of salt and pure water. Has your room become more or less orderly since the last time you expended energy to straighten it up?

Disorder can be discussed in quantitative terms; its measure is a quantity called **entropy.** Greater entropy implies greater disorder in any system. Not all energy conversion processes result in the same ability to do work. Various amounts of useful energy may be lost as the original forms are converted to unusable forms of energy—the heat associated with disorder. In the universe as a whole, or in any closed system, the amount of entropy increases; this is the principle of degradation of energy, also known as the **second law of thermodynamics.** Other ways of stating the second law will appear in this chapter.

CHEMICAL EQUILIBRIUM

In principle, all chemical reactions can run both forward and backward. For example, if compound A can be converted into compound B (A → B), then B can in principle be converted into A (B → A)—although at given concentrations of A and B only one of these directions will be favored. Think of the overall reaction as a result of competition between forward and reverse reactions. Increasing the concentration of the reactants (A) speeds up the forward reaction; increasing the concentration of the products (B) favors the reverse. At some point the forward and reverse reactions take place at the same rate. At this point no further change in the system is observable, although individual molecules are still forming and breaking apart. This balance between forward and reverse reactions is known as **chemical equilibrium.**

When a reaction goes more than halfway to completion, we say that the reaction A → B is a **spontaneous reaction;** in this case the reaction B → A is not spontaneous. A spontaneous reaction is one that, given enough time, goes largely to completion by itself, without the addition of energy. In fact, as spontaneous reactions proceed, they release energy (called free energy). If a reaction runs spontaneously in one direction (from reactant A to product B, for example), then the reverse reaction (from B to A) requires a steady supply of energy to drive it (Figure 6.2). For example, starch slowly but spontaneously breaks down in water, producing the disaccharide maltose; maltose, however, does not form starch spontaneously.

Entire chemical pathways also have a spontaneous direction. The complete oxidation of glucose in the processes that provide energy for cellular work is spontaneous. On the other hand, the synthesis of glucose from carbon dioxide and water by plants is

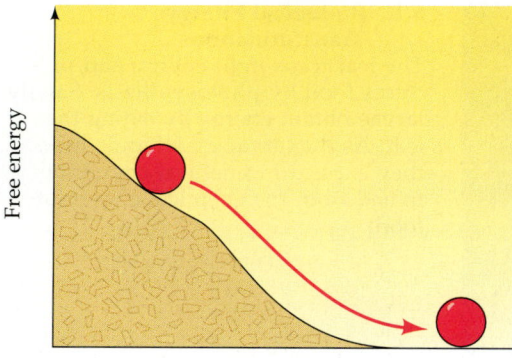

6.2 A Spontaneous Reaction
The reactants in a spontaneous reaction are like a ball on the side of a hill. The products form as the "reactant ball" rolls down the hill with a release of free energy. But the ball will not roll *up* the hill spontaneously; it would require work (the input of energy) to move the ball back up the hill.

a nonspontaneous reaction that must be driven by energy from the absorption of light in photosynthesis. (The important processes of fermentation, cellular respiration, and photosynthesis are discussed at length in Chapters 7 and 8).

Spontaneity has nothing to do with time. A reaction may be extremely slow, yet still be spontaneous. The burning of a newspaper and the slow browning of pages in old library files are both spontaneous processes, but they have different time scales. A spontaneous reaction moves toward equilibrium, using up reactants and making products. Given enough time, every spontaneous reaction eventually reaches equilibrium concentrations of reactants and products. By supplying energy to reactions, cells *prevent* the attainment of equilibrium as long as they live. In this way, cells keep their chemical composition different from that of the environment around them.

The Equilibrium Constant

Each reaction has a specific point of equilibrium—the point at which certain concentrations of reactants and products have been reached. Consider the following example. Every living cell contains glucose 1-phosphate, and its conversion to glucose 6-phosphate, accelerated by a particular enzyme, is a common event in cells. (Later in this chapter we explain how enzymes work.) In our example, the necessary enzyme is added to a solution of glucose 1-phosphate, at an initial concentration of 0.02 M (0.02 molar; see Chapter 2). As the reaction comes to equilibrium, the concentration of the product, glucose 6-phosphate, rises from 0 to 0.019 M, while the concentration of glucose 1-phosphate falls to 0.001 M (Figure 6.3). The reaction proceeds until equilibrium is reached at these concentrations. From then on, the reverse re-

action to glucose 1-phosphate is progressing at the same rate as the forward reaction to glucose 6-phosphate. At equilibrium, then, the forward reaction has gone 95 percent of the way to completion—the forward reaction is a spontaneous reaction. This result is obtained every time the experiment is run under the same conditions: at 25°C and pH 7.

The **equilibrium constant,** K_{eq}, is defined as the ratio of the concentrations of products and reactants *at equilibrium:*

$$K_{eq} = \frac{[\text{product}]}{[\text{reactant}]}$$

where the brackets indicate concentrations in, for example, moles per liter. By convention, reaction products are always shown in the numerator and reactants in the denominator. In our example,

$$K_{eq} = \frac{[\text{glucose 6-phosphate}]}{[\text{glucose 1-phosphate}]}$$

$$= 0.019 \ M/0.001 \ M = 19$$

Suppose that we run the experiment again, this time starting with only the product—with, say, 0.02 M glucose 6-phosphate and no glucose 1-phosphate. A reaction occurs, with equilibrium being reached at 0.001 M glucose 1-phosphate and 0.019 M glucose 6-phosphate—the same as before. As this result shows, the starting proportions of reactants and products do not affect equilibrium concentrations. Equilibrium concentrations are defined by K_{eq}—that is, the equilibrium constant *is constant* for a given chemical reaction as long as other conditions remain unchanged.

In general, for a reaction of the type A $\rightleftharpoons$ B, the equilibrium constant K_{eq} is defined by the equation

$$K_{eq} = \frac{[B]}{[A]}$$

where [B] and [A] are the concentrations of product B and reactant A, in moles per liter, *at equilibrium.* For the more complex reaction C + D $\rightleftharpoons$ E + F, the equilibrium constant is

$$K_{eq} = \frac{[E][F]}{[C][D]}$$

For the ionization of acetic acid in water at 25°C,

$$CH_3COOH \rightarrow CH_3COO^- + H^+$$

$$K_{eq} = \frac{[CH_3COO^-][H^+]}{[CH_3COOH]} = 2 \times 10^{-5} \ M$$

The smallness of this equilibrium constant tells us that the ionization of acetic acid is quite limited—less than one-half of one percent of the acetic acid molecules are ionized, and the ionization is not a spontaneous reaction. Acetic acid is a weak acid. When a strong acid such as hydrochloric acid dissolves in

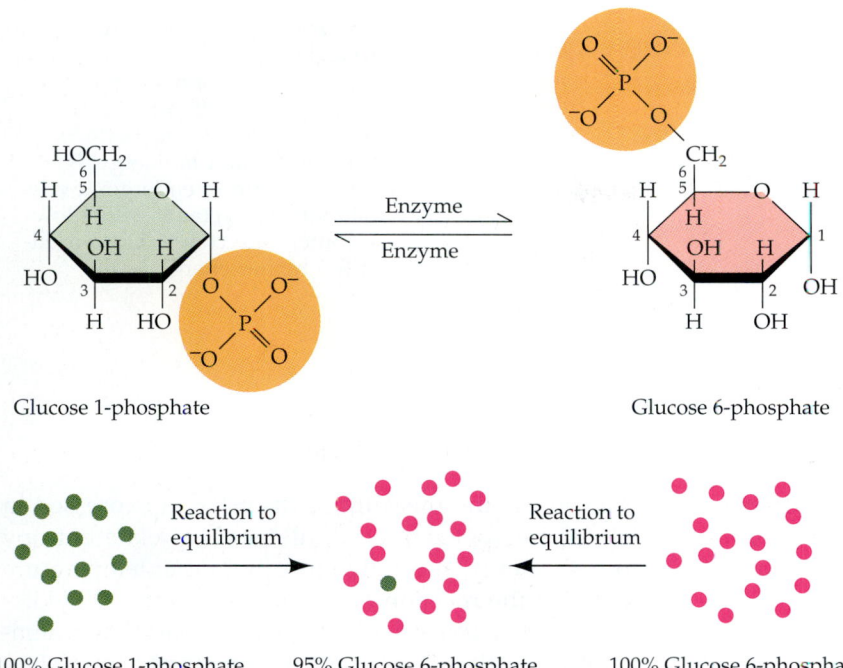

Glucose 1-phosphate   Glucose 6-phosphate

6.3 Concentration at Equilibrium
An enzyme speeds both the conversion of glucose 1-phosphate to glucose 6-phosphate and the reverse reaction. At 25°C and a pH of 7, there will always be 95 percent glucose 6-phosphate and 5 percent glucose 1-phosphate at equilibrium, no matter what the starting percentages of the two compounds are. Notice from the structural formulas that the compounds get their names from the position of attachment of the phosphate group (orange) to the carbon ring.

100% Glucose 1-phosphate   95% Glucose 6-phosphate / 5% Glucose 1-phosphate   100% Glucose 6-phosphate

water, almost all the molecules ionize; thus the equilibrium constant is large, showing this reaction is spontaneous.

Table 6.1 gives examples of equilibrium constants. Notice that the values vary widely. A high value of K_{eq} means that the reaction goes far toward completion; a very low value means that the reaction scarcely goes at all—in fact, the reverse reaction predominates. The table also gives a related measure of a reaction's spontaneity, its free energy change, which we discuss in the next section.

Free Energy and Equilibria

What determines the point of chemical equilibrium? What distinguishes a spontaneous reaction from one that can proceed only with a considerable input of

energy? The second question answers the first. A spontaneous reaction, one that goes far toward completion, is a reaction that gives off a great deal of **free energy,** energy that can be used to do work. Such a reaction is said to be **exergonic.** A reaction with a tiny equilibrium constant, on the other hand, is **endergonic** because it can be made to go only by the addition of free energy (Figure 6.4). Free energy is symbolized by G (for "Gibbs free energy," named after the nineteenth-century Yale thermodynamicist Josiah Willard Gibbs). It cannot be measured absolutely; however, the *change* in free energy, ΔG, of a reaction can be determined readily. It is related directly to the value of K_{eq} for the reaction: The greater the value of K_{eq}, the more free energy is given off. Values of ΔG are given in Table 6.1.

In the universe as a whole, or in any closed sys-

TABLE 6.1
Equilibrium Constants and Standard Free Energy Changes of Selected Reactions

REACTION[a]	EQUILIBRIUM CONSTANT	STANDARD FREE ENERGY CHANGE (KCAL/MOL)
Acetic acid + H_2O → acetate + H_3O^+	0.00002	+6.3
Malate → fumarate + H_2O	0.28	+0.75
Fructose 6-phosphate → glucose 6-phosphate	2.0	−0.4
Glucose 1-phosphate → glucose 6-phosphate	19	−1.7
Glucose 6-phosphate + H_2O → glucose + phosphate	260	−3.3
Sucrose + H_2O → glucose + fructose	140,000	−7.0

[a]The reactions are arranged from top to bottom in order of increasing tendency to go to completion as written ("go to the right").

Exergonic reaction
(spontaneous; energy-releasing)

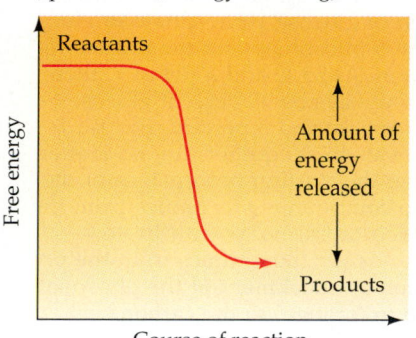

Endergonic reaction
(not spontaneous; energy-requiring)

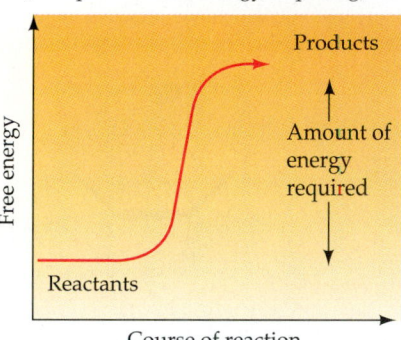

6.4 Exergonic and Endergonic Reactions

In an exergonic reaction, energy is *released* as reactants with a relatively high energy content form products with a lower amount of energy. Energy is *required* for an endergonic reaction, in which reactants with a low energy content are converted to products with a higher energy level.

tem, the quantity of free energy is always decreasing and entropy is always increasing. (This is another way of stating the second law of thermodynamics.) In a spontaneous reaction, the reactants possess more free energy than do the products. A reaction that goes nearly to completion ("to the right" as written) has a large, negative ΔG, indicating that it releases a large amount of free energy. In an exergonic reaction such as the conversion of glucose 1-phosphate to glucose 6-phosphate, ΔG is a negative number (in this example, $\Delta G = -1.7$ kcal/mol). Recall that equilibrium for this reaction has a product-to-reactant ratio of 19:1—that is, the reaction goes nearly to completion. A large, positive ΔG means that the reaction hardly proceeds at all as written; if the products are present, such a reaction runs backward ("to the left") to near completion. A ΔG value near zero is characteristic of a readily reversible reaction: Reactants and products have almost the same free energies.

Free Energy, Heat, and Entropy

We have seen how free energy is related to chemical equilibrium: Where equilibrium is attained, free energy is at a minimum. This relationship makes it clear that ΔG measures the useful chemical energy obtainable from a reaction. As an exergonic reaction proceeds, free energy is released and may be used to do chemical work. The cell harvests free energy from exergonic reactions such as the oxidation of foodstuffs, or from sunlight, and uses the free energy to drive vital endergonic reactions (such as those of photosynthesis).

Free energy is related to two other forms of energy: heat and a form associated with the entropy of the system. Both forms can be discussed in the context of spontaneous reactions: As a spontaneous reaction proceeds, entropy increases and heat is usually released. In any chemical reaction or physical process, each of these three forms of energy—free energy, heat, and the energy associated with entropy—may change.

Entropy, the measure of disorder, is expressed in kilocalories per degree (kcal/deg). To relate entropy to free energy, multiply entropy by the temperature at which the reaction occurs to obtain energy in kilocalories. For this calculation we use the Kelvin temperature scale, or absolute temperature. (In Kelvin units, 0 K is equal to about $-273°C$. Since one Kelvin unit is equivalent to one Celsius degree, to convert from Celsius to Kelvin, simply add 273 to the Celsius temperature.) As an example, the increase in entropy as 1 mole of ice melts to form water is 5.26 kcal/deg. When this reaction occurs at the freezing point (0°C, or 273 K), the energy lost to disorder is 5.26 kcal/deg $\times$ 273 K = 1436 kcal.

The change in free energy (ΔG) of any reaction is defined in terms of the change in heat (ΔH) and the change in entropy (ΔS):

$$\Delta G = \Delta H - T\Delta S$$

where T is the absolute temperature at which the reaction is occurring. The $T\Delta S$ term shows that the effect of a change in disorder is greater at high temperatures than at low. (Multiplying a given value of ΔS by the larger number that represents a higher temperature gives a larger energy change.)

Consider the relative importance of these factors in the following example: The combustion of 1 mole of glucose gives off 673 kcal of heat and the disorder increases by 0.0433 kcal/deg. At 25°C (298 K), $T\Delta S = 298$ K (0.0433 kcal/deg) = 12.9 kcal, which can be rounded off to 13 kcal. Both heat and disorder contribute to the spontaneity of glucose combustion. Thus for the complete oxidation of 1 mole of glucose, the change in free energy is calculated as follows:

$$\Delta G = -673 \text{ kcal} - 13 \text{ kcal} = -686 \text{ kcal}$$

Although the entropy factor (13 kcal/mol) is much smaller than the heat factor (673 kcal/mol), it does contribute to the total change in free energy. Some other spontaneous reactions have large negative ΔG values because their changes in entropy are large.

Recall that the change in free energy determines

the equilibrium constant for the reaction. Each chemical reaction is characterized by a particular equilibrium constant (K_{eq}) and by another constant, which we discuss in the next section.

REACTION RATES

When we know a reaction's change in free energy (ΔG), we know where the equilibrium of the reaction lies. The more negative ΔG is, the further the reaction proceeds toward completion. ΔG does not tell us anything, however, about the **rate of a reaction**—the speed at which it moves toward equilibrium.

Getting Over the Energy Barrier

A key to understanding reaction rates lies in recognizing that there is an energy barrier between reactants and products. Think about a butane lighter. The burning of the fuel (that is, the reaction of butane with oxygen to release carbon dioxide and water vapor) is obviously exergonic—once started, the reaction goes to completion, which occurs when all the butane has been burned. Since burning butane liberates free energy as light and heat, you might expect this reaction to proceed on its own. If you simply allow butane to flow and mix with air, however, nothing happens. To start the burning of butane, you have to provide a spark. Even though the burning of butane is spontaneous because it releases energy, the need for a spark to start the reaction shows that there is some sort of energy barrier between reactants and products.

In general, reactions proceed only after they are pushed over the energy barrier by bits of energy (such as the heat from a spark), which are called **activation energy** (E_a; Figure 6.5a). Recall the ball rolling down the hill in Figure 6.2. The ball has more free energy at the top of the hill because somebody expended energy carrying or throwing it up there. As the ball rolls down the hill, the reaction is exergonic: The ball is losing energy. Rolling the ball back up the hill requires energy and is thus an endergonic reaction, which cannot happen spontaneously.

Suppose now that the ball on the hillside is in a little depression (Figure 6.5b). Rolling down the hill is still an exergonic process, but to start the ball rolling, a small amount of activation energy must first be exerted. In a chemical reaction, the energy barrier—the "hump" over which reactants must "roll" before they can proceed spontaneously to form products—is energy needed to change reactants into intermediate molecular forms called transition-state species. Transition-state species have higher free energies than either the products or the reactants. A transition-state species corresponds to a ball that has

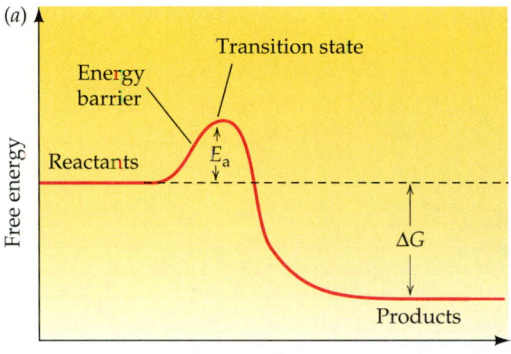

(a)

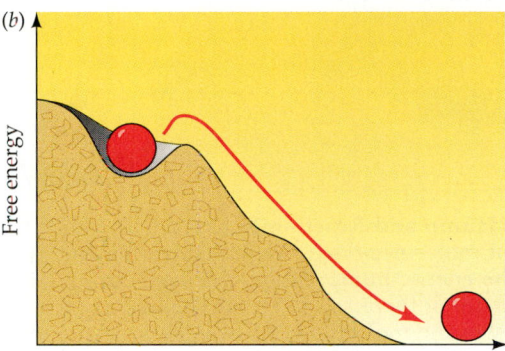

(b)

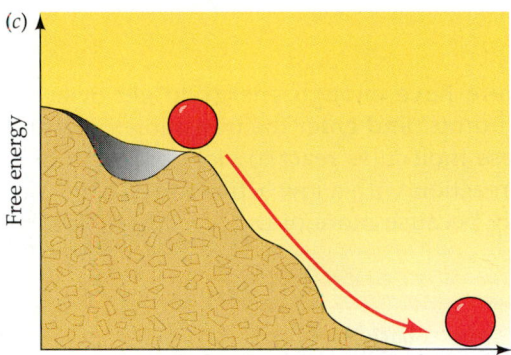

(c)

6.5 Reactions Need Activation Energy to Get Them Going
(a) If supplied with activation energy (E_a), the reactants surmount an energy barrier. Then an energy-releasing reaction proceeds spontaneously, releasing free energy (ΔG). (b) If a ball on the side of a hill is in a depression, it requires a push (activation energy) to get it out of the depression before it can roll to the bottom. (c) Here the reactant ball on the hill has received an input of activation energy and is poised (activated) for a spontaneous journey down to the product level, releasing free energy as it goes.

just been rolled up from the depression (Figure 6.5c). The activation energy that starts a reaction is recovered during the ensuing "downhill" phase of the reaction, so it is not a part of the drop in free energy, ΔG (see Figure 6.5a).

In any situation, some molecules have more energy than others. Picture a mixture of reactant molecules with various energies. Some of the molecules

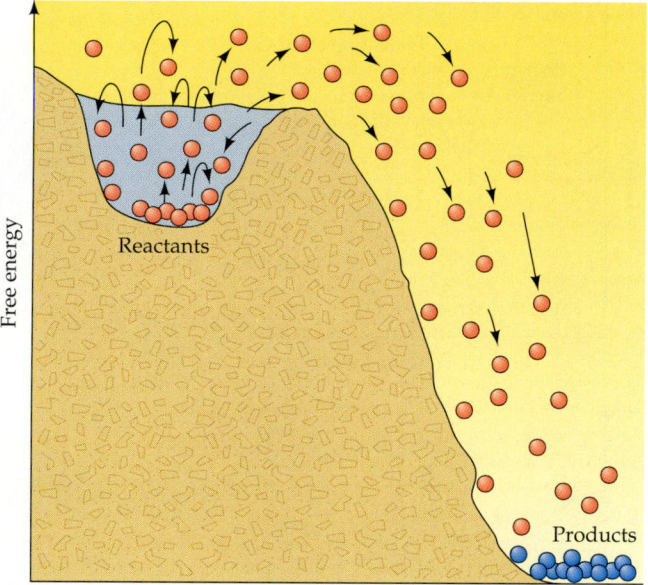

6.6   Energy Hump and Reaction Rate
Some reactant molecules have enough activation energy to get over the energy hump; others fall back into the depression, remaining there as reactants. The higher the hump, the fewer the molecules that can get over it, and the lower the reaction rate.

in the mixture have enough energy to get over the activation "hump" and enter the transition state, and some of these molecules react to yield products (Figure 6.6). A reaction with a low activation energy goes more rapidly because more of the reactant molecules

have enough energy to get over the hump. When activation energy is high, the reaction is slow unless more energy is provided, usually as heat. If the system is heated, all the reactant molecules become more energetic, and since more of them thus have energy in excess of the required activation energy, the reaction speeds up.

Biochemical reactions in our own mouths confront an energy barrier. Our saliva acts on the starch in food. As we saw in Chapter 3, starch is a polymer consisting of many glucose monomers. Starch reacts with water in the presence of saliva. The water cleaves the starch polymer into oligomers by breaking some of the bonds connecting the glucose units (Figure 6.7a). Starch in pure water, however, is quite stable because the activation energy of the reaction is high. (When we say "stable" we mean that a reaction does not proceed at all or proceeds so slowly that it would take a very long time to detect any change.) For the reaction to proceed, water and starch molecules must first collide; then the bonds connecting the glucose units must stretch and break. New bonds must then form between the broken ends and either H or OH derived from water. As shown in Figure 6.7b, there is an intermediate stage in which the bond between two glucose units is longer than normal and new bonds are in the process of forming. This is the transition-state species; it would be at the top of the energy barrier in a diagram like Figure 6.5. What is the difference between saliva and pure water that causes starch to break down in saliva but not in water? We'll explain in a moment.

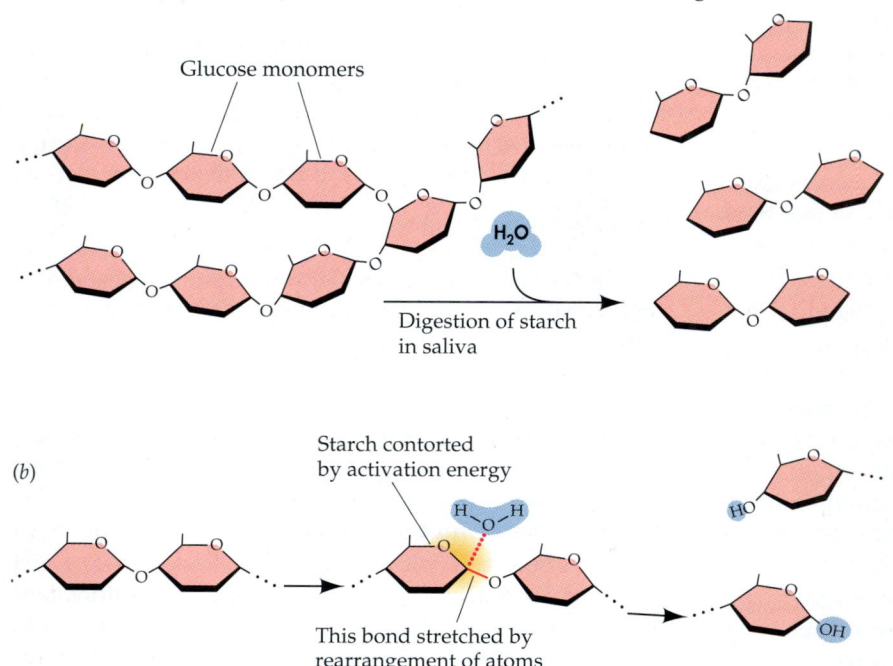

(a)   **Starch**   **Oligomers**

Glucose monomers

H_2O

Digestion of starch
in saliva

(b)

Starch contorted
by activation energy

This bond stretched by
rearrangement of atoms

6.7   The Breakdown of Starch
(a) In saliva, long starch molecules are digested into oligomers. (b) Before the bonds between glucose monomers in starch can break, an activated intermediate stage (center) must form, as. In the yellow area, activation energy contorts the starch molecule, stressing an otherwise stable bond and preparing the molecule for reaction with water. On the right, the reaction has been completed with the breaking of the stressed bond and the addition of atoms from a water molecule.

Rate Constants

Any chemical reaction proceeds at a rate directly proportional to the concentration of the reactants. The rate—in, for example, micromoles per minute—simply equals the reactant concentration times a **rate constant,** k, related to activation energy. At any given temperature, each specific reaction has its own characteristic rate constant. For the simple case,

$$A \xrightarrow{k} B$$

the rate of formation of product B is equal to k times the concentration of A, or $r_B = k[A]$, where r_B is the rate of formation of B. The rate constant is affected by temperature: As the temperature increases, so does the rate constant k. Recall from the previous section that increased temperature makes the reactant molecules more energetic, enabling more of them to get over the E_a hump and thus speeding up the reaction.

When starch dissolves in pure water, the rate of its conversion to smaller carbohydrate molecules is extremely slow, as we have already noted. Even when the starch solution is heated to the boiling point, essentially none of the water and starch molecules gain enough energy to exceed the activation energy for the reaction. Nonetheless, we know that starch does get digested at a significant rate in the presence of saliva (which is fortunate, for we would starve to death if all our digestive processes took place as slowly as the spontaneous breakdown of starch in pure water). How can a reaction such as that shown in Figure 6.7 be sped up by the presence of something such as saliva? The answer is that saliva is a catalyst.

THE HIGHLY SPECIFIC CATALYSIS OF ENZYMES

Although its function is to speed up a reaction, a **catalyst** does not cause anything to take place that would not take place eventually without it, and it does not become part of the reaction products. A catalyst merely lowers the activation energy of the reaction, allowing equilibrium to be approached at a faster rate. Most nonbiological catalysts are nonspecific. Platinum black (a soft powder), for example, catalyzes virtually any reaction in which molecular hydrogen is a reactant because it weakens the bond between the atoms in the H_2 molecule. By contrast, most biological catalysts, called **enzymes,** are highly specific. An enzyme usually catalyzes only a single chemical reaction or, at most, a very few closely related reactions. Saliva contains an enzyme that catalyzes the breakdown of starch, but that enzyme does

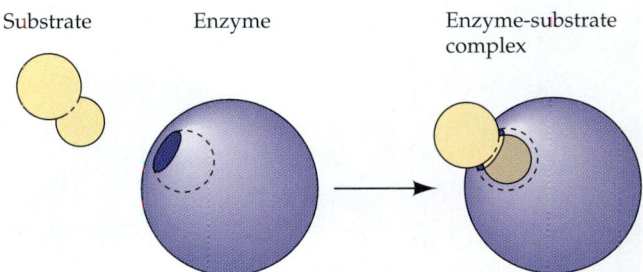

6.8 Enzyme and Substrate
An enzyme is a protein catalyst with an active site capable of binding a particular type of substrate molecule, forming an enzyme–substrate complex.

not catalyze the breakdown of fats, of proteins, or even of other polymers of glucose.

The molecules that are acted on catalytically are the enzyme's **substrates.** Substrate molecules bind themselves to the surface of the enzyme at the enzyme's **active site,** where catalysis takes place (Figure 6.8). The binding of a substrate to the active site forms an **enzyme–substrate complex** held together by one or more means, such as hydrogen bonding, ionic attraction, or covalent bonding. The enzyme–substrate complex may form product and free enzyme:

$$E + S \rightleftharpoons ES \rightleftharpoons E + P$$

where E is the enzyme, S is the substrate, P is the product, and ES the enzyme–substrate complex. An enzyme present in saliva acts on its substrate, starch, forming oligosaccharides, the product. Note that E, the free enzyme, is in the same chemical form at the end of the reaction as at the beginning. During the reaction it may change chemically, but by the end of the reaction it is restored to its initial form.

An enzyme gets its specificity from the exact structure of its active site. The tertiary structure of the enzyme lysozyme is shown in Figure 6.9. Lysozyme is an enzyme that protects the animals that produce it by destroying invading bacteria, which it accomplishes by cleaving certain polysaccharide chains in the cell walls of the bacteria. Lysozyme is found in tears and other bodily secretions and is particularly abundant in the whites of bird eggs. In Figure 6.9, the active site of lysozyme appears as a large indentation filled with the substrate (shown in green). The substrate fits precisely into the active site, whereas other molecules—with different shapes or different chemical groups on their surfaces—cannot form a complex with the enzyme.

Some enzymes change shape after binding with their substrate (Figure 6.10). These shape changes improve the alignment and result in an **induced fit** between the enzyme and the substrate. The enzyme α-amylase, which is present in saliva and specific for

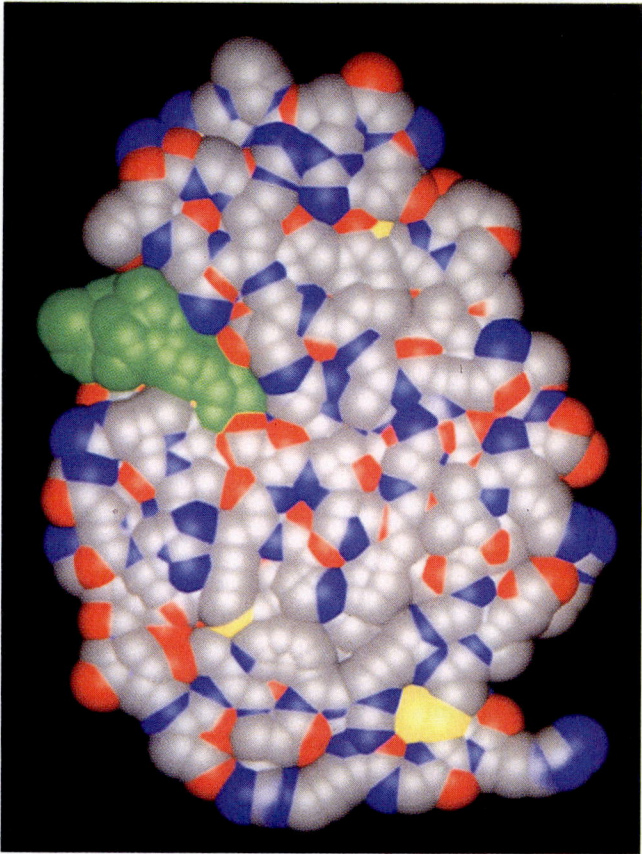

6.9 Tertiary Structure of Lysozyme
A substrate, shown in green, is bound to a lysozyme molecule. In the enzyme, the carbons are shown in gray, oxygens in red, nitrogens in blue, and sulfurs in yellow. Hydrogen atoms have been omitted. Lysozyme attaches precisely to the polysaccharide substrate, stressing particular bonds and allowing the usually stable polymer to be broken.

digesting starch to oligosaccharides, undergoes such a change as it catalyzes the reaction shown in Figure 6.7.

When an enzyme lowers the activation energy, both the forward and the reverse reactions speed up, so the enzyme-catalyzed reaction proceeds toward equilibrium more rapidly than the uncatalyzed one. Remember that the final equilibrium is the same with or without the enzyme. Adding an enzyme to a reaction does not change the difference in free energy (ΔG) between the reactants and the products; it changes only the activation energy (Figure 6.11), and consequently the rate constant.

Substrate Concentration and Reaction Rate

For a reaction of the type A → B, the rate of the uncatalyzed reaction is directly proportional to the concentration of A. Most enzyme-catalyzed reactions, on the other hand, generate plots like the one in Figure 6.12. At first the reaction rate increases as the substrate concentration increases, but then it levels off. Further increases in the substrate concentration do not increase the reaction rate. Since the concentration of the enzyme is usually much lower than that of the substrate, what we are seeing is a saturation phenomenon like the ones that occur in facilitated diffusion and active transport across membranes (see Figure 5.10). As more substrate is added,

6.10 Some Enzymes Change Shape When Substrate Binds
The deep cleft on the left side of the enzyme hexokinase divides the molecule into upper (darker shading) and lower lobes and contains the site where the substrate, glucose, binds.

(a) Hexokinase with empty active site

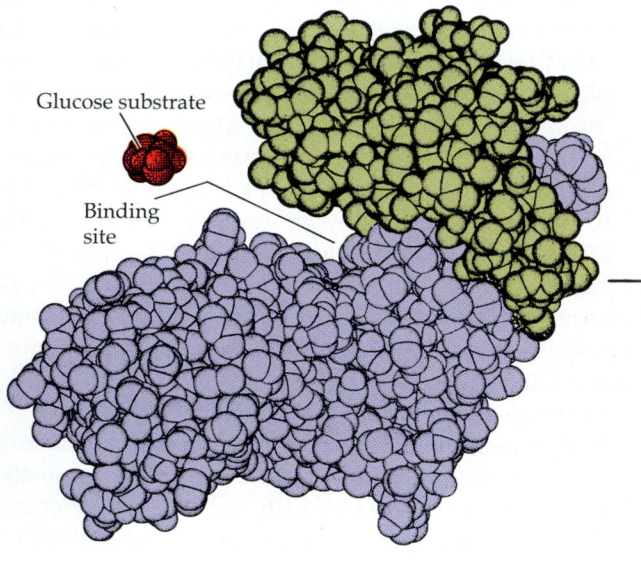

Glucose substrate

Binding site

(b) Binding of substrate in active site induces movement of lobes of hexokinase molecule, closing the binding site cleft

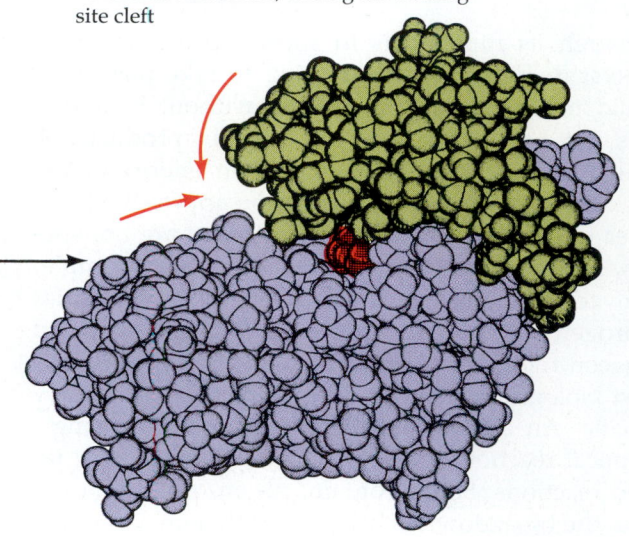

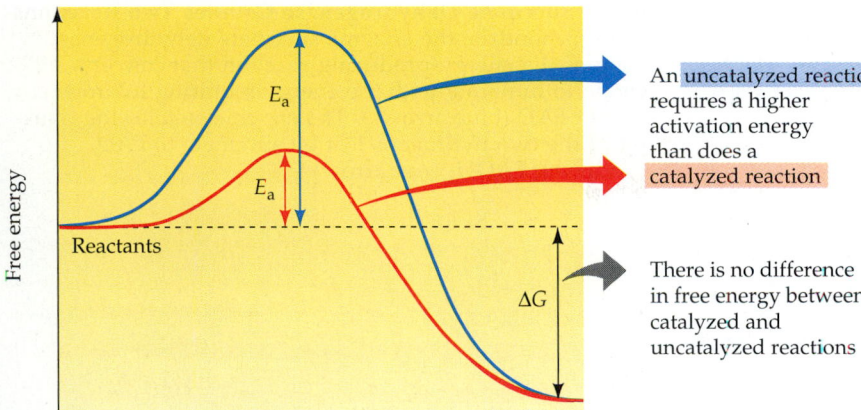

6.11 Enzymes Lower the Activation Energy
An enzyme-catalyzed reaction (red curve) has a lower activation energy (E_a) than does an uncatalyzed reaction (blue curve). There is no difference in the amount of free energy released (ΔG).

An uncatalyzed reaction requires a higher activation energy than does a catalyzed reaction

There is no difference in free energy between catalyzed and uncatalyzed reactions

more of the enzyme molecules are tied up as enzyme–substrate complexes. When all the enzyme molecules are bound to substrate molecules, nothing is gained by adding more substrate because no enzyme molecules are available to act as catalysts.

The study of the rates of enzyme-catalyzed reactions is called *enzyme kinetics.* As we will see later in this chapter, some graphs of rate as a function of substrate concentration are quite different from that in Figure 6.12. Such graphs tell us a great deal about the nature of the enzyme-catalyzed reaction.

Coupling of Reactions

Some of the most important reactions in living organisms are not spontaneous but proceed because specific enzymes **couple** them with other reactions that *are* spontaneous. Consider, for example, a pair of coupled reactions that occur in mitochondria. The first reaction, which converts succinate to fumarate, is highly spontaneous, with a large drop in free energy. The second reaction, the hydrogenation of FAD

to FADH$_2$, is nonspontaneous and requires a large input of free energy. The catalyst that couples these two reactions is the enzyme succinate dehydrogenase (Figure 6.13).

When the first reaction takes place in a mitochondrion, the two hydrogen atoms that are removed from succinate are transferred to a molecule of a carrier substance, FAD (flavin adenine dinucleotide). Succinate dehydrogenase couples the exergonic reaction to the endergonic one by ensuring that hydrogen atoms liberated by succinate are used to make FADH$_2$. One site on the enzyme surface binds succinate; another site nearby binds FAD. Every time a succinate ion reacts with succinate dehydrogenase to form fumarate, much of the free energy that is released by this highly exergonic process is immediately trapped and used to synthesize FADH$_2$. FADH$_2$ acts as a carrier of the hydrogen and the chemical free energy until another enzyme couples the exergonic dehydrogenation of FADH$_2$ with the endergonic hydrogenation of some other compound (see Chapter 7).

We have already seen, in Chapter 5, other examples of coupled reactions (see Figures 5.14 and 5.15). In animals the sodium–potassium pump (for primary active transport) is an enzyme that couples the exergonic breakdown of ATP with the endergonic pumping of Na$^+$ and K$^+$ against their concentration differences. In secondary active transport, another protein couples the exergonic influx of Na$^+$ with the endergonic influx of glucose. We will see in Chapter 40 how the contractile proteins of muscle couple the exergonic breakdown of ATP with the performance of mechanical work against a load. In metabolic pathways, there is another type of coupling, in which successive enzyme-catalyzed steps share compounds. A reaction A + B ⇌ C may be endergonic but still proceed rapidly if the next step, C + D ⇌ E, is so exergonic that the overall reaction (A + B + D ⇌ E) is exergonic.

These examples illustrate an important generalization: Coupled reactions are the major means of carrying out energy-requiring reactions in cells.

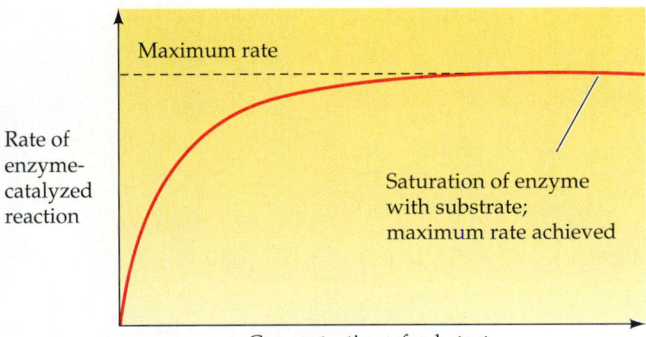

Maximum rate

Rate of enzyme-catalyzed reaction

Saturation of enzyme with substrate; maximum rate achieved

Concentration of substrate

6.12 Enzymes and Reaction Rate
At the maximum reaction rate, all enzyme molecules are tied up with substrate molecules; at this point, adding more substrate would not make the reaction go any faster.

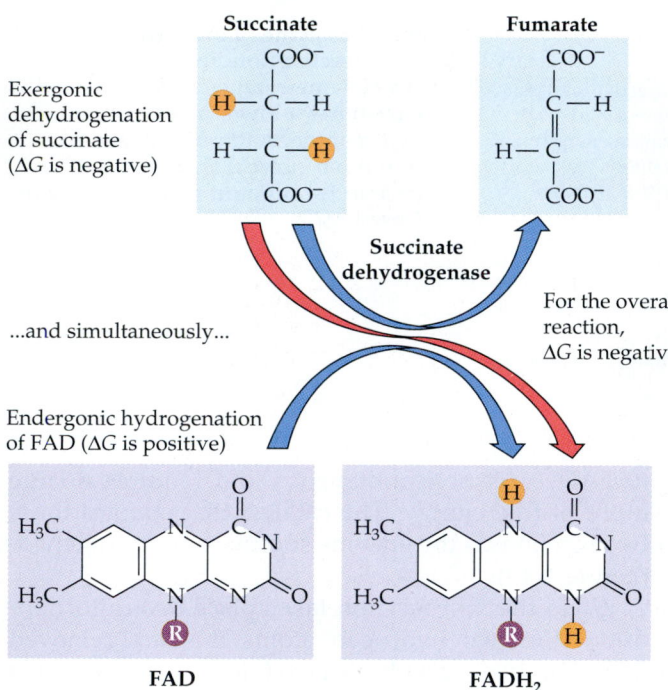

Exergonic dehydrogenation of succinate (ΔG is negative)

Succinate

Fumarate

Succinate dehydrogenase

...and simultaneously...

For the overall reaction, ΔG is negative

Endergonic hydrogenation of FAD (ΔG is positive)

FAD

FADH$_2$

6.13 Succinate Dehydrogenase Couples Two Reactions

In mitochondria, the enzyme succinate dehydrogenase couples the energy-producing reaction that converts succinate to fumarate with the energy-requiring hydrogenation of FAD (blue arrows). The enzyme enables the transfer of the two hydrogens lost by succinate to FAD, producing FADH$_2$ (red arrow).

MOLECULAR STRUCTURE OF ENZYMES

Until the 1960s biochemists knew little about the behavior of enzymes at the molecular level. It was generally agreed that the substrates of enzymes bind to active sites on the surface of the enzyme molecule, but the actual structure of an active site was not understood. The remarkable ability of an enzyme to select exactly the right substrate was explained by the assumption that the binding of the substrate to the site depends on a precise interlocking of molecular shapes. In 1894 the great German chemist Emil Fischer compared the fit between an enzyme and substrate to that of a lock and key. Fischer's model persisted for more than half a century with only indirect evidence to support it.

The first direct evidence came in 1965, when David Phillips and his colleagues at the Royal Institution in London succeeded in crystallizing the enzyme lysozyme and, using the techniques of X-ray crystallography, determined its structure. Since then, the structures of many other enzymes have been determined by such X-ray diffraction studies. Computers are now programmed to draw proteins from X-ray crystallographic data. This work has revealed a great deal about how the enzyme molecule is designed, how it works, and how it is controlled. Small enzymes consist of a single folded polypeptide chain; large en-

6.14 Some Enzymes Cleave Proteins

The hypothetical polypeptide in this figure is eight amino acids long; its backbone is shaded gray and the names of amino acids are indicated at the bottom of the figure. Arrows point to the peptide linkages cleaved by four different enzymes.

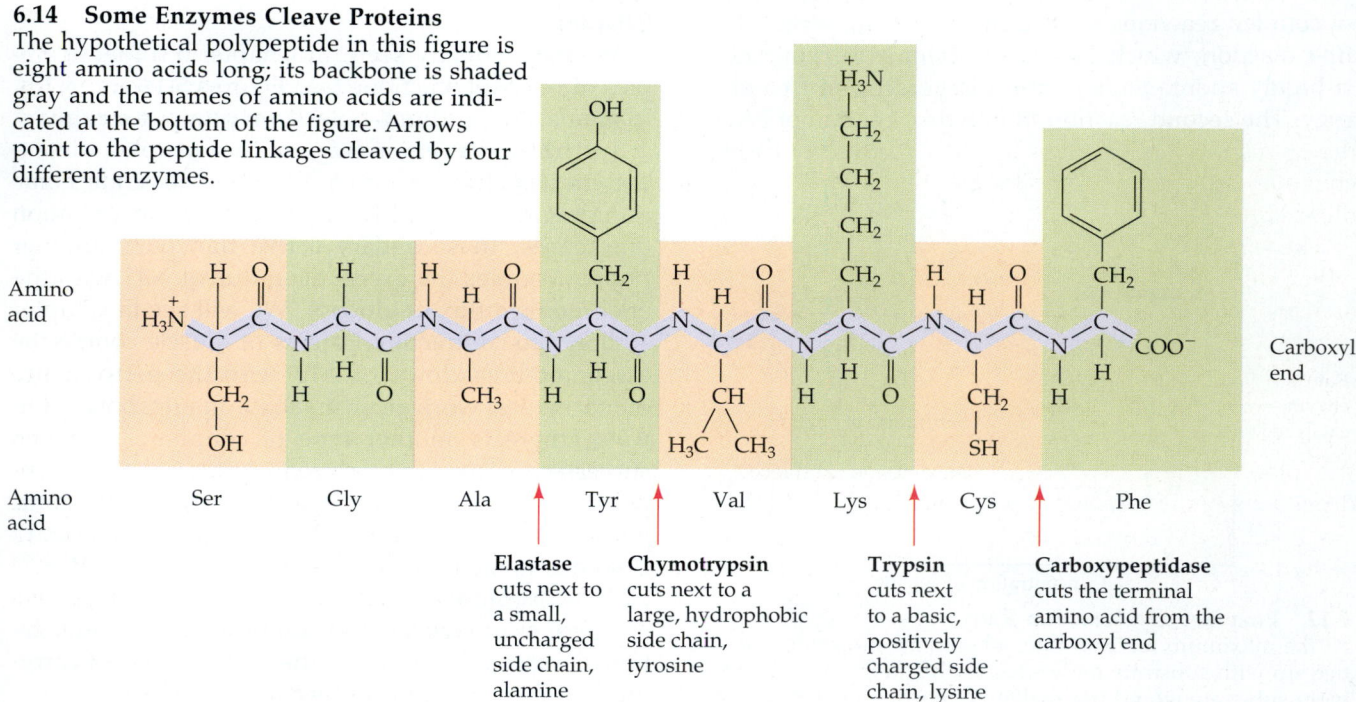

Amino acid

Carboxyl end

Amino acid

| Ser | Gly | Ala | Tyr | Val | Lys | Cys | Phe |

Elastase cuts next to a small, uncharged side chain, alamine

Chymotrypsin cuts next to a large, hydrophobic side chain, tyrosine

Trypsin cuts next to a basic, positively charged side chain, lysine

Carboxypeptidase cuts the terminal amino acid from the carboxyl end

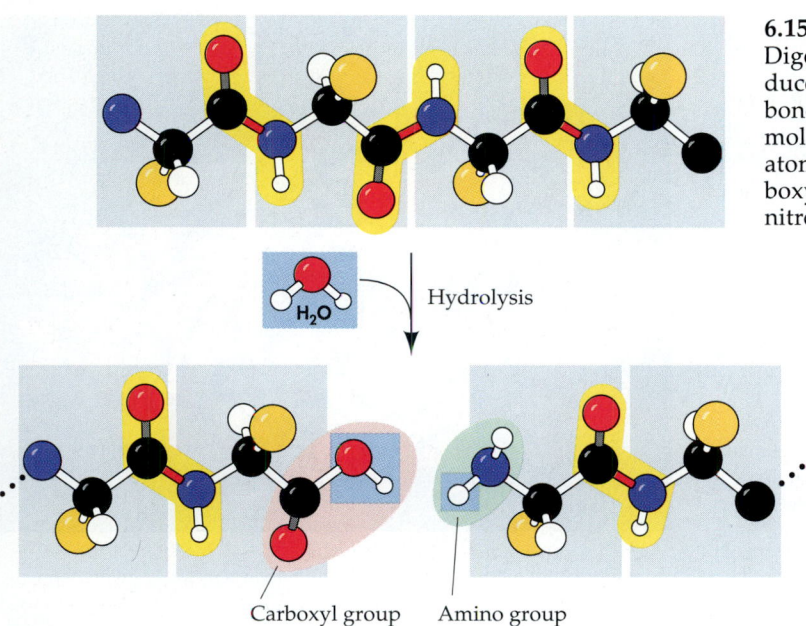

6.15 Hydrolysis of Peptide Linkages
Digestive enzymes such as the four introduced in Figure 6.14 hydrolyze peptide bonds. In the hydrolysis of a peptide, a water molecule donates an –OH group to the carbon atom in the peptide linkage, forming a carboxyl group. The H from water goes to the nitrogen atom, creating an amino group.

Hydrolysis

H_2O

Carboxyl group   Amino group

zymes may contain several, often identical, polypeptide chains. Frequently the active site contains a metal ion that enhances the reaction, or a small, nonprotein molecule may be attached to the active site. These metals and small molecules are called prosthetic groups and will be discussed later in this chapter.

Structures and Actions of Protein-Digesting Enzymes

Different enzymes may behave generally in the same way and yet have very specific differences in behavior. Consider four enzymes that allow animals to digest proteins: carboxypeptidase, chymotrypsin, trypsin, and elastase. All four break the peptide linkages that connect amino acids in polypeptide chains, but each attacks only a very specific linkage. Carboxypeptidase snips one amino acid at a time from the carboxyl (COOH) end of a chain; the other three enzymes cleave chains at particular places in the middle, as shown in Figure 6.14.

The explanation for this fastidious specificity, which underlies the entire biochemistry of living organisms, lies in the architecture of the enzyme molecules. As Fischer suggested, the structure of the active site is molded to fit the substrate molecule. The binding of the substrate to the active site of the enzyme depends on the same forces that maintain the folded tertiary structure of the enzyme itself: hydrogen bonds, the electrostatic attraction and repulsion of charged chemical groups, and the interaction of hydrophobic groups. Substrates may also be covalently bonded to enzymes. (These forces are described in Chapter 2.)

Protein-digesting enzymes **hydrolyze** a peptide bond—that is, they break the polypeptide chain by

adding a water molecule across the peptide bond (Figure 6.15). Carboxypeptidase lowers the energy barrier for this reaction, thus sharply increasing the reaction rate. Because of the specificity of the enzyme, only certain peptide bonds are hydrolyzed at this rapid rate—only those carboxyl-terminal bonds next to bulky hydrophobic side chains that fit comfortably into a hydrophobic pocket in the active site of carboxypeptidase. The other protein-digesting enzymes have different active sites that bind different side chains on the substrate.

Prosthetic Groups and Coenzymes

Although some enzymes consist entirely of one or more polypeptide chains, others possess a tightly bound nonprotein portion called a **prosthetic group.** This may be a single metal ion, a metal ion contained in a small organic molecule, or a **coenzyme,** a complex organic molecule required in some way for the action of one or more enzymes (Figure 6.16). Not all coenzymes are bound as a prosthetic group, however; some are separate and move from enzyme molecule to enzyme molecule.

Some coenzymes assist the catalytic activities of enzymes by accepting or donating electrons or hydrogen atoms (see Chapter 7). Other coenzymes alter the structure of a substrate in such a way that the substrate becomes more reactive. In animals, coenzymes of these two types often are produced from vitamins in the diet. Another group of coenzymes transfers phosphate groups, along with a great deal of free energy, from molecule to molecule. Metal ions attached to enzyme proteins generally function by binding the substrate to the enzyme or by withdrawing electrons from the substrate (Figure 6.17).

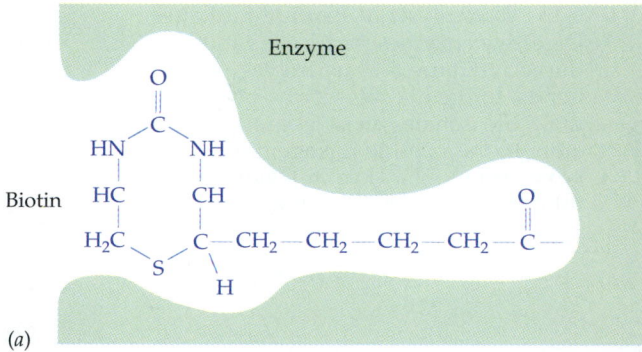

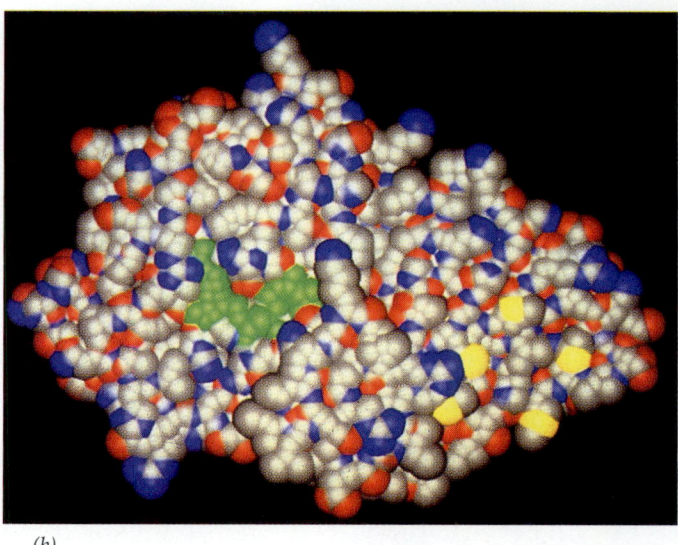

(b)

6.16   Coenzymes

Some enzymes require coenzymes in order to function. (a) The coenzyme biotin bonds covalently to any of several different enzymes that carry a carboxyl group. Because it bonds to its enzyme, this coenzyme is a prosthetic group. (b) The coenzyme NAD (nicotinamide adenine dinucleotide), shown in green, bonds to the enzyme G3PD (glyceraldehyde 3-phosphate dehydrogenase). This illustration shows more realistically than the representation in (a) the relative sizes of enzyme and coenzyme. Hydrogen atoms have been omitted in this drawing, which is colored in the same way as Figure 6.9.

REGULATION OF ENZYME ACTIVITY

Various substances, some occurring naturally in cells and others artificial, act upon enzymes to increase or decrease the rates of enzyme-catalyzed reactions. Those that occur naturally regulate metabolism; the

artificial ones are used either to treat disease or to study how enzymes work. Some substances, called **inhibitors,** that inhibit enzyme-catalyzed reactions produce irreversible effects. Other inhibitors produce reversible effects—that is, these inhibitors can become unbound. Enzymes consisting of multiple subunits are subject to another type of control called allosteric regulation. We will discuss all these types

(a)

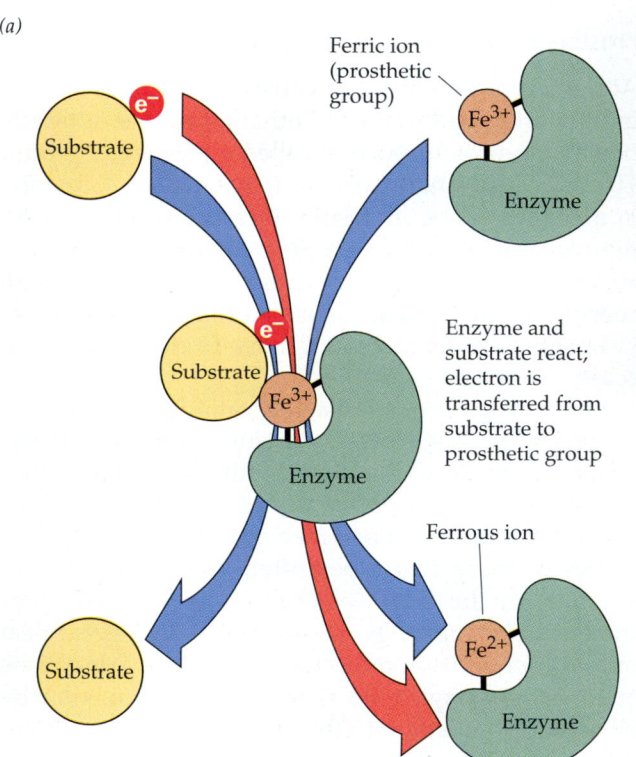

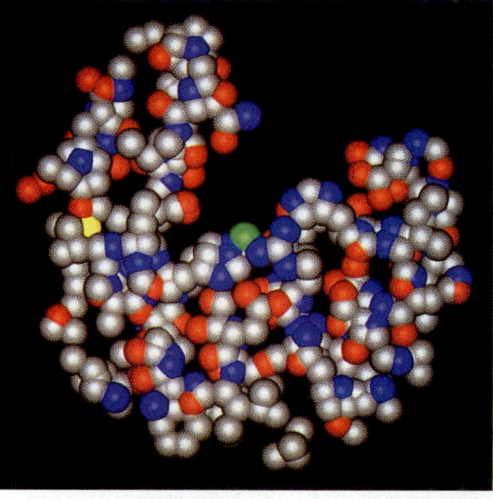

(b)

6.17   Metal Ions as Prosthetic Groups

(a) A ferric ion attached to an enzyme as a prosthetic group. In this reaction the ferric ion (Fe^{3+}) withdraws the electron from the substrate and becomes a ferrous ion (Fe^{2+}); the substrate moves on, altered by the loss of an electron. (b) A tiny part of the enzyme thermolysin, with a zinc ion bound as a prosthetic group (green). Close to the zinc ion is part of a carboxyl group of a glutamic acid. This carboxyl group and the zinc ion collaborate in binding the substrate. The side chain of the glutamic acid having this carboxyl group is thus the part of the enzyme molecule most important for catalysis.

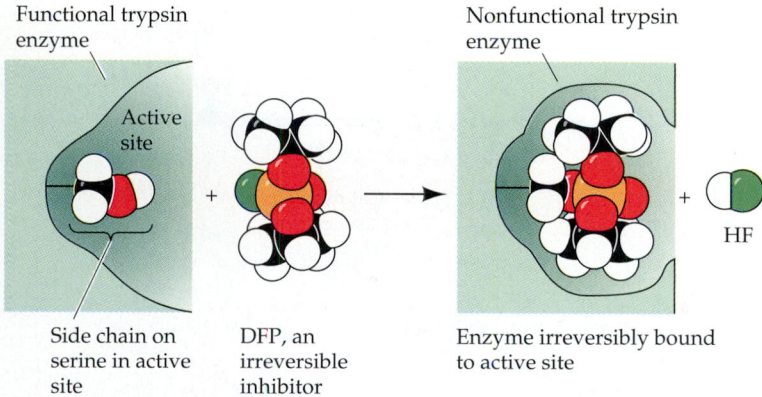

Functional trypsin enzyme

Active site

Side chain on serine in active site

DFP, an irreversible inhibitor

Nonfunctional trypsin enzyme

Enzyme irreversibly bound to active site

HF

6.18 Irreversible Inhibition
DFP disables the digestive enzyme trypsin by reacting with a hydroxyl group belonging to the amino acid serine. After this reaction, trypsin can no longer act on peptide linkages as described in Figure 6.14.

of regulation and conclude this section with a look at the effects of pH and temperature on enzyme activity.

Irreversible Inhibition

Some inhibitors irreversibly modify certain side chains at active sites, ruining the enzymes by destroying their capacity to function as catalysts. An example of such an **irreversible inhibitor** is DFP (diisopropylphosphorofluoridate). DFP reacts with a hydroxyl group belonging to the amino acid serine (see Table 3.1) at the active site of the enzyme trypsin, preventing the use of this side chain in the catalytic mechanism (Figure 6.18).

DFP is an irreversible inhibitor not only for the protein-digesting enzyme trypsin but also for many other enzymes whose active sites contain serine. Among these is acetylcholinesterase, an enzyme that is essential for the orderly propagation of signals from one nerve cell to another (see Chapter 38). Because of their effect on acetylcholinesterase, DFP and other similar compounds are classified as nerve gases.

Reversible Inhibition

Not all inhibitory action is irreversible. Some inhibitor molecules are similar enough to a particular enzyme's natural substrate to bind to the active site, yet different enough that the enzyme catalyzes no chemical reaction. When such a molecule is bound to the enzyme, the natural substrate cannot enter the active site; thus the intruder effectively wastes the enzyme's time, inhibiting its catalytic action. These are called **competitive inhibitors** because they compete with the natural substrate for the active site and block it (Figure 6.19a). The blockage is reversible, however; a competitive inhibitor may become unbound, leaving the active site unchanged. If enough natural substrate molecules are present, they can compete successfully with inhibitor molecules for empty active sites.

Consider the enzyme succinate dehydrogenase, which is subject to competitive inhibition. Recall that this enzyme, found in all mitochondria, removes two hydrogen atoms from succinate to produce fumarate (see Figure 6.13); it then transfers the hydrogens to another molecule, as shown in Figure 6.19b. The other molecules shown in the figure are competitive inhibitors of succinate dehydrogenase. They resemble succinate enough that the enzyme is fooled into binding them. Having bound them, however, the enzyme can do nothing more with them, because the inhibitors are the wrong size and shape or have key chemical groups in the wrong places. The enzyme molecule cannot bind a succinate molecule until the inhibitor molecule has moved out of the active site.

Dissociation of the inhibitor does occur because binding of a competitive inhibitor is reversible, *as is binding of the substrate*. For example, when the competitive inhibitor malonate is added to a solution containing succinate and succinate dehydrogenase, the reaction of succinate to fumarate is slowed. The effect of malonate can be overcome, however, if enough succinate is added. The relative concentrations of substrate and inhibitor determine which of these is more likely to bind to the active site.

Inhibitors that do not react with the active site are called **noncompetitive inhibitors.** Noncompetitive inhibitors bind to the enzyme at a site away from the active site. Their binding causes a conformational change in the protein that alters the active site (Figure 6.19c). The active site still binds substrate molecules, but the rate of product formation is reduced. Noncompetitive inhibitors can become unbound, so their effects are reversible. Because they do not bind to the active site, their effects do not change as substrate concentration changes.

(a) Competitive inhibition

Competitive inhibitor in active site

Active site

Natural substrate

Enzyme

The enzyme molecule's function is disabled as long as the inhibitor remains bound but if the inhibitor becomes unbound, a substrate molecule may bind to the active site.

(b) Competitive inhibition of succinate dehydrogenase

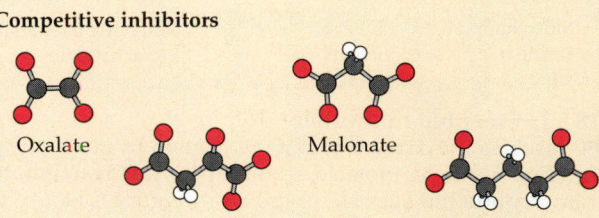

Succinate (substrate) + A ⇌ Fumarate + AH_2

Catalyzed by succinate dehydrogenase

Competitive inhibitors

Oxalate

Malonate

Oxaloacetate

Glutarate

The above series of molecules of increasing length compete with succinate for the enzyme's active site. The similarity that fits them all to the same active site is the presence of two negatively charged carboxyl groups, one at each end of the molecule.

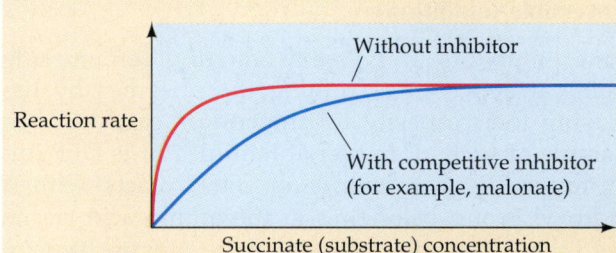

Reaction rate

Without inhibitor

With competitive inhibitor (for example, malonate)

Succinate (substrate) concentration

In the absence of a competitive inhibitor, the enzyme increases reaction rate more than it does when one is present. As substrate concentrations increase, however, competitive inhibition becomes less effective and, eventually, completely ineffective.

(c) Noncompetitive inhibition

Substrate

Enzyme

Noncompetitive inhibitor

Enzyme

A noncompetitive inhibitor may not prevent the substrate from binding to the active site, but it modifies the active site.

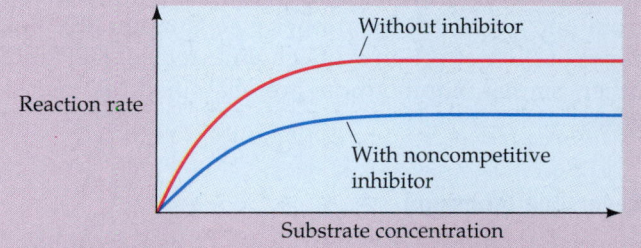

Reaction rate

Without inhibitor

With noncompetitive inhibitor

Substrate concentration

The modification of the active site slows down the rate at which the enzyme catalyzes the reaction. Even high concentrations of substrate do not overcome the effect of the inhibitor; compare with graph in (b).

6.19   Reversible Inhibition

(a) A competitive inhibitor, in competition with the substrate, can bind the active site and thus can disable the enzyme. This binding, like that of the substrate, is reversible. (b) Succinate dehydrogenase (see Figure 6.13) is an example of an enzyme that is subject to competitive inhibition. Competitive inhibitors resemble the substrate chemically. (c) A noncompetitive inhibitor binds at a site away from the active site. The effect may not disable the enzyme but may slow down the reaction.

Allosteric Enzymes

Many important enzymes are larger and more complex than the ones we have discussed so far, which are individual polypeptides. These complex enzymes have quaternary structures (see Chapter 3) consisting of two or more polypeptide subunits, each with a molecular weight in the tens of thousands.

The activity of these complex enzymes is controlled by molecules, called **effectors,** that may have no similarity either to the reactants or to the products of the reaction being catalyzed. Effectors operate by binding to a site on the enzyme other than the active site. Binding at this **allosteric site** enhances or diminishes reactions at the active site; effectors thus can be activators or inhibitors of enzymes. Because many effector–substrate pairs differ structurally, this phenomenon is called allostery, meaning "different shape." Enzymes subject to allosteric control are called **allosteric enzymes;** all have two or more subunits.

Allosteric enzymes and single-subunit enzymes differ greatly in their effects on reaction rates when the substrate concentration is low. Graphs of reaction rates plotted against substrate concentration show this difference. For an enzyme with a single subunit, the plot looks like that in Figure 6.20*a* (see also Figure 6.12). The reaction rate first increases very sharply with increasing substrate concentration, then tapers off to a constant maximum rate as the supply of enzyme becomes saturated with substrate. The plot for an allosteric enzyme is radically different (Figure 6.20*b*), with a sigmoidal (S-shaped) appearance. The increase in reaction rate with increasing substrate concentration is slight at low substrate concentrations, but there is a range over which the reaction rate is extremely sensitive to relatively small changes in the substrate concentration. Because of this sensitivity, allosteric enzymes are important in fine-tuning the activities of a cell. We can understand this behavior in terms of the structure of an allosteric enzyme.

Mechanism of Allosteric Effects

An allosteric enzyme has not only more than one subunit, but more than one *type* of subunit: The **catalytic subunit** has an active site that binds the enzyme's substrate; the **regulatory subunit** has one or more allosteric sites that bind specific effector molecules. A molecule of an allosteric enzyme usually consists of two or more catalytic subunits and two or more regulatory subunits. An allosteric enzyme exists in two or more distinct forms with different catalytic efficiencies, and these forms are in equilibrium with each other. In the simple cases we will examine, the **active form** has full catalytic activity, whereas the **inactive form** is totally without activity (Figure 6.21). In the active form, the active sites on the catalytic subunits can bind substrate and convert it to product. In the inactive form, the active sites have been distorted in such a way that they cannot bind substrate; however, the allosteric sites are able to bind an effector, which we will consider in this example to be an inhibitor. The regulatory subunits of the active form of the enzyme have deformed allosteric sites and cannot bind effector. When neither substrate nor inhibitor is present, the active and inactive forms convert rapidly back and forth in equilibrium (column 1 in Figure 6.21), the equilibrium constant being characteristic of the given enzyme.

What happens when inhibitor or substrate is present? If substrate is present (top of column 2 in Figure 6.21), some of the substrate binds to the active sites of active enzyme molecules; while the enzyme–substrate complex exists, those enzyme molecules cannot be converted to the inactive form. (The presence of a substrate molecule in either active site prevents the enzyme molecule from being converted to the inactive form.) Inactive molecules, however, are being converted to active ones at the same rate as before, so an increase in the concentration of active enzyme results from the presence of substrate! In addition, because each active enzyme molecule (in this example) has two active sites, one enzyme molecule can bind two substrate molecules and simultaneously catalyze reactions of both of them. This explains the upward curvature at the lower left of a plot of reaction rate versus substrate concentration for an alloste-

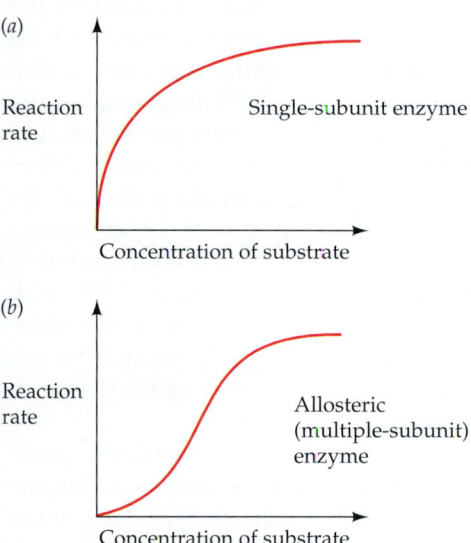

6.20 Allostery and Reaction Rate
How the rate of an enzyme-catalyzed reaction changes with increasing substrate concentration depends on whether the enzyme consists of one or more than one polypeptide subunit.

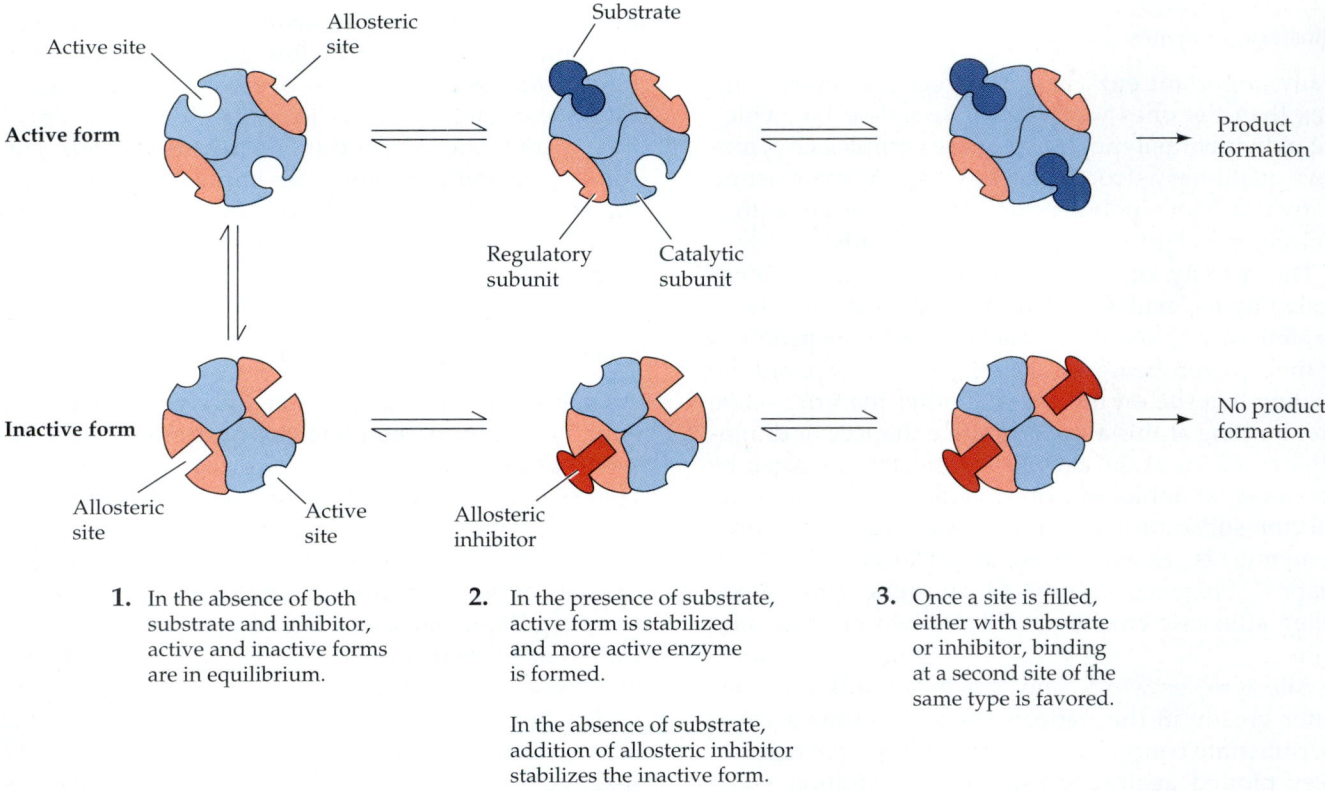

Active site
Allosteric site
Substrate

Active form

Regulatory subunit
Catalytic subunit

Product formation

Inactive form

Allosteric site
Active site
Allosteric inhibitor

No product formation

1. In the absence of both substrate and inhibitor, active and inactive forms are in equilibrium.

2. In the presence of substrate, active form is stabilized and more active enzyme is formed.

   In the absence of substrate, addition of allosteric inhibitor stabilizes the inactive form.

3. Once a site is filled, either with substrate or inhibitor, binding at a second site of the same type is favored.

6.21 Allosteric Regulation of Enzymes
The hypothetical enzyme shown here has four subunits, two catalytic (blue) and the other two regulatory. When the enzyme is in its active form, the active sites on the catalytic subunits can accept substrate. When the enzyme is in its inactive form, the allosteric sites on the regulatory subunits can accept inhibitor.

ric enzyme (see Figure 6.20): Increasing the substrate concentration increases the availability of active enzyme and of active sites and hence rapidly accelerates the reaction rate.

If inhibitor is present (bottom of column 2 in Figure 6.21), the concentration of active enzyme *decreases* and the reaction is thus inhibited. The inhibitor binds to the allosteric site of the *inactive* form of the enzyme, preventing the conversion of the inactive to the active form; but the conversion of the active to the inactive enzyme is not affected by the inhibitor. The overall effect is to decrease the concentration of the active form and thus inhibit the enzymatic reaction. An allosteric activator works in a similar way, except that it binds to the regulatory subunit of the *active* form of the enzyme and holds it in the active configuration. Note that allosteric inhibitors and activators do not modify the structure of the enzyme; rather, they interfere with its conversion to another form with which it is normally in equilibrium.

This mechanism of allosteric inhibition and activation was proposed by the French molecular biologist Jacques Monod and his colleagues in 1965. Yet another of Monod's many contributions to biology is described in Chapter 12.

Control of Metabolism through Allosteric Effects

An organism's metabolism is the sum total of the biochemical reactions that take place within it. These reactions proceed along **metabolic pathways,** which are sequences of enzyme-catalyzed reactions. In the sequences, the product of one reaction is the substrate for the next. Some pathways synthesize, step-by-step, the important chemical building blocks from which macromolecules are built; others trap energy from the environment; still others have different functions. Some metabolic pathways are branched. In a branched pathway, one or more of the intermediate substances are acted upon by more than one enzyme and thus are sent through more than one metabolic branch.

At the branching points where metabolic pathways diverge, **regulatory enzymes** catalyze reactions. These regulatory enzymes are like switches. What flips such a switch? The end product of a branch pathway may block the initial step in that branch pathway, reducing the formation of the end product (Figure 6.22). This illustrates the principle of **negative feedback,** also evident, for example, in thermostats on furnaces. (Negative feedback is discussed in detail

6.22 Feedback in Metabolic Pathways

The reactions C to D and C to H are the first committed steps in the branch pathways leading to end products G and J, respectively. These end products can block the first committed steps by acting as allosteric inhibitors of the enzymes catalyzing the reactions. Thus, if levels of products G and J build up beyond what the cell needs for other reactions, the extra molecules provide the negative feedback that turns off the synthesis of G and J.

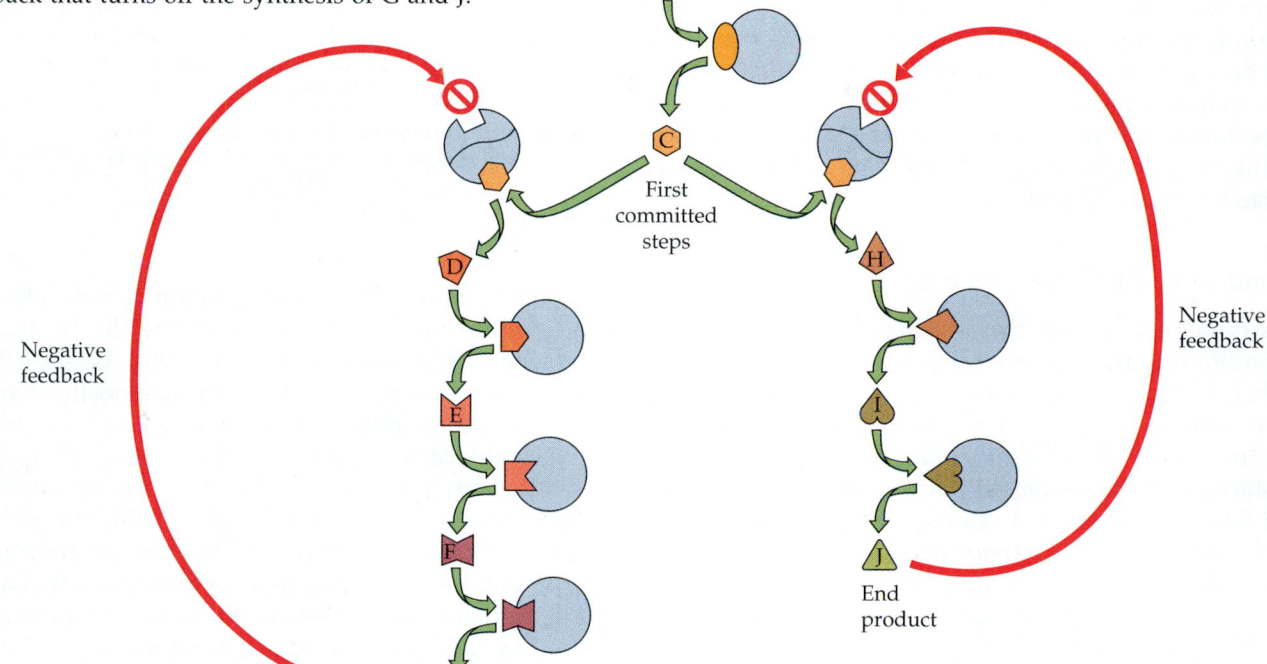

6.23 Concerted Feedback Inhibition

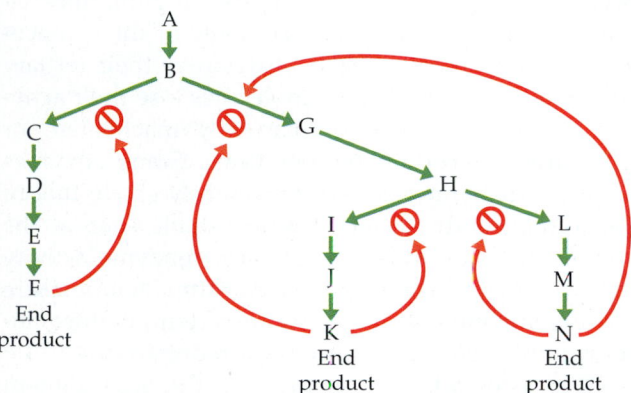

A variety of feedback controls come into play in branching metabolic pathways. Inhibiting the B → C reaction, for example, shunts reactant B into the B → G step. In some cases in which a single compound (H) gives rise to more than one end product (K and N), both end products can join in concerted inhibition of the enzyme catalyzing the first reaction committed to the production of H (B → G); each end product may also inhibit its own branch pathway.

in Chapters 35 and 36.) The end product of a particular pathway typically is an allosteric inhibitor of the regulatory enzyme catalyzing the first **committed step** in its own synthesis—that is, the earliest step in the branched pathway that leads only to the synthesis of that end product and no other. The first committed steps in metabolic pathways are particularly effective points for feedback control. For instance, inhibition of the B-to-C step in Figure 6.23 shunts all the reactants into the other branch of the pathway, whereas inhibition of the C-to-D reaction, one step later, leads only to a possibly harmful and certainly wasteful buildup of substance C. Living things generally do not accumulate unneeded intermediates.

When two different end products, produced by different branches of a pathway, are both present in excessive concentrations, they may act together to inhibit an earlier branch-point enzyme—one that catalyzes the first committed step for formation of these two products (see the B-to-G step in Figure 6.23). **Concerted feedback inhibition** like this results in further efficiency: In this example intermediates G and H do not build up as they would if only the steps to I and L were inhibited by their individual end products. Concerted feedback inhibition requires that the enzyme have two allosteric sites, both of which must be bound to inhibitors to stop the enzyme's activity.

Allosteric regulation is very effective: It allows rapid adjustment to short-term changes in metabolism or in the environment. The activities of enzyme molecules are adjusted by their interactions with small molecules, the end products. If a particular enzyme is not needed, might it not be a good idea simply to stop making it until it is needed? Wouldn't it be advantageous to be able to regulate *production* as well enzyme *activity*? This is indeed the case, and the regulation of enzyme synthesis plays an important role in controlling development and metabolism (see Chapters 12 and 17).

Sensitivity of Enzymes to the Environment

Enzymes enable cells to perform reactions under mild conditions, unlike the extremes of temperature and pH employed by chemists in the laboratory. Enzymes are extremely sensitive to changes in the medium around them. For example, the rates of most enzyme-catalyzed reactions depend on the pH of the medium in which they occur. Each enzyme is most active at a particular pH; its activity decreases as the solution is made more acidic or more basic (Figure 6.24).

Several factors contribute to this effect. One is the ionization of carboxyl, amino, and other groups on either the substrate or the enzyme. Carboxyl groups (—COOH) ionize to become negatively charged carboxylate ions (—COO$^-$) in neutral or basic solutions. Similarly, amino groups (—NH$_2$) accept H$^+$ ions in neutral or acidic solutions, becoming positively charged ammonium ions (—NH$_3^+$). This means, for

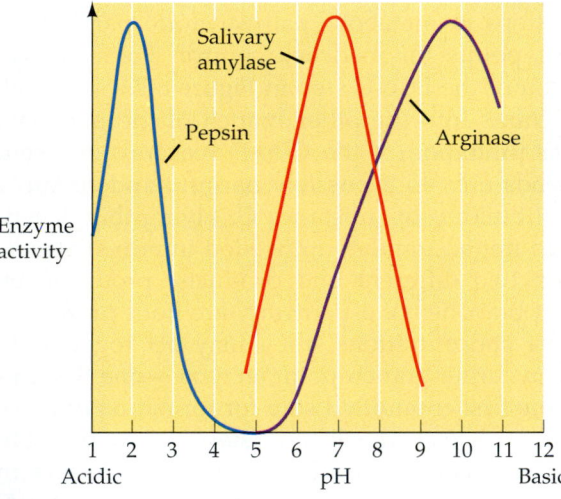

6.24   pH Affects Enzyme Activity
Each enzyme catalyzes reactions most efficiently at a particular pH, as shown by the peaks of the activity curves for three enzymes.

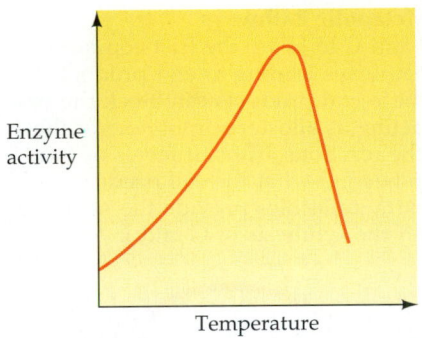

6.25   Temperature Affects Enzyme Activity
An enzyme is most active at a particular temperature.

example, that in a neutral solution a molecule with an amino group is attracted electrically to another molecule that has a carboxyl group because both groups are ionized and they have opposite charges. The attraction does not occur in acidic solution (in which the carboxyl group is not ionized) or in basic solution (in which the amino group is not ionized). Different enzymes function best at different pH values. Evolution has matched enzymes to their environment—for example, digestive enzymes that act in the stomach work best at the very low pH values that prevail in the stomach after a meal.

Temperature also has a profound effect on enzyme activity (Figure 6.25). At low temperatures, warming increases the rate of an enzyme-catalyzed reaction because at higher temperatures a greater fraction of the reactant molecules have enough energy to provide the activation energy of the reaction. Temperatures that are too high, however, inactivate enzymes because at high temperatures enzyme molecules vibrate and twist so rapidly that some of their noncovalent bonds break. The heat destroys their tertiary structure, and the enzyme molecules lose their activity. Enzymes become permanently inactivated, or **denatured,** at certain temperatures. Some enzymes denature at temperatures only slightly above that of the human body; a very few are stable even at the boiling point of water. A graph of enzyme activity versus temperature peaks at the **optimal temperature** for the enzyme. Above the optimal temperature, inactivation of enzyme molecules predominates.

Organisms adapt to changes in the environment in many ways, one of which is based on groups of enzymes, called **isozymes,** that catalyze the same reaction but have differing physical properties. Isozymes may be chemically similar to one another (made from different combinations of the same subunits, for example) or totally unrelated. Within a given set, different isozymes may have different optimal temperatures. An example is the enzyme acetylcholinesterase in the rainbow trout. If a rainbow trout is transferred from relatively warm water to near-freezing water (2°C), the fish produces an isozyme of acetylcholinesterase that is different from the acetylcholinesterase isozyme produced at the higher temperature. The new isozyme has a lower optimal temperature than does the previously formed one, which helps the fish to perform normally in the colder water.

ENZYMES, RIBOZYMES, AND ABZYMES

This chapter is about enzymes: biological catalysts that are proteins. However, enzymes may not have been the first catalytic macromolecules to evolve. As we will see in Chapters 13 and 18, the first biological catalysts may have been RNA molecules. Catalytic RNAs, or **ribozymes**, still function today. Also, by immunizing mice with synthetic analogs of substrates, we can cause them to produce novel antibodies (antibodies are another class of proteins; see Chapter 16) that have modest catalytic activity. Catalytic antibodies, or abzymes, have potential as "designer" catalysts.

In Chapters 7 and 8 we will see how enzymes that catalyze the reactions of two crucial sets of pathways—cellular respiration and photosynthesis—provide cells and organisms with the energy they need to live, grow, and reproduce. Enzymes and other biological catalysts will appear again and again as we continue our examination of the living world.

SUMMARY of Main Ideas about Enzymes, Catalysis, and Energy

Enzymes are biological catalysts that speed up chemical reactions in cells.

An enzyme's active site binds its substrate.
  Review Figure 6.8

The tertiary structure of some enzymes changes as they bind their substrates.
  Review Figure 6.10

Inhibitors block the functioning of enzymes.
  Review Figures 6.18 and 6.19

The rate of an enzyme-catalyzed reaction is affected by the concentrations of substrates and by the temperature and pH of the medium.
  Review Figures 6.12, 6.24, and 6.25

Forward and reverse reactions reach a point of chemical equilibrium at which they both proceed at the same rate.
  Review Figure 6.3

Every chemical reaction has its own equilibrium constant, K_{eq}, which is related to the change in free energy, ΔG.
  Review Table 6.1

Enzymes speed up reactions by lowering the energy barrier between reactants and products.
  Review Figures 6.5, 6.6, and 6.11

Exergonic reactions are spontaneous, but endergonic reactions proceed only if free energy is provided.
  Review Figure 6.4.

Many energy-requiring reactions in cells proceed because some enzymes couple exergonic and endergonic reactions.
  Review Figure 6.13

Prosthetic groups, such as coenzymes or metal ions, help some enzymes catalyze reactions.

Allosteric (multiple-subunit) enzymes and single-subunit enzymes behave differently.
  Review Figures 6.20 and 6.21

Feedback regulation is one type of control for regulating metabolic pathways.
  Review Figures 6.22 and 6.23

Two laws of thermodynamics, describing energy in any closed system, underlie what happens in a chemical reaction.

The first law of thermodynamics states that the quantity of energy in the system remains constant: energy cannot be created or destroyed.

The second law of thermodynamics states that the quantity of free (usable) energy in the system decreases and the quantity of entropy (disorder) increases.

Free energy, heat, temperature, and entropy are related by the equation

$$\Delta G = \Delta H - T\Delta S$$

SELF-QUIZ

1. Which statement about energy is incorrect?
   a. It can neither be created nor destroyed.
   b. It is the capacity to do work.
   c. All its conversions are fully reversible.
   d. In the universe as a whole, the amount of free energy decreases.
   e. In the universe as a whole, the amount of entropy increases.

2. Which statement about thermodynamics is incorrect?
   a. Free energy is given off in an exergonic reaction.
   b. Free energy can be used to do work.
   c. A spontaneous reaction is exergonic.
   d. Free energy tends always to a minimum.
   e. Entropy tends always to a minimum.

3. In a chemical reaction,
   a. the rate depends on the equilibrium constant.
   b. the rate depends on the activation energy.
   c. the entropy change depends on the activation energy.
   d. the activation energy depends on the equilibrium constant.
   e. the change in free energy depends on the activation energy.

4. Which statement is *not* true of enzymes?
   a. They consist of proteins, with or without a nonprotein part.
   b. They change the rate constant of the catalyzed reaction.
   c. They change the equilibrium constant of the catalyzed reaction.
   d. They are sensitive to heat.
   e. They are sensitive to pH.

5. The active site of an enzyme
   a. never changes shape.
   b. forms no chemical bonds with substrates.
   c. determines, by its structure, the specificity of the enzyme.
   d. looks like a lump projecting from the surface of the enzyme.
   e. changes the equilibrium constant of the reaction.

6. A prosthetic group
   a. is a tightly bound, nonprotein part of an enzyme.
   b. is composed of protein.
   c. does not participate in chemical reactions.
   d. is present in all enzymes.
   e. is an artificial enzyme.

7. The rate of an enzyme-catalyzed reaction
   a. is constant under all conditions.
   b. decreases with an increase in substrate concentration.
   c. cannot be measured.
   d. depends on the equilibrium constant.
   e. can be reduced by inhibitors.

8. Which statement is *not* true of enzyme inhibitors?
   a. A competitive inhibitor binds the active site of the enzyme.
   b. An allosteric inhibitor binds a site on the active form of the enzyme.
   c. A noncompetitive inhibitor binds elsewhere than the active site.
   d. Noncompetitive inhibition cannot be completely overcome by adding more substrate.
   e. Competitive inhibition can be completely overcome by adding more substrate.

9. Which statement is *not* true of feedback inhibition of enzymes?
   a. It is exerted through allosteric effects.
   b. It is directed at the enzyme catalyzing the first committed step in a branch of a pathway.
   c. Concerted feedback inhibition is based on two or more end products.
   d. It acts very slowly.
   e. It is an example of negative feedback.

10. Which statement is *not* true of temperature effects?
    a. Raising the temperature may reduce the activity of an enzyme.
    b. Raising the temperature may increase the activity of an enzyme.
    c. Raising the temperature may denature an enzyme.
    d. Some enzymes are stable at the boiling point of water.
    e. The isozymes of an enzyme have the same optimal temperature.

FOR STUDY

1. How is it possible for endergonic reactions to occur in organisms?

2. Consider two proteins: One is an enzyme dissolved in the cytosol; the other is an ion channel in a membrane. Contrast the structures of the two proteins, indicating at least two important differences.

3. Plot free energy versus the course of a reaction for an endergonic reaction and for an exergonic reaction. Include the activation energy in both plots. Label E_a and ΔG in both graphs.

4. Consider an enzyme that is subject to allosteric regulation. If a competitive inhibitor (not an allosteric inhibitor) is added to a solution of such an enzyme, the ratio of enzyme molecules in the active form to those in the inactive form increases. Explain this observation.

READINGS

Dressler, D. and H. Potter. 1991. *Discovering Enzymes*. W. H. Freeman, New York. How enzymes are studied; nicely illustrated.

Harold, F. 1986. *The Vital Force: A Study of Bioenergetics*. W. H. Freeman, New York. A detailed introduction to the study of energy and life.

Karplus, M. and J. A. MacCammon. 1986. "The Dynamics of Proteins." *Scientific American*, April. This article will correct any misconception of proteins as rigid molecules; it describes the constant, rapid changes in local shape that underlie the functioning of proteins.

Koshland, D. E., Jr. 1973. "Protein Shape and Biological Control." *Scientific American*, October. This paper shows that the ability of proteins to change shape in specific circumstances underlies the control and coordination of biological processes.

Newsholme, E. A. and C. Start. 1973. *Regulation of Metabolism*. John Wiley & Sons, New York. A rigorous treatment of regulation of enzyme activity, with emphasis on feedback control.

Stryer, L. 1995. *Biochemistry*, 4th Edition. W. H. Freeman, New York. Good discussion of the structure of proteins.

7

Pathways that Release Energy in Cells

Releasing Energy for a Novel Purpose
This sea walnut uses some of its energy to generate its own light in the deep ocean.

Cells convert energy from one form to another as they carry out the business of life. Plants, as we will see in the next chapter, convert light energy into chemical energy—food for themselves and for the animals that eat them. Some organisms convert the chemical energy of food into light, as does this small deep-sea animal, called a sea walnut. Instead of using chemical energy from food only to grow or reproduce, the sea walnut illuminates its environment. Why does it spend energy in such an apparently frivolous way? In fact, nobody knows for sure—it has been suggested that the animal uses its light to frighten off potential predators, but not many biologists believe this. The function of light emission by sea walnuts remains a mystery. We do know a lot about how the energy of food is made available, however.

GLYCOLYSIS, CELLULAR RESPIRATION, AND FERMENTATION

Organisms must have energy to live. How do they get it? Organisms draw their energy from four biochemical processes that power the machinery of life: photosynthesis, glycolysis, cellular respiration, and fermentation (Figure 7.1). These are the biochemical pathways. These processes consist of metabolic pathways made up of many small chemical steps. In photosynthesis, plants and some other autotrophs use light energy to synthesize food compounds, as we will discuss in Chapter 8. In this chapter we are concerned with the extraction of energy from food molecules as practiced by both heterotrophs *and* autotrophs.

Glycolysis prepares food for cellular respiration. A series of preparatory reactions convert food molecules into a compound in cellular respiration. The glycolytic pathway is also the first part of fermentation. Whether the process that follows glycolysis is cellular respiration or the rest of fermentation depends on the type of organism extracting the energy and on whether the environment is **aerobic** (containing oxygen gas, O_2) or **anaerobic** (lacking oxygen gas). Glycolysis takes place in either case. Fermentation, which occurs primarily in cells where the oxygen supply is limited or depleted, converts food molecules to waste products such as lactic acid or ethanol, releasing some energy. Cellular respiration, which requires oxygen, releases much more energy from a given amount of food than does fermentation. The pathways that release energy in cells are regulated by a system of allosteric feedback control.

The combined operation of glycolysis and cellular respiration is the biological equivalent of burning the sugar glucose. When glucose is burned with a match, it yields carbon dioxide, water, and energy in the form of heat and light. In cells, glucose is broken

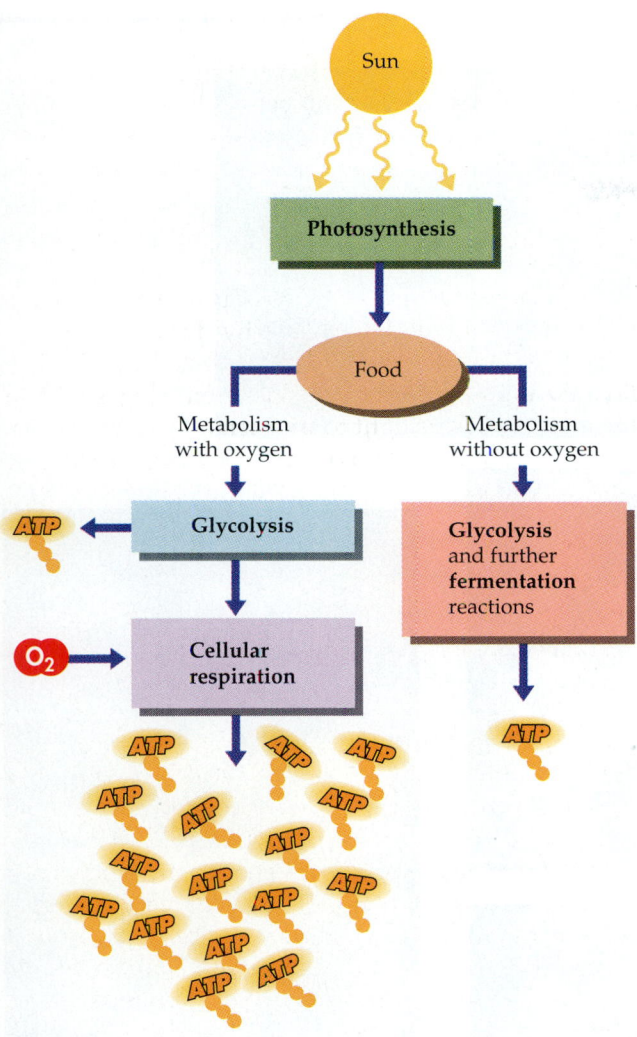

ATP = Energy for work

7.1   Energy for Life
Organisms obtain energy through four types of metabolism. In photosynthesis, autotrophic organisms use light energy to synthesize food compounds. Heterotrophic organisms process these food compounds by glycolysis, fermentation, and cellular respiration. Fermentation is particularly important to cells in which oxygen content is low or depleted, but it releases much less energy from a given amount of food than does cellular respiration. Glycolysis precedes cellular respiration and is the first part of fermentation.

down to the same products, but much of the released energy, rather than being lost as light and large amounts of heat, is trapped in the energy-storage compound **ATP** (adenosine triphosphate). In this chapter we show how cells use enzymes to "burn" glucose and then harness the released energy as ATP—the "spendable energy" referred to earlier.

The complete biological "combustion" of glucose takes place as coupled reactions. To understand these events we must first learn the biochemistry of ATP, which is the cell's principal compound for short-term energy storage and for energy transfers within the cell.

ATP

All living cells rely on the ATP molecule for the short-term storage of energy and the transfer or application of energy to do work. Bioluminescence is a visually dramatic example of the use of ATP (Box 7.A). ATP is a sort of energy currency. How do cells spend, make, and save this currency in the course of their activities?

Spending, Making, and Saving ATP

To "spend" a molecule of its ATP, a cell must break the molecule, releasing the energy of one of its bonds. Many different enzymes can catalyze the breakdown of ATP, whose structure is shown in Figure 7.2. The breakdown (hydrolysis) of ATP yields **ADP** (adenosine diphosphate) and an inorganic phosphate ion. This breakdown is an exergonic reaction, yielding approximately 12 kcal of free energy per mole of ATP under biological conditions. This is enough energy to drive typical endergonic reactions in the cell.

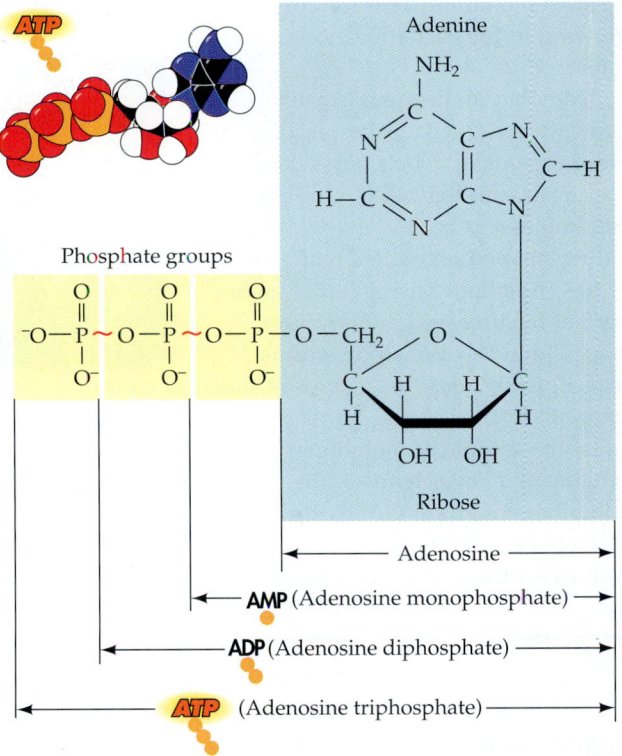

7.2   Structure of ATP
An ATP molecule consists of the base adenine bonded to ribose (a sugar), with three phosphate groups bonded to another carbon on the ribose. The compounds ADP and AMP are like ATP but have one and two fewer phosphate groups, respectively. Adenosine consists of adenine bonded to ribose, without phosphate groups. When the high-energy bonds (color) between the phosphate groups are broken, energy is released.

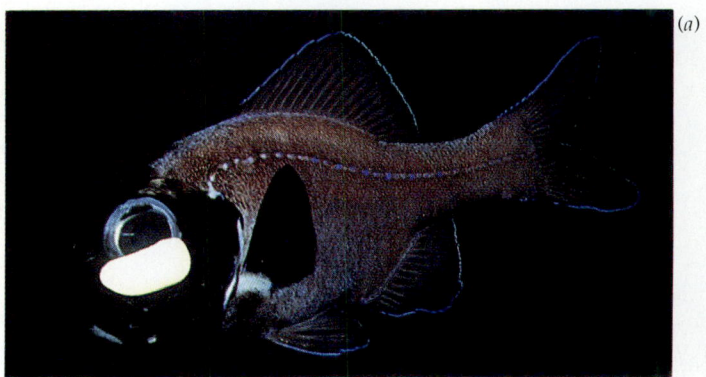

(a)

BOX 7.A

Some Organisms Use ATP to Make Light

Bioluminescence—the production of light by living organisms—always requires ATP as an energy source. Although we know that a firefly uses its bioluminescence to communicate with fireflies of the opposite sex, we are in the dark as to the role of light in most bioluminescent species. Bioluminescence has evolved independently in many kinds of organisms, from bacteria to vertebrates. The comb jelly shown at the beginning of this chapter is a bioluminescent ctenophore. The glowing mushrooms in *(b)* are a bioluminescent fungus growing in the Costa Rican rainforest.

Many bioluminescent organisms live within the cells or tissues of other organisms. Their hosts then appear to emit light. For example, bioluminescent bacteria populate the kidney-shaped organ below the eye of the "flashlight" fish in *(a)*. Another type of bioluminescent bacterium lives within the nematode worms that, in turn, infest the tissues of the caterpillars in *(c)*.

In an example of bioluminescence "engineered" by scientists, the insertion of a gene from a firefly into tobacco plants produced a dazzling display when the plants were watered

(b)

(c)

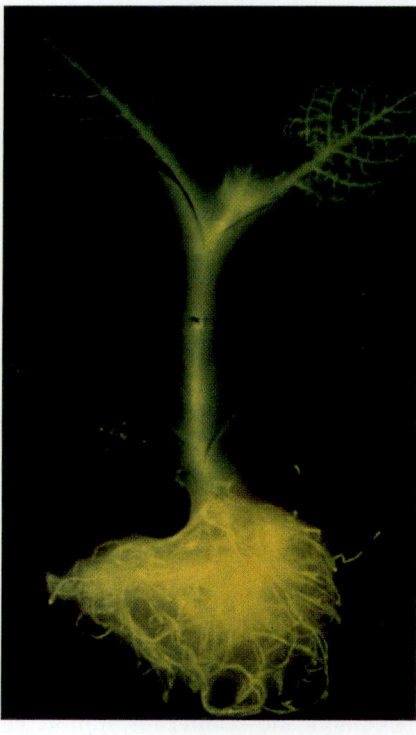

(d)

with an appropriate substrate *(d)*. The plant's ATP provides the energy for the reaction catalyzed by an enzyme encoded in the firefly gene. This triumph of recombinant DNA technology (Chapter 14) affords a powerful tool for studying such diverse topics as development, gene expression, cellular energetics, and the movement of proteins within

---

ADP, which possesses less free energy than does ATP, can combine with a phosphate ion to make a new molecule of ATP if enough energy is provided by an exergonic reaction. The formation of ATP from ADP and a phosphate ion is endergonic and consumes as much free energy as is released by the breakdown of ATP. Many different enzyme-catalyzed reactions in the cell can provide the energy to convert ADP to ATP. The main ones in eukaryotes, however, are the reactions of cellular respiration, in which the maximum amount of energy is released from food molecules and trapped as the stored energy of ATP. Later in this chapter we will see in detail exactly how cells produce ATP.

Figure 7.3 summarizes how ATP traps and releases energy. An exergonic reaction is coupled with the formation of ATP, and at a later time or in another part of the cell, an endergonic reaction is coupled

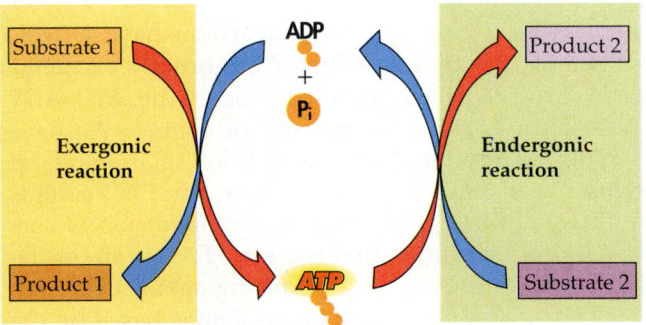

7.3 Formation and Use of ATP
Energy released by an exergonic reaction can be contained when it is used to meld ADP and an inorganic phosphate ion (P_i) into a molecule of ATP. The ATP molecule can then drive an endergonic reaction by splitting into ADP and P_i, releasing energy in the process.

with the splitting of the ATP to ADP and a phosphate ion.

An active cell requires millions of molecules of ATP per second to drive its biochemical machinery. Even so, the cell diverts some of its ATP into synthesizing long-term energy-storage compounds. For this purpose, plants synthesize starch, a long-chain polymer of glucose (see Chapter 3), and sometimes fats. Animals store energy in glycogen (another glucose polymer) and fats. Of course, any large molecule synthesized by the cell (a protein, for example) is a storehouse of energy, but energy storage may not be its primary function. ATP can be considered the circulating currency of energy exchange in living organisms, and starches and fats the savings accounts. When animals need energy, they draw on their deposits of fat and carbohydrates. They break down these deposits into carbon dioxide and water, forming ATP from the energy released in the process. Similarly, plants draw on their stored fats or on deposits of starch, which they convert to glucose; plants then break the glucose down to carbon dioxide and water while forming ATP.

The Energy Content of ATP

To understand the respiratory pathways, you need to have a feeling for why such large changes in free energy accompany the formation and hydrolysis of ATP. ATP is a phosphate ester whose hydrolysis releases much more free energy than most other esters release when hydrolyzed. (An **ester** is an organic compound produced by the reaction of an alcohol with an acid.)

For the hydrolysis of ATP to ADP and phosphate, the change in free energy, ΔG, is about -12 kcal/mol at the temperature, pH, and substrate concentrations typical of living cells. (The "standard" ΔG for ATP hydrolysis is generally given as -7.3 kcal/mol, but that value is valid only at pH 7 and with ATP, ADP,

and phosphate present at concentrations of 1 *M*—conditions that differ greatly from those found in cells.) The hydrolysis of most other phosphate esters produces considerably less than half as much free energy as the hydrolysis of ATP. By transferring phosphate groups, ATP can "prime" other compounds for future chemical reactions.

Part of the unusually large free energy of the hydrolysis of ATP comes from the large number of negative charges near each other on its neighboring phosphate groups. When ATP is hydrolyzed, the charges are spread over two molecules—ADP and inorganic phosphate—which can get far apart; these products are thus more stable than the ATP molecule with its cluster of negative charges. The hydrolysis of ADP to AMP (adenosine monophosphate) and inorganic phosphate liberates an even slightly greater amount of free energy than does that of ATP to ADP, although this energy is not often harnessed for work. Hydrolyzing the last phosphate group (that is, converting AMP to adenosine) does not spread out the negative charges any further, so the free-energy change is low, similar to that for any other ester hydrolysis. Because of their larger free energies of hydrolysis, the first and second bonds broken in ATP are sometimes called high-energy bonds, although "high-energy" refers to the energy of hydrolysis and not to any intrinsic energy of the bond itself. The high-energy bond is sometimes symbolized by $\sim$, and ATP can be written A—P—P$\sim$P, where P represents an entire phosphate group. The phosphate ion, HPO_4^{2-}, is often abbreviated P_i, meaning inorganic phosphate.

As one might expect, given the large amount of free energy released in the hydrolysis of ATP, the reverse reaction—making ATP from ADP and P_i—requires a substantial input of energy. The two major mechanisms of ATP production arise from reactions in which molecules transfer electrons or hydrogen atoms to other molecules. How do cells manage these transfers?

THE TRANSFER OF HYDROGEN ATOMS AND ELECTRONS

Certain pathways of energy metabolism release hydrogen atoms that must be captured and passed on to other reactions. A hydrogen atom consists of a proton and an electron. The transfer of electrons is an oxidation–reduction reaction, or **redox reaction**. The *gain* of one or more electrons by an atom, ion, or molecule is called **reduction**. The *loss* of one or more electrons is called **oxidation**. Although oxidation and reduction are always defined in terms of traffic in *electrons*, we must also think in these terms when hydrogen atoms (not hydrogen ions) are gained or lost, because transfers of hydrogen atoms

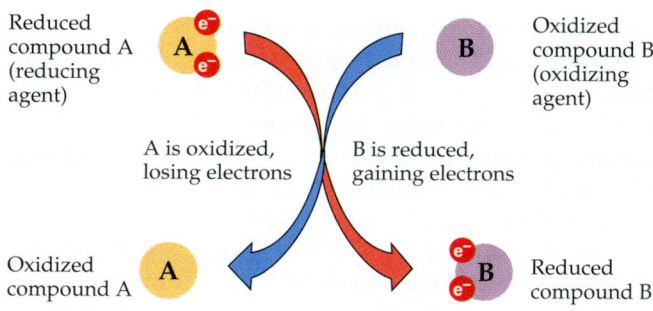

7.4   Oxidation and Reduction Are Coupled
As compound A is oxidized, compound B is reduced. In the process A loses electrons and B gains electrons. In this redox reaction, A is the reducing agent because it donates electrons and B is the oxidizing agent because it accepts electrons.

involve transfers of electrons. Thus when a molecule loses hydrogen atoms, the molecule becomes oxidized.

Oxidation and reduction *always* occur together: As one material is oxidized, the electrons it loses are transferred to another material, reducing it. In a redox reaction, we call the reactant that becomes reduced an **oxidizing agent** and the one that becomes oxidized a **reducing agent** (Figure 7.4). An oxidizing agent accepts electrons; a reducing agent gives up electrons. In the process of oxidizing the reducing agent, the oxidizing agent itself becomes reduced. Conversely, the reducing agent becomes oxidized as it reduces the oxidizing agent. Energy is transferred in the reaction, with energy originally present in the reducing agent becoming associated with the reduced product. ΔG is negative as long as the overall redox reaction is spontaneous. As we will see, some of the key reactions of cellular respiration are highly exergonic redox reactions.

At some early stage in evolution, organisms began to form reducing agents and oxidizing agents. Natural selection favored the use of certain of these

agents (the ones whose redox reactions have suitable values of ΔG) as a system for the orderly exchange of electrons, analogous to the use of the ATP–ADP system for the orderly transfer of energy. We have already encountered an example of such agents at work in cells: In Chapter 6 we saw that FAD accepts hydrogens during the respiratory conversion of succinate to fumarate; in that reaction, FAD is an oxidizing agent and $FADH_2$ is a reducing agent.

The main electron banking system is based on the compound **NAD** (nicotinamide adenine dinucleotide; Figure 7.5), which exists in two chemically distinct forms: one oxidized (NAD^+) and the other reduced ($NADH + H^+$). The function of NAD is to carry hydrogen atoms (with their high-energy electrons) and free energy from compounds being oxidized and to give up hydrogen atoms (with their electrons) and free energy to compounds being reduced (Figure 7.6). The reduction

$$NAD^+ + 2 H \rightarrow NADH + H^+$$

is accompanied by a free energy increase of 52.4 kcal/mol if oxygen gas is the final oxidizing agent:

$$NADH + H^+ + \frac{1}{2} O_2 \rightarrow NAD^+ + H_2O$$

$$\Delta G = -52.4 \text{ kcal/mol}$$

(Note that the oxidizing agent appears here as "½ O_2" instead of "O." This is to emphasize that it is oxygen gas, O_2, that takes part in the reaction.) In the same way that ATP can be thought of as a means of packaging free energy in bundles of about 12 kcal/mol, NAD can be thought of as a means of packaging approximately 50-kcal/mol bundles.

Various energy carriers are chemically related. Be-

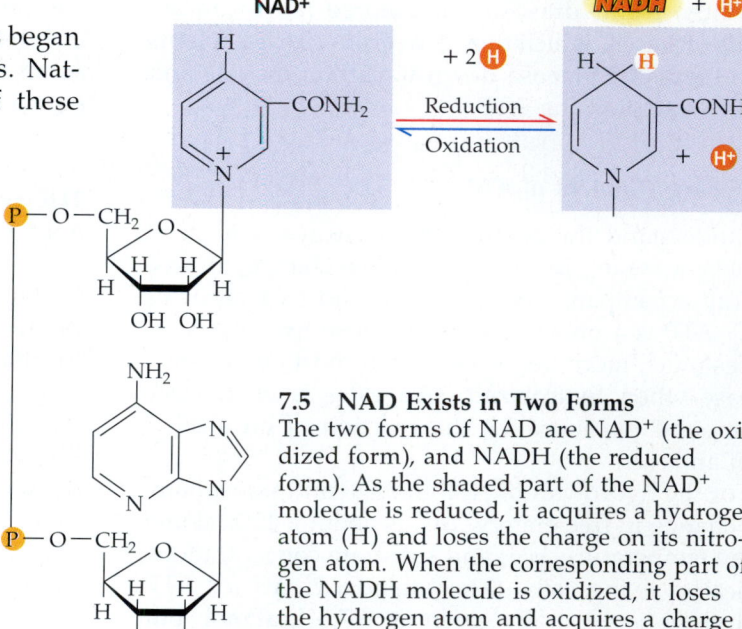

7.5   NAD Exists in Two Forms
The two forms of NAD are NAD^+ (the oxidized form), and NADH (the reduced form). As the shaded part of the NAD^+ molecule is reduced, it acquires a hydrogen atom (H) and loses the charge on its nitrogen atom. When the corresponding part of the NADH molecule is oxidized, it loses the hydrogen atom and acquires a charge on its nitrogen atom.

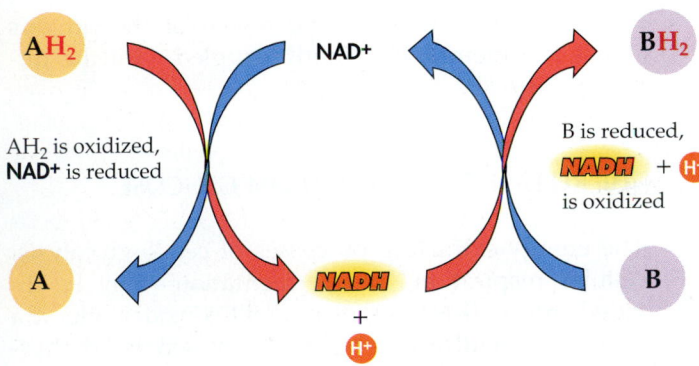

7.6 NAD Is an Energy Carrier

As compound AH_2 is oxidized, it releases its hydrogen atoms with their electrons to NAD^+, reducing NAD^+ to $NADH + H^+$. Elsewhere $NADH + H^+$ may reduce compound B to BH_2, at which time NADH is oxidized to NAD^+. Thanks to its ability to carry electrons and free energy, NAD is a major and universal energy intermediary in cells.

cause part of the NAD molecule looks very much like a molecule of ATP (see Figures 7.2 and 7.5), it is easy to imagine that several energy carriers made up of adenine, ribose, phosphates, and other groups have evolved over time from a common (and less efficient) precursor molecule.

The structures of NAD and some other carrier molecules include compounds that we humans need but cannot synthesize for ourselves; these are classified as vitamins. Nicotinamide, which is part of NAD, forms directly from nicotinic acid, or niacin, a member of the vitamin B complex. Another member of this same vitamin complex is riboflavin, which is part of a carrier called **FAD** (flavin adenine dinucleotide), which we will encounter frequently. We need only small amounts of vitamins because these carrier molecules are recycled through the metabolic machinery. Vitamins are discussed more fully in Chapter 43.

HOW DO CELLS PRODUCE ATP?

In many contexts in this and the next chapter we will be discussing the production of ATP by cells. Two basic mechanisms accomplish the production of ATP: substrate-level phosphorylation and the chemiosmotic mechanism. Every living cell produces at least part of its ATP by **substrate-level phosphorylation**, the enzyme-catalyzed transfer of phosphate groups from donor molecules to ADP molecules, driven by

energy obtained from oxidation. Consider, for example, the following pair of reactions in glycolysis. One intermediate in the glycolytic pathway, glyceraldehyde 3-phosphate, reacts with P_i and NAD^+, becoming 1,3-bisphosphoglycerate. In this enzyme-catalyzed reaction, an aldehyde is oxidized to a carboxylic acid, with NAD^+ acting as the oxidizing agent. The oxidation provides so much energy that the newly added phosphate group is linked to the rest of the molecule by a bond with even higher energy than the high-energy bond of ATP (Figure 7.7a). A second enzyme catalyzes the transfer of this phosphate group from 1,3-bisphosphoglycerate to ADP, forming ATP (Figure 7.7b). Both reactions are exergonic, even though a substantial amount of energy is consumed in the formation of ATP.

Free-floating enzymes catalyze phosphorylation at the substrate level. By contrast, the **chemiosmotic mechanism** requires the participation of molecules that are embedded in membranes. In the chemiosmotic mechanism, protons are pumped across a membrane, effectively charging a "battery." The pump causes a difference in proton concentration (pH) across the membrane. Because the protons are positively charged and the membrane has a low permeability to anions, a difference in electric charge also builds up across the membrane. One side of the membrane is then electrically negative to the other side. The proton concentration gradient and the charge difference together constitute a **proton-motive force** that tends to drive the protons back across the membrane, just as the charge on a battery drives the flow of electrons, discharging the battery. The dis-

(a) Oxidation of substrate   (b) Transfer of phosphate to ADP

7.7 Enzymes Catalyze Substrate-Level Phosphorylation

(a) Enzyme I catalyzes the oxidation of a substrate; in this example, glyceraldehyde 3-phosphate is oxidized, becoming 1,3-bisphosphoglycerate. 1,3-Bisphosphoglycerate contains a phosphate group linked to the molecule by a high-energy bond. (b) A second enzyme catalyzes the transfer of this phosphate group to ADP, forming ATP.

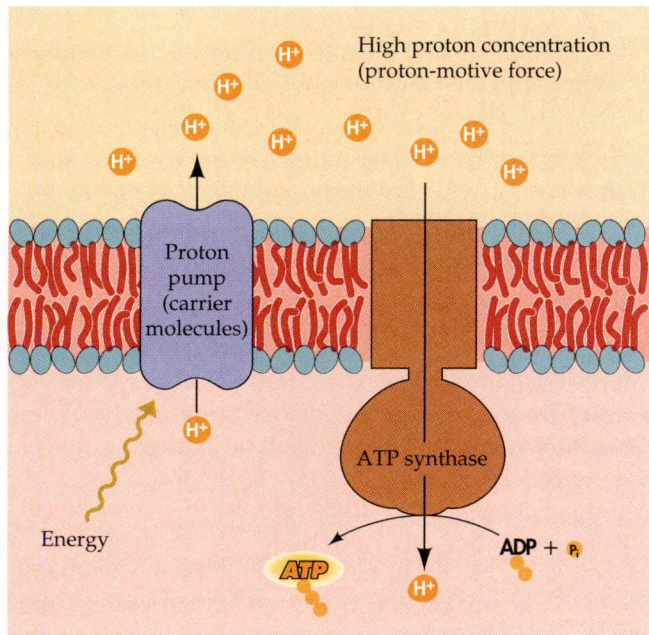

High proton concentration
(proton-motive force)

Proton
pump
(carrier
molecules)

ATP synthase

Energy

ADP + P$_i$

ATP

7.8 Membranes Support Chemiosmotic ATP Production

The energy-driven transport of electrons from one carrier in a membrane to another pumps protons across the membrane, establishing a proton-motive force that tends to drive the protons back. Protons return across the membrane only by using ATP synthases. These membrane proteins operate only when ADP is present; they couple energy from the discharge of the proton-motive force with the formation of ATP from ADP and P$_i$.

charge of the proton-motive force is prevented by the impermeability of most of the membrane to protons. Protons can return across the membrane only by passing through **ATP synthases**, membrane protein complexes that permit protons to flow only if ADP is present. The energy released by the discharge of the proton-motive force is coupled with the endergonic formation of ATP.

What pumps the protons across the membrane in the first place? High-energy electrons are passed from one carrier in the membrane to another. The individual redox reactions are exergonic, and some of the energy is used to pump protons against the gradient of pH and electric charge (Figure 7.8). In cellular respiration, the electrons derive from food molecules; in photosynthesis, the electron transfers are initiated by the energy of sunlight. We will consider these electron transfers in more detail later in this chapter and in Chapter 8.

The chemiosmotic mechanism plays several key roles in the living world. It is responsible for most of the production of ATP in cellular respiration; it participates in the trapping of light energy in ATP formation in photosynthesis (Chapter 8); it provides the energy for driving the propeller-like motion of bacterial flagella (Chapter 4). These three systems differ in detail, but each includes the pumping of protons

across a membrane and the return of the protons across the membrane, tightly coupled with the formation of ATP.

THE RELEASE OF ENERGY FROM GLUCOSE

The energy-extracting processes of cells—glycolysis, cellular respiration, and fermentation—may be divided into pathways that we will consider one at a time. These pathways also tend to be separated physically in the cell (Figure 7.9); they evolved separately, and perhaps at different times.

Glycolysis consists of the glycolytic pathway. This near-universal process was probably the first energy-releasing pathway to evolve; if any earlier pathway existed, it has disappeared from Earth. Today virtually all living cells, even the most evolutionarily ancient, use glycolysis. It is this pathway in which glucose is metabolized to **pyruvate** (pyruvic acid). The glycolytic pathway contains an oxidative step in which an electron carrier, NAD$^+$, becomes reduced, acquiring electrons. Each molecule of glucose processed through glycolysis produces a net yield of two molecules of ATP. The major products of glycolysis are ATP (which the cell will use to drive endergonic reactions), pyruvate, and the two electrons acquired by NAD. Both the pyruvate and the electrons must be processed further.

Cellular respiration is made of up two pathways: the citric acid cycle and the respiratory chain. In eukaryotes in the presence of oxygen and in some bacteria, the pyruvate from glycolysis is oxidized in a cyclical series of respiratory reactions known as the **citric acid cycle** (also called the Krebs cycle or the tricarboxylic acid cycle). In eukaryotes, the reactions of the citric acid cycle are catalyzed by enzymes present in the liquid matrix inside the mitochondrion. Several steps release the carbon atoms of pyruvate (originally the carbon atoms of glucose) as carbon dioxide (CO_2) molecules and transfer more electrons to carriers. The products of the citric acid cycle are, then, carbon dioxide, which must be eliminated in some way from the organism, and many more stored electrons (along with accompanying hydrogen nuclei) than are produced in glycolysis. As we are about to see, more stored electrons means a greater ultimate harvest of ATP.

Hydrogen is an outstanding fuel. When it reacts with oxygen, a great deal of free energy is released; better still, the "waste" product of this reaction—water—is no problem either to the environment or to any organism that produces it. In both glycolysis and the citric acid cycle, hydrogen atoms are acquired by the molecules they reduce. Through these pathways most of the energy originally present as the covalent bonds of glucose becomes associated with reduced NAD (NADH + H$^+$).

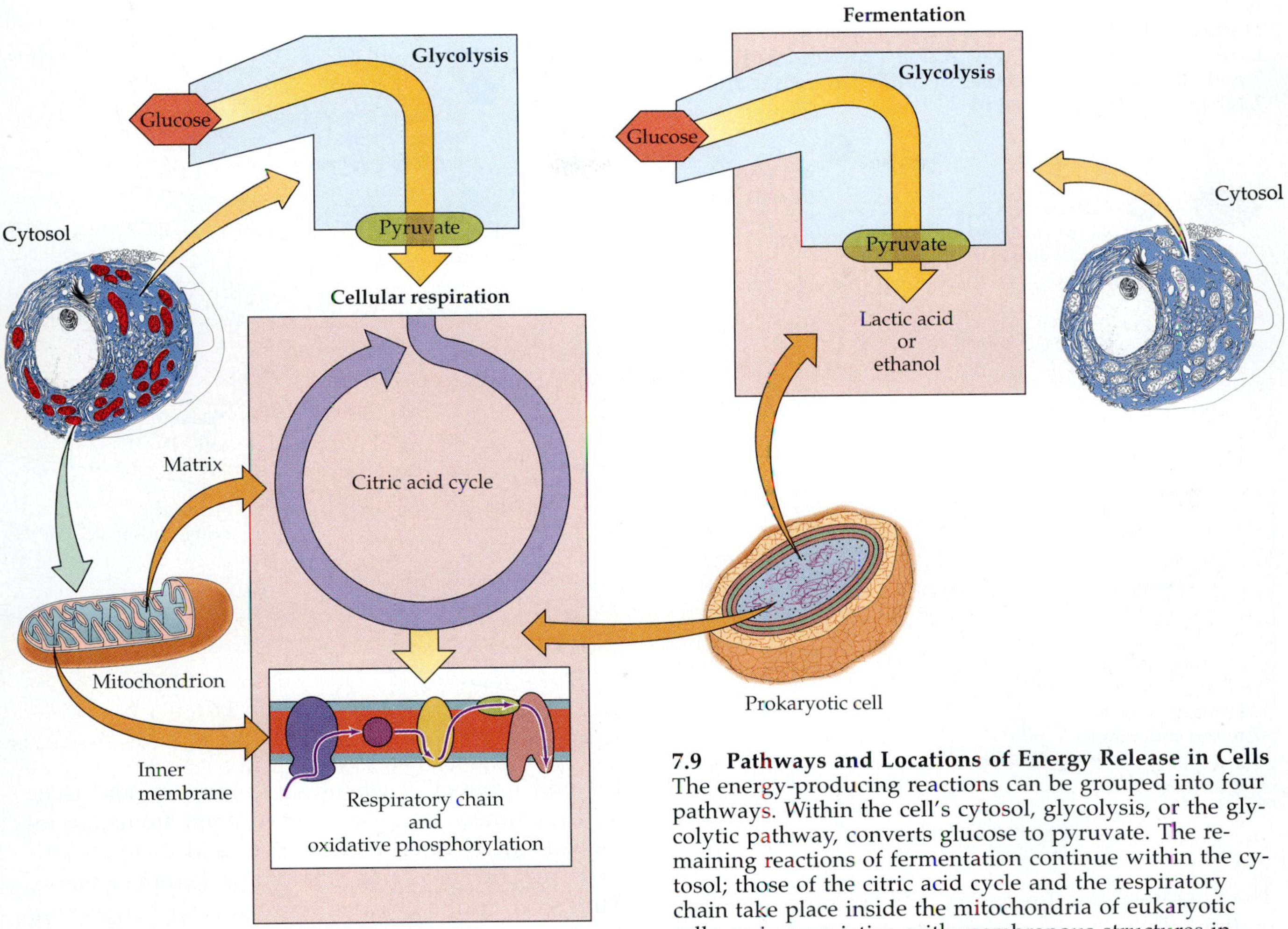

Glycolysis

Glucose

Pyruvate

Cytosol

Cellular respiration

Matrix

Citric acid cycle

Mitochondrion

Inner
membrane

Respiratory chain
and
oxidative phosphorylation

Fermentation

Glycolysis

Glucose

Pyruvate

Lactic acid
or
ethanol

Cytosol

Prokaryotic cell

7.9 Pathways and Locations of Energy Release in Cells
The energy-producing reactions can be grouped into four
pathways. Within the cell's cytosol, glycolysis, or the glycolytic pathway, converts glucose to pyruvate. The remaining reactions of fermentation continue within the cytosol; those of the citric acid cycle and the respiratory
chain take place inside the mitochondria of eukaryotic
cells, or in association with membranous structures in
some prokaryotes.

The principal role of the **respiratory chain** is to
release energy from reduced NAD in such a way that
it may be used to form ATP. This pathway is a series
of successive redox reactions in which hydrogen
atoms—or, in the later steps, electrons derived from
hydrogen atoms—are passed from one type of membrane carrier to another and finally are allowed to
react with oxygen gas to produce water. In eukaryotes, the carriers (and associated enzymes) are bound
to the folds of the inner mitochondrial membranes,
the **cristae** (Figure 7.10 and Chapter 4). In both prokaryotes and eukaryotes, the transfer of electrons
along the respiratory chain drives a chemiosmotic
mechanism (see Figure 7.8). This is the way in which
the vast majority of the ATP in animals is formed.

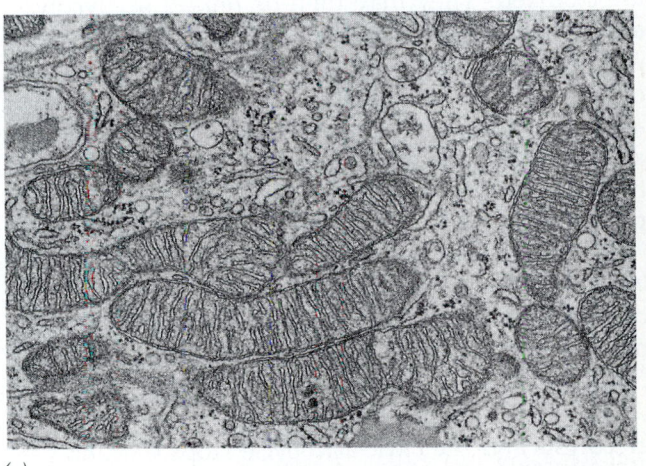

(a)

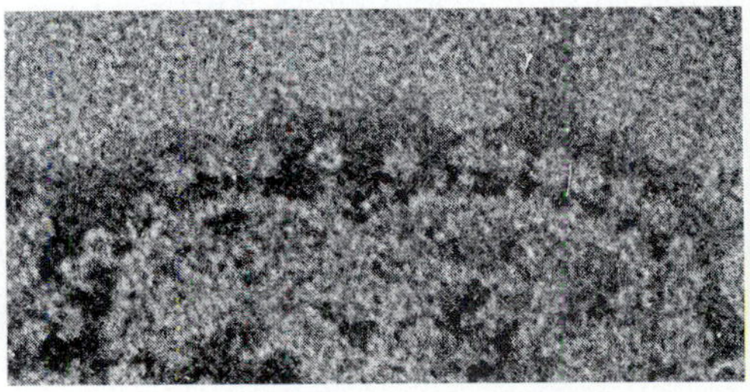

7.10 Cellular Powerhouses
(a) Numerous cristae, arising from the inner of the two mitochondrial
membranes, reach into the matrix of these mitochondria. The cristae are
the sites of the ATP-producing reactions of cellular respiration. (b) A
high-magnification view of the inner mitochondrial membrane. Spherical "knobs" project into the mitochondrial matrix; these protein knobs
catalyze the synthesis of ATP.

(b)

7.11 Glycolysis Converts Glucose to Pyruvate

Starting with hexokinase, ten enzymes catalyze ten reactions in turn. Along the way, ATP is produced (reactions 7 and 10), and two molecules of NAD$^+$ are reduced to 2 NADH + 2 H$^+$ (reaction 6).

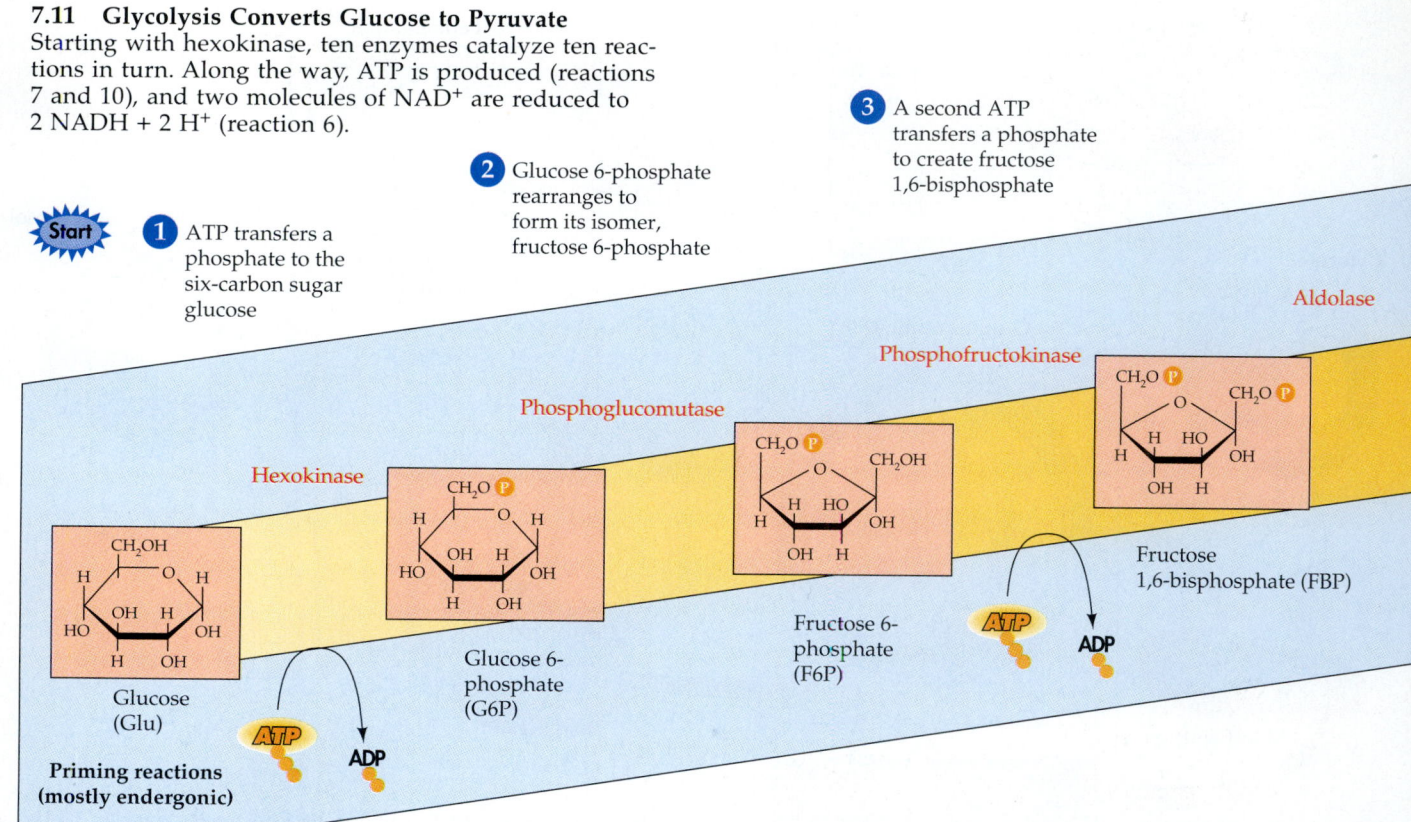

1 ATP transfers a phosphate to the six-carbon sugar glucose

2 Glucose 6-phosphate rearranges to form its isomer, fructose 6-phosphate

3 A second ATP transfers a phosphate to create fructose 1,6-bisphosphate

Hexokinase

Phosphoglucomutase

Phosphofructokinase

Aldolase

Glucose (Glu)

Glucose 6-phosphate (G6P)

Fructose 6-phosphate (F6P)

Fructose 1,6-bisphosphate (FBP)

Priming reactions (mostly endergonic)

The chemiosmotic formation of ATP during the operation of the respiratory chain is called **oxidative phosphorylation**.

As the energy is released, the reduced NAD (NADH + H$^+$) and other agents of electron transfer are oxidized. They may then be reused in glycolysis and the citric acid cycle, steadily draining off hydrogen atoms and allowing those pathways to continue operating. This oxidation of NADH + H$^+$ is thus another consequence of the respiratory chain. Overall, the inputs to the respiratory chain are stored hydrogen atoms and oxygen gas (O$_2$), and the outputs are water and stored energy in the form of ATP.

If we are deprived of oxygen for too long, we die because the respiratory chain cannot function. Without oxygen molecules as receptors, the carriers in our mitochondrial cristae are unable to jettison the hydrogen atoms or electrons bound to them. Soon no oxidized carriers are available; when that happens, glycolysis and the citric acid cycle stop. With no glycolysis, no citric acid cycle, and no respiratory chain activity, we have insufficient ATP. Without ATP, our cells cannot maintain their structure and metabolism, and we die.

Our muscle cells have an alternative way to rid themselves of the hydrogen atoms produced during glycolysis: The hydrogens are passed right back to the end product of glycolysis (pyruvate), and lactic acid is formed. Because the hydrogen atoms are removed, the electron carriers that held them are oxidized and can be used again to process more glucose. Thus even in the absence of oxygen, glycolysis continues (sometimes at an increased rate), without the activity of the citric acid cycle or the respiratory chain, and ATP continues to be produced. The cells that have in their cytosol the enzymatic machinery necessary for this reaction are thereby enabled to function for a time in the absence of oxygen. Eventually, however, the concentration of lactic acid in muscles reaches a toxic level. This anaerobic production of ATP, which releases only a small part of the energy for organisms that require oxygen, is called **fermentation**. For some organisms that live entirely without oxygen, fermentation is the sole pathway that can combine with glycolysis to release energy. We will examine fermentation in more detail later in this chapter.

GLYCOLYSIS

A molecule of glucose taken in by a cell enters the glycolytic pathway, which consists of ten reactions that gradually convert the six-carbon glucose molecule into two molecules of the three-carbon compound pyruvic acid (Figure 7.11). These reactions are accompanied by the *net* formation of two molecules of ATP and by the reduction of two molecules of

4 The fructose ring opens, and the six-carbon fructose 1,6-bisphosphate breaks into two different three-carbon sugar phosphates, DAP and G3P

5 Dihydroxyacetone phosphate rearranges to form its isomer, glyceraldehyde 3-phosphate (G3P)

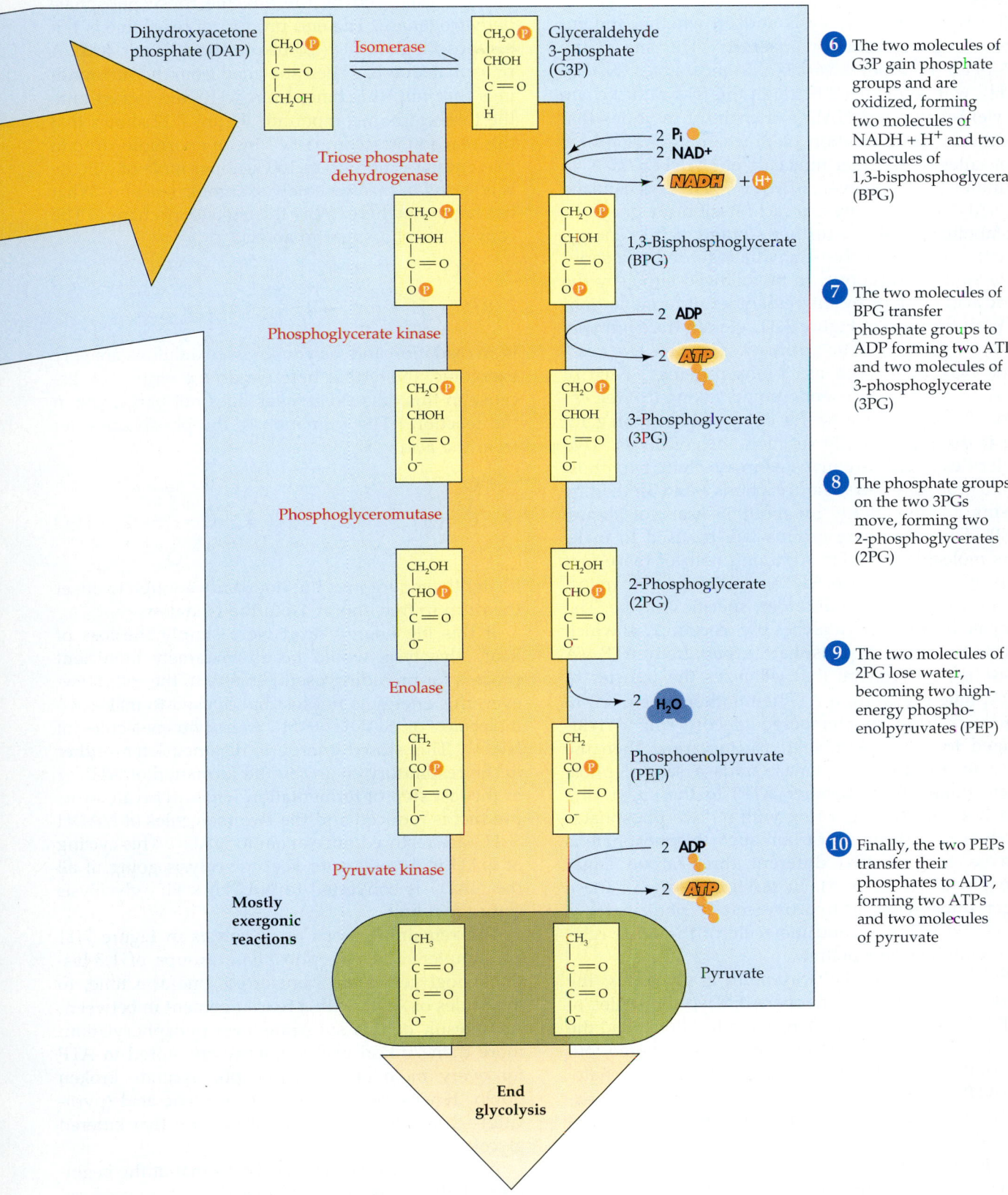

6 The two molecules of G3P gain phosphate groups and are oxidized, forming two molecules of NADH + H$^+$ and two molecules of 1,3-bisphosphoglycerate (BPG)

7 The two molecules of BPG transfer phosphate groups to ADP forming two ATPs and two molecules of 3-phosphoglycerate (3PG)

8 The phosphate groups on the two 3PGs move, forming two 2-phosphoglycerates (2PG)

9 The two molecules of 2PG lose water, becoming two high-energy phospho-enolpyruvates (PEP)

10 Finally, the two PEPs transfer their phosphates to ADP, forming two ATPs and two molecules of pyruvate

NAD^+ to two molecules of $NADH + H^+$. At the end of the pathway, energy is located in ATP, and four hydrogen atoms are passed on in a reducing agent. The fate of the pyruvic acid depends on the type of cell carrying out glycolysis and on whether the environment is aerobic or anaerobic. The fate of the $NADH + H^+$, too, is variable. In most cases, $NADH + H^+$ will be oxidized through the respiratory chain to yield water and NAD^+—a chain of reactions that results in the formation of much more ATP (three molecules of ATP per molecule of $NADH + H^+$). In fermentation, however, $NADH + H^+$ is reoxidized to NAD^+ either by pyruvic acid itself or by one of its metabolites, with no further storage of free energy. In either case, glycolysis may be regarded as a series of *preparatory* reactions, to be followed either by the citric acid cycle or by the remainder of fermentation.

With the help of Figure 7.11, we can trace our way through the glycolytic pathway. The first five reactions may be viewed as "pump-priming." Each of these five reactions is endergonic, taking up free energy; the cell is *investing* free energy rather than gaining it during the early reactions of glycolysis. Two molecules of ATP are invested in attaching two phosphate groups to the sugar (reactions 1 and 3), thereby raising its free energy by about 15 kcal/mol (Figure 7.12). The phosphate groups will be used to make new molecules of ATP. Although both of these first steps of glycolysis use ATP as one of the substrates, each is catalyzed by a different, specific enzyme. The enzyme hexokinase catalyzes the reaction 1, in which glucose receives a phosphate group from ATP. (A *kinase* is any enzyme that catalyzes the transfer of phosphate group from ATP to another substrate.) In reaction 2, the six-membered glucose ring is rearranged to a five-membered fructose ring; then the enzyme phosphofructokinase adds a second phosphate (taken from another ATP) to the sugar ring (reaction 3). The sugar ring with its two phosphates is opened, and the six-carbon sugar bisphosphate is cleaved to give two different three-carbon sugar phosphates (reaction 4). In reaction 5, one of these sugar phosphates (dihydroxyacetone phosphate) is converted into a second molecule of the other (glyceraldehyde 3-phosphate).

By this time, the halfway point in glycolysis, the following things have happened: Two molecules of ATP have been *used* in the priming reactions, and the six-carbon glucose molecule has been converted into two molecules of a three-carbon sugar phosphate. No ATP has been gained, and nothing has been oxidized; in short, it looks as if we are going determinedly backward.

Now, however, the pump is primed and things begin to happen rapidly. In what follows, remember that each step occurs twice for each glucose molecule going through glycolysis, because each molecule has

by now been split into two molecules of three-carbon sugar phosphate, both of which go through the remaining steps of glycolysis. Reaction 6 is a two-step reaction catalyzed by the enzyme triose phosphate dehydrogenase. The end product of reaction 6 is 1,3-bisphosphoglycerate (1,3-bisphosphoglyceric acid). A phosphate ion has been snatched from the surroundings (but not, this time, from ATP) and tacked onto the three-carbon compound. Figure 7.12 shows that this reaction is accompanied by an enormous drop in free energy—more than 100 kcal per mole of glucose is released in this extremely exergonic reaction. What has happened here? Why the big energy change? The conversion of a sugar to an acid

$$R-\overset{\overset{\displaystyle O}{\|}}{C}-H + (O) \rightarrow R-\overset{\overset{\displaystyle O}{\|}}{C}-OH$$

is an oxidation and, as you know, oxidations are very exergonic. (Note that here we do *not* write $\frac{1}{2}O_2$ because in this case oxygen gas does not participate in the reaction.) The formation of the phosphate ester from the acid

$$R-\overset{\overset{\displaystyle O}{\|}}{C}-OH + HPO_4^{2-} \rightarrow R-\overset{\overset{\displaystyle O}{\|}}{C}-O-\overset{\overset{\displaystyle O}{\|}}{\underset{\underset{\displaystyle O^-}{|}}{P}}-O^- + H_2O$$

is slightly endergonic, but not nearly enough to offset the drop in free energy from the oxidation.

If this big energy drop were simply the loss of heat, glycolysis would be an extremely inefficient process for providing useful energy to the cell. However, this energy is not lost but is used to make two molecules of $NADH + H^+$ from two molecules of NAD^+. This stored energy is regained later—either in the respiratory chain, by the formation of ATP, or in the last step of fermentation when pyruvate or its product is reduced and the two molecules of $NADH + H^+$ are restored once again to NAD^+. This cycling of NAD is necessary to keep glycolysis going; if all the NAD^+ is converted to $NADH + H^+$, glycolysis comes to a halt.

The remaining steps of glycolysis in Figure 7.11 are simpler. The two phosphate groups of 1,3-bisphosphoglycerate are transferred, one at a time, to molecules of ADP, with a rearrangement in between. As a result of this substrate-level phosphorylation, more than 20 kcal of free energy are stored in ATP for every mole of 1,3-bisphosphoglycerate broken down. Finally, we are left with pyruvic acid (pyruvate)—2 mol for each mole of glucose that entered glycolysis.

A review of the reactions shows that at the beginning of glycolysis two molecules of ATP are used per molecule of glucose, but that ultimately four are produced (two for each of the two 1,3-bisphosphoglycer-

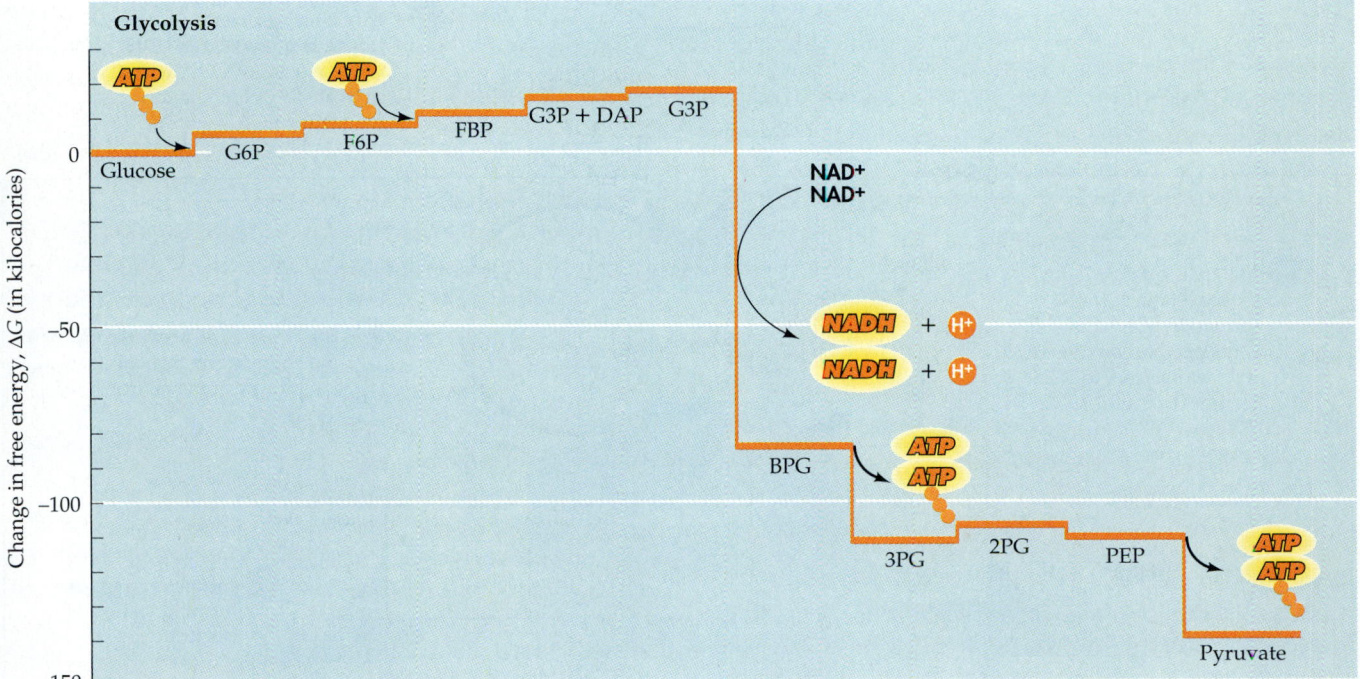

7.12 Free Energy Changes during Glycolysis
Each reaction of glycolysis changes the free energy available, as shown by the differing energy levels of the series of reactants and products from glucose to pyruvate. Note at the upper left that the investment of each of two ATPs in the priming reactions raises the free energy content of the sugar. Key energy-releasing reactions account for the largest drops in free energy. The quantities of NADH + H^+ and ATP in the diagram reflect those produced from one molecule of glucose by the reactions of glycolysis. The net gain is 2 ATP, and 2 NADH + 2 H^+ are released.

ates)—a net gain of two ATP molecules and two NADH + H^+. Under anaerobic conditions, the total usable energy yield from the metabolism of glucose is usually two ATP molecules. The NADH + H^+ is rapidly recycled to NAD^+ by fermentation for reuse by the triose phosphate dehydrogenase of glycolysis. In the presence of oxygen, on the other hand, eukaryotes and some bacteria are able to reap far more energy by the further metabolism of pyruvate and by reoxidizing the reduced NAD of glycolysis through the respiratory chain.

[By now, you might be wondering why we are using words like *pyruvate* and *pyruvic acid* interchangeably. At pH values commonly found in cells, the ionized form—pyruvate—is present rather than the acid—pyruvic acid. Similarly, all carboxylic acids are present as ions (the *-ate* forms) at these pHs. Thus on grounds of chemical accuracy and simplicity, it is better to name the negative ion than the acid. However, custom often prevails over accuracy, and the acids are often named instead; for example, nobody seems to want to change the name *citric acid cycle* to the more accurate *citrate cycle*.]

THE BEGINNING OF CELLULAR RESPIRATION: THE CITRIC ACID CYCLE

The end product of glycolysis, pyruvate, is the starting point of two different pathways: the citric acid cycle of cellular respiration, and the fermentation pathway (see Figure 7.9). In this section we take a look at the citric acid cycle, in which pyruvate is incinerated to carbon dioxide (CO_2).

Figure 7.12 shows that the metabolism of glucose to pyruvate is accompanied by a drop in free energy of about 140 kcal/mol. About one-third of this energy is captured in the formation of ATP and reduced NAD. Oxidizing the pyruvate yields additional free energy for biological work. The citric acid cycle takes pyruvate and breaks it down to CO_2, using the hydrogen atoms to reduce carrier molecules and to pass chemical free energy to those carriers. The reduced carriers are later oxidized in the respiratory chain, which we discuss in the next section; and an enormous amount of free energy is transferred from the reduced carriers to ATP in the process. The principal inputs to the citric acid cycle are pyruvic acid, water, and oxidized electron carriers; the principal outputs are carbon dioxide and reduced electron carriers:

pyruvic acid ($C_3H_4O_3$) + 3 H_2O + 5 carrier →
3 CO_2 + 5 carrier · (2 H)

Overall, then, for each molecule of pyruvate, during the citric acid cycle three carbons are removed as CO_2 and five pairs of hydrogen atoms are used to reduce carrier molecules, with the simultaneous storage of

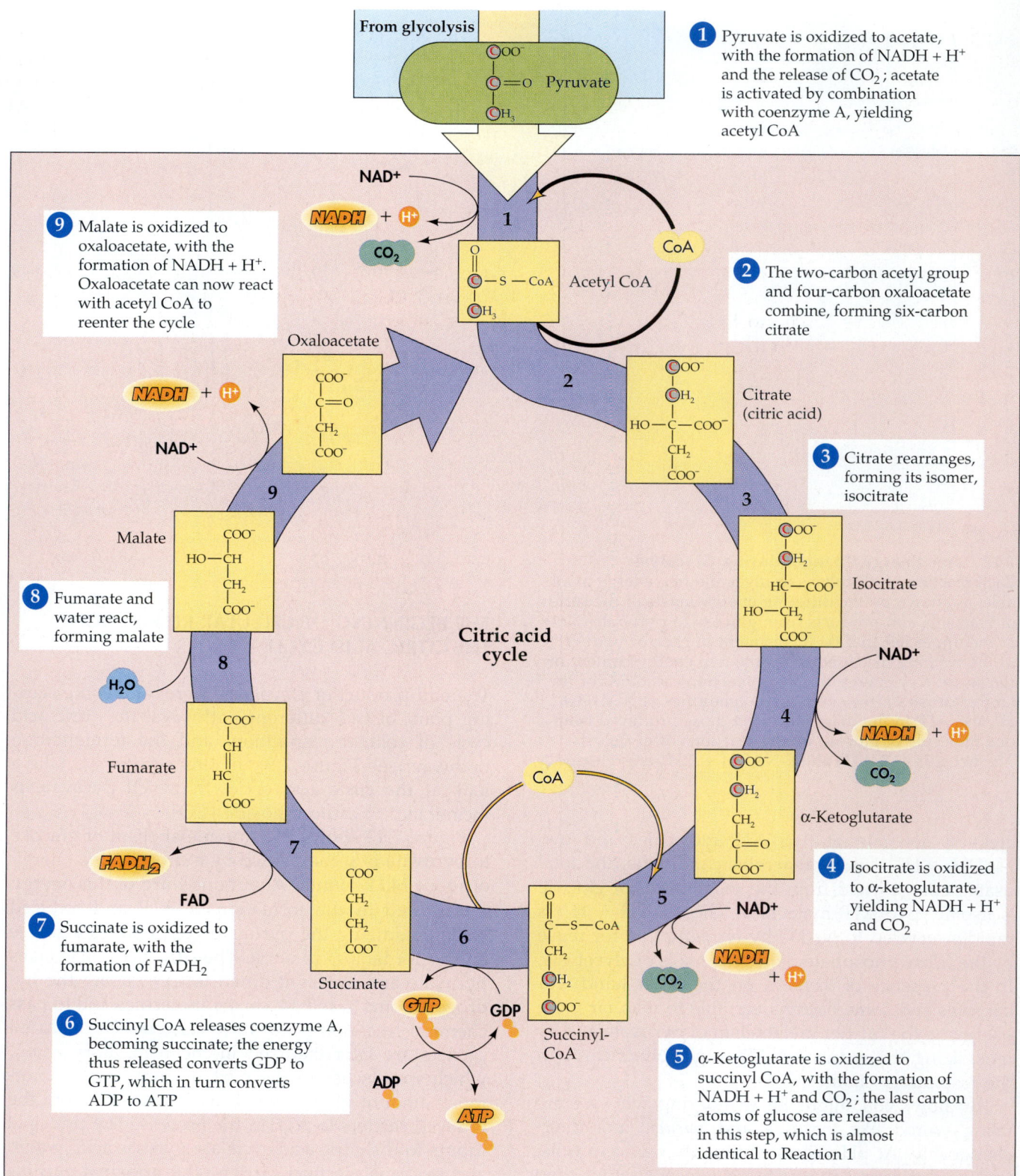

From glycolysis

Pyruvate

1 Pyruvate is oxidized to acetate, with the formation of NADH + H⁺ and the release of CO_2; acetate is activated by combination with coenzyme A, yielding acetyl CoA

Acetyl CoA

CoA

2 The two-carbon acetyl group and four-carbon oxaloacetate combine, forming six-carbon citrate

9 Malate is oxidized to oxaloacetate, with the formation of NADH + H⁺. Oxaloacetate can now react with acetyl CoA to reenter the cycle

Oxaloacetate

Citrate (citric acid)

3 Citrate rearranges, forming its isomer, isocitrate

Malate

Isocitrate

8 Fumarate and water react, forming malate

Citric acid cycle

α-Ketoglutarate

Fumarate

4 Isocitrate is oxidized to α-ketoglutarate, yielding NADH + H⁺ and CO_2

7 Succinate is oxidized to fumarate, with the formation of $FADH_2$

Succinate

CoA

Succinyl-CoA

6 Succinyl CoA releases coenzyme A, becoming succinate; the energy thus released converts GDP to GTP, which in turn converts ADP to ATP

5 α-Ketoglutarate is oxidized to succinyl CoA, with the formation of NADH + H⁺ and CO_2; the last carbon atoms of glucose are released in this step, which is almost identical to Reaction 1

7.13 The Citric Acid Cycle
The first reaction produces the two-carbon acetyl CoA. Notice that the two carbons from acetyl CoA are traced with color through reaction 5, after which they may be at either end of the molecule (note the symmetry of succinate and fumarate). Reactions 1, 4, 5, 7, and 9 accomplish the major overall effect of the cycle—the storing of energy—by passing electrons to the carrier molecule NAD. Reaction 6 also stores energy.

energy. The energy-storing reactions of the cycle are a major reason for its existence.

At the beginning of the citric acid cycle, pyruvate is oxidized (yielding useful free energy and CO_2) and converted into an activated form of acetic acid, CH_3COOH, called **acetyl CoA** (acetyl coenzyme A). Then acetyl CoA, with two carbon atoms in its acetate group, reacts with a four-carbon acid (oxaloacetate) to form the six-carbon compound citric acid (citrate). The remainder of the cycle consists of a series of enzyme-catalyzed reactions in which citric acid is degraded, leading to the production of useful free energy from redox reactions, to the release of two of the carbons as CO_2, and to the production of a new four-carbon molecule of oxaloacetate from the other four carbons. This new oxaloacetate can react with a second acetyl CoA, producing a second molecule of citrate, and so forth. Acetyl CoA comes into the cycle from pyruvate, CO_2 goes out, the rest of the compounds in the cycle are used and replaced, and energy from redox reactions is stored. As we describe the citric acid cycle in detail, concentrate on how it is maintained in a steady state—that is, with material entering and leaving and with intermediate compounds like succinate and malate turning over constantly but without changing concentration. (The concept of steady state is important; be sure it is clear to you.)

As you read the next paragraphs, you can follow the sequence of steps in the citric acid cycle in Figure 7.13. The product of reaction 1, acetyl CoA, is 7.5 kcal/mol higher in energy than simple acetate. (Acetyl CoA can donate acetate to acceptors such as oxaloacetate much as ATP can donate phosphate to various acceptors.) There are three steps in this first reaction: (1) pyruvate is oxidized to acetate, with the release of CO_2; (2) part of the energy from this oxidation is saved by reducing NAD^+ to $NADH + H^+$; and (3) some of the remaining energy is stored temporarily by combining the acetate with CoA. An analogous three-step reaction occurs in glycolysis when glyceraldehyde 3-phosphate is converted to 1,3-bisphosphoglycerate (see Figure 7.11, reaction 6). In that reaction, a sugar is oxidized to an acid, some of the energy released by oxidation is stored in $NADH + H^+$, and some of the remaining energy is preserved in a second phosphate bond in the molecule. A good metabolic idea is likely to appear more than once; we will see this one again, later in the citric acid cycle. As you might guess, a complex set of steps such as those in the reaction from pyruvate to acetyl CoA requires more than one type of catalytic protein. In fact, this reaction is catalyzed by the *pyruvate dehydrogenase complex*, which consists of 72 subunits—24 each of three different types of protein, for a total molecular weight of 4.6 million. This complex is an impressive example of biological organization.

The energy temporarily stored in acetyl CoA helps to drive the formation of citrate from oxaloacetate (reaction 2). During this reaction, the coenzyme molecule falls away, to be recycled and bound to another acetate by the pyruvate dehydrogenase complex. Citrate is rearranged to isocitrate (reaction 3). In reaction 4, a CO_2 molecule and two hydrogen atoms are removed in the conversion of isocitrate to α-ketoglutarate. As Figure 7.14 indicates, this reaction produces a large drop in free energy. The released energy is stored in $NADH + H^+$ and can be recovered later in the respiratory chain, when the $NADH + H^+$ is reoxidized.

Like reaction 1 (the oxidation of pyruvate to acetyl CoA), reaction 5 of the citric acid cycle is complex. The α-ketoglutarate molecule is oxidized to succinate, CO_2 is given off, some of the oxidation energy is stored in $NADH + H^+$, and some is preserved temporarily by combining succinate with CoA. This temporarily stored energy is saved in reaction 6, in which GTP (guanosine triphosphate) is first made and then used to make ATP—another example of substrate-level phosphorylation. A smaller amount of free energy is released in reaction 7, when two hydrogens are transferred to an enzyme containing FAD (an oxidizing agent similar to NAD^+ that we discussed earlier in this chapter); one more NAD^+ reduction (reaction 9) occurs after a molecular rearrangement (reaction 8). The oxaloacetate that remains after all these reactions is ready to combine with another acetyl CoA molecule and go around the cycle again. Bear in mind that the citric acid cycle operates twice for each glucose molecule that enters glycolysis.

CONTINUATION OF CELLULAR RESPIRATION: THE RESPIRATORY CHAIN

Without oxidizing agents to be reduced and act as electron carriers, the oxidative steps of glycolysis and the citric acid cycle could not occur. Two such agents—the crucial molecules NAD^+ and FAD—are regenerated by the respiratory chain. The reaction of substrate and oxidizing agent can be represented as follows

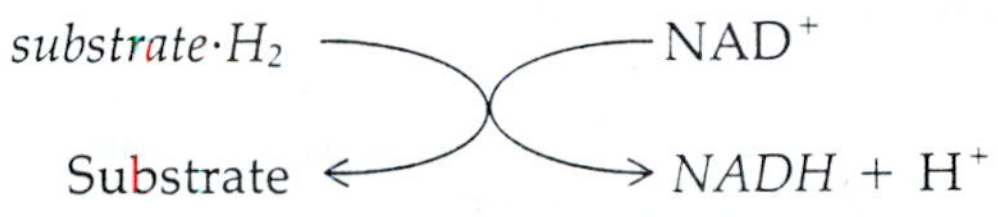

$$substrate \cdot H_2 \longrightarrow NAD^+$$
$$Substrate \longleftarrow NADH + H^+$$

with the hydrogen and its high-energy electrons being passed from the originally reduced substrate (such as malate in the citric acid cycle) to the oxidizing agent NAD^+. We see that the presence of the oxidizing agent is critical, because without it to accept electrons the substrates could not be oxidized, and there would be no respiratory metabolism.

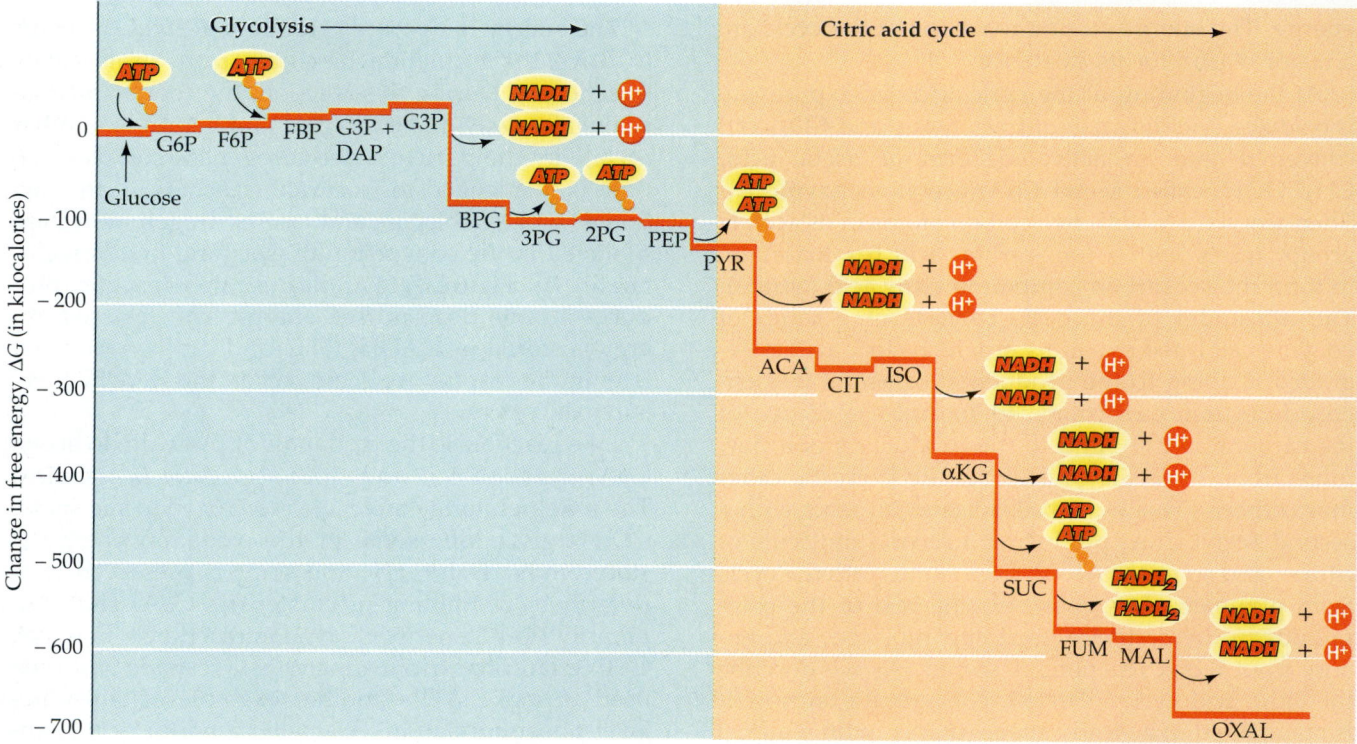

7.14 The Citric Acid Cycle Releases Much More Free Energy Than Glycolysis Does
Electron carriers (NAD in glycolysis; NAD and FAD in the citric acid cycle) are reduced and ATP is generated by reactions coupled to reactions producing major drops in free energy.

But what about all that NADH? If this reaction continued, it appears that all the cell's NAD$^+$ would become reduced, leaving none to act as an oxidizing agent. Fortunately, there is something in most cells—a specific oxidizing agent—that can reoxidize the NADH + H$^+$. This agent is a carrier called ubiquinone (Q). Q acts as follows to oxidize NADH + H$^+$:

$$NADH + H^+ \quad \diagdown \qquad \diagup \quad Q$$
$$NAD^+ \quad \diagup \qquad \diagdown \quad QH_2$$

NAD$^+$ is once again available, so glycolysis and the citric acid cycle may continue. But won't we run out of oxidized Q now? No, because there is another carrier, **cytochrome c**, a small protein that can reoxidize the QH$_2$. Does this respiratory chain have an end? Yes—cytochrome c is reoxidized by molecular oxygen, the final oxidizing agent:

$$cyt\ c\ (red) \quad \diagdown \qquad \diagup \quad \tfrac{1}{2}\,O_2$$
$$cyt\ c\ (ox) \quad \diagup \qquad \diagdown \quad H_2O$$

This is very satisfactory indeed. Oxygen gas is abundant in most places where life is found on Earth today, so there is no worry about running out of oxidizing agent. In addition, the "waste" product of the respiratory chain, water, is nontoxic and presents no disposal problem. The two hydrogens in the water, by the way, may be thought of as the hydrogens abstracted from a substrate back in the citric acid cycle or in glycolysis. They have been passed from one carrier to another and finally used to reduce molecular oxygen, reoxidizing cytochrome c and allowing the various respiratory pathways to continue.

The respiratory chain is more complicated than we have just indicated, for the chain also contains three large protein complexes through which electrons are passed (Figure 7.15). Between NADH + H$^+$ and Q lies **NADH-Q reductase**, a complex of 25 polypeptide subunits, with a total molecular weight of 850,000. **Cytochrome reductase**, with 9 subunits and a molecular weight of 250,000, lies between Q and cytochrome c. **Cytochrome oxidase**, with 8 subunits and a molecular weight of 160,000, lies between cytochrome c and oxygen. Different subunits within each of the complexes bear different electron carriers, so electrons are transported *within* each complex. All the components of the respiratory chain are proteins or are attached to proteins, except for Q, which is a smaller molecule.

Why should the respiratory chain have so many links? Why, for example, do we not just use the following single step?

$$NADH + H^+ + \tfrac{1}{2}\,O_2 \rightarrow NAD^+ + H_2O$$

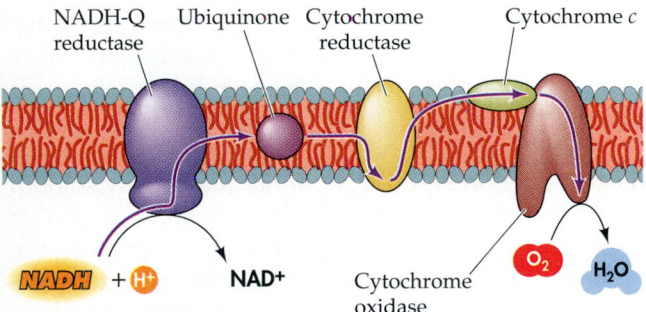

7.15 The Oxidation of NADH + H⁺
Oxidation of NADH + H⁺ by the respiratory chain produces a great deal of ATP. As traced by the purple arrows, electrons from NADH + H⁺ are passed through a series of carrier molecules in the inner mitochondrial membrane (or the plasma membrane of an aerobic prokaryote), releasing enough energy to produce ATP from ADP + P_i along the way. The carriers gain free energy and become reduced as electrons are passed to them and release free energy when they are oxidized, passing the electrons to the next carrier in the chain.

Would this not accomplish the same thing, and more efficiently? To begin with, there is no enzyme that will catalyze the direct oxidation of NADH by oxygen. More fundamentally, this would be an untamable reaction. It would be terrifically exergonic—rather like setting off a stick of dynamite in the cell. There is no biochemical way to harvest that burst of energy efficiently and put it to physiological use (no metabolic reaction is so endergonic as to consume a significant fraction of that energy in a single step).

Instead, evolution has led to the lengthy chain we observe today: a *series* of reactions, each releasing a smaller, relatively manageable amount of energy. Electron transport within each of the three protein complexes results, as we shall see, in the formation of ATP. Thus the vast energy supply originally contained in glucose and other foods is finally tucked into the cellular energy currency, ATP. For each pair of electrons passed along the respiratory chain from NADH + H⁺ to oxygen, three molecules of ATP are formed.

The carriers of the respiratory chain (including those contained in the three protein complexes) differ as to how they change upon reduction. NAD^+, for example, accepts one proton and two electrons, leaving the proton from the other hydrogen atom to float free: NADH + H⁺. Others, including Q, bind both protons and both electrons in becoming, for example, QH_2. The remainder of the chain, however, is only an electron-transport process. Electrons, but not protons, are passed from Q to cytochrome *c*. The cytochromes contain iron atoms that in their oxidized (ferric) states are Fe^{3+} and in their reduced (ferrous) states Fe^{2+}. The iron atoms are held in place by **heme groups** like that found in hemoglobin.

Electrons pour into the pool of Q molecules from the NADH + H⁺ pathway, or they can come from another source: the succinate-to-fumarate reaction of the citric acid cycle (see Figure 7.13, reaction 7). Another protein complex, **succinate-Q reductase**, links the oxidation of succinate to the reduction of Q (Figure 7.16). The enzyme that constitutes the first part

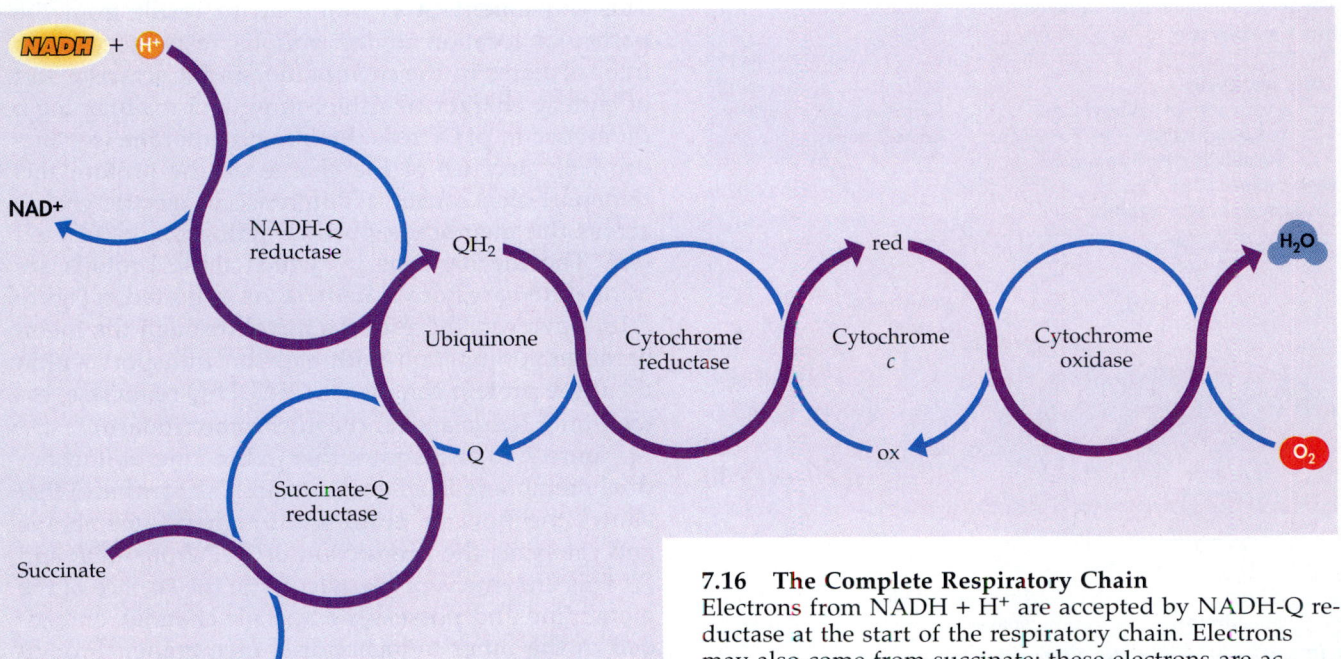

7.16 The Complete Respiratory Chain
Electrons from NADH + H⁺ are accepted by NADH-Q reductase at the start of the respiratory chain. Electrons may also come from succinate; these electrons are accepted by succinate-Q reductase, rather than by NADH-Q reductase. For every succinate oxidized by the chain, two molecules of ATP are eventually produced.

of succinate-Q reductase has attached to it an FAD carrier molecule, which is reduced by succinate to $FADH_2$. Later, hydrogen atoms are transferred to the Q molecules. No ATP is generated in the succinate-to-Q branch of the respiratory chain. Hence the pathway from the oxidation of succinate forms only two ATP molecules, compared with the three obtained when NAD^+ is the first oxidizing agent.

Oxidative Phosphorylation and Mitochondrial Structure

For many years, scientists struggled to understand how the operation of the respiratory chain caused oxidative phosphorylation—the formation of ATP in the mitochondrion. The problem was solved in 1961, when the British biochemist Peter Mitchell proposed the chemiosmotic theory. This elegant model illustrates once again the intimate relationship between structure and function in biology, so let's begin by reviewing the placement of the various components of respiratory metabolism within the cell.

The reactions of glycolysis are older than those of the citric acid cycle, having evolved before the most ancient of today's prokaryotes. It is not surprising, therefore, that the enzymes for glycolysis float free in the cytosol of the cell or are bound to the cytoskeleton. They are not enclosed within any organelle and are even found in most cells that lack organelles. By contrast, the enzymes of the citric acid cycle and the respiratory chain are isolated in the mitochondria.

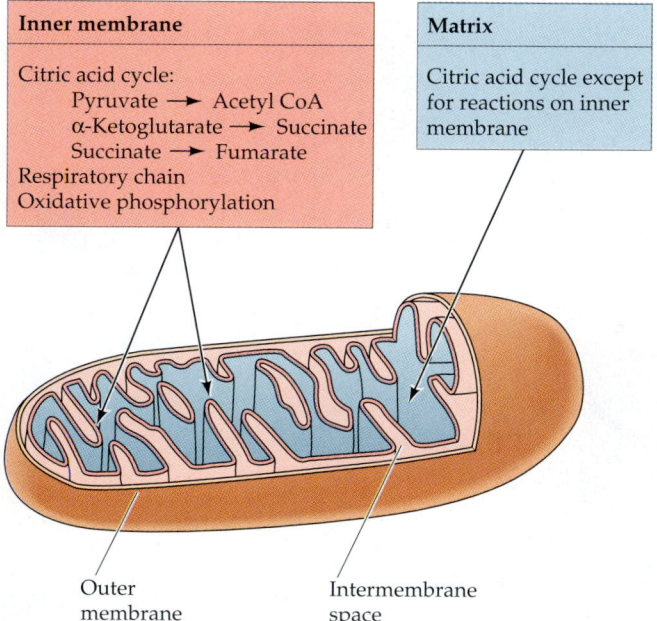

Inner membrane

Citric acid cycle:
  Pyruvate → Acetyl CoA
  α-Ketoglutarate → Succinate
  Succinate → Fumarate
Respiratory chain
Oxidative phosphorylation

Matrix

Citric acid cycle except for reactions on inner membrane

Outer membrane   Intermembrane space

7.17 Reactions in the Mitochondrion
Most of the important reactions of cellular respiration in eukaryotic cells take place in the mitochondrion's matrix or inner membrane.

(Some aerobic bacteria also carry out these reactions. Although they lack mitochondria, the bacteria have membrane systems with which these enzymes are closely associated.)

A typical mitochondrion from a mammalian cell is diagrammed in Figure 7.17. It has a relatively smooth outer membrane and an inner membrane that is folded back and forth deep into the interior of the organelle, giving the inner membrane an enormous surface area in relation to the volume that it encloses. That enclosed volume is filled with a protein-rich fluid, the mitochondrial matrix. The enzymes of the citric acid cycle are dissolved in the mitochondrial matrix, with three exceptions: succinate dehydrogenase, which catalyzes reaction 7, and the two enormous complexes that catalyze reactions 1 and 5 (see Figure 7.13). These enzymes are buried in the inner membrane (Box 7.B). The carriers and enzymes of the respiratory chain (other than cytochrome c) are also embedded in the inner mitochondrial membrane. Cytochrome c is an extrinsic protein (see Chapter 5) and lies in the space between the inner and outer mitochondrial membranes, loosely attached to the inner membrane. Ubiquinone, a small, nonprotein molecule, is free to move within the hydrophobic interior of the phospholipid bilayer of the inner membrane (Figure 7.18).

Mitchell proposed, and then showed, that operation of the respiratory chain results in the transport of hydrogen ions, against their concentration difference, through the inner membrane of the mitochondrion from inside to outside ("outside" being the space between the two mitochondrial membranes). This movement of H^+ appears to result from the particular location of the various respiratory chain intermediates in the membrane, and it acts as a sort of battery charger by establishing and maintaining a difference in pH across the inner membrane (see Figure 7.8). Because of the charge on the proton, this transport also causes a difference in electric charge across the membrane, further aiding the battery effect. The mechanisms by which these protons are transported are not yet known. As indicated in Figure 7.18, however, the protons travel through the membrane in conjunction with electron transport within the three protein complexes (NADH-Q reductase, cytochrome reductase, and cytochrome oxidase).

Figure 7.18 also shows that in the inner mitochondrial membrane is an enzyme (an ATP synthase) that allows the flow of protons through the membrane and catalyzes the production of ATP from ADP and P_i. This enzyme is perpendicular to the surface of the membrane and possesses a specific channel, embedded in the inner mitochondrial membrane, through which the excess protons on the outside of the membrane may flow back into the matrix. The ATP synthase part of this enzyme, which sticks out of the

BOX 7.B

Dissecting the Mitochondrion

We have said that certain enzymes of the citric acid cycle are in the mitochondrial matrix, whereas others are embedded in mitochondrial membranes. How was this learned? Think for a moment about how you might determine where a mitochondrial enzyme resides. Then read on, pausing after each step to see whether you can guess what comes next.

To begin with, you will need a centrifuge. As described at the end of Chapter 4, the centrifuge can be used to isolate a sample of mitochondria for study. After a relatively pure sample of mitochondria has been obtained, the matrix must be separated from the surrounding membranes. Think about osmosis (see Chapter 6). Think about lysis in a hypotonic solution (a membrane-bounded structure swells until it bursts). The next step is to transfer the mitochondria into a small volume of hypotonic solution (distilled water will do). Water will rush into the mitochondria, causing them to burst.

The result is a suspension consist-

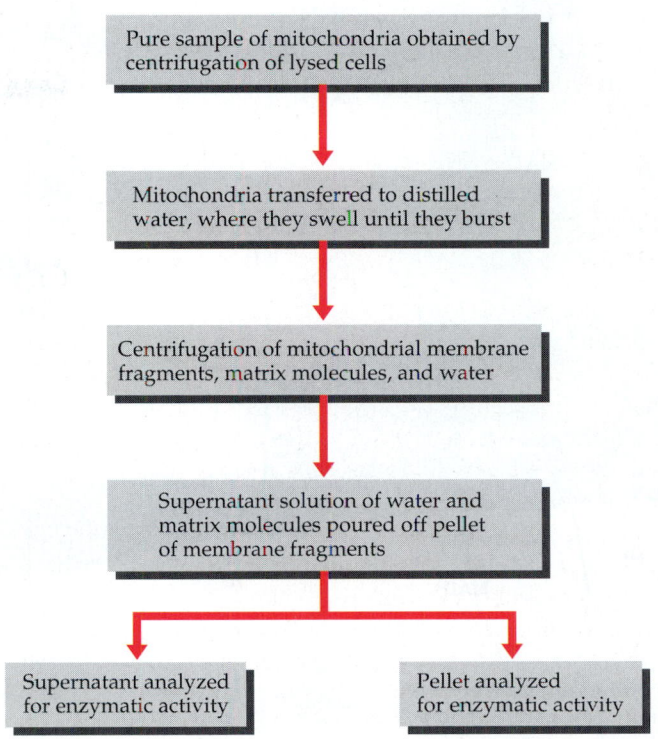

Pure sample of mitochondria obtained by centrifugation of lysed cells

Mitochondria transferred to distilled water, where they swell until they burst

Centrifugation of mitochondrial membrane fragments, matrix molecules, and water

Supernatant solution of water and matrix molecules poured off pellet of membrane fragments

Supernatant analyzed for enzymatic activity

Pellet analyzed for enzymatic activity

ing of the water, the mitochondrial matrix, and fragments of the membranes. What next? Think first. Centrifuge this suspension. The membranes sink to the bottom of the centrifuge tube, forming a pellet; the matrix remains in solution. Pour off the solution into a second tube; keep the membranous pellet in the first.

Now all that remains is to see which enzyme activities are in each

tube. If a particular enzyme (malate dehydrogenase from the citric acid cycle, for example) is found only in the tube containing the solution, we may reasonably conclude that it was originally present in the matrix. If another enzyme, such as succinate dehydrogenase, is found only in the first tube (which contains the pellet), it was most likely contained in the membrane.

inner mitochondrial membrane and is visible in electron micrographs, looks like a large knob.

The Chemiosmotic Mechanism of Mitochondria Summarized

To summarize the chemiosmotic mechanism: The flow of electrons through the respiratory chain transfers protons from the inside to the outside of the inner mitochondrial membrane, leading to an accumulation of protons on the outside. By the laws of diffusion (see Chapter 5), these excess protons tend to move spontaneously back into the matrix, which they can do only by passing through the channel-like ATP synthase molecules. This process provides the

conditions necessary for ATP production. Also, as protons diffuse away from an area of their high concentration, energy is released.

According to the chemiosmotic model, one would expect that the mitochondrion could be "fooled" into making more ATP by the following clever trick. A sample of isolated mitochondria is maintained in a solution at pH 8 (slightly basic) until it is fully adjusted; then suddenly the mitochondria are transferred into a second solution at pH 4 (fairly acidic) containing ADP and P_i. This transfer should lead to an excess of protons on the outside of the inner membrane, from where they should be able to proceed through the proposed ATP synthase channels, causing a burst of ATP production. This phenomenon

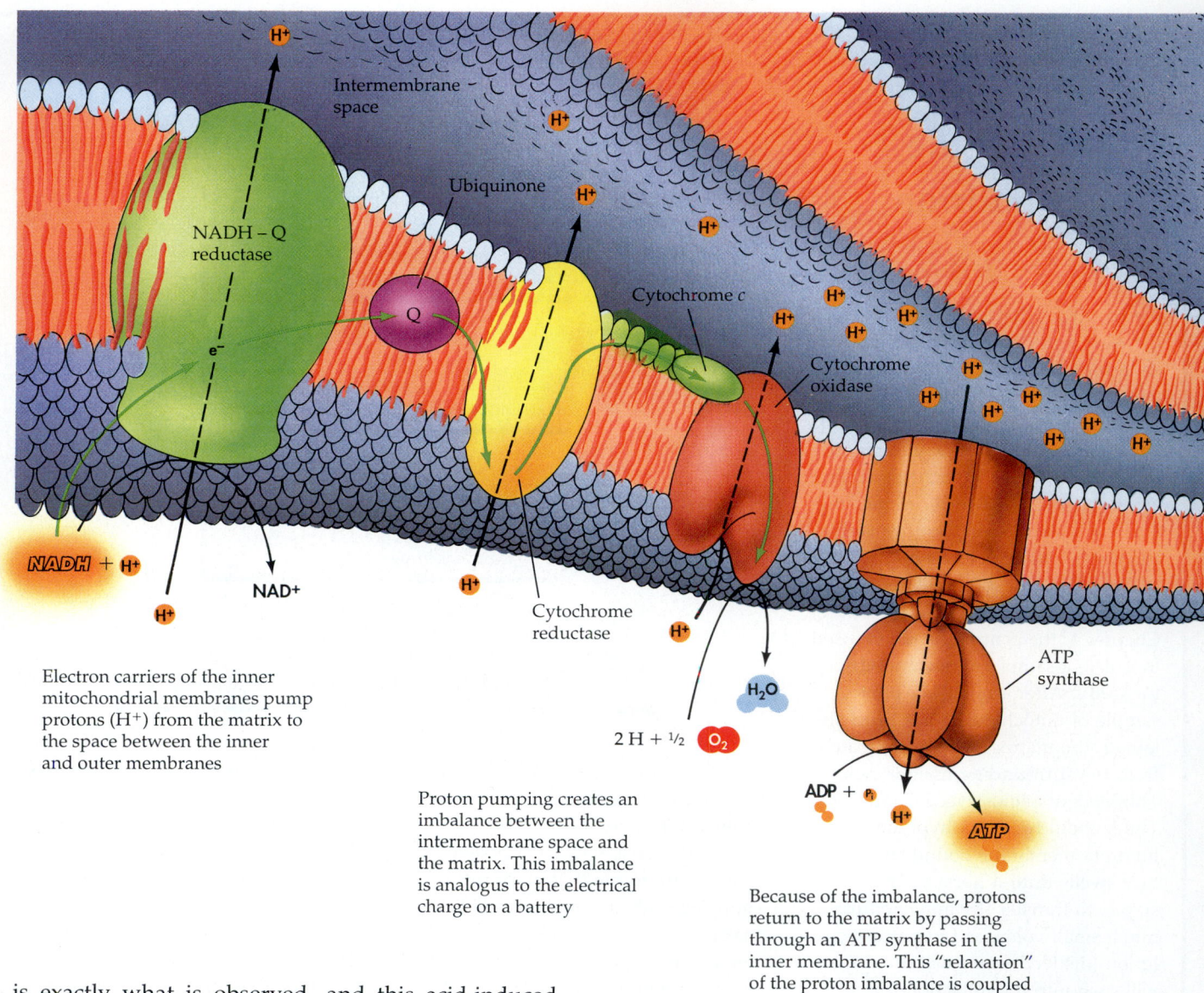

Intermembrane space

NADH – Q reductase

Ubiquinone

Cytochrome c

Cytochrome oxidase

Q

e⁻

Cytochrome reductase

$NADH + H^+$

NAD^+

ATP synthase

H_2O

$2 H + \frac{1}{2} O_2$

$ADP + P_i$

H^+

ATP

Electron carriers of the inner mitochondrial membranes pump protons (H⁺) from the matrix to the space between the inner and outer membranes

Proton pumping creates an imbalance between the intermembrane space and the matrix. This imbalance is analogus to the electrical charge on a battery

Because of the imbalance, protons return to the matrix by passing through an ATP synthase in the inner membrane. This "relaxation" of the proton imbalance is coupled with the formation of ATP in the matrix

is exactly what is observed, and this acid-induced ATP production by isolated mitochondria stands as one of the stronger pieces of evidence favoring the chemiosmotic theory as the explanation for how oxidative phosphorylation proceeds in cells.

For the chemiosmotic mechanism to work, the inner membrane of the mitochondrion must be quite impermeable to protons; otherwise, the protons would leak back across the membrane as fast as they were pumped out by the respiratory chain. Cellular respiration would race along, and no ATP would form. The mechanism works correctly only because the respiratory chain and ATP formation are strictly coupled. Many years ago, biochemists discovered compounds that, when added to mitochondria or respiring tissue, uncouple respiratory metabolism from ATP formation. These **respiratory uncouplers** carry protons back across the membrane, discharging the proton-motive force. As a result, food is metabolized, but all the released energy is lost as heat, rather than being trapped in ATP. One naturally oc-

curring respiratory uncoupler plays an important role in regulating the temperature of some mammals: In "brown fat," uncoupling of respiration raises the body temperature (see Chapter 35).

The normal coupling of respiratory metabolism and ATP formation affords an important opportunity for the regulation of cellular respiration. Protons cannot return to the mitochondrial matrix unless ADP is available for conversion to ATP. When a cell's ATP level is high, there is little ADP, and this shortage stops the flow of protons. The excessive buildup of the proton-motive force stops the respiratory chain, and thus the citric acid cycle and glycolysis. On the other hand, if a cell is working hard and thus using lots of ATP, the ADP level rises, and respiratory metabolism speeds up. This phenomenon is called respiratory control.

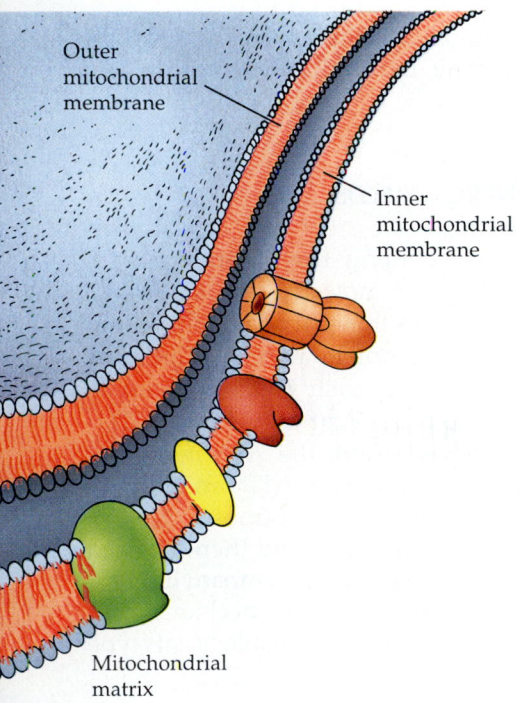

Outer mitochondrial membrane

Inner mitochondrial membrane

Mitochondrial matrix

7.18 A Chemiosmotic Mechanism Produces ATP
As electrons pass along the series of carriers in the respiratory chain, protons are pumped from the mitochondrial matrix to the intermembrane space. As the protons return to the matrix through an ATP synthase, ATP forms, as shown in Figure 7.8.

FERMENTATION

Suppose that the supply of oxygen to a respiring cell is cut off, perhaps by drowning or by extreme exertion. As we can see in Figure 7.16, the first consequence of an insufficient supply of O_2 is that the cell cannot reoxidize cytochrome c, so all of that compound is soon in the reduced form. When this happens, there is no oxidizing agent to reoxidize QH_2, and soon all the Q is in the reduced form. So it goes, until the entire respiratory chain is reduced. By this point, there remains no NAD^+ and no oxidized FAD; therefore, the oxidative steps in glycolysis and the citric acid cycle also stop. If the cell has no other way to obtain energy from its food, it will die.

If, however, the cell is one—such as a muscle cell—that has the necessary enzymes, it will switch to fermentation. This process has two defining characteristics. First, a fermentative reaction uses NADH + H^+ to reduce pyruvate or one of its metabolites, with the important consequence that NADH is oxidized, regenerating NAD^+. Once the cell has some NAD^+, it can carry more glucose through glycolysis (that is, through the early steps of fermentation). The amount

of NAD^+ obtained from the fermentative step is just enough to take a comparable amount of glucose through glycolysis, with none left over to carry the pyruvate into the citric acid cycle. Instead, this newly produced pyruvate is also fermented, producing more NAD^+ to oxidize more glucose, and so forth. This illustrates the second characteristic of fermentation: By allowing glycolysis to continue, fermentation enables a sustained production of ATP—only as much as can be obtained from substrate-level phosphorylation in glycolysis (not the much greater yield obtainable with the citric acid cycle, the respiratory chain, and oxidative phosphorylation), but enough to keep the cell going.

In fact, when cells capable of fermentation become anaerobic, the rate of glycolysis speeds up tenfold or even more. Thus a substantial rate of ATP production is maintained, although the efficiency in terms of ATP molecules per glucose molecule is greatly reduced. Some bacteria of the genus *Clostridium*, while growing anaerobically in the presence of glucose, grow and multiply as rapidly as the fastest-growing aerobic bacteria. This rapid growth is made possible by the fact that the *Clostridium* bacteria are running the glycolytic reactions much more rapidly than the aerobes do.

Figure 7.19 shows the inputs and outputs of a particular type of fermentation, called lactic acid fermentation (because its end product is lactic acid, or lactate). Lactic acid fermentation takes place in many microorganisms and in our muscle cells. Unlike muscle cells, however, nerve cells (neurons) are incapable of fermentation because they lack the enzyme that reduces pyruvate to lactate. For this reason, in the absence of oxygen our nervous system (including the brain) is rapidly destroyed and is the first part of the body to die.

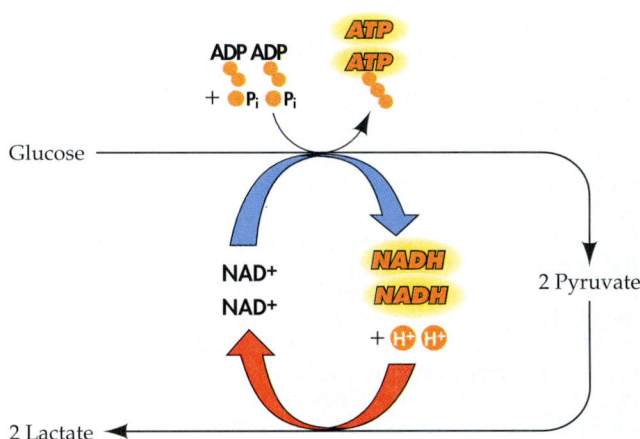

Glucose

ADP ADP
+ P_i P_i

ATP
ATP

NADH
NADH
+ H^+ H^+

NAD^+
NAD^+

2 Pyruvate

2 Lactate

7.19 Lactic Acid Fermentation
Glycolysis produces pyruvate from glucose, as well as ATP and NADH + H^+. In lactic acid fermentation, pyruvate is then reduced to lactic acid (lactate) using NADH + H^+ as the reducing agent.

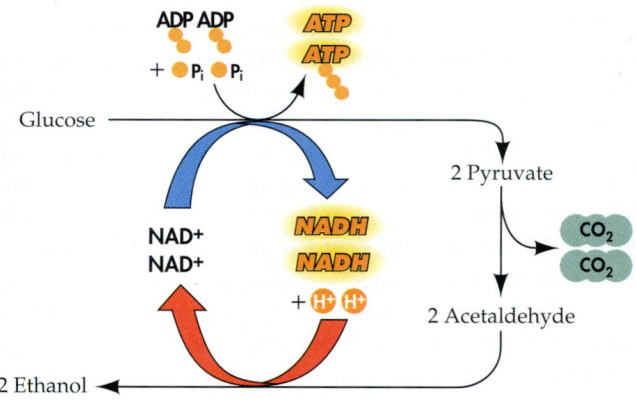

7.20 The Basis for the Brewing Industry
In alcoholic fermentation, pyruvate from glycolysis is converted to acetaldehyde, with the release of CO_2. The NADH + H^+ from glycolysis acts as a reducing agent, reducing acetaldehyde to ethanol.

Different forms of fermentation are observed in other kinds of organisms. Certain yeasts and some plant cells in anaerobic conditions carry on a process called **alcoholic fermentation** (Figure 7.20). Carbon dioxide is removed from pyruvate during alcoholic fermentation, leaving the compound acetaldehyde. This acetaldehyde is then reduced by NADH + H^+ to produce ethyl alcohol (ethanol). Remember that recycling NAD allows the fermenting cell to produce ATP by glycolysis. The brewing industry relies on alcoholic fermentation to produce wine and beer.

As noted earlier, some organisms carry on no energy metabolism other than fermentation. Some of these organisms are confined to totally anaerobic environments; others can carry on fermentation if they find themselves in the presence of oxygen. And several bacteria carry on cellular respiration—not fermentation—without using oxygen gas as an electron

7.21 Cellular Respiration Yields More Energy Than Glycolysis Does
Glycolysis yields two molecules of ATP for every glucose molecule entering the pathway. The ensuing citric acid cycle and respiratory chain produce an additional 34 ATP molecules for every glucose molecule. The source of most of these ATP molecules is the oxidation of reduced carriers (produced in glycolysis and the citric acid cycle) by the respiratory chain. We get three molecules of ATP for each NAD^+ regenerated by the respiratory chain and two molecules of ATP for each FAD. Thus the total gross yield of ATP from one molecule of glucose taken through glycolysis and respiration is 38. However, we must subtract two from that gross, for a net yield of 36 ATP. This is because the inner mitochondrial membrane is impermeable to NADH, and a "toll" of one ATP must be paid for each NADH (produced in glycolysis) that is shuttled into the mitochondrial matrix. The 36 molecules of ATP from the oxidation of glucose through glycolysis combined with cellular respiration still far exceeds the net of two molecules of ATP from fermentation.

acceptor. Instead, to oxidize their cytochromes these bacteria reduce nitrate ions (NO_3^-) to nitrite ions (NO_2^-).

COMPARATIVE ENERGY YIELDS

The total yield of stored energy from fermentation is two molecules of ATP per molecule of glucose oxidized. The maximum yield that can be obtained from glycolysis followed by complete aerobic respiration of a molecule of glucose is much greater—about 36 molecules of ATP (Figure 7.21). (Study Figures 7.11, 7.13, and 7.18 to review where the ATP molecules come from.) Why is so much more ATP produced by aerobic respiration? Because carriers (mostly NAD^+) are reduced in the citric acid cycle and then oxidized by the respiratory chain, with the accompanying production of ATP by the chemiosmotic mechanism. In an aerobic environment, a species capable of this type

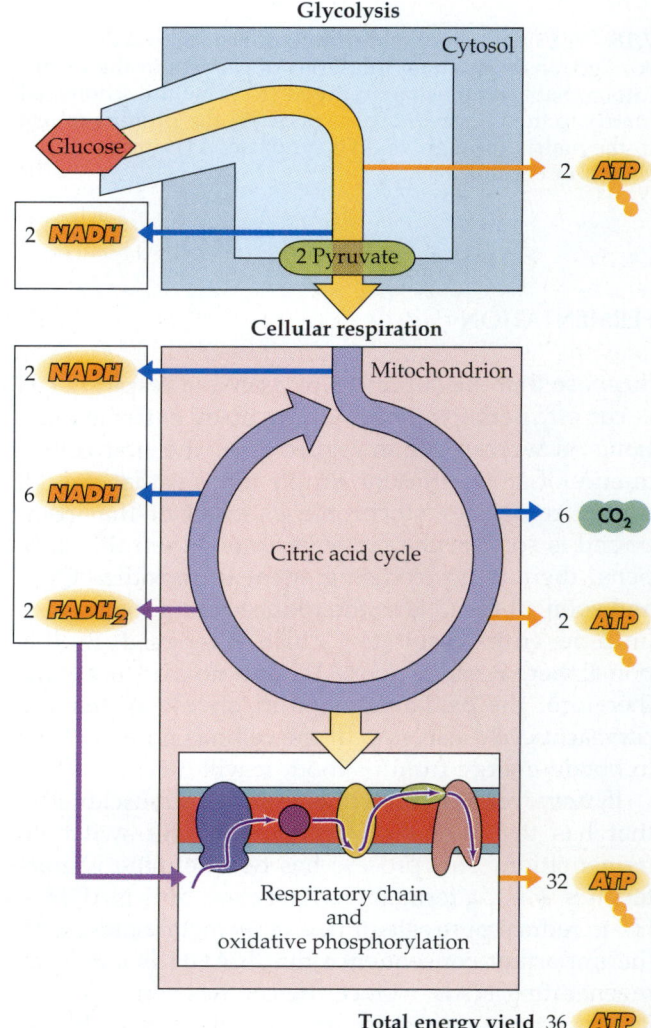

of metabolism will be at an advantage (in terms of energy availability per glucose molecule) over one that is limited to fermentation.

If glucose is simply burned, the reaction is

Glucose + 6 O_2 →
$$6 CO_2 + 6 H_2O - 686 \text{ kcal/mol}$$

with all of the 686 kcal of energy being released as heat and light. The complete biological "combustion"—respiration—of glucose could be represented by the same overall reaction, with the key difference that 36 molecules of ATP are formed for each molecule of glucose used. If each mole of ATP stores 12 kcal of energy (the actual amount varies as a function of the concentrations of ATP, ADP, and P_i in the cell), then 432 kcal (36×12 kcal) are stored to drive nonspontaneous reactions later, instead of being lost as heat.

CONNECTIONS WITH OTHER PATHWAYS

The respiratory pathways do not operate in isolation from the rest of metabolism. Rather, there is an interchange, with traffic flowing in both directions. For example, materials other than glucose can serve as the starting materials for respiratory ATP production. Other monosaccharides may be used, after being converted to glucose. Polymers such as starch and glycogen are **digested** (hydrolyzed) to glucose and subsequently metabolized to yield ATP. Fats are first digested to yield glycerol and fatty acids (see Chapter 3); the glycerol is then readily converted to glyceraldehyde 3-phosphate (an intermediate in glycolysis), and the fatty acids are broken down to form acetyl CoA (an early intermediate in the citric acid cycle).

Each of these reactions also operates in reverse. Thus, in the synthesis of fats, fatty acids form from acetyl CoA and glycerol forms from glyceraldehyde 3-phosphate. This occurs only when the cell has an adequate energy supply; otherwise the acetyl CoA would be needed strictly for the citric acid cycle and ATP formation. With an abundant supply of starting material (food, such as glucose), however, the cell can divert some acetyl CoA to fatty acid production and some glyceraldehyde 3-phosphate to glycerol formation. The fat that forms on our bodies, adding baggage, is a result of this diversion.

Some intermediates of the citric acid cycle are used in the synthesis of various important cellular constituents. Succinyl CoA (succinyl coenzyme A) is a starting point for chlorophyll synthesis, and α-ketoglutarate is a key starting material for amino acid (and hence protein) production. Other amino acids are formed from oxaloacetate. (Still other amino acids derive from pyruvate, which is not an intermediate in the cycle.) Acetyl CoA has many different fates:

In addition to its role in fatty acid production, it is a building block for various pigments, plant growth substances, rubber, and the steroid hormones of animals—among other functions.

The number of possible uses for acetyl CoA also presents a problem: If too many molecules of citric acid cycle intermediates are withdrawn from the cycle for use in other pathways, the oxaloacetate concentration could be lowered so much that there would no longer be enough to react with incoming acetyl CoA to keep the cycle going. This problem is avoided by a number of replenishing reactions, which bring in material from other parts of metabolism or which bypass some of the steps of the citric acid cycle, keeping atoms in the pathway. One such reaction bypasses the first step of the citric acid cycle (in which pyruvate is converted to acetyl CoA with the loss of a carbon atom as CO_2.) In the alternative reaction, pyruvate combines with a CO_2 molecule, forming oxaloacetate, the four-carbon substance at the end of the cycle. Thus the pool of citric acid cycle intermediates is increased, making up for materials that are lost from other parts of the cycle. We will refer to this combination again in the next section, which discusses the regulation of respiratory metabolism.

FEEDBACK REGULATION

Whereas fermentation produces only two molecules of ATP for every glucose molecule, passing the pyruvate from glycolysis to the citric acid cycle and respiratory chain yields 36 ATPs per glucose molecule. Thus an aerobically respiring organism obtains 18 times more energy per mole of glucose oxidized than one that is respiring anaerobically. In other words, when a yeast cell switches from aerobic respiration to anaerobic fermentation at low concentrations of oxygen, it must metabolize glucose 18 times faster to obtain the same amount of energy. But as soon as aerobic respiration begins again, glycolysis in the yeast cell slows down. The amount of glucose used is only as much as is needed for energy production under the existing conditions, anaerobic or aerobic. This phenomenon is called the **Pasteur effect**, after its discoverer, Louis Pasteur. What is the mechanism that slows down glycolysis when the respiratory chain begins to operate?

The mechanism by which glycolysis, the citric acid cycle, and the respiratory chain are regulated is allosteric control of the enzymes (see Chapter 6). Some products of later reactions, if they are in oversupply, can suppress the action of enzymes that catalyze earlier reactions. On the other hand, an excess of the products of one branch of a synthetic chain can speed up reactions in another branch, diverting raw materials away from their own synthesis (Figure 7.22).

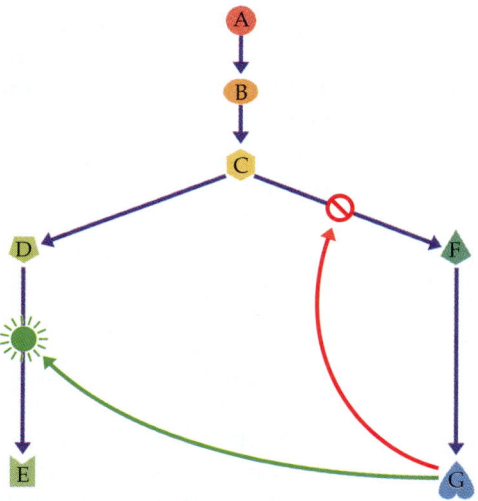

7.22   Allosteric Regulation
Compound G inhibits the enzyme for the conversion of C to F, blocking that reaction and ultimately its own synthesis, demonstrating negative feedback by allosteric regulation. Compound G also provides positive feedback to the enzyme catalyzing the step from D to E, changing that enzyme to a form that will catalyze the reaction.

These negative and positive feedback control mechanisms are used at many points in the energy-extracting processes and are summarized in Figure 7.23.

The main control point in glycolysis is the conversion of fructose 6-phosphate to fructose 1,6-bisphosphate by the enzyme phosphofructokinase. This enzyme is allosterically inhibited by ATP and activated by ADP or AMP. The enzyme is also inhibited by citrate (see Figure 7.23), for reasons that will become clear shortly. As long as fermentation proceeds, yielding a relatively small amount of ATP, phosphofructokinase operates at full efficiency. But when aerobic respiration begins producing ATP 18 times faster than before, the excess ATP allosterically inhibits the conversion of fructose 6-phosphate, and the rate of glucose utilization drops.

Pyruvate occupies a key position in the network diagrammed in Figure 7.23. In fermentation it is reduced to lactate, which can be either returned as pyruvate or used to resynthesize glucose for storage. Under aerobic conditions, pyruvate is converted to acetyl CoA, which enters the citric acid cycle by combining with oxaloacetate. Finally, as we noted in the last section, pyruvate can be used to produce more oxaloacetate by reaction with CO_2. The pathway pyruvate takes depends on conditions and needs in the cell. In the pyruvate-to-lactate conversion of fermentation, pyruvate is reduced by the NADH + H[+] produced in glycolysis. However, the affinity of the respiratory chain for NADH is much greater than that of the enzyme that forms lactate. (Recall that in the respiratory chain NADH-Q reductase oxidizes

NADH + H[+]; see Figure 7.15.) Thus if the respiratory chain is operating, it steals all the available NADH, turning fermentation off.

At a second control point for pyruvate reactions, acetyl CoA regulates oxaloacetate production. If enough oxaloacetate is present to keep the citric acid cycle going as fast as acetyl CoA is produced, the concentration of acetyl CoA remains low. If too little oxaloacetate is available, acetyl CoA builds up, activating the enzyme that produces oxaloacetate and restoring the level of oxaloacetate needed for the operation of the citric acid cycle. Acetyl CoA is an **allosteric activator** for the reaction.

Thus the concentration of acetyl CoA determines the balance point between two competing reactions: one that uses oxaloacetate in turning the citric acid cycle and another that makes oxaloacetate if it is in short supply. ADP is an opposing **allosteric inhibitor** of this same oxaloacetate-producing enzyme. If the cell is low in ATP, it is high in ADP; by inhibiting the oxaloacetate-producing enzyme, ADP directs more pyruvate to become citrate, causing the citric acid cycle and respiratory chain to operate more rapidly and form more ATP.

The main control point for the citric acid cycle is the conversion of isocitrate to α-ketoglutarate. ATP and NADH are feedback inhibitors of this reaction; ADP and NAD[+] are activators. If too much ATP is accumulating, or if NADH + H[+] is being produced faster than it can be used by the respiratory chain, the isocitrate reaction is almost completely blocked and the citric acid cycle is essentially shut down. This would lead to a pileup of large amounts of isocitrate and citrate, except that the conversion of acetyl CoA to citrate is also slowed by ATP and NADH + H[+]. The negative effects of halting the isocitrate reaction are thus spread backward through the chain of reactions. A certain excess of citrate does accumulate, however, and this excess acts as a negative feedback inhibitor to slow the fructose 6-phosphate reaction early in glycolysis. Consequently, if the citric acid cycle has been slowed down because of an excess of ATP (and not because of a lack of oxygen), glycolysis is shut down as well. Both processes resume when the ATP level falls and they are needed. Allosteric control keeps the process in balance.

Another control point in Figure 7.23 involves a method for storing excess acetyl CoA by using it to synthesize fatty acids. Excess citrate is an allosteric activator for one of the enzymes in the pathway for making fatty acids. If too much ATP is being made and the citric acid cycle is shut down, the accumulation of citrate switches acetyl CoA to the synthesis of fatty acids for storage. These may be metabolized later to produce more acetyl CoA.

Allosteric control of this sort is one of the most impressive examples of the tight organization that

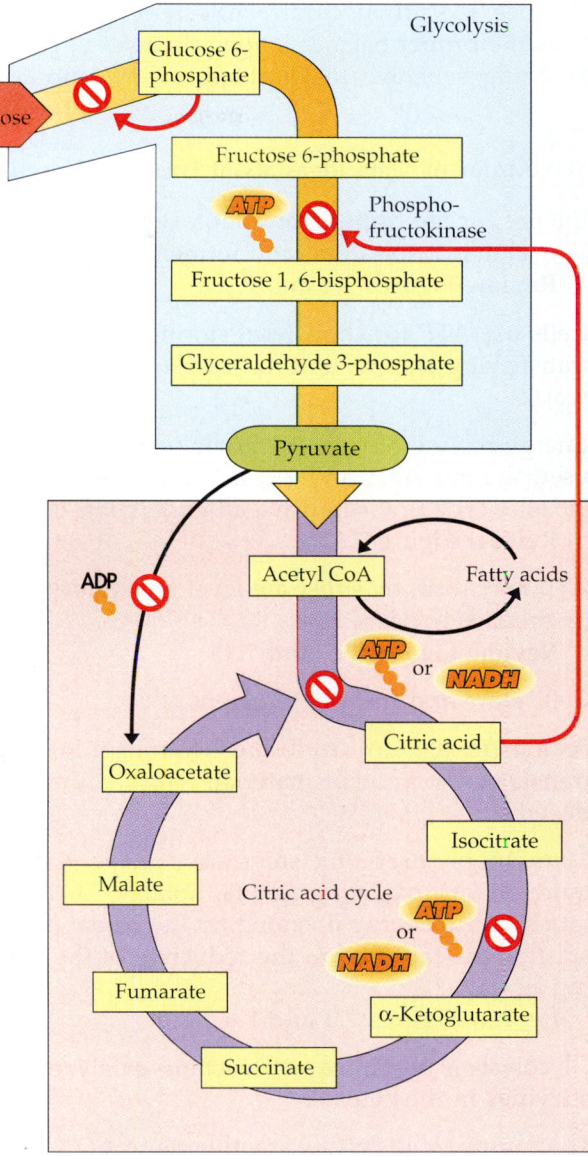

7.23 Feedback Regulation of Glycolysis and the Citric Acid Cycle
Positive and negative allosteric regulation control glycolysis and the citric acid cycle. Because it provides both negative feedback on the enzyme catalyzing the third step in glycolysis and positive feedback for the synthesis of fatty acids from acetyl CoA, citric acid is much like compound G in Figure 7.22. Note that feedback controls glycolysis and the citric acid cycle at crucial early steps in the pathways, increasing the efficiency of the pathways and preventing the excessive buildup of intermediates. Note also that the compounds inhibiting or activating enzymes are often the energy-carrying compounds themselves—ATP, ADP, NAD$^+$, NADH, and so forth. If too much ATP accumulates, for example, it inhibits a key reaction and thus slows down ATP production. If much ATP has been consumed, resulting in the formation of ADP and P_i, the ADP activates enzymes in the pathway to stimulate the production of more ATP.

can evolve through natural selection, when selection pressure favors efficient operation in the competition among organisms for limited resources. Each of the feedback controls regulates a part or various parts of the energy-releasing pathways and keeps them operating in harmony and balance. It is unnecessary (and therefore inefficient and disadvantageous) to run the glycolytic mechanism too fast if it is supplemented by the more energy-efficient processes of the citric acid cycle and respiratory chain. In terms of energy production, it is wasteful to produce more acetyl CoA if there is insufficient oxaloacetate to handle it in the citric acid cycle. It is also senseless to shunt too much pyruvate into making oxaloacetate and to neglect production of the fuel acetyl CoA because, in the step that keeps the citric acid cycle turning, a two-carbon molecule of acetyl CoA must react with a four-carbon molecule of oxaloacetate to produce six-carbon citrate. Allosteric control maintains the proper balance among the uses of pyruvate by being sensitive to shortages or oversupplies of

acetyl CoA. All the other allosteric feedback controls help make the system more efficient and hence contribute to the success of the individual that carries them.

SUMMARY of Main Ideas about Energy Release in Cells

Energy for life comes from photosynthesis, glycolysis, cellular respiration, and fermentation.
    Review Figures 7.1 and 7.9

Cells use ATP for short-term storing of energy, for transferring energy, and for releasing energy to do work.

Energy released by an exergonic reaction may be used to form ATP, and the splitting of ATP provides energy that can drive an endergonic reaction.
    Review Figure 7.3

ATP may form by substrate-level phosphorylation or chemiosmotically.
    Review Figures 7.7 and 7.18

Cells store and transfer electrons.

As a material is oxidized, the electrons it loses are transferred to another material, which is thereby reduced.

Many highly exergonic steps are oxidations that require specific oxidizing agents, notably, NAD^+; much of the energy liberated by the oxidation of a substrate is captured in the reduction of the oxidizing agent.
    Review Figures 7.4 and 7.6

Glycolysis is a pathway of reactions catalyzed by enzymes in the cytosol.

The inputs to glycolysis are glucose, NAD^+, and $ADP + P_i$. Its outputs are pyruvate, $NADH + H^+$, and ATP.
    Review Figure 7.11

Glycolysis provides starting materials for cellular respiration and fermentation; it also releases energy.
    Review Figures 7.9 and 7.12

Cellular respiration proceeds through the citric acid cycle and the respiratory chain in the mitochondria of eukaryotes and in membrane systems in certain bacteria.
    Review figure 7.17

The inputs to the citric acid cycle are pyruvate, NAD^+, FAD, and $ADP + P_i$. Its outputs are CO_2, $NADH + H^+$, $FADH_2$, and ATP.
    Review Figure 7.13

Reduced electron carriers from glycolysis and the citric acid cycle are reoxidized by the respiratory chain.

The inputs to the respiratory chain are oxygen, $NADH + H^+$ (or $FADH_2$), and $ADP + P_i$. Its outputs are ATP, NAD^+ (or FAD), and water.
    Review Figure 7.18

Many organisms derive their energy supply from fermentation, which proceeds in the absence of oxygen gas.

Fermentation oxidizes the $NADH + H^+$ produced in glycolysis, allowing glycolysis to continue.
    Review Figures 7.19 and 7.20

For each molecule of glucose used, fermentation yields 2 molecules of ATP. In contrast, glycolysis combined with the citric acid cycle and the respiratory chain yields up to 36 molecules of ATP per molecule of glucose.
    Review Figures 7.19, 7.20, and 7.21

Allosteric feedback controls regulate the web of reactions that constitute energy metabolism.
    Review Figure 7.23

SELF-QUIZ

1. Which statement about ATP is *not* true?
    a. It is formed only under aerobic conditions.
    b. It is used as an energy currency by all cells.
    c. Its formation from ADP and phosphate is an endergonic reaction.
    d. It provides the energy for many different biochemical reactions.
    e. Some ATP is used to drive the synthesis of storage compounds.

2. Oxidation and reduction
    a. entail the gain or loss of proteins.
    b. are defined as the loss of electrons.
    c. are both endergonic reactions.
    d. always occur together.
    e. proceed only under aerobic conditions.

3. NAD$^+$
   a. is a type of organelle.
   b. is a protein.
   c. is an oxidizing agent.
   d. is a reducing agent.
   e. is formed only under aerobic conditions.

4. Glycolysis
   a. takes place in the mitochondrion.
   b. produces no ATP.
   c. has no connection with the respiratory chain.
   d. is the same thing as fermentation.
   e. reduces two molecules of NAD for every glucose molecule processed.

5. Fermentation
   a. takes place in the mitochondrion.
   b. takes place in all animal cells.
   c. does not require O_2.
   d. requires lactic acid.
   e. prevents glycolysis from taking place.

6. Which statement is *not* true of pyruvate?
   a. It is the end product of glycolysis.
   b. It gets reduced during fermentation.
   c. It feeds into the citric acid cycle.
   d. It is a protein.
   e. It contains three carbon atoms.

7. The citric acid cycle
   a. takes place in the mitochondrion.
   b. produces no ATP.
   c. has no connection with the respiratory chain.
   d. is the same as fermentation.
   e. reduces two molecules of NAD for every glucose molecule processed.

8. Which statement is *not* true of the respiratory chain?
   a. It takes place in the mitochondrion.
   b. It uses O_2 as an oxidizing agent.
   c. It leads to the production of ATP.
   d. It regenerates oxidizing agents for glycolysis and the citric acid cycle.
   e. It operates simultaneously with fermentation.

9. Which statement is *not* true of the chemiosmotic mechanism?
   a. Protons are pumped across a membrane.
   b. Protons return through the membrane by way of a channel protein.
   c. ATP is required for the protons to return.
   d. Proton pumping is associated with the respiratory chain.
   e. The membrane in question is the inner mitochondrial membrane.

10. Which statement is *not* true of oxidative phosphorylation?
    a. It is the formation of ATP during the operation of the respiratory chain.
    b. It is brought about by the chemiosmotic mechanism.
    c. It requires aerobic conditions.
    d. In eukaryotes, it takes place in mitochondria.
    e. Its functions can be served equally well by fermentation.

FOR STUDY

1. Trace the sequence of chemical changes that occurs in mammalian brain tissue when the oxygen supply is cut off. (The first change is that the cytochrome oxidase system becomes totally reduced, because electrons can still flow from cytochrome c but there is no oxygen to accept electrons from cytochrome oxidase. What are the remaining steps?)

2. Trace the sequence of chemical changes that occurs in mammalian muscle tissue when the oxygen supply is cut off. (The first change is exactly the same as that in Study Question 1.)

3. Some cells that use the citric acid cycle and the respiratory chain can also thrive by using fermentation under anaerobic conditions. Given the lower yield of ATP (per molecule of glucose) in fermentation, why can these cells function so efficiently under anaerobic conditions?

4. Describe the mechanisms by which the rates of glycolysis and of aerobic respiration are kept in balance with one another.

READINGS

Alberts, B., D. Bray, J. Lewis, M. Raff, K. Roberts and J. D. Watson. 1994. *Molecular Biology of the Cell*, 3rd Edition. Garland Publishing, New York. Chapter 7 develops the themes introduced in this chapter; Chapter 3 is also useful as an introduction.

Hinkle, P. C. and R. E. McCarty. 1978. "How Cells Make ATP." *Scientific American*, March. Discussion of the chemiosmotic mechanism, in which ATP is formed by protons passing back through a membrane after being pumped out by the respiratory chain.

Lodish, H., D. Baltimore, A. Berk, L. Zipursky, P. Matsudaira and J. Darnell. 1995. *Molecular Cell Biology*, 3rd Edition. Scientific American Books, New York. Chapter 17 gives an excellent more detailed treatment of the topics in this chapter.

Stryer, L. 1995. *Biochemistry*, 4th Edition. W. H. Freeman, New York. Although more advanced than this chapter, the section on glycolysis and respiration is straightforward and does not demand an advanced knowledge of chemistry.

Voet, D. and J. G. Voet. 1990. *Biochemistry*. John Wiley & Sons, New York. A general textbook with a full discussion of energy, enzymes, and catalysis. Outstanding illustrations.

Sunlight: The Source of Our Energy

8
Photosynthesis

We are the creatures of the sun. Its light is the source—direct or indirect—of the free energy that powers life on Earth. This is the important message about photosynthesis, which you may already know. As biologists, however, we might express a more refined appreciation of how we are able to use the sun's energy.

As biologists we can say, "We are creatures of the chloroplasts." This abundant organelle, which gives green plants their color, captures the sun's energy for us. Inside plants, within stacks of connected, inner membranes wrapped in two outer membranes, chemical reactions convert and store the energy that we draw upon. What are the reactants and what are the products? What sorts of molecules embedded in the inner membranes or floating in their vicinity participate in the reactions? What, exactly, happens inside this organelle that feeds the world? These are the central topics of this chapter, but before taking them up, let's consider photosynthesis in a more general way.

SUNLIGHT AND LIFE ON EARTH

Our own dependence on the sun, although absolute, is indirect. Like the other animals, the fungi, many protists, and most monerans, we depend upon a ready supply of partially reduced, carbon-containing compounds as a food source. From such compounds we get all the free energy that keeps us alive and functioning. From them, too, we obtain the carbon atoms used in every organic molecule in our bodies. In a word, we are **heterotrophs**: We need to feed upon something else. In a world populated exclusively by heterotrophs, all life would grind to an end as the food gradually disappeared.

Our world owes the continued existence of life to the presence of **autotrophs**—organisms that do not need previously formed organic substances from their environment. For autotrophs, an energy source (such as light) and an inorganic carbon source (such as carbon dioxide gas) suffice as a diet. From these simple ingredients, autotrophs make the reduced carbon compounds from which their bodies are built and their food needs met. By feeding on autotrophs, the heterotrophs of the world meet their needs for energy and matter. The principal autotrophs are photosynthetic organisms that use visible light as their energy source. From light, carbon dioxide, and water, they begin the chemistry that sustains almost the entire biosphere.

Photosynthesis is the conversion of light energy to chemical energy by living things. Organisms that conduct photosynthesis—plants and photosynthetic protists and monerans—stand at the gateway to the

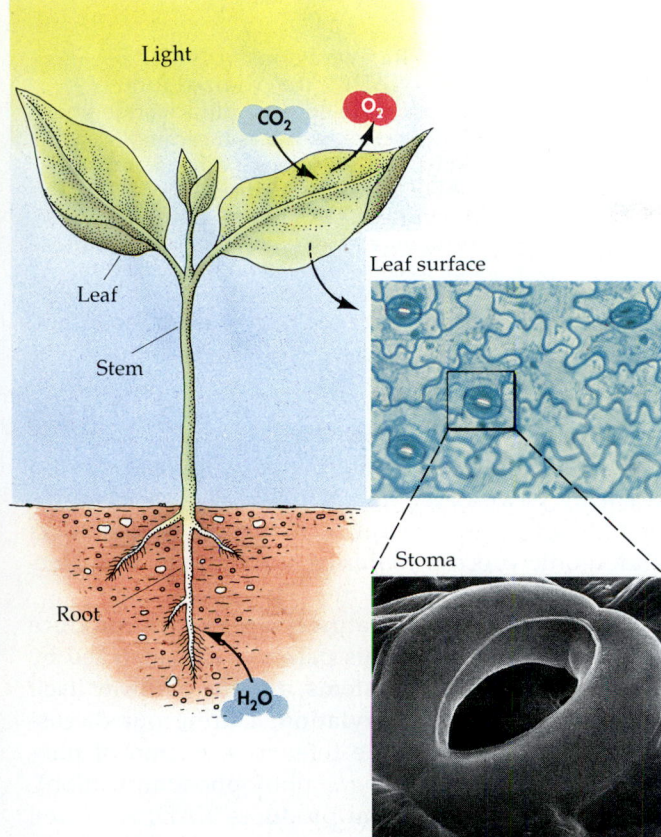

8.1 Ingredients for Photosynthesis
A typical terrestrial plant uses light from the sun, water from the soil, and carbon dioxide from the atmosphere to form organic compounds by photosynthesis. Carbon dioxide enters the leaves through openings called stomata. The top micrograph shows several stomata, magnified about 100 times; the stoma on the far right was closed when the photograph was taken. Below is a scanning electron micrograph of a single open stoma.

living world, at the interface where inorganic becomes organic, where nonlife becomes life. The worldwide extent of photosynthetic activity is stunning: Each year, tens of billions of tons of carbon atoms are taken from carbon dioxide and incorporated into molecules of sugars, amino acids, and other compounds.

EARLY STUDIES OF PHOTOSYNTHESIS

By the beginning of the nineteenth century, scientists understood the broad outlines of photosynthesis. Photosynthesis was known to use three principal ingredients—water, carbon dioxide (CO_2), and light—and to produce not only food but also oxygen gas (O_2). Scientists had learned that the water for photosynthesis comes primarily from the soil (for plants living on land) and must travel from the roots to the leaves; that carbon dioxide is taken in from the atmosphere through tiny openings, called **stomata**

(singular, *stoma*), in the leaves (Figure 8.1); and that light is absolutely necessary for the production of oxygen and food. The last of the important early discoveries, made during the first decade of the nineteenth century, was that carbon dioxide uptake and oxygen release are closely related and that both depend on light action. By 1804, scientists could summarize photosynthesis in plants by writing

$$CO_2 + H_2O + \text{light energy} \rightarrow \text{sugar} + O_2$$

but many details of the process remained hidden.

It was almost a century and a half before it was determined whether the oxygen released during photosynthesis comes from the carbon dioxide or from the water. The direct demonstration depended on one of the first uses of an isotopic tracer (see Chapter 2) in biological research. In the experiments, two groups of green plants were allowed to carry on photosynthesis. Plants in the first group were supplied with water containing the heavy-oxygen isotope ^{18}O and with CO_2 containing only the common isotope ^{16}O; plants in the second group were supplied with CO_2 labeled with ^{18}O and water containing only the common isotope. Oxygen gas was collected from each group of plants, and it was found that O_2 containing ^{18}O was produced in abundance by the plants given ^{18}O-labeled water but not by plants given labeled CO_2. From these results, scientists concluded that the oxygen gas produced during photosynthesis comes from water (Figure 8.2).

With the information about where the O_2 comes from in hand, and taking into account the number of

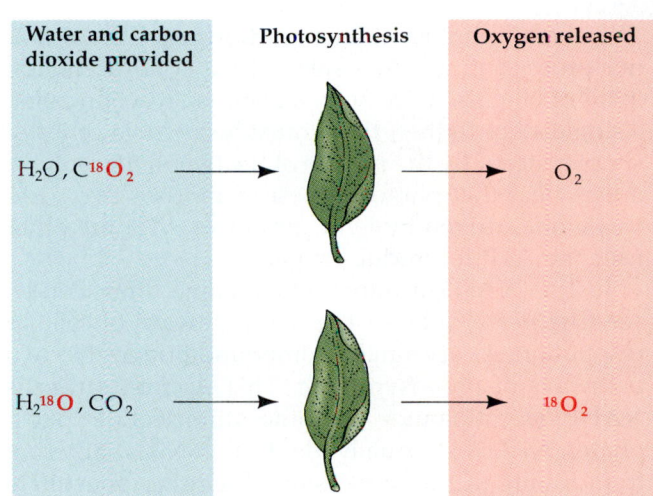

8.2 Water Produces the Oxygen Liberated during Photosynthesis
Experimenters gave some plants isotope-labeled carbon dioxide, $C^{18}O_2$, and unlabeled water (top); they gave other plants isotope-labeled water, $H_2^{18}O$ and unlabeled CO_2 (bottom). Because only plants in the bottom group released isotope-labeled oxygen gas, $^{18}O_2$, we know that water is the source of the oxygen.

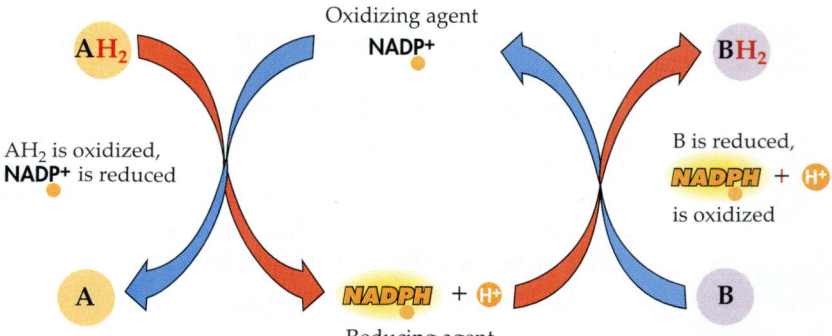

8.3 The Two Forms of NADP
When NADP$^+$, the oxidized form of nicotinamide adenine dinucleotide phosphate, reacts with a reduced molecule (AH$_2$), AH$_2$ becomes oxidized and NADP$^+$ becomes reduced to NADPH + H$^+$. In turn, NADPH + H$^+$, the reduced form, may react with an oxidized molecule such as B, becoming oxidized to NADP$^+$ and reducing B to BH$_2$. The red arrows trace the path of electrons in the reactions.

CO_2 molecules needed to form a simple sugar such as glucose, we may now rewrite the overall equation for photosynthesis as

$$6\ CO_2 + 12\ H_2O \rightarrow C_6H_{12}O_6 + 6\ O_2 + 6\ H_2O$$

Water appears on both sides of the equation because water is used as a reactant (the 12 molecules on the left) and released as a product (the 6 new ones on the right). Note that this equation is essentially the reverse of the overall equation for cellular respiration, given in Chapter 7.

Although the O_2 is in one sense a waste product, it is vital to all oxygen-requiring organisms. The photosynthetic production of oxygen by green plants is an important source of atmospheric oxygen, which most organisms—including the green plants themselves—require in order to complete their respiratory chains and thus obtain the energy to live.

THE PATHWAYS OF PHOTOSYNTHESIS

The overall photosynthetic reaction just shown cannot proceed in a single step. There is no precedent in all of chemistry for such a complex reaction being a single step. Rather, there must be a whole series of simpler steps. By the middle of the twentieth century, it was clear that photosynthesis comprises two pathways: one, driven by light, produces ATP; the other uses the ATP to produce sugar.

Just as NAD (nicotinamide adenine dinucleotide; see Chapter 7) bridges the two pathways of cellular respiration, a very similar compound bridges the two pathways of photosynthesis. This electron carrier is **NADP** (nicotinamide adenine dinucleotide phosphate). NADP is virtually identical to NAD, differing from it only in the possession of another phosphate group attached to the ribose portion of the molecule. Whereas NAD participates in metabolic breakdown reactions and energy transfers, NADP participates in synthetic reactions that require energy and reducing power. Like NAD, NADP exists in two forms. One (NADP$^+$) is an oxidizing agent, whereas the other (NADPH + H$^+$) is a reducing agent (Figure 8.3).

NADPH + H$^+$ is an intermediary for energy and reducing power. ATP and NADPH + H$^+$ are carriers of reducing power because reduction is always an endergonic process requiring both energy and electrons.

One photosynthetic pathway uses light to produce ATP. The reactions of this pathway are catalyzed by enzymes called **photosystems**, and the pathway itself is called **photophosphorylation**. During our discussion of this pathway, we refer to a version of photophosphorylation (*noncyclic* photophosphorylation), carried out by plants, that produces NADPH as well as ATP. Before discussing the second pathway of photosynthesis (in which ATP is used to produce sugar) in detail, we will briefly describe another version of photophosphorylation (*cyclic* photophosphorylation), which produces only ATP.

The NADPH + H$^+$ and ATP produced in the first pathway of photosynthesis are used in the second pathway, where reactions trap CO_2 and reduce the resulting acid to sugar. These sugar-producing reactions constitute the Calvin–Benson cycle (Figure 8.4), also known as the photosynthetic carbon reduction cycle, or simply the dark reactions (because none of them uses light, as does photophosphorylation). The reactions of both pathways proceed within the chloroplast, but, as we will see, they reside in different parts of the organelle. *Both* pathways stop in the dark because ATP synthesis and NADP$^+$ reduction require light. The rate of each set of reactions is dependent upon that of the other. They are tied together by the exchange of ATP and ADP and of NADP$^+$ and NADPH.

LIGHT AND PIGMENTS

The living world makes marvelous use of light. In photosynthesis, light is a source of *energy*; in most other light-related phenomena, it is involved in the transmission of *information*. Many of these phenomena will be described in later chapters. In them, we will find that light can be modulated in many ways to carry information: Its *brightness* may be varied, as creasing wavelength and of decreasing energy per photon. Visible light fits into this **electromagnetic spectrum** between ultraviolet and infrared radiation (Figure 8.6). Although considerable attention has been devoted to the apparent paradox of light being simultaneously a wave phenomenon and a particle phenomenon, this is nothing to be concerned about in our study of photosynthesis.

The speed of light in a vacuum is one of the universal constants of nature. In a vacuum light travels at 3×10^{10} centimeters per second (186,000 miles per second), a value symbolized as c. In air, glass, water, and other media, light travels slightly more slowly. Let's consider light as a long train of waves moving in a straight line and see what the train would look like to a stationary observer. Successive peaks of the waves pass the observer with a uniform **frequency** (ν) determined by the wavelength and the speed of light. The exact relationship is $\nu = c/\lambda$, where ν (the Greek letter *nu*) is the frequency; c is the speed of light; and λ (Greek *lambda*) is the wavelength. Often ν is expressed in hertz (Hz), c in centimeters per second (cm/sec), and λ in nanometers (nm). (One nanometer equals 10^{-9} meter or 10^{-7} centimeter; see the conversion table inside the back cover.)

Humans perceive light as having distinct colors (the reason for this will be explained in Chapter 39). The colors relate to the wavelengths of the light, as shown in Figure 8.6. Most of us can see electromagnetic radiation with wavelengths from 400 to 700 nm. At 400 nm we are at the blue end of the visible spectrum, whereas 700 nm is the red end. Wavelengths in the range from about 100 to 400 nm are ultraviolet radiation; those immediately above 700 are referred to as infrared.

The amount of energy, E, contained in a single photon is directly proportional to its frequency. The constant of proportionality that describes this relationship, h, is named Planck's constant after Max Planck, who first introduced the concept of the photon. With this information we can write the equation

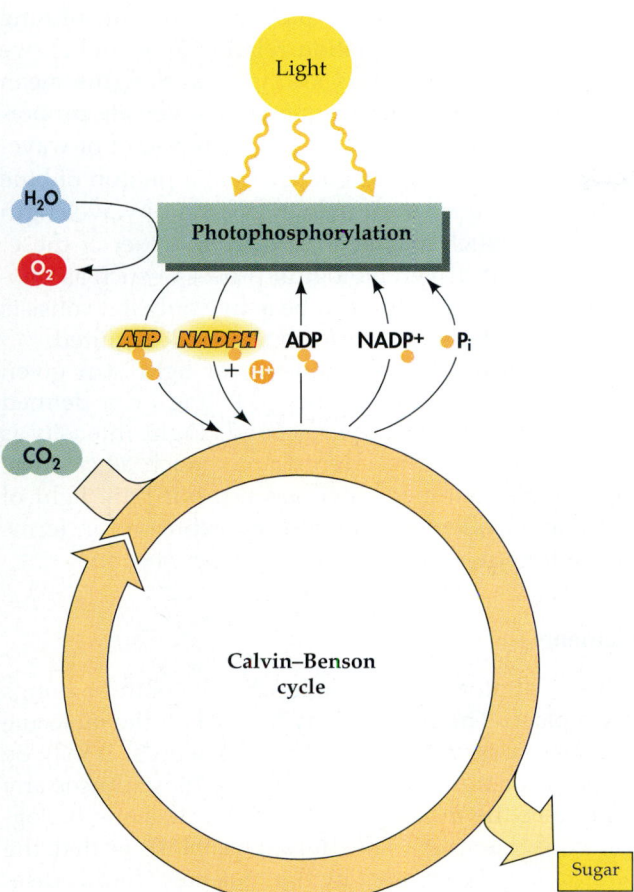

8.4 An Overview of Photosynthesis
Light energy and water are used in the first pathway, photophosphorylation, to produce ATP, NADPH + H⁺, and (in many organisms), O₂. Carbon dioxide and the ATP plus NADPH + H⁺ produced during photophosphorylation are used in the Calvin–Benson cycle (the second pathway) to produce sugars and other food molecules.

may its *color*, and it may be presented for various *durations*—continuously or in short, long, or variable periods. Some of the material to be covered will be more meaningful if we first learn to deal in a quantitative way with the brightness, color, and energy content of light.

Basic Physics of Light

Light is a form of radiant energy. It comes in discrete packets called **photons**. Light also behaves as if it were propagated in waves. The **wavelength** is the distance from the peak of one wave to the peak of the next (Figure 8.5). Different colors result from different wavelengths. Light and other forms of radiant energy—cosmic rays, gamma rays, X rays, ultraviolet radiation, infrared radiation, microwaves, and radio waves—are **electromagnetic radiation**. We have listed these forms of radiation here in order of in-

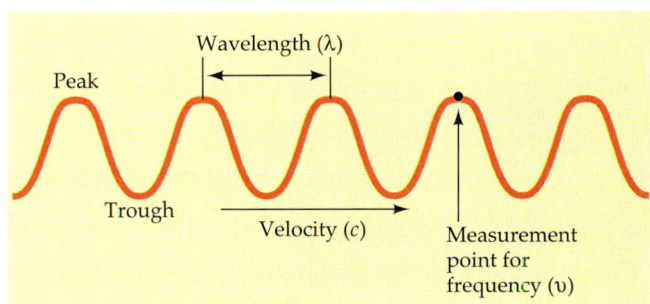

8.5 Light Has Wavelike Properties
The wavelength, λ, is the distance between the peaks of successive waves; and the frequency, ν, is the number of peaks passing an observation point in a second. The velocity with which the train of waves moves is c.

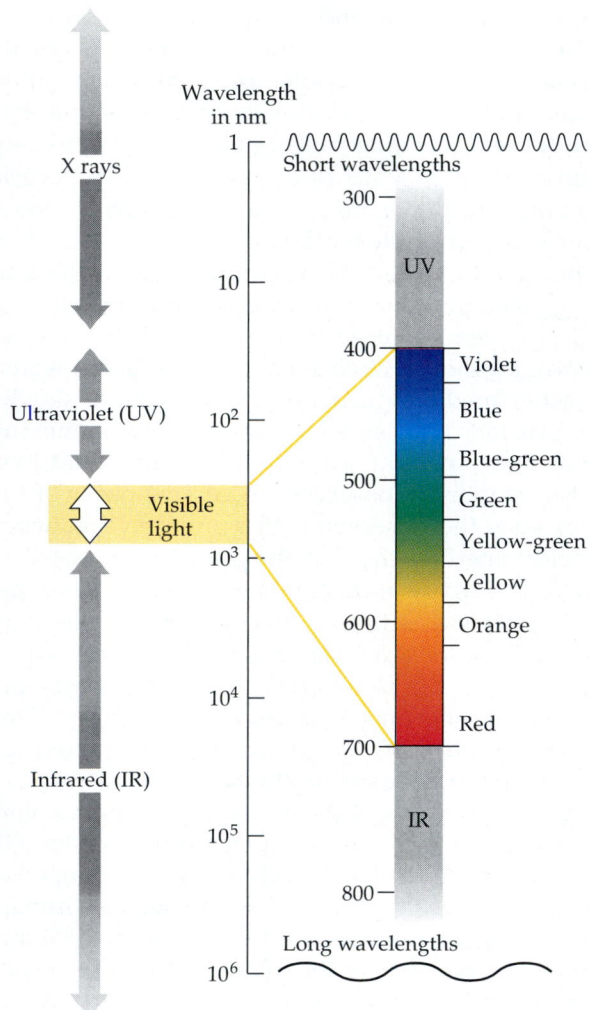

8.6 The Electromagnetic Spectrum
Wavelengths of electromagnetic radiation can be arranged on a scale called the electromagnetic spectrum. A portion of the spectrum in the vicinity of light visible to humans is represented here. Visible light comprises wavelengths between about 400 and 700 nm, although not everyone can see over this entire range. Ultraviolet radiation extends from the short-wavelength end of the visible spectrum, infrared radiation from the long-wavelength end.

$E = h\nu$, where ν is the frequency in Hz. Substituting c/λ for ν (from the equation relating λ, ν, and c), we see that $E = hc/\lambda$. Thus shorter wavelengths mean greater energies—that is, energy is inversely proportional to wavelength. A photon of red light of wavelength 660 nm has less energy than a photon of blue light at 430 nm; an ultraviolet photon of wavelength 284 nm is much more energetic than either of these. For any light-driven biological process—such as photosynthesis—a photon can be active only if it consists of enough energy to perform the work required.

The brightness, or **intensity**, of light at a given point is the amount of energy falling on a defined area—such as 1 cm^2—per second. Light intensity is usually expressed in energy units (such as calories) per square centimeter per second, but pure light of a single wavelength may also be expressed in terms of photons per square centimeter per second.

Pigments

When a photon meets a molecule, one of three things takes place. The photon may bounce off the molecule or it may pass through it. In other words, it may be reflected or transmitted. Neither of these causes any change in the molecule, and neither has any biological consequences. The third possibility is that the photon may be *absorbed* by the molecule. In this case, the photon simply disappears. Its energy, however, cannot disappear, because energy is neither created nor destroyed. The molecule acquires the energy of the absorbed photon and is thereby raised from a **ground state** (lower energy) to an **excited state** (higher energy). The difference in energy between this excited state and the ground state is precisely equal to the energy of the absorbed photon. The

8.7 Exciting a Molecule
(a) When a molecule, initially in the ground state, absorbs a photon, the molecule is raised to an excited state possessing more energy. (b) The absorption of the photon "boosts" one of the molecule's electrons to an orbital farther from the nucleus.

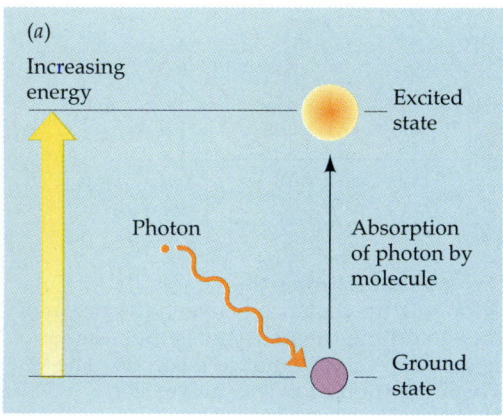

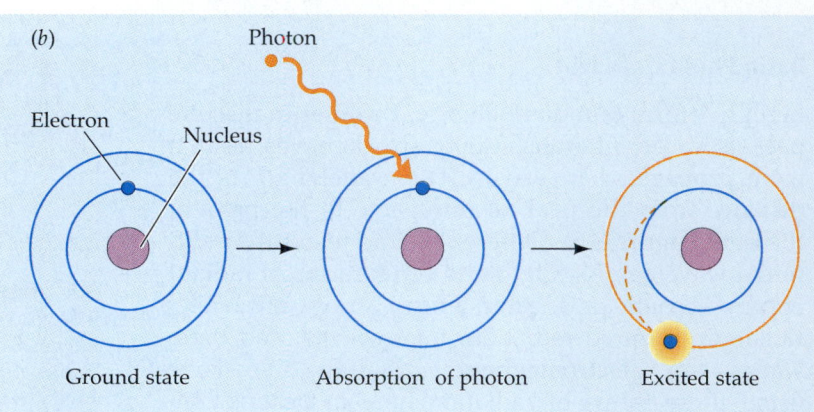

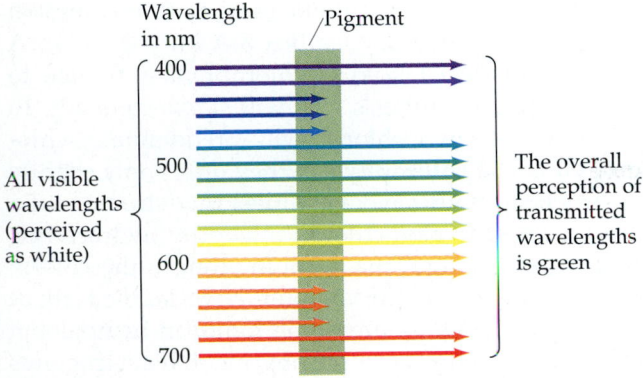

8.8 Why a Leaf Looks Green
The pigment chlorophyll (vertical bar), present in the leaves of all green plants, absorbs photons from specific wavelengths of visible light (short arrows). We see the combination of wavelengths that are not absorbed (long arrows) as the characteristic color of the pigment.

increase in energy boosts one of the electrons in the molecule into an orbital farther from the nucleus; in a sense, this electron is now held less firmly by the molecule (Figure 8.7). We will see the chemical consequence of this later in this chapter.

All molecules absorb electromagnetic radiation. The specific wavelengths absorbed by a particular molecule are characteristic of that type of molecule. Some molecules cannot absorb wavelengths in the *visible* region; those that can are called **pigments**.

When a beam of white light (light containing visible light of all wavelengths) falls on a pigment, certain wavelengths of the light are absorbed. The remaining wavelengths, which are reflected or transmitted, make the pigment appear to us to be colored. If, for example, a pigment absorbs both blue and red light, as does the pigment chlorophyll, what we see is the remaining light—primarily, green (Figure 8.8). The fact that chlorophyll absorbs light in both the blue and the red region of the spectrum indicates that it has two excited states of differing energy levels, both close enough to the ground state to be reached with the energy of photons of visible light.

Absorption Spectra and Action Spectra

A given type of molecule can absorb radiant energy of only certain wavelengths. If we plot a compound's absorption of light as a function of the wavelength of the light, the result is an **absorption spectrum** (Figure 8.9). Absorption spectra are good "fingerprints" of compounds; sometimes an absorption spectrum contains enough information to enable us to identify an unknown compound. The fact that the peaks in an absorption spectrum are smoothly rounded, rather than sharp spikes, tells us that a given excited state is not an extremely narrow range of energies. Rather, it consists of a substantial family of energy sublevels, differing by tiny increments of energy much smaller than those contained in a pho-

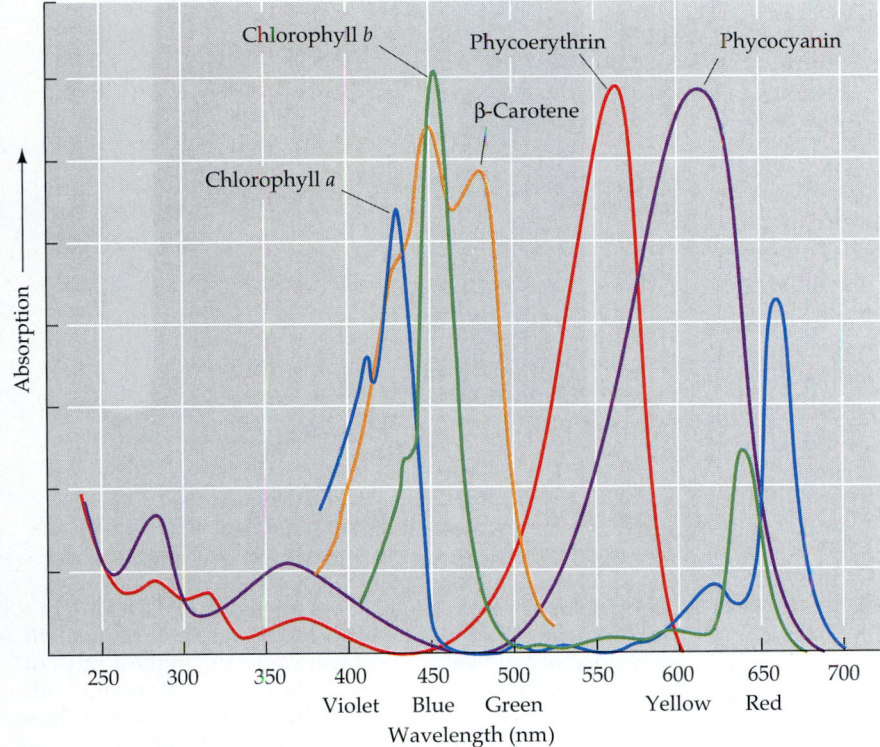

8.9 Photosynthetic Pigments Have Distinctive Absorption Spectra
Because these pigments, all of which participate in photosynthesis, absorb photons most strongly at different wavelengths, photosynthesis uses most of the visible spectrum. Notice how much of the visible spectrum would go to waste if chlorophyll *a* were the only pigment absorbing light for photosynthesis.

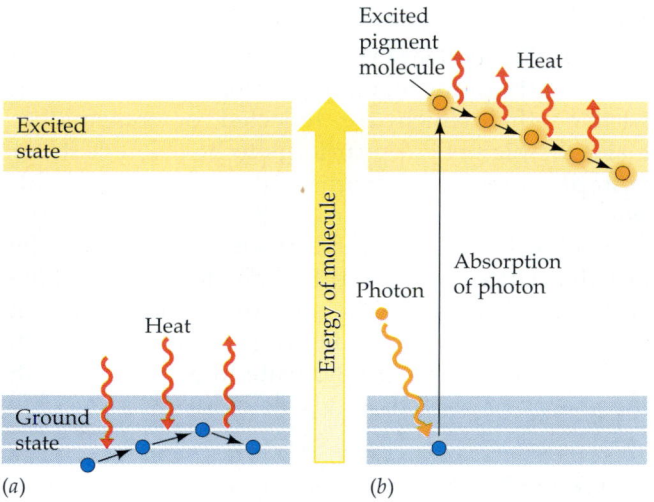

8.10 A Pigment's Energy Sublevels
(a) The ground state of a pigment consists of several sublevels of slightly different energy. When a molecule in the ground state rises from one sublevel to the next, it absorbs a tiny amount of heat; when the molecule falls to the next lower sublevel, it gives off heat. A molecule may be raised from any of these sublevels to an excited state. *(b)* All excited states also consist of energy sublevels. After reaching an excited state by absorbing a photon, a pigment molecule may give off minute amounts of heat as it drops from one sublevel to the next within the excited state.

ton of visible light. As molecules move from one of these sublevels to another they absorb or release small amounts of heat (Figure 8.10).

Light may be analyzed for its biological effectiveness, the magnitude of its effect on a particular activity such as photosynthesis. We may plot the effectiveness of light as a function of wavelength. The resulting graph is an **action spectrum.** Figure 8.11 shows the action spectrum for photosynthesis by a freshwater plant. As you can see, all wavelengths of visible light are at least somewhat effective in causing photosynthesis, although some are more effective than others. Because light must be absorbed in order to produce a chemical or biological effect, action spectra are helpful in determining what pigment or pigments are used in a particular photobiological process such as photosynthesis; that is, we should be able to find which pigment or pigments have absorption spectra that match the action spectrum of the process.

The Photosynthetic Pigments

Certain pigments are important in biological reactions, and we will discuss them as they appear in the book. Here we discuss pigments, found in leaves and in other parts of photosynthetic organisms, that play roles in photosynthesis. Of these, the most important are the **chlorophylls**. Chlorophylls occur universally in the plant kingdom, in photosynthetic protists, and

in virtually all photosynthetic bacteria (the exception being the halobacteria; see Box 8.A on page 174). A mutant individual lacking chlorophyll is unable to perform photosynthesis and will starve to death. In green plants, two chlorophylls predominate, **chlorophyll *a*** and **chlorophyll *b***; they differ only slightly in structure. Both have a complex ring structure of a type referred to as a chlorin: a lengthy hydrocarbon "tail," and a central magnesium atom in the chlorin ring (Figure 8.12). (In Chapter 7 we learned about porphyrins, such as the heme found in hemoglobin and the cytochromes; porphyrins have structures very similar to those of chlorins.)

We saw in Figure 8.9 that the chlorophylls absorb blue and red wavelengths, which are near the two ends of the visible spectrum. Thus if *only* chlorophyll pigments were active in photosynthesis, much of the visible spectrum would go unused. However, all photosynthetic organisms possess **accessory pigments** that absorb photons intermediate in energy between the red and the blue wavelengths and then transfer a portion of the energy to chlorophyll to use in photosynthesis. Among these accessory pigments are **carotenoids** such as β-carotene (see Chapter 3); the carotenoids absorb photons in the blue and blue-green wavelengths and appear deep yellow. The

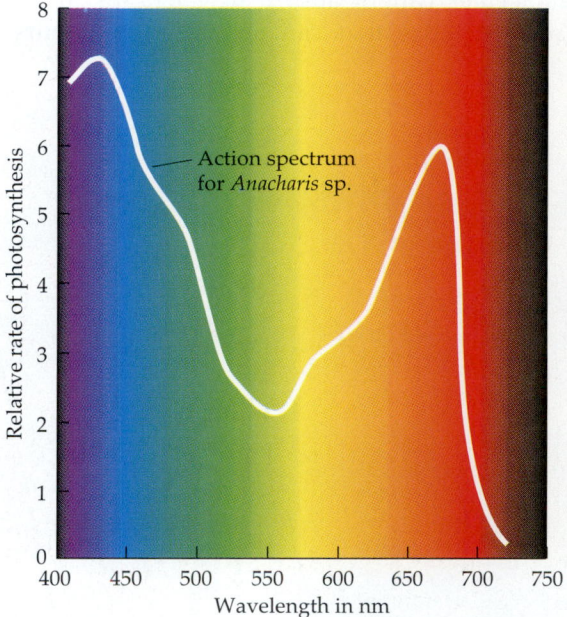

8.11 Action Spectrum of Photosynthesis
An action spectrum plots the biological effectiveness of wavelengths of radiation against the wavelength. Here the rate of photosynthesis in the freshwater plant *Anacharis* is plotted against wavelengths of visible light. You can see that wavelengths in the blue and orange-red regions of the visible spectrum cause the highest rates of photosynthesis. If we compare this action spectrum with the absorption spectra of specific pigments, such as those in Figure 8.9, we can identify which pigments are responsible for the process.

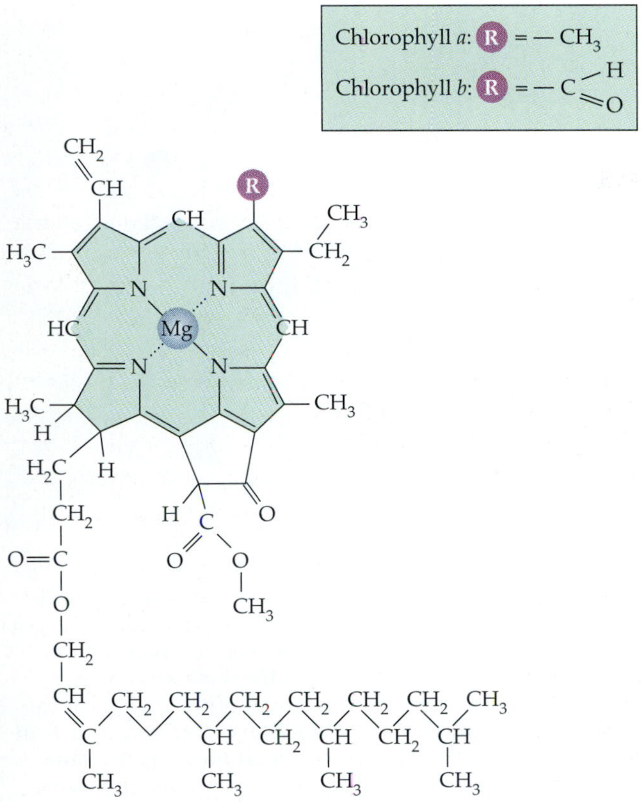

Chlorophyll a: $R = -CH_3$

Chlorophyll b: $R = -C{\overset{H}{\underset{O}{\diagdown}}}$

8.12 A Molecule of Chlorophyll
Chlorophyll consists of a chlorin ring with magnesium (shaded area), plus a hydrocarbon "tail." Chlorophyll *a* and chlorophyll *b* differ only in the groups attached to the position on the chlorin ring designated with an R (for "residue").

phycobilins (phycocyanin and phycoerythrin), which are found in red algae and in cyanobacteria (contributing to their respective colors), absorb various yellow-green, yellow, and orange wavelengths (see Figure 8.9). Such accessory pigments, in collaboration with the chlorophylls, constitute an energy-absorbing "antenna" covering much of the visible spectrum.

In the energy-absorbing antenna, any pigment molecule with a suitable absorption spectrum may absorb an incoming photon and become excited. The excitation passes from one pigment molecule to another in the antenna, moving to those pigments absorbing longer wavelengths (lower energies) of light. Thus the excitation must end up in the one pigment molecule in the antenna that absorbs the longest wavelength—the molecule that occupies the **reaction center** of the antenna. The reaction center is the part of the antenna that converts light absorption into chemical energy. The pigment molecule in the reaction center is always a molecule of chlorophyll *a*. There are many other chlorophyll *a* molecules in the antenna, but all of them absorb light at shorter wavelengths than does the molecule in the reaction center.

PHOTOPHOSPHORYLATION

Using the Excited Chlorophyll Molecule

A pigment molecule enters an excited state when it absorbs a photon (see Figure 8.7). The molecule usually does not stay in the excited state for very long. One means of returning to the ground state is by **fluorescence**. In this process, the boosted electron falls back from its higher orbital to the original, lower one. This process is accompanied by a loss of energy, which is given off as another photon (Figure 8.13*a*). The molecule absorbs one photon and within approximately 10^{-9} seconds emits another photon of longer wavelength than the one absorbed. The emitted light is fluorescence. If energy is simply absorbed and then rapidly returned as a photon of light, there can be no chemical or biological consequences.

For biological work to be done, something must happen to transfer energy in some other way during the billionth of a second before a photon is emitted as fluorescence. Photosynthesis conserves energy by using the excited chlorophyll molecule in the reaction center as a reducing agent (Figure 8.13*b*). Ground state chlorophyll (symbolized as Chl) is not much of a reducing agent, but excited chlorophyll (Chl*) is a good one. To understand the reducing capability of Chl*, recall that in an excited molecule, one of the electrons is zipping about in an orbital farther from its nucleus than it was before. Less tightly held, this electron can be passed on in a redox reaction to an oxidizing agent. Thus Chl* (but not Chl) can react with an oxidizing agent A in a reaction like this:

$$Chl^* + A \rightarrow Chl^+ + A^-$$

This, then, is the first biochemical consequence of light absorption by chlorophyll in the chloroplast: The chlorophyll becomes a reducing agent and participates in a redox reaction that would not have occurred in the dark. As we are about to see, the further adventures of that electron (the one passed from chlorophyll to A) produce ATP and a stable reducing agent (NADPH), both of which are required in the Calvin–Benson cycle.

Noncyclic Photophosphorylation: Formation of ATP and NADPH

The NADPH-producing version of photophosphorylation we have been discussing is called **noncyclic photophosphorylation**. With its appearance, the evolution of life on Earth made a crucial advance because noncyclic photophosphorylation uses light energy not only to form ATP and NADPH + H$^+$, but also to release O_2.

In noncyclic photophosphorylation, electrons from water replenish chlorophyll molecules that have given up electrons. These electrons are transferred to oxidizing agents and ultimately to NADP$^+$, reducing

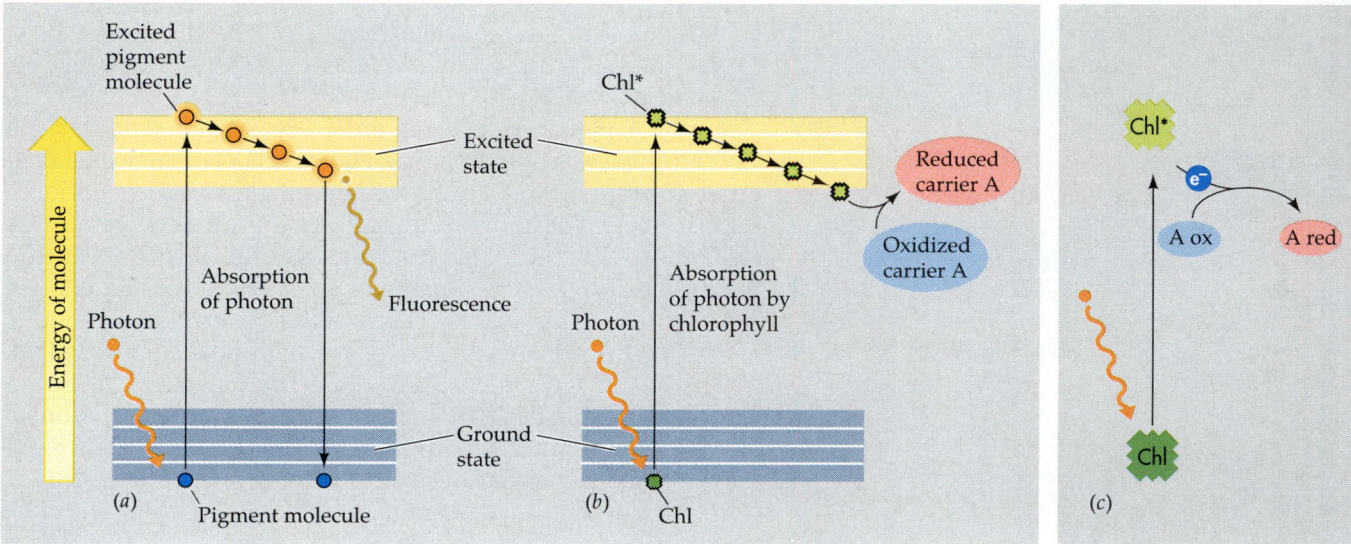

8.13 Excitation, Fluorescence, Redox
(a) When a pigment molecule absorbs a photon, boosting an electron to a higher orbital, the pigment moves to an excited state from its ground state. Although the molecule may then pass from one energy sublevel to the next, it spends very little time in the excited state. An excited molecule may return to the ground state by fluorescence, in which the electron falls back to its original lower orbital and a photon is emitted. Alternatively, an excited pigment molecule may pass the excitation to another pig- ment molecule, such as chlorophyll. *(b)* When a ground-state chlorophyll molecule (Chl) becomes excited (Chl*), it may become a reducing agent; the electron boosted to a higher orbital may pass to an oxidized electron carrier (A), reducing the carrier. Thus much of the energy of the excited state is preserved rather than being lost, as it is in fluorescence. *(c)* This diagram, which presents the same information as *(b)*, illustrates the conventions to be used in Figures 8.14 and 8.15.

it to NADPH + H$^+$. The original source of the electrons is a plentiful one: water. As the electrons are passed from water to, ultimately, NADP, they go through a series of electron carriers. These spontaneous redox reactions are exergonic, and some of the free energy released is used ultimately to form ATP.

Noncyclic photophosphorylation requires the participation of two distinct molecules of chlorophyll—actually, two separate sets of chlorophyll molecules, in separate energy-absorbing antennas of pigment molecules. One of these sets, called **photosystem I**, is used to make a reducing agent strong enough to reduce NADP$^+$ to NADPH + H$^+$. The reaction center of the antenna for photosystem I contains a chlorophyll *a* molecule in a form called P$_{700}$ because it can absorb light of wavelength 700 nm. **Photosystem II**, the other set of pigment molecules, takes electrons from water and passes them up to the series of redox carriers involved in the conversion of ADP + P$_i$ to ATP. The reaction center of the antenna for photosystem II contains a chlorophyll *a* molecule in a form called P$_{680}$ because it absorbs maximally at 680 nm. Thus photosystem II requires somewhat more energetic photons than does photosystem I. To keep noncyclic photophosphorylation going, both photosystems I and II must constantly be absorbing light, thereby boosting electrons to higher orbitals from which they may be captured by specific oxidizing agents.

We can follow the noncyclic pathway from water to NADP in Figure 8.14. Photosystem II (P$_{680}$) absorbs photons, sending electrons from P$_{680}$ to an oxidizing agent (pheophytin-I) and causing P$_{680}$ to become oxidized to P$_{680}^+$. Electrons from the oxidation of water, which forms H$^+$ ions and O$_2$, are passed to P$_{680}^+$ of photosystem II, reducing it once again to P$_{680}$, which can absorb photons, and so on. The electron donated by photosystem II to its oxidizing agent passes through a series of exergonic redox reactions, storing energy that is later used to form ATP. In photosystem I, P$_{700}$ absorbs photons, becoming excited to P$_{700}^*$, which then reduces its own oxidizing agent (ferredoxin) while being oxidized to P$_{700}^+$. Then P$_{700}^+$ is returned to the ground state by accepting electrons passed through the chain from photosystem II. Now photosystem I is accounted for, and we must consider only the electrons from photosystem I and the protons from the original oxidation of water at the beginning of the scheme. These are used in the last step of noncyclic photophosphorylation, in which two electrons and two protons (from two operations of the noncyclic scheme) are used to reduce a molecule of NADP$^+$ to NADPH + H$^+$.

In sum, noncyclic photophosphorylation uses a molecule of water, four photons (two each absorbed by photosystems I and II), one molecule each of NADP$^+$ and ADP, and one P$_i$ ion; from them it produces one molecule each of NADPH + H$^+$ and ATP,

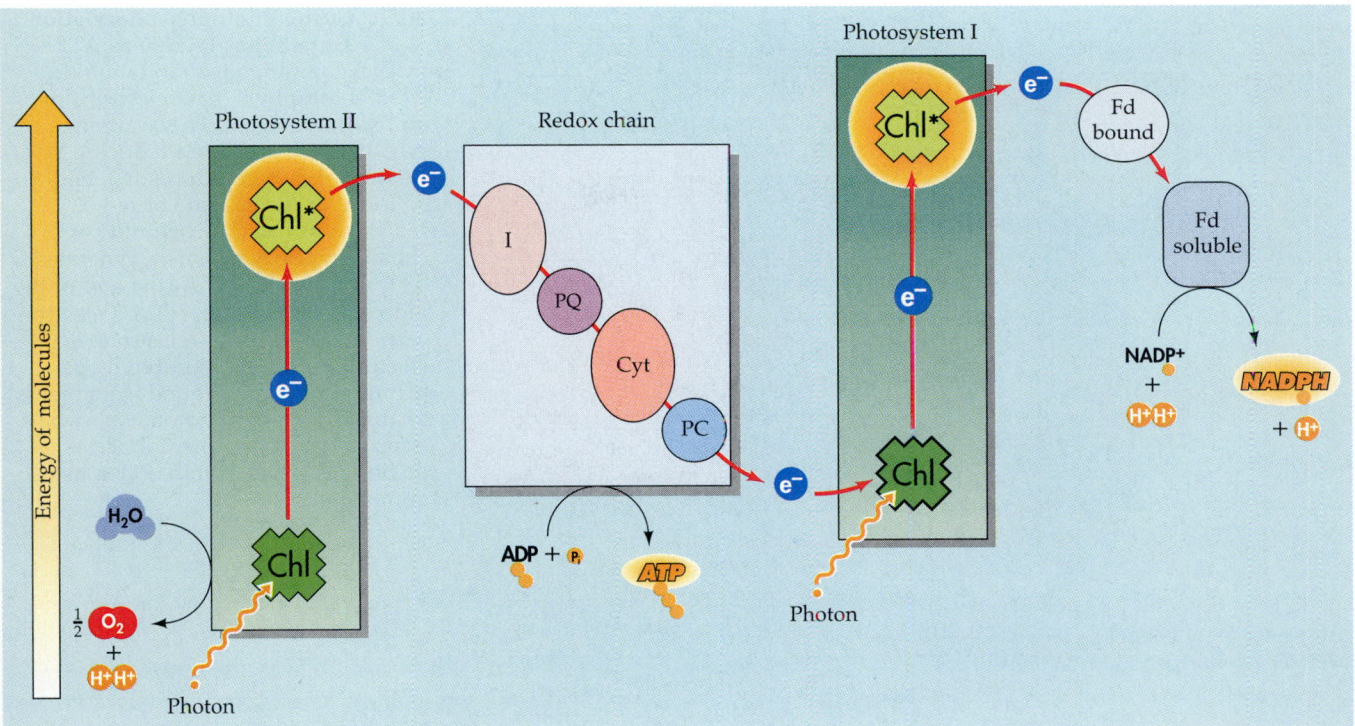

8.14 Noncyclic Photophosphorylation Uses Two Photosystems
Photosystems I and II—each containing chlorophyll—operate to keep noncyclic photophosphorylation going. ATP is produced by the redox chain between the photosystems, NADPH + H⁺ is produced by the redox reactions that follow the electron's passing through photosystem I, and O_2 is produced as a by-product of the breakdown of water. Abbreviations: Cyt, cytochrome; Fd, ferredoxin; I, pheophytin-I; PC, plastocyanin; PQ, plastoquinone.

and one-half molecule of oxygen (review Figure 8.14). A substantial fraction of the light energy absorbed in noncyclic photophosphorylation is lost as heat, but another significant fraction is trapped in ATP and NADPH + H⁺.

Cyclic Photophosphorylation: Formation of ATP but Not NADPH

Photophosphorylation that produces only ATP is called **cyclic photophosphorylation** because the electron passed from an excited chlorophyll molecule at the outset cycles back to the chlorophyll molecule at the end of the chain of reactions. Water, which supplies electrons to restore chlorophyll molecules to the ground state in noncyclic photophosphorylation, does not enter these reactions; thus no O_2 is produced from them.

Before cyclic photophosphorylation begins, P_{700}, the reaction center chlorophyll of photosystem I, is in the ground state. It absorbs a photon and becomes the reducing agent $P_{700}*$. The $P_{700}*$ then reacts with oxidized ferredoxin (Fd_{ox}) to produce reduced ferredoxin (Fd_{red}). The reaction is spontaneous—that is, it is exergonic, releasing free energy.

In noncyclic photophosphorylation, this Fd_{red} re-

duces NADP⁺ to form NADPH + H; but Fd_{red} is a good enough reducing agent to pass its added electron to *another* oxidizing agent, plastoquinone (PQ, a small organic molecule). This is what happens in cyclic photophosphorylation (Figure 8.15), which occurs in some organisms when the ratio of NADPH + H⁺ to NADP⁺ in the chloroplast is high. The Fd_{red} reduces PQ (which is part of the electron transport chain that connects photosystems I and II; see Figure 8.14) in the reaction $Fd_{red} + PQ_{ox} \rightarrow Fd_{ox} + PQ_{red}$. Acting as if the electron passed to it came from pheophytin-I (as happens in noncyclic photophosphorylation), PQ_{red} passes the electron to a cytochrome complex (Cyt). The electron continues down the chain until it completes its cycle by returning to P_{700}. This cycle is a series of redox reactions, each exergonic, and the released energy is stored in a form that ultimately can be used to produce ATP.

Remember that when $P_{700}*$ passed its electron on to Fd, we were left with a molecule of positively charged P_{700}^+ (having lost an electron, the chlorophyll has one unbalanced positive charge). In due course, P_{700}^+ interacts with a reducing agent that donates an electron, converting it back to uncharged P_{700}. This reducing agent, plastocyanin (PC), is the last member of the electron transport chain in Figure

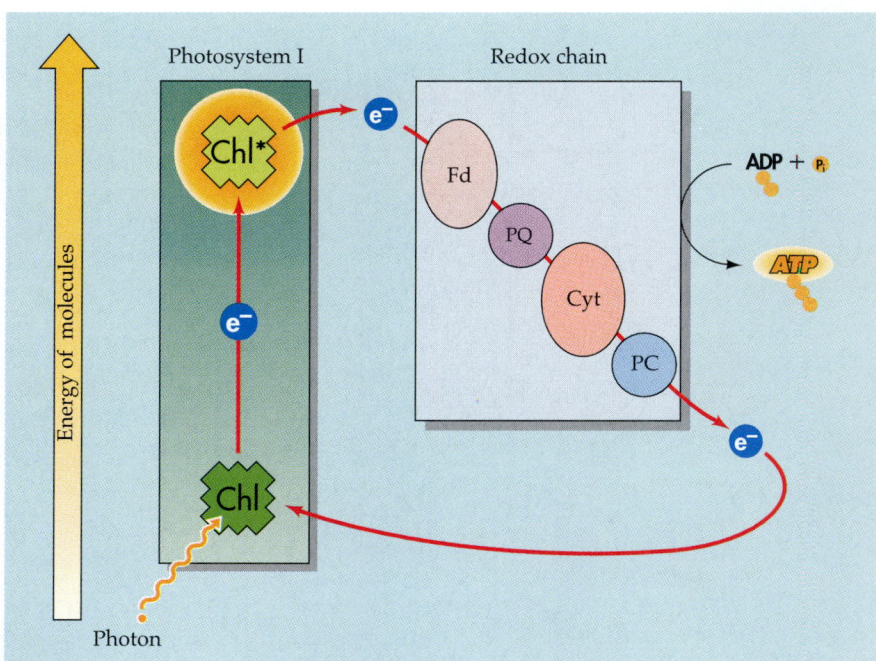

8.15   Cyclic Photophosphorylation Traps Light Energy as ATP
In cyclic photophosphorylation, excited chlorophylls pass electrons to an oxidizing agent, Fd, becoming positively charged and reducing Fd. Reduced Fd then reduces PQ, and so forth, in the cascade of redox reactions from Fd through PC. A chemiosmotic mechanism generates ATP from ADP + P_i. At the end of the redox chain, the last reduced electron carrier (PC_{red}) passes electrons to electron-deficient chlorophylls, returning them to a ground state ready to absorb another photon. Abbreviations: Cyt, cytochrome; Fd, ferredoxin; PC, plastocyanin; PQ, plastoquinone.

8.15. By the time the electron (passed from P_{700}* and on through the redox chain) comes back to P_{700}^+ and reduces it, all the energy from the original photon has been released. In each of the redox reactions, some free energy is lost, until all of the original energy has been converted to heat *except* for that used to form ATP.

Comparing the cartoon in Figure 8.16 with Figure 8.15 may help make the concept of cyclic photophosphorylation clearer to you.

The Mechanism and Location of ATP Formation

How does electron transport in the photophosphorylation pathways form ATP? In Chapter 7 we considered the **chemiosmotic mechanism** for ATP formation in the mitochondrion. The chemiosmotic mechanism also operates in photophosphorylation. In chloroplasts, as in mitochondria, protons (H^+ ions) are transported across a membrane, resulting in a difference in pH and in electric charge across the membrane. In the mitochondrion, protons are pumped from the matrix, across the internal membrane, and into the space between the inner and outer mitochondrial membranes (see Figure 7.18). In the chloroplast, the electron carriers are located in the thylakoid membranes (see Figure 4.16). The electron carriers are oriented so that protons move into the interior of the thylakoid, so the inside becomes acidic with respect to the outside. This difference in pH leads to the passive movement of protons back out of the thylakoid, through protein channels in the membrane. These proteins are ATP synthases, enzymes that catalyze the formation of ATP; they are

activated by the movement of protons through the channels, just as in mitochondria (Figure 8.17).

The hypothesis that this chemiosmotic mechanism is responsible for the formation of ATP in chloroplasts was tested by Andre Jagendorf (of Cornell University) and Ernest Uribe (now at Washington State University) in the following way. Chloroplast thylakoids were isolated from spinach leaves and then kept in the dark, so that there would be no light energy to drive the production of ATP. The thylakoids were moved from a neutral solution to one with a low pH so that by diffusion the interiors of the thylakoids would become acidic. Then they were transferred to a solution that had a higher pH, so that the interiors of the thylakoids would be more acidic than the outsides—mimicking the situation created by light-driven pumping of protons into the interiors. This

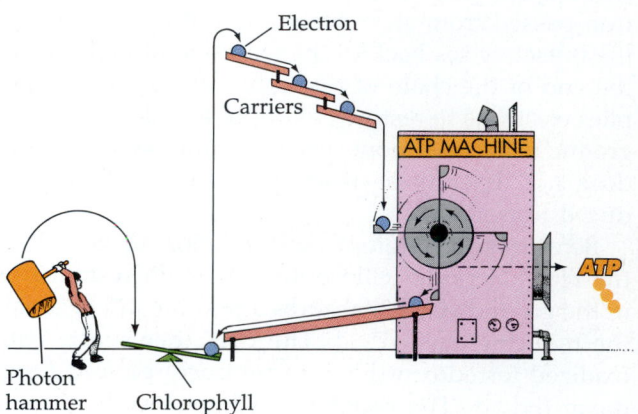

8.16   Cyclic Photophosphorylation Cycles Electrons

8.17 Chloroplasts Form ATP Chemiosmotically

Protons (H^+) pumped across the thylakoid membrane from the stroma during noncyclic photophosphorylation make the interior of the thylakoid more acidic than the outside. Driven by this pH difference, the protons then return to the stroma through ATP synthase channels, activating the synthases to catalyze the formation of ATP from ADP + P_i. These reactions take place in several places on the thylakoid membrane simultaneously. Compare this chemiosmotic model with the one in Figure 7.18 that explains the activities of the inner mitochondrial membrane.

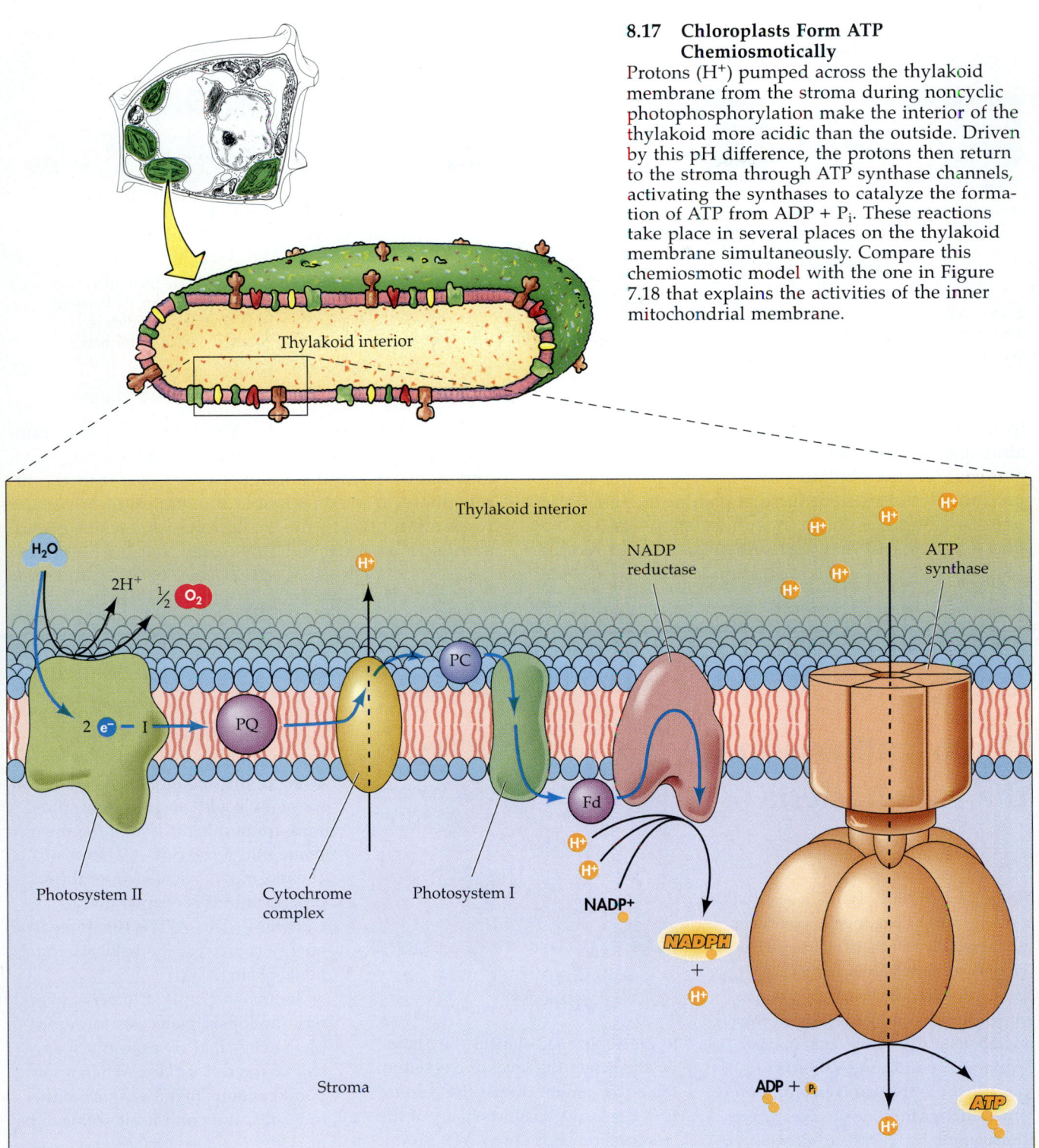

final step immediately resulted in the formation of ATP, even though no light was available to serve as the energy source (Figure 8.18). This is precisely the result predicted by the chemiosmotic model. (A very similar experiment, using mitochondria, pH changes, and ATP formation, was described in Chapter 7.)

Box 8.A describes an unusual use of chemiosmotic mechanisms to form ATP from sunlight.

Noncyclic and Cyclic Photophosphorylation Revisited

Photosystem I evolved before photosystem II; thus, cyclic photophosphorylation evolved before noncyclic photophosphorylation. Early in evolutionary history, photosynthetic bacteria used photosystem I and cyclic photophosphorylation to make ATP. This

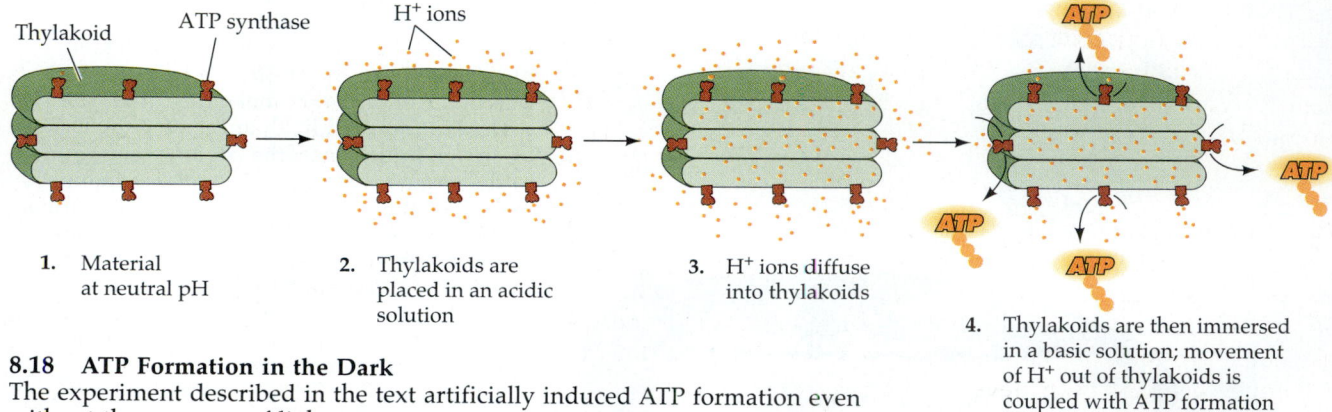

1. Material
    at neutral pH

2. Thylakoids are
    placed in an acidic
    solution

3. H$^+$ ions diffuse
    into thylakoids

4. Thylakoids are then immersed
    in a basic solution; movement
    of H$^+$ out of thylakoids is
    coupled with ATP formation

8.18   ATP Formation in the Dark
The experiment described in the text artificially induced ATP formation even
without the presence of light energy.

form of photosynthesis evolved long before Earth's atmosphere contained significant quantities of oxygen gas. Nearly 3 billion years ago the cyanobacteria produced photosystem II, thus gaining the ability to perform noncyclic photophosphorylation—and to extract electrons from water and use them to reduce NADP$^+$. Over hundreds of millions of years, noncyclic photophosphorylation by cyanobacteria, algae, and plants poured enough oxygen gas into the atmosphere to make possible the evolution of cellular respiration. Today the evolutionarily ancient photosynthetic bacteria still have only cyclic photophos-

BOX 8.A

Photosynthesis in the Halobacteria

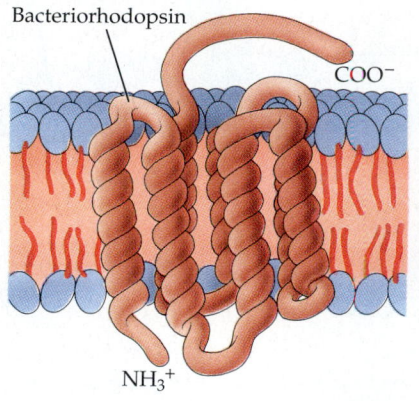

From time to time we discover that some group of organisms conducts its metabolic affairs in ways that previously were totally unexpected. In 1971, for example, biologists were surprised by discoveries about the metabolism of certain bacteria that live in salt ponds and salt lakes, places where the salt concentration is much higher than in the ocean. Because sunlight reaches to depths where these "salt-lovers" live, the surprise was not that these **halobacteria** trap sunlight energy. The surprise was that they do so without using chlorophyll.

These halobacteria species do not always use their unique light-trapping equipment. They are heterotrophs that get ATP by using oxygen

to metabolize substances they take in. But when the level of oxygen in the environment drops, the respiratory electron transport chains of the halobacteria shut down ATP production. Under these conditions, halobacteria turn on their light-trapping equipment.

The light-trapping equipment includes **retinal**, a carotenoid pigment. Retinal combines with a protein to form **bacteriorhodopsin**, which is incorporated into the plasma membrane. Each bacteriorhodopsin mole-

cule is organized in seven helical regions roughly perpendicular to the plane of the membrane. When one of these molecules absorbs light, protons are pumped through the membrane out of the cell. ATP forms by a chemiosmotic mechanism of the sort described in this chapter and in Chapter 7. This ATP is the immediate energy source for the halobacterium's metabolism.

Retinal is purple. Clusters of bacteriorhodopsin molecules form purple patches that cover as much as half of the cell surface. When a salt pond contains high densities of the tiny cells, the pond itself seems to be purple.

As you progress through this book, you will hear of retinal again. This pigment is also found in the vertebrate eye, where it plays a key role in vision. It is striking that retinal is used in two such different processes—vision and photophosphorylation—in organisms so widely separated in ancestry.

phorylation. Cyanobacteria, algae, and plants, which perform mostly noncyclic photophosphorylation, still perform cyclic photophosphorylation to produce ATP when the ratio of NADPH + H$^+$ to NADP$^+$ in their chloroplasts is high.

THE CALVIN–BENSON CYCLE

Real progress in understanding the dark reactions—the second main pathway of photosynthesis—came only after World War II. Satisfactory experimental techniques had not been developed before then. A group of scientists at the University of California, Berkeley, led by Melvin Calvin and including Andrew Benson and James Bassham, broke the problem wide open. The problem, as posed by the Berkeley group, was to learn the biochemical steps between the uptake of CO_2 and the appearance of the first complex carbohydrates in the chloroplast. Its solution depended on three advances in technique: (1) the discovery and availability of a radioactive carbon isotope, ^{14}C; (2) the development of a technique—paper partition chromatography—for the rapid separation of individual compounds from complex solutions; and (3) the development of autoradiography, a technique for locating colorless but radioactive compounds on a paper chromatogram. These advances are described in Box 8.B.

Armed with ^{14}C, paper partition chromatography,

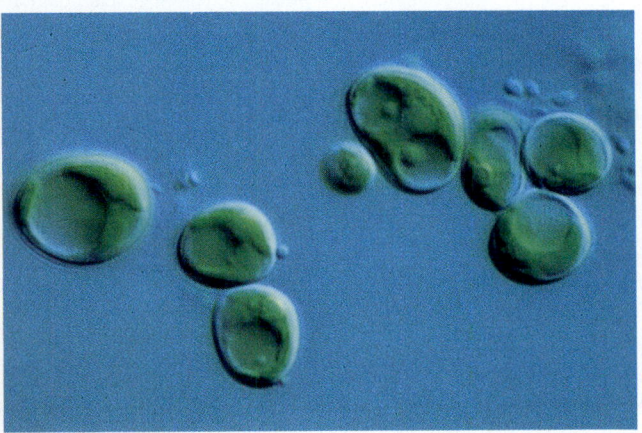

8.19 Algae Used in Experiments on Photosynthesis
The alga *Chlorella pyrenoidosa* was one of several used in studies by Calvin's group; this image is magnified about 1,600 times. Single-celled algae were chosen over leafy plants because their enzymes could be inactivated very quickly by ethanol, which penetrated the cell walls rapidly. This allowed the researchers to stop the reactions in the cells at a chosen time after treating the cells with radioactive $^{14}CO_2$ in dissolved form. Algae also could be grown continuously in cultures, providing a ready supply of research material with little variation.

and autoradiography, the Berkeley group set out to investigate how photosynthetic organisms metabolize CO_2. They worked mostly with unicellular aquatic algae, such as the green alga *Chlorella* (Figure 8.19). Algae were grown in dense suspensions in a flattened flask (called a "lollipop" because of its shape) between two bright lights, ensuring a rapid rate of photosynthesis (Figure 8.20*a*). To start an experiment, a solution containing dissolved $^{14}CO_2$ was suddenly squirted into the lollipop. At a carefully measured time after this squirt, a sample of the culture was rapidly drained into a container of boiling ethanol. The ethanol performed two functions: It killed the algae, stopping photosynthesis immediately, and it extracted the ^{14}C-containing intermediates of $^{14}CO_2$ metabolism from the algae (along with many other compounds). A sample of this ethanolic extract was spotted on filter paper for paper chromatography followed by autoradiography. Typical results of a 30-second exposure to $^{14}CO_2$ are shown in Figure 8.20*b*. As you can see, many biochemical reactions had taken place during that short interval.

The First Stable Product of Carbon Dioxide Fixation

During 30 seconds of continuous exposure to $^{14}CO_2$, many different products were formed in Calvin's lollipop. To determine which of them formed first, the experiment had to be repeated several times, using ever-shorter exposures. Even after exposures of less than two seconds, half a dozen or more labeled compounds appeared in the autoradiographs. One, however, was produced most rapidly and in greatest abundance: **3PG** (3-phosphoglycerate, also called 3-phosphoglyceric acid, or PGA), which we have already encountered as an intermediate in glycolysis. 3PG is the first stable product of CO_2 fixation. The Berkeley group isolated the individual carbon atoms from this 3PG and found that the carbon of the carboxyl group was much more intensely radioactive than the other two carbon atoms (Figure 8.21). (A single atom either is or is not radioactive. What we mean by "more intensely radioactive" is that in a *population* of molecules, the fraction of carboxyl carbons labeled with radioactivity was greater than the fractions of radioactive carbons in the other two carbon positions.)

The Berkeley group drew two important conclusions from finding the higher fraction of radioactive carboxyl carbons. First, the heavy labeling showed that the carboxyl carbon is obtained directly from CO_2. The existence of label in the other carbon atoms led to the second conclusion: Some kind of *cyclical* process is involved, a process by which 3PG is made by adding CO_2 to another compound that is itself produced from photosynthetic 3PG.

BOX 8.B

Tools that Cracked the Calvin–Benson Cycle

Melvin Calvin's group at Berkeley needed a way to keep track of the carbon atom from CO_2 taken up during photosynthesis in order to solve one of the mysteries of the dark reactions: What happens to the carbon after the CO_2 molecule reacts? It was the program to produce the first atomic bomb that indirectly provided the solution when the radioactive carbon isotope ^{14}C was made available to scientists. With this carbon isotope, samples of CO_2 could be prepared in which some of the carbon atoms were ^{14}C rather than the stable isotope ^{12}C found in nature.

Paper partition chromatography

Any compound incorporating the radioactive material would also be radioactive.

Next, Calvin's group needed an improved method of separating complicated mixtures into their individual components. Living things contain thousands of different chemical components. If we want to study just one of them, we must separate it from all the thousands of others. A powerful new separation tool, called **paper partition chromatography**, became available shortly before the Berkeley group began its work.

In paper chromatography, a drop of a solution containing the compounds to be separated is placed near one end of a strip of filter paper (as shown in part *a* of the figure below). The paper is then lowered into a container with a suitable mixture of organic solvents (liquids such as chloroform and alcohols) until the end of the paper is submerged. The solvent works its way up through the paper by capillary action. As the solvent climbs, so do the compounds to be separated. For now, let us say that

the compounds in the mixture are pigments, so that we can see what happens to them. By the time the solvent has moved several centimeters up the paper, one can usually see that the mixture of pigments has separated because the different pigments are visible as colored spots at various distances up the paper.

If repeated, such experiments show that a given compound always travels the same relative distance. If we take the distance (from the starting spot) moved by a particular compound and divide it by the distance (from the same point) moved by the solvent, that ratio is the "front ratio" (R_F) of that compound under those conditions of temperature and solvent composition. Part *b* of the figure shows how we determine the R_F. Once calculated, the R_F of a compound can be used to identify it.

Some compounds are difficult to separate by paper partition chromatography. In such cases, **two-dimensional chromatography** sometimes

(a)
Paper strip
End of strip placed in solvent; solvent begins to rise
Solvent
Mixture of compounds to be separated

(b)
Final position of solvent
Final position of compound
Distance traveled by solvent = A
Distance traveled by compound = B
Initial position of compound

$$R_F = \frac{B}{A}$$

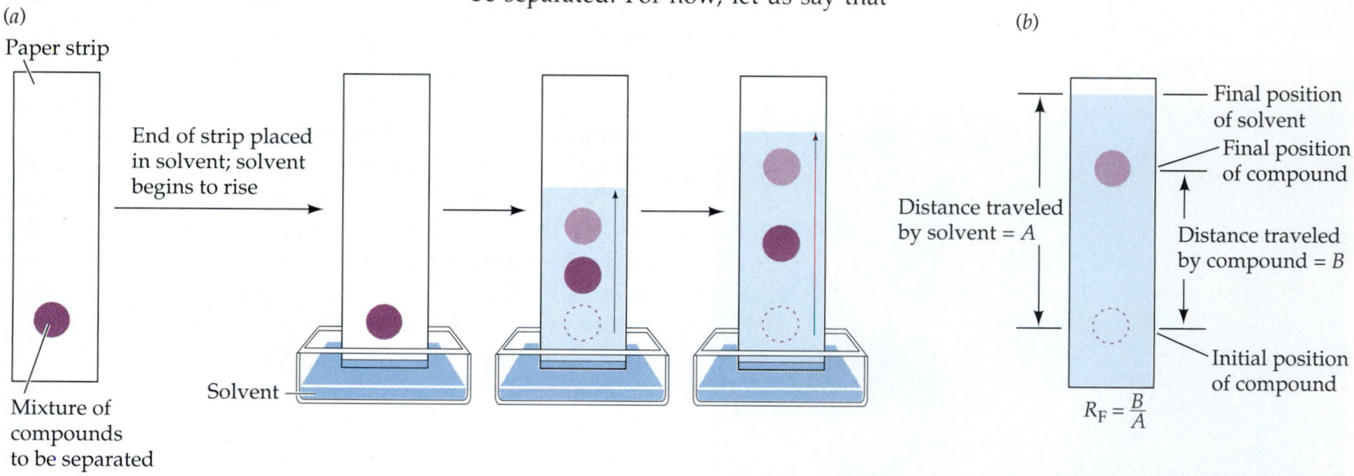

What Is the Carbon Dioxide Acceptor?

What is this compound, obtained from the metabolism of 3PG, that binds with CO_2 to make more 3PG? Given the structure of 3PG, it was reasonable to expect that the mysterious CO_2 acceptor would be a *two*-carbon compound that could react with CO_2 to become a *three*-carbon compound—3PG—with the CO_2 becoming a carboxyl group. If this idea were

correct, it should have been possible for the Berkeley group to find, on their chromatograms, a compound with only two carbon atoms, both of which were radioactive after a lollipop experiment. They did *not* find such a compound, and this made the problem more difficult. The "obvious" answer—that the CO_2 acceptor is a two-carbon compound—was wrong. Where would they go from here?

Two-dimensional chromatography

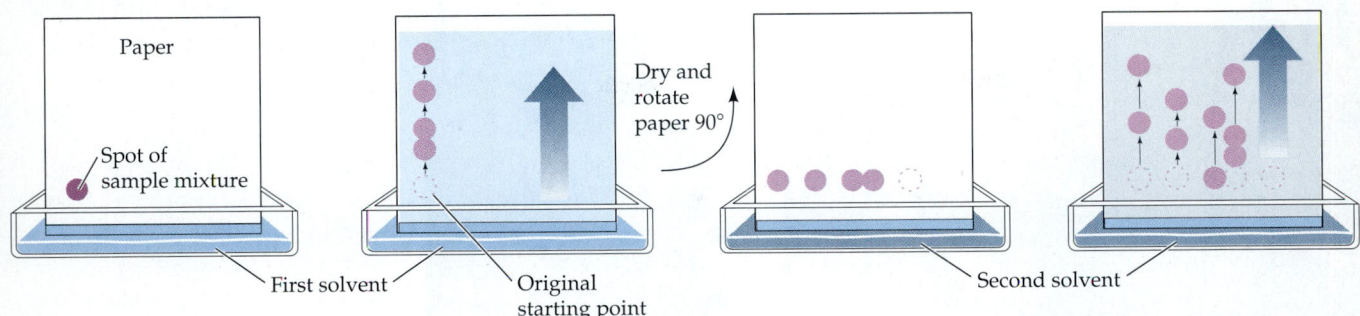

works. In this technique, the sample mixture is applied as a single spot near one corner of a square of filter paper. Paper chromatography is run as before, using one solvent and achieving a partial separation of the mixture along one edge of the paper. The paper is then dried and turned so that the row of partially separated compounds is now along the bottom edge. Chromatography is run again with a different solvent. If the properties of the second solvent are markedly different from those of the first, the second run will separate compounds that stayed together in the first run. The result looks like a scatter plot. Because each compound has a certain R_F in each of the solvents,

the precise pattern can be repeated again and again.

The third thing Calvin's group needed was a method for locating tiny amounts of radioactive material visually. Suppose now that you are doing paper chromatography runs not with a mixture of pigments, but with a mixture of radioactive and nonradioactive compounds. The radioactivity comes from ^{14}C incorporated into precursors as described at the beginning of this box. You could take advantage of the fact that radioactive emissions, like light, expose photographic film. The method that does this, **autoradiography**, works as follows: The paper chromatogram (the filter-paper sheet on which chro-

matography has been performed) has spots of ^{14}C-containing materials at unknown places. This chromatogram is taken into a darkroom, covered either with a type of photographic film or with a liquid photographic emulsion, and kept in the dark for a suitable length of time, during which the radioactive decay of ^{14}C releases particles that expose the film. Later the film is developed, revealing dark spots (composed of silver grains in the film; see Box 2.A) that correspond to the places on the chromatogram where there were accumulations of ^{14}C-containing materials. The R_F values of the radioactive compounds can thus be determined from the positions of the dark spots.

Autoradiography

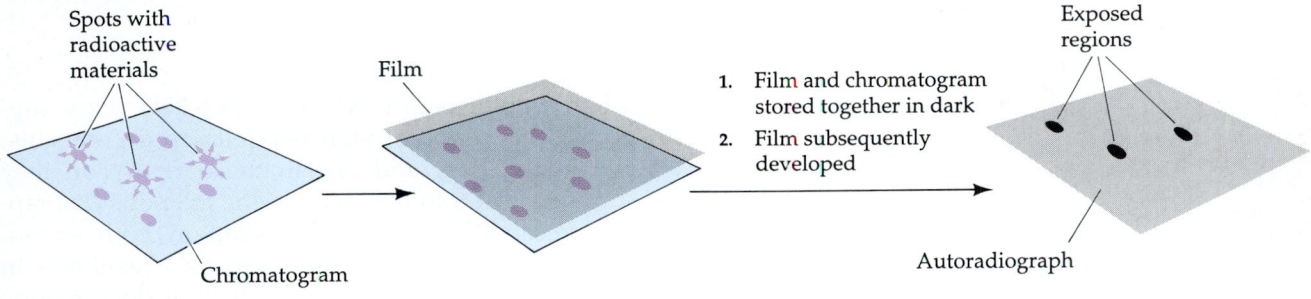

At this point in the investigation, concentrating on a photosynthetic *cycle* became useful. Consider a tentative cycle of the sort shown in Figure 8.22a: CO_2 reacts with X, the CO_2 acceptor, to produce 3PG. From the 3PG, photosynthetic organisms make products (things like glucose) and more X; this new X can react with another molecule of CO_2 and keep the cycle going. But what is X, and what are the other intermediates in the cycle?

It was observed that 3PG was the only acid phosphate produced in significant amount, whereas many kinds of *sugar* phosphates appeared on the chromatograms. On this basis, the Berkeley group guessed that the first thing to happen to 3PG is its conversion to a three-carbon sugar phosphate (glyceraldehyde 3-phosphate, which we will call G3P). Such a reaction is a reduction, and since reductions are highly endergonic, the Berkeley group proposed that this re-

(a)

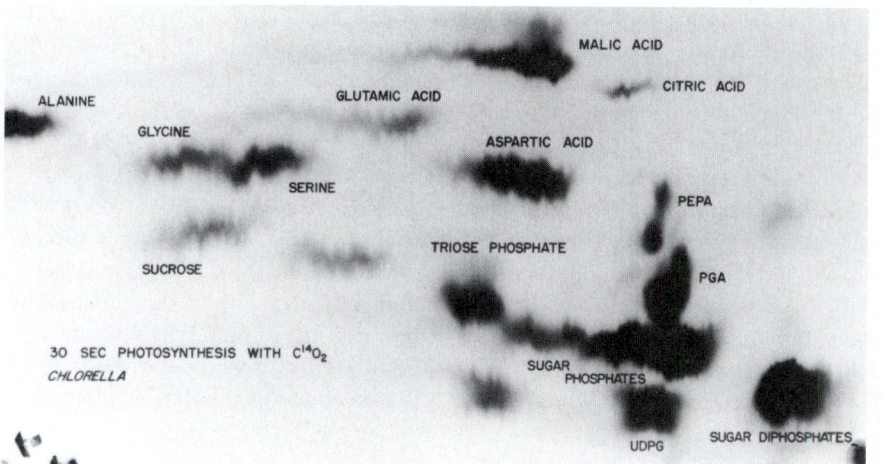

(b)

8.20   A Lollipop and Its Products

(a) The lollipop used in experiments on photosynthesis. The thin flask was filled with a suspension of algae and illuminated from both sides. After injection of $^{14}CO_2$, a sample was drained from the flask into boiling ethanol. (b) A chromatogram showing products of algal photosynthesis. The dark spots are compounds containing ^{14}C—all formed in the 30 seconds following injection of $^{14}CO_2$. The spot labeled PGA (an older notation) corresponds to the position of 3-phosphoglycerate (3PG), a compound we will discuss in the next section.

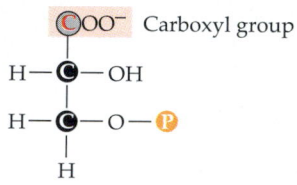

COO⁻ Carboxyl group

H—C—OH

H—C—O—(P)

H

3-Phosphoglycerate

8.21   3PG Is the First Stable Product of CO_2 Fixation

In experiments in which an algal sample from a lollipop was killed with ethanol only a few seconds after the introduction of $^{14}CO_2$, it was found that most 3-phosphoglycerate molecules had a ^{14}C in the carboxyl group, as indicated by the red symbol; smaller numbers of 3PG molecules had the label in the other two carbon atoms. The heavy labeling in the carbon of the carboxyl group indicated that the carbon was obtained directly from $^{14}CO_2$.

action would require ATP (Figure 8.22b). They supported this proposal with two lollipop experiments, one in the light and one in the dark. When they supplied $^{14}CO_2$ to the algae in the lollipop, 3PG rapidly became radioactively labeled in both experiments, but G3P became radioactively labeled only in the light, showing that the reduction does require energy—presumably in the form of ATP generated in the light.

At this point an extremely clever suggestion was made: If we assume that Figure 8.22b accurately models what occurs in the chloroplast, then it should be possible to regulate this cycle in the laboratory by two simple means. First, turning off the light would specifically block the step from 3PG to G3P, the sugar phosphate, because the necessary ATP and NADPH + H^+ can only be produced with an input of energy; in photosynthesis, the energy source is light. There-

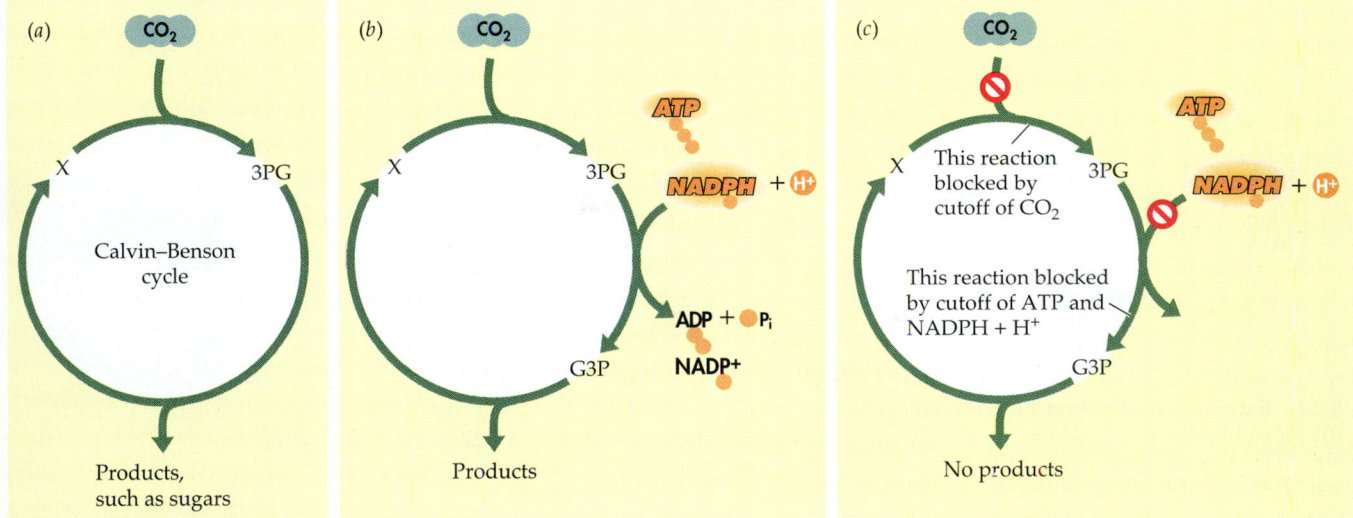

8.22 Manipulating the Dark Reactions of Photosynthesis

(a) After it became apparent that CO_2 combines with some molecule to form 3PG and that a cyclical process is involved, a pathway could be devised in which a molecule of compound X combines with CO_2 to form 3PG and in which further reactions regenerate X. (b) Ensuing speculation and experimentation led to the proposition that 3PG was reduced to a sugar phosphate, glyceraldehyde 3-phosphate (G3P); this endergonic reaction, a reduction, would require ATP, as shown here. (c) In the proposed model, a cutoff of CO_2 blocks the formation of 3PG from compound X and CO_2; a cutoff of ATP (by turning off the light) blocks the formation of G3P from 3PG.

are the changes we would expect to see *if* our model (Figure 8.22c) is correct. Similar reasoning should convince you that when the CO_2 supply is cut off (with the lights on), the concentration of X rises and that of 3PG falls. With no CO_2 available, X is no longer used up and 3PG is no longer formed, but 3PG can still be reduced to G3P.

Having devised this model, the Berkeley group proceeded to study the effects of changes in light intensity and CO_2 supply on the concentrations of all the major radioactive compounds found in the

fore ATP should be made in the chloroplast only when the light is on. Second, the reaction from X to 3PG could easily be blocked by cutting off the supply of CO_2 to the lollipop (Figure 8.22c).

Assume that photosynthesis is proceeding at a steady pace, so the concentrations of 3PG and the CO_2 acceptor X in the cells are constant. The lights are on, of course, and there is plenty of CO_2. Suddenly the investigator turns off the lights, thus blocking the cycle as proposed in Figure 8.22c: What change in 3PG concentration occurs immediately? What change in the concentration of X occurs immediately? *Stop.* Think about this before you read on.

When the light is turned off, no more ATP is made. Without ATP, the reaction from 3PG to the three-carbon sugar G3P cannot take place. Therefore, 3PG is no longer being used up. However, there is nothing to stop the formation of 3PG from incoming CO_2 and X, as long as there is any X around. Therefore, the immediate consequence of turning off the light is an *increase* in the level of 3PG. On the other hand, X continues to be used up, because its reaction with CO_2 does not depend on light or ATP; but the formation of X slows down because 3PG can no longer be reduced and ultimately form new X. Therefore, the concentration of X *decreases* (Figure 8.23). These

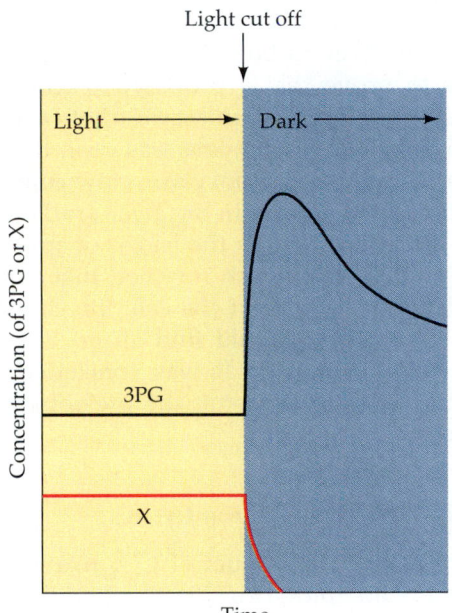

8.23 Changes in the Dark

When the light to a lollipop is turned off, 3PG accumulates as compound X is combined with CO_2. The concentration of X, on the other hand, falls when the light is turned off. X is not replenished because 3PG is not converted to G3P to continue the cycle (light is necessary to provide the ATP for the endergonic 3PG → G3P reaction).

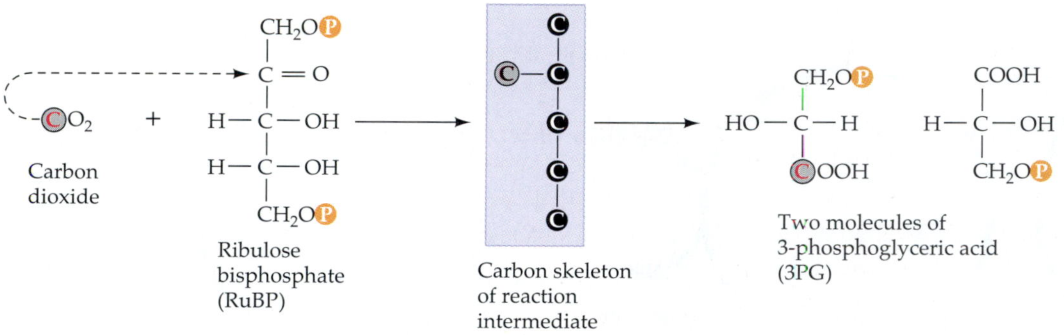

8.24   RuBP Is the Carbon Dioxide Acceptor
Ribulose bisphosphate (RuBP) is the CO_2-accepting compound X in Figures 8.22 and 8.23. The combination of CO_2 and RuBP forms a reaction intermediate, which then splits into two molecules of 3PG. The fate of the carbon atom in CO_2 is traced in red.

lollipop experiments. The first thing they noticed was that only one compound underwent the concentration changes proposed for 3PG—and that was 3PG itself. This result showed that their model was likely to be accurate. But would any compound behave in the way predicted for the mysterious compound X, the CO_2 acceptor? Yes—and only one: a five-carbon sugar phosphate called **RuBP** (ribulose bisphosphate). It seemed, then, that instead of the originally proposed reaction (in which CO_2 was thought to react with a two-carbon sugar to form 3PG), there must be a reaction in which CO_2, with its single carbon, combines with the five-carbon RuBP to give two molecules of the three-carbon 3PG (Figure 8.24).

The best way to prove that a proposed reaction does in fact take place is to find an *enzyme* that catalyzes it. In this case, such an enzyme was soon discovered. The enzyme, RuBP carboxylase, now commonly called **rubisco**, was found in the algae studied by the Berkeley group and also in the leaves of spinach. Best of all, studies of spinach revealed that rubisco is found in only one part of the cell: the chloroplast—exactly where one should find an enzyme concerned with photosynthesis. It was concluded, then, that RuBP is the CO_2 acceptor, the previously unknown compound X.

Filling the Gaps in the Calvin–Benson Cycle

Having discovered the first product of CO_2 fixation (3PG) and the CO_2 acceptor (RuBP), the Berkeley group proceeded to work out the remaining reactions of the cycle. They found some relatively complicated steps between G3P and RuBP; among the intermediates are sugar phosphates with four, five, six, and seven carbon atoms. All the proposed intermediates have been found in chloroplasts, as have all the necessary enzymes. It was also discovered that ATP is needed at one more point in the Calvin–Benson cycle:

in the step producing RuBP (ribulose *bis*phosphate) from RuMP (ribulose *mono*phosphate). This additional ATP requirement makes it even easier to understand why turning off the light drastically reduces the concentration of RuBP (X in Figure 8.23).

For now, you need learn only the material in Figure 8.25, which summarizes the key features of the Calvin–Benson cycle. In the chloroplast, most of the enzymes that catalyze the reactions of this pathway are dissolved in the stroma (see Figure 4.16), and this is where the reactions take place. RuBP reacts with CO_2 to form 3PG. 3PG is then reduced—in a reaction requiring ATP as well as hydrogens provided by $NADPH + H^+$—to a three-carbon sugar phosphate (G3P). What follows this step is a complex sequence of reactions with two principal outcomes: the formation of more RuBP, and the release of products such as glucose. The production of one molecule of glucose ($C_6H_{12}O_6$) requires the Calvin–Benson cycle to operate six times on successive CO_2 molecules. Just as the respiration of one mole of glucose *yields* 686 kcal of energy (see Chapter 7), it *requires* 686 kcal to make one mole of glucose from CO_2.

Figure 8.26 gives a general summary of photosynthesis. The glucose produced in photosynthesis is subsequently used to make other compounds besides sugars. The carbon of glucose is incorporated into amino acids, lipids, and the building blocks of the nucleic acids. The products of the Calvin–Benson cycle are of crucial importance to the entire biosphere, for they serve as the food for all of life. Their covalent bonds represent the total energy yield from the harvesting of light by plants. Most of this stored energy is released by the plants themselves in their own glycolysis and cellular respiration. However, much plant matter ends up being consumed by animals. Glycolysis and cellular respiration in the animals then releases free energy from the plant matter for use in the animal cells.

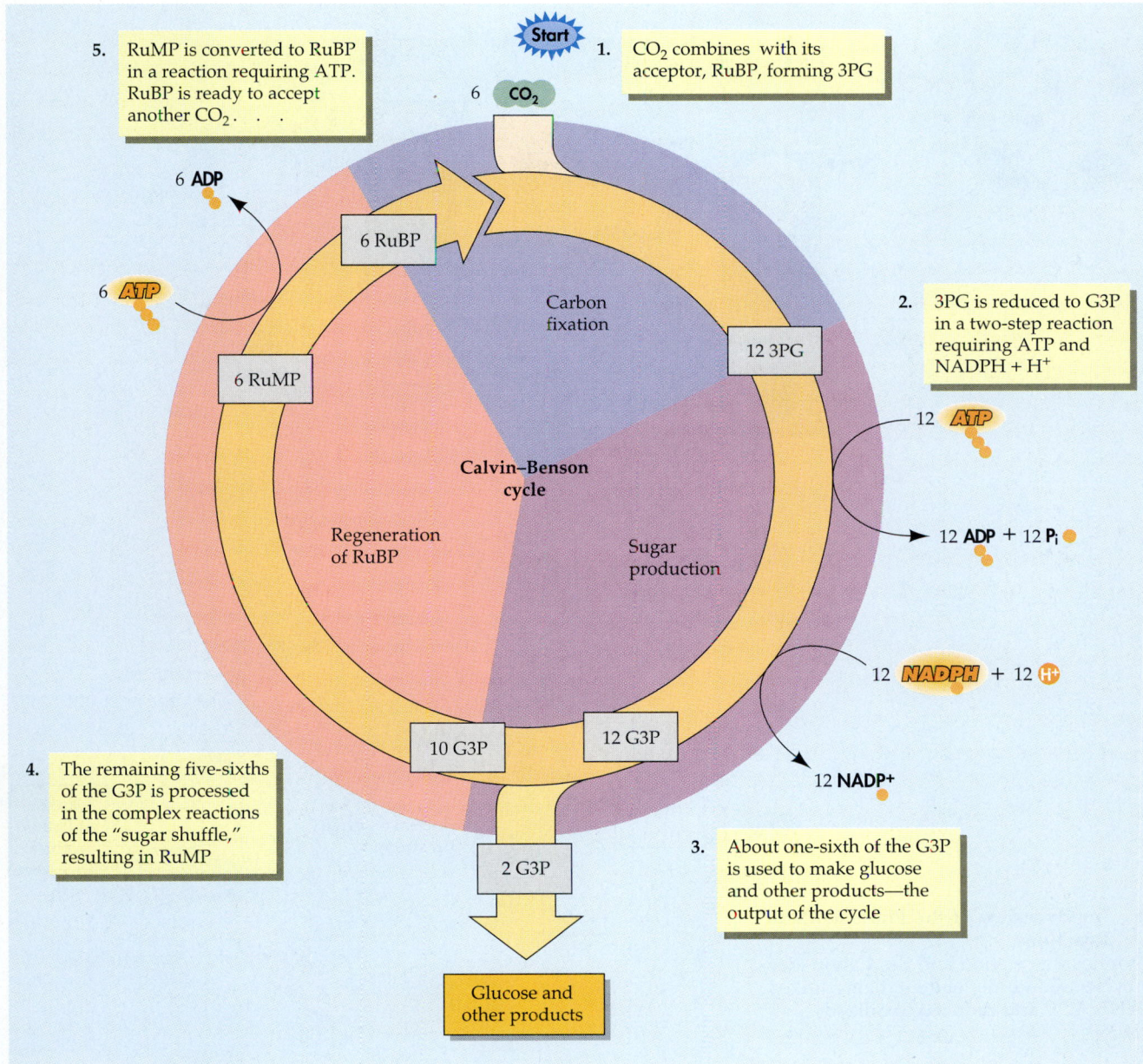

5. RuMP is converted to RuBP in a reaction requiring ATP. RuBP is ready to accept another CO_2. . .

Start

6 CO_2

1. CO_2 combines with its acceptor, RuBP, forming 3PG

6 **ADP**

6 **ATP**

6 RuBP

6 RuMP

Carbon fixation

Calvin–Benson cycle

12 3PG

2. 3PG is reduced to G3P in a two-step reaction requiring ATP and NADPH + H^+

12 **ATP**

12 **ADP** + 12 **P_i**

Regeneration of RuBP

Sugar production

12 **NADPH** + 12 **H^+**

4. The remaining five-sixths of the G3P is processed in the complex reactions of the "sugar shuffle," resulting in RuMP

10 G3P

12 G3P

12 **NADP+**

2 G3P

3. About one-sixth of the G3P is used to make glucose and other products—the output of the cycle

Glucose and other products

8.25 The Calvin–Benson Cycle
The Calvin-Benson cycle, sometimes called the dark reactions of photosynthesis, uses CO_2 to produce glucose and other organic molecules that contain the carbon and energy necessary for the many and varied processes of life. This diagram shows only the key steps; the values given are those necessary to make one molecule of glucose, which requires six "turns" of the cycle.

PHOTORESPIRATION

The substrate specificity of the enzyme rubisco is not limited to CO_2. Rubisco also catalyzes a reaction of RuBP with oxygen. (*Rubisco* stands for *ribulose bisphosphate carboxylase/oxygenase*.) The oxygenase reaction is favored when CO_2 levels are low or O_2 levels are high. One of the products of this reaction is glycolate, a two-carbon compound that leaves the chloroplast and diffuses into organelles called microbodies (see Chapter 4). In the microbody, glycolate is oxidized (in an oxygen-requiring reaction); later the product undergoes reactions in mitochondria leading to the release of CO_2. The rate of the overall process (from RuBP and O_2 to the release of CO_2) is roughly proportional to the light intensity. Because of its dependence on light and because it takes up oxygen and releases carbon dioxide, this process is called **photorespiration** (Figure 8.27).

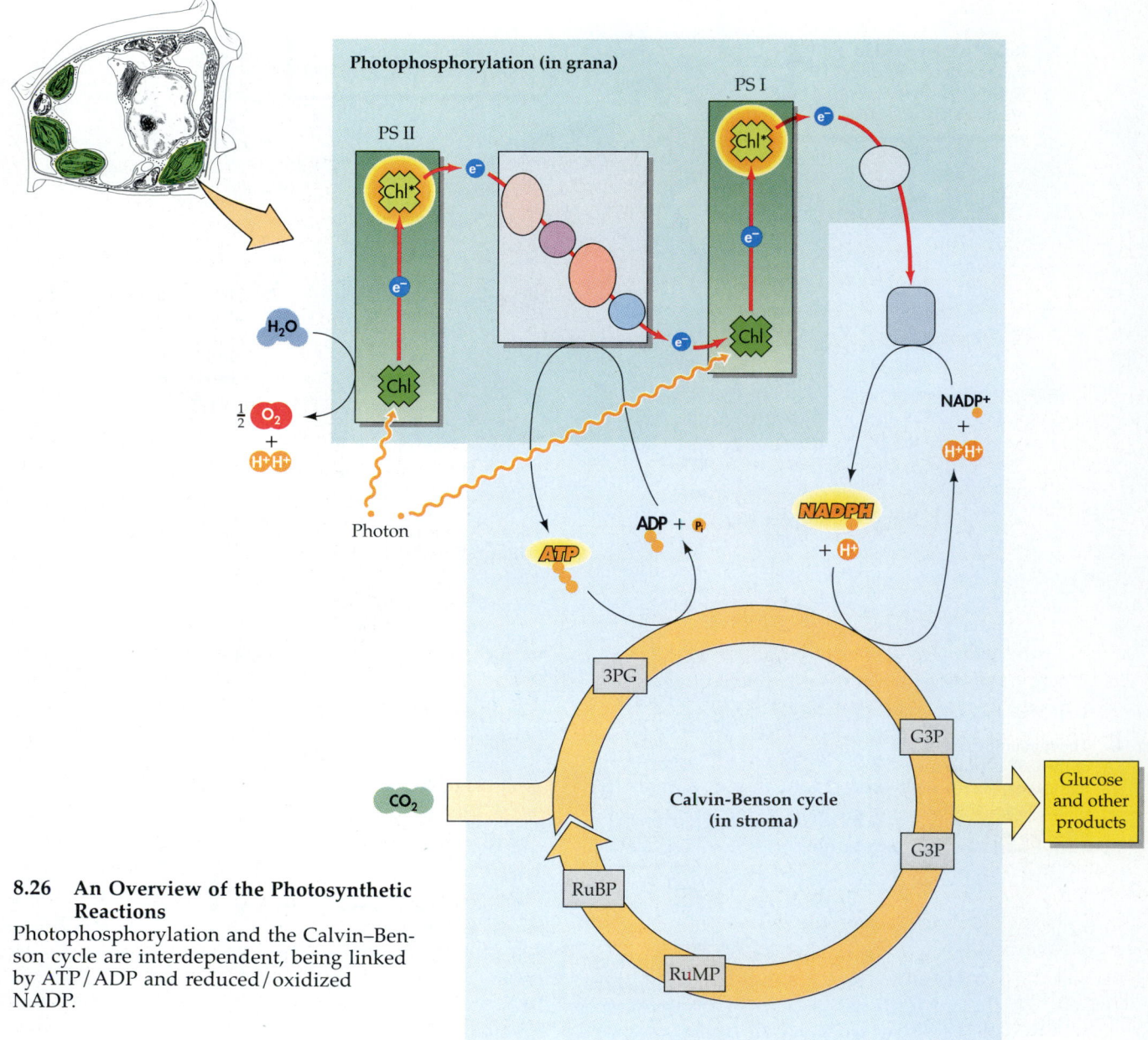

8.26   An Overview of the Photosynthetic Reactions
Photophosphorylation and the Calvin–Benson cycle are interdependent, being linked by ATP/ADP and reduced/oxidized NADP.

You can see that photorespiration interferes with photosynthesis; in fact, it apparently reverses it—but without resulting in ATP formation as does cellular respiration. The role of photorespiration in the life of the plant is unknown; it may simply be wasteful, with no positive role. With this in mind, many scientists are attempting to develop a gene that codes for a form of rubisco that recognizes only CO_2 as its substrate, and to insert that gene into crop plants. (See Chapter 14 for a discussion of this recombinant DNA technology.)

It seems odd, though, that rubisco, the most abundant single protein in the living world, should apparently function less than optimally. Most types of plants photorespire away a substantial fraction of the CO_2 initially fixed in photosynthesis. But, as we are about to see, some plants have minimized photorespiration, thus maximizing the efficiency of their photosynthesis.

ALTERNATE MODES OF CARBON DIOXIDE FIXATION

The discoveries of the Berkeley group led to the expectation that the exposure of a plant to both light and $^{14}CO_2$ would always lead to the appearance of 3-phospho[^{14}C]glycerate as the first labeled product of CO_2 fixation. Thus scientists were surprised when they learned that such treatment of chloroplasts extracted from sugarcane leaves leads instead to the formation of four-carbon acids as the first ^{14}C-con-

(a)

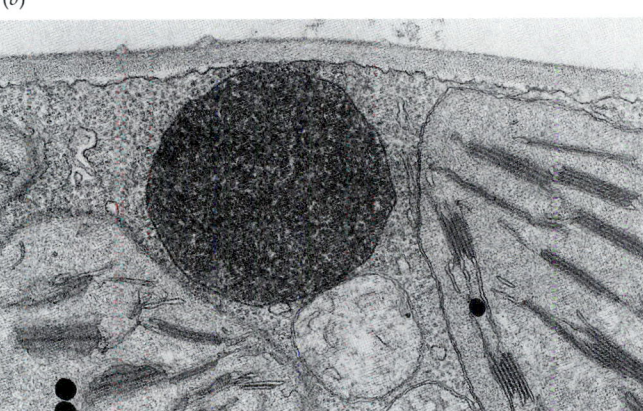

(b)

8.27 Chloroplasts, Microbodies, and Photorespiration
(a) In the chloroplast, the enzyme rubisco may catalyze a reaction between RuBP and O_2; further reactions lead to the reduced two-carbon compound glycolate (CC_{red}). Glycolate leaves the chloroplast and diffuses into a microbody, where it is oxidized. Later reactions lead to the formation of CO_2, which diffuses out of the cell. With the uptake of O_2 and the release of CO_2, the overall result re- sembles respiration, and because the process occurs in rough proportion to light intensity, it has been dubbed "photorespiration." (b) The dark, round object is a microbody in a mesophyll cell of a tobacco leaf. Portions of adjacent chloroplasts are visible to the lower left and far right, a mitochondrion is to the immediate lower right, and the plasma membrane and cell wall are above the microbody in this view.

taining products. Subsequently it was shown that many plants follow this pattern in fixing CO_2. Known as **C_4 plants** because the first products of their CO_2 fixation are four-carbon compounds, these plants perform the normal Calvin–Benson cycle, but they have an additional early step that traps CO_2 without losing carbon to photorespiration, greatly increasing the overall photosynthetic yield. Because this step functions even at low levels of CO_2 within the leaf, C_4 plants very effectively optimize photosynthesis.

C_4 plants live in environments where water is not always available. To prevent excessive water loss, the leaves keep their stomata closed much of the time. This leads to a depletion, by photosynthesis, of CO_2 within the leaf. However, because their leaves contain the enzyme **PEP carboxylase** (phosphoenolpyruvate carboxylase), C_4 plants have a means of compensating for this depletion. PEP carboxylase catalyzes the reaction of PEP (phosphoenolpyruvate, a three-carbon acid) with CO_2 to yield the four-carbon compound oxaloacetate as the first product of CO_2 fixation. PEP carboxylase has a much greater affinity for CO_2 than does rubisco and, because it lacks an oxygenase function, PEP does not support photorespiration, so C_4 plants can trap CO_2 even when that gas is present in a much reduced concentration. (You may recall that PEP is a late intermediate in glycolysis—see Figure 7.11—and that oxaloacetate is the last intermediate in the citric acid cycle—see Figure 7.13. Evolution has led to the use of certain compounds in a number of different ways in living things.)

The leaf anatomy of C_4 plants differs from that of **C_3 plants** (plants that produce the three-carbon com- pound 3PG in the first step of the Calvin–Benson cycle). C_3 plants have only one type of cell capable of photosynthesis (Figure 8.28a), but the leaves of C_4 plants have two classes of photosynthetic cells, each with a distinctive type of chloroplast. The cells are arranged as shown in Figure 8.28b, with a photosynthetic **mesophyll** layer surrounding an inner layer of **bundle sheath cells**, which are also photosynthetic. Like those of C_3 plants, the mesophyll cells of C_4 plants contain chloroplasts filled with grana that trap carbon dioxide for photosynthesis. From this trapped CO_2, C_4 plants produce four-carbon compounds that diffuse into the bundle sheath cells. Once in the bundle sheath cells these compounds are decarboxylated (a carboxyl group is removed) to release CO_2, which is recaptured by rubisco and used in the Calvin–Benson cycle—the cycle that in C_3 plants takes place entirely within the mesophyll cells (Figure 8.29). The chloroplasts in the bundle sheath cells of C_4 plants lack well-developed grana but typically have substantial starch grains deposited in them because they, rather than the mesophyll chloroplasts, are the sites where sugars are finally formed and starches are stored.

What this system does is pump CO_2 from a region where its concentration is low (the intercellular spaces within the leaf) to one where it is relatively more abundant (the bundle sheath layer). At the same time, C_4 photosynthesis bypasses photorespiration, thus retaining more of the carbon that is fixed in photosynthesis. PEP carboxylase in the mesophyll chloroplasts can take up CO_2 when rubisco cannot, and, because it lacks the oxygenase activity of ru-

(a)   Arrangement of cells in a C₃ leaf

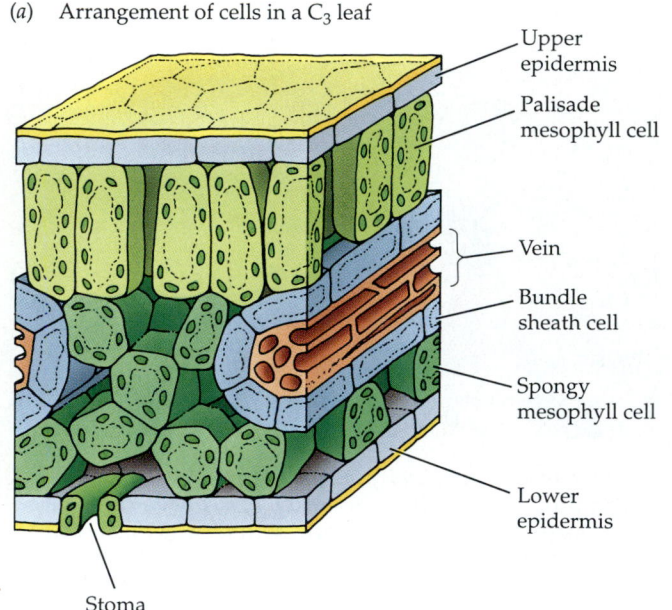

Upper
epidermis

Palisade
mesophyll cell

Vein

Bundle
sheath cell

Spongy
mesophyll cell

Lower
epidermis

Stoma

(b)   Arrangement of cells in a C₄ leaf

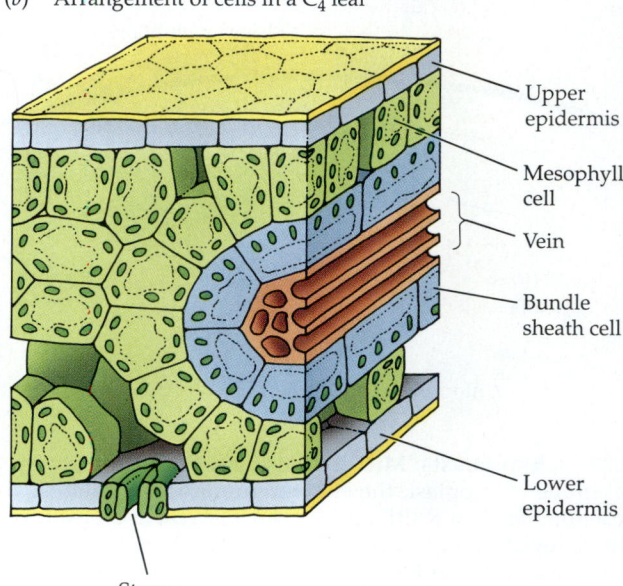

Upper
epidermis

Mesophyll
cell

Vein

Bundle
sheath cell

Lower
epidermis

Stoma

8.28   C₄ Plants Differ from C₃ Plants in Leaf Anatomy
(a) In the leaf of a C₃ plant, the bundle sheath cells surrounding the vascular elements of a vein are relatively small; the upper part of the leaf is filled with upright palisade mesophyll cells; and loosely arranged spongy mesophyll cells allow gases to circulate in the lower layers within the leaf. Both mesophyll layers carry on photosynthesis, but the bundle sheath cells do not. (b) In a C₄ leaf, the bundle sheath cells are usually larger and contain prominent chloroplasts toward their outer edges; uniform mesophyll cells surround the entire vascular bundle. This arrangement facilitates the incorporation of carbon from CO₂ into four-carbon compounds by the mesophyll cells and the passage of these carbon-containing compounds to the bundle sheath cells, where the reactions of the Calvin–Benson cycle take place. The mesophyll serves as a CO₂ pump.

bisco, does not cause photorespiration. The temporary products (four-carbon acids) are then loaded into the bundle sheath, allowing the release of sufficient CO₂ to keep the rubisco in the bundle sheath busy. The O₂ level in the bundle sheath is not so high, however, as to cause photorespiration. Table 8.1 compares C₃ and C₄ photosynthesis.

A related but distinguishable system called **Crassulacean acid metabolism** (CAM) functions in certain other plants that face frequent water shortages. Many of these are members of the family Crassulaceae, which includes the ice plants and some other succulent plants. Many cacti also perform CAM. Because of the way in which the stomata of CAM plants are regulated, these plants have access to atmospheric CO₂ only at night. By day their stomata are closed, preventing water loss, and no CO₂ can enter the leaf. Using PEP carboxylase to trap CO₂ at night allows these plants to store great quantities of CO₂ in the form of carboxyl groups of four-carbon acids. By day, behind closed stomata, the CO₂ is released within the leaves; it is recaptured by rubisco, and photosynthesis then proceeds by way of the Calvin–Benson cycle. The difference between this system and C₄ photosynthesis is that here the PEP comes from the respiratory breakdown of sugars at night; in C₄ photosynthesis, PEP is produced photosynthetically in a light-requiring reaction by day.

TABLE 8.1
Comparison of Photosynthesis in C₃ and C₄ Plants

VARIABLE	C₃ PLANTS	C₄ PLANTS
Photorespiration	Extensive	Minimal
Perform Calvin–Benson cycle	Yes	Yes
Primary CO₂ acceptor	RuBP	PEP
CO₂-fixing enzyme	Rubisco (RuBP carboxylase)	PEP carboxylase
First product of CO₂ fixation	3PG (3-carbon compound)	Oxaloacetate (4-carbon compound)
Affinity of carboxylase for CO₂	Moderate	High
Leaf anatomy: photosynthetic cells	Mesophyll	Mesophyll + bundle sheath
Classes of chloroplasts	One	Two

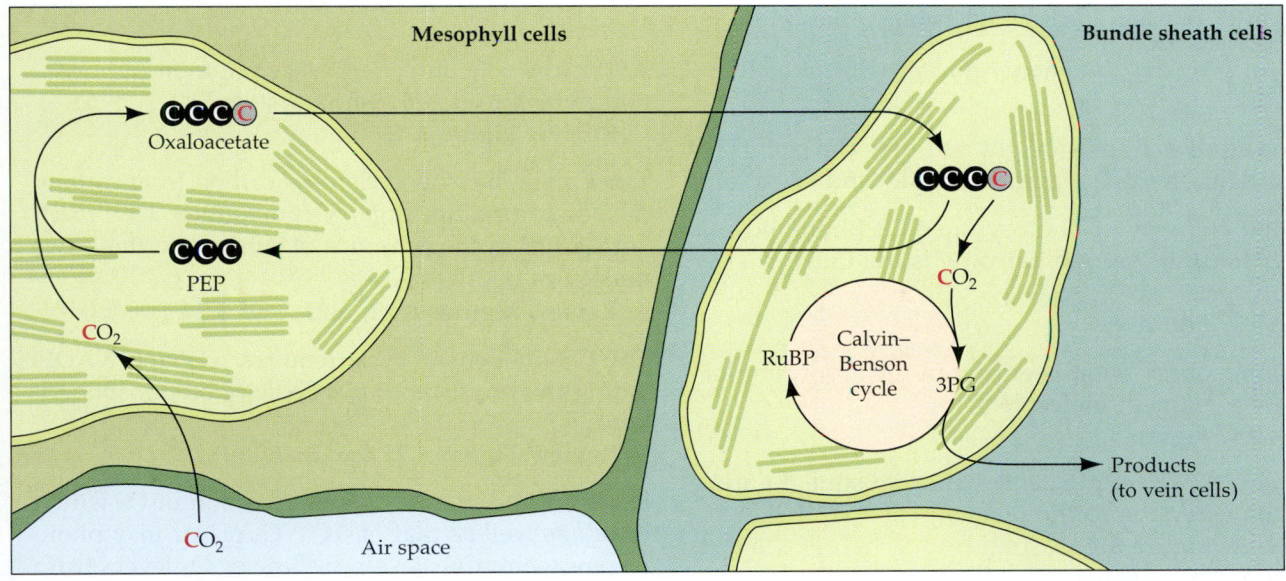

Mesophyll cells

Oxaloacetate

PEP

CO_2

CO_2

Air space

Bundle sheath cells

CO_2

RuBP

Calvin–Benson cycle

3PG

Products (to vein cells)

(a)

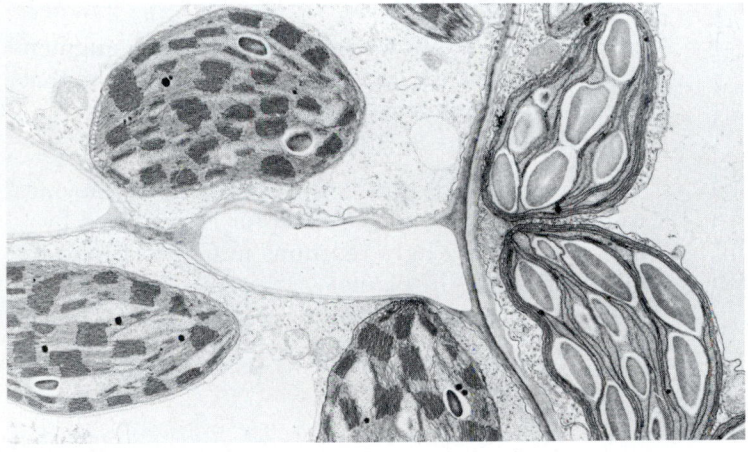

(b)

8.29 C₄ Photosynthesis

(a) In C_4 photosynthesis, mesophyll cells in the leaf take up CO_2 and incorporate the carbon atom into four-carbon compounds. The four-carbon compounds diffuse into adjacent bundle sheath cells, where they are decarboxylated, releasing CO_2. The enzyme rubisco picks up this CO_2, and the usual Calvin–Benson cycle of C_3 photosynthesis ensues. (b) Portions of two mesophyll cells (left) and two chloroplasts in a single bundle sheath cell (right) from the leaf of a C_4 plant. Note the numerous grana and few starch grains in the chloroplasts of the mesophyll cells; in the chloroplasts of the bundle sheath cell, where the Calvin–Benson cycle forms the products of photosynthesis, there are many large, oval starch granules but very few membranes organized into grana.

PHOTOSYNTHESIS AND CELLULAR RESPIRATION

In plants, cellular respiration takes place both in the light and in the dark, whereas photosynthesis takes place only in the light. The site of glycolysis is the cytosol, that of respiration is the mitochondria, and that of photosynthesis is the chloroplasts. Thus photosynthesis and respiration can proceed simultaneously but in different organelles.

For a plant to live, it must photosynthesize more than it respires, giving it a net gain of carbon dioxide and energy from the environment. Accordingly, the plant world, along with photosynthetic bacteria and protists, exports food—and oxygen—to the animal kingdom and to all other nonphotosynthetic organisms (with the exception of a few types of bacteria). Animals require both food and oxygen; they return carbon dioxide that plants may use in photosynthesis. Thus both carbon dioxide and oxygen have natural cycles.

Photosynthesis and respiration have important similarities. In eukaryotes, both processes reside in specialized organelles that have complex systems of internal membranes. ATP synthesis in both processes relies on the chemiosmotic mechanism, involving the pumping of protons through a membrane. Another key feature of both respiration and photosynthesis is electron transport, that is, the passing of electrons from carrier to carrier in a series of exergonic redox reactions. In respiration, the carriers receive electrons from high-energy food molecules and pass them ultimately to oxygen, forming water. On the other hand, photosynthesis requires an input of light energy to make chlorophyll, a reducing agent strong enough to initiate the transfer of electrons. In photosynthesis, water is the source of the electrons, and oxygen is released from water in a very early step. The electrons from water end up in NADPH and, finally, in food molecules.

SUMMARY of Main Ideas about Photosynthesis

The sun provides the energy for virtually all biological work.

Photosynthesis traps sunlight and uses the converted light energy to synthesize ATP and sugars.

The first pathway of photosynthesis is photophosphorylation; the second pathway is the Calvin–Benson cycle.
   Review Figure 8.4

Photophosphorylation begins with the absorption of light by a pigment molecule.
   Review Figure 8.8

Several chloroplast pigments help trap sunlight and pass the excitation to the pigment cholorophyll.
   Review Figure 8.9

In photophosphorylation, an excited chlorophyll molecule passes an electron and energy to an oxidizing agent; the electron then passes through a series of carriers, and some of the energy thus released is used to form ATP chemiosmotically.
   Review Figures 8.13 and 8.18

Noncyclic photophosphorylation produces NADPH + H$^+$ as well as ATP; it also produces O$_2$ by the breakdown of water.

Two chlorophyll-containing photosystems participate in noncyclic photophosphorylation.
   Review Figure 8.14

ATP is the sole output of cyclic photophosphorylation, which uses only photosystem I.
   Review Figure 8.15

The Calvin–Benson cycle uses carbon from carbon dioxide to produce organic compounds. CO$_2$ reacts with RuBP in the presence of rubisco, forming two molecules of 3PG.
   Review Figures 8.22, 8.23, and 8.24

The Calvin–Benson cycle requires ATP and NADPH + H$^+$, the products of noncyclic photophosphorylation.
   Review Figure 8.25 and 8.26

Because rubisco catalyzes the reaction of O$_2$ with RuBP, as well as that of CO$_2$, C$_3$ plants may photorespire when CO$_2$ levels are low or O$_2$ levels are high.
   Review Figure 8.27

C$_4$ plants rarely photorespire because they augment the Calvin–Benson cycle with further reactions that permit photosynthesis even at low CO$_2$ levels.
   Review Figure 8.28 and 8.29 and Table 8.1

C$_4$ photosynthesis and CAM use the same enzyme to trap CO$_2$, but the CO$_2$ acceptor is produced by energy-requiring light reactions in C$_4$ plants and by glycolysis in CAM plants.

SELF-QUIZ

1. Which statement about light is *not* true?
   a. Its velocity in a vacuum is constant.
   b. It is a form of energy.
   c. The energy of a photon is directly proportional to its wavelength.
   d. A photon of blue light has more energy than one of red light.
   e. Different colors correspond to different frequencies.

2. Which statement about light is true?
   a. An absorption spectrum is a plot of biological effectiveness versus wavelength.
   b. An absorption spectrum may be a good means of identifying a pigment.
   c. Light need not be absorbed to produce a biological effect.
   d. A given kind of molecule can occupy any energy level.

   e. A pigment loses energy as it absorbs a photon.

3. Which statement is *not* true of chlorophylls?
   a. They absorb light near both ends of the visible spectrum.
   b. They can accept energy from other pigments, such as carotenoids.
   c. Excited chlorophyll (Chl*) can either reduce another substance or fluoresce.
   d. Excited chlorophyll is an oxidizing agent.
   e. They contain magnesium.

4. In cyclic photophosphorylation
   a. oxygen gas is released.
   b. ATP is formed.
   c. water donates electrons and protons.
   d. NADPH + H$^+$ is formed.
   e. CO$_2$ reacts with RuBP.

5. Which does *not* happen in noncyclic photophosphorylation?
   a. Oxygen gas is released.
   b. ATP is formed.
   c. Water donates electrons and protons.
   d. NADPH + H$^+$ is formed.
   e. CO$_2$ reacts with RuBP.

6. In the chloroplast
   a. light leads to the pumping of protons out of the thylakoids.
   b. ATP is formed when protons are pumped into the thylakoids.
   c. light causes the stroma to become more acidic than the thylakoids.
   d. protons return passively to the stroma through protein channels.
   e. proton pumping requires ATP.

7. Which is *not* true of the Calvin–Benson cycle?
   a. CO$_2$ reacts with RuBP to form 3PG.

b. RuBP is formed by the metabolism of 3PG.

c. ATP and NADPH + H$^+$ are formed when 3PG is reduced.

d. The concentration of 3PG rises if the light is switched off.

e. Rubisco catalyzes the reaction of CO_2 and RuBP.

8. In C$_4$ photosynthesis
   a. 3PG is the first product of CO_2 fixation.
   b. rubisco catalyzes the first step in the pathway.
   c. four-carbon acids are formed by PEP carboxylase in the bundle sheath.

d. photosynthesis continues at lower CO_2 levels than in C$_3$ plants.

e. CO_2 released from RuBP is transferred to PEP.

9. C$_4$ photosynthesis and the acid metabolism in Crassulaceae differ in that
   a. only C$_4$ photosynthesis uses PEP carboxylase.
   b. CO_2 is trapped by night in Crassulaceae and by day in C$_4$ plants.

c. four-carbon acids are formed only in C$_4$ photosynthesis.
d. only Crassulaceae commonly grow in dry or salty environments.
e. only C$_4$ photosynthesis helps conserve water.

10. Photorespiration
    a. takes place only in C$_4$ plants.
    b. includes reactions carried out in microbodies.
    c. increases the yield of photosynthesis.
    d. is catalyzed by PEP carboxylase.
    e. is independent of light intensity.

FOR STUDY

1. Both photophosphorylation and the Calvin–Benson cycle stop when the light is turned off. Which specific reaction stops first? Which stops next? Continue answering the question "Which stops next?" until you have explained why both pathways have stopped.

2. In what principal ways are the reactions of photophosphorylation similar to the respiratory chain and oxidative phosphorylation discussed in Chapter 7? Differentiate

between cyclic and noncyclic photophosphorylation in terms of (1) the *products* and (2) the *source* of electrons for reduction of oxidized chlorophyll.

3. The development of what three experimental techniques made it possible to elucidate the Calvin–Benson cycle? How were these techniques used in the investigation?

4. If water labeled with ^{18}O is added to a suspension of photosynthesizing chloroplasts, which of the following compounds will first become labeled with ^{18}O: ATP, NADPH, O_2, or 3PG? If water labeled with 3H is added to a suspension of photosynthesizing chloroplasts, which of those compounds will first become radioactive? If CO_2 labeled with ^{14}C is added to a suspension of photosynthesizing chloroplasts, which of those compounds will first become radioactive?

READINGS

Alberts, B., D. Bray, J. Lewis, M. Raff, K. Roberts and J. D. Watson. 1994. *Molecular Biology of the Cell*, 3rd Edition. Garland Publishing, New York. Chapter 14 on energy conversion contains a good discussion of photosynthesis.

Bjorkman, O. and J. Berry. 1973. "High-Efficiency Photosynthesis." *Scientific American*, October. A discussion of C$_4$ photosynthesis and what it means to the plants in which it is found.

Clayton, R. K. 1980. *Photosynthesis*. Cambridge University Press, New York. An advanced general treatment of photosynthesis by a prominent photobiologist.

Govindjee and W. J. Coleman. 1990. "How Plants Make Oxygen." *Scientific American*, February. A "clock" in photosystem II that splits water into oxygen gas, protons, and electrons.

Hall, D. O. and K. K. Rao. 1987. *Photosynthesis*, 4th Edition. Edward Arnold, New York. An intermediate-level treatment of all the major topics in photosynthesis and excellent bibliography, all in 100 pages.

Stryer, L. 1995. *Biochemistry*, 4th Edition. W. H. Freeman, New York. Chapter 22 gives an advanced but clear treatment of topics in photosynthesis.

Voet, D. and J. G. Voet. 1990. *Biochemistry*. John Wiley & Sons, New York. Chapter 22 discusses photosynthesis.

Weinberg, C. J. and R. H. Williams. 1990. "Energy from the Sun." *Scientific American*, September. Photosynthesis and biomass technology, along with other solar-derived technologies such as wind and solar-thermal, are considered as sources of energy for industrial and other uses.

Youvan, D. C. and B. L. Marrs. 1987. "Molecular Mechanisms of Photosynthesis." *Scientific American*, June. A difficult but interesting article on events in the first fraction of a millisecond of photosynthesis in a bacterium. Part of the article is better read after reading Part Two of this book.

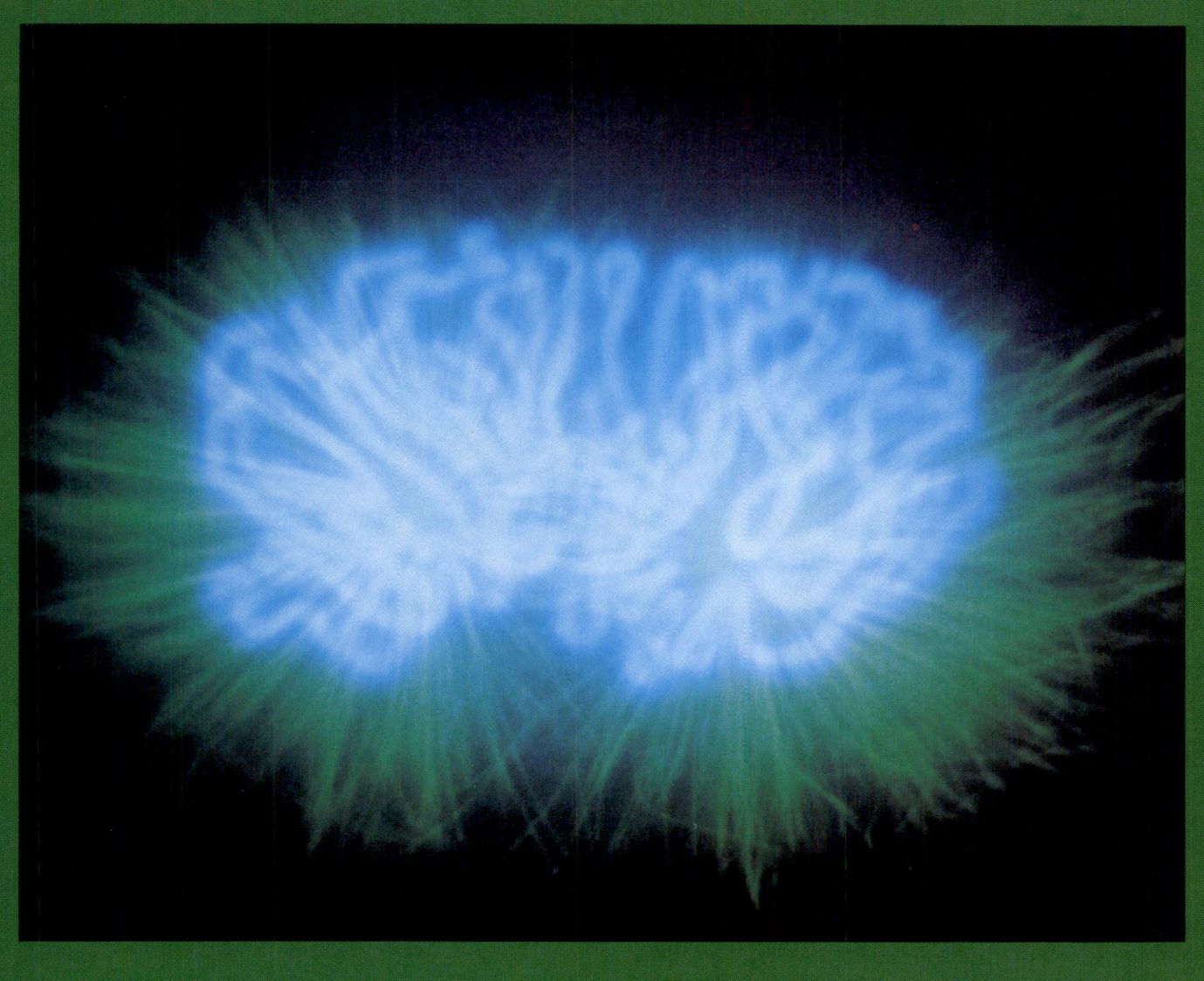

PART TWO
Information and Heredity

More than one hundred trillion (10^{14}) cells make up an adult human. As a fertilized human egg—a single cell—develops into a university student, its nucleus gives rise to over one hundred trillion nuclei, each containing basically the same genetic information as did the fertilized egg. An intricate mechanism first copies the genetic material in the nucleus and then partitions it into two daughter nuclei so that each gets one complete copy of the genetic information.

Part Two of this textbook deals with information and heredity. Multicellular organisms use their information to develop from a single cell, and each cell uses some of the information to build macromolecules and organelles. Each cell of an organism contains a full set of the information needed to build the whole organism, but a given cell acts on only the information that is relevant to it. The information is contained in the structures of nucleic acid molecules and is processed by enzymes. The structures of the enzymes, in turn, are encoded in the nucleic acids.

Heredity is the passing of information from one generation to the next. The orderly expression of hereditary information is required for development. This is especially evident in the development of the immune system, an important defense against disease. Recombinant DNA technology, an application of our knowledge about information and heredity, is providing us with new approaches to the treatment of many diseases.

Our early successes in understanding the molecular aspects of information and heredity came from studies of the genetics and molecular biology of prokaryotes. One entire chapter in Part Two (Chapter 12) is devoted to prokaryotes, and these fascinating organisms appear from time to time elsewhere in this part of the book. However, eukaryotic molecular biology is now making remarkable strides, and eukaryotes will occupy most of our attention. This chapter focuses almost exclusively on eukaryotic cells, although the last section addresses cell division in prokaryotes.

THE DIVISIONS OF EUKARYOTIC CELLS

The "intricate mechanism" that copies the genetic material for cell division is **mitosis**, which produces exact copies of a nucleus. This biological copying machine turns out the nuclei of the many cells of an organism's adult body.

A second mechanism for nuclear division is **meiosis**. This mechanism produces four daughter nuclei, each with only *half* the genetic information contained in the original cell, and each differing from the others in the exact information contained. When organisms reproduce sexually, pairs of such cells

9
Chromosomes and Cell Division

combine. Each sexual union of cells may produce new genetic combinations. Because of meiosis, offspring of the same parents differ.

The division of a eukaryotic cell consists of two steps: first the division of the nucleus (by mitosis or meiosis), then the division of the cytoplasm. Between divisions—that is, for most of its life—a eukaryotic cell is in a condition called **interphase**. What determines whether a cell in interphase will proceed to divide? How does mitosis lead to exact copies, and how does meiosis lead to diversity of products? Why do we need both exact copies *and* diverse products? Why do most organisms have sex in their life cycles? In this chapter we will learn the details of mitosis, meiosis, and interphase, as well as their biological consequences, which are of the utmost importance for heredity, development, and evolution.

EUKARYOTIC CHROMOSOMES AND CHROMATIN

All human cells, other than eggs and sperm, contain two full sets of genetic information, one from the mother and the other one from the father. Eggs and sperm, however, contain only a single set. Any particular egg or sperm in your body contains some information from your mother and some from your father. The genetic information consists of molecules of DNA packaged as **chromosomes** in the nucleus (see Chapter 4). Through a microscope, the nucleus appears relatively featureless (except for the nucleolus) during most of the life of a cell; the chromosomes cannot be seen (Figure 9.1).

The basic unit of the eukaryotic chromosome is a gigantic, linear, double-stranded molecule of DNA complexed with many proteins. During many stages of a eukaryotic cell's life cycle, each chromosome contains only one such DNA molecule. At other times, however, the DNA molecule doubles; the chromosome then comprises two joined **chromatids**, each made up of one DNA molecule complexed with proteins. At the particular times when chromosomes are visible in microscopes, the chromatids are joined in a specific, small region of the chromosome called the **centromere** (Figure 9.2). As we will see, centromeres direct the movement of chromosomes when a nucleus divides. A body that has a single centromere, whether it contains one or two DNA molecules, is properly called a *chromosome*.

The complex of DNA and proteins in a eukaryotic chromosome is referred to as **chromatin**. The DNA carries the genetic information, and the proteins organize the chromosome physically and regulate the activities of the DNA. Chromatin changes dramatically during mitosis and meiosis. During interphase, the chromatin of a chromosome is strung out so thinly that the chromosome cannot be seen clearly as a defined body under the light microscope. During most of mitosis and meiosis, however, the chromatin is coiled and compacted to a high degree, so that the chromosome appears as a bulky object (Figure 9.2*a*). This alternation of forms of chromatin relates to what the chromatin is doing during interphase and division. The genetic material is duplicated before each mitosis. Mitosis then separates this duplicated material into two new nuclei. This separation is accomplished more easily if the DNA is neatly arranged in compact units rather than being tangled up like a plate of spaghetti. During interphase, however, the DNA must direct the growth and other activities of

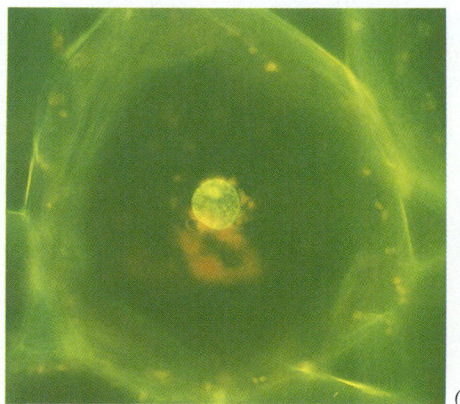

(a)

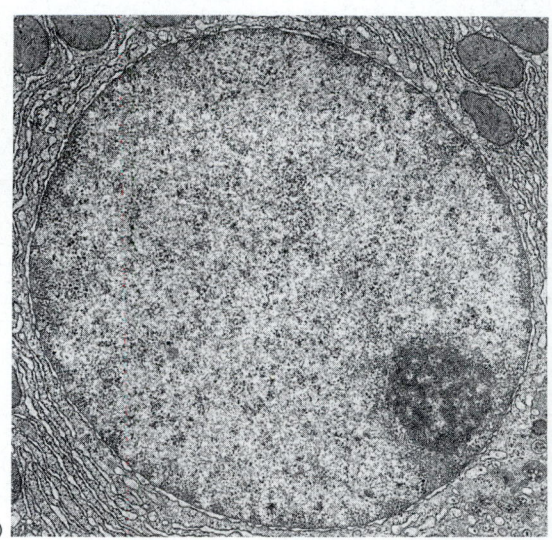

(b)

9.1 Nuclei
(a) The bright object in the middle of this spinach cell is a nucleus, as resolved through a light microscope. The smaller round spot inside the nucleus is a nucleolus, the site of ribosome assembly. (b) This view of the nucleus of an animal cell required an electron microscope. The two membranes of the nuclear envelope are distinct, as are a number of pores in the envelope. The nucleolus is at the lower right.

(a)

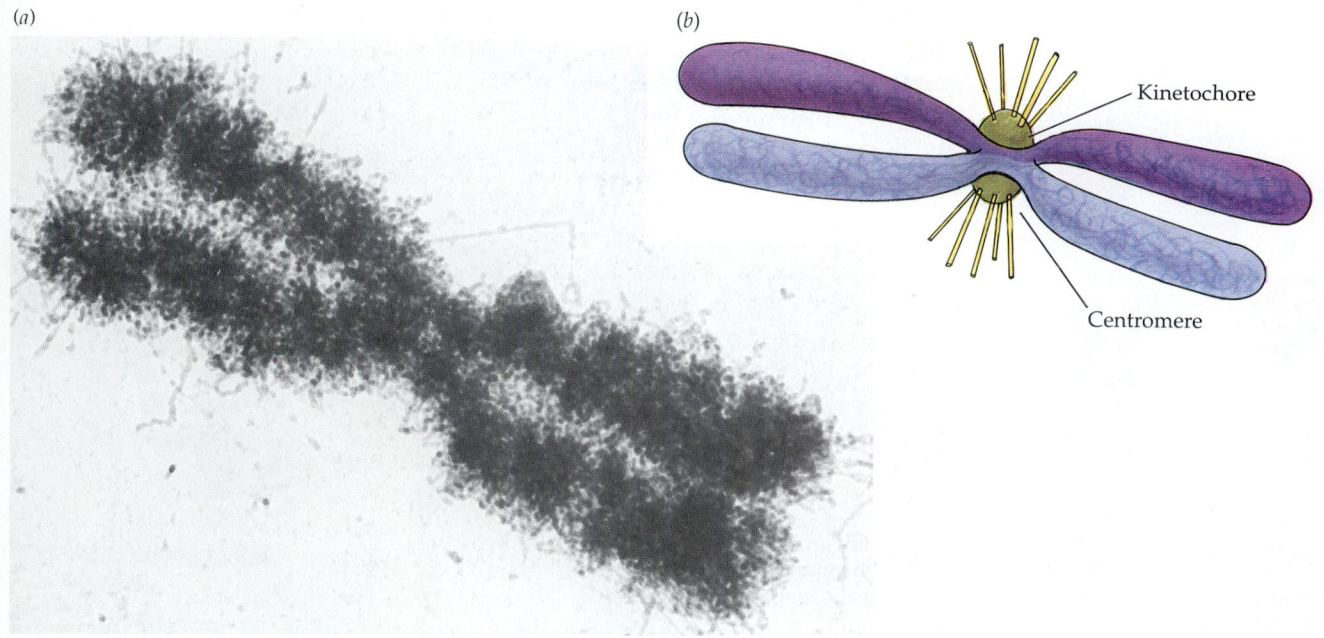

(b)

Kinetochore

Centromere

9.2 Chromosomes, Chromatids, and Chromatin
(a) A human chromosome in which the centromere is visible as a pinched-in region in the center. At the stage of the cell cycle captured here, the chromosome consists of two chromatids lying side by side. Individual fibers of chromatin are visible at the chromosome's edges. (b) A diagrammatic representation of the chromosome, with the two chromatids shown in different shades of blue. (Kinetochores will be described later in this chapter.)

the cell. As we will see in Chapters 11 and 13, DNA does this by interacting with enzymes while unwound and exposed.

Chromatin proteins associate closely with the DNA in chromosomes. Chromosomes contain large quantities of five classes of proteins, all of which are known as **histones**. Small for protein molecules, histones have a positive charge at pH levels found in the cell. The positive charge is a result of their par-

ticular amino acid compositions. Histone molecules join together to produce complexes around which the DNA is wound. Eight histone molecules, two of each of four of the histone classes, unite to form a core or spool shaped so that the DNA molecule fits snugly in a coil around it. The fifth class of histone (H1) appears to fit on the outside of the DNA, perhaps "clamping" it to the histone core. Strong evidence indicates that chromatin consists of a great number of these beadlike units, or **nucleosomes**, connected by a DNA thread (Figure 9.3).

A chromatid has a single DNA molecule running through vast numbers of nucleosomes. The many nucleosomes of a mitotic chromosome may pack together and coil as shown in Figure 9.4. During both

9.3 DNA Plus Histones Form Nucleosomes
(a) The DNA double helix coils around a central core of eight histone molecules to make a nucleosome. Another histone (H1) clamps the DNA to the core. (b) The "beads" of these chromatin fibers are nucleosomes. The "threads" connecting them are DNA.

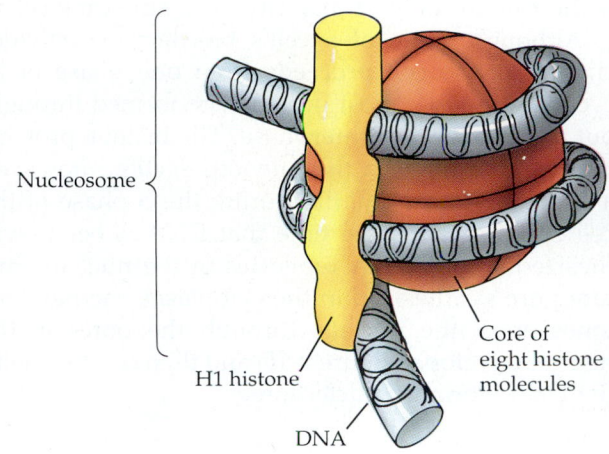

Nucleosome

H1 histone

DNA

Core of eight histone molecules

(a)

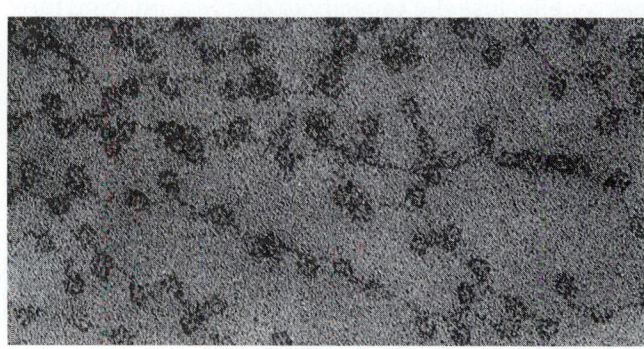

(b)

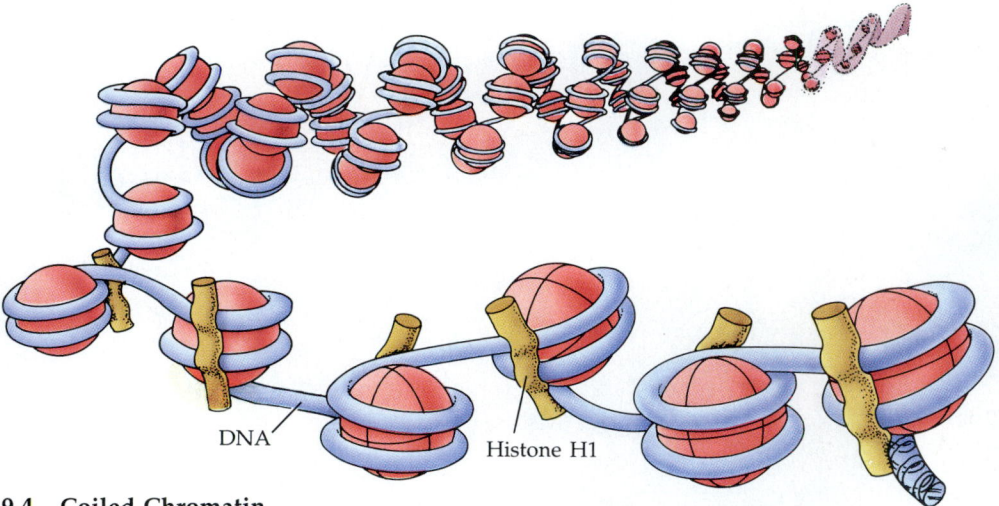

DNA

Histone H1

9.4 Coiled Chromatin
Nucleosomes are packed into a coil that twists into another, larger coil—and so forth—to produce condensed, supercoiled chromatin fibers such as those seen around the edges of the chromatids in Figure 9.2a. See also Figure 9.5.

mitosis and meiosis, the chromatin becomes ever more coiled and condensed, with further folding of the chromatin continuing up until the chromosomes separate (Figure 9.5). A diverse group of acidic proteins are also present in small quantities in chromosomes. The roles of these proteins will be considered in Chapter 13.

Although we know less about the organization of interphase chromatin than about that of mitotic chromatin, we do know that interphase chromatin has nucleosomes that are spaced at the same intervals as in supercoiled chromatin. During interphase, DNA thus remains associated with histone molecules while it replicates and directs synthesis of RNA. Current research is investigating the possibility that the structure of the nucleosomes changes as the cell proceeds with its interphase activities.

THE CELL CYCLE

A cell lives and functions until it divides or dies—or, if it is a sex cell, until it fuses with another sex cell. Some cells, such as red blood cells, muscle cells, and nerve cells, lose the capacity to divide as they mature; cells of certain other types rarely divide. And then there are cancerous cells, which, having escaped from the normal controls on division, divide rapidly and inappropriately. Most cells, however, have some probability of dividing, and some are specialized for rapid division. Thus for many kinds of cells we may speak of a **cell cycle** that has mitosis as one phase and interphase as the other (Figure 9.6). A given cell lives for one turn of the cycle and becomes two cells.

For life as a whole, the cycle repeats again and again as a constant source of renewal. The cell cycle, even for tissues engaged in rapid growth, consists mainly of interphase. Examining any collection of cells, such as a root tip or a slice of liver, reveals that most of the cells are in interphase most of the time. Only a small percentage of the cells are in mitosis at any given moment; this fact can be confirmed, in certain cultures of cells, by watching a single cell through its entire cell cycle.

The cell's DNA replicates during the **S phase** of interphase (the S stands for synthesis). The gap between the S phase and the onset of mitosis is referred to as Gap 2, or **G2**. Another gap phase—**G1**—separates the end of mitosis from the onset of the next S phase. If a cell is not going to divide, it may remain in G1 for weeks or even for many years until it dies—it seemingly will not waste effort replicating its genetic material. (There are some exceptions in which cells that will not divide do synthesize DNA and are thus stuck in G2, but continuation of the G1 phase is the rule in the vast majority of nondividing cells.)

Although some of a cell's biochemical activities change as the cell proceeds from one phase of its cycle to the next, most proteins are formed throughout all subphases of interphase. The histone proteins that we discussed in the previous section, however, are synthesized primarily during the S phase of the cell cycle, at the same time that DNA is being synthesized. While DNA replicates in the nucleus, histones are synthesized in the cytoplasm; the new histones enter the nucleus through the pores in the nuclear envelope (Figure 9.1b) and then combine with the DNA, forming nucleosomes.

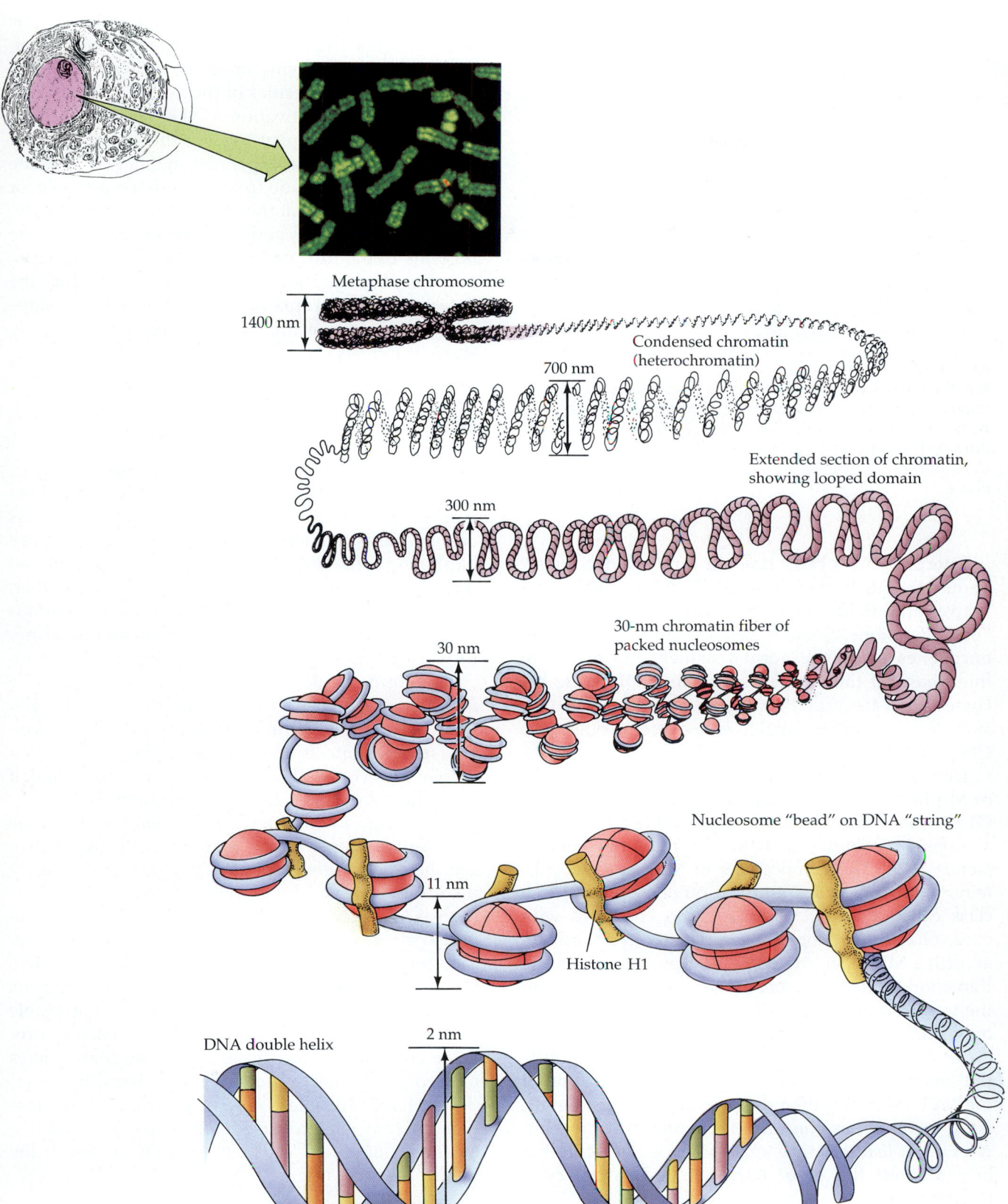

Metaphase chromosome

1400 nm

Condensed chromatin
(heterochromatin)

700 nm

Extended section of chromatin,
showing looped domain

300 nm

30-nm chromatin fiber of
packed nucleosomes

30 nm

Nucleosome "bead" on DNA "string"

11 nm

Histone H1

DNA double helix

2 nm

9.5 How DNA Packs into a Metaphase Chromosome

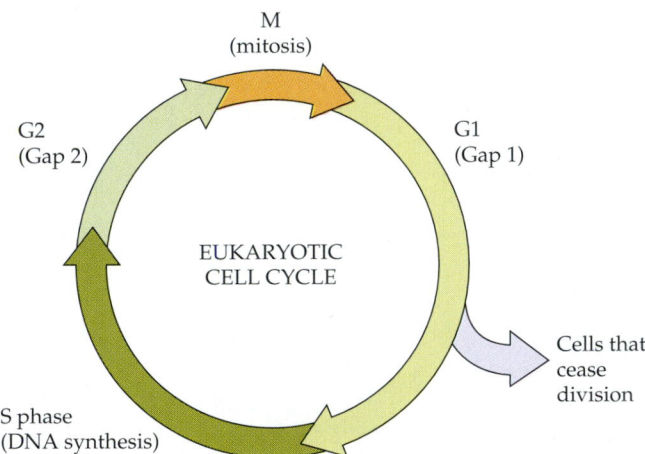

9.6   The Cell Cycle
A cell's life history is made up of a short mitosis and a longer interphase (green arrows). Interphase has three subphases in cells that divide. DNA is synthesized only during the short S phase between the G1 and G2 phases. Cells that do not divide are usually arrested in the G1 phase.

Interphase is a busy time—a time of "decisions." Should the nucleus replicate its DNA? Should it enter mitosis and divide? Some forms of cancer result from "bad" decisions. Some genes of a type called proto-oncogenes direct the normal sequence of events in interphase. If they mutate (become modified) to become what are called oncogenes, the normal decisions are no longer made, and cancer results (see Chapter 13).

How are appropriate decisions to enter the S phase or M phase (mitosis) made? These transitions—from G1 to S, and from G2 to M—require the activation of a protein complex called the maturation-promoting factor, MPF. The components of MPF are two proteins: one called cdc2 and the other belonging to a class called **cyclins**. When a cyclin combines with cdc2, enzymes act on the complex, converting it to an active MPF. One type of cyclin participates in the transition to the S phase; another type participates in the transition to M. In both cases, the active MPF not only brings about the appropriate phase of the cell cycle, but also activates other enzymes that degrade the cyclin, resulting eventually in the inactivation of the MPF itself! But then the *other* type of cyclin gradually increases in concentration sufficiently to bind with cdc2, leading once again to an active MPF that brings about the other transition in the cell cycle (Figure 9.7).

How does MPF produce its results? MPF has enzymatic activity that causes the addition of phosphate groups to other enzymes, making them active; the activity of these other enzymes is to add phosphate groups to still other enzymes, thus making *them* active. Each activation step multiplies the original effect, since each enzyme molecule catalyzes the activation of many molecules of the next kind of enzyme in this **cascade** of activation. (Other such cascades of phosphate transfer play roles in the action of many animal hormones, as explained in Chapter 36.) The products of the cascade from MPF are responsible for the observed effects at the cellular level. For example, one of the enzymes adds phosphate groups to the proteins of the nuclear lamina (see Chapter 4), causing the disintegration of the lamina and thus the detachment of chromatin from the nuclear envelope *and* the breakup of the nuclear envelope in an early stage of mitosis.

MITOSIS

In mitosis, a single nucleus gives rise to two nuclei that are genetically identical to each other and to the parent nucleus. Mitosis is a process of continuous change. Resist the temptation to think of it as a series of photographic slides in which one scene is replaced directly by another, distinctly different one. Rather, it is like a movie showing continuous action. For our discussion, however, it is convenient to look at mitosis as a series of important frames selected at intervals from the movie. Look at Figure 9.8 just one frame at a time, as we call for it in the text. After you have been through the entire story in the text, you can review mitosis by looking at Figure 9.8 as a whole; but you will find it easier to understand if you take it in smaller bites the first time through.

Let us begin with interphase, when the nucleus is between divisions. At the beginning of Figure 9.8 we see the nuclear envelope, the nucleoli, and a barely discernible tangle of chromatin. Immediately before mitosis (at the interphase–prophase transition), there is also a pair of **centrosomes** lying near the nucleus. The centrosomes are regions of the cell that help orchestrate chromosomal movement; these regions are not bounded by membranes and are not visible as discrete objects. In many organisms, each centrosome contains a pair of centrioles. The centrosomes of seed plants and some other organisms, however, do not have centrioles. Where present, each of these pairs of centrioles consists of one "parent" and one smaller "daughter" centriole at right angles to the parent centriole (Figure 9.9).

Development of the Chromosomes and Spindle

The appearance of the nucleus changes in the next frame of Figure 9.8, as the cell enters **prophase**, the

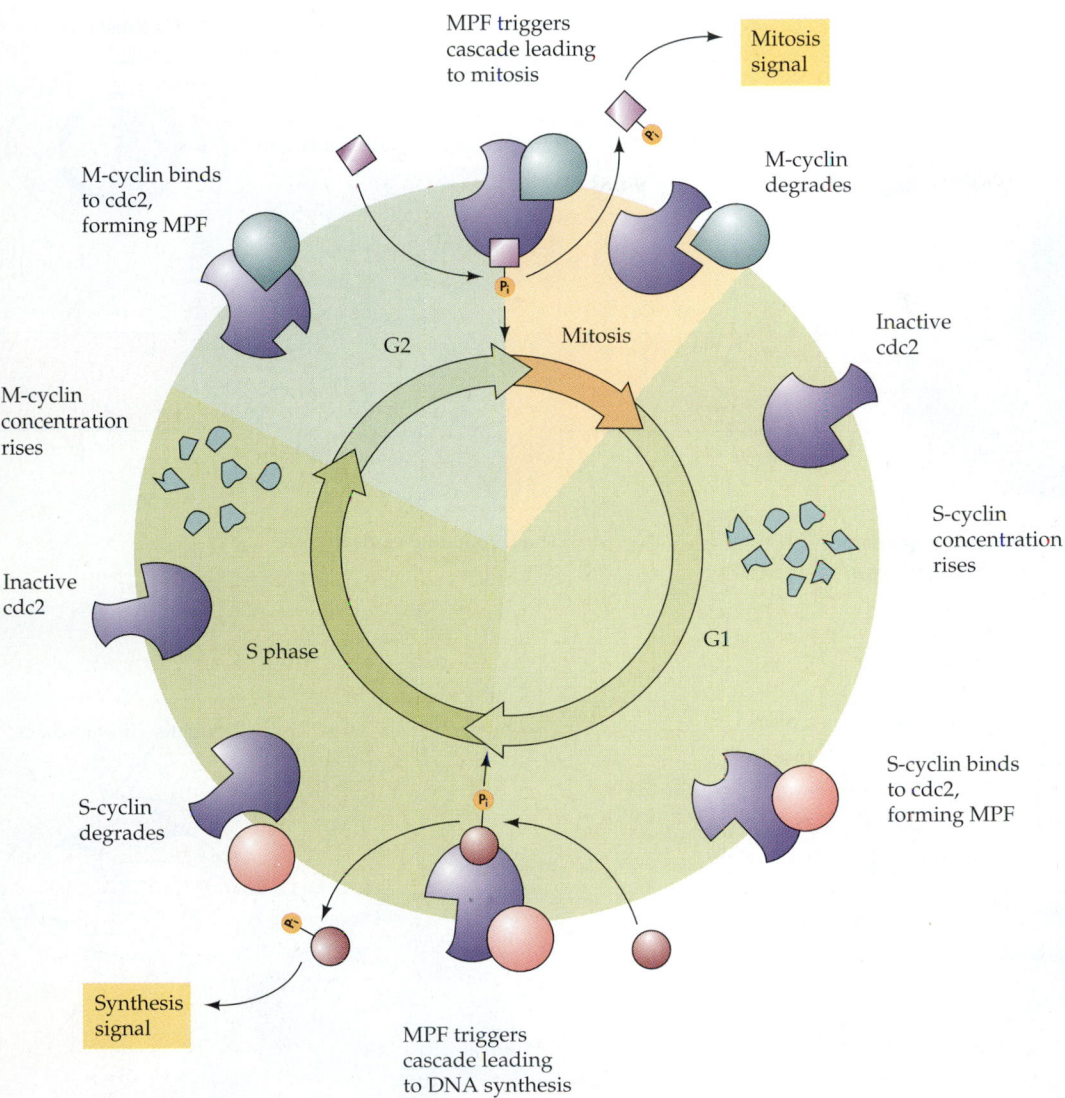

MPF triggers cascade leading to mitosis

Mitosis signal

M-cyclin binds to cdc2, forming MPF

M-cyclin degrades

M-cyclin concentration rises

Inactive cdc2

G2

Mitosis

Inactive cdc2

S-cyclin concentration rises

S phase

G1

S-cyclin degrades

S-cyclin binds to cdc2, forming MPF

Synthesis signal

MPF triggers cascade leading to DNA synthesis

9.7   Cyclins Trigger Decisions in the Cell Cycle
The signals for a cell to enter the M or S phase are triggered when the protein cdc2 binds to a cyclin protein (M-cyclin or S-cyclin, respectively), producing a protein complex called MPF (maturation-promoting factor). MPF enzyme activity in turn activates other enzymes in a cascade effect.

beginning of mitosis. The nucleolar material disperses. The centrosomes, with or without pairs of centrioles, move away from each other toward opposite ends of the cell. Each centrosome then serves as a **mitotic center** that organizes microtubules. In animal cells, some of the microtubules point away from the nuclear region and form starlike groupings called **asters**. Other microtubules, called **polar microtubules**, run between the mitotic centers and make up the developing **spindle** (Figure 9.10a). The spindle is actually two *half spindles*: Each polar microtubule runs from one mitotic center to the middle of the spindle, where it overlaps with polar microtubules of

the other half spindle (Figure 9.10b). Polar microtubules are unstable, constantly forming and falling apart until they contact polar microtubules from the other half spindle, at which point they become more stable.

The chromatin also changes during prophase. The extremely long, thin fibers take on a more orderly form as a result of coiling, supercoiling, and compacting (review Figure 9.5). At this level of magnification and at this stage of the nuclear cycle, each chromosome is seen to consist of two chromatids held tightly together over much of their length. The two chromatids of a single chromosome are identical

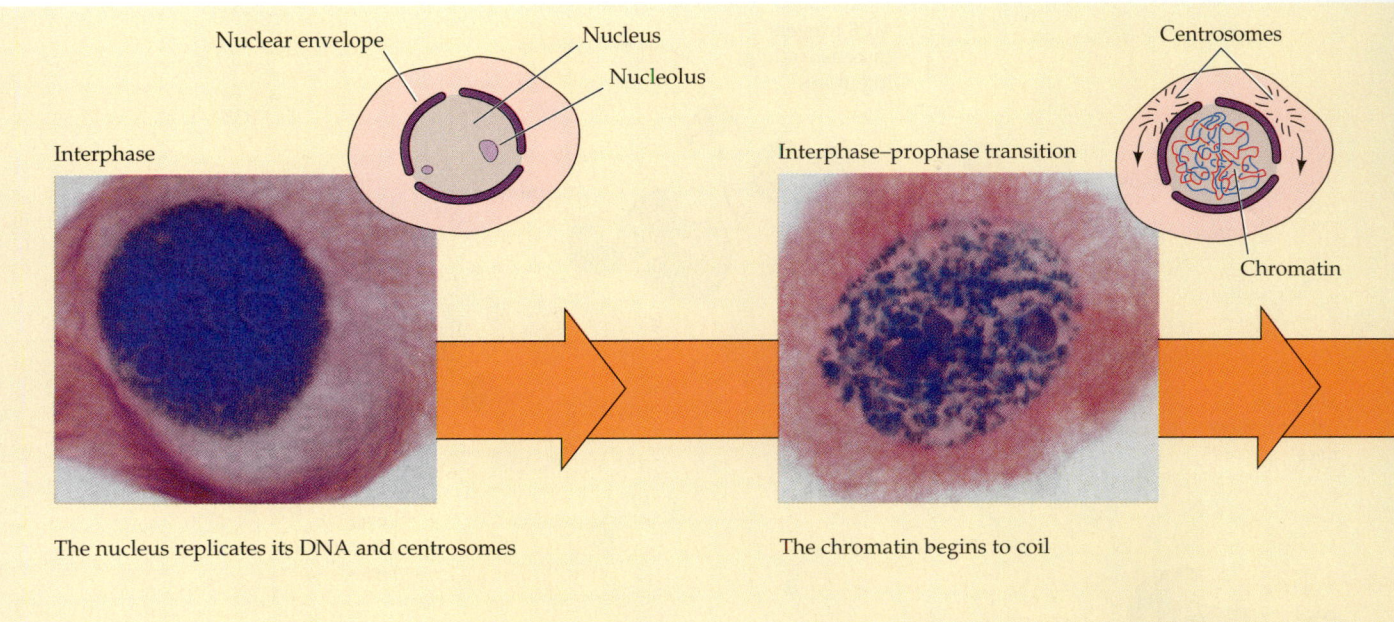

Interphase

The nucleus replicates its DNA and centrosomes

Interphase–prophase transition

The chromatin begins to coil

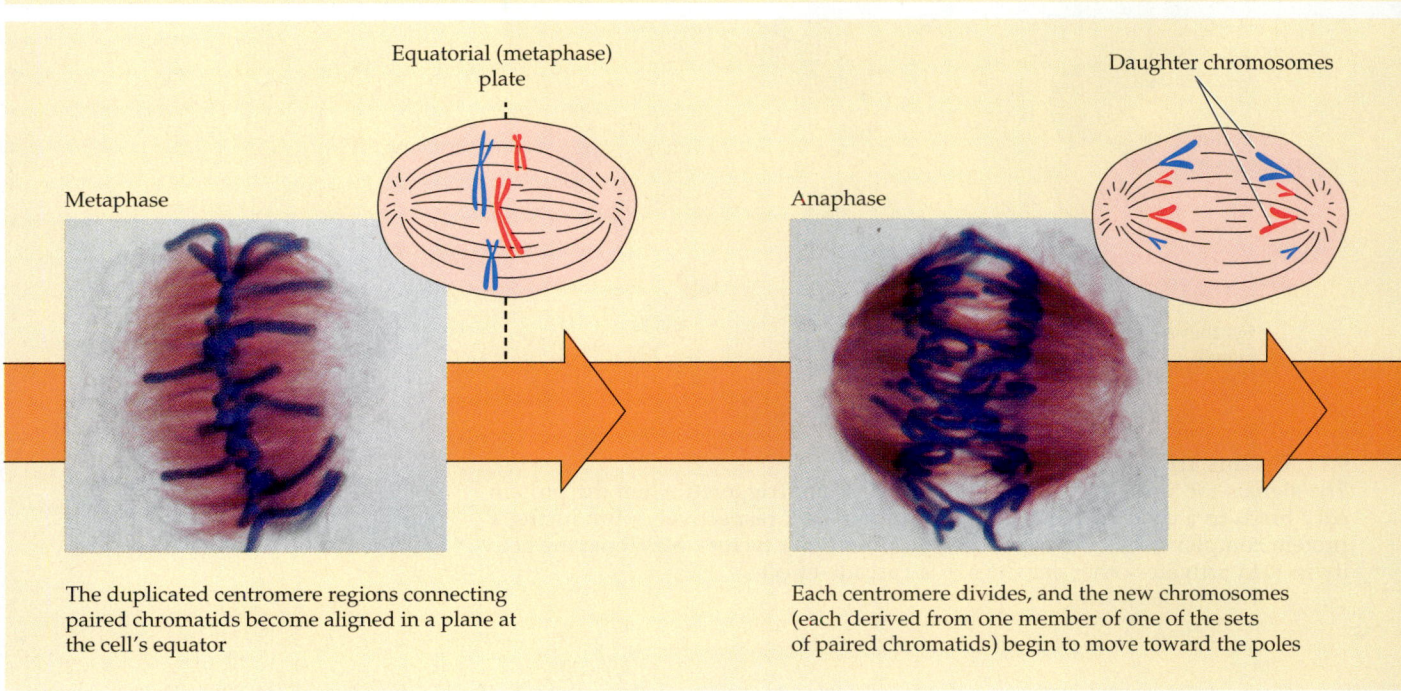

Metaphase

The duplicated centromere regions connecting paired chromatids become aligned in a plane at the cell's equator

Anaphase

Each centromere divides, and the new chromosomes (each derived from one member of one of the sets of paired chromatids) begin to move toward the poles

in structure, chemistry, and the hereditary information they carry because one chromatid, formed during the S phase of the previous interphase, is a replica of the other. Within the region of tight binding of the chromatids lies the centromere, which must be present in order for the chromatids to become associated with microtubules of the spindle. Very late in prophase, specialized three-layered structures called **kinetochores** develop in the centromere region, one on each chromatid (see Figure 9.2).

Dancing Chromosomes

The somewhat condensed chromosomes start to move at the end of prophase—the beginning of **prometaphase**, the next frame in Figure 9.8. The nuclear envelope suddenly disintegrates into membranous sacs (it takes only 20–30 seconds). This disintegration is caused by the activation of one of the MPFs, a process beginning with the accumulation of a cyclin during G2 of interphase. Groups of microtubules,

Chromatids of chromosomes

Developing spindle

Aster

Prophase

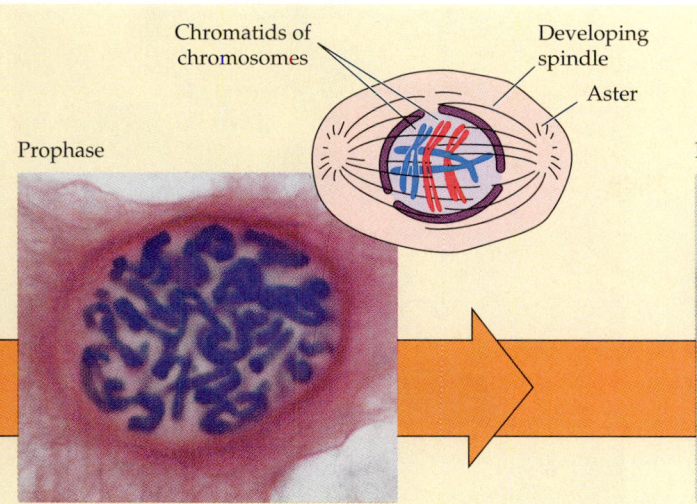

The chromatin continues to coil and supercoil, making the chromatin more and more compact. The chromosomes consist of identical, paired chromatids

Prometaphase

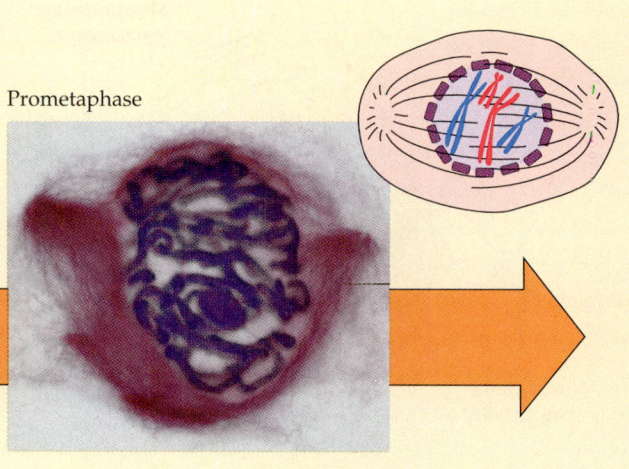

The nuclear envelope breaks down. Kinetochore microtubules appear and interact with the polar microtubules of the spindle, resulting in movement of the chromosomes

Telophase

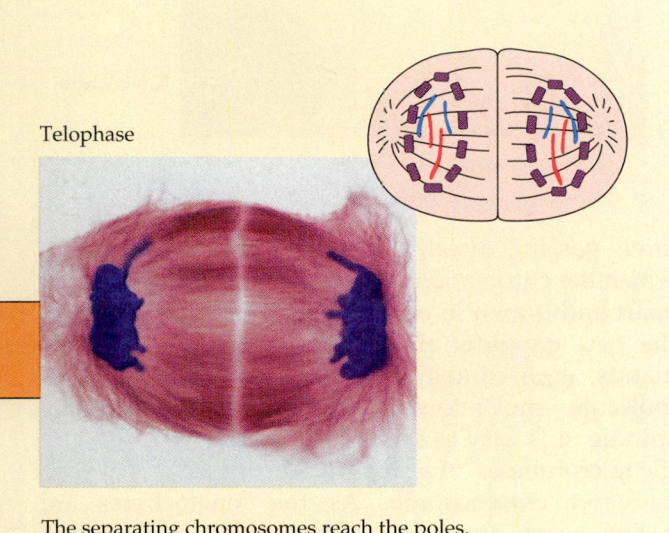

The separating chromosomes reach the poles. Telophase passes into the next interphase as the nuclear envelopes and nucleoli re-form and the chromatin becomes diffuse

9.8 Mitosis

Mitosis results in two nuclei, genetically identical to one another and to the nucleus from which they formed. These photomicrographs are of plant nuclei, which lack centrioles and asters. The red dye stains microtubules and thus the spindle; the blue dye stains the chromosomes. In plants, the first steps toward division of the cell itself cause changes in the telophase cell that disrupt staining, causing the white line seen in the photomicrograph. The diagrams are of corresponding phases in animal cells, in order to introduce the structures not found in plants. In the diagrams, the chromosomes are stylized to emphasize the fates of the individual chromatids.

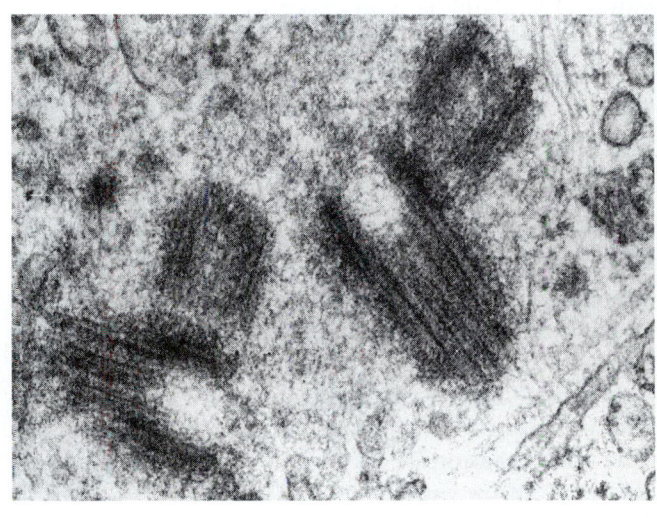

9.9 Centrioles

At a right angle to each parent centriole in this animal cell is a daughter centriole that is about one-half as long. Early in nuclear division one parent–daughter pair migrates to one side of the nucleus, the other to the opposite side. Centrioles consist mostly of microtubules.

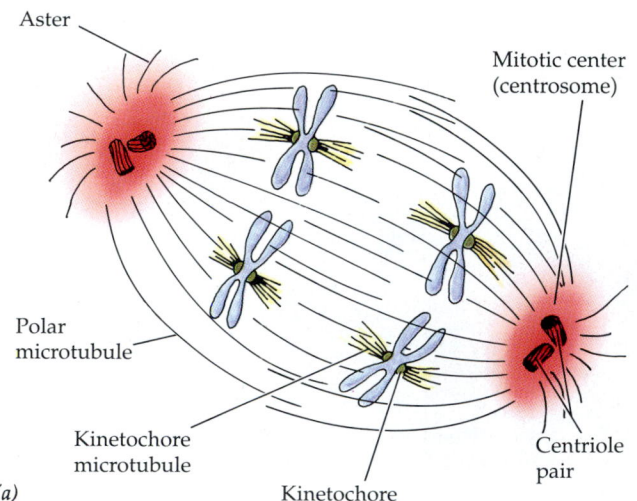

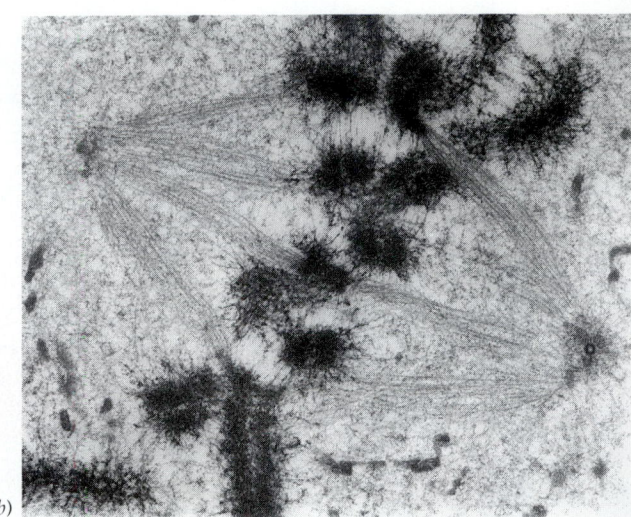

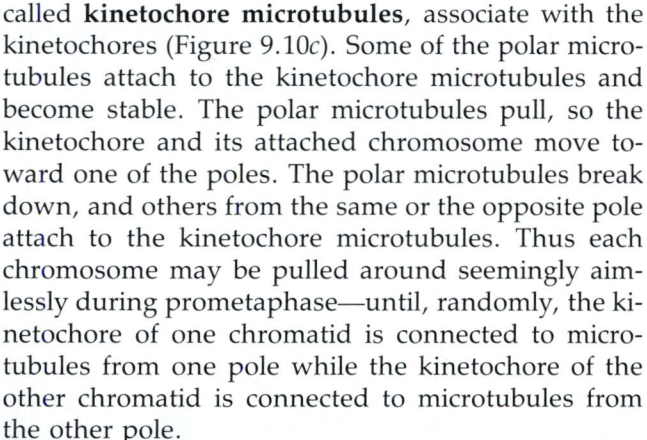

Kinetochore
microtubules

Kinetochore

(a)

(b)

(c)

9.10   The Mitotic Spindle Consists of Microtubules
(a) During prometaphase, polar microtubules extend from
each pole of the spindle apparatus. Kinetochore microtubules attach to the kinetochores in the centromeres of
the chromosomes and to polar microtubules. (b) Polar
microtubules extending from the poles are visible in this
electron micrograph of metaphase. The large dark objects
in the middle are chromosomes. (c) Kinetochore microtubules extend down from the top of this electron micrograph to the kinetochore, which is seen as a dark, three-
layered "plate."

called **kinetochore microtubules**, associate with the
kinetochores (Figure 9.10c). Some of the polar microtubules attach to the kinetochore microtubules and
become stable. The polar microtubules pull, so the
kinetochore and its attached chromosome move toward one of the poles. The polar microtubules break
down, and others from the same or the opposite pole
attach to the kinetochore microtubules. Thus each
chromosome may be pulled around seemingly aim-
lessly during prometaphase—until, randomly, the kinetochore of one chromatid is connected to microtubules from one pole while the kinetochore of the
other chromatid is connected to microtubules from
the other pole.

When this happens to a pair of kinetochores, the
microtubules attached to them stop falling apart, perhaps because of the tension established by the opposing pulls from the two poles. The polar microtubules pull in such a way that the kinetochores
approach a region halfway between the ends of the
spindle. This region, which may be thought of as an
invisible plane perpendicular to the long axis of the
spindle, is called the **equatorial plate**, or metaphase
plate.

The cell reaches **metaphase** when all the kinetochores arrive at the equatorial plate (see Figure 9.8).
The condensation of chromatin that began with prophase continues until the end of metaphase. At this
time, the centromeres divide. Metaphase is usually

brief, passing directly into **anaphase**, the phase in
which the chromatids of each chromosome are pulled
apart and drawn to opposite ends of the spindle. As
the new **daughter chromosomes**—the former chromatids, each containing one double-stranded DNA
molecule—move toward the opposite poles of the
spindle, it is easy to see that the motion is caused by
the microtubules "tugging" at the kinetochore of each
daughter chromosome. As the kinetochores are
pulled apart, the arms of the chromosomes drag
along passively. The mechanism of the tugging is not
fully understood. The microtubules are probably
shortening, and the kinetochore microtubules may
be sliding along the polar microtubules.

Also during anaphase, the poles of the spindle are
often pushed farther apart by the action of some of
the polar microtubules, thus contributing to the separation of the daughter chromosomes. The polar microtubules contain dynein, a protein also associated
with the microtubules of cilia and flagella (Chapter
4). Presumably, then, the movement of the poles is
produced in a manner similar to that in which eukaryotic flagella and cilia are made to beat.

Amazingly little energy is expended in moving a
chromosome during anaphase. The hydrolysis of 20
ATP molecules is enough to move a chromosome
from the equatorial plate to the pole.

The End of Mitosis

When the chromosomes stop moving at the end of anaphase, the cell enters **telophase** (the final frame in Figure 9.8). Two identical collections of chromosomes, which carry identical sets of hereditary instructions, are at the opposite ends of the spindle, which begins to break down (as do the asters, if present). A new nuclear envelope forms around each group of chromosomes. The chromosomes begin to uncoil, and continue uncoiling until they become the diffuse tangle of chromatin characteristic of interphase. The nucleolus or nucleoli reappear at specific sites on specific chromosomes. When these changes are complete, telophase—and mitosis—is at an end, and each of the daughter nuclei enters another interphase.

In interphase, the DNA duplicates and new chromatids form, so that each chromosome consists of two chromatids. The duplication of DNA is a major topic and is discussed in Chapter 11. Centrioles, if present, replicate during interphase: The two paired centrioles first separate, and then each acts as a parent for the formation of a new daughter centriole at right angles to it (see Figure 9.9).

Mitosis is beautifully precise. Its result is the formation of two nuclei *identical to each other* and to the parent nucleus in chromosomal makeup, and hence in genetic constitution.

Cytokinesis

Mitosis refers only to the division of the nucleus; it is not always immediately accompanied by the division of the rest of the cell, called **cytokinesis.** Generally, however, cytokinesis follows immediately upon mitosis. Animal cells, which lack cell walls, usually divide by a furrowing of the membrane, as if an invisible thread were tightening between the two parts (Figure 9.11a). The "invisible" thread consists of microfilaments of actin and myosin (Chapter 4), two proteins that interact here to produce a contraction, just as they do in muscles (Chapter 40).

Plant cells must divide differently because they have cell walls. As the spindle breaks down after mitosis, membranous vesicles derived from the Golgi apparatus appear in the equatorial region roughly midway between the two daughter nuclei of a dividing plant cell. With the help of microtubules, the vesicles begin to form a cell plate, that is, the beginning of a new cell wall (Figure 9.11b).

After cytokinesis, both daughter cells contain all the components of a complete cell. Mitosis ensures the precise distribution of chromosomes. Organelles such as ribosomes, mitochondria, and chloroplasts need not be distributed equally between daughter cells, as long as many of each are present in both cells; accordingly, there is no mechanism comparable to mitosis to provide for their equal allocation to daughter cells. Although centrioles, where present, were once thought to organize the mitotic spindle, scientists now speculate that the association of the

(a)

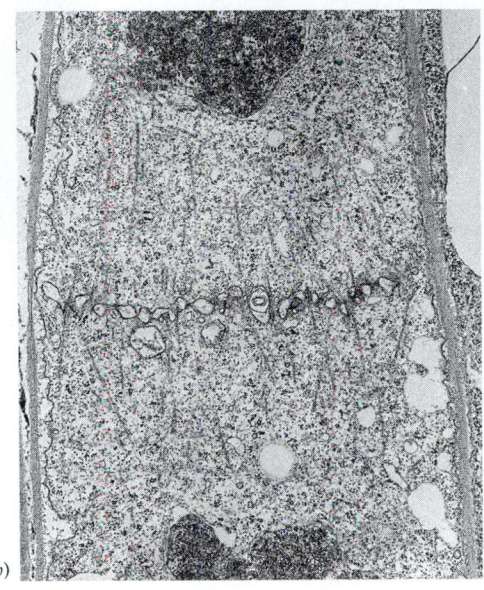

(b)

9.11 Cytokinesis
(a) A sea urchin egg has just completed cytokinesis at the end of the first division in its development into an embryo. The division furrow has completely separated the cytoplasm of one daughter cell from the other, although their surfaces remain in contact. Tiny, hairlike microvilli cover the surfaces of both cells. (b) The horizontal row of vesicles in this dividing plant cell in late telophase will join to form a cell plate between the cell above and the cell below. Microtubules, visible above and below in the cytoplasm, run between the vesicles.

centrioles with mitotic centers simply ensures that centrioles, like chromosomes, are distributed equally to the daughter cells.

SEX AND REPRODUCTION

As the cell cycle repeats itself, a single cell can give rise to a vast number of others. The cell could be a unicellular organism reproducing with each cycle, or a cell that divides to produce a multicellular organism. The multicellular organism, in turn, may be able to reproduce itself by releasing one or more of its cells, derived from mitosis and cytokinesis, as a spore *or* by having a multicellular piece break away and grow on its own (Figure 9.12). The multicellular organism reproducing by releasing cells and the unicellular organism are examples of **asexual reproduction**, sometimes called vegetative reproduction. Asexual reproduction is based on mitotic division of the nucleus and, accordingly, produces offspring that are genetically identical to the parent. It is a rapid and effective means of making new individuals and is widely practiced in nature.

A drawback of asexual reproduction is its very uniformity, which leads to the production of a **clone** of genetically identical progeny. Although the clone may be well adapted to its existing environment, it may be at great risk should conditions change. In contrast, organisms that produce genetically different offspring are more successful when the environment varies unpredictably in time and space, because at least *some* of their genetically diverse offspring may

be individuals able to meet the different challenges of a changing environment.

Diversity is fostered by **sexual reproduction**. Genetic information from two separate cells combines. In the reproduction of most animal species, the two cells are contributed by two separate parents. Each parent provides a sex cell, or **gamete**. Each gamete is **haploid**, meaning that it contains a single set of chromosomes; the number of chromosomes in such a single set is denoted by n. The two gametes—often identifiable as a female egg and a male sperm—fuse to produce a single cell, the **zygote** or fertilized egg. This fusion is called **fertilization**. Its consequence, the zygote, contains genetic information from both gametes and, hence, from both parents. The zygote also has *two* sets of chromosomes; it is said to be **diploid**, denoted by $2n$. In many species, including all animals, the zygote develops by mitotic divisions into a multicellular adult. Because the zygotic nucleus is diploid, all the body cells produced by mitosis are also diploid ($2n$). Sexual life cycles are summarized in Figure 9.13.

What happens when the adult from this zygote reproduces? If the gametes were produced by mitosis in this diploid parent, then they too would be diploid. Thus after fusion of two diploid ($2n$) gametes, the next-generation zygote would be $4n$. This is not a tenable situation because subsequent generations would contain more and more chromosomes. There must be a *reduction* step in the sexual life cycle, that is, a special type of nuclear division that reduces the chromosome number from diploid to haploid. This form of division in sexually reproducing organisms is meiosis.

Meiosis in animal cells directly produces haploid gametes. However, in plants and some fungi, meiosis gives rise to haploid **spores**, which undergo mitosis, producing multicellular haploid bodies (find the correct path in Figure 9.13). Particular cells in these haploid bodies ultimately give rise, by mitosis, to haploid gametes, and the life cycle continues. Details of some life cycles are considered in Chapters 23 to 27. The simplest possible sexual life cycle is one in which two haploid gametes fuse to give one diploid ($2n$) zygote, which immediately undergoes meiosis, yielding a new set of haploid (n) gametes. Embellishments on this scheme consist mainly of the addition of mitotic divisions leading to multicellularity in the haploid phase, the diploid phase, or both.

Keep in mind that sex and reproduction are *not* the same thing. Sex is the combining of genetic material from two cells. Reproduction is the formation of new individuals, whether of unicellular or multicellular organisms. Sex and reproduction can be widely separated in time in a unicellular species, a phenomenon illustrated in the life cycle of the protist *Paramecium* (see Chapter 23).

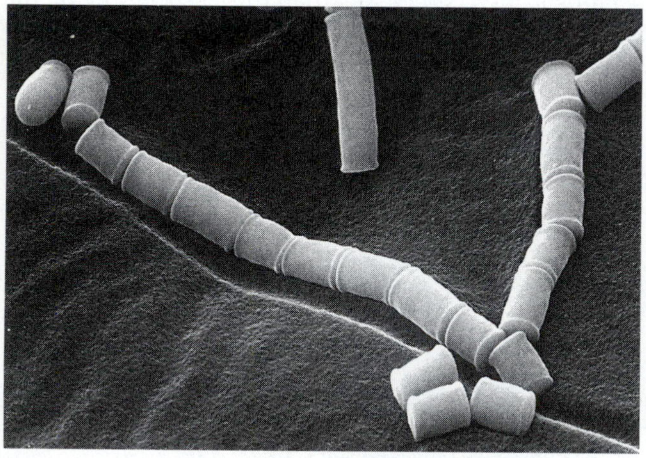

9.12 Asexual Reproduction
These spool-shaped cells are asexual spores formed by a fungus. Each spore contains a nucleus produced by a mitotic division. A spore and the fungal body that will grow from it following germination are the same genetically as the parent that fragmented to produce the spores. The general shape of the parent can be guessed from spores still in contact end-to-end.

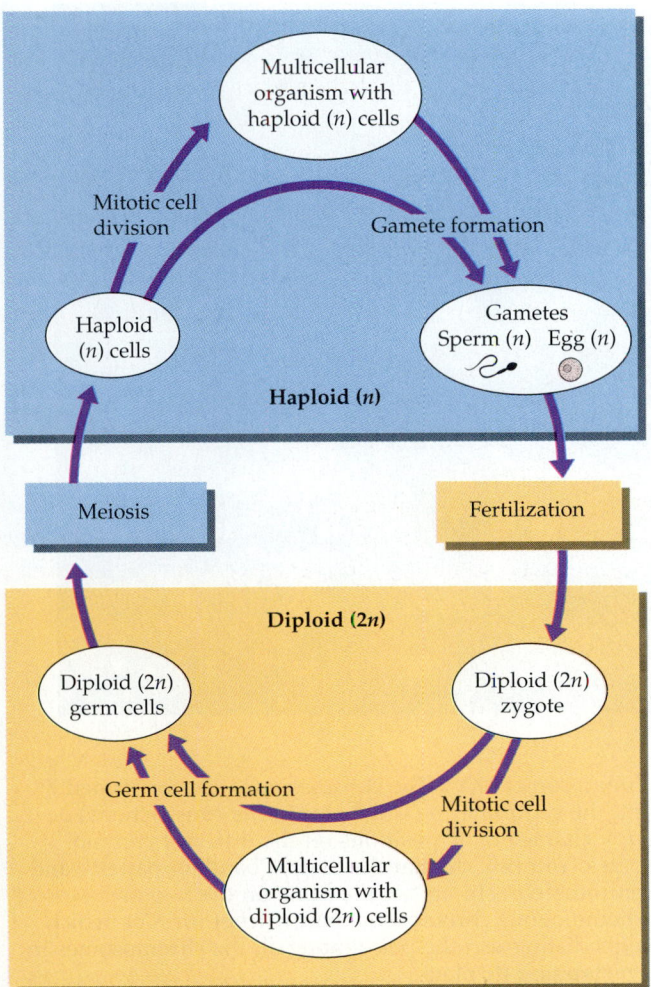

9.13 Fertilization and Meiosis Alternate in Sexual Reproduction
Haploid (*n*) cells or organisms alternate with diploid (2*n*) cells or organisms. A zygote, formed during fertilization, may differentiate into a germ cell, or it may form a multicellular organism that eventually produces germ cells. Whatever their origins, germ cells form haploid cells through meiosis. A haploid cell may differentiate into a gamete (sperm or egg), or it may form a multicellular organism that eventually produces gametes. Different organisms follow different paths around this cycle, and they may spend very different proportions of their life in different stages. Each variation is called a life cycle.

The essence of sexual reproduction is the selection of half of a parent's diploid chromosome set to make a haploid gamete, followed by the fusion of two such haploid gametes to produce a diploid cell containing genetic information from the two gametes. Both of these steps contribute to a shuffling of genetic information in the population, so that usually no two individuals have exactly the same genetic constitution. This result is the opposite of the situation with asexual reproduction. The diversity provided by sexual reproduction opened up enormous opportunities

for evolution. Although both asexual and sexual modes of reproduction have existed for billions of years, there are many more species of sexually reproducing organisms than of asexually reproducing organisms.

THE KARYOTYPE

When nuclei are in metaphase of mitosis, the centromeres are spread out on the equatorial plate. At this time it is often possible to count and characterize the individual chromosomes. This process is relatively simple in some organisms, thanks to techniques that can capture cells in metaphase and spread out the chromosomes. A photograph of the entire set of chromosomes can then be made, and the images of the individual chromosomes can be cut out and pasted together in an orderly arrangement. Such a rearranged photograph reveals the number, forms, and types of chromosomes in a cell, all of which constitute its **karyotype** (Figure 9.14).

Individual chromosomes can be recognized by their length, the positions of their centromeres, and characteristic banding when they are stained and observed at high magnification. When the cell is diploid, the karyotype consists of pairs of chromosomes—23 pairs for a total of 46 chromosomes in our species, and more or fewer pairs in other diploid species (Table 9.1). In each recognizable pair of chro-

TABLE 9.1
Numbers of Pairs of Chromosomes in Different Species of Plants and Animals

COMMON NAME	SPECIES	NUMBER OF CHROMOSOME PAIRS
Mosquito	*Culex pipiens*	3
Housefly	*Musca domestica*	6
Garden onion	*Allium cepa*	8
Toad	*Bufo americanus*	11
Rice	*Oryza sativa*	12
Frog	*Rana pipiens*	13
Alligator	*Alligator mississipiensis*	16
Cat	*Felis domesticus*	19
House mouse	*Mus musculus*	20
Rhesus monkey	*Macaca mulatta*	21
Wheat	*Triticum aestivum*	21
Human	*Homo sapiens*	23
Potato	*Solanum tuberosum*	24
Cattle	*Bos taurus*	30
Donkey	*Equus asinus*	31
Horse	*Equus caballus*	32
Dog	*Canis familiaris*	39
Chicken	*Gallus domesticus*	≈39
Carp	*Cyprinus carpio*	52

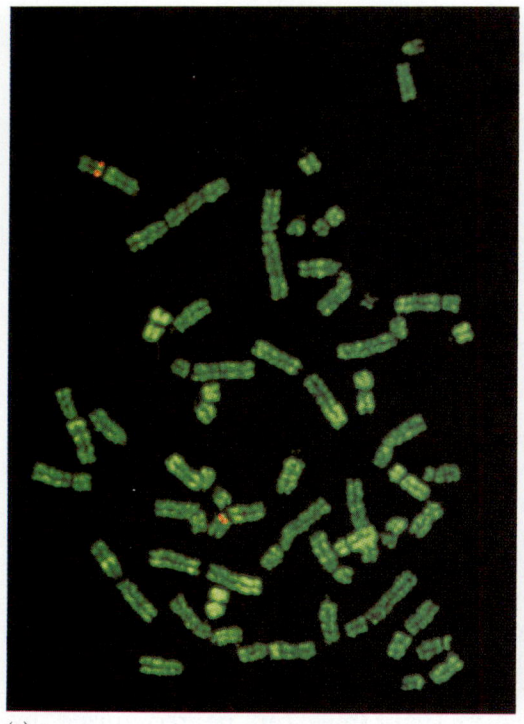

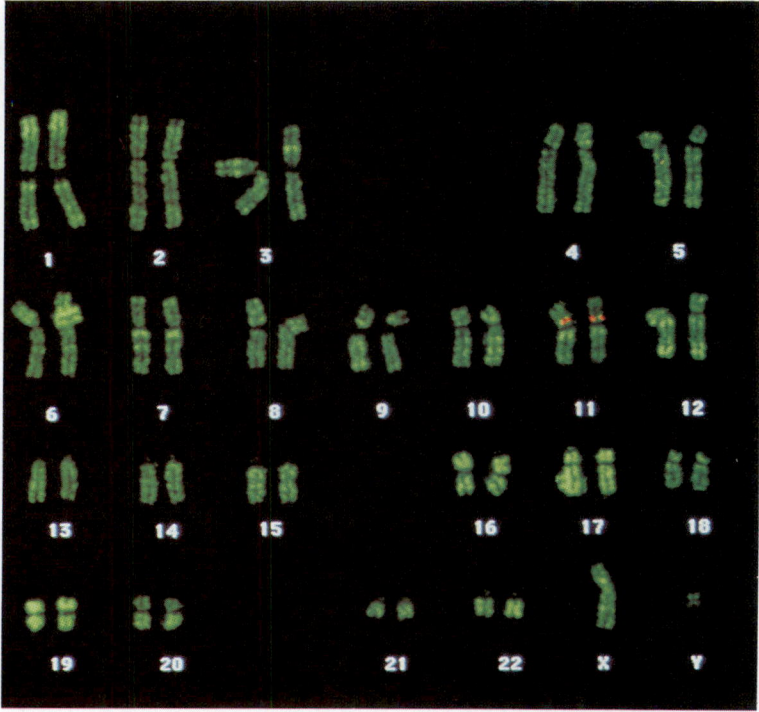

(a)   *(b)*

9.14   Human Cells Have 46 Chromosomes
Chromosomes of a human male. *(a)* The chromosomes are spread out because immersion in hypotonic solution bloated and then ruptured the cell, which was in metaphase of mitosis. You should be able to count 46 chromosomes. *(b)* A karyotype has been arranged from the metaphase spread of chromosomes. There are 23 pairs of homologous chromosomes, including a pair of sex chromosomes (XY). The chromosomes appear striped, and the centromeres (which appear as constrictions) occupy characteristic positions on the different chromosomes. You can see that the length, banding pattern, and centromere positions are the same on the two members of a homologous chromosome pair (except for XY), which helps distinguish the pair among all the chromosomes in a metaphase display.

mosomes, one chromosome comes from one parent and one from the other. The members of such a **homologous pair**, called homologs, are identical in size and appearance (with the exception of so-called sex chromosomes in some species; see Chapter 10), and they bear corresponding, though generally not identical, types of genetic information. Haploid cells contain only one of the homologs from each pair of chromosomes. Thus when haploid gametes fuse in fertilization, the resulting diploid zygote ends up with two homologs of each type.

MEIOSIS

Meiosis is the mechanism that reduces the diploid number of chromosomes to the haploid number for sexual reproduction. To understand the process and its specific details, it is useful to keep in mind the overall functions of meiosis: (1) to reduce the chromosome number from diploid to haploid, (2) to ensure that each of the four products has a complete set of chromosomes, and (3) to promote genetic diversity among the products. Pay particular attention to the fact that, although two divisions occur during meiosis, the DNA is replicated only once.

Two unique features characterize the first meiotic division, **meiosis I**. The first feature is that homologous chromosomes pair along their entire lengths, a process called **synapsis** that lasts from prophase to the end of metaphase of meiosis I. The second key feature is that homologous chromosomes separate during meiosis I. The individual chromosomes, each consisting of two joined chromatids, remain intact until the end of metaphase of **meiosis II**, the second meiotic division. In the discussion that follows, refer to Figure 9.17 (on pages 206 and 207) to help you visualize each step.

The First Meiotic Division

Meiosis I is preceded by an interphase during which each chromosome is replicated, so that each chromosome then consists of two sister chromatids. Meiosis I begins with a long **prophase I** (the first four frames of Figure 9.17), marked by a number of important changes. Very early in prophase I, the homologous chromosomes synapse; they are already

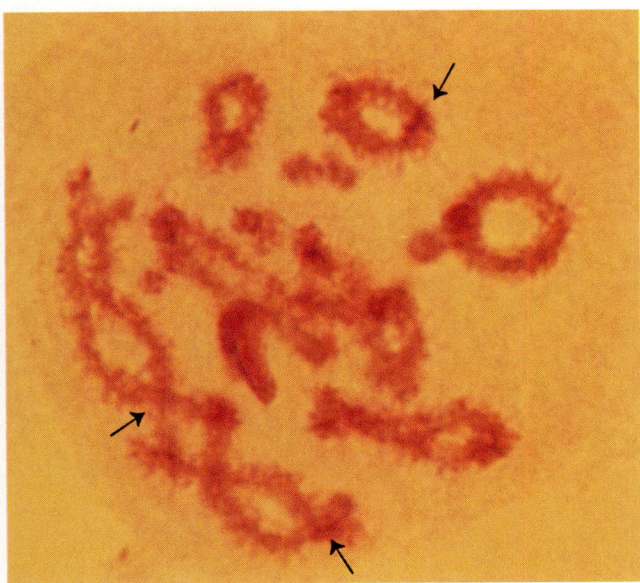

9.15   Chiasmata
Chiasmata—locations where segments of chromatids are being exchanged—are visible near the middles of some of the chromatids, and near the ends of others. Three of the many chiasmata in this micrograph are indicated with arrows.

tightly joined by the time they can be seen clearly under the light microscope. Throughout prophase I and metaphase I, the chromatin continues to coil and compact progressively, so that the chromosomes appear ever thicker and smoother.

Partway through prophase I, the homologous chromosomes seem to *repel* each other, especially near the centromeres, but they are held together by physical attachments (Figure 9.15). The regions having these attachments take on an X-shaped appearance and are called **chiasmata** (singular: *chiasma*, meaning "cross" in Greek). A chiasma reflects an exchange of material between chromatids on homologous chromosomes—what geneticists call crossing over (Figure 9.16). We will have a great deal to say about crossing over and its genetic consequences in coming chapters. Although the chromosomes exchange material shortly after synapsis begins, the chiasmata do not become visible until later, when the homologs are repelling each other.

Prophase I is followed by **prometaphase I** (not pictured in Figure 9.17), during which the nuclear envelope and the nucleoli disappear. A spindle forms, and microtubules become attached to the kinetochores of the chromosomes. In meiosis I, there is only one kinetochore per chromosome, not one per chromatid as in mitosis.

By **metaphase I**, the kinetochores have become connected to the poles, all the chromosomes have moved to the equatorial plate, and the homologous chromosomes are about to be pulled apart. Up to this point, they have been held together by chiasmata; it is this connection that provides the tension needed to stabilize the polar microtubules of the spindle. They separate in **anaphase I**, when individual chromosomes, each still consisting of *two* chromatids, are pulled to the poles, one homolog of a pair going to one pole and the other to the opposite pole (see Figure 9.17). (Note that this process differs from the separation of *chromatids* during mitotic anaphase.) Each of the two daughter nuclei from this division is haploid—that is, it contains only one set of chromosomes, compared to the two sets of chromosomes that were present in the original diploid nucleus. However, because they consist of two chromatids rather than just one, each of these chromosomes has twice the mass of a chromosome at the end of a mitotic division.

In some species, but not in others, there is a **telophase I**, with the reappearance of nuclear envelopes and so forth. When there is a telophase I, it is followed by an **interkinesis** phase similar to mitotic interphase. During interkinesis the chromatin is somewhat, but not completely, uncoiled. There is no replication of the genetic material because each chromosome already consists of two chromatids. In contrast to mitotic interphase, the sister chromatids are generally not genetically identical, because crossing over in prophase I has scrambled the original chromatids to some degree.

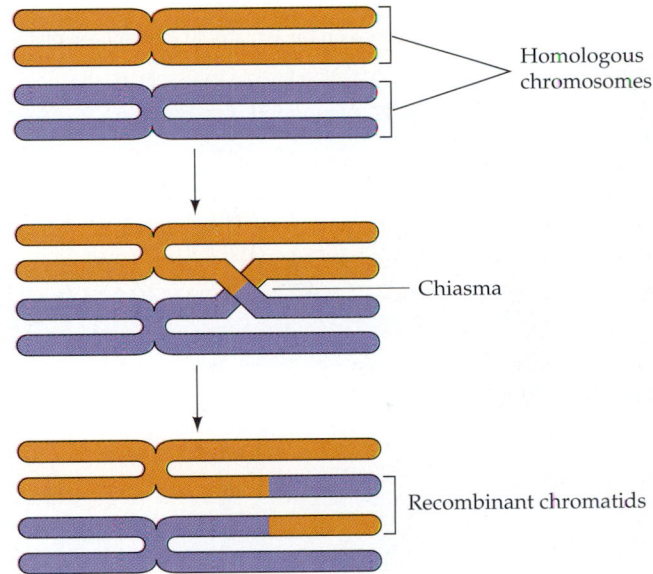

9.16   Crossing Over Forms Genetically Diverse Chromosomes
Early in prophase I, two chromatids of different homologs often cross over, break, and rejoin, so that each homolog has some DNA from the other. The products of crossing over are recombinant chromatids. This recombination can have important genetic and evolutionary consequences.

Meoisis I

Early prophase I

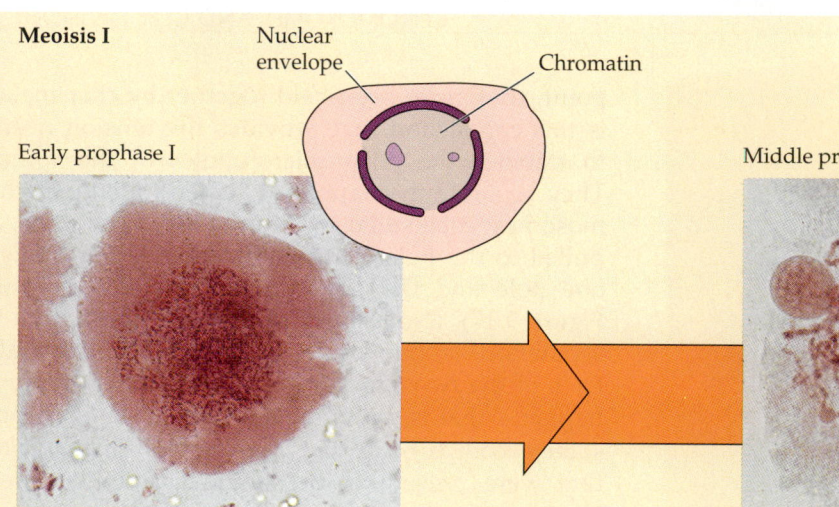

Nuclear envelope · Chromatin · Centrosomes

Middle prophase I

The chromatin begins to condense following interphase

Synapsis aligns homologs, and chromosomes shorten

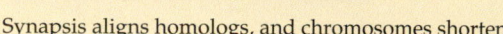

Metaphase I

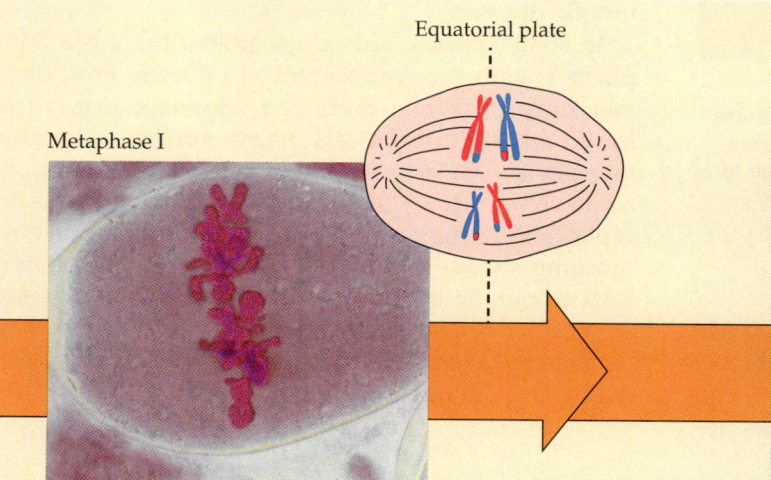

Equatorial plate

Anaphase I

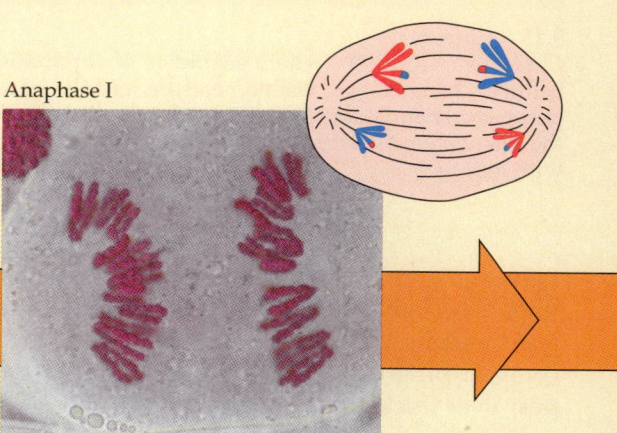

The chromosomes line up on the equatorial (metaphase plate)

The homologous chromosomes (each with two chromatids) move to opposite poles of the cell

Metaphase II

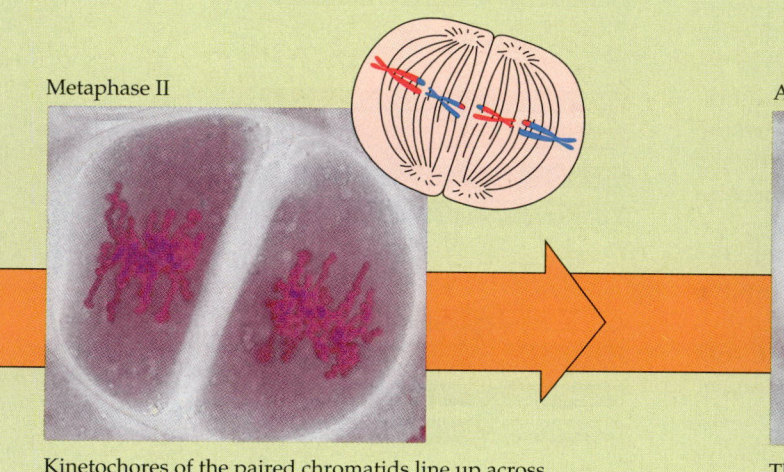

Anaphase II

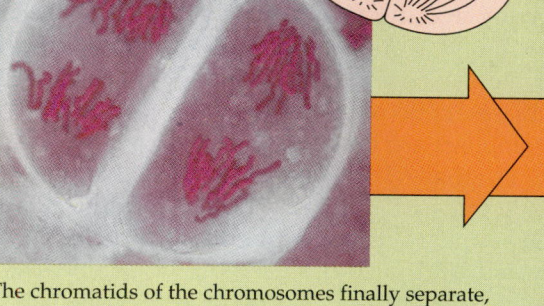

Kinetochores of the paired chromatids line up across the equator of each cell

The chromatids of the chromosomes finally separate, becoming chromosomes in their own right, and are pulled to opposite poles

9.17 Meiosis

In meiosis, two sets of chromosomes are divided among four cells, each of which then has half as many chromosomes as the original cell. This happens as a result of two successive nuclear divisions. The photomicrographs shown here are of meiosis in the male reproductive organ of a lily. As in Figure 9.8 (mitosis), the diagrams are of meiosis in an animal.

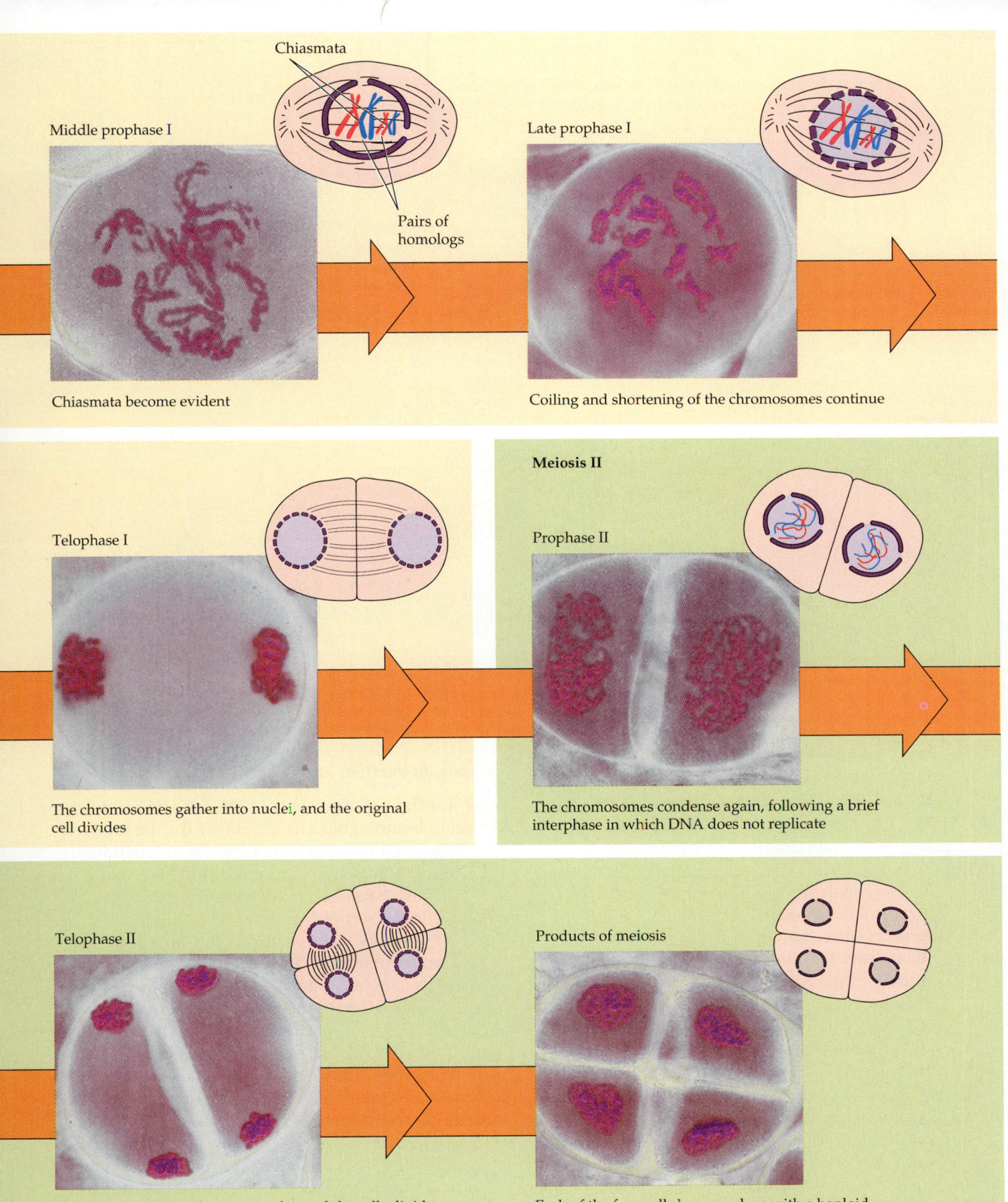

Middle prophase I

Chiasmata

Pairs of homologs

Chiasmata become evident

Late prophase I

Coiling and shortening of the chromosomes continue

Telophase I

The chromosomes gather into nuclei, and the original cell divides

Meiosis II

Prophase II

The chromosomes condense again, following a brief interphase in which DNA does not replicate

Telophase II

The chromosomes gather into nuclei, and the cells divide

Products of meiosis

Each of the four cells has a nucleus with a haploid number of chromosomes. Each of the four cells shown here will now develop into a pollen grain

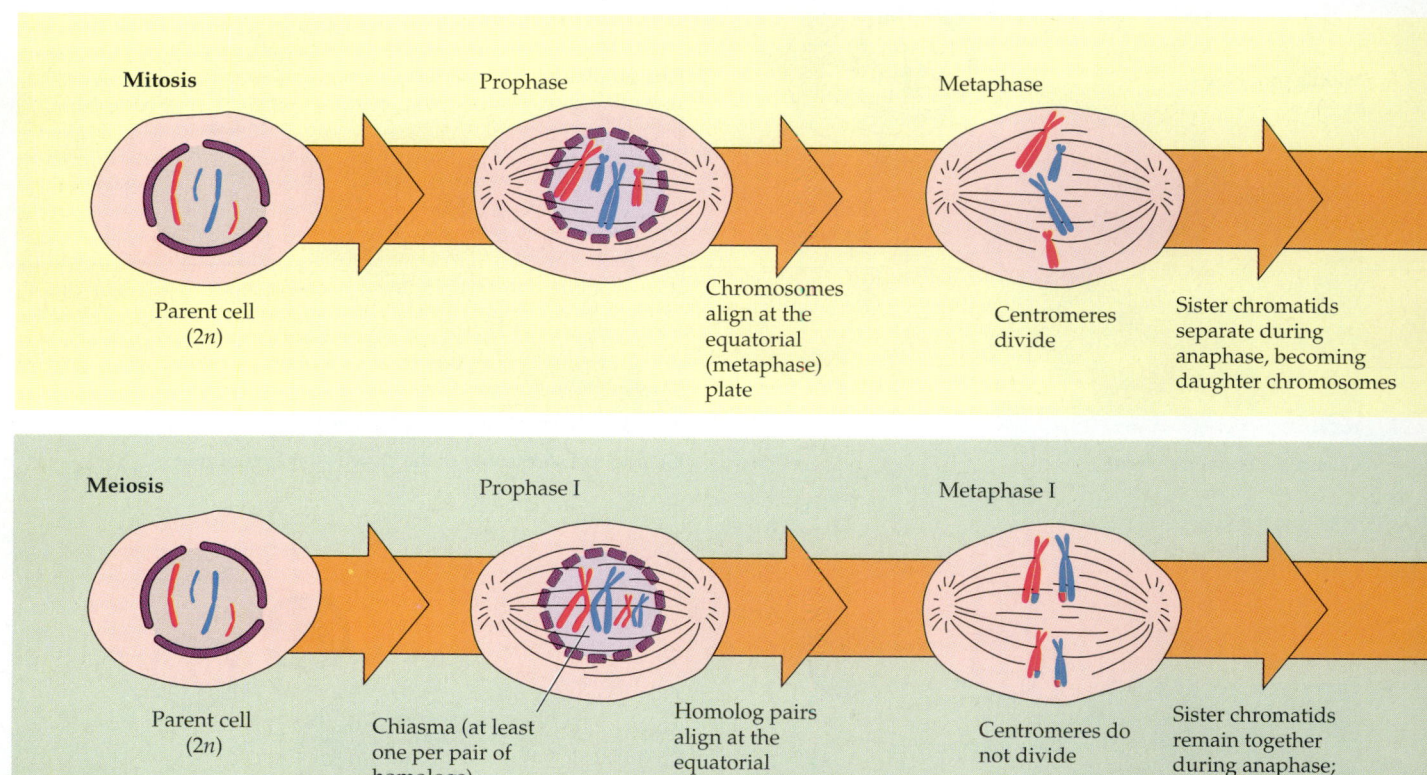

Mitosis

Prophase

Metaphase

Parent cell
(2n)

Chromosomes
align at the
equatorial
(metaphase)
plate

Centromeres
divide

Sister chromatids
separate during
anaphase, becoming
daughter chromosomes

Meiosis

Prophase I

Metaphase I

Parent cell
(2n)

Chiasma (at least
one per pair of
homologs)

Homolog pairs
align at the
equatorial
(metaphase)
plate

Centromeres do
not divide

Sister chromatids
remain together
during anaphase;
homologs separate

The Second Meiotic Division

Meiosis II is similar to mitosis. In each nucleus produced by meiosis I, the chromosomes line up at equatorial plates in metaphase II; the chromatids, each having a centromere, separate; and new daughter chromosomes (consisting now of single chromatids) move to the poles in anaphase II. There are three major differences between meiosis II and mitosis: (1) DNA replicates before mitosis but not before meiosis II; (2) in mitosis the chromatids making up a given chromosome are identical, whereas in meiosis II they differ over part of their length if they participated in crossing over in prophase of meiosis I; (3) the number of chromosomes on the equatorial plate of each of the two nuclei is n in meiosis II rather than $2n$ as in the single mitotic nucleus.

Figure 9.18 compares mitosis and meiosis. The final result of meiosis is four nuclei: Each nucleus is haploid and each has a single full set of chromosomes that differs from other such sets in its exact genetic composition. The differences, to repeat a very important point, result from crossing over during prophase I and from the separation of maternal and paternal chromosomes during anaphase I.

Synapsis, Reduction, and Diversity

What are the consequences of the synapsis and separation of homologous chromosomes during meiosis? In *mitosis*, each chromosome behaves independently of its homolog; its two chromatids are sent to opposite poles at anaphase. If we start a mitotic division with x chromosomes, we end up with x in each daughter nucleus (and each chromosome then consists of one chromatid, or double-stranded molecule of DNA). In *meiosis*, synapsis organizes things so that chromosomes of maternal origin pair with their paternal homologs. Then the separation during meiotic anaphase I ensures that each pole receives one chromosome member from each pair of homologous chromosomes. (Remember that each chromosome still consists of two chromatids.) For example, at the end of meiosis I in humans, each daughter nucleus contains 23 of the original 46 chromosomes—one member of each homologous pair. In this way, the chromosome number is decreased from diploid to haploid; in this way, too, meiosis I guarantees that each daughter nucleus gets a full set of chromosomes, for it must get one of each pair of homologous chromosomes.

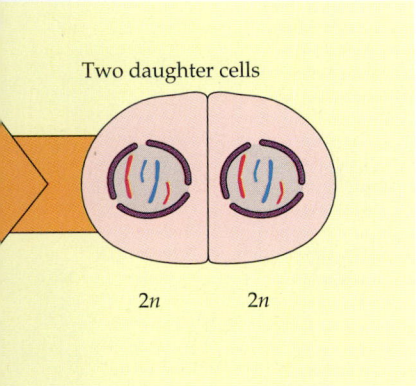

Two daughter cells

2n   2n

9.18 Mitosis and Meiosis Compared

Mitosis is a mechanism for constancy; the parent nucleus produces two identical daughter nuclei. Meiosis is a mechanism for diversity; the parent nucleus produces four daughter nuclei, each different from the parent nucleus and from its sister nuclei. The distinctive features of meiosis are synapsis and the failure of the centromeres to divide at the end of metaphase I.

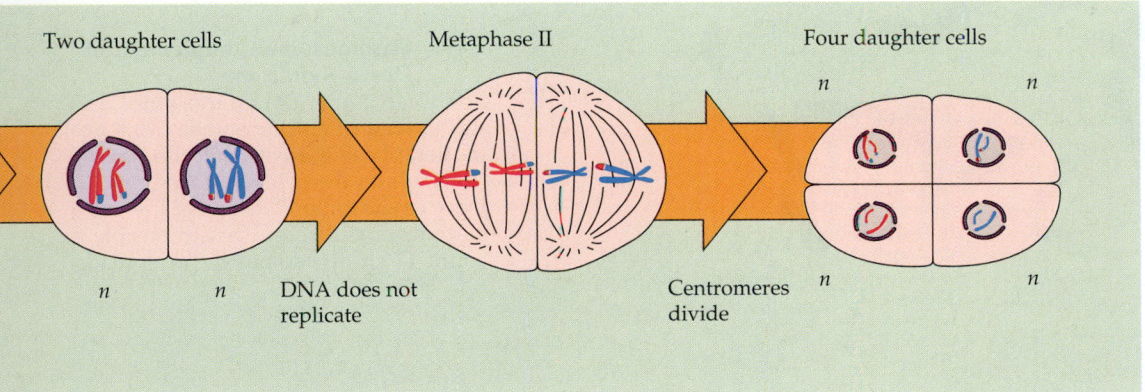

Two daughter cells   Metaphase II   Four daughter cells

n   n   DNA does not replicate   Centromeres divide   n   n

The products of meiosis I become genetically diverse for two reasons. First, synapsis during prophase I allows the maternal chromosome to interact with the paternal one; if there is crossing over, the recombinant chromatids contain some genetic material from each parent. Second, it is a matter of pure chance which member of a pair of chromosomes goes to which daughter cell at anaphase I. If there are two pairs of chromosomes in the diploid parent nucleus, a particular daughter nucleus could get paternal chromosome 1 and maternal chromosome 2, or paternal 2 and maternal 1, or both maternals, or both paternals. It all depends on the random way in which the chromosomes line up at metaphase I. Note that of the four possible chromosome combinations just described, two produce daughter nuclei that are essentially the same as one of the parental types (except for any material exchanged by crossing over). With three pairs, only two of the eight possible chromosome combinations are essentially the same as one of the parents. You can see that the probability of getting back the original parental combinations decreases rapidly as the number of chromosome pairs increases; most species of diploid organisms do, indeed, have more than two pairs (see Table 9.1).

Meiotic Errors and Their Consequences

A pair of homologous chromosomes may fail to separate during meiosis I, or sister chromatids may fail to separate during meiosis II or mitosis. This phenomenon, called **nondisjunction**, results in the production of aneuploid cells. **Aneuploidy** is a condition in which one or more chromosomes or pieces of chromosomes are lacking or are present in excess. If, for example, the chromosome-21 pair fails to separate during the formation of a human egg, allowing both members of the pair to go to one pole during anaphase I, then the resulting egg contains either two copies of chromosome 21 or none at all. If an egg with two of these chromosomes is fertilized by a normal sperm, the resulting zygote and infant has three copies of the chromosome: it is **trisomic** for chromosome 21 (Figure 9.19). As a result of carrying the extra chromosome 21, such a child demonstrates the symptoms of Down syndrome: impaired intelligence, characteristic abnormalities of the hands, tongue, and eyelids, and an increased susceptibility to diseases such as leukemia.

Other abnormal events also lead to aneuploidy. In a process called **translocation**, a piece of a chromo-

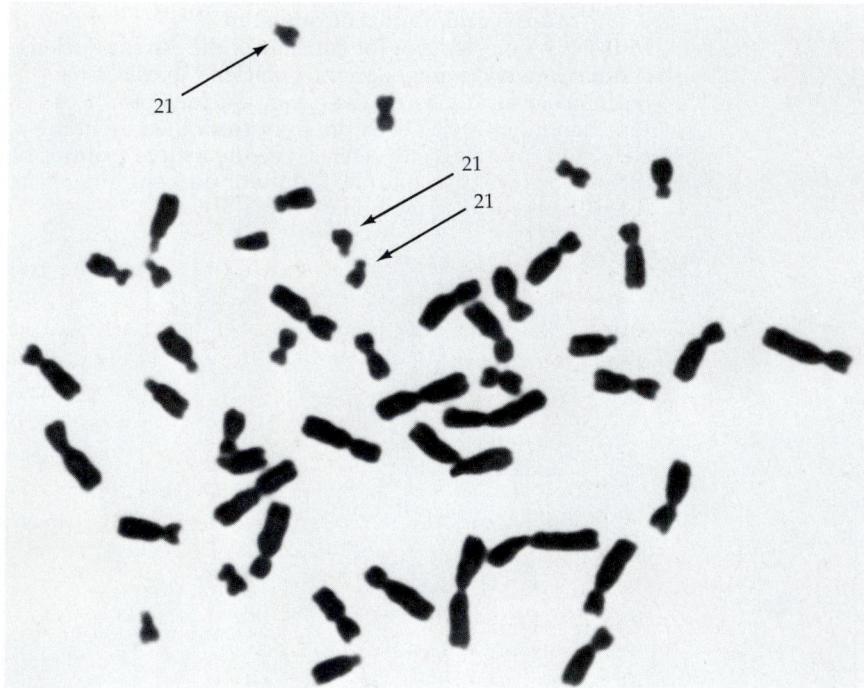

9.19 Chromosome Complement of Down Syndrome
The three copies of chromosome 21 in this spread of chromosomes from a cell in metaphase indicate that the person from whom the cell was taken has Down syndrome. Because of the extra chromosome 21, you should be able to count 47 chromosomes in the spread.

some may break away and become attached to another chromosome. For example, a particular large part of one chromosome 21 may be translocated to another chromosome. Individuals who inherit this translocated piece along with two normal chromosomes 21 also have Down syndrome.

Other human disorders result from particular chromosomal abnormalities. Sex chromosome aneuploidy causes such disorders as Turner syndrome and Klinefelter syndrome, which we discuss in Chapter 10 in connection with sex determination. Deletion of a portion of chromosome 5 results in cri-du-chat syndrome, so named because the afflicted infant's cry sounds like that of a cat. This syndrome includes severe mental retardation.

Trisomies (and the corresponding monosomies, where there is only one chromosome instead of two) are surprisingly common in human zygotes, but most of the embryos that develop from such zygotes do not survive to birth. Trisomies for chromosomes 13, 15, and 18 greatly reduce probabilities of surviving to birth and all lead to death before the age of one year; trisomies and monosomies for other chromosomes are lethal to the embryo. At least one-fifth of all recognized pregnancies self-terminate (miscarry) during the first two months, largely because of such trisomies and monosomies. (The actual fraction of self-terminated pregnancies is certainly much higher, because the earliest terminations usually go unrecognized.)

PLOIDY, MITOSIS, AND MEIOSIS

We have seen that both diploid and haploid nuclei divide by mitosis. Multicellular diploid and multicellular haploid individuals develop from single-celled beginnings by mitotic divisions. Mitosis may proceed in diploid organisms even when a chromosome from one of the haploid sets is missing or when there is an extra copy of one of the chromosomes (as in Down syndrome). There are also circumstances in which triploid (3*n*), tetraploid (4*n*), and higher-order polyploid nuclei are formed. Each of these **ploidy levels** represents an increase in the number of complete sets of chromosomes present. If by accident or (in some organisms) by design, the nucleus has one or more extra full sets of chromosomes (that is, if it is triploid, tetraploid, or of still higher ploidy), this condition in itself does not prevent mitosis. In mitosis, each chromosome behaves independently of the others.

In meiosis, by contrast, chromosomes synapse to begin division; if even one chromosome has no homolog, then anaphase I cannot send representatives of that chromosome to both poles. A diploid nucleus can undergo normal meiosis; a haploid one cannot. A tetraploid nucleus has an even number of each kind of chromosome, so it is possible for each chromosome to pair with its homolog, but a triploid nucleus cannot undergo normal meiosis because one-third of the chromosomes would lack partners. The requirement of an even number of chromosomes for longer than it is thick. The bacterium itself is about 1 μm (1,000 nm) in diameter and about 4 μm long. Thus the space into which the long thread of DNA is packed in the bacterial nucleoid is very small relative to the length of the DNA molecule. It is not surprising that the molecule usually appears in electron micrographs as a hopeless tangle of fibers.

When bacterial cells are gently lysed (broken open; see Chapter 5) to release their contents, the chromosome sometimes comes untangled and spreads out to its full length. Several techniques have shown that the bacterial chromosome is a closed circle rather than the linear structure found in eukaryotes. Circular chromosomes are probably to be found in all prokaryotes, as well as in some viruses and in the chloroplasts and mitochondria of eukaryotic cells. The bacterial chromosome is attached to the plasma membrane. When a new DNA molecule forms from the old one, it too attaches to the membrane. As the cell elongates during growth, the two attachment points separate so that when the new wall and membrane material form at fission, the two chromosomes are included in separate daughter cells (Figure 9.21).

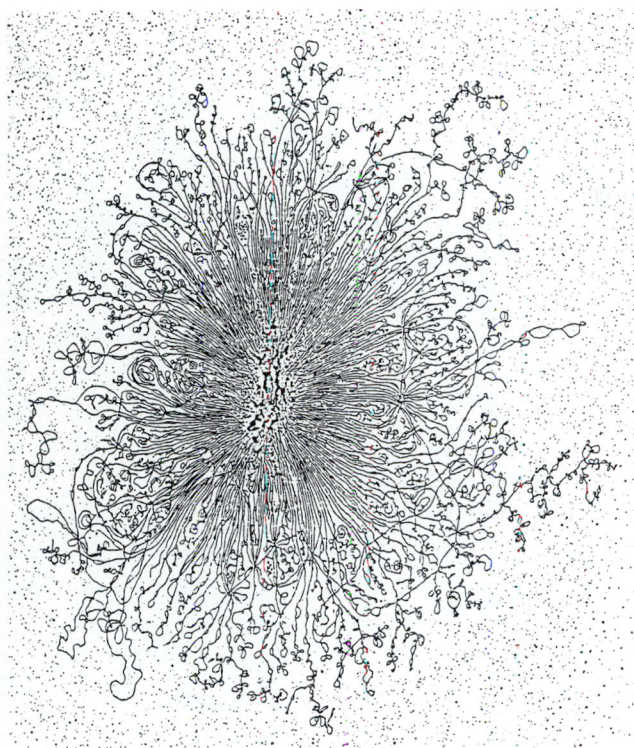

9.20 The Bacterial Chromosome Is a Circle
The long, looping fibers of DNA from this cell of the bacterium *Escherichia coli* are all part of one continuous circular chromosome.

meiosis has important consequences for the fertility of triploid, tetraploid, and other chromosomally unusual organisms that may be produced by intentional breeding or by natural accidents.

CELL DIVISION IN PROKARYOTES

In this chapter we have considered the structures and events of the division of the *eukaryotic* nucleus. Prokaryotic cells, by definition, lack nuclei and hence do not employ mitosis or meiosis in connection with their cell divisions. Still, prokaryotic division, called fission, must include an orderly distribution of genetic information to its daughter cells. Let's consider briefly how this is accomplished.

A prokaryote carries its genetic information on a chromosome that differs in composition and structure from the eukaryotic chromosome. The chromosome of a prokaryote is made of DNA, with protein components only temporarily bound to it and not part of the long-term structure. In the bacterium *Escherichia coli*, the main chromosome is a single circular molecule of DNA about 1.6 million nm (1.6 mm) long (Figure 9.20). The molecule is half a million times

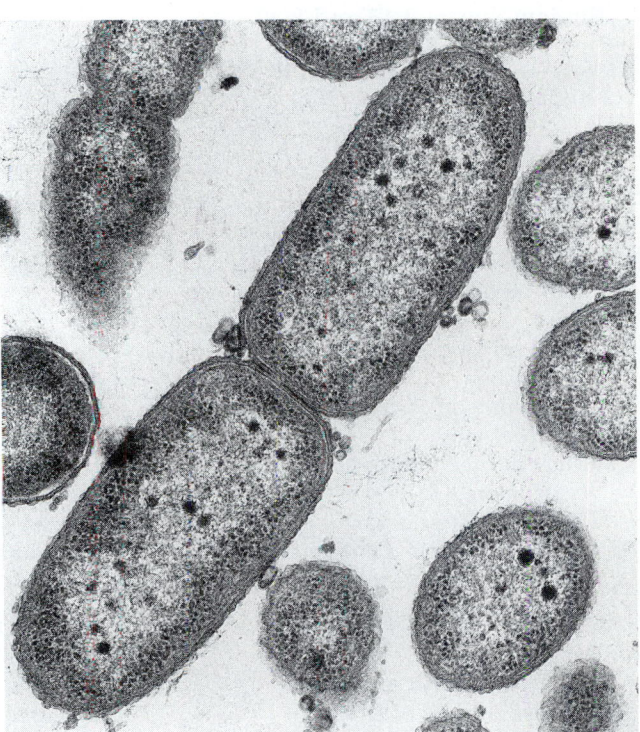

9.21 Bacterial Fission
These two cells of the bacterium *Pseudomonas aeruginosa* have almost completed fission. Plasma membranes have completely formed—separating the cytoplasm of one cell from that of the other—and only a small gap of cell wall remains to be completed. Each cell contains a complete chromosome in the light-toned nucleoid visible in the center of the cells.

SUMMARY of Main Ideas about Mitosis and Meiosis

Mitosis and meiosis are the processes by which the nuclei of eukaryotic cells divide.

Located within the nucleus, a eukaryotic chromosome carries genetic information in a continuous DNA molecule that wraps around aggregates of histone proteins to form nucleosomes.
   Review Figures 9.3, 9.4, and 9.5

Microtubular spindles move chromosomes during nuclear divisions.
   Review Figure 9.10

Mitotic division of a diploid nucleus produces two diploid daughter nuclei, each genetically and chromosomally identical to the parental nucleus.
   Review Figure 9.8

In the cell cycle, mitosis alternates with interphase, which consists of a synthesis phase and two gap phases.
   Review Figures 9.6 and 9.7

Meiotic division of a diploid nucleus produces four genetically diverse haploid daughter nuclei for sexual reproduction.
   Review Figure 9.13 and 9.17

Meiosis is a pair of nuclear divisions with no intervening synthesis phase.

During the first meiotic division, homologous chromosomes, each consisting of two chromatids, synapse and then separate into daughter nuclei.
   Review Figure 9.17 and 9.18

During the second meiotic division, chromatids separate into daughter nuclei.
   Review Figures 9.17 and 9.18

Crossing over during meiosis forms recombinant chromatids.
   Review Figure 9.16

After a nucleus divides by mitosis or meiosis, the rest of the cell may divide by cytokinesis.
   Review Figure 9.11

The karyotype of an organism identifies its chromosomal makeup; the human karyotype consists of 23 pairs of homologous chromosomes.
   Review Figure 9.14

Asexually reproduced offspring are genetically identical to the parent.

A prokaryotic cell, which has no nucleus, divides by fission after the DNA of its chromosome has replicated and separated to opposite areas of the plasma membrane.

SELF-QUIZ

1. Which of the following statements is *not* true of eukaryotic chromosomes?
   a. They sometimes consist of two chromatids.
   b. They sometimes consist of a single chromatid.
   c. They normally possess a single centromere.
   d. They consist of chromatin.
   e. They are always clearly visible as defined bodies under the light microscope.

2. Nucleosomes
   a. are made of chromosomes.
   b. consist entirely of DNA.
   c. consist of DNA wound around a histone core.
   d. are present only during mitosis.
   e. are present only during interphase.

3. Which of the following statements is *not* true of the cell cycle?

   a. It consists of mitosis and interphase.
   b. The cell's DNA replicates during G1.
   c. A cell can remain in G1 for weeks or much longer.
   d. Most proteins are formed throughout all subphases of interphase.
   e. Histones are synthesized primarily during the S phase.

4. Which of the following statements is *not* true of mitosis?
   a. A single nucleus gives rise to two identical daughter nuclei.
   b. The daughter nuclei are genetically identical to the parent nucleus.
   c. The centromeres divide at the onset of anaphase.
   d. Homologous chromosomes synapse in prophase.
   e. Mitotic centers organize the microtubules of the spindle fibers.

5. Which of the following statements is true of cytokinesis?
   a. A cell plate is formed in cytokinesis of animal cells.
   b. Furrowing of the membrane initiates cytokinesis in plant cells.
   c. Cytokinesis generally follows immediately upon mitosis.
   d. Actin and myosin are important in cytokinesis in plant cells.
   e. Cytokinesis is the division of the nucleus.

6. In sexual reproduction
   a. gametes are usually haploid.
   b. gametes are usually diploid.
   c. the zygote is usually haploid.
   d. the chromosome number is reduced during mitosis.
   e. spores are formed during fertilization.

7. In meiosis
  a. meiosis II reduces the chromosome number from diploid to haploid.
  b. DNA replicates between meiosis I and II.
  c. the chromatids making up a chromosome in meiosis II are identical.
  d. each chromosome in prophase I consists of four chromatids.
  e. homologous chromosomes are separated from one another in anaphase I.

8. In meiosis
  a. a single nucleus gives rise to two identical daughter nuclei.
  b. the daughter nuclei are genetically identical to the parent nucleus.
  c. the centromeres divide at the onset of anaphase I.
  d. homologous chromosomes synapse in prophase I.
  e. no spindle forms.

9. Which of the following statements is *not* true of aneuploidy?
  a. It results from chromosomal nondisjunction.
  b. It does not happen in humans.
  c. An individual with an extra chromosome is trisomic.
  d. Trisomies are common in human zygotes.
  e. A piece of one chromosome may translocate to another chromosome.

10. In prokaryotes
  a. there are no meiotic divisions.
  b. mitosis proceeds as in eukaryotes.
  c. the genetic information is not carried in chromosomes.
  d. the chromosomes are identical to those of eukaryotes.
  e. cell division follows division of the nucleus.

FOR STUDY

1. How does a nucleus in the G2 phase of the cell cycle differ from one in the G1 phase?

2. What is a chromatid? When does a chromatid become a chromosome?

3. Compare and contrast mitosis (and subsequent cytokinesis) in animals and plants.

4. Suggest two ways in which one might, with the help of a microscope, determine the relative durations of the various phases of mitosis.

5. Contrast mitotic prophase and meiotic prophase I. Contrast mitotic anaphase and meiotic anaphase I.

READINGS

All introductory genetics texts contain chapters on meiosis and how this process distributes genetic information.

Alberts, B., D. Bray, J. Lewis, M. Raff, K. Roberts and J. D. Watson. 1994. *Molecular Biology of the Cell*, 3rd Edition. Garland Publishing, New York. An outstanding book in which to pursue the topics of this chapter in greater detail. Chapters 8, 17, 18, and 20 have definitive modern treatments of the nucleus, mitosis, the cell cycle, meiosis, and more.

Glover, D. M., C. Gonzalez and J. W. Raff. 1993. "The Centrosome." *Scientific American*, June. New findings on the structure and function of the organelle that directs the assembly of the cytoskeleton and controls cell division—when it is present.

Mazia, D. 1961. "How Cells Divide." *Scientific American*, September. A classical description of mitosis by a leading researcher of cell division.

Mazia, D. 1974. "The Cell Cycle." *Scientific American*, January. This article discusses the four major stages of the cell cycle.

McIntosh, J. R. and K. L. McDonald. 1989. "The Mitotic Spindle." *Scientific American*, October. A fascinating description of the growth, disassembly, and interactions of the components of the spindle.

Mitchison, J. M. 1972. *The Biology of the Cell Cycle*. Cambridge University Press, New York. One of the few basic texts available on cell division and the cell cycle.

Murray, A. W. and M. W. Kirschner. 1991. "What Controls the Cell Cycle." *Scientific American*, March. A discussion of how cyclins, cdc2, and the maturation promoting factor (MPF) regulate mitosis and meiosis.

Sloboda, R. D. 1980. "The Role of Microtubules in Cell Structure and Cell Division." *American Scientist*, May/June. Includes a discussion of the spindle apparatus.

Mendel's Research Material
Studies of garden peas cracked the secrets of inheritance.

10

Mendelian Genetics and Beyond

Y ou have seen how the nuclei of eukaryotic cells divide mitotically and meiotically. You know that by these processes cells pass copies of their genetic information to their descendants. By thinking about meiosis and sexual reproduction, you can account for the fact that offspring of the same parents may differ. But what if no one understood cell division—especially the complexities of meiosis? How could there be any understanding of how traits pass from parents to offspring without such background information? The answer is the subject of this chapter.

The term **Mendelian genetics** refers to certain basic inheritance patterns. The term honors the Austrian monk Gregor Johann Mendel (1822–1884), the person who first made rigorous, quantitative observations of the patterns of inheritance and proposed plausible mechanisms to explain them. In organisms that reproduce sexually and have more than one chromosome (and orderly meiosis), many traits pass from parent to offspring in accord with these patterns. When Mendel began his work with the garden pea in his monastery garden, little was known about the sex lives of plants or the consequences of sexual reproduction for the inheritance of traits. Chromosomes, mitosis, and meiosis were unknown.

MENDEL'S DISCOVERIES

Some observations that Mendel found useful in his studies had been made in the late eighteenth century by a German botanist, Josef Gottlieb Kölreuter. Kölreuter studied many plants by cross-pollinating them. He attempted many crosses between plants, produced many **hybrids** (the offspring of genetically different parents), and learned a great deal about pollination (see Chapter 33). In some instances he confirmed the common observation that hybrids are intermediate between their parents with respect to obvious traits such as size, color, and flower shape; more important, he found and emphasized that in some cases the hybrids are not intermediate but closely resemble just one of the parents.

Kölreuter also studied the offspring from **reciprocal crosses**. These are crosses made in both directions; that is, in one set of crosses, males with one form of a trait that we will call "A" are crossed with females having the "a" form of the same trait, while in a complementary set of crosses "a" males and "A" females are the parents. In an example of reciprocal crosses of plants, pollen (which carries the male sperm) from a plant with trait "A" is placed on the female organ—from which the sperm can travel to the eggs—of a plant with trait "a" in one set of crosses. In the reciprocal cross, pollen from "a" plants is placed on the female organs of "A" plants. (In many plant species the same individuals have both male and female reproductive organs; each plant may then reproduce as a male, as a female, or as both—which makes such plants excellent material for genetic studies.) In Kölreuter's experience, reciprocal crosses always gave identical results.

This was the state of knowledge in genetics when Mendel began his work. In one sense, the time was ripe for his discoveries, for it had recently been shown that one female gamete combines with one male gamete to bring about fertilization. On the other hand, the role of the chromosomes as bearers of genetic information was unknown, and mitosis and meiosis were yet to be discovered. Mendel himself was well qualified to make the big step forward. Although in 1850 he had failed an examination for a teaching certificate in natural science, he later undertook intensive studies in physics, chemistry, mathematics, and various aspects of biology at the University of Vienna. His work in physics and mathematics is probably what led to his applying experimental and quantitative methods to the study of heredity—and these were the key ingredients in his success.

Mendel worked out the basic principles of the heredity of plants and animals over a period of about nine years, the work culminating in a public lecture in 1865 and a detailed written account in 1866. However, his theory was not accepted. In fact, it was ignored. Perhaps the chief difficulty was that the physical basis of his theory was not understood until the discovery of meiosis, some years later. The most prominent biologists at the time Mendel published his results simply were not in the habit of thinking in mathematical terms, even the simple terms used by Mendel. Mendel's paper on plant hybridization appeared in a journal that was received by 120 libraries, and he sent reprinted copies (of which he had obtained 40) to several distinguished scholars. We know that at least one of these scholars died years later without even having opened the pages of the Mendel reprint. Whatever the reasons, Mendel's pioneering paper had no discernible influence on the scientific world for more than 30 years.

Then, in 1900, Mendel's discoveries burst into prominence as a result of independent experiments by the Dutchman Hugo de Vries, the German Karl Correns, and the Austrian Erich von Tschermak. Each of these scientists carried out crossing experiments and obtained quantitative data about the progeny; each published his principal findings in 1900; each cited Mendel's 1866 paper. At last the time was ripe for biologists to appreciate the significance of what these four geneticists had discovered—largely because meiosis had by then been described. That Mendel made his discoveries *prior* to the discovery of meiosis was due in part to the methods of experimentation he used.

Mendel's Strategy

Mendel chose the garden pea for his studies because of its ease of cultivation, the feasibility of controlled pollination, and the availability of varieties with differing traits. He controlled pollination by moving pollen from one plant to another; thus he knew the parentage of the offspring in his experiments. If untouched, the peas Mendel studied naturally self-pollinate—that is, the female organs of flowers receive pollen from the male organs of the same flowers—and he made use of this natural phenomenon in some of his experiments.

Mendel began by examining varieties of peas in a search for heritable traits (traits that can be passed from parent to offspsring) suitable for study. A suitable trait would be one that was "true-breeding." To be considered true-breeding, peas with white flowers, when crossed with one another, would have to give rise *only* to progeny with white flowers; tall plants bred to tall plants would have to produce only tall progeny. The suitable traits were also ones that had well-defined, contrasting alternatives that could be obtained in true-breeding form, such as purple flowers versus white flowers. For most of his work, Mendel concentrated on the seven pairs of contrasting traits shown in Figure 10.1. Before performing a given cross, he made sure that each potential parent was from a true-breeding strain; this was an essential point in his analysis of his experimental results.

Mendel then placed pollen he collected from one parental strain onto the stigma (female organ) of flowers of the other strain. The plants providing and receiving the pollen were the parental generation, designated **P**. In due course, seeds formed and were planted. The resulting new plants constituted the first filial generation, F_1. Mendel and his assistants examined each F_1 plant to see which traits it bore and then recorded the number of F_1 plants expressing each trait. In some experiments the F_1 plants were allowed to self-pollinate and produce a second filial generation, or F_2. Again, each F_2 plant was characterized and counted. Mendel performed other crosses in which the F_2 was produced by crossing F_1 hybrids with one of the true-breeding parental strains.

Always, each type of progeny was counted and recorded. This attention to quantitative detail was a unique advance in experimental biology; it allowed Mendel to make numerical comparisons and ultimately to develop a hypothesis, or model—a proposed explanation for the numbers he observed. In sum, Mendel devised a well-organized plan of research, pursued it faithfully and carefully, recorded great amounts of quantitative data, and analyzed the numbers he recorded to explain the relative proportions of the different kinds of progeny. His 1866 paper stands to this day as a model of clarity. His results

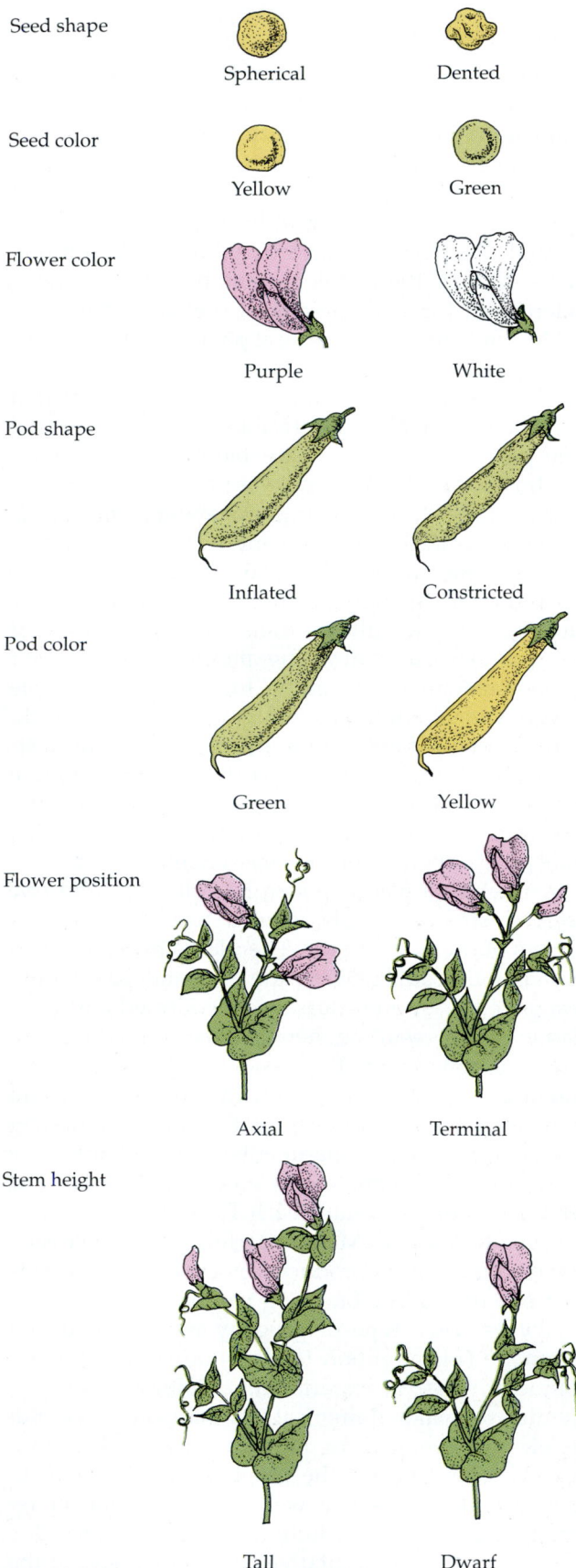

Seed shape

Spherical   Dented

Seed color

Yellow   Green

Flower color

Purple   White

Pod shape

Inflated   Constricted

Pod color

Green   Yellow

Flower position

Axial   Terminal

Stem height

Tall   Dwarf

10.1   Inherited Traits Studied by Mendel
Mendel's work on the genetics of peas focused on these seven pairs of traits. He isolated each as a true-breeding trait before he began his studies of various crosses.

and the conclusions to which they led are the subject of the next few sections.

Experiment 1

"Experiment 1" in Mendel's paper included a **monohybrid cross**, one in which the parents were both hybrids for a single trait. He took pollen from plants of a true-breeding strain with dented seeds and placed it on the stigmas of flowers of a true-breeding, spherical-seeded strain. He also performed the reciprocal cross, placing pollen from the spherical-seeded strain on the stigmas of flowers of the dented-seeded strain. In both cases, all the F_1 seeds that were produced were spherical—it was as if the dented trait had disappeared completely. The following spring Mendel grew 253 F_1 plants from these spherical seeds, each of which was allowed to self-pollinate—a monohybrid cross—to produce F_2 seeds. In all, there were 7,324 F_2 seeds, of which 5,474 were spherical and 1,850 dented (Figure 10.2).

Mendel observed that the spherical seed trait was **dominant** because it was expressed over the dented seed trait, which he called **recessive**. In each of the other six pairs of traits studied by Mendel, one proved to be dominant over the other. When he crossed plants differing in any of these traits, only one of each pair of traits was evident in the F_1 generation. However, the trait that was not seen in the F_1 *reappeared* in the F_2. Most important, the ratio of the two traits in the F_2 was always the same: approx-

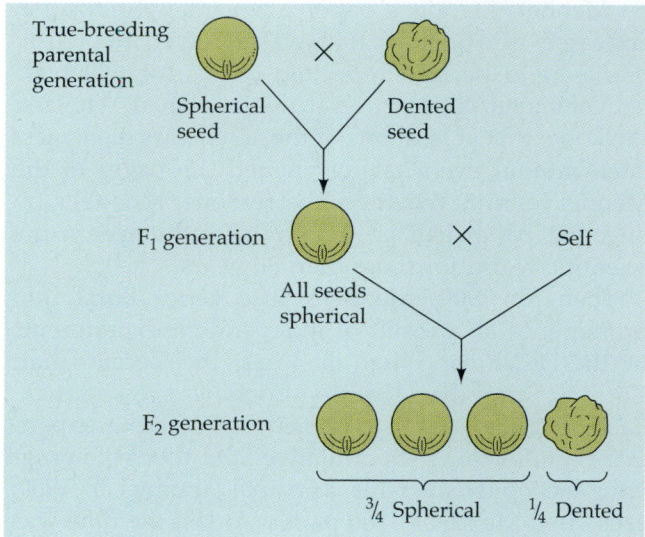

True-breeding parental generation

Spherical seed   ×   Dented seed

F_1 generation

All seeds spherical   ×   Self

F_2 generation

¾ Spherical   ¼ Dented

10.2   Mendel's Experiment 1
Mendel crossed plants of true-breeding varieties of peas with different seed shapes. Plants grown from the spherical F_1 seeds and then self-pollinated produced F_2 in which about ¼ of the seeds were dented and ¾ were spherical. The pattern was the same regardless of which variety contributed the pollen in the parental generation.

TABLE 10.1
Mendel's Results from Monohybrid Crosses

P			F₂			
DOMINANT	×	RECESSIVE	DOMINANT	RECESSIVE	TOTAL	RATIO
Spherical	×	Dented seeds	5,474	1,850	7,324	2.96:1
Yellow	×	Green seeds	6,022	2,001	8,023	3.01:1
Purple	×	White flowers	705	224	929	3.15:1
Inflated	×	Constricted pods	882	299	1,181	2.95:1
Green	×	Yellow pods	428	152	580	2.82:1
Axial	×	Terminal flowers	651	207	858	3.14:1
Tall	×	Dwarf stems	787	277	1,064	2.84:1

imately 3:1—that is, three-fourths of the F_2 showed the dominant trait and one-fourth showed the recessive trait (Table 10.1). In his Experiment 1, the ratio was 5474:1850 = 2.96:1. Reciprocal crosses in the parental generation gave similar outcomes in the F_2.

Terminology for Mendelian Genetics

All by themselves the results from Experiment 1 disproved the widely believed theory that inheritance is a "blending" phenomenon. According to the blending theory, Mendel's F_1 seeds should have had an appearance intermediate between those of the two parents—they should have been slightly dented. Furthermore, the blending theory offered no explanation for the reappearance of the dented trait in the F_2 seeds after its apparent absence in the F_1 seeds. From his results Mendel proposed a **particulate theory**, in which the hereditary carriers are present as discrete units that retain their integrity in the presence of other units.

As he wrestled mathematically with his data, Mendel reached the conclusion that each pea has two such units for each character, one derived from each parent. Each gamete contains one unit, and the resulting zygote (and each cell of the adult that develops from it) contains two. This conclusion is the core of his model of inheritance. Mendel's "unit" is now called a **gene**.

Mendel reasoned that in Experiment 1, the spherical-seeded parent had a pair of genes of the same type, which we will call S, and the parent with dented seeds had two s genes. The SS parent produced gametes each containing a single S, and the ss parent produced gametes each with a single s. Each member of the F_1 generation had an S from one parent and an s from the other; an F_1 could thus be described as Ss. We say that S is dominant over s because s is not evident when both genes are present.

The physical appearance of a character is its **phenotype**. Mendel correctly supposed the phenotype to

be the result of the **genotype**, or genetic constitution, of the organism showing the phenotype. In Experiment 1 we are dealing with two phenotypes (spherical seeds and dented seeds) and three genotypes: The dented-seed phenotype is produced only by the genotype ss, whereas the spherical-seed phenotype may be produced by the genotypes SS and Ss. The different forms of a gene (S and s in this case) are called **alleles**. Individuals that breed true for a character contain two copies of the same allele. For example, a strain of true-breeding peas with dented seeds must have the genotype ss—if S were present, the plants would produce spherical seeds. We say individuals that produce dented seeds are **homozygous** for the allele s, meaning that they have two copies of the same allele. Some peas with spherical seeds—the ones with the genotype SS—are also homozygous. However, other spherical-seeded plants are **heterozygous** because they have two different alleles of the gene in question; these plants have the genotype Ss. To illustrate these terms with a more complex example, one in which there are three gene pairs, an individual with the genotype $AABbcc$ is homozygous for two genes—because it has two A alleles and two c alleles—but heterozygous for the gene with alleles B and b. An individual that is homozygous for a character is called a homozygote; a heterozygote is heterozygous for the character in question.

Segregation of Alleles

How does Mendel's model explain the composition of the F_2 generation in Experiment 1? Consider first the F_1, which has the spherical-seeded phenotype and the genotype Ss. According to the model, when any F_1 individual produces gametes, the alleles **segregate**, or separate, so that each gamete receives only *one* member of the pair of genes. Half the gametes contain the S allele and half the s allele. The random combination of these gametes produces the F_2 generation (Figure 10.3). Three different F_2 genotypes

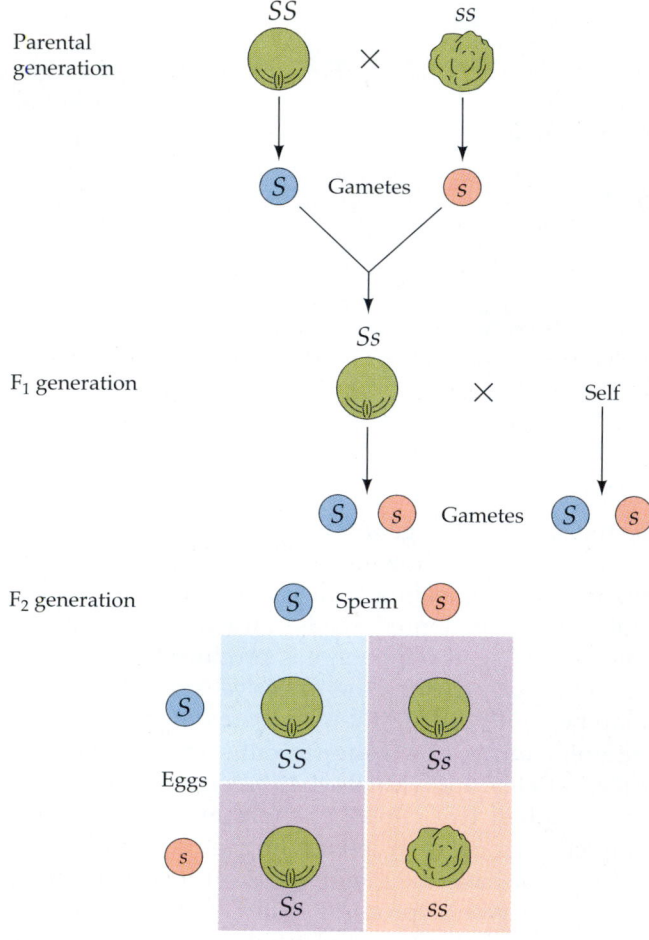

Parental generation

Gametes

F₁ generation

Self

Gametes

F₂ generation

Sperm

Eggs

Punnett square

10.3 Mendel's Explanation of Experiment 1
Mendel concluded that heredity depends on factors from each parent, and that these factors do not blend in the offspring. The drawing shows a modern version of Mendel's explanation of the experiment in Figure 10.2. A parent homozygous for the allele for spherical seeds is crossed with a parent homozygous for the allele for dented seeds. Each parent makes gametes of only one kind, either S or s, and these combine at fertilization to form plants that all have the genotype Ss and the spherical-seeded phenotype. When the F₁ plants self-pollinate they produce two kinds of eggs, S and s, and the same two types of male sex cells. These combine randomly in four different ways to form F₂ plants, as shown in the box at the bottom of the figure. Three of these combinations produce genotypes that determine the spherical-seeded phenotype and one produces the genotype for the dented-seed phenotype, resulting in the observed ratio of 3:1. The box in which the F₂ plants are displayed is called a Punnett square; it is a convenient device for keeping track of all the ways gametes can combine at fertilization.

two ways: by using the "Punnett square," devised in 1905 by the British geneticist Reginald Crundall Punnett, or by using simple probability calculations. The Punnett square is illustrated in Figure 10.3 and other figures in this chapter; probabilities are discussed in Box 10.A. By either method (they amount to the same thing), it becomes apparent that self-pollination of the F₁ genotype Ss will give the three F₂ genotypes in the expected ratio 1 SS:2 Ss:1 ss. Because S is dominant and s recessive, the ratio of *phenotypes* is 3 spherical (SS and Ss) to 1 dented (ss), just as Mendel observed.

Mendel did not live to see his theory placed on a sound physical footing based on chromosomes and DNA. Genes are now known to be portions of the DNA molecules in chromosomes. More specifically, a gene is a portion of the DNA that resides at a particular **locus**, or site, within the chromosome and that encodes a particular function. Remember that the cells in a multicellular organism have the same genotype because they are all derived by mitosis from a single cell, the zygote (Chapter 9). Each diploid cell has two homologous chromosomes of each type, and therefore has two alleles at each locus. Consistent with Mendel's model, if the two homologous chromosomes have copies of the same allele at a given locus, the cell and the organism are homozygous at that locus; if the homologs have differing alleles, the cell and the organism are heterozygous at that locus. Because meiosis reduces the number of chromosomes per cell, each gamete contains only one member of each homologous pair of chromosomes and, hence, only one allele at any given locus. If you visualize S and s as occupying specific, homologous sites on a pair of homologous chromosomes in an F₁ individual, you will see how they would be inherited through successive generations, by way of meiosis and the random fusion of gametes.

On the basis of monohybrid crosses such as that of Experiment 1, Mendel proposed his first law, called the **law of segregation**, which says that *alleles segregate from one another during the formation of gametes.* Mendel did not know about chromosomes or meiosis, but we do and we can picture the alleles segregating as chromosomes separate into gametes in meiosis (Figure 10.4).

The Test Cross

Mendel's theory adequately explains the ratios of phenotypes observed in F₁ and F₂ generations obtained from crosses of differing, true-breeding strains. To be regarded as fully satisfactory, however, the theory must also be able to predict—accurately—the outcome of other kinds of experiments. One such challenge was posed by Mendel himself. According to his theory, ⅔ of the F₂ spherical seeds from Ex-

are possible: SS, Ss (which is the same thing as sS), and ss. Our quantitative way of looking at things may lead us to wonder about what proportions of these genotypes we might expect to observe in the F₂ progeny. The expected frequencies of these three genotypes in our example may be determined in either of

BOX 10.A

Elements of Probability

Many people find it easiest to solve genetics problems using probability calculations, perhaps because the basic underlying considerations are a familiar part of daily life. When we flip a coin, for example, we expect it to have an equal probability of landing "heads" or "tails." When we roll a fair die, we expect to have equal chances of getting any of the numbers from one to six. We properly bet more money on a coin's giving at least one heads in a pair of tosses than we do on it coming up heads in a single toss—yet if we are even slightly sophisticated, we recognize that on a given toss, the probability of heads is independent of what happened in all the previous tosses. (For a fair coin, a run of 10 straight heads implies nothing about the next toss.

No "law of averages" increases the likelihood that the next toss will come up tails, and no "momentum" makes an eleventh occurrence of heads any more likely. On the eleventh toss, the odds are still 50:50.)

The basic conventions of probability are simple: If an event is absolutely certain to happen, its probability is 1. If it cannot happen, its probability is 0. Otherwise, its probability lies between 0 and 1. A coin toss results in heads half the time, and the probability of heads is ½—as is the probability of tails. If *two* coins (a penny and a dime, say) are tossed, each acts independently of the other. What, then, is the probability of both coins coming up heads? Half the time, the penny comes up heads; of that fraction, half the time the dime also comes up heads. Therefore, the joint probability of two heads is half of one-half, or $\frac{1}{2} \times \frac{1}{2} = \frac{1}{4}$. To find the joint probability of *independent* events, *multiply* the probabilities of the individual events.

To apply this to the crosses we have been discussing, we need only deal with gamete formation and random fertilization. A homozygote can produce only one type of gamete, so, for example, an *SS* individual has a probability equal to 1 of producing gametes with the genotype *S*. The heterozygote *Ss* produces *S* gametes with a probability of ½, and *s* gametes with a probability of ½ as well. Consider, now, the F_2 progeny of the cross of Figure 10.3. They are obtained by self-pollinating F_1 hybrids of genotype *Ss*. The probability that an F_2 plant is *SS* must be $\frac{1}{2} \times \frac{1}{2} = \frac{1}{4}$—there is a 50:50 chance of the sperm's being *S*, and this is independent of the 50:50 chance of the egg's being *S*. Similarly, the probability of *ss* offspring is $\frac{1}{2} \times \frac{1}{2} = \frac{1}{4}$. The probability of getting *S* from the sperm and *s* from the egg is also ¼, but the same genotype can also result from *s* in the sperm and *S* in the egg, with a probability of ¼. Thus the probability that an F_2 plant is a heterozygote is $\frac{1}{4} + \frac{1}{4} = \frac{1}{2}$. All three of the genotypes are expected in the ratio ¼ *SS*:½ *Ss*:¼ *ss*—hence the 1:2:1 ratio of genotypes and the 3:1 ratio of phenotypes seen in Figure 10.3.

---

periment 1 should be heterozygous, each carrying both *S* and *s* alleles (recall the genotype ratio 1:2:1). Therefore, if all the spherical seeds were allowed to grow into F_2 adults and self-pollinate, the ⅔ of the plants that were heterozygous would produce seeds of which about ¾ would be spherical and ¼ dented (recall the phenotype ratio 3:1). The other ⅓ of the F_2 plants, being *SS* homozygotes, would produce only spherical seeds (Figure 10.5). This result is what Mendel observed. It is impossible to know the genotype of a plant displaying a dominant phenotype simply by looking at it, but looking at the phenotypes of progeny obtained by self-fertilizing the plant gives us the answer.

The **test cross** is another way to test whether a given individual showing a dominant trait is homozygous or heterozygous. In a test cross, the individual in question is crossed with an individual known to be homozygous for the recessive trait—an easy individual to identify because its only phenotype is the recessive one. For the gene that we have been considering, the recessive homozygote for the test cross is *ss*. The individual being tested may be described initially as *S*— because we do not yet know the identity of the second allele. If the individual being tested is homozygous dominant (here, *SS*), all offspring of the test cross will be *Ss* and show the dominant character (spherical seeds). If, however, the tested individual is heterozygous (*Ss*), then approximately ½ of the offspring of the test cross will show the dominant trait, but the other ½ will be homozygous recessive (Figure 10.6). These are exactly the results that are obtained; thus Mendel's model predicts accurately the results of such test crosses.

Independent Assortment of Alleles

What happens if a cross is made between two parents that differ at two or more loci? When a double heterozygote (for example, *AaBb*) makes gametes, do the alleles of maternal origin go together to one gamete

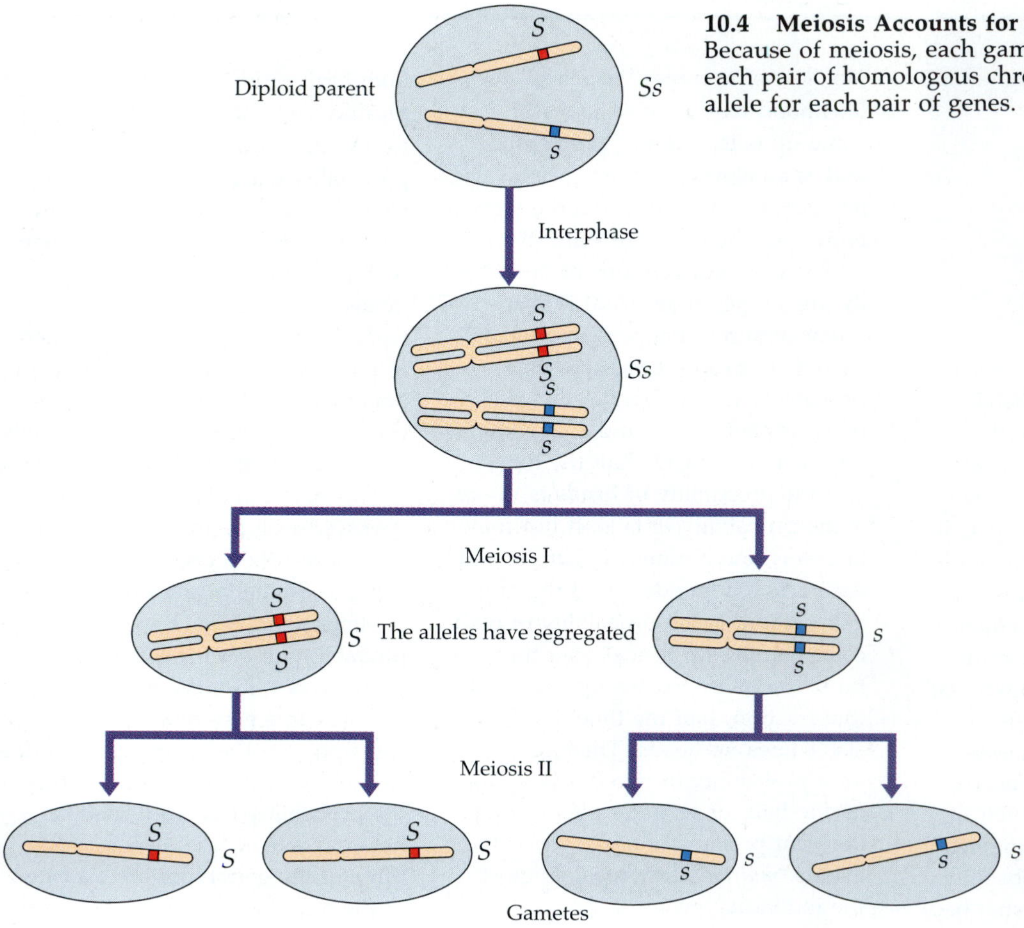

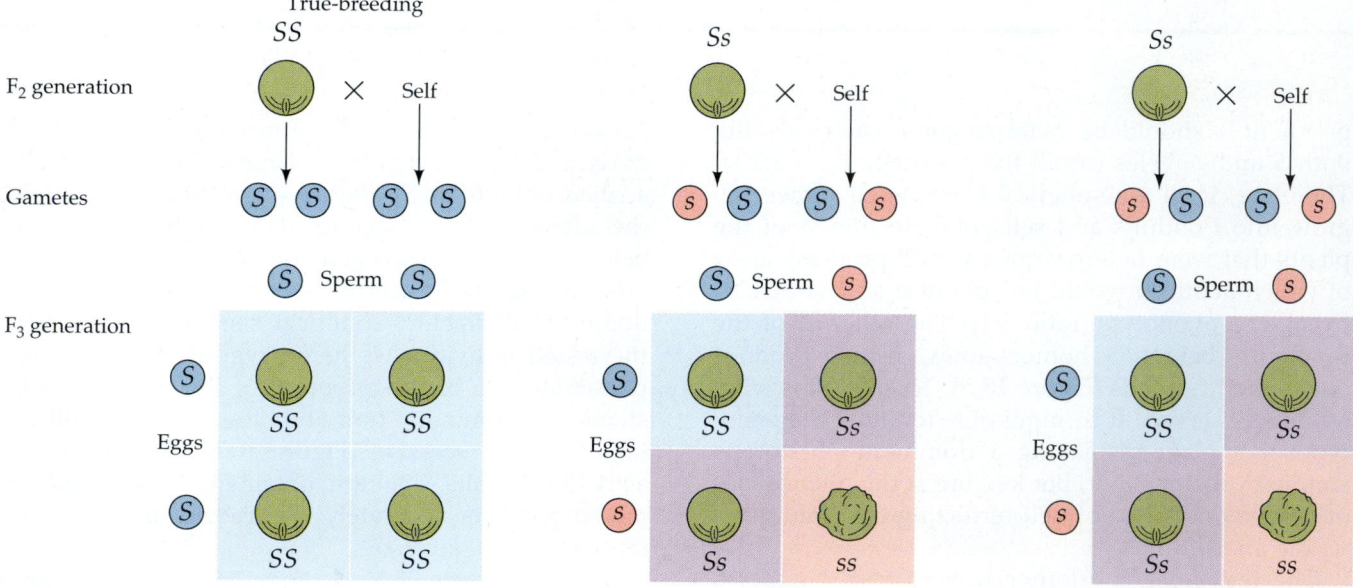

10.4 Meiosis Accounts for the Segregation of Alleles
Because of meiosis, each gamete contains *one* member of each pair of homologous chromosomes, and thus *one* allele for each pair of genes.

10.5 On to the F₃ Generation
According to Mendel's ideas, the spherical-seeded F₂ plants in Figure 10.3 are of two kinds. One-third of them (the *SS* homozygotes), upon self-pollination, will produce only spherical-seeded offspring. Two-thirds of them (the *Ss* heterozygotes) will, like the original F₁ plants in Figure 10.2, produce spherical- and dented-seeded plants in a 3:1 ratio.

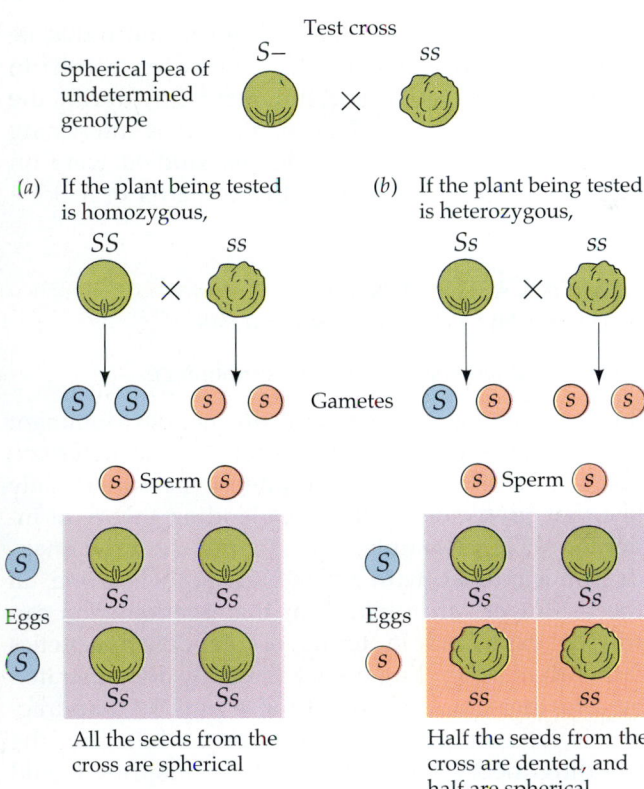

Test cross

Spherical pea of undetermined genotype $S-$ × ss

(a) If the plant being tested is homozygous,

SS × ss

Gametes

Sperm

Eggs

All the seeds from the cross are spherical

(b) If the plant being tested is heterozygous,

Ss × ss

Sperm

Eggs

Half the seeds from the cross are dented, and half are spherical

10.6 Homozygous or Heterozygous? Try a Test Cross

A plant with a dominant phenotype may be homozygous or heterozygous. Its genotype can be deduced by observing the phenotypes of progeny produced by crossing it with a homozygous recessive plant—that is, by making a test cross. (a) If all progeny show the dominant phenotype, the plant being tested must have been homozygous for the dominant allele. (b) If half the progeny have the dominant phenotype and half have the recessive phenotype, the plant in question must have been heterozygous.

and those of paternal origin to another gamete? Or does a single gamete receive some maternal and some paternal alleles? To answer these questions Mendel performed a series of **dihybrid crosses**: crosses made between parents identically heterozygous at two loci.

In these experiments Mendel began with peas that differed for two characters of the seeds: seed shape and seed color. One true-breeding strain produced only spherical, yellow seeds and the other strain produced only dented, green ones. Plants of the first strain can be designated $SSYY$, indicating that they are homozygous both for the S allele at the seed-shape locus and for the Y allele at the seed-color

locus. The second doubly homozygous strain is $ssyy$. The doubly heterozygous F_1 offspring from a cross between these two strains are $SsYy$. Because the S and Y alleles are dominant, these F_1 seeds would all be yellow and spherical.

Now what about the dihybrid cross? There are two ways in which these doubly heterozygous plants might produce gametes, as Mendel saw it (remember that he had never heard of chromosomes, let alone of meiosis). First, if the alleles maintained the associations they had in the original parents, then only two types of gametes would be produced: SY and sy; and the F_2 progeny resulting from self-pollination of the F_1 plants would consist of three times as many plants bearing spherical, yellow seeds as ones with dented, green seeds. Were such results to be obtained, there would be no reason to suppose that seed shape and seed color were really regulated by two different genes, because spherical seeds would always be yellow, and dented ones green.

The second possibility is that the segregation of S from s is *independent* of the segregation of Y from y during the production of gametes. In this case, four kinds of gametes would be produced, and in equal numbers: SY, Sy, sY, and sy. When these gametes combined at random, they would produce an F_2 of nine different genotypes. The progeny can have any of three possible genotypes for shape (SS, Ss, or ss)

BOX 10.B

Probabilities in the Dihybrid Cross

If F_1 plants heterozygous for two independent traits self-pollinate, the resulting F_2 plants express four phe-

notypes. The proportions of these phenotypes are easily determined by probabilities. The probability of a seed's being yellow is ¾ (see Box 10.A); by the same reasoning, the probability of a seed's being spherical is also ¾. The two traits are determined by separate genes and are independent of one another, so the joint probability of a seed being both yellow and spherical is ¾ × ¾ = ⁹⁄₁₆. For the dented, yellow members of

the F_2, the probability of yellow is again ¾; the probability of dented seeds is ½ × ½ = ¼. The joint probability of a seed being both yellow and dented is, then, ¾ × ¼ = ³⁄₁₆. The same probability applies, for similar reasons, to the spherical, green F_2 seeds. Finally, the probability of F_2 seeds being both dented and green must be ¼ × ¼ = ¹⁄₁₆. Looking at all four phenotypes, we see they are expected in the ratio of 9:3:3:1.

and any of three for color (*YY*, *Yy*, or *yy*). These nine genotypes would produce just four phenotypes (spherical, yellow; spherical, green; dented, yellow; dented, green). By using either a Punnett square or simple probability calculations, we can show that these four phenotypes would be expected to occur in a ratio of 9:3:3:1.

Mendel's dihybrid crosses produced the results predicted by the second possibility. Four different phenotypes appeared in a ratio of about 9:3:3:1 in the F₂, rather than only the two parental types as predicted from the first possibility (Figure 10.7; Box 10.B). The parental traits appeared in new combinations in two of the phenotypic classes (spherical, green and dented, yellow). These are called **recombinant phenotypes**. These results led Mendel to the formulation of what is now known as Mendel's second law: *Alleles of different genes assort independently of one another during gamete formation.* This **law of independent assortment** is not as universal as the law of segregation because it applies only to genes that lie

on separate chromosomes, and not to those that lie on the same chromosome. It is, however, correct to say that *chromosomes* assort independently during the formation of gametes (Figure 10.8). It is interesting to note that all the genes Mendel studied were on different chromosomes; was that a matter of luck?

GENETICS AFTER MENDEL: ALLELES AND THEIR INTERACTIONS

Incomplete Dominance and Codominance

Some genes have alleles that are neither dominant nor recessive to each other. Instead, the heterozygotes show an intermediate phenotype superficially like that predicted by the old blending theory of inheritance. For example, if a true-breeding red snapdragon is crossed with a true-breeding white one, all the F₁ flowers are pink. That this phenomenon can still be explained in terms of Mendelian genetics rather than a blending theory is readily demonstrated by a further cross. If one of these pink F₁ snapdragons is crossed with a true-breeding white one, the blending theory predicts that all the offspring would be a still-lighter pink. In fact, approximately ½ of the offspring are white and ½ the same pink as the original F₁. Suppose now that the F₁ pink snapdragons are self-pollinated. The resulting F₂ plants are distributed in a ratio of 1 red : 2 pink : 1 white (Figure 10.9). Clearly the hereditary particles—the genes—have not blended, but they are readily sorted out in their original forms.

We can understand these results in terms of the Mendelian model. All we need to do in cases like this is recognize that the heterozygotes show a phenotype intermediate between those of the two homozygotes. Genes code for the production of specific proteins, many of which are enzymes. Different alleles at a locus code for alternative forms of a protein that differ in structure and, when the protein is an enzyme, often have different degrees of catalytic activity. In the snapdragon example, one allele codes for an enzyme that catalyzes a reaction leading to the formation of a red pigment in the flowers. The alternative allele codes for an altered protein lacking catalytic activity for pigment production. Plants homozygous for this alternative allele cannot synthesize red pigment, and their flowers are white. Heterozygous plants, with only one allele for the functional enzyme, produce just enough red pigment so that their flowers are pink. Homozygous plants with two alleles for the functional enzyme produce more red pigment, resulting in red flowers. When a heterozygous phenotype is intermediate, as in this example, the gene is said to be governed by **incomplete dominance**.

There are more examples of incomplete dominance

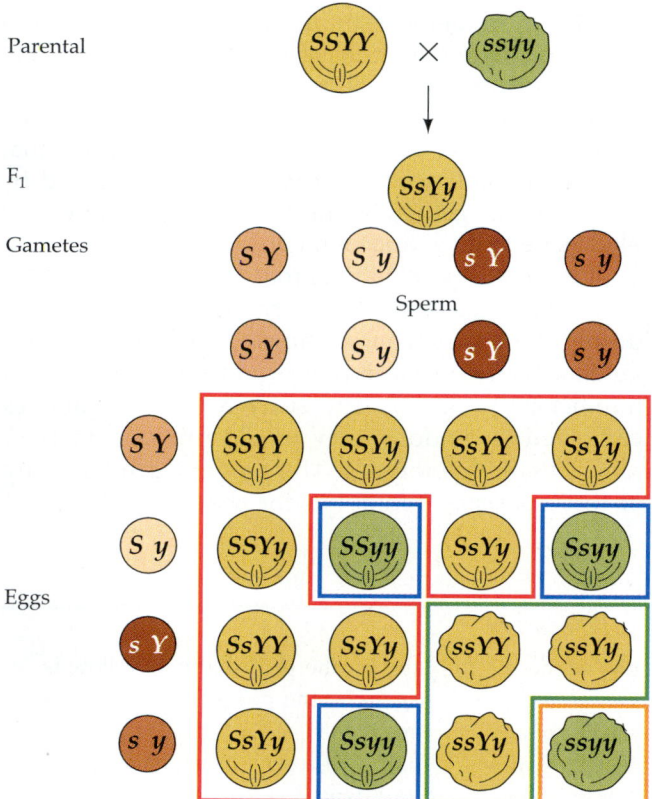

10.7 Independent Assortment
Plants heterozygous for two genes (*SsYy*) make four kinds of gametes in equal proportions. Random combination of these gametes produces equal numbers of the 4 × 4 = 16 combinations displayed in the boxes; these 16 combinations result in nine different genotypes. Because *S* and *Y* are dominant over *s* and *y*, respectively, the nine genotypes determine four phenotypes (indicated by four different outline colors in the Punnett square) in the ratio of 9:3:3:1.

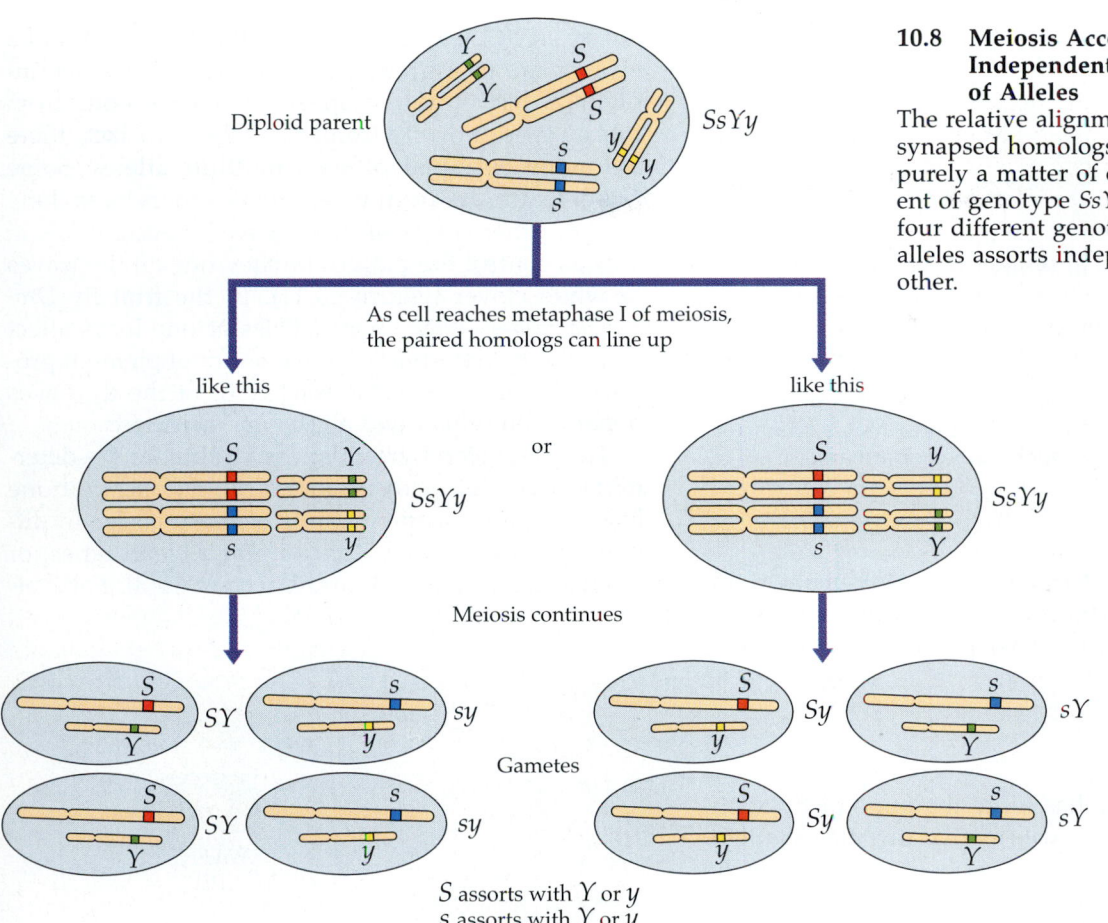

10.8 Meiosis Accounts for Independent Assortment of Alleles

The relative alignment of two pairs of synapsed homologs at metaphase I is purely a matter of chance. Thus a parent of genotype *SsYy* forms gametes of four different genotypes—each pair of alleles assorts independently of the other.

than of complete dominance. Thus an unusual feature of Mendel's report is that all seven of the examples he described (see Table 10.1) are characterized by complete dominance. For dominance to be complete, a single copy of the dominant allele must produce enough of its protein product to give the maximum phenotypic response. For example, just one copy of the dominant allele *T* at one of the loci studied by Mendel leads to the production of enough of a growth-promoting chemical so that the *Tt* heterozygotes are as tall as the homozygous dominant plants (*TT*)—the second copy of *T* causes no further growth of the stem. The homozygous recessive plants (*tt*) are much shorter because the allele *t* does not lead to the production of the growth promoter.

10.9 Incomplete Dominance Follows Mendel's Laws

Heterozygous snapdragons produce pink flowers because the allele for red flowers is incompletely dominant over the allele for white ones. When true-breeding red and white parents cross, all plants in the F₁ generation are pink. When these F₁ plants self-pollinate, they produce F₂ offspring that are white, pink, and red in a ratio of 1:2:1. A test cross, diagrammed at the right, confirms that pink snapdragons are heterozygous; see Figure 10.6 for the reasoning.

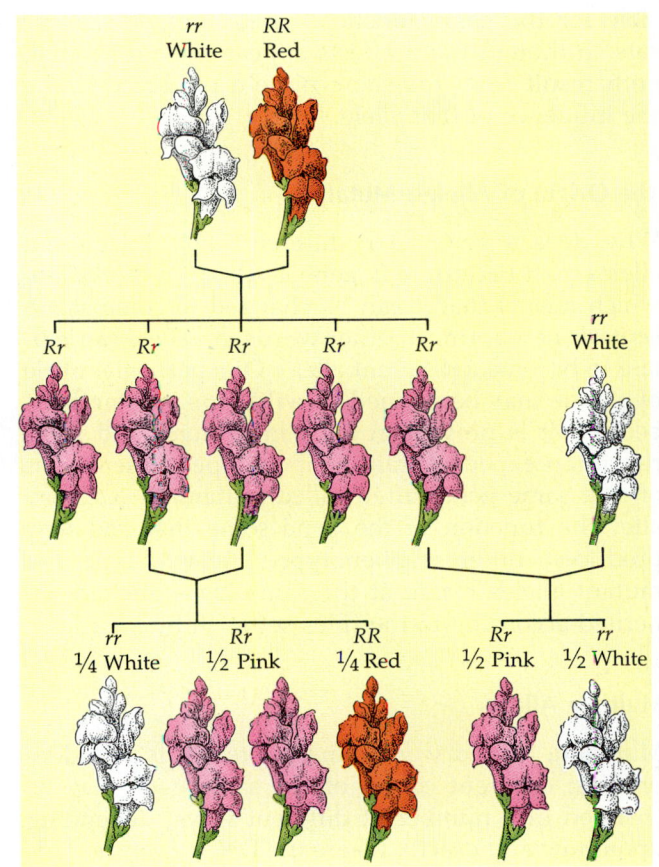

V^hV^h V^hV^f V^fV^f

10.10   Codominance in White Clover
White clover leaves have characteristic patterns of
chevrons and colored areas, all genetically determined.
The leaves on the left are homozygotes of genotype V^hV^h;
those on the right are homozygotes of genotype V^fV^f.
Since V^hV^f heterozygotes show both the chevron that is
characteristic of V^h and the colored area that is characteristic of V^f, the V^h and V^f alleles are codominant.

Sometimes two alleles at a locus produce two different phenotypes, both of which appear in heterozygotes (Figure 10.10). This phenomenon is called
codominance.

Pleiotropy

When a single allele has more than one distinguishable phenotypic effect, we say that the allele is **pleio-
tropic**. The most familiar example of pleiotropy is the
allele responsible for the coloration pattern (light
body, darker extremities) of Siamese cats, discussed
later in this chapter. The same allele is also responsible for the characteristic crossed eyes of Siamese
cats. Although these effects appear to be unrelated,
both result from the same protein produced under
the influence of that allele.

The Origin of Alleles: Mutation

Why does a gene have different alleles? Different
alleles exist because any gene is subject to **mutation**,
which means that it can be changed to some *stable,
heritable* new form. In other words, an allele can mutate to become a different allele. One particular allele
of a gene may be defined as **wild-type**, or standard,
because it is present in most individuals and gives
rise to an expected trait or phenotype. Other forms
of that same gene, often called **mutant** alleles, may
alter the function of the gene somewhat and may
produce a different phenotype. The wild-type and
mutant alleles reside at the same locus and are inherited according to the rules set forth by Mendel.

Multiple Alleles

Mutation, to be discussed in Chapter 11, is a random
process; different copies of the same gene may be
changed in a number of different ways, depending
upon how and exactly where the DNA changes. This

implies that there may be more than two alleles of a
given gene in a group of individuals. (Any one individual has only two alleles, of course—one from
the mother and one from the father.) In fact, there
are many examples of such **multiple alleles**. Some
clover leaves are plain green, while others have chevrons of other colors on their leaves. Seven alleles at
a locus control the pattern of chevrons on the leaves
of white clover (Figure 10.11). In the fruit fly *Drosophila melanogaster*, many alleles at one locus affect
eye color by determining the amount of pigment produced (Table 10.2). The exact color of the fly's eyes
depends on which two alleles are inherited.

The ABO blood group system in humans is determined by a set of three alleles (I^A, I^B, and i) at one
locus. Different combinations of these alleles in different people produce four different blood types, or
phenotypes: A, B, AB, and O (Table 10.3). Early at-

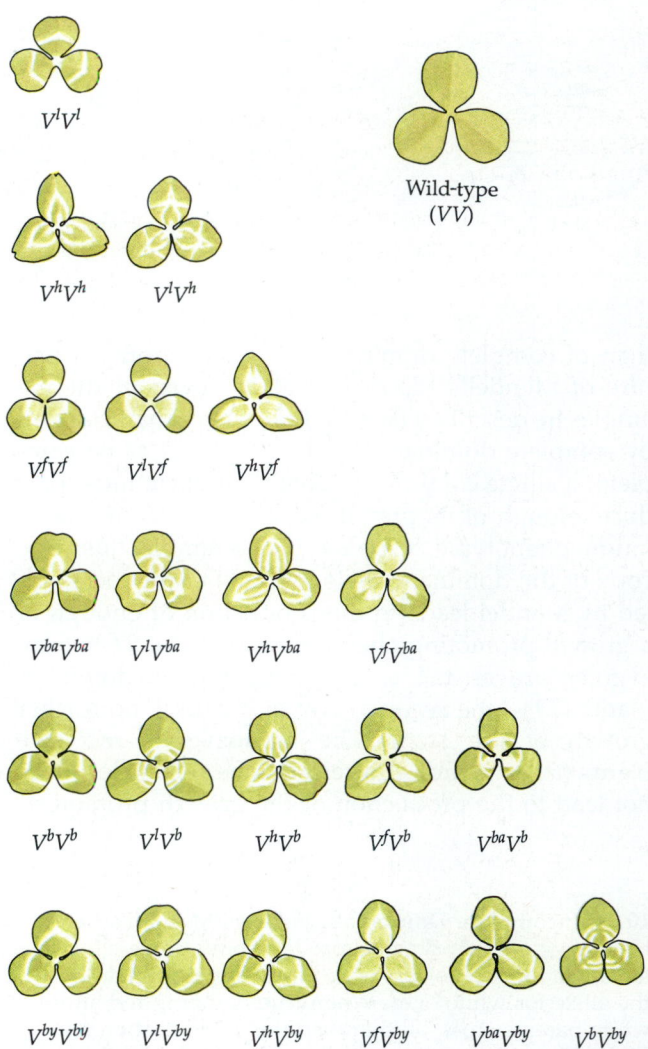

10.11   Multiple Alleles in White Clover
Seven alleles at the same locus determine the pattern of
chevrons and colored areas on white clover leaves. Many
of these alleles show codominance in heterozygotes.

TABLE 10.2
Multiple Alleles for Eye Color in *Drosophila melanogaster*

GENOTYPE	PHENOTYPE	DEGREE OF PIGMENTATION OF THE EYE
w^+w^+	wild-type (dull red)	0.6800
$w^{col}w^{col}$	colored	0.1636
$w^{sat}w^{sat}$	satsuma	0.1404
w^ww^{col}		0.1114
$w^{co}w^{co}$	coral	0.0798
w^ww^w	wine	0.0650
$w^{a3}w^{a3}$	apricot-3	0.0632
$w^{ch}w^{ch}$	cherry	0.0410
w^ew^e	eosin	0.0324
$w^{bl}w^{bl}$	blood	0.0310
w^aw^a	apricot	0.0197
w^tw^t	tinged	0.0062
ww	white	0.0044

The Rh factor, so named because it was first found in rhesus monkeys, is another substance on the surface of red blood cells. In most human populations, almost 100 percent of the individuals have the Rh factor, and their blood is said to be Rh^+ (Rh-positive). Among Caucasians, however, only 83 percent are Rh^+; the others lack the Rh factor, and their blood is called Rh^- (Rh-negative). Like the ABO blood types, the Rh factor is genetically determined. A single locus with at least eight multiple alleles is responsible. Certain dominant alleles cause the production of the Rh factor; Rh^- individuals are homozygous recessives.

Another system of multiple alleles is illustrated by the scallops in Study Question 1 at the end of the chapter. The question of how differing alleles may be maintained in a population through time will be examined in Chapter 19.

FOCUS ON CHROMOSOMES

Linkage

In the immediate aftermath of the rediscovery of Mendel's laws, the second law—independent assortment—was considered to be generally applicable. However, some investigators, including Punnett (the inventor of the square), began to observe strange deviations from the expected 9:3:3:1 ratio in some dihybrid crosses. In particular, they sometimes observed an apparent excess of parental phenotypes and a shortage of recombinant phenotypes among the F_2 progeny. Suppose that the original cross was between the genotypes *AABB* and *aabb*. If alleles at the *A* locus assorted independently of alleles at the *B* locus, the F_2 should consist of 9/16 individuals with the dominant phenotypes for *A* and *B*, 3/16 individuals dominant only for *A* (*A—bb*), 3/16 individuals dominant only for *B* (*aaB—*), and 1/16 double recessive homozygotes (*aabb*). What Punnett and others observed instead were large excesses of *aabb* over the 1/16 expected.

These results become understandable when we assume that the two loci are on the *same chromosome*—

tempts at blood transfusion—made before these blood types were understood—often killed the patient. Around the turn of the century, however, the Austrian scientist Karl Landsteiner mixed blood cells and serum (blood from which cells have been removed) from different individuals. He found that only certain combinations of blood types are compatible. In other combinations, the red blood cells form clumps because of the presence in the serum of specific proteins, called antibodies (see Chapter 16), that react with foreign, or "nonself," cells and macromolecules (Figure 10.12). When transfusions are given, a perfect matchup of blood types between donor and patient is best. For example, if the patient has red blood cells of type A, then the blood donor should have type A cells as well. Certain combinations other than perfect matchups are also usually successful, as indicated in Table 10.3.

Blood type

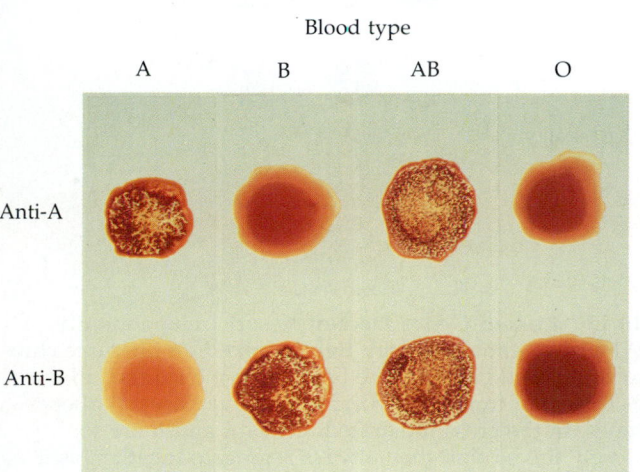

10.12 ABO Blood Reactions Are Important in Transfusions
Cells of blood types A, B, AB, and O were mixed with anti-A or anti-B antibodies. Red blood cells that react with antibodies clump together (speckled appearance in the photograph); red blood cells that do not react with antibody remain evenly dispersed. Note that anti-A reacts with A and AB cells but not with B or O. Which blood types do anti-B antibodies react with? As you look down the columns, note that each of the types, when mixed separately with anti-A and with anti-B, gives a unique pair of results; this is the basic method by which blood is typed.

TABLE 10.3
The ABO Blood System

BLOOD TYPE	GENOTYPE	REACTION WITH ANTI-A SERUM	REACTION WITH ANTI-B SERUM	TYPE OF DONOR BLOOD ACCEPTED
A	$I^A I^A$ or $I^A i$	Clumping of red blood cells	No clumping	A or O
B	$I^B I^B$ or $I^B i$	No clumping	Clumping of red blood cells	B or O
AB	$I^A I^B$	Clumping of red blood cells	Clumping of red blood cells	A, B, AB, or O
O	ii	No clumping	No clumping	O

that is, they are linked together. After all, since the number of genes in a cell far exceeds the number of chromosomes, each chromosome must contain many genes. (The human genome consists of perhaps 50,000 genes, distributed over 23 pairs of chromosomes.) Suppose, now, that the A and B loci are on the same chromosome. To remind ourselves that the genes are linked in this fashion, let us write the genotypes differently: One parent is $\overline{AB}\,\overline{AB}$ and the other $\overline{ab}\,\overline{ab}$. The former produces gametes that are of one type, $\overline{AB}$; the latter produces $\overline{ab}$ gametes. Thus the genotype of the F_1 is $\overline{AB}\,\overline{ab}$. Now, the key difference between dihybrid crosses with linkage and those without is in the formation of gametes by the F_1. Without linkage, as we have seen, four types of gametes are produced in equal frequency (AB, Ab, aB, and ab). *With* linkage, however, most of the gametes must be either $\overline{AB}$ or $\overline{ab}$ because the two loci are physically "tied together" on the same chromosome—they are part of the same DNA molecule. Instead of the 9:3:3:1 ratio of four F_2 phenotypes produced from a dihybrid cross without linkage (see Figure 10.7), a dihybrid cross with linkage yields only parental phenotypes in the F_2 (Figure 10.13). (As we will see, a few recombinant gametes will appear when genes are less closely linked than in our example. They result from crossing over between loci; see Chapter 9.)

The full set of loci on a given chromosome constitutes a **linkage group**. The number of linkage groups in a species, determined by experiments such as the dihybrid cross described here, should equal the number of homologous chromosome pairs, as determined by microscopic examination of nuclei undergoing meiosis or mitosis.

Sex Determination

In Kölreuter's experience, and later in Mendel's, reciprocal crosses apparently always gave identical results. This is because in diploid organisms, chromosomes come in pairs. One member of each chromosome pair derives from each parent; it does

not matter, for example, whether a dominant allele was contributed by the mother or by the father. But this is not always the case; sometimes the parental origin of a chromosome does matter. To understand the types of inheritance in which parental origin is important, we must consider the ways in which sex is determined in different species.

In maize, a plant much studied by geneticists, every diploid adult has both male and female structures. These two types of tissue are genetically identical, just as roots and leaves are genetically identical.

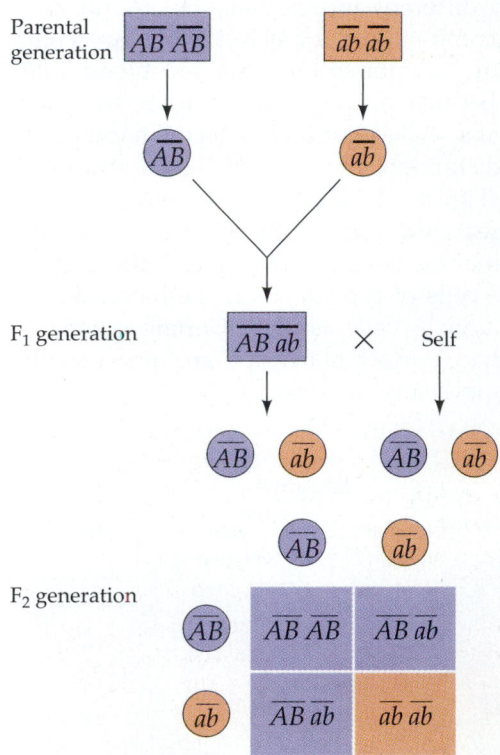

10.13 Linked Genes Do Not Assort Independently
When two genes are very tightly linked on the same chromosome (indicated by the line above the letters), all the F_2 offspring from a dihybrid cross have parental phenotypes. In the cross illustrated, genes A and B are so closely linked that they always segregate together, as if they were a single gene.

Plants such as maize, and animals such as earthworms, which produce both male and female gametes in the same organism, are said to be **monoecious** (from the Greek for "single house"). Some plants, such as date palms and oak trees, and most animals are **dioecious**, meaning some of the individuals can produce only male gametes and the others can produce only female gametes. In most dioecious organisms, sex is determined by differences in the chromosomes; but such determination operates in a bewildering variety of ways (Figure 10.14). The sex of a honeybee, for example, depends on whether it develops from a fertilized or an unfertilized egg. A fertilized egg is diploid and gives rise to a female bee—either a worker or a queen, depending on the diet during larval life. An unfertilized egg is haploid and gives rise to a male drone.

In many other animals, including ourselves, sex is determined by a single **sex chromosome** or by a pair of them. Both males and females have two copies of each of the rest of the chromosomes, which are called **autosomes**. The sex chromosome that is present in different numbers is the **X chromosome**. For example, female grasshoppers have two X chromosomes, whereas males have only one. These females form eggs containing one copy of each autosome and one X chromosome. The males form two types of sperm. Half contain an X chromosome and one copy of each autosome, and the other half contain only autosomes. This is a natural consequence of meiosis in the two sexes. In females, the two X chromosomes synapse in prophase I; one goes to each of the daughter nuclei from meiosis I. Males have but a single X chromosome, so there is no synapsis. Thus half the sperm end up containing an X chromosome, but the others get none. Female grasshoppers are described as being XX (ignoring the autosomes) and males as XO (pronounced "ex-oh"). When an X-bearing sperm fertilizes an egg, the zygote is XX and develops into a female. When a sperm without an X fertilizes an egg, the zygote is XO and develops into a male. This chromosomal mechanism ensures that the two sexes are produced in approximately equal numbers. No such mechanism for numerical equality of the sexes exists in the diploid–haploid system of bees.

Female mammals have two X chromosomes and males have one (Figure 10.14b). However, male mammals also have a sex chromosome that is not found in females: the **Y chromosome**. Females may be represented as XX and males as XY. The males produce two kinds of gametes, each with a complete set of autosomes but differing with respect to their sex chromosomes: Half the gametes carry an X chromosome and the rest carry a Y. When an X-bearing sperm fertilizes an egg, the resulting XX zygote is female; when a Y-bearing sperm fertilizes an egg, the XY zygote is male.

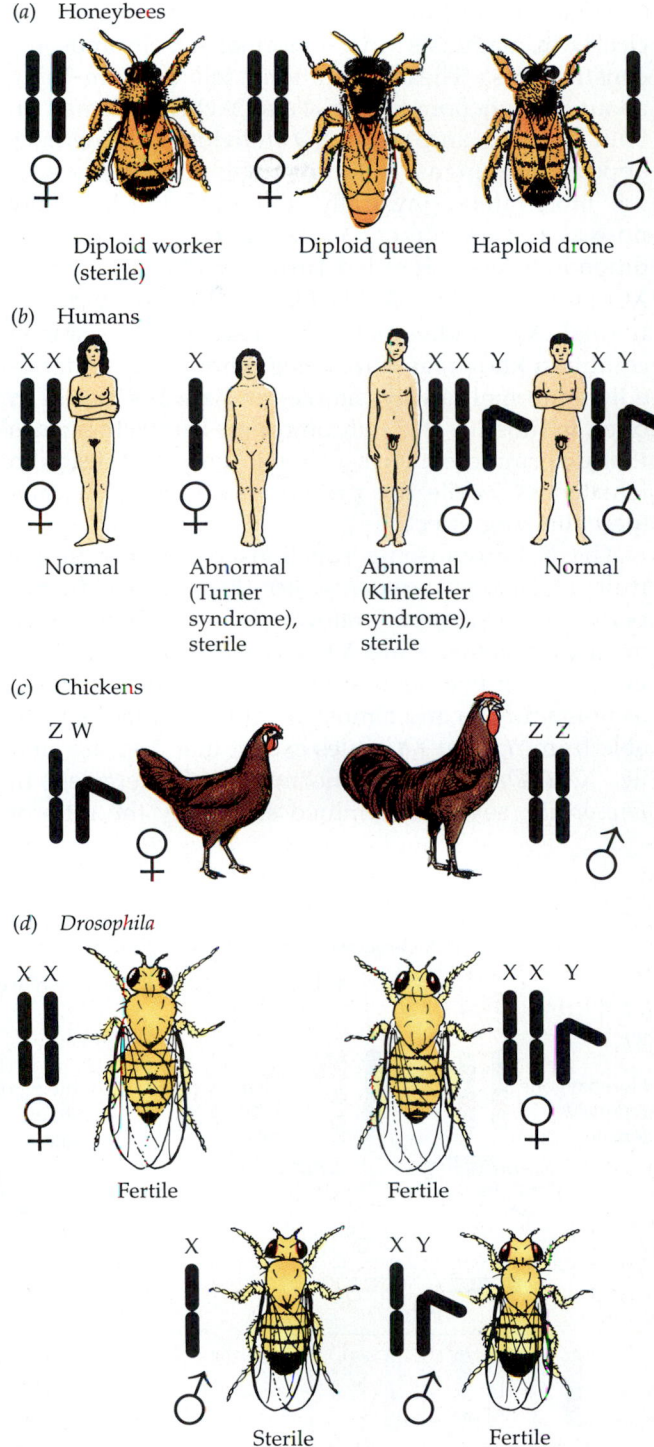

10.14 Sex Is Determined in Different Ways
(a) In honeybees, fertilized eggs develop into diploid females and unfertilized eggs develop into haploid males. In other animals, sex is determined by special sex chromosomes. (b) Normal human females (left) carry two X chromosomes; normal males (right) carry one X and one Y chromosome. Persons who have some other number of sex chromosomes may develop abnormally. (c) In birds, it is the males that carry two identical sex chromosomes (ZZ) and females that have differing ones (ZW). (d) Drosophila (fruit fly) females have two X chromosomes and may also have a Y chromosome; males have an X chromosome and, if they are fertile, a Y chromosome.

Some subtle but important differences show up clearly in mammals with abnormal sex chromosome constitutions. These conditions tell us something about the functions of the X and Y chromosomes. In both humans and mice, XO individuals sometimes appear. In humans, XO individuals are females who are moderately physically abnormal but mentally normal and are almost always sterile. The XO condition in humans is called Turner syndrome. In mice, XO individuals are fertile females that are virtually normal. XXY individuals also arise. XXY humans (a condition known as Klinefelter syndrome) are decidedly abnormal, always sterile, and always males. In brief, in humans the Y chromosome carries the genes that determine maleness. The absence of Y leads to femaleness, while the presence of Y has a definite masculinizing effect.

The Y chromosome functions differently in the fruit fly *Drosophila melanogaster* (Figure 10.14*d*). Superficially, *Drosophila* follows the same pattern as mammals: Females are XX and males are XY. However, XO individuals are males (rather than females as in mammals) and almost always are indistinguishable from normal XY males except that they are sterile. XXY *Drosophila* are normal, fertile females. In *Drosophila*, sex is determined strictly by the ratio of X chromosomes to autosome sets. If there is one X chromosome for each set of autosomes, the individual is a female; if there is only one X chromosome for the two sets of autosomes, the individual is a male. The Y chromosome plays no sex-determining role in *Drosophila*, but it is needed for male fertility.

In birds, moths, and butterflies, males are XX and females are XY. To avoid confusion, this is usually expressed as ZZ (male) and ZW (female) (Figure 10.14*c*). In these organisms, it is the female that produces two types of gametes. Thus the egg determines the sex of the offspring, rather than the sperm as in humans and fruit flies.

Sex Linkage

How does the existence of sex chromosomes affect patterns of inheritance? In *Drosophila* and in humans, the Y chromosome carries few known genes, whereas a substantial number of genes affecting a great variety of traits are carried on the X chromosome, which leads to an important deviation from the usual Mendelian ratios for the inheritance of genes located on the X chromosome. Any such gene is present in two copies in females, but in only one copy in males. Therefore, females may be heterozygous for genes that are on the X chromosome, but males will always be **hemizygous** for these genes—they will have only

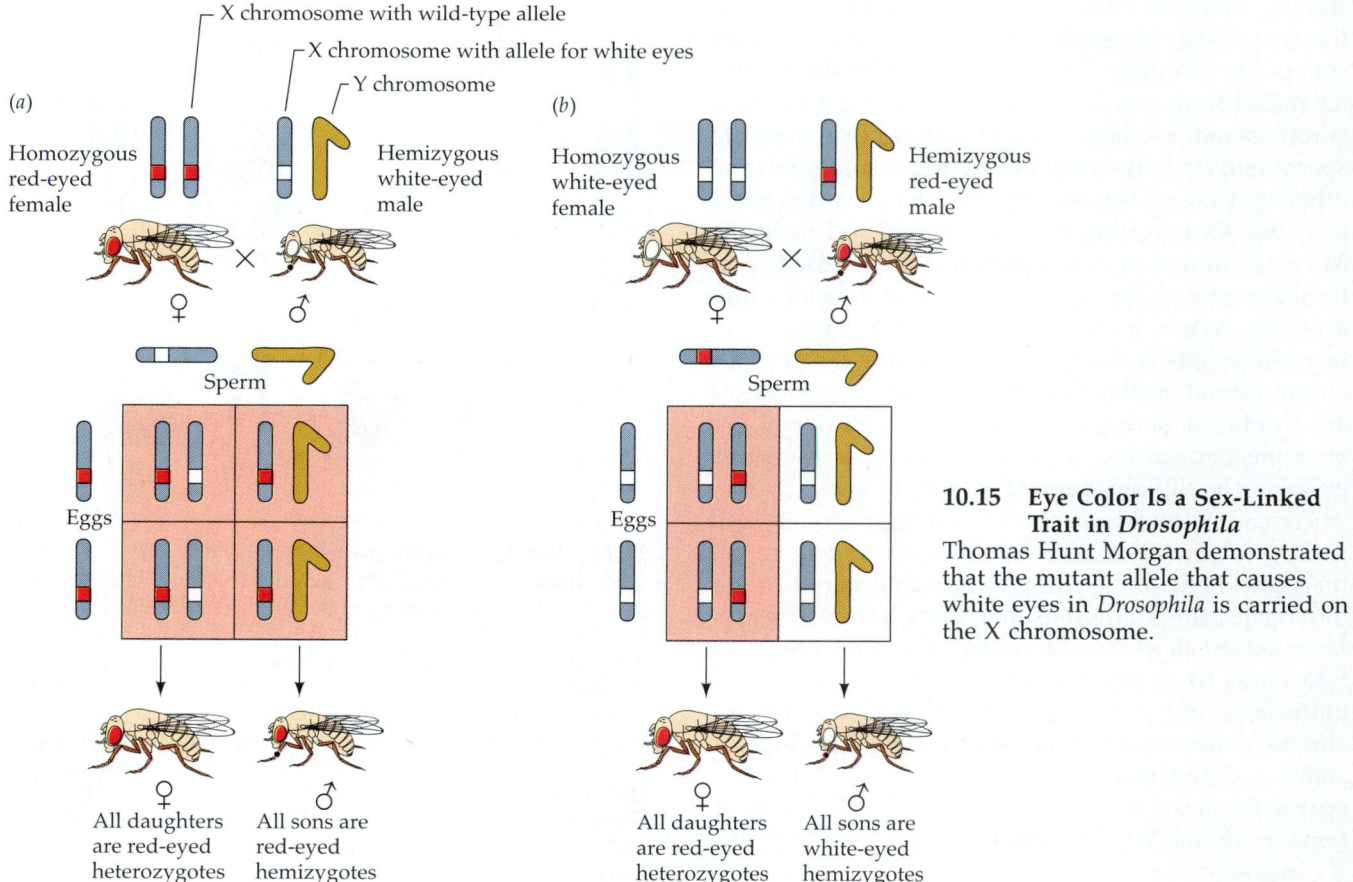

10.15   Eye Color Is a Sex-Linked Trait in *Drosophila*
Thomas Hunt Morgan demonstrated that the mutant allele that causes white eyes in *Drosophila* is carried on the X chromosome.

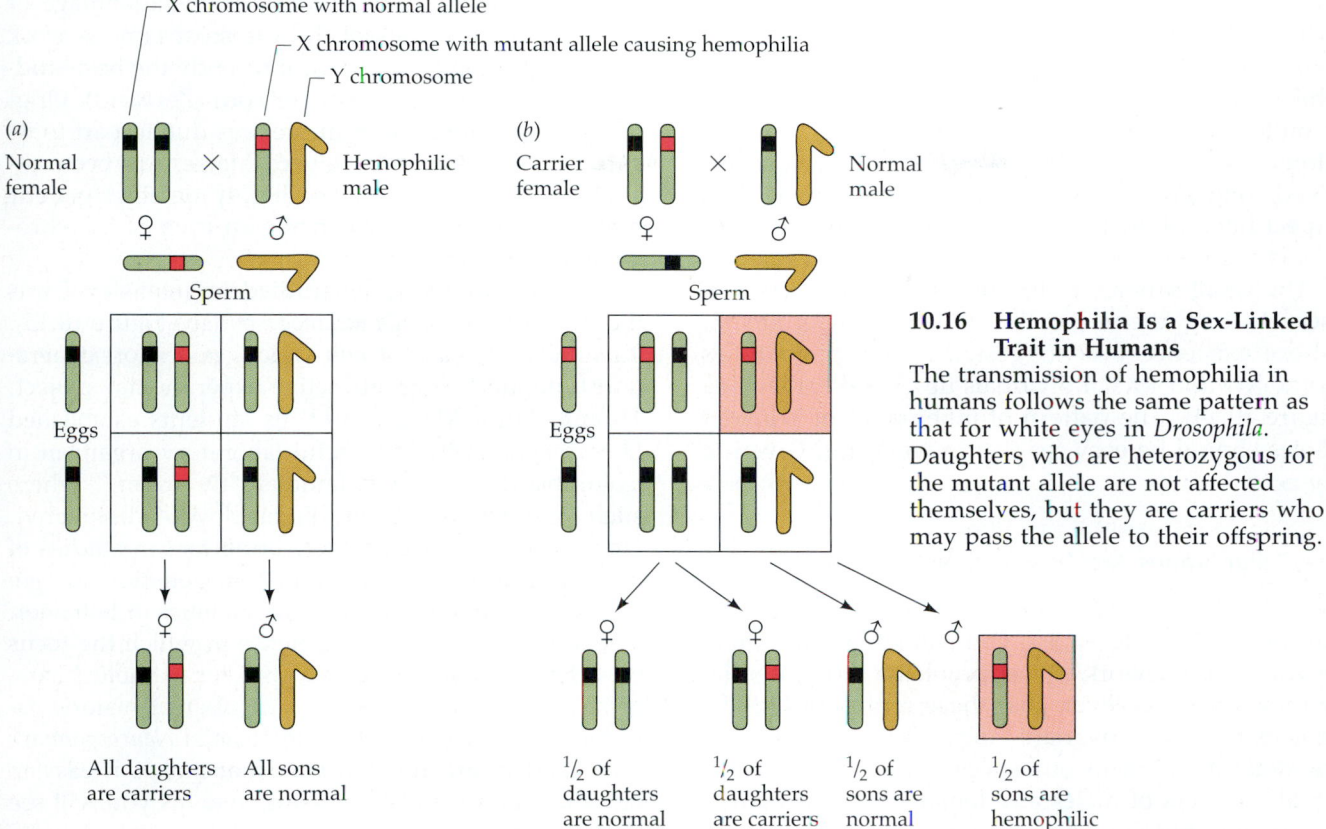

X chromosome with normal allele

X chromosome with mutant allele causing hemophilia

Y chromosome

(a)
Normal
female × Hemophilic
male

(b)
Carrier
female × Normal
male

Sperm

Sperm

Eggs

Eggs

10.16 Hemophilia Is a Sex-Linked Trait in Humans
The transmission of hemophilia in humans follows the same pattern as that for white eyes in *Drosophila*. Daughters who are heterozygous for the mutant allele are not affected themselves, but they are carriers who may pass the allele to their offspring.

All daughters
are carriers

All sons
are normal

$^1/_2$ of
daughters
are normal

$^1/_2$ of
daughters
are carriers

$^1/_2$ of
sons are
normal

$^1/_2$ of
sons are
hemophilic

one of each. It is useful here, as it is in many instances when studying genetics, to think of loci whose alleles govern easily observable phenotypes as **markers** of the chromosomes on which they are located. Reciprocal crosses of parents that have different markers on their sex chromosomes do not give identical results; this is a sharp deviation from the inheritance of markers on autosomes.

The first and still one of the best examples of **sex-linked inheritance**—inheritance of traits governed by loci on the sex chromosomes—is that of eye color in *Drosophila*. The wild-type eye color of these flies is red. In 1910, Thomas Hunt Morgan discovered a mutation that causes white eyes. He experimented by crossing flies of the wild-type and mutant phenotypes. His results demonstrated that the eye-color locus is on the X chromosome. When homozygous red-eyed females were crossed with (hemizygous) white-eyed males, all the sons and daughters had red eyes, because red is dominant over white and all the progeny had inherited a wild-type X chromosome from their mothers (Figure 10.15a). However, in the reciprocal cross, in which a white-eyed female was mated to a red-eyed male, all the sons were white-eyed and all the daughters red-eyed (Figure 10.15b). The sons from the reciprocal cross inherited their only X chromosome from their white-eyed mother; the Y chromosome they inherited from their father

does not carry the eye-color locus. The daughters, on the other hand, got an X chromosome with the white allele from their mother and an X chromosome bearing the red allele from their father; they were therefore red-eyed heterozygotes. When Morgan mated these same heterozygous females with red-eyed males, he observed that half their sons had white eyes, but all their daughters had red eyes.

The human X chromosome carries many loci. The alleles at these loci follow the same pattern of inheritance as those for white eyes in *Drosophila*. One X-chromosome locus, for example, has an allele that causes hemophilia, a hereditary disorder characterized by the failure of blood to clot properly; victims suffer from excessive and often fatal bleeding. Hemophilia appears in individuals who are homozygous for a mutant recessive allele. A hemophilic man married to a homozygous normal woman will not produce any hemophilic children. The sons inherit a single, normal X from their mother and will neither have the disease nor transmit it to their children. The daughters get an X chromosome bearing a normal allele from their mother and one bearing the allele for hemophilia from their father. Because hemophilia is recessive, however, the daughters will not be hemophilic (Figure 10.16a). They will, however, be heterozygous *carriers*. Such carriers of an X-linked trait will transmit the disease to half their sons and the carrier role to half their daughters (Figure 10.16b). What parental genotypes would produce a female hemophiliac? Her father would have to be a hemophiliac, and her mother a carrier. Because hemophilia is quite rare, two such people are unlikely to meet. Moreover, until recently hemophilic males rarely survived long enough to reproduce. One might thus expect hemophilic females to be extremely rare, and in fact very few have ever been found.

The small human Y chromosome carries very few loci. Among them are the maleness determiners, whose existence was suggested by the phenotypes of the XO and XXY individuals described earlier (see Figure 10.14). The pattern of inheritance of Y-linked alleles should be easy for you to work out. Give it a try now.

Mendelian Ratios Are Averages, Not Absolutes

You have now been introduced to the basic Mendelian ratios: 3:1, 1:1, 9:3:3:1; you will figure out others as you do homework or test problems. It is essential to remember, however, that these represent highest probabilities, not invariant rules. The X-Y system of sex determination in our species results in roughly equal numbers of males and females in a substantial population, but you know that a given family of four children may not consist of two girls and two boys. It is not unusual for four children to be of the same sex; in fact, one family in eight who have four children will have all boys or all girls. How do we know if this deviation is simply the result of chance?

When we are trying to understand the genetic basis for the results of a cross, it is important to know how much deviation from a predicted ratio we can reasonably expect due to normal chance. Statistical methods have been devised for determining whether the observed deviation from an expected ratio can be attributed to chance variation or whether the deviation is large enough to suggest that the observed ratio is caused by something specific. These methods take several factors into consideration, but one of the most important is **sample size**. If we expect to find a 3:1 ratio between two phenotypes and we look at a sample of only 8 progeny, we should not be surprised to find 7 individuals of one phenotype and 1 of the other (rather than the expected 6 and 2). It would be surprising, however, in a sample of 80 to find 70 of one phenotype and 10 of the other—you would question whether that really represented a 3:1 ratio. In experiments in genetics—and in quantitative biology in general—sample sizes should be large so that the data are easier to evaluate with confidence.

Special Organisms for Special Studies

Prokaryotic and eukaryotic organisms of many kinds have been used in genetic studies. A few species have been used many times because of one advantage or another. Gregor Mendel did his most famous work with the garden pea, but until recently the best-studied higher plant was maize, or corn (Zea mays). Originally, this emphasis on maize was due in part to its great agricultural importance. Maize has been examined so thoroughly that highly detailed **genetic maps** locating particular genes on each of the chromosomes are available.

The first animal to be studied in great detail was the fruit fly, Drosophila melanogaster (see Figure 10.15). Its small size, ease of cultivation, and short generation time made it an attractive experimental subject. Thomas Hunt Morgan and his students established Drosophila as a highly useful laboratory organism in Columbia University's famous "fly room," where such phenomena as sex linkage were discovered. Drosophila remains extremely important in studies of chromosome structure, population genetics, the genetics of development, and the genetics of behavior.

There was a period in genetics in which the focus was the common salmon-colored bread mold Neurospora crassa. It was used in a number of historic experiments. The products of meiosis in Neurospora are organized in an unusual way that makes it easy to visualize the results of crossing over, as you will see in the next section.

More recently, molecular geneticists and developmental biologists have directed heavy attention to two other organisms. We discuss research on the tiny worm Caenorhabditis elegans in Chapter 17 and the small plant Arabidopsis thaliana, considered by many to be a weed, in Chapters 33 and 34.

Meiosis in *Neurospora*

The life cycle of Neurospora, like those of other fungi, is complex. Neurospora grows from haploid spores that divide mitotically to produce a feltlike mat of long strands called hyphae. Eventually, as the end result of a sexual process, haploid nuclei from two individuals unite to produce zygotes in a specialized fruiting structure. As soon as a diploid nucleus is formed, it undergoes meiosis. Thus the zygote itself constitutes a greatly reduced diploid generation.

Meiosis in Neurospora is a tidy process that packages all of the nuclei produced by the divisions of a single zygote in a long, thin sac called an ascus (plural: asci). The four haploid nuclei then divide once again by mitosis. The eight nuclei produced by this sequence of events are incorporated into eight spores, all neatly lined up within the ascus (Figure 10.17). Because the ascus is so thin, the nuclei cannot pass one another as the divisions proceed, so the pairs of spores can easily be identified with the meiotic division that produced them. This makes Neurospora an especially useful organism in which to examine segregation, assortment, and recombination

10.17 *Neurospora* Packs Its Haploid Spores into an Ascus

A rosette of asci of *Neurospora crassa*, resulting from a cross of a spore-color mutant with a wild-type strain. Each ascus contains eight spores; their arrangements reflect different segregation patterns.

of genetic markers. Accordingly, we use it to illustrate the material in the next section.

Recombination in Eukaryotes

We have seen that each chromosome has many loci and that all the loci on one chromosome are linked to each other. If homologous chromosomes did not undergo crossing over when paired (see Chapter 9), all the markers on a given chromosome would all segregate together as a unit. A geneticist would have no way of knowing that linked loci are actually different loci. Mendel's second law (independent assortment of alleles of different loci) would apply only to loci on different chromosomes.

What actually happens is more complex and therefore more interesting. Markers located at different places on the same chromosome do sometimes separate from one another as the result of crossing over. The farther apart two markers are on a chromosome, the greater the likelihood that they will separate and recombine. Geneticists use **recombination frequencies** (the observed frequencies in the offspring of marker combinations different from those of the parents) to generate genetic maps that indicate the arrangement of markers along the chromosome (Box 10.C).

Genetic markers on the same chromosome pair recombine by **crossing over**, which results from the physical exchange of corresponding genetic segments between two homologous chromosomes during prophase I of meiosis. In other words, recombination occurs at the stage when homologous chromosomes are paired. Recall that the DNA has duplicated by this stage, and each chromosome consists of two chromatids. Thus crossing over occurs at the four-

strand stage. The exchange event at any point along the length of the chromosome involves only two of the four chromatids, one from each member of the chromosome pair (Figure 10.18). The lengths of chromosome are exchanged reciprocally, so both chromatids involved in crossing over are recombinant (that is, each chromatid contains genes from both parents); and no genes are created or destroyed. The points at which the chromatids break in the exchange seem to correspond perfectly (at the level of base pairs in the DNA), so that the amount of material donated by a chromatid exactly equals the amount it receives.

At any point along the paired chromosomes, only two of the four chromatids participate in crossing over; but other crossovers may occur at other points. These other crossovers may involve the same pair of

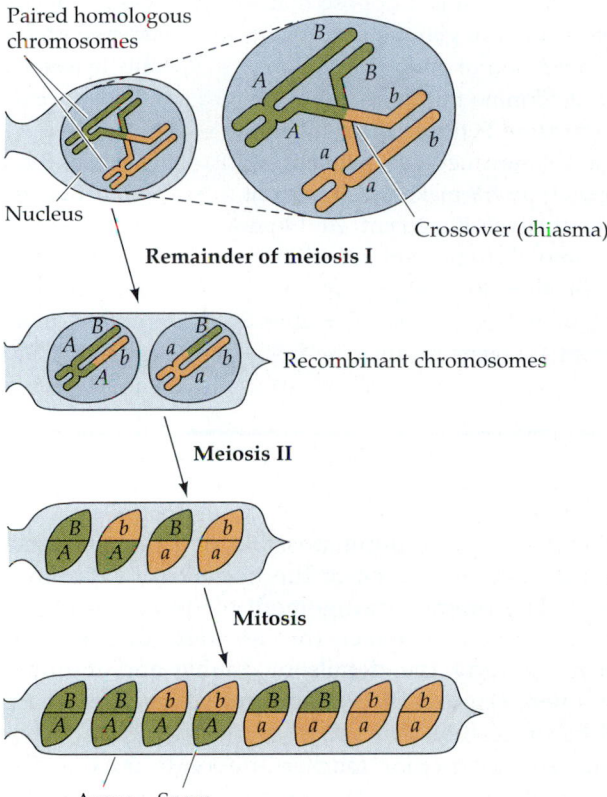

Prophase I of meiosis

Paired homologous chromosomes

Nucleus

Crossover (chiasma)

Remainder of meiosis I

Recombinant chromosomes

Meiosis II

Mitosis

Ascus Spore

10.18 Crossing Over in *Neurospora*

Neurospora retains the products of meiosis (spores) in a single package (the ascus). This diagram shows only one of *Neurospora*'s several chromosome pairs, beginning with the diploid nucleus during prophase I of meiosis. A single crossover between genes *A* and *B* forms two recombinant chromatids. In this particular ascus, at the end of meiosis II the nuclei with recombinant chromosomes (carrying *Ab* and *aB*) lie in the middle of the ascus, while those with parental chromosomes (carrying *AB* and *ab*) lie at the ends. A mitotic division follows meiosis II, increasing the number of spores in the ascus to eight. Note that the recombinant chromosomes remain at the middle of the ascus.

BOX 10.C

Gene Mapping in Eukaryotes

Neurospora is an excellent organism for illustrating the principles of genetic mapping. In mapping experiments, we do not make use of the orderly packaging of spores in asci; rather, spores are collected at random and examined. Remember that this organism is haploid for most of its life cycle; haploid genotypes thus characterize strains. Suppose that we cross a strain of genotype *AB* with another strain of genotype *ab*. We then determine the **frequency of recombination** between the two markers as follows. Let us say that spores of genotype *AB* make up 40 percent of the total, *ab* 40 percent, *Ab* 10 percent, and *aB* 10 percent. Of all these spores, 40 + 40 = 80 percent are of the parental genotypes, 10 + 10 = 20 percent are recombinant. The fre-

quency of recombination between the two markers is thus 20 percent.

To determine the linear sequence and spacing of markers on the chromosome, we can perform a three-factor cross. If *Neurospora* strains that are *ABC* and *abc* are crossed, the following classes of spore genotypes might be observed:

1. *ABC*   38.0 percent
2. *abc*   40.2 percent
3. *Abc*   7.2 percent
4. *aBC*   6.6 percent
5. *ABc*   3.1 percent
6. *abC*   3.7 percent
7. *AbC*   0.5 percent
8. *aBc*   0.7 percent

Classes 1 and 2 are parental types. Classes 3 and 4 are single recombinants between *A* and *B*. Classes 5 and 6 are single recombinants between *B* and *C*. Note that all four of the classes 3 through 6 are also recombinants between *A* and *C*. Classes 7 and 8 are recombinant between *A* and *B* and between *B* and *C*, but not between *A* and *C*; they represent double crossover types. Note that 7 and 8 are the least frequent classes, which is what we

would expect of double crossover types. The two crossovers cancel each other in recombining *A* and *C*, leaving them in the parental arrangement relative to each other.

To compute map distances, we add the crossovers in all the classes that are recombinant between a given pair of markers. The "distance" between *A* and *B* is thus the sum of classes 3, 4, 7, and 8: 7.2 + 6.6 + 0.5 + 0.7 = 15 percent recombination. The distance between *B* and *C* (8 percent recombination) is obtained by summing classes 5, 6, 7, and 8. The distance between *A* and *C* is the sum of classes 3 through 6, plus two times class 7, plus two times class 8 (because each of these last two classes contains two crossovers between *A* and *C*): so the distance between *A* and *C* = 7.2 + 6.6 + 3.1 + 3.7 + 2(0.5) + 2(0.7) = 23 percent recombination. We can thus draw a map locating the three markers, showing their map "distances," as follows:

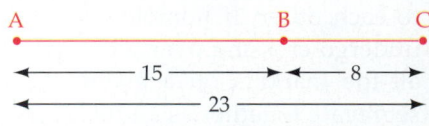

chromatids or any other possible pair that includes one member from each of the homologous chromosomes. The precise arrangement of spores in the ascus of *Neurospora* makes this an ideal organism in which to study the details of double and multiple crossovers (Figure 10.19). The probability that there will be more than one crossover in the segment between two particular markers depends on the distance between them—the greater the distance, the more likely crossover events will take place.

Cytogenetics

By making experimental crosses and calculating the recombination frequencies, geneticists can show that certain genes are associated in a linkage group, in a specific order (Box 10.C). Such a linkage group is logical, but to what extent does it actually correspond with the physical structure of a chromosome as seen

under the microscope? To establish a relationship between a genetic linkage group and a chromosome, the cytogeneticist (a person who studies the microscopic appearance of chromosomes in relation to genetics) tries to find an individual in whom the normal linkage relationships are changed. The cytogeneticist then examines that individual's cells under the microscope, looking for a corresponding visible change in one or more chromosomes.

In the tissues of most species, the chromosomes are too small for an observer to see any except the most gross changes. One exception is the giant chromosomes in the salivary glands of the larvae of *Drosophila*. Called **polytene chromosomes**, they have replicated their DNA many times without cytokinesis, so that many copies of each DNA molecule lie side by side to form thick, snakelike structures that can be seen clearly even with a low-magnification lens (Figure 10.20). Condensed thickenings, or chro-

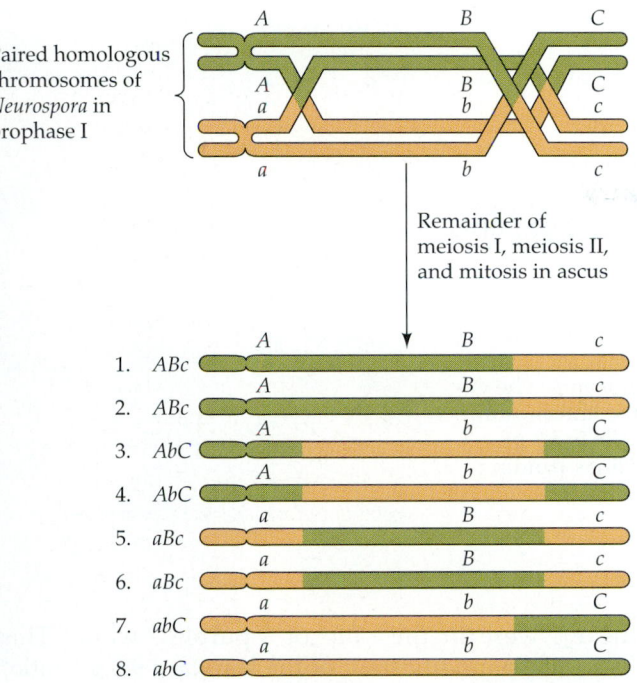

Paired homologous chromosomes of *Neurospora* in prophase I

Remainder of meiosis I, meiosis II, and mitosis in ascus

1. *ABc*
2. *ABc*
3. *AbC*
4. *AbC*
5. *aBc*
6. *aBc*
7. *abC*
8. *abC*

Chromosomes in the 8 spores in ascus

10.19 There Can Be Multiple Crossovers
A chromosome carrying alleles *A*, *B*, and *C* synapses with its homolog, which carries *a*, *b*, and *c*. The chromosomes cross over at three points. The crossovers are faithfully recorded in the order of spores, and all eight spores have recombinant chromosomes.

momeres, along these chromosomes are paired with sister chromomeres on the parallel strands of the polytene chromosome so the chromosomes appear to have a pattern of transverse bands. (These bands, visible in interphase in polytene chromosomes, are not the same as the bands seen by special staining of chromosomes in mitosis.) Each polytene chromosome has a characteristic pattern of bands. The two homologous polytene chromosomes are closely synapsed, causing the two chromosomes to look like a single structure. The thickness, spacing, and sharpness or diffuseness of the bands are so characteristic that an experienced cytogeneticist can often tell at a glance if the order or position of a group of bands has been changed.

One type of chromosomal change is an **inversion**. Inversions can be detected by genetic mapping as a change in the linkage relations of markers on a chromosome. If the normal order of the markers is *ABCDEFGHI*, an inversion may have the order *ABCDGFEHI*. When the polytene chromosomes of such individuals are examined, the order of a group of bands in one of the chromosomes also appears reversed. A linkage group with an inversion in it can be correlated with a chromosome whose bands are inverted, and the geneticist can infer that certain

genes are located in certain regions of the visibly altered chromosome.

Together with inversions, other chromosomal changes observable in polytene chromosomes, such as the **deletion**, or loss, of genetic material, can be useful in correlating the recombination map with the physical chromosome. Where there is genetic evidence of a deletion, a small group of bands (or even a single band) often is missing from one of the chromosomes. From the study of a great many chromosomal changes in *Drosophila*, it is well demonstrated that the order of genes deduced by the two methods agrees, but the distances do not. You can see this for yourself in Figure 10.21. Because of the exceedingly complex folding of DNA in a eukaryotic chromosome, one can never be sure that the microscopically observed length of a segment of chromosome is a good reflection of the length of DNA in that segment, nor is recombination-map distance necessarily a reliable reflection of the length of DNA. Despite these limitations, the combination of cytogenetics and recombination analysis has been successful in probing the composition of chromosomes of eukaryotes.

INTERACTIONS OF GENES WITH OTHER GENES AND WITH THE ENVIRONMENT

Thus far we have treated the phenotype of an organism, with respect to a given trait, as a simple result of its genotype, and we have implied that a single trait results from the alleles of a single locus. In fact, several loci may interact to determine a trait's phenotype. To complicate things further, the physical environment may interact with the genetic constitution of an individual in determining the phenotype.

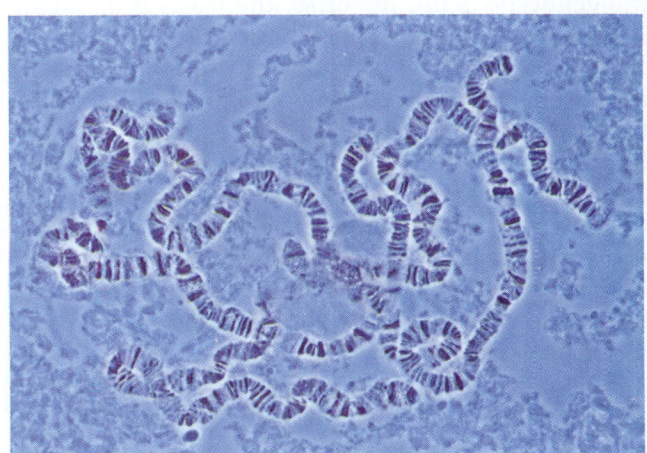

10.20 Polytene Chromosomes are Huge
A complete set of banded polytene chromosomes in a cell from the salivary gland of a *Drosophila* larva.

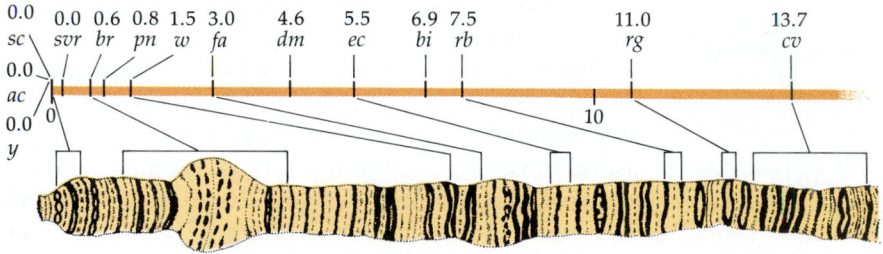

10.21 Map of One End of a Polytene X Chromosome
From the genetic map at the top you can read distances (in map units) between loci (italic letters) determined by recombination studies. Notice that these studies place the three loci at the left at the same point on the chromosome. The cytological map below gives, for 8 of the 14 loci, positions along the chromosome determined from studies of polytene chromosomes. Notice that the *order* of these 8 loci is the same on both maps. Notice also, by using the lines running between the maps, how different the *distances* on the two maps are.

Epistasis

When a particular trait is the result of a series of chemical reactions, each controlled by a different locus, the gene that acts at the earliest step in the series may, in one of its allelic forms, mask the expression of one or all of the other loci. This phenomenon of **epistasis**, in which one gene alters the effect of another, is illustrated by several loci that determine coat color in mice. The wild-type color is agouti, a grayish pattern resulting from bands on the individual hairs. The dominant allele *B* determines that the hairs will have bands and thus that the color will be agouti, whereas the homozygous recessive genotype *bb* results in unbanded hairs. Another, unlinked locus affects an early step in the formation of hair pigments. The dominant allele *A* at this locus allows normal color development, but *aa* blocks all pigment production and results in an all-white albino. As a result, *aa* is said to be epistatic over the *B* locus. Whether the genotype is *BB*, *Bb*, or *bb*, the result is an albino if the alleles of the other locus are both *a* (Figure 10.22). If a mouse with genotype *AABB* (and thus the agouti phenotype) is crossed with an albino of genotype *aabb*, the F₁ is *AaBb* and of the agouti phenotype. If the F₁ mice are crossed with each other to produce an F₂, the epistasis of *aa* will result in an expected phenotypic ratio of 9 agouti:3 black:4 albino. Can you show why this is so? The underlying ratio is the usual 9:3:3:1 for a dihybrid cross with unlinked genes, but be sure to look closely at each genotype and watch out for epistasis.

Epistasis can work in both directions between two genes, as first observed by William Bateson and Reginald Punnett. They performed a cross between two sweet pea plants (not the edible peas studied by Mendel). Each parent had white flowers. To the astonishment of Bateson and Punnett, the F₁ all had purple flowers! The F₂, obtained by self-pollinating

the F₁, were in the ratio of 9 purple:7 white. This looks like a modification of the standard 9:3:3:1 ratio, with the last three groups lumped into one. Let's try, as did Bateson and Punnett, to figure it out. First, because this looks like a dihybrid cross ratio, we assume that two different loci are involved. We recall that both dominant alleles are present in 9/16 of the offspring of a dihybrid cross (see Figure 10.7), and we notice that this is the proportion of the F₂ that are purple. We therefore decide that each individual in this group has at least one copy of each dominant allele and write *A—B—* as their genotype. Because only 9/16 are purple, it must be that having a dominant allele for only one of the genes will *not* produce a color. Thus the genotypes *A—bb* and *aaB—* fail to give purple flowers.

Here's how we may represent the whole experiment: If the original parents were *AAbb* and *aaBB*, both would have been white, as observed; all the F₁ would have been *AaBb* and purple, also as observed. You should now work out the genotypes of the F₂ and then convert them to phenotypes, remembering that all the genotypes give white flowers *except* the ones that are *A—B—*. If you do this exercise carefully, you will obtain the observed 9:7 ratio. We may say that *aa* is epistatic to *B* and *bb* is epistatic to *A*, in that both of these doubly recessive genotypes alter the expression of the dominant allele at the other locus, thereby determining that the phenotype will be white. Another way to describe this kind of situation is to say that the two loci are complementary. Complementary loci are mutually dependent, the expression of each being dependent upon the alleles of the other.

The epistatic action of complementary genes may be explained as follows: The dominant alleles *A* and *B* in this example code for the production of enzymes that catalyze two separate reactions in the production

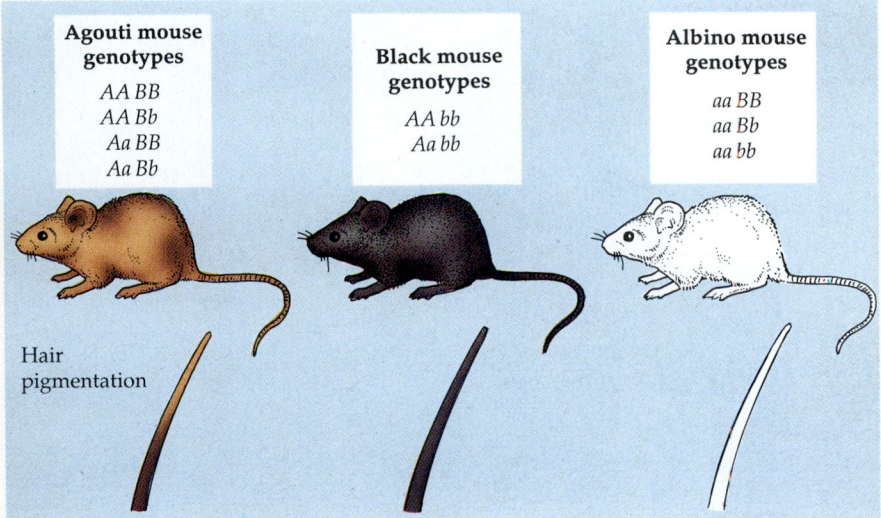

10.22   Genes May Interact Epistatically
Mice that have at least one dominant allele at each locus are agouti. Mice with genotype *aa* are albino regardless of their genotype for the other locus, because the *aa* genotype blocks all pigment production. Mice with *bb* genotypes are black unless they also are *aa* (which makes them albino).

of a purple pigment. In order for the pigment to be produced, both reactions must take place. If a plant is homozygous for either *a* or *b*, the corresponding reaction will not occur, no purple pigment will form, and the flowers will be white.

Quantitative Inheritance and Environmental Effects

Individual heritable traits are often found to be controlled by many genes, each contributing to the final outcome. As a result, variation in such traits is **continuous** rather than, as in the examples we have been considering, **discontinuous**. In the experiment of Bateson and Punnett, for example, the sweet pea flowers were either white or purple; variation was discontinuous. But many traits that are under genetic control—such as height and other aspects of size, or skin color—vary continuously. We may think of these continuously varying traits as being controlled by multiple **polygenes**: loci whose alleles increase or decrease the observed character (Figure 10.23). Polygenes affecting a particular quantitative trait are common on many chromosomes. One of Mendel's wise decisions was to deal only with discontinuous variation, which is relatively simple. Had he, like Charles Darwin and others, concentrated on continuous variation, we might still not know the basic rules of heredity!

Humans differ with respect to the amount of a dark pigment, melanin, in the skin. There is great variation in the amount of melanin among different people, but much of this variation is determined by

alleles at just four (possibly three) loci. None of the alleles at these loci demonstrates dominance. Of course, skin color is not entirely determined by the genotype, since exposure to sunlight can cause the production of more melanin—that is, tanning.

Such environmental variables as light, temperature, and nutrition can sharply affect the translation of a genotype into a phenotype. A familiar example is the Siamese cat (Figure 10.24). This handsome animal normally has darker fur on its ears, nose, paws, and tail than on the rest of its body. These darkened parts are ones that have a somewhat lower temperature. A few simple experiments show that the Siamese cat has a genotype that results in dark fur, but only at temperatures somewhat below the general body temperature. If some dark fur is removed from the tail and the cat is kept at higher-than-usual temperatures, the new fur that grows in is light. Conversely, removal of light fur from the back, followed by local chilling of the area, causes the spot to fill with dark fur.

Genotype and environment interact to determine the phenotype of an organism. It is sometimes possible to determine the proportion of individuals in a group with a given genotype that actually show the expected phenotype. This proportion is called the **penetrance** of the genotype. The environment may also affect the **expressivity** of the genotype, that is, the degree to which it is expressed. For an example of environmental effects on expressivity, consider Siamese cats that are kept indoors and outdoors in different climates.

Uncertainty over how much of the observed vari-

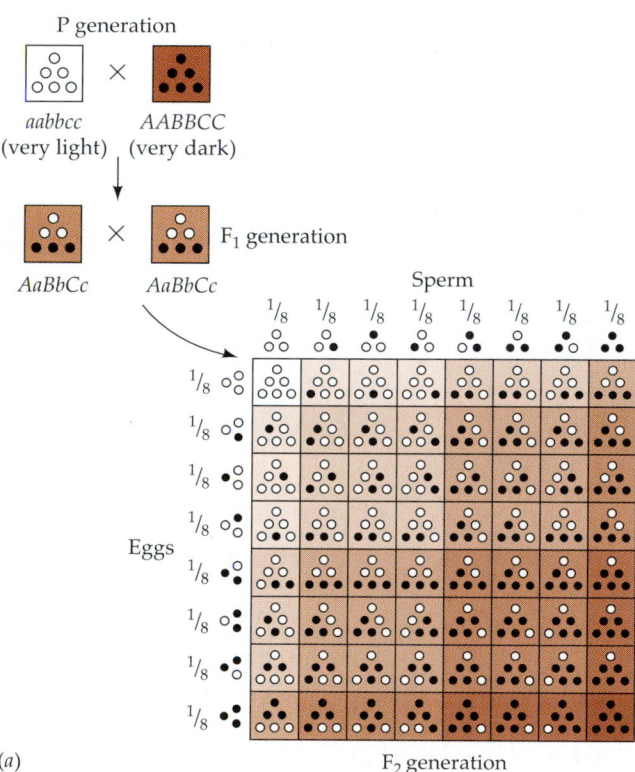

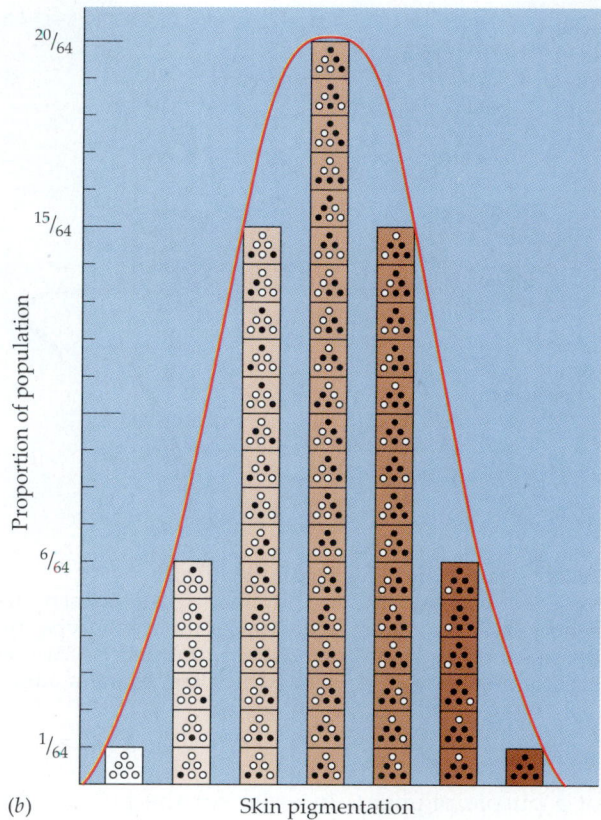

10.23 Polygenes Determine Human Skin Pigmentation
A model of polygenic inheritance based on three genes. The alleles *A*, *B*, and *C* contribute dark pigment to the skin, but the alleles *a*, *b*, and *c* do not. The more *A*, *B*, and *C* alleles an individual possesses, the darker that person's skin will be. The pattern of inheritance of the alleles is shown in *(a)*. The black circles represent the alleles *A*, *B*, and *C*; the white circles represent the alleles *a*, *b*, and *c*. The frequencies of the phenotypes in *(a)* are graphed in *(b)*. If both members of a couple have intermediate pigmentation (*AaBbCc*, for example), they are unlikely to have children with either very light or very dark skin.

ation is due to the environment and how much to the effects of the several polygenes complicates the analysis of quantitative inheritance. A useful approach that avoids this difficulty is to study identical twins. Since these individuals are genetically identical, any differences between such twins must be attributed to environmental effects.

The phenotype of an organism depends on its total genetic makeup and on its environment. Some of the interactive effects will become more obvious when we focus, in Chapter 12, on the regulation of gene expression.

NON-MENDELIAN INHERITANCE

You have studied the basic patterns of Mendelian inheritance in terms of chromosomal behavior. Does all inheritance in eukaryotes conform to the Mendelian pattern? Consider the four-o'clock plant shown in Figure 10.25. This particular four-o'clock shows three different patterns of chlorophyll distribution in three different parts of the shoot. One branch is all

10.24 The Environment Affects the Phenotype
This Siamese cat has dark fur on its extremities, where temperature is below the general body temperature.

All-green branch

All-white branch

Main shoot is variegated

10.25 A Variegated Four-o'Clock with Green and White Branches
This is the plant used for the experiment summarized in Table 10.4. Pollen is transferred from a flower on one part of the plant to a flower on another part.

white; it has no chlorophyll. Another branch is entirely green. The rest of the shoot is variegated—that is, there is chlorophyll in some patches of cells and none in others. Each part of the plant has flowers, so we can do the following experiment. Take pollen (which produces male gametes) from some flowers and transfer it to the female parts of flowers on another part of the plant, thus performing a cross. Table 10.4 shows the nine possible crosses and their outcomes. Study the table for a moment before reading further and try to discern the basic pattern of inheritance. Do you see the surprising feature? Look again. What should become obvious is that the phenotype of the "father" (the parent producing the pollen) is irrelevant to the outcome! Only the "mother" (the parent producing the egg) seems to play a role in determining the phenotype of the offspring.

TABLE 10.4		
Results of the Four-o'Clock Experiments		
PHENOTYPE OF BRANCH WITH FEMALE PARENT	PHENOTYPE OF BRANCH WITH MALE PARENT	PHENOTYPE OF PROGENY
White	White	White
White	Green	White
White	Variegated	White
Green	White	Green
Green	Green	Green
Green	Variegated	Green
Variegated	White	Variegated, white, or green
Variegated	Green	Variegated, white, or green
Variegated	Variegated	Variegated, white, or green

This pattern of inheritance, in which the progeny's phenotype is unaffected by the father, is found in all sorts of eukaryotic organisms and is referred to as **maternal inheritance**. How does it work? The essence of Mendelian inheritance is that information carried on chromosomes is partitioned with great precision during meiosis, but eukaryotic cells have other self-reproducing entities besides the nuclear chromosomes. Chloroplasts and mitochondria carry some genetic information in small circular chromosomes (see Chapter 4). The DNA of these organelles is subject to mutation just as is the DNA in the chromosomes of the nucleus, so we may speak of alleles of non-nuclear genes. These genes are not inherited in the same way as nuclear chromosomal genes are because the eggs of most species contain large amounts of cytoplasm, but the sperm contain hardly any. Generally speaking, all the mitochondria in a zygote come from the cytoplasm of its mother's egg, even though half the zygote's nuclear chromosomes come from its father. In plant zygotes all the chloroplasts come from the maternal cytoplasm. Hence any particle that is inherited through the cytoplasm is said to be maternally inherited. In such cases, reciprocal crosses give quite different results, as we saw in Table 10.4.

SUMMARY of Main Ideas about Mendelian Genetics

By crossing different strains of pea plants and counting each type of progeny, Mendel discovered some characteristic ratios of inheritance.

F_1 progeny of a cross between a dominant homozygote and a recessive homozygote express only the dominant trait.

Both parental phenotypes reappear in the F_2 produced by a monohybrid cross, in the ratio 3:1.

The F_2 consists of three genotypes in the ratio 1:2:1.
Review Figures 10.2, 10.3, 10.4, and Table 10.1

A monohybrid cross with incomplete dominance results in a 1:2:1 phenotype ratio in the F_2.
Review Figure 10.10

A test cross shows whether an individual expressing a dominant phenotype is homozygous or heterozygous.
Review Figure 10.6

Heterozygous organisms display both traits when the two alleles are codominant.
Review Figure 10.10

Some loci are represented by multiple alleles; different alleles arise by mutation.
Review Figures 10.11, 10.12, and Tables 10.2 and 10.3

In a dihybrid cross with unlinked genes, alleles of the two loci assort independently at meiosis.
Review Figure 10.8

Dihybrid crosses with unlinked genes yield a 9:3:3:1 ratio of four phenotypes in the F_2.
Review Figure 10.7

If the markers in a dihybrid cross are linked, there is a much higher proportion of parental phenotypes in the F_2 and fewer—or no—recombinant individuals.
Review Figure 10.13

Crossing over between chromatids produces recombination.
Review Figures 10.18 and 10.19

From the frequency of recombination, taken as a measure of the distance between genes, mapping techniques reveal the linear order of genes on a chromosome.

The presence or absence of certain chromosomes determines sex in many organisms.

In humans and *Drosophila*, XX determines female, and XY male.
Review Figure 10.14

Traits governed by loci on the X or the Y chromosome are not inherited in the same ratios as those observed for autosomal markers.

Reciprocal crosses for sex-linked markers give nonidentical results.
Review Figures 10.15 and 10.16

Some genes modify the expression of other genes.
Review Figure 10.22

Interactions of multiple genes control many traits.
Review Figure 10.23

The environment may influence both the penetrance and the expressivity of a genotype.

Some traits are coded for by DNA contained in the mitochondria or chloroplasts, which are inherited maternally.

For traits that are inherited maternally, reciprocal crosses do not give identical results.
Review Figure 10.25 and Table 10.4

SELF-QUIZ

1. Which statement is *not* true for Mendel's cross of *TT* peas with *tt* peas?
   a. Each parent can produce only one type of gamete.
   b. F_1 individuals produce gametes of two types, each gamete being *T* or *t*.
   c. Three genotypes are observed in the F_2 generation.
   d. Three phenotypes are observed in the F_2 generation.
   e. This is an example of a monohybrid cross.

2. The phenotype of an individual
   a. depends at least in part on the genotype.
   b. is either homozygous or heterozygous.
   c. determines the genotype.
   d. is the genetic constitution of the organism.
   e. is either monohybrid or dihybrid.

3. Which of the following statements is *not* true of alleles?
   a. They are different forms of the same gene.
   b. There may be several at one locus.
   c. One may be dominant over another.
   d. They may show incomplete dominance.
   e. They occupy different loci on the same chromosome.

4. Which statement is *not* true of an individual that is homozygous for an allele?
   a. Each of its cells possesses two copies of that allele.
   b. Each of its gametes contains one copy of that allele.
   c. It is true-breeding with respect to that allele.
   d. Its parents were necessarily homozygous for that allele.
   e. It can pass that allele to its offspring.

5. Which of the following statements is *not* true of a test cross?
   a. It tests whether an unknown individual is homozygous or heterozygous.
   b. The test individual is crossed with a homozygous recessive individual.
   c. If the test individual is heterozygous, the progeny will have a 1:1 ratio.
   d. If the test individual is homozygous, the progeny will have a 3:1 ratio.
   e. Test cross results are consistent with Mendel's model of inheritance.

6. Linked genes
   a. must be immediately adjacent to one another on a chromosome.
   b. have alleles that assort independently of one another.
   c. never show crossing over.
   d. are on the same chromosome.
   e. always have multiple alleles.

7. In the F_2 generation of a dihybrid cross
   a. four phenotypes appear in the ratio 9:3:3:1 if the loci are linked.
   b. four phenotypes appear in the ratio 9:3:3:1 if the loci are unlinked.
   c. two phenotypes appear in the ratio 3:1 if the loci are unlinked.
   d. three phenotypes appear in the ratio 1:2:1 if the loci are unlinked.
   e. two phenotypes appear in the ratio 1:1 whether or not the loci are linked.

8. The sex of a honeybee is determined by
   a. ploidy, the male being haploid.
   b. X and Y chromosomes, the male being XY.
   c. X and Y chromosomes, the male being XX.
   d. the number of X chromosomes, the male being XO.
   e. Z and W chromosomes, the male being ZZ.

9. In epistasis
   a. nothing changes from generation to generation.
   b. one gene alters the effect of another.
   c. a portion of a chromosome is deleted.
   d. a portion of a chromosome is inverted.
   e. the behavior of two genes is entirely independent.

10. Individual heritable traits
   a. are always determined by dominant and recessive alleles.
   b. always vary discontinuously.
   c. can sometimes be controlled by many genes.
   d. were first studied in this century.
   e. do not exist outside the laboratory.

FOR STUDY

1. Utilizing the Punnett squares below, show that for typical dominant and recessive autosomal traits, it does not matter which parent contributes the dominant allele and which the recessive allele. Cross true-breeding tall plants (TT) with true-breeding dwarf plants (tt).

   Tall Female × Dwarf Male   Dwarf Female × Tall Male

   Male gametes   Male gametes

   Female gametes   Female gametes

2. Show diagrammatically what occurs when the F_1 offspring of the cross in Question 2 self-pollinate.

   Male gametes

   Female gametes

3. A new student of genetics suspects that a particular recessive trait in fruit flies (dumpy wings) is sex-linked. A single mating between a fly having dumpy wings (dp; female) and a fly with wild-type wings (Dp; male) produces 3 dumpy-winged females and 2 wild-type males. On the basis of these data, is the trait sex-linked or autosomal? What were the genotypes of the parents? Explain how these conclusions can be reached on the basis of so few data.

4. The sex of fishes is determined by the same X–Y system as in humans and *Drosophila*. An allele of one locus on the Y chromosome of the fish *Lebistes* causes a pigmented spot to appear on the dorsal fin. A male fish with a spotted dorsal fin is mated with a female fish with an unspotted fin. Describe the phenotypes of the F_1 and the F_2 from this cross.

5. In *Drosophila melanogaster*, the recessive allele p, when homozygous, determines pink eyes. Pp or PP results in wild-type eye color. Another gene, on another chromosome, has a recessive allele, sw, that produces short wings when homozygous. Consider a cross between females of genotype $PPSwSw$ and males of genotype $ppswsw$. Describe the phenotypes and genotypes of the F_1 generation and of the F_2 generation produced by allowing the F_1 to mate with one another.

6. On the same chromosome of *Drosophila melanogaster* that carries the p (pink eyes) locus, there is another locus that affects the wings. Homozygous recessives, $byby$, have blistery wings, while the dominant allele By produces wild-type wings. The p and by loci are very close together on the chromosome; that is, the two loci are tightly linked. In answering these questions, assume that no crossing over occurs.
   a. For the cross $PPByBy \times ppbyby$, give the phenotypes and genotypes of the F_1 and of the F_2 produced by F_1 interbreeding.
   b. For the cross $PPbyby \times ppByBy$, give the phenotypes and genotypes of the F_1 and of the F_2.
   c. For the cross of Question 7b, what further phenotype(s) would appear in the F_2 generation if crossing over occurred?
   d. Draw a nucleus undergoing meiosis, at the stage in which the crossing over (Question 7c) occurred. In which generation (P, F_1, or F_2) did this crossing over take place?

7. Consider the following cross of *Drosophila melanogaster* with alleles as described in Question 6. Males with genotype $Ppswsw$ are crossed with females of genotype $ppSwsw$. Describe the phenotypes and genotypes of the F_1 generation.

8. In the Blue Andalusian fowl, a single pair of alleles controls the color of the feathers. Three colors are observed: blue, black, and splashed white. Crosses among these three types yield the following results:

Parents	Progeny
Black × blue	Blue and black (1:1)
Black × splashed white	Blue
Blue × splashed white	Blue and splashed white (1:1)
Black × black	Black
Splashed white × splashed white	Splashed white

   a. What progeny would result from the cross blue × blue?
   b. If you wanted to sell eggs, all of which would yield blue fowl, how should you proceed?

9. In *Drosophila melanogaster*, white (*w*), eosin (*w^e*), and wild-type red (*w^+*) are multiple alleles of a single locus for eye color. This locus is on the X chromosome. An eosin-eyed female is crossed with a male with wild-type eyes. All the female progeny are red-eyed; half the male offspring have eosin (pale orange) eyes, and half have white eyes.
   a. What is the order of dominance of these alleles?
   b. What are the genotypes of the parents and progeny?

10. Color blindness is a recessive trait. Two people with normal vision have two sons, one color-blind and one with normal vision. If the couple also has daughters, what proportion of them will have normal vision? Explain.

11. A mouse with an agouti coat is mated with an albino mouse of genotype *aabb*. Half the offspring are albino, one-quarter are black, and one-quarter are agouti. What are the genotypes of the agouti parents and of the various kinds of offspring? (Hint: see the section "Epistasis.")

12. Sweet peas (genotype *aaBB*) with white flowers are crossed with sweet peas with purple flowers. Of the progeny, half have purple flowers and half have white flowers. What can you say about the genotype of the purple-flowered parent? (Hint: This is another problem dealing with epistasis.)

13. The photograph shows the shells of 15 bay scallops, *Argopecten irradians*. These scallops are hermaphroditic—that is, a single individual can reproduce sexually, as did the pea plants of the F₁ generation in Mendel's experiments. Three color schemes are evident: yellow, orange, and black and white. The color-determining locus has three alleles. The top row shows a yellow scallop and a representative sample of its offspring, the middle row shows a black-and-white scallop and its offspring, and the bottom row shows an orange scallop and its offspring. Assign a suitable symbol to each of the three alleles participating in color control; then determine the genotype of each of the three parent individuals and tell what you can about the genotypes of the different offspring. Explain your results carefully.

READINGS

Cooper, N. G. 1995. *The Human Genome Project: Deciphering the Blueprint of Heredity.* University Science Books, Mill Valley, CA. This unique book follows an introduction to the ideas of classical and molecular genetics with a complete discussion of the purpose, approach, technology, pitfalls, and implications of the Human Genome Project.

Griffiths, A. J. F., J. H. Miller, D. T. Suzuki, R. C. Lewontin and W. M. Gelbart. 1993. *An Introduction to Genetic Analysis,* 5th Edition. W. H. Freeman, New York. An excellent textbook of modern genetics. Chapters 2 and 3 are particularly relevant to this chapter; Chapters 4 and 5 are also useful.

Mange, E. J. and A. P. Mange. 1994. *Basic Human Genetics.* Sinauer Associates, Sunderland, MA. Genetics, especially chromosomal inheritance, can be studied using humans as examples; this book does so at an introductory level.

Russell, P. J. and J. M. Nickerson. 1992. *Genetics,* 3rd Edition. Harper/Collins, New York. A well-balanced treatment of a broad range of topics in genetics. Highly recommended.

Sapienza, C. 1990. "Parental Imprinting of Genes." *Scientific American,* October. When reciprocal crosses aren't equivalent.

Stern, C. and E. R. Sherwood (Eds.). 1966. *The Origin of Genetics: A Mendel Source Book.* W. H. Freeman, New York. A collection of the writings of researchers at the dawn of the science of genetics, including translations of Mendel's papers and letters. The last two articles discuss the likelihood that Mendel fudged his data.

Sturtevant, A. H. and G. W. Beadle. 1962. *An Introduction to Genetics.* Dover, New York. First published by W. B. Saunders in 1939. Though old, this text holds up as a fine introduction to formal chromosome genetics.

Gregor Mendel described the basic patterns of inheritance in plants and animals and devised a powerful explanation for the mechanisms underlying these patterns. The second of these accomplishments is most impressive because Mendel had no way of knowing the physical basis for his proposed mechanisms. He never knew what a gene is, in chemical terms, nor did he know about the behavior of chromosomes. Hence he could not have known how genes are copied between generations or how new alleles arise. He could not have had the slightest inkling of how a gene works—that is, how the genotype produces a phenotype.

During the same years Mendel was analyzing how the characteristics of pea plants were inherited, a Swiss chemist, Friedrich Miescher, was at work trying to identify the chemical composition of cell nuclei. Miescher focused on a material that has a high ratio of nuclear to cytoplasmic volume: pus cells from bandages discarded from the wounds of soldiers. Miescher found that the nuclei contained large amounts of protein and of a previously undescribed compound that we now call DNA.

Neither Mendel's nor Miescher's work was understood or appreciated for the remainder of the nineteenth century, and genetics and DNA did not meet until 1944. During the half-century since then, geneticists, biochemists, biophysicists, and molecular biologists have developed a detailed picture of the chemistry and functioning of the genetic material. In this and the next two chapters, we will try to show you how the main questions of molecular genetics have been studied.

WHAT IS THE GENE?

During the first half of the twentieth century, the hereditary material was generally assumed to be protein. The impressive chemical diversity of proteins made this assumption seem reasonable. Also, some proteins—notably enzymes and antibodies—show great specificity. By contrast, nucleic acids were known to have only a few components and seemed too simple to carry the complex information expected in the genetic material. The recognition that the gene is not a protein, but rather deoxyribonucleic acid, or DNA, was slow in coming and depended on the interaction of several types of research.

The Transforming Principle

The history of biology is filled with incidents in which research on some specific topic has—with or without answering the question originally under investigation—contributed richly to another, apparently un-

The Double Helix of DNA

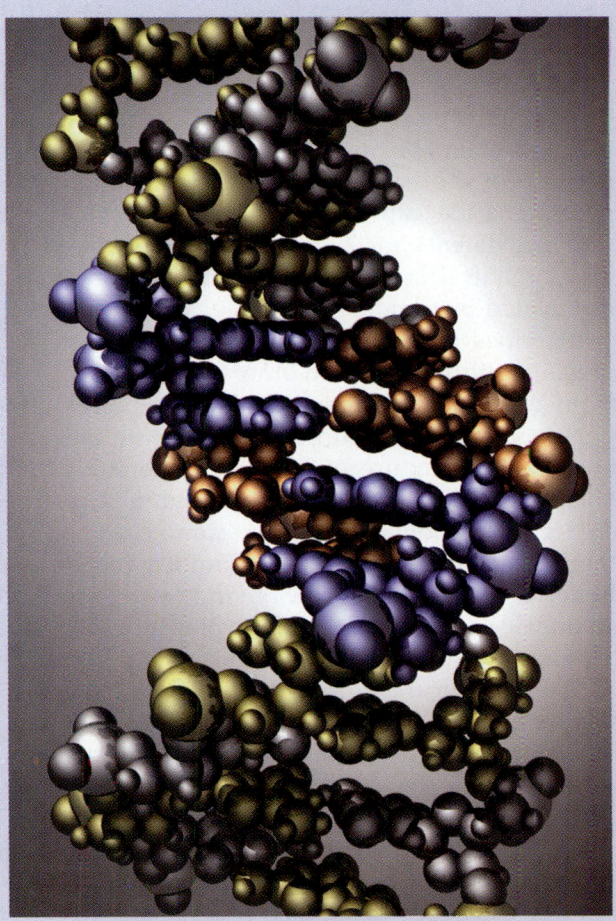

11

Nucleic Acids as the Genetic Material

related area. Such a case is the work of Frederick Griffith, an English physician. In the 1920s Griffith was studying the disease-causing behavior of the bacterium *Streptococcus pneumoniae*, or pneumococcus, one of the agents that produce pneumonia in humans. He identified two strains of pneumococcus, designated S and R because the former produces shiny, smooth (S) colonies when grown in the laboratory, whereas the colonies of the latter are rough (R) in appearance. When the S strain was injected into mice, they died within a day, and the hearts of the dead mice were found to be teeming with the deadly bacteria. When the R strain was injected instead, the mice did not become diseased. In other words, the S strain is virulent (disease-causing) and the R strain is nonvirulent. This difference was eventually shown to be due to a difference in the chemical makeup of the bacterial surface: The S strain has a polysaccharide capsule that protects the bacterium from the defense mechanisms of the host (see Chapter 16). The R strain lacks this capsule and can be inactivated by a mouse's defenses.

In hopes of developing a vaccine against pneumonia, Griffith inoculated other mice with heat-killed S pneumococci. Neither heat-killed S nor living R pneumococci produced infection. Griffith inoculated mice with a mixture of living R bacteria and heat-killed S bacteria. To his astonishment, all these mice died of pneumonia. When he examined blood from the hearts of these mice, he found it full of living bacteria, many of them belonging to the virulent S strain! He concluded that, in the presence of the dead S pneumococci, some of the inoculated R pneumococci had been transformed into virulent organisms (Figure 11.1).

Did transformation of the bacteria depend upon something the mouse did? No. It was soon shown that the same transformation occurs when living R and heat-killed S bacteria are simply incubated together in a test tube. Next it was discovered that a cell-free extract of heat-killed S cells also transforms R cells. (A cell-free extract contains all the contents of ruptured cells, but no intact cells.) This result demonstrated that some substance—called at the time a chemical **transforming principle**—from the dead S pneumococci can cause a permanent change in the affected R cells. (Remember that great numbers of *living* bacteria of the S type were always found in the mice that died as a result of being inoculated with

the mixture of heat-killed S and living R bacteria.) From these observations scientists concluded that the transforming principle carried heritable information; thus it could be thought of as genetic material. We now know that any genetic trait in pneumococci or in several other types of bacteria can be passed, by way of transforming principles, from one bacterium to another.

The Transforming Principle is DNA

A crucial step in the history of biology was the identification of the transforming principle, accomplished over a period of several years by Oswald T. Avery and his colleagues at what is now Rockefeller University. They treated samples of the transforming principle in a variety of ways to destroy different types of substances—proteins, nucleic acids, carbohydrates, lipids—and then tested the treated samples to see if they had retained transforming activity. The answer was always the same: If the DNA in the

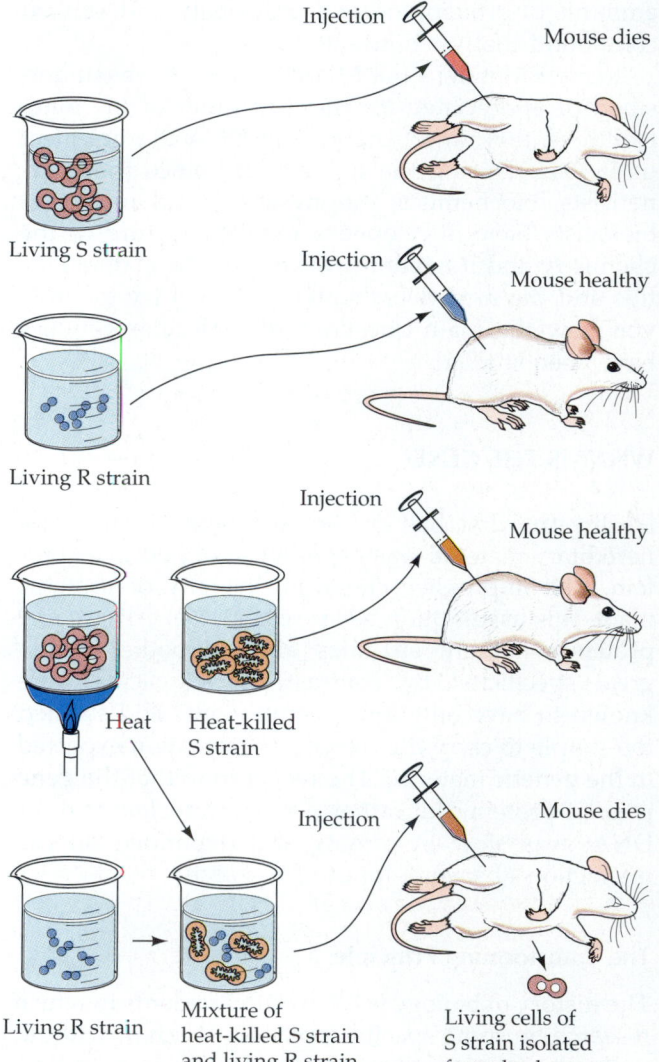

11.1   Genetic Transformation of Nonvirulent R Pneumococci
Griffith's experiments demonstrated that some factor in the virulent S strain could transform nonvirulent R-strain bacteria into a lethal form, even when the S-strain bacteria had been killed by high temperatures.

Injection — Mouse dies
Living S strain

Injection — Mouse healthy
Living R strain

Injection — Mouse healthy
Heat   Heat-killed S strain

Injection — Mouse dies
Living R strain   Mixture of heat-killed S strain and living R strain   Living cells of S strain isolated from dead mouse

sample was destroyed, transforming activity was lost; everything else was dispensable. As a final step, Avery, with Colin MacLeod and MacLyn McCarty, isolated virtually pure DNA from a sample of pneumococcal transforming principle and showed that it was highly active in causing bacterial transformation. Their work, published in 1944, was a milestone in establishing that DNA is the genetic material in cells; at the time, however, it did not receive the attention it deserved, and scientists in other labs continued their attempts to identify the hereditary material.

The Genetic Material of a Virus

A report published in 1952 by Alfred D. Hershey and Martha Chase of the Carnegie Laboratory of Genetics had a much greater immediate impact than did Avery's 1944 paper. The Hershey–Chase experiment was carried out with a virus that infects bacteria. This virus, called T2 bacteriophage, consists of a DNA core packed within a protein coat (Figure 11.2*a*). The virus is thus made of the two materials that were, at the time, the leading candidates for the genetic material.

When a T2 bacteriophage attacks a bacterial cell, part—but not all—of the virus enters the cell; about 20 minutes later the bacterial cell lyses (breaks apart), releasing 200 to 1,000 new T2s. Hershey and Chase set out to determine which part of the virus—protein or DNA—is the hereditary material that gets inside the bacterial cell. The idea was to trace these two components during the life cycle of the virus, so Hershey and Chase labeled each with a specific radioactive tracer. All proteins contain some sulfur (in the amino acids cysteine and methionine), an element that is not present in DNA, and sulfur has a radioactive isotope, ^{35}S. The deoxyribose–phosphate "backbone" of DNA, on the other hand, is rich in phosphorus (see Chapter 3), an element not present in proteins—and phosphorus also has a radioactive isotope, ^{32}P. Thus Hershey and Chase grew one batch of T2 in a bacterial culture in the presence of ^{32}P, so that all the viral DNA was labeled with ^{32}P. Similarly, the proteins of another batch of T2 were labeled with ^{35}S (Figure 11.2*b*).

In separate experimental runs, Hershey and Chase combined radioactive viruses containing either ^{32}P or ^{35}S with bacteria (Figure 11.2*c*). After a few minutes, the mixtures were swirled vigorously in a kitchen blender, which (without bursting the bacteria) stripped away the parts of the virus coats that had not penetrated the bacteria. The bacteria were then separated from the rest of the material in a centrifuge. It was found that more than three-fourths of the ^{35}S (and thus the protein) had separated from the bacteria, and that two-thirds or more of the ^{32}P (and thus the DNA) had settled to the bottom of the centrifuge tube along with the bacteria. Although the numbers were not as clear-cut as one might like, these results suggested that the DNA was transferred to the bacteria while the protein remained outside.

Confirmation of this tentative conclusion came when other batches of bacteria and labeled T2 were incubated together for longer periods, allowing a progeny generation of viruses to be collected. When this was done, Hershey and Chase found that the resulting T2 progeny contained less than 1 percent of the original ^{35}S but about one-third of the original ^{32}P—and thus, presumably, one-third of the DNA. This result showed that T2 injects the DNA from its head into a bacterium while the external protein structures remain outside the bacterial cell. Because DNA was carried over in the virus from generation to generation, whereas protein was not, a logical conclusion was that the hereditary information of the viruses is contained in the DNA. The Hershey–Chase experiment convinced most scientists that DNA is indeed the carrier of hereditary information.

NUCLEIC ACID STRUCTURE

Once scientists agreed that the genetic material is DNA, they wanted to learn just what the DNA molecule looks like. In its structure they hoped to find clues to two questions: how the molecule is replicated between nuclear divisions, and how it causes the synthesis of specific proteins. Both expectations were fulfilled.

Evidence from X-Ray Crystallography and Biochemistry

The structure of DNA was deciphered only after many types of experimental evidence and theoretical considerations were put together. The most crucial "hard" evidence was obtained by X-ray crystallography. The positions of atoms in a crystalline substance can be inferred by the pattern of diffraction of X rays passed through the crystal, but even today this is not an easy task when the substance is of enormous molecular weight. In the early 1950s, even a highly talented X-ray crystallographer could (and did) look at the best available images from DNA preparations and fail to see what they meant. Nonetheless, the attempt to characterize DNA would have been impossible without the crystallographs prepared by the English chemist Rosalind Franklin. Franklin's work, in turn, depended upon the success of the English biophysicist Maurice Wilkins in preparing very uniformly oriented DNA fibers, which made far more manageable samples for crystallography than had previous samples.

(a)   **The virus:** T2 bacteriophage

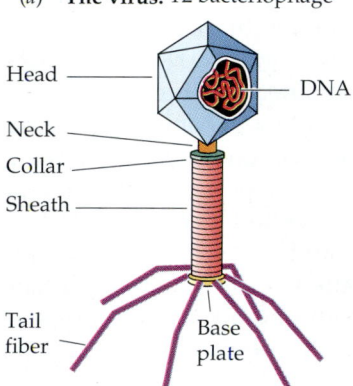

Head —————

DNA

Neck —————

Collar —————

Sheath —————

Tail
fiber —————

Base
plate

11.2   T2 and the Hershey-Chase Experiment

(a) The external structures of the bacteriophage T2 consist entirely of protein. This cutaway view shows a strand of DNA within the head. (b) What made the Hershey–Chase experiment possible was the production of two sets of uniquely tagged T2 bacteriophage—one for the DNA and one for the protein coat. (c) Because progeny generations of viruses incorporated the radioactively tagged DNA but not the radioactively tagged proteins, the experiment demonstrated that DNA, not protein, is the hereditary material.

(b)   **Preparation of specifically labeled viruses**

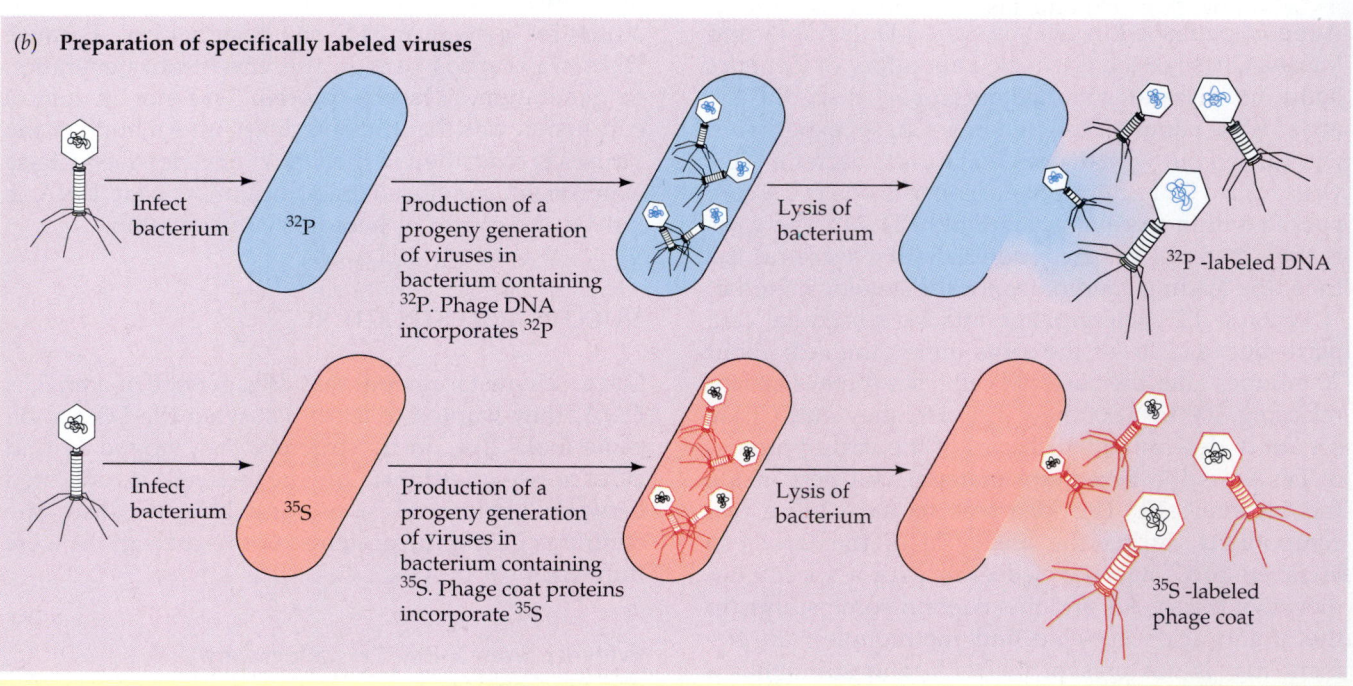

Infect bacterium → ^{32}P Production of a progeny generation of viruses in bacterium containing ^{32}P. Phage DNA incorporates ^{32}P → Lysis of bacterium → ^{32}P -labeled DNA

Infect bacterium → ^{35}S Production of a progeny generation of viruses in bacterium containing ^{35}S. Phage coat proteins incorporate ^{35}S → Lysis of bacterium → ^{35}S -labeled phage coat

(c)   **The experiment**

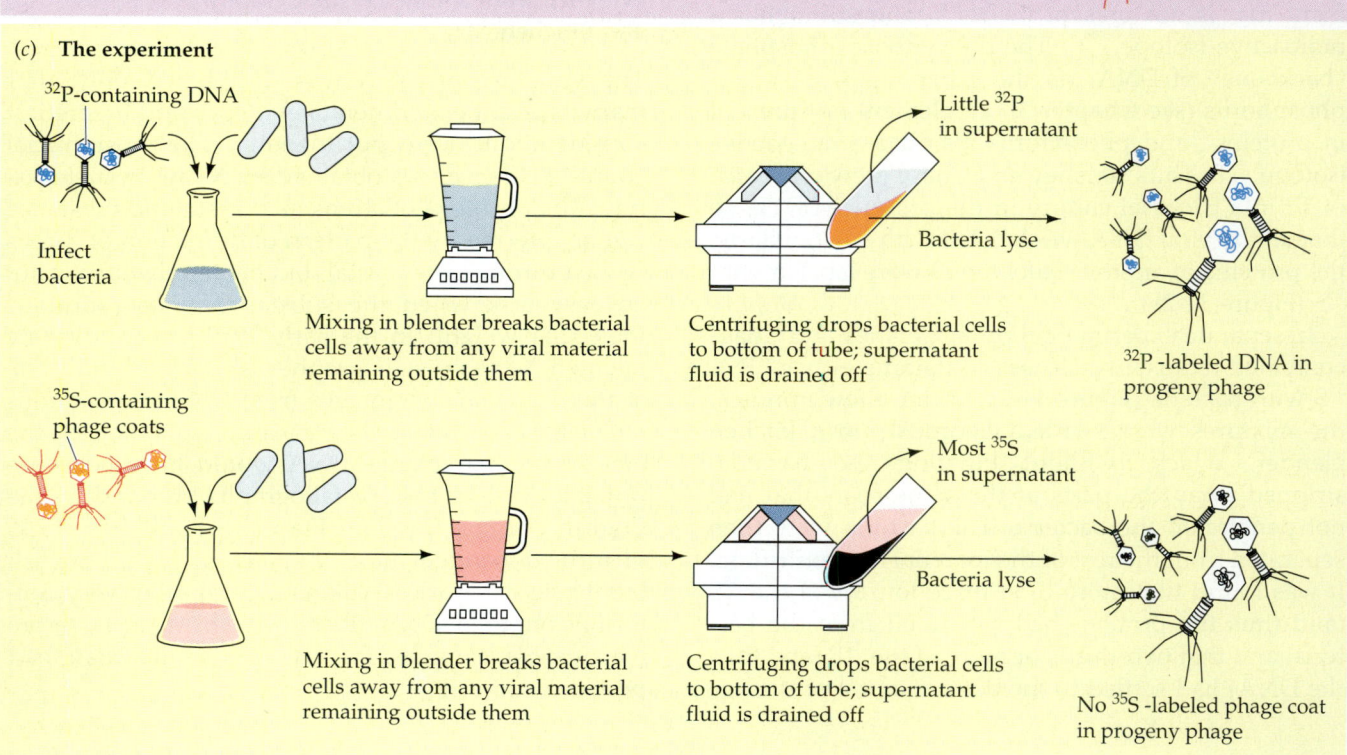

^{32}P-containing DNA

Infect bacteria

Mixing in blender breaks bacterial cells away from any viral material remaining outside them

Centrifuging drops bacterial cells to bottom of tube; supernatant fluid is drained off

Little ^{32}P in supernatant

Bacteria lyse

^{32}P -labeled DNA in progeny phage

^{35}S-containing phage coats

Mixing in blender breaks bacterial cells away from any viral material remaining outside them

Centrifuging drops bacterial cells to bottom of tube; supernatant fluid is drained off

Most ^{35}S in supernatant

Bacteria lyse

No ^{35}S -labeled phage coat in progeny phage

The chemical composition of DNA also provided important clues about its structure. Biochemists knew the chemical structures of the four monomers, or nucleotides, of DNA (see Chapter 3), as well as how one nucleotide is joined to another to form a polynucleotide chain. Recall that a nucleotide of DNA consists of a molecule of the sugar deoxyribose, a phosphate group, and a nitrogen-containing base (see Figures 3.22 and 3.23). The only differences between the four nucleotides of DNA are in their nitrogenous bases: the purines **adenine** and **guanine** and the pyrimidines **cytosine** and **thymine**. In 1950 Erwin Chargaff at Columbia University reported observations of major importance. He and his colleagues had found that DNA from many different species—and from different sources within a single organism—exhibits certain regularities. In any DNA the following rules hold: The amount of adenine equals the amount of thymine, and the amount of guanine equals the amount of cytosine. As a result, the total amount of purines equals the total amount of pyrimidines. The structure of DNA could scarcely have been worked out without this information, yet its significance was overlooked for at least three years.

Watson, Crick, and the Double Helix

Solving the puzzle was accelerated by the technique of model building: assembling three-dimensional representations of possible molecular structures. This technique, originally exploited in structural studies by the American chemist Linus Pauling, was used by the English physicist Francis Crick and the American geneticist James D. Watson, then both at the Cavendish Laboratory of Cambridge University. Watson and Crick attempted to combine all that had been learned so far about DNA structure into a single, coherent model. The crystallographers' results convinced Watson and Crick that the DNA molecule is **helical** (cylindrically spiral) and provided the values of certain distances within the helix. The results of density measurements and model building suggested that there are two polynucleotide chains in the molecule. The modeling studies also led to the conclusion that the two chains in DNA run in opposite directions, that is, that they are **antiparallel**.

Crick and Watson attempted several models. Late in February of 1953, they built the one that established the general structure of DNA. There have been minor amendments to their first published structure, but the principal features remain unchanged.

Key Elements of DNA Structure

Four features summarize the molecular architecture of DNA: *The molecule is (1) a double-stranded helix, (2) of uniform diameter, (3) twisting to the right (that is,*

twisting in the same direction as the threads on most screws), with (4) the two strands running in opposite directions. As you know, the two strands are polynucleotide chains. The sugar–phosphate backbones of the chains coil around the outside of the helix, with the nitrogenous bases pointing toward the center. The two chains are held together by hydrogen bonding between specifically paired bases. As implied by Chargaff's studies, adenine (A) pairs with thymine (T) by forming two hydrogen bonds, and guanine (G) pairs with cytosine (C) by forming three hydrogen bonds. Because the A–T and G–C pairs, like rungs of a ladder, are of equal length and fit identically into the double helix, the diameter of the helix is uniform (Figure 11.3a). A pair of purines would cause a swelling in the molecule and a pair of pyrimidines would cause a constriction, but DNA never includes such pairs. Every base pair consists of one purine (A or G) and one pyrimidine (C or T). Two grooves, one broad (the major groove) and one narrow (the minor groove), spiral around the outside of the DNA molecule (Figure 11.3b).

We say the two DNA strands run in opposite directions, but what does this mean? The direction of a polynucleotide can be defined by looking at the linkages between adjacent nucleotides. (These linkages are called phosphodiester bonds.) To avoid confusion with the carbon and nitrogen atoms in the ring structure of the bases (which are numbered 1, 2, 3, . . .), a prime symbol is placed after a number that refers to a carbon atom of a sugar: 1', 2', 3', and so on. In the sugar–phosphate backbone of DNA, the phosphate groups connect to the 3' carbon of one deoxyribose molecule and the 5' carbon of the next, linking successive sugars together (Figure 11.4). Thus the two ends of a polynucleotide differ. Polynucleotides have a free (not connecting to another nucleotide) 5' phosphate ($-OPO_3^{3-}$) group at one end—the **5' end**—and a free 3' hydroxyl ($-OH$) group at the other—the **3' end**—just as polypeptides have a free amino group at one end and a free carboxyl group at the other end. The 5' end of one strand in a DNA double helix is paired with the 3' end of the other strand, and vice versa; that is, the strands run in opposite directions (see Figure 11.3a).

Alternative Structures for DNA

We've been talking about DNA as if its shape never varied—that is, as if it were always a right-handed double helix, with two grooves of unequal width spiraling up its side. This is the form of most DNA. Over the past 40 years, however, some minor variations on this theme have been discovered, and in 1979, a strikingly different structure was observed in some samples of DNA. These molecules twist to the *left* rather than to the right, and they have only one

(a)

(b)

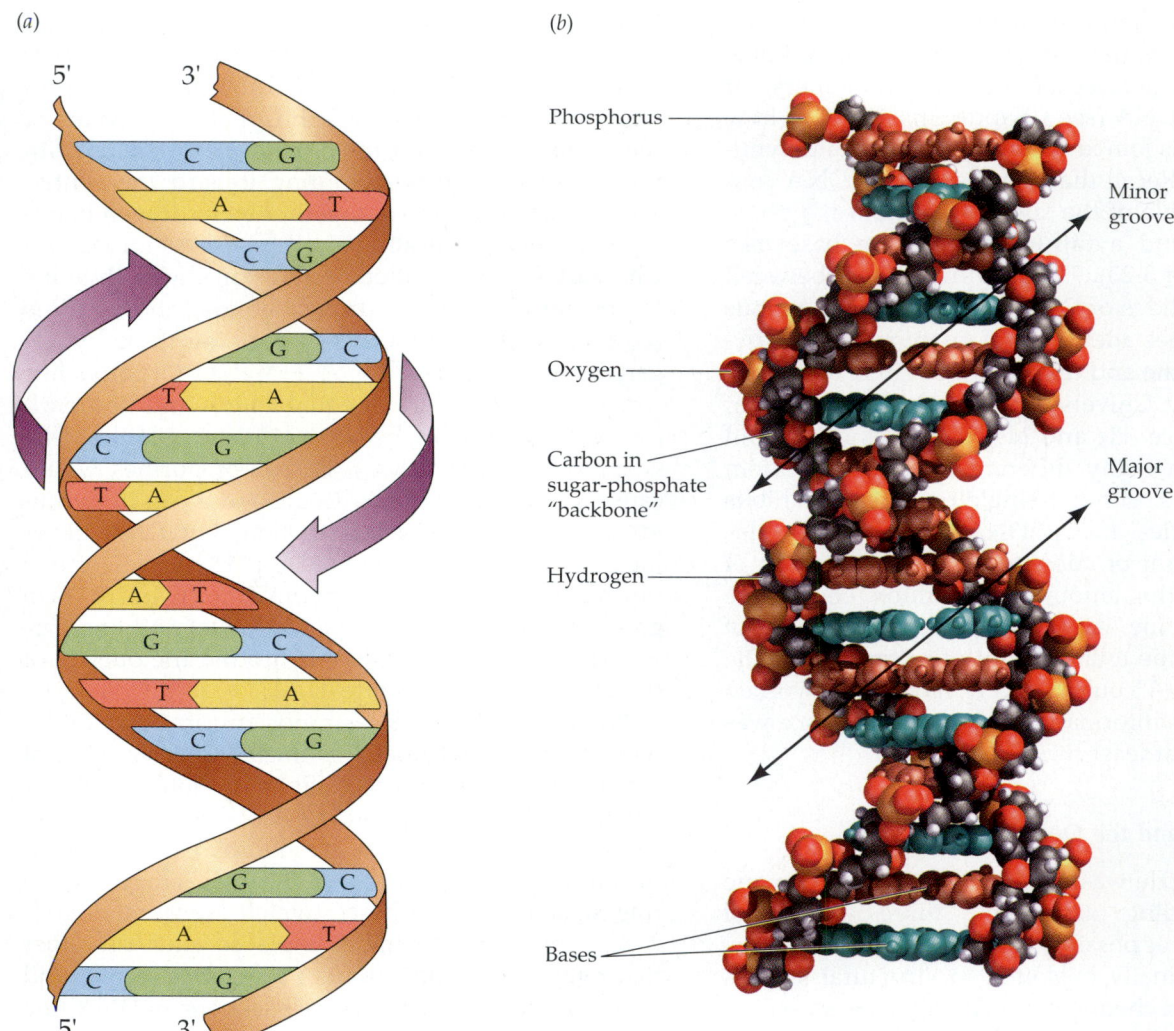

11.3 DNA Is a Double Helix

(a) Watson and Crick proposed that DNA is a double-helical molecule. The brown bands represent the two sugar–phosphate chains, with pairs of bases forming horizontal connections between the chains. The two chains run in opposite directions. (b) Biochemists can now pinpoint the position of every atom in a DNA macromolecule. To see that the essential features of the original Watson–Crick model have been verified, follow with your eyes the double-helical ribbons of sugar–phosphate groups and note the horizontal rungs of the bases.

groove. This form of DNA is referred to as **Z-DNA** because its sugar–phosphate backbones follow a zig-zag course rather than the smooth spiral of normal DNA backbones (Figure 11.5). The structure of Z-DNA is a left-handed double helix. It was first observed when scientists synthesized short DNA molecules that had alternating purine and pyrimidine bases on each strand. Then researchers discovered that short stretches of Z-DNA appear naturally in the DNA of living organisms, for example, in the chromosomes of the fruit fly *Drosophila melanogaster*.

Whether Z-DNA has a role distinct from that of normal DNA is not yet known. One possibility, which is mere speculation at this time, is that the

sharply different shape of Z-DNA makes it recognizable to one or more proteins that play regulatory roles.

Structure of RNA

To understand how DNA functions, you need to know about RNA. RNA (ribonucleic acid) is a polynucleotide similar to DNA (see Figure 3.25) but different in three ways: (1) RNA generally consists of only one polynucleotide strand (thus Chargaff's equalities, G = C and A = T, are true only for DNA and not for RNA); (2) the sugar molecule found in ribonucleotides is ribose, rather than deoxyribose as

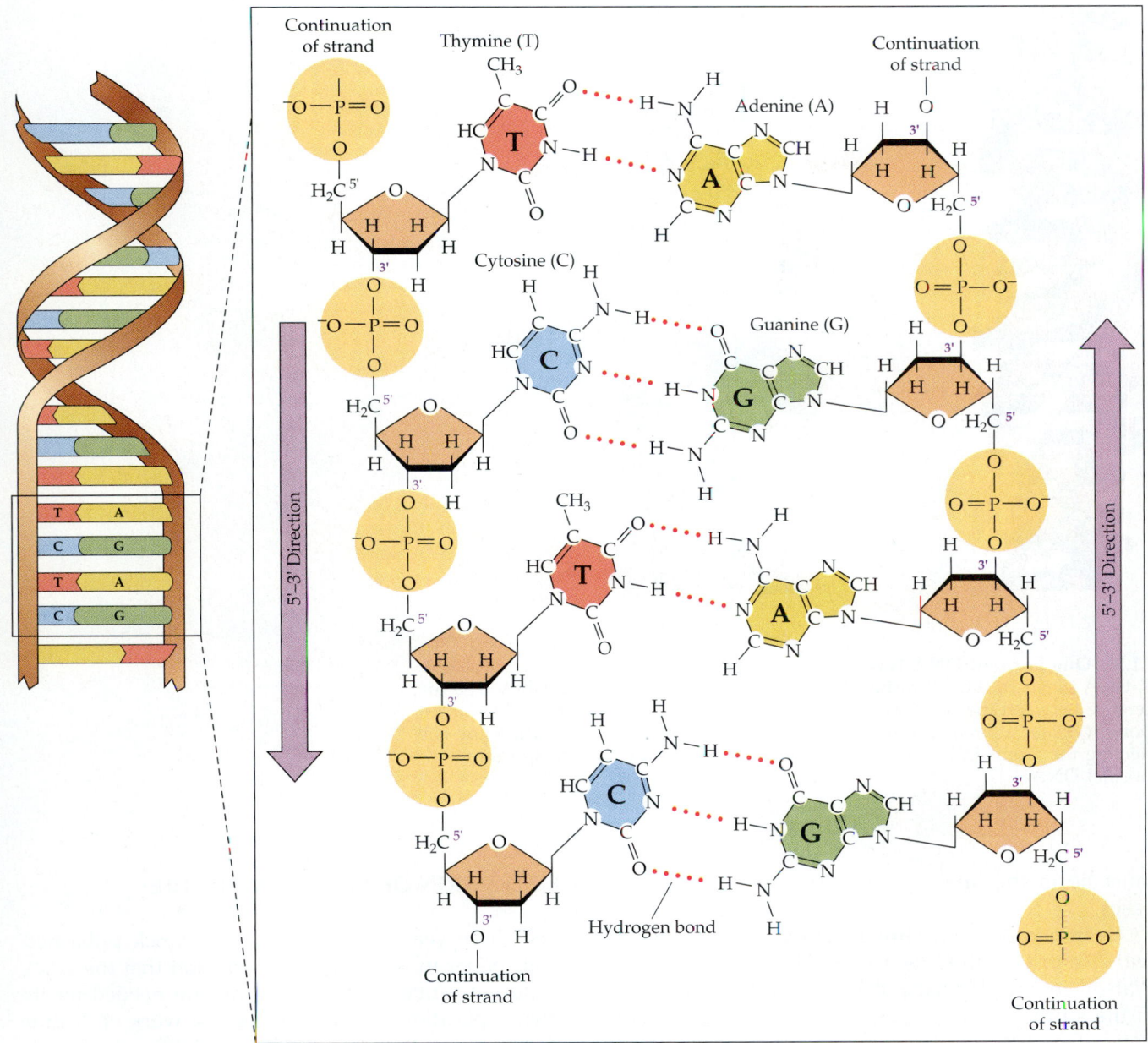

11.4 Base Pairing in DNA Is Complementary
In the sugar–phosphate backbone of DNA, each phosphate group links the 3' carbon of one sugar to the 5' carbon of the next sugar along the backbone. This asymmetry gives each DNA strand a 5' "head" and a 3' "tail." Complementary strands line up head-to-tail. Pairs of complementary bases form hydrogen bonds that hold the two strands of a DNA double helix together. T–A pairs form two hydrogen bonds; G–C pairs form three hydrogen bonds.

strand of RNA may pair between complementary bases. As we will see in the discussion of transfer RNA later in this chapter, when RNA bases do pair, uracil is like thymine in that it is complementary to adenine.

Implications of the Double-Helical Structure of DNA

Watson and Crick's double-helical model gave hints about how DNA carries out its biological role. First, the molecule is, in a sense, boring—it runs on and on, nucleotide pair after nucleotide pair, with no kinks and no bulges. Such a molecule can carry and convey information in only one way: The information

in DNA; and (3) although three of the nitrogenous bases (adenine, guanine, and cytosine) found in ribonucleotides are identical with those bases found in deoxyribonucleotides, the fourth base in RNA is uracil (U), which is similar to thymine but lacks the methyl (—CH₃) group. Different stretches of a single

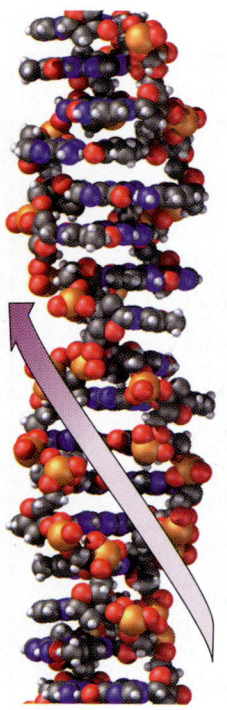
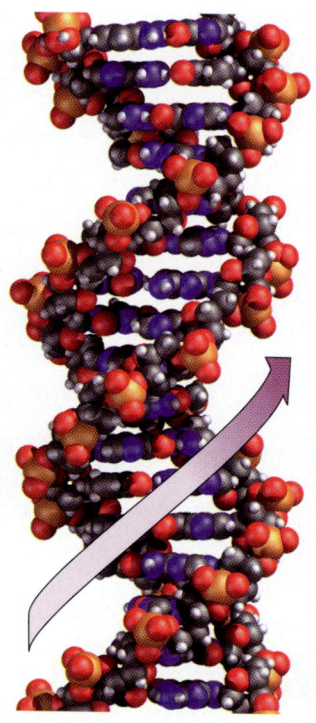

Z-DNA   Normal DNA

11.5   One Form of DNA Has a Reverse Twist
Z-DNA twists to the left rather than to the right as does normal DNA. The difference between the two forms is evident if you compare the phosphate backbones (yellow atoms) on the two models; the zigzag (Z) pattern of the line connecting the atoms in Z-DNA is clearly different from the smooth spiral in normal DNA.

must lie in the linear sequence of the nitrogenous bases.

An implication of **complementary base pairing**, A with T and G with C, was pointed out by Crick and Watson in the original publication of their findings in the journal *Nature* (1953): "It has not escaped our notice that the specific pairing we have postulated immediately suggests a possible copying mechanism for the genetic material." Each strand of the double helix is complementary to the other—that is, at each point, the base on one strand is complementary with the base on the other strand. If the spirals unwound, each strand could serve as a guide for the synthesis of a new one; thus each single strand of the parent double helix could produce a new double-stranded molecule identical to the original. All the information needed to construct new DNA molecules is present in a single strand of DNA, because that information is the sequence of bases. The double-helical structure of DNA also suggested an explanation for some mutations: They might simply be changes in the linear sequence of nucleotide pairs.

In sum, *gene function, gene replication, and gene mutation could all be accounted for by the double-helical structure of DNA.*

REPLICATION OF THE DNA MOLECULE

Just three years after Watson and Crick published their paper in *Nature*, their prediction that the DNA molecule contains all the information needed for its own replication was proved by the work of Arthur Kornberg at Washington University in St. Louis. Kornberg showed that DNA can replicate in a test tube with no cells present. All that is required is a mixture containing DNA, a specific enzyme (which he called DNA polymerase), and a mixture of the four precursors: the nucleoside triphosphates deoxy-ATP, deoxy-CTP, deoxy-GTP, and deoxy-TTP (Figure 11.6). If any one of the four nucleoside triphosphates is omitted from the reaction mixture, DNA does not replicate itself. The intact DNA serves as a template for the reaction—a guide to the exact placement of nucleotides in the new strand. Where there is a T in the template, there must be an A in the new strand, and so forth.

Two other models of how the double helix might replicate were suggested after the original paper by Watson and Crick. In the model of conservative replication (Figure 11.7a), the original double helix would serve somehow as a template but would either

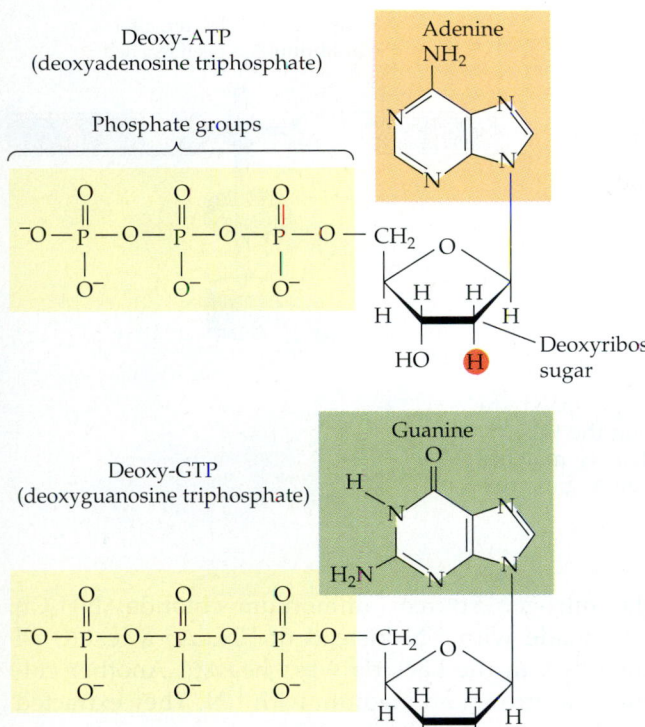

11.6 Building Blocks of DNA
The four deoxyribonucleoside triphosphates that form DNA differ only in their nitrogenous bases.

11.7 Three Models for DNA Replication All Obeyed Base Pairing Rules

In each model, the stretches of original DNA are blue and newly synthesized DNA is orange. (a) Conservative replication would preserve the original molecule and generate an entirely new molecule. (b) Dispersive replication would produce two molecules with old and new DNA interspersed along each strand. (c) Semiconservative replication would also produce molecules with both old and new DNA, but each molecule would contain one old strand and one new one.

be reconstituted or, perhaps, would never unwind at all. Thus the new molecule would contain none of the atoms of the original. According to the model of dispersive replication (Figure 11.7b), fragments of the original molecule would serve as templates, assembling two molecules, each containing old and new parts, perhaps at random. In **semiconservative replication** (the model proposed by Watson and Crick; Figure 11.7c), the original two strands would separate, and each would function as the template for a new partner. Each molecule produced by semicon-

Original DNA
double helix

DNA molecules
after one
round of
replication

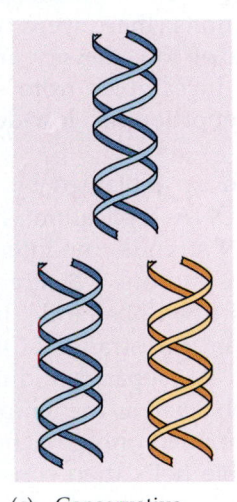

(a) Conservative
replication

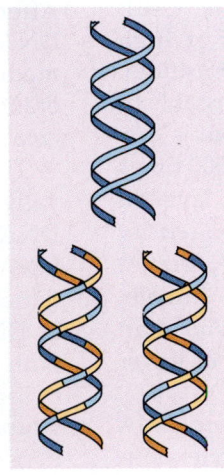

(b) Dispersive
replication

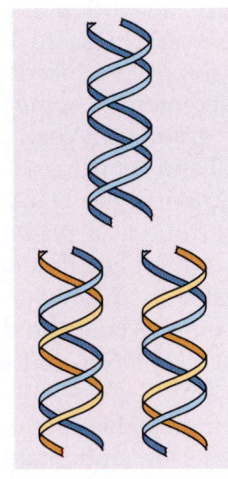

(c) Semiconservative
replication

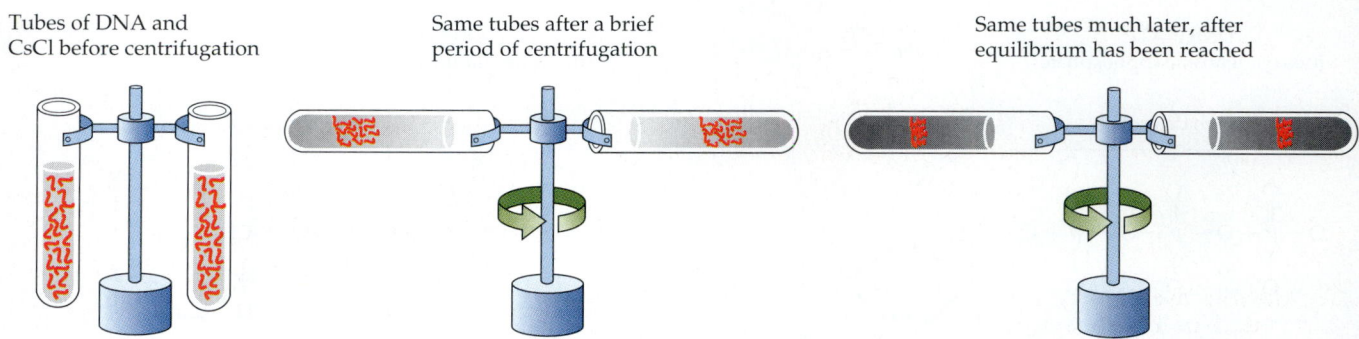

Tubes of DNA and CsCl before centrifugation

Same tubes after a brief period of centrifugation

Same tubes much later, after equilibrium has been reached

11.8   Density Gradient Centrifugation
When a solution of cesium chloride is centrifuged at extremely high speed, the cesium ions tend to sink slightly, forming a density gradient along the tube. Another substance in the tube will float at the point where its density matches that of the gradient. In this illustration, the red DNA molecules aggregate in a single band.

servative replication would therefore consist of one old and one new strand. After a short time, experimental work confirmed the model of semiconservative replaction.

Demonstration of Semiconservative Replication

A clever experiment by Matthew Meselson and Franklin Stahl convinced the scientific community that semiconservative replication is the correct model. Working at the California Institute of Technology in 1957, they devised a simple way to distinguish old strands of DNA from new ones. The key was to use a "heavy" isotope of nitrogen. Heavy nitrogen (^{15}N) is a rare, nonradioactive isotope that makes molecules more dense than chemically identical molecules containing the common isotope ^{14}N. To study DNA of different densities (that is, DNA containing ^{15}N versus DNA containing ^{14}N), Meselson, Stahl, and Jerry Vinograd invented a new centrifugation procedure that allowed them to determine the density of DNA from a specific sample. At a certain molarity, a solution of cesium chloride (CsCl) has a density very close to that of DNA; at high gravitational forces produced in an ultracentrifuge, cesium ions sediment to some extent, thus establishing a density gradient. When a DNA sample is dissolved in CsCl and centrifuged at about 100,000 times the force of gravity, the DNA gathers in a layer in the centrifuge tube at a position where the density of the CsCl solution equals that of the DNA (Figure 11.8). To cluster in this way, DNA that is initially lower in the tube, where the density is greater than its own, must rise; DNA that is in a region of lower density must sink.

After developing this method of measuring DNA densities, Meselson and Stahl could begin experimenting. They grew a culture of the bacterium *Escherichia coli* for 17 generations on a medium in which the nitrogen source (ammonium chloride, NH$_4$Cl) was made with ^{15}N instead of ^{14}N. As a result, all the DNA in the bacteria was "heavy." Another culture was grown on medium with ^{14}N. They extracted DNA from both cultures. When these extracts were combined and centrifuged with CsCl, two separate DNA bands formed, showing that this method would work for separating DNA samples of slightly different densities.

Meselson and Stahl then conducted their main experiment. They grew another culture on ^{15}N medium and *transferred* it to normal ^{14}N medium. *E. coli* reproduces every 20 minutes, and Meselson and Stahl collected some of the bacteria from each generation after the transfer. They extracted DNA from the samples. After DNA was duplicated and the cells divided to produce each new generation, the DNA banding in the density gradient was different from the original banding. Initially, the DNA was uniformly labeled with ^{15}N and hence was relatively dense. After one generation, when the DNA had been duplicated once, all the DNA was of an intermediate density. After two generations, there were two equally large DNA bands: one of low density and one of intermediate density. In samples from subsequent generations, the proportion of low-density DNA increased steadily.

These data can be explained by the semiconservative model of DNA replication. The high-density DNA had two ^{15}N strands, the intermediate-density DNA had one ^{15}N and one ^{14}N strand, and the low-density DNA had two ^{14}N strands. In the first round of DNA replication, the strands of the double helix, both heavy with ^{15}N, separated; during the process of separation, each acted as the template for a second strand, which contained only ^{14}N and hence was less dense. Each double helix then consisted of one ^{15}N and one ^{14}N strand and was of intermediate density. In the second replication, the ^{14}N-containing strands

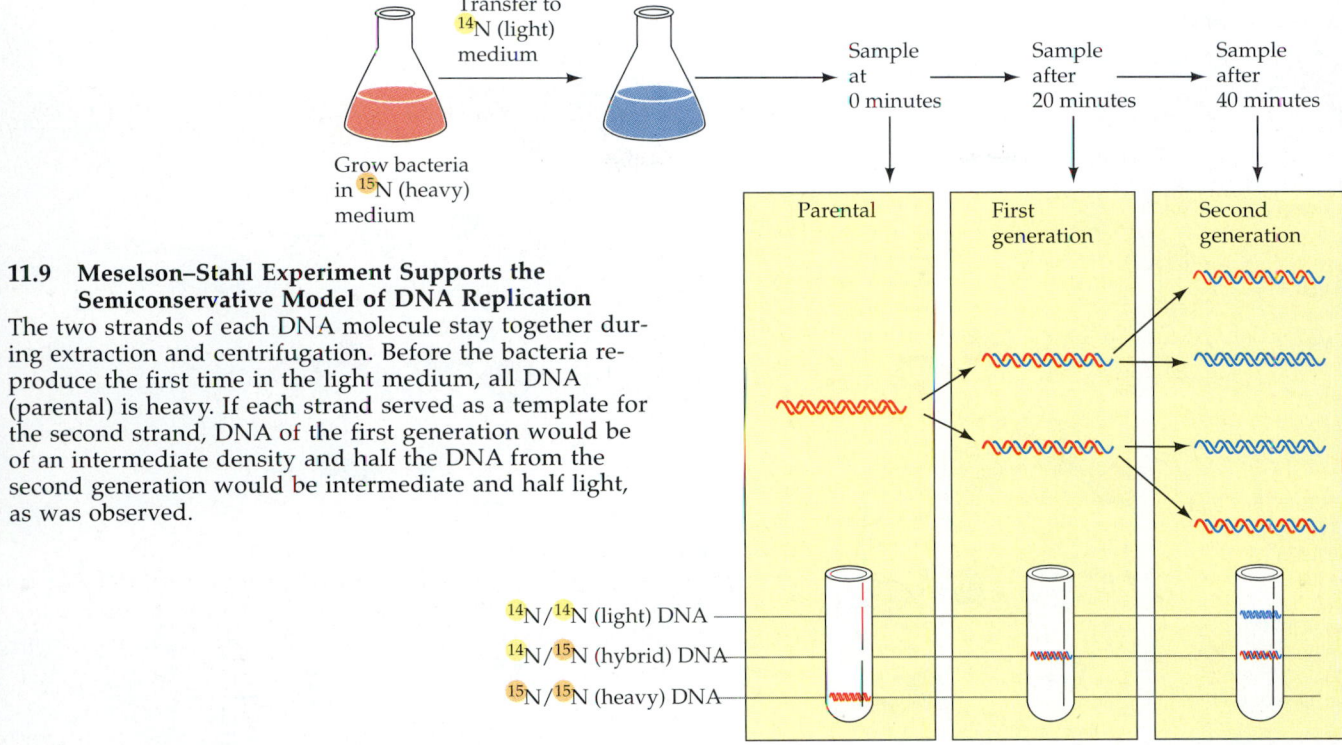

11.9 Meselson–Stahl Experiment Supports the Semiconservative Model of DNA Replication
The two strands of each DNA molecule stay together during extraction and centrifugation. Before the bacteria reproduce the first time in the light medium, all DNA (parental) is heavy. If each strand served as a template for the second strand, DNA of the first generation would be of an intermediate density and half the DNA from the second generation would be intermediate and half light, as was observed.

^{14}N/^{14}N (light) DNA
^{14}N/^{15}N (hybrid) DNA
^{15}N/^{15}N (heavy) DNA

directed the synthesis of partners with ^{14}N, creating low-density DNA, and the ^{15}N strands got new ^{14}N partners (Figure 11.9).

If the DNA had replicated in accord with either of the other models, the results would have been quite different. Under the conservative model, after one generation there would have been two bands—one for heavy DNA (^{15}N–^{15}N) and the other for light (^{14}N–^{14}N); no DNA of intermediate density would have formed at any time. If dispersive replication had taken place, the first round of replication would have produced DNA of intermediate density, but the density of all the DNA would have decreased with each subsequent replication. The crucial observation proving the semiconservative model was that intermediate-density DNA (^{15}N–^{14}N) appeared in the first generation and continued to appear in subsequent generations.

Replicating an Antiparallel Double Helix

How is semiconservative DNA replication accomplished? That it depends upon enzymes should not surprise you. We now know that there are different types of DNA polymerases, with different functions, and that the replication process is intricate. Kornberg's basic observations, including the need for a DNA template and for a mixture of nucleoside triphosphates, still hold. He also showed that nucleotides are always added to the growing chain at the same end: the 3′ end, the end at which the DNA

strand has a free hydroxyl group on the 3′ carbon of its terminal deoxyribose (Figure 11.10; see also Figure 11.4). This hydroxyl group reacts with a phosphate group on the 5′ carbon of the deoxyribose of a deoxyribonucleoside triphosphate (see Figure 11.6), and thus the chain grows. Bonds linking the phosphate groups of the deoxyribonucleoside triphosphate break and thereby release the energy for this reaction. Two of the phosphate groups diffuse away, and one becomes part of the sugar–phosphate backbone of the growing DNA molecule.

For double-stranded DNA to replicate, the strands must unwind and separate from each other. Only then can they function as templates for the synthesis of new, complementary strands. The unwinding results in a **replication fork**—a moving, Y-shaped structure that is the region where new DNA strands are being synthesized. Recall that the two original strands are antiparallel, with the 3′ end of one strand paired with the 5′ end of the other. As the replication fork moves along the parent DNA molecule, an enzyme, DNA polymerase III, catalyzes the replication of both strands. How can this be accomplished, given that new nucleotides are added only at the 3′ end of a polynucleotide chain?

One parent strand is being exposed beginning at its 3′ end, which presents no problem—its complementary strand is synthesized continuously as the replication fork proceeds. This daughter strand is called the **leading strand**. The other daughter strand, the **lagging strand**, is produced in discontinuous

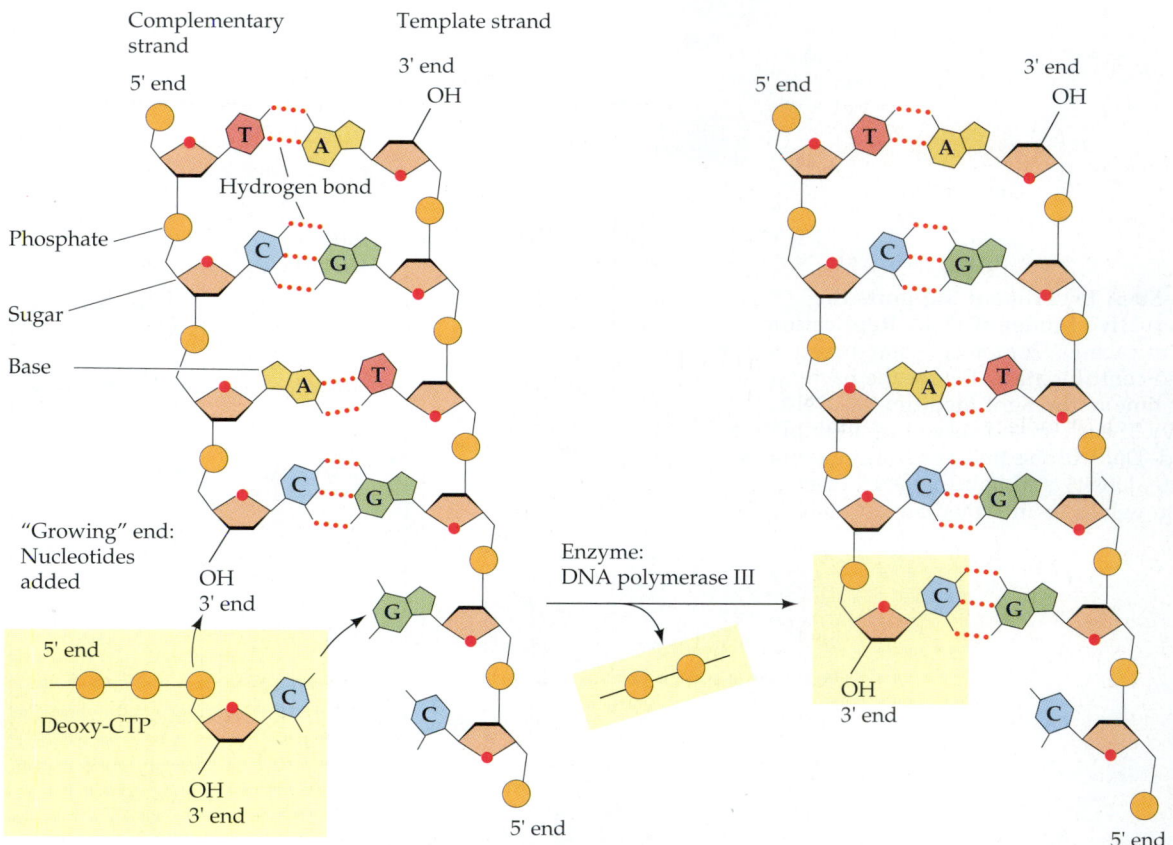

11.10 A Replicating DNA Strand Grows from 5' to 3'
A DNA strand, with its 3' end at the top and its 5' end at the bottom, is the template for the synthesis of the complementary strand at the far left. The new strand has its 5' end at the top and its 3' end at the bottom—that is, it is antiparallel to the template strand. DNA polymerase III adds the next deoxyribonucleotide, with the base C, at the free —OH group at the 3' end of the growing chain. At the right are the same template and the growing strand, now one base longer.

spurts (100 to 200 nucleotides at a time in eukaryotes; 1,000 to 2,000 at a time in prokaryotes). The discontinuous stretches are synthesized just as the leading strand is, by adding the 5' end of a nucleotide to the 3' end of the daughter strand, but the stretches are synthesized in the opposite direction with respect to the replication fork. These stretches of new DNA for the lagging strand are called **Okazaki fragments** after their discoverer, the Japanese biochemist Reiji Okazaki. While the leading strand grows continuously "forward," the lagging strand grows in shorter, "backward" stretches with gaps between them (Figure 11.11). The gaps between the Okazaki fragments are then filled in by DNA polymerase I, and another enzyme, **DNA ligase**, links the fragments.

Working together, two DNA polymerases, DNA ligase, and several other proteins (Box 11.A) do the complex job of DNA synthesis with a speed and accuracy that are almost unimaginable. In *E. coli*, the complex makes new DNA at a rate in excess of 1,000 base pairs per second and makes mistakes in fewer than one base in 10^8–10^{12}.

On a bacterial chromosome, DNA synthesis begins at just one point, the **origin of replication**. Each chromosome of a eukaryote, on the other hand, has many origins of replication. DNA synthesis may thus proceed simultaneously in many areas of a single eukaryotic chromosome. Synthesis proceeds in both directions from an origin of replication as two replication forks move away from it (Figure 11.12).

PROOFREADING AND DNA REPAIR

DNA must be faithfully replicated and maintained. The price of failure is great—it may even be death.

BOX 11.A

Collaboration of Proteins at the Replication Fork

The replication of a DNA molecule is an amazingly complex process. We have already described the general problem of synthesizing both a leading and a lagging strand at the same time—but this is only the most obvious difficulty. DNA replication includes all the following steps: (1) unwinding the two parent strands, (2) providing a primer for the synthesis of a new strand, (3) elongating each of the daughter strands, (4) filling in the gaps between the Okazaki fragments of the lagging strand, (5) connecting the completed Okazaki fragments, and (6) editing the newly synthesized strands for accuracy of replication. Each of these steps requires one or more specific proteins, many of which are enzymes.

The unwinding of the double helix is mediated by two related enzymes called **helicases**, one of which attaches to each of the parental DNA strands. Energy to separate the strands comes from the hydrolysis of ATP. The separated strands would tend to interact with each other because of their complementarity, but this is prevented by the attachment of **single-stranded DNA-binding proteins** to each of the separated strands. These binding proteins hold the single strands of DNA in a configuration that binds readily to DNA polymerase III. A pair of DNA polymerase III molecules at the replication fork catalyzes elongation of the leading and lagging strands. The discontinuous production of the lagging strand, however, results in a repeated need for a new primer to start the synthesis of the next Okazaki fragment. The primer is a short single strand of RNA rather than of DNA; it is formed, complementary to the template DNA strand, by an enzyme called a **primase**, which is one of several polypeptides bound together in an aggregate called a **primosome**. DNA polymerase III ex-

tends the primer. DNA polymerase I later replaces the RNA primer segments with DNA segments. Finally, each newly completed Okazaki fragment is linked to the completed portion of the lagging strand in a reaction catalyzed by DNA ligase.

Besides catalyzing elongation of the leading and lagging strands, DNA polymerase III plays another crucial role in DNA replication: It checks the accuracy of its own work. After adding a monomer to a strand, DNA polymerase III tests the new base pair to see that it is complementary to the nucleotide in the template strand. If an incorrect nucleotide has been inserted, the DNA polymerase excises the erroneously selected nucleotide and tries again. As a result, DNA replication is startlingly faithful, with an error rate on the order of one wrong nucleotide per billion—even though the error rate before the proofreading process is on the order of one wrong nucleotide per 10,000.

Even this description of DNA replication is simplified. In *E. coli*, for example, more than 30 polypeptides participate. The largest of the proteins, DNA polymerase III, has a molecular weight of 760,000 and consists of seven or eight polypeptide subunits.

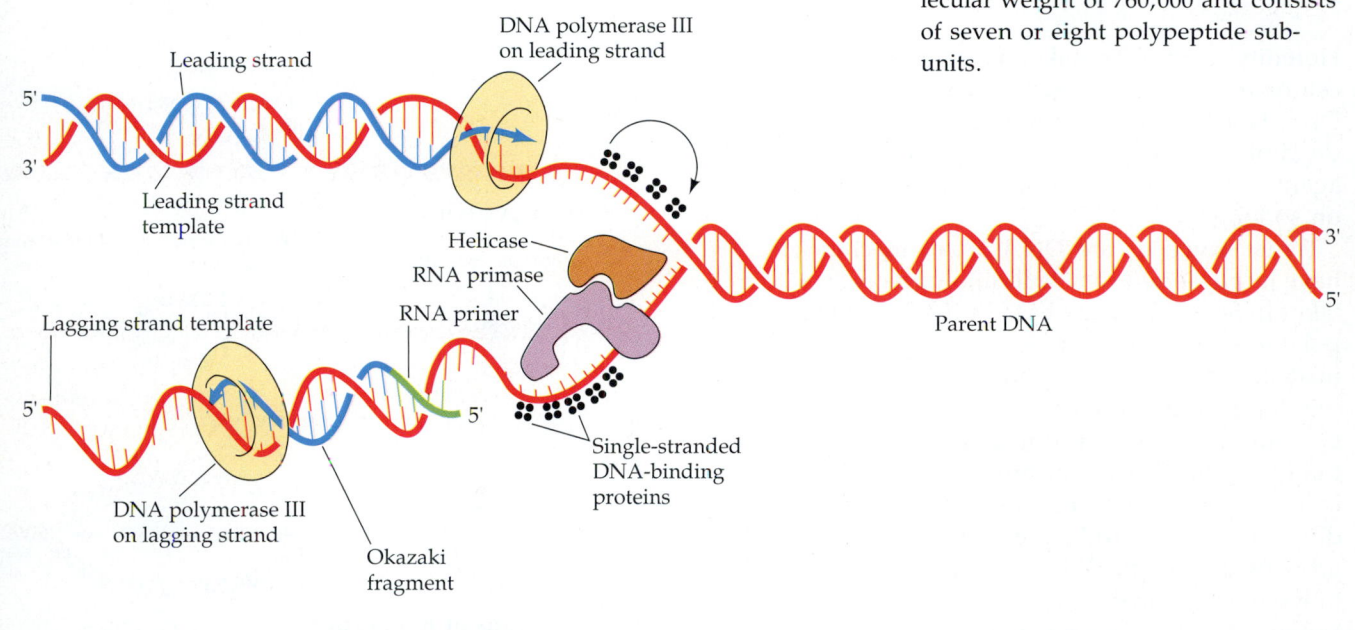

Leading strand

DNA polymerase III on leading strand

5'

3'

Leading strand template

Lagging strand template

DNA polymerase III on lagging strand

Helicase

RNA primase

RNA primer

5'

5'

Okazaki fragment

Single-stranded DNA-binding proteins

Parent DNA

3'

5'

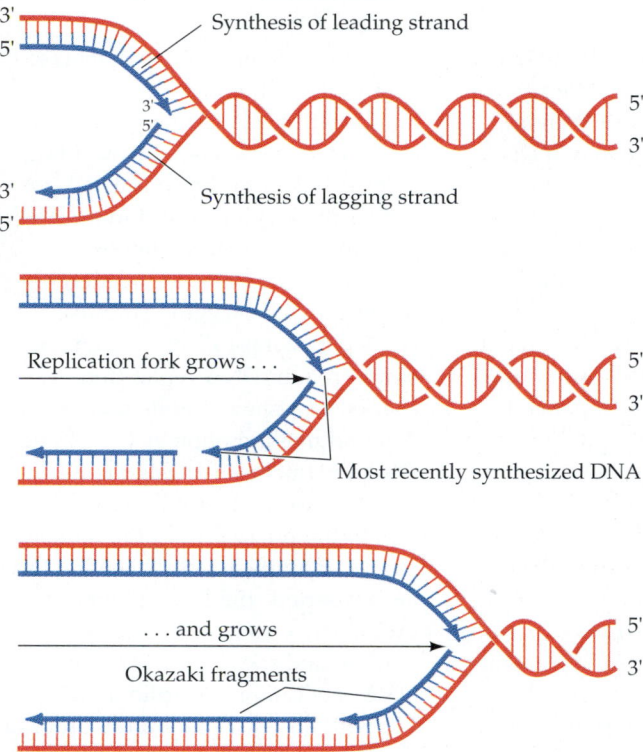

11.11 DNA Is Synthesized at a Replication Fork
As the original DNA strands unwind and separate, both daughter strands are synthesized in the 5'-to-3' direction, although their template strands are antiparallel. The leading strand is synthesized continuously, but the lagging strand is synthesized in Okazaki fragments that are joined later. Abbreviated here, eukaryotic Okazaki fragments are hundreds and prokaryotic fragments thousands of nucleotides long.

Heredity itself is at stake, as is the functioning of a cell or multicellular organism. Yet the replication of DNA is *not* perfectly accurate, and the DNA of non-dividing cells is subject to damage by environmental agents. In the face of these threats, why has life gone on so long?

What saved us are DNA repair mechanisms. We have just asserted that replication proceeds with mistakes in fewer than one base in 10^8–10^{12}. In fact, DNA polymerases initially make a significant number of mistakes in assembling polynucleotide strands; in *E. coli*, for example, the error rate would result in flaws in approximately one out of every ten genes each time the cell divides. In humans, about 1,000 genes in every cell would be affected each time the cell divided. After introducing a new nucleotide into a growing polynucleotide strand, however, the DNA polymerases carry out a "proofreading" function. When a DNA polymerase "notices" a mispairing of bases, it removes the improperly introduced nucleotide and tries again. Odds are that the polymerase will be successful in inserting the correct monomer

the second time because the error rate is about 1 in 10^4 base pairs. This repair mechanism greatly lowers the overall error rate for replication.

What if a DNA molecule becomes damaged during the often long life of a cell? Some cells live and play important roles for many years, even though their DNA is constantly at risk from such hazards as high-energy radiation, chemicals that induce mutations, and random spontaneous chemical reactions. Cells owe their lives to the many DNA repair mechanisms that have evolved. An important class of repair mechanisms that includes the proofreading function just described is called excision repair. Certain enzymes "inspect" the cell's DNA. When they find mispaired bases, chemically modified bases, or spots in which one strand has more bases than the other (with the result that one or more bases of one strand form an unpaired loop), these enzymes make a cut in the defective strand. Another enzyme cuts away the bases adjacent to and including the offending base, and DNA polymerase and DNA ligase synthesize a new, usually correct piece to replace the excised one. Our dependence on such repair mechanisms is underscored by our susceptibility to a number of diseases that arise from DNA-repair defects. One example is the skin disease xeroderma pigmentosum. People with this disease lack a mechanism that normally repairs damage caused by the ultraviolet radiation in sunlight. Without this mechanism, a person exposed to sunlight invariably develops skin cancer.

Normal, undamaged DNA is vital to life. But what exactly is DNA for? What do genes do? What do they control?

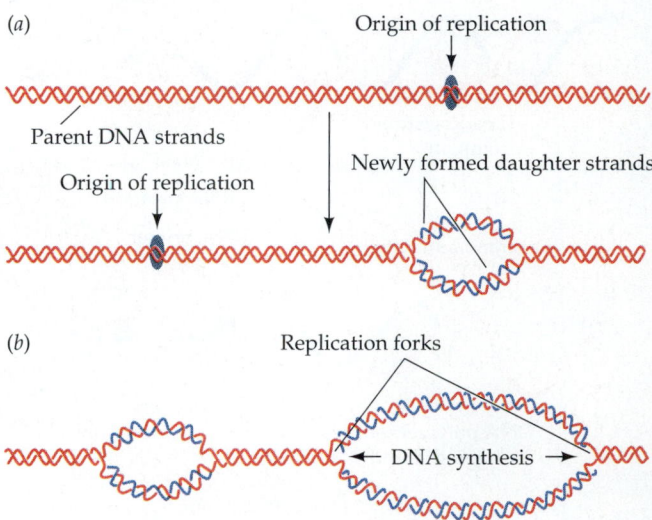

11.12 Origins of Replication
DNA synthesis can begin at many origins in a eukaryotic chromosome. (*a*) The parent strands separate at an origin of replication. (*b*) Synthesis proceeds in both directions from an origin of replication.

WHAT DO GENES CONTROL?

There are many steps between genotype and phenotype. Genes cannot, all by themselves, directly produce a phenotypic result such as the color of an eye or a flower, the shape of a seed, or a cleft chin, any more than a thermostat without a furnace can heat a house. What are the steps between the genes on the chromosomes and the phenotype of the organism? The first hints came early in this century from the work of the English physician Archibald Garrod. Alkaptonuria is a hereditary disease in which the patient's urine turns black when exposed to air. Garrod recognized that this symptom showed that the biochemistry of the affected individual was different from that of other persons. He suggested in 1908 that alkaptonuria and some other hereditary diseases are consequences of "inborn errors of metabolism." He proposed that what makes the urine dark is a defect in an enzyme. Garrod studied the pattern of inheritance of alkaptonuria and reasoned that it affects individuals who are homozygous for the recessive allele of a particular gene (see Chapter 10), which in normal individuals codes for an enzyme needed for metabolism of the amino acid tyrosine. His proposals were the first plausible approach for explaining how genes are expressed. However, like Mendel's explanation of inheritance in the garden pea, Garrod's proposals were too advanced for their time and sat almost unnoticed for over 30 years.

A series of experiments performed by George W. Beadle and Edward L. Tatum at the California Institute of Technology in the 1940s confirmed and extended Garrod's ideas. Beadle and Tatum experimented with the bread mold *Neurospora crassa*. *Neurospora* can be grown on a simple, completely defined medium (that is, one in which all the ingredients are known) containing inorganic ions, a simple source of nitrogen (such as ammonium chloride), an organic source of energy and carbon (such as glucose), and a single vitamin (biotin; see Chapter 43). From this minimal medium, the enzymes of wild-type *Neurospora* can catalyze the metabolic reactions needed to make all the chemical constituents of its cells. Beadle and Tatum reasoned that mutations might alter the enzymes so that they could no longer do their jobs. In that case, mutants of *Neurospora* might be found that could not make certain compounds they needed; such mutants would grow only on media to which those compounds were added. Mutants of this type have since been named **auxotrophs** ("increased eaters"), in contrast to the wild-type **prototrophs** ("original eaters") that constituted the original *Neurospora* population. Prototrophs grow on minimal medium, whereas auxotrophs require specific additional nutrients.

Beadle and Tatum irradiated cells of *Neurospora* with X rays to increase the frequency of mutations and then isolated some nutritional mutants. These auxotrophs did not grow on the minimal medium that supported the growth of the wild-type strain, but they did grow on a complex medium enriched with amino acids (the monomers of proteins), purines and pyrimidines (the nitrogenous bases of nucleic acids), and vitamins. Beadle and Tatum tested these mutants to determine the simplest nutritional supplements that would support their growth (Figure 11.13). Among the collection of mutants were individual strains that required a specific amino acid or vitamin. In almost every case, the nutritional requirement was simple: Only a single compound had to be added to the minimal medium to support the growth of any given mutant. This result supported the idea that mutations have simple effects—and, perhaps, that each mutation causes a defect in only one enzyme in the metabolic pathway leading to the synthesis of the required nutrient.

The auxotrophs identified in this way could be divided into classes on the basis of the nutritional supplements that would support their growth. For example, all mutants that did not grow on minimal medium but grew on minimal medium supplemented with the amino acid arginine were classified as *arg* mutants. Other sets of mutants were found that required adenine, or proline, or vitamin B₁, and so forth. Within a group of mutants with the same nutritional requirement, mapping studies established that some of the individual mutations were at different loci on a chromosome or were on different chromosomes, indicating that different genes can govern a common biosynthetic pathway. For example, Beadle and Tatum found no fewer than 15 different *arg* mutants. These mutants were then grown in the presence of various suspected intermediates in the synthetic metabolic pathway for arginine. Different mutants were able to grow on different intermediates, as well as on arginine-supplemented medium (Figure 11.14).

Growing mutants on different intermediates helped to determine the biochemical steps by which various amino acids and other compounds are synthesized in *Neurospora*. Much more important, however, this work led to the formulation of the one-gene, one-enzyme theory. According to this theory, the function of a gene is to control the production of a single, specific enzyme. This proposal strongly influenced the subsequent development of the sciences of genetics and molecular biology. Garrod had pointed in the same direction more than three decades earlier, but only now were other scientists prepared to act on the suggestion.

The one-gene, one-enzyme hypothesis was re-

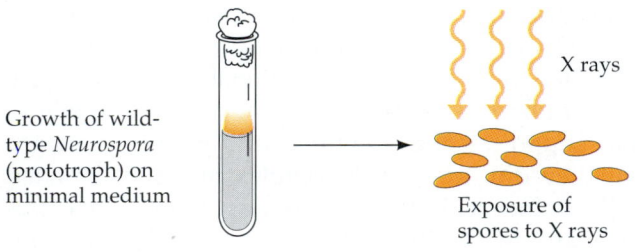

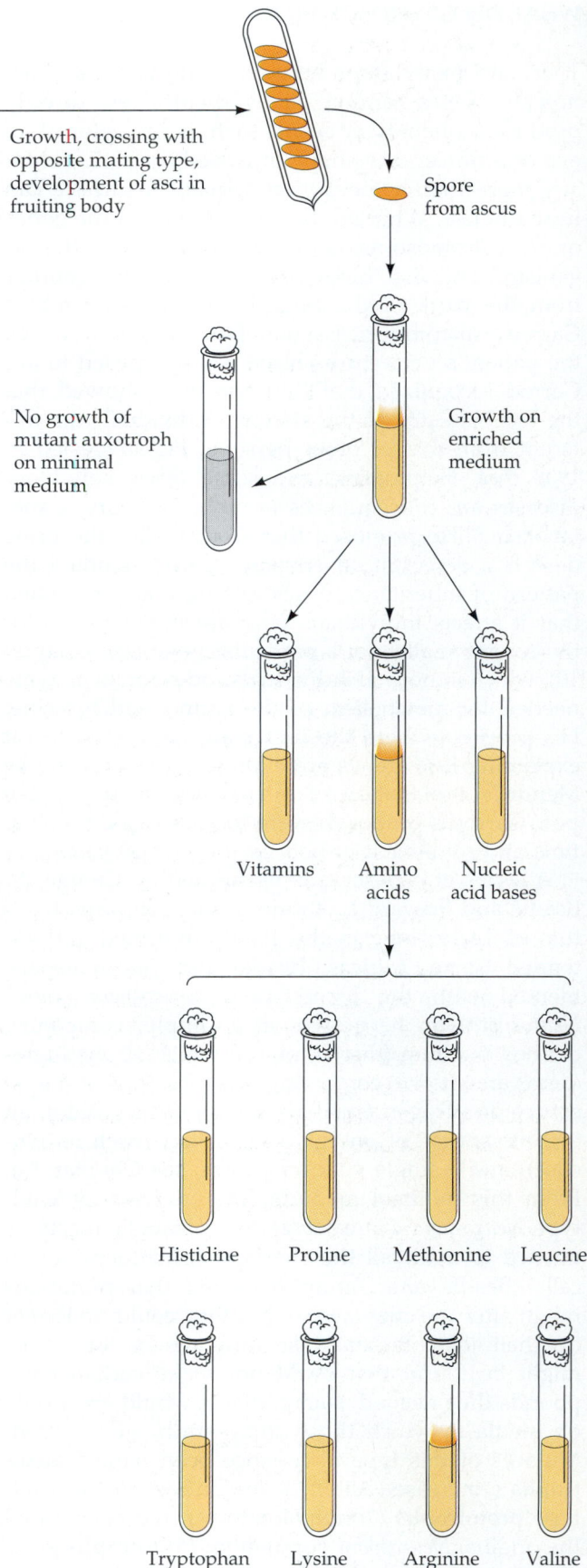

11.13 Identifying Nutritional Mutations of *Neurospora*
Offspring (spores) of cultures that have been irradiated in order to increase their mutation rate are individually placed on an enriched medium. Each spore forms a fungal mat that contains only one kind of haploid nucleus. Part of each mat is placed on minimal medium to test whether it is a mutant auxotroph. The auxotroph's nutritional requirement (in this instance, arginine) is determined by finding which substance supports its growth (indicated here by an orange "glow"). Because each spore is haploid, by the end of the procedure the phenotype reveals the genotype—an important reason for using *Neurospora*.

fined in the mid-1950s as a result of work on sickle-cell anemia. This serious disease is the consequence of a recessive allele that, when homozygous, results in defective red blood cells. Where oxygen is abundant, as in the lungs, the cells are normal in structure and function. But at the low oxygen levels characteristic of working muscles, the red blood cells collapse into the shape of a sickle (Figure 11.15). Linus Pauling, at the California Institute of Technology, speculated that the disease results from a defect in hemoglobin, a protein that fills red blood cells and carries oxygen. Human hemoglobin is a tetramer, consisting of two each of two different polypeptide chains. After Pauling's suggestion, it was shown that one of the two kinds of polypeptides differs by one amino acid between normal and sickle-cell hemoglobin. This result suggested the more satisfying one-gene, one-polypeptide theory: The function of a gene is to control the production of a single, specific polypeptide. Much later, it was discovered that some genes code for forms of RNA that do not get translated into polypeptides.

FROM DNA TO PROTEIN

How does DNA function? We have learned about its structure and how it replicates. Next we need to know what it does. Beadle and Tatum demonstrated that genes are responsible for the production of proteins. But how does a gene specify the synthesis of a protein?

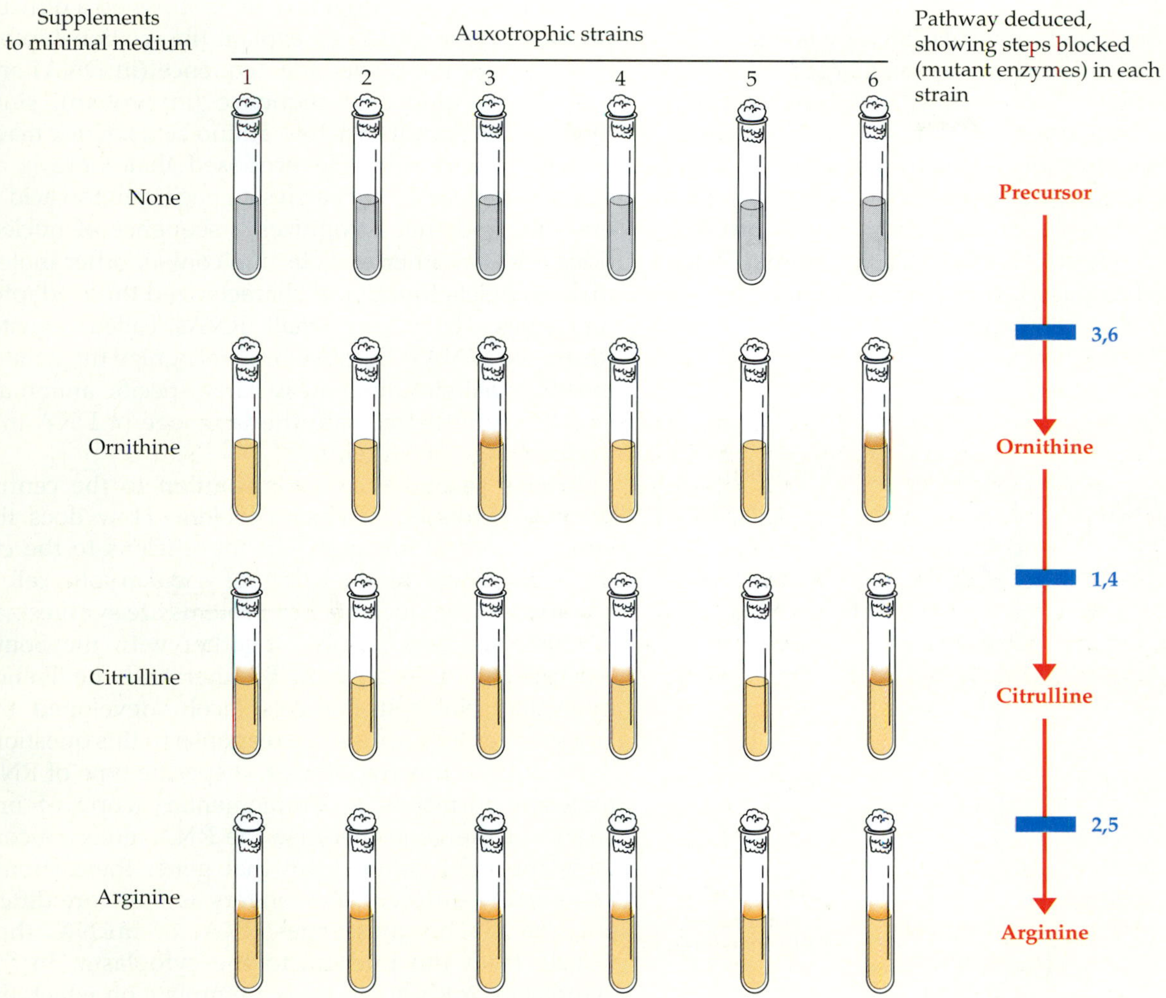

Supplements to minimal medium	Auxotrophic strains						Pathway deduced, showing steps blocked (mutant enzymes) in each strain
	1	2	3	4	5	6	

None — Precursor

3,6

Ornithine — Ornithine

1,4

Citrulline — Citrulline

2,5

Arginine — Arginine

11.14 Dissecting a Biochemical Pathway

You have just isolated six arginine-requiring auxotrophic mutants of *Neurospora*. You know that ornithine and citrulline are chemically related to arginine, and you think that they may be intermediates in its synthesis. Let's use these six mutants to deduce the biosynthetic pathway to arginine. All six strains grow when supplied with arginine (fourth row of cultures) but not on unsupplemented minimal media (top row). Strains 2 and 5 grow only on media supplemented with arginine. What does this suggest? Because these strains cannot grow on either ornithine or citrulline, their mutations must interfere with the synthesis of arginine itself—the final step. Strains 1 and 4 grow when supplied with citrulline but not with ornithine. What does this suggest? Their mutations must block the synthesis of citrulline but permit the conversion of citrulline to arginine. Thus part of the pathway must be citrulline → arginine. Note that strains 1 and 4 do not grow when supplemented with ornithine, suggesting that if ornithine is an intermediate, then it must occur before this genetic block. What can you infer from the behavior of strains 3 and 6? We leave that for you to decide. The simplest pathway that is consistent with all these observations is shown at the right.

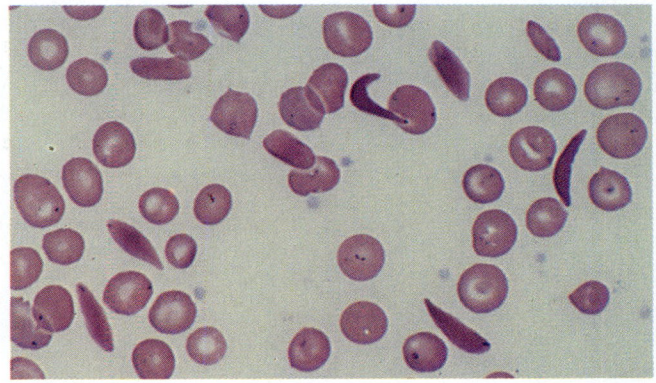

11.15 Sickled Blood Cells

Most of these human red blood cells are normal: They are flattened and roughly circular, with concave centers. Some of the cells, recognizable by their shape, are sickled. This change in shape results from a single amino acid substitution in one of the polypeptides of the protein hemoglobin. The sickled cells are fragile and are eliminated in the spleen, with anemia the result.

The Central Dogma of Molecular Biology

The **central dogma** of molecular biology is one of the most important concepts to have emerged in the attempt to explain how genes make polypeptide chains. The central dogma is, simply, that DNA codes for the production of RNA (transcription), RNA codes for the production of protein (translation), and protein does *not* code for the production of protein, RNA, or DNA (Figure 11.16). In Crick's words, "once 'information' has passed into protein *it cannot get out again*."

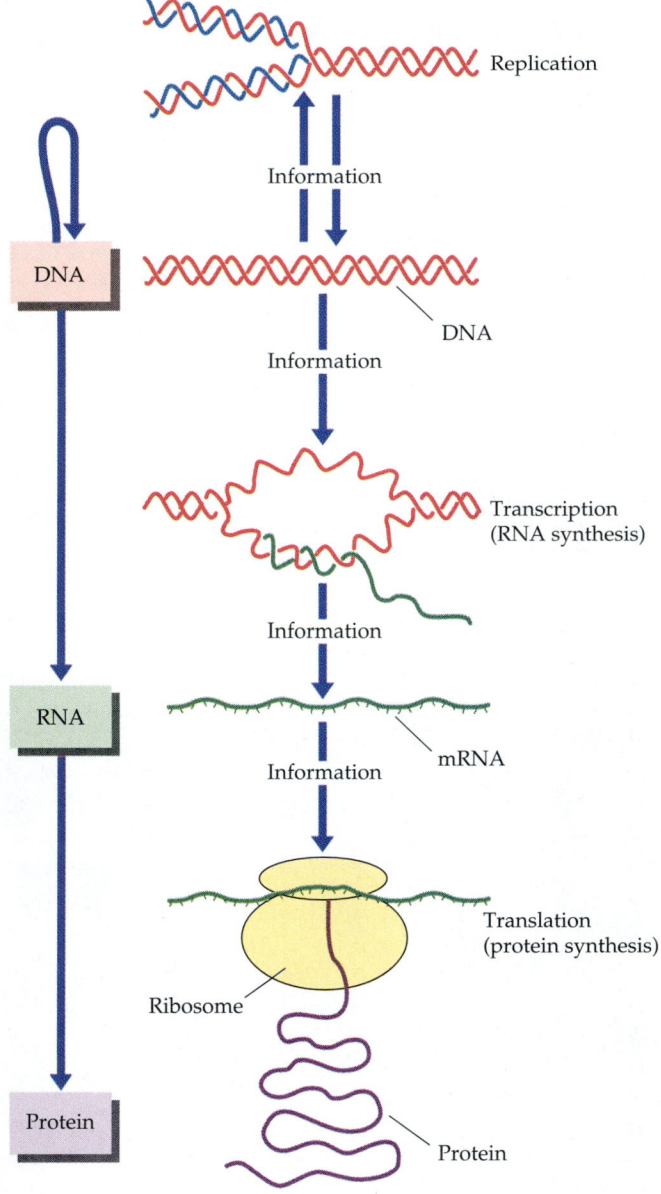

11.16 The Central Dogma
Information coded in the sequence of base pairs in DNA is replicated in new molecules of DNA and passed to molecules of RNA. Information in RNA is passed to proteins.

Crick contributed two key ideas to the development of the central dogma. The first solved a difficult problem: How could one explain the relationship between a specific nucleotide sequence (in DNA) and a specific amino acid sequence (in protein), since nucleotides do not attach to amino acids. Crick made a clever suggestion: He proposed that there is an adaptor molecule that carries a specific amino acid at one end and that recognizes a sequence of nucleotides with its other end. In due course, other molecular biologists found and characterized these adaptor molecules. They are small RNAs called transfer RNAs, or **tRNAs**. Because they recognize the genetic message *and* simultaneously carry specific amino acids, tRNAs can translate the language of DNA into the language of proteins.

Crick's second major contribution to the central dogma addressed another problem: How does the genetic information get from the nucleus to the cytoplasm? (Most of the DNA of a eukaryotic cell is confined to the nucleus, but proteins are synthesized in the cytoplasm.) Crick, together with the South African geneticist Sydney Brenner and the French molecular biologist François Jacob, developed the "messenger hypothesis" in response to this question. According to this hypothesis, a specific type of RNA molecule forms as a complementary copy of one strand of the gene. Because the RNA molecule contains the information from that gene, there should be as many different messengers as there are different genes. This messenger RNA, or **mRNA**, then travels from the nucleus to the cytoplasm. In the cytoplasm, mRNA serves as a template on which the tRNA adaptors line up to bring amino acids in the proper order into a growing polypeptide chain in the process called **translation** (see Figure 11.16).

Summarizing the main features of the central dogma, the messenger hypothesis, and the adaptor hypothesis, we may say that *a given gene is transcribed to produce a messenger RNA complementary to one of the DNA strands*, and that *transfer RNA molecules translate the sequence of bases in the mRNA into the appropriate sequence of amino acids*.

RNA Viruses and the Central Dogma

According to the central dogma of molecular biology, DNA codes for RNA and RNA codes for protein. All cellular organisms have DNA as their hereditary material. Only among viruses (which are not cellular) is a variation on the central dogma found. Many viruses, such as the tobacco mosaic virus, have RNA rather than DNA as their nucleic acid. Heinz Fraenkel-Conrat of the University of California at Berkeley separated the protein and RNA fractions of the tobacco mosaic virus and then recombined them to obtain active virus particles. When he took RNA from one mutant strain of this virus and combined it with protein from another, the resulting viruses replicated to produce more virus particles like the first (the RNA-donating) strain. Thus he showed that RNA is the genetic material of the tobacco mosaic virus. RNA itself is the template for the synthesis of the next generation of viral RNA and viral proteins. In this virus, DNA is left out of the information flow.

Rous sarcoma virus is an RNA virus that causes a cancer in chickens. The virus enters a chicken cell and subsequently causes the cell to make a DNA "transcript" of the viral RNA, the reverse of the usual process. The afflicted cell does not burst, but it changes permanently in shape, metabolism, and growth habit. The new DNA becomes part of the hereditary apparatus of the infected chicken cell. In 1964, Howard Temin of the University of Wisconsin discovered that the virus carries an enzyme for the manufacture of DNA, using viral RNA as the informational template. The enzyme was named **reverse transcriptase** because it transcribes DNA from RNA rather than RNA from DNA. Viruses that employ reverse transcriptase are known as **retroviruses**; one of the most studied retroviruses is the human immunodeficiency virus (HIV), which causes AIDS (see Chapter 16). The central dogma requires slight modification to account for the flow of information in retroviruses and their hosts. However, it is still true that information does not flow from protein back to the nucleic acids.

Transcription

The formation of a specific RNA under the control of a specific DNA is called **transcription**. Transcription requires the enzyme **RNA polymerase**, the appropriate ribonucleoside triphosphates (ATP, GTP, CTP, and UTP), and the DNA template. Only *one* of the DNA strands—the **template strand**—is transcribed. The other, complementary DNA strand remains untranscribed. The DNA molecule must partially unwind, as in DNA replication, to expose the bases on the template strand that will be transcribed.

RNA polymerase catalyzes the continuous transcription of DNA in only one direction. This fact was shown in an ingenious experiment using an artificial nucleoside triphosphate (cordycepin triphosphate) that has an unusual nucleoside portion (3'-deoxy-adenosine, or cordycepin; Figure 11.17). Cordycepin triphosphate lacks a hydroxyl group at the 3' position of its sugar; thus only its 5' end (and not its 3' end) can be attached in an RNA strand. The nucleotides in a strand of RNA, as in DNA, are covalently bonded such that the 3' end of one attaches to the 5' end of the next, and so on. If mRNA grew by adding nucleotides to its 5' end, cordycepin triphosphate could not form a bond with it. Thus, the presence of this

Cordycepin triphosphate

11.17 Cordycepin Triphosphate
This molecule is very similar to adenosine triphosphate (ATP; see Figure 7.2) but it lacks an —OH group at its 3' carbon end (blue oval). The incorporation of cordycepin triphosphate into mRNA blocks elongation of the strand because it has no attachment site for the next nucleotide.

compound among the available nucleoside triphosphates would have no effect on transcription. When added to a transcription reaction mixture, however, cordycepin triphosphate actually *does* strongly inhibit mRNA formation. Since this result makes sense only if the 3' position of the RNA molecule is the growing point, this experiment clearly shows that mRNA grows from the 5' end to the 3' end.

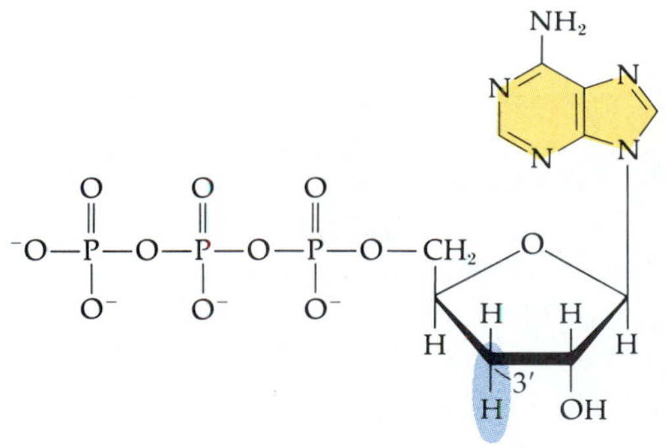

Formation of mRNA is inhibited because RNA polymerase mistakes cordycepin triphosphate for normal adenosine triphosphate (ATP), and attaches the 5' carbon of cordycepin triphosphate to the 3' end of the growing RNA chain. After cordycepin triphosphate is attached, the RNA chain cannot react with another nucleoside triphosphate because cordycepin triphosphate has no hydroxyl group on its 3' end. Once cordycepin triphosphate has joined a chain, then, no other nucleotides can be added, mRNA elongation ceases, and only short strands of mRNA with terminal cordycepin molecules can be recovered.

When DNA polymerase catalyzes the replication of DNA, the two strands of the parent molecule are unwound, and each strand becomes paired with a new strand. In transcription, DNA is unwound, but it must then be *rewound*. The DNA strand that is to be transcribed must be partly separated from its partner so that it may serve as a template for mRNA synthesis, but the RNA transcript peels away as it is formed, allowing the DNA that has already been transcribed to be rewound (Figure 11.18).

Transcription of genes begins at regions called **initiation sites** that tell the RNA polymerase where to attach and which strand to copy. Similarly, particular base sequences in the DNA specify the termination of transcription. The transcription of DNA is under precise control: Particular genes are transcribed in

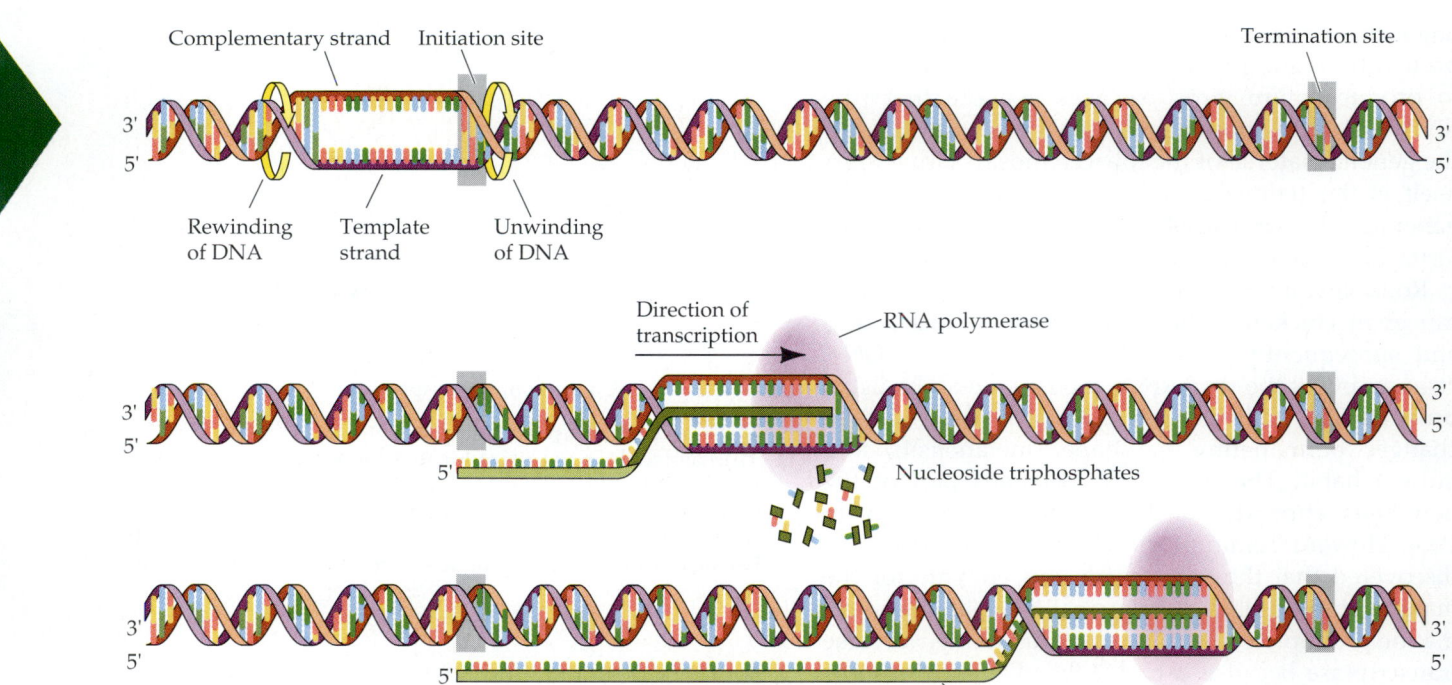

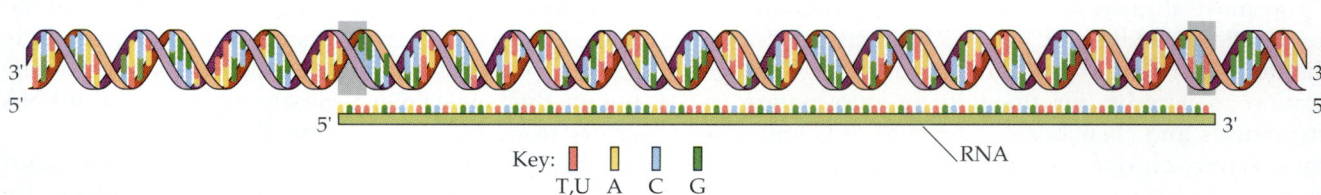

Key: T,U A C G

11.18   DNA Is Transcribed into RNA
The DNA double helix unwinds to give RNA polymerase, moving in the 5′-to-3′ direction along the template strand, access to the nucleotide sequence. As the growing RNA transcript is released from the template, the two DNA strands rewind. RNA transcripts are made from only one strand of a DNA double helix. The base-pairing rules are similar to those for DNA: adenine with uracil and guanine with cytosine. The RNA polymerase, which is much larger than shown here, actually covers about 50 base pairs.

certain cells at specific times, whereas other genes are transcribed at other times. This intriguing regulatory process is discussed in Chapter 12.

It is not just mRNA that is produced by transcription. The same process is used in the synthesis of tRNA and of ribosomal RNA, or **rRNA**, which constitutes a major fraction of the ribosome. These other forms of ribonucleic acid are coded for by specific regions of the DNA. In prokaryotes, most of the DNA acts as a template for the production of mRNA, tRNA, or rRNA. The situation in eukaryotes is more complicated, as will be explained later in this chapter and in Chapter 13.

Transfer RNA

You can think of the genetic information transcribed in an mRNA molecule as a series of three-letter "words." Each sequence of three nucleotides (the "letters") along the chain specifies a particular amino acid. The three-letter "word" is called a **codon**. Let's see how a codon is related to the amino acid for which it codes. As predicted by Crick, the codon and the amino acid are related by way of an adaptor—a specific type of tRNA. For each of the 20 amino acids, there is at least one specific tRNA molecule.

A tRNA molecule is small, consisting of only about 75 to 80 nucleotides. Robert Holley of Cornell University was the first to work out the complete nucleotide sequence of a particular tRNA species. He noticed that several regions, apparently separate when the molecule was viewed stretched out, had complementary base sequences. It seemed that these regions could come together by folding and then could stabilize the fold with complementary base pairing. Some years later it became possible to deter-

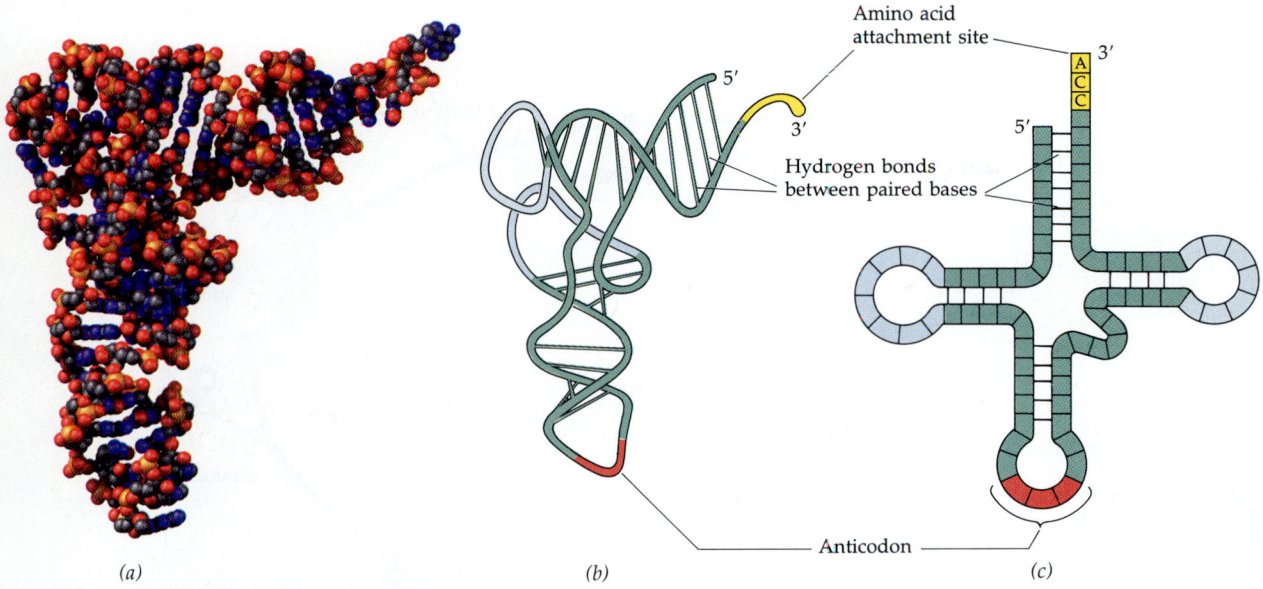

Amino acid
attachment site

Hydrogen bonds
between paired bases

Anticodon

(a)   (b)   (c)

11.19   Transfer RNA: Crick's "Adaptors"
Molecules of tRNA "read" the genetic code. (a) This computer-generated space-filling representation shows the three-dimensional structure of a tRNA, which is then drawn in (b). This three-dimensional shape, brought about by the four regions of base pairing, is diagrammed for clarity in (c). Notice that the region of tRNA that binds to the amino acid is far from the anticodon that interacts with the mRNA.

mine the three-dimensional structures of tRNAs, and it was found that such pairing does occur, giving all tRNAs certain shapes in common. At one end of every tRNA molecule is a site to which the amino acid attaches. At the opposite end is a group of three bases, called the **anticodon**, that constitutes the point of contact with mRNA (Figure 11.19). Each tRNA species has a unique anticodon, allowing it to unite by complementary base pairing with only one codon. This is the key to the specificity of translation.

The three-dimensional shape of tRNAs allows them to combine specifically with the binding sites on ribosomes. The structure of tRNA molecules relates clearly to their functions: They carry amino acids, associate with mRNA molecules, and interact with ribosomes.

How does a tRNA molecule combine with the correct amino acid? A family of **activating enzymes**, known more formally as aminoacyl-tRNA synthases, accomplishes this task. Each activating enzyme is specific for one amino acid. It must also find its appropriate tRNA, which it does by recognizing short sequences of bases on the tRNA, away from the anticodon. The enzyme reacts first with a molecule of amino acid and a molecule of ATP, producing a high-energy AMP-amino acid that remains bound to the enzyme. The high energy results from the breaking of the bonds in the ATP. The enzyme then catalyzes a shifting of the amino acid from the AMP (adenosine

monophosphate) to the 3'-terminal nucleotide of the tRNA, where it is held by a relatively high-energy bond. The activating enzyme finally releases this **charged tRNA** (tRNA with its attached amino acid), which can then charge another tRNA molecule (Figure 11.20). The high-energy bond in the charged tRNA provides the energy for the synthesis of a peptide bond joining adjacent amino acids.

The Ribosome

Ribosomes are required for translation of the genetic information into a polypeptide chain. Each ribosome consists of two subunits, a heavy (large) one and a light (small) one (Figure 11.21a). In eukaryotes, the heavy subunit consists of three different molecules of rRNA (ribosomal RNA) and about 45 different protein molecules, arranged in a precise pattern. The light subunit in eukaryotes consists of one rRNA molecule and 33 different protein molecules. The ribosomes of prokaryotes are somewhat smaller than those of eukaryotes. Mitochondria and chloroplasts also contain ribosomes, some of which are even smaller than those of prokaryotes. When not active in the translation of mRNA, the ribosomes exist as separated subunits.

Each ribosome has two tRNA-binding sites that participate in translation. The ribosome also binds to the mRNA that it is translating.

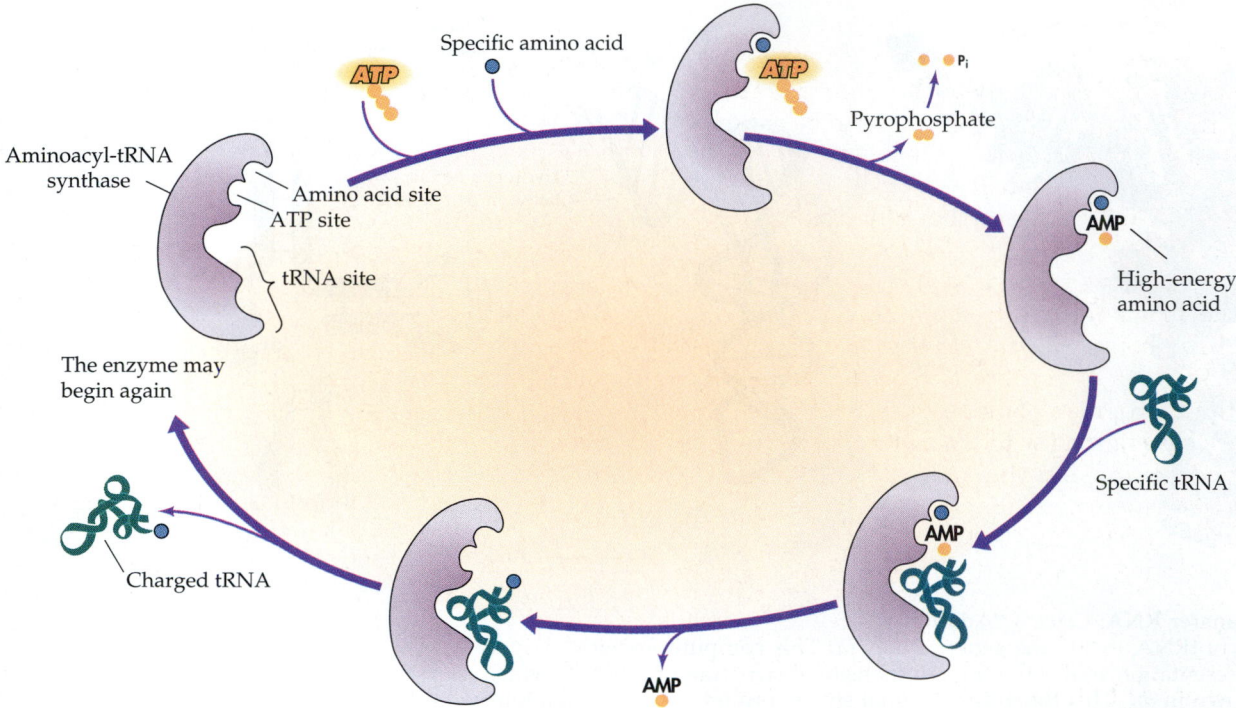

11.20   Charging a tRNA Molecule

An activating enzyme, aminoacyl-tRNA synthase, has a three-part active site that recognizes three smaller molecules: a specific amino acid, ATP, and a specific tRNA. The enzyme activates the amino acid, catalyzing a reaction with ATP in which high-energy AMP amino acid and a pyrophosphate ion are formed. The enzyme then catalyzes a reaction of the activated amino acid with the correct tRNA, producing a charged tRNA molecule with a high-energy bond that provides energy for synthesizing a peptide bond. The enzyme is free to charge another tRNA.

11.21   Translation of Genetic Information

(a) Each ribosome consists of a light and a heavy subunit, which separate when they are not in use. (b) An initiation complex. After the initiation complex forms, the heavy subunit joins it, completing the ribosome. (c) The polypeptide chain elongates as the mRNA is translated. (d) Translation terminates when the ribosome encounters a stop signal on the mRNA. The completed polypeptide, final tRNA, and ribosomal subunits all dissociate from the mRNA. Circled three-letter abbreviations represent amino acids (see Table 3.1).

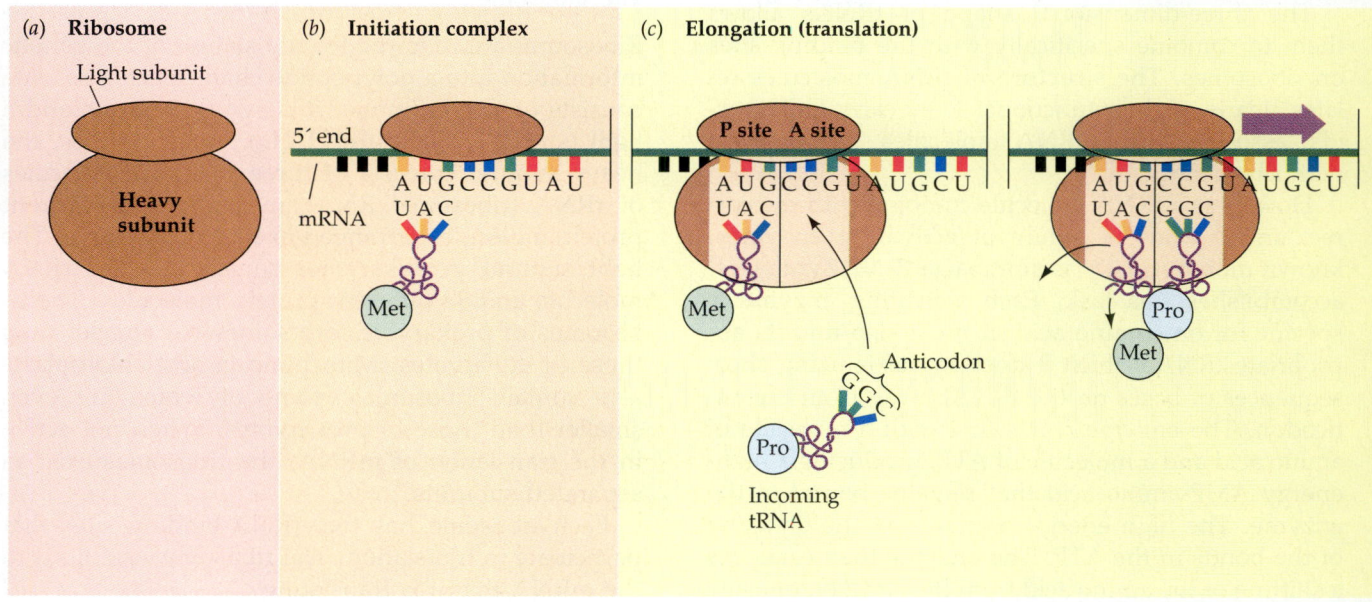

Translation

We have been working our way through the steps by which the sequence of bases in the template strand of a DNA molecule specifies the sequence of amino acids in a protein. We are now at the last step: translation, the RNA-directed assembly of a protein. The translation of mRNA begins with the formation of an initiation complex, which consists of a charged tRNA bearing the first amino acid and a light ribosomal subunit, both bound to the starting point on the mRNA chain. The anticodon of the charged tRNA binds to the appropriate point on the mRNA by complementary base pairing with the first codon. Hydrogen bonds form, linking the codon and the anticodon. The light ribosomal subunit binds to the same point (Figure 11.21b). The heavy subunit of the ribosome then joins the complex (Figure 11.21c).

There are two RNA-binding sites on the ribosome: the A site (which accepts a tRNA molecule bearing one amino acid) and the P site (which will carry a tRNA molecule bearing a growing polypeptide chain). The first charged tRNA now lies in the P site of the ribosome, and the A site is over the second codon. A charged tRNA whose anticodon is complementary to the second codon enters the open A site. The first amino acid joins the amino acid on the tRNA in the A site, with the peptide linkage forming in such a way that the first amino acid is the N-terminus of the new protein (see Chapter 3), while the second amino acid remains attached to its tRNA by its carboxyl group (—COOH). The first tRNA, having released its amino acid, dissociates from the complex, returning to the cytosol to become charged with another amino acid of the same kind. The second tRNA,

now bearing a dipeptide, shifts to the P site of the ribosome, which moves along the mRNA by another codon. The process continues—the latest charged tRNA enters the open A site, picks up the growing polypeptide chain from the one in the P site, and then moves to the newly vacated P site—until a stop codon enters the A site and terminates translation (Figure 11.21d). (Stop codons are described later in this chapter, in the section on the genetic code.) The newly synthesized protein separates from the ribosome. The N-terminus of the new protein is the amino acid corresponding to the first codon on the mRNA; the C-terminus is the last amino acid to join the chain.

Several ribosomes can work simultaneously at translating a single mRNA molecule to produce a number of molecules of the protein at the same time. As soon as the first ribosome has moved far enough from the initiation point, a second initiation complex can form, then a third, and so on. The first ribosome to initiate translation is the first to finish translating the message and be released. The assemblage of a thread of mRNA with its beadlike ribosomes and their growing polypeptide chains is called a polyribosome, or **polysome**. Cells that are actively synthesizing proteins contain large numbers of polysomes and fewer free ribosomes or ribosomal subunits. Figure 11.22 shows a polysome in action.

A given ribosome is not specifically adapted to produce just one kind of protein. It was once thought that there were specific ribosomes for each type of protein produced in a cell, but that is not the case. A ribosome can combine with any mRNA and any tRNAs and thus can be used to make different polypeptide products. The mRNA contains the informa-

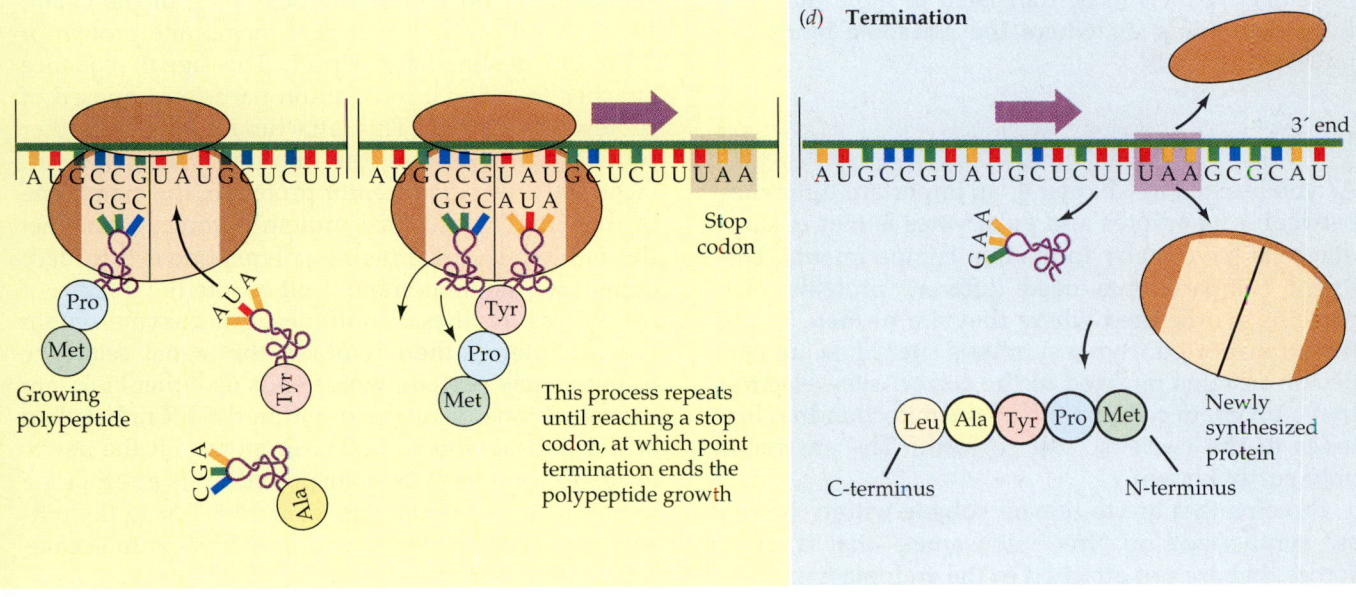

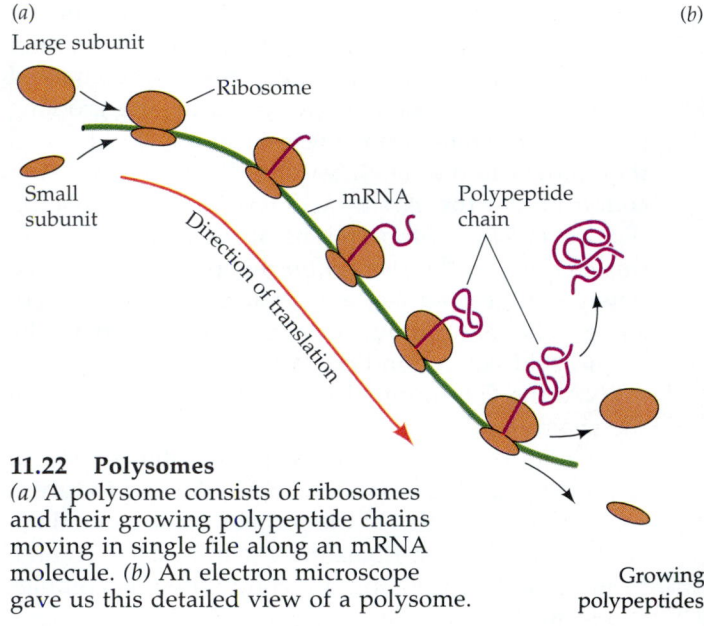

(a)
Large subunit

Ribosome

Small subunit

Direction of translation

mRNA

Polypeptide chain

11.22   Polysomes
(a) A polysome consists of ribosomes and their growing polypeptide chains moving in single file along an mRNA molecule. *(b)* An electron microscope gave us this detailed view of a polysome.

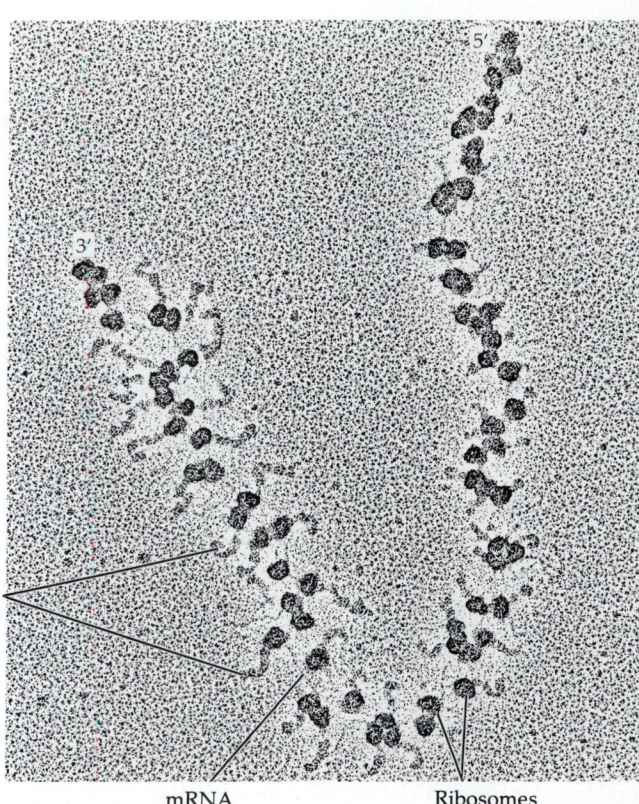

(b)

Growing polypeptides

mRNA   Ribosomes

tion that specifies the polypeptide sequence. The ribosome is simply the molecular machine that accomplishes the task. Its structure allows it to hold the mRNA and tRNAs in the right positions, thus allowing the growing polypeptide to be assembled efficiently.

We have simplified this account of translation in at least two ways. First, we have not mentioned the specific proteins and small molecules that play roles in the polypeptide elongation and release processes. Second, we have described protein synthesis as it occurs in prokaryotes. In eukaryotes, things are more complex (see Chapter 13); in particular, several steps come between transcription and translation.

In at least one virus certain genes overlap, allowing some mRNAs to be translated in more than one way, depending on where the ribosome binds the mRNA (Box 11.B).

The Role of the Endoplasmic Reticulum

As you learned in Chapter 4, an important difference between prokaryotes and eukaryotes is that eukaryotic cells have many individual compartments. Different compartments need different proteins. Are proteins synthesized where they are needed, or are they transported from a synthesis site? How are particular proteins targeted to the correct site—electron transport chain components to the mitochondria, histones to the nucleus, and so forth? The answer is only partly known.

Proteins that are to remain soluble within the cell are synthesized on "free" ribosomes—that is, ribosomes that are not attached to the endoplasmic retic-

ulum (ER). On the other hand, proteins that are to become parts of membranes, or are to be exported from the cell, or are to end up in lysosomes or peroxisomes, are generally synthesized on the ribosomes of the rough ER. All protein synthesis, however, *begins* on free ribosomes. The first few amino acids of a polypeptide chain determine whether production of the protein will be completed on the rough ER or on free ribosomes.

If a specific sequence of amino acids, the **signal sequence**, is present at the beginning of the chain, the finished product will be a membrane protein or a protein destined for export. The signal sequence attaches to a signal recognition particle composed of protein and RNA. This attachment blocks further protein synthesis until the ribosome can become attached to a specific receptor protein in the membrane of the ER. The receptor protein becomes a channel through which the growing polypeptide is extruded, either into the membrane itself or into the interior of the ER, as synthesis continues. An enzyme within the ER interior then removes the signal sequence from the new protein, which ends up either built into the membrane or retained within the ER rather than in the cytosol (Figure 11.23). From the ER the newly formed protein can be transported to its appropriate location—to other cellular membranes or to the outside of the cell—without mixing with other molecules in the cytoplasm.

BOX 11.B

Making the Most of Your DNA

A virus named φX174, one of the smallest bacteriophages, is made up of a few types of protein molecules plus a small, circular molecule of DNA. This viral DNA must code for nine types of proteins, including some that are not part of the mature phage but that are needed during viral replication or release. We know the lower limit for the length of a DNA base sequence that can code for a protein. Scientists puzzled over φX174 because its total DNA content (about 5,400 nucleotides) appears to be too low by 10–15 percent to code for all nine proteins. This problem was resolved in a most unexpected way in 1977, when the English

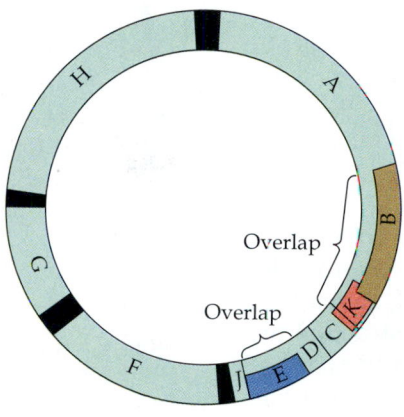

The circle represents the chromosome of the φX174 virus; the letters indicate the proteins coded for by the genes in the colored regions. Note that some regions coding for one protein also include the information for other proteins; for example, the region coding for protein A also codes for protein B and part of protein K. The few short stretches of the chromosome that do not code for proteins are shown in black.

biochemist Frederick Sanger and his colleagues reported the complete structure, nucleotide by nucleotide, of the DNA molecule of φX174. With the nucleotide map spread out before

them, and with their knowledge of the primary structures of the proteins, these scientists were able to discover that the genes coding for two of the proteins are embedded within two of the other genes, as indicated in the figure. Putting it another way, the same stretch of DNA can participate in coding for two entirely different proteins. The mRNA transcribed from such a shared stretch of DNA contains codes for both proteins. Which protein is produced depends on where translation begins—that is, on which of two initiation sites becomes bound to a ribosome. When you read about frameshift mutations in this chapter, you might recall the story of φX174 and marvel that such an improbable thing as shared DNA should ever have come about. The phenomenon of overlapping genes does not appear to be common; it may have evolved in φX174 because of the limitation on the size of the DNA molecule imposed by the small protein coat.

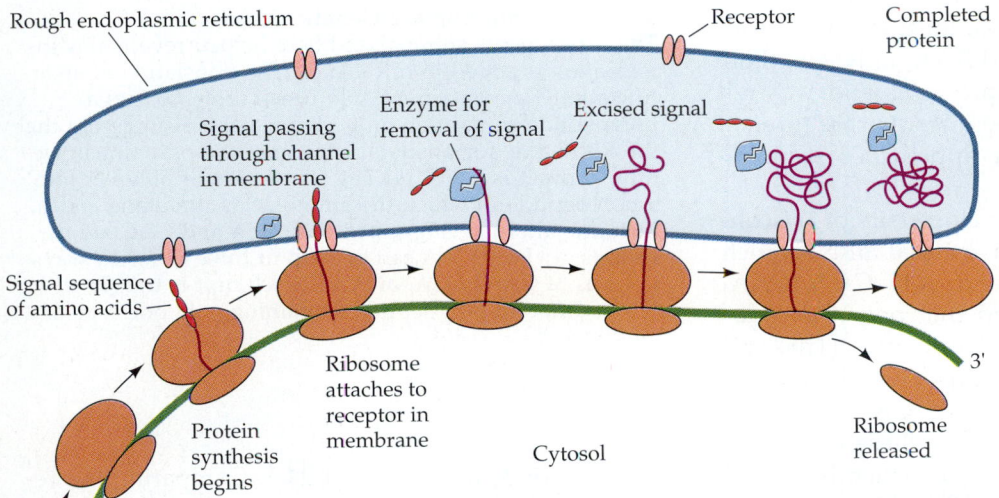

11.23   A Signal Sequence Moves Polypeptide Elongation into the ER
Synthesis of a protein to be exported from the cell begins free in the cytoplasm, as ribosomal subunits and mRNA come together and the first few amino acids, including a signal sequence, are linked. Synthesis continues only after a signal–receptor interaction creates a channel through which the polypeptide will elongate into the interior of the endoplasmic reticulum. When the protein is complete, it is released to the ER interior, and the ribosomal subunits separate from the membrane and from one another.

The Genetic Code

Which mRNA codons translate into which amino acids? Molecular biologists broke the code in which genetic information is stored in the early 1960s. The problem seemed overwhelming at the outset: How could more than 20 "code words" be written with an "alphabet" consisting of only four "letters"? How, in other words, could four bases code for 20 or so different amino acids? It was not yet possible to determine the base sequence in a nucleic acid, so scientists could not simply compare the primary structure (the amino acid sequence) of a protein with the base sequence of the appropriate DNA or mRNA molecule. However, there were ways of getting partial information about the code, even though nucleic acid chemistry was not yet very far advanced.

Marshall W. Nirenberg and J. H. Matthaei, at the National Institutes of Health, made the first breakthrough when they realized that they could use a very simple artificial polymer instead of a complex, natural mRNA as a messenger. They could then see what the simple, artificial messenger coded for. In 1961, Nirenberg and Matthaei published the first of their papers on polypeptide synthesis directed by artificial mRNAs. Nirenberg had prepared an artificial mRNA in which all the bases were uracil; the molecule was called poly U. When poly U was added to a reaction mixture containing ribosomes, amino acids, activating enzymes, tRNAs, and other factors, a polypeptide formed. This polypeptide contained only one kind of amino acid: phenylalanine (Phe). Poly U coded for poly Phe! Accordingly, it appeared that UUU was the mRNA code word—the codon—for phenylalanine (Figure 11.24a). Following up on this success, Nirenberg and Matthaei easily showed that CCC codes for proline and AAA for lysine. (Poly G presented some chemical problems and was not tested initially.) UUU, CCC, and AAA were three of the easiest codons; different approaches were required to work out the rest.

Har Gobind Khorana, at the University of Wisconsin, painstakingly synthesized artificial mRNAs such as poly CA (CACACA. . .) and poly CAA (CAACAACAA. . .). Khorana found that poly CA codes for a polypeptide consisting of threonine (Thr) and histidine (His), in alternation (His–Thr–His–Thr. . .; Figure 11.24b). There are thus two possible codons in poly CA, CAC and ACA. One of these must code for His and the other for Thr—but which is which? The answer came from the results with poly CAA, which produces three different polypeptides: poly Thr; poly Gln (polyglutamine); and poly Asn (polyasparagine). To understand this, we must know that an artificial messenger can be read beginning at any point in the chain; there is no specific initiator region. Thus poly CAA can be read as a polymer of CAA, of

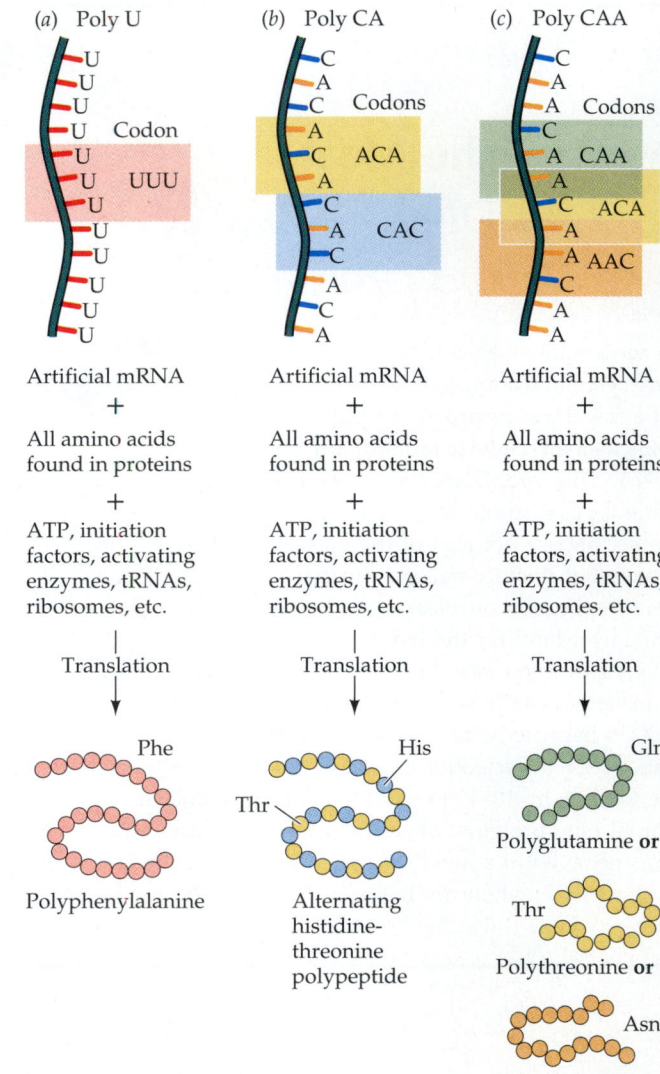

11.24   Deciphering the Genetic Code
The experiments summarized here helped reveal how information is coded in mRNA. (a) The translation of an artificial mRNA containing only uracil (poly U) into a polypeptide containing only phenylalanine suggested that the RNA code for phenylalanine contains only uracil; we now know it is UUU. (b) The translation of poly CA into a polypeptide in which the amino acids threonine and histidine alternate suggested that ACA and CAC are the codons. (c) Poly CAA can be read in three different ways: in units of CAA, ACA, or AAC. Each unit is translated into a different polypeptide containing only one type of amino acid.

ACA, or of AAC (Figure 11.24c). Comparing the results of the poly CA and poly CAA experiments, you should be able to figure out which code word is for Thr and which is for His.

Nirenberg discovered something that led to the decoding of the remaining words in the code book in 1964 and 1965. He found that simple "mRNAs," only three monomers long and each amounting to a

Second letter

	U	C	A	G	
U	UUU UUC Phenyl-alanine UUA UUG Leucine	UCU UCC UCA UCG Serine	UAU UAC Tyrosine UAA Stop codon UAG Stop codon	UGU UGC Cysteine UGA Stop codon UGG Tryptophan	U C A G
C	CUU CUC CUA CUG Leucine	CCU CCC CCA CCG Proline	CAU CAC Histidine CAA CAG Glutamine	CGU CGC CGA CGG Arginine	U C A G
A	AUU AUC Isoleucine AUA AUG Methionine; initiation codon	ACU ACC ACA ACG Threonine	AAU AAC Asparagine AAA AAG Lysine	AGU AGC Serine AGA AGG Arginine	U C A G
G	GUU GUC GUA GUG Valine	GCU GCC GCA GCG Alanine	GAU GAC Aspartic acid GAA GAG Glutamic acid	GGU GGC GGA GGG Glycine	U C A G

First letter (left margin) · Third letter (right margin)

11.25 The Universal Genetic Code
Genetic information is encoded in mRNA in three-letter units—codons—made up of the bases uracil (U), cytosine (C), adenine (A), and guanine (G). To decode a codon, find its first letter in the left column, then read across the top to its second letter, then read down the right column to its third letter. The amino acid the codon specifies is in the corresponding row. For example, AUG codes for methionine, and GUA codes for valine.

codon, can bind to ribosomes and that the resulting complex can then cause the binding of the corresponding charged tRNA. Thus, for example, simple UUU causes phenylalanyl-tRNA charged with phenylalanine to bind to the ribosome. After this discovery, complete deciphering of the code book was relatively simple. To find the "translation" of a codon, Nirenberg could use a sample of that codon as an artificial mRNA and see which amino acid became bound.

The complete genetic code is shown in Figure 11.25. Notice that there are many more RNA codons than there are different amino acids in proteins. Combinations of the four "letters" (the bases) give 64 different three-letter codons, yet these determine only 20 amino acids. Three of the codons (UAA, UAG, UGA) are **stop codons**, or chain terminators; when the translation machinery reaches one of these codons, translation stops and the polypeptide is released from the ribosome–mRNA–tRNA complex. AUG, which codes for methionine, is also the **start codon**, which acts as the initiation signal for tran-

scription. Still, 61 codons are far more than enough to code for 20 amino acids—and indeed there are repeats. Thus we say that the code is **degenerate**, which means that an amino acid may be represented by more than one codon. The degeneracy is not evenly divided among the amino acids. For example, methionine and tryptophan are represented by only one codon each, whereas leucine is represented by six different codons.

The term *degeneracy* should not be confused with *ambiguity*. To say that the code was ambiguous would mean that a single codon could specify either of two (or more) different amino acids. There would be doubt whether to put in, say, leucine or something else. The genetic code is not ambiguous. Degeneracy in the code means that there is more than one unequivocal way to say, "Put leucine here." In other words, a given amino acid may be coded for by more than one codon, but a codon can code for only one amino acid.

The code appears to be universal, applying to all the species on our planet. The code must be an ancient one that has been maintained intact throughout the evolution of living things. Only one exception is known: Within mitochondria the code differs slightly but detectably from that in prokaryotes and elsewhere in eukaryotic cells. The code is not even quite the same in the mitochondria of different eukaryotes. The significance of this difference is not yet clear.

You should remember that the codons in Figure 11.25 are mRNA codons. The master words on the DNA strand that was transcribed to produce the mRNA are complementary to these codons, so, for example, AAA in the template DNA strand corresponds to phenylalanine (which is coded for by the mRNA codon UUU). Does this code really work? Final proof was obtained by synthesizing artificial DNA of known sequence, introducing it into bacteria, and inducing the bacteria to produce the specific protein coded for by that DNA. We can now program bacteria to synthesize proteins no organism ever made before.

MUTATIONS

Accurate DNA replication, transcription, and translation all depend upon the reliable pairing of complementary bases. Errors occur, though infrequently, in all three. In particular, errors in the replication of DNA during the production of the gametes are crucial to evolution. If there were no mutations—heritable changes in the genetic information—there would be no evolution. Minute changes in the genetic material often lead to easily observable changes in outward form and function. The detection of a mutation depends on our ability to observe these phenotypic effects. Some effects of mutation in humans are obvious—dwarfism, for instance, or the presence of more than five fingers on each hand. A mutant genotype in a microorganism may be obvious if, for example, it results in a change in color or in nutritional requirements, as we have discussed for *Neurospora*. Other mutations, however, may be virtually unobservable. In humans, for example, there is a mutation that drastically lowers the level of an enzyme called glucose 6-phosphate dehydrogenase that is present in many tissues, including red blood cells. The red blood cells of a person carrying the mutant gene are abnormally sensitive to an antimalarial drug called primaquine; when such people are treated with this drug, their red blood cells rupture, causing serious medical problems. People with the normal allele have no such problem. Before the drug came into use, no one was aware that such a mutation existed. Similarly, distinguishing a mutant bacterium from a normal one may be a very subtle matter, dependent on what tools are available.

Some mutations cause their phenotypes only un-

der certain restrictive conditions and are not detectable under other conditions. We call organisms that carry such mutations conditional mutants. Many conditional mutants are temperature-sensitive, unable to grow at some restrictive temperature, such as 37°C, but able to grow normally at a lower, permissive temperature, such as 30°C. The mutant allele in such an organism may code for an enzyme with an unstable tertiary structure that is altered at the restrictive temperature.

All mutations are alterations in the nucleotide sequence in DNA. We divide mutations into two categories based on the extent of the alteration. **Point mutations** are mutations of single genes. One allele becomes another because of small alterations in the sequence or number of nucleotides—even as small as the substitution of one nucleotide for another. **Chromosomal mutations**, introduced in Chapter 10, are more extensive alterations. Chromosomal mutations may change the position or direction of a DNA segment without actually removing any genetic information, or they may cause a segment of DNA to be irretrievably lost. Both point mutations and chromosomal mutations are heritable.

Point Mutations

Many point mutations consist of the substitution of one base for another in the DNA and hence in the mRNA. Often base-substitution mutations change the genetic message so that one amino acid substitutes for another in the protein. Figure 11.15 shows the sickled blood cells that result from one such mutation. A **missense mutation** such as this may sometimes cause the protein not to function, but often the effect is only to reduce the functional efficiency of the protein. Individuals carrying missense mutations may survive even though the affected protein is essential to life. In the course of evolution, some missense mutations even improve functional efficiency.

Nonsense mutations, another type of base-substitution mutation, are more often disruptive than are missense mutations. A nonsense mutation is one in which the base substitution results in the formation of a chain-terminator codon, such as UAG, in the mRNA product. A nonsense mutation results in a shortened protein product, since translation does not proceed beyond the point where the mutation occurred.

Not all point mutations are base substitutions. Single base pairs may be inserted into or deleted from DNA. Such mutations are known as **frame-shift mutations** because they interfere with the decoding of the genetic message by throwing it out of register. Think again of codons as three-letter words, each corresponding to a particular amino acid. Translation proceeds codon by codon; if a base is added to the message or subtracted from it, translation proceeds perfectly until it comes to the one-base insertion or deletion. From that point on, the three-letter words in the message are one letter out of register. In other words, such mutations shift the "reading frame" of the genetic message. Frame-shift mutations almost always lead to the production of completely nonfunctional proteins.

Consider the following example to see how a frame-shift mutation works. If a template strand of DNA reads from 3' to 5' as follows:

3'—AGATACGTGCTGCAT—5'

it is transcribed to yield an mRNA with the following sequence from 5' to 3':

5'—UCUAUGCACGACGUA—3'

which can be divided up into the following codons:

. . . UCU AUG CAC GAC GUA . . .

which, as you can determine from Figure 11.25, translates to the following amino acid sequence (from N-terminus to C-terminus):

. . . serine methionine histidine aspartic acid valine . . .

Now suppose that, through a deletion, the DNA strand loses the fifth base in the sequence above, an A, so it reads as follows:

3'—AGATCGTGCTGCAT—5'

Transcription yields the following mRNA sequence:

5'—UCUAGCACGACGUA—3'

or

. . . UCU AGC ACG ACG UA . . .

which, in turn, is translated to the amino-acid sequence

. . . serine serine threonine threonine . . .

No wonder frame-shift mutations are so disruptive. An organism carrying such a mutant gene can survive only if the affected gene product is not essential to the cellular machinery, or if the organism carries another copy of the gene in its normal form.

Certain chemicals, called **mutagens**, can induce mutations. Among these are base analogs: purines or pyrimidines not found in natural DNA but enough like the natural bases that they can be incorporated into DNA. Base analogs are mutagenic presumably because they are more likely than natural DNA bases to mispair. For example, 5-bromouracil is very similar to thymine and it is easily incorporated into DNA in place of thymine (Figure 11.26). But 5-bromouracil is much more likely than thymine to engage in an abnormal pairing with guanine, and it therefore induces

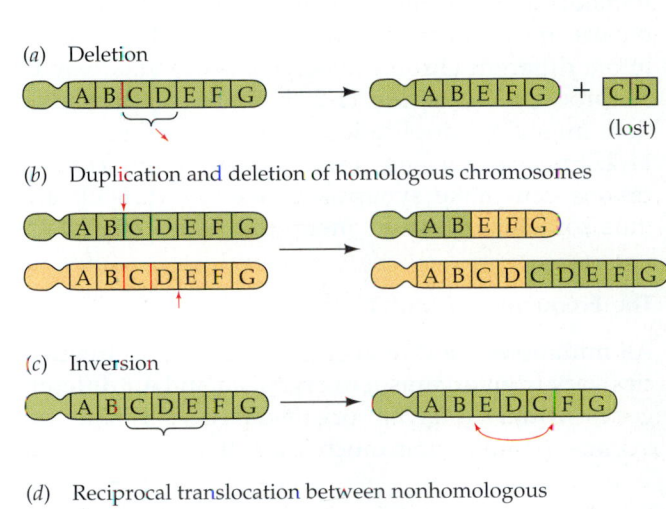

11.26   Base Analogs Are Similar to Bases
Enzymes that replicate DNA cannot distinguish between the base analog 5-bromouracil and the base thymine because the two molecules are so similar in shape. Once incorporated in a DNA strand, 5-bromouracil may rearrange itself to a form that resembles cytosine and then pair with guanine.

mutations of A–T to G–C and G–C to A–T. Thus 5-bromouracil is a potent chemical mutagen.

Chromosomal Mutations

Genetic strands can break and rejoin, grossly disrupting the sequence of genetic information. There are four types of such chromosomal mutations: deletions, duplications, inversions, and reciprocal translations (Figure 11.27).

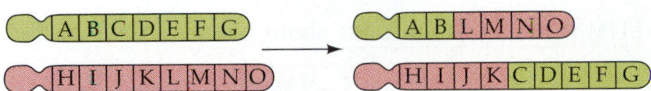

11.27   Chromosomal Mutations
Chromosomes may break during replication and parts of chromosomes may rejoin. Letters on the colored chromosomes represent segments along the length of the chromosome. (*a*) A deletion results when a chromosome breaks in two places and a segment is lost when the parts rejoin. (*b*) A duplication and a deletion result when homologous chromosomes break at different points and swap segments. (*c*) An inversion results when a broken segment is reinserted in reverse order. (*d*) A reciprocal translocation results when two nonhomologous chromosomes exchange segments.

Deletions remove part of the genetic material. Like frame-shift point mutations, they cause death unless they affect unnecessary genes or are masked by the presence, in the same cell, of normal copies of the deleted genes. It is easy to imagine one mechanism that could produce deletions: A DNA molecule might break at two points and the two end pieces might rejoin, leaving out the DNA between the breaks.

Another mechanism by which deletion mutations might arise would lead simultaneously to the production of a second kind of chromosomal mutation: a **duplication**. Duplication would come about if homologous chromosomes broke at different positions and then reconnected to the wrong partners. One of the two molecules produced by this mechanism would lack a segment of DNA, and the other would have two tandem copies of the information that was deleted from the first.

Breaking and rejoining can also lead to **inversion**—the removal of a segment of DNA and its reinsertion into the same location, but "flipped" end for end so that it runs in the opposite direction. If an inversion includes part of a segment of DNA that codes for a protein, the resulting protein will be drastically altered and almost certainly nonfunctional.

The fourth type of chromosomal mutation, called **translocation**, results when a segment of DNA breaks, moves from a chromosome, and is inserted into a different chromosome. Translocations may be reciprocal, as in Figure 11.27*d*; notice that the mutation involving duplication and deletion in Figure 11.27*b* is also a *non*reciprocal translocation. Translocations can make synapsis in meiosis difficult and thus sometimes lead to aneuploidy (see Chapter 9).

The Frequency of Mutations

All mutations are rare events, but mutation frequencies vary from organism to organism and for different genes within a given organism. Usually the frequency of mutation is much lower than one mutation per 10^4 genes per DNA duplication, and sometimes the frequency is as low as one mutation per 10^9 genes per duplication. Most mutations are point mutations in which one nucleotide is substituted for another during the synthesis of a new DNA strand.

THE ORIGIN OF NEW GENES

Most mutations harm the organism that carries them. Some mutations are neutral (they have no effect on the organism's ability to survive or produce offspring). Once in a while, however, a mutation improves an organism's adaptation to its environment or becomes favorable when environmental variables change. Duplication mutations may be the source of "extra" genes. Most of the more complex creatures living on Earth have more DNA and therefore more genes than the simpler creatures do. Humans, for example, have 1,000 times more genetic material than bacteria have.

How do new genes arise? If whole genes were sometimes duplicated by the mechanism described in the previous section, the bearer of the duplication would have a surplus of genetic information that might be turned to good use. Subsequent mutations in one of the two copies of the gene might not have an adverse effect on survival because the other copy of the gene would continue to produce functional protein. The extra gene might mutate over and over again without ill effect because its function would be fulfilled by the original copy. If the random accumulation of mutations in the extra gene led to the production of some useful protein (for example, an enzyme with an altered specificity for the substrates it binds, allowing it to catalyze different—but related—reactions), natural selection would tend to perpetuate the existence of this new gene. New copies of genes also arise through the activity of transposable elements, which are discussed in Chapters 12 and 13.

SUMMARY of Main Ideas about Nucleic Acids as the Genetic Material

DNA is the genetic material of all cellular organisms and many viruses.

Genetic information can be transferred from dead bacteria to genetically different live bacteria by transformation. The transforming principle is DNA.
Review Figure 11.1

The Hershey–Chase experiment gave compelling evidence that DNA is the genetic material.
Review Figure 11.2

DNA consists of two polynucleotide chains forming a double helix.

DNA chains are held together by hydrogen bonding between their nitrogenous bases. Base pairing is complementary: A–T, T–A, G–C, C–G.

Normal DNA has a right-handed helical structure with two grooves.
Review Figures 11.3 and 11.4

RNA is usually a single polynucleotide chain.

DNA replication, catalyzed by DNA polymerases, is semiconservative.
Review Figures 11.7 and 11.9

Because the two antiparallel DNA strands are synthesized in the 5′-to-3′ direction, the lagging strand must be synthesized discontinuously.
Review Figures 11.10 and 11.11

DNA polymerases introduce occasional mistakes, but the proofreading function of these enzymes corrects most of the mistakes.

Study of blocks in biochemical pathways showed that many genes encode single polypeptides.
Review Figures 11.13 and 11.14

DNA codes for RNA and RNA codes for protein.
Review Figure 11.6

In retroviruses, RNA serves as a template for DNA synthesis in the host cell as carried out by reverse transcriptase.

Transcription forms an RNA molecule with a base sequence complementary to that of the template strand of the DNA.
Review Figure 11.18

In translation, the information transcribed into mRNA is used to make a polypeptide from amino acids.

tRNAs bind amino acids.
Review Figures 11.19 and 11.20

A ribosome binds to mRNA, and binds charged tRNAs at two other sites.
Review Figure 11.21

Several ribosomes can simultaneously translate the same mRNA.
Review Figure 11.22

Most amino acids are specified by more than one codon. There are also a start codon and three stop codons.
Review Figure 11.25

Mutations, which are rare, change the genetic information.

Point mutations change single nucleotide pairs.

Chromosomal mutations affect larger stretches of a DNA molecule.
Review Figure 11.27

SELF-QUIZ

1. Griffith's studies of *Streptococcus pneumoniae*
   a. proved that DNA is the genetic material of bacteria.
   b. proved that DNA is the genetic material of bacteriophages.
   c. demonstrated the phenomenon of bacterial transformation.
   d. proved that bacteria reproduce sexually.
   e. proved that protein is not the genetic material.

2. In the Hershey–Chase experiment
   a. DNA from parent bacteriophage appeared in progeny bacteriophages.
   b. most of the phage DNA never entered the bacteria.
   c. more than three-fourths of the phage protein appeared in progeny phages.
   d. DNA was labeled with radioactive sulfur.
   e. DNA formed the coat of the bacteriophage.

3. Which statement about complementary base pairing is *not* true?
   a. It plays a role in DNA replication.
   b. In DNA, T pairs with A.
   c. Purines pair with purines, and pyrimidines pair with pyrimidines.
   d. In DNA, C pairs with G.
   e. The base pairs are of equal length.

4. In semiconservative replication of DNA
   a. the original double helix remains intact and a new double helix forms.
   b. the strands of the double helix separate and act as templates for new strands.
   c. polymerization is catalyzed by RNA polymerase.
   d. polymerization is catalyzed by a double-helical enzyme.
   e. DNA is synthesized from amino acids.

5. Which of the following does not occur during DNA replication?
   a. Unwinding of the parent double helix
   b. Formation of short pieces that are united by DNA ligase
   c. Complementary base pairing
   d. Use of a primer
   e. Polymerization in the 3′-to-5′ direction

6. Transcription
   a. produces only mRNA.
   b. requires ribosomes.
   c. requires tRNAs.
   d. produces RNA growing from the 5′ end to the 3′ end.
   e. takes place only in eukaryotes.

7. Which statement is *not* true of translation?
   a. It is RNA-directed polypeptide synthesis.
   b. An mRNA molecule can be translated by only one ribosome at a time.
   c. The same genetic code is in effect in all organisms.
   d. Any ribosome can be used in the translation of any mRNA.
   e. There are both start and stop codons.

8. Which statement is *false*?
   a. Transfer RNA functions in translation.
   b. Ribosomal RNA functions in translation.
   c. RNAs are produced in transcription.
   d. Messenger RNAs are produced on ribosomes.
   e. DNA codes for mRNA, tRNA, and rRNA.

9. The genetic code
   a. is different for prokaryotes and eukaryotes.
   b. has changed during the course of recent evolution.
   c. has 64 codons that code for amino acids.
   d. is degenerate.
   e. is ambiguous.

10. A mutation that results in the codon UAG where there had been UGG
    a. is a nonsense mutation.
    b. is a missense mutation.
    c. is a frame-shift mutation.
    d. is a large-scale mutation.
    e. is unlikely to have a significant effect.

FOR STUDY

1. The genetic code is described as degenerate. What does this mean? How is it possible that a point mutation, consisting of the replacement of a single nitrogenous base in DNA by a different base, might not result in an error in protein production?

2. Suppose that Meselson and Stahl had continued their experiment on DNA replication for another 10 bacterial generations. Would there still have been any ^{14}N–^{15}N DNA present? Would it still have appeared in the centrifuge tube? Explain.

3. Look back at Khorana's experiment with poly CA and poly CAA, in which the codons for histidine and threonine were determined. Using the genetic code (Figure 11.25) as a guide, deduce what results Khorana would have obtained had he used poly UG and poly UGG as artificial messengers. (In fact, very few such artificial messengers would have given useful results.) For an example of what could happen, consider poly CG and poly CGG. Using poly C as the messenger, a mixed polypeptide of arginine and alanine (−Arg−Ala−Arg−Ala−) would be obtained; poly CGG would give three polypeptides: polyarginine, polyalanine, and polyglycine. Can any codons be determined from only these data? Explain.

4. What causes transcription to start? to stop? What causes translation to start? to stop?

READINGS

Felsenfeld, G. 1985. "DNA." *Scientific American*, October. A well-illustrated description of DNA structure and function.

Griffiths, A. J. F., J. H. Miller, D. T. Suzuki, R. C. Lewontin and W. M. Gelbart. 1993. *An Introduction to Genetic Analysis*, 5th Edition. W. H. Freeman, New York. An excellent textbook of modern genetics. Chapters 11, 12 and 13 are particularly relevant.

Judson, H. F. 1979. *The Eighth Day of Creation*. Simon and Schuster, New York. A sparkling history of molecular biology, with the best available description of the events surrounding the discovery of the structure of DNA.

Radman, M. and R. Wagner. 1988. "The High Fidelity of DNA Duplication." *Scientific American*, August. How error avoidance and error correction work. Why don't they work even better?

Smith, M. 1979. "The First Complete Nucleotide Sequencing of an Organism's DNA." *American Scientist*, vol. 67, pages 57–67. How nucleotide sequences in DNA are worked out and the interesting discovery of overlapping genes in the bacteriophage φX174.

Stent, G. S. and R. Calendar. 1978. *Molecular Genetics*, 2nd Edition. W. H. Freeman, New York. A brilliant technical and historical introduction to molecular genetics and the role of DNA.

Upton, A. C. 1982. "The Biological Effects of Low-Level Ionizing Radiation." *Scientific American*, February. How radiation leads to mutations.

Watson, J. D. 1968. *The Double Helix*. Atheneum, New York. A captivating and, to some, infuriating book in which Watson describes the events leading to the discovery of DNA structure.

Watson, J. D., N. H. Hopkins, J. W. Roberts, J. A. Steitz and A. M. Weiner. 1987. *Molecular Biology of the Gene*, 4th Edition. Two volumes. Benjamin/Cummings, Menlo Park, CA. See especially Chapters 3, 9, 10, and 12 to 15 of Volume I.

You are looking at *Escherichia coli*, the best-understood prokaryote that ever lived. We know its molecular biology more intimately than that of any other organism. *E. coli* was the "host" species in 1973, when biologists first succeeded in "transplanting" a gene from one species into another species and in seeing that gene expressed—transcribed and translated—in its new host. Since then *E. coli* has hosted many kinds of foreign or synthetic genes of interest to biomedical scientists. For even longer, however, biologists have poked and probed *E. coli*, seeking to understand basic genetic and molecular biological principles.

Prokaryotes such as *E. coli* and bacteriophages (bacterial viruses) have often been easier subjects than eukaryotes for experimental study. Molecular biologists working with bacteria and viruses in the 1950s and 1960s discovered most of the principles described in Chapter 11. Perhaps these principles would still be hidden from us if work had been limited to garden peas, *Neurospora*, corn, and fruit flies. What are the advantages of working with bacteria and viruses? First, it is easier to work with small amounts of DNA. A typical bacterium contains about $\frac{1}{1000}$ as much DNA as a single human cell, and a typical bacteriophage contains about $\frac{1}{100}$ as much DNA as a bacterium. Second, data on large numbers of organisms can be obtained easily from prokaryotes, but not from most eukaryotes. A single milliliter of medium can contain more than 10^9 *E. coli* cells or 10^{11} bacteriophages and cost less than a penny. In addition, a culture of *E. coli* can be grown under conditions that allow it to double every 20 minutes. By contrast, 10^9 mice would cost more than 10^9 dollars and would require a cage that would cover about 3 square miles. Growth of a generation of mice takes about 3 months instead of 20 minutes.

To be of use to geneticists and molecular biologists, bacteria and viruses must be genetically variable and, preferably, have some form of sex life. That bacteria mutate was demonstrated by Salvador Luria and Max Delbrück in 1943. (Remember that mutation is the source of genetic variation.) Three years later, Joshua and Esther Lederberg and Edward Tatum proved that some bacteria can engage in a form of sexual reproduction. The only similarity between this "sexual reproduction" and sexual reproduction in eukaryotes is that genetic information is transferred from one individual to another.

The ease of growing and handling bacteria and their viruses permitted the explosion of molecular biology that came shortly thereafter (you read about some of these discoveries in Chapter 11). The relative biological simplicity of bacteria and bacteriophages contributed immeasurably to the discoveries about the genetic material, the replication of DNA, and the mechanisms of gene expression. Later these bacteria

The Prettiest *E. coli* You Ever Saw
This transmission electron micrograph of *Escherichia coli* was artificially colored to illuminate the bacteria's interior components.

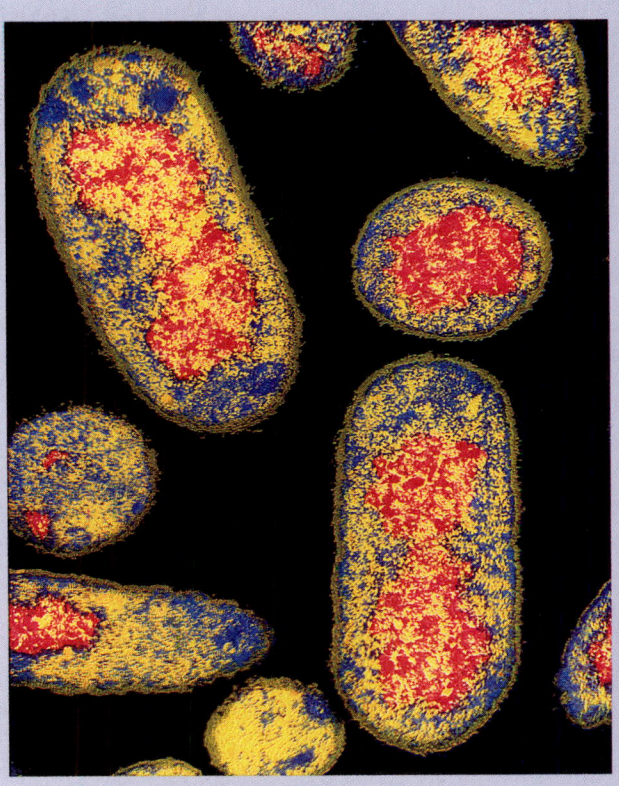

12

Molecular Genetics of Prokaryotes

and bacteriophages were the first subjects of recombinant DNA technology (see Chapter 14). Questions of interest to all biologists continue to be studied in prokaryotes, and prokaryotes continue to be important tools for biotechnology and for research on eukaryotes.

MUTATIONS IN BACTERIA AND BACTERIOPHAGES

E. coli—or any other bacterial species—reproduces by the division of single cells into two identical offspring. A single cell gives rise to a clone—a population of genetically identical individuals. As long as conditions remain favorable, a population of *E. coli* can double every 20 minutes. Researchers grow *E. coli* on the surface of a solid medium containing a sugar, minerals, a nitrogen source such as ammonium chloride, and a solidifying agent such as agar. Aseptic conditions are maintained so that there are no other microorganisms to compete with the bacterial cells on the medium. Each bacterium gives rise to a small, rapidly growing colony. If enough cells (10^7–10^8) are used, the resulting 10^7–10^8 colonies grow until they merge, forming what is called a bacterial lawn. If the culture begins with a pure type, such as the strain called *E. coli* K, the entire lawn will be of the same type.

We can do an experiment to demonstrate that bacterial mutants arise. First we mix a large sample of *E. coli* K with a suspension of a bacteriophage, such as the one called T4, and pour the mixture over growth medium in a petri plate. Wherever the virus finds a bacterial cell, it attaches to it, infects it, and eventually causes it to burst, killing the bacterium and releasing many new viruses. These viruses, in turn, attack neighboring cells. Soon visible **plaques**, or circular clearings, begin to appear in the lawn wherever the viruses have killed bacteria (Figure 12.1). The plaques, which are caused by the virus-induced bursting, or lysis, of bacteria, grow and grow. As you scan several such plates, however, here and there you will find a bacterial colony growing in the midst of a plaque, in spite of the surrounding hordes of viruses. Each of these colonies has arisen from a mutant bacterium that is resistant to the virus. We call these bacteria *E. coli* K/4 (pronounced "K-bar-4") because they are resistant to phage T4. We know that resistance is a heritable trait because the mutant bacteria give rise to colonies of cells that are similarly resistant to T4.

The phages also mutate. If you prepare plates inoculated simultaneously with *E. coli* K/4 and phage T4, you expect to see no plaques because K/4 is resistant to T4. Occasional plaques *are* found, however. These plaques must arise from mutations, but what

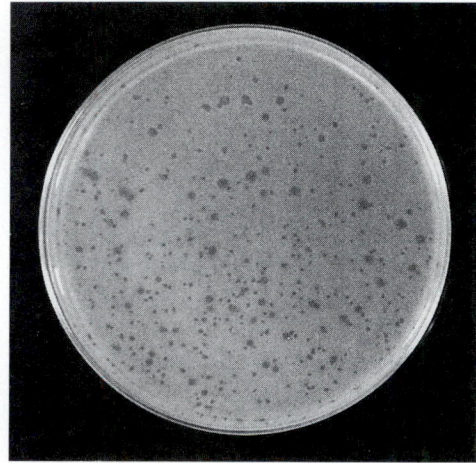

12.1   Bacteriophages Clear Bacterial Lawns
The dark circles are clear plaques in an opaque lawn of *E. coli*. Plaques form where bacteriophages have lysed bacterial cells.

have mutated—the bacteria or the viruses? Give this question a moment's thought. This time it must be the T4 that have mutated. A mutation of one *E. coli* K/4 back to the wild-type K would not result in plaque formation because only that single cell would be infected by the phage and burst; the neighboring bacteria would still be resistant and would still form an even lawn. However, a mutant phage T4 can infect a K/4 cell and, through its progeny phages, lead to the formation of a plaque. Such a phage is designated T4*h* because it has changed with regard to its host—the type of bacterial cell it can infect. The mutation of the phage, like that of the bacterium, is heritable, as the ability of the progeny phages to lyse the K/4 and form a growing plaque shows.

We have just seen how bacteria and bacteriophages are grown in the laboratory and what the consequences of certain mutations are. What we have seen, by the way, is an example of evolution on a small scale. *E. coli* K mutate to K/4 at a low rate all the time in nature, and bacteriophage T4 mutate to T4*h*. The mutants normally do not take over the entire population but exist in low frequency as members of the bacterial or phage population. However, when the environment favors one genotype in a population over others, the proportions of the different genotypes in the population change. Here, for example, T4 kills *E. coli* K, but not K/4, so K/4 soon becomes predominant.

BACTERIAL CONJUGATION

The existence and heritability of mutations in bacteria and their viruses made these microbes attractive subjects for genetic investigations. But if their reproduction were solely asexual, bacteria and their phages would not be useful for genetic analysis. Some form of exchange of genetic information between individuals is necessary. Luckily, such exchange does occur. Genetic recombination—a sexual phenomenon—is a rare event in *E. coli*; it was demonstrated by Joshua and Esther Lederberg and Edward Tatum in 1946.

The Lederbergs and Tatum used two nutrient-requiring (auxotrophic) strains of *E. coli* K12 as parents. Because strain I required the amino acid methionine and the vitamin biotin for growth, its genotype is symbolized as *met⁻ bio⁻*. Strain II requires neither of these substances but cannot grow without the amino acids threonine and leucine. Considering all four factors, we note that strain I is *met⁻bio⁻thr⁺leu⁺* and strain II *met⁺bio⁺thr⁻leu⁻*. These two mutant strains were mixed and cultured together for several hours on a medium supplemented with methionine, biotin, threonine, and leucine, so that both could grow. The bacteria were then removed from the medium by centrifugation, washed, and transferred to minimal medium. Neither parental strain could grow on this medium because of their nutritional requirements. However, a few colonies *did* appear on the plates. Because they grew in the minimal medium, these colonies must have consisted of bacteria that were *met⁺bio⁺thr⁺leu⁺*—that is, they were prototrophic

(Figure 12.2). These colonies appeared at a rate of approximately 1 for every 10 million cells put on the plates.

From what you have learned thus far, try to formulate at least three possible explanations for the appearance of these prototrophic colonies. One possibility is mutation, but this hypothesis can be rejected for the following reasons: The observed colonies would have had to arise from *double* mutations because each parental strain started with two defective alleles. Given the range of single-mutation frequencies, one would expect such double mutations to occur in, at most, 1 out of every 10^{12} cells—100,000 times less frequently than was actually observed. Neither parental strain, when grown alone under the same conditions, was ever observed to give rise spontaneously to prototrophic colonies.

A second possibility is transformation, the incorporation of DNA from dead bacteria into live ones, as first described by Griffith (see Figure 11.1). This explanation was ruled out by Bernard D. Davis of the U.S. Public Health Service. Davis conducted his experiment with a U-shaped tube, the two arms of which were separated by a very fine filter. The pores in the filter were large enough for DNA to pass through, but small enough to prevent the passage of bacteria. He placed a culture of strain I *E. coli* K12 in one arm and a culture of strain II in the other. Then he applied alternating pressure and suction so that the growth medium was flushed back and forth. The flushing mixed the solutions from the two arms without mixing the bacteria. DNA from dead bacteria *would* have been able to pass through the filter, but

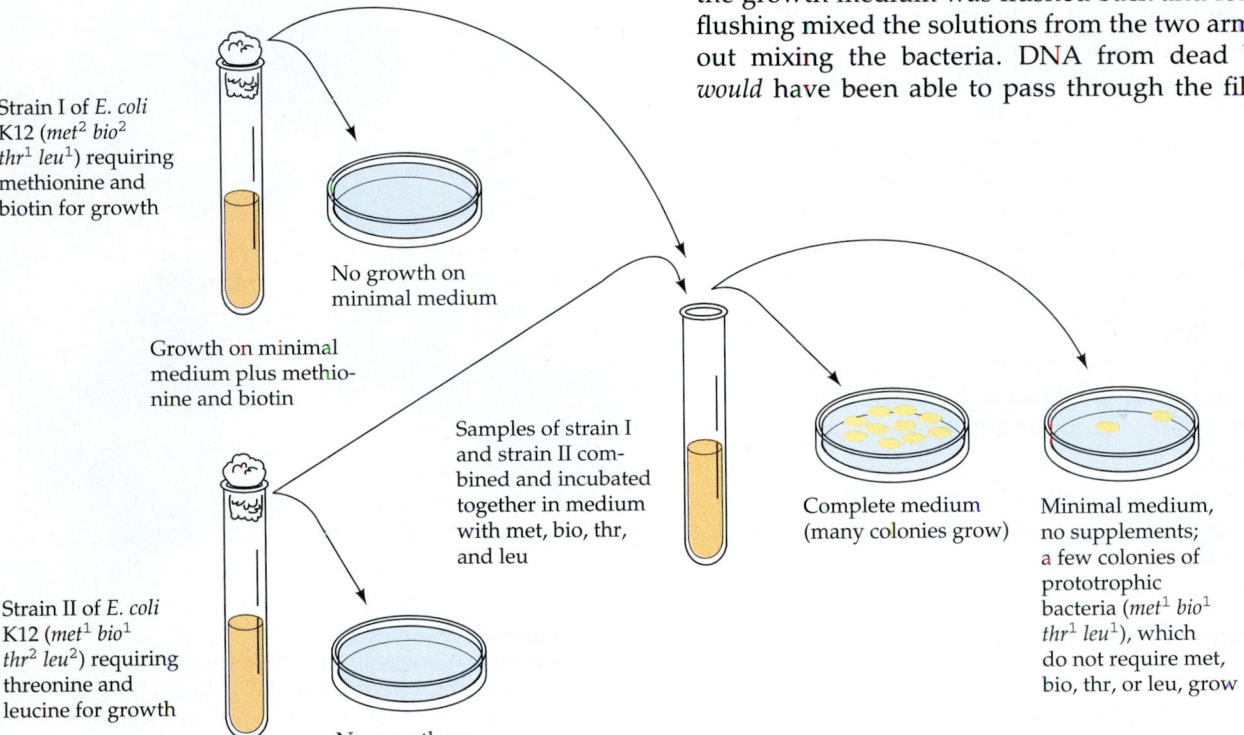

Strain I of *E. coli* K12 (*met² bio² thr¹ leu¹*) requiring methionine and biotin for growth

No growth on minimal medium

Growth on minimal medium plus methionine and biotin

Samples of strain I and strain II combined and incubated together in medium with met, bio, thr, and leu

Strain II of *E. coli* K12 (*met¹ bio¹ thr² leu²*) requiring threonine and leucine for growth

No growth on minimal medium

Growth on minimal medium plus threonine and leucine

Complete medium (many colonies grow)

Minimal medium, no supplements; a few colonies of prototrophic bacteria (*met¹ bio¹ thr¹ leu¹*), which do not require met, bio, thr, or leu, grow

12.2 New Prototrophic Colonies Appear
After growing together, a mixture of complementary auxotrophic strains contains a few cells that can give rise to new prototrophic colonies.

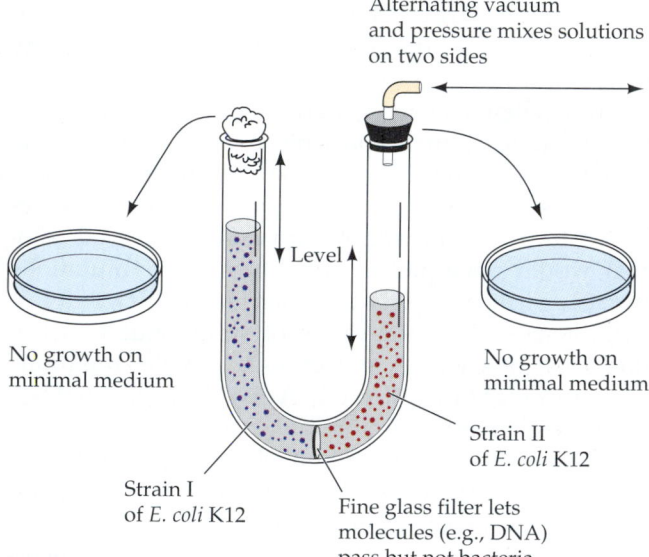

Alternating vacuum and pressure mixes solutions on two sides

Level

No growth on minimal medium

No growth on minimal medium

Strain I of *E. coli* K12

Strain II of *E. coli* K12

Fine glass filter lets molecules (e.g., DNA) pass but not bacteria

12.3   The Davis U-Tube Experiment
Because no prototrophic bacterial cell was recovered from either side of the U-tube, transformation was ruled out as a possible explanation for the appearance of prototrophic colonies in the Lederberg–Tatum experiment (Figure 12.2). Davis's results suggested that the appearance of prototrophic cells requires physical contact between cells of the parental strains.

Davis found *no* wild-type bacteria on either side of the filter (Figure 12.3). These results showed that the phenomenon observed by the Lederbergs and Tatum requires physical contact between cells of the two strains, not just incorporation of DNA.

The third possibility is that the bacteria had **conjugated** in pairs, allowing their genetic material to mix and recombine to produce prototrophic colonies from cells containing *met*$^+$ and *bio*$^+$ alleles from strain II and *thr*$^+$ and *leu*$^+$ alleles from strain I. This explanation was confirmed by other experiments, in which two cells of differing genotype mated, and one cell— the recipient—received DNA that included the two wild-type alleles for the loci in the recipient. Recombination then created a genotype with four wild-type alleles. The physical contact required for conjugation was later observed under the electron microscope (Figure 12.4).

What sort of a process brings about the recombination of genes after bacteria conjugate? We will learn about this shortly.

Isolating Specific Bacterial Mutants

Throughout this chapter we will consider experiments that use bacteria and phages with various specific genotypes, as we have just done in the conjugation experiment. How can one obtain a strain with a particular genotype, such as *met*$^-$*bio*$^-$*thr*$^+$*leu*$^+$?

To isolate bacteria carrying a particular mutation—

say *met*$^-$—we start with a strain carrying the wild-type allele for which mutations are desired. We then subject these bacteria to procedures that increase the mutation rate, such as irradiation with ultraviolet or X rays, or the addition of a chemical mutagen. Now the search begins. First, we let the bacteria in the culture grow and increase their numbers by keeping them in a medium that includes the compound that will be needed by the desired mutant strain (in our example, the medium must include methionine). The overwhelming majority of the cells in the culture are unchanged; these wild-type cells must be eliminated.

There is more than one way to eliminate the wild-type cells, but we will describe one invented by Bernard Davis. He knew that the antibiotic penicillin kills only growing bacteria. Therefore, his method was to put a mixed culture of many wild-type and a few mutant bacteria into a medium *lacking* the nutrient for which the desired mutants were auxotrophic (again, methionine in our example), and add penicillin. In this experiment, the cells that do not need methionine grow rapidly—and commit "penicillin suicide." Because they grow, they die. The desired mutants, on the other hand, fail to grow (because the needed nutrient is unavailable), so they avoid damage by the penicillin. These mutants are then trans-

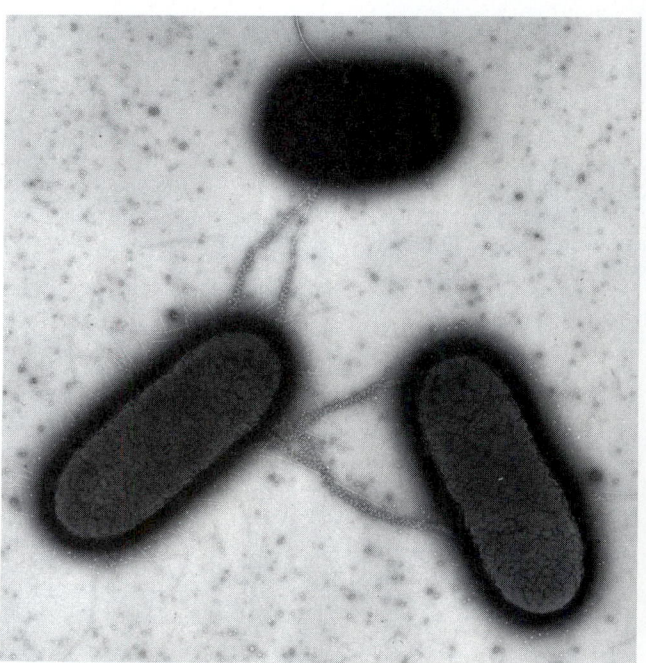

12.4   Bacteria Conjugate
In this electron micrograph of *E. coli*, the "male" cell on the left is connected to two "female" cells by thin tubes called F-pili. In this instance, two F-pili are in contact with each female cell. The tiny "beads" on the pili are bacteriophages that attach specifically to F-pili, making the pili more visible. After cells are joined by F-pili, they are drawn into closer contact, and DNA is transferred from one cell to the other.

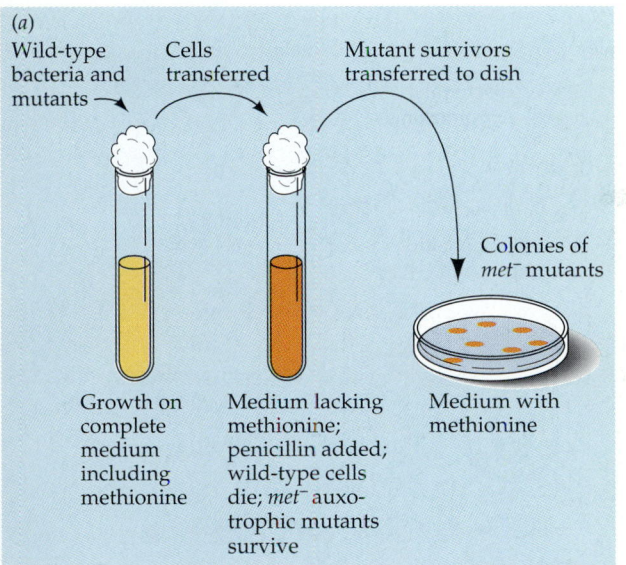

(a)
Wild-type bacteria and mutants → Cells transferred → Mutant survivors transferred to dish

Colonies of *met⁻* mutants

Growth on complete medium including methionine

Medium lacking methionine; penicillin added; wild-type cells die; *met⁻* auxotrophic mutants survive

Medium with methionine

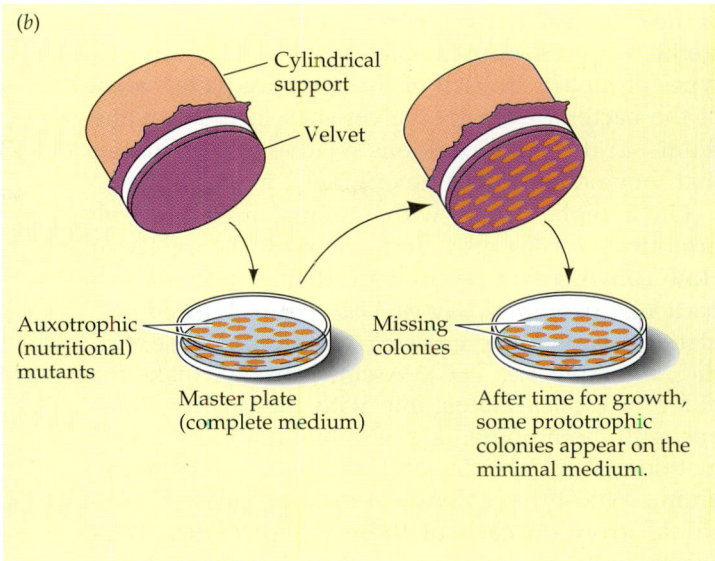

(b)
Cylindrical support
Velvet

Auxotrophic (nutritional) mutants

Master plate (complete medium)

Missing colonies

After time for growth, some prototrophic colonies appear on the minimal medium.

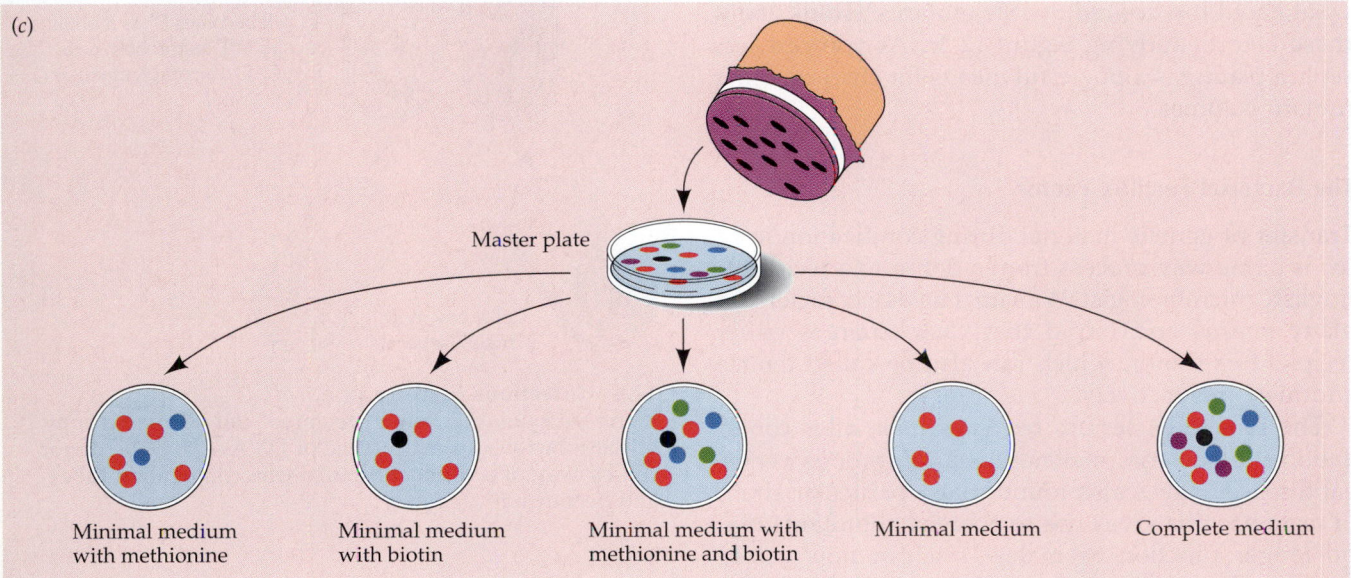

(c)
Master plate

Minimal medium with methionine

Minimal medium with biotin

Minimal medium with methionine and biotin

Minimal medium

Complete medium

12.5 Isolating and Identifying Auxotrophic Mutants
(a) Penicillin kills growing wild-type cells; the nongrowing, methionine-requiring auxotrophic mutants survive. (b) In the replica plating process, the velvet replicates the spatial pattern, permitting identification of auxotrophic colonies present on the master plate but missing from the replica plate with minimal medium. (c) The colonies shown in green on the master plate grow only on the plate containing both methionine and biotin (middle) and are therefore *met⁻bio⁻*. Can you identify the nutritional requirements of each remaining colony on the master plate?

ferred to medium that has no penicillin but does contain methionine, so that they may grow and form colonies (Figure 12.5a).

A particular methionine mutant can then be chosen and used as a parental strain for selecting a *met⁻ bio⁻* (*thr⁺leu⁺*) double mutant by the same general approach as was used to select the single (*met⁻*) mutant. In this case, of course, methionine must be present in the medium at all times, and the absence of biotin is used to protect the *met⁻ bio⁻* double mutants from penicillin-induced death. By selecting for the second mutation, one obtains the desired double mutant (*met⁻bio⁻*).

How does one identify the progeny of various recombinant types after performing a cross between two strains of bacteria? One method is **replica plating**, a technique invented by Joshua and Esther Led-

erberg. A small sample (about 0.1 milliliter) of a mixed suspension, presumably including the desired genotype as well as others, is spread on a plate with complete medium and allowed to produce colonies. A sterilized piece of velvet, mounted on a cylindrical support that fits easily into a petri plate, is now pressed gently against the medium. Its fuzzy surface picks up substantial numbers of bacteria from each of the colonies. The velvet is next pressed against the sterile surfaces of new plates containing different types of media. In each of these replica plates, some of the bacteria from the velvet stick to the agar medium—in the same positions relative to one another that they occupied on the original, "master" plate.

On a replica plate with minimal medium, only prototrophic cells grow into colonies (Figure 12.5b). How can this fact be used to identify bacteria of a particular genotype, say *met⁻bio⁻*, from a mixed population obtained by crossing two strains (one *met⁻ bio⁺*, the other *met⁺bio⁻*)? Assume that we make five different replica plates: one with minimal medium, one with methionine, one with biotin, one with both methionine and biotin, and one with complete medium. Wild-type colonies from the master plate would grow on each of these replica plates. What about single mutants (*met⁻bio⁺* and *met⁺bio⁻*)? And what about the desired double mutants? Think about these before studying Figure 12.5c. As you can see, replica plating is a powerful means for characterizing mutant colonies.

The Bacterial Fertility Factor

Transfer of genetic material during conjugation in *E. coli* is a one-way process from a donor to a recipient. English microbiologist William Hayes characterized many strains and found that each strain is either recipient or donor, which can also be called female and male, respectively.

The female bacterium becomes male after conjugation—in bacteria, maleness is an infectious venereal disease! Hayes also found that a particular strain of male bacteria gives rise to occasional mutants that no longer function as males—but can now act as females. Hayes rationalized these observations by proposing that maleness in bacteria is due to the presence of a fertility factor, called **F**. Males possess the factor and are F⁺; females, lacking the factor, are F⁻. In a cross of F⁺ × F⁻, a copy of the F factor is transferred to the female, thus rendering it F⁺, while the original male remains F⁺ (Figure 12.6).

The F factor is an extra piece of DNA that can replicate itself and persist in the cell population as if it were a second chromosome independent of the normal bacterial chromosome. Males can change into females simply by losing the F factor through mutation. Genes on the F factor direct a number of processes, among which is the formation on the surface of the male bacterium of long, thin, hairlike projections called **F-pili** (singular: *pilus* = hair). These are tubes with ends that attach to the surface of female cells (see Figure 12.4). Initial contact is made by an F-pilus, and subsequently a mating contact is made that allows DNA to be transferred.

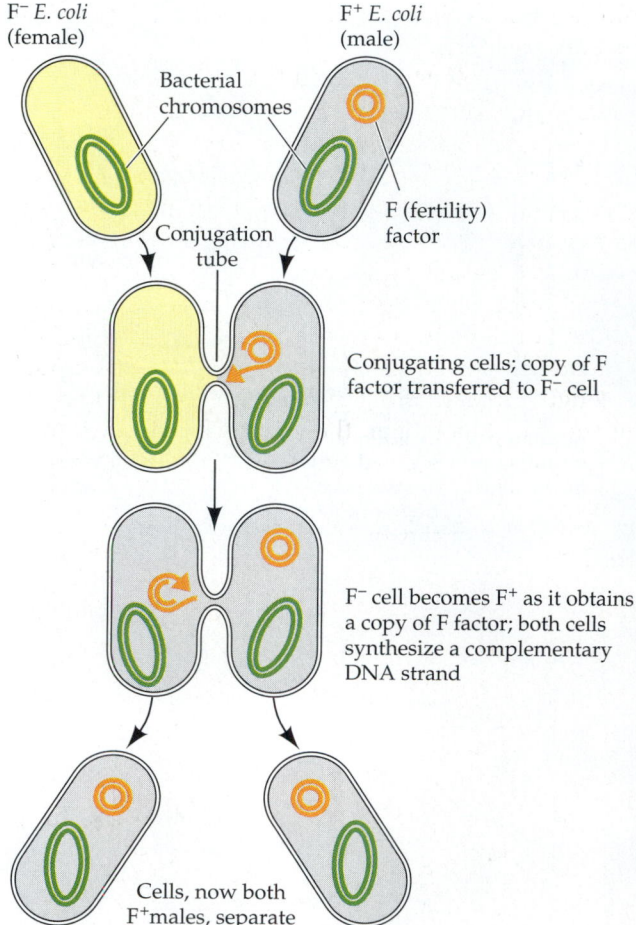

12.6   Infectious Fertility in *E. coli*
During conjugation, the F⁻ recipient cell receives a copy of the F factor—an extra piece of DNA—from the donor cell by way of a connecting tube (the conjugation tube) and becomes F⁺.

Transfer of Male Genetic Elements

The discovery that genetic recombination could follow conjugation opened the possibility of mapping the genetic material of bacteria. However, early attempts at mapping were complicated by the fact that very few recombinant offspring arose from F⁺ × F⁻ crosses, thus making it difficult to obtain reliable quantitative data. The situation changed with the discovery of certain mutant male strains. Recombinant offspring *were* obtained when these males were used in crosses. These strains were called **Hfr** mutants, for *High frequency of recombination*. Hfr males, unlike ordinary F⁺ males, do not generally transfer their F factor to the female. Also, they transfer only certain marker genes with high frequency, transferring other markers no more frequently than ordinary F⁺ males do. We know now that in Hfr strains the F factor is actually incorporated into the bacterium's chromosome. Work in 1955 by the French biologists Elie Wollman and François Jacob explained these observations.

Jacob and Wollman showed that the markers from an Hfr male enter the female one at a time. In their most dramatic experiments, they used the technique of **interrupted mating**. They mixed Hfr and F⁻ bacteria at high concentration to initiate conjugation; at various times thereafter they diluted samples of the mixture and agitated them in a kitchen blender for two minutes. Such agitation separates conjugating bacteria but does not damage them. The number of Hfr markers passed to the females depended upon the length of time allowed before conjugation was interrupted—the longer the conjugation, the more markers were transferred (Figure 12.7a). The markers always entered in a particular order from any particular Hfr strain. The Hfr mutant almost never transferred the F factor.

Jacob and Wollman recognized immediately that this interrupted mating technique provided a simple way to map the chromosome. They prepared different mutant strains and crossed pairs of strains; then they interrupted successive samples from the crosses. Because the markers are transferred in a particular sequence, the length of mating time required before a particular marker is transferred and thus available to appear in recombinant progeny is a measure of its location on the chromosome (Figure 12.7b).

Different Hfr mutants have different genetic maps (Table 12.1). If you examine the table, however, you may be able to spot a regularity in the different maps. Jacob and Wollman noticed that although different markers are transferred first in different Hfr strains, the maps are always consistent in that a marker *B* that lies between markers *A* and *C* always does so in every Hfr strain. That is, the starting points may vary, but the *order* of genes remains constant. Even when genes are in a reversed order, *B* is still between *A* and *C*. The simplest conclusion is that the bacterial chromosome is *circular* (Figure 12.8). If you break the circle in different places and convert the results into linear form, you can see how the maps are generated.

From these and other experiments, Jacob and Wollman concluded that (1) the *E. coli* chromosome is circular, (2) Hfr males have the F factor incorporated into their chromosome, (3) the location where the F factor is inserted varies, giving rise to different Hfr strains, (4) the inserted F factor marks the point at which the chromosome "opens" as conjugation begins, and (5) one end—always the same one—of the opened chromosome leads the way into the female. The piece of chromosome continues to move through the conjugation tube until mating is inter-

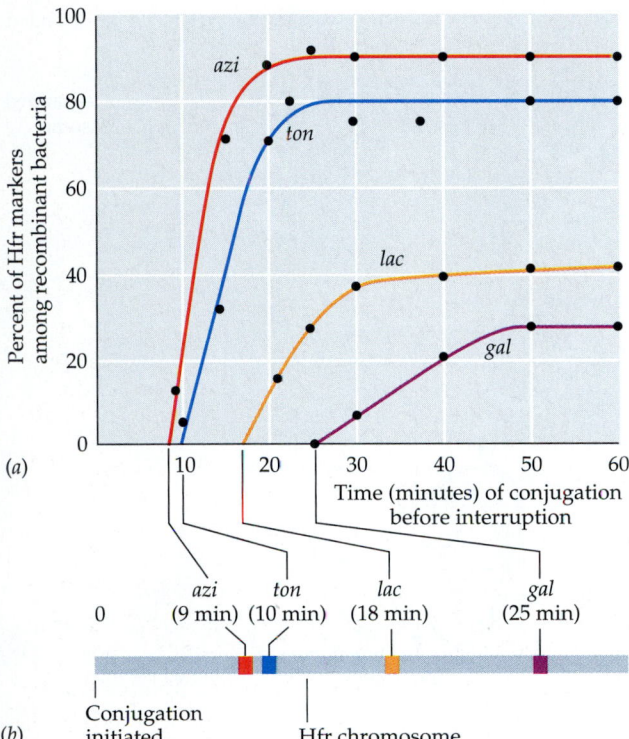

(a)

(b) Conjugation initiated

Hfr chromosome

azi (9 min) ton (10 min) lac (18 min) gal (25 min)

12.7 Chromosome Maps from Interrupted Matings

Chromosome maps of *E. coli* were constructed by interrupting conjugating cells at various times and counting recombinants from the matings. (a) The map in (b) was constructed from data plotted on this graph. In the particular Hfr strain used, recombinants carrying the *azi* allele were first detected from matings interrupted 10 minutes after mixing, as were recombinants carrying the *ton* allele. By extrapolation from other points on the graph, it was determined that the *azi* gene was transferred at 9 minutes, the *ton* gene at 10 minutes. Timings for the *lac* and *gal* genes were established in the same way. (b) The units on this chromosome map are minutes rather than recombination frequencies.

TABLE 12.1
Sequences of Markers Transferred by Various Hfr Strains

ORDER OF ENTRY	H	1	2	3	4
			Hfr STRAIN		
1	T	L	pro	ade	B₁
2	L	T	T₁	lac	ilu
3	azi	B₁	azi	pro	mal
4	T₁	ilu	L	T₁	trp
5	pro	mal	T	azi	gal
6	lac	trp	B₁	L	ade
7	ade	gal	ilu	T	lac
8	gal	ade	mal	B₁	pro
9	trp	lac	trp	ilu	T₁
10	mal	pro	gal	mal	azi
11	ilu	T₁	ade	trp	L
12	B₁	azi	lac	gal	T

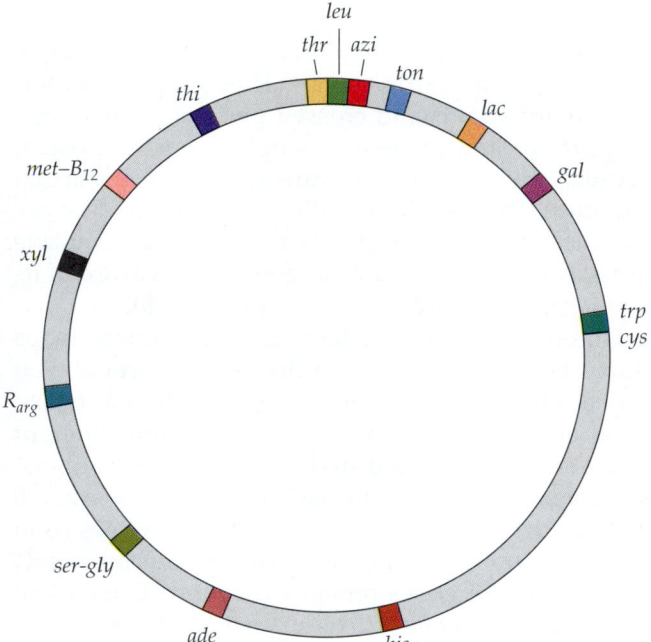

12.8   *E. coli*'s Chromosome Is Circular
This map of the *E. coli* chromosome summarizes data
from interrupted conjugation experiments with many Hfr
strains (see Figure 12.7 and Table 12.1). The three-letter
abbreviation for each gene is derived from its phenotype:
*leu*cine-requiring, *azi*de-resistant, *xyl*ose-utilizing, and so
on. This early version of the map shows only a few
genes. By now, well over half of the estimated 3,000 *E.
coli* genes have been mapped.

rupted naturally or otherwise. At the very end of the
opened chromosome lies the portion of the F factor
that determines maleness.

What moves from the Hfr to the F⁻ is just one
strand of the double-stranded Hfr chromosome.
Transfer is initiated when one strand within the F
factor is nicked (Figure 12.9). The 5′ end of the nicked
strand begins to unravel from the chromosome and
moves to the F⁻. Meanwhile, the transferred strand
is replaced in the Hfr by DNA synthesis at the 3′ end
of the nick, using the intact circular strand as the
template. Thus the male still contains a double-
stranded set of DNA sequences even after donating
a fair amount of DNA to the F⁻. As it enters the F⁻,
the nicked DNA strand replicates, becoming double-
stranded. Markers on this piece of DNA will give rise
to recombinant bacteria only if they become incorporated into the F⁻ chromosome by crossing over
(Figure 12.10). About half of the transferred Hfr
markers get incorporated in this way; the others are
lost as the cell divides.

Sexduction

Sometimes the F factor of an Hfr male separates from
the chromosome. In the separation process, the F

factor may carry with it a bit of the chromosome. Any
genes thus captured by the F factor are transferred
to the F⁻ recipient when conjugation occurs. The
process in which genes are carried by the autonomous F factor into the F⁻ is called **sexduction**. The
modified F factor is called an F′ (F-prime) factor (Figure 12.11).

An F or F′ factor, like the bacterial chromosome,
is a circular DNA molecule. Genes carried by the F′
factor are allelic to genes on the main chromosome
in the recipient cell. Thus cells harboring an F′ factor
may contain more than one allele of a particular gene.
Such cells may be used to study whether there is
dominance among the alleles of a gene present on
the F′ factor—a question that usually cannot be investigated in these normally haploid bacteria.

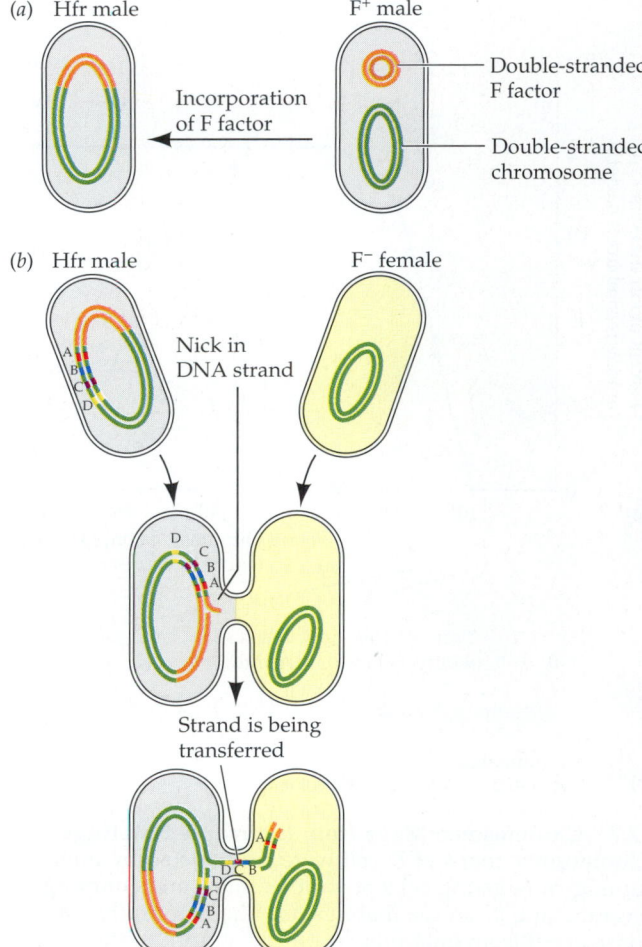

12.9   Origin and Behavior of Hfr Strains
(*a*) A cell becomes an Hfr male when the F factor of an F⁺
cell is incorporated into its chromosome. (*b*) During conjugation, the Hfr male's chromosome opens within the inserted F factor and one strand of the DNA double helix is
transferred to the recipient cell. Because most of the F factor is the last DNA to be transferred, the recipient cell
usually does not become a male, since the complete chromosome is rarely transferred.

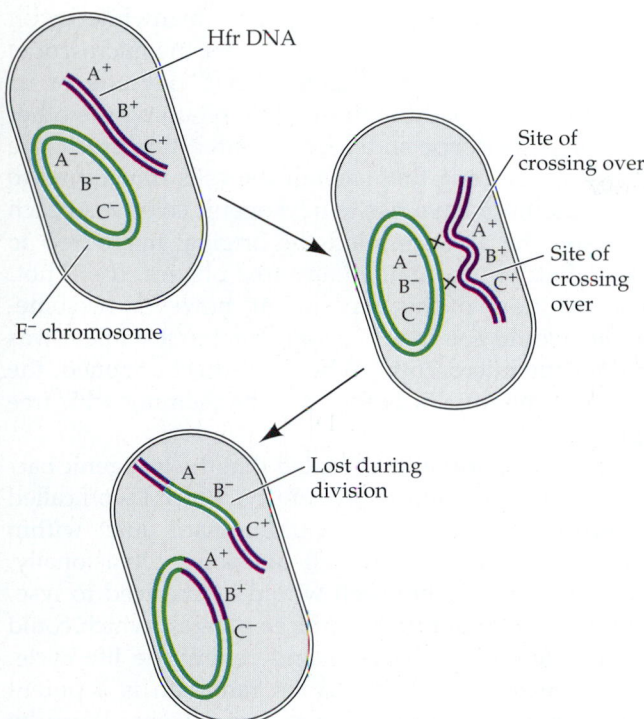

12.10   Recombination Following Conjugation
DNA from an Hfr donor may become incorporated into the recipient cell's chromosome through crossing over. Only part of the donor chromosome here was incorporated—the part containing A^+ and B^+. The $A^+B^+C^-$ sequence becomes a permanent part of the recipient genotype, and the reciprocal $A^-B^-C^+$ segment is lost.

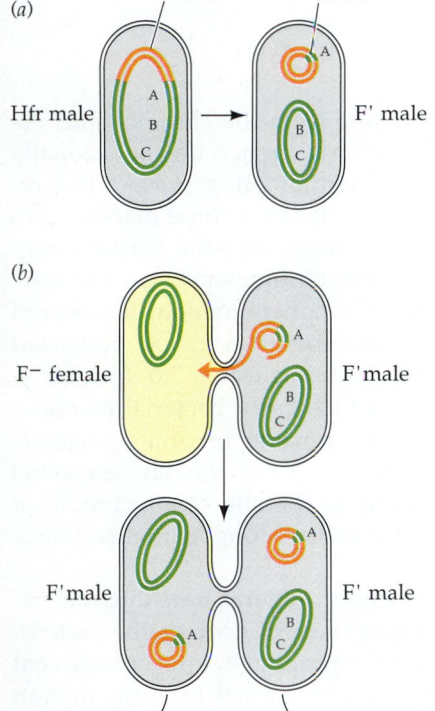

12.11   Sexduction
(a) An F factor in an Hfr cell may carry some genes with it as it leaves the cell's chromosome, changing the cell into an F' cell. (b) A recipient that conjugates with an F' cell receives any genes carried by the F' factor by sexduction and becomes an F' cell.

BACTERIOPHAGES

Recombination in Phages

Genetic recombination in phages was demonstrated in 1946, the same year the Lederbergs and Tatum revealed the sex life of bacteria. Here too, mutant phenotypes were needed as markers. The alleles of some phage genes affect the appearance of plaques (see Figure 12.1), and we have already noted that mutations can change the ability of a phage to infect certain hosts. To understand the basis for genetic recombination in phages, recall that phages reproduce by injecting their genetic material into a host bacterium, which then supports the synthesis of a large number of progeny phages.

Alfred Hershey and Raquel Rotman, at Washington University in St. Louis, performed a series of experiments in which *E. coli* were simultaneously infected by *two* different mutant strains of the bacteriophage T2. In their first experiment, one of the phage strains was genotypically h^+r and the other was hr^+. (We need not worry here about what the phenotypes were; just note that h^+ and h are alleles at one locus and r^+ and r are alleles at another locus.) We would expect the addition of these phages to a culture of *E. coli* to produce substantial numbers of phages of both parental types. Hershey and Rotman found not only the parental types, however, but also many phages of genotypes h^+r^+ and hr—that is, recombinant phages (Figure 12.12). As more markers in such phage crosses were studied, a map began to take form. In due course, it was learned that the phage has a single, circular chromosome.

Lysogeny and the Disappearing Phages

Up to now we have treated bacteriophages as if their life cycle was always **lytic**: a cycle in which a phage infects a bacterial cell, the phage replicates, the cell lyses, and many new phages are released to renew the cycle. With some bacteria and some phages, however, infection does not always result in lysis of the bacteria. The phages seem to disappear from the culture, leaving the bacteria immune to further attack by the same strain of phage. In such cultures, however, a few free phages are sometimes detected. Bacteria harboring phages that are not lytic are called **lysogenic**. When lysogenic bacteria are combined with other bacteria that are sensitive to the phages, they cause the sensitive cells to lyse.

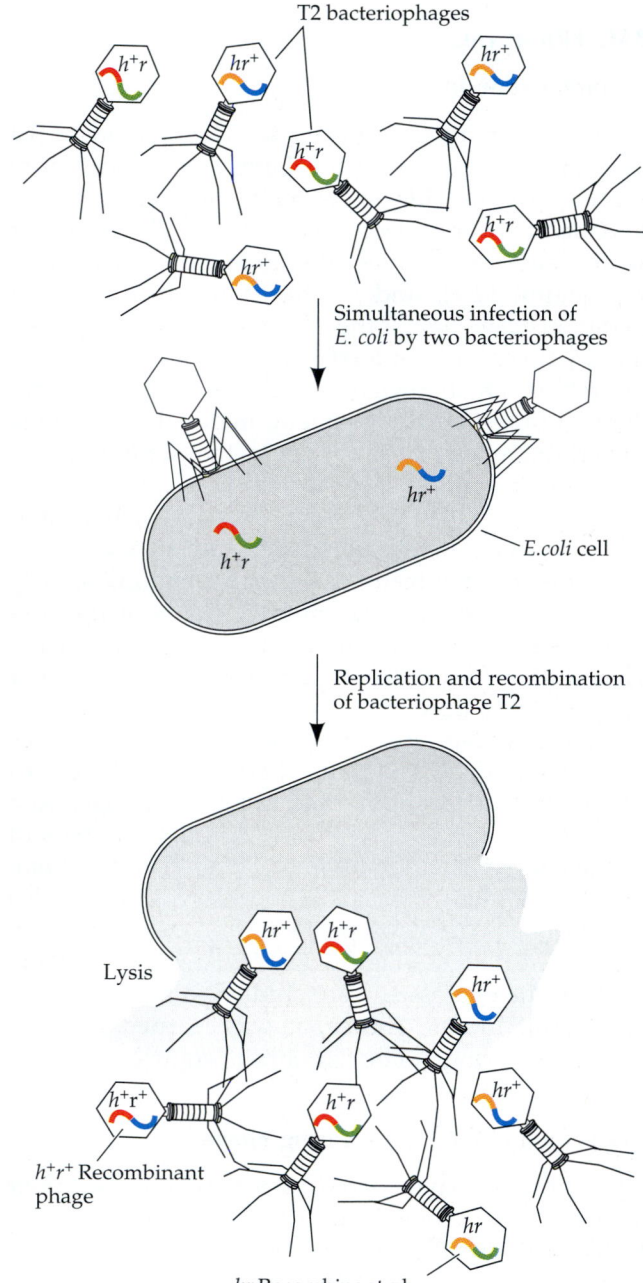

12.12 Bacteriophage Genetic Recombination
Bacteriophage strains can be crossed by infecting a culture of susceptible bacteria with both strains simultaneously.

To see where the free phages come from in a culture of lysogenic bacteria, the French microbiologist André Lwoff performed the following delicate experiment with the large bacterium species *Bacillus megaterium*. From a lysogenic culture of *B. megaterium*, Lwoff isolated a single bacterial cell, mounting it in a drop of medium on a microscope slide. He watched patiently until the cell divided and then removed one of the daughter cells with a micropipette. This cell was transferred to an agar medium to see whether

its offspring would be lysogenic. Meanwhile Lwoff watched the daughter cell still under the microscope. When it divided, he again farmed out one of its daughters to solid medium while retaining the other on the microscope slide. He repeated this procedure 19 times, finding that each of the cells transferred to solid medium gave rise to a lysogenic colony. At each transfer, he also sampled the original microdrop to see whether it contained any free phages. It did not. In repetitions of this experiment, however, he sometimes would see the *B. megaterium* burst while it was under the microscope. Whenever this happened, the drop of medium was found to be teeming with free phages.

These experiments revealed that the lysogenic bacteria contain a noninfective entity, which Lwoff called a **prophage**. Prophages could remain quiet within bacteria through many cell divisions. Occasionally, however, a lysogenic cell would be induced to lyse, releasing a large number of free phages, which could then infect other bacteria and renew the life cycle. Lwoff learned that ultraviolet radiation is a potent inducer of the production of free phages. Work by many investigators established finally that the prophage is a molecule of phage DNA that has been incorporated into the bacterial chromosome. Notice the similarity between prophage behavior and that of the F factor that inserts into the chromosome to give rise to an Hfr male. Just as the F factor sometimes leaves the chromosome, so may the prophage—whereupon the phage DNA is activated to multiply rapidly, to make many new phages, and to lyse the bacterium. The lytic and lysogenic cycles are contrasted in Figure 12.13.

Transduction

The prophage can escape from the chromosome. As you might expect, bacterial genes are occasionally taken along by the departing phage DNA. The resulting phages can then introduce these markers into other bacteria that they infect, and the markers may be incorporated into the chromosomes of the new hosts (Figure 12.14*a*). This phenomenon, discovered in 1956 by Joshua Lederberg, is called **restricted transduction** (here *transduce* means "to transfer"). Transducing phages, which carry bacterial markers, cause newly infected bacteria to become lysogenic. In restricted transduction, only the chromosomal genes that are adjacent to the site of attachment of the prophage may be taken along with the phage DNA.

A related phenomenon, **general transduction**, results from the incorporation of part of the *bacterial* chromosome, *without* the prophage, into a phage coat (Figure 12.14*b*). The resulting particle, even though it lacks any phage genes, can infect another bacter-

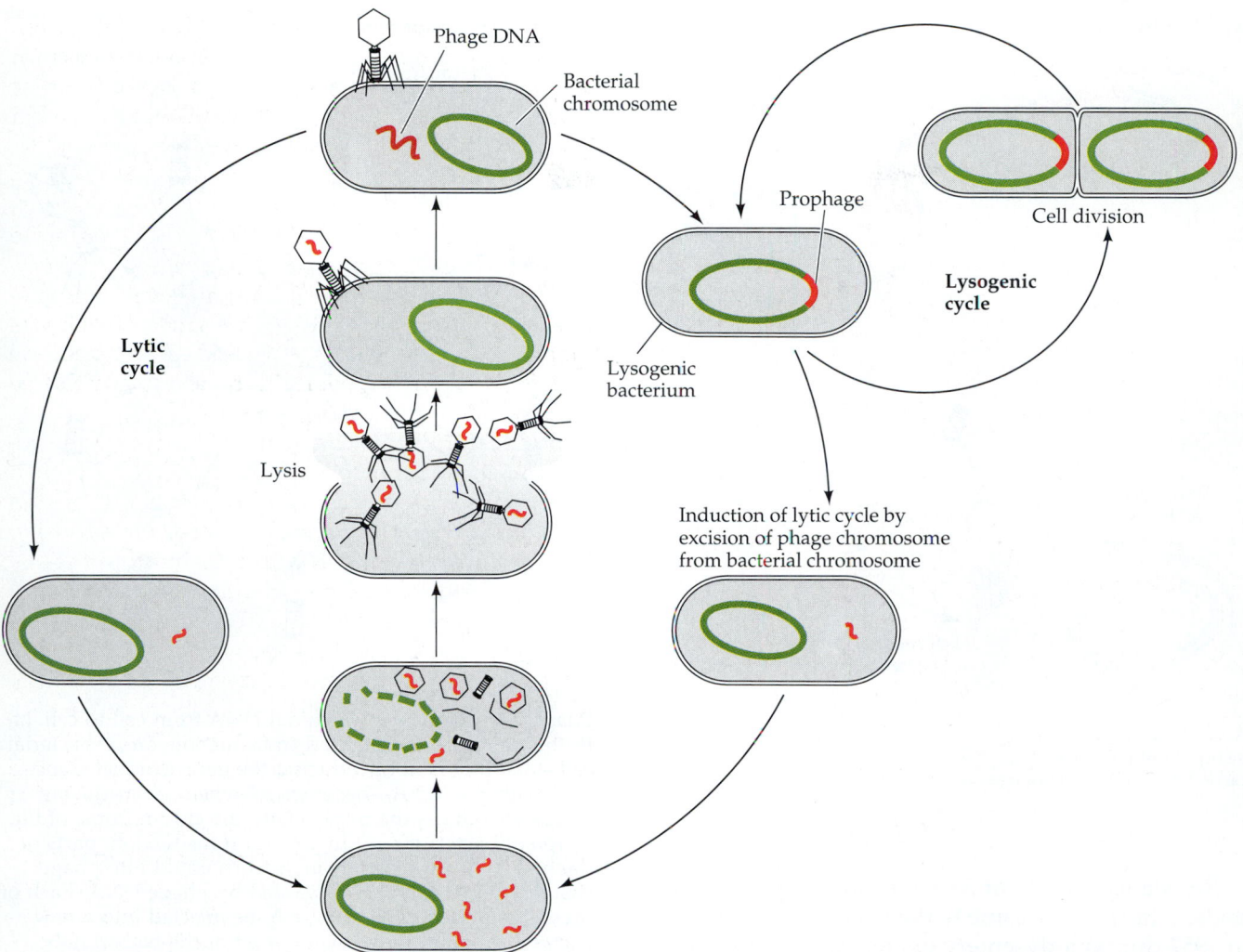

Phage DNA

Bacterial chromosome

Prophage

Cell division

Lysogenic cycle

Lysogenic bacterium

Lytic cycle

Lysis

Induction of lytic cycle by excision of phage chromosome from bacterial chromosome

12.13 Lytic and Lysogenic Phage Cycles
Phage DNA injected into a host bacterium may be incorporated into the host's chromosome, becoming a noninfective prophage (lysogenic cycle), or it may remain free (lytic cycle). The lytic cycle, which produces new phages and lyses the host cell, may be repeated until all host cells are lysed. In the lysogenic cycle, the prophage is replicated as part of the host's chromosome; but if a prophage separates from the host's chromosome, it becomes a lytic phage.

ium, injecting the piece of DNA from its former host. A bacterium thus infected with a piece of foreign bacterial DNA does *not* become lysogenic, nor does it form new phages and burst as in the lytic cycle—it has not really been infected by a phage. It simply contains extrachromosomal bacterial DNA, as if it had conjugated with an Hfr cell. If crossing over takes place between the host chromosome and the transduced DNA, the transfer of markers is completed. In contrast with restricted transduction, general transduction can move any part of the bacterial chromosome. There is no limitation on what chromosomal markers might become enclosed in a phage coat. The phage coat is big enough to house several adjacent bacterial genes. General transduction therefore is another powerful tool for mapping the bacterial chromosome—with viral assistance.

EPISOMES AND PLASMIDS

The F factor and viral prophages are examples of **episomes**. Episomes are nonessential genetic elements that can exist in either of two states: independently replicating within a cell, or integrated into the main chromosome. Episomes cannot arise by mutation; they must be obtained by infection from outside the bacterium. The infection can come from a virus or from another bacterium. As we have seen, episomes may be used as vehicles for transferring genetic markers from one bacterium to another. Other nonessential genetic elements, which exist only as free, independently replicating circles of DNA that cannot be incorporated into the bacterial chromosome, are called **plasmids**. (An episome is simply a plasmid that has the possibility of becoming part of the chromosome.)

(a) **Restricted transduction**

A⁻ cell becomes A⁺ with
incorporation of marker
carried by transducing phage

(b) **General transduction**

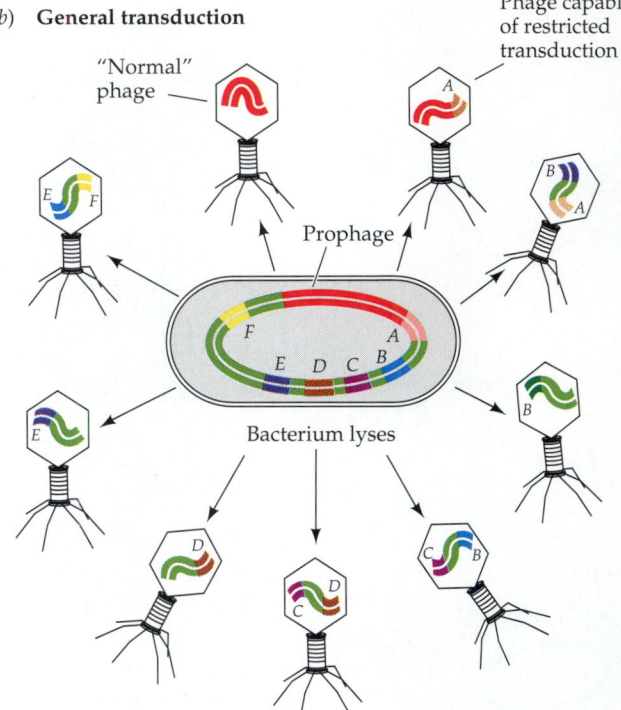

12.14 Transduction

Phages may transduce bacterial DNA from cell to cell. *(a)* In this example of restricted transduction, an A^- bacterial cell—one that is not producing the gene product associated with marker A—becomes A^+ when a transducing phage introduces the marker into the chromosome of the recipient bacterium. *(b)* In general transduction, parts of the bacterial chromosome are incorporated into phage coats without being accompanied by phage DNA. Each of these bits of bacterial DNA may be injected into a new bacterial cell and may become part of the bacterium's DNA by recombination.

Resistance factors, or **R factors**, are important plasmids. R factors first came to the attention of biologists in 1957 during a dysentery epidemic in Japan, when it was discovered that some strains of the dysentery bacterium *Shigella* were resistant to several antibiotics. Researchers found that resistance to the entire spectrum of antibiotics could be transferred by conjugation even when no markers on the main chromosome were transferred. Also, F⁻ cells could serve as donors, indicating that the genes for antibiotic resistance were not carried by the F factor. Eventually it was shown that the genes were carried on plasmids. Each of these plasmids (the R factors) carries one or more genes conferring resistance to particular antibiotics. As far as biologists can determine, R factors appeared long before antibiotics were discovered and used, but they seem to have become more abundant in modern times. Can you propose a hypothesis to explain why R factors might be more widespread now than they were in the past?

Bacteria do not require plasmids to live. Thus in order for a particular type of plasmid to be maintained within a population of bacteria, it must have an origin of replication (see Chapter 11). That is, the plasmid must be a **replicon**, capable of independent replication, so that it divides at roughly the same rate as the bacterium. Otherwise, it is simply diluted out of the population.

TRANSPOSABLE ELEMENTS

As we have seen, plasmids, episomes, and even phage coats can transport genes from one bacterial cell to another. Another type of "gene transport" within the individual cell relies on segments of chromosomal or plasmid DNA called **transposable elements**. Copies of transposable elements can be inserted at other points in the same or other DNA molecules, often producing multiple physiological effects resulting from the disruption of the genes into which the transposable elements are inserted (Figure 12.15*a*).

The first transposable elements to be discovered in prokaryotes were large pieces of DNA, typically 1,000 to 2,000 base pairs long, found in many places in the *E. coli* chromosome. The sequence of a transposable element can replicate independently of the rest of the chromosome. The copy then inserts itself at other, seemingly random places in the chromosome. The genes encoding the enzymes necessary for this insertion are found within the transposable

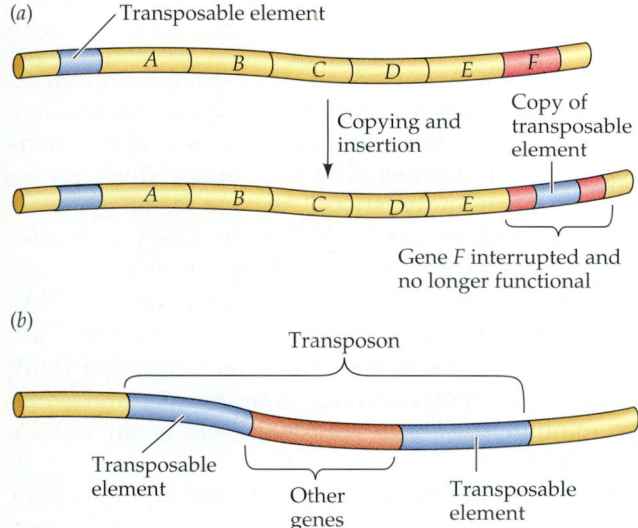

(a) Transposable element

Copying and insertion

Copy of transposable element

Gene *F* interrupted and no longer functional

(b)

Transposon

Transposable element

Other genes

Transposable element

12.15 Transposable Elements
(a) If a transposable element appears in the middle of another gene, that gene can no longer be transcribed to yield an appropriate mRNA; thus the interrupted gene cannot function. *(b)* A transposon consists of two transposable elements flanking another gene or genes; the entire transposon is copied and inserted as a unit.

element itself. Many transposable elements discovered later were longer (about 5,000 base pairs) and carried one or more additional genes. These longer elements with additional genes are called **transposons** (Figure 12.15*b*).

Initially, transposable elements were called jumping genes, but the frequency of their insertion is usually very low. Transposition is closely regulated—random insertion would often lead to the inactivation of an essential gene and the death of the cell. The process by which transposable elements move is complex and incompletely understood. In one step, the DNA at the new site is cut by an enzyme, transposase, that is encoded by the transposable element. The enzyme makes a staggered cut, and the copy of the transposable element inserts itself between the ends. When the gaps are repaired by DNA polymerase, a short sequence of chromosomal DNA at one end of the inserted sequence appears again, in duplicate, at the other end; the duplication results from the original staggered cut.

Transposable elements have contributed to the evolution of plasmids. The plasmids called R factors originally gained their genes for antibiotic resistance through the activity of transposable elements; one piece of evidence for this conclusion is that each resistance gene in an R factor is part of a transposon. Transposons on the F factor and on the bacterial chromosome interact to direct the insertion of the F factor into the chromosome in the development of an Hfr male.

CONTROL OF TRANSCRIPTION IN PROKARYOTES

Let us now consider how the activities of prokaryotic genes are *regulated*. As a normal inhabitant of the human gut, *Escherichia coli* has to adjust to sudden changes in its chemical environment. Its host may present it with one foodstuff one hour and another the next. For example, the bacteria may suddenly be deluged with milk, the main carbohydrate of which is lactose. This sugar is a β-galactoside—a disaccharide containing galactose β-linked to glucose (see Chapter 3). Before lactose can be of any use to the bacteria, it must first be taken into their cells by an enzyme called β-galactoside permease. Then it must be hydrolyzed to glucose and galactose by another enzyme, β-galactosidase. A third enzyme, thiogalactoside transacetylase, is also required for lactose metabolism. When *E. coli* is grown in a medium that does not contain lactose or other β-galactosides, the levels of all three of these enzymes within the bacterial cell are extremely low—the cell does not waste energy and material making the unneeded enzymes. If, however, the environment changes so that lactose is the predominant sugar and very little glucose is present, the synthesis of all three of these enzymes begins promptly and their levels may rise more than a thousandfold. Regulation of enzyme synthesis by the genes that code for them thus promotes efficiency in the cell.

Compounds that evoke the synthesis of an enzyme (as does lactose in this example) are called **inducers**. The enzymes that are evoked are called **inducible enzymes**, whereas enzymes that are made all the time at a constant rate are called **constitutive enzymes**. If lactose is removed from *E. coli*'s environment, synthesis of the three enzymes stops almost immediately. The enzyme molecules that have already been formed do not disappear; they are merely diluted during subsequent growth and reproduction until their concentration falls to the original low level within each bacterium.

The blueprints for the synthesis of these three enzymes are called **structural genes**, indicating that they specify the primary structure (that is, the amino acid sequence) of a protein molecule. When Jacob, Wollman, and Monod mapped the particular structural loci coding for enzymes that metabolize lactose, they discovered that all three lie close together in a region that covers only about 1 percent of the *E. coli* chromosome.

It is no coincidence that these three genes lie next to one another. The information from them is transcribed into a single, continuous molecule of mRNA. A molecule that contains transcripts of more than one gene is called a **polycistronic messenger**. Because this particular polycistronic messenger governs the synthesis of all three lactose-metabolizing enzymes, either all or none of the enzymes are made, depending on whether their common message—their mRNA—is present in the cell.

Processing a Polycistronic Messenger

How can a single mRNA molecule make three different polypeptides? The answer is that the polycistronic mRNA contains punctuation marks to specify the end of one polypeptide chain and the start of the next. A molecule of tRNA is always attached to a growing polypeptide chain, but a finished molecule of protein does not contain any tRNA. This indicates that the last step in prokaryotic protein synthesis must involve not only the termination of the polypeptide chain but also the removal of the terminal tRNA. The termination signal is encoded in the mRNA. Three codons of the genetic code (UAA, UAG, and UGA) mean "terminate translation," and one of them must be present at the end of each structural gene.

It is easy to see how a polycistronic messenger can give rise to one polypeptide for each structural gene. A ribosome begins at one end of the message, translates until it comes to the termination signal of the first structural gene transcript, and then releases the first polypeptide. The ribosome itself may remain bound and start translating the second structural gene at the next initiation site and, when it finishes, release the second polypeptide, and so on. When the ribosome has translated all the structural gene transcripts on the mRNA, it is released (Figure 12.16).

Promoters

Some genes are transcribed more often than others. In Chapter 11 we said that RNA polymerase attaches to DNA and starts transcribing, but we did not mention where it attaches. The polymerase does not attach itself randomly; special regions for attachment, called **promoters**, are built into the DNA molecule. There is one promoter for each structural gene or set of structural genes to be transcribed into mRNA. Promoters serve as a punctuation, telling the RNA polymerase where to start and which strand of DNA to read. A promoter and one or more structural genes are enough to specify the synthesis of an mRNA molecule.

Not all promoters are identical. One promoter may bind RNA polymerase very effectively and therefore trigger frequent transcription of its structural genes; in other words, it competes effectively for the available RNA polymerase. Another promoter may bind the polymerase poorly, and its structural genes are rarely transcribed. The efficiency of the promoter sets a limit on how often each structural gene can be transcribed. An enzyme that is needed in large amounts is encoded by a structural gene whose promoter is efficient, but the synthesis of an enzyme that is needed only in tiny amounts is controlled by an inefficient promoter.

What about enzymes, such as those that metabolize lactose, that bacteria need in large amounts at some times but not at all at others? The genes coding for these enzymes must contain a very efficient promoter so that the maximum rate of mRNA synthesis is high, but there must also be a way to stop mRNA synthesis when the enzymes are not needed.

12.16   Translation of Polycistronic mRNA
A polycistronic mRNA codes for several polypeptides. Each structural gene along the mRNA codes for a different polypeptide, and these genes are "punctuated" by termination signals that signal the ribosome to release the completed polypeptide. The ribosome is released when it has translated all the structural gene transcripts.

Operons

Prokaryotes have evolved a mechanism that meets these two needs: they place an obstacle between the

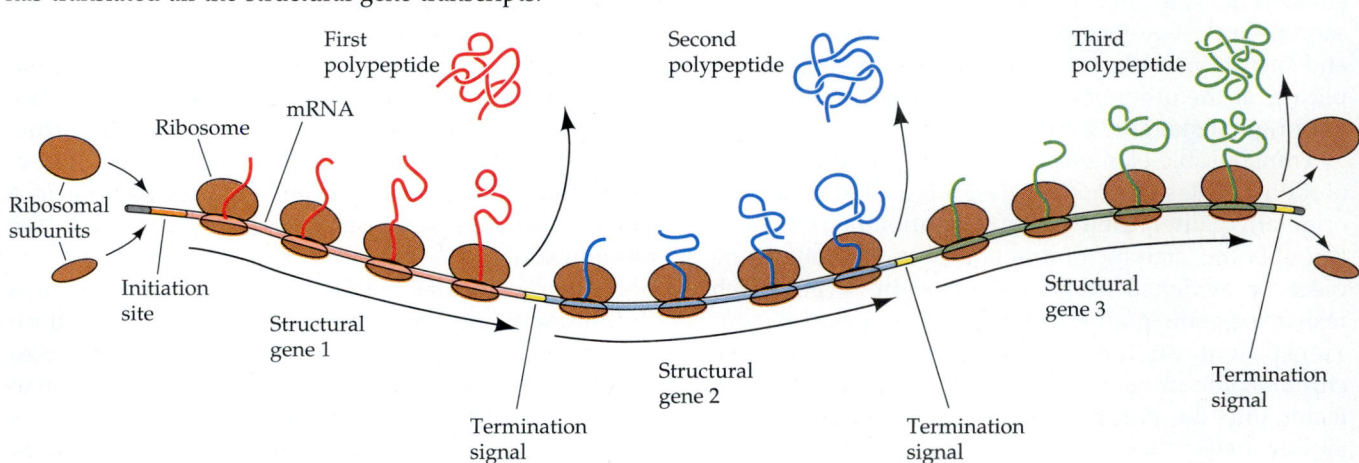

promoter and its structural genes. A short stretch of DNA called the **operator** can bind a special type of protein molecule, the **repressor**, creating the obstacle. When the repressor (the protein) is bound to the operator (the DNA), it blocks the transcription of mRNA (Figure 12.17). When the repressor is not attached to the operator, messenger synthesis proceeds rapidly. The whole unit, consisting of closely linked structural genes and the stretches of DNA that control their transcription, is called an **operon**. An operon always consists of two binding sites on the DNA molecule—a promoter and an operator—and one or more structural genes.

E. coli has three different ways of controlling transcription of operons. Two depend on interactions with the operator, and the third depends on interaction with the promoter. The two operator controls differ: one induces transcription, and the other represses it. We will now consider each of the three control systems in turn.

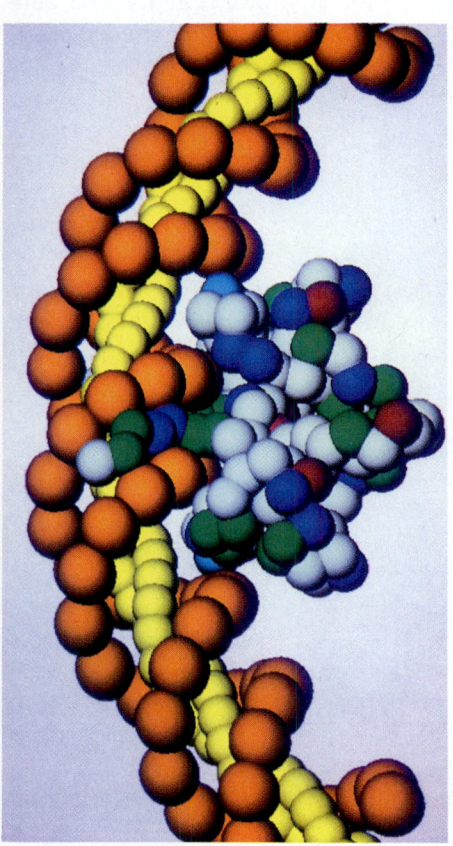

12.17 Repressor Bound to Operator Blocks Transcription
The yellow and orange spheres denote parts of the DNA molecule, which contains the operator. A portion of the repressor has already bound to the minor groove of the operator region of the DNA, and the remaining portion is about to bind to the major groove. The amino acids of the repressor are colored according to the following conventions: pale blue, hydrophobic; green, hydrophilic; red, positive charge; blue, negative charge.

Operator–Repressor Control That Induces Transcription: The *lac* Operon

The operon that controls and contains the genes for the three lactose-metabolizing enzymes is called the *lac* operon. As just explained, RNA polymerase binds the promoter, and the repressor binds to the operator. How is the operon controlled? The key lies in the repressor and its binding to the operator. The repressor is able to bind not only to its specific operator but also to inducers. Inducers of the *lac* operon are molecules of lactose and certain other β-galactosides. Binding of the inducer changes the shape of the repressor (by allosteric modification; see Chapter 6), and the change in shape makes the repressor fall off the operator.

For example, when lactose (an inducer) is added to a culture of E. coli, it enters the cell and promptly combines with the *lac* operon's repressor, changing the repressor's shape and causing it to detach from the operator. RNA polymerase can then bind to the promoter and start transcribing the structural genes of the *lac* operon. The mRNA transcribed from these genes is translated by ribosomes, which synthesize the enzyme products of the operon.

What happens if the concentration of lactose drops? As lactose concentration decreases, the inducer (lactose) molecules separate from the repressor, the repressor quickly becomes bound to the operator, and transcription of the *lac* operon stops. Translation stops soon thereafter, because the mRNA that is already present quickly degrades. The inducer, which is the target of the enzyme products of the operon, regulates the binding of the repressor to the operator.

Repressor proteins are coded for by **regulatory genes**—genes that control the activity of structural genes. The regulatory gene that codes for the repressor of the *lac* operon is called the i gene (for "inducibility"). The i gene happens to lie close to the operon that it controls, but many other regulatory genes are distant from their operons. Like all genes, the i gene itself has a promoter, which can be designated p_i. This promoter is very inefficient, allowing the production of just enough mRNA to synthesize about ten molecules of repressor per cell per generation. There is no operator between p_i and the i gene. Therefore, the repressor of the *lac* operon is constitutive, that is, it is made at a constant rate not subject to environmental control. Figure 12.18 shows the sequence of the regulatory gene and the *lac* operon, and Figure 12.19 outlines how the *lac* operon is regulated.

Let us review the important features of inducible systems such as the *lac* operon. The unregulated condition of the *lac* operon is one of being turned *on*. Control is exerted by a regulatory protein—the repressor—that turns the operon *off*. Some genes, such

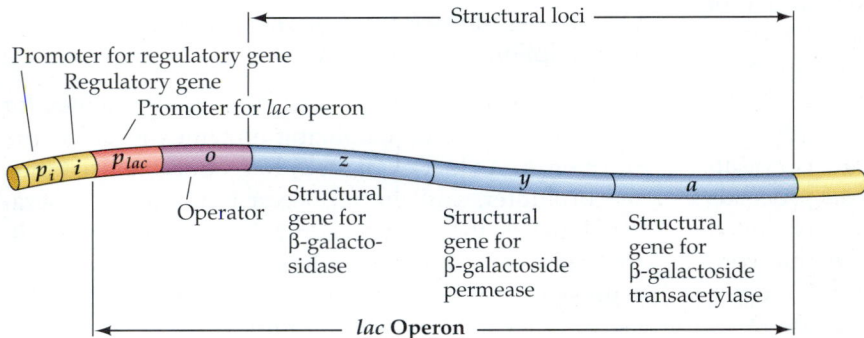

Structural loci

Promoter for regulatory gene
Regulatory gene
Promoter for *lac* operon

p_i i p_{lac} o z y a

Operator
Structural gene for β-galactosidase
Structural gene for β-galactoside permease
Structural gene for β-galactoside transacetylase

lac Operon

12.18 The *lac* Operon of *E. coli* and Its Regulator
An operon is a segment of DNA that includes structural genes along with sequences that regulate the transcription of those genes. The *lac* operon includes a promoter, an operator, and three structural genes. The regulatory gene (*i*) encodes a repressor protein that controls the operon.

as *i*, produce proteins whose sole function is to regulate the expression of other genes, and certain other DNA sequences (namely, operators and promoters) do not code for any proteins. Promoters are not even transcribed.

Operator–Repressor Control That Represses Transcription: The Tryptophan Operon

We saw that *E. coli* benefits from having an inducible system for lactose metabolism. Only when lactose is present does the system switch on. Equally valuable to a bacterium is the ability to switch off the synthesis of certain enzymes in response to something in the environment. For example, if the amino acid tryptophan, an essential nutrient, is present in ample concentration, it is advantageous to stop making the enzymes for tryptophan synthesis. When the formation of an enzyme is turned off in response to such a biochemical cue, the enzyme is said to be **repressible**.

Monod realized that repressible systems, such as the one for tryptophan synthesis, could work by mechanisms similar to those of inducible systems, such as the *lac* operon. In repressible systems, the repressor cannot shut off its operon unless it first unites with a **corepressor**, which may be either the nutrient itself (tryptophan in this case) or an analog of it. If the nutrient is absent, the operon is transcribed at a maximum rate. If the nutrient is present, the operon is turned off (Figure 12.20).

The difference between inducible and repressible systems is small but significant. In inducible systems, a substance in the environment (the inducer) interacts with the regulatory-gene product (the repressor), rendering it *incapable* of binding to the operator and thus incapable of blocking transcription. In repressible systems, a substance in the environment (the

corepressor) interacts with the regulatory-gene product to make it *capable* of binding to the operator and blocking transcription. Although the effects of the substances are exactly opposite, the systems as a whole are strikingly similar.

In both the inducible lactose system and the repressible tryptophan system, the regulatory molecule functions by binding the operator. Let us next consider an example of control by binding the *promoter*.

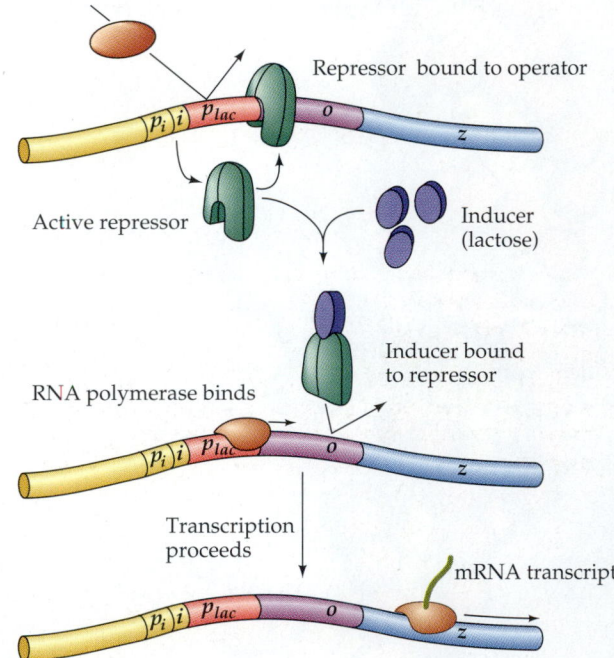

RNA polymerase can't bind; transcription blocked

Repressor bound to operator

p_i i p_{lac} o z

Active repressor

Inducer (lactose)

Inducer bound to repressor

RNA polymerase binds

p_i i p_{lac} o z

Transcription proceeds

mRNA transcript

p_i i p_{lac} o z

12.19 The *lac* Operon: Transcription Induced by Removal of Repressor
In an *E. coli* cell growing in the absence of lactose, the repressor protein coded for by gene *i* prevents transcription by binding to the operator. Lactose induces transcription by binding to the repressor, which cannot then bind to the operator. As long as the operator remains free of repressor, RNA polymerase that recognizes the promoter can transcribe the operon. Refer to Figure 12.18 for an explanation of the colors in the gene sequence.

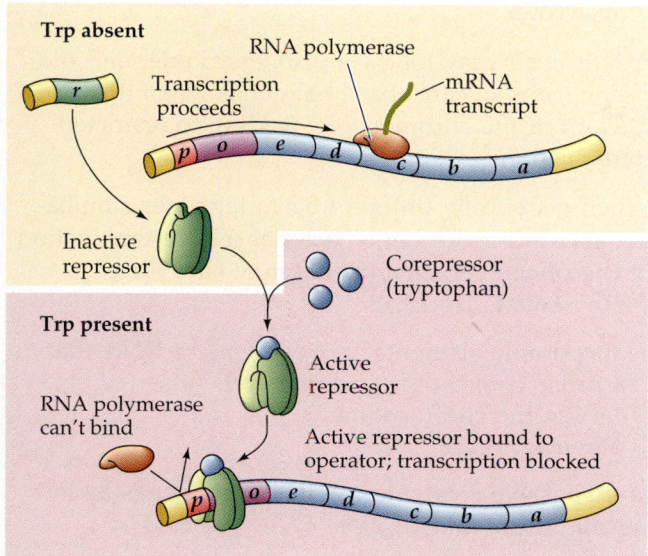

12.20 The Tryptophan Operon: Transcription Repressed by Binding of Repressor
In an *E. coli* cell growing in the absence of tryptophan (Trp), regulatory gene *r* produces an inactive repressor that does not bind to the operator of the tryptophan operon. RNA polymerase can thus transcribe the operon's structural genes into mRNAs that are translated into enzymes of the tryptophan pathway. When tryptophan is present, it converts the inactive repressor into an active one, which does bind to the operator, thus blocking RNA polymerase from transcribing the structural genes and preventing synthesis of the enzymes of the tryptophan pathway. Because tryptophan activates an otherwise inactive repressor, it is called a corepressor.

Control by Increasing Promoter Efficiency

A bacterial cell has the means to increase the transcription of certain relevant genes when it needs a new energy source. *E. coli* prefers to get its energy from glucose in the environment. When glucose is unavailable, *E. coli* must get energy from another source, such as lactose or certain other sugars or even amino acids. The alternative energy source must be catabolized (degraded) by reactions requiring catabolic enzymes. There are regulatory molecules that enhance the transcription of operons containing the genes for these enzymes. The mechanism of this effect is entirely different from the two operator–repressor mechanisms just discussed, which turn the operator on or off. This third type of mechanism makes the promoter function more efficiently.

Suppose that a bacterial cell lacks a glucose supply but has access to another food that can be catabolized to yield energy. In operons containing genes for catabolic enzymes, the promoters bind RNA polymerase in a series of steps (Figure 12.21). First, a special protein (abbreviated **CRP**, for *cAMP receptor protein*) binds the low-molecular-weight compound adeno-

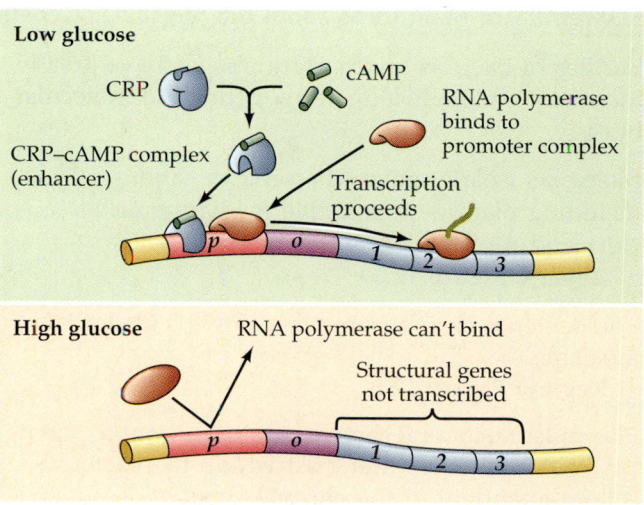

12.21 Transcription Enhanced at the Promoter Site
This operon's structural genes encode enzymes that break down a food source other than glucose. When supplies of glucose are low, a receptor protein (CRP) and cAMP form a complex that binds to the promoter and activates it, allowing transcription of structural genes that encode enzymes for catabolizing the alternative energy source. A cell that contains ample glucose and does not require energy from other sources contains little cAMP and little CRP–cAMP; in such a cell, the structural genes are not transcribed and the catabolic enzymes are not formed.

sine 3',5'-cyclic monophosphate, better known as cAMP. Next, the CRP–cAMP complex binds close to the binding site of the RNA polymerase and enhances the binding of the polymerase 50-fold. The promoters of the *lac* operon and many other genes responsible for sugar metabolism are activated in this way.

When glucose becomes abundant in the medium, breakdown of the alternative food molecules is not needed, so the cell diminishes or abolishes the synthesis of the corresponding catabolic enzymes. Glucose lowers the concentration of cAMP by a mechanism that is not yet understood but is sometimes called catabolite repression.

Preventing versus Enhancing Transcription

The inducible and repressible systems—the two operator–repressor systems—are examples of *negative* control of transcription because the regulatory molecule (the repressor) in each case *prevents* transcription. The promoter system is an example of *positive* control of transcription because the regulatory molecule (the CRP–cAMP complex) *enhances* transcription.

As we will see in Chapter 13, the regulation of gene expression in eukaryotes is more intricate than in prokaryotes. Transcription in eukaryotes is subject primarily to positive control.

SUMMARY of Main Ideas about the Molecular Genetics of Prokaryotes

Studies of bacteria and bacteriophages have greatly increased our knowledge of genetics and molecular biology.

Biologists isolate mutant bacteria in various ways, including plating on selective media, penicillin suicide, and replica plating.
    Review Figure 12.5

The bacterial chromosome is a single, circular DNA molecule.
    Review Figure 12.8

Plasmids, such as R factors, are nonessential, circular DNA molecules that exist within the bacterial cell independent of the chromosome.

Episomes, such as the F factor and prophages, are plasmids that can also be incorporated into the chromosome.

Genetic recombination in bacteria may follow conjugation, transduction, or transformation.

In conjugation, DNA is transferred from a male bacterium to a female bacterium while the two cells are physically attached.
    Review Figure 12.6

Some strains of male bacteria donate DNA much more frequently than others.
    Review Figure 12.9

In transduction, DNA is transferred from one bacterium to another by a bacteriophage.
    Review Figure 12.14

In transformation, living cells take up DNA from their environment.

All known bacteriophages have lytic life cycles; some also have lysogenic cycles.

During a lysogenic cycle, the phage DNA is maintained as a prophage incorporated into the bacterial chromosome.
    Review Figure 12.13

A lytic cycle may follow a lysogenic cycle, and the prophage may carry bacterial genes when it separates from the chromosome, leading to restricted transduction.

When genetically different bacteriophages simultaneously infect the same bacterial cell, recombination of the phage genetic material may take place.
    Review Figure 12.12

Transposable elements are segments of DNA that can cause copies of themselves to be inserted elsewhere in the chromosome.
    Review Figure 12.15

Transposable elements can inactivate genes by inserting into them.

Structural genes are expressed by being transcribed into an mRNA and translated into a protein; regulatory genes control the expression of other genes.

Much of the genetic material of bacteria is organized into operons, each consisting of one or more structural genes, an operator, and a promoter.
    Review Figure 12.18

Interactions with the operator control some operons.

Transcription of structural genes is prevented when the operator is bound by a repressor.

Operator–repressor interactions underlie repressible systems, such as the tryptophan operon, as well as inducible ones, such as the lactose operon.
    Review Figures 12.19 and 12.20

Interactions with the promoter underlie control of some catabolic enzymes.

Binding of the CRP–cAMP complex to the promoter greatly enhances binding of RNA polymerase. Glucose availability regulates the concentration of cAMP.
    Review Figure 12.21

SELF-QUIZ

1. In bacterial conjugation
   a. each cell donates DNA to the other.
   b. a bacteriophage carries DNA between bacterial cells.
   c. one partner possesses a fertility factor.
   d. the two parent bacteria merge like sperm and egg.
   e. all the progeny are recombinant.

2. Which statement is *not* true of the bacterial fertility factor?
   a. It is a plasmid.
   b. It confers "maleness" on the cell in which it resides.
   c. It can be transferred to a female cell, making it male.
   d. It has thin projections called F-pili.
   e. It can become part of the bacterial chromosome.

3. Hfr mutants
   a. are female bacteria that are highly efficient recipients of genes.
   b. rarely transfer all the markers on the chromosome.
   c. keep their F factor separate from the chromosome at all times.
   d. are unable to conjugate with other bacteria.
   e. transfer markers in random order.

4. Lysogenic bacteria
   a. lack a prophage.
   b. are accompanied by free phages when growing in culture.
   c. lyse immediately.
   d. cannot release their phages.
   e. are susceptible to further attack by the same strain of phage.

5. Which statement is *not* true of transduction?
   a. The viral DNA is an episome.
   b. Transduction is a useful tool for mapping a bacterial chromosome.
   c. In restricted transduction, the newly infected cell becomes lysogenic.
   d. Transduction results in genetic recombination.
   e. To carry bacterial markers, the viral coat must contain viral DNA.

6. Plasmids
   a. are circular protein molecules.
   b. are required by bacteria.
   c. are tiny bacteria.
   d. may confer resistance to antibiotics.
   e. are a form of transposable element.

7. Which statement is *not* true of a transposable element?
   a. It can be copied to another DNA molecule.
   b. It can be copied to the same DNA molecule.
   c. It is typically 100 to 500 base pairs long.
   d. It may be part of a plasmid.
   e. It encodes the enzyme transposase.

8. In an inducible operon
   a. an outside agent switches on enzyme synthesis.
   b. a corepressor unites with the repressor.
   c. an inducer affects the rate at which repressor is made.
   d. the regulatory gene lacks a promoter.
   e. the control mechanism is positive.

9. The promoter
   a. is the region that binds the repressor.
   b. is the region that binds RNA polymerase.
   c. is the gene that codes for the repressor.
   d. is a structural gene.
   e. is an operon.

10. The CRP–cAMP system
    a. produces many catabolites.
    b. requires ribosomes.
    c. operates by an operator–repressor mechanism.
    d. is a form of positive control of transcription.
    e. relies on operators.

FOR STUDY

1. Viruses sometimes carry DNA from one cell to another by transduction. Sometimes a segment of bacterial DNA is incorporated into a phage protein coat without any phage DNA. These particles can infect a new host. Would the new host become lysogenic if the phage originally came from a lysogenic host? Why or why not?

2. For studies of metabolism in a particular species of bacteria, you need to isolate the following mutant strains: a histidine auxotroph (a strain, *his⁻*, that cannot synthesize the amino acid histidine) and a tryptophan auxotroph (*trp⁻*, unable to synthesize the amino acid tryptophan). After irradiating a culture of the bacteria with ultraviolet light to increase the mutation rate, you expect to find some *his⁻* and *trp⁻* auxotrophs in the culture. Describe all the steps you would take in order to increase the percentages of *his⁻* and *trp⁻* auxotrophs, using the "penicillin suicide" technique.

3. You are provided with two strains of *Escherichia coli*. One, an Hfr strain that is sensitive to streptomycin, carries the markers A^+, B^+, and C^+. The other is an F^- strain that is resistant to streptomycin and carries the markers A^-, B^-, and C^-. You mix the two cultures. After 20, 30, and 40 minutes you take samples of the mixed culture and swirl them vigorously in a blender. Next you add streptomycin to the swirled cultures. You examine surviving bacteria by replica plating. Some of the bacteria from the 20-minute sample are B^+; in the 30-minute sample there are both B^+ and C^+ bacteria; but A^+ bacteria are found only in the 40-minute sample. What can you say about the arrangement of the A, B, and C loci on the bacterial chromosome? Explain your answer fully.

4. You have isolated three strains of *E. coli*, which you name I, II, and III. You attempt to cross these strains, and you find that recombinant progeny are obtained when I and II are mixed or when II and III are mixed, but not when I and III are mixed. By diluting a suspension of II and plating it out on solid medium, you isolate a number of separate clones. You find that almost all these clones can conjugate with strain I to produce recombinant offspring. One of the clones derived from strain II, however, lacks the ability to conjugate with strain I. Characterize strains I, II, and III and the nonconjugating clone of strain II in terms of the fertility (F) factor.

5. In the lactose operon of *E. coli*, repressor molecules are coded for by the regulatory gene. The repressor molecules are made in very small quantities and at a constant rate per cell. Would you surmise that the promoter for these repressor molecules is efficient or inefficient? Is synthesis of the repressor constitutive, or is it under environmental control?

6. A key characteristic of a repressible enzyme system is that the repressor molecule must react with a corepressor (typically, the end product of a pathway) before it can combine with the operator of an operon to shut the operon off. How is this different from an inducible enzyme system?

READINGS

Darnell, J. E., Jr. 1985. "RNA." *Scientific American*, October. A discussion of aspects of transcription and of gene regulation in prokaryotes and eukaryotes.

Griffiths, A. J. F., J. H. Miller, D. T. Suzuki, R. C. Lewontin and W. M. Gelbart. 1993. *An Introduction to Genetic Analysis*, 5th Edition. W. H. Freeman, New York. An up-to-date revision of one of the field's classic textbooks.

Judson, H. F. 1979. *The Eighth Day of Creation*. Simon and Schuster, New York. A constantly fascinating history of molecular biology, with much attention to the regulation of gene expression. For a lay audience.

Nomura, M. 1984. "The Control of Ribosome Synthesis." *Scientific American*, February. A discussio of how ribosomes are assembled and the roles of operons in regulating ribosome production in bacteria.

Stent, G. S. and R. Calendar. 1978. *Molecular Genetics*, 2nd Edition. W. H. Freeman, New York. Technical and historical information charmingly presented.

Your life depends on the continual delivery of oxygen to your living tissues. This job is accomplished by your red blood cells with their cargo of hemoglobin molecules. Each red blood cell functions for about four months and then is destroyed, to be replaced by a new one. The replacement rate to keep you going is 100 billion red blood cells per day! DNA deep within your bones directs these specialized cells to form and mature, by which time they are quite unlike other cells. Red blood cells are simple, membranous bags that lack compartments and are full of hemoglobin. They are simple, but each one must be exactly right. The plasma membrane must have the right proteins, in the right proportions, to govern the movement of oxygen and other substances into and out of the cell.

The DNA deep within your bones must direct not only the perfect assembling of each red blood cell, but also the production of hundreds of thousands of perfect hemoglobin molecules for each one of today's hundred billion new red blood cells. Each hemoglobin molecule requires four perfect polypeptides, coded for by two different genes, plus many enzymes, coded for by other genes, to do the assembling.

Why are these genes expressed only in the special places where their products are needed? What keeps other genes in the DNA deep within your bones switched off, so that their products never appear there? In this chapter you will learn about the structure of the eukaryotic gene and see how that structure permits the control of gene expression.

EUKARYOTES AND EUKARYOTIC CELLS

Most eukaryotic cells are much larger and more internally complex than prokaryotic cells. In particular, eukaryotic cells typically contain an array of membrane-bounded organelles specialized for various functions. Eukaryotic cells contain much more DNA than do prokaryotic cells. For example, most mammals have about 1,000 times as much DNA per cell as does the bacterium *Escherichia coli*. DNA is packaged differently in eukaryotic cells than it is in prokaryotic cells: The eukaryotic chromosome is organized into nucleosomes (see Chapter 9). For these and other reasons, gene expression and the patterns of inheritance in eukaryotes differ from the same phenomena in prokaryotes.

Eukaryotes evolved as unicellular organisms, and only unicellular eukaryotes existed for a very long time before multicellularity evolved. Single-celled eukaryotes had some real advantages in a world full of prokaryotes. Their internal organelles allowed them to gain efficiency by separating various activities into special compartments. The eukaryotes were the first

DNA in Your Body Directs Production of a Hundred Billion New Red Blood Cells—Every Day

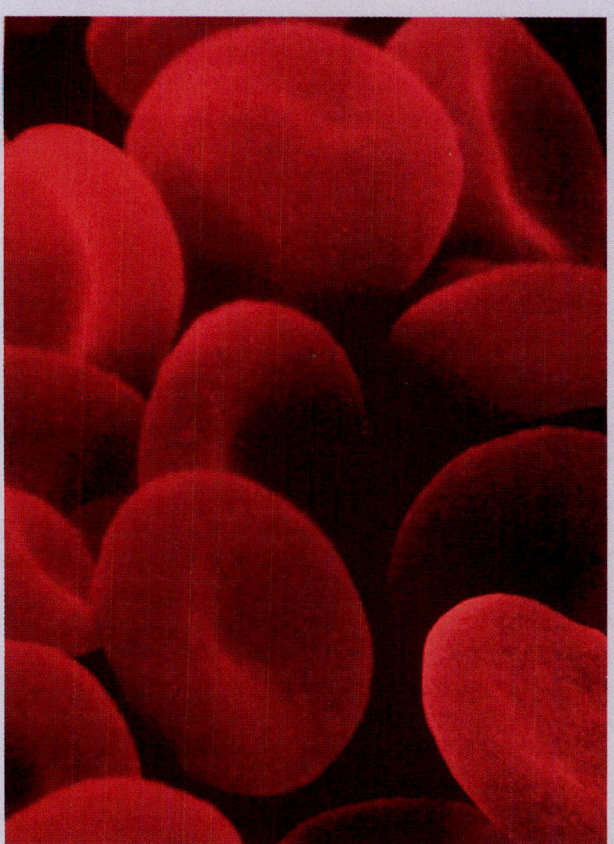

13

Molecular Genetics of Eukaryotes

organisms with sex as we usually understand it—that is, with equal samples from the genomes of two individuals making up the genome of the offspring, and with sexual processes typically taking place in the life of each individual. In prokaryotes, sexual processes occur very rarely, and the "male" usually contributes only a small fraction of its genome to the offspring. Sexual reproduction afforded eukaryotes a greater ability to produce offspring that could be successful in heterogeneous (mixed) and changing environments.

Many eukaryotes today are multicellular. There is usually a division of labor among the cells; therefore a multicellular organism has various types of cells that contain different proteins and are capable of performing different specialized functions. The human body has at least 200 different cell types, differing in a few major proteins and many minor proteins. During the development of the multicellular body, different genes are expressed at different times, or in different specific tissues (see Chapter 17). Such differential gene expression is important even in the development of the tiny but complex bodies of unicellular eukaryotes. How is gene expression managed? Or, in other words, how are eukaryotic genes turned on and off?

Before we can address this question, we must examine the structure of the eukaryotic gene itself, as well as the complex series of steps from gene to protein product in eukaryotes. Eukaryotic genes tend to differ from those of most prokaryotes in that they have stretches of DNA that are not expressed in polypeptide products. That is, most eukaryotic genes are "split" by the presence of noncoding DNA in the midst of the coding sequences. As we shall see, this characteristic requires processing at the molecular level that is not found in prokaryotes.

What is known about eukaryotic gene structure and expression has been revealed by application of the techniques of molecular biology. Because understanding one of these techniques in particular—nucleic acid hybridization—makes it easy to learn about some of the important discoveries, we will begin by considering this technique.

HYBRIDIZATION OF NUCLEIC ACIDS

Nucleic acid hybridization depends on the association, through complementary base pairing, of single-stranded nucleic acids. If we carefully heat a sample of DNA, the hydrogen bonds forming the base pairs are destroyed and the two strands of each double helix separate—we say that the DNA **denatures**. If we then lower the temperature slowly, the complementary strands join again, or **reanneal**, to form double-stranded DNA, with each base pair obeying the A–T, G–C pairing rules (Figure 13.1a,b). To make this procedure work efficiently, one must enzymatically or mechanically "cut" the DNA into short segments a few hundred bases long and carefully regulate the temperature and salt concentration in the test solution.

Suppose, now, that we denature a sample of DNA with heat and then combine it with a sample of RNA that has been transcribed from part of that DNA. Because the RNA is complementary to the one strand of DNA that coded for it (the template strand), it may anneal with that DNA strand to form a hybrid (Figure 13.1c). As the temperature is lowered, the RNA transcript and the other DNA strand compete to anneal with the template strand. The reannealing of the DNA strands can be prevented by immobilizing the denatured DNA on a nitrocellulose filter before it cools, which keeps the separated DNA strands from coming together, thereby favoring RNA binding. (The immobilized DNA is still accessible for hybridization with nucleic acids in solution.) The tendency for DNA strands to reanneal can also be reduced by outnumbering the DNA strands with RNA strands.

EUKARYOTIC GENE STRUCTURE

Now that we understand how nucleic acid hybrids are formed, we are ready to consider the eukaryotic gene and how it differs from the prokaryotic gene. The structure of the eukaryotic gene has been studied by comparing the DNA with its RNA transcripts. One way of making this comparison is as follows: We denature a sample of DNA and then add corresponding mRNA, such as would be found in the cytoplasm. From the resulting hybridization we obtain uniform, double-stranded DNA–mRNA hybrid structures associated with *single-stranded*, looped structures. The loops are displaced, noncoding DNA strands. This method revealed in 1977 that double-stranded hybrid regions are studded with both single- and double-stranded loops. What could this mean?

The mRNA is a faithful transcript of the information required for protein synthesis. After hybridization, all the mRNA is bound, through complementary base pairing, with the appropriate region of the single-stranded DNA. However, there is some DNA *in the middle of the gene* that is not represented in the mRNA. That is, some DNA sequences are not represented in the information that ends up in the mRNA that encodes the protein product. In fact, most (but not all) vertebrate genes as well as many other eukaryotic genes contain such intervening sequences, called **introns**: segments of DNA that do not encode any part of the polypeptide product of the gene. Later in this chapter we will learn that not all the transcribed RNA gets into the cytoplasm to be

(a) Upon being carefully heated, the two polynucleotide strands of a DNA molecule denature (separate)

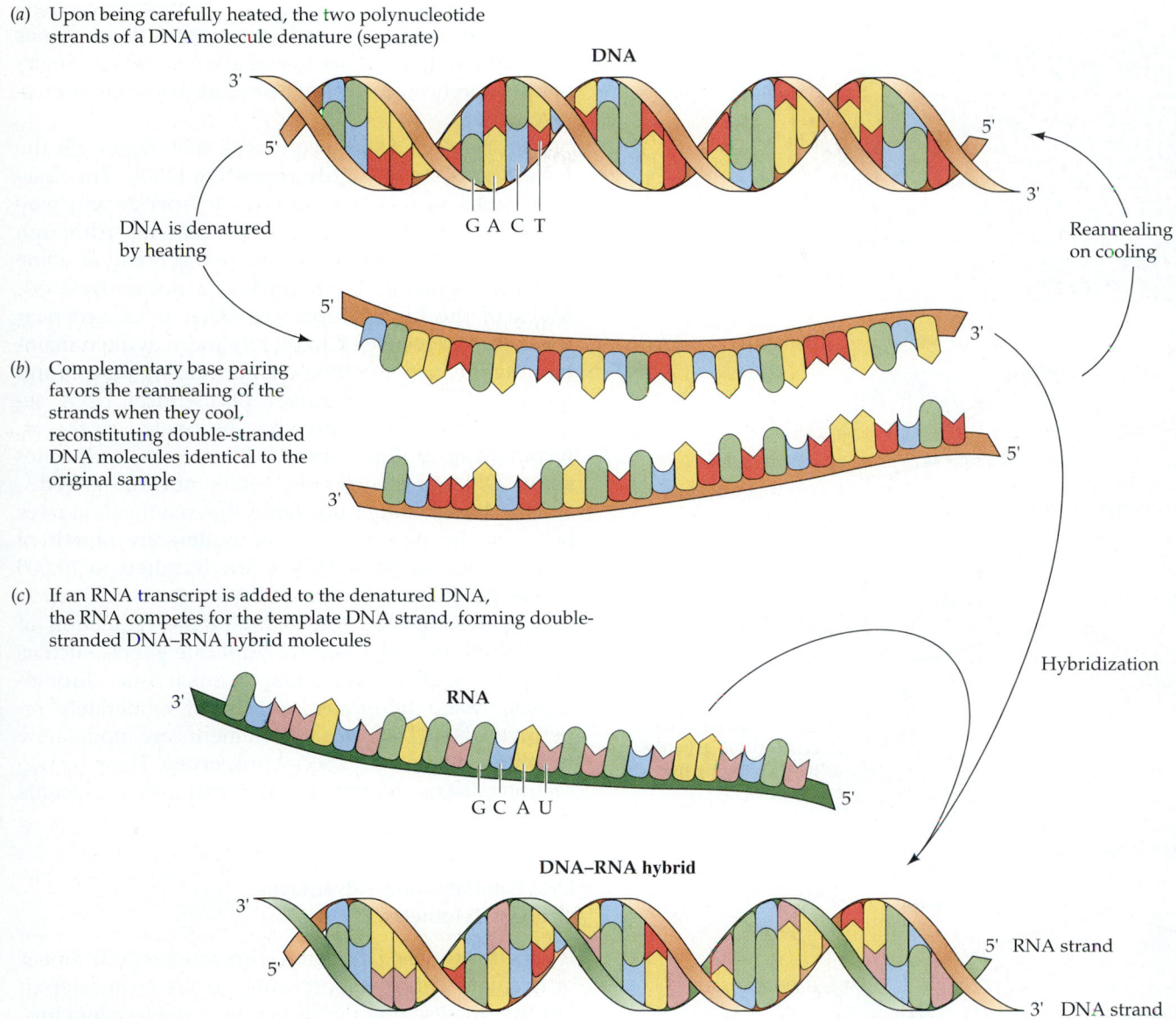

DNA

DNA is denatured by heating

G A C T

Reannealing on cooling

(b) Complementary base pairing favors the reannealing of the strands when they cool, reconstituting double-stranded DNA molecules identical to the original sample

(c) If an RNA transcript is added to the denatured DNA, the RNA competes for the template DNA strand, forming double-stranded DNA–RNA hybrid molecules

RNA

G C A U

Hybridization

DNA–RNA hybrid

5' RNA strand

3' DNA strand

13.1 Nucleic Acid Hybridization
Denatured strands of DNA anneal with complementary sequences of either DNA or RNA.

mRNA. The parts of the gene that *are* represented in the mRNA product are called **exons** because they are *ex*pressed regions of the gene (Figures 13.2 and 13.3).

Introns do not scramble the sequence that codes for a polypeptide; they just reside in the middle of it. The base sequence of the exons, taken in order, is exactly complementary with that of the mature mRNA product. The introns simply separate the coding sequence for the protein into parts. Exons and introns are found in all groups of eukaryotes and even in a few prokaryotes. We are still seeking to understand the significance of introns to the organisms that possess them. Later in this chapter we will describe the posttranscriptional events that remove the transcripts of introns.

REPETITIVE DNA IN EUKARYOTES

What else can we learn by using nucleic acid hybridization? If we denature eukaryotic DNA and then let the complementary strands reanneal, we can collect data on how long it takes to renature. From these data we observe that some parts of the genome anneal only very slowly, whereas other segments quickly find partners. Why should one DNA sequence anneal quickly and another slowly? The answer is that there are multiple copies of some, but not all, stretches of DNA. If a particular single strand of DNA has, say, a few hundred complementary segments with which it can anneal, it will be able to find a partner much more rapidly than one for which only a single acceptable partner exists.

(a)

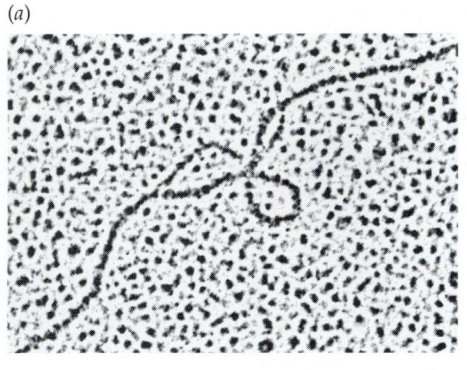

(b)

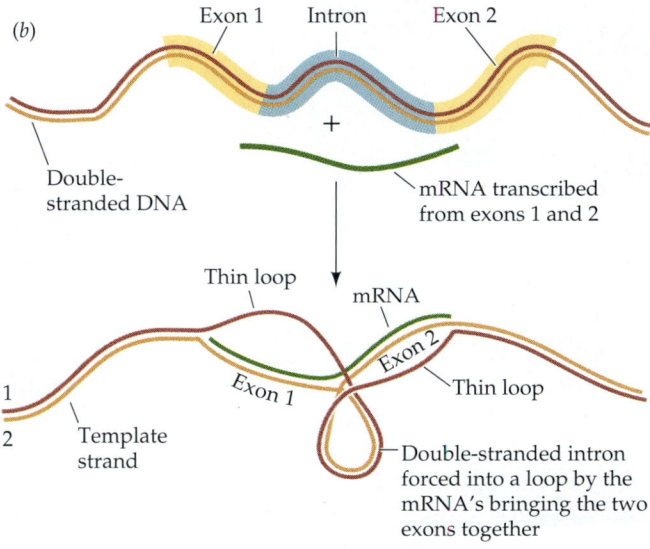

Exon 1   Intron   Exon 2

Double-
stranded DNA

+

mRNA transcribed
from exons 1 and 2

Thin loop

mRNA

1

2   Template
        strand

Exon 1

Exon 2

Thin loop

Double-stranded intron
forced into a loop by the
mRNA's bringing the two
exons together

(c) With no introns:

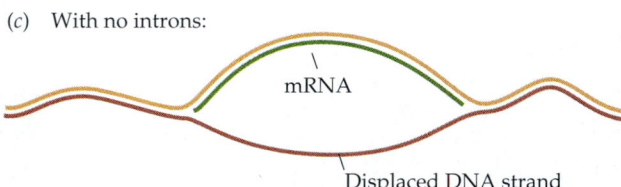

mRNA

Displaced DNA strand

13.2 Evidence for Extra DNA in the Eukaryotic Gene
Mouse DNA was partially denatured and mixed with
mRNA transcribed from one of the genes in the DNA.
(a) Examination of the resulting mixture by electron microscopy revealed thick nucleic acid, bearing thick loops
(here one points downward) and thinner loops. (b) A diagrammatic interpretation of (a). The mRNA hybridized
with the template DNA strand, forming double strands
and thin loops; these thin loops are the unpaired, complementary strand of DNA. The double-stranded DNA from
the intron reannealed because it had no counterpart in
the mRNA. The mRNA joined the two exons, forcing the
intron DNA into a thick loop. (c) Hybridization pattern
observed when no introns are present, as in prokaryotic
DNA.

Researchers applying this reannealing technique,
called liquid hybridization, discovered that there are
three classes of eukaryotic DNA. The class that reanneals the slowest consists of **single-copy sequences**—
genes that, like prokaryotic genes, have only one

copy in each **genome**, or haploid set of chromosomes.
Single-copy sequences code for most of the enzymes
and structural proteins in eukaryotes. Some single-
copy sequences form long spacers between successive genes.

The class of DNA sequences that reanneals the
fastest consists of **highly repetitive DNA**. This fraction varies widely from species to species and may
make up a third or more of the genome. Although
there are half a million copies per genome of some
of these segments, their function is not understood.
Much of the highly repetitive DNA is located near
the centromeres (see Chapter 9) and may help maintain the integrity of chromosomes during mitosis and
meiosis. The large number of identical DNA sequences at a centromere may be related to the attachment of multiple spindle fibers to each chromosome. DNA of this class is usually not transcribed.

The class of eukaryotic DNA that reanneals at rates
between the two extremes is **moderately repetitive
DNA**, which is present in a few hundred to 10,000
copies per genome. Some of these moderately repetitive genes may be important in the regulation of
development, and some are duplicate genes, such as
those for rRNA. The ends of eukaryotic chromosomes, called **telomeres**, consist of moderately repetitive DNA. Transposable elements are moderately
repetitive genes with special properties. The next two
sections discuss telomeres and transposable elements
in detail.

**Disadvantages and Advantages
of Ends: Telomeres**

Is it better for the DNA in a chromosome to be linear
or circular? This question cannot really be answered,
but the possession of ends is a real problem for chromosomes when it comes to DNA replication. The
problem is that the monomers of DNA cannot simply
string together from scratch; they can be added only
to the 3' end of a previously existing nucleic acid

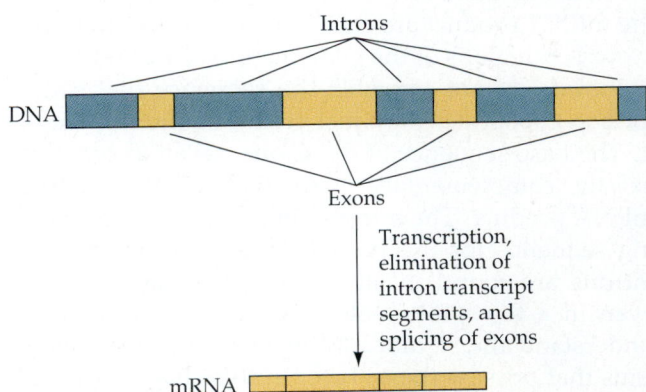

Introns

DNA

Exons

Transcription,
elimination of
intron transcript
segments, and
splicing of exons

mRNA

13.3 mRNA Is Encoded Only in Exons

strand. That is, they require a **primer**. In both prokaryotes and eukaryotes primers consist of RNA laid down using the other strand of the existing DNA as a template. At a later stage in DNA replication, a growing strand of DNA approaches the other (5') end of the primer and eventually replaces the primer, so the entire new molecule consists of DNA, with no inserted stretches of RNA (see Box 11.A).

Recall from Chapter 11 that replication proceeds differently on the two strands of a DNA molecule. Both new strands form in the 5'-to-3' direction, but one (the leading strand) grows continuously from one end to the other, while the other (the lagging strand) grows as a series of Okazaki fragments. With the circular prokaryotic chromosome, as both DNA strands extend continuously around the whole circular chromosome, production of a complete series of Okazaki fragments is not a problem—there is always some DNA at the 5' end of a primer, ready to replace it.

Now think about replication at an end of a linear eukaryotic chromosome. The leading strand can grow without incident to the very end. But how does the last Okazaki fragment for the end of the lagging strand form? Replication must begin with an RNA primer at the 5' end of the forming strand, but there is nothing beyond the primer in the 5' direction to replace the RNA. Thus the new daughter chromosome in a eukaryote lacks a bit of double-stranded DNA, a bit of genetic information, at each end. The ends of the chromosome have been clipped off. This clipping problem threatens to shorten all eukaryotic chromosomes; why don't they all disappear over the course of many replications? The answer is that a repeated DNA sequence forms a telomere at each end of the chromosome. This sequence of moderately repetitive DNA can double back on itself and thus serve as the primer for completing the lagging strand as the chromosome is replicated.

This activity is essential for the preservation of eukaryotic chromosomes, as the requirements for making a viable artificial chromosome illustrate. We can make artificial chromosomes that, when inserted into yeast cells, behave like normal chromosomes—they are replicated, they are distributed normally in mitosis, and their genes are transcribed and translated. For this process to work, however, the artificial chromosome must contain three features: a centromere (to allow attachment to mitotic and meiotic spindles); an origin of replication (to allow replication); and a telomere at each end (to protect the chromosome from shortening). The rest of the artificial chromosome may consist of any DNA sequences we wish to put there.

The telomeres may serve another important function in the behavior of eukaryotic chromosomes. By attaching to the nuclear envelope, telomeres may help initiate the process of synapsis early in meiosis (see Chapter 9).

Transposable Elements in Eukaryotes

Eukaryotes have **transposable elements** that are moderately repetitive segments of DNA. Because these transposable elements insert themselves into different parts of chromosomes, they have been called "jumping genes" (see Chapter 12). The first evidence for transposable elements came from studies on maize, a eukaryote, conducted by Barbara McClintock at Cold Spring Harbor Laboratory.

The method by which transposable elements are copied in eukaryotes differs from that in prokaryotes. The copying mechanism in eukaryotes requires an RNA intermediate. Recall that in prokaryotes the DNA of the transposable elements can simply copy itself. The transposable elements of both prokaryotes and eukaryotes are always parts of chromosomes; unlike plasmids, they do not function as independent pieces of DNA.

Thus far, scientists have learned very little about the roles of transposable elements in the life of a cell. Although the protein products of some eukaryotic transposable elements have been identified, the cellular functions of these products have not been determined. Transposable elements may, in effect, be parasites that simply replicate themselves; on the other hand, they may play important roles in the survival of the organisms in whose chromosomes they reside.

Transposable elements can act as mutators—that is, by jumping into genes, they can eliminate or change the functions of those genes (see Chapter 12). Transposable elements can bring about deletions, insertions, transpositions, and inversions; they also may be a source of duplications, in which multiple copies of genes are created. Some genes may be inactivated by the insertion of transposable elements; other genes may be placed in new relative positions, affecting their transcription. For some genes, transposable element insertions constitute more than 99 percent of all mutations.

Recall from Chapter 4 the endosymbiotic theory of the origin of chloroplasts and mitochondria, which proposes that chloroplasts and mitochondria are the descendants of once free-living prokaryotes. Transposable elements seem to have played a part in this process. Chloroplasts and mitochondria possess DNA, and some parts of these organelles are encoded by genes on this extranuclear DNA. Other parts of the organelles are coded for by nuclear genes, a finding that might appear to weaken the endosymbiotic theory. It has recently been shown, however, that in the course of evolution genes of some organisms have been transposed to the nuclei from both chloroplasts and mitochondria. The insertion of transposable elements, and the subsequent loss of the originals of these genes from the chloroplasts and mitochondria, can therefore be used to counter this argument against the endosymbiotic theory. Thus, because of our relatively new knowledge about transposable elements, a finding once considered to be evidence *against* the endosymbiotic theory can now be used as evidence *for* the theory.

How did transposable elements arise? Their source is still unknown, but there are interesting hints at a relationship between retroviruses (tumor viruses that use reverse transcriptase to transcribe their RNA to DNA) and the transposable elements of eukaryotes, as we will see when we discuss cancer-causing genes in Chapter 15. It is possible that some transposable elements arose from retroviruses.

GENE DUPLICATION AND GENE FAMILIES

Some genes have just two or only a few copies. These copies arise, evolutionarily, in various ways. In the previous section we noted that transposable elements are a source of gene duplication. Another source is unequal crossing over (see Chapter 10), in which mispaired chromosomes cross over in such a way as to put both copies of a gene on the same chromosome. A set of duplicated genes is called a **gene family**. Members of a gene family may reside on different chromosomes, or they may be bunched tightly on a single chromosome.

Once more than one copy of a gene exists, the copies may evolve differently. One copy must retain the original function, or the organism may not survive. As long as one copy does this, however, the others may change slightly, extensively, or not at all.

An evolutionarily ancient gene family found in vertebrates codes for the globin proteins. Globins are required for the binding and transport of oxygen; some of them are components of hemoglobin, the oxygen-carrying pigment of red blood cells, and one is found in myoglobin, a related protein that binds

and stores oxygen within muscle fibers (see Chapter 41). Each molecule of hemoglobin is a tetramer. The four globin polypeptides that make up the tetramer in adult humans are two of one type (α) and two of another (β). The hemoglobin tetramers of human fetuses also contain two types of globin polypeptides, both of which differ from those found in adult hemoglobin. Still other α-like and β-like polypeptides are found in the earliest embryonic stages.

All these globin polypeptides are coded for by globin genes descended from a single ancestral globin gene. This fact has been ascertained from similarities in the amino acid sequences of the polypeptides and from the locations of introns within the genes. Like the members of other gene families, the globin genes differ from one another more in their introns than in their exons; the exons are probably conserved with little variation because their gene products perform essential functions, whereas the introns have no products.

The genes for human α-like globins lie in a tight cluster on one of our chromosomes, and the genes for β-like globins lie in a cluster on a different chromosome. In both clusters there are additional stretches of DNA, closely similar in base sequence to the globin genes, that are not expressed. Such apparently nonfunctional genes are called **pseudogenes** (Figure 13.4). How were pseudogenes discovered, if they have no function? Nucleic acid hybridization revealed the existence of pseudogenes because they hybridize to a significant extent with adult globin mRNA (as do all other members of the globin gene family).

Most pseudogenes are probably duplicate genes that changed so much during evolution that they no longer function. The changes may have inactivated promoters, caused nonsense mutations in exons, or eliminated the clipping out of an intron. Some pseudogenes, however, did not arise by gene duplication; they were derived by reverse transcription of the mRNA. Such "processed pseudogenes" lack introns and are found away from the rest of a gene cluster. Whatever functions the pseudogenes had in the past

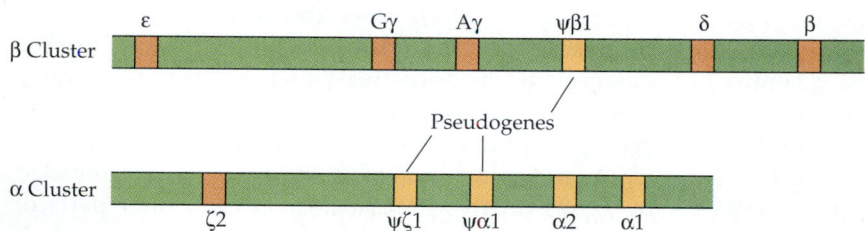

13.4 Gene Families May Include Pseudogenes
The human α-like and β-like globin gene families are organized into clusters. Each gene cluster includes both functional genes and pseudogenes; pseudogenes are prefixed by the Greek letter psi (ψ).

(when they were not pseudogenes) are now performed by other genes in the family. In many gene families, pseudogenes outnumber functional genes, often by several fold. As we mentioned already, different active members of the human globin gene family serve at different times in development. There are other gene families in which different active members function in different tissues or under different environmental conditions instead of at different times. The gene families for rRNA, tRNAs, and histones are examples in which duplication is used to meet the demand for large amounts of the gene product.

RNA PROCESSING IN EUKARYOTES

The RNA transcript of a eukaryotic gene contains both introns and exons. How do we get from this product to an mRNA, and from there to the protein encoded by the gene? The original product of transcription of a eukaryotic gene is a heterogeneous nuclear RNA, or **hnRNA**, so called because of the great range of sizes of RNAs of this class. As we are about to see, a great deal more must be done to the hnRNA to produce a mature mRNA that is ready to be translated.

Capping and Tailing RNA

An early step in the processing of an hnRNA molecule is the addition of a **cap**—a modified molecule of guanosine triphosphate (GTP)—at the 5' end of the hnRNA. This modified G cap is retained during the processing of mRNA and facilitates the binding of the mRNA to a ribosome for translation (see Chapter 11); it may also help protect the mRNA from degradation.

At the other end of some hnRNA molecules, the 3' end, a string of 100 to 200 adenine nucleotides, called a **poly A tail**, is added by a process called polyadenylation. (Poly A tails constitute 5 to 20 percent of the length of mature mRNA molecules produced from hnRNA.) Neither the modified G cap nor the poly A tail is coded for in the DNA; both are added as part of the early processing of hnRNA.

The poly A tail is thought to protect the modified hnRNA and the mRNA from degradation; some evidence suggests that the tail is needed for translation as well. Somewhat less than one-third of the hnRNA molecules in mammalian cells have poly A tails, and about 70 percent of the resulting mRNA molecules have the tails. We do not yet know why some RNAs get poly A tails and others do not. For those hnRNAs that do get poly A tails, however, the poly A is essential in order for the hnRNA to mature into mRNA.

Splicing RNA

The next step in the processing of eukaryotic RNA is the removal of the regions coded for by introns in the DNA. If these RNA regions, which are also called introns, were not removed, a nonfunctional mRNA or an improper protein would be produced. A process called **RNA splicing** removes introns and splices exons together. Figure 13.5 illustrates how splicing of an hnRNA transcript 7,700 nucleotides long results in a mature mRNA of only 1,872 nucleotides. Capping, tailing (polyadenylation), the removal of seven introns, and the splicing of eight exons do the job.

In the mRNA splicing reaction, a loop forms, extruding the intron and bringing the adjacent exons together. How do the exons become linked? At the boundaries between introns and exons there are **consensus sequences**—short stretches of DNA that appear, with little variation, in many different genes. Eukaryotic nuclei contain molecules of small nuclear RNA, or **snRNA**, that contain regions complementary to the consensus sequences. To accomplish the splicing, an snRNA combines with a set of proteins to produce a small nuclear ribonucleoprotein particle, or **snRNP**. One of the snRNPs recognizes and binds the consensus sequence at one end of the intron; a second, different snRNP recognizes and binds the consensus sequence at the other end of the intron; other snRNPs recognize and bind a sequence in the intron itself. Together, six different snRNPs constitute a "splicing machine" called a spliceosome. The spliceosome joins the exons and releases the introns (Figure 13.6).

Splicing mechanisms differ among RNA classes: mRNAs, rRNAs, and tRNAs are all spliced in different ways. Molecular biologists were startled to learn in 1981 that in the protist *Tetrahymena thermophila*, the RNA precursor of rRNA can catalyze the splicing of its own intron. That is, the RNA, in the absence of any protein, is catalytic. Another case of a catalytic RNA (not involving RNA splicing) has been discovered, this time in *E. coli*. The existence of RNAs with catalytic powers may help explain evolution at the dawn of life (see Chapter 18).

The Stability of mRNAs

In eukaryotes, RNA exits the nucleus through pores in the nuclear envelope (see Figure 4.10). The transport of RNA from the nucleus to the cytoplasm is mediated by carrier proteins, but the details of this mechanism are not yet known. Mature mRNA in the cytoplasm may be relatively stable, lasting for hours or days. In prokaryotes, however, an mRNA molecule usually lasts for only a few minutes following transcription—its life is so short that translation be-

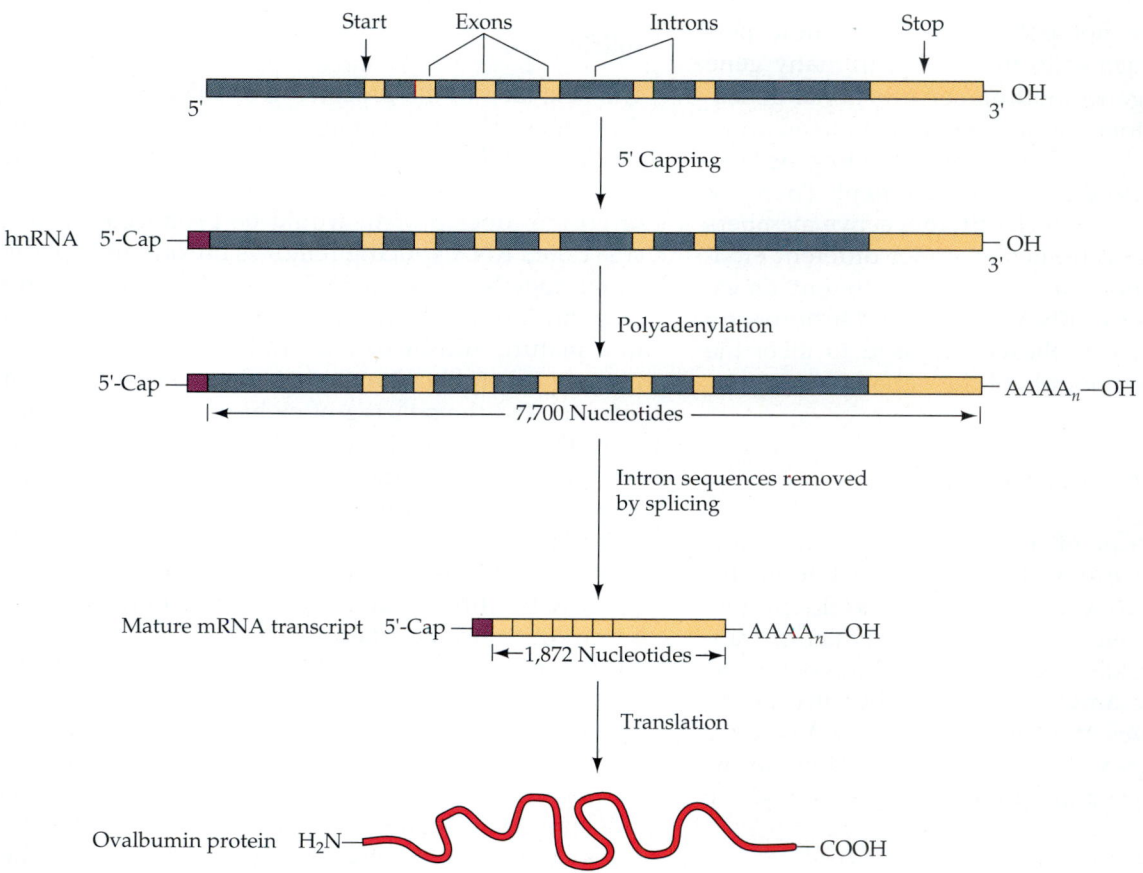

13.5 Eukaryotic Cells Process hnRNA
The heterogenous nuclear RNA (hnRNA) of the gene that codes for ovalbumin, the major protein in egg whites, is capped at the 5' end and has a poly A tail added at the 3' end. It is 7,700 nucleotides long and contains seven introns and eight exons. Splicing removes over three-quarters of the nucleotides and results in a mature mRNA that can be translated to yield ovalbumin.

gins before the mRNA is completely formed. As we will see, different eukaryotic RNAs may differ from one another in their stability. The stability of mRNAs plays an important role in the development of animals and plants.

CONTROL OF GENE EXPRESSION IN EUKARYOTES

For development to proceed normally, the right genes must be expressed at just the right times and in just the right cells. Indeed, the expression of eukaryotic genes is precisely regulated. The modes of this regulation are many. In a few cases, gene regulation depends on *changes in the DNA itself*—genes are actually rearranged on the chromosomes (an example is given at the end of this chapter). *Transcription* in eukaryotes is subject to complex mechanisms of regulation, which have a surprising variety. *Translation* may be regulated by a variety of means. Even the polypeptide products of translation may require

further processing before they become biologically active, so we can also speak of *posttranslational control* of the expression of some genes.

We begin with some cases in which gene expression is under *transcriptional* control. There are at least three ways in which transcription may be controlled: genes may be inactivated, specific genes may be amplified, or—most frequently—specific genes may be transcribed selectively.

Transcriptional Control: Gene Inactivation

As mitosis or meiosis concludes, chromosomes uncoil, but not completely (see Chapter 9). During interphase, one portion of the chromatin—the **euchromatin**—is diffuse and thus does not stain. The **heterochromatin**, however, retains its coiling and continues to be stainable by the dyes that stain mitotic chromosomes. Heterochromatin generally is not transcribed, and any genes that it contains are thus inactivated.

How such inactivation controls gene expression is

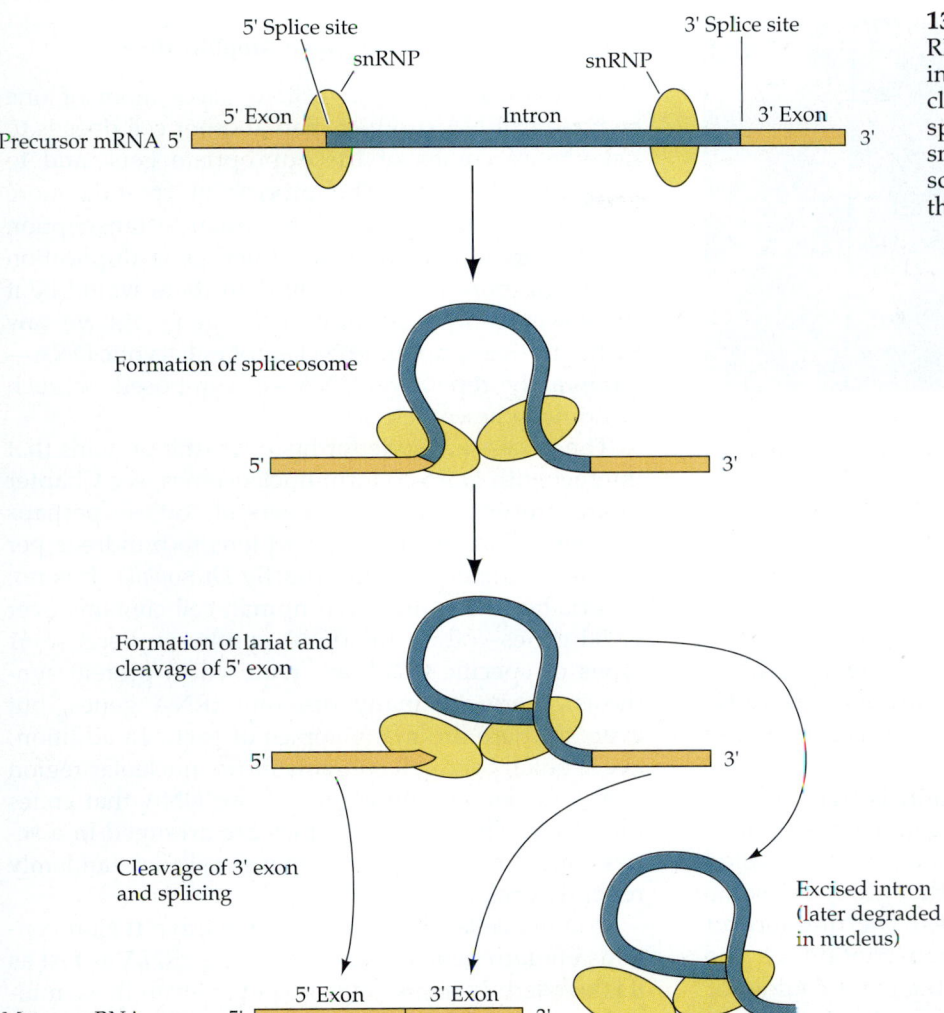

5' Splice site
snRNP
3' Splice site
snRNP

5' Exon
Intron
3' Exon

Precursor mRNA 5'
3'

Formation of spliceosome

5'
3'

Formation of lariat and cleavage of 5' exon

5'
3'

Cleavage of 3' exon and splicing

Excised intron (later degraded in nucleus)

5' Exon
3' Exon

Mature mRNA 5'
3'

13.6 snRNPs Splice RNA
RNA splicing depends on the binding of several small nuclear ribonucleoprotein particles to form the spliceosome. (Only two of the snRNPs are shown here.) The transcript of the intron forms a "lariat" that is cleaved away.

easy to understand if you think about the X and Y chromosomes of mammals. The normal female mammal has two X chromosomes; the normal male has one X and one Y. The Y chromosome has few, if any, genes that are also present on the X chromosome, and the Y appears to be transcriptionally inactive in most cells. Hence there is a 100 percent difference between females and males in the dosage of X-chromosome genes. Why is this not a case of aneuploidy involving a rather large chromosome? Aneuploidy for an autosome of comparable size is invariably lethal. Why then is not one sex or the other grossly deformed or completely inviable?

The answer was found in 1961 by Mary Lyon and Liane Russell, working independently. Lyon suggested that one of the X chromosomes in each cell of a normal female mammal is inactivated early in embryonic life and remains inactive ever after. The choice of which X in any pair of X chromosomes remains active is random. Because many cells are ultimately produced from each cell in which the choice is made, female mammals contain patches of

tissue in which one or the other X is active. This interpretation is supported by genetic, biochemical, and cytological evidence. Interphase cells of normal female mammals have a single, stainable nuclear body called a **Barr body** (after its discoverer, Murray Llewellyn Barr) that is not present in males. The Barr body is the inactive X chromosome, condensed into heterochromatin (Figure 13.7). The cells of women who have only one X chromosome, like those of normal men, contain *no* Barr bodies. Other women, who have a chromosomal constitution of XXX, have cells with *two* Barr bodies; there are even XXXXY males who have three Barr bodies in each cell. We may thus infer that interphase cells of each individual, male or female, contain a *single* active X chromosome, making the dosage of *expressed* X-chromosome genes constant and the same in both sexes.

In individual chromosomes, the presence of limited regions of transcriptionally active euchromatin may sometimes be observed. This activity is most obvious in polytene chromosomes (see Chapter 10), the giant chromosomes found in insect salivary

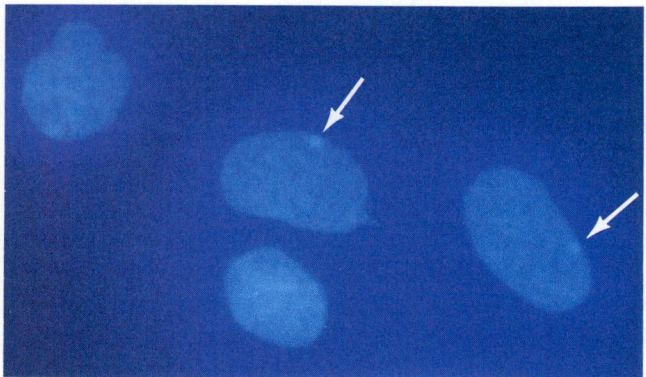

13.7 The Barr Body
The spots marked with arrows in these nuclei of human female cells are Barr bodies. A Barr body is the condensed, inactive member of the pair of X chromosomes in the cell.

glands. In some preparations of polytene chromosomes, puffs are visible. These **chromosome puffs** are regions of maximally extended chromatin, whose DNA is being transcribed (Figure 13.8).

Forming part of heterochromatin is not the only way that genes are inactivated. Another mechanism of gene inactivation is a chemical modification, called **DNA methylation**, that adds methyl groups to some cytosine residues in certain genes. The presence of the methyl groups prevents transcription of the genes. In humans and chickens, the DNA coding for globin synthesis is unmethylated in developing red blood cells. In cells that do not need to produce globin, however, the cytosine residues of the globin genes are highly methylated, so no globin is produced.

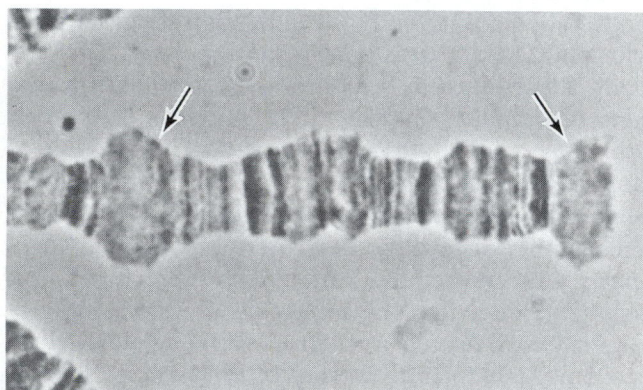

13.8 Chromosome Puffs
Arrows point to two puffs in this chromosome from a salivary gland cell of an insect larva. Puffs reveal regions where DNA is being transcribed to RNA. As the development of a larva proceeds, different regions of a chromosome puff when different gene products are needed.

Transcriptional Control: Gene Amplification

One obvious way for a cell to make more of one enzyme or RNA product than another cell does is to have more copies of the appropriate gene and to transcribe them all. The process of creating more genes of one kind in order to enhance transcription is called **gene amplification**. Such gene duplication results in more DNA per cell than there would be if there were only one copy of the gene. As we saw earlier in this chapter, one class of eukaryotic DNA—moderately repetitive DNA—is composed of such multiple gene copies.

The genes that code for histones (the proteins that interact with DNA to form nucleosomes; see Chapter 9) are present in great numbers of copies—perhaps 500 per cell in sea urchins and tens to hundreds per cell in mammals and the fruit fly *Drosophila*. It is not surprising, then, that each human cell contains over 2,000 genes coding for tRNA synthesis. Because 61 types of specific tRNA are required in protein synthesis, there are many different tRNA genes, but even so there are many copies of each. In addition, every eukaryotic cell contains in the nucleolar region of the nucleus many copies of the DNA that codes for rRNA. The multiple copies are arranged in a series, one after another in what is called a tandemly repetitive region.

In most cells, the tandemly repetitive region contains enough gene copies to produce rRNA as fast as it is needed. In some cells, however, even these multiple genes are apparently insufficient. The germ cells in amphibians that are destined to become eggs, for example, store large amounts of ribosomes for the early development of the embryo. During egg formation, each of these cells, called oocytes, multiplies the number of copies of rRNA genes until there are about a thousand nucleoli, containing more than a million rRNA genes in all, floating free in its nucleoplasm. Each of these nucleoli consists of DNA with repeating segments coding for rRNA, and each is transcribed to furnish the tremendous amount of rRNA that is stored in a mature oocyte for use by the embryo. This amplification ensures that the oocyte cytoplasm has enough rRNA to sustain rapid development during the entire period from fertilization to the formation of the gastrula (see Chapter 17), a structure consisting of hundreds of thousands of cells.

In frog oocytes, each haploid set of chromosomes contains about 500 copies of the rRNA genes before gene amplification; afterward, there are nearly a million. Evidently no genes in these cells except those responsible for rRNA synthesis are amplified. It is thought that genes are amplified in the oocytes of many animals and, under certain conditions, in vascular plants.

Figure 13.9 shows active rRNA genes from an amphibian nucleolus. The axial strands, or connecting threads, are nuclear DNA that codes for rRNA. The fuzzy-looking "triangles" attached to the DNA are many strands of rRNA in the making. Each partial rRNA molecule is attached to the DNA by an RNA polymerase molecule. Many polymerases can transcribe the DNA simultaneously. The apex of each triangle is the point at which RNA synthesis starts, so the RNA strands protruding from this region are very short. As transcription proceeds along the DNA, the RNA strands become longer and longer as more and more of the DNA coding for rRNA is read. Many of these fuzzy triangles are repeated along the DNA of the nucleolar organizer, showing us that the rRNA-coding DNA sequence itself is repetitive. Notice, too, that between the triangles there is quite a bit of silent "spacer," DNA that does not seem to be transcribed. The function of these spacers is unknown, but there is evidence suggesting that they serve as a "loading zone" for the RNA-polymerase proteins that transcribe the genes.

Transcriptional Control: Selective Gene Transcription

Very few genes become amplified. A much more common type of control simply switches the transcription of individual single-copy genes on or off. In some cases **selective transcription** of genes is mediated by eukaryotic steroid hormones (see Chapter 36). Insects with polytene chromosomes are good subjects for experimental studies of such hormonal control because transcription switched on by the hormones is accompanied by the formation of easily observable chromosome puffs (see Figure 13.8). The insect hormone ecdysone has three different types of specific effects on particular genes. Transcription of some genes, as indicated by puff formation, begins within four hours after treatment with ecdysone. Other genes form puffs several hours later—the later puffs seem to depend on both ecdysone and the protein products of earlier puffs. In addition, certain other genes stop producing puffs when ecdysone is present. Thus a complex repertoire of transcriptional events is under hormonal control.

Recall that transcriptional control in prokaryotes relies on operons, which are subject to either negative or positive control. Eukaryotic genes do not have operons, and the control of their transcription is almost always positive. In the negative control systems of prokaryotes, genes are expressed unless they are turned *off* by regulatory proteins. The positively regulated eukaryotic gene is turned *on* by proteins. Eukaryotes have much more DNA than do prokaryotes, so the danger that regulatory proteins will bind to inappropriate sites is much greater. This danger is reduced, however, by requiring the protein to bind at *multiple* sites in order to initiate the transcription of a single gene. As a result, the **promoter** for a eukaryotic gene is more complex than a prokaryotic promoter (see Chapter 12). In both eukaryotes and prokaryotes, the promoter is the stretch of DNA to which RNA polymerase binds to initiate transcription.

Unlike prokaryotes, which have only one type of RNA polymerase, eukaryotes have *three* RNA polymerases. RNA polymerase I transcribes the DNA that encodes rRNA; not surprisingly, this is the most abundant RNA polymerase in a eukaryotic cell. RNA polymerase II transcribes the structural genes that encode mRNAs and thus has the greatest diversity of products. RNA polymerase III transcribes the DNA that encodes tRNAs and some other small RNA species. We focus on RNA polymerase II for the rest of this discussion.

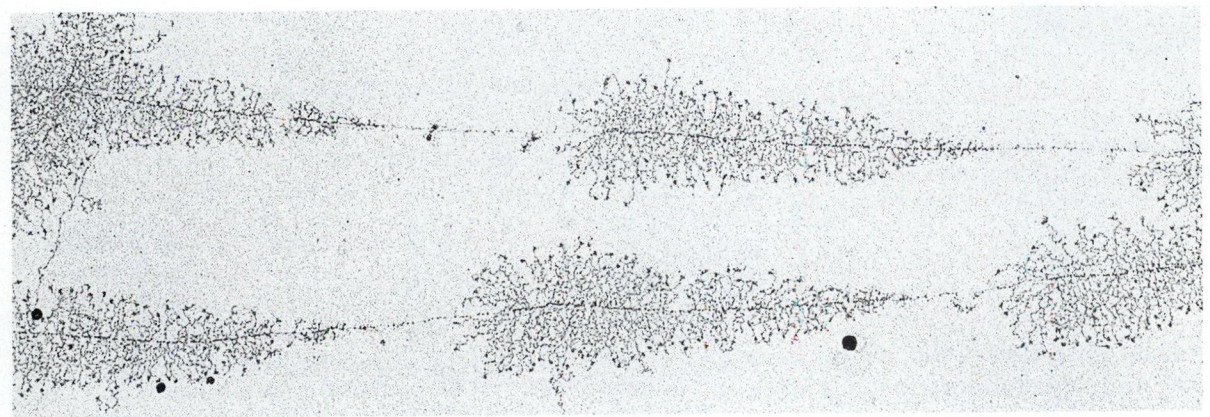

13.9 Transcription in the Nucleolus
Elongating strands of rRNA transcripts form arrowhead-shaped regions, each centered on a strand of DNA that codes for the rRNA.

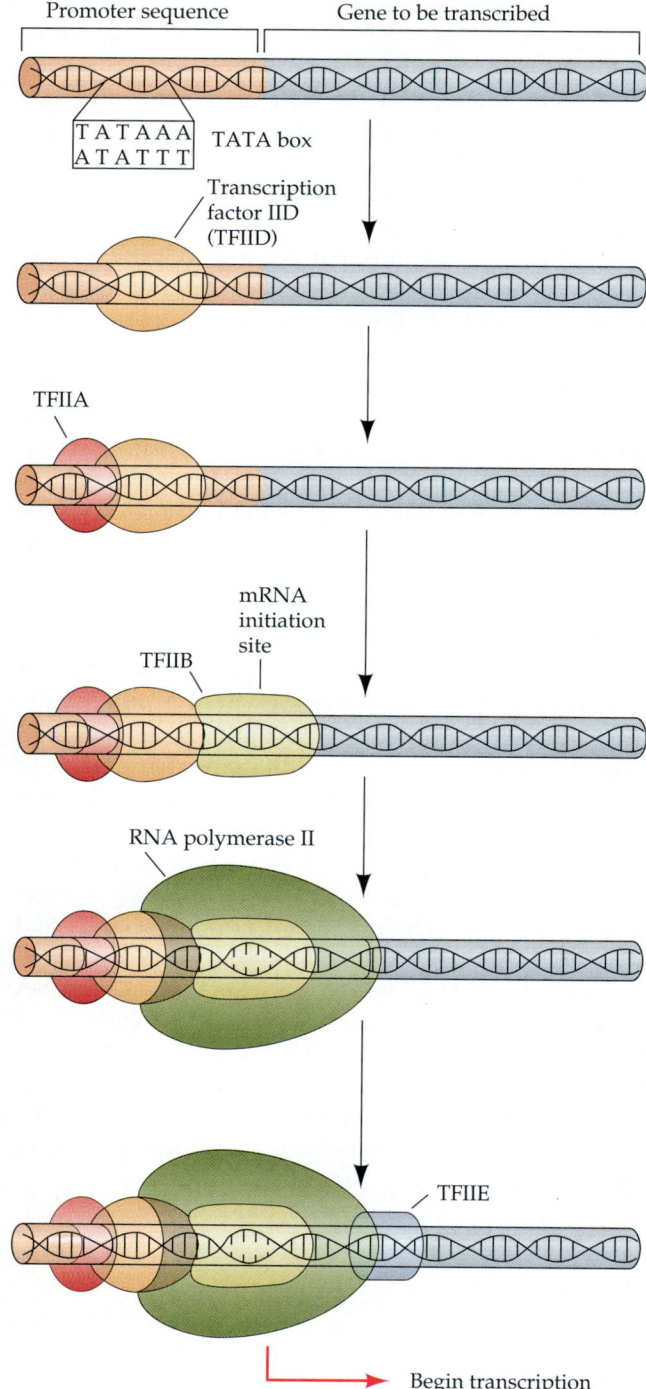

13.10 Forming a Transcription Complex
Interactions among the TATA box, four transcription factors, and RNA polymerase II lead to the formation of the transcription complex.

such sequence is the TATA box, an eight-base-pair sequence consisting only of T–A pairs. The TATA box is found in many eukaryotic promoters about 25 base pairs before the starting point for transcription. One of the transcription factors binds to the TATA box. Next, other transcription factors bind to other sequences on the promoter. Then RNA polymerase II joins the growing transcription complex, which is completed by the addition of yet another transcription factor (Figure 13.10). Still other DNA sequences, each requiring a different transcription factor, precede the TATA box on the chromosome (Figure 13.11). Transcription begins only after all the sequences have bound their transcription factors. These DNA sequences and their transcription factors have two principal roles in the regulation of transcription: The TATA box helps pinpoint the exact starting point for transcription, and the other sequences determine the efficiency of the promoter.

Some sequences, such as the TATA box, are common to the promoters of many genes and are recognized by transcription factors found in all the cells of an organism. Other sequences are specific to only a few genes and are recognized by transcription factors found only in certain tissues; these play important parts in differentiation.

Another important type of regulation is brought about by **enhancers**, DNA sequences that bind transcription factors and stimulate specific promoters, thus enhancing the transcription of specific genes. Enhancers can act at greater distances from the regulated genes than do promoters. In fact, enhancers may lie far away, in either direction along the sequence, from the genes they regulate. Many enhancers are specific to particular cell types and are inactive in others.

How do transcription factors and other DNA-binding proteins recognize and interact with specific sequences of bases in DNA? Such proteins have do-

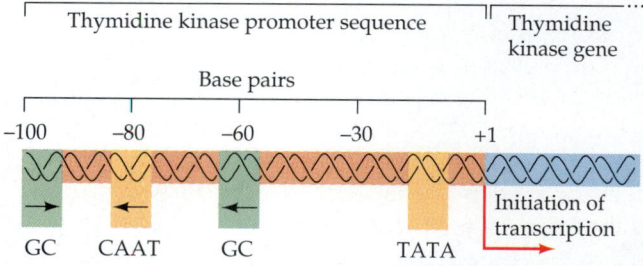

13.11 DNA Modules in a Eukaryotic Promoter
The promoter for the gene that encodes the enzyme thymidine kinase is typical of eukaryotic promoters. It contains the TATA box and three other DNA sequences, two of them identical (GC) but oriented in opposite directions. Transcription factors bind to the sequences to initiate transcription.

RNA polymerase II by itself cannot simply bind to the chromosome and initiate transcription. Rather, it can bind and act only after various regulatory proteins, or **transcription factors**, have assembled on the chromosome. Each transcription factor recognizes and binds to a particular sequence of the DNA. One

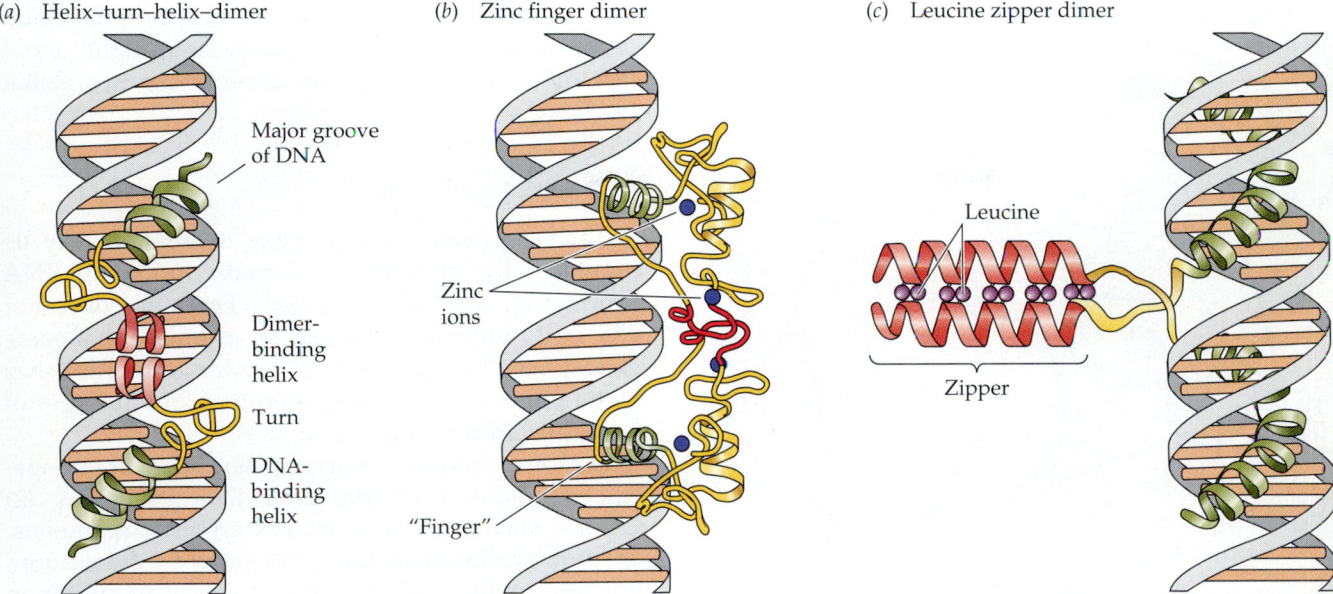

(a) Helix–turn–helix–dimer
(b) Zinc finger dimer
(c) Leucine zipper dimer

Major groove of DNA

Dimer-binding helix

Turn

DNA-binding helix

Zinc ions

"Finger"

Leucine

Zipper

mains that fit into the major groove of a DNA molecule and other parts that bind these domains together in a way that holds the protein tightly to the DNA (Figure 13.12).

Translational Control

We might guess that eukaryotes do *not* control the expression of their genes at the translational level because each of their mRNAs codes for only one polypeptide (suggesting that the production of a given polypeptide could be adequately regulated by mechanisms acting on transcription). Our guess would be wrong, however; translation of some genes *is* regulated. This level of control is a very rapid one and acts closest to the formation of the polypeptide product. Let's consider some examples of translational control of differentiation.

Several different mechanisms provide translational control. One involves a hormone acting on mRNA. As mammals prepare to lactate—to produce milk—a hormone, prolactin, acts on the mammary gland as the final trigger for milk production. The primary effect of this hormone is a dramatic increase in the translation of mRNA for casein, a major milk protein. Prolactin increases the longevity of casein mRNA, allowing it to be translated 25 more times than it is in the absence of the hormone.

A second mechanism of translational control—the capping of mRNA—is evident in the oocytes of sea urchins and certain moths. As already noted, most mRNAs become capped with a modified G unit during RNA processing. Uncapped messages are not translated. Stored mRNA in a tobacco hornworm oocyte, for example, has the G portion of the cap,

13.12 Transcription Factors Bind DNA
The structures of transcription factor proteins favor DNA binding; structural motifs include these examples. (a) Helix-turn-helix proteins are dimers with two pairs of α-helical regions that fit in the major groove of DNA and bind specific base sequences. (b) Many transcription factors have "zinc fingers"—regions, held in shape by zinc ions, that protrude precisely into the major groove. (c) Some protein dimers that bind DNA are held together by "leucine zippers" in which multiple copies of the amino acid leucine in the two polypeptide strands attract one another strongly by hydrophobic bonding. In these renditions, the red areas indicate the portion of the molecule that joins the monomers together in a dimer; green regions are where the transcription factor dimers bind to the DNA; and yellow indicates the linker regions that hold the red and green regions in the correct relative positions.

but the G has not been modified; hence these uncapped messages cannot be translated. When a tobacco hornworm egg is fertilized, the uncapped message is modified to complete the cap, and the mRNAs can then be translated.

In another example, an elegant set of controls ensures that hemoglobin is synthesized efficiently and in appropriate quantities in developing red blood cells. Hemoglobin is a moderately complex molecule that consists of two α-globin chains, two β-globin chains (both types of globins are proteins), and four smaller heme molecules, one for each globin chain. Thus the complete hemoglobin molecule consists of three distinct types of components, in a ratio of 2:2:4 (Figure 13.13). The components are usually synthesized in just this ratio; if their production gets out of balance, severe illnesses beset the organism. Three separate mechanisms maintain the ratio. First, any excess of heme results in feedback inhibition (see Chapter 6) of heme synthesis, thus reducing the imbalance. Second, excess heme increases the transla-

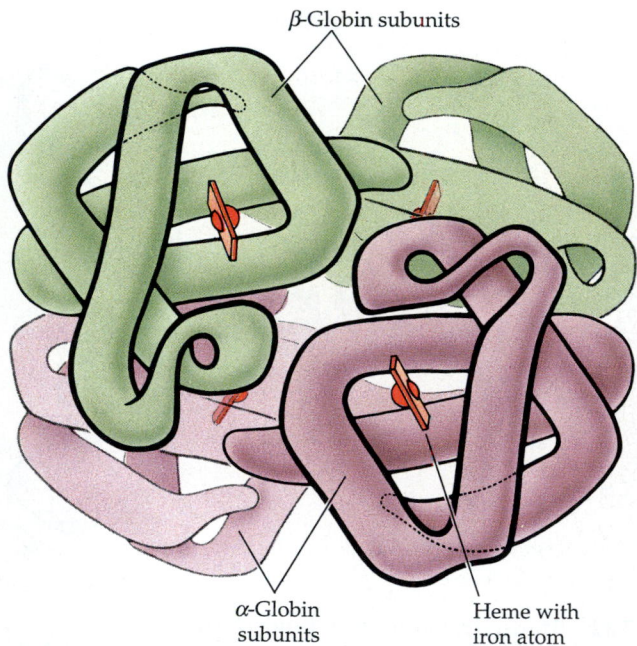

13.13 Hemoglobin Consists of Three Types of Molecules
The hemoglobin molecule is made of two α-globin (purple) and two β-globin (green) polypeptides, and each of these globins contains a heme molecule (red).

tion of globin messengers. Third, an appropriate ratio of α-globin to β-globin chains is brought about through control of the translation of the two globin mRNAs.

Posttranslational Control

We have considered how gene expression may be regulated by the control of transcription, of RNA processing, and of translation. The story does not end here, however, because the expression of some genes may be modified even *after* translation has taken place. Here is a brief summary of four types of **posttranslational control**.

(1) Some proteins are specifically *inactivated* by specific degradation shortly after their formation. (2) Others, such as insulin and certain other hormones, are not produced in an active form by translation, but are made active by later chemical modification (Figure 13.14). (3) Some proteins have to be inserted into particular compartments of the cell, such as mitochondria, and others must be directed through the endoplasmic reticulum and inserted into the plasma membrane before they become active. Proteins destined to associate with or pass through particular membranes contain **leader sequences** of amino acids at their N-terminal (NH₂) ends. A given leader se-

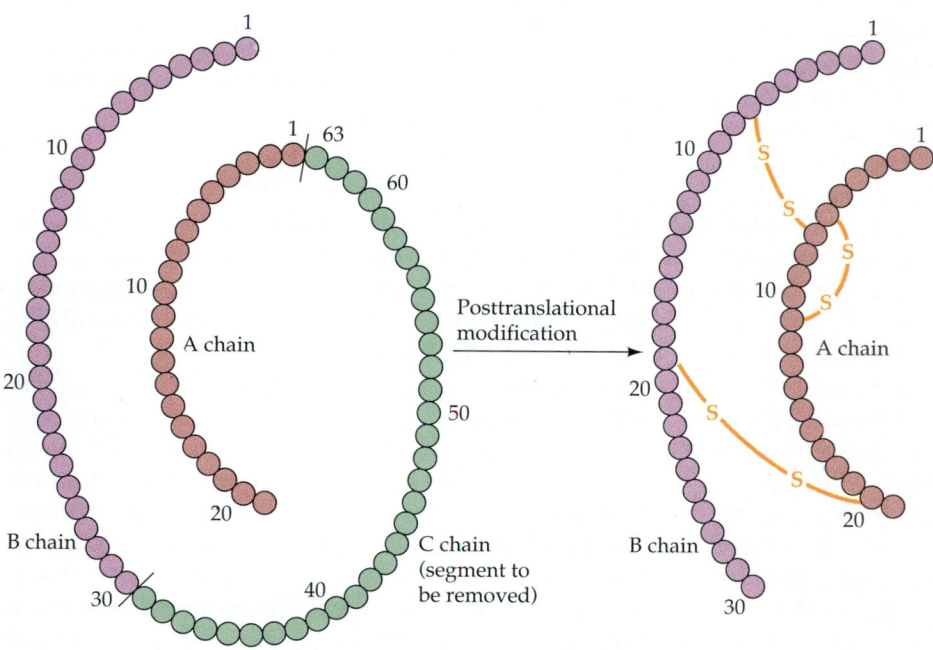

13.14 Posttranslational Events in Insulin Synthesis
The translation product of the mRNA that codes for the hormone insulin is a larger molecule, proinsulin, that does not act as a hormone. *After* translation, part of proinsulin is removed, leaving the active polypeptide hormone, insulin.

13.15 Some Proteins Are Inactive Until They Reach Their Destination

After being synthesized in the cytoplasm, the inactive forms of certain proteins move to their destinations, where, with final modifications, they become active molecules. The genes are finally expressed only where their products are needed. Leader sequences of amino acids guide the proteins across the membranes to their destinations. This example shows two proteins, one destined for the mitochondrial matrix and the other for the intermembrane space.

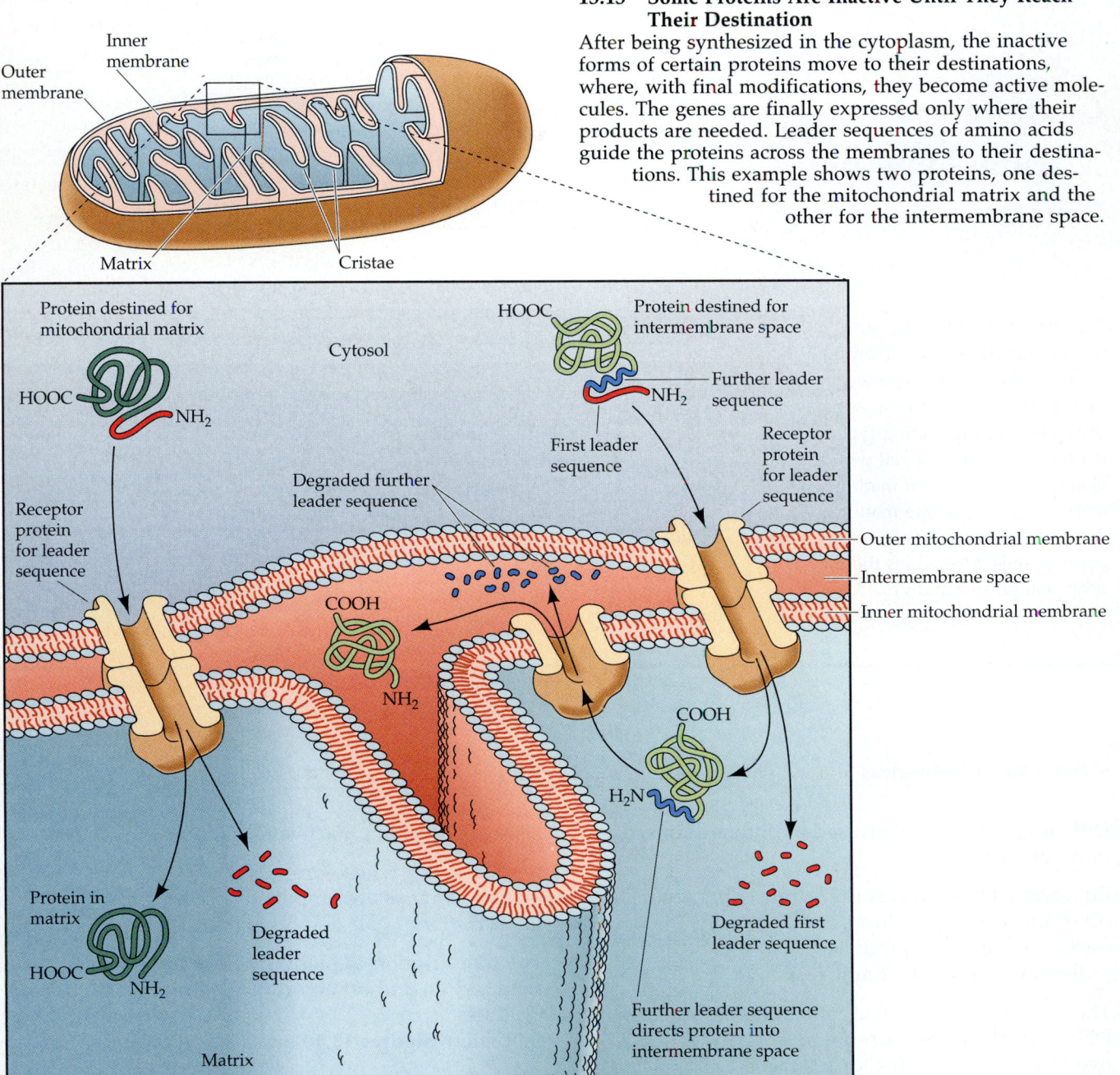

quence is recognized by a specific recognition protein in the appropriate membrane, allowing transit through that membrane. Proteases then remove the leader sequence, leaving the final, active protein as a product (Figure 13.15). (4) Finally, some proteins are inactive until they are incorporated into larger, compound structures. The proteins tubulin and actin, which form microtubules and microfilaments, respectively (see Chapter 4), do not become active until these structures have formed.

This brief overview of posttranslational control completes our consideration of molecular events in gene expression. In looking at the many mechanisms for controlling gene expression in eukaryotes, we gave no details about the very first mechanism we mentioned: rearrangement of genes on chromosomes. Box 13.A describes a dramatic example of such gene swapping. In Chapter 17 we will consider specific examples of the control of gene expression during animal development.

BOX 13.A

Cassettes and the Mating Type of Yeasts

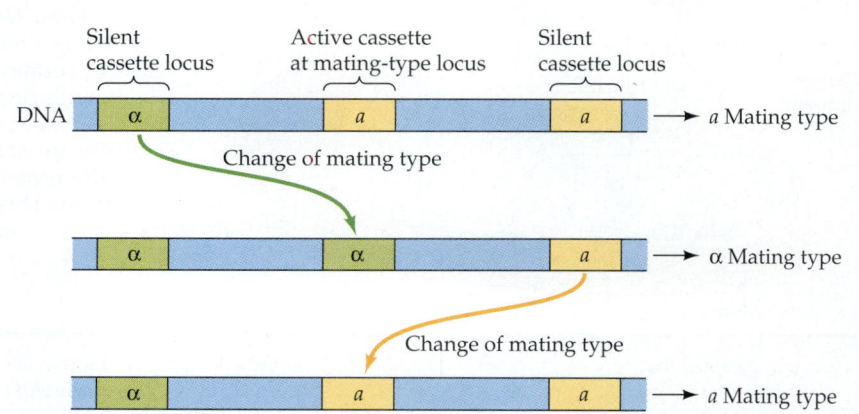

Silent cassette locus   Active cassette at mating-type locus   Silent cassette locus

DNA | α | | a | | a | → *a* Mating type

Change of mating type

| α | | α | | a | → α Mating type

Change of mating type

| α | | a | | a | → *a* Mating type

Yeasts have two mating types, *a* and α. The mating type is determined by a locus—the mating-type locus—on one of the chromosomes. A cell with the *a* allele at the mating-type locus is of mating type *a*; a cell with the α allele at the locus is of mating type α. In some yeasts, the mating type may change with almost every generation of cells. How does the mating type change so rapidly?

The mechanism has been likened to a cassette recorder, with the mating-type locus being the tape deck into which a cassette is inserted and "played." The yeast cell keeps two cassettes—unexpressed copies of the *a* and α alleles—at other loci on the chromosome that bears the mating-type locus, as shown here. From time to time the allele in the mating-type locus is removed, and one of the two unexpressed loci is copied to the mating-type locus. If this newly inserted cassette differs from the previous occupant of the mating-type locus, the mating type changes, because yeasts are unicellular and are haploid throughout most of their life cycle.

SUMMARY of Main Ideas about Gene Expression in Eukaryotes

Different genes are expressed at different times or in different tissues.

Eukaryotic DNA and its RNA transcripts are made up of introns that are removed and exons that code for the polypeptide product.
Review Figures 13.2 and 13.3

The three classes of eukaryotic DNA are single-copy DNA, which codes for proteins; moderately repetitive DNA, which codes for telomeres and transposable elements; and highly repetitive DNA, whose function is unknown.

The hnRNA produced by the transcription of a gene is capped and tailed, and introns are spliced from it; by these modifications it becomes mRNA.
Review Figures 13.5 and 13.6

The expression of structural genes is controlled in many different ways and at different points in the sequence from DNA to RNA to protein.

Euchromatin can be transcribed; heterochromatin is generally not transcribed.

DNA methylation inactivates genes.

Certain genes whose products are needed in enormous quantities are amplified before the time for transcription.

The transcription of many single-copy genes is switched on and off according to the need for their products.
Review Figures 13.10 and 13.12

After some mRNAs are produced, translational controls block synthesis of their proteins until the proteins are needed.

Posttranslational control results in a final polypeptide that differs from the polypeptide product of translation.
Review Figures 13.14 and 13.15

Transposable elements modify genes and chromosomes.

Duplication of genes in the course of evolution has given rise to families of genes.
Review Figure 13.4

SELF-QUIZ

1. Which statement is *not* true of nucleic acid hybridization?
   a. It depends upon complementary base pairing.
   b. A DNA strand can hybridize with another DNA strand.
   c. An RNA strand can hybridize with a DNA strand.
   d. A polypeptide can hybridize with a DNA strand.
   e. Double-stranded DNA denatures at high temperatures.

2. Which statement is *not* true of introns?
   a. Their name is short for "intervening sequence."
   b. They do not encode any part of the polypeptide product of the gene.
   c. They are found in all vertebrate genes.
   d. They are transcribed.
   e. They interrupt, but do not scramble, the coding sequence for a polypeptide.

3. In regard to repetitive DNA
   a. much highly repetitive DNA lies near the centromeres.
   b. highly repetitive DNA reanneals most slowly of the three classes of DNA.
   c. highly repetitive DNA is transcribed often and rapidly.
   d. single-copy DNA is rare in eukaryotes.
   e. transposable elements are single-copy genes.

4. Capping of hnRNA
   a. takes place at its 3′ end.
   b. facilitates binding of the mRNA to a ribosome.
   c. is coded for in the DNA.
   d. prevents its translation.
   e. consists of the addition of a poly A tail.

5. Which statement is *not* true of RNA splicing?
   a. It removes introns.
   b. It is performed by small nuclear ribonucleoprotein particles (snRNPs).
   c. There are different splicing mechanisms for mRNAs, rRNAs, and tRNAs.
   d. It is directed by consensus sequences.
   e. It lengthens the RNA molecule.

6. Which genes are *not* commonly present in many copies in eukaryotes?
   a. Those that encode histones
   b. Those that encode mRNAs
   c. Those that encode tRNAs
   d. Those that encode rRNA
   e. Those that are present in a tandemly repetitive region

7. Which statement is *not* true of selective transcription in eukaryotes?
   a. Different classes of RNA polymerase transcribe different parts of the genome.

   b. Transcription requires transcription factors.
   c. Genes are transcribed in groups called operons.
   d. Control is almost always by positive regulation.
   e. The promoter is more complex than in prokaryotes.

8. Transcription factors in eukaryotes
   a. consist of DNA.
   b. consist of RNA.
   c. include such sequences as the TATA box.
   d. allow the binding of RNA polymerase to the promoter.
   e. cause operons to be transcribed.

9. Translational control
   a. is not observed in eukaryotes.
   b. is a slower form of regulation than is transcriptional control.
   c. occurs by only one mechanism.
   d. requires that mRNAs be uncapped.
   e. ensures that hemoglobin is synthesized in appropriate quantity.

10. Which statement is *not* true of telomeres?
    a. They contain repetitive DNA.
    b. They appear at the ends of eukaryotic chromosomes.
    c. They are the sites at which spindle fibers attach.
    d. They protect chromosomes from shortening during replication.
    e. They may play a role in synapsis during meiosis.

FOR STUDY

1. In rats a gene 1,440 base pairs long codes for an enzyme made up of 192 amino acid units. Discuss this apparent discrepancy.

2. Describe the steps in the production of mature, translatable mRNA from a eukaryotic gene that contains introns.

3. How can nucleic acid hybridization techniques be used to determine whether a gene possesses introns?

4. Describe the origin and development of gene families such as the one containing genes that code for globins.

5. Prepare a list of the possible ways in which transcription and translation may be regulated in eukaryotes. Contrast this with the situation in prokaryotes.

READINGS

Cech, T. R. 1986. "RNA as an Enzyme." *Scientific American*, November. A description of exciting discovery of the catalytic activity of certain RNAs and its roles in the molecular biology of eukaryotes. These findings help us to understand the origin of life.

Chambon, P. 1981. "Split Genes." *Scientific American*, May. Introns and exons—their origin and how they are handled.

Donelson, J. E. and M. J. Turner. 1985. "How the Trypanosome Changes Its Coat." *Scientific American*, February. Trypanosomes evade the host's immune system by constantly switching on new genes that code for different surface antigens.

Griffiths, A. J. F., J. H. Miller, D. T. Suzuki, R. C. Lewontin and W. M. Gelbart. 1993. An Introduction to Genetic Analysis, 5th edition. W. H. Freeman, New York. An up-to-date revision of one of the field's classic textbooks. See especially Chapter 17.

Grunstein, M. 1992. "Histones as Regulators of Genes." *Scientific American*, October. Histones not only organize the nucleosomes, but they regulate transcription.

McKnight, S. L. 1991. "Molecular Zippers in Gene Regulation." *Scientific American*, April. A description of an important class of DNA-binding proteins, some of which are transcription factors.

Ptashne, M. 1989. "How gene activators work." *Scientific American*, January. How genes are turned on and off in eukaryotic cells.

Rennie, J. 1993. "DNA's new twists." *Scientific American*, March. Transposable elements and other features.

Rhodes, D. and A. Klug. 1993. "Zinc Fingers." *Scientific American*, February. How zinc fingers help some proteins bind to DNA and regulate transcription.

Steitz, J. A. 1988. "Snurps." *Scientific American*, June. How spliceosomes remove intron transcripts.

Varmus, H. 1987. "Reverse Transcription." *Scientific American*, September. Reverse transcription is not confined to the retroviruses. This article describes reverse transcription in eukaryotes, stressing its likely relevance to the emergence of DNA as the genetic material.

Suppose you could "teach" bacteria or other unicellular organisms to produce important chemicals normally produced only by humans. Might that be medically useful? Suppose that you could teach bacteria to clean up oil spills in the ocean. Might that be a powerful tool for environmental protection? Suppose that you could teach important crop plants to make their own fertilizer. Wouldn't *that* be useful as we attempt to feed the ever-expanding human population?

The first item on this "wish list" is already a reality, the second has been achieved to a very limited extent, and the third is being vigorously pursued. In each case, the "teaching" process consists of providing an organism with genes from another organism that is capable of doing something the organism receiving the genes can't do.

This process brings together techniques from molecular biology, microbial genetics, and biochemistry. It has become known as **recombinant DNA technology**, popularized as "genetic engineering" and "cloning." By recombinant DNA, we mean DNA made up of connected segments from mixed sources—perhaps from two different species, or perhaps a combination of natural and synthetic DNA. An example of combining DNA from different species is the insertion of a gene from a human into the DNA of a yeast. This feat can convert the yeast cells into "factories" producing the protein product of the human gene. An early example of combining natural and synthetic DNA was the use of bacteria to produce the human hormone somatostatin, a 14-amino acid polypeptide whose sequence was known. Biologists at the City of Hope Medical Center and at the University of California, San Francisco, synthesized a completely artificial stretch of DNA, part of which was designed to code for the amino acid sequence of somatostatin. This DNA was first inserted into a plasmid that had been isolated from a bacterial cell. The scientists then introduced the recombinant plasmid into a culture of *Escherichia coli*, where the plasmid replicated, producing multiple copies of itself in each bacterial cell. These *E. coli* could then be induced to synthesize human somatostatin. Somatostatin may be useful in treating pancreatitis (a disease of the pancreas), acromegaly (abnormal enlargement of bones in the hands, feet, and face), and insulin-dependent diabetes. A short time ago somatostatin could not be considered for medical use because almost none of it was available from natural sources. Recombinant DNA technology, however, has made possible the large-scale production of somatostatin, opening vistas for treatment of these diseases.

As a tool, recombinant DNA technology has revolutionized much of experimental biology. Most recent advances in understanding how the genes of eukaryotes are regulated and organized have come

Brewing Medicine
Hepatitis vaccine produced by recombinant DNA technology "brews" in an 800-liter fermenter.

14

Recombinant DNA Technology

through the application of recombinant DNA technology. Recombinant DNA technology is also revolutionizing agriculture, medicine, and other areas of the chemical industry, as well as forensics and the battle against environmental pollution. In this chapter we will consider the basic techniques of this technology and some of its applications in the laboratory and beyond.

THE PILLARS OF RECOMBINANT DNA TECHNOLOGY

Scientists realized that chemical reactions used in living cells for one purpose may be applied in the laboratory for other, novel purposes. Recombinant DNA technology is based on this realization, and on the recognition of the properties of certain enzymes and of DNA itself. Naturally occurring enzymes that cleave DNA, help it grow, and repair it are numerous and diverse, and many of them are now used in the laboratory to manipulate and combine DNA molecules from different sources.

The nucleic acid base-pairing rules underlie many of the fundamental processes of molecular biology. The mechanisms of DNA replication, transcription, and translation all rely on complementary base pairing. Similarly, all the key techniques of recombinant DNA technology—sequencing, splicing, locating, and identifying DNA fragments—make use of the complementary pairing of A with T (or U) and of G with C.

CLEAVING AND SPLICING DNA

The basic operations of cleaving and splicing DNA are good examples of how scientists use enzymes and complementary base pairing creatively. The enzymes are used in the laboratory to achieve different overall purposes than they would in the living cell.

Restriction Endonucleases

All organisms must have mechanisms to deal with their enemies. As we saw in Chapter 12, bacteria are attacked by bacteriophages that inject their genetic material into their hosts. Eventually the phage genetic material may be replicated by the enzyme systems of the host. Some bacteria defend themselves against such invasions by producing enzymes called **restriction endonucleases** that can cleave double-stranded DNA molecules—such as those injected by many phages—into smaller, noninfectious fragments (Figure 14.1). There are many such enzymes, each of which cleaves DNA at a specific site defined by a *sequence of bases* and called a **recognition site**. The

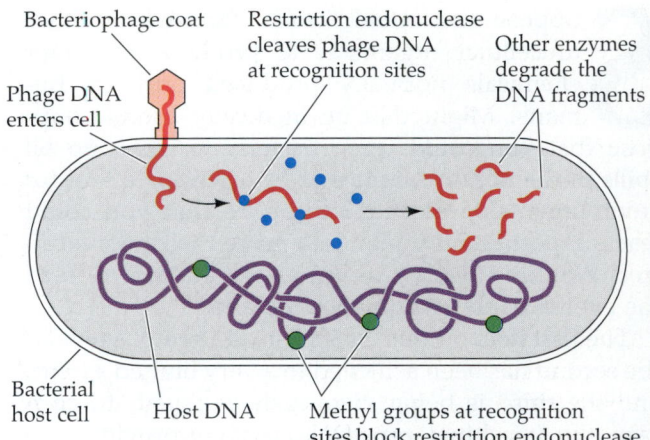

14.1 Bacteria Fight Phages with Restriction Endonucleases
Bacteria produce restriction endonucleases that break up phage DNA. Other enzymes protect the bacteria's own DNA from being cleaved; they do this by methylating the host DNA.

DNA of the bacterial host is not cleaved by its own restriction endonucleases because of the activity of specific methylases, enzymes that add methyl (—CH_3) groups to certain of the bases within the recognition sites. The methylation of the bases makes the recognition sites unrecognizable to the restriction endonucleases, thus preventing cleavage of the host DNA, but the unmethylated phage DNA is efficiently degraded.

A specific sequence of bases defines each recognition site. For example, the restriction endonuclease *Eco*RI (named after its source, *E. coli*) cuts DNA only where it encounters these paired sequences in the double helix:

5'. . .GAATTC. . .3'

3'. . .CTTAAG. . .5'

Other restriction endonucleases recognize different base sequences. The sequence recognized by *Eco*RI occurs on the average about once in 4,000 base pairs—about once per four prokaryote genes. This restriction endonuclease can chop a typical big piece of DNA into smaller pieces containing, on the average, just a few genes. Remember that "on the average" does not mean that it cuts at regular intervals along all stretches of DNA. The *Eco*RI recognition sequence does not occur even once in the 40,000-base-pair sequence of the DNA of phage T7—a characteristic that is crucial for the survival of T7, because *E. coli* is its host. Fortunately for *E. coli*, the DNA of other phages does contain this recognition sequence, which prevents *E. coli* from being overrun by phages.

Different restriction enzymes that recognize different recognition sites may cut the same sample of

DNA. Several hundred restriction endonucleases have been extracted from various bacteria, and many are available for recombinant DNA research. Thus cutting a sample of DNA in many different, specific places is an easy task in the laboratory. We can use restriction endonucleases as "knives" for genetic "surgery."

Sticky Ends and DNA Splicing

An important property of some restriction endonucleases is that they make staggered cuts in the DNA rather than cutting both strands at the same point. For example, *Eco*RI cuts DNA as shown at the top of Figure 14.2*a* (note that the cut is within the recognition sequence given in the previous section). After the two cuts are made, the two strands are held together by only four base pairs. The hydrogen bonds of those base pairs are too weak to persist at warm temperatures (room temperature or above), so the pieces separate. As a result, there are single-stranded tails at the site of each cut. These tails are called **sticky ends** because they have a specific base sequence that is capable of binding (by complementary base pairing, at low temperature) with complementary sticky ends.

After a piece of DNA has been cut by a restriction endonuclease, it is possible for the complementary sticky ends to rejoin. The original ends can join, or an end may pair with another fragment. If more than one recognition site for a given restriction endonuclease is present in a sample, the enzyme can make a number of cut pieces, all with the *same* sequences in their sticky ends. When the temperature is lowered, the pieces reassociate at random. At the lower temperature, base pairs are more stable and four base pairs may hold the two pieces of DNA together. The new associations are unstable, however, because they are maintained by only a few hydrogen bonds.

The joined sticky ends can be permanently united by a second enzyme, DNA ligase, which makes the joining very stable. The usual function of DNA ligase in the cell, as mentioned briefly in Chapter 11, is to unite the Okazaki fragments of the lagging strand during DNA replication (see Box 11.A). DNA ligase also mends breaks in polynucleotide chains, thus helping in DNA repair.

A piece of DNA can be inserted into a plasmid as shown in Figure 14.3, as long as both the plasmid and the source for the DNA piece contain recognition sites for the same restriction endonuclease. The DNA to be inserted is cleaved from within its molecule by cutting both ends with a particular restriction endonuclease. The circular plasmid is cleaved with the same endonuclease, transforming it into a linear molecule with sticky ends. The sticky ends of the piece of DNA join the sticky ends of the cleaved plasmid, and DNA ligase seals the joining, regenerating a circular plasmid. The plasmid now contains the inserted

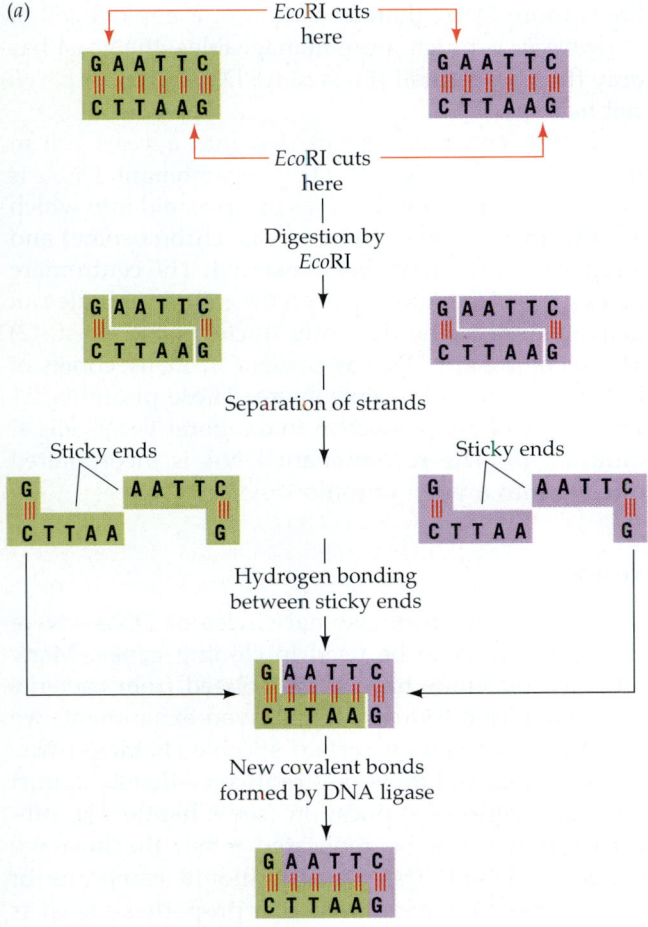

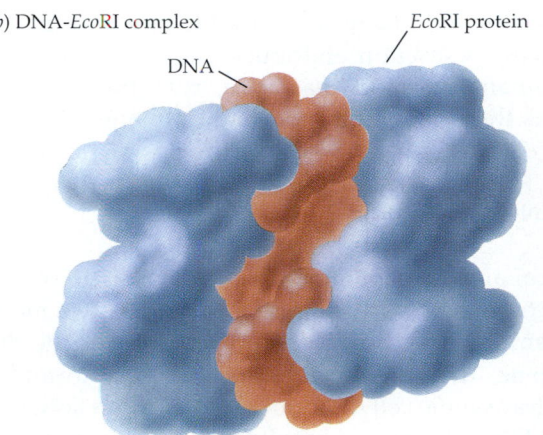

14.2   Cutting and Splicing DNA
(*a*) Some restriction endonucleases sever one strand of the double helix at one point and the other strand at a different point. The separated pieces have single-stranded sticky ends capable of combining with the sticky ends of complementary single strands. Newly joined pieces are stabilized by the action of DNA ligase. (*b*) The binding of the restriction endonuclease *Eco*RI to its recognition site on a DNA molecule.

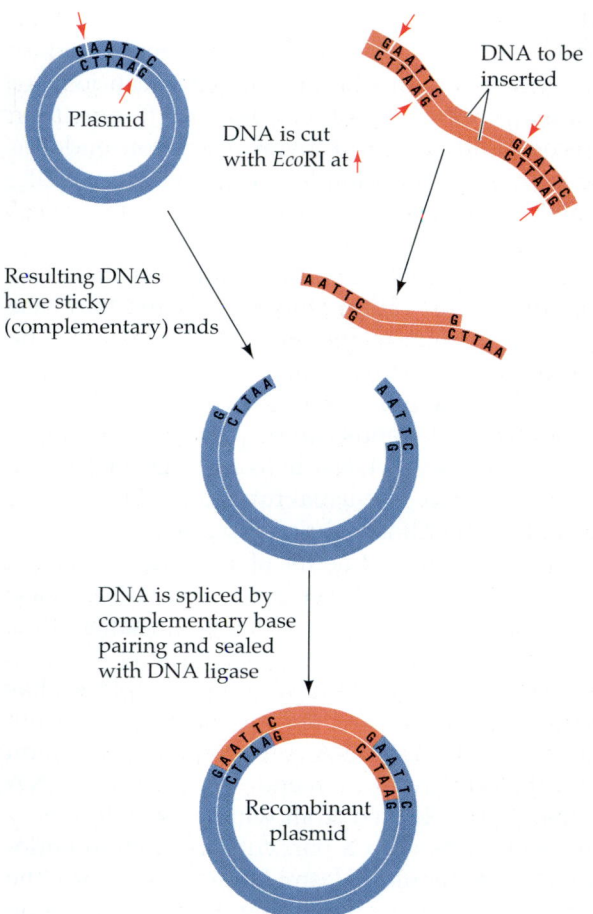

14.3 Insertion of a DNA Sample into a Plasmid
DNA from an outside source is inserted into a plasmid in the middle of a recognition site (red arrows) for the restriction endonuclease.

DNA—in the middle of what used to be a recognition site for the restriction endonuclease. Notice that the recombinant plasmid has two recognition sites, whereas the original plasmid had only one.

CLONING GENES

A typical aim of recombinant DNA work is to obtain many copies of a particular gene. To do so, we make **transgenic** bacterial or yeast cells containing the desired gene, then allow the transgenic cells to multiply. A transgenic cell or organism is one that contains foreign DNA integrated into its own genetic material. To ensure that the foreign gene, natural or synthetic, gets integrated, we must first insert it into a suitable **cloning vector**—such as a virus or a plasmid—before introducing it into the bacterial or yeast host. A stretch of DNA that is introduced by itself into a cell does not get replicated and will eventually be degraded; a cloning vector (and any gene contained in it), however, *does* get replicated.

Host Cells

The early successes of recombinant DNA technology were achieved with bacteria as hosts. However, bacteria are not ideal organisms for processing and studying eukaryotic genes. By now, cells of some eukaryotes have also been used as hosts.

Yeasts are now the most commonly used host cells for recombinant DNA studies of eukaryotic genes. Because it is widely employed in such studies, *Saccharomyces cerevisiae* (baker's or brewer's yeast), is rapidly becoming the eukaryote best understood at the molecular level, just as *Escherichia coli* is the best-understood prokaryote. Unlike most other eukaryotes, yeasts are unicellular; however, yeasts are typical eukaryotes in most respects, including their genetic mechanisms. The DNA of *S. cerevisiae* is organized into 17 chromosomes (in the haploid phase) that can be separated in the laboratory using a new technique that will be described later in this chapter.

Yeasts share three of the greatest advantages of *E. coli* for work in molecular biology. Because they are tiny and unicellular, they are easy to grow in vast numbers in small volumes of medium. They multiply almost as rapidly as *E. coli*. (The division time of a typical culture of *E. coli* is in the range of 20 to 60 minutes and that of *S. cerevisiae* is 2 to 8 hours.) Finally, although a mammalian cell has about 1,000 times more DNA than does *E. coli*, a haploid cell of *S. cerevisiae* is much more manageable—the yeast has only three and a half times more DNA than an *E. coli* cell has.

Foreign genes can be carried into a yeast cell in three different ways: (1) The recombinant DNA is present in one or a few copies of a plasmid into which a centromere (cloned from a yeast chromosome) and a replication site have been inserted. The centromere makes the plasmids stable, so the mitotic spindle can deliver them to the daughter nuclei of the yeast. (2) The recombinant DNA is present in many copies of a plasmid without a centromere. These plasmids are unstable and are passed on to daughter yeast cells at random. (3) The recombinant DNA is incorporated directly into a yeast chromosome.

Vectors

Plasmids—extrachromosomal circles of DNA—were the first vectors to be used in cloning genes. Many different plasmids have been isolated from bacterial and eukaryotic sources. For a given experiment, we select a plasmid with certain specific characteristics. First, the plasmid must be a **replicon**—that is, it must have an origin of replication (see Chapter 11); otherwise it will not be replicated when the host cell divides. Second, the plasmid should carry one or more genes conferring particular properties—such as resistance to specific antibiotics—that may be used for selection purposes. Third, the plasmid should have only one recognition site where the restriction endonuclease will cut it. If it has no such site, it cannot be opened for the insertion of the new genes; if it has two or more, the restriction endonuclease and DNA ligase may form many diverse products rather than just one. With a single site, the probability of achieving the desired insertion is higher but still not great (Figure 14.4).

Plasmids are efficient vectors for cloning small DNA fragments, but they multiply more slowly—and are less stable—when they contain large DNA fragments. Nowadays much recombinant DNA research uses viruses, in part because they are stable for cloning larger DNA fragments. Genes from other sources are inserted into the DNA or RNA of the viruses used as vectors. Viral vectors are commonly used for studies of eukaryotic DNA. The recombinant viral chromosome produced for such a study must be able to fit into the viral coat. Thus a nonessential piece of the chromosome may be "edited out" before the foreign gene is inserted so that the new recombinant will be small enough to be packaged into the viral coat for delivery to the cell.

Inserting Foreign DNA into Host Cells

Although some viral cloning vectors can infect host cells directly, most vectors require assistance in inserting their DNA into host cells.

As you will recall from Chapters 11 and 12, bacteria are capable of **transformation**—they can contact and take up isolated DNA from other bacteria. This ability provided researchers with the method for introducing plasmid vectors into bacteria. Bacteria are mixed with a solution containing plasmids and, under the appropriate conditions, some of the bacteria take up the plasmid DNA.

A similar process in eukaryotic cells, **transfection**—the uptake, incorporation, and expression of foreign DNA—often includes other steps. The cell walls must be removed before DNA can enter plant or fungal cells. The walls can be digested by appropriate enzymes, such as an enzyme found in snail guts that the snail uses to digest its meals. The resulting plant or fungal cell, without a wall, is called a **protoplast**. Plant and yeast protoplasts can be transfected with plasmids.

DNA may be inserted into eukaryotic cells in other, more drastic ways. In one method, **micropro-**

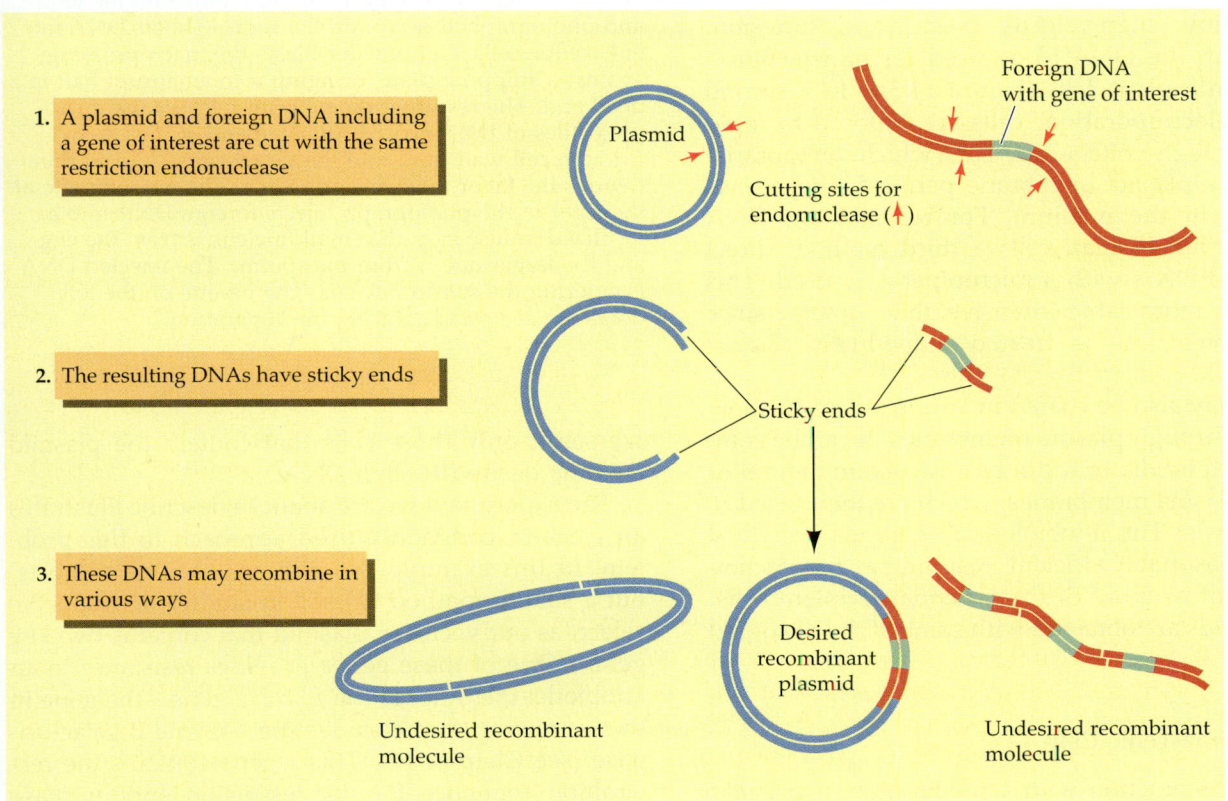

1. A plasmid and foreign DNA including a gene of interest are cut with the same restriction endonuclease

Plasmid

Foreign DNA with gene of interest

Cutting sites for endonuclease (↓)

2. The resulting DNAs have sticky ends

Sticky ends

3. These DNAs may recombine in various ways

Undesired recombinant molecule

Desired recombinant plasmid

Undesired recombinant molecule

14.4 Insertion into Plasmids Gives Multiple Products
These are only three of the many ways in which the plasmid and foreign DNA can combine. If the plasmid had more than one cutting site for the restriction endonuclease, the range of combinations would be much greater. As we will soon see, there is an elegant way to isolate the desired recombinant plasmid.

(a)

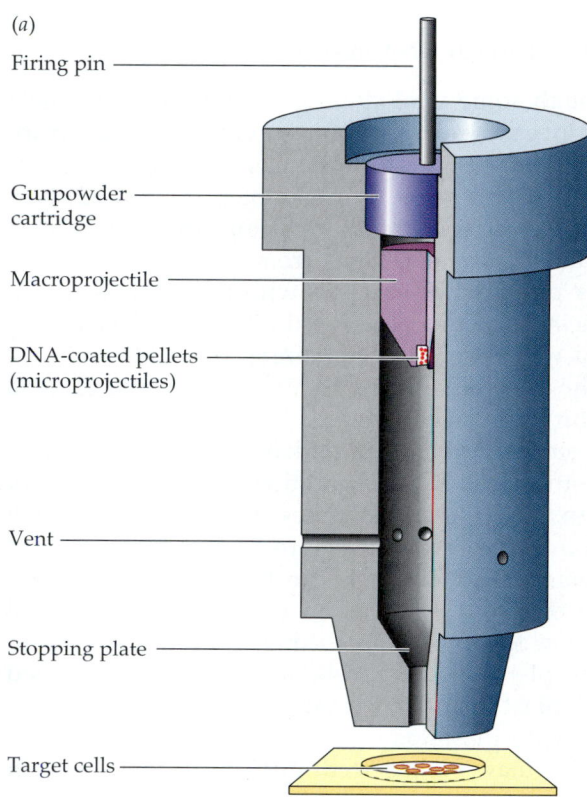

Firing pin

Gunpowder cartridge

Macroprojectile

DNA-coated pellets (microprojectiles)

Vent

Stopping plate

Target cells

(b)

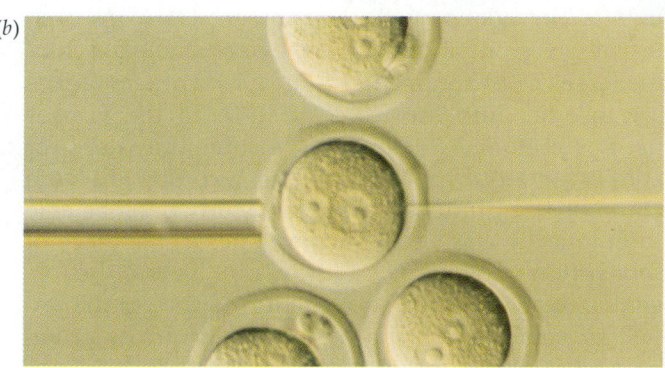

14.5 Inserting DNA into Cells

(a) This "DNA particle gun"—diagrammed on the left and photographed in use on the right—shoots DNA into eukaryotic cells. Gunpowder blasts the macroprojectile against a stopping plate, bringing it to an abrupt halt in a dead end. The resulting momentum releases the micro-projectiles of DNA-coated tungsten and carries them through cell walls and into the target cells. *(b)* And then there is the labor-intensive approach. The micropipette at the right of the photograph injects foreign DNA into a fertilized mouse egg (the small nucleus is from the egg and the larger one is from the sperm). The injected DNA is entering the sperm nucleus. The pipette on the left holds the cell steady during the "operation."

jectiles—tiny, high-velocity particles of tungsten, coated with the DNA to be used for transfection—are shot into plant cells (Figure 14.5*a*). In a second method, **electroporation**, cells are exposed to rapid pulses of high-voltage current, which temporarily renders the plasma membrane permeable to macro-molecules in the medium. For recombinant DNA studies of mammalian cells a third method, direct injection of DNA with a micropipette, is used. This method is more labor-intensive than others, since each recipient cell is treated individually (Figure 14.5*b*).

DNA may also be coated in various ways to allow it to pass through plasma membranes. It can be com-plexed with lipids, or it can be enclosed in natural or artificial plasma membranes, which are then fused to the host cells. The nonbiological methods described here are reasonably efficient, resulting in transfection of 1 percent or more of the potential recipient cells. Higher yields are obtained with some of the biological methods.

Selecting Transgenic Cells

Following interaction with a preparation of plasmid vectors, a population of host yeast or bacterial cells is heterogeneous, since only a small percentage of the cells have taken up the plasmid. Also, only a few of the plasmids that have moved into host cells con-tain the DNA sequence we wish to clone. How can we select only those cells that contain the plasmid *with* the desired foreign DNA?

The experiment we are about to describe illustrates an elegant, commonly used approach to this prob-lem. In this example, we will use bacteria as hosts, but a similar method is used in studies of yeast. We select, as our vector, a plasmid that contains two key genes. One of these genes provides resistance to an antibiotic; the other is the *E. coli z* gene—the gene in the *lac* operon that encodes the enzyme β-galactosi-dase (see Chapter 12). This *z* gene contains the rec-ognition sequence for the restriction endonuclease we will use. The following description is diagrammed in Figure 14.6.

When we add the endonuclease to the plasmid preparation, it cuts the plasmid within the *z* gene, leaving sticky ends. We then add the foreign DNA,

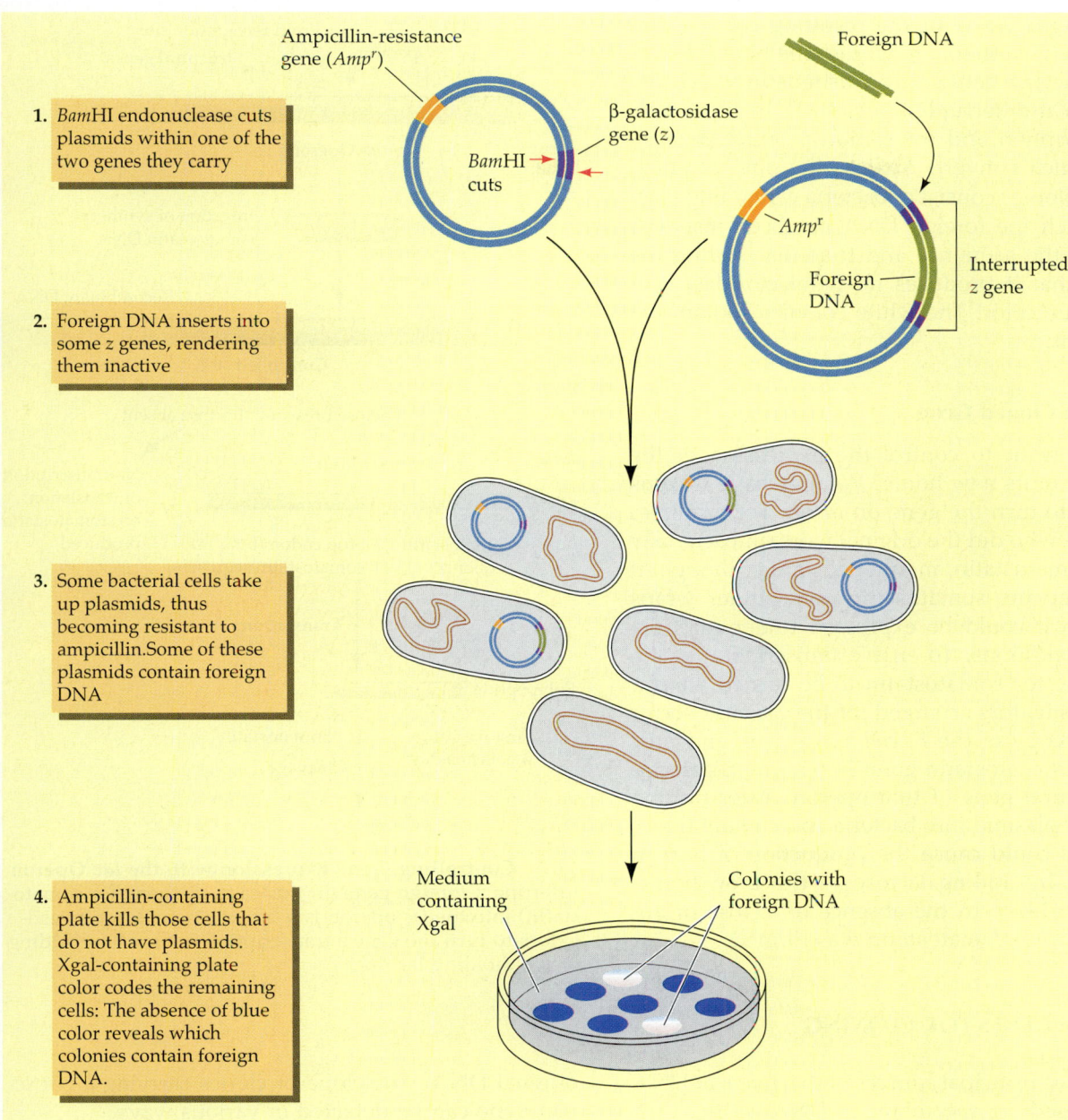

1. *Bam*HI endonuclease cuts plasmids within one of the two genes they carry

Ampicillin-resistance gene (*Amp*ʳ)

Foreign DNA

β-galactosidase gene (z)

*Bam*HI cuts

*Amp*ʳ

Foreign DNA

Interrupted z gene

2. Foreign DNA inserts into some z genes, rendering them inactive

3. Some bacterial cells take up plasmids, thus becoming resistant to ampicillin.Some of these plasmids contain foreign DNA

4. Ampicillin-containing plate kills those cells that do not have plasmids. Xgal-containing plate color codes the remaining cells: The absence of blue color reveals which colonies contain foreign DNA.

Medium containing Xgal

Colonies with foreign DNA

14.6   Color Coding the Colonies We Want
Inserting genes that confer antibiotic resistance into plasmids allows us to use ampicillin to eliminate those bacterial cells that do not contain the plasmid. To discover which plasmid-containing cells also took up the foreign DNA, we place the cells in Xgal. Those cells with no foreign DNA produce β-galactosidase, which turns the Xgal blue. In colonies of cells that took up the foreign DNA, however, β-galactosidase transcription is interrupted and no blue color is seen.

which has been cut with the same endonuclease, and allow the sticky ends of this DNA and of the plasmids to recombine. This process gives us a heterogeneous population of plasmids, some containing the foreign DNA. Combination of the plasmid population with host cells leads to transformation, giving a hetero-

geneous population of bacteria, some containing a plasmid.

Which cells contain plasmids? Which plasmids contain the foreign DNA? We add the antibiotic to which the gene in the plasmid confers resistance. This treatment kills all the bacteria that do not contain the plasmid. But which of the surviving bacteria contain the plasmids with the foreign DNA? β-Galactosidase, the product of the plasmid's z gene, catalyzes the hydrolysis of a number of β-galactosides, including a colorless synthetic compound called Xgal (pronounced "ex-gal"). The products of hydrolysis of Xgal include an intensely blue compound. Therefore, if we grow the antibiotic-resistant cells on a medium containing Xgal, some of the resulting colonies are deep blue, while others are white. The blue colonies consist of bacteria containing plasmids in which the z gene is an uninterrupted sequence. In these bacteria, transcription and translation produce β-galactosidase, which converts Xgal to the blue product. The white colonies consist of bacteria containing plasmids in which the foreign DNA has been inserted, interrupting the sequence and thus inactivating the z gene, so that it produces no β-galactosidase and hence no blue color! The white colonies contain the cells we want.

Controlling a Cloned Gene

What if we want to control the expression of the foreign gene in its new home? We may have reasons for wanting to turn the gene on or off. For example, the scientists who did the original work on bacterially produced somatostatin, mentioned earlier, thought it safest to keep the somatostatin gene under wraps except when it could be expressed under carefully controlled conditions. To enable transcription of the inserted gene for somatostatin to be turned on and off, the investigators arranged for the plasmid vector to include a copy of the E. coli lac operon, and they inserted the somatostatin gene into z, the β-galactosidase structural gene of that operon. After putting the modified plasmid into bacteria and cloning them, the scientists could cause the production of human somatostatin by adding lactose to the growth medium (Figure 14.7). In the absence of lactose or another inducer, no somatostatin was formed.

SOURCES OF GENES FOR CLONING

There are three principal sources of the genes or DNA fragments used in recombinant DNA work. One source is pieces of chromosomes inserted into vectors; these DNA-vector units are maintained as gene libraries. A second source is complementary DNA, obtained by reverse transcription from specific RNAs. The third source is laboratory synthesis of specific polynucleotide sequences.

Gene Libraries and Shotgunning

DNA from a desired source can be isolated and broken into small fragments, usually by restriction endonucleases. These fragments can then be incorporated into bacteriophage DNA or plasmids that are, in turn, cloned in bacteria. This technique, called **shotgunning** because the DNA is fragmented at random rather than into specific desired pieces, produces a collection of clones called a **gene library** (Figure 14.8). Each clone carries a fragment of the original DNA, and the library as a whole carries all of the

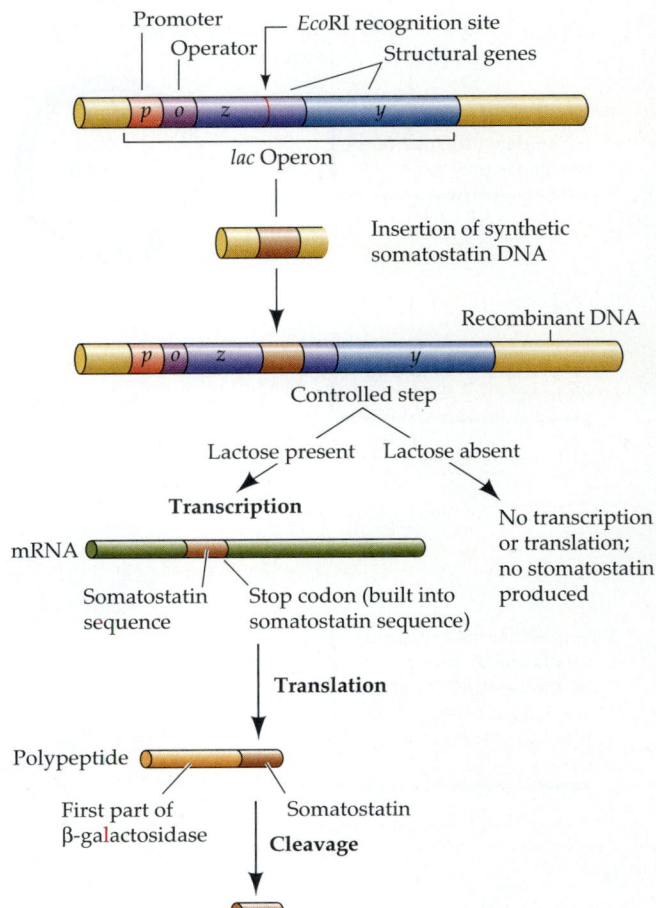

14.7 Controlling Gene Expression with the lac Operon Inserting a foreign gene (in this case, the gene for somatostatin) into the lac operon is a "trick" that allows investigators to turn the gene's transcription on or off by adding or withholding lactose.

original DNA. The clone or clones carrying a particular gene can be detected in various ways.

Complementary DNA

If a specific RNA, such as a particular mRNA, is available, one can make a complementary DNA, or **cDNA**, by using the RNA as a template. Although the starting amount of specific mRNA is usually small, the cDNA can be cloned.

The steps that produce cDNA are illustrated in Figure 14.9. Recall that most eukaryotic mRNAs have a poly A tail—a string of adenine, or A, residues at their 3' end. The first step in cDNA production is to allow an mRNA to hybridize with a molecule (called oligo-dT) consisting of a string of T residues. After the hybrid forms, the oligo-dT can serve as a primer and the mRNA can serve as a template for the enzyme **reverse transcriptase**—the enzyme that retroviruses use to synthesize DNA from RNA templates in host cells. If the primer and reverse transcriptase

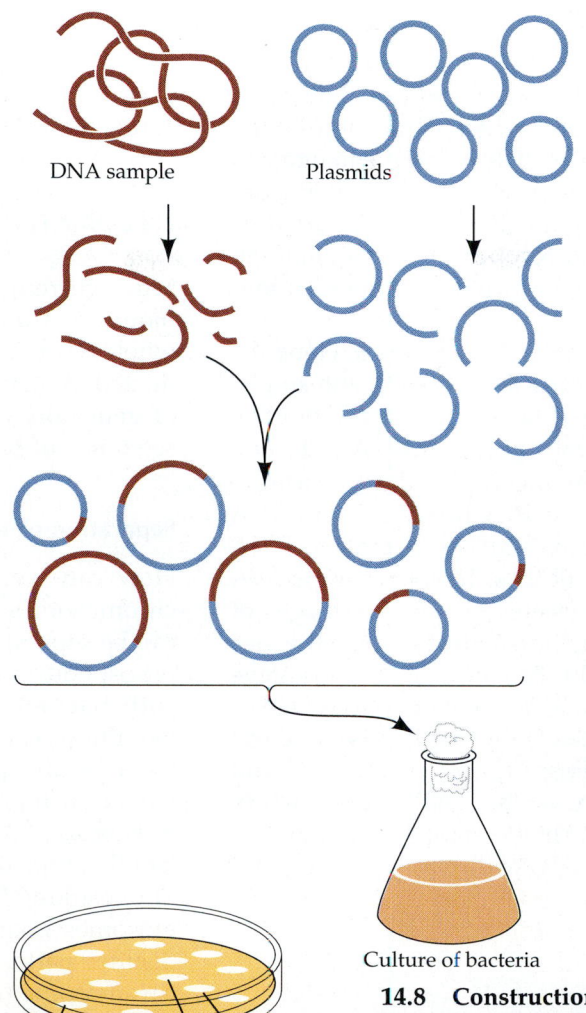

1. A DNA sample and plasmids are cleaved with the same restriction endonuclease

2. DNA fragments and opened plasmids are mixed and spliced with DNA ligase

3. A mixture of different plasmids results

4. Plasmids are mixed with bacteria and placed on a nutrient medium

5. Colonies contain clones of different fragments of original DNA

6. Individual recombinant colonies are isolated and each is maintained as a pure culture; each such culture is a "volume" in the gene library

DNA sample

Plasmids

Culture of bacteria

Individual recombinant cultures

14.8 Construction of a Gene Library
Bacteria are transformed by DNA inserted into plasmids or other vectors. Each recombinant bacterium gives rise to a colony containing part of the original DNA sample. Cultures of these colonies can be analyzed to determine which genes they contain.

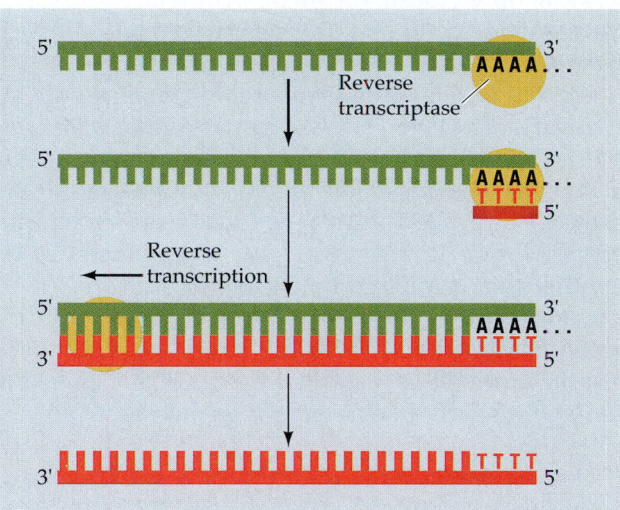

1. mRNA template with 3' poly A tail is combined with reverse transcriptase enzyme

2. A short oligo-dT primer is added and allowed to hybridize with the poly A tail

3. Reverse transcriptase synthesizes cDNA using the mRNA template and deoxyribonucleoside triphosphate substrates, creating a DNA–RNA hybrid

4. When synthesis is completed, the mRNA is removed, leaving single-stranded cDNA

14.9 Synthesis of Complementary DNA
Synthesis of single-stranded cDNA requires the enzyme reverse transcriptase, an mRNA template with a poly A tail on the 3' end, a short primer chain of oligo-dT, and the four deoxyribonucleoside triphosphates.

are given a source of DNA precursors (the four deoxyribonucleoside triphosphates; see Figure 11.6), they will synthesize a strand of cDNA complementary to the mRNA. After the cDNA strand is removed from the mRNA (by increasing the pH of the solution, thus denaturing the cDNA-mRNA hybrids and degrading the mRNA), it can be used for cloning, hybridization, or other experiments.

If the next step in a research project is to clone the cDNA by inserting it into a DNA vector, the single-stranded cDNA must first be converted into double-stranded DNA, using DNA polymerase. A collection of complementary DNAs incorporated into bacteriophage vectors is called a **cDNA library**. Such cDNA libraries contain only *expressed* DNA, that is, DNA that is represented by mRNAs. In comparing cDNA libraries from different tissues or different stages of development, one can gain insight into which genes are being turned on or off. By contrast, the advantage of a *gene* library is that is *does* include DNA that is not transcribed, as well as DNA that is processed out of the mRNAs (including introns, promoters, and regulatory signals such as the TATA box). "Chromosome walking," which we discuss later in this chapter, can be done only with gene libraries, not with cDNA libraries.

Synthetic DNA

When we know the amino acid sequence of the desired polypeptide product, we can get the DNA we need for cloning by synthesizing it directly. This process works well for small DNA molecules. This approach was used by the group that first cloned a DNA sequence coding for somatostatin. Commercially made instruments for automated DNA synthesis are now available.

How do we design a small, synthetic gene? Using the genetic "code book" (see Figure 11.25), we can figure out the appropriate base sequence for the synthetic gene. Using this sequence as the starting point, we can add other sequences to serve specific purposes. What else might we add? For instance, how can we ensure that translation begins and ends at the right places? We can add codons for initiation and termination of translation. How can we prepare the synthetic gene for insertion into a cloning vector? We can add to the ends of our synthetic gene appropriate recognition sequences for the restriction endonuclease that we will use to splice the synthetic DNA into the cloning vector. Other refinements are also possible.

Besides being used to synthesize small polypeptides such as somatostatin, synthetic DNA has been put to two other principal uses. One use, to be discussed in the next section, is as "probes" for detecting specific DNA fragments. The other use is in "directed" mutagenesis—the production of specific alterations in DNA regions with known sequences.

EXPLORING DNA ORGANIZATION

Much current research uses recombinant DNA techniques to study the organization of DNA, including whole chromosomes. These studies may be conducted at many levels, from examining the sequence of genes on a chromosome down to examining the sequence of bases in a DNA fragment.

Separation of Intact Chromosomes

How can we obtain the largest DNA molecules—chromosomes—in pure form? Human chromosomes can be separated by an elegant automated technique. A suspension of chromosomes (chromosomes mixed with but not dissolved in a liquid) is stained with two fluorescent dyes. Each chromosome takes up the two dyes in a particular ratio. The suspension is then passed at high speed through a fine tube, where it is exposed to laser beams of two different wavelengths, each of which is absorbed by one of the dyes. The resulting fluorescence from the dyes in the chromosomes is analyzed. Each chromosome exhibits a distinct pattern of fluorescence. When the desired chromosome (recognized by its fluorescence pattern) passes the observation point, the drop in which it is contained is given a tiny electric charge. As the drop falls away from the nozzle of the apparatus, it is attracted by a charged plate and falls into a tube, while the other, uncharged drops fall into another tube (Figure 14.10). In this way, one type of chromosome can be isolated from all others in the suspension. This system can be adjusted so that any one of the 22 autosomes, or the X or Y chromosome, is collected—at rates in the hundreds per second.

Separation and Purification of DNA Fragments

DNA fragments differing in length can be separated by **gel electrophoresis** (Figure 14.11a). A mixture of negatively charged DNA fragments is placed in a porous gel, and an electric field with a positive charge is applied across the gel. The negative charge on each DNA fragment is proportional to its length because each phosphate group in the chain is negatively charged. As the fragments move through the field, attracted by the positively charged electrode, the gel resists their movement with different intensity, depending on their length. Longer fragments are retarded more strongly than small fragments. The smallest fragments move the fastest and therefore travel the farthest across the gel. Each DNA fragment thus stops at a different point on the gel, and the

14.10 Automated Separation of Chromosomes

This apparatus rapidly separates specific human chromosomes based on how each chromosome stains with two fluorescent dyes. Droplets containing the desired chromosome are identified by their stain, electrically charged, and deflected into a collecting tube.

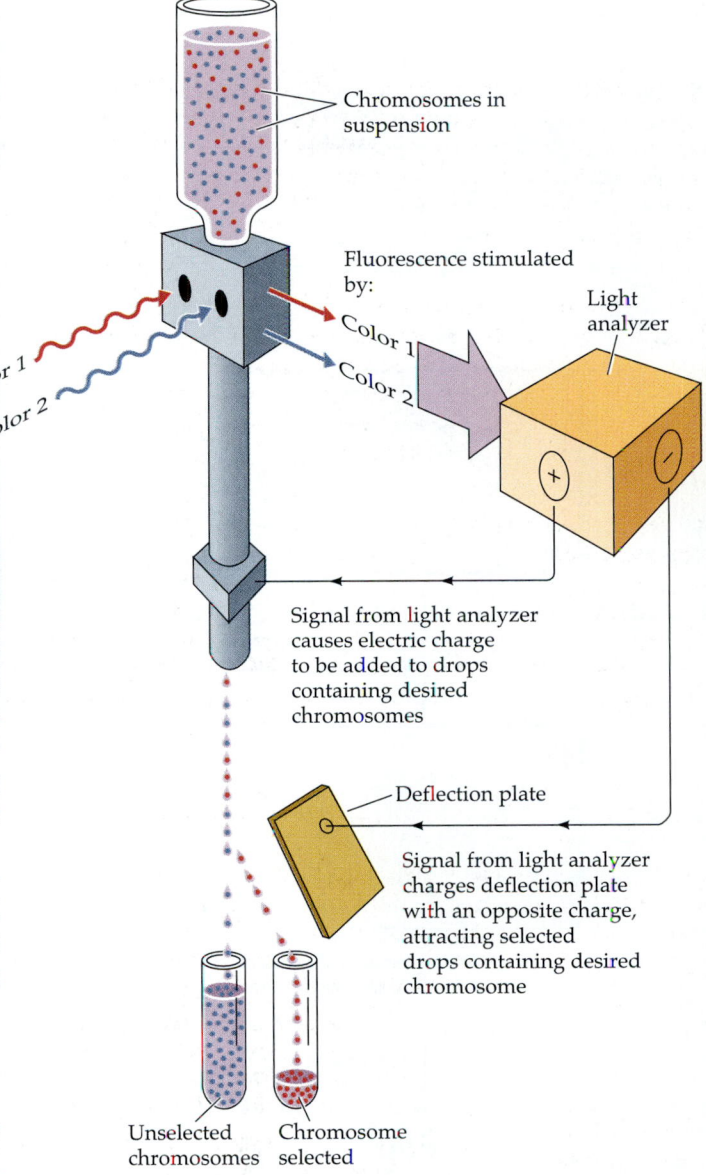

Chromosomes in suspension

Laser color 1

Laser color 2

Fluorescence stimulated by:

Color 1

Color 2

Light analyzer

Signal from light analyzer causes electric charge to be added to drops containing desired chromosomes

Deflection plate

Signal from light analyzer charges deflection plate with an opposite charge, attracting selected drops containing desired chromosome

Unselected chromosomes

Chromosome selected

separated materials can then be removed from the gel in their pure form. Different samples may be "run" side by side in different "lanes" in the electric field. DNA fragments of known length or molecular weight are often run next to experimental mixtures to provide a size reference. After the samples are run, the DNA on the gel plate can be examined by covering the plate with a fluorescent dye that stains the DNA and then placing the plate under ultraviolet light (Figure 14.11b).

Electrophoresis is a very useful technique in molecular biology, biochemistry, and cell biology. All sorts of molecules, particularly proteins and nucleic acids, can be separated by virtue of their differing sizes and electric charges.

Detection of Specific DNA Fragments

Electrophoresis separates DNA fragments of different sizes but does not itself show which of the separated fragments contains a particular piece of DNA. Detection of a particular piece depends on complementary base pairing with a suitable **probe**—a strand of DNA or RNA known to have the base sequence complementary to the piece of DNA sought. For example, a specific mRNA can be used as a probe to locate the gene from which it was transcribed. Such hybridization experiments cannot be done in the gel, however; first the DNA must be transferred to a nitrocellulose filter that binds and immobilizes single-stranded DNA.

A technique called **Southern blotting** is often used in such a search for a specific DNA fragment. A mixture of DNA fragments including the one of interest is separated by electrophoresis. The electrophoresis gel is then soaked in an alkaline solution, which breaks the hydrogen bonds of the DNA fragments, separating the strands. The gel is then "blotted" with a sheet of nitrocellulose, in a setup similar to that shown in Figure 14.12, to transfer some of the

DNA to the sheet. The sheet is removed and heated, which "fixes" the DNA so that it is immobilized on the nitrocellulose in its original position. Then the nitrocellulose filter is soaked in a solution containing the probe (mRNA or cDNA that has been labeled, radioactively or otherwise). When probe molecules meet DNA strands to which they are complementary, they are trapped by complementary base pairing, forming double-stranded nucleic acid molecules. These trapped molecules remain attached to the sheet, while the other probe molecules are washed away. The positions of the desired fragments on the sheet may then be detected by finding the bound, labeled probe.

Southern blotting was named after its inventor, Scottish molecular biologist E. M. Southern. Related techniques have been named Northern blotting (RNA blotting) and Western blotting (protein blotting).

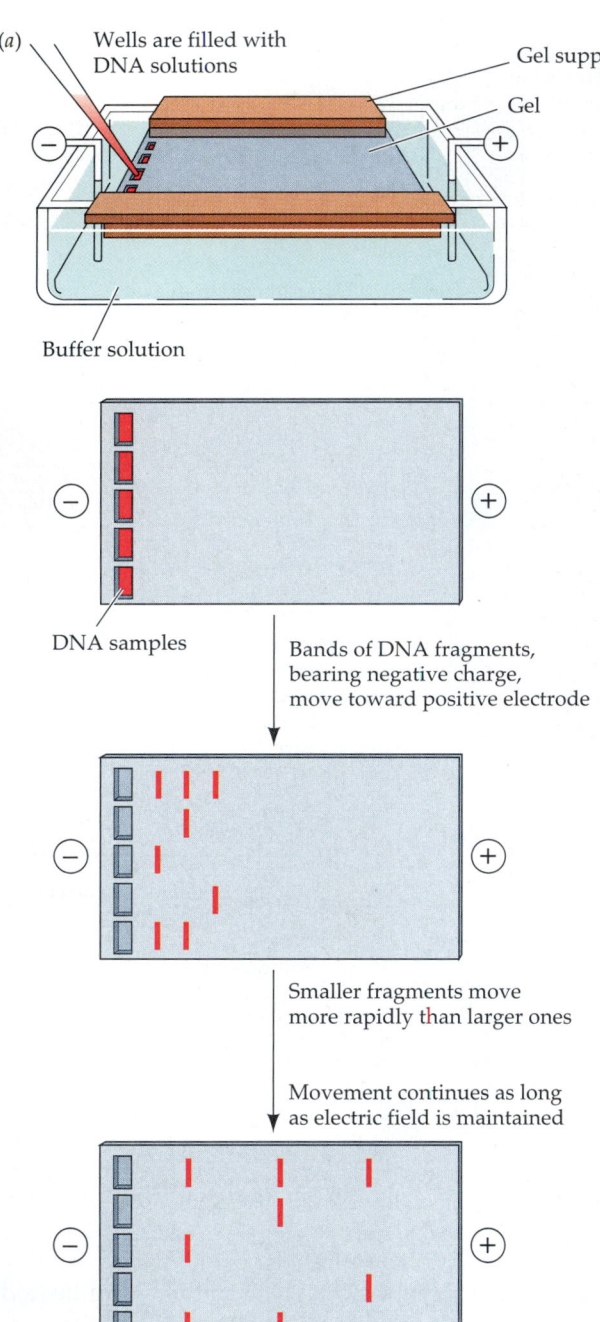

(a)
Wells are filled with
DNA solutions

Gel support

Gel

Buffer solution

DNA samples

Bands of DNA fragments,
bearing negative charge,
move toward positive electrode

Smaller fragments move
more rapidly than larger ones

Movement continues as long
as electric field is maintained

**14.11   Separating Macromolecules by
Gel Electrophoresis**
(a) One version of a setup for separating nucleic acid
fragments or other macromolecules by gel electrophoresis. Samples are injected by pipette into wells in a horizontal gel slab, and an electric field is applied. DNA fragments move at different rates determined by their sizes.
(b) Examining a stained slab under ultraviolet light. The
pink bands contain the separated DNA fragments.

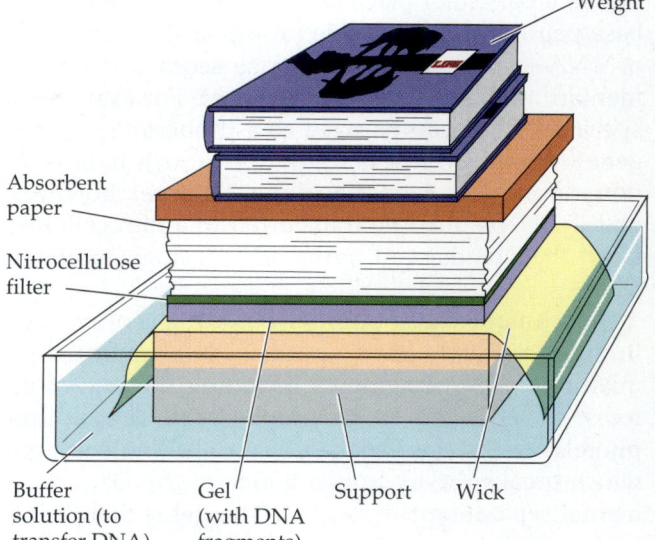

Weight

Absorbent
paper

Nitrocellulose
filter

Buffer
solution (to
transfer DNA)

Gel
(with DNA
fragments)

Support

Wick

14.12   Southern Blotting
A buffer solution moves through a wick, an electrophoresis gel, a nitrocellulose filter, and into a stack of absorbent
paper that acts as a blotter. As the buffer moves upward,
it transfers DNA strands from the gel to the nitrocellulose
filter, which immobilizes the DNA strands.

Autoradiography

Radioactive materials release energy or subatomic particles that can darken a photographic emulsion. This phenomenon is the basis of autoradiography, a technique illustrated in Box 2.A and described in Box 8.B. Autoradiography can be combined with some of the techniques we have just described to help scientists locate particular DNA fragments. It plays an important part in the technique we describe next.

Localization of Genes on Chromosomes

Radioactive probes (RNA or cDNA) can be used to determine on which chromosome a gene or other specific DNA segment lies and even where it lies on the chromosome. A microscope slide is prepared with cells in metaphase of nuclear division (Chapter 9). The slide is treated with weak base to separate the DNA strands slightly—to break the hydrogen bonds and expose single-stranded DNA for hydrogen bonding. The probe is then poured onto the slide. The probe binds only to the DNA sequence to which it is complementary, and the unbound probe is washed away. The slide is stained to reveal the bands on the chromosomes, and autoradiography shows where the probe has bound in relation to the banding pattern. This technique, summarized in Figure 14.13, is called **in situ hybridization** (*in situ* means "in place").

Restriction Mapping

For some research purposes we may want to determine the actual sequence of bases in samples of DNA (gene fragments, genes, or chromosomes). Before we can begin that project, however, we must obtain a preliminary, less detailed map of a DNA molecule. Restriction endonucleases, with their highly specific recognition sequences, can be used to subdivide a DNA molecule into fragments that can then be ordered to form a **restriction map** (Figure 14.14). Sometimes the base sequences of the individual fragments themselves can be determined, a step on the way to sequencing the entire molecule.

Chromosome Walking

Chromosomes are too large for direct DNA sequencing, and they are too large to clone using viruses or plasmids as vectors. To study a chromosome, then, we must first cleave it into a large number of smaller fragments. The chromosome can be treated with one or more restriction endonucleases in such a way that cleavage is incomplete, so individual molecules in the sample are cleaved in different places and the differ-

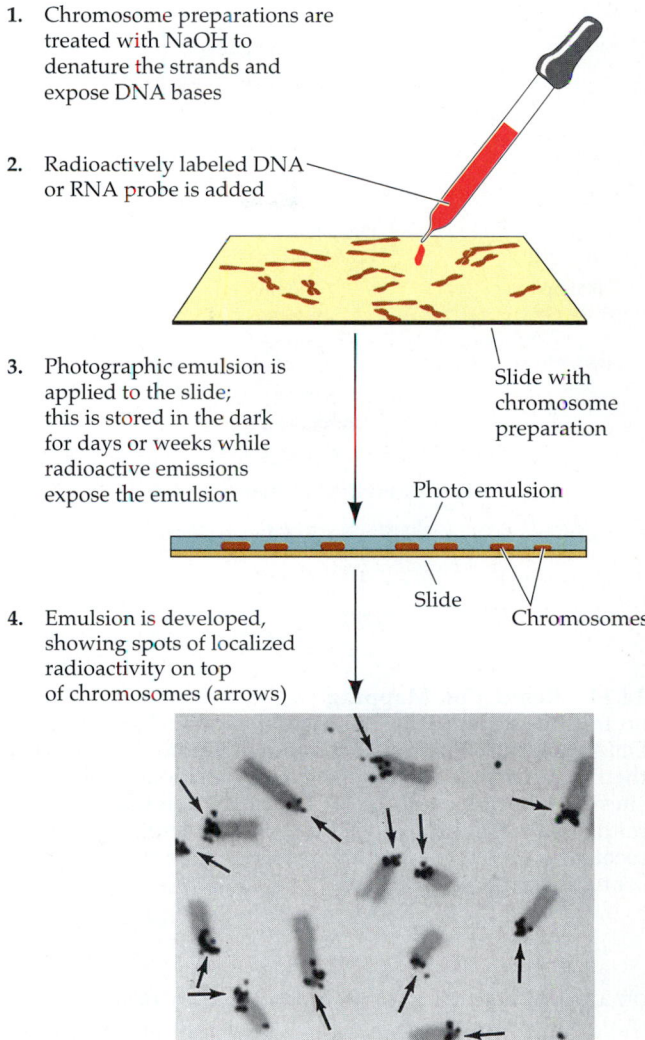

1. Chromosome preparations are treated with NaOH to denature the strands and expose DNA bases

2. Radioactively labeled DNA or RNA probe is added

Slide with chromosome preparation

3. Photographic emulsion is applied to the slide; this is stored in the dark for days or weeks while radioactive emissions expose the emulsion

Photo emulsion

Slide

Chromosomes

4. Emulsion is developed, showing spots of localized radioactivity on top of chromosomes (arrows)

14.13 In Situ Hybridization
A radioactive DNA or RNA probe locates specific genes on specific chromosomes. Autoradiography reveals where the probe has bound in relation to the banding pattern of the chromosome.

ent fragments overlap one another. After we determine the base sequence in each fragment, we need to know the order of the fragments themselves if we are to know the base sequence of the whole chromosome. How do we determine the order of these smaller fragments in the original chromosome?

An easy and elegant approach, called **chromosome walking**, is illustrated in Figure 14.15. First two samples of the original DNA are cleaved with different restriction endonucleases. The fragments from each sample are then cloned, creating two gene libraries. A clone from the first library is made single-stranded and used as a probe of the second gene library; the probe hybridizes with a fragment only if the fragment contains a base sequence complementary to the probe

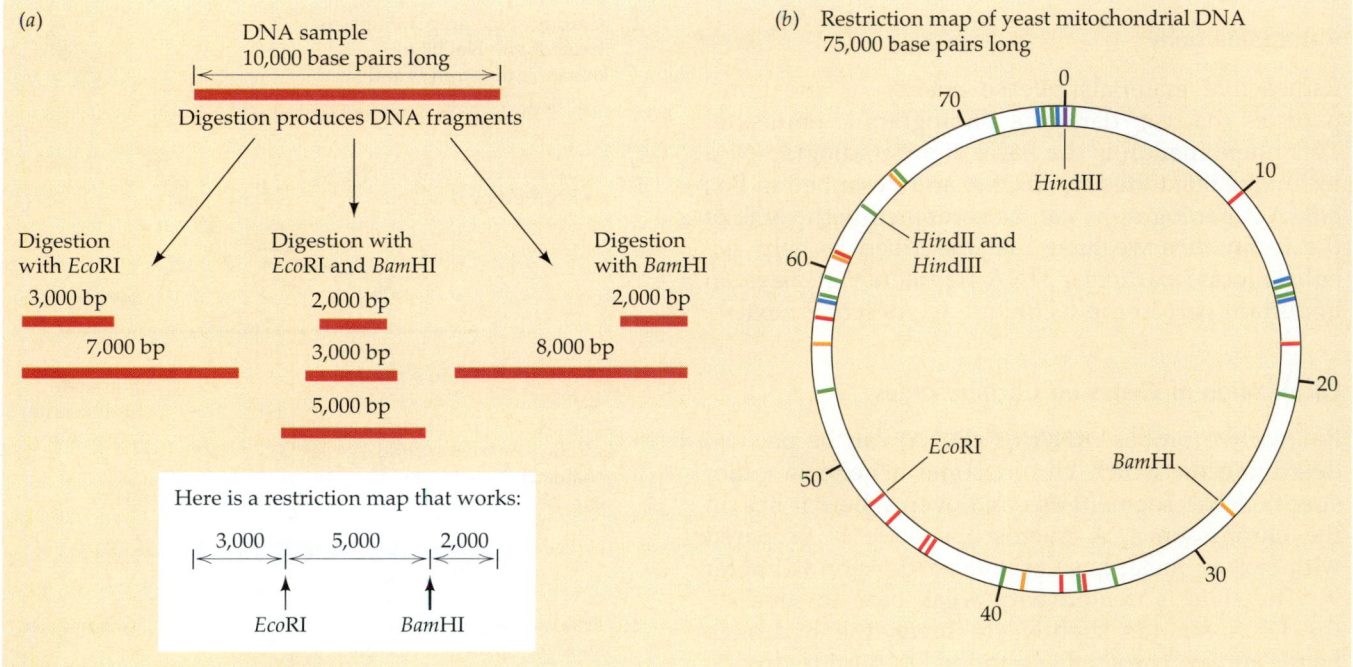

14.14   Restriction Mapping
(a) In three experiments, restriction endonucleases cut samples of the same DNA molecule. The red bars summarize the data obtained. By piecing together the fragments, we can determine the positions of the endonuclease recognition sites in the original DNA. (b) The restriction map of DNA isolated from yeast mitochondria. Cutting sites for four restriction endonucleases are shown. Try to reconstruct the experimental data that gave rise to this map. (Numbers represent thousands of base pairs.)

on one of its strands. If hybridization occurs, then the other strand of the hybrid fragment must contain the same sequence as the probe—that fragment overlaps the probe fragment. The cloned fragment from the second library is then used as a probe of the *first* library to identify a *third* overlapping fragment, the third fragment is used as a probe of the *second* library, and so forth. By identifying successive overlapping fragments, chromosome walking reveals the order of the fragments in the original DNA.

Sequencing DNA

The discovery of restriction endonucleases opened possibilities for determining the base sequence of a fragment of DNA. By using various restriction enzymes, one could cut a sample of DNA into multiple fragments. Hybridization techniques enabled biologists to determine which genes are associated with which fragments and to construct a partial map of the DNA sample. But there was still the problem of determining the base sequences within these smaller pieces. In the mid-1970s, the British biochemist Frederick Sanger (who in 1953 had determined the first protein primary structure), Allan Maxam and Walter Gilbert (both of Harvard University), and others de-

vised techniques for the rapid sequencing of DNA. A number of methods are in current use; one is described in Box 14.A.

Why should we want to determine the base sequence of a DNA sample? One reason is that this knowledge, along with our knowledge of the genetic code, enables us to determine the primary structure of the gene's protein product. Although we could analyze the protein directly, it turns out to be easier to determine the primary structure of a protein by analyzing the corresponding DNA.

Most important, knowing base sequences may help us determine how regulatory sites (such as promoters) function. Also, closely related genes may have quite different functions, and sequencing helps us understand how they differ. In addition, we can use DNA sequencing to identify precise molecular defects, as in genetic diseases. Gene cloning coupled with DNA sequencing has provided important information about the insertion of transposable elements in eukaryotes as well as in prokaryotes. DNA sequencing can provide information of interest to evolutionary biology as well. By comparing the base sequences of homologous genes from different organisms, we can extend our knowledge of the evolutionary relationships among species.

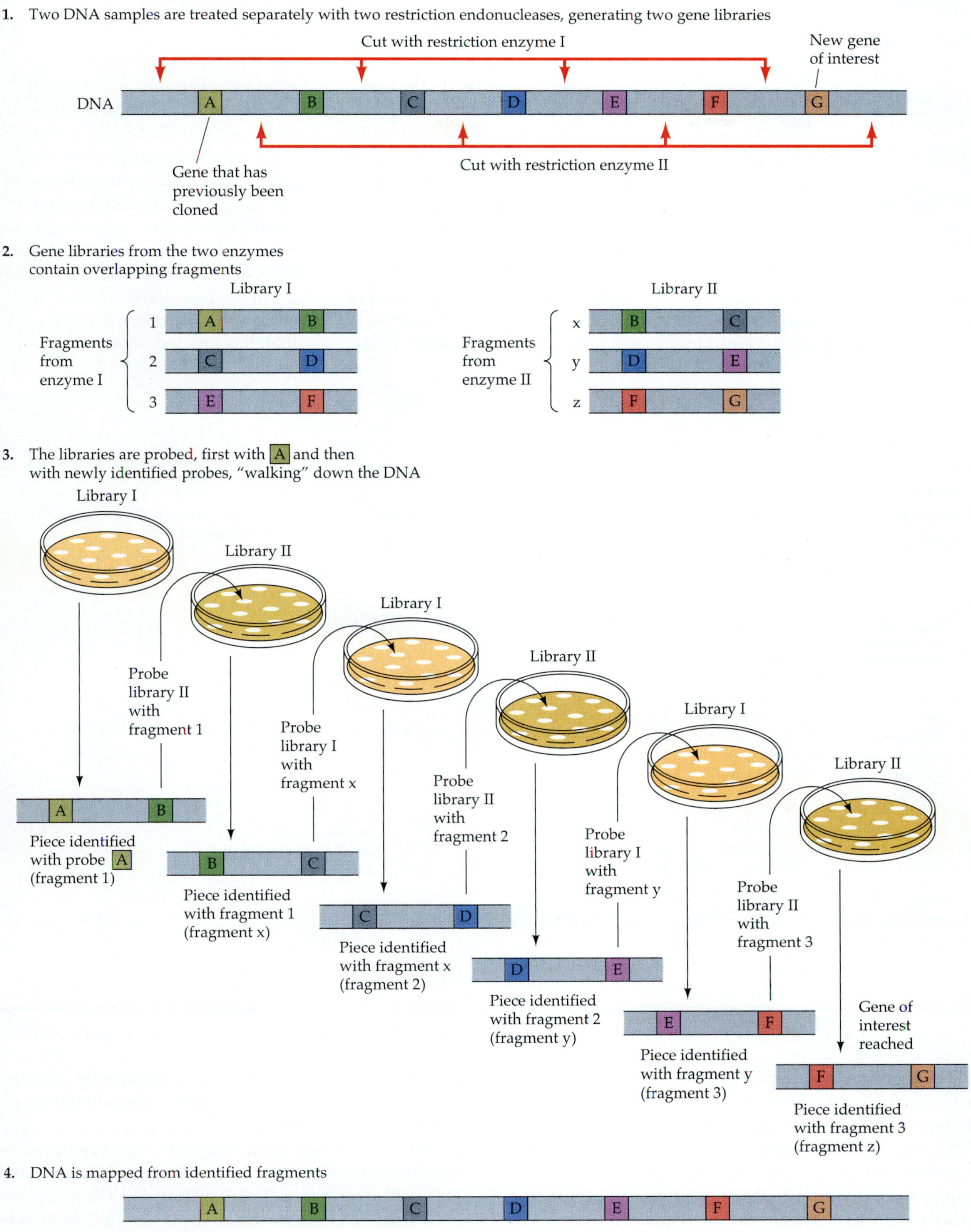

1. Two DNA samples are treated separately with two restriction endonucleases, generating two gene libraries

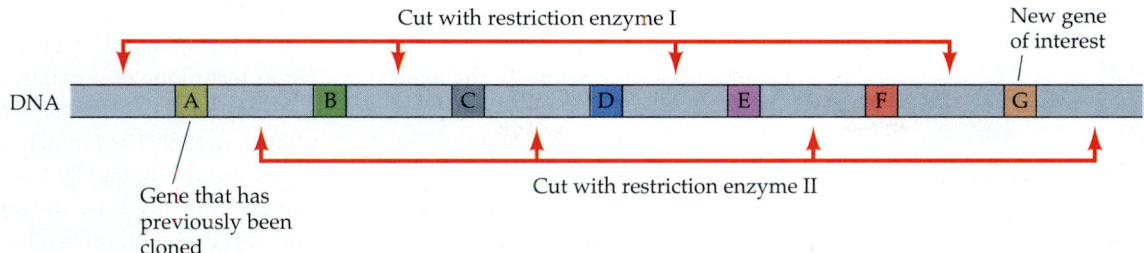

2. Gene libraries from the two enzymes contain overlapping fragments

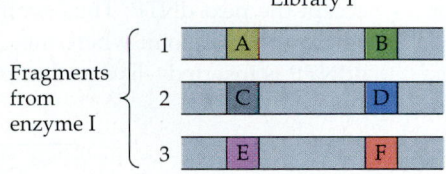

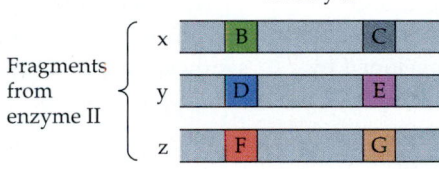

3. The libraries are probed, first with A and then with newly identified probes, "walking" down the DNA

Library I

Library II

Library I

Library II

Library I

Library II

Probe library II with fragment 1

Probe library I with fragment x

Probe library II with fragment 2

Probe library I with fragment y

Probe library II with fragment 3

Gene of interest reached

Piece identified with probe A (fragment 1)

Piece identified with fragment 1 (fragment x)

Piece identified with fragment x (fragment 2)

Piece identified with fragment 2 (fragment y)

Piece identified with fragment y (fragment 3)

Piece identified with fragment 3 (fragment z)

4. DNA is mapped from identified fragments

14.15 Chromosome Walking
The "volumes" in a gene library (see Figure 14.8) can be put in the proper order—their order in the original DNA molecule—by chromosome walking.

BOX 14.A

Determining the Base Sequence of DNA

One of the simplest, fastest, and most accurate methods for sequencing DNA samples was developed by Leroy Hood and his colleagues. This method has similarities to the experiment described in Chapter 11 (in the section entitled "Transcription") in which the compound cordycepin triphosphate is used to interrupt RNA synthesis. This experiment demonstrated that RNA synthesis, like DNA synthesis, proceeds from the 5' to the 3' end of the molecule.

The nucleoside triphosphates normally used as substrates for DNA synthesis are the 2'-deoxyribonucleoside triphosphates (dNTPs). The Hood technique also makes use of 2',3'-dideoxyribonucleoside triphosphates, or ddNTPs (shown in figure (a)). Suppose that a DNA strand is being synthesized, by the addition of one dNTP after another. If a ddNTP is picked up instead, it joins the growing chain; but, since it lacks a free hydroxyl group at C3', it cannot accept the next dNTP. Thus synthesis stops at the point where the ddNTP is inserted—like cordycepin

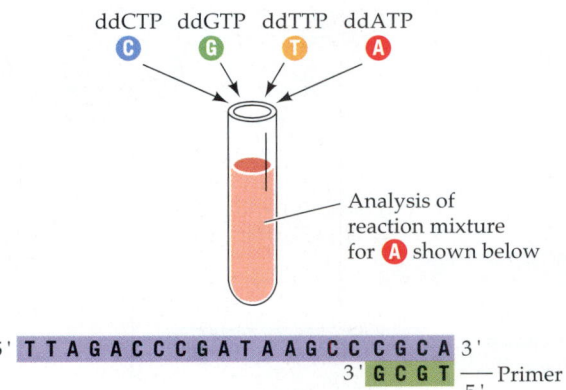

Ribonucleoside triphosphate (NTP)   Base (A,U,G or C)   HO   OH

Deoxyribonucleoside triphosphate (dNTP)   Base (A,T,G or C)   HO   H   2'

Dideoxyribonucleoside triphosphate (ddNTP)   Base (A,T,G or C)   H 3'   H 2'

1. Single-stranded DNA sequence to be determined

5' T T A G A C C C G A T A A G C C C G C A 3'

2. A sample of this unknown DNA is combined with primer, DNA polymerase, 4 dNTPs, and 4 ddNTPs each bound to a fluorescent dye, and synthesis begins. The results are illustrated here by what binds to a T in the unknown strand. If dATP is picked up from the mixture, synthesis continues. If ddATP is picked up, synthesis stops. A series of fragments of different lengths is made, each ending with a ddNTP

ddCTP   ddGTP   ddTTP   ddATP
C   G   T   A

Analysis of reaction mixture for A shown below

5' T T A G A C C C G A T A A G C C C G C A 3'
3' G C G T 5' —— Primer

GENE COPIES BY THE BILLION

Biologists often want to obtain particular pieces of DNA—a particular gene, for example—in quantity sufficient for biochemical studies. A powerful technique, the **polymerase chain reaction**, has made this process of DNA amplification relatively easy. First described in 1984, the polymerase chain reaction has become one of the most widely used techniques in molecular biology. It can produce billions of copies of a single piece of a DNA molecule—the target sequence—in a few hours. By contrast, the conventional techniques of cloning recombinant DNA require days to weeks.

The polymerase chain reaction is a cyclic process in which the following sequence of steps is repeated over and over again. Double-stranded DNA is heat-denatured into single strands, primers for DNA synthesis are added to the 3' ends of the target DNA sequence on the separated strands, and DNA polymerase catalyzes the production of new complementary strands. A single cycle doubles the amount of triphosphate, the ddNTP terminates the chain.

In this technique for sequencing DNA, single-stranded DNA to be sequenced is combined with DNA polymerase (to synthesize the complementary strand), a primer, the four dNTPs (dATP, dGTP, dCTP, and dTTP)—and small amounts of each of the four ddNTPs (ddATP, ddGTP, ddCTP, or ddTTP), each attached to a fluorescent molecule emitting a different color of light (see part (b) of the figure). The reaction mixture soon contains a DNA mixture made up of the unknown single strand and shorter complementary strands. The complementary strands, each terminating in a ddNTP, are of various lengths.

For example, each time a T is reached on the template strand, the growing complementary strand adds, at random, either dATP or ddATP. If ddATP is added, chain growth terminates at that point. By chance, some of the replicating strands grow to greater lengths than others before coming to a ddATP stopping point. The other ddNTPs also produce strands of various lengths. The strands can then be displayed by gel electrophoresis, resulting in a series of colored bands. The color of each band tells us which ddNTP terminated the strands in that band. Because we thus know which ddNTP gave rise to each strand, we also know which base is last in the strands of each set. Given that these strands are complementary to the original sample, we can then determine the exact sequence of bases in the original DNA sample, as shown in the figure.

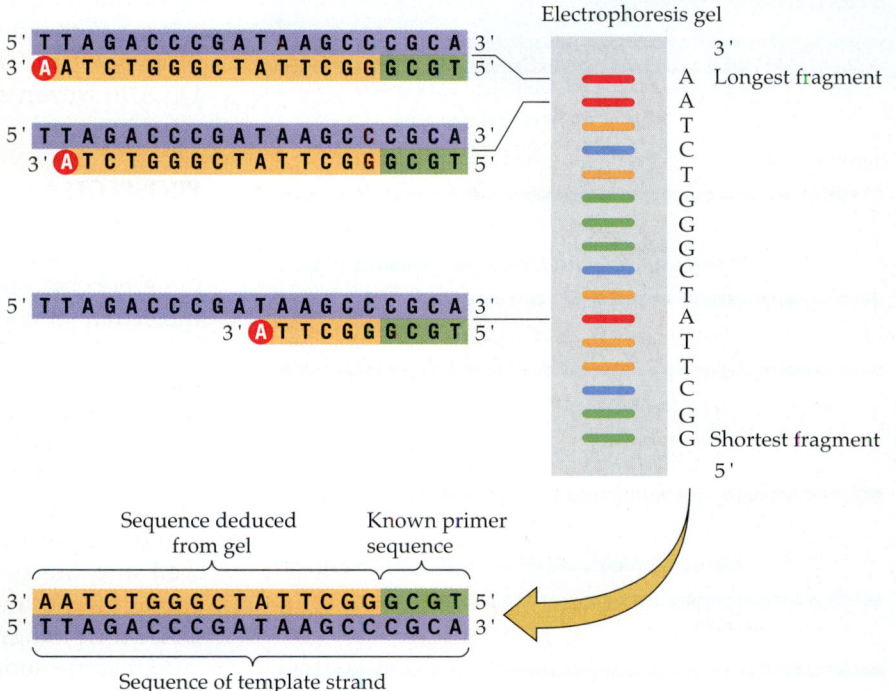

3. The resulting strands of various lengths are separated by gel electrophoresis, which can detect length differences as short as one base. Each strand fluoresces a color that identifies the ddNTP that terminated the strand

4. The sequence of the newly synthesized strand of DNA can now be deduced and converted to the sequence of the template strand

the DNA sequence in the reaction mixture and leaves the new DNA in the double-stranded state (Figure 14.16). Each cycle takes only one to a few minutes.

To use the polymerase chain reaction, a scientist must know a sequence of about 20 bases at the 3' end of the target sequence on each DNA strand. Knowing that sequence, the scientist can create the two oligonucleotides that are complementary to the sequence to use as primers for DNA synthesis. The primers are crucial: Whether the starting DNA sample is ample or limited, pure or crude, long or short, the primers indicate where DNA synthesis is to take place. The primers give the polymerase chain reaction its versatility.

Separating the strands of a DNA molecule requires a temperature of about 94°C. Therefore, the reaction mixture must be heated to that temperature in each cycle of the polymerase chain reaction. At first this was a problem, because the heating destroyed the DNA polymerase that catalyzed the polymerase chain reaction. Now, however, the polymerase chain reaction is run with a temperature-resistant DNA

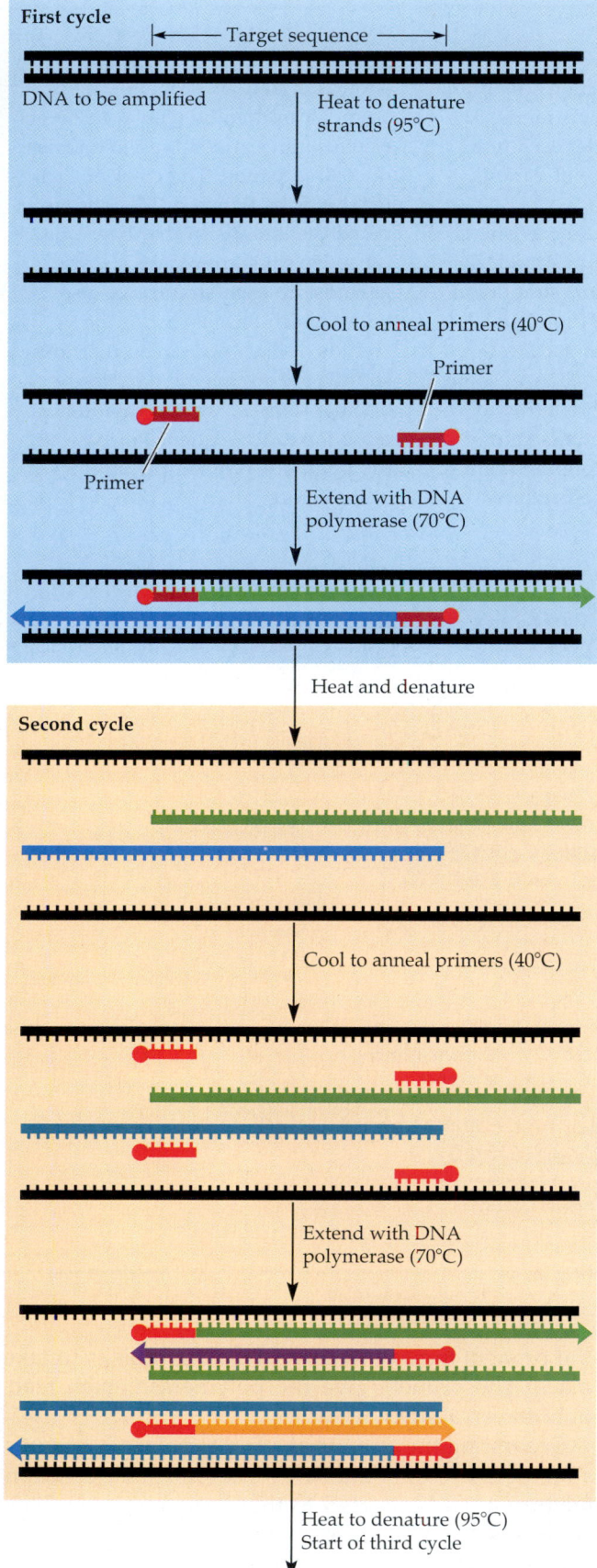

First cycle

Target sequence

DNA to be amplified

Heat to denature strands (95°C)

Cool to anneal primers (40°C)

Primer

Primer

Extend with DNA polymerase (70°C)

Heat and denature

Second cycle

Cool to anneal primers (40°C)

Extend with DNA polymerase (70°C)

Heat to denature (95°C)
Start of third cycle

14.16 The Polymerase Chain Reaction

Because it is used so often and in so many laboratories, this technique was named "molecule of the year" by the journal Science in 1990. In the presence of appropriate primers, a heat-resistant DNA polymerase copies both strands of the desired DNA sequence; by alternating heating (to separate DNA strands) and synthesis phases, a PCR machine can rapidly make enormous numbers of copies.

polymerase from the bacterium *Thermus aquaticus*, which lives in hot springs. This polymerase survives the high temperatures used to separate the DNA strands. The whole process has been automated, and the compact "PCR machine" (polymerase chain reaction machine) is found in thousands of laboratories all over the world.

The amplification of DNA samples by the polymerase chain reaction has found many uses, ranging from the diagnosis of hereditary diseases in human fetuses to the identification of semen samples in criminological investigations and the study of ancient DNA in frozen samples from the last Ice Age.

PROSPECTS

The techniques of recombinant DNA technology have become standard tools for investigators of the molecular biology of both prokaryotes and eukaryotes and in the pharmaceutical industry as well as the clinical laboratory. These techniques are being used to turn selected strains of bacterial, yeast, plant, and animal cells into factories for the production of important polypeptides. Thus we can supply patients with gene products they cannot make for themselves; surely this technology will be a major tool of medicine well into the twenty-first century. We will consider some of the medical applications of recombinant DNA technology in Chapter 15.

Concerns about safety were expressed in the early days of recombinant DNA research, and at one point the research was called to a temporary halt. The worst of the fears have abated, however, and the technology is now widely agreed to be safe as long as certain reasonable precautions are observed. These precautions, set forth as guidelines by the U.S. National Institutes of Health, include precisely defined containment provisions (methods for keeping experimental organisms from escaping from the laboratory) and rules for handling different biological materials. Most experiments are thought to pose no hazard. Some are considered risky and accordingly are conducted under more restrictive conditions. When first issued, the guidelines specifically restricted recombinant DNA work with the genes of cancer viruses. But gradually, as various concerns have been put to rest, the guidelines have been relaxed. Relaxation of the guidelines has led to major advances in our understanding of the cancer-producing oncogenes (see Chapter 15).

Plant Agricultural Biotechnology

Earth's human population is increasing by more than 100 million individuals each year. In less than four decades the population will reach nine billion. To feed all those people we will have to triple world food production. Most humans depend on only a few crop plants for essential calories—rice, wheat, maize, sorghum, and several others (such as potatoes) that dominate in certain local agricultures. Because only a few genetic strains of each of these crops are planted in a given year, the food supply of hundreds of millions of people is in constant peril from mutations in disease-causing organisms that could overcome the plants' resistance and destroy entire harvests. Equally important threats are changes in weather patterns, or political and economic disruptions that affect supplies of essential agricultural chemicals, fertilizers, or fossil fuels.

Other than the overriding need to contain population growth, the greatest potential for improving human welfare with modern biological technology lies in the search for economically feasible crop plants that are higher-yielding; more nutritious; disease-resistant; drought-, salt-, and pollution-tolerant; and otherwise able to meet the challenges of our over-extended planetary resources.

Plant breeding is a form of "genetic engineering" that has been used for a long time; it has been practiced intensively in the twentieth century. By the judicious crossing of existing strains of plants, we have been able to increase crop yields about 1 percent per year for the past century. Among the greatest triumphs of the plant breeders were the hybridization of corn (in the 1930s) and the "Green Revolution" (in the 1950s and 1960s, during which improved strains of wheat and rice were used to increase food production in many parts of the world). Although plant breeding will continue to play a key role in the development of agriculture, it has three significant limitations. (1) Plant breeding is a slow process, and it requires many acres of land. (2) It is nonspecific in that the entire genomes of the parents participate in a cross (the genome of an organism is all the genetic information the organism contains); thus unwanted genes may appear in the progeny along with the desirable ones for which the breeder is selecting. (3) The breeder can work only with strains (and, more rarely, species) that can interbreed with one another.

The addition of recombinant DNA technology to the tools of the breeder addresses each of those three limitations. First, recombinant DNA techniques are often applied to many millions of independent cells at a time, all within a single flask, thus eliminating the space problem; and the cells, unlike whole plants, multiply many times a day. Second, individual genes can be transferred from one plant to another without dragging along other, undesired, genes. Third, gene transfer is not restricted to closely related plants; in fact, a plant may be given genes from *any* organism. When we know the specific biochemical basis for a change that we want to make in a plant, recombinant DNA technology is extraordinarily useful.

For both conventional plant breeding and recombinant DNA work, there is a continuing and growing need for a source of suitable genes, sometimes referred to as **germ plasm**, to introduce into existing species. One goal of conservationists and biologists is to maintain an adequate global supply of germ plasm, as seeds in repositories and as plants in nature, to ensure genetic diversity in nature and a continuing supply of tools for the breeder and the biologist. A lost species is a lost treasury of germ plasm. Recombinant DNA, perhaps as gene libraries, could supplement a gene repository for germ plasm.

The cloning vector commonly used in recombinant DNA work with plants is a plasmid found in *Agrobacterium tumefaciens*, a bacterium. *A. tumefaciens* is a pathogen that causes the plant disease crown gall, which is characterized by large tumors (Figure 14.17a). The bacterium contains a large plasmid, called Ti (for Tumor-inducing). Part of the Ti plasmid is T-DNA, a transposon (see Chapter 12) that produces copies of itself in the chromosomes of infected plant cells. This transposon is the key to gene cloning in plants. The gene to be cloned is inserted into the transposon in a Ti plasmid, the plasmid is inserted (by transformation) into *A. tumefaciens*, the bacterium infects the plant, and the gene is copied into the plant's chromosomes along with the rest of the transposon (Figure 14.17b,c). Since the recombinant DNA is found only in tumor cells, this might seem like a nonheritable change in the plant's genome. However, tumor cells can be isolated and grown in culture, eventually giving rise to a complete, new, normal plant (Figure 14.18). If a plant grows from a tumor cell containing recombinant DNA, each of its cells is transgenic, as explained earlier in this chapter.

Scientists have produced transgenic crop plants that are resistant to herbicides or to insects and others that are resistant to viral diseases (Figure 14.19). Still other transgenic plants produce an enzyme that breaks down a plant growth substance that normally causes senescence (aging; see Chapter 34) and thus spoilage of harvested fruit.

Transgenic bacteria will also play growing roles in agriculture. For example, nitrogen fixation—the conversion of atmospheric nitrogen gas to ammonium ions usable as a nitrogen source for plants—is crucial

(a)

14.17   Crown Gall and *Agrobacterium tumefaciens*
(a) A. tumefaciens causes tumors—crown gall—on infected plants, such as this geranium. New shoots are forming within the gall, which lies at the base of the geranium stem. *(b)* A. tumefaciens contains the Ti plasmid which, in turn, contains a transposon called T-DNA. Copying the transposon to a chromosome of an infected plant is a key step in the development of the disease. *(c)* A scientist can clone a desired gene by inserting it into the transposon of a modified Ti plasmid, transforming *A. tumefaciens* with the plasmid, and infecting plants with the bacteria.

(b)

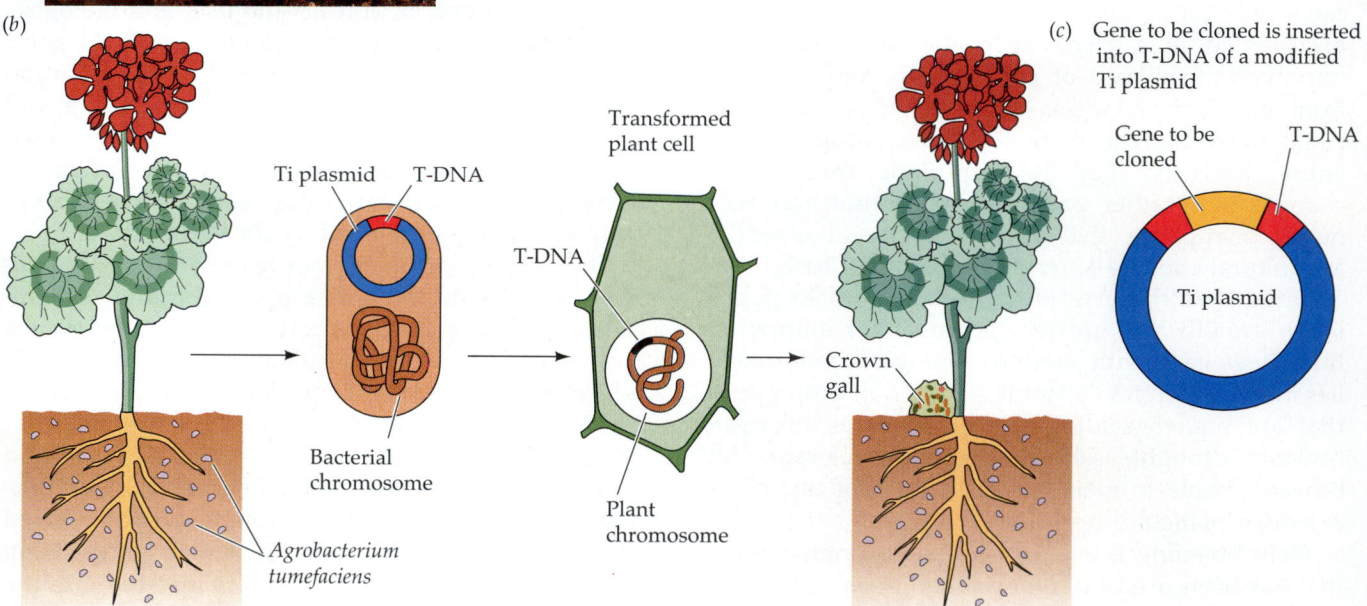

14.18   Multiplying a Clone
This rosette of transgenic cotton plantlets is developing from a culture of crown gall tumor tissue. The plantlets contain a gene they picked up from the *Agrobacterium tumefaciens* that caused the tumor.

to agriculture and to life on Earth (see Chapter 32). Much current research is focusing on modifying both the nitrogen-fixing bacteria and the plants that harbor them in order to improve the efficiency of nitrogen fixation. In another example, genetically altered bacteria have already been produced that, when present on plant surfaces, prevent frost formation (Figure 14.20).

Recombinant DNA Technology and the Environment

With disturbing frequency, great quantities of oil have been accidentally released from tankers into surrounding waters, causing severe damage to sensitive environments (Figure 14.21). Genetically engineered bacteria, equipped with DNA that codes for enzymes that cleave hydrocarbons and other constituents of oil, are being developed to combat such

14.19 Genetically Engineered Resistance to a Virus
When tested in the field, transgenic tomato plants (right) treated to resist infection by the cucumber mosaic virus produced marketable fruit. The control plant (left) was stunted by viral infection and produced no tomatoes.

spills. Several biotechnology companies are producing these and other bacterial strains modified to deal with other types of organic pollutants, such as sewage and dioxins.

At the other end of the petroleum industry, genetically engineered bacteria will be useful in the production of chemicals to enhance yields from oil drilling. Oil yield in the United States continues to decline, and products of recombinant DNA technology represent an approach to increasing yields from existing and newly discovered reservoirs.

Yeasts and other microorganisms can concentrate metals such as nickel, gold, and plutonium from dilute solutions. Thus they can be used both in combating pollution and in increasing the recovery of important metals from waste. Genetic manipulation of bacteria and yeast for metal recovery and for the separation of metals from their ores is being actively pursued.

Release of transgenic organisms into the environment in large numbers must be done with care. Generally, bacteria prepared for release into the soil, lakes, or oceans will be designed to do their work effectively but not to persist long in the environment. In spite of such care, debates on the safety and merits of this kind of work should continue. We must always consider the environmental impact and health concerns as well as the technological and economic advantages.

Genome Projects and Modern Medicine

There are hundreds of human hereditary diseases, and a new one is discovered every two or three days.

14.20 Fighting Frost with Bacteria
Strawberry blossoms and *Pseudomonas syringae* bacteria in water at –2.8°C. The tube on the left contains normal *P. syringae* and is frozen solid. The *P. syringae* on the right, however, lack a protein on their surface that provides a "template" for water molecules to align as ice—the gene that codes for the protein has been deleted. Ice does not form readily in the presence of these transgenic bacteria. Spraying such bacteria on crops may provide protection from freezing.

14.21 An Environmental Disaster
This oil slick resulted from a spill in the Delaware River near Philadelphia. Biotechnologists are working on methods to deal biologically with such spills.

It has been estimated that each person is heterozygous for no fewer than 30 recessive genetic diseases. Recombinant DNA technology affords new methods—often the first methods available—for combating these diseases. Several **genome projects** are already under way. These are major efforts to *map* and *sequence* the human genome and, in part or in whole, the genomes of certain other species. Mapping precedes sequencing and in itself yields much valuable information. The development of effective gene therapies (see Chapter 15) will depend on the mapping and characterization of the genes responsible for the diseases, and this provides major motivation for supporting human genome projects. Genes responsible for cystic fibrosis and other diseases have already been mapped and cloned and can now be studied to learn about the gene products, the diseases themselves, and their possible treatment.

Genome projects require the gathering of vast amounts of data, including the locations of genes, their functions, their base sequences, and more. Processing—and simply storing—so much information requires massive computing resources, and the analysis of it all will depend heavily on advances in computing theory. Biology and computer science will be full partners in this venture. While computer science develops new algorithms and parallel processing techniques, biology and biotechnology will develop new physical and biochemical techniques for DNA manipulation and analysis.

The return on this investment will be magnificent, including a revolution in the diagnosis and treatment of hereditary diseases. Even if we sequence every base in the human genome, however, we still will not know what all the DNA sequences *do*—or how their products do what they do. An understanding of genetic regulation in eukaryotes requires knowledge of much more than base sequences. But without the mapping and sequencing data from genome projects, these greater problems will remain unsolved.

SUMMARY of Main Ideas about Recombinant DNA Technology

Recombinant DNA technology serves research in most areas of present-day biology; in addition, its applications are important in agriculture, medicine, forensics, environmental protection, and the chemical industry.

Foreign DNA can be inserted into a vector (a virus or plasmid) with the assistance of restriction endonucleases and other enzymes.
   Review Figures 14.2, 14.3, and 14.4

The modified vectors are inserted into suitable hosts for cloning.
   Review Figures 14.5 and 14.6

Bacteria are widely used as hosts, but eukaryotic hosts, particularly yeasts, are increasingly used for transfection with eukaryotic genes.

Gene libraries, cDNA production, and direct chemical synthesis are sources of DNA for cloning.
   Review Figures 14.8 and 14.9

DNA fragments are commonly separated by gel electrophoresis and then identified by Southern blotting.
   Review Figures 14.11 and 14.12

Specific DNA fragments can be amplified rapidly by means of the polymerase chain reaction.
   Review Figure 14.16

DNA fragments and chromosomes may be analyzed by restriction mapping, chromosome walking, and DNA sequencing.
   Review Figures 14.14 and 14.15

SELF-QUIZ

1. Restriction endonucleases
   a. play no role in bacteria.
   b. cleave single-stranded DNA molecules.
   c. cleave DNA at highly specific recognition sites.
   d. are inserted into bacteria by bacteriophages.
   e. add methyl groups to specific DNA base sequences.

2. "Sticky ends"
   a. are double-stranded ends of DNA fragments.
   b. are complementary to other specific sticky ends.
   c. rejoin best at elevated temperatures.
   d. are removed by restriction endonucleases.
   e. are identical for all restriction endonucleases.

3. Which statement is *not* true of DNA ligase?
   a. It is an enzyme.
   b. It is a normal constituent of cells.
   c. It can unite sticky ends in recombinant DNA work.
   d. It functions in the normal replication of DNA.
   e. It mends breaks in polypeptide chains.

4. Which feature is undesirable in a plasmid for cloning a gene?
   a. Possession of an origin of replication
   b. Possession of genes conferring resistance to antibiotics
   c. Possession of recognition sites for multiple restriction endonucleases
   d. Possession of multiple recognition sites for the endonuclease to be used
   e. Possession of genes other than the one to be cloned.

5. Transfection can be accomplished by
   a. using high-velocity particles of tungsten coated with DNA.
   b. Southern blotting.
   c. gel electrophoresis.
   d. the polymerase chain reaction.
   e. heating the material to denature it.

6. Complementary DNA
   a. is produced from ribonucleoside triphosphates.
   b. is produced using oligo-dU.
   c. is produced by reverse transcription.
   d. requires no primer.
   e. requires no template.

7. Southern blotting
   a. is used to detect a specific DNA fragment.
   b. is used to detect a specific RNA fragment.
   c. is used to detect a specific polypeptide fragment.
   d. is a technique for separating nucleic acid fragments.
   e. is used to separate chromosomes.

8. Restriction mapping
   a. is a useful tool for separating DNA fragments.
   b. is a useful tool for subdividing a DNA molecule into manageable fragments.
   c. cannot be used on prokaryotic DNA.

   d. can be used to produce large quantities of specific DNA.
   e. is an expensive and controversial procedure.

9. The polymerase chain reaction
   a. is a method for sequencing DNA.
   b. is used to detect a specific DNA fragment.
   c. is used to produce large quantities of specific DNA.
   d. is used to map genes.
   e. is used to transcribe specific genes.

10. Genome projects
   a. are a matter for the distant future.
   b. all focus on the human genome.
   c. will tell us what all our genes do.
   d. have already yielded medically useful results.
   e. will probably all be carried out in one carefully chosen university.

FOR STUDY

1. Using examples from this chapter, describe how molecular biologists have found new uses in recombinant DNA technology for enzymes produced by bacteria.

2. Make a thorough list of the phenomena and techniques discussed in this chapter that depend on complementary base pairing.

3. You have attempted to insert a copy of a particular gene into a plasmid, specifically placing your gene in the middle of a gene conferring resistance to the antibiotic streptomycin. The plasmid also has a gene conferring resistance to the antibiotic aureomycin. You have transformed bacteria (sensitive to both antibiotics) with your plasmid suspension. Describe the procedures you would use to select the bacteria that have taken up the plasmid. What additional steps would be required to select the bacteria that have taken up copies of the plasmid containing your gene?

4. Discuss (a) what you see as important positive features of a human genome project and (b) what you consider to be negative features of such a project.

READINGS

Gasser, C. S. and R. T. Fraley. 1992. "Transgenic Crops." *Scientific American*, June. A brief overview of some major projects for the improvement of crops through biotechnology.

Gilbert, W. 1991. "Toward a Paradigm Shift in Biology." *Nature*, January 10. A discussion of the impact of recombinant DNA technology on pure research in biology.

Griffiths, A. J. F., J. H. Miller, D. T. Suzuki, R. C. Lewontin and W. M. Gelbart. 1993. *An Introduction to Genetic Analysis*, 5th Edition. W. H. Freeman, New York. An outstanding genetics textbook, thoroughly up to date.

Murray, A. W. and J. W. Szostak. 1987. "Artificial Chromosomes." *Scientific American*, November. Tools for cloning human genes in yeast and for the investigation of chromosomal behavior during mitosis and meiosis.

Neufeld, P. J. and N. Colman. 1990. "When Science Takes the Witness Stand." *Scientific American*, May. Ethical issues in the use of DNA evidence.

Watson, J. D., J. Tooze and D. T. Kurtz. 1989. *Recombinant DNA: A Short Course*, 2nd Edition. Scientific American Books, New York. Begins at the elementary level but ends up presenting a great deal of molecular biology. A short, intense book, but readable.

Weinberg, R. A. 1985. "The Molecules of Life." *Scientific American*, October. Introductory chapter to a special issue on the molecules of life; gives a good overview of the role of recombinant DNA technology in various aspects of molecular biology.

Weintraub, H. M. 1990. "Antisense RNA and DNA." *Scientific American*, January. Deactivation of specific genes—a powerful research tool, perhaps someday a medical tool as well.

Searching for Genetic Causes—and Cures
Andrew Slay (seated) suffers from spinal muscular atrophy, a severe neuromuscular disease caused by a defect in a single gene. He is with Dr. T. Conrad Gilliam, a geneticist who has conducted a long search to find the gene responsible. Once the gene is located, modern genetic techniques may be able to correct the defect and cure the disease.

15

Genetic Disease and Modern Medicine

What happens when a gene doesn't work? If a person carries alleles that fail to code correctly for a particular protein, the effect may be medically insignificant—blue eyes instead of brown eyes, for example. If the missing protein catalyzes a step in a pathway such as cellular respiration, however, the zygote itself is not viable, and there will *be* no person. Between these extremes lie many inherited genetic diseases, such as sickle-cell anemia and cystic fibrosis. Our species is subject to more than 4,000 known heritable metabolic defects, most of which block production of necessary enzymes or other proteins. Many of these defects are recessive traits, whose alleles are unable to code for the normal proteins. It has been estimated that each of us is heterozygous—a carrier—for 30 or more of these recessive disorders.

Does all genetic ill health result from having defective alleles? The answer is no. We discussed another type of genetic ill health, chromosomal aberrations, in Chapters 9 and 10. Conditions such as Down syndrome result from either inappropriate numbers of chromosomes or inappropriate chromosomal structures. These conditions are inherited.

Is all genetic ill health inherited? Again, the answer is no. Cancer is not usually inherited, but it *is* a genetic disease because it results from effects on certain important genes that we will discuss later in this chapter. The change may occur in any part of the body, and in some cases, it may be expressed in many parts of the body.

Is there any hope that cancer and other genetic diseases can be defeated? Modern biomedical science has made considerable strides since gaining the ability to locate and clone genes whose defective alleles are responsible for genetic diseases. In the next few years we should see major advances in our ability to treat and prevent some of these diseases. Let's begin by considering a few of the better-known human genetic disorders.

SOME INHERITED DISEASES

Human populations were isolated from one another for long periods of evolutionary time. As will be explained in Chapter 19, an allele that remains rare in one isolated population may, with time, become much less rare in another population. Some defective alleles responsible for inherited diseases became less rare in certain human groups. The probability of your carrying a defective allele for certain diseases thus depends on your ancestry.

Cystic fibrosis, caused by an autosomal recessive allele (see Chapter 10), is a major killer of young people; most of its victims die during their early to mid twenties. Among Caucasians, cystic fibrosis is the most common potentially lethal autosomal recessive disease; it is also a significant threat to African-Americans. Cystic fibrosis affects 1 in 2,500 newborn infants, and about 1 in 20 people in the United States is heterozygous for the recessive allele that causes it. The symptoms of cystic fibrosis include severe respiratory difficulties resulting from thick mucus in the lungs, as well as liver, pancreatic, and digestive failure. At present, the median survival of victims of cystic fibrosis is about 29 years, but because of advances in our understanding and treatment of the disease, a person born *today* with cystic fibrosis will probably survive about 40 years even if there is no further progress in treatment. Current treatments include thumping of the chest to help clear the lungs of mucus, antibiotics to protect against infections, a special diet, and other treatments. By methods that we will discuss in this chapter, biomedical scientists recently isolated the defective gene that causes cystic fibrosis. The gene encodes a chloride channel in the cell membrane; cells of persons homozygous for the defective allele are unable to export chloride ions. Because the cells thus contain excess salt, they import water osmotically from the mucus that surrounds them, making the mucus thicker and harder to sweep away by ciliary action (see Chapter 40).

Tay-Sachs disease, another example of autosomal recessive inheritance, is most common among Ashkenazic Jews, appearing in the United States once in every 3,600 births in this group, a rate of occurrence that is about 100 times higher than in the remainder of the American population. Tay-Sachs disease results from a failure to produce a specific enzyme, hexosaminidase A. Without this enzyme, the nervous system degenerates. Children with Tay-Sachs disease appear normal for their first few months, but they usually lose their eyesight by about the age of one year, and they rarely survive more than five years.

Like victims of Tay-Sachs disease, infants with **phenylketonuria** seem normal at birth. Homozygotes for the autosomal recessive allele responsible for phenylketonuria cannot metabolize the amino acid phenylalanine. The principal consequence is severe mental retardation, and untreated victims usually do not live more than 30 years. Most states require screening of newborn infants for phenylketonuria because the condition, if caught immediately, is treatable and retardation can be prevented: Phenylketonuric infants kept on a diet low in phenylalanine develop normally.

Sickle-cell anemia is also caused by a defective allele of an autosomal gene. This disease is most common among people whose ancestors came from tropical areas or the Mediterranean. Among African-Americans, about 1 percent are homozygous for the defective allele and have the disease, and about 14

percent are heterozygous carriers who do not develop the disease. Homozygotes produce an abnormal hemoglobin, which leads to defects in the red blood cells; the molecular basis of the disease was discussed in Chapter 11 (see Figure 11.15).

Duchenne muscular dystrophy is an X-linked recessive trait that appears in the first three years of life. Muscular deterioration becomes more and more severe as the disease progresses; victims generally die in their twenties. About 1 in every 3,500 males born in the United States has the disease. The gene associated with the disease has been isolated and cloned. It is the largest known human gene (over a million base pairs) and encodes a protein found only in the plasma membranes of skeletal muscle cells.

The most common form of inherited mental retardation is the **fragile-X syndrome**, an X-linked recessive trait. About 1 in 1,500 males and 1 in 2,500 females are affected. The defective alleles that cause the syndrome are always carried on an abnormal chromosome that is easily identified under the microscope. Near one tip, the abnormal chromosome has a tight constriction that tends to break during preparation for microscopy (Figure 15.1). Although the basic pattern of inheritance of fragile-X syndrome is that of an X-linked recessive trait, there are strange departures from this behavior. Not all individuals with a fragile-X chromosome are mentally retarded. The gene responsible for fragile-X syndrome was

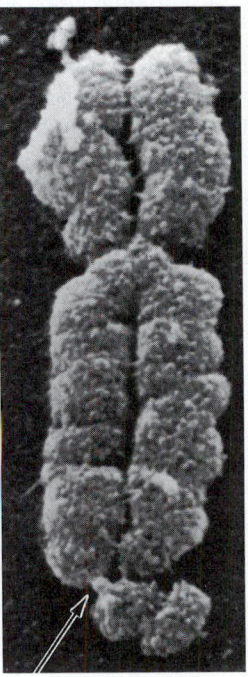

15.1 A Fragile-X Chromosome at Metaphase
The extreme constriction at the lower tip of this metaphase chromosome is the site of the fragile-X defect, clearly visible under the scanning electron microscope.

cloned and sequenced in 1991, revealing a surprise we describe in the next section.

Hemophilia, in which the blood does not clot normally, is an X-linked recessive trait. The failure of the blood to clot results from its failure to produce a protein that participates in the clotting process. Some hemophiliacs are at risk of death from even minor cuts, because they cannot stop bleeding. Treatment is primarily blood transfusion and the provision of a blood clotting factor. At the end of 1992, a blood clotting factor produced by recombinant DNA technology was approved for general medical use.

Not all hereditary diseases are the products of recessive alleles. **Huntington's disease** is an example of an autosomal dominant trait. It leads inevitably to premature death because the nervous system degenerates. Because the trait is dominant, on average half the children of an affected individual receive this allele. Traditionally, a person with a parent with Huntington's disease had to face a difficult personal decision about undertaking parenthood because the symptoms do not typically appear until age 35–50—well past the age when an affected individual is most likely to become a parent. Recently, however, scientists developed an effective way for people at risk for this disease to discover whether they carry the defective dominant allele. But this solution also raises difficult questions. Take some time to consider what they might be. Still more recently, biomedical scientists have isolated the gene responsible for Huntington's disease.

What about autosomal dominant alleles that act before reproductive age? If the resulting phenotype is lethal, we would expect the allele to disappear from the population in short order because an individual carrying it would not survive to pass the gene on. The known autosomal dominant disorders whose symptoms appear at an early age are indeed far from lethal. They include achondroplastic dwarfism (in which bone development is drastically inhibited) and polydactyly (excess fingers or toes).

Among the many other hereditary disorders, some are merely mild handicaps, such as red–green color blindness. Others are more severe. But an individual who carries the allele for a severe hereditary disorder and is expected to display its symptoms may instead show no signs of the disease. This phenomenon occurs with the fragile-X syndrome, as we will see in the next section.

TRIPLET REPEATS AND THE FRAGILITY OF SOME HUMAN GENES

About one-fifth of all males with a fragile-X chromosome are phenotypically normal. Most daughters of these phenotypically normal males are normal, too; but many of the sons of those daughters are mentally retarded. In any family in which fragile-X syndrome appears, the later generations tend to demonstrate earlier onset and more severe symptoms of the disease. It is almost as if the defective allele itself is changing—and getting worse. And that is in fact what is happening.

The gene responsible for fragile-X syndrome contains a repeated triplet of bases, CGG, at one point in its sequence. The CGG triplet repeats 6 to 54 times in normal alleles; the most common allele has 29 CGG repeats. In the alleles of mentally retarded individuals with fragile-X syndrome, this region has expanded to 200 to 1,300 repeats of the CGG triplet! What about the males and their daughters who show no symptoms? The alleles they carry have 52 to 200 repeats. Such an allele is referred to as a **premutation** because the number of repeats in these alleles may increase, especially when a woman passes the allele to her children (Figure 15.2).

Following the discovery of this unexpected expansion of a **triplet repeat**, a search began for the same sort of phenomenon in other genetic diseases. Research revealed that the same thing happens in myotonic dystrophy, in which the normal allele has 5 CTG

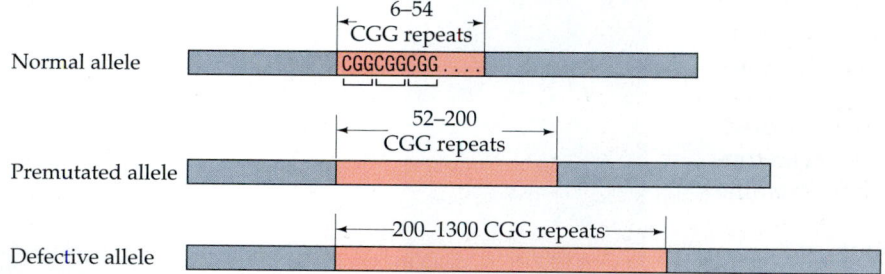

15.2 The CGG Triplet Region Expands in the Fragile-X Gene
The defective alleles that cause fragile-X syndrome have many more triplet repeats than do normal alleles. Individuals who carry the premutation do not have fragile-X syndrome; however, the number of repeats in premutated alleles is likely to increase in the next generation.

repeats, mildly affected individuals have alleles with about 50 repeats (the premutation condition), and the most severely affected individuals have alleles with up to 1,000 repeats. With Kennedy's disease, the normal allele has 21 CAG repeats, whereas alleles of affected persons have 40 to 52 repeats. Huntington's disease also results from an expansion of a region of CAG triplets. In Huntington's disease, in contrast to the fragile-X syndrome, expansion of triplet repeats happens more commonly when the gene is inherited from the father.

Triplet repeats are not limited to disease genes. Many other human genes contain triplet repeats, as do some genes of all animal species that have been studied since the original discovery of triplet repeats. What may be unique, or at least unusual, about human genes is that their triplet repeats expand readily. Genes of other animals do not appear to be as unstable in this regard.

Because they are so common, we assume that triplet repeats play important roles in normal alleles. We do not know, however, what these roles are, nor do we know why triplet repeats often are found within the protein-coding segment of a gene but may, as in the myotonic dystrophy gene and the fragile-X gene, be found outside the protein-coding segment. Triplet repeats are a subject of intense research interest. Their totally unexpected discovery is an early example of the surprises that can be unearthed by the Human Genome Project (see Chapter 14).

DEALING WITH GENETIC DISEASE: AN OVERVIEW

How do we deal with a genetic disease? We approach it in a series of steps, some established long ago and others discovered only recently. The final steps are yet to be defined.

First, we must characterize the symptoms of the disease—this is the realm of the physician. Epidemiologists (those who study the frequency and distribution of diseases within populations) often take the next steps, tracing the disease to discover whether it strikes most often at a particular group of people or in a particular place. Epidemiological studies can yield startling results; for example, the discovery that in the regions of Africa where sickle-cell anemia is most common, malaria is also very common. This discovery led in turn to the discovery that individuals who are *heterozygous* for the defective allele that causes sickle-cell anemia are more resistant to malaria than are homozygotes for the normal allele. This information helps us to understand why the sickle-cell allele remains common even though many persons homozygous for it do not live long enough to pass their genes on to offspring: In its

heterozygous state, it appears to confer an advantage on the individuals who carry it—if they live in an area where malaria is prevalent.

Armed with epidemiological data, medical geneticists can study affected families and develop detailed pedigrees to try to define the pattern of inheritance of the disease—is it autosomal recessive, X-linked recessive, or some other pattern? Examining pedigrees gives much valuable information, but it gets us only so far. Eventually we must make a jump from the Mendelian level of analysis to the molecular level.

New techniques enable us to identify the chromosomes on which disease genes reside. Recombinant DNA technology and related techniques, some of which we discussed in Chapter 14, then allow us to pinpoint the gene at a specific locus on the chromosome. We can clone the gene and compare the DNA of patients afflicted with the disease with that of people free of the disease. A current goal is to offset defective alleles by outfitting the patient with a functional allele—although there are ethical questions about technologies of this type. To further complicate things, we have discovered that many genetic diseases result from lesions in more than one gene. This chapter, however, focuses primarily on the single-gene case, which does adequately account for some of the most prominent genetic diseases.

Let's consider these steps in more detail. As you read, watch for the conceptual breakthroughs that opened new prospects for dealing with genetic disease. Most of the techniques we describe will be superseded as research progresses. We try to show only enough detail to enable you to see where we stand today in the battle against genetic disease.

FINDING A DEFECTIVE GENE: MAPPING HUMAN CHROMOSOMES

Mapping genes in humans opens new possibilities for diagnosing and treating human genetic diseases. Striking technical advances, along with an emerging willingness to pay a great deal of money to invent and implement them, gave rise to the Human Genome Project.

Mapping the human genome is an impressive and difficult undertaking. Until only very recently, two formidable barriers obstructed any program to map human genes. The first is the long generation time in humans, together with the relatively small number of offspring produced by most couples. The second barrier is the moral unacceptability of performing breeding experiments of the type that were crucial to mapping the genes of maize, bacteria, viruses, and the fruit fly *Drosophila melanogaster*. Although breeding experiments will presumably always be unacceptable in human studies, recent technological advances are helping to overcome some of the difficulty in mapping the human genome. As we will see in the next few sections, the traditional approach—pedigree analysis—is now supplemented by new, sophisticated techniques that enable scientists to identify much more precisely where particular genes are located.

Because of the distinctive character of sex-linked inheritance, the first human genes to be associated with a particular chromosome were all found on the X chromosome, beginning in 1911 with the discovery of a gene for color blindness. It was another 59 years before a human gene was located on an autosome.

Pedigree Analysis

Until recently the mapping of human genes relied on existing data from family trees and mostly amounted to simply assigning a locus to a sex chromosome or autosome. Known as **pedigree analysis**, this process documents the transmission of a particular genetic characteristic over two or more generations and reveals whether the pattern is dominant or recessive. Although human pedigrees include data on relatively few individuals and few generations, we can sometimes obtain useful results by comparing the pedigrees of different families exhibiting the hereditary trait of interest. Such pedigrees are our closest acceptable approximation to the breeding experiments performed with plants and experimental animals.

Some of the symbols commonly used in pedigree charts are shown in Figure 15.3. Figure 15.4 shows three examples of pedigrees; before reading the caption, try to determine the pattern of inheritance—for example, autosomal recessive or sex-linked dominant—demonstrated by each chart.

Pedigree analysis has also been used to test for the linkage of pairs of traits. By noting the relationships of individuals affected by one, the other, or both traits in a family, we can determine whether the traits are linked or unlinked and may even obtain a rough estimate of the map distance between linked loci.

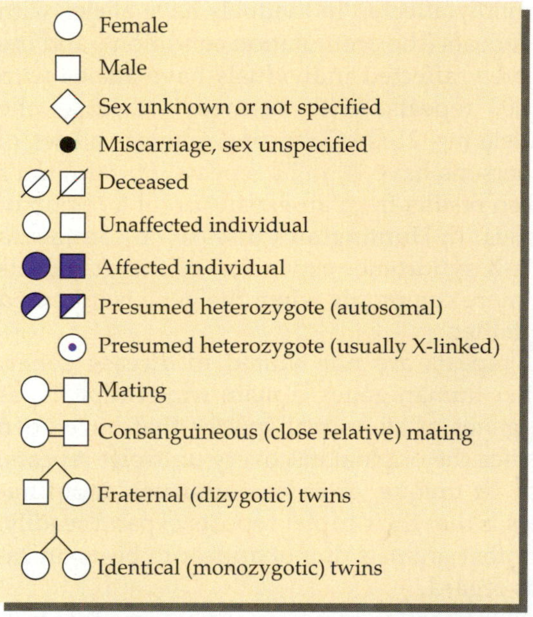

Somatic-Cell Genetics

Once we know that a gene associated with a disease is carried on an autosome instead of on a sex chromosome, the next step in mapping is to determine *which* autosome. One of several new techniques used to map human genes is **somatic-cell hybridization**. This method, which involves creating and analyzing hybrid cells that contain the chromosomes of two different species, helps researchers determine which chromosome houses a particular gene.

Somatic-cell hybridization relies on a remarkable phenomenon: Somatic cells (that is, cells other than gametes) isolated from two animal species and cul-

15.3   Symbols for Pedigree Charts
These symbols are used to characterize individuals and relationships in pedigree charts.

tured together on a suitable medium, occasionally fuse to form a hybrid cell with two nuclei, one from each species! Biologists have been able to increase the probability of such fusion by adding certain chemicals, such as polyethylene glycol, to the culture medium.

After a hybrid cell forms, the two nuclei may fuse to create a nucleus containing the chromosomes of both species. Mitosis in hybrid cells is sometimes abnormal, with successive mitoses leading to the disappearance of more and more chromosomes of one of the species. As a result, some daughter cells have nuclei containing all the chromosomes of one species and only one or a few chromosomes of the second species (Figure 15.5). In mouse–human hybrid cells, it is most often the human chromosomes that are lost.

At this point, you may be able to guess how mouse–human hybrid cells can be used to identify the particular chromosomes on which certain genes are located. Researchers can observe several colonies of hybrid cells, testing them for two characteristics: (1) their ability to form the products of a particular gene and (2) their possession of different human chromosomes (based on the characteristic appearances of the individual chromosomes; see Figure 9.14). A given gene product will appear in all cells that have a particular chromosome and in no cells that lack that chromosome (see Study Question 3). Using this technique, scientists have successfully mapped a few hundred human genes to specific chromosomes. This effort has been enhanced by the

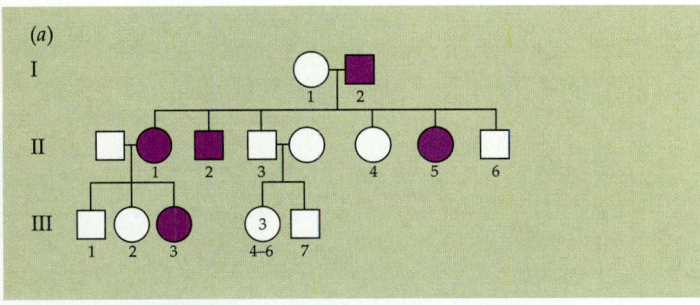

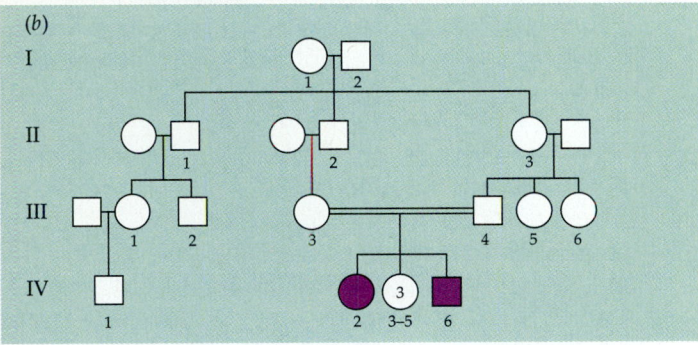

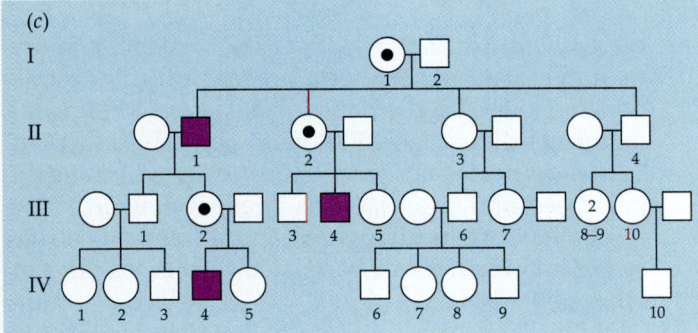

15.4 Patterns of Inheritance

In these pedigrees, roman numerals identify generations; arabic numerals identify individuals within each generation. The symbols are explained in Figure 15.3; numbers inside of symbols mean multiple individuals of that discription (e.g., three sisters in generation IV). (a) Inheritance of an autosomal dominant trait. Individual I-2 is a male heterozygous for the trait; if he were homozygous, all of the progeny in generation II would show the trait. None of the affected individuals in generations II or III can be homozygotes, because each has received a normal recessive allele from one parent. This is the pattern of inheritance for Huntington's disease. (b) Inheritance of an autosomal recessive trait. If the trait were dominant rather than recessive, either III-3 or III-4 would have to show the trait in order for IV-2 and IV-6 to have inherited it. III-3 and III-4 are first cousins—their mating is represented by the double horizontal line. There is no indication of sex linkage. This is the pattern of inheritance for cystic fibrosis. (c) Inheritance of a sex-linked recessive trait. The affected individuals—II-1, III-4, and IV-4—are all males. The carriers are all females. This is the pattern of inheritance for hemophilia.

availability of data from pedigree analysis. Once a gene has been localized on a particular chromosome by somatic-cell genetic techniques, we know that any other genes linked to that gene (as previously discovered by pedigree analysis) must reside on the same chromosome.

Another technique of somatic-cell genetics is chromosome-mediated gene transfer. With this method, instead of combining cells from two species, the researcher combines somatic cells of a recipient species (such as a mouse) with *chromosomes* isolated from metaphase cells of a donor species (such as *Homo sapiens*). When recipient cells are incubated with isolated chromosomes, only some parts of some chromosomes are taken up by the cells, and only a small fraction of what is taken up becomes associated with

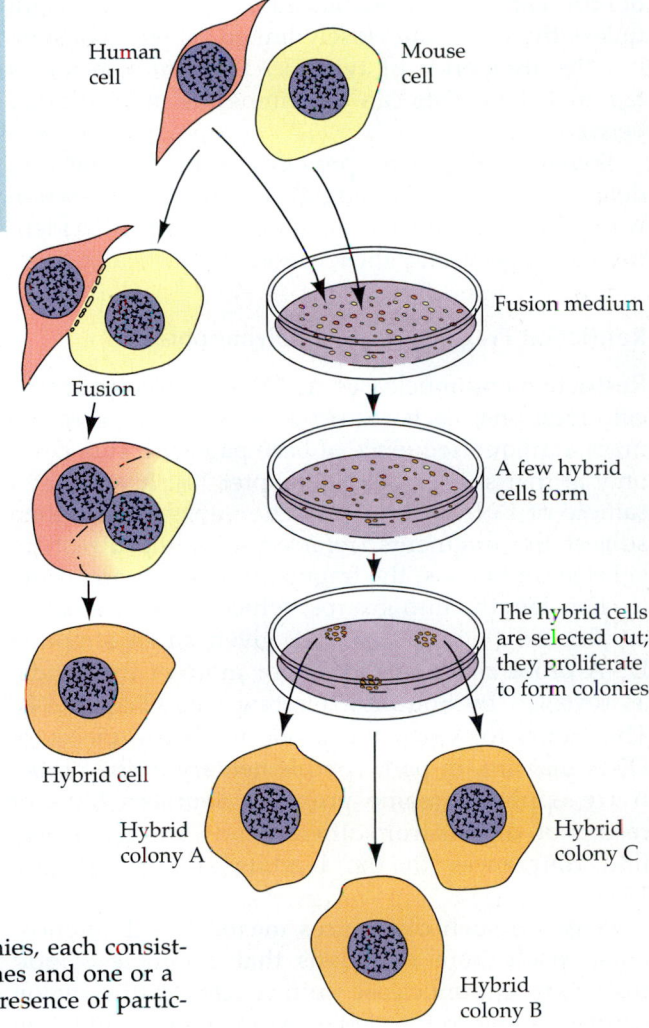

15.5 Mouse and Human Cells Can Fuse
The final result of this hybridization process is a series of colonies, each consisting of many cells with a full complement of mouse chromosomes and one or a few human chromosomes. Each colony may be tested for the presence of particular genes and chromosomes.

(a)

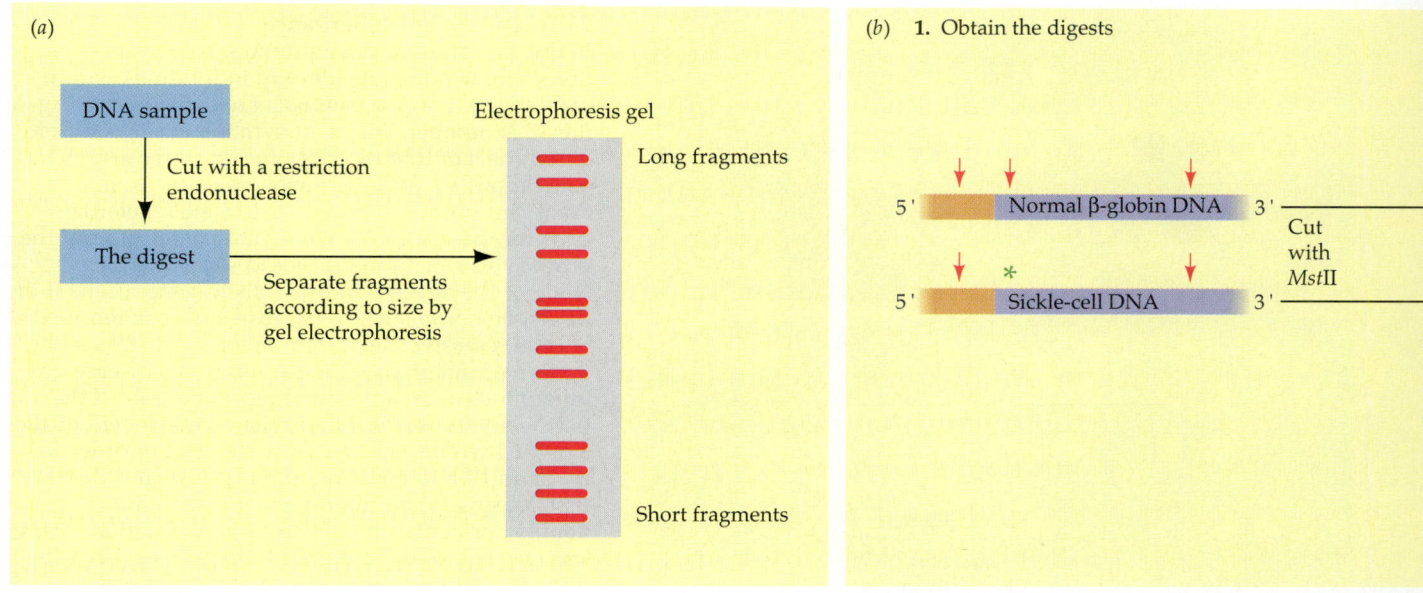

(b) **1. Obtain the digests**

the host chromosomes of the mouse and is thus stabilized. The chances that two particular genes of the donor will both be stabilized are extremely slight unless the genes are closely linked. Thus by observing the frequency of two loci that are transferred together, biologists can determine the map distance between them.

Somatic-cell genetics provides useful, but not very delicate, tools for mapping genes. In the next section we will see how restriction endonucleases help identify more specifically the location of a particular gene.

Restriction Fragment Length Polymorphisms

Restriction endonucleases cut DNA molecules at specific locations; each restriction endonuclease recognizes a unique sequence of base pairs and cuts DNA only at that sequence (see Chapter 14). If we treat a sample of DNA with a restriction enzyme and then subject the fragments (referred to as the digest) to gel electrophoresis, the fragments separate according to their lengths into groups, which appear as bands on the gel (Figure 15.6a). A given single-stranded DNA probe may bind to one or more of the bands, as revealed by Southern blotting (see Chapter 14). The bands to which the probe binds are pieces of DNA that are, in part, complementary to the probe. If we apply the same probe to Southern blots of restriction digests from different individuals, we may find differences in the binding patterns (Figure 15.6b).

What do such differences mean? The differences must result from mutations that eliminate or add restriction endonuclease cutting sites in the region corresponding to the probe. A single point mutation

can easily cause such a change. The existence of more than one band pattern for a probe is called a restriction fragment length polymorphism, or **RFLP** (pronounced "rifflip"). A RFLP band pattern is inherited in Mendelian fashion and can be followed through a pedigree—and it can serve as a genetic marker. Many RFLPs have been discovered and are now available as landmarks that can be related to the positions of other genes.

How can we use RFLPs to help determine the location of a gene? We can study pedigrees for evidence that the gene of interest and certain RFLP band patterns are inherited together, implying that the gene and the markers lie in the same part of a chromosome (Figure 15.7). RFLP mapping enabled scientists to determine that the cystic fibrosis gene is located on a portion of the long arm of human chromosome 7. But that portion was still huge—about 1.5 million base pairs long. How can we find the gene in a "haystack" that big?

Narrowing the Search

Chromosome walking (see Chapter 14) is a suitable technique for finding a gene in a relatively small stretch of DNA. For larger stretches—such as the 1.5-million-base-pair stretch containing the cystic fibrosis gene—an analogous procedure called chromosome jumping was developed. Using this technique, scientists established a series of starting points for chromosome walks and narrowed the location of the cystic fibrosis gene to a much smaller part of chromosome 7.

Additional experiments narrowed the search even further. Eventually the start and stop signals were

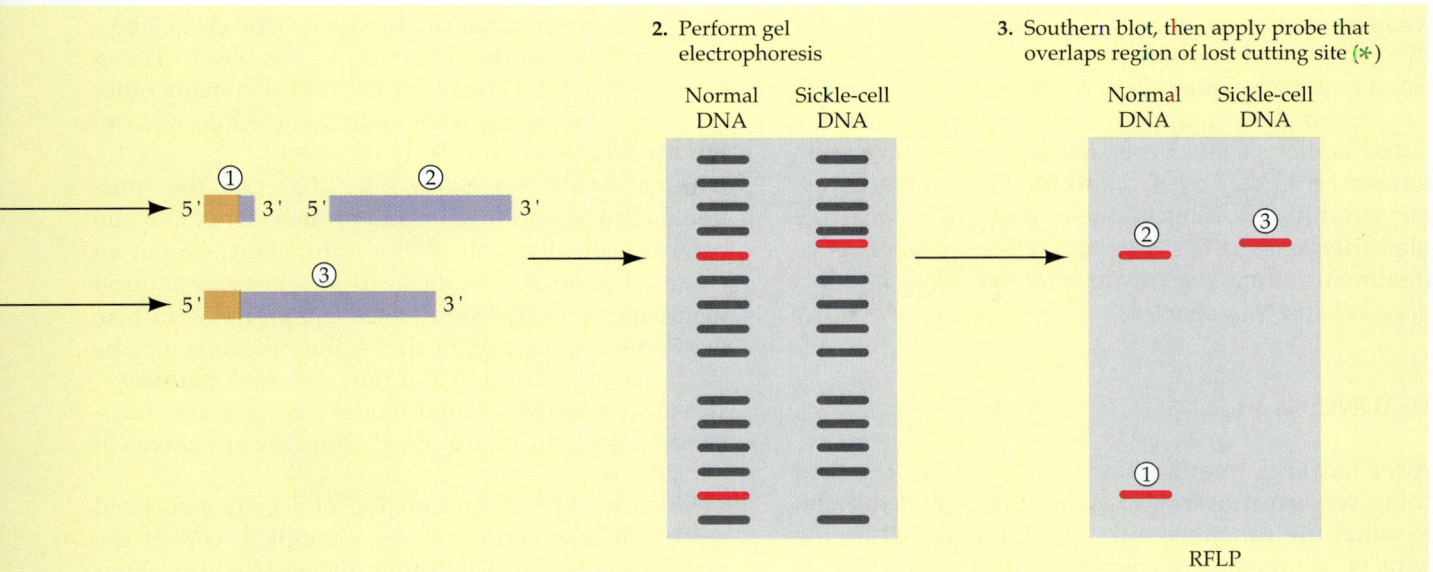

2. Perform gel electrophoresis

3. Southern blot, then apply probe that overlaps region of lost cutting site (✷)

15.6 Probing Restriction Digests

(a) DNA fragments cut by a restriction endonuclease can be put in order by size using gel electrophoresis. (b) Probing restriction digests may reveal that DNA from two individuals differs, as in this comparison of DNAs from two homozygous humans, one having the allele that codes for normal β-globin polypeptide (a subunit of the hemoglobin molecule) and the other having the sickle-cell allele. The restriction endonuclease *Mst*II has three cutting sites (red arrows) in or near the β-globin gene, but one of these sites is eliminated (green asterisk) by the mutation that forms the sickle-cell allele. A probe overlapping that site reveals the RFLP in this region.

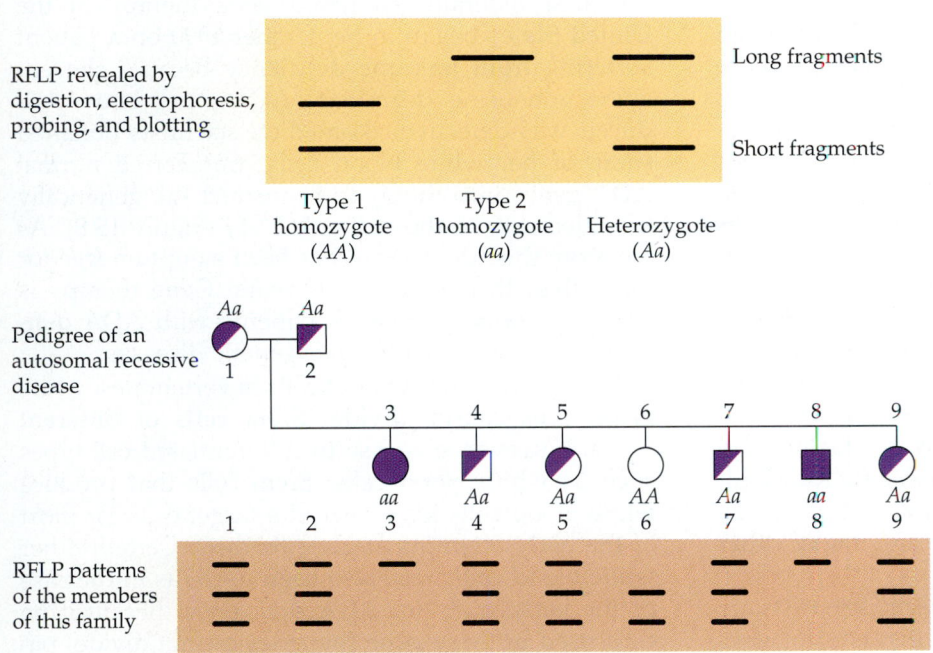

15.7 RFLP Mapping

In the hypothetical pedigree shown here, the inheritance of the RFLP shown at the top center and that of an autosomal recessive disease are closely linked. Affected individuals have the genotype *aa*. Note the fragment pattern from child number 6, which suggests that he has genotype *AA*. This RFLP can serve as a marker in mapping the locus of the disease gene.

located, and ultimately the nucleotide sequence of the gene was determined. Thanks to these sophisticated new techniques, the gene for cystic fibrosis has been identified, mapped, and sequenced. As we noted earlier, a few other disease genes have been located by similar means, and the number is increasing steadily. Locating the gene that causes a particular disease is the first step to developing an effective treatment. What do we do after we have found a gene related to a disease?

DEALING WITH A DEFECTIVE GENE

After locating a gene related to a disease, the next thing we want to know is what the gene *does*—that is, what the function of the protein encoded by the gene is. Ultimately, of course, we want to learn what to do about the defective allele—how to cure the disease. Recent advances are helping us toward both of these goals. We can study specific gene function in laboratory animals called "knockout mice" that are genetically engineered to contain the defective allele. We can also try to get around the effects of a defective allele by inserting functioning alleles into the cells of persons with the disease.

Knockout Mice

Since it is usually impossible to use humans as the laboratory subjects when studying the functions of defective alleles, what researchers needed was an **animal model** for the specific disease. The ideal animal would show the symptoms of the disease, would be useful in biochemical and molecular biological studies, and could be the test organism for possible therapies. Unfortunately for us, nonhuman animals do not usually get the same genetic diseases humans do. Intensive laboratory work, however, resulted recently in the development of a variety of transgenic rodents. Particularly useful in studies of human diseases are the so-called **knockout mice**— mice in which defective alleles have been inserted to replace the functioning alleles, which were "knocked out." Many of the genes studied in this fashion are associated with specific human diseases. Knockout mice are now available as models for cystic fibrosis, sickle-cell anemia, atherosclerosis, and many other human diseases.

With knockout mice as their subjects, scientists are now able to study human genetic diseases in ways that would be difficult or impossible in work with humans. Using the techniques of biochemistry, molecular biology, physiology, and pharmacology, we can explore how the defective alleles lead to the expression of a disease and we can test potential drugs and other treatments. Work with knockout mice

helped determine that the product of the cystic fibrosis gene is a chloride channel, as we saw earlier in this chapter. Knockout mice are useful in many other types of investigations as well; in principle, we can modify any gene and study its effect.

Surprisingly, knockout mice that lack the functional allele encoding a crucial protein sometimes still behave normally. This ability to function without an essential gene may indicate that otherwise unused members of a gene family (see Chapter 13) can take over when necessary, or that some functions may be accomplished through entirely different pathways. To be an effective animal model for a disease, however, a knockout mouse must show the symptoms of the disease.

Once we know the function of a gene associated with a disease, how can we attempt to correct the mistake made by the defective allele? As we are about to see, there is a promising strategy for dealing with this problem.

Gene Therapy

Perhaps the most obvious thing to do when a person lacks a functional allele is to provide one. If a person cannot make an essential protein because he or she is homozygous for a defective recessive allele, can we simply give the person a supply of cells with correct genetic instructions for making the protein? Such **gene therapy** is under intensive investigation. The first federally approved gene therapy in the United States began in September 1990 on a patient suffering from immune deficiency because the enzyme adenosine deaminase (ADA) her body produced was defective. Biomedical scientists obtained some of her white blood cells, transferred normal ADA genes into them, and returned the genetically modified cells to the patient's body (Figure 15.8). As we write this, the patient has been symptom-free for more than three and a half years. Gene therapy is being performed on other patients with ADA deficiency, as well as with other genetic diseases.

Like many other types of cells in vertebrates, white blood cells cannot divide. **Stem cells** of different types, however, give rise to differentiated cell types such as white blood cells. Stem cells that produce white blood cells have been the target cells for most of the early attempts at gene therapy. Certain other white blood cells have also been used, as in the case of the early trials on ADA deficiency. Because the cells used in these gene therapies cannot divide, patients require repeated treatments, including isolation of their cells, genetic engineering of the cells, and return of the cells to the body. Stem cells inserted with the functional allele offer the possibility of one-step gene therapy because they should continue to generate new cells with the functional allele.

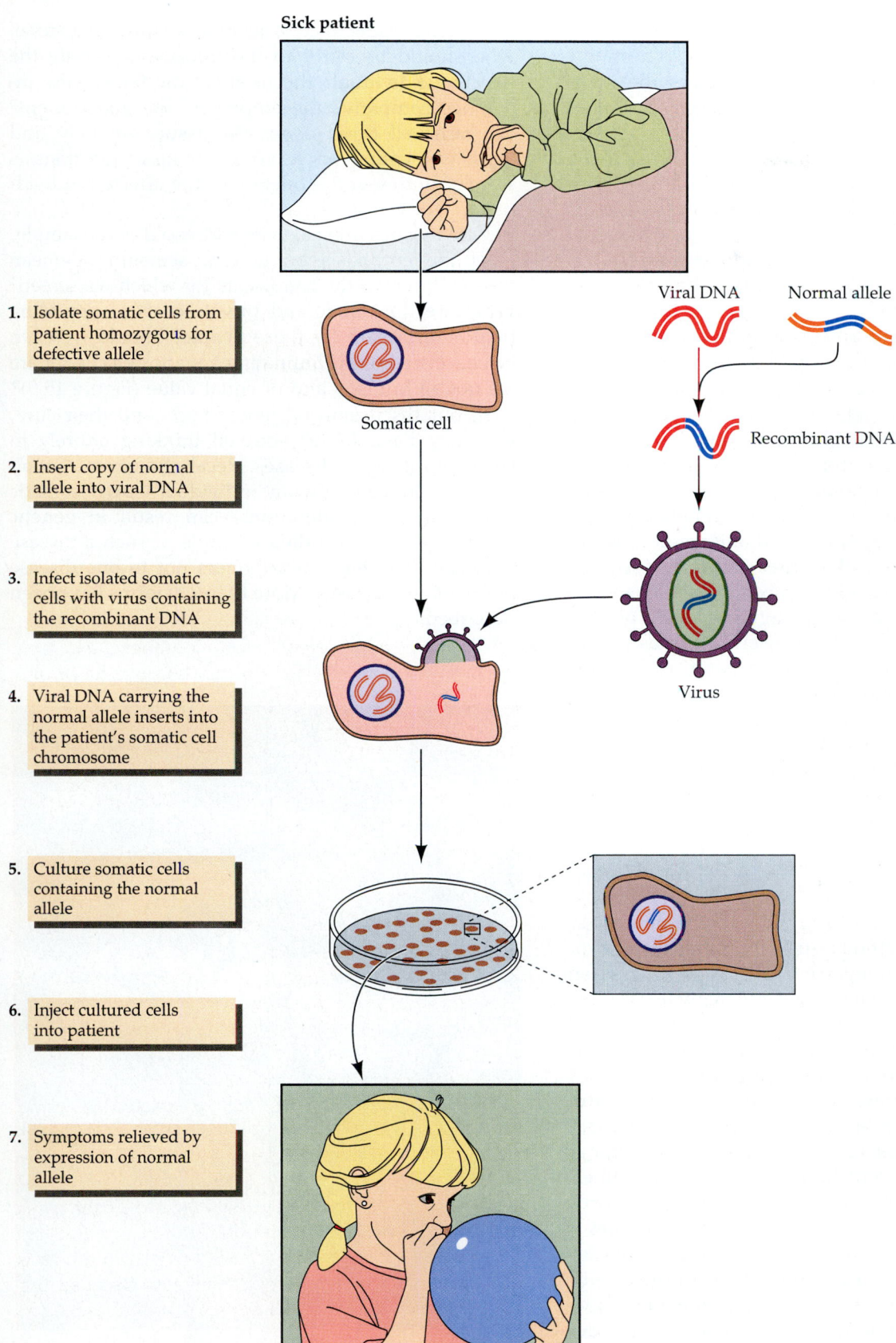

Sick patient

Viral DNA   Normal allele

Recombinant DNA

Somatic cell

Virus

Well patient

1. Isolate somatic cells from patient homozygous for defective allele

2. Insert copy of normal allele into viral DNA

3. Infect isolated somatic cells with virus containing the recombinant DNA

4. Viral DNA carrying the normal allele inserts into the patient's somatic cell chromosome

5. Culture somatic cells containing the normal allele

6. Inject cultured cells into patient

7. Symptoms relieved by expression of normal allele

15.8   Gene Therapy
Recombinant DNA technology provides techniques for inserting a functional gene into a patient's somatic cells, rendering the patient capable of producing the protein encoded in that allele.

How likely are we to practice this specific form of gene therapy—isolating, modifying, and returning a patient's cells—on a large scale? Unless the intermediate recombinant DNA technological steps can be automated, this approach to treating genetic disease appears to be too labor-intensive, too costly, and too dependent on sophisticated laboratory equipment and techniques for large-scale use. It has been suggested that, if these problems are resolved, the techniques are more likely to be applied to the treatment of cancer or infectious diseases than to single-gene genetic defects.

In any case, there are many ethical concerns associated with gene therapy. For example, should such therapy be limited to the modification of somatic cells, such as stem cells and white blood cells, or should it be extended to germ cells (gametes) in order to ensure that future generations do not inherit a genetic disease? The general opinion at this time inclines to the view that we should *not* genetically modify germ cells. The argument is that if we do this, we will be tempted to make other changes—inserting genes for increased height, for example—that are less nobly motivated and too strongly resemble playing God. Similar concerns about inserting genes for enhanced intelligence, on the other hand, can be dismissed as far beyond the realm of possibility for a long time to come, since no such genes have been identified. Some of these ethical considerations apply not only to gene therapy but also to the widespread practice of screening fetuses and newborn babies for genetic defects.

Screening for Harmful Alleles

During pregnancy, cells from the growing fetus may be sampled by methods such as amniocentesis or chorionic villi sampling (see Chapter 37) and then tested for harmful alleles. There are many tests, some focusing on chromosomes, others on proteins, still others on biochemical reactions; most recently it has become possible to test the DNA directly. RFLPs tightly linked to known disease genes are a powerful tool for such screening, as is the polymerase chain reaction (see Chapter 14). The goal of such screening, of course, is to prevent human suffering. The ill effects of phenylketonuria, as mentioned earlier in this chapter, can be prevented by an appropriate diet if the presence of the defective allele responsible for the condition is detected in the infant. In the case of some other diseases, abortion may be the only active step that can be undertaken in response to the detection of a disease-causing allele. This solution, although acceptable to many, is abominable to others. This is not the only ethical issue to arise from genetic screening.

At least as challenging is the issue of privacy. If genetic screening is conducted as a matter of course, who should be party to the information—only the affected individual? the mother? the father? the insurance company? the employer? the government? We must all think about this issue; we may find ourselves voting on it. *None* of these questions is simple, not even the one about the affected individual.

Other issues must also be addressed. For example, does the emphasis on genetic screening demean those who have the conditions for which we screen? What are the public and private costs of providing treatment or care for these people? Should we even be concerned with eliminating genetic defects, for are we not all humans and of equal value (Figure 15.9)?

In our discussion of gene defects and their cure, you may have found yourself thinking entirely in terms of inherited diseases. Recall, however, that alterations in genes in an individual's own somatic cells—*uninherited* alterations—can result in genetic disease. The best-studied example of such a disease is cancer. The term *cancer* refers not to one disease but to many diseases. More than 200 forms have been described.

15.9 Life with a Genetic Defect
Many children are born each year with Down syndrome, a genetic defect that results in mental retardation and other symptoms. Despite their handicap, such individuals enjoy the activities of life and often become productive members of society.

CANCER

Each year about 6.5 million new cases of cancer are diagnosed worldwide. What do these cases have in common? All forms of **cancer** differ from other diseases in two ways. First, a cancerous cell loses control over its own division. Cells that have been transformed to the cancerous state divide rapidly and inappropriately, ultimately forming tumors (large masses of cells). **Benign tumors**, which are not cancers, remain localized where they form, but **malignant tumors**—cancers—invade surrounding tissues and spread to other parts of the body. This spreading, called **metastasis**, is the second defining characteristic of cancer.

Metastasis proceeds in two stages: First the cancer cells extend into surrounding tissues; then they enter either the bloodstream or the lymphatic system (Figure 15.10). Cancer cells metastasizing by way of the lymph are slowed by lymph nodes, where they must pause before proceeding to the next node. The removal of a series of lymph nodes and the ducts between them can often end the disease in such cases. For example, a mastectomy (removal of breast tissue) to treat breast cancer often includes the removal of lymph nodes. Metastasis through the bloodstream is another matter. Less common than metastasis through the lymphatic system, it is rapid and often fatal.

Different forms of cancer affect different parts of the body. **Carcinomas**—cancers that arise in surface tissues such as the skin and the lining of the gut—account for more than 90 percent of all human cancers. Lung cancer, breast cancer, colon cancer, and liver cancer are all carcinomas. (Among these, the first three account for more than half of all cancer deaths in Europe and North America; liver cancer is the most common fatal tumor in Africa and Asia.) **Sarcomas** are cancers of tissues such as bone, blood vessels, and muscle. **Lymphomas** and **leukemias** affect the cells that give rise to the white and red blood cells, respectively. (White blood cells of several types participate in the body's immune system, and red blood cells carry oxygen throughout the body.)

Where does a cancer begin? That is, what triggers the *transformation of a normal cell to a cancer cell?*

Genes, Viruses, and Cancer

The first cancer-causing agent to be identified was a virus that produces sarcomas in chickens. Since that discovery, many other viruses have been demonstrated to cause cancers in animals. The ability of certain viruses to cause cancer in humans became apparent only many years later, and it appears likely that viruses cause only a small fraction of human cancer.

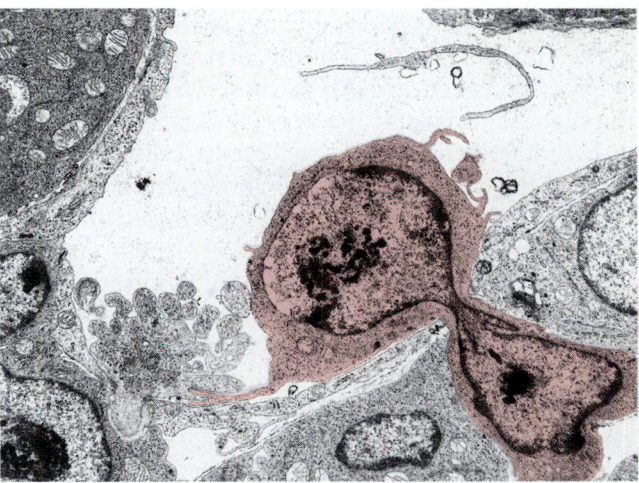

15.10 The Spread of Cancer
A cancer cell (shaded in color) squeezes into a small blood vessel through a cell in the vessel's wall. Cancer spreads through animal bodies as the malignant cells invade tissues and enter the bloodstream or lymphatic system. The blood or lymph then transports the cancer cells throughout the body, starting new tumors in various organs.

Because many cancers are diseases of old age, it was proposed that cancers result from an accumulation of mutations in a single cell. What could account for these mutations? Chemical **carcinogens** in the diet or in the immediate environment are substances that can cause cancer. Polluted air and tobacco smoke contain carcinogens. All known carcinogens react directly or indirectly with DNA to cause mutations. Ultraviolet radiation, X rays, and other forms of high-energy radiation also cause mutations and thus cancer. Excessive exposure to sunlight, with its substantial content of ultraviolet rays, can lead to cancers of the skin. Skin cancers are particularly common among fair-skinned persons and albinos, as well as persons whose enzymes for repairing damage to DNA are defective.

Chemical carcinogens, radiation, and tumor-inducing viruses all cause mutations or otherwise modify the DNA content of a nucleus. What, then, is the nature of these changes? Since cancerous transformation results from changes in genes required for normal growth, we we must first understand what normal growth is.

Growth Factors and Cancer-Related Genes

An animal consists of masses of cells of many types. The cells divide at different times and at different rates; some do not divide at all after a certain developmental stage. Cell division is tightly orchestrated in a healthy animal.

How is cell division regulated in the various parts of the animal body? Part of the answer lies in a group of proteins called **growth factors**, which circulate in the blood and trigger the normal division of cells. A growth factor acts only on certain cells—its target cells. What keeps it from triggering division in any cell it happens to meet? Each growth factor can be bound only by a unique receptor protein, and the receptor proteins are embedded in the plasma membranes of the target cells. After binding its growth factor, a receptor protein becomes active as an enzyme, catalyzing reactions that participate in cell division. These reactions are not triggered unless the growth factor is present.

Growth factors and receptor proteins are coded for by genes that may change and cause cancer. Cancers originate in the activities of cancer-producing alleles, or **oncogenes**: dominant alleles that *arise from the normal recessive alleles that encode, among other things, a growth factor and certain receptor proteins*. The normal alleles are called **proto-oncogenes**. Proto-oncogenes are absolutely essential to the normal development of the cell. It is the *mutant* form, the oncogene, that causes cancer. The question of how oncogenes produce cancer is an area of intense study. There is no single, simple answer. Ultimately the oncogene leads to unregulated cell multiplication. The regulation of cell division, so important to normal cells, is absent in cancerous cells.

Proto-oncogenes can lead to cancers by various mechanisms, three of which we describe here. One mechanism is simple mutation. A single point mutation can cause certain proto-oncogenes to become oncogenes and induce tumor formation. A second mechanism is overproduction. Multiple copies of the proto-oncogene may form through gene amplification (see Chapter 13) or because normal mechanisms of gene regulation fail. In a third mechanism the proto-oncogene moves (by transposition or by chromosomal translocation; see Chapter 11) to a new chromosomal site near the promoter of a very active gene, resulting in continual transcription of the proto-oncogene.

A different type of mutation—a recessive mutation of a dominant **tumor-suppressor gene**—initiates most human cancers. Tumor-suppressor genes encode proteins that inhibit cell division, thus suppressing the formation of tumors. The best-studied tumor-suppressor gene, *p53*, encodes a protein that halts the cell cycle at a point before division, allowing a cell to repair any damaged DNA before it divides. Mutations of *p53* are evident in tumors in half of all cancer cases. The first clear example of a cancer associated with mutation of a tumor-suppressor gene (in this case not *p53*) was retinoblastoma, a tumor of the eye most often observed in children.

Transformation to the cancerous state requires multiple mutations in a single cell. Different types of cancer result from different sequences, but usually more than one tumor-suppressor gene must mutate, and often a proto-oncogene mutates to an oncogene (Figure 15.11). The requirement for multiple mutations, which generally take a long time, is consistent with the observation that many cancers are diseases of old age. Not all the changes need arise by mutation, however; as explained in the previous section, viruses are a source of genetic change, and of cancer.

Many cancers in mammals and birds are triggered by retroviruses (see Chapter 11). Recall that these viruses have RNA, not DNA, as their genetic material. To replicate within cells of their hosts, retroviruses must have their RNA copied to DNA by reverse transcription (Figure 15.12). Reverse transcription is accomplished by an enzyme, reverse transcriptase, that is coded for by a viral gene. The DNA reverse transcript of viral RNA is inserted into a host chromosome, where it may trigger the transformation of the host cell into a cancerous cell.

Some retroviruses that lack oncogenes can cause tumors in animals, but this process typically requires many months. It has been discovered in such cases that the viral DNA was inserted at a locus very close to a proto-oncogene in the host chromosome. Under these circumstances cancer formation by oncogenes depends on viral sequences to induce gene activity in the host.

Retroviruses that are highly cancer-producing (oncogenic) contain, in addition to the three genes necessary for their own reproduction, an oncogene that causes cancerous growth and behavior in the infected cell. The promoters of viral oncogenes are strong and cause frequent transcription. How did oncogenic viruses get their oncogenes?

Origins of Oncogenic Viruses and Transposable Elements

The oncogenes of retroviruses most likely arose from animal proto-oncogenes by the incorporation of an RNA transcript of a proto-oncogene into a viral coat along with part of the viral RNA genome (a process similar to transduction, which was described in Chapter 12). This theory of the origin of retroviral oncogenes is supported by the observation that a retroviral oncogene is similar in base sequence to the suspected "parental" proto-oncogene, except that the proto-oncogene has introns, whereas the retroviral oncogene is uninterrupted. The oncogene differs further from the proto-oncogene in that it has been shortened at either the N-terminal or C-terminal end and has experienced one or more point mutations.

It has also been suggested that some (but not all) transposable elements may have arisen from retroviruses that became immobilized. This suggestion is

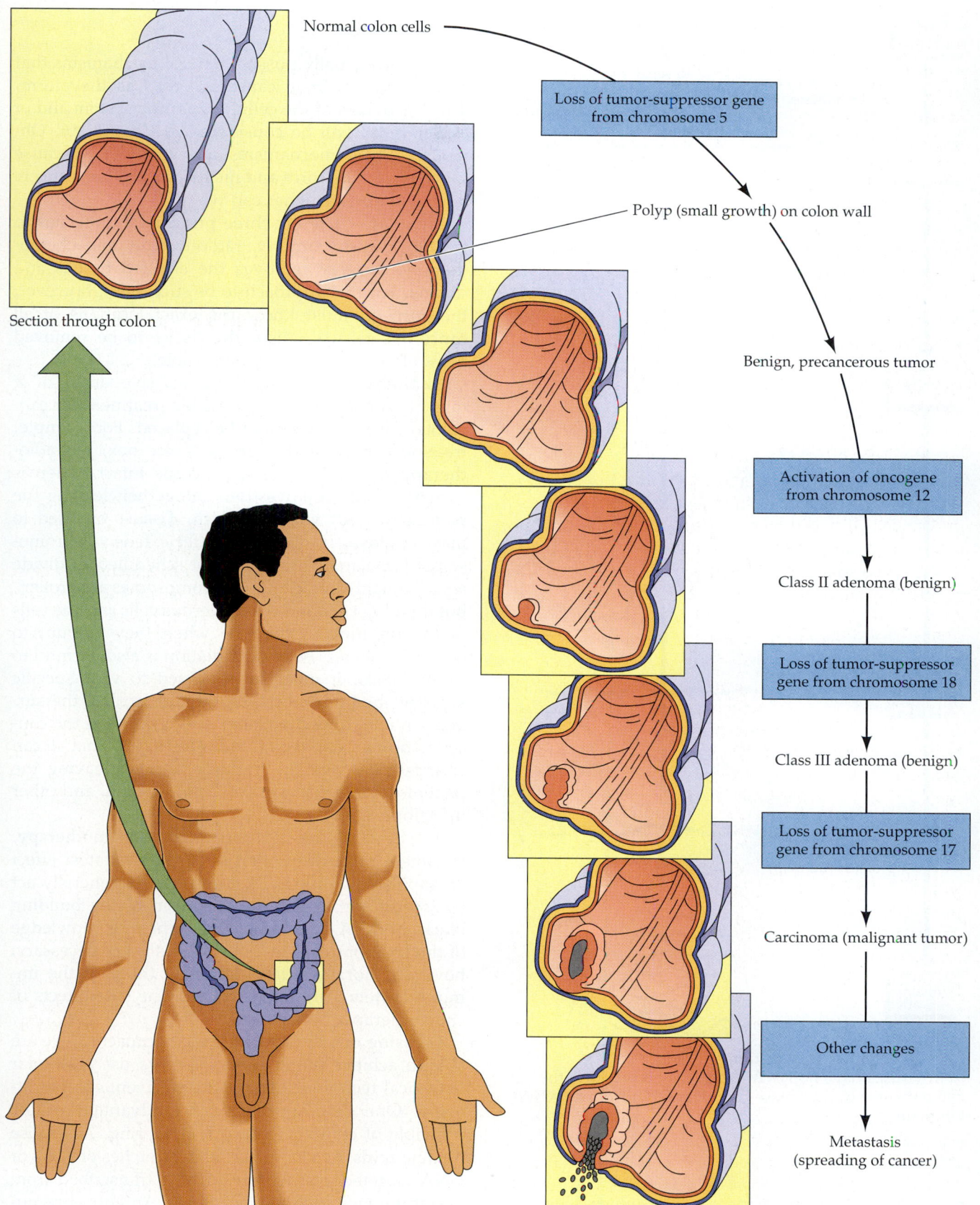

Normal colon cells

Loss of tumor-suppressor gene from chromosome 5

Polyp (small growth) on colon wall

Benign, precancerous tumor

Section through colon

Activation of oncogene from chromosome 12

Class II adenoma (benign)

Loss of tumor-suppressor gene from chromosome 18

Class III adenoma (benign)

Loss of tumor-suppressor gene from chromosome 17

Carcinoma (malignant tumor)

Other changes

Metastasis (spreading of cancer)

15.11   Multiple Steps May Transform a Normal Cell into a Cancerous Cell
These stages in the development of normal tissue into a cancer of the colon (the large intestine) reflect a series of mutations required to transform a normal cell into a cancerous one.

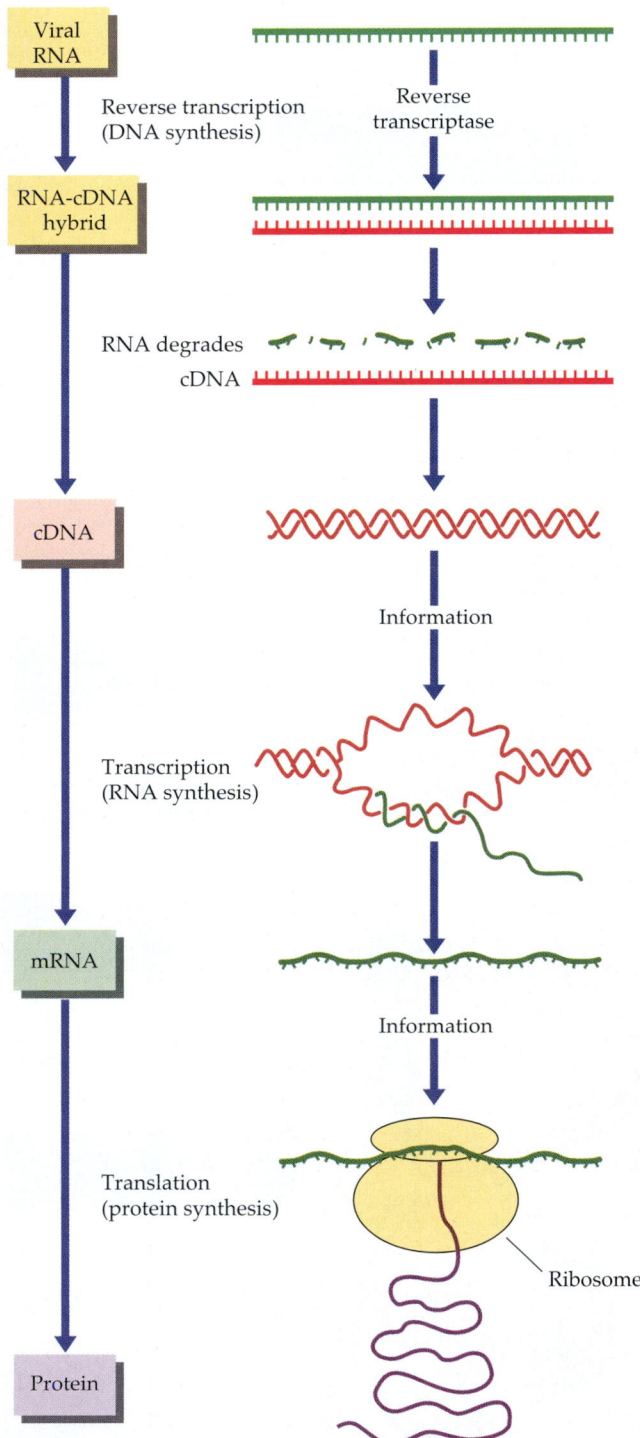

Viral RNA

Reverse transcription (DNA synthesis)

Reverse transcriptase

RNA-cDNA hybrid

RNA degrades

cDNA

cDNA

Information

Transcription (RNA synthesis)

mRNA

Information

Translation (protein synthesis)

Ribosome

Protein

15.12 Reverse Transcription
Retroviruses have no DNA; when they reproduce, the enzyme reverse transcriptase produces DNA from an RNA template.

based on DNA-sequencing studies, which have revealed strong similarities between the base sequences of the ends of certain retroviruses and those of certain transposable elements, indicating a possible common origin.

Treating Cancer

Our bodies usually possess natural mechanisms that detect and eliminate cancerous cells as they form. These activities of the cellular immune system and of "killer cells" will be considered in Chapter 16. Obviously these mechanisms are imperfect, because cancers often do form and metastasize. When cancers arise and spread, what can we do?

There are currently three principal approaches to treating cancer: surgery, radiotherapy, and chemotherapy. Surgery removes the affected tissues and organs. For this approach to be successful, however, the surgeon must know the exact locations of all cancerous tissues. Also, the tissue to be removed must obviously not be irreplaceable.

Radiotherapy—exposure to massive doses of X rays or gamma rays—is a possible treatment for cancerous tissues that cannot be replaced. For example, treating cancer of the larynx (voice box) by radiotherapy may leave the vocal cords intact, whereas surgery would remove them altogether, leaving the patient without normal speech. Tissues exposed to massive doses of radiation suffer extensive chromosomal breakage; some cells not scheduled to divide again remain alive after the chromosomes are broken, but dividing cells such as cancerous cells and the cells of the immune system die when they attempt to undergo mitosis. Because radiation is also harmful to normal cells, it must be restricted to very specific areas of the body; thus the radiologist, like the surgeon, must know the extent and location of the cancer. Another danger of radiotherapy is that it can seriously damage the immune system, leaving the patient defenseless against bacterial, viral, and other infections.

The third major cancer treatment is **chemotherapy**, treatment with drugs that have their greatest effect on rapidly dividing cells. These drugs generally act by interfering with the metabolism of the building blocks of DNA. In chemotherapy, precise knowledge of the locations of cancerous tissues is not necessary; however, normal cells—especially those of the immune system—are also affected. The side effects of chemotherapy are often severe.

By using one or more of these approaches, we are able to achieve many cures. Can we use molecular biological techniques to open new avenues of treatment? One new approach takes advantage of the principle of complementary base pairing. **Antisense nucleic acids** are single-stranded stretches of RNA or DNA targeted against the mRNAs transcribed from harmful genes such as oncogenes. An antisense nucleic acid is complementary at each nucleotide to its target mRNA and therefore forms hydrogen bonds that join the two molecules (Figure 15.13). In this "duplex" condition the mRNA is inactive; it cannot

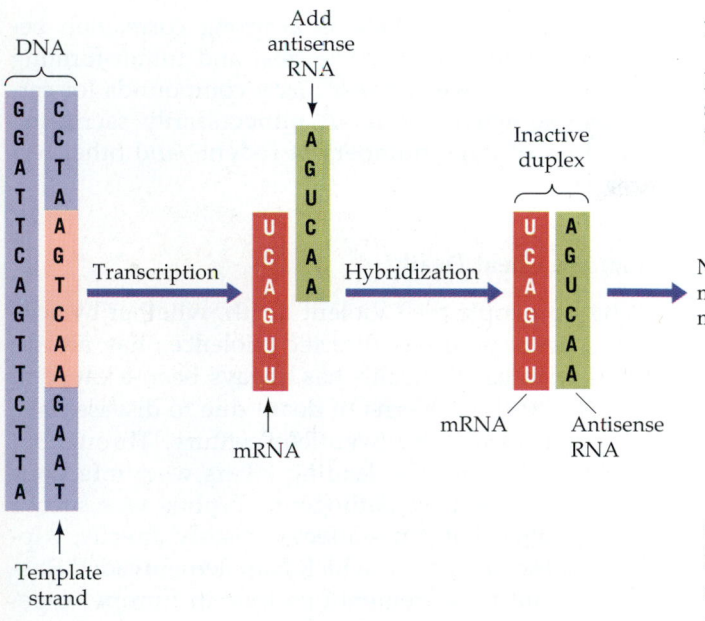

DNA

Template strand

Add antisense RNA

Transcription

mRNA

Hybridization

Inactive duplex

mRNA   Antisense RNA

Not translated; may be degraded or modified by enzymes

15.13 Antisense RNA Inactivates the Corresponding mRNA

In one type of cancer therapy, targeted single-stranded stretches of RNA are injected into cells, inactivating the transcription of oncogene mRNAs.

be translated by ribosomes, and it may be degraded or modified by enzymes.

Antisense RNA injected into a cell has succeeded in inactivating a previously injected mRNA, but wouldn't it be more effective to "teach" a cell how to make its own antisense RNA so that repeated injections would not be necessary? We could, for example, insert a gene encoding a particular antisense RNA into a suitable RNA virus and then use the virus to insert its RNA into the patient's cells. This approach has worked, both against RNAs that had themselves been inserted by viral vectors and against genes normally present in the cell. There is hope that antisense nucleic acids may eventually be used to combat viral infections, including those caused by oncogenic viruses. Another, longer-term possibility is the clinical use of antisense nucleic acids to inactivate cellular oncogenes and thus bring cancers under control. One of the most difficult aspects of such a therapy will be to design an antisense nucleic acid sufficiently specific to inactivate the oncogene without impairing the action of the normal allele—the proto-oncogene that is so important in the normal growth of a cell.

There is hope, too, that immunological techniques, such as vaccination against the few known oncogenic viruses, may reduce the incidence of cancer. What other preventive measures might we take to reduce the risk of developing cancer in the first place rather than having to fight a cancer after it is started?

Preventing Cancer

The single most effective way to minimize the risk of cancer is never to smoke—or, if you smoke, to stop immediately and forever. Avoiding second-hand smoke—smoke exhaled by smokers in your vicinity—is also highly advisable. The relationship between smoking and cancer has been established beyond question.

Other cancer-preventive measures are less firmly established. However, minimizing exposure to carcinogens—both chemical carcinogens and mutagenic forms of irradiation, such as ultraviolet radiation and X rays—makes obvious sense. Although a deep tan was once a seemingly important trophy of summer, we now know that there is no such thing as a "healthy" tan. Various commercially available "sunblocks" prevent ultraviolet radiation from reaching the body surface. Some of these compounds may themselves prove to be carcinogenic, however; therefore, the safest approach to sunlight is avoidance—wearing protective clothing and minimizing direct exposure, especially at midday. Avoidance is also the best approach to chemical carcinogens in the diet, in the home, and in the workplace. But how do we know which compounds are carcinogenic?

The Ames Test

The most widely employed test for chemical carcinogens is based on the observation that every known carcinogen either is itself a mutagen or is converted to a mutagen by the liver. In the **Ames test**, named for its inventor Bruce Ames, the suspected carcinogenic compound is combined with a suspension of ground-up liver cells and mutant bacterial cells that cannot grow in the absence of, say, a particular amino acid. We then look for the appearance of bacteria that can grow in the absence of that amino acid (Figure 15.14). Such bacteria must result from an alteration that reverses the original mutation; thus their presence indicates that the compound added to the suspension is carcinogenic.

In contrast to other tests, such as direct observation of a compound's ability to cause tumors when applied to mice or rats, the Ames test is simple, fast,

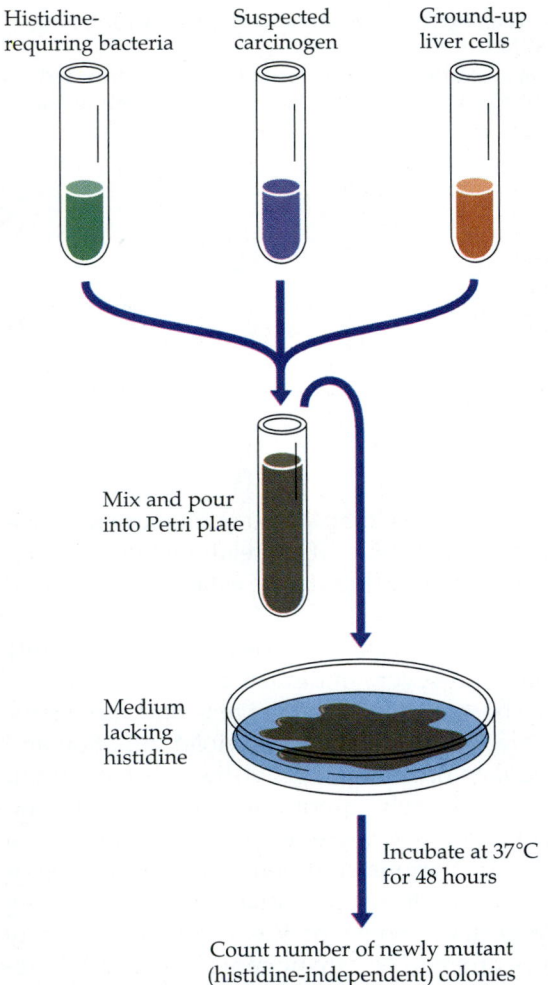

Histidine-requiring bacteria

Suspected carcinogen

Ground-up liver cells

Mix and pour into Petri plate

Medium lacking histidine

Incubate at 37°C for 48 hours

Count number of newly mutant (histidine-independent) colonies

15.14 The Ames Test
The Ames test for a suspected carcinogen's ability to cause mutations works because liver cells convert carcinogens to mutagens. In looking for "back mutations" in an already-mutant strain of bacteria, we recognize mutagenized cells by their ability to produce colonies on a medium on which the original strain could not grow.

and inexpensive. There is a strong correlation between activity in the Ames test and tumor-forming activity. Thus we can test many compounds for carcinogenic activity without unnecessarily sacrificing the lives of large numbers of rodents and other test animals.

Can We Cheat Death?

Why do people die? Violent death, whether by accident, war, or other directed violence, has always been with us. Ill health has always been a cause of death, but the patterns of death due to disease have changed during the twentieth century. Throughout recorded history, the leading killers were infectious diseases caused by pathogens. Typhus was such a potent killer that it has affected history directly, wiping out the army with which Napoleon invaded Russia. Plague took tremendous tolls in Europe, especially during the ninth and fourteenth centuries. In the twentieth century, advances in medicine and, even more important, in public health measures led to a startling reduction in infectious disease. By the 1960s the leading killers were *noninfectious* diseases, notably heart disease and cancer. Can we combat heart disease and cancer as effectively as we have dealt with infectious diseases? That remains to be seen. Success against noninfectious diseases, however, would still not mean that we have conquered death. There will always be a "number-one killer." Only its identity will change.

Consider a specific example: A new disease has killed almost two million people within the decade since its discovery. That disease is AIDS—acquired immunodeficiency syndrome. AIDS kills indirectly, by destroying the body's immune system—a major protection against other diseases. We will consider the immune system and AIDS in the next chapter.

SUMMARY of Main Ideas about Genetic Disease and Modern Medicine

Chromosomal aberrations cause one type of genetic ill health.
Review Figures 9.19 and 10.16

Defective alleles that are recessive cause many genetic diseases; dominant defective alleles cause only a few.

Failure to produce a particular protein is the defect responsible for many genetic diseases.

The number of triplet repeats in some disease-causing alleles is much greater than the number in the corresponding normal alleles.
Review Figure 15.2

Current efforts to map human genes will help in the diagnosis and treatment of genetic diseases.

Pedigree analysis reveals whether the inheritance pattern of a disease is dominant or recessive and whether the defective allele is carried on an autosome or on a sex chromosome.
Review Figures 15.3 and 15.4

To determine which chromosome carries a certain gene, researchers use somatic-cell hybridization.
Review Figure 15.5

Mutations in recognition sites for restriction endonucleases change the band pattern obtained by gel

electrophoresis of restriction digests, giving rise to restriction fragment length polymorphisms (RFLPs); RFLPs are convenient guideposts for locating genes on chromosomes.
   Review Figures 15.6 and 15.7

After a gene has been isolated, its functions can be studied in appropriately designed knockout mice.

Gene therapy is one way to deal with a defective gene, but it is unlikely to be used on a large scale.
   Review Figure 15.8

Although most cancers are not inherited, cancer is a genetic disease because it appears when a cell's genetic control of its own division fails.

Cancer metastasizes in the bloodstream or lymphatic system.

Cancers begin when the DNA of genes required for normal growth changes, transforming normal cells into cancer cells.

Transformation to the cancerous state requires multiple mutations that may include the conversion of proto-oncogenes to oncogenes and the inactivation of tumor-suppressor genes.
   Review Figure 15.11

About half of all cancers contain cells with mutations in the tumor-suppressor gene *p53*.

The principal approaches to treating cancer are surgery, radiotherapy, and chemotherapy.

The most effective way to prevent cancer is to avoid exposure to carcinogens.

SELF-QUIZ

1. Major symptoms of cystic fibrosis are caused by
   a. fibrous materials that collect in cysts.
   b. cysts that clog up the fibrosis organ.
   c. defective transport of calcium across the plasma membrane.
   d. defective transport of chloride across the plasma membrane.
   e. defective transport of sodium across the plasma membrane.

2. Phenylketonuria can be treated by
   a. dietary restrictions.
   b. tissue transplantation.
   c. depletion of an excessive substance.
   d. surgical repair of tissue.
   e. replacment of a missing gene product.

3. Fragile-X syndrome is an X-linked disorder
   a. that occurs with the same frequency in males and females.
   b. that occurs much more frequently in females than in males.
   c. that is extremely rare in all populations.
   d. whose expression within a family varies between generations.
   e. with a uniform phenotype.

4. The type of inheritance in which most affected persons (about ½ males and ½ females) have normal parents (and sometimes the parents are related to each other) is

   a. Y-linked.
   b. X-linked dominant.
   c. X-linked recessive.
   d. autosomal dominant.
   e. autosomal recessive.

5. The assignment of human autosomal genes to specific chromosomes
   a. was initially less difficult than assigning X-linked genes.
   b. was initially just as difficult as assigning X-linked genes.
   c. was initially more difficult than assigning X-linked genes.
   d. has been possible since 1900.
   e. will probably be entirely completed by 1999.

6. Researchers in somatic cell genetics fuse cells from different species for the purpose of
   a. amplifying certain genes by the polymerase chain reaction.
   b. producing transgenic animals.
   c. assigning genes to specific chromosomes.
   d. gene cloning.
   e. alleviating genetic deficiencies.

7. Restriction fragment length polymorphisms (RFLPs) are variations in the lengths of
   a. proteins produced by restrictions of mutation.
   b. time needed to produce certain types of mutations.
   c. restriction enzymes, due to new mutations.
   d. RNA segments produced by a restriction enzyme.

   e. DNA segments produced by a restriction enzyme.

8. Knockout mice
   a. readily contract communicable diseases.
   b. are exceptionally healthy.
   c. do not contract the genetic diseases humans do.
   d. have had defective alleles inserted in place of functional alleles.
   e. have not yet found uses in biomedical research.

9. Gene therapy
   a. has not yet been tested on human patients.
   b. can include isolating, modifying, and returning a patient's own cells.
   c. is likely to be widely used—and soon—in treating genetic disorders.
   d. could not be used to cure a genetic disease.
   e. is likely to focus on the germ cells rather than somatic cells.

10. A proto-oncogene is
   a. a transcription factor produced by an oncogene.
   b. an abnormal gene that prevents cells from dividing.
   c. a normal gene that regulates cell division.
   d. one of the few genes normally found in red blood cells.
   e. one of the few genes normally found in a retrovirus.

FOR STUDY

1. Genetic defects may be produced by either dominant or recessive alleles. Of the known autosomal dominant disorders, those whose symptoms appear at an early age are relatively benign. Why should this be so?

2. It seems possible that genes associated with disorders may often contain regions of triple repeats. Suggest how you might make use of this idea to search for new dise. DNA segments produced by a restriction enzyme.

3. Mouse–human cell hybridization gave rise to five hybrid colonies. The table at the right shows the human genes and human chromosomes that were present (plus sign) in each colony. From these data, what can you say as to which genes are located on which chromosomes?

4. Many cancers are diseases of old age. Account for this on the basis of what you have learned about the transformation of a healthy cells to the cancerous state.

5. If someone has a genetic disease, who has the right to know? The person's spouse or fiance? The person's parents or children? The person's employer? The company that insures the person? List arguments *for* and *against* informing each of these people or groups.

| | | HYBRID CELL LINES | | | | |
		A	B	C	D	E
Human genes	1	+	−	−	+	−
	2	−	+	−	+	−
	3	+	−	−	+	−
	4	+	+	+	−	−
Human chromosomes	1	−	+	−	+	−
	2	+	−	−	+	+
	3	−	−	−	+	+

From Griffiths et al., *An Introduction to Genetic Analysis*, 5th Edition, p. 168.

READINGS

Capecchi, M. R. 1994. "Targeted gene replacement." *Scientific American*, March. The story of "knockout mice."

Croce, C. M. and G. Klein. 1985. "Chromosome Translocations and Human Cancer." *Scientific American*, March. A clear treatment of oncogenes and their activation.

Culotta, E. and D. E. Koshland, Jr. 1993. "Molecule of the Year: p53 Sweeps Through Cancer Research." *Science*, vol. 262, pages 1958–1961. A brief look at the best-studied tumor-suppressor gene.

Golde, D. W. 1991. "The Stem Cell." *Scientific American*, December. The cell that gives rise to blood cells and immune systems. Applications in treatment of cancer and immune defects.

Griffiths, A. J. F., J. H. Miller, D. T. Suzuki, R. C. Lewontin and W. M. Gelbart. 1993. *An Introduction to Genetic Analysis*, 5th Edition. W. H. Freeman, New York. Chapters 6 and 15 are particularly relevant; Chapter 22 is also useful.

Hunter, T. 1984. "The Proteins of Oncogenes." *Scientific American*, August. A detailed treatment of several specific oncogenes and their products.

Lawn, R. M. and G. A. Vehar. 1986. "The Molecular Genetics of Hemophilia." *Scientific American*, March. The use of recombinant DNA technology and bacteria to produce a blood-clotting protein that may save the lives of hemophiliacs.

Liotta, L. A. 1992. "Cancer Cell Invasion and Metastasis." *Scientific American*, February. How cancer spreads in the body, and prospects for treatment.

Mange, E. J. and A. P. Mange. 1993. *Basic Human Genetics*. Sinauer Associates, Sunderland, MA. Genetics, especially chromosomal inheritance, can be studied using humans as examples; this book does so at an introductory level.

Miller, J. A. 1990. "Genes That Protect against Cancer." *BioScience*, vol. 40, pages 563–566. The roles of anti-oncogenes, also known as tumor-suppressor genes.

Patterson, D. 1987. "The Causes of Down Syndrome." *Scientific American*, August. Identification and mapping of genes responsible for the most common cause of mental retardation.

Rennie, John. 1994. "Grading the Gene Tests." *Scientific American*, June. A good start for a discussion of ethical issues in screening for hereditary diseases.

Verma, I. M. 1990. "Gene Therapy." *Scientific American*, November. A discussion of how healthy alleles are introduced to correct heritable disorders.

Weinberg, R. A. 1988. "Finding the Anti-Oncogene." *Scientific American*, September. Isolation of a gene that prevents a cell from proliferating out of control.

Weintraub, H. M. 1990. "Antisense RNA and DNA." *Scientific American*, January. Antisense nucleic acids as tools for basic research and as tools against viruses and, maybe, cancer.

White, R. and J.-M. Lalouel. 1988. "Chromosome Mapping with DNA Markers." *Scientific American*, February. A description of the powerful tool known as restriction-fragment length polymorphism.

By the year 2000 as many as 40 million people, worldwide, will have become infected with HIV, the human immunodeficiency virus, which causes AIDS (*a*cquired *i*mmune *d*eficiency *s*yndrome). That number about equals the combined total populations of the states of New York, Michigan, and Pennsylvania. The number is an estimate by the World Health Organization, made in mid-1993, by which time over 13 million people had already become infected. Of these 13 million, over 2 million had developed AIDS, and of those 2 million, most had died. Who are these millions of people? They are women, men, children, and new-born infants. About half of those infected with HIV contracted the infection between the ages of 15 and 24. Most of them became infected through heterosexual intercourse.

AIDS does not kill people directly. The death of an AIDS patient is usually the result of infection by a disease-producing organism that is normally present in all of us, but that the AIDS victim cannot tolerate. Why are AIDS patients at extreme risk from infections that rarely trouble most other people? You may have read or heard that the development of AIDS is strongly associated with a reduction in numbers of certain types of blood cells. What does this have to do with the disastrous loss of ability to deal with infection?

In this chapter we focus on the mechanisms by which organisms combat infection and disease. The environment is alive with microorganisms capable of producing disease in other organisms. All organisms have defenses that help them cope with microorganisms and foreign macromolecules. After considering what these mechanisms are and how they work, we conclude with an examination of how they fail in persons infected with AIDS.

NONSPECIFIC DEFENSES AGAINST PATHOGENS

Consider the challenges faced by a potential pathogen (a disease-causing organism such as a bacterium, virus, fungus, protist, or animal parasite) as it approaches the body of an animal. There are hurdles that the pathogen must overcome: It must arrive at the body surface of a potential host, enter the host, multiply inside the host, and finally, prepare to infect the next host. Failure to overcome any one of these hurdles ends the reproductive career of a pathogenic organism. Animals have defenses that stop many different pathogens from invading their bodies. Because these initial defenses give general protection against different pathogens, they are called **nonspecific defenses**.

An Artist's Response
The work of American artist Keith Haring is dedicated to the victims of AIDS.

16

Defenses against Disease

Pathogens arrive at a potential host by way of airborne droplets (as from a sneeze), food or drink, an animal that bites, direct contact with an infected individual, or contact with some pathogen-carrying object in the environment. Many of the most massive improvements in public health have come with the control of sewage or with campaigns against insects, ticks, and other animals that carry pathogens; these measures prevent pathogens from reaching us. "Entering the host" means different things to different pathogens. Some pathogens simply multiply on the surface of a mucous membrane that lines, for example, the throat or intestine of the host. Other pathogens penetrate and multiply within these surface cells, and still others pass into deeper tissues of the body or into the bloodstream.

Skin is a primary nonspecific defense against invasion. Bacteria and viruses rarely penetrate healthy, unbroken skin. Damaged skin or other surface tissue, however, is another matter. The sensitivity of surface tissue accounts in part for the greatly increased risk of infection by HIV in a person who already has a sexually transmitted disease. It may also partially explain why a woman can become infected by HIV during heterosexual intercourse much more easily than a man can, since the vaginal lining is more frequently damaged by intercourse than is the skin of the penis.

In addition to skin, we have other defenses against invasion by pathogens. One of these is our **normal flora**: the bacteria and fungi that live and reproduce in great numbers on our surfaces without causing disease (Figure 16.1). These natural occupants compete with pathogens for locations and nutrients, and some of them produce inhibitor compounds that are toxic to potential pathogens.

The one type of healthy surface that *is* penetrable by bacteria is the mucous membrane, a type of mucus-secreting tissue found in parts of the visual, respiratory, digestive, and urogenital systems. However, these areas of the body have other defense mechanisms to discourage penetration by pathogens. Secretions such as tears, nasal drips, and saliva possess an enzyme called **lysozyme** that attacks the cell walls of many bacteria. Mucus in our noses traps most of the microorganisms in the air we breathe, and most of those that get past this filter end up trapped in mucus deeper in the respiratory tract. They are removed from the respiratory tract by the beating of cilia in the respiratory passageway, which moves a sheet of mucus and the debris it contains up toward the nose and mouth. Another effective means of removing microorganisms from the respiratory tract is the sneeze.

Pathogens that travel as far into a person as the digestive tract (stomach, small intestine, and large intestine) are met by other defenses. The stomach is

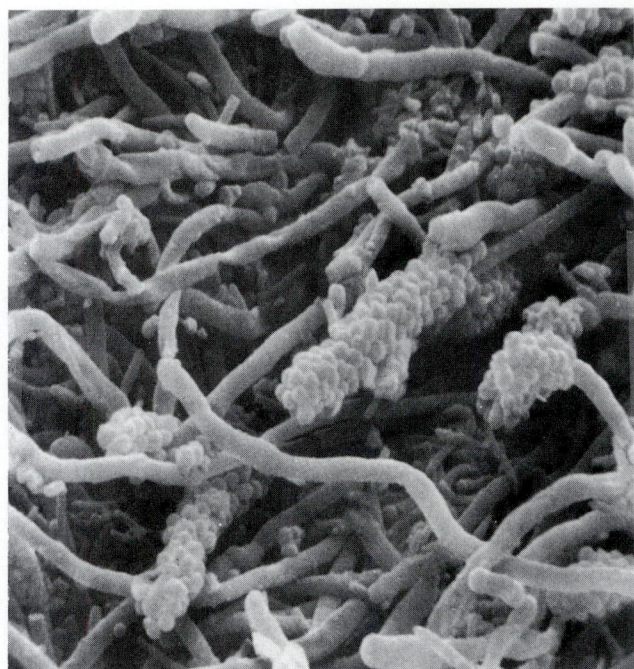

16.1 Normal Flora Gone Rampant
The human mouth harbors a wide variety of microorganisms, most of which cause no damage under normal conditions. When bacteria accumulate on the surfaces of teeth, the result is called plaque. The presence of plaque contributes to tooth decay. This electron micrograph shows plaque on a tooth three days after the person stopped brushing.

an unfriendly environment for most bacteria because of the hydrochloric acid that is secreted into it. The lining of the small intestine cannot be penetrated by bacteria, and some pathogens are killed by bile salts secreted into this part of the tract. The large intestine harbors many bacteria, which multiply freely; however, these are usually removed quickly with the feces. (The digestive system is described fully in Chapter 43.)

If Pathogens Evade the Blocks to Entry

Pathogens that manage to penetrate the surface cells of an animal's body encounter still other defenses. These defenses fall into two categories: nonspecific and specific (the specific defenses are the immune system). One of the nonspecific mechanisms is a "battle" for iron that takes place between some pathogens and the host. Both require iron for their metabolism, but they utilize it in different forms. Pathogen and host produce different substances to trap the iron, and competition may be intense. If iron is in limited supply, the pathogen seldom wins the battle.

The nonspecific defenses of the animal host also include antimicrobial proteins in its tissues and body fluids. Two important types of antimicrobial proteins are complement proteins and interferons, which we will discuss in more detail later in the chapter.

In your body, and in the bodies of all other vertebrates (animals with backbones), blood circulates through a system of vessels, pumped by a heart. Blood contains a mixture of different types of cells. As we will see, the circulating behavior and other activities of certain types of blood cells give vertebrate animals powerful defenses against pathogens. (The circulatory system is described fully in Chapter 42.)

White blood cells called **phagocytes** provide an extremely important nonspecific defense against pathogens that penetrate the surface of the host. Some phagocytes adhere to certain tissues; others travel freely in the circulatory system. Pathogens become attached to the membrane of a phagocyte (Figure 16.2). The phagocyte ingests the pathogens by endocytosis. Once inside an endocytic vesicle in a phagocyte, pathogens are destroyed by enzymes from lysosomes that fuse with the vesicle (see Figure 4.22). A single phagocyte can ingest 5 to 25 bacteria before it dies from the accumulation of toxic breakdown products. Even when phagocytes do not destroy all the invaders, they usually reduce the number of pathogens to the point where other defenses can finish the job. So important is the role of the phagocytes that if their functioning is impaired by disease, the animal usually soon dies of infection.

A class of small white blood cells, known as **natural killer cells**, can initiate the lysis of some tumor cells and some normal cells that are infected by a virus. It is possible that the natural killer cells seek out cancer cells that appear in the body.

Another important nonspecific defense is **inflammation**. The body employs this characteristic, highly generalized response in dealing with infections, mechanical injuries, and burns. The damaged cells themselves cause the inflammation by releasing various substances. You have experienced the symptoms of inflammation: redness and swelling, with heat and pain. The redness and heat result from dilation of blood vessels in the infected or injured area. The blood capillaries (the smallest vessels) become leaky, allowing some blood plasma (described later in this chapter) and phagocytes to escape into the tissue, causing swelling. The pain results from increased pressure (from the leakage) and from the action of some leaked enzymes. Certain of the plasma proteins and the phagocytes are responsible for most of the healing aspects of inflammation. The heat may also play a healing role if it raises the temperature beyond that at which the pathogen that triggered the inflammation can multiply effectively.

Viral Diseases and Interferon

When you have a viral disease, such as influenza, you are unlikely to develop another viral disease at the same time. An apparent explanation for this phenomenon was provided in 1957 by Alick Isaacs and Jean Lindemann of the National Institute for Medical Research in London. They found that inoculating the cells of a developing chick with influenza virus causes the cells to produce small amounts of an antimicrobial protein called **interferon** that increases the resistance of neighboring cells to infection by influenza or *other* viruses. Interferons have been found in many vertebrates and are one of the body's lines of nonspecific defense against viral infection.

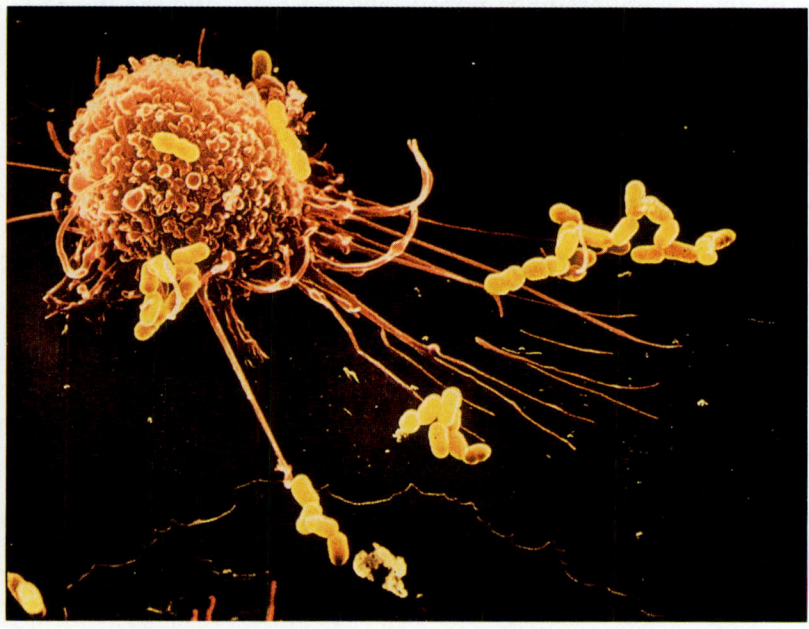

16.2 A Phagocyte and Its Bacterial Prey Some bacteria (appearing yellow in this artificially colored scanning electron micrograph) have become attached to the surface of a phagocyte in the human bloodstream. Many of these bacteria will be taken into the phagocyte and destroyed before they can multiply and damage the human host. The long protuberances of the phagocyte probably help it move and adhere to other cells.

Interferons differ from species to species, and each vertebrate species produces at least three different interferons. All interferons are glycoproteins (proteins with a carbohydrate component) consisting of about 160 amino acid units. By binding to receptors in the membranes of cells, interferons inhibit the ability of the viruses to replicate. Interferons have been the subject of intensive research because of their possible applications in medicine—for example, the treatment of influenza and the common cold.

Nonspecific Defenses of Plants

Plants also have a variety of mechanisms, both mechanical and chemical, by which they resist or even actively oppose infection by pathogens. The outer surfaces of plants are protected by tissues such as the epidermis or cork. If pathogens get past these barriers, the differences between the defense systems of plants and animals become apparent. Animals generally *repair* tissues that have been infected—they heal, through appropriate developmental pathways. Plants, on the other hand, do not make repairs. Instead, they develop in ways that seal off the damaged tissue so that the rest of the plant does not become infected. Trees seal off damaged tissue by producing new wood that differs in orientation and chemical composition from the previously deposited wood. Some of the new cells also contain substances that resist the growth of microorganisms and hence tend to protect the rest of the plant.

The healing mechanism just described is primarily mechanical. Many plants have chemical defenses as well. For example, infection of one of these plants by certain fungi stimulates the plant to produce substances that are toxic to the fungi. Some of these toxic substances can act against some bacteria as well. Their antifungal activity is nonspecific—that is, the substances can destroy many species of fungi in addition to the one that originally triggered their production. Physical injuries, viral infections, and even certain chemical compounds can also induce the production of these substances.

SPECIFIC DEFENSES: THE IMMUNE SYSTEM

Our nonspecific defenses are numerous and effective, but some invaders nevertheless elude the nonspecific defenses and must be dealt with by defenses targeted against specific threats. The destruction of specific pathogens is an important function of an animal's **immune system**. The immune system recognizes and attacks specific invaders, such as bacteria and viruses. After responding to a particular type of pathogen once, the immune system can usually respond more rapidly and powerfully to the same

threat in the future. Thus the functions of an immune system are to *recognize*, *selectively eliminate*, and *remember* foreign invaders.

An animal with a defective immune system can die from infection by even "harmless" bacteria. Some microorganisms routinely carried in or on an animal's body without harm are potentially pathogenic and will cause disease if the host's immune system is stressed in some way.

The immune system is made up of cells that travel in the body's fluids. Blood is a fluid tissue. About 55 to 65 percent of a human's blood is the yellowish liquid matrix called **plasma**; the remainder consists of red blood cells, white blood cells, and platelets. Plasma is mostly water, but contains many other important substances.

As we will see in Chapter 42, some of the circulating components of blood are returned to the heart not by veins but by the lymphatic system (Figure 16.3). **Lymph**, a blood filtrate that accumulates in the spaces outside the blood capillaries, contains water, solutes, and white blood cells that have left the capillaries, but no red blood cells. The lymph is collected in vessels called lymph ducts and routed back into blood vessels near the heart.

White blood cells are larger and far less numerous in the blood than red blood cells (Figure 16.4). White blood cells have nuclei and are colorless. Like the cancer cell in Figure 15.10, they can move through tissues by squeezing through junctions between the cells that make up the walls of blood capillaries. In response to invading pathogens, the number of white blood cells in the blood and lymph may rise sharply, providing medical professionals with a useful clue for detecting an infection.

The most abundant types of white blood cells are the phagocytes (which, as you already know, are important as nonspecific defenses) and the **lymphocytes**. Two groups of lymphocytes, the **B cells** and **T cells**, together with specialized cells that arise from them, are the important cells of the immune system. Both B cells and T cells originate from cells in the bone marrow. The precursors of T cells migrate to an organ called the thymus and develop their unique properties there, becoming mature T cells. The B cells migrate from the bone marrow to the outer regions of the body, circulate in the blood and lymph vessels, and pass through the lymph nodes and spleen. B and T cells look the same under the light microscope, but they have quite different functions in immune responses.

RESPONSES OF THE IMMUNE SYSTEM

Foreign organisms and substances that invade the animal body and escape the internal, nonspecific de-

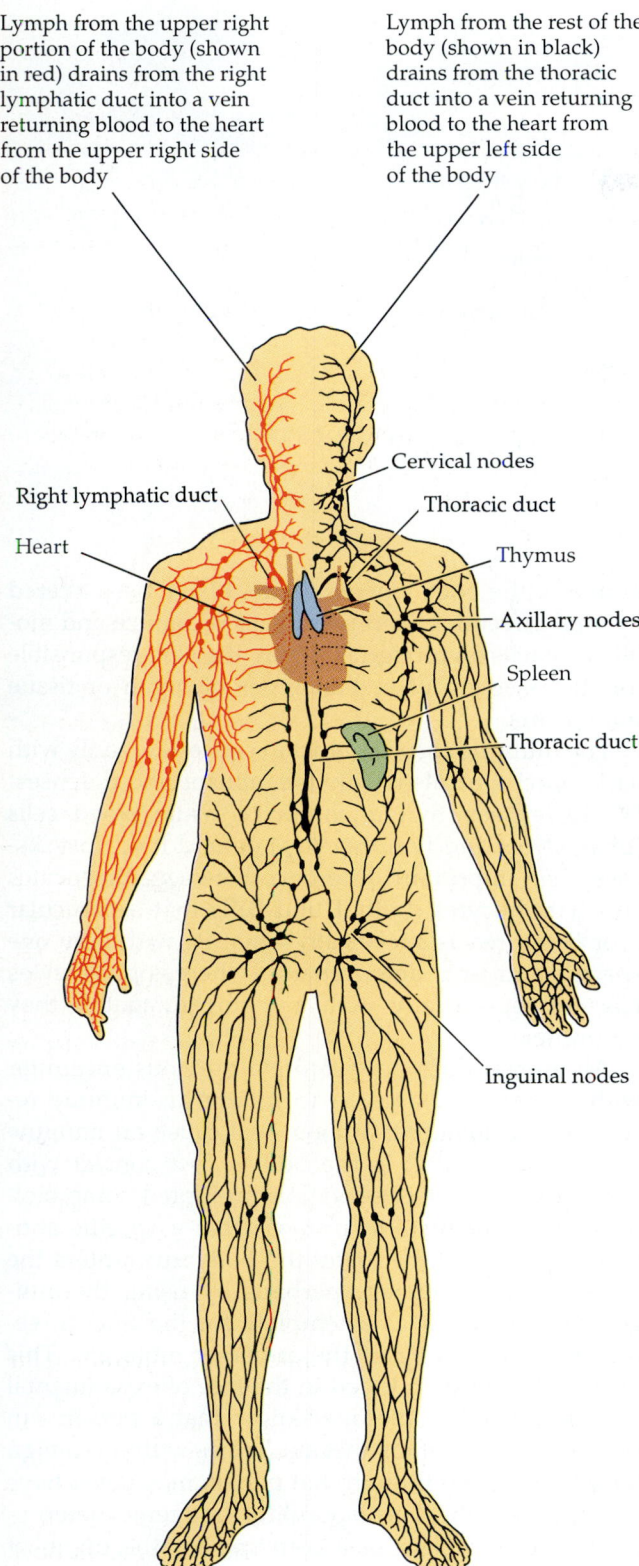

Lymph from the upper right portion of the body (shown in red) drains from the right lymphatic duct into a vein returning blood to the heart from the upper right side of the body

Lymph from the rest of the body (shown in black) drains from the thoracic duct into a vein returning blood to the heart from the upper left side of the body

Right lymphatic duct

Heart

Cervical nodes

Thoracic duct

Thymus

Axillary nodes

Spleen

Thoracic duct

Inguinal nodes

16.3   The Human Lymphatic System
A network of ducts collects lymph from the body's tissues and carries it toward the heart, where it mixes with blood to be pumped back to the tissues. There are major lymph nodes in the neck, armpits, and groin. What we call "swollen glands" in the neck are cervical lymph nodes in which invading bacteria are trapped.

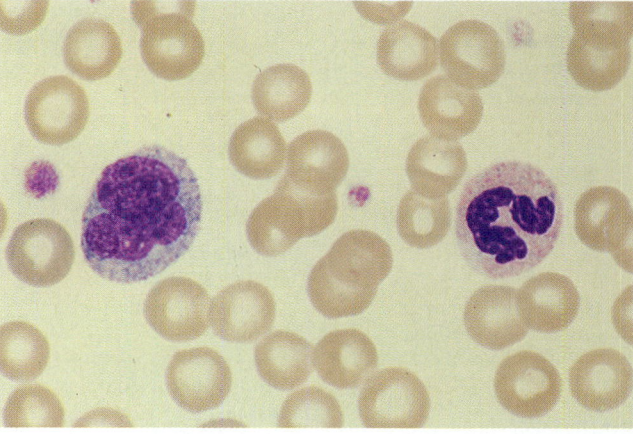

16.4   Blood Cells
The large, stained cell to the left is a white blood cell called a monocyte. The other stained cell is a white blood cell called a neutrophil. Notice that the numerous red blood cells in this blood smear are unstained because red blood cells do not have nuclei.

fenses come up against the immune system (Table 16.1). The immune system has two responses against invaders. Protein molecules produced by the immune system attack bacteria and viruses that are in fluids of the animal body but outside the cells. This is the **humoral immune response** (from the Latin *humor*, "fluid"). Certain cells of the immune system mount an attack called the **cellular immune response** against viruses that have become established within cells of the animal body and against fungi and microscopic animals.

The two responses operate in concert—simultaneously, cooperatively, and sharing mechanisms. Both responses aim at the same specific areas on the surfaces of the foreign bodies that invade an animal. We call the foreign bodies (organisms or molecules) **antigens**; the specific sites on antigens that the immune system attacks are **antigenic determinants**.

Highly specific protein molecules called **antibodies** carry out the humoral immune response against invaders in the fluids. An animal produces a vast diversity of antibodies that, among them, can react with virtually any conceivable antigen in the bloodstream or lymph. Some antibodies travel free in the blood and lymph; others exist as integral membrane proteins on B cells. Where do antibodies come from? Some of an animal's B cells differentiate to become **plasma cells**. Plasma cells then produce antibodies.

Each antibody recognizes and binds to a particular site—an antigenic determinant—on one or more antigens that invade an animal's body (Figure 16.5). An antigenic determinant is a specific chemical grouping that may be present on many different molecules. A large antigen such as a whole cell may have many

TABLE 16.1
Animal defenses against disease

NONSPECIFIC DEFENSES		SPECIFIC DEFENSES (IMMUNE SYSTEM)
BLOCK INVADERS	ATTACK ALL SUCCESSFUL INVADERS	ATTACK SPECIFIC SUCCESSFUL INVADERS
Skin	Competition for iron	Humoral immune
Normal flora	Antimicrobial proteins	response
Mucous membranes	(complement and	Cellular immune response
Protective secretions	interferons)	
	Phagocytes	
	Natural killer (NK) cells	
	Inflammation	

different antigenic determinants on its surface, each capable of binding a specific antibody.

The cellular immune response is directed against a multicellular antigen or an antigen that has become established within a cell of the animal. The animal cell is able to display the invader's antigenic determinants on its surface; we will explain how this happens later in the chapter. The cellular immune response, in contrast to the humoral response, does not use antibodies. Instead, it is carried out by T cells that roam through the bloodstream and lymph. The T cells have **T-cell receptors**: surface glycoproteins that recognize and bind to antigenic determinants.

Like antibodies, T-cell receptors have specific molecular configurations that bind to specific antigenic determinants. Once bound to a determinant, each particular type of T cell initiates characteristic activity. Some T cells recognize and help destroy foreign cells or any of the body's own cells that have been altered by viral infections. Because T cells recognize and mobilize attacks on foreign material, they are responsible for the rejection of certain types of organ or tissue transplants.

The immune responses act in concert not only with each other but also with the nonspecific defenses. We have seen that an animal's white blood cells (phagocytes and lymphocytes) defend it against disease. The important difference between phagocytes and lymphocytes (both T and B) is that a particular T or B cell reacts specifically—that is, with only one specific antigenic determinant—whereas phagocytes react nonspecifically with any foreign matter they encounter.

An animal does not require a previous encounter with a particular antigen to mount an immune response. An invading antigen will cause an immune response even if it is the body's first contact with that antigen. This observation prompted some biologists and chemists to propose that a specific antibody must be formed *after* the body encounters the corresponding antigen, perhaps by using the antigenic determinant as a template for the final three-dimensional folding of the antibody molecule. This idea had to be abandoned in the face of experimental results, however. We now know that a person can produce *millions* of distinct antibodies without foreign templates—even though that person may never have encountered the corresponding antigenic determinants. The problem of accounting for the origin of

16.5 Each Antibody Matches an Antigenic Determinant
Three antigens are depicted in tan: a virus (*a*) and two distinct globular proteins (*b* and *c*). Each has on its surface antigenic determinants that are recognized by specific antibodies. An antibody recognizes and binds to its antigenic determinant wherever it is; for example, the antibody depicted in purple locates its unique determinant both on the virus (*a*) and on protein (*b*).

such a tremendous diversity of specific proteins will be considered later in this chapter.

Immunological Memory and Immunization

The effectiveness of the immune system is enhanced by **immunological memory**. The first time a vertebrate animal is exposed to a particular antigen, there is a time lag (usually several days) before the number of antibody molecules and the number of T cells circulating in the bloodstream catch up with the number of invaders. But for years afterward, sometimes for life, the immune system "remembers" that particular antigen and remains capable of responding more quickly than it did on the first encounter. Whereas the first exposure to the antigen results in some response, a second exposure causes a much greater, longer-sustained, and more rapid production of antibodies and T cells (Figure 16.6).

The ability of the human body to remember a specific antigen explains why **immunization** has virtually wiped out such deadly diseases as smallpox, diphtheria, and polio in medically sophisticated countries. (Smallpox, in fact, has been eliminated worldwide from the spectrum of infectious diseases affecting humans, thanks to a concentrated international effort by the World Health Organization. As far as we know, the only remaining smallpox viruses on Earth are those kept in some laboratories.) **Vaccination** means injecting a small amount of virus or bacteria or their proteins (usually treated to make them harmless) into the body. Later, if the same or very similar disease organisms attack, the body's cells are prepared because they have already been exposed to the antigen. They recognize the antigen and quickly overwhelm the invaders with a massive production of lymphocytes and antibodies.

Clonal Selection and Its Consequences

Each person possesses an enormous number of different B cells and T cells, apparently capable of dealing with practically any antigen ever likely to be encountered. How does this diversity arise? You may also wonder why some of our antibodies and T cells do not attack and destroy the components of our own bodies. An individual can mount an immune response against another person's proteins, yet it rarely mounts one against its own. The immune system can distinguish **self** (one's own antigens) from **nonself** (those from outside the body). The versatility of immune responses, immunological memory, and the recognition of self can all be explained satisfactorily by a particular theory of the origin of specific antibodies.

In 1954 the Danish immunologist Niels K. Jerne proposed a new view of the relationship between antigen and antibody. His idea was that the antigen does not itself specify the structures of the antibodies formed against it; instead, in Jerne's view, those an-

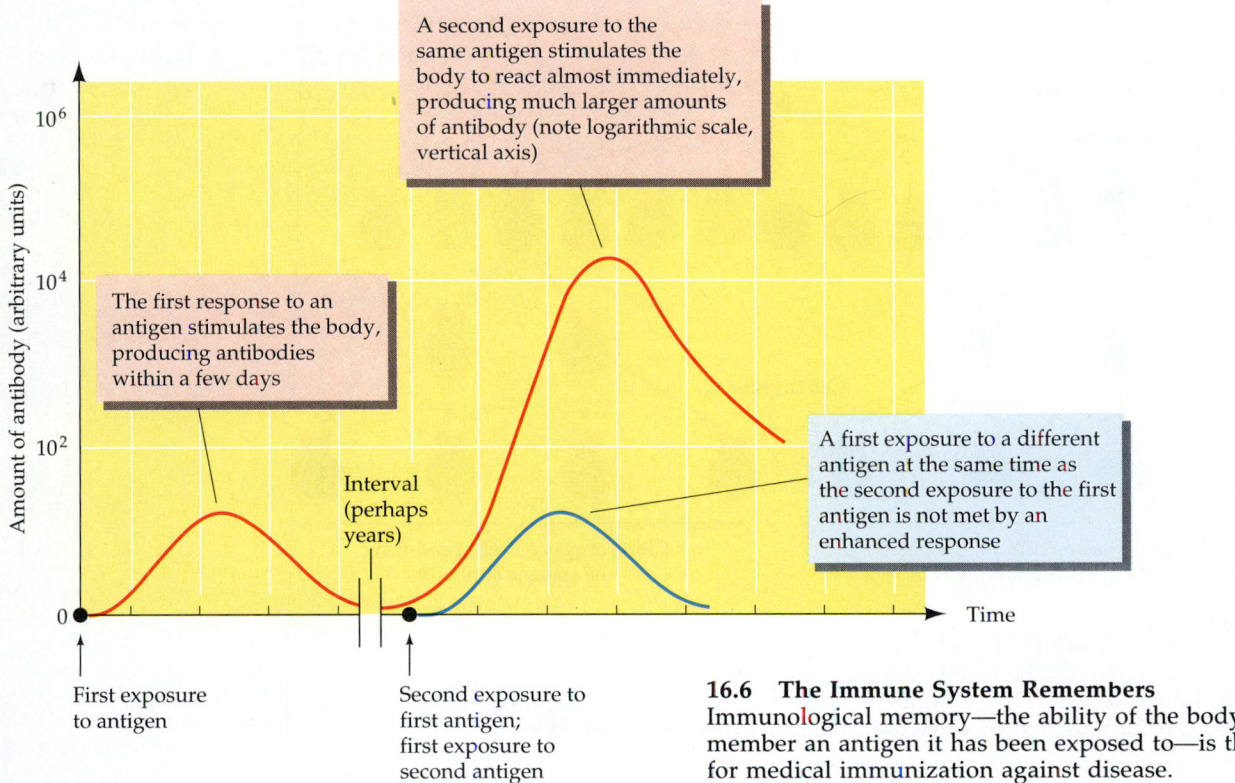

16.6 The Immune System Remembers
Immunological memory—the ability of the body to remember an antigen it has been exposed to—is the basis for medical immunization against disease.

tibodies are already being produced in small quantity, and the antigen specifically stimulates the cells that are making those particular antibodies to increase production. (Recall that antibodies are produced by plasma cells, which are differentiated B cells). Jerne hypothesized that there must be a population of different B cells corresponding to each of the antigenic determinants to which the organism can respond. Jerne's model was improved and extended by the Australian immunologist MacFarlane Burnet, who named it the **clonal selection theory**. According to the theory, the individual animal contains an enormous variety of different B cells, each type able to produce only one kind of antibody. Recent molecular work has shown that the DNA sequences encoding antibodies are arranged in slightly different ways in different B cells, so different B cells in a single animal have slightly different genotypes. After an antigen enters the body, it encounters B cells that can recognize its antigenic determinants. As a consequence of this meeting, each of these B cells begins to multiply, giving rise to a large clone of plasma cells, each of which secretes antibody of the same specificity (Figure 16.7).

The clonal selection theory accounts nicely for the body's ability to respond rapidly to any of a vast number of different antigens. In the extreme case, even a single B cell might be sufficient for an immunological response by the body, provided it encounters the antigen and then proliferates into a large enough clone rapidly enough to deal with the invasion. Clonal selection accounts for the proliferation of both B and T cells.

The clonal selection theory also explains two other phenomena: recognition of self (discussed in the next section), and immunological memory, evidence of which was illustrated in Figure 16.6. According to the clonal selection theory, an activated lymphocyte produces two types of daughter cells. The ones that carry out the attack on the antigen are **effector cells**—either plasma cells that produce antibodies, or T cells that bind antigenic determinants. The others, called **memory cells**, are long-lived cells that retain the ability to start dividing on short notice to produce more effector and more memory cells. Effector cells live only a few days, but memory cells may survive for decades. When the body first encounters a particular antigen, one or more types of lymphocytes become activated and divide to produce clones of effector and memory cells. (Why more than one type? Because the antigen may possess more than one antigenic determinant.) The effector cells destroy the invaders at hand and then die, but one or more clones of different memory cells have now been added to the immune system. Thus if the animal encounters the same antigen a second time, it can respond more rapidly and more massively, as Figure 16.6 shows.

Self, Nonself, and Tolerance

In addition to explaining immunological memory, the clonal selection theory may explain the recognition of self. Given the great array of different lymphocytes directed against particular antigens, how is it that a healthy animal apparently does not produce self-destructive immune responses? There appear to be

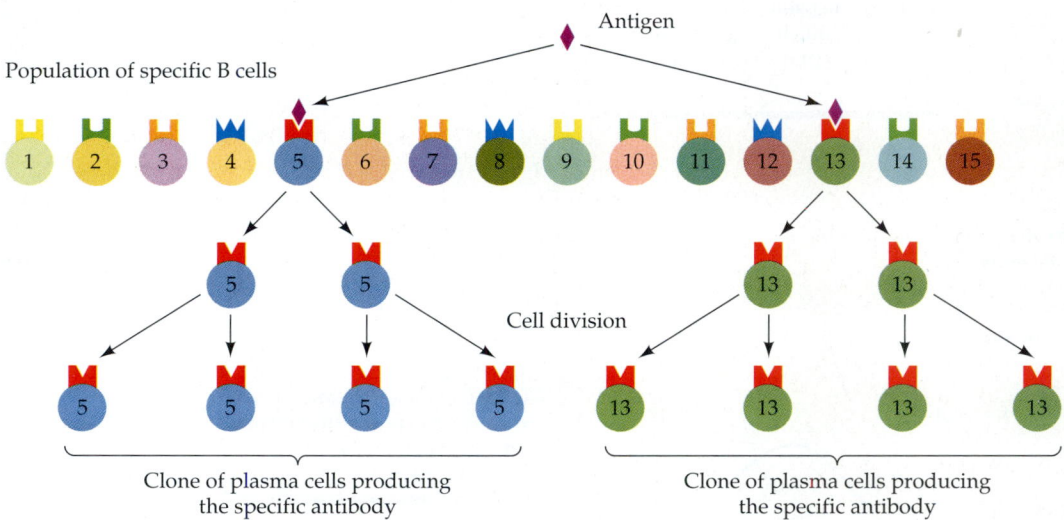

16.7 Clonal Selection
An animal produces many kinds of B cells, each genetically unique and specific for a particular antigen. Of the B cells shown here, only two—5 and 13 (red)— recognize the antigen. The antigen stimulates these B cells to proliferate, forming clones of plasma cells that produce the particular antibody specified by the ancestral cell's genotype.

two mechanisms of self-tolerance: clonal deletion and suppressor T cells.

Lymphocytes that have not yet fully differentiated are not capable of attacking antigens. When "anti-self" lymphocytes in this undifferentiated state encounter corresponding self antigens, the antiself lymphocytes are either inactivated or eliminated; thus, no clones of antiself lymphocytes normally appear in the bloodstream. This phenomenon is referred to as **clonal deletion**.

The second mechanism that may account for self-tolerance depends upon a class of lymphocytes known as **suppressor T cells**. These antigen-specific cells inhibit the activities of effector T and B cells. At least some aspects of self-tolerance probably are the result of the inactivation of anti-self lymphocytes by specific suppressor T cells.

In 1945 Ray D. Owen at the California Institute of Technology observed that some *nonidentical* twin cattle contained some of each other's red blood cells, even though their blood cells were of differing types and might have been expected to cause immune responses resulting in their elimination. Four years later Burnet suggested that the blood cells had passed between the animals before the differentiation of immune specificities was complete and thus were regarded as "self" when the recognition of self developed. Burnet further proposed that, if this were true, one should be able to inject foreign antigen into an animal early in its development and cause that animal henceforth to recognize that antigen as "self." Inducing **immunological tolerance** in this way was demonstrated in 1953 by the English immunologist Peter B. Medawar, who used two strains of mice, each so highly inbred that they were almost a clone. Medawar injected cells from adult mice of one strain into newborn mice of another strain. Other newborn mice of the second strain served as uninjected controls. Eight to ten weeks later, he tested for tolerance in the treated and untreated mice of the second strain by grafting skin from the first strain onto them. The untreated mice rejected the grafts but the treated mice accepted them (Figure 16.8). Medawar thus discovered that immunological tolerance to an antigen can be induced by exposure to the antigen early in development.

The establishment of tolerance, whether by the production of appropriate suppressor T cells or by clonal deletion, must be repeated throughout the life of the animal because lymphocytes are produced constantly. Continued exposure to self-antigen helps to maintain tolerance. If for some reason an animal stops producing a given protein—that is, a self-antigen—a clone of lymphocytes directed against that protein may become established. Then if the protein is synthesized again later, it may cause a full-fledged immune response, resulting in an **autoimmune dis-**ease (such as rheumatoid arthritis, in which an immune system attacks part of the body in which it resides).

Development of Plasma Cells

Let's consider in more detail the *mechanisms* of the immune responses. When a B or T cell is activated by an antigen, it proliferates, with both effector and memory cells being produced. The activation of a B cell begins with the binding of particular antigens to the antibodies carried on the B-cell surface. A particular type of T cell, called a helper T cell, must be present; the helper T cell also binds to the antigen. Then, cellular division and differentiation lead to the formation of plasma cells (the effector cells) and memory cells. As plasma cells develop, the number of ribosomes and the amount of endoplasmic reticulum in their cytoplasm increase greatly. These increases prepare the cells for synthesizing large amounts of antibodies for secretion (Figure 16.9). All the plasma cells arising from a given B cell produce antibodies of specificity identical to that of the receptors that bound antigen to the parent B cell.

For many years, scientists wanted to obtain cultures of plasma cells all producing the same antibody so that they could prepare large quantities of pure antibody. Methods were finally discovered for producing antibodies from single clones of cells (Box 16.A), but what *are* the molecules we call antibodies?

IMMUNOGLOBULINS: AGENTS OF THE HUMORAL RESPONSE

Antibodies are also called **immunoglobulins**. We generally use the latter term when we turn from the larger picture of how the humoral immune response attacks antigens in the body fluids to look more closely at the molecules that do the attacking. The chemical structure of the most common form of immunoglobulins was worked out by Gerald M. Edelman (Rockefeller University) and Rodney M. Porter (Oxford University), who found that the basic immunoglobulin molecule is a tetramer consisting of four polypeptides. The polypeptides are two identical "light" chains and two identical "heavy" chains. Disulfide bonds hold the chains together. Each chain consists of a **constant region** and a **variable region** (Figure 16.10). The constant regions of both light and heavy chains are similar from one immunoglobulin to another. On the other hand, the amino acid sequence of the variable region (considering the variable ends of the heavy and light chains together) is unique in each of the millions of different types of immunoglobulins that can mount a humoral immune response. Thus the variable regions of a light and a

Strain A experimental mouse

Lymphoid cells from strain A mouse are injected into strain B mice. Strain B control mice are not injected

8 to 10 weeks later, treated and untreated strain B mice mature into adults and skin grafts from strain A, B, and C mice are implanted

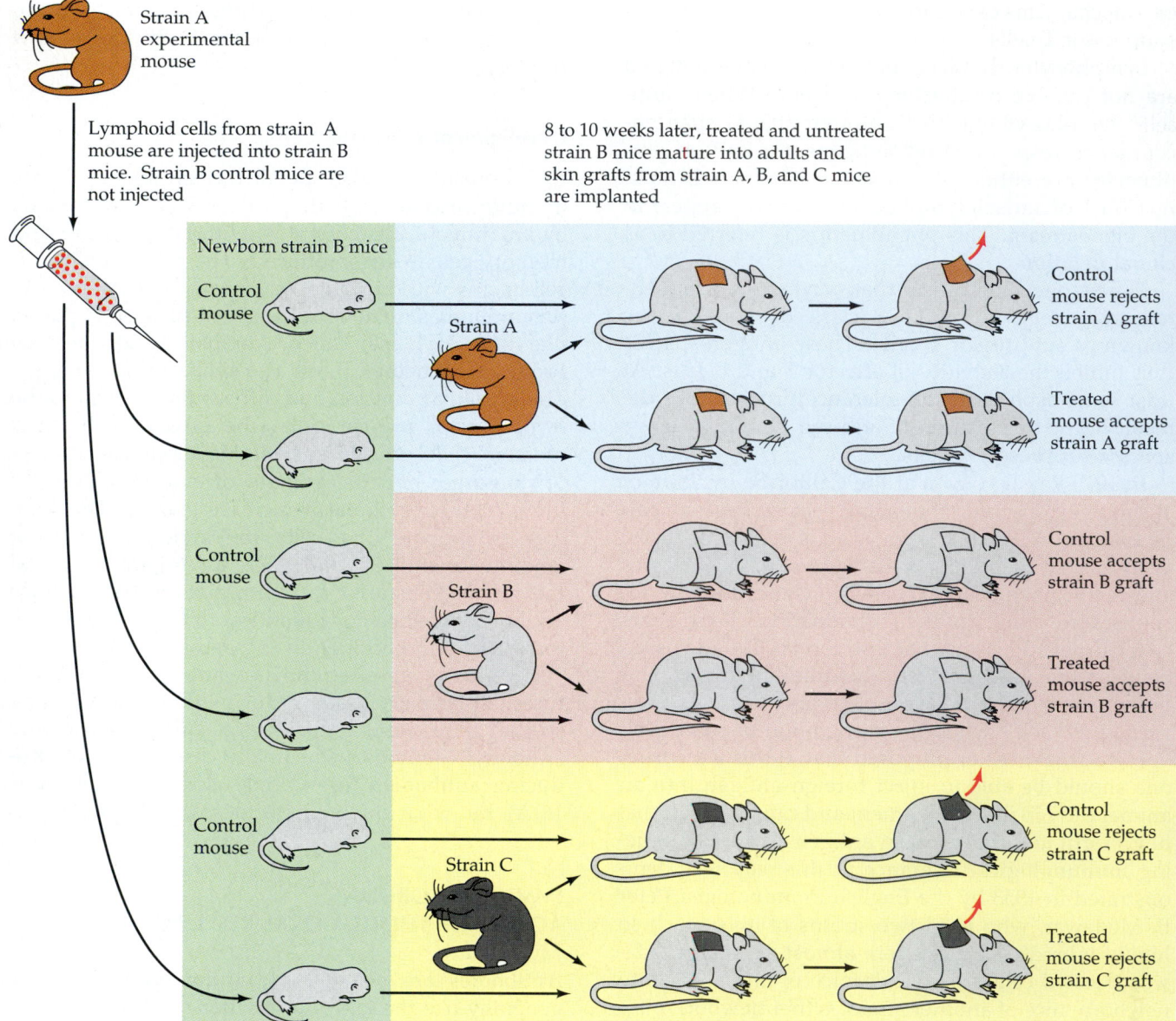

Newborn strain B mice

Control mouse

Strain A

Control mouse rejects strain A graft

Treated mouse accepts strain A graft

Control mouse

Strain B

Control mouse accepts strain B graft

Treated mouse accepts strain B graft

Control mouse

Strain C

Control mouse rejects strain C graft

Treated mouse rejects strain C graft

16.8   Making Nonself Seem Like Self

The ability of adult mice to recognize and reject grafts of foreign skin can be overcome. Adult mice of strain B that were injected shortly after birth with lymphoid cells from strain A tolerate grafts from strain A or strain B, but reject grafts from other strains, such as strain C. The control mice—adults of strain B raised from uninjected newborn mice—accept grafts only from other strain B mice, rejecting skin from strain A as well as from strain C. What is recognized as "self" and "nonself" thus depends partly on when it is first encountered.

heavy chain combine to form, on each of the immunoglobulin's "arms," a highly specific, three-dimensional structure similar to the active site of an enzyme. This characteristic part of a particular immunoglobulin molecule is what binds with a particular, unique antigenic determinant.

Although the variable regions are responsible for the *specificity* of an immunoglobulin, the constant regions are equally important, for it is the constant regions that determine the type of action to be taken

in eliminating the antigen, as we will see in the next section. In particular, the constant regions of the heavy chains determine whether the antibody remains part of the cell's plasma membrane or is secreted into the bloodstream. The two halves of an antibody, each consisting of one light and one heavy chain, are identical, so each of the two arms can combine with an identical antigen, leading sometimes to the formation of a large complex of antigen and antibody molecules (Figure 16.11).

16.9   A Plasma Cell
Note the prominent nucleus (recognizable by the double membrane), the cytoplasm crowded with rough endoplasmic reticulum, and an extensive Golgi complex—all structural features of a cell actively synthesizing and exporting proteins.

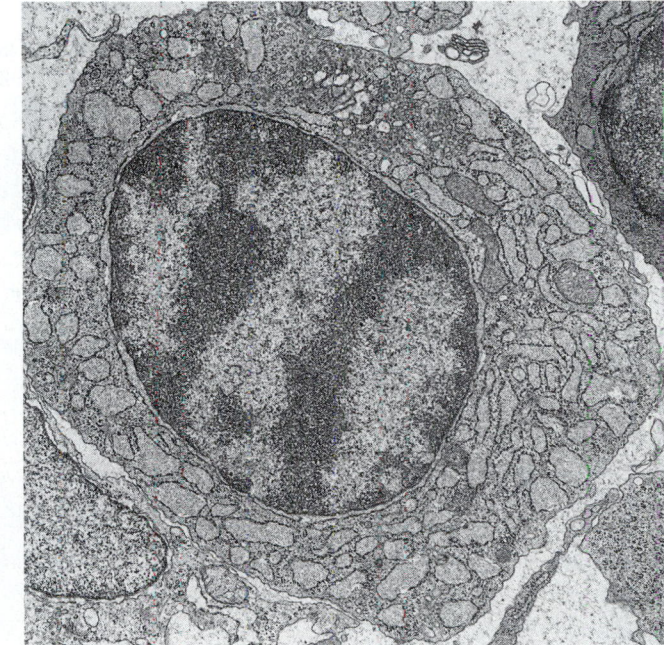

(a)

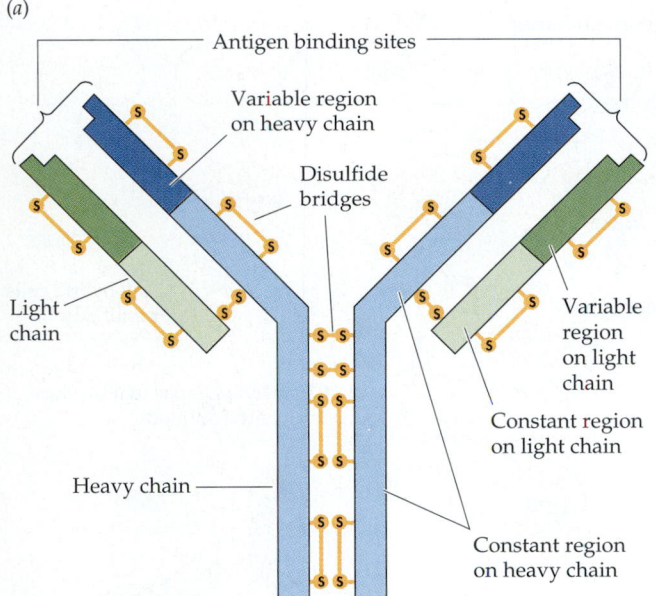

Antigen binding sites

Variable region on heavy chain

Disulfide bridges

Light chain

Heavy chain

Variable region on light chain

Constant region on light chain

Constant region on heavy chain

(b)

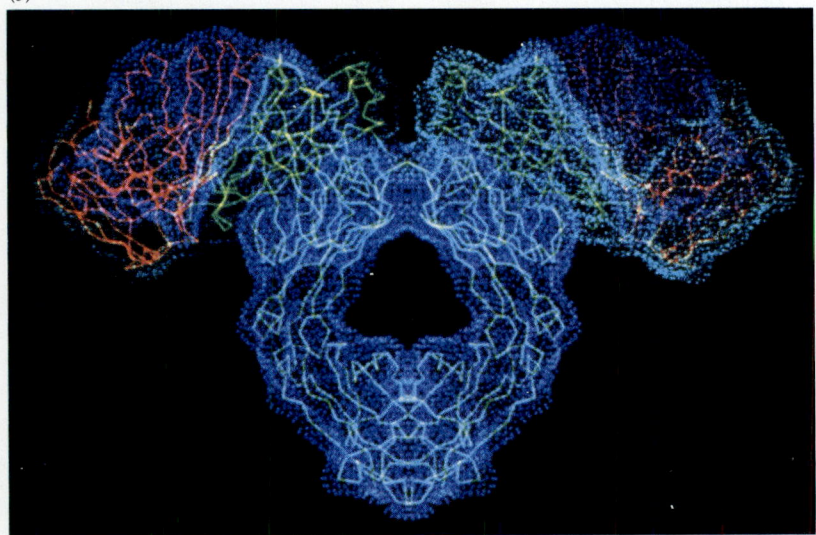

16.10   Structure of Immunoglobulins
(a) Disulfide bridges hold the four polypeptide subunits of an immunoglobulin together. The variable regions recognize and bind antigens. (b) An immunoglobulin molecule in roughly the same orientation as (a), drawn by a computer. Green indicates the light chains, mixed with red for the variable regions and yellow for the constant regions. Dark blue indicates the heavy chains, mixed with red for the variable regions and light blue for the constant regions.

BOX 16.A

Monoclonal Antibodies

Because most antigens carry many different antigenic determinants, animals usually produce a complex mixture of antibodies. Scientists found it virtually impossible to separate the individual antibody types for chemical study. In the mid-1970s, however, Cesar Milstein (an Argentine biochemist living in Cambridge, England) and a colleague from Switzerland, Georges Köhler, made an important breakthrough. They knew that a single lymphocyte produces only a single species of antibody. In principle, all one needed to do was to cause a single lymphocyte to multiply in pure culture to get a large population of cells, all dedicated to the production of the same antibody. However, the antibody-producing cells cannot be cultured. On the other hand, cancerous tumors of plasma cells, called myelomas, grow rapidly in culture. The cells of a given tumor all produce the same antibody, but they produce far too little of the antibody to be useful sources. Milstein and Köhler made use of both cell types—normal lymphocytes and myeloma cells—to produce hybrid cells (**hybridomas**) that made specific normal antibodies in quantity and that, like the myeloma cells, could proliferate rapidly and indefinitely in culture.

Clones of hybrid cells are made as follows. An animal is inoculated with an antigen to trigger specific lymphocyte proliferation. Later, the spleen is dissected out and lymphocytes are collected from it. (The spleen, like the lymph nodes and certain other lymphoid tissues associated with the gut, is a site of lymphocyte accumulation and maturation.) These lym-

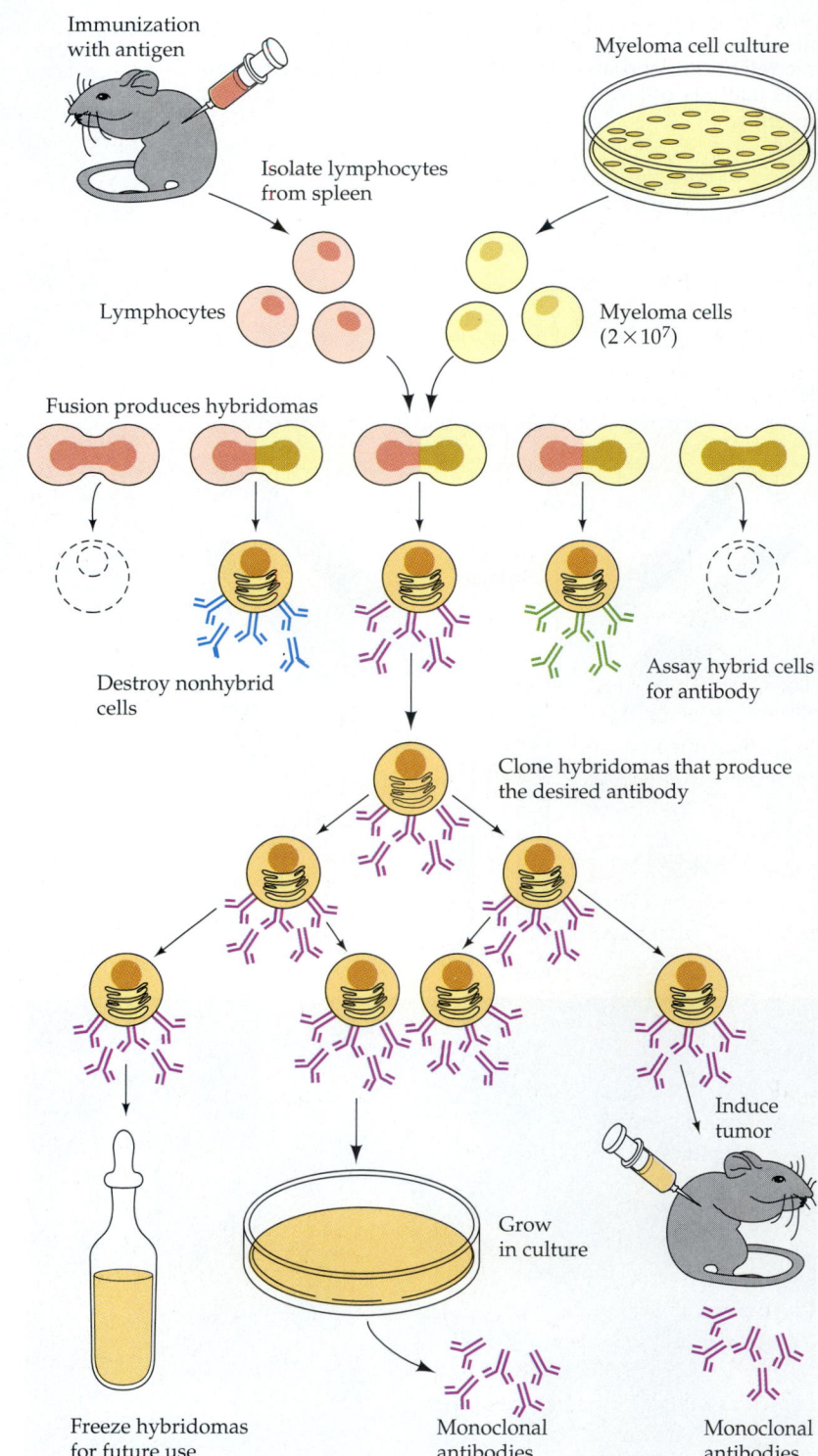

Immunization with antigen

Myeloma cell culture

Isolate lymphocytes from spleen

Lymphocytes

Myeloma cells (2×10^7)

Fusion produces hybridomas

Destroy nonhybrid cells

Assay hybrid cells for antibody

Clone hybridomas that produce the desired antibody

Induce tumor

Grow in culture

Freeze hybridomas for future use

Monoclonal antibodies

Monoclonal antibodies

phocytes are combined under appropriate conditions with myeloma cells from a single tumor. Some lymphocytes fuse with myeloma cells, giving rise to hybridomas. The cell mixture is then treated to destroy all nonhybrid cells. The hybridomas are grown in a suitable medium so that each one forms a clone. Individual clones are tested, and the ones that produce the desired antibodies—specific for one antigenic determinant— are selected. These clones produce **monoclonal antibodies** (uniform antibodies from a single clone of cells) in large quantities, either from a mass culture or following transfer into an animal where they can grow as a tumor. The hybridomas may also be frozen for storage.

Monoclonal antibodies are ideal for the study of specific antibody chemistry, and they have been used to further our knowledge of cell membranes as well as for specialized laboratory procedures such as tissue typing for grafts and transplants. Monoclonal antibodies have many practical applications. One possibility is passive immunization—inoculation with specific antibody rather than with an antigen that causes the patient to develop his or her own antibody (as most vaccines are designed to do). Monoclonal antibodies can also be used for detecting specific cancers; diagnostic kits using monoclonal antibodies for colon cancer are already available. Antitumor monoclonal antibodies have been successfully employed in immunotherapy for tumors. In yet another use, poisons directed against tumors can be attached to specific monoclonal antibodies as a form of cancer treatment. Developmental biologists (Chapter 17) have used monoclonal antibodies to show that the cell membranes of different kinds of cells in an animal contain different molecules and that the molecules change as the animal develops.

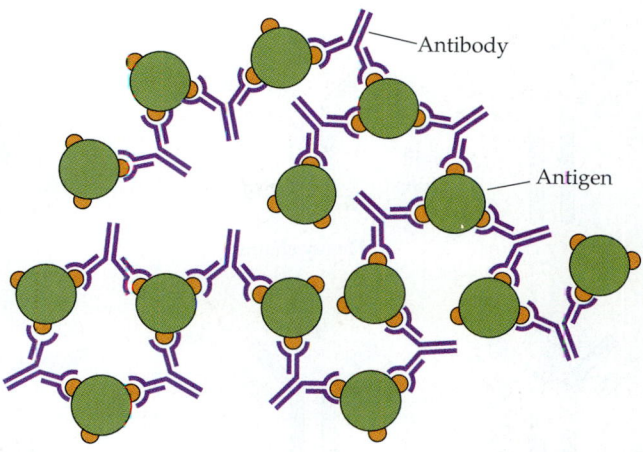

16.11 Antibody–Antigen Complex
An antibody has two sites that can bind to different molecules of antigen, and more than one antibody may bind to the same antigen molecule. Large antibody–antigen complexes may precipitate from the blood; they play a role in many types of clinical laboratory tests.

There are five immunoglobulin classes based on differences in the constant regions of the heavy chains (Figure 16.12). One, called immunoglobulin M, or IgM, is always the first antibody product of a plasma cell. The cell may later switch over to the production of other classes of immunoglobulins—but with the same antibody specificity. The four other classes—IgA, IgD, IgE, and IgG—play different roles in the immune system.

IgG molecules, which have the γ heavy-chain constant region, make up about 85 percent of the total immunoglobulin content of the bloodstream. They consist of a single immunoglobulin unit (two identical heavy chains and two identical light chains) and are produced in greatest quantity during a secondary immune response (see Figure 16.6). IgG defends the body in several ways. For example, some IgG molecules that have bound antigens become attached by their heavy chains to a type of phagocyte called macrophages. This IgG–macrophage union makes it easier for the phagocytes to ingest the antigens (Figure 16.13). Another major function of IgG is to activate antimicrobial proteins, a potent set of nonspecific defenses collectively called the complement system (discussed in the next section); this also enhances phagocytosis.

The bulk of the antibodies produced at the beginning of a primary immune response are IgM molecules. They differ from IgG in being composed of five immunoglobulin units (see Figure 16.12). Because they have more binding sites, IgM molecules are more active than IgG molecules in activating the complement system and promoting the phagocytosis of antibody-coated cells.

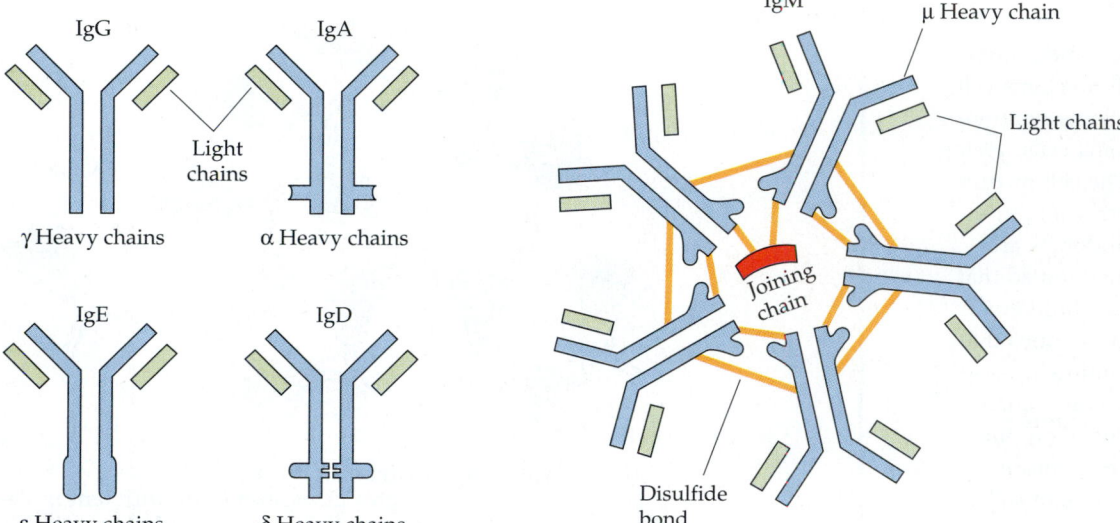

16.12 Classes of Immunoglobulins
Each class of immunoglobulin has its own type of heavy chain, identified by a Greek letter. IgM, unlike other antibodies, is made up of five immunoglobulin subunits. Two of the five units are bonded to a single joining chain.

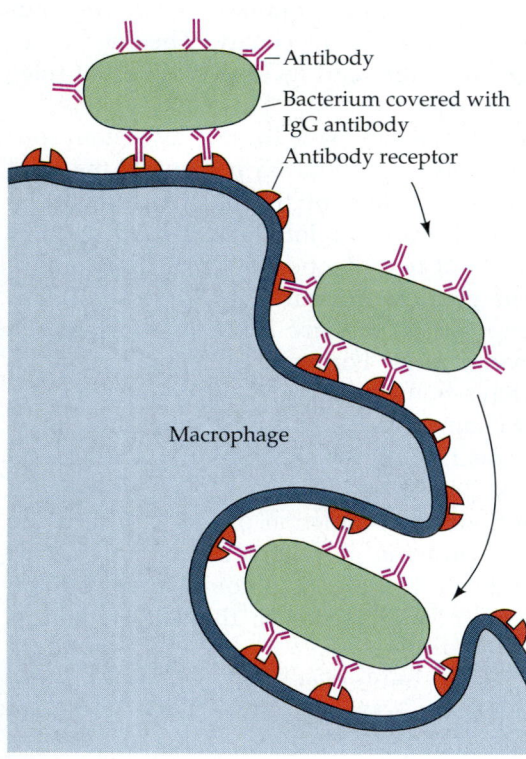

16.13 IgG Antibodies Promote Phagocytosis
Phagocytes called macrophages have receptors on their cell surfaces for part of the IgG molecule. Thus when a bacterium has reacted with IgG, the resulting complex binds readily to macrophages, activating phagocytosis (ingestion of the antigen by the macrophage).

Only small amounts of IgD antibody travel free in the bloodstream. The major role of IgD is to serve, as IgM does, as membrane receptors on B cells.

IgE antibodies take part in inflammation and allergic reactions. IgE helps kill worm pathogens such as those that cause the disease schistosomiasis, which affects some 200 million people in Africa and South America. Where inflammation occurs, IgE may participate in bringing white blood cells, components of the complement system, and other factors into the inflamed region. For most of us, the effect of IgE is most apparent when we suffer allergies. IgE molecules bind to antigenic determinants on the substances that provoke the allergy (the allergens), and they also bind to receptor sites on the surfaces of cells called mast cells. The mast cell–IgE–allergen complex stimulates the release of histamine and other compounds, leading in turn to inflammation. Hives, hay fever, eczema, and asthma are all common allergic reactions (Figure 16.14).

Body secretions such as saliva, tears, milk, and gastric fluids all contain immunoglobulins, specifically IgA. IgA molecules are transported across epithelial cells to join the secreted fluids. IgA exists as both monomers and dimers; Figure 16.12 shows only the monomeric unit.

ANTIBODIES AND NONSPECIFIC DEFENSES WORKING TOGETHER

Recall that antimicrobial proteins were listed at the beginning of the chapter among the nonspecific defenses. Vertebrate blood contains about 20 different antimicrobial proteins that make up the **complement system**. These proteins, in different combinations, provide three types of defenses. Their most impres-

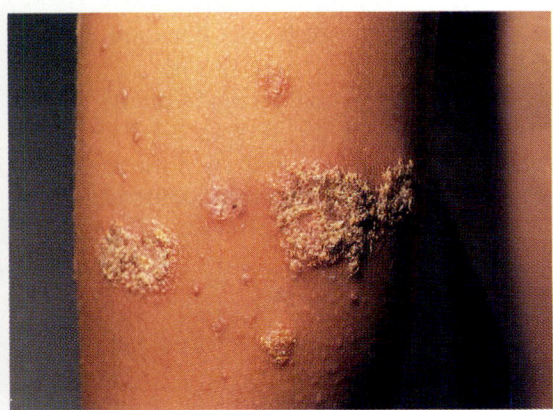

16.14   An Allergic Reaction
Eczema is a common, noncontagious allergic reaction characterized by itching, redness, and the appearance of crusted lesions.

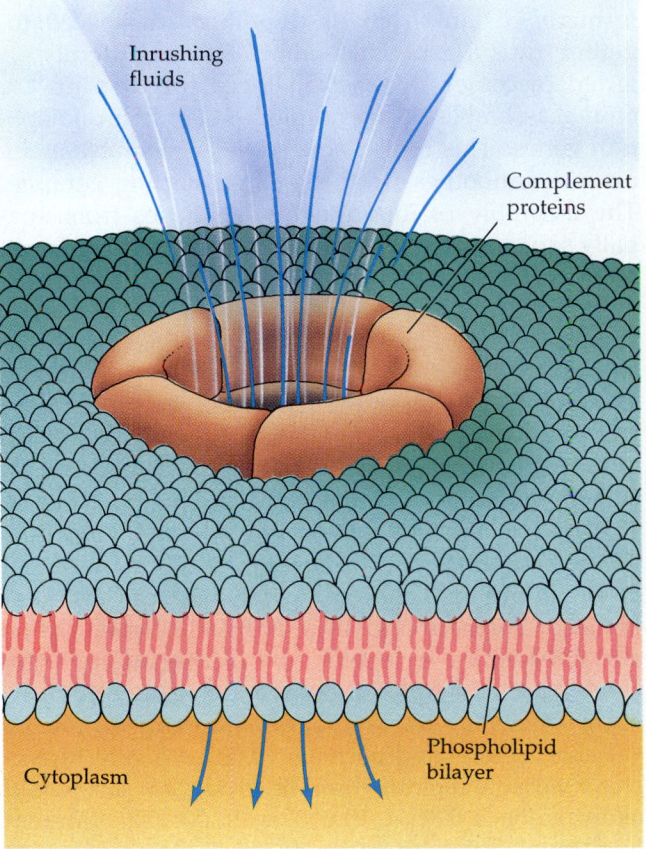

16.15   A Pore in a Lytic Complex Destroys a Foreign Cell
The end result of a cascade of reactions is a precisely arranged group of proteins extending through the phospholipid bilayer of the plasma membrane of a foreign cell. A pore in the complex makes the affected cell leaky; fluids rushing into the cell then cause it to burst.

sive defense, carried out with the help of antibodies, is to lyse (burst) foreign cells—bacteria, for example. When IgG antibodies bind to antigenic determinants on the surface of a foreign cell, this binding may bring about the binding of the first of the complement proteins to the cell surface. What follows is a cascade of reactions, with different complement proteins acting upon one another in succession. The final product of the complement cascade is a lytic complex—a doughnut-shaped structure in the membrane of the foreign cell that renders the membrane leaky, allowing fluids to enter the cell rapidly, causing lysis (bursting) of the foreign cell (Figure 16.15).

In the second type of defense, complement proteins help phagocytes destroy foreign microorganisms. The phagocytes can recognize foreign cells more easily after complement proteins attach to the foreign cells. The third defensive activity of the complement system is to attract phagocytes to sites of infection.

The complement system is *nonspecific* in its action. The specific recognition that is the first step toward forming a lytic complex is accomplished by the antibodies that bind to cell-surface antigens of the invading cells, not to the complement proteins themselves. Hereditary deficiencies in one or another of the complement proteins can cause characteristic diseases, mostly infections and hypersensitivity diseases.

THE ORIGIN OF ANTIBODY DIVERSITY

A newborn mammal possesses a full set of genetic information for immunoglobulin synthesis. At each of the loci coding for the heavy and light chains it has an allele from the mother and one from the father. Throughout the animal's life, each of its cells

begins with the same full set. However, the genomes of B cells become modified during development in such a way that each cell eventually can produce one—and only one—type of antibody. Different B cells develop different antibody specificities. A question that vexed immunologists and geneticists for decades was how a single organism could produce so many different specific immunoglobulins—perhaps a million, or even a billion, antibody specificities. Research in recent years has effectively answered this question.

The most surprising part of the answer is that functional immunoglobulin genes are assembled from DNA segments that initially are spatially separate. Every cell has hundreds of DNA segments potentially capable of participating in the synthesis of the variable regions, the parts of the antibody molecule conferring immunological specificity. B-cell precursors differ from all other cells in that these DNA segments are *rearranged* during B-cell development. Pieces of the DNA are deleted, and DNA segments formerly distant from one another are spliced together; thus a gene is assembled from randomly selected pieces of DNA. Each B-cell precursor in the animal assembles its own unique set of immunoglobulin genes. This remarkable process generates many diverse antibodies from the same starting genome. The assembly of immunoglobulin genes from spatially separate DNA segments was first demonstrated by Susumu Tonegawa, in Switzerland, in 1976.

In both humans and mice, the DNA segments coding for immunoglobulin heavy chains are on one chromosome and those for light chains are on others. The variable region of the light chain is coded for by two families of DNA segments, and the variable region of the heavy chain is coded for by three families. (We discussed gene families in Chapter 13; see Figure 13.4.) Gene families are assembled randomly; thus the diversity afforded by several hundred DNA segments is multiplied by the combination of different families. Furthermore, since light and heavy chains are synthesized independently of one another, the combination of light and heavy chains introduces more diversity. If, say, 1,000 different light-chain variable regions and perhaps 10,000 different heavy-chain variable regions could be produced, this would allow some $1,000 \times 10,000 = 10,000,000$ different immunoglobulins to be produced.

Are we forgetting another contribution to antibody diversity? B-cells are diploid, so each contains *two* chromosomes with the heavy-chain genes and two of each of the others with light-chain genes. So shouldn't each B-cell produce two types of heavy chains and two types of light chains—and thus as many as four different types of antibodies? If this were true, the clonal selection theory, which depends upon a given cell producing a single antibody species, would be in shambles. The theory holds, however, because only one homologous chromosome from each relevant pair produces an mRNA that is translated into the corresponding antibody chain. (The underlying mechanism is not yet well understood.)

Although the diploidy of B cells does not contribute to antibody diversity, there is one more factor that does: *mutation.* Mutation occurs frequently in the genes of the variable region during B-cell development. It has been estimated that such mutations increase the total number of different antibody specificities by 10 to 100 times or more.

How a B Cell Produces a Particular Heavy Chain

To see how DNA rearrangement generates antibody diversity, let's consider how the heavy chain of IgM is produced. B cells produce this antibody, which then inserts itself into the plasma membrane of the B cells.

The locus governing heavy-chain synthesis is on chromosome 12 of mice and on chromosome 14 of humans. In mice, the locus is arranged as shown in Figure 16.16, with a long stretch of DNA occupied by a family of 100 or more V (variable) segments. Humans have about 300 V segments. In a given B cell, only *one* of these segments is used to produce part of the variable region of the heavy chain; the remaining V segments are discarded or rendered inactive. At a distance of many nucleotides from the V segments is a family of 10 or more D (diversity) segments. Again, only one of these is used to produce part of the variable region of the heavy chain of a given B cell, as is only one J (joining) segment from the family of four such segments (in mice) lying yet farther along the chromosome. This combination of one each of V, D, and J segments forms a complete variable region for a functional gene. Still farther along the chromosome, and separated from the suite of J segments, is a family of eight segments, one of which codes for the constant region of the heavy chain. Light chains are produced from similar families of DNA segments, but without D segments.

How does order emerge from this seeming chaos of DNA segments? Two important steps impose order. First, substantial chunks of DNA are deleted

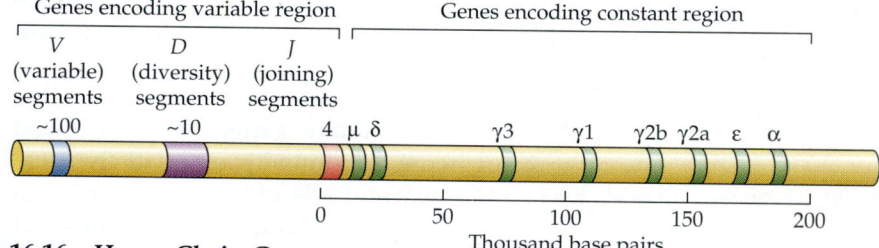

16.16 Heavy-Chain Genes
Immunoglobulin heavy chains are encoded by a gene that has many segments. The variable region for the heavy chain of a particular antibody is encoded by one V segment, one D segment, and one J segment. Each of these segments is taken from a pool of like segments. The constant region is selected from another pool of segments.

from the chromosome during the rearrangement of the segments. As a result of these deletions, a particular *D* segment is joined directly to a particular *J* segment, and then the *D* segment is joined to one of the *V* segments; thus a single "new" sequence, consisting of one *V*, one *D*, and one *J* segment, can now code for the variable region of the heavy chain. All the progeny of this cell constitute a clone having this same sequence for the variable region. Different B cells result from DNA being deleted in different places on the chromosome, leading to different variable-region sequences. In mice—which have about 100 *V* segments, 10 or more *D* segments, and 4 *J* segments to choose from—about $100 \times 10 \times 4 = 4,000$ different heavy-chain variable regions are possible. This estimate increases when we take into account the variation introduced by mutation.

The second step in organizing the synthesis of an immunoglobulin chain follows transcription. Splicing of the RNA transcript (see Chapter 13) removes the product of an intron that includes any *J* segments lying between the selected *J* segment and the first constant-region segment. Splicing also removes the products of introns contained in both the *V* segment and the *C* (constant-region) segment (Figure 16.17). The result is an mRNA that can be translated, directly yielding the heavy chain of the cell's specific antibody.

Note that two distinct types of nucleic acid splicing contribute to the formation of an antibody. *DNA* splicing, before transcription, joins the *V*, *D*, and *J* segments. *RNA* splicing, after transcription, joins the *J* segment to the constant region.

The Constant Region and Class Switching

Early in its life a plasma cell produces IgM molecules that are responsible for the specific recognition of a particular antigenic determinant. At this time, the constant region of the antibody's heavy chain is encoded by the first constant-region segment, the μ segment (see Figure 16.16). During an immunological response—later in the life of the plasma cell—another deletion may occur in the plasma cell's DNA, positioning the heavy-chain variable-region gene (consisting of the same *V*, *D*, and *J* segments) next to a constant-region segment farther down the original DNA, such as the γ, ε, or α constant region (Figure 16.18).

Such a deletion—called **class switching**—results in the production of an antibody with a different *function* but the same *antigen specificity*. The new antibody has the same variable regions of the light and heavy chains but a different constant region of the heavy chain. This antibody falls into one of the four other immunoglobulin classes (IgA, IgD, IgE, or IgG; see Figure 16.12), depending on which of the constant-region segments is adjacent to the variable-region gene. Once class switching has occurred, the plasma cell cannot go back to making the previous immu-

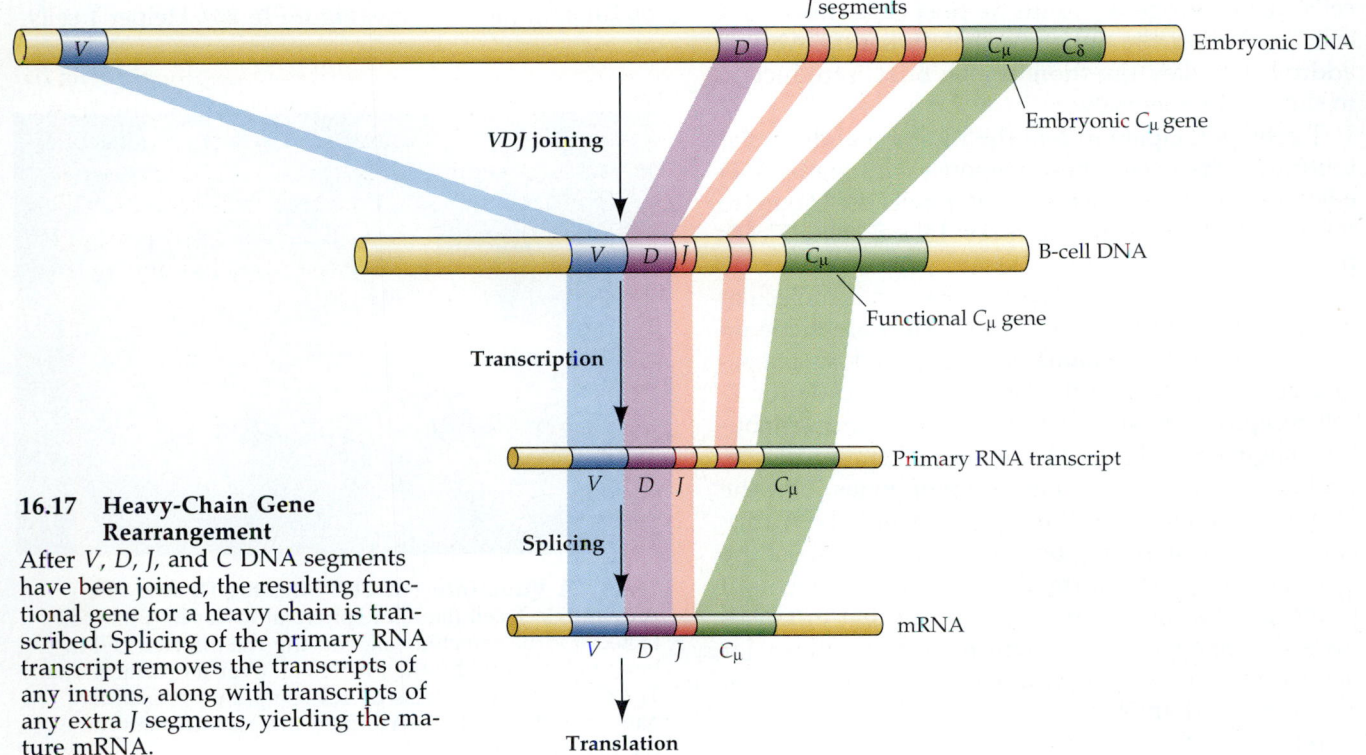

16.17 Heavy-Chain Gene Rearrangement
After *V*, *D*, *J*, and *C* DNA segments have been joined, the resulting functional gene for a heavy chain is transcribed. Splicing of the primary RNA transcript removes the transcripts of any introns, along with transcripts of any extra *J* segments, yielding the mature mRNA.

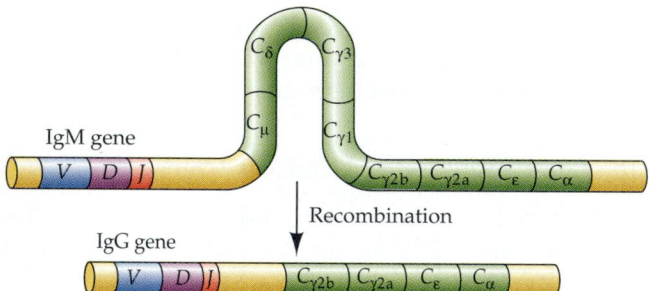

16.18 Class Switching
The functional gene produced by joining *V*, *D*, *J*, and *C* segments may later be modified, causing a different *C* segment to be transcribed. This modification, known as class switching, is accomplished by deleting part of the constant region. Shown here is class switching from an IgM gene to an IgG gene.

noglobulin class—that part of the DNA has been deleted. On the other hand, if additional constant-region segments are still present, another class switch may occur.

T CELLS: AGENTS OF BOTH RESPONSES

Thus far we have been concerned primarily with the humoral immune response, whose effector molecules are the antibodies secreted by plasma cells that develop from activated B cells. T cells are the effectors of the cellular immune response, which is directed against multicellular invaders and against antigens that have become established *within* the animal's cells. If the antigens are inside host cells, how do T cells encounter the antigenic determinants? We will address that key question after a brief introduction to the T cells themselves.

T cells participate in *both* the humoral and cellular immune responses. These responses are highly specific, and we will consider how T cells contribute to the specificity of both responses. Like B cells, T cells possess specific receptors.

T-cell receptors are glycoproteins with molecular weights about half that of an IgG. They are made up of two polypeptide chains, each encoded by a separate gene (Figure 16.19). The genes that code for T-cell receptors are similar to those for immunoglobulins, suggesting that both are derived from a single, evolutionarily more ancient group of genes. Like the immunoglobulins, T-cell receptors include both variable and constant regions and are assembled by *V–D–J* joining. Once formed, the receptors are bound to the plasma membrane of the T cell that produces them. In the next sections, which introduce the major histocompatibility complex, we discuss how T-cell receptors bind antigens.

When T cells are activated by contact with a spe-

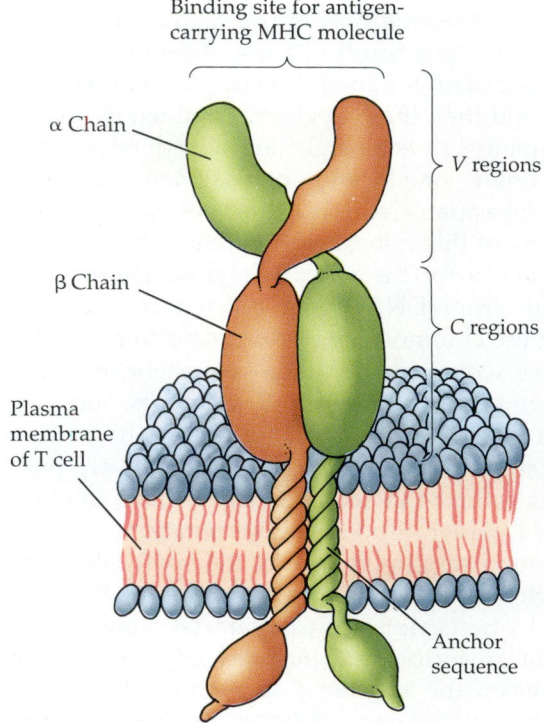

16.19 A T-Cell Receptor
A T-cell receptor consists of two polypeptide chains—an α chain and a β chain. Each polypeptide chain possesses a hydrophobic region that anchors the chain in the phospholipid bilayer of the T-cell plasma membrane.

cific antigenic determinant, they develop and give rise to several distinct types of effector cells. Cytotoxic T cells, or **T$_C$ cells**, are one type of effector cell. T$_C$ cells recognize virus-infected cells and kill them by causing them to lyse (Figure 16.20). Helper T cells,

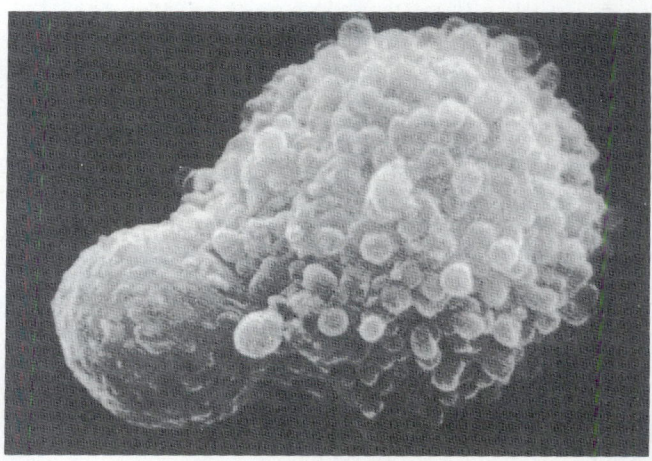

16.20 A Virus-Infected Cell Bites the Dust
A cytotoxic T cell (smaller sphere at lower left) has contacted a virus-infected cell, causing the infected cell to lyse, as indicated by the blisters all over its surface. The T$_C$ cell induced the lysis by releasing a protein, aptly named perforin, that perforated the plasma membrane of the infected cell.

or **T$_H$ cells**, assist both the cellular and humoral immune systems. As mentioned already, a T$_H$ cell of appropriate specificity must bind an antigen before a B cell can be activated by it. Still other T cells differentiate into suppressor T cells, or **T$_S$ cells**. These regulatory cells inhibit the responses of both B and T cells to antigens. They probably play an important role in immune tolerance and in the acceptance of self antigens. Now that we are familiar with the major types of T cells, we can return to the seemingly difficult question of how T cells meet the antigenic determinants if the antigens themselves are inside host cells.

The Major Histocompatibility Complex

The key to the interactions of T cells and antigenic determinants, and to the interactions of B cells and the various classes of T cells, lies in a tight cluster of loci called the major histocompatibility complex, or **MHC**. These loci code for specific proteins on the surfaces of cells. Because of the number of MHC genes and the number of their alleles, different animals of the same species are highly likely to have different MHC genotypes—and that difference is what leads to the rejection of organ transplants.

Similarities in structure and base sequences between MHC genes and the genes coding for antibodies suggest that the MHC genes may be descended from the same ancestral genes as are those for antibodies and T-cell receptors. Major aspects of the immune systems may thus be woven together by a common evolutionary thread.

There are three classes of MHC loci. Class I MHC loci code for proteins (antigens) that are present on the surface of every cell in the animal. These proteins function in antiviral T-cell immunity. The products of class II loci are found only on the surfaces of B cells, T cells, and certain phagocytes called macrophages (Figure 16.21). It is this class of MHC products that is primarily responsible for the interaction of T$_H$ cells, macrophages, and B cells in antibody responses. Class III MHC loci code for some of the proteins of the complement system that interact with antigen–antibody complexes and result in the lysis of foreign cells (see Figure 16.15). Now let's see how class I and class II MHC products help T cells interact with antigenic determinants.

T Cells and the Humoral Immune Response

A macrophage ingests an antigen and breaks it into antigen fragments, each with one or more antigenic determinants. The fragments are called **processed antigen**. The products of class II MHC loci bind processed antigen, carry it to the surface of the cell, and display it on the outside of the cell where it is avail-

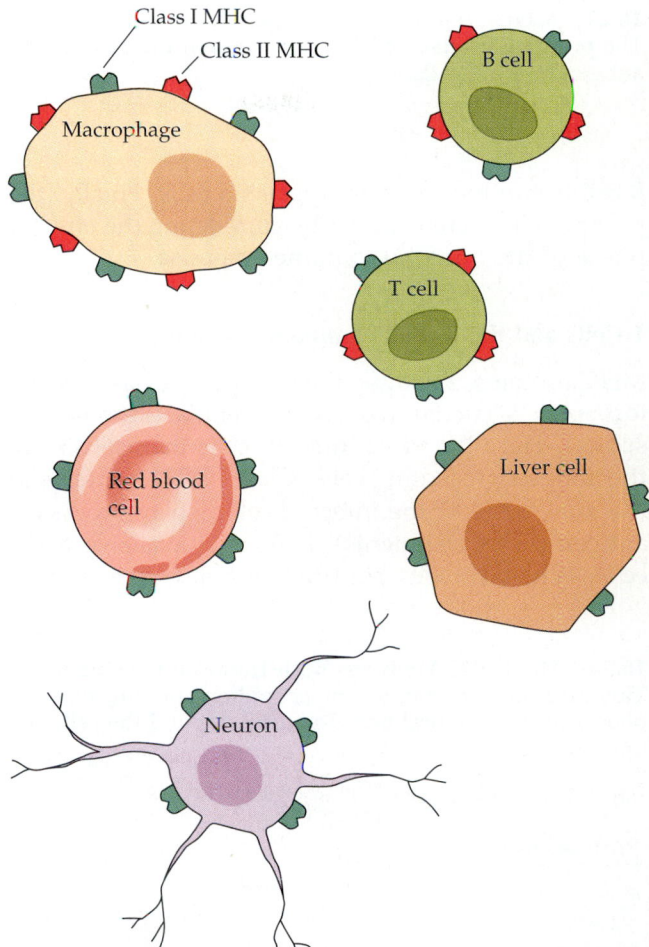

16.21 MHC Products in Plasma Membranes
Class I MHC proteins can be found on every cell in an animal; class II MHC proteins are found only on B cells, T cells, and macrophages.

able to T$_H$ cells (Figure 16.22). Because the MHC products *present* antigen, the macrophages are thus referred to as **antigen-presenting cells**.

A T$_H$ cell can bind to processed antigen only if the T-cell receptors correspond to the displayed antigenic determinant—and only if the processed antigen is carried by an MHC protein. The T-cell receptor binds both antigenic determinant and MHC protein. When binding is complete, the macrophage releases substances that activate the T$_H$ cell, causing it to produce a clone of differentiated cells capable of interacting with B cells. The steps to this point constitute the activation phase of the response. Next comes the effector phase, in which B cells are activated.

B cells, too, are antigen-presenting cells. B cells take up antigen bound to their receptors, process it, and display it on class II MHC products. An activated T$_H$ cell binds only if it recognizes both the displayed antigenic determinant and the MHC protein. The bound T$_H$ cell releases helping signals that cause the

16.22   MHC Products and Antigen-Presenting Cells

The products of class II MHC loci present the processed antigen to the T cells.

B cell to produce a clone of plasma cells. Finally, the plasma cells secrete antibody, completing the effector phase of the humoral immune response.

T Cells and the Cellular Immune Response

MHC proteins, this time the products of class I MHC loci, play a similar role in the cellular immune response. Here the virus-infected cells themselves are the antigen-presenting cells. Class I MHC proteins in the membrane of the infected cell display processed antigen (virus fragments), making it available to T_C cells. When T_C cells bind the complex of processed

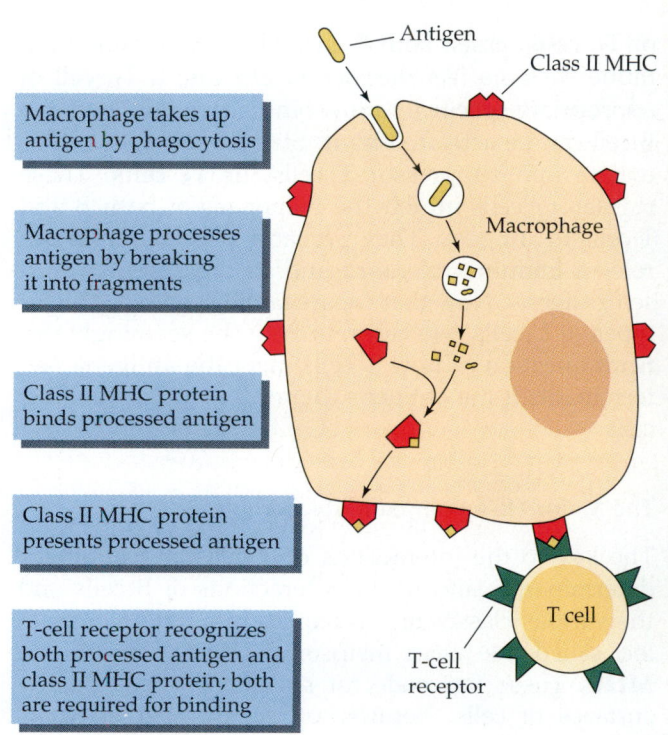

Antigen

Class II MHC

Macrophage takes up antigen by phagocytosis

Macrophage

Macrophage processes antigen by breaking it into fragments

Class II MHC protein binds processed antigen

Class II MHC protein presents processed antigen

T-cell receptor recognizes both processed antigen and class II MHC protein; both are required for binding

T cell

T-cell receptor

16.23   The MHC Mediates Both Immune Responses

This diagram summarizes the activation and effector phases of the humoral immune response and the cellular immune response.

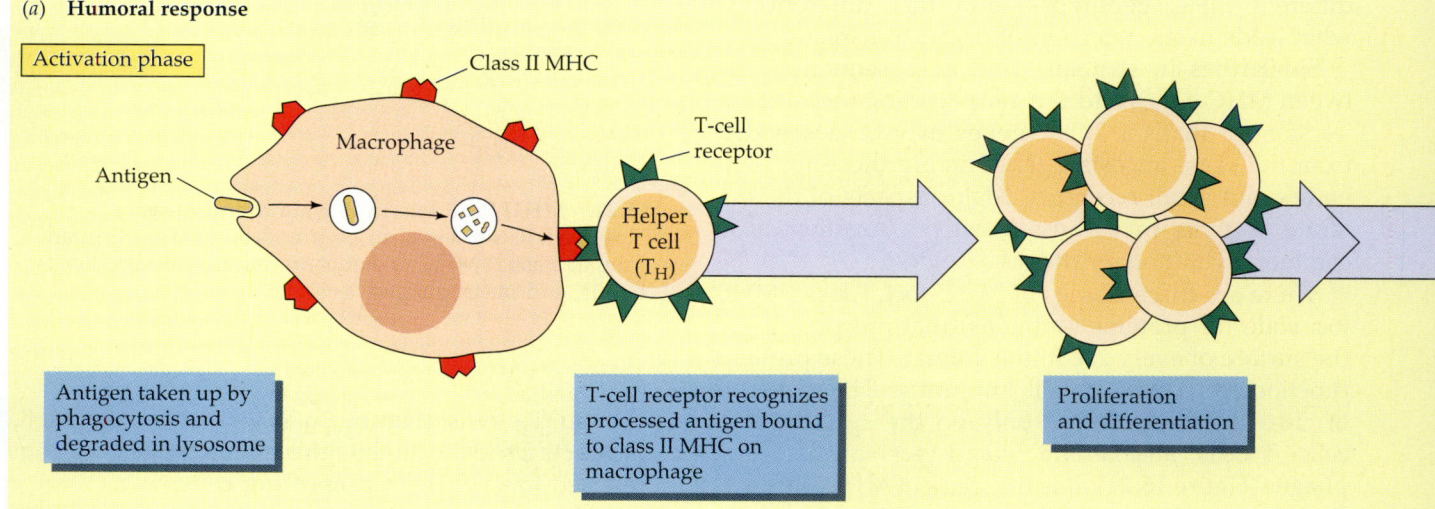

(a) **Humoral response**

Activation phase

Class II MHC

Macrophage

T-cell receptor

Antigen

Helper T cell (T_H)

Antigen taken up by phagocytosis and degraded in lysosome

T-cell receptor recognizes processed antigen bound to class II MHC on macrophage

Proliferation and differentiation

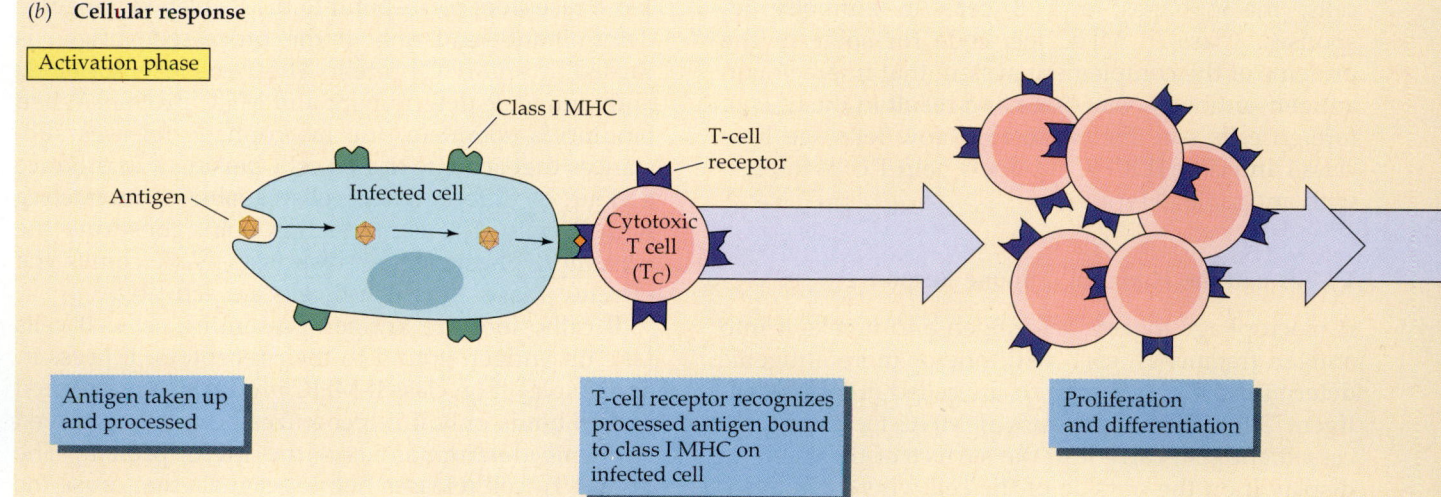

(b) **Cellular response**

Activation phase

Class I MHC

T-cell receptor

Antigen

Infected cell

Cytotoxic T cell (T_C)

Antigen taken up and processed

T-cell receptor recognizes processed antigen bound to class I MHC on infected cell

Proliferation and differentiation

antigen and class I MHC protein, they become activated. In the effector phase of the cellular immune response, T-cell receptors on the activated T_C cells recognize processed viral antigen displayed by class I MHC proteins on the surface of virus-infected cells. The T_C cells then produce lytic signals (perforin; see Figure 16.20), causing the infected cell to lyse. Because T-cell receptors recognize self MHC products, they rid the body of its own virus-infected cells.

The roles of T cells in the humoral and cellular immune systems are similar (Figure 16.23). Both use MHC proteins on the surfaces of antigen-presenting cells, and both have well defined activation and effector phases. Next we consider another crucial interaction between T cells and the MHC.

The MHC and Tolerance of Self

The MHC plays a key role in establishing tolerance to self, without which an animal would be destroyed by its own immune system. Developing T cells undergo "testing" in the thymus. One test asks, Can this cell recognize the body's MHC proteins? A T cell unable to recognize self MHC would be useless to the animal because it could not participate in any immune reactions. Such a cell fails the test and dies within about three days. The second and more crucial test asks, Does this cell bind to self MHC protein *and* to one of the body's own antigens? A T cell that satisfied both of these criteria would be harmful or lethal to the animal; it fails the test and is destroyed immediately.

The T cells that survive this pair of tests are those that recognize the animal's own MHC proteins and that ignore the animal's own antigens. Such T cells are the ones that can do the work of the cellular immune system. They get their diplomas and mature into either T_C cells, T_H cells, or T_S cells.

Transplants

A major side effect of the MHC antigens became important with the development of organ transplant surgery, sometimes with devastating results. Because

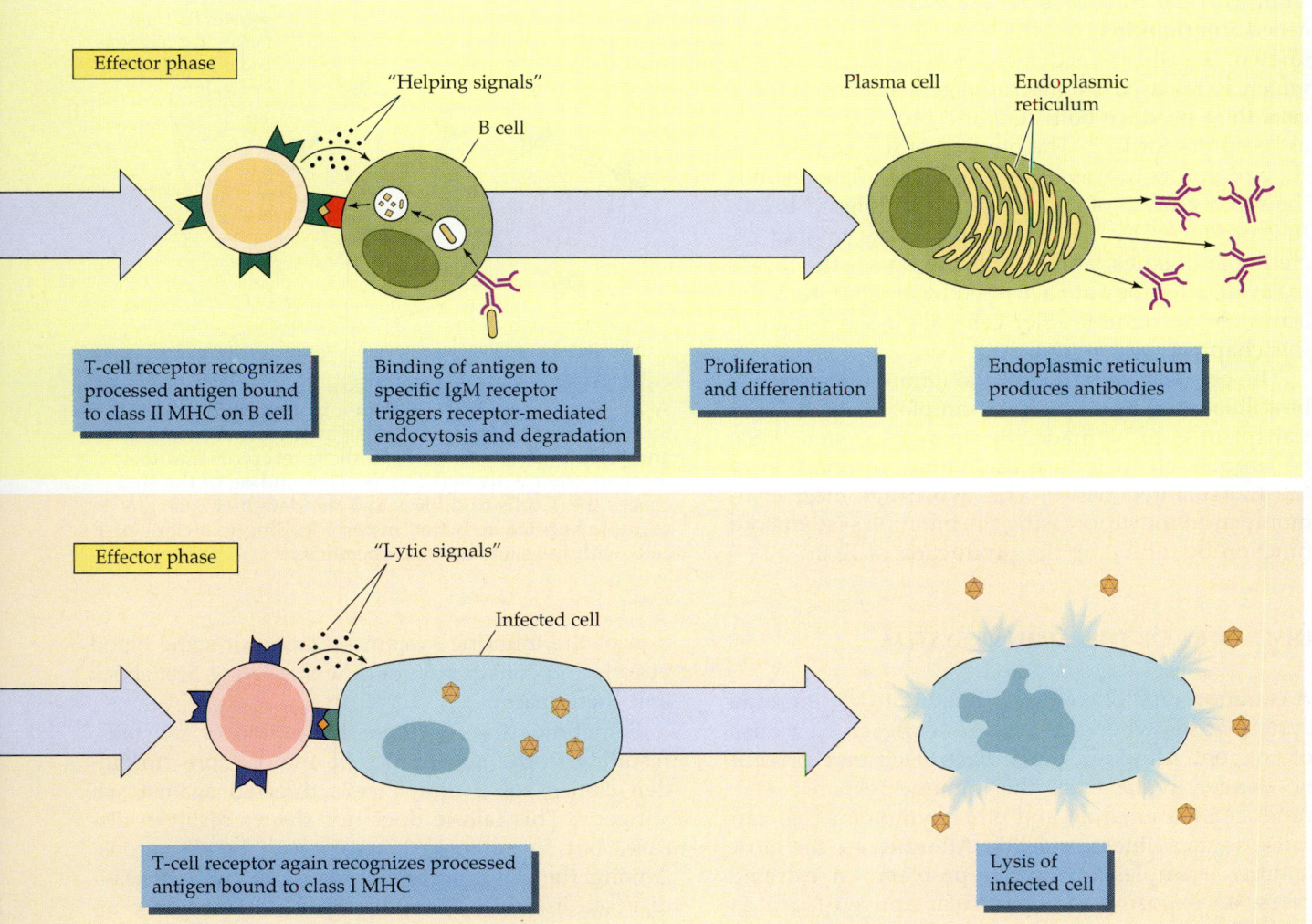

Effector phase

"Helping signals"

B cell

Plasma cell

Endoplasmic reticulum

| T-cell receptor recognizes processed antigen bound to class II MHC on B cell | Binding of antigen to specific IgM receptor triggers receptor-mediated endocytosis and degradation | Proliferation and differentiation | Endoplasmic reticulum produces antibodies |

Effector phase

"Lytic signals"

Infected cell

| T-cell receptor again recognizes processed antigen bound to class I MHC | Lysis of infected cell |

the proteins produced by the MHC are specific to each individual, they act as antigens if transplanted into another individual. An organ or a piece of skin transplanted from one person to another is recognized as nonself and soon provokes an immune response; the tissue is then killed, or "rejected," by the cellular immune system. But if the transplant is performed immediately after birth or if it comes from a genetically identical person (an identical twin), the material is recognized as self and is not rejected.

Physicians can overcome the rejection problem for a while by treating the patient with drugs, such as cyclosporin, that suppress the immune system. This technique, however, compromises the ability of patients to defend themselves against bacteria and viruses. Cyclosporin and some other immunosuppressants interfere with communication between cells of the immune system. How do the cells of the immune system normally communicate with one another?

Interleukins

Communication in the immune system consists of signals that pass from macrophages to T cells, and from T_H cells to B cells. These signals are proteins called **interleukins**, of which more than 10 are now known. T cells are activated by IL-1 (interleukin-1), which is released by macrophages. The activated T cells then produce both IL-2 and proteins that serve as receptors for IL-2. The binding of IL-2 to a T cell's IL-2 receptors causes the cell to divide. The result is the rapid growth of a clone of T cells (Figure 16.24). IL-2 produced by T_H cells helps B cells to start secreting antibodies, and it probably causes the B cells to divide after they are activated by antigen. IL-2 also activates the natural killer cells discussed earlier in this chapter.

The central role of IL-2 in the immune response is best illustrated by a medical example. When a tissue transplant is to be made, the immune system must be suppressed to reduce the danger of rejection of the transplanted tissue. The two drugs most commonly used for suppressing the immune system both function by inhibiting the production of IL-2.

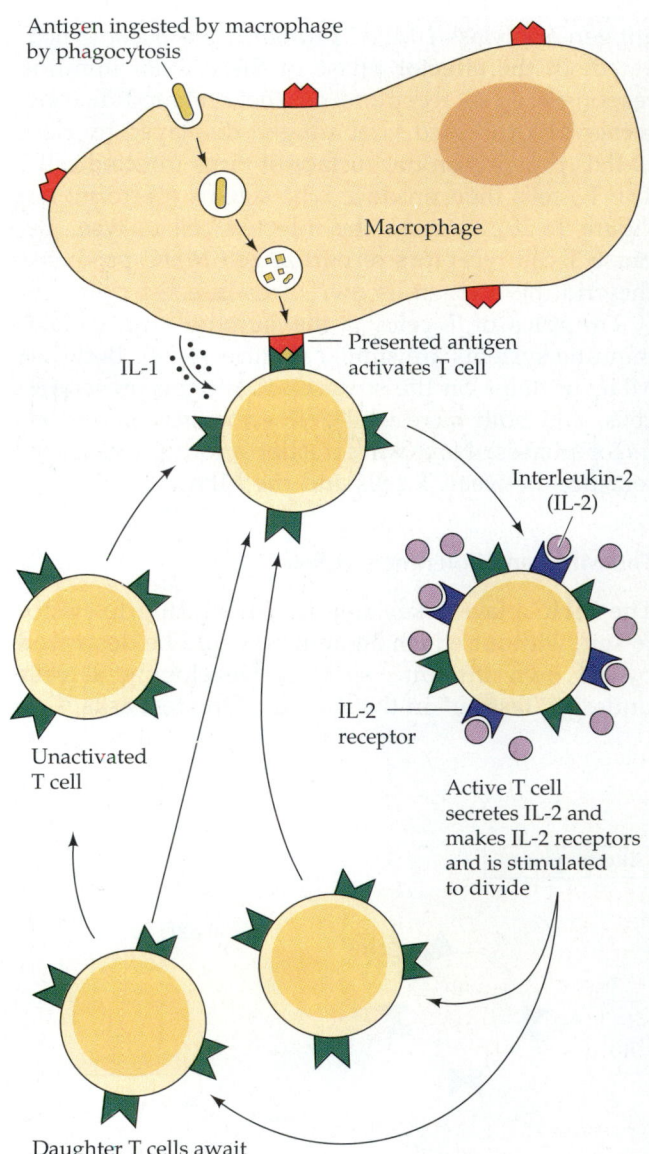

16.24 The Action of Interleukin-2
A T cell, activated by IL-1 from a macrophage, makes IL-2 receptors that become part of its own plasma membrane and secretes IL-2 that binds to these receptors and to those of other activated T cells. The binding of the IL-2 causes the T cells to divide, and the daughter cells are ready to become activated in turn, leading to a clone of T cells with the same antigen specificity.

DISORDERS OF THE IMMUNE SYSTEM

A common type of condition relating to the immune system arises when the system overreacts to a dose of antigen. Although the antigen itself may present no danger to the host, the immune response may produce inflammation and other symptoms that can cause serious illness or death. **Allergies** are the most familiar examples of such a problem. In extreme cases, an exposure to a particular antigen (such as the toxin of a bee sting) may lead to a fatal overreac-

tion of the immune system—dilation of some blood vessels and constriction of others, causing first shock and then death.

Sometimes the immune recognition of self fails, resulting in the appearance of one or more "forbidden clones" of B and T cells directed against self antigens. This failure does not always result in disease, but in some instances it can be disastrous. Among the **autoimmune diseases** of our species—diseases in which components of the body are attacked by its own immune system—are rheumatic

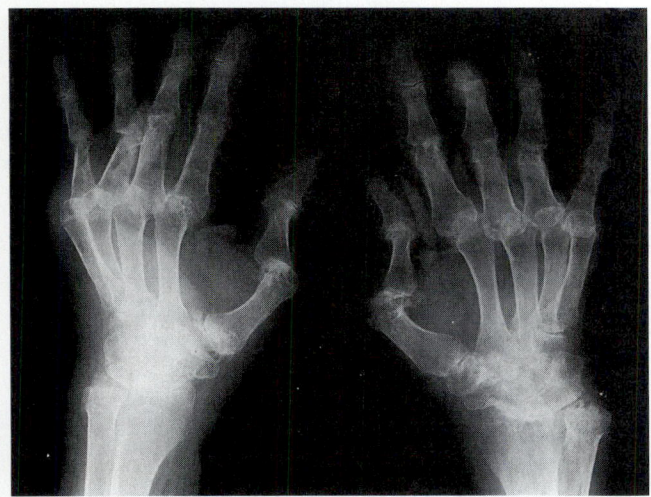

16.25 Rheumatoid Arthritis
This X ray shows the bones in the hands of a 50-year-old man with a 20-year history of rheumatoid arthritis. This autoimmune disease affects the joints, destroying the cartilage with an inflammatory reaction and causing pain and the severe deformities seen here.

fever, rheumatoid arthritis (Figure 16.25), ulcerative colitis, myasthenia gravis (muscle weakness), and several others. Many medical scientists believe that multiple sclerosis results from an abnormality in a type of T cell. The abnormal T cells mount an immune attack on the myelin sheath, an insulating material that surrounds many nerve cells (see Chapter 38). When the myelin sheath is damaged, the result is a severely debilitating loss of nerve function, including blindness and loss of motor control.

Immune Deficiency Disorders

There are various immune deficiency disorders, such as those in which B cells are never formed and others in which B cells lose the ability to give rise to plasma cells. In either case, the affected individual is unable to produce antibodies and thus lacks a major line of defense against microbial pathogens.

Because of its essential roles in both antibody responses and cellular immune responses, the T_H cell is perhaps the most central of all the components of the immune system—the worst one to lose to an immune deficiency disorder. The disease responsible for the epidemic discussed at the beginning of this chapter, acquired immune deficiency syndrome, or **AIDS**, homes in on the T_H cells. At the onset of the infection, the number of HIV (*h*uman *i*mmunodeficiency *v*irus) particles in the blood soars until the immune system mounts a response that clears most—but, unfortunately, not all—of the HIV particles. Some of the particles infect and remain latent in the T_H cells, often for many years. As more cells become infected, the immune system is weakened.

Eventually the HIV count in the blood soars again and the immune system fails completely.

HIV is a retrovirus—a virus with RNA as its genetic material, capable of inserting its own genome into the genome of its animal host. The structure of HIV is shown in Figure 16.26. A central core, with a protein coat (p24 capsule protein), contains two identical molecules of RNA, as well as certain enzymes. An envelope, derived from the plasma membrane of the cell in which the virus was formed, surrounds the core. The envelope is studded with an envelope protein (gp120, where gp stands for glycoprotein) that enables the virus to infect its target T_H cell.

HIV attacks host cells at a membrane protein called CD4, which is found primarily on T_H cells and macrophages. CD4 acts as the receptor for the viral envelope protein gp120. The binding of gp120 to CD4 starts a complex series of events (Figure 16.27). When HIV infects a cell, the viral core is admitted into the cell; then the core "uncoats" itself, releasing its contents. Among the enzymes in the core is **reverse transcriptase**, which catalyzes the formation of a double-stranded DNA molecule encoding the same information as the viral RNA (see Chapters 13 and 14). Another enzyme catalyzes the destruction of RNA molecules transcribed from the host cell's own DNA. The DNA transcript enters the nucleus of the host cell and is spliced into a chromosome, much as bacteriophage DNA may become incorporated into a bacterial chromosome. Another HIV enzyme, called **integrase**, catalyzes this splicing. The DNA transcript of the HIV RNA thus becomes a permanent part of the chromosome, replicating with it at each division of a host cell.

A DNA transcript of the HIV RNA, once incorporated into the genome of a T_H cell, may remain there latent for days, or even for a decade or more. The latent period ends if the HIV-infected T_H cell becomes activated. Then the viral DNA is transcribed, yielding many molecules of viral RNA, some of which are translated, forming the enzymes and structural proteins of a new generation of viruses. Other viral RNA molecules are incorporated directly into the new viruses as their genetic material. Formation of new viruses may be slow; the viruses bud from the infected cell, surrounding themselves with modified plasma membrane from the host. More-rapid virus production leads to lysis of the host cell. Several viral genes control the rate of production of individual proteins and of whole viruses. One viral gene, responsible for the antigenic properties of the envelope protein, has a high mutation rate, causing the antigen to change, making HIV a moving target for what is left of the host's immune system—and complicating efforts to develop a vaccine against AIDS.

We do not yet know the mechanisms that lead to

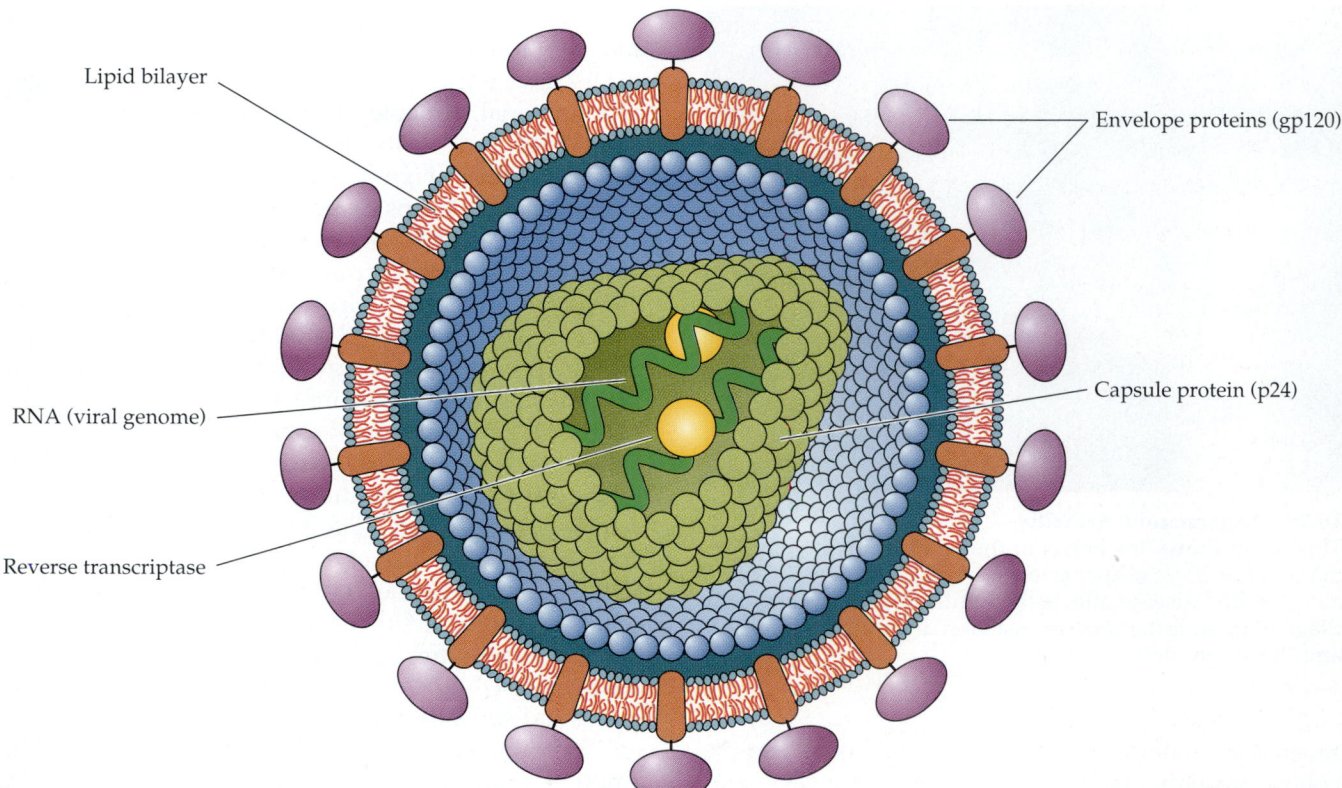

Lipid bilayer

Envelope proteins (gp120)

Capsule protein (p24)

RNA (viral genome)

Reverse transcriptase

16.26   Structure of HIV, the AIDS Virus
HIV has a core containing RNA and various proteins, including the enzyme reverse transcriptase. Another protein (shown in blue) surrounds the core. The surface of the virus is complex: A phospholipid bilayer, studded with knobs of envelope protein, covers the protein layer surrounding the core.

gradual, selective depletion of the T_H cells. As a consequence of this depletion, however, the host's immune system becomes unable to function. AIDS patients usually die of "opportunistic" infections, diseases caused by bacteria and fungi that are almost always eliminated by the immune systems of uninfected individuals. **AIDS-related complex**, a less severe form of the disease, has milder symptoms but appears to develop into full-blown AIDS in most patients. Thus far, HIV infection appears to lead to death in all cases.

In the United States, the highest incidence of AIDS occurs among drug addicts (from the use of shared, contaminated needles) and male homosexuals. AIDS can be transmitted by blood transfusions, by sexual activity (either homosexual or heterosexual), and from mother to fetus. HIV is not transmitted by mosquitoes or other insect vectors, or by kissing or casual contact. On a worldwide basis, the most common mode of transmission is heterosexual intercourse. At this writing, most of the HIV-infected people in the world live in Africa (Figure 16.28), but the virus is spreading fastest among people living in south and

southeast Asia. By the turn of the century, residents of developing countries will bear 90 percent of new HIV infections as the AIDS epidemic rages on throughout the world. There is not likely to be an effective vaccine or cure by that time. Millions of people will have died before the AIDS epidemic ends. In the meantime, medical and biological research on the subject is proceeding with great intensity.

**Prospects for an AIDS Cure—
or an AIDS Vaccine**

Prospects for a *cure* for AIDS are exceedingly dim at this time. What would an AIDS cure entail? It would include the detection and elimination of every HIV-infected cell in the body. Given the long latent period of infection, this task appears overwhelming. If a cure seems far off, what then?

Many investigators are seeking a vaccine to forestall HIV infection. You have probably seen newspaper or television news reports of some of these studies, and you will certainly see many more. Here again, however, there are major stumbling blocks. One of the greatest problems is the exceptional genetic variation of HIV. New strains are constantly appearing, and a promising new vaccine against one strain may afford no protection whatsoever against new or even other old strains. What next?

There is substantial agreement that we can learn

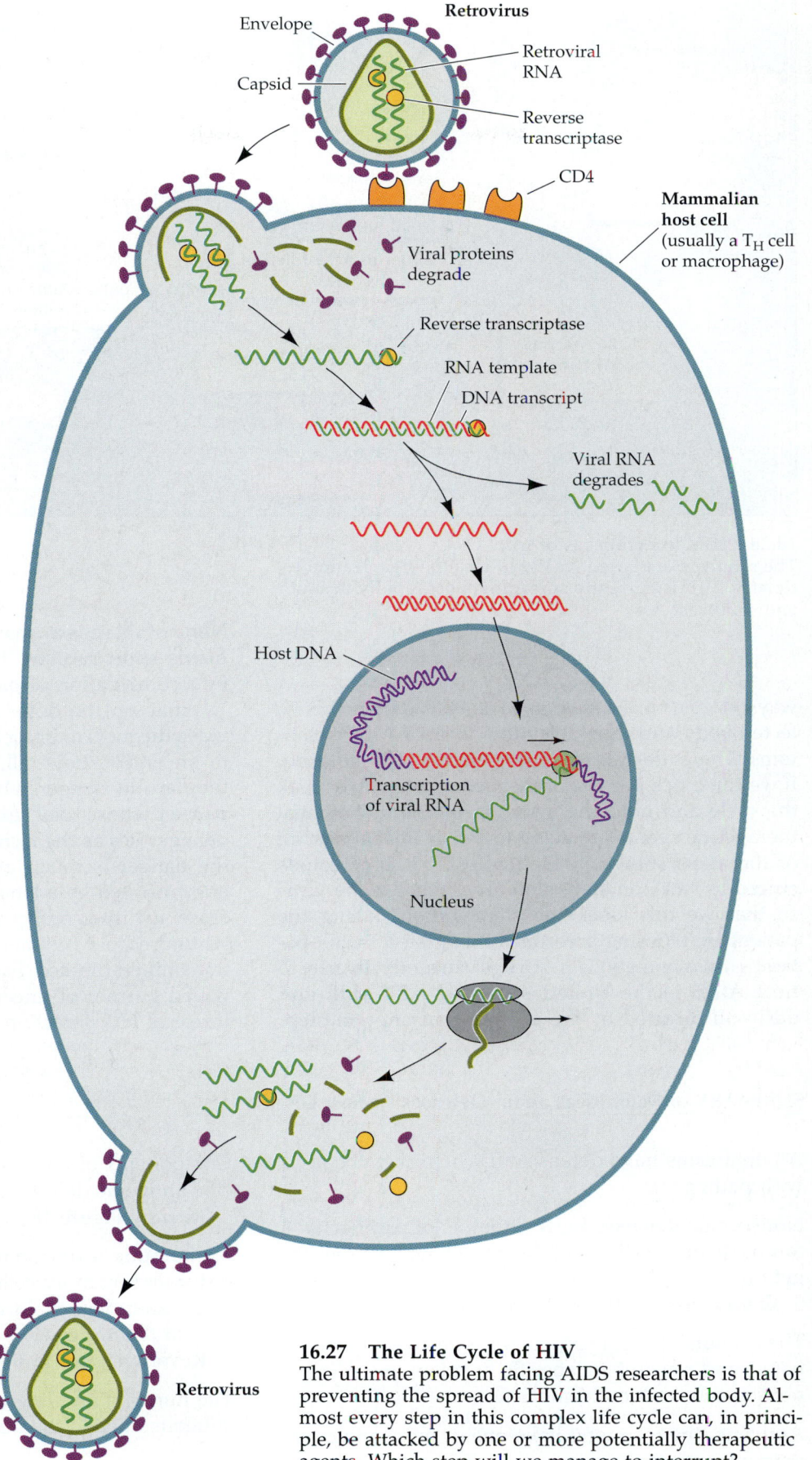

1. Retrovirus attaches to host cell at membrane protein CD4

2. The viral core is uncoated as it enters the host cell by endocytosis

3. Viral RNA uses reverse transcriptase to make complementary DNA

4. Viral RNA degrades

5. Single-stranded reverse transcript synthesizes second complementary DNA strand

6. DNA enters cell nucleus and is integrated into host chromosome, forming a provirus

7. Proviral DNA transcribes viral RNA, which is exported to cytoplasm

8. Viral RNA is translated

9. Viral proteins, new capsids, and envelopes are assembled

10. Assembled virus buds from the cell membrane and releases virus particles

11. Virus matures

16.27 The Life Cycle of HIV
The ultimate problem facing AIDS researchers is that of preventing the spread of HIV in the infected body. Almost every step in this complex life cycle can, in principle, be attacked by one or more potentially therapeutic agents. Which step will we manage to interrupt?

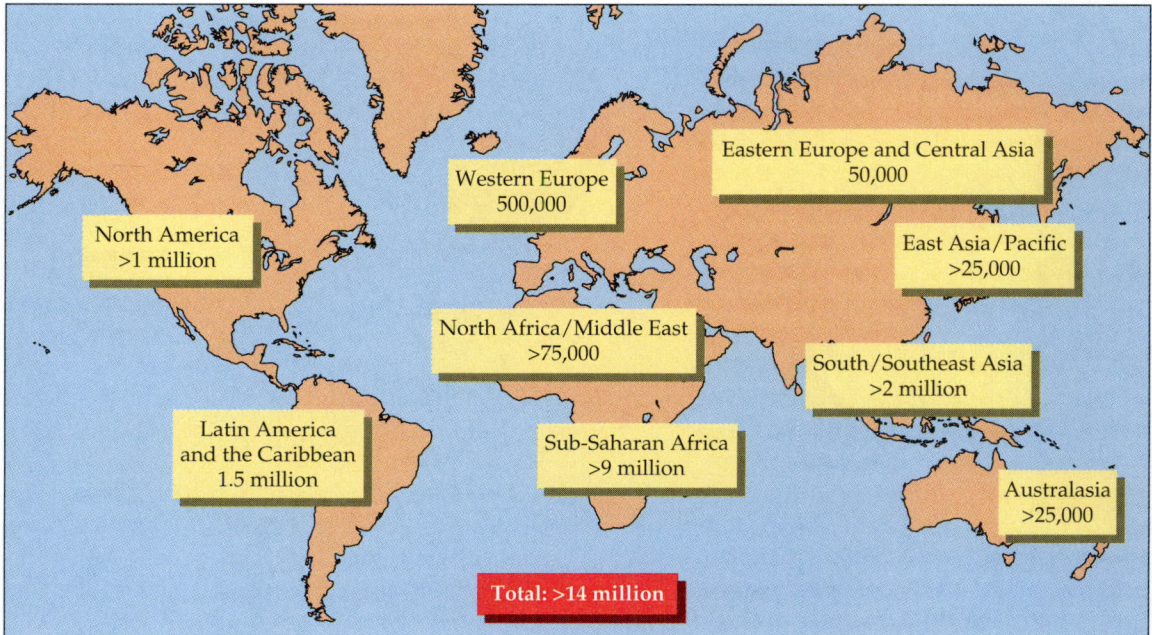

North America
>1 million

Western Europe
500,000

Eastern Europe and Central Asia
50,000

East Asia/Pacific
>25,000

North Africa/Middle East
>75,000

South/Southeast Asia
>2 million

Latin America
and the Caribbean
1.5 million

Sub-Saharan Africa
>9 million

Australasia
>25,000

Total: >14 million

16.28   HIV Infections as of Mid-1994
The numbers are based on World Health Organization estimates. HIV is currently spreading most rapidly in south
and southeast Asia.

ways to *control the replication* of HIV within the infected body and thus delay the onset of AIDS symptoms. There are at least 16 steps in the HIV life cycle.
If we can block just *one* of them completely, we break
the cycle and hold the infection in check. Potential
therapeutic agents are being tested against almost all
of the steps shown in Figure 16.27. It is of course
crucial to work on steps that are unique to the virus
so that we can block the step without killing the
patient by blocking a corresponding step in the patient's own metabolism. Drugs currently in use to
treat AIDS in the United States are AZT, ddI, and
ddC—all directed at the reverse transcription step.

None of these is a "magic bullet" against the disease.
In the short run, we hope to delay HIV replication
by a combination of partially effective treatments.
    What can be done until biomedical science pro-
vides the tools to bring the worldwide AIDS epidemic
to an end? Above all, people must recognize that
they are in danger whenever they have sex with a
partner whose total sexual history is not known. The
danger rises as the number of sex partners rises, and
the danger is much greater if sexual intercourse is
not protected by a latex condom. The danger of heterosexual intercourse transmitting HIV rises ten- to
a hundredfold if either partner has another sexually
transmitted disease. Face it: *Every* penetrative sex act
with a partner of uncertain HIV status carries a real
threat of HIV infection, and *no cure is in sight*.

SUMMARY of Main Ideas about Defenses Against Disease

All organisms have defenses that help them cope
with pathogens.

Nonspecific defenses block attempts of pathogens
to invade our bodies and destroy pathogens that do
get in.
    Review Figure 16.2 and Table 16.1

The immune system recognizes, selectively eliminates, and remembers specific invaders, including
nonself cells and macromolecules.

Antigens (the invaders) have one or more antigenic
determinants.

Components of the immune system recognize specific antigenic determinants.
    Review Figure 16.5

An immune response includes the formation of effector and memory cells; memory cells allow a
larger and more rapid response to a second exposure to the antigen.
    Review Figure 16.6

The humoral immune response is directed against
antigens outside the host cells.

The humoral immune response relies on antibodies (immunoglobulins), each of which recognizes and binds one specific antigenic determinant.

B cells give rise to plasma cells, the effector cells that produce antibodies.

The cellular immune response is directed against multicellular pathogens and against viruses that have established themselves in host cells.

The cellular immune response relies on T-cell receptors, each of which recognizes and binds one specific antigenic determinant.
   Review Figure 16.19

Effector T cells include cytotoxic (T_C), helper (T_H), and suppressor (T_S) T cells.

Specific B and T cells become activated when exposed to specific antigenic determinants; they multiply and differentiate to give rise to clones of cells of identical specificity.
   Review Figure 16.7

Continued exposure to self antigens beginning at or before birth results in the deletion or inactivation (by suppressor T cells) of cells that would otherwise produce antiself antibodies.

The immunoglobulin molecule is a tetramer of two identical light-chain polypeptides and two identical heavy-chain polypeptides.
   Review Figure 16.10

Both light and heavy chains contain constant and variable regions.

Immunoglobulin specificity is determined by the variable regions.

Different heavy chains determine the type of effects produced by different antibody classes.

Antibody diversity results from the chance selection of different regions of a chromosome in constructing a single gene coding for the immunoglobulin molecule, as well as from frequent mutation of some of the genes.
   Review Figures 16.16 and 16.17

Genes of the major histocompatibility complex (MHC) encode cell-surface proteins that present antigen to helper T cells and effector cells of the cellular immune system.
   Review Figures 16.21, 16.22, and 16.23

Histocompatibility antigens also are the basis for the rejection of nonself tissue transplants.

Cells of the immune system communicate by releasing interleukins.
   Review Figure 16.24

HIV, a retrovirus that attacks helper T cells, causes AIDS.
   Review Figure 16.26

Reverse transcriptase, carried by the HIV virus, catalyzes the formation of a DNA transcript of the viral RNA.

The DNA transcript of HIV RNA is spliced into the host genome and may remain latent for years before being expressed and producing new viruses, eventually crippling the host's immune system.
   Review Figure 16.27

There is no immediate prospect of either a cure or a vaccine for AIDS.

SELF-QUIZ

1. Which statement is *not* true of phagocytes?
   a. Some travel in the circulatory system.
   b. They ingest microorganisms by endocytosis.
   c. A single one can ingest 5 to 25 bacteria before it dies.
   d. Although they are important, an animal can do perfectly well without them.
   e. Lysosomes play an important role in their function.

2. Immunoglobulins
   a. help antibodies do their job.
   b. recognize and bind antigenic determinants.
   c. are among the most important genes in an animal.
   d. are the chief participants in nonspecific defense mechanisms.
   e. are a specialized class of white blood cells.

3. Which statement is *not* true of an antigenic determinant?
   a. It is a specific chemical grouping.
   b. It may be part of many different molecules.
   c. It is the part of an antigen to which an antibody binds.
   d. It may be part of a cell.
   e. A single antigen has only one antigenic determinant on its surface.

4. T-cell receptors
   a. are the primary receptors for the humoral immune system.
   b. are carbohydrates.
   c. cannot function unless the animal has previously encountered the antigen.
   d. are produced by plasma cells.
   e. are important in combatting viral infections.

5. According to the clonal selection theory
   a. an antibody changes its shape according to the antigen it meets.
   b. an individual animal contains only one type of B cell.

c. the animal contains many types of B cells, each producing one kind of antibody.

d. each B cell produces many types of antibodies.

e. no clones of antiself lymphocytes appear in the bloodstream.

6. Immunological tolerance
   a. depends on repeated exposure throughout the life of the animal.
   b. develops late in life and is usually life-threatening.
   c. disappears at birth.
   d. results from the activities of the complement system.
   e. results from DNA splicing.

7. The extraordinary diversity of antibodies results in part from
   a. the action of monoclonal antibodies.
   b. the splicing of RNA molecules.
   c. the action of suppressor T cells.
   d. the splicing of gene segments.
   e. their remarkable nonspecificity.

8. Which of the following play no role in the antibody response?
   a. Helper T cells
   b. Interleukins
   c. Macrophages
   d. Reverse transcriptase
   e. Products of class II MHC loci

9. The major histocompatibility complex
   a. codes for specific proteins found on the surfaces of cells.
   b. plays no role in T-cell immunity.
   c. plays no role in antibody responses.
   d. plays no role in skin graft rejection.
   e. is coded for by a single locus with multiple alleles.

10. Which of the following plays no role in AIDS?
    a. Integrase
    b. Reverse transcriptase
    c. Transcription
    d. Translation
    e. Transfection

FOR STUDY

1. Describe the part of an antibody molecule that interacts with an antigenic determinant. How is it similar to the active site of an enzyme? How does it differ from the active site of an enzyme?

2. Contrast immunoglobulins and T-cell receptors, with respect to structure and function.

3. Discuss the diversity of antibody specificities in an individual in relation to the diversity of enzymes. Does every cell in an animal contain genetic information for all the organism's enzymes? Does every cell contain genetic information for all the organism's immunoglobulins?

4. Describe and contrast two ways in which DNA splicing plays roles in the immune response.

5. Discuss the roles of monoclonal antibodies in medicine and in biological research.

READINGS

Ada, G. L. and G. Nossal. 1987. "The Clonal-Selection Theory." *Scientific American*, August. A fascinating historical account of the development of the central concept of immunology.

Atkinson, M. A. and N. K. Maclaren. 1990. "What Causes Diabetes?" *Scientific American*, July. An explanation of the origin of diabetes, in terms of an autoimmune response, that may lead to the development of preventive therapies for insulin-dependent diabetes.

Boon, T. 1993. "Teaching the Immune System to Fight Cancer." *Scientific American*, March. Therapies designed to cause a patient's own T-cells to attack tumors.

Cohen, I. R. 1988. "The Self, the World, and Autoimmunity." *Scientific American*, April. The nature of autoimmune diseases, and an approach to their prevention.

Edelman, G. M. 1970. "The Structure and Function of Antibodies." *Scientific American*, August. The amino acid sequence of an antibody dictates its unique characteristics and determines its ability to interact with an antigen.

Engelhard, V. H. 1994. "How Cells Process Antigens." *Scientific American*, August. Clearly explains the roles of the MHC molecules and other key players.

Golde, D. W. 1991. "The Stem Cell." *Scientific American*, December. The cell that gives rise to blood cells and immune systems—applications in cancer treatment and immune defects.

Golub, E. S. and D. R. Green. 1991. Immunology: A Synthesis, 2nd Edition. Sinauer Associates, Sunderland, MA. An outstanding textbook of immunology, with special attention to the experimental basis for what we know.

Haynes, B. F. 1993. "Scientific and Social Issues of Human Immunodeficiency Virus Vaccine Development." *Science*, vol. 260, pages 1279–1286. The technical—as well as the social and ethical—difficulties in conquering AIDS are enormous.

Lerner, R. A. and A. Tramontano. 1988. "Catalytic Antibodies." *Scientific American*, March. A powerful new tool for biotechnology, combining the talents of enzymes and antibodies, which may also be useful in augmenting the capabilities of the immune system.

Marrach, P. and J. Kappler. 1986. "The T Cell and Its Receptor." *Scientific American*, February. A detailed consideration of the key actors in the cellular immune system.

Smith, K. A. 1990. "Interleukin-2." *Scientific American*, March. A clear description of the role of interleukin-2 in the expansion of a clone of T cells.

Tonegawa, S. 1985. "The Molecules of the Immune System." *Scientific American*, October. A beautifully illustrated account of the structures of antibodies and T-cell receptors and of how they are formed.

von Boehmer, H. and P. Kisielow. 1991. "How the Immune System Learns about Self." *Scientific American*, October. Experiments with transgenic mice established the clonal deletion theory.

The baby octopus in the photograph, like its still-unhatched sisters and brothers, has an intricate body. If it is lucky enough to survive, it will grow a hundredfold or more in size, and its adult form will differ from the hatchling you see here. The differences, however, will not be nearly as great as the differences between this newborn animal and the single cell it developed from.

A fertilized egg, or zygote, is a single cell and often very tiny; but within the miniscule structure is the astounding potential to create an entire organism. The human zygote, for example, is only about one-tenth of a millimeter in diameter, or one-fifth the diameter of the period at the end of this sentence. Small as it is, the zygote gives rise to a precisely shaped, extremely complex adult. What happens between the single-celled and adult states? For one thing, an amazing number of new cells forms—more than a hundred trillion (10^{14}) of them. Are these cells all alike? Scarcely. The human body consists of more than 200 types of cells, each with one or more important roles. The body changes form significantly as various tissues take up new positions. Myriad intricate steps make up the development of the body.

THE STUDY OF ANIMAL DEVELOPMENT

Development is a process of progressive change during which an organism successively takes on the forms of the several stages of its life cycle. In its early stages of development an animal (or plant) is called an **embryo**. Sometimes the embryo is contained within a protective structure such as an eggshell or a uterus. An embryo does not actively feed because it obtains its food directly or indirectly (by way of the egg, for example) from its mother. Although we focus on the embryo in this chapter, development is a process that continues through all stages, ceasing only with the death of the animal.

The goal of the developmental biologist is to understand the steps from zygote to adult and the molecular mechanisms that underlie them. The task is daunting—just look at the incredible diversity of animal body types on Earth and imagine the different paths that must lead to that diversity! Early twentieth-century work on development offered hope as important generalizations emerged from the study of the first several hours of development in organisms as diverse as sea urchins, frogs, and flies. But after more studies, the patterns of development of some animals appeared to differ from those of others, and the apparent lack of "rules" of development was discouraging.

Today, however, we find ourselves in a true "golden age" of developmental biology. Some of the pieces are coming together. A few of the gaps be-

Life Emerges from an Octopus Egg

17

Animal Development

17.1 The Cast of Characters
Some of the species scientists use extensively in research on animal development include sea urchins, frogs, and chickens. Here the contrast between adult organisms and their embryo or larval stages is clear.

tween different patterns of development have been bridged, and, as is true throughout biology, discoveries in one field have unexpected application in other fields. The work of geneticists, developmentalists, endocrinologists (those who study hormone action), biochemists, neurobiologists, and others has come to a common focus. Seemingly unrelated studies of roundworms, fruit fly larvae, and mice contribute to one another and to the understanding of human development. Current research has uncovered a stunning unity in certain developmental processes at the cellular and molecular levels, in spite of the equally stunning diversity at the level of the entire organism. Our ignorance still exceeds our knowledge, however.

In our study of animal development, we begin with an an overview of the *unity* that exists across animal groups in the earliest stages of development. We will see that, by cell division and cell movement, a zygote first becomes a solid sphere of cells. The

sphere then develops into a hollow ball of cells. Groups of cells begin to infold, forming a gut. Organs begin to form. Once we have the sequence—the unity—of these earliest stages, we will consider the important developmental processes of determination, differentiation, and pattern formation as we take a closer look at specific points in the development of several animal species.

CLEAVAGE

Some species have been particularly useful subjects in studies of animal development. Foremost among them are frogs, chickens, sea urchins, fruit flies, and a tiny roundworm (Figure 17.1). A sea urchin, a frog, a bird, and a roundworm all start life in the same way: as dividing cells. As we explore the unity in the development of these animals, however, our attention is drawn, even at the first cell divisions, to differences.

Becoming Multicellular

When an egg is fertilized, the resulting zygote nucleus is activated, and DNA replication and mitosis commence. The activation of the nucleus begins the process of cleavage, in which the zygote divides mitotically and gives rise, over a period of hours, to hundreds of cells, and eventually to thousands. In most embryos, initially all the cells divide at the same time in each round of division: First there is a single cell, then there are 2, then 4, 8, 16, and so forth.

From species to species, differences in cleavage, including the arrangement of the daughter cells, or **blastomeres**, depend especially on the amount and distribution of **yolk**—nutrients stored in the egg. The sea urchin, for example, has a small egg (0.15 mm in diameter) with only a small amount of yolk that is distributed uniformly. In such an egg, the blastomeres separate completely from one another as they are formed; division proceeds nearly simultaneously from both ends, or poles, of the cell (Figure 17.2a).

By comparison, a frog egg is rather larger (0.5–1 mm), with the yolk concentrated in one half (the vegetal hemisphere) and pigment concentrated in the other (the animal hemisphere). After fertilization, there is a substantial redistribution of the cytoplasmic contents of the zygote, including the movement of some pigmented material. As a result, a **gray crescent** forms on one side of the zygote, opposite the point of sperm entry and near the boundary between the animal and vegetal hemispheres. The gray crescent contains less pigment than does the rest of the animal hemisphere. As we will see, the gray crescent is of great significance in later development. The blastomeres of a frog zygote divide completely, but the division begins at the point farthest from the vegetal hemisphere. This point is called the **animal pole**; the point farthest from the animal pole is called the **vegetal pole**. The plane of the first cleavage passes through the animal pole, through the middle of the gray crescent on one side of the zygote, through the site of sperm entry on the other, and eventually through the vegetal pole (Figure 17.2b).

A bird egg, which is larger than that of a frog and

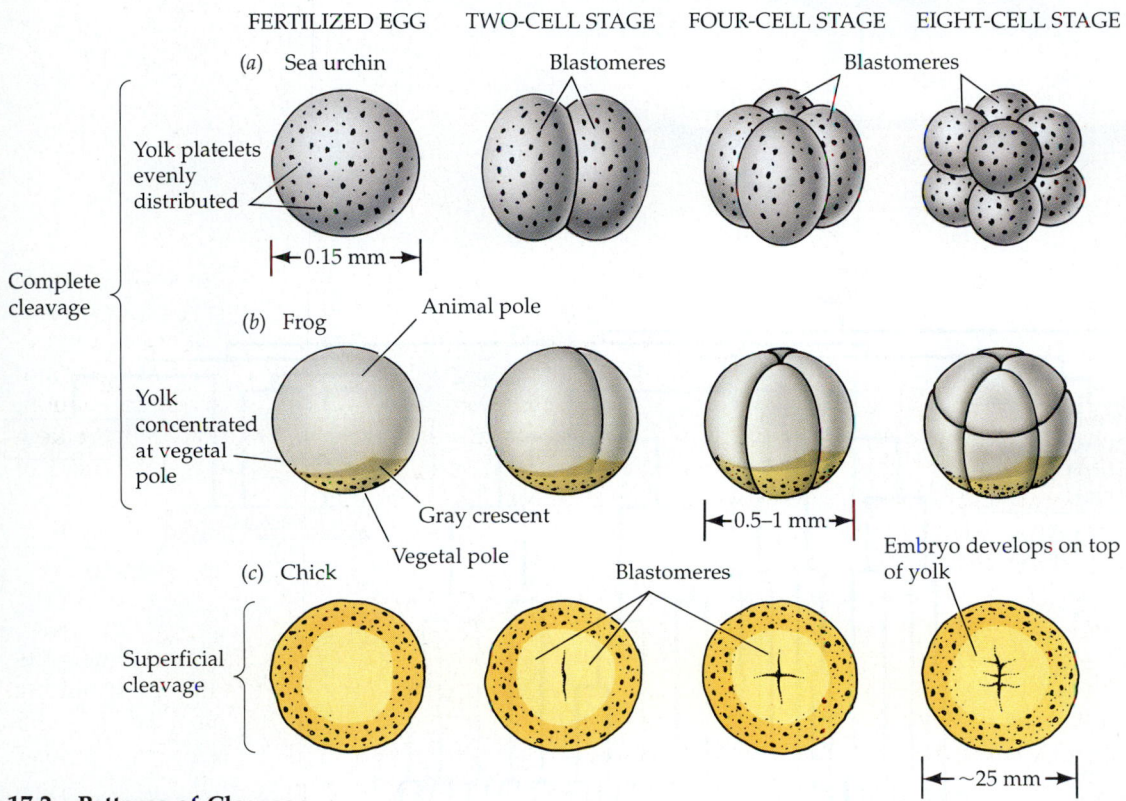

FERTILIZED EGG   TWO-CELL STAGE   FOUR-CELL STAGE   EIGHT-CELL STAGE

(a) Sea urchin

Yolk platelets evenly distributed

Blastomeres

Blastomeres

0.15 mm

Complete cleavage

(b) Frog

Animal pole

Yolk concentrated at vegetal pole

Gray crescent

Vegetal pole

0.5–1 mm

(c) Chick

Embryo develops on top of yolk

Blastomeres

Superficial cleavage

~25 mm

17.2 Patterns of Cleavage

Patterns of early embryonic development reflect differences in the way egg cytoplasm is organized. (a) A sea urchin egg is small, with little yolk (black dots). All of the first three cell divisions are complete, cutting through the entire egg. Both of the first two divisions divide the cell lengthwise; the third cuts across the middle of the cells in a plane perpendicular to that of the first two cleavages. After three divisions, then, the sea urchin embryo consists of eight approximately equal-size cells arranged in two layers. (b) A frog egg is larger than a sea urchin's, with yolk concentrated toward the vegetal pole. All cell divisions are complete: the first two begin near the animal pole; the third is transverse, cutting across the cells near the animal pole. As a result, the cells of the eight-cell stage differ in size, in the amount of yolk they contain, and in their proximity to the gray crescent. (c) A chicken egg is mainly yolk with only a small mass of cytoplasm. The first few divisions of the cytoplasm (shown here from above) do not extend through the yolk; they yield a thin, flat embryo on the surface.

contains much more yolk, is markedly different. Yolk occupies the bulk of the cell, with the yolk-poor cytoplasm confined to the surface or to one end of the cell. Cell division in such zygotes is incomplete: Following each mitosis, the nuclei are separated completely, but initially the daughter cells are not separated completely by plasma membranes. As cleavage continues, however, more and more cells with complete plasma membranes are formed. The embryo develops initially as a disc-shaped mass on top of the yolk (Figure 17.2c).

In some animals, such as roundworms and clams, sister blastomeres become committed to different developmental roles already when there are eight or fewer cells. The parts of the embryo that will be derived from each blastomere are fixed after the second or third division—in a few species, even after the first division. If a blastomere is lost from an embryo of this type, the corresponding portion of the animal is not produced. This developmental pattern is called **mosaic development**, with each blastomere contributing a specific set of "tiles" to the final "mosaic" of the organism. As a consequence, in such embryos the future fates of cells produced at each early division are predictable. Mosaic development is dramatically illustrated in a tiny roundworm, *Caenorhabditis elegans* (Figure 17.3).

By contrast to *C. elegans*, sea urchin and vertebrate embryos are characterized by **regulative development**, in which the loss of some cells during cleavage does not affect development because the remaining cells compensate for the loss. In humans, for example, the separation of one embryo into two masses at the 64-cell stage or beyond, and the subsequent de-

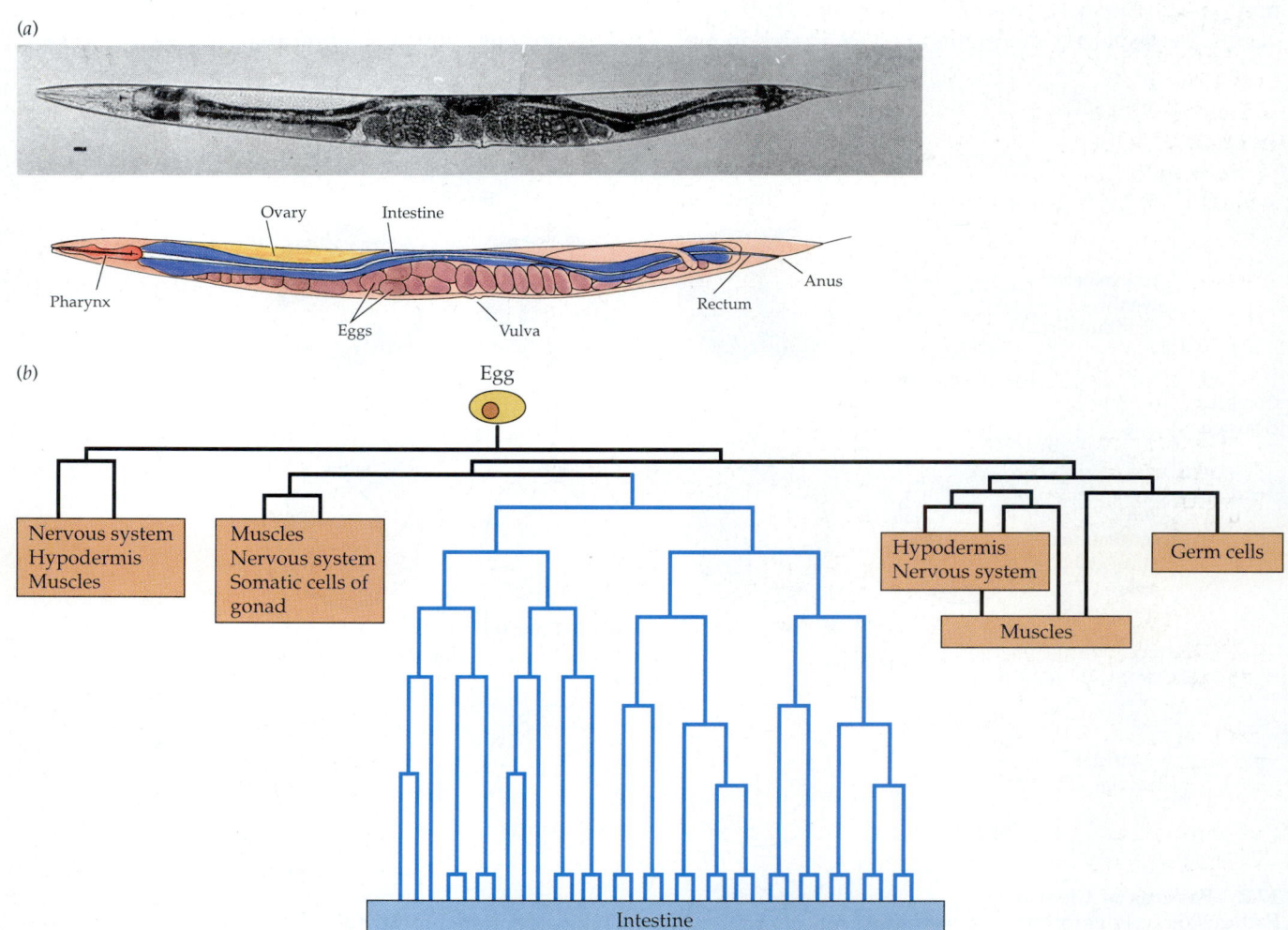

17.3 Development of *C. elegans*
Caenorhabditis elegans is a 1-millimeter-long nematode, or roundworm. There are two sexes: male and the hermaphrodite shown here, which has both male and female reproductive structures. The hermaphrodite reaches its adult stage just three and a half days after fertilization of the egg. *(a)* Because its internal structures are visible and its anatomy is relatively simple, the transparent, colorless worm is a useful organism for tracking cellular development. *(b)* It has been possible to trace all divisions from a single cell to the 959 cells found in the fully developed adult. Each fork in this tree represents the mitotic division of a cell. Here we focus on those cells that give rise to the intestine. At the eight-cell embryonic stage, a single cell is can be identified as the source of all future intestinal cells.

velopment of each, produces identical twins. Even in regulative development, however, there is a point beyond which a loss of cells does affect the outcome. By the same token, the development of primarily mosaic embryos includes some steps in which substitutions are possible.

In spite of the differences just discussed, a number of features of cleavage—the series of mitotic divisions that divides the zygote cytoplasm into many smaller cells—are common to most animals. First, mitosis is usually quite rapid during this stage. For a given organism, the blastomeres proliferate more rapidly than do any other cells. Second, there is little or no increase in overall volume during cleavage, so the blastomeres become smaller and smaller with each division; the entire ball of cells remains approximately the same size as the fertilized egg. Third, the ratio of nuclear volume to cytoplasmic volume in the embryo increases steadily throughout cleavage because the original supply of cytoplasm (from the egg) is being shared by ever more nuclei. Finally, during cleavage massive synthesis of DNA and associated chromosomal proteins takes place.

Formation of the Blastula

Throughout cleavage, the embryo retains roughly the same external spherical form. As cleavage proceeds in sea urchins, frogs, birds, and many other animals, the sphere becomes a hollow structure, the **blastula**. Its cavity, the **blastocoel**, forms because the blastomeres selectively detach from one another on their innermost surfaces during early cleavage, leaving a fluid-filled space in the center. The blastomeres themselves constitute a sheet of cells, the **blastoderm**. In sea urchins the blastoderm is only one cell thick; in

other animals, such as frogs, it may be several cells thick (Figure 17.4).

The developing blastula is a dynamic structure; each division of blastomeres results in changed contacts among the cells. Within the blastula, however, the contents of the embryo are distributed much as they were in the original zygote; in the frog embryo, for example, the yolk is still more concentrated near the vegetal pole. Only in the subsequent gastrula stage of development do massive rearrangements of cells and materials begin within the embryo. How does the embryo proceed next to lay the groundwork for the production of specific structures such as a digestive tube (a gut)?

GASTRULATION

In all animals, cells from the surface of the blastula move to the interior to form layers. Cells redistribute themselves to transform the blastula into a **gastrula**, an embryo with a gut connected to the outside world by either an anus or a mouth. During gastrula formation, called **gastrulation**, some surface cells move into the interior, resulting in a two- or three-layered embryo. These layers are called germ layers because they give rise to distinct tissues and organs (Table 17.1). The blastoderm **invaginates** (bends inward) to form a pocket of an inner germ layer, the **endoderm**, leaving an outer germ layer, the **ectoderm**. In a few animals such as jellyfish, a two-layered embryo consisting only of ectoderm and endoderm develops into a two-layered adult with little cellular diversity. Adults of most species, however, develop from a three-layered embryo: The third germ layer, the **mesoderm**, forms between ectoderm and endoderm. In

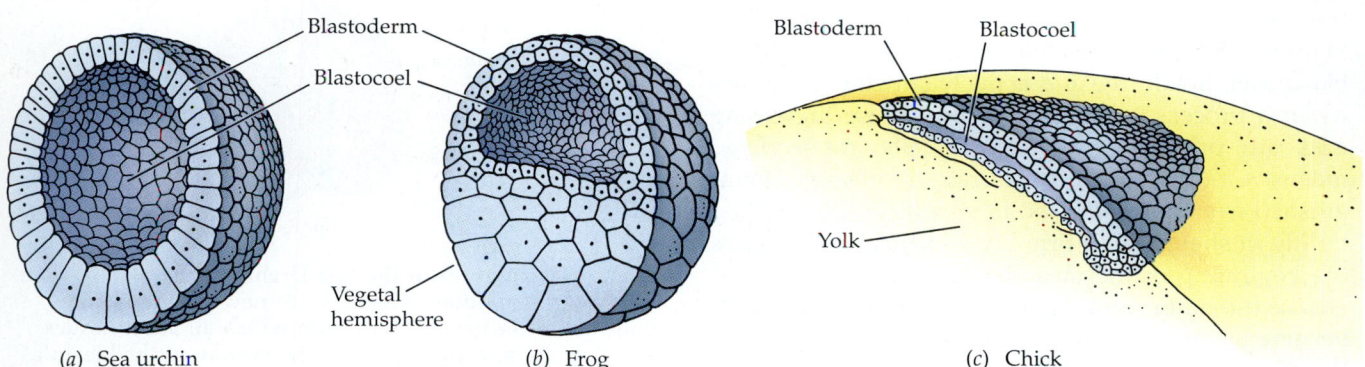

(a) Sea urchin   (b) Frog   (c) Chick

17.4 Blastulas Reflect Patterns of Cleavage
From the beginnings shown in Figure 17.2, continued cell divisions produce hollow blastulas, cut open here to show the blastocoel. (a) In sea urchins and organisms with similar eggs, the blastoderm is a single sheet of cells, and the blastocoel they enclose is spherical. (b) In frogs, the blastoderm is many cells thick in the vegetal hemisphere and the blastocoel is flattened on one side. (c) The blastocoel of birds is a lens-shaped cavity defined by a thin blastoderm layer below and a slightly thicker one above.

TABLE 17.1
Fates of Embryonic Germ Layers in Vertebrates[a]

GERM LAYER	FATE
Ectoderm	Brain and nervous system; lens of eye; inner ear; lining of mouth and of nasal canal; epidermis of skin; hair and nails; sweat glands, oil glands, milk secretory glands
Mesoderm	Skeletal system: bones, cartilage, notochord; gonads; muscle; outer coverings of internal organs; dermis of skin; circulatory system: heart, blood vessels, blood cells; kidneys
Endoderm	Inner linings of: gut, respiratory tract (including lungs), liver, pancreas, thyroid, and urinary bladder

[a]The final structures are complex, containing cells from more than one germ layer. Interactions among tissues are usually important in determining the composition and structure of an organ.

the illustrations throughout this chapter, endoderm cells are shown in yellow, mesoderm cells in pink, and ectoderm cells in blue.

Despite many variations in detail, some general features are common to the gastrulation of all animals. The rate of mitosis during gastrulation is much slower than it was during cleavage. There are massive movements of cells, giving rise to adjacent external and internal tissues: ectoderm, mesoderm, and endoderm. As development proceeds, different genes are activated in cells from the different germ layers, so that different gene products are formed.

Gastrulation in sea urchins results from several types of cell movements. Certain cells of a sea urchin blastula first become columnar, flattening one end of the embryo slightly. Then, as shown at the top of Figure 17.5, the central columnar cells bulge into the blastocoel, break free and actively crawl into the cavity in a process called **ingression**. These ingressing cells are **primary mesenchyme** cells. (*Mesenchyme* means a loosely organized array of cells, as distinguished from an *epithelium*, in which cells are packed tightly in sheets.) The primary mesenchyme cells are the beginning of the mesoderm. Next, the flattened end of the embryo invaginates as the columnar cells become wedge-shaped and buckle inward to produce the endoderm pocket. The ectoderm remains on the outside.

Invagination during gastrulation forms a new cavity, the **archenteron** or "primitive intestine," that opens to the exterior through the **blastopore**. As the archenteron lengthens, more cells move into the blastocoel from the archenteron's tip in a second round of ingression. These **secondary mesenchyme** cells

first form fine extensions, called filopodia, that extend through the blastocoel and attach to the inner surface of the ectoderm opposite the blastopore. Contraction of these filopodia lengthens the archenteron further, until the tip of the archenteron approaches the area of the future mouth. At this stage, the secondary mesenchyme cells detach from the endoderm. Migrating groups of mesenchyme cells eventually form a continuous mesoderm layer. The migration of the primary and secondary mesenchyme

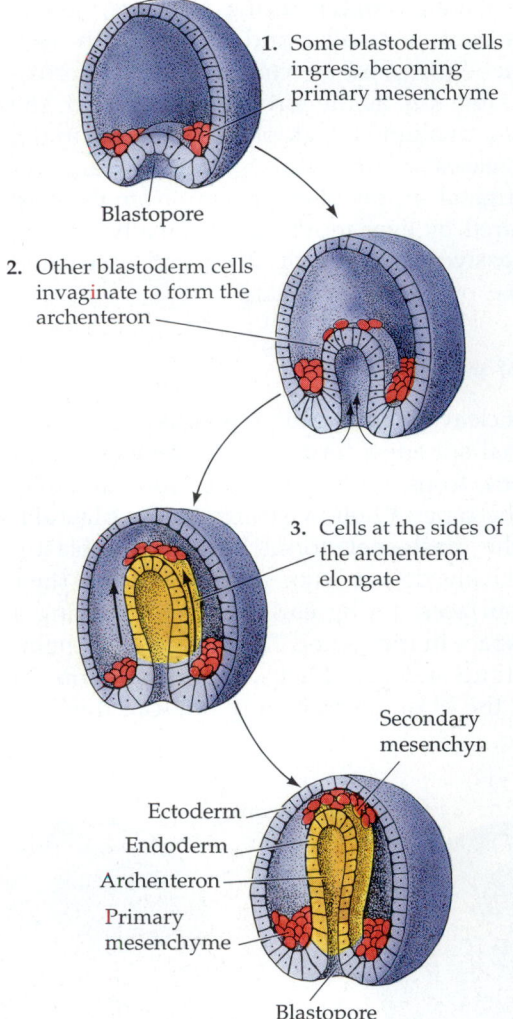

1. Some blastoderm cells ingress, becoming primary mesenchyme

Blastopore

2. Other blastoderm cells invaginate to form the archenteron

3. Cells at the sides of the archenteron elongate

Secondary mesenchyn

Ectoderm
Endoderm
Archenteron
Primary mesenchyme

Blastopore

17.5 Gastrulation in the Sea Urchin
During gastrulation, cells move to new positions and form the three germ layers from which all adult tissues develop. First, certain cells (at the bottom of the blastula in these drawings) leave the surface layer and crawl into the blastocoel, forming the primary mesenchyme (pink). Other surface cells buckle inward, and a new cavity—the archenteron—develops. Cells that line the archenteron become the endoderm layer (yellow), while cells that remain outside the blastopore become the ectoderm layer (blue). Groups of secondary mesenchyme cells (pink) ingress from the tip of the archenteron, eventually joining other mesenchyme cells to form the mesoderm.

cells is guided in part by fibers of an extracellular protein, fibronectin, laid down by ectodermal cells. The mesenchyme cells selectively attach to the fibronectin fibers and crawl along them.

The archenteron becomes the digestive cavity of the embryo, and its opening to the outside—the blastopore—becomes the anus of the sea urchin. The mouth develops later, from a perforation that forms at the other end. Animals such as sea urchins and vertebrates, in which the mouth develops from a second opening distant from the blastopore, are called deuterostomes (from the Greek for "mouth second"). Other animals, such as earthworms and insects, in which the mouth develops from the blastopore and the *anus* is formed by a second opening, are called protostomes ("mouth first"). The deuterostome–protostome distinction is one of the bases we use in classifying animals (see Chapter 26).

Gastrulation in Embryos with Much Yolk

In frogs, substantial quantities of yolk influence gastrulation and the formation of the three germ layers. Cells at one side of a frog embryo change shape and initiate invagination. Movement of cells starts just below the center of the gray crescent. Recall that the gray crescent is a product of cytoplasmic reorganization in the frog zygote. The first site of invagination marks the **dorsal lip** of the blastopore. Then two things happen. First, the cells at the animal end of the embryo begin to increase their surface area and to expand, a process that continues throughout gastrulation. As this expansion proceeds, the cells then turn inward at the dorsal lip of the blastopore and "flow" into the interior of the embryo (Figure 17.6). This inward turning of an expanding sheet of cells is called **involution**. The archenteron forms as involution proceeds. The dorsal lip of the blastopore extends, eventually forming a complete circle, with cells involuting at its dorsal (upper), lateral (side), and ventral (lower) lips. Within the circular blastopore, a "yolk plug" consisting of food-laden cells from the vegetal hemisphere of the embryo can be seen. The cells involuting over the lip of the blastopore give rise to both endoderm and mesoderm layers.

The formation of the mesoderm is complex, involving several types of cell movement. Some mesodermal precursors crawl along fibronectin fibers, as in sea urchin embryos. The major force moving future mesoderm to the interior and away from the blastopore, however, is convergent extension, a pattern of movement in which cells become narrower, elongate, and move between one another like cars rearranging themselves in heavy traffic. During convergent extension a strip of mesoderm differentiates to form the **notochord**, a stiff, supportive rod of cartilage. Later in the frog's development, the support-

ive function of the notochord is taken over by the vertebrae.

Gastrulation in Birds and Mammals

Because of their massive yolk content, bird eggs exhibit a pattern of gastrulation that differs from both sea urchin and frog eggs. Starting with the disc-shaped blastula (Figure 17.2c), surface cells lying on either side of the embryo migrate toward the center line and then forward, forming a depression called the **primitive streak** that is analogous to the blastopores of sea urchins and frogs. As migrating cells reach the primitive streak, they separate from one another and ingress through it (Figure 17.7). Once inside, they come together again, forming internal sheets that will become the mesoderm and endoderm. In this manner, the gut-forming cells of the endoderm are brought inside the embryo, the skin- and nerve-forming cells of the ectoderm remain external, and the mesoderm, which forms most of the organs, is brought between them. Interactions among cells of these germ layers then begin to determine the fates of specific cells in different regions of the embryo (see Table 17.1).

Gastrulation in mammals is somewhat similar to that in birds—and to that in reptiles, from which both birds and mammals evolved. Were it not for the shared evolutionary origins of birds and mammals, the primitive streak and other birdlike features of mammalian gastrulation would be surprising, because mammalian eggs are small and lacking in yolk in comparison with bird eggs. Other aspects of early mammalian development differ from the corresponding stages in birds and, in fact, in all other animals.

One striking feature of mammalian development is its slow pace. By contrast, the worm *Caenorhabditis elegans* reaches its adult form in three and a half days after fertilization; after two days a frog embryo has become clearly recognizable as a tadpole. But by two days a human embryo has reached only the 2-cell stage, and by three days only the 8-cell stage. At the 8-cell stage the mammalian embryo becomes more compact, and tight junctions (see Chapter 5) form between the outer cells of the mass, isolating the inside of the mass from the external environment. (Recall that tight junctions allow no gaps between adjacent cells.) Following additional cleavages, some cells form an inner cell mass enclosed by a surrounding layer called the trophoblast. Thus, the mammalian blastula, or **blastocyst,** differs in several respects from the blastulas of other animals (Figure 17.8). For example, in mammals the embryo develops within and is protected and nourished by the body of the mother. The blastocyst becomes implanted in the maternal uterus, where the trophoblast gives rise to the placenta (see Chapter 37), while certain cells of the

1. Gastrulation begins when cells just below the center of the gray crescent invaginate to form the dorsal lip of the future blastopore

2. Cells at the animal pole spread out, pushing surface cells below them toward and across the dorsal lip. Those cells involute into the interior of the embryo where they form the endoderm and mesoderm

3. This involution creates the archenteron and destroys the blastocoel. The dorsal lip forms a circle, with cells on both its dorsal and ventral surfaces; the yolk plug is visible through the blastopore

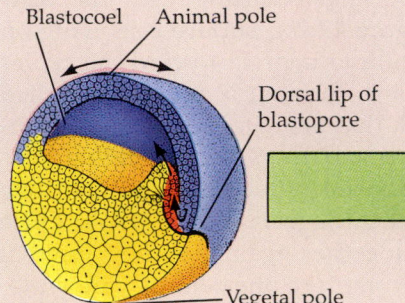
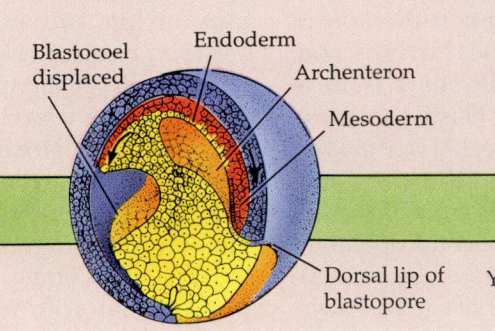
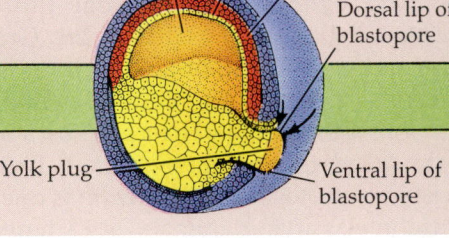

17.6   Gastrulation in the Frog
The developmental sequence begun in Figures 17.2*b* and 17.4*b* continues with gastrulation. The color conventions for the endoderm, ectoderm, and mesoderm remain as in previous figures; green is added, identifying cells that will form the nervous system.

inner cell mass give rise to protective membranes that are not part of the embryo. The disc-shaped remainder of the inner cell mass gives rise to the embryo itself by steps including a gastrulation process somewhat reminiscent of that of birds.

As we have seen, the patterns of gastrulation differ in detail from group to group of animals. However,

the result in all cases is an embryo with a gut and two or three germ layers. The next step is the development of a nervous system.

NEURULATION

As an animal develops from gastrula to adult, many specialized organs and organ systems are formed, so it is not surprising that developmental patterns differ from one organ to another. One organ system is the central nervous system—the brain and spinal cord—which develops as a tube from a sheet of cells. At the beginning, this sheet is an external layer of the embryo; at the end, the tube is internal. The formation of such a tube in the development of the central nervous system is referred to as **neurulation**. We will consider neurulation in the frog embryo.

Recall that the late gastrula of a frog already has three tissue layers: ectoderm, endoderm, and mesoderm. At this stage, the embryo has anterior (front) and posterior (rear) ends, and the blastopore is positioned at the extreme rear. On the dorsal side of the embryo, the ectoderm begins to thicken, forming a flattened **neural plate** (Figure 17.9). The edges of the neural plate thicken and move upward to form neural folds. In the center of the plate, a neural groove forms and deepens as the folds begin to roll toward each other. As this is going on, the embryo as a whole elongates its anterior-to-posterior axis. The neural folds continue to roll toward one another

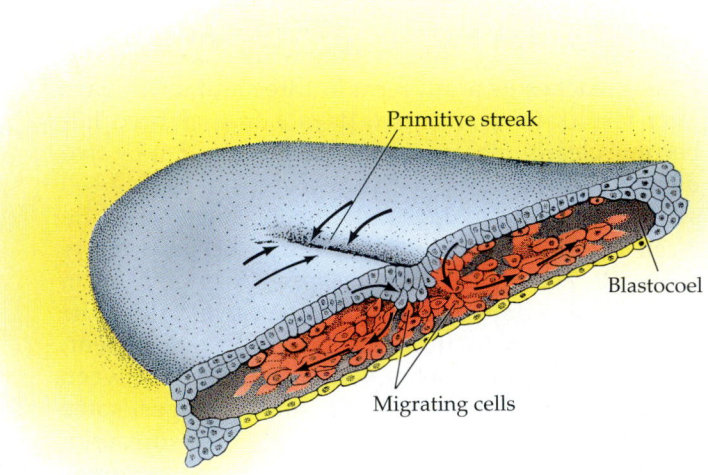

17.7   Gastrulation in Birds
During gastrulation in a bird embryo (shown from above), cells from the blastula's surface migrate toward the primitive streak and then separate from the surface layer and crawl into the blastocoel. Once in the blastocoel, the migrating cells come together in two sheets to form mesoderm (pink) and endoderm (yellow), while the cells that remain on the surface form ectoderm (blue).

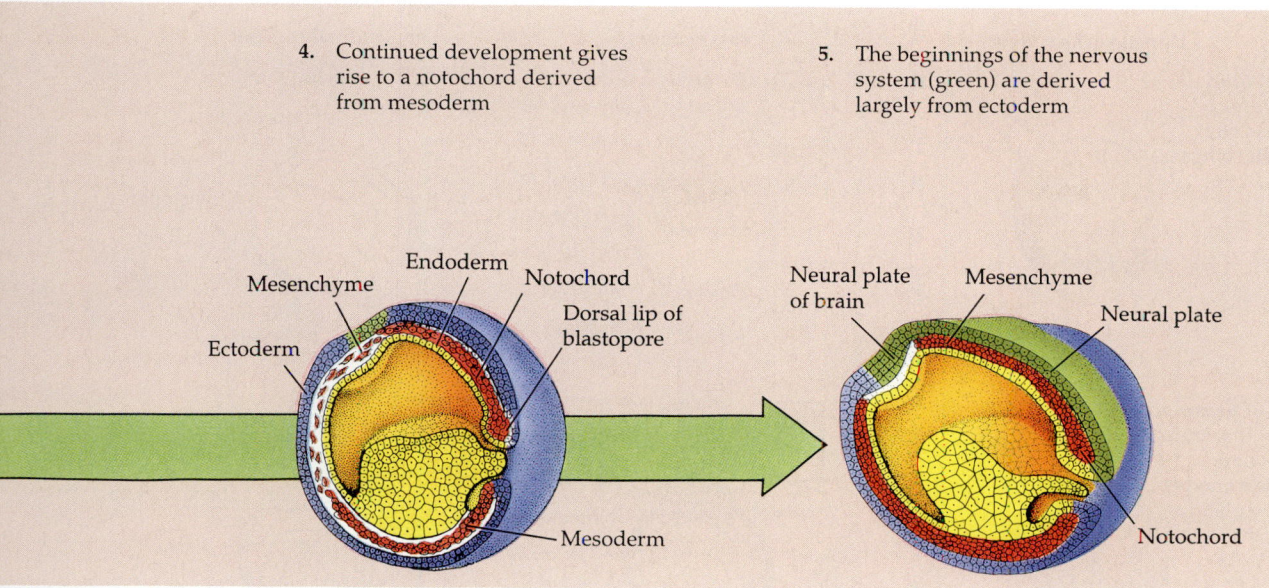

4. Continued development gives rise to a notochord derived from mesoderm

5. The beginnings of the nervous system (green) are derived largely from ectoderm

Mesenchyme
Endoderm
Notochord
Dorsal lip of blastopore
Ectoderm
Mesoderm

Neural plate of brain
Mesenchyme
Neural plate
Notochord

and finally, late in neurulation, fuse to form a narrow, hollow cylinder. This cylinder, the **neural tube**, becomes detached from the overlying ectoderm of the embryonic surface. Thus cells that once were part of a surface sheet are now incorporated into an internal tube (see the cross sections in Figure 17.9).

Many important events take place during neurulation, but the defining event is the formation of the neural tube. This type of process, in which a sheet of ectoderm from the surface becomes a tube inside the embryo, is repeated in many developmental contexts.

The entire nervous system derives from ectoderm,

and the anterior end of the neural tube ultimately becomes the brain. While the ectoderm is undergoing these dramatic changes, what is happening to the embryonic frog's mesoderm?

Late in gastrulation, as you saw in Figure 17.6, the notochord forms from mesoderm, providing support for the developing embryo. This cartilaginous rod continues to develop during neurulation, as shown in the longitudinal sections in Figure 17.9. After the notochord forms, other portions of the mesoderm to either side of the notochord and neural tube condense into blocks of cells called **somites** (Figure 17.10). Some somite cells give rise to muscles, others give rise to bones—the vertebrae. The somites develop soonest in the anterior end of the embryo and then sequentially toward the posterior end.

LATER STAGES OF DEVELOPMENT

Although we have seen differences in the ways frog, sea urchin, and chick embryos pass through their earliest stages of development, our main purpose so far in this chapter has been to show that various animals have much in common during the stages of cleavage, gastrulation, and neurulation. After neurulation, cells continue to divide, change shape, and move as embryos progress toward adulthood, but the later stages of development differ profoundly among animal groups.

Growth

Growth—irreversible increase in size—is often extensive. This increase in size results from cell multipli-

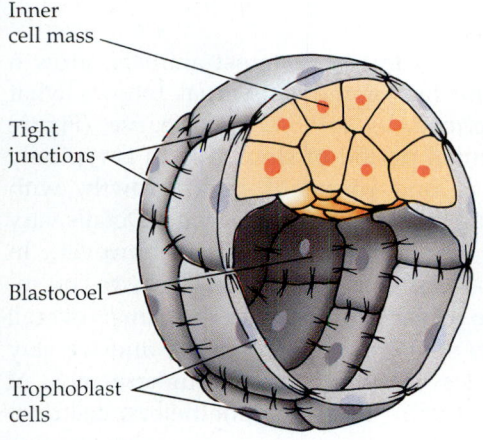

Inner cell mass
Tight junctions
Blastocoel
Trophoblast cells

17.8 The Distinctive Blastocyst of a Mammal
The mammalian blastula, called a blastocyst, features an inner cell mass, some of whose cells give rise to the embryo after the blastocyst implants in the wall of the mother's uterus. Gastrulation follows; it resembles gastrulation in birds, despite the absence of a large mass of yolk.

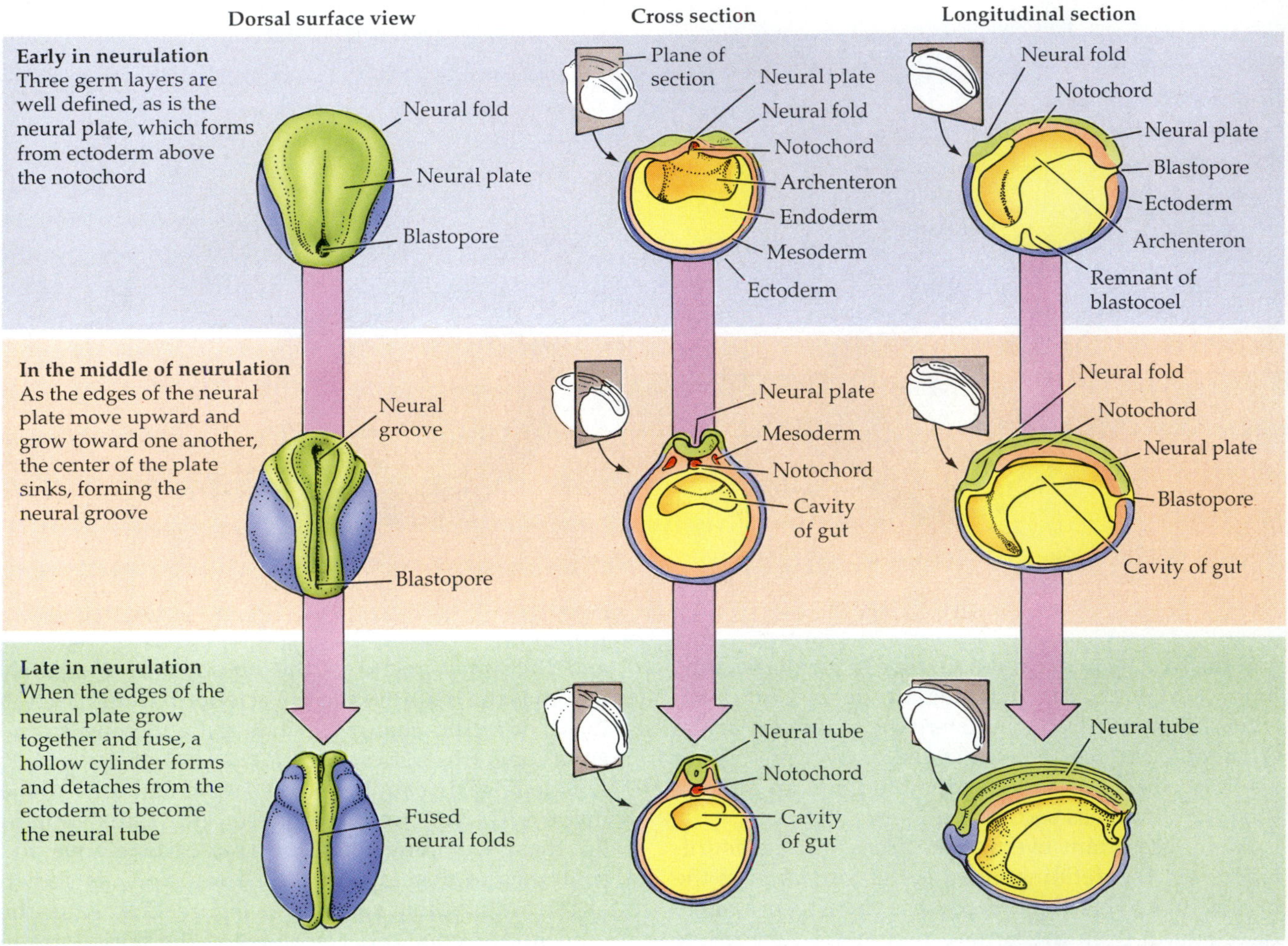

Dorsal surface view	Cross section	Longitudinal section

Early in neurulation
Three germ layers are well defined, as is the neural plate, which forms from ectoderm above the notochord

Neural fold
Neural plate
Blastopore

Plane of section
Neural plate
Neural fold
Notochord
Archenteron
Endoderm
Mesoderm
Ectoderm

Neural fold
Notochord
Neural plate
Blastopore
Ectoderm
Archenteron
Remnant of blastocoel

In the middle of neurulation
As the edges of the neural plate move upward and grow toward one another, the center of the plate sinks, forming the neural groove

Neural groove
Blastopore

Neural plate
Mesoderm
Notochord
Cavity of gut

Neural fold
Notochord
Neural plate
Blastopore
Cavity of gut

Late in neurulation
When the edges of the neural plate grow together and fuse, a hollow cylinder forms and detaches from the ectoderm to become the neural tube

Fused neural folds

Neural tube
Notochord
Cavity of gut

Neural tube

17.9 Neurulation in the Frog
Continuing from Figure 17.6, these drawings outline the development of the frog's neural tube. The three views across the top show a late gastrula. The longitudinal sections (right column) show the development of the notochord and its position relative to the neural tube.

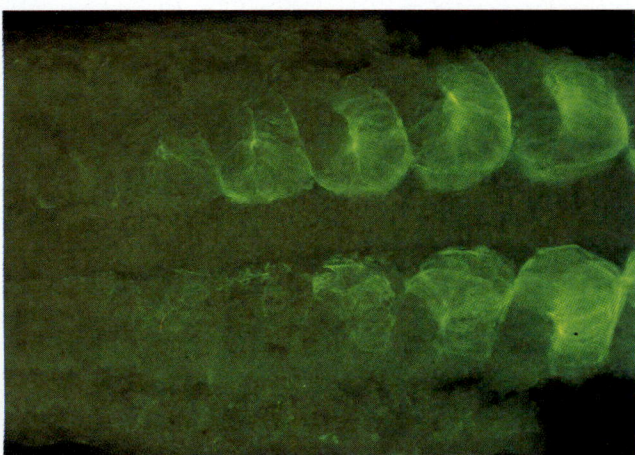

17.10 Somites Give Rise to Muscle and Bone
Somites in this amphibian embryo look like green blocks because of a fluorescent stain.

cation and cell expansion. In most animals, growth throughout the life of the individual follows what may be described as an S-shaped curve (Figure 17.11a). An initial period of slow growth is typically followed by a long phase of rapid growth, with growth slowing markedly at some stage. Details vary considerably among animal groups, however. In many groups—lobsters, for example—growth continues until the organism dies. In humans, overall growth ceases sometime after puberty, and we stay at a more or less constant size throughout most of our adult life (Figure 17.11c). Nonetheless, cell division continues at a rapid pace in some tissues throughout our lives, replacing cells such as those sloughed off by our skin (more than 1 gram a day) and intestinal lining, as well as the millions of blood cells turned over each minute (see Chapter 16).

Returning to the lobster, we see another departure from the pattern of Figure 17.11a, because the growth

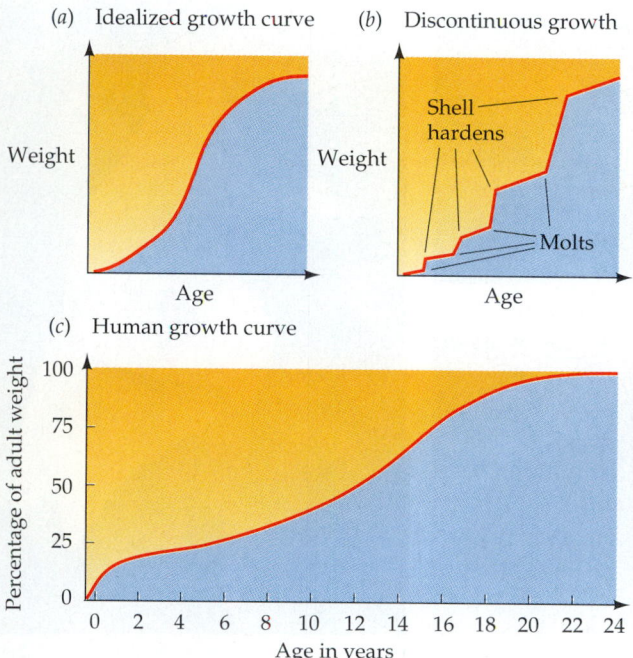

(a) Idealized growth curve

Weight

Age

(b) Discontinuous growth

Weight

Shell hardens

Molts

Age

(c) Human growth curve

Percentage of adult weight: 100 75 50 25 0

Age in years: 0 2 4 6 8 10 12 14 16 18 20 22 24

17.11 Growth Patterns in Animals
All types of animals increase in weight as they grow older, but their growth rates differ, as these growth curves show. The steeper the curve, the faster the growth rate. (a) Idealized S-shaped growth curve characteristic of many species. There is some growth at every age, but the maximum growth rate is during the middle of the life cycle. (b) Discontinuous growth characteristic of hard-shelled animals that must molt to grow. A growth spurt follows each molt, before the new shell hardens. (c) A variant of the S-shaped curve shows the growth pattern for humans. A decade of gradual, preadolescent growth precedes a growth spurt during puberty; within a few years, the final adult weight is approached.

of a lobster is *discontinuous*. Because of its rigid external skeleton, the individual must molt (shed its skeleton) in order to grow. Accordingly, a lobster grows in spurts following successive molts (Figure 17.11b).

Larval Development and Metamorphosis

Many animals go through a larval stage in development. A **larva** is an immature form of an animal, differing in appearance from the adult. Examples of larvae include the caterpillar stage of a butterfly and the tadpole of a frog. A tadpole must change dramatically to become a four-legged adult frog, a radical rearrangement of structures called **metamorphosis**. Most larvae differ strikingly from adults of the same species.

During metamorphosis, new adult structures must be formed and old larval structures lost. Many changes must happen between the tadpole and the frog. The gut is shortened, corresponding to the transition from a vegetarian tadpole to a carnivorous

adult. The brain is remodeled to allow binocular (two-eyed) vision. Limbs appear and the tail disappears, corresponding to the transition from a swimming tadpole to a jumping frog. The loss of old tissues and old parts is accomplished by programmed cell death; that is, certain cells, such as those constituting the tadpole's tail, are purposely killed in developing organisms. Cell death plays a part not only in metamorphosis, but also in the embryonic development of most species, thus contributing in a general way to the development of form in animals.

The overall pattern of development in butterflies, moths, and many other insects is familiar to you (Figure 17.12). From a fertilized egg there develops a creeping larva that feeds voraciously, growing through a series of molts (the larval stages between molts are called **instars**). Some specialized cells, arranged in clusters called **imaginal discs**, remain undifferentiated throughout larval growth but later give rise to the tissues of the adult. The final instar stops feeding and then surrounds itself with a cocoon and transforms into a **pupa**. In the pupa, tremendous changes take place (Figure 17.13). Some larval cells die, others are reprogrammed to make different products characteristic of the adult, and the imaginal discs differentiate into new adult structures. Such a major revision between larva and adult is referred to as **complete metamorphosis**. In sharp contrast is the **gradual metamorphosis** characteristic of other insects, including grasshoppers and cockroaches. In gradual metamorphosis, the instars between molts are known as nymphs (or, when aquatic, naiads) and resemble miniature adults in many physical features.

LOOKING CLOSER AT DEVELOPMENT

With the understanding that animals develop from zygotes through cleavage, gastrulation, and neurulation stages, we can now examine three processes that reveal a great deal about how cells, and the animals themselves, change during development. These processes are determination, differentiation, and pattern formation.

Table 17.1 showed that cells of the three germ layers have fates. Some ectoderm cells, for example, will become the outermost layers of the animal's skin (the epidermis). How did developmental biologists learn what is summarized in Table 17.1? What is the experimental evidence?

Staining specific cells of the early embryo and observing which cells of older embryos contained the stain enabled biologists to determine which adult structures develop from certain parts of the blastula and early gastrula. For instance, the shaded area of the frog blastoderm shown in Figure 17.14 has the fate of becoming (that is, it is destined to become)

(a)

(b)

(c)

(d)

(e)

17.12 Complete Metamorphosis in a Moth
The comet-tail moth from Madagascar undergoes complete metamorphosis. (a) Eggs and first instar larvae (one in the process of hatching). (b) Third instar. (c) Fifth instar. (d) Pupa (removed from cocoon). The sweeping disposal of old tissues and the development of adult structures from imaginal discs in the pupa leads to the designation "complete." (e) Adult male moth, approximately 30 minutes after emergence from the pupa. Its long tails are not yet fully expanded.

part of the skin of the tadpole larva if left in place. If we cut out a piece from this region and transplant it to another place on an early gastrula, however, the type of tissue it becomes is determined by its *new* location, as Figure 17.14 shows. The developmental potential of blastoderm cells—that is, their range of possible development—is thus greater than their fate, which is limited to what normally develops.

Does developing embryonic tissue retain its broad developmental potential? Generally speaking, no. The developmental potential of cells becomes restricted fairly early. If taken from a region fated to develop into brain, for example, late gastrula tissue becomes brain tissue even if transplanted to parts of an early gastrula destined to become other structures. The tissue of the late gastrula is thus said to be determined: Its fate has been sealed, regardless of its surroundings. By contrast, the younger experimental tissue in Figure 17.14 has not yet become determined.

Determination is not something that is visible under the microscope—cells do not change appearance when they become determined. Changes in biochemistry, structure, and function constitute the *differentiation* of cells. Determination precedes differentiation. **Determination** means commitment; the final realization of this commitment is differentiation which we discuss next. We will then examine two mechanisms by which cells become determined.

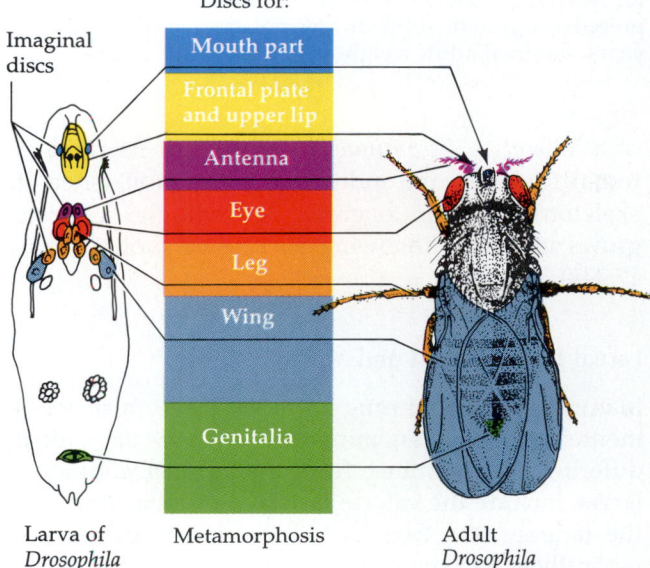

Discs for:

Imaginal discs

Mouth part

Frontal plate and upper lip

Antenna

Eye

Leg

Wing

Genitalia

Larva of *Drosophila*

Metamorphosis

Adult *Drosophila*

17.13 Complete Metamorphosis in Fruit Flies
In insects that undergo complete metamorphosis, such as the fruit fly, the embryo hatches from its egg into a soft-bodied larva that feeds for some time before entering the pupal stage. In the pupa, most of the larval tissues die and are reabsorbed, providing building blocks for subsequent development. The remaining larval tissues are specialized imaginal discs, which differentiate to form the organs of the adult insect.

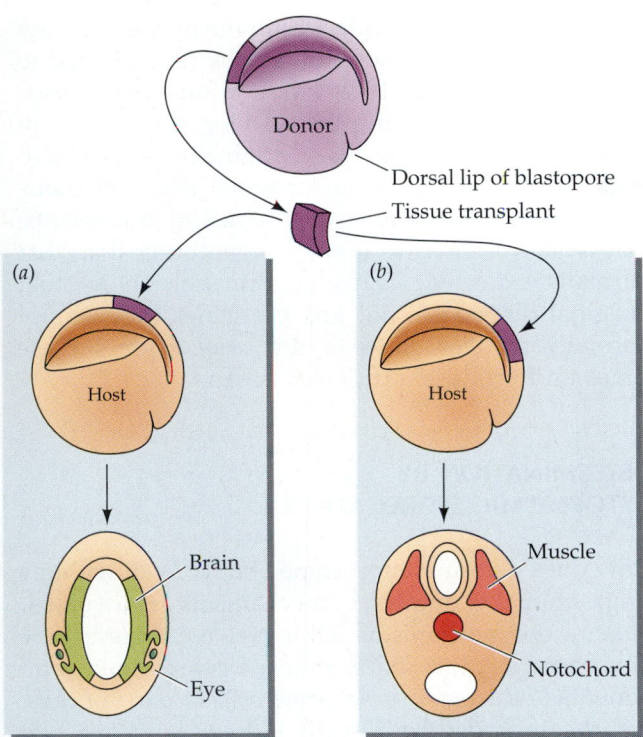

17.14 Developmental Potency in Early Gastrulas
Tissues in early gastrulas retain the capacity to develop in alternate ways, as demonstrated in transplantation experiments with frogs. Tissue destined to become part of a tadpole's skin is cut from an early gastrula and transplanted to another gastrula (the host). (a) When transplanted to a region of the host destined to become brain tissue, the donor tissue also develops into brain tissue. (b) When transplanted to a mesodermal region of the host, transplanted tissue becomes mesoderm and ultimately muscle and notochord. The wide developmental potential of early gastrulas is lost by the late gastrula stage.

DIFFERENTIATION

The process by which a cell achieves its fate—by which it acquires its final structure and physiological function—is **differentiation**. Because the cells of a multicellular organism arise by mitotic divisions of a single-celled zygote, they are for the most part genetically identical. In the absence of all mutation, all the cells in an organism have the same hereditary makeup; yet the adult organism is composed of many distinct types of cells. This apparent contradiciton results from the close regulation of the expression of various parts of the genome. Most of what we know about differentiation comes in large part from recombinant DNA technology (see Chapter 14).

The zygote is **totipotent**: It has the ability to give rise to every type of cell in the adult body. Its genetic "library" is complete, containing instructions for all the structures and functions that will arise throughout the life cycle. Later in the development of animals (and probably to a lesser extent in plants), the cellular

descendants of the zygote lose their totipotency and become determined—that is, committed to form only certain parts of the embryo. When a cell achieves its determined fate, it is said to have differentiated. The mechanisms of differentiation relate primarily to changes in the transcription and translation of genetic information (see Chapter 13).

Is Differentiation Irreversible?

An early explanation of the mechanisms of differentiation was that the cell nucleus undergoes irreversible genetic changes in the course of development. It was suggested that chromosomal material is lost, or that some of it is irreversibly inactivated.

Differentiation is clearly irreversible in certain types of cells. The mammalian red blood cell, which loses its nucleus during development, is an example. Another is the tracheid, a water-conducting cell in vascular plants. The development of a tracheid culminates in the death of the cell, leaving only the pitted cell walls that were formed while the cell was alive (see Chapter 29). In these two extreme cases, the irreversibility of differentiation can be explained by the absence of a nucleus. Generalizing about mature cells that retain functional nuclei is more difficult. Most biologists tend to think of plant differentiation as reversible and of animal differentiation as irreversible, but this is not a hard-and-fast rule. A lobster can regenerate a missing claw, but a cat cannot regenerate a missing paw. Why is differentiation reversible in some cells but not in others? At some stage of development do changes within the nucleus permanently commit a cell to specialization?

At the Institute of Cancer Research in Philadelphia in the 1950s, Robert Briggs and Thomas J. King performed experiments to see whether genetic material was preserved or was permanently inactivated or lost during normal development. To find out whether the nuclei of frog blastulas had lost the ability to do what the zygote nucleus could do, they carried out a series of meticulous transplants. First they removed the nucleus from an unfertilized egg (thus forming what is called an enucleated egg). Then, with a very fine glass tube, they punctured a cell of a blastula and drew up part of its contents, including the nucleus, which they then injected into the enucleated egg, and the egg was activated. More than 80 percent of these operations resulted in the formation, from the egg and its new nucleus, of a normal blastula; of these blastulas, more than half developed into normal tadpoles and, ultimately, adult frogs. These experiments showed that no information has been lost from the nucleus by the time the blastula has formed. On the other hand, Briggs and King found that when the nuclei were derived from older embryonic stages, fewer larvae developed (Figure 17.15).

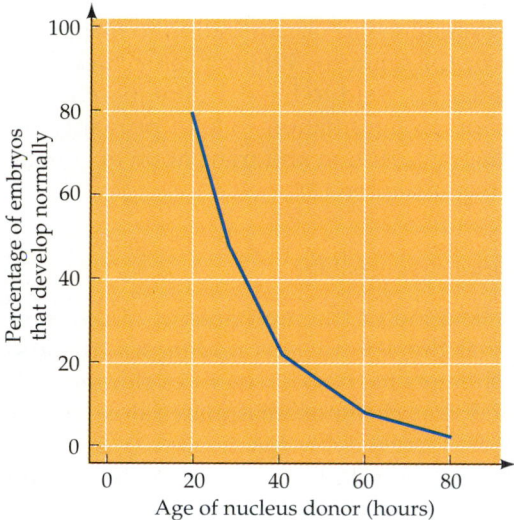

17.15   Nuclei Lose Potency with Age
Nuclei from older frog embryos transplanted into enucleated eggs are less likely to direct successful development of an embryo than are younger nuclei.

This work was carried further by John B. Gurdon and his associates at Oxford University, who performed similar transplants using nuclei from gastrulas, swimming tadpoles, and even adult frogs. Nuclei from differentiated adult cells were transplanted into enucleated eggs, the eggs were raised to the blastula stage, nuclei were isolated in turn from these blastulas, and these nuclei were used for further transplants. When placed in an enucleated egg, a nucleus obtained by such serial transplants occasionally was able to direct development to a tadpole stage, complete with brain, gut, blood, heart, and other parts. Work of this sort convinced many developmental biologists that the loss of particular genes is not the cause of differentiation, and that genes no longer expressed in certain cells are still present and can be expressed if they are placed in a "younger" environment such as an enucleated egg. In the next section we consider another example of development that does not result from the loss of genes.

Transdetermination of Imaginal Discs

The fates of the imaginal discs of insects are determined long before metamorphosis. If transplanted from one larva to another, an imaginal disc still develops into the same type of organ (a wing, for example, or an antenna) that it would have if left undisturbed, and that organ is formed wherever the imaginal disc is placed in the host body.

If transplanted into an *adult* insect, an imaginal disc remains undifferentiated (because the hormonal signal for its development is lacking), but the cells of the disc continue to divide within the new host. Later these transplanted disc cells may be transplanted to other adult insects or to larvae. If returned to a larva, the disc cells *almost* always develop into the adult organ for which they were originally determined. Occasionally, however, an imaginal disc will **transdetermine** in the course of a series of transplants; that is, it will develop into an organ other than that normally expected. Transdetermination shows that imaginal discs have not lost the genes they do not normally express, since a disc *may* express these genes and produce a different organ.

DETERMINATION BY CYTOPLASMIC SEGREGATION

How does determination come about? Even within a single animal, a number of mechanisms are involved. Most of the mechanisms fall into two categories, the first based on the segregation of cytoplasmic components of the egg into separate cells or parts of cells, and the second based on the influences of one part of the embryo on another part. We will consider cytoplasmic segregation first, beginning with its role in distinguishing one end of an animal from the other.

Polarity in the Egg and Zygote

Polarity, the difference of one end from the other, is obvious in development. Our heads are distinct from our feet, and the distal ends of our arms (wrists and fingers) differ from the proximal ends (shoulders) of our arms. An animal's polarity develops early, even in the egg itself. Yolk may be distributed asymmetrically in the egg and the embryo, and other chemical substances may be confined to specific parts of the cell or may be more concentrated at one pole than at the other. In some animals, the original polar distribution of materials in the egg's cytoplasm changes as a result of fertilization, yielding a new polar distribution in the zygote. As cleavage proceeds, the resulting blastomeres contain unequal amounts of the materials that were not distributed uniformly in the zygote. As we learned from the work of Briggs and King and of Gurdon, cell nuclei do not always undergo irreversible changes during early development; thus we can explain some embryological events on the basis of the *cytoplasmic* differences in blastomeres.

Even as apparently simple a structure as a sea urchin egg has polarity. As the gastrula forms (see Figure 17.5), one can see a slight difference in blastomere size; the cells in the vegetal half—the ones that ingress—are somewhat larger. A striking differ-

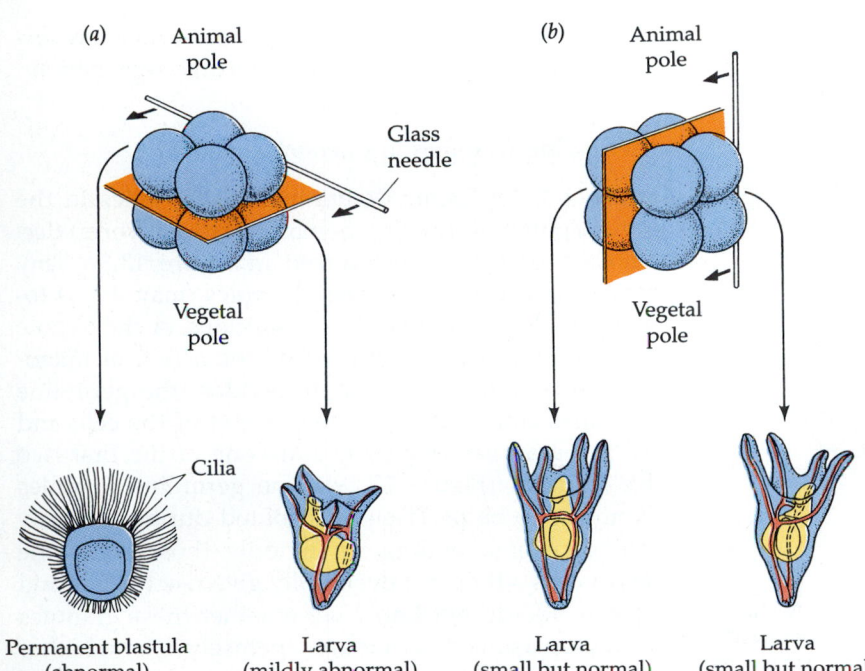

(a) Animal pole

Glass needle

Vegetal pole

Cilia

Permanent blastula (abnormal)

Larva (mildly abnormal)

(b) Animal pole

Vegetal pole

Larva (small but normal)

Larva (small but normal)

17.16 Early Asymmetry in the Embryo
The animal and vegetal halves of very young sea urchin embryos differ in their developmental potential. The vegetal half alone cannot direct development at all, and completely normal development requires cells from both halves.

ence between blastomeres can be demonstrated well before that, however. The Swedish biologist Sven Hörstadius showed in the 1930s that the development of sea urchin embryos that have been divided in half at the eight-cell stage depends on how the separation is performed. If the embryo is split into "left" and "right" halves, with each half containing cells from both the animal and the vegetal pole, normal-shaped but dwarfed larvae develop from the halves (Figure 17.16). If, however, the cut separates the four cells at the animal pole from the four at the vegetal pole, the result is different. The animal half develops into an abnormal blastula with large cilia at one end and cannot form a larva, whereas the vegetal half develops into a small, somewhat misshapen, larva with an oversized gut. For fully normal development, factors from both the animal and vegetal halves of the embryo are necessary. By further experiments on eggs, Hörstadius showed that this unequal distribution of material between the animal and vegetal halves is already present in the unfertilized egg.

When a frog egg becomes fertilized, some components of the cytoplasm are again restricted to certain parts of the egg, resulting in—among other things—the formation of the gray crescent. In the early part of this century, Hans Spemann found that the gray crescent contains materials essential for normal embryonic development. In normal cleavage, the first cell division usually divides the gray crescent about equally between the daughter cells (Figure 17.17a). These two blastomeres, if separated, give rise to normal embryos. This process is regulative devel-

opment, as described earlier. Sometimes, however, the zygote divides in such a way that one of the blastomeres contains the entire gray crescent and the other none of this material (Figure 17.17b). If the two cells of an embryo that has divided this way are separated, the half with the gray crescent develops normally but the half lacking the gray crescent forms only an unorganized cellular mass. These experiments by Hörstadius and by Spemann were among many that established that the unequal distribution of materials in the egg cytoplasm plays a role in directing embryonic development.

Cytoplasmic Factors in Polarity in *Drosophila*

In the eggs and larvae of the fruit fly *Drosophila melanogaster*, polarity is based on the distribution of more than a dozen mRNAs and proteins. These **cytoplasmic determinants** are products of specific genes in the mother and are distributed to the eggs, often in a nonuniform manner. They determine the dorsoventral (back–belly) and anteroposterior (head–tail) axes of the embryo. The discovery that this part of fruit fly development is determined by cytoplasmic segregation was due to the striking appearance of the mutant larvae produced when the determinants are distributed abnormally. For example, larvae formed by females homozygous for the *bicaudal D* mutation consist solely of two hind ends, joined at the middle and possessing no head (Figure 17.18).

We know that cytoplasmic determinants specify these axes from the results of experiments in which cytoplasm was transferred from one egg to another.

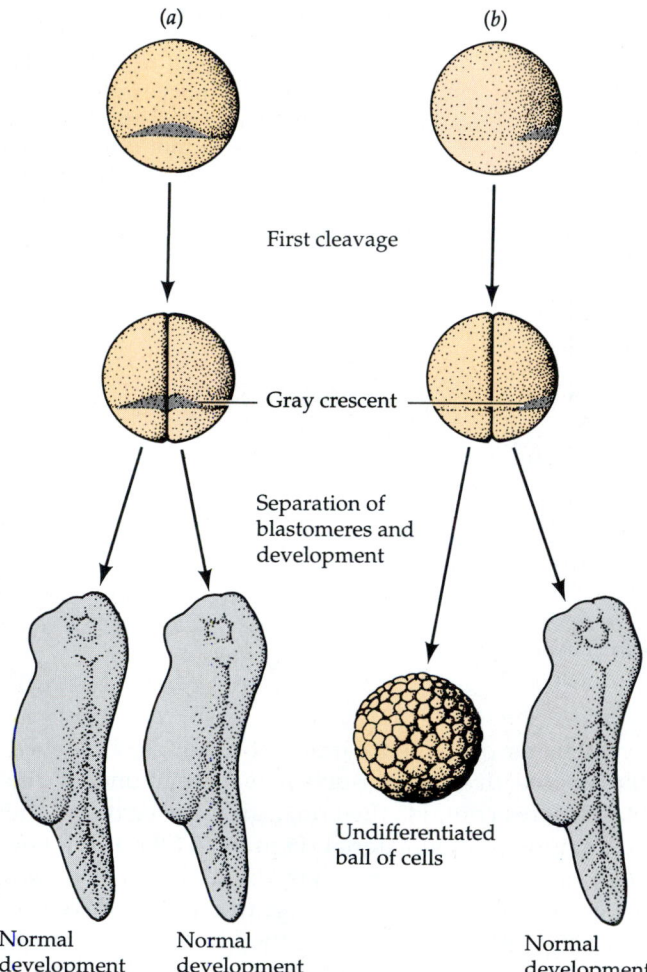

(a) *(b)*

First cleavage

Gray crescent

Separation of
blastomeres and
development

Undifferentiated
ball of cells

Normal
development

Normal
development

Normal
development

17.17 Developmental Importance of the Gray Crescent
The asymmetric distribution of materials in the egg establishes the egg's polarity and determines its developmental architecture. Without the material contained in the gray crescent, a blastomere can form only an undifferentated cell mass.

Females homozygous for the *bicoid* mutation produce larvae with no head and no thorax. However, if eggs of homozygous-mutant *bicoid* females are inoculated at the anterior end with cytoplasm from the anterior region of a wild-type egg, the treated eggs develop into normal larvae—with heads developing from the part of the egg that receives the wild-type cytoplasm. On the other hand, removal of 5 percent or more of the cytoplasm from the anterior of a wild-type egg results in an abnormal larva that looks like a *bicoid* mutant larva.

Another gene, *nanos*, plays a comparable role in the development of the posterior end of the larva. Eggs from homozygous-mutant *nanos* females develop into larvae with missing abdominal segments; injecting the mutant eggs with cytoplasm from the posterior region of wild-type eggs, however, allows

normal development. Let's consider one more example of determination by cytoplasmic segregation.

Germ-Line Granules in *Caenorhabditis*

Various cytoplasmic determinants play roles in the development of the tiny nematode (roundworm) *Caenorhabditis elegans* (introduced in Figure 17.3). Tiny particles called **germ-line granules** may be cytoplasmic determinants. Their positions in the zygote and embryo are determined by the action of microfilaments. Before the zygote divides, the germ-line granules collect at the posterior end of the cell, and all the granules appear in only one of the first two blastomeres (Figure 17.19). The germ-line granules continue to be precisely distributed during the early cell divisions, ending up in only those cells—the germ cells—that will eventually give rise to eggs and sperm. We do not know yet whether these granules are cytoplasmic determinants themselves, or whether they are simply distributed together with the "real" cytoplasmic determinant.

We have seen several examples of determination by cytoplasmic segregation. Now we will examine another important general mechanism of determination, in which certain tissues induce the determination of other tissues.

(a) Wild-type

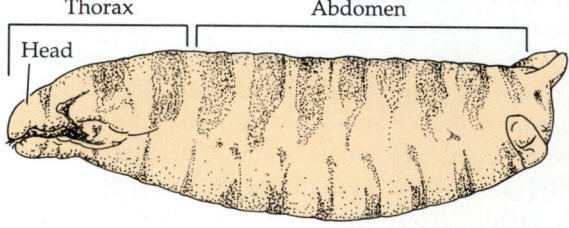

Thorax   Abdomen

Head

(b) *bicaudal*

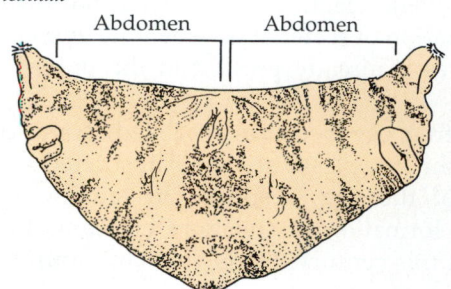

Abdomen   Abdomen

17.18 The *bicaudal* Mutation
The anteroposterior axis of *Drosophila* larvae arises from the interaction of several cytoplasmic determinants. *(a)* A larva produced by a wild-type female. *(b)* A larva produced by a female homozygous for the bicaudal D allele. The larva has no head or middle—it consists of two hind ends.

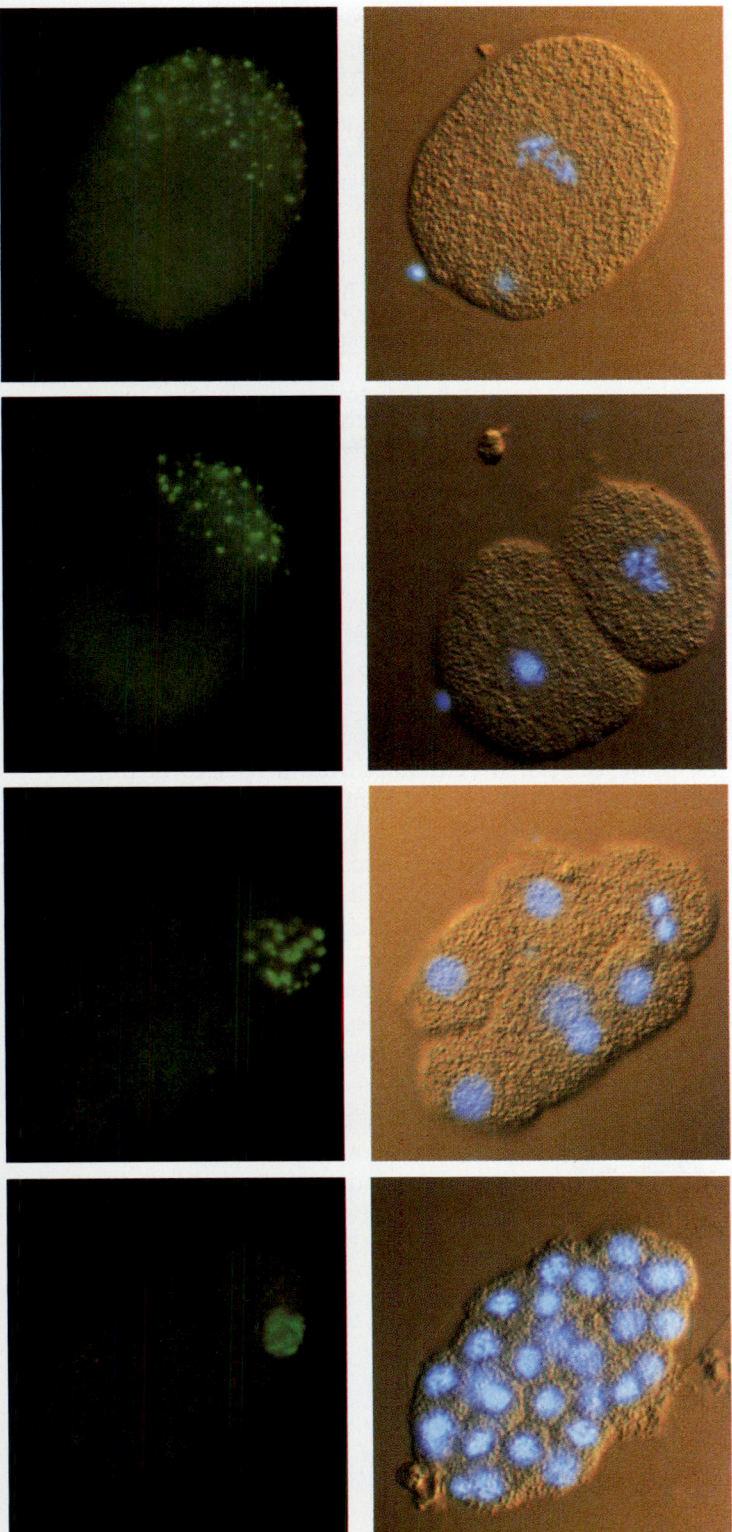

17.19 Distribution of Germ-Line Granules in Caenorhabditis
Micrographs show a development embryo of *Caenorhabditis elegans*. In the left column, the germ-line granules (bright spots, stained with antibodies) move to the poterior end of the embryo. Eventually the granules are confined to the cell that gives rise to gametes. By contrast, the nuclei (right column; stained blue) of the same embryo are distributed evenly among its cells.

DETERMINATION BY EMBRYONIC INDUCTION

Experimental work has clearly established that the fates of particular tissues are determined by interactions with other specific tissues in the embryo. In the developing embryo there are many such instances of **induction**, in which one tissue causes an adjacent tissue to develop in a different manner.

Induction and the Organizer

In 1924 Hilde Mangold and Hans Spemann, working with newt embryos, performed a classic demonstration of induction. First Mangold transplanted a piece of the dorsal lip of the blastopore from an early gastrula of a lightly pigmented donor species to a particular place on the surface of an early gastrula of a heavily pigmented host species. She could follow the fate of the transplanted piece because it was lighter in color than the surrounding host tissue. As predicted from the known fate of the surrounding tissue—to become mesoderm—the graft itself developed into principally mesodermal products. But something unexpected also happened: In the region of the graft, an extra neural plate appeared, containing neural folds made from host tissue (Figure 17.20). The procedure was repeated many times, and in some instances the extra neural plate continued to develop into a secondary embryo attached to the main one! The grafted dorsal lip had induced the nearby host tissue to develop along completely different lines than it would have without the graft.

The dorsal lip of the blastopore not only has the fate of becoming part of the notochordal mesoderm, but in addition it induces any ectoderm that it contacts to organize into a neural tube. With the formation of the neural tube, the principal axes of the embryo—the anteroposterior, dorsoventral, and left–right axes—are formed. For this reason Spemann called the dorsal lip the **embryonic organizer**.

The development of the lens in the vertebrate eye is another classic example of induction. Lens formation is shown in Figure 17.21: The developing forebrain bulges out at both sides to form the **optic vesicles**, which expand until they come in contact with the ectoderm of the head. The head ectoderm in the region of contact with the optic vesicles thickens, forming a **lens placode** The lens placode invaginates, folds over on itself, and ultimately detaches from the surface to produce a structure that will develop into the lens. If the growing optic vesicle is cut away before it contacts the surface ectoderm, no lens forms in the head region from which the optic vesicle has been removed. An impermeable barrier placed between the optic vesicle and the ectoderm also prevents the lens from forming. These observations sug-

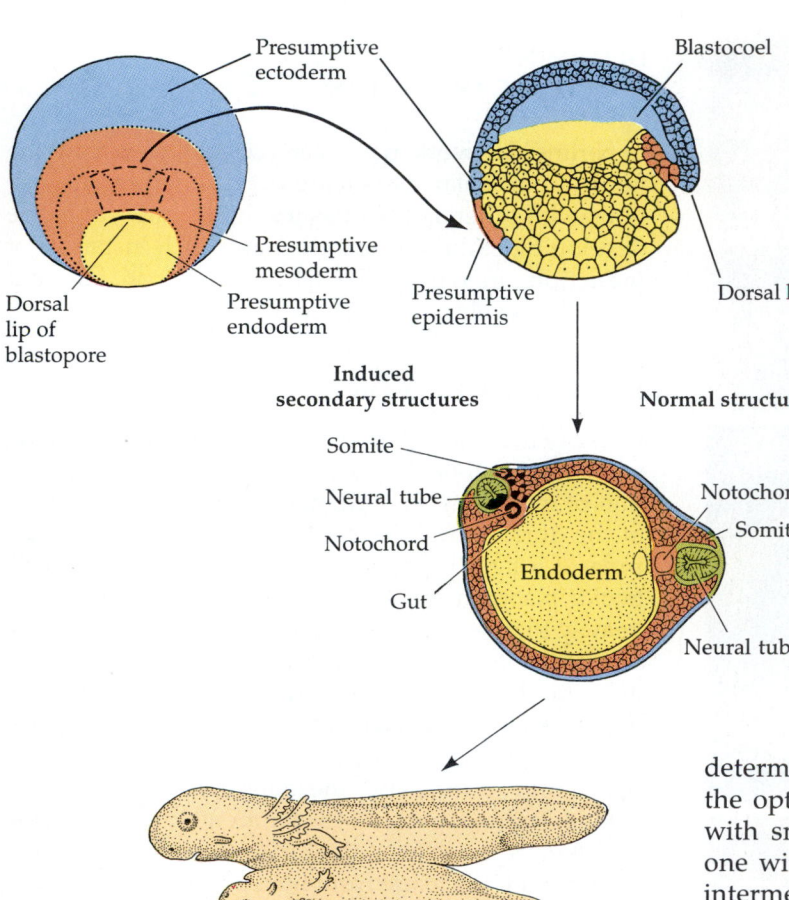

Induced secondary structures

Normal structures

Presumptive ectoderm

Blastocoel

Dorsal lip of blastopore

Presumptive mesoderm

Presumptive endoderm

Presumptive epidermis

Dorsal lip

Somite

Neural tube

Notochord

Gut

Endoderm

Notochord

Somite

Neural tube

17.20 Embryonic Induction
Mangold transplanted a donor's dorsal lip (presumptive mesodermal tissue) to an ectodermal region in the host gastrula. The donor and host organisms were differently pigmented, so she could distinguish graft tissue from host tissue. She expected the graft to become mesodermal tissue and it did. Unexpectedly, she saw that host ectoderm near the graft (that would normally have developed into epidermis) developed into a second neural plate, and in some instances into a second embryo. Mangold concluded that the dorsal lip induced nearby host ectoderm to alter its development.

gest that ectoderm begins to develop into a lens when it receives a signal—an **inducer**—from its contact with an optic vesicle.

The interaction of tissues in eye development is a two-way street: There is a "dialogue" between the developing optic vesicle and the ectoderm. The lens determines the size of the optic cup that forms from the optic vesicle. If ectoderm from a species of frog with small eyes is grafted over the optic vesicle of one with large eyes, both lens and optic cup are of intermediate size. The lens also induces the ectoderm over it to develop into a cornea. Thus an entire chain of inductive interactions participates in development of the parts required to make an organ such as the eye. Induction triggers a sequence of gene expression in the responding cells. Tissues do not induce themselves; rather, different tissues interact and induce each other.

One of the most resistant problems in developmental biology has been that of determining the specific chemical nature of the inducers. In some cases,

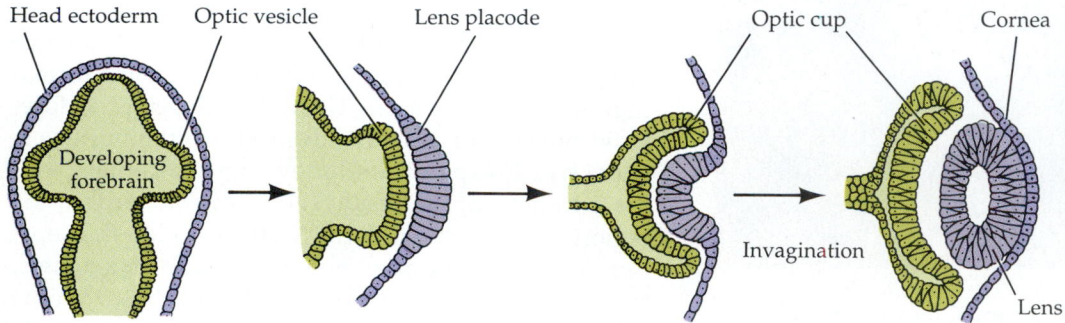

Head ectoderm   Optic vesicle   Lens placode   Optic cup   Cornea

Developing forebrain

Invagination

Lens

17.21 Inducers in the Vertebrate Eye
The vertebrate eye develops as inducers take their turns. Studies on frogs revealed that the optic vesicle induces overlying ectoderm to form placode tissue that, in turn, induces the formation of an optic cup. The optic cup then induces the placode to invaginate and form the eye's lens. Once the developing lens has separated from and become covered by surface ectoderm, it induces this ectoderm to form a cornea.

specific diffusible proteins may be involved; the inducer that acts earliest in frog gastrulas appears to be a growth factor. In other cases, however, insoluble extracellular materials such as collagen and other proteins may be involved in induction. Generally speaking, induction is a phenomenon confined to embryonic tissues; however, it continues in the production of certain white blood cells in the adult immune system.

Induction at the Cellular Level

The tiny worm *Caenorhabditis elegans* is an excellent organism for studying the mechanisms of induction because, as we saw in Figure 17.3, different parts of the worm's body form from different cell lineages in the developing organism. The hermaphroditic form (containing both male and female reproductive organs) of *C. elegans* lays eggs through a pore called the vulva on the ventral surface of the worm. A single cell, called the anchor cell, induces the vulva to form. If the anchor cell is destroyed by laser surgery, no vulva forms. The anchor cell controls the fates of six cells on the animal's ventral surface. Each of these cells has three possible fates. By the manipulation of

two genetic switches, a given cell becomes either a primary vulval precursor, a secondary vulval precursor, or simply part of the worm's surface, the epidermis.

The anchor cell produces an inducer. Cells that receive enough of the inducer become vulval precursors; those slightly farther from the anchor cell become epidermis. The first switch, controlled by the inducer from the anchor cell, determines whether a cell takes the "track" toward becoming part of the vulva or the track toward becoming epidermis. Now the cell closest to the anchor cell, having received the most inducer, becomes the primary vulval precursor and produces its own inducer, which acts on the two neighboring cells and directs them to become secondary vulval precursors. Thus the primary vulval precursor cell controls a second switch, determining whether a vulval precursor will take the primary track or the secondary track. The two inducers control the activation or inactivation of specific genes in the responding cells (Figure 17.22). These genes are explained in detail in Box 17.A.

There is an important lesson to draw from this example. Much of development is controlled by switches that allow a cell to proceed down either of

17.22 Two Switches Determine Vulva Formation in *C. elegans*
Two secreted proteins act as inducers. The primary inducer, produced by the anchor cell, activates a gene whose products determine that cells will develop as vulval precursors rather than epidermal cells; the second inducer activates another gene, thus determining that cells will develop as secondary precursors.

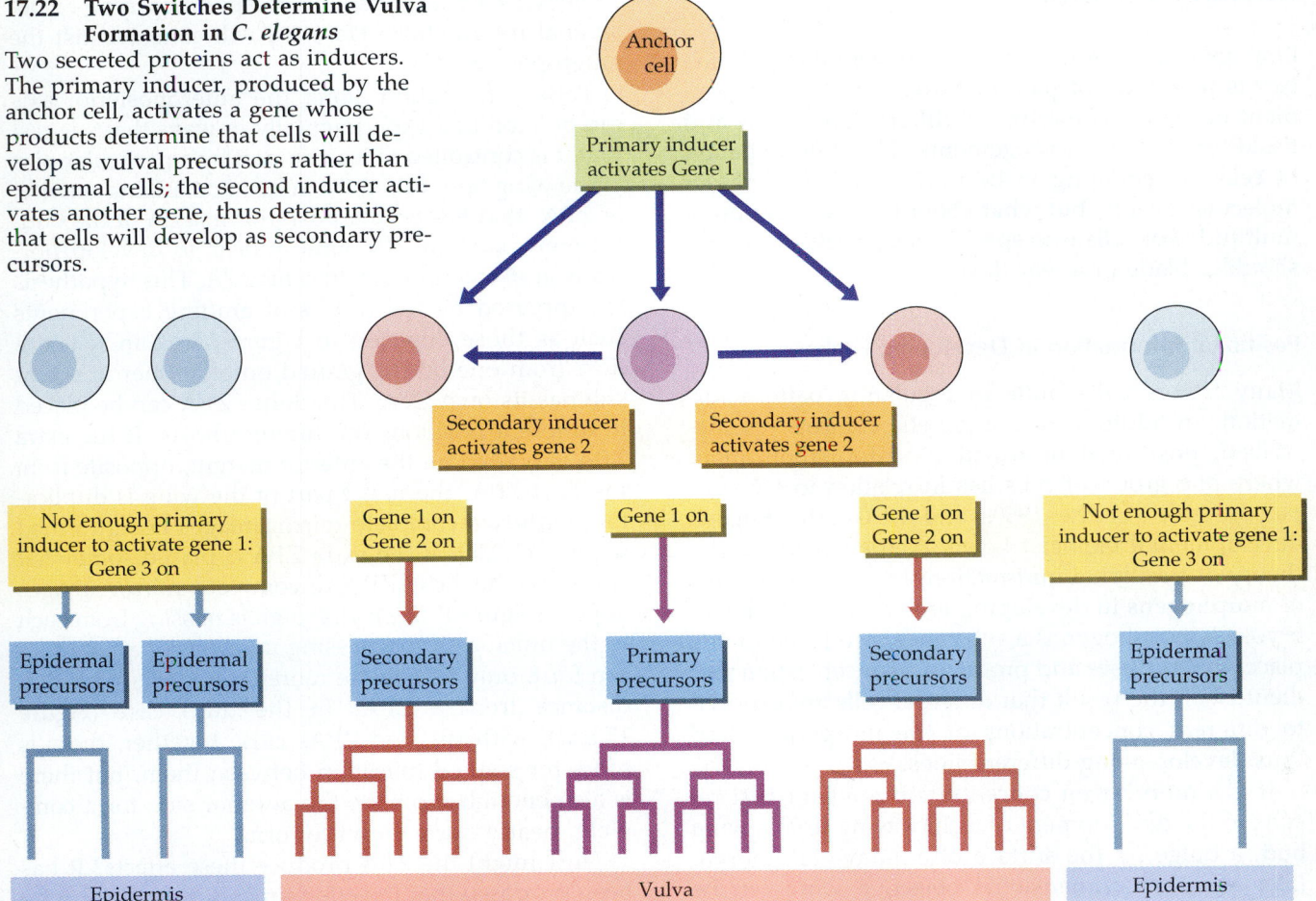

BOX 17.A

Genes Interact in Induction

Let us see how several genes interact to produce the vulva of *Caenorhabditis elegans*. In the anchor cell, a gene called *lin-3* is transcribed and translated to produce the first inducer, a protein similar to a growth factor (see Chapter 15). In the target cells, another gene (called *let-23*) is responsible for producing a receptor for this inducer. The receptor is a protein that extends through the plasma membrane of the cell. The part of the receptor that lies outside the cell recognizes the inducer; when it binds the inducer, the part on the inside of the cell picks up a phosphate group from ATP. The phosphorylated receptor then binds an adapter protein produced by another gene, *sem-5*. Next, this complex interacts with another protein, a G-protein (itself encoded by another gene, *let-60*). Upon meeting the adaptor–receptor complex, the G-protein acquires a phosphate group from GTP (hence the name G-protein). The G-protein thus gains the ability to direct another gene to produce the second inducer—which, in turn, activates yet another gene, *lin-12*, in the cells that become the secondary vulval precursors.

Many events at plasma membranes are controlled by receptor proteins that interact with ATP and with G-proteins. This type of membrane activity is described in more detail in our discussion of animal hormones (see Chapter 36).

two alternative tracks. One challenge for the developmentalist is to find these switches and determine how they work. Another challenge is to account for the larger-scale phenomenon of pattern formation.

PATTERN FORMATION

One area of current research in developmental biology is the study of **pattern formation**, the development of organs consisting of differentiated cells and tissues in ordered arrangements. The differentiation of cells is beginning to be understood in terms of molecular events, but what about the organization of multitudes of cells into specific body parts, such as a shoulder blade or a tear duct?

Positional Information in Developing Limbs

Many factors collaborate in regulating pattern formation. In addition to the genetic controls just described, **positional information**—information about where one group of cells lies in relation to others—plays a role. In the 1960s and 1970s, the English developmental biologist Lewis Wolpert developed a theory of positional information, based on gradients of **morphogens** in developing limb buds in chick embryos. A morphogen is a substance, produced in one place, that diffuses and produces a concentration gradient, with the result that different cells are exposed to different concentrations of the morphogen and thus develop along different lines.

It is a morphogen concentration gradient that results in the development of a chick wing from a wing bud, a bulge on the surface of a 3-day-old embryo. Like any three-dimensional object, a wing can be described in terms of three perpendicular axes. We are interested here in the anteroposterior axis of the wing—the axis that corresponds to the axis of the body running from the head to the tail of the chick. (The proximodistal axis runs from the base of the limb to its tip, and the dorsoventral axis from back to belly.) Each axis has a corresponding type of positional information. Here we will consider just the anteroposterior axis.

Pattern formation along the anteroposterior axis can be modified experimentally in ways that suggest that it is controlled at least in part by a particular part of the wing bud, called the zone of polarizing activity, or **ZPA**, that lies on the posterior margin of the bud. Different parts of the limb appear to develop normally at *specific distances from the ZPA*. This hypothesis is supported by the results of grafting experiments such as those depicted in Figure 17.23, in which a ZPA from one bud is grafted onto another bud that still has its own ZPA. The donor ZPA can be placed in different positions on different hosts. If the extra ZPA is grafted on the anterior margin, opposite from the host ZPA, the distal part of the wing is duplicated, with two complete mirror images being formed (Figure 17.23b). If the extra ZPA is placed somewhat closer to the host ZPA, incomplete mirror images appear (Figure 17.23c): One digit is missing from each of the units, as if the missing units are of a type that can form only if they are more than some minimum distance from a ZPA. In the third case (Figure 17.23d), with the two ZPAs close together, there is room for some duplication between them, but there is also enough room on the anterior side for a complete, nearly normal unit to form.

How might the ZPA produce these effects? It has been proposed that the effects result from the activity

(a)

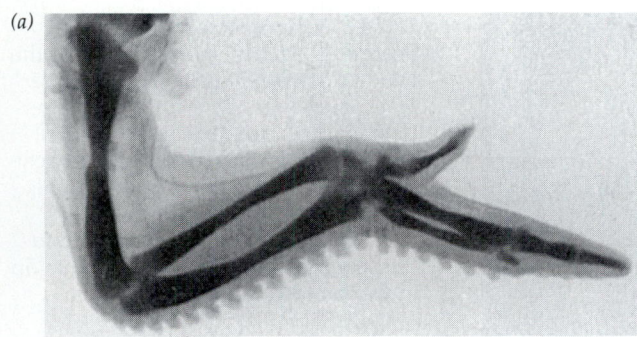

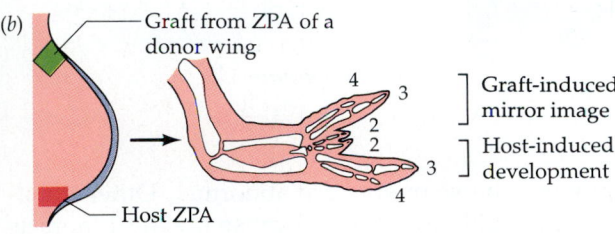

(b)

Graft from ZPA of a donor wing

Graft-induced mirror image

Host-induced development

Host ZPA

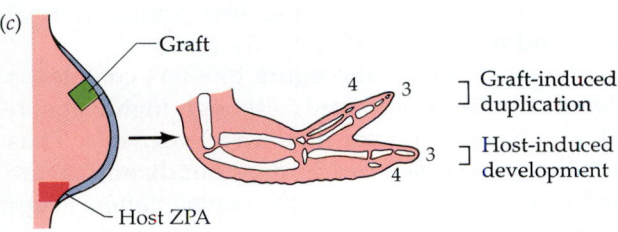

(c)

Graft

Graft-induced duplication

Host-induced development

Host ZPA

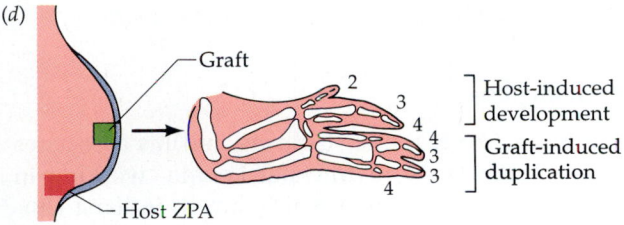

(d)

Graft

Host-induced development

Graft-induced duplication

Host ZPA

17.23 The ZPA Provides Positional Information
(a) By the age of 9.5 days, the embryonic chick wings have developed to the stage shown in this preparation that has been stained to reveal a full set of wing bones. (b–d) A zone of polarizing activity (ZPA) is located on the posterior margin of the chick wing bud—the area indicated as "Host ZPA." In each of the experiments diagrammed here, the ZPA from one bud was grafted onto another wing bud which still had its own ZPA. (b) ZPA grafted on the anterior margin causes mirror-image duplication of the distal part of the wing—digits 4, 3, and 2. (c) A ZPA grafted closer to the host ZPA causes duplication of digits 4 and 3, but no digit 2 develops. (d) A ZPA grafted still closer to the host ZPA allows a nearly normal set of digits 2, 3, and 4 to develop on the anterior side but also results in partial duplication of digits 3 and 4 between the two ZPAs.

of an apparent morphogen, called retinoic acid, produced by the ZPA. Several pieces of experimental evidence support this theory. Quite recently, however, contradictory evidence has convinced many workers that retinoic acid plays a role other than that of the ZPA morphogen. We still lack a satisfactory explanation for the fascinating effects shown in Figure 17.23.

Establishing Body Segmentation

Developmental biologists have also studied pattern formation using *Drosophila* fruit flies as experimental subjects. Insects (and many other animals) develop a highly modular body composed of different types of modules, called segments. Complex interactions of different sets of genes underlie the pattern for-

mation of segmented bodies. Unlike the body segments of segmented worms such as earthworms, the segments of the *Drosophila* body are different from one another. The *Drosophila* adult has a head (one segment), three different thoracic segments, eight abdominal segments, and a genital segment at the posterior end. Thirteen segments in the *Drosophila* larva correspond to these adult segments.

Key genes act sequentially to organize a developing *Drosophila* larva. An overall framework of anteroposterior and dorsoventral axes is laid down initially by the activity of the genes that produce the cytoplasmic determinants referred to earlier (see Figure 17.18). As we saw, mutations in those genes result in the duplication or deletion of body parts, such as anterior and posterior halves of the embryo. The number, boundaries, and polarity of the larval segments are determined by proteins encoded by several more genes, the **segmentation genes**. Three classes of segmentation genes participate, one after the other, to regulate finer and finer details of the segmentation pattern. First, **gap genes** organize large areas along the anteroposterior axis. Mutations in gap genes result in the omission of several larval segments. Second, **pair-rule genes** divide the embryo into two-segment-long units. Mutations in pair-rule genes result in embryos missing every second segment. Third, **segment-polarity genes** determine the boundaries and anteroposterior organization of the segments themselves. Mutations in segment-polarity genes result in segments in which some posterior structures are replaced by reversed (mirror-image) anterior structures.

Finally, after the basic pattern of segmentation has been established by the segmentation genes, differences between the segments are mediated by the activities of **homeotic genes**. These genes are expressed in different combinations along the length of the body and tell each segment what to become.

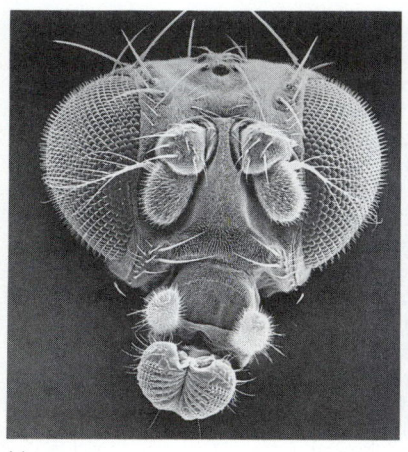

(a)

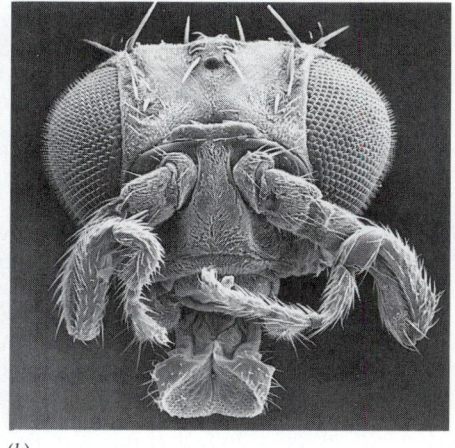

(b)

17.24   A Homeotic Mutation
Scanning electron micrographs of the heads of two *Drosophila melanogaster*. (a) A wild-type fly, with normal antennae. (b) An *antennapedia* mutant, with roughly normal legs in the positions usually occupied by antennae. This abnormality results from a homeotic mutation—a drastic mutation causing one structure to develop in the place of another.

Homeotic Mutations

Our present understanding of the genetics of pattern formation began with the discovery of dramatic mutations, called homeotic mutations, in *Drosophila*. Instead of a normal body part, the insect with a homeotic mutation has another part characteristic of another body segment. Two bizarre examples are the *antennapedia* mutant, in which legs grow in the place of antennae (Figure 17.24), and the *ophthalmoptera* mutant, in which wings grow in the place of eyes. Homeotic genes fall into a few tight clusters. Of these, the best characterized is the **bithorax complex**. The eight or more genes of the bithorax complex control development of the abdomen and posterior thorax of the fly. Development of the head and anterior thorax is controlled by another homeotic gene cluster, the **antennapedia complex**. The functions of the two complexes interact substantially.

Because many mutations in the bithorax complex are so severe they prevent development past the early larval stages, they can be studied only in larvae right after they hatch out of the egg. Figure 17.25 shows that if mutations have deleted all or part of the bithorax complex, the larvae have thirteen seg-

ments but the segments are abnormal. Other mutations in the bithorax complex cause the third thoracic segment of the adult to develop exactly like the second, resulting in a fly that has two pairs of normal wings and no halteres (Figure 17.26).

As we just saw, if the entire bithorax complex is deleted, the larva produced, although highly abnormal, still has the normal number of segments. Thus the bithorax complex clearly does not determine the number of segments. It is the segmentation genes that determine the number and polarity of segments. In the course of normal development, the homeotic genes give each segment its distinctive character.

The Homeobox

In the early 1980s Walter Gehring and his associates William McGinnis and Michael Levine, working in Switzerland, and Thomas Kaufman, at Indiana University, undertook a study of the antennapedia complex, using the techniques of recombinant DNA technology. They set out to isolate and clone the *antennapedia* (*Antp*) gene, a member of the antennapedia complex, from *Drosophila*. As part of this study, they prepared a DNA that could hybridize to this *Antp* gene. Surprisingly, the *Antp* DNA hybridized with both the *Antp* gene and a nearby segmentation gene (the *fushi tarazu* gene, *ftz*, from the Japanese for

17.25   *bithorax* Mutations in *Drosophila* Larvae
Genotypes at loci of the bithorax complex govern how segments develop in *Drosophila* larvae. The wild-type larva shown on the left has a head (H), three thoracic segments (T1–T3), eight abdominal segments (A1–A8), and one genital segment (G). The larva in the center lacks the entire bithorax complex. It has a normal head and first thoracic segment, but the other segments have developed like normal second thoracic segments. The larva on the right lacks part of the bithorax complex; all its segments are like normal eighth abdominal segments.

17.26 *bithorax* **Mutations in Adult** *Drosophila*
Because of two mutations in its bithorax complex, this
fly's third thoracic segment developed as if it were a second thoracic segment. In wild-type *Drosophila* the second
thoracic segment gives rise to wings and legs in the adult,
while the third thoracic segment produces a pair of legs
and a pair of small, winglike structures called halteres.

"too few segments"). The *Antp* and *ftz* genes must
have DNA sequences of close similarity, because part
of each gene hybridizes with the same DNA. Further
hybridization studies demonstrated that the same
shared stretch of DNA is also found in the *bicoid* gene
of the bithorax complex, in some other parts of the
Drosophila genome, and in genes in other insect species. In fact, this important sequence of 180 base pairs
of DNA, called the **homeobox**, has now been shown
to be part of a few genes of many animals and plants.

What does the homeobox, which is present in almost all eukaryotes, do? The homeobox sequence
codes for a region of 60 amino acids—the homeodo-
main—that is part of some proteins. These proteins
return to the nucleus and bind to DNA, regulating
the transcription of other genes. A computerized
search of published sequences of DNA from numerous species revealed a similarity between the homeobox and parts of certain regulatory genes in yeast—
genes that produce proteins that also bind to specific
DNA sequences. Some genes with homeoboxes are
expressed only at certain times and in certain tissues
as development proceeds, as would be expected if
these proteins regulate development.

What are we to make of the presence of the homeobox in such diverse species as humans, fruit flies,
frogs, *C. elegans*, and tomatoes—and of its presence
in several genes in the same organism? The ubiqui-
tous presence of the homeobox suggests that both
the antennapedia and bithorax complexes may have
arisen from a single ancestral gene. Further, it implies
that a single gene in some ancient organism may have
been the evolutionary progenitor of what is now a
widespread controlling system for development.

YOU'VE SEEN THE PIECES; NOW LET'S BUILD A FLY

One of the most striking and important observations
about development in *Drosophila*—and in other animals—is that it results from a *sequence* of changes,
each change triggering the next. The sequence is a
transcriptionally controlled cascade. To see this more
clearly, let's "build" a *Drosophila* larva step by step,
beginning with the unfertilized egg.

Before the egg is fertilized, mRNA for the *bicoid*
protein is localized at the end destined to become the
anterior end of the animal. The egg is fertilized and
laid, and cleavage begins. At the same time, the *bicoid*
mRNA is translated, forming *bicoid* protein that diffuses away from the anterior end, establishing a gradient of the protein (Figure 17.27a). Another morphogen, the *nanos* protein, diffuses from the posterior
end, forming a gradient in the other direction. Thus,
each nucleus in the developing blastula is exposed to
a different concentration of *bicoid* protein and to a
different ratio of *bicoid* protein to *nanos* protein (Figure
17.27b). What do these morphogens do?

The two morphogens regulate the expression of
the gap genes. Both morphogens are transcription
factors (see Chapter 13)—they enhance or repress the
expression of the gap genes. High concentrations of
bicoid protein turn on the gap gene called *hunchback*,
but the *bicoid* protein also turns off another gap gene
called *Krüppel*. The pattern of gap gene activity resulting from morphogen actions is shown in Figure
17.28. What is the function of the gap genes?

The proteins encoded by the gap genes are another
set of transcription factors that control the expression
of the pair-rule genes. Many of these in turn encode
transcription factors that control the expression of the
segment-polarity genes, giving rise to a complex,
striped pattern that foreshadows the segmented body
plan of *Drosophila*. By this point, each nucleus of the
blastula is exposed to a distinctive set of transcription
factors. The body pattern of the larva is established
even before gastrulation begins, although no segmentation is visible yet. When the segments do appear, why aren't they all alike?

The homeotic genes give the different segments
their different properties. Each homeotic gene is expressed over a characteristic portion of the embryo.
Developmental biologists recently discovered that the
six homeotic genes are arranged on the same chromosome—in the same order as the order of their
function from the anterior to the posterior end of the
larva! Each homeotic gene encodes a transcription
factor, and each of these transcription factors has a
homeobox as part of its primary structure.

And that is how genes and their products specify
the structure of a *Drosophila* larva by a transcriptionally controlled cascade. Does this have any relevance

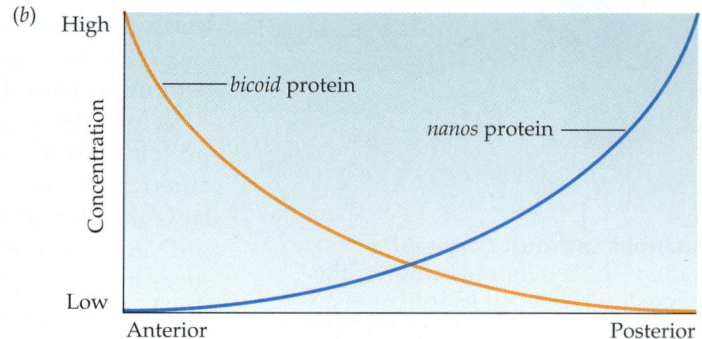

17.27 *bicoid* Protein Forms a Gradient
Translation of mRNAs at the ends of the *Drosophila* larva leads to gradients of the morphogen products. *(a)* The concentration of *bicoid* protein is highest at the embryo's anterior end—bright yellow in this photograph. The colorized gradient moves from orange to red as *bicoid* concentration decreases into the dark posterior end. *(b)* Morphogen gradients originating at the two ends of the embryo expose each cell to a unique environment.

to the development of other organisms—notably humans? In fact, it *is* relevant. Recent experiments have shown that mice and humans (the two best-studied mammals) have clusters of homeobox genes. Thirty-eight genes are divided into four clusters, each located on a different chromosome. As in *Drosophila*, these homeobox genes are arranged in the same order on each chromosome as the order of their expression from anterior to posterior of the developing animal. These genes are expressed in particular segments of the animal, just as in the "simple" fruit fly model. Thus there *are* some rules of development that apply to much of the animal kingdom. What might this and other findings of developmental biology imply for our understanding of the mechanisms of evolution?

DEVELOPMENTAL BIOLOGY AND EVOLUTION

As you are about to discover in Part Three, our current understanding of evolution is heavily dependent on contributions from the field of population genetics. More recently, developmental genetics has also begun to make key contributions to the study of evolution. In particular, homeotic genes are believed to contribute to "macroevolution"—the larger jumps involved in the appearance of new species and, long ago, of major groups of animals and other organisms. Figure 17.29 offers a speculative view of how the progressive addition of homeotic gene functions could account for some aspects of insect evolution. This model is meant simply to illustrate how homeotic genes may contribute to evolution and is not to be taken literally as a description of what actually happened.

17.28 Gap Genes in Action
Interactions of proteins coded by gap genes define domains of the larval body in *Drosophila*. In this larva, *hunchback* (orange) and *Krüppel* (green) proteins overlap, forming a boundary (yellow) between two domains.

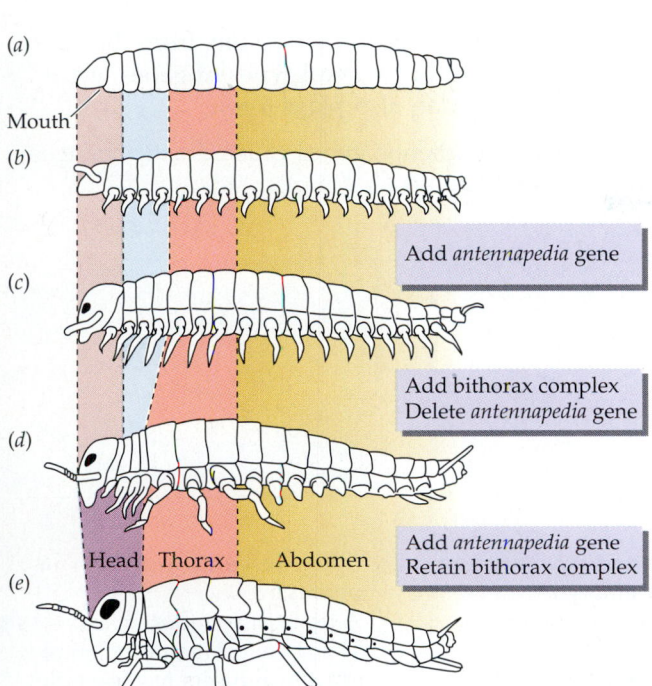

(a)

Mouth

(b)

Add *antennapedia* gene

(c)

Add bithorax complex
Delete *antennapedia* gene

(d)

Head | Thorax | Abdomen

Add *antennapedia* gene
Retain bithorax complex

(e)

17.29   How Homeotic Genes May Have Contributed to Insect Evolution
This hypothetical model illustrates how the addition of homeotic gene functions may have led progressively from (a) an earthwormlike animal to (e) the winged insects of today. The intermediate stages correspond roughly to animal groups alive today—(b) onychophorans, (c) centipedes, and (d) wingless insects (see Chapter 26).

Some developmental biologists are concerned with evolutionary questions, ranging from how the first embryos came into being to how macroevolution occurs. Let us, too, turn next to the study of evolution: its mechanisms, its consequences, and its history.

SUMMARY of Main Ideas About Animal Development

An animal develops by progressive changes from the zygote, through the embryonic and subsequent stages of its life cycle, until it dies.

The zygote divides mitotically, giving rise to thousands of cells by cleavage. The amount and distribution of yolk varies among animal groups, leading to different patterns of cleavage.
   Review Figure 17.2

Cleavage results in the formation of a hollow blastula.
   Review Figure 17.4

Redistribution of cells transforms the blastula into a gastrula, a two- or three-layered embryo with a gut.

The pattern of gastrulation varies among animal groups.
   Review Figures 17.5, 17.6, and 17.7

A gastrula has an outer layer (ectoderm) and an inner layer (endoderm); in most animals there is an intervening layer (mesoderm).

The ectoderm, mesoderm, and endoderm each give rise to particular parts of the adult animal.
   Review Table 17.1

The nervous system is one of the first embryonic organ systems to develop.

In neurulation, part of the surface ectoderm folds and becomes a neural tube within the embryo.
   Review Figure 17.9

Early products of the mesoderm are the notochord (a supporting rod); later products are the somites (which develop into muscles and the backbone).

Many animals grow throughout their lives; some develop continuously, and others develop discontinuously because they have external skeletons.
   Review Figure 17.11

Some animals, such as butterflies and frogs, undergo metamorphosis, in which a juvenile form reorients radically while developing into the adult form.

As the embryo develops, cells become determined—committed to developing into particular parts of the embryo and of the adult animal.

When a cell is determined, its fate is specified; its developmental flexibility is restricted.

At early stages a cell may have a particular fate, but the fate may change if the cell is transplanted to another part of an embryo.

Determination precedes differentiation.

Differentiation is the process in which cells change in structure, chemical makeup, and function. One source of differentiation is the segregation of some components of the egg cytoplasm into separate cells of the developing embryo.
   Review Figure 17.19

Chemical inducers are another source of differentiation.
Review Figures 17.21 and 17.22

Pattern formation is the process by which body parts become delineated and organized in predictable arrangements.

Pattern formation may result from the application of information about where different groups of cells lie in relation to one another.
Review Figure 17.23

Body segmentation in insects results from the successive expression of several types of genes; cytoplasmic segregation also plays a role.

Mutations in homeotic genes cause drastic changes in pattern formation.

SELF-QUIZ

1. Which statement is *not* true of cleavage?
   a. The blastomeres are produced by mitosis.
   b. Cleavage patterns depend in part on the distribution of yolk.
   c. The first few divisions often occur simultaneously.
   d. The first cleavage results in a cell toward the animal pole and one toward the vegetal pole.
   e. In eggs with a lot of yolk, daughter cells are not completely separated.

2. A blastula
   a. is a solid ball of cells.
   b. is surrounded by a sheet of cells called the blastocoel.
   c. develops over a few hours of cleavage.
   d. has a lower ratio of nuclear to cytoplasmic volume than did the zygote.
   e. is much larger than the fertilized egg.

3. Gastrulation
   a. is identical in all animal species.
   b. always results from ingression of cells at the vegetal pole.
   c. always produces a primitive streak.
   d. always results in a roughly spherical gastrula.
   e. is a process in which a two- or three-layered embryo is formed.

4. Which statement is *not* true of the dorsal lip of the amphibian blastopore?
   a. It is the first site of invagination as the blastula is forming.

   b. Cells turn in here and flow into the interior of the embryo.
   c. It serves as the embryonic organizer.
   d. It spreads, eventually forming a complete circle.
   e. Cells involuting here give rise to both endoderm and mesoderm.

5. Which statement is *not* true of cytoplasmic determinants in *Drosophila*?
   a. They specify the dorsoventral and anteroposterior axes of the embryo.
   b. Their positions in the embryo are determined by microfilament action.
   c. They are products of specific genes in the mother insect.
   d. They often produce striking effects in larvae.
   e. They have been studied by the transfer of cytoplasm from egg to egg.

6. The organizer of an amphibian embryo
   a. is a homeotic mutation.
   b. is the homeobox.
   c. induces adjacent ectoderm to organize into a neural tube.
   d. is the ventral lip of the blastopore.
   e. is a product of imaginal discs.

7. Which statement is *not* true of embryonic induction?
   a. One tissue induces an adjacent tissue to develop in a certain way.
   b. It triggers a sequence of gene expression in target cells.

   c. It may be either instructive or permissive.
   d. A tissue may induce itself.
   e. The chemical identification of specific inducers has been difficult.

8. In establishing body segmentation in *Drosophila* larvae,
   a. the first steps are specified by homeotic genes.
   b. mutations in pair-rule genes result in embryos missing every other segment.
   c. mutations in gap genes result in the insertion of extra segments.
   d. segment-polarity genes determine the dorsoventral axes of segments.
   e. segmentation is the same as in earthworms.

9. Homeotic mutations
   a. are often so severe that they can be studied only in larvae.
   b. cause subtle changes in the forms of larvae or adults.
   c. occur only in prokaryotes.
   d. do not affect the animal's DNA.
   e. are confined to the apical ectodermal ridge.

10. Which statement is *not* true of the homeobox?
    a. It is transcribed and translated.
    b. It is found only in animals.
    c. Proteins containing the homeodomain bind to DNA.
    d. It is a stretch of DNA shared by many genes.
    e. Its activities often relate to body segmentation.

FOR STUDY

1. Discuss the differences—and the reasons for the differences—in the gastrulas of a sea urchin, a frog, and a chicken.

2. Consider a cell in the lens of a frog eye. Trace its developmental history back to a particular part of the zygote. Describe all the ways in which other cells or tissues have interacted with it.

3. During development, the developmental potential of a tissue becomes ever more limited until, in the normal course of events, the developmental potential is the same as the original fate. On the basis of what you have learned in this chapter and in Chapter 13, discuss possible mechanisms for the progressive limitation of the developmental potential.

4. How was it possible for biologists to obtain such a complete accounting of all the cells in the roundworm *Caenorhabditis elegans*? Why can't we reason directly from studies of *C. elegans* to comparable problems in our own species?

READINGS

De Robertis, E. M., G. Oliver and C. V. E. Wright. 1990. "Homeobox Genes and the Vertebrate Body Plan." *Scientific American*, July. Extends material in this chapter to amphibian development; good discussion of evolution of anatomy.

Gehring, W. J. 1985. "The Molecular Basis of Development." *Scientific American*, October. A lucid account of homeotic mutations and the homeobox.

Gilbert, S. F. 1994. *Developmental Biology*, 4th Edition. Sinauer Associates, Sunderland, MA. An exceptionally well-balanced treatment of developmental biology, covering both molecular/cellular concepts and embryology. Gives a feeling for the history of the discipline.

Goodman, C. S. and M. J. Bastiani. 1984. "How Embryonic Nerve Cells Recognize One Another." *Scientific American*, December. How do brains develop their specific "wiring"? This article describes experimental methods for dealing with this question.

Holliday, R. 1989. "A Different Kind of Inheritance." *Scientific American*, June. Genes can be methylated in one cell, and the pattern of methylation is inherited by the products of cell division. Thus a pattern of gene activity is transmitted from one cell generation to the next.

Hynes, R. O. 1986. "Fibronectins." *Scientific American*, June. Proteins that guide migrating cells during development, and their possible role in cancer.

Stent, G. S. and D. A. Weisblat. 1982. "The Development of a Simple Nervous System." *Scientific American*, January. Details of a specific pattern of development in the leech.

PART THREE
Evolutionary Processes

Buried under layers of younger rocks, Earth's oldest rocks contain no fossils, indicating that for many years after its formation there was no life on Earth. All organisms consist of cells and, as far as we know, all cells come from preexisting cells. The first cells, however, must have come from noncells. How did this come about? Under what conditions did life originate on Earth? Biologists have made exciting recent progress in determining the physical conditions on Earth when life arose and in deducing the stages by which nonlife evolved into protolife (the immediate precursor of life) and, eventually, into life. These investigations are the subject of this chapter.

In Parts One and Two you learned that the chemical reactions essential for life proceed within the cells of organisms because of the activity of catalysts. The functioning of organisms requires that the rates of these chemical reactions be regulated so that the substances needed by the organisms are produced at the right times and in the right amounts. What catalyzed the chemical reactions of protolife and early life, and how were the rates of those reactions regulated?

The concept that organisms reproduce themselves also was central to the discussion at many points in Parts One and Two. The basis of biology is that cells and multicellular organisms replicate themselves. How could something that is self-replicating have been started before there were cells? What was the earliest self-replicating unit and how did it reproduce?

When life first appeared, biological evolution—changes in the genetic compositions of organisms through time—also began. Biological evolution over more than 3 billion years has resulted in the millions of species of organisms living today and the many millions more that lived in the past but that became extinct. What are the mechanisms of evolutionary change? How did the first organisms become divided into millions of different types? How do biologists determine which organisms gave rise to which descendants? How do we understand biological evolution and its varied products? Part Three answers these questions.

IS LIFE EVOLVING FROM NONLIFE TODAY?

Until the last century people believed that new life appeared regularly—for example, that flies and maggots arose from rotting meat and barnyard manure, lice from sweat, glowworms from rotting logs, eels and fish from sea mud, and frogs and mice from moist earth. For more than 2,000 years, **spontaneous generation**—the formation of living organisms from nonliving matter—was accepted by most people as a continuing, obvious fact.

Stratified Rocks Help Scientists Tell Time
Rocks are laid down in discernible strata over time. In Coal Canyon, Arizona, wind and water action have exposed such layers in a striking scene.

18

The Origin of Life on Earth

411

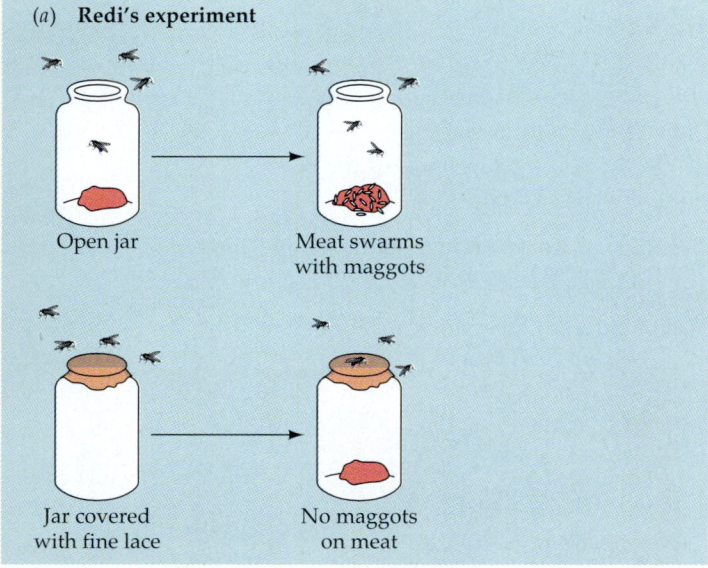

(a) **Redi's experiment**

Open jar → Meat swarms with maggots

Jar covered with fine lace → No maggots on meat

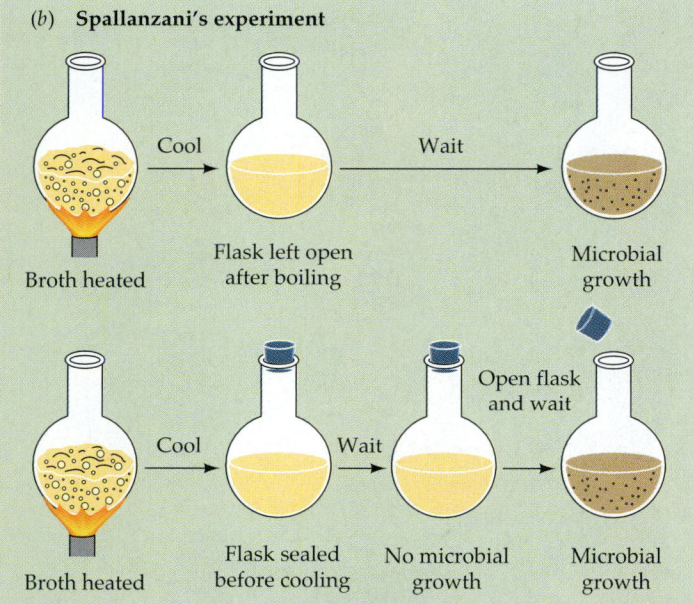

(b) **Spallanzani's experiment**

Broth heated — Cool → Flask left open after boiling — Wait → Microbial growth

Broth heated — Cool → Flask sealed before cooling — Wait → No microbial growth → Open flask and wait → Microbial growth

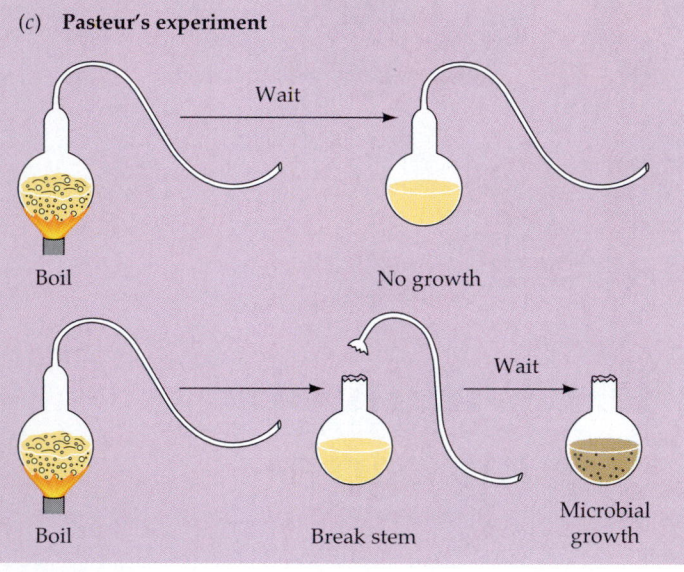

(c) **Pasteur's experiment**

Boil — Wait → No growth

Boil → Break stem — Wait → Microbial growth

18.1 Tests of Spontaneous Generation

Experiments by several scientists eventually ruled out prevailing theories of the spontaneous generation of life, but it took two centuries.

As far as we know, before 1668 no one had performed an experiment to determine, for example, whether maggots really arose spontaneously from decaying meat. In that year Francesco Redi, an Italian physician, demonstrated that maggots in meat are the larvae of flies and that if meat is protected so that adult flies cannot lay their eggs on it, no maggots appear (Figure 18.1a).

In the latter part of the eighteenth century, the Italian biologist Lazzaro Spallanzani showed that if nutrient media are placed in sealed containers after being (as we say today) adequately sterilized, they remain devoid of life (Figure 18.1b). He failed to convince his contemporaries, however, partly because others performed the same experiments with less care and obtained different results. Also, some people argued that Spallanzani's techniques not only killed the microorganisms already present but also rendered the air unfit for the generation and growth of new ones.

In 1862 the great French scientist Louis Pasteur finally convinced most people that spontaneous generation does not occur. Pasteur performed a series of meticulous experiments showing that microorganisms come only from other microorganisms and that a genuinely sterile solution remains lifeless indefinitely unless contaminated by living creatures. His most elegant experiment relied on swan-necked flasks that were open to the air (ruling out the "spoiled air" objection raised against Spallanzani). Pasteur filled the flasks with nutrient medium, heated them to kill any microorganisms present, then cooled them slowly. The shape of the necks kept any new organisms from falling into the medium. No new growth appeared in the flasks (Figure 18.1c). As a result of these experiments, the aphorism *omne vivum e vivo* ("all life from life") became widely accepted.

WHEN DID LIFE ON EARTH ARISE?

Pasteur's experiments suggested that life is not evolving from nonlife today, but they shed no light on when or how life did arise. Life that arose on Earth can be no older than Earth itself, but how old is Earth? Speculations about the age and origin of Earth extend back to the earliest written records. Hindu Brahmans believe that Earth has always existed—that it is eternal. In the Western world, however, it has long been believed that Earth had a beginning—and a relatively recent one at that. In 1650 Irish Archbishop James Ussher, estimating from his study of Hebrew texts, calculated that the creation of Earth took place in 4004 B.C. Although his estimate was not universally accepted, until the nineteenth century most people in the Western world shared Bishop Ussher's view that Earth was relatively young.

During the nineteenth century, however, scientific evidence accumulated suggesting a much older age of Earth. Lord Kelvin, one of the great British physicists of the nineteenth century, estimated that Earth is about 20 million years old. He did this by calculating how much heat would have been generated as Earth contracted from a large mass of gases to its current small solid state. Geologists also became convinced that Earth was much older than most people believed, but they could not say how old. Their evidence came primarily from the remains of organisms that were found in sedimentary rocks.

The geologists' guiding concept was the **principle of superposition**—that as rocks form by the piling up of sediments, younger rocks are deposited on top of older ones (Figure 18.2). **Fossils**—remains of organisms that lived when the sediments were accumulating—were preserved within the rocks. Geologists could see slight differences among similar organisms as they compared the layers, or **strata**, of older and younger rocks in the same place. They also found remains of familiar organisms at widely separated locations. By assuming that rocks at different locations that contained a particular type of fossil were likely to be of approximately the same age, they inferred that widely separated sedimentary deposits, each containing the same type of fossil, had been laid down at approximately the same time. By making such comparisons among many locations, and always considering the superposition of strata, geologists had determined the general order of events in the history of life long before they knew the actual times that these events occurred. This method, however, gave only relative ages.

At the end of the nineteenth century, scientists discovered radioactivity. Out of this discovery came a way to date rocks and the fossils they contain because as rocks form, radioactive isotopes of uranium, thorium, rubidium, potassium, and other elements—in proportion to their presence in the environment—are incorporated into the rocks. Each type of radioactive isotope then begins to decay at its own constant rate, becoming by this process a stable isotope. These rates of decay can serve as **radiometric clocks** because the absolute ages of rocks can be calculated from the proportions of radioactive and stable isotopes present. Uranium-238, for example, spontaneously decays into lead at a slow but precisely known rate. By knowing this constant rate and by comparing the amount of ^{238}U still present in a rock with the amount of lead derived from its decay, geochemists can estimate the age of the rock with less than a 5 percent error.

Radioisotopes that decay rapidly are useful for dating more recent rocks and fossils; those that decay slowly can be used to date older ones. The earliest known fossils have been radiometrically dated at 3.8 billion years old (Figure 18.3), so we know that life arose in the very ancient past.

Another way to date the origin of life is to estimate the age of the genetic code. We can make such an estimate by comparing the sequences of tRNA among

18.2 Younger Rocks Lie on Top of Older Rocks
The positions of the rocks in the Grand Canyon of the Colorado River reveal their *relative* ages, but radiometric clocks were needed to estimate the *absolute* ages given here.

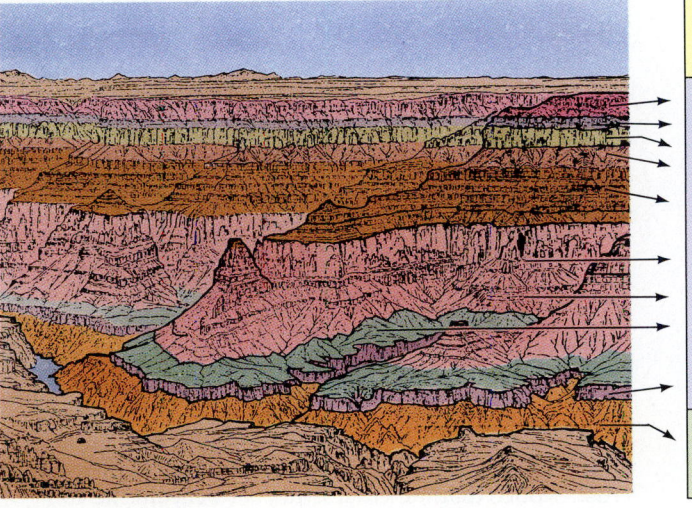

Rock formation	Approximate age in millions of years	Era
Kaibab limestone	250	
Toroweap limestone	255	
Coconino sandstone	260	
Hermit shale	265	
Supai sandstone	285	
Redwall limestone	335	Paleozoic
Muav limestone	515	
Bright Angel shale	530	
Tapeats sandstone	545	
Zoroaster granite and Vishnu schist	1,700–2,000	Precambrian

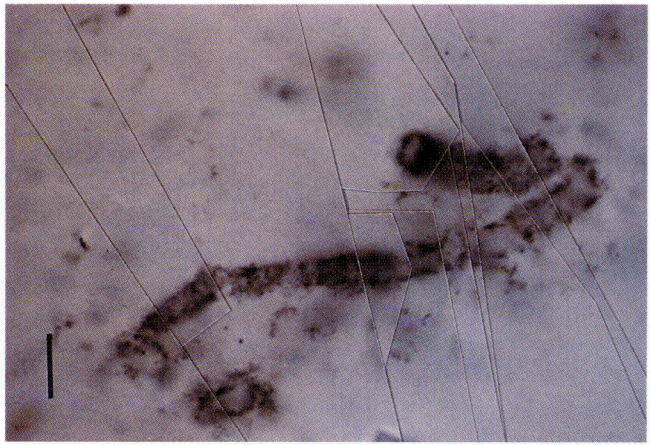

18.3   The Oldest Known Fossil Cells
This fossilized structure, found in rocks from Western Australia, is believed to be a filamentous bacteria. Such fossil cells appear in rocks formed as early 3.8 billion years ago.

different organisms. The more dissimilar the alignment of base sequences of different organisms, the longer they probably have been evolving as separate lineages. The patterns of similarities of tRNAs suggest that the genetic code originated before the separation of the eubacteria and archaebacteria, which is estimated from dated fossils to have happened about 3.7 billion years ago. Putting all of this evidence together, we estimate that the genetic code is about 3.8 billion years old.

HOW DID LIFE EVOLVE FROM NONLIFE?

Fossils and absolute dating both are important in answering the question of *when* life first arose, but they cannot tell us *how* life arose. To understand how life arose, we need to know the physical conditions that prevailed on Earth before there was life and when life arose. Those conditions determined which types of chemical reactions took place on Earth. If we know what types of reactions took place, we can suggest how those reactions could have led to the appearance of life.

Scientists now propose that between 10 and 20 billion years ago a mighty explosion occurred. The matter of the universe, which had been highly concentrated, began to spread apart rapidly. The "big bang" sent gases hurtling in all directions. Eventually clouds of gases collapsed upon themselves through gravitational attraction, forming the galaxies, which are great clusters of hundreds of billions of stars.

Somewhat less than 5 billion years ago, toward the outer edge of our galaxy (the Milky Way), our solar system (the sun, Earth, and our sister planets) took form. Most of the planets probably formed by grav-

itational attraction and the aggregation of cold dust particles. As Earth slowly grew by this process, the weight of the outer layers compressed the interior of the planet. The resulting pressures, combined with the energy from radioactive decay, heated the interior until it melted. Within this viscous liquid, the settling of the heavier elements produced an iron and nickel core with a radius of approximately 3,400 km. Around the core lies a mantle of dense silicate materials that is 3,000 km thick. Over the mantle is a lighter crust, more than 40 km thick under the continents but thinning to as little as 5 km in some places under the oceans (Figure 18.4).

Conditions on Earth at the Time of Life's Origin

Earth's mantle and crust released carbon dioxide, nitrogen, and other heavier gases. These gases were held by Earth's gravitational field, and gradually formed a new atmosphere. Earth accumulated an atmosphere consisting mostly of methane (CH_4), carbon dioxide (CO_2), ammonia (NH_3), hydrogen (H_2), nitrogen (N_2), and water vapor (H_2O). As Earth cooled, some of the water vapor escaping from inside the planet condensed into seas.

No free oxygen was present in this early atmosphere because oxygen is a very reactive element. It reacted with hydrogen to form water and with components of Earth's crust and atmosphere to form iron oxides, silicates, carbon dioxide, and carbon mon-

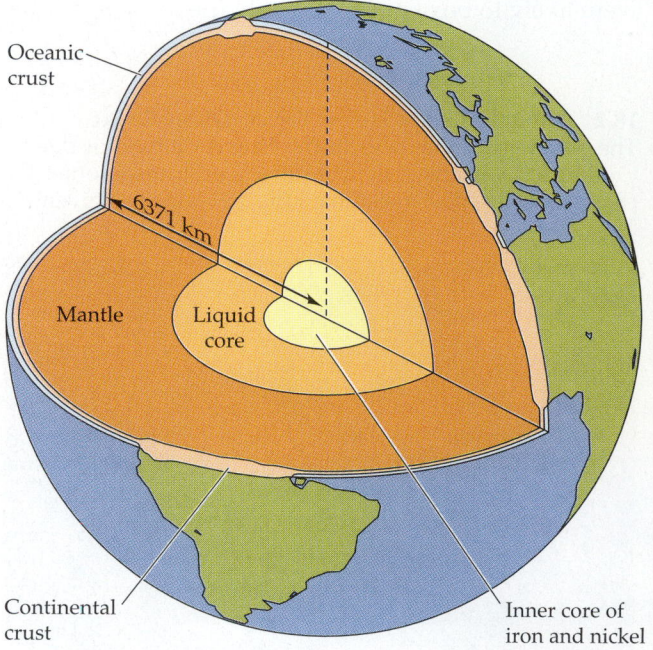

18.4   A Section through the Planet
Earth has a solid central core surrounded by a liquid core and a dense mantle, which is covered in turn by a solid crust. Notice that the crust is thicker beneath the continents than under the oceans.

BOX 18.A

A Concise Scenario for the Origin and Early Evolution of Life

1. The solar system condenses from a dust cloud; Earth and other planets form.

2. Earth's original atmosphere (mostly hydrogen) is lost. Heat from radioactive decay and gravitational compression melts Earth's interior.

3. Gases escape from the hot interior, producing a reducing atmosphere consisting mostly of nitrogen, ammonia, water vapor, carbon dioxide, and methane. There is no free oxygen.

4. Water vapor condenses into seas. Ultraviolet light, radioactive decay, volcanic heat, and lightning provide the energy to produce organic compounds from atmospheric gases.

5. Protolife evolves the ability to use RNAs similar to today's ribozymes as templates for protein synthesis and as catalysts for chemical reactions.

6. Protolife evolves DNA as the genetic material after hydrophobic environments are provided within membrane-enclosed spaces. **Life has evolved.** Copying errors (mutations) yield variations upon which natural selection acts. There is still no free oxygen.

1 BILLION YEARS have elapsed

7. Simple prokaryotic organisms evolve varied synthetic and respiratory pathways, using a variety of molecules as substrates for their energy. One of these pathways uses glucose as the energy source for the synthesis of ATP. (This pathway is later used by all eukaryotes.) Several photosynthetic pathways evolve, but none of them liberates oxygen gas.

2 BILLION YEARS have elapsed

8. Some cyanobacteria evolve photosynthetic pathways that liberate large quantities of oxygen.

9. Oxygen, poisonous to anaerobic organisms, accumulates in the atmosphere. Some prokaryotes evolve the ability to combine oxygen with their metabolic products—aerobic respiration begins.

10. As oxygen builds up, some is converted to ozone, which accumulates in the upper atmosphere. This ozone shields Earth from ultraviolet radiation, allowing life to invade shallow waters and the land.

11. Respiring organisms, using efficient oxidation reactions as their energy source for ATP synthesis, evolve the cytochromes of the terminal respiratory chain.

3 BILLION YEARS have elapsed

12. Eukaryotic cells evolve. Endosymbiotic cyanobacteria become chloroplasts. Other endosymbiotic bacteria become mitochondria. Multicellular organisms evolve from these efficient eukaryotic cells.

oxide. For more than 2 billion years, all oxygen was bound up with other elements, and Earth was only a **reducing environment**. (See Chapter 7 for a discussion of oxidation and reduction.) As a setting for chemical reactions, then, early Earth differed fundamentally from present-day Earth, which has an atmosphere that contains large quantities of free oxygen.

An Overview of Earth's History

Earth's early history and the evolution of life are summarized in Box 18.A. Because numbers in the billions are difficult to comprehend, we present the history of Earth as a hypothetical 30-day month in Figure 18.5. Each "day" on this geological calendar represents approximately 150 million years. The fossil record and the broad pattern of the history of life on Earth that it reveals will be treated in detail in Chapter 28. For the moment, you need remember only that life first appeared long ago, that organisms have continued to evolve ever since life arose, and that scientists knew that life originated billions of years ago before they had testable ideas about how it came about.

Scientists have been trying, during the twentieth century, to reconstruct a probable sequence of events by which life could have evolved from nonlife. The first step in the reconstruction is to describe the environment in which the events took place. Reconstruction of those conditions helps us understand how life may have arisen.

Beginning the Sequence of Events: The Synthesis of Organic Compounds

What sort of chemical reactions would have occurred spontaneously in Earth's early reducing environ-

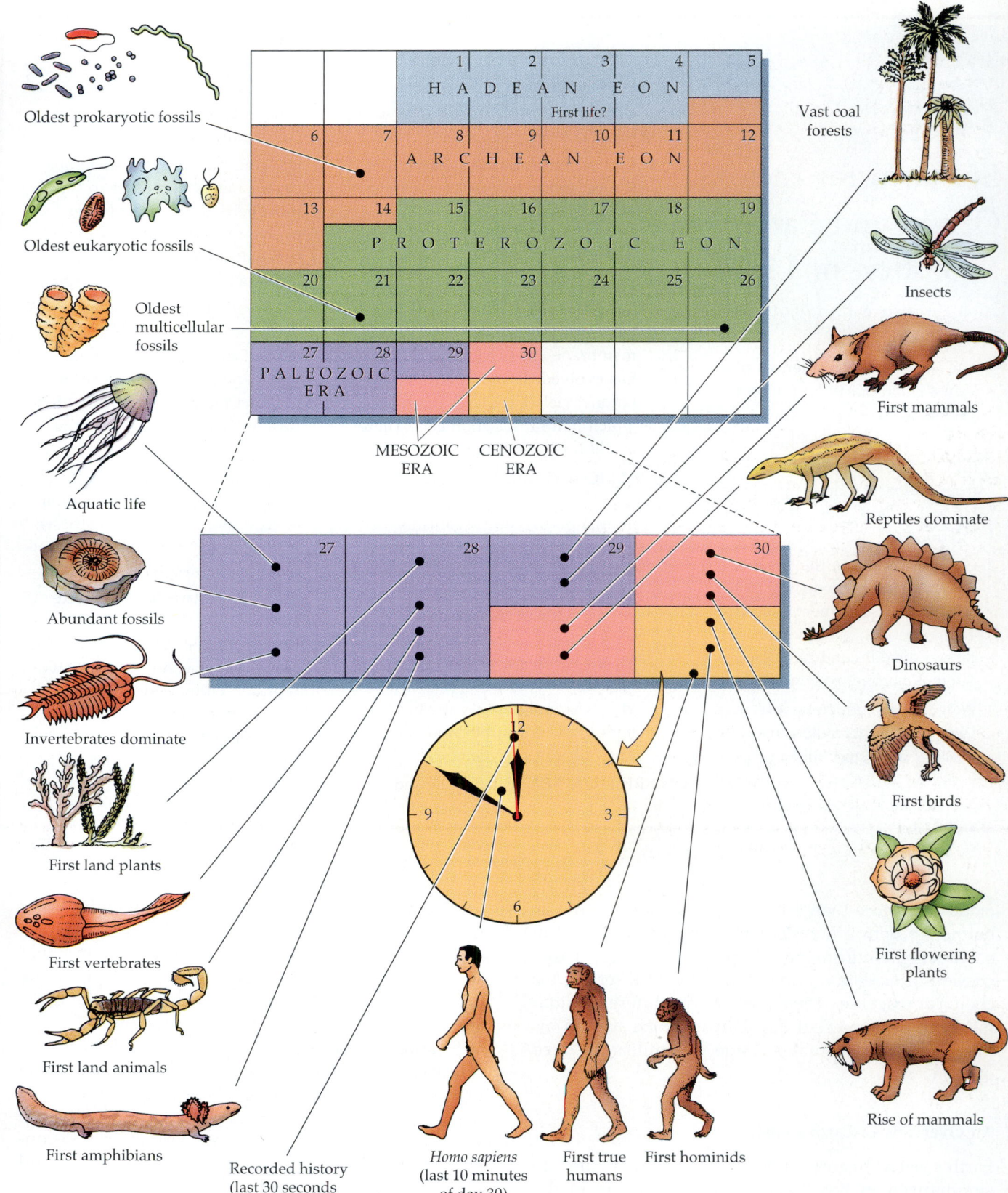

Oldest prokaryotic fossils

Oldest eukaryotic fossils

Oldest multicellular fossils

Aquatic life

Abundant fossils

Invertebrates dominate

First land plants

First vertebrates

First land animals

First amphibians

Recorded history (last 30 seconds of day 30)

Homo sapiens (last 10 minutes of day 30)

First true humans

First hominids

Vast coal forests

Insects

First mammals

Reptiles dominate

Dinosaurs

First birds

First flowering plants

Rise of mammals

1 HADEAN EON 2 3 4 5
First life?
6 7 8 9 10 11 12 ARCHEAN EON
13 14 15 16 17 18 19 PROTEROZOIC EON
20 21 22 23 24 25 26
27 28 29 30 PALEOZOIC ERA
MESOZOIC ERA   CENOZOIC ERA

18.5 Life's Calendar
Major periods and events in the history of life on Earth are represented in this calendar, on which a "day" lasts about 150 million years. On this scale, *Homo sapiens* evolved in the last 10 minutes of day 30, and recorded history is confined to the final 30 seconds.

ment? Could such reactions have been the first step toward the origin of life? In the 1950s, to investigate these questions, Stanley Miller studied chemical reactions proceeding under conditions simulating those of early Earth. He established a reducing atmosphere of hydrogen, ammonia, methane, nitrogen, carbon dioxide, and water vapor and passed these gases over a spark to simulate lightning (Figure 18.6). Within a few hours the system contained numerous simple organic compounds (Figure 18.7a). In water, these compounds were rapidly converted into amino acids, simple acids, and other compounds (Figure 18.7b). Because the molecules in Figure 18.7b are relatively stable, they quickly accumulate in solution.

Comparable results are obtained under a variety of conditions, provided that free oxygen is absent—that is, if the environment is a reducing one. Thus, once Earth cooled enough for water to condense and form oceans, molecules of many kinds formed spontaneously, and they probably accumulated until they reached relatively high concentrations. Life emerged from nonlife in such an anaerobic "primordial soup."

The important large molecules of which organisms are composed—polysaccharides, proteins, and nucleic acids—are polymers formed by the combination of subunits called monomers. The **polymerization** reactions that generate these large molecules belong to a class of reactions called condensations or dehydrations (see Chapter 3). During a condensation reaction, water is formed, one hydrogen atom coming from one molecule and the other hydrogen atom and the oxygen atom coming from another molecule. Large molecules are assembled through repeated condensations of monomers, and each condensation reaction requires energy. The molecules of organisms that direct the synthesis of other molecules identical to themselves are polymers.

Which of the molecules formed on the prebiotic ("before life") Earth were most likely capable of catalyzing their own replication? Nucleic acids, the basis of today's genetic code, are clearly well suited for this purpose, and the two types of monomers (purines and pyrimidines) that are linked together to form the nucleic acid polymers are formed under conditions similar to those believed to have prevailed on the early Earth. The key process in the formation of these nucleic acid bases is the polymerization of hydrogen cyanide (HCN), one of the molecules formed in Miller's experiments. This polymerization, by the steps shown in Figure 18.8a, yields large amounts of the tetramer diaminomaleonitrile. Diaminomaleonitrile can rearrange and react again with hydrogen cyanide to form the purine base adenine (Figure 18.8b). Under slightly different conditions, diaminomaleonitrile first reacts with formamide generated by the hydrolysis of hydrogen cyanide and

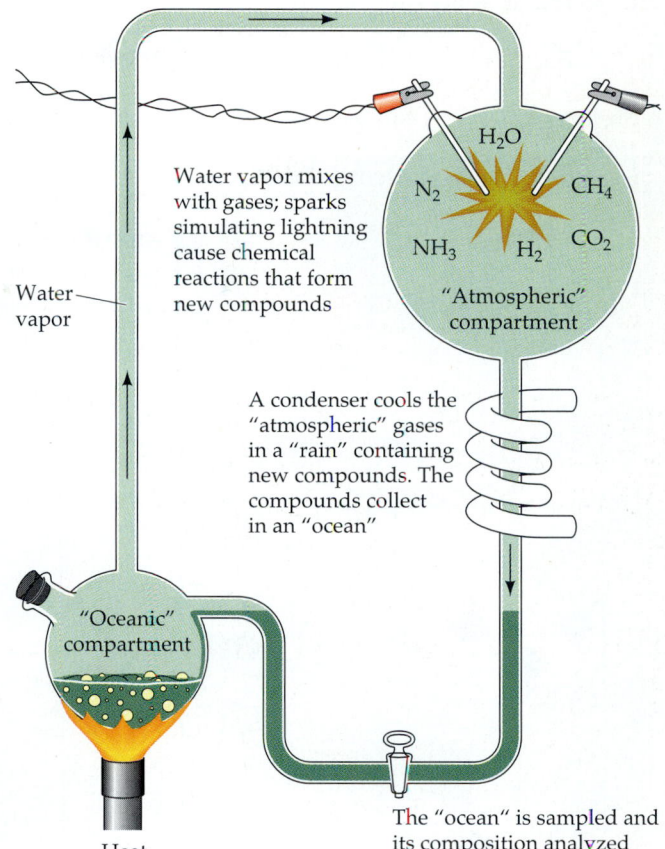

Water vapor mixes with gases; sparks simulating lightning cause chemical reactions that form new compounds

H₂O
N₂ CH₄
NH₃ H₂ CO₂
"Atmospheric" compartment

Water vapor

A condenser cools the "atmospheric" gases in a "rain" containing new compounds. The compounds collect in an "ocean"

"Oceanic" compartment

Heat

The "ocean" is sampled and its composition analyzed

18.6 Synthesis of Molecules in an Experimental Atmosphere
The apparatus shown here is similar to the one used by Stanley Miller to determine which molecules are spontaneously produced in a reducing atmosphere similar to the one that existed on early Earth. Samples of products were withdrawn both from the fluid at the bottom of the apparatus and from the "oceanic" boiler in which the fluid collected in solution.

then reacts with hydrogen cyanide to give the other important purine base, guanine. Other similar reactions yield the pyrimidine bases.

Sugars, another basic component of living cells, are generated by the polymerization of formaldehyde, another molecule formed in Miller's experiments (see Figure 18.7a). Solutions of formaldehyde spontaneously polymerize to form several different five- and six-carbon sugars, but these sugars are unstable in aqueous solutions and break down to alcohols and carboxylic acids. Three- and four-carbon sugars are very stable, however, and can accumulate for hundreds of years in aqueous solutions. Over millions of years, these organic molecules would have accumulated in the oceans. They would have reached even higher concentrations in drying bodies of water, where they would have continued to polymerize.

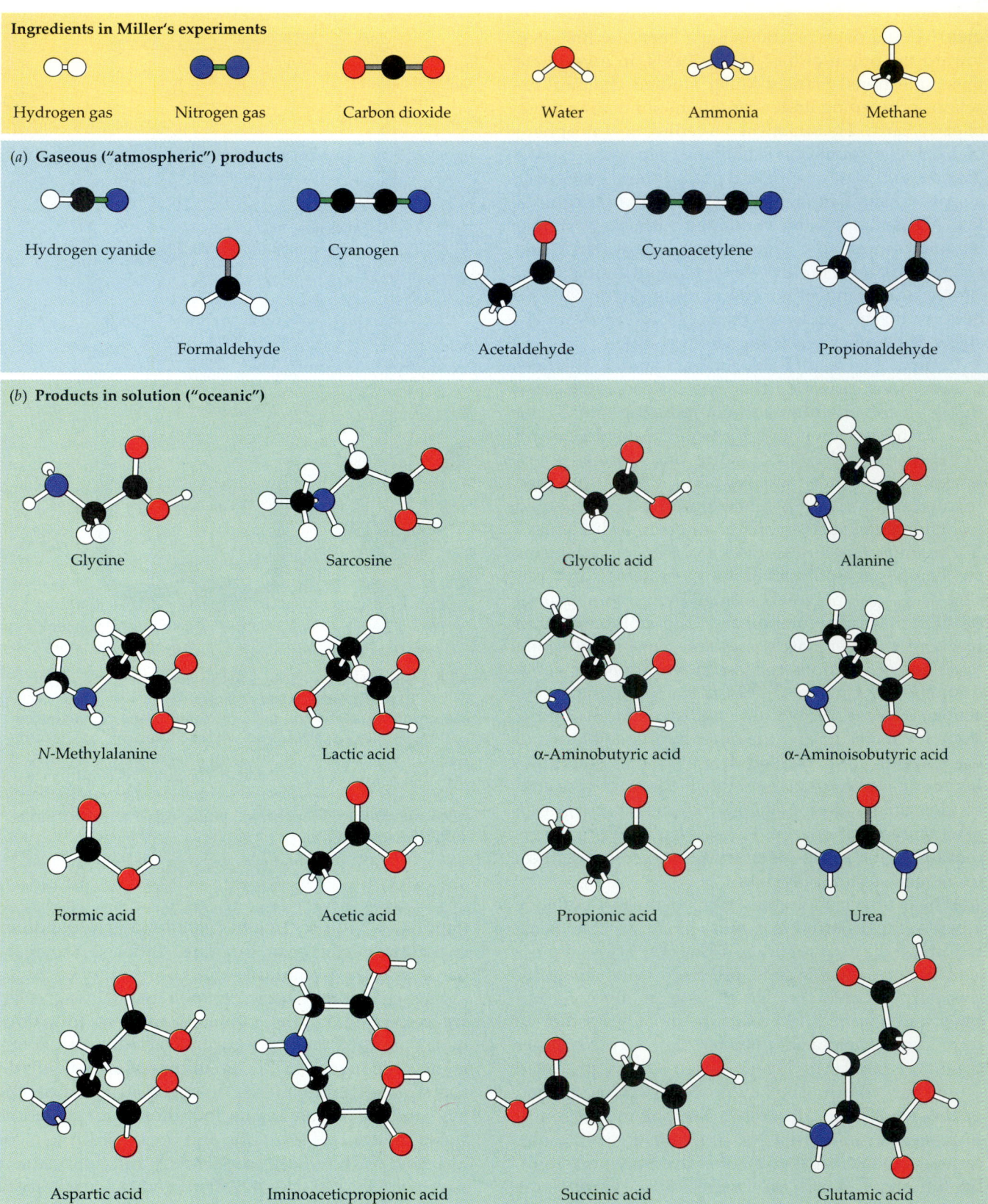

Ingredients in Miller's experiments

Hydrogen gas   Nitrogen gas   Carbon dioxide   Water   Ammonia   Methane

(a) **Gaseous ("atmospheric") products**

Hydrogen cyanide   Cyanogen   Cyanoacetylene

Formaldehyde   Acetaldehyde   Propionaldehyde

(b) **Products in solution ("oceanic")**

Glycine   Sarcosine   Glycolic acid   Alanine

N-Methylalanine   Lactic acid   α-Aminobutyric acid   α-Aminoisobutyric acid

Formic acid   Acetic acid   Propionic acid   Urea

Aspartic acid   Iminoaceticpropionic acid   Succinic acid   Glutamic acid

18.7   Many Molecules Were Synthesized in Miller's Experiments
Different small organic molecules formed in the "atmospheric" (a) and
"oceanic" (b) compartments of Miller's apparatus. Most of these organic and
amino acids are important components of contemporary living organisms.

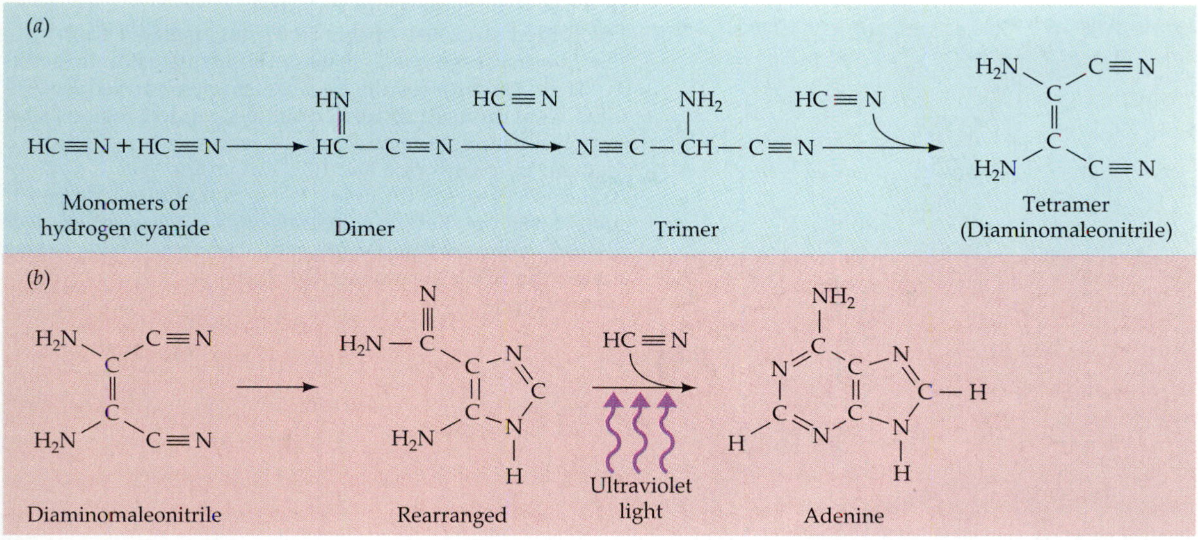

(a)

(b)

18.8 From Simple Organic Molecules to an Amino Acid
(a) Hydrogen cyanide readily forms dimers. Adding a third hydrogen cyanide
molecule generates the trimer aminomaleonitrile, and adding a fourth forms
the tetramer diaminomaleonitrile. (b) Diaminomaleonitrile rearranges itself into
another tetramer, diaminomalconitrite. Adenine forms in reactions of diamino-
maleonitrite and hydrogen cyanide activated by ultraviolet light.

Continuing the Sequence: More Polymerization

Monomers that form readily under prebiotic conditions can polymerize by phosphorylation (the addition of a phosphate group). Phosphorylated monomers are stable enough to accumulate in solution but
reactive enough to polymerize further. The first biologically active polymers of nucleic acid bases may
have been compounds formed from three- and four-
carbon sugars. Such polymers could have helped polymerize specific amino acids into peptides, as RNA
does today. They could also have self-replicated by
forming complementary double-stranded molecules,
just as pentose-based nucleic acid polymers do today.
At some point, by as yet unidentified processes,
polymers based on small sugars were replaced by
those based on larger sugars, such as ribose, the
sugar that is part of RNA.

RNA: The First Biological Catalyst

The enzymes that control the types and rates of reactions within organisms are proteins. As you
learned in Chapter 11, proteins are synthesized by a
process that begins with transcription of information
from DNA to an RNA molecule having a base sequence complementary to that of one strand of the
DNA. This information is then translated into mRNA
and is eventually used to synthesize a specific polypeptide from an array of amino acids. Amino acids
are brought to a ribosome by specific tRNA molecules
and are attached sequentially to the growing poly-

peptide. Until recently, biologists believed that nucleic acids lacked catalytic abilities, and that only enzymes could catalyze chemical reactions. If so, how
could catalysis evolve if proteins needed nucleic acids
for their formation but nucleic acids needed proteins
to catalyze their own replication?

Inability to solve this "chicken-and-egg" dilemma
held up research on the origin of life for several
decades. The discovery that provided a solution to
the dilemma came in 1981 from researchers working
with the unicellular protist *Tetrahymena thermophila*.
The scientists were studying the excision of introns
and the splicing together of exons. To isolate the
catalysts required for the reaction, the investigators
established two cell-free systems. One contained
RNA molecules from which the introns were to be
excised and proteins to serve as catalysts; the other,
the control system, contained only RNA molecules.
As expected, the introns were excised in the RNA–
protein system, but contrary to expectations, excision
and splicing also took place in the control. The investigators found that the intron itself—a 400-nucleo-
tide sequence of RNA—carried out the excision and
splicing. A *Tetrahymena* ribosome contains several
molecules of RNA along with a variety of proteins.
Experiments have demonstrated that the RNA rather
than the ribosomal protein is the catalyst of protein
synthesis. RNAs that catalyze reactions are called
ribozymes (Figure 18.9). Ribozymes have now been
found in many organisms, strongly suggesting that
the first biological catalysts were RNAs.

The current system of cellular reactions, based

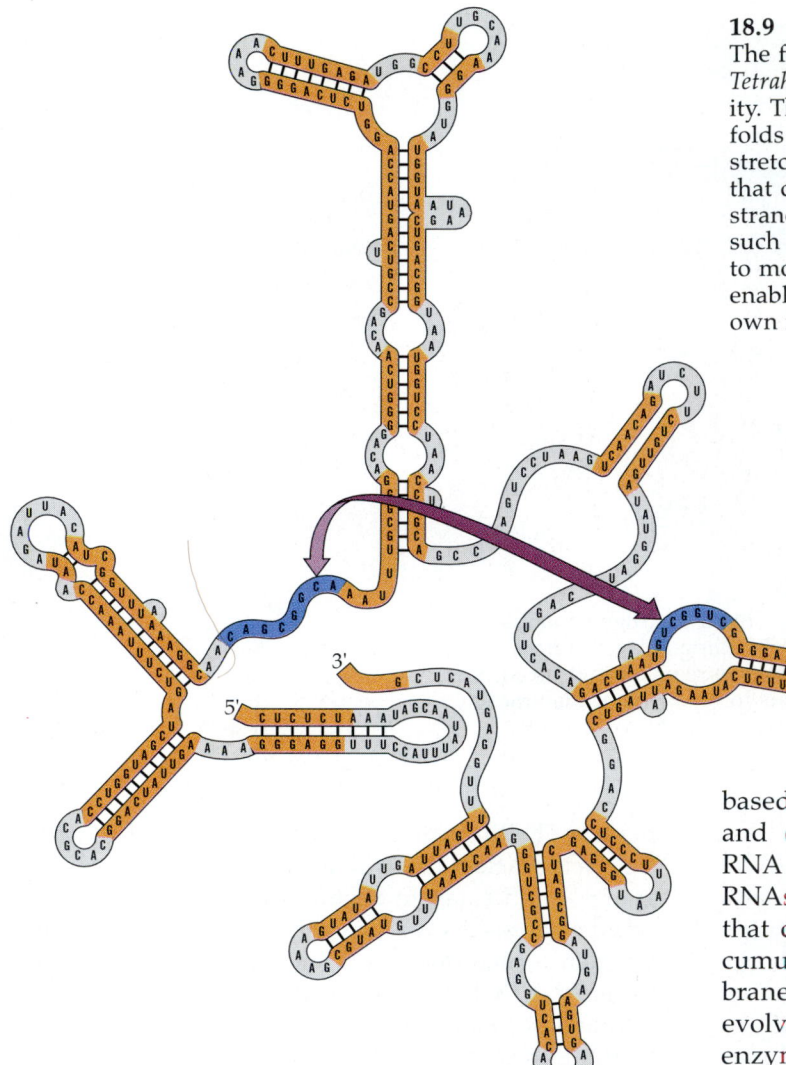

18.9 A Ribozyme from a Protist

The folded structure of this ribozyme, isolated from *Tetrahymena thermophila*, is essential for its catalytic activity. This structure results because single-stranded RNA folds back on itself to form double-stranded regions. Two stretches of the molecule with sequences of nucleotides that can pair with one another, can join to create double-stranded segments (orange). Long-range interactions—such as the one between the regions shown in blue—lead to more complex three-dimensional shapes. Those shapes enable the RNA to catalyze the excision of some of its own introns.

upon the "central dogma" (DNA → RNA → protein), evolved from gradual changes in much simpler processes. The first information-carrying molecules were short strands of RNA that replicated themselves without the help of enzymes. Evidence that RNAs can replicate themselves came first from experiments conducted by Manfred Eigen in the late 1970s. Eigen added RNA molecules to solutions containing monomers for making more RNA and found that sequences of 5 to 10 nucleotides were formed. If he added a simple inorganic molecule such as zinc, much longer sequences were copied.

These experiments showed that RNA could act on itself, but they did not demonstrate that RNA could act on other molecules as true enzymes do. Within two years, however, studies of ribonuclease P—a tRNA-processing enzyme that contains RNA and a protein in a single package—showed that the RNA alone can cut the pre-tRNA molecule at the correct spot, whereas the protein cannot do so. Thus many scientists now believe that the first genetic code was based on RNA that catalyzed both its own replication and other chemical reactions. Beginning with an RNA molecule that catalyzed its own replication, RNAs would have diversified into a variety of forms that could catalyze other processes, such as the accumulation of lipidlike molecules to form cell membranes and the synthesis of proteins. After proteins evolved, however, they would have taken over most enzymatic functions because they are better catalysts than is RNA.

The shapes of single-stranded RNA molecules depend upon the sequences of nucleotides because their folding is determined by hydrogen bonds between complementary sequences of bases. To replicate, these different RNAs would have competed with one another for monomers. Some RNA molecules would have been better at replicating in certain environments because they had base sequences that produced the most stable configurations under the particular conditions of temperature and salinity. With time these molecules would have come to dominate the populations of RNA in the corresponding environments.

In the "RNA world," in which RNA was the only genetic information, ribozymes would have needed to fold into an RNA polymerase that used RNA as a template and to unfold and act as templates for other replicase molecules. Investigators have produced RNA molecules with these capabilities by starting with completely random-sequence RNA. These ribozymes ligate (bind together) two RNA molecules that are aligned on a template by catalyzing the attachment of a 3'-end hydroxyl on an adjacent 5'-end triphosphate. This reaction is similar to that employed by the enzymes that synthesize RNA. In addition, "evolution" of these RNA molecules in the test tube improved their ligation activity and led to ribozymes with reaction rates 7 million times faster than the uncatalyzed reaction rate! Therefore, ribozymes may have evolved quite rapidly when conditions on Earth became suitable for the formation of nucleic acids.

From Ribozymes to Cells

How proto-living units might have evolved into cells was first suggested by the experiments of the Russian scientist Alexander Oparin, who spent much of his career studying complex solutions. Oparin observed that if he shook a mixture of a large protein and a polysaccharide, the drops that formed were divided into two "phases." Their interiors, which were primarily protein and polysaccharide, with some water, were surrounded by an aqueous solution containing low concentrations of proteins and polysaccharides. These drops, known as **coacervates**, are quite stable and will form in solutions of many different types of polymers.

Coacervates have other properties relevant to the origin of life. Many substances, when added to a coacervate preparation, are preferentially concentrated within the drops. Lipids coat the boundaries of drops with membranelike structures that strengthen the drops and help contol the rates of passage of materials into and out of the drops. Coacervate drops that contain enzyme molecules exhibit a "metabolism": They can absorb substrates, catalyze reactions, and let the products diffuse back out into the solution (Figure 18.10). Oparin even succeeded in making chlorophyll-containing coacervate drops that absorbed an oxidized dye from the solution, used light energy to reduce it, and returned the reduced dye to the medium.

Oparin regarded complex coacervate drops as possible precursors to cells because they provided the physical framework within which metabolic reactions took place. Because drops in which chemical reactions were better controlled would have survived longer than drops with more poorly controlled reactions, refinements of metabolic processes by the use of enzymes could have evolved. Billions of droplets would have competed with one another for substrate materials from the environment. They would have evolved better abilities to replicate internal reactions and would have engaged in an early form of reproduction by passing on at least rough copies of their molecules to daughter droplets when they divided. At some point, such droplets may have accumulated enough RNA to display the properties we now associate with life.

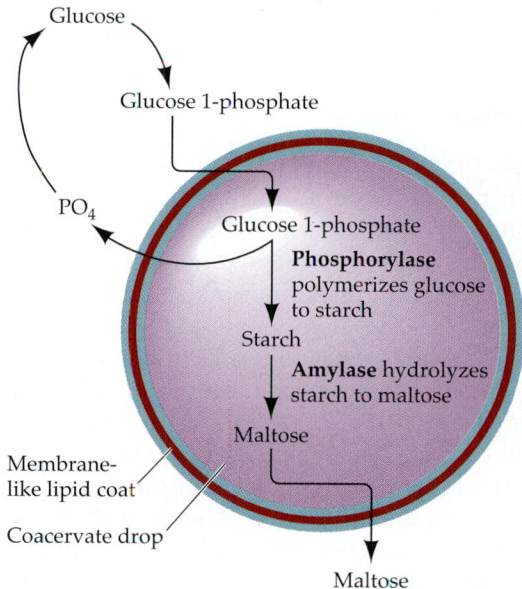

18.10 "Metabolism" of Coacervate Drops
This hypothetical coacervate drop has a lipid coat that surrounds the drop with a membranelike structure. The drop absorbs glucose 1-phosphate from the surrounding medium. Within the drop the enzyme phosphorylase polymerizes glucose to starch, and the enzyme amylase hydrolyzes starch to maltose.

The Evolution of DNA

If the genetic code was first embodied in and transmitted by RNA, then RNA must have provided the template upon which DNA was synthesized. Unlike RNA, DNA is stable only in hydrophobic environments, which in aqueous environments are found only within lipid-based cell membranes. Therefore, DNA probably did not evolve until RNA-based life became enclosed in membrane-bounded cells. Because DNA is a more stable storage location for genetic information than is RNA, however, once the appropriate hydrophobic environments were available, DNA probably evolved rapidly, replacing RNA as the genetic code for most organisms. RNAs would then have assumed their current roles as intermediaries in the translation of genetic information.

METABOLISM OF EARLY ORGANISMS

As you saw in Box 18.A, Earth's atmosphere lacked oxygen at the time when life first appeared. Box 18.A also shows that Earth's atmosphere was a reducing one for about a billion years after life evolved. Therefore, the earliest organisms must have been obligate anaerobes—organisms that obtain their energy without using oxygen. No traces of the metabolic pathways of those early organisms have been preserved, but we can make reasoned guesses about stages in their evolution by studying the metabolism, especially photosynthesis, of living anaerobic bacteria.

Anaerobic Photosynthesis

The bacteria that survive today in environments lacking oxygen must remain in an oxygen-free environment because oxygen is poisonous to them. Three major types of anaerobic photosynthetic eubacteria—green sulfur bacteria, purple sulfur bacteria, and purple nonsulfur bacteria—live today in sediments that lack oxygen. These eubacteria all contain types of chlorophyll, called bacteriochlorophyll *a*, *c*, and *d*. Anaerobic photosynthetic bacteria also contain red and yellow carotenoids, which absorb light of wavelengths that are not absorbed by chlorophyll and pass the energy along to chlorophyll. The photosynthetic system of these bacteria is embedded in membrane complexes called thylakoids (see Chapter 4). The thylakoids possess the electron transport chains by which captured solar energy is passed along and used to generate ATP (see Chapter 8). Photosynthetic bacteria, which were probably similar in their metabolism to today's forms, were so abundant about 3.4 billion years ago that their partly decomposed remains formed extensive deposits of carbon resembling the coal deposits produced by vascular plants 3 billion years later.

To reduce carbon dioxide (CO_2), a photosynthetic cell needs a source of hydrogen atoms. Many bacteria use light energy to generate ATP and NADPH + H^+, but which waste product they liberate depends upon the source of hydrogen atoms they use. The green and purple sulfur bacteria obtain their hydrogen atoms from hydrogen sulfide (H_2S) and generate sulfur as a waste product (Figure 18.11). The purple nonsulfur bacteria obtain hydrogen atoms from organic compounds such as ethanol, lactic acid, or pyruvic acid, or directly from hydrogen gas. In some environments today, the hydrogen is provided by other bacteria as the end products of their fermentations. Under the anaerobic conditions of early Earth, hydrogen sulfide and other compounds containing hydrogen would have been more abundant than they are today because atmospheric oxygen quickly oxidizes them into water, carbon dioxide, and sulfur oxides.

On the early Earth, volcanoes poured large quantities of hydrogen, nitrogen, carbon dioxide, methane, and hydrogen sulfide into the atmosphere. In the waters, compounds such as ammonia, nitrates, carbon dioxide, sulfates, and phosphates also circulated. Under these conditions, a variety of anaerobic bacteria evolved and thrived. The cycles of these elements cannot be completed even today without the involvement of anaerobic bacteria (see Chapter 49). Thus even though anaerobic bacteria survive only in environments lacking oxygen, life as we know it

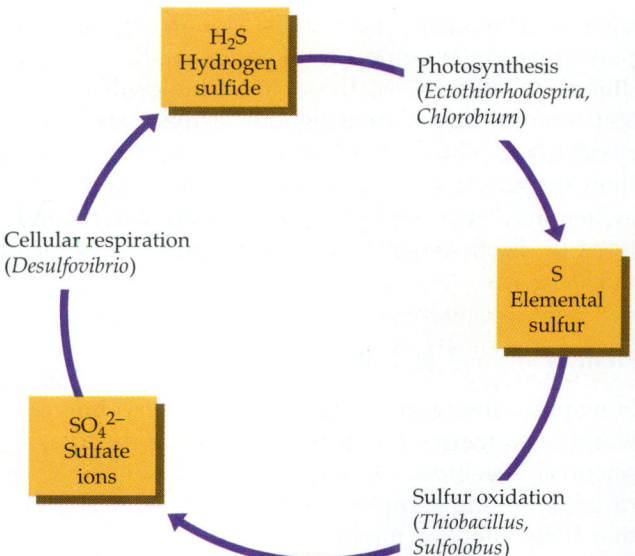

18.11   Oxidizing Sulfur
Several species of eubacteria and archaebacteria that use sulfur-containing compounds as oxidizing and reducing agents may function together to carry out sulfur cycles, one of which is shown here. Representative genera able to conduct the processes are given in parentheses. Many such cycles are found in nature.

would not have evolved without their metabolic activities, and it still depends on them today.

Aerobic Photosynthesis: Source of the Atmosphere's Free Oxygen

The most recent step in the evolution of photosynthesis, which originated more than one billion years ago, is acquisition of the ability to use water as the source of hydrogen. This ability appeared first in certain sulfur bacteria that evolved into cyanobacteria (see Chapter 22). The ability to split water was doubtless the cause of the extraordinary success of the cyanobacteria, a success that changed the planet and life on it. The oxygen they liberated opened the way for the evolution of oxidation reactions as the energy source for the synthesis of ATP, and thus for the evolution of the full respiratory chain of reactions now carried out by all aerobic cells. The evolution of life irrevocably changed the nature of our planet. Life created the free oxygen of our atmosphere, and it removed most of the carbon dioxide from the atmosphere by transferring it into sediments.

HOW PROBABLE WAS THE EVOLUTION OF LIFE?

Scientists have been able to gather information that provides many insights into the origin of life on Earth. In laboratory experiments they have studied

chemical reactions under conditions similar to those believed to have prevailed on the early Earth. Under laboratory conditions, they have witnessed the "evolution" of ribozymes with catalytic activity from random-sequence RNA.

Taken together, this information suggests that the evolution of life was highly probable under the conditions that prevailed on Earth several billion years ago. The molecules on which life is based form spontaneously under such conditions, and the organization of those molecules into larger units also proceeds spontaneously. Much remains to be learned about the early evolution of the complex metabolism of living organisms, but techniques are available for doing so. Good understanding of details of events that happened billions of years ago is now within our grasp.

If the origin of life was once probable, why is new life not still being assembled from nonliving matter on today's Earth? Simple biological molecules released into today's environment are quickly consumed by already living things. They cannot accumulate to the densities that characterized the "primordial soup," even in anaerobic environments. In aerobic environments these molecules are quickly oxidized to other forms and would not accumulate even if they were not consumed. Generation of life from nonlife on Earth did happen, but it was an event of the remote past. Once it had evolved, life prevented other life from arising from nonlife.

SUMMARY of Main Ideas About the Origin of Life

Life does not arise from nonliving matter on Earth today.
   Review Figure 18.1

Using known rates of decay of radioactive elements, scientists can date ancient rocks and the fossils they contain, an important part of generating an overview of the history of Earth.

Life on Earth arose by chemical evolution under conditions much different from today's.

Earth had a reducing atmosphere when life evolved.

Many types of organic molecules are synthesized in laboratory experiments under conditions that mimic the early reducing atmosphere.
   Review Figures 18.6 and 18.7

Polymerization reactions generated the large molecules of which organisms are composed.
   Review Figure 18.8

The first genetic material may have been RNA that had both a catalytic function and an information-transfer function.

Some RNAs—called ribozymes—still have catalytic functions today.
   Review Figure 18.9

DNA may have evolved after RNA-based proto-living systems became surrounded by membranes that provided hydrophobic environments in which DNA is stable.

The earliest protocells may have been similar to coacervates.
   Review Figure 18.10

For more than 2 billion years, the only organisms were anaerobic bacteria.
   Review Box 18.A and Figure 18.5

Cyanobacteria, which evolved the ability to split water to obtain hydrogen, created the free oxygen of Earth's atmosphere.

SELF-QUIZ

1. The atmosphere of early Earth consisted largely of
   a. water vapor.
   b. hydrogen.
   c. carbon dioxide.
   d. nitrogen.
   e. all of the above.

2. Pasteur's experiments convinced most people that spontaneous generation of life did not happen because
   a. he was extremely meticulous.
   b. he used very fine mesh screens to cover his flasks.
   c. he did not boil his flasks for a long time.
   d. his swan-necked flasks ruled out the "spoiled air" objection to Spallanzani's experiments.
   e. by the time he performed his experiments, many people no longer believed in spontaneous generation.

3. The principle of superposition is that
   a. lighter elements rest on top of denser ones.
   b. lighter rocks rest on top of denser ones.
   c. organisms are preserved in rocks if they land on top of hard places.
   d. younger rocks lie on top of older ones.
   e. as rocks age, they are thrust upward by earthquakes.

4. The best method of obtaining good estimates of the absolute ages of ancient events is
   a. quantitatively measuring rates of formation of sedimentary rocks.
   b. measuring the past salinity of the oceans as recorded in sedimentary rocks.
   c. applying the principle of superposition.
   d. taking accurate measurements of fossil morphology.
   e. determining the radioactive isotope composition of ancient deposits.

5. To determine which molecules might have formed spontaneously on early Earth, Stanley Miller used an apparatus with an atmosphere containing
   a. oxygen, hydrogen, and nitrogen.
   b. oxygen, hydrogen, ammonia, and water vapor.
   c. oxygen, hydrogen, and methane.
   d. hydrogen, oxygen, and carbon dioxide.
   e. hydrogen, ammonia, methane, and water vapor.

6. Most biologists think that RNA was the first genetic material because

a. amino acids were produced in Stanley Miller's apparatus.
b. DNA is the universal genetic material of eukaryotes.
c. the existence of ribozymes suggests that early cells could have used RNA to catalyze chemical reactions and transfer information.
d. RNA is simpler than DNA.
e. DNA is not stable in hydrophobic environments.

7. Biologists believe that the current DNA–RNA–protein system is the result of a long period of evolution because
   a. the transcription of DNA to mRNA and translation of mRNA into proteins consists of many steps.
   b. DNA replication is complicated but relatively error-free.
   c. the current system is very complex and precise.
   d. evidence indicates that RNA preceded DNA as the genetic material.
   e. all of the above.

8. The question whose answer enabled research on the origin of catalysis to proceed rapidly was,
   a. How could complex life have evolved on such a young Earth?
   b. How could the precise duplication of DNA have evolved?

c. How could catalysis evolve given that RNA needs proteins for its synthesis and proteins need RNA for their synthesis?
d. How could eukaryotes evolve from prokaryotes?
e. How did the first cells form?

9. The key process in the formation of nucleic acid bases is
   a. the polymerization of hydrogen cyanide.
   b. the polymerization of formaldehyde.
   c. the spontaneous formation of monomers.
   d. the spontaneous formation of proteins.
   e. the polymerization of proteins.

10. Metabolism of living prokaryotes provides important insights into the chemical processes used by early organisms because
    a. many prokaryotes live in environments similar to those in which life first evolved.
    b. prokaryotes are simpler to study and hence are better known than are eukaryotes.
    c. many bacteria are obligate aerobes.
    d. many bacteria use oxygen as their oxidizing agent.
    e. fermentation evolved before aerobic respiration.

FOR STUDY

1. Why is determining the composition of Earth's early atmosphere a key step in inferring how life arose?

2. Why is the ability of ribozymes to catalyze their own synthesis and the synthesis of proteins so important for understanding the origin of life?

3. On the one hand, scientists are confident that life no longer arises from nonliving matter. On the other hand, most biologists believe that life did arise on this planet, billions of years ago, from nonliving matter. How can scientists hold both of these beliefs?

4. Why do biologists believe that the evolution of life was highly probable on the early Earth?

5. How might each of the following have been involved in the evolution of coacervate drops?
   a. coating of drop boundaries with lipids
   b. wave action in bodies of water
   c. catalysts within the drops

READINGS

Badash, L. 1989. "The Age-of-the-Earth Debate." *Scientific American*, August. The story of the controversies and discoveries that have changed the generally accepted age of Earth by 4.5 billion years over the past three centuries.

Bernal, J. D. 1967. *The Origin of Life.* World Publishing Company, Cleveland, OH. A clear, but dated, consideration of technical and philosophical issues. Includes reprints of original papers by Oparin and Haldane.

Cech, T. R. 1986. "RNA as an Enzyme." *Scientific American*, November. A well-illustrated discussion of the discovery of the catalytic abilities of RNA and their implications for the origin and early evolution of life.

Dyson, F. J. 1985. *The Origins of Life*. Cambridge University Press, New York. A concise argument in favor of multiple origins of life.

Eigen, M., W. Gardiner, P. Schuster and R. Winkler-Oswatitch. 1981. "The Origin of Genetic Information." *Scientific American,* April. A molecular biological view of the origin of genes.

Horgan, J. 1991. "In the Beginning. . .". *Scientific American,* February. Scientists are having a hard time agreeing on when, where, and—most important—how life first emerged on Earth.

Life: Origin and Evolution. 1979. W. H. Freeman, San Francisco. (A Scientific American Book.) A collection of articles on many aspects of the origin of life.

Loomis, W. F. 1988. *Four Billion Years*. Sinauer Associates, Sunderland, MA. A very readable book on the evolution of genes and organisms, concentrating on the first billion years.

Margulis, L. 1984. *Early Life*. Jones and Bartlett, Boston. An engaging account of the earliest organisms and how they evolved. Good treatment of the symbiont theory of the origin of eukaryotes by its principal proponent.

Schopf, J. W. and C. Klein (Eds.). 1992. *The Proterozoic Biosphere*. Cambridge University Press, New York. An excellent source of information on interactions between life and early Earth.

Weinberg, S. 1993. *The First Three Minutes*. Basic Books, New York. A stimulating, but demanding, account of the origin of Earth for those interested enough to make the investment.

Woese, C. R. 1984. "The Origin of Life." Carolina Biological Supply Company, Burlington, NC. A discussion of why old theories of the origins of life are incorrect.

They Are Not All the Same
When observed closely, the individuals in a population of red and green macaws vary a great deal.

19

The Mechanisms of Evolution

We are aware that no two people (unless they are identical twins) look exactly alike. We also recognize our pets as distinct individuals. But we have great difficulty in seeing differences among individuals of most other species of organisms. The brilliant red and green macaws feeding on a clay cliff in the Peruvian jungle may all appear identical to the untrained eye; scientists who study them closely, however, realize that each is unique. The colored feathers display many slight variations in pattern, and the black-and-white feathers that surround the birds' eyes form patterns that, like human fingerprints, are unique to the individual. Members of many groups, particularly among behaviorally sophisticated animals such as vertebrates, readily recognize one another and adjust their behavior accordingly.

Differences among individuals in local populations, even if they are subtle, are the raw material upon which evolutionary mechanisms act to produce the striking variability revealed by the multitude of organisms living on Earth today. A good fossil record can reveal much about when and how the forms of organisms changed. Fossils may also provide clues about the reasons for those changes, but they provide only indirect evidence of the causes of evolutionary change. To obtain direct evidence we must study evolutionary changes happening today. The study of variability is at the heart of investigations into the mechanisms of evolution.

In this chapter we discuss the agents of evolution and the short-term studies designed to investigate them. By testing hypotheses observationally and experimentally we can answer key questions about the processes guiding evolutionary changes. In later chapters we consider how we use this information to explain longer-term features of the evolutionary record.

Although ideas about evolution have been put forth for centuries, until the last one hundred years none of the hypotheses about the causes of evolutionary change were tested. Charles Darwin stated his hypotheses more than a century ago (see Chapter 1), but he did not understand the underlying genetic basis of evolution. This essential ingredient was provided by Gregor Mendel (see Chapter 10). Mendel's and Darwin's insights, melded in the course of the twentieth century, form the basis for most current evolutionary hypotheses and the studies designed to test them.

WHAT IS EVOLUTION?

The fossil record shows that organisms have changed over time. The basic hereditary and metabolic mechanisms shared by nearly all organisms show that all

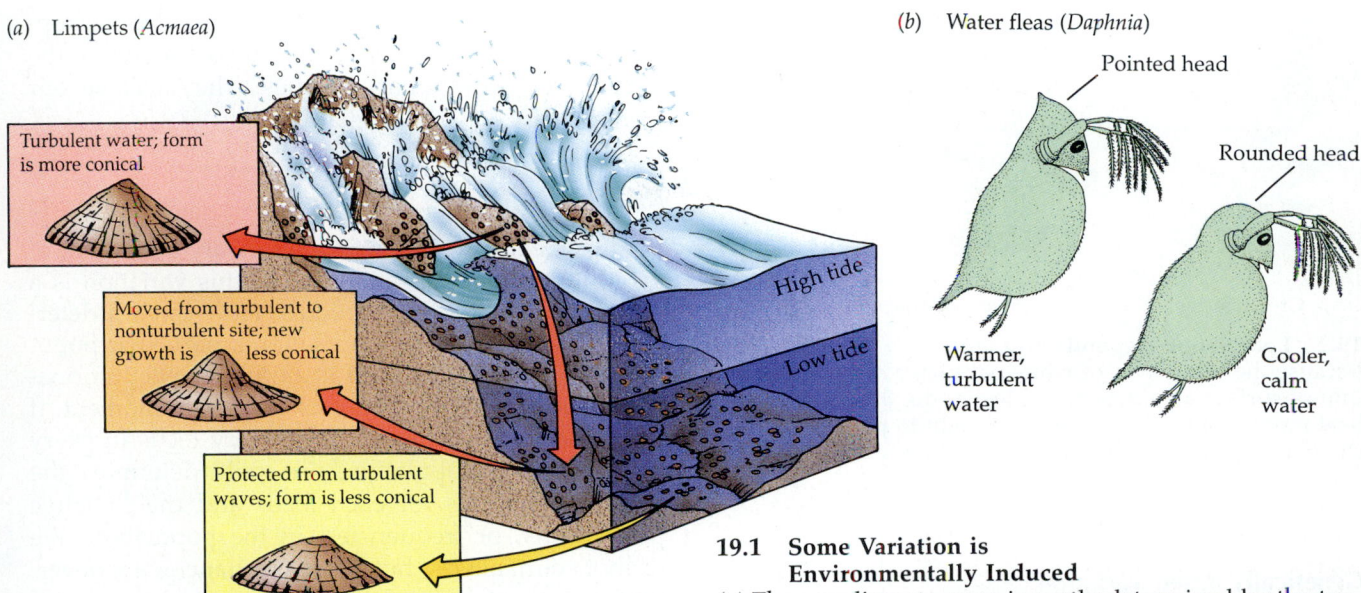

(a) Limpets (*Acmaea*)

Turbulent water; form is more conical

Moved from turbulent to nonturbulent site; new growth is less conical

Protected from turbulent waves; form is less conical

High tide

Low tide

(b) Water fleas (*Daphnia*)

Pointed head

Rounded head

Warmer, turbulent water

Cooler, calm water

19.1 Some Variation is Environmentally Induced

(a) The way limpets grow is partly determined by the turbulence of the water in which they live. (b) Head shape in water fleas depends on both water temperature and water turbulence. Such capacity to respond to outside change enables organisms to adjust their shapes to the various environments they encounter during their lives.

living things are descended from a common ancestor. In other words, life exhibits a reproductive continuity from its beginnings to the present. Each of us is connected to the earliest life by an unbroken strand of millions of generations of life. Change has accompanied the passing of these generations. Indeed, because offspring are not exactly like their parents, descendants may be quite different from their distant ancestors. The change over time in the genetic composition of populations is called **biological evolution**. How and why does biological evolution happen?

Organisms evolve because they interact with their environments in complex ways that influence the likelihood that individuals survive and reproduce. Biological evolution requires variation among a population's individuals in traits that are **heritable** — that is, traits that can be passed on to offspring. The genetic constitution governing a heritable trait is called its **genotype** (see Chapter 10). A population changes genetically—it evolves—when individuals with different genotypes survive or reproduce at different rates. Change may come about because the heritable traits influence the ability of individuals to obtain mates or food or to avoid hazards. The agents of evolutionary change act upon the physical expression of a genotype, its **phenotype**. Not all phenotypic variation, however, is governed by genotypes. Some of the variation observed within populations is genetically determined but some is not.

Environmentally Induced Variation

The shapes and sizes of many marine animals that live attached to a substrate (such as a rock) are affected by water temperature, concentration of nutrients (particularly calcium), competition with neigh-

bors, and turbulence of the water. For example, limpets (*Acmaea*) growing high in the intertidal zone (the shore exposed at low tide), where they experience heavy wave action, are more cone-shaped than are limpets of the same species growing in the subtidal zone, where they are protected from wave action. We know that this difference is not genetic because individuals from high in the intertidal zone add new growth to their shells to produce a flatter, subtidal shape when they are transplanted to the subtidal zone (Figure 19.1a). Similarly, if a water flea, *Daphnia cucullata*, grows in cool or calm water, it develops a rounded head, but if it originally lives in, or is moved to, warm or turbulent water, it develops a pointed "helmet" (Figure 19.1b).

The cells of the leaves on a tree or shrub are normally genetically identical. Yet leaves on the same tree often differ in shape and size. Leaves closer to the top of an oak tree, where they receive more wind and sunlight, may be more deeply lobed than leaves lower down on the same tree (Figure 19.2). These within-plant variations, however, are not passed on to offspring. What *is* passed on is the ability to form various types of leaves from the same genotype in response to different environmental conditions.

Environmentally induced variation is often important in biology. Researchers may manipulate such variation experimentally and study the induced changes to gain understanding of how organisms develop and adjust to their environments. But to investigate evolutionary questions, biologists must work with variation in traits that are heritable.

Grown in sun   Grown in shade

19.2   Leaf Shape Depends on Light
Because the sun leaves of white oaks have more edge per
unit of surface area than shade leaves do, they dissipate
heat more rapidly and allow more light to pass to the
shade leaves, which are lower on the tree.

Genetically Based Variation

High levels of genetic variation characterize nearly
all natural populations. This fact has been revealed
over and over again for thousands of years by people
attempting to produce desirable traits in plants and
animals. For example, artificial selection for different
traits in a common European wild mustard produced
many important crop plants (Figure 19.3). Plant and
animal breeders can achieve such results only if the
original population has genetic variation for the traits
of interest. The almost universal success of breeders
indicates that genetic variation is common, but it does
not tell us how much there is.

To understand evolution, we need to know more
precisely how much genetic variation populations
have, the sources of that genetic variation, and how
genetic variation is maintained and expressed in populations in space and over time. We also need to
know the agents that change the genetic variation in
populations, how they act, and their relative importance in affecting the direction of evolutionary
changes.

THE STRUCTURE OF POPULATIONS

The appropriate unit for defining and measuring genetic variation is a **population**: a group of organisms
of the same species occupying a particular geographic
region. The size of the geographic region we recognize depends on the objectives of a particular investigation. For some purposes, a population needs to
be defined as the group of individuals in a particular
place that mate with each other. A population in this
sense may be part of a larger population—a geographic population—that extends over so great an
area that its members in some places cannot mate
with those in others. A locally interbreeding group
within a geographic population may be called a sub-

population, a **deme**, or a Mendelian population. Populations whose members interbreed are often the
subjects of evolutionary studies. They will be our
primary concern in this chapter.

Measures of Genetic Variation

Because genetic variation is a necessary ingredient of
evolutionary change, quantifying this variation is a
part of many biological studies. How can we determine how much genetic variation a particular population has? The **gene pool** is all the genetic information present in a population at any given moment. If
we could count every allele at every locus in every
organism in a population, we could determine the
number of alleles for each locus and their relative
proportions, or frequencies, in the population. We
cannot count alleles, but **allele frequencies** are nevertheless central to the study of genetic variation in
populations. Biologists estimate allele frequencies using simple calculations. Once you understand how
to calculate allele frequencies, you will find it easy to
think about the amount of genetic variation in populations.

Assume that among all the members of a population there are only two alleles for a given locus. For
convenience, we will label one allele A and the other
a. (Whether or not allele A is dominant over a does
not matter at this point.) Two different alleles may
combine to form three different genotypes: AA, Aa,
and aa. The relative frequencies of alleles A and a can
be calculated as follows: In a population consisting
of N diploid individuals, let X be the number of
individuals that are homozygous for the A allele
(AA), let Y be the number that are heterozygous (Aa),
and let Z be the number that are homozygous for the
a allele (aa). Note that $X + Y + Z = N$, the total
number of individuals in the population, and that
the total number of alleles present in the population
is $2N$. Each AA individual has two A alleles and each
Aa individual has one A allele. Therefore, the total
number of A alleles in the population is $2X + Y$, and
the total number of a alleles in the population is
$2Z + Y$. If p represents the frequency of the A allele
and q the frequency of a, then

$$p = \frac{2X + Y}{2N} \text{ and } q = \frac{2Z + Y}{2N}$$

To see how this works, let's calculate allele frequencies in two populations, each consisting of 200 diploid
individuals. Population 1 has mostly homozygotes
(90 AA, 40 Aa, and 70 aa), whereas population 2 has
mostly heterozygotes (45 AA, 130 Aa, and 25 aa). In
population 1, where $X = 90$, $Y = 40$, and $Z = 70$,

$$p = \frac{2X + Y}{2N} = \frac{180 + 40}{400} = 0.55$$

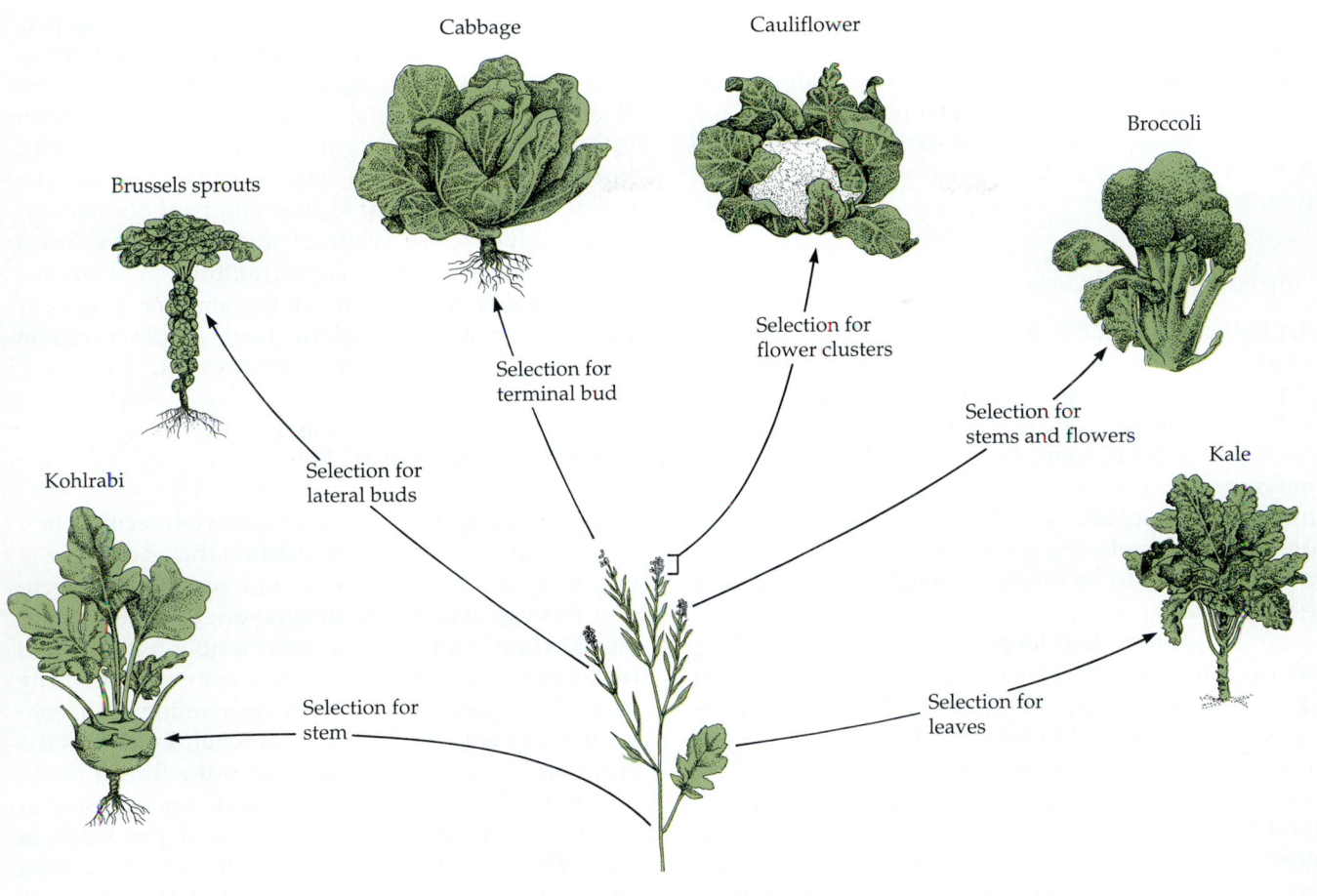

Cabbage

Cauliflower

Broccoli

Brussels sprouts

Selection for
flower clusters

Selection for
terminal bud

Selection for
stems and flowers

Selection for
lateral buds

Kale

Kohlrabi

Selection for
stem

Selection for
leaves

Brassica oleracea
(a common wild mustard)

19.3 Many Vegetables from One Species
European agriculturalists selected individual wild mustard plants (*Brassica oler-acea*) that produced unusually large leaves, stems, buds, or flowers. Using these individuals as parents and crossing their offspring repeatedly for many generations, these agriculturalists were able to produce all the different crop plants shown here.

and

$$q = \frac{2Z + Y}{2N} = \frac{140 + 40}{400} = 0.45$$

In population 2, where $X = 45$, $Y = 130$, and $Z = 25$,

$$p = \frac{2X + Y}{2N} = \frac{90 + 130}{400} = 0.55$$

and

$$q = \frac{2Z + Y}{2N} = \frac{50 + 130}{400} = 0.45$$

These calculations demonstrate two important points. First, notice that for each population $p + q = 1$. Frequencies are measures that range from 0 to 1; the sum of all allele frequencies at a locus must equal 1. If there is only one allele in a population, its fre-

quency is 1. If an allele is missing from a population, its frequency is 0, and the locus in that population is represented by one or more other alleles. Because $p + q = 1$, $q = 1 - p$, which means that when there are two alleles at a locus in a population, we can calculate the frequency of one allele and then easily obtain the second frequency by subtraction.

The second thing to notice from the calculations is that the two populations—one consisting mostly of homozygotes and the other mostly of heterozygotes—have exactly the same allele frequencies for *A* and *a*. Therefore, they have the same gene pool for this locus. However, the alleles in the gene pool are distributed differently among genotypes. These distribution patterns, together with information about allele frequencies, describe the **genetic structure** of a population.

Although we began our calculations with *numbers* of genotypes, for many purposes, genotypes, like alleles, are best thought of as frequencies. In population 1 of our example the genotype frequencies, which we calculate as the number of individuals with the genotype divided by the total number of individuals in the population, are 0.45 *AA*, 0.20 *Aa*, and 0.35 *aa*. What are the genotype frequencies of population 2?

Polymorphic Populations

An individual that has two different alleles at a given locus is said to be heterozygous for that locus. If a population has two or more alleles for a given locus, and the frequency of the most common allele is not greater than 95 percent, the population is said to be **polymorphic** for that locus. Polymorphism may be transitional—occurring only during the time when one allele is replacing another—but most polymorphism appears to be relatively stable, persisting over long periods.

Although they had long been interested in doing so, evolutionists were unable to obtain many data on genetic variation within natural populations until the development of molecular techniques in the 1950s. The DNA of eukaryotes is transcribed into RNA, which is, in turn, translated into proteins. Proteins (and DNA) from different individuals can be distinguished on the basis of their rates of migration on a gel in an electric field, a process called electrophoresis (see Chapter 14). Samples are placed on the edge of a sheet of jellylike material and exposed to an electric field. The rate at which a sample migrates through the gel is determined by both the charge on and the size of its molecules: smaller molecules move faster than larger ones.

Results from the hundreds of species whose proteins have been analyzed electrophoretically show that genetic variation is nearly universal. All groups of animals have genetic variation, but groups differ in the amount of variation they contain (Table 19.1). Vertebrates are usually less variable than invertebrates, and species that form small, local populations within which close relatives interbreed have lower levels of variation than larger, outbred populations. But whatever the reason for the amount of genetic variation, most populations have ample variation upon which evolutionary agents can act.

THE HARDY–WEINBERG RULE

Evolution is change in the genetic composition of a population over time. A population that is not changing, that has the same allele and genotype frequencies from generation to generation, is said to be at **equilibrium**. How can we determine if a population is changing or is at equilibrium without measuring its allele frequencies over many generations? The major method is to use a statistical result known as the Hardy–Weinberg rule, named after the British mathematician G. H. Hardy and the German biologist W. Weinberg, who each derived it independently in 1908. The rule specifies the conditions a population must meet to be at equilibrium and the genotype frequencies that will be found in such a population. By comparing the frequencies we observe in real populations with those specified by the Hardy–Weinberg rule, we can detect changes and direct our attention to the most likely causes of a change.

We discuss the conditions for the Hardy–Weinberg

TABLE 19.1
Genetic Variation among Some Groups of Animals

	NUMBER OF SPECIES	AVERAGE NUMBER OF LOCI ASSAYED PER SPECIES	PROPORTION OF HETEROZYGOUS LOCI PER POPULATION	PER INDIVIDUAL
Insects				
Drosophila	28	24	0.529	0.150
Wasps	6	15	0.243	0.062
Others	4	18	0.531	0.151
Marine invertebrates	9	26	0.587	0.147
Marine snails	5	17	0.175	0.083
Land snails	5	18	0.437	0.150
Fish	14	21	0.306	0.078
Amphibians	11	22	0.336	0.082
Reptiles	9	21	0.231	0.047
Birds	4	19	0.145	0.042
Rodents	26	26	0.202	0.054
Large mammals	4	40	0.233	0.037

equilibrium in detail in the sections that follow. To meet the conditions, a population must be very large and be made up of sexually reproducing diploid individuals. All individuals in the population must survive and reproduce equally well, and mating must combine genotypes at random. The rule says that if all these conditions hold, the frequencies of alleles at a locus will remain constant from generation to generation and the frequencies of the genotypes will also remain constant after one generation. When there are two alleles at the same locus, allele and genotype frequencies are related as follows:

$$p^2_{(AA)} + 2pq_{(Aa)} + q^2_{(aa)} = 1$$

To see that this relationship is true, consider an example in which the frequency (p) of A alleles is 0.6. Because individuals select mates without regard to genotype, gametes carrying A or a combine at random—that is, as predicted by the frequencies p and q. The probability that any given sperm or egg in this example bears an A allele rather than an a allele is 0.6. In other words, 6 out of 10 random selections of a sperm or an egg will bear an A allele. Because $q = 1 - p$, as we saw earlier in this chapter, the probability of an a allele is $1 - 0.6 = 0.4$.

To obtain the probability of two A-bearing gametes coming together at fertilization, we multiply the two independent probabilities of drawing them: $p \times p = p^2 = 0.36$. Therefore, 0.36, or 36 percent, of the offspring in the next generation will have the AA genotype. Similarly, the probability of bringing together two a-bearing gametes is $q \times q = q^2 = 0.16$, so 16 percent of the next generation will have the aa genotype. As Figure 19.4 shows, there are two ways of producing a heterozygote: An A sperm may combine with an a egg, or an a sperm may combine with an A egg. In each case, the probability is $p \times q$. Consequently, the overall probability of obtaining a heterozygote is $2pq$. What percentage of the next generation will be heterozygotes?

It is easy now to show that the allele frequencies p and q remain constant for each generation. Notice that $p^2 + pq$ represents the total of the A alleles. The fraction that this frequency constitutes of all alleles is

$$\frac{p^2 + pq}{p^2 + 2pq + q^2} = \frac{p(p + q)}{(p + q)(p + q)} = \frac{p}{p + q} =$$

$$\frac{p}{p + (1 - p)} = p$$

Similarly, the frequency of a in the next generation will be

$$\frac{q^2 + pq}{p^2 + 2pq + q^2} = \frac{q(p + q)}{(p + q)(p + q)} = \frac{q}{p + q} =$$

$$\frac{q}{(1 - q) + q} = q$$

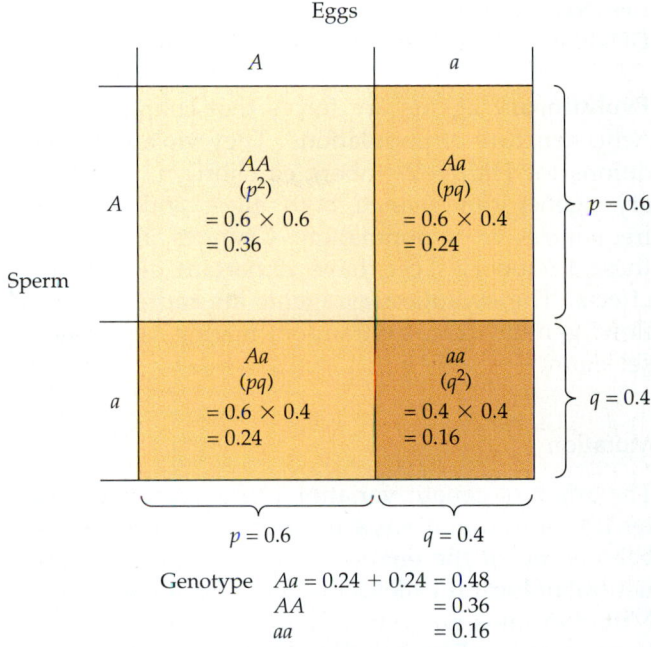

19.4 Calculating Hardy–Weinberg Genotype Frequencies
The areas within the squares are proportional to the expected frequencies of possible matings if mating is random with respect to genotype and the frequency (p) of A alleles is 0.6 and the frequency (q) of a alleles is 0.4.

Thus the original allele frequencies are unchanged.

The most important message of the Hardy–Weinberg rule is that allele frequencies remain the same unless some agent acts to change them. The rule also shows us exactly what distribution of genotypes to expect for a population at equilibrium at any value of p and q.

You may already have recognized that populations in nature rarely meet the conditions of the Hardy–Weinberg rule. Why, then, is the rule considered so important for the study of evolution? The answer is that without it we cannot tell whether evolutionary agents are operating. If individuals in a population mate randomly and no other agents are operating to change allele frequencies, then the genotype frequencies we actually observe in the population will approximate those we calculate from the Hardy–Weinberg formula. However, if the frequencies of genotypes deviate significantly from the expected Hardy–Weinberg values, we have evidence that some agent of evolution is in action.

In other words, if the genotype frequencies in a population do not fit Hardy–Weinberg frequencies, we know that something of evolutionary interest is influencing the population. Without knowing the rule, there would be no way of assessing whether observed genotype frequencies are interesting or even surprising.

CHANGING THE GENETIC STRUCTURE OF POPULATIONS

Evolutionary agents are forces that change the genetic structure of populations. They violate the conditions for Hardy–Weinberg equilibrium. Evolutionary agents can change both allele and genotype frequencies in a population. Changes in either of these frequencies can have important evolutionary effects. The evolutionary agents are mutation, gene flow, genetic drift, nonrandom mating, and natural selection.

Mutation

The origin of genetic variation is mutation (see Chapter 10). Most mutations are harmful or neutral to their bearers, but if the environment changes, previously neutral or harmful alleles may become advantageous. Mutation rates are very low for most loci that have been studied. Rates as high as one mutation in a thousand zygotes per generation are rare; one in a million is more typical. Nonetheless, these rates are sufficient to create considerable genetic variation over long time spans. In addition, mutations can restore to populations alleles that other evolutionary agents remove. Thus mutations both create and help maintain variation within populations.

One condition for Hardy–Weinberg equilibrium is that there be no mutation. Although this condition is never strictly met, the rate at which mutations arise at single loci is so low that mutations will result in only very small deviations from Hardy–Weinberg expectations. If large deviations are found, it is appropriate to dismiss mutation as the cause and to look for evidence of other evolutionary agents. Repeated mutations can bring the frequencies of rare alleles up to levels as high as several percent. Earlier in the chapter we chose 5 percent as the frequency for defining polymorphism in order to exclude rare alleles that may be present only because of repeated mutations.

Gene Flow

Because few populations are completely isolated from other populations of the same species, usually some migration between populations takes place. **Gene flow** refers to migration *followed by breeding* in the new location. For a population to be in Hardy–Weinberg equilibrium, there must be no gene flow between it and other populations. Gene flow ranges from extremely low to very high, depending on the number of migrating individuals and their genotypes. Immigrants (individuals entering a population) may add new alleles to the pool of a population or may change the frequencies of alleles already present. Emigrants (individuals leaving a population) may completely remove alleles or may reduce the frequencies of alleles when they leave a population. For a population to be in Hardy–Weinberg equilibrium, there must be no migration among its subpopulations.

Genetic Drift

On average, half the times a coin is tossed it will come up heads; for any particular toss the coin is equally likely to come up heads or tails. A ratio of 8 heads to 2 tails is not uncommon in 10 tosses of a coin. A ratio of 80 heads to 20 tails in 100 tosses is much less likely. If you tossed a fair coin 100 times or more, you would find that the number of tosses yielding heads approximately equalled the number of tosses yielding tails. To demonstrate the principle of probability that guides the outcome of coin tossing, you need to toss a coin many times. Otherwise you would be demonstrating the effects of chance.

Chance alters allele frequencies in small populations. Such alteration is called **genetic drift**. This is the reason that a population must be very large to be in Hardy–Weinberg equilibrium. If only a few individuals or a few gametes are drawn at random to form the next generation, the alleles they carry are not likely to be in the same proportions as alleles in the gene pool from which they were drawn. This phenomenon is illustrated in Figure 19.5, which shows allele frequencies as proportions of red and yellow beans. Most of the beans that survive to germinate the next generation in this example are red, so the new population has a much higher frequency of red beans than the previous generation had.

In very small populations, genetic drift may be strong enough to influence the direction of change of allele frequencies even when other evolutionary agents are pushing the frequencies in a different direction. Harmful alleles, for example, may increase because of genetic drift, and rare advantageous alleles may be lost. Two important causes of genetic drift are bottlenecks and founder effects.

BOTTLENECKS. Even organisms that normally have large populations may pass through occasional periods when only a small number of individuals survive. During these population **bottlenecks**, genetic variation can be lost by chance. For example, suppose we have performed a cross of $Aa \times Aa$ individuals of a species of *Drosophila* to produce an offspring population in which $p = q = 0.5$ and in which the genotype frequencies are 0.25 AA, 0.50 Aa, and 0.25 aa. If we randomly select four individuals from among the offspring to form the next generation, the allele frequencies in this small sample may differ markedly from $p = q = 0.5$. If, for example, we happen by chance to draw two AA homozygotes and two het-

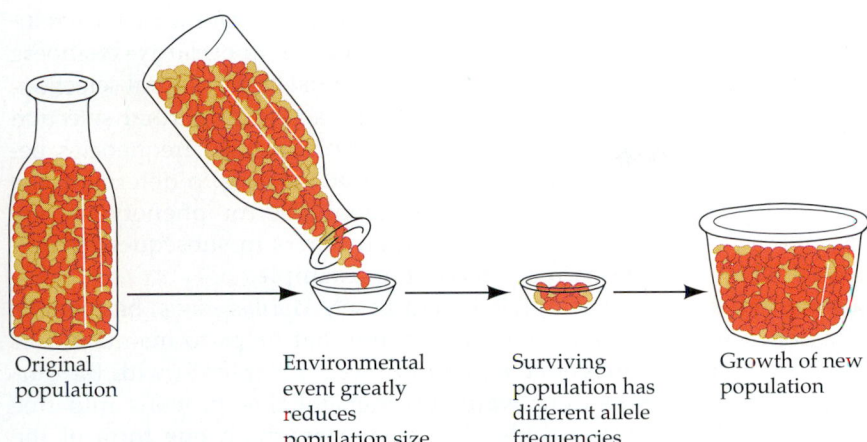

19.5 Genetic Drift
As this imaginary population passes through a bottleneck that reduces the population's size, the frequencies of yellow and red alleles change because of random sampling.

Original population

Environmental event greatly reduces population size

Surviving population has different allele frequencies

Growth of new population

erozygotes (*Aa*), the genotype frequencies in this "surviving population" are $p = 0.75$ and $q = 0.25$. If we begin a large evolutionary study by replicating this experiment 1,000 times, one of the two alleles will be missing entirely from about eight of the 1,000 "surviving populations."

Populations in nature pass through bottlenecks for many different reasons. Predators may reduce populations of their prey to very small sizes. During the 1890's, hunting reduced northern elephant seals to about 20 animals in a single population on the coast of Mexico. The actual breeding population may have been even smaller because only a few males mate with all the females and father the offspring in any generation (Figure 19.6). By analyzing small samples of tissue collected from northern elephant seals on the California coast, scientists determined that these seals have less genetic variation than any other seal

that has been studied. The investigators examined 24 proteins from each animal electrophoretically (see Chapter 14), looking for evidence of genetic variation among the seals at the loci encoding the proteins. They found no evidence of variation in any of the 24 proteins. By contrast, the southern elephant seal, whose numbers were not severely reduced by hunting, has much more genetic variation.

FOUNDER EFFECTS. When a species expands into new regions, populations may be started by a small number of pioneering individuals. These pioneers are not likely to have all the alleles found in their source population. Even if they do, the allele frequencies are likely to differ from those in the source population. The situation is equivalent to that for a large population reduced by a bottleneck, but rather than a small surviving population, there is a small founding pop-

19.6 A Species with Low Genetic Variation
In this northern elephant seal breeding colony on Año Nuevo Island, California, a few males control the beaches on which most females have their pups. Because these few males sire most of the offspring, the size of the breeding population is smaller than the whole population. This nonrandom mating, together with a bottleneck that occurred when the seals were overhunted in the late nineteenth century, resulted in a population with very little genetic variation.

ulation. This type of genetic drift is called a **founder effect**. Because many plant species reproduce sexually by self-fertilization, a new population may be started by a single seed—an extreme example of a founder effect.

The inhabitants of the Faroe Islands, north of Scotland, demonstrate the founder effect. These islands were settled by Danes around A.D. 825—about 40 generations ago. Most of the few additional immigrants who came after the population was founded also were Danes. The size of the initial colonizing group is unknown, but since 1769, when the first census was taken, the population has fluctuated between 4,000 and 5,000 people except during times of major epidemics. The human population of the Faroes is divided into seven groups among which there has been, for centuries, a relatively steady but low level of migration. All island groups differ strikingly from present-day Danish populations—a founder effect—and populations from different islands have drifted apart genetically from one another.

Nonrandom Mating

One Hardy–Weinberg condition specifies that individuals do not choose mates on the basis of their genotypes—that is, that mating must be random. Often, however, individuals with certain genotypes mate more often with individuals of either the same or different genotypes than would be expected on a random basis. When such **assortative mating** takes place, in the next generation homozygous genotypes are overrepresented and heterozygous genotypes are underrepresented in comparison with Hardy–Weinberg expectations. Self-fertilization, another form of nonrandom mating, is common in many groups of organisms, especially plants. Selfing tends to reduce the frequencies of heterozygous individuals in populations below Hardy–Weinberg expectations. Under assortative mating and self-fertilization, genotype frequencies change but allele frequencies remain the same. Nonrandom mating can alter allele frequencies, however, if some individuals are more successful in mating than others. We consider this situation, which is called sexual selection, in Chapter 46.

Natural Selection

Not all individuals survive and reproduce equally well in a particular environment. Therefore, some individuals contribute more offspring to the next generation than do other individuals. Individuals vary for heritable traits that determine the success of their reproductive efforts. The differential contribution of offspring resulting from different heritable traits was called **natural selection** by Charles Darwin. When genotype differences cause differential reproductive

success, natural selection is operating as an evolutionary agent. One condition for Hardy–Weinberg equilibrium is that there must be no natural selection. Biologists investigate the action of natural selection by comparing allele (or phenotype) frequencies between generations and attempting to determine the reasons why some genotypes (or phenotypes) are better represented than others in subsequent generations. Let's look at an example.

The edible blue mussel, *Mytilus edulis*, has genetic variation for an enzyme that helps to maintain cells in osmotic equilibrium (see Chapter 5) with the surrounding water by metabolizing proteins into free amino acids. The allele governing one form of the enzyme is at very high frequencies in populations growing in highly saline (salty) seawater. However, mussels with this form of the enzyme do not survive well in low-salinity estuaries, where the osmotic environment is different. In estuaries, the frequency of the allele for this form of the enzyme is reduced within each generation by the higher death rate of the mussels that possess it. This is an example of natural selection. Each spring the frequency of this allele rises again in the estuaries when large numbers of larvae, produced by adults living in high-salinity environments, invade and settle. Because estuarine populations are a small fraction of the total population, they are repeatedly swamped by migrating offspring from individuals growing in marine environments, but the newcomers are eliminated from the population. Thus the operation of one evolutionary agent—natural selection—counteracts another evolutionary agent—gene flow (Figure 19.7).

Natural selection is the only evolutionary agent

19.7 Natural Selection Counters Gene Flow
The white organisms are young blue mussels; most of them are adapted to highly saline seawater, but they have settled—and now dominate the population—in an estuary with lower-salinity water. They will not survive or reproduce well there. Most of the large adults shown here have a genotype that is adapted to estuaries.

that adapts populations to their environments. The changes in allele frequencies produced by natural selection result from the relative abilities of individuals to survive and reproduce in the environments in which they live. The study of natural selection is thus central to evolutionary investigations.

SELECTION OPERATING ON MORE THAN ONE LOCUS. Up to this point we have been considering how evolution changes single loci in a population. When we consider more than a single locus, the useful Hardy–Weinberg rule does not apply. Natural selection usually operates simultaneously on many traits or on traits governed jointly by more than one locus. Consider how sexual recombination multiplies the opportunities for natural selection to operate in this way. In asexually reproducing organisms, the daughter cells resulting from the mitotic division of a single cell normally contain identical genotypes at all loci. In a population of organisms producing new individuals in this asexual way—without any combination of genetic material from different individuals—every new individual is genetically identical to its parent unless there has been a mutation. When organisms exchange genetic material during sexual reproduction, however, the offspring differ from their parents because chromosomes assort randomly during meiosis and because fertilization brings together material from two different cells.

Sexual recombination generates an endless variety of genotypic combinations that increases the *evolutionary potential* of populations. Because it increases

the variation among the offspring produced by an individual, sexual recombination improves the chance that some of the offspring will be successful in the varying and often unpredictable environments they will encounter. Sexual recombination does not influence the frequency of different alleles or the rate of evolutionary change, however. Instead it generates new combinations of genetic material upon which selection can act.

Depending upon which traits are favored in a population, natural selection can produce any one of several quite different results. In the example in Figure 19.8, the variable trait is size, a trait likely to be controlled by many loci. If many genetic and environmental factors contribute to size—and there is no selection—then the distribution of sizes in a population should approximate the bell-shaped curve shown in the upper parts of the figure. If both the smallest and the largest individuals contribute relatively fewer offspring to the next generation than those closer to the center, **stabilizing selection** is operating (Figure 19.8*a*). Natural selection frequently acts in this way, and by doing so may counter increases in variation brought about by mutation or

19.8 Natural Selection Operates on a Variable Trait
In these examples, body size is being studied—a trait controlled by several gene loci. The phenotypes favored by natural selection are shown in yellow under the bell-shaped curves that plot the distributions of phenotypes in a population before selection (top row). The lower graphs show the effects of natural selection on phenotype distribution, and on the shape of the original curve.

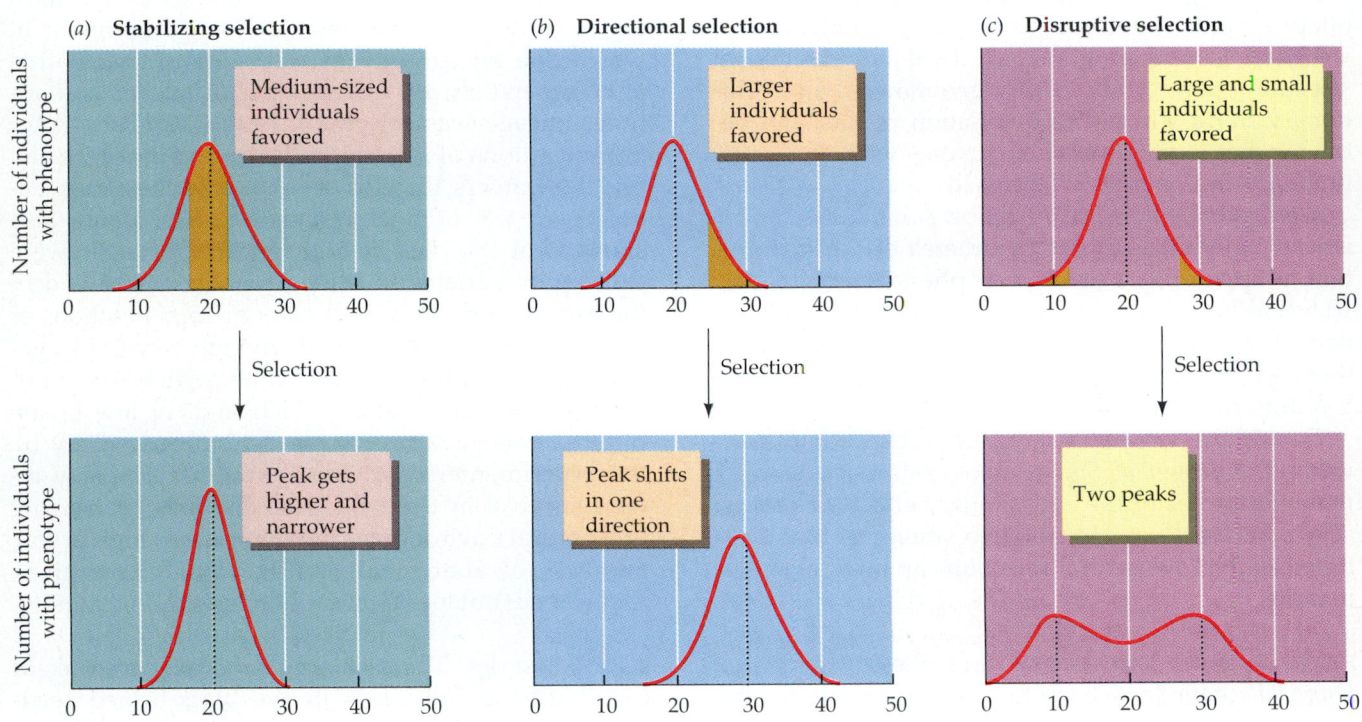

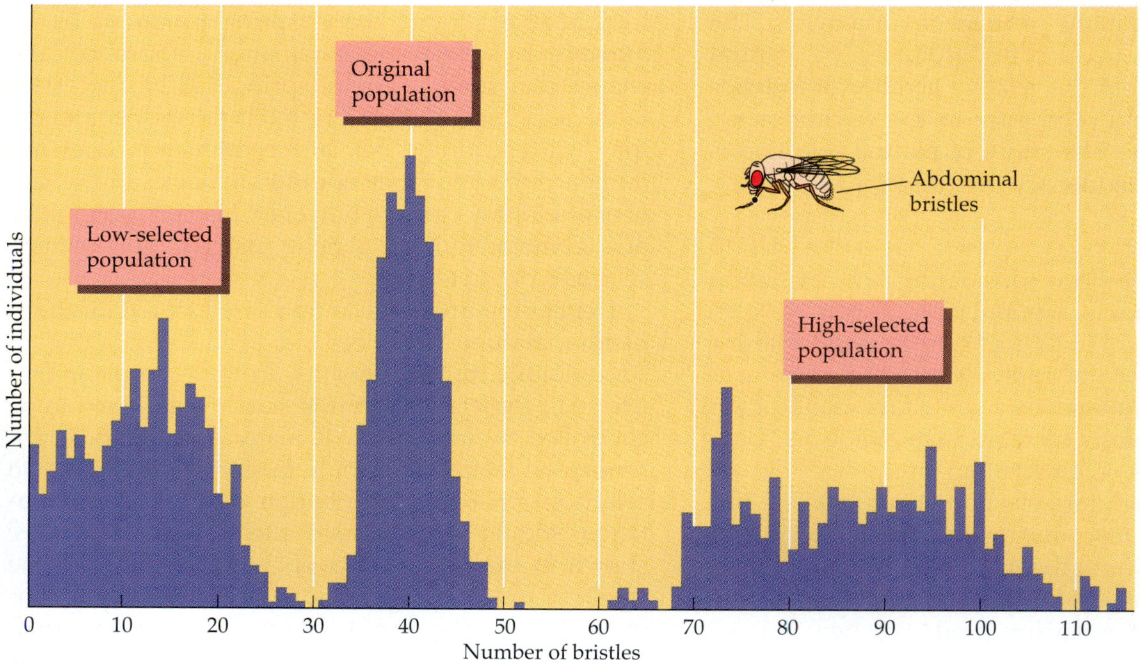

19.9   Bristle Number Can Evolve Rapidly in *Drosophila*
In the experimental data graphed above, after 35 generations of artificial selection, neither the population selected for low abdominal bristle number nor the high-selected population had any overlap with the original population in number of bristles. All three populations, however, maintained a great deal of variation in this trait.

migration. We know from the fossil record that most populations do not usually evolve rapidly. The slow rate of evolution may be largely an effect of stabilizing selection.

If individuals at one extreme of the size distribution—the larger ones, for example—contribute more offspring than other individuals do, then **directional selection** is operating (Figure 19.8*b*). If directional selection operates over many generations, an evolutionary trend within the population results. But because what is favored often changes with time, evolutionary trends may be reversed.

Disruptive selection, selection simultaneously favoring individuals at both extremes of the distribution, is apparently a much rarer phenomenon (Figure 19.8*c*). When disruptive selection operates, individuals at the extremes contribute more offspring than those in the center, producing two peaks in the distribution of a trait.

During the 135 years since the publication of Darwin's *The Origin of Species*, biologists have studied many examples of the evolutionary effects of natural selection. Next we describe two situations that illustrate clearly how natural selection can influence populations.

FRUIT FLIES IN THE LABORATORY. Fruit flies of the genus *Drosophila* have long been favored organisms for the study of selection because they are easy to rear in captivity and have short generation times. By selecting lineages of fruit flies for high or low number of bristles, investigators have been able to conduct intensive studies of the response to natural selection of the numbers of bristles on the bodies of the flies. Figure 19.9 shows the results of an experiment in which one lineage of flies was selected, generation after generation, for low number of bristles and another lineage was selected for high numbers. After 35 generations of selection, flies in both lineages had bristle numbers that fell well outside the range observed in the original population. The figure also shows that the flies in both selected lineages were still highly variable in bristle number at the end of the experiment. That is, even after 35 generations of intense selection, there was enough remaining genetic variation for still further changes in bristle number. The environment in which high or low bristle number was "adaptive" was determined strictly by the experimenters, but bristle number probably affects survival in the wild. Had the flies been living in a natural environment in which either high or low numbers of abdominal bristles would have given them an advantage, they could have evolved rapidly.

AFRICAN FINCHES. The strikingly bimodal (two-peaked) distribution of bill sizes in the black-bellied seedcracker *Pyrenestes ostrinus*, a west African finch, illustrates how disruptive selection can adapt populations in nature. Seeds of two species of sedges (marsh plants) are the most abundant food source for the finches during parts of the year. Birds with large bills readily crack the hard seeds of the sedge *Scleria verrucosa*; birds with small bills crack those seeds only with difficulty. Small-billed birds prefer to feed on the soft seeds of the sedge *Scleria goossensii*, and they feed more efficiently on those seeds than do birds with larger bills. Young finches whose bills deviate markedly from the two predominant bill sizes do not survive as well as finches whose bills are close to one of the two sizes represented by the distribution peaks (Figure 19.10). Because there are few abundant food sources in the environment and because the seeds of the two sedges do not overlap in hardness, birds with intermediate-sized bills are inefficient in utilizing ei-

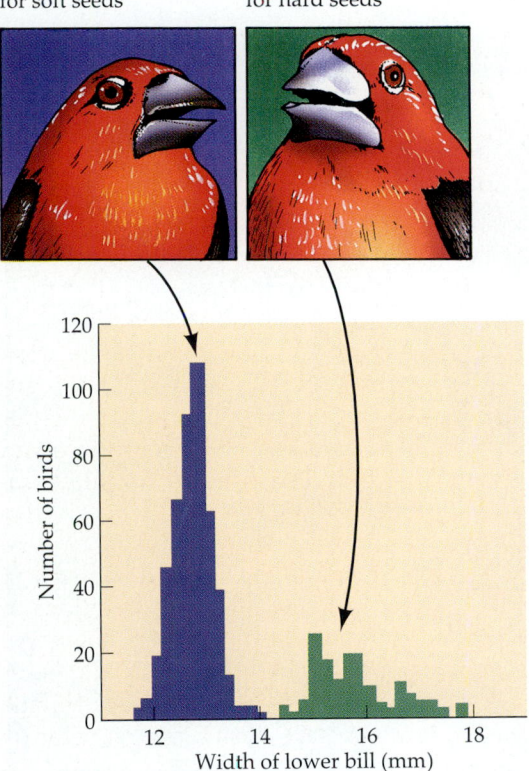

19.10 Natural Selection Alters Bill Sizes
Bill size among black-bellied seedcrackers of western Africa is distributed in a bimodal pattern. Large-billed birds can crack the hard seeds of one type of sedge plant, whereas smaller-billed birds feed more efficiently on the soft seeds of another type. Birds with intermediate-sized bills, being unable to use seeds of either plant efficiently, survive poorly.

ther of the principal food sources. Disruptive selection therefore maintains a bill-size distribution with two peaks.

Fitness

A central concept in natural selection is **fitness**. The fitness of a genotype or phenotype is its reproductive contribution to subsequent generations *relative* to the contribution of other genotypes or phenotypes. The word *relative* is critical; the absolute number of offspring produced by an individual does not influence allele frequencies. Changes in *absolute* numbers of offspring are responsible for increases and decreases in the *size* of a population, but it is the relative success among genotypes within a population that leads to changes in allele frequencies—that is, to evolution.

An individual may influence its own fitness in two ways. First, it may produce its own offspring, contributing to its individual fitness. Second, it may help the survival of relatives that have the same alleles by descent from a common ancestor. This is called kin selection. Together individual fitness and kin selection determine the inclusive fitness of the individual. Among species that either are solitary or reproduce in groups no larger than a pair and its offspring, individual fitness strongly dominates inclusive fitness. Among highly social species, such as social insects, some birds, and many primates, kin selection also may be very important. We will return to this topic in Chapter 46.

GENETIC VARIATION AND EVOLUTION

Genetic drift, stabilizing selection, and directional selection all tend to reduce genetic variation within populations. But we have seen that there is considerable genetic variation within most groups of organisms. Why isn't the genetic variation of a species lost over time? To answer this question we need to know how genetic variation is distributed and how it is maintained.

Distribution of Genetic Variation

Some of the genetic variation among populations of many species is correlated with the geographic distribution of the populations (Figure 19.11). For example, plant populations may vary geographically in the chemicals they synthesize to defend themselves against herbivores. Some individuals of the clover *Trifolium repens* are cyanogenic—they produce the poisonous chemical cyanide. Cyanogenic individuals are less appealing to herbivores—particularly mice and slugs—than are acyanogenic individuals (those who do not produce cyanide). However, clover

(a)   (b)

19.11 Humans Are Highly Variable
(a) These women of the Masai tribe of Kenya, Africa, differ among themselves, but they are distinct from (b) these Kuna Indian women from the San Blas Islands, off the Caribbean coast of Panama.

plants with cyanide are more likely to be killed by frost, because freezing damages cell membranes and releases the toxic cyanide into the plant's own tissues. Thus in populations of *Trifolium repens*, the frequency of cyanide-producing individuals increases gradually from north to south and from east to west across Europe (Figure 19.12). Cyanogenic plants typify clover populations only in areas where the winters are mild. Chemical variation is maintained in the overall population of *Trifolium repens* because herbivores graze acyanogenic individuals heavily in areas with mild winters, but cold temperatures prevent plants from using cyanide as a defense in areas having more severe winters.

Gradual geographic changes in phenotypes and genotypes, as illustrated here by *Trifolium repens*, are called **clines**. Clines are widespread among most groups of organisms. In some regions, however, frequencies of certain traits change abruptly, creating **step clines**. Color patterns in the rat snake *Elaphe obsoleta* are a good example (Figure 19.13). In this species the color differences are complex and striking. Single color patterns dominate extensive regions, and each change from one pattern to the next is abrupt. Because there are no obvious environmental changes in the regions where color patterns change, the causes of the abrupt changes are unknown.

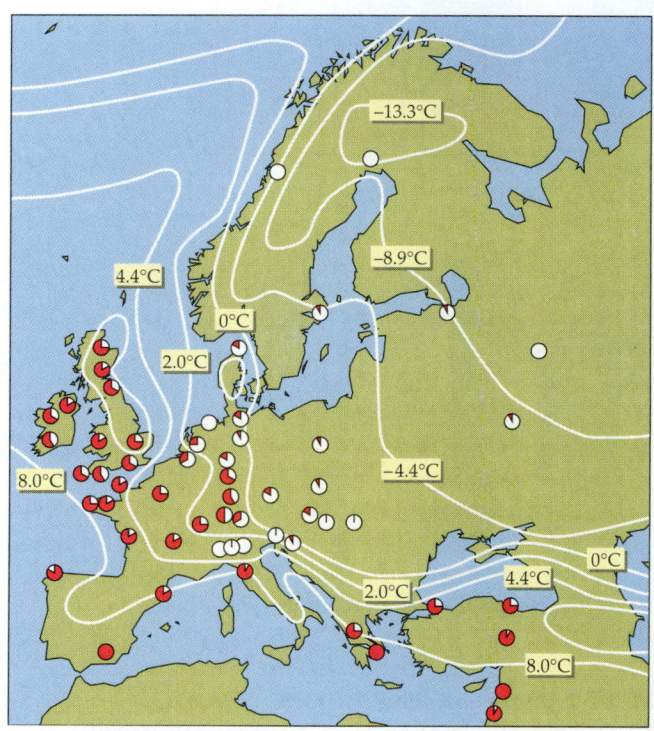

19.12 Geographic Variation in Cyanogenic Clovers
The frequencies of cyanide-producing individuals in populations of white clover (*Trifolium repens*) are represented by the proportion of each circle that is blackened. The isotherms (lines connecting points with equal temperatures) plot January mean temperatures.

19.13 Step Clines in the Rat Snake
There are no obvious environmental changes at the boundaries where the color patterns of these snakes change.

Maintenance of Genetic Variation

Natural selection often preserves variation by favoring different traits in different areas, as illustrated by clines and step clines. In such cases, much of the variation in a large population is preserved as differences between subpopulations. Natural selection also preserves variation as polymorphisms—genetic differences within a local population—which we discussed earlier in this chapter. For example, polymorphism is maintained when the success of a genotype (or phenotype) depends upon its frequency relative to other genotypes (or phenotypes). A fish that lives in Lake Tanganyika in east central Africa provides an example. The mouth of this scale-eating fish, *Perissodus microlepis*, opens either to the right or to the left as a result of an asymmetrical jaw joint (Figure 19.14). *Perissodus* approaches its prey (another fish) from behind and dashes in to bite off several scales from its flank. "Right-mouthed" individuals always attack from the victim's left; "left-mouthed" individuals always attack from the victim's right. The distorted mouth enlarges the area of teeth in contact with the prey's flank, but only if the scale-eater attacks from the appropriate side.

Prey fish are alert to approaching scale-eaters, so attacks are more likely to be successful if the prey must watch both flanks. Guarding by the prey favors equal numbers of right-mouthed and left-mouthed *Perissodus* because if one morph were more common than the other, prey fish would pay more attention to potential attacks from that flank. Over an 11-year period in which the fish in Lake Tanganyika were studied, natural selection maintained the polymorphism, keeping the two morphs of *Perissodus* at about equal frequencies.

Natural selection also preserves genetic variation when it favors different traits at different times, as is evident when the selective agents are predators. The speed at which a fish can dart is influenced by its total number of vertebrae (segments of the backbone) and the ratio of trunk vertebrae to tail vertebrae, which are genetically determined characteristics. No single ratio of the two types of vertebrae results in the maximum swimming speed at all sizes.

Among three-spined sticklebacks, small individuals (7.4 to 7.8 mm long) can achieve the greatest burst speed if they have a high ratio of trunk to tail vertebrae. Among larger fish (8.3 to 9.0 mm long), individuals with a lower ratio of trunk to tail vertebrae can achieve higher burst speeds. Because burst speed is important for escaping from predators, three-spined sticklebacks with higher trunk-to-tail ratios survive better when they are small, whereas individuals with lower ratios survive better when they grow

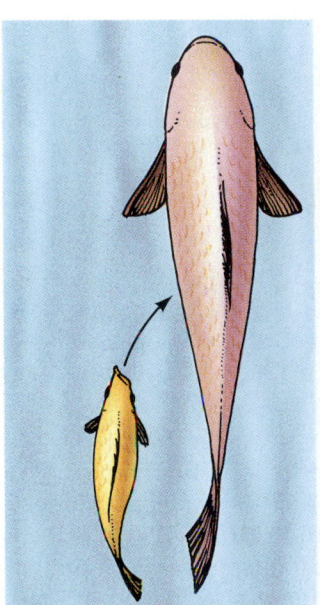
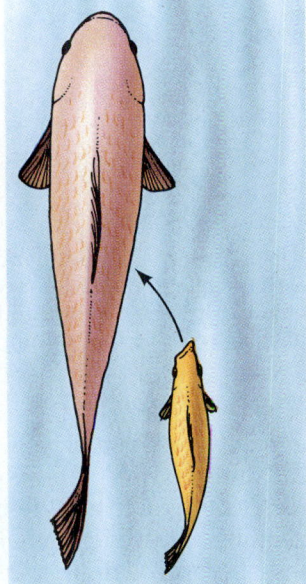

"Right mouthed" *Perissodus* attack prey from the left rear side

"Left mouthed" *Perissodus* attack prey from the right rear side

19.14 Balanced Polymorphism in "Mouthedness"
Natural selection maintains equal frequencies of left-mouthed and right-mouthed individuals of a scale-eating fish (*Perissodus microlepis*) in Lake Tanganyika.

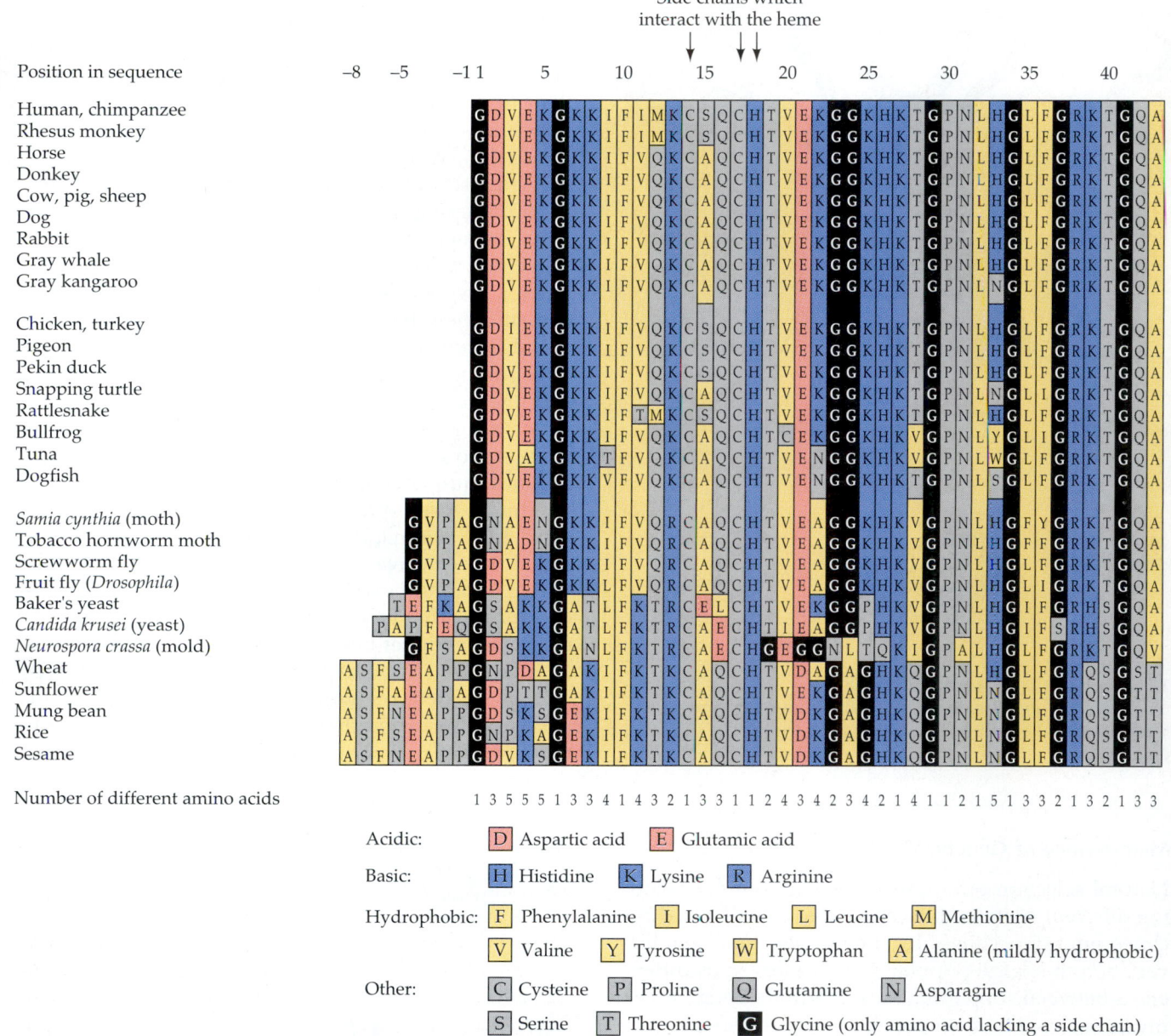

Side chains which
interact with the heme

19.15   Amino Acid Sequences of Cytochromes *c*
Invariant positions in the cytochromes *c* of these species have 1's at the bottom of their columns. Conservation of charge at a position is shown by the constancy of color in its column.

larger. If predation rates were the same from year to year, one optimal ratio would result in the highest overall survival to adulthood. In Holden Lake, British Columbia, however, where biologists have studied these fish, predation rates vary within and among years. During some years individuals with higher trunk-to-tail ratios survive to adulthood better than individuals with lower ratios; in other years the reverse is true. As a result there is considerable genetic variation for the total number and the ratios of trunk and tail vertebrae among three-spined sticklebacks in the lake.

Adaptive and Adaptively Neutral Genetic Variation

Natural selection often maintains genetic variation, but not all genetic variation affects the fitness of organisms; that is, some variation is adaptively neutral. For example, some amino acid substitutions caused by mutations do not affect the functioning of the molecules of which they are a part. Such substitutions may be lost, or their frequencies may increase over time, untouched by natural selection.

Modern molecular techniques enable us to mea-

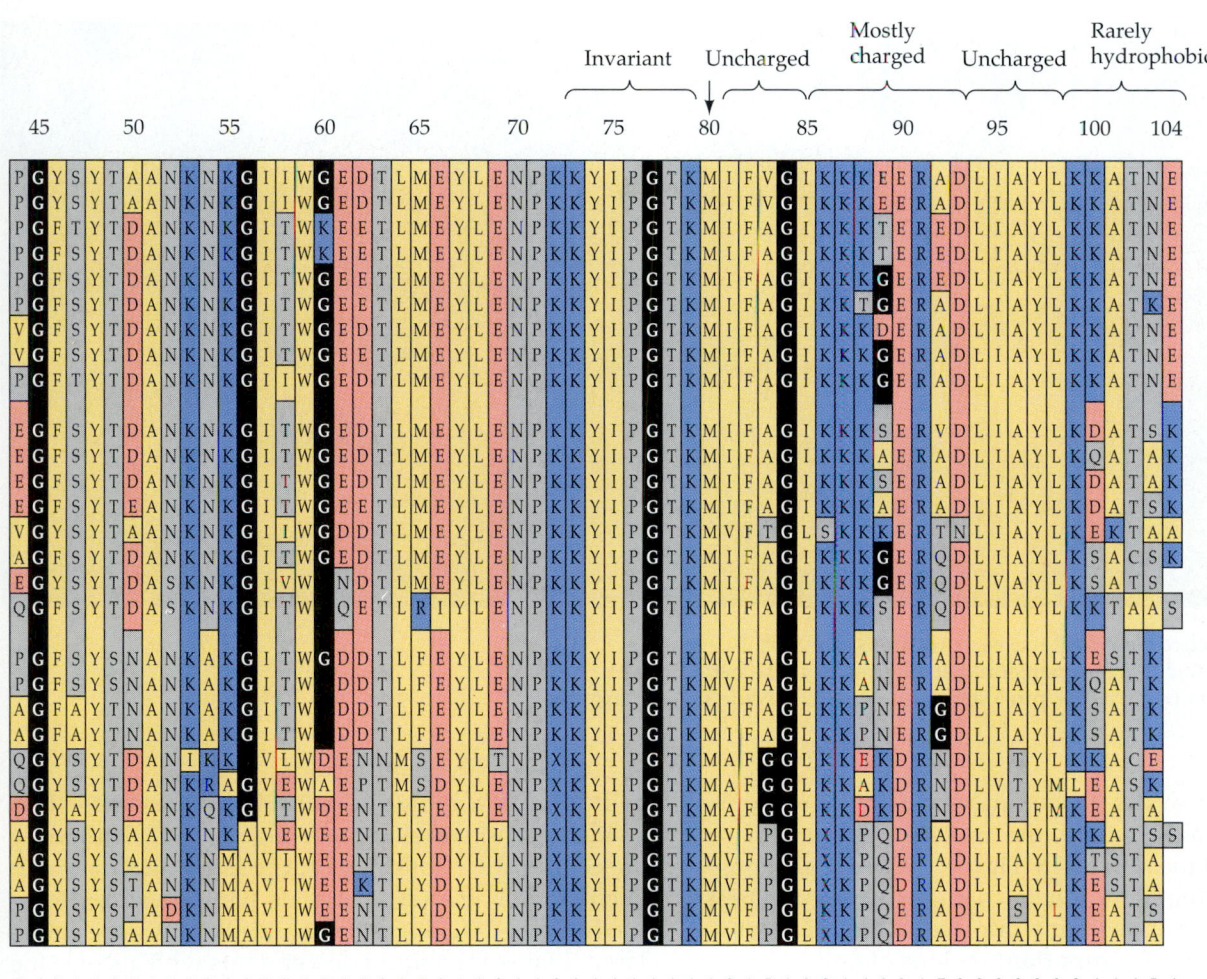

sure this form of variation and provide the means by which to distinguish adaptive from neutral variation. We illustrate this technique by examining a well-studied protein, cytochrome *c*. Cytochrome *c* is one component of the electron transport chain of mitochondria (see Chapter 7). Together with other proteins of the citric acid cycle and respiratory chain, cytochrome *c* is part of the common heritage of all eukaryotic organisms.

Cytochrome *c* amino acid sequences are known for nearly 100 species of organisms, ranging from yeasts to humans. Biologists' attempts to understand the significance of similarities and differences in these sequences are aided by information about the three-dimensional structure of cytochrome *c*, which has been determined for several species. The amino acid sequences for cytochrome *c* from 33 species are compared in Figure 19.15. The great sequence similarity among all species indicates that all cytochromes *c* are variations on a common theme. Notice that 33 positions along the chain are occupied by the same amino acid in every species. There have undoubtedly been many mutations in DNA at these positions, but it appears that mutations at those positions have re-

peatedly been weeded out by natural selection as eukaryotes evolved. The side chains of the invariant amino acids at positions 14, 17, 18, and 80 (marked by arrows on the figure) all interact with the iron-containing heme group. Evidently any changes in these amino acids adversely affect the functioning of the vital heme group. Most of the remaining positions show little variation, but there are seven positions with six or more different amino acids each, indicating that the functional requirements of those positions are nonspecific.

The figure shows that the positions having only a few variants are consistent in their charges. For instance, region 80 to 85 is always hydrophobic, or at least free of charged groups. Region 86 to 93 is mostly charged, region 94 to 98 is never charged, and region 99 to 104 is rarely hydrophobic. These charge consistencies suggest that all these cytochrome *c* molecules are folded in the same way and that conserving charge is necessary to preserve the three-dimensional structure of the molecules.

This interpretation is confirmed by the three-dimensional structure of rice and tuna cytochromes *c*, which are folded in exactly the same way (Figure

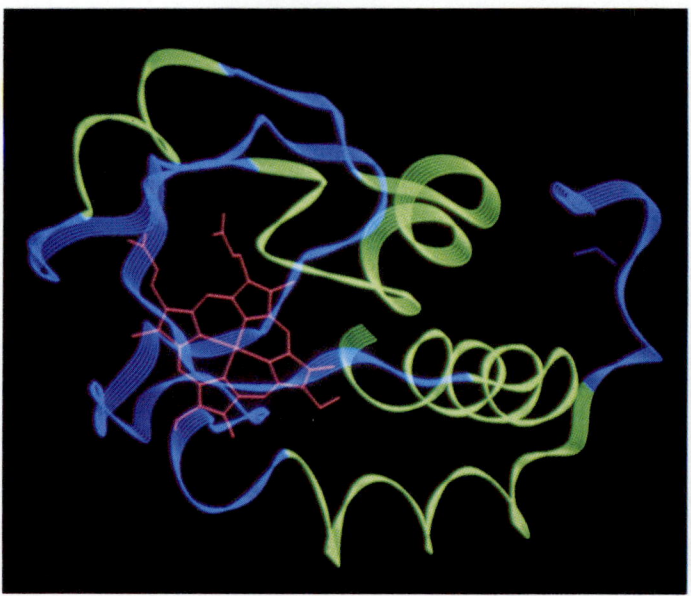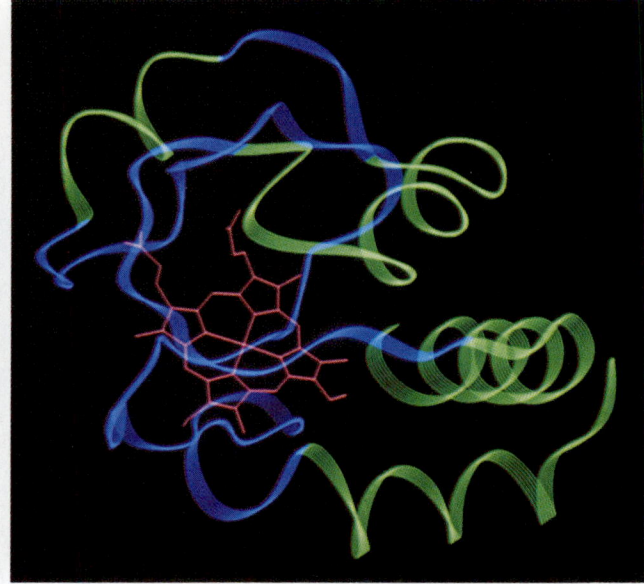

19.16 Cytochrome *c* Molecules Fold the Same Way
Note the similar three-dimensional structures of these cytochromes *c* from rice
(left) and tuna (right). The pink structure is the molecule's heme group.

19.16). Most of the glycine molecules in cytochrome *c* are in places where the molecule makes sharp bends. Glycine (black in Figure 19.15) is the only amino acid that lacks a side chain, making it the best choice for these positions. Molecular biologists now detect functionally important regions of molecules by searching for and studying positions that lack variation because constancy usually indicates an area in which the action of natural selection has been strong.

In contrast, substitutions that are functionally neutral may simply persist when they arise. If so, these substitutions may accumulate slowly over time at the rate at which they arise. The cytochromes *c* of eukaryotes have diverged from one another at the rate of about three amino acid substitutions per 100 million years (Figure 19.17), suggesting that among different species, the differences in cytochromes *c* at many of the variable positions are the result of slow accumulations of neutral amino acid substitutions.

INTERPRETING LONG-TERM EVOLUTION

The short-term changes in populations that we have been discussing in this chapter are called **microevolution**. Microevolutionary studies are an important part of evolutionary biology because short-term changes can be observed directly and can be manipulated experimentally. Although studies of short-term changes reveal much about evolution, by themselves they cannot provide a complete explanation of the long-term changes that are often called **macro-**

evolution. We could measure the shapes of the legs of horses and correlate the shape with their running speed and their ability to escape from predators. But such studies would not tell us why the small, four-toed, forest-dwelling horse *Hyracotherium* gave rise to large, plains-inhabiting descendants with a single toe (hoof).

Patterns of macroevolutionary changes can be strongly influenced by events that occur so infrequently or so slowly that they are unlikely to be observed during microevolutionary studies. The evolution of horses was influenced by long-term climatic changes that caused forests to shrink and grasslands to expand over large parts of Earth. Also, how evolutionary agents act may change over time; even among the descendants of a single ancestral species, different lineages may be evolving in different directions, as happened during the evolution of horses. Therefore, we cannot interpret the past simply by extending the results of short-term experiments and observations backward in time. Additional types of evidence must be gathered if we wish to understand the course of evolution over billions of years. The task of gathering some of this evidence is challenging because we must infer the nature of events in the remote past from traces we can measure today. How scientists approach this challenge will be described in the next two chapters. Finally, in Chapter 28, after we have described the major groups of organisms more fully, we will synthesize this information to provide a picture of patterns of macroevolution and their causes.

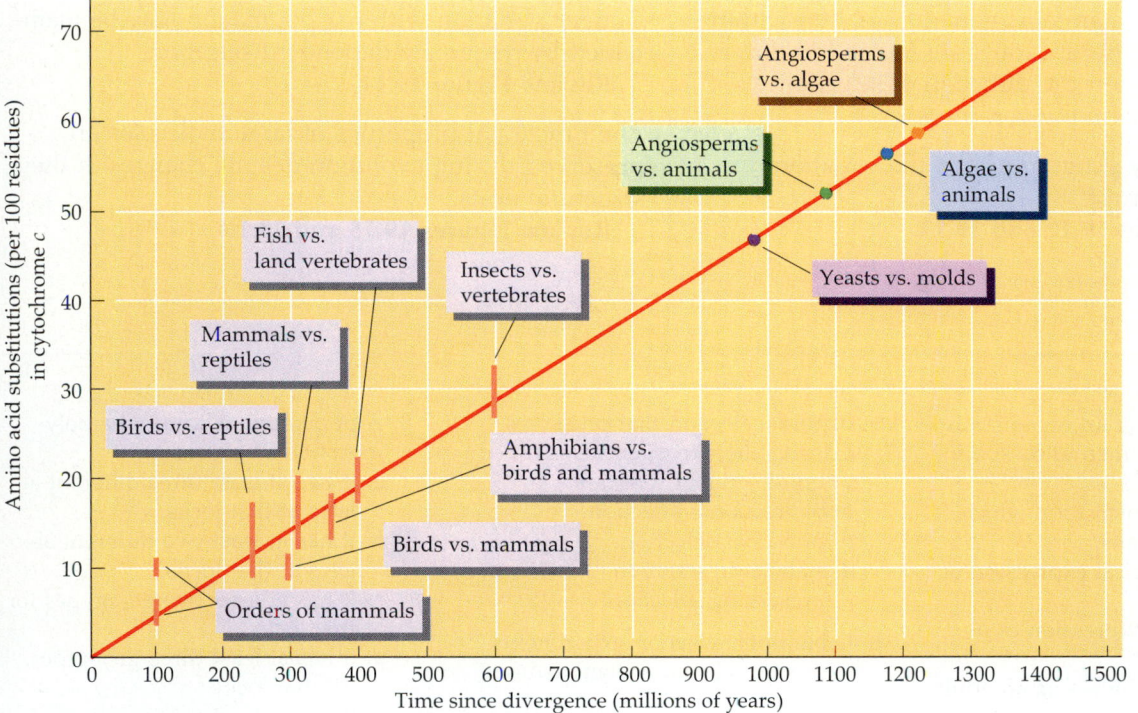

19.17 Cytochrome *c* Molecules Evolved Slowly
This plot of average rates of amino acid substitutions does not show the many short-term changes in evolutionary rates within and among the taxa. The extent of intrataxon variation is shown for animals by the vertical bars, which plot mean deviations from the average for the taxa being compared.

SUMMARY of Main Ideas about the Mechanisms of Evolution

Biological evolution results from evolutionary agents acting over millions of years.

For a population to evolve, its members must possess genetic variation, which is the raw material upon which agents of evolution act.

Allele frequencies measure the amount of genetic variation in a population. Genotype frequencies show how a population's genetic variation is distributed among its members.

A population evolves when individuals having different genotypes survive or reproduce at different rates.

In a population at Hardy-Weinberg equilibrium, allele and genotype frequencies remain the same from generation to generation.
   Review Figure 19.4

Biologists can determine if some agent of evolution is acting on a population by comparing its genotype frequencies with Hardy–Weinberg equilibrium frequencies.

Changes in allele frequencies and genotype frequencies within populations are caused by mutations, gene flow, genetic drift, nonrandom mating, and natural selection.

The origin of genetic variation is mutation.

Gene flow is migration from one population to another, followed by breeding in the new location.

Genetic drift alters allele frequencies primarily in small populations.
   Review Figure 19.5

Natural selection may preserve allele frequencies or cause them to change over time.

Natural selection is the only agent of evolution that adapts populations to their environments.

Stabilizing selection, directional selection, and disruptive selection change the distributions of phenotypes governed by more than one locus.
   Review Figure 19. 8

Natural selection maintains genetic variation within a species when different traits are favored in different places and when the direction of selection changes over time in a given place.

Species may vary geographically; the variation can be gradual or abrupt.
Review Figures 19.12 and 19.13

Genetic variation within a population may be maintained by frequency-dependent selection.
Review Figure 19.14

The functional properties of large molecules are preserved during evolution despite changes in their structural units.
Review Figures 19.15 and 19.16

SELF-QUIZ

1. The phenotype of an organism is
   a. the type specimen of its species in a museum.
   b. its genetic constitution which governs its traits.
   c. the chronological expression of its genes.
   d. the physical expression of its genotype.
   e. the form it achieves as an adult.

2. The appropriate unit for defining and measuring genetic variation is
   a. the cell.
   b. the individual.
   c. the population.
   d. the community.
   e. the ecosystem.

3. Which of the following is *not* true of allele frequencies?
   a. The sum of any set of allele frequencies is always 1.
   b. If there are two alleles at a locus and we know the frequency of one of them, we can obtain the frequency of the other by subtraction.
   c. If an allele is missing from a population, its frequency is 0.
   d. If two populations have the same gene pool for a locus, they will have the same proportion of homozygotes at that locus.
   e. If there is only one allele at a locus, its frequency is 1.

4. In a population at Hardy–Weinberg equilibrium in which the frequency, p, of A alleles is 0.3, the expected frequency of Aa individuals is
   a. 0.21.
   b. 0.42.
   c. 0.63.
   d. 0.18.
   e. 0.36.

5. Natural selection that preserves existing allele frequencies is called
   a. unidirectional selection.
   b. bidirectional selection.
   c. prevalent selection.
   d. stabilizing selection.
   e. preserving selection.

6. Laboratory selection experiments with fruit flies have demonstrated that
   a. bristle number is not genetically controlled.
   b. bristle number is not genetically controlled but changes in bristle number are caused by the environment in which the fly is raised.
   c. bristle number is genetically controlled but there is little variation on which natural selection can act.
   d. bristle number is genetically controlled but selection cannot result in flies having more bristles than originally present in the population.
   e. bristle number is genetically controlled and selection can result in flies with bristle numbers higher than present in any flies before selection was started.

7. Disruptive selection maintains bill-size variability in the African seedcracker because
   a. it is difficult to form bills of intermediate shapes.
   b. the two major food sources of the finches differ markedly in size and hardness.
   c. males use their large bills in displays.
   d. migrants introduce different bill sizes into the population each year.
   e. older birds need larger bills than younger birds.

8. A population is said to be polymorphic for a locus if
   a. it has at least three different alleles at that locus.
   b. it has at least two different alleles at that locus.
   c. it has at least two genotypes for that locus.
   d. it has at least three genotypes for that locus.
   e. it has at least two genotypes for that locus, the rarest of which sum to a frequency greater than 5 percent.

9. A cline is defined as
   a. the distribution of an organism across a slope.
   b. an abrupt change in frequencies of certain traits over time.
   c. a gradual change in frequencies of certain traits over time.
   d. an abrupt change in frequencies of certain traits over space.
   e. a gradual change in frequencies of certain traits over space.

10. Certain positions in the cytochrome c molecule are occupied by the same amino acids in all species because
   a. there have been no mutations at those positions.
   b. there have been no mutations in hydrophobic regions.
   c. there have been no mutations in charged regions.
   d. all mutations at those positions were deleterious and were weeded out by natural selection.
   e. there has been no selection affecting those positions.

FOR STUDY

1. During the past 45 years, more than 200 species of insects that attack crop plants have become highly resistant to DDT and other pesticides. Using your recently acquired knowledge of evolutionary processes, explain the rapid and widespread evolution of resistance. Propose ways of using pesticides that would slow down the rate of evolution of resistance. Now that DDT has been banned in the United States, what do you expect to happen to levels of resistance to DDT among insect populations? Justify your answer.

2. In nature, mating among individuals in a population is never truly random and natural selection is seldom totally absent. Why, then, does it make sense to use the Hardy–Weinberg model, which is based on assumptions known to be generally false? Can you think of other models in science that also are based on false assumptions? How are such models used?

3. An investigator is studying populations of house mice living in barns and sheds on a large farm. Each building has a population of between 25 and 50 mice. Populations in different buildings have strikingly different frequencies of alleles determining coat color and tail length. By marking most individuals, the investigator determines that mice only rarely move between buildings. He interprets his study as providing evidence for random genetic drift. Could other agents of evolution plausibly account for the pattern? If so, how could they be distinguished?

4. As far as we know, natural selection cannot adapt organisms to future events. Yet many organisms exhibit responses in advance of natural events. For example, many mammals go into hibernation while it is still quite warm. Similarly, many birds leave the temperate zone for their southern wintering grounds long before winter has arrived. How can these "anticipatory" behaviors evolve?

5. Many people believe that species, like individual organisms, have life cycles. They believe that species are born by some process of speciation, undergo growth and expansion, and inevitably die out as a result of "species old age." Is there any agent of evolution that could cause such a species life cycle? If not, how do you explain the high rates of extinction of species in nature?

READINGS

Crow, J. F. 1986. *Basic Concepts in Population, Quantitative, and Evolutionary Genetics.* W. H. Freeman, New York. An excellent introduction to all aspects of modern population genetics.

Dickerson, R. E. 1972. "Cytochrome *c*: The Structure and History of an Ancient Protein." *Scientific American,* April. Discussion of the evidence for the 1.3-billion-year history of cytochrome *c*, the best-known protein.

Endler, J. A. 1986. *Natural Selection in the Wild.* Princeton University Press, Princeton, NJ. A thorough review of the problems and successes in measuring natural selection in nature.

Futuyma, D. J. 1986. *Evolutionary Biology,* 2nd Edition. Sinauer Associates, Sunderland, MA. A comprehensive review of all aspects of evolutionary biology.

Grant, P. R. 1991. "Natural Selection and Darwin's Finches." *Scientific American,* October. The finches of the Galapagos—the classic example of how natural selection works over millions of years—have now been observed to evolve in real time. This overview is by one of the foremost current researchers in the field.

Li, W.-H. and D. Graur. 1991. *Fundamentals of Molecular Evolution.* Sinauer Associates, Sunderland, MA. An accessible introduction to the dynamics of evolutionary change at the molecular level, this book describes the effects of various molecular mechanisms on genes—a driving force behind the evolutionary process.

Weiner, J. 1994. *The Beak of the Finch: A Story of Evolution in Our Time.* Alfred Knopf, New York. A popular account of the work of two biologists (Peter and Rosemary Grant) who have spent years recording the evolution of beak size among Galapagos finches. As its subtitle indicates, the thrust of the book is to show examples of evolution and speciation that have taken place in time spans that scientists can observe without having to rely on reconstructions of an incomplete fossil record.

Wilson, E. O. and W. H. Bossert. 1971. *A Primer of Population Biology.* Sinauer Associates, Sunderland, MA. A brief, self-teaching textbook designed to form a bridge between elementary textbooks and more advanced treatments of population genetics, evolutionary theory, and ecology.

A Whole New World
These tiny leaf beetles represent only five of the many species that were unknown to scientists before recent research on tropical forest tree canopies.

20

Species and Their Formation

If you were to walk through a tropical forest, you would probably be unaware of most of the organisms in the tops of the trees above you. The singing of birds, chattering of monkeys, and calling of cicadas and crickets would reach your ears, but many other small organisms cannot be heard. Only when biologists began propelling into the tops of trees a chemical mist that causes the smaller animals to fall to the ground did they become aware of the incredible richness of species (primarily insects) living high above the ground in tropical forests. Most of the individuals now being collected by this technique belong to previously unknown species. So many new species are being discovered by this and similar techniques that some biologists estimate that the number of species of organisms on Earth today may be as high as 30 to 50 million even though only 1.5 million have been described and named.

You already know that all these species, as well as those that lived in the past, are believed to be descendants of a single ancestral species. But what are species and how did these millions of species form? How does one species become two? What factors stimulate such splitting? How much time does species formation take? These and other related questions are the subject of this chapter.

WHAT ARE SPECIES?

Species means "kind." Many plants and animals, especially ones of the temperate zone, were classified as species by their appearances, or **morphologies**, more than 200 years ago by the Swedish biologist Carolus Linnaeus and his students. Among organisms in which males and females look very different, the two sexes were sometimes called different species. As soon as such individuals were discovered to be males and females of the same population, however, they were placed together in the same species.

An influential definition of species that incorporated the concept of shared reproduction was proposed by Ernst Mayr in 1940. He stated, "Species are groups of actually or potentially interbreeding natural populations which are reproductively isolated from other such groups." All parts of this definition identify important aspects of Mayr's **biological species concept**. The "groups" in this definition are collections of the local populations, or demes, that we discussed in Chapter 19. The words "actually or potentially" assert that, even if some members of a species are not in the same place and hence are unable to mate, they should not be placed in separate species if they would be likely to mate if they were together. The word *natural* is an important part of the definition because only *in nature* does the exchange of genes, which occurs within species, affect evolu-

20.1 Captivity Lowers Barriers to Interbreeding
Although horses and zebras do not interbreed in the wild, they have done so on this ranch in Kenya. The resulting hybrids, known as zebroids, display traits of both parental species.

tionary processes; the interbreeding of two different species that may occur in captivity does not (Figure 20.1).

Mayr viewed species as evolutionary units that are evolving separately from other such units, but he also included within one species geographically separated populations that are not exchanging genes, if they were judged to be capable of interbreeding. Other evolutionary biologists prefer not to unite within a single species populations that are evolving separately, even if they might interbreed if they were together.

Many species that were classified morphologically, when nothing was known about their reproductive behavior, fit Mayr's definition. This is not surprising because the members of an evolutionary unit share genes inherited from common ancestors. As a result, these individuals are likely to share many of the same alleles coding for their morphological traits, so their morphologies will be similar.

THE COHESIVENESS OF SPECIES

Species are cohesive units. Someone knowledgeable about a group of organisms, such as orchids or lizards, usually can determine the species of an individual simply by examining it superficially. The patterns of morphological similarities that unite groups of organisms and separate them from other groups

are familiar to all of you. For example, you can easily recognize as members of the same species redwinged blackbirds from New York and red-winged blackbirds from California (Figure 20.2). The standard field guides to birds, mammals, insects, and flowers are possible only because most species are cohesive units that change in appearance only gradually over large geographic distances.

Even though familiar, however, these patterns are quite surprising. Redwings from New York do not exchange genes with redwings from California. The large geographic range of the redwing is divided into many well-separated areas within which most individuals remain throughout their lives. Species within which self-fertilization is the rule, even though some sexual recombination occurs, also show cohesion. What maintains the cohesion of widespread or partially self-fertilizing species?

The most important factor promoting cohesion is **gene flow**, the migration of individuals resulting in their reproduction some distance from where they were born. As we saw in Chapter 19, even a very low rate of gene flow among populations can maintain the cohesion of a species because it prevents populations from diverging as a result of genetic drift or adaptation to local conditions. If natural selection favors similar organisms in different places, the effects of gene flow are reinforced because migrating individuals do not differ greatly genetically from individuals in the populations they join.

20.2 Redwings Are Redwings Everywhere
Both of these birds are obviously red-winged blackbirds, even though one (a) lives in California and the other (b) lives in New York.

(a)

(b)

HOW DO NEW SPECIES ARISE?

Evolution creates two patterns across time and space—vertical evolution and speciation. **Vertical evolution,** also called **anagenesis,** is change in a single lineage through time. With sufficient time, the changes may be so great that the descendants are given another species name even though no "new" species has formed. Anagenetic changes are a common feature of the fossil record. **Speciation** is the process by which one evolutionary unit splits into two units, which thereafter evolve as distinct lineages. Although Charles Darwin entitled his book *The Origin of Species*, he did not extensively discuss how a single species splits into two or more daughter species. Rather, he was concerned principally with demonstrating vertical evolution, that species are altered by natural selection over time.

The critical process in the formation of two species from one ancestral species is the separation of the gene pool belonging to the ancestral species into two separate gene pools. Then, within each isolated gene pool, allele and gene frequencies may change as a result of the action of evolutionary agents. If sufficient differences have accumulated during the period

of isolation, the two populations may not exchange genes when they again come together. Gene flow among populations may be interrupted in several ways, each of which characterizes a mode of speciation. These modes are the focus of the next three sections.

Allopatric Speciation

Often one species evolves into two daughter species because a physical barrier subdivides its populations into two segments. Such barriers include climate, water gaps for terrestrial organisms, dry land for aquatic organisms, and mountains. Barriers can form when continents drift, sea levels rise, or climates change. Alternatively, some members of a population may cross an existing barrier and establish a new population. Speciation that results when a population is divided by a barrier is known as **allopatric speciation** (from the Greek *allo* = "different" and *patris* = "country"), or **geographic speciation.**

Natural selection adapts populations to their environments (see Chapter 19). If the environments on the two sides of the physical barrier differ, different selective pressures may cause the populations to diverge genetically. Genetic drift may also bring about genetic changes. The two populations may also start with different mixtures of alleles (the founder effect; see Chapter 19). Differences that accumulate while a barrier is in place may become so large that the populations will fail to establish gene flow if the barrier later breaks down—that is, they will have come to be different species.

The unusual finches of the Galapagos Islands, 1,000 kilometers off the coast of Ecuador, demonstrate the importance of geographic isolation. Darwin's finches (as they are usually called because Darwin was the first scientist to study them) arose on the Galapagos Islands by speciation from a single South American species that colonized the islands. Today there are 14 species of Galapagos finches, all of which differ strikingly from the probable mainland ancestor (Figure 20.3). The islands of the Galapagos archipelago are sufficiently isolated from one another that the finches seldom migrate between them. Also, environmental conditions differ among islands: Some are relatively flat and arid; others have forested mountain slopes. Therefore, populations of finches on different islands have differentiated enough that when occasional immigrants arrive from other islands, they either do not breed with the residents or, if they do, their offspring do not survive as well as those produced by pairs composed of island residents. The genetic distinctiveness of different populations is thus maintained.

A barrier's effectiveness at preventing gene flow depends on the size and mobility of species. What is a firm barrier for a terrestrial snail may be no barrier at all to a butterfly or a bird. Populations of wind-pollinated plants are totally isolated at the maximum distance pollen may be blown by the wind, but individual plants are effectively isolated at much shorter distances. Among animal-pollinated plants, the extent of the barrier is the distance that pollinators travel while carrying pollen (see Chapter 34). Even animals with great powers of dispersal are often reluctant to cross narrow strips of unsuitable habitat. For animals that cannot swim or fly, narrow water-filled gaps may be effective barriers.

Peripheral populations of a species, situated at the outskirts of its range, are often geographically isolated from the main population. Because conditions at the edges of ranges are likely to differ from those toward the center of the range, isolated peripheral populations are likely to evolve differences from the main population. However, many peripheral populations do not remain isolated long enough to become new species because conditions at the margins of ranges are often unfavorable for their survival. Isolated peripheral populations thus have high extinction rates.

Parapatric Speciation

Parapatric speciation (from the Greek *para* = "beside") is the development of reproductive isolation among adjacent members of a population in the absence of a geographic barrier. Parapatric speciation occurs much less often than allopatric speciation because gene flow usually prevents differentiation between populations in contact. Occasionally, however, a species boundary forms where there is a marked change in environment, as occurs between populations of plants growing at the boundaries of different soil types.

Unusually abrupt changes in soil are created by mining activities that leave rubble (tailings) with high concentrations of heavy metals that are detrimental to plant growth. The soils developing on the tailings at the Goginian lead mine near Aberystwyth, Wales, for example, are highly contaminated with lead, but where the tailings end, they suddenly give way to normal rich pastureland (Figure 20.4). The pasture grass *Agrostis tenuis* is common to both sites, but there is a sharp gradient in lead tolerance among plants less than 20 meters apart. Plants on the mine tailings grow well under lead concentrations that would be lethal to plants growing just a few meters away. Nearly complete reproductive isolation exists between plants on contaminated and normal soil because they flower at different times. These two populations have not yet been designated as separate species, but reproductive isolation between them has already evolved, demonstrating that gene flow can slow or stop even in the absence of a distinct physical barrier.

Sympatric Speciation

In **sympatric speciation** (from the Greek *sym* = "with"), a gene pool becomes subdivided even though members of the daughter species are not physically separated during the speciation process. One means of sympatric speciation is **polyploidy**, a multiplication of the number of chromosomes (see Chapter 9). Polyploidy can arise by the duplication of the chromosomes of a single species or by the combination of chromosomes from two different species. Polyploid individuals of either type usually cannot exchange genes successfully with members of the parent populations because the polyploid individuals have different numbers of chromosomes than their parents have. Therefore, chromosomes from such matings cannot pair properly during the first metaphase of meiosis. As a result, the zygotes usually fail to develop properly. Thus, the matings either result in no offspring, or, if offspring are produced, they die before they mature.

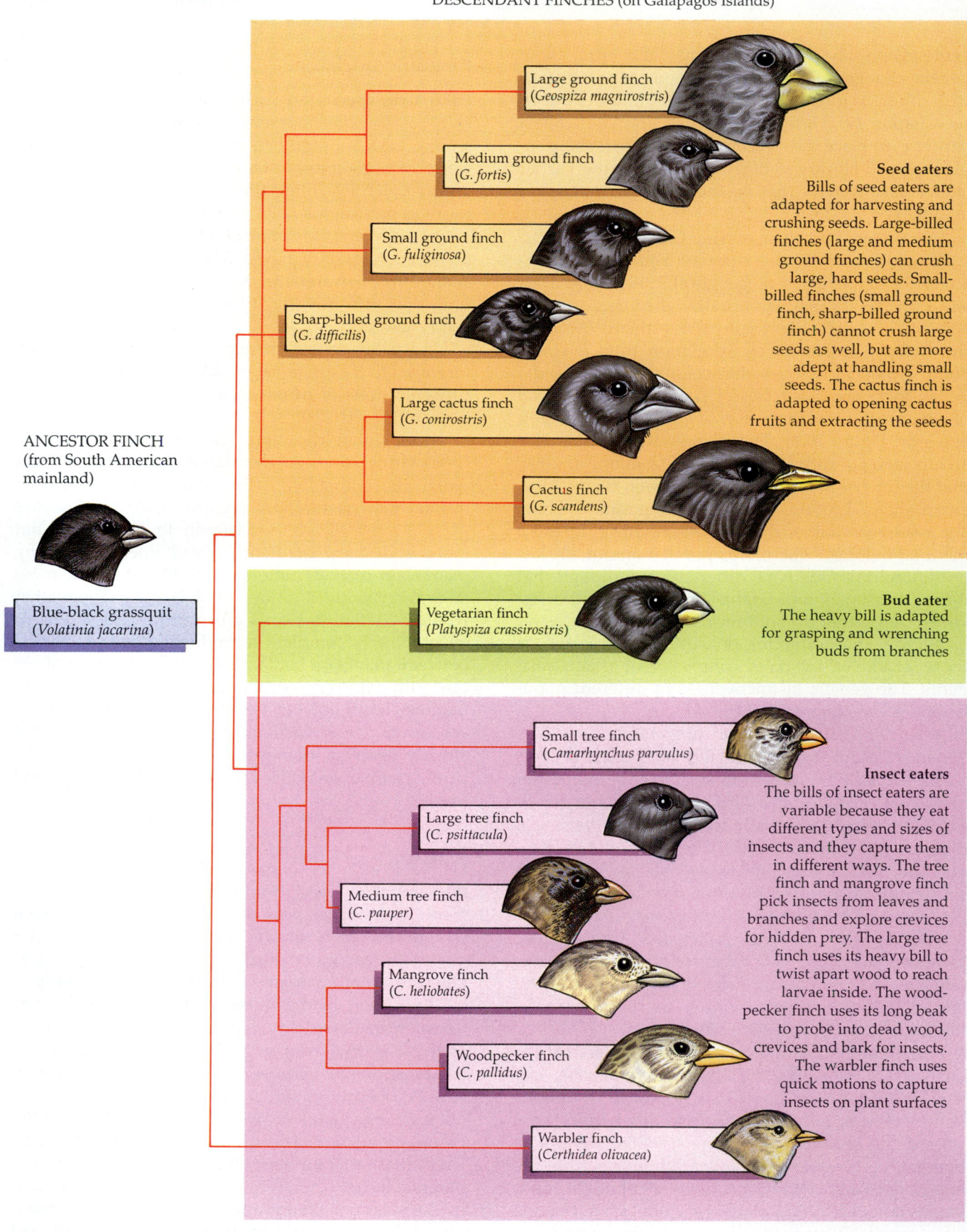

20.3 Evolution among Galapagos Finches
The descendants of the blue-black grassquits that colonized the Galapagos Islands several million years ago evolved into 14 species whose members are variously adapted to feed on seeds, buds, and insects.

450

20.4 Parapatric Speciation
Agrostis tenuis individuals growing on tailings of this Welsh mine are reproductively isolated from nearby individuals growing on uncontaminated.

Polyploidy can create a new species within one generation if the polyploid individuals can mate among themselves or self-fertilize. Plants can accomplish this task much more easily than animals can because plants of many species can reproduce by self-fertilization as well as by outcrossing (crossing with a relatively unrelated individual). If the polyploidy arises in several offspring of a single parent, the siblings can fertilize one another. Animals that have speciated by polyploidy either are parthenogenetic—that is, they consist of females that produce young from unfertilized eggs (see Chapter 37)—or they survived the initial generations probably by means of matings among siblings.

Speciation by polyploidy has been very important in the evolution of flowering plants. Botanists estimate that more than half of all species of flowering plants are polyploids. Most of these arose as a result of hybridization between two species, followed by self-fertilization. The importance of these processes and the speed with which they can produce new species are illustrated by salsifies (*Tragopogon*), members of the sunflower family. Salsifies are weedy plants that thrive in disturbed areas around towns. People have inadvertently spread them around the world from their ancestral ranges, which are in Eurasia.

Three diploid species of salsify were introduced into North America early in this century: *T. porrifolius*, *T. pratensis*, and *T. dubius*. Two tetraploid hybrids ("tetraploid" implies four sets of chromosomes)—*T. mirus* and *T. miscellus*—between species of the original three were first reported in 1950. Both hybrids have spread since their discovery and today are more widespread than their diploid parents (Figure 20.5).

Studies of their cells have revealed that both hybrid species have been formed more than once. Some populations of *T. miscellus*, for example, have the chloroplast genome of *T. pratensis*, whereas other populations have the chloroplast genome of *T. dubius*. Differences in the ribosomal genes of local populations of *T. miscellus* show that this species has evolved independently at least three times. Scientists seldom know the dates and locations of species formation so well. The success of newly formed hybrid species of salsifies illustrates why so many species of flowering plants originated as polyploids.

Among animals, sympatric speciation rarely happens by polyploidy, but it can result from the way species exploit resources. For example, many parasites and insect herbivores (plant eaters) attack only one or a few host species. Many insects find their mates on their host plants and subsequently mate and lay their eggs there. Individuals that colonize a new type of host plant will thus mate with each other—eventually forming new species—rather than with individuals still feeding on the original host. Sympatric speciation via plant host selection has happened among flies of the genus *Rhagoletis*. These flies attack fruits of trees and breed only when the fruits are mature. In North America, *Rhagoletis* species that attack native hawthorns occasionally spread to domestic fruit trees, where they often become host races (species adapted to a specific host). In 1864, for example, *R. pomonella* invaded apples in the Hudson River Valley of New York; it has since spread over most of the apple-growing regions of North America. In the 1960s a second host race of *R. pomonella* evolved on cherries in Door County, Wisconsin. The three races now live sympatrically in Door County, where they are reproductively isolated because the fruits of cherries, apples, and hawthorns do not ma-

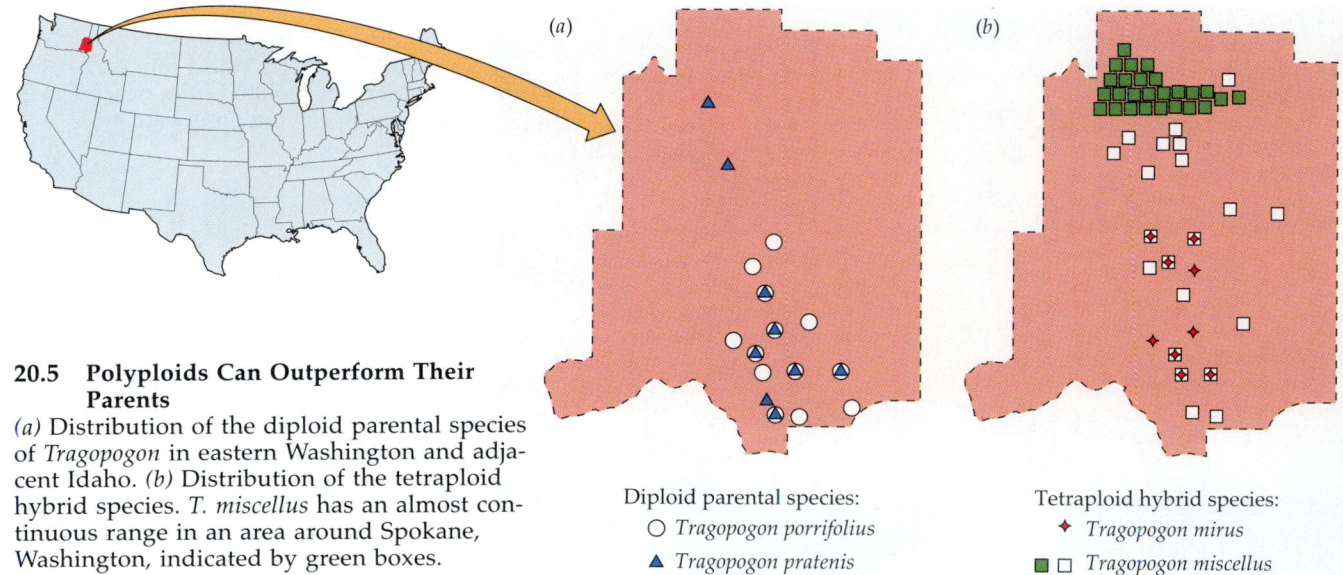

20.5 Polyploids Can Outperform Their Parents
(a) Distribution of the diploid parental species of *Tragopogon* in eastern Washington and adjacent Idaho. (b) Distribution of the tetraploid hybrid species. *T. miscellus* has an almost continuous range in an area around Spokane, Washington, indicated by green boxes.

Diploid parental species:
○ *Tragopogon porrifolius*
▲ *Tragopogon pratenis*

Tetraploid hybrid species:
✦ *Tragopogon mirus*
■ □ *Tragopogon miscellus*

ture at the same times. The extreme species richness among insects (see Chapters 26 and 48) may have resulted in part from sympatric speciation by means of host plant specialization.

How Important Is Each Speciation Mode?

The three modes of speciation—allopatric, parapatric, and sympatric—differ in their relative importance. We know that sympatric speciation has been very common among plants because there are so many polyploid plant species. The few examples of sympatric speciation that are known among animals have happened among small animals such as insects. Allopatric speciation is clearly the dominant mode of speciation among large animals. Because most animals do not self-fertilize or form polyploid populations, they lack traits conducive to providing reproductive isolation in sympatry. Parapatric distributions are common among animals, but they are probably caused by allopatric speciation followed by range expansion that puts the species in contact, rather than by parapatric speciation.

REPRODUCTIVE ISOLATING MECHANISMS

Any trait of an organism that prevents individuals of two different populations from producing fertile hybrids is called a **reproductive isolating mechanism**. Any two populations that are not producing fertile hybrids are reproductively isolated from each other, but they do not necessarily have isolating mechanisms. For example, although a geographic barrier may separate two populations, it is not an isolating mechanism because it is not a property of the organisms.

Reproductive isolating mechanisms fall into two categories depending on whether they take effect before or after fertilization. **Prezygotic mechanisms** lower the probability that hybrid zygotes will be formed, either by preventing or reducing mating between individuals of different species or by preventing fertilization of eggs if individuals of two species have mated. **Postzygotic mechanisms** prevent zygotes resulting from hybrid matings from developing into viable, fertile adults. Isolating mechanisms of both types prevent successful interbreeding between two populations. The most important isolating mechanisms are described here and summarized in Table 20.1.

GEOGRAPHIC ISOLATION. Individuals of different species may select different places in the environment in which to live. As a result, they may never come into contact during their respective mating seasons and are thus reproductively isolated by location.

TEMPORAL ISOLATION. Many organisms have mating periods that are as short as a few hours or days. If the mating periods of two species do not coincide, they will be reproductively isolated by time.

BEHAVIORAL ISOLATION. Mating behavior is stimulated in many animal species by courtship pheromones or scents (see Chapter 46). Males that do not perform the behavior appropriate for the species may not be accepted as mates by receptive females.

MECHANICAL ISOLATION. Differences in the sizes and shapes of reproductive organs may prevent the union of gametes from different species.

TABLE 20.1
Reproductive Isolating Mechanisms

TYPE OF MECHANISM	HOW IT ESTABLISHES A BARRIER
Prezygotic mechanisms	
Geographic	Individuals of the two populations do not come together because they live in different places
Temporal	Mating or flowering occur at different times of day or different seasons
Behavioral	Individuals of the other population are not acceptable as mates
Mechanical	Structural differences prevent individuals from mating
Chemical	Male and female gametes do not attract one another
Postzygotic mechanisms	
Hybrid inviability	Hybrid zygotes do not develop or they fail to reach sexual maturity
Hybrid sterility	Hybrids do not produce functional gametes
Hybrid weakness	Offspring of hybrids have reduced survival or reproductive rates

CHEMICAL ISOLATION. Sperm of one species may not be attracted to the eggs of another species because the eggs do not release the appropriate attractive chemicals, or the sperm may be unable to penetrate the egg because it is chemically incompatible.

HYBRID INVIABILITY AND INFERTILITY. If sperm of one species fertilize eggs of another species, the zygotes formed may be abnormal, or the hybrids may mature normally but be infertile when they in turn attempt to reproduce. For example, the offspring of matings between horses and donkeys—mules—are vigorous, but mules are sterile, thus producing no descendants (Figure 20.6). Polyploidy, which we discussed in the previous section, functions as a postzygotic isolating mechanism because chromosomes of polyploids cannot pair properly with those of their parents.

HYBRID ZONES

Sometimes when contact is reestablished between formerly geographically isolated populations, so few differences have accumulated that the individuals interbreed freely with members of the other population and produce offspring that are as successful as those resulting from matings within the population. The offspring of parents from genetically dissimilar populations are called **hybrids** (see Chapter 10). If successful hybrids spread through both populations and reproduce with other individuals, the gene pools combine quickly, and no new species result from the period of isolation.

Alternatively, rather than the combining of gene pools throughout the two populations, interbreeding may occur only where the two populations come into contact, resulting in a **hybrid zone.** For example, the ranges of two species of salamanders, *Plethodon jordani* and *P. glutinosus*, are continuous throughout the southern Appalachian Mountains, but they hybridize

20.6 Sturdy but Sterile
Mules are widely used as pack animals because of their stamina. For that purpose, their infertility is unimportant.

(a)

(b)

(c)

20.7   A Hybrid Zone Moves Upward
Plethodon jordani (a) and *Plethodon glutinosus (b)* mate and
produce hybrid offspring *(c)* in several locations in the
southern Appalachians. As the climate changes, the hybrid zone shifts to a higher elevation.

in a few places (Figure 20.7). One hybrid zone occurs
between the elevations of 685 and 945 meters in the
Nantahala Mountains of North Carolina. These salamanders are terrestrial and highly sedentary; individuals generally move less than a few meters
throughout their lives. *P. glutinosus* survives better
in warmer and drier sites (lower elevations), *P. jordani*
in cooler and wetter sites (higher elevations). The
hybrid zone is shifting gradually to a higher elevation, probably as a result of the poorer survival of *P.
jordani* individuals because the climate of the region
is gradually becoming warmer and drier. The hybrid
zone remains narrow, both because the salamanders
are highly sedentary and because hybrids survive
well only in a narrow transition zone between sites
favored by each of the parent species.

To determine what happens when formerly separate populations come together, studies ideally begin
when contact is first established. An opportunity to
observe the formation of a hybrid zone is provided
by blue and snow geese (Figure 20.8). These geese
breed in Arctic North America and spend the winter
in the southern United States. Birds with white plumage (snow geese) dominate breeding populations
in the West; birds with dark plumage (blue geese)
dominate in the East. Historical evidence shows that
the two color forms were almost completely sepa-

rated geographically until the third decade of this
century. The recent hybrid zone is due to a change
in the winter feeding ranges of the birds. Birds of
both types now winter in large flocks in the rice-
growing regions of inland Texas and Louisiana. The
geese select mates while on the wintering grounds,
and pairs migrate to nest on the breeding grounds
from which the female came. Interbreeding is becoming common between the two forms; thus a hybrid
zone is developing in a small region of the Canadian
Arctic. Biologists are monitoring the spread of this
hybrid zone to determine whether isolating mechanisms are developing or whether the zone will continue to spread.

20.8   A New Hybrid Zone Forms
Blue and snow geese are forming mixed pairs because the
species now winter together in Louisiana rice fields.

(a)

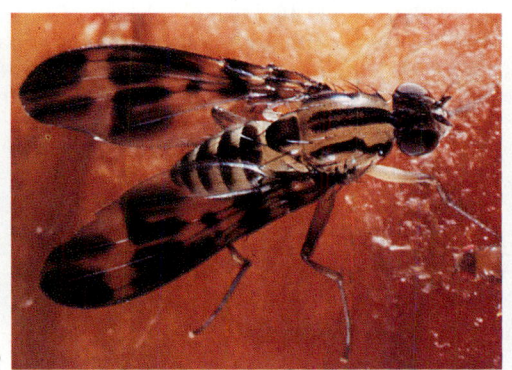

(b)

(c)

20.9 Morphologically Different, Genetically Similar Although these fruit flies, a small sample of the hundreds of species found only on the Hawaiian Islands, are extremely variable in appearance, they are genetically almost identical. *(a) Drosophila sylvestris. (b) D. conspicua. (c) D. balioptera.*

HOW MUCH DO SPECIES DIFFER GENETICALLY?

Sympatric species can maintain their distinctness even if they have not diverged very much genetically. Small genetic differences can cause morphological and physiological differences that lower the success of hybrids or prevent individuals from being acceptable as mates to members of the other species. Evidence that the genetic changes accompanying speciation may be small is provided by the many species of *Drosophila* on the Hawaiian Islands. Flies of all species of Hawaiian *Drosophila* share nearly all of their alleles. Most morphological differences *among* the species are based upon variability already present *within* each of the species. All of the hundreds of species of this genus that have evolved in Hawaii during the past 40 million years, even those that have diverged morphologically, are relatively similar genetically (Figure 20.9).

Other research confirms that the differences among species generally are similar in type to the differences within species. There is no compelling evidence that evolution of the types of differences that separate most closely related species requires any mechanisms or processes other than those known to operate within species.

HOW LONG DOES SPECIATION TAKE?

There is no general rule about how much time speciation requires. A population may speciate by polyploidy within one breeding season. At the other extreme, some populations that have been isolated for millions of years remain reproductively compatible. American and European sycamores have been iso-

lated from one another for at least 20 million years. They are, nonetheless, morphologically very similar (Figure 20.10), and they can form fertile hybrids. Other examples cover virtually the entire range between these two extremes. For example, a strain of *Drosophila paulistorum* could produce fertile hybrids with other strains when it was first collected. After being isolated in the laboratory for only a few years, however, its matings with other strains were sterile. The several species of cichlid fishes that are found only in Lake Nabugabo in Africa and that are reproductively isolated are another example of relatively rapid speciation. These species must have formed quickly because Lake Nabugabo has existed as an independent lake for only 4,000 years. Species may form quickly or very slowly, but what factors affect the rate of species formation?

Behavior and Speciation Rates

Animals with complex behavior probably speciate more quickly than simpler ones. An important characteristic of behaviorally complex organisms is that they make sophisticated discriminations among potential mating partners. Not only do they distinguish members of their own species from members of other

20.10 Geographically Separated, Morphologically Similar
Although they have been separated on different continents for at least 20 million years, (a) American sycamores (*Platanus occidentalis*) and (b) European sycamores (*P. hispanica*) are similar in appearance and can form fertile hybrids.

(a)

(b)

species, but they also make subtle discriminations within species on the basis of size, shape, appearance, and behavior. These discriminations may be based on the quality of the genes of the potential partner, the quality of parental care likely to be given, or both. Such behavioral discrimination can greatly amplify differences in fitness already caused by the interactions of the pool of potential partners with the physical environment, prey, or predators. Therefore, mate selection—a form of nonrandom mating—is probably a major cause of rapid evolution in general and of reproductive isolation between species.

Ecology and Speciation Rates

The fossil record shows that some lineages give rise to many more species than do others. The large, hooved mammals of Africa show how evolutionary agents may influence the rate of accumulation of differences that affect speciation rates. These mammals are abundant on African savannas (grassy plains) today, and we have a good fossil record of them going back millions of years. A great deal is known about feeding behavior, habitat specificity, population size, and social organization—in short, the ecology—of the living species.

Different lineages of these large mammals varied markedly in their rate of speciation. Speciation rates could be correlated with birth rates because animals with higher birth rates produce more offspring per generation upon which natural selection can act. No pattern among the birth rates of current members of these lineages supports this hypothesis, however. Speciation rates *are* correlated with diet: Grazers (which eat grass and other nonwoody plants) and browsers (which eat branches and leaves of woody plants) speciate faster than omnivores (which eat both plants and animals) and anteaters (Figure 20.11).

The grazers and browsers require large expanses of open grasslands and woodlands, respectively. In Africa, these resources disappeared from and reappeared in large areas during periods of climatic change, isolating populations and causing both high extinction rates and high rates of differentiation among populations between these isolated regions. Omnivores and anteaters, on the other hand, maintained more continuous populations during these climatic changes, so gene flow continued and reproductive isolation was not established.

This evolutionary pattern is consistent with the action of natural selection within lineages, but natural selection cannot be proposed as an explanation from short-term studies of living populations alone. Only by combining information about the ecology of living species with data from the fossil record can we understand patterns of speciation and long-term evolution of these mammals. This example also shows how some groups may show stasis, long periods during which they change little, while other lineages in the same environment are evolving rapidly.

Evolutionary Radiation and Speciation Rates

The fossil record and current distributions of organisms together reveal that at certain times in some lineages, speciation rates have been especially high, giving rise to a large number of daughter species. Such **evolutionary radiations** have occurred on all continents, but the resulting species are much more obvious on island archipelagos. Because many organisms disperse poorly across large water-filled gaps, islands lack many plant and animal groups found on the mainland. This situation creates ecological opportunities that may stimulate rapid evolutionary changes. Because the water barriers among the islands restrict gene flow, populations on different is-

1   Wildebeest
2   Topi
3   Warthog
4   Kob
5   Duiker
6   Patas monkey
7   Kudu
8   Giraffe
9   Springbok
10   African buffalo
11   Klipspringer
12   Impala
13   Spring hare
14   Eland
15   African bush pig
16   Duiker
17   Hamadryas baboon
18   Aardvark

20.11   Speciation Rates in Some African Mammals
Grazers and browsers have speciated more rapidly than omnivores and anteaters. (Animals belonging to two different genera are both commonly referred to as duikers—a phenomenon discussed further in Chapter 21. Thus duikers appear as both browsers and omnivores.)

lands can evolve adaptations to their local environments. Together these two factors make it likely that when individuals do migrate between islands, they will not be able to interbreed with individuals in the places where they arrive.

The most remarkable evolutionary radiations have occurred in the Hawaiian archipelago, the most isolated islands in the world. The Hawaiian Islands lie 4,000 km from the nearest major land mass and 1,600 km from the nearest group of islands. The islands are arranged in a line of decreasing age—the youngest islands to the southeast, the oldest to the northwest. The native biota of the Hawaiian islands includes 1,000 species of flowering plants, 10,000 species of insects, 1,000 land snails, and more than 100 birds. However, there were no amphibians, no terrestrial reptiles, and only one native mammal—a bat—until humans introduced additional species. The 10,000 known native species of insects on Hawaii are believed to have evolved from only about 400 immigrant species; only seven immigrant species are believed to be necessary to account for all the native Hawaiian land birds.

More than 90 percent of all plant species on the Hawaiian islands are **endemic**—that is, they are found nowhere else. Several groups of flowering plants have more diverse forms and life histories on the islands and live in a wider variety of habitats than do their close relatives on the mainland. An outstanding example is the group of sunflowers called silverswords and tarweeds (genera *Argyroxiphium*, *Dubautia*, and *Wilkesia*). The 28 species in the silversword group are believed to be derived from a single species of tarweed from the Pacific coast of North America. Whereas all mainland tarweeds are small, upright, nonwoody plants (herbs), Hawaiian silversword species include prostrate and upright herbs, shrubs, trees, and vines (Figure 20.12). They occupy nearly all the habitats of the islands, from sea level to above timberline in the mountains. Despite their extraordinary diversification, however, the silverswords have differentiated very little genetically.

The island silverswords are more diverse in size and shape than the mainland tarweeds because the colonizers arrived on islands that had very few plant species. In particular, there were few trees and

(a)

(b)

(c)

20.12   Rapid Evolution among Hawaiian Plants
Silverswords and tarweeds in three closely related genera of the sunflower family (Asteraceae) are believed to have descended from a single herbaceous ancestor that colonized Hawaii from the Pacific coast of North America. (a) *Argyoxiphium sandwichense.* (b) *Wilkesia gymnoxiphium.* (c) *Dubautia laxa.* Their rapid evolution makes them appear more distantly related than they actually are.

shrubs because such large-seeded plants only rarely migrate to oceanic islands; many island trees and shrubs have evolved from nonwoody ancestors. On the mainland, however, tarweeds have lived in ecological communities that contain tree and shrub lineages older than their own. Therefore, opportunities to exploit the tree way of life were already preempted.

An even more spectacular evolutionary radiation took place among the cichlid fishes of Lake Victoria, East Africa. The more than 300 species of cichlids in Lake Victoria are believed to be descendants of a single species that colonized Lake Victoria from other, older East African lakes. Using DNA sequences from the cytochrome *b* gene as a molecular clock, scientists calculated that the radiation of Lake Victoria cichlids was accomplished in about 200,000 years. Among the many types of cichlids in the lake are generalized bottom feeders, thick-lipped insect eaters, snail and clam crushers, and fish eaters (Figure 20.13). Scientists do not know how gene flow was disrupted hundreds of times between cichlid populations long enough to produce reproductive isolation within a single lake.

Some biologists believe that most evolutionary change takes place at the time of speciation. They suggest that the isolation of small populations, together with random changes in allele frequencies, disrupts the integrated functioning of the genome and allows for rapid responses to agents of evolution. According to this view, once the speciation process has been completed, the better-integrated new genotypes resist change, leading to **stasis**—long periods of little change that are interrupted only by the next round of speciation. The fossil record reveals examples of both stasis and long-term gradual evolution, but the record is too incomplete to tell us how often these patterns happen.

THE SIGNIFICANCE OF SPECIATION

The result of speciation processes operating over billions of years is a world in which life is organized into millions of species, each adapted to live in a particular place and to use environmental resources in a particular way. The world would be very different if speciation had been a rare event in the history of life. How the millions of species are distributed over the surface of Earth and organized into ecological communities will be a major focus of Part Seven of this book. Remember, however, that the differences among species usually arise as traits that adapt populations to their environments, not as devices for reproductive isolation. Some of the traits that adapt populations to their environments may also function as isolating mechanisms if the populations come together again, but this result is an incidental by-product of evolution. When and where speciation will happen is not evident from studies of microevolutionary changes because such studies cannot predict when individuals will cross existing barriers or where or when new barriers will be imposed and because many differences between diverging populations may be caused initially by genetic drift.

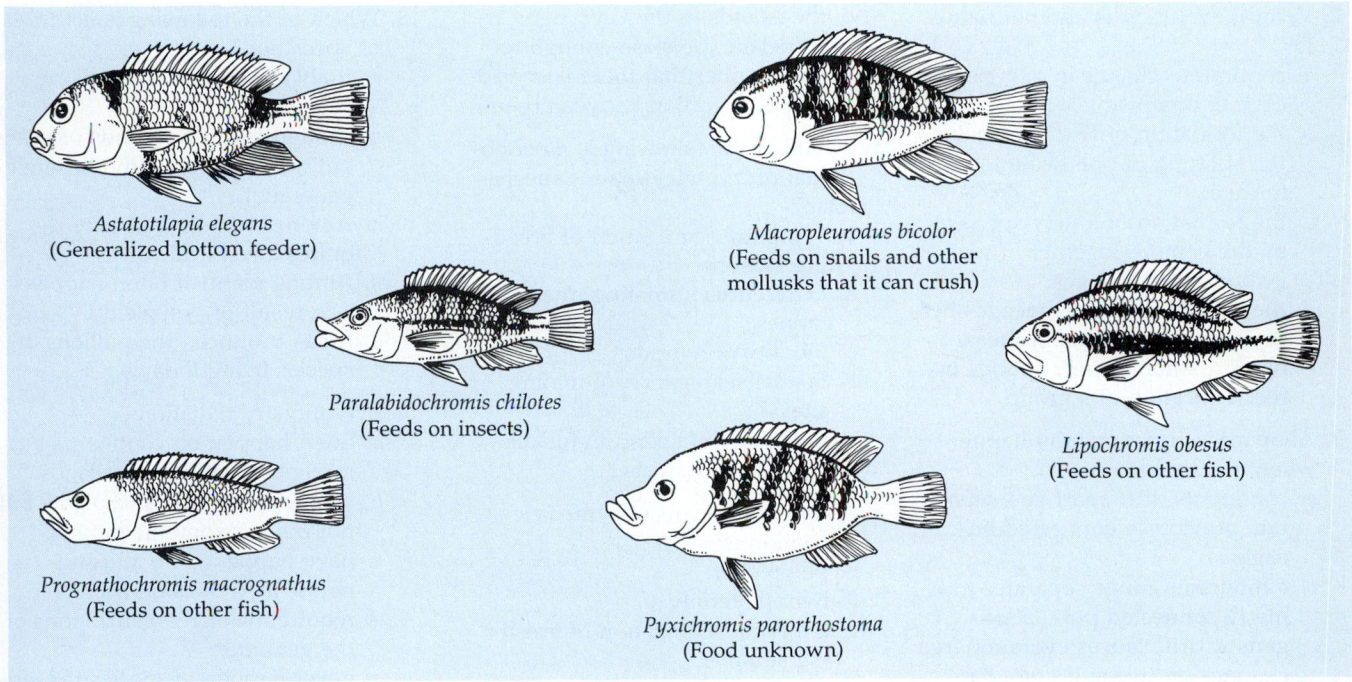

Astatotilapia elegans
(Generalized bottom feeder)

Macropleurodus bicolor
(Feeds on snails and other
mollusks that it can crush)

Paralabidochromis chilotes
(Feeds on insects)

Lipochromis obesus
(Feeds on other fish)

Prognathochromis macrognathus
(Feeds on other fish)

Pyxichromis parorthostoma
(Food unknown)

20.13 Lake Victoria Cichlids Have Radiated Rapidly
All of these fish species evolved from a common ancestor within the past
200,000 years. Their most striking differences adapt them for feeding on different types of food.

SUMMARY of Main Ideas About Speciation

Speciation—the splitting of one species into two separate species—is a central evolutionary process.

A species is a group whose members can mate to produce fertile offspring but are isolated reproductively from other groups in nature. Species are cohesive units even when they occasionally hybridize with one another.

There are three modes of speciation: allopatric, parapatric, and sympatric. Allopatric speciation is the most important means of speciation among animals.
Review Figures 20.3

Species may be formed quickly by a multiplication of chromosome numbers because polyploid off-spring are sterile in crosses with members of the parent species. Polyploidy has been a major factor in plant speciation.
Review Figure 20.5

Polyploidy is rare among animals.

Reproductive isolation may be due to prezygotic or postzygotic isolating mechanisms or both.
Review Table 20.1

As a result of speciation, new lineages are created that henceforth evolve genetically independently.

Evolutionary radiation may permit populations to adapt rapidly to new environments, resulting in many subsequent speciation events.
Review Figures 20.11, 20.12, and 20.13

SELF-QUIZ

1. A biological species is
   a. a group of actually interbreeding natural populations that is reproductively isolated from other such groups.
   b. a group of potentially interbreeding natural populations that is reproductively isolated from other such groups.
   c. a group of actually or potentially interbreeding natural populations that is reproductively isolated from other such groups.
   d. a group of actually or potentially interbreeding natural populations that is reproductively connected to other such groups.
   e. a group of actually interbreeding natural populations that is reproductively connected to other such groups.

2. Vertical evolution is another name for
   a. continuous change in a single lineage of organisms.
   b. the formation of two species by the splitting of one evolutionary lineage.
   c. the formation of a new species by the coming together of two evolutionary lineages.
   d. the reduction of two lineages by the extinction of one of them.
   e. the formation of new species by reclassification of a group.

3. Allopatric speciation may happen when
   a. continents drift apart and separate previously connected lineages.
   b. a mountain range separates formerly connected populations.
   c. genetic drift causes evolutionary changes on two sides of a barrier.
   d. the range of a species is separated by loss of intermediate habitat.
   e. all of the above.

4. Finches speciated on the Galapagos Islands because
   a. the Galapagos islands are a long way from the mainland.
   b. the Galapagos Islands are very arid.
   c. the Galapagos Islands are small.
   d. the islands in the Galapagos archipelago are sufficiently isolated from one another that there is little migration between them.
   e. the islands in the Galapagos archipelago are close enough to one another that there is considerable migration between them.

5. Which of the following is not a potential prezygotic isolating mechanism?
   a. Temporal segregation of breeding seasons
   b. Differences in mating pheromones
   c. Sterility of hybrids
   d. Spatial segregation of mating sites
   e. Inviability of sperm in female reproductive tracts

6. A common means of sympatric speciation is:
   a. polyploidy.
   b. hybrid sterility.
   c. temporal segregation of breeding seasons.
   d. spatial segregation of mating sites.
   e. imposition of a geographical barrier.

7. Sympatric species are often very similar in appearance because
   a. appearances are often not influenced by natural selection.
   b. genetic changes accompanying speciation are often small.
   c. genetic changes accompanying speciation are usually large.
   d. speciation usually requires major reorganization of the genome.
   e. the traits that differ among species are not the same as the traits that differ among individuals within species.

8. Which of the following is not true of speciation?
   a. It always takes thousands of years.
   b. It often takes thousands of years but may happen within a single generation.
   c. Among animals it usually requires a physical barrier.
   d. Among plants it often happens as a result of polyploidy.
   e. It has produced the millions of species living today.

9. Evolutionary radiations
   a. often happen on continents but rarely on island archipelagos.
   b. characterize birds and plants but not other taxonomic groups.
   c. have happened on all continents, as well as on islands.
   d. require major reorganizations of the genome.
   e. never happen in species-rich environments.

10. Speciation is often rapid within lineages in which species have complex behavior because
   a. individuals of such species make very fine discriminations among potential mating partners.
   b. such species have short generation times.
   c. such species have high reproductive rates.
   d. such species have complex relationships with their environments.
   e. none of the above.

FOR STUDY

1. Gene exchange between populations is prevented by geographic isolation, by behavioral responses before mating (for example, females rejecting courting males of other species), and by mechanisms that function after mating has occurred (for example, hybrid sterility). All of these are commonly called isolating mechanisms. In what ways are the three types very different? If you were to apply different names to them, which one would you call an isolating mechanism? Why?

2. The blue goose of North America has distinct color morphs—a blue one and a white one. As we have seen, they can interbreed, and matings between the two color types are common. On their breeding grounds in northern Canada, however, blue individuals mate with blue individuals much more frequently than would be expected by chance. Suppose that 75 percent of all mated pairs consisted of two individuals of the same morph, what would you conclude about speciation processes in these geese? If 95 percent of pairs were of the same morph? If 100 percent of pairs were of the same morph?

3. Although many species of butterflies are divided into local populations among which there is little gene flow, these butterflies often show relatively little geographic variation. Describe the studies you would carry out to determine what maintains this morphological similarity.

4. Distinguish among the following terms: allopatric speciation, parapatric speciation, and sympatric speciation. For each of the three statements below, indicate which type of speciation is implied:

a. This process occurs most commonly in nature as a result of polyploidy.

b. The present sizes of national parks and wildlife preserves may be too small to allow this type of speciation to occur among organisms restricted to those areas.

c. This process generally occurs in species inhabiting areas where sharp environmental contrasts exist.

5. Evolutionary radiations are common and easily studied on oceanic islands. In what types of mainland situations would you expect to find major evolutionary radiations? Why?

READINGS

Atchley, W. R. and D. S. Woodruff (Eds.). 1981. *Evolution and Speciation: Essays in Honor of M. D. J. White.* Cambridge University Press, New York. A rich collection of essays on many aspects of speciation.

Bush, G. L. 1975. "Modes of Animal Speciation." *Annual Review of Ecology and Systematics,* vol. 6, pages 339–364. A thorough review of the types of speciation and the evidence for them by a strong proponent of sympatric speciation. Contains an extensive bibliography.

Endler, J. T. 1977. *Geographic Variation, Species, and Clines.* Princeton University Press, Princeton, NJ. A theoretical analysis of how sympatric and parapatric speciation might occur.

Mayr, E. 1970. *Populations, Species, and Evolution.* Harvard University Press, Cambridge, MA. An abridged version of the most thorough work on speciation theory as applied to animals.

Otte, D. and J. A. Endler (Eds.). 1989. *Speciation and Its Consequences.* Sinauer Associates, Sunderland, MA. A comprehensive collection of essays on concepts, methods, and consequences of speciation. Includes general treatments and analyses of specific cases.

Stebbins, G. L. 1950. *Variation and Evolution in Plants.* Columbia University Press, New York. Although now dated, this is one of the great classics on speciation, full of many examples of speciation among plants.

An Escape Mechanism
This lizard may shed its tail in an attempt to escape from the hawk. The physiological mechanisms by which it achieves this are different from the mechanisms a salamander uses.

21

Systematics and Reconstructing Phylogenies

When attacked by predators, such as owls or snakes, some salamanders and lizards can disengage their tails. The separated tail then whips back and forth dramatically while the animal remains motionless. Predators are attracted by the moving tail and eat it rather than the whole animal.

Although both salamanders and lizards can detach, or autotomize, their tails, they do not use the same mechanism to do so. Lizards achieve tail autotomy by a breakage within vertebrae, whereas in salamanders the breakage is between vertebrae. In addition, tail autotomy has evolved independently several times among salamanders, each group of which uses a different combination of structural adaptations. If scientists did not know the structural details of tail autotomy, they would probably conclude, incorrectly, that tail autotomy had evolved only once. How can multiple origins of traits such as tail autotomy be distinguished from a single origin? How do we determine evolutionary histories and relationships among organisms? How can we express those relationships in classification systems that can be used as a foundation for further studies of those organisms?

Systematics, the science that addresses these questions, has ancient roots. People have long been interested in the richness of the living world and have attempted to understand its origins and maintenance. At first, the study of biological diversity was motivated by purely practical reasons: to determine which plants and animals might be useful sources of food, medicine, and other products—a motivation that continues today. During the seventeenth and eighteenth centuries, the study of nature was stimulated in Western culture by the desire to reveal the hidden order and harmony thought to have been provided by God. Because God was assumed to have had a plan, it followed that the diversity of living organisms must obey some general laws and that classifying the organisms might reveal these laws, at least in part. These early attempts at classification led to the complex systems we use today.

In this chapter we consider the goals and methods of modern systematics, show how biologists use classification systems to express evolutionary relationships—the **phylogeny** of organisms—among living and fossil organisms, and illustrate how knowledge of evolutionary relationships is used to solve other biological problems.

THE IMPORTANCE OF CLASSIFICATIONS

Classification systems improve our ability to explain relationships among things. For biologists, this capability is especially important when we attempt to

21.1 Many Different Plants Are Called Bluebells
(a) These North Dakotan flowers, a species of *Campanula*, are usually called bluebells. (b) In Virginia, these *Mertensia* are also called bluebells. (c) This English bluebell is *Endymion nonscriptus*. None of these plants is closely related to the others.

reconstruct the evolutionary pathways that have produced the diversity of organisms living today. Classification systems can also be an aid to memory. It is impossible to remember the characteristics of many different things unless we can group them into categories based on shared characteristics. Such classifications improve our predictive powers. If we know, for example, that females of all known mammalian species have mammary glands with which they produce milk for their offspring, we can be quite certain that a newly discovered animal that has a uniquely mammalian combination of other traits, such as hair and a constant, high body temperature, will also have this method of feeding its offspring, even if the first individuals we happen to find are males that lack functional mammary glands.

Classification systems provide unique names for organisms. If the names are changed, the systems provide means of tracing the changes. Common names, even if they exist (most organisms have no common names), are very unreliable and often confusing. For example, plants called "bluebells" are found in England, Scotland, Texas, and the Rocky Mountains—but none of the bluebells in any of those places is closely related evolutionarily to the bluebells in any of the other places (Figure 21.1). Fish called pickerel are prized for eating in central Ontario, Canada, whereas to the south, in the Great Lakes region, pickerels are regarded as undesirable for the table. The inconsistency is due to the fact that around the Great Lakes the name pickerel is applied to a fish species that is called the grass pike in central Ontario. In neither location is that species regarded as good eating. A different species is called pickerel in central Ontario. These cases emphasize the need for formal, unique names for organisms.

Recognizing and interpreting similarities and differences among organisms is easier if the organisms are classified into groups that are ordered and ranked. Any group of organisms treated as a unit in a classification system is called a **taxon** (plural: taxa). **Taxonomy** is the theory and practice of classifying organisms. **Systematics**, a larger field of which taxonomy is a part, is the scientific study of the diversity of organisms. Systematists use classification systems to help understand how organisms evolved as they did and to communicate their knowledge to other scientists.

THE HIERARCHY OF THE LINNAEAN SYSTEM

The biological classification system that is used today is based on the work of the great Swedish biologist Carolus Linnaeus. In the Linnaean system, each species is assigned two names, one identifying the species itself and the other the genus to which the species belongs (Figure 21.2). A **genus** (plural: genera; adjectival form: generic) is a group of closely related species. In many cases the name of the taxonomist who first proposed the species name is added at the end. Thus, *Homo sapiens* Linnaeus is the name of the modern human species. *Homo* is the genus to which the species belongs, and *sapiens* identifies the species; Linnaeus proposed the species name *sapiens*. You can think of the generic name *Homo* as being equivalent to your surname and the specific name *sapiens* as being equivalent to your first name. This two-name system, referred to as **binomial nomenclature**, is universally employed in biology. The generic name is always capitalized; the species name is not. Both names are always italicized, whereas common names are not. Reference to more than one species in a genus is expressed with the abbreviation spp. after the generic name (for example, *Drosophila* spp.); the abbreviation sp. is used after a generic name if the identity of the species is uncertain. Rather than repeating a generic name when it is used several times in the same discussion, biologists often spell it out only once and abbreviate it to the initial letter thereafter (for example, *D. melanogaster* is the abbreviated form of *Drosophila melanogaster*).

In the Linnaean system, species are grouped into higher taxonomic categories. The number and limits of these categories are based on taxonomists' judgments of evolutionary uniqueness among organisms. Thus, a higher taxonomic category may have only a single species if that species has evolved independently from all other species for a very long time. Some genera, on the other hand, contain hundreds of species.

Theoretically, genera and higher taxonomic categories could be delineated by the length of time since the taxa last shared a common ancestor. For example, genera might be separated by, say, 15 million years of independent evolution; higher taxonomic categories by still longer times. Usually, however, we do not know how long various lineages have been evolving separately, so this method of establishing higher taxonomic categories is not common.

The category (taxon) above the genus in the Linnaean system is the **family**. The names of animal families end in the suffix -idae. Thus Formicidae is the family that contains all ant species, and Hominidae contains humans, a few of our fossil relatives, and chimpanzees and gorillas. Family names are

AVES ACCIPITRES. Vultur.

I. ACCIPITRES.

Roſtrum e Mandibula ſuperiore denticulum utrinque exſerens.

40. **VULTUR.** *Roſtrum* rectum, apice aduncum. *Caput* impenne, antice nuda cute. *Lingua* bifida.

Gryphus. 1. V. maximus, caruncula verticali longitudine capitis. †
Vultur Gryps Gryphus. *Klein. av.* 45.
Cuntur. *Raj. av.* 11.
Habitare fertur Chili.
Rara avis in terris; mihi ignota; videſis Kleinium.

Harpyja. 2. V. occipite ſubcriſtato. †
Yzquauhtli *Hern. mex. pp.* 34.
Aquilæ criſtatæ genus. *Raj. av.* 161.
Habitat in Mexico.
Magnitudo *arietis.* Collum, Dorſum, Cauda, Criſta *ſurrecta nigra;* Subtus *nigro candidoque fulvo mixta. Homines etiam adoritur. Hernand.*
Cfr. Aquila mexicana coronata. Roſtrum *Vulturis.* Oculi *membrana nictitante. Sub ingluvie pennæ albæ, quas iratus dimittit usque ad pedum digitos. Alæ Caudaque ſubtus albo nigroque punctatæ colore Tigridis. Erectus conſidet.* Pennas occipitis ſæpius erigit in formam coronæ. *Fertur unico ictu cranium hominis iratus findere. Viſus Madriti in Vivario Regio a Z. Hallman.*

Papa. 3. V. naribus carunculatis, vertice colloque denudato.
Vultur elegans. *Edv. av.* 2 *t.* 2.
Vultur. *Alb av.* 2. *p.* 4. *t.* 4.
Habitat in India *occidentali.*
Obſ. Vultur. *Alb. av.* 3. *p.* 1. *t.* 1. *an femina hujus?*
Caput & Collum *quaſi excoriata retrahere poteſt intra vaginam cutis plumoſæ colli inferioris.*

Aura. 4. V. fuſcogriſeus, remigibus nigris, roſtro albo.
Tzopilotle ſ. Aura. *Hernand. mex.* 331.

21.2 Part of Linnaeus's Classification of Vultures
This page from the tenth edition of *Systema Naturae* (1758) shows the descriptions Linnaeus provided for the species he named.

based on the name of a member genus. Formicidae is based on *Formica*, and Hominidae is based on *Homo*. Plant classification follows the same procedures except that the suffix -aceae is used instead of -idae. Thus Rosaceae is the family that includes the genus of roses (*Rosa*) and its immediate relatives. Unlike generic and species names, family names are not italicized, but they are capitalized.

Families are, in turn, grouped into **orders**, and orders are grouped into **classes**. Classes of animals and protists are grouped into **phyla** (singular: phylum). (The phyla of plants, fungi, and bacteria were formerly called divisions.) Phyla are grouped into **kingdoms**. The hierarchical units of this classification system, as applied to the blackburnian warbler (*Dendroica fusca*) and the moss rose (*Rosa gallica*) are shown in Figure 21.3.

CLASSIFICATION SYSTEMS AND EVOLUTION

Biological classification systems are designed to express relationships among organisms. Different systems express different relationships and thus are based on different features. If, for instance, we were interested in a system that would help us decide what plants and animals were desirable as food, we might devise a classification based on tastiness, ease of capture, and the type of edible parts each organism possessed. Early Hindu classifications of plants were designed according to such criteria. Hindu classifications of animals were made from several points of view, such as method of reproduction, habitat, mode of life, usefulness to people, and number of senses. One ancient system divided animals into (1) those with placentas, (2) those formed from eggs, (3) those that generated spontaneously, and (4) those born of vegetable matter. Another system divided animals into (1) those born of moisture and heat, (2) those that bear live young, (3) those that lay eggs, and (4) those that burst forth from the ground.

We do not use such systems today, but they served the needs of the people who developed them. It is inappropriate to ask whether those classifications, or any others, including contemporary ones, are right or wrong. Classification systems can be judged only in terms of their utility and consistency with their stated goals. To evaluate any classification system we must first ask, What is it trying to accomplish? Then we can ask, How well does it accomplish its objectives?

Reclassifying Organisms

Many organisms were given species names by Linnaeus and his followers before evolution became widely accepted as biology's central concept. These workers, who had only the morphological traits of organisms to work with, classified each species they described by placing it in a genus, a family, an order, a class, a phylum, and a kingdom. As evolutionary relationships began to be understood, a process of reclassifying organisms to reflect these relationships began, and this process continues today. Many organisms that still carry "Linnaeus" after their species names are now in different genera, families, orders, classes, phyla, and even kingdoms than the ones in which they were originally placed by Linnaeus. Other organisms are still classified exactly as they were originally.

In general, systematists are slowly reclassifying organisms whose descendants are so different from their ancestors that they are more similar to organisms in other lineages than to other descendants of the same ancestor. For example, we now know that

birds and crocodilians (a group of reptiles that includes crocodiles and alligators) are more closely related to each other than crocodilians are to snakes and lizards. The classification used until recently, however, placed crocodilians with snakes, lizards, and turtles, and separated birds into their own group (Figure 21.4*a*). This classification emphasizes the morphological similarity of crocodilians to snakes, lizards, and turtles, and the great differences between birds and those animals, but it is not in accord with their evolutionary relationships. In a classification based on evolutionary relationships, birds are grouped with crocodilians and their ancestors into a single taxon separate from snakes, lizards, and turtles (Figure 21.4*c*). Many classifications based on morphological similarities that are inconsistent with evolutionary relationships are still in use today.

Cladistic Classification

Classifications based solely on strong evidence of evolutionary relationships among organisms are called **cladistic classifications**. The goal of cladistic systematics is to determine the evolutionary histories of organisms and then to express those relationships in treelike diagrams. Such diagrams, called evolutionary trees, traditionally were constructed from morphological data, but sometimes these data were not consistent with evolutionary relationships. Chapter 15 introduced you to pedigrees as graphic summaries of relationships between ancestors and descendants in human families. Evolutionary trees are rather like pedigrees of lineages of organisms, except that they are often constructed with the ancestor at the bottom rather than at the top. The base of the "trunk" of the tree represents the point in the past when the lineage consisted of only the ancestor. Cladistic systematics gets its name from **clade**, which is the entire portion of a phylogeny that is descended from a **common ancestor** (a single ancestral species). An evolutionary tree constructed using only evidence of evolutionary relationships is called a **cladogram**. A cladogram illustrates evolutionary relationships by showing points at which lineages diverged from common ancestral forms.

Let us return to an example already mentioned to see how all this works. Biologists used cladistic methods to reconstruct the phylogeny of birds, reptiles, and mammals. They found that the mammal lineage was the first to branch off from the lineage of the common ancestor. As you can see from the cladogram in Figure 21.4*b*, the next phylogenetic event was the evolution of a common ancestor of birds and crocodilians that differed from the common ancestor of lizards, snakes, and turtles. Notice that the clado-

Kingdom	Plantae (plants)		
± 275,000 species			
Phylum	Tracheophyta (vascular plants)		
± 250,000 species			
Class	Angiospermae (flowering plants)		
± 235,000 species			
Order	Rosales (roses and their allies)		
± 18,000 species			
Family	Rosaceae		
± 3,500 species			
Genus	*Rosa*		
± 500 species			
Species	*Rosa gallica*		
Moss rose			

Less specific ↑   More specific ↓

21.3   Hierarchy in the Linnaean System
The moss rose and the blackburnian warbler
as they are classified under the Linnaean system.

gram is a single, higher-order clade consisting of all the descendants of the common ancestor.

What can cladistic systematists use as strong evidence of the branching points in evolutionary trees? What evidence leads them to summarize their findings by drawing a new branch on the cladogram they are constructing? The most important evidence comes from identifying ancestral and derived traits. A trait shared with a common ancestor is called an **ancestral trait**: most mammals have four limbs, having inherited this trait from a common ancestor. A trait that differs from the ancestral trait in a lineage is called a **derived trait**: most mammalian limbs terminate with five digits (e.g., fingers), but in the hooved mammals, the number of digits is one. The state of having a single digit has been determined to be derived from the five-digit state.

Distinguishing derived traits from ancestral traits is difficult because traits may diverge (become dissimilar), making ancestral states unrecognizable. For example, horses and bats are both four-limbed mammals who inherited the four-limbed trait from a common ancestor; the forelegs of a horse, however, bear very little resemblance to the wings of a bat. Alternatively, the same trait may arise more than once during evolution. These traits may then converge, resulting in a similar appearance even though they are not derived from a common ancestor.

Any two traits descended from a common ancestral structure are said to be **homologous**. **General homologous traits** are shared by many organisms—for example, the four-limbed state of most vertebrates, including rats, mice, and dogs. The four-limbed condition appears to have evolved only once, and it did so long ago, so the trait is homologous among vertebrates. **Special homologous traits** are shared by only a few species. They arose more recently during evolution than general homologous traits did. Rats and mice, but not dogs, have long, continuously growing incisor teeth. Such incisors

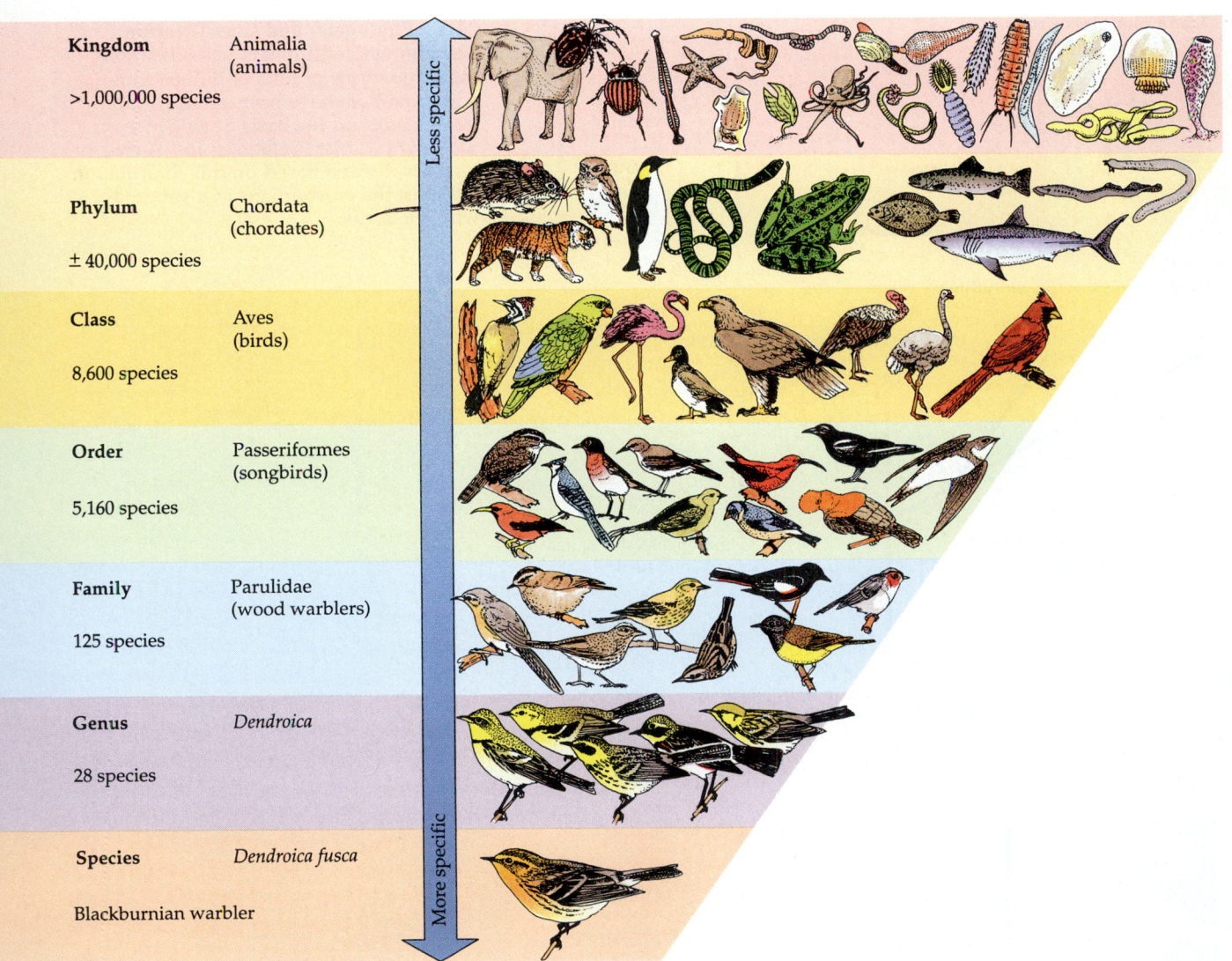

Kingdom	Animalia (animals)	Less specific ↑
>1,000,000 species		
Phylum	Chordata (chordates)	
± 40,000 species		
Class	Aves (birds)	
8,600 species		
Order	Passeriformes (songbirds)	
5,160 species		
Family	Parulidae (wood warblers)	
125 species		
Genus	*Dendroica*	
28 species		
Species	*Dendroica fusca*	More specific ↓
Blackburnian warbler		

suggest that rats and mice are more closely related to one another than either is to dogs. On the other hand, because nearly all vertebrates have four limbs, that trait is of no use for determining relationships among vertebrates.

Although homologous traits come from common ancestral traits, they may not have similar appearances or functions because, over time, such traits can diverge until they are very different. Consider, for example, the leaves of plants. Several lines of evidence, especially details of their structure and development, indicate that the leaves of vascular plants have been modified in many ways from their original light-trapping function to become protective spines, tendrils, and brightly colored structures that attract pollinators (Figure 21.5). Because all these structures are modified leaves, they are homologs of one another. Like structural traits, behavioral traits can be homologous, because behavior also evolves.

Not all resemblances are products of a common ancestry. If a trait evolves more than once, so that it is possessed by more than one species although it was not found in their most recent common ancestor, it exhibits **homoplasy**. Homoplasy can result when structures that were formerly very different come to resemble one another because they have been modified by natural selection to perform similar functions. We call this convergent evolution. For example, the structures that aid plants in climbing over other plants have evolved from stipules, leaflets, leaves, and inflorescences (Figure 21.6).

Parallel evolution also can produce homoplasy when the same character evolves independently in different lineages. As we pointed out at the beginning of the chapter, tail autotomy has evolved independently several times among salamanders. Biologists know this because the details of the mechanisms differ and because autotomy is found among species on different branches of the salamander evolutionary tree (Figure 21.7). If all tail autotomy were homolo-

(a) **Traditional classification**

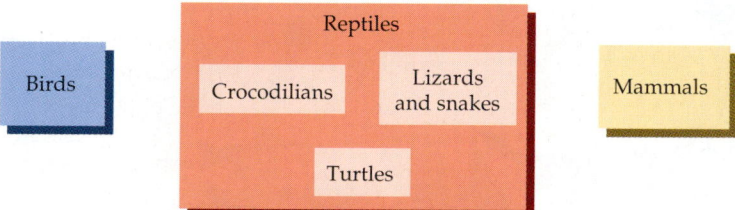

(b) **Cladogram showing evolutionary relationships**

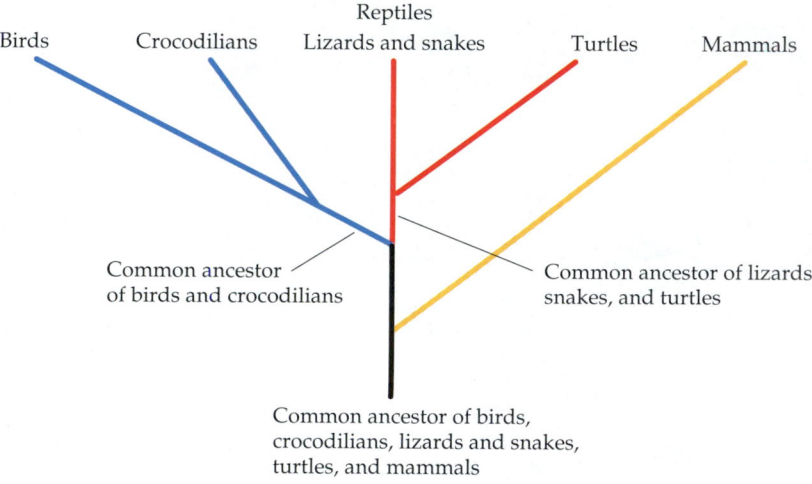

(c) **Cladistic classification**

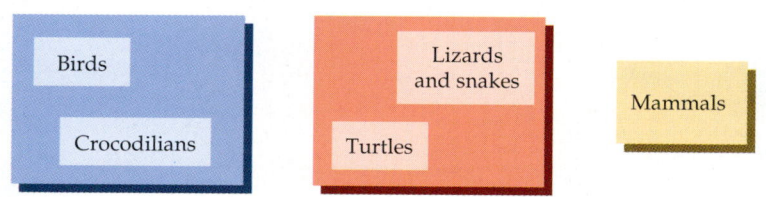

21.4 Phylogeny and Classification

The animals traditionally classified into the three major groups of birds, reptiles, and mammals have been grouped into three clades on the basis of evidence of their evolutionary relationships. A cladistic classification based on this information includes the crocodilians in a subgroup with the birds.

(a)

(b)

21.5 Homologous Structures Derived from Leaves

(a) The orange spines of the barrel cactus (*Ferrocactus acanthodes*) and (b) the colorful bracts of *Heliconia rostrata* are both modified leaves.

Passiflora rubra

Bignonia capreolata

Tendril derived from stipule

Normal stipule

Tendrils derived from 3-lobed leaves

Tendrils derived from leaf

Hooks derived from stems of flower heads

Clematis afoliata

Uncaria gambir

21.6 Convergent Climbing Structures
Convergent evolution rather than common ancestry is responsible for the similarity of the tendrils in these four plant species. All of the tendrils serve the same purpose: to help the plant climb over other plants. Therefore, these tendrils are homoplasies rather than homologies.

gous, the mechanism by which it occurs would probably be the same in all species.

Homoplasy is widespread in evolution. Its existence makes attempts to determine true phylogenies difficult because, in the absence of extensive knowledge of the development of the traits, biologists are likely to assume that converged traits are truly homologous. Workers need a means by which to identify which forms of traits are ancestral and which are derived and a means by which to distinguish homologies from homoplasies.

A method for reconstructing phylogenies that serves both of these needs was developed by the German entomologist Willi Hennig in the 1950s. Hennig suggested that if two species possess the same trait, systematists should *provisionally* (that is, until proven otherwise) assume that the trait is homologous in the two species. Hennig also proposed that general homology could be distinguished from special homology as follows. A general homologous trait is one that is found not only in one or more species of a group whose phylogeny is being reconstructed, but also appears *outside* this group in what is known as an **outgroup**. An outgroup is a taxon that is closely related to the group whose phylogeny is being reconstructed but that branched off from the lineage of the group below its base on the evolutionary tree. For example, the presence of four limbs in mammals is considered a general homologous trait because members of the outgroup—animals closely related to mammals—also have four limbs. Traits found only *within* the group, on the other hand, are special homologous traits.

In Hennig's system, the members of the group whose phylogeny is being reconstructed are then clustered according to the number of special homol-

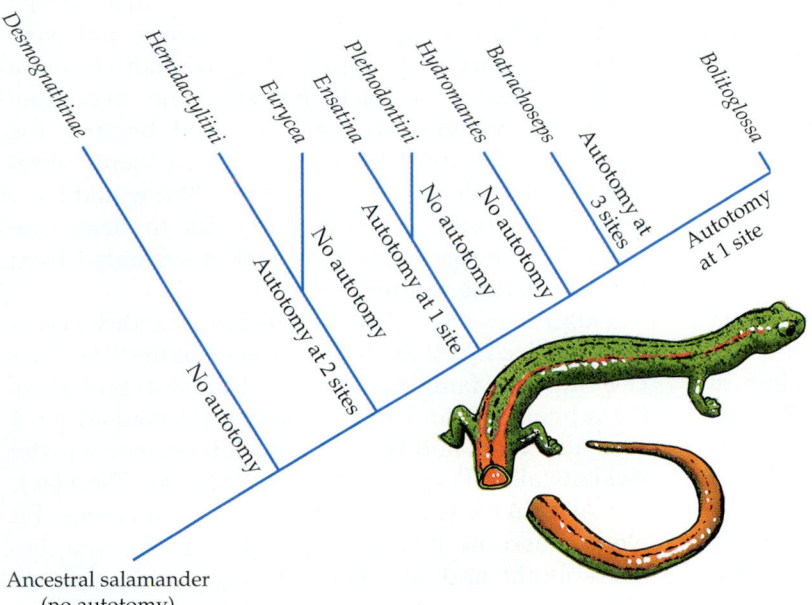

Desmognathinae

Hemidactyliini

Eurycea

Ensatina

Plethodontini

Hydromantes

Batrachoseps

Bolitoglossa

Autotomy at 3 sites

Autotomy at 1 site

No autotomy

No autotomy

Autotomy at 1 site

No autotomy

Autotomy at 2 sites

No autotomy

Ancestral salamander (no autotomy)

21.7 Salamander Tails Break in Different Places
This cladogram of a salamander lineage indicates that the tail separates in different places in different species. Each method of tail separation is a unique derived trait modified from an ancestral trait in which the tail does not detach.

TABLE 21.1
Derived Traits among Some Vertebrates

| | DERIVED TRAIT | | | | | |
TAXON	JAWS	CLAWS OR NAILS	LUNGS	FEATHERS	FUR	MAMMARY GLANDS
Hagfish (outgroup)	–	–	–	–	–	–
Perch	+	–	–	–	–	–
Pigeon	+	+	+	+	–	–
Chimpanzee	+	+	+	–	+	+
Salamander	+	–	+	–	–	–
Lizard	+	+	+	–	–	–
Mouse	+	+	+	–	+	+

A plus sign indicates the trait is present, a minus sign that it is absent.

ogous traits they share. Species that share a recent common ancestor should share many homologous traits. They should share few homoplastic traits, however, because homoplastic traits appear independently of evolutionary relationships. Therefore, as more and more traits are measured, the data are increasingly likely to support a single phylogenetic pattern, and biologists are thus more likely to be able to distinguish between homologies and homoplasies. A few of the traits originally assumed to be homologies may turn out to be homoplasies, but the best way to determine the true status of shared traits is to assume that they are homologous until additional evidence suggests that they are not.

To see how a cladogram is constructed, consider seven vertebrate animals—hagfish, perch, pigeon, chimpanzee, salamander, lizard, and mouse. We will assume initially that a given derived trait evolved only once during the evolution of the lineage and that no derived traits were lost from any of the descendant groups. The traits we have selected are either present (+) or absent (−). For reasons that will become evident in Chapter 27, hagfishes are believed to be more distantly related to the other vertebrates than they are to each other. Therefore, we will choose hagfishes as the outgroup for our analysis. Derived traits are those that have been acquired by other members of the lineage since they separated from hagfishes. The taxa and the traits we will consider are shown in Table 21.1.

Cladistic methods infer the branching points in evolutionary trees by counting the number of derived traits descendants have in common. The chimpanzee and mouse share five derived traits, suggesting that they have a very recent common ancestor. The pigeon has four derived traits, of which it shares three with the lizard, chimpanzee, and mouse. The lizard has three derived traits, all of which it shares with

the chimpanzee, mouse, and pigeon. The salamander has only two derived traits, and the perch has only one.

Using this information, we can construct a cladogram. The taxon with no derived traits, the hagfish, is the outgroup, and the animals that share the most derived traits—the chimpanzee and the mouse—are assumed to have the most recent common ancestor. Notice that the pigeon has one derived trait—feathers—that is not shared with any other taxon, suggesting that feathers evolved after the lineage leading to birds separated from that of other animals in the lineage. We can infer that this separation came after jaws, claws, and lungs evolved because birds share those derived traits with several taxa that lack feathers. The most probable cladogram for these taxa is shown in Figure 21.8.

This cladogram was easy to construct because the traits we used fulfilled the assumptions that derived traits appeared only once in the lineage and were never lost once they appeared. If we had chosen a snake instead of a lizard, however, the second assumption would have been violated because the ancestors of snakes had limbs, which were subsequently lost, along with their claws. We would have needed to examine additional traits to determine when the lineage leading to snakes separated from the one leading to other taxa.

Outgroups have also been used to identify ancestral and derived traits in lineages of butterflies. Species in two families, the brush-footed butterflies (Nymphalidae) and the monarchs (Danaidae) have four functional and two very small legs, whereas the swallowtails (Papilionidae), the sulfurs (Pieridae), and all other butterflies have six functional legs. Biologists assume that having six legs is ancestral because moths and all other orders of insects have six functional legs. The four-legged trait in monarchs

Hagfish   Perch   Salamander   Lizard   Pigeon   Mouse   Chimp

Feathers

Fur;
mammary
glands

Claws
or nails

Lungs

Jaws

21.8   A Probable Cladogram of Seven Vertebrates
This cladogram can be constructed from the information given in Table 21.1. The hagfish is designated as the outgroup for this exercise, with the other groups displaying derived traits the hagfish does not have. A particular derived trait appears after each branch point.

and brush-footed butterflies is thus inferred to be a derived trait of butterflies descended from six-legged ancestors (Figure 21.9). If other special shared traits also unite brush-footed butterflies and monarchs, we would conclude that these two groups of butterflies share a more recent common ancestor than they do with any other groups of butterflies.

Many traits must be analyzed to reconstruct a phylogeny, and systematists use various methods to combine information from the different traits. Each method is based on specific operating rules that are, in effect, provisional assumptions about how evolution proceeds. A simple method—the one we used in our vertebrate example— makes two assumptions: (1) that the evolution of traits is irreversible—that is, an ancestral trait can change into a derived one, but the reverse change does not happen—and (2) that each trait can change only once within a lineage.

(a)

21.9   Six Legs is the Ancestral Number
Having six functional legs is the ancestral trait among butterflies; some species, however, have four legs—a derived trait. (a) *Pieris protodice* has six legs. (b) The monarch butterfly, *Danaus plexippus*, displays the derived trait of four legs.

(b)

To see how this method works, consider the imaginary data in Table 21.2. If we have information on only the first four traits (1 through 4) each of which can exist in only one of two forms (0 or 1) in the five species (labeled A through E), we can construct the following unique cladogram using the two rules just stated.

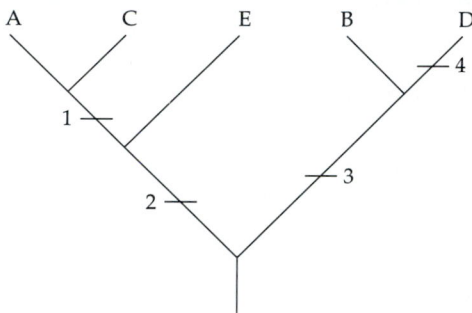

TABLE **21.2**
Ancestral and Derived Traits among Five Imaginary Species

SPECIES	TRAIT				
	1	2	3	4	5
A	1	1	0	0	0
B	0	0	1	0	1
C	1	1	0	0	1
D	0	0	1	1	0
E	0	1	0	0	1

The ancestral form of each trait is coded 0 and the derived state is coded 1.

If we then obtain information on trait 5, however, we discover that it defines a group (BCE) that shares a common ancestor. This evidence contradicts the cladogram we constructed using the first four traits. Thus, one of the two assumptions must have been violated, but more data must be gathered to determine which one it was.

Such contradictions arise whenever traits appear and are lost multiple times during the evolution of a lineage. A good fossil record often can show what really happened. As you know, a fossil is any recognizable structure from an organism, or any impression from such a structure, that has been preserved. A good fossil record helps reveal ancestral traits. For example, the excellent fossil record of horses shows that modern horses evolved from ancestors that had five toes (Figure 21.10).

If suitable fossils are unavailable, other operating rules must be employed. For example, we could relax our restrictions and allow a derived trait to be lost or to evolve more than once. However, a cladogram that postulates fewer changes in the same trait is more likely to be accurate than one that requires more changes, because reversals and multiple origins of traits are relatively rare events. Therefore, systematists often employ **parsimony** in reconstructing a phylogeny. Parsimony involves arranging the organisms such that the number of changes in traits that must be postulated to account for the inferred lineage is minimized. Parsimony is used as an operating rule in many types of biological investigations, including

21.10 Some Ancient Horses
The fossil record gives evidence that modern horses, with their single-toed limbs (hooves), evolved from ancestors with five and three toes. This artist's reconstruction of the North American Great Plains of 12 million years ago depicts several species of three-toed horses; only *Pliohippus*, grazing under the large elm tree in the background, had the one-toed trait of modern horses.

biogeography (see Chapter 50). It is useful not because evolutionary changes are necessarily parsimonious, but because it is generally wiser not to adopt more complicated explanations when simpler ones are capable of explaining the known facts. More-complicated explanations are accepted only when evidence requires them.

TRAITS USED IN RECONSTRUCTING PHYLOGENIES

As we have just seen, many different traits must be measured if we wish to distinguish homologies from homoplasies. Because organisms differ in many ways, the number of traits that systematists use to reconstruct phylogenies is very large. Some traits are readily preserved in fossils; others, such as behavior and molecular structure, rarely survive fossilization processes. Nevertheless, by using cladistic techniques, systematists can take into consideration behavioral and molecular traits as well as structural traits.

Structure and Behavior

An important source of information for systematics is **gross morphology**, that is, sizes and shapes of body parts. These traits are useful because they are under genetic control and are relatively stable, but they do change over evolutionary time. Because living organisms have been studied for centuries, we have a wealth of morphological data, and we have wonderful museum and herbarium collections of or-

ganisms whose traits, including molecular ones, can be measured. This is also the type of information most readily available from fossils. Sophisticated methods are now available for measuring and analyzing morphology and for estimating the amount of morphological variation among individuals, populations, and species.

The early developmental stages of many organisms reveal similarities with other organisms that are lost by the time adulthood is reached. For example, the larvae of the marine creatures called sea squirts have a supporting rod in their backs—the notochord—which disappears as they develop into adults. Many other animals—all the animals called vertebrates—also have such a structure. Because larval sea squirts share this and other structures with vertebrates, they are judged to be more closely related to vertebrates than would be suspected by examining adult sea squirts (Figure 21.11). Larval morphology does not, however, always reflect evolutionary ancestry. Many larvae have been highly modified by evolution for their particular existence. Butterflies, for example, are not closely related to animals that superficially resemble their caterpillars. Therefore, care must be taken when using different stages of organisms to reconstruct phylogenies.

Living organisms often reveal their relationships by similarities in their behavior. This information is most useful for detecting relationships among closely related organisms; in other words, most shared behavioral traits are special homologous traits. For example, the German student of animal behavior Konrad Lorenz showed that homologous courtship behavior patterns support other evidence that species

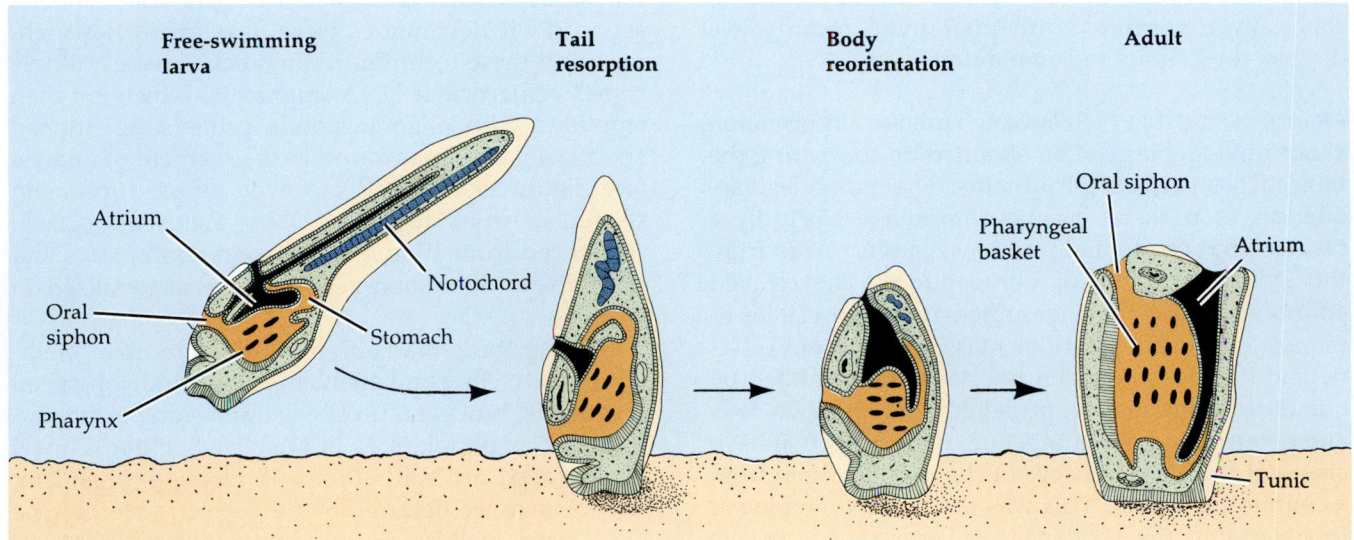

21.11 A Larva Reveals Evolutionary Relationships
Sea squirt larvae have a well-developed notochord (blue) that reveals their relationship with other chordates.

(a)   (b)

21.12 Homologous Courtship Displays among Ducks
(a) The North American ruddy duck *Oxyura jamaicensis* and (b) the Asian
white-headed duck *O. leucocephala*, although geographically isolated and possessing different plumages, are in fact closely related, as the similar courtship
displays of males of both species indicate. The raised-tail posture attracts the
brown female.

of ducks with quite different plumages are nonetheless very closely related (Figure 21.12). Many of these
species of ducks, despite substantial differences in
their plumages, can mate and produce fertile hybrid
offspring—evidence that they are also genetically
similar.

Molecular Traits

Like the sizes and shapes of their body parts, the
molecules of organisms are heritable characteristics
that may come to differ greatly between lineages over
evolutionary time. Molecular traits are thus also valuable for systematists. Among the most important molecular traits of organisms are the structures of their
nucleic acids (DNA and RNA) and proteins. Because
they are so important for reconstructing phylogenies,
especially for groups with poor fossil records, we
discuss these traits in some detail.

PROTEIN STRUCTURE.   Relatively precise information
about phylogenies can be obtained by comparing the
molecular structures of proteins. The amino acid sequences of proteins can be determined easily by a
process that sequentially removes amino acids from
the amino terminus of polypeptides. The cleaved
amino acids are then identified using special techniques (chromatography or mass spectroscopy). Genetic differences between two taxa are estimated by
obtaining homologous proteins from the two taxa
and determining the number of amino acids that have
changed since the lineages of the taxa diverged from
a common ancestor. This was the method employed
to determine the sequences of amino acids in the
cytochromes *c* in Figure 19.15. As we saw, that information revealed a great deal about how natural
selection influenced the evolution of cytochrome *c*.

NUCLEIC ACID STRUCTURE.   The structure of DNA provides excellent evidence of evolutionary relationships
among organisms. Cells of eukaryotes have genes in
their nuclei and mitochondria. Plant cells also have
genes in their chloroplasts. The chloroplast genome
(cpDNA), which is used extensively in phylogenetic
studies of plants, consists of a circular, double-
stranded DNA molecule. This molecule is evolutionarily highly stable. All land plants have nearly the
same complement of 100 chloroplast genes that code
transfer RNAs, some ribosomal RNA subunits, and
some proteins, particularly those involved in photosynthesis. Mitochondrial DNA, which is very similar
to cpDNA, has been used extensively for evolutionary studies of animals.

Deriving useful information from DNA has been
much easier since 1985 when polymerase chain reaction (PCR) techniques were developed. Researchers using these techniques can quickly make multiple
copies of particular DNA fragments, which are then
amplified with oligonucleotide primers constructed
specifically for genes found in the segment of interest
(see Figure 14.16). PCR can even be used to obtain
sequences from degraded DNA. Genes have been
sequenced from 17-million-year-old fossil plants and
from 25- to 30-million-year-old insects fossilized in
amber!

Despite the remarkable recent advances in methods for extracting and analyzing DNA from long dead
organisms, however, prospects for cloning extinct organisms, as popularized in Michael Crichton's novel
Jurassic Park, are still remote. To clone an organism,
scientists must determine the correct order of the
DNA from small fragments; if the organism is extinct,
doing so is impossible, even if the fragments could
be sequenced. In addition, synthesizing a DNA complement of more than 100 million base pairs, the

probable size of a dinosaur genome, would be a task far beyond current and foreseeable capabilities.

Chloroplast DNA has been used to reveal phylogenetic relationships among plants in the sunflower family (Asteraceae). The Asteraceae, one of the largest families of flowering plants, comprises about 1,100 genera and 20,000 species. Fossil evidence suggests that the family originated about 30 million years ago, but because it soon underwent rapid radiation, relationships within the family could not be determined from fossil evidence. Evidence from analyzing DNA structure, however, has been useful in determining these relationships.

The bases in the 151-kilobase cpDNAs (that is, cpDNAs consisting of 151,000 base pairs) of nearly all flowering plants occur in the same order. However, a single 22-kilobase inversion of the normal pattern is found in 57 genera representing nearly all major lineages of the Asteraceae. This inversion is not found among members of other families allied to the Asteraceae, or in members of the Barnadesiinae, one lineage within the sunflower family. The absence of this inversion among families related to the sunflowers suggests that it is a general homologous trait. Because cpDNA is so evolutionarily stable, cladistic criteria indicate that the presence of this inversion is probably a derived trait and that its absence among the Barnadesiinae is an ancestral condition. Therefore, the lineage leading to the Barnadesiinae probably separated from the lineage leading to the rest of the family early in the evolution of the sunflowers (Figure 21.13).

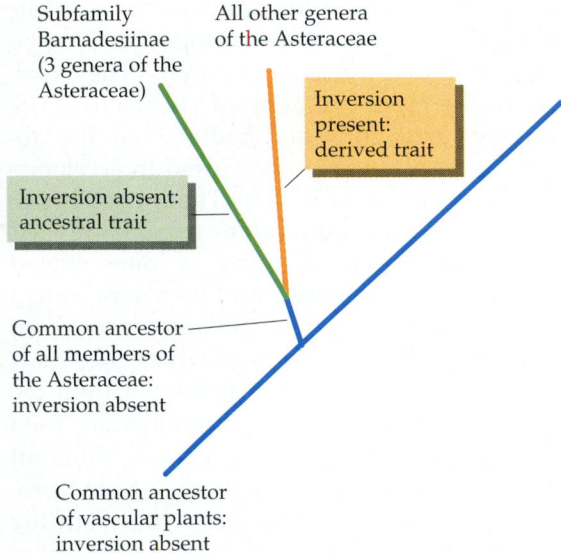

21.13 Chloroplast DNA and Sunflower Phylogeny
An early branch point in the evolutionary tree of the sunflowers, family Asteraceae, is marked by the presence of a distinctive inversion in cpDNA. This inversion is present in most lineages of the family but is absent among members of the Barnadesiinae.

TABLE 21.3
Genetic Differences among Some Vertebrates as Estimated by DNA Hybridization

TAXA COMPARED	PERCENTAGE DIFFERENCE IN DNA SEQUENCES
Human/chimpanzee	1.6
Human/gibbon	3.5
Human/rhesus monkey	5.5
Human/galago (nocturnal primate)	28.0
House mouse/Norway rat	20.0
Cow/sheep	7.5
Cow/pig	20.0

DNAs can be compared, even if the precise sequences of their bases are not known, by a process called **DNA hybridization**. As you may recall from Chapter 14, the two strands of the DNA double helix can be separated by heating them, but they reanneal when cooled. If DNAs from two species are mixed and heated, they form hybrid double helices when they cool. Because of differences in the base sequences of the two DNAs, however, they do not match up well. Less heat is required to separate these hybrid helices than to separate helices composed of DNA from a single species. The degree of mismatching of the DNA is related to its thermal stability in a very consistent way. With about 1 percent base-pair mismatching, the temperature at which 50 percent of the helices dissociate is lowered by 1 percent. DNA sequences differing by more than about 20 percent of their base pairs thus do not form stable duplexes, so this method can be used to compare only relatively similar species.

DNA hybridization reveals that the DNAs of humans and chimpanzees are much more similar than would be expected given the considerable morphological differences between the two species (Table 21.3). This similarity indicates that humans and chimpanzees share a common ancestor more recently than previously thought.

A long-standing debate among biologists over whether the giant panda of China was more closely related to bears or to raccoons was resolved by DNA hybridization data, which clearly indicate that the giant panda is a bear (Figure 21.14). The unusual (that is, nonbearlike) features of the giant panda are recent adaptations to its specialized diet: bamboo.

Combinations of Traits

Although molecules are increasingly used in analyses of evolutionary relationships among organisms, many types of data continue to be valuable to sys-

21.14 The Giant Panda Is a Bear
In the above cladogram, drawn slightly differently from previous cladograms, the horizontal line at the far left represents the lineage of a common ancestor shared by the taxa identified on the right. The DNA hybridization data used to construct the cladogram indicate that the giant panda of Asia is a bear. The lesser panda (above), also of Asia, was long thought to be closely related to the giant panda (below), but it is actually more closely related to raccoons and their allies of the Americas.

tematists. Indeed, by including data from many sources, systematists are able to reach more powerful conclusions than they could if they used only one type of information. For example, a major uncertainty about relationships among invertebrates has been resolved by an analysis that combines data from 141 morphological traits, some adult and some larval, with comparisons of rRNA subunit sequences. This analysis strongly supports the view that annelids and arthropods, despite the similarities in their body segmentation (see Chapter 26), are not closely related. Rather, these two groups are members of lineages that separated early during invertebrate evolution.

EVALUATING DIFFERENT METHODS

Many different evolutionary trees may be proposed for any complex lineage. Determining which one is the best reconstruction of the phylogeny is an extremely difficult task. The fossil record is incomplete, and often we cannot distinguish homologies from homoplasies with certainty.

Testing the accuracy of the methods used helps biologists choose among evolutionary trees. Simple organisms with short life cycles are useful in testing methods of reconstructing phylogenies, just as they are excellent subjects for studying basic genetic mechanisms.

An imaginative method for using microorganisms to test cladistic methods was developed in 1992 by researchers at the University of Texas. They took advantage of the rapid evolution of viruses to create an evolutionary radiation (see Chapter 20) in the laboratory. A chemical mutagen was used to accelerate evolution in bacteriophage T7. The investigators tracked changes in the nucleotides over many generations and developed an array of nine related phage "species." The species could be distinguished by the nucleotide fragment lengths obtained when their DNA was cut by restriction enzymes. The exact phylogeny of the nine species was known from the records of the experiment. The investigators then reconstructed this phylogeny using five different methods. All five cladistic methods they tested produced a generally correct evolutionary tree, but the parsimony method yielded an inferred tree that matched almost exactly the known phylogeny. This does not mean that parsimony always yields the best reconstruction, but the experiment increases our confidence in the value of using parsimony as an operating rule.

THE FUTURE OF SYSTEMATICS

Although molecular methods are very important for reconstructing phylogenies, information from many sources will always be valuable. The fossil record, for example, which reveals when lineages diverged and began their independent evolutionary histories, is necessary to provide absolute dating for evolutionary events. Fossils provide important evidence that allows us to distinguish ancestral from derived traits. In addition, no single method is suitable for all time frames and for all types of organisms. Therefore, the range of data used in classification is likely to increase rather than decrease in the future. Because systematics integrates activities from many different biological disciplines, a systematist needs to have a command of molecular techniques, natural history, and computer programming.

OTHER USES OF PHYLOGENIES

Typically, phylogenies are reconstructed in efforts to determine evolutionary relationships among organisms. Nevertheless, phylogenies can be used to answer many other types of biological questions. Indeed, it is difficult to think of any biological problem whose solution would not be assisted by a reliable evolutionary tree.

For example, phylogenetic data can be used to answer questions about the ways in which organisms achieved their current distribution patterns. Data on blood groups from human populations in many places around the world were used to construct a "gene substitution tree" (Figure 21.15a). This tree was then used to infer the probable migration patterns of humans as they colonized Earth (Figure 21.15b). Similarly, anthropologists and linguists use cladistic methods to analyze the evolution of human languages.

Many biological statements are basically phylogenetic statements. Any statement claiming an association between a trait and a group of organisms is a claim about when during a lineage the trait first arose

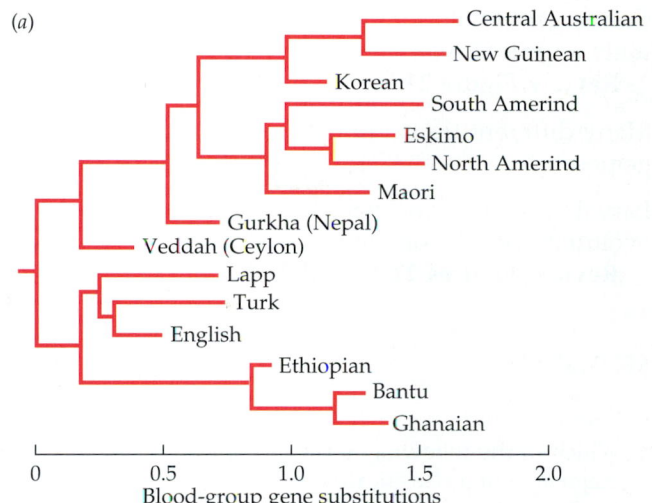

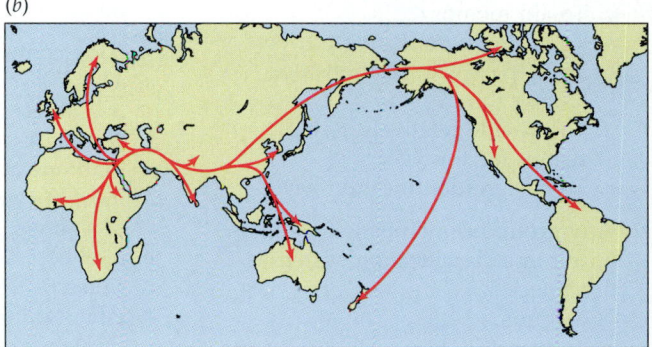

21.15   Blood Types and Migration Patterns
An evolutionary "tree" of human blood groups (a) was used to reconstruct these possible routes by which humans may have populated Earth (b).

and the fate of the trait since its first appearance. For example, the statement that DNA is the genetic material for all eukaryotes is a claim that organisms evolved DNA as their genetic material before eukaryotic cells appeared—that DNA is an ancestral and a general homologous trait among eukaryotes—and that DNA has been maintained as the genetic material in the subsequent evolution of all surviving eukaryote lineages.

SUMMARY of Main Ideas about Systematics and Phylogenies

Biological nomenclature assigns to each organism a unique combination of a generic and a specific name.

The classification of a species is its genus, family, order, class, phylum, and kingdom.
   Review Figure 21.3

Cladistic classifications are based on evidence of evolutionary relationships.
   Review Figures 21.4, 21.7, and 21.8

Reconstructing a phylogeny requires distinguishing between ancestral and derived traits within a lineage, as well as between homologous and homoplastic traits.

Divergent evolution may make homologous traits appear dissimilar.
   Review Figure 21.6

Convergent evolution may make nonhomologous traits appear similar.
Review Figure 21.5

Many different traits are used to reconstruct phylogenies.

Larval structures and behavior sometimes reveal evolutionary relationships.
Review Figures 21.11 and 21.12

The structures of nucleic acids are important taxonomic tools.
Review Figure 21.13

The most accurate method for reconstructing a phylogeny can be determined experimentally.

SELF-QUIZ

1. Which of the following is *not* a major role of a classification system?
   a. To aid memory
   b. To improve predictive powers
   c. To help explain relationships among things
   d. To provide relatively stable names for things
   e. To design identification keys.

2. Any group of organisms treated as a unit in a classification system is
   a. a species.
   b. a genus.
   c. a taxon.
   d. a clade.
   e. a phylogen.

3. A genus is
   a. a group of closely related species.
   b. a group of genera.
   c. a group of similar genotypes.
   d. a taxonomic unit larger than a family.
   e. a taxonomic unit smaller than a species.

4. Outgroups are used in cladistic analyses to
   a. distinguish homoplasies from homologies.
   b. distinguish homoplasies from convergence.
   c. distinguish between general homologies and special homologies.

   d. determine relationships between closely related taxa.
   e. reduce the size of the lineage being analyzed.

5. A clade contains
   a. all—and only—the descendants of a single ancestor.
   b. all the descendants of more than one ancestor.
   c. most but not all the descendants of a single ancestor.
   d. members of two or more lineages.
   e. a few of the descendants of a single ancestor.

6. Which of the following is *not* a way of identifying ancestral traits?
   a. Determining which traits are found among fossil ancestors
   b. Using an outgroup in which the trait is also found
   c. Using a number of outgroups having the trait
   d. Determining how many species in the lineage share the trait today
   e. Experimentally creating a known lineage

7. Traits that evolve very slowly are useful for determining relationships at the level of
   a. phyla.
   b. genera.
   c. orders.
   d. families.
   e. species.

8. Homologous traits are
   a. similar in function.
   b. similar in structure.
   c. similar in structure but derived from different ancestral structures.
   d. derived from a common ancestor whether or not they have the same function today.
   e. derived from different ancestral structures and have dissimilar structures.

9. The genes that are most extensively used to determine evolutionary relationships among plants are
   a. nuclear genes.
   b. chloroplast genes.
   c. mitochondrial genes.
   d. genes in flowers.
   e. genes in roots.

10. Which of the following is *not* a way in which phylogenies are used?
    a. To establish evolutionary relationships.
    b. To determine how rapidly traits evolve.
    c. To determine historical patterns of movement of organisms.
    d. To help identify unknown organisms.
    e. To infer evolutionary trends.

FOR STUDY

1. The great blue heron (*Ardea herodias*) is found over most of North America. The very similar grey heron (*Ardea cinerea*) ranges over most of Europe and Asia. These two herons currently are treated as different species. A colleague argues that they should be lumped into one species. What facts should you consider in evaluating your colleague's suggestion?

2. Why are systematists so concerned with identifying lineages descended from a single ancestor?

3. How are fossils used to identify ancestral and derived traits of organisms?

4. A student of the evolution of frogs has performed DNA hybridization experiments among about 25 percent of all frog species. As a result of these experiments she proposes a new classification of frogs that differs strikingly from the traditionally accepted one. Should frog taxonomists immediately accept this new classification? Why or why not?

5. Linnaeus developed his system of classification before Darwin proposed his theory of evolution by natural selection, and most classifications of organisms were developed by people who were not evolutionists. Yet most of these classifications are still used today, with minor modifications, by evolutionists. Why?

READINGS

Brooks, D. R. and D. H. McLennan. 1991. *Phylogeny, Ecology, and Behavior. A Research Program in Comparative Biology.* University of Chicago Press, Chicago. An excellent book that shows how rich insights can be derived by integrating phylogenetic analyses with studies of ecology and behavior. Includes a chapter describing the tools used to reconstruct phylogenies.

Eldredge, N. and J. Cracraft. 1980. *Phylogenetic Patterns and the Evolutionary Process.* Columbia University Press, New York. A good sampling of various perspectives on concepts and practices in systematics.

Ferguson, A. 1980. *Biochemical Systematics and Evolution.* Blackie & Son, Glasgow. A concise introduction to the use of biochemical techniques at all levels of taxonomy.

Harvey, P. H. and M. D. Pagel. 1991. *The Comparative Method in Evolutionary Biology.* Oxford University Press, Oxford. A good discussion of methods of reconstructing phylogenies and the use of comparative approaches in evolutionary studies.

Hillis, D. M. and C. Moritz (Eds.). 1990. *Molecular Systematics.* Sinauer Associates, Sunderland, MA. A description of methods used currently in molecular systematics; designed to guide beginners through a molecular systematic study. Chapter 11, "Phylogenetic Reconstruction," is especially useful as an overview.

Mayr, E. 1982. *The Growth of Biological Thought.* Belknap Press, Cambridge, MA. A good treatment of the history of thinking about systematics, biological diversity, and evolution.

Pääbo, S. 1993. "Ancient DNA." *Scientific American,* November. DNA from organisms that died thousands of years ago can be partially reproduced. Although deciphering is incomplete, the study of reconstituted DNA fragments allows revealing comparisons to be made between modern and ancient species.

Ridley, M. 1986. *Evolution and Classification: The Reformation of Cladism.* Longman, London. A critical evaluation of current schools of thought in systematics that provides clear statements of the goals and methods of all approaches.

Wiley, E. O. 1981. *Phylogenetics: The Theory and Practice of Phylogenetic Systematics.* Wiley, New York. A clear overview of the phylogenetic approach to evolution and classification.

The Evolution
of Diversity

The most abundant organisms on Earth are so small that we cannot see them with the naked eye. These organisms, the members of the kingdoms Eubacteria and Archaebacteria, are the most successful of all creatures on Earth if success is measured by numbers of individuals. The bacteria in one person's mouth, for example, outnumber all the humans who have ever lived. Although small, bacteria play many critical roles in the biosphere, interacting in one way or another with every other living thing. Some bacteria perform key steps in the cycling of nitrogen, sulfur, and carbon; others trap energy from the sun or from inorganic chemical sources; and some help animals digest their food. The members of these two kingdoms outdo all other groups in metabolic diversity, and they occupy more—and more extreme—habitats than any other groups. Their effects on our environment are diverse and profound. How can such tiny and seemingly simple organisms be such successes in a world full of complex and intelligent animals such as us?

Viruses are even smaller than bacteria. Although small and structurally simple, they are not evolutionarily ancient. Rather, they are believed to have arisen from the plant, animal, and bacterial groups that they infect. We learned about some features of a few viruses in Chapters 4 and 12; here we take a broad look at the variety of viral types.

In the first six chapters of Part Four we celebrate the diversity of the living world—the products of evolution. This chapter focuses on the two bacterial kingdoms and on viruses. Chapters 23 to 27 deal with the kingdoms Protista, Fungi, Plantae, and Animalia.

GENERAL BIOLOGY OF THE BACTERIA

The bacterial kingdoms, which consist of all organisms whose cells are prokaryotic, have the most ancient origins of any group present today. The earliest bacterial fossils date back at least 3.5 billion years, and as we saw in Chapter 18, these ancient traces indicate that there was considerable diversity among the prokaryotes even during the Archean eon. The prokaryotes reigned supreme on an otherwise sterile Earth for more than 2 billion years, adapting to new environments and to changes in existing ones. Bacteria have spread to every conceivable habitat on the planet, including the insides of other organisms. By any standard, they must be judged enormously successful creatures. We begin our study of the bacteria by examining the remarkable diversity of ways in which they obtain their energy.

Metabolic Diversity in the Bacteria

The long evolutionary history of bacteria has led to extraordinary diversity of their metabolic "lifestyles."

The Most Numerous Organisms
These rod-shaped members of the kingdom Eubacteria, invisible to the naked eye, are flourishing on the tip of a hypodermic needle.

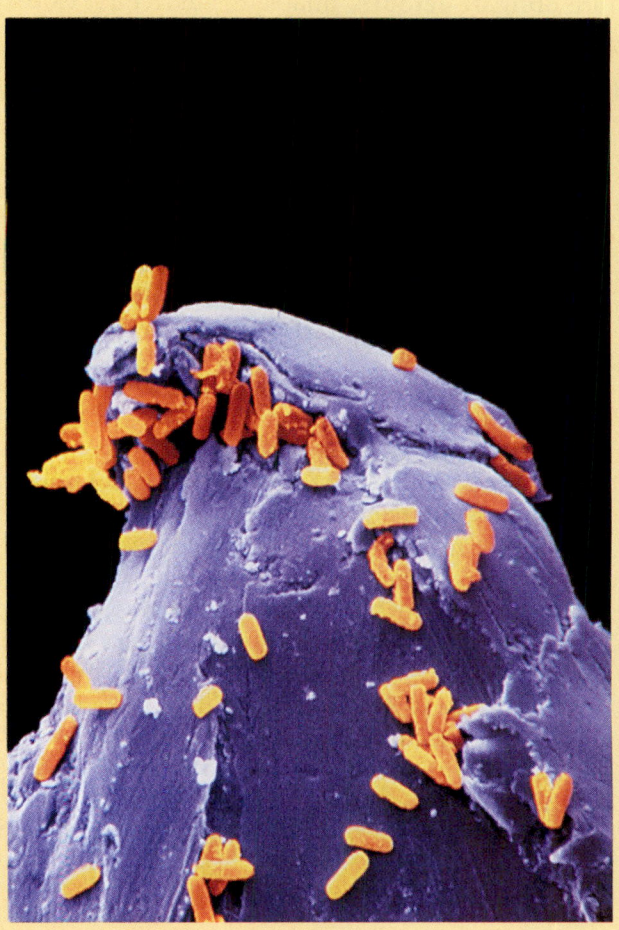

22

Bacteria and Viruses

Organisms that can shift their metabolism between anaerobic and aerobic modes are called **facultative anaerobes**. Some facultative anaerobes can conduct only anaerobic metabolism (fermentation) but are not damaged by oxygen when it is present. Although many types of prokaryotes are facultative anaerobes, alternating between fermentation and cellular respiration as conditions dictate, others can live only by fermentation because they are poisoned by oxygen gas. These oxygen-sensitive fermenters are called **obligate anaerobes**. At the other extreme, some bacteria are **obligate aerobes**, unable to survive for extended periods in the *absence* of oxygen.

Some bacteria carry out respiratory electron transport without using oxygen as an electron acceptor. These forms use oxidized inorganic ions such as nitrate, nitrite, or sulfate as electron acceptors. Among these organisms are the denitrifiers, bacteria that return nitrogen to the atmosphere, completing the cycle of nitrogen in nature (see Chapter 32).

Biologists recognize four broad nutritional categories in the bacteria. In the first category are the **photoautotrophs**, which are photosynthetic. Photoautotrophs use light as their source of energy and carbon dioxide as their source of carbon. One group, the cyanobacteria, like the photosynthetic eukaryotes, performs photosynthesis with chlorophyll *a* as the key pigment and produces oxygen as a byproduct of noncyclic photophosphorylation.

By contrast the other photosynthetic bacteria use bacteriochlorophyll as their key photosynthetic pigment, and they do not release oxygen gas. These photosynthesizers produce particles of pure sulfur instead because hydrogen sulfide (H_2S) rather than H_2O is the electron donor for photophosphorylation (Figure 22.1a). Bacteriochlorophyll absorbs light of longer wavelength than the chlorophyll used by all other photosynthesizing organisms. As a result, bacteria using this pigment can grow in water beneath fairly dense layers of algae because light of the wavelengths they can use is not appreciably absorbed by the algae (Figure 22.1b).

The second nutritional category of bacteria is the **photoheterotrophs**. These bacteria use light as their source of energy but must obtain their carbon atoms from organic compounds made by other organisms. The photoheterotrophs are also known as the purple nonsulfur bacteria. They use such compounds as carbohydrates, fatty acids, and alcohols as their organic "food."

Chemoautotrophs, of the third category, obtain their energy by oxidizing inorganic substances, and they use some of that energy to fix carbon dioxide in reactions analogous to those of the photosynthetic carbon reduction cycle. The chemoautotrophs (Figure 22.2) include the nitrifiers, which oxidize ammonia or nitrite ions to form nitrate ions that are taken up by plants, as well as other bacteria that oxidize hydrogen gas, hydrogen sulfide, sulfur, and other materials. Scientists exploring the ocean bottom near the Galapagos Islands in 1977 discovered a spectacular example of chemoautotrophy. They found an entire ecosystem based on chemoautotrophic bacteria that are eaten by a large community of crabs, mollusks, and giant worms (pogonophorans; see Figure 26.23)—all at a depth of 2,500 meters, far below any hint of light from the sun but in the immediate neighborhood of volcanic vents in the ocean floor. The bacteria obtain energy by oxidizing hydrogen sulfide released from the vents.

Finally, there are the **chemoheterotrophs**, which typically obtain both energy and carbon atoms from one or more organic compounds. Most bacteria are chemoheterotrophs—as are all animals, fungi, and many protists.

Nitrogen fixation, in which atmospheric nitrogen

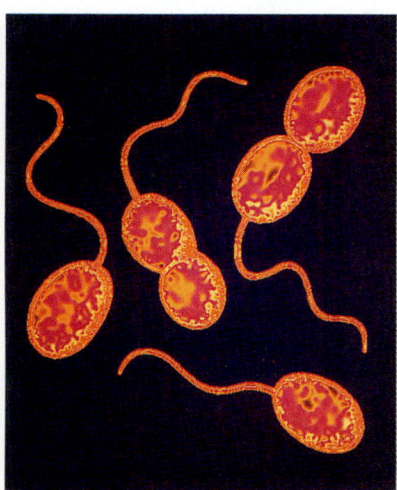

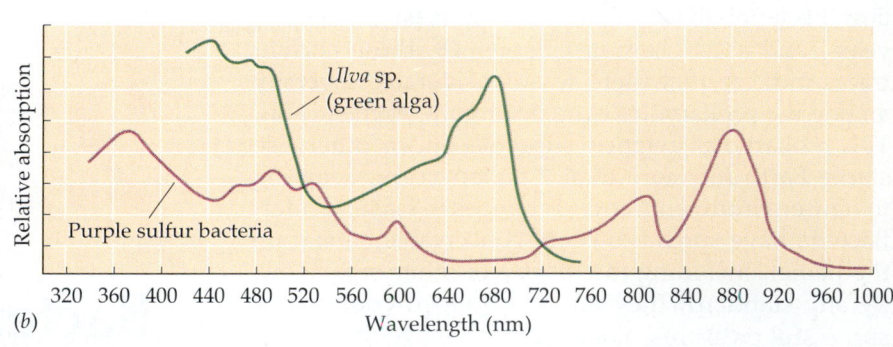

(b)

(a)

22.1 Some Bacteria Photosynthesize
(a) Cells of *Thiocystis*, a purple sulfur bacterium, store granules of sulfur that they produce via anaerobic photosynthesis. Two of the cells shown here are in the process of dividing. (b) Notice that the alga does not absorb any light wavelengths longer than 750 nm. Purple sulfur bacteria can conduct photosynthesis beneath the alga, using the longer wavelengths.

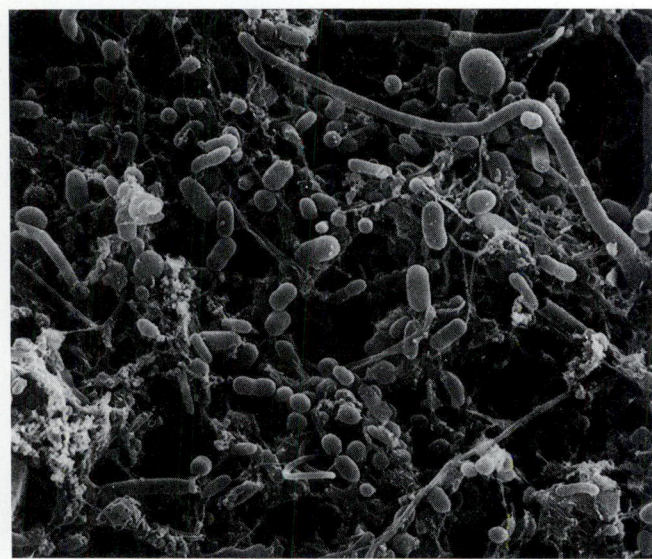

22.2 Some Bacteria Are Chemoautotrophic
All of these bacteria are chemoautotrophs that live near a hot water vent along the Galapagos Rift in the eastern Pacific, where their fixing of carbon supports an entire community of organisms that thrives in total darkness.

gas is converted into chemical forms usable by bacteria and other living things, is another important metabolic adaptation. This vital process is carried out by a wide variety of bacteria but by no other organisms.

Dividing the Bacteria into Kingdoms and Phyla

The bacteria of today are astoundingly diverse. They probably represent the current products of many independent evolutionary lines that have been separate for billions of years. That is, although they have a common prokaryotic heritage, different groups of bacteria have been following their separate evolutionary paths for most of the history of life. In addition, because each of these evolutionary lines has spread over the surface of Earth and has adapted to many or most of the possible environmental challenges, diversity is great within each line, even compared with the diversity within the other kingdoms. The result is a taxonomist's nightmare.

There are at least three possible motivations for classification schemes: to reveal evolutionary relationships, to help identify unknown organisms, and to illustrate diversity. The favored approach of many biologists is to group organisms by evolutionary relationship, providing where possible a natural classification. The standard taxonomic reference for bacteria, *Bergey's Manual of Systematic Bacteriology*, groups species in a way that facilitates the identification of unknown organisms, taking into account as much

evidence from current research as possible. Classification schemes designed to reveal the diversity of the living world, on the other hand, generally are not excessively concerned with incorporating all the newest evidence. In this book we take a mixed approach: We let evolutionary relationships determine the assignment of bacteria to kingdoms, but at the level of phylum and below we attempt primarily to show the extent of bacterial diversity. (Some bacteriologists use the term "division" rather than "phylum.")

Until recently, most biologists lumped all bacteria into a single kingdom called the kingdom Monera. Now, however, we follow the lead of Carl Woese, of the University of Illinois, who has defined two separate kingdoms of bacteria, the **Archaebacteria** and the **Eubacteria**. As we will see, archaebacteria differ strongly from all other bacteria—in some ways more strongly than bacteria in general differ from eukaryotes! Except for recognizing two bacterial kingdoms, we generally follow *Bergey's Manual* in this chapter.

The condition of the bacterial cell wall is a key characteristic that taxonomists use in the classification of bacteria. Some bacteria have no cell walls, and of those that do, some are classified as "gram-positive," others as "gram-negative," for reasons we explain shortly. The kingdom Eubacteria is thus divided into three phyla: **gram-negative bacteria**, **gram-positive bacteria**, and **mycoplasmas** (bacteria lacking cell walls). The cell walls of archaebacteria are chemically unrelated to those of the gram-negative and gram-positive bacteria, so these organisms are grouped into phyla according to the habitats in which they thrive. The kingdom Archaebacteria is thus divided into three phyla: **thermoacidophiles**, **methanogens**, and **strict halophiles** (Table 22.1). As you will see, the Archaebacteria live in some of the strangest places on Earth.

Although both archaebacteria and eubacteria are prokaryotes, remember that archaebacteria stand far apart from the other bacteria.

Prokaryotes versus Eukaryotes

The architectures of prokaryotic and eukaryotic cells were compared in Chapter 4 (you may wish to review Figures 4.2 to 4.6 and Boxes 4.A and 4.B). The basic unit of bacteria is the prokaryotic cell, which contains a full complement of genetic and protein-synthesizing systems, including DNA, RNA, and all the enzymes needed to transcribe and translate the genetic information into protein. The cell also contains at least one system for generating the ATP it needs.

The prokaryotic cell differs from the eukaryotic cell in two crucial ways. First, the organization and replication of the genetic material differs. The DNA of the prokaryotic cell is not organized within a membrane-bounded nucleus, and it is not complexed with

TABLE 22.1
Groups within the Bacteria

GROUP	REPRESENTATIVE GENERA
Kingdom Archaebacteria	
Thermoacidophiles	*Sulfolobus*
Methanogens	*Methanopyrus*
Strict halophiles	*Halobacterium*
Kingdom Eubacteria	
Gram-negative bacteria	
Gliding bacteria	*Beggiatoa, Thiothrix*
Spirochetes	*Borrelia, Treponema*
Curved and spiral bacteria	*Bdellovibrio, Spirillum*
Gram-negative rods	*Escherichia, Rhizobium, Salmonella*
Gram-negative cocci	*Neisseria, Nitrosococcus*
Rickettsias and chlamydias	*Rickettsia, Chlamydia*
Cyanobacteria	*Anabaena, Anacystis, Oscillatoria*
Gram-positive bacteria	
Gram-positive rods	*Bacillus, Clostridium*
Gram-positive cocci	*Staphylococcus, Streptococcus*
Actinomycetes	*Actinomyces, Corynebacterium, Streptomyces*
Mycoplasmas	*Mycoplasma*

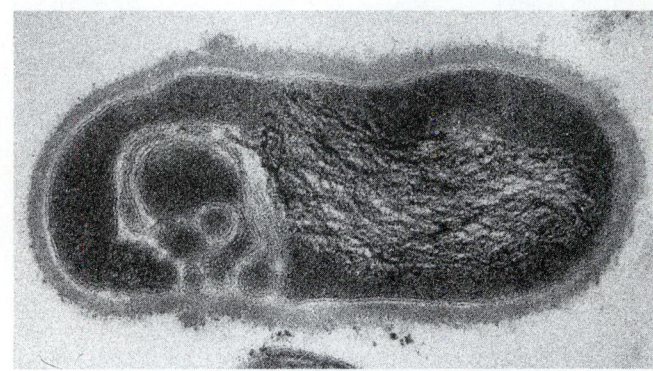

22.3   Some Bacteria Have Internal Membranes
A large mesosome is continuous with the plasma membrane on the left in this cell of *Corynebacterium parvum*. A large, fibrous mass of DNA attached to the mesosome fills most of the remainder of the cell. Unlike eukaryotic organelles, the mesosome is not a separate, membrane-bounded compartment.

flagellum consists of a single fibril made of the protein flagellin, in contrast to the flagellum of eukaryotes, which usually contains a circle of nine pairs of microtubules surrounding two central microtubules, all made of the protein tubulin. The bacterial flagellum (see Figure 4.5) rotates about its base, rather than beating, as a eukaryotic flagellum or cilium does.

Structural Characteristics of Bacteria and Bacterial Associations

Most bacteria have a thick and relatively stiff cell wall. If the wall contains a great deal of peptidoglycan (a

histones to form chromatin, as in eukaryotes. DNA molecules in prokaryotes are circular. The elaborate mechanism of mitosis is missing; prokaryotic cells divide by their own elaborate method, fission, after replicating their DNA. Second, prokaryotes have none of the membrane-bounded cytoplasmic organelles that eukaryotes have—mitochondria, chloroplasts, Golgi apparatus, endoplasmic reticulum—but the cytoplasm of a prokaryotic cell may contain a variety of mesosomes (see Figure 4.4) and photosynthetic membrane systems not found in eukaryotes.

The absence of characteristic eukaryotic organelles should not be construed as a total absence of internal membranes and other internal structures from the prokaryotic cell. Membranous mesosomes frequently associate with new cell walls during cell division, and in electron micrographs a bacterial cell's DNA is often seen attached to a mesosome (Figure 22.3). Many aerobic bacteria have respiratory enzymes that are bound to elaborate internal membrane systems, and photosynthetic bacteria have extensive and highly organized internal membranes laden with photosynthetic pigment systems (see Figures 4.3 and 22.14).

Although many prokaryotes are not motile, some can move by means of flagella—whiplike filaments that extend singly or in tufts from one or both ends of the cell, or all around it (Figure 22.4). A bacterial

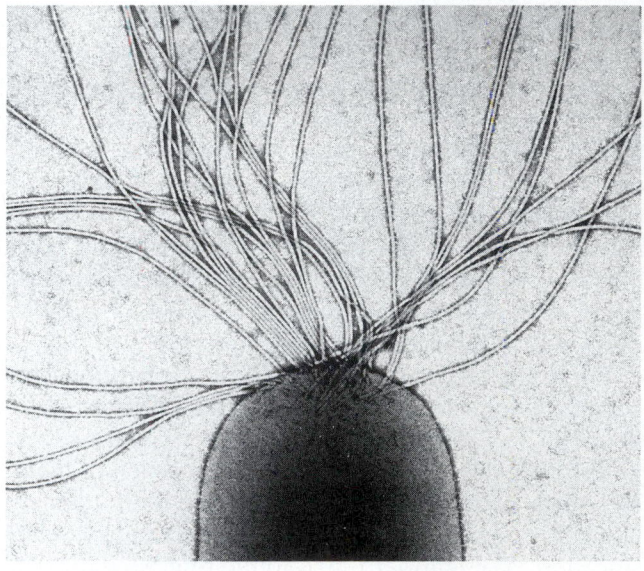

22.4   Some Bacteria Have Flagella
A tuft of flagella at one end of *Spirillum graniferum*; the other end bears a similar tuft. Other species, such as *Thiocystis* (see Figure 22.1*a*), have a single flagellum.

polymer of amino sugars), it takes on a blue to purple color when treated with the Gram stain (Box 22.A); bacteria with such walls are called **gram-positive**. Bacteria whose cells contain less peptidoglycan and appear pink to red following Gram staining are called **gram-negative**. This difference is useful in classifying eubacteria, but the cell walls of different bacteria also vary in other characteristics. As already mentioned, mycoplasmas are eubacteria that have no cell walls. Archaebacteria have walls, but their walls contain no peptidoglycan. Peptidoglycan is a substance unique to eubacteria; its absence from the walls of archae-

bacteria indicates a key difference between the two bacterial kingdoms, resulting from the separation of these two groups at the beginning of evolutionary history.

Three shapes are particularly common among the bacteria, as shown in Figure 4.6 and in various figures in this chapter. These are spheres, rods, and curved or spiral forms. A spherical bacterium is a **coccus** (plural: cocci); a rod-shaped bacterium is a **bacillus** (plural: bacilli). Cocci may live singly or may associate in two- or three-dimensional arrays as chains, plates, or blocks of cells. Bacilli and spiral forms may be

BOX 22.A

The Gram Stain and Bacterial Cell Walls

In 1884 Hans Christian Gram, a Danish physician, developed an uncomplicated staining process that has lasted into our high-technology era as the single most common tool in the study of bacteria. The **Gram stain** separates most types of bacteria into two distinct groups: gram-positive and gram-negative. A smear of bacteria on a microscope slide is soaked in a violet dye and treated with iodine; it is then washed with alcohol and

counterstained with safranine (a red dye). Gram-positive bacteria retain the violet dye and appear blue to purple, as shown here in the upper micrograph. The alcohol washes the violet stain off gram-negative bacteria; these bacteria then pick up the safranine counterstain and appear pink to red, as shown in the lower micrograph. Gram-staining characteristics are a crucial consideration in grouping some kinds of bacteria and are important in determining the identity of bacteria in an unknown sample. Mycoplasmas, which lack walls, and Archaebacteria, which lack peptidoglycan, are not stained at all by the Gram stain.

The different staining reactions probably relate to differences in the amount and accessibility of peptidoglycan in the cell walls of bacteria. These electron micrographs and asso-

ciated sketches show a thick layer of peptidoglycan outside the plasma membrane on gram-positive cells (top); the gram-negative cell wall typically has only one-fifth as much peptidoglycan, and outside the peptidoglycan layer the cell is surrounded by a second, outer membrane quite distinct in chemical makeup from the plasma membrane (bottom). The consequences of these different features are endless and relate to the disease-causing characteristics of some bacteria. Indeed, the bacterial cell wall is a favorite target in medical combat against diseases that are caused by bacteria because it has no counterpart in eukaryotic cells. Antibiotics and other agents that specifically destroy peptidoglycan-containing cell walls tend to have little, if any, effect on the cells of humans and other eukaryotes.

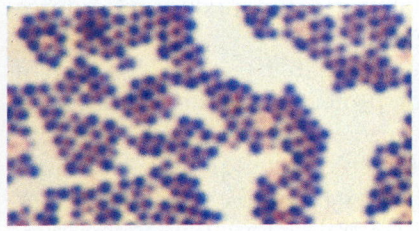

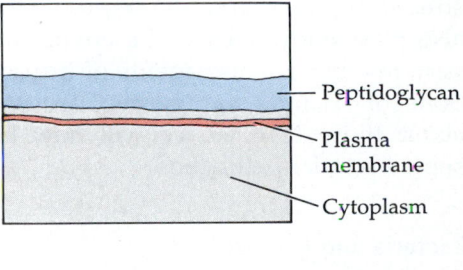

Peptidoglycan
Plasma membrane
Cytoplasm

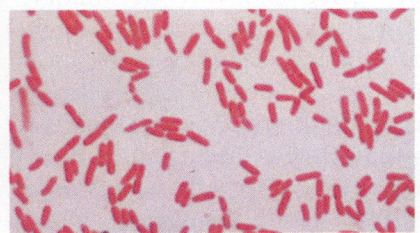
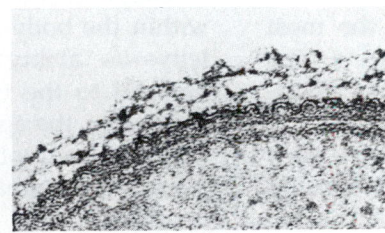
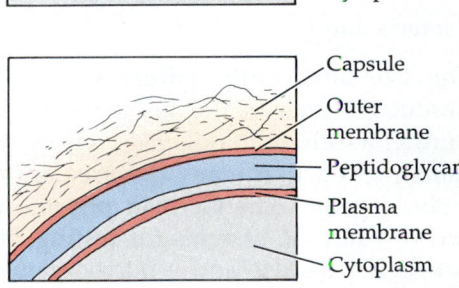

Capsule
Outer membrane
Peptidoglycan
Plasma membrane
Cytoplasm

single or may form chains. Note that associations such as chains do not signify multicellularity because each cell is fully viable and independent. Associations arise as cells adhere to one another after reproducing by fission, with each becoming two cells. Some bacteria associate in chains that become enclosed within delicate tubular sheaths. These associations are called filaments. All the cells of a filament divide simultaneously.

Some bacteria have other structural features. For example, some attach to their substrate by stalks that may be an extension of the cell wall or a product secreted outside the cell.

Reproduction and Resting

Most bacteria reproduce by fission or by producing spores; both processes are asexual, or **vegetative**. As you will recall, however, there are also sexual processes—transformation, conjugation, and transduction—that allow the exchange of genetic information between some bacteria (see Chapters 11 and 12).

Other bacteria produce **endospores**. These spores are resting structures, not reproductive ones. When environmental conditions become extremely hot, cold, dry, or otherwise harsh, the bacterium produces an endospore. It replicates its DNA and encapsulates one copy, along with some of its cytoplasm, in a tough cell wall (Figure 22.5). The parent cell then breaks down, releasing the endospore. This is not a reproductive process; the endospore merely replaces the parent cell. The endospore can survive the harsh environmental conditions because it is dormant—that is, its normal activity is suspended. Later, if it encounters favorable conditions, the endospore germinates—that is, it becomes metabolically active and divides, forming new cells like the parent. Some endospores can germinate even after more than a thousand years of dormancy.

Bacteria differ from one another in many ways—structurally, metabolically, reproductively—and they also play many roles in the environment. Although very few bacteria are agents of disease, popular notions of bacteria as "germs" arouse our curiosity about those few, so we will now briefly consider some bacterial pathogens.

Bacteria and Disease

The late nineteenth century was one of the most productive eras in the history of medicine, a time during which bacteriologists, chemists, and physicians proved that many diseases are caused by microbial agents. The German physician Robert Koch laid down a set of rules for testing the relationship between a disease and a microorganism. According to Koch, the disease in question could be attributed

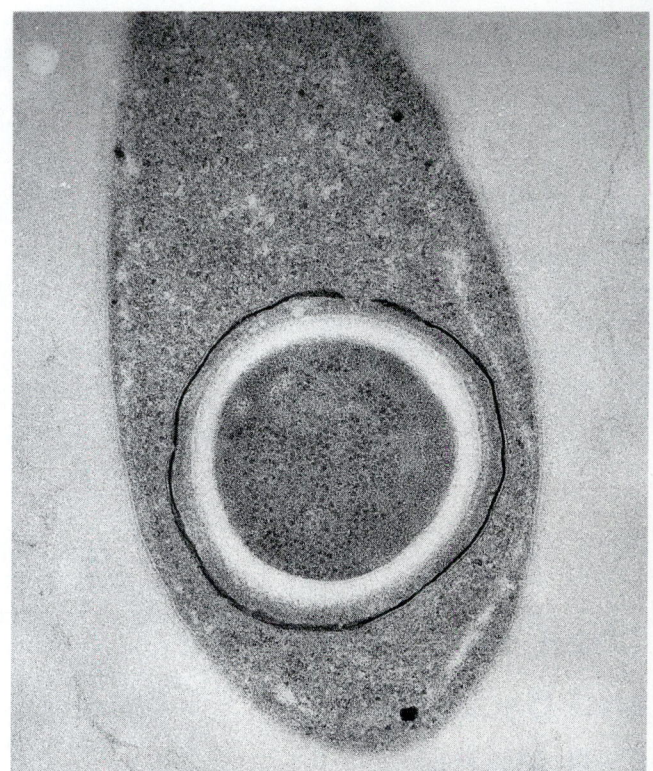

22.5 A Structure for Waiting Out Bad Times
Clostridium tetani, the bacterium that causes tetanus, produces endospores as resistant resting structures. This endospore—the round, thick-walled structure—lies near one end of the cell.

to a particular microorganism if (1) the microorganism could always be found in diseased individuals, (2) the microorganism taken from the host could be grown in pure culture, (3) a sample of the culture produced the disease when injected into a new, healthy host, and (4) the newly infected host yielded a new, pure culture of microorganisms identical to that obtained in step 2. These rules—called **Koch's postulates**—are still important in modern procedures for investigating new diseases.

In Chapter 16 we considered the immune system and other modes of protection against diseases of microbial origin. We also examined, briefly, the problems faced by a pathogenic bacterium in establishing itself in a host. The consequences of an infection for the host depend on several factors. One is the **invasiveness** of the pathogen—its ability to multiply within the body of the host. Another is its **toxigenicity**—its ability to produce chemical substances harmful to the tissues of the host. *Corynebacterium diphtheriae*, the agent of diphtheria, has low invasiveness and multiplies only in the throat, but its toxigenicity is so great that the entire body that it infects is affected. By contrast, *Bacillus anthracis*, which causes anthrax (a disease primarily of cattle and sheep), has low toxigenicity but an invasiveness so great that the entire bloodstream ultimately teems with the bacteria.

Remember that in spite of our frequent mention of human pathogens, only a small minority of the known bacterial species are agents of disease. Many more species play positive roles in our lives and in the biosphere—participating in the digestive processes of animals, in the processing of nitrogen and sulfur in soils, as decomposers in all ecosystems, and as key participants in many of our own industrial and agricultural processes.

PHYLA OF THE ARCHAEBACTERIA

The kingdom Archaebacteria consists of a few prokaryotic genera that live in habitats notable for such characteristics as extreme salinity (salt content), low oxygen concentration, high temperature, or high or low pH (Figure 22.6a). On the face of it, the archaebacteria do not seem to belong together as a group. However, they do share characteristics, such as the definitive lack of peptidoglycan in their walls and the possession of lipids of distinctive composition. Also, the base sequences of their ribosomal RNAs show great similarities. By the criterion of rRNA base sequence, archaebacteria differ as thoroughly from eubacteria as either group does from the eukaryotic kingdoms. Biologists had classified the members of the kingdom Archaebacteria in widely separated groups of bacteria until the late 1970s, when Carl Woese and others, using techniques of molecular biology, recognized their close relationship and ancient origin and grouped them together as the Archaebac-

teria (the prefix is from the Greek *archaios*, "ancient"). The kingdom can be clearly divided into three phyla: the thermoacidophiles, the methanogens, and the strict halophiles.

Thermoacidophiles

Some archaebacteria are both thermophilic (heat-loving) and acidophilic (acid-loving). *Sulfolobus* is a typical genus of such thermoacidophiles. Bacteria of this genus live in hot sulfur springs at temperatures of 70 to 75°C. They die of "cold" at 55°C (131°F). Hot sulfur springs are also extremely acidic. *Sulfolobus* grows best in the pH range from 2 to 3, but it readily tolerates pH values as low as 0.9. Thermoacidophiles that have been tested maintain an internal pH near 7 (neutral) in spite of the acidity of their environment. Thermoacidophiles thus thrive where very few other organisms can even survive.

Methanogens

Ten species of prokaryotes, previously assigned to unrelated bacterial groups, share the property of producing methane (CH_4) by reducing carbon dioxide. All these methanogens are obligate anaerobes, and methane production is the key step in their energy metabolism. By comparing rRNA base sequences, Woese established that all methanogens are closely related. Methanogens release approximately 2 billion tons of methane gas into Earth's atmosphere each year, accounting for all the methane in our air, including that associated with mammalian flatulence (the passing of gas). Approximately one-third of the methane production comes from methanogens in the

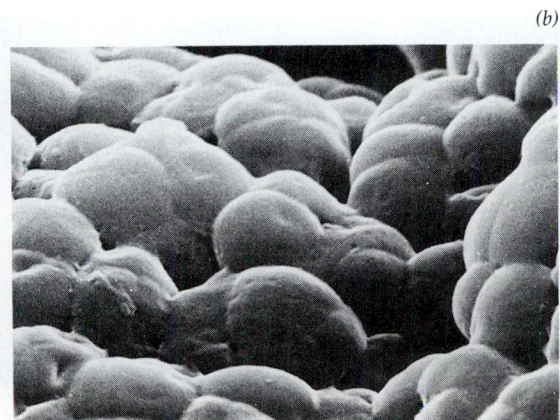

22.6 Archaebacteria
(a) Masses of archaebacteria form a mat around a steaming hot spring at Yellowstone National Park. (b) Other archaebacteria, such as these methane-producing bacteria, live anaerobically in the digestive tracts of animals.

guts of grazing herbivores such as cows (Figure 22.6b; Chapter 43). One methanogen, *Methanopyrus*, lives on the ocean bottom near blazing volcanic vents. *Methanopyrus* can survive and grow at 110°C; it is the current record-holder for temperature tolerance. It grows best at 98°C and not at all at temperatures below 84°C.

Strict Halophiles

The strict halophiles live exclusively in extremely salty environments. Because they contain pink carotenoids, they can be seen easily under some circumstances. Halophiles grow in the Dead Sea and in brines of all types—pickled fish may sometimes show reddish-pink spots that are colonies of halophilic bacteria. Photographs of salt flats taken from orbiting satellites show a distinct pink tinge resulting from the presence of vast numbers of *Halobacterium* and its halophilic relatives. Few other organisms can live in the saltiest of the homes that the strict halophiles occupy; most would "dry" to death, losing too much water by osmosis to the hypertonic (more concentrated) environment. Strict halophiles have been found in lakes with pH values as high as 11.5—the most alkaline environment inhibited by living organisms, almost as alkaline as household ammonia.

PHYLA OF THE EUBACTERIA

Gram-Negative Bacteria

Most bacteria are gram-negative; this phylum takes up more than three-fourths of *Bergey's Manual*. Gram-negative bacteria are highly diverse in form and metabolism. We will describe only a few of the groups of this phylum.

GLIDING BACTERIA. The gliding bacteria are filaments or rods whose movement resembles gliding. Members of the genus *Beggiatoa* (Figure 22.7a) are an example. The physical basis of the gliding movement is not yet known. The cells may use a secreted slime in their locomotion; they leave slimy trails behind them as they move over soil or dead organic matter (Box 22.B).

Gliding bacteria form remarkable structures called fruiting bodies. A group of cells aggregates to make one of these structures. The fruiting bodies of some species are simple globes more than 1 mm in diameter; those of other species are more complex, branched structures (Figure 22.7b). Within the fruiting bodies of some species, single cells transform themselves into thick-walled spores that can resist harsh environmental conditions; in other fruiting bodies, whole clusters of cells form a cyst that is resistant to drying. Both spores and cysts can germinate under favorable conditions to yield the next generation of typical gliding cells.

SPIROCHETES. The spirochetes are characterized by unique structures called **axial filaments**, composed of flagella, running along the cell body between a thin, flexible cell wall and an outer envelope (Figure 22.8a). The cell body is a long cylinder coiled into a spiral. The flagella constituting the axial filaments begin at either end of the cell and overlap in the middle. The axial filaments are thought to be respon-

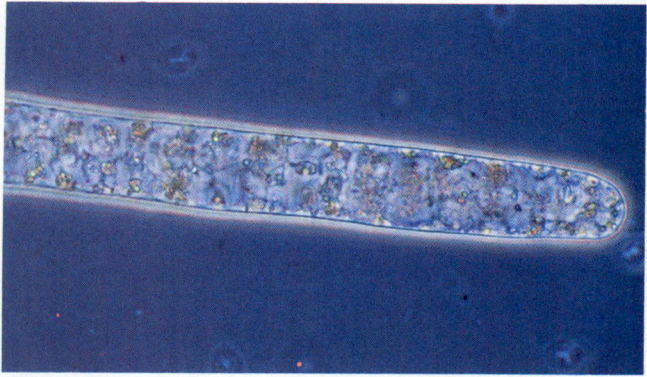

(a)

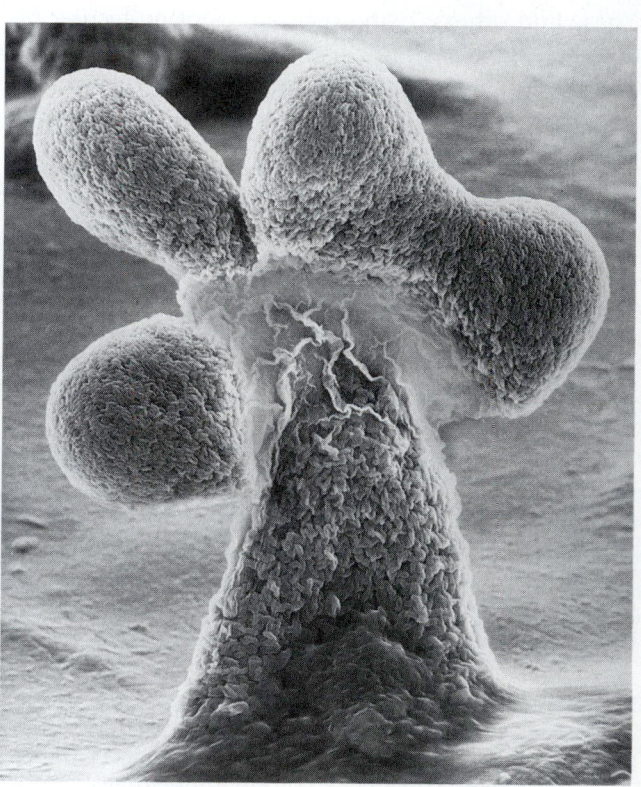

(b)

22.7 Gliding Bacteria
(a) This filament of *Beggiatoa* was isolated from ocean mud. (b) Individual bacteria aggregated to make up the stalk and knobs of this fruiting body of *Stigmatella aurantiaca*.

BOX 22.B

Gliding through the Soil

Bacteriologists collect *Beggiatoa* at night by scraping up samples of dirt. Why do this by night rather than by day? These gliding bacteria live deeper in the soil by day, so it is easier to collect them in the top layer at

night. Why do they move toward the surface by night and head for deeper layers in the daytime? They are aerobic bacteria and need oxygen to metabolize, but they obtain their energy by oxidizing reduced sulfur-containing compounds such as hydrogen sulfide. These sulfur compounds are produced deep in the soil, away from oxygen which is present in air. Because oxygen itself oxidizes the sulfur compounds, *Beggiatoa* must grow in a rather narrow zone—a zone of compromise between being deep enough to have a supply of the sul-

fur compounds but high enough to get enough oxygen. Another group of bacteria (the cyanobacteria, described later in this chapter) also contribute to the oxygen supply, however, by performing photosynthesis. During the daylight hours, cyanobacteria produce oxygen photosynthetically and thus increase the oxygen supply in the upper soil layers. At night, when the cyanobacteria do not photosynthesize and thus produce no oxygen, *Beggiatoa* must glide toward the surface of the soil to get enough oxygen.

---

sible for the motility (movement) of these organisms, and there are typical basal rings where the flagella are attached to the cell wall, but the precise mechanism by which they work is unknown.

Many spirochetes live in humans as parasites including *Treponema pallidum*, the organism that causes the venereal disease syphilis (Figure 22.8*b*). Others live free in mud or water.

CURVED AND SPIRAL BACTERIA. The curved and spiral bacteria have diverse properties, with little in common other than a curved form and a gram-negative reaction. Unlike spirochetes, the spiral bacteria in this group do not have an axial filament. Some, notably members of the genus *Spirillum*, live free in fresh or salt water, whereas others are parasites. Some species cause diseases in animals. Several strains of *Campy-*

lobacter fetus, for example, are increasingly recognized as causes of intestinal inflammation and other damage to humans (Figure 22.9). *Bdellovibrio* is a particularly interesting genus in this group. Some species of *Bdellovibrio* penetrate and reproduce within other bacteria—one bacterium parasitizing another.

GRAM-NEGATIVE RODS. The gram-negative rods, as a group, demonstrate both the advantage and weakness of the classification scheme employed here. The advantage is that such groups are readily recognized in the laboratory by their shape and staining properties. The disadvantage is that the group is so large and diverse that it can scarcely be a "natural" group with close evolutionary ties among its members. Some gram-negative rods are aerobic, others are facultative anaerobes, and still others are obligate anaerobes. Nitrogen-fixing genera such as *Rhizobium* (see Figure 32.10), which plays an important role in plant nutrition, are included, as are *Nitrobacter, Thio-*

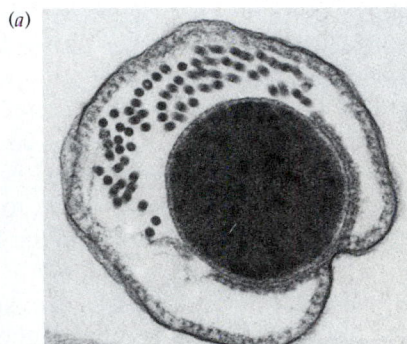
(a)

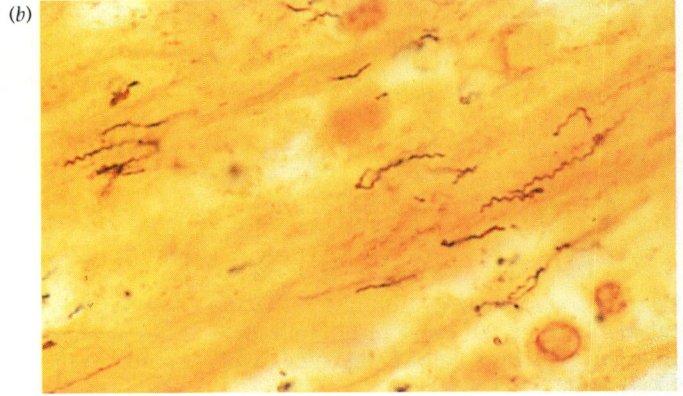
(b)

22.8 Spirochetes
(a) A spirochete from the gut of a termite, seen in cross section. The axial filaments are the dots inside the outer envelope but outside the dense cell body. (b) Corkscrew-shaped cells of *Treponema pallidum* in fetal tissue. Transmitted by sexual contact or from mother to fetus, *T. pallidum* causes syphilis.

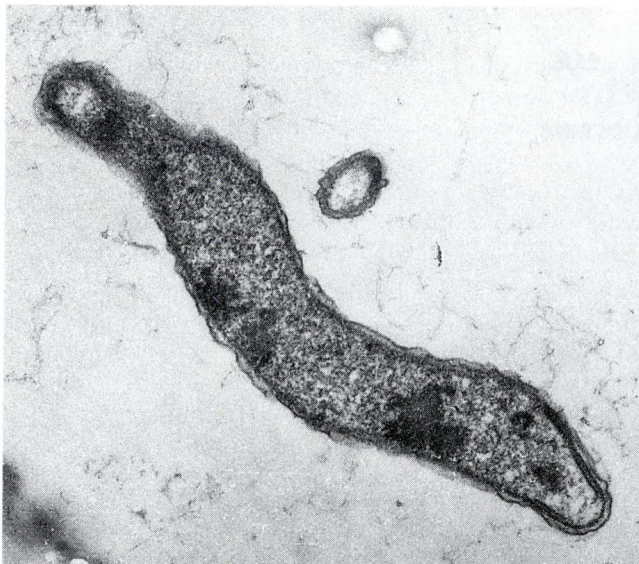

22.9   A Spiral Bacterium
Various strains of the spiral *Campylobacter fetus* cause sheep and cattle to abort their fetuses and cause intestinal inflammation in humans.

22.11   Crown Gall
This massive growth on a white oak trunk is crown gall, a plant disease caused by *Agrobacterium tumefaciens*.

bacillus, and other anaerobes that use nitrogen or sulfur compounds instead of oxygen for respiration.

Escherichia coli, probably the most studied organism of all, is a gram-negative rod (see Box 4.B). So, too, are many of the most famous human pathogens, such as *Yersinia pestis* (the cause of plague), *Shigella dysenteriae* (dysentery), *Vibrio cholerae* (cholera), and *Salmonella typhimurium* (a common agent of food poisoning). A bacterium from this group is shown in Figure 22.10.

Certain gram-negative rods invade animal cells, where they survive and cause diseases. For example,

Yersinia pseudotuberculosis, the agent of guinea-pig plague, invades the intestinal cells of guinea pigs and other mammals. Its ability to invade mammalian cells results from the possession of a single gene, called *inv*, that codes for a single large protein. The role of *inv* was revealed when biologists using recombinant DNA techniques successfully transferred this gene from *Y. pseudotuberculosis* to *E. coli*, with the result that the recipient *E. coli* cells were able to invade mammalian cells.

Most plant diseases are caused by fungi; viruses cause others; but about 200 plant diseases are of bacterial origin. **Crown gall**, with its characteristic tumors (Figure 22.11), is one of the most striking. The causal agent of crown gall is *Agrobacterium tumefaciens*, a gram-negative rod. *A. tumefaciens* harbors a plasmid containing the genes responsible for the crown gall disease. The plasmid is used in recombinant DNA studies as a vehicle for inserting genes into new plant hosts (see Chapter 14).

GRAM-NEGATIVE COCCI.   Cocci are spherical bacterial cells. Some are gram-negative, others gram-positive. Some of the gram-negative cocci use oxides of nitrogen or sulfur as terminal electron acceptors for cellular respiration; among them are the nitrifiers (see Chapter 32). Because it causes the venereal disease gonorrhea, *Neisseria gonorrhoeae* is the most well-known coccus (Figure 22.12). *N. gonorrhoeae* changes

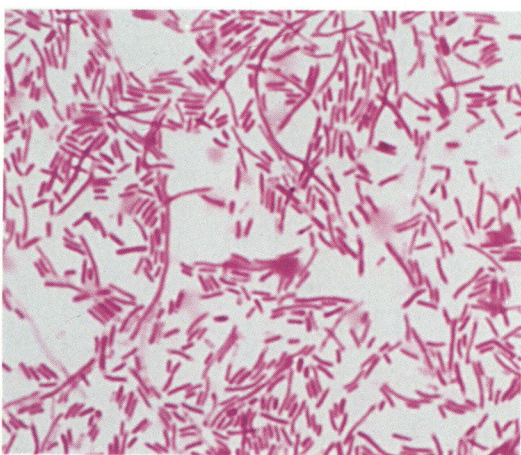

22.10   Gram-Negative Rods
Salmonella typhi, the cause of typhoid fever, is a gram-negative rod. The pink color is the "negative" response to the Gram stain.

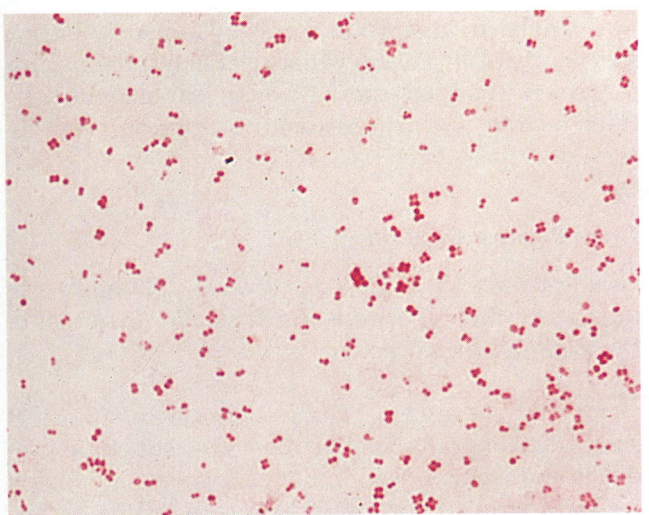

22.12   A Gram-Negative Coccus
Neisseria gonorrhoeae, which causes gonorrhea, often aggregates in groups of two to four cells.

its antigenic determinants frequently by means of transposable elements, making it an elusive target for the human immune system. The mechanism is similar to the cassette mechanism used by yeasts to change their mating type (see Box 13.A). *N. meningitidis* causes a frequently fatal infection of the linings of the nervous system. Other genera of gram-negative cocci, such as *Acinetobacter* and *Moraxella*, also cause infections.

RICKETTSIAS AND CHLAMYDIAS.   The rickettsias and related organisms were once grouped with viruses be-

cause they are so small and because they reproduce only within the cells of other organisms. The rickettsias, extremely small parasites approximately 1 μm in length and 0.3 μm in diameter, have cell walls and chemical characteristics similar to those of some gram-negative bacteria. With one exception, rickettsias have never been grown outside living cells. Rickettsias are agents of several serious diseases in humans, notably Rocky Mountain spotted fever (which is actually more common in the southeastern United States than in the Rocky Mountains) and typhus (Figure 22.13*a*). Rickettsias are frequently carried by arthropods, particularly fleas and ticks, but they do not seem to cause any disease symptoms in their arthropod hosts.

Chlamydias are often grouped with the rickettsias because of their size (0.2 to 1.5 μm in diameter) and because they too can live only as parasites within the cells of other organisms. These tiny spheres are unique prokaryotes because of their complex reproductive cycle, which involves two different types of cells (Figure 22.13*b*). In humans, various strains of chlamydias cause eye infections (especially trachoma), venereal disease, and some forms of pneumonia.

CYANOBACTERIA.   The cyanobacteria (blue-green bacteria) are very independent nutritionally. They perform photosynthesis by using chlorophyll *a* and liberating oxygen gas, they carry out fermentation under anaerobic conditions, and many fix nitrogen (see Chapter 32 for a discussion of nitrogen fixation). Cyanobacteria require only water, nitrogen gas, oxygen, a few mineral elements, light, and carbon dioxide. De-

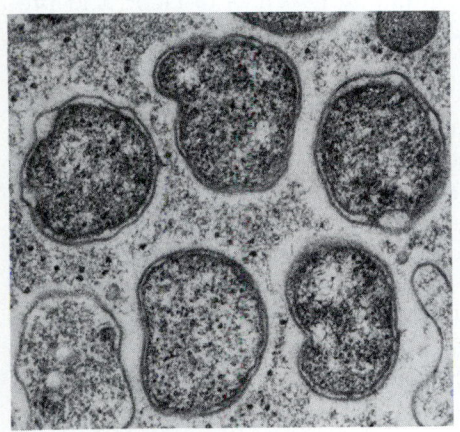
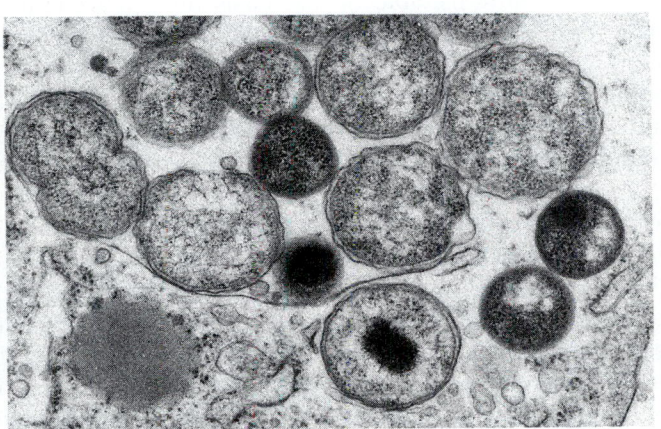

(a)   (b)

22.13   Rickettsias and Chlamydias
(a) A cluster of tiny rickettsias in the cytoplasm of an infected animal cell. The species shown here is *Rickettsia prowazekii*, the cause of typhus; it is carried from person to person by lice. (b) The small, dense elementary bodies and the larger, thin-walled initial bodies are the two major phases of the life cycle of *Chlamydia psittaci*. The elementary bodies are taken into a cell by phagocytosis and then develop into initial bodies; the initial bodies, which cannot infect other cells, grow and divide (note the dividing initial body at the left). Finally, initial bodies reorganize into elementary bodies, which are liberated by rupture of the host cell.

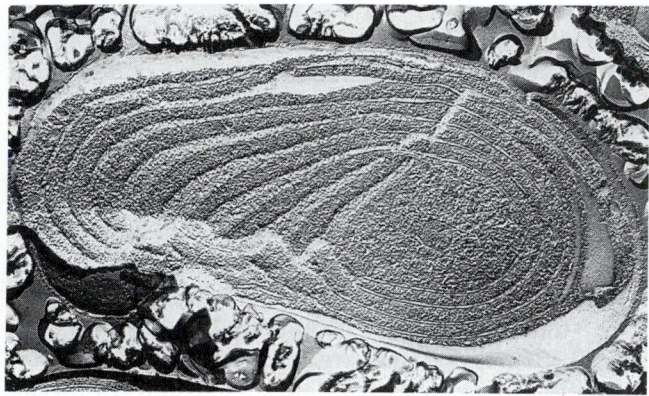

22.14 Thylakoids in Cyanobacteria
A cell of a cyanobacterium prepared by the freeze-etch method to emphasize the extensive system of internal membranes. These photosynthetic membranes are present through most of the cytoplasm.

spite their ability to carry out the type of photosynthesis otherwise characteristic only of eukaryotic photosynthesizers, they are true prokaryotes. They contain none of the familiar membrane-bounded organelles of eukaryotic cells, they do not have a discrete nucleus, their chromosomes lack histones, and their cell walls contain peptidoglycan. Even so, cyanobacteria contain elaborate and highly organized internal membrane systems, the photosynthetic lamellae, or thylakoids (Figure 22.14). They are also the only prokaryotic photosynthesizers that contain chlorophyll *a*, with the exception of one other cyanobacteriumlike genus, *Prochloron*.

The cyanobacteria form a closely related, homogeneous, logical grouping. They are placed in a group by themselves because of their mode of photosynthesis; they are also distinguishable on the basis of rRNA sequence. Lest we forget the complexity of prokaryote taxonomy, however, we should mention that a few leading microbiologists classify the cyanobacteria as a subgroup of the gliding bacteria because mobile cyanobacteria use the gliding type of locomotion.

Cyanobacteria can associate in colonies or can live free as single cells. Depending on the species and on growth conditions, colonies of cyanobacteria may range from flat sheets one cell thick to spherical balls of cells. Some filamentous colonies show differentiation into three cell types: vegetative cells, spores, and **heterocysts**. Heterocysts are specialized for nitrogen fixation; the enzyme nitrogenase gives them this ability. All the known cyanobacteria with heterocysts fix nitrogen. Heterocysts also have a role in reproduction: When filaments break apart to reproduce, the heterocyst may serve as a breaking point. Figures 22.15*a* and *b* show heterocysts within filaments.

Cyanobacteria reproduce only by fission. There are viruses that can infect cyanobacteria and then transfer genetic material from one organism to another by transduction, but true sexuality has never been observed in these bacteria.

Gram-Positive Bacteria

Gram-positive bacteria are diverse, although the number of their species is much smaller than that of the gram-negative bacteria.

GRAM-POSITIVE RODS. There are two principal subgroups of gram-positive rods. One subgroup produces endospores. Recall that endospores are highly resistant structures containing a copy of the bacterium's DNA, some ribosomes and other cytoplasmic constituents, a thick peptidoglycan coat, and an outer spore coat (see Figure 22.5). Members of this endospore-forming group include the many species of *Bacillus*, including *B. anthracis*, the agent of anthrax in sheep and humans; *B. thuringiensis*, a species used in commercial preparations to kill moth and butterfly larvae that are plant pests; and *B. subtilis* (Figure 22.16*a*). The genus *Clostridium*, also in this subgroup, includes *C. denitrificans*, a free-living bacterium that plays an important role in Earth's nitrogen cycle; two producers of potent toxins, *C. botulinum* (the agent of various forms of botulism) and *C. tetani* (tetanus); and *C. perfringens* (which causes food poisoning and gas gangrene). The toxins produced by *C. botulinum* are among the most poisonous ever discovered—the lethal dose for humans is about one-millionth of a gram (1 μg).

The members of the other subgroup of gram-positive rods do not form endospores. These include bacteria of the genera *Lactobacillus* (a lactic acid producer; Figure 22.16*b*), *Listeria*, and others.

GRAM-POSITIVE COCCI. There are many gram-positive cocci. Cells of the genus *Staphylococcus* (Figure 22.17*a*) called staphylococci, are about 1 μm in diameter. Staphylococci are abundant on the human body surface and are responsible for boils and many other skin problems. *S. aureus* is the best-known human pathogen; it is found in 20 to 40 percent of normal adults (and in 50 to 70 percent of hospitalized adults) and can cause respiratory, intestinal, and wound infections, in addition to skin diseases. Staphylococci produce toxins that are a major cause of food poisoning and that cause toxic shock syndrome.

Streptococcus is another important genus of gram-positive cocci. Cells of this group are known as streptococci. Streptococci divide along a single axis and stay together after fission, forming chains (Figure 22.17*b*). *S. mutans* is an acid-producing oral species

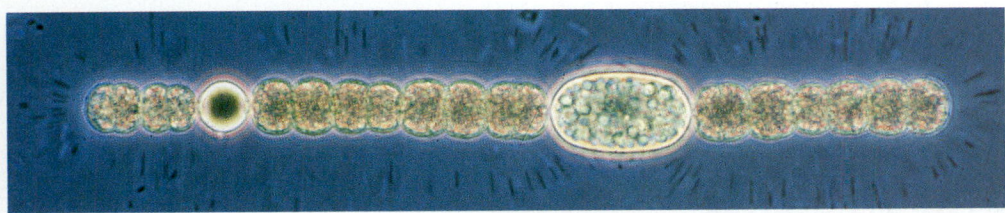

(a)

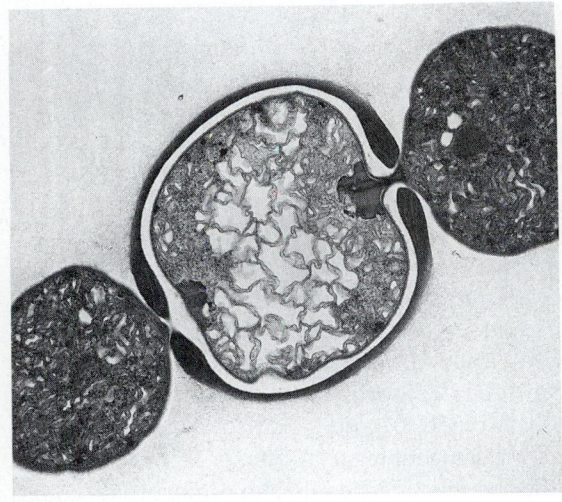

(b)

(c)

22.15 Cyanobacteria

(a) *Anabaena* is a genus of filamentous cyanobacteria. The filament shown here has a heterocyst (the thick-walled, circular cell) and a resting spore (the elongated, thick cell near the other end). (b) The spherical cell in the center of this electron micrograph is a heterocyst. A thin neck attaches it to each of the other two cells. A thick wall separates the cytoplasm of the nitrogen-fixing heterocyst from the surrounding environment. (c) Cyanobacteria appear in enormous numbers in some environments. This tidewater pond has experienced eutrophication: Phosphorus and other nutrients generated by human activity have accumulated in the pond, feeding an immense green mat composed of various species of unicellular cyanobacteria.

that can erode tooth enamel. There is no major organ system in the human body that is not subject to one form of streptococcal infection or another. The first demonstration that DNA is the hereditary material was performed by Avery, MacLeod, and McCarty, using cultures of *S. pneumoniae*, formerly called *Diplococcus pneumoniae* (see Figure 11.1).

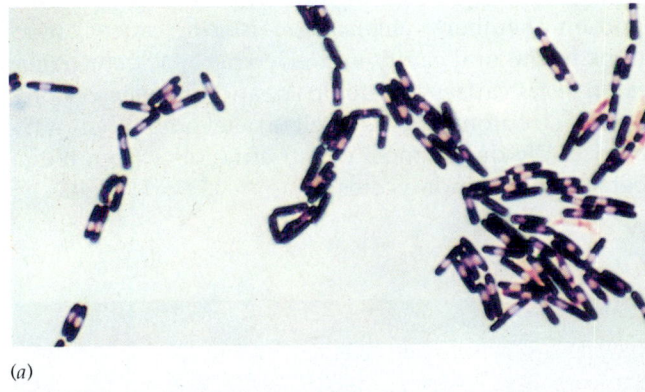

(a)

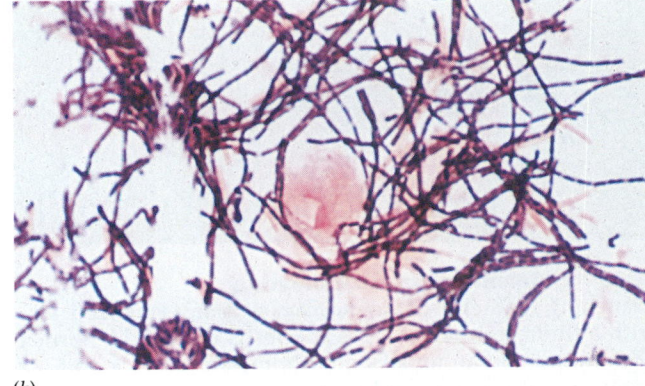

(b)

22.16 Gram-Positive Rods

(a) In response to unfavorable changes in their environment, these gram-positive *Bacillus subtilis* are developing endospores (the clear areas in the middle of the otherwise deeply-stained cells). (b) Gram-positive rods of *Lactobacillus* do not form endospores.

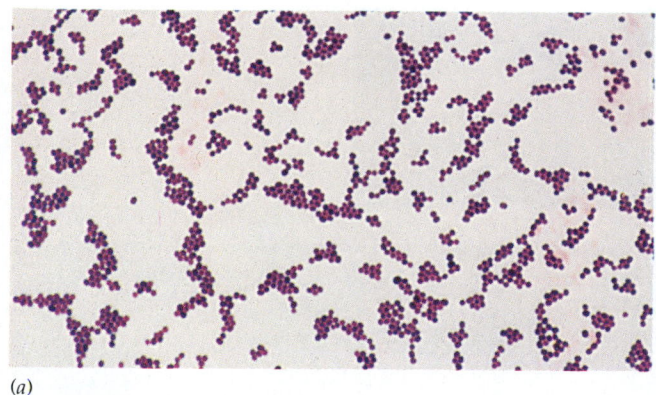

(a)

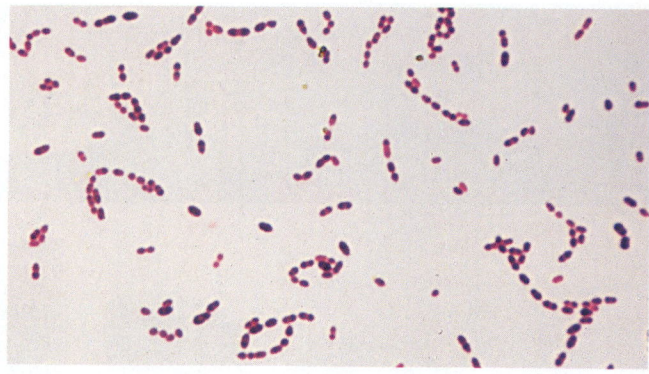

(b)

22.17   Gram-Positive Cocci

(a) "Grape clusters" are the usual arrangement of gram-positive staphylococci such as these *Staphylococcus aureus* cells. (b) Streptococci usually aggregate in chains of two or more cells, as shown here by *Streptococcus pyogenes*.

ACTINOMYCETES. Actinomycetes develop an elaborately branched system of filaments (Figure 22.18). These bacteria closely resemble the filamentous bodies of fungi and were, in fact, once classified as fungi.

Some actinomycetes reproduce by forming chains of spores at the tips of the filaments. In the species that do not form spores, the branched, filamentous growth ceases and the structure breaks up into typical cocci or rods, which then reproduce by fission.

The actinomycetes include several medically important members: *Actinomyces israelii*, causes infections in the oral cavity and elsewhere, *Mycobacterium tuberculosis* causes tuberculosis, and *Streptomyces* produces streptomycin as well as several other antibiotics. We derive most of our antibiotics from members of the actinomycetes.

Mycoplasmas

The third phylum of the kingdom Eubacteria consists of bacteria that lack cell walls. These bacteria, the mycoplasmas, are the smallest cellular creatures ever discovered—they are even smaller than rickettsias and chlamydias (Figure 22.19). Mycoplasmas have diameters between 0.1 and 0.2 μm. The mycoplasmas are small in another crucial sense: They have less than half as much DNA as do the other prokaryotes. It has been speculated that the amount of DNA in a mycoplasma may be the minimum amount required to code for the absolutely essential properties of a cell.

Most mycoplasmas are parasites found within the cells of animals and plants. As intracellular parasites, they are not subjected to the osmotic challenges faced by free-living bacteria, so they need no cell walls. As a result, they take on irregular shapes, and they cannot be killed with penicillin, which kills other bacteria by interfering with wall synthesis.

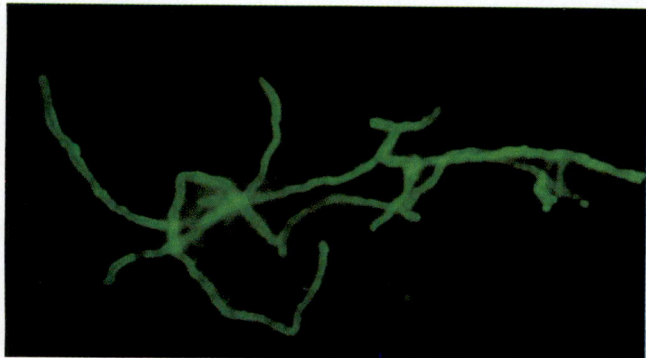

22.18   Filaments of an Actinomycete

Branching filaments of *Actinomyces israelii*, visualized with a fluorescent stain. This species is part of the normal flora in the human tonsils, mouth, intestinal tract, and lungs but will invade body tissues and cause severe abscesses when afforded the opportunity.

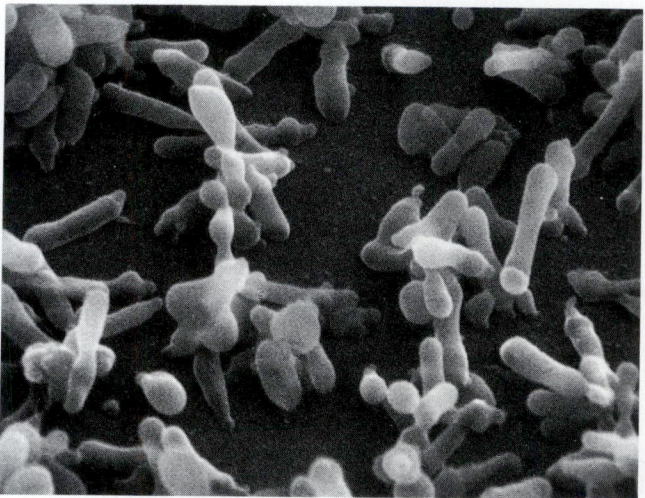

22.19   The Tiniest Bacteria

Cells of *Mycoplasma gallisepticum*. Lacking cell walls and containing only about half as much DNA as other bacterial cells, mycoplasmas are the smallest known bacteria.

VIRUSES

Discovery of the Viruses

Most viruses are much smaller than most bacteria (Table 22.2). Viruses have become well understood only within the last half century, but the first step on this path of discovery was taken by the Russian botanist Dmitri Ivanovsky in 1892. In studying tobacco mosaic disease, Ivanovsky tried to isolate the causal agent of the disease by passing an infectious extract of diseased tobacco leaves through a fine porcelain filter, a technique that had been used previously to isolate disease-causing bacteria. To his surprise, the liquid that passed through the filter still caused tobacco mosaic disease. Other workers soon showed that similar filterable agents, so tiny that they cannot be seen under the light microscope, cause several plant and animal diseases. They discovered that alcohol, which kills cultures of bacteria, does not destroy the tiny agents' ability to cause disease.

In 1935 Wendell Stanley, of what is now Rockefeller University, was the first to succeed in crystallizing viruses. The crystalline viral preparation became infectious again when it was dissolved. It was soon shown that crystallized viral preparations consist of protein and nucleic acid. Finally, direct observation of viruses with electron microscopes confirmed how much they differ from bacteria and other organisms.

Viral Structure

Unlike the organisms that make up the six taxonomic kingdoms of the living world, viruses are **acellular**; that is, they are not cells and do not consist of cells. Unlike cellular creatures, viruses do not metabolize energy—they neither produce ATP nor conduct fermentation, cellular respiration, or photosynthesis.

TABLE 22.2
Common Sizes of Microorganisms

MICROORGANISM	TYPE	TYPICAL SIZE RANGE (CUBIC MICROMETERS)
Protists	Eukaryote	5,000–50,000
Photosynthetic bacteria	Prokaryote	5–50
Spirochetes	Prokaryote	0.1–2
Mycoplasmas	Prokaryote	0.01–0.1
Pox viruses	Virus	0.01
Influenza virus	Virus	0.0005
Polio virus	Virus	0.00001

Modified from R. Y. Stanier, E. A. Adelberg, and J. Ingraham. 1976. *The Microbial World*, 4th Edition.

Whole viruses never arise directly from preexisting viruses. They are *obligate intracellular parasites*; that is, they develop and reproduce only within the cells of specific hosts. The cells of animals, plants, fungi, protists, and bacteria serve as hosts to viruses. Viruses outside host cells exist as individual particles called **virions**. The virion, the basic unit of a virus, consists of a central core of either DNA or RNA (but not both) surrounded by a **capsid**, or coat, which is composed of one or more proteins (Figure 22.20). These proteins are assembled so as to give the virion a characteristic shape. Many animal viruses also acquire a membrane consisting of lipids and proteins as they bud through host cell membranes in the course of viral reproduction, and many bacterial viruses have specialized "tails" made of protein. The complex architecture of HIV-I, the AIDS virus, is shown in Figure 16.26.

Reproduction of Viruses

Viruses reproduce by taking over their host cell's metabolism; the viral nucleic acid directs the production of new viruses from host materials. Animal viruses begin the process by attaching to the plasma membrane of the host cell. They are then taken up by endocytosis, which traps them within a membranous vesicle inside the cell. The membrane of the vesicle breaks down, and the host cell digests the protein capsid. At this point, the viral nucleic acid takes charge. The host cell replicates the viral nucleic acid and synthesizes new capsid protein as directed by the viral nucleic acid. New capsids and new viral nucleic acid combine spontaneously, and in due course, the host cell releases the new virions. Animal viruses usually escape from the host cell by budding through virus-modified areas of the plasma membrane. During this process the completed virions acquire a membrane similar to that of the host cell (Figure 22.21).

Plant viruses must pass through a cell wall as well as through the host plasma membrane. They accomplish this penetration through their association with **vectors**, which are intermediate carriers of disease from one organism to another. Infection of a plant usually results from attack by a virion-laden insect vector. The insect uses its proboscis (snout) to penetrate the cell wall, allowing the virions to move via the proboscis from the insect into the plant. Plant viruses, such as tobacco mosaic virus, can be introduced artificially without insect vectors: First we mechanically bruise a leaf or other part; then we expose the bruised plant surface to a suspension of virions. Bacteriophages (bacterial viruses) which also must penetrate cell walls, are often equipped with tail assemblies that inject their nucleic acid through the cell wall into the host bacterium. Virions escape from

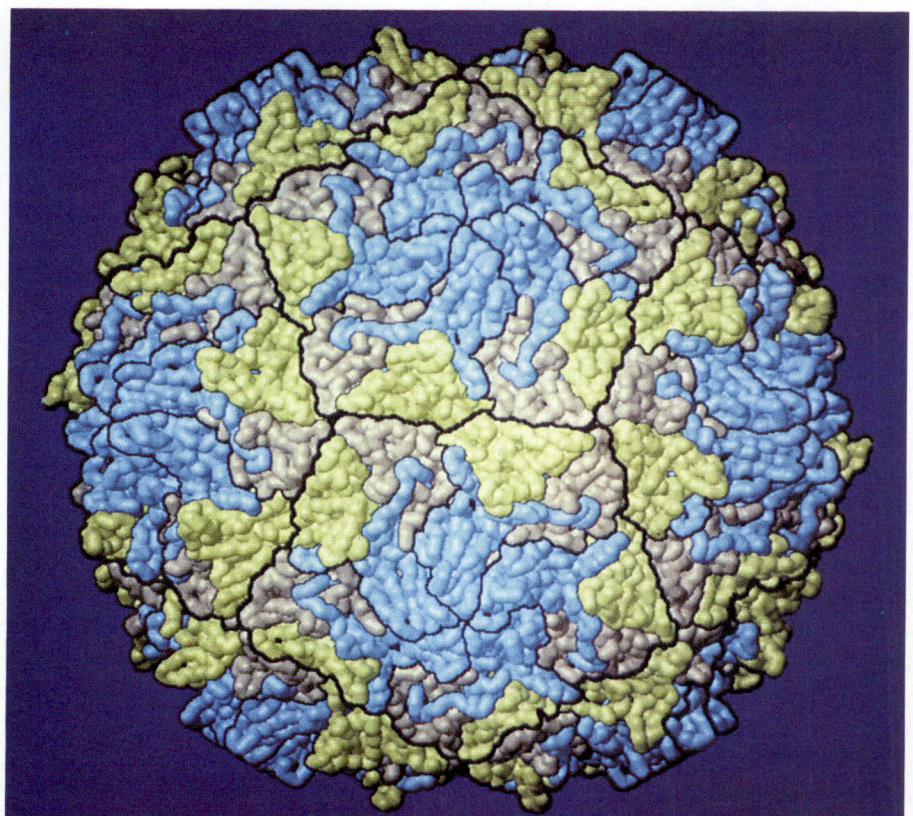

22.20   A Virion
The capsid of a poliovirus virion, as drawn by computer. There are three major proteins in the capsid, each shown in a different color. The proteins are organized into building blocks, outlined by the wavy black lines.

plant or bacterial cells by lysing (breaking apart) the host cell, rather than by budding. Some bacteriophages have lytic life cycles (cycles that result in rapid lysis of host cells); others have lysogenic life cycles, in which the viral and host nucleic acids replicate at the same time, and the virus may be present as a "silent" provirus for many bacterial cell generations.

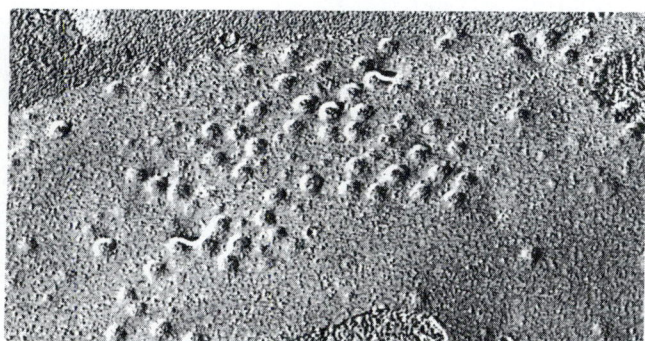

22.21   Buds of an Animal Virus
The numerous small bumps on the curved surface of this cell are buds of Sindbis virus (named after the village in Egypt where it was first isolated) in the plasma membrane. At this stage the capsids are acquiring membranous envelopes that completely surround each virion when the process is complete. These membrane envelopes make the first contact when these virions infect new host cells.

Bacteriophage life cycles were described in detail in Chapter 12.

How does a virion recognize a suitable host? Some bacteriophages use a specific interaction between the proteins of the bacteriophage tail and of the host cell wall. Membrane-surrounded animal viruses probably depend on the membrane to recognize suitable cells: Because the membrane was obtained from the previous host, it can readily fuse with the plasma membrane of a new host cell. It is not known how other virions recognize their host cells.

Classification of Viruses

A common way to classify viruses separates them by whether they have DNA or RNA and then by whether their nucleic acid is single- or double-stranded. Some of the RNA viruses have more than one molecule of RNA, and the DNA of one virus family is circular. Further levels of classification depend upon factors such as the overall shape of the virus and the symmetry of the capsid. Most capsids may be categorized as **helical** (coiled like a spring; Figure 22.22*a*), **icosahedral** (a regular solid with 20 faces; Figure 22.22*b*), or **binal** (having a polyhedral, or many-faced, head with a helical tail; Figure 22.22*c*). Another level of classification is based on the presence or absence of a membranous envelope around

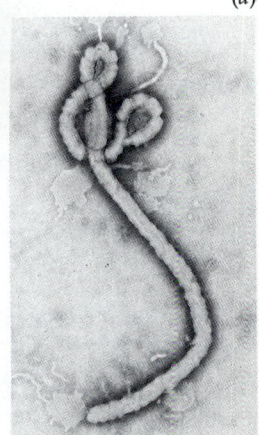

22.22   Viruses Come in Different Shapes
(a) Tobacco mosaic virus consists of an inner helix of
RNA covered with a helical array of protein molecules.
This computer model corresponds to about one-seventh
of the entire, long virus. (b) Adenoviruses have an icosahedral capsid as an outer shell. Inside this 20-faced structure is a spherical mass of other proteins and DNA. (c)
These T2 bacteriophages illustrate the binal form of capsid. (d) Not all virions are regular in shape. Wormlike
virions of Ebola virus infect humans, causing hemor-
rhages.

the virion; still further subdivision relies on capsid
size. Table 22.3 shows only the major levels of virus
classification and some examples; technical names of
the taxonomic groups are omitted.

The distribution of viruses in terms of host organisms is puzzling. Viral diseases of flowering plants
are very common, but such diseases are rare in the
cone-bearing seed plants, ferns, algae, and fungi.
Almost all vertebrates are susceptible to viral infections, but among invertebrates, such infections are
common only in arthropods. A group of viruses
called **arboviruses** (short for arthropod-borne viruses) causes serious diseases, such as encephalitis,
in humans and other mammals. Arboviruses are
transmitted to the mammalian host through a bite
(certain arboviruses are carried by mosquitoes, for
example). Although carried within the arthropod
vector's cells, arboviruses apparently do not affect
the insect host severely; they affect only the bitten
and infected mammal.

Viroids: RNA without a Capsid

Pure viral nucleic acids can produce viral infections
under laboratory conditions, although only ineffi-
ciently. Might there be infectious agents in nature
that consist of nucleic acid without a protein capsid?
In 1971, Theodore Diener of the U.S. Department of
Agriculture reported the isolation of agents of this
type, called **viroids**. Viroids are single-stranded RNA
molecules consisting of 270 to 380 nucleotides. They
are one-thousandth the size of the smallest viruses.
All the viroids studied thus far have substantial regions of internal complementarity, so they fold into

double-stranded rods. These rods are most abundant
in the nuclei of infected cells.

Viroids have been found only in plant cells. There
they produce a variety of diseases. Two mechanisms
are known by which viroids may be transmitted from
plant to plant. If two plants (one infected and one
uninfected) are injured and their wounded surfaces
come into contact, viroids may be transmitted from
the infected plant to the uninfected one. The other
mechanism of transfer operates from generation to
generation. If a pollen grain or an ovule produced by
an infected plant contains viroids, these will infect
the plant produced after this gamete unites with another in fertilization.

There is no evidence that viroids are translated to
synthesize proteins, and it is not known how they
cause disease. Viroids are replicated by the enzymes
of their plant hosts. Similarities in base sequences
between viroids and transposable genetic elements
strongly suggest that viroids evolved from transposable elements (see Chapters 12 and 13).

Scrapie-Associated Fibrils: Infectious Proteins?

A class of protein fibrils, called **scrapie-associated
fibrils**, or prions, has been associated with certain
degenerative diseases of mammalian central nervous
systems. These fibrils consist entirely of protein, with
no evident nucleic acid component. The fibrils are
associated with scrapie, a disease of sheep and goats,
and may be the infective agent of the disease. Such
fibrils have also been identified in connection with
two similar diseases of the human central nervous

TABLE 22.3
A Classification Scheme for Some Animal Viruses

	NUCLEIC ACID			VIRION		
VIRUS GROUP	MOL. WT. (millions)	TYPE	STRANDS	SHAPE	SIZE (nm)	NOTES
Families of viruses affecting both vertebrates and other hosts						
Poxviridae	160–200	DNA	2	Brick-shaped	300×240×100	Pox viruses
Parvoviridae	1.2–1.8	DNA	1	Icosahedral	20	Hosts include rats and insects
Reoviridae	15	RNA	2	Icosahedral	50–80	Vertebrate, insect, and plant hosts
Rhabdoviridae	4	RNA	1	Bullet-shaped	175 × 70	Rabies, vesicular stomatitis
Families of viruses of vertebrates						
Herpetoviridae	100–200	DNA	2	Icosahedral	150	Herpes
Adenoviridae	20–29	DNA	2	Icosahedral	70–80	Adenovirus
Papovaviridae	3–5	DNA	2	Icosahedral	45–55	Papillomas
Retroviridae	10–12	RNA	1	Spherical	100–200	Tumor viruses
Paramyxoviridae	7	RNA	1	Spherical	100–300	Measles, Newcastle disease
Orthomyxoviridae	5	RNA	1	Spherical	80–120	Influenza
Togaviridae	4	RNA	1	Spherical	40–60	Rubella, hog cholera, arboviruses
Coronaviridae	?	RNA	1	Spherical	80–120	
Arenaviridae	3.5	RNA	1	Spherical	85–120	
Picornaviridae	2.6–2.8	RNA	1	Icosahedral	20–30	Digestive and respiratory diseases
Bunyaviridae	6	RNA	1	Spherical	90–100	

system, kuru and Creutzfeldt–Jakob disease. One investigator of scrapie-associated fibrils, Stanley Prusiner of the University of California, San Francisco, has suggested a possible relationship between the fibrils and Alzheimer's disease, a severe dementia that is most common in the elderly.

The mechanism of action of scrapie-associated fibrils is unknown. Genes coding for the known parts of their amino acid sequences exist in the chromosomes of both infected and uninfected animals. These genes may be proviruses (see Chapter 12); further research will surely produce a detailed picture of the relationship between these genes and scrapie-associated fibrils.

SUMMARY of Main Ideas about Bacteria

Members of the kingdoms Archaebacteria and Eubacteria are all those organisms whose cells are prokaryotic.

Bacteria are obligate aerobes, obligate anaerobes, or facultative anaerobes depending on whether they obtain energy by cellular respiration, fermentation, or both.

As classified by the energy and carbon sources they can use, bacteria are photoautotrophs, photoheterotrophs, chemoautotrophs, or chemoheterotrophs.

Archaebacteria, all of which lack peptidoglycan in their cell walls, are classified by their environments into three phyla: thermoacidophiles, methanogens, and strict halophiles.

Eubacteria are classified by their Gram stain reaction and by presence or absence of a cell wall into three phyla: gram-negative bacteria, gram-positive bacteria, and mycoplasmas.
Review Table 22.1

SUMMARY of Main Ideas about Viruses

Viruses were first identified as disease-causing agents small enough to pass through filters that trap bacteria.

Viruses are acellular and are obligate intracellular parasites. A virus consists of RNA or DNA and protein.

Viruses reproduce by taking over the metabolism of their host cells.

Viruses are classified by their hosts and their nucleic acids.
Review Table 22.3

SELF-QUIZ

1. Most bacteria
   a. are agents of disease.
   b. lack ribosomes.
   c. evolved from the most ancient protists.
   d. lack a cell wall.
   e. are chemoheterotrophs.

2. All photosynthetic bacteria
   a. use chlorophyll *a* as their photosynthetic pigment.
   b. use bacteriochlorophyll as their photosynthetic pigment.
   c. release oxygen gas.
   d. produce particles of sulfur.
   e. are photoautotrophs.

3. Gram-negative bacteria
   a. appear blue to purple following Gram staining.
   b. are the most abundant of the bacterial groups.
   c. are all either rods or cocci.
   d. contain no peptidoglycan in their walls.
   e. are all photosynthetic.

4. Endospores
   a. are produced by viruses.
   b. are reproductive structures.
   c. are very delicate and easily killed.
   d. are resting structures.
   e. lack cell walls.

5. Archaebacteria
   a. are gram-negative.
   b. lack peptidoglycan in their cell walls.
   c. survive only at moderate temperatures and near neutrality.
   d. all produce methane.
   e. show little similarity among their subgroups.

6. Actinomycetes
   a. are important producers of antibiotics.
   b. belong to the kingdom Fungi.
   c. are never pathogenic to humans.
   d. are gram-negative.
   e. are the smallest known bacteria.

7. Which statement is *not* true of mycoplasmas?
   a. They lack cell walls.
   b. They are the smallest known cellular organisms.
   c. They contain the same amount of DNA as do other prokaryotes.
   d. They cannot be killed with penicillin.
   e. Most live within plant or animal cells.

8. Viruses
   a. are cellular in structure.
   b. produce ATP.
   c. are obligate parasites.
   d. all have DNA as their genetic material.
   e. undergo mitosis.

9. In viral reproduction,
   a. capsid protein structure is encoded in the host's DNA.
   b. new viruses are produced from host materials.
   c. the host cell must die before new viruses can be made.
   d. viral core nucleic acid is encoded in the host's DNA.
   e. new capsids are synthesized around new viral nucleic acid molecules.

10. Which statement is *not* true of viroids?
    a. They are smaller than viruses.
    b. They consist solely of RNA.
    c. They produce diseases in plants.
    d. They are translated to synthesize proteins.
    e. Their nucleic acid folds into double-stranded rods.

FOR STUDY

1. Contrast the biology of animal viruses, plant viruses, and bacteriophages.

2. Differentiate among the members of the following sets of related terms.
   a. prokaryotic/eukaryotic

   b. obligate anaerobe/facultative anaerobe/obligate aerobe
   c. photoautotroph/photoheterotroph/chemoautotroph/chemoheterotroph
   d. gram-positive/gram-negative

3. For each of the types of organism listed below, give a single characteristic that may be used to differ-

   entiate it from the related organism(s) in parentheses.
   a. spirochetes (spiral bacteria)
   b. *Bacillus* (*Lactobacillus*)
   c. mycoplasmas (free-living bacteria)
   d. cyanobacteria (other photoautotrophic bacteria)

4. Until fairly recently, the cyanobacteria were called blue-green algae and were not grouped with the bacteria. Suggest several reasons for this (abandoned) tendency to separate the bacteria and cyanobacteria. Why are the cyanobacteria now grouped with the bacteria?

5. The rickettsias were once grouped with the viruses. Why? Now they are grouped with the bacteria. Why?

READINGS

Balows, A., H. G. Trüper, M. Dworkin, W. Harder and K.-H. Schleifer (Eds.). 1991. *The Prokaryotes*. Three volumes. Springer-Verlag, New York. The ultimate reference on the bacteria: ecophysiology, isolation, identification, applications.

Brock, T. D., M. T. Madigan, J. M. Martinko and J. Parker. 1994. *Biology of the Microorganisms*, 7th Edition. Prentice-Hall, Englewood Cliffs, NJ. An excellent general textbook, including a chapter on viruses.

Carmichael, W. W. 1994. "The Toxins of Cyanobacteria." *Scientific American*, January. Some of their products can kill cattle, but these and other products may find pharmaceutical use.

Fischetti, V. A. 1991. "Streptococcal M Protein." *Scientific American*, June. How rheumatic fever and strep throat bacteria evade the body's defenses.

Hirsch, M. S. and J. C. Kaplan. 1987. "Antiviral Therapy." *Scientific American*, April. Approaches to killing viruses without damaging their host cells.

Hogle, J. M., M. Chow and D. J. Filman. 1987. "The Structure of Poliovirus." *Scientific American*, March. The relationship of viral structure and function is considered. The authors also speculate on viral evolution.

Koch, A. L. 1990. "Growth and Form of the Bacterial Cell Wall." *American Scientist*, vol. 78, pages 327–341. How does this cell wall, a single peptidoglycan molecule, allow the cell to grow but keep it from bursting?

McEvedy, C. 1988. "The Bubonic Plague." *Scientific American*, February. The bubonic plague shaped world history, and it hasn't disappeared yet.

Shapiro, J. A. 1988. "Bacteria as Multicellular Organisms." *Scientific American*, June. The behavior of highly regular bacterial colonies.

Stanier, R. Y., J. L. Ingraham, M. L. Wheelis and P. R. Painter. 1986. *The Microbial World*, 5th Edition. Prentice-Hall, Englewood Cliffs, NJ. A wide-ranging, authoritative text.

Tiollais, P. and M.-A. Buendia. 1991. "Hepatitis B Virus." *Scientific American*, April. Virus structure, epidemiology, vaccines, genetics, and replication.

Woese, C. 1981. "Archaebacteria." *Scientific American*, June. An early treatment of the subject by their leading student.

iatoms are recognizable by their remarkable shells—intricate constructions made
up of a glassy material and consisting of
two pieces fitting together like the top and
bottom of a petri plate. Some diatoms are motile,
although they have no flagella, legs, or other locomotor organs. They aren't animals, because they are
photosynthetic; they are clearly not plants, because
they do not produce embryos protected by surrounding tissues. Are they bacteria? No: Microscopic examination shows that they are eukaryotic. Like their
fellow unicellular protists, diatoms are tiny but complex. How can single cells show the complexity of
form and function found in the kingdom Protista?

The origin of the eukaryotic cell was one of the
greatest events in evolutionary history. In this chapter we celebrate the origin and early diversification
of the eukaryotes and the complexity achieved in
some single cells; that is, we celebrate the kingdom
Protista. This kingdom is a great evolutionary grab
bag, defined largely by exclusion. Thus the most precise definition of the protists, as we are using the
term, is *all eukaryotes that are not plants, fungi, or animals*. All protists are eukaryotic, and all evolved from
bacteria. The remaining three kingdoms—Plantae,
Fungi, and Animalia—evolved from protists.

Most protists are unicellular, but many are multicellular. Some biologists reserve the term Protista for
unicellular eukaryotes and nothing else. These biologists assign the organisms called multicellular protists in this book variously to the other three eukaryotic kingdoms. We find, however, that a classification
scheme that excludes multicellular organisms from
the kingdom Protista creates more problems than it
solves. In our scheme, there are many multicellular
protists.

Dividing the protists into phyla presents problems, and the taxonomy of this kingdom is an area
of intense research. There are several sources of this
interest. The marvelous diversity of body forms and
metabolic lifestyles would seem to be reason enough
for a fascination with these organisms, but questions
about the apparently multiple origins of protists from
the bacteria and about the origins of the multicellular
kingdoms from the Protista stimulate further interest,
and new tools of molecular biology make it possible
to explore evolutionary relationships in greater detail
and with greater confidence than ever before.

PROTISTA AND THE
OTHER EUKARYOTIC KINGDOMS

Part of the difficulty in placing certain organisms in
the appropriate kingdom is a natural consequence of
the evolutionary origin of the other eukaryotic kingdoms from the kingdom Protista. The other eukary-

Diatoms: Photogenic Protists

23
Protists

otic organisms—the fungi, plants, and animals— arose from protists in various ways. Deciding just where to draw the lines between the Protista and the other eukaryotic kingdoms is difficult. Different protists tend to resemble animals, plants, or fungi to one extent or another. The animallike protists are referred to as **protozoans**, the plantlike protists as **algae**. There are also **funguslike protists**—the slime molds, chytrids, and water molds.

In this book, we tend to assign eukaryotic organisms to the kingdom Protista if they are unicellular or colonial, although this is not a hard-and-fast rule. We separate protists from fungi using several criteria: presence of flagella (no fungi have flagella), gametes (some protists have visibly different male and female gametes; fungi do not), lack of a dikaryotic stage which fungi have; see Chapter 24), and chemistry of the cell wall (the cell walls of all fungi consist primarily of chitin). Development is what sets algae apart from plants: Whereas plants develop from embryos protected by tissues of the parent plant, algae develop from a single cell without such protection. The separation between protists and animals is relatively easy: An organism is an animal if it is a multicellular heterotroph with ingestive metabolism.

GENERAL BIOLOGY OF THE PROTISTS

Most protists are aquatic. Some live in marine environments, others in fresh water, and still others in the body fluids of other organisms. The slime molds inhabit damp soil and the moist, decaying bark of rotting trees.

Protists are strikingly diverse in their metabolism, perhaps second only to the bacteria. Nutritionally, some are autotrophs, whereas others are absorptive heterotrophs, and still others are ingestive heterotrophs. Some switch with ease between the autotrophic and heterotrophic modes of nutrition.

One protist phylum consists entirely of nonmotile organisms, but the other phyla include cells that move by amoeboid motion, by ciliary action, or by means of flagella. Most unicellular protists are tiny, but the multicellular plantlike protists include the giant kelps, which are among the longest organisms in existence.

Vesicles

Unicellular organisms tend to be of microscopic size, as we explained in Chapter 1. An important reason that cells are small is that they need enough membrane surface area to support the exchange of materials required for the life of the cell. The size unicellular protists can achieve is limited by their surface-to-volume ratio. Many relatively large unicellular protists minimize this problem by having membrane-

bounded vesicles of various types that increase their effective surface area. For example, many freshwater protists address their osmotic problems as follows.

Members of several of the protist phyla have **contractile vacuoles** that help them cope with their hypotonic environments. Because these organisms have a more negative osmotic potential than their freshwater environment does, they constantly take in water by osmosis. Excess water collects in the contractile vacuole and is then pushed out (Figure 23.1).

A beautifully simple experiment confirms that bailing out water is the principal function of the contractile vacuole. First we observe cells under the microscope and note the rate at which the vacuoles are contracting—they look like little eyes winking. Then we place other cells of the same type in solutions of differing osmotic potential. The less negative the osmotic potential of the surrounding solution, the more hypertonic the cells are and the faster the water rushes into them, causing the contractile vacuoles to pump more rapidly. Conversely, the contractile vacuoles stop pumping if the solute concentration of the medium is increased so that it is isotonic with the cells.

A second important type of vesicle found in many protists is the **food vacuole**. Protists such as *Paramecium* engulf solid food, forming food vacuoles within which the food is digested (Figure 23.2). Smaller vesicles containing digested food pinch away from the food vesicle and enter the organism's cytoplasm. These tiny vesicles provide a large surface area across which the products of digestion may be absorbed by the rest of the cell.

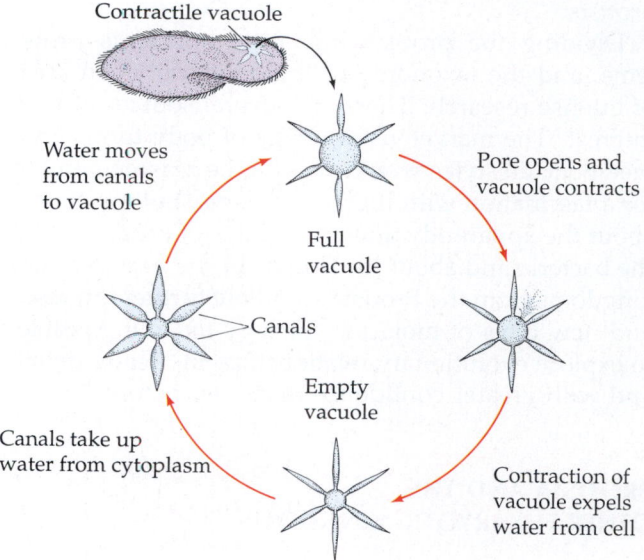

23.1 Contractile Vacuoles Bail Out Excess Water
A tiny pore in the contractile vacuole near a protist's surface opens when the vacuole is full, connecting it with the outside world; then the vacuole quickly contracts, expelling its contents.

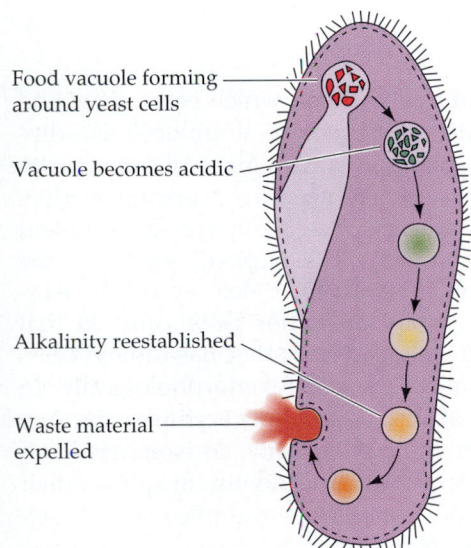

23.2 An Experiment with *Paramecium* Food Vacuoles
A biologist stained yeast cells with Congo red, a pH-sensitive dye, and fed them to a population of Paramecium. The paramecia formed food vacuoles. Changing colors revealed what was going on within the vacuoles. First, the dye quickly turned blue-green, indicating that the food vacuoles were becoming acidic; the increased acid concentration helped digest the yeast cells. As digestion came to an end and the products began to move through the vacuole membrane into the cytoplasm, the pH increased and the dye returned to its red color. Eventually the food vacuoles merged with the plasma membrane and expelled the digestive wastes to the environment.

The Cell Surface

A few protists, such as some amoebas (members of the phylum Rhizopoda), are surrounded only by a plasma membrane, but most have stiffer surfaces that maintain the cell's structural integrity. Many algae and funguslike protists have cell walls; these walls are often complex in structure.

The protozoans have a variety of surface coverings. Among these are "shells" either secreted by the organism itself, as foraminiferans (protists in the phylum Foraminifera) do, or made of bits of sand and thickenings immediately beneath the plasma membrane, as in some other amoebas (Figure 23.3). Among other roles, the plasma membrane often plays a part in the cell's sensitivity to its environment.

Sensitivity to the Environment

Many protists sense environmental stimuli and adjust their behavior appropriately. We expect such behavior of animals, but how can unicellular organisms show this complexity? The answer is not simple, but it often includes the presence of specialized organelles that carry out stimulus–response functions. Unicellular protists sense several environmental stimuli, including touch, temperature, light, and chemicals.

The cell requires separate systems to detect the several stimuli.

Endosymbiosis

In Chapter 4 we introduced the concept of endosymbiosis (organisms living together, one inside the other). As one of the most bizarre examples, we selected the protist *Mixotricha paradoxa*, which has a variety of bacteria living inside it and on its surface, but also is an endosymbiont itself, inhabiting termites. Endosymbiosis is very common among the protists, and in some instances the endosymbionts are also protists. All radiolarians (protists in the phylum Actinopoda), for example, harbor photosynthetic protists as endosymbionts (Figure 23.4). As a result, the radiolarians appear greenish or yellowish, depending on the type of endosymbiont they harbor. This arrangement is beneficial to the radiolarian, for it can make use of the food produced by its photo-

(a)

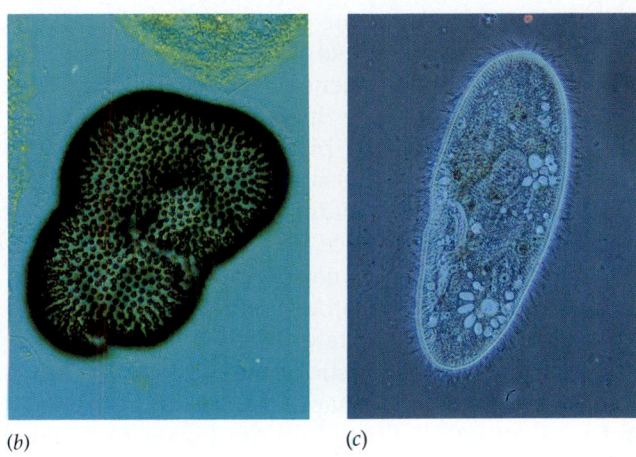
(b) (c)

23.3 Diversity in Protozoan Cell Surfaces
(a) Foraminiferan shells are made of protein hardened with calcium carbonate. (b) This shelled amoeba, *Difflugia*, constructed its shell by cementing sand grains together. (c) Spirals of protein make the surface of this *Paramecium*—known as its pellicle—flexible but resilient.

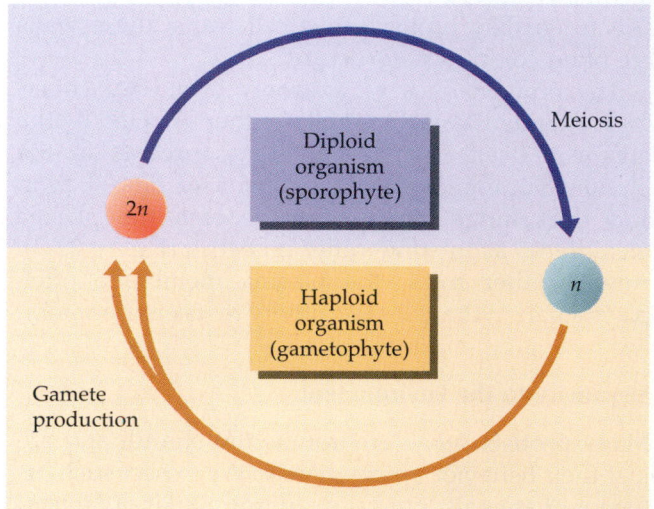

23.4 Protists within Protists
Photosynthetic algae living as endosymbionts within these radiolarians provide food for the radiolarians, as well as part of the pigmentation seen through the glassy skeletons. Both the algae and the radiolarians are protists.

synthesizing guest. In return the guest may make use of metabolites made by the host, or it may simply receive physical protection. Alternatively, the guest may actually be a victim, exploited for its photosynthetic products while receiving no benefit itself.

Endosymbiosis is important in the lives of many protists. This and other phenomena have contributed to the great success of the kingdom Protista, a kingdom that flourished for hundreds of millions of years before the first multicellular species evolved. Another source of this success was the remarkable diversity of reproductive strategies practiced by protists.

Reproduction

Although most protists indulge in both asexual and sexual reproduction, some groups lack sexual reproduction. As we will see, some asexually reproducing protists do engage in genetic recombination, even though it does not relate directly to reproduction.

Asexual reproductive processes in the kingdom Protista include binary fission (simple splitting of the cell), multiple fission, budding (the outgrowth of a new cell from the surface of an old one), and the formation of spores. Sexual reproduction also takes various forms. In the protozoans, reduction from the diploid to the haploid state differs from the formal meiosis that is characteristic of the multicellular kingdoms. Reduction to the haploid state does take place and is eventually followed by the fusion of haploid cells to produce a diploid zygote. In many algae and funguslike protists, both diploid and haploid cells undergo mitosis, giving rise to an alternation of generations.

Alternation of Generations

In **alternation of generations**, which some algae and funguslike protists demonstrate, a multicellular, diploid, spore-producing organism gives rise to a multicellular, haploid, gamete-producing organism; then fusion of two gametes once again creates a diploid organism (Figure 23.5). The haploid organism, the diploid organism, or both may also reproduce asexually. The two organisms differ genetically, in that one has haploid cells and the other has diploid cells, but they may or may not differ morphologically. In **heteromorphic alternation of generations**, the two organisms differ morphologically; in **isomorphic alternation of generations**, they do not, in spite of their genetic difference. We will see examples of both heteromorphic and isomorphic alternation of generations as we consider the phyla of protists.

Gametes are not generally produced directly by meiosis in multicellular protists, fungi, or plants. Instead, specialized cells of the diploid organism, called **sporocytes**, divide meiotically to produce four spores. The spores may eventually germinate and divide mitotically to produce multicellular haploid organisms. These haploid organisms constitute the **gametophyte** generation, so called because they produce gametes—by *mitosis* and cytokinesis. Unlike spores, gametes can produce new organisms only by fusing with other gametes. The fusion of two gametes produces a diploid zygote, which then undergoes mitotic divisions to produce a diploid organism: the **sporophyte** generation. The sporocytes of the sporophyte generation undergo meiosis at some point and produce haploid spores, starting the cycle anew.

23.5 Alternation of Generations
A diploid generation that produces spores alternates with a haploid generation that produces gametes.

TABLE 23.1
Classification of Protozoans

PHYLUM	COMMON NAME	FORM	LOCOMOTION	EXAMPLES
Zoomastigophora	Zooflagellates	Unicellular, some colonial	One or more flagella	*Trypanosoma, Trichonympha*
Rhizopoda	Amoebas	Unicellular, no definite shape	Pseudopods	*Amoeba, Entamoeba*
Actinopoda	Actinopods	Unicellular	Pseudopods	Radiolarians, heliozoans
Foraminifera	Foraminiferans	Unicellular	Pseudopods	Foraminiferans
Apicomplexa	Amoeboid parasites	Unicellular	None	*Plasmodium*
Ciliophora	Ciliates	Unicellular	Cilia	*Paramecium, Blepharisma, Vorticella*

PROTOZOANS

All protozoans are unicellular. Most ingest their food by endocytosis. Their diversity, which includes many of the most abundant or commonly observed protist phyla, is summarized in Table 23.1.

Phylum Zoomastigophora

Because all members of the phylum Zoomastigophora possess one or more flagella, they are called flagellates—or zooflagellates, to distinguish them from plantlike flagellates. The phylum gets its name (from the Greek for "whip bearer") from the whiplike motion with which the flagella provide power. The Zoomastigophora constitute the largest protist phylum, numbering more than 10,000 species. This phylum is probably also the most evolutionarily ancient of the protist phyla; it may be ancestral to some or all of the others. Zooflagellates reproduce vegetatively by mitosis and cytokinesis—the simplest and most direct way.

Some free-living zooflagellate species make their living by preying on other protists. An impressive variety of other zooflagellates live as internal parasites on animals, including humans (Figure 23.6).

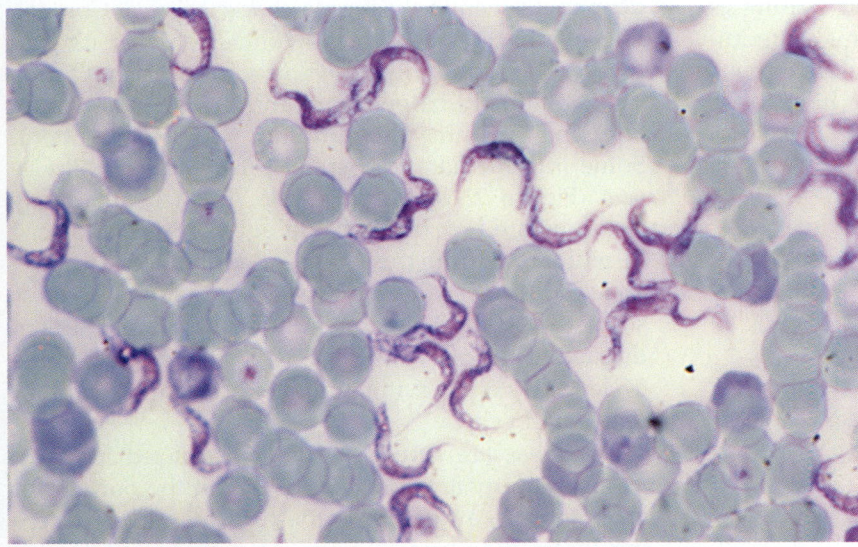

23.6 A Zooflagellate
Trypanosoma gambiense is a parasitic zooflagellate, prevalent in Africa, that causes sleeping sickness in mammals. In trypanosomes the flagellum runs along one edge of the cell as part of a structure called the undulating membrane.

Within the guts of certain wood-eating roaches and termites live an array of huge zooflagellates possessing some of the most bizarre and complicated body forms found anywhere among the protists (see Chapter 4).

One group of zooflagellates, the Choanoflagellida, is thought to be ancestral to the sponges, the most ancient of the surviving phyla of animals (see Chapter 26). Sponges are colonial, rather than truly multicellular, and the Choanoflagellida bear a striking resemblance to the most characteristic type of cell found in the sponges (see Figure 26.7).

Some zooflagellates are human pathogens. The vector (intermediate host) for sleeping sickness—which is caused by the parasitic zooflagellate *Trypanosoma* (see Figure 23.6) and is one of the most dreaded diseases of Africa—is an insect, the tsetse fly. Carrying its deadly cargo, the tsetse fly bites livestock, wild animals, and even humans, infecting all of them with *Trypanosoma*. *Trypanosoma* then multiplies in the mammalian bloodstream and produces toxic substances. When these parasites invade the nervous system, the symptoms of sleeping sickness appear—and are followed by death. Other disease-causing zooflagellates include *Giardia lamblia*, which contaminates water supplies and causes the intestinal disease giardiasis, and *Trichomonas vaginalis*, which causes a common but usually mild venereal disease.

Phylum Rhizopoda

The Rhizopoda—amoebas and their relatives—are protists that form pseudopods, extensions of their constantly changing body mass (Figure 23.7). Amoebas have often been portrayed in popular writing as simple blobs—the simplest form of "animal" life imaginable. Superficial examination of a typical amoeba shows how such an impression might have been

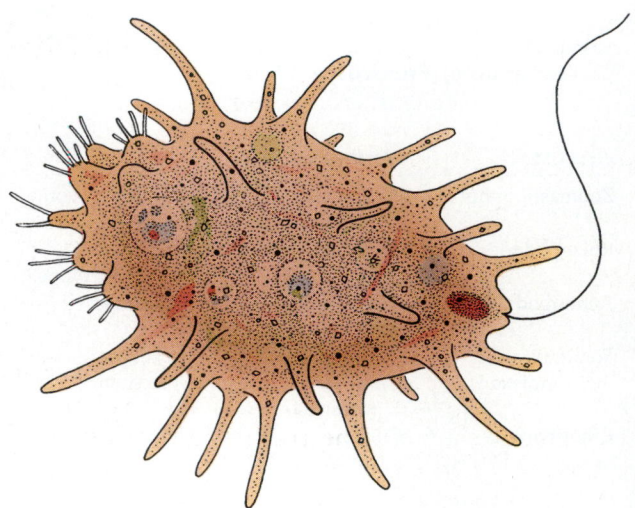

23.8 An Intermediate Form
A cell of *Mastigamoeba aspera* displays both the pseudopods characteristic of Rhizopoda and the flagellum characteristic of Zoomastigophora. This species is arbitrarily placed in the Rhizopoda.

obtained. An amoeba consists of a single cell with no definite shape. It feeds on small organisms and particles of organic matter by phagocytosis, engulfing them with its pseudopods. Particles of food are sealed off in food vacuoles within the cytoplasm of the amoeba. The material is then slowly digested and assimilated into the main body of the organism. Pseudopods are also the organs of locomotion. (The mechanism of amoeboid motion will be discussed in Chapter 40.)

The amoeba is not an extremely ancient organism, despite its seemingly simple characteristics. Compelling evidence points to the conclusion that its simplicity is a secondarily derived condition in evolution. The phylum Rhizopoda apparently originated from ancestors within the phylum Zoomastigophora. Some intermediate forms still exist. *Mastigamoeba aspera* (Figure 23.8), for example, is such an exactly intermediate link that it could equally well be placed in either phylum, the Zoomastigophora or the Rhizopoda. Amoebas of the free-living genus *Naegleria*, some of which can enter humans and cause a fatal disease of the nervous system, have a two-stage life cycle, one stage having amoeboid cells and the other flagellated cells.

Amoebas are actually rather specialized forms of protists. Many are adapted for life on the bottoms of lakes, ponds, and other bodies of water. Their creeping locomotion and their manner of engulfing food particles fit them for life close to a relatively rich supply of sedentary organisms or organic particles. Other amoebas are even more specialized. All amoebas are animallike, existing as predators, parasites, or scavengers. None are photosynthetic. Some amoe-

23.7 An Amoeba
This *Amoeba proteus* has several pseudopods.

bas are shelled, living in casings of sand grains glued together or in shells secreted by the organism itself (see Figure 23.3*b*).

Phylum Actinopoda

Once classified among the Rhizopoda, the Actinopoda are recognizable by their thin pseudopods, which are reinforced by microtubules. The pseudopods play at least three roles. They greatly increase the surface area of the cell for exchange of materials with the environment, they help the cell float in its marine or freshwater environment, and they are the cell's feeding organs. The pseudopods trap smaller organisms, often taking up the prey by endocytosis and transporting it to the main cell body.

Radiolarians, actinopods that are exclusively marine, are perhaps the most beautiful of all microorganisms (Figure 23.9*a*). Almost all radiolarian species secrete glassy skeletons from which needlelike pseudopods project. Part of the skeleton is a central capsule within the cytoplasm. The skeletons of the different species are as varied as snowflakes, and many have elaborate geometric designs. A few radiolarians are among the largest of the protists, with skeletons measuring several millimeters across. Innumerable radiolarian skeletons, some as old as 700 million years, form the sediment under some seas in the tropics.

Heliozoans are actinopods lacking a central capsule (Figure 23.9*b*). Most heliozoans live in fresh water.

Phylum Foraminifera

Foraminiferans are marine creatures that secrete shells of calcium carbonate (see Figure 23.3*a*). Their long, threadlike, branched pseudopods reach out through numerous microscopic pores in the shells and interconnect to create a sticky net, which the foraminiferan uses to catch smaller plankton (free-floating microscopic organisms). After foraminiferans reproduce—by mitosis and cytokinesis—the daughter cells abandon the parent shell and make new shells of their own. The discarded skeletons of infinite legions of ancient foraminiferans make up extensive limestone deposits in various parts of the world, forming a covering hundreds to thousands of meters deep over millions of square kilometers of ocean bottom. Foraminiferan skeletons also make up the sand of some beaches. A single gram of such sand may contain as many as 50,000 foraminiferan shells.

The shells of individual foraminiferan species have distinctive shapes and are easily preserved as fossils in marine sediments. Each geological period has its own distinctive foraminiferan species. For this reason, and because they are so abundant, the remains of foraminiferans are especially valuable as indicators in the classification and dating of sedimentary rocks, as well as in oil prospecting.

Phylum Apicomplexa

The Apicomplexa are exclusively parasitic protozoans, thus named because the apical end of their spore contains a mass of organelles. These organelles help the apicomplexan spore invade its host tissue. Unlike many other protists, apicomplexans lack contractile vacuoles. Because their rigid cell walls limit expansion, they do not take in excess water. Apicomplexans generally have an indefinite body form like that of an amoeba, but this shape in no way indicates a relationship to the Rhizopoda. Rather, this body form has evolved over and over again in parasitic

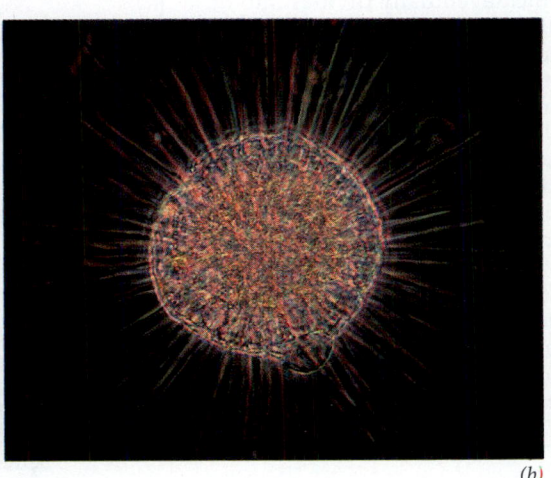

(a)   (b)

23.9 Actinopods
(a) Radiolarians display a glassy skeleton of delicate intricacy.
(b) *Actinophaerium* is a heliozoan with long pseudopods.

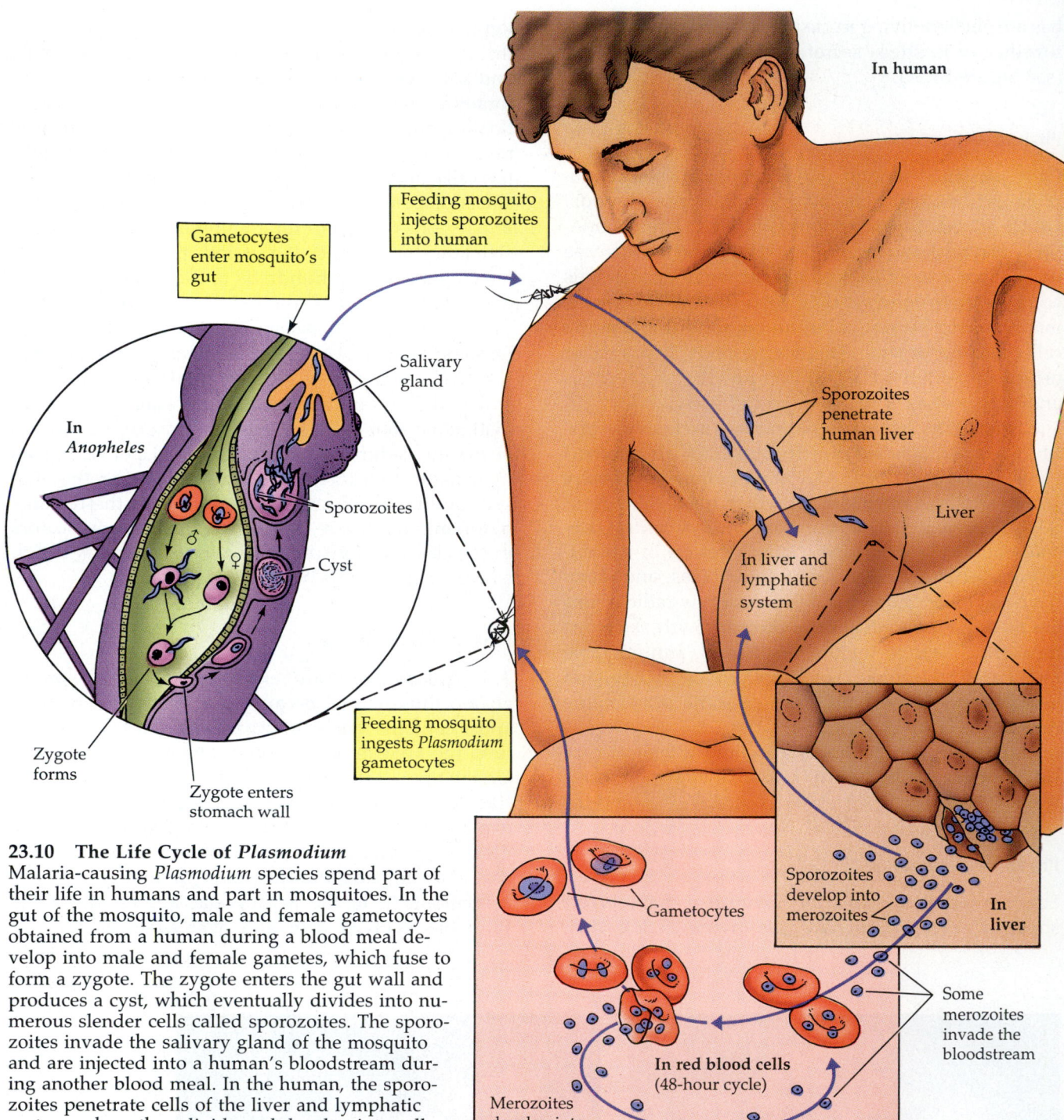

Labels in figure:
- Gametocytes enter mosquito's gut
- Feeding mosquito injects sporozoites into human
- In human
- Salivary gland
- In *Anopheles*
- Sporozoites
- Sporozoites penetrate human liver
- Cyst
- Liver
- In liver and lymphatic system
- Zygote forms
- Zygote enters stomach wall
- Feeding mosquito ingests *Plasmodium* gametocytes
- Sporozoites develop into merozoites
- In liver
- Gametocytes
- Some merozoites invade the bloodstream
- In red blood cells (48-hour cycle)
- Merozoites develop into gametocytes

23.10 The Life Cycle of *Plasmodium*

Malaria-causing *Plasmodium* species spend part of their life in humans and part in mosquitoes. In the gut of the mosquito, male and female gametocytes obtained from a human during a blood meal develop into male and female gametes, which fuse to form a zygote. The zygote enters the gut wall and produces a cyst, which eventually divides into numerous slender cells called sporozoites. The sporozoites invade the salivary gland of the mosquito and are injected into a human's bloodstream during another blood meal. In the human, the sporozoites penetrate cells of the liver and lymphatic system, where they divide and develop into cells of another stage, called merozoites. Merozoites may, in turn, invade fresh cells of the human liver or lymphatic system, where they divide and develop into another generation of merozoites; or they may enter red blood cells, where they may divide, grow, lyse cells, and invade fresh red blood cells on a 48-hour cycle. Eventually some merozoites inside red blood cells develop into male and female gametocytes, ready to be picked up by a hungry mosquito to start the life cycle again.

protists. The form has appeared, for example, even in parasitic dinoflagellates, a group of algae whose nonparasitic relatives have highly distinctive, regular body forms. Like many animal obligate parasites, apicomplexans have elaborate life cycles featuring asexual and sexual reproduction by a series of very dissimilar life stages. Often these stages are associated with two types of host organisms.

The best-known apicomplexans are the malaria parasites of the genus *Plasmodium*, a highly specialized group of organisms that spend part of their life cycle within human red blood cells (Figure 23.10). Malaria continues to be a major problem in some tropical countries although it has been almost eliminated from the United States; indeed, malaria is one of the world's most serious diseases in terms of number of people infected. Female mosquitoes of the genus *Anopheles* transmit *Plasmodium* to humans. *Plasmodium* enters the human circulatory system when an infected *Anopheles* mosquito penetrates the human skin in search of blood. The parasite cells find their way to the liver and the lymphatic system, change their form, multiply, and reenter the bloodstream, attacking red blood cells. The attackers multiply in a red blood cell for approximately two days, producing up to 36 new *Plasmodium* cells each. The victimized cell then bursts, releasing a new swarm of parasites to attack other red blood cells.

If another *Anopheles* bites the victim, some of the parasitic *Plasmodium* cells are taken into the mosquito along with the blood, thus infecting the mosquito. The parasites attack cells of the mosquito's gut, reproduce, and move into the salivary glands, from which they can be passed to another human host. The *Plasmodium* life cycle that spreads malaria is best broken by the removal of stagnant water, in which mosquitoes breed. The use of insecticides to reduce the *Anopheles* population can be effective, but possible ecological, economic, and health risks should be considered.

Phylum Ciliophora

Because the Ciliophora characteristically have hairlike cilia, they have the common name ciliates. This protozoan phylum ranks with the Zoomastigophora in diversity and ecological importance (Figure 23.11). All ciliates are heterotrophic, and they are much more specialized in body form than most flagellates and other protists. Ciliates are also characterized by the possession of two types of nuclei, a large **macronucleus** and, within the same cell, from one to as many

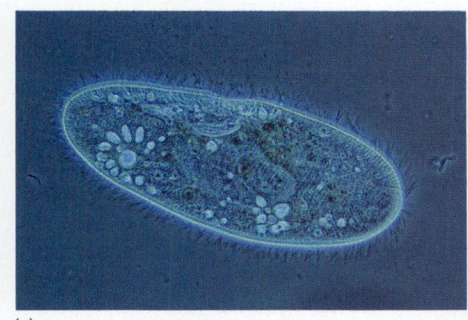

(a)

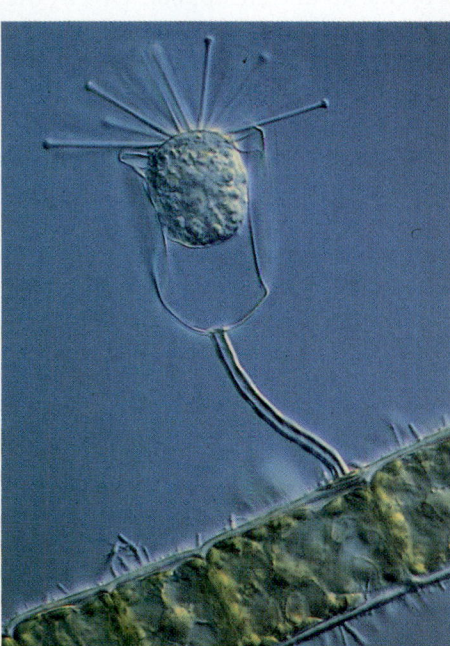

(c)

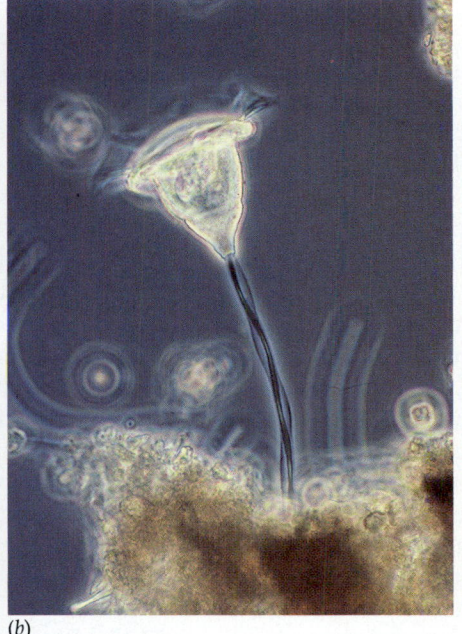

(b)

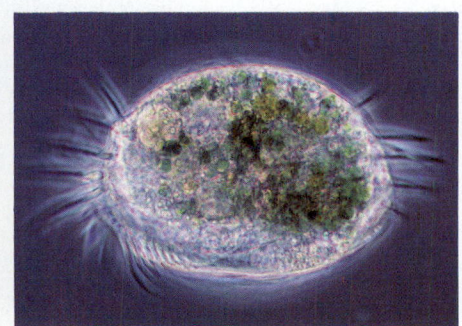

(d)

23.11 Diversity in the Ciliates
(a) *Paramecium caudatum*, a free-swimming organism, belongs to the ciliophoran subgroup called holotrichs, which have many cilia of uniform length. (b) Members of the peritrich subgroup of Ciliophora have cilia on their mouthparts and usually attach to their substrate, as demonstrated by this *Vorticella*. (c) Tentacles replace cilia as suctorians, another subgroup, develop. Suctorians, such as this *Paracineta*, attach to the substrate by a stalk. (d) The organelles of *Euplotes* are developed for specific uses. This ciliate "walks" on fused cilia called cirri that project from its body. Other cilia in *Euplotes* are fused into flat sheets that sweep in food particles; this individual has just fed on green algae.

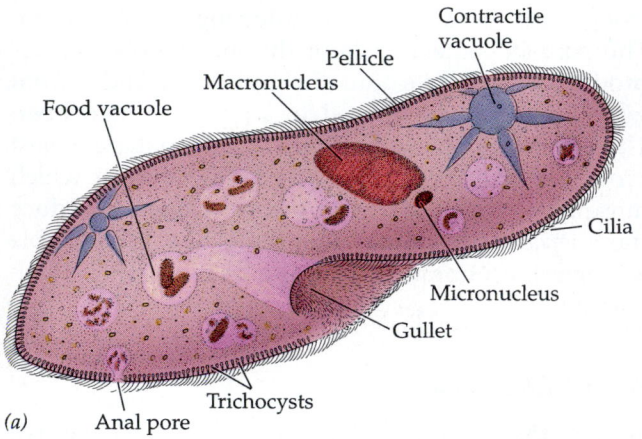

(a)

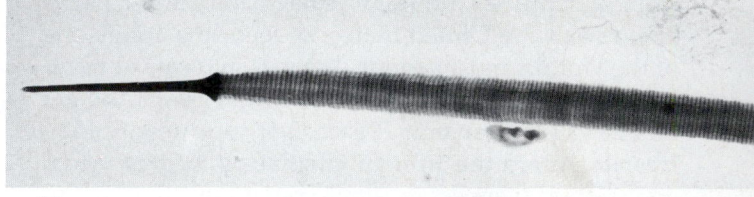

23.12 Anatomy of *Paramecium*

(a) The major structures of a typical paramecium. (b) A trichocyst discharged from beneath the pellicle of a paramecium has a sharp point and a straight filament. This image is magnified about 10,000 times.

(b)

as 80 **micronuclei**. The micronuclei, which are typical eukaryotic nuclei, are essential for genetic recombination. The macronucleus contains many copies of the genetic information, packaged in units containing very few loci each; the macronuclear DNA is transcribed and translated to control the life of the cell. Although we do not know how this system of macro- and micronuclei came into being, we know something about the behavior of these nuclei, which we will discuss in the next section.

A CLOSER LOOK AT ONE CILIATE. *Paramecium*, a frequently studied genus, exemplifies the complex structure and behavior of ciliates (Figure 23.12*a*). The slipper-shaped cell is covered by an elaborate **pellicle**, a structure composed principally of an outer membrane and an inner layer of closely packed, kidney-shaped structures (called alveoli) that embrace the cilia. Defensive organelles called trichocysts are also present as a layer of the pellicle. A microscopic explosion expels the trichocysts in a few milliseconds, and they emerge as sharp darts, driven forward at the tip of a long, expanding shaft (Figure 23.12*b*).

The cilia provide a form of locomotion that is generally more precise than that made possible by flagella or pseudopods. A paramecium can direct the beat of its cilia to propel itself either forward or backward in a spiraling manner (Figure 23.13). A paramecium can back off swiftly when it encounters a barrier or a negative stimulus. Some of these large ciliates hold the speed record for the kingdom Protista—faster than 2 millimeters per second. A few of

the cilia of a paramecium are sensory in function; they are somehow able to transmit stimuli back through the remainder of the cytoplasm, thereby coordinating the rapid movements of the entire organism.

Paramecia reproduce by cell division. The micronuclei divide mitotically, but the macronucleus simply pinches apart to yield two daughter macronuclei. Paramecia also have an elaborate sexual behavior called **conjugation** (Figure 23.14). Two paramecia line up tightly against each other and fuse in the oral region of the body. Extensive reorganization and exchange of nuclear material occurs during the next several hours, as follows. In each cell, all the micronuclei except one degenerate. The remaining micronucleus in each cell divides meiotically, a process that reduces the chromosome number from diploid to haploid. Each of the four meiotic products in each cell divides mitotically. Then all but one of the eight resulting haploid nuclei break down, and the last one divides mitotically, producing two haploid nuclei. The macronuclei break up and disappear. One of the two haploid nuclei in each cell remains in its "home" cell, and the other migrates to the partner cell, where it fuses with its counterpart.

The exchange is fully reciprocal—each of the two paramecia gives and receives an equal amount of DNA. Afterward the two organisms separate and go their own ways, each genetically "refreshed" by the

23.13 *Paramecium* "Swims" with Its Cilia

Beating its cilia in coordinated waves that progress from one end of the cell to the other, a paramecium can move in either direction with respect to the long axis of the cell; this one is moving from left to right. The cell also rotates in a spiral as it travels.

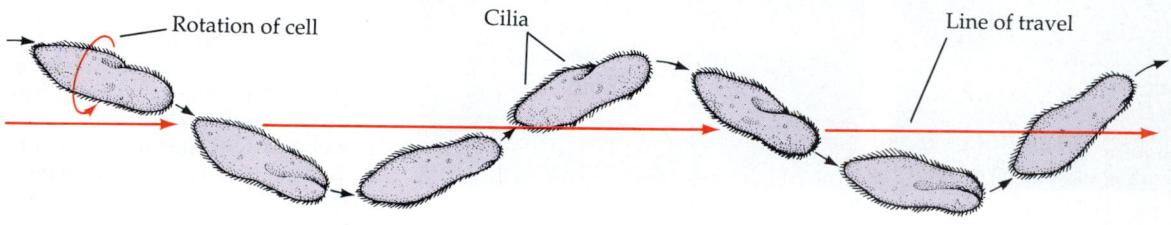

23.14 Paramecia Achieve Genetic Recombination by Conjugating
These two *Paramecium* individuals are conjugating. Their cells fuse and exchange micronuclei, thereby permitting genetic recombination. After conjugation, the cells separate and continue their lives as two individuals.

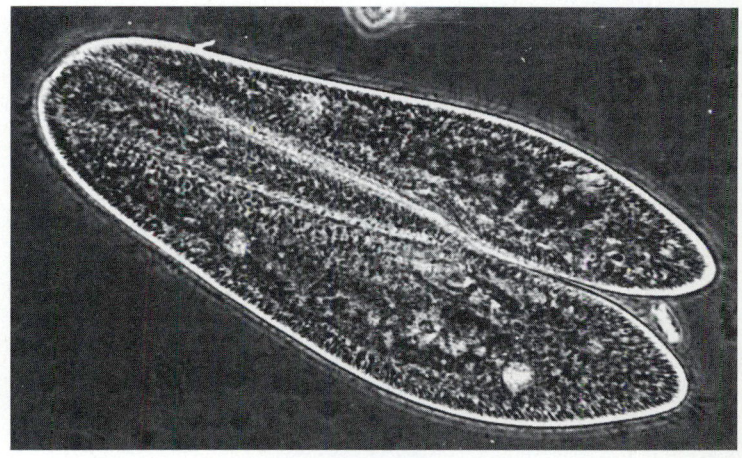

recombination that occurred during conjugation. The new, recombined diploid nucleus of each cell divides mitotically, producing two diploid nuclei, one that is destined to be the new macronucleus and a second that gives rise to the appropriate number of micronuclei by further mitotic divisions.

Conjugation in *Paramecium* is a *sexual* process of genetic recombination, but it is not a *reproductive* process. The same two cells that begin the process are there at the end, and no new cells are created. As a rule, each clone of paramecia must periodically conjugate. Laborious experimentation has shown that if some species are not permitted to conjugate, the asexual clones can live through no more than approximately 350 cell divisions before they die out.

CYTOPLASMIC ORGANIZATION IN THE CILIATES. Most ciliates possess all the traits of *Paramecium*. Some, however, are notable for the exceptional degree of development of their individual organelle systems. Certain ciliates, for example, have the equivalent of legs. Fused cilia called cirri move in an independent, but coordinated, fashion and enable the organism to "walk" over surfaces (see Figure 23.11*d*). Nervelike **neurofibrils** leading to individual cirri coordinate this locomotion. The coordination is lost if the neurofibrils are experimentally cut. Many types of ciliates possess **myonemes**; musclelike fibers within the cytoplasm. The contraction of myonemes causes a rapid retraction of the stalk in ciliates such as *Vorticella* when the organism is disturbed.

What may be the ultimate cytoplasmic organization is displayed by highly specialized ciliates that live in the digestive tracts of cows and many other hoofed mammals. These ciliates possess not only myonemes, neurofibrils, and elaborately fused cilia, but also a cytoplasmic "skeleton" and a "gut" complete with "mouth," "esophagus," and "anus" (Figure 23.15).

When examining the intricate structures of these organisms and many other protists, we must pause

and remember that we are looking at only one cell. Structural complexity in multicellular organisms—fungi, animals, and plants—is based on the diversity and coordination of cell types. These protists, on the other hand, owe their complexity to the diversity and coordination of organelles within a single cell.

FUNGUSLIKE PROTISTS

Unlike true fungi, the funguslike protists have flagella (although the flagella may be present in only certain stages of their life cycles) and do *not* have a dikaryotic stage in their life cycle (see Chapter 24) or chitin in their cell walls. Table 23.2 summarizes the phyla of funguslike protists.

Two groups of funguslike protists are so similar at first glance that they have sometimes been grouped in a single phylum. These so-called acellular and cel-

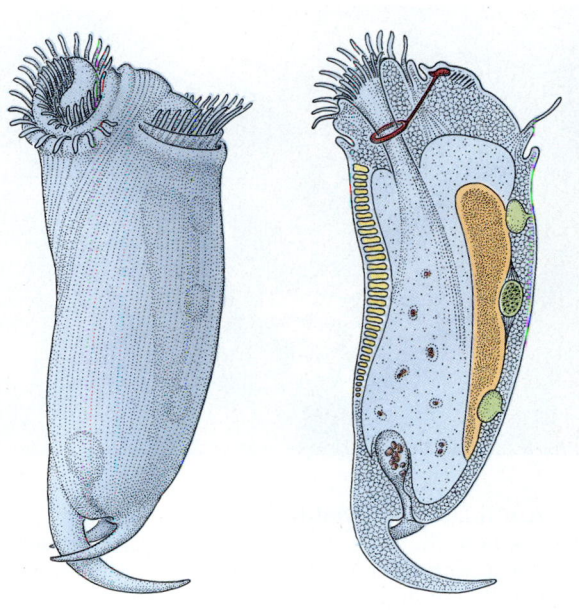

23.15 An Exceptional Ciliate
Surface and cutaway views of *Diplodinium dentatum*. The "mouth" is at the top, the "esophagus" (surrounded by a contractile ring that can close it) is immediately below the mouth, and the "anus" is toward the bottom. The plates comprising the "skeleton" are stacked on the left.

TABLE **23.2**
Classification of Funguslike Protists

PHYLUM	COMMON NAME	FORM	LOCOMOTION	EXAMPLES
Myxomycota	Acellular slime molds	Single cells and coenocytes	Amoeboid	*Physarum*
Acrasiomycota	Cellular slime molds	Single cells and aggregates	Amoeboid	*Dictyostelium*
Protomycota	Chytrids and hyphochytrids	Unicellular or mycelium	None	*Cladochytrium, Rhizidiomyces*
Oomycota	Water molds and downy mildews	Coenocytic mycelium	None	*Saprolegnia, Achlya, Phytophthora*

lular slime molds are so different, however, that some biologists even classify them in different kingdoms. The two groups of slime molds share some characteristics. Both are motile, both ingest particulate food by endocytosis, and both form spores on erect fruiting bodies. They undergo striking changes in organization during their life cycles, and one stage consists of isolated cells engaging in absorptive nutrition. Some slime molds may attain areas of 1 m or more in diameter while in their less-aggregated stage. Such a large slime mold may weigh more than 50 g. Slime molds of both types favor cool, moist habitats, primarily in forests. They range from colorless to brilliant yellow and orange.

Phylum Myxomycota

If the nucleus of an amoeba began rapid mitotic division accompanied by a tremendous increase in cy-toplasm and organelles, the resulting organism might resemble the vegetative phase of the *acellular* slime molds (phylum Myxomycota). During most of its life history, an acellular slime mold is a wall-less mass of cytoplasm with numerous diploid nuclei. This mass streams over its substrate in a remarkable network of strands called a **plasmodium** (Figure 23.16*a*). A plasmodium of a myxomycete is an example of a **coenocyte**, a body in which many nuclei are enclosed in a single plasma membrane. The outer cytoplasm of the plasmodium (closest to the environment), which is normally in a state less fluid than that of the interior, provides some structural rigidity.

Myxomycetes such as *Physarum* (a popular research subject) provide a dramatic example of **cytoplasmic streaming**. The outer cytoplasmic region becomes more fluid in places, and there is a rush of cytoplasm into those areas, stretching the plasmodium in one direction or another. This streaming

(a)

23.16   Acellular Slime Molds
(a) Plasmodia of the yellow slime mold *Physarum* cover a rock in Nova Scotia. (b) The fruiting structures (sporangiophore and sporangia) of *Physarum polycephalum*.

(b)

somehow reverses its direction every few minutes as cytoplasm rushes into a new area and drains away from an older one. As the plasmodium spreads over its substrate in this manner, it engulfs food particles—predominantly bacteria, yeasts, spores of fungi, and other small organisms, as well as decaying animal and plant remains. Sometimes an entire wave of plasmodium moves across the substrate, leaving strands behind. A contractile protein called myxomyosin participates in the streaming mechanism. Minute fibers—microfilaments—mediate the movement in conjunction with a myosinlike molecule.

An acellular slime mold can grow almost indefinitely in its plasmodial stage, as long as the food supply is adequate and other conditions, such as moisture and pH, are favorable. If conditions become unsuitable, however, one of two things can happen. The plasmodium can form a resistant structure, an irregular mass of hardened cell-like components called a **sclerotium**, which rapidly becomes a plasmodium again upon restoration of favorable conditions; or the plasmodium can transform itself into spore-bearing fruiting structures (Figure 23.16b). Rising from heaped masses of plasmodium, these stalked or branched fruiting structures—called **sporangiophores**—derive their rigidity from the deposition of cellulose or chitin at the surfaces of their component cells.

The nuclei of the plasmodium are diploid, and they divide by meiosis during the development of the sporangiophore. One or more knobs—variously colored and shaped **sporangia**—develop on the end of the stalk. Within a sporangium, haploid nuclei become surrounded by walls and form spores. Eventually, as the sporangiophore dries, it sheds its spores. The spores germinate into wall-less, flagellated, haploid cells called **swarm cells** or swarmers, which can either divide mitotically to produce more haploid swarm cells or function as gametes. Swarm cells can manage on their own and can become walled and resistant cysts when conditions are un-

favorable. When conditions improve again, the cysts release flagellated swarm cells. Two swarmers can fuse to form a diploid zygote, which divides by mitosis, but without wall formation between the nuclei, and thus forms a new, coenocytic plasmodium.

Phylum Acrasiomycota

The phylum Acrasiomycota encompasses the *cellular* slime molds. Whereas the plasmodium is the basic vegetative unit of acellular slime molds, an amoeboid cell is the vegetative unit of the cellular slime molds. Large numbers of cells called **myxamoebas**, which have single haploid nuclei, engulf bacteria and other food particles and reproduce by mitosis and fission. This simple developmental stage can persist indefinitely (as long as food and moisture are available) as swarms of independent, isolated cells.

When conditions become unfavorable, however, cellular slime molds aggregate and form fruiting structures, as do their acellular counterparts. The apparently independent myxamoebas aggregate into an irregular mass called a **pseudoplasmodium** (Figure 23.17a). Unlike the true plasmodium of the acellular slime molds, this structure is not simply a giant sheet of cytoplasm; the individual myxamoebas retain their plasma membranes and, therefore, their identity. The pseudoplasmodium is not a coenocyte. The chemical signal that causes the myxamoebas to aggregate into a pseudoplasmodium is 3′,5′-cyclic adenosine monophosphate (cAMP), a "messenger" that plays many important roles in chemical signaling in animals (see Chapter 36). A pseudoplasmodium may migrate over its substrate for several hours before ultimately constructing a delicate, stalked fruiting structure (Figure 23.17b). Cells at the top of the fruiting structure de-

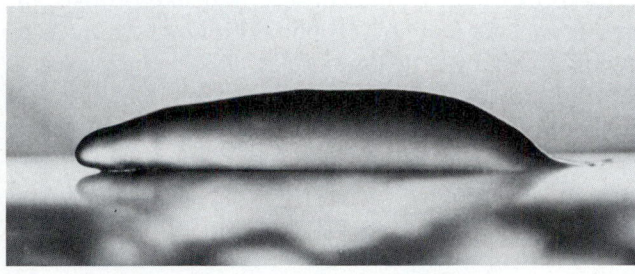

(a)

(b)

23.17 A Cellular Slime Mold
(a) A pseudoplasmodium of *Dictyostelium discoideum* migrating over the substrate. (b) Fruiting structures of *D. discoideum* in various stages of development.

velop into thick-walled spores. The spores are released; later, under favorable conditions, they germinate, releasing myxamoebas.

The cycle from myxamoebas through a pseudoplasmodium and spores to new myxamoebas is asexual. There is also a sexual cycle, in which two myxamoebas (possibly of different mating types; see Chapter 24) fuse. The product of this fusion develops into a spherical structure that ultimately germinates, releasing new myxamoebas.

Phylum Protomycota

Chytrids and hyphochytrids, the two groups that constitute the phylum Protomycota, are aquatic microorganisms sometimes classified as fungi. We place them among the protists because they possess flagellated cells, but they resemble fungi in that their cell walls consist primarily of chitin.

All chytrids and hyphochytrids are either parasitic (on organisms such as algae, mosquito larvae, and nematodes) or **saprobic**, obtaining their nutrients by breaking down dead organic matter. They live in freshwater habitats or in moist soil. Some of the Protomycota are unicellular, while others have **mycelia**—masses of filaments—made up of branching chains of cells. Hyphochytrids and chytrids differ in some structural and metabolic details. Chytrids reproduce both sexually and asexually, but no sexual stages have been observed in the hyphochytrids.

Allomyces, a well-studied genus of the chytrids, displays an isomorphic alternation of generations; that is, it has haploid and diploid phases of the life cycle that are indistinguishable except on the bases of their chromosome number and of their reproductive products. The diploid generation of *Allomyces* produces numerous diploid flagellate spores, called **zoospores**, by mitosis and cytokinesis. (Many funguslike protists and algae reproduce vegetatively from zoospores.) Haploid zoospores are also produced, by meiotic divisions and cytokinesis, in discrete sporangia (spore cases), which are cell-like units defined by distinct cross walls. Diploid zoospores produce new diploid organisms; haploid zoospores produce haploid organisms. The haploid organism produces gametes at the tips of **hyphae** (filaments). A single haploid organism can produce both male and female gametes, with the male gametes at the very ends of the hyphae and the female gametes just below, in specialized, walled-off structures known as **gametangia** (gamete cases) (Figure 23.18). Both female and male gametes have flagella. The motile female gamete produces a chemical attractant (pheromone; see Chapter 46) that attracts the swimming male gamete.

One hyphochytrid, *Rhizidiomyces apophysatus*, parasitizes funguslike protists of the phylum Oomycota

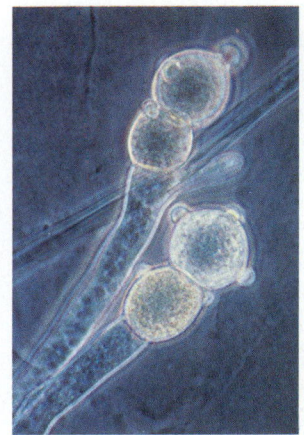

23.18 Reproductive Structures of a Chytrid
Gametangia (which contain female gametes) of *Allomyces*. The small rounded structures attached to the gametangia are male gametes.

(water molds; see the next section). Upon contacting water mold cells, the zoospores of *R. apophysatus* lose their flagella and develop germination tubes that extend into the host tissue. Ultimately a branching mass of **rhizoids** (rootlike structures) spreads throughout the infected water mold. The hyphochytrid forms a sporangium, and a tube develops, through which a mass of the hyphochytrid's cytoplasm and nuclei is released. The mass cleaves, freeing a large number of zoospores. The zoospores swim away, some coming in contact with water molds and repeating the life cycle.

Phylum Oomycota

The phylum Oomycota consists in large part of the water molds and their funguslike terrestrial relatives, such as the downy mildews. If you have seen a whitish, cottony mold growing on dead fish or dead insects in water, it was probably the mycelium of a water mold of the common genus *Saprolegnia* (Figure 23.19). *Saprolegnia* itself is a common target of parasitism by the hyphochytrid *Rhizidiomyces*, described in the previous section. These funguslike protists are coenocytic: They have hyphae with no cross walls to divide them into discrete cells. Their cytoplasm is continuous throughout the mycelium, and there is no single structural unit with a single nucleus, except in certain reproductive stages. A distinguishing feature of the oomycetes is that they have flagellated reproductive cells. They are diploid throughout most of their life cycle and have cellulose in their cell walls.

The water molds, such as *Saprolegnia*, are all aquatic and saprobic. Some other members of the phylum Oomycota are terrestrial. Although most terrestrial oomycetes are harmless or helpful saprobes, a few are serious plant parasites attacking crops such

23.19   A Water Mold
The mycelium of *Saprolegnia* radiates from the carcass of a salamander.

as avocados, grapes, and potatoes. The mold *Phytophthora infestans*, for example, is the causal agent of late blight of potatoes, which brought about the great Irish potato famine of 1845 to 1847. *P. infestans* destroyed the entire Irish potato crop in a matter of days in 1846. Among the consequences of the famine were a million deaths from starvation and the emigration of about 2 million people, mostly to the United States. *Albugo*, another member of the Oomycota, is a well-known parasitic genus that causes a mealy blight on sweet potato leaves, morning glories, and numerous other plants. An obligate parasite, *Albugo* has never been grown on any medium other than its plant host.

ALGAE

Algae are photosynthetic, carrying out probably 50 to 60 percent of all the photosynthesis on Earth with the kingdom Plantae accounting for most of the rest. The overall contribution of cyanobacteria and other photosynthetic bacteria is smaller, although it is locally important in some aquatic ecosystems. Algae differ from plants in that the zygote of an alga is on its own; the parent gives the zygote no protection. A plant zygote, on the other hand, grows into a multicellular embryo that is protected by parental tissue.

Algae exhibit a remarkable range of growth forms. Some are unicellular; others are filaments composed either of distinct cells or of multinucleate structures without cross walls (coenocytes). Still others—including the algae commonly known as seaweeds—are multicellular and intricately branched or arranged in multicellular, leaflike extensions. The bodies of a few types of algae are even subdivided into tissues and organs. Certain algal phyla—for example, the phy-

lum Chlorophyta (the green algae)—include representatives of almost all these growth forms.

Algal life cycles also show extreme variation, but all algae except members of the phylum Rhodophyta (red algae) have forms with flagellated motile cells in at least one stage of their life cycle. Some are unicellular and motile throughout most of their existence (for example, the dinoflagellates—in the protist phylum Pyrrophyta).

Table 23.3 summarizes the classification of algae.

Phylum Pyrrophyta

A distinctive mixture of photosynthetic and accessory pigments gives the chloroplasts of Pyrrophyta, all of which are unicellular, a golden-brown color. The dinoflagellates, the major group within the Pyrrophyta, are of great ecological, evolutionary, and morphological interest. Dinoflagellates are probably second in importance only to diatoms (members of the algal phylum Chrysophyta, the topic of the next section) as primary photosynthetic producers of organic matter in the oceans. Many dinoflagellates are endosymbionts, living within the cells of other organisms, including various invertebrates and even other marine protists. Dinoflagellates are particularly common endosymbionts in corals, to whose growth they contribute mightily by photosynthesis. Some dinoflagellates are nonphotosynthetic and live as parasites within other marine organisms.

Dinoflagellates are distinctive cells (Figure 23.20*a*). They have two flagella, one in an equatorial groove around the cell, the other starting at the same point as the first and passing down a longitudinal groove before extending free into the surrounding medium. Most dinoflagellates are marine organisms. Some dinoflagellates reproduce in enormous numbers in warm and somewhat stagnant waters. The result can be a "red tide," so called because of the reddish color of the sea that results from fluorescence of the chlorophyll in the dinoflagellates (Figure 23.20*b*). During a red tide, the concentration of dinoflagellates may reach 60 million cells per liter of ocean water. Certain species produce a potent nerve toxin; a red tide made up of these species can kill great numbers of fish. A particularly severe red tide in the Gulf of Mexico in the summer of 1971 killed tons of fish along the west coast of Florida. The genus *Gonyaulax* produces a potent toxin that can accumulate in shellfish in amounts that, although not fatal to the shellfish, may kill a person who eats it.

Many dinoflagellates are bioluminescent (see Box 7.A). In complete darkness, cultures of these organisms emit a faint glow. If one suddenly disturbs a culture physically, by stirring it or bubbling air through it, the organisms each emit a number of bright flashes, perhaps a thousandfold brighter than

TABLE 23.3
Classification of Algae

PHYLUM	COMMON NAME	FORM	LOCOMOTION	REPRESENTATIVE GENERA
Pyrrophyta	Dinoflagellates	Unicellular	Two flagella	*Gonyaulax, Ceratium, Noctiluca*
Chrysophyta	Diatoms	Usually unicellular	Usually none	*Diatomia, Fragillaria, Ochromonas, Synura*
Euglenophyta	Flagellates	Unicellular	Two flagella	*Euglena*
Phaeophyta	Brown algae	Multicellular	Two flagella on reproductive cells	*Macrocystis, Fucus*
Rhodophyta	Red algae	Multicellular or unicellular	None	*Chrondrus*, coralline algae
Chlorophyta	Green algae	Unicellular, colonial, or multicellular	Most have flagella at some stage	*Chlorella, Ulva, Acetabularia*

the dim glow of an undisturbed culture. The flashing then rapidly subsides. A ship passing through a tropical ocean containing a rich growth of these species produces a bow wave and a wake that glow eerily as billions of these dinoflagellates discharge their light systems.

Phylum Chrysophyta

Diatoms and their relatives make up the phylum Chrysophyta. Some species are single-celled; others are filamentous. Many have sufficient carotenoids in

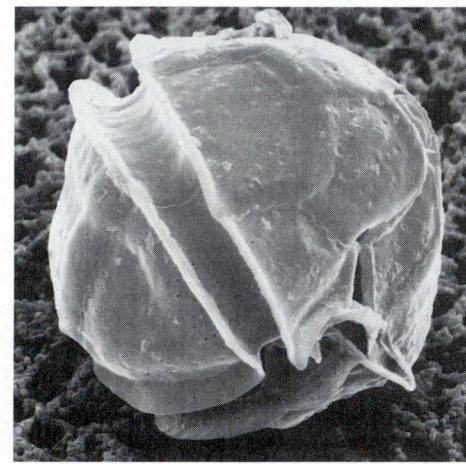

23.20 Dinoflagellates
(a) A single *Gonyaulax tamarensis*. Each groove in the side of the cell normally contains a flagellum, but the flagella are not visible here. (b) In astronomical numbers, this species causes a toxic red tide, as seen here off the coast of Baja California.

(a)

(b)

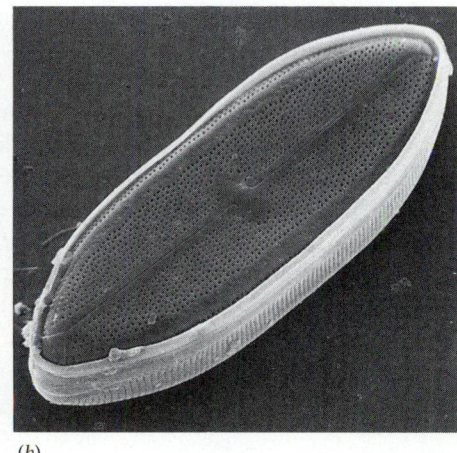

(b)

(a)

23.21 Diatom Diversity
(a) Diatoms exhibit a splendid variety of species-specific forms. (b) *Neidium iridus*. The dark and light areas of this scanning electron micrograph emphasize the distinct two-piece construction of the cell wall.

their chloroplasts to give them a yellow or brownish color. All make chrysolaminarin (a carbohydrate) and oils as photosynthetic storage products.

Architectural magnificence on a microscopic scale is the hallmark of the diatoms (Figure 23.21a). Despite their remarkable morphological diversity, however, all diatoms are symmetrical—either bilaterally (division along only one plane results in identical halves) or radially (division along any plane results in identical halves).

Many diatoms deposit silicon in their cell walls. The cell wall of some species is constructed in two pieces, with the walls of the top overlapping the walls of the bottom like the top and bottom of a petri plate (Figure 23.21b). The silicon-impregnated walls have

intricate, unique patterns; in fact, the taxonomy of these marine or freshwater organisms is based entirely on their wall patterns.

Diatoms reproduce both sexually and asexually. Asexual reproduction is by cell division and is somewhat constrained by the silica-containing cell wall. Both the top and the bottom of the "petri plate" become tops of new "plates" without changing appreciably in size; as a result, the new cells made from former bottoms are smaller than the parent cells. If the process continued indefinitely, one cell line would simply vanish. Sexual reproduction largely solves the problem. Gametes form, shed their cell walls, and fuse. The resulting zygote then increases substantially in size before a new wall is laid down (Figure 23.22).

23.22 Diatom Reproduction
Silicon-impregnated cell walls, shown edge-on in this diagram, are two-part "petri plates." In asexual reproduction by mitosis and cytokinesis (vertical sequence on the right), the parts separate, each becoming the top of a new "plate." In the process, the offspring cells from the bottom parts become progressively smaller. Zygotes produced by sexual reproduction (cycle at left) grow and lay down new full-size cell skeletons.

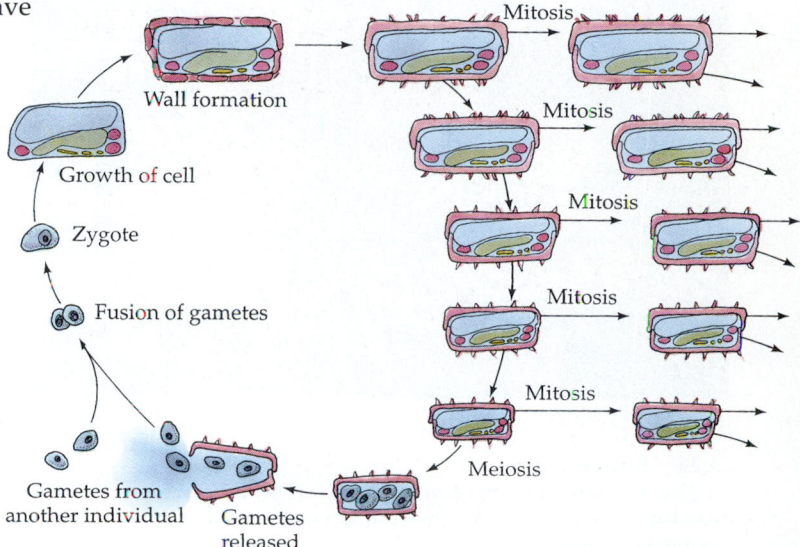

Mitosis

Wall formation

Mitosis

Growth of cell

Mitosis

Zygote

Mitosis

Fusion of gametes

Mitosis

Meiosis

Gametes from another individual

Gametes released

Diatoms are everywhere in the marine environment and are frequently present in great numbers, making them the leading primary photosynthetic producers in the oceans. Diatoms are also common in fresh water. Because the walls of dead diatom cells resist decomposition, certain sedimentary rocks are composed almost entirely of these silica-containing skeletons that sank to the sea floor. Diatomaceous earth, which is obtained from such rocks, has many industrial uses—from insulation and filtration to metal polishing.

Phylum Euglenophyta

The 800 species of the phylum Euglenophyta are sometimes grouped with the Zoomastigophora because they resemble the protozoans in some respects, notably in lacking a cell wall. Like the Zoomastigophora, the Euglenophyta are unicellular flagellates, but most members of the Euglenophyta carry on photosynthesis, as do the other algae. Figure 23.23 depicts a cell of the genus *Euglena*. Like most other members of the phylum, this common freshwater organism has a complex cell plan. It propels itself through the water with one of its two flagella, which sometimes doubles as an anchor to hold the organism in place. The flagellum provides power by means of a wavy motion that spreads from base to tip. *Euglena* reproduces vegetatively by mitosis and cytokinesis.

Euglena has very flexible nutritional requirements. In sunlight it is fully autotrophic, using its chloroplasts to synthesize organic compounds through photosynthesis. When kept in the dark, the organism loses its photosynthetic pigment and begins to feed

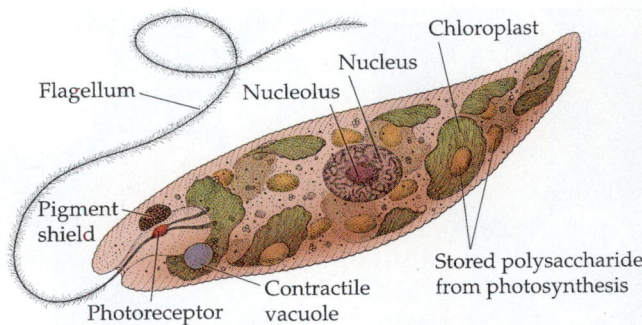

23.23   *Euglena* Is a Photosynthetic Flagellate
Several species of *Euglena* are among the best-known flagellates. Photosynthetic chloroplasts are prominent features in a typical *Euglena* cell.

exclusively on dead organic material floating in the water around it. Such a "bleached" cell of *Euglena* resynthesizes its photosynthetic pigment when returned to the light and hence becomes autotrophic again. *Euglena* cells treated with certain antibiotics or mutagens lose their photosynthetic pigment completely; neither they nor their descendants are ever autotrophs again. Those descendants, however, function perfectly well as heterotrophs.

Phylum Phaeophyta

All members of the phylum Phaeophyta, commonly called brown algae, are multicellular, composed either of branched filaments or of leaflike growths called **thalli** (Figure 23.24). The brown algae obtain their namesake color from the carotenoid fucoxanthin, which is abundant in the plastids. The combi-

(a)

(b)

23.24   Brown Algae
(a) This species of *Hormosira* is an intertidal brown alga growing in Australia. (b) The filamentous brown alga *Ectocarpus*, as seen through a light microscope.

nation of this yellow-orange pigment with the green of chlorophylls *a* and *c* yields a dirty brown color.

The Phaeophyta include the largest of the algae. Giant kelps, such as those of the genus *Macrocystis*, may be up to 60 meters long (see Figure 29.2). The brown algae are almost exclusively marine. Some float in the open ocean; the most famous example is the genus *Sargassum*, which forms dense mats of vegetation in the Sargasso Sea in the mid-Atlantic. Most brown algae, however, are attached to rocks near the shore. A few thrive only where they are regularly exposed to heavy surf; a notable example is the sea palm *Postelsia palmaeformis* of the Pacific coast. The attached forms all develop a specialized structure, called a **holdfast**, that literally glues them to the rocks (Box 23.A). Some brown algae may differentiate extensively into stemlike stalks and leaflike blades, and some develop gas-filled cavities or bladders. For biochemical reasons that are only poorly understood, these gas cavities often contain as much as 5 percent carbon monoxide—a concentration high enough to kill a human.

In addition to organ differentiation, the larger brown algae also exhibit considerable tissue differentiation. Most of the giant kelps have photosynthetic filaments only in the outermost regions of the stalks and blades. Inside the photosynthetic region lie filaments of long cells that closely resemble the food-conducting tissue of plants (see Chapter 30). Called trumpet cells because they have flaring ends, they are tubes that rapidly conduct the products of photosynthesis through the body of the alga.

The brown algae exemplify the extraordinary diversity found among the algae. One genus of simple brown algae is *Ectocarpus* (see Figure 23.24*b*). Its branched filaments, a few centimeters in length, commonly grow on shells and stones. The gametophyte and sporophyte phases of its alternation of generations can be distinguished only by chromosome number or reproductive products (zoospores or gametes). Thus the generations are isomorphic. By contrast, some kelps of the genus *Laminaria* and some other brown algae show a more complex heteromorphic alternation of generations. The large and obvious generation of these species is the sporophyte. Meiosis in special fertile regions of the leaflike fronds produces haploid zoospores. These germinate to form a tiny, filamentous gametophyte. The gametophytes produce either eggs or sperm. The genus *Fucus* carries gametophyte reduction still further: It has no multicellular haploid phase, only a multi*nucleate* haploid phase. The gametes themselves are formed directly by meiosis.

The cell walls of brown algae may contain as much as 25 percent alginic acid, a gummy polymer of sugar acids. The substance cements cells and filaments together and also provides good holdfast glue. It is used commercially as an emulsifier in ice cream, cosmetics, and other products.

Phylum Rhodophyta

Almost all Rhodophyta (red algae) are multicellular (Figure 23.25). The characteristic color of the red algae is a result of the pigment phycoerythrin, which is found in relatively large amounts in the chloroplasts of many species. In addition to phycoerythrin, red algae contain phycocyanin, carotenoids, and chlorophyll. The red algae include species that grow in the shallowest tide pools, as well as the algae found deepest in the ocean (as deep as 170 meters where nutrient conditions are right and the water is clear

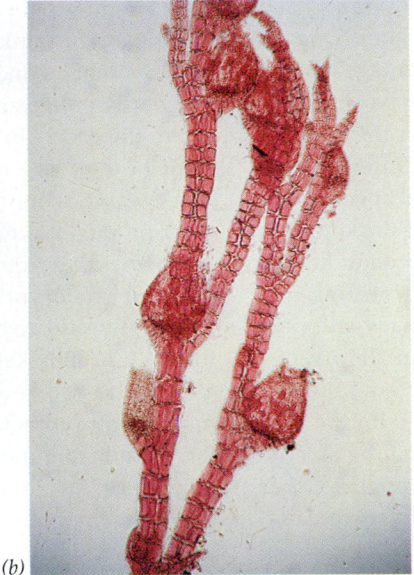

23.25 Red Algae
(*a*) Dulse is a large, edible red alga. (*b*) A species of *Polysiphonia*, with both vegetative and reproductive structures, seen under the light microscope. Both these algae get their red color largely from the pigment phycoerythrin.

BOX 23.A

Algae in a Turbulent Environment

One of the most interesting environments, because of both its animal and its algal life, is the intertidal zone of a rocky shore (*a*). Here the resident organisms are subjected to frequent, dramatic changes in water supply, availability of oxygen, temperature, and other factors, not the least of which is an intermittent pounding by surf. Multicellular algae that grow in this environment are equipped with firm holdfasts (*b*) that anchor them against the surging water. Although seaweeds are common in the rocky intertidal zone, they rarely grow on sandy beaches, where there are no stable attachment sites.

Two algal growth forms are evident in a rocky intertidal environment, one featuring upright growth and the other a prostrate, fleshy, crustlike growth. The upright algae may be more subject to grazing by animals, but they can seize a photosynthetic advantage by overgrowing and shading the prostrate forms. The prostrate algae, on the other hand, have the reciprocal advantage and disadvantage: They may be shaded by the upright algae, but they are less subject to grazing. Different individuals of the same algal species may have one or the other of these two general forms, a phenomenon known as **heteromorphy**. Where grazing pressure varies substantially and predictably, there may be an alternation between the two patterns of growth (upright and prostrate), with the prostrate form predominating at times of high grazing pressure. Other heteromorphic algae, subject to vari-

(*a*)

Algae growing in the intertidal zone on an exposed rocky shore take a tremendous pounding by the surf. (*a*) Sea palm (*Postelsia palmaeformis*), a brown alga growing along the California coast. (*b*) The tough, branched holdfast that anchors the sea palm dominates this photograph.

(*b*)

able but unpredictable grazing pressure, produce both forms at all times.

One might think of this violent environment as one where only a few hardy organisms can exist clinging to the rocks. In fact, however, some of the rocky intertidal areas where wave action is heaviest are the most productive communities on Earth, yielding from two to ten times as much photosynthetic product as the most productive tropical rainforests. Shrubby kelps in such areas carry up to two and a half times as much photosynthetic surface as do the plant communities of tropical rainforests. What factors make this environment supportive of such in-

tense production? The heavy pounding of the surf contributes in at least four ways: (1) It constantly replaces the nutrient-laden water in contact with the algae, so the supply of mineral elements is always at a maximum. (2) It keeps certain important consumers, such as sea urchins, from attacking and eating the algae. (3) It dislodges some of the less tenacious organisms from the rocks, making it possible for shrubby kelps, sea palms, and other well-adapted algae to take their places and form dense "forests." (4) It keeps the fronds of the algae in constant motion, so that none are shaded for long, maximizing photosynthesis.

enough to permit the penetration of light). Very few red algae inhabit fresh water. Most grow attached to a substrate by a holdfast. In a sense the red algae, along with several other groups of algae, are misnamed. They have the capacity to change the relative amounts of their various photosynthetic pigments depending upon the light conditions where they are growing. Thus the leaflike *Chondrus crispus*, a common North Atlantic red alga, may appear bright green when it is growing at or near the surface of the water and deep red when growing at greater depths. The pigmentation—the ratio of pigments present—depends to a remarkable degree upon the intensity of the light that reaches the alga. In deep water, where the light is dimmest, the algae accumulate large amounts of phycoerythrin, an accessory photosynthetic pigment (see Figure 8.9). The algae in deeper water have as much chlorophyll as the green ones near the surface, but the accumulated phycoerythrin makes them look red. Variation in pigmentation depending on light intensity is known as **chromatic adaptation**.

In addition to being the only algae with phycoerythrin and phycocyanin among their pigments, the red algae have two other unique characteristics: They store the products of photosynthesis in the form of floridean starch, which is composed of very small, branched chains of approximately 15 glucose units; and they produce no motile, flagellated cells at any stage in their life cycle. The male gametes are naked and slightly amoeboid, and the female gametes are completely immobile.

Some red algal species enhance the formation of coral reefs. They share with the coral animals the biochemical machinery for depositing calcium carbonate both in and around their cell walls. After the death of the algal cells, the calcium carbonate persists, sometimes forming substantial rocky masses.

Some red algae also produce large amounts of mucilaginous polysaccharide substances, which are based mostly on the sugar galactose with a sulfate group attached. This material readily forms solid gels and is the source of agar, a substance widely used in the laboratory for making a solid aqueous medium on which tissue cultures and many microorganisms may be grown.

Some marine red algae are parasitic upon other red algae. The hosts are photosynthetic, but the parasites are often colorless and nonphotosynthetic, deriving their nutrition from the host. Lynda Goff and Annette Coleman of Brown University discovered that the parasitic red alga *Choreocolax* inserts its nuclei into cells of its host red alga, *Polysiphonia*. This is the first example found of regular introduction of parasite nuclei into living host cells. Apparently parasite genes are expressed in the host cytoplasm, diverting the host's metabolism.

Phylum Chlorophyta

It is thought that the plant kingdom evolved from one or more representatives of the phylum Chlorophyta, commonly known as green algae. These algae have uniform pigmentation and a characteristic storage product. The Chlorophyta and the Euglenophyta are the only protist groups that contain the full complement of photosynthetic pigments also characteristic of the kingdom Plantae. Chlorophyll *a* predominates, and a major pigment is chlorophyll *b*, which none of the other algae have. The carotenoids found in these groups, predominantly β-carotene and certain xanthophylls (carotenoids with one or more hydroxyl groups), are likewise those characteristic of plants. The principal photosynthetic storage product, like that of the plant kingdom, is long, straight, or branched chains of glucose that together make up starch.

BODY SHAPE AND CELLULARITY IN THE CHLOROPHYTA. We find in the green algae an incredible variety in shape and construction of the algal body. *Chlorella* is an example of the simplest type: unicellular and flagellated. Surprisingly large and well-formed colonies of cells are found in such freshwater groups as the genus *Volvox* (Figure 23.26a). The cells are not differentiated into tissues and organs as in the plants and animals, but the colonies show vividly how the preliminary step of this great evolutionary development might have been taken.

The intermediate stages between the one-celled state and the extreme colonial state of *Volvox* are preserved in loosely colonial forms, such as *Gonium* and *Pandorina*. By contrast, *Oedogonium* is multicellular and filamentous, with each cell having only one nucleus. *Cladophora* is multicellular, but each cell is multinucleate. *Bryopsis* is tubular and coenocytic, forming cross walls only with the formation of reproductive structures. *Acetabularia* (see Figure 4.12) is a single, giant uninucleate cell with remarkable morphology, becoming multinucleate only at the end of the reproductive stage. *Ulva lactuca* is a membranous sheet two cells thick; the unusual appearance of *U. lactuca* justifies its common name: sea lettuce (Figure 23.26b). Finally, there are the remarkable unicellular desmids with their elaborately sculptured cell walls (Figure 23.26c).

LIFE CYCLES IN THE CHLOROPHYTA. The life cycles of green algae are also diverse. We will examine two of these in detail, beginning with the life cycle of the sea lettuce *Ulva lactuca*, (Figure 23.27). The diploid sporophyte of this common seashore alga is a "leaf" a few centimeters in diameter. Specialized cells (sporocytes) differentiate and undergo meiosis and cytokinesis, producing motile haploid spores (zoospores).

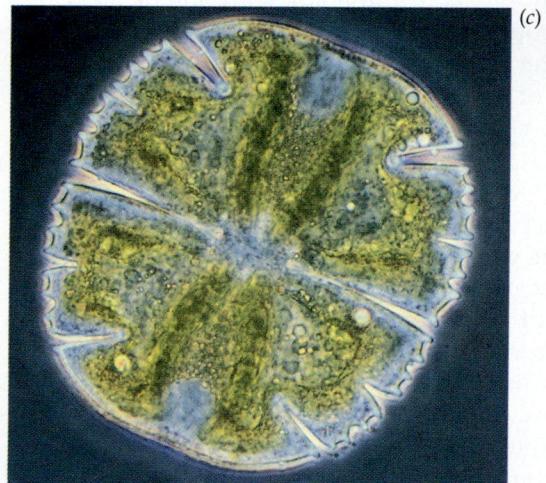

23.26 Green Algae
(a) Colonies of *Volvox* showing the precise spacing of cells and containing a number of daughter colonies. (b) A stand of sea lettuce, *Ulva lactuca*, submerged in a tide pool. (c) A microscopic desmid of the genus *Micrasterias*. A narrow, nucleus-housing isthmus joins two elaborate semicells—halves of the unicellular organism. A single, large, ornate chloroplast fills much of the volume of each semicell.

23.27 The Life Cycle of *Ulva*
The diploid sporophytes and haploid gametophytes look very much alike and give this green alga its common name, sea lettuce.

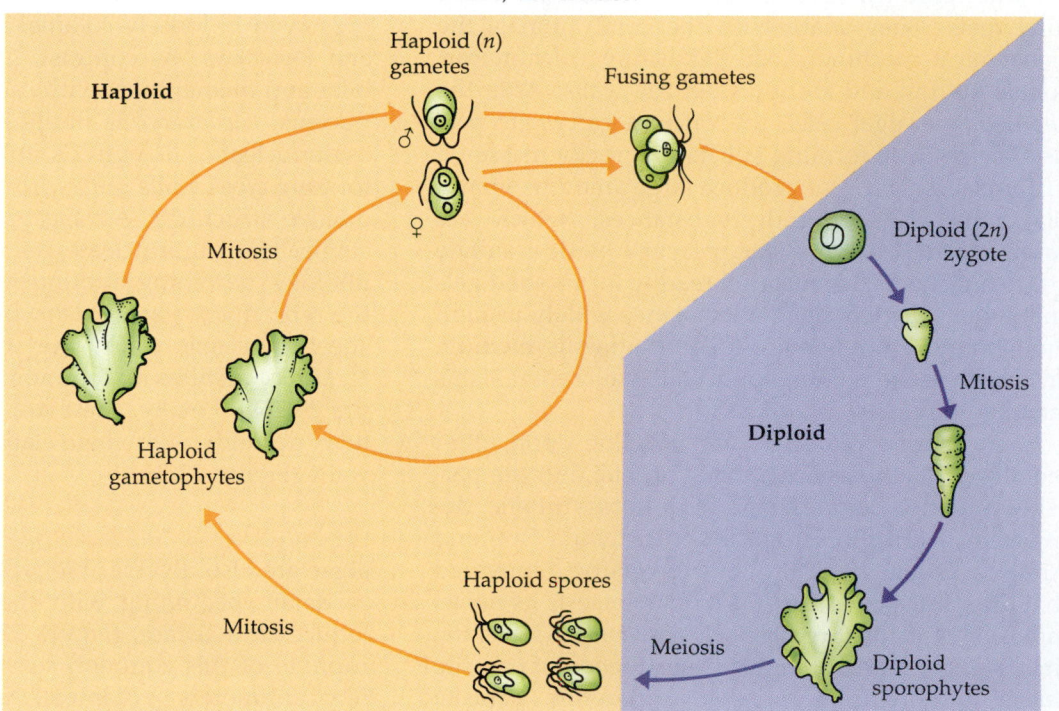

These swim away, each propelled by four flagella, and some eventually find a suitable place to settle. The spores then lose their flagella and begin to divide mitotically, producing a thin filament that develops into a broad sheet only two cells thick. The haploid gametophytes thus produced look just like the diploid sporophytes.

Each spore contains genetic information for just one mating type (see Chapter 24); and a given gametophyte can produce only male or female gametes, never both. The gametes arise mitotically within single cells (gametangia), rather than within any specialized multicellular structure, as in mosses and vascular plants (see Chapter 25). Both types of gametes bear two flagella (in contrast to the four flagella of a haploid spore) and hence are motile. In most species of *Ulva* the female and male gametes are indistinguishable structurally, making those species **isogamous**—having gametes of identical appearance. Some other algae and funguslike protists are also isogamous. (Yet other algae, including some species of *Ulva*, are **anisogamous**—having female gametes distinctly larger than the male gametes.) Sperm and egg come together and unite, losing their flagella as the zygote forms and settles. After resting briefly, the zygote begins mitotic division, forming a new sporophyte. Gametes that fail to find partners can settle down on a favorable substrate, lose their flagella, undergo mitosis, and produce new gametophytes directly; in other words, the gametes can also function as zoospores. Few species other than *Ulva* have motile gametes that can also function as zoospores.

A life cycle like that of *Ulva* is **isomorphic**: Sporophyte and gametophyte generations are identical in structure. By contrast, in the funguslike protists and many other algae, the generations are heteromorphic (see Box 23.A). In one variation of the heteromorphic cycle—the **haplontic** life cycle (Figure 23.28)—multicellular gametophytes produce gametes that fuse to form a zygote. The zygote functions directly as a sporocyte, undergoing meiosis to produce spores, which in turn produce new gametophytes. In the entire haplontic life cycle, only one cell—the zygote—is diploid. The filamentous green algae of the genus *Ulothrix* are examples of haplontic algae.

Other algae have a **diplontic** life cycle like that of many animals. In the diplontic life cycle meiosis of sporocytes produces gametes directly; these fuse, and the zygote divides mitotically to form a new multicellular sporophyte. In such organisms, every cell except the gametes is diploid. Between these two extremes are algae whose gametophyte and sporophyte generations are both multicellular, but that have one phase (usually the sporophyte) that is much larger and more prominent than the other.

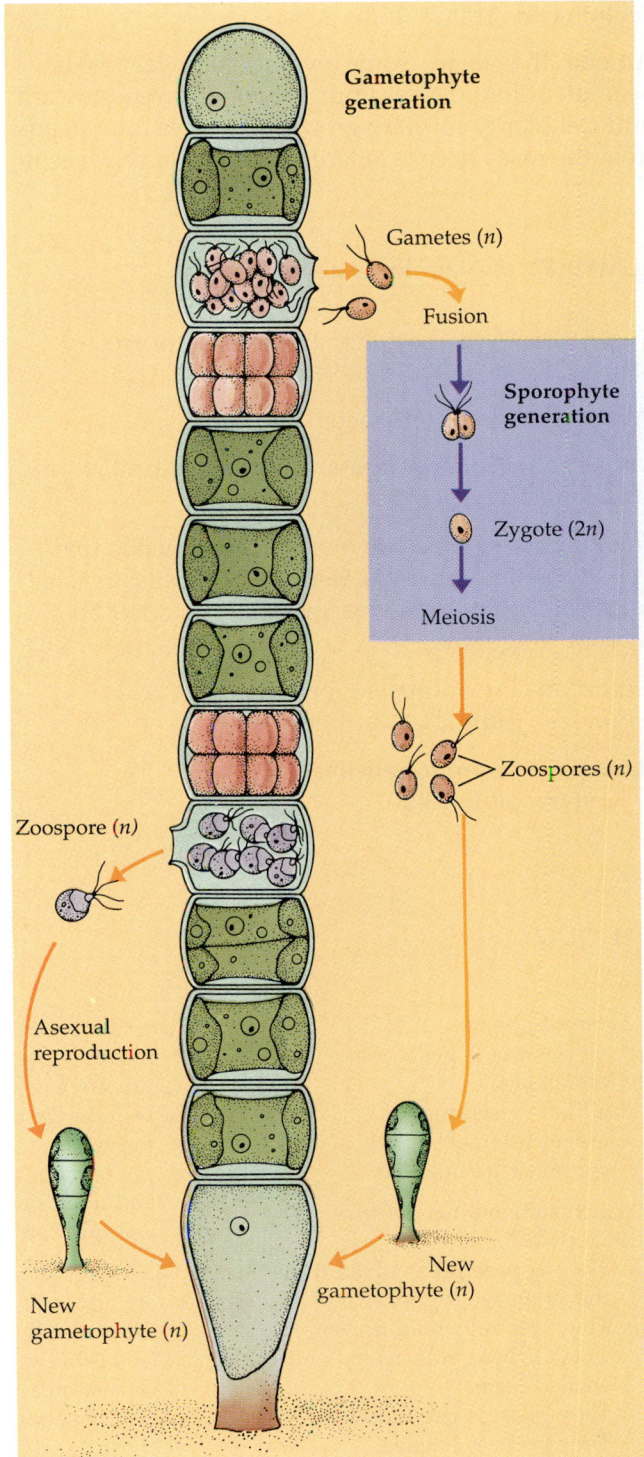

23.28 A Haplontic Life Cycle
In the life cycle of *Ulothrix*, a multicellular gametophyte generation alternates with a unicellular sporophyte generation. At times some gametophyte cells divide mitotically, forming zoospores that develop into multicellular gametophytes. Some cells of a gametophyte divide mitotically to form gametes, which fuse in pairs to form zygotes. A zygote undergoes meiosis, producing haploid spores that develop into the next gametophyte generation.

Evolutionary Trends in the Algae

Among the algae, some evolutionary trends have been identified. For example, there is a trend toward multicellularity in some groups. Another trend leads from the mass release of large numbers of tiny, seemingly undifferentiated gametes (isogamy) toward increased protection of a single, large egg (anisogamy). For example, the female gamete of the green alga *Oedogonium* is large and immobile, and the male gamete is free-swimming and flagellated—an extreme case of anisogamy known as oogamy.

SUMMARY of Main Ideas about Protists

In contrast to the bacteria, all protists are eukaryotic.

Protists evolved from bacterial ancestors.

Unlike fungi, plants, and animals, most protists are unicellular.

The kingdom Protista is more heterogeneous than the other kingdoms; it is defined by exclusion as all eukaryotes that are not assigned to the other eukaryotic kingdoms.

Protozoans are animallike protists.
Review Table 23.1

Some protists are funguslike.
Review Table 23.2

Algae are plantlike protists.
Review Table 23.3

Many protists have highly differentiated bodies, even though they consist of only a single cell.

Alternation of generations is a feature of some protist life cycles.
Review Figures 23.5, 23.7, and 23.28

Photosynthetic protists play a major role in the energy balance of the living world; saprobic protists are among the important decomposers.

Some parasitic protists are agents of disease in plants and animals.
Review Figure 23.10

The multicellular kingdoms all evolved from protists.

SELF-QUIZ

1. Zoomastigophora
   a. all possess flagella.
   b. are all algae.
   c. all have pseudopods.
   d. are all colonial.
   e. are never pathogenic.

2. Which statement is *not* true of the Rhizopoda?
   a. They apparently evolved from the Zoomastigophora.
   b. They use amoeboid movement.
   c. They include both naked and shelled forms.
   d. They possess pseudopods.
   e. They possess flagella.

3. The Apicomplexa
   a. possess flagella.
   b. being amoeboid, are closely related to the Rhizopoda.
   c. are all parasitic.
   d. are among the funguslike protists.
   e. include the trypanosomes that cause sleeping sickness.

4. The Ciliophora
   a. move by means of flagella.
   b. use amoeboid movement.
   c. include *Plasmodium*, the agent of malaria.
   d. possess both a macronucleus and micronuclei.
   e. are autotrophic.

5. The Myxomycota
   a. are also called the acellular slime molds.
   b. lack fruiting bodies.
   c. consist of large numbers of myxamoebas.
   d. consist at times of a mass called a pseudoplasmodium.
   e. possess flagella.

6. The Acrasiomycota
   a. are also called the acellular slime molds.
   b. lack fruiting bodies.
   c. form a plasmodium that is a coenocyte.
   d. use cAMP as a "messenger" to signal aggregation.
   e. possess flagella.

7. Which statement is *not* true of the algae?
   a. They differ from plants in lacking protected embryos.
   b. They are photosynthetic autotrophs.
   c. They have chitin in their cell walls.
   d. They include both unicellular and multicellular forms.
   e. Their life cycles show extreme variation.

8. Which statement is *not* true of the Phaeophyta?
   a. They are all multicellular.
   b. They use the same photosynthetic pigments as do plants.
   c. They are almost exclusively marine.
   d. A few are among the largest organisms on Earth.
   e. Some have extensive tissue differentiation.

9. The Rhodophyta
   a. are mostly unicellular.
   b. are mostly marine.
   c. owe their red color to a special form of chlorophyll.
   d. have flagella on their gametes.
   e. are all heterotrophic.

10. Which statement is *not* true of the Chlorophyta?
    a. They use the same photosynthetic pigments as do plants.
    b. Some are unicellular.
    c. Some are multicellular.
    d. All are microscopic in size.
    e. They display a great diversity of life cycles.

FOR STUDY

1. For each type of organism given below, give a single characteristic that may be used to differentiate it from the other, related organism(s) in parentheses.
   a. Foraminifera (radiolarians)
   b. *Vorticella* (*Paramecium*)
   c. *Euglena* (*Volvox*)
   d. *Trypanosoma* (*Giardia*)
   e. amoeba (foraminiferan)
   f. *Physarum* (*Dictyostelium*)

2. For each of the groups listed below, give at least two characteristics used to distinguish the group from others.
   a. Ciliophora
   b. Apicomplexa
   c. Rhizopoda
   d. Zoomastigophora

3. What is a major role in the world played by the photosynthetic protists? What is a major role in the world played by the saprobic protists?

4. Giant seaweed (mostly brown algae) have "floats" that aid in keeping their fronds suspended at or near the surface of the water. Why is it important that the fronds be suspended?

5. Justify the placement of *Euglena* in the Euglenophyta. Why might it just as well be placed in the Chlorophyta? Why are the Euglenophyta, some Chlorophyta, and the Zoomastigophora often included together in a group called the Mastigophora?

6. Why are algal pigments so much more diverse than those of plants?

READINGS

Brusca, R. C. and G. J. Brusca. 1990. *Invertebrates*. Sinauer Associates, Sunderland, MA. Chapter 5, on the protozoans, covers many algae as well—showing once again that different authorities follow different taxonomic schemes for the kingdom Protista.

Donelson, J. E. and M. J. Turner. 1985. "How the Trypanosome Changes Its Coat." *Scientific American*, February. Trypanosomes evade the host's immune system by constantly switching on new genes that code for different surface antigens.

Farmer, J. N. 1980. *The Protozoa*. Mosby, St. Louis. A full treatment of the animallike protists.

Godson, G. N. 1985. "Molecular Approaches to Malaria Vaccines." *Scientific American*, May. The life cycle and molecular biology of a problem pathogen.

Lee, J. J., S. H. Hutner and E. C. Bovee (Eds.). 1985. *An Illustrated Guide to the Protozoa*. Allen Press, Lawrence, KS. A publication of the Society of Protozoologists.

Margulis, L. 1993. *Symbiosis in Cell Evolution*, 2nd Edition. W. H. Freeman, New York. It is interesting to review the endosymbiotic theory as you study the protists, which may illustrate some of the stages proposed by Margulis.

Saffo, M. B. 1987. "New Light on Seaweeds." *BioScience*, October. Experimental observations on the vertical distribution of algae.

Vidal, G. 1984. "The Oldest Eukaryotic Cells." *Scientific American*, February. A look at ancient marine protists.

Releasing Carbon That Other Organisms Will Use
The red fungus is obtaining nutrients from the log, thus taking part in the decomposition of the fallen tree.

24

Fungi

Earth would be a messy place without fungi. They are at work in forests and garbage dumps, breaking down the remains of dead organisms and even substances such as plastic. For half a billion years or so, the ability of fungi to decompose substances has been important for life on Earth chiefly because, by breaking down carbon compounds, the fungi return carbon and other elements to the environment, where they can be used again by living organisms.

Already crucial for the continuation of life as we know it, the fungi may be about to take on an even more important decomposition task. This task is necessary because our species has been loading toxic substances into the environment for some time. Might the removal of these substances be a job for the fungi? Indeed it might, and many biologists are exploring the ability of different fungi to break down pesticides and other toxic substances into harmless products of metabolism.

GENERAL BIOLOGY OF THE FUNGI

Some members of the kingdom Fungi are the most important degraders of dead organic matter because they are superbly adapted for absorptive nutrition. Other fungi are parasites. Parasitic fungi attack all eukaryotes, including other fungi (Figure 24.1). Fungi are found in all environments. Many are exquisitely constructed, and their life cycles are extremely complex. Whether some organisms should be classified as fungi or as protists has challenged taxonomists because both kingdoms encompass great diversity in anatomy and life cycles. To clarify the separation into kingdoms, we view the kingdom Fungi as encompassing *heterotrophic organisms* with *absorptive nutrition*. The fungi, as we define them, are **saprobes** (organisms living on dead matter), parasites, or **mutualists** (organisms living in mutually beneficial symbiosis with other organisms). Fungi form spores, and none of their cells ever possess flagella. Fungi reproduce sexually by **conjugation**, in which filaments of different mating types fuse. The cell walls of all fungi consist of the polysaccharide **chitin**, which is also found in the cell walls of certain funguslike protists as well as in the skeletons of arthropods. Most fungi are multicellular.

These criteria enable us to distinguish between the fungi and some protists. The slime molds consist of one protist phylum (Acrasiomycota) that takes up food by phagocytosis rather than by absorption and another (Myxomycota) that has cells with flagella. The other funguslike protists (phyla Protomycota and Oomycota) also have flagellated cells, and the Oomycota have cellulose, rather than chitin, in their cell walls.

(a)

(b)

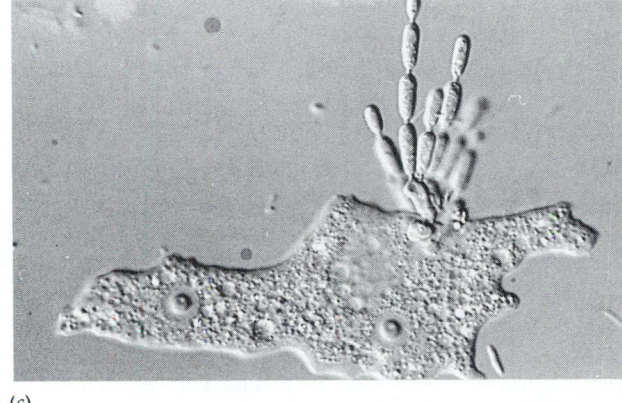

(c)

We consider the kingdom Fungi to consist of a single phylum, the Eumycota, or true fungi (Table 24.1). We distinguish the classes of the Eumycota on the basis of the methods and structures they use for sexual reproduction and, to a lesser extent, on criteria such as the presence or absence of cross walls separating their cells. The Eumycota are divided into four classes: Zygomycetes (conjugating fungi); Ascomycetes (sac fungi); Basidiomycetes (club fungi); and Deuteromycetes (imperfect fungi), which do not form sexual structures by which they might be identified as members of one of the other three classes.

The Fungal Body

The body of a multicellular fungus is called a **mycelium**. Its cells are organized into rapidly growing individual filaments called **hyphae**. Certain hyphae, the **rhizoids**, anchor saprobic fungi to their substrate (the dead organism upon which they feed). Parasitic fungi may have modified hyphae, called **haustoria**, that penetrate the cells of the host organism. The total hyphal growth of a mycelium (not the growth of an individual hypha) may exceed one kilometer per day. The hyphae may be highly dispersed or may

24.1 Parasitic Fungi Attack Living Organisms
(a) The gray masses on this ear of corn are the parasitic fungus *Ustilago maydis*, commonly called corn smut. (b) The tropical fungus whose fruiting body is growing out of the carcass of this ant has developed from a spore ingested by the ant. The spores of this fungus must be ingested by insects before they will germinate and develop. The growing fungus absorbs organic and inorganic nutrients from the ant's body, eventually killing it, after which the fruiting body produces a new crop of spores. (c) An amoeba (below) being parasitized by a fungus of the genus *Amoebophilus* (which means "amoeba-lover").

TABLE 24.1
Classification of Fungi

PHYLUM	CLASS	COMMON NAME	FEATURES	EXAMPLES
Eumycota (80,000 species)	Zygomycetes	Conjugating fungi	No hyphal cross walls, usually no fleshy fruiting body	*Rhizopus*
	Ascomycetes	Sac fungi	Ascus, perforated cross walls	*Neurospora*, yeast
	Basidiomycetes	Club fungi	Basidium, complete cross walls	*Puccinia*, mushrooms
	Deuteromycetes	Imperfect fungi	No known sexual stages	*Arthrobotrys* (traps nematodes)

be clumped into a cottony mass. Sometimes the mycelium is organized into elaborate fruiting bodies such as mushrooms.

The attack of a parasitic fungus on a plant illustrates the roles of some fungal structures (Figure 24.2). The hyphae of a fungus invade a leaf through pores called stomata (see Chapter 8) or through wounds. Once inside, the hyphae form a mycelium. Haustoria grow into living plant cells, absorbing the nutrients within the cells. Eventually fruiting bodies form, either within the plant body or on its surface.

Fungi and Their Environment

The tubular hyphae of a fungus give it a unique relationship with its physical environment: The fun-

24.3 A Kitchen Nuisance
The green mold *Penicillium digitatum* on an orange.

gal mycelium has an enormous surface-to-volume ratio compared with those of most other large multicellular organisms. This large surface-to-volume ratio is a marvelous adaptation for absorptive nutrition, in which nutrients are absorbed across the cell surfaces. Throughout the mycelium, except in fruiting bodies, all the cells are very close to their environmental food source.

Another characteristic of fungi is their tolerance for highly hypertonic environments (those with more negative osmotic potentials; see Chapter 5). Many fungi are more resistant than are bacteria to damage in hypertonic surroundings. For example, jelly in the refrigerator will not become a growth medium for bacteria because it is too cold, but it may eventually harbor mold colonies. You have probably seen the green mold *Penicillium* growing on oranges in the refrigerator (Figure 24.3). The refrigerator example illustrates another trait of many fungi: tolerance of temperature extremes. Many fungi tolerate temperatures as low as 5 or 6 degrees below freezing, and some tolerate high temperatures of 50°C or more.

Nutrition of Fungi

All fungi are heterotrophs that obtain food by absorption. Many are saprobes, obtaining their energy, carbon, and nitrogen directly from dead organic matter by the action of enzymes. Many others are parasites. Other fungi form mutualistic associations with other organisms. Saprobic fungi, along with the bacteria, are the major decomposers of the biosphere, contributing to decay and thus to the recycling of the elements used by living things. In the forest, for example, the invisible mycelia of fungi obtain nutrients from fallen trees, thus decomposing the trees (see the illustration at the beginning of this chapter).

Because saprobic fungi will grow on artificial media, we can determine their exact nutritional require-

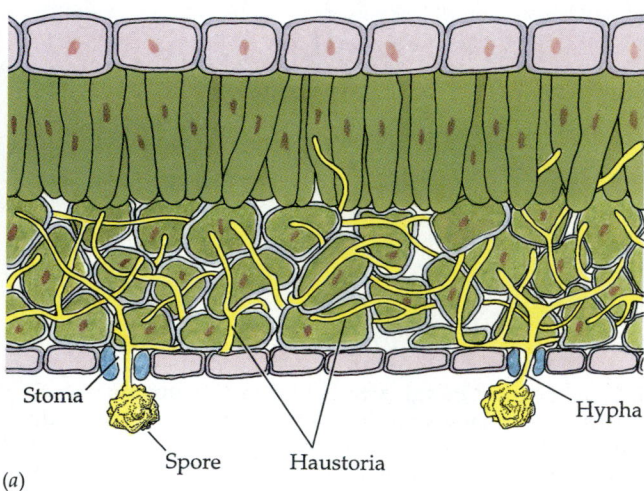

Stoma

Hypha

Spore   Haustoria

(a)

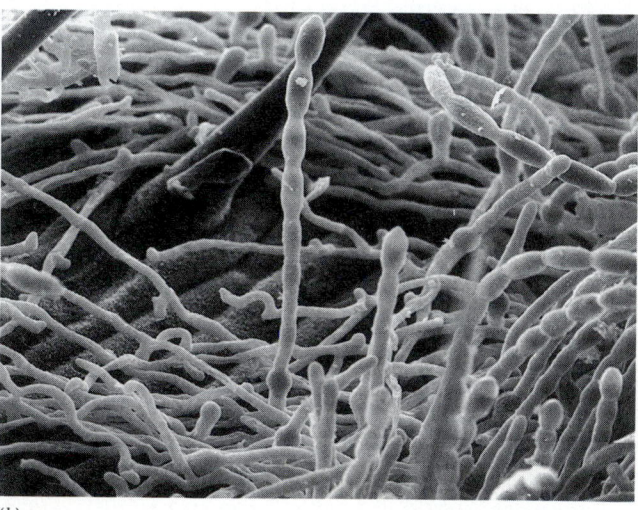

(b)

24.2 A Fungus Attacks a Leaf
(a) Fungal spores germinate on the surface of the leaf. Elongating hyphae pass through stomata into the interior of the leaf, elongate further, and branch. Some hyphae develop into haustoria that penetrate cells within the leaf. *(b)* The surface of the grass *Stenotaphrum secundatum* (dark) infected with the fungus *Erysiphe graminis* (light).

ments. Sugars are the favored sources of carbon, and most fungi obtain nitrogen from proteins or protein breakdown products. Many fungi can use nitrate (NO_3^-) or ammonium (NH_4^+) ions as their sole source of nitrogen. No known fungus gets its nitrogen directly from nitrogen gas, as can some bacteria and plant–bacteria associations (see Chapter 32). Vitamins play a role in fungal nutrition. Most fungi are unable to synthesize their own thiamin (vitamin B_1) or biotin (another B vitamin) and hence must absorb them from their environment. Other compounds that are vitamins for animals can be made by fungi. Like all other organisms, fungi also require some mineral elements.

Nutrition in the parasitic fungi is particularly interesting. **Facultative** parasites can grow parasitically but can also grow by themselves on defined artificial media. Biologists can work out the exact nutritional requirements by varying the composition of the growth medium. **Obligate** parasites cannot be grown on any available defined medium; they can grow only on their specific hosts, usually plants (see Figure 24.1a). Obligate parasites include various mildews and rusts. Because their growth is so limited, they must have unusual nutritional requirements. Biologists are thus very interested in learning more about them.

Some fungi have adaptations allowing them to function as active predators, trapping nearby microscopic protists or animals, from which they obtain nitrogen and energy. The most common approach is to secrete sticky substances from the hyphae so that passing organisms stick tightly to them. Fungal haustoria then quickly invade the prey, growing and branching within it, spreading through its body, absorbing nutrients, and eventually killing it.

A more dramatic adaptation for predation is the **constricting ring** formed by some species of *Arthrobotrys*, *Dactylaria*, and *Dactylella* (Figure 24.4). All these fungi grow in soil; when nematodes (tiny roundworms) are present, the fungi form three-celled rings with a diameter that just fits a nematode. A nematode crawling through one of these rings stimulates it, causing the cells of the ring to swell and trap the worm. Fungal hyphae quickly invade and digest the unlucky victim. R. G. Thorn and G. L. Barron of the University of Guelph reported that 11 of 27 species of wood-decaying mushrooms tested have the ability to supplement their nitrogen supply by trapping nematodes.

Certain highly specific associations between fungi and other organisms have nutritional consequences for the fungal partner. **Lichens** are associations of a fungus with either a unicellular alga or a cyanobacterium; the combination functions effectively as a plant. In the union the fungus draws nutrition from the photosynthetic bacterium or alga. **Mycorrhizae** are associations between fungi and the roots of plants (Box 24.A). In such associations the fungus is fed by the plant but provides minerals (primarily phosphorus) to the plant root, so the plant's nutrition is also promoted. Seed germination in most orchid species depends on the presence of a specific mutualistic fungus, which itself derives nutrients from the seed and seedling of the orchid.

Perhaps the most striking fungal associations are with insects. Some leaf-cutting ants "farm" fungi, feeding the fungi and later harvesting and eating

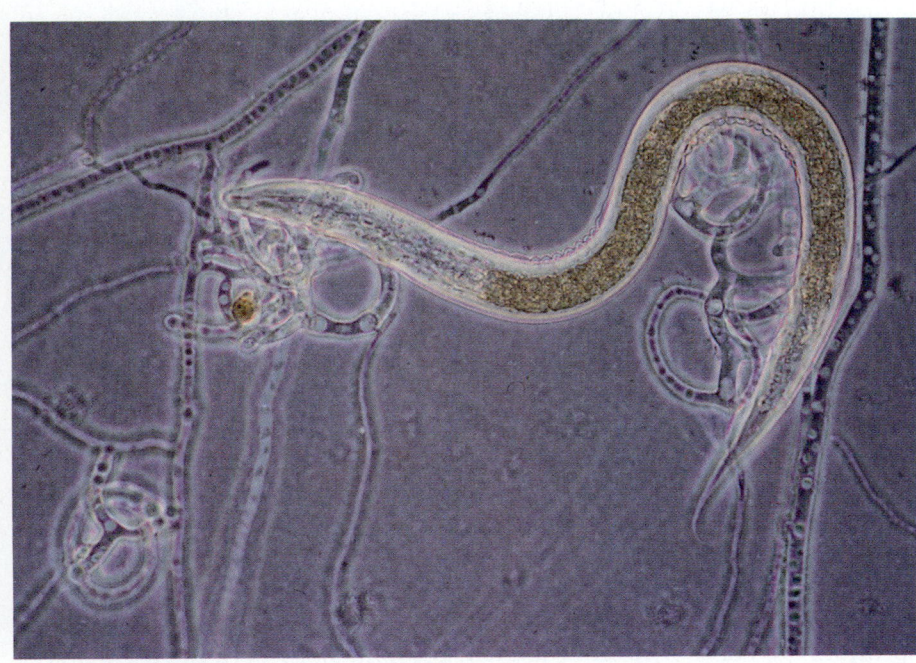

24.4 Some Fungi Are Predators
A nematode (roundworm) trapped in a sticky loop of the soil-dwelling fungus *Arthrobotrys conoides*; at the lower left is an empty loop.

BOX 24.A

Plant Roots and Fungi

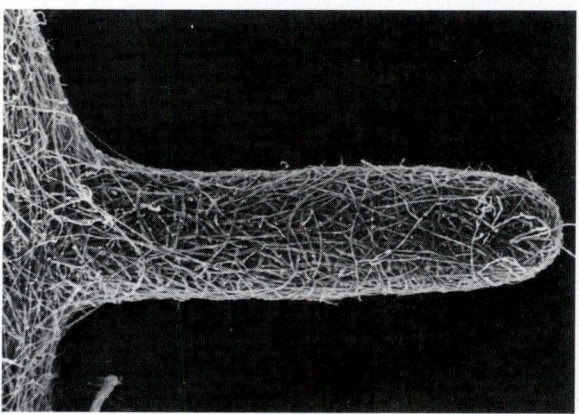

Hyphae of the fungus *Pisolithus tinctorius* cover this eucalyptus root, forming a mycorrhiza.

Many plants, including almost all tree species, depend on a mutually beneficial symbiotic association with fungi for an adequate supply of water and mineral elements. Unassisted, the root hairs of such plants do not absorb enough of these materials to sustain maximum growth. The roots, however, become infected with a fungus, which wraps around them and may invade their cells, forming a mycorrhiza. Such roots characteristically branch extensively and become swollen and club-shaped, as shown here. The hyphae of the fungus attached to the root increase the surface area for the absorption of water and minerals, and the mass of the mycorrhiza, like a sponge, holds water efficiently in the neighborhood of the root.

In most families of flowering plants are some species that form mycorrhizae, as do liverworts, ferns, club mosses, and gymnosperms (see Chapter 25). Fossils of mycorrhizal structures more than 300 million years old have been found. Certain plants that live in nitrogen-poor habitats, such as cranberry bushes and

orchids, invariably have mycorrhizae. Orchid seeds will not germinate in nature unless they are already infected by the fungus that will form their mycorrhizae. Plants without chlorophyll, such as the parasite dodder (see Figure 32.13), always have mycorrhizae, which are often shared with the roots of green, photosynthetic plants.

The symbiotic fungus–plant association of the mycorrhiza is important to both partners. The fungus obtains important organic compounds, such as sugars and amino acids, from the plant. In return, the fungus greatly increases the absorption of water and minerals (especially phosphorus) by the plant. The fungus also often provides certain growth hormones (including auxin and cytokinins; see Chapter 34), and it protects the plant

against attack by microorganisms. Plants with active mycorrhizae typically are a deeper green and may resist drought and temperature extremes better than plants of the same species with little mycorrhizal development. Attempts to introduce some plant species to new areas have failed until a bit of soil from the native area (presumably containing the fungus necessary to establish mycorrhizae) was added.

The partnership between plant and fungus results in a plant better adapted for life on land. K. A. Pirozynzki and D. W. Malloch, of the Canadian Department of Agriculture, have suggested that the evolution of this symbiotic association was the single most important step leading to the colonization of the terrestrial environment by living things.

them (Figure 24.5). The ants collect leaves and flower petals and chew them into small bits, upon which they "plant" bits of fungal mycelium. The ants even "weed" these gardens by removing other fungal species. The species of fungus cultivated by the ants are found nowhere other than in these gardens.

Scale insects live in association with the fungus *Septobasidium*. The fungus spreads over a colony of the insects, infecting some of them and thus parasitizing them—without killing them. As new insects hatch from the eggs within the colony, some of them become infected. They take the fungus along as they establish new colonies. The fungus protects the col-

ony against drying and against some predators but also draws its nutrition from the insects.

Reproduction in Fungi

Both asexual and sexual reproduction are common in the fungi. Asexual reproduction takes several forms. One form is the production of haploid spores within structures called sporangia. The second form is the production of naked spores at the tips of hyphae; these spores, called **conidia** (from the Greek *konis*, "dust"), are not produced in sporangia. The third form of asexual reproduction, performed by unicel-

24.5 Ants Farm a Fungus
Workers of the Costa Rican ant species *Atta cephalotes* add a cut piece of leaf to their fungal garden. The fungus is the white material. In response to this care, the fungus grows and serves as food for the ants.

lular fungi, is cell division—either a relatively equal division or an asymmetric division in which a tiny bud is produced. The fourth form of asexual reproduction, seen in some multicellular fungi, is simple breakage of the mycelium.

Sexual reproduction in many fungi features an interesting twist. In addition to a distinction between male and female structures, there is a genetically determined distinction between two or more **mating types:** Individuals of the same mating type cannot mate with one another, even though they can each function either as a female or as a male. This distinction prevents self-fertilization. Individuals of different mating types differ genetically from one another but are visually indistinguishable.

The nuclei of fungi are haploid except in the zygotes formed in sexual reproduction, which are diploid. Zygotes undergo meiosis, producing haploid spores. Haploid fungal spores, whether produced sexually or asexually, divide mitotically to produce multicellular individuals. Fungal cells may have a configuration other than the familiar haploid and diploid—that is, dikaryotic.

Dikaryon Formation in Fungi

Sexual reproduction begins in some fungi in an unusual way: The cytoplasm of two individuals fuses long before their nuclei fuse, so *two genetically different haploid nuclei exist within the same cell*. The bread mold *Neurospora crassa* exhibits such reproduction. *Neurospora* has two mating types, *A* and *a*. Each mating type can produce both female and male structures. A haploid individual of *Neurospora* consists of hyphae

entwined into a mycelium. This haploid organism will mate only with a haploid individual of the other mating type.

How does the mating take place? Hyphae from individuals of different mating types grow toward one another, forming mating structures that eventually touch (top of Figure 24.6). At the points of contact, enzymes digest the cell walls, allowing some cytoplasm and the nucleus of one cell to invade the other. The recipient mating cell then contains two haploid nuclei, one from each parent. The recipient cell develops a hypha consisting of cells that each have two nuclei—one of each mating type. This hypha is called a **dikaryon** (having two nuclei). Because the two nuclei in each cell of a dikaryon differ genetically, the hypha is also called a **heterokaryon** (having *different* nuclei).

The heterokaryotic hypha develops into a heterokaryotic mycelium consisting of undifferentiated regions and specialized fruiting structures. Within the fruiting structures, many pod-shaped sexual sporangia called **asci** (singular: *ascus*) are produced, each containing two dissimilar nuclei—one from each parent. Ultimately, these pairs of nuclei fuse within the asci, giving rise to zygotes long after the original "mating" of the *Neurospora* hyphae. The zygote—the only diploid structure in the entire life cycle of *Neurospora*—undergoes meiosis, producing four haploid nuclei. Each of the four products of meiosis divides mitotically; then cytokinesis results in a total of eight spores. The spores are shed by the ascus and germinate to form new haploid mycelia, each of a specific mating type. Spores may also form asexually (see Figure 24.6).

The reproduction of *Neurospora* displays several unusual features. First, there are no gamete cells, only gamete nuclei. Second, there is never any true diploid tissue, although for a long period during development the genes of both parents are present in the dikaryon and can be expressed. In effect, the cells are neither diploid (2*n*) nor haploid (*n*); rather, they are (*n* + *n*). A harmful recessive mutation in one nucleus may be compensated for by a normal allele on the same chromosome in the other nucleus. Dikaryosis is perhaps the most significant of the genetic peculiarities of the fungi. Finally, although these organisms grow in moist places, the gamete nuclei are not motile and are not released into the environment; therefore, liquid water is not required for fertilization. (As we will see, some members of the plant kingdom require liquid water for the meeting of gametes, as do some aquatic animals.)

Multiple Hosts in a Fungal Life Cycle

Some fungi are very specific about what host organism they use as a source of nutrition, and some even

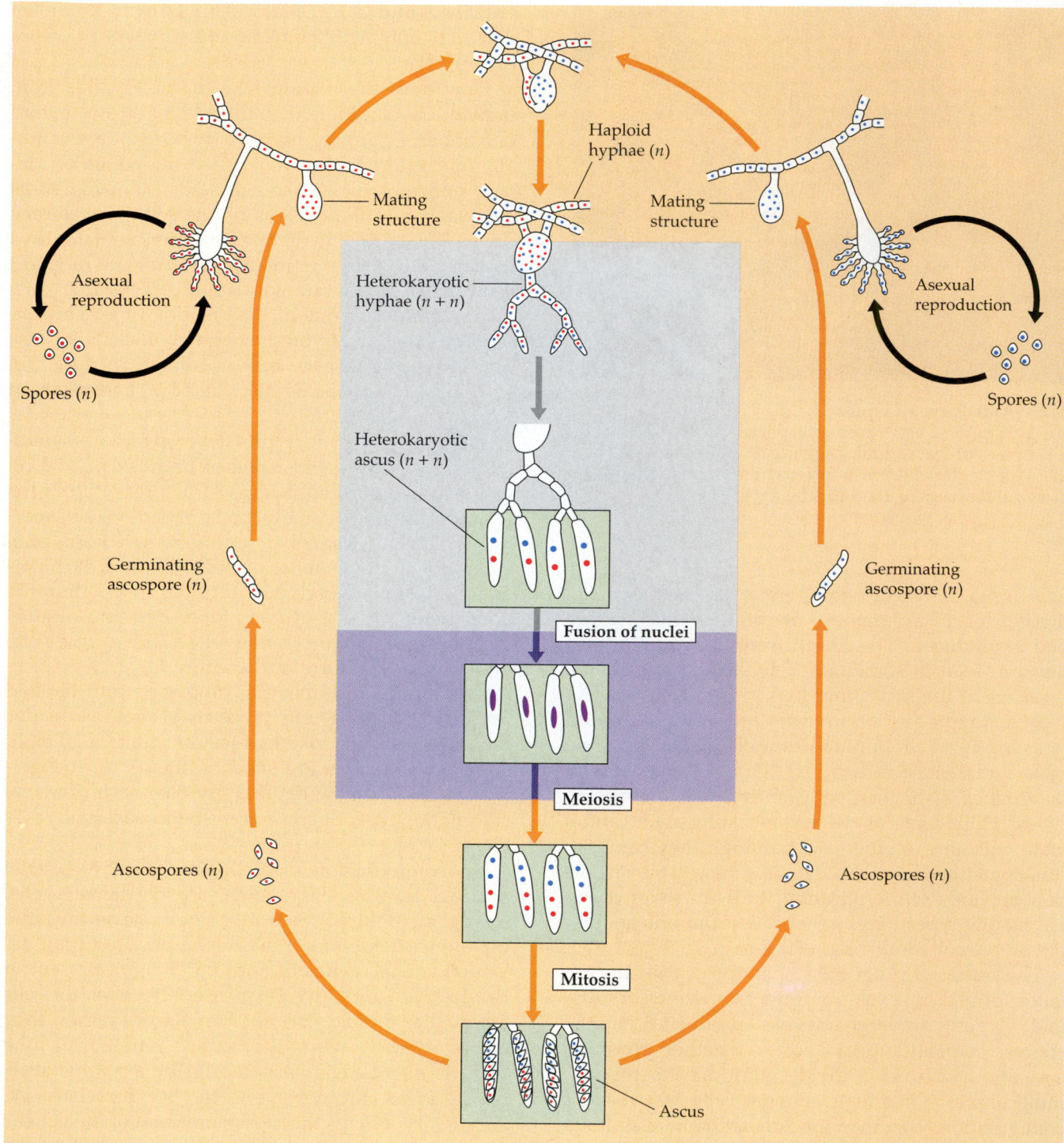

24.6 Life Cycle of *Neurospora*
The two hyphae are of different mating types, as indicated by the different-colored (red and blue) nuclei. The sexual phase is shown in the center of the diagram; the zygote, shown with a purple nucleus, is the only diploid structure in the entire cycle. Spores can also form asexually (black arrows).

use different hosts for different stages of their life cycles. (Remember that a host is an organism that harbors a symbiotic organism—generally a parasite—and provides it with nutrition.) One of the most striking examples of a fungal life cycle involving two different hosts is the complicated life cycle of the black stem rust of wheat, *Puccinia graminis*, the agent of a major agricultural disease. In the epidemic of 1935, *P. graminis* caused a loss of approximately one-fourth of the wheat crop in Canada and the United States.

The life cycle of *P. graminis* illustrates three principal features of fungal life cycles: the utilization of two different hosts, the extent of dikaryosis, and the sheer complexity of some cycles. During the summer dikaryotic ($n + n$) hyphae of *P. graminis* proliferate in the stem and leaf tissues of wheat plants, drawing their nutrition from the wheat plants and damaging them severely. These dikaryotic hyphae produce extensive amounts of summer spores called **uredospores**. These one-celled, orange spores are scattered by the wind and infect other wheat plants, on which the summer hyphae then proliferate. Like the hyphae, the uredospores are dikaryotic. This part of the life cycle is shown at the "ten o'clock" position in Figure 24.7. Dark brown winter spores called **teliospores** begin to appear on the hyphae in late summer. Each teliospore consists of two dikaryotic cells. Both cells have thick walls and are resistant to freezing. The teliospores remain dormant until spring. Then the two haploid nuclei in each cell fuse to form two cells, each diploid ($2n$). These are the only truly dip-

loid cells in the entire life cycle. The teliospore germinates once the nuclei have fused. Each of the two cells develops into a reproductive structure, and each of the reproductive structures divides meiotically to produce four haploid **basidiospores** (discussed in this chapter under Class Basidiomycetes).

The basidiospores are of two different mating types: plus (+) and minus (−). They are carried by the wind and, if they land on a leaf of the common barberry plant, they germinate and produce haploid hyphae (+ or −, depending on the mating type of the germinating basidiospore). These hyphae invade the barberry leaf. The hyphae form flask-shaped structures on the upper surfaces of the leaf, within

24.7 Some Fungal Life Cycles Are Very Complex
Wheat rusts have complex life cycles punctuated by cell fusion, nuclear fusion, and meiosis and divided between two host plants. Because nuclear fusion is delayed until after cell fusion, rusts have a dikaryotic ($n + n$) phase in their life cycles (gray arrows) in addition to the usual haploid (n; orange arrows) and diploid ($2n$; blue arrows) generations. The photograph at the upper left shows the formation of uredospores on wheat. The lower photograph is a cross section of a barberry leaf; on the leaf's upper surface are pycniospore-producing spermagonia of *Puccinia graminis*, while aeciospores are forming on the leaf's underside.

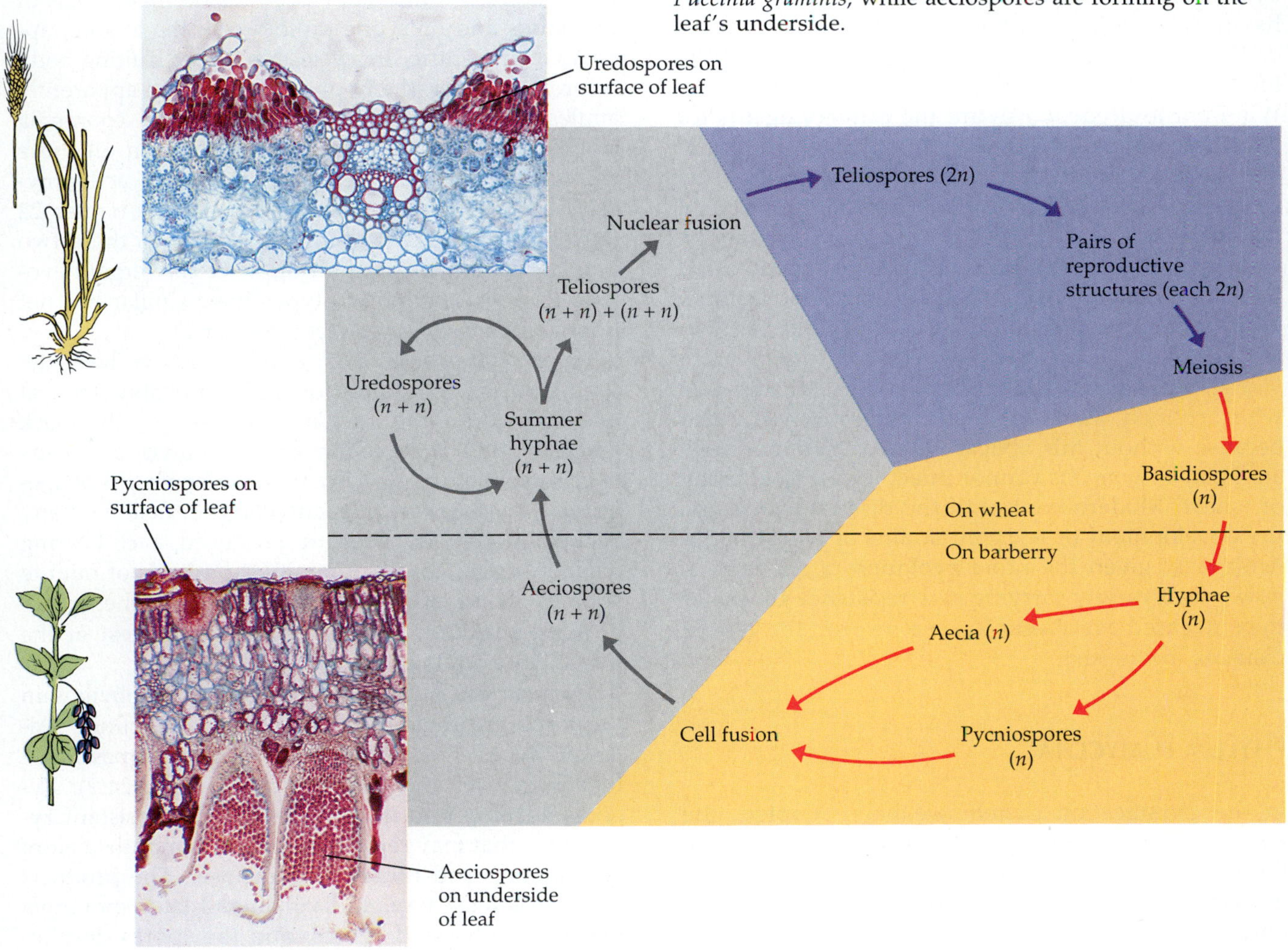

which some of the hyphae pinch off tiny, colorless, haploid **pycniospores** from their tips. The hyphae also form another type of structure, called an **aecium**, near the lower surface of the barberry leaf. Thus the leaf contains "flasks" on the upper surface and aecia near the lower one, and these are connected by the hyphae. Insects attracted to a sweetish liquid produced in the flasks carry pycniospores from one flask to another.

A pycniospore of one mating type fuses with a receptive hypha of the other type within a flask. The nucleus of the pycniospore repeatedly divides mitotically, and the products move through the hyphae into immature aecia, where they produce dikaryotic cells with two haploid nuclei, one of each mating type. These cells develop into **aeciospores**, each containing two unlike nuclei. The aeciospores are scattered by the wind, and some germinate on wheat plants, continuing the life cycle. When the dikaryotic aeciospores germinate, they produce the summer hyphae with two nuclei in each cell. In the stages from the aeciospore to the teliospore, *P. graminis* is dikaryotic; that is, the individual cells contain nuclei from both "parents." Overwintered teliospores, at first dikaryotic, become diploid. Basidiospores are produced by meiosis and cytokinesis, so they are haploid, each having a mating type of + or −.

The different types of spores produced during the life cycle of *P. graminis* play very different roles. Wind-borne uredospores are the primary agents for spreading the rust from wheat plant to wheat plant and to other fields. Resistant teliospores allow the rust to survive the harsh winter but contribute little to the spreading of the rust. Basidiospores spread the rust from wheat to barberry plants. Pycniospores and receptive hyphae initiate the sexual cycle of the rust. Finally, aeciospores spread the rust from barberry to wheat plants.

The traditional means of combatting *P. graminis* in wheat country was to remove barberry from the area because without this obligate (necessary) intermediate host, *P. graminis* cannot infect a new generation of wheat. Modern control of the disease focuses on the development of resistant strains of wheat, a difficult task, given the rapid evolution of the rust. A new wheat variety carrying new resistance genes has to be released every year to keep up with the genetic changes in the rust.

PHYLUM EUMYCOTA

In the classification system we have adopted, the kingdom Fungi consists of only one phylum, the Eumycota, which is divided into four classes: Zygomycetes, Ascomycetes, Basidiomycetes, and Deuteromycetes.

24.8 A Zygomycete
This small forest of filamentous structures is made up of sporangiophores of the fungus *Phycomyces*. The stalks end in tiny, rounded sporangia.

Class Zygomycetes

The conjugating fungi, or zygomycetes, have hyphae without cross walls. They produce no motile cells of any kind, and only one diploid cell, the zygote, appears in the entire life cycle. No fleshy fruiting body is formed; rather, the hyphae spread in an apparently aimless fashion, with occasional stalked sporangia reaching up into the air (Figure 24.8). In all these regards, the zygomycetes resemble the actinomycetes, a group of bacteria discussed in Chapter 22 (see Figure 22.19). The similarity between these two types of organisms is an example of convergent evolution, in which the two types have similar but not homologous structures (see Chapter 21).

About 600 species of zygomycetes have been described. A zygomycete you have probably seen at one time or another is *Rhizopus stolonifer*, the black bread mold. The mycelium of a zygomycete spreads over its substrate, growing forward by means of specialized hyphae. In vegetative reproduction, many stalked sporangiophores are produced, each bearing a single sporangium containing hundreds of minute spores. As in other filamentous fungi, the spore-forming structure is separated from the rest of the hypha by a wall.

Because the conjugation process is so obvious in zygomycetes, they are called conjugating fungi. Adjacent hyphae of two different mating types grow together, fuse, and form zygotes (Figure 24.9). Zygotes develop into thick-walled, highly resistant **zygospores** that may remain dormant for months before germinating and undergoing meiosis. The products of meiosis form haploid hyphae and later sporangia that release haploid spores, and the spores develop

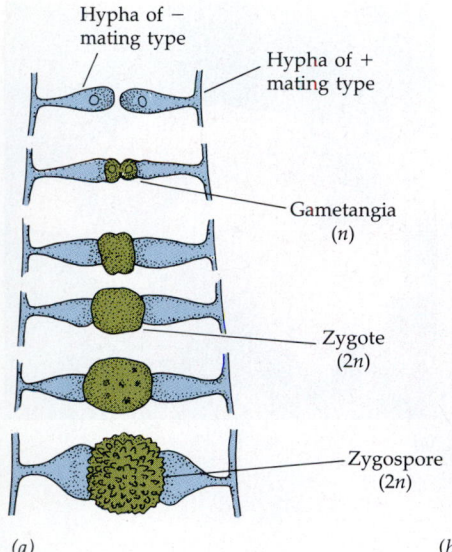

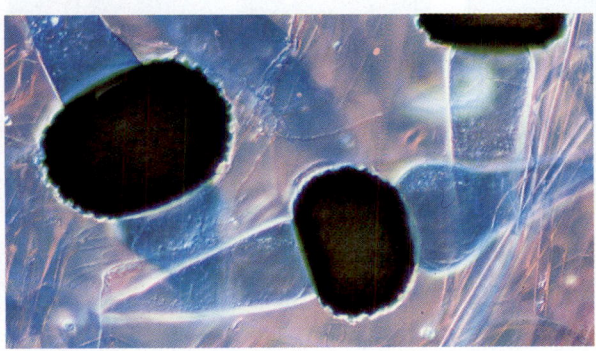

24.9 Conjugation in a Zygomycete

(a) When two hyphae of differing mating type grow side by side, they produce branches that grow toward each other. The tips of these branches develop into gametangia, structures that produce gametes. The gametangia—and then the gametes—fuse, and the resulting zygote develops into a resistant zygospore. (b) Conjugation in the black bread mold *Rhizopus stolonifer* produced these zygospores.

Labels in figure (a): Hypha of − mating type; Hypha of + mating type; Gametangia (n); Zygote ($2n$); Zygospore ($2n$)

into a new generation of hyphae. What causes the hyphae to grow together and fuse? The two hyphae release pheromones (chemical attractants) that direct this conjugation process. (Recall that a pheromone directs gamete attraction in *Allomyces* as well; see Chapter 23.)

Class Ascomycetes

The ascomycetes, or sac fungi, are a large and diverse group of fungi distinguished by the production of **asci** (Figure 24.10; see also Figure 24.6). The ascus is the characteristic sexual reproductive structure of the ascomycetes. The two key events in sexual reproduction—the fusion of two haploid nuclei and meiosis—take place within this single cell, the ascus. Ascomycete hyphae are segmented by cross walls (unlike zygomycete hyphae), but a pore in each cross wall permits movement of cytoplasm and organelles (including the nuclei) from one cell to the next.

The approximately 30,000 known species of ascomycetes can be divided into two broad groups, depending on whether the asci are contained within a specialized fruiting structure. Species that have a fruiting structure, the **perithecium**, are collectively called euascomycetes (true ascomycetes); those without perithecia are called hemiascomycetes (half ascomycetes).

The sexual cycle of euascomycetes includes the formation of a dikaryon stage (see Figure 24.6). When the hyphae of two compatible mating types fuse, a heterokaryon forms. Nuclei pass from one mycelium to the other, and then the introduced nuclei divide simultaneously with the host nuclei. Eventually asci form. Only with the formation of asci do the nuclei finally fuse. Both nuclear fusion and the subsequent meiosis of the zygote take place within individual asci. The meiotic products form **ascospores** that are ultimately shed by the ascus to begin the new gametophyte generation.

Ascus formation itself is a peculiar process. A hook (crosier) forms at the tip of the heterokaryotic hypha, and the paired nuclei come to lie on either side of it. Both nuclei then undergo mitosis simultaneously, with their mitotic axes parallel to the axis of the hypha. New walls are laid down and the two nuclei in the top of the hook finally fuse. Meiosis begins ascospore formation.

THE HEMIASCOMYCETES. In general, the hemiascomycetes are very small, and many species are unicellu-

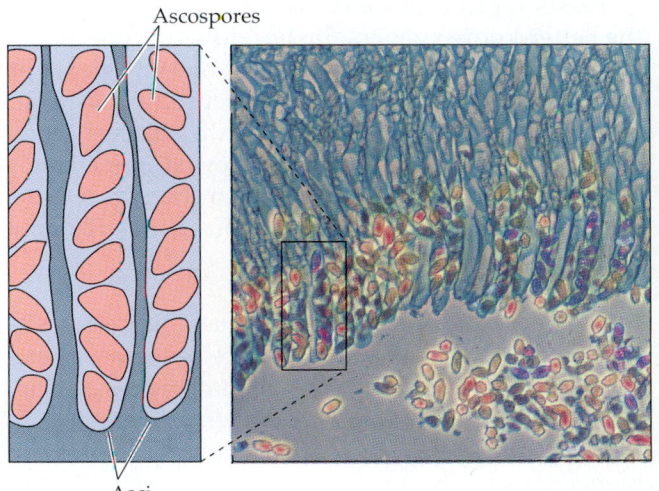

24.10 Asci

Asci and ascospores of the black morel *Morchella elata*. Each ascus contains eight ascospores—the products of meiosis followed by a single mitotic division and cytokinesis.

Labels in figure: Ascospores; Asci

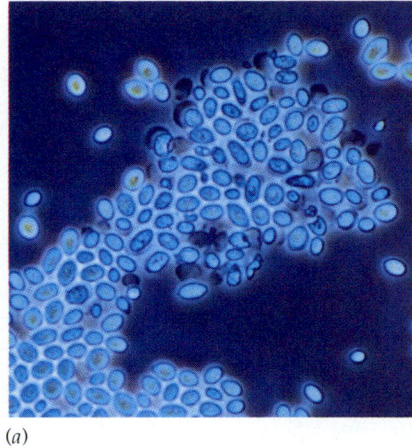

(a)

(b)

(c)

24.11 Some Ascomycetes
(a) Cells of baker's yeast, *Saccharomyces cerevisiae*. Some are budding. These hemiascomycetes are facultative anaerobes—organisms that can grow in either the presence or the absence of free oxygen. (b) Two yellow morels, *Morchella esculenta*. Morels are characterized by their netlike caps and subtle flavor. (c) These orange cups, *Aleuria aurantia*, are representative cup fungi. Morels and cup fungi are euascomycetes.

lar. Perhaps the best known are the yeasts, especially baker's or brewer's yeast (*Saccharomyces cerevisiae*; Figure 24.11a). The yeasts are among the most important domesticated fungi. *S. cerevisiae* metabolizes glucose obtained from its environment to ethanol and carbon dioxide. Carbon dioxide bubbles form in bread dough and give baked bread its light texture. Although baked away in bread making, the ethanol and carbon dioxide are both retained in beer. Other yeasts live on fruits such as figs and grapes and play an important role in the making of wine.

Yeasts reproduce asexually either by fission or, in the better-known genera, by **budding** (the outgrowth of a new cell from the surface of an old one). The single cells are haploid. Sexual reproduction takes place only occasionally between two adjacent, compatible cells; nuclear fusion is followed immediately by meiosis and a single mitosis, so the entire structure becomes an ascus. Yeasts have no dikaryon stage.

THE EUASCOMYCETES. Among the euascomycetes, the other major group of sac fungi, are several common molds, including *Neurospora*, the pink molds (see Figure 24.6). *Neurospora* are found everywhere.

Many euascomycetes are serious parasites on higher plants. Chestnut blight and Dutch elm disease are caused by euascomycetes. The powdery mildews are a group that infects cereal grains, lilacs, and roses, among many other plants. They can be a serious problem to grape growers, and a great deal of research has been done on ways to control these agricultural pests.

Two particularly delicious euascomycete fruiting structures are morels (Figure 24.11b) and truffles. Truffles grow in a mutualistic association with the roots of some species of oaks. People traditionally used pigs to find truffles because some truffles secrete a substance with an odor similar to a pig's sex-attraction substance. Unfortunately, pigs also eat truffles, so dogs are now the usual truffle hunters.

The euascomycetes also include the cup fungi (Figure 24.11c). In these organisms the fruiting structures are cup-shaped and can be as large as several centimeters across. The inner surfaces of the cups are covered with a mixture of sterile filaments and asci, and they produce huge numbers of spores. Although these fleshy structures appear to be composed of distinct tissue layers, microscopic examination shows that their basic organization is still filamentous—a tightly woven mycelium. Such fruiting structures are formed only by the dikaryotic mycelium.

The euascomycetes reproduce asexually by means of conidia (Figure 24.12) that form at the tips of spe-

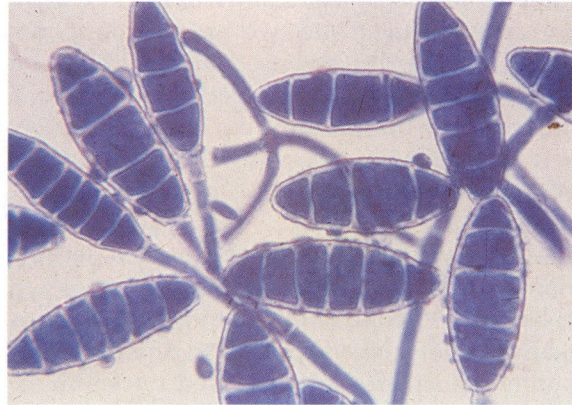

24.12 Conidia
The large oval shapes are lines of four to six conidia at the tips of hyphae of an ascomycete. Each conidium can develop into a multicellular haploid individual, completing the asexual reproductive cycle.

cialized hyphae. These small chains of conidia are produced by the millions and are sufficiently resistant to survive for weeks in nature. The conidia are what give molds their characteristic colors.

Class Basidiomycetes

About 25,000 species of club fungi, or basidiomycetes, have been described. Some basidiomycetes produce the most spectacular fruiting structures found anywhere among the fungi. These are the puffballs (which may be over half a meter in diameter), mushrooms of all kinds (more than 3,250 species, including the familiar *Agaricus campestris* and the poisonous toadstools), and the giant bracket fungi often encountered on trees and fallen logs in a damp forest (Figure 24.13). Bracket fungi do great damage to cut lumber and to stands of timber. Some basidiomycetes are among the most damaging plant pathogens, including wheat rust (*Puccinia graminis;* see Figure 24.7) and the smut fungi (see Figure 24.1*a*) that parasitize cereal grains. In sharp contrast, other basidiomycetes contribute to the well-being of plants as fungal partners in mycorrhizae (see Box 24.A).

Basidiomycete hyphae are characteristically completely **septate** (walled off). (Recall that the zygomycetes have hyphae without walls and that the ascomycetes have walls with large pores.) The **basidium** (plural: basidia), a swollen cell at the tip of a hypha, is the characteristic sexual reproductive structure of the basidiomycetes. It is the site of nuclear fusion and meiosis (Figure 24.14). Thus the basidium plays the same role in the basidiomycetes as the ascus does in the ascomycetes. After nuclear fusion takes place in the basidium, the resulting diploid cell undergoes meiosis, and the four products are extruded from the basidium, forming haploid **basidiospores** on tiny stalks. The basidiospores are scattered (see Figure 24.13*a*) and then germinate, giving rise to haploid hyphae. As these hyphae grow, haploid hyphae of different mating types meet and fuse, forming dikaryotic hyphae, each cell of which contains two nuclei, one from each parental hypha. The dikaryotic mycelium grows and eventually produces fruiting structures.

The elaborate fruiting structure of a fleshy basidiomycete, such as the gill mushroom drawn in Figure 24.14, is topped by a cap, or pileus, which has gills on its underside. Basidia develop in enormous numbers between the gills. The basidia discharge their spores into the air spaces between adjacent gills, and the spores sift down into air currents for dispersal and germination as new haploid mycelia. The exact pattern of the gills and the spore color are criteria for distinguishing mushroom species. If a mature cap is placed gill side down on a piece of paper for a few hours in a quiet place, the ejected basidiospores settle from between the gills, leaving on the paper an elegant replica of the gill pattern and visible evidence of spore color.

(a)

(b)

(c)

24.13 Basidiomycete Fruiting Structures
Although some only persist a few days or weeks, the fruiting structures of the class Basidiomycetes are probably the most familiar structures produced by fungi. (*a*) *Calostoma cinnabarina*, a puffball species from the rainforests of Costa Rica. When raindrops hit them, these puffballs release clouds of spores for dispersal. (*b*) A deadly species of *Galerina*, a poisonous mushroom. (*c*) This bracket fungus, *Polyporus sulphureus*—the "chicken mushroom"—is parasitizing a tree.

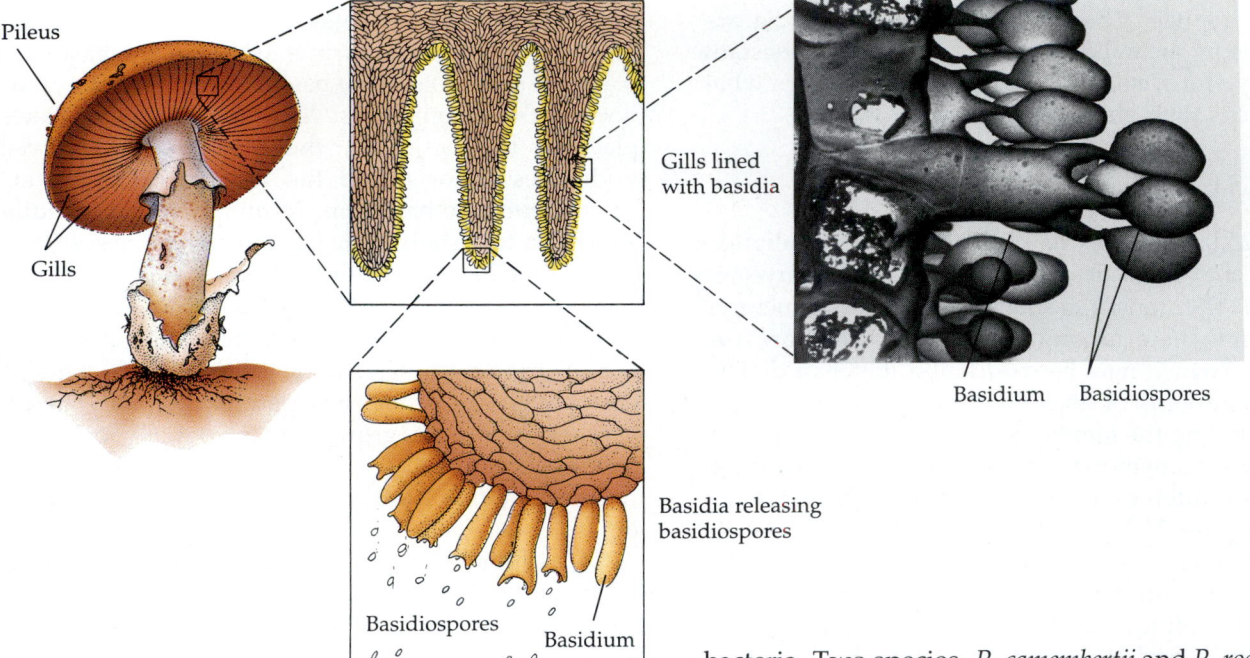

Pileus

Gills

Gills lined
with basidia

Basidia releasing
basidiospores

Basidiospores

Basidium

Basidium   Basidiospores

24.14   Anatomy of a Mushroom's Fruiting Structures
Two nuclei fuse in a basidium; meiosis and further development lead to the production of four or eight narrow
projections through which nuclei and other organelles
squeeze, forming basidiospores. The gills on the underside of a mushroom's cap are lined with basidia, from
which basidiospores are shed into the air.

Class Deuteromycetes

Mechanisms of sexual reproduction readily distinguish the three preceding classes of the Eumycota
(Zygomycetes, Ascomycetes, and Basidiomycetes).
However, many fungi, both saprobes and parasites,
lack sexual stages entirely; presumably these stages
have been lost in evolution. Classifying these fungi
with any of the three major classes is thus difficult.
True fungi that cannot be classified taxonomically are
simply dumped into the orphanage known as the
imperfect fungi, or deuteromycetes. At present,
about 25,000 species are classified as imperfect fungi.
Among them are the pathogens that cause athlete's
foot and ringworm.

If sexual structures are found on a fungus classified as a deuteromycete, the fungus is reassigned to
the appropriate class. That happened, for example,
with a fungus that produces plant growth hormones
called gibberellins (see Chapter 34). Originally classified as the deuteromycete *Fusarium moniliforme*, this
fungus was later found to produce asci, whereupon
it was renamed *Gibberella fujikuroi* and transferred to
the class Ascomycetes.

Penicillium (see Figure 24.3) is a deuteromycete genus of green molds that produce the antibiotic penicillin, presumably for defense against competing

bacteria. Two species, *P. camembertii* and *P. roquefortii*,
are the organisms responsible for the characteristic
flavors of Camembert and Roquefort cheeses. Not
surprisingly, people who are hypersensitive to penicillin may react violently if they eat one of these
cheeses. Another deuteromycete genus of importance in some diets is *Aspergillus*, the genus of brown
molds. *A. tamarii* acts on soybeans in the production
of soy sauce, and *A. oryzae* is used in brewing the
Japanese alcoholic beverage saki.

LICHENS

A lichen is not an organism; it is a meshwork of two
radically different organisms: a fungus and a photosynthetic microorganism. Together the organisms
constituting a lichen survive some of the harshest
environments on Earth (Figure 24.15). In spite of this
hardiness, lichens are very sensitive to air pollution
because they are unable to excrete toxic substances
that they absorb. Hence they are not common in
industrial cities. Lichens are good biological indicators of air pollution because of their sensitivity.

The fungal components of most lichens are ascomycetes, but some are basidiomycetes or imperfect
fungi (only one zygomycete serving as the fungal
component of a lichen has been reported). The photosynthetic component may be either a cyanobacterium or a green alga. Relatively little experimental
work has been done with lichens, perhaps because
they grow so slowly—typically less than one centimeter per year. Thus only recently have workers been
able to culture the fungal and photosynthetic partners separately and then to reconstruct a lichen from
the two.

There are about 20,000 "species" of lichens. They

24.15 Lichens in Frigid Environments
(a) This fractured sandstone rock from the cold desert of
Antarctica shows the black, white, and green layers of a
lichen that grows around crystals under the rock's hard
surface crust. The black layer consists of fungal hyphae
and algal cells; the white layer is mostly colorless fungal
hyphae; the green layer consists of hyphae and algae. No
signs of life are visible on the outer surface of this environment. (b) Many types of lichens growing on an
Alaskan tundra.

(a)

(b)

are found in all sorts of exposed habitats: tree bark,
open soil, or bare rock. Reindeer "moss" (actually
not a moss at all, but the lichen *Cladonia subtenuis*)
covers vast areas in arctic, subarctic, and boreal regions, where it is an important part of the diets of
reindeer and other large mammals. Lichens come in
various forms and colors. Crustose (crustlike) lichens
look like colored powder dusted over their substrate;
foliose (leafy) and fruticose (shrubby) lichens may
appear quite complex (Figure 24.16).

The most widely held interpretation of the lichen
relationship is that it is a type of mutually beneficial
symbiosis. Hyphae of the fungal mycelium are tightly
pressed against the photosynthetic cells of the alga
or cyanobacterium and sometimes even invade them.
The bacterial or algal cells not only survive these
indignities but continue their growth and photosyn-

(b)

(a)

24.16 Lichen Body Forms
Lichens fall into three principal classes based on their
form. (a) Crustose lichens such as the orange, white, and
gray species in this photograph often grow on otherwise
bare rock, as shown here, or on tree bark. (b) A wet foliose lichen growing on a dead twig. (c) A fruticose lichen
growing on a wooden fence.

(c)

24.17   Lichen Anatomy

(a) Soredia of a fruticose lichen. Each soredium consists of a photosynthetic cell surrounded by fungal hyphae. Soredia detach readily from the parent lichen and travel in air currents, founding new lichens when they settle in a suitable environment. (b) Cross section showing layers of a lichen. Algal cells are the round, green objects; fungal hyphae are orange.

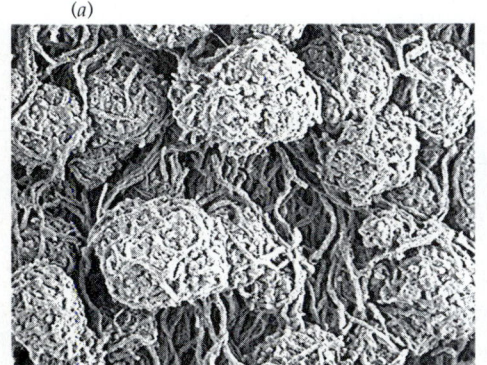

(a)

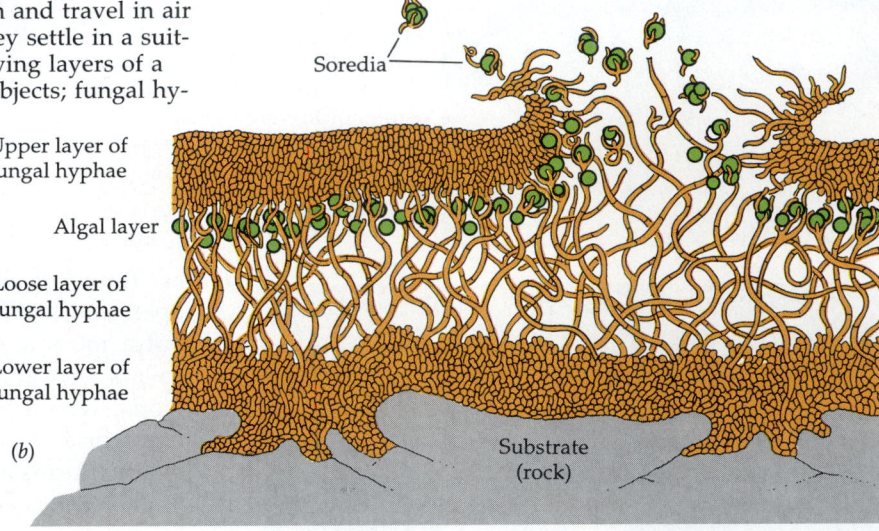

Soredia

Upper layer of fungal hyphae

Algal layer

Loose layer of fungal hyphae

Lower layer of fungal hyphae

(b)

Substrate (rock)

thesis. In fact, algal cells in a lichen "leak" photosynthetic products at a greater rate than do similar cells growing on their own. On the other hand, photosynthetic cells from lichens grow more rapidly on their own than when combined with a fungus. On this ground, we may consider lichen fungi as parasitic upon their photosynthetic partners.

Lichens can reproduce simply by fragmentation of the vegetative body, which is called the **thallus**, or else by specialized structures called **soredia**. Soredia consist of one or a few photosynthetic cells surrounded by fungal hyphae (Figure 24.17a). The soredia become detached, move in air currents, and upon arriving at a favorable location, develop into a new lichen. If the fungal partner is an ascomycete or a basidiomycete, it may go through its sexual cycle, producing either ascospores or basidiospores. When these are discharged, however, they disperse alone, unaccompanied by the photosynthetic partner, and thus may not be capable of reestablishing the lichen association. Nevertheless, many lichens produce characteristic fruiting structures in which the asci or basidia are located.

Visible in a cross section of a typical foliose lichen (Figure 24.17b) are a tight upper region of fungal

hyphae alone, a layer of cyanobacteria or algae, a looser hyphal layer, and finally hyphal rhizoids that attach the whole structure to its substrate. The meshwork has properties that enable it to hold water fairly tenaciously. Some nutrients for the photosynthetic cells arrive through the fungal hyphae, the meshwork provides a suitably moist environment for the photosynthetic cells, and the fungi derive fixed carbon from the photosynthesis of the algal or cyanobacterial cells.

Lichens are often the first colonists on new areas of bare rock. They satisfy most of their needs from the air and from rainwater, augmented by the absorption of some minerals from their rocky substrate. A lichen begins to grow shortly after a rain, as it begins to dry. As it grows, the lichen acidifies its environment slightly, and this acid contributes to the slow breakdown of rocks, an early step in soil formation. After further drying the lichen's photosynthesis ceases. The water content of the lichen may drop to less than 10 percent of its dry weight, at which point the lichen becomes highly insensitive to extremes of temperature. The flora of Antarctica features more than 100 times as many species of lichens as of plants.

SUMMARY of Main Ideas about Fungi

Fungi are the principal degraders of dead organic matter in the biosphere.

Some fungi are important pathogens, but others are beneficial, including those used in baking, brewing, and producing various foods and antibiotics.

Fungi are heterotrophic, usually multicellular, organisms with absorptive nutrition.

Some fungi are saprobes, some are mutualists, and some are parasites.

Some parasitic fungi are facultative parasites; others are obligate parasites.

Some fungi form mutualistic associations with algae or bacteria (in lichens) or with plant roots (in mycorrhizae).

The fungal body is composed of hyphae, massed to form a mycelium.

Fungi possess chitinous cell walls and lack flagellated cells.

Fungi reproduce asexually by means of spores formed within sporangia, by conidia formed at the tips of hyphae, or by budding or fragmentation.
Review Figures 24.8, 24.11a, and 24.12

Fungi reproduce sexually when hyphae of different mating types meet and fuse.
Review Figures 24.6 and 24.9

In addition to the haploid and diploid states, many fungi demonstrate a third nuclear condition: the dikaryotic, or (*n* + *n*), state.
Review Figures 24.6 and 24.7

The kingdom Fungi consists of the phylum Eumycota (true fungi).

The classes of the phylum Eumycota differ in their reproductive structures, spore formation, and, less importantly, the cross walls (if any) of their hyphae.
Review Table 24.1

Lichens are mutualistic combinations of a fungus, usually an ascomycete, with a green alga or a cyanobacterium.
Review Figure 24.17

Lichens are found in some of the seemingly most inhospitable environments on the planet.

SELF-QUIZ

1. Which statement is *not* true of fungi?
   a. A multicellular fungus has a body called a mycelium.
   b. Hyphae are composed of individual mycelia.
   c. Many tolerate highly hypertonic environments.
   d. Many tolerate low temperatures.
   e. They are anchored to their substrate by rhizoids.

2. The absorptive nutrition of fungi is aided by
   a. heterokaryon formation.
   b. spore formation.
   c. the fact that they are all parasites.
   d. their large surface-to-volume ratio.
   e. their possession of chloroplasts.

3. Which statement about fungal nutrition is *not* true?
   a. Some fungi are active predators.
   b. Some fungi form mutualistic associations with other organisms.
   c. All fungi require mineral nutrients.
   d. Fungi can make some of the compounds that are vitamins for animals.
   e. Facultative parasites can grow only on their specific hosts.

4. Which statement is *not* true of heterokaryosis?
   a. The cytoplasm of two cells fuses before their nuclei fuse.
   b. The two haploid nuclei are genetically different.
   c. The two nuclei are of the same mating type.
   d. The heterokaryotic stage ends when the two nuclei fuse.
   e. Not all fungi have a heterokaryotic stage.

5. Reproductive structures consisting of a photosynthetic cell surrounded by fungal hyphae are called
   a. ascospores.
   b. basidiospores.
   c. conidia.
   d. soredia.
   e. gametes.

6. The zygomycetes
   a. have hyphae without cross walls.
   b. produce motile gametes.
   c. form fleshy fruiting bodies.
   d. are haploid throughout their life cycle.
   e. have structures homologous to those of the actinomycetes.

7. Which statement is *not* true of the ascomycetes?
   a. They include the yeasts.
   b. They form reproductive structures called asci.
   c. Their hyphae are segmented by cross walls.

   d. Many of their species have a heterokaryotic stage.
   e. All have fruiting structures called perithecia.

8. The basidiomycetes
   a. often produce fleshy fruiting structures.
   b. have hyphae without walls.
   c. have no sexual stage.
   d. never produce large fruiting structures.
   e. form diploid basidiospores.

9. The deuteromycetes
   a. have distinctive sexual stages.
   b. are all parasitic.
   c. include some commercially important species.
   d. include the sac fungi.
   e. are never components of lichens.

10. Which statement is *not* true of lichens?
    a. They can reproduce by fragmentation of their vegetative body.
    b. They are often the first colonists in a new area.
    c. They render their environment more basic (alkaline).
    d. They contribute to soil formation.
    e. They may contain less than 10 percent water by weight.

FOR STUDY

1. You are shown an object that looks superficially like a pale green mushroom. Describe at least three criteria (including anatomical and chemical ones) that would enable you to tell whether the object is a piece of a plant or a piece of a fungus.

2. Differentiate between the members of each of the following pairs of related terms.
   a. hypha/mycelium
   b. euascomycete/hemiascomycete
   c. ascus/basidium
   d. rhizoids/haustoria

3. For each type of organism listed below, give a single characteristic that may be used to differentiate it from the other, related organism(s) in parentheses.
   a. Zygomycetes (Ascomycetes)
   b. Basidiomycetes (Deuteromycetes)
   c. Ascomycetes (Basidiomycetes)
   d. baker's yeast (*Neurospora crassa*)

4. Many fungi are dikaryotic during part of their life cycle. Why are dikaryons described as ($n + n$) instead of $2n$?

5. If all the fungi on Earth were suddenly to die, how would the surviving organisms be affected? Be thorough and specific in your answer.

READINGS

Alexopoulos, C. J. and C. W. Mims. 1979. *Introductory Mycology,* 3rd Edition. John Wiley & Sons, New York. Still a leading textbook on the biology of fungi.

Moore-Landecker, E. 1991. *Fundamentals of Fungi.* Prentice-Hall, Englewood Cliffs, NJ. The most up-to-date introduction to the kingdom.

Newhouse, J. R. 1990. "Chestnut Blight." *Scientific American,* July. A new biological method for controlling the fungus responsible for a disease that virtually eliminated an important tree species.

Raven, P. H., R. F. Evert and S. Eichhorn. 1991. *Biology of Plants,* 5th Edition. Worth, New York. Includes an excellent discussion of the fungi.

What a thrill it would be to see the face of Earth at different stages of its evolution. Successive photographs from space would show the drifting of the continents. At the surface we would see scenes scarcely imaginable to us today. For more than four billion years the terrestrial environment was basically a mass of rock. Its appearance changed relatively little until plants colonized the land and slowly spread over its surface.

Earth did not take on a green tint until less than half a billion years ago, long after the ancestors of today's plants had invaded the land at some time in the Paleozoic era (see Table 28.1). The earliest land plants were small, but their metabolic activities helped convert native rock to soil that could support the needs of their successors. Evolution led rapidly (in geological terms) to larger and larger plants, and in the Carboniferous period (345 to 290 million years ago) great forests were widespread. We would not recognize these forests, for few of their trees were like those that we know today.

During the tens of millions of years since, these trees were replaced by others more familiar to us—pinelike conifers and their relatives and, in the last 100 million years, the broad-leaved forms of modern vegetation. Today Earth is a patchwork of widely differing environments, supporting differing communities of plants, animals, fungi, protists, and bacteria. What do we notice when we look at any of these environments?

Consciously or not, we see its *plants*—or, in a very harsh environment, its relative *lack* of plants.

THE PLANT KINGDOM

Broadly defined, a plant is a *multicellular photosynthetic eukaryote*. The definition is not precise because a few organisms that can be regarded only as plants are not photosynthetic. These parasitic species, such as Indian pipe (Figure 25.1), are clearly related to photosynthetic plants (they have leaves, roots, flowers, and so on) but possess adaptations that provide them with alternative modes of nutrition. The definition casts an enormous net over a wide range of organisms that, while all multicellular and photosynthetic, differ in size, cellular organization, photosynthetic and associated pigments, and cell wall chemistry. At present we prefer to narrow the definition of a plant to exclude algae, which we discussed in Chapter 23.

According to this narrower definition, plants are multicellular, photosynthetic eukaryotes that have the following additional properties. *They develop from embryos protected by tissues of the parent plant.* (This characteristic is definitive; an older classification scheme used the term Embryophyta to refer to pre-

What's Missing?

25

Plants

cisely those organisms that we include here in the kingdom Plantae.) Their cells have walls that contain cellulose as the major strengthening polysaccharide. Their chloroplasts contain chlorophylls *a* and *b* and a limited array of specific carotenoids (see Chapter 8). Their major storage carbohydrate is starch. Their life cycles have important features in common.

Alternation of Generations in Plants

A universal feature of the life cycles of plants is the alternation of generations. If we consider the plant life cycle to begin with a single cell, the diploid zygote, then the first phase of the cycle features the formation, by mitosis and cytokinesis, of a multicellular embryo and eventually the mature diploid plant (see Figure 23.5). This multicellular, diploid plant is the **sporophyte**. (Sporophyte means "spore plant.") Cells contained in **sporangia** (singular: *sporangium*, "spore reservoir") on the sporophyte undergo meiosis to produce haploid, unicellular spores. By mitosis and cytokinesis a spore forms a haploid plant. This multicellular, haploid plant, the **gametophyte** ("gamete plant"), produces haploid gametes. The fusion of two gametes (syngamy, or fertilization) results in the formation of a diploid cell, the zygote, and the cycle begins again.

The sporophyte generation extends from the zygote through the adult, multicellular, diploid plant; the gametophyte generation extends from the spore through the adult, multicellular, haploid plant to the gamete. The transitions between the phases are accomplished by fertilization and meiosis.

The gametophyte and sporophyte of any plant look totally different, unlike the generations of some

25.1 Stretching the Definition
Indian pipe is not photosynthetic, but it is a plant in all other respects.

algae, such as *Ulva*, which are indistinguishable to the eye (see Figure 23.27). In plants and in *Ulva*, the sporophyte and gametophyte differ genetically: One has diploid cells, the other haploid cells.

Classification of Plants

We may divide the kingdom Plantae into smaller, more closely related groups in several ways. One popular approach is to divide the plant kingdom into 12 phyla of equal taxonomic status. We will group some of them (Table 25.1) to emphasize the evolu-

---

TABLE 25.1
Classification of Plants[a]

PHYLUM	COMMON NAME	CHARACTERISTICS
Nonvascular Plants		
Hepaticophyta	Liverworts	No filamentous stage
Anthocerophyta	Hornworts	Embedded archegonia
Bryophyta	Mosses	Filamentous stage
Vascular Plants		
Lycophyta	Club mosses	Simple leaves in spirals; no seeds
Sphenophyta	Horsetails	Simple leaves in whorls; no seeds
Psilophyta	Whisk ferns	No true leaves; no seeds
Pterophyta	Ferns	Complex leaves; no seeds
Cycadophyta	Cycads	Fern-like leaves, swimming sperm; seeds in cones
Ginkgophyta	Ginkgo	Deciduous, fan-shaped leaves; seeds
Gnetophyta		Vessels in wood; seeds
Coniferophyta	Conifers	Many have seeds in cones; needlelike leaves
Anthophyta	Flowering plants	Double fertilization; seeds in fruit

[a] No extinct groups are included in this classification.

tionary trends that you will be learning. Three phyla (mosses, liverworts, and hornworts) were once considered classes of a single larger phylum, of which the most familiar examples are mosses. We use the term **nonvascular plants** to refer collectively to these three phyla. All members of the other nine phyla possess well-developed systems that transport materials throughout the plant body. We call these nine phyla, collectively, the **vascular plants**.

Plant Colonizers of the Land

Plants and their ancestors pioneered and developed the terrestrial environment. That environment differs dramatically from the aquatic environment in several ways. The density of the plant body is much closer to that of water than of air, so aquatic plants are buoyant in water and are supported against gravity. A multicellular plant on land must either have a support system against gravity or else sprawl unsupported on the ground. It also must have the means to transport water and minerals from the soil to its aerial (raised) parts or else live where water is abundant enough to bathe all of its parts regularly. How did such organisms arise from aquatic ancestors to thrive in such a challenging environment?

Origins of the Plant Kingdom

It is widely agreed that the plant kingdom arose from the green algae (the protist phylum Chlorophyta). Which group of green algae was ancestral to the plants is not certain, although recent evidence suggests the stoneworts, characterized by multicellular sex organs and complex body patterns, as likely candidates. The characteristics of green algae that make them the most likely ancestors of the plants include the possession of photosynthetic and accessory pigments in their plastids that are similar to those of plants, the use of starch as their principal storage carbohydrate, and the presence of cellulose as the principal component of their cell walls.

In addition, some green algae (not all) show the same oogamous type of reproduction—which features a large, stationary egg—that is characteristic of sexual reproduction in plants. Like plants, a few green algae have a life cycle with both a multicellular gametophyte and a multicellular sporophyte generation. Furthermore, there are green algae with bulky, three-dimensional bodies like plants, although most green algae are unicellular, filamentous, or two-dimensional. The chloroplasts of plants and many green algae have thylakoids that are arranged into grana. Some green algae produce new cell walls after cell division using a mechanism similar to that in plants. No other phylum of algal protists shares so many traits with the Plantae.

Although we do not know what the ancestors of

plants looked like, many evolutionary biologists believe that those ancestral green algae lived at the margins of ponds or marshes, ringing them with a green mat. What kinds of plants evolved from that mat?

Two Groups within the Plant Kingdom

A green algal ancestor gave rise to the first nonvascular plants, which were, in turn, ancestral to all of today's plants. Later the first vascular plants arose from a nonvascular plant ancestor (Figure 25.2).

The nonvascular plants—mosses, liverworts, and hornworts—never evolved into large plants. They

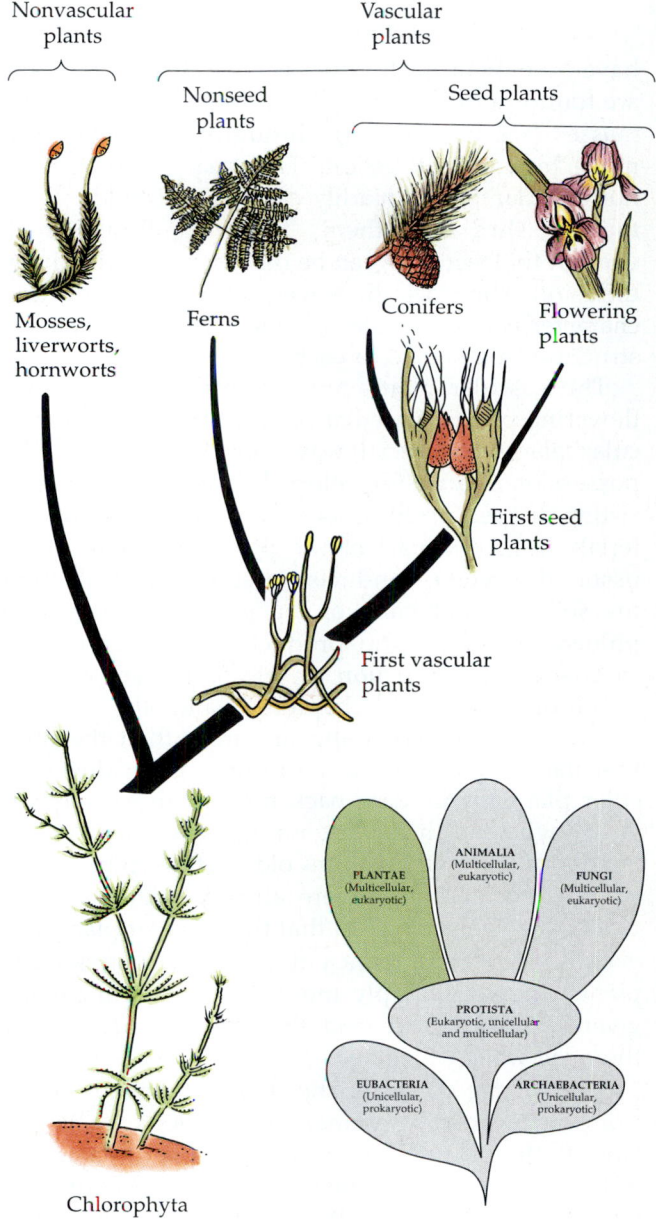

25.2 From a Green Alga to Plants
The green algae (protist phylum Chlorophyta) gave rise to both the nonvascular and the vascular plants.

25.3 Some Nonvascular Plants Form Mats
Moss colonizing an old lava bed in Iceland.

have little or no water-transporting tissue, yet some are found in dry environments. Many grow in dense masses (see Figure 25.3), through which water can move by capillary action. The "leafy" structure of nonvascular plants readily catches and holds water that splashes onto them. These plants are small enough that minerals can be distributed internally by diffusion. They lack the leaves, stems, and roots that characterize the vascular plants, although they have structures analogous to each.

The vascular plants are the ferns, conifers, and flowering plants. Vascular plants differ from nonvascular plants in crucial ways, one of which is the possession of a well-developed **vascular system** consisting of specialized tissues for the transport of materials from one part of the plant to another. One tissue, the **xylem**, conducts water and minerals from the soil to aerial parts of the plant; the other, the **phloem**, conducts the products of photosynthesis from sites of production or release to sites of utilization or storage (see Chapters 29 and 30).

The vascular plants appear earlier than the nonvascular plants in the fossil record. The oldest vascular plant fossils date back more than 410 million years, whereas the oldest nonvascular plant fossils are about 350 million years old, dating from a time when vascular plants were already widely distributed. This does not mean that the vascular plants are evolutionarily more ancient than the nonvascular plants; they are simply more likely to form fossils, given their structure and the chemical makeup of their cell walls.

Most of the characteristics that distinguish plants from algae are evolutionary adaptations to life on land. Both nonvascular and vascular plants have protective coverings that prevent drying, and both have means of taking up water from the soil (vascular plants have roots; nonvascular plants have rhizoids). Support against gravity is derived from the turgor (internal fluid pressure; see Chapter 29) of plant cells,

by a woody stem in some vascular plants, and by thickened cell walls in nonvascular plants. Unlike the algae, plants form embryos, young sporophytes contained within a protective structure. We will examine the adaptations of the vascular plants, but let's concentrate first on the nonvascular plants.

NONVASCULAR PLANTS

Most nonvascular plants are small and grow in dense mats in moist habitats (Figure 25.3). The largest of these plants are about 20 centimeters tall, and most are only a few centimeters tall or long. Why have larger nonvascular plants never evolved? The probable answer is that they never evolved an efficient system for conducting water and minerals from the soil to distant parts of the plant body.

Most nonvascular plants live on the soil or on other plants, but some grow on bare rock, dead and fallen tree trunks, and even on the buildings in which we live and work. These small plants have been here much longer than we and perhaps a quarter of a billion years longer than the flowering plants. Nonvascular plants are widely distributed over six continents and exist very locally on the coast of the seventh (Antarctica). They are very successful plants, well adapted to their niches, which are virtually all terrestrial. Although a few nonvascular plants live in fresh water, these aquatic forms are descended from terrestrial ones. There are no marine nonvascular plants. All nonvascular plants have a waxy covering that retards water loss, and their embryos are protected within layers of maternal tissue.

The Nonvascular Plant Life Cycle

The life cycles of nonvascular plants differ sharply from those of other plants and of their algal ancestors in that the conspicuous green plant that we recognize is the gametophyte, whereas the familiar ferns and seed plants are sporophytes. The nonvascular gametophyte is photosynthetic and thus nutritionally independent, whereas the sporophyte may or may not be photosynthetic but is *always* dependent upon the gametophyte and remains permanently attached to it.

The sporophyte produces unicellular, haploid spores as products of meiosis. A spore germinates, giving rise to a multicellular, haploid gametophyte whose cells contain chloroplasts and are thus photosynthetic. Eventually, gametes form within specialized sex organs. The **archegonium** is a multicellular, flask-shaped female sex organ with a long neck and a swollen base (Figure 25.4*a*). The base contains a single egg. The **antheridium** is a male sex organ in which sperm, each bearing two flagella, are produced in large numbers (Figure 25.4*b*). Once released, the sperm must swim to a nearby archegonium on the same or a neighboring plant. The sperm are aided in this task by chemical attractants released by the egg or the archegonium. Before sperm can enter the archegonium, certain cells in the neck of the archegonium must break down, leaving a water-filled canal through which the sperm swim to complete their journey. Note the dependence on liquid water for all of these events.

On arrival at the egg, one of the sperm nuclei fuses with the egg nucleus to form the zygote. Mitotic divisions of the zygote produce a multicellular, diploid sporophyte embryo. The base of the archegonium grows to protect the embryo during its early growth. Eventually the developing sporophyte elongates sufficiently to break out of the archegonium, but it remains connected to the gametophyte by a "foot" that is embedded in the parent tissue and absorbs water and nutrients from it. The sporophyte remains attached to the gametophyte throughout its life. The sporophyte produces a **capsule**, or sporangium, within which meiotic divisions produce spores and thus the next gametophyte generation.

The structure and pattern of elongation of the sporophyte differ among the three phyla of nonvascular plants—the liverworts, hornworts, and mosses. The evolutionary relationship of the three phyla is uncertain.

Phylum Hepaticophyta: Liverworts

The gametophytes of some liverworts are "leafy" and prostrate. The simplest liverwort gametophytes, however, are flat plates of cells, a centimeter or so in length, that produce antheridia or archegonia on their upper surfaces and rhizoids on the lower. Liverwort sporophytes are shorter than those of mosses and hornworts, rarely exceeding a few millimeters. The sporophyte has a stalk that connects capsule and foot. The stalk elongates and thus raises the capsule

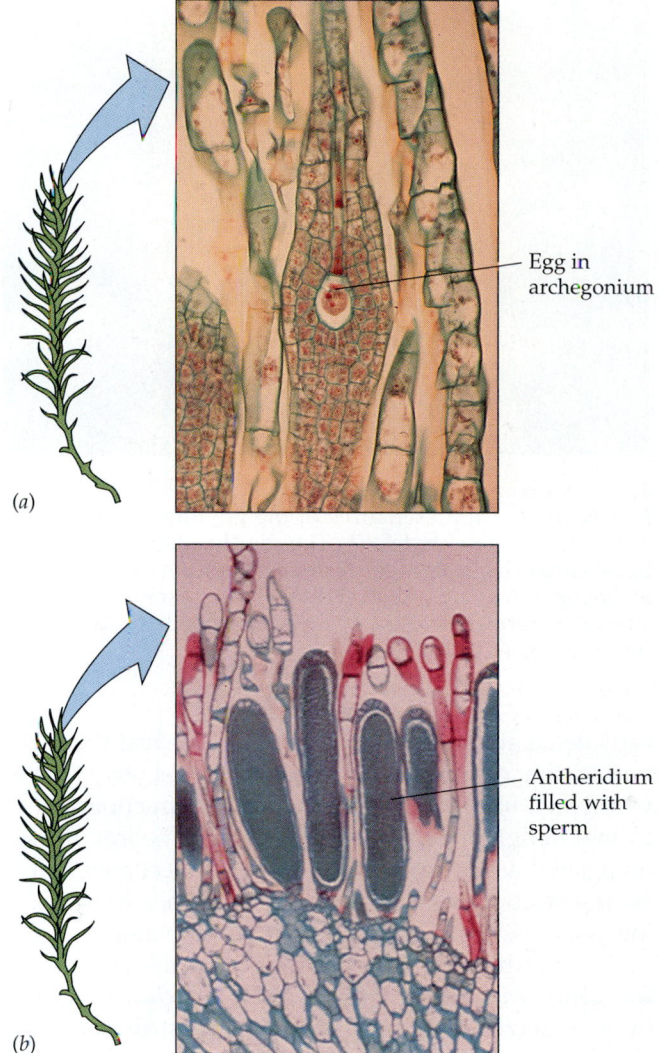

(a)

(b)

Egg in archegonium

Antheridium filled with sperm

25.4 Mosses Have Sex Organs
(*a*) Archegonia of the moss *Mnium* (phylum Bryophyta). The large egg cell in the center of the archegonium looks like an eye. Archegonia develop at the tip of a gametophyte. It is in the archegonium that the egg will be fertilized and begin development into a sporophyte. (*b*) Antheridia, also located at the tip of a gametophyte of *Mnium*. These male organs contain a large number of sperm. When released, the sperm must locate an archegonium and swim down its neck to the egg.

above ground level, favoring dispersal of spores when they are released. The elongation of the sporophyte occurs broadly over the length of the stalk, which has no specific growing zone. There are no stomata (pores allowing gas exchange between the atmosphere and the plant's interior) on the liverwort sporophyte.

The capsules of liverworts are simple: a globular capsule wall surrounding a mass of spores. In some species of liverworts spores are not released by the sporophyte until the surrounding capsule wall rots. In other liverworts, however, the spores are disseminated by structures called **elaters** located within the

(a)

(b)

(c)

25.5   A Liverwort

Marchantia is a representative of the phylum Hepatico-phyta. (a) Gametophytes. (b) The disc-headed structures bear antheridia; the finger-headed structures bear archegonia. (c) Cups filled with gemmae—small, lens-shaped outgrowths of the body, each capable of developing into a new plant.

capsule. Elaters are long cells with a helical thickening of the cell wall. As an elater loses water, the whole cell shrinks longitudinally to a fraction of its former length, thus compressing the helical thickening like a spring. When the stress becomes sufficient, the compressed "spring" snaps back to its resting position, throwing spores in all directions.

Among the most familiar liverworts are species of the genus *Marchantia* (Figure 25.5). *Marchantia* is easily recognized by the characteristic structures on which its male and female gametophytes bear their antheridia and archegonia. Like most liverworts, *Marchantia* reproduces vegetatively by simple fragmentation of the gametophyte. *Marchantia* and some other liverworts and mosses also reproduce vegetatively by means of **gemmae** (singular: *gemma*), lens-shaped clumps of cells loosely held in structures called gemma cups.

Phylum Anthocerophyta: Hornworts

The hornworts—so named because their sporophytes look like little horns—appear at first glance to be liverworts with very simple gametophytes (Figure 25.6a). These gametophytes are flat plates of cells, a few cells thick. However, hornworts have two characteristics that distinguish them from liverworts and mosses. First, the archegonia are embedded in the gametophytic tissue instead of being borne on stalks. Second, of all the nonvascular plant sporophytes, those of the Anthocerophyta come closest to being capable of indefinite growth. The stalk of either the liverwort or moss sporophyte stops growing as the capsule matures, so elongation of the sporophyte is

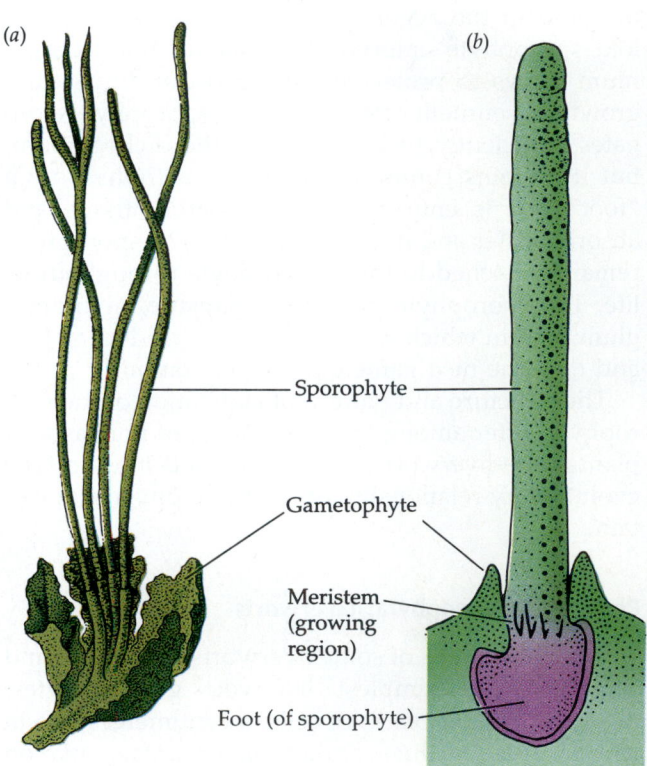

(a)

(b)

Sporophyte

Gametophyte

Meristem (growing region)

Foot (of sporophyte)

25.6   Hornworts and Their Growth

(a) The hornwort *Anthoceros* (phylum Anthocerophyta), drawn slightly larger than actual size. (b) The sporophyte of a hornwort grows from its basal end.

strictly limited. In a hornwort such as *Anthoceros*, however, there is no stalk, but a basal region of the capsule remains capable of indefinite cell division, continuously producing new spore-bearing tissue above (Figure 25.6b). Sporophytes of some hornworts growing in mild and continuously moist conditions can become as tall as 20 centimeters, making them the tallest known nonvascular plants.

Whereas the photosynthetic cells in other plants have many small chloroplasts, each cell of a hornwort has only a single, large chloroplast. Hornworts have internal cavities filled with a mucilage, often populated by cyanobacteria that convert atmospheric nitrogen gas into a nutrient form usable by the host plant (see Chapter 32).

Phylum Bryophyta: Mosses

The most familiar nonvascular plants are the mosses. There are more species of mosses than of liverworts and hornworts combined, and these hardy little plants are found in almost every terrestrial environment.

The moss gametophyte that develops following spore germination is a branched, filamentous plant, or **protonema** (plural: protonemata), that looks much like a filamentous green alga and is unique to this phylum. Some of the filaments contain chloroplasts and are photosynthetic; others are nonphotosynthetic and anchor the protonema to the substrate. The nonphotosynthetic filaments, called **rhizoids**, are the nonvascular plant counterpart of the root hairs of vascular plants. After a period of growth, cells close to the tips of the photosynthetic filaments divide rapidly in three dimensions to form buds. The buds eventually differentiate a distinct apex and produce the familiar leafy moss plant with the "leaves" spirally arranged.

These leafy shoots produce antheridia or archegonia at their tips (Figure 25.7a). The antheridia release sperm that travel through liquid water to the archegonia, where they fertilize the eggs. Sporophyte development in most mosses follows a precise pattern, resulting ultimately in the formation of an absorptive foot, a stalk, and, at the tip, a swollen capsule. In contrast to hornworts, which grow from the base (see Figure 25.6), the moss sporophyte grows at its apical end. Cells at the tip of the stalk divide,

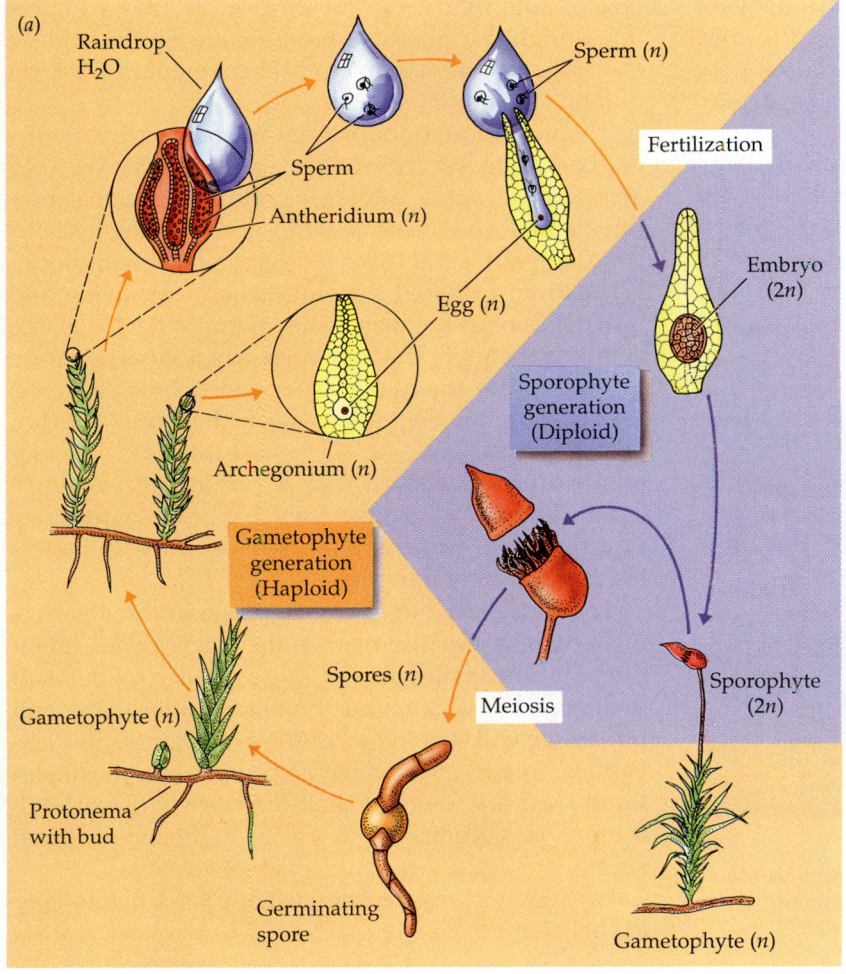

25.7 Moss Life Cycle
(a) The familiar "leafy" moss plant is the haploid gametophyte. Fertilization in mosses requires water so that sperm can swim to eggs. The sporophyte is attached to and nutritionally dependent on the gametophyte. (b) The teeth of this moss capsule help expel the thousands of spores.

supporting elongation of the structure and giving rise to the capsule. Unlike liverworts, moss sporophytes often possess stomata, allowing gas exchange with the atmosphere.

The archegonial tissue also grows rapidly as the stalk elongates, for a time keeping pace with the rapidly expanding sporophyte. Eventually, however, it is outgrown and is torn apart around its middle. The top portion of the archegonium frequently persists on the top of the rapidly elevating capsule as a little pointed cap, the **calyptra**. The top of the capsule is shed after meiosis and development of numerous mature spores within. Groups of cells just below the lid form a series of teeth surrounding the opening. Highly responsive to humidity, the teeth arch into the mass of spores when the atmosphere is dry and then out again as it becomes moist (Figure 25.7b). The spores are thus dispersed when the surrounding air is moist, that is, when conditions favor their subsequent germination.

Only a few mosses depart from this pattern of capsule development. A familiar exception is the genus *Sphagnum*, which occurs in tremendous quantities in northern bogs and tundra extending well into the Arctic (Figure 25.8) and whose global biomass probably exceeds that of all other mosses combined. Species in this genus have a very simple capsule with an air chamber in it. Air pressure builds up in this chamber, eventually causing the capsule lid to pop open, dispersing the spores with an audible explosion.

Many mosses contain a type of cell, called a hydroid, that dies and leaves a tiny channel through which water may travel. The hydroid is the likely progenitor of the characteristic water-conducting cell of the vascular plants. The possession of hydroids

and of a limited system for transport of foods by some mosses shows that the term *nonvascular* plant is not perfect.

VASCULAR PLANTS

The vascular plants are an extraordinarily large and diverse group, yet they can be said to have been launched by a single evolutionary event. Sometime during the Paleozoic era, probably well before the Silurian period, the sporophyte generation of a now long-extinct green alga produced a new cell type, the **tracheid**. The tracheid is the principal water-conducting element of the xylem in all vascular plants except the angiosperms (flowering plants); even in angiosperms the tracheid persists along with a more specialized and efficient system of vessels and fibers.

The evolutionary appearance of a tissue composed of tracheids had two important consequences. First, it provided a pathway for long-distance transport of water and mineral nutrients from a source of supply to regions of need. Second, it provided something almost completely lacking—and unnecessary—in the largely aquatic algae: rigid structural support. Support is important in a terrestrial environment because land plants tend to grow upward as they compete for sunlight for photosynthesis. Thus the tracheid set the stage for the complete and permanent invasion of land by plants.

The evolutionary descendants of the early colonizers belong to several distinct plant phyla. We may assort those phyla into two groups: those that produce seeds and those that do not. The life cycle of vascular plants that lack seeds is remarkably uniform. Haploid and diploid generations are free-living and independent at maturity. The sporophyte is the large and obvious plant that one normally notices in nature (in contrast to the nonvascular sporophyte, which is attached to and dependent upon the gametophyte). Gametophytes of the non-seed-forming vascular plants are rarely more than a centimeter or two in length and are short-lived, whereas the sporophyte of a tree fern, for example, may be 15 or 20 meters tall and may live for years.

The most prominent resting stage in the life cycle of a seedless vascular plant is the single-celled spore. This feature makes its life cycle similar to those of the fungi, the algae, and the nonvascular plants but not, as we will see, to that of the seed plants. Seedless vascular plants must have an aqueous environment for at least one stage of their life cycle because fertilization is accomplished by a motile, flagellated sperm.

We will discuss the life cycle of seed plants later in this chapter.

25.8 *Sphagnum* Forms Peat Bogs
The moss *Sphagnum* is responsible for the formation of peat, which is used for fuel. This peat bog in Ireland is being harvested for commercial use.

25.9 An Ancient Forest

A little over 300 million years ago, a forest grew in a setting similar to tropical river delta habitats of today. A dense forest of lycopods 10 to 20 meters high grew along the low natural levee forming the edge of the river. A few tree ferns approximately 10 meters high were present. A small fern grew on the trunk of a tree fern. On the ground, scattered clumps of weak-stemmed seed ferns grew, as did tangled clumps of herbaceous sphenopods. Beyond the levee a swampy area supported groups of larger, spruce-tree shaped sphenopods. Farther in the distance giant gymnosperms—up to 40 meters tall—towered over the forest.

Evolution of the Vascular Plants

The green algal ancestors of the plant kingdom successfully invaded the terrestrial environment between 400 and 500 million years ago. The evolution of a water-impermeable cuticle (waxy outer covering) and of protective layers for the gamete-bearing structures helped to make the invasion permanent, as did the initial absence of herbivores (plant-eating animals). By the late Silurian period (see Table 28.1) vascular plants were being preserved as fossils that we can study today. Several remarkable developments occurred during the Devonian period, 400 to 345 million years ago. Three groups of non-seed-producing vascular plants that still exist made their first appearances during that period: the lycopods (club mosses), horsetails, and ferns. The proliferation of these plants made the terrestrial environment ever more hospitable to animals; amphibians and insects arrived soon after the plants became established. Fossil remains about 360 million years old provide the first evidence of seed plants.

Trees of various kinds, which appeared in the Devonian, came into their own in the Carboniferous period (345 to 290 million years ago). Mighty forests of lycopods up to 40 meters tall, horsetails, and tree ferns flourished in the tropical swamps of what would become North America and Europe (Figure 25.9). In the subsequent Permian period the continents came ponderously together to form a single, gigantic land mass, called Pangaea. The continental interior become warmer and dryer, but late in the period glaciation was extensive. The 200-million-year reign of the lycopod–fern forests came to an end, to be replaced by forests of gymnosperms (nonflowering seed plants) that ruled throughout the Triassic and Jurassic periods. These forests changed with time as the gymnosperm groups evolved. Gymnosperm forests dominated during the era in which the continents drifted apart and dinosaurs strode the Earth.

The oldest evidence of angiosperms (flowering plants) dates into the Cretaceous period, about 120 million years ago. The angiosperms radiated almost explosively and, over a period of about 55 million years, became the dominant plant life of the planet.

Ancient Vascular Plants

The first vascular plants belonged to a now-extinct phylum (Rhyniophyta). The rhyniophytes appear to have been the only vascular plants in the Silurian period of the Paleozoic era, and they dominated the landscape. Early versions of the structural features of all the other vascular plant phyla appeared in the plants of that time. These shared features strengthen the case for the origin of all vascular plants from a common nonvascular plant ancestor.

In 1917 the British paleobotanists Robert Kidston and William H. Lang reported well-preserved fossils of vascular plants embedded in Devonian rocks near Rhynie, Scotland. Their preservation was remarkable, considering that the rocks were over 395 million years old. These fossil plants had a simple vascular system consisting of tracheids and phloem. Flattened scales on the stems of some of the plants lacked vascular tissue and thus were not comparable with the true leaves of any other vascular plants. The plants were without roots, apparently anchored in the soil by horizontal portions of stem (**rhizomes**) that bore rhizoids. These rhizomes also bore aerial branches. Sporangia—homologous with the nonvascular plant capsule—were found at the tips of the

stems. Branching was dichotomous; that is, the shoot apex divided to produce two equivalent new branches, each pair diverging at approximately the same angle from the original stem (Figure 25.10). Scattered fragments of such plants had been found earlier, but never in such profusion or so well preserved as those discovered near Rhynie by Kidston and Lang.

The presence of tracheids indicated that these plants, named *Rhynia* after the site of their discovery, were vascular plants. But were they sporophytes or gametophytes? Close inspection of thin sections of fossil sporangia revealed that the spores were in groups of four. In almost all living seedless vascular plants (with no evidence to the contrary from fossil forms), the four products of a meiotic division and cytokinesis remain attached to one another during their development into spores. The spores separate only when they are mature, and even after separation their walls reveal the exact geometry of how they were attached. Thus a group of four closely packed spores is found only immediately after meiosis, and a plant that produces such a group of four must be a diploid sporophyte.

Although the Rhyniophyta apparently were ancestral to the other vascular plant phyla, the rhyniophytes themselves are long gone. None of their fossils appear anywhere after the Devonian period.

Further Developments in the Ancient Vascular Plants

Within a few tens of millions of years, during the Devonian period, three new phyla of vascular plants appeared on the scene, arising from rhyniophyte ancestors. These new groups featured advances over the rhyniophytes, including one or more of the following: true roots, true leaves, and a differentiation between two types of spores.

THE ORIGIN OF ROOTS. *Rhynia* and its close relatives lacked true roots. They had only rhizoids, arising from a prostrate rhizome, with which to gather water and minerals. How, then, did subsequent groups of vascular plants come to have the complex roots we see today?

A French botanist, E. A. O. Lignier, proposed an attractive hypothesis in 1903 that is still widely accepted today. Lignier argued that the ancestors of the first vascular plants branched dichotomously. This accounts for the dichotomous branching observed in the rhyniophytes themselves. Lignier also suggested that such a branch could bend and penetrate the soil, branching there (Figure 25.11). The underground portion could anchor the plant firmly, and even in this primitive condition it could absorb water and minerals. The subsequent discovery of fossil plants

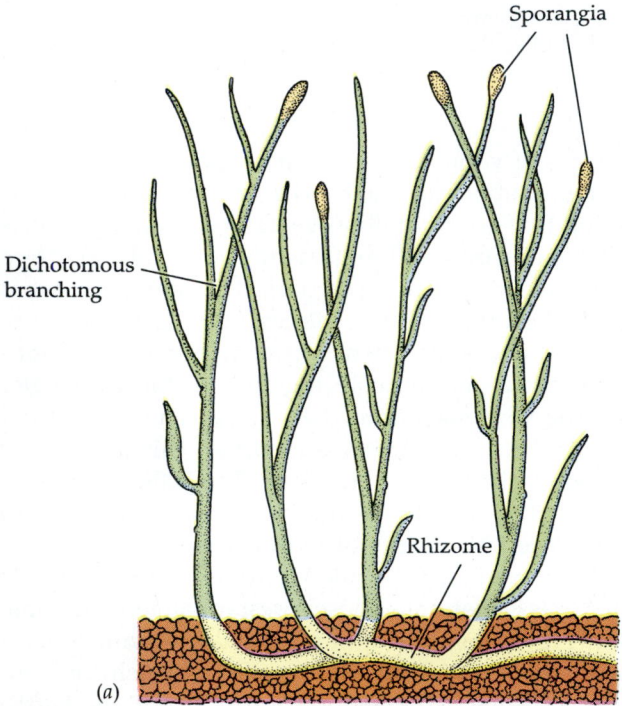

25.10 The First Plants Lacked Roots and Leaves
An extinct plant in the genus *Rhynia* (phylum Rhyniophyta), named after Rhynie, Scotland, near where its fossil remains were discovered. The horizontal underground stem is called a rhizome; the aerial shoots were less than 50 centimeters in height, and some were topped by sporangia.

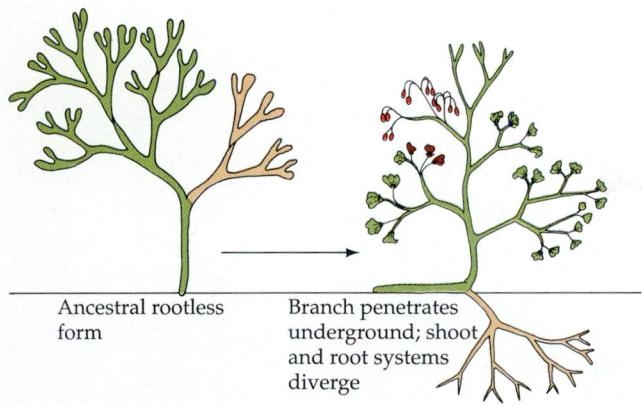

25.11 Is This How Roots Evolved?
According to Lignier's hypothesis, branches from ancestral rootless plants could have penetrated the soil, where they gradually evolved into a root system.

from the Devonian period, all having horizontal stems with both underground and aerial branches, supported Lignier's hypothesis.

The underground branches, in an environment sharply different from that above the ground, were subjected to very different selection pressures during the succeeding millions of years. Thus the two parts of the plant axis (the shoot and root systems) diverged in structure and came to have distinct internal and external anatomies (see Chapter 29). We believe, however, that the root and shoot systems of vascular plants are homologous—in fact, that they were once part of the same organ.

THE ORIGIN OF TRUE LEAVES. Thus far, we have used the term *leaf* rather loosely. We spoke of "leafy" mosses; we also commented on the absence of "true leaves" in rhyniophytes. In the strictest sense, a **leaf** is a flattened photosynthetic structure emerging laterally from a main axis or stem and possessing true vascular tissue. This tight definition allows a closer look at true leaves in the vascular plants, which shows us that there are two different types of leaves, probably of different evolutionary origins.

The first type of leaf is usually small and only rarely has more than a single vascular strand, at least in plants alive today. Plants in two phyla have such *simple* leaves: the Lycophyta (club mosses) and the Sphenophyta (horsetails, or scouring rushes), of which only a few genera survive. The evolutionary origin of this type of leaf is thought to be the progressive development of vascular tissue within small, scalelike outgrowths of the stem (Figure 25.12a). The principal characteristic of this type of leaf is that its vascular strand departs from the vascular system of the stem in such a way that the structure of the stem's vascular system is scarcely disturbed. This was true even in the fossil lycopod and horsetail trees of the

Carboniferous period, many of which had leaves several centimeters long.

The other type of leaf is encountered only in ferns and in seed plants. This larger, more *complex* type of leaf is thought to have arisen from the flattening of a dichotomously branching stem system, with the development of extensive photosynthetic tissue between the branch members (Figure 25.12b). Another feature of the complex leaf is that its vascular system

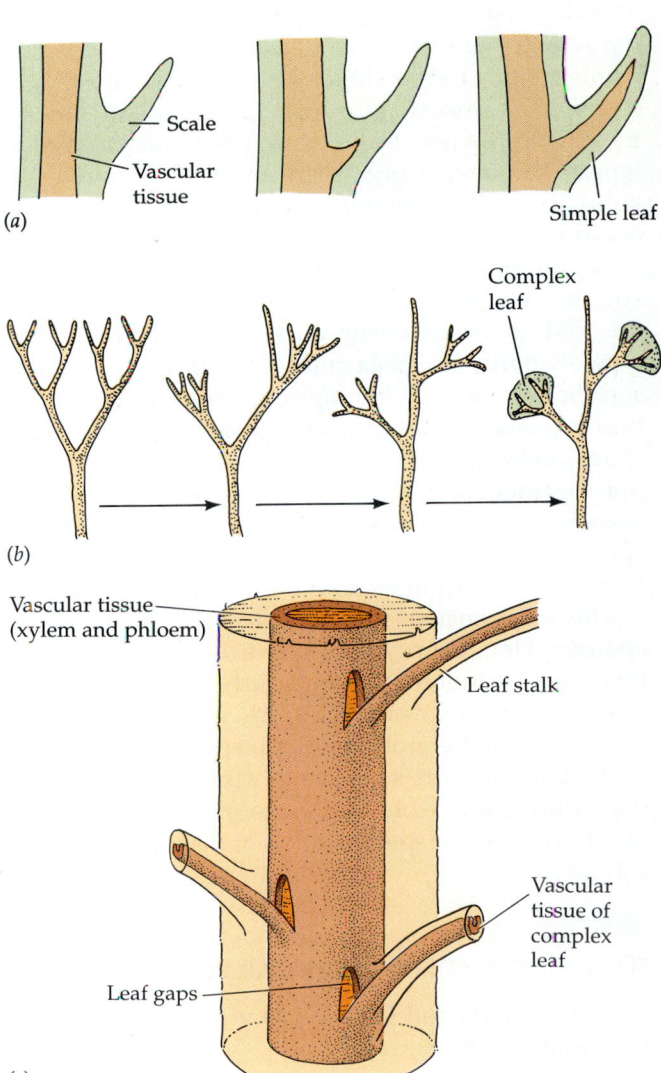

25.12 Evolution of Leaves
(a) Simple leaves might have evolved from scales such as those on the stems of some rhyniophytes. The diagram shows progression from a scale without vascular tissue (left) to one with some vascular tissue (center), to a true leaf (right). All three types appear in fossil plants. (b) Complex leaves may have originated as a branching stem system (left) and become progressively reduced (left center) and flattened (right center). Flat plates of photosynthetic tissue developed between small end branches (right). The end branches evolved into the veins of leaves. (c) Where vascular tissue departs from a stem into a complex leaf, there is a gap in the stem tissue immediately over the junction.

creates a major alteration in the architecture of the stem vascular system where it departs for the leaf base (Figure 25.12c). The complex leaf probably evolved several times, in different phyla of vascular plants.

HOMOSPORY AND HETEROSPORY.   In the most-ancient vascular plants both the gametophyte and sporophyte are free-living and photosynthetic. Spores produced by the sporophytes are of a single type, and they develop into a single type of gametophyte, bearing both female and male reproductive organs. Such plants, which bear a single type of spore, are said to be **homosporous** (Figure 25.13a). The sex organs on the gametophytes of homosporous plants are of two types. The female organ is a multicellular archegonium, typically containing a single egg. The male organ is an antheridium, containing many sperm.

A different system, with two distinct types of spores, evolved somewhat later. Plants of this type are said to be **heterosporous** (Figure 25.13b). One type of spore, the **megaspore**, develops into a larger, specifically female gametophyte (megagametophyte) that produces only eggs. The other type, the **microspore**, develops into a smaller, male gametophyte (microgametophyte) that produces only sperm. Megaspores are produced in small numbers in megasporangia on the sporophyte, and microspores in large numbers in microsporangia.

The most ancient vascular plants were all homosporous. Heterospory evidently evolved a number of times, independently, in the early evolution of the vascular plants descended from the rhyniophytes. The fact that heterospory evolved repeatedly suggests that it affords selective advantages. As we will see, subsequent evolution in the plant kingdom featured ever greater specialization of the heterosporous condition.

SURVIVING SEEDLESS VASCULAR PLANTS

Phyla Lycophyta and Sphenophyta: Club Mosses and Horsetails

The Lycophyta, or club mosses, and the Sphenophyta, or horsetails (also called scouring rushes because silica deposits found in the cell walls made them useful for cleaning), are represented by relatively few surviving species. Both have true roots, and both bear only simple leaves. Both include homosporous species and heterosporous species. Like all seedless vascular plants, both have a heteromorphic alternation of generations with a large, independent sporophyte and a small, independent gametophyte. Here, however, the resemblance ends.

Whereas in club mosses the leaves are arranged spirally on the stem (Figure 25.14a), in the horsetails

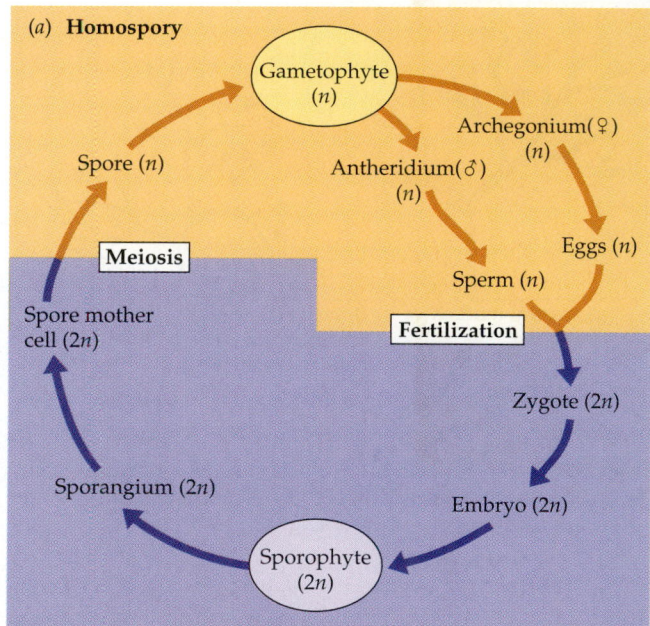

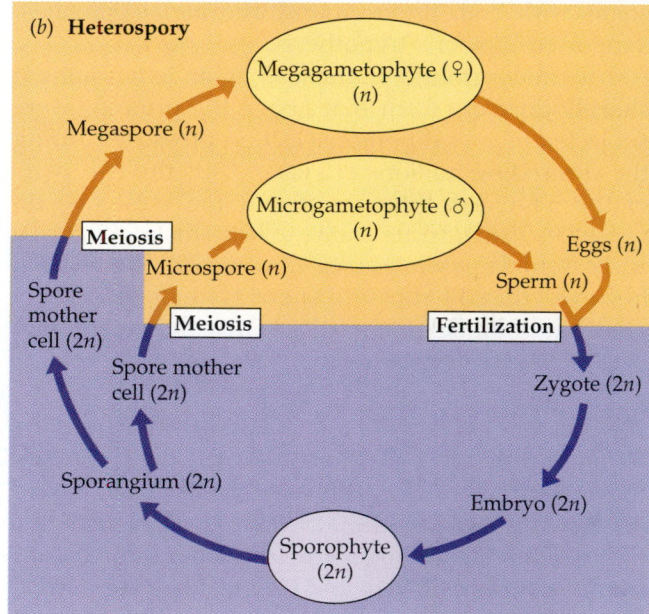

25.13   Homospory and Heterospory

they form distinct whorls (Figure 25.15a). The sporangia in club mosses appear in conelike structures called **strobili** (singular: *strobilus*) and are tucked in the upper angle between a specialized leaf and the stem (Figure 25.14b); the sporangia of horsetails are curved back toward the stem on the ends of short stalks (sporangiophores; Figure 25.15b). Growth in club mosses comes entirely from groups of dividing cells at the tips of the stems, whereas growth in horsetails comes to a large extent from discs of dividing cells just above each whorl of leaves, so each segment of the stem grows from its base. This difference in growth mechanism is dramatic, and it has

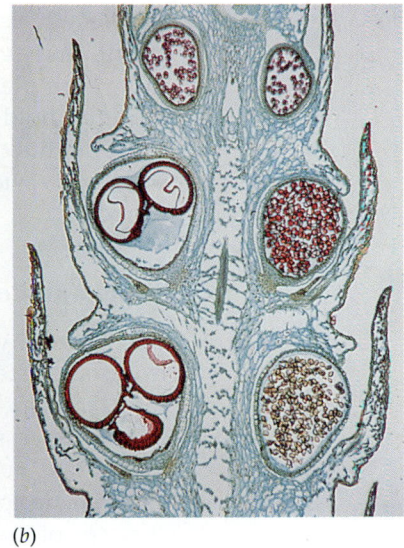

(a)   (b)

25.14 Club Mosses
(a) *Lycopodium obscurum*, a club moss (phylum Lycophyta), with conelike strobili at the tips. Club mosses have simple leaves arranged spirally on their stems. (b) Thin section through a strobilus of *Selaginella*, another club moss.

a counterpart in the more familiar flowering plants: Grasses grow as horsetails do, while many other flowering plants grow apically, as mosses do.

Although only minor elements of the vegetation of today, these two phyla appear to have been the dominant vegetation during the Carboniferous period. One type of coal, called cannel coal, is formed almost entirely from fossilized spores of a tree lycopod named *Lepidodendron;* this finding is an indication of the abundance of this genus in the forests of that time.

Phylum Psilophyta: Two Simple Plants

There once was some disagreement as to whether rhyniophytes are entirely extinct. The confusion arose because of the existence today of two genera of rootless, spore-bearing plants, *Psilotum* and *Tmesipteris*. Are they the living relics of the rhyniophytes? If so, an enormous hole in the geologic record—between the rhyniophytes, which apparently became extinct over 300 million years ago, and *Psilotum* and *Tmesipteris*, which are modern plants—remains unexplained. Most botanists now treat these two genera as their own phylum, the Psilophyta, or whisk ferns. Although *Psilotum nudum* (Figure 25.16) has only minute scales instead of true leaves, plants of the genus *Tmesipteris* have flattened photosynthetic organs with well-developed vascular tissue.

(a)

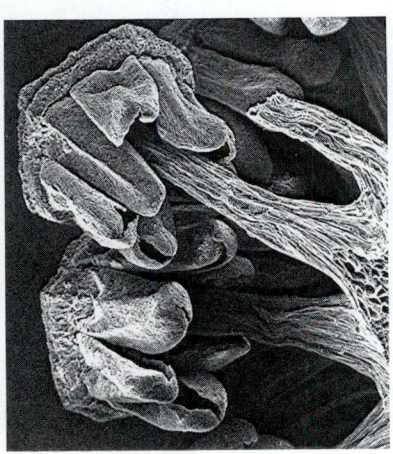

(b)

25.15 Horsetails
(a) Vegetative and sterile shoots of the horsetail *Equisetum telmateia* (phylum Sphenophyta). Leaves form in spaced whorls at nodes on the stems of the vegetative shoots; a few of the candle-shaped fertile shoots are visible toward the lower right. (b) Sporangia and sporangiophores of the horsetail *Equisetum arvense*.

25.16   A Modern Whisk Fern that Looks Like a Rhyniophyte
Aerial branches of *Psilotum nudum*, a plant once considered by some to be a surviving rhyniophyte and by others to be a fern. It is now included in the phylum Psilophyta.

The whisk ferns are probably highly specialized plants that evolved fairly recently from anatomically more complex ancestors, although one school of thought holds that they are evolutionarily ancient descendants of anatomically simple ancestors. Today *Psilotum* is widely distributed in the tropics and subtropics.

Plants with Complex Leaves

The sporophytes of the ferns and seed plants have true roots, stems, and leaves. Their leaves are typically large and have branching vascular strands; some species have small leaves as a result of evolutionary reduction, but they have more than one vascular strand. The sporangia of these plants are on the lower surfaces of leaves or modified leaves or, more rarely, on the margins of leaves. Fern and seed plant sporophytes are independent and much larger than the gametophytes. The gametophytes of ferns are independent, but those of gymnosperms and angiosperms are dependent upon the sporophytes.

Phylum Pterophyta: Ferns

The true ferns constitute the phylum Pterophyta, which first appeared during the Devonian period and today consists of about 12,000 species (Figure 25.17). Ferns are characterized by fronds (large leaves with complex vasculature), by the absence of seeds, and by a requirement for water for the transport of the male gametes to the female gametes.

Most ferns inhabit shaded, moist woodlands and

(a)

(b)

(c)

25.17   Fern Fronds Take Many Forms
(a) Fronds of the cinnamon fern *Osmunda cinnamomea* (phylum Pterophyta). (b) Tiny fronds of the water fern *Marsilea*. (c) "Fiddleheads" (developing fronds) of a fern; these structures will unfurl and expand to give rise to the complex adult frond.

25.18 Fern Sori Contain Sporangia
Sori, each with many spore-producing sporangia, on the underside of a frond of the western sword fern *Polystichum munitum*.

swamps. Some, the tree ferns, reach heights of up to 20 meters. Tree ferns lack the rigidity of woody plants, and thus do not grow in sites exposed directly to strong winds but rather in ravines or beneath trees in forests.

During its development, the fern frond unfurls from a tightly coiled "fiddlehead" (see Figure 25.17c). Some fern leaves become climbing organs and may grow to be as much as 30 meters long. The sporangia are found on the undersurfaces of the leaves, sometimes covering the whole undersurface and sometimes only at the edges; in some species the sporangia are clustered in groups called **sori** (singular: *sorus*; Figure 25.18).

Devonian fossil beds have yielded ferns with some characteristics that are rhyniophyte and some that resemble those of other vascular plant phyla. For example, like a modern fern, the genus *Protopteridium* had flattened branch systems with extensive photosynthetic tissue between the branches, but like a rhyniophyte, it bore terminal sporangia and lacked true roots (Figure 25.19). During late Paleozoic times, the ferns underwent considerable evolutionary experimentation in the structure of their leaves and particularly in the arrangement of their vascular tissue.

The Fern Life Cycle

The undersides or edges of fern fronds carry sporangia in which cells undergo meiosis to form haploid spores (Figure 25.20). Once shed, spores often travel great distances and eventually germinate to form small, independent gametophytes. These gametophytes produce antheridia and archegonia, although not necessarily at the same time or on the same gametophyte. Sperm swim through water to archegonia, often on other gametophytes, where they unite with an egg. The resulting zygote develops into a

new sporophyte embryo. The young sporophyte sprouts a root and can thus grow independently of the gametophyte. In the alternating generations of a fern, the gametophyte is small, delicate, and short-lived, but the sporophytes can be very large and can sometimes survive for hundreds of years.

Most ferns are homosporous. However, two orders of aquatic ferns, the Marsileales and Salviniales, have evolved heterospory. Male and female spores of these plants (which germinate to produce male and female gametophytes, respectively) are produced in different sporangia, and the male spores are always much smaller and greater in number.

A few genera of ferns produce a tuberous, fleshy gametophyte instead of the characteristic flattened, photosynthetic structure described earlier. These tuberous gametophytes depend on a mutualistic fungus for nutrition; in some genera, even the sporophyte embryo must become associated with the fungus before extensive development can proceed.

THE SEED PLANTS

The most recent group to appear in the evolution of the plant kingdom is the seed plants: the **gymnosperms** (such as pines and their relatives) and the **angiosperms** (flowering plants). In seed plants the gametophyte generation is reduced even further than it is in ferns. The haploid gametophyte develops partly or entirely while attached to and nutritionally

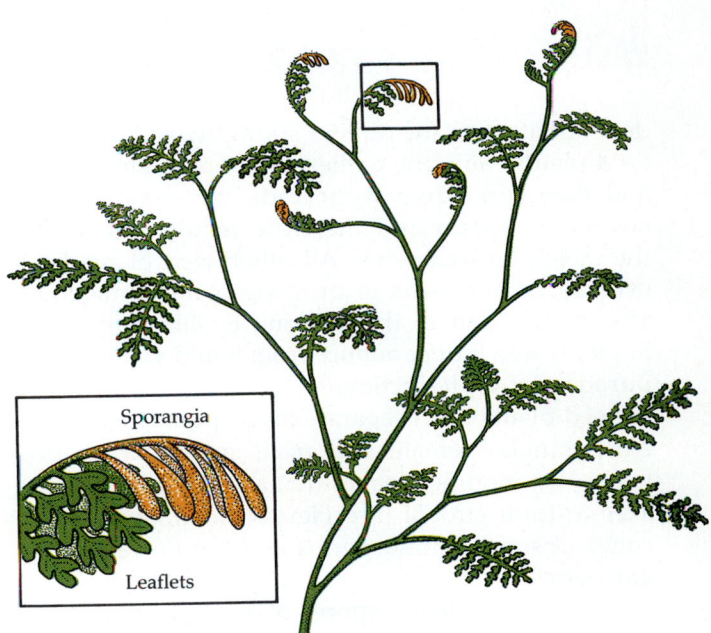

Sporangia

Leaflets

25.19 A Fossil Fern
During their evolutionary history, ferns exhibited combinations of structures. This fossil fern (*Protopteridium*, from the Devonian period) lacked true roots and had branches with both leaflets and terminal sporangia.

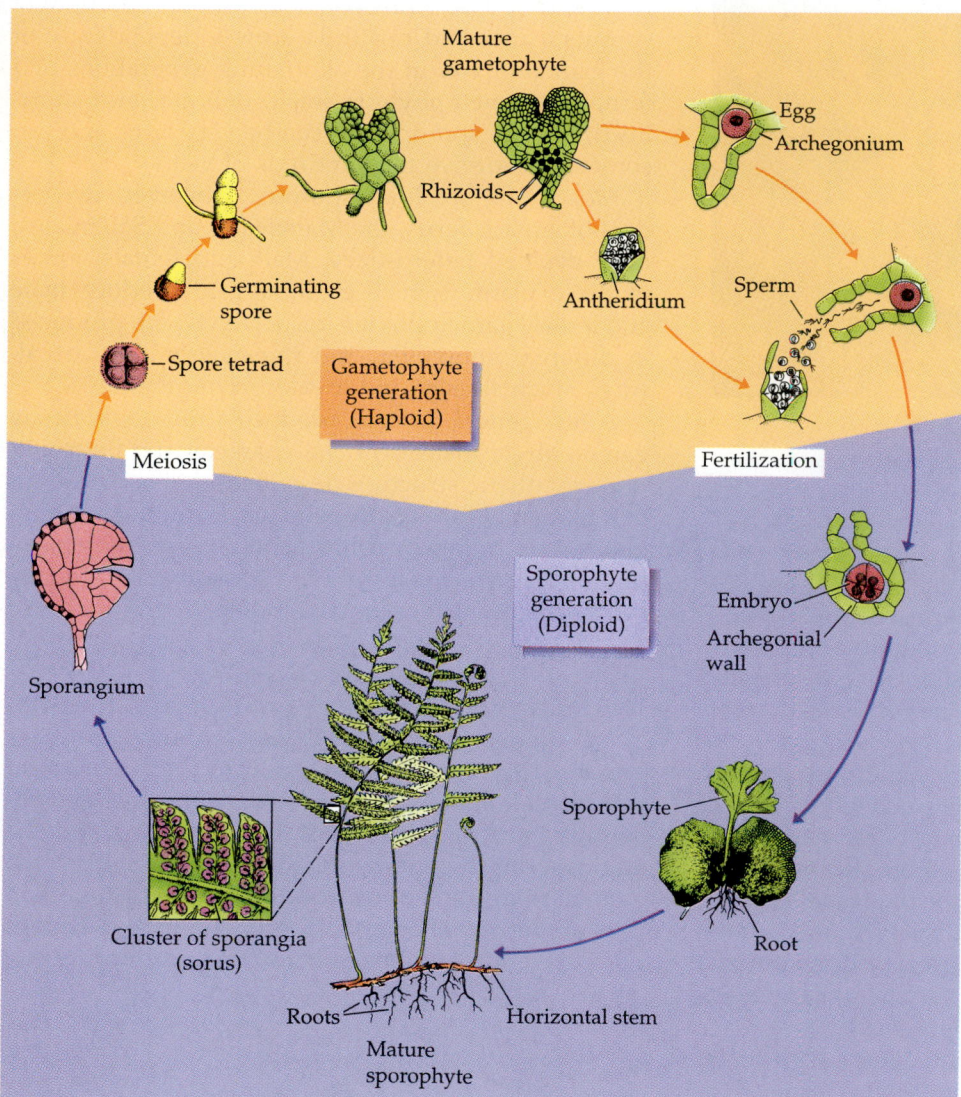

25.20 Life Cycle of a Fern
The most familiar stage in the fern life cycle is the mature, diploid sporophyte. Following meiosis, haploid spores emerge from sporangia on the undersides of the sporophyte's fronds. When the spores germinate, they form small, heart-shaped gametophytes. Eggs in archegonia on gametophytes are fertilized by swimming sperm from antheridia. The embryo that develops from the diploid zygote in the archegonium eventually sends out roots and develops into a sporophyte.

dependent upon the diploid sporophyte. Among the seed plants, only the earliest types of gymnosperms and their few survivors (cycads, for example) had swimming sperm and therefore required water for the meeting of gametes. All other seed plants have evolved other means of bringing gametes together. The culmination of this striking evolutionary trend in plants was independence from liquid water for the purposes of reproduction.

Seed plants form separate megasporangia and microsporangia—female and male sporangia, respectively—on modified leaves that are grouped on short axes to form strobili (see Figure 25.14), such as the cones of some gymnosperms and the flowers of angiosperms.

As in other plants, spores of seed plants are produced by meiosis within the sporangia, but in this case they are not shed. Instead, the gametophytes develop within the sporangia and depend upon them for food and water. In most species only one of the meiotic products in a megasporangium survives. The

surviving haploid nucleus divides mitotically, and the resulting cells (without walls) divide again to produce a small multicellular female gametophyte. In the angiosperms, female gametophytes do not normally include more than eight nuclei. The female gametophyte of a seed plant is retained within the megasporangium, where it matures, is fertilized, and undergoes the early development of the next sporophyte generation.

Within the microsporangium the meiotic products develop into microspores that undergo one or a few cell divisions to form male gametophytes, the familiar **pollen grains** (Figure 25.21; see also Figure 34.2). Distributed by wind, an insect, a bird, or a plant breeder, a pollen grain that reaches the appropriate surface of a sporophyte, near the female gametophyte, develops further. It produces a slender **pollen tube** that grows and digests its way through the sporophyte tissue toward the female gametophyte.

When the tip of the male gametophyte's pollen tube reaches the female gametophyte, either sperm

25.21 Wind Carries Gymnosperm Pollen
The wind carries pollen grains from a male (above) to a female strobilus.

nuclei or a sperm cell is released from the tube, and fertilization occurs. The resulting diploid zygote divides repeatedly, forming a young sporophyte that develops to an embryonic stage at which the entire system becomes temporarily suspended (often referred to as a dormant stage). The end product at this stage is a **seed**. A seed may contain tissues from three generations. The seed coat develops from tissue of the diploid parent sporophyte. Within the seed coat is a layer of haploid female gametophyte tissue from the next generation (this tissue is fairly extensive in most gymnosperm seeds, but its place is taken by a tissue called endosperm, which we will discuss shortly, in angiosperm seeds). In the center of the seed package, the third generation, in the form of the embryo of the new sporophyte, is found.

The embryos of seedless plants develop directly into sporophytes, which either survive or die, depending on environmental conditions; there is no resting stage in the life cycle. By contrast, the multicellular seed of a gymnosperm or angiosperm is a well-protected resting stage for the embryo. Layers of cells enclose the embryo, and the seeds of some species may remain viable (capable of growth and development) for many years, germinating when conditions are favorable for the growth of the sporophyte. When the young sporophyte begins to grow, it draws on food reserves in the seed. During the dormant stage the seed coat protects the embryo from drying and may also protect against potential predators that would otherwise eat the embryo or

the food reserves. Many seeds have structural adaptations that promote dispersal by the wind or, more often, by animals. The possession of seeds is a major reason for the enormous evolutionary success of seed plants, which are the prominent elements of Earth's modern land flora in most areas.

THE GYMNOSPERMS

Although there are probably fewer than 750 species of living gymnosperms, these plants are second only to the angiosperms (flowering plants) in their dominance of the land masses. There are four phyla of gymnosperms living today (Figure 25.22). The cycads (phylum Cycadophyta) are palmlike plants of the tropics, growing as tall as 20 meters. Ginkgos (phylum Ginkgophyta), which were common during the Mesozoic era, are represented today by a single genus, the maidenhair tree. The phylum Gnetophyta consists of three very different genera that share a characteristic type of cell (the vessel element) in their xylem tissue that is found in no other group of plants except the angiosperms. One member of the phylum is *Welwitschia,* a long-lived desert plant with just two straplike leaves that sprawl on the sand and can become as long as 3 meters. Far and away the most abundant of the gymnosperms are the conifers (phylum Coniferophyta), cone-bearing plants such as pines.

All living gymnosperms have active secondary growth (growth in diameter of stems and roots; see Chapter 29), and all but the Gnetophyta have only tracheids as water-conducting and support cells in the xylem. Despite this water transport and support system (which might seem suboptimal compared with that of angiosperms), the gymnosperms include some of the tallest trees known. The coastal redwoods of California are the tallest gymnosperms; the largest are well over 100 meters tall. Xylem produced by gymnosperms is the principal resource of the lumber industry.

Fossil Gymnosperms

The earliest fossil evidence of gymnosperms is found in Devonian rocks. The early gymnosperm story is especially interesting, illustrating the difficulties under which students of fossil plants frequently labor. Years ago a relatively rare and poorly preserved plant fossil, *Archaeopteris,* was described. The fossil appeared to many workers to be an ancient heterosporous fernlike plant similar to *Protopteridium* (see Figure 25.19). Another fairly common Devonian fossil, *Callixylon,* remained a puzzle. It consisted only of well-petrified logs, some wider than a meter in diameter and longer than 20 meters. Among the deli-

(a)

(b)

(d)

(c)

25.22 Gymnosperms

(a) This bread palm belongs to the cycads, the most ancient phylum of gymnosperms (phylum Cycadophyta). Many cycads have growth forms that resemble both ferns and palms. (b) Characteristic broad leaves of the maidenhair tree, *Ginkgo biloba* (phylum Ginkgophyta); the leaves turn brilliant gold in the autumn. (c) *Welwitschia mirabilis* growing in the Namib Desert of Africa. Two huge, straplike leaves grow throughout the life of the plant, breaking and splitting as they grow. Seeds are produced on the cone-bearing branches. *Welwitschia* belongs to the Gnetophyta, the gymnosperm phylum most closely related to the flowering plants. (d) A dramatic conifer, the giant sequoia *Sequoiadendron giganteum* (phylum Coniferophyta), growing in Yosemite National Park, California.

cate and herbaceous rhyniophytes it seemed entirely out of place. Its wood was composed of pitted tracheids, and it obviously had possessed a highly effective mechanism for secondary growth. Charles Beck of the University of Michigan resolved the puzzle when, in 1960, he found fronds of *Archaeopteris* clearly attached to and part of *Callixylon* logs. Because the name *Archaeopteris* had been published first, the name *Callixylon* was dropped, and the "rare" *Archaeopteris* was suddenly recognized as being a common Devonian plant. It had both rhyniophyte and fernlike characteristics, but its woody tissue, based on tracheids, was clearly that of a gymnosperm.

By the Carboniferous period, several new lines of

gymnosperms had evolved, including one now extinct phylum, the seed ferns, that possessed fernlike foliage but had characteristic gymnosperm seeds attached to the leaf margins. The first true conifers appeared at approximately the same time. Either they were not dominant trees or they did not grow where conditions were right for fossilization. During the Permian period, however, the conifers and cycads came into their own. Gymnosperms dominated the forests until less than 80 million years ago; they even dominate some present-day forests.

Phylum Coniferophyta: Conifers

The great Douglas fir and cedar forests of the northwestern United States and the massive forests of pine, fir, and spruce that clothe the northern continental regions and upper slopes of mountain ranges rank among the great vegetation formations of the world (the "boreal forest"; see Figure 50.15). All these trees belong to one phylum of gymnosperms, Coniferophyta—the conifers, or cone-bearers. All conifers are heterosporous. Male and female spores are produced in separate male and female cones.

The Gymnosperm Life Cycle

We will use the life cycle of a pine to illustrate reproduction in gymnosperms (Figure 25.23). The production of male gametophytes as pollen grains frees the plant once and for all from its dependence upon external liquid water for fertilization. Instead of water, wind assists conifer pollen grains in their first stage of travel to the female gametophyte. The pollen tube provides the means for the last stage of travel by growing and digesting its way through maternal sporophytic tissue and eventually releasing a sperm cell near the egg. The megasporangium, which will form the female gametophyte containing eggs within archegonia, is enclosed in a special layer of sporophytic tissue, called the **integument**, that will eventually develop into the seed coat. The integument, its enclosed megasporangium, and the tissue attaching it to the maternal sporophyte forms the **ovule**. The pollen grain travels through the small opening in the integument at the apex of the ovule, the **micropyle**.

The word *gymnosperm* means, literally, "naked-seeded." Conifer ovules (which develop into seeds upon fertilization) are borne exposed on the upper surfaces of cone scales without fruit tissue to protect the seeds. The only protection from the environment is that the scales are tightly pressed against each other within the cone. Some pines have such tightly closed female cones that only fire suffices to split them open and release the seeds. One example is the lodgepole pine, which covers vast fire-ravaged areas in the Rocky Mountains and elsewhere.

About half of the conifer species have fruitlike, fleshy tissues associated with their seeds; examples are the "berries" of juniper and yew. Animals may eat these tissues and then disperse the seeds, often carrying them considerable distances, which may spread the conifer population. True fruits are one of the characteristics of yet another plant phylum, the one that is dominant today: Anthophyta.

THE ANGIOSPERMS: FLOWERING PLANTS

The phylum Anthophyta consists of the flowering plants, also known as the angiosperms. This highly diverse phylum includes about 275,000 species. In other chapters, when we mention plants in discussing processes such as long-distance transport in the xylem and phloem, or the chemical regulation of development, generally we are referring to the angiosperms. The angiosperms represent the current extreme of an evolutionary trend that runs throughout the vascular plants, in which the *sporophyte* generation becomes *larger* and *more independent* of the gametophyte, while the *gametophyte* generation becomes *smaller* and *more dependent* upon the sporophyte. Angiosperms differ from other plants in several ways, although exceptions exist.

Double fertilization is the single most reliable distinguishing characteristic of the angiosperms. *Two male gametes*, contained within a single male gametophyte, participate in fertilization events within the female gametophyte of an angiosperm. One sperm nucleus combines with the egg to produce a diploid zygote, the first cell of the sporophyte generation. The other sperm nucleus usually combines with two other haploid nuclei of the female gametophyte to form a triploid ($3n$) nucleus. This, in turn, divides to form a triploid tissue, the **endosperm**, that nourishes the embryonic sporophyte during its early development. Double fertilization occurs in all present-day angiosperms and in one other plant, *Ephedra* (a member of the gymnosperm phylum Gnetophyta). William Friedman, at the University of Georgia, discovered in 1990 that *Ephedra* has double fertilization. Thus although double fertilization is the most reliable character for recognizing a plant as an angiosperm, it is not absolutely reliable. We are not sure when and how double fertilization evolved because there is no fossil evidence on this point.

A second consistent characteristic of angiosperms is the possession of specialized water-transporting cells called **vessel elements** in the xylem, but these are also found in a few gymnosperms and ferns, and even in a club moss and a horsetail. Another distinctive cell in angiosperm xylem is the **fiber**, which plays an important role in supporting the plant body. The name *angiosperm* refers to the diagnostic character that the seeds of these plants are enclosed in a modified leaf called a carpel. Of course, the most memorable diagnostic feature of angiosperms is that they have flowers.

The Flower

If you examine any familiar flower, you will notice that each part has one or more veins and, in the case of the outer parts, looks somewhat like a leaf. These parts are in fact modified for the flower function from leaves. Thus the terms that we have already used for other plants apply here. A generalized flower (for which there is no exact counterpart in nature) is shown in Figure 25.24. In the flower, modified leaves

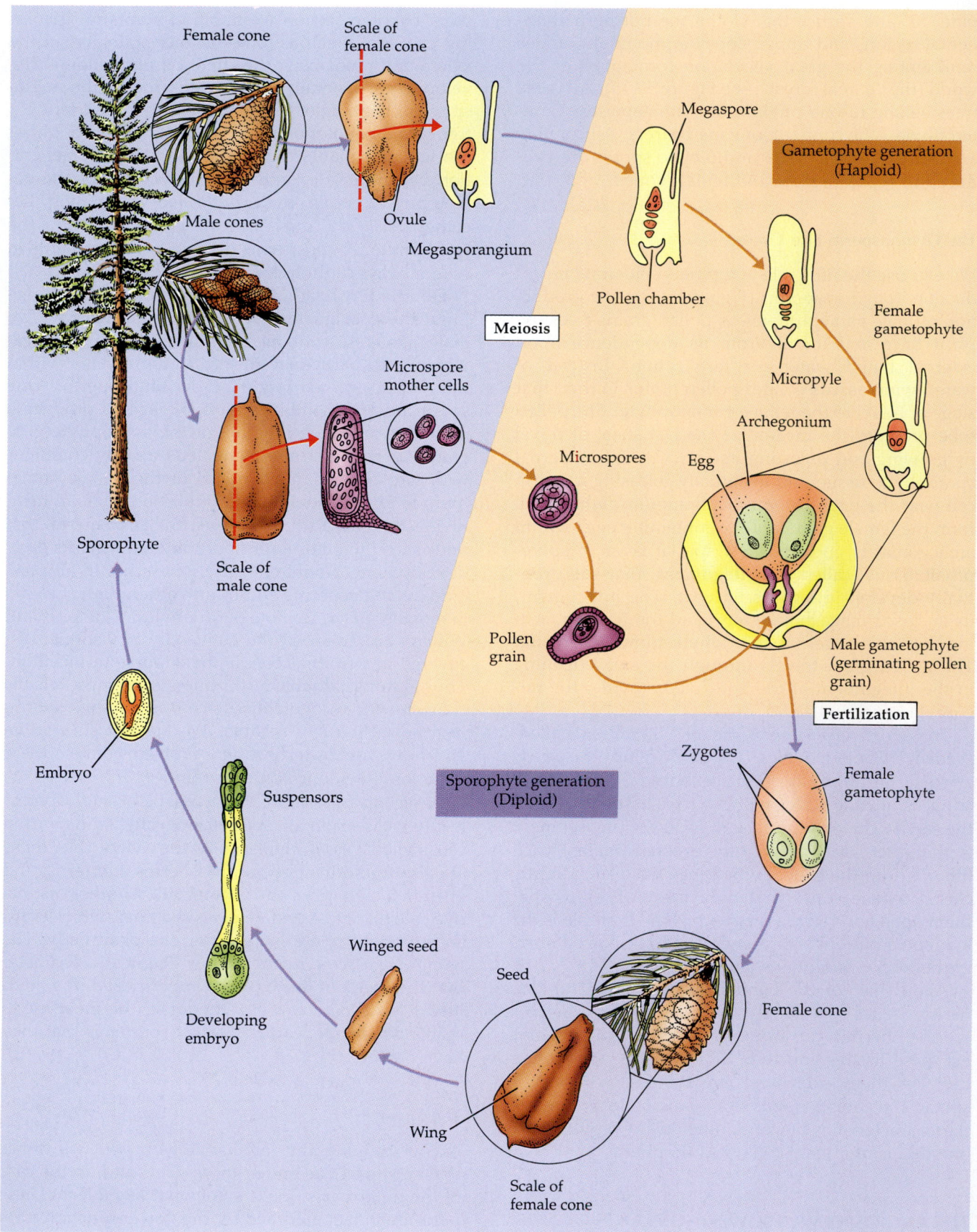

25.23   Life Cycle of a Pine
The seed protects the embryo. The sporophyte is enormous, but the gametophytes are tiny. Note that the same plant has both pollen-producing male cones
and egg-producing female cones.

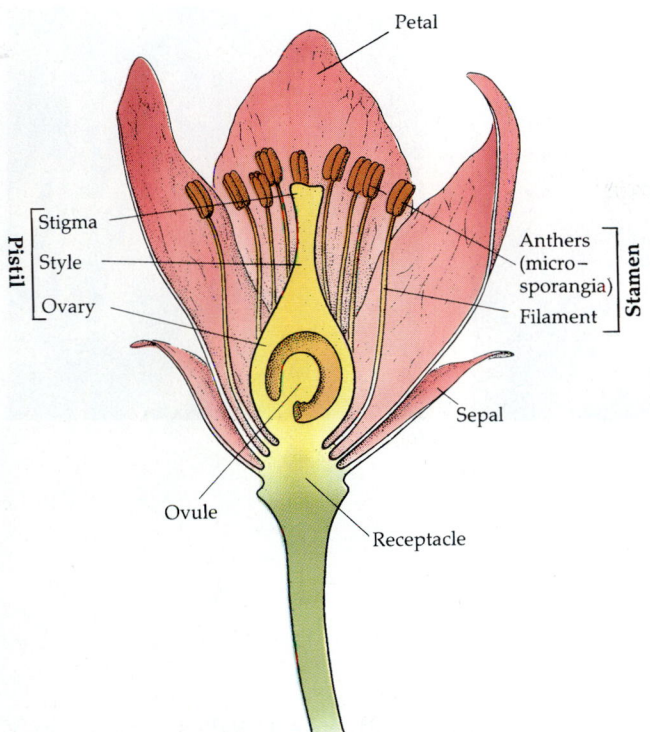

25.24 Structure of a Generalized Flower
Not all flowers possess all the structures shown here, but they must possess stamens, pistils, or both in order to play their role in sexual reproduction. This flower, possessing both, is a perfect flower.

bearing microsporangia are called **stamens**, each composed of a filament and an **anther** that contains pollen-producing sporangia; those bearing megasporangia are called **carpels**. A structure composed of one or more fused carpels is called a **pistil**. The swollen base of the pistil, containing one or more ovules, is an **ovary**; the apical stalk of the pistil is a **style**; and the terminal surface that receives the pollen is called a **stigma**.

In addition, there often are a number of specialized sterile (non-spore-bearing) leaves, the inner ones called **petals** (collectively, the **corolla**), and the outer the **sepals** (collectively, the **calyx**). The corolla and calyx, which can be quite showy, often play roles in attracting animal pollinators to the flower. From base to apex, the sepals, petals, stamens, and carpels are arranged in whorls and attached to a central receptacle.

The flower in Figure 25.24 has both megasporangia and microsporangia and is said to be **perfect**, meaning that it contains both functional female and functional male parts. Many angiosperms produce two types of flowers, one type with only megasporangia and the other with only microsporangia; consequently, either the stamens or the carpels are nonfunctional or absent in a given flower, and the flower is referred to as **imperfect**. Species such as corn or birch that have both female and male flowers on the

same plant are said to be **monoecious** (meaning "one-housed"—but, it must be added, one house with separate rooms). The sexes are completely separated in some other species of angiosperms, such as willows and date palms; in these species, a given plant produces either male or female sporangia, but never both. Such species are said to be **dioecious** ("two-housed"). In other words, there are truly female plants and truly male plants.

In the generalized flower of Figure 25.24, we illustrated distinct petals and sepals arranged in distinct whorls, although in nature sometimes the petals and sepals are arranged in a continuous spiral and are indistinguishable. Such appendages are called **tepals**. In other cases petals, sepals, or tepals are completely absent.

Flowers may be single or grouped together to form an **inflorescence**. Different families of flowering plants have their own, characteristic types of inflorescences, such as the umbels of the carrot family, the heads of the aster family, the spikes of many grasses, and others (Figure 25.25).

Evolution of the Flower

Botanists disagree about which type of flower is evolutionarily the most ancient. One of the earliest types has many tepals (or sepals and petals), carpels, and stamens, all spirally arranged (Figure 25.26a). Evolutionary change within the angiosperms included some striking modifications from this early condition: reduction in the number of each type of organ, differentiation of petals from sepals, stabilization of each type of organ to a fixed number, arrangement in whorls, and, finally, change in symmetry from radial (as in a lily) to bilateral (as in a sweet pea or orchid), often accompanied by an extensive fusion of parts (Figure 25.26b). A great variety of corolla types have emerged in the course of evolution, as you will realize if you think of some of the flowers you recognize.

The first carpels to evolve were modified leaves, folded but incompletely closed and thus intermediate between the scales of the gymnosperms and the true carpels of the angiosperms. In the groups of angiosperms that evolved later, the carpels fused and became progressively more buried in receptacle tissue (Figure 25.27a); in the flowers of the latest groups to evolve, the other flower parts are attached at the very top of the ovary rather than at the bottom. The stamens of the most-ancient flowers may have been leaflike (Figure 25.27b), little resembling those of the generalized flower in Figure 25.24.

Why do so many flowers have pistils with long styles and anthers with long filaments? Natural selection has favored length in both of these structures probably because length increases the likelihood of successful pollination. Long filaments may bring the

(a)   (b)   (c)

25.25   Inflorescences
(a) The inflorescences of Queen Anne's Lace (*Daucus carota*) are umbels. Each umbel bears flowers on stalks that arise from a common center. (b) Sunflowers are members of the aster family; their inflorescences are heads. In a head, each of the long, petallike structures is a ray flower; the central portion of the head consists of dozens to hundreds of disc flowers. (c) Grasses such as these foxtails have inflorescences called spikes.

(a)

(b)

25.26   Flower Form and Evolution
(a) A flower of *Magnolia grandiflora*, showing major features of early flowers: radial symmetry with the individual tepals, carpels, and stamens separate, numerous, and attached at their bases in a spiral arrangement. (b) The orchid *Paphiopedilum villosum* has a bilaterally symmetrical structure that evolved much later than the form of the magnolia flower in (a). One of the three petals evolved into the complex lower "lip." Inside, the stamen and pistil are fused, and there is a single anther.

anthers in contact with insect bodies, or they may put the anthers where they catch the wind better. Similar arguments apply to long styles. A long style may serve another purpose as well. If several pollen grains land on one stigma, a pollen tube will start growing from each grain toward the ovary. If there are more pollen grains than ovules, there is a "race" for the ovules. The race down the style can be viewed as "mate selection" by the plant holding that style. Pollen has played another, major role in evolution, as we are about to see.

Pollen and the Coevolution of Angiosperms and Animals

Most gymnosperms are wind-pollinated, but most angiosperms are animal-pollinated. This distinction has been true probably ever since the first angiosperms appeared. Animals visit flowers to obtain nectar or pollen and in the process often end up carrying pollen from one flower to another, or from one plant to another. Thus in its quest for food, the animal contributes to the genetic diversity of the plant population. Insects—especially bees—are among the most important pollinators, but birds and some species of bats also play major roles.

For more than 100 million years, angiosperms and their animal pollinators have coevolved in the terrestrial environment. The animals have affected the evolution of the plants, and the plants have affected the evolution of the animals. Pollination by just one or a very few committed animal species provides a plant species with a reliable mechanism for transferring pollen from one to another of its members. Flower structure has become incredibly diverse under selective pressure.

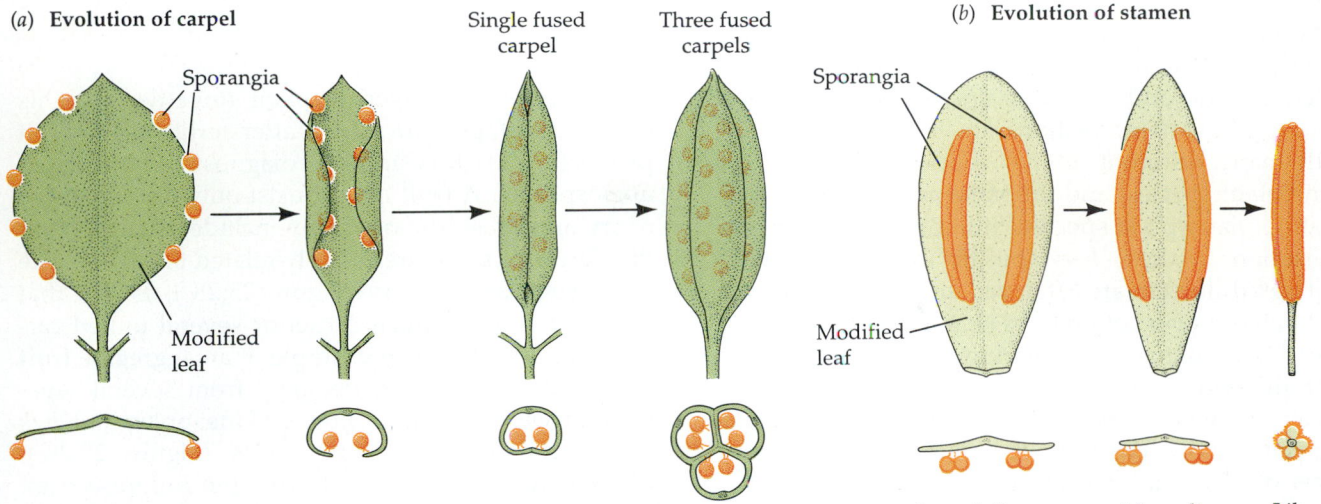

(a) Evolution of carpel

Sporangia

Modified leaf

Single fused carpel

Three fused carpels

(b) Evolution of stamen

Sporangia

Modified leaf

Austrobaileya sp.   Magnolia   Lily

25.27 Evolution of Carpels and Stamens
(a) The carpel began as a modified leaf with sporangia (shown here in orange) along its edges. In the course of evolution, leaf edges curled inward and finally fused. At the end of the sequence, three carpels have fused to form a three-chambered ovary. (Below each stage is a cross section.) (b) Three modern plants show the major stages in stamen evolution. The leaflike portion of the structure was progressively reduced until only the microsporangia (orange) remained.

(a)

(b)

(c)

(d)

25.28 Fruits Come in Many Forms and Flavors
(a) A simple fruit: sour cherry. (b) An aggregate fruit: blackberry. (c) A multiple fruit: pineapple. (d) An accessory fruit: apple.

Some of the products of coevolution are highly specific; for example, some yucca species are pollinated by one and only one species of moth. Most plant–pollinator interactions are much less specific, with many different animal species pollinating the same plant species, and the same animal species pollinating many plant species. Still, there has been specialization in these less specific interactions. Bird-pollinated flowers are often red and odorless; insect-pollinated flowers often have characteristic odors and may have conspicuous markings ("nectar guides") that are evident only in the ultraviolet region of the spectrum, where insects have better vision than in the red region. We treat coevolution and other aspects of plant–animal interactions in more detail in Chapter 48.

The Fruit

The ovary of a flowering plant (together with its seeds) develops into a fruit after fertilization. Fruit production is thus another diagnostic character of angiosperms. A fruit may consist only of the mature ovary and its seeds, or it may include other parts of the flower or structures closely related to it. A **simple fruit**, such as a cherry (Figure 25.28*a*), is one that develops from a single carpel or several united carpels. A raspberry is an example of an **aggregate fruit** (Figure 25.28*b*)—one developing from several separate carpels of a single flower. Pineapples and figs are examples of **multiple fruits** (Figure 25.28*c*), formed from a cluster of flowers (an inflorescence). Fruits derived from parts in addition to the carpel

25.29 Life Cycle of an Angiosperm

Pollen grains develop from microspores in the inflorescences at the top of a corn plant. Megagametophytes develop in ovules in lower inflorescences. When a pollen grain lands on the stigma of a pistil, a pollen tube grows through the pistil until it reaches the megagametophyte.

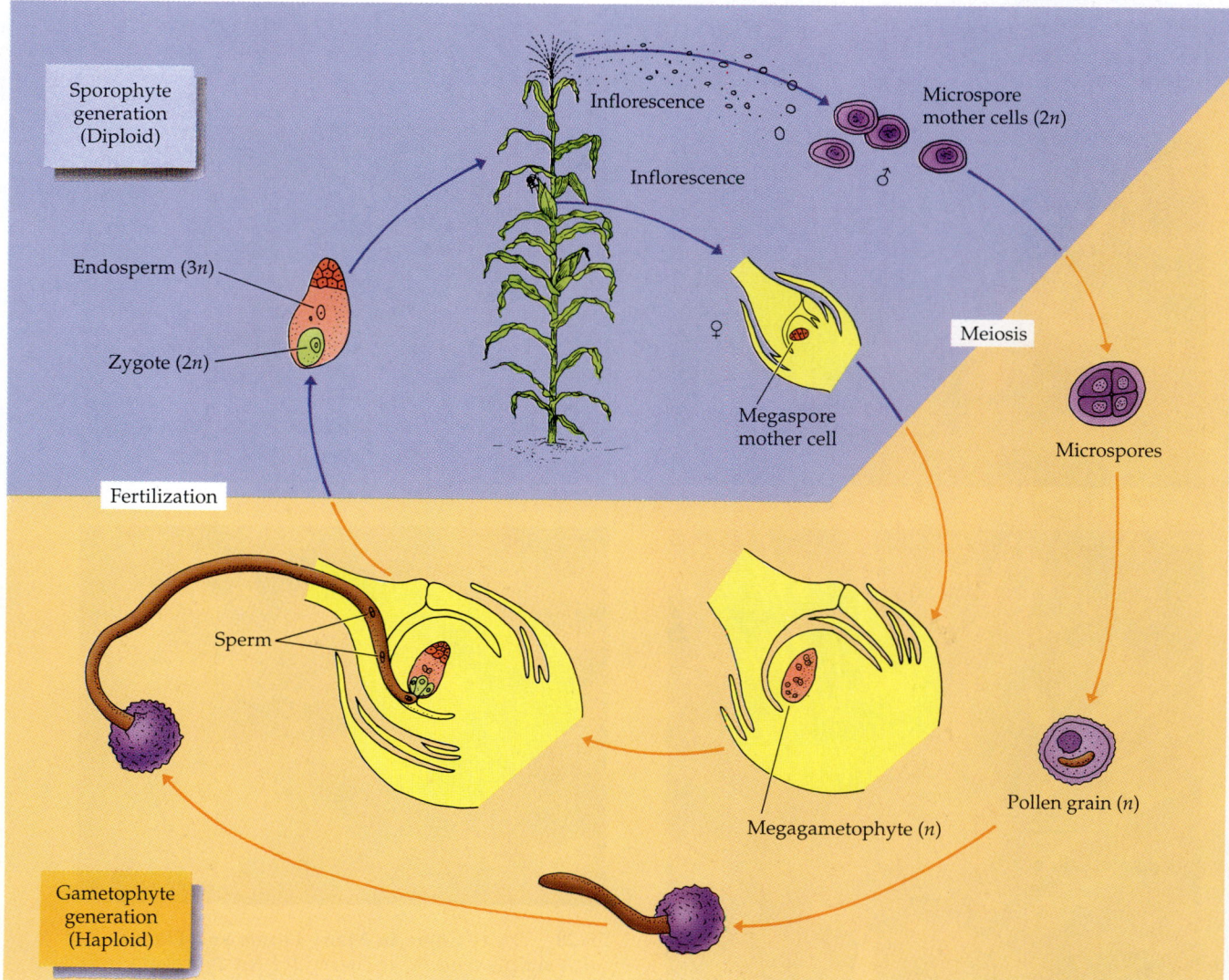

and seeds are called **accessory fruits** (Figure 25.28*d*); examples are apples, squash, and bananas. The development, ripening, and dispersal of fruits will be considered in Chapter 33.

The Angiosperm Life Cycle

The life cycle of the angiosperms will be considered in detail in Chapter 34. As summarized in Figure 25.29, it has similarities with and differences from that of the gymnosperms (see Figure 25.23). Like gymnosperms, angiosperms are heterosporous. The female gametophyte is even more reduced than in the gymnosperms. The ovules are contained within carpels, rather than being exposed on the surfaces of scales as in gymnosperms. The male gametophytes are, again, pollen grains.

The ovule develops into a seed, containing the products of the double fertilization that characterizes angiosperms. The triploid endosperm serves as storage tissue for starch or lipid reserves, storage proteins, and other reserve substances. The diploid zygote develops into an embryo consisting of an embryonic axis and one or two **cotyledons**. Also called seed leaves, cotyledons have different fates in different plants. In many, they serve as absorptive organs that take up and digest the endosperm. In others, they enlarge and become photosynthetic when the seed germinates. Often they play both of these roles.

Classes of Angiosperms

The angiosperms are divided into the class Monocotyledones (the monocots; Figure 25.30) and the class Dicotyledones (the dicots; Figure 25.31). The names derive from the presence of a single embryonic cotyledon in the monocots and of two cotyledons in the dicots. There are, however, other major differences between the two classes (see Figure 29.5). These include differences in leaf vein patterns, arrangement of vascular tissue in the stem and root, numbers of flower parts, and the presence or absence of secondary growth (produced by a cambium; see Chapter 29). The cotyledons of some, but not all, dicots store the reserves originally present in the endosperm.

The monocots include grasses, cattails, lilies, orchids, and palm trees. The dicots include the vast number of familiar seed plants: most of the herbs, weeds, vines, trees, and shrubs. Examples are oaks, willows, violets, sunflowers, and chrysanthemums.

(a)

(b)

(c)

25.30 Monocots
Monocots include popular garden flowers such as lilies and orchids (see Figure 25.26*b*). (*a*) Date palms are an important food source in many parts of the world. Palms are among the few monocot trees. (*b*) Grasses, such as this wheat, seen in the summer, are monocots. (*c*) The lotus-of-the-Nile grows in water.

(a)

(b)

(c)

25.31 Dicots
(a) The cactus family is a large group of dicots, with about 1,500 species in the Americas. This claret cup cactus takes its name from its scarlet flowers. (b) Snapdragons, *Anterrhinum majus. (c)* These wood roses from Yellowstone National Park are members of the family Rosaceae, as are the familiar roses from your local florist. Figures 25.25 and 25.26a show other dicots.

Origin and Evolution of the Angiosperms

How did the angiosperms arise? Modern cladistic analyses (see Chapter 21) have settled this once vexing question. It is widely agreed that the angiosperms and two groups of gymnosperms, the Gnetophyta and the long-extinct cycadeoids, a subgroup of the cycads, arose from a single ancestral species that gave rise to no other groups. A close relationship between the angiosperms and the Gnetophyta was long sus-

pected, primarily on the grounds that some Gnetophyta have vessel elements, which characterize the angiosperms. In 1990 this theory was strengthened by the light-microscopic confirmation of double fertilization in *Ephedra*, a member of the Gnetophyta. The cycadeoids, which became extinct at about the same time as did the dinosaurs, shared several important characteristics with the Gnetophyta and the angiosperms. The reproductive organs of one of the cycadeoids, although clearly a gymnosperm structure with naked seeds, had suggestive similarities to the flower of *Magnolia*.

The next great area of controversy is likely to be the question, Which were the first angiosperms? Two candidates are the magnolia family (see Figure 25.26a) and another family, the Chloranthaceae, whose flowers are much simpler in anatomy than those of the magnolias. One complication in the search for the earliest fossil angiosperms is that we are unlikely to be able to tell whether an ancient fossil plant practiced double fertilization.

The first angiosperms, according to fossil evidence, were probably trees. The gymnosperms from which they evolved must also have been trees. The history of vascular plant evolution up to that point featured a progression from herbaceous (nonwoody) early vascular plants to larger, woody forms. During angiosperm evolution, herbaceous forms appeared once again. Today the predominant ground cover consists of herbaceous angiosperms. The evolution of small, nonwoody plants from tall, woody ones is still a matter of research and debate.

25.32 Herbs and Tundra
Here in the tundra of Denali National Park, Alaska, the landscape is dominated by herbs.

Where Herbaceous Plants Predominate

When a piece of ground becomes available for new colonization—perhaps as a result of fire, or the appearance of a new sand bar, or clearing by humans—the first colonizers are generally herbs. Their seeds may have been present in a dormant state in the soil, or they may have been brought in by wind or animals. The rapidly growing herbs produce seeds that germinate and contribute to the growth of the herb population. Before the land was cleared, the foliage of shrubs and trees may have shaded the ground, making little light available for photosynthesis by small herbs. When the taller plants are removed, the herbs have their day in the sun. Later, as the land is modified by these herbaceous plants, other, larger forms appear (or reappear) and generally take over as succession proceeds (see Chapter 49).

Although not dominating the scene, some herbaceous angiosperms do well in forests. They succeed through good timing, performing their growth, photosynthesis, and reproduction early in the season before the foliage develops fully on the trees above. Once the leafy canopy of the forest becomes dense, the shoots of these herbs die back, leaving underground organs (rhizomes or tubers) with stored food for the beginning of the following year's growth.

Herbs predominate in at least one extreme environment: tundras (Figure 25.32). Here the ability of some herbs to store photosynthate in underground organs stands them in particularly good stead, because they leave no aboveground parts to be damaged during the long cold season. The cold is so intense that shrubs and trees are unable to survive, but the dormant underground parts of herbs succeed.

SUMMARY of Main Ideas about Plants

Plants are multicellular, photosynthetic eukaryotes that develop from embryos protected by parental tissue.

Plants evolved from a green algal ancestor.
Review Figure 25.2

All plants have a life cycle in which a sporophyte generation alternates with a gametophyte generation differing in size and structure.

Major trends in the evolution of the plant kingdom include:
- growing independence from liquid water for reproduction
- increasing size and independence of the sporophyte generation
- a shift from homospory to heterospory
- increasing plant size

In the course of their evolution, many angiosperm lineages partly reversed the trend toward larger size as some descendants of larger plants became small herbs.

The nonvascular plants include the liverworts, hornworts, and mosses.

Nonvascular plants either lack vascular tissues completely or have only a rudimentary system of water- and food-conducting cells.

Nonvascular plants lack true leaves and roots.

The nonvascular sporophyte generation is smaller than the gametophyte generation and depends upon the gametophyte for water and nutrition.
Review Figure 25.7

The vascular plants have tracheids and other specialized cells that together constitute vascular tissue.

Xylem is a vascular tissue for conducting water and minerals.

Phloem is a vascular tissue for conducting foods.

Club mosses and horsetails have simple leaves and little vascular tissue; leaves with more complex vasculature are characteristic of all other phyla of vascular plants.
   Review Figure 25.12

Ferns have a dominant sporophyte generation.
   Review Figure 25.20

Gymnosperms and angiosperms possess seeds; ferns do not.

Nonvascular plants and ferns depend on liquid water for fertilization.

The pollen of seed plants has evolved mechanisms for dispersal, so liquid water is not needed for the transfer of male gametes.

Gymnosperms have a life cycle in which naked seeds are produced on the scales of cones.
   Review Figure 25.23

Angiosperms are distinguished by the production of flowers and fruits.
   Review Figures 25.24 and 25.27

The vascular tissues of angiosperms contain cell types (such as vessel elements and fibers) rarely found elsewhere in the plant kingdom.

Double fertilization, producing a zygote and an endosperm, is unique to flowering plants and one member of the Gnetophyta.
   Review Figure 25.29

The two angiosperm classes—the monocots and the dicots—differ in vein patterns in leaves, number of cotyledons, and numbers of flower parts.

SELF-QUIZ

1. Plants differ from algae in that
   a. only plants are photosynthetic.
   b. only plants are multicellular.
   c. only plants possess chlorophyll.
   d. only plants have multicellular embryos protected by the parent.
   e. only plants are eukaryotic.

2. Which statement is *not* true of the alternation of generations in plants?
   a. It is heteromorphic.
   b. Meiosis occurs in sporangia.
   c. Gametes are always produced by meiosis.
   d. The zygote is the first cell of the sporophyte generation.
   e. The gametophyte and sporophyte differ genetically.

3. Which statement is *not* evidence for the origin of plants from the green algae?
   a. Some green algae have multicellular sporophytes and multicellular gametophytes.
   b. Both plants and green algae have cellulose in their cell walls.
   c. The two groups have the same photosynthetic and accessory pigments.
   d. Both produce starch as their principal storage carbohydrate.
   e. All green algae produce large, stationary eggs.

4. The nonvascular plants
   a. lack a sporophyte generation.
   b. grow in dense masses, allowing capillary movement of water.
   c. possess xylem and phloem.
   d. possess true leaves.
   e. possess true roots.

5. The rhyniophytes
   a. possessed vessel elements.
   b. possessed true roots.
   c. possessed sporangia at the tips of stems.
   d. possessed leaves.
   e. lacked branching stems.

6. Club mosses and horsetails
   a. have larger gametophytes than sporophytes.
   b. possess small leaves.
   c. are represented today primarily by trees.
   d. have never been a dominant part of the vegetation.
   e. produce only simple fruits.

7. Which statement is *not* true of ferns?
   a. The sporophyte is larger than the gametophyte.
   b. Most are heterosporous.
   c. The young sporophyte can grow independently of the gametophyte.
   d. The frond is a large leaf.
   e. The gametophytes produce archegonia and antheridia.

8. The gymnosperms
   a. dominate all land masses today.
   b. have never dominated land masses.
   c. have active secondary growth.
   d. all have vessel elements.
   e. lack sporangia.

9. Which statement is *not* true of flowers?
   a. Pollen is produced in the anthers.
   b. Pollen is received on the stigma.
   c. An inflorescence is a cluster of flowers.
   d. A species having female and male flowers on the same plant is dioecious.
   e. A flower with both mega- and microsporangia is said to be perfect.

10. Which statement is *not* true of fruits?
   a. They develop from ovaries.
   b. They may include other parts of the flower.
   c. A multiple fruit develops from several carpels of a single flower.
   d. They are produced only by angiosperms.
   e. A cherry is a simple fruit.

FOR STUDY

1. Mosses and ferns share a common trait that makes water droplets a necessity for sexual reproduction. What is this trait?

2. Ferns display a dominant sporophyte stage (with large fronds). Describe the major advance in anatomy that enables most ferns to grow much larger than mosses.

3. What features distinguish club mosses from horsetails? What features distinguish these groups from rhyniophytes and psilophytes? from ferns?

4. Suggest an explanation for the great success of the angiosperms in occupying terrestrial habitats.

5. Contrast simple leaves with complex leaves in terms of structure, evolutionary origin, and occurrence among plants.

6. In many locales large gymnosperms predominate over large angiosperms. Under what conditions might gymnosperms have the advantage, and why?

READINGS

Burnham, C. R. 1988. "The Restoration of the American Chestnut." *American Scientist,* vol. 76, pages 478–487. If you read the article on chestnut blight recommended in the previous chapter, you may be encouraged to see that there are methods, based on Mendelian genetics, to rebuild the native population of this important tree species.

Cox, P. A. 1993. "Water-Pollinated Plants." *Scientific American,* October. Some strategies for coping with a tricky environment.

Crosson, P. R. and N. J. Rosenberg. 1989. "Strategies for Agriculture." *Scientific American,* September. Discusses approaches to increasing yields for an expanding population, but points out that social and economic changes will also be required.

Friedman, W. E. 1990. "Double Fertilization in *Ephedra,* a Nonflowering Seed Plant: Its Bearing on the Origin of An-giosperms." *Science,* vol. 247, pages 951–954. A breakthrough in understanding the origin of the flowering plants.

Graham, L. E. 1985. "The Origin of the Life Cycle of Land Plants." *American Scientist,* vol. 73, pages 178–186. How plants made it to the terrestrial environment.

Heyler, D. and C. M. Poplin. 1988. "The Fossils of Montçeau-les-Mines." *Scientific American,* September. Plants and animals of the Carboniferous period, discovered in a rich fossil lode in central France. Presents a detailed picture of life in a time long past.

Hinman, C. W. 1986. "Potential New Crops." *Scientific American,* July. Prospects for food and other materials from such plants as jojoba, buffalo gourd, and others.

Niklas, K. J. 1986. "Computer-Simulated Plant Evolution." *Scientific American,* March. An examination of some hypotheses concerning plant evolution. The hypotheses were modeled on computers, generating testable suggestions. The article presumes no knowledge of computers.

Niklas, K. J. 1987. "Aerodynamics of Wind Pollination." *Scientific American,* July. Do the mechanics of pollination seem improbable to you? Wind-pollinated plants have many adaptations that favor successful pollination.

Raven, P. H., R. F. Evert and S. Eichhorn. 1991. *Biology of Plants,* 5th Edition. Worth, New York. An excellent general botany textbook.

Strange Animals in a Watery World
Beneath the waters of the Sea of Cortez a red and yellow coral frames a colorful flatworm—two families of animals most of us recognize only dimly.

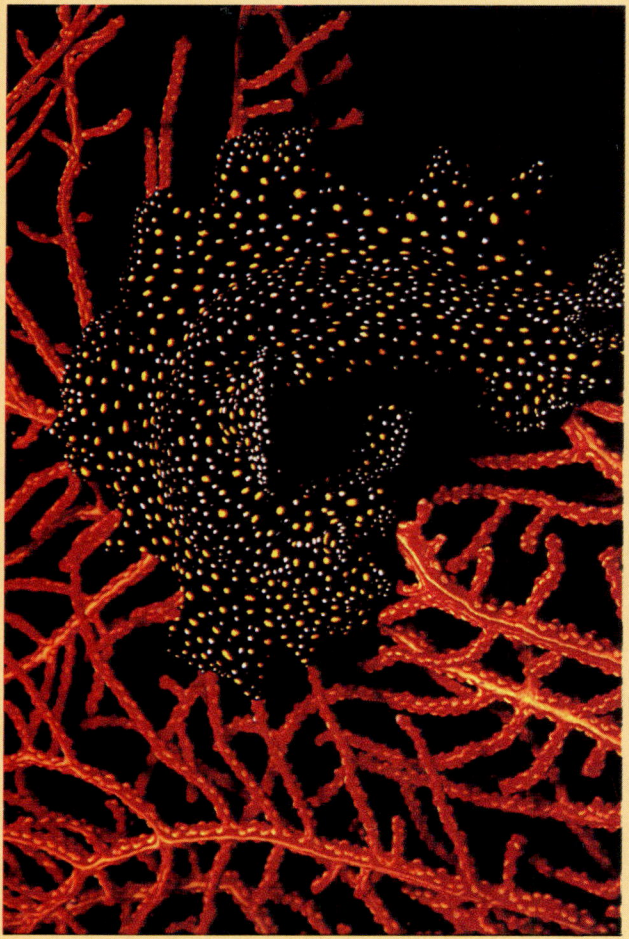

26

Sponges and Protostomate Animals

Even though animals—members of the kingdom Animalia—are among the most conspicuous living things in the world around us, and even though, as members of this kingdom, we have a special interest in its other members, the task of identifying all the animal species has barely begun. This taxonomic task is enormous because animals are so diverse. Most of the 30 million or so species of organisms are animals. Humans tend to be most aware of other large animals that share our terrestrial environment, but these are only a few of Earth's myriad animal species. A brief excursion into a coral reef would bring us into contact with animals of many other families—some of which we might not even recognize as animals.

Animals require a variety of complex organic molecules as sources of energy, and they obtain these molecules by active expenditure of energy. This energy is used either to move the animals through their environment or to cause the environment and the food it contains to move to the animals. The food animals eat includes most other members of the animal kingdom as well as members of all the other kingdoms. Much of the diversity of animal sizes and shapes evolved as animals acquired the ability to capture and eat many different kinds of food.

Because they lack rigid cell walls, animals cannot use high osmotic pressures to control the exchange of materials with their environments, as plants do. Animals have evolved other mechanisms for regulating their internal composition. The need to find food has given a strong selective advantage to structures that provide detailed information about the environment and to structures able to receive and coordinate this information. Consequently, most animals are behaviorally much more complex than are plants.

We devote two chapters to the kingdom Animalia, dividing animals into two major groups believed to represent lineages that have been evolving separately since the Cambrian period. The lineages differ fundamentally in their early embryological development. The word *protostomate* (Greek for "mouth first") in this chapter title and the word *deuterostomate* ("mouth second") in the next chapter title identify these developmental types. Sponges are also animals, but their developmental patterns differ from those of all other animals. The developmental patterns of sponges, protostomes, and deuterostomes will be described shortly.

Appreciation of animal structure and functioning can best be achieved through first-hand experience in the field and laboratory. The accounts in this chapter and the next, however, serve as an orientation to the major groups of animals, their similarities and differences, and the evolutionary pathways that resulted in the current number and variety of animal species.

Animals accomplish their diverse activities within the constraints of basic structural plans or designs. The entire animal, its organ systems, and the integrated functioning of its parts are known as its **body plan**. Body plans reflect and provide clues to the evolutionary history of animal lineages. Consequently, we use them as a way to organize our treatment of animal groups in this and the next chapter. As you will see, animals in many lineages evolved increasing body complexity, but many simple animals remain common today in a world with many complex species.

HOW ARE ANIMALS CLASSIFIED?

Biologists try to classify animals in a way that reflects their evolutionary relationships. Clues to these relationships are found in the fossil record, in patterns of embryological development, and in the comparative morphology and physiology of living and fossil animals. Reconstructing the phylogenies of animals is difficult because fossils of most animal phyla appear suddenly, near the beginning of the Cambrian period, and because animal evolution is replete with the convergence of traits. Therefore, biologists must look at a variety of traits in their attempts to reconstruct animal phylogenies.

Body Symmetry

A fundamental aspect of an animal's body plan is its overall shape, that is, its **symmetry**. A symmetrical animal can be divided along at least one plane into similar halves. Animals that have no plane of symmetry are said to be **asymmetrical** (Figure 26.1*a*). Many sponges are asymmetrical. Most animals, however, have some kind of symmetry. The simplest form is **spherical symmetry**, in which body parts radiate out from a central point. An infinite number of planes passing through the central point can divide a spher-

ically symmetrical organism into similar halves. Spherical symmetry is widespread among protists (for example, radiolarians; see Figure 23.4), but most animals possess other forms of symmetry.

A body with **radial symmetry** has one main axis around which body parts are arranged (Figure 26.1*b*). A perfectly radially symmetrical animal can be divided into similar halves by any plane that passes through the main axis. Some simple sponges and a few other animals have such symmetry, but most radially symmetrical animals are modified so that only two planes can divide them into similar halves. These animals are said to have **biradial symmetry**. Three animal phyla—Cnidaria, Ctenophora, and Echinodermata—are composed primarily of radially or biradially symmetrical animals.

A **bilaterally symmetrical** animal can be divided into mirror images only by a cut through the midline of its body from the front (anterior) to the back (posterior) (Figure 26.1*c*). The side of a bilaterally symmetrical animal without a mouth is its **dorsal** surface; the side with a mouth is its **ventral** surface. Bilateral symmetry is a common characteristic of animals that move through their environments. Such symmetry is strongly correlated with the development of sense organs and central nervous tissues at the anterior end of the animal, a process known as **cephalization**. Cephalization may have been selected for among motile animals because, in many of them, the anterior end is the part that encounters new environments first.

Developmental Pattern

The early development of embryos is often **evolutionarily conservative**: it evolves very slowly. For this reason, development reveals a great deal about evolutionary relationships among animals. Patterns of early development have been used extensively to identify major lineages of animals. During development from a single-celled zygote to a multicellular

(*a*) Asymmetrical

(*b*) Radial

(*c*) Bilateral

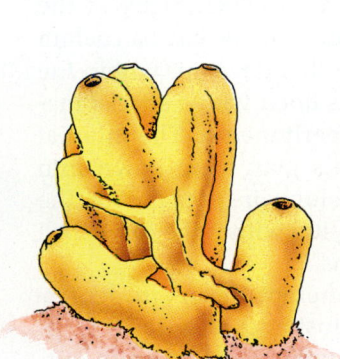

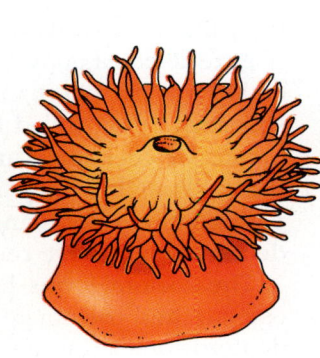

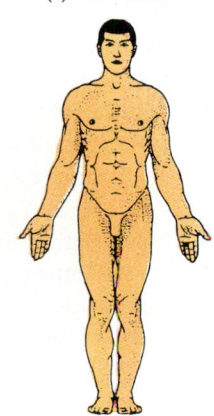

26.1 Body Symmetry
(*a*) A sponge is asymmetrical. (*b*) Radial symmetry: A sea anemone. (*c*) A human has bilateral body symmetry.

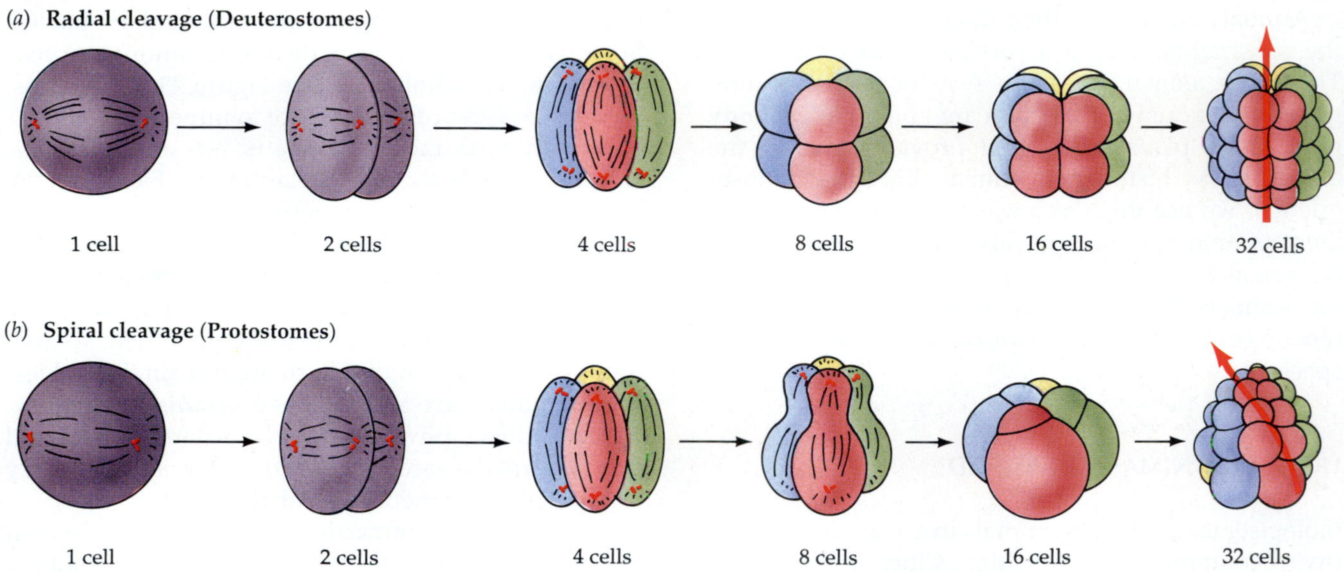

(a) Radial cleavage (Deuterostomes)

1 cell → 2 cells → 4 cells → 8 cells → 16 cells → 32 cells

(b) Spiral cleavage (Protostomes)

1 cell → 2 cells → 4 cells → 8 cells → 16 cells → 32 cells

26.2 Egg Cleavage Patterns
(a) Radial cleavage produces equal-sized cells that lie directly above one another. (b) In spiral cleavage the plane of division (last panel; red arrow) is oblique and the cells are arranged in a spiral. Because spiral cleavage is unequal, some cells are larger than others.

adult, animals form a number of cell layers (review Chapter 17, especially Figures 17.4 through 17.9). These layers behave as units during early embryonic development and give rise to different tissues and organs in the adult animal. The embryos of **diploblastic** animals have only two cell layers: the outer ectoderm and the inner endoderm. The embryos of **triploblastic** animals have a third layer, called mesoderm, which lies between the ectoderm and endoderm. Mesoderm derives in most species from the endoderm, although in a few species it derives from the ectoderm.

On the basis of differences in their early developmental patterns, animals other than sponges, cnidarians, and ctenophores can be divided into two major lineages. In the **protostome** lineage, cleavage of the fertilized egg is determinate; that is, if the egg is allowed to undergo a few cell divisions and the cells are then separated, each cell develops into only a partial embryo (which part depends on the cell's original position in the blastula). Cleavage of the fertilized egg in the other lineage, the **deuterostomes**, typically is **indeterminate**: cells separated after several cell divisions can still develop into complete embryos.

The pattern of egg cleavage is another trait that distinguishes protostomes from deuterostomes. In deuterostomes cleavage is radial: Cells divide along a plane either parallel or perpendicular to the long axis of the fertilized egg (Figure 26.2a). Cleavage in protostomes is spiral: The plane of division is oblique

to the long axis of the egg, causing the cells to be arranged in a spiral (Figure 26.2b). Finally, the mouth of the embryo of a deuterostome originates at some distance from the blastopore (see Chapter 17), which becomes the anus. Among protostomes, however, the mouth arises from the blastopore or very nearby.

Body Cavities

Animals can be divided into several groups that differ in the type of body cavity they have and its development. Body cavity formation is another important example of how embryology gives clues about evolutionary relationships among animals. The **acoelomates** lack an enclosed body cavity (Figure 26.3a). Their only internal cavity is the digestive cavity; the space between the gut and the body wall is filled with masses of cells that are collectively called mesenchyme. The tubular, fluid-filled cavity that houses the proboscis of a ribbon worm is called a **rhynchocoel** (see Figure 26.20). The **pseudocoelomates** have a **pseudocoel** (Figure 26.3b), a body cavity derived from the blastocoel (the first cavity formed inside a proliferating ball of embryonic cells). The pseudocoel provides a liquid-filled space in which many of the body organs float. **Coelomate** animals have a **coelom** (Figure 26.3c), a body cavity that develops within the embryonic mesoderm and is lined with a special mesodermal lining called the **peritoneum**. The internal organs of coelomate animals may hang down into the coelom, but they are slung in pouches of the peritoneum rather than floating within the cavity. All deuterostomes have coeloms.

The coelom forms by different means in deuterostomes and coelomate protostomes (Figure 26.4). In deuterostomes the coelom typically arises by an outpocketing of the embryonic gut. These pockets ulti-

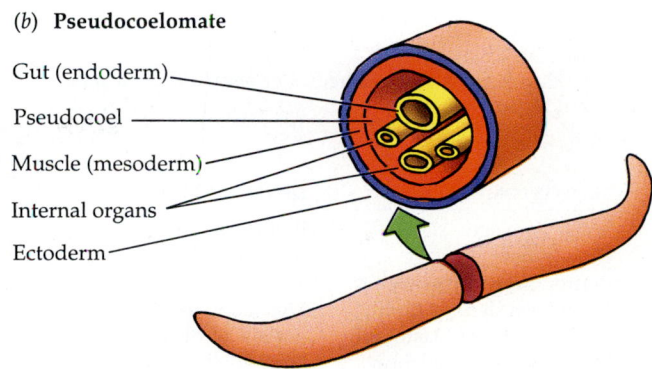

(a) **Acoelomate**

Gut (endoderm)
Mesenchyme
Muscle layer (mesoderm)
Ectoderm

(b) **Pseudocoelomate**

Gut (endoderm)
Pseudocoel
Muscle (mesoderm)
Internal organs
Ectoderm

(c) **Coelomate**

Gut (endoderm)
Internal organ
Peritoneum (mesoderm)
Coelom
Muscle (mesoderm)
Ectoderm

26.3   Animal Body Cavities
The three major types of animal body cavities differ in the types of cells lining them. Following the conventions adopted in Chapter 17, tissues derived from ectoderm are colored blue, those from mesoderm are pink, and those from endoderm are yellow. *(a)* Acoelomates such as flatworms do not have enclosed body cavities. *(b)* A roundworm is a pseudocoelomate animal. *(c)* Segmented worms (such as common earthworms) are coelomate.

mately pinch off and come to lie in the blastocoel. In coelomate protostomes, mesoderm develops within the embryo from a single cell, often near the blastopore, which migrates into the blastocoel. A split then forms in the mesoderm, creating the coelomic cavities. The early embryological differences between protostomes and deuterostomes, summarized in Ta-

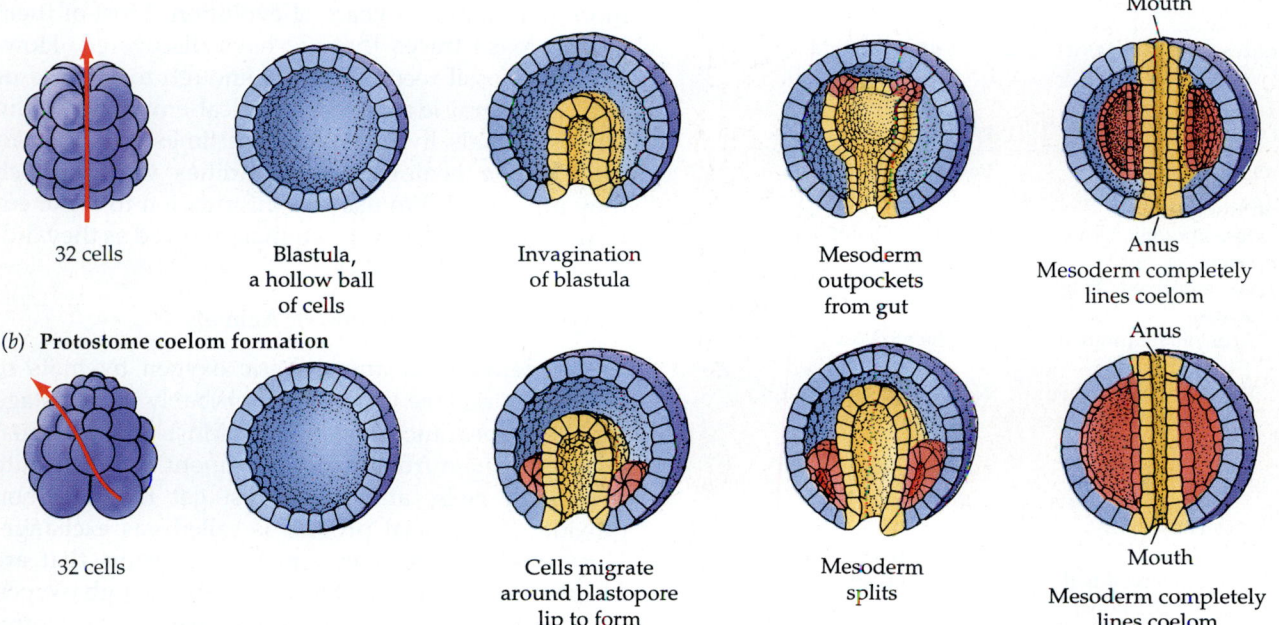

(a) **Deuterostome coelom formation**

32 cells | Blastula, a hollow ball of cells | Invagination of blastula | Mesoderm outpockets from gut | Mouth / Anus / Mesoderm completely lines coelom

(b) **Protostome coelom formation**

32 cells | | Cells migrate around blastopore lip to form mesoderm | Mesoderm splits | Anus / Mouth / Mesoderm completely lines coelom

26.4   Coeloms Form by Pocketing or Splitting
These sequences continue the developmental processes shown in Figure 26.2 to the stage of coelom formation. *(a)* Among deuterostomes the coelom typically arises by outpocketing. *(b)* The coelom of coelomate protostomes arises from a split in the mesoderm.

TABLE 26.1
Developmental Differences between Protostomes and Deuterostomes

PROTOSTOMES	DEUTEROSTOMES
Spiral cleavage	Radial cleavage
Determinate cleavage	Indeterminate cleavage
Blastopore becomes mouth	Blastopore becomes anus
Mesoderm derives from cells on lip of blastopore	Mesoderm derives from walls of developing gut
Mesoderm splits to form coelom	Mesoderm usually outpockets to form coelom

ble 26.1, indicate that they are lineages that separated early in the evolution of animals.

Body cavities are of great functional significance to animals, as we will soon see. The fundamental developmental patterns that produce different types of cavities are useful, together with other traits, for comparing the body plans and grades (levels of structural complexity) of animal phyla. An overview of the structural complexities of animals is given in Table 26.2. We also use these traits to help understand the way of life of animal groups and patterns of evolution within the phyla.

TABLE 26.2
Grades of Complexity among Animal Phyla

GRADE	PHYLA
Without true tissues	Porifera, Placozoa
With true tissues	
Two embryonic tissue layers	Cnidaria, Ctenophora
Three embryonic tissue layers	
Acoelomate (no body cavity)	Platyhelminthes
Rhynchocoel-bearing (tubular, fluid-filled cavity)	Nemertea
Pseudocoelomate (blastocoel persists as body cavity)	Loricifera, Nematoda, Rotifera
Coelomate (mesodermal body cavity develops)	Pogonophora, Annelida, Onychophora, Tardigrada, Chelicerata, Crustacea, Uniramia, Mollusca, Phoronida, Chaetognatha, Brachiopoda, Bryozoa, Echinodermata, Hemichordata, Chordata

THE ORIGINS OF ANIMALS

Animals probably arose evolutionarily from ancestral colonial protists as a result of division of labor among their aggregated cells. Division of labor probably evolved because an undifferentiated mass of cells can exchange materials with its environment only relatively slowly, and because some functions are performed better by specialized cells. Within the ancestral colonies of cells—perhaps similar to those still existing in *Volvox* and other colonial flagellated protists (see Figure 23.26a)—some cells became specialized for movement, others for nutrition, and still others differentiated into gametes. Once division of labor began, the units continued to differentiate, all the while improving their coordination with other working groups of cells. These coordinated groups of cells evolved into the larger and more-complex organisms that we now call animals.

Multicellular animals may have arisen from the protists at least three times. The sponges (phylum Porifera), cnidarians and ctenophores (phyla Cnidaria and Ctenophora), and flatworms (phylum Platyhelminthes) may represent three separate evolutionary lines. The small phylum Placozoa may also have evolved independently from protists. The other animal phyla probably evolved from a flatworm or flatwormlike ancestor. Possible evolutionary relationships among animals are shown in Figure 26.5. New information continues to modify and refine our understanding of these relationships.

Millions of species of animals have arisen during more than a billion years of evolution. Most of them left no fossil traces that we have discovered. However, the fossil record is good enough to provide us with a general idea of the physical environments in which animals lived at different times and to characterize the ecological communities within which they interacted. We use this information to make educated guesses as to why animals evolved as they did.

Oxygen and the Evolution of Animals

The generation of atmospheric oxygen by mats of cyanobacteria (see Chapter 18) probably set the stage for the appearance of animals. Most animals must take up oxygen from the environment and distribute it to their cells, and they must get rid of carbon dioxide; this crucial process is called **gas exchange**. Very small, unicellular, aquatic organisms that are adapted to stagnant water can obtain enough oxygen by simple diffusion even when oxygen concentrations are very low. Larger unicellular organisms, however, have lower surface-to-volume ratios. In order to obtain enough oxygen by simple diffusion, environmental concentrations of oxygen must be higher than those that can support the small prokar-

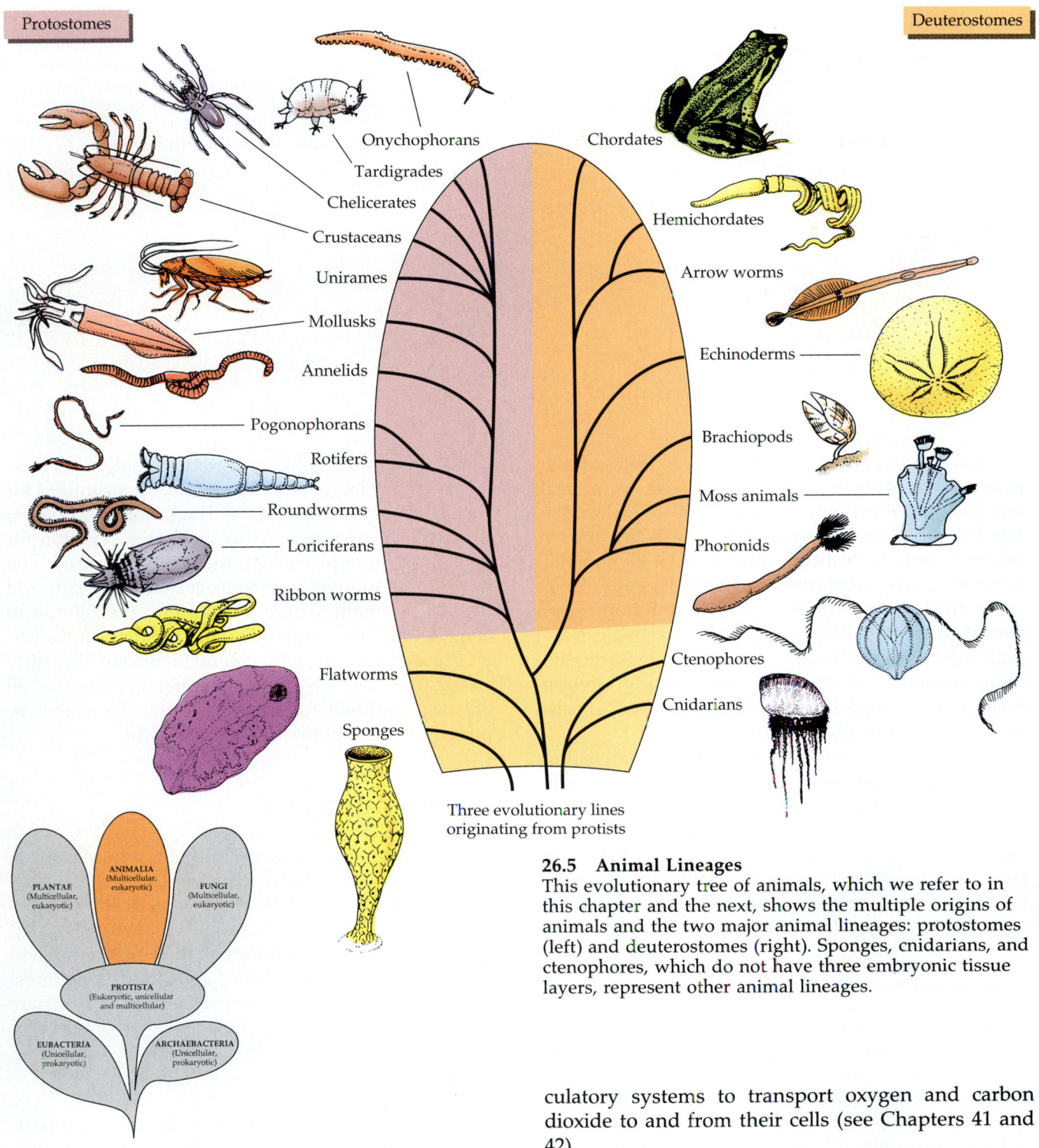

Protostomes

Deuterostomes

Onychophorans
Tardigrades
Chelicerates
Crustaceans
Unirames
Mollusks
Annelids
Pogonophorans
Rotifers
Roundworms
Loriciferans
Ribbon worms
Flatworms
Sponges

Chordates
Hemichordates
Arrow worms
Echinoderms
Brachiopods
Moss animals
Phoronids
Ctenophores
Cnidarians

Three evolutionary lines
originating from protists

ANIMALIA
(Multicellular,
eukaryotic)

PLANTAE
(Multicellular,
eukaryotic)

FUNGI
(Multicellular,
eukaryotic)

PROTISTA
(Eukaryotic, unicellular
and multicellular)

EUBACTERIA
(Unicellular,
prokaryotic)

ARCHAEBACTERIA
(Unicellular,
prokaryotic)

26.5 Animal Lineages
This evolutionary tree of animals, which we refer to in
this chapter and the next, shows the multiple origins of
animals and the two major animal lineages: protostomes
(left) and deuterostomes (right). Sponges, cnidarians, and
ctenophores, which do not have three embryonic tissue
layers, represent other animal lineages.

yotic cells. Whereas bacteria can thrive on 1 percent
of current atmospheric oxygen levels, large eukaryotic cells require oxygen levels that are at least 2 to 3
percent of current atmospheric concentrations. Small
multicellular animals with a high surface-to-volume
ratio can exchange gases through their body surfaces.
Larger animals have evolved specialized structures
such as gills and lungs to exchange gases, and cir-

culatory systems to transport oxygen and carbon
dioxide to and from their cells (see Chapters 41 and
42).

About 1,500 mya (million years ago), oxygen concentrations became high enough for large eukaryotic
cells to flourish and diversify (Figure 26.6). By 700
mya, protist communities were rich in species and
ecological types. Photosynthesizing species were
pursued by ciliated predatory forms. Different systems of photosynthesis were used in different environments, and a variety of compounds for storing
energy not needed for immediate metabolism had
evolved.

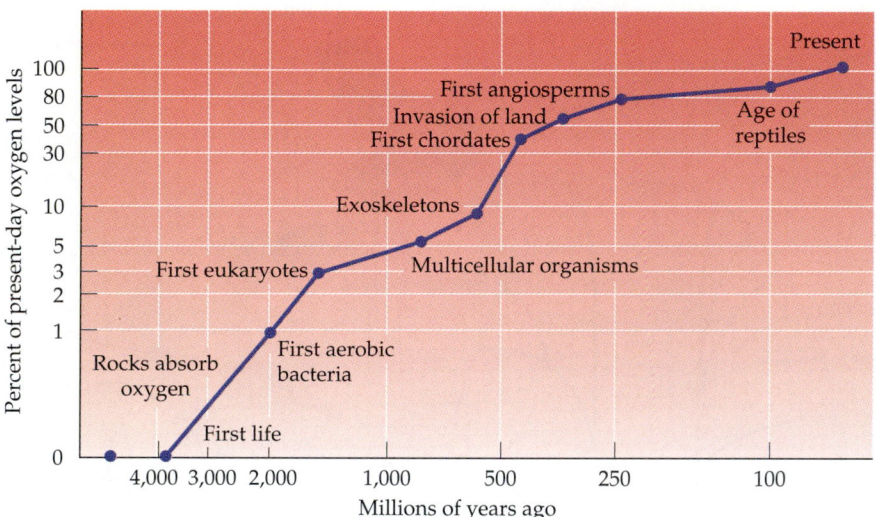

26.6 Large Cells Need More Oxygen
Hundreds of millions of years of evolution of life elapsed before atmospheric oxygen levels became high enough to support large eukaryotes and multicellular organisms. (Both axes are on logarithmic scales.)

Those early ecological communities were simpler than current ones in that all organisms were small and they did not engage in sexual reproduction. The fact that no part of a small organism is far from its external surface facilitates the transfer of materials from the environment to all parts of its body. However, small size makes it more difficult for an organism to prevent the movement of undesirable materials into its cells. In general, small organisms are more strongly influenced by the physical environment than are large animals. Small organisms are also highly vulnerable to larger predators.

Some 700 to 570 mya, it is likely that a combination of changes in the physical environment (especially oxygen concentrations) and interactions between predators and their prey favored increasing sizes of organisms. The steady increase in atmospheric concentrations of oxygen made it possible for larger animals to evolve because they could obtain enough oxygen to maintain the metabolism of all their cells.

Early Animal Evolution

When the first animals evolved, the mats of cyanobacteria that until then had been the primary photosynthesizers were being replaced by algae, both floating in the water and on the bottoms of shallow seas. Therefore, the primary food supplies available to the earliest animals were algae floating in the water (**phytoplankton**; animals floating in the water are called **zooplankton**), the extensive algal mats covering the shallow sea bottoms, protists, and bacteria.

The earliest animals were probably colonies of flagellated cells that fed on bacteria and protists. Some of these animals developed specialized cells with stinging tentacles that allowed the capture of larger prey. Others evolved morphological and behavioral adaptations for grazing in the algal mats. Both of these changes favored the evolution of larger animals

that could move about. Not surprisingly, the presence of these larger animals created opportunities for carnivores that fed on them. The predators, in turn, may well have generated in their prey selection for shells and other protection, the ability to burrow, the use of safe refuges such as caves and crevices, and faster movement. These themes in the evolution of animals continue today. For the moment, however, let's examine how they operated during the early evolution of the invertebrates, remembering that all life was confined to the seas during this extensive period of the evolution of life on Earth.

SIMPLE AGGREGATIONS

Animals probably arose from protists whose cells remained together after division, forming a multicellular colony. It is, in fact, difficult to distinguish between a protist colony and some simple multicellular animals that have little differentiation or coordination among their cells. The living animals that are most similar to the probable ancestral colonial protists are the sponges.

Phylum Porifera

The sponges (phylum Porifera, Latin for "pore bearers") are **sessile**: they live attached to the substrate and do not move about. All sponges, even large ones, have very simple body plans. The body of a sponge consists of a loose aggregation of cells built around a water canal system. A sponge has no mouth or digestive cavity, no muscles, and no nervous system. In fact, there are no organs in the usual sense of the word. A sponge is so loosely organized that even if it is strained through a filter and completely disassociated, its cells can reassociate into a new sponge.

Throughout most of its life an individual sponge

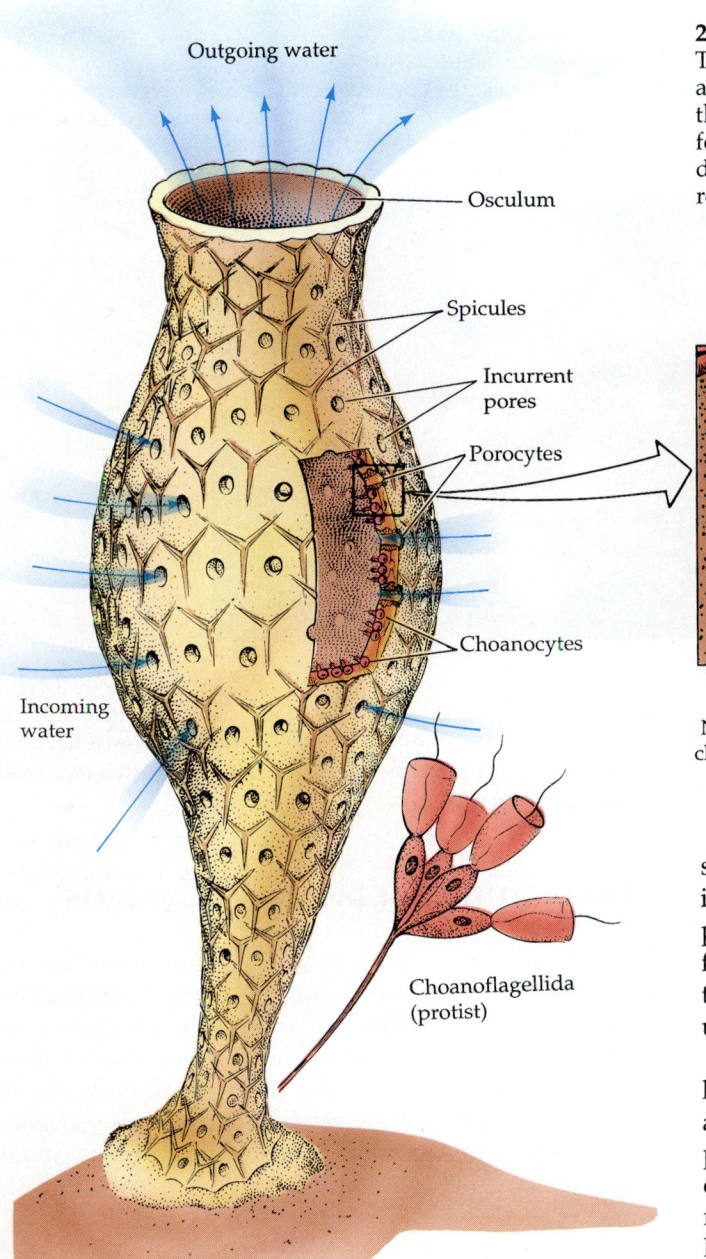

Outgoing water

Osculum

Spicules

Incurrent pores

Porocytes

Choanocytes

Incoming water

Choanoflagellida (protist)

26.7 Sponge Body Plan

The flow of water though the sponge is shown by blue arrows. The enlargement shows the detailed structure of the body wall studded with choanocytes—specialized feeding cells whose resemblance to cells of the protist order Choanoflagellida suggest there may an evolutionary relationship.

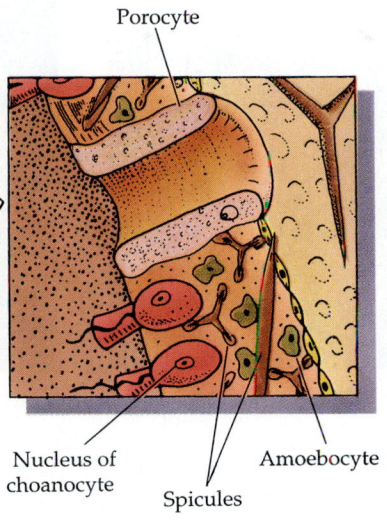

Porocyte

Nucleus of choanocyte

Spicules

Amoebocyte

special epidermal cells (in simple sponges) or through intercellular pores (in complex sponges). The water passes into small chambers within the body where food particles are captured by the choanocytes, and then exits through one or more larger openings (Figure 26.7).

Between the thin epidermis and the choanocytes lies a layer of cells, some of which are amoebalike and move about in the body. Also present is a supporting skeleton composed of the protein spongin, either in the form of spines or spicules or as an elastic network of spongin fibers. The spicular skeletons of larger sponges may be very complex, as is evident in the sponge skeletons you can buy in a hardware store.

Unlike most other animals, sponges have no distinct body tissues, no cavity between the layers of cells, and no recognizable body symmetry. Nonetheless, sponges come in a wide variety of sizes and shapes (Figure 26.8). The different ways that water moves have probably influenced the evolution of sponge morphologies, as have the advantages to sponges if they do not repeatedly filter the same water. Sponges living in intertidal or shallow subtidal environments, where they are subjected to strong wave action, hug the substrate. They spread laterally and have multiple pores scattered over their body surface. Many sponges living in calm waters are simple, with a single large opening on top of the body. Water is taken in through pores on the sides of the body and expelled upward through the large opening

is immobile, attached to the substrate. It feeds by drawing water into itself and filtering out the small organisms and nutrient particles that flow past the walls of its inner cavity. Unique feeding cells called **choanocytes** line the inside of the water canals; these cells have a collar consisting of cytoplasmic extensions that surround a flagellum. Similar cells are found among flagellates of the protist order Choanoflagellida, suggesting an evolutionary relationship between the two groups. Changes in body plans during sponge evolution led to better capture of suspended food from the water. The flagellated choanocytes set up water currents. Water flows into the animal either by way of small pores that perforate

26.8   Sponges Differ in Size and Shape
(a) Glass sponges (class Hexactinellida), such as this *Euplectella aspergillum*, are named for their glasslike spicules. (b) The forms of spicules are clear in this close-up of *Leucosolenia* (class Calcarea). (c) The brown volcano sponge (class Demospongiae) is typical of many simple marine sponges.

(a)

(b)

(c)

on top. Sponges living in flowing water do not need to exert much energy to move water through their bodies. Most of them are flattened and are oriented at right angles to the direction of current flow, so they intercept water as it flows past them. Water is drawn out of the top pores of such tall, thin sponges just as air is drawn out of a tall chimney by the wind.

Sponges reproduce both sexually and asexually. In most species a single individual produces both eggs and sperm. Water currents carry sperm from one individual to another. Asexual reproduction is by budding and other processes that produce fragments able to develop into new sponges. Most of the 10,000 species of sponges are marine animals, but about 50 species are found in fresh water.

Phylum Placozoa

Placozoans are small animals, less than 3 millimeters in diameter. Their bodies consist of no more than a few thousand cells and only four cell types. Placozoans lack any kind of symmetry, and they have no body cavity or distinct tissues or organs (Figure 26.9). Their body plan is a flat plate consisting of two layers of flagellated cells that enclose a fluid-filled area with fibrous cells. For a long time biologists thought that placozoans were larvae of some type of sponge or cnidarian, but in the late 1960s placozoans were observed to achieve sexual maturity and produce gametes. Although most often observed in aquaria, placozoans are also widely distributed in shallow tropical ocean waters. They feed by taking bacteria or small protists into their own cells, or by secreting enzymes onto their protist prey, which are then digested outside the placozoan's body. The prey's re-

mains are then absorbed by the placozoan's cells. Only two species have been described in this unusual group.

THE EVOLUTION OF DIPLOBLASTIC ANIMALS

Animals in all phyla other than Porifera and Placozoa have distinct cell layers and symmetrical bodies. The members of two animal phyla, the Cnidaria and Ctenophora, have only two cell layers; that is, they are **diploblastic**. They have no enclosed body cavity and are radially symmetrical. These animals probably represent a lineage that evolved from protists independently of all other animals.

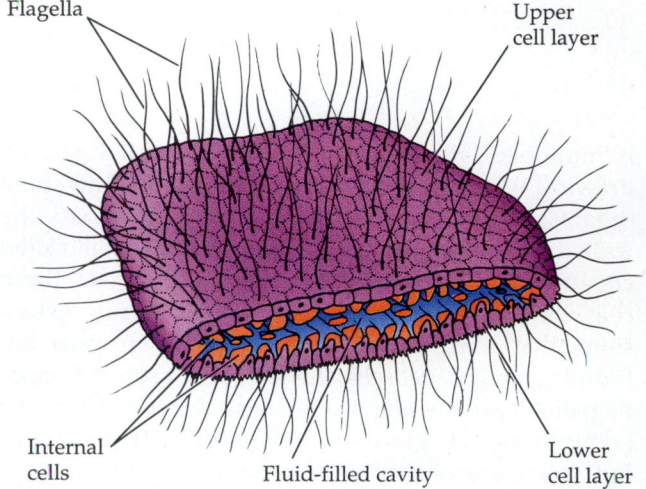

26.9   Placozoans Lack Symmetry
This cross section illustrates the simple structure of a placozoan.

Portuguese man-of-war

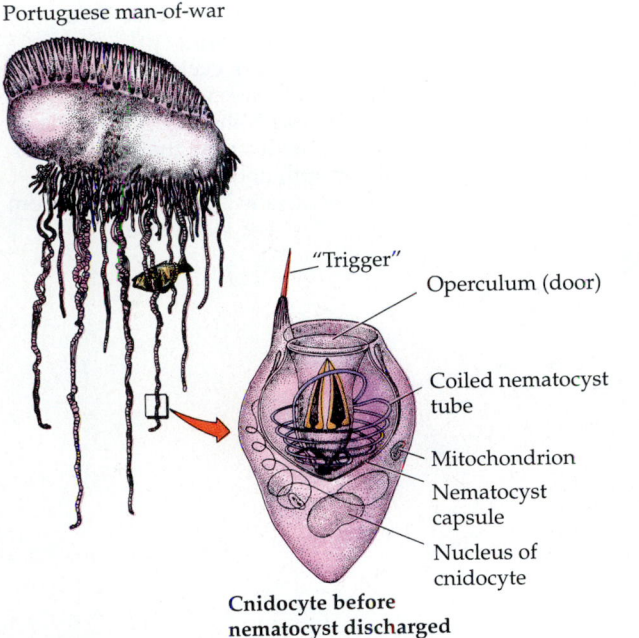

Cnidocyte before
nematocyst discharged

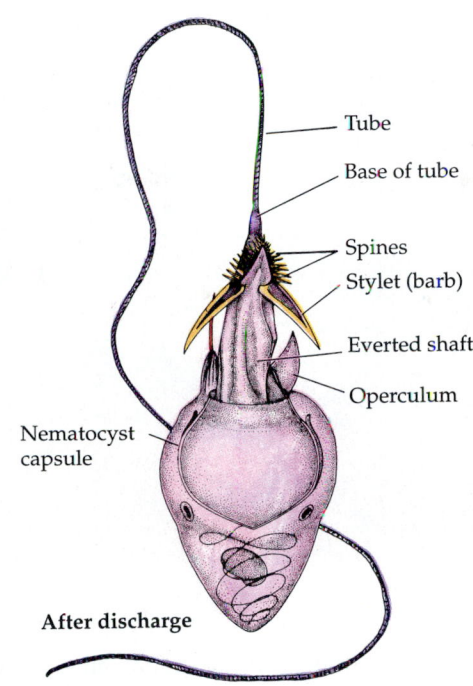

After discharge

26.10 Nematocysts Are Potent Weapons

Possessing a large number of nematocysts, cnidarians such as this Portuguese man-of-war can subdue and eat very large prey. Nematocysts remain coiled inside feeding cells called cnidocytes until their discharge is triggered by the presence of potential prey. Once discharged, stylets and spines on the nematocyst anchor it to the prey.

Phylum Cnidaria

Cnidarians appeared early in evolutionary history and radiated into many different species; they may have constituted more than half of the late Precambrian animal species. Modern cnidarian species—jellyfish, sea anemones, corals, and hydroids—number roughly 10,000. All but a few species are marine.

A key feature of cnidarians is their cnidocytes, specialized feeding cells containing stinging structures called **nematocysts** that can discharge toxins into their prey (Figure 26.10). We suspect that cnidocytes evolved through a symbiotic relationship with protists that lived within the bodies of the ancestors of cnidarians because some protists (for example, some dinoflagellates) have similar structures. Cnidocytes, which are borne on tentacles, allow cnidarians to capture large prey. The nematocysts paralyze and help hold prey, and are responsible for the sting that some jellyfish and other cnidarians can inflict on human swimmers. At the extreme, the tropical Pacific sea wasp (genus *Chironex*) and the Portuguese man-of-war (genus *Physalia*), can cause fatal injuries.

After being captured, a cnidarian's food is transported to its mouth by the tentacles. The mouth is connected to a dead-end sac called the **gastrovascular cavity**, which functions in digestion, circulation, and gas exchange. The same opening serves as both mouth and anus in cnidarians. This digestive apparatus and the row of tentacles armed with nematocysts that surrounds the mouth enable the animal to capture and swallow much larger food particles than sponges can. Cnidarians also have epithelial cells with muscle fibers whose contractions enable the organism to move, and nerve nets that integrate body activities.

CNIDARIAN LIFE CYCLES. Most cnidarian species have two distinct stages in their life cycles (Figure 26.11). The **polyp** is a cylindrical stalk with the mouth and tentacles at one end opposite a site of attachment to the substrate. This stage is usually asexual; individual polyps may reproduce by budding, thereby forming a colony. The **medusa** (plural: medusae) is the familiar, free-swimming, sexual stage shaped like a bell or an umbrella. It typically floats with its mouth and tentacles facing downward. Medusae produce eggs and sperm and release them into the water. When fertilized, an egg develops into a free-swimming, ciliated larva called a **planula** (plural: planulae) that eventually settles to the bottom and transforms into a polyp.

The cnidarian body plan is radially symmetrical and has only two cell layers. A middle cell layer eventually develops from the ectoderm, but it never produces complex internal organs like those of the triploblastic animal phyla. Although the polyp and medusa appear very different, they actually share a similar body plan. A medusa is essentially a polyp without a stalk. Conversely, a polyp is a medusa that is attached to the substrate by a stalk. Most of the outward differences between polyps and medusae

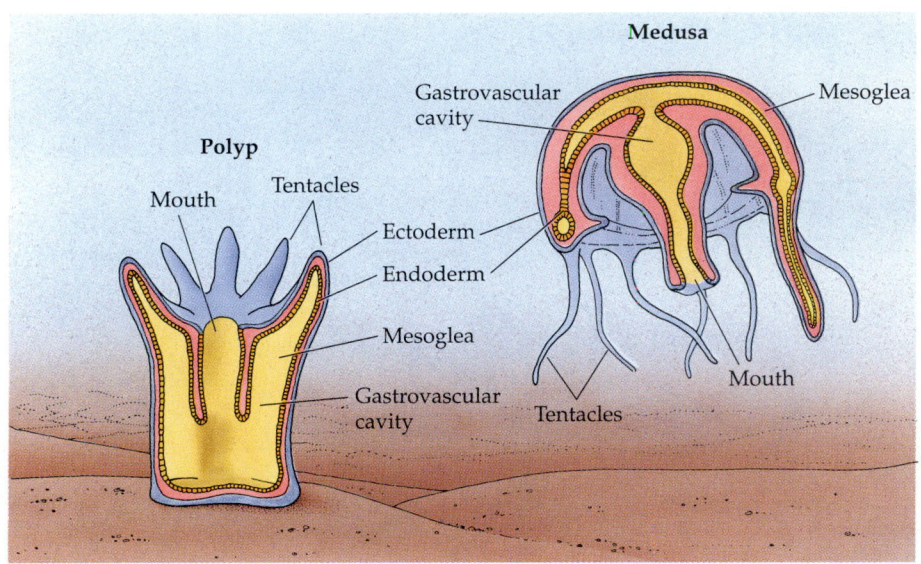

Polyp

Mouth

Tentacles

Ectoderm

Endoderm

Mesoglea

Gastrovascular
cavity

Medusa

Gastrovascular
cavity

Mesoglea

Mouth

Tentacles

**26.11   Cnidarians Have Two
Body Forms**
During the life cycle of a cnidarian,
the usually sessile, asexual polyp alternates with the free-swimming,
sexual medusa. As the positions of
the mouth and tentacles indicate,
the medusa is "upside down" from
the polyp—or vice versa.

are due to the **mesoglea,** a middle body layer composed of jellylike material that is largely devoid of
cells and has a very low metabolic rate. In polyps the
mesoglea is usually thin; in medusae it is very thick,
constituting the bulk of the animal.

HYDROZOANS. In the class Hydrozoa—the group containing the only freshwater cnidarians—the polyp
usually dominates the life cycle. A few species have

solitary polyps, but most hydrozoans are colonial. A
single planula eventually gives rise to many polyps,
all interconnected and sharing a continuous gastrovascular cavity (Figure 26.12). Within a colony, the
polyps often differentiate. Some polyps have tentacles with many nematocysts; they capture prey for
the colony. Others lack tentacles and are unable to
feed, but are specialized for the production of medusae. Still others are fingerlike and are adapted to

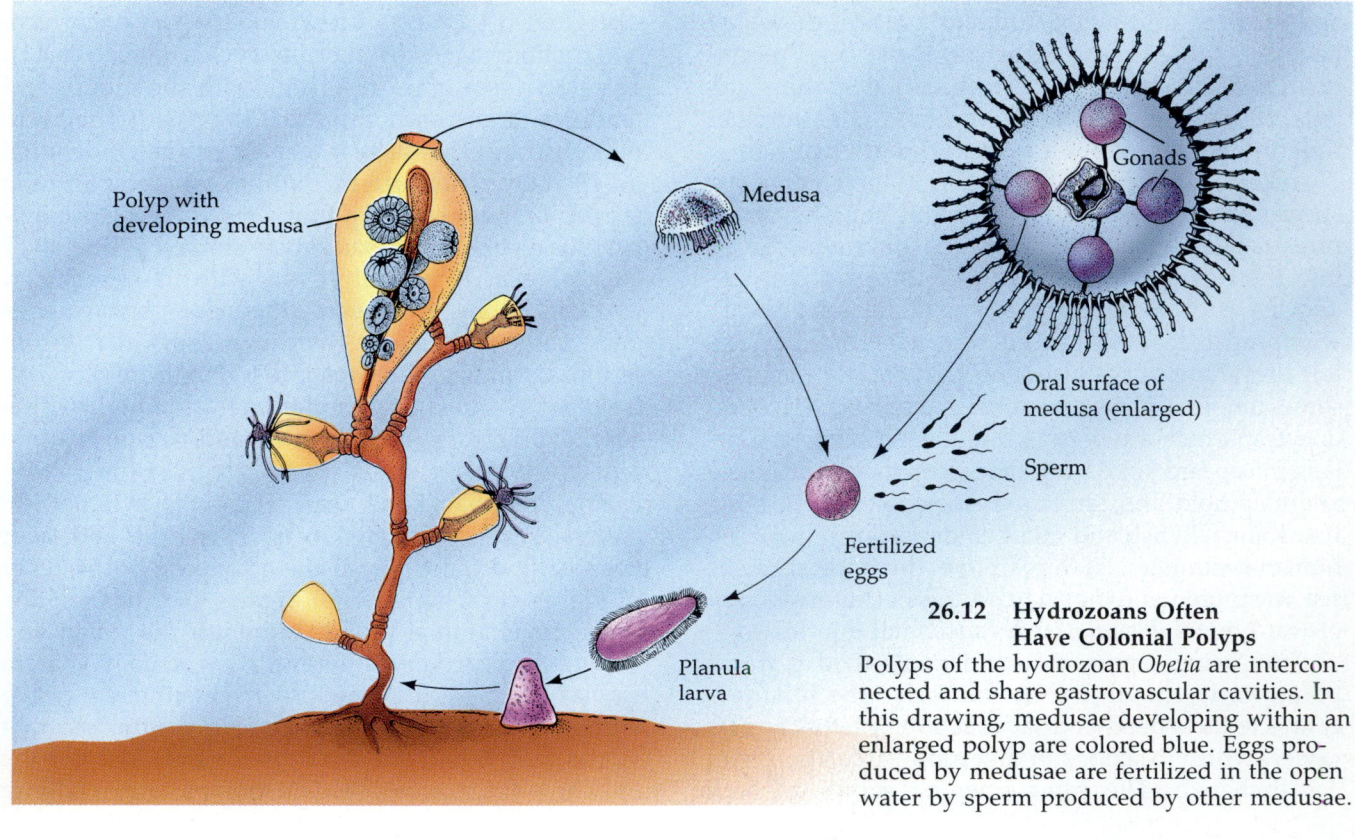

Polyp with
developing medusa

Medusa

Gonads

Oral surface of
medusa (enlarged)

Sperm

Fertilized
eggs

Planula
larva

**26.12   Hydrozoans Often
Have Colonial Polyps**
Polyps of the hydrozoan *Obelia* are interconnected and share gastrovascular cavities. In
this drawing, medusae developing within an
enlarged polyp are colored blue. Eggs produced by medusae are fertilized in the open
water by sperm produced by other medusae.

defend the colony. All of these polyp types, however, are ultimately derived from a single, sexually produced planula.

The siphonophores are free-floating hydrozoans in which medusae and polyps combine to form the most complex of cnidarian colonies. Individual medusae are modified for specific functions: to act as gas-filled floats, to move the colony through the water by jet propulsion, or to defend the colony. And, of course, there are also feeding polyps and reproductive medusae.

SCYPHOZOANS. The several hundred species of the class Scyphozoa are all marine. Some are as large as two and a half meters in diameter. The mesoglea of their medusae is very thick and firm, giving rise to their common name, jellyfish. The medusa typically has the form of an inverted cup, and the tentacles with nematocysts extend downward from the margin of the cup. Contraction of muscles ringing the cup margin expels water from the cup. When the muscles relax, the cup expands and again fills with water. This muscular contraction cycle allows scyphozoan medusae to actively swim through the water. Food captured by the tentacles is passed to the mouth and distributed to one of four gastric pouches, where enzymes begin digesting the food.

The medusa, rather than the polyp, dominates the life cycle of scyphozoans. Gonads (sex organs) develop in tissues close to the gastric pouches. An individual medusa is female or male, and releases eggs or sperm into the open sea. The fertilized egg develops into a small, heavily ciliated planula that quickly settles on a substrate and changes into a small polyp. This polyp feeds and grows and may produce additional polyps by budding. After a period of growth, the polyp begins to bud off small medusae by transverse division of its body column (Figure 26.13). These small medusae feed, grow, and transform into adult medusae. Thus a polyp that grows from a single fertilized egg is capable of producing many genetically identical medusae that will eventually reproduce sexually.

ANTHOZOANS. The roughly 6,000 species of sea anemones and corals that constitute the class Anthozoa are all marine. Unlike other cnidarians, the medusa has been completely eliminated from the anthozoan life cycle. The polyp produces eggs and sperm, and the fertilized egg develops into a planula that metamorphoses directly into another polyp. Many species can also reproduce asexually, by budding or fission. Like all other cnidarians, anthozoans are carnivores that capture prey with nematocyst-studded tentacles. However, the digestive cavity of anthozoans is more complex than that of other cnidarians. It is partitioned by sheets, called mesenteries, that increase

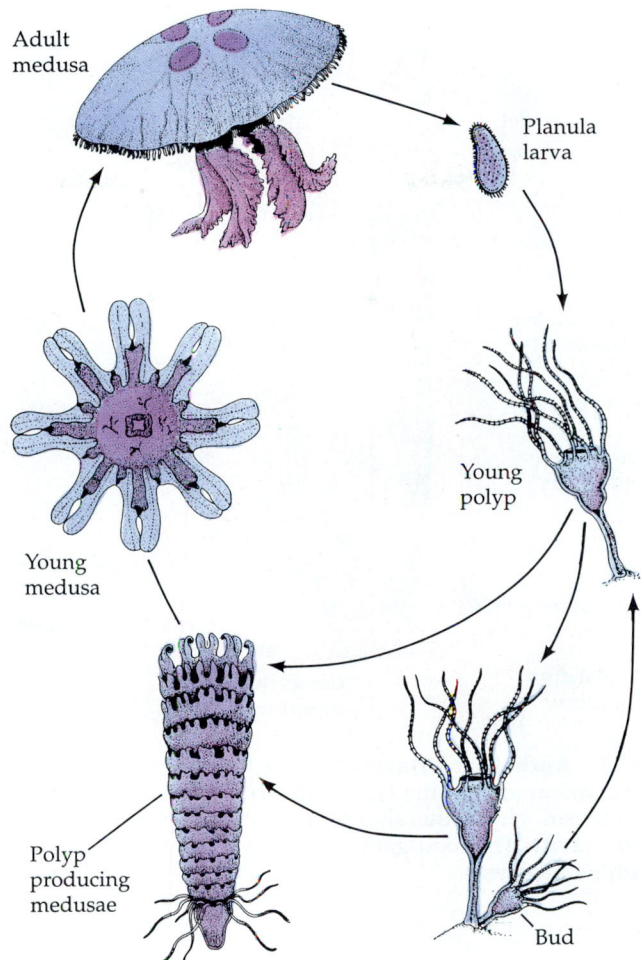

26.13 Medusae Dominate Scyphozoan Life Cycles
Scyphozoan medusae are the familiar jellyfish of coastal waters. The small polyps soon produce medusae.

the surface area available for secreting digestive enzymes and absorbing nutrients (Figure 26.14). Gonads also develop on these mesenteries.

Sea anemones are solitary anthozoans that lack specialized protective coverings. They are widespread in both warm and cold ocean waters. Many sea anemones are able to crawl slowly on their pedal discs; a few species can swim. Corals, by contrast, are usually sessile and colonial. The polyps of most corals secrete an organic matrix upon which calcium carbonate, the eventual skeleton of the colony, is deposited. The forms of coral skeletons are species-specific and highly diverse (Figure 26.15a). The common names of coral groups—horn corals, brain corals, staghorn corals, organ pipe corals, sea fans, sea whips, among others—accurately convey their appearances.

As a coral colony grows, old polyps die and leave their calcareous skeletons intact. The living members form a layer on top of a growing reef of skeletal remains. Reef-building corals are restricted to clear, warm waters. They are especially abundant in the

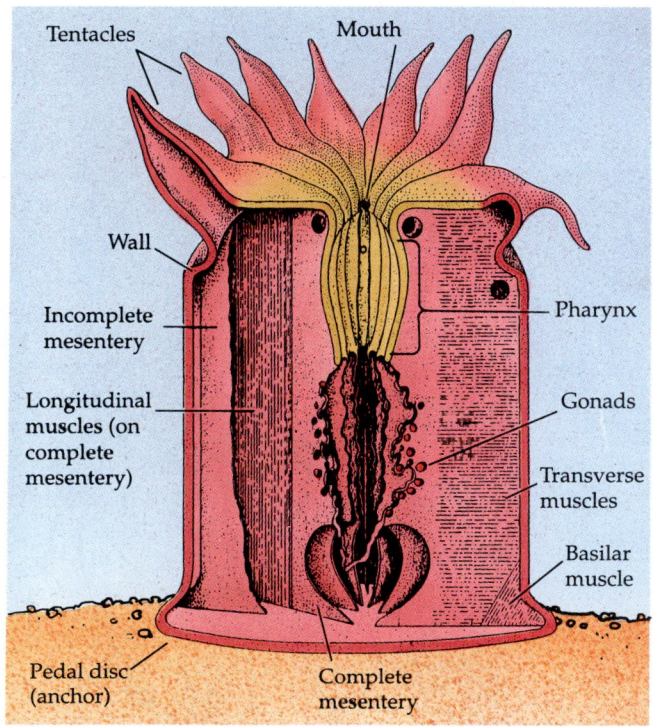

26.14 Anthozoans Have Complex Polyps
A sea anemone has the typical muscular structure of an anthozoan. Numerous sheets (mesenteries) partition the body cavity. The food-gathering tentacles are studded with nematocysts.

Indo-Pacific region, where they form chains of islands and reefs. The Great Barrier Reef along the northeast coast of Australia is a coral formation more than 2,000 km long and up to 150 km wide. A continuous coral reef hundreds of miles long in the Red Sea has been calculated to contain more material than all the buildings in the major cities of North America combined.

Corals flourish in nutrient-poor, clear, tropical waters. For a long time scientists wondered how corals obtained enough nutrients to grow as rapidly as they do. The answer is that highly modified dinoflagellates live symbiotically within the cells of the corals and, by their photosynthesis, provide carbohydrates to their hosts and help with calcium deposition. In turn, the dinoflagellates within the coral's tissues are protected from predators. This symbiotic relationship explains why reef-forming corals are restricted to surface waters, where light levels are high enough to allow photosynthesis (Figure 26.15b).

KEYS TO CNIDARIAN DIVERSITY. Cnidarians became the dominant marine organisms late in the Precambrian period, more than 600 mya, and they remain important components of marine ecological communities today (Figure 26.16). The cnidarian body plan combines a low metabolic rate with the ability to capture relatively large prey. The bulk of the body of medusae and many polyps is made up of the largely inert mesoglea. As a result, even a large cnidarian such as a sea anemone requires relatively little food and can fast for weeks or months.

(a)

(b)

26.15 Corals
(a) Many different species of corals grow together on this reef in Fiji. (b) The ends of the branches of staghorn coral are spread, maximizing the amount of sunlight the photosynthetic dinoflagellates living symbiotically inside their cells receive.

26.16 Diversity among Cnidarians
(a) The structure of the polyps of the North Atlantic coastal hydrozoan *Gonothyraea loveni* is visible here. (b) This sea nettle jellyfish, *Chrysaora melanaster*, illustrates the complexity of some scyphozoan medusae. (c) The nematocyst-studded tentacles of the stubby rose anemone *Tealia coriacea* (class Anthozoa) are poised to capture large prey carried to the animal by water movement.

(a)   (b)

(c)

Nematocysts allow cnidarians to subdue prey that are much more active and structurally more complex than the cnidarians themselves. Nonetheless, many cnidarians eat microscopic prey, and ancestral cnidarians (as well as some living species) probably consumed bacteria and protists. Many corals and other cnidarians have symbiotic associations with dinoflagellates that enable them to grow where food is scarce. These traits allow cnidarians to survive in environments where encounter rates with prey are much lower than required to sustain animals with higher metabolic needs. Cnidarians have persisted with few modifications for hundreds of millions of years.

Phylum Ctenophora

The ctenophores, or comb jellies, constitute the diploblastic phylum Ctenophora. Although the body plans of ctenophores and cnidarians are superficially similar, there are substantial differences. Both have two cell layers separated by a thick, gelatinous mesoglea, radial symmetry, and feeding tentacles. However, ctenophores (Figure 26.17) have a gut that opens through two anal pores opposite the mouth. These pores assist the mouth in voiding wastes from the gut. Ctenophores have eight rows of ciliated plates, called ctenes; they move by beating the cilia rather than by muscular contractions. Ctenophoran tentacles are solid and lack nematocysts; instead, the tentacles are covered with sticky filaments to which prey adhere. After capturing prey, the tentacles are retracted and bring the food to the mouth. In some species, the entire surface of the body is coated with a sticky mucus that captures prey. All of the 100 known species of ctenophores are carnivorous marine animals.

Most ctenophores cannot capture prey as large relative to their own bodies as can some cnidarians, but their sticky tentacles, which dangle in the water, accumulate large amounts of planktonic prey. Like cnidarians, ctenophores have low metabolic rates because they are composed primarily of inert mesoglea. They are common carnivores in the open sea, where prey are often scarce.

Ctenophore life cycles are simple. Gametes from gonads located on the walls of the gastrovascular cavity are liberated into the cavity and then discharged through the mouth or through pores. Fertilization takes place in the open ocean. In nearly all species the fertilized egg develops directly into a miniature ctenophore that gradually grows into an adult.

Anal pores

Gut

Ciliated plates
(ctenes)

Pharynx

Tentacle

Mouth

Sticky
filaments

(a)

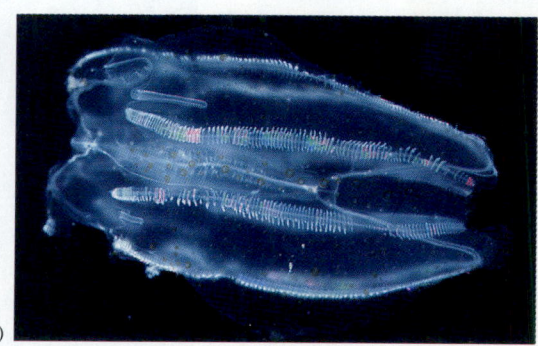

(b)

26.17   Comb Jellies Feed with Tentacles
(a) The sticky filaments to which prey adhere cover the long tentacles of this comb jelly. *(b)* The comb jelly *Leucothea* has much shorter tentacles.

THE EVOLUTION OF BILATERAL SYMMETRY

Bilateral symmetry probably first arose in simple organisms—flattened masses of cells crawling over a substrate, feeding as they went, just as placozoans do today. Sponges and placozoans are no more complex than the earliest of these simple animals probably were. An early lineage that may have evolved from such simple flat organisms is the phylum containing the flatworms.

Phylum Platyhelminthes

Flatworms (phylum Platyhelminthes) are triploblastic, bilaterally symmetrical animals whose internal organs, though simple, are more complex than those of cnidarians and ctenophores. Flatworms have no enclosed body cavity, they lack organs for transporting oxygen to internal tissues, and they have only simple organs for excreting metabolic wastes. This body plan dictates that each cell must be near a body surface in order to respire. The flattened body form makes such positioning of the cells possible. The digestive tract of a flatworm, if there is one, consists of a mouth opening into a dead-end sac. However, the sac is often highly branched, forming intricate patterns that increase the surface area available for absorption of nutrients. All living flatworms feed on animal tissues, some as carnivores and parasites, others as scavengers. Motile flatworms glide over surfaces powered by broad layers of cilia. This form of movement is very slow, but it is sufficient for small scavenging animals.

The flatworms probably most similar to the ancestral forms are the turbellarians (class Turbellaria): small, free-living marine and freshwater animals (a few live in moist terrestrial habitats). Freshwater turbellarians of the genus *Dugesia*, better known as planarians, are the most familiar flatworms. At one end

they have a head with chemoreceptor organs, two simple eyes, and a tiny brain composed of anterior thickenings of the longitudinal nerve cords.

Although the earliest flatworms were free-living, the flatworm body plan was readily adapted to a parasitic existence (Figure 26.18). A likely evolutionary transition was from feeding on carrion, to invading the bodies of dying hosts, to invading and consuming parts of living, healthy hosts or absorbing nutrients through the body surface. Most of the 25,000 species of living flatworms—the tapeworms (class Cestoda) and flukes (class Trematoda)—are parasitic. These worms inhabit the bodies of many vertebrates and cause some serious human diseases. Other flukes (class Monogenea) are external parasites of fishes and other aquatic vertebrates.

Parasites live in nutrient-rich environments in which food is delivered to them, but they face other challenges. To complete their life cycles, parasites must overcome the defenses of their hosts. They die when their host dies, and they must disperse offspring to new hosts. Some parasitic flatworms void their eggs with the host's feces. Later, these eggs are ingested directly by other host individuals. Most parasitic species, however, have complex life cycles involving two or more hosts and several larval stages (Figure 26.19).

Complex life cycles probably evolve because they increase the likelihood that other host individuals will be infected. Individuals of one host species, for example, may carry parasites between individuals of two other host species. An intermediate host may provide a good feeding environment for the parasite, allowing it to develop and reproduce within an organism that is likely to be eaten by the primary host. These very advantages for the parasites, however, provide humans with opportunities to reduce infections by interrupting a transmission stage in the parasite's life cycle.

(a)

(b)

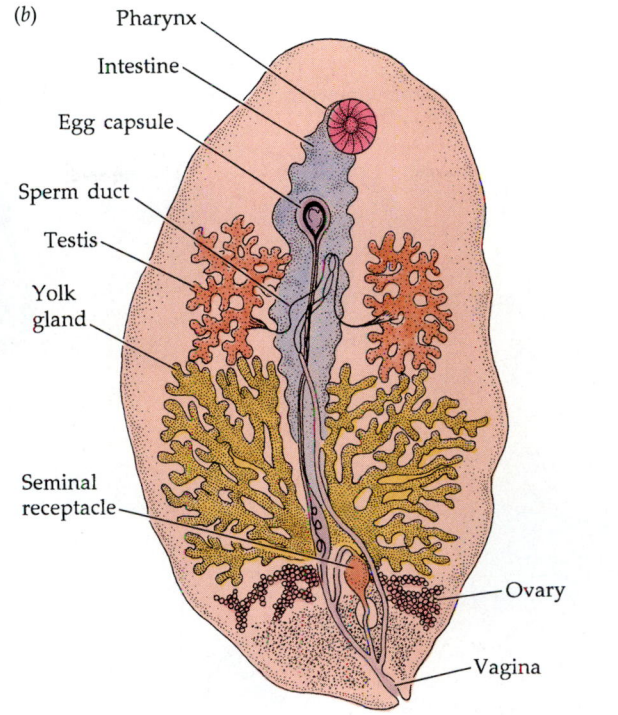

Pharynx

Intestine

Egg capsule

Sperm duct

Testis

Yolk gland

Seminal receptacle

Ovary

Vagina

26.18 Flatworms Live Freely and Parasitically
(a) *Prostheceraeus bellostriatus* is a free-living marine flatworm of the Pacific Coast of North America. (b) The flatworm *Syndesmis* lives parasitically in the gut of sea urchins. As is typical of internal parasites, its body is filled primarily with sexual organs. (c) The parasitic sheep liver fluke *Fasciola hepatica* is filled with highly branched large gonads.

Phylum Nemertea

Ribbon worms (phylum Nemertea) are triploblastic, dorsoventrally flattened animals with nervous and excretory systems similar to those of flatworms. Unlike flatworms, however, they have a complete digestive tract with a mouth at one end and an anus at the other. Food items move in one direction through the digestive tract of a ribbon worm and are acted upon by digestive enzymes—a more efficient system than one in which food remains must be ejected through the mouth.

In the body of almost all ribbon worms is a fluid-filled area called the rhynchocoel, within which floats a hollow, muscular proboscis. The proboscis is the feeding organ, and it may extend much of the length of the worm. Contraction of the muscles surrounding the rhynchocoel causes the proboscis to be ejected explosively through an anterior opening (Figure 26.20). Thus the rhynchocoel transmits forces that move the proboscis rapidly, although it does not move the rest of the animal. Small ribbon worms move by beating their cilia. Larger ones employ waves of contraction of body muscles to move on the surface of sediments or to burrow. Movement by both of these methods is slow.

The proboscis of most ribbon worms is armed with a stylet that pierces the prey. Paralysis-causing toxins produced by the proboscis are discharged into the wound made by the stylet. Some species lack a stylet and capture prey by wrapping the muscular proboscis around it. The proboscis is then withdrawn into the rhynchocoel by means of a retractor muscle, and the ribbon worm takes the prey into its mouth. In addition to capturing prey, the rhynchocoel helps the ribbon worm to burrow. The proboscis is pushed into the substrate and when it contracts and fattens, the

(c)

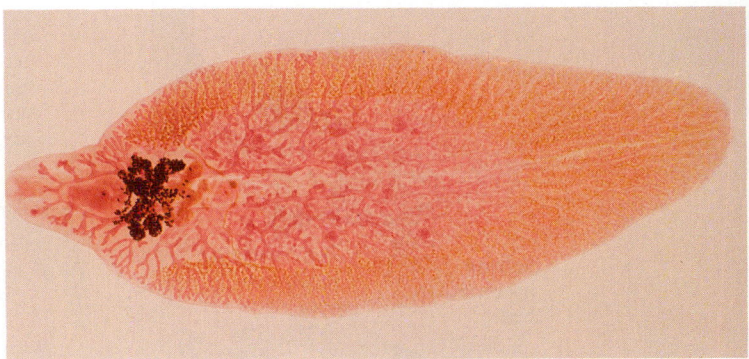

worm is pulled forward. Similar burrowing mechanisms are found in other phyla.

The approximately 900 species of ribbon worms are nearly all marine, but some are found in freshwater and a few live in moist terrestrial tropical environments. All ribbon worms are carnivores, feeding mostly on arthropods and small worms of several different phyla.

THE DEVELOPMENT OF BODY CAVITIES

Rapid movement is advantageous for prey and for the predators that pursue them. Fast-moving prey and predators evolved in the early Cambrian period (600 mya). Animals with fluid-filled body cavities can move more rapidly than animals lacking them can. Such cavities function as skeletons that transfer forces generated by contracting muscles from one part of the body to another (see Chapter 40).

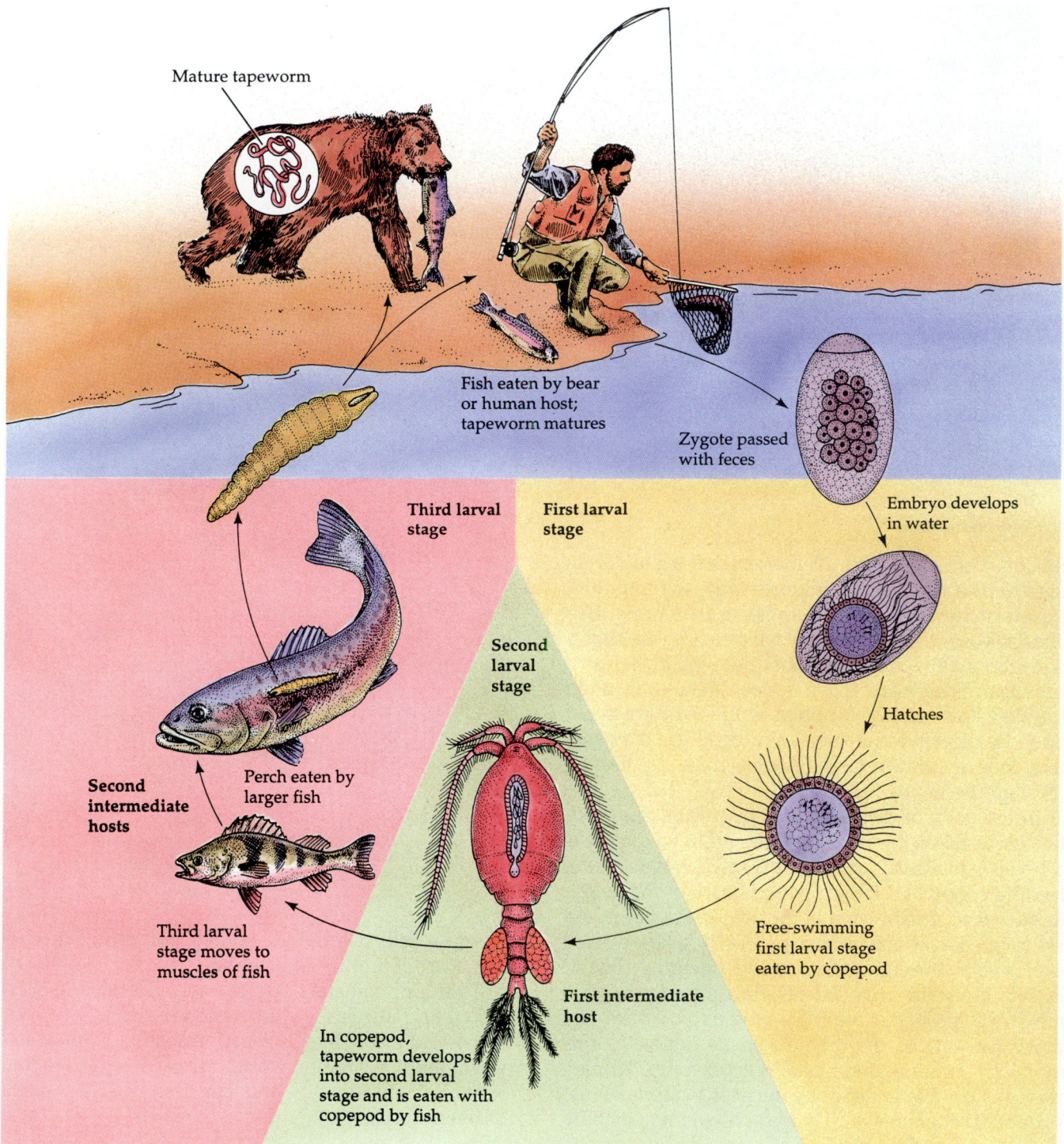

26.19   Returning to a Host by a Complex Route
The broad fish tapeworm *Diphyllobothrium latum* must
pass through the bodies of a copepod (a type of crustacean) and a fish before it can reinfect its primary host, a
mammal. Such complex life cycles assist the flatworm's
recolonization of hosts, but they also offer opportunities
for humans to break the cycle with hygienic measures.

Hydrostatic skeletons facilitate movement because
they are incompressible. When muscles surrounding
part of a fluid-filled body cavity contract, the fluid
must move to another part of the cavity. If the body
tissues around the cavity are flexible, fluids moving
from one region will cause the expansion of another
region. Fluids can thus move specific body parts or
even the whole animal, provided that temporary attachments can be made to the substrate. In the re-

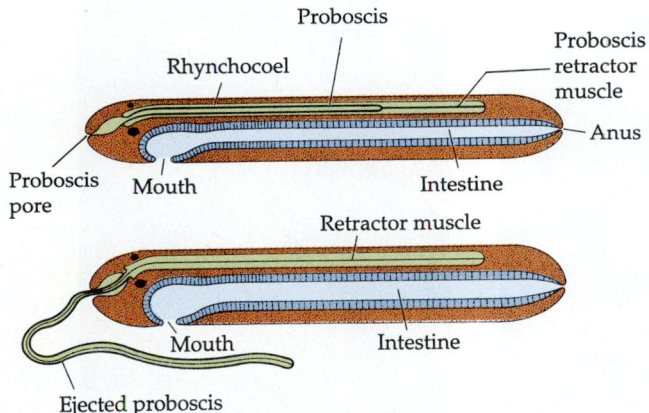

26.20 The Proboscis is the Feeding Organ of a Ribbon Worm
Floating in a cavity called the rhyncocoel (brown), the proboscis (green) can be moved rapidly. The worm, however, moves slowly.

maining animal phyla we will encounter many variations of fluid-filled body cavities used to change the shapes of organisms and move them around. Indeed, the types and numbers of body cavities provide a key to the lives of these animals and the degree to which they can control their shapes and movements.

Many lineages of wormlike (vermiform) animals evolved from flatwormlike ancestors because the flatworm body form enables animals to move efficiently through muddy and sandy marine sediments. Worms may be defined as bilaterally symmetrical, legless, soft-bodied animals that are at least several times longer than they are wide. By these criteria, the organisms of 16 animal phyla qualify as worms. Among these are the loriciferans, a new phylum (Loricifera) discovered in 1983. Living at great depths and clinging so tightly to ocean sediments that separating the animal from the sand is difficult, these tiny worms escaped human detection for centuries.

The evolutionary relationships of the vermiform phyla are obscure because their soft-bodied ancestors left few fossil traces. Therefore, we discuss these animals largely in terms of their body plans and lifestyles.

PSEUDOCOELOMATE ANIMALS

Early in the development of most animals a blastocoel develops within the blastula (see Chapter 17). In some animals the blastocoel persists into adulthood as a pseudocoel, a fluid-filled cavity in which body organs float (see Figure 26.3b). A pseudocoel can function as a hydrostatic skeleton. Of the many phyla of pseudocoelomate animals, we will discuss only the two that have many species: Nematoda and Rotifera.

Phylum Nematoda

Pseudocoelomate roundworms constitute the phylum Nematoda. A thick, multilayered cuticle secreted by the underlying epidermis gives the roundworm body its shape (Figure 26.21a). As a roundworm grows, it sheds and resecretes its cuticle four times. Because the cross-sectional shape of the body is round, a roundworm has a relatively small amount of body surface for exchanging oxygen and other materials with the environment. Exchange does take place through the cuticle, but it also takes place through the intestine, a layer that is only one cell thick. Materials are moved through the gut by rhythmic contraction of a highly muscular pharynx at the anterior end.

Roundworms are one of the most abundant and universally distributed of all animal groups. We unintentionally eat and drink enormous numbers of roundworms in our lifetimes. A single rotting apple from the ground of an orchard was examined and found to contain 90,000 roundworms. One square meter of mud taken from off the coast of the Netherlands yielded 4,420,000 individuals. The topsoil of rich farmland has up to 3 billion roundworms per acre. Countless numbers live as scavengers in the upper layers of the soil, in the bottoms of lakes and streams, and as parasites in the bodies of most kinds of plants and animals. The largest known roundworm, which reaches a length of nine meters, is found in the placenta of female sperm whales. About 20,000 species have been described, but the actual number of living species may exceed one million!

The diets of roundworms are as varied as their habits. Many roundworms live parasitically within their hosts. Many are predators, preying on protists and other small animals (including other roundworms). The role of some roundworms as parasites of people, cats, dogs, cows, sheep, and economically important plants has stimulated much research, mostly directed at controlling their proliferation.

The structure of parasitic roundworms does not differ much from that of free-living species (Figure 26.21a,b), but the life cycles of many parasitic species have evolved complexities that facilitate transfer of individuals among hosts. *Trichinella spiralis*, the parasite that causes the disease trichinosis, has a relatively simple life cycle. The larvae of *Trichinella* form cysts in the muscles of their mammalian hosts (Figure 26.21c). If they are present in great numbers, these cysts cause severe pain or death. *Trichinella* is transmitted to a mammal when it eats the flesh of an infected individual. The cysts are activated and the larvae leave them, bore through the host's intestinal wall, and are carried in the bloodstream to muscles, where they feed and again form cysts. Thus the intermediate host is likely to be another mammal (usu-

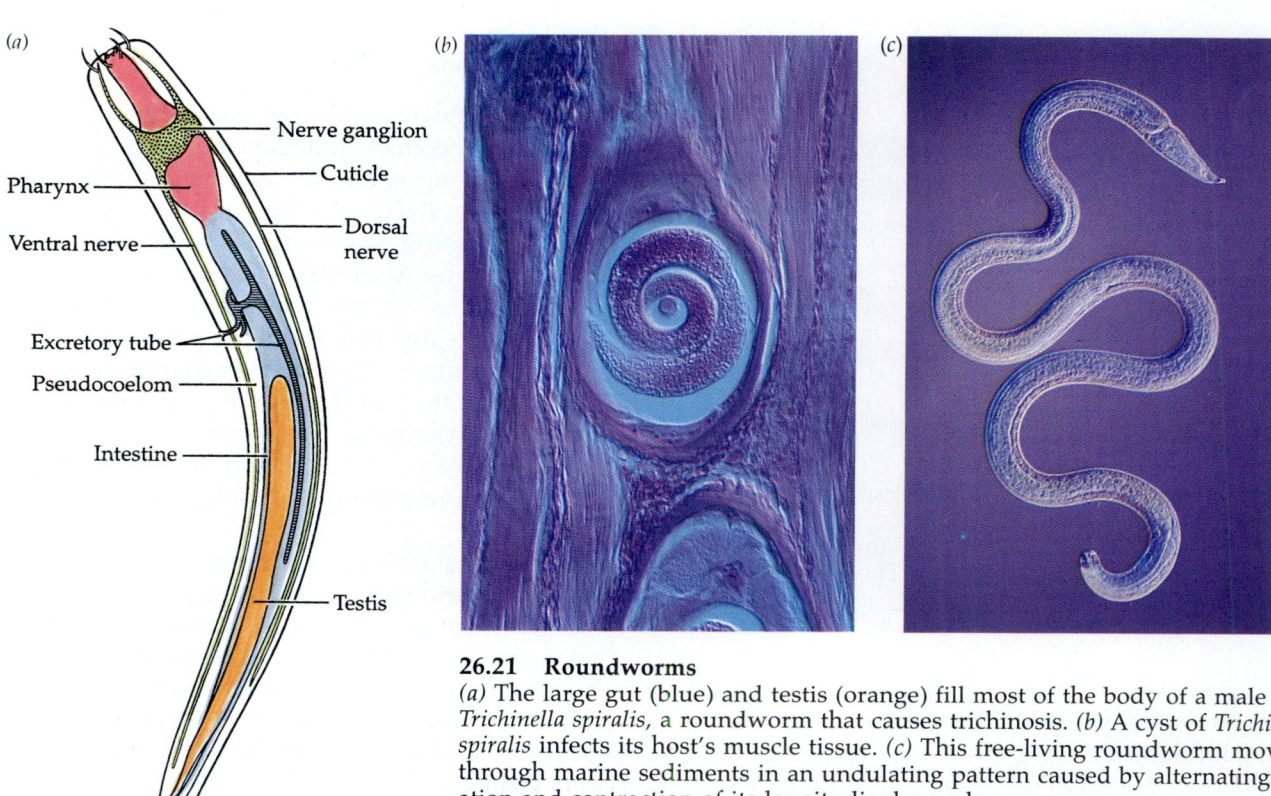

(a)

Nerve ganglion

Cuticle

Pharynx

Dorsal nerve

Ventral nerve

Excretory tube

Pseudocoelom

Intestine

Testis

Anus

(b)

(c)

26.21 Roundworms
(a) The large gut (blue) and testis (orange) fill most of the body of a male *Trichinella spiralis*, a roundworm that causes trichinosis. *(b)* A cyst of *Trichinella spiralis* infects its host's muscle tissue. *(c)* This free-living roundworm moves through marine sediments in an undulating pattern caused by alternating relaxation and contraction of its longitudinal muscles.

ally a pig in the case of human infections). No special stage in the *Trichinella* life cycle lives in an alternative host. Other roundworm life cycles, however, are as complex as that of flatworms such as the broad fish tapeworm (see Figure 26.19).

Phylum Rotifera

The second most abundant and widespread group of pseudocoelomate animals is the phylum Rotifera. Rotifers are triploblastic, bilateral, unsegmented animals. Most rotifers are tiny (50 to 500 μm long), smaller than some ciliate protists, but they have a highly developed organ structure (Figure 26.22). A complete gut passes from an anterior mouth to a posterior anus, and the pseudocoel functions as a hydrostatic skeleton. Most rotifers are active, but they propel themselves through the water by means of rapidly beating cilia rather than by muscular contraction. This type of movement is effective for them because rotifers are so small.

The most distinctive organs of rotifers are the ones used to collect and process food. A conspicuous,

ciliated organ (the corona) surmounts the head of many species. Coodinated beating of cilia provides the force for locomotion and also sweeps particles of organic matter from the water into the mouth and down to a complicated skeletal structure (the mastax), which has teeth that grind the food. By contracting the muscles surrounding the pseudocoel, a few rotifer species that prey on protists and small animals can evert the mastax through the mouth and seize small objects.

Some rotifers are marine, but most of the 1,800 known species live in fresh water. A small number, loosely referred to as terrestrial, rest on the surfaces of mosses and lichens in a desiccated, inactive state until it rains. When rain falls, they absorb water, become active, and swim about and feed in the films of water that temporarily cover the plants. Most rotifers are short-lived; typical life spans are between one and two weeks.

COELOMATE ANIMALS

Although the evolution of body cavities provided many new movement capabilities, control over body shape is crude if the cavity has muscles on only one side, as a pseudocoel does. A coelom, which is surrounded by muscles, allows better control over movements of the fluids it contains. Even with a coelom, however, control over movement is limited

(a)

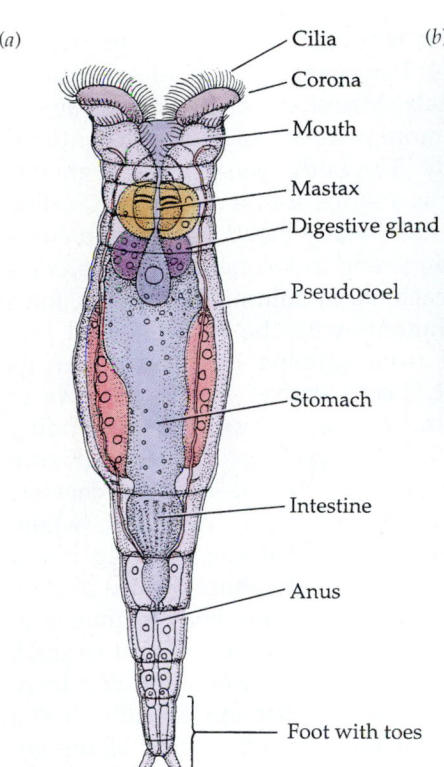

- Cilia
- Corona
- Mouth
- Mastax
- Digestive gland
- Pseudocoel
- Stomach
- Intestine
- Anus
- Foot with toes

(b)

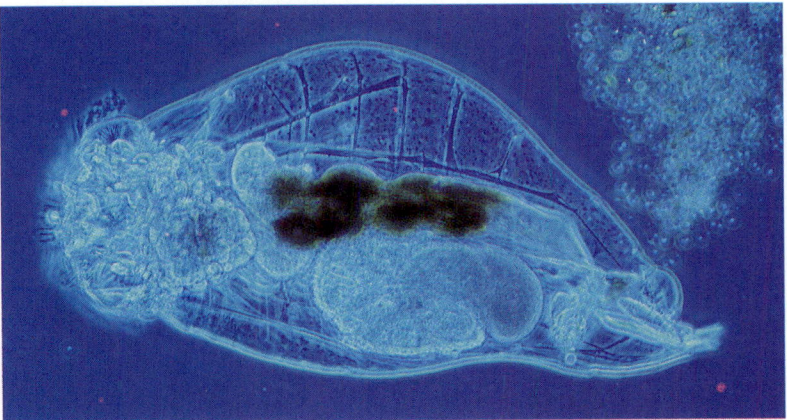

26.22 Rotifers
(a) Philodina roseola, a free-living rotifer, reflects the general structure of many species in this phylum. *(b)* The internal anatomy of the small rotifer *Epiphanes senta* is clear in this micrograph.

in an animal that has only a single large body cavity. If the cavity is separated into compartments, however, localized changes in shape produced by contractions of circular and longitudinal muscles in the individual segments are possible. Thus the animal can change the shape of each segment independently of the others. Segmentation of the coelom probably evolved a number of different times in both protostomes and deuterostomes.

Phylum Pogonophora: Losing the Gut

Many early protostomes were small animals with thin body coverings. Gases and waste products were routinely exchanged through the body wall. Marine waters and sediments contain abundant food particles in the form of bacteria and dissolved organic matter that also can be taken in directly through the body wall. One lineage of protostomes, the phylum Pogonophora, evolved into burrowing forms with a crown of tentacles through which gases are exchanged, and they entirely lost their digestive tracts. Pogonophorans were not discovered until this century because they live only in deep water, often thousands of meters below the surface (Figure 26.23). In these deep oceanic sediments, pogonophorans are abundant, reaching densities of many thousands per square meter. About 145 species have been described.

The coelom of a pogonophoran consists of an anterior compartment, into which the tentacles can be withdrawn, and a long, subdivided cavity extending much of the length of its body. Experiments using radioactively labeled molecules have shown that pogonophorans take up dissolved organic matter at high rates from sediments in which they live.

The largest and most remarkable pogonophorans reach up to 2 meters in length and live near deep-ocean hydrothermal vents—openings in the sea floor through which hot, sulfur-rich water pours. The tissues of these species are filled with bacteria that fix carbon by oxidizing hydrogen sulfide. The pogonophorans either consume these bacteria directly or live

26.23 Pogonophorans
The red bodies of these deep-sea-vent pogonophorans project from the long white tubes in which they live.

on their metabolic by-products, while maintaining the bacteria in optimal environments near the vents. Hydrothermal vent ecosystems, which are not based upon light as their energy source, have attracted considerable interest because they may be the type of environment in which life on Earth originated (see Chapter 18).

Phylum Annelida: Many Subdivisions of the Coelom

A subdivided body cavity allows an animal to alter the shape of its body in complex ways and to control its movements more precisely. The annelids (phylum Annelida), a diverse group of worms, have such a subdivided body cavity (Figure 26.24). The approximately 15,000 described species of annelids live in marine, freshwater, and terrestrial environments. Each body segment of a typical annelid has a pair of cavities. The bulging waves that move up and down the length of the body are made possible by its segmentation. A separate nerve center controls each segment, but the centers are connected by nerve cords that coordinate their functioning. The coelom in each segment is isolated from those in other segments. Because most annelids lack a rigid external protective covering, the flexible outer body wall plays a key role in locomotion. The thin body wall also serves as a general surface for gas exchange in most species. However, this thin, permeable body surface restricts annelids to moist environments because they lose body water rapidly.

POLYCHAETES. More than half of all annelid species are placed in the class Polychaeta. Nearly all polychaetes are marine animals. Most have one or more pairs of eyes and one or more pairs of tentacles at the anterior end of their body. The body wall in most segments extends laterally as a series of thin outgrowths, called **parapodia**, that have many blood vessels and function in gas exchange and in locomotion. Stiff bristles protrude from each parapodium; the bristles form temporary attachments with the substrate and prevent the animal from slipping backward when its muscles contract. Many species live in burrows in soft sediments and capture prey from surrounding water with elaborate feathery tentacles (Figure 26.25a). The sexes are separate in most polychaetes. Gametes are often released into the water, where fertilization occurs. A fertilized egg develops into a ciliated larva known as a **trochophore** (Figure 26.25b). As a trochophore feeds, it forms body segments at its rear end; it eventually metamorphoses into a small adult worm. In many ways the formation of a large complex organism from similar but slightly altered repeating segmental units is reminiscent of the colonial development of structure by siphonophores.

OLIGOCHAETES. More than 90 percent of the approximately 3,000 described species of oligochaetes (class Oligochaeta) live in freshwater or terrestrial habitats. Oligochaetes have no parapodia, eyes, or anterior tentacles, and they have relatively few bristles (setae). Earthworms—the most familiar oligochaetes— are scavengers and ingestors of soil from which they

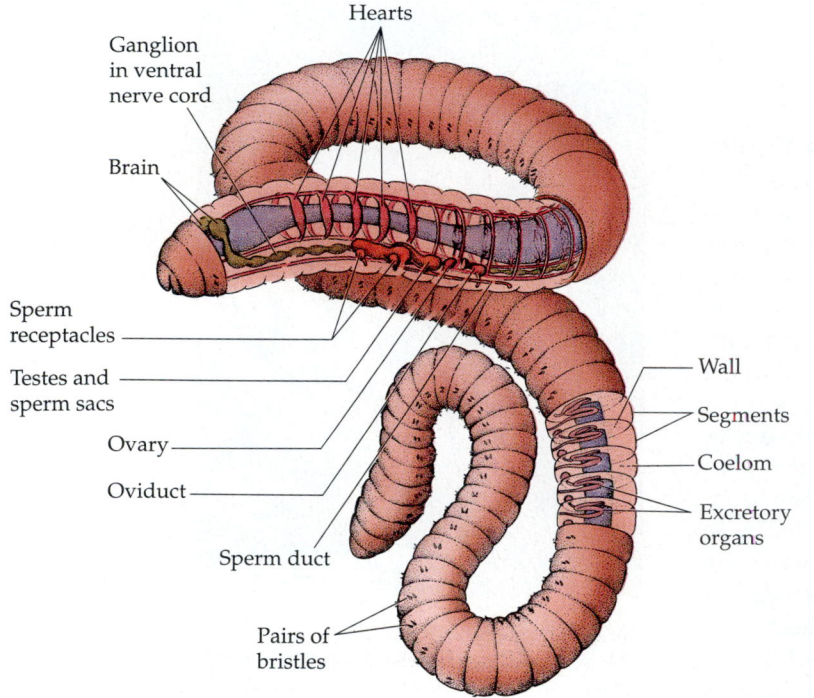

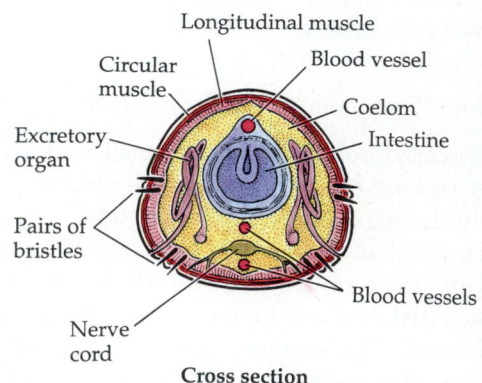

Cross section

26.24 Many Body Segments

The segmented structure of annelids is apparent both externally and internally. Most organs of an earthworm are highly segmented.

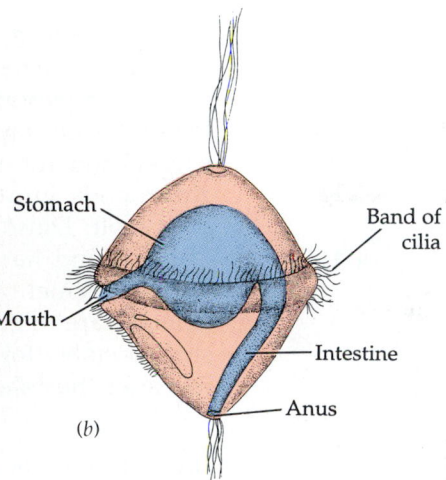

Stomach

Band of cilia

Mouth

Intestine

Anus

(a)

(b)

(c)

(d)

26.25 Annelids
(a) The Christmas tree worm (*Spirobranchus grandis*, class Polychaeta) of the Caribbean has striking feeding tentacles. (b) A trochophore larva. (c) Individual earthworms (*Lumbricus*, class Oligochaeta) are hermaphroditic. When they copulate, as they are doing here, each individual both donates and receives sperm. (d) This freshwater leech (class Hirudinea) has conspicuous anterior and posterior suckers.

extract food particles. Unlike polychaetes, all oligochaetes are **hermaphroditic**: each individual has both male and female sex organs. Sperm are exchanged simultaneously between two copulating individuals (Figure 26.25c). Eggs are laid in a cocoon outside the adult's body. The cocoon is shed and, when their development is complete, miniature worms emerge and begin independent life.

LEECHES. The leeches (class Hirudinea) probably evolved from oligochaetelike ancestors. Most species live in freshwater or terrestrial habitats and, like oligochaetes, lack parapodia and tentacles. Leeches are hermaphroditic; each individual serves as a sperm donor and a sperm recipient during copulation. Unlike that of other annelids, the coelom of leeches is not divided into compartments, and the coelomic space is largely filled with mesenchyme tissue. Therefore, the movement of leeches differs radically from that of other annelids. Groups of segments at each end of a leech are modified to form suckers, which serve as temporary anchors (Figure 26.25d). With its posterior anchor attached, the leech extends its body by contracting its circular muscles. The anterior sucker is then attached, the posterior one detached, and the leech shortens itself by contracting its longitudinal muscles.

Most leeches are external parasites of other animals. The mouth has three toothed jaws, with which the leech makes an incision in its host to feed on its blood. An anticoagulant secreted into the wound keeps the host's blood flowing. Leeches were formerly widely employed in medicine for bloodletting. Even today they are used to reduce fluid pressures in tissues damaged by, for example, a snake bite, and to eliminate pools of coagulated blood.

THE EVOLUTION OF EXTERNAL SKELETONS

Fluid-filled body cavities acted upon by surrounding muscles provide the skeletal support for most of the animals we have discussed so far. The body plans of these animals vary according to the size and extent of the cavities, how the cavities are lined, and whether they are subdivided. Most of these animals have external coverings that are relatively thin and flexible and allow gas exchange (although some, such as roundworms, have tough cuticles). In Precambrian times, the body covering in several flatwormlike lineages became thickened by incorporating of layers of protein and a strong, flexible, modified polysaccharide called **chitin**. Following this change, which was initially probably protective in function, the body covering acquired support and locomotor functions— it became an **exoskeleton**.

How can an animal with a rigid exoskeleton and no cilia move? One possibility is to have appendages that are moved by muscles. The lineages of animals that acquired such appendages evolved into the phyla collectively called **arthropods**. Arthropods can move rapidly because they evolved jointed appendages. The musculature of species with exoskeletons is adapted to the special needs of armored existence. Arthropod muscles operate particular segments of the body and the appendages attached to them (Figure 26.26a). Arthropod appendages serve many functions, including walking, swimming, food capture

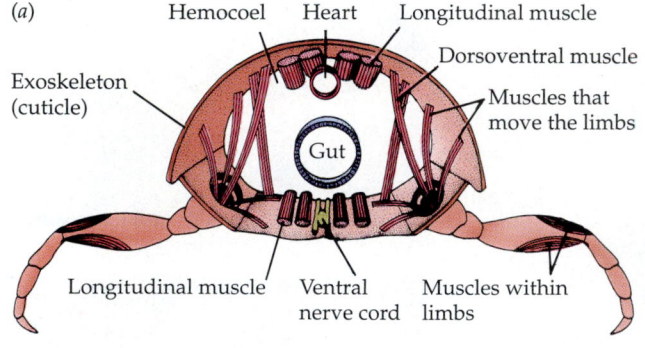

(a) Hemocoel   Heart   Longitudinal muscle
Dorsoventral muscle
Exoskeleton (cuticle)
Muscles that move the limbs
Gut
Longitudinal muscle   Ventral nerve cord   Muscles within limbs

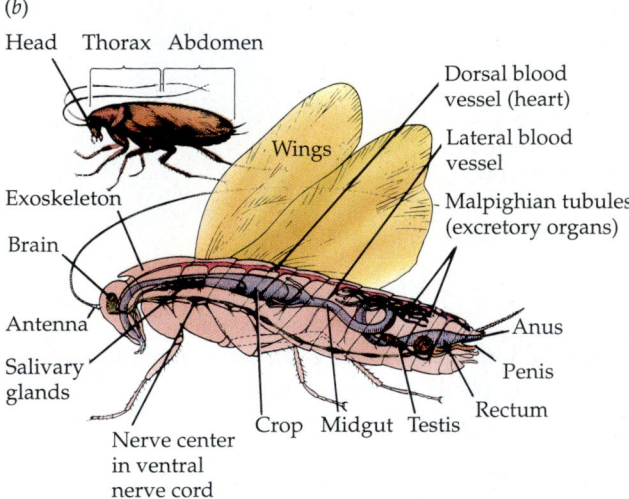

(b)
Head   Thorax   Abdomen
Dorsal blood vessel (heart)
Wings
Lateral blood vessel
Exoskeleton
Malpighian tubules (excretory organs)
Brain
Antenna
Anus
Salivary glands
Penis
Rectum
Crop   Midgut   Testis
Nerve center in ventral nerve cord

26.26   External and Jointed
(a) This cross section through a segment of a generalized individual shows the structure of an arthropod, which is characterized by an exoskeleton with jointed appendages. (b) The body plan of this insect differs in many details from that of other arthropods, but the basic theme of segmented body with modified appendages is general to most arthropod lineages.

and manipulation, copulation, and sensory perception (Figure 26.6b). This division of labor among body regions afforded a versatility to each individual arthropod species not found in its less-differentiated ancestors.

The development of an exoskeleton affected the animals that acquired it in other ways. First, the coelom lost its primary function as a hydrostatic skeleton. During the evolution of the dominant arthropod lineages, the coelom became much reduced. The major body cavity became the hemocoel, filled with blood that directly bathes the animal's organs. Second, because the exoskeleton slows gas exchange, arthropods evolved new means of taking up oxygen and releasing carbon dioxide. For example, many arthropods evolved gills, structures specialized for gas exchange.

Third, arthropods had to develop a new mechanism for growth because the rigid, nonliving exo-

skeleton prevented the animal from gradually increasing in size. The mechanism that evolved is **molting**: a periodic shedding of the exoskeleton followed by the rapid hardening of a new and larger exoskeleton formed from cells under the old one. Arthropods grow in spurts immediately following each molt. During this growth the new exoskeleton expands and hardens. While this process is underway, movement is difficult or impossible and the animals are highly vulnerable to predators. Soft-shelled crabs, for example, are arthropods captured just after they shed their old exoskeletons.

Diversification of the Arthropods

The evolution of an exoskeleton and its accessories had yet another profound influence on arthropod evolution. Encasement within armor does more than just protect an animal from predators. It also provides support for walking on dry land, and, if it has special waterproofing, it keeps the animal from drying out quickly in air. Arthropods were, in short, excellent candidates to invade the land, and they did so—repeatedly. Several lineages of arthropods colonized the land, but all of the other groups are overshadowed in numbers and diversity by the insects. The great majority of not only arthropods but of all animal species are insects.

Ancient Arthropod Lineages

A small phylum of about 75 species of inconspicuous tropical animals, the onychophorans (phylum Onychophora), are similar to what many people think were ancestors of the earliest arthropods (Figure 26.27a). Onychophorans are relatively soft bodied and they use their body cavities as hydrostatic skeletons. They are, in effect, worms with soft, fleshy, unjointed, claw-bearing legs formed by outgrowths of the body. Their tracheas are mostly unbranched. Their body is covered by a thin, flexible cuticle that contains chitin.

The water bears (phylum Tardigrada) are similar to onycophorans in many ways (Figure 26.27b). They also have fleshy, unjointed legs, and they use their fluid-filled body cavity as a hydrostatic skeleton. Unlike onychophorans, however, tardigrades are extremely small, and they lack a circulatory system and gas-exchange organs. The 550 extant species live in marine sands and gravels and on temporary water films on plants. When these films dry out, water bears also lose water and shrink to small barrel-shaped objects that can survive at least a decade in a dehydrated state. Onychophorans and water bears are probably not ancestors of other arthropod lineages, but they may be similar to many early arthropods.

(a)

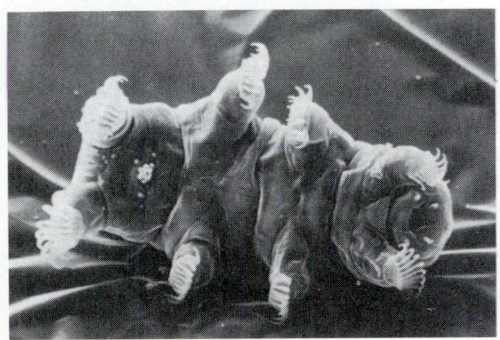

(b)

26.27 Arthropods with Unjointed Legs
(a) *Peripatus* (phylum Onychophora) has unjointed legs
and uses its body cavity as a hydrostatic skeleton. (b) The
appendages and general anatomy of water bears (phylum
Tardigrada) superficially resemble those of onychophorans.

A once-dominant line of arthropods, the trilobites
(phylum Trilobita) flourished in the seas of the Cambrian and Ordovician periods but were extinct by the
close of the Paleozoic era. Trilobites were heavily
armored and their body segmentation and appendages followed a relatively simple, repetitive plan (Figure 26.28). Why trilobites declined in abundance and
eventually became extinct is unknown.

26.28 A Trilobite
The relatively simple, repetitive segments of the now-extinct trilobites are illustrated here by fossils of *Dalmanites
limulurus*, a species that lived during the Silurian period.

Phylum Chelicerata

The bodies of all chelicerates (phylum Chelicerata)
are divided into two major regions, the anterior of
which bears two pairs of appendages modified to
form mouth parts and four pairs of walking legs. The
63,000 described species are usually placed in three
classes, only one of which contains many species.

  The pycnogonids (class Pycnogonida), or sea spiders, are a small group of marine species that are
seldom seen except by marine biologists (Figure
26.29a). The class Merostomata contains a single or-

(a)

(b)

26.29 Minor Chelicerates
(a) Although they are not spiders, it is easy to see why
sea spiders (class Pycnogonida) were given their common
name. (b) This spawning aggregation of horseshoe crabs
(class Merostomata) was photographed on a sandy beach
in Delaware.

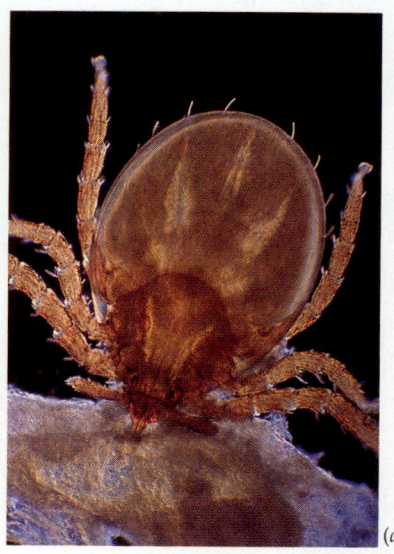

26.30 Arachnid Diversity
(a) Scorpions, such as *Uroctonus mondax* from California, are nocturnal predators. *(b)* Wolf spiders are active daytime predators with large eyes and good vision. *(c)* Harvestmen, often called daddy longlegs, are scavengers. *(d)* Ticks are blood-sucking external parasites on vertebrates. This wood tick, *Ixodes ricinus*, is piercing the skin of its human host.

der: the Xiphosura, or horseshoe crabs. These marine animals, which have changed very little during a long fossil history, have a large, horseshoe-shaped covering over most of the body. They are common in shallow waters along the eastern coasts of North America and Southeast Asia, where they scavenge and prey on bottom-dwelling invertebrates. Periodically they crawl into the intertidal zone to mate and lay eggs (Figure 26.29*b*).

The arachnids (class Arachnida) are relatives of horseshoe crabs that invaded the land, where they diversified into many species. Most arachnids have simple life cycles in which miniature adults hatch from eggs and begin independent lives almost immediately. Some species, however, have more complex life cycles. Others retain their eggs during development and give birth to live young. The most diverse and ecologically important groups are the scorpions, harvestmen, spiders, mites, and ticks (Figure 26.30).

Spiders have evolved into effective predators. Some have excellent vision that enables them to chase and seize their prey. Others spin elaborate silken webs to snare prey. The webs of different groups of spiders are strikingly varied and enable the spiders to position their snares in many different environments. Spiders also use silk to construct safety lines during climbing, and for homes, mating structures, protection for developing young, and dispersal. The threads are produced by modified abdominal appendages that are connected to internal glands that secrete the proteins of which silk is constructed.

Phylum Crustacea

The crustaceans (phylum Crustacea) are the dominant arthropods of the oceans. The members of one group alone, the copepods, are so numerous in plankton communities that they may well be the most abundant of all animals. Most of the 40,000 species of crustaceans have a body that is divided into three

SPONGES AND PROTOSTOMATE ANIMALS

(a)

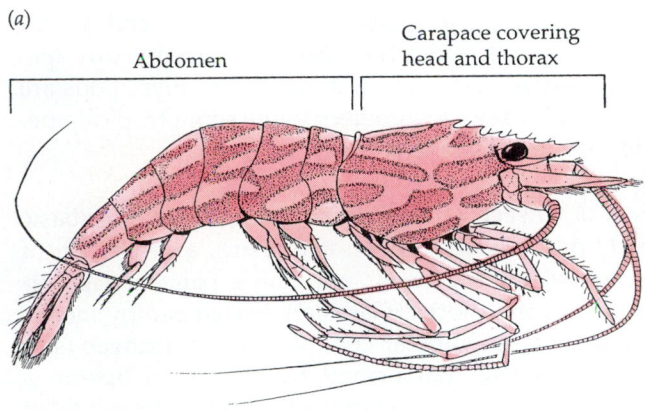

Abdomen | Carapace covering head and thorax

(b)

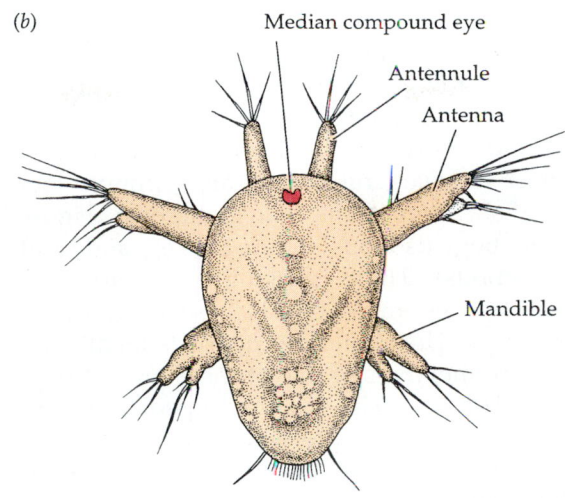

Median compound eye

Antennule

Antenna

Mandible

26.31 Crustacean Structure
(a) The bodies of crustaceans are divided into three regions, each of which bears appendages. *(b)* A nauplius larva has one compound eye and three pairs of appendages.

regions: head, thorax, and abdomen. The segments of the head are fused together, and the head bears five pairs of appendages. Each of the multiple thoracic and abdominal segments usually bears one pair. In many species, a fold of the exoskeleton, the carapace, extends dorsally and laterally back from the

head to cover and protect some of the other segments (Figure 26.31*a*).

The sexes are separate in nearly all crustaceans; males and females come together to copulate. The fertilized eggs of most crustacean species are attached to the outside of the female's body, where they are brooded during their early development. At hatching, the young of some species are released as larvae; those of other species are released as juveniles similar

(a)

(b)

26.32 Crustacean Diversity
(a) This California spiny lobster, *Panulirus interruptus*, is a decapod crustacean. *(b)* The sow bug *Ligia occidentalis* is an intertidal isopod found on the California coast. *(c)* *Euchaeta* is a typical planktonic copepod. *(d)* The appendages of this gooseneck barnacle (*Lepas anatifera*) are protruded from its shell and in the feeding position.

(c)

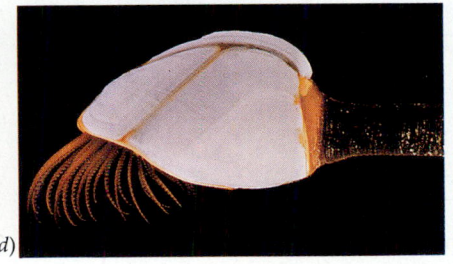

(d)

in form to the adults. Some species, however, release fertilized eggs freely into the water or attach the eggs to an object in the environment. The typical crustacean larva, called a **nauplius**, has three pairs of appendages and one compound eye (Figure 26.31*b*).

The most familiar crustaceans are shrimps, lobsters, crayfish, and crabs (all decapods; Figure 26.32*a*); sow bugs (isopods; Figure 26.32*b*); and sand fleas (amphipods). There is also a wide variety of other small species, many of which superficially resemble shrimps. The abundant copepods mentioned previously are members of one of these groups (Figure 26.32*c*). The barnacles (class Cirripedia) are unusual crustaceans that are sessile as adults (Figure 26.32*d*). With their unique calcareous shells, barnacles superficially resemble mollusks, but as the zoologist Louis Agassiz remarked more than a century ago, a barnacle is "nothing more than a little shrimp-like animal, standing on its head in a limestone house and kicking food into its mouth."

Phylum Uniramia

The body of a unirame (phylum Uniramia) is divided into two or three regions. The anterior regions have few segments, but the posterior region, the abdomen, has many segments. The phylum Uniramia consists of two subphyla of primarily terrestrial animals—Myriapoda and Insecta—which have elaborate systems of channels to bring oxygen to the cells of internal organs.

MYRIAPODS. Animals in the subphylum Myriapoda—centipedes, millipedes, and two other groups of inconspicuous animals—have two body regions, a head and a trunk. Centipedes and millipedes have well-formed heads and long, flexible, segmented trunks that bear many pairs of legs (Figure 26.33). Centipedes are predators of insects and other small animals. Millipedes are scavengers and plant eaters.

More than 3,000 species of centipedes and 10,000 species of millipedes have been described; many species remain unknown. Although most myriapods are less than a few centimeters long, some tropical species are ten times that long.

INSECTS. Insects (subphylum Insecta) have three basic body parts (head, thorax, abdomen), a single pair of antennae on the head, and three pairs of legs attached to the thorax. Gases are exchanged by means of air sacs and tubular channels called **tracheae** (singular: trachea) that extend from external openings inward to tissues throughout the body. The adults of most flying insects have two pairs of stiff, membranous wings attached to the thorax—except for flies, which have only one pair, and beetles, in which the forewings form heavy, hardened wing covers.

The one and a half million species of insects that have been described are believed to be only a small fraction of the total number living on Earth today (Figure 26.34). Insects are found in nearly all terrestrial and freshwater habitats, and they utilize as food nearly all species of plants and many species of animals. Some are internal parasites of plants and animals; others suck blood externally or consume surface body tissues. Insects effectively transmit many viral, bacterial, and protist diseases among plants and animals. Very few insect species live in the oceans. In freshwater environments, on the other hand, insects are sometimes the dominant animals, burrowing through substrates or extracting suspended prey from the water.

Most insects have the full number of adult segments when they hatch from their eggs, but species differ strikingly in their state of maturity at hatching and the processes by which they achieve adulthood. Wingless insects (class Apterygota), probably most similar to insect ancestors, have **simple development**. They hatch from the egg looking very similar to the adults, and they mature mostly by increasing

(a)

(b)

26.33 Myriapods
(*a*) Centipedes such as *Scolopendra heros* have powerful jaws for capturing active prey. (*b*) Millipedes, which are scavengers and plant eaters, have smaller jaws and legs.

(a)

(b)

(c)

(d)

(e)

(f)

(g)

(h)

26.34 A Diversity of Insects

(a) This firebrat, *Thermobia domestica* is a typical member of the apterygote order Thysanura. (b) Unlike most other insects, this adult mayfly (order Ephemeroptera) cannot fold its wings over its back. Representatives of some of the largest insect orders are (c) a broad-winged katydid (order Orthoptera), (d) a mating pair of harlequin bugs (order Hemiptera), (e) a predaceous diving beetle (order Coleoptera), (f) a swallowtail butterfly (order Lepidoptera), (g) a robber fly (order Diptera), and (h) a bumblebee (order Hymenoptera).

in size. Development in the winged insects (class Pterygota) is more complex. The hatchlings are less similar in form to adults, and they undergo substantial changes at each molt in the process of becoming an adult. The immature stages of insects between molts are called **instars**. If changes between its instars are gradual, an insect is said to have **incomplete metamorphosis**. If dramatic changes occur between one instar and the next, an insect is said to have **complete metamorphosis** (see Figures 17.12 and 17.13).

Entomologists, the scientists who study insects, divide the winged insects into about 28 different orders; Figure 26.34 shows some of them. We can make sense out of this bewildering variety by recognizing three major adaptive types. Members of one lineage cannot fold their wings back against the body. Although they are often excellent flyers, they require a great deal of open space in which to maneuver. The only surviving members of these insects are the orders Odonata (dragonflies and damselflies) and Ephemeroptera (mayflies). These insects all have aquatic larvae that metamorphose into flying adults after they crawl out of the water. Dragonflies and damselflies are active predators as adults, but adult mayflies lack functional digestive tracts and live only long enough to mate and lay eggs.

A second major evolutionary lineage includes the orders Orthoptera (grasshoppers, crickets, roaches, mantids, walking sticks), Isoptera (termites), Plecoptera (stone flies), Dermaptera (earwigs), Thysanoptera (thrips), Hemiptera (true bugs), and Homoptera (aphids, cicadas, and leafhoppers). These insects have incomplete metamorphosis. Hatchlings are sufficiently similar in form to adults to be recognizable. They acquire adult organ systems, such as wings and compound eyes, gradually through a number of juvenile instars.

Insects belonging to the third lineage have complete metamorphosis. With complete metamorphosis, different life stages are specialized for living in different environments and using different food sources. In many species the larvae are adapted for feeding and growing and the adults are specialized for reproduction and dispersal. The adults of some species do not feed at all, living only long enough to mate, disperse, and lay eggs. In many species whose adults do feed, the adults use different food resources than the larvae do. The most familiar example of complete metamorphosis is the caterpillar that changes into a butterfly, but other insects, including beetles and flies, have similar transformations. Their larvae are wormlike, with or without legs. The larvae transform into the adult form during a specialized "inactive" phase, the **pupa**, in which many larval tissues are broken down and the adult form develops. About 85 percent of all winged insects have complete metamorphosis. Familiar examples are the orders Neuroptera (lacewings and their relatives), Coleoptera (beetles), Trichoptera (caddisflies), Lepidoptera (butterflies and moths), Diptera (flies), and Hymenoptera (sawflies, bees, wasps, and ants).

Because they can fold their wings over their backs when not in use, insects belonging to the latter two adaptive types are able to fly from one place to another and then, upon landing, tuck their wings out of the way and crawl into crevices and other tight places. Several orders, including the Mallophaga (biting lice), Anopleura (sucking lice), and Siphonaptera (fleas), are parasitic. Although descended from flying insects, these orders have lost the ability to fly.

Why have the insects undergone such incredible evolutionary diversification? Insects may have originated from a centipedelike ancestor at least as far back as the Devonian period, more than 350 mya. With this early start, they were able to exploit the newly formed forests and other ancient forms of land vegetation. The terrestrial environments penetrated by insects were like a new planet, an ecological world comparable in size and with more complexity than the surrounding seas, but one with relatively few species of terrestrial animals. By Carboniferous times, a great diversity of insect types already swarmed over the land, and winged insects—the first animals to fly—had appeared. Thus it is not surprising that there are more species of insects than there are of all other animals put together.

CALCIFIED PROTECTION

Several animal lineages evolved calcified body coverings. These organisms, some of which we have already discussed, probably evolved from flatworm-like ancestors. The most spectacular calcified shells, however, evolved in the phylum Mollusca—the mollusks.

Phylum Mollusca

As a group, the mollusks (phylum Mollusca) underwent one of the most remarkable animal evolutionary radiations. The earliest known mollusks were worm-like animals with calcified spicules on their dorsal surface, a rasping feeding structure known as the **radula** (plural: radulae) at the anterior end, and gills at the posterior end (Figure 26.35). The ancient mollusks may have grazed on algae, much like some of their modern descendants. Selective pressures from predators are what probably favored further development of the protective spicules into a series of plates or a single, caplike shell, both of which provide better protection.

The remarkable evolutionary radiation of mollusks

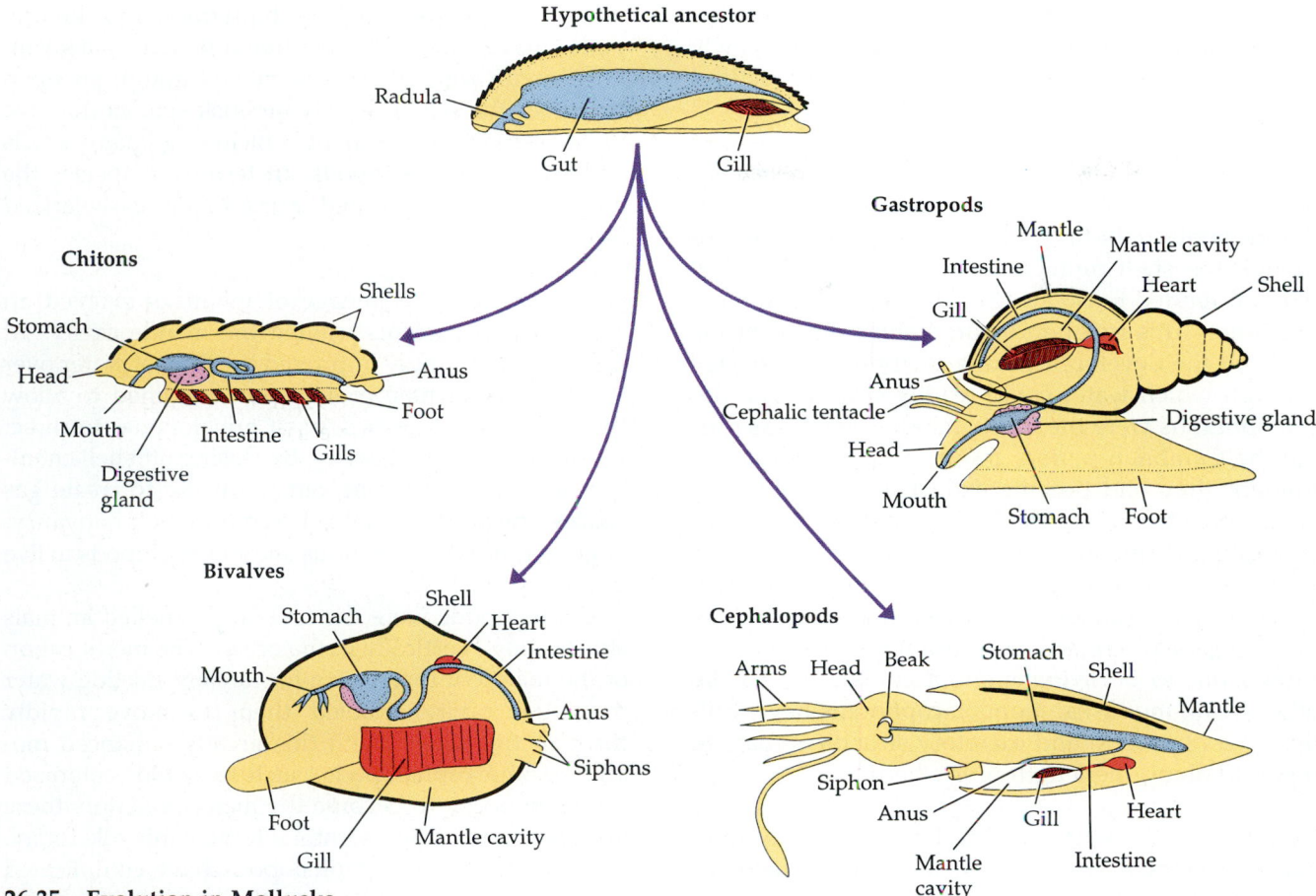

26.35 Evolution in Mollusks
Modifications of the hypothetical ancestral features
yielded a diverse array of molluscan body plans.

was based on a body plan with three structural components: the foot, the mantle, and the visceral mass.
Animals that appear very different, including snails,
clams, and squids, are all built from these components. Using the presence of these structures as the
defining characteristic of mollusks, invertebrate zoologists have placed 100,000 species of animals into
this single phylum.

The **foot** is a large, muscular structure that originally was the molluscan organ of locomotion as well
as the support for internal organs. In the lineage
leading to squids and octopuses, the foot has been
modified to form arms and tentacles borne upon a
head with complicated sense organs. In other groups,
such as clams, it has been transformed into a burrowing organ. In some lineages it is greatly reduced.

The **mantle** is a sheet of specialized tissue that
covers the internal organs as if it were a body wall.
The mantle secretes the shell, which provides external protection in most mollusks, but which has been
modified to function as internal support in slugs and
squids and is almost lost in octopuses. The featherlike
gills of mollusks are located between the mantle and

the **visceral mass**—the major internal organs. When
the cilia on the gills flex, they create a flow of oxygenated water over the gills.

The coelom of mollusks is much reduced, but the
open circulatory system has large, fluid-filled cavities
that are major components of the hydrostatic skeleton. The radula was originally an organ for scraping
algae from rocks, a function it retains in many living
mollusks. In some, however, the radula has been
modified for different functions: in some species it
serves as a drill, in others as a poison dart. Mollusks
range in size from snails that are only 1 millimeter
high to giant squids more than 18 meters long—the
largest known protostomes.

CHITONS. The molluscan lineage in which the spicules
united to form plates gave rise to the chitons (class
Polyplacophora). Chitons are the living mollusks
most similar to probable molluscan ancestors (Figure
26.36a). The chiton body is symmetrical, and the internal organs, particularly the digestive and nervous
systems, are relatively simple. Development proceeds through a trochophore larva almost indistinguishable from that of annelids. Most chitons are
marine herbivores that scrape algae from rocks with
their sharp radulae. An adult chiton spends most of

its life glued tightly to rock surfaces by its large, muscular, mucus-covered foot. It can move slowly by means of rippling waves of muscular contractions in its foot.

MONOPLACOPHORANS. A lineage in which the spicules became united into a single caplike shell gave rise to the monoplacophorans (class Monoplacophora). Although the shell protected the gills and other body parts, it tended to isolate the gills from oxygen-bearing water. This isolation favored enlargement of the gills and the development of a cavity under the shell through which water could readily circulate. Monoplacophorans were the most abundant mollusks during the Cambrian period, 600 mya. Although it was thought they had become extinct many millions of years ago, in 1952, off the Pacific coast of Costa Rica, an oceanographic vessel dredged up 10 specimens of an unusual little mollusk with a cap-shaped shell. Finding these animals, which were placed in the genus *Neopilina*, created a sensation because they turned out to be living monoplacophorans. Unlike other living mollusks, monoplacophorans have multiple gills, muscles, and excretory structures that are repeated along the length of the body.

BIVALVES AND GASTROPODS. One lineage of early mollusks developed a two-part shell that extended over the sides of the body as well as the top, giving rise to the bivalves (class Bivalvia), the familiar clams, oysters, scallops, mussels, and other important edible shellfish, together with a host of similar, less-familiar forms (Figure 26.36b). The name *bivalve* derives from the structure of the shell of these animals, which has two major pieces connected by a flexible hinge. Bivalves are largely sedentary and have greatly reduced heads. The foot is compressed and is used by many clams to burrow into mud and sand. Food is filtered from the water by the large gills, which are also the main sites of gas exchange.

Another lineage of early mollusks became the gastropods (class Gastropoda). Most gastropods are motile, using their large foot to move slowly across the substrate or to burrow through it. The shell and visceral mass of a gastropod larva undergo a 180° counterclockwise torsion relative to the foot during development. The result is that the digestive tract and nervous system are twisted so that the anus, the opening of the mantle, and the gills are moved to the front of the body, just behind and over the head. The gills of gastropods are the primary sites of gas exchange. In some species, they are also feeding devices.

Gastropods are the most diverse and widely distributed of the molluscan classes (Figure 26.36c, d, and e). Some species can crawl, including a rich variety of snails, whelks, limpets, slugs, abalones, and the often brilliantly ornamented nudibranchs. Still other gastropods—the sea butterflies and heteropods—have a modified foot that functions as a swimming organ with which they move through the open waters of the sea. The only mollusk species that live in terrestrial environments, including many snails and slugs, are gastropods. In terrestrial species the mantle cavity is modified into a highly vascularized lung.

CEPHALOPODS. One lineage of mollusks evolved an exit tube for currents leaving the mantle cavity. At first this tube simply improved the flow of water over the gills; subsequently it became modified to allow the early cephalopods (class Cephalopoda) to direct water from the shell cavity. By closing off shell chambers and then pumping out the water to create gas spaces, the animals could also control their buoyancy. Together, these adaptations allow cephalopods to live in open water.

Cephalopods were the first large, shelled animals able to move vertically in the ocean. The modification of the mantle into a device for forcibly ejecting water from the cavity enabled them to move rapidly through the water. With this greatly enhanced mobility, some cephalopods, such as squid, colonized the open ocean to become the major predators there (Figure 26.36f). They continue to play this role today. As active predators, cephalopods have complicated sense organs, especially eyes that are comparable to those of vertebrates in their ability to resolve images (see Chapter 39).

Cephalopods appeared near the beginning of the Cambrian period, and by the mid-Devonian period a wide variety of types were present. Increases in size and reductions in external hard parts characterize the subsequent evolution of many lineages. The cephalopod foot is closely associated with the large, branched head bearing tentacles and a siphon. The large, muscular mantle is a solid external supporting structure. The gills hang within the mantle cavity. A beak enables cephalopods to capture and subdue a variety of prey.

The earliest cephalopod shells were divided by partitions penetrated by tubes through which liquids could be removed. As fluid moves out of a chamber, gas diffuses into it. Changes in gas content of the chambers allow the animal to adjust its buoyancy so that it can maintain any position in the water. Of these early cephalopods, only the nautiloids (genus *Nautilus*) survive (Figure 26.36g). Another group, the abundant and diverse ammonites used by geologists to identify geological layers, became extinct.

Nautiloids can control their buoyancy, but the bulky, chambered shell is an impediment to rapid locomotion. In other cephalopod lineages the shell was reduced and other mechanisms for controlling buoyancy were substituted. For example, many species employ mechanisms of "chemical lift" by secret-

26.36 Mollusk Diversity
(a) *Tonicella lineata* is a chiton (class Polyplacophora) common in the intertidal zone of the Pacific coast of North America. (b) Scallops such as the bay scallop *Aequipecten irradians*, are unusual among bivalves because they can swim by rapidly opening and closing their valves. Notice the eyes on the mantle margin. (c) The mantle of the chestnut cowry (*Cypraea spadicea*) projects over its coiled shell. (d) Terrestrial and marine slugs have evolutionarily lost their shells; this shell-less sea slug, *Dirona albolineata*, is very conspicuous. (e) The land snail *Monadenia infumata*, is typical of many terrestrial gastropods. (f) Octopuses are active predators. This one is searching for prey on a coral head off the coast of Mexico. (g) *Nautilus pompilius* is one of a few surviving species of the rich nautiloid evolutionary radiation.

ing body fluids that are less dense than seawater into coelomic spaces or into vacuoles in the various muscles. Still other cephalopods abandoned control of their buoyancy. They maintain themselves in the water by swimming actively, or they live on the sea floor, as octopuses do.

THEMES IN PROTOSTOME EVOLUTION

Most protostome evolution took place in the oceans. Because water provides good support, early animals used body fluids as the basis for their support. When acted upon by surrounding circular and longitudinal muscles, these fluid-filled spaces function as skeletons that enable large animals to move. Subdivisions of the body cavity allow better control of movement and permit different parts of the body to be moved independently of one another. Thus some protostome lineages gradually evolved abilities to change their shapes in complex ways and to move with great speed on and through substrates or in the water.

Predation may have been the major selective pressure for the development of hard, external body coverings. Such coverings evolved in many invertebrate phyla. Originally protective in function, they became key elements in the development of new systems of locomotion. Locomotory abilities permitted prey to escape more readily from predators but also allowed predators to pursue their prey more effectively. The evolution of protostomes has been and continues to be a complex saga of arms races among predators and prey.

During much of protostome evolution, the only food in the water was dissolved organic matter and very small organisms. Consequently, many different lineages of animals evolved feeding structures designed to extract small prey from water and structures for moving water through or over their prey-collecting devices. Because water flows readily, bringing food with it, sessile lifestyles also evolved repeatedly during protostome evolution. Most surviving protostome phyla have at least some sessile members. A sessile animal gains access to local resources but forfeits access to more distant resources. A sessile animal is exposed to physical agents and predators from which it cannot escape by movement.

Sessile animals cannot come together to mate; instead, most rely on the fertilizaton of gametes that have been ejected into the water. Some species eject both eggs and sperm into the water. Others retain their eggs within their bodies and extrude only their sperm, which are carried by the water to other individuals. Species whose adults are sessile often have motile larvae, many of which have complicated mechanisms for locating suitable sites on which to settle. Many colonial sessile protostomes are able to grow in the direction of better resources or into sites offering better protection.

A frequent consequence of a sessile existence is intense competition for space that provides access to light and other resources. Such competition is intense among plants in most terrestrial environments. In the sea, especially in shallow waters, animals also compete directly for space. They have evolved mechanisms for overgrowing one another and for engaging in toxic warfare where they come into contact (Figure 26.37).

Colonial organization enables animals to compete more effectively for space because colonies are better

---

TABLE 26.3
General Characteristics of Protostome Phyla

PHYLUM	SYMMETRY	BODY CAVITY	DIGESTIVE TRACT	CIRCULATORY SYSTEM
Porifera	Asymmetrical, radial	None	None	None
Placozoa	Asymmetrical	None	None	None
Cnidaria	Radial, biradial	None	Dead-end sac	None
Ctenophora	Biradial	None	Dead-end sac	None
Platyhelminthes	Bilateral	None	Dead-end sac	None
Nemertea	Bilateral	Rhynchocoel	Complete	Closed
Nematoda	Bilateral	Pseudocoel	Complete	None
Rotifera	Bilateral	Pseudocoel	Complete	None
Pogonophora	Bilateral	Coelom	None	None
Annelida	Bilateral	Coelom	Complete	Closed or open
Onychophora	Bilateral	Coelom	Complete	Closed or open
Tardigrada	Bilateral	Coelom	Complete	Closed or open
Chelicerata	Bilateral	Coelom	Complete	Closed or open
Crustacea	Bilateral	Coelom	Complete	Closed or open
Uniramia	Bilateral	Coelom	Complete	Closed or open
Mollusca	Bilateral	Reduced coelom	Complete	Open except in cephalopods

26.37 Toxic Warfare between Sea Anemones
When two sea anemones come into contact they attack one another by developing special tentacles that produce toxins.

at overgrowing neighbors than single individuals are. Individual members of colonies, if they are directly connected, also can share resources. The ability to share resources enables some individuals to specialize for particular functions, such as reproduction, defense, or feeding. The nonfeeding individuals can derive their nutrition from their feeding associates.

Although we have concentrated on the evolution of greater complexity in protostome lineages, many lineages that remained simple still survive today. Cnidarians are common in the oceans; roundworms are abundant in most aquatic and terrestrial environments. Parasitic lineages have repeatedly evolved simpler body plans—but more complex life cycles—than those of their nonparasitic ancestors.

All of the phyla of protostome animals had evolved by the Cambrian period, 600 mya, but extinction and diversification within those lineages continue to the present. The characteristics of the protostome phyla are summarized in Table 26.3. Many of the evolutionary trends demonstrated by protostomes also dominated the evolution of deuterostomes, the lineage that led to the chordates, the group to which we belong. Hard external body coverings evolved and were later abandoned by many lineages. After considering the evolution of deuterostomes in the next chapter, we will return to the major themes in animal evolution and describe patterns that characterize all groups of animals.

SUMMARY of Main Ideas about Sponges and Protostomate Animals

Members of the kingdom Animalia are multicellular, heterotrophic eukaryotes with ingestive nutrition.

Most animals have a recognizable symmetry, either radial or bilateral.
   Review Figure 26.1

Protostomes and deuterostomes differ in early embryological development and type of body cavity.
   Review Table 26.1 and Figures 26.2, 26.3, and 26.4

Both sponges and cnidarians are simple animals. Cnidarians, however, with their nematocyst-studded tentacles, can capture prey that is larger and more complex than themselves.
   Review Figures 26.7 and 26.10

Most cnidarian life cycles have a sessile polyp and a free-swimming, sexual, medusa stage.
   Review Figures 26.11 through 26.14

Flatwormlike animals were probably the ancestors of most animal phyla other than Porifera, Cnidaria, Ctenophora, and Placozoa.

Some flatworms are free-living but most live as internal parasites of other animals.
   Review Figures 26.18 and 26.19

The first animals were small, simple organisms lacking systems for internal transport of materials and having only body fluids as the basis for their support.

Some animals evolved digestive systems with openings at both ends so that food moved through the gut in only one direction.

Systems for internal transport of oxygen, carbon dioxide, food, and metabolic wastes were part of the greater complexity of some evolving animal groups.
   Review Figure 26.21

Subdivision of the body cavity and evolution of hard, external coverings gave evolving animal groups better control of movement, better protection from predators, and the ability to burrow into and walk on top of substrates.
   Review Figures 26.24 and 26.25

The arthropod lineages evolved jointed appendages.

There are more species of arthropods than any other animal group; most of these arthropod species are insects.

Insects evolved wings and became the first animals to fly.
   Review Figure 26.26

The highly varied mollusks all have a body plan consisting of a foot, mantle, and visceral mass that are highly modified in the different groups, allowing them to occupy a wide range of habitats.
Review Figure 26.35

Most protosome evolution took place in the oceans, but several lineages invaded the land to become dominant organisms there.

Although many lineages became more compex, simple protostomes have survived with little modification for millions of years.

SELF-QUIZ

1. The body plan of an animal is
   a. its general structure.
   b. the functional interrelationship of its parts.
   c. its general form and the functional interrelationship of its parts.
   d. its general form and its evolutionary history.
   e. the functional interrelationship of its parts and its evolutionary history.

2. A bilaterally symmetrical animal can be divided into mirror images by
   a. any cut through the midline of its body.
   b. any cut from its anterior to its posterior end.
   c. any cut from its dorsal to its ventral surface.
   d. only a cut through the midline of its body from its anterior to its posterior end.
   e. only a cut through the midline of its body from its dorsal to its ventral surface.

3. In protostomes, cleavage of the fertilized egg is
   a. delayed while the egg continues to mature.
   b. determinate; that is, cells separated after a few divisions develop into only partial embryos.
   c. indeterminate; that is, cells separated after a few divisions develop into complete embryos.
   d. triploblastic.
   e. diploblastic.

4. In which ways did early animal communities differ from modern ones?
   a. All organisms were small and engaged in sexual reproduction.
   b. Some organisms were large but none of them engaged in sexual reproduction.
   c. Some organisms were large but all of them lived on the surface of the substrate.
   d. All organisms were small; they lived both on the surface and below it.
   e. All organisms were small and lacked sexual reproduction

5. The sponge body plan is characterized by
   a. a mouth and digestive cavity but no muscles or nerves.
   b. muscles and nerves but no mouth or digestive cavity.
   c. a mouth, digestive cavity, and spicules.
   d. muscles and spicules but no digestive cavity or nerves.
   e. no mouth, digestive cavity, muscles, or nerves

6. The phyla of diploblastic animals are
   a. Porifera and Cnidaria.
   b. Cnidaria and Ctenophora.
   c. Cnidaria and Platyhelminthes.
   d. Ctenophora and Platyhelminthes.
   e. Porifera and Ctenophora.

7. The abundance of cnidarians is probably due to
   a. their ability to live in both salt and fresh water.
   b. their ability to move rapidly in the water column.
   c. their ability to capture and consume large numbers of small prey.
   d. their low metabolic rates and their ability to capture large prey.
   e. their ability to capture large prey and to move rapidly.

8. Which of the following protostome phyla are coelomate?
   a. Rotifera, Pogonophora, Annelida, Chelicerata, Uniramia, and Crustacea
   b. Pogonophora, Annelida, Chelicerata, Uniramia, Crustacea, and Nematoda
   c. Pogonophora, Annelida, Chelicerata, Uniramia, Crustacea, and Mollusca
   d. Rotifera, Annelida, Chelicerata, Uniramia, Crustacea, and Mollusca
   e. Nematoda, Rotifera, Pogonophora, and Annelida

9. Insects that hatch from eggs into juveniles similar to the adults are said to have
   a. instars.
   b. neopterous development.
   c. simple development.
   d. incomplete metamorphosis.
   e. complete metamorphosis.

10. Which of the following is *not* part of the molluscan body plan?
    a. Mantle
    b. Foot
    c. Radula
    d. Visceral mass
    e. Jointed skeleton

11. Many lineages of protostomes evolved feeding structures designed to extract small prey from the water because
    a. during much of protostome evolution the only food available was dissolved organic matter and very small organisms.
    b. during much of protostome evolution small animals were more abundant than large animals.
    c. large animals were available as food but were difficult to capture.
    d. to be successful in competition for space, protostomes had to feed on small prey.
    e. water flowed naturally over their feeding structures, so early protostomes did not have to work to get food.

FOR STUDY

1. Differentiate among the members of the following sets of related terms.
   a. radial symmetry/bilateral symmetry
   b. protostome/deuterostome
   c. indeterminate cleavage/determinate cleavage
   d. spiral cleavage/radial cleavage
   e. incomplete metamorphosis/complete metamorphosis
   f. coelomate/pseudocoelomate/acoelomate

2. For each type of organism listed below, give a single trait that may be used to distinguish it from the type of organism in parentheses.
   a. cnidarians (sponges)
   b. gastropods (all other mollusks)
   c. polychaetes (other annelids)
   d. ribbon worms (roundworms)

3. Segmentation has arisen a number of times during protostomate evolution. What advantages does segmentation provide? Given these advantages, why do so many unsegmented animals survive?

4. Many animals extract food from the surrounding medium. What protostomate phyla contain animals that extract suspended food from the water column?

5. Name the phyla in which some species form large colonies of attached individuals. What advantages does colonial living provide to animals?

6. Discuss the structures animals have evolved to enable them to capture large prey.

7. A major factor influencing the evolution of animals was predation. What major animal features appear to have evolved in response to predation? How do they help their bearers avoid becoming a meal for some other animal?

READINGS

Barnes, R. S. K., P. Calow and P. J. W. Olive. 1993. *The Invertebrates. A New Synthesis*. Blackwell Scientific Publications, Oxford. A thorough general textbook with a strong functional approach to the invertebrates.

Barrington, E. J. W. 1979. *Invertebrate Structure and Function*, 2nd Edition. Halsted-Wiley, New York. Engaging coverage of invertebrates, with an emphasis on morphology.

Borror, D. J., C. A. Triplehorn and N. F. Johnson. 1989. *An Introduction to the Study of Insects*, 6th Edition. Saunders, Philadelphia. A clearly written introduction to the richness of the insects.

Brusca, R. C. and G. J. Brusca. 1990. *Invertebrates*. Sinauer Associates, Sunderland, MA. A thorough account of the invertebrates that provides detailed treatments of body plans and includes excellent discussions of phylogenies.

Cloudsley-Thompson, J. L. 1976. *Insects and History*. St. Martins Press, New York. A vivid and readable account of instances in which insects have affected human society; special emphasis on insects as carriers of disease organisms.

Kozloff, E. N. 1990. *Invertebrates*. Saunders, Philadelphia. An excellent reference book on all invertebrate taxa.

Morris, S. C. and H. B. Whittington. 1979. "The Animals of the Burgess Shale." *Scientific American*, July. A brief overview of the rich collection of fossils from this important site.

Noble, E. R. and G. A. Noble. 1982. *Parasitology: The Biology of Animal Parasites*, 5th Edition, Lea & Febiger, Philadelphia. A general text covering the major parasitic protists and worms that affect animals, including people.

Ruppert, E. E. and R. D. Barnes. 1994. *Invertebrate Zoology*, 7th Edition. Saunders, Philadelphia. One of the best general textbooks, covering all groups from the animallike protists to the invertebrate chordates.

Frogs that "Fly"

27

Deuterostomate Animals

Flying frogs have evolved from nonflying ancestors several times. Although they evolved their "flying" ability independently, flyers usually have enlarged hands and feet, fully webbed fingers and toes, and accessory skin flaps on the outer edges of their arms and legs. Flying frogs really glide rather than fly, but they can glide at angles less than 45 degrees to the horizontal and they can change directions during their descent. Some frogs that lack the structural traits of flying frogs drop from trees, but these frogs must descend at angles much greater than 45 degrees and can change direction very little. They may be said to parachute rather than glide. Parachuting has evolved many times among frogs that make daily vertical movements between feeding, breeding, and resting sites. Flying frogs can glide farther and maneuver better through dense vegetation than parachuting frogs can.

All frogs that take to the air begin their descent with fully extended legs, but they bend their legs somewhat as they pick up speed. By bending its limbs, a parachuting frog increases both its stability during the descent and the likelihood that it will land upright. A frog can glide only if it has the flyer morphology and if it bends its legs. Both the structural and behavioral changes must evolve in a lineage before a frog can glide and turn while gliding. Combinations of behavioral and structural changes comparable to those that permit some frogs to "fly" have evolved among deuterostome lineages. Elaborate behavior has been especially important in the evolution of the lineages leading to the vertebrates, the broad group to which humans belong.

The deuterostome lineage is characterized by indeterminate cleavage in the early embryo, formation of the mesoderm from outpocketing of the embryonic gut, and a blastopore that becomes the anus (see Table 26.1). We do not know what the first deuterostomes were like because the original lineage had already split into several lineages before deuterostomes began to be preserved in the fossil record. All deuterostomes are triploblastic and have well developed coelomic body cavities (Table 27.1), but their body plans vary considerably. There are far fewer species of deuterostomes than of protostomes.

TRIPARTITE DEUTEROSTOMES

The lophophorates (from the Greek for "crest bearers") share one feature with protostomate animals—the mouth often forms from or near the blastopore—but they are probably deuterostomes. The body of a lophophorate is divided into three parts: the prosome (anterior), mesosome (middle), and metasome (posterior). In most species each region has a separate coelomic compartment: the protocoel, mesocoel, and

TABLE 27.1
General Characteristics of Deuterostomate Animal Phyla[a]

PHYLUM	SYMMETRY	CIRCULATORY SYSTEM
Lophophorate phyla: Phoronida, Bryozoa, Brachiopoda	Bilateral	None in most
Hemichordata	Bilateral	Closed
Chaetognatha	Bilateral	None
Echinodermata	Biradial	Open or none
Chordata	Bilateral	Closed

[a]Members of all deuterostomate phyla have a coelom and a complete digestive tract.

metacoel, respectively. Typically, these animals secrete a tough outer covering. All lophophorates have a U-shaped gut, with the anus located close to the mouth but outside the tentacles.

The most conspicuous feature of these animals is the **lophophore,** a circular or U-shaped ridge around the mouth that bears one or two rows of ciliated, hollow tentacles (Figure 27.1). This large and complex structure, which is both a food-collecting organ and a surface for gas exchange, is held in position and moved by contraction of muscles surrounding the mesocoel. Adult animals of all species in these groups are sessile and use the tentacles and cilia of the lophophore to capture phytoplankton and zooplankton.

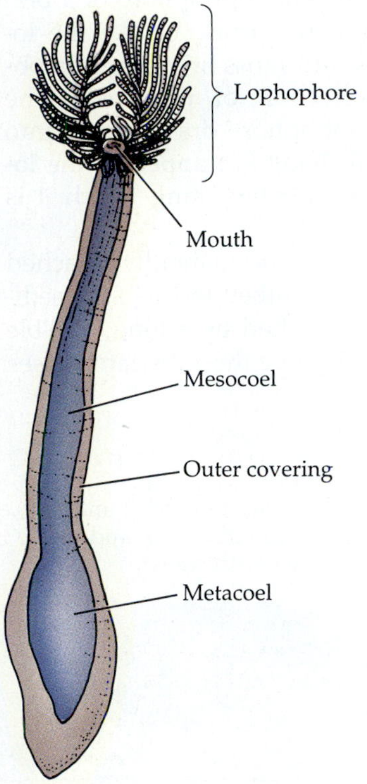

Lophophorates are subdivided into three phyla: Phoronida, Bryozoa, and Brachiopoda. Nearly all members are marine; a few live in fresh water. About 4,500 living species are known, but many times that number of species existed during the Paleozoic and Mesozoic eras. Fossils of these animals provide important records of past ecological communities.

Phylum Phoronida

There are only about 15 species of phoronids (phylum Phoronida), sedentary worms that live in muddy or sandy sediments or attached to rocky substrates. Phoronids are found in waters ranging from intertidal zones to about 400 meters deep. They live in chitinous tubes, which they secrete. Their most conspicuous external feature is the lophophore (Figure 27.2a). Cilia drive water into the top of the lophophore. Water exits through the narrow spaces between the tentacles. Suspended food particles are caught and transported by ciliary action to the food groove and into the mouth.

Phylum Bryozoa (Ectoprocta)

Moss animals (phylum Bryozoa or Ectoprocta) are colonial lophophorates with a "house" secreted by the body wall. They are called moss animals because of the plantlike appearance of their colonies (Figure 27.2b), which are similar in many ways to coral polyps. A colony consists of many individuals connected by strands of tissues along which materials can be moved. Most moss animals are marine, but a few live in fresh water. Moss animals are the only lophophorates able to extend and retract their lophophores, and they can rock and rotate the extended lophophore as it feeds to increase contact with prey in the water (Figure 27.3).

27.1 Lophophore Artistry
The lophophore dominates the anatomy of this phoronid, *Phoronis architecta*.

27.2   Lophophorate Animals
(a) The lophophores of these phoronids (*Phoronis* sp.) from Monterey, California, are extended in the feeding position. (b) A colony of freshwater moss animals, *Cristatella mucedo*. Notice the delicate structure of the lophophores.

A colony of moss animals is created by the asexual reproduction of its founding member. Colonies of different species have different shapes. One colony may contain up to 2 million individuals. In some species, individual colony members are specialized in anatomy and function; some are adapted solely for feeding, others for reproduction, defense, or support. Moss animals reproduce sexually by releasing sperm into the external environment, where they are collected from the surrounding water by other individuals. Eggs are fertilized internally, and developing embryos are brooded before exiting as larvae that seek suitable sites for attachment.

Phylum Brachiopoda

Brachiopods, or lamp shells (phylum Brachiopoda), are solitary, marine, lophophorate animals that superficially resemble bivalve mollusks (Figure 27.4). Brachiopods possess a mantle and a calcareous shell with two opposing valves, which can be pulled shut to protect the soft body inside. The shell differs from that of bivalves in that the two halves are dorsal and ventral rather than lateral. The lophophore of a brachiopod is extended into two arms. Unlike the lophophores of phoronids and moss animals, it is positioned within the protection of the shell. The beating of cilia on the lophophore draws water into the slightly opened shell. Food is trapped in the lophophore and directed to a ridge along which it is transferred to the mouth.

Nearly all brachiopods are permanently attached to a solid substrate or are embedded in soft sediments. Most species are attached by a long, flexible stalk that holds the animal above the substrate. Gases

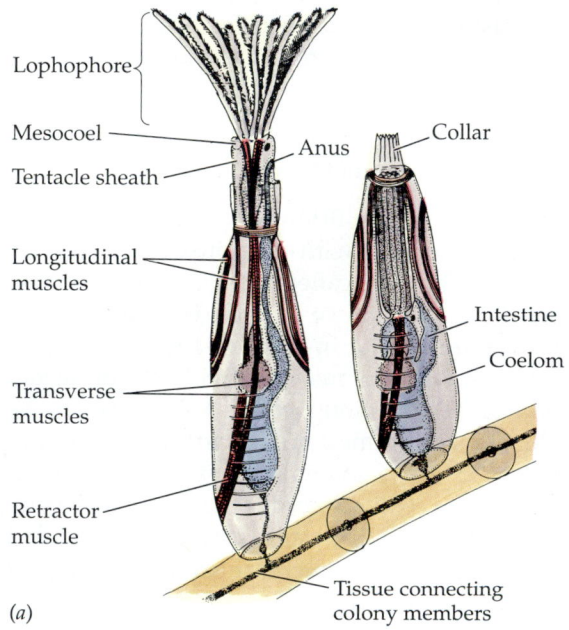

Lophophore
Mesocoel
Tentacle sheath
Anus
Collar
Longitudinal muscles
Intestine
Coelom
Transverse muscles
Retractor muscle
Tissue connecting colony members

(a)

27.3   The Lophophore in Action
(a) Moss animals use various muscles to extend and retract their lophophores. (b) They can also rock and rotate the lophophore to increase contact with prey.

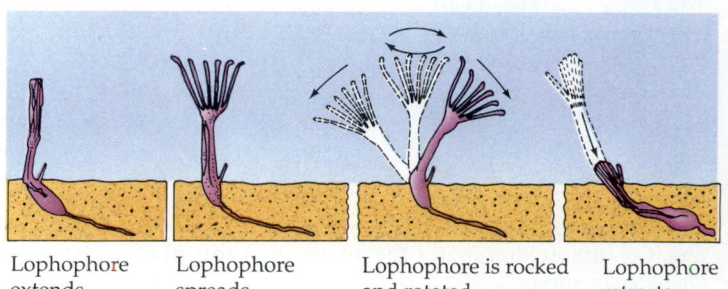

Lophophore extends

Lophophore spreads

Lophophore is rocked and rotated

Lophophore retracts

(b)

27.4 Brachiopods
The stalk of this North Pacific brachiopod, *Laqueus* sp., is barely visible to the left of the shell. The lophophore is visible between the valves of the shell.

are exchanged across nonspecialized body surfaces, especially the tentacles of the lophophore. Most brachiopods release their gametes into the water, where fertilization takes place. The larvae, which resemble the adults, remain in the water column for only a few days before they settle and metamorphose into adults.

Brachiopods reached their peak abundance and diversity in Paleozoic and Mesozoic times. More than 26,000 fossil species have been described. Only about 350 species survive, but they are common in some marine environments.

INNOVATIONS IN FEEDING

Evolution in one lineage of deuterostomes resulted in several modifications of the lophophore and the coelomic cavity within it. These modifications provided new ways of capturing and handling food. All living representatives of this lineage are wormlike animals that live buried in marine sands or muds, under rocks, or attached to algae.

Phylum Hemichordata

The hemichordates (phylum Hemichordata) have a tripartite body plan. The three regions of the body—proboscis, collar, and trunk—appear to be homologous to the prosome, mesosome, and metasome of lophophorate animals. The animals in the two highly divergent clades within this phylum—Pterobranchia and Enteropneusta—are adapted to different ways of capturing food.

The ten living species of pterobranchs (class Pterobranchia) seem to have changed relatively little from the ancestors of their lineage. They are sedentary animals up to 12 millimeters long that live in tubes

secreted by the proboscis. Some species are solitary; others form colonies of individuals joined together (Figure 27.5*a*). Behind the proboscis is a collar with one to nine pairs of arms bearing long tentacles that capture prey and permit gas exchange. The digestive tract is U-shaped, with the anus situated next to the tentacles. The proboscis encloses a coelomic cavity that has a pair of openings to the exterior through which excretory wastes leave the body.

In the other hemichordate lineage, the lophophore was apparently lost and the proboscis grew larger and became a digging organ. The survivors of this lineage are the 70 species of acorn worms (class Enteropneusta). These animals live in burrows in muddy and sandy sediments (Figure 27.5*b*). The enlarged proboscis is coated with a sticky mucus that traps prey and is then conveyed by ciliary action to the mouth. In the esophagus, the food-laden mucus is compacted into a ropelike mass that is moved through the digestive tract by ciliary action. Behind the mouth is a pharynx that opens to the outside through a number of slits, more than 100 pairs in some species. These openings allow water to exit;

(a)

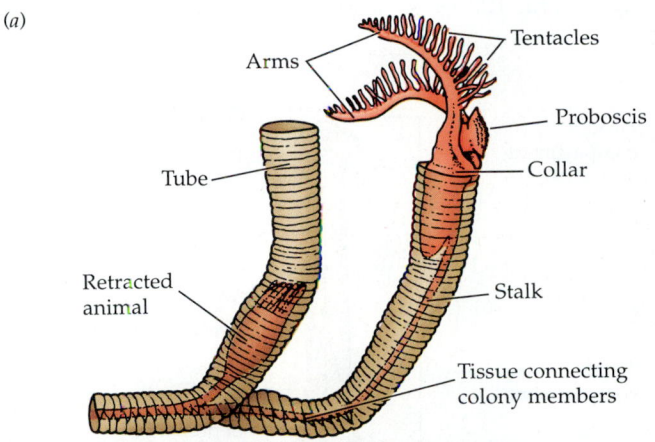

(b)

27.5 Hemichordates
(*a*) *Rhabdopleura* is a colonial pterobranch. Two members of a colony are depicted here; the animal on the left is retracted into the tube secreted by its proboscis. (*b*) Acorn worms (class Enteropneusta) have lost the lophophore. The proboscis (upper right) of this *Saccoglossus kowaleskii* is modified for digging; this individual was extracted from its burrow.

highly vascularized tissue surrounding them serves as a gas-exchange apparatus. An acorn worm breathes with the anterior end of its gut by pumping water into its mouth and out through the gill slits.

ACTIVE FOOD SEEKERS

As we saw earlier, adult lophophorates attach to a substrate and extract prey from the water either by creating currents that bring prey to the lophophore or by extending the lophophore into moving currents. However, in one lineage of deuterostomes, the arrow worms, the tripartite body plan characteristic of lophophorate animals was modified for active dispersal and pursuit of prey in the open water.

Phylum Chaetognatha

Arrow worms (phylum Chaetognatha) are tripartite, streamlined, bilaterally symmetrical animals (Figure 27.6). Most swim in the open sea, but a few live on the seafloor. Arrow worms have a long fossil history. Their abundance as fossils indicates that they were already common more than 500 mya (million years ago). The 100 or so known species of arrow worms

are small marine carnivores, all less than 12 centimeters long. They are so small that their requirements for gas exchange and excretion can be met by diffusion through the body surface. Arrow worms lack a circulatory system. Wastes and nutrients are moved around the body in the coelomic fluid, which is propelled by cilia that line the coelom. The body plan is based on a coelom that is divided into head, trunk, and tail compartments. There is no distinct larval stage. Miniature adults hatch directly from eggs released into the water.

Arrow worms are among the dominant planktonic predators in open oceans. They typically lie motionless in the water until movement of the water signals the approach of prey, which range from small protists to young fish as large as the arrow worms themselves. The arrow worm then darts forward and grasps the prey with the stiff spines adjacent to its mouth. Arrow worms are stabilized in the water by one or two pairs of lateral fins and a caudal fin. Arrow worms are not powerful enough to swim against strong water currents, but some species undertake daily vertical migrations of up to several hundred meters, moving to deeper water during the day and back to the surface at night. (Similar vertical migrations are undertaken by many other small oceanic animals, probably because surface waters are dangerous during the day when predators that hunt using vision are active.)

CALCIFYING THE SKELETON

Bilateral symmetry is characteristic of animals that move about actively. Radial symmetry is found primarily among sedentary or slow-moving animals. In one lineage, which originated with a Precambrian deuterostome, radial symmetry evolved in association with a complex body plan. The result was one of the most striking evolutionary radiations, which, surprisingly, combines radial symmetry with locomotion.

Phylum Echinodermata

The remarkable lineage in which radial symmetry combined with locomotion gave rise to the echinoderms (phylum Echinodermata), some of the most unusual living organisms and the most structurally complex of all radially symmetrical animals. One of the two main changes during the evolution of the echinoderms was calcification of an internal skeleton, which gave protection against predators. The calcified plates of the early ancestors were enlarged and thickened until they fused inside the entire body. These plates are covered by thin layers of skin and some muscles.

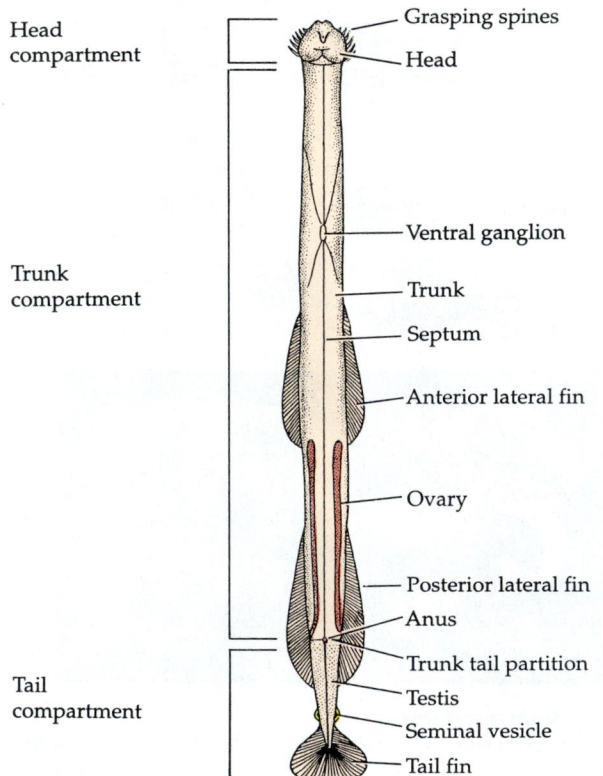

27.6 An Arrow Worm
Arrow worms (phylum Chaetognatha) have a tripartite body plan. The fins and grasping spines are powerful adaptations for a predatory life.

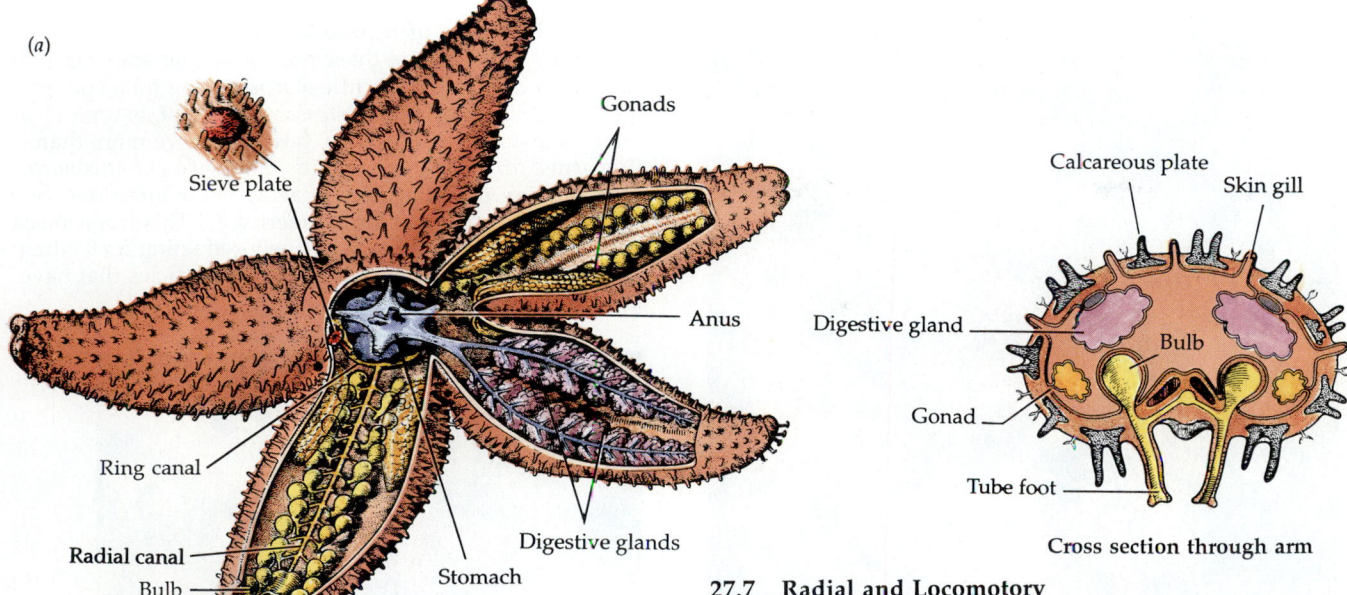

(a)

Sieve plate

Gonads

Anus

Ring canal

Radial canal

Bulb

Tube foot

Digestive glands

Stomach

Adult sea star

Calcareous plate

Skin gill

Digestive gland

Bulb

Gonad

Tube foot

Cross section through arm

27.7 Radial and Locomotory

(a) This cutaway view of a sea star and the cross section through one of its arms reveal its internal organs and organ systems. *(b)* The sea star larva moves through the water by beating its cilia.

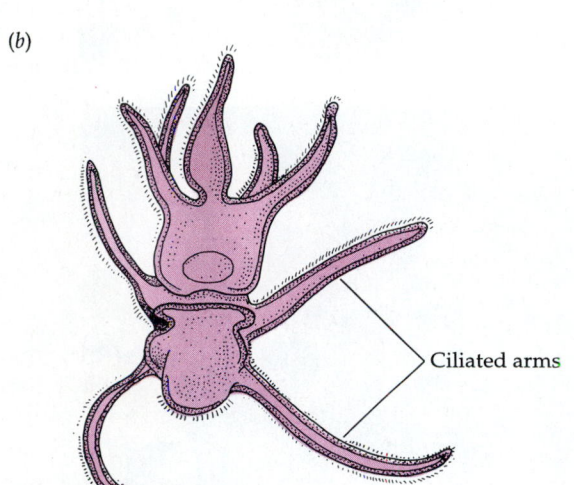

(b)

Ciliated arms

Sea star larva

The second major change was the development of the **water vascular system,** a series of seawater channels and spaces derived by enlargement and extension of the mesocoel, one of the three coelomic cavities of the ancestral forms. The water vascular system is a network of hydraulic canals leading to extensions called **tube feet** that function in gas exchange, locomotion, and feeding. Seawater enters the water vascular system through a sievelike pore. In most species a calcified canal leads from the pore to another canal that rings the esophagus. From the ring canal other canals radiate, extending through the arms (in species that have arms) and connecting with the tube feet (Figure 27.7*a*).

Echinoderms have an extensive fossil record.

About 23 classes have been described, of which only six survive today. There are about 7,000 species of modern echinoderms, but an additional 13,000 species, probably only a small fraction of those that actually lived, have been described from their fossil remains. Nearly all living species have a bilaterally symmetrical, ciliated larva (Figure 27.7*b*) that feeds for a while as a planktonic organism before settling and transforming into a radially symmetrical adult. Living echinoderms are divisible into two lineages: subphylum Pelmatozoa and subphylum Eleutherozoa. The two groups differ in the forms of their water vascular systems and the number of arms they have.

PELMATOZOANS. Sea lilies and feather stars (class Crinoidea) are the only surviving pelmatozoan lineage. Sea lilies were abundant 300 to 500 mya, but only about 80 species survive today. Sea lilies attach to the substrate by means of a flexible stalk consisting of a stack of calcareous discs. The main body of the animal, attached to the stalk, is a cup-shaped structure that contains a tubular digestive system. From five to several hundred arms, usually in multiples of five, extend outward from the cup. Jointed calcareous plates cover the arms, enabling them to bend. A groove runs down the center of each arm. On both sides of the groove are tube feet covered with mucus-secreting glands. A sea lily feeds by orienting its arms in passing water currents. Food particles strike the tube feet, are transferred by them to the groove, and are carried to the mouth by ciliary action. The tube feet of sea lilies are also used for gas exchange and elimination of nitrogenous wastes.

(a)

(c)

27.8   A Diversity of Echinoderms

(a) The flexible arms of these golden feather stars are clearly visible. (b) This brittle star is resting on a sponge. (c) The rainbow sea star, *Orthasterias koehleri*, is typical of many sea stars; some species, however, have more than five arms. (d) Purple sea urchins, *Strongylocentrotus purpuratus*, are important grazers of algae in the intertidal zone of the Pacific coast of North America. (e) This tropical sea cucumber, *Cucumaria* sp., is actively retracting its feeding tentacles so that it can digest the food particles that have adhered to them.

(b)

(d)

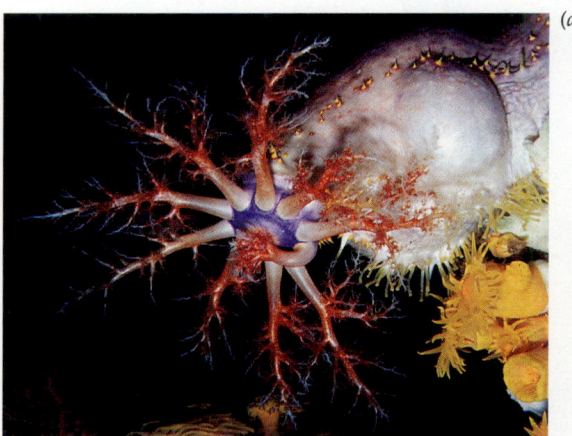

(e)

Feather stars (Figure 27.8a) are similar to sea lilies, but they have flexible appendages with which they grasp the substrate while they are feeding and resting. Feather stars can walk on the tips of their arms or swim by rhythmically beating their arms. Feather stars feed in much the same manner as sea lilies. About 600 species of feather stars have been described.

ELEUTHEROZOANS.   Most surviving echinoderms are members of the subphylum Eleutherozoa. The most-familiar echinoderms are the sea stars (class Asteroidea; Figure 27.8b). Many sea stars prey on polychaetes, gastropods, bivalves, and fishes. Each tube foot of a sea star is a small adhesive organ consisting of an internal bulb connected by a muscular tube to an external sucker. The tube foot is moved by hydraulic expansion and contraction. When the bulb and the circular muscles of the tube contract, the tube foot elongates. The tube foot adheres to a surface by secreting a sticky substance around the sucker. With hundreds of tube feet acting simultaneously, a sea star can exert an enormous and continuous force. It can grasp a clam in its arms, anchor the arms with its tube feet, and, by steady contraction of the muscles in the arms, gradually exhaust the clam's muscles and pull the shell apart. Tube feet serve as organs of locomotion, and because of their thin walls they also are important sites for gas exchange.

Sea stars that feed on bivalves are able to push their stomach out through their mouth and then through the narrow space between the shells of a bivalve. The stomach secretes digestive enzymes into the soft parts of the bivalve, and the animal is digested and consumed. Some species feed on smaller prey or suspended particles and do not extrude their stomachs. Sea stars are important predators in many marine environments, such as coral reefs and rocky intertidal zones, where they strongly influence the species composition of animal communities.

The brittle stars (class Ophiuroidea) are most similar in structure to sea stars. The flexible arms (Figure 27.8c) are composed of jointed hard plates. Although the arms thrash during locomotion, in burrowing species and in young individuals of most species the tube feet also play a locomotor role. Brittle stars generally have five arms, but each arm may branch a number of times. Unlike most other members of the phylum, brittle stars have only one opening to the digestive tract. Most of the 2,000 species of brittle stars are deposit feeders. They ingest particles from the surfaces of sediments, assimilate the organic material from them, and eject the remainder through their mouths. Some species remove suspended food particles from the water, and others capture small animals.

The remaining three classes of echinoderms lack arms. The sea daisies (class Concentricycloidea) were not discovered until 1986, and little is known about them. They have tiny discoid bodies with a ring of marginal spines but no arms. Sea daisies are the only echinoderms in which the water vascular system has two ring canals and in which the tube feet are arranged in a circle around the edge rather than along grooves radiating from the center. Sea daisies are found on rotting wood in deep ocean waters. They apparently feed on bacteria, which they digest and absorb either through a membrane that covers the oral surface or via a shallow, saclike stomach.

Sea urchins (class Echinoidea; Figure 27.8d) are armless, hemispherical animals that are covered with spines attached to the underlying skeleton via ball-and-socket joints. They resemble sea stars with their "arms" folded and fused over their backs. The spines come in various sizes and shapes, and a few produce highly toxic substances. The skeletal plates of a sea urchin are bound tightly together. Many sea urchins consume algae, scraping them from the rocks with a complex rasping structure. Others feed on small organic debris collected by their tube feet or spines.

Sea cucumbers (class Holothuroidea; Figure 27.8e) resemble stretched, flexible sea urchins, but lacking spines and having greatly reduced skeletal plates. Tube feet located on either side of five grooves along the body of the animal are used primarily for attaching to the substrate rather than for moving. Some species lack tube feet entirely, except those at the anterior end. The anterior tube feet are modified into large, feathery tentacles that can be protruded around the mouth. The tentacles are coated with a sticky substance to which prey or the surrounding substrate adhere. Periodically, the sea cucumber sticks its tentacles into its mouth, wipes them off, and then digests the adhered material.

EVOLUTION OF THE CHORDATE PHARYNX

One deuterostome lineage evolved a unique way of exploiting the abundant food provided by marine phytoplankton and zooplankton. The structures that enlarged in this lineage to become a device for removing plankton from the water are the **pharyngeal slits,** which originally functioned as sites for exchanging gas and eliminating water. The requirement for effective gas exchange—a large surface area—also serves well for capturing prey; the lophophore, for example, serves this dual function in many lophophorate animals. Enlargement of the pharyngeal slits eventually led to remarkable evolutionary developments that produced animals unlike members of any other animal phyla.

Phylum Chordata

The evolutionary lineage leading to the chordates (phylum Chordata) lost the lophophore and proboscis, replacing them with enlarged pharyngeal slits as a feeding device. Chordates are bilaterally symmetrical animals that have pharyngeal slits at some stage in their development. The main features of their body plan are an internal skeleton; a dorsal, hollow nervous system; a ventral heart; and a tail that extends beyond the anus. Although not all skeletons of chordates are the same, all species have a dorsal supporting rod, the **notochord,** at some stage during their development. In some species, the notochord is lost during metamorphosis to the adult stage. In other species, it is replaced by other skeletal structures having the same function.

27.9 Tunicates: Pharyngeal Basket Specialists
The pharyngeal basket of this sea squirt, *Ciona intestinalis*, occupies most of its body cavity.

The tunicates (subphylum Urochordata) may be similar to the ancestors of all chordates. The 2,500 species of tunicates are all marine animals, most of which are attached to the substrate as adults. Swimming, tadpolelike larvae are the dispersal stage of the tunicate life cycle. These larvae reveal the close evolutionary relationships between tunicates and other chordates. In addition to its pharyngeal slits, a tunicate larva has a dorsal, hollow nerve cord and a notochord. Muscles attach to the notochord, which provides relatively rigid support. After a short time as a member of the plankton, the larva settles on the seafloor and transforms into a sessile adult that feeds by extracting plankton from the water with its pharynx, which is enlarged into a pharyngeal basket.

More than 90 percent of known species of tunicates are sea squirts (class Ascidiacea). The baglike bodies of the adults are surrounded by a tough tunic, composed of protein and a complex polysaccharide, secreted by the epidermal cells. Much of the body is occupied by the large pharyngeal basket lined with cilia, whose beating moves water through the animal (Figure 27.9). The cilia also move the thin layer of mucus that lines the basket and to which food particles adhere. Water enters the body through an anterior opening, passes through the pharyngeal basket into a chamber that is enclosed by the tunic, and out through another opening well removed from the site where the water entered. Some sea squirts are solitary, but others produce large colonies by asexual budding from a single founder. In some colonial species individuals have their own entrances, but they share a single exit.

The larvaceans (class Appendicularia) become reproductively mature and complete their life cycles as planktonic organisms. These tunicates never settle on the bottom; rather they swim in the water, filtering prey through screens made of mucopolysaccharides. What began as a dispersal stage became a new life-

style in these animals. There are only a few species of larvaceans, but they are widespread in the world's oceans.

The 25 species of lancelets (subphylum Cephalochordata) are small, fishlike animals that live partly buried in soft marine sediments and extract small prey from the water with their pharyngeal baskets (Figure 27.10). The notochord of a lancelet extends the entire length of the body throughout its life.

SUCKING MUD: THE RISE OF THE VERTEBRATES

The pharyngeal basket is an efficient tool for extracting prey from surrounding water. Because of its many exit openings, the basket is also effective in mud, where many inedible particles are ingested along with the food. In the late Cambrian, more than 500 mya, some early chordates evolved improved structures for extracting food from mud and sand. A jointed, dorsal **vertebral column** replaced the notochord as the primary support in these chordates, which are called **vertebrates.** Vertebrates also evolved external armor, which enabled them to live above the substrate, where predators, principally arthropods at that time, were abundant, rather than having to burrow to escape.

Key to the vertebrate body plan is the vertebral column, which typically supports a skull in front and

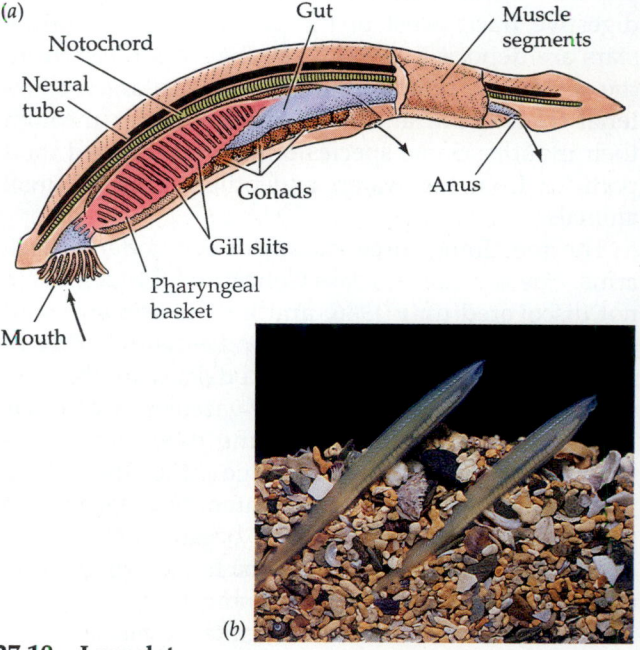

27.10 Lancelet
(a) The internal structure of a lancelet. Note the large pharyngeal basket (red) with gill slits. (b) The anterior ends of two adult lancelets (*Branchiostoma lanceolatum*) protrude from shell gravel.

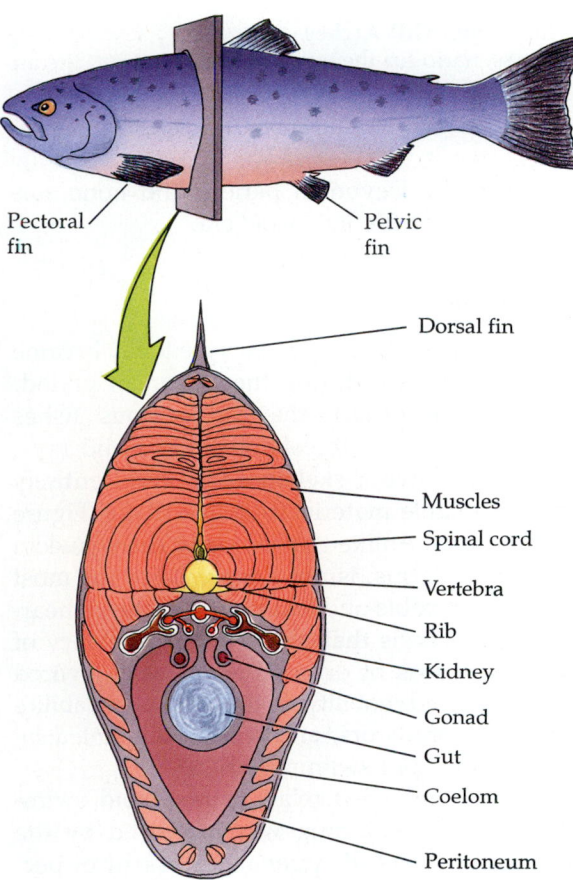

Pectoral fin

Pelvic fin

Dorsal fin

Muscles
Spinal cord
Vertebra
Rib
Kidney
Gonad
Gut
Coelom
Peritoneum

27.11 The Vertebrate Body Plan
A cross section of a bony fish demonstrates the key elements of the vertebrate body plan: internal skeleton, dorsal nervous system (spinal cord), organs suspended in the coelom, and segmented muscles.

27.12 Agnathans
Sea lampreys, *Petromyzon marinus*, have attached to a carp with their large, sucking mouths and are rasping away at its flesh.

two pairs of appendages (Figure 27.11). The skull and appendages evolved after the vertebral column. Vertebrates have a large coelom in which the body organs are suspended, but it serves neither as a hydrostatic skeleton nor as a gas-exchange structure. A well-developed circulatory system, driven by contractions of a ventral heart, delivers oxygen to internal organs.

The early vertebrates swam over the bottom, sucking mud as they went. They were jawless fishes (class Agnatha), called ostracoderms, that were typically between 6 and 30 centimeters long. The name *ostracoderm*, meaning "shell-skinned," refers to their bony external armor. With their heavy armor, these fishes could swim only slowly, but swimming above the substrate was easier than having to burrow through it, as all previous sediment-feeders had done. This new mobility may have been the major breakthrough that enabled vertebrates to exploit their environments in new ways.

One of those ways was to attach to dead, rotting flesh and use the pharynx to create a suction to pull fluids and partly decomposed tissues into the mouth.

Modern fishes that feed in this way are the lampreys and hagfishes (Figure 27.12), the only jawless fishes to have survived beyond the Devonian period. These fishes have tough scaly skin instead of external armor. Hagfishes ingest the tissues of dead animals; lampreys suck the blood of living fishes or eat the flesh of dying fishes. The round mouth is a sucking organ with which the animals attach to their prey and rasp at the flesh. All hagfishes are marine, but lampreys live in both fresh and salt water.

Jaws: A Key Evolutionary Novelty

In the Devonian period an immense variety of new kinds of fishes evolved in the seas and fresh waters. Many of these were jawless, but in one lineage jaws evolved from some of the cartilaginous or bony arches that supported the gill region (Figure 27.13). The advantage of a jaw is that it allows the fish to grasp a living, relatively large prey while consuming its tissues. Further development of the jaws and teeth led to the ability to chew both soft and hard body parts of prey. Although many intermediates must have existed between jawless fishes and the fully jawed ancestors of modern fishes, it is not difficult to imagine how each stage would have functioned better than those that preceded it.

The most important early jawed fishes were the heavily armored placoderms (class Placodermi). Some of these fish evolved elaborate fins and sleek body forms that must have improved their ability to maneuver in open water. A few became huge and, together with squids, were probably the most important predators in the Devonian oceans. Despite their early abundance, most placoderms disappeared

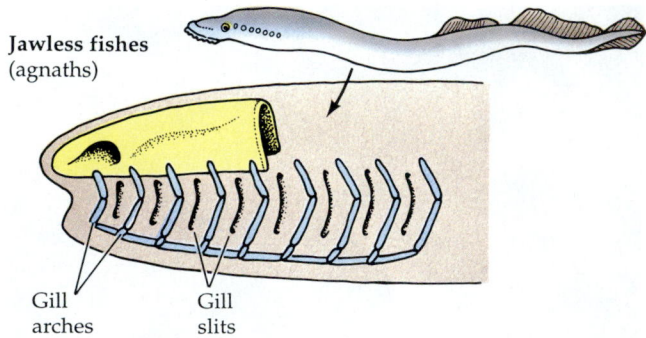

Jawless fishes
(agnaths)

Gill
arches

Gill
slits

Early jawed fishes
(placoderms)

Jaw (from
gill arches)

Gill
slit

Modern jawed fishes
(cartilaginous and bony fishes)

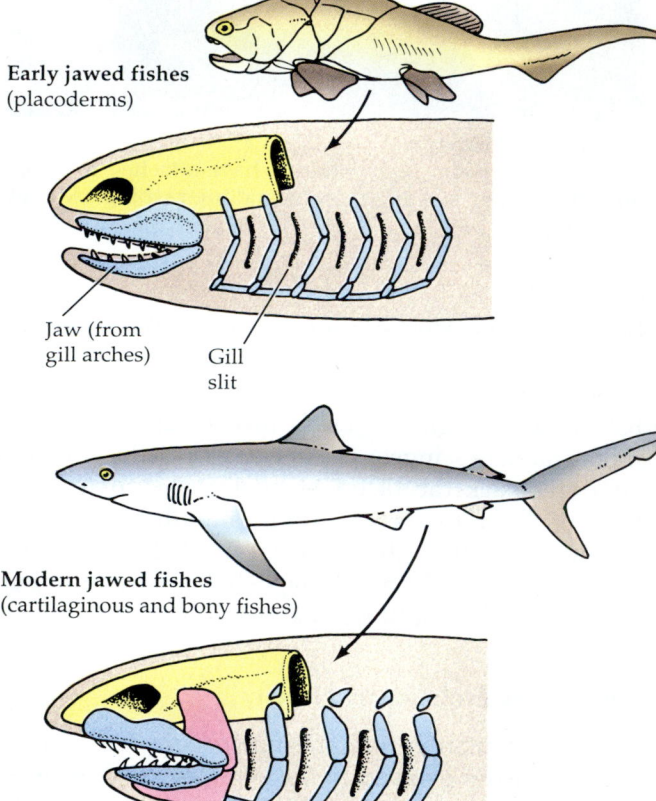

27.13 Jaws from Gill Arches
A probable scenario for the evolution of jaws from the anterior gill arches of fishes.

by the end of the Devonian period, and none survived to the end of the Paleozoic era.

Fins and Mobility

Two other groups of fishes that survive today became numerically important during the Devonian period. Members of one group—the cartilaginous fishes (class Chondrichthyes): the sharks, skates and rays, and chimaeras—have a skeleton composed entirely of a firm but pliable material called **cartilage** (Figure 27.14). Their skin, unlike that of many early jawless fishes and placoderms, is not armored. For the most part the skin is flexible and leathery, sometimes bearing bristly projections that give it the consistency of sandpaper. The loss of external armor was favored because it increased mobility, hence improving ability to escape from predators, and was accompanied by the evolution of rapid swimming.

Most sharks and their relatives are rapid swimmers. Control of swimming was improved by the evolution of two pairs of appendages, a pair of pectoral fins just behind the gill slits, and a pair of pelvic fins just in front of the anal region that stabilize the fish as it moves. These are the two pairs of limbs that are a key feature of the vertebrate body plan (see Figure 27.11). Sharks move forward by means of their tail and pelvic fins. Skates and rays swim by means

27.14 Cartilaginous Fishes
(a) Most sharks, such as the seven-gill shark *Notorhynchus cepedianus*, are active predators living in open waters. (b) Skates and rays, represented here by the bullseye electric ray *Diplobatis ommata*, have their mouths on the ventral surface of their body and feed on the ocean bottom.

(a)

(b)

of the undulating movements of their pectoral fins, which are greatly enlarged.

Most sharks are predators, but others evolved to feed by straining plankton. The world's largest fish, the whale shark (*Rhincodon typhus*), which may grow to more than 15 meters long and weigh more than 9,000 kilograms, is a plankton strainer. Most skates and rays live on the ocean floor and feed on mollusks and other invertebrates buried in the sediments. Chimaeras feed on mollusks whose shells they crack with their hard, flat teeth. The cartilaginous fishes originated in the sea and only rarely penetrate fresh water.

Mobility and Buoyancy

Most of the early evolution of fishes took place in the oceans, but one lineage evolved in fresh water. These fishes have internal skeletons of bone rather than cartilage, giving them their common name, bony fishes (class Osteichthyes). In fresh water, oxygen is often in short supply. The pharyngeal slits in bony fishes open into a single chamber covered by a hard flap. Movement of the flap improves the flow of water over the gills and brings more oxygen in contact with the gas-exchange surfaces. Early bony fishes also evolved lunglike sacs that supplemented the gills in respiration. As we will see in the next section, this evolutionary step was important for another lineage of fishes that colonized the land.

In most fishes the lungs evolved into **swim bladders,** organs of buoyancy that help keep the fish suspended in water. By adjusting the amount of gas in its swim bladder, a fish can control the depth in the water column at which it is stable. Only a small group of fishes, the lungfishes, still uses the lungs for the original purpose of respiration.

The external armor of bony fishes is greatly reduced, but most species are covered with flat, smooth, thin scales that provide some protection without weighing down the fish. With their light skeletons and their swim bladders, some bony fishes recolonized the seas to become major players in marine ecological communities.

Among the more than 20,000 species of bony fishes living today is a remarkable diversity of sizes, shapes, and lifestyles (Figure 27.15). The smallest bony fish is a goby that is only about 1 centimeter long as an adult. The largest are ocean sunfishes that weigh up to 900 kilograms. Fishes are adapted to exploit nearly all types of food sources available in fresh and salt water. In the oceans they filter plankton from the water, rasp algae from rocks, eat corals and other colonial invertebrates, dig invertebrates from soft sediments, and prey upon all other vertebrates except large whales and dolphins. In fresh water they eat plankton, devour insects of all aquatic orders,

harvest fruits that fall into the water in flooded forests, and prey upon other aquatic vertebrates. Many live buried in soft sediments, where they grab passing prey or from which they emerge at night to feed in the water column above. Many are solitary, but others form large schools in open water. Many fishes perform complicated behaviors by which they maintain schools, build nests, court and choose mates, and care for their young. They are a group in which behavior has stimulated many evolutionary changes.

With their fins and swim bladders, fishes can readily control their position in open water. Their eggs, however, tend to sink. Most fishes attach their eggs to plants or to the substrate. Some species with very small eggs discharge them directly into surface waters, where they are buoyant enough to complete their development before they sink very far. Most fishes, however, move to food-rich shallow waters to spawn, which is why coastal waters and estuaries are so important in the life cycles of many species. Some fishes, such as salmon, actually abandon salt water for breeding, ascending rivers to spawn in streams and freshwater lakes. Conversely, some species, such as eels, that live most of their lives in fresh water migrate to the sea to spawn.

BREATHING AIR AND EXPLORING THE LAND

The evolution of lunglike sacs in the early bony fishes set the stage for the invasion of land by some of their descendants. Early bony fishes probably used their lungs to supplement the gills when oxygen levels in the water were low. This ability would also have allowed them to leave the water temporarily when pursued by predators unable to breath air. With their unjointed fins, however, bony fishes were unable to do more than flop around on land as most fishes do today if placed out of water. Evolution of joints in the fins enabled fishes to move over land to find new bodies of water when those in which they lived dried up or became overpopulated. Later, these fishes began to use terrestrial food sources and became more fully adapted to life on land.

A group of bony fishes commonly thought to be ancestors of forms that invaded the land were the crossopterygians, or lobe-finned fishes (subclass Crossopterygii). Lobe-fins flourished from the Devonian period into Mesozoic times. They were once thought to have become extinct about 25 million years ago, but in 1939 a lobe-fin was caught by a commercial fisherman off the east coast of Africa. Since that time, several dozen specimens of this extraordinary fish, *Latimeria chalumnae*, have been collected. *Latimeria*, a predator on other fishes, reaches a length of about 1.5 meters and weighs up to 82 kilograms (Figure 27.16a).

The skeleton of *Latimeria* is composed mostly of cartilage, and its swim bladder contains fat rather than gases. *Latimeria* has a distinctive class of immunoglobulin genes that differ from those of bony fishes and terrestrial vertebrates but are similar to those of sharks.

In and Out of the Water: The Amphibians

During the Devonian period, the amphibians (class Amphibia) arose from ancestors that had lungs with which they could breathe air. Amphibians had another breathing advantage: thin skins that allowed respiration. In addition, the stubby fins of their ancestors evolved into the walking legs of amphibians. The design of these legs has remained largely unchanged throughout the evolution of terrestrial vertebrates. (Figure 27.16*b*). Devonian predecessors of amphibians gradually evolved to be able to live on swampy land and, eventually, on drier land. They were probably able to crawl from one pond or stream to another by pulling themselves along on their fin-like legs, as do some modern species of catfishes and other fishes. The earliest amphibians, whose bodies were fishlike, merely evolved modifications of this locomotor ability.

About 4,500 species of amphibians live on Earth today, many fewer than those known only from fossils. Living amphibians belong to three classes (Figure 27.17): the wormlike, tropical, burrowing caecilians (order Gymnophiona); frogs and toads (Order Anura = tailless); and salamanders (order Urodela or Caudata = tailed). Most species of frogs and toads live in tropical and warm temperate regions, al-

27.15 A Diversity of Bony Fishes
(a) The lungfish *Protopterus aethiopicus* survives the frequent East African droughts by breathing air. *(b)* The Volga sturgeon *Huso huso*, the major source of Russian caviar, is a survivor of an ancient lineage of bony fishes. *(c)* Salmonids such as the rainbow trout *Salmo gairdneri* are commercially important fishes of the northern temperate zone. *(d)* Angelfishes like this emperor angelfish, *Pomacanthus imperator*, are dominant fishes on tropical coral reefs. *(e)* Perches are widely distributed in marine and fresh waters. The yellow perch, *Perca flavescens*, is typical of many of these species.

(c)

(a)

(d)

(b)

(e)

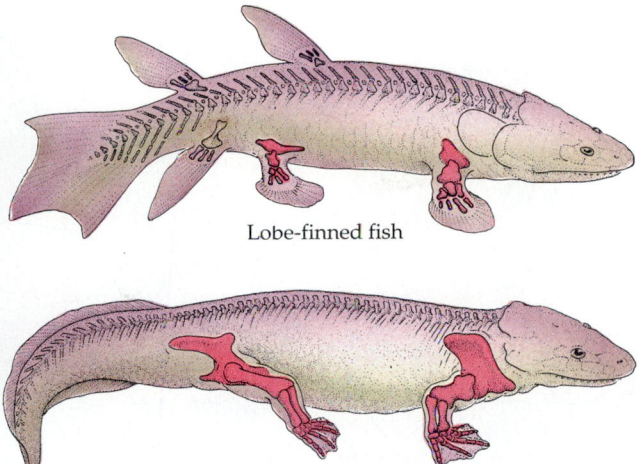

Lobe-finned fish

Early amphibian *(b)*

27.16   The Transition from Sea to Land
(a) Latimeria is the sole survivor of a lineage, thought to
have become extinct, from which land invaders may have
evolved. *(b)* Only a modest enlargement of the bones of a
fish with fins like those of *Latimeria* is needed to convert
them into walking legs.

though a few are found at high latitudes. Salamanders are more diverse in temperate regions, but many
species are found in the tropics, particularly in the
mountains of Middle America, where cool, moist
conditions prevail.

Most amphibians live in water at some time in
their lives. In the typical life cycle, part or all of the
adult life is spent on land, usually in a moist habitat,
but adults return to fresh water to lay their eggs
(Figure 27.18). An amphibian egg must remain moist
in order to develop because it is surrounded by a
delicate envelope through which it loses water readily if the surroundings are dry. The egg of most species gives rise to a larva that lives in water before
metamorphosing into a terrestrial adult.

There are interesting variations on this life cycle.
Some amphibians are entirely aquatic, never leaving
the water at any stage in their lives. Others are en-

(a)

(c)

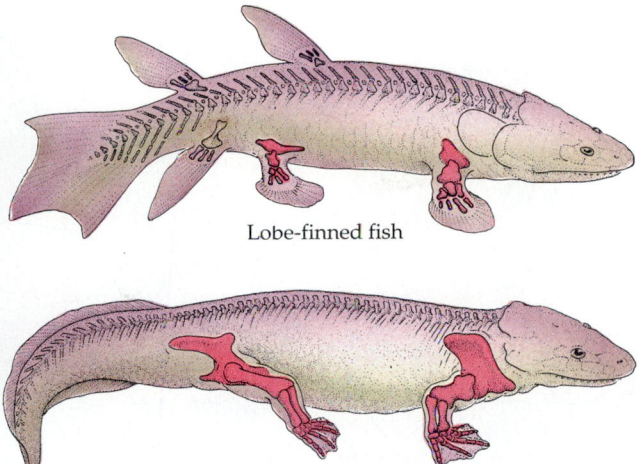

(b)

27.17   Amphibians
(a) Gymnophis mexicana, a burrowing Central American
caecilian, superficially looks more like a worm than an
amphibian. *(b)* The bright blue skin of this South American "poison arrow frog" contains toxins that humans
have used to render hunting implements more lethal.
(c) The bright colors of *Tylototriton verrucosus*, a Chinese
salamander, indicate to at least some visually hunting
predators that it is poisonous.

27.18 In and Out of the Water
Most stages in the life cycle of many temperate-zone frogs take place in the water. (Many tropical frogs, by contrast, lay their eggs in wet places on land.)

tirely terrestrial, laying their eggs in moist places on land. In many lungless salamanders that live in rotting logs or in the soil, gas exchange takes place entirely through the skin and lining of the mouth. However, all terrestrial species are confined to moist environments because amphibians lose water through their skins when exposed to dry air, although some toads have tough skins that enable them to live for long periods of time in dry places.

Amphibians are the focus of much attention be-cause populations of many species are declining rapidly throughout the world. For example, the golden toad is disappearing from the Monteverde Cloud Forest Reserve in Costa Rica, a reserve established primarily to protect this rare species. The reasons for the declines are not known, but biologists are monitoring amphibian populations closely to learn more about the causes of their difficulties and to determine the implications of amphibian declines for other organisms.

Colonization of the Land

Two morphological changes allowed vertebrates to exploit the full range of terrestrial habitats. The first was an egg whose shell is relatively impermeable to water but is permeable to gases and which thus can be deposited in relatively dry places. The second was a suite of traits that included a tough skin impermeable to water and kidneys that could excrete concentrated urine.

Amniotes evolved both of these morphological changes and, as a result, were the first vertebrates to become common over much of the terrestrial surface of Earth. The amniote egg has a leathery or brittle calcium-filled shell that retards evaporation of the fluids inside. Within the shell and surrounding the embryo are three membranes that protect the embryo from desiccation and assist it in excretion and respiration. The egg also supplies the embryo with large quantities of food—yolk—that permit it to attain a relatively advanced state of development before it hatches and must feed itself (Figure 27.19).

An early amniote lineage, the reptiles (class Reptilia), arose from early tetrapods in the Carboniferous period, some 300 mya. About 6,000 species of reptiles live today. Most reptiles do not care for their offspring after the eggs hatch. In fact, most of them desert their eggs after laying them. In some species eggs do not develop shells and are retained inside the female's body until they hatch. Still others have evolved placentas, organs that nourish the developing embryos (see Chapter 37).

The skin of a reptile is covered with horny scales containing keratin that greatly reduce loss of water from the body surface and is thus unavailable as an organ of gas exchange. In reptiles gases are exchanged almost entirely by the lungs, which are much larger than those of amphibians. A reptile forces air into and out of its lungs by bellowslike movements of the ribs. The reptilian heart is divided into chambers, which separate oxygenated from unoxygenated blood. With this heart reptiles can generate higher blood pressures than amphibians can. Reptiles can sustain higher levels of muscular activity than amphibians can, although they tire much more rapidly than do birds or mammals.

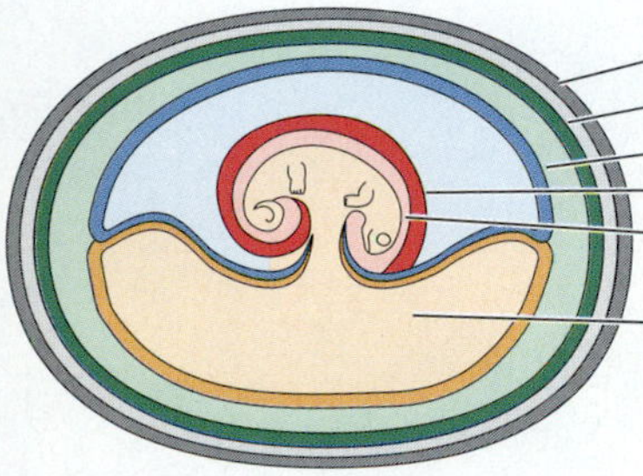

Modern Reptiles

There are three subclasses of modern reptiles. Turtles and tortoises (subclass Chelonia; (Figure 27.20*a*) have an armor of bony plates, both dorsally and ventrally, forming a shell into which the head and limbs can be withdrawn. Most turtles live in lakes and ponds, but tortoises are terrestrial, and sea turtles spend their entire lives at sea except when they come ashore to lay eggs. Most turtles and tortoises are primarily herbivorous, eating a variety of aquatic and terrestrial plants, but some species are strongly carnivorous.

Only two species of the reptilian subclass Sphenodontida survive: the tuataras that live only on a few islands off the coast of New Zealand (Figure 27.20*b*). Tuataras superficially resemble lizards but differ from them in certain internal anatomical features.

The third reptilian subclass, Squamata, includes the lizards, amphisbaenas, and snakes (Figures 27.20*c* and 27.20*d*). Most lizards are insectivores, but some are herbivores, and still others prey on vertebrates. Some species of monitors that live in the East Indies are the largest lizards, growing as long as 3 m. Most lizards walk on four limbs, but some are limbless. All snakes are legless; they probably evolved from burrowing lizards. All snakes are carnivores that can swallow objects much larger than their own diameter. Three groups of snakes have evolved poison glands and inject venom into their prey with their teeth. Snakes range in size up to pythons more than 10 meters long.

Dinosaurs and Their Relatives

During the Mesozoic era, one amniote lineage, the thecodonts (class Archosauria) split from other reptiles and underwent an extraordinary diversification. One thecodont lineage gave rise to crocodilians (subclass Crocodylia—crocodiles, caimans, gavials, and alligators—which are confined to tropical and warm temperate environments (Figure 27.21). Crocodilians spend much of their time in water, but they build nests on land or on floating piles of vegetation. The eggs, which are warmed by heat generated by the decaying organic matter of the nest, typically are tended by the female until they hatch. All crocodilians are carnivorous; they prey upon vertebrates of all classes, including large mammals.

Another thecodont lineage led to the dinosaurs, the prevalent large terrestrial animals for millions of years and the largest terrestrial animals ever to inhabit Earth. Some of the largest dinosaurs weighed up to 100 tons and were agile and fast-moving (see

Shell
Chorion
Allantois
Amnion
Amniotic cavity
Yolk sac

27.19 An Egg for Dry Places
The evolution of the amniote egg—with its shell, embryo-enveloping membranes (amnion, chorion, and allantois), and embryo-nourishing yolk sac—was a major step in the colonization of the terrestrial environment.

(a)

(b)

(c)

(d)

27.20 Reptilian Diversity
(a) The green sea turtle, *Chelonia mydas*, is widely distributed in tropical oceans. (b) The Madagascar chameleon, *Camaelo pantheri*, has a long tail with which it can grasp branches and large eyes that move independently in their sockets. (c) The tuatara, *Sphenodon punctatus*, looks like a typical lizard, but it is one of only two survivors of a lineage that separated from lizards long ago. (d) An eyelash viper (*Bothrops schlegeli*) swallows a lizard.

27.21 Crocodilians
(a) Most crocodilians are tropical, but alligators live in warm temperate environments. They are found in Asia and, like this *Alligator mississippiensis*, in the southeastern United States. (b) A saltwater crocodile, *Crocodillus porosus*, from the Northwest Territory of Australia.

(a)

(b)

BOX 27.A

The Four-Minute Mile

Many mammals can run much faster, yet we humans are proud to have achieved a four-minute mile. Terrestrial vertebrates did not achieve such speeds easily. Amphibians and reptiles fill and empty their lungs using some of the same muscles they use for walking. In addition, because the limbs protrude laterally, their movement generates a strong lateral force that bends the body from side to side. Recent studies have shown that these animals cannot breathe while they walk or run. Therefore, they can operate aerobically only briefly. Because they depend upon anaerobic glycolysis while running, they tire rapidly.

In the lineage leading to dinosaurs and birds and in the lineage leading to mammals, the legs assumed more vertical positions, which reduced the lateral forces on the body during locomotion. Special ventilatory muscles that can operate independently of locomotory muscles also evolved. These muscles are visible in living birds and mammals. We can infer their existence in dinosaurs from the structure of the vertebral column and the capability of many dinosaurs for bounding, bipedal (using two legs) locomotion. The ability to breathe and run simultaneously, a capability we take for granted, was a major innovation in the evolution of terrestrial vertebrates.

A future four-minute miler?

Figure 28.12). The ability to move actively on land was not achieved easily. The first terrestrial vertebrates probably moved only very slowly, much more slowly than their aquatic relatives. The reason is that they apparently could not walk and breathe at the same time. Not until evolution of the lineages leading to the mammals, dinosaurs, and birds did special muscles evolve enabling the lungs to be filled and emptied while the limbs moved (Box 27.A). This ability enabled its bearers to maintain steady, high levels of activity, which generated enough heat to result in high body temperatures.

Zoologists sometimes refer to birds (subclass Aves) as "feathered dinosaurs." That phrase embodies an important truth because birds are descendants of a lineage of dinosaurs. Early birds that evolved in the Mesozoic era, such as *Archaeopteryx* (Figure 27.22), were intermediate between dinosaurs and modern birds in many ways. Although *Archaeopteryx* was covered with feathers and had well-developed wings, the fingers of its forearms were not much reduced and it had a long tail. This early bird may have been

27.22 A Mesozoic Bird
An artist's recreation of *Archaeopteryx* shows features resembling both its dinosaur relatives and the modern birds that would come later.

(a) Anatomy of a feather

Vane

Rachis

Barbule

Barb

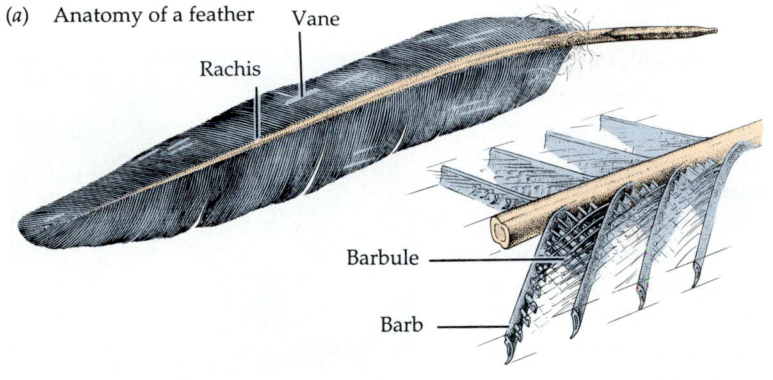

(b) Arrangement of feathers in wing

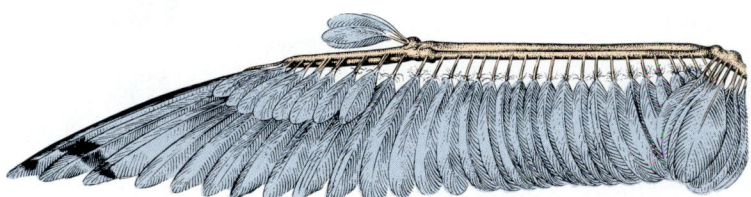

(c) Feather tracts

Capital

Humeral

Alar

Spinal

Femoral

Crural

Caudal

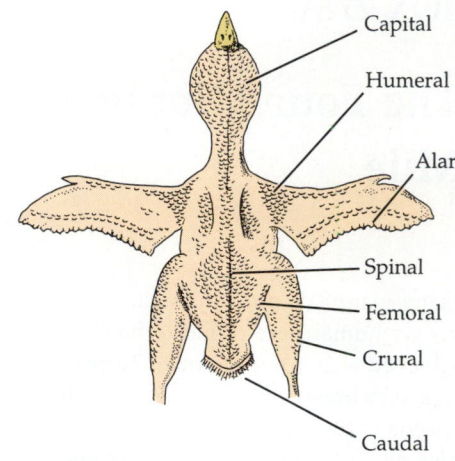

27.23 Feathers and Wings
(a) A feather is sturdy and light, with a complicated anatomy. (b) Large feathers create flying surfaces. (c) The distinct feather tracts on the dorsal surface of a bird.

a weak, flapping flier, relying much on gliding. Its brain case and breastbone were not significantly modified from those of the dinosaurs.

The characteristic feature of birds is their feathers, which are highly modified scales (Figure 27.23a). The flying surface of the wing is created by large quills that arise from the forearm and from reduced, stubby fingers (Figure 27.23b). Other strong feathers sprout like a fan from the shortened tail and serve as stabilizers during flight. Still other feathers, the contour feathers and down feathers, which arise from well-defined tracts (Figure 27.23c), cover the body like a garment and insulate it to control loss of body heat.

The bones of birds are hollow but have internal struts, so they are light but strong. The sternum (breastbone) forms a large, vertical keel to which the breast muscles are attached. These muscles pull the wings downward during the main propulsive movement in flight. Flight is metabolically expensive; a flying bird consumes energy at a very high rate. Bird lungs operate with a pattern of air flow that allows a complete exchange of respiratory gases (see Chapter 41).

Because birds have high metabolic rates, they generate large amounts of heat. They can maintain high body temperatures because they control the rate of heat loss using their feathers, which may be held close to the body or elevated to alter the amount of trapped air they contain and hence the amount of insulation they provide. A bird's metabolic rate is so high that it consumes about eight times the amount of energy per day as a lizard of the same weight! The brain of a bird is relatively large in proportion to body

size, primarily because the cerebellum, the center of sight and muscular coordination, is enlarged. The beaks of modern birds lack teeth.

Most birds lay their eggs in a nest, where they are incubated by the body heat of an adult. The high body temperatures of birds result in short incubation periods, less than two weeks in many small species. The megapodes of Australia and New Guinea incubate their eggs in a pile of rotting organic matter, taking advantage of the heat from decomposition to keep the eggs warm. Nestlings that hatch at a relatively helpless stage and are fed for some time by their parents are referred to as **altricial. Precocial** offspring can run about and feed themselves shortly after hatching. Adults attend even most precocial offspring for a while, warning them of and protecting them from predators, guiding them to good foraging places, protecting them from bad weather, and, in some cases, feeding them.

Because birds invest heavily in parental care by laying large, yolk-laden eggs, incubating the eggs, and guarding and feeding the nestlings, they have lower reproductive rates than nearly all groups of animals we have discussed. Despite their low reproductive rates, birds have evolved very rapidly, another indication of the importance of behavior as a stimulus for evolutionary change.

As a group, birds eat almost all types of animal and plant material. A few aquatic species have bills modified for filtering small food particles from the water. In terrestrial environments, insects are the most important food for birds. About 60 percent of all species of birds belong to the order Passeriformes

(a)

(b)

(c)

(d)

27.24   Birds: Feathered Dinosaurs
(a) Like the macaroni penguin, most penguins live in harsh Antarctic environments, although one species breeds as far north as the Equator. (b) Parrots are most abundant and diverse in Australia and New Guinea. The crimson rosella, a common bird in the eucalyptus forests of temperate Australia, feeds primarily on the seed capsules of eucalyptus trees. (c) The American robin is a well-known member of the order Passeriformes—the perching birds, or song birds. (d) The superb starling, *Spreo superbus*, of East Africa is also a passeriform song bird, but is actually a rather poor songster.

(the perching birds), the majority of which are primarily or exclusively insectivorous. In addition, birds eat fruits and seeds, nectar and pollen, leaves and buds, carrion, and other vertebrates. Birds are major predators of flying insects during the day; some species exploit that food source even at night. By eating the fruits and seeds of vascular plants in many terrestrial ecosystems, birds perform a vital function: seed dispersal.

As adults, birds range in size from the 2-gram bee hummingbird of the West Indies to 150-kilogram ostriches. Some birds of Madagascar and New Zealand known from fossils were even larger, but they were exterminated by early people when they first reached those islands. Although there are more than 8,600 species of living birds, more than in any other vertebrate group except fishes, birds are less diverse structurally than are other vertebrates, probably because of the constraints imposed by flying (Figure 27.24).

THE ORIGIN OF MAMMALS

Mammals (class Mammalia) appeared in the early part of the Mesozoic era, branching from the now extinct order Therapsida. Small mammals coexisted with reptiles and dinosaurs for 150 to 200 million years, but when the large reptiles and dinosaurs disappeared at the close of the Mesozoic era, mammals increased dramatically in number, diversity, and size.

Skeletal simplification accompanied the evolution of mammals from their therapsid ancestors. During mammalian evolution, most lower-jaw bones moved back to the middle ear, leaving a single bone in the lower jaw, and the number of bones in the skull decreased. The bulk of the limbs and the bony girdles from which they are suspended was reduced, and the limbs became oriented beneath the body, as they are in dinosaurs and birds, rather than poking out to the side and then down, as in reptiles. (The significance of this limb orientation is explained in Box

Reptile   Therapsid   Mammal

27.25 Legs under the Body
The limbs of reptiles extend from the sides of their bodies. Therapsids had mammallike thoracic and pelvic girdles that permitted their legs to be positioned underneath the body; mammals continued this trend so that their limbs are even more vertically oriented.

27.A.) Fossils of later therapsids suggest that their legs were positioned underneath the body. Thus early mammals represent a continuation of changes that were already under way (Figure 27.25).

The skeletal features we have been discussing are readily fossilized. The important soft parts of mammals, however, were seldom preserved in the fossil record. Key mammalian features such as mammary glands, sweat glands, hair, and a four-chambered heart may have evolved among the later therapsids, but the existing record does not tell us when this happened. Mammals are unique among animals in suckling their young with a nutritive fluid (milk) secreted by mammary glands. Mammalian eggs are fertilized within the body of the female, and the embryos undergo some development within the uterus prior to being born. Mammals have a protective and insulating covering called hair, which is luxuriant in some species but almost entirely absent in other species, such as whales and dolphins. Instead of hair, the latter have thick layers of fat (blubber) under their skins as insulation.

Mammals have far fewer, but more highly differentiated teeth than reptiles do. Differences in the number, type, and arrangement of teeth in mammals reflect their varied diets. Understanding the relationships between the condition of the teeth and diet among living mammals enables us to infer most features of the diets of extinct groups. Mammals range in size from tiny shrews weighing only about 2 grams to the blue whale, which measures up to 31 meters long and weighs up to 160,000 kilograms, the largest animal ever to live on Earth.

The approximately 4,000 species of living mammals are placed into three major groups: monotremes, marsupials, and eutherians. The subclass Prototheria contains a single order, the Monotremata, represented by two families and only three species, which are found in Australia and New Guinea. These mammals, the duck-billed platypus and the spiny anteaters, or echidnas, differ from other mammals in

laying eggs and in possessing some reptilelike anatomical features (Figure 27.26a). They nurse their young on milk, but there are no nipples on the mammary glands. Rather the milk oozes out and is lapped off the fur by the offspring.

The other two groups of mammals are in the subclass Theria. Females of one group, the order Marsupialia, containing about 240 species, have a ventral pouch or folds in which the young are carried and fed (Figure 27.26b). Gestation (pregnancy) in marsupials is short; the young are born tiny but with well-developed forelimbs, with which they climb to the pouch. Once the offspring has left the uterus, female marsupials may become sexually receptive again. They can then carry a fertilized egg capable of initiating development and replacing the offspring in the pouch should something happen to it. The marsupial mode of reproduction is adapted to relatively harsh and uncertain conditions, where adults must travel long distances to find food and where droughts may cause young offspring to die.

At one time marsupials were widely distributed on the southern continents, but today the majority of species are restricted to the Australian region, with a modest representation in South America (Figure 27.26c). One species, the Virginia opossum, ranges north into the United States. Marsupials radiated into virtually all mammalian lifestyles except that no species are marine or can fly. The largest living marsupial is the red kangaroo of Australia, which weighs up to 90 kilograms, but much larger marsupials existed in Australia until recently. These large marsupials were probably exterminated by people soon after they reached Australia about 40,000 years ago.

Most living mammals are **eutherians** (order Eutheria; sometimes they are called placentals, but this is not a good name because some marsupials also have placentas). Eutherians are more highly developed at birth than are marsupials, and no external pouch houses them after birth. The nearly 4,000 species of eutherians are placed into 16 orders (Figure 27.27),

(a)

(b)

(c)

27.26 Monotremes and Marsupials
(a) The echidna, or spiny anteater, is one of three surviving species of monotremes. It is common and widespread in Australia. *(b)* Australia's kangaroos are perhaps the most familiar marsupials. Here a female red kangaroo (*Macropus rufus*) carries her offspring in the distinctive pouch. *(c)* The mouse opossum *Marmosa murina* is a South American marsupial from the Amazon basin.

(a)

(b)

(c)

(d)

27.27 Eutherian Diversity
(a) The golden-mantled ground squirrel (order Rodentia) is one of many species of small, diurnal (active during the day) rodents of the western North American mountains. *(b)* With their powers of echolocation (see Chapter 39), many bats can locate and capture prey even in complete darkness. This big brown bat, *Eptesicus fuscus* (order Chiroptera), is about to capture a large moth. *(c)* Cats (order Carnivora) ambush their prey and capture them after short chases. The massive legs of the African lion are not suited to long-distance running. *(d)* Large hooved mammals (order Artiodactyla) are important herbivores over much of Earth. Sheep, such as these bighorns, are primarily found in mountainous regions in the Northern Hemisphere.

631

the largest of which is the Rodentia (rodents), with about 1,700 species. The second largest order, the Chiroptera (bats), has about 850 species, followed by the order Insectivora (moles and shrews), with slightly more than 400 species. The largest mammals are marine, but some terrestrial mammals, such as elephants and rhinoceroses, weigh more than several thousand kilograms.

Eutherians are extremely varied in form and ecology. Along with insects, they are the most important grazers in terrestrial ecosystems. They have exerted strong selective pressures favoring the evolution of many features of terrestrial plants, such as spines, tough leaves, and growth forms.

A summary of all animal phyla is given in Table 27.2.

HUMAN EVOLUTION

One mammalian lineage that has undergone extensive recent evolutionary radiation is the primates, the group to which we belong. Primates probably descended from small, arboreal (tree-inhabiting) insectivores some time in the Cretaceous period. The major traits that distinguish primates from other mammals are their adaptations to arboreal life. These adaptations include dexterous hands with opposable thumbs that can grasp branches and manipulate food, nails rather than claws, eyes on the front of the face that provide good depth perception, and very small litters (usually one) of offspring that receive extended parental care.

The primate lineage split into two main branches early in its evolutionary history (Figure 27.28). One

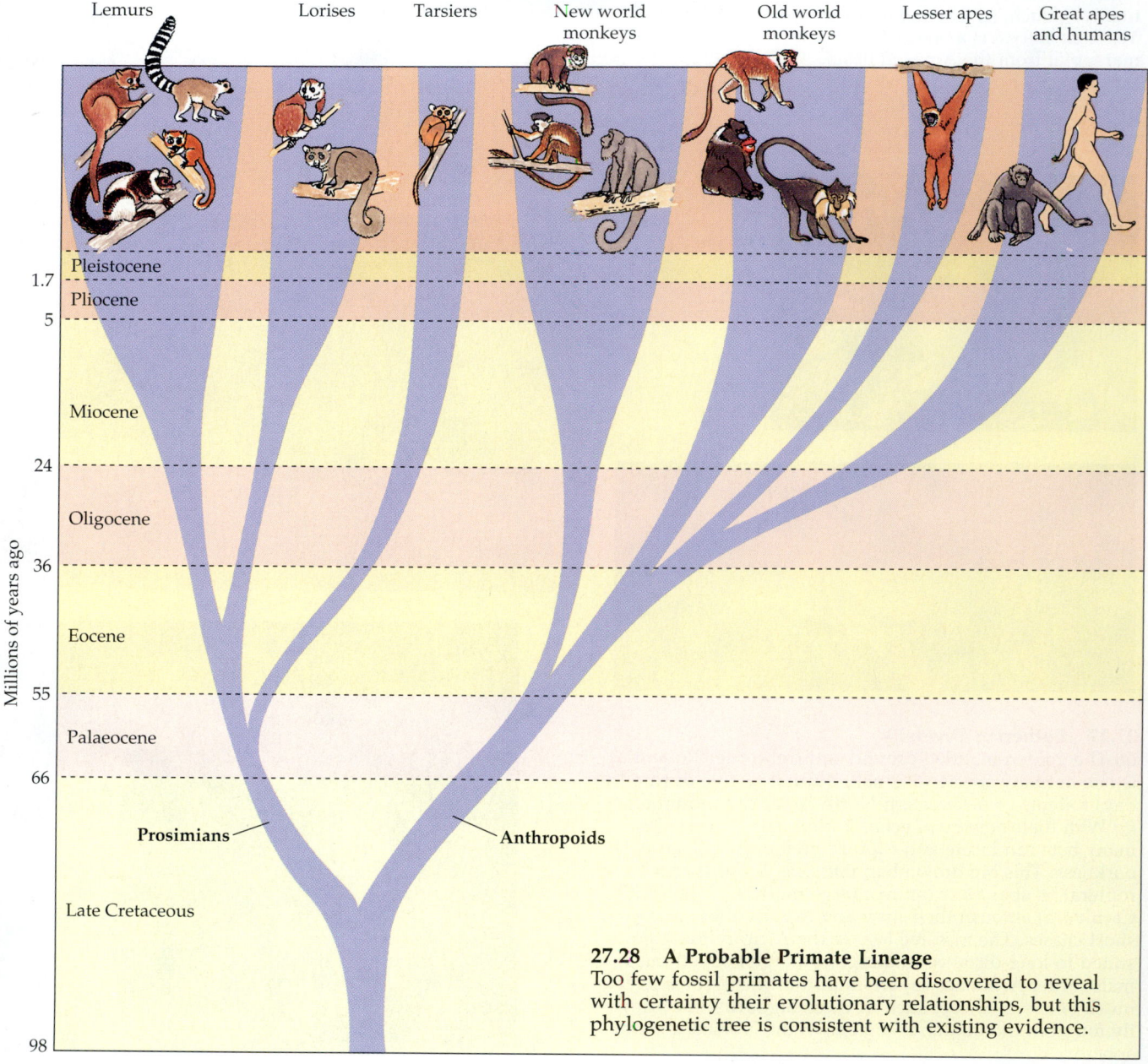

27.28 A Probable Primate Lineage
Too few fossil primates have been discovered to reveal with certainty their evolutionary relationships, but this phylogenetic tree is consistent with existing evidence.

TABLE 27.2
Summary of Living Members of the Kingdom Animalia

PHYLUM	NUMBER OF LIVING SPECIES DESCRIBED	SUBGROUPS
Protostomes		
Porifera: Sponges	10,000	
Cnidaria: Cnidarians	10,000	Hydrozoa: Hydras and hydroids Scyphozoa: Jellyfish Anthozoa: Corals, sea anemones
Ctenophora: Comb jellies	100	
Platyhelminthes: Flatworms	25,000	Turbellaria: Free-living flatworms Trematoda: Flukes (all parasitic) Cestoda: Tapeworms (all parasitic) Monogenea: Ectoparasites of fish
Nemertea: Ribbon worms	900	
Nematoda: Roundworms	20,000	
Rotifera: Rotifers	1,800	
Pogonophora: Pogonophorans	145	
Annelida: Segmented worms	15,000	Polychaeta: Polychaetes (all marine) Oligochaeta: Earthworms, freshwater worms Hirudinea: Leeches
Onychophora: Onychophorans	75	
Tardigrada: Water bears	400	
Chelicerata: Chelicerates	63,000	Pycnogonida: Sea spiders Merostomata: Horseshoe crabs Arachnida: Scorpions, harvestmen, spiders, mites, ticks
Crustacea: Crustaceans	40,000	Crabs, shrimp, lobsters, barnacles, copepods
Uniramia: Unirames	1,500,000	Myriapoda: Centipedes, millipedes Insecta: Insects
Mollusca: Mollusks	100,000	Polyplacophora: Chitons Monoplacophora: Monoplacophorans Bivalvia: Clams, oysters, mussels Gastropoda: Snails, slugs, limpets Cephalopoda: Squids, octopuses, nautiloids
Deuterostomes		
Phoronida: Phoronids	15	
Bryozoa (Ectoprocta): Moss animals	4,000	
Brachiopoda: Lamp shells	350	More than 26,000 fossil species described
Hemichordata: Hemichordates	100	Pterobranchia: Pterobranchs Enteropneusta: Acorn worms
Chaetognatha: Arrow worms	100	
Echinodermata: Echinoderms	7,000	Crinoidea: Sea lilies, feather stars Asteroidea: Sea stars Ophiuroidea: Brittle stars Concentricycloidea: Sea daisies Echinoidea: Sea urchins Holothuroidea: Sea cucumbers
Chordata: Chordates	40,000	Urochordata: Tunicates Cephalochordata: Lancelets Agnatha: Lampreys, hagfishes Placodermi: Placoderms Chondrichthyes: Cartilaginous fishes Osteichthyes: Bony fishes Amphibia: Amphibians Reptilia: Reptiles Archosauria: Dinosaurs, crocodilians, birds Mammalia: Mammals

(a)

(b)

(c)

lineage gave rise to the prosimians—lemurs, tarsiers, pottos, and lorises (Figure 27.29). Prosimians were formerly found on all continents, but today they are restricted to Africa, tropical Asia, and Madagascar. All mainland species are arboreal and nocturnal, but on Madagascar, where there has been a remarkable prosimian radiation, there are also diurnal and terrestrial species. Until the recent arrival of humans, there were no other primates on Madagascar.

The anthropoids—monkeys, apes, and humans—evolved from an early primate stock about 55 million years ago in Africa or Asia. New World monkeys have been evolving separately from Old World monkeys long enough that they could have reached South America from Africa when those two continents were still connected. Perhaps because tropical America has

27.29 Prosimians
(a) The sifaka lemur, *Propithecus verreauxi*, is one of many lemur species of Madagascar, where they are part of a unique assemblage of plants and animals. *(b) Loris tardigradis*, the slender loris, of southern India. *(b)* In the rainforests of Borneo, this tarsier (*Tarsius bancanus*) seems otherworldly to our eyes.

(a)

(b)

27.30 Monkeys
(a) Golden lion tamarins (*Leontopithecus rosalia*) are New World monkeys, living in the trees of the coastal Brazilian rainforest. *(b)* Some Old World species, such as these Japanese macaques (*Macaca fuscata*) live and travel in groups.

27.31 Apes
(a) The gibbons, genus *Hylobates*, are the smallest of the apes. This white-handed gibbon is from Laos. (b) Intelligent and endangered, orangutans face massive habitat destruction in their native Indonesia. (c) A mother chimpanzee, *Pan troglodytes*, carries her offspring on her back. Her "knuckle walk" position is characteristic of apes when they move on the ground. (d) A family of gorillas in Rwanda.

been heavily forested for a long time, all New World monkeys are arboreal (Figure 27.30a). Many of them have long, prehensile (adapted for grasping) tails with which they can hold on to branches. Many Old World primates are arboreal, but there are terrestrial Old World species, some of which live and travel in large groups (Figure 27.30b). No Old World primates have prehensile tails.

About 35 mya the lineage leading to the modern apes separated from other Old World primates. The first apes were arboreal, but some species came to live in drier habitats with scattered trees where they obtained most of their food from the ground. Jaw bones of apes that lived between 15 and 8 mya have been found in Africa, the Near East, and Asia. These apes, genus *Ramapithecus*, have features suggesting that they were the beginning of the lineage leading to humans. Like us, ramapithecines had short muzzles and small canines. Their chewing teeth are worn down flat, indicating that they chewed from side to side as we do rather than up and down as chimpanzees and gorillas do.

The four living genera of apes—gorillas (*Gorilla*), chimpanzees (*Pan*), orangutans (*Pongo*), and gibbons (*Hylobates*)—are restricted to tropical Africa and Asia (Figure 27.31). Several lines of evidence suggest that gibbons and orangutans are not as closely related to us as chimpanzees and gorillas are, but which of the latter two apes is our closer relative (Table 27.3)? The three possible cladograms are shown in Figure 27.32. Which one is the most plausible depends on the relative weights given to different kinds of evidence.

TABLE 27.3
Which Ape Is the Closest Relative of Humans?

CHARACTERISTIC	HUMAN	CHIMPANZEE	GORILLA	FAVORED CLADOGRAM[a]
Limb length	Arms shorter than legs	Legs shorter than arms	Legs shorter than arms	3
Canine teeth	Small	Large	Large	?
Thumbs	Long	Short	Short	3
Head hair	Long	Short	Short	?
Calf muscles	Large	Small	Small	?
Buttocks	Fat	Thin	Thin	?
Number of chromosomes	46	48	48	3
Structure of chromosomes 5 and 12	Like other primates	Different from other primates	Different from other primates	3
Fluorescence of chromosomes Y and 13	Same as gorilla	Like other primates	Same as human	1
Alpha hemoglobin chain, compared to human	Same as chimpanzee	Same as human	One amino acid different	2
GM factor in blood	Same variability as chimpanzee	Same variability as human	Not variable	2
Sequence of amino acids in glycogen	Generally like other primates	Same as gorilla	Same as chimpanzee	3

[a]See Figure 27.32.

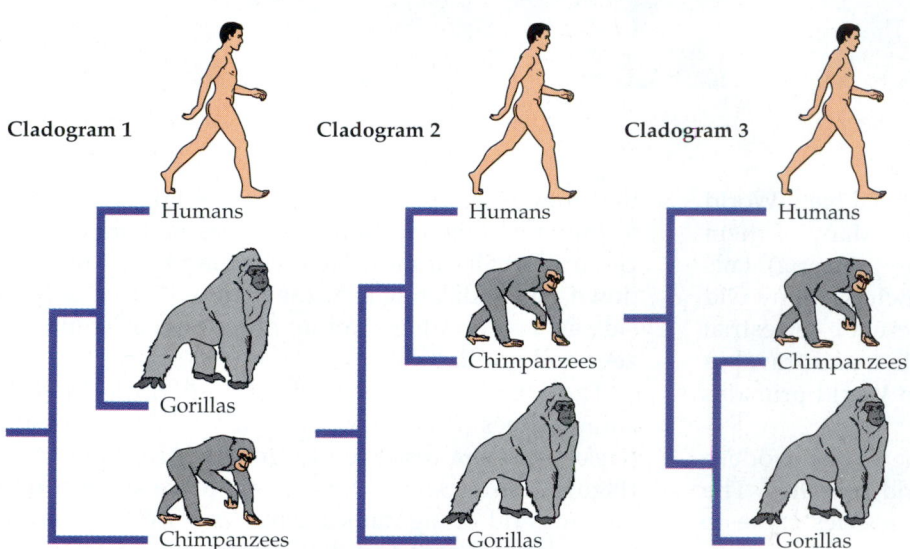

27.32 Which Cladogram Is Correct?
These three cladograms are all compatible with existing evidence. Which one is preferred depends on which traits are given more weight.

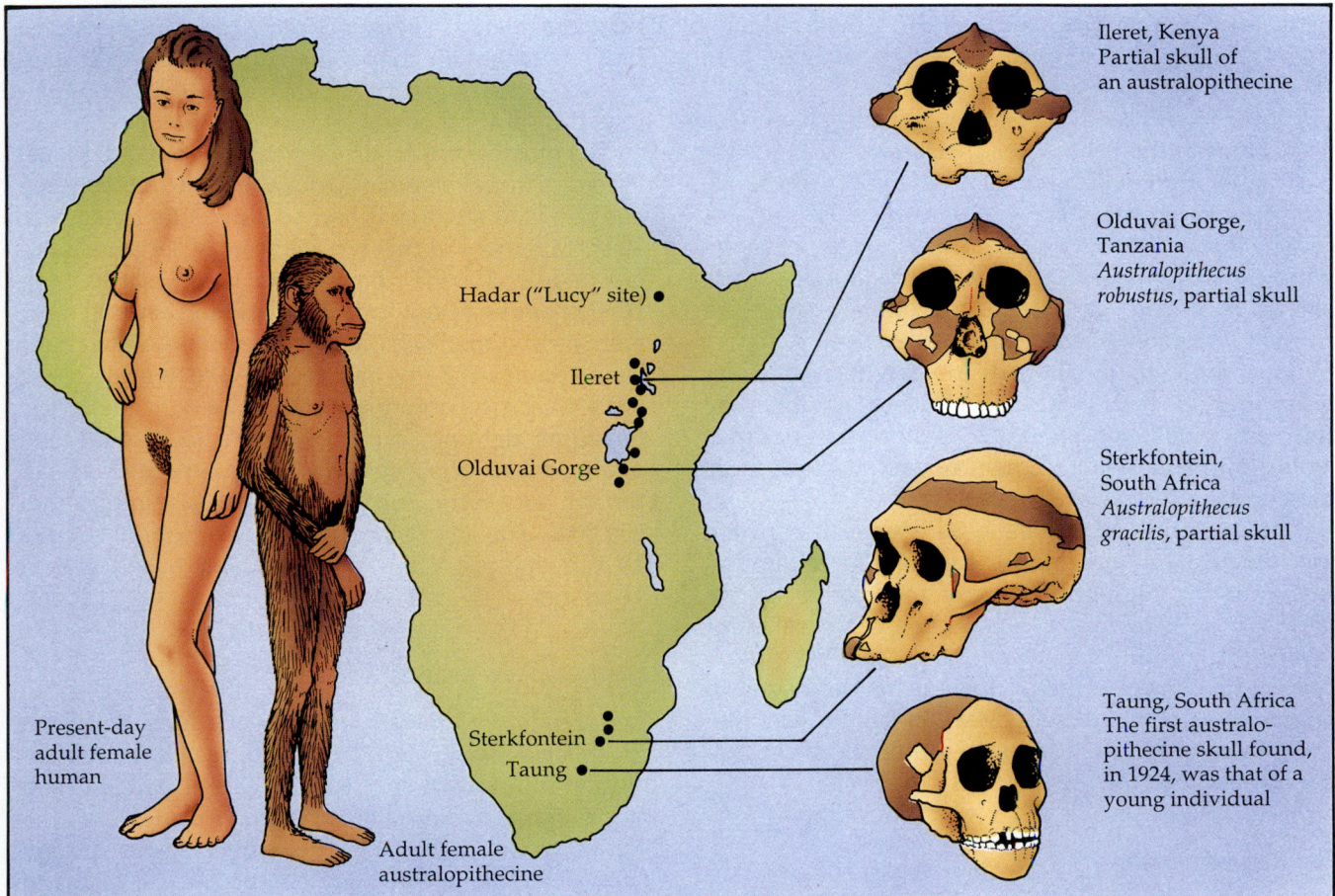

Ileret, Kenya
Partial skull of
an australopithecine

Olduvai Gorge,
Tanzania
*Australopithecus
robustus*, partial skull

Sterkfontein,
South Africa
*Australopithecus
gracilis*, partial skull

Taung, South Africa
The first australo-
pithecine skull found,
in 1924, was that of a
young individual

Hadar ("Lucy" site)

Ileret

Olduvai Gorge

Sterkfontein

Taung

Present-day
adult female
human

Adult female
australopithecine

27.33 Australopithecine Fossils
Dots indicate the sites in eastern Africa where australopithecine fossils have
been found. Very few such fossilized remains are complete; the illustrations of
some of the partial skulls excavated show the actual bone that was found in
lighter color; darker areas were reconstructed with modern materials.

Evidence from bones, teeth, and soft parts of the
body suggested that chimpanzees and gorillas are
more closely related to one another than either of
them is to humans (cladogram 3), but evidence from
studies of chromosomes and molecules suggests
otherwise.

Some experts believe that *Ramapithecus* was the
direct ancestor of modern apes and humans; others
believe that *Ramapithecus* is a member of only the
hominid (human) lineage. But there is no doubt that
members of another lineage, the australopithecines,
are direct human ancestors. The australopithecines
had distinct morphological adaptations for **bipedalism,** locomotion in which the body is held erect and
moved exclusively by movements of the hind legs
beneath it. *Australopithecus africanus*, which had a
broad, bowl-shaped pelvis similar to that of modern
humans, was presumably already well adapted for
walking upright. Bipedal locomotion frees the hands
to manipulate objects and to carry them while walking. It also elevates the eyes, enabling the animal to

see over tall vegetation to spot predators and prey.
Both advantages were probably important for early
australopithecines.

The first australopithecine skull was found in
South Africa in 1924; since then fragments have been
found in other sites in Africa (Figure 27.33). The
oldest and most complete fossil skeleton of an australopithecine, approximately 3.5 million years old,
was discovered in Ethiopia. That individual, a young
female known to the world as Lucy, attracted a great
deal of attention because she was so complete and
well preserved. Lucy has been assigned to the species
Australopithecus afarensis, the most likely ancestor of
later hominids. All the evidence from different parts
of her skeleton suggests that Lucy, who was only
about one meter tall, walked upright.

From *Australopithecus afarensis* ancestors, a number
of species of australopithecines evolved. Several million years ago, two distinct types of australopithecines lived together over much of eastern Africa. The
more robust type (about 40 kilograms) is represented

by at least two species, both of which died out suddenly about 1.5 million years ago. The smaller (25 to 30 kilograms), more slender *A. africanus* is just as old, but it is much rarer as a fossil, suggesting that it was less common than the other species.

Because they were less agile, members of the robust species probably stayed relatively close to trees, to which they retreated at night and when predators were near. Members of the small *A. africanus* were able to run faster, and because of their smaller size needed less food per day to survive. They probably lived in more-open, drier savannas where food was less abundant than in the moister areas inhabited by the more robust species. Their small size and greater agility probably enabled them to exploit these more dangerous and less productive areas.

The evolution of the hand as a precise grasping instrument was valuable for carefully selecting and manipulating high-quality plant food items. In savannas, the growing regions of grasses close to or under the ground are prime food items. The more precise its control of hand movement, the better an animal is able to pull on grasses in such a way that the growing region is harvested rather than being left behind because the stem breaks off above it.

The Rise of *Homo*

Many experts believe that a population of *Australopithecus africanus* or a similar species gave rise to the genus *Homo* about 2.5 million years ago; then early members of *Homo* lived at the same time as australopithecines for perhaps half a million years. Two major changes accompanied the evolution of *Homo* from *Australopithecus*: an increase in body size and a striking increase in brain size to about double that of the late australopithecines. The oldest fossil remains of members of the genus *Homo*, *H. habilis*, discovered in the Olduvai Gorge, Tanzania are estimated to be 2 million years old. Tools used by these early hominids to obtain food were found with the fossils. Other fossils of *H. habilis* have been found in Kenya, Ethiopia, and South Africa, indicating that the species had a wide range in Africa.

Homo habilis lived in relatively dry areas, where for much of the year the main food reserves are subterranean roots, bulbs, and tubers. Exploitation of these food resources requires an ability to dig into hard, dry soils, something that cannot be done with the unaided hand. Moreover, although these underground storage organs of plants are good sources of carbohydrates, they are much lower in proteins than are leaves, so *H. habilis* would have needed to supplement its diet with animal foods. Roots can be harvested in large quantities in a relatively short time by an individual with a simple digging tool, and they would have been harvestable by *H. habilis* women

carrying infants. Therefore, male *H. habilis* may have had more time for cooperative hunting of large, dangerous animal prey than if their diet had been dominated by leafy foods.

The only other known extinct species of our genus, *Homo erectus*, evolved in Africa about 1.6 million years ago. Soon thereafter it had spread as far as eastern Asia. Members of this species were as large as modern people, but their bones were somewhat heavier. *H. erectus* used fire for cooking and for hunting large animals and made characteristic stone tools that have been found in many parts of the Old World. These tools were probably used for a variety of purposes, including digging, capturing animals, cleaning and cutting meat, scraping hides, and cutting wood. Although *H. erectus* survived in Eurasia until about 250,000 years ago, it was replaced in tropical regions by our species, *Homo sapiens*, about half a million years ago.

Homo sapiens Evolves

The trends we have observed in the transition from *Australopithecus* to *H. erectus* continued with the evolution of our own species. The earliest humans had larger brains and smaller teeth than members of the earlier species of *Homo*. These changes were probably favored by an increasingly complex social life. The ability of group members to communicate with one another was valuable in cooperative hunting and gathering, for sharing information about the location and use of food sources, and for improving one's status in the complex social interactions that must have characterized those societies just as they do ours today.

Several types of *H. sapiens* existed during the mid-Pleistocene epoch. All were skilled hunters of large mammals, even though plants continued to be important components of their diets. During this period another distinctly human trait emerged: religious practice and a concept of life after death. These beliefs are revealed by the fact that deceased individuals were buried with tools and clothing in their graves, presumably for their existence in the next world.

One type of *Homo sapiens*, generally known as Neanderthal because it was first discovered in the Neander Valley in Germany, was widespread in Europe and Asia between about 75,000 and 30,000 years ago. Neanderthals were short, stocky, and powerfully built people whose massive skulls housed brains somewhat larger than our own. They manufactured a variety of tools and were highly efficient at hunting large mammals, which were ambushed and subdued in close combat by several individuals attacking together. For a short time their range overlapped that of Cro-Magnon people, a more modern form of *H. sapiens*, but then they abruptly disappeared. Many

27.34 Hunting, Pastoralism, and Agriculture
(a) Cro-Magnon people depended on hunting to obtain food. As this painting from the caves of France shows, Cro-Magnons often depicted animals with arrows and spears flying toward them; these paintings may have served as part of religious rituals designed to increase success during hunts. (b) Pastoralism—the raising of domestic animals to provide food and other products—displaced hunting in many societies. This Fulani herder occupies dry season temporary quarters in Burkina Faso, Africa. (c) Agricultural practices increase food production, feeding an ever-increasing human population. Agricultural development has totally transformed the landscape in these hills above Port-au-Prince, Haiti.

(a)

scientists believe that the Neanderthals were exterminated by the Cro-Magnons, just as *H. habilis* may have been exterminated by *H. erectus*.

Cro-Magnon people made and used a variety of sophisticated tools. They created the remarkable paintings of large mammals, many of them showing scenes of hunting, that have been discovered in caves in various parts of Europe (Figure 27.34a). The animals depicted were characteristic of the cold steppes and grasslands that occupied much of Europe during periods of glacial expansion. Cro-Magnon people spread across Asia, reaching North America perhaps as early as 20,000 years ago, although the date of their arrival in the New World is uncertain. As they spread rapidly southward through North and South America, they may have exterminated, by overhunting, populations of many species of large mammals that had been abundant on those continents.

(b)

The Evolution of Language and Culture

As our ancestors evolved larger brains, they also increased their behavioral capabilities, especially the capacity for language. Most animal communication consists of a limited number of signals, nearly all of which pertain to immediate circumstances and are associated with changed emotional states induced by those circumstances. (The language of honeybees is unusual in that it contains a symbolic component referring to events distant in both space and time; see Chapter 45.) Human language is far richer in its symbolic character than are any other animal utterances. Our words can refer to past and future times and to distant places. We are capable of learning thousands of words, many of them referring to abstract concepts. We can rearrange those words to form sentences with complex meanings.

The expanded mental abilities of humans are largely responsible for the development of **culture,** the process by which knowledge and traditions are passed along from one generation to another by teaching and observation. Culture can change rapidly

(c)

because genetic changes are not necessary for a cultural trait to spread through a population. The primary disadvantage of culture is that each generation must be taught its norms.

The designs of the tools and other implements associated with human fossils, as well as the cave paintings these people created, reveal cultural traditions. Cultural learning greatly facilitated the spread of the domestication of plants and animals and the resultant conversion of most human societies from ones in which food was obtained by hunting and gathering to ones in which **pastoralism** (herding large animals) and **agriculture** dominated (Figure 27.34b,c). The agricultural revolution, in turn, led to an increasingly sedentary life, the development of cities, greatly expanded food supplies, and rapid growth of the human population. Twenty-five thousand years ago only 3 million people lived on Earth. Ten thousand years ago Earth housed only about 5 million people. As agriculture spread during the subsequent 5,000 years, however, the human population increased rapidly to about 100 million. Rapid increase in the human population has continued to the present unabated.

Agriculture developed in the Middle East approximately 11,000 years ago, and from there it spread rapidly to the northwest across Europe, finally reaching the British Isles about 4,000 years ago. The plants and animals domesticated by these people were cereal grains such as wheat and barley; legumes such as beans, lentils, and peas; and woody plants such as grapes and olives. Others, such as rye, cabbage, celery, and carrots, were domesticated as agriculturalists spread across Europe. Cattle, sheep, goats, horses, pigs, dogs, cats, and chickens were the most important domesticated animals.

Agriculture developed independently in eastern Asia, contributing to our modern diets such plants as soybeans, rice, citrus fruits, and mangoes. There was some exchange, even in early times, among agricultural centers in the Old World, but when people crossed the cold and barren Bering Land Bridge into the New World, they apparently brought no domesticated plants with them. These people subsequently developed rich and varied agricultural systems based on important crops such as corn, tomatoes, kidney and lima beans, peanuts, potatoes, chili peppers, and squashes.

The Human Lineage

When and how people spread around Earth from the tropics has been the subject of much controversy. The study of mitochondrial DNA from contemporary human populations in Africa, Asia, Europe, Australia, and New Guinea has provided valuable information. Mitochondrial DNA is particularly useful for determining lineages over short time spans because it accumulates random changes about 10 times faster than chromosomal DNA does and because all mitochondria are of maternal origin, so changes in nucleotide sequences result only from ticking of the molecular clock and not from sexual recombination.

The most surprising result of these investigations is the suggestion that the human evolutionary tree may have a single ancestor, sometimes called "mitochondrial Eve." Despite this name, the data do not suggest that all modern people are descended from a single female. Rather, they indicate that the ancestral population was small enough that, perhaps as a result of genetic drift, only one set of mitochondrial genes was transmitted to later generations.

Although there is general agreement that all humans are descended from a small ancestral population, there is less agreement about the area of its origin. Some calculations using the mitochondrial DNA clock suggest that the human lineage began about 200,000 years ago in Africa. African populations have at least twice as much genetic variation as people from other continents, which suggests that humans have a longer history in Africa than elsewhere. However, some recently discovered Asian fossils appear to be as old as the earliest human fossils from Africa. A number of laboratories are actively studying human molecular data, so the uncertainties may be resolved in the near future.

THEMES IN DEUTEROSTOME EVOLUTION

In several important ways, deuterostome evolution paralleled protostome evolution. Both lineages exploited the abundant food supplies buried in soft marine substrates, attached to rocks, or suspended in water columns. Because of the ease with which water can be moved, many groups of both lineages developed elaborate structures for extracting prey from water (Figure 27.35a). In both lineages, a coelomic cavity evolved and subsequently became divided into compartments that allowed better control of body shape and movement. Both lineages evolved locomotor abilities. Some members of both groups evolved mechanisms for controlling their buoyancy in water, using gas-filled internal spaces whose contents can be adjusted to control the depth at which the animal is stable (Figure 27.35b). Similar planktonic larval stages evolved in marine members of many protostome and deuterostome phyla; these all fed on tiny planktonic organisms during dispersal in the open water.

Both protostomes and deuterostomes colonized the land, the former via beaches, the latter via fresh water, but the consequences were very different. The jointed external skeletons of arthropods, although they provide excellent support and protection in air, are not suitable for large animals. In addition, an

Protostomate animals

(a) Filtering devices

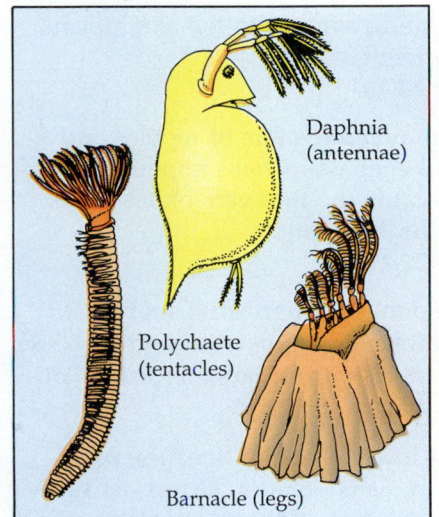

Daphnia (antennae)

Polychaete (tentacles)

Barnacle (legs)

(b) Buoyancy controls

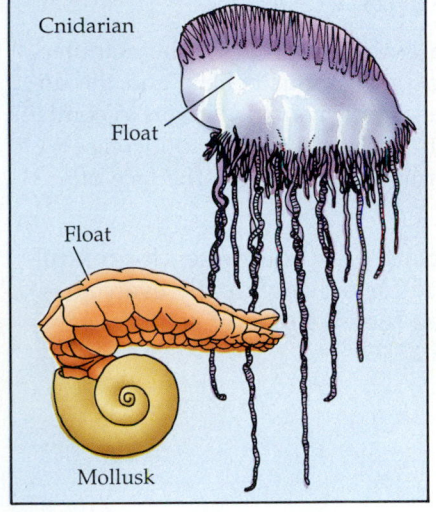

Cnidarian

Float

Float

Mollusk

Deuterostomate animals

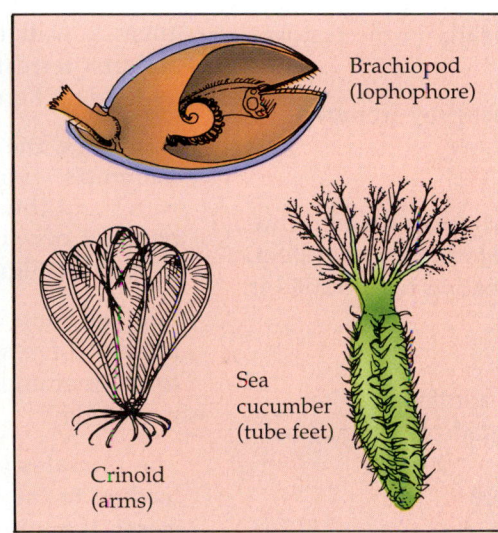

Brachiopod (lophophore)

Sea cucumber (tube feet)

Crinoid (arms)

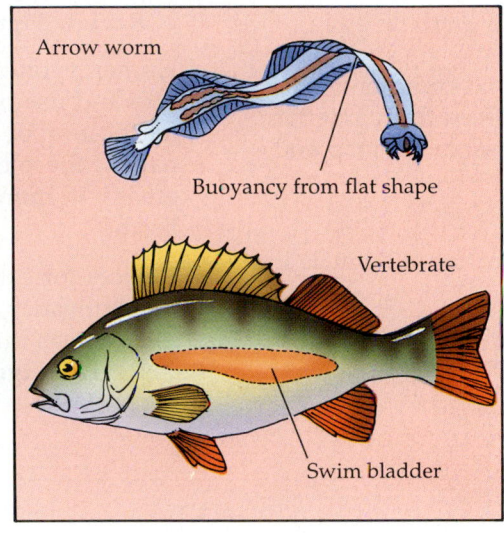

Arrow worm

Buoyancy from flat shape

Vertebrate

Swim bladder

27.35   Parallel Evolution
Devices for (a) filtering food from the water and (b) maintaining buoyancy in the water evolved in both protostomate and deuterostomate animal lineages.

arthropod must shed its skin and become temporarily vulnerable in order to grow. The internal, jointed skeletons of vertebrates, however, permit growth to large size without any temporary vulnerable stages. Consequently, although arthropods are abundant and diverse on land, vertebrates are the only very large terrestrial animals to have evolved on Earth.

Terrestrial lineages of vertebrates recolonized aquatic environments a number of times. Interestingly, suspension feeding reevolved in several of these lineages. For example, the largest living mammals, the baleen whales (the toothless whales, including blue, humpback, and right whales), feed upon relatively small prey that they extract from the water with large plates in their mouths.

Unlike the oceans, where the dominant photosynthetic plants are unicellular algae, most photosynthesis on land is carried out by vascular plants. Dom-

inant plants in most terrestrial environments are large, complex organisms that provide the primary physical structure of those ecological communities. Although plants are suitable sources of energy for animals, most plant tissues are difficult to digest. Herbivores must ingest large quantities of fibers and defensive chemicals along with the energy-rich molecules they need. Because larger animals can exist on food of poorer quality than that tolerated by small animals, a common pattern in herbivore evolution is a steady increase in body size. This pattern is striking in the evolution of reptiles and in the later evolution of mammalian herbivores. The evolution of large herbivores, in turn, favored the evolution of larger carnivores able to attack and overpower them. This evolutionary trend may have come to a temporary halt because of the invention of weapons by a moderately sized, omnivorous primate—the human.

SUMMARY of Main Ideas about Deuterostomate Animals

Deuterostome evolution began early in the history of life on Earth.

Members of several phyla extract prey from the water with a U-shaped lophophore.
Review Figures 27.1 and 27.3

Echinoderms, which have a body plan based on radial symmetry and a unique water-vascular system, move slowly on the substrate, eating food buried in sediments or attached to rocks.
Review Figure 27.7

Simple chordates are sessile as adults and filter prey from the water with enlarged pharyngeal baskets.
Review Figures 27.9 and 27.10

Vertebrates evolved jointed internal skeletons that enabled them to swim rapidly.
Review Figure 27.11

Early vertebrates lacked jaws and could only suck up their food. Jaws, which evolved from anterior gill arches, enabled their possessors to grasp and chew their prey.
Review Figure 27.13

Jawed fishes rapidly became dominant animals in both marine and fresh waters.

Fishes that had evolved jointed fins that enabled

them to walk on land gave rise to the amphibians, the first terrestrial vertebrates.
Review Figure 27.16

Amniotes, the common ancestors of reptiles and mammals, evolved eggs with shells impermeable to water and thus became the first vertebrates independent of water for breeding.
Review Figure 27.19

Reptiles were the dominant terrestrial animals for millions of years. Another amniote lineage gave rise to the mammals, the dominant large terrestrial animals today.

The primates split into two major lineages, one leading to the prosimians (lemurs, lorises, and tarsiers), the other leading to the anthropoids (monkeys, apes, and humans).
Review Figure 27.28

Human evolution is characterized by the manufacture and use of tools, religious beliefs, and the domestication of plants and animals. Taken in combination, these traits enabled humans to increase greatly in number and to transform the face of Earth.

Devices for extracting prey from water, for controlling buoyancy, and for moving rapidly evolved many times during the evolution of animals.
Review Figure 27.35

SELF-QUIZ

1. Which of the following are deuterostomate phyla with a tripartite body plan?
   a. Rotifera, Phoronida, Ectoprocta, and Brachiopoda
   b. Phoronida, Ectoprocta, Brachiopoda, and Hemichordata
   c. Phoronida, Ectoprocta, Hemichordata, and Chordata
   d. Echinodermata, Ectoprocta, Brachiopoda, and Chordata
   e. Phoronida, Ectoprocta, Hemichordata, and Echinodermata

2. The structure used by brachiopods to capture food is a
   a. pharyngeal gill basket.
   b. proboscis.
   c. lophophore.
   d. mucous net.
   e. radula.

3. The water vascular system of echinoderms is a
   a. series of seawater channels derived by enlargement and extension of a coelomic cavity.
   b. series of seawater channels derived by enlargement and extension of the pharyngeal cavity.
   c. series of channels derived by enlargement and extension of a coelomic cavity and filled with coelomic fluid.
   d. series of channels derived by enlargement and extension of a coelomic cavity and filled with freshwater.
   e. series of channels that can be filled to different degrees with water to enable the animal to control its buoyancy.

4. The pharyngeal gill slits of chordates originally functioned as sites for
   a. taking up oxygen only.
   b. releasing carbon dioxide only.
   c. taking up oxygen and releasing carbon dioxide.
   d. removing small prey from the water.
   e. forcibly expelling water to move the animal.

5. The key to the vertebrate body plan is
   a. a pharyngeal gill basket.
   b. a vertebral column to which internal organs are attached.
   c. a vertebral column to which two pairs of appendages are attached.
   d. a vertebral column to which a pharyngeal gill basket is attached.

e. a pharyngeal gill basket and two pairs of appendages.

6. Which of the following fishes do *not* have a cartilaginous skeleton?
   a. Chimaeras
   b. Lungfishes
   c. Sharks
   d. Skates
   e. Rays

7. In most fishes, lunglike sacs evolved into
   a. pharyngeal gill slits.
   b. true lungs.
   c. coelomic cavities.
   d. swim bladders.
   e. none of the above.

8. Most amphibians return to water to lay their eggs because
   a. water is isotonic to egg fluids.
   b. adults must be in water while they guard their eggs.
   c. there are fewer predators in water than on land.
   d. amphibians need water to produce their eggs.

e. amphibian eggs quickly lose water and dry out if their surroundings are dry.

9. The horny scales that cover the skin of reptiles prevents them from
   a. using their skin as an organ of gas exchange.
   b. sustaining high levels of metabolic activity.
   c. laying their eggs in water.
   d. flying.
   e. crawling into small spaces.

10. Which of the following is *not* true of the feathers of birds?
    a. They are highly modified reptilian scales.
    b. They insulate the body.
    c. They arise from well-defined tracts.
    d. They help birds fly.
    e. They are important sites of gas exchange.

11. Monotremes differ from other mammals by
    a. not producing milk.

b. lacking body hairs.
c. laying eggs.
d. living in Australia.
e. having a pouch in which the young are raised.

12. Bipedalism is believed to have evolved in the human lineage because
    a. bipedal locomotion is more efficient than quadrupedal locomotion.
    b. bipedal locomotion is more efficient than quadrupedal locomotion and it frees the hands to manipulate objects.
    c. bipedal locomotion is less efficient than quadrupedal locomotion but it frees the hands to manipulate objects.
    d. bipedal locomotion is less efficient than quadrupedal locomotion but bipedal animals can run faster.
    e. bipedal locomotion is less efficient than quadrupedal locomotion but natural selection does not act to improve efficiency.

FOR STUDY

1. In what animal phyla has the ability to fly evolved? How do structures used for flying differ among these animals?

2. Extracting suspended food from the water column is a common mode of foraging among animals.

Which groups contain species that extract prey from the air? Why is this mode of obtaining food so much less common than extracting prey from water?

3. Compare the buoyancy systems of cephalopods and fishes.

4. Why does possession of an external skeleton limit the size of a terrestrial animal more than possession of an internal skeleton does?

5. Large size confers benefits and poses certain risks. What are these risks and benefits?

READINGS

Alexander, R. M. 1975. *The Chordates.* Cambridge University Press, New York. A comprehensive and readable account of the biology of the chordates.

Bakker, R. T. 1975. "Dinosaur Renaissance." *Scientific American*, April. A discussion of the relationships between birds and dinosaurs; presents evidence that dinosaurs were warm-blooded.

Bond, C. E. 1979. *Biology of Fishes.* Saunders, Philadelphia. A leading text for courses on ichthyology (fish biology).

Carroll, R. C. 1987. *Vertebrate Paleontology and Evolution.* W. H. Freeman, New York. A thorough account of the fascinating evolutionary history of the vertebrates.

Colbert, E. H. 1980. *Evolution of the Vertebrates: A History of the Backboned Animals Through Time,* 3rd Edition. Wiley-Interscience, New York. A thoughtful discussion of the origins and evolutionary radiations of the vertebrate groups.

Gill, F. B. 1995. *Ornithology,* 2nd Edition. W. H. Freeman, New York. A technically accurate and readable introduction to bird biology for students at any level.

Langston, W., Jr. 1981. "Pterosaurs." *Scientific American*, February. An account of the largest animals ever to fly. Includes notes on evolutionary relationships among birds and reptiles.

Pough, F. H., J. B. Heiser and W. N. McFarland. 1989. *Vertebrate Life.* Macmillan, New York. An excellent treatment of the evolution and ecology of the vertebrates.

Vaughan, T. A. 1978. *Mammalogy,* 2nd Edition. Saunders, Philadelphia. The leading textbook on mammals; good coverage of the orders of mammals and general aspects of mammalian biology.

Willson, M. F. 1984. *Vertebrate Natural History.* Saunders, Philadelphia. A thorough treatment of all aspects of the lives of vertebrates.

Volcanic Eruptions May Exterminate Life over Limited Areas

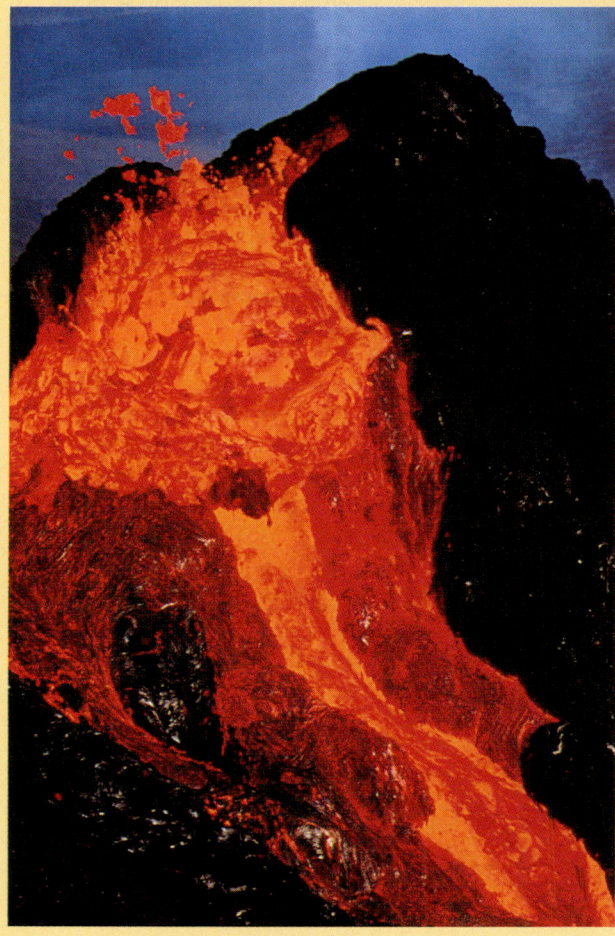

28

Patterns in the Evolution of Life

On the morning of August 27, 1883, Krakatau, an island the size of Manhattan located in the Sunda Strait between Sumatra and Java, was devastated by a series of volcanic eruptions. The most violent one created an airwave that traveled around the world seven times. The eruption was heard as far away as Singapore and the city of Perth in Australia. Tidal waves caused by the eruption hit the shores of Java and Sumatra, demolishing towns and villages and killing 40,000 people. All life on Krakatau was destroyed. Only a remnant of the former island, covered by a hot layer of pumice 40 meters thick, remained above sea level.

The 1883 explosion at Krakatau was not the largest in recorded history. The 1815 eruption of Tambora on the Indonesian island of Sumbawa was five times greater. An even larger eruption on Sumatra 75,000 years ago created 65-kilometer-long Lake Toba. Great volcanic eruptions have probably happened repeatedly throughout the history of Earth, but most volcanic eruptions are small. Large eruptions are much less frequent than small ones.

Creating another type of impact, small meteorites collide with Earth more often than large ones do. At least 30 meteorites between the sizes of baseballs and soccer balls hit Earth each year, but collisions with large meteorites are rare. The largest meteorite to hit the United States weighed 5,000 kilograms and fell in Norton County, Kansas, on February 18, 1948. A gigantic prehistoric meteorite formed Canyon Diablo in Arizona; one 3,200 meters in diameter created the Chubb Crater at Ungava, Quebec.

Volcanic eruptions and falling meteorites typically exterminate life over only relatively small areas. However, the largest catastrophic events, which happen less than once every million years, can affect life regionally or globally. What are those events and how much have they influenced the history of life? If the living world as we know it has been strongly molded by such events, studies of microevolutionary patterns and their causes can provide only a partial picture of the evolution of life on Earth.

In this chapter we examine the evidence for major disruptions in the history of life. We also look at how life on Earth changed between major catastrophic events to find out which features of organisms and which lineages evolved rapidly and which ones changed very little. To understand the patterns of change, we must think in time frames spanning many millions of years and imagine events and conditions very different from those we now observe. The geological past is, to us, a foreign land. The Earth of the distant past is a foreign planet in which continents were not where they are today and climates were different. One of the remarkable achievements of modern science has been the development of techniques that enable us to infer past conditions and to date them with some precision.

HOW EARTH HAS CHANGED

Earth's history has been profoundly influenced by internal processes, such as the activity of volcanoes and the compression and shifting of Earth's crust, and by external events, such as meteorite falls. Before considering these events, we must establish some mileposts in Earth's history to which we can refer when we talk about events in the remote past.

In Chapter 18 we learned that radioactivity provides a way to date rocks precisely because in successive, equal periods of time, the same fraction of the remaining radioactive material of any radioisotope decays, becoming the corresponding stable isotope. For example, in 14.3 days, one-half of any sample of phosphorus-32 (^{32}P), a radioactive isotope of phosphorus, decays. During the next 14.3 days, one-half of the remaining half decays, leaving only one-fourth of the original sample of ^{32}P. After 42.9 days, three half-lives have passed, so only one-eighth (that is, ½ × ½ × ½) of the original radioactive material remains, and so forth. Each radioisotope has a characteristic half-life. Tritium (^{3}H), for example, has a half-life of 12.3 years, and carbon-14 (^{14}C) has a half-

life of about 5,700 years. Some radioisotopes have much longer half-lives: The half-life of potassium-40 (^{40}K) is 1.3 billion years; that of uranium-238 (^{238}U) is about 10 billion years

How is the regularity of radioactive decay used to date rocks? We can use ^{14}C to age fossils contained in fairly young rocks; we then infer the ages of sedimentary rocks from the ages of the fossils embedded in them. The half-life of ^{14}C is long enough to allow us to determine the time of death of anything that has died within the last 15,000 years and has left remains that contain carbon. The ratio of radioactive ^{14}C to nonradioactive ^{12}C in a living creature is always the same as that in the environment because carbon is constantly being exchanged between the environment and organisms. The production of new ^{14}C in the upper atmosphere (by the reaction of neutrons with ^{14}N) just balances the natural radioactive decay of ^{14}C, so a steady state exists. However, as soon as a tree or any other living thing dies, it ceases to equilibrate its carbon compounds with the rest of the world. Its decaying ^{14}C is not replenished from outside, and the ratio of ^{14}C to ^{12}C decreases. By measuring what fraction of the total carbon in a specimen is ^{14}C, we can easily calculate how much time has elapsed since it died. Some radiocarbon dates of archaeological objects are shown in Figure 28.1.

Geologists divide Earth's history into four **eons:**

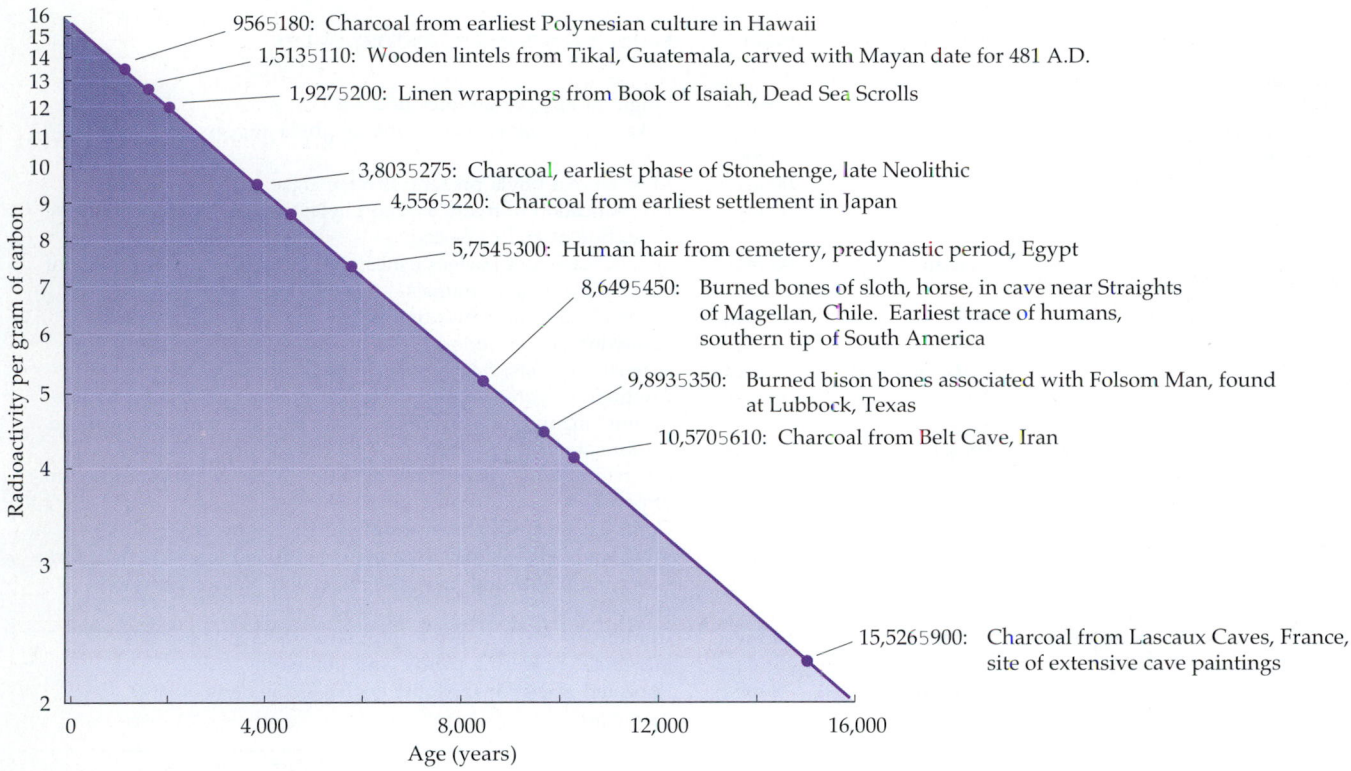

28.1 Radioactive Clocks Tell Time
Sometimes, as with the wooden lintels from Tikal, Guatemala, radioactively determined dates are confirmed by evidence from cultural or other sources.

the Hadean eon, the Archean eon, the Proterozoic eon, and the Phanerozoic eon. The Phanerozoic eon is subdivided into **eras**, which are further subdivided into **periods** (Table 28.1). The boundaries between these divisions were originally based on major differences in the fossils contained in successive layers of rocks. The relative ages of the rocks were thus known, but their absolute ages were unknown. When radioactivity was discovered and understood, the dates in the table could be estimated. By using a variety of radioactive isotopes, whose different half-lives span the full range of Earth's history, geologists were able to date the boundaries.

The fossils of organisms, not the rocks themselves, provided the information geologists first used to order the sequence of events, because evolving life provides a record that can be ordered through time. There is no directional evolution of rocks; a rock of a particular type can be formed at any time. Thus the names of the eons and eras are based on fossils—primarily those of animals—not on the rocks. This is why eras bear the suffix *zoic*, from the Greek word for animals.

Major Changes from Internal Processes

The maps and globes that adorn our walls, shelves, and books, give an impression of a static Earth. It is easy for us to believe that the continents have always been where they are, but this conclusion would be quite incorrect. Earth's crust consists of solid plates approximately 40 km thick that float on a liquid mantle. The plates move because the sea floor is spreading along ocean ridges where material from the mantle rises and pushes the plates aside. Where plates come together they either move sideways along fault lines or one plate moves under the other, creating mountain ranges. The movement of the plates and the continents they contain has had enormous effects on climate, sea level, and the distribution of organisms.

In the late Cambrian period, 550 million years ago (mya), there were six continents, all located at equatorial latitudes. By a process known as **continental drift**, all of them united into a single large continent, called Pangaea, during Permian times (Figure 28.2*a*). During several periods in the Paleozoic era, sea levels

TABLE 28.1
Earth's Geological History

EON	ERA	PERIOD	ONSET[a]	MAJOR EVENTS IN THE HISTORY OF LIFE
Hadean			4.5 bya	
Archean			3.8 bya	Origin of life; prokaryotes flourish
Proterozoic			2.5 bya	Eukaryotes evolve; several animal phyla appear
Phanerozoic			600 mya	
	Paleozoic	Cambrian	600 mya	Most animal phyla present; diverse algae
		Ordovician	500 mya	Diversification of many animal phyla; first jawless fishes. **Mass extinction** at end of period
		Silurian	440 mya	Diversification of jawless fishes; first bony fishes; colonization of land by plants and animals
		Devonian	400 mya	Diversification of fishes; first insects and amphibians. **Mass extinction** late in period
		Carboniferous	345 mya	Extensive forests; first reptiles; insects radiate
		Permian	290 mya	Continents aggregate into Pangaea; reptiles radiate; amphibians decline; many types of insects. **Mass extinction** at end of period
	Mesozoic	Triassic	245 mya	Continents begin to drift; early dinosaurs; first mammals; diversification of marine invertebrates. **Mass extinction** near end of period
		Jurassic	195 mya	Continents drifting; diverse dinosaurs; first birds
		Cretaceous	138 mya	Most continents widely separated; continued dinosaur radiation; flowering plants and mammals diversify. **Mass extinction** at end of period
	Cenozoic	Tertiary	66 mya	Continents nearing current positions; radiations of birds, mammals, flowering plants, and pollinating insects
		Quaternary	2 mya	Repeated glaciations; people evolve; extinctions of large mammals

[a]bya = billion years ago; mya = million years ago

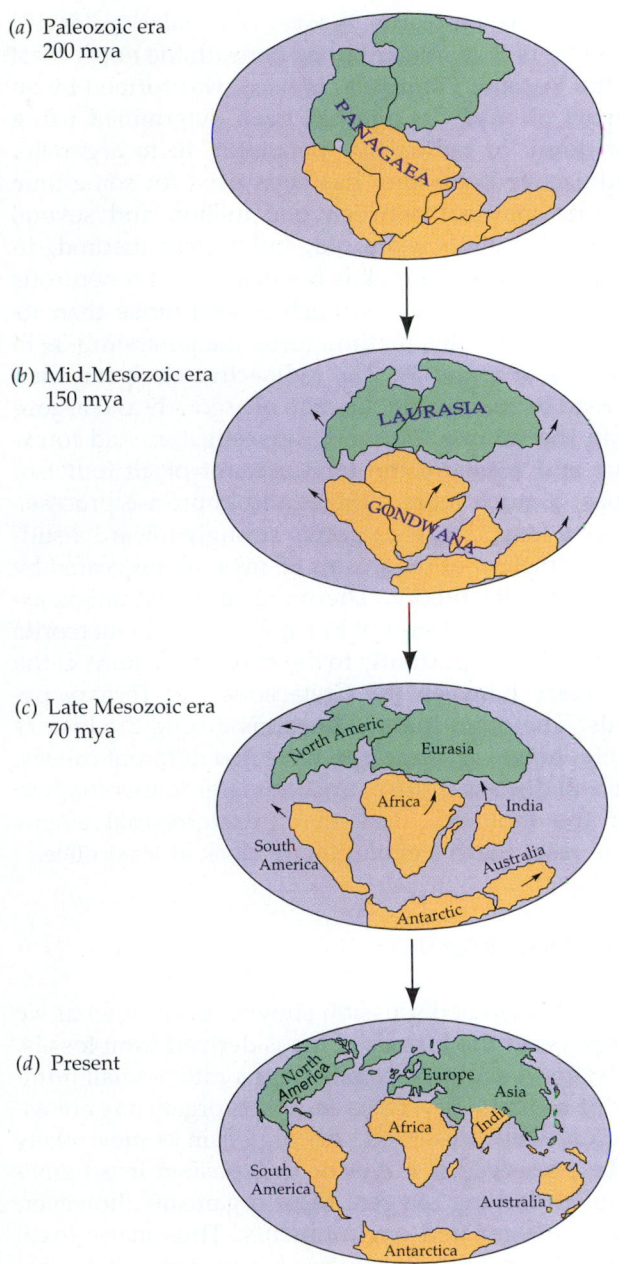

(a) Paleozoic era
200 mya

(b) Mid-Mesozoic era
150 mya

(c) Late Mesozoic era
70 mya

(d) Present

28.2 The Continents Drift
(a) Earth in the Paleozoic era, about 200 mya. The super-continent Pangaea is alone in a great ocean. The two colors and outlines of present-day continents show the breakup and drifting that would happen in the future. (b,c) Earth in the Mesozoic era. Pangaea split into the northern and southern land masses of Laurasia and Gondwanaland (or Gondwana). During the Mesozoic era, Gondwanaland began to break up into several land masses. (d) The present position of the continents.

rose and shallow seas spread over parts of the continents. There were also periods when sea levels dropped and parts of the continental shelves were exposed, and there were periods when both events happened.

Pangaea began to break up during the Permian period. By the late Triassic period, what is now North America and Eurasia formed a single large northern continent, Laurasia. The southern land masses were united into another continent, Gondwanaland. These two continents were separated by a gradually widening oceanic channel. By the mid-Cretaceous period, Gondwanaland had begun to break up into Africa, South America, and a third land mass consisting of what are now Australia, Antarctica, and India (Figure 28.2b).

By the late Cretaceous period, the Indian plate had separated from the rest of Gondwanaland and begun its northward drift, eventually to collide with Asia (Figure 28.2c). The Himalaya Mountains formed as a result of this impact. South America and Africa continued to drift apart; Australia separated from Antarctica and slowly moved to its present position much closer to the equator. North America and Eurasia also drifted apart, but they were later rejoined at the present Bering Sea. About 3 mya, the Isthmus of Panama arose, thereby connecting North and South America for the first time in more than 250 million years.

Through much of its history, Earth's climate was considerably warmer than it is today, and temperatures declined more slowly toward the poles. At other times, however, Earth was much colder than it is today. Large areas were covered with glaciers during the late Proterozoic, the Carboniferous, the Permian, and the Quaternary, but these cold periods were separated by very long periods of milder climates. Because we live in one of the colder periods in the history of Earth, it is difficult for us to imagine the mild climates characteristic of high latitudes during much of the history of life.

Usually climates change slowly, but there have been major climatic shifts over periods as short as 5,000 to 10,000 years, primarily as a result of changes in Earth's orbit around the sun. A few shifts appear to have been even more rapid than that. For example, during one interglacial period, the Antarctic Ocean changed from being ice-covered to being nearly ice-free in less than 100 years. Climates have sometimes changed rapidly enough that extinctions due to them appear "instantaneous" in the fossil record.

Changes from External Events

You may have noticed from Table 28.1 that the history of life includes several **mass extinctions**. These were periods when relatively more species became extinct than during intervening periods. The hypothesis that the mass extinction about 66 mya might have been caused by the collision of Earth with a large meteorite was proposed in 1979 by Luis Alvarez and

28.3   Evidence of a Meteorite-Caused Mass Extinction? A thin band rich in iridium—a metal common in meteorites but rare on Earth—marks the boundary between rocks deposited in the Cretaceous and Tertiary periods.

several of his colleages at the University of California, Berkeley. These scientists based their hypothesis on the finding of abnormally high concentrations of iridium, an element in the platinum group, in a thin layer separating rocks deposited during the Cretaceous period from those of the Tertiary period (Figure 28.3). Iridium is abundant in some meteorites but is exceedingly rare on Earth's surface. To account for the estimated amount of iridium in this layer, Alvarez postulated that a meteorite 10 kilometers in diameter collided with Earth at a speed of 72,000 kilometers per hour. The force of such an impact would have ignited massive fires, created great tidal waves, and sent up an immense dust cloud that encircled Earth, blocking the sun and cooling the planet. The settling dust would have formed the iridium-rich layer.

This hypothesis was greeted with much skepticism when it was first proposed. Many paleontologists (scientists who study fossils) believed that the mass extinction had taken place over a period of millions of years, not instantaneously as the meteorite theory demanded. Vulcanologists pointed out that some volcanoes emit substantial quantities of iridium. Therefore, a period of intense volcanic activity could have produced the iridium layer. And why, some critics asked, had the impact crater not been discovered?

This controversy stimulated much activity. Some scientists tried to locate the site of the supposed impact; others worked to improve the precision with which events of that age could be dated; still others tried to determine more closely the speed with which extinctions had occurred at the Cretaceous–Tertiary boundary. Progress on all three fronts has tended to

support the meteorite theory. A circular crater 180 kilometers in diameter buried beneath the north coast of the Yucatan Peninsula, Mexico, was formed by an impact 65 mya. Its age has been determined using the decay of radioactive potassium-40 to argon-40. Radioactive potassium has been used for some time to date material between one million and several hundred million years old, but a new method, in which a sample of rock is bombarded with neutrons in a nuclear reactor, is much more precise than its predecessors. This method turns the potassium-39 in the rock to argon-39. The radioactive clock can then be read by measuring the ratio of argon-39 and argon-40 in the sample. Formerly, investigators had to extract and measure the total amount of all four isotopes, a much more difficult and imprecise process.

New fossil evidence points strongly toward a sudden extinction of organisms 65 mya, as suggested by the meteorite theory. Therefore, most scientists accept that the collision of Earth with a large meteorite contributed importantly to the mass extinctions at the boundary between the Cretaceous and Tertiary periods. The other mass extinctions during the history of life, however, appear to have had different causes. We will discuss those extinctions later in this chapter. For the moment, note that extraterrestrial events have reset Earth's evolutionary clock at least once.

THE FOSSIL RECORD

As the previous discussion shows, much of what we know about the history of life is derived from fossils. Most fossil evidence comes from a rather small number of sedimentary rocks in which organisms are especially well preserved. An organism is most likely to be preserved if it dies or is deposited in an environment lacking oxygen. Most organisms, however, live in oxygenated environments. Thus many fossil assemblages are collections of organisms that were transported by wind or water to sites without oxygen. Occasionally, however, organisms are preserved where they lived. In such cases—especially those of cool, anaerobic swamps, where conditions for preservation were excellent—we obtain a picture of whole organismal communities.

The Completeness of the Fossil Record

About 300,000 species of fossil organisms have been described, and the number is growing steadily. However, this number is only a tiny fraction of the species that have ever lived. We do not know how many species lived in the past, but we have ways of making reasonable estimates. Of the present-day **biota**—the species in all groups (archaebacteria, bacteria, protists, fungi, plants, and animals)—approximately 1.5 million species have already been described. The actual number of living species is probably at least 10 million (and possibly as high as 50 million) because most species of insects, the richest animal group, have not yet been described. Thus the number of known fossil species is less than 2 percent of the probable total of living species.

Because life has existed on Earth for 3.5 billion years, and because species last, on average, less than 10 million years, the total number of species that lived in the past must have been many times the number that are alive today. Earth's biota has been replaced many times during geological history. If at any moment in the past the number of species was no greater than at present (or even if it was substantially less), the total number of species over evolutionary time would be much greater than the current number.

The sample of fossils, although small in relation to the total number of extinct species, is not uniformly poor. The record is especially good for the phyla of marine animals that have hard skeletons. Among the nine major phyla that have hard-shelled members, approximately 200,000 species have been described from fossils, roughly twice the number of living marine species in these same groups. Paleontologists lean heavily on these groups in their inter-

pretations of the evolution of life in the past. Insects, although much rarer as fossils, are still well represented in the fossil record (Figure 28.4). Fossil insects belonging to about 1,265 families and all 30 of the common orders having living species have been identified.

The incompleteness of the fossil record can mislead our interpretations of what happened. Most described fossils come from a relatively small number of sites, and an organism that evolved elsewhere may have been fossilized at one of these sites. When such a fossil is found, it gives the impression that the organism evolved very rapidly from one of the species that already lived there when, in fact, it may have evolved slowly elsewhere and moved to the site. For example, horses evolved slowly over millions of years in North America. Many different lineages arose and died out. Ancestors of horses crossed the Bering Land Bridge into Asia several times, the last one only several million years ago (Figure 28.5). Evidence of each crossing appears suddenly in the Asian fossil record as a major new form of organism. If we lacked fossil evidence of horse evolution in North America, we might conclude that horses evolved very rapidly somewhere in Asia.

Patterns in the Fossil Record

The general richness of species in lineages of protists, plants, and animals over geological time is shown as "spindle diagrams" in Figure 28.6. The changing width of a spindle shows how the number of species in a phylum changed through time, but the figure does not show that many different lineages arose, prospered, and died out within the major phyla. Note from the figure that most phyla have been in existence for at least 500 million years and that, in general, animal phyla are older than plant phyla.

LIFE IN THE REMOTE PAST

To understand and interpret the broad patterns in the evolution of life, we need an overview of what life on Earth was like during each geological era; we also need to know when and why the course of evolution was disrupted by major catastrophic events.

Species-Rich Precambrian Life

Most kingdoms of eukaryotic organisms—protists, fungi, and animals, but not plants—evolved prior to the Cambrian period, but the known fossil record for that ancient time is fragmentary. The record for animals is the most complete. The best fossil assemblage

28.4 A Fossil Beetle
Trapped in the sap of a tree in the Dominican Republic about 25 mya, this ground beetle is exquisitely preserved in the amber that formed from the sap. Details of its external anatomy are clearly visible.

Bering bridge

To all parts of old world

Homeland of the horses

Central American bridge

Hyracotherium (50 mya)

Mesohippus (25 mya)

Hipparion (8 mya)

Pliohippus (4 mya)

Equus (recent)

28.5 Horses Have a Complex Evolutionary History
The earliest horses evolved in North America; many lineages arose and died out, and ancestors of several of these lineages crossed into Asia over the Bering land bridge and into South American over the Central American land bridge.

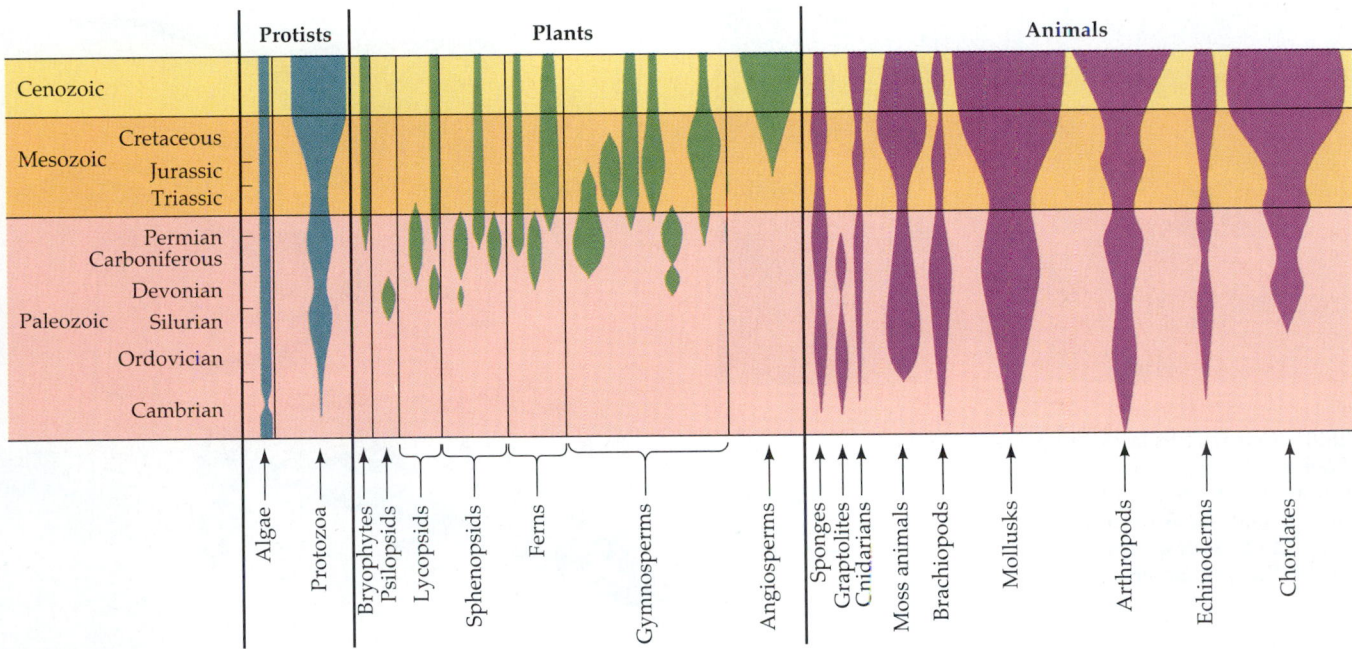

		Protists	Plants	Animals

Cenozoic

Mesozoic — Cretaceous, Jurassic, Triassic

Paleozoic — Permian, Carboniferous, Devonian, Silurian, Ordovician, Cambrian

Algae, Protozoa, Bryophytes, Psilopsids, Lycopsids, Sphenopsids, Ferns, Gymnosperms, Angiosperms, Sponges, Graptolites, Cnidarians, Moss animals, Brachiopods, Mollusks, Arthropods, Echinoderms, Chordates

28.6 Broad Features of the Fossil Record
The times of appearance in the fossil record of the major phyla of protists, plants, and animals. The approximate number of fossil species in each group at any time is indicated by the width of the spindles.

of Precambrian animals, all soft-bodied invertebrates, is known as the Ediacaran fauna, named after the Australian site where it was discovered (Figure 28.7). (The term *fauna* refers to all of the species of animals living in a particular area. The corresponding term for plants is *flora*.)

Among the Ediacaran animals are forms believed by some biologists to represent early annelids, arthropods, echinoderms, and cnidarians, but they are very different from later members of those phyla. They may instead represent a separate radiation of animal groups that have no living descendants. About two-thirds of the species that have been identified in the Ediacaran fauna are thought to be cnidarians. Scientists have discovered fossils of worms and mollusks slightly younger than the Ediacaran fauna that are probably the immediate ancestors of many of the animals found in early Cambrian deposits. These younger worms and mollusks do not appear to be direct descendants of Ediacaran animals.

The fossil record shows that the volume of organisms increased dramatically in late Precambrian times. The shallow Precambrian seas teemed with life. Protists and small multicellular animals fed on

(a)

28.7 Ediacaran Animals
Fossils excavated at Ediacara in southern Australia are 600 milllion years old and illustrate the diversity of life that evolved in Precambrian times. *(a) Spriggina floundersi. (b) Mawsonites.*

(b)

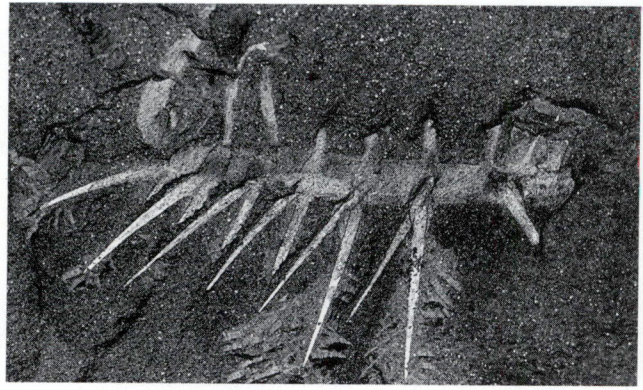

(a)

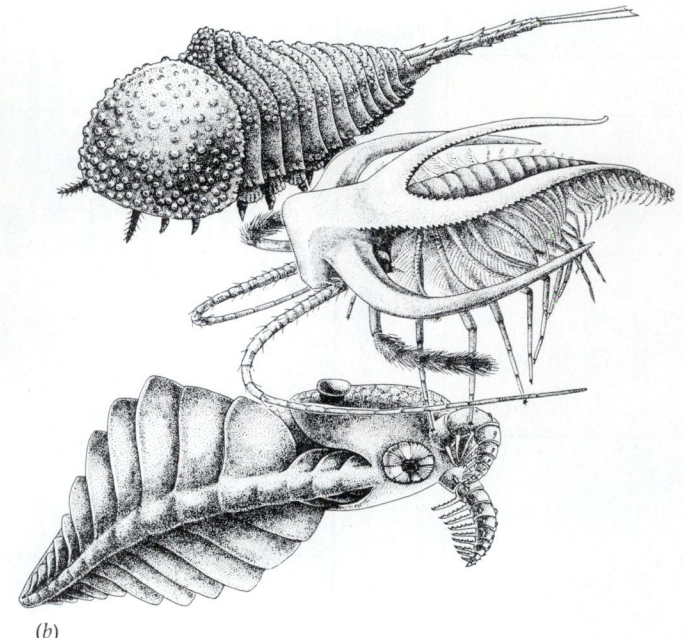

(b)

28.8 Extinct Phyla in the Burgess Shale
These animals, which lived in mid-Cambrian times, do not appear to be members of any lineage that survived to the present. (a) A fossil *Hallucigenia*. (b) An artist's conception of several other animals of the many represented in the Burgess Shale.

floating algae. Living plankton and plankton remains were devoured by animals that filtered food items from the water or ingested surface sediments and digested the organic remains in them.

Even Richer Paleozoic Life

By the mid-Cambrian period, about 530 mya, all animal phyla that have species living today were already present, as revealed by the exceptionally well preserved fossils in the Burgess Shale in British Columbia. Many species belonging to about 10 extinct phyla are known only from this fossil bed (Figure 28.8). Animals with many different body plans evolved in the Cambrian period, but most of them failed to survive past Cambrian times. The evolution of hard skeletons in representatives of so many phyla between 600 and 530 mya suggests that predation became very intense during this period.

During the Cambrian period, a small group of wormlike animals evolved fanlike tentacles covered with cilia. Coordinated beating of these cilia created a stream of water through the tentacles. The cilia captured food material in the water and transferred it to the mouth. Another change was the enclosure of the tentacles within a shell, forcing the entire current of water to pass through the tentacles. This adaptation improved filtering efficiency and protected the tentacles from predators. These changes so improved feeding efficiency that organisms possessing them (the early brachiopods) became dominant animals in the early Cambrian seas.

During the Ordovician period (500 to 440 mya) many animal phyla radiated, creating a great profusion of classes and orders. Abundant sponges filtered food from the water much as today's sponges do.

Cnidarians, also filter feeders, built large reefs but were different from modern corals, which first appeared in the mid-Ordovician period. There was a great increase in other kinds of animals that filter small prey from the water, such as bivalve mollusks and echinoderms. Floating graptolites, members of a now extinct phylum, were abundant (Figure 28.9). Trilobites (see Figure 26.28) were the dominant arthropods, but they also became extinct. Ancestors of club mosses and horsetails colonized wet terrestrial environments, but they were still relatively small. At the end of the Ordovician, when sea levels dropped considerably and much of the continental shelf was exposed, numerous families and orders of marine

28.9 Graptolites
Graptolites were abundant in oceans of the Ordovician period (see Figure 28.6). These specimens were fossilized in black shale.

28.10 A Devonian Marine Community
During the Devonian period present-day New York State was an underwater reef. This reconstruction shows how part of the reef may have appeared; notice the trilobites and the unusual corals and ammonite cephalopods.

animals became extinct, probably because the environments in which they lived were greatly reduced in area.

During the Silurian period (440 to 400 mya), graptolites declined, corals proliferated, and the colonization of land continued. Nevertheless, late-Silurian marine communities probably looked much the same as those of the Ordovician period. Early vertebrates diversified during this period, and the first terrestrial arthropods—scorpions and millipedes—appeared.

Rates of evolutionary change accelerated in many groups of organisms during the Devonian period (400 to 345 mya). There was great evolutionary radiation of corals, trilobites, and shelled squidlike cephalopods (Figure 28.10). Fishes diversified as jawed forms replaced jawless ones, and the heavy armor characteristic of most earlier fishes gave way to the less rigid outer coverings of modern fishes. Terrestrial communities also changed markedly during the Devonian period. Land plants became common, and some reached the size of trees. They were mostly club mosses and horsetails, along with some tree ferns. Toward the end of the period the first gymnosperms appeared. A springtail from this period is the first known fossil of an insect. The first fishlike amphibians began to occupy the land.

Extensive forests grew during the Carboniferous period (345 to 290 mya), as was shown in Figure 25.9. The compressed remains of trees that grew in swampy forests where they fell into deep, anaerobic mud that preserved them from biological degradation is the coal we now mine for energy. Carboniferous beds are rich in fossils, many of which retain traces of the fine details of their structure (Figure 28.11). The diversity of terrestrial animals increased greatly in the Carboniferous period. Snails, scorpions, cen-

tipedes, and insects were present in great abundance and variety. Amphibians became better adapted to terrestrial existence. Some of them were large animals more than 5 meters long, quite unlike any surviving today. From one amphibian stock, the first reptiles evolved late in the period.

Permian period (220 to 245 mya) deposits contain representatives of most orders of insects, including dragonflies with wingspreads that measured 2 feet, the largest insects that ever lived. By the end of the period reptiles greatly outnumbered amphibians. These reptiles included a variety of terrestrial forms, as well as species that were major predators in marine and freshwater environments. In fresh waters, the

28.11 A Carboniferous Fossil
A fossilized tree fern excavated in France. Like the fossil fuels we burn today, this fossil is a remnant of the massive forests of the Carboniferous period.

28.12 Mesozoic Dinosaurs
Dinosaurs of the Mesozoic era still capture our imagination. This painting illustrates some of the large species from the Jurassic and Cretaceous periods.

Permian period was a time of extensive radiation of bony fishes. Ammonites (cephalopod mollusks) proliferated in the oceans, but their radiation was abruptly terminated at the end of the period when the greatest of all mass extinctions occurred.

At the end of the Permian period about 95 percent of all species, both terrestrial and marine, became extinct. Trilobites were extinguished completely. The number of species in such groups as anthozoan corals, ammonites, ostracods, crinoids, and brachiopods declined sharply. On land, many of the species of trees that dominated the great coal-forming forests also became extinct.

What caused the mass extinction at the end of the Permian? We have no evidence of massive volcanic activity or collisions of large meteorites with Earth. The Permian extinctions appear to have happened slowly, perhaps over more than 10 million years. The probable cause was the coalescing of the continents into the supercontinent Pangaea (see Figure 28.2*a*). The interior of Pangaea, far removed from the oceans, experienced very harsh climates. The largest glaciers in Earth's history formed, and the massive transfer of moisture from the oceans to the land caused sea levels to drop, drying out many of the shallow seas where most marine organisms lived.

Provincialization during the Mesozoic Era

As the Mesozoic era started, the few surviving organisms found themselves in a relatively empty world. As Pangaea separated into individual continents, glaciers melted and the oceans rose and reflooded the continental shelves, forming huge, shallow inland seas. Life again diversified, but lineages

different from the major ones of the Permian period came to dominate Earth during the Mesozoic era. Brachiopods were replaced by mollusks. The trees that dominated the great coal-forming forests were replaced by cycadeoids, cycads, ginkgos, and conifers.

Earth's biota, which had until that time been relatively homogeneous, became increasingly **provincialized**—that is, individual continents acquired distinctive terrestrial floras and faunas, and the biotas of continental shelves also diverged regionally. The provincialization that began during the Mesozoic influences the geography of life even today (see Chapter 50).

During the Triassic period (244 to 195 mya) the ammonites underwent a second great radiation, modern corals proliferated, and some invertebrate groups, such as bivalve mollusks, increased in diversity. Many invertebrate groups that previously had been restricted to living on surfaces of bottom sediments evolved burrowing forms. On land, gymnosperms and seed ferns became the dominant woody plants. The great radiation of thecodonts, which eventually gave rise to dinosaurs, crocodilians, and birds, began (Figure 28.12). Late in the Triassic period a therapsid lineage evolved into the first mammals.

The beginning of the Jurassic period (194 to 138 mya) is marked by a mass extinction that eliminated most of the ammonites and many other marine invertebrates. The ammonites underwent yet another radiation, and many invertebrate groups evolved sturdier shells, probably because of increased predation pressures from crabs, gastropods, and fishes. Bony fishes began the great radiation that culminated in their dominance of the seas. Frogs, salamanders, and lizards first appeared during the Jurassic. Flying reptiles evolved, and dinosaurs radiated into bipedal predators and large quadrupedal herbivores. One lineage of dinosaurs gave rise to the first birds in the mid-Jurassic period. Several groups of mammals also evolved.

During the Cretaceous period (138 to 66 mya) marine invertebrates increased in variety and number of species, and pressures from marine predators increased. Many species of crabs with powerful claws evolved, and new forms of carnivorous marine snails able to drill holes in shells (species similar to modern oyster drills, cone shells, and tritons) began to fill the Cretaceous seas. Skates, rays, and bony fishes with powerful teeth capable of crushing mollusk shells also evolved, and large, powerful marine reptiles, the placodonts, fed heavily on clams. The shells of clams evolved increasing thickness during the Cretaceous, but predators were so effective that clams disappeared from the surfaces of most marine sediments. The survivors in those environments were species that burrowed into the substrate, where they were more difficult to capture.

On land, dinosaurs continued to diversify. The first snakes appeared during the Cretaceous, but their lineage did not radiate until much later. By the end of the period, many groups of modern mammals had evolved, but these species were generally small in size. Early in the Cretaceous period, possibly somewhat earlier, flowering plants evolved from gymnosperm ancestors and began the radiation that led to their current dominance on land.

Another mass extinction took place at the end of the Cretaceous period. On land, all vertebrates larger than about 25 kilograms in body weight apparently became extinct. In the seas, many planktonic organisms and bottom-dwelling invertebrates became extinct. As we discussed earlier in the chapter, this extinction was probably caused by a large meteorite that collided with Earth off the Yucatan Peninsula.

The Cenozoic Era

The Cenozoic era (66 mya to the present) is characterized by an extensive radiation of mammals, but other taxa were undergoing important changes as well. The richness of the fossil record has made it possible to subdivide the Cenozoic era into a number of periods and epochs.

In the Tertiary period (66 to 2 mya) the angiosperms diversified extensively and came to dominate world forests, except in cool regions. In the middle of the Tertiary, when the climate became considerably drier and cooler, many lineages of angiosperms

evolved herbaceous (nonwoody) forms, and parklands and grasslands spread over much of Earth. By the beginning of the Cenozoic era, invertebrate faunas were already modern in most respects. It is among the vertebrates that evolutionary change during the Tertiary periods was most rapid. Living groups of reptiles, such as snakes and lizards, underwent extensive radiations during this period. In fact, snakes have undergone speciation as rapidly as any taxon during the past few million years.

The main radiation of birds took place in the early Tertiary period. By the Eocene epoch (55 mya), all modern orders were present. There are few differences between the birds of the Miocene epoch (25 mya) and those of today. Mammals were common at the beginning of the Tertiary period, but the largest were no bigger than small dogs. The differentiation of most orders of mammals took place early in the Tertiary. Grazing mammals proliferated during times when Earth was drier and herbaceous vegetation dominated large expanses of the continents, as it does today. The groups of mammals that have undergone the most extensive diversification in recent times are the bats, rodents, hoofed mammals, and primates.

The present geological period, the Quaternary, began with the Pleistocene epoch about 2 mya. The Pleistocene was a time of drastic cooling and climatic fluctuations. There were four major and several minor glacial episodes during which Earth became much cooler and distributions of animals and plants shifted toward the equator. The last glaciers retreated from temperate latitudes fewer than 10,000 years ago. Organisms are still adjusting to these changes. Many high-latitude ecological communities have occupied their current locations no more than a few thousand years. Interestingly, Pleistocene climatic fluctuations resulted in little extinction of species. However, many large birds and mammals became extinct in North and South America and in Australia, coincident with the arrival of people on those continents. These were the first human-caused extinctions. How our species is affecting biological diversity today will be discussed in Chapter 51.

THE TIMING OF EVOLUTIONARY CHANGE

This overview of how life evolved as continents shifted locations and at least one large meteorite struck Earth inspires some questions. How fast has evolution proceeded? Has the rate been constant over time or among lineages? How much have extinction rates varied? How and when did evolutionary novelties arise? Scientists have made enough progress in studying evolution that at least tentative answers can be given to some of the major questions.

Rates of Evolutionary Change

Rates of morphological change have not been constant during the history of life, but how variable are they? Were there times when many or most lineages were changing rapidly and other times when most lineages were changing slowly? For several reasons, answering these questions is difficult. First, the fossil record is incomplete in both space and time. As we have seen, the sudden appearance of a group of organisms in the fossil record could signal either rapid evolution or migration of the organisms from some other area. Second, fossils preserve primarily hard body parts. They reveal almost nothing of changes in the sizes and shapes of soft organs or molecules. None of the changes in the structure of cytochrome *c* that we discussed in Chapter 19, for example, can be determined from fossils. The history of the molecule has been deduced by comparing the cytochromes *c* of living species.

Despite these difficulties, the fossil evidence suggests that many species experienced times of **stasis**, during which they changed very little over long periods. For example, as we have already seen, many marine lineages evolved very slowly during the Silurian period. The horseshoe crabs that lived 300 million years ago are almost identical in appearance to those living today (see Figure 26.29*b*). The chambered nautiluses of the late Cretaceous are indistinguishable from living species. Such "living fossils" are found today in environments that have changed relatively little for millennia. These environments tend to be harsh in one way or another. The sandy coastlines that are the homes of horseshoe crabs experience temperature and salinity extremes that are lethal to many other organisms. Chambered nautiluses spend their days in deep, dark ocean waters, ascending to feed in food-rich surface waters only under the protective cover of darkness. Their intricate shells provide little protection against today's visually hunting fishes.

Periods of stasis may be broken by times during which morphological changes are rapid, but why does a slowly changing lineage suddenly begin to change rapidly? The general answer is that changes, either in the physical or biological environment, create conditions that favor new traits. The sizes of spines on the three-spined stickleback (*Gasterosteus aculeatus*) have changed rapidly a number of times. This marine fish, which is widespread and not more than 75 millimeters long, has repeatedly invaded fresh water throughout its long evolutionary history. Sticklebacks in all marine and most freshwater populations have well-developed pelvic girdles with prominent spines that make it difficult for other fishes to swallow them. Large predatory insects, however, can readily grasp the stickleback's spines.

These insects prey selectively on stickleback individuals with the largest spines. When sticklebacks invade freshwater habitats where predatory fish are absent but predatory insects are present, the lineages rapidly evolve smaller spines. Populations with reduced spines are found primarily in young lakes that were covered by ice during the last glaciation (Figure 28.13). These lakes lack large predatory fishes.

The extensive fossil record of sticklebacks shows that spine reduction evolved many times in different populations that invaded fresh water. In addition, molecular data reveal that each freshwater population is most closely related to an adjacent marine population, not to other freshwater populations.

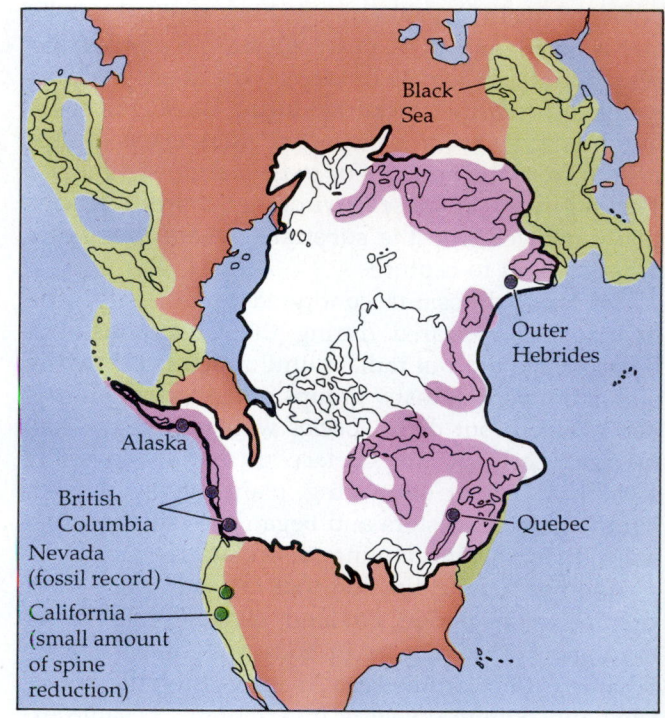

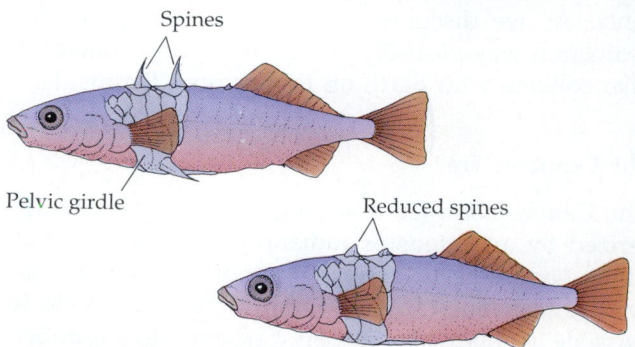

28.13 Natural Selection Acts on Spines
In this map, the region of the Northern Hemisphere that was once covered by Pleistocene glaciers is outlined in black. The current range of sticklebacks includes formerly glaciated areas (lavender) and unglaciated areas (green). Places where sticklebacks are known to have reduced spines are indicated by circles.

28.14 Sudden Appearances
These mammals appear suddenly as fossils in rocks that formed about 53 million years ago in the Bighorn Basin of Wyoming and Montana.

Therefore, spine reduction has evolved rapidly many times in different places in response to the same ecological situation: the absence of predatory fish.

Some paleontologists believe that evolutionary changes are often very rapid at the time of speciation. Such a pattern is difficult to detect in the fossil record, but the hypothesis can be tested by comparing rates of change in structures among lineages living in the same environments but speciating at different rates. If evolutionary changes are generally most rapid at the time of speciation, lineages in which speciation rates are high should show higher rates of morphological change than lineages with slower speciation rates. One such test, performed using two lineages of freshwater fishes, gave a negative result. Minnows in the genus *Notropis* speciated much more rapidly than sunfishes in the genus *Lepomis*, but rates of morphological evolution have been about the same in the two genera over the same time period. More tests are needed to determine whether these fishes are unusual or are typical of many evolving lineages.

Sometimes climates change so abruptly that some organisms cannot evolve rapidly enough to adapt to the new conditions. One such rapid climatic change happened about 55 million years ago, when the temperature of the deep sea rose from 10°C to 18°C in just 2,000 years as a result of a major change in ocean circulation. Associated with this temperature shift was a change in the oxygen isotope ratio in seawater carbonates. The oceans exchange carbon with the atmosphere in the form of carbon dioxide, which is taken up by photosynthetic plants. Animals eating those plants preserve the same isotope ratios in their bones. The oxygen isotope ratios changed suddenly in the teeth and bones of mammals living 53 mya in Montana and Wyoming. The change coincided exactly with the disappearance of a fauna of archaic mammals and its replacement by a new fauna of mammals (Figure 28.14). Although abrupt warming was clearly the driving environmental change, how it exerted its effects or where the new mammals came from is unknown.

Is a pattern of long periods of evolutionary stasis punctuated by times of rapid morphological change compatible with the microevolutionary mechanisms we discussed in Chapter 19, or must new mechanisms be postulated? We might first ask, Are the rapid rates of evolution too rapid to be accounted for by known microevolutionary mechanisms? The answer to this question is clearly no.

One measure of the rate of evolution of a character is the **darwin**, which is defined as a change of magnitude *e* (the base of natural logarithms, 2.718) per million years. Artificial selection in the laboratory can change characters at rates higher than 60,000 darwins for short periods. The skeletons of house sparrows that were introduced into the United States from Europe during the last century diverged from their ancestors at rates between 50 and 300 darwins. By contrast, the highest rate of change that has been measured in characters among rapidly evolving fossil mammals is 12 darwins for short time periods. Rates of increase in sizes of dinosaurs were less than one darwin. Clearly, the known rates of evolution are compatible with microevolutionary mechanisms, but that does not prove that they were caused by microevolutionary mechanisms.

A more powerful argument that new mechanisms are not needed to explain uneven evolutionary rates comes from population genetics. Population genetic theory predicts that evolution by natural selection should be punctuated because if natural selection is strong enough to overcome the effects of genetic drift, it should result in major changes in a phenotypic character in less than 50,000 generations. This rate of change appears very rapid in the geological record, where a thousand years is a short time interval. Thus, variable rates of evolution, although they do not tell us much about the factors that determine them, are to be expected given what we know about microevolutionary mechanisms.

Extinction Rates

More than 99 percent of the species that have ever lived are extinct. However, we know relatively little about causes of extinction, except for species that have become extinct in historical times. Extinctions have occurred throughout the history of life, but extinction rates have changed dramatically. There have been periods in which some groups had high extinction rates while others were proliferating. Paleontologists distinguish between normal or background extinction rates and rates characterizing mass extinctions (Figure 28.15).

Each mass extinction changed the flora and fauna of the next period by eliminating some types of organisms, thereby increasing the relative abundance of others. For example, among planktonic Foraminifera, important marine protists, the only survivors of mass extinction were relatively simple species with broad geographical ranges. Among the mollusks of the Atlantic coastal plain of North America, species with broad geographic ranges were less likely to become extinct during normal periods than were species with small geographic ranges. By contrast, during the late Cretaceous mass extinction, *groups of closely related species* with large geographic ranges survived better than those with small ranges, even if the

28.15 The Six Mass Extinctions

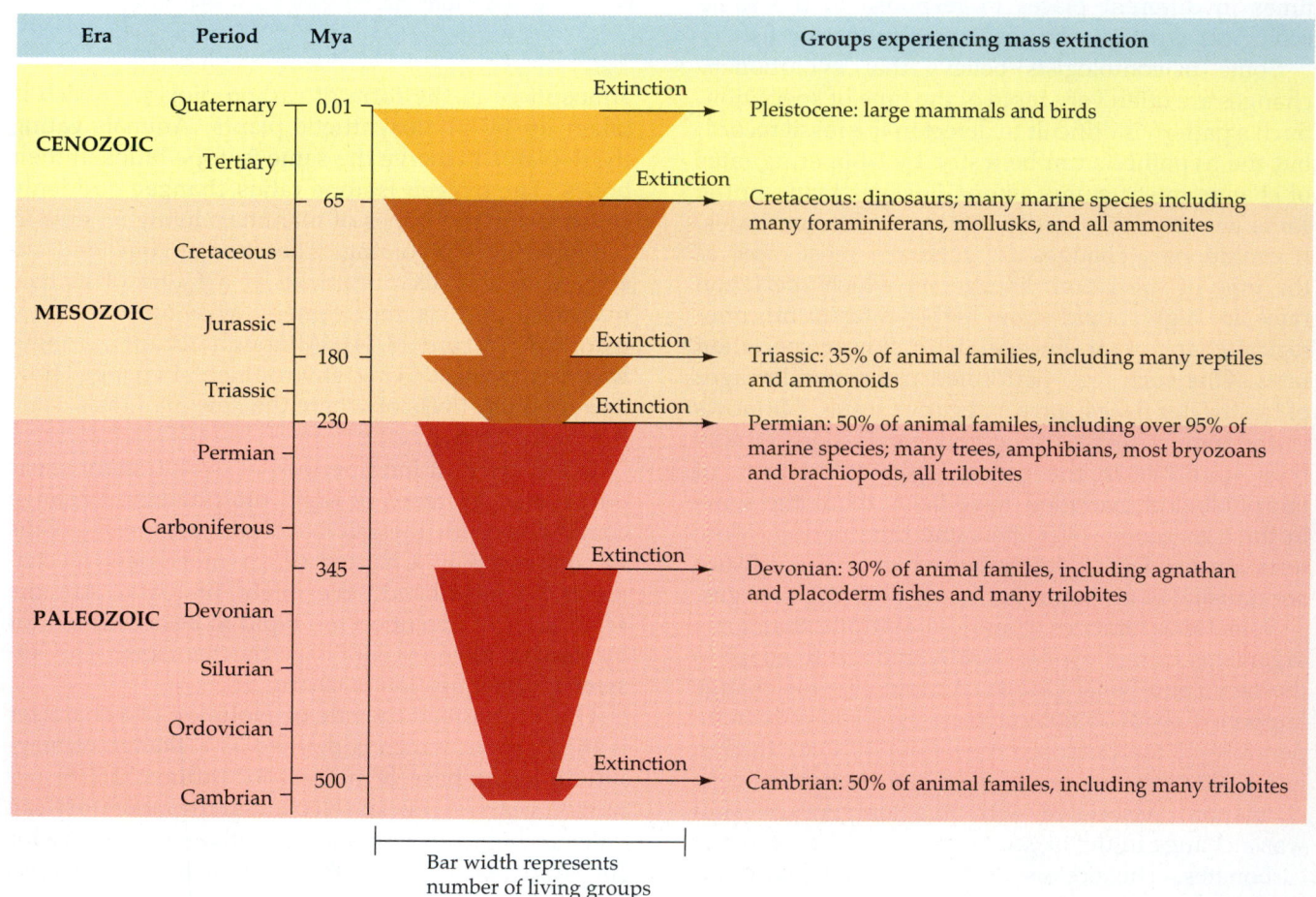

Era	Period	Mya	Groups experiencing mass extinction
CENOZOIC	Quaternary	0.01	**Extinction** → Pleistocene: large mammals and birds
	Tertiary		
MESOZOIC		65	**Extinction** → Cretaceous: dinosaurs; many marine species including many foraminiferans, mollusks, and all ammonites
	Cretaceous		
	Jurassic		
		180	**Extinction** → Triassic: 35% of animal families, including many reptiles and ammonoids
	Triassic	230	**Extinction** → Permian: 50% of animal families, including over 95% of marine species; many trees, amphibians, most bryozoans and brachiopods, all trilobites
PALEOZOIC	Permian		
	Carboniferous		
		345	**Extinction** → Devonian: 30% of animal families, including agnathan and placoderm fishes and many trilobites
	Devonian		
	Silurian		
	Ordovician		
		500	**Extinction** → Cambrian: 50% of animal families, including many trilobites
	Cambrian		

Bar width represents number of living groups

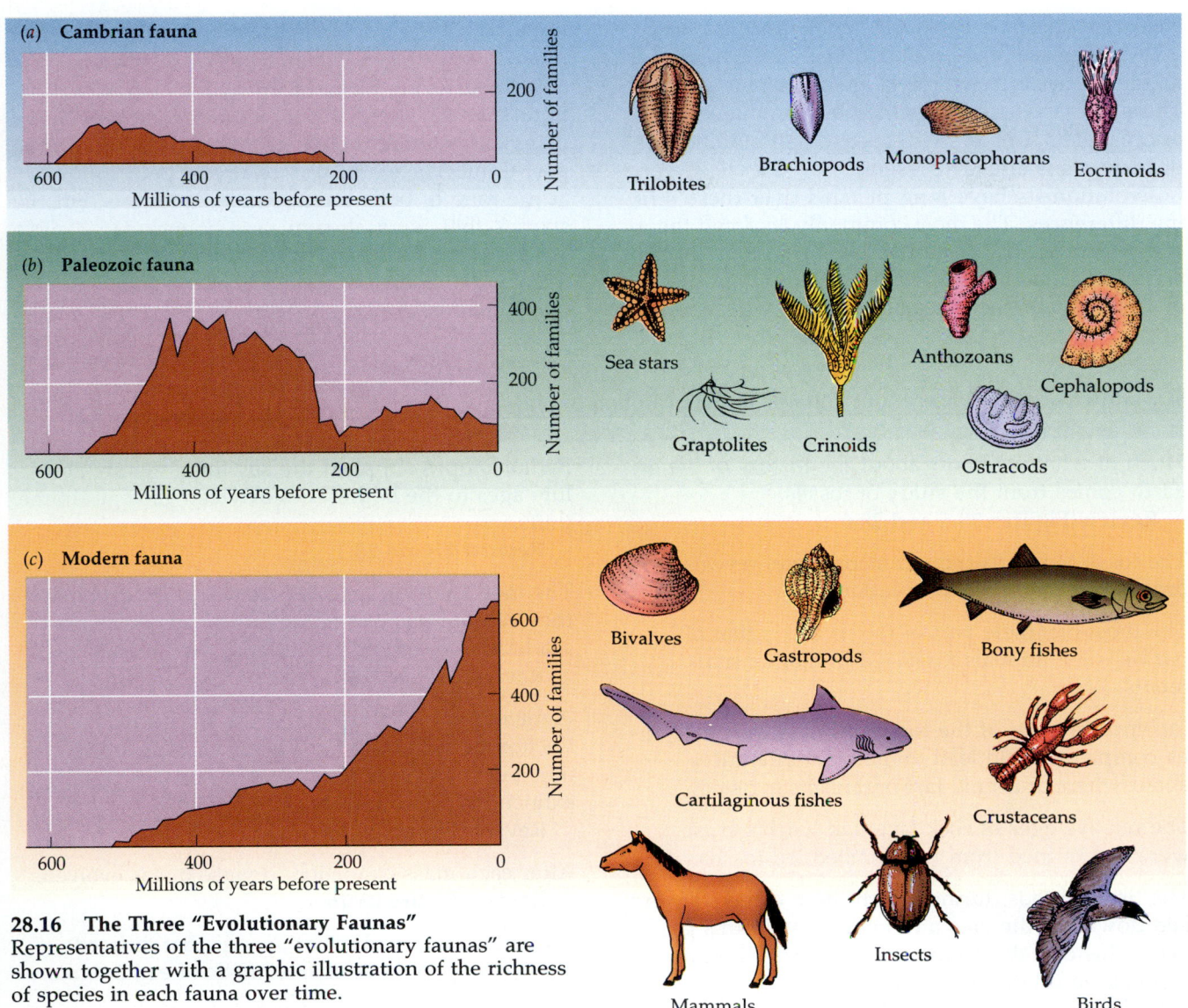

(a) Cambrian fauna

Number of families

Trilobites Brachiopods Monoplacophorans Eocrinoids

(b) Paleozoic fauna

Number of families

Sea stars Anthozoans Cephalopods

Graptolites Crinoids Ostracods

(c) Modern fauna

Number of families

Bivalves Gastropods Bony fishes

Cartilaginous fishes Crustaceans

Insects

Mammals Birds

28.16   The Three "Evolutionary Faunas"
Representatives of the three "evolutionary faunas" are shown together with a graphic illustration of the richness of species in each fauna over time.

individual species in them had small ranges. Similar patterns are found in other molluscan groups elsewhere, suggesting that traits favoring long-term survival during normal times are often different from those that favor survival during times of mass extinctions.

At the end of the Cretaceous period, extinction rates on land were much higher among large vertebrates than among small ones. The same was true during the Pleistocene mass extinction, when extinction rates were high only among large mammals and birds. In addition, as we have seen, during some mass extinctions marine organisms were heavily hit while terrestrial organisms survived well. Other extinctions affected organisms in both environments.

Rates of evolutionary change have been highly variable. During some periods many lineages of organisms were rapidly evolving; during other periods few lineages were changing rapidly. The fossil record is not complete enough, however, to reveal whether most periods of rapid evolutionary change accompany times of speciation.

Origins of Evolutionary Novelties

Three times during the history of life, many new evolutionary lineages originated (Figure 28.16). The first such event, known as the Cambrian explosion, took place about half a billion years ago. The second, about 60 million years later, resulted in the Paleozoic fauna. The great Permian extinctions 300 million years later were followed by the third event, the Triassic explosion, which led to our modern fauna. These explosions all resulted in the evolution of many new species, but qualitatively they differed.

During the Cambrian explosion all the major groups of present-day organisms appeared, along with a number of phyla that subsequently became extinct. As many as 100 different phyla of organisms may have evolved in the Cambrian, of which only 30 survive today. The Paleozoic and Triassic explosions greatly increased the number of families, genera, and species, but no new phyla of organisms evolved. These later explosions resulted in many new species of organisms, but all these species had modifications of body plans already present.

Evolutionists have long puzzled over these striking differences. The most commonly accepted theory is that because the Cambrian explosion took place in a world that contained few species of organisms, all of them small, the ecological setting was favorable for the evolution of many new body plans and different ways of life. Many types of organisms were able to survive initially in this world, but as competition intensified and new types of predators evolved, many forms were unable to persist. In addition, the post-Cambrian world was relatively poor in species at the time of both evolutionary explosions, but the species that were already present included a wide array of body plans and ways of life. Therefore, new major innovations were less likely to evolve in this world than in the Cambrian period.

SUMMARY of Main Ideas about Patterns in the Evolution of Life

Much of what we know about the history of life on Earth comes from the study of fossils.
Review Figures 28.5 and 28.6

The fossil record, although incomplete, reveals broad patterns in life's evolution.

The most complete part of the record is that of hard-shelled animals fossilized in marine sediments.

Incomplete parts of the fossil record, if assumed to be complete, could lead to incorrect inferences about where and how fast organisms evolved.

The relative ages of rock layers in Earth's crust were determined from their embedded fossils.

The time periods during which the rock layers were laid down are the eras and periods of Earth's geological history; the boundaries between these units were based on differences between their fossil biotas.
Review Table 28.1

Radioisotopes supplied the key for assigning absolute ages to the boundaries between geological time units.
Review Figure 28.1

The movement of Earth's plates and the continents they contain has had enormous effects on climates, sea levels, and the distributions of organisms.
Review Figure 28.2

A meteorite probably caused the abrupt mass extinction between the Cretaceous and Tertiary periods, but the causes of other, less abrupt mass extinctions are unknown.
Review Figure 28.15

After each mass extinction, diverse biotas evolved.
Review Figure 28.16

Each new biota had major species that differed markedly from those characteristic of earlier biotas.

Changes in biotas were associated with changes in climate.

SELF-QUIZ

1. The number of species of fossil organisms that has been described is about
   a. 50,000.
   b. 100,000.
   c. 200,000.
   d. 300,000.
   e. 500,000.

2. Radioactive carbon can be used to date the ages of fossil organisms because
   a. all organisms contain many carbon compounds.
   b. radioactive carbon has a regular rate of decay to nonradioactive carbon.
   c. the ratio of radioactive to nonradioactive carbon in living organisms is always the same as that in the atmosphere.
   d. the production of new radioactive carbon in the atmosphere just balances the natural radioactive decay of ^{14}C.
   e. all of the above.

3. About two-thirds of the species that have been identified in the Ediacaran fauna are thought to be
   a. algae.
   b. protists.
   c. echinoderms.
   d. cnidarians.
   e. ferns.

4. The total of all species of organisms living in a region is known as its
   a. biota.
   b. flora.
   c. fauna.
   d. flora and fauna.
   e. diversity.

5. The coal beds we now mine for energy are the remains of
   a. trees that grew in swamps during the Carboniferous period.
   b. trees that grew in swamps during the Devonian period.
   c. trees that grew in swamps during the Permian period.

d. herbaceous plants that grew in swamps during the Carboniferous period.

e. none of the above.

6. The cause of the mass extinction at the end of the Mesozoic period was
a. continental drift.
b. collision of Earth with a large meteorite.
c. changes in Earth's orbit.
d. massive glaciation.
e. changes in the salinity of the oceans.

7. The times during the history of life when many new evolutionary lineages appeared were
a. Precambrian, Cambrian, and Triassic.
b. Precambrian, Cambrian, and Tertiary.
c. Cambrian, Paleozoic, and Triassic.

d. Cambrian, Triassic, and Devonian.
e. Paleozoic, Triassic, and Tertiary.

8. Most scientists believe that the collision of Earth with a large meteorite was a major contributor to the mass extinction at the boundary between the Cretaceous and Tertiary periods because
a. there is an iridium-rich layer at the boundary of rocks from these two periods.
b. the probable site of collision has been found off the Yucatan Peninsula.
c. mass extinction at the end of the Cretaceous happened very suddenly.
d. new methods have allowed scientists to date the iridium layer very precisely.
e. all of the above.

9. We know that organisms can evolve very rapidly because
a. the fossil record reveals periods of very rapid evolutionary change.
b. theoretical models of evolutionary change show that rapid change can be produced by natural selection.
c. rapid evolutionary changes have been produced under artificial selection.
d. rapid evolutionary changes have been measured in natural populations of organisms during the past century.
e. all of the above

10. In which of the following periods was there *no* mass extinction?
a. the end of the Cambrian period.
b. the end of the Devonian period.
c. the end of the Permian period.
d. the end of the Triassic Period.
e. the end of the Silurian period.

FOR STUDY

1. Some lineages of organisms evolved to contain large numbers of species, whereas other lineages produced only a few species. Is it meaningful to consider the former as more successful than the latter? What does the word *success* really mean in evolution? How does your answer influence your thinking about *Homo sapiens*, the only surviving respresentative of the Hominidae—a family that never had many species in it?

2. If extinction rates in groups of organisms are relatively constant over long time spans, as they sometimes are, then recently evolved species should be no better adapted to their environments than older species are. Does this observation contradict the belief that natural selection adapts organisms to their environments?

3. Why is it useful to be able to date past events absolutely as well as relatively?

4. In a study of lungfish evolution, fossil lungfishes and modern specimens were assigned scores based on the number of derived versus ancestral traits they possessed. On

this scale, an organism that has only ancestral traits receives a score of 0. An organism that has only derived traits receives a score of 100. The rate of appearance of

these traits in lungfishes is plotted in the two graphs below. What does this pattern suggest about evolutionary radiation among lungfishes?

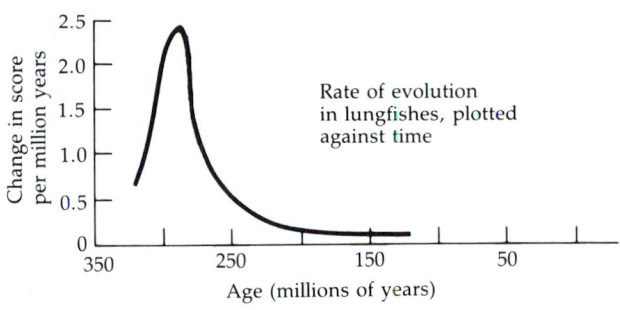

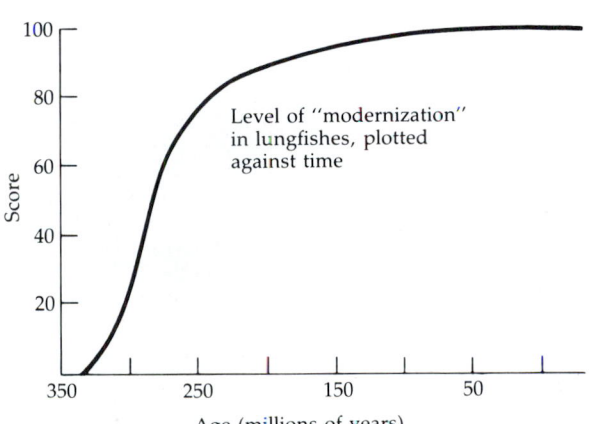

5. What factors favor increases in body size? How could *average* body size among species in a lineage decrease even if natural selection is favoring large body sizes in most of the species in the lineage?

6. Does the uneven pace of evolution negate the belief that natural selection is the principal agent that adapts organisms to their environments? What does the theory of natural selection, by itself, predict about rates of evolutionary change?

READINGS

Bonner, J. T. 1988. *The Evolution of Complexity by Means of Natural Selection.* Princeton University Press, Princeton, NJ. An excellent treatment of broad patterns in the evolutionary record and the developmental and physiological bases for them.

Futuyma, D. J. 1986. *Evolutionary Biology,* 2nd Edition. Sinauer Associates, Sunderland, MA. The best general treatment of evolution and its mechanisms.

Gates, D. M. 1993. *Climate Change and Its Biological Consequences.* Sinauer Associates, Sunderland, MA. An accessible book with extensive discussions of Earth's past climates and the methods scientists use to recreate climatic history.

Gould, S. J. 1989. *Wonderful Life: The Burgess Shale and the Nature of History.* W. W. Norton, New York. An engaging account of the remarkable fauna, containing representatives of many phyla of animals that left no modern descendants, in the Burgess Shale. Explores the implications of this fauna for our view of the history of life.

Morris, S. C. and H. B. Whittington. 1979. "The Animals of the Burgess Shale." *Scientific American,* July. A short account of the finds in one of these rich and important fossil beds.

Raup, D. M. and S. M. Stanley. 1978. *Principles of Paleontology,* 2nd Edition. W. H. Freeman, New York. An excellent account of the history of life and how it is studied.

Simpson, G. G. 1944. *Tempo and Mode in Evolution.* Columbia University Press, New York. A dated but classic book providing clear pictures of the long-term patterns of evolutionary change.

Stanley, S. M. 1979. *Macroevolution: Pattern and Process.* W. H. Freeman, New York. An argument that macroevolutionary mechanisms are different from microevolutionary mechanisms.

Stanley, S. M. 1987. *Extinction.* Scientific American Library, New York. A beautifully illustrated account of the fossil record and the phenomenon of extinction.

Ward, P. D. 1992. *On Methuselah's Trail: Living Fossils and the Great Extinctions.* W. H. Freeman, New York. A delightful personal account by a paleontologist who has studied both "living fossils" and true fossils embedded in rocks formed at the times of the great extinctions.

White, M. E. 1986. *The Greening of Gondwana.* Reed Books, Sydney. A beautifully illustrated account of the geological history of Australia and the unusual plants that evolved there.

PART FIVE

The Biology of
Vascular Plants

Leaves are conspicuous parts of most flowering plants. As major sites of photosynthesis, leaves are the food sources for the rest of the plant and, in fact, for much of the living world. We all recognize most leaves as such when we see them. Their basic form—flattened, and with veins—helps them perform their nutritive function. The reactions of photosynthesis proceed in chloroplasts within certain leaf cells, and the photosynthetic products are distributed to other parts by a plumbing system that includes the leaf's veins. Some raw materials travel to the photosynthesizing cells through the veins; other materials travel from pores in the leaf surface through air spaces within the leaf. The structure of the leaf lets the photosynthesizing cells get the light they need; the flattened shape, coupled with the orientation of the leaf, minimizes internal shading. The structure of the leaf thus supports its photosynthetic function. What about the rest of the plant? The stems and roots of flowering plants also have interesting and complex structures, with structure and function well matched.

Part Five deals with plant structure and function. There are many aspects of plant function to consider. Plants—even the tallest trees—transport water from the soil to their tops, and they transport the products of photosynthesis from the leaves to the roots and other parts. Plants interact with their environment, both living and nonliving. They defend themselves against bacteria, fungi, animals, and other plants. Some plants can cope with hostile environments such as deserts, salt marshes, or sites polluted by mining and other human activities. Plants must obtain nutrients—not only the raw materials of photosynthesis, but also mineral elements such as potassium and calcium. Plants respond to environmental cues as they develop. They produce chemical signals that cause structural and functional changes appropriate to the environmental cues. Among the most important changes are those that lead to reproduction. Because we can understand plant function only in terms of the underlying structure, this chapter focuses on the structure of the plant body.

FLOWERING PLANTS

Recall that flowering plants are vascular plants characterized by double fertilization, by endosperm, and by seeds enclosed in modified leaves; their xylem contains vessel elements and fibers. If you have not been thinking about plants for a while, you might want to review the sections entitled "Two Groups within the Plant Kingdom" and "The Angiosperms: Flowering Plants" in Chapter 25 before reading the rest of Chapter 29.

Frosted Sites of Photosynthesis

29

The Flowering Plant Body

Flowering plants consist of a few important organs whose life-supporting functions can be understood in terms of their large-scale structure, as well as the microscopic structure of their component cells. The cells are grouped into tissues, and the tissues are grouped into organs. In this chapter we will present some anatomical features common to many flowering plants. As always in biology, it is important to remember that there are differences between organisms of the same species as well as between species. Let us begin, then, by looking at four important or familiar species: coconut palm, red maple, rice, and soybean.

FOUR EXAMPLES

Coconut Palm

In some cultures the coconut palm (*Cocos nucifera*; Figure 29.1) is called the Tree of Life because every aboveground part of the plant has value to humans. People use the stem (the trunk) of this tropical coastal lowland tree as lumber. They dry the sap from its trunk for use as a sugar, or they ferment it to drink. They use the leaves to thatch their homes and to make hats and baskets. They eat the apical bud at the top of the trunk in salads. The coconut fruit serves many purposes. The hard shell can be used as a container or burnt as fuel; the fibrous middle layer, or coir, of the fruit wall can be made into mats and rope. The seed of the coconut palm has both a liquid endosperm (coconut milk) and a solid endosperm (coconut meat). Because the refreshing and delicious milk contains no bacteria or other pathogens, it is a particularly important drink wherever the water is not fit for drinking. Millions of people get most of their protein from coconut meat. Much coconut meat is dried and marketed as copra, from which coconut oil is pressed. Coconut oil is the most widely used vegetable oil in the world; it is used in the manufacture of a range of products from hydraulic brake fluid to synthetic rubber and, although nutritionally poor, as food. Ground copra serves as fertilizer and as food for livestock.

The trunk of a coconut palm differs in three basic ways from the trunks of many other familiar trees. The most striking difference is that it bears no branches, and all the leaves are borne in a cluster at the top of the trunk. Second, the coconut trunk tapers little from the base of the tree to the top—even the youngest part of the trunk is essentially as thick as the base. We will discuss this phenomenon later in the chapter. Third, a cross section of the trunk reveals no annual rings.

Each coconut palm tree has separate male and female flowers; both are small and inconspicuous. The male flowers have six stamens. The leaves of the coconut palm are large and made up of numerous long, narrow leaflets, each having veins running parallel to one another.

Red Maple

One of the most familiar native trees in the eastern United States is the red, or scarlet, maple (*Acer rub-*

(a)

(b)

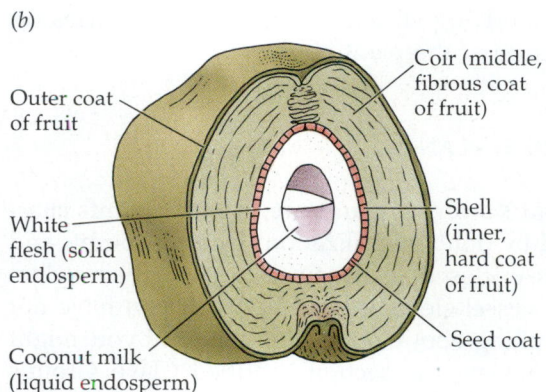

Outer coat of fruit

Coir (middle, fibrous coat of fruit)

White flesh (solid endosperm)

Shell (inner, hard coat of fruit)

Coconut milk (liquid endosperm)

Seed coat

29.1 Coconut Palm
(*a*) A coconut plantation in the South Pacific. Palms are among the only monocot trees. (*b*) A cross section of the coconut's fruit.

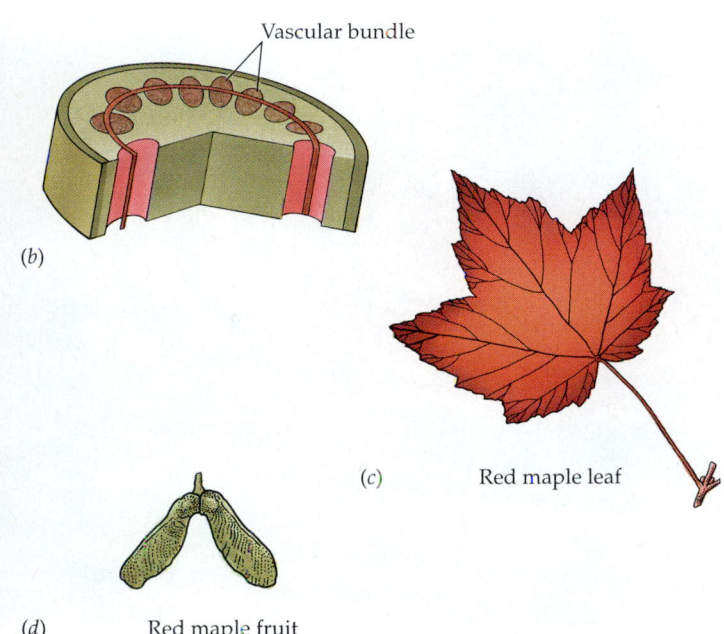

(a)

Vascular bundle

(b)

(c) Red maple leaf

(d) Red maple fruit

29.2 Red Maple
(a) A red maple tree in autumn. (b) Vascular bundles in the stem of a young maple, diagrammed in cross section. (c) A leaf of red maple. (d) The characteristic winged fruit of the maple family.

rum; Figure 29.2). Unlike the coconut palm, the red maple does not provide us with a great variety of useful products, but it enriches us by its beauty. Not only is it abundant in forests, but we admire it in parks and as a street tree growing to 10 to 30 meters tall. We use its wood as lumber, although the sugar maple is a more important commercial source of maple wood.

Microscopic examination of a very young maple stem reveals vascular bundles of water- and food-conducting cells, arranged in a cylindrical pattern. Like the palm, the mature maple tree has a thick, massive trunk, but a cross section of the trunk shows that the wood is made up of many annual rings. The roots, too, are woody. The red maple leaf—the symbol of Canada—consists of a single blade with three to five lobes, with veins that radiate from a single focal point. These leaves are among the brilliant contributors to the fall colors of eastern forests. The scarlet flowers have four sepals, four petals, eight stamens, and one pistil. The distinctive, winged fruit of the maple family contains two seeds.

Rice

More than half of the world's human population derives the bulk of its food energy from the seeds of a single plant: rice (*Oryza sativa*; Figure 29.3). Rice is particularly important in the diets of people in the Far East, where it has been cultivated for nearly 5,000 years. People use rice straw in many ways, such as thatching for roofs, food and bedding for livestock, and clothing. Rice hulls also have many uses, ranging from fertilizer to fuel.

Rice is a fast-growing plant, yielding more than one crop per year. Some rice is fed to livestock, but most is eaten by humans. When milled for human consumption, rice is an incomplete food because milling removes the bran that contains B vitamins. Even unmilled rice is a poor source of protein; thus, it should be eaten with other, supplementary foods such as soybeans or fish. Most rice varieties are grown submerged in water for the bulk of the growing season. Fish are often raised in rice paddies, where they serve as supplemental food for humans, add fertilizer to the paddy, and control the mosquito population.

The rice plant looks much like other cereal grain plants. The leaves are long, narrow, flat, and more than half a meter in length, with veins running parallel to one another along the length of the leaf. Rice stems do not thicken and become woody as do the stems of trees and shrubs. Rice flowers have six stamens and one ovary. The vascular bundles in the rice stem are scattered, rather than lying in a ring as in the red maple stem.

(a)

(b)

29.3   Rice
(a) Terraced rice paddies in Bali, Indonesia. *(b)* A rice plant.

Soybean

Soybeans (*Glycine max*; Figure 29.4) were first grown in China thousands of years ago, but today the United States is the largest single producer. Soybeans are featured in many foods and sauces. They also yield a commercially important oil, used in the manufacture of adhesives, paints, inks, and plastics. After oil has been squeezed from the seeds, the residue may be fed to livestock or made into soy flour. Soybean stems may be used for straw.

29.4   Soybean
Their leaves dominate this farmer's field of soybeans in summer.

The soybean plant stands from less than a meter to more than 2 meters in height. Soybean leaves have three lobes, with veins radiating in a netlike pattern. The vascular bundles of the young soybean stem, like those of the red maple, are arranged in a cylindrical pattern. The flowers are small, either white or blue, and consist of five sepals, five petals, ten stamens, and one pistil. Soybean plants tend to be drought-resistant because they have richly branching root systems, which often extend more than 1.5 meters below the soil surface.

CLASSES OF FLOWERING PLANTS

Comparison of the features of these four plants—coconut palm, red maple, rice, and soybean—suggests at least two ways in which they may be classified. We may divide them into trees (coconut palm, red maple) and herbs (rice, soybean) based on their growth plan, or we may divide them into monocots (coconut palm, rice) and dicots (red maple, soybean) based on one clearly distinguishing character (possession of one or two seed leaves—cotyledons—in their embryo) as well as on several important anatomical characters (Figure 29.5). Monocots are generally narrow-leaved flowering plants such as grasses (including rice), lilies, orchids, and palms. Dicots are broad-leaved flowering plants such as soybeans, roses, sunflowers, and maples. As we learned in Chapter 25, the monocots and dicots are the two classes, Monocotyledones and Dicotyledones, that make up the phylum Anthophyta—flowering plants (angiosperms). We'll consider other parts of plants in addition to cotyledons, but first let's consider some overall organizing concepts.

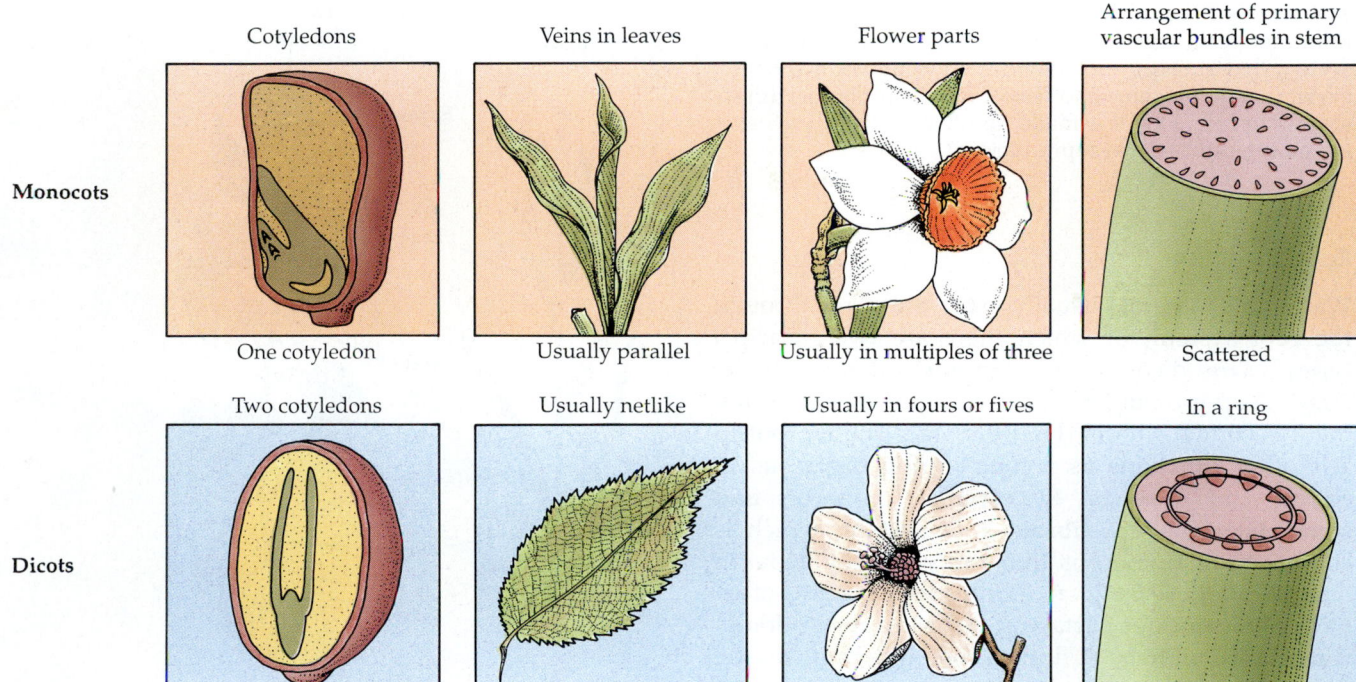

| | Cotyledons | Veins in leaves | Flower parts | Arrangement of primary vascular bundles in stem |

Monocots — One cotyledon — Usually parallel — Usually in multiples of three — Scattered

Dicots — Two cotyledons — Usually netlike — Usually in fours or fives — In a ring

29.5 Monocots versus Dicots

AN OVERVIEW OF THE PLANT BODY

As the plant body grows, it may lose parts, and it forms new parts that may grow at different rates. Each branch of a plant may be thought of as a unit in many ways independent of the other branches. A branch of a plant does not bear the same relationship to the remainder of the body as does an arm to the remainder of the human body. Each branch lives out its own history, and branches grow independently, exploring different parts of the surrounding environment. Branches may respond differently to gravity, some growing more or less vertically and others horizontally. Leaves are units of another sort, produced in fresh batches to take over the constant function of feeding the plant. Often the shapes of different leaves of the same plant differ depending on their differing local environments. Leaves are much shorter-lived than are branches. Branch roots are semi-independent units.

The partial independence of plant parts results in a decentralization of control systems. Branches experiencing different local environments send differing reports to the rest of the plant. If a plant has more than one stem, each one may receive a different report as to the availability of water and minerals because each is served by different roots. In spite of this decentralization, the plant functions as a coherent unit.

Animals grow all over; that is, all parts of the body grow as the individual develops from embryo to adult. Most plant growth, by contrast, is in specific regions of active cell division and cell expansion. The regions of cell division are called **meristems**. Meristems at the tips of the root and stem produce the plant body by dividing to make the cells that compose the parts of the plant. All plant organs arise from cell divisions in the meristems, followed by cell expansion. As you read this chapter, notice the emphasis on the activities of meristems.

ORGANS OF THE PLANT BODY

The bodies of most vascular plants are divided into three principal organs: the **leaves**, the **stem**, and the **root system** (Figure 29.6). A stem and its leaves, taken together, are called a shoot. The **shoot system** of a plant consists of all stems and all leaves. Broadly speaking, the leaves are the chief organs of photosynthesis. The stem holds and displays the leaves to the sun, maximizing the photosynthetic yield, and provides transport connections between the roots and leaves. The points where leaves attach to the stem are called **nodes**, and the stem regions between nodes are **internodes** (see Figure 29.6). Roots anchor the plant in place, and their extreme branching and fine form adapt them to absorb water and mineral nutrients from the soil.

Each of the principal organs can best be understood in terms of its function and its structure. By structure we mean both gross form and microscopic anatomy—the component tissues as well as their arrangement.

29.6 Body Plans of Monocots and Dicots

Both monocots and dicots absorb water through a root system that anchors and provides nutrients for a shoot system made of stems and leaves in which photosynthesis takes place. Flowers, made up of specialized leaves, are adapted for sexual reproduction.

Roots

Water and minerals usually enter the plant through the root system, of which there are two principal types. Many dicots have a **taproot system**: a single, large, deep-growing root accompanied by less prominent secondary roots (Figure 29.7a). The taproot itself often functions as a food-storing organ, as in carrots and radishes. By contrast, monocots and some dicots have a **fibrous root system**, which is composed of numerous thin roots roughly equal in diameter (Figure 29.7b–e). Fibrous root systems often have a tremendous surface area for the absorption of water and minerals. A fibrous root system holds soil

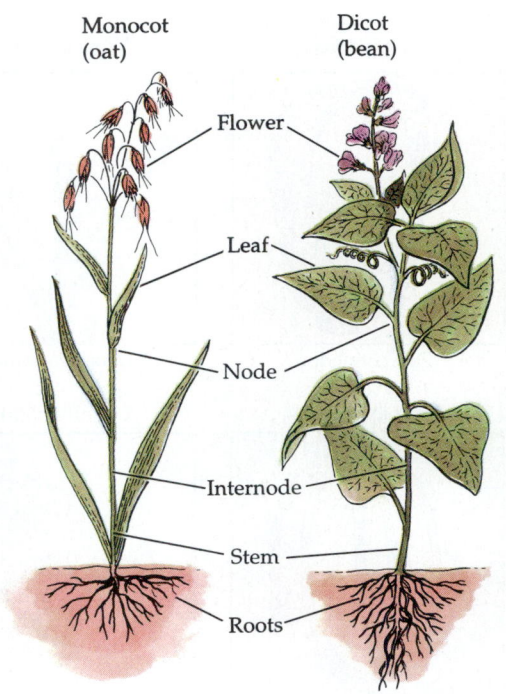

Monocot (oat)   Dicot (bean)

Flower
Leaf
Node
Internode
Stem
Roots

(a)

(b)

29.7 Root Systems

The taproot system of a dandelion (a) contrasts with the fibrous root system of grasses (b). Fibrous root systems are diverse (c–e), with forms adapted to the different environments in which they grow.

(c)   (d)   (e)

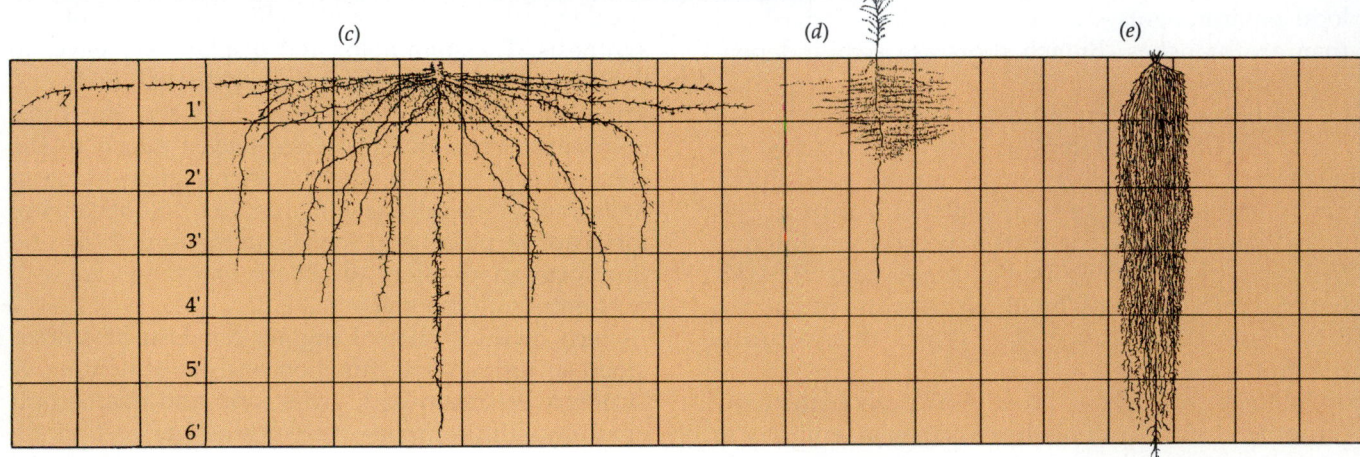

1'
2'
3'
4'
5'
6'

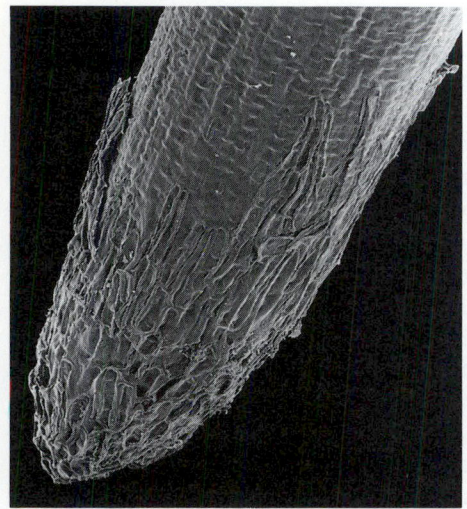

29.8 The Root Cap Protects the Root Tip
As the root grows through the soil, the root cap wears off and is replaced by cells from the root apical meristem.

very well, giving grasses with such systems a protective role on steep hillsides where runoff from rain could cause erosion.

A tissue composed of rapidly dividing cells is located at the tip of the root proper, just behind the root cap. This tissue is the **root apical meristem**, which produces all the cells that contribute to growth in the length of the root. Some of the daughter cells

from the root apical meristem are contributed to the **root cap** that protects the delicate growing region of the root as it pushes through the soil (Figure 29.8). Cells of the root cap are often damaged or scraped away and must therefore be replaced constantly. The root cap is also the structure that detects the pull of gravity and thus causes the root to grow downward. Most root cells—those that are produced at the other end of the meristem—elongate. Following elongation, these cells differentiate, giving rise to the various tissues of the mature root.

Some plants have roots that arise from points along the stem, or from the leaves. Known as **adventitious roots**, they also form (in many species) when a piece of shoot is cut from the plant. Adventitious rooting enables the cutting to establish itself in the soil. Some plants—corn, for example—use adventitious roots as props to help support the young shoot.

Stems

Unlike most roots, a stem may be green and capable of photosynthesis. A stem bears leaves at its nodes, and where each leaf meets the stem there is a **lateral bud**, which develops into a branch if it becomes active (Figure 29.9a). A branch is also a stem. The branching patterns of plants are highly variable, depending upon the species, environmental conditions, and a gardener's pruning activities.

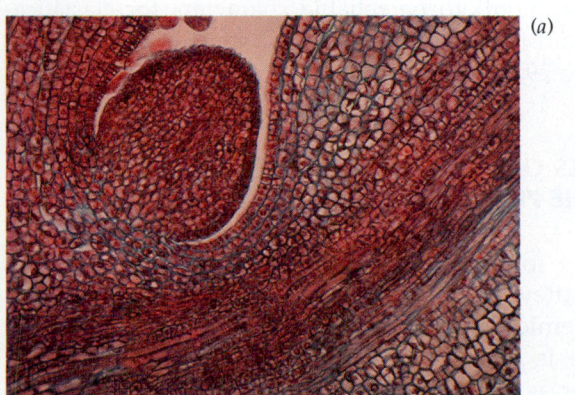

(a)

29.9 A Selection of Stems
(a) Microscopic view of a lateral bud developing at the junction between leaf and stem of a lilac; the bud contains vascular tissue and may develop into a branch. (b) A potato is a modified stem called a tuber; the sprouts that grow from its eyes are branches. (c) Runners (red) of beach strawberry are horizontal stems; such a stem produces roots at intervals, providing a local water supply and allowing rooted portions to live independently if the runner is cut.

(c)

(b)

Some stems are highly modified. The potato **tuber**, for example, is actually a portion of the stem. Its eyes (Figure 29.9*b*) contain lateral buds, and a sprouting potato is just a branching stem. The runners of strawberry plants and Bermuda grass are horizontal stems from which roots grow at frequent intervals (Figure 29.9*c*). If the links between the rooted portions are broken, independent plants can develop from each side of the break. This is a form of asexual reproduction, which we will discuss in Chapter 33.

Stems bear buds—embryonic shoots—of various types. We have already mentioned lateral buds, which give rise to branches. At the tip of each stem or branch is an **apical bud** containing a **shoot apical meristem**, which produces the cells for the growth and development of the stem. The shoot apical meristem also produces **leaf primordia**, which expand to become leaves in the apical bud (see the upper micrograph in Figure 29.16). At times that vary from species to species, buds are formed that develop into flowers.

Shoot systems have various forms, in which branches take on different relationships to the plant as a whole (Figure 29.10). Some shoots branch underground, and their branches emerge from the soil looking like separate plants.

Leaves

In most plants the leaves are responsible for most of the photosynthesis, producing food for the plant and releasing oxygen gas. Leaves also carry out metabolic reactions that make nitrogen available to the plant for the synthesis of proteins and nucleic acids (see Chapter 32). Leaves are important food-storage organs in some species; in others—the succulents—the leaves store water. The thorns of cacti are modified leaves. Certain leaves of poinsettias, dogwood, and some other plants are brightly colored and help attract pollinating animals to the often less-striking flowers. Many plants, such as peas and squash, have tendrils—modified leaves that support the plant by wrapping around other plants. A less obvious but often crucial function of leaves is to shade neighboring plants. Like all other organisms, plants compete; if a plant can reduce the photosynthetic capability of

its neighbors by intercepting sunlight, it can obtain a greater share of the available water and mineral nutrients. Finally, as we will see in Chapter 33, the "timer" by which some plants measure the length of the night is located in the leaves.

Leaves are marvelously adapted to serve as light-gathering, photosynthetic organs. Typically, the **blade** of a leaf is flat, and during the daytime it is held by its stalk, or **petiole**, at an angle almost perpendicular to the rays of the sun. Some leaves track the sun, moving so that they constantly face it. If leaves were thicker, the outer layers of cells would absorb so much of the light that the interior layers of cells would be too dark and thus would be unable to photosynthesize.

The different leaves of a single plant may have quite different shapes. The form of a leaf results from a combination of genetic, environmental, and developmental influences. Most species, however, bear leaves of some broadly defined type (Figure 29.11). A leaf may be **simple**, consisting of a single blade, or **compound**, in which blades, or leaflets, are arranged along an axis or radiate from a central point. In a simple leaf, or in a leaflet of a compound leaf, the veins may be parallel to one another or in a netlike arrangement. The general development of a specific leaf pattern is programmed in the individual's genes and is expressed by differential growth of the leaf veins and of the tissue between the veins. As a result plant taxonomists have often found leaf forms (outline, margins, tips, bases, and patterns of arrangement) to be reliable characters for classification and identification. At least some of the forms in Figure 29.11 probably look familiar to you.

LEVELS OF ORGANIZATION IN THE PLANT BODY

Newly formed cells expand to their final size, and then they differentiate, that is, become structurally or chemically specialized for particular functions. A tissue is an organized group of cells, working together as a functional unit. Simple tissues are composed of a single type of cell; compound tissues are composed of several cell types. Plant tissues are or-

29.10 Types of Shoot Systems
Do you recognize some of these stem types? You can probably find most of them among the weeds on your campus.

Leaf shapes

Margins

Apices and bases

Arrangements on stem

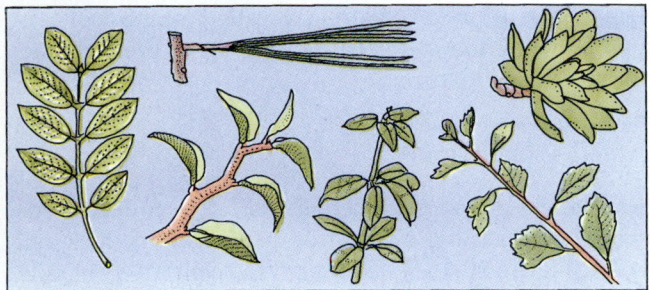

Parts and types

29.11   The Diversity of Leaves

ganized into three tissue systems that extend throughout the body of the plant, from organ to organ. These three systems are the vascular tissue system (xylem and phloem), which conducts materials from one part of the body to another; the dermal tissue system, which protects the body surface; and the ground tissue system, which plays many roles, including producing and storing food materials.

To understand the structures and functions of the tissue systems, we must know the nature of their building blocks. Some cells are alive when functional; others function only after their living parts have died and disintegrated. Some cells develop chemical capabilities not demonstrated by other cells. Several cell types differ dramatically in the structure of their cell walls. We will first consider the types of cells that make up the plant body and then see how aggregations of cells form functioning tissues and tissue systems.

PLANT CELLS

Living plant cells have all the essential organelles common to eukaryotes (see Chapter 4). In addition, they have some structures and organelles not shared by cells of the other kingdoms. Some plant cells contain chloroplasts and microbodies. Many plant cells contain large central vacuoles. Every plant cell is surrounded by a cellulose-containing cell wall.

Cell Walls

The division of a plant cell is completed when cell walls form, separating the daughter cells. The first barrier to form is the **middle lamella** (Figure 29.12a). The formation of this layer is followed by the secretion of structural materials, including cellulose, by the newly separated cells. Each daughter cell, as it expands to its final size, secretes more cellulose and other polysaccharides to complete formation of the **primary wall** (Figure 29.12b).

Once cell expansion stops, a plant cell may deposit more polysaccharides and other materials—such as lignin, characteristic of wood, or suberin, characteristic of cork—in one or more layers internal to the primary wall. These layers collectively form the **secondary wall** (Figure 29.12c), which often serves supporting or waterproofing roles.

Although the cell wall lies outside the plasma membrane of the cell, it is not a chemically inactive region. Chemical reactions in the wall play an important role in cell expansion. Cell walls may thicken or be sculpted or perforated as part of the differentiation into various cell types.

Except where the secondary wall is waterproofed, the structure is porous to water and to most small

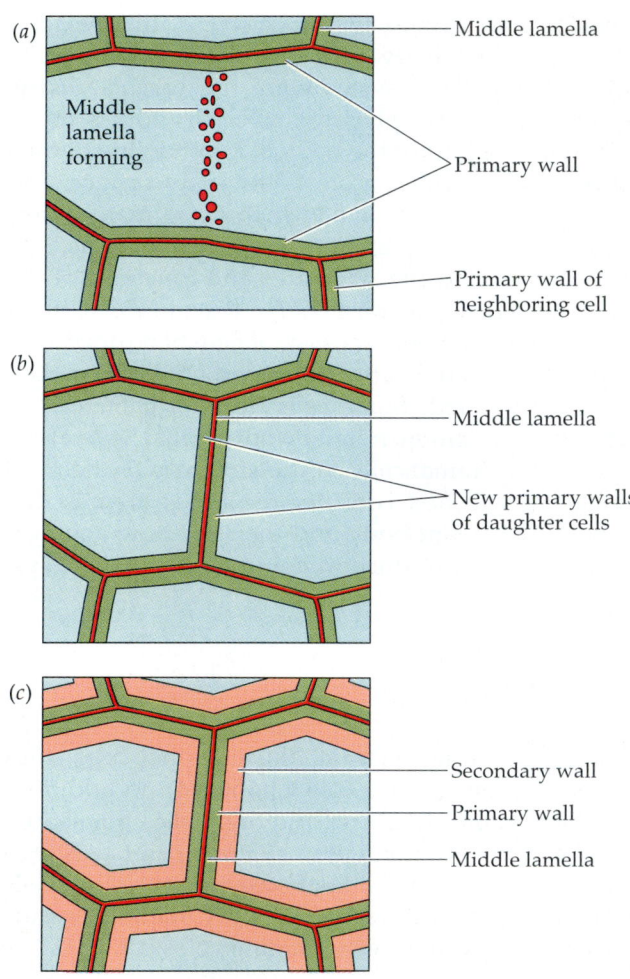

(a)

Middle lamella forming

Middle lamella

Primary wall

Primary wall of neighboring cell

(b)

Middle lamella

New primary walls of daughter cells

(c)

Secondary wall

Primary wall

Middle lamella

29.12   Cell Wall Formation
The middle lamella is the first wall layer to form. Each daughter cell secretes a primary wall. Once cell expansion stops, the cell may secrete more layers, forming secondary walls.

molecules. Water and dissolved materials can move directly from cell to cell without passing into the cell wall space because plant cells have structures called **pit pairs** connected by strands of cytoplasm called plasmodesmata (see Chapter 4). A pit is a thinning in the primary wall of a cell at a place where the secondary wall either is absent or is separated from the primary wall by a space. Where there is a pit in the wall of one cell, there is usually a corresponding pit in the adjacent cell's wall; together they are a pit pair. Plasmodesmata pass through pit pairs and the middle lamella between them, allowing molecules with molecular weights of about 850 daltons or less to pass freely from one cell to the other.

Parenchyma Cells

The most numerous cells in the young plant body are the **parenchyma** cells (Figure 29.13a). Parenchyma

cells are alive when they perform their functions in the plant. They usually have thin walls, consisting only of a primary wall and the shared middle lamella. Many parenchyma cells have shapes similar to those of soap bubbles crowded into a limited space—figures with 14 faces. They are not elongated or otherwise asymmetrical. Most have large central vacuoles.

Many parenchyma cells store various substances, such as starch or lipids. In the cytoplasm of these cells, starch is often stored in specialized plastids called leucoplasts (see Chapter 4). Lipids may be stored as oil droplets, also in the cytoplasm. Other parenchyma cells appear to serve as packing material and play a vital role in supporting the stem. Leaves have a particularly important type of parenchyma cell that is specialized for photosynthesis and is equipped with abundant chloroplasts. Some other parenchyma cells—but not these photosynthetic cells—retain the capacity to divide and hence may give rise to new meristems, as when a branch root forms within a region of parenchyma cells inside a taproot.

Sclerenchyma Cells

Sclerenchyma cells function when dead. A heavily thickened secondary wall performs their function: support. There are two types of sclerenchyma cells: elongated **fibers** and variously shaped **sclereids**. Fibers, often organized into bundles, provide a relatively rigid support both in wood and in other parts of the plant (Figure 29.13b). The bark of trees owes much of its mechanical strength to long fibers. Sclereids may pack together very densely, as in a nut's shell or other types of seed coats (Figure 29.13c). Isolated clumps of sclereids, called stone cells, in pears and some other fruits give them their characteristic gritty texture.

Collenchyma Cells

Another type of supporting cell, the **collenchyma** cell, remains alive even after laying down thick cell walls (Figure 29.13d). Collenchyma cells are generally elongated. In these cells the primary wall thickens and no secondary wall forms. Collenchyma provides support to petioles, nonwoody shoots, and growing organs. Tissue made of collenchyma cells, although resistant to bending, is more flexible than is sclerenchyma; stems and leaf petioles strengthened by collenchyma can sway in the wind without snapping as they might if they were strengthened by sclerenchyma.

Water-Conducting Cells of the Xylem

The xylem of vascular plants contains cells called tracheary elements, which die before they assume

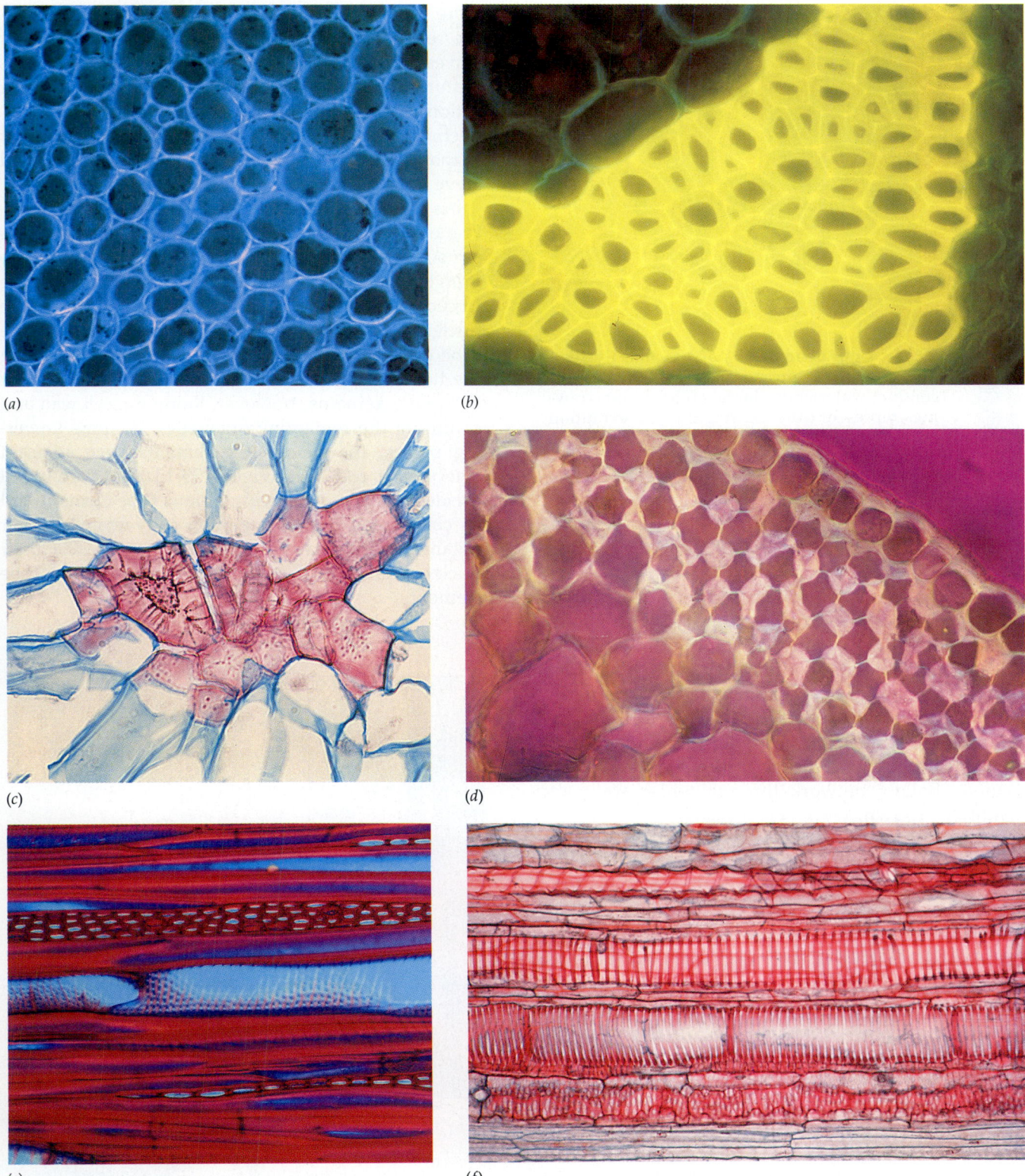

(a)

(b)

(c)

(d)

(e)

(f)

29.13 Plant Cells

(a) Parenchyma cells in the leaf of a primrose plant; note the uniform cell walls. (b) Sclerenchyma: Fibers in a broad bean pod. A stain causes the heavily thickened walls to fluoresce a brilliant yellow. (c) Sclerenchyma: Thick-walled sclereids; these extremely thick secondary cell walls are laid down in layers. They provide support and a hard texture to structures such as nuts and seeds. (d) Collenchyma cells make up the five outer cell layers of this spinach leaf vein. They are recognizable because their cell walls are very thick at the corners of the cells and thin elsewhere. (e) Tracheids appear deep red in this micrograph of bassswood; note the complexity of the cell walls. (f) Vessel elements in the stem of a squash. The secondary walls are stained red; note the different patterns of thickening, including rings and spirals. Which cells in this figure function when they are alive and which only when they are dead?

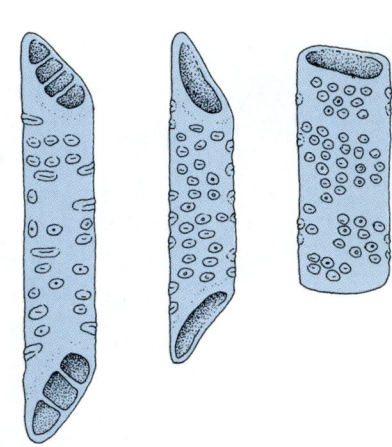

29.14   Conducting Cells of Vascular Systems
The xylem of angiosperms contains vessels that conduct water and minerals. These four drawings represent different stages in the evolution of the vessel element; the one on the left is the most ancient, and the one at the right is the most recently evolved.

their ultimate function of transporting water and dissolved minerals. The tracheary elements of gymnosperms and angiosperms differ significantly. The tracheary elements of gymnosperms are **tracheids**— spindle-shaped cells interconnected by numerous pits in their cell walls (Figure 29.13*e*). Because the cell contents—the nucleus and cytoplasm—disintegrate upon death, a group of dead tracheids forms a continuous hollow network through which water can readily be drawn.

Flowering plants evolved a water-conducting system made up of vessels. The individual cells, called **vessel elements**, also die before they become functional. Vessel elements are generally larger in diameter than tracheids; they are laid down end-to-end and lose all or part of their end walls, so that each vessel is a continuous hollow tube consisting of many vessel elements and providing a clear pipeline for water conduction (Figure 29.13*f*). In the course of angiosperm evolution, vessel elements have become shorter, and their end walls have become less and less obliquely oriented and less obstructed (Figure 29.14). The xylem of many angiosperms also includes tracheids.

Sieve Tube Elements

The phloem, in contrast to xylem, consists primarily of living cells. In flowering plants the characteristic cell of the phloem is the **sieve tube element** (Figure 29.15*a*). Like vessel elements, these cells meet end-to-end and form long sieve tubes, which transport foods from their sources to tissues that consume or

store them. In plants with mature leaves, for example, excess products of photosynthesis move from leaves to tissues in roots. As sieve tube elements mature during their development, a chemical drilling action expands small holes in the end walls, connecting the contents of neighboring cells. The result is that the end walls look like sieves and are called **sieve plates** (Figure 29.15*b*).

As the holes in the sieve plates expand, the membrane around the central vacuole disappears, allowing some of the cytosol and the vacuole's contents to mingle and form a single fluid; this mixture can be forced from cell to cell along the sieve tube. The nucleus and some of the other organelles in the sieve tube element also break down and thus do not clog the holes of the sieve. A "fixed," stationary layer of cytoplasm remains, however, lining the cell wall and confining the remaining organelles. In some flowering plants, the sieve tube elements have adjacent **companion cells** (see Figure 29.15*a*). A parent cell divides, thereby producing a sieve tube element and its companion cell. Companion cells retain all their organelles and may, through the activities of their nuclei, regulate the performance of the sieve tube elements.

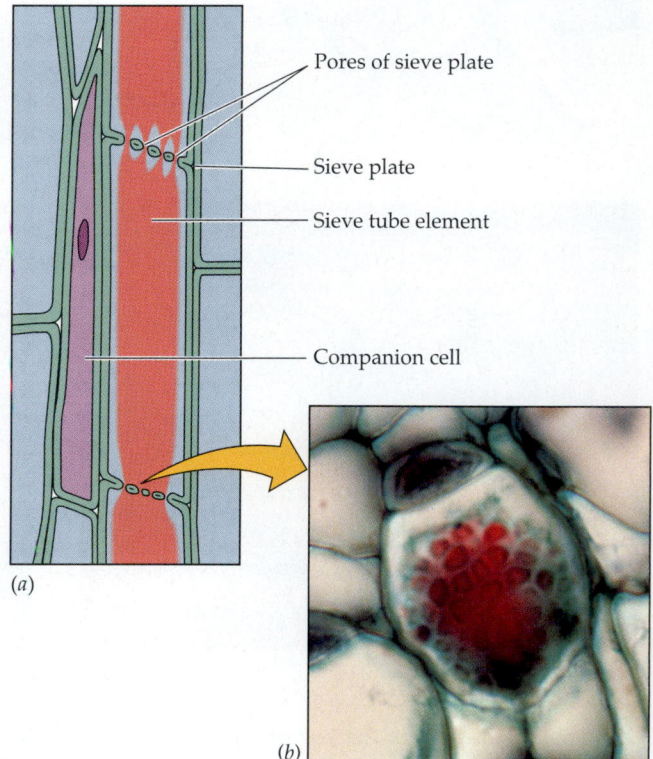

Pores of sieve plate

Sieve plate

Sieve tube element

Companion cell

(a)

(b)

29.15   Sieve Tube Elements
(a) Sieve tube elements are usually accompanied by companion cells. *(b)* Micrograph of a sieve plate at the end of a sieve tube element. Phloem sap passes through the holes in sieve plates from one sieve tube element to the next.

PLANT TISSUES AND TISSUE SYSTEMS

Parenchyma cells make up parenchyma tissue, a simple tissue—that is, one composed of only one type of cell. Sclerenchyma and collenchyma are other simple tissues. Cells of various types also combine to form complex tissues. Xylem and phloem are complex tissues, composed of more than one type of cell. All xylem contains parenchyma cells, which store food. The xylem of angiosperms contains vessel elements, as well as thick-walled sclerenchyma fibers that provide considerable mechanical strength to the xylem. In most gymnosperms, tracheids serve both in water conduction and in support because vessels and fibers are absent. In addition, old xylem that is no longer active in transport becomes compacted at the center of the tree trunk and continues to contribute support for the tree. As a result of its cellular complexity, xylem can perform a variety of functions, including transport, support, and storage. The phloem of angiosperms includes sieve tube elements, companion cells, fibers, sclereids, and parenchyma cells.

The **vascular tissue system**, which includes the xylem and phloem, is the conductive, or "plumbing," system of the plant. All living cells require a source of energy and chemical building blocks. As already mentioned, the phloem transports food from the sites of production (called sources; commonly the leaves) to sites of utilization or storage (called sinks) elsewhere in the plant. The xylem distributes water and mineral ions taken up by the roots to the stem and leaves.

The **dermal tissue system** is the outer covering of the plant. All parts of the young plant body are covered by an **epidermis**, either a single layer of cells or several layers. The shoot epidermis secretes a layer of wax, the **cuticle**, that helps retard water loss. The protective covering of the stems and roots of older woody plants is the **periderm**, which is composed of cork and other tissues that will be discussed later in this chapter.

The **ground tissue system** makes up the rest of a plant and consists primarily of parenchyma tissue, often supplemented by collenchyma or sclerenchyma tissue. The ground tissues function primarily in storage, support, and photosynthesis. Let's look at how the tissue systems are organized in the different organs of a flowering plant, as well as how this organization develops as the plant grows.

GROWTH AND MERISTEMS

At the tip of each shoot or branch is a shoot apical meristem, and at the tip of each root is a root apical meristem. Growth from the apical meristems is called **primary growth**. These meristems give rise to the entire body of many plants (Figure 29.16).

Other plants develop what we commonly refer to as wood and bark. These complex tissues are derived from other meristems. One, called **vascular cambium**, is a cylindrical tissue consisting primarily of vertically elongated cells that divide frequently, producing derivative cells both to the inside of the vascular cambium layer, forming new xylem, and to the outside, forming new phloem. As trees grow in diameter, the outermost layers of the stem are sloughed off. Without the activity of **cork cambium**, which in a tree forms continuously in the bark, this sloughing off would expose the tree to potential damage, including excessive water loss or invasion by microorganisms. Cork cambium produces new cells, primarily in the outward direction. The walls of these cells become impregnated with the waxy substance suberin, thus augmenting the dermal tissue system. Growth in the diameter of stems and roots, produced by the vascular and cork cambia, is called **secondary growth**. It is the source of wood and bark.

In some plants, meristems may remain active for years—even centuries. Such plants grow in size, or at least in diameter, throughout their lifetimes. This phenomenon is known as **indeterminate growth**. **Determinate growth**, which stops at some point, is characteristic of most animals, as well as some plant parts, such as leaves, flowers, and fruits. The life cycles of plants fall into three categories: annual, biennial, and perennial. **Annuals**, such as many food crops, live less than a year. **Biennials**—carrots and cabbage, for example—grow for all or part of one year and live on into a second year, during which they flower, set seed, and die. **Perennials**, such as oak trees, live for a few to many years.

THE MERISTEMS AND THEIR PRODUCTS

The Young Root

Cell divisions in the root apical meristem produce both the protective root cap and the other primary tissues of the growing root. When a meristematic cell divides, the products initially take up no more volume between them than did the dividing cell. One of the products of each cell division develops into another meristematic cell the size of its parent, while the other product develops differently. The products above the apical meristem—away from the root cap—constitute three cylindrical primary meristems that give rise to the three tissue systems of the root. The innermost primary meristem, the **procambium**, gives rise to the vascular tissue system; the **ground meristem** gives rise to the ground tissue system; and the outermost, the **protoderm**, gives rise to the dermal tissue system (Figure 29.17). The apical and primary

29.16 Meristems and the Plant Body
The root apical meristem and shoot apical meristem, shown in the insets, give rise to the plant body, as do the lateral bud meristems. The vascular cambium and cork cambia thicken the stem and root.

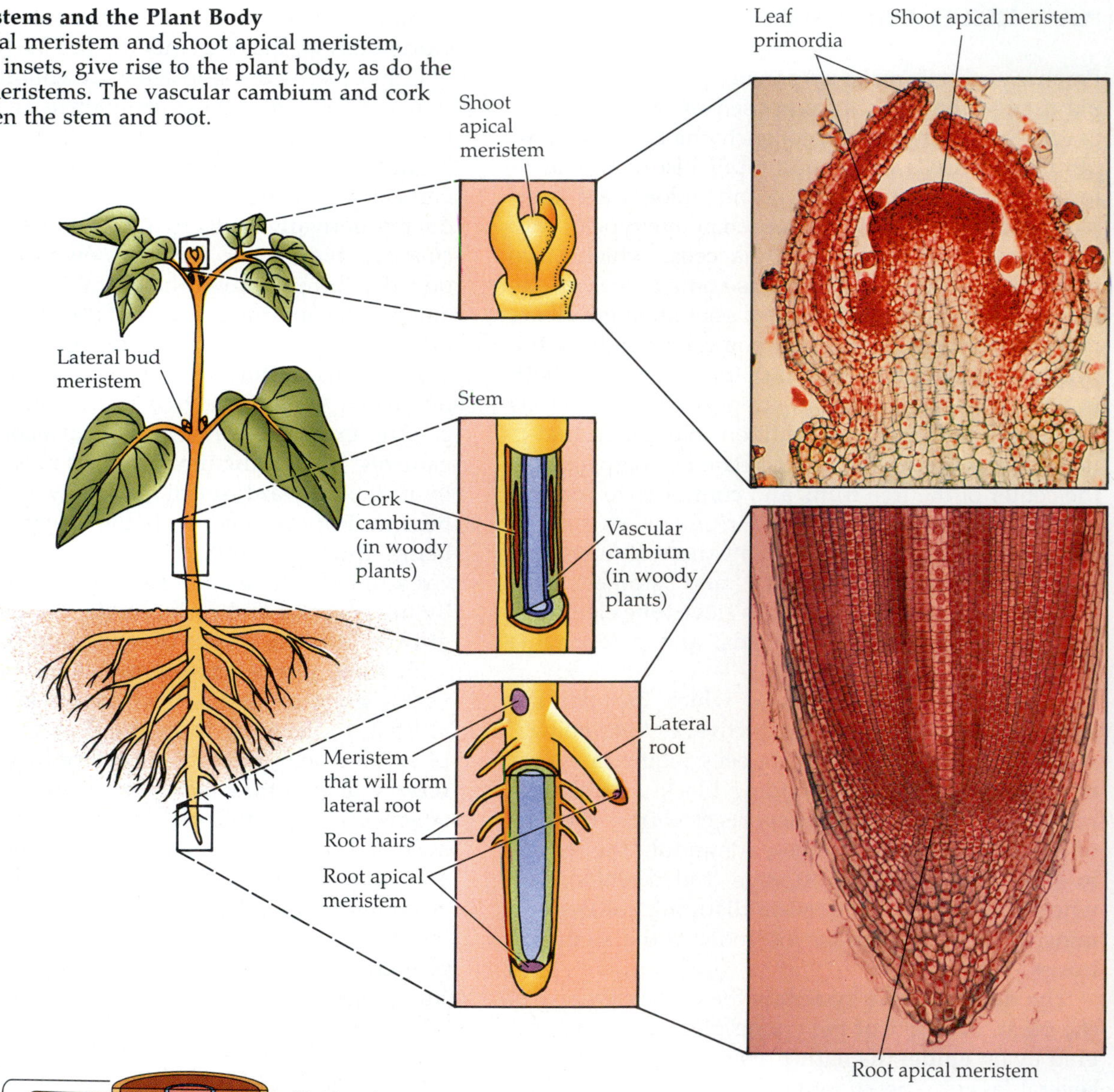

Shoot apical meristem

Lateral bud meristem

Stem

Cork cambium (in woody plants)

Vascular cambium (in woody plants)

Meristem that will form lateral root

Lateral root

Root hairs

Root apical meristem

Leaf primordia

Shoot apical meristem

Root apical meristem

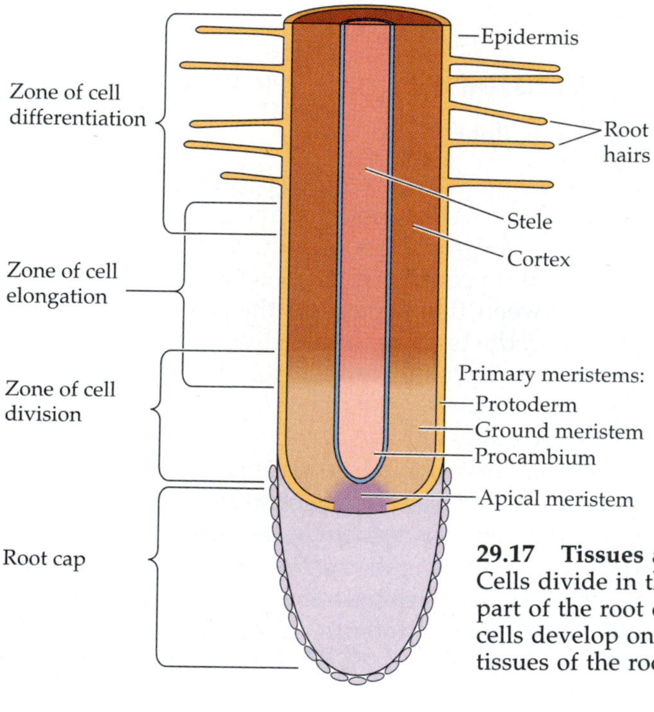

Zone of cell differentiation

Zone of cell elongation

Zone of cell division

Root cap

Epidermis

Root hairs

Stele

Cortex

Primary meristems:
Protoderm
Ground meristem
Procambium

Apical meristem

meristems are the zone of cell division, the source of all the cells of the root's primary tissues. Just above this zone, where the cells are somewhat older, is the zone of cell elongation, in which cells are elongating and thus causing the root to reach farther into the soil. Where the cells are older yet, and where they are differentiating—taking on specialized forms and functions—is the zone of cell differentiation. These three zones grade imperceptibly into one another; there is extensive cell division even as far up as the

29.17 Tissues and Regions of the Root Tip
Cells divide in the root apical meristem. Some of the daughter cells become part of the root cap, which is constantly being eroded away, but most daughter cells develop on the side away from the tip and differentiate into the primary tissues of the root.

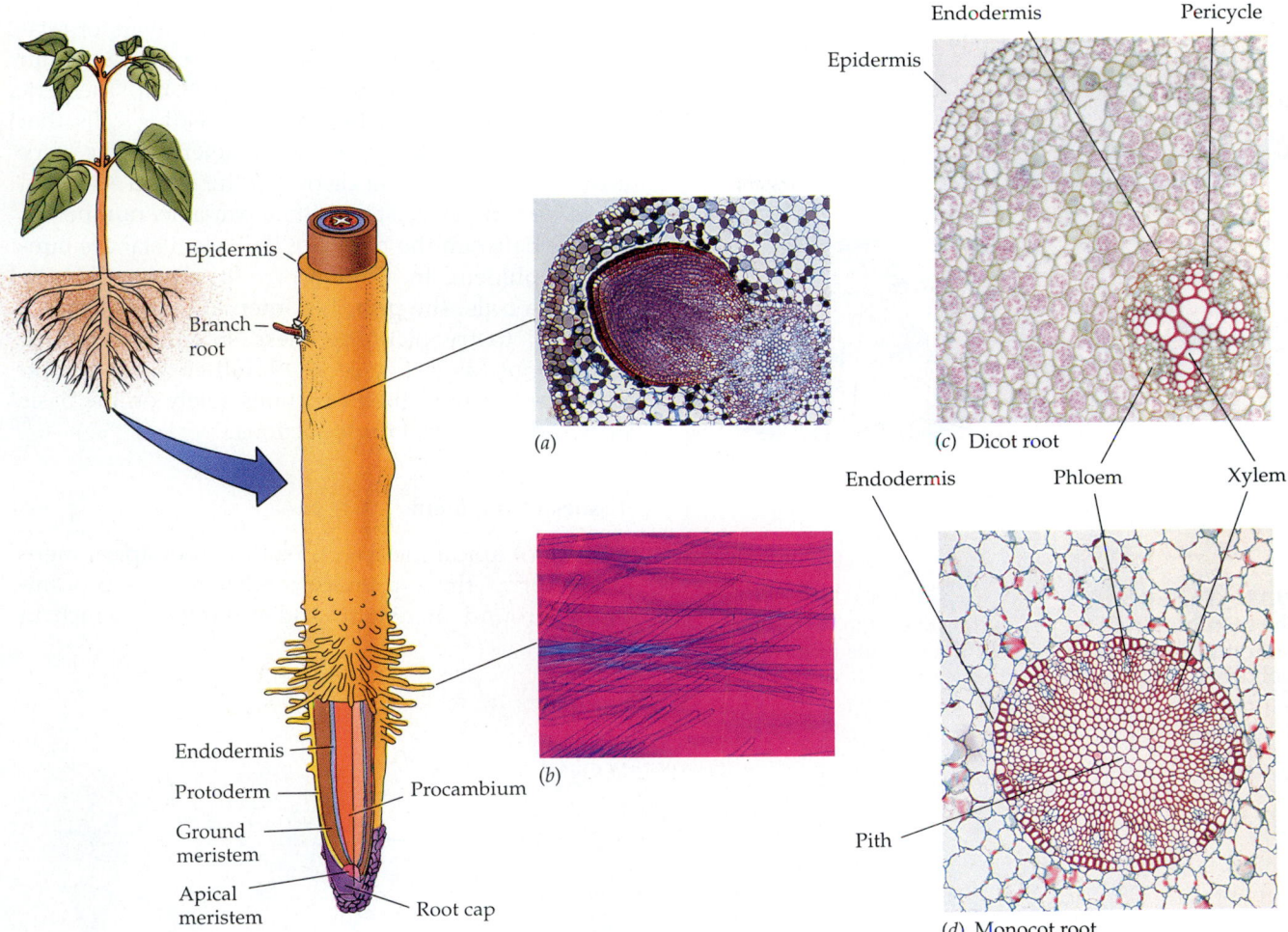

Epidermis

Branch root

Endodermis

Protoderm

Ground meristem

Apical meristem

Procambium

Root cap

(a)

(b)

Endodermis

Epidermis

Pericycle

(c) Dicot root

Endodermis

Phloem

Xylem

Pith

(d) Monocot root

29.18 Root Anatomy

The drawing at the left shows a generalized root structure. (a) A branching root tip. Cells in the pericycle divide and the products differentiate, forming the tissues of a branch root. (b) Root hairs, viewed under polarized light. (c, d) The primary root tissues of a dicot and a monocot. The monocot (an orchid) has a central pith region; the dicot (ranunculus) does not.

zone of cell differentiation, and some cells differentiate even in the zone of cell division.

The protoderm gives rise to the outer layer of root cells, the epidermis, which is adapted for protection and for the absorption of mineral ions and water (Figure 29.18). Epidermal cells are flattened, and many of them produce amazingly long, delicate **root hairs** that vastly increase the surface area of the root (Figure 29.18b). It has been estimated that a mature rye plant has a total root surface of more than 1,500 square kilometers (600 square miles), all contained within about 6 liters of soil. Root hairs grow out among the soil particles, probing nooks and crannies and taking up water and minerals.

Internal to the root's epidermis is a region of ground tissue many cells in thickness, called the cor-

tex (see Figure 29.17). The cells of the cortex are relatively unspecialized and often function in food storage. In many plants, but especially in trees, epidermal and sometimes cortical cells form an association with a fungus. This association, called a **mycorrhiza**, increases the absorption of minerals and water by the plant (see Box 24.A). Some plant species have poorly developed root hairs or no root hairs. These plants cannot survive unless they develop mycorrhizae that help in mineral absorption.

Proceeding inward, we come to the **endodermis** of the root, a single layer of cells that is the innermost cell layer of the cortex. Endodermal cells differ markedly in structure from the rest of the cortical cells; parts of their walls contain suberin, a waxy substance that forms a waterproof seal. The endodermal cells control the access of water and dissolved substances to the inner, vascular tissues (see Figure 30.6). Elsewhere in the root, water can pass freely through cell walls and between cells.

Once past the endodermis, we enter the domain produced by the procambium. This domain, the vascular cylinder or **stele**, consists of three tissues: the pericycle, the xylem, and the phloem (Figure 29.19).

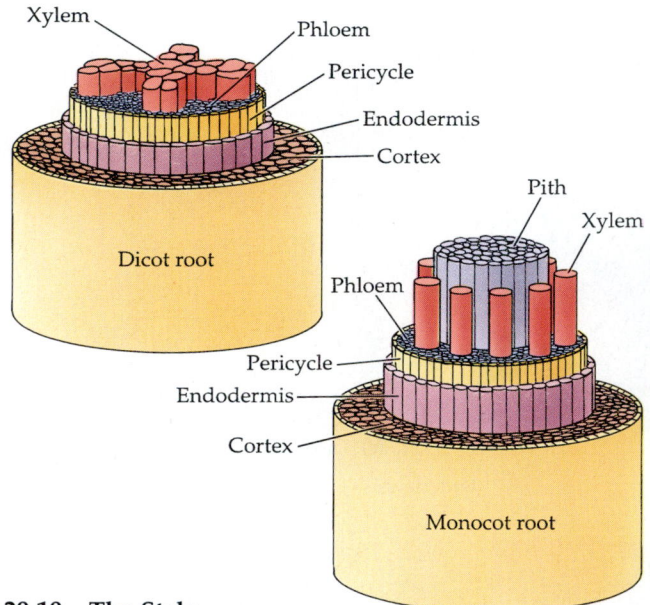

29.19 The Stele
The distribution of tissues in the stele—the region internal to the endodermis—differs in the roots of dicots and monocots.

The **pericycle** consists of one or more layers of relatively undifferentiated cells. It is the tissue within which branch roots arise (see Figure 29.18a); the pericycle also provides a few of the dividing cells that enable the root to grow in diameter. At the very center of the root of a dicot lies the xylem—seen in cross section as a star with a variable number of points. Between the points of the xylem star are bundles of phloem. In a monocot root, a region of parenchyma cells, the **pith**, lies internal to the xylem. It is useful to try picturing these structures in three dimensions, as in Figure 29.19, rather than attempting to understand their functions solely on the basis of two-dimensional cross sections.

Tissues of the Stem

The shoot apical meristem, like the root apical meristem, forms three primary meristems: the procambium, ground meristem, and protoderm, which in

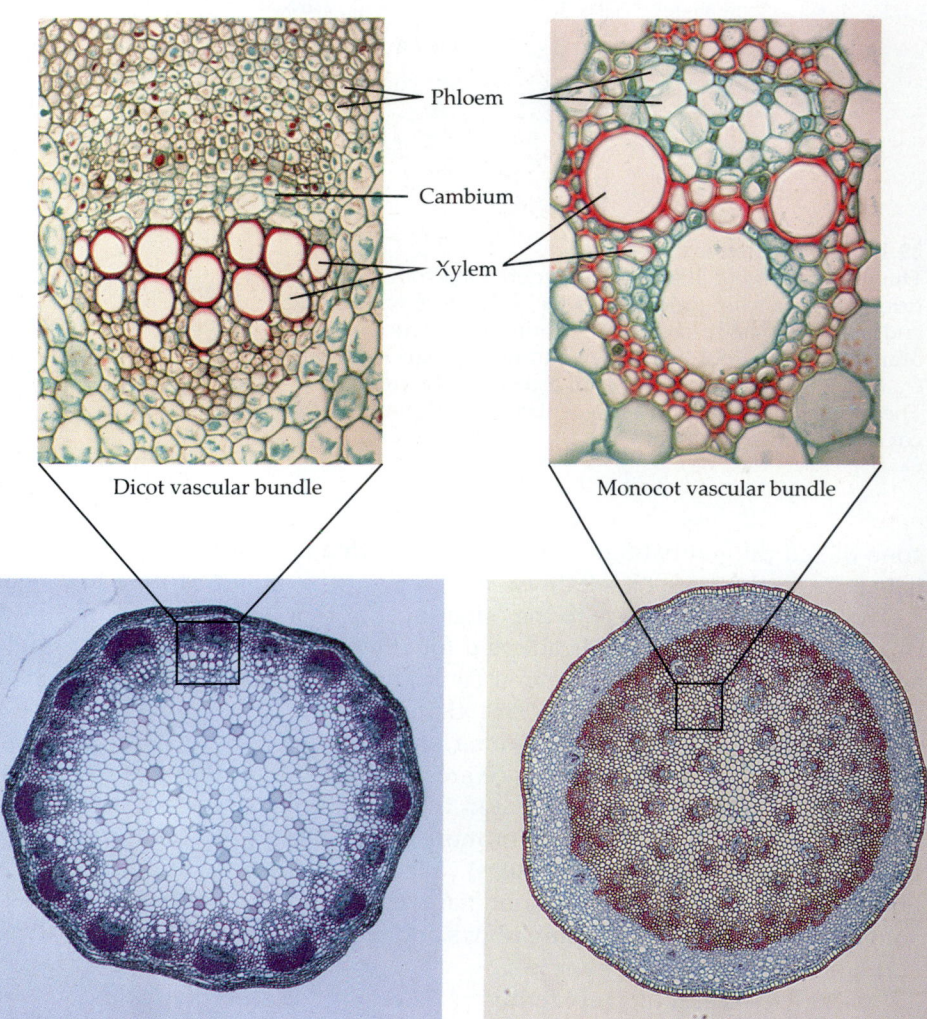

Dicot vascular bundle

Monocot vascular bundle

29.20 Vascular Bundles in Stems
The vascular tissues in stems are organized into bundles. (a) In dicots the vascular bundles are arranged in a circle with pith in the center and cortex outside the ring, as in this young sunflower stem. (b) This cross section shows the scattered arrangement of bundles typical of monocot stems. In both monocots and dicots, the bundles are oriented so that xylem is toward the center of the stem and phloem is to the outside.

(a) Dicot

(b) Monocot

turn give rise to the three tissue systems. Leaves arise from leaf primordia that form as cells divide on the sides of shoot apical meristems (see Figure 29.16). The growing stem has no cap analogous to the root cap, but the leaf primordia can act as a protective covering. Dicot stems extend in a region of elongation below the shoot apical meristem. Grasses and some other monocots, however, elongate at the bases of internodes and leaves, where there is some meristematic tissue. Lawn and range grasses can grow back after mowing or grazing because they grow from basal meristems close to the soil surface.

The plumbing of angiosperm stems differs from that of roots. In roots, the vascular tissue lies in the middle, with the xylem at or near the very center. The vascular tissue of a young stem, however, is divided into discrete **vascular bundles**. Vascular bundles generally form a cylinder in the dicots (Figure 29.20a) but are seemingly scattered throughout the cross section of the stem in the monocots (Figure 29.20b). Each vascular bundle contains both xylem and phloem.

The stem contains other important tissues in addition to the vascular tissues. Internal to the vascular bundles of dicots is a storage tissue, pith, and to the outside lies a similar storage tissue, the cortex. The cortex may contain strengthening collenchyma cells with thickened walls (Figure 29.21a). In many monocots the pith is hollowed out (Figure 29.21b). The pith, the cortex, and the regions between the vascular bundles in dicots—called pith rays—constitute the ground tissue system of the stem. The outermost cell layer of the young stem is the epidermis, which functions primarily to minimize the loss of water from the cells within.

Secondary Growth of Stems and Roots

Some stems and roots show little or no growth in diameter, remaining slender, but many others, all dicots, undergo considerable thickening. This thickening is of great importance and interest because it gives rise to wood and bark, as well as making the support of large trees possible. Secondary growth results from the activity of two meristematic tissues, vascular cambium and cork cambium (see Figure 29.16). Vascular cambia consist of cells that divide to produce new—secondary—xylem and phloem cells, while cork cambia produce mainly waxy-walled cork cells.

Initially, the vascular cambium is a single layer of cells between the primary xylem and the primary phloem. The root or stem increases in diameter when the cells of the vascular cambium divide, producing secondary xylem cells toward the inside of the root or stem and secondary phloem cells toward the outside (Figure 29.22a). In a stem, cells of the pith rays between the vascular bundles also divide, forming a continuous cylinder of vascular cambium running the length of the stem. This cylinder in turn gives rise to complete cylinders of secondary xylem—wood—and secondary phloem—bark (Figure 29.22b).

As the vascular cambium produces secondary xylem and phloem, its principal products are vessel elements and supportive fibers in the xylem, and sieve tube elements, companion cells, and fibers in the phloem. Not all xylem and phloem cells are adapted for transport or support; some store materials in the stem or root. Living cells such as these storage cells must be connected to the sieve tubes of the phloem, or they would starve to death. The con-

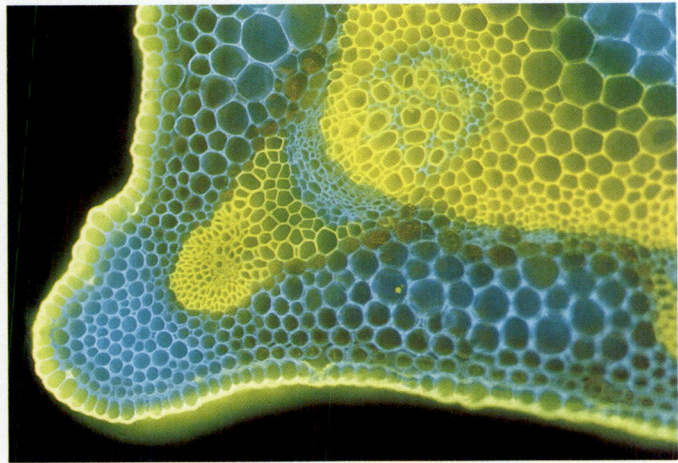

(a)

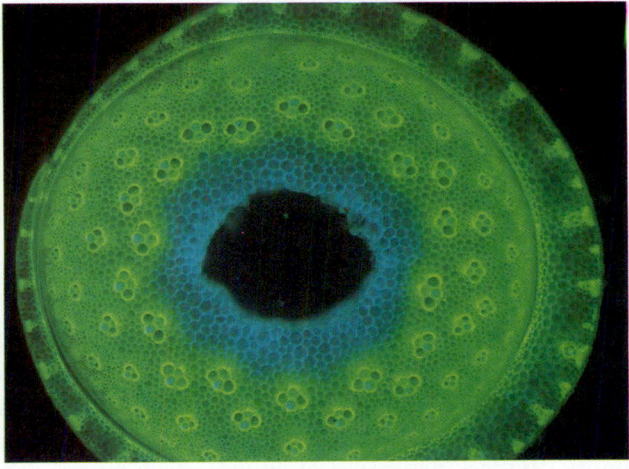

(b)

29.21 Other Stem Structures
(a) The stem of a broad bean resists bending but is not brittle. Collenchyma, the tissue in bright blue in the projection to the lower left of this stem cross section, pro- vides flexible support. The next tissue toward the interior, shown in gold, consists of phloem fibers that are stiffer than the collenchyma. (b) This bamboo looks like a "typical" monocot, except for its hollowed-out pith.

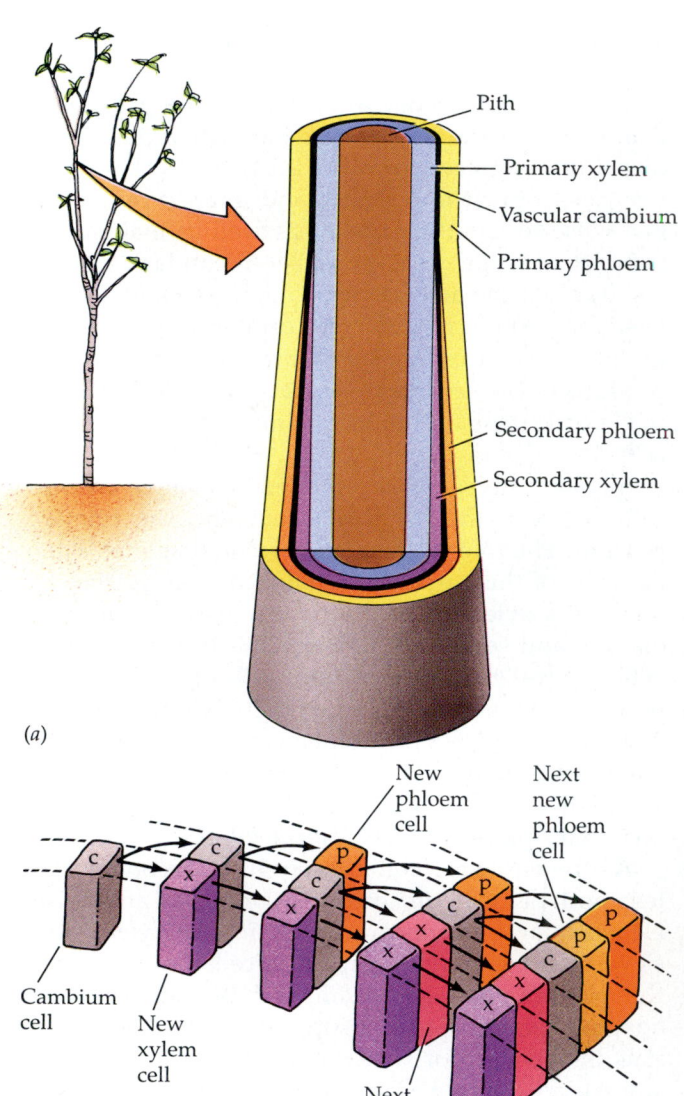

(a)

Pith
Primary xylem
Vascular cambium
Primary phloem

Secondary phloem
Secondary xylem

New phloem cell
Next new phloem cell

Cambium cell
New xylem cell
Next new xylem cell

(b)

29.22 Vascular Cambium Thickens Stems and Roots
Stems and roots grow thicker because a thin layer of cells, the vascular cambium, remains meristematic—capable of dividing. (a) This longitudinal section of a woody stem shows the vascular cambium thickening the stem by producing secondary xylem and secondary phloem. (b) When a vascular cambial cell (gray) divides, it produces either a new xylem cell toward the inside of the stem or root, or a new phloem cell toward the outside. Older xylem and phloem cells are pushed farther from the cambium with each division of the cambium.

nections are provided by **vascular rays**, which are composed of cells derived from the vascular cambium. The rays, laid down progressively as the cambium divides, are rows of living parenchyma cells running perpendicular to the xylem vessels and phloem sieve tubes (Figure 29.23). As the root or stem continues to increase in diameter, new vascular rays are initiated so that this storage and transport tissue

continues to meet the needs of the bark and of the living cells in the xylem. The cambium itself increases in circumference with the growth of the root or stem, for if it did not, it would split. The vascular cambium grows by the division of some of its cells in a plane at right angles to the plane that gives rise to secondary xylem and phloem. The products of each of these divisions lie within the vascular cambium itself.

Many dicots have vascular cambia and cork cambia and thus undergo secondary growth. In the rare cases in which monocots form thickened stems—palm trees, for example—a greater girth is achieved by quite a different mechanism.

WOOD. Most trees in temperate-zone forests have **annual rings** (Figure 29.24), which result from changing environmental conditions during the growing season. In the springtime, when water is relatively plentiful, the tracheids or vessel elements produced by the vascular cambium tend to be large in diameter and thin-walled. As water becomes less available during the summer, narrower cells with thicker walls

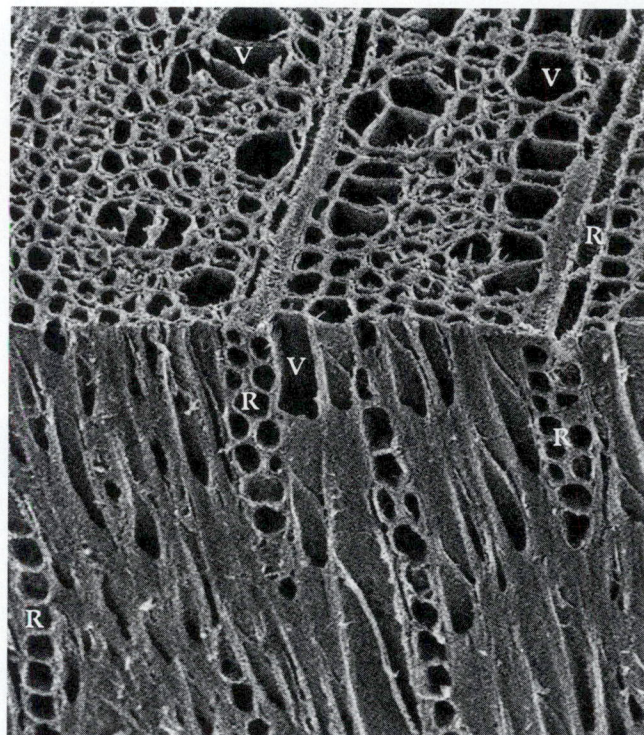

29.23 Vascular Rays
Wood of tulip poplar, showing that the orientation of xylem vessels (V) is perpendicular to that of vascular rays (R). The longitudinal section of the stem (lower half of the micrograph) shows the xylem vessels as long vertical tubes and the ray cells as circles. The cross section (top), which is perpendicular to the longitudinal section, shows the xylem vessels as circles and the rays as tubes. Rays transport food horizontally from the phloem to storage cells; xylem vessels conduct water vertically.

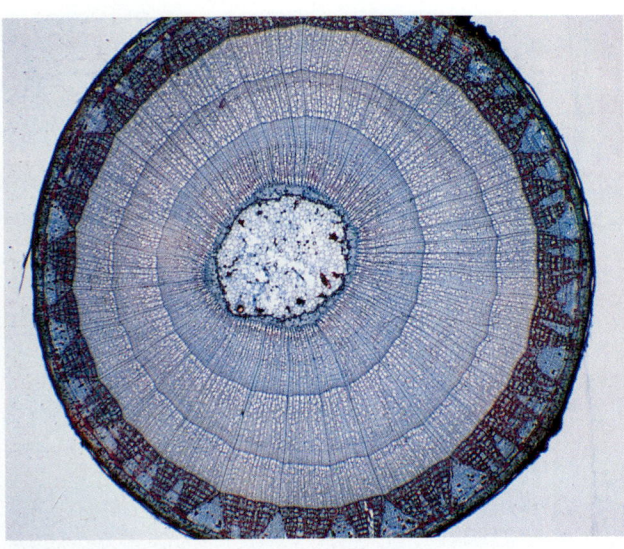

29.24 Annual Rings
Rings of xylem vessels are the most noticeable feature of this cross section from a three-year-old basswood stem.

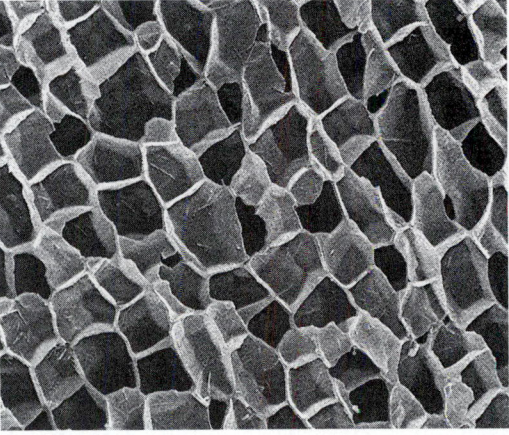

29.25 Cork
Commercial cork from the cork oak is seen in this scanning electron micrograph.

are produced; this summer wood is darker and perhaps more dense. Thus each year is usually recorded in a tree trunk by a clearly visible annual ring consisting of one light and one dark layer. Trees in the wet tropics do not lay down such obvious, regular rings.

The difference between old and new regions also contributes to the appearance of wood. As a tree grows in diameter, the xylem toward the center becomes clogged with resins and ceases to conduct water and minerals. This heartwood is darker; the sapwood—that portion that is actively conducting all water and minerals in the tree—is lighter and more porous. Knots—which we find attractive in knotty pine but regard as a defect in structural timbers—are branches: As a trunk grows, a branch extending out of it becomes buried in new wood and appears as a knot when the trunk is cut lengthwise.

PERIDERM. Obviously, as secondary growth continues, something has to give. Expansion of the vascular tissues stretches and breaks the epidermis and cortex, which are ultimately lost. Derivatives of the phloem then become the outermost tissue of the stem. Woody roots behave similarly. Because the epidermis is specialized in part for the retention of water, how does the plant cope if this tissue is shed? Before layers of epidermal cells are broken away, cells lying near the surface begin to divide and produce layers of cork, a tissue composed of cells with thickened, waterproof walls (Figure 29.25). The dividing cells, derived from the phloem, form a cork cambium. Sometimes cells are also produced to the inside by the cork cambium; these cells constitute what is known as the phelloderm. Cork is waterproofed by

suberin. The cork soon becomes the outermost tissue of the stem or root. Cork, cork cambium, and phelloderm—if present—make up the periderm of the secondary body. As the vascular cambium continues to produce secondary vascular tissue, the corky layers are in turn lost, but a similar process of cell division in the underlying phloem gives rise to new corky layers.

As periderm forms, there is still a need for gas exchange with the environment. Carbon dioxide must be released and oxygen must be taken up for cellular respiration. **Lenticels** are spongy regions that allow such gas exchange (Figure 29.26).

Leaf Anatomy

Figure 29.27*a* shows a typical dicot leaf in cross section. Generally a leaf has two zones of photosynthesizing tissues referred to as **mesophyll**, meaning "middle of the leaf." The upper layer or layers of mesophyll consist of roughly cylindrical cells. This zone is referred to as palisade mesophyll. The lower layer or layers consist of irregularly shaped cells called spongy mesophyll. Within the mesophyll is a great deal of air space through which carbon dioxide can diffuse to surround all photosynthesizing cells. Vascular tissue branches extensively in the leaf, forming a network of **veins**. These veins extend to within a few cell diameters of all the cells of the leaf, ensuring that the mesophyll cells are well supplied with water. The products of photosynthesis are loaded into the phloem of the veins for export to the rest of the plant (Figure 29.27*b*).

Covering the entire leaf is a layer of nonphotosynthetic cells constituting the epidermis. To retard water loss, the epidermal cells and their overlying waxy cuticle must be highly impermeable, but this

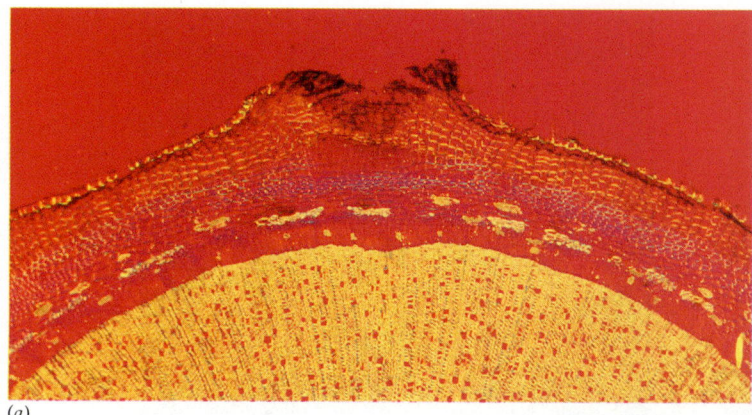

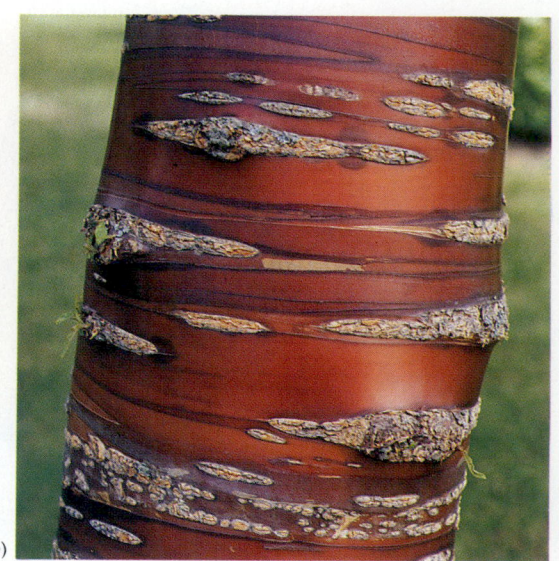

(a)

(b)

29.26   Lenticels Allow Gas Exchange through the Periderm
(a) The region at the top that appears broken open is a lenticel in a year-old elder twig; note the spongy tissue that constitutes the lenticel. (b) The rough areas on the trunk of this Chinese plum tree are lenticels. Most tree species have lenticels much smaller than these.

impermeability poses a problem: While keeping water within the leaf, the epidermis keeps carbon dioxide, the raw material of photosynthesis, out. The problem of balancing water retention and carbon dioxide availability is solved by an elegant regulatory system that will be discussed in more detail in Chapter 30. This system is based on pairs of **guard cells**, modified epidermal cells that change shape, thereby opening or closing pores called **stomata** (singular: stoma) between the guard cells (Figure 29.27c; see also Figure 30.10). When the stomata are open, carbon dioxide can enter, but water can be lost.

In Chapter 8 we described C$_4$ plants, which can fix carbon dioxide efficiently even when the carbon dioxide supply falls to a level at which the photosynthesis of C$_3$ plants is inefficient. One adaptation that helps C$_4$ plants do this is a modified leaf anatomy, as shown in Figure 8.28. Notice that the photosynthetic cells in the C$_4$ leaf are grouped around the veins in concentric layers: an outer mesophyll layer and an inner bundle sheath. These layers each contain different types of chloroplasts, leading to the biochemical division of labor described in Chapter 8.

SUPPORT IN A TERRESTRIAL ENVIRONMENT

Water buoys up aquatic plants, but terrestrial plants must either sprawl on the ground or somehow be supported against gravity. There are two principal types of support for terrestrial plants, one based on the osmotic properties of cells and the other based on tissues stiffened by specialized cell walls. One type of support is the pressure potential (sometimes called turgor) of the cells in the body. A small plant can maintain an erect posture if its cells are turgid,

but it collapses—wilts—if the pressure potential falls too low. Think about the difference between a wilted plant and a turgid one; the distinction dramatically illustrates the role of pressure potential in supporting the body. Support by the pressure potential is often augmented by the second type of support, the presence of strengthening tissues such as collenchyma and sclerenchyma. Collenchyma is more flexible than sclerenchyma, which provides a more rigid, stronger support.

The most important support found in many plants is wood—a mass of secondary xylem. Wood is such a strong yet lightweight material that we have used it in buildings, furniture, and other structures for millennia. All dicot trees are supported by their woody stems. Not all wood is the same, however. Let's consider some of the special adaptations of secondary xylem.

Reaction Wood

As the branches of growing trees grow longer and heavier, why don't they simply sag to the ground? This problem is averted by means of a gravity-induced asymmetry in wood structure: Specialized **reaction wood**, differing from normal wood, keeps the limb straight. Angiosperms and gymnosperms have different kinds of reaction wood, and in different places. In gymnosperms, **compression wood** forms on the lower side of a branch. It is prestressed—that is, it is laid down under compressive stress—and expands, thus tending to push the branch upward (Figure 29.28a). Compression wood contains thicker and shorter tracheids, with more lignin and less cellulose in their walls than normal wood has. By contrast, the reaction wood of angiosperms, called **ten-**

(a)

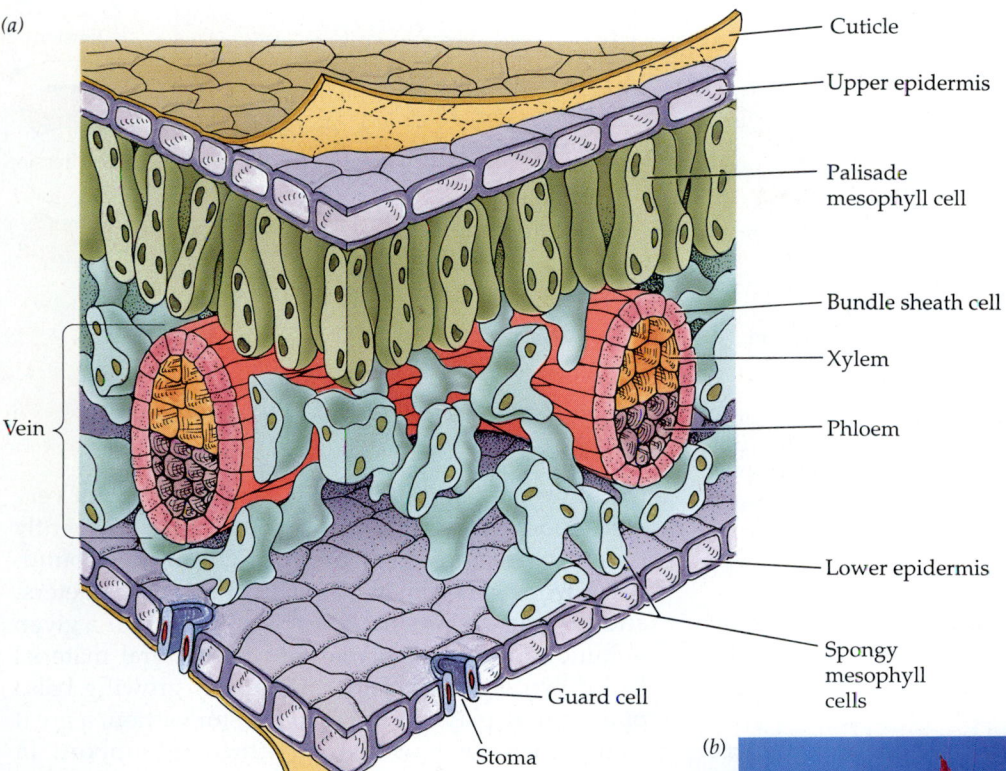

- Cuticle
- Upper epidermis
- Palisade mesophyll cell
- Bundle sheath cell
- Xylem
- Phloem
- Lower epidermis
- Spongy mesophyll cells
- Guard cell
- Stoma
- Cuticle
- Vein

29.27 The Dicot Leaf
(a) Cross section of a dicot leaf. (b) The network of fine veins in this Japanese maple leaf (*Acer palmatum*) carries water to the mesophyll cells and carries photosynthetic products away from them. (c) The lower epidermis of a dicot leaf, stained. The small, heavily stained, paired cells are guard cells; the gaps between them are stomata, through which carbon dioxide enters the leaf.

(b)

(c)

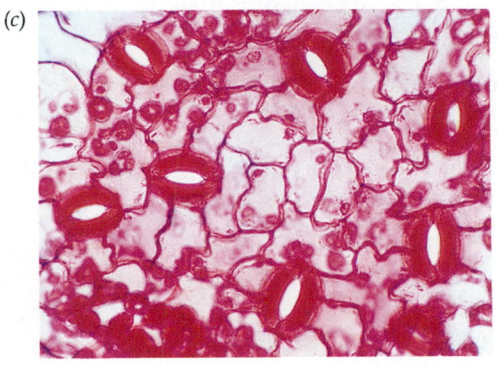

sion wood, is formed on the upper side of the branch. It is laid down under tension and shrinks, thus tending to pull the branch upward, or at least to resist downward bending (Figure 29.28b). In tension wood the fibers have more heavily thickened walls, containing less lignin and more cellulose, and there are fewer and smaller vessels than in normal wood.

That a gravitational stimulus, not the sagging of the branch, determines which type of reaction wood forms is illustrated by an experiment performed by the Australian plant physiologist A. B. Wardrop (Figure 29.28c). Wardrop bent the trunk of a young angiosperm sapling into a circle and allowed reaction wood to develop. Reaction wood formed on the top side of the top of the loop, where it was under tension, and also on the top side of the bottom of the loop, where it was under compression.

Trees grown indoors tend to be much more spindly than their outdoor counterparts, apparently because indoor trees are not subjected to buffeting by wind. They develop a firmer trunk if they are simply

29.28   Reaction Wood
Reaction wood reduces the tendency of branches to sag. (a) Compression wood, the reaction wood of gymnosperms, is heavily lignified wood that forms on the lower sides of branches. (b) Tension wood, the reaction wood of angiosperms, forms on the upper sides of branches. (c) An angiosperm sapling was bent into a loop and tied in place. Tension wood formed on the tops of both horizontal regions; as shown in the cutouts, one region of tension wood was under tension, but the other was actually under compressive stress. This result indicates that the stimulus to tension wood formation is gravitational (tension wood forms on the upper side) rather than a response to the stress itself.

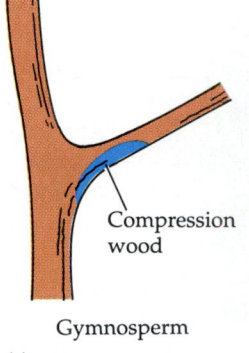

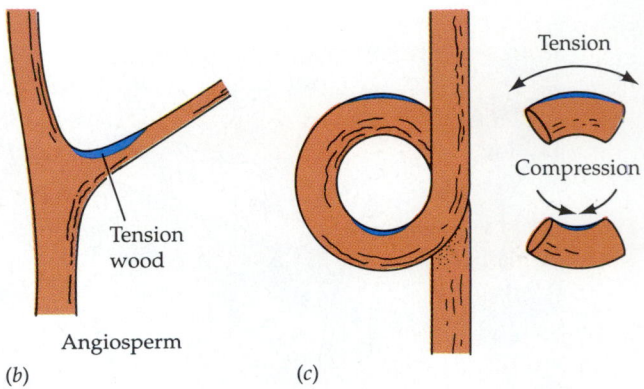

shaken or pounded with a padded mallet from time to time. The change in wood deposition caused by such treatments may be akin to reaction wood formation.

The variety in the structure and development of woods has resulted in different economic uses for different woods, uses that parallel the functions of the woods.

Quonset Huts and Marble Palaces

People put up Quonset huts for cheap, relatively short-term shelter. They build palaces of marble if they want to create a monument for the ages. We find similar contrasts if we compare the wood of trees growing under very different conditions. Consider balsa and mahogany: one is extremely light and soft,

the other dark and hard. Like most species with very soft woods, balsa is a fast-growing tree, frequently found in areas recently burnt or cut to the ground. Balsa wood has cells with relatively large diameters, and the wood fibers have very thin walls. For a given volume of wood, the amount of structural material laid down is slight. Thus the rapidly growing balsa plant can display its foliage to the sun without a great commitment of resources to structural support. In short, balsa is the botanical equivalent of a Quonset hut.

Mahogany grows extremely slowly. It is very sturdy. Its fine-textured wood has tiny cells, and the wood fibers have thick walls. In contrast to balsa, mahogany wood is "expensive" to form—it has much more dry weight per volume. As a result of its hardness, mahogany wood can support the plant as a long-lived tree in rainforests. Mahogany contains impregnating materials that darken the wood and help render it resistant to fungal attack. It is the botanical equivalent of a palace.

SUMMARY of Main Ideas about the Flowering Plant Body

Plants grow from tissues called meristems that retain the capacity of cell division.

Stems and roots grow from apical meristems.
   Review Figure 29.16

The meristems that thicken stems and roots are the vascular cambium and the cork cambia.
   Review Figures 29.16 and 29.22

The shoot and root systems make up the plant body.
   Review Figures 29.7, 29.9, 29.10, 29.17, 29.18, and 29.19

Secondary growth thickens the stem, producing wood and periderm.
   Review Figures 29.22, 29.23, 29.24, 29.25, and 29.26

Leaves exchange materials with the rest of the plant by means of vascular tissues in the veins, and they exchange gases with the environment by means of stomata.
   Review Figure 29.27

The two classes of flowering plants, monocots and dicots, differ in plant body characteristics.
   Review Figures 29.5 and 29.6

Flowering plants have cells specialized for support, conduction of water and minerals, conduction of food materials, photosynthesis, storage, and cell division.
   Review Figures 29.13, 29.14, and 29.15

Flowering plants have three tissue systems: the vascular tissue system (xylem and phloem), the dermal tissue system (epidermis and periderm), and the ground tissue system (parenchyma, collenchyma, and sclerenchyma cells).

Wood provides support for many plants; reaction wood provides support for branches, counteracting the tendency of the branches to sag.
Review Figure 29.28

SELF-QUIZ

1. Which of the following is *not* a difference between monocots and dicots?
   a. Dicots more frequently have broad leaves.
   b. Monocots commonly have flower parts in multiples of three.
   c. Monocot stems do not generally undergo secondary thickening.
   d. The vascular bundles of monocots are commonly arranged as a cylinder.
   e. Dicot embryos commonly have two cotyledons.

2. Roots
   a. always form a fibrous root system that holds the soil.
   b. possess a root cap at their tip.
   c. form branches from lateral buds.
   d. are commonly photosynthetic.
   e. do not show secondary growth.

3. The plant cell wall
   a. lies immediately inside the plasma membrane.
   b. is an impenetrable barrier between cells.
   c. is always waterproofed with either lignin or suberin.
   d. consists of a primary wall and secondary wall, separated by a middle lamella.
   e. contains cellulose and other polysaccharides.

4. Which statement about parenchyma cells is *not* true?
   a. They are alive when they perform their functions.
   b. They typically lack a secondary wall.
   c. They often function as storage depots.
   d. They are the most numerous cells in the primary plant body.
   e. They are found only in stems and roots.

5. Tracheids and vessel elements
   a. die before they become functional.
   b. are important constituents of all bryophytes and tracheophytes.
   c. have walls consisting of middle lamella and primary wall.
   d. are always accompanied by companion cells.
   e. are found only in the secondary plant body.

6. Which statement is *not* true of sieve tube elements?
   a. Their end walls are called sieve plates.
   b. They die before they become functional.
   c. They link end-to-end, forming sieve tubes.
   d. They form the system for translocation of foods.
   e. They lose the membrane that surrounds their central vacuole.

7. The pericycle
   a. separates the stele from the cortex.
   b. is the tissue within which branch roots arise.
   c. consists of highly differentiated cells.
   d. forms a star-shaped structure at the very center of the root.
   e. is waterproofed by Casparian strips.

8. Secondary growth of stems and roots
   a. is brought about by the apical meristems.
   b. is common in both monocots and dicots.
   c. is brought about by vascular cambia and cork cambia.
   d. produces only xylem and phloem.
   e. is brought about by vascular rays.

9. Periderm
   a. contains lenticels that allow for gas exchange.
   b. is produced during primary growth.
   c. is permanent; once formed it lasts as long as the plant does.
   d. is the innermost part of the plant.
   e. contains vascular bundles.

10. Which statement about leaf anatomy is *not* true?
   a. Stomata are controlled by paired guard cells.
   b. The cuticle is secreted by the epidermis.
   c. The veins contain xylem and phloem.
   d. The cells of the mesophyll are packed together, minimizing air space.
   e. C_3 and C_4 plants differ in leaf anatomy.

FOR STUDY

1. When a young oak was 5 meters tall, a thoughtless person carved his initials in its trunk at a height of 1.5 meters above the ground. Today that tree is 10 meters tall. How high above the ground are those initials? Explain your answer in terms of the manner of plant growth.

2. Consider a newly formed sieve tube element in the secondary phloem of an oak tree. What kind of cell divided to produce the sieve tube element? What kind of cell divided to produce that parent cell? Keep tracing back in this manner until you arrive at a cell in the apical meristem.

3. Distinguish between sclerenchyma cells and collenchyma cells in terms of structure and function.

4. Distinguish between primary and secondary growth. Do all angiosperms undergo secondary growth? Explain.

5. What anatomical features make it possible for a plant to retain water as it grows? Describe the tissues and how and when they form.

READINGS

Esau, K. 1977. *Anatomy of Seed Plants,* 2nd Edition. John Wiley, New York. A comprehensive treatment; particularly good on secondary growth.

Feldman, L. J. 1988. "The Habits of Roots." *BioScience,* vol. 38, pages 612–618. Considers many aspects of the biology of roots, including structure, competition, associations with soil microorganisms, and others.

Mangelsdorf, P. C. 1986. "The Origin of Corn." *Scientific American,* August. What was the ancestry of this popular vegetable? The gross anatomy of some possible ancestors is compared.

Niklas, K. J. 1989. "The Cellular Mechanics of Plants." *American Scientist,* vol. 77, pages 344–349. This fine article, subtitled "How Plants Stand Up," details how cell walls and other aspects of stem architecture enable terrestrial plants to stand erect.

Raven, P. H., R. F. Evert and S. Eichhorn. 1991. *Biology of Plants,* 5th Edition. Worth, New York. An excellent general botany textbook.

Swaminathan, M. S. 1984. "Rice." *Scientific American,* January. This article deals primarily with ways to increase the yield of rice, but it also describes the structure of the plant.

Wilson, B. F. and R. R. Archer. 1979. "Tree Design: Some Biological Solutions to Mechanical Problems." *BioScience,* vol. 29, pages 293–298. As trees grow, they constantly redesign themselves. This article examines aspects of this process, such as reaction wood formation, from an engineering viewpoint.

Life first arose and flourished in the oceans, but the vascular plants arose on land. Plants were the first eukaryotes to face the challenges of life out of water. The mechanisms that terrestrial organisms require for taking in and conserving water differ from those of aquatic organisms. The cells and tissues of aquatic organisms are bathed in the water and minerals that they require, but a terrestrial plant must have a transport system to distribute water and minerals throughout its body. And because leaves are the sites of food production, all plants except the smallest ones need a system to transport food from the leaves to other parts of the body.

Terrestrial organisms also need ways to support their bodies. In a watery environment a giant kelp—a marine algal protist—can spread out like an enormous tree because of the buoying action of the surrounding water. On land, large organisms can resist the pull of gravity only with the help of rigid materials such as wood or bone. As you learned in the previous chapter, the pressure of water in the tissues provides much of the support for nonwoody terrestrial plants, which wilt in the absence of water. And, of course, wood is the most important source of support for many large terrestrial plants. The cellular anatomy that gave us an understanding of the support systems of terrestrial plants is also central to an understanding of how water, minerals, and food are transported within the bodies of plants.

UPTAKE AND TRANSPORT OF WATER AND MINERALS

Terrestrial plants obtain both water and mineral nutrients from the soil, usually by way of their roots. You know that leaves are loaded with chloroplasts and that water is one of the ingredients for food production by photosynthesis. How do leaves, high in a tree, obtain water from the soil? What are the mechanisms by which water and minerals enter the plant body through the dermal tissue of the root, pass through the ground tissue, enter the stele, and ascend as sap in the xylem? Because neither water nor minerals can move through the plant into the xylem without crossing at least one plasma membrane, we will first consider two membrane phenomena, osmosis and the uptake of minerals.

Osmosis

Osmosis, the movement of a solvent such as water through a membrane in accordance with the laws of diffusion, was discussed in Chapter 5. Recall that the **osmotic potential** of a solution results from the presence of dissolved solutes. The greater the solute con-

How Do Water and Minerals Get Up There?
Wood not only supports these California coast redwoods but is also the tissue through which water and minerals are transported more than 100 meters into the air.

30
Transport in Plants

centration, the more negative the osmotic potential and hence the greater the tendency of water to move into that solution from another solution of lower solute concentration. The two solutions must be separated by a differentially permeable membrane (permeable to water but impermeable to the solute). Solutions—or cells—with identical osmotic potentials are isotonic; if two solutions differ in osmotic potential, the one with the less negative osmotic potential is hypotonic to the other. Recall, too, that osmosis is a passive process—ATP is not required.

Unlike animal cells, plant cells are surrounded by a relatively rigid cell wall. After a certain amount of water enters a plant cell, the entry of more water is resisted by an opposed **pressure potential**, sometimes called turgor pressure, owing to the rigidity of the wall. As more and more water enters, the pressure potential becomes greater and greater. The pressure potential is analogous to the air pressure in an automobile tire; it is a real pressure that can be measured with a pressure gauge. Cells with walls do not burst when placed in distilled water because of the rigidity of the walls. Water enters by osmosis until the pressure potential exactly balances the osmotic potential. At this point, the cell is quite turgid—that is, it has a high pressure potential.

The overall tendency of a solution to take up water from pure water is called the **water potential**. The water potential is the sum of the (negative) osmotic potential and the (positive) pressure potential. For distilled water under no applied pressure, all three of these potentials are defined as equal to zero. In all cases in which water moves between two cells, or between a cell and its environment, or between two solutions separated by a membrane, the following rule of osmosis applies: *Water always moves toward the region of more negative water potential* (Figure 30.1).

Osmotic phenomena are of great importance. The turgor of most plants is maintained by the pressure potentials of their cells; if the pressure potential is lost, a plant wilts (Figure 30.2). The movement of water within a plant follows a gradient of water potential, and as we will see, the flow of phloem sap through the sieve tubes is driven by a gradient in pressure potential.

Uptake of Minerals

Mineral nutrient ions are taken up across plasma membranes with the help of proteins. (You may wish to review the section "Crossing the Membrane Barrier" in Chapter 5.) Some of these proteins are carriers for the facilitated diffusion of particular ions. Facilitated diffusion does not require ATP. The concentrations of some ions in the soil solution are lower than those required inside the plant, however. Thus the plant must take up these ions against a concentration

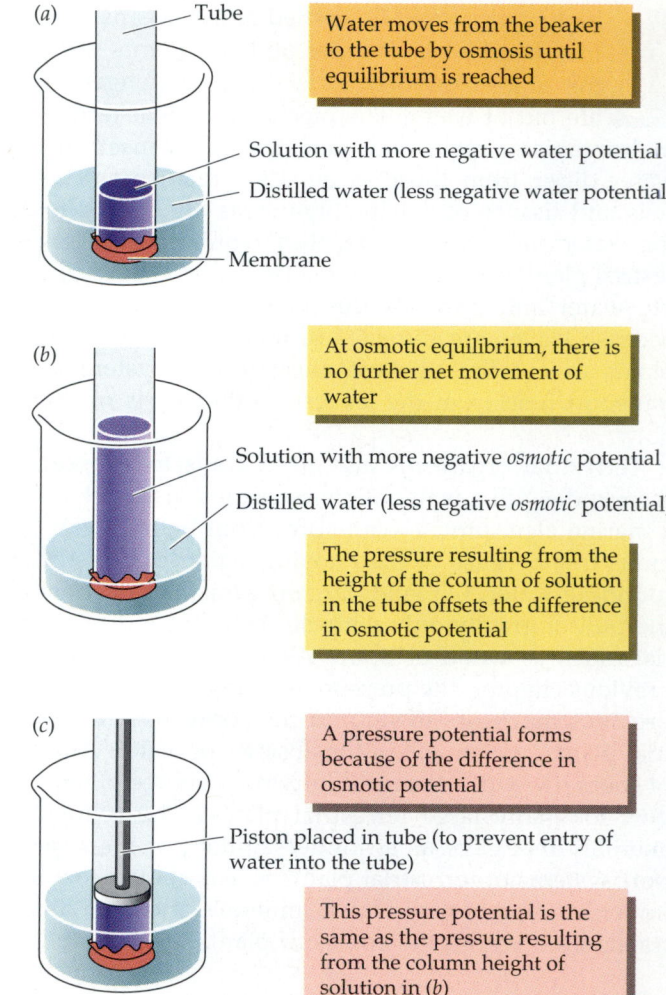

(a) Tube — Water moves from the beaker to the tube by osmosis until equilibrium is reached

Solution with more negative water potential
Distilled water (less negative water potential)
Membrane

(b) At osmotic equilibrium, there is no further net movement of water

Solution with more negative *osmotic* potential
Distilled water (less negative *osmotic* potential)

The pressure resulting from the height of the column of solution in the tube offsets the difference in osmotic potential

(c) A pressure potential forms because of the difference in osmotic potential

Piston placed in tube (to prevent entry of water into the tube)

This pressure potential is the same as the pressure resulting from the column height of solution in (b)

30.1 Water Potential, Osmotic Potential, and Pressure Potential
(a) The solution in the tube has a negative osmotic potential owing to the presence of dissolved solutes; the pressure potential = 0; thus the water potential is negative. The beaker contains distilled water (water potential = 0). *(b)* Water moves through the membrane into the tube because of the difference in water potentials. *(c)* If the water potentials are made equal by applying pressure with a piston, water does not enter the tube.

gradient. Such active transport is an energy-requiring process, and it depends upon cellular respiration as a source of ATP. Active transport, too, requires specific carrier proteins.

Plants do not have a sodium–potassium pump for active transport. Rather, plants have a **proton pump** that uses energy obtained from ATP to push protons out of the cell against a proton concentration gradient (Figure 30.3a). Because protons (H^+) are positively charged, their accumulation on one side of a membrane has two results. First, the region outside the membrane becomes positively charged with respect to the region inside. Second, there is a proton con-

(a)

(b)

30.2 Turgor in Plants
(a) This coleus plant remains turgid as long as the pressure potential of its cells is high. (b) When cells lose too much water, their pressure potential drops and the plant wilts.

centration difference. Each of these results has consequences for the movement of other ions. Because of the charge difference across the membrane, the movement of positively charged ions such as K^+ into the cell through their membrane channels is enhanced because these positive ions are moving into a region of negative charge (Figure 30.3b). The proton concentration difference can drive a form of secondary active transport in which negatively charged ions such as Cl^- are moved into the cell against a concentration gradient by a symport that couples their movement with that of H^+ (Figure 30.3c). In all, there is a vigorous traffic of ions across plant membranes. How do biologists measure these ion movements?

A technique called **patch clamping** allows us to monitor the flow of ions through just a few carrier proteins—or even just one—at a time. First we remove the cell wall, by digesting it with enzymes, to expose the plasma membrane. Then we immobilize the naked cell by pulling it part way into a very fine glass micropipette (Figure 30.4a). Next, we press a still finer glass micropipette against the exposed plasma membrane and apply a slight suction so that a tiny patch of the membrane is effectively isolated from the rest of the surface (Figure 30.4b). Once a tight seal has been made, we can proceed in one of various ways. In one approach, by pulling very carefully, we can tear the patch away from the rest of the

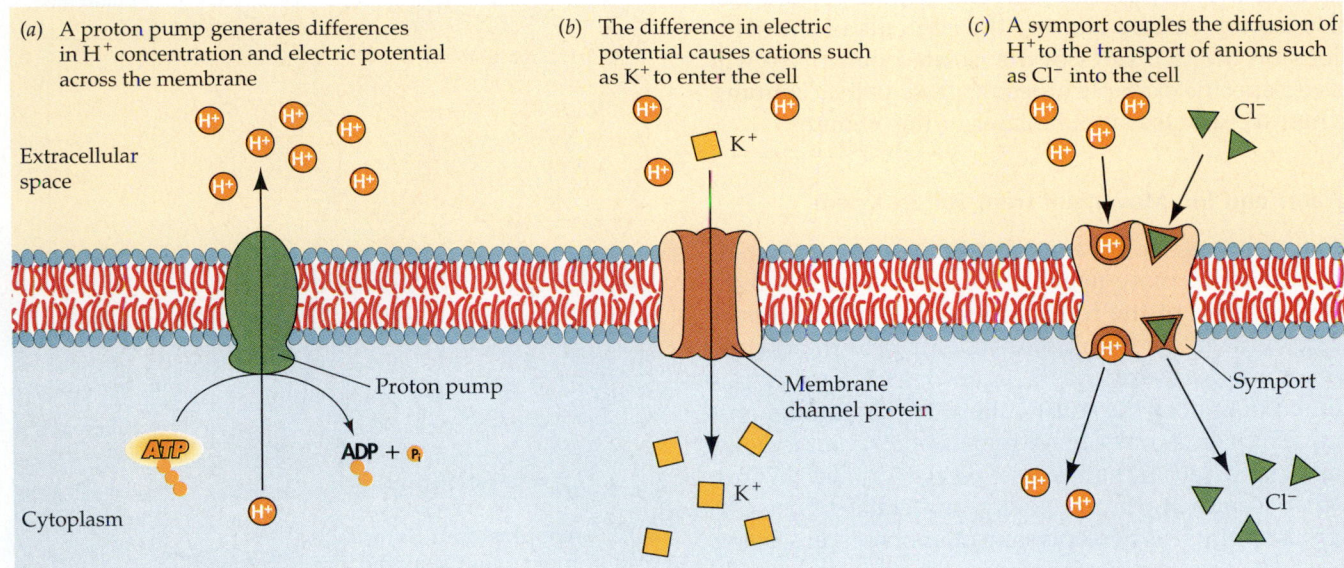

(a) A proton pump generates differences in H^+ concentration and electric potential across the membrane

(b) The difference in electric potential causes cations such as K^+ to enter the cell

(c) A symport couples the diffusion of H^+ to the transport of anions such as Cl^- into the cell

Extracellular space

Cytoplasm

Proton pump

ADP + P$_i$

Membrane channel protein

Symport

30.3 The Proton Pump and Its Effects
(a) The buildup of hydrogen ions transported across the cell membrane by the proton pump triggers the movement of both cations (b) and anions (c).

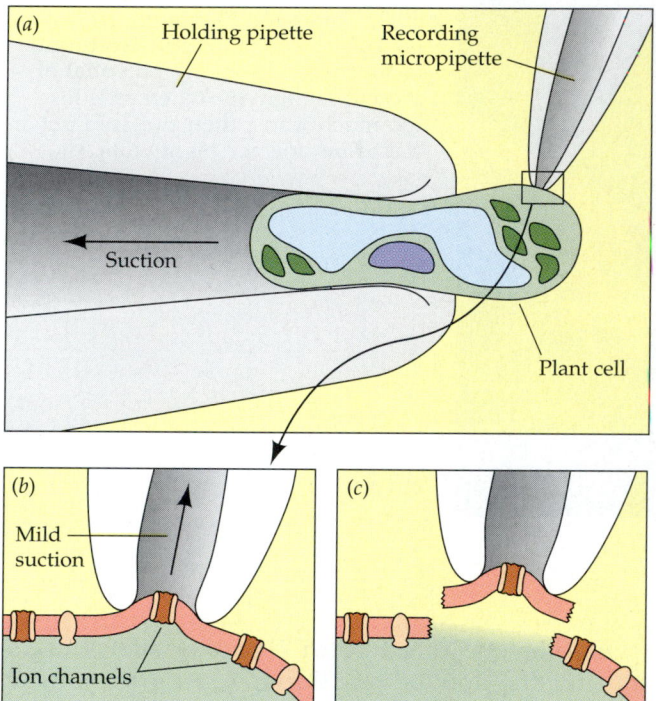

30.4 Patch Clamping
(a) The naked cell, lacking a wall, is trapped in the tip of a holding micropipette. (b) A part of the membrane is sucked into a smaller, recording micropipette. (c) The part of the plasma membrane has been torn away. It contains only a single ion channel, which can be studied by measuring electric currents across it as ions flow.

membrane and study the flow of ions through the carrier or carriers contained in just that patch (Figure 30.4c). Because ions are electrically charged, we can measure their movement through the patch by recording the tiny electric current that flows. We can also experiment by altering the contents of the solutions on the two sides of the isolated patch. We will give a specific example of results from patch clamping when we discuss stomata later in this chapter.

Water and Ion Movement from Soil to Xylem

Water moves along a gradient of water potential, toward ever more-negative regions. Water moves into the stele of the root because the water potential is more negative within the stele than in the cortex. The cortex, in turn, has a more negative water potential than does the soil solution. Minerals enter and move in plants in various ways. Where water is flowing, dissolved minerals are carried along. Where water is moving more slowly, minerals diffuse. At certain points, where plasma membranes are being crossed, some minerals are sped along by active transport.

Water and minerals from the soil may pass through the dermal and ground tissues to the stele

via two plant parts: the apoplast and the symplast. Plant cells are surrounded by cell walls that lie outside the plasma membrane, and intercellular spaces are common in many tissues. The walls and intercellular spaces together constitute the **apoplast** (from the Greek for "away from living material"). The apoplast is a continuous meshwork through which water and dissolved substances can flow or diffuse without ever having to cross a membrane. Movement of materials through the apoplast is thus unregulated. The remainder of the plant body is the **symplast** (from the Greek for "together with living material")—that is, the plant body enclosed by membranes, the continuous meshwork consisting of the living cells, connected by plasmodesmata (Figure 30.5). The selectively permeable plasma membranes of the cells control access to the symplast, so movement of water and dissolved substances into the symplast is tightly regulated.

Water and minerals can pass from the soil solution through the apoplast to the inner border of the cortex. As you may recall from Chapter 29, to enter the stele, water and minerals must pass through the endodermis (the inner cell layer of the cortex; see Figure 29.18). What distinguishes the endodermis from the rest of the ground tissues is the presence of **Casparian strips**. These waxy, suberin-containing structures line the endodermal cells at their tops, bottoms, and

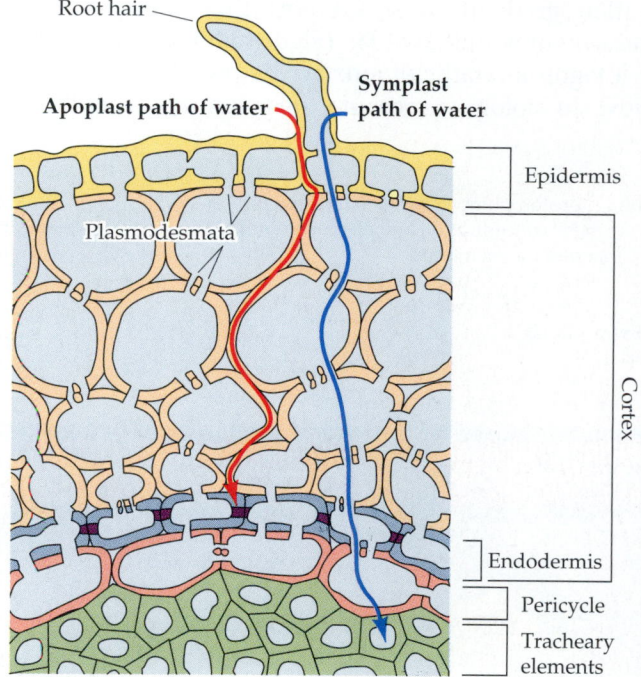

30.5 Apoplast and Symplast
Water and minerals spread through the apoplast, which consists of cell walls and intercellular spaces. At some point, water and minerals must enter the symplast—the remainder of the plant body—or they will be unable to pass into the stele.

Paths of water halted by
Casparian strips

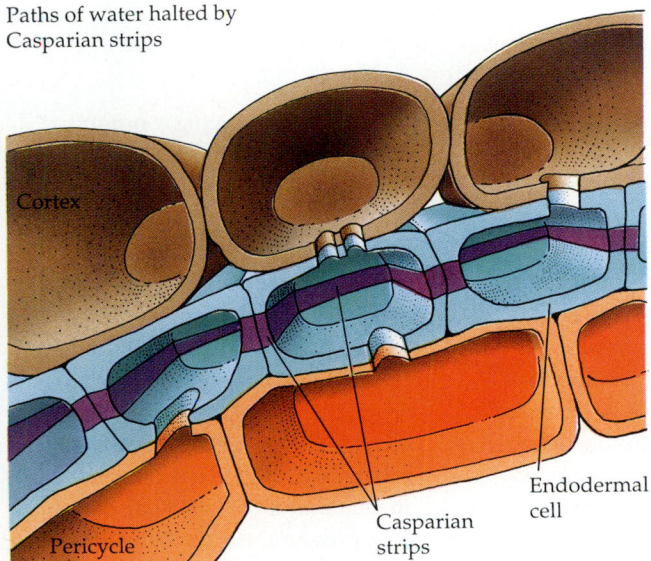

Cortex

Pericycle

Casparian
strips

Endodermal
cell

30.6 Casparian Strips
Casparian strips prevent water in the apoplast from passing between the endodermal cells and into the stele. Water must first enter the living endodermal cells; by entering the symplast, it can evade the Casparian strips.

sides, acting as a gasket that prevents water and ions
from moving between them (Figure 30.6). The endodermis thus completely separates the apoplast of
the cortex from the apoplast of the stele. The Casparian strips do not obstruct the outer or inner faces
of the endodermal cells. Accordingly, water and ions
can enter the stele only by way of the symplast—that
is, by entering and passing through the cytoplasm of
the endodermal cells. Thus transport proteins in
membranes between the apoplast and symplast determine which minerals pass, and at what rates.

Once they have passed the endodermal barrier,
water and minerals leave the symplast. Parenchyma
cells in the pericycle or xylem help minerals move
back into the apoplast. Some of these parenchyma
cells, called **transfer cells**, are structurally modified
for transporting mineral ions from their cytoplasm
(part of the symplast) into their cell walls (part of the
apoplast). The wall that receives the transported ions
has many knobby growths extending into the transfer
cell, increasing the surface area of the plasma membrane, the number of transport proteins, and thus
the rate of transport. Transfer cells also have many
mitochondria that produce the ATP needed to power
the active transport of mineral ions. As mineral ions
move into the solution in the walls, the water potential of the wall solution becomes more negative; thus
water moves out of the cells into the wall solution by
osmosis. Active transport of ions moves the ions directly, and water follows passively. The end result is
that water and minerals end up in the xylem, where
they constitute the sap.

Ascent of Sap in Xylem: The Magnitude of the Problem

The water and minerals in the xylem must be transported to the entire shoot system, all the way to the
highest leaves and apical buds. Before we consider
the mechanisms underlying this transport, we should
know what needs to be explained: How much sap is
transported, and how high must it go?

In answer to the first question, consider the following example: A single maple tree 15 meters tall
was estimated to have some 177,000 leaves, with a
total leaf surface area of 675 square meters. On a
summer day, that tree lost 220 liters of water *per hour*
to the atmosphere by evaporation from the leaves.
To prevent wilting, xylem transport in that tree
needed to provide 220 liters of water to the leaves
every hour.

How high must the xylem sap be transported? The
question may be rephrased: How tall are the tallest
trees? The tallest gymnosperms, the coast redwoods—*Sequoia sempervirens*—exceed 110 meters in
height, as do the tallest angiosperms, the Australian
Eucalyptus regnans. Any successful explanation of
transport in the xylem must account for transport
over these great distances.

Early Models of Transport in the Xylem

Some of the earliest models to explain the rise of sap
in the xylem were based on a hypothetical pumping
action by living cells in the stem. Experiments published in 1893 by the German botanist Eduard Strasburger definitively ruled out such models (Box 30.A).

Another early suggestion was a model based on
capillary action, the rising of watery solutions in very
thin tubes or in woven materials like paper. At first
glance this theory seems reasonable. The diameters
of vessel elements and tracheids are tiny, and the
narrower the tube, the higher water will rise by capillary action. However, the diameters of tracheids are
only small enough to support a capillary column of
about 40 centimeters—shorter than many shrubs, let
alone a giant eucalyptus towering over us to a height
greater than the length of a football field.

Root Pressure

After the capillary model was questioned, some plant
physiologists turned to a model based on root pressure—a pressure exerted by the root tissues that
would force liquid up the xylem. The basis for root
pressure is a higher solute concentration, and accordingly a more negative water potential, in the
xylem sap than in the soil solution. This negative
potential draws water into the stele; once there, the
water has nowhere to go but up.

BOX 30.A

There Are No Pressure Pumps in the Xylem

Eduard Strasburger was the leading plant cytologist of the late nineteenth century. He was one of those who established that the nucleus is the carrier of hereditary information, and he is best remembered today for pioneering work that led to the discovery of meiosis. He also performed some of the first important experiments that led to our current understanding of water movement in the xylem. His contemporaries generally believed that living cells in the xylem played a key role, probably by acting as pressure pumps, pushing the sap upwards.

Strasburger worked with trees about 20 meters in height. He sawed them through at their bases and plunged the cut ends into buckets containing solutions of poisons such as picric acid. The solutions rose through the trunks, as was readily evident from the progressive death of the bark higher and higher up. When the solutions reached the leaves, the leaves died, too, and the solutions were no longer transported; the liquid levels in the buckets stopped dropping at that point. This simple experiment established three important points: (1) Living, "pumping" cells are not responsible for the upward movement of the solutions, for the solutions themselves killed all living cells with which they came in contact. (2) The leaves play a crucial role in causing the transport. As long as they were alive, the solutions continued to be transported upward; when the leaves died, transport ceased. (3) Transport in these experiments, which covered distances of 20 meters and more, was not caused by root pressure, for the trunks had been completely separated from the roots. It is interesting, if disappointing, that theories of xylem transport based on root pressure were entertained for some years following Strasburger's definitive experiments.

There is good evidence for root pressure—for example, the phenomenon of **guttation**, in which liquid water is forced out through openings in leaves (Figure 30.7). Guttation occurs only under conditions of high atmospheric humidity and plentiful water in the soil. Root pressure is also the source of the sap that oozes from the cut stumps of some plants, such as *Coleus*, when their tops are removed. However, root pressure cannot account for the ascent of sap in trees. Root pressures seldom exceed one or two times atmospheric pressure, and they are actually less at times when transport in the xylem is most rapid. If root pressure were driving sap up the xylem, we should observe a positive pressure potential in the xylem at all times. In fact, as we are about to see, the xylem sap is under a tension—a negative pressure potential—when it is ascending. Furthermore, as Strasburger had already shown, materials can be transported upward in the xylem even when the roots have been removed.

The Evaporation–Cohesion–Tension Mechanism

To understand how sap rises in the xylem, even to the tops of the tallest trees, we must begin by looking at the final step in the process of water movement from soil to root to leaf and out to the atmosphere. At the end of the line, water evaporates from the moist walls of mesophyll cells, diffuses through the air spaces of the leaf, and finally leaves as water vapor through the open stomata.

The evaporation of water from mesophyll cells makes their water potential more negative—effectively, it increases the solute concentration of the cells—so more water enters osmotically from the nearest tiny vein. The removal of water from the xylem of the veins establishes a tension, or pull, on the entire column of water contained within the xylem, so the column is drawn upward all the way from the roots. The ability of water to be pulled

30.7 Guttation
Root pressure forces water through openings in the tips of this strawberry leaf.

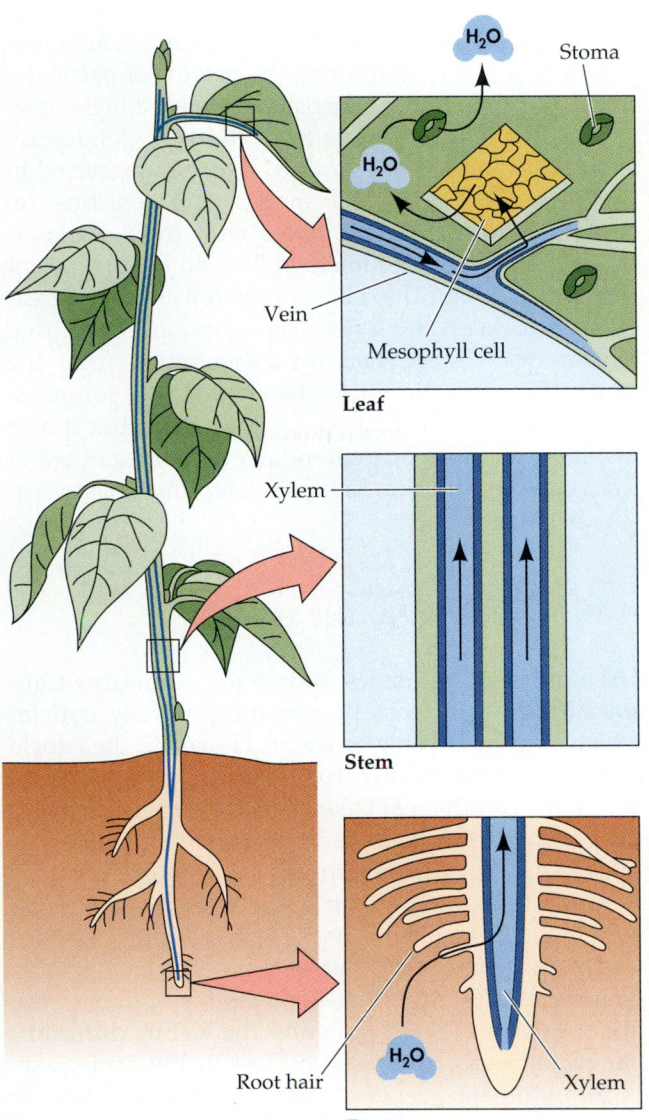

Leaf

H₂O
Stoma
H₂O
Vein
Mesophyll cell

Stem

Xylem

Root

Root hair
H₂O
Xylem

1. Water diffuses out of stoma

2. Water evaporates from mesophyll cell

3. Water from vein enters mesophyll cell by osmosis

4. Tension pulls water column upward and outward in xylem of vein in leaf

5. Tension pulls water column upward in xylem of stem

6. Tension pulls water column upward in xylem of root

7. Water molecules form a cohesive column

8. Water moves into stele by osmosis

30.8 Water Transport in Plants
Evaporation, cohesion, and tension account for the movement of water from the soil to the atmosphere.

upward through a tiny tube results from the remarkable cohesiveness of water—the tendency of water molecules to adhere to one another through hydrogen bonding (see Chapter 2). The narrower the tube, the greater the tension the water column can withstand without breaking. As the water column in the xylem is pulled upward, more water enters the xylem in the root by osmosis from surrounding cells.

In summary, the key elements of water transport in the xylem are *evaporation* from the moist cells in the leaves and a resulting *tension* in the remainder of the xylem's water owing to its *cohesion*, which pulls up more water to replace that which has been lost (Figure 30.8). All this requires no work on the part of the plant. At each step water moves passively to a region with a more strongly negative water potential. Dry air has the most negative water potential; the soil solution has the least negative water poten-

tial; xylem sap has a water potential more negative than that of cells in the root but less negative than that of mesophyll cells in the leaf.

Mineral ions contained in the xylem sap rise passively as the solution ascends from root to leaf. In this way the nutritional needs of the shoot are met. Some of the mineral elements brought to the leaves are subsequently redistributed to other parts of the plant by way of the phloem, but the initial delivery from the roots is through the xylem.

The evaporative loss of water from the shoot is called **transpiration**. In addition to promoting the transport of minerals, transpiration contributes to temperature regulation. As water evaporates from mesophyll cells, heat is taken up from the cells, and the leaf temperature drops. This cooling effect may be important in enabling plants to live in certain environments.

Measuring Tension in the Xylem Sap

The evaporation–cohesion–tension model can be true only if the column of solution in the xylem is under tension. The most elegant demonstrations of this tension, and of its adequacy to account for the ascent of sap in the tallest trees, were performed in the early 1960s by Per Scholander of the Scripps Institution of Oceanography in La Jolla, California. Scholander measured tension in stems with a device called a pressure bomb.

The principle of the pressure bomb is as follows: Consider a stem in which the xylem sap is under tension. If the stem is cut, the sap pulls away from the cut, into the stem. Now the tissue is placed in a cylinder—the bomb—in which the pressure may be raised. The cut surface remains outside the bomb. As pressure is applied, the xylem sap is forced back to the cut surface. When the sap first becomes visible again at the cut surface, the pressure in the bomb is recorded. This pressure is the same as the tension that was originally present in the xylem (Figure 30.9).

Scholander used the pressure bomb to study dozens of plant species, from diverse habitats, growing under a variety of conditions. In all cases in which the xylem sap was ascending, it was found to be under tension. The tension disappeared in some of the plants at night. In developing vines, the xylem sap was not under tension until leaves formed. Once leaves appeared, transport in the xylem began, and tensions were recorded.

Suppose you wanted to measure tensions in the xylem at various heights in a large tree, such as a Douglas fir more than 80 meters tall, to confirm that the tensions were sufficient to account for the rate at which sap was moving up the trunk. How would

you get stem samples for measurement? Scholander surveyed a tree to determine the heights of particular twigs and then had a sharpshooter with a high-powered rifle shoot the twigs from the tree. As quickly as the twigs fell to the ground, they were inserted in the pressure bomb and their xylem tensions recorded. Scholander had twigs shot from a tree at heights of 27 and 79 meters at four different times of day. At each hour the difference in tensions was great enough to keep the xylem sap ascending, and that tension was established by transpiration from the leaves. Transpiration provides the impetus for transport of water and minerals in the xylem, but it also results in the loss of tremendous quantities of water from the plant. How do plants keep this loss to reasonable levels?

TRANSPIRATION AND THE STOMATA

The epidermis of leaves and stems minimizes transpirational water loss by secreting a waxy **cuticle**, which is impermeable to water. However, the cuticle is also impermeable to carbon dioxide, posing a problem: How can the leaf balance its need to retain water with its need to obtain carbon dioxide for photosynthesis? An elegant compromise has evolved, in the form of stomata (singular: stoma). A **stoma** is a gap in the epidermis; its opening and closing are controlled by a pair of specialized epidermal cells called **guard cells** (Figure 30.10a). When the stomata are open, carbon dioxide can enter the leaf by diffusion, but water vapor may also be lost in the same way. Closed stomata prevent water loss but also exclude carbon dioxide from the leaf. Most plants compromise by opening the stomata only when the light intensity is sufficient to maintain a good rate of photosynthesis. At night, when darkness precludes photosynthesis, the stomata remain closed; no carbon dioxide is needed at this time, and water is conserved. Even during the daytime, the stomata close if water is being lost at too great a rate.

The stoma and guard cells in Figure 30.10a are typical of dicots. Monocots typically have specialized epidermal cells associated with their guard cells. The principle of operation, however, is the same for both monocot and dicot stomata.

The mechanism by which stomata open and close is now at least partially understood. The guard cells control the size of the stomatal opening. When the stomata are about to open, potassium ions (K^+) are actively transported into the guard cells from the surrounding epidermis. (The redistribution of potassium can be visualized by means of an instrument called an electron microprobe, and it can be measured by patch clamping.) The accumulation of potassium ions makes the water potential of the guard cells

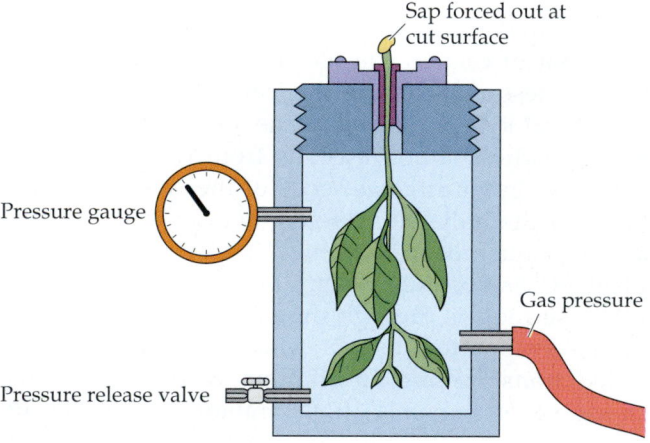

30.9 A Pressure Bomb
By applying just enough pressure so that xylem sap is pushed back to the cut surface of a plant sample, a scientist can determine the tension on the sap in the living plant.

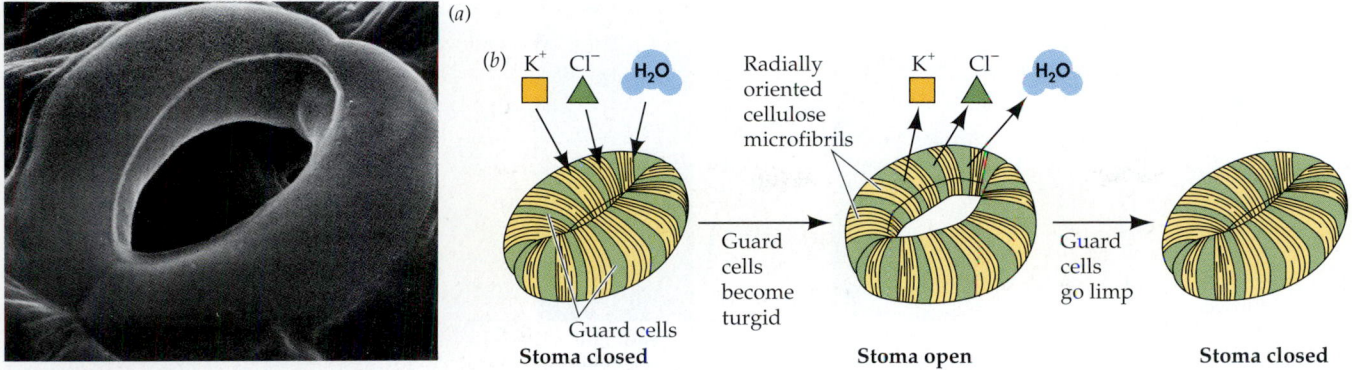

30.10 Stomata

(a) Mesophyll cells and air space inside this dicot leaf are visible through the gaping stoma between the two sausage-shaped guard cells. (b) Potassium ions are actively transported into the guard cells; their resulting negative water potential causes them to take up water and stretch so that a gap—the stomatal opening—appears between them. As K^+ then diffuses passively out of the guard cells, water follows by osmosis, the guard cells go limp, and the stoma closes. Negatively charged ions traveling with K^+ maintain electrical balance and contribute to the changes in osmotic potential that affect the guard cells.

more negative. Water enters the guard cells by osmosis, making them more turgid and stretching them in such a way that a gap, the stoma, appears between them. The pattern of stretching is controlled by the orientation of cellulose microfibrils in the walls of the guard cells. The stoma closes by the reverse process: Potassium ions diffuse passively out of the guard cells, water follows by osmosis, turgidity is lost, and the guard cells collapse together and seal off the stoma (Figure 30.10*b*). Negatively charged chloride and organic ions also move along with the potassium ions, maintaining electrical balance and contributing to the change in osmotic potential of the guard cells.

What controls the movement of potassium into and out of guard cells? The control system is complex, with more than one type of sensor system. For one thing, the level of carbon dioxide in the spaces inside the leaf is monitored; a low level favors opening of the stomata, thus allowing an increased carbon dioxide level and an enhanced rate of photosynthesis. On the other hand, certain cells monitor their own water potentials. If they are too dry—that is, if their water potential is too negative—they release a substance called abscisic acid. According to one hypothesis, abscisic acid then acts on the guard cells, causing them to release potassium ions, thus closing the stomata and preventing further drying of the leaf. (Some scientists think that abscisic acid serves only to keep the stomata closed, rather than causing the closure; further experiments on the timing of abscisic acid production are needed to resolve this point.)

Light also controls the opening of the stomata, which makes sense in view of the fact that most plants conduct photosynthesis only in the light; we would expect stomata to be closed in the dark, thus preventing unnecessary water loss. It was recently discovered that, under certain conditions, brief exposures to blue light cause guard cells to acidify their environment—that is, to pump out protons. Patch clamping experiments revealed further details about the relationship between blue light and proton pumping (Figure 30.11). The proton pump enables the guard cell to take up K^+ and Cl^- ions as described earlier in this chapter (see Figure 30.3).

CRASSULACEAN ACID METABOLISM AND THE STOMATAL CYCLE

Most plants change their stomatal openings on a schedule shown by the blue curve of Figure 30.12: The stomata are typically open for much of the day and closed at night (they may, however, close during very hot days). But not all plants follow this pattern. Many **succulent plants**—fleshy plants that live in dry areas or near the ocean—are members of the flow-

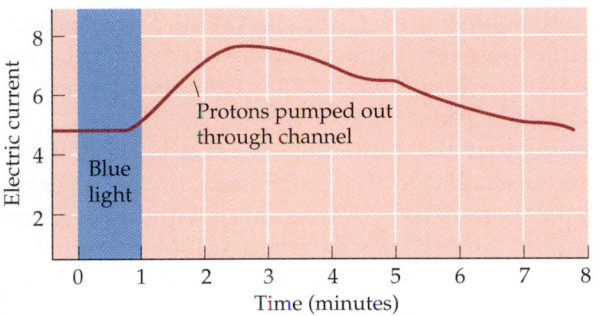

30.11 Patch Clamping Reveals Light-Induced Proton Pumping

A tracing of the tiny electric current that results from the flow of protons through a single channel in the plasma membrane of a guard cell. A brief exposure to blue light against a background of constant, dim red light causes protons to flow out of the cell for a few minutes.

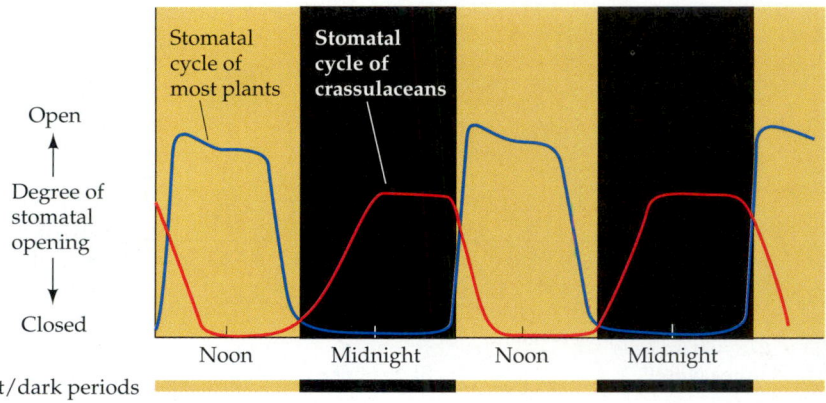

30.12   Stomatal Cycles
Most plants open their stomata during the daytime. Plants of the family Crassulaceae, such as the rock dudleya in the photograph, have evolved means to reverse this stomatal cycle. Crassulacean stomata open during the night.

ering plant family Crassulaceae. The crassulaceans have some unusual biochemical and behavioral features. One that was particularly surprising to its discoverers is their "backward" stomatal cycle: Their stomata are open at night and closed by day (red curve in Figure 30.12). It was then discovered that crassulacean leaf tissues become acidic at night and more neutral in the daytime.

The mystery was resolved with the following discoveries. At night, while the stomata are open, carbon dioxide diffuses freely into the leaf and reacts in the mesophyll cells with phosphoenolpyruvic acid to produce organic acids such as malic acid and aspartic acid (see Chapter 8). These acids accumulate to high concentrations in the vacuoles of the mesophyll cells. At daybreak the stomata close, thus preventing water loss. Throughout the day, the organic acids are broken down to release the carbon dioxide once again—behind closed stomata. Because the carbon dioxide cannot diffuse out of the plant, it is available for photosynthesis. This set of chemical reactions, discussed in Chapter 8, is referred to as crassulacean acid metabolism, or **CAM**. CAM and the accompanying stomatal behavior were subsequently observed in species of many other plant families besides the Crassulaceae.

Notice that the formation of organic acids is absolutely essential to the functioning of the reversed stomatal pattern of CAM plants. Without the acid formation, carbon dioxide could still be admitted to the leaf at night and saved for daytime. However, it could build up in the intercellular spaces of the leaf only to the same level—0.03 percent of the atmosphere—as in the surrounding air. This amount would be used up by the Calvin–Benson cycle of photosynthesis in a matter of minutes in the daytime. Instead, during the night a CAM plant makes the carbon dioxide into organic acids as fast as it comes

in, thus allowing more carbon dioxide to enter. Acid formation in effect fills the leaf with carbon dioxide. CAM is well adapted to environments where water is scarce: A leaf with its stomata open at night loses much less water (because the environment is cooler then) than does a leaf with its stomata open by day.

In both CAM and non-CAM plants, the carbon dioxide is converted to the products of photosynthesis. How does the plant deliver these products to the parts of the plant that do not perform photosynthesis?

TRANSLOCATION OF SUBSTANCES IN THE PHLOEM

How substances in the phloem move from sources, such as leaves, to sinks, such as the root system, remains a topic of interest in plant physiology. Sugars, amino acids, some minerals, and a variety of other substances are translocated in the phloem. Any model to explain this translocation must account for a few important facts: (1) Translocation stops if the phloem tissue is killed by heating or other methods; thus the mechanism must be different from that of transport in the xylem. (2) Translocation often proceeds in both directions—up and down the stem or petiole—simultaneously. This bidirectional transport may be explained in terms of neighboring sieve tubes conducting in opposite directions, with each sieve tube transporting all its contents in a single direction. (3) Translocation is inhibited by compounds that inhibit cellular respiration and thus limit the ATP supply.

To answer some of the most pressing questions about translocation, plant physiologists needed to obtain samples of pure phloem sap from individual sieve tube elements. This task was simplified when

30.13 Phloem Sap Gets Around
Aphids—the white organisms with "sculpted" abdomens—are drilling into a plant to obtain phloem sap. A drop of the sap has formed at the anus of one of the aphids, and an ant is about to collect it.

scientists recognized that a common garden pest, the aphid, feeds by drilling into a sieve tube. An aphid inserts its stylet, or feeding organ, into a stem until the stylet enters a sieve tube. Within the sieve tubes, the pressure is much greater than in the surrounding plant tissues, so phloem sap is forced up the stylet and into the aphid's digestive tract. So great is the pressure that sugary liquid is forced out the insect's anus (Figure 30.13). At times, ants collect this sugary discharge as food, and some species of ants actually "farm" colonies of aphids, moving them from place to place and protecting them from enemies.

Plant physiologists use the aphid somewhat differently. When liquid appears on the aphid's abdomen, indicating a connection with a sieve tube, the physiologist freezes the aphid and cuts its body away from the stylet. Phloem sap continues to exude from the cut stylet, where it may be collected for analysis. Study of the sap gives accurate information about the chemical composition of the contents of a single sieve tube element over time. From that information one can infer such things as translocation rates. Data obtained by this and other means led to the general adoption of the pressure flow model as an explanation for transport in the phloem.

The Pressure Flow Model

Phloem sap flows, under pressure, through the sieve tubes. Two important steps in the flow are the active, ATP-requiring transport of sugars and other solutes *into* the sieve tubes in source areas and the *removal* of the solutes by active transport where the sieve tubes enter sinks. According to the **pressure flow model**, the solute concentration at the source end of a sieve tube is higher than at the sink end, so water

has a greater tendency to enter the sieve tube by osmosis at the source end. In turn, this entry of water causes a greater pressure potential at the source end, so the entire fluid content of the sieve tube is, in effect, squeezed toward the sink end of the tube (Figure 30.14). This mechanism was first proposed more than half a century ago, but some of its features are still debated.

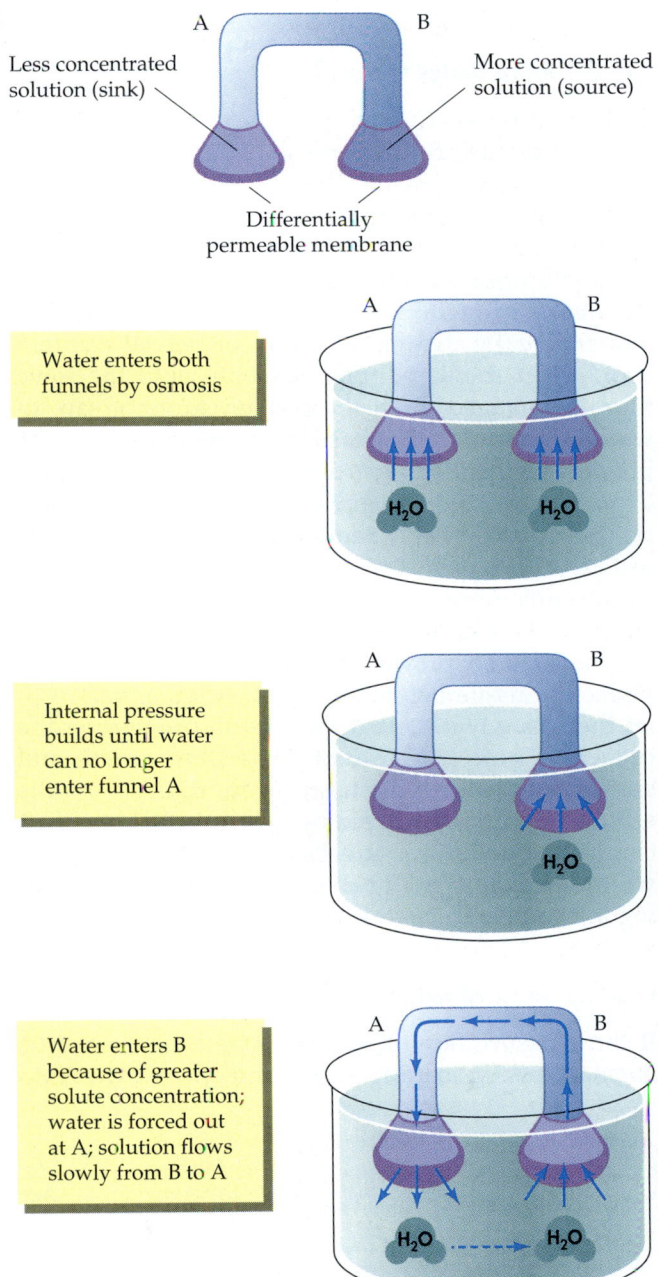

A — Less concentrated solution (sink)

B — More concentrated solution (source)

Differentially permeable membrane

Water enters both funnels by osmosis

Internal pressure builds until water can no longer enter funnel A

Water enters B because of greater solute concentration; water is forced out at A; solution flows slowly from B to A

30.14   The Pressure Flow Model
This experimental demonstration of the pressure flow model describes how pressure potential and water potential combine to drive sugars and other solutes from the source (B) to the sink (A). Phloem sap may flow through sieve tubes in this manner.

Other mechanisms have been proposed to account for translocation in sieve tubes. Some have been disproved, and none of the rest have been supported by a weight of evidence comparable to that for the pressure flow model. The pressure flow model depends on two things: The sieve plates must be unclogged, so that bulk flow from one sieve tube element to the next is possible, and there must be an effective method for loading sucrose and other solutes into the phloem in source tissues and removing them in sink tissues. Are these conditions met?

Are the Sieve Plates Clogged or Free?

Early electron microscopic studies of phloem samples cut from plants produced results that seemed to contradict the pressure flow model. The pores in the sieve plates always appeared to be plugged with masses of a fibrous protein, suggesting that phloem sap could not flow freely. But what is the function of the fibrous protein? One possibility is that this protein is usually distributed more or less at random throughout the sieve tube elements until the sieve tube is damaged; then the sudden surge of sap toward the cut surface carries the protein into the pores, blocking the pores and effectively caulking the leak. That is, the protein does *not* block the pores unless the phloem is damaged. How might this be tested? How could we cut samples of phloem for microscopic observation without causing the sap to surge to the cut surface?

One way to prevent the surge of the sap is to freeze the tissue before cutting it. Another way is to let the tissue wilt so that there is no pressure in the phloem. These and related ideas were tested, and sure enough, when the tissue is cut in this way, the sieve plates are unclogged by protein! Thus, the first condition of the pressure flow model is met. Now, what about the need for an effective method for loading and unloading solutes?

Loading and Unloading of the Phloem

If the pressure flow model is correct, there must be mechanisms for loading sugars and other solutes into the phloem from source regions and for unloading them into sink regions. One mechanism of phloem loading has been demonstrated in a number of plants. Sugars and other solutes to be transported are passed from cell to cell through the symplast in the mesophyll. After these substances reach cells adjacent to the ends of leaf veins, they leave the mesophyll cells and enter the apoplast, sometimes with the help of transfer cells. Then specific sugars, amino acids, some mineral elements, and a few other compounds are actively transported into cells of the phloem, thus reentering the symplast (Figure 30.15).

Passage through the apoplast selects substances to be accumulated for translocation because substances can enter the phloem only upon passing through a differentially permeable membrane. In many plants, the cells thus loaded are companion cells (see Chapter 29), which then transfer the solutes to the adjacent sieve tube elements. Loading of the phloem with solutes results in a very negative water potential in the sieve tubes; thus water enters by osmosis from the surrounding tissue and maintains a high pressure potential within the sieve tubes. As Figure 30.15 shows, sucrose movement from the mesophyll to the sieve tube elements takes place entirely within the symplast in some species; that is, transfer of solutes from symplast to apoplast and back again is not a universal feature of phloem loading.

A form of secondary active transport (see Chapter 5) loads sucrose into the companion cells and sieve tube elements. Sucrose is carried through the plasma membrane from apoplast to symplast by a sucrose–proton symport—the entry of sucrose and of protons is strictly coupled. For this symport to work, there must be a high concentration of protons in the apoplast; the protons are supplied by a primary active transport system, the proton pump. The protons then "relax" back into the cell through the symport, bringing sucrose with them (Figure 30.16).

In sink regions, the transported solutes are actively transported out of the sieve tubes and into the surrounding tissues. This unloading serves two purposes: It helps to maintain the gradient of osmotic potential and hence of pressure potential in the sieve tubes, and it promotes the buildup of sugars and

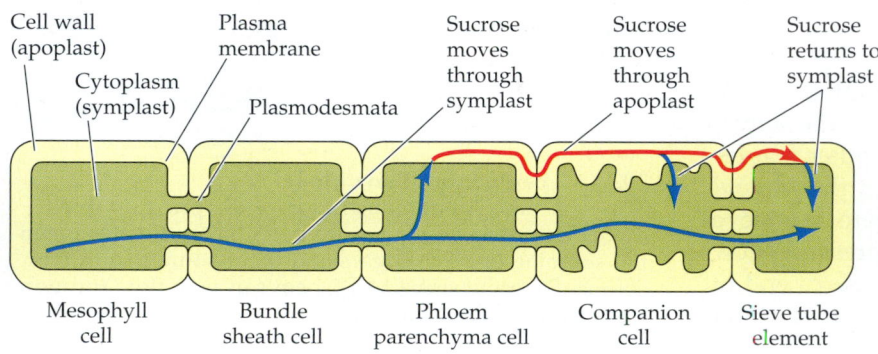

Cell wall (apoplast)   Plasma membrane   Sucrose moves through symplast   Sucrose moves through apoplast   Sucrose returns to symplast

Cytoplasm (symplast)   Plasmodesmata

Mesophyll cell   Bundle sheath cell   Phloem parenchyma cell   Companion cell   Sieve tube element

30.15 Solutes May Enter Sieve Tubes Via the Apoplast
Chloroplasts in the mesophyll produce sucrose, which moves primarily through the symplast (blue arrow) as it passes from cell to cell on its way to the sieve tube elements. In many species, however, sucrose exits into the apoplast (red arrow), from which it is loaded into the companion cells or sieve tube elements.

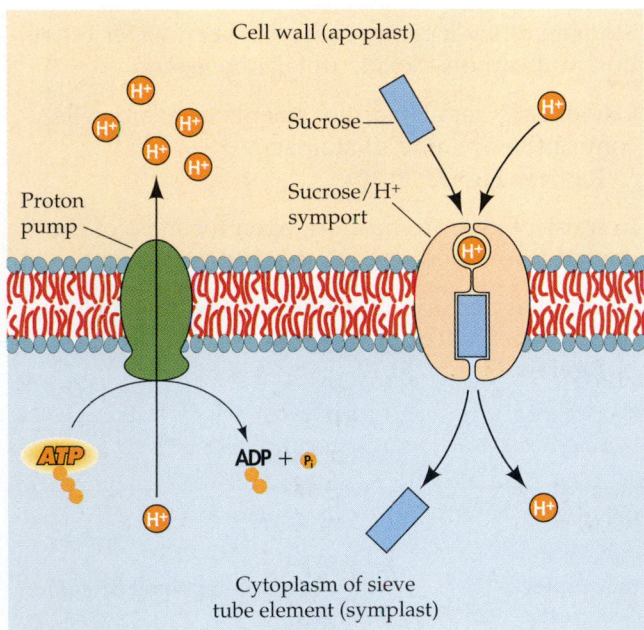

30.16 Sucrose–Proton Symport Loads the Phloem
The proton pump builds a gradient of proton concentration, and the relaxation of the gradient through the sucrose–proton symport carries sucrose, against its own concentration gradient, into the companion cells and sieve tube elements.

30.17 Collecting Maple Sap
Sap from the blue tap flows through tubes to a central location for collection.

starch to high concentrations in storage regions, such as developing fruits and seeds.

Sucrose in the Xylem

The xylem sap, as well as the phloem sap, occasionally may contain sugars. In sugar maples and many other deciduous trees and shrubs of the temperate zones, excess photosynthate produced in late summer and early fall is stored as starch in living xylem cells of the trunk and twigs. Later, in early spring, the starch is digested into sugars that appear initially in the xylem sap, which may be collected and concentrated into syrup (Figure 30.17). The activities of plants may vary with the time of year, and patterns of transport and storage may change accordingly.

SUMMARY of Main Ideas about Transport in Plants

Water and solutes move through tissues in the apoplast and symplast.

In the root, water and solutes may pass between the cortex and stele only in the symplast.

Casparian strips in the endodermis block water and solute movement in the apoplast.
   Review Figures 30.5 and 30.6

Water moves between cells by osmosis.

The water potential of a cell or solution is the sum of the osmotic potential and the pressure potential.

Water always moves toward the cell or solution with a more negative water potential.
   Review Figure 30.1

A proton pump mediates the active transport of many solutes across membranes in plants.
   Review Figures 30.3 and 30.16

Patch clamping is a powerful method for studying the movement of ions through membranes.
   Review Figures 30.4 and 30.11

Water and minerals are transported through tracheids or vessel elements in the xylem.

Root pressure plays a minor role in the transport of water and minerals in the xylem.

Evaporation from moist-walled cells in the leaf lowers the water potential of those cells and thus pulls water—held together by its cohesiveness—up through the xylem from the root.
   Review Figure 30.8

Evaporation of water cools the leaves.

Products of photosynthesis, and some minerals, are translocated through sieve tubes in the phloem. The difference in solute concentration between sources and sinks allows a difference in pressure potential along the sieve tubes, resulting in bulk flow.
Review Figure 30.14

Solutes are actively loaded into the phloem in source regions and unloaded in sink regions.
Review Figure 30.15

Stomata allow a compromise between water retention and carbon dioxide uptake by leaves.

Osmotically regulated movements of guard cells control the opening of stomata.
Review Figure 30.10

In most plants, stomata are open for most of the daylight hours but closed at night.

Plants that perform crassulacean acid metabolism have a reversed stomatal cycle.
Review Figure 30.12

SELF-QUIZ

1. Osmosis
   a. requires ATP.
   b. results in the bursting of plant cells placed in pure water.
   c. can cause a cell to become turgid.
   d. is independent of solute concentrations.
   e. continues until the pressure potential equals the water potential.

2. Water potential
   a. is the difference between the osmotic potential and the pressure potential.
   b. is analogous to the air pressure in an automobile tire.
   c. is the movement of a solvent through a membrane.
   d. determines the direction of water movement between cells.
   e. is defined as 1.0 for pure water under no applied pressure.

3. Which statement about proton pumping across the plasma membranes of plants is *not* true?
   a. It requires ATP
   b. The region inside the membrane becomes positively charged with respect to the region outside.
   c. It enhances the movement of K^+ ions into the cell.
   d. It pushes protons out of the cell against a proton concentration gradient.
   e. It can drive the secondary active transport of negatively charged ions.

4. Patch clamping
   a. is a mechanism that protects plants against insect pests.
   b. is a mechanism for sealing leaks in the plasma membrane.
   c. can be performed with the cell wall intact.
   d. can be used as a remedy for mineral deficiencies.
   e. can be used to monitor the flow of ions through carrier proteins.

5. Which of the following statements is *not* true?
   a. The symplast is a meshwork consisting of the (connected) living cells.
   b. Water can enter the stele without entering the symplast.
   c. The Casparian strips prevent water from moving between endodermal cells.
   d. The endodermis is a cell layer in the cortex.
   e. Water can move freely in the apoplast without entering cells.

6. Which of the following is *not* part of the evaporation–cohesion–tension model?
   a. Water evaporates from the walls of mesophyll cells.
   b. Removal of water from the xylem exerts a pull on the water column.
   c. Water is remarkably cohesive.
   d. The wider a tube, the greater the tension its water column can withstand.
   e. At each step, water moves to a region with a more strongly negative water potential.

7. Stomata
   a. control the opening of guard cells.
   b. release less water to the environment than do other parts of the epidermis.
   c. are usually most abundant on the upper epidermis of a leaf.
   d. are covered by a waxy cuticle.
   e. close when water is being lost at too great a rate.

8. Plants that perform crassulacean acid metabolism
   a. incorporate carbon dioxide into organic acids.
   b. have leaves that become more acidic during the daytime hours.
   c. close their stomata at night.
   d. are also called C_4 plants.
   e. must live in environments where water is plentiful.

9. Which statement is *not* true of phloem transport?
   a. It takes place in sieve tubes.
   b. It depends upon mechanisms for loading solutes into the phloem in sources.
   c. It stops if the phloem is killed by heat.
   d. In sinks, solutes are actively transported into sieve tube elements.
   e. A high pressure potential is maintained in the sieve tubes.

10. The fibrous protein in sieve tube elements
    a. clogs the sieve plates at all times.
    b. never clogs the sieve plates.
    c. serves no known function.
    d. may caulk leaks when a plant is damaged.
    e. provides the motive force for transport in the phloem.

FOR STUDY

1. Epidermal cells protect against excess water loss. How do they perform this function?

2. Phloem transports material from sources to sinks. What is meant by a source and a sink? Give examples of each.

3. Contrast the transport of organic substances through the phloem with the transport of water and minerals through the xylem. Touch on mechanisms and on overall direction.

4. Transpiration exerts a powerful pulling force on the water column in the xylem. When would you expect transpiration to proceed most rapidly? Why? Describe the source of the pulling force.

5. Plants that can perform crassulacean acid metabolism (CAM) are adapted to environments in which water is available in limited supply—they open their stomata only at night. Could a non-CAM plant, such as a pea plant, enjoy a similar advantage if it opened its stomata only at night? Explain.

READINGS

Neher, E. and B. Sakmann. 1992. "The Patch Clamp Technique." *Scientific American*, March. This article, by the Nobel laureates who introduced patch clamping, deals with animal cells rather than plants, but it gives a good account of the technique.

Raven, P. H., R. F. Evert and S. Eichhorn. 1991. *Biology of Plants*, 5th Edition. Worth, New York. A sound general botany textbook.

Salisbury, F. B. and C. W. Ross. 1992. *Plant Physiology*, 4th Edition. Wadsworth, Belmont, CA. An authoritative textbook with excellent chapters on transport and translocation.

Taiz, L. and E. Zeiger. 1991. *Plant Physiology*. Benjamin/Cummings, Redwood City, CA. Chapters 3, 4, 6, and 7 give more advanced treatments of the topics presented in this chapter.

A Beetle Disarms Laticifers
This beetle is inactivating a milkweed's defense system by cutting supply lines.

31

Environmental Challenges to Plants

Whate we see today is only one frame of a long movie of interactions among organisms on Earth. Some of these interactions represent adaptations, visible in this single frame, that are being challenged and perhaps further refined. Let's begin our discussion of the challenges that the environment poses to plants by looking at the current status of a deadly "game" between a plant and one of its insect enemies.

Milkweeds are latex-producing, or laticiferous, plants. When damaged, a milkweed releases copious amounts of a white, rubbery, toxic liquid—latex—from tubes called laticifers. Latex has long been suspected to deter insects from eating the plant because laticiferous plants are not attacked by insects that feed on neighboring plants of other species. This observed behavior is consistent with, but does not prove, the hypothesis that the latex keeps the insects at bay. Stronger support for the hypothesis was afforded in 1987 by David Dussourd, now of the University of Maryland, and Thomas Eisner of Cornell University, who studied field populations of *Labidomera clivicollis*, a beetle that is one of the few insects that feed on *Asclepias syriaca*, the field milkweed.

The two zoologists observed a remarkable prefeeding behavior: The beetles cut a few veins in the leaves before settling down to dine. In the undamaged plant the latex is under pressure, so cutting the veins, with their adjacent laticifers, causes massive leakage and depressurizes the system. By cutting a few veins, the beetles interrupt the latex supply to a downstream portion of the leaf. The beetles then move to the relatively latex-free portion and eat their fill. Some other insects that do not feed on undamaged milkweeds will eat parts of leaves that have had the latex supply cut off. When presented simultaneously with leaves on undamaged plants and leaf parts that have had their laticifers cut, *L. clivicollis* and other insects that share this vein-cutting behavior select the relatively latex-free leaf parts.

Does this behavior of the beetles negate the adaptational value of latex protection? Not at all. There are still great numbers of potential insect pests that are effectively deterred by the latex. And this is just one frame of the movie. It may be that, over time, milkweed plants producing higher concentrations of toxins will be selected by virtue of their ability to kill beetles that cut the laticifers.

THREATS TO PLANT LIFE

All plants, especially terrestrial ones, face environmental challenges, as we indicated in Chapters 25 and 30. Some environments, however, pose exceptional problems and thus drastically limit the kinds

31.1 Desert Annuals Evade Drought
Seeds of desert plants often lie dormant for long periods awaiting conditions appropriate for germination. When they do germinate, they grow and reproduce rapidly before the short wet season passes. They cover the desert landscape with color for only a few weeks, since water is inadequate at other times.

of plants that can live in them. The most challenging physical environments include ones that are very dry (deserts), ones at the other extreme that are water-logged and thus limit the availability of oxygen, ones that are dangerously salty, and ones that contain high concentrations of toxic substances such as heavy metals. This chapter focuses in part on how some plants manage to thrive in such environments.

The biological environment is also a threat to all plants. It includes herbivores like *Labidomera clivicollis* that consume plants, as well as pathogenic fungi, bacteria, and viruses. The defenses of plants against these biological threats are the other major focus of the chapter.

DRY ENVIRONMENTS

Water for plants and other organisms is often in short supply in the terrestrial environment. Some terrestrial habitats, such as deserts, intensify this challenge, and many plants that inhabit particularly dry areas have one or more structural adaptations that

allow them to conserve water. Plants adapted to dry environments are called **xerophytes**.

Some desert plants have no special *structural* adaptations for water conservation other than those found in almost all flowering plants. Instead they have an alternative *strategy*. Simply put, these plants carry out their entire life cycle—from seed to seed—during a brief period in which the surrounding desert soil is sufficiently moist (Figure 31.1). Through the long dry periods that intervene, only the seeds remain alive, until enough moisture is present to trigger the next life cycle. These desert annuals simply evade the periods of drought. Plants that remain active during the dry periods must have special adaptations that enable them to survive.

Special Adaptations of Leaves to Dry Environments

The secretion of a heavier layer of cuticle over the leaf epidermis to retard water loss is a common adaptation to dry environments. An even more common adaptation is a dense covering of epidermal hairs. Some species have stomata only in sunken cavities below the leaf surface, which reduces the drying effects of air currents; often these stomatal cavities contain hairs as well (Figure 31.2). Ice plants and their relatives have fleshy leaves in which water may be stored. Others, such as ocotillo, produce leaves only when water is abundant, shedding them as the soil dries out (Figure 31.3). Cacti and similar

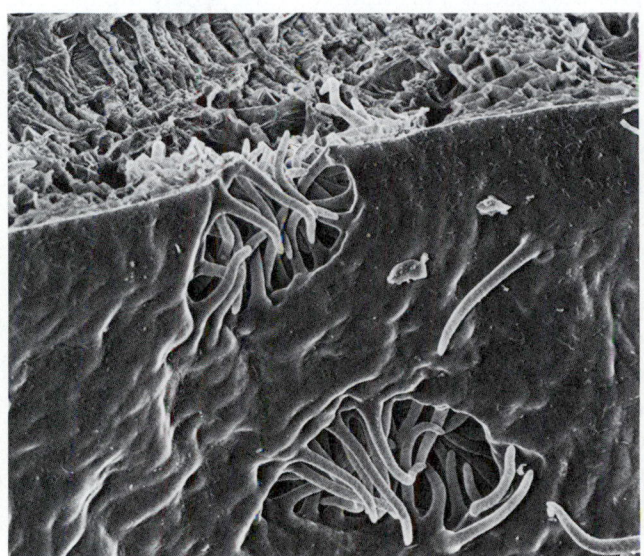

31.2 Stomatal Crypts
Stomata in some water-conserving leaves are in sunken pits called stomatal crypts. The hairs covering these two crypts presumably trap moist air. (Above the cut edge near the top of the photo, a section of the leaf's interior can be seen.)

(a) (b)

31.3 Opportune Leaf Production
Plants in hot, dry environments lose great amounts of water through their leaves. The ocotillo, which lives in the lower deserts of the southwestern United States and northern Mexico, has leaves only when water is available in the soil. During dry periods, the thorny, leafless stems of an ocotillo appear almost dead (a), but when water is on hand, leaves develop rapidly (b) and provide the plant with photosynthetic products.

plants have spines rather than typical leaves, and photosynthesis is confined to the fleshy stems. The spines may reflect incident radiation or perhaps dissipate heat. Corn and some related grasses have leaves that roll up during dry periods, thus reducing the leaf surface area through which water is lost. Some trees that grow in arid regions have leaves that hang vertically at all times, thus evading the midday sun. Characteristic examples are some eucalyptuses (Figure 31.4).

Xerophytic adaptations of leaves minimize water loss by the plant. However, such adaptations simultaneously minimize the uptake of carbon dioxide and thus limit photosynthesis. In consequence, most xerophytes grow slowly, but they utilize water more efficiently than do other plants; that is, they fix more grams of carbon by photosynthesis per gram of water lost to transpiration than other plants do.

Other Adaptations to a Limited Water Supply

Roots may also be adapted to environments low in water. The Atacama Desert in northern Chile often goes several years without receiving any measurable rainfall. The landscape there is almost barren save

for a substantial number of surprisingly large mesquite trees of the genus *Prosopis*. How do these trees obtain water? They have taproots that grow to very great depths, sufficient to reach underground water supplies (Figure 31.5). These trees also obtain water from condensation on their leaves.

A more common adaptation of desert plants is to have root systems that grow each rainy season but die back during dry periods. Cacti, on the other hand, have shallow but extensive fibrous root systems that effectively trap water at the surface of the soil following even light rains.

Xerophytes and other plants receiving inadequate water may accumulate the amino acid proline to substantial concentrations in their vacuoles. As a consequence, the osmotic potential and water potential of the cells become more negative, thus tending to extract more water from the soil.

As we have seen, there are many ways in which some plants eke out an existence in environments with very little water. What happens if there is too much water?

WHERE WATER IS PLENTIFUL AND OXYGEN IS SCARCE

When soils become waterlogged, the availability of oxygen from the soil declines. Most plants cannot tolerate this situation for long. Some species, however, are adapted to life in a swampy habitat. Their

31.4 Shade at Midday
Because eucalyptus leaves hang vertically, their flat surfaces are not presented directly to the midday sun. This adaptation minimizes heating as well as water loss.

31.5 Mining Water with Deep Taproots
This California desert is not as arid as the Chilean Atacama, but *Prosopis juliflora*, a mesquite, still reaches far down in the sand dunes for its water supply.

roots grow slowly and hence do not penetrate deeply. With an oxygen level too low to support aerobic respiration, the roots carry on alcoholic fermentation (see Chapter 7), which provides a supply of ATP for the activities of the root system.

The root systems of some plants adapted to swampy environments have **pneumatophores**, extensions that grow out of the water and up into the air (Figure 31.6). Oxygen diffusing into pneumatophores aerates the submerged parts of the root system.

Submerged or partially submerged aquatic plants often have large air spaces in the leaf parenchyma and in the petioles. Tissue with such air spaces is called **aerenchyma** (Figure 31.7). Aerenchyma stores and permits the diffusion of oxygen, and it provides buoyancy.

Thus far we have considered water supply—either too little or too much—as a limiting factor in plant growth. Are there other substances that make an environment hospitable or inhospitable to plant growth?

SALINE ENVIRONMENTS

On a world scale, no toxic substance restricts plant growth more than does salt (sodium chloride). Sa-

31.6 Coming Up for Air
The roots of the mangroves in this tidal swamp obtain oxygen through their pneumatophores—extensions growing out of the water, under which the remainder of the root system is buried.

31.7 Aerenchyma Lets Oxygen Get to Submerged Tissues
This scanning electron micrograph, a cross section of a petiole of the yellow water lily, shows a vascular bundle and a number of open channels of the aerenchyma. The channels are lined by cells, including branched ones that send projections into the channels. Because aerenchyma has far fewer cells than does a comparable volume of most petiole tissue, less respiratory metabolism is carried on and the need for oxygen is much reduced.

line—salty—habitats support, at best, sparse vegetation. The **halophytes**, plants adapted to such habitats, belong to a wide variety of flowering plant groups. Saline environments themselves are diverse, ranging from hot, dry, salty deserts to moist, cool, salty marshes. Along the seashore are salty environments created by ocean spray. The ocean itself is a saline environment, as are river estuaries, where fresh and salt water meet and mingle. The salinization of agricultural land is an increasing world problem; where crops are irrigated, sodium ions in the water accumulate in the soil to ever greater concentrations. Biologists in Israel and elsewhere have had some success in breeding crops that can be watered with seawater or diluted seawater.

Saline environments pose an osmotic problem. Because of a high salt concentration, the environment has an unusually large negative water potential. To obtain water from such an environment, resident plants must have an even more negative water potential than that of plants in nonsaline environments; otherwise, the plants lose water and wilt. A second problem related to the saline environment is the potential toxicity of high concentrations of certain ions, notably sodium. Chloride ions may also be toxic at high concentrations. How can halophytes cope with a highly saline environment, while nonhalophytes cannot?

Salt Accumulation and Salt Glands

Most halophytes share one adaptation: They accumulate sodium and, usually, chloride ions and they transport these ions to the leaves. Nonhalophytes accumulate relatively little sodium, even when placed in a saline environment; of the sodium that is absorbed by their roots, very little is transported to the shoot. The increased salt concentration in halophytes makes their water potential more negative, so they can take up water from the saline environment. We still do not know how halophytes are able to tolerate such high internal sodium and chloride concentrations without being poisoned.

Some halophytes have other adaptations to life in saline environments. For example, some have salt glands in their leaves. These glands excrete salt, which collects on the leaf surface until it is removed by rain or wind (Figure 31.8). This adaptation, which reduces the danger of poisoning by accumulated salt, is found both in some desert plants, such as tamarisk, and in some mangroves growing in seawater in the tropics.

Salt glands can play multiple roles, as in the arid-zone shrub *Atriplex halimus*. This shrub has glands that secrete salt into small bladders on the leaves where, by increasing the gradient in water potential, it helps the leaves obtain water from the roots. At the same time, by making the water potential of the leaves more negative, the salt reduces the transpirational loss of water to the atmosphere.

These adaptations are specific to halophytes. Several other adaptations are shared by halophytes and xerophytes.

31.8 Secreting Salt
This salty mangrove has used special glands to secrete salt, which now appears as crystals on the leaves.

Adaptations Common to Halophytes and Xerophytes

Many halophytes accumulate the amino acid proline in their cell vacuoles. Unlike sodium, proline is relatively nontoxic. As in xerophytes, the accumulated proline makes the water potential more negative.

Succulence—the possession of fleshy, water-storing leaves—is an adaptation to dry environments. The same adaptation is common among halophytes, as might be expected, since saline environments also make water uptake difficult for plants. Succulence characterizes many halophytes occupying salt marshes. There the salt concentration in the soil solution may change throughout the day; while the tide is out, for instance, evaporation increases the salt concentration. Succulence may offer a reserve of water for the plant during the period of maximum salinity; when the salinity drops as the tide comes in, the leaf's store of water is replenished.

Other general adaptations to a saline environment are of the same sorts observed in xerophytes. These include high root-to-shoot ratios, sunken stomata, reduced leaf areas, and thick cuticles.

A Versatile Halophyte

The halophyte *Triglochin maritima* is unusual because it can adjust to a wide range of environmental salinities. Recent work in Toronto, Canada, has established that *T. maritima* plants change in many ways when they are shifted to environments with lower or higher salt concentrations. Researchers watered some plants with seawater and some with diluted seawater. The plants watered with pure seawater produced much smaller leaves, with smaller cells, than those watered with diluted seawater, but they retained their leaves longer. The rate of leaf production is the same regardless of salinity, but when the environment is changed, the pattern of leaf production changes accordingly.

These morphological differences in *T. maritima* leaves are accompanied by physiological changes as well. The leaf cells of the plants grown on undiluted seawater contained much higher concentrations of sodium and chloride ions as well as of proline. The rates of photosynthesis in the leaves of the plants watered with undiluted seawater also were higher.

Salt is not the only toxic solute in soils. Some other solutes are, in fact, more toxic than salt when presented at the same concentration.

HABITATS LADEN WITH HEAVY METALS

High concentrations of heavy metals, such as copper, lead, nickel, and zinc, poison most plants, even

31.9 Plant Life on a Mine Tailing
Although high concentrations of copper kill most plants, Bermuda grass is colonizing this copper mine tailing in Copperhill, Tennessee.

though plants require some heavy metals at low concentrations. Some sites are naturally rich in heavy metals as a result of normal geological processes. Acid rain leads to the release of toxic aluminum ions in the soil. Other human activities, notably the mining of metallic ores, leave localized areas—known as tailings—with substantial concentrations of heavy metals and low concentrations of nutrients. Such sites are hostile to most plants, and seeds falling on them generally do not produce adult plants.

Mine tailings rich in heavy metals, however, generally are not completely barren (Figure 31.9). They may support healthy plant populations that differ genetically from populations of the same species on the surrounding normal soils. How can these plants survive? Within some species, a few individuals may have genotypes that allow them to survive in soils rich in heavy metals. Those individuals may grow poorly on such soils compared with their potential for growth on more normal soils but may nevertheless survive. Or, because most plants cannot survive in such habitats and hence the competition may be sharply reduced, the few plants growing in them may even thrive.

Initially, some plants were thought to tolerate heavy metals by excluding them: By not taking up the metal ions, the plant could avoid being poisoned. Measurements have shown, however, that tolerant plants growing on mine tailings do take up the heavy metals, accumulating them to concentrations that would kill most plants. Thus the tolerant plants must have a mechanism for dealing with the heavy metals they take up.

The British biologist D. Jowett made an interesting discovery about plants that tolerate heavy metals. In

31.10   Serpentine Barrens
The rocky serpentine barrens in the foreground support only a sparse vegetation of goldenrod and asters, contrasted with the green, forested hills in the background.

Wales and Scotland, bent grass (*Agrostis*) grows near many mines (see Figure 20.4). From mine to mine, the heavy metals in the soil differ. Jowett obtained samples of bent grass from several such sites and tested their ability to grow in various solutions, each containing only one of the heavy metals. In general, the plants tolerated a particular heavy metal—the one most abundant in their habitat—but were sensitive to other heavy metals. That is, their tolerance was for one or two heavy metals only, rather than for the heavy metals as a group.

Tolerant populations can evolve and colonize an area surprisingly rapidly. The bent grass population around a particular copper mine in Wales is resistant to copper and relatively abundant, even though the copper-rich soil dates from mining done late in the nineteenth century, only a century ago. If populations can evolve and cope with toxic soils, can they deal as well with soils in which nutrients are in short supply or in improper balance?

SERPENTINE SOILS

One unproductive soil type that is found in many parts of the world is derived from rock called **serpentine**. Calcium is in short supply in serpentine soils, as are some other essential plant nutrients. Magnesium is present in greater concentration than calcium, reducing plant growth. Chromium, nickel, and certain other heavy metals may be abundant. These factors make serpentine soils inhospitable to many plants. The vegetation on most serpentine soils differs dramatically from that on immediately adjacent nonserpentine soils, and the serpentine vegetation generally is more sparse and less diverse (Figure 31.10).

The shortage of calcium and the higher magnesium concentration probably are the principal challenges facing potential colonizers of a serpentine soil. A number of species have physiological adaptations to meet the challenges successfully. Different species exhibit striking differences in their response to calcium supply in the soil. Biologists divided some serpentine soil into several samples, adjusted the calcium level of each, and grew jewel flower (*Streptanthus glandulosis*), which grows on serpentine, and tomato—a crop plant intolerant of serpentine—on each sample. The growth of the tomato plants was sharply dependent on calcium concentration, while that of the jewel flower was remarkably insensitive to it (Figure 31.11). Serpentine plants such as jewel flower can absorb calcium efficiently even from soils highly deficient in that element. These serpentine plants may also be able either to exclude excess magnesium or to tolerate high internal magnesium concentrations.

Having considered plant adaptations to their physical environment—water, oxygen, salt, and toxic substances—we will now turn to the biological environment. Specifically, we will consider how plants interact with the animals that eat them, and with fungi, bacteria, and viruses that produce plant diseases.

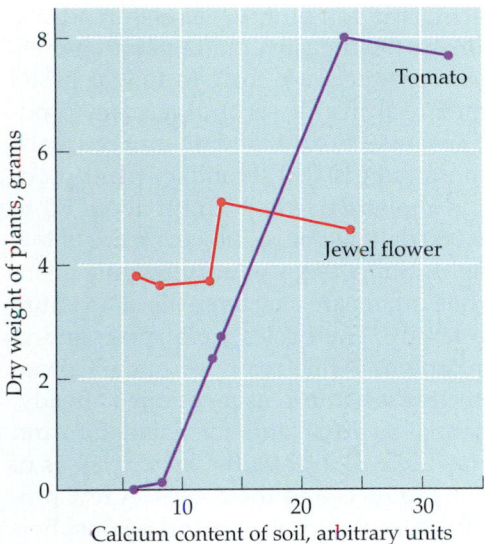

31.11 Differing Responses to Calcium Supply
Adding calcium to serpentine soil permits growth of a
crop plant (tomato) but has no effect on the growth of a
serpentine plant (the jewel flower).

PLANTS AND HERBIVORES

Herbivores—animals that eat plants—depend on
plants for energy and nutrients. Plants have defense
mechanisms to protect them against herbivores, as
we will see, but first let's consider some examples in
which herbivores have a positive effect on the plants
that they eat.

Grazing and Plant Productivity

Consider the phenomenon of grazing, in which an
animal predator eats part of its prey, such as the
leaves of plants, without killing the prey organism,
which has the potential to grow back. What are the
consequences of grazing? Is it detrimental to the
plants, or are they somehow adapted to their place
in the food chain of nature? In fact, certain plants
and their predators evolved together, each acting as
the agent of natural selection on the other. Because
of this coevolution, grazing increases photosynthetic
production in certain plant species.

The removal of some leaves from a plant typically
increases the rate of photosynthesis of the remaining
leaves. This phenomenon probably is the result of
several factors. First, nitrogen obtained from the soil
by the roots no longer needs to be divided among so
many leaves. Second, the transport of sugars and
other photosynthetic products from the leaves may
be enhanced because the demand for those products
in the sinks—such as roots—is undiminished, while
the sources—leaves—have been decreased. A third
and particularly significant factor, especially in

grasses, is an increase in the availability of light to
the younger, more active leaves or leaf parts. The
removal of older, dying—or even dead—leaves and
leaf parts by a grazer decreases the shading of
younger leaves, and unlike most other plants, which
grow from their shoot and leaf tips, grasses grow
from the base of the shoot and leaf.

Some grazed plants continue to grow until much
later in the season than ungrazed but otherwise similar plants do. This longer growing season results in
part because the removal of apical buds by the grazers stimulates lateral buds to become active, thus
producing a more heavily branched plant. In addition, leaves on ungrazed plants may die earlier in the
growing season than leaves on grazed plants.

A clear case of increased productivity due to grazing was reported in 1987 by Ken Paige of the University of Utah and Thomas Whitham of Northern
Arizona University. Mule deer and elk graze many
plants, including one called scarlet gilia. The grazing
removes about 95 percent of the aboveground part
of each scarlet gilia. However, each plant quickly
regrows not one but four replacement stems. The
cropped (grazed) plants produce three times as many
fruits by the end of the growing season as do ungrazed plants (Figure 31.12). Paige and Whitham

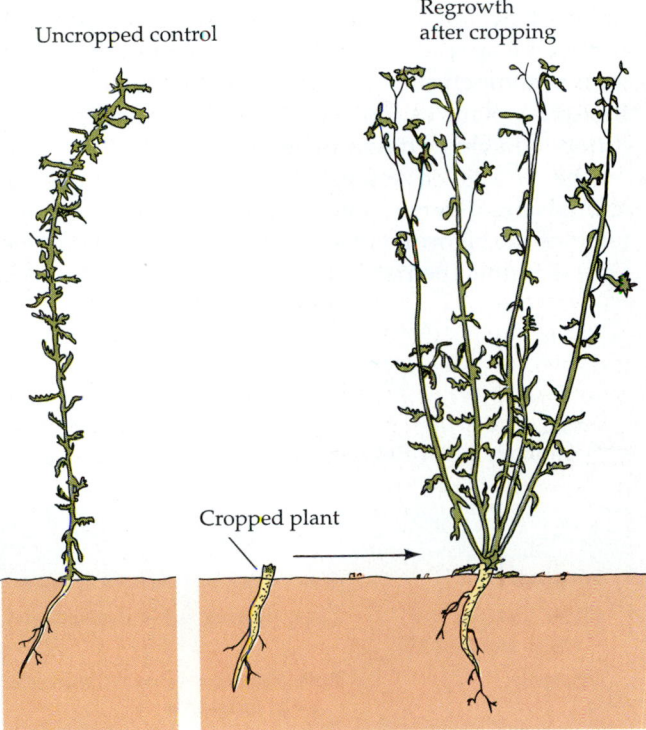

31.12 Overcompensation for Being Eaten
A scarlet gilia was cropped to the point indicated. It then
grew four new stems and produced almost three times as
many offspring as did uncropped plants like the one on
the right.

cropped some scarlet gilia in the laboratory; these plants, too, produced more fruits than did uncropped plants. Not only does the productivity of cropped plants increase, but we may also conclude that grazing by herbivores increases the fitness of scarlet gilia, as the cropped plants pass their genes on to more surviving offspring than do uncropped plants.

Some other plants also profit from moderate herbivory. In addition to the increase in its productivity, a plant benefits by attracting animals that spread its pollen or that eat its fruit and thus distribute its seeds through their feces. However, resisting attack by fungi and herbivorous animals and inhibiting the growth of neighboring plants are also to the advantage of a plant.

Chemical Defenses against Herbivores

Plants attract, resist, and inhibit other organisms often by producing special chemicals known as **secondary products**. (Primary products are substances such as proteins, nucleic acids, carbohydrates, and lipids, that are produced and used by all living things.) Although different kinds of organisms share a biochemical heritage of primary products, they may also differ as radically in chemical content as in external appearance. Animals and fungi, for example, need various enzymes to digest their food, a need not shared by most plants. The plant kingdom is noteworthy for its profusion of secondary products that serve special functions. These compounds help plants compensate for being unable to move. Although a plant cannot flee its herbivorous enemies, it may be able to defend itself chemically.

The effects of defensive secondary products on animals are diverse. Some act upon the nervous systems of herbivorous insects, mollusks, or mammals. Others mimic the natural hormones of animals, causing some insect larvae to fail to develop into adults. Still others damage the digestive tracts of herbivores. Some secondary products are toxic to fungal pests. We make commercial use of secondary plant products as fungicides, insecticides, and pharmaceuticals.

There are more than 10,000 secondary plant products, ranging in molecular weight from about 70 to more than 400,000 daltons; most, however, are of low molecular weight. Some are produced by only a single species, while others are characteristic of an entire genus or even family. Their roles are diverse, and in most cases unknown. While many secondary products have protective functions, as mentioned already, others are essential as attractants for pollinators and seed dispersers. Table 31.1 gives the major classes of secondary plant products and their roles. A few proteins and amino acids also protect plants against herbivores (Box 31.A). In the next section we will look at a specific example of an insecticidal secondary product.

A Versatile Secondary Product

Some plants produce canavanine, an amino acid that is not found in proteins but that is closely similar to the amino acid arginine, which is found in almost all proteins (Figure 31.13). Canavanine has recently been found to have two important roles in plants that produce it in significant quantity. The first role is as a nitrogen-storing compound in seeds. The second role is a defensive one and is based on the similarity of canavanine to arginine. Many insect larvae that consume canavanine-containing plant tissue are poisoned: The canavanine is mistakenly incorporated into the insect's proteins in some of the places where the DNA has coded for arginine. Canavanine is different enough in structure from arginine that some of the proteins end up with modified tertiary struc-

TABLE 31.1
Secondary Plant Products

CLASS	SOME ROLES
Alkaloids	Affect herbivore nervous systems
Other nitrogen and sulfur compounds	Cause cancers, nerve damage, and pain in herbivores
Phenolics	Taste obnoxious to herbivores; affect herbivore nervous systems; act as fungicides
Quinones	Inhibit growth of competing plants
Terpenes	Act as fungicides and insecticides; attract pollinators
Steroids	Mimic animal hormones; prevent normal development of insect herbivores
Flavonoids	Attract pollinators and animals that disperse seeds

BOX 31.A

A Protein for Defense against Insects

A group of scientists at the ARCO Plant Cell Research Institute and the University of Wisconsin recently studied the abilities of seeds of wild and domesticated common beans (*Phaseolus vulgaris*) to resist attack by two species of bean weevils. The investigation began with the observation that some wild bean seeds show high resistance to the weevils, whereas no cultivated bean seeds show such resistance. The scientists discovered that all weevil-resistant bean seeds contain a specific seed protein, arcelin. This protein has never been found in cultivated bean seeds. Therefore, the scientists hypothesized that arcelin is responsible for the resistance of some seeds to predation by the weevils.

Because other differences between wild and cultivated beans might have been responsible for the resistance, two series of experiments were performed to test the relationship between resistance and the possession of arcelin. In one series, cultivated and wild bean plants were crossed. The progeny seeds of such crosses showed an absolute correlation between the presence of arcelin and resistance to weevils. In the other series of experiments, the scientists worked with "artificial" bean seeds made by removing seed coats of cultivated beans and grinding the remainder of the seeds into flour. Different concentrations of arcelin were added to different batches, and the flour was molded into artificial seeds. Bean weevils were then allowed to attack the artificial seeds. The more arcelin the artificial seeds contained, the more resistant they were to weevils.

The scientists then proceeded to prepare, clone, and sequence a cDNA (see complementary DNA, Chapter 14) for arcelin so that they could compare the structure of arcelin with that of other, possibly related proteins that may confer insect resistance on seeds of beans and other legumes. The goal of this ongoing work is to introduce genes for arcelin or other resistance-conferring proteins into agriculturally important crops such as beans. In preliminary tests, this group also showed that arcelin in cooked beans is not harmful to rats—a first step toward demonstrating whether arcelin is safe in food for humans.

---

tures and hence reduced biological activities. The defects in protein structure and function lead in turn to developmental abnormalities in the insect.

A few insect larvae are able to eat canavanine-containing plant tissue and still develop normally. How can this be? In these larvae the enzyme that charges the tRNA specific for arginine discriminates accurately between arginine and canavanine. The canavanine they ingest is thus not incorporated into the proteins they form. The corresponding enzyme in the susceptible larvae discriminates much less effectively between those two amino acids, so canavanine is frequently substituted for arginine.

As we have seen, some plants have potent weapons for their struggle with herbivores. Can plants also deal with smaller invaders—the pathogenic fungi, bacteria, and viruses?

PROTECTION AGAINST FUNGI, BACTERIA, AND VIRUSES

Plants resist infection by pathogens by a variety of mechanical and chemical means. The outer surfaces of plants are protected by tissues such as the epidermis or cork, and these tissues are generally covered by cutin, suberin, or waxes. If pathogens pass these barriers, differences between the defense systems of plants and animals (see Chapter 16) become apparent. Animals generally repair tissues that have been infected; they heal, through appropriate developmental pathways. Plants, on the other hand, do not make repairs. Instead, they develop in ways that seal off the damaged tissue so that the rest of the plant

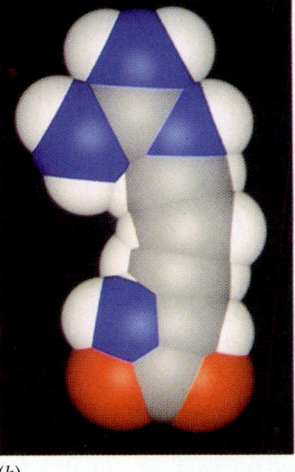

(a)   (b)

31.13 A Toxic Secondary Product and Its Analog
(a) Computer model of canavanine, a nitrogen-storing compound produced by some plants. (b) The amino acid arginine.

BOX 31.B

Take Two Aspirin and Call Me in the Morning

Aspirin—acetylsalicylic acid—is one of the best-selling drugs in the world and has been for many years. People commonly take aspirin to reduce fever and pain. In our bodies aspirin is hydrolyzed to produce the substance that actually causes the effects, salicylic acid. Since ancient times, people in Asia, Europe, and the Americas have used willow leaves and bark to relieve pain and fever. The active ingredient contained in willow (*Salix*) is salicylic acid, and it now appears that all plants contain at least some salicylic acid. This compound appears to play a number of roles in the plants themselves—notably a role in disease resistance.

The acquired resistance that sometimes follows the hypersensitive reaction is accompanied by the synthesis of pathogenesis-related proteins (PR proteins). Treatment of plants with salicylic acid or with aspirin leads to the production of PR proteins and to a resistance to pathogens. Salicylic acid treatment provides substantial protection against tobacco mosaic virus (a well-studied plant pathogen) and some other viruses.

Scientists strongly suspect that salicylic acid serves as a signal for disease resistance—that microbial infection in one part of a plant leads to the export of salicylic acid to other parts of the plant, where it causes the production of PR proteins before the infection can spread. The PR proteins would, according to this hypothesis, limit the extent of the infection.

Although, of course, the "medicinal" effect of salicylic acid in plants is not the same as that of aspirin in humans, temperature changes are part of the response in both. If we have a fever, aspirin lowers our body temperature. In some plant species salicylic acid or aspirin *causes* a "fever." In skunk cabbage, jack-in-the-pulpit, and other members of the family Araceae, for example, production of salicylic acid leads to a dramatic increase in the temperature of part of the plant, and the higher temperature leads to the release of an odorous compound that attracts insect pollinators. In a few plants salicylic acid has yet another effect: It induces flowering. It is not yet known whether these various effects—protection against disease, induction of elevated temperatures, and induction of flowering—are related or entirely unrelated. In any case, salicylic acid plays key roles in the lives of plants.

---

does not become infected. Trees seal off damaged tissue by producing new wood different in orientation and chemical composition from the previously deposited wood. Some of the new cells also contain substances that resist the growth of microorganisms and hence tend to protect the rest of the plant.

Sealing off damaged tissue is primarily a mechanical mechanism for healing. Many plants have chemical defenses as well. Certain fungi and bacteria, when they infect one of these plants, stimulate the host to produce substances called **phytoalexins**. These substances are not present until the plant is infected, but within hours of the onset of infection they are produced in the infected area and in immediately neighboring cells. Phytoalexins are toxic to many fungi and bacteria. Their antimicrobial activity is nonspecific: Phytoalexins can destroy many species of fungi and bacteria in addition to the one that originally triggered their production. Physical injuries, viral infections, and even certain chemicals can also induce the production of phytoalexins.

Some plants that are resistant to fungal, bacterial, or viral diseases owe this resistance to the **hypersensitive reaction**. In this reaction, cells around the site of microbial infection produce phytoalexins and other chemicals and then die, leaving a necrotic lesion—a "dead spot"—that contains what is left of the microbial invasion. The remainder of the plant remains free of the infecting microbe. The hypersensitive reaction can impart long-lasting resistance to subsequent attacks by pathogens. One of the chemicals that may contribute to this long-term disease resistance may surprise you; see Box 31.B.

There has been a great deal of recent excitement about the discovery of a new tool in the fight against fungal, viral, and bacterial pests: plant **disease resistance genes**. Some of these genes, which afford resistance to specific pathogen strains, have been identified and sequenced. The two most important questions about these genes are, how do they work? and how can we best transfer them to crop plants?

As we have seen, many plants produce toxic chemicals that protect them from herbivores and from pathogenic microbes. Why don't these secondary products kill the plants that produce them?

Plants that produce toxic secondary products generally use one of the following measures to protect themselves: (1) keeping the toxic material isolated in a special compartment, (2) producing the toxic substance only after the plant's cells have been damaged, or (3) using modified enzymes or modified receptors that do not recognize the toxic substance. The first method is the most common.

Plants using the first method store their poison in vacuoles if it is water-soluble. If hydrophobic, the poison is stored in laticifers (see the beginning of this chapter) or in waxes on the epidermal surface. The storage keeps the toxic substance away from mitochondria, chloroplasts, and other parts of the plant's own metabolic machinery.

Some plants store the precursors of toxic substances in one compartment, such as the epidermis, and store the enzymes that convert the precursors to the active poison in another compartment, such as the mesophyll. These plants use the second protective measure of producing the toxic substance only after being damaged. When an herbivore chews part of the plant, cells rupture and the enzymes come in contact with the precursors, releasing the toxic product that repels the herbivore. The only part of the plant that is damaged by the toxic material is that which was already damaged by the herbivore. Plants that respond to attack by producing cyanide—a potent inhibitor of cellular respiration in all organisms that respire—are among those using this protective measure.

The third protective measure is used by the canavanine-producing plants described earlier. These plants produce a tRNA-charging enzyme for arginine that does not bind canavanine. Some herbivores can evade being poisoned by canavanine because their enzymes, like that of the canavanine-producing plants, do not use canavanine by mistake.

Not all plants use protective chemicals to defend themselves against herbivores or pathogens. Should we encourage such defenses in the plants that we cultivate? That is, should we be breeding crop plants that make their own pesticides?

A TRADEOFF: PROTECTION OR TASTE?

A plant with sturdy chemical defenses may taste bad, make us sick, or even kill us. Not surprisingly, we have bred our food plants to minimize toxicity and obnoxious tastes—that is, to make the plants contain very little in the way of chemical defenses. As a result, we must take steps to save our relatively defenseless crops. A current goal of agricultural biotechnology is to develop crop plants that produce their own useful pesticides that are not harmful or offensive to us.

SUMMARY of Main Ideas about Environmental Challenges to Plants

Xerophytes are plants adapted to environments where water is scarce.
Review Figures 31.2, 31.3, 31.4, and 31.5

Some plants are adapted to environments that are waterlogged and provide little oxygen.
Review Figures 31.6 and 31.7

Halophytes are plants adapted to saline environments.
Review Figure 31.8

Some plants are resistant to the effects of normally toxic materials such as heavy metals.

Plants produce thousands of secondary products, some of which defend the plants against herbivores.
Review Table 31.1

SELF-QUIZ

1. Which statement about latex is *not* true?
   a. It is sometimes contained in laticifers.
   b. It is typically white in color.
   c. It is often toxic to insects.
   d. It is a rubbery solid.
   e. Milkweeds produce it.

2. Which of the following is *not* an adaptation to dry environments?
   a. A less negative osmotic potential in the vacuoles
   b. Hairy leaves
   c. A heavier cuticle over the leaf epidermis
   d. Sunken stomata
   e. Root systems that grow each rainy season and die back when it is dry

3. Some plants adapted to swampy environments meet their need for oxygen for their roots by means of a specialized tissue called
   a. parenchyma.
   b. aerenchyma.
   c. collenchyma.
   d. sclerenchyma.
   e. chlorenchyma.

4. Halophytes
   a. all accumulate proline in their vacuoles.
   b. have less negative osmotic potentials than other plants.
   c. are often succulent.
   d. have low root-to-shoot ratios.
   e. rarely accumulate sodium.

5. Which of the following is *not* a commonly toxic heavy metal?
   a. Copper
   b. Lead
   c. Nickel
   d. Potassium
   e. Zinc

6. Plants tolerant to heavy metals commonly
   a. grow poorly where the soil contains heavy metals.
   b. do not take up the heavy metal ions.
   c. are tolerant to all heavy metals.
   d. are slow to colonize an area rich in heavy metals.
   e. weigh more than plants that are sensitive to heavy metals.

7. Herbivory
   a. is predation by plants on animals.

   b. always reduces plant growth.
   c. usually increases the rate of photosynthesis in the remaining leaves.
   d. reduces the rate of transport of photosynthetic products from the remaining leaves.
   e. always is lethal to the grazed plant.

8. Which statement is *not* true of secondary plant products?
   a. Some attract pollinators.
   b. Some are poisonous to herbivores.
   c. Most are proteins or nucleic acids.
   d. Most are stored in vacuoles.
   e. Some mimic the hormones of animals.

9. Which of the following is *not* a common defense against bacteria, fungi, and viruses?

   a. New wood formation
   b. Phytoalexins
   c. A waxy covering
   d. The hypersensitive reaction
   e. Mycorrhizae

10. Plants sometimes protect themselves from their own toxic secondary products by
   a. producing special enzymes that destroy the toxic substances.
   b. storing precursors of the toxic substances in one compartment and the enzymes that convert precursors to toxic products in another compartment.
   c. storing the toxic substances in mitochondria or chloroplasts.
   d. distributing the toxic substances to all cells of the plant.
   e. performing crassulacean acid metabolism.

FOR STUDY

1. How might plant adaptations affect the evolution of herbivores? How might adaptations of herbivores affect plant evolution?

2. The stomata of the common oleander (*Nerium oleander*) are sunk in crypts in its leaves. Whether or not you know what an oleander is, you should be able to describe an important feature of its natural habitat.

3. Explain thoroughly why halophytes often use the same mechanisms for coping with their challenging environment as xerophytes do for theirs.

4. In ancient times, people used less sophisticated methods for mining than we use today. Thus ancient mines often yield substantial profits to modern-day miners who find and work them. Based on material in this chapter, how might you try to find an ancient mine site?

5. We mentioned the possibility of designing crop plants that produce their own pesticides. In Chapter 33 you will read about designing crop plants capable of detoxifying weed-killers, so that crops grow after farmers have destroyed competing vegetation. Discuss the likely usefulness and possible drawbacks of such applications of recombinant DNA technology.

READINGS

Barrett, S. C. H. 1987. "Mimicry in Plants." *Scientific American,* September. A discussion of how some plants use camouflage to avoid predation, and some weeds survive by mimicking crops so that humans will select them.

Dussourd, D. E. and T. Eisner. 1987. "Vein-Cutting Behavior: Insect Counterploy to the Latex Defense of Plants." *Science,* vol. 237, pages 898–901. Describes the phenomenon discussed at the beginning of this chapter. Clear, readable account of field observations and sound experimentation.

Goulding, M. 1993. "Flooded Forests of the Amazon." *Scientific American,* March. Describes the adaptations of plants and other organisms in a seriously threatened environment.

Lewin, R. 1987. "On the Benefits of Being Eaten." *Science,* vol. 236, pages 519–520. Describes the work on scarlet gilia cited (under "Grazing and Plant Productivity") in this chapter. Discusses other examples of cropping.

Rosenthal, G. A. 1986. "The Chemical Defenses of Higher Plants." *Scientific American,* January. Plants employ many chemicals that repel or poison

herbivores or that retard the growth of herbivorous insects; some herbivores use these plant-derived compounds for their own defense.

Shigo, A. L. 1985. "Compartmentalization of Decay in Trees." *Scientific American,* April. A description of how trees defend themselves by sealing off the damage done to them.

Taiz, L. and E. Zeiger. 1991. *Plant Physiology.* Benjamin/Cummings, Redwood City, CA. A good general textbook. Chapters 13 and 14 focus on the topics of this chapter.

Pitcher plants live in wet, boggy regions with acidic soil. Their pitcher-shaped leaves collect small amounts of rainwater. Insects are attracted into these pitchers either by bright colors or by scent. Once an insect enters the pitcher, the stiff, downward-pointing hairs that line the plant prevent it from ever leaving. The insect eventually dies and is digested by a combination of enzymes and bacteria in the water; the plant uses the nutrients—especially nitrogen—from the insect protein to thrive in an environment where it is very difficult to obtain enough nitrogen from the soil.

Why do plants need nitrogen? The answer is simple if we recall the chemical structures of amino acids—and hence proteins—and nucleic acids. These vital components of all living things contain nitrogen, as do chlorophyll and many other important biochemical compounds. If a plant cannot get enough nitrogen, it cannot synthesize these compounds at a rate adequate to keep itself healthy.

Nitrogen deficiency is the most common mineral deficiency of plants; the visible symptoms of nitrogen deficiency include uniform yellowing, or chlorosis, of leaves because chlorophyll, which is responsible for the green color of leaves, contains nitrogen. Thus without nitrogen there is no chlorophyll, and without chlorophyll the leaves turn yellow.

ACQUIRING NUTRIENTS

Every living thing needs raw materials from its environment. These **nutrients** include the ingredients of macromolecules: carbon, hydrogen, oxygen, and nitrogen. Carbon and oxygen enter the living world through photosynthesis carried out by plants and by some bacteria and protists; these organisms obtain carbon and oxygen from atmospheric carbon dioxide. The principal source of hydrogen is water, usually taken up from the soil solution by plants. For hydrogen, too, photosynthesis is the gateway to the living world. Nitrogen, which constitutes about four-fifths of the atmosphere, exists as the virtually inert gas N_2 (dinitrogen). A large amount of energy is required to break the triple covalent bond linking the two nitrogen atoms in a molecule of nitrogen gas and to obtain a reasonably reactive form from which amino acids and other nitrogen-containing organic compounds may be synthesized. Movement of dinitrogen into organisms begins with processing by some highly specialized bacteria in the soil. The bacteria fix (convert into a form usable by other organisms) and oxidize dinitrogen, yielding materials that can be taken up by plants. The plants, in turn, provide organic nitrogen and carbon to animals, fungi, and many microorganisms.

A Meat-Eating Plant

32

Plant Nutrition

The proteins of organisms contain sulfur, and their nucleic acids contain phosphorus. There is magnesium in chlorophyll, and iron in many important compounds, such as the cytochromes. Within the soil, minerals dissolve in water, forming a solution that contacts the roots of plants. Plants take up most of these **mineral nutrients** from the soil solution in ionic form.

Autotrophs and Heterotrophs

The plant kingdom provides carbon, oxygen, hydrogen, and nitrogen to the rest of the living world. Plants and some protists and bacteria are autotrophs; that is, they make their own organic food from simple *inorganic* nutrients—carbon dioxide, water, nitrate or ammonium ions containing nitrogen, and a few soluble minerals (Figure 32.1). Organisms that require at least one of their raw materials in the form of *organic* compounds are called heterotrophs; herbivores depend directly and carnivores depend indirectly on autotrophs as their source of nutrition.

Most autotrophs are **photosynthetic;** that is, they use light as the source of energy for synthesizing organic compounds from inorganic raw materials. Some autotrophs, however, are **chemosynthetic,** deriving their energy not from light but from reduced inorganic substances such as hydrogen sulfide (H_2S) in their environment. All chemosynthesizers are bacteria. The activities of chemosynthetic bacteria that fix nitrogen are vital to the nutrition of plants. But how does a plant get to the bacterial products—or to its nutrients in general?

How Does a Sessile Organism Find Nutrients?

An organism that is sessile (stationary) must exploit energy that is somehow brought to it. Most sessile animals depend primarily upon the movement of water to bring energy in the form of food to them, but a plant's supply of energy arrives at the speed of light! A plant's supply of essential materials, however, is strictly local, and the plant may deplete its local environment of water and minerals as it develops. How does a plant cope with such a problem? One answer is to extend itself by growing into new resources—growth is a plant's version of locomotion. Roots mine the soil. They grow to reach new sources of minerals and become more elaborate to obtain more water. Growth of leaves helps a plant secure light and carbon dioxide. A plant may compete with other plants for light by outgrowing them, both capturing more light for itself and also preventing the growth of its neighbors by shading them.

As it grows, the plant, or even a single root, must deal with environmental diversity. Animal droppings give high local concentrations of nitrogen. A particle

32.1   What Do Plants Need?
Plants require only light plus carbon dioxide, water, nitrate or ammonium ions, and several essential minerals. These parsley plants are growing on nothing more than a solution containing these ingredients. This is hydroponic culture.

of calcium carbonate in the soil may make a tiny area alkaline, while dead organic matter may make a nearby area acidic. Does the root take up whatever materials it encounters?

Bulk Ingestion versus Selective Uptake of Nutrients

Animals ingest their meals, taking in unneeded and sometimes toxic materials along with needed nutrients; what an animal ingests is not determined by its needs. Part of what animals ingest must eventually be disposed of as waste products, such as urea. Plants, by contrast, do not urinate or produce wastes in other obvious ways. Instead, they control their uptake of most substances, matching the uptake rates to their biochemical needs. The major waste products released to the environment by plants are carbon dioxide or, during active photosynthesis, oxygen gas.

TABLE 32.1
Elements Required by Plants

ELEMENT	SOURCE	ABSORBED FORM	MAJOR FUNCTIONS
Nonmineral elements			
Carbon (C)	Atmosphere	CO_2	In all organic molecules
Oxygen (O)	Atmosphere	CO_2	In most organic molecules
Hydrogen (H)	Soil	H_2O	In most organic molecules
Nitrogen (N)	Soil	NH_4^+ and NO_3^-	In proteins, nucleic acids, etc.
Mineral nutrients			
Macronutrients			
Phosphorus (P)	Soil	$H_2PO_4^-$	In nucleic acids, ATP, phospholipids, etc.
Potassium (K)	Soil	K^+	Enzyme activation; water balance; ion balance
Sulfur (S)	Soil	SO_4^{2-}	In proteins, coenzymes
Calcium (Ca)	Soil	Ca^{2+}	Affects the cytoskeleton, membranes, and many enzymes; second messenger
Magnesium (Mg)	Soil	Mg^{2+}	In chlorophyll; required by many enzymes; stabilizes ribosomes
Micronutrients			
Iron (Fe)	Soil	Fe^{3+}	In active site of many redox enzymes and electron carriers; needed for chlorophyll synthesis
Chlorine (Cl)	Soil	Cl^-	Photosynthesis; ion balance
Manganese (Mn)	Soil	Mn^{2+}	Activates many enzymes
Boron (B)	Soil	$H_2BO_3^-$, HBO_3^{2-}	May be needed for carbohydrate transport (poorly understood)
Zinc (Zn)	Soil	Zn^{2+}	Enzyme activation; auxin synthesis
Copper (Cu)	Soil	Cu^{2+}	In active site of many redox enzymes and electron carriers
Molybdenum (Mo)	Soil	MoO_4^{3-}	Nitrogen fixation; nitrate reduction

This is not to say that plants do not take up toxic substances from the soil. They do, at times, as discussed in Chapter 31. However, plants exert a more systematic control over what can enter their bodies than do animals. What important minerals do get admitted, and what are their roles?

WHICH NUTRIENTS ARE ESSENTIAL?

Table 32.1 lists the mineral elements essential for plants. They all come from the soil solution and derive ultimately from rock. The criteria for calling something an **essential element** are the following: (1) The element must be necessary for normal growth and reproduction. (2) The element cannot be replaceable by another element. (3) The requirement must be direct—that is, not the result of an indirect effect, such as the need to relieve toxicity caused by another substance.

Before a plant that is deficient in an essential element dies, it usually displays characteristic **deficiency symptoms.** Table 32.2 describes the symptoms of some common mineral deficiencies, and the opening figure in this chapter shows an example. Such symptoms help horticulturists diagnose nutrient deficiencies in plants.

Several essential elements fulfill multiple roles—some structural, others catalytic. Magnesium, as we have mentioned, is a constituent of the chlorophyll molecule and hence is essential to photosynthesis. It is also required as a cofactor by numerous enzymes in cellular respiration and other metabolic pathways. Iron is a constituent of many molecules, including some proteins, that participate in oxidation–reduction reactions. Phosphorus, usually in phosphate groups, is found in many compounds, particularly in pathways of energy metabolism such as photosynthesis and glycolysis. The transfer of phosphate groups is important in many energy-storing and energy-releasing reactions, notably those that use or produce ATP. Other roles of phosphate groups include the activation and inactivation of enzymes.

Plant tissues contain high concentrations of potassium, which plays a major role in maintaining electric neutrality of cells. Potassium ions (K^+) balance the negative charges of ionized carboxyl groups (—COO^-) of organic acids. Potassium also helps move water from cell to cell. There are no "pumps" for the active transport of water, yet water must be moved from place to place—for example, into and out of the guard cells surrounding stomata (see Chapter 30). Plants and animals achieve water movement by actively transporting K^+ from one cell to

TABLE 32.2
Some Mineral Deficiencies

DEFICIENCY	SYMPTOMS
Calcium	Growing points die back; young leaves are yellow and crinkly
Iron	Young leaves are white or yellow with green veins
Magnesium	Older leaves have yellow in stripes between veins
Manganese	Younger leaves are pale with stripes of dead patches
Nitrogen	Oldest leaves turn yellow and die prematurely; plant is stunted
Phosphorus	Plant is dark green to purple and stunted
Potassium	Older leaves have dead edges
Sulfur	Young leaves are yellow to white with yellow veins
Zinc	Older leaves have many dead spots

thought they had excluded. Some minerals are required in such tiny amounts that there may be enough in a seed to feed the embryo and the resultant plant throughout its entire lifetime and leave enough in the next seed to get the next generation well started. There was enough chloride on dust particles and water droplets in the air in Berkeley, California (where some of the work on essential elements was performed), for example, to provide the infinitesimal amounts needed to keep experimental plants growing. The essentiality of chlorine thus was not established until 1954, after special air filters had been installed in the laboratory. Simply touching a plant may give it a significant dose of chlorine in the form of chloride ions from sweat.

Only rarely are new essential elements reported now; either the list is virtually complete, or more

another. Chloride ions (Cl^-) follow the K^+ passively, maintaining electrical balance. Movement of these ions changes the water potential of the cells, and water then moves passively to maintain osmotic balance.

Calcium plays many roles in plants. Its function in the processing of hormonal and environmental cues is the subject of great current interest (the analogous function of Ca^{2+} in animal cells is discussed in Chapter 36). Calcium also affects membranes and cytoskeleton activity, participates in spindle formation for mitosis and meiosis, and is a constituent of the middle lamella of cell walls.

The essential minerals in Table 32.1 are divided into two categories: the macronutrients and the micronutrients. Plant tissues need **macronutrients** in concentrations of at least 1 milligram per gram of their dry matter, and they need **micronutrients** in concentrations of less than 100 micrograms per gram of their dry matter. (Dry matter, or dry weight, is what remains after all the water has been removed from a tissue sample.) The essential elements differ in that some, such as nitrogen, may move around within the plant, while others, such as iron, are not redistributed. *All* these elements are essential to the life of all plants. Other elements may be essential to some plants—and perhaps to all.

Plant physiologists identified most of the essential elements by the technique outlined in Figure 32.2. An element is considered essential if a plant does not grow, flower, or produce viable seed when deprived of that element. The technique is limited by the possibility that some elements thought to be absent from the solutions are present. Some of the chemicals used in early experiments were so impure that they provided micronutrients that the first investigators

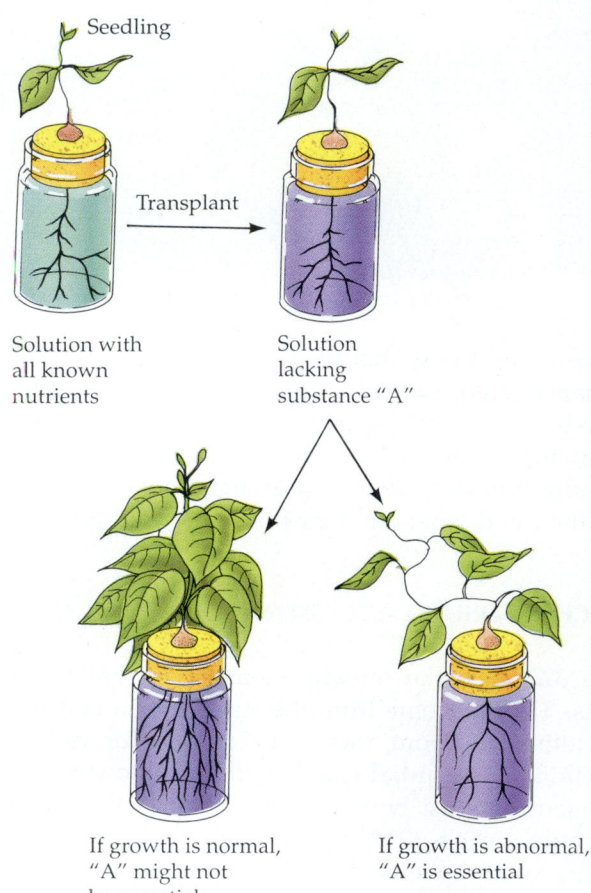

32.2 Identifying Essential Plant Nutrients
The plant physiologist transplants a seedling to a solution lacking only one of the ingredients thought to be essential for growth (substance "A" in this example). If the plant grows and reproduces normally after being transplanted, the missing ingredient is assumed to be nonessential. The experimental environment must be rigorously controlled because some essential nutrients are needed in only tiny amounts that may be present as contaminants of other materials or on objects.

likely, we will need more sophisticated techniques to find others. A new essential element likely to be announced soon—the first since chlorine—is nickel, which was shown in 1984 to be essential for legumes. Some minerals are essential for certain plants but apparently not for others. Continued research may show that only legumes need nickel or that all plants do.

Where does the plant find these essential minerals? How does the plant get the environment to yield the minerals to it?

TABLE 32.3
Soil Particles

SOIL TYPE	PARTICLE SIZE (mm)
Coarse sand	0.2–2.0
Fine sand	0.02–0.2
Silt	0.002–0.02
Clay	<0.002

SOILS

Soils are of great importance to plants, and plant interactions with the soil are complex. Plants obtain their mineral nutrients from the soil or the water in which they grow. Water for terrestrial plants also comes from the soil, as does the supply of oxygen for the roots; soil also provides mechanical support for plants on land. Soil harbors bacteria that perform chemical reactions leading to products required for plant growth; on the other hand, soil may also contain organisms harmful to plants.

Soils have living and nonliving components. The living components include plant roots, as well as populations of bacteria and fungi (Figure 32.3). The nonliving portion of the soil includes rock fragments ranging in size from large boulders to particles 2 μm and less in diameter (Table 32.3). As we will see in the next section, the **clay particles** play a special role in plant nutrition. Soils also contain minerals, water, gases, and organic matter from animals, plants, fungi, and bacteria. Soils change constantly because of both nonhuman natural causes—such as rain, high

and low temperatures, and the activities of plants and animals—and human activities, farming in particular. Soils from different parts of the world differ dramatically in their chemical composition and physical structure because the temperature, water supply, and other factors during their formation differ from place to place.

The structure of any soil changes with depth, revealing a soil profile. Although soils differ greatly, virtually all soils consist of two or more **horizons**—recognizable horizontal layers—lying on top of one another (Figure 32.4). Mineral nutrients tend to be leached—that is, dissolved in rain or irrigation water and carried to deeper horizons. Other processes also move materials down—or up—in the soil. Soil sci-

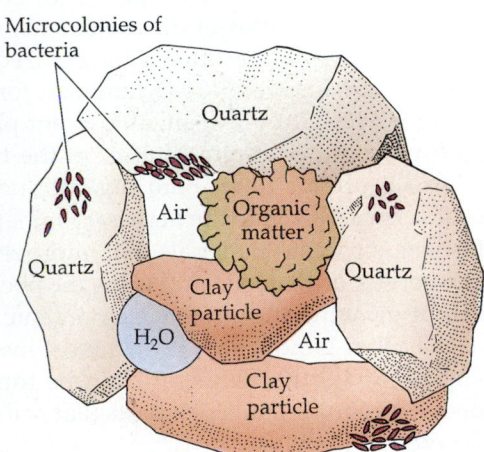

32.3 The Complexity of Soil
Soil consists of more than inorganic particles such as clay and quartz. It contains living organisms such as the bacteria shown here. Air and water are present in pores in soil crumbs like this one.

32.4 A Soil's Profile
The A, B, and C horizons can sometimes be seen in road cuts such as this one in Australia. The upper layers developed from the bedrock. The dark upper layer is home to most of the living organisms in the soil.

entists recognize three major zones in the profile of a typical soil. The A horizon is the zone from which minerals have been depleted by leaching. Most of the organic matter in the soil is in the A horizon, as are most roots, earthworms, soil insects, nematodes, and soil protists. Successful agriculture depends on the presence of a suitable A horizon. The B horizon is the zone of infiltration and accumulation of materials leached from above, and the C horizon is the original parent material from which the soil is derived. Some deep-growing roots extend into the B horizon, but roots rarely enter the C horizon.

Soils and Plant Nutrition

The supply of minerals to plants depends upon the presence of clay particles, which have a net negative charge. Many of the minerals that are important for plant nutrition, such as potassium, magnesium, and calcium, exist in soil as positive ions chemically attached to clay particles. To become available to plants, the positive ions must be detached from the clay particles, and this is accomplished by reactions with protons (hydrogen ions, H^+), which are released into the soil by roots or by the ionization of carbonic acid (H_2CO_3). (Carbonic acid is almost universally present in soils because it forms whenever CO_2 from respiring roots or from the atmosphere dissolves in water, according to the reaction $CO_2 + H_2O \rightarrow H_2CO_3$.)

The clay particles get their net negative charge from negatively charged ions that are permanently attached to them. Positively charged ions in solution associate reversibly with these attached negative ions. Protons then trade places with ions such as potassium (K^+) on the clay particles, thus putting the nutrients back into the soil solution. This trading of places is called **ion exchange** (Figure 32.5). The fertility of a soil is determined primarily by its ability to provide nutrients such as potassium, magnesium, and calcium in this manner.

Clay particles effectively hold and exchange positively charged ions, but there is no comparable exchanger of negatively charged ions. As a result, important negative ions such as phosphate, nitrate, and sulfate—the primary sources of phosphorus, nitrogen, and sulfur, respectively—leach rapidly from soil, whereas positive ions tend to be retained in the A horizon.

Fertilizers and Lime

Agricultural soils often require fertilizers because irrigation leaches mineral nutrients from the soil and the harvesting of crops removes the nutrients that the plants took up from the soil during growth. Crop yields fall if too much of any element is removed.

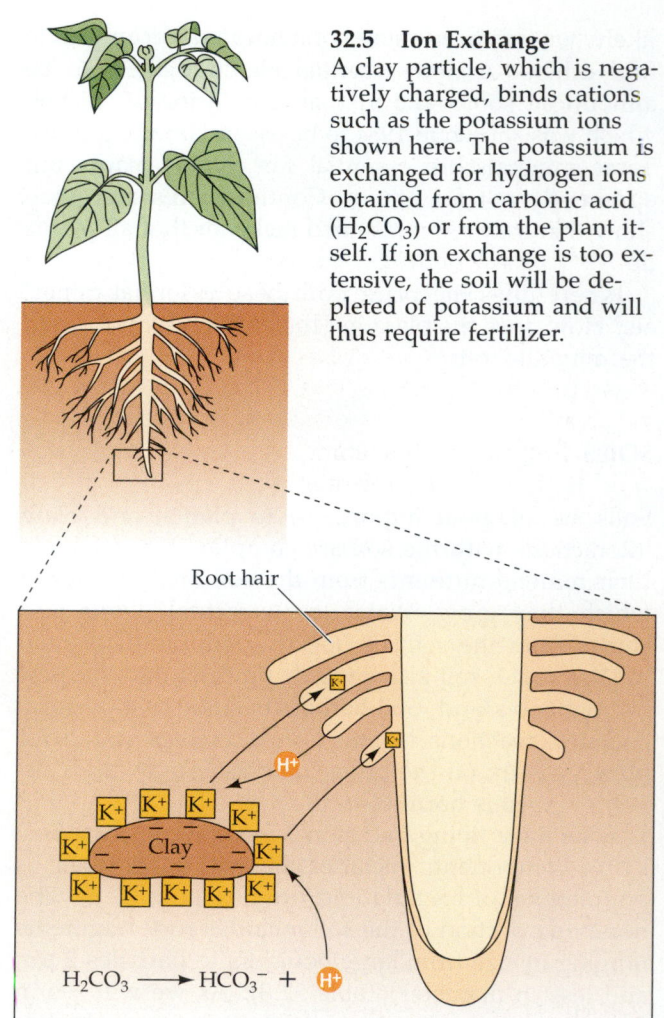

32.5 Ion Exchange
A clay particle, which is negatively charged, binds cations such as the potassium ions shown here. The potassium is exchanged for hydrogen ions obtained from carbonic acid (H_2CO_3) or from the plant itself. If ion exchange is too extensive, the soil will be depleted of potassium and will thus require fertilizer.

Minerals may be replaced by organic fertilizers, such as rotted manure, or inorganic fertilizers of various types. The three elements most commonly added to agricultural soils are nitrogen (N), phosphorus (P), and potassium (K). The ratios of these elements vary among fertilizers, which are often characterized by their N–P–K percentages. A 5–10–5 fertilizer, for example, contains 5 percent nitrogen, 10 percent phosphate, and 5 percent potassium. Sulfur, in the form of sulfate, also is often added. Both organic and inorganic fertilizers can provide the necessary minerals. Organic fertilizers contain materials that improve the physical properties of the soil, providing air pockets for gases, root growth, and drainage. Inorganic fertilizers, on the other hand, provide an almost instantaneous supply of soil nutrients and can be formulated to meet the requirements of a particular soil and a particular crop.

The availability of nutrient ions, whether naturally present in the soil or added as fertilizer, is altered by changes in soil pH. Rainfall and the decomposition of organic substances in the soil lower the pH. Such acidification of the soil can be reversed by **liming**—

the application of calcium-containing material (usually calcium carbonate or calcium hydroxide). This practice is older than agricultural history. It is easy to guess how we learned the use of fertilizer; it didn't take much insight to notice improved plant growth around animal feces. Perhaps a similar observation about limestone, or chalk, or oyster shells—all sources of calcium—led to the practice of liming. Whatever its ancient source, the addition of Ca^{2+} allows H^+ to be released from soil particles by ion exchange and leached away, raising the soil pH (see Chapter 2 for a review of pH). Liming also increases the availability of Ca^{2+} to plants, which require it as a macronutrient. Sometimes a soil is not acidic enough; in this case, a farmer can add sulfate ions to decrease the pH. Iron and some other elements are more available at a slightly acid pH.

Spraying leaves with a nutrient solution is another effective way to deliver some essential elements to growing plants. Plants take up more copper, iron, and manganese when these elements are applied as foliar (leaf) sprays rather than as soil fertilizer. By adjusting the concentrations of nutrient ions and the pH to optimize uptake and to minimize toxicity and by controlling time of spraying to avoid "burning," one can achieve excellent results. Foliar application of mineral nutrients is increasingly used in wheat production, but fertilizer is still delivered most commonly by way of the soil.

Soil Formation

The type of soil in a given area depends on the rock from which it formed, the climate, the topography (features of the landscape), the organisms living there, and the length of time that soil-forming processes have been acting. Rocks are broken down in part by **mechanical weathering,** the physical breakdown—with no accompanying chemical changes—of materials by wetting, drying, and freezing. The most important parts of soil formation, however, include **chemical weathering,** the chemical alteration of at least some of the materials in the rocks. The key process is the formation of clay. Both the physical and the chemical properties of soils depend on the amount and kind of clay particles they contain. Just grinding up rocks does not produce a clay that swells and shrinks and is chemically active. The rock must be chemically changed as well. The initial step in the chemical weathering of most soil minerals is hydrolysis, as illustrated for feldspar, a common soil mineral, in the following formula:

$$\langle Si, Al, O \rangle K^+ + H^+OH^- \rightarrow \langle Si, Al, O \rangle H^+ + K^+OH^-$$

| FELDSPAR | WATER | HYDROLYZED FELDSPAR | POTASSIUM HYDROXIDE |

Two examples illustrate the diversity of soil-forming processes. In wet tropical regions, where rainfall and temperatures are high and the soils are usually moist, water moves rapidly downward through the soil. Silica and soluble nutrients are quickly leached, leaving insoluble iron and aluminum compounds in the A horizon. These are often oxidized, giving bright reddish colors to the soils. This type of soil formation is known as **laterization.** The resulting soil is very poor in nutrients. In semiarid regions, where water from rainfall evaporates rapidly, there is no net movement of water downward through the soil. Instead, water penetrates for a distance, stops, and then is drawn back up by the roots of plants and by evaporation from the soil surface. Under these conditions the soil remains rich in mineral nutrients, and a hard layer of calcium carbonate often forms in the B horizon. The B horizon may be so hard that it prevents deeper penetration of the soil by plant roots. These soils are very fertile, however, when supplied with additional water and nitrogen. Much of the success of irrigated agriculture depends on the high nutrient content of arid-zone soils.

Effects of Plants on Soils

How soil forms in a particular place also depends on the types of plants growing there. Plant litter is a major source of carbon-rich materials that break down to form **humus**—dark-colored organic material, each particle of which is too small to be recognizable with the naked eye.

Soils rich in exchangeable positive ions of mineral nutrients tend to support plants that extract large quantities of nutrients for incorporation into their tissues. When these tissues die and decompose, they produce a rich, alkaline humus called mull. Plants growing on nutrient-poor soils extract fewer nutrients and form tissues that yield a poor, acidic humus known as mor. The mor produced by conifers is particularly resistant to decay and may accumulate in thick layers on the surface of the soil.

Soils age. Young soils support rapidly increasing amounts of vegetation (biomass) that contribute materials for humus, which thus increases at the same rapid rate; the green and yellow lines in Figure 32.6 plot these increases for a hypothetical example. The increase in biomass comes at the expense of nutrients, which decline rapidly, as exemplified by the drop in calcium carbonate in Figure 32.6. The fraction of clay in the soil gradually increases, decreasing the availability of water for plant growth. The loss of mineral nutrients from the soil also contributes to the long-term decline in biomass.

Biologists were slow to recognize the importance of the long-term changes in soils because nearly all the soils of the northern temperate zone, where most

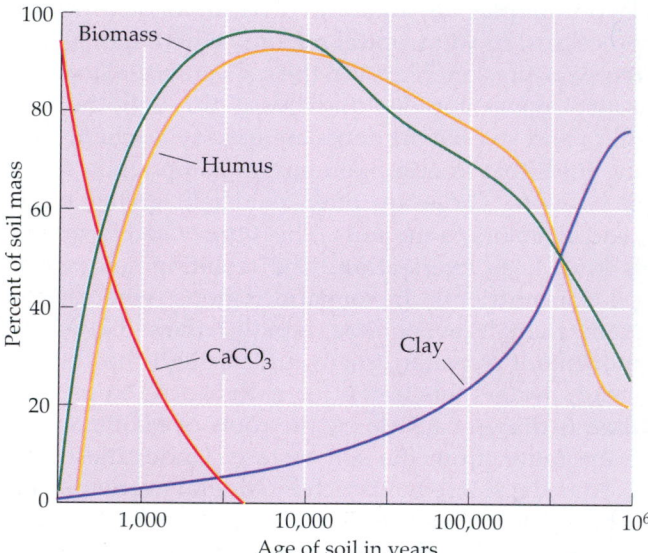

32.6   Aging of Soils
Weathering processes gradually remove most exchangeable nutrient ions from a soil, leaving an infertile residue. The pattern plotted for calcium carbonate ($CaCO_3$) is typical of many soil nutrients. In this hypothetical example, the rich humus and the biomass it supports decrease over time while the fraction of clay in the soil increases.

soil scientists live and work, are very young, dating from the last glacial period only a few thousand years ago. In large areas of the tropics and subtropics, however, especially away from areas of recent mountain formation, soils are ancient and have few remaining nutrients. The right-hand edge of Figure 32.6 suggests the makeup of such soils, which are unsuitable for agriculture unless heavily fertilized.

Nitrogen is the essential element most often required as fertilizer. In the next few sections we will look at how nitrogen is made available and at how certain soil bacteria participate in this process.

NITROGEN FIXATION

Plants cannot use nitrogen gas (dinitrogen: N_2) directly as a nutrient. Making nitrogen from N_2 available to plants takes a great deal of energy because N_2 is a highly unreactive substance. A few species of bacteria can convert it into a more useful form. These prokaryotic organisms, the nitrogen fixers, convert N_2 to ammonia (NH_3). Some of them must live in intimate association with specific eukaryotes before they develop functional nitrogen-fixing machinery. There are relatively few kinds of nitrogen fixers, and what few there are have a small biomass relative to the mass of other organisms on Earth. Without the nitrogen fixers, however, other organisms would not survive! This elite group of prokaryotes is just as essential in the biosphere as are the photosynthetic autotrophs.

Organisms fix approximately 90 million tons of atmospheric dinitrogen per year. Tens of millions of tons are fixed industrially, by a method called the Haber process. A smaller amount of nitrogen is fixed in the atmosphere by nonbiological means such as lightning, volcanic eruption, and forest fires; the products thus formed are brought to Earth by rainwater. By far the greatest share of total world nitrogen fixation, however, is that performed biologically by nitrogen-fixing organisms.

Nitrogen-fixing species are widely scattered in the kingdom Eubacteria. One group of microorganisms fixes nitrogen only in close association with the roots of certain seed plants; the best known of these bacteria belong to the genus *Rhizobium*. Some *Rhizobium* species live free in the soil, where they do not fix nitrogen. Others live in nodules on the roots of plants in the legume family, which includes peas, soybeans, alfalfa, and many tropical shrubs and trees (Figure 32.7). These nodule-inhabiting *Rhizobium* species do fix nitrogen. Some cyanobacteria fix nitrogen in association with fungi in lichens or with ferns, cycads, or bryophytes. Finally, the filamentous bacteria called actinomycetes fix nitrogen in association with root nodules on shrub species such as alder. How were the nitrogen-fixing roles of bacteria discovered?

The ancient Chinese, Greeks, and Romans, and probably members of other early civilizations, recognized that plants such as clover, alfalfa, and peas improve the soil in which they are grown. Two Ger-

32.7   Root Nodules
These large, round, tumorlike nodules are developing from a yellow wax bean root—the structure to the right. The nodules house nitrogen-fixing bacteria.

32.8 Equipped to Colonize
A retreating glacier in Alaska left this area of bare rock exposed. *Dryas drummondii*, the low-lying plant seen here in bloom, was one of the first plants to colonize the bare area. It and other rapid colonizers growing here share the characteristic of having nitrogen-fixing root nodules.

man chemists, Hellriegel and Wilfarth, first showed in 1888 that the root nodules on these plants are caused by bacteria and are sites of nitrogen fixation. These particular plant-infecting bacteria all belong to the genus *Rhizobium*, and the various species of *Rhizobium* show a fairly high specificity for the species of legume they nodulate.

In the oceans, various photosynthetic bacteria, including cyanobacteria, fix nitrogen; in fresh water, cyanobacteria are the principal nitrogen fixers. On land, free-living soil bacteria make some contribution to nitrogen fixation, but it is bacteria in the root nodules of plants that produce most of the fixed nitrogen. Unlike various free-living nitrogen fixers that fix what they need for their own uses and release the fixed nitrogen only upon their deaths, bacteria in root nodules release up to 90 percent of the nitrogen they fix to the plant and excrete some amino acids into the

soil, making nitrogen immediately available to other organisms. Some farmers alternate their crops, planting clover or alfalfa occasionally to increase the useful nitrogen content of the soil.

Root nodules permit some *non*leguminous plants to be pioneers, to occupy environments having few or no other plants (Figure 32.8; see also Figure 48.25). Shrubs such as alder thrive in mountainous areas, with their roots grasping chunks of rock in the talus (debris) slopes below the cliffs. The western mountain lilac *Ceanothus* grows well in extremely gravelly soils that have little or no organic matter and hence no fixed nitrogen. Eastern sweet gale *Myrica* flourishes on almost pure sand; its growth is dependent on nitrogen from nodules. Pioneer plants, with their bacterial partners, make initially barren habitats available to other plants and to the animals that depend on them.

How does biological nitrogen fixation work?

Chemistry of Nitrogen Fixation

Nitrogen fixation progressively reduces the dinitrogen molecule by adding pairs of hydrogen atoms to cleave the three bonds between the nitrogen atoms:

$$N \equiv N \xrightarrow{2H} HN = NH \xrightarrow{2H} H_2N - NH_2 \xrightarrow{2H} 2\ NH_3$$
DINITROGEN   AMMONIA

Throughout this series of reactions, the reactants are firmly bound to the surface of a single enzyme called **nitrogenase** (Figure 32.9). The reactions require a strong reducing agent (see Chapter 7) to transfer hydrogen atoms to dinitrogen and the intermediate products, as well as a great deal of energy, which is supplied by ATP. Depending on the species of nitrogen fixer, either respiration or photosynthesis may provide the necessary reducing agent and ATP.

Nitrogenase is extremely sensitive to oxygen—so

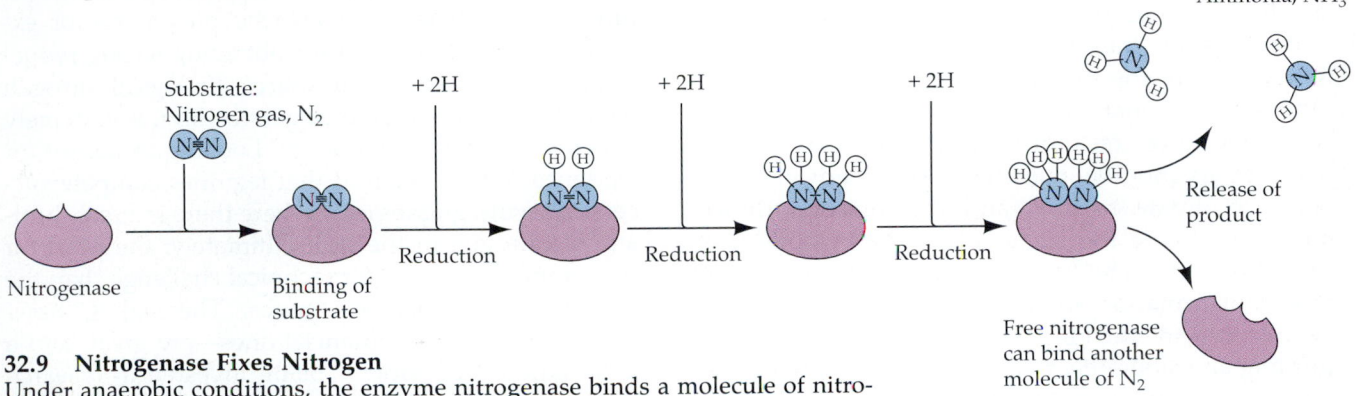

32.9 Nitrogenase Fixes Nitrogen
Under anaerobic conditions, the enzyme nitrogenase binds a molecule of nitrogen gas, and the nitrogen is reduced by the addition of three successive pairs of hydrogen atoms. The final products—two molecules of ammonia—are released, freeing the nitrogenase to bind another dinitrogen molecule.

much so that its discovery was delayed because investigators had not thought to seek it under anaerobic conditions, which are inconvenient to establish in the laboratory. Because nitrogenase cannot function in the presence of oxygen, it is not surprising that many nitrogen fixers are anaerobes. Free-living *Rhizobium* species respire aerobically, but they do not fix nitrogen under these conditions. Legumes respire aerobically, but their *Rhizobium*-containing, nitrogen-fixing root nodules maintain an anaerobic internal environment. Aerobic nitrogen fixers must decrease their internal oxygen levels drastically in order for the process to work. One means for doing this—the production of a special type of hemoglobin—will be described in the next section.

Symbiotic Nitrogen Fixation

The legume nodule provides an excellent example of symbiosis, in which two different organisms live in physical contact and, in association, do things that neither organism can do separately. In the form of symbiosis called mutualism, both organisms benefit from the relationship. Neither free-living *Rhizobium* species nor uninfected legumes can fix nitrogen. Only when the two are closely associated in root nodules does the reaction take place.

The establishment of this symbiosis between *Rhizobium* and a legume requires a complex series of steps with active contributions by both the bacteria and the root. First the root releases chemical signals that attract the *Rhizobium* to the vicinity of the root. The bacteria, in turn, produce substances that cause cell divisions in the root cortex, leading to the formation of a primary nodule meristem. Next comes the infection of the plant by the bacteria. The bacteria first attach to root hairs that project from the epidermal cells and then produce one or more growth substances that cause the cell walls of the root hairs to invaginate—fold inward. With help from the Golgi apparatus in the cells, the invagination proceeds inward through several cells as an infection thread; the bacteria in the thread continue to divide, although slowly (Figure 32.10).

At this stage the bacteria are still outside the plant in a sense, for the thread is lined with cellulose and other cell-wall materials. The thread grows into the cortex tissue of the root until it encounters cells in the primary nodule meristem. The meristem cells begin to divide rapidly, and the infection thread bursts, releasing the bacteria into the cytoplasm of these host cells. The bacteria now undergo a remarkable transformation, increasing about tenfold in size, developing an outside membranous envelope, and forming an elaborately folded internal membrane. At this stage the infecting bacteria are called bacteroids. The bacteroids are, in effect, organelles for nitrogen fixation; it has been suggested that, in the course of further evolution, such bacteroids may in the future give rise to permanent nitrogen-fixing organelles in some plant species.

The final step before the fixation of nitrogen can begin is that the plant produces leghemoglobin, which surrounds the bacteroids. Hemoglobin is an oxygen-carrying pigment that one seldom associates with plants, but some nodules contain enough of its close relative leghemoglobin to be bright pink when viewed in cross section. The leghemoglobin traps oxygen, keeping it away from the oxygen-sensitive bacteroids and nitrogenase.

The Need to Augment Biological Nitrogen Fixation

Bacterial nitrogen fixation does not suffice to support the needs of agriculture. Native Americans used to plant dead fish along with corn so that the decaying fish would release fixed nitrogen that the developing corn could use. Industrial nitrogen fixation is becoming ever more important to world agriculture because of the degradation of soils and the need to feed a rapidly expanding population. Research on biological nitrogen fixation is being vigorously pursued, with commercial applications very much in mind (for an example see Box 32.A).

An alternative to the current process for industrial nitrogen fixation is urgently needed because of the cost of energy and other economic factors associated with it. At present, manufacturing nitrogen-containing fertilizer takes more energy than any other aspect of crop production in the United States. One line of investigation centers on recombinant DNA technology as a means of "teaching" new plants to produce nitrogenase. Workers in many industrial and academic laboratories are working on the insertion of bacterial genes coding for nitrogenase into plasmids, and on the incorporation of such plasmids into the cells of angiosperms, particularly crop plants. However, developing crops that can fix their own nitrogen will take more than just the insertion of genes for nitrogenase; there must also be provisions for excluding free oxygen and for obtaining strong reducing agents and an energy source. Biological nitrogen fixation, like industrial nitrogen fixation, is extremely expensive in terms of energy. Looking to nature for evidence of this, we find that legumes compete successfully with grasses only where there is a real shortage of nitrogen in the soil. Ultimately, the need for ATP represents a greater technical challenge than the insertion of nitrogenase genes. The stakes, however—especially the financial ones—are great, and a great amount of effort is being invested in research along these lines. This is, in fact, one of the busiest areas in the burgeoning field of biotechnology.

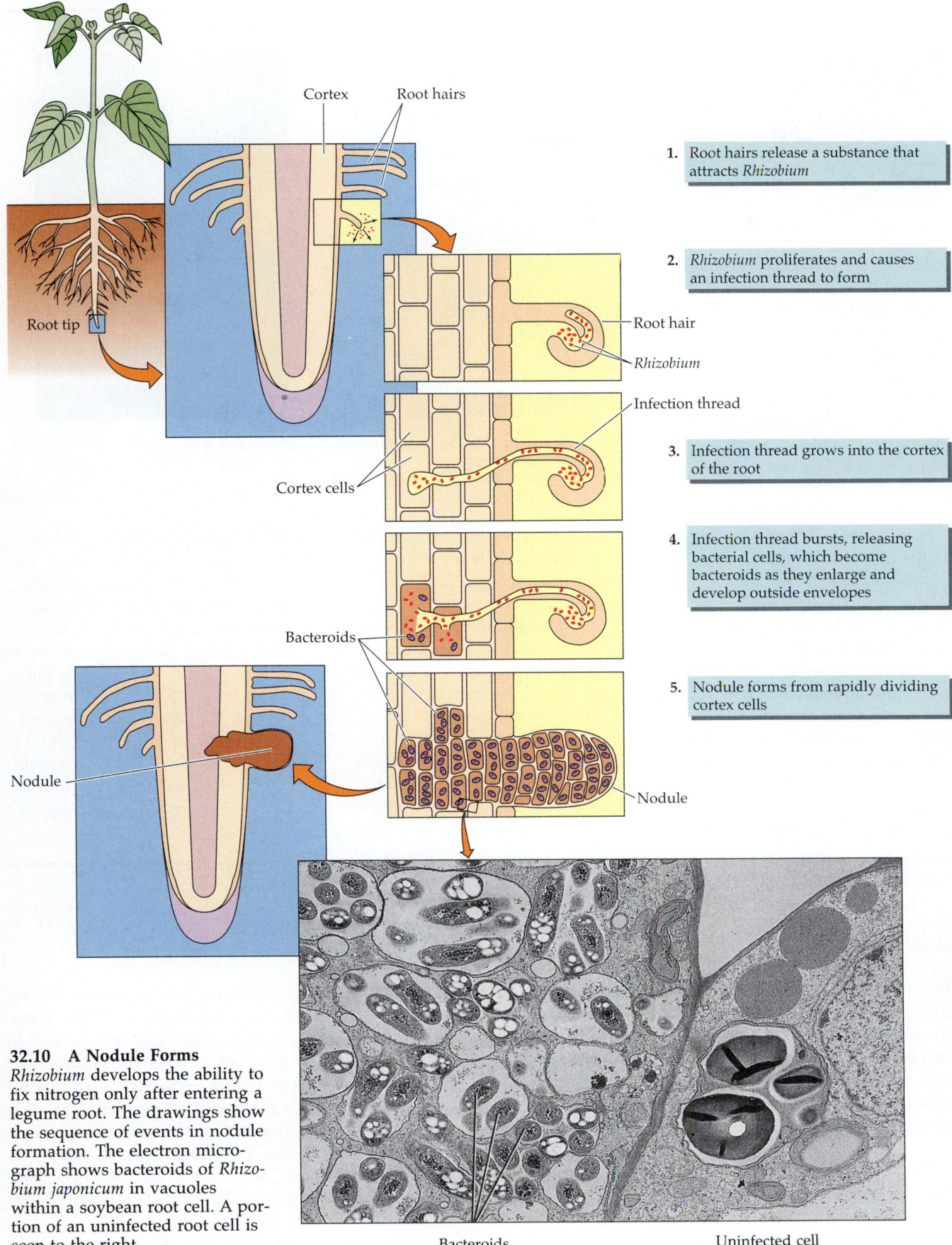

1. Root hairs release a substance that attracts *Rhizobium*

2. *Rhizobium* proliferates and causes an infection thread to form

Root hair

Rhizobium

Infection thread

3. Infection thread grows into the cortex of the root

4. Infection thread bursts, releasing bacterial cells, which become bacteroids as they enlarge and develop outside envelopes

5. Nodule forms from rapidly dividing cortex cells

Cortex   Root hairs

Root tip

Cortex cells

Bacteroids

Nodule

Nodule

Bacteroids   Uninfected cell

32.10   A Nodule Forms
Rhizobium develops the ability to fix nitrogen only after entering a legume root. The drawings show the sequence of events in nodule formation. The electron micrograph shows bacteroids of *Rhizobium japonicum* in vacuoles within a soybean root cell. A portion of an uninfected root cell is seen to the right.

BOX 32.A

Biotechnology in a Plant

When should nitrogen-containing fertilizer be added to a crop? If too much fertilizer is added during the growing season, the farmer is wasting money, and leaching of nitrates from the overfertilized soil leads to serious pollution of the groundwater. If too little fertilizer is added, the crop's yield—and the farmer's income—is low. The best judge of a plant's nitrogen status is the plant itself, and one crop plant has been "taught" how to report its nitrogen status to scientists.

Aladar Szalay of the University of Alberta and Thomas Baldwin of Texas A & M University isolated genes from *Vibrio harveyi*, a luminescent marine bacterium. The genes code for the production of luciferase, the enzyme that catalyzes the light-producing reaction. These scientists then inserted the luciferase genes at a key spot—between the gene coding for nitrogenase production and the promoter of the nitrogenase gene—in

The photo on the left, shot under normal light, shows root nodules of a soybean plant that carries the luciferase genes. The

the chromosome of *Rhizobium japonicum*, a bacterium that participates in the nitrogen-fixing root nodules of soybean plants.

The nitrogenase genes in soybean root nodules are inactive when the plants are getting enough nitrogen from the soil. If more nitrogen is needed, the nitrogenase promoter activates the nitrogenase gene, allowing the nodules to fix their own nitrogen directly. The insertion of *V. harveyi* luciferase genes between the *R. japonicum* nitrogenase gene and its

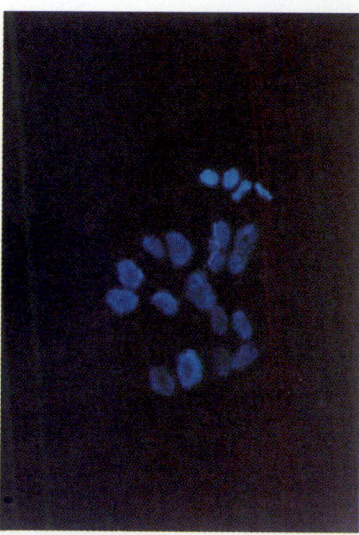

right-hand photo is the same scene, shot in the dark in the presence of decanal fumes.

promoter has the following result: When more nitrogen is needed, the nitrogenase promoter activates the neighboring genes, including those that code for luciferase. One can determine when luciferase is present by adding a suitable substrate, in this case a volatile substance known as decanal. Decanal vapor is absorbed directly by cells. If decanal and luciferase are both present, light is emitted. Thus when the modified *R. japonicum* is fixing nitrogen, it glows.

DENITRIFICATION

Because nitrogen fixation decreases the amount of nitrogen gas in the atmosphere, we should mention, if only briefly, the opposite process, called **denitrification**. Some normally aerobic bacteria, mostly species of the genera *Bacillus* and *Pseudomonas*, can use nitrate (NO_3^-) as a terminal electron acceptor in place of oxygen if they are kept under anaerobic conditions:

$$2\,NO_3^- + 10\,e^- + 12\,H^+ \rightarrow N_2 + 6\,H_2O$$

These bacteria are extremely common and, as the equation shows, they return dinitrogen to the atmosphere. Nature's nitrogen cycle is discussed in Chapter 49.

NITRIFICATION

Fixed nitrogen released into the soil by nitrogen fixers is primarily in the form of ammonia (NH_3) and ammonium ions (NH_4^+). Although ammonia is toxic to plants, ammonium ions can be taken up safely at low concentrations. Most plants, however, grow better with nitrate (NO_3^-) than with ammonium ions as a source of nitrogen because of the toxicity of ammonium ions. The form of nitrogen taken up is also affected by soil pH: Nitrate ions are taken up preferentially under more acidic conditions, ammonium ions under more basic ones.

Where do nitrate ions in the soil come from? Ammonia is oxidized to nitrate by the process of **nitri-**

fication. This process is carried out in the soil by bacteria. Bacteria of two genera, *Nitrosomonas* and *Nitrosococcus*, are capable of converting ammonia to nitrite ions (NO_2^-), and *Nitrobacter* bacteria oxidize nitrite to nitrate. These three genera of prokaryotes constitute a critical ecological link, taking the products of other crucial bacteria, the nitrogen fixers, and converting them into a form more available to plants, and hence to the rest of the biosphere.

What do these bacteria get in return for being so ecologically helpful? Actually, they carry on nitrification for their own selfish ends. These three genera are chemosynthetic autotrophs; that is, their chemosynthesis is powered by the energy released by oxidation of ammonia or nitrite. For example, by passing the electrons from nitrite through an electron transport chain, *Nitrobacter* can make ATP, and using some of this ATP, it can also make NADH. With the ATP and NADH, the bacterium can convert carbon dioxide and water to glucose and other foods. In short, the nitrifiers base their entire biochemistry— their entire lives—on the oxidation of ammonia or nitrite ions. *Nitrobacter* can convert 6 molecules of carbon dioxide to 1 molecule of glucose for every 78 nitrite ions that they oxidize—not terribly efficient, but efficient enough to keep *Nitrobacter* living, growing, and reproducing.

NITRATE REDUCTION

We have seen dinitrogen *reduced* to ammonia in nitrogen fixation and ammonia *oxidized* to nitrate in nitrification. We will now see that plants *reduce* the nitrate they take up all the way back to ammonia before using it further to manufacture amino acids (Figure 32.11). The reactions of **nitrate reduction** are carried on by the plant's own enzymes. The later steps, from nitrite to ammonia, take place in the chloroplasts; this conversion is not a part of photosynthesis. The final products of nitrate reduction are amino acids, from which the plant's proteins and all its other nitrogen-containing compounds are formed.

Nitrogen metabolism, in bacteria and in plants, is complex. It is also of great importance. Nitrogen atoms constitute approximately 1 to 5 percent of the dry weight of a leaf, and nitrogen-containing compounds constitute 5 to 30 percent of the plant's total dry weight. The nitrogen content of animals is even higher, and all the nitrogen in the animal world arrives by way of the plant kingdom. As we are about to see, plants also play an important part in delivering sulfur to animals.

SULFUR METABOLISM

All living things require sulfur, which is a constituent of two amino acids, cysteine and methionine, and hence of almost all proteins. Sulfur is also a component of other biologically crucial compounds, such as coenzyme A (see Chapter 7). Animals must obtain their cysteine and methionine from plants, but plants can start with sulfate ions (SO_4^{2-}) obtained from the soil or from a liquid environment. Interestingly, all of the most abundant elements in plants are taken up from the environment in their most oxidized forms—sulfur as sulfate, carbon as carbon dioxide, nitrogen as nitrate, phosphorus as phosphate, and hydrogen as water. In plants, sulfate is reduced and incorporated into cysteine; from this amino acid all the other sulfur-containing compounds in the plant are made. These important processes—sulfate reduction and the utilization of cysteine—are closely analogous to the reduction of nitrate to ammonia and the subsequent utilization of ammonia by plants.

Numerous bacteria base their metabolism on the modification of sulfur-containing ions and compounds in their environment. One group, for example, performs reactions analogous to nitrification. Just as the nitrifiers oxidize ammonia to nitrate, the chemosynthetic sulfur bacteria oxidize hydrogen sulfide

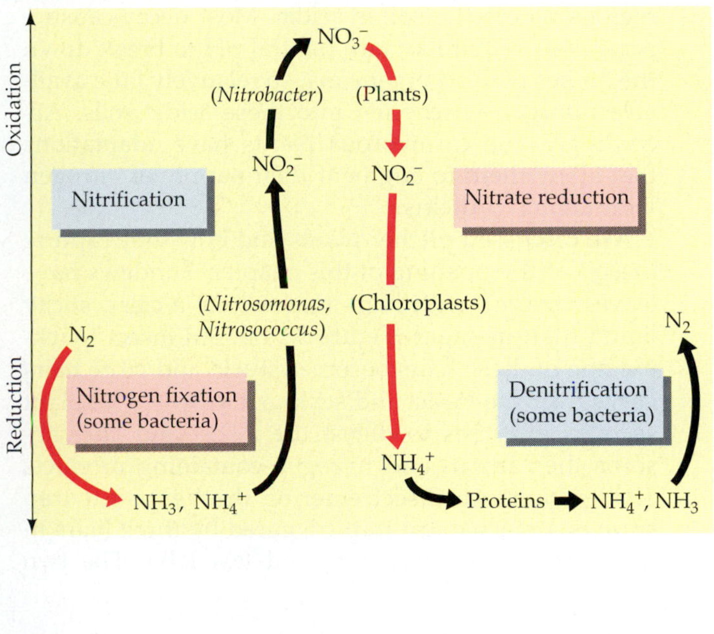

32.11 The Path of Nitrogen
Bacteria fix nitrogen from the atmosphere and conduct nitrification. Plants reduce nitrates back to ammonia, the form in which nitrogen is incorporated into proteins. Finally, some denitrifying bacteria can oxidize ammonia back to nitrogen gas, which returns to the atmosphere. Red arrows identify steps in which nitrogen compounds become reduced.

(H₂S) to sulfate, using the energy thus released to make ATP and fix carbon dioxide.

Thus far in this chapter we have considered the mineral nutrition of plants. As you already know, another crucial aspect of plant nutrition is photosynthesis—the principal source of energy and carbon for plants and for the biosphere as a whole. Not all plants, however, are photosynthetic autotrophs. A few, in the course of their evolution, have lost the ability to feed themselves by photosynthesis. How do these plants get their energy and carbon?

HETEROTROPHIC SEED PLANTS

A few plants are parasites that obtain their food directly from the living bodies of other plants (Figure 32.12). Perhaps the most familiar parasitic plants are the mistletoes and dodders. Mistletoes are green and carry on some photosynthesis, but they parasitize other plants for water and mineral nutrients and may derive photosynthetic products from them as well. Another parasitic plant, the Indian pipe (see Figure 25.1), once was thought to obtain its food from dead organic matter; it is now known to get its nutrients,

32.13 A Carnivorous Sundew
A sundew has trapped an insect on its sticky hairs. Secreted enzymes will digest the carcass externally.

32.12 Parasitic Dodder
Orange-brown tendrils of dodder wrap around other plants. This parasitic plant obtains water, sugars, and other nutrients from its host through tiny, rootlike protuberances that penetrate the surface of the host.

with the help of fungi, from nearby actively photosynthesizing plants. Hence it too is a parasite.

Some other heterotrophic plants are the 450 or so carnivorous species—those that augment their nitrogen and phosphorus supply by capturing and digesting flies and other insects. The best-known plant carnivores are Venus's-flytrap (genus *Dionaea*), sundews (genus *Drosera*), and pitcher plants (genus *Sarracenia*). These plants are normally found in boggy regions where the soil is acidic. Most decay-causing organisms require a more neutral pH to break down the bodies of dead organisms, so relatively little available nitrogen is recycled into these acidic soils. Accordingly, the carnivorous plants have adaptations that allow them to augment their supply of nitrogen with animal proteins.

We discussed pitcher plants and how they capture insects at the opening of this chapter. Sundews have leaves covered with hairs that secrete a clear, sticky liquid high in sugar (Figure 32.13). An insect touching one of these hairs becomes stuck, and more hairs curve over the insect and stick to it as well. The plant secretes enzymes to digest the insect and later absorbs the carbon- and nitrogen-containing products of digestion. An insect entering the Venus's-flytrap springs a mechanical trap triggered by three hairs in the center of a partially closed leaf lobe. The two halves of the leaves close, and spiny outgrowths at the margins of the leaves interlock to imprison the insect. Enzymes secreted by the plant digest the trapped insect. None of the carnivorous plants must feed on insects; they grow adequately without insects, but in nature, they grow faster and are a darker green when insects are available to them. The extra supply of nitrogen is used to make more proteins and chlorophyll, as well as other nitrogen-containing compounds.

SUMMARY of Main Ideas about Plant Nutrition

Nearly all plant species are photosynthetic autotrophs.

A few plant species are parasitic heterotrophs that feed on other plants.

Carnivorous plant species are autotrophs that supplement their nitrogen supply by feeding on insects.

The only nutrients required by most plants are carbon dioxide, water, nitrate or ammonium ions, and several mineral salts.
    Review Tables 32.1 and 32.2

Clay particles hold and exchange positively charged mineral nutrients, making them available to plants.
    Review Figure 32.5

A few species of soil bacteria are responsible for all nitrogen fixation, the conversion of nitrogen gas to usable nitrogen compounds.
    Review Figure 32.9

Some nitrogen-fixing bacteria live free in the soil; others, such as some species of *Rhizobium*, live symbiotically within the roots of plants.
    Review Figure 32.10

Ammonium ions are the main product of nitrogen fixation in the soil, yet most plants preferentially take up nitrate ions under most conditions.

Nitrate ions are available because other species of soil bacteria carry on nitrification.

The nitrifying bacteria are chemosynthetic autotrophs.

Once in the plant, nitrate is reduced to ammonia, which is incorporated into organic compounds.
    Review Figure 32.11

Sulfate ions, produced in the soil by bacteria that oxidize hydrogen sulfide, are reduced within the plant.

SELF-QUIZ

1. Which of the following is *not* an essential mineral element for plants?
    a. Potassium
    b. Magnesium
    c. Calcium
    d. Lead
    e. Phosphorus

2. Fertilizers
    a. are often characterized by their N–P–O percentages.
    b. are not required if crops are removed frequently enough.
    c. restore needed mineral nutrients to the soil.
    d. are needed to provide carbon, hydrogen, and oxygen to plants.
    e. are needed to destroy soil pests.

3. Which of the following is *not* an important step in soil formation?
    a. The removal of bacteria
    b. Mechanical weathering

    c. Chemical weathering
    d. Clay formation
    e. Hydrolysis of soil minerals

4. Laterization
    a. results in a very productive soil.
    b. often takes place in mine tailings.
    c. produces a soil rich in copper, lead, nickel, and zinc.
    d. produces a soil rich in chromium and poor in calcium.
    e. produces a soil rich in insoluble iron and aluminum compounds.

5. Nitrogen fixation
    a. is performed only by plants.
    b. is the oxidation of nitrogen gas.
    c. is catalyzed by the enzyme nitrogenase.
    d. is a single-step chemical reaction.
    e. is possible because N_2 is a highly reactive substance.

6. Nitrification
    a. is performed only by plants.
    b. is the reduction of ammonium ions to nitrite and nitrate ions.
    c. is the reduction of nitrate ions to nitrogen gas.
    d. is catalyzed by the enzyme nitrogenase.
    e. is performed by certain bacteria in the soil.

7. Nitrate reduction
    a. is performed by plants.
    b. takes place in mitochondria.
    c. is catalyzed by the enzyme nitrogenase.
    d. includes the reduction of nitrite ions to nitrate ions.
    e. is known as the Haber process.

8. Which of the following statements about sulfur is *not* true?
   a. All living things require it.
   b. It is a component of DNA and RNA.
   c. It is a constituent of two amino acids.
   d. Its metabolism is similar to the metabolism of nitrogen.
   e. Many bacteria base their metabolism on reactions of sulfur-containing ions.

9. Which of the following is a parasite?
   a. Venus's-flytrap
   b. Pitcher plant
   c. Sundew
   d. Dodder
   e. Tobacco

10. All heterotrophic seed plants
    a. are parasites.
    b. are carnivores.
    c. are incapable of photosynthesis.
    d. derive their nutrition from animals.
    e. develop from multicellular embryos.

FOR STUDY

1. Methods for determining whether a particular element is essential have been known for over a century. Since the methods are so old and well established, why is it that the essentiality of some elements was discovered only recently?

2. If a Venus's-flytrap were to be deprived of soil sulfates and hence made unable to synthesize the amino acids cysteine and methionine, would it die from lack of proteins? Discuss.

3. Soils are dynamic systems. What changes might result when land is subjected to heavy irrigation for agriculture, after being relatively dry for many years? What changes in the soil might result when a virgin, deciduous forest is replaced by crops that are harvested each year?

4. What is the significance of nitrification to plants? to the bacteria that carry on nitrification?

READINGS

Brill, W. J. 1981. "Agricultural Microbiology." *Scientific American*, September. A look at how microbes can be used to support plant growth.

Epstein, E. 1984. "Rhizostats: Controlling the Ionic Environment of Roots." *BioScience*, November. Computers, roots, and plant nutrition.

Power, J. F. and R. F. Follett. 1987. "Monoculture." *Scientific American*, March. There is an increasing tendency to grow the same crop year after year. This article discusses how this practice affects soils, and whether it is good or bad in general.

Raven, P. H., R. F. Evert and S. Eichhorn. 1991. *Biology of Plants*, 5th Edition. Worth, New York. An excellent general botany textbook.

Taiz, L. and E. Zeiger. 1991. *Plant Physiology*. Benjamin/Cummings, Redwood City, CA. An authoritative textbook with good chapters on mineral nutrition and related topics.

What causes the different parts of a plant to take on distinctive forms? Why do roots and shoots differ in structure? What causes roots to form? Hormones, which are chemical signals, play important roles in determining patterns of plant development. Recall, for example, that plants need not start as seeds; a piece of shoot (a cutting) can form roots and give rise to an intact plant as, in the extreme case, can a small piece of tissue or even a single cell. Gardeners commonly induce cuttings to root. How do they do it? How do they change the course of development in this way?

The gardener, or a biologist, or you, can do what we did to produce this photograph—we treated the cut stems with a hormone called auxin. A tiny amount of this hormone causes some stem cells to develop not into stem structures but, instead, into root apical meristems and thus into new roots. By modifying the amount of auxin applied, we can control the number and lengths of new roots. This simple experiment shows the basis of our discussion in this chapter—that hormones profoundly affect plant development.

WHAT REGULATES PLANT DEVELOPMENT?

The development of a plant—the progressive changes that take place throughout its life—is regulated in complex ways. There are four major players in regulation: the *environment, hormones,* the pigment *phytochrome,* and the plant's *genome.*

Environmental cues, such as light or temperature changes, trigger some important developmental events, such as flowering and the onset and end of dormancy. Hormones and phytochrome mediate the effects of environmental cues.

Hormones—compounds produced in small quantity in one part of an organism and then moved to other parts where they produce effects—mediate many developmental phenomena, such as stem growth and autumn leaf fall. The plant hormones include **auxin, gibberellins, cytokinins, abscisic acid,** and **ethylene.** Each is produced in one or more specific parts of the plant's body. Each plays multiple regulatory roles, affecting several different aspects of development (Table 33.1).

Like the hormones, phytochrome regulates many processes. Unlike the hormones, it does not travel from one part of the plant to another but acts within the cells that produce it. Light (an environmental cue) acts directly on phytochrome, which in turn regulates developmental processes such as the many changes accompanying the growth of a young plant out of the soil and into the light.

What Made the Difference?
The cutting with more roots was treated with auxin, a hormone discussed in this chapter. The other cutting is an untreated control.

33

Regulation of Plant Development

TABLE 33.1
Plant Hormones

| | | | | ACTIVITY | | |
HORMONE	SITE OF PRODUCTION	SEED DORMANCY	SEED GERMINATION	SEEDLING GROWTH	APICAL DOMINANCE	LEAF ABSCISSION
Gibberellins	Embryo, young leaves, root and shoot apices	Breaks	Promotes	Promotes cell division and expansion	—	—
Auxin	Embryo, young leaves, shoot apical meristem	—	—	Promotes cell expansion	Inhibits lateral buds	Inhibits
Cytokinins	Roots	—	—	Promotes cell division	Promotes lateral buds	Inhibits
Ethylene	Ripening fruit, senescing tissue, stem nodes	—	—	—	—	Promotes
Abscisic acid	Root cap, older leaves, stem	Imposes	Inhibits	—	—	—

No matter what cues direct development, ultimately the plant's genome determines how the plant and its parts develop. The genome is the master plan of the plant, but it is interpreted differently depending on the status of the environment. The genome encodes phytochrome and the enzymes that catalyze the formation of the hormones and mediate some of their actions; it is also the target for some hormone actions. For several decades hormones and phytochrome were the focus of most work on plant development, but recent advances in molecular genetics now allow us to focus on underlying processes.

AN OVERVIEW OF DEVELOPMENT

Let's now review the life history of a flowering plant, from seed to death, focusing on how the developmental events are regulated. Keep in mind that as plants develop, the environment, hormones, and phytochrome affect three fundamental processes: cell division, cell expansion, and cell differentiation. Try to envision how the division, expansion, and differentiation of cells contribute to different developmental phenomena.

From Seed to Seedling

Consider a seed. All developmental activity may be suspended in this seed; in other words, it may be **dormant.** Cells in dormant seeds do not divide, expand, or differentiate. Seed dormancy may be broken by one of several physical mechanisms—mechanical abrasion, fire, leaching of inhibitors by water—described later in this chapter. As the seed **germinates** (begins to develop), it first imbibes (takes up) water. The growing embryo must then obtain building blocks—monomers—by digesting the polymeric reserve foods stored either in the cotyledons or in the endosperm. The embryos of some plant species secrete gibberellins that direct the mobilization of the reserves.

If the seed germinates underground, it must elongate rapidly and cope with life in the absence of light. Phytochrome controls this stage and ends it when the shoot reaches the light. The growth of the seedling, both in darkness and light, is also regulated by auxin and gibberellin, and auxin is known to regulate tissue and organ formation. Thus auxin affects cell differentiation. Other information regulating these phenomena comes from cytokinins (Figure 33.1).

Reproductive Development

Eventually the plant flowers. Flowering may be initiated when the plant reaches an appropriate age or size. Some plant species, however, flower at particular times of the year, meaning that the plant must sense the appropriate date. These plants are photoperiodic (see Chapter 34); they measure the length of the night (shorter in the summer, longer in the winter) with great precision. Although we don't know *how* it works, we do know a lot about photoperiodism. We know that the plant measures the length of

	WINTER DORMANCY	FLOWERING	FRUIT DEVELOPMENT
	Breaks	Stimulates in some plants	Promotes
	—	—	Promotes
	—	—	—
	—	—	Promotes ripening
	Imposes	—	—

mains a mystery, but it seems likely that a "flowering hormone"—named florigen, but not yet discovered—travels from the leaf to the point of flower formation. Perhaps information travels in some other form, but this form has not been discovered either.

After flowers have formed, hormones, including auxin and gibberellin, play further roles. Hormones and other substances control the growth of a pollen tube down the style of a pistil. Following fertilization, a fruit develops, controlled in several ways by gibberellin and auxin (Figure 33.2). The ripening of the fruit is also under chemical control, commonly by the gaseous hormone ethylene.

Dormancy, Senescence, and Death

Some perennials have buds that enter a state of winter dormancy during the cold season. (Perennial plants are those that grow for a number of years.) Abscisic acid helps maintain such dormancy. Finally, the plant **senescences** (deteriorates because of aging) and dies. Death, which may be under environmental control, follows senescent changes that are controlled by regulators such as ethylene (Figure 33.3).

The elements of this brief overview will be considered in detail in this chapter and the next. Let's begin with regulation at the start of the life cycle.

SEED DORMANCY AND GERMINATION

Some seeds are, in effect, instant plants, because all they need for germination is water. Many other species have seeds whose germination is regulated in more complex ways because they are initially dor-

darkness, that there is a biological clock—itself a mystery, and that light absorbed by phytochrome can affect the time-measuring process.

Once a leaf has determined, by measuring the night length, that it is time for the plant to flower, that information must be transported to the places where flowers will form. How this comes about re-

33.1 From Seed to Seedling
Environmental factors, hormones, and phytochrome regulate the first stages of plant growth.

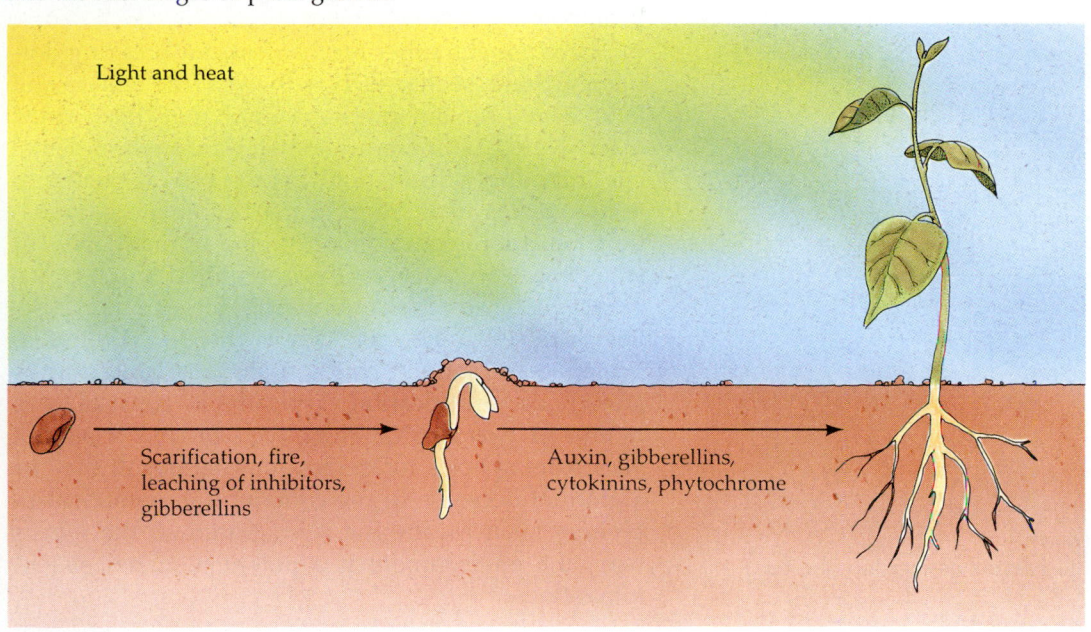

Light and heat

Scarification, fire, leaching of inhibitors, gibberellins

Auxin, gibberellins, cytokinins, phytochrome

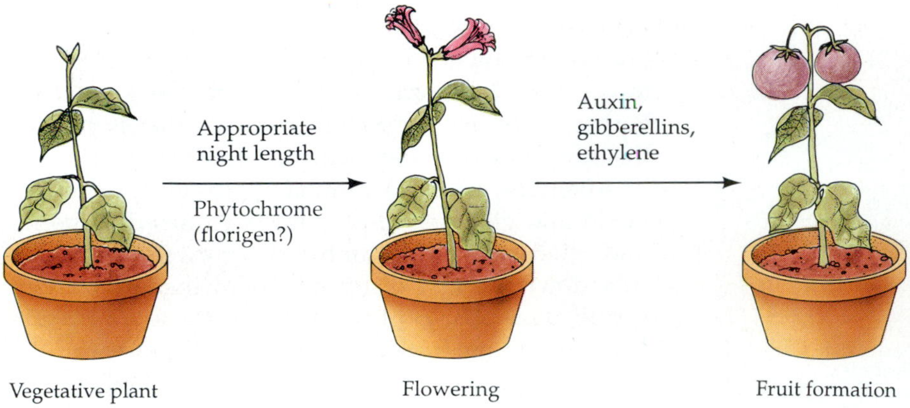

33.2   Flowering and Fruit Formation
Environmental factors, hormones, and phytochrome regulate plant reproduction.

mant. Seed dormancy may last for weeks, months, years, or even decades! The mechanisms of dormancy are numerous and diverse, but three principal strategies dominate: exclusion of water or oxygen from the embryo, mechanical restraint of the embryo, and chemical inhibition of embryo development.

Some seeds exclude water or, sometimes oxygen, by having an impermeable seed coat. The breaking of dormancy in such seeds depends on the abrasion of the seed coat, perhaps by the seed's tumbling across the ground or through creek beds, by its passing through the digestive tracts of birds, or by other means. Such modification of the seed coat is called **scarification.** The environment regulates germination of these seeds by providing or withholding the means of scarification.

The seed coat may also impose dormancy by the simple mechanical restraint of the embryo; if the embryo cannot expand, the seed cannot germinate. In the laboratory we can promote germination of such a seed by simply cutting away part of the coat or by partially dissolving it with strong acid. In nature, soil microorganisms probably play a major role in soft-

ening seed coats of this type, and the action of digestive enzymes in the guts of birds or other animals is also important. Another agent of scarification to release mechanical restraint is fire, which causes significantly increased germination in some natural habitats. Fire can also melt wax in seed coats, making water available to the embryo (Figure 33.4; see also Figure 47.16).

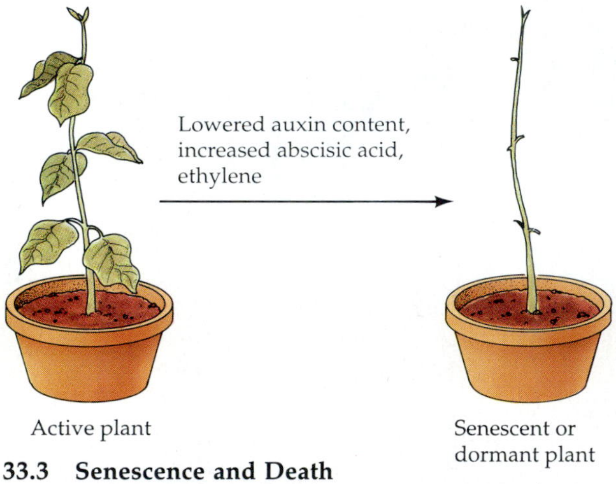

The action of chemical germination inhibitors is another mechanism of seed dormancy. As long as the concentration of inhibitor is high, the seed remains dormant. One means of reducing the level of inhibitor is by leaching, that is, prolonged exposure to water. Another is the scorching of the seeds by fire, which breaks down some inhibitors. Usually inhibitors of germination are already present in the dry seed, but in a few cases they are produced only after the seed has begun to take up water. The most common chemical inhibitor of seed germination is abscisic acid. In some seeds the level of abscisic acid or other inhibitors does not decline during germination; rather, the effect of the inhibitor is overcome by gradually increasing the concentrations of growth promoters.

There are still other mechanisms for breaking dormancy. Some seeds, such as those of the tomato and lima bean, remain dormant until they have dried extensively. Temperature can be an environmental cue initiating germination. Even a brief exposure to temperatures near freezing may stimulate germination, but more commonly a period of many days or weeks of low temperature is required to end dormancy. In agriculture and forestry, it is common practice to refrigerate seeds such as those of conifers to hasten germination. This refrigeration procedure is known as **stratification;** typically, it consists of a month or two at 5°C. The effects of cold treatment vary from species to species, but one result may be a gradual decrease in the content of germination inhibitors such as abscisic acid. Although the means vary, the environment is a potent regulator of germination.

33.3   Senescence and Death
Environmental factors and hormones regulate the final stages of plant growth.

33.4 Fire and Seed Germination
This fireweed germinated and flourished after a great fire in Denali National Park, Alaska.

Adaptive Advantages of Seed Dormancy

Why might regulating the germination of seeds be a good thing? For many species, dormancy results in germination at a favorable time. Seeds that require a long cold period for germination commonly germinate in the spring, when water is usually abundant—germination in the dry days of late summer could be risky. Some other seeds will not germinate until a certain amount of time has passed, regardless of how they are treated. This period of **afterripening** prevents germination while the seed of a cereal grain, for example, is still attached to the parent plant, and it tends to favor dispersal of the seed.

Regulating germination may increase the likelihood of a seed's germinating in the right place. For example, some cypress trees grow in standing water, and their seeds germinate only if leached extensively by water (Figure 33.5). Many weeds must have their seed coats damaged before they will germinate, and other weed seeds will not germinate unless they have been exposed to light. Either type of weed germinates best in disturbed soils. You may have noticed how a freshly cultivated patch of soil quickly teems with weeds that are then free from competition with other plants. Seeds that must be scorched by fire in order to germinate also avoid competition; they germinate only when the area has been largely cleared by fire. Light-requiring seeds, which germinate only at or near the surface of the soil, are generally tiny seeds with few food reserves. The germination of some seeds is inhibited by light; these germinate only when well buried. Light-inhibited seeds are generally large and well stocked with nutrients.

Seed dormancy helps annual species counter the effects of year-to-year variations in the environment.

Some seeds remain dormant throughout an unfavorable year, and other seeds germinate at different times during the year. Seed dormancy can also contribute to the dispersal of a plant species. Seeds that remain dormant until they have passed through the guts of birds or other animals will likely be carried some distance before they are deposited. Seeds carried by birds in their digestive tracts can give rise to the first plants on newly formed volcanic islands, for example.

33.5 Leaching of Germination Inhibitors
The seeds of *Nyssa aquatica* germinate only after being leached by water, which increases the chances that they germinate in a situation suitable for their growth.

33.6 The Radicle Emerges
The tip of this lima bean's radicle has just broken through its protective sheath. The appearance of the radicle is one of the first externally visible events in seed germination.

Seed Germination

Imbibition—the uptake of water—is the first step in the germination of a seed. Typically, the seed is dry prior to the start of germination. Its water potential (see Chapter 30) is very negative, and water can be taken up readily if the seed coat allows it. The magnitude of the water potential is demonstrated by the force exerted by seeds expanding in water. Cocklebur seeds that are imbibing can exert a pressure of up to 1,000 atmospheres against a restraining force. As a seed takes up water, it undergoes metabolic changes: Certain preformed enzymes become activated, RNA and then new enzymes are synthesized, the rate of cellular respiration increases, and other metabolic pathways become activated. Interestingly, there is no DNA synthesis and no cell division during these early stages of seed germination. Emergent growth arises solely from expansion of small, preformed cells. DNA is synthesized only after the radicle—the embryonic root—begins to grow and poke out beyond the seed coat (Figure 33.6).

Mobilizing Food Reserves

Until the young plant (the **seedling**) becomes able to carry on photosynthesis, it depends on built-in reserves from the seed to meet its needs for energy and materials. The principal reserve of energy and carbon in the seeds of some species is the carbohydrate starch. More species, however, store lipids—fats or oils—as reserves in their seeds. Typically, the endosperm of the seed holds amino acid reserves in the form of proteins, rather than as free amino acids. Plant species differ in what their reserves are.

The giant molecules of starch, lipids, and proteins must be digested into monomers before they can enter the cells of the embryo to be used as building blocks and energy sources. Starch is a polymer of glucose, the starting point for glycolysis and cellular respiration. Starch is digested to glucose. The lipids can be digested to release glycerol and fatty acids, both of which yield energy through cellular respiration. Glycerol and fatty acids can also be converted to glucose, which permits fat-storing plants to make all the building blocks they need for growth. Lipid-storing seeds can pack more energy in a smaller space than starch-storing seeds can because lipids contain more calories per unit of weight than does starch. Some species that store lipids in their seeds store starch in their roots or tubers, where space is not a concern. Proteins are digested as well. The growing embryo can break down the proteins to obtain the amino acids it needs to assemble into its own myriad proteins. Plant species differ in their patterns of digestion of reserve polymers.

Germinating barley and other cereal seeds digest proteins and starch as follows: As the embryo becomes active, it secretes gibberellins. These diffuse through the endosperm to a surrounding tissue called the **aleurone layer,** which lies inside the seed coat. The gibberellins trigger a crucial series of events in the aleurone layer. First, protein-containing bodies called aleurone grains break down, releasing amino acids. The aleurone layer then uses the amino acids in the assembly of digestive enzymes, including amylases (starch-degrading enzymes), proteases (protein-degrading enzymes), and ribonucleases (RNA-degrading enzymes). These enzymes, along with certain others already present in the aleurone layer, are next secreted into the endosperm, where they catalyze the release of sugars and amino acid monomers from reserve polymers for use by the growing embryo (Figure 33.7).

We will now consider each of the major plant hormones, beginning with the gibberellins. Here is a study hint for the rest of this chapter. We have seen in a brief overview that information about hormones can be organized by considering the development of each part of the plant one after another. Now, to take a closer look at the hormones, we will organize the information by considering each hormone in turn. Try outlining the material in the first way, showing in detail how the development of roots, stems, buds, and leaves are regulated. In general, this is an excellent way to study: Look for a different way to organize the material presented in lecture or in the book.

GIBBERELLINS

We just encountered the gibberellins in the mechanism by which the germinating seeds of barley and

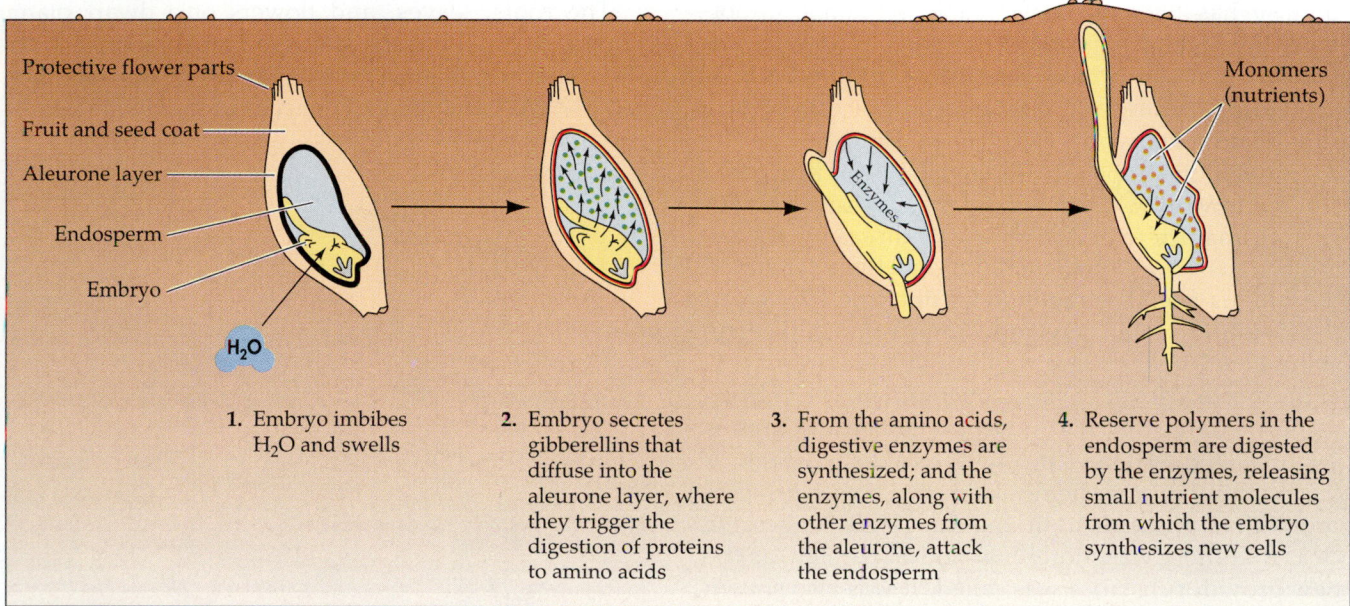

1. Embryo imbibes H_2O and swells

2. Embryo secretes gibberellins that diffuse into the aleurone layer, where they trigger the digestion of proteins to amino acids

3. From the amino acids, digestive enzymes are synthesized; and the enzymes, along with other enzymes from the aleurone, attack the endosperm

4. Reserve polymers in the endosperm are digested by the enzymes, releasing small nutrient molecules from which the embryo synthesizes new cells

Protective flower parts
Fruit and seed coat
Aleurone layer
Endosperm
Embryo
Monomers (nutrients)

33.7 Embryos Mobilize Polymer Reserves

Seed germination in cereal grasses consists of a cascade of processes. Gibberellin signals the conversion of reserve polymers into monomers that can be used by the developing embryo.

other cereals convert their reserve proteins and starch into soluble monomers (see Figure 33.7). This is an elegant sequential mechanism, in which gibberellin functions as a *signal* to give the embryo access to the nutrients it needs. Gibberellins produce a wide variety of effects on plant development in addition to this triggering of digestive enzyme synthesis.

Discovery of the Gibberellins

The gibberellins are a large family of closely related compounds (Figure 33.8*a*,*b*), some found in plants and others in a pathogenic (disease-causing) fungus, where they were first discovered. The discovery of the gibberellins followed a crooked path, beginning with a book dictated in 1809 by a Japanese farmer named Konishi. In it, he described the symptoms of *ine bakanae-byo*, the "foolish seedling" disease of rice. Seedlings affected by the disease grow more rapidly than healthy plants, but the rapid growth gives rise to spindly plants that die before producing seed. The

33.8 Plant Hormones

Chemical structures of some representative compounds.

(*a*) Gibberellin A_1 (Important in stem growth)

(*b*) Gibberellin A_3 (Commercially available)

(*c*) Auxin (Indoleacetic acid)

(*d*) Ethylene (The "senescence hormone")

(*e*) Kinetin (A cytokinin discovered in aged DNA)

(*f*) Zeatin (A naturally occuring cytokinin in plants)

(*g*) Abscisic acid (The "stress hormone")

disease has been of considerable economic importance in several parts of the world. In 1898 it was learned that the *bakanae* disease, including both its growth-promoting and toxic effects, was caused by a fungus now known as *Gibberella fujikuroi.*

In 1925 the Japanese biologist Eiichi Kurosawa studied how *G. fujikuroi* caused the excessive spindly growth characteristic of the *bakanae* disease. He grew the fungus on a liquid medium and then separated the fungus from the medium by filtering. He heated the medium to kill any remaining fungus. The resulting heat-treated filtrate still stimulated the growth of uninfected rice seedlings. He found no such effects using medium that had never contained the fungus (Figure 33.9). Thus, Kurosawa established that *G. fujikuroi* produces a chemical substance with growth-promoting properties. In the late 1930s it became clear that there was more than one gibberellin, as the new growth substance was called. It was also shown that the toxic effects of the fungus were caused by another, inhibitory substance.

Were the gibberellins simply exotic products of an obscure fungus, or did they play a more general role in the growth of plants? Bernard O. Phinney of the University of California, Los Angeles, partially answered this question in 1956, when he reported the spectacular growth-promoting effect of gibberellins on certain dwarf strains of corn. These plants were known to be genetic dwarfs; each phenotype was produced when a particular recessive allele was present in the homozygous condition (see Chapter 10 if you need to review these genetic terms). Gibberellin applied to nondwarf—tall—corn seedlings had virtually no effect, but gibberellin applied to the dwarfs caused them to grow as tall as their normal relatives. (A comparable effect of gibberellin on dwarf mustard plants is shown in Figure 33.10.) This result suggested to Phinney that (1) gibberellins are normal constituents of corn and perhaps of all plants, and (2) the dwarfs are short because they cannot produce their own gibberellin. According to these hypotheses, nondwarf plants manufacture enough gibberellins to promote their full growth. Phinney and other scientists tested extracts from numerous plant species to see if they promoted growth in dwarf corn and they found that many such extracts did. These findings provided direct evidence that plants that are not genetic dwarfs contain gibberellinlike substances.

33.9 Kurosawa's Experiment
Kurosawa demonstrated that *bakanae* disease is caused by substances produced by *Gibberella fujikuroi.* Fungus-free medium previously used to culture the fungus made rice seedlings grow rapidly, as if they had the *bakanae* disease. As a control, Kurosawa determined that medium in which the fungus had never been cultured did not cause disease symptoms.

The roots, leaves, and flowers of a dwarf plant appear normal, but the stems are much shorter than their counterparts on other plants. What does this tell us? We know that all the cells of the dwarf plants

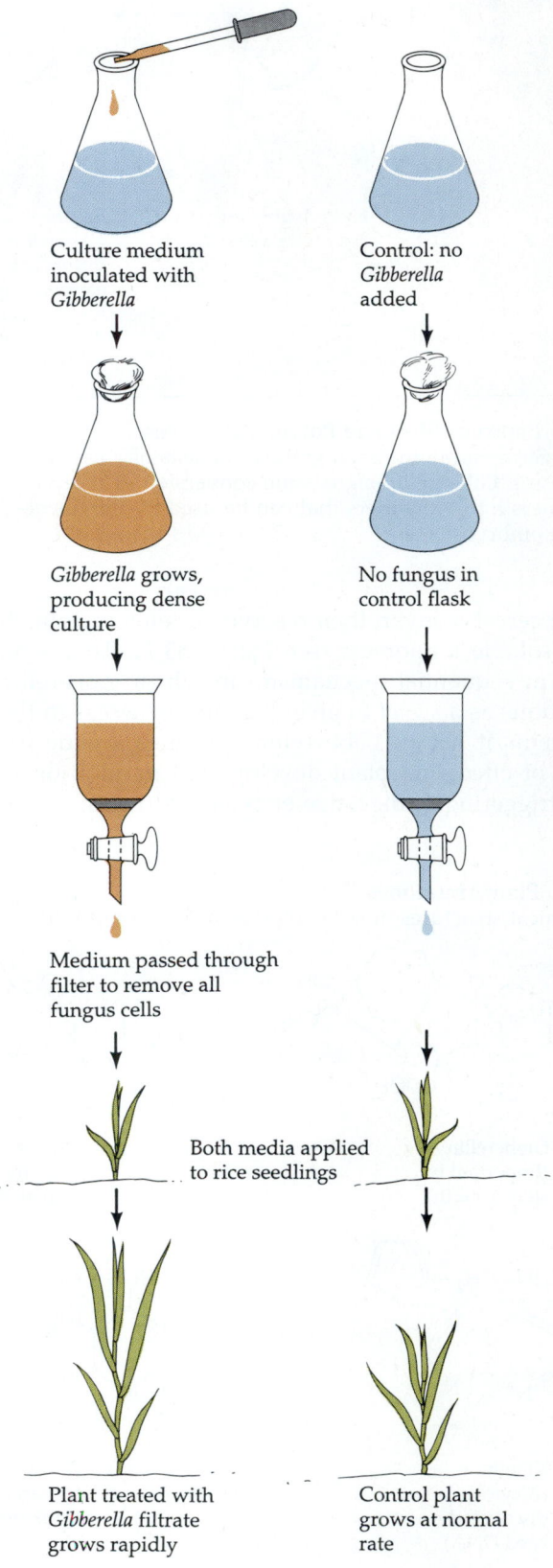

Culture medium inoculated with *Gibberella*

Control: no *Gibberella* added

Gibberella grows, producing dense culture

No fungus in control flask

Medium passed through filter to remove all fungus cells

Both media applied to rice seedlings

Plant treated with *Gibberella* filtrate grows rapidly

Control plant grows at normal rate

33.10   Gibberellin's Effect on a Dwarf Plant
In this experiment, a tiny amount of gibberellin was
added to the dwarf mustard plant (*Brassica rapa*) on the
right; on the left is an untreated control plant. After 22
days, the plant on the right reached a size more typical of
a nondwarf plant, while the control remained dwarf.

relatives. In one experiment, removal of seeds from
very young seeded grapes prevented their normal
growth, suggesting that the seeds are sources of a
fruit growth regulator. It was then shown that spraying young seedless grapes with a gibberellin solution
causes them to grow as large as seeded varieties.
Subsequent biochemical studies showed that the developing seeds produce gibberellin, which diffuses
out into the immature fruit tissue.

Some biennial plants respond dramatically to an
increased level of gibberellin. Biennial plants grow
vegetatively in their first year and flower and die in
their second year. In their second year, in response
either to the increasing length of days or to the winter
cold period, the apical meristems of biennials produce elongated shoots that eventually bear flowers.
This elongation is called **bolting.** When the plant
senses the appropriate environmental cue—longer
days or a sufficient winter chilling—it produces more
gibberellins, raising the gibberellin concentration to
a level that causes the shoot to bolt. Plants of some
biennial species will bolt if sprayed with a gibberellin
solution even though they have not experienced the
environmental cue (Figure 33.11).

have the same genome and all parts of the dwarf
plants contain much less gibberellin than do the organs of other plants. From this we may infer that
stem elongation *requires* gibberellin or the products
of gibberellin action; on the other hand, gibberellin
plays no comparable role in the development of
roots, leaves, and flowers.

Why So Many Gibberellins?

We do not yet know how many different gibberellins
exist. Each year brings reports of more, with several
dozen now having been characterized. Some gibberellins are produced in the root system, others in
young leaves. For many years plant physiologists
were puzzled by the existence of such a great number
of different gibberellins. Recent work, however, has
led to the conclusion that only one gibberellin, gibberellin A_1, actually controls stem elongation; the
other gibberellins found in stems are simply intermediates in the production of gibberellin A_1. As we
will see in the next section, gibberellins affect processes other than stem elongation, but we do not yet
know which gibberellin has any other particular effect.

Other Activities of Gibberellins

Gibberellins and other hormones regulate the growth
of fruits. It has long been known that seedless varieties of grapes form smaller fruit than their seeded

33.11   Bolting
Spraying with gibberellin causes cabbage and some other plants to bolt.
While untreated control plants retain
their compact leafy heads, the internodes of treated plants elongate dramatically, resulting in towering shoots.

No gibberellin   With gibberellin

Gibberellins also cause fruit to grow from unfertilized flowers, promote seed germination in lettuce and some other species, and in the spring help bring buds out of winter dormancy. Hormones usually have multiple effects within the plant and often interact with one another in regulating developmental processes. In controlling stem elongation, for example, gibberellins interact with another hormone, auxin.

AUXIN

If you pinch off the apical bud at the top of a bean plant, lateral buds that were once inactive become active. Similarly, pruning a shrub causes an increase in branching. If you cut off the blade of a leaf but leave its petiole attached to the plant, the petiole drops off sooner than it would if the leaf were intact. If a plant is kept indoors, its shoot system grows toward a bright window. What these diverse responses of shoot systems have in common is that they are mediated by a plant hormone called auxin—or **indoleacetic acid** in chemical terms (see Figure 33.8c).

Discovery of Auxin

The discovery of auxin and its numerous physiological activities traces back to work done in the 1880s by Charles Darwin and his son Francis, who were interested in plant movements. One type of movement they studied was **phototropism**, the growth of plant structures toward light (as in shoots) or away from it (as in roots).

An obvious question that they asked was, What part of the plant senses the light? To answer this question, the Darwins worked with canary grass seedlings grown in the dark. The dark-grown grass seedling has a coleoptile, or leaf sheath, covering the immature shoot. To find the light-receptive region of the coleoptile, the Darwins tried "blindfolding" it in various places and then illuminating it from one side (Figure 33.12). The coleoptile grew toward the light whenever its tip was exposed. If the top millimeter or more of the coleoptile was covered, however, there was no phototropic response. Thus the tip contains the photoreceptor. The bending, however, takes place in a growing region a few millimeters below the tip. Therefore, the Darwins reasoned, some type of message must travel within the coleoptile from the tip to the growing region.

Others later demonstrated that the message is a chemical substance; it cannot pass through a barrier impermeable to chemicals but does pass through certain nonliving materials, such as gelatin. The tip of the coleoptile produces a hormone that moves down

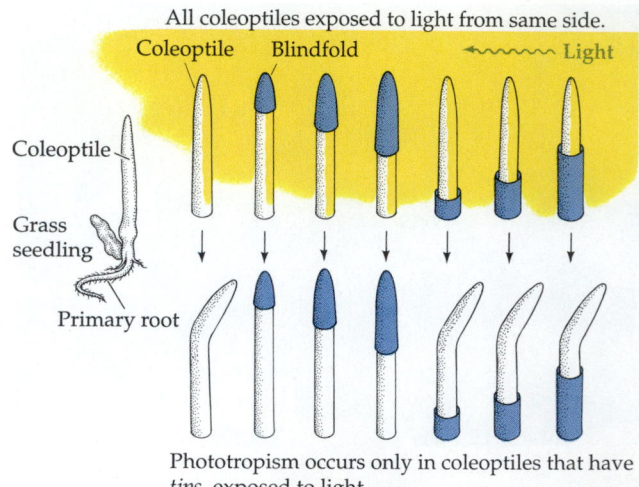

All coleoptiles exposed to light from same side.

Phototropism occurs only in coleoptiles that have *tips* exposed to light.

33.12   The Darwins' Experiment
The top drawings show some of the ways in which Charles and Francis Darwin "blindfolded" dark-grown grass seedlings; the lower drawings show what they observed in each case. Coleoptiles responded to light only when the top millimeter or so was exposed; they responded as they grew by bending toward the light a few millimeters below the tip. These observations suggested that the plant's "eye" (that which senses light) is in the tip, and that it sends a message from the tip to the region of bending.

the coleoptile to the growing region. If the tip is removed, the growth of the coleoptile is sharply inhibited; if the tip is then carefully replaced, growth resumes, even if the tip and base are separated by a thin layer of gelatin. The hormone moves down from the tip but does not move from one side of the coleoptile to the other. If the tip of an oat coleoptile is cut off and replaced so that it covers only one side of the cut end of the shoot, the coleoptile curves as the cells on the side below the replaced tip grow more rapidly than those on the other side.

With this information as a beginning, the Dutch botanist Frits W. Went succeeded, where many had failed, in isolating the hormone from oat coleoptiles. Went removed coleoptile tips and placed their cut surfaces on a block of gelatin, hoping the hormone would diffuse into the gelatin. Then he placed pieces of the gelatin block on decapitated coleoptiles—positioned to cover only one side, just as coleoptile tips had been placed in some of the earlier experiments. As they grew, the coleoptiles curved toward the side away from the gelatin. This curvature demonstrated that the hormone had indeed diffused into the gelatin block from the isolated coleoptile tips (Figure 33.13). The hormone had at last been isolated from the plant. It was later named auxin; still later, it was shown to be indoleacetic acid. Went's historic experiment was performed in 1926—the very year Kurosawa published his classic account of the isolation of a growth

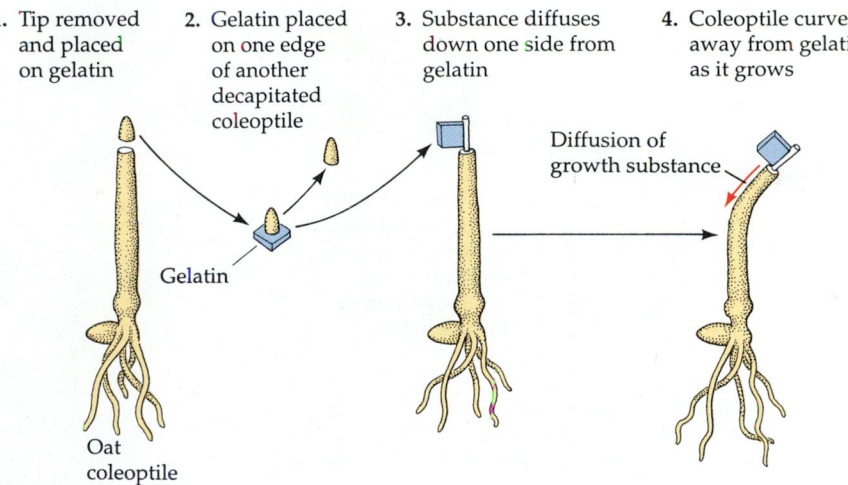

1. Tip removed and placed on gelatin

2. Gelatin placed on one edge of another decapitated coleoptile

3. Substance diffuses down one side from gelatin

4. Coleoptile curves away from gelatin as it grows

Gelatin

Diffusion of growth substance

Oat coleoptile

33.13 Went's Experiment
Went isolated the growth substance from coleoptile tips in a small block of gelatin. The gelatin block caused another coleoptile to bend, confirming that the block contained the growth substance. The substance is now called auxin.

substance from the fungus *Gibberella fujikuroi*, an accomplishment closely analogous to Went's.

Auxin Transport

Auxin was studied in a number of ways once it had been isolated from the plant. Early experiments showed that its movement through certain tissues is strictly polar, that is, unidirectional along a line from apex to base. By inverting the setups in half of the experiments, scientists determined that the apex-to-base direction of auxin movement has nothing to do with gravity; the polarity of this movement is a totally biological matter. Many plant parts show at least partial polarity of auxin transport. For example, auxin moves in leaf petioles from the blade end toward the stem end.

Phototropism and Gravitropism

The *lateral* movement of auxin in the apex affects the direction of plant growth. When light strikes a coleoptile from one side, auxin at the tip moves toward the shaded side. The imbalance thus established is maintained down the coleoptile, so that in the growing region below there is more auxin on the shaded side, causing the unequal growth that results in a coleoptile bent toward the light. This is phototropism (Figure 33.14*a*).

Similarly, but even in the dark, auxin moves to the lower side of a shoot that has been tipped over, causing more rapid growth in the lower side and, hence, an upward bend of the shoot. This phenomenon is **gravitropism,** the growth of a plant part in a direction determined by gravity (Figure 33.14*b*). The upward gravitropic response of shoots is defined as negative; the gravitropism of roots, which bend downward, is positive.

Auxin and Vegetative Development

Like the gibberellins, auxin has many roles in the plant. Cuttings from the shoots of some plants can produce roots and thus grow into entire new plants. For this to happen, certain undifferentiated cells in the *interior* of the shoot, originally destined to function only in food storage, must set off on an entirely new mission: to differentiate and become organized into the meristem of a root. These changes are very similar to those in the pericycle of a root when a branch root forms (see Chapter 29). Shoot cuttings of many species can be stimulated to grow roots profusely by dipping the cut surfaces into an auxin solution.

The effect of auxin on leaf **abscission,** the separation of old leaves from stems, is quite different. If the blade of the leaf is excised, the petiole abscises more rapidly than if the leaf had remained intact (Figure 33.15*a*). If the cut surface is treated with an auxin solution, however, the petiole remains attached to the plant, often longer than an intact leaf would have (Figure 33.15*b*). It appears that the time of abscission of leaves in nature is determined in part by a decrease in the movement of auxin, produced in the blade, through the petiole.

Auxin maintains **apical dominance,** the tendency of some plants to grow a single main stem with minimal branching. This phenomenon can be shown by an experiment with dark-grown pea seedlings. If the plant remains intact, the stem elongates and the lat-

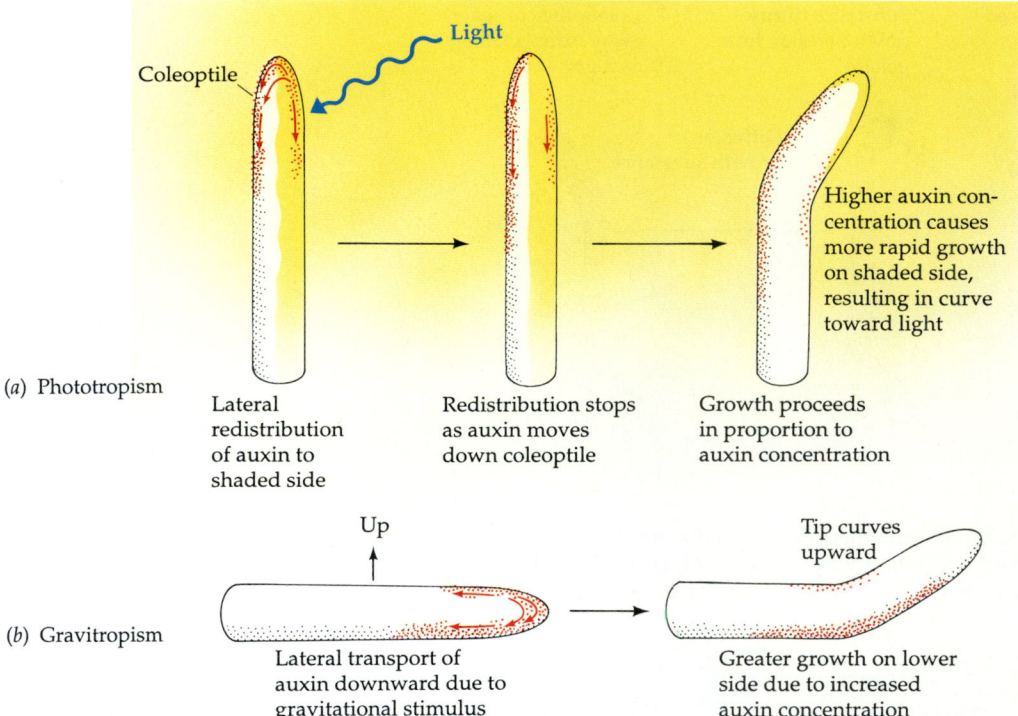

(a) Phototropism

Lateral
redistribution
of auxin to
shaded side

Redistribution stops
as auxin moves
down coleoptile

Growth proceeds
in proportion to
auxin concentration

Higher auxin con-
centration causes
more rapid growth
on shaded side,
resulting in curve
toward light

(b) Gravitropism

Lateral transport of
auxin downward due to
gravitational stimulus

Greater growth on lower
side due to increased
auxin concentration

33.14 Plants Respond to Light and Gravity
Auxin in the coleoptile tip moves toward the shaded side, beginning the
phototropic response. Auxin accumulates on the lower side of a horizontal
coleoptile, beginning the gravitropic response.

eral buds remain inactive. Removal of the apical
bud—the major site of auxin production—causes the
lateral buds to grow out vigorously, but this growth
is prevented if the cut surface of the stem is treated
with an auxin solution (Figure 33.16). Apical buds of
branches exert apical dominance, with their own lateral buds being inactive unless the apex is removed.
Note that in the experiments on leaves and stems
that we have discussed, removal of a particular part
of the plant produces an effect—abscission or loss of
apical dominance—and that the effect is prevented
by treatment with auxin. These results are consistent
with other data showing that the excised part of the
leaf or stem is an auxin source and that auxin in the
intact plant helps maintain apical dominance and delays the abscission of leaves.

Many synthetic auxins—chemical analogs of indoleacetic acid—have been produced and studied.
One of them, 2,4-dichlorophenoxyacetic acid (2,4-D),
has the striking property of being lethal to dicots at
concentrations that are harmless to monocots. This
property made 2,4-D a widely used **selective herbicide** that could be sprayed on a lawn or a cereal crop
to kill the dicots, thus eliminating most of the weeds.
Because 2,4-D takes a long time to break down, however, it pollutes the environment, so scientists are
seeking new approaches to selective weed killing
(Box 33.A).

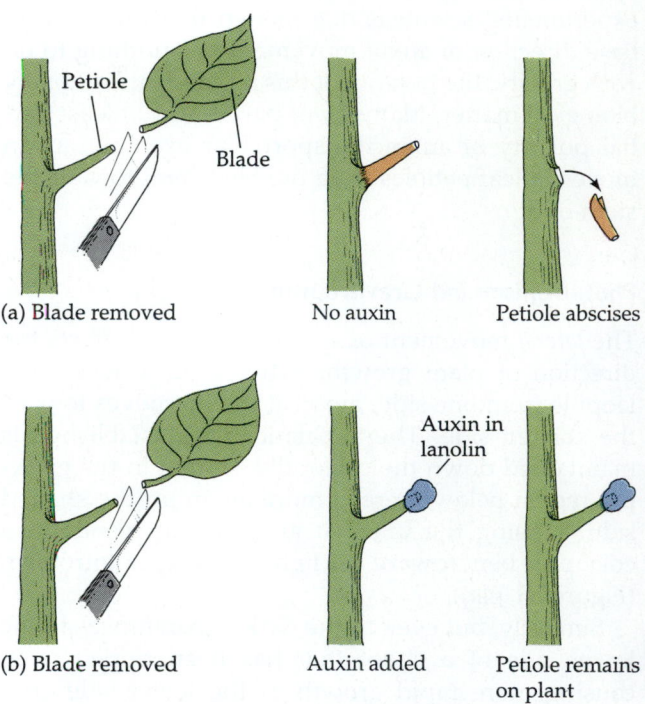

(a) Blade removed — No auxin — Petiole abscises

(b) Blade removed — Auxin added — Petiole remains on plant

33.15 Auxin Delays Leaf Abscission
The leaf blade is a source of auxin throughout the growing season.

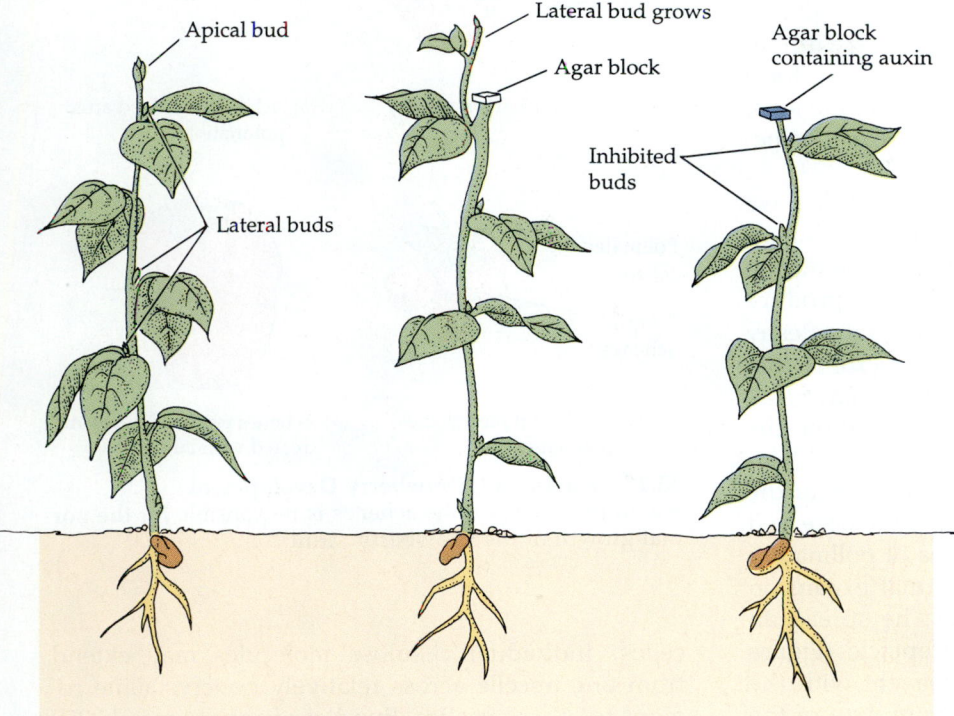

(a) In intact plants lateral buds are inhibited by apical bud

Apical bud

Lateral buds

(b) Decapitate plant. Agar block (no auxin) applied to stump. Buds grow out

Lateral bud grows

Agar block

(c) Agar block containing auxin applied at time of decapitation. Buds inhibited

Agar block containing auxin

Inhibited buds

33.16 Auxin and Apical Dominance
Apical dominance results from auxin produced by the apical bud, as suggested by these experimental results.

BOX 33.A

Strategies for Killing Weeds

The first commercial herbicides—weed killers—were substances that killed any plant they touched. Today we use such general herbicides when it is not necessary to save other vegetation; for example, products containing 3-aminotriazole (3AT) are used to kill stands of poison oak or poison ivy. We cannot use such an all-out, nonspecific approach if we wish to kill certain plants but retain others.

The next approach was the use of selective herbicides such as 2,4-D. Although 2,4-D is an effective herbicide, its use is controversial because it pollutes the environment.

The advent of recombinant DNA technology opened entirely new avenues of herbicide research. One approach is to transfer genes that confer resistance to herbicides into plants we want to be unaffected by the herbicides. The concept is simple and attractive: First, develop a highly effective herbicide that will kill all the plants in a field; second, genetically engineer the chosen crop plants so that they are resistant to that herbicide; finally, spray the field and watch the weeds die while the crop, protected by its inserted gene, prospers. Biotechnology companies view this as a promising method for weed control, although some scientists fear its environmental and health consequences.

Scientists at Calgene, a biotechnology company in Davis, California, recently reported the results of such an experiment. Their goal was to render selected plants insensitive to bromoxynil, a potent inhibitor of photosynthesis. The Calgene scientists discovered that a particular soil bacterium, *Klebsiella ozaenae*, converts bromoxynil to an inactive product. From this bacterium they isolated the gene that codes for the enzyme that inactivates bromoxynil. They then inserted the *Klebsiella* gene into tobacco plants. To make sure the gene would be expressed in the tobacco plants, they inserted it along with and under the control of a light-activated promoter. The promoter normally controls the gene that codes for rubisco, the enzyme that fixes carbon dioxide in photosynthesis. The experiment was a success: Transgenic plants containing the bacterial gene grew vigorously after being sprayed with bromoxynil solutions that killed control tobacco plants.

Auxin and Fruit Development

Although fruit development normally depends on prior fertilization of the egg, in many species treatment of an unfertilized ovary with auxin or gibberellin causes **parthenocarpy**—fruit formation without fertilization of the egg. Parthenocarpic fruit form spontaneously in some plants, including dandelions, seedless grapes, and the cultivated varieties of pineapple and banana.

The strawberry is an unusual "fruit." What we commonly call the fruit is actually a modified stem, or receptacle, with the tiny, dry "seeds" being the true fruits, called achenes. The achenes produce auxin, and the auxin induces the growth of the fleshy receptacle, as the French botanist Jean-Pierre Nitsch demonstrated (Figure 33.17). When he removed all the achenes within three weeks after pollination, the stem tissue did not develop into a "strawberry" (Figure 33.17b). If he pollinated only one to three of the many pistils and kept the others virgin, he observed localized receptacle growth in the area of pollination (Figure 33.17c). He could induce normal expansion if, after removing all of the achenes, he spread an auxin-containing paste over the receptacle (Figure 33.17d). These three results are consistent with the hypothesis that the achenes cause the growth of the receptacle by producing auxin.

Plant hormones control other aspects of fruit physiology and of the development and senescence of flower parts as well. These activities illustrate again the great diversity of important roles that hormones play.

Are There Master Reactions?

Each of the known plant hormones causes a variety of responses that are often seemingly unrelated, thus raising one of the major questions of plant physiology: Do all the effects of a particular hormone arise from a single mechanism—a common master chemical reaction? We know so little about how the plant hormones act at the molecular level that this question cannot yet be answered. For one hormone—auxin—we are beginning to gain some insight into its central mechanisms, if indeed there is more than one molecular mechanism. To appreciate this mechanism, we must first briefly consider the architecture of the plant cell wall.

Cell Walls and Growth

The principal strengthening component of the plant cell wall is cellulose, a large polymer of glucose. In the wall, cellulose molecules tend to associate with one another, forming crystalline regions called mi-

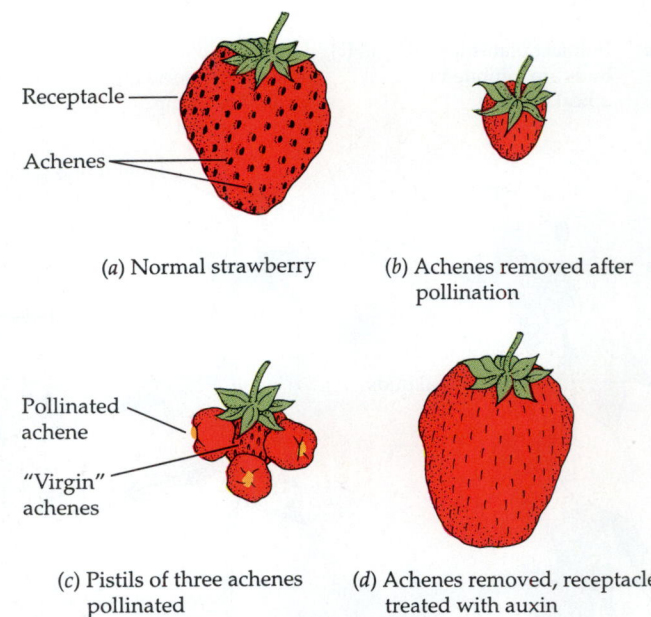

(a) Normal strawberry

(b) Achenes removed after pollination

(c) Pistils of three achenes pollinated

(d) Achenes removed, receptacle treated with auxin

33.17 Auxin and Strawberry Development
Auxin produced by the achenes is responsible for the normal growth of the strawberry "fruit."

celles. Individual cellulose molecules may extend from one micelle across relatively noncrystalline regions to other micelles. Bundles of approximately 250 cellulose molecules, including many micelles, constitute microfibrils visible with an electron microscope. What makes the cell wall rigid is a network of cellulose microfibrils connected by bridges of other, smaller polysaccharides (Figure 33.18a). Peter Albersheim and his colleagues at the University of Colorado proposed a model for the molecular architecture of the cell wall, showing how the other polysaccharides may interconnect the cellulose microfibrils (Figure 33.18b).

The growth of a plant cell is driven primarily by the uptake of water, which enters the cytoplasm of the cell and its vacuole (see Figure 4.24). As the vacuole expands, the cell grows rapidly, with the vacuole often making up more than 90 percent of the volume of a mature cell. As the vacuole expands, it presses the cytoplasm against the cell wall, and the wall resists this force. For the cell to grow, its wall must loosen and be stretched. As the wall stretches, it should become thinner; however, because new polysaccharides are deposited throughout the wall and new cellulose microfibrils are deposited at the inner surface of the wall, the cell wall maintains its thickness. Thus the cellulose microfibrils in the outermost part of the wall are the oldest, and those in the innermost part the youngest.

The wall plays a key role in controlling the growth rate of a plant cell. How does the plant determine the behavior of its cell walls?

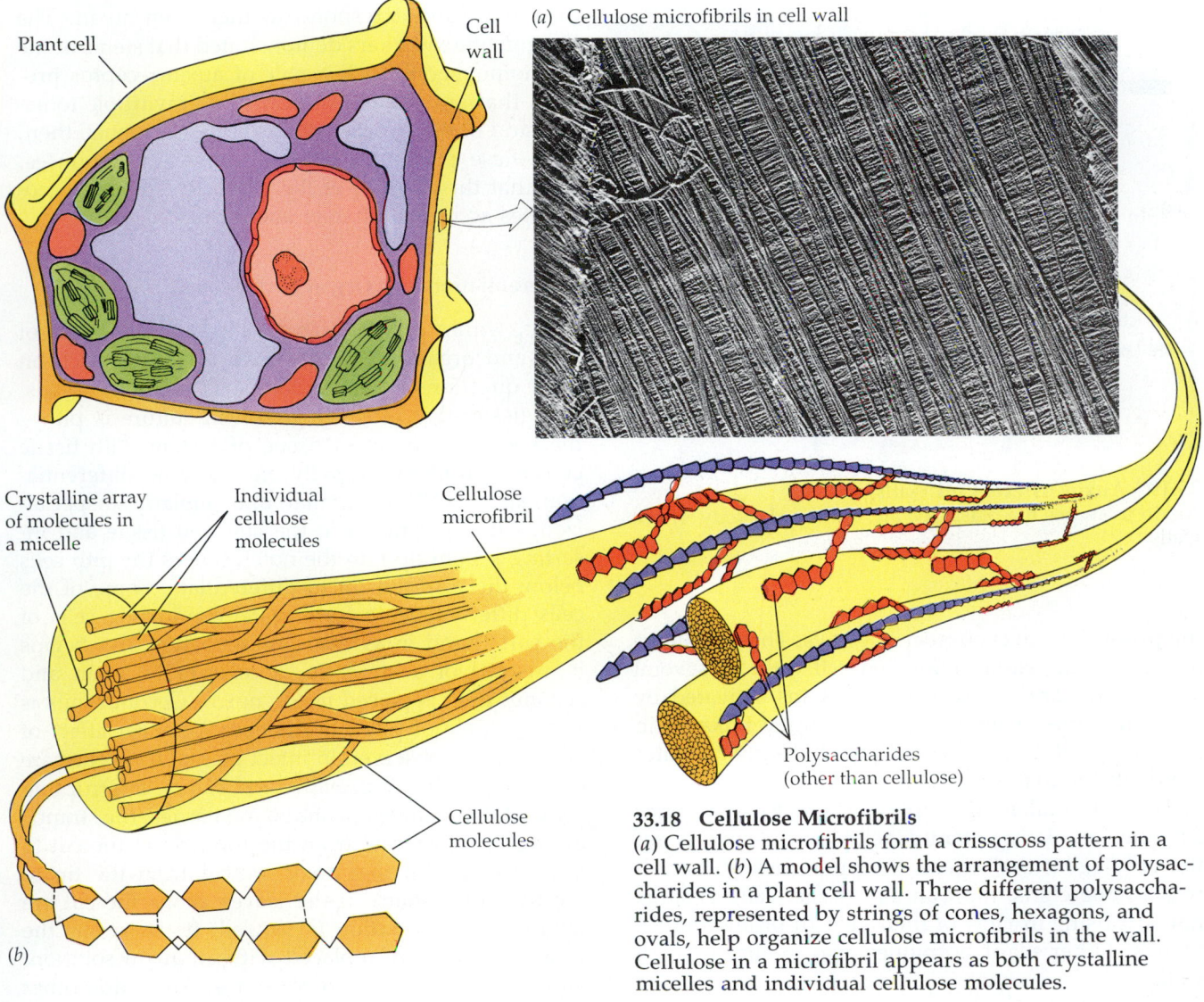

(a) Cellulose microfibrils in cell wall

Plant cell

Cell wall

Crystalline array of molecules in a micelle

Individual cellulose molecules

Cellulose microfibril

Cellulose molecules

Polysaccharides (other than cellulose)

(b)

33.18 Cellulose Microfibrils
(a) Cellulose microfibrils form a crisscross pattern in a cell wall. (b) A model shows the arrangement of polysaccharides in a plant cell wall. Three different polysaccharides, represented by strings of cones, hexagons, and ovals, help organize cellulose microfibrils in the wall. Cellulose in a microfibril appears as both crystalline micelles and individual cellulose molecules.

Auxin and the Cell Wall

Auxin can loosen cell walls—make them more stretchable—as the Dutch physiologist A. J. N. Heyn first demonstrated half a century ago. Heyn hung segments of oat coleoptiles on pins and hung weights on the ends of the segments, causing them to bend. Then he removed the weights, allowing the segments to bend back. Recovery was incomplete; that is, some of the bending was not reversible. Heyn called the reversible bending **elasticity** and the irreversible bending **plasticity** (Figure 33.19). Pretreating the coleoptile segments with auxin significantly increased their plasticity; it loosened the wall. This result suggested that auxin-induced cell expansion might result from just such a loosening effect.

It was later shown that auxin itself does not loosen cell walls upon contact. Rather, there is an intervening step; auxin may cause the release of a "wall-loosening factor" from the cytoplasm. Work in the

1970s in the United States and in Europe indicated that the wall-loosening factor was simply hydrogen ions (protons, H^+). Acidifying the growth medium (that is, adding H^+) causes segments of stems or coleoptiles to grow as rapidly as segments treated with auxin, and treating coleoptile segments with auxin causes acidification of the medium. Treatments that block acidification by auxin also block auxin-induced growth.

It was proposed that hydrogen ions, secreted into the cell wall as a result of auxin action, activate some enzyme or enzymes in the wall. These enzymes might digest specific linkages connecting the cellulose microfibrils, or they might alter bonds in the matrix in which the microfibrils reside. The end result in either case would be a temporary loosening of the wall. In 1992, a group led by Daniel Cosgrove at Pennsylvania State University isolated two interest-

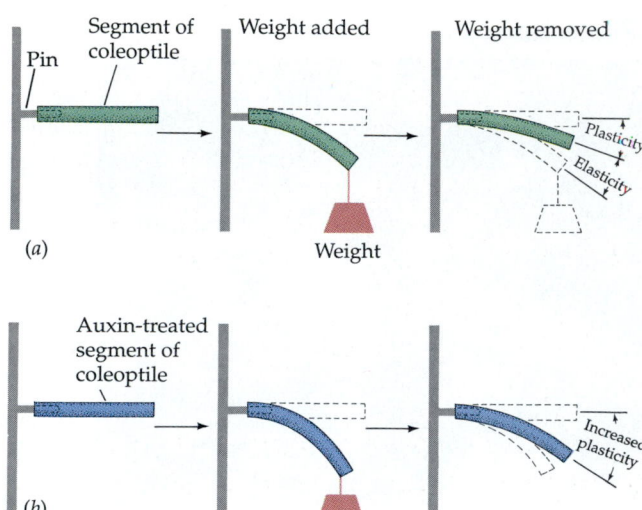

33.19   Auxin Affects Cell Walls
Auxin increases the plasticity, but not the elasticity, of cell walls.

ing proteins from cucumber cell walls. These proteins cause the extension of isolated cell walls of several species and appear to be the proteins activated by hydrogen ions. They seem not to digest linkages in polysaccharides such as cellulose, but rather to alter bonds in the matrix.

The cell wall is an important site for the major activities regulating plant development. Auxin is not the only agent that affects the properties of the cell wall. Gibberellins, too, can loosen the wall, which is not surprising in view of their growth-promoting activities. Other plant hormones also modify the cell wall.

Auxin Receptors

We will see in Chapter 36 that animal hormones begin to act only after binding with specific receptor proteins. Does the action of plant hormones also require their recognition by receptor proteins? Recently, it has been shown that several proteins can bind various plant hormones. It has not been shown, however, that these proteins function in the living plant as receptors that mediate the effects of the regulators.

In 1989 Glenn Hicks, David Rayle, and Terri Lomax, at Oregon State University, provided the first solid evidence for a connection between such apparent receptor proteins and an auxin-related plant response. They were working with the *diageotropica* (*dgt*) mutation of tomato. Plants homozygous for the *dgt* mutation fail to show normal gravitropism; instead of growing upright, they simply sprawl on the ground. They also show other symptoms indicating

that they cannot respond to their own auxin. The Oregon State workers demonstrated that stems of the *dgt* homozygotes lack a pair of auxin-receptor proteins that are present in normally gravitropic tomatoes and in many other plant species. It seems, then, that these proteins participate in auxin responses, and that the absence of these proteins in *dgt* homozygotes accounts for their aberrant growth.

Differentiation and Organ Formation

What, within a plant, signals the different types of cells and organs to form? Much of the research on these questions has been done with cultured tissues. One tissue that is easily grown in culture is pith—the spongy, innermost tissue of a stem. Pith tissue cultures proliferate rapidly but show no differentiation; all the cells are similar, and similarly unspecialized. Cutting a notch in the cultured tissue and inserting a stem tip into the notch causes the pith cells below the inserted tip to differentiate. Some of the cells differentiate to form water-conducting cells of the sort found in xylem. Differentiation also begins if, instead of a stem tip, a mixture of auxin and coconut milk is placed in the notch. Coconut milk is a rich source of plant hormones. A similar effect of auxin can be observed in intact plants. If notches are cut in the stems of *Coleus blumei* plants, interrupting some of the strands of conducting tissues, the strands gradually regenerate from the top side of the cut to the lower (recall that auxin moves from the tip to the base of a stem). If the leaves above the cut are removed, regeneration is slowed; if, however, the missing leaves are replaced with an auxin solution, new conductive tissue regenerates. Auxin and other plant hormones signal the formation of specific cell types.

Other work with cultured tissues has helped clarify which hormones control organ formation. Undifferentiated cultures of tobacco pith form roots when treated with an appropriate concentration of auxin. Another group of hormones—the cytokinins—causes buds and then shoots to form in such cultures. The pattern of organ formation depends on the ratio of auxin to cytokinin in the medium. A high proportion of auxin favors roots and a high proportion of cytokinins favors buds, but both processes are most active when both hormones are present.

CYTOKININS

The cytokinins, which have a variety of effects besides stimulating bud formation, promote cell division in cultured tissues, an activity that led to their discovery.

Discovery of the Cytokinins

After studying cell division for many years, Folke Skoog, at the University of Wisconsin, and his associate Carlos Miller reasoned that, since a plant hormone (auxin) is what regulates cell *expansion*, a plant hormone must also regulate cell *division*. Because cell division requires DNA replication, Skoog and Miller suspected that the hypothetical hormone regulates the metabolism of nucleic acids and that it might even consist of DNA itself.

In 1955 they and other members of Skoog's group were studying the effects of various compounds on the rate of cell division in cultures of carrot root tissue. They took an old bottle of herring sperm DNA (the only DNA on hand) off the shelf and added a bit to the medium in which some of the cultures were growing. They were gratified to observe that these cultures began to proliferate much more rapidly than the controls. Determined to discover what component of the DNA preparation was the active material, they purchased some fresh herring sperm DNA, but they were disappointed to find that it was quite inactive. Deciding that the only difference between the two samples of DNA was age, they tried to "age" the new DNA by heating it in a sterilizer. This worked! The heated DNA preparation caused rapid cell division in the carrot cultures. Careful chemical work revealed that a single substance in the old or sterilized DNA preparations was the active material. Skoog's group named this substance kinetin (see Figure 33.8e) and suggested that it might be just one of a family of compounds, which are now called cytokinins.

For several years they and other investigators tried in vain to find kinetin in plant tissues. What they did find were two closely related compounds called zeatin (see Figure 33.8f) and isopentenyl adenine. These two are naturally occurring cytokinins; kinetin, on the other hand, may be considered a synthetic, since it has not been isolated from plant tissue.

Other Activities of the Cytokinins

Cytokinins are believed to form primarily in the roots and to move to other parts of the plant. Adding an appropriate combination of auxin and cytokinin to the medium yields rapid growth of plant tissues. Cytokinins can cause certain light-requiring seeds to germinate when the seeds are kept in constant darkness. Cytokinins usually inhibit the elongation of stems, but they cause lateral swelling of stems and roots; the fleshy roots of radishes are an extreme example. Cytokinins stimulate lateral buds to grow into branches; thus the balance between auxin and cytokinin levels controls the bushiness of a plant.

Cytokinins increase the expansion of cut pieces of leaf tissue, so they may regulate normal leaf expansion. Cytokinins also delay the senescence of leaves. If leaf blades are detached from a plant and floated on water or a nutrient solution, they quickly turn yellow and show other signs of senescence. If instead they are floated on a solution containing a cytokinin, they remain green and senesce much more slowly. Cytokinins apparently regulate the redistribution of biologically active materials from one part of a plant to another. When one of a pair of leaves opposite each other on the stem of a bean plant is treated with a cytokinin, the treated leaf remains dark green and healthy. The untreated leaf opposite it, on the other hand, turns completely yellow and senesces rapidly as a result of its loss of nutrients to the treated leaf.

ETHYLENE

Whereas the cytokinins oppose or delay senescence, another plant hormone promotes it. This hormone is the gas ethylene, $H_2C{=}CH_2$, which is sometimes called the senescence hormone (see Figure 33.8d). Ethylene can be produced by all parts of the plant, and like all plant hormones, it exerts a number of effects. Back when streets were lit by gas rather than by electricity, leaves on trees near street lamps abscised earlier than those on trees farther from the lamps. We now know that it was ethylene, a combustion product of the illuminating gas, that caused the abscission. Auxin delays leaf abscission, but ethylene strongly promotes it; thus a balance of auxin and ethylene controls abscission (Figure 33.20).

Another effect of ethylene that is related to senescence is the ripening of fruit. The old saying, "One rotten apple spoils the barrel," is true. That rotten apple is a rich source of ethylene, which triggers the ripening and subsequent rotting of the others in the barrel. As the fruit ripens, it loses chlorophyll and its cell walls break down. Ethylene produced in the fruit tissue promotes both processes. Ethylene also causes an increase in its own production. Thus, once ripening begins, more and more ethylene is formed, and because it is a gas, ethylene diffuses readily throughout the fruit and even to neighboring fruit on the same or other plants.

Another gas, carbon dioxide, antagonizes the effects of ethylene. Commercial shippers and storers of fruit can thus precisely control the ripening of their wares. They hasten ripening by adding ethylene to the storage chambers; they slow ripening by adding carbon dioxide. This use of ethylene is the single most important use of a plant hormone in agriculture and commerce.

In some places autumn is a time of striking change

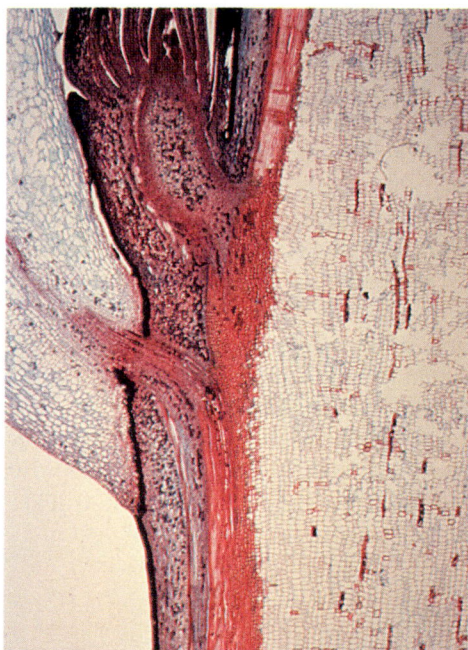

33.20 When a Leaf Is About to Fall
The petiole to the left is about to abscise from the stem by breaking away at the abscission layer (the black band). There is a bud in the axil of the leaf.

33.21 Leaf Senescence
Where leaves would be a liability under winter conditions, they senesce and die in autumn. Leaves are senescing in this forest in Vermont; only a skeleton of branches will remain to face the elements.

in the colors of the leaves of deciduous trees and shrubs. The display of colors is followed by the falling of the leaves, which have, in effect, passed through "old age" and died (Figure 33.21). Equally dramatic, in a different way, is the aging and death of entire plants, especially when they grow in great fields. Many crop plants grow vigorously throughout most of the year but then, after flowering and setting fruit, die together by the thousands. These examples show that senescence consists of irreversible, deteriorative changes controlled by internal factors.

Is senescence simply an undesirable but unavoidable fact of life, or does it play a useful role? Leaf senescence and the subsequent abscission are of real importance for the survival of the plant. Leaves senesce and abscise at the end of the growing season, shortly before the onset of the severe conditions of winter. Many species of plants have delicate leaves that could be a liability during a typical winter in the temperate zone: The temperature would be too low for efficient photosynthesis (and the ground might be frozen, making water unavailable), yet water could still be lost from the stomata in the leaves; damage from freezing would render the leaf unable to function normally during the next growing season in any case. Before the leaves die and are shed, their proteins are hydrolyzed to yield amino acids, which are then exported from the leaves to the stems—an important form of resource conservation. Controlled leaf abscission thus costs the plant little and benefits

it greatly. In other parts of the world, plants shed their leaves during the harsh dry season and grow them during the wet periods, which are more favorable for growth.

What about the senescence and death of the entire plant that follows flowering and seed setting in some species? This process appears to be an adaptation for producing more offspring—by pumping so much energy (food) and so many nutrients into the seeds that the parent essentially starves itself to death.

Although primarily associated with senescence, ethylene is active at other stages of plant development as well. The stems of many dicot seedlings that have not yet seen light—as they grow upward in soil during germination—often form an apical hook (Figure 33.22). The apical hook is maintained through an asymmetric production of ethylene gas, which inhibits the elongation of cells on the inner surface of the hook. Ethylene inhibits stem elongation in general, promotes lateral swelling of stems (as do the cytokinins), and causes stems to lose their sensitivity to gravitropic stimulation.

ABSCISIC ACID

Abscisic acid is another hormone with multiple effects in the living plant (see Figure 33.8g). This com-

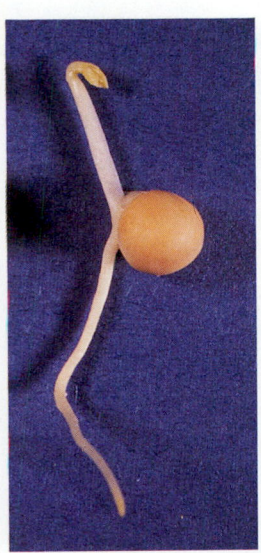

33.22 The Apical Hook of a Dicot
Asymmetric production of ethylene is responsible for the apical hook of this young pea seedling, which was grown in the dark.

pound inhibits stem elongation and is generally present in high concentrations in dormant buds and some dormant seeds. As we saw in Chapter 30, it also regulates gas and water vapor exchange between leaves and the atmosphere, through its effects on the stomata in the leaf surface. Abscisic acid is sometimes referred to as the stress hormone of plants because it accumulates when plants are deprived of water and because of its possible role in maintaining winter dormancy of buds.

In temperate zones the shoots of perennial plants do not grow constantly and in all seasons. At some time in the year the terminal buds of temperate-zone perennials become inactive, and growth ceases until the next spring. This dormancy minimizes damage to the plant during a harsh winter. Buds on a typical deciduous tree of the northern temperate zone undergo a number of changes well in advance of winter. These changes include the formation of thickened, overlapping **bud scales** that are covered with wax, which helps waterproof the bud contents—the leaf primordia and the growing point of the stem (Figure 33.23). The winter bud often contains an insulating material consisting of modified, cottony leaves.

Elsewhere in the plant are other changes. Leaves abscise, and the scars produced where leaves were formerly attached to the stems are sealed with a corky material. Lateral growth of the trunk ceases, and the solute concentrations in the transport systems increase, lowering the freezing point of the sap. These are several of the changes that constitute winter dormancy. In at least some species, some of these changes appear to be associated with an increased concentration of abscisic acid in the buds. Both the onset and termination of winter dormancy are precisely controlled.

ENVIRONMENTAL CUES

An environmental cue, the length of the night, determines the onset of winter dormancy. As summer wears on, the days get shorter—that is, the nights become longer. Leaves have a mechanism for measuring the length of the night, as we will see in the next section. This is a marvelous way to determine the season of the year. If a plant determined the season by the temperature, it could be fooled by a winter warm spell or by unseasonable cold weather in the summer. The length of the night, on the other hand, is determined by Earth's rotation around the sun and does not vary; plants use this accurate indicator to time several aspects of growth and development.

Length of night is one of several environmental cues detected by plants, or by individual parts such as leaves. Light—its presence or absence, its intensity, and its duration—provides various cues. Temperature, too, provides important environmental cues, both by its value at any particular time and by the distribution of warmer and colder stretches over a period of time. The plant "reads" an environmental cue and then "interprets" it, often by stepping up or decreasing its production of hormones.

LIGHT AND PHYTOCHROME

Light regulates many aspects of plant development. For example, some seeds will not germinate in darkness but do so readily after even a brief exposure to light. Studies have shown that blue and red light are highly effective in promoting germination, whereas green light is not. Of particular importance to plants is the fact that far-red light *reverses* the effect of a prior exposure to red light! Far-red is a very deep red, bordering on the limit of human vision and cen-

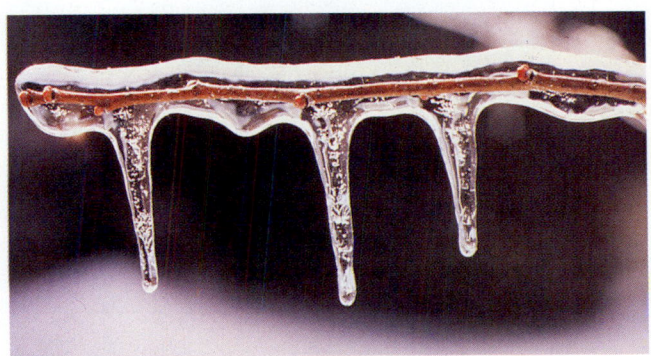

33.23 Winter Dormancy
As winter approaches, many deciduous plants cease growth, cover their buds with scales, and shed their leaves—all changes that aid survival in harsh conditions. This winter-dormant twig is coated with ice.

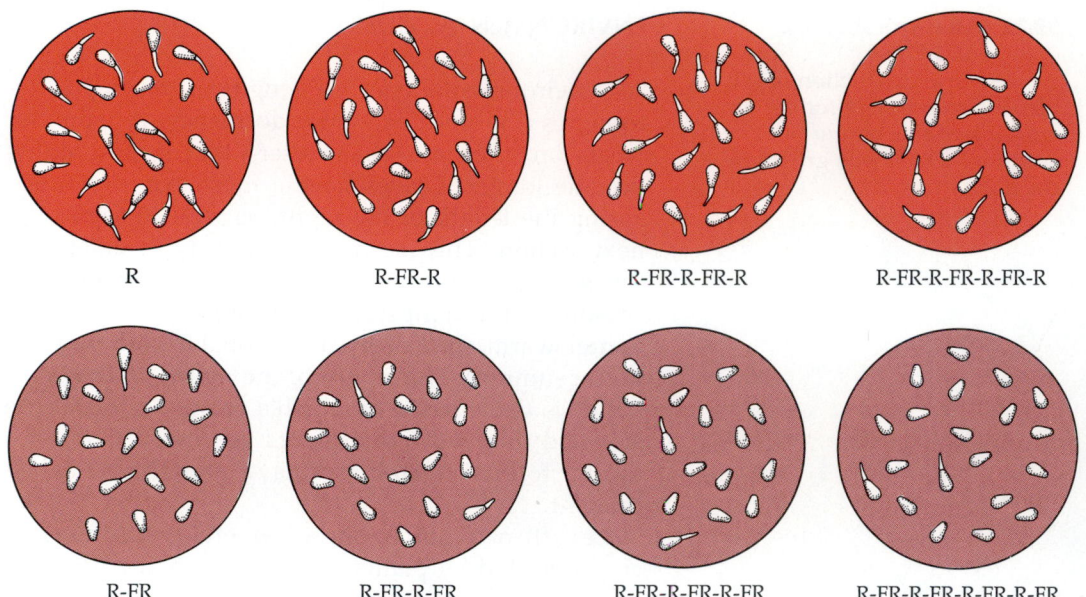

R   R-FR-R   R-FR-R-FR-R   R-FR-R-FR-R-FR-R

R-FR   R-FR-R-FR   R-FR-R-FR-R-FR   R-FR-R-FR-R-FR-R-FR

33.24   Sensitivity of Seeds to Light
Lettuce seeds were exposed to alternating periods of red light for 1 minute and
far-red light for 4 minutes. Seeds germinated if the final exposure was to red
(top), and remained dormant if the final exposure was to far-red (bottom). In
each case the final exposure reversed the effect of the preceding exposure to the
other wavelength of light.

tered upon a wavelength of 730 nm; red wavelengths
are around 660 nm. If exposed to brief, alternating
periods of red and far-red light in close succession,
seeds respond only to the final exposure: If red, they
germinate; if far-red, they remain dormant (Figure
33.24). This reversibility of the effects of red and far-
red light regulates many other aspects of plant development.

The basis for the red and far-red effects resides in

a bluish pigment—a protein called **phytochrome**—
which exists in plants in two interconvertible forms.
The pigment is blue because it absorbs red and far-
red light (Figure 33.25), and reflects blue light. Red
light converts phytochrome into one form; far-red
converts the phytochrome back into the other form
(Figure 33.26). Light drives the interconversion of the
two forms of phytochrome, both in the test tube and
in the living plant. The form that absorbs principally
red light is called P_r. Upon absorption of a photon of
red light, a molecule of P_r is converted into P_{fr}, the
far-red absorbing form. P_{fr} has a number of important
biological effects. As we have just seen, one of them
is to initiate germination in certain seeds.

33.25   Absorption Spectra of Phytochrome
P_r absorbs most strongly at 660 nm, in the red region of
the spectrum; P_{fr} absorbs most strongly at 730 nm, in the
far-red.

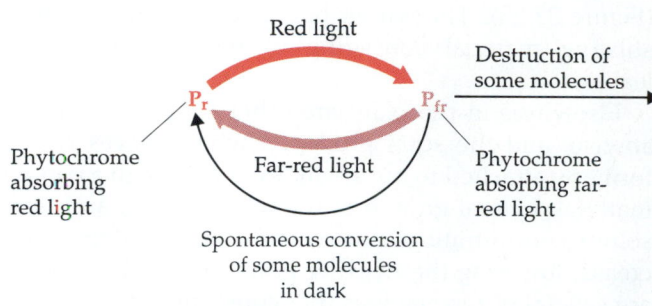

Red light

Destruction of
some molecules

P_r P_{fr}

Phytochrome
absorbing
red light

Far-red light

Phytochrome
absorbing far-
red light

Spontaneous conversion
of some molecules
in dark

33.26   Behavior of Phytochrome
Phytochrome can exist either as P_r, which absorbs red
light, or as P_{fr}, which absorbs far-red light. Each form is
converted to the other by light of the appropriate color. In
the living plant, some P_{fr} is spontaneously converted to
P_r, and some P_{fr} is destroyed.

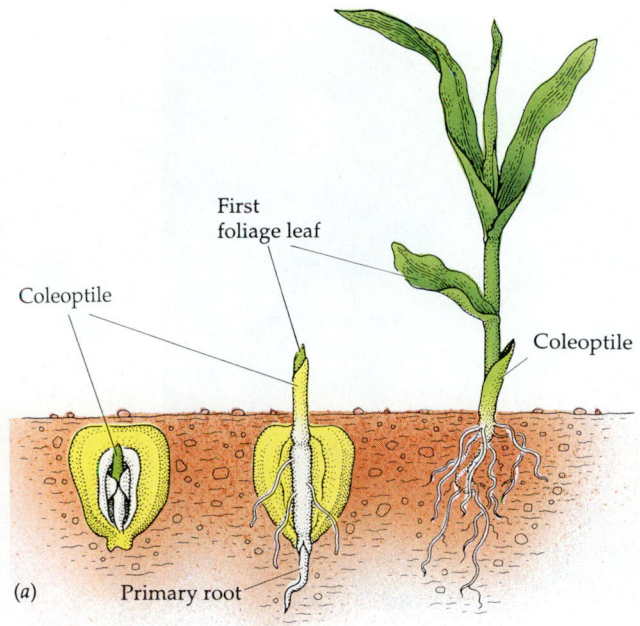

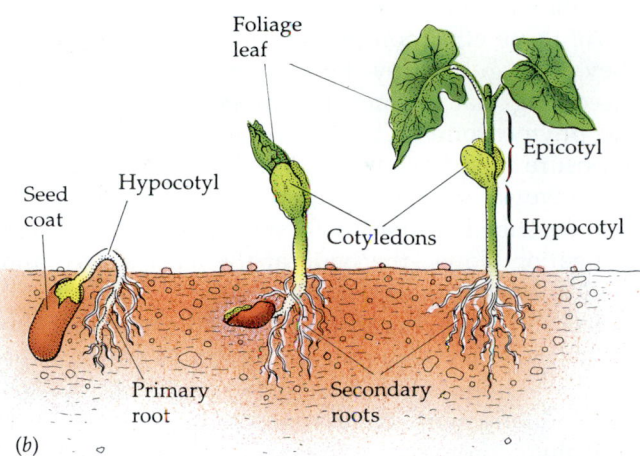

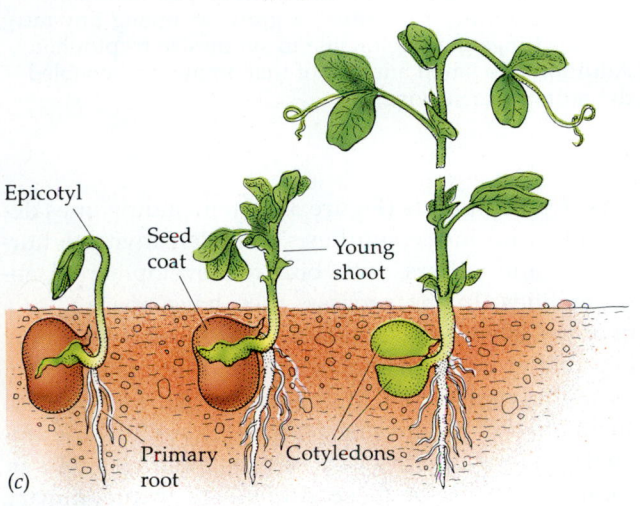

33.27 Patterns of Early Shoot Development
Shoots are protected as they penetrate the soil during germination. (a) A coleoptile covers the early shoot of corn and other monocots. After the shoot emerges from the soil, it pierces the coleoptile and grows out. (b) The shoot apex of most dicots, such as the bean shown here, is protected by the cotyledons as the upper part of the plant is pulled through the soil by the elongating hypocotyl. When the epicotyl elongates, the first foliage leaves emerge. (c) The cotyledons of other dicots, such as peas, remain in the soil. The shoot apex is pulled up as the bent epicotyl elongates.

We do not yet know how P_{fr} produces its many effects. Some of a plant's phytochrome may be included in membranes, where it may regulate how ions move into and out of cells and organelles. P_{fr} may function as a channel through which ions move until the conversion to P_r closes the channel. This and other ideas are currently undergoing active investigation.

Depending on whether any of a seedling's tissue is exposed to light, phytochrome helps regulate its early growth. The radicle, the embryonic root, is the first portion of the seedling to escape the seed coat. The shoot emerges later. Frequently seeds germinate below the soil surface, yielding a young plant that is **etiolated**—pale as a result of being kept in darkness. The seedling must reach the surface and begin photosynthesis before its food reserves are expended and it starves. Plants have evolved a variety of ways to cope with this problem. Etiolated flowering plants, for example, do not form chlorophyll. They turn

green only upon exposure to light, thereby conserving precious resources, for chlorophyll would be of no use in the dark. Only when light is available does it "pay" to expend metabolic energy on the production of chlorophyll. An etiolated shoot elongates rapidly to hasten its arrival at the soil surface, where photosynthesis quickly begins.

Early shoot development varies among the flowering plants (Figure 33.27). In some monocots, such as grasses, the shoot is initially protected by a leaf sheath, the coleoptile. The developing shoot later grows out through the coleoptile. Dicots lack this protective structure. In most dicots, the hypocotyl elongates and the cotyledons are carried to the surface, where they become the first important photosynthetic structures. In other dicots, such as peas, the cotyledons remain below the soil surface, and tissue above them, the epicotyl, grows up through the soil. In all three cases, elongation—whether of the coleoptile, the hypocotyl, or the epicotyl—proceeds much more rapidly in the dark than in the light.

The shoot of etiolated dicot seedlings curves back into a hook at the apex during part of their development, in response to ethylene production (see Figure 33.22). The hook protects the tender apical bud from being forced directly upward to batter its way through the soil. The leaves of an etiolated seedling do not expand. An underground leaf cannot photosynthesize, and dragging an expanded leaf through the soil would be a real problem.

All these etiolation phenomena are adaptive, and they are regulated by the pigment phytochrome. In a seedling that has never been exposed to light, all the phytochrome is in the red-absorbing (P_r) form. Exposure to light converts P_r to P_{fr}, and the P_{fr} initiates reversal of the etiolation phenomena we have just described: Chlorophyll synthesis begins, shoot elongation slows, the hook at the apex opens, and the leaves start to expand.

Some of these responses, notably the onset of chlorophyll synthesis, require activation of parts of the plant's genome. Recall that we identified the genome, along with environmental cues, hormones, and phytochrome, as a major determinant of development.

STUDYING THE PLANT GENOME

How do we get at the genome to study it? The most common approach is to obtain many developmental mutants and to analyze and modify them. That, however, is much more easily said than done. For years, biologists were much more successful using such approaches with bacteria and viruses than with plants and animals. Among the advantages of bacteria (particularly *E. coli*) for such studies are their small size and short generation times. But what would help the plant biologist? As is often the case, a series of seemingly unrelated observations led to the realization that a little weed, related to mustard, had the right properties to be the plant biologist's *E. coli*.

Arabidopsis thaliana is a small herb that produces tiny seeds. Thus hundreds of seeds can be planted in a single petri plate, facilitating searches for mutants. The generation time is short—about a month and a half from seed to flower—and a single plant can produce thousands of seeds. The genome is very small for a plant—approximately the same size as that of the fruit fly *Drosophila melanogaster*. It is easy to perform techniques such as transformation on *Arabidopsis* seedlings.

Research groups have isolated many interesting

33.28 Wild-Type and Mutant *Arabidopsis thaliana*
The large wild-type seedling is growing among tiny mutant seedlings that are unable to synthesize tryptophan. Addition of a small amount of that amino acid enabled the mutants to survive.

Arabidopsis mutants (Figure 33.28), including ones deficient in hormone synthesis or in sensitivity to hormones and others with bizarre developmental abnormalities. In the process, they have opened up a new era of research in plant development. It is now possible to perform analyses similar to those used in studying animal development (see Chapter 17). Not surprisingly, we are discovering that plants and animals have a number of developmental elements in common. Many of these studies are in preliminary stages, however, and we have taken a conservative approach in this text by limiting discussion to processes that have been clearly demonstrated and avoiding extension of animal developmental reasoning to areas less thoroughly studied in plants. We are optimistic, though, about prospects for future work on such things as morphogens and pattern formation, which have been studied in animal development, by working with mutant forms of plants. Homeotic mutations—mutations that change organ identities—are found in plants such as *Arabidopsis* as well as in animals, as we shall see in the next chapter.

SUMMARY of Main Ideas about the Regulation of Plant Development

Environmental cues, hormones, phytochrome, and the genome regulate plant development.

Seed germination is followed by seedling growth, flowering, fruit formation, senescence, and death.

Seeds may be dormant for a time before germination.

Seeds imbibe water to germinate and then mobilize food reserves.
Review Figure 33.7

If a seed germinates underground, the seedling grows rapidly until it reaches the surface.
Review Figure 33.1

A growing plant orients itself with respect to light and gravity (phototropism and gravitropism).
Review Figure 33.14

Photoperiodic plants flower at particular times of the year.
Review Figure 33.2

Following fertilization, a fruit develops and then ripens.

The environment and hormones trigger a plant's senescence and death.
Review Figure 33.3

Some perennial plants become dormant in the winter.
Review Figure 33.21

Plants have several hormones, each producing a variety of responses.
Review Table 33.1 and Figure 33.8

Auxin and the gibberellins regulate shoot development and many other phenomena.
Review Figures 33.7, 33.10, 33.11, 33.14, 33.16, 33.17, and 33.19

Abscisic acid is the "stress hormone."

Ethylene is the "senescence hormone."

Cytokinins are important from seed germination to senescence.

Normal development depends on a proper dynamic balance of all the hormones.

Phytochrome is a protein—not a hormone—that regulates many developmental phenomena.

Phytochrome exists in two forms, P_r and P_{fr}, that are interconverted by red and far-red light.
Review Figures 33.25 and 33.26

Red light triggers seed germination by converting P_r to P_{fr}.
Review Figure 33.24

Much current work in plant development focuses on the genome.

SELF-QUIZ

1. Which of the following is *not* an advantage of seed dormancy?
   a. It makes the seed more likely to be digested by birds that disperse it.
   b. It counters the effects of year-to-year variations in the environment.
   c. It increases the likelihood of a seed germinating in the right place.
   d. It favors dispersal of the seed.
   e. It may result in germination at a favorable time of year.

2. Which of the following does *not* participate in seed germination?
   a. Imbibition of water
   b. Metabolic changes
   c. Growth of the radicle
   d. Mobilization of food reserves
   e. Extensive mitotic divisions

3. To mobilize its food reserves, a germinating barley seed
   a. becomes dormant.
   b. undergoes apomixis.
   c. secretes gibberellins into its endosperm.
   d. converts glycerol and fatty acids into lipids.
   e. embryo takes up proteins from the endosperm.

4. The gibberellins
   a. are responsible for phototropism and gravitropism.
   b. are gases at room temperature.
   c. are produced only by fungi.
   d. cause bolting in some biennial plants.
   e. inhibit the synthesis of digestive enzymes by barley seeds.

5. In coleoptile tissue, the transport of auxin
   a. is from base to tip.
   b. is from tip to base.
   c. can be toward either the tip or the base, depending on the orientation of the coleoptile with respect to gravity.
   d. is by simple diffusion, with no preferred direction.
   e. does not occur because auxin is used where it is made.

6. Which process is *not* directly affected by auxin?
   a. Apical dominance
   b. Leaf abscission
   c. Synthesis of digestive enzymes by barley seeds
   d. Root initiation
   e. Parthenocarpic fruit development

7. Plant cell walls
   a. are strengthened primarily by proteins.
   b. often make up more than 90 percent of the total volume of an expanded cell.
   c. can be loosened by an increase in pH.
   d. become thinner and thinner as the cell grows longer and longer.
   e. are made more plastic by treatment with auxin.

8. Which statement is *not* true of the cytokinins?
   a. They promote bud formation in tissue cultures.
   b. They delay the senescence of leaves.
   c. They usually promote the elongation of stems.
   d. They cause certain light-requiring seeds to germinate in the dark.
   e. They stimulate the development of branches from lateral buds.

9. Ethylene
   a. is antagonized by carbon dioxide.
   b. is liquid at room temperature.
   c. delays the ripening of fruits.
   d. generally promotes stem elongation.
   e. inhibits the swelling of stems, in opposition to cytokinin effects.
10. Phytochrome
    a. is a nucleic acid.
    b. exists in two forms interconvertible by light.
    c. is a red or far-red colored pigment.
    d. is sometimes called the "stress hormone."
    e. is the photoreceptor for phototropism.

FOR STUDY

1. How may it be advantageous for some species to have seeds whose dormancy is broken by fire?

2. Cocklebur fruits contain two seeds each, and the two seeds are kept dormant by two different mechanisms. How may this be advantageous to cockleburs?

3. Corn stunt virus causes a great reduction in the growth rate of infected corn plants, so the diseased plants take on a dwarfed form. Since their appearance reminds you of the genetically dwarfed corn studied by Phinney, you suspect that the virus may inhibit the synthesis of gibberellins by the corn plants. Describe two experiments you might conduct to test this hypothesis, only one of which should require chemical measurement.

4. Whereas relatively low concentrations of auxin promote the elongation of segments cut from young plant stems, higher concentrations generally inhibit growth, as shown in the figure. In some plants, the inhibitory effects of high auxin concentrations appear to be secondary: High auxin concentrations cause the synthesis of ethylene, which is what causes the growth inhibition. Cobalt ions inhibit ethylene synthesis. How do you think the addition of cobalt ions to the solutions in which the stem segments grew would affect the appearance of the graph?

5. When carbon dioxide is applied to plants, it has effects opposite to those of ethylene. Carbon dioxide is sometimes added to the atmosphere around fruit being shipped to other parts of the country. What do you suppose is the purpose of this procedure?

READINGS

Bewley, J. D. and M. Black. 1994. *Seeds: Physiology of Development and Germination*, 2nd Edition. Plenum, New York. Lots of more advanced information; includes a chapter on agricultural and industrial aspects.

Evans, M. L., R. Moore and K.-H. Hasenstein. 1986. "How Roots Respond to Gravity." *Scientific American*, December. Classical and modern experimentation on the mechanisms of gravitropism.

Mandoli, D. F. and W. R. Briggs. 1984. "Fiber Optics in Plants." *Scientific American*, August. How plants guide light to regions of high phytochrome concentration. Includes an excellent description of the light environment in a wheat field.

Moses, P. B. and N.-H. Chua. 1988. "Light Switches for Plant Genes." *Scientific American*, April. How is light absorption by phytochrome transduced into developmental effects? Some stretches of DNA respond to phytochrome by turning on specific genes.

Raven, P. H., R. F. Evert and S. Eichhorn. 1991. *Biology of Plants*, 5th Edition. Worth, New York. A well-balanced general botany textbook.

Salisbury, F. B. and C. W. Ross. 1992. *Plant Physiology*, 4th Edition. Wadsworth, Belmont, CA. A sound textbook with excellent chapters on plant hormones and development.

Taiz, L. and E. Zeiger. 1991. *Plant Physiology*. Benjamin/Cummings, Redwood City, CA. Chapters 15 through 20 of this authoritative textbook deal with plant hormones, phytochrome, and other aspects of development.

What are all these flowers for—decoration? A flower is a sexual reproductive structure, containing either female or male organs or both. The female and male organs produce, respectively, eggs and sperm. Petals produce neither eggs nor sperm, but the showy petals of many plants attract animals that carry pollen from plant to plant, assisting in cross-fertilization. Thus the petals are part of a plant's way of getting sperm and eggs together.

Biologists are still seeking answers to some important questions about flowering and reproduction. We want to know just how environmental clues lead to flower formation. The plants in this photo are of several species, each having flowers with characteristic structures. The kinds of flower parts are pretty much the same, but their numbers, shapes, colors, and arrangements are different—we want to know how flowers develop in general, and in the specific ways characteristic of different species. In this chapter we deal with several aspects of plant reproduction, including some for which we are still in search of answers.

MANY WAYS TO REPRODUCE

Plants have many ways of reproducing themselves—and with humans helping, there are even more ways. Flowers are sex organs. It is thus no surprise that almost all flowering plants reproduce sexually. But many reproduce asexually as well; some reproduce asexually most of the time. Which is the better way to reproduce? Most of the answers to this question relate to genetic diversity or genetic recombination because sexual reproduction produces new genetic combinations. The details of sexual reproduction differ among different species of flowering plants. In our discussion of Mendel's work (see Chapter 10), we saw that some plants can reproduce sexually either by cross-pollinating or self-pollinating. Self-pollination is possible because, as we explained in Chapter 25, in many species each individual has both male and female sex organs.

Both sexual and asexual reproduction are important in agriculture. Annual crops, including wheat, rice, millet, and corn—the great grain crops, all of which are grasses—as well as plants in other families, such as soybeans and safflower, are grown from seed—that is, sexually. Other crops begin asexually by slips, grafts, or other means. Orange trees, which have been under cultivation for centuries, are grown from seed except for one type, the navel orange. This plant has apparently arisen only once in history. Early in the nineteenth century, on a plantation on the Brazilian coast, one seed gave rise to one tree that had aberrant flowers. Parts of the flowers

What Are All the Flowers For?

34

Plant Reproduction

aborted, and seedless fruits were formed. Every navel orange in the world comes from a navel orange tree derived asexually from another, which came from another, and another, and so on back to that original Brazilian tree. Navel oranges must be propagated asexually. Strawberries need not be, because they are capable of setting seed. Nonetheless, asexual propagation of strawberries is common because vast numbers of plants that are genetically identical to one particularly desirable plant can be produced from strawberry runners.

We will treat asexual reproduction in greater detail at the end of this chapter. We begin, however, by considering sexual reproduction.

SEXUAL REPRODUCTION

Sexual reproduction provides genetic diversity through recombination. Meiosis and mating shuffle genes into new combinations, giving a population a variety of genotypes in each generation. This genetic diversity may serve well as the environment changes or as the population expands into new environments. The adaptability resulting from genetic diversity is why sexual reproduction is often thought of as being "better" than asexual reproduction.

As you know, the flower is an angiosperm's device for sexual reproduction. Figure 25.24 shows the structure of a "typical" flower. (You may also find it useful to refer back to pages 563–569 in Chapter 25.) A complete flower consists of four groups of modified leaves: the carpels, stamens, petals, and sepals. To review briefly, the female organs, bearing megasporangia, are the carpels. Enclosed within the carpels are one or more ovules. Each ovule consists of the megasporangium it encloses, its protective layers of surrounding integuments, and its stalk. There is an opening in the integuments, the micropyle, through which the pollen tube grows on its way to the megasporangium. A pistil, consisting of one or more fused carpels, bears a pollen-receptive stigma at the tip of its elongated style. The male organs, bearing microsporangia, are the stamens. Each stamen consists of a filament bearing an anther with four microsporangia.

In addition to these sexual parts, many flowers have petals and sepals arranged in whorls around the spore-bearing carpels and stamens. The petals, often colored, constitute the corolla. Below them the sepals, often green, constitute the calyx. All the parts of a flower are borne on a stem tip, the receptacle.

Gametophytes

(Before reading this section, you may wish to review the section at the beginning of Chapter 25 entitled "Alternation of Generations.")

Gametophytes, the gamete-producing generation, develop from haploid spores in sporangia. Female gametophytes (megagametophytes), which are called **embryo sacs**, develop in megasporangia. Male gametophytes, the **pollen grains**, develop in microsporangia.

The megasporocyte, a cell within the ovule's megasporangium, divides meiotically to produce four haploid megaspores. All but one of the megaspores then degenerate. The surviving megaspore undergoes mitotic divisions, usually producing eight nuclei, all initially contained within a single cell—three nuclei at one end, three at the other, and two in the middle. Subsequent cell wall formation leads to an elliptical, seven-celled megagametophyte with a total of eight nuclei. At the end of the megagametophyte nearest the micropyle are three tiny cells: the **egg** and two cells called **synergids**. At the opposite end are three antipodal cells, and in the large central cell are two **polar nuclei**. Note that the embryo sac (the female gametophyte or megagametophyte) is the entire seven-celled, eight-nucleate structure. Follow the arrows on the left-hand side of Figure 34.1 to review the development of the embryo sac.

The male gametophyte, or pollen grain, consists of fewer cells than the female. As the right-hand side of Figure 34.1 illustrates, development of the pollen grain begins when the microsporocytes divide meiotically within the four microsporangia, which are fused into two anthers. Each resulting microspore normally undergoes one mitotic division within the spore wall before the anthers open and shed the pollen. Further development of the pollen grain, which we will describe shortly, is delayed until the pollen arrives at the stigma of a pistil. The transfer of pollen from the anther to the stigma is referred to as **pollination**.

Pollination

Gymnosperms and angiosperms evolved independence from water as a medium for gamete travel and fertilization—a freedom not shared by other plant groups. The sperm nuclei of gymnosperms and angiosperms travel within a pollen grain (Figure 34.2), and the pollen tube provides a route to the ovary. But how do pollen grains travel from an anther to a stigma? The mechanisms that have evolved for pollen transport are many and various. In one, pollination is accomplished before the flower bud opens, as in peas and their relatives—resulting in the self-fertilization described in Chapter 10. In self-pollination, pollen is transferred by the direct contact of anther and stigma within the same flower.

Wind is the vehicle for pollen transport in many species, especially grasses (Figure 34.3). Wind-pollinated flowers have sticky or featherlike stigmas, and they produce pollen grains in great numbers. Some

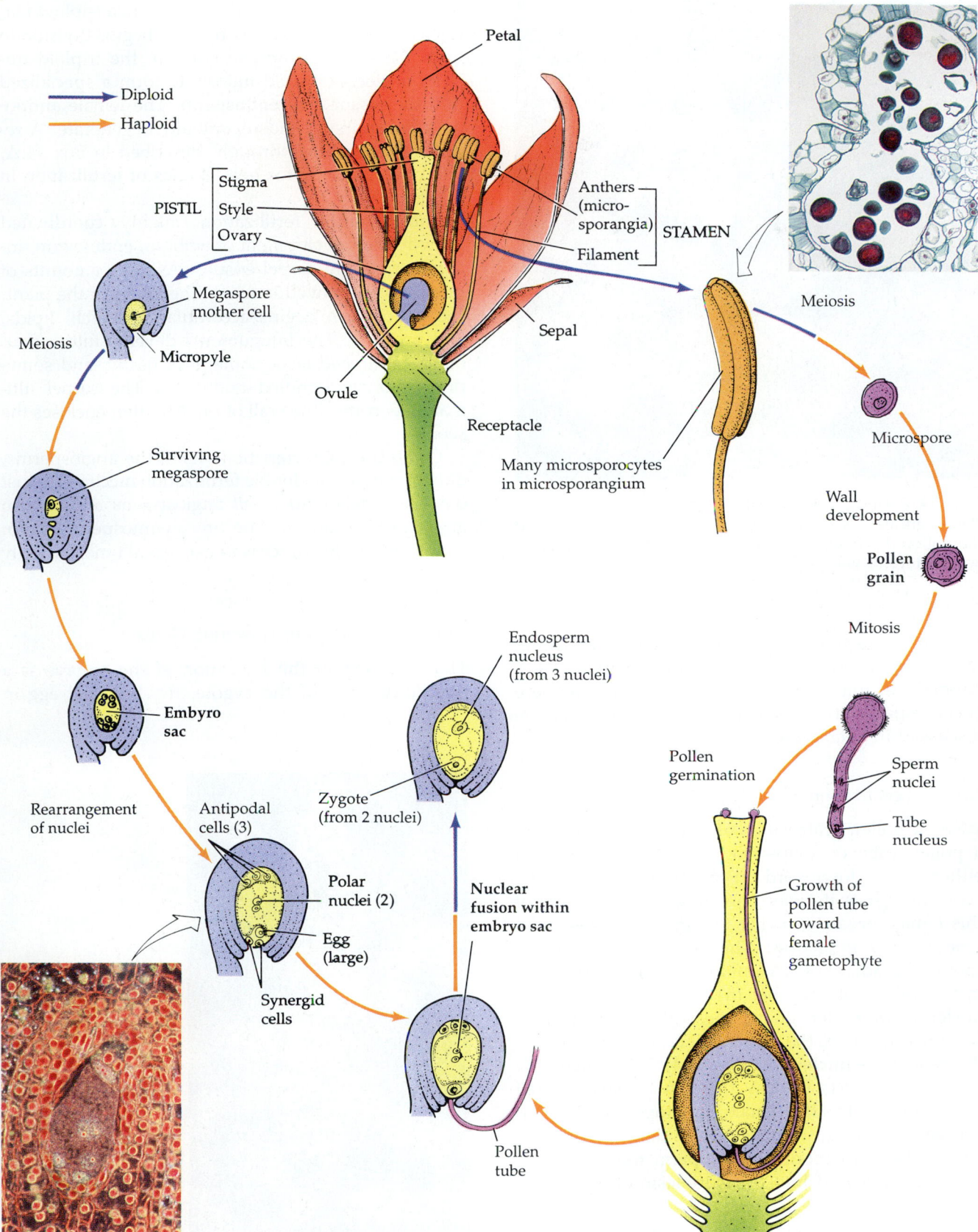

Diploid
Haploid

Petal

Stigma
Style — PISTIL
Ovary

Anthers
(micro-
sporangia) — STAMEN
Filament

Megaspore
mother cell

Micropyle

Meiosis

Ovule

Sepal

Receptacle

Many microsporocytes
in microsporangium

Meiosis

Microspore

Wall
development

Pollen
grain

Mitosis

Surviving
megaspore

Embryo
sac

Rearrangement
of nuclei

Antipodal
cells (3)

Polar
nuclei (2)

Egg
(large)

Synergid
cells

Endosperm
nucleus
(from 3 nuclei)

Zygote
(from 2 nuclei)

Nuclear
fusion within
embryo sac

Pollen
germination

Sperm
nuclei

Tube
nucleus

Growth of
pollen tube
toward
female
gametophyte

Pollen
tube

34.1   Development of Gametophytes and Nuclear Fusion
The embryo sac is the female gametophyte; its development in an ideal flower is illustrated on the left. The
pollen grain, whose development is illustrated on the
right, is the male gametophyte; the micrograph shows
several developing pollen grains in the anthers. Male and
female nuclei fuse within the embryo sac (bottom center).

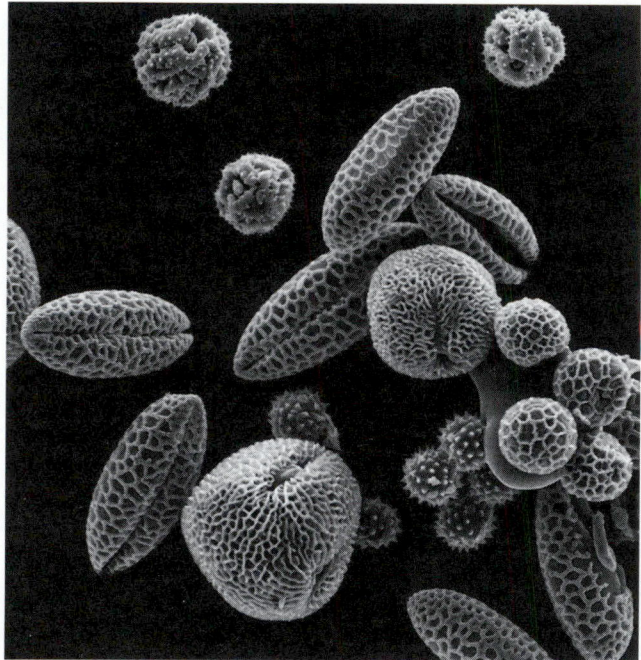

34.2 A Pollen Grain Sampler
Pollen grains of geranium, tiger lily, phlox, marigold, and dandelion. Each species has a characteristic size, shape, and pattern of wall sculpturing.

aquatic angiosperms are pollinated by water action, with water carrying pollen grains from plant to plant. Animals are important pollinators; the mutually beneficial aspects of such plant–animal associations are discussed in Chapter 48.

Double Fertilization

When a pollen grain lands on the stigma of a pistil, a pollen tube develops from the pollen grain and either grows downward on the inner surface of the style or digests its way down the spongy tissue of this female organ, growing millimeters or even centimeters in the process. The pollen tube follows a chemical gradient of calcium ions or other substances in the style until it reaches the micropyle. Of the two nuclei in the pollen grain, one is the **tube nucleus**, close to the tip of the pollen tube, and the other is the **generative nucleus** (Figure 34.4). The pollen tube eventually digests its way through megasporangial tissue and reaches the female gametophyte. The generative nucleus meanwhile has undergone one mitotic division to produce two **sperm nuclei**, *both* of which are released into the cytoplasm of one of the synergids.

From this synergid, which degenerates, each sperm nucleus enters a different cell. One sperm nucleus enters the egg cell and fuses with its nucleus, producing the diploid zygote. The other sperm nucleus enters the central cell of the embryo sac and

fuses with the two polar nuclei to form a triploid ($3n$) nucleus. While the zygote nucleus begins division to form the new sporophyte embryo, the triploid nucleus undergoes rapid mitosis to form a specialized nutritive tissue, the **endosperm**. The female antipodal cells and synergids eventually degenerate. A recent research breakthrough, described in Box 34.A, will open the way for new studies of fertilization in plants.

Shortly after fertilization, highly coordinated growth and development of embryo, endosperm, integuments, and carpel ensues. As large amounts of nutrients are moved in from other parts of the plant, the endosperm begins accumulating starch, lipids, and proteins. The integuments develop into a double-layered seed coat, sometimes fleshy, and sometimes heavily lignified and hard. The carpel ultimately becomes the wall of the fruit that encloses the seed.

Of all the characteristic traits of the angiosperms, only one trait, the double fertilization mechanism just described, is found in *all* angiosperms and *only* in angiosperms—and in one tiny gymnosperm group. The origin of this process in geological time is wholly unknown.

Embryo Formation in Flowering Plants

The first step in the formation of the embryo is a mitotic division of the zygote, the fertilized egg in

34.3 Wind Pollination
A flowering inflorescence of a grass. The numerous anthers all point away from the stalk and stand free of the plant, promoting dispersal of the pollen by wind.

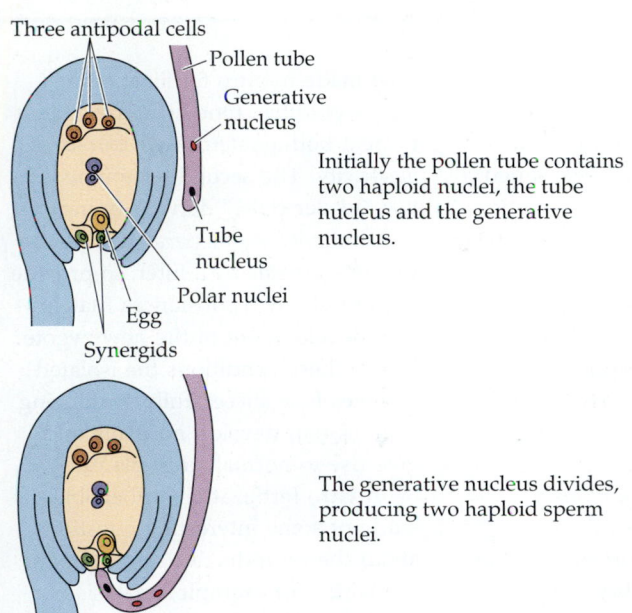

Three antipodal cells
Pollen tube
Generative nucleus
Tube nucleus
Polar nuclei
Egg
Synergids

Initially the pollen tube contains two haploid nuclei, the tube nucleus and the generative nucleus.

The generative nucleus divides, producing two haploid sperm nuclei.

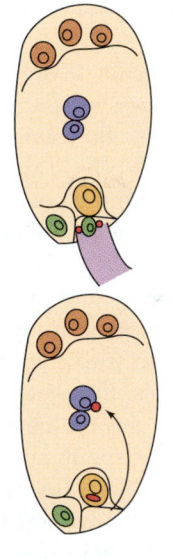

The sperm nuclei enter the cytoplasm of a synergid.

The synergid breaks down; one sperm nucleus fertilizes the egg, forming the zygote, the first cell of the 2*n* sporophyte generation. The other unites with the two polar nuclei, forming the first cell of the 3*n* endosperm.

34.4 Pollen Nuclei and Double Fertilization
The sperm nuclei contribute to the formation of the diploid zygote and the triploid endosperm. Double fertilization is a characteristic feature of angiosperm reproduction.

the embryo sac, giving rise to two daughter cells. Even at this stage the two cells face different fates. An asymmetric (uneven) distribution of contents within the zygote causes one end to produce the embryo proper and the other end to produce an early supporting structure, the **suspensor** (Figure 34.5). Polarity has been established, as has the longitudinal axis of the new plant. A filamentous suspensor and a globular embryo are distinguishable after just four mitotic divisions. The suspensor soon ceases to elongate, and, as development continues, the first organs take form within the embryo.

In dicots (monocots are somewhat different) the embryo soon takes on a characteristic heart-shaped form as the cotyledons start to grow. Further elon-

gation of the cotyledons and of the main axis of the embryo gives rise to what is called the torpedo stage, during which some of the internal tissues begin to differentiate. The elongated region below the cotyledons is called the hypocotyl. At the top of the hypocotyl, between the cotyledons, is the shoot apex; at the other end is a root apex. Each of these apical regions contains an apical meristem whose dividing cells give rise to the organs of the mature plant. In many species, such as peas and peanuts, the cotyledons absorb the food reserves from the surrounding endosperm and grow very large in relation to the rest of the embryo (Figure 34.6*a*). In others, including the castor bean, the cotyledons remain thin (Figure 34.6*b*); they will draw on the reserves in the endo-

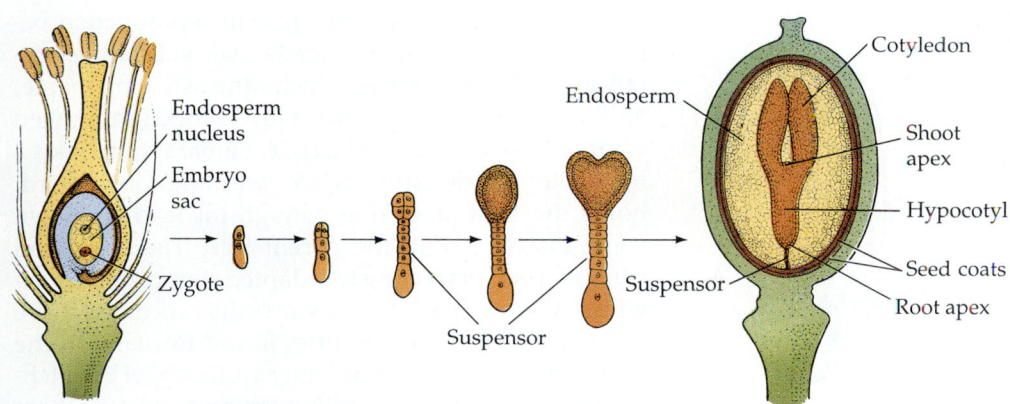

Endosperm nucleus
Embryo sac
Zygote
Suspensor
Endosperm
Suspensor
Cotyledon
Shoot apex
Hypocotyl
Seed coats
Root apex

34.5 Early Development of a Dicot
The zygote nucleus divides mitotically, one daughter cell giving rise to the embryo proper and the other to the suspensor. The embryo develops through intermediate stages to form the torpedo stage (far right). The tissues surrounding the embryo sac develop into seed coats.

BOX 34.A

In Vitro Fertilization: Test-Tube Plants

Hundreds of children have been conceived by means of in vitro fertilization, the combination of egg and sperm in a petri plate rather than in the mother's body. Can we do the same thing with plant gametes—make test-tube plants? It sounds simple, but in vitro fertilization in plants did not succeed fully until 1993.

What made this achievement so difficult? In animals, sperm and eggs are set free and can unite readily. In plants the gametes are not free of other cells. The sperm are contained within the pollen tube, and osmotic treatments are needed to free them from the pollen. The egg presents an even greater problem because it is contained in the embryo sac. Treatment with enzymes that digest cell walls, followed by microdissection, can release the egg. However, the freed sperm and egg appear to be incapable of fusing spontaneously. What is wrong? One possibility is that fertilization depends on activities of other cells in the embryo sac; another possibility is that the egg nucleus is damaged by the enzymes used to remove the egg cell wall.

A combination of techniques has now made in vitro fertilization in plants possible. One key technique is to treat both gametes with bursts of electricity. The second technique is to use "feeder cells," derived from normal embryos and separated from the treated gametes by a filter, to provide as yet unknown substances that support development of the new zygote. Under these conditions the isolated gametes fuse successfully, producing zygotes that develop normally and give rise to normal adult plants.

In vitro fertilization will enable the study of some interesting questions about the reproductive development of plants. For example, analysis of the products of the cultured feeder cells will shed light on the role of the synergids in normal, in vivo fertilization in plants.

---

sperm as needed when the seeds germinate. In either case, the endosperm is the maternal plant's contribution to the nutrition of the next generation (Box 34.B).

Why does the embryo stop developing within the seed? In the late stages of embryonic development the seed loses water—sometimes as much as 95 percent of its original water content. In its dried state, the embryo is incapable of further development. It remains in this dormant state until the conditions are right for germination. (Recall from Chapter 33 that a necessary first step in seed germination is the massive imbibition of water.)

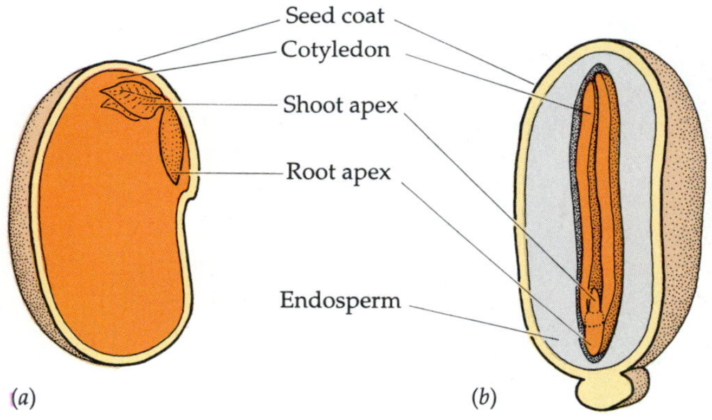

Fruit

After fertilization, the ovary wall of a flowering plant—together with its seeds—develops into a fruit. A fruit may consist of only the mature ovary and its seeds, or it may include other parts of the flower or structures that are closely related to it. The major variations on this theme are illustrated in Figure 25.28.

Some fruits play a major role in reproduction because they help disperse seeds over substantial distances. A number of trees, including ash, elm, maple, and tree of heaven, produce a dry, winged fruit called a samara (see Figure 29.2d). A samara spins like a helicopter blade and, while whirling downward, holds the fruit aloft long enough for it to be blown some distance from the parent tree. The dandelion fruit is also marvelously adapted for dispersal by wind. Water disperses some fruits; coconuts have been spread in this way from island to island in the Pacific (Figure 34.7a). Still other fruits travel by hitching rides with animals—either inside or outside them (Figure 34.7b). Burdocks, for example, have hooks that adhere to animal fur, and many other plants have prickled, barbed, hairy, or sticky fruits. Fleshy fruits such as berries provide food for mammals or

(a)   (b)

34.6   Variety in Dicot Seeds
In some dicots the cotyledons absorb much of the endosperm and fill most of the seed (a). In others the endosperm remains separate and the cotyledons remain thin (b).

(a)   (b)

34.7   Dispersal of Fruits
(a) These coconuts are germinating where they washed ashore on a Tahitian beach. (b) Many sticky fruits, such as the burrs of the common burdock, hitch rides on animals, traveling far from their parent plants.

BOX 34.B

Maternal "Care" in Plants

The mammalian mother has a special relationship with her embryonic and newborn offspring. The mammalian embryo receives its nourishment by way of the maternal bloodstream, and it is subjected to hormonal influences from the mother, but not from the father. Maternal *plants* also influence their offspring in important ways. Of course, the offspring receive half their genetic endowment by way of the egg, but the maternal parent has other, specific effects, especially relating to the size of seeds produced, because the seeds develop from and within tissues of the maternal flower.

Barbara A. Schaal, of Washington University, St. Louis, has studied maternal effects in the Texas bluebonnet, *Lupinus texensis*. She demonstrated that variation in seed mass in these plants could be attributed to responses of the maternal plant to environmental variation. Differences in seed mass in turn affected the subsequent performance of the offspring, as the figure shows. A higher percentage of larger seeds germinated, and they germinated earlier. Larger seeds gave rise to larger seedlings. The seedlings from larger seeds grew more leaves and survived better during the first month of life after germination. By six weeks of age, differences in size and survivorship between seedlings from larger and smaller seeds were no longer evident. However, between the ages of 80 and 130 days, the period of most intense reproduction in the bluebonnet, plants from larger seeds pro-

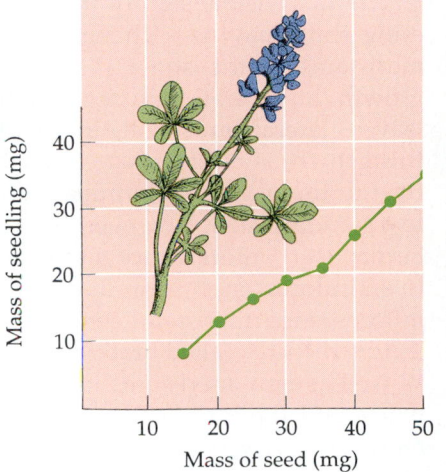

duced more ovules than did those from smaller seeds. Because the size of the seed is determined more by the maternal parent than by the genotype of the embryo, Schaal concluded that maternal effects were important influences on the development and reproduction of offspring.

birds; their seeds travel safely through the animal's digestive tract or are regurgitated, in either case being deposited some distance from the parent plant.

Having discussed mechanisms for dispersing seeds and pollen, we should make it clear that most seeds, and most pollen grains, end up close to their sources. Long-range dispersal is the exception, not the rule. In cases of extensive dispersal, paternal genes usually travel farther than maternal genes because, although both maternal and paternal genes travel with the seed, only paternal genes travel with the pollen grain.

We have traced the sexual life cycle from the flower to the fruit to the dispersal of seeds. We discussed seed germination and vegetative development of the seedling in Chapter 33. Now let's complete the sexual life cycle by considering the transition from the vegetative to the flowering state, and how this transition is regulated.

TRANSITION TO THE FLOWERING STATE

Flowering may terminate, repeatedly interrupt, or accompany vegetative growth. The transition to the flowering state often marks the end of vegetative growth for a plant. If we view a plant as something produced by a seed for the purpose of bearing more seeds, then the act of flowering is one of the supreme events in a plant's life.

The first visible sign of the transition to the flowering state may be a change in one or more apical meristems in the shoot system. During vegetative growth an apical meristem continues to produce leaves, lateral buds, and internodes (Figure 34.8*a*); this growth is *indeterminate* (see Chapter 33), in contrast to the usually determinate growth of an animal to a standard size. However, if a vegetative meristem becomes an **inflorescence meristem** that will give rise to an inflorescence, it produces other structures. The inflorescence meristem generally produces smaller leafy structures called **bracts** separated by internodes, as well as new meristems in the angles between the bracts and the internodes (Figure 34.8*b*). These new meristems may be inflorescence meristems or **floral meristems**, which give rise to the flowers themselves (Figure 34.8*c*). Each floral meristem in turn typically produces four consecutive whorls of organs—the sepals, petals, stamens, and carpels—separated by very short internodes. In contrast to vegetative meristems and some inflorescence meristems, floral meristems are responsible for *determinate* growth—the limited growth of the flower.

How does the floral meristem give rise, in short order, to whorls of four different organs? This problem is not unlike that of accounting for the construction of a fruit fly larva (see Chapter 17). Recent work

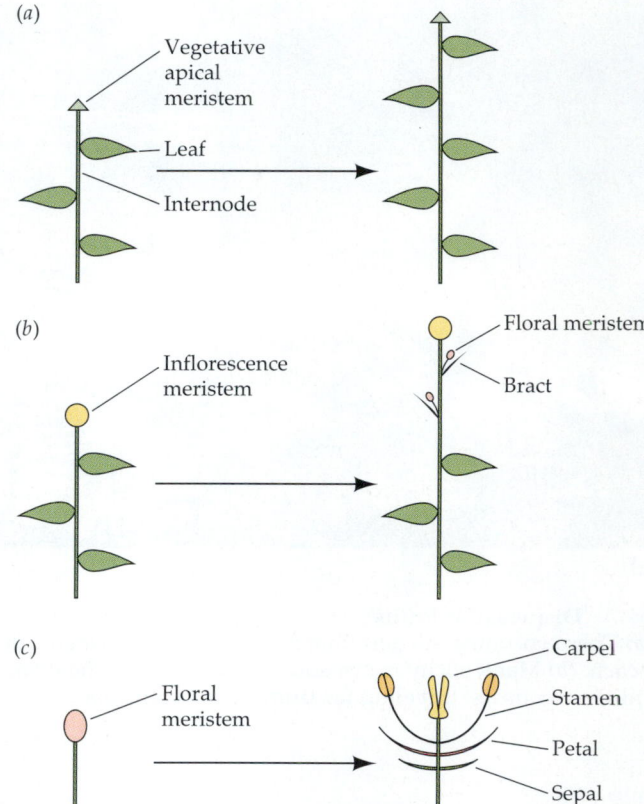

34.8 Flowering and the Apical Meristem
A vegetatively growing apical meristem continues to produce leaves and internodes. Inflorescence and floral meristems each give rise to a characteristic set of products.

underscores some of the similarities. Most strikingly, a group of **homeotic genes**—organ-identity genes—work in concert to specify the successive whorls. We recognize the presence of homeotic genes because mutations in these genes, called homeotic mutations, lead to major alterations in flower structure. Table 34.1 gives a sense of some of these mutations. Do you see a pattern in the alterations? These mutations, along with analysis of the homeotic genes and their products, are leading us to a preliminary understanding of how normal flowers develop. Earlier work, which we will consider next, helped to explain how the transition to the flowering state is initiated.

Flowering is triggered in very different ways in different plant species. Environmental cues include seasonal changes in the lengths of days and nights and seasonal temperature changes.

PHOTOPERIODIC CONTROL OF FLOWERING

In 1920 W. W. Garner and H. A. Allard of the U.S. Department of Agriculture studied the behavior of a newly discovered mutant tobacco plant. The mutant

TABLE 34.1
Homeotic Mutations in Flower Development

	PHENOTYPE			
GENOTYPE	WHORL 1	WHORL 2	WHORL 3	WHORL 4
Wild type	Sepals	Petals	Stamens	Carpels
Mutant A	Carpels	Stamens	Stamens	Carpels
Mutant B	Sepals	Sepals	Carpels	Carpels
Mutant C	Sepals	Petals	Petals	Sepals

was named Maryland Mammoth because of its large leaves and exceptional height (and where it was found). The other plants in the field flowered, but the Maryland Mammoth continued to grow. Garner and Allard took cuttings of the Maryland Mammoth into their greenhouse, and the plants that grew from the cuttings finally flowered in December. Garner and Allard also noticed that some soybean plants all flowered at about the same time, in late summer, even though they had been planted at different times in the spring.

Garner and Allard guessed that both of these observations had something to do with the seasons.

They tested a number of likely seasonal variables, such as temperature, but the key variable proved to be the length of day. They moved plants between light and dark rooms at different times to vary the length of day artificially and were able to establish a direct link between flowering and day length. The **critical day length** for Maryland Mammoth tobacco proved to be 14 hours (Figure 34.9). The plants did

34.9 Day Length and Flowering
By varying the length of day, Garner and Allard showed that Maryland Mammoth tobacco flowers only when days are shorter than 14 hours; that is, the critical day length is 14 hours.

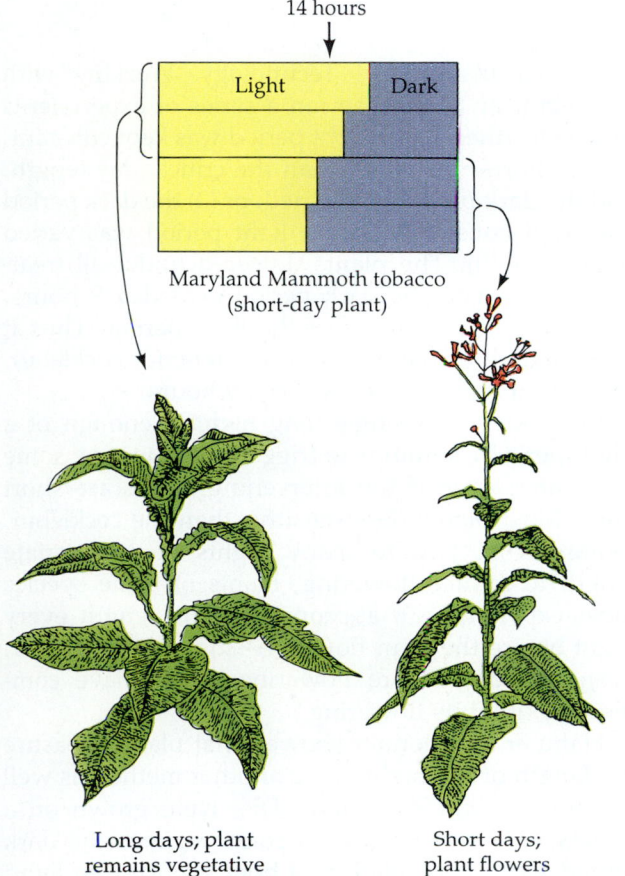

Maryland Mammoth tobacco
(short-day plant)

Long days; plant remains vegetative   Short days; plant flowers

Henbane, *Hyoscyamus niger*
(long-day plant)

Long days; plant flowers   Short days; plant remains vegetative

not flower if the light period was longer than 14 hours each day, but flowering commenced after the days became shorter than that. This phenomenon of control by lengths of days and nights is called **photoperiodism**. Soybeans and Maryland Mammoth tobacco are **short-day plants** (SDPs). Spinach and clover are examples of **long-day plants** (LDPs), which flower only when the day is longer than a critical minimum. Generally LDPs are triggered to flower in midsummer and SDPs in late summer, or sometimes in the spring.

It is a historical accident that we use the terms short-day plant and long-day plant. The SDPs could as well have been called long-night plants and the LDPs short-night plants, because the natural day has a fixed length of 24 hours. Do plants measure the length of day or the length of night? The answer will be given presently.

Other Patterns of Photoperiodism

Some plants require more-complex photoperiodic signals in order to flower. One group, the short-long-day plants, must first experience short days and then long ones. Accordingly, because they pass first through the short days of early spring and then through ever longer ones, they flower during the long days before midsummer. Another group, the long-short-day plants, cannot flower until the long days of summer have been followed by shorter ones, so they bloom only in the fall. Long-short-day plants will not bloom in the spring, in spite of its short days, nor will a short-long-day plant flower in late summer.

Other effects besides flowering are also under photoperiodic control. We have learned, for example, that the onset of winter dormancy is triggered by short days. (Animals, too, show a variety of photoperiodic behaviors; in aphids, for example, long days favor the development of sexually reproducing females, whereas females that reproduce asexually develop when days are short.)

It is important to note that the flowering of some angiosperms is not photoperiodic. In fact, there are more **day-neutral** plants than there are short-day and long-day plants. Some plants are photoperiodically sensitive only when young and become day-neutral as they grow older. Others require specific combinations of day length and other factors to flower.

Importance of Night Length

The terms short-day plant and long-day plant became entrenched before it was learned that plants actually measure the length of night, or of darkness. This fact was demonstrated by Karl Hamner of the University of California at Los Angeles and James Bonner of the

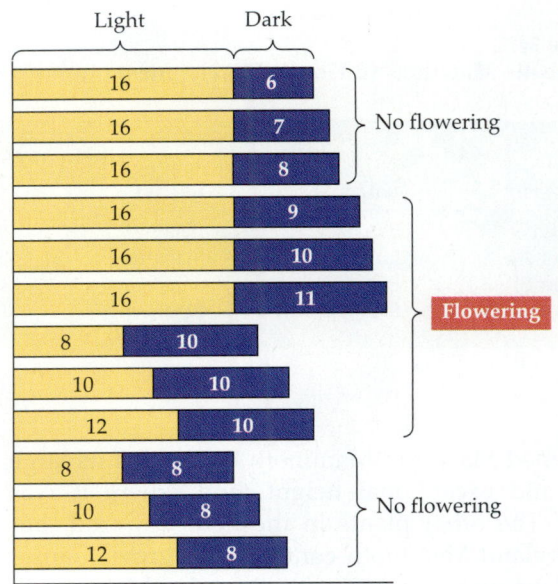

34.10 Night Length and Flowering
In the experiments symbolized by the six upper bars, plants were exposed to 16 hours of light, followed by dark periods of various duration. Only plants given 9 or more hours of dark flowered. In the 6 experiments indicated at the bottom, plants were exposed to various light periods followed by 8 or 10 hours of dark. Only plants given 10 hours of dark flowered. Experiments like these showed that short-day plants really should be called long-night plants.

California Institute of Technology. Working with cocklebur, an SDP, they ran a series of experiments in which either (1) the light period was kept constant, either shorter or longer than the critical day length, and the dark period was varied, or (2) the dark period was kept constant and the light period was varied (Figure 34.10). The plants flowered under all treatments in which the dark period exceeded 9 hours, regardless of the length of the light period. Thus it is the length of the *night* that matters; for cocklebur, the critical night length is about 9 hours.

In cocklebur, a single long night is enough of a photoperiodic stimulus to trigger full flowering some days later, even if the intervening nights are short ones. Most plants, less sensitive than the cocklebur, require from two to many nights of appropriate length to induce flowering. Plants of some species must experience an appropriate night length every night before they can flower. A single shorter night, even one day before flowering would have commenced, inhibits flowering.

Hamner and Bonner showed that plants measure the length of the night using another method as well (Figure 34.11a). SDPs and LDPs were grown on a variety of light regimes. In some regimes the dark period was interrupted by a brief exposure to light;

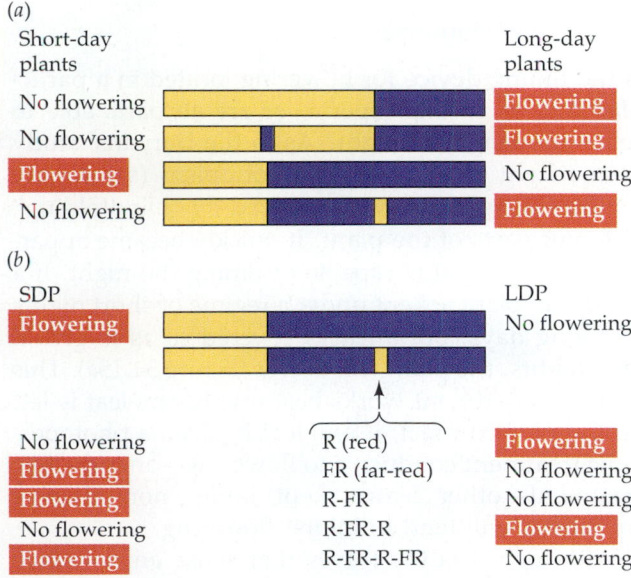

(a)

Short-day plants

Long-day plants

No flowering | Flowering
No flowering | Flowering
Flowering | No flowering
No flowering | Flowering

(b)

SDP | LDP

Flowering | No flowering

No flowering	R (red)	Flowering
Flowering	FR (far-red)	No flowering
Flowering	R-FR	No flowering
No flowering	R-FR-R	Flowering
Flowering	R-FR-R-FR	No flowering

34.11 Interrupted Days and Nights
(a) Short-day plants require long, uninterrupted nights to flower. Long-day plants flower when the night is short; interrupting their long day has no effect but interrupting a long night with a brief period of light induces flowering. (b) When plants are exposed to red (R) and far-red (FR) light in alternation, the final treatment determines the effect of the light interruption, suggesting that phytochrome participates in photoperiodic responses.

in others, the light period was interrupted briefly by darkness. Interruptions of the light period by darkness had no effect on the flowering of either short-day or long-day plants. Even very brief interruptions of the dark period, however, completely nullified the effect of a long night. An SDP flowered only if the long nights were uninterrupted. An LDP experiencing long nights flowered if these were broken by exposure to light. Thus a plant must have a timing mechanism that measures the length of a continuous dark period and uses the result to trigger flowering or to remain vegetative. Despite much study, the nature of the timing mechanism is still unknown.

Phytochrome seems to participate in the photoperiodic timing mechanism. In the interrupted-night experiments, the most effective wavelengths of light were red (Figure 34.11b). The effect of a red-light interruption of the night was fully reversed by a subsequent exposure to far-red light. It was once thought that the timing mechanism might simply be the slow conversion of phytochrome during the night from the P_{fr} form—produced during the light hours—to the P_r form. But this suggestion is inconsistent with most of the experimental observations and must be wrong. Phytochrome must be only a photoreceptor. The time-keeping role is played by a biological clock.

Circadian Rhythms and the Biological Clock

It is abundantly clear that organisms have a way of measuring time and that they are well adapted to the 24-hour day–night cycle of our planet. Some sort of biological clock resides within the cells of all eukaryotes, and the major outward manifestations of this clock are known as **circadian rhythms** (Latin *circa*, = "about" and *dies*, = "day"). Plants provide innumerable examples of approximately 24-hour cycles. The leaflets of a plant such as clover or the tropical tree *Albizia* normally hang down and fold at night and rise and expand during the day. Flowers of many plants show similar "sleep" movements, closing at night and opening during the day. They continue to open and close on an approximately 24-hour cycle even when the light and dark periods are experimentally modified (Figure 34.12).

The circadian rhythms of protists, animals, fungi, and plants share some important characteristics. First, the **period** is remarkably insensitive to temperature, although the **amplitude** of the fluctuation may be drastically reduced by lowering the temperature (Figure 34.13 explains these terms). Second, circadian rhythms are highly persistent; they continue even in an environment in which there is no alternation of light and dark. Third, circadian rhythms can be **entrained**, within limits, by light–dark cycles that differ from 24 hours. That is, the period an organism expresses can be made to coincide with that of the light–dark regime. The period in nature is approximately 24 hours. If an *Albizia* tree, for example, were to be

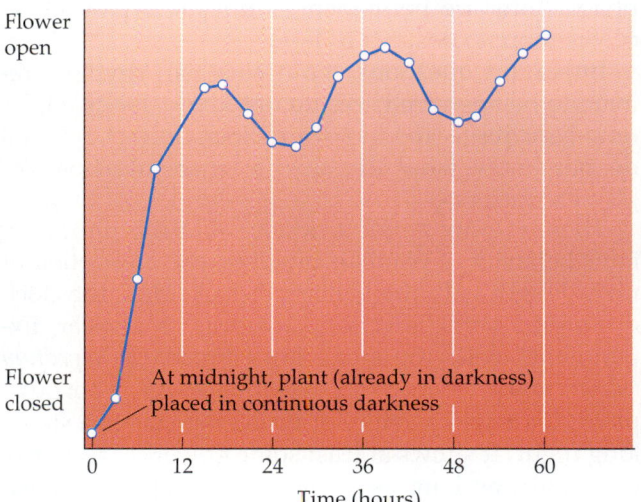

34.12 Sleep Movements of Flowers
Kalanchoe flowers close at night. In the middle of the night (time = 0), when the flowers were completely closed, biologists transferred a *Kalanchoe* plant to a dark box and kept it there for another 60 hours. During that time the flowers continued to open and partially close on a 24-hour cycle.

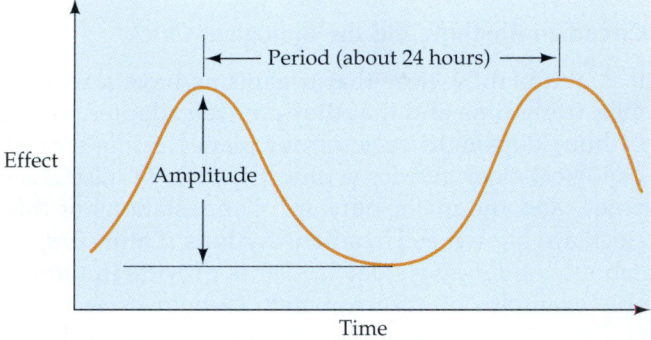

34.13 Features of Circadian Rhythms
Circadian rhythms are characterized on the basis of time, measured in periods of about 24 hours, and on the basis of the magnitude of the rhythmic effect, measured by the cycle's amplitude.

placed under artificial light on a day–night cycle totaling exactly 24 hours, the rhythm expressed would show a period of exactly 24 hours. If, however, an experimenter used a day–night cycle of, say, 22 hours, then the rhythm would be entrained to a 22-hour period.

If light–dark cycles can entrain circadian rhythms, it follows that light alone should also be able to shift the rhythm. If an organism is maintained under constant darkness, with its circadian rhythm expressed on the approximately 24-hour period, a brief exposure to light can make the next peak of activity appear either later or earlier than one would have predicted, depending on when the exposure is given. Moreover, the organism does not then return to its old schedule if kept in darkness. If the first peak is delayed by 6 hours, the subsequent peaks are all 6 hours late. Such phase shifts are permanent—until the organism receives more exposures to light.

Important questions about circadian rhythms remain to be answered. We do not know, for example, how light resets the biological clock. In fact, we still do not know what the clock's biochemical or biophysical basis is.

There is now ample evidence that the photoperiodic behavior of plants is based on the interaction of night length with the biological clock. How the clock is coupled with flowering, however, is unclear. Experiments with the small lawn weed *Chenopodium rubrum* provided one sort of evidence. As a short-day plant, *Chenopodium* will flower in response to a single long night. It shows at least some flowering response to a night as long as 96 hours. A single red flash during this extremely long night can either enhance or inhibit this response, depending on precisely when it is administered (Figure 34.14). The effect of the red flash oscillates on an approximately daily basis (like the clock), but just how the underlying daily rhythm relates to flowering remains to be learned.

A Flowering Hormone?

Is the timing device for flowering located in a particular part of an angiosperm, or are all parts able to sense the length of night? As in the Darwins' study of the light receptor for phototropism (see Figure 33.12), this question was resolved by "blindfolding" different parts of the plant. It quickly became apparent that each leaf is capable of timing the night. If a short-day plant is kept under a regime of short nights and long days, but a leaf is covered so as to give it long nights, the plant will flower (Figure 34.15a). This type of experiment works best if only one leaf is left on the plant. In fact, if one leaf is given a photoperiodic treatment conducive to flowering—an inductive treatment—other leaves kept under noninductive conditions will tend to inhibit flowering.

Although it is the leaves that sense an inductive night period, the flowers form elsewhere on the plant. Thus a message must be sent from the leaf to the site of flower formation. Three lines of evidence suggest that this message is a chemical substance—a flowering hormone. First, if a photoperiodically induced leaf is removed from the plant shortly after the inductive night period, the plant does not flower. If, however, the induced leaf remains attached for several hours, the plant flowers. This result suggests that something—the hypothetical hormone—must be synthesized in the leaf in response to the inductive night, then move out of the leaf to induce flowering.

The second line of evidence for the existence of a flowering hormone comes from grafting experiments. If two cocklebur plants are grafted together, and if one plant is given inductive long nights and its graft

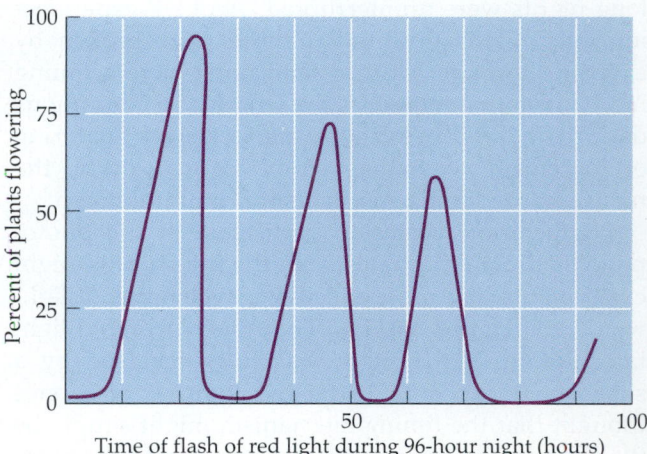

34.14 A Flowering Rhythm
The short-day plant *Chenopodium rubrum* flowers in response to a single 96-hour "night." If single flashes of red light are given during such a 96-hour dark period, they either enhance or inhibit the flowering response. The peaks of enhancement are on about a 24-hour cycle initiated by the light-to-dark transition that started the night.

(a) Cockleburs on short nights/long days

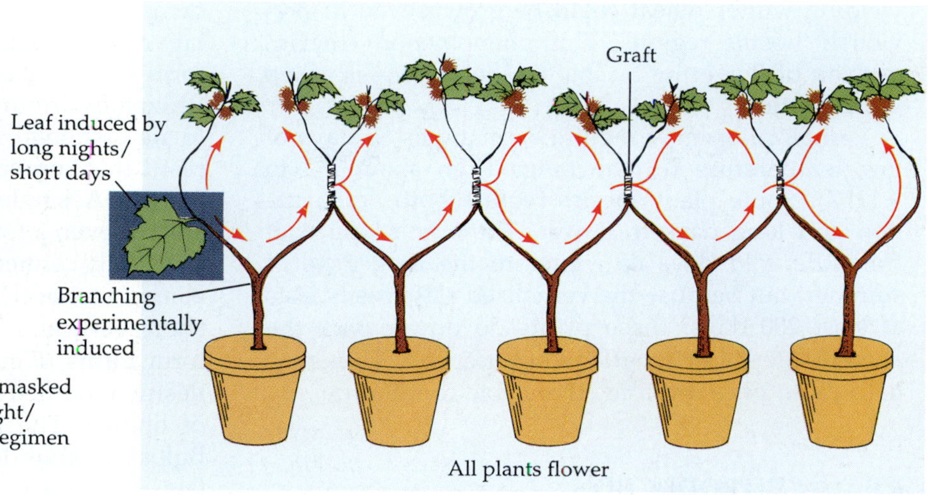

(b) Grafted cockleburs on short nights/long days

34.15 Evidence for a Flowering Hormone
(a) Cocklebur, a short-day plant, will not flower if kept under long days and short nights. If even one leaf is masked for part of the day, however—thus shifting that leaf to short days and long nights—the plant will flower; note the burrs. Because the flowers are formed far from the induced leaf, some substance probably carries the flowering message from the leaf. (b) Five cocklebur plants grafted together and kept under long days and short nights, with most leaves removed. If a leaf on a plant at one end of the chain is subjected to long nights, all of the plants will flower. Arrows indicate the routes of the hypothetical flowering hormone from the induced leaf.

partner is given noninductive short nights, both plants flower (Figure 34.15b). Grafting experiments also provided the third line of evidence for a flowering hormone. Jan A. D. Zeevaart, a plant physiologist now at Michigan State University, exposed a single leaf of the SDP *Perilla* to a short-night/long-day regime, inducing the plant to flower. Then he detached this leaf and grafted it onto another, noninduced, *Perilla* plant—which responded by flowering. The same leaf grafted onto successive hosts caused each of them to flower in turn. As long as three months after the leaf was exposed to the short-night/long-day regime, it could still cause plants to flower.

The Search for Florigen

Experiments such as Zeevaart's suggest that the photoperiodic induction of a leaf causes a more or less permanent change in it, inducing it to start and to continue producing a flowering hormone that is transported to other parts of the plant, switching those target parts to the reproductive state. So reasonable is this idea that biologists have named the hormone, even though after decades of active searching the compound has yet to be isolated and characterized. The hormone is called **florigen**. The direct demonstration of florigen activity remains a cherished goal of plant physiologists. Gibberellins regulate the flowering of many species, especially of long-day plants, but these hormones do not have the properties of florigen.

As a final teaser, we will describe an experiment that suggests that the florigen of short-day plants is identical with that of long-day plants, even though SDPs produce it only under long nights and LDPs only under short nights. An SDP and an LDP were grafted together, and both flowered, as long as the photoperiodic conditions were inductive for one of the partners. Either the SDP or the LDP could be the one induced, but both would always flower. These results suggest the transfer of a flowering-inducing hormone, the elusive florigen, from one plant to the other.

VERNALIZATION AND FLOWERING

In both wheat and rye, we distinguish two categories of flowering behavior. Spring wheat, for example, is sown in the spring and flowers the same year. It is an annual plant. Winter wheat is biennial and must be sown in the fall; it flowers in its second summer. If winter wheat is not exposed to cold after its first year, it will not flower normally the next year. The implications of this finding were of great agricultural interest in Russia because winter wheat is a better producer than spring wheat, but it cannot be grown in parts of Russia because the winters there are too cold for its survival. A number of studies performed in Russia during the early 1900s demonstrated that if seeds of winter wheat were premoistened and prechilled, they would develop and flower normally the same year when sown in the spring. Thus, high-yielding winter wheat could be grown even in previously hostile regions. This phenomenon—the induction of flowering by low temperatures—is called **vernalization**.

Vernalization may require as many as 50 days of low temperature (in the range from about −2 to +12°C). Some plant species require both vernalization and long days to flower. There is a long wait from the cold days of winter to the long days of summer, but because the vernalized state easily lasts at least 200 days, these plants do flower once the appropriate night length is experienced. Thus vernalization, once induced, is a stable condition.

ASEXUAL REPRODUCTION

Although sexual reproduction takes up the bulk of the space in this chapter's discussion of how plants reproduce, asexual reproduction is responsible for an important fraction of the new plant individuals appearing on Earth. This suggests another answer to the question asked at the beginning of this chapter: In some circumstances, asexual reproduction is better. For example, think about genetic recombination. We have already noted that when a plant self-fertilizes there are fewer opportunities for genetic recombination than there are with cross-fertilization. With self-fertilization, the only genetic variation that can be arranged into new combinations is that possessed by the single parental plant. A plant that is heterozygous for a locus can produce among its progeny both kinds of homozygotes for that locus plus the heterozygote, but it cannot produce any progeny that carry alleles that it does not itself possess. Asexual reproduction goes a step further: It eliminates genetic recombination. When a plant reproduces asexually, it produces a clone of progeny with genotypes identical to its own. If a clone is highly adapted to its environment, the many copies of that genotype that may be formed by asexual reproduction may spread throughout that environment. This ability to exploit a particular environment is an advantage of asexual reproduction.

Asexual Reproduction in Nature

We call stems, leaves, and roots vegetative parts, distinguishing them from the reproductive parts. The modification of a vegetative part of a plant is what makes **vegetative reproduction** possible. The stem is the part modified in many cases. Strawberries and some grasses produce **stolons**, horizontal stems that form roots at intervals and establish potentially independent daughter plants (see Figure 29.9c). Other stolons are branches that sag to the ground and put out roots. The rapid multiplication of water hyacinths

demonstrates the effectiveness of stolons for vegetative reproduction. Some plants, such as potatoes, form **tubers**, the fleshy tips of underground stems. **Rhizomes** are underground stems that can give rise to new shoots. Bamboo is a striking example of a plant that reproduces vegetatively by means of rhizomes. A single bamboo plant can give rise to a stand—even a forest—of plants constituting a single, physically connected entity. Whereas stolons and rhizomes are horizontal stems, bulbs and corms are short, vertical, underground stems. Lilies and onions form **bulbs** (Figure 34.16a), short stems with many fleshy, modified leaves, such as the familiar "scales" of onions. The leaves make up most of the bulb. Bulbs are thus large buds that store food and can later give rise to new plants. Crocuses, gladioli, and many other plants produce **corms**, underground stems that function very much as bulbs do. Corms are conical and consist primarily of stem tissue; they lack the fleshy scales characteristic of bulbs.

Not all vegetative reproduction arises from modified stems. Leaves may also be the source of new plantlets, as in the succulent plants of the genus *Kalanchoe* (Figure 34.16b). Many kinds of angiosperms, ranging from grasses to trees such as aspens and poplars, form interconnected, genetically homogeneous populations by means of **root suckers**—horizontal roots. What appears to be a whole stand of aspen trees, for example, is a clone derived from a single tree by root suckers (see Figure 47.9).

Plants that reproduce vegetatively often grow in physically unstable environments such as eroding hillsides. Plants with stolons or rhizomes, such as beach grasses, rushes, and sand verbena, are common pioneers on coastal sand dunes. Rapid vegetative reproduction enables the plants, once introduced, not only to multiply but also to survive burial by the shifting sand; in turn, the dunes are stabilized by the extensive network of rhizomes or stolons.

Dandelions and some other plants reproduce by **apomixis**, the asexual production of seeds. As you learned in Chapter 9, meiosis reduces the number of chromosomes in gametes, and fertilization restores the sporophytic number of chromosomes in the zygote. Some plants can skip over *both* meiosis and fertilization and still produce seeds. Apomixis produces seeds within the female gametophyte without the mingling and segregation of chromosomes and without the union of gametes. The ovule develops into a seed, and the ovary wall develops into a fruit. An apomictic embryo has the sporophytic number of chromosomes. The result of apomixis is a fruit with seeds genetically identical to the parent plant.

Interestingly, apomixis sometimes requires pollination. In some apomictic species a sperm nucleus must combine with the polar nuclei in order for the endosperm to form. In other apomictic species, the

(b)

34.16 Vegetative Reproduction
(a) The short stem is visible at the bottom of this sectioned daffodil bulb. White storage leaves grow from the stem; the yellow parts contain flower buds. (b) The plantlets forming on the margin of this *Kalanchoe* leaf will fall to the ground and start independent lives.

(a)

pollen provides the signals for embryo and endosperm formation, although neither sperm nucleus participates in fertilization. Pollination and fertilization are not the same thing.

Asexual Reproduction in Agriculture

Farmers take advantage of some of these natural forms of vegetative reproduction. Farmers and scientists have also added new types of asexual reproduction by manipulating plants. One of the oldest methods of vegetative reproduction used in agriculture consists simply of making cuttings, or **slips**, of stems, inserting them in soil, and waiting for them to form roots and thus become autonomous plants. Rooting is sometimes hastened by treating the slips with a plant hormone, auxin, as described in Chapter 33.

Agriculturists reproduce many woody plants by **grafting**—attaching a piece of one plant to the stem or root of another plant. The part of the resulting plant that comes from the root-bearing "host" is called the **stock**; the part grafted on is called the **scion**. Figure 34.17 shows three types of grafts. In order for a graft to "take," the cambium of the scion must become associated with the cambium of the stock (see Chapter 29). The cambia of scion and stock both form masses of wound tissue. If the two masses meet and fuse, the resulting continuous cambium can produce xylem and phloem, allowing transport of water and minerals to the scion and of photosynthate to the stock. Grafts are most often successful when

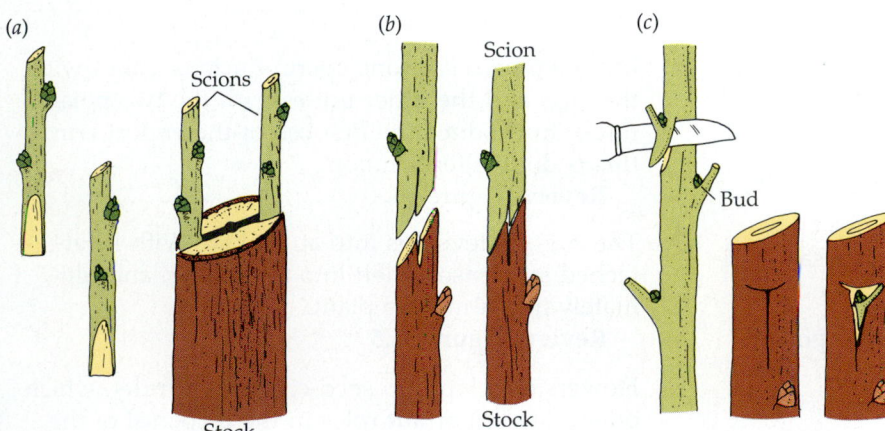

34.17 Grafting
The scions are shown in green and the stocks in brown. (a) Cleft grafting; the scions are placed so that their vascular cambia are aligned with the vascular cambium in the stock. (b) Whip grafting. (c) Budding.

the stock and scion belong to the same or closely related species. Grafting techniques are of great importance in agriculture; most fruit grown for the market in the United States is produced on trees grown from grafts.

There are many reasons for grafting plants for fruit production. The most common is the desire to combine a hardy root system with a shoot system that produces the best-tasting fruit. This motive is illustrated by the story of the wine grape *Vitis vinifera*. In 1863, plant lice of the genus *Phylloxera* inflicted great damage in French vineyards. The roots of vines on more than 2.5 million acres were destroyed. The problem was solved by importing great numbers of *V. vinifera* plants, which have *Phylloxera*-resistant root systems, from California. These plants were used as stocks to which French vines were grafted as scions. Thus the fine French grapes could be grown using roots resistant to the lice. The battle continues, however; in recent years, a new strain of *Phylloxera* has been damaging the grape vines in California.

Scientists in universities and industrial laboratories have been developing new ways to produce valuable plant materials. For example, gene splicing can provide plants with capabilities they previously lacked (see Chapter 14). By causing cells of different sorts to fuse, one can obtain plants with exciting new combinations of properties. By cloning—making genetically identical copies of—small bits of tissue, one can obtain large numbers of equally desirable plants. A problem remains: How can one efficiently take such small, delicate materials and get them to grow in the field? Plants in nature solved this problem long ago. The product of sexual and apomictic reproduction in flowering plants is a compact package, protectively wrapped, containing an embryonic member of the next generation along with a supply of the nutrients it needs to begin its independent existence. This package is the seed. What was needed as a tool of plant biotechnology, and what has actually been de-

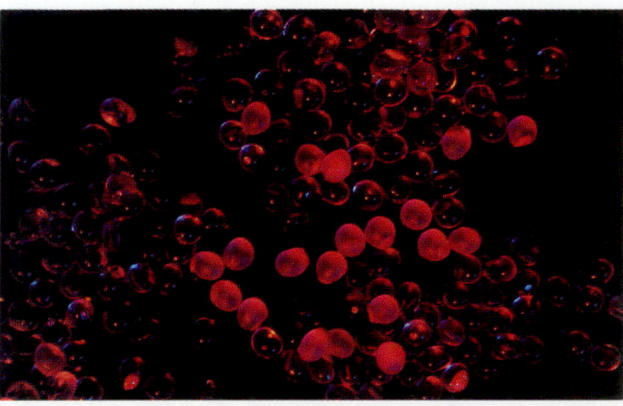

34.18 Artificial Seeds
These water-soluble capsules, developed by Plant Genetics, Inc., house somatic embryos along with nutrients and other chemicals.

veloped, is an artificial seed, containing the product of laboratory invention (Figure 34.18).

Artificial seeds contain a multicellular "somatic embryo." This is not a sexually produced embryo, but an embryolike product of mitotic divisions in tissue culture. Individual cells or small clusters of cells isolated from the body of a suitable parent plant may develop in liquid culture into structures similar to normal embryos derived from zygotes. So that the somatic embryo does not dry out, and so that it may be stored and transported before planting, it is embedded in a water-soluble gel; then the combined embryo and gel are encapsulated in a protective plastic coat. The coat and gel dissolve away after the artificial seed is planted. Other materials may be added to the gel, among them suitable inorganic nutrients, fungicides, and pesticides. Such scientifically designed artificial seeds should be used more and more often as the remaining problems are solved and methods are perfected for the mass production of these tiny packages.

SUMMARY of Main Ideas about Plant Reproduction

In the sexual cycle of angiosperms, a gametophyte (gamete-producing) generation alternates with a sporophyte (spore-producing) generation.

The gametophytes develop in the flowers of the sporophytes.

The mature female gametophyte (embryo sac) typically contains eight nuclei in a total of seven cells.

The entire male gametophyte travels as a pollen grain.
 Review Figure 34.1

In the embryo sac, one sperm nucleus unites with the egg, and the other unites with the two polar nuclei to produce the first cell of the endosperm; this is double fertilization.
 Review Figure 34.4

The zygote develops into an embryo with its attached suspensor, then into a seedling, and ultimately into a mature plant.
 Review Figure 34.5

Flowers develop into seed-containing fruits, which often play important roles in the dispersal of the species.

For a vegetatively growing plant to flower, an apical meristem in the shoot system must change.
Review Figure 34.8

Some angiosperms have photoperiodic flowering behavior.

Short-day plants flower only when the length of the nightly dark period exceeds a critical night length.
Review Figures 34.9, 34.10, and 34.11

Long-day plants flower only when the length of the nightly dark period is shorter than a critical night length.

Some angiosperms have more complex photoperiodic requirements than short-day or long-day plants have, but most are day-neutral.

The mechanism of photoperiodic control appears to include a biological clock and phytochrome.
Review Figures 34.12, 34.13, and 34.14

In some species, vernalization is required for the plants to flower.

There is evidence for a flowering hormone, florigen, but the substance has yet to be convincingly isolated from any plant.
Review Figure 34.15

Some angiosperms reproduce asexually.

Stolons, tubers, rhizomes, bulbs, corms, or root suckers are means by which plants may reproduce vegetatively.

Some species produce seeds by apomixis.

SELF-QUIZ

1. Which of the following does *not* participate in asexual reproduction?
   a. Stolon
   b. Rhizome
   c. Fertilization
   d. Tuber
   e. Apomixis

2. Apomixis includes
   a. sexual reproduction.
   b. meiosis.
   c. fertilization.
   d. a diploid embryo.
   e. no production of a seed.

3. Sexual reproduction in angiosperms
   a. is by way of apomixis.
   b. requires the presence of petals.
   c. can be accomplished by grafting.
   d. gives rise to genetically diverse offspring.
   e. cannot result from self-pollination.

4. The typical angiosperm female gametophyte
   a. is called a megaspore.
   b. has 8 nuclei.
   c. has 8 cells.
   d. is called a pollen grain.
   e. is carried to the male gametophyte by wind or animals.

5. Pollination in angiosperms
   a. never requires water.
   b. never occurs within a single flower.
   c. always requires help by animal pollinators.
   d. is also called fertilization.
   e. makes most angiosperms independent of water for reproduction.

6. Which statement is *not* true of double fertilization?
   a. It is found in all angiosperms.
   b. It is found in no plants other than angiosperms.
   c. One of its products is a triploid nucleus.
   d. One sperm nucleus fuses with the egg nucleus.
   e. One sperm nucleus fuses with two polar nuclei.

7. The suspensor
   a. gives rise to the embryo.
   b. is heart-shaped in dicots.
   c. separates the two cotyledons of dicots.
   d. ceases to elongate early in embryo development.
   e. is larger than the embryo.

8. Which statement is *not* true of photoperiodism?
   a. It is related to the biological clock.
   b. Phytochrome plays a role in the timing process.

c. It is based on measurement of the length of the night.
   d. Most plant species are day-neutral.
   e. It is limited to the plant kingdom.

9. Although florigen has never been isolated, we think it exists because
   a. night length is measured in the leaves, but flowering occurs elsewhere.
   b. it is produced in the roots and transported to the shoot system.
   c. it is produced in the coleoptile tip and transported to the base.
   d. we think that gibberellin and florigen are the same compound.
   e. it may be activated by prolonged (more than a month) chilling.

10. Which statement is *not* true of vernalization?
   a. It may require more than a month of low temperature.
   b. The vernalized state generally lasts for about a week.
   c. Vernalization makes it possible to have two winter wheat crops each year.
   d. It is accomplished by subjecting moistened seeds to chilling.
   e. It was of interest to Russian scientists because of their native climate.

FOR STUDY

1. For a crop plant that reproduces both sexually and asexually, which method of reproduction might the farmer prefer?

2. Thompson seedless grapes are produced by vines that are triploid. Think about the consequences of this chromosomal condition for meiosis in the flowers.

Why are these grapes seedless? Describe the role played by the flower in fruit formation when no seeds are being formed. How do you suppose Thompson seedless grape plants are propagated?

4. Poinsettias are popular ornamental plants that typically bloom just before Christmas. Their flowering is

photoperiodically controlled. Are they long-day or short-day plants? Explain.

5. You plan to induce the flowering of a crop of long-day plants in the field by using artificial light. Is it necessary to keep the lights on continuously from sundown until the critical night length is reached? Explain.

READINGS

Barrett, S. C. H. 1987. "Mimicry in Plants." *Scientific American*, September. Some plants mimic insects, encouraging the mimicked insects to act as pollinators.

Cleland, C. E. 1978. "The Flowering Enigma." *BioScience*, April. Florigen eluded physiologists when this article was published, and it still does.

Cox, P. A. 1993. "Water-Pollinated Plants." *Scientific American*, October. An unusual environment calls for different pollination strategies, and it provides unusual opportunities for research.

Goodman, B. 1993. "A 'Shotgun Wedding' Finally Produces Test-Tube Plants." *Science*, vol. 261, page 430. A brief account of the discovery of in vitro fertilization in plants.

Handel, S. N. and A. J. Beattie. 1990. "Seed Dispersal by Ants." *Scientific American*, August. Some plants produce seeds bearing specialized fat bodies that are eaten by ants after the ants carry the seeds to their nests. This and other mechanisms are discussed.

Niklas, K. J. 1987. "Aerodynamics of Wind Pollination." *Scientific American*, July. Do the mechanics of pollination seem improbable to you? Wind-pollinated plants have many adaptations that favor successful pollination.

Raven, P. H., R. F. Evert and S. Eichhorn. 1991. *Biology of Plants*, 5th Edition. Worth, New York. A well-balanced general botany textbook.

Salisbury, F. B. and C. W. Ross. 1992. *Plant Physiology*, 4th Edition. Wadsworth, Belmont, CA. An authoritative textbook with good chapters relating to reproductive development.

Taiz, L. and E. Zeiger. 1991. *Plant Physiology*. Benjamin/Cummings, Redwood City, CA. Chapter 21 deals with the control of flowering.

PART SIX
The Biology
of Animals

Animals live in amazing places, including the most extreme environments on our planet. The adaptations that make these lifestyles possible can be fascinating. Consider, for example, emperor penguins, the largest penguin species. They spend most of their lives feeding at sea, but they breed and raise their young far from the open sea during the bitter cold winter season of Antarctica. As the winter ice shelf forms off Antarctica, emperor penguins gather at its edge. They then walk over 100 kilometers inland across the ice, to breed in one of the coldest, most inhospitable places on Earth. After the female lays an egg, she walks back to the sea to feed; the male incubates the egg, then protects and feeds the chick until the female returns with her body fat replenished. The female then takes over the feeding of the chick, and the male walks back to the sea, having fasted for more than four months.

In Part Six you will learn about adaptations that allow birds to fly at altitudes at which humans cannot exist without technological assistance, seals to remain underwater for more than half an hour, mice to live without water in the hottest deserts, and bats to capture flying insects in total darkness. The explanations of how animals achieve such feats are found in physiology, the study of how animals work. Unusual adaptations are not just the spice of physiology; they are extensions of basic physiological mechanisms and therefore help us to understand the principles of normal physiological functions in humans and other animals.

Earlier in this book we learned how cells work. Because animals consist of cells, how cells function—including their energy metabolism, their needs for nutrients, their production of waste products such as carbon dioxide, and their osmotic balance—can be extended to whole animals. For the simplest animals, sponges and cnidarians, this extension is straightforward, since most of those species live in the sea and their bodies are only two cell layers thick. Seawater contains nutrients, it has a suitable composition of salts, and it provides a stable physical environment. Each cell of a sponge or a jellyfish is in direct contact with the environment and functions almost as an autonomous unit. It receives its nutrients directly from and releases its wastes directly into the seawater. With aquatic animals that are larger and more complex than sponges and cnidarians, not every cell of their bodies can be in direct contact with the external environment. With terrestrial animals, the cells of their bodies cannot be in direct contact with their external environment, air, because it would dry and kill them. For these reasons, most cells of most animals are served by an internal environment consisting of extracellular fluids. That internal environment provides appropriate physical conditions for the cells of the animal, supplies all the nutrients they need, and removes all their wastes.

A Long Walk to a Cold Breeding Ground
Emperor penguins migrate between the sea and their breeding grounds. The distance may be more than 100 kilometers.

35

Physiology, Homeostasis, and Temperature Regulation

HOMEOSTASIS

Most of animal physiology focuses on how the internal environment is maintained in a condition that meets the needs of the cells. The internal environment is obviously not as vast as the sea. Its nutrient content can be rapidly exhausted, and its physical conditions are altered by the metabolic activities of the cells it serves. Animal organs and organ systems keep various aspects of the internal environment at a steady state, a condition called **homeostasis**. Gas-exchange organs provide oxygen to and remove carbon dioxide from the extracellular fluids, digestive organs supply nutrients, and excretory systems eliminate wastes.

Homeostasis is an essential feature of complex animals, and it has enabled animals to adapt to nearly every environment on Earth. It has also enabled biochemical systems to become more efficient by being adapted to function over narrow ranges of physical parameters. If an organ fails to function properly, homeostasis of the internal environment is lost, and as a result cells become sick and die. The sick cells are not just those of the organ that functions improperly, but the cells of all other organs as well. Loss of homeostasis is therefore a serious problem that makes itself worse. To avoid loss of homeostasis, the activities of organs must be controlled and regulated in response to changes in the internal environment.

Control and regulation require information; hence the organ systems of information—the endocrine system and the nervous system—must be included in our discussions of every physiological function. For that reason, we treat the endocrine and nervous systems early in this part of the book. Subsequent chapters deal with the organ systems that are responsible for homeostasis of various aspects of the internal environment.

ORGANS AND ORGAN SYSTEMS

The diversity of adaptations that enable animals to live in just about any environment presents us with a bewildering number of details that can make the study of physiology seem daunting. Thus we begin our study with a road map of the organs, organ systems, and physiological functions of at least one species. That species might as well be *Homo sapiens*, but the road map also applies to most other vertebrates and to some invertebrates as well.

The Structure of Organs

Organs are made of tissues, and a tissue consists of cells with similar structure and function. Biologists who study cells and tissues recognize many types of cells but group them into only four general types of tissues: epithelial, connective, muscle, and nervous.

Most **epithelial tissues** are sheets of tightly connected cells such as those that cover the body surface and those that line various hollow organs of the body, including the digestive tract and the lungs. Some epithelial cells have secretory functions—for example, those that secrete mucus, digestive enzymes, or sweat. Other epithelial cells have cilia and help substances move over surfaces or through tubes. Since epithelial cells create boundaries between the inside and the outside of the body and between body compartments, they frequently have absorptive and transport functions. An epithelium can be stratified, as is the skin, which consists of many layers of cells, or it can be simple, as is the lining of the gut, which consists of a single layer of cells. We'll encounter epithelial tissues in our discussions of the linings and tubules of reproductive systems (see Chapter 37), the linings of gas-exchange systems (see Chapter 41), the linings of digestive tracts (see Chapter 43), and the tubules of excretory systems (see Chapter 44).

Connective tissues support and reinforce other tissues. Unlike epithelial tissues, which consist of densely packed, tightly connected populations of cells, most connective tissues consist of a dispersed population of cells embedded in an extracellular matrix. The properties of the matrix differ in different types of connective tissues. The connective tissue in skin contains many elastic fibers that can be stretched and then return to their original position. The connective tissues that connect muscles to bone and bones to one another have many collagen fibers with high tensile strength. Bone is a connective tissue in which the extracellular matrix has been hardened by mineral deposition (see Chapter 40). Two other major types of connective tissue are adipose tissue (fat cells) and the cellular components of the blood (see Chapter 42).

Muscle tissue consists of cells that can contract and therefore cause movements of organs, limbs, or most other parts of the body. Muscles are the most important **effectors** of the body; they enable it to do things (see Chapter 40). Since muscle tissues play important roles in most organs and organ systems, we'll encounter them in many places in Part Six.

Nervous tissue (see Chapters 38, 39, and 40) enables animals to deal with information. The cells of nervous tissue, **neurons**, are extremely diverse. Some respond to specific types of stimuli, such as light, sound, pressure, or certain molecules, by generating electric signals in their membranes. These electric signals can be conducted via long extensions of the cell to other parts of the body, where they are passed on to other neurons, muscle cells, or secretory cells. Because neurons are involved in controlling the activities of most organ systems, we will run across them frequently in Part Six.

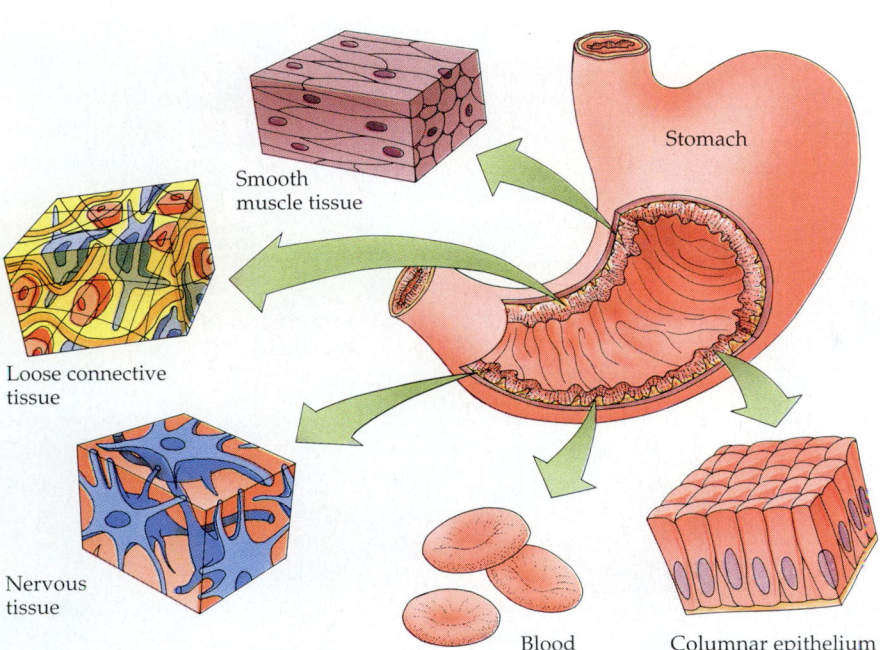

35.1 One Organ Contains Many Tissue Types
The human stomach—a section of the gut—contains many tissue types that enable it to perform its functions within the digestive system (see Figure 35.5a).

Smooth muscle tissue

Loose connective tissue

Nervous tissue

Stomach

Blood

Columnar epithelium

Organs are usually made up of more than one tissue type, and most organs include all tissue types. The gut—the organ that digests food and absorbs nutrients—is a good example (Figure 35.1). The gut is lined internally with a single layer of columnar epithelial cells. Some secrete mucus or enzymes and others mainly absorb nutrients. Beneath the gut lining is connective tissue, within which are glands and blood vessels. Concentric layers of muscle tissue move food through the gut and mix it with the secretions of the epithelial cells. Neurons extend between the layers of other tissues to control both the secretions and the movements of the gut.

An individual organ, such as the gut, and other organs with complementary functions may be the parts of an organ system. The major organ systems of the body are outlined in Figures 35.2 through 35.7. We'll discuss each of these systems in the chapters that follow.

Organs for Information and Control

The principal organ system that processes information and uses that information to control the physiology and behavior of the animal is the nervous system (Figure 35.2a). It consists of the brain and spinal cord (the central nervous system) along with peripheral nerves that conduct electric signals from sensors to the central nervous system and conduct signals from the central nervous system to effectors, which are either muscle tissue or secretory tissue. The sensors of the nervous system are diverse; they include eyes, ears, organs of taste and smell, and cells sensitive to temperature, touch, pressure, stretch, and pain.

The endocrine system (Figure 35.2b) also processes information and controls the functions of organs, but its messages are distributed mostly in the blood to the entire body as chemical signals called hormones. The principal organs of the endocrine system are ductless glands that secrete specific hormones into the blood. In addition, many other tissues contain individual cells that secrete hormones. There are strong interactions between the nervous system and

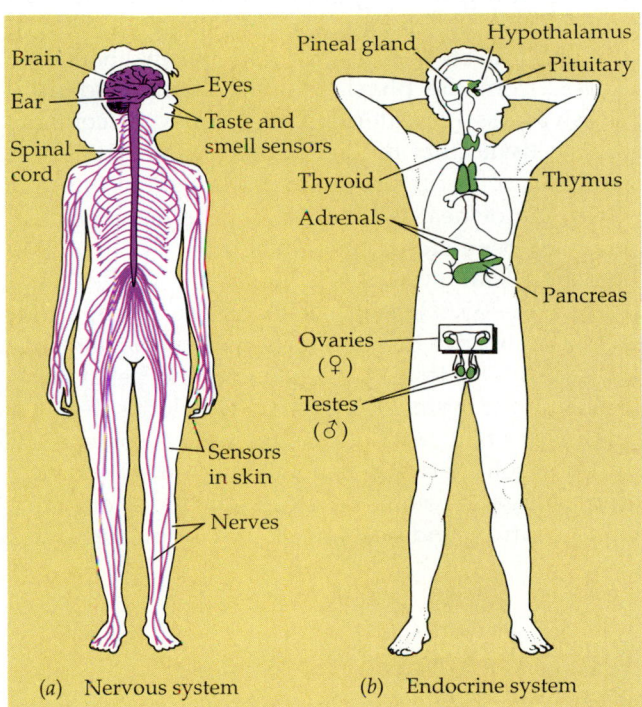

Brain
Ear
Spinal cord
Eyes
Taste and smell sensors
Sensors in skin
Nerves

Pineal gland
Hypothalamus
Pituitary
Thyroid
Thymus
Adrenals
Pancreas
Ovaries (♀)
Testes (♂)

(a) Nervous system   (b) Endocrine system

35.2 Organs of Information and Control

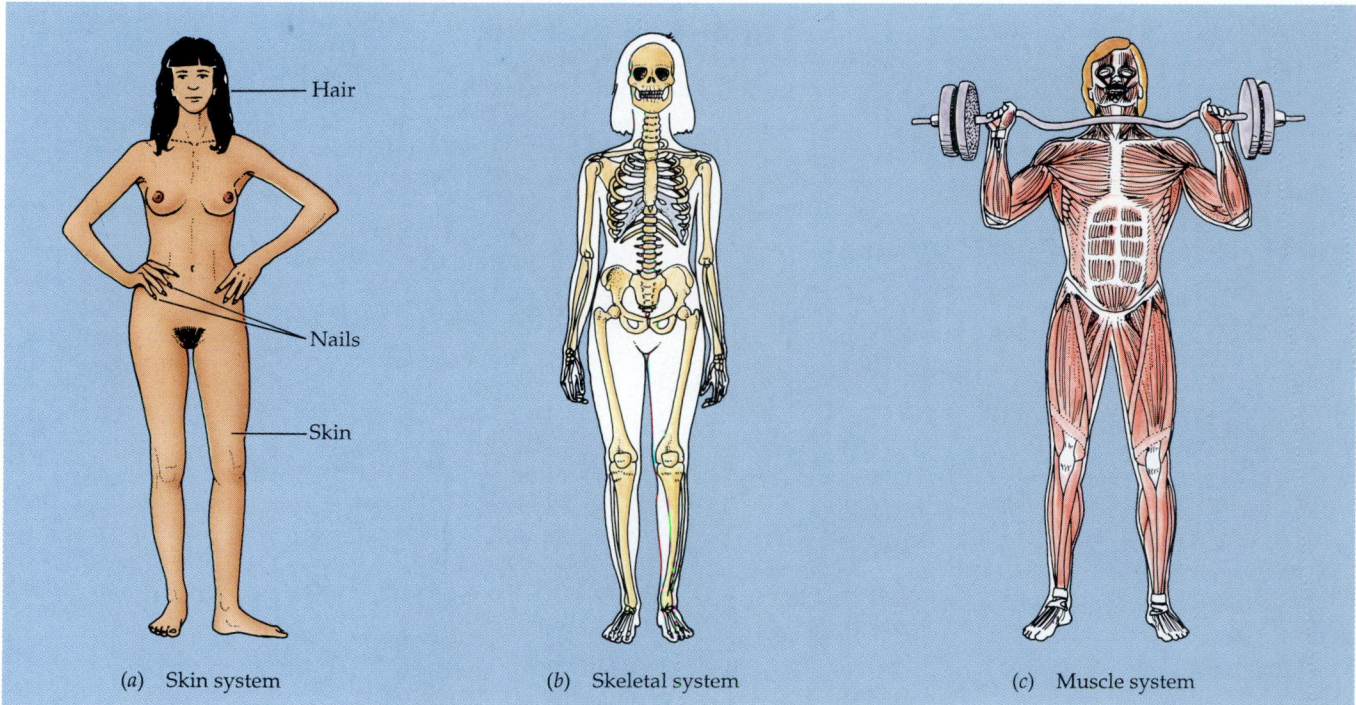

(a) Skin system   (b) Skeletal system   (c) Muscle system

35.3 Organs of Protection, Support, and Movement

the endocrine system. Cells in the brain produce hormones that control parts of the endocrine system. In turn, there are cells in the brain that respond to the hormones produced by endocrine glands.

Organs for Protection, Support, and Movement

The largest organ of the body is the skin, along with its special elaborations, hair and nails (Figure 35.3a). The skin protects the body from organisms that cause disease, from the physical environment, and from excessive loss of water. Because the skin contains nervous tissue that is sensitive to various stimuli, it is a major sense organ. The skin is also an effector: As a route of heat exchange with the environment, it helps to regulate body temperature.

The skeletal system supports and protects the body (Figure 35.3b). The skeleton is also an important effector: It forms the supports and the levers that muscles pull on to cause movement and behavior.

The muscle system (Figure 35.3c) includes the skeletal muscles that are under our conscious control and cause all voluntary movements, the muscles of the internal organ systems that are not under our conscious control, and the muscles that constitute the heart.

Organs of Reproduction

The male and female reproductive systems consist of gonads (testes and ovaries, respectively), which produce sex cells (gametes). Additional organs deliver

the sex cells to the site where fertilization takes place (Figure 35.4). The female reproductive system includes the uterus, the organ that supports the development of the embryo. The female's mammary glands provide nutrients for the infant. In the gonads of both sexes are tissues that secrete hormones that play roles in sexual development and reproduction.

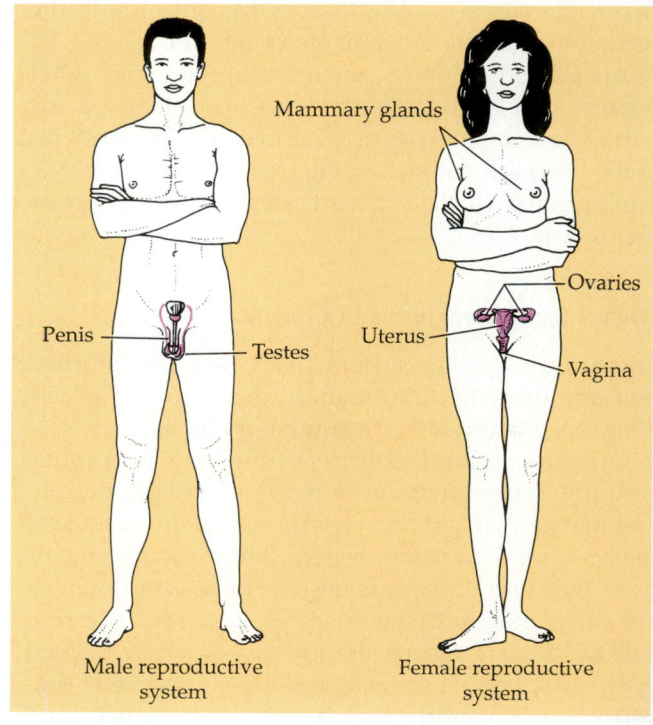

Male reproductive system   Female reproductive system

35.4 Organs of Reproduction

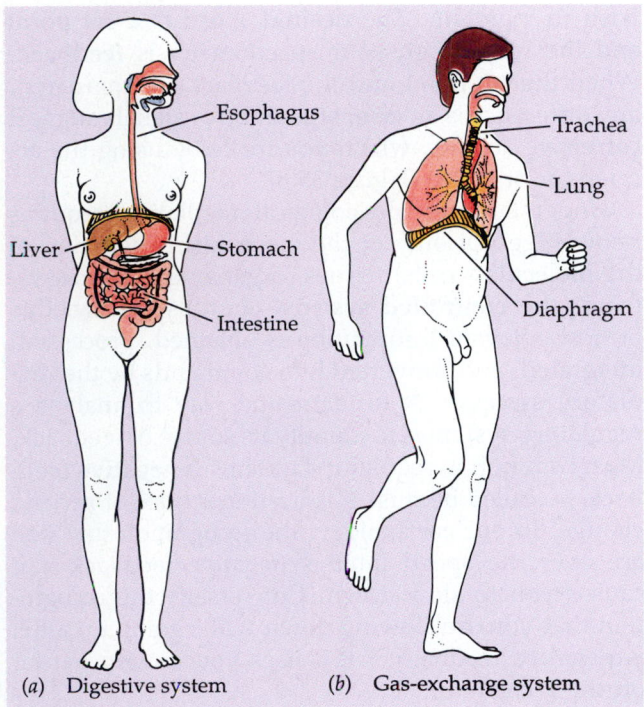

(a) Digestive system   (b) Gas-exchange system

35.5 Organs of Nutrition

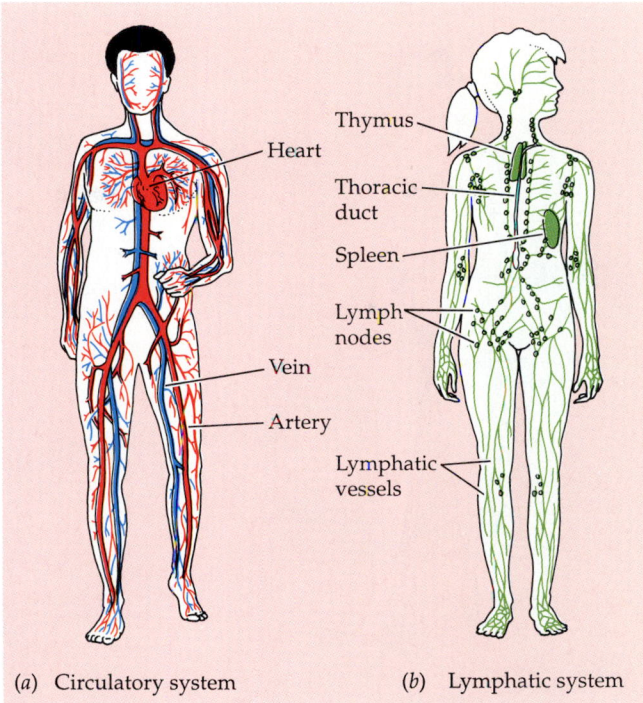

(a) Circulatory system   (b) Lymphatic system

35.6 Organs of Transport

Organs of Nutrition

The digestive system is largely a continuous tubular structure that extends from mouth to anus (Figure 35.5a). This tube, also called the gut, is divided into different segments that serve different functions in the processing and digestion of food and the absorption of nutrients. Glands associated with the gut deliver into it digestive enzymes and other molecules that break down complex food molecules. The lower gut stores and periodically eliminates solid wastes and water.

The gas-exchange system, also called the respiratory system, provides oxygen, which is essential for cellular respiration (Figure 35.5b). Carbon dioxide, a waste product of cellular respiration, is eliminated by the gas-exchange system. The gas-exchange organs of humans are lungs, which are a system of progressively dividing airways leading to tiny but numerous sacs with membranous gas-exchange surfaces that have a very large combined surface area. The muscles that move air into and out of the lungs are another component of the gas-exchange system.

Organs of Transport

Oxygen must be transported from the lungs to the tissues of the body, and carbon dioxide must be transported from the tissues to the lungs. These gases are transported by the circulatory system, which includes a pump (the heart), a system of blood vessels (veins and arteries), and blood (Figure 35.6a). The circulatory system also transports nutrients from the

gut, delivers nitrogenous wastes to the excretory system, transports hormones, transports heat, and generates mechanical forces. Blood is made up of cellular components in a liquid medium called plasma. The blood plasma is virtually continuous with the extracellular fluids that are the internal environment of the body.

The lymphatic system is another transport system consisting of a set of vessels that extend throughout the body, but, unlike the circulatory system, it does not include a pump and its vessels do not form a complete circuit (Figure 35.6b). The lymphatic system picks up extracellular fluid and delivers it to the blood circulatory system.

Organs of Excretion

Urine forms in the kidneys. Urine includes nitrogenous wastes from the metabolism of proteins and nucleic acids as well as excess salts and other substances that the body excretes. The kidneys play crucial roles in maintaining the correct water content of the body and the correct salt composition of the extracellular fluids. Urine passes to a bladder for storage until it is released to the exterior through the urethra (Figure 35.7).

In the chapters that follow we will explore physiological functions and the organ systems that accomplish them. In the remainder of this chapter we will discuss general principles of homeostasis, which have applications to all animal physiology. We'll illustrate these principles by considering in detail the

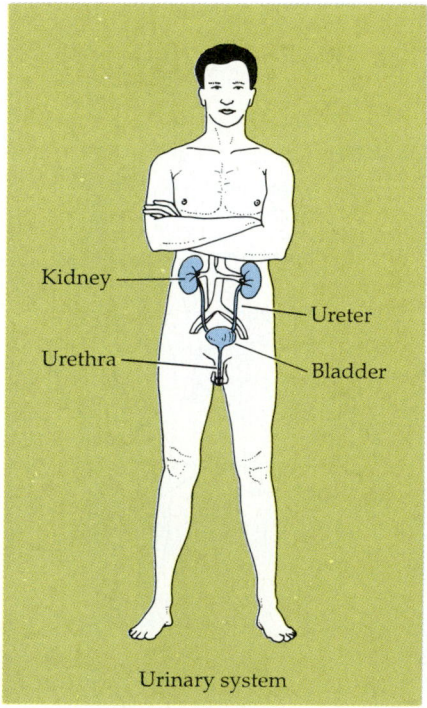

35.7   Organs of Excretion and Water/Salt Balance

regulation of body temperature. Temperature is an important physical parameter of the internal environment. It can be perturbed by the activities of cells and by changes in the outside environment. Animals have evolved a number of adaptations for dealing with changes in temperature.

GENERAL PRINCIPLES OF HOMEOSTASIS

Homeostasis refers to a constant state of internal conditions in the body. Homeostasis depends on the functions of the organs and organ systems of the body; these functions must be controlled and regulated to achieve a relatively constant internal environment. The terms *control* and *regulation* might seem interchangeable, but their meanings differ. Control implies the ability to *change* the rate of a reaction or process. Regulation—the more sophisticated and more specific physiological concept—refers to *maintaining* a variable within specific levels or limits. You control the speed of your car by using the accelerator and brake, but you regulate it by using the accelerator and brake to maintain a particular speed.

Set Points and Feedback

Regulation requires, in addition to control mechanisms, the ability to obtain and use information. You can regulate the speed of a car only if you know the speed at which you are traveling and the speed you wish to maintain. The desired speed is a **set point** and the reading on your speedometer is **feedback**. When the set point and the feedback are compared, any difference is an **error signal**. Error signals suggest corrective actions, which you make by using the accelerator or brake (Figure 35.8).

Understanding physiological regulation requires knowledge not only of the mechanisms of action of the molecules, cells, tissues, organs, and organ systems—the **controlled systems**—but also knowledge of how relevant information is obtained, processed, integrated, and converted into commands by the regulatory systems. A fundamental way to analyze a regulatory system is to identify its source of feedback. Most common in regulatory systems is **negative feedback**, so called because it is used to reduce or reverse change. In our car analogy, the recognition that you are over the speed limit is negative feedback if it causes you to slow down. Conversely, the recognition that you are slowing down while going up a hill is negative feedback if it causes you to step harder on the accelerator.

Regulatory Systems

To understand the features of a regulatory system, consider a thermostat—a relevant analogy, since we will be looking at the biological thermostats of vertebrate animals later in this chapter. The thermostat that is part of the heating–cooling system of a house is a regulatory system. It has upper and lower set points that you can adjust, and it receives feedback from a sensor. The circuitry of the thermostat converts any differences between the set points and the sensor into signals that activate the controlled systems—the furnace and the air conditioner. When room temperature rises above the upper set point, the thermostat activates the air conditioner, thus reducing room temperature below the set point. When room temperature falls below the lower set point, the thermostat activates the furnace, thus raising room temperature toward that set point. Hence the sensor of room temperature provides negative feedback (Figure 35.9).

Negative feedback makes good sense for physiological regulatory systems, so you may wonder if there is any such thing as positive feedback in physiology. Although not as common as negative feedback, it does exist. Rather than returning a system to a set point, **positive feedback** amplifies a response. One example is sexual behavior, in which a little stimulation can cause more behavior, which causes more stimulation, and so on. Positive feedback is not used by regulatory systems that maintain stability!

Feedforward information is another feature of regulatory systems. The function of feedforward is to change the set point. Seeing a deer ahead on the road

35.8 Control, Regulation, and Feedback
As you drive a car, the posted speed limit is your set point and the speedometer gives you feedback information. Comparing the speed limit to the speedometer reading gives you error signals that you convert into corrective actions by using the brakes and the accelerator to regulate the car's speed. The sight of a deer in the road ahead is feed*forward* information.

when you are driving is an example of feedforward; this information takes precedence over the posted speed limit, and you change your set point to a slower speed. If you want the temperature of your house to be lower at night than during the day, you can add a clock to the thermostat to provide feedforward information about time of day.

These general considerations about control, regulation, and regulatory systems help to organize our thinking about physiological systems, but the physiological systems can be far more complex than the thermostat and driving analogies. In some systems we do not even know the nature of the feedback. For most people, body weight is regulated, although it might not be at the level we prefer. Without consciously counting calories, the brain controls hunger so that food intake matches energy expenditure. We do not understand what information the brain uses to achieve this remarkable feat of regulation, but there are some interesting hypotheses and active research on the problem.

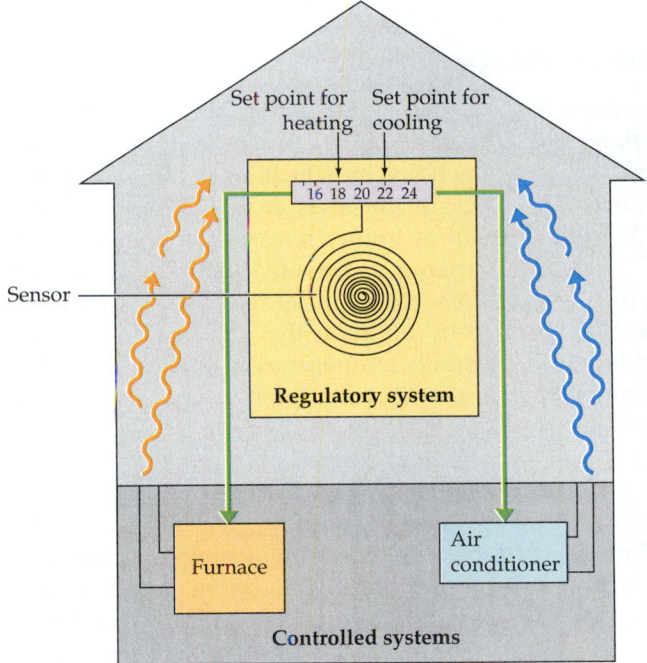

35.9 A Thermostat Regulates House Temperature
Changes in room temperature cause a sensor to move relative to set points on the thermostat and activate the furnace or the air conditioner. Room temperature, as detected by the sensor, is feedback to the regulatory system.

One regulatory system that we understand well is the system that regulates body temperature. Let's begin by asking two related questions: Why do organisms need to thermoregulate? What is the effect of temperature on living systems?

THE EFFECTS OF TEMPERATURE ON LIVING SYSTEMS

Over the face of Earth, temperatures vary enormously—from boiling hot springs to the frigid Antarctic plateau. Because heat always moves from a warmer object to a cooler object, any change in the temperature of the environment causes a change in the temperature of an organism in that environment, unless the organism does something to regulate its temperature. Living cells are restricted to a narrow range of temperatures. If cells cool to below 0°C, ice crystals can form within them; this can fatally damage their structures. Some cells are adapted to prevent freezing and others to survive freezing, but generally cells must remain above 0°C to stay alive. The upper temperature limit is around 45°C for most cells. Some specialized algae can grow in hot springs at 70°C, and some bacteria can live at near 100°C, but in general, proteins begin to denature as temperatures rise above 45°C. As proteins denature, they lose their functional properties. Most cellular functions are limited to the range of temperatures between 0 and 45°C, which are considered the thermal limits for life.

The Q_{10} Concept

Even within the range of 0 to 45°C, temperature changes can create problems for animals. Like the biochemical reactions of which they are made up, most physiological processes are temperature-sensitive, going faster at higher temperatures (see Figure 6.27). The temperature sensitivity of a reaction or process can be described in terms of the $\mathbf{Q_{10}}$, a quotient calculated by dividing the rate of a process or reaction at a certain temperature, R_T, by the rate of that process or reaction at a temperature 10°C lower, R_{T-10}:

$$Q_{10} = \frac{R_T}{R_{T-10}}$$

The Q_{10} can be measured for a simple enzymatic reaction or for a complex physiological phenomenon. If a reaction or process is not temperature-sensitive, it has a Q_{10} of 1. Most biological Q_{10}'s are between 2 and 3, which means the reaction rates double or triple as the temperature increases by 10°C (Figure 35.10).

Changes in temperature can be particularly disruptive to an animal's functioning because all the component reactions in the animal do not have the

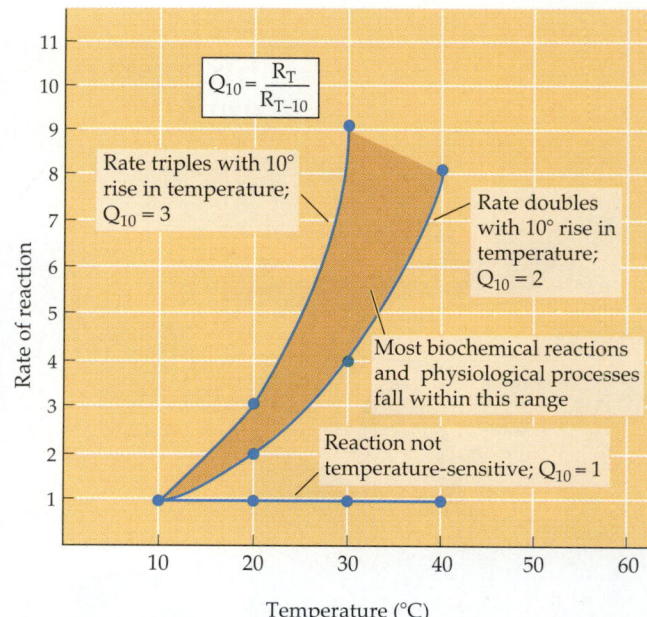

35.10 $\mathbf{Q_{10}}$ **and Reaction Rate**
The larger the Q_{10}, the faster the reaction rate rises as the temperature rises.

same Q_{10}. Individual reactions with different Q_{10}'s are linked together in complex networks that carry out physiological processes. Temperature changes shift the rates of some of the reactions more than those of others, thus disrupting the balance and integration that the processes require. To maintain homeostasis, organisms must be able to compensate for or prevent changes in temperature.

Metabolic Compensation

The body temperatures of some animals are tightly coupled to the temperature of the environment. Think of a fish in a pond in a highly seasonal environment. As the temperature of the pond water changes from 4°C in midwinter to 24°C in midsummer, the body temperature of the fish does the same. If we bring such a fish into the laboratory in the summer and measure its metabolic rate (the sum total of the energy turnover of its cells) at different water temperatures, we might calculate a Q_{10} of 2 and plot our data as shown by the red line in Figure 35.11. We predict from our graph that in winter, when the temperature is 4°C, the fish's metabolic rate will be only one-fourth of what it was in the summer. We then return the fish to its pond. When we bring the fish back to the laboratory in the winter and repeat the measurements, we find, as the blue line shows, that its metabolic rate at 4°C is not as low as we predicted, but is almost the same as it was at 24°C in the summer. If we repeat the measurement over a range of temperatures, we find that the fish's meta-

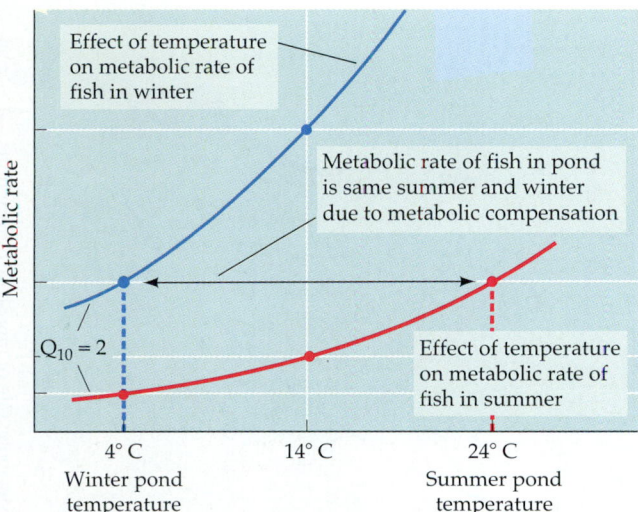

Effect of temperature on metabolic rate of fish in winter

Metabolic rate of fish in pond is same summer and winter due to metabolic compensation

$Q_{10} = 2$

Effect of temperature on metabolic rate of fish in summer

Metabolic rate

4° C
Winter pond temperature

14° C

24° C
Summer pond temperature

35.11 Metabolic Compensation for Seasonal Differences
When the metabolic rate of a fish is measured in summer and in winter, it is shown to be temperature-sensitive. At normal environmental temperatures in nature, however, the fish may have the same metabolic rate summer and winter. Metabolic compensation acclimatizes a fish to its changing environment.

bolic rate is always higher than the rate we measured at the same temperature in the summer-acclimatized fish. **Acclimatization** is the process of physiological and biochemical change that an animal undergoes in response to seasonal changes in climate.

The reason for the difference between our prediction from the summer data and our measurements on the winter fish is that seasonal acclimatization in the fish has produced **metabolic compensation**. Metabolic compensation readjusts the biochemical machinery to counter the effects of temperature. What might account for such a change? Look again at Figure 6.25, which suggests a hypothesis. If the fish we are studying have duplicate enzymes that operate at different optimal temperatures, they may compensate metabolically by catalyzing reactions with one set of enzymes in the summer and another set in the winter. The end result of such readjustment is that metabolic functions are much less sensitive to seasonal changes in temperature than they are to shorter-term thermal fluctuations.

THERMOREGULATORY ADAPTATIONS

Organisms have evolved numerous behavioral and physiological mechanisms for maintaining optimal body temperatures. The best way to discuss these thermoregulatory adaptations is to look at a thermoregulatory classification of the animals that possess them. Animals are commonly described as cold-

blooded or warm-blooded. Taken at face value these terms could lead to some ridiculous errors. For example, desert reptiles could be classified as warm-blooded during the day and cold-blooded at night, many insects might be classified as warm-blooded during flight and cold-blooded at rest, and hibernating mammals would be very cold-blooded during most of the winter.

Biologists prefer a different set of terms. A **homeotherm** is an animal that regulates its body temperature at a constant level; a **poikilotherm** is an animal whose temperature changes. This system of classification says something about the biology of the animals, but it also presents problems. Should a fish in the deep ocean, where the temperature changes very little, be called a homeotherm? Should a hibernating mammal that allows its body temperature to drop to nearly the temperature of its environment be called a poikilotherm? The problem posed by the hibernator has been set aside by coining a third category, the **heterotherm**: an animal that regulates its body temperature at a constant level *some* of the time. Homeotherm, poikilotherm, and heterotherm are useful descriptive terms.

Another set of terms classifies animals on the basis of thermoregulatory mechanisms. **Ectotherms** depend largely on external sources of heat, such as solar radiation, to maintain their body temperatures above the environmental temperature. **Endotherms** can regulate their body temperature by producing heat metabolically or by mobilizing active mechanisms of heat loss. Mammals and birds behave as endotherms; animals of all other species behave as ectotherms most of the time.

Laboratory Studies of Ectotherms and Endotherms

Let's choose a small lizard to represent ectotherms and a mouse of the same body size as the lizard to represent endotherms. In the laboratory we put each animal in a small metabolism chamber that enables us to measure the body temperature of the animal and its metabolic rate as we change the temperature of the chamber from 0 to 35°C. The results obtained from the two species are very different (Figure 35.12). The body temperature of the lizard always equilibrates with that of the chamber, whereas the body temperature of the mouse remains at 37°C. The metabolic rate of the lizard increases with temperature. Below about 27°C the metabolic rate of the mouse increases as chamber temperature decreases (notice that you must read the graph right to left to see this). It seems the lizard cannot regulate its body temperature or metabolism independently of environmental temperature. The mouse regulates its body temperature by altering its rate of metabolic heat production.

(a)

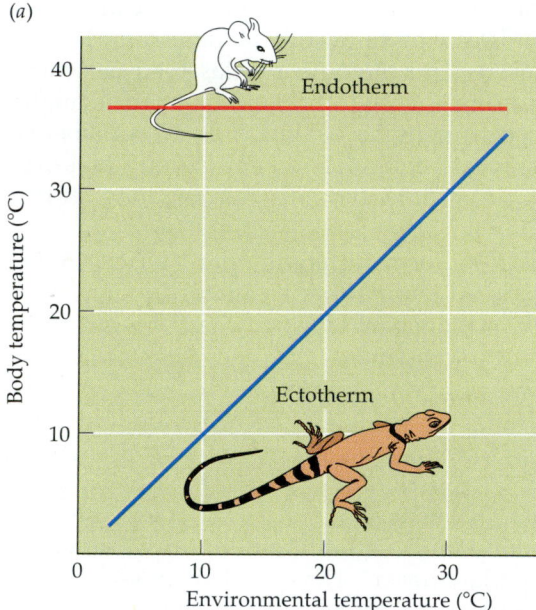

(b)

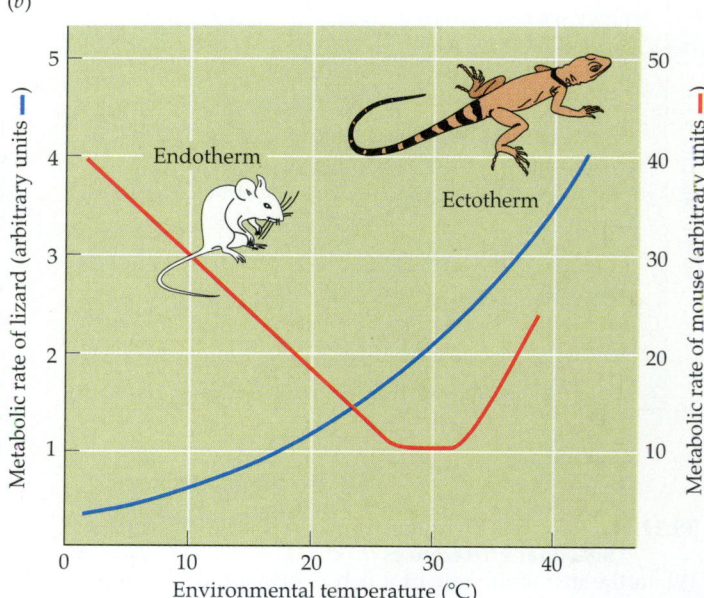

35.12   Effects of Environmental Temperature Differ
(a) The body temperatures of a lizard (blue line) and a
mouse (red line) respond differently to changes in environmental temperature. (b) The reason for the difference
is that, as the environmental temperature falls, the rate of
metabolic heat production in the lizard decreases,
whereas that of the mouse increases.

Field Study of an Ectotherm

A logical next step is to test in nature our laboratory
conclusion that the lizard cannot regulate its body
temperature. We can do this by implanting a capsule
containing a radio telemeter in the lizard's body and
then releasing the lizard in its desert environment.
The radio telemeter measures the lizard's body temperature and then converts it to a radio signal that
can be heard through a portable radio. We can thus
observe the body temperature of the lizard as it goes
about its normal behavior. Our prediction is that the
temperature of the lizard will follow the temperature
of the environment, which can change more than
40°C in a few hours.

The results of the experiment, however, differ
strikingly from this prediction. At night the temperature in the desert may drop close to freezing, but
the temperature of the lizard remains stable at 16°C.
This is not difficult to explain; the lizard spends the
night in a burrow where the soil temperature is a
constant 16°C. Early in the morning, soon after sunrise, the lizard emerges from its burrow. The air temperature is still quite cool, but the body temperature
of the lizard rises to 35°C in less than 30 minutes.
The lizard achieves this rapid rise in temperature by
basking on a rock with maximum exposure to the
sun. As its dark skin absorbs solar radiation, its body
temperature rises considerably above the surround-

ing air temperature. By altering its exposure to the
sun, the lizard maintains its body temperature at
around 35°C all morning as it seeks food, avoids
predators, and interacts with potential mates or competitors. By noon the air temperature near the surface
of the desert has risen to 50°C, but the lizard's body
temperature remains around 35°C. It is now staying
mostly in shade, frequently up in bushes where there
is a cooling breeze. As afternoon progresses, air temperature declines, and the lizard again spends more
of its time in the sun and on hot rocks to maintain
its body temperature around 35°C. The lizard returns
to its burrow just before sunset, and its body temperature rapidly drops to 16°C. Figure 35.13 reviews
the patterns of the lizard's behavior and the temperature changes over the course of a day.

This field experiment shows that the lizard can
regulate its body temperature quite well by behavioral mechanisms rather than by metabolic mechanisms. The deficiency in our laboratory experiment
was that the lizard in the chamber could not use its
thermoregulatory behavior. If we give a lizard access
to a thermal gradient in the laboratory, it is capable
of regulating its body temperature by selecting the
right place on the gradient. If only a hot place and a
cold place are available, it will shuttle back and forth.
It will maintain a different body temperature during
the night than during the day, and if it is infected
with pathogenic bacteria it will give itself a fever by
selecting higher temperatures on the gradient (see
Box 35.A on page 795).

The lesson to be learned from the discrepancies
between the results from the laboratory and field
experiments on the lizard are encapsulated by a quo-
tation from the German embryologist Hans Spe-

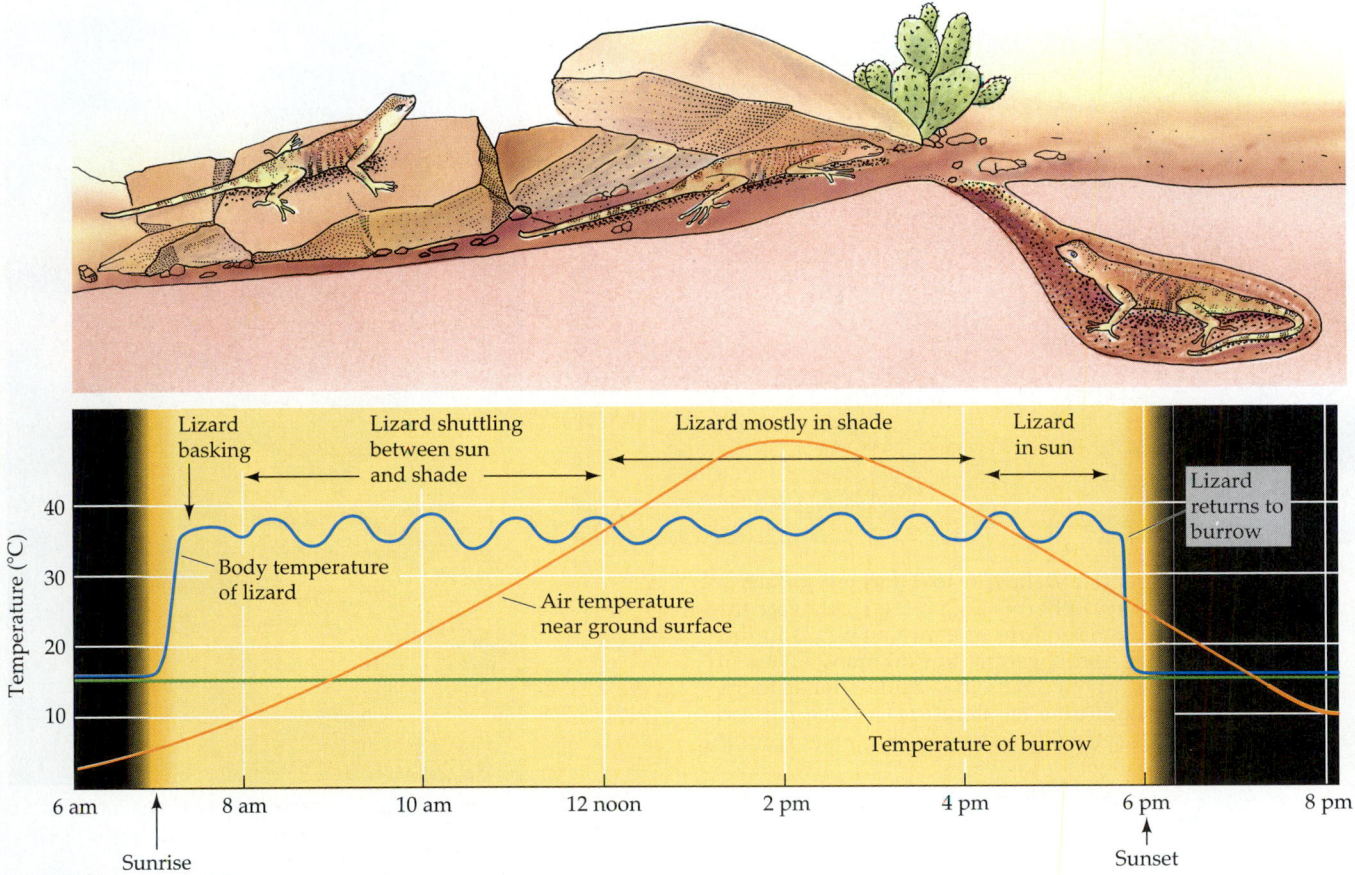

35.13 A Lizard Can Regulate Its Body Temperature Behaviorally

mann: "An experiment is like a conversation with an animal, but the animal must be permitted to answer in its own language."

Behavioral Thermoregulation

Behavioral thermoregulation is not the exclusive domain of ectotherms. It is also the first line of defense for endotherms. When the option is available, most animals select thermal microenvironments that are best for them. They may change their posture, orient to the sun, move between sun and shade, and move between still air and moving air, as demonstrated by the lizard in our field experiment. Examples of more complex thermoregulatory behavior are nest construction and social behavior such as huddling. In humans, the selection of clothing is quite important. Behavioral thermoregulation is widespread in the animal kingdom (Figure 35.14).

Control of Blood Flow to the Skin

Physiological thermoregulation is not the exclusive domain of endotherms. Ectotherms exhibit various physiological thermoregulatory adaptations. Both ectotherms and endotherms can alter the rate of heat

exchange between their bodies and their environments by controlling the flow of blood to the skin. For example, when a person's body temperature rises as a result of exercise, blood flow to the skin increases, and the skin surface gets quite warm. The extra heat brought from the body core to the skin by the blood is lost to the environment, thus tending to bring body temperature back to normal. By contrast, when a person is exposed to cold, the blood vessels supplying the skin constrict, decreasing blood flow and heat transport to the skin, thus reducing heat loss to the environment.

The control of blood flow to the skin is an important adaptation for ectotherms like the marine iguana of the Galapagos Islands. The Galapagos are volcanic islands on the equator, bathed by cold oceanic currents. Marine iguanas are reptiles that bask on black lava rocks near the ocean and swim in the sea, where they feed on submarine algae. When the iguanas cool to the temperature of the sea, they are slower and more vulnerable to predators, and probably incapable of efficient digestion. They therefore alternate between feeding in the sea and basking on the rocks. It is advantageous for iguanas to retain body heat as long as possible while swimming and to warm up as fast as possible when basking. They adjust their cool-

(a)

(b)

35.14   Endotherms Use Behavior to Thermoregulate
Humans and other endotherms adjust their behavior to
the environmental temperature in many ways. (a) In the
extreme cold of the Arctic, people put on many layers of
insulating clothing. The ice huts they build shelter them
from even colder nighttime temperatures. (b) An African
elephant showers itself with water to bring relief from the
heat.

ing and heating rates by changing the flow of blood
to the skin.

Blood vessels to the skin constrict when an iguana
is in the ocean and dilate when it is basking. In
addition, an iguana's heart rate is slower when it is
swimming than when it is basking. Slowed heart rate
and constricted vessels when swimming mean that
less blood is being pumped through the skin, and
therefore less heat is being transported from deep in
the iguana's body to its skin to be lost to the water.
Faster heart rate and dilated vessels when the iguana
is basking increase the transport of heat from the
skin to the rest of its body. Of course, basking on
black rocks under the equatorial sun can be too much
of a good thing. When an iguana reaches an optimal
body temperature, it lifts its body off the rocks and
orients itself to minimize the surface area of its body
that is directly exposed to solar radiation. Thus, the
marine iguana uses both physiological and behavioral
mechanisms to regulate its body temperature.

Metabolic Heat Production

The use of metabolic heat production to maintain a
body temperature above that of the environment is
surprisingly common among ectotherms. For example, the powerful flight muscles of many insects, such
as dragonflies, moths, bees, and beetles, must reach
a fairly high temperature (35 to 40°C) before the in-

sects can fly, and they must maintain these high
temperatures during flight, even at air temperatures
around 0°C. Such insects use the flight muscles themselves to produce the required heat. These muscles
are about 20 percent efficient; that is, about 20 percent
of the energy they consume goes into useful work
and 80 percent is lost as heat. Flight thus produces
an enormous amount of heat, which keeps body temperature elevated. To reach flight temperature from
resting temperature, the insects contract their flight
muscles isometrically: The muscles that contract alternately during flight to produce the wingbeats contract simultaneously during warm-up. Even though
the muscles are contracting and producing heat, the
wings do not move (Figure 35.15). During warm-up
some bees and moths appear to be shivering because
the wings show movements of small amplitude.

The heat-producing ability of insects can be quite
remarkable. It enables moths to fly at night when air
temperatures are low and solar basking is not possible. The heat-producing ability of a species of scarab
beetle that lives in the mountains north of Los Angeles, California, has made it possible for these beetles to have an unusual mating behavior. The beetles
spend most of their life cycle in the soil, except for
mating, at which time females and males emerge
from the soil and the males fly in search of females.
What is unusual is that they engage in this behavior
in winter, at night, during snowstorms. The drop in

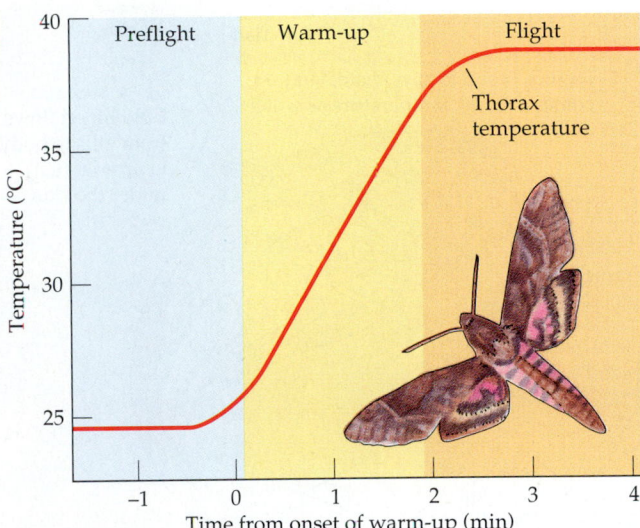

35.15   A Moth's Preflight Warm-Up
Prior to "takeoff," insects such as the sphinx moth contract the flight muscles in their thoraxes to generate heat and warm the muscles up to the temperature required for flight.

barometric pressure associated with a storm probably triggers the emergence from the soil. These beetles were long considered to be very rare because very few entomologists look for beetles in the mountains, in winter, at night, during snowstorms. Presumably the same is true for potential predators!

Honeybees regulate temperature as a group. They live in large colonies consisting mostly of female worker bees that maintain the hive and rear young that are hatched from eggs laid by the single queen bee in the colony. During winter, honeybee workers combine their individual heat-producing abilities to regulate the temperature of the brood. They cluster in the area of the hive where the brood is located and adjust their joint metabolic heat production and density of clustering so that the brood temperature remains remarkably constant, at about 34°C, even as outside air temperature drops below freezing.

Some reptiles use metabolic heat production to raise body temperature above air temperature. The female Indian python protects her eggs by coiling her body around them. If air temperature falls, she uses isometric contractions of her body wall muscles to generate heat. Like the use of flight muscles by insects, this adaptation of the python is analogous to shivering in mammals. The python is able to maintain the temperature of her body—and therefore that of her eggs—considerably above air temperature.

Biological Heat Exchangers

If heat that active muscles produce is not rapidly lost to the environment, it can be used to raise body temperature above the temperature of the surrounding air or water. It is particularly difficult for fish to slow the loss of body heat to the environment because blood pumped from the heart comes into close contact with water flowing over the thin gill membranes before it travels through the body. Therefore, any heat transferred to the blood from active muscles is lost rapidly to the environment. It is therefore surprising to find that some large, rapidly swimming fishes, such as bluefin tuna and great white and mako sharks, can maintain temperature differences as great as 10 to 15°C between their bodies and the surrounding water (Figure 35.16). The heat comes from their powerful swimming muscles, of course, but the ability to conserve that heat is due to remarkable arrangements of the blood vessels.

In the usual fish circulatory system, oxygenated blood from the gills collects in a large, dorsal vessel, the aorta, which travels through the center of the fish, distributing blood to all organs and muscles. "Hot" fish such as bluefin tuna have smaller central dorsal aortas. Most of their oxygenated blood is transported in large vessels just under the skin (Fig-

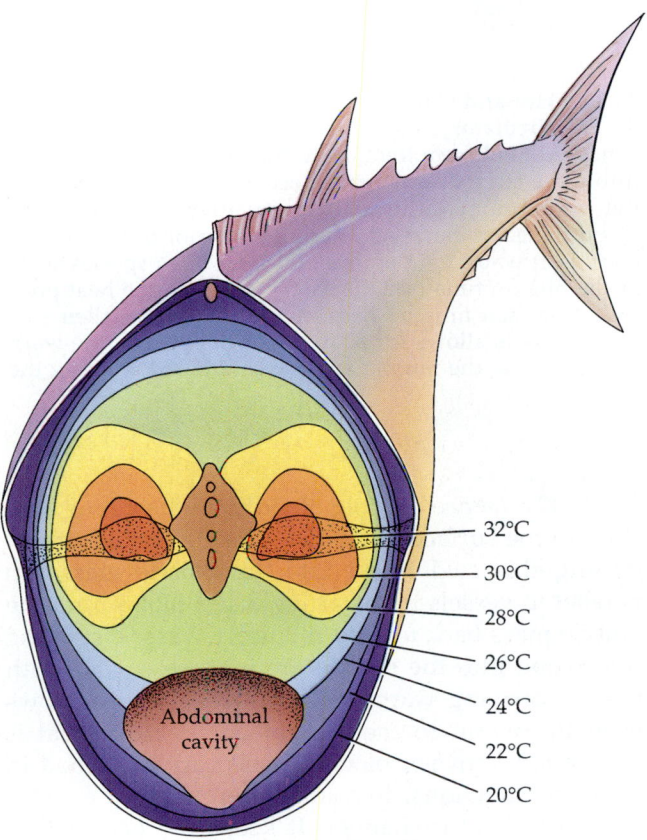

Water temperature 20°C

35.16   Cold Water, Warm Muscles
In an actively swimming fish such as the bluefin tuna, the muscles that power swimming generate heat that keeps the fish's internal body temperature much higher than that of the surrounding water.

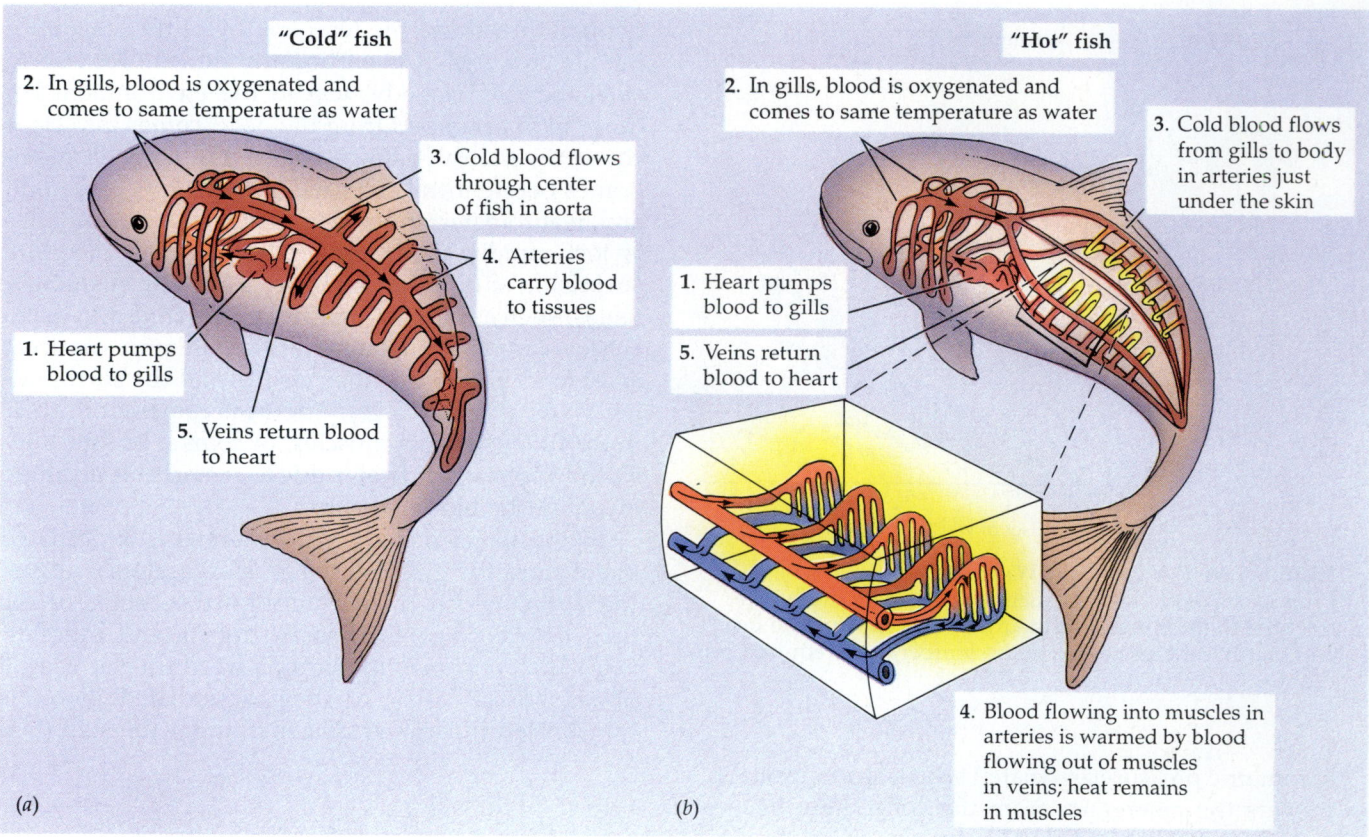

35.17   Hot and Cold Fish
(a) The circulatory systems of most fish conduct the oxygenated blood from the gills to the organs of the fish through a large dorsal aorta. Because the blood comes into equilibrium with water temperature in the gills, the blood the aorta carries through the interior of the fish's body is at water temperature. (b) "Hot" fish species such as the bluefin tuna (see Figure 35.16) retain the heat produced by their muscles because the anatomy of their blood vessels allows for heat exchange between the warm blood leaving the muscle and the cold blood entering the muscle.

ure 35.17). Hence the cold blood from the gills is kept close to the surface of the fish. Smaller vessels transporting this cold blood into the muscle mass run parallel to vessels transporting warm blood from the muscle mass back toward the heart. Vessels carrying cold blood into the muscle are in close contact with vessels carrying warm blood away, and heat flows from the warm to the cold blood. Because heat is exchanged between blood vessels carrying blood in opposite directions, this adaptation is called a countercurrent heat exchanger. It keeps the heat within the muscle mass, enabling the fish to have an internal body temperature considerably above the water temperature. Why is it advantageous for the fish to be warm? Each 10-degree rise in muscle temperature increases its sustainable power output almost threefold!

THERMOREGULATION IN ENDOTHERMS

An endotherm responds to changes in the temperature of its environment primarily by changing its metabolic rate. Within a narrow range of environmental temperatures called the **thermoneutral zone**, the metabolic rate of the endotherm is low and independent of temperature. The metabolic rate of a resting animal at a temperature within the thermoneutral zone is called the **basal metabolic rate**. It is usually measured on animals that are quiet but awake and that are not using energy in the digestive processes or for reproduction. A resting animal consumes energy at the basal metabolic rate just to carry out all of its metabolic functions other than thermoregulation.

The basal metabolic rate of an endotherm is about six times greater than the metabolic rate of a similarly sized ectotherm at rest and at the same body temperature (see Figure 35.12b). A gram of mouse tissue consumes energy at a much higher rate than does a gram of lizard tissue when both tissues are at 37°C. This difference is due to a basic change in cell metabolism that accompanied the evolution of endotherms from their ectothermic ancestors. The higher level of heat production by endotherms makes it easier for them to maintain a temperature difference between the body and the environment.

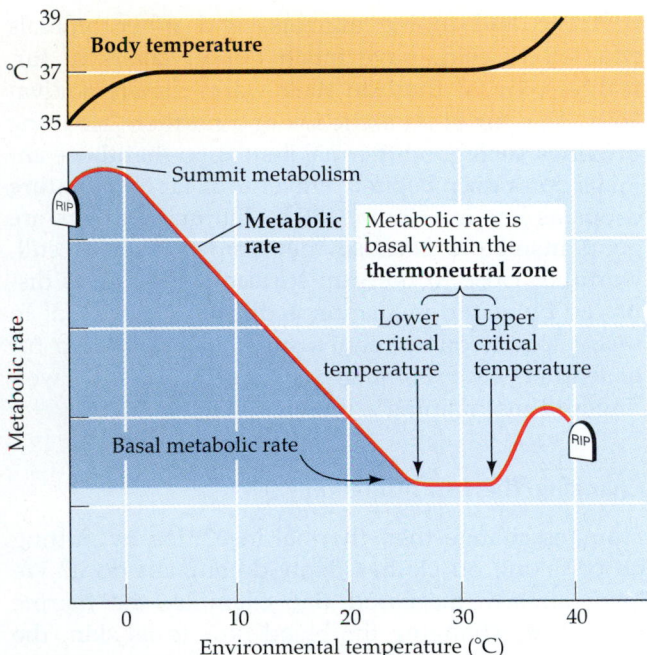

35.18 Environmental Temperature and Mammalian Metabolic Rates

Within a mammal's thermoneutral zone its metabolic rate is low and constant; the animal thermoregulates behaviorally and by changing its thermal insulation. Above the upper critical temperature the animal must expend energy to lose heat by panting or sweating, which makes its metabolic rate increase. Below the lower critical temperature, the animal produces metabolic heat to compensate for increased heat loss to the environment, as indicated by the dark-shaded area under the plotted line.

Active Heat Production and Heat Loss

The thermoneutral zone is bounded by a lower critical temperature and an upper critical temperature. Below the lower critical temperature an endotherm's metabolic rate increases as environmental temperature declines because the animal must produce more and more heat to maintain a constant body temperature as heat loss to the environment increases. As the environment gets colder, eventually the animal reaches its summit metabolism, or maximum possible thermoregulatory heat production. If the environmental temperature falls still lower, the animal's body temperature will begin to drop. On the other hand, when the environmental temperature goes above the upper critical temperature, the animal pants or sweats. Since these active heat loss responses require an increased expenditure of energy, the metabolic rate rises. A graph of metabolic rate as a function of environmental temperature illustrates the thermoregulatory responses of an endotherm (Figure 35.18).

Mammals use two mechanisms—shivering and nonshivering heat production—to create heat for thermoregulation. Birds use only shivering heat production. Shivering uses the contractile machinery of skeletal muscles to consume ATP without causing observable behavior. The muscles pull against each other so that little movement other than a tremor results. All of the energy from the conversion of ATP to ADP in this process is released as heat. Most nonshivering heat production occurs in specialized tissue called **brown fat**. It looks brown because of its abundant mitochondria and rich blood supply (Figure 35.19). In brown fat cells a protein called thermogenin

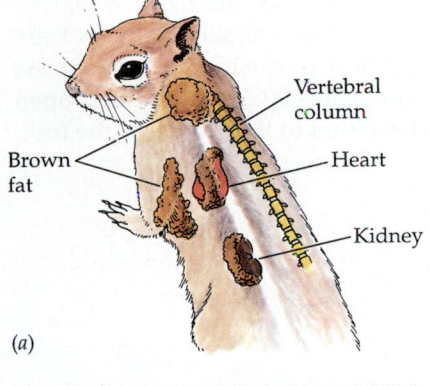

(a)

35.19 Brown Fat: A Heat-Producing Tissue

In many mammals, brown fat produces heat. (a) In a ground squirrel, brown fat occurs in specific anatomical locations. (b) White fat viewed through a light microscope. Each cell is filled with a globule of lipid and has few organelles. The tissue has few blood vessels. (c) Brown fat viewed through a light microscope at the same magnification reveals cells with many intracellular structures and multiple droplets of lipid. Numerous capillaries run through the tissue. (d) An electron micrograph of brown fat shows the tight packing of mitochondria in a brown fat cell. A portion of a lipid droplet is visible in the upper right; in the upper left is part of a blood capillary.

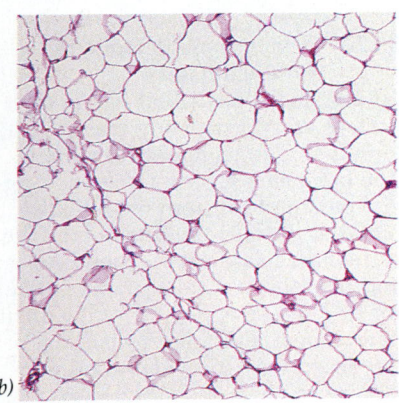

(b)

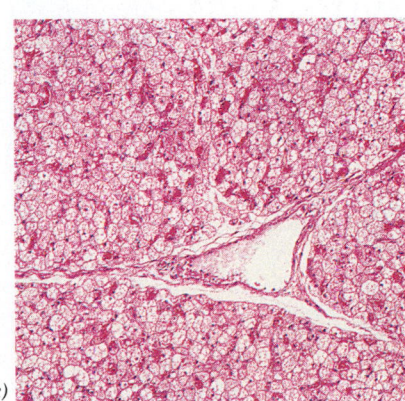

(c)

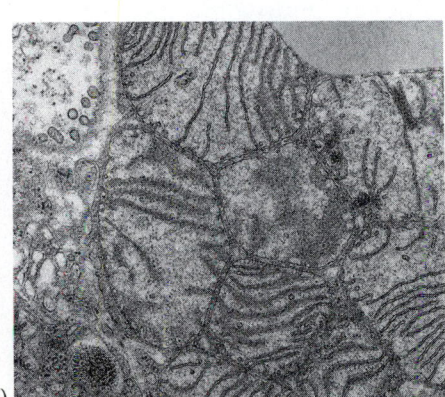

(d)

uncouples oxidative phosphorylation. Thus metabolic fuels are consumed to produce heat without the production of ATP. Brown fat is especially abundant in newborn infants of many mammalian species, in some adult mammals that are highly acclimatized to cold, and in mammals that hibernate.

Living in the Cold

The coldest habitats on Earth are in the Arctic, the Antarctic, and at the tops of high mountains. Many birds and mammals, but no reptiles or amphibians, live in the coldest habitats. The ability to produce a substantial amount of heat metabolically has enabled endotherms to exploit formidable, frigid environments. Most tropical species of birds and mammals, however, would not fare well in those environments. What adaptations besides endothermy characterize species that live in the cold?

The most important adaptations of endotherms to cold environments are those that reduce their heat loss to the environment. Since most heat loss is from the body surface, many cold-climate species have smaller surface areas than their warm-climate cousins, even when their body masses are the same. Rounder body shapes and shorter appendages reduce the surface area-to-volume ratios of some cold-climate species; compare, for example, the desert jackrabbit and the arctic hare (Figure 35.20). Another means of decreasing heat loss is to increase thermal insulation. Fur, feathers, and layers of fat decrease the loss of heat from an endotherm to the environment. You can experience the effectiveness of thermal insulation by comparing what it feels like to sit on a cold stone or metal bench while wearing only cotton shorts with what it feels like after you put a feather pillow or a wool blanket between you and the bench.

Arctic and alpine animals, and other animals adapted to cold, have much thicker layers of fur, feathers, or fat than do their warm-climate equivalents. The fur of an arctic fox or a northern sled dog provides such good thermal insulation that those animals don't even begin to shiver until air temperature drops as low as −20 to −30°C. Fur and feathers are good insulators because they trap a layer of still, warm air close to the skin surface. If that air is displaced by water, insulation is drastically reduced. In many species oil secretions spread through their fur or feathers by grooming is critical for resisting wetting and maintaining a high level of insulation.

Changing Thermal Insulation

Humans change their thermal insulation by putting on or taking off clothes. How do animals do it? We have already discussed one example, the marine iguana. By changing the blood flow to its skin, the marine iguana increases or decreases the exchange of heat between the environment and its body—it changes its thermal insulation. Increasing or decreasing blood flow to the skin is an important thermoregulatory adaptation for endotherms as well. In a hot environment, your skin feels hot because of the high rate of blood flow through it, but when you are sitting in an overly air-conditioned theater, your hands, feet, and other body surfaces feel cold as blood flow to those areas decreases. The wolf has an elegant mechanism for decreasing heat loss from its feet without the risk of freezing them. As long as the wolf's foot temperature is more than a few degrees above freezing, certain blood vessels in its foot are constricted and blood flow to the foot is minimal. As foot temperature approaches 0°C, these vessels open and allow more warm blood to flow through the foot, thus keeping it from freezing.

(a)

(b)

35.20 Adaptations to Hot and Cold Climates
(a) The desert jackrabbit has a large surface area for its body mass, largely due to its long extremities. The large ears serve as heat exchangers, passing heat from the rabbit's blood to the surrounding air. (b) The arctic hare has shorter extremities and therefore a smaller surface area for its body size. The fur of the arctic hare, longer and thicker than that of the desert hare, provides good insulation.

For highly insulated arctic animals and for many large mammals from all climates, getting rid of excess heat can be a serious problem, especially during exercise. Arctic species usually have a place on the body surface, such as the abdomen, that has only a thin layer of fur and can act as a window for heat loss. Large mammals, such as elephants, rhinoceroses, and water buffalo, have little or no fur and seek places where they can wallow in water when the air temperature is too high. Having water in contact with the skin greatly increases heat loss because water has a much greater capacity for absorbing heat than does air.

Evaporative Water Loss

The evaporation of water is a very effective means of dissipating heat. A gram of water absorbs about 580 calories when it evaporates. However, water is heavy, animals do not carry an excess supply of it, and hot environments tend to be arid places where water is a scarce resource. Therefore, evaporation of water by sweating or panting is usually a last resort for animals adapted to hot environments. Sweating and panting are active processes that require the expenditure of metabolic energy. That's why the metabolic rate increases when the upper critical temperature is exceeded (see Figure 35.18). A sweating or panting animal is producing heat in the process of dissipating heat. This can be a losing battle. Animals can survive in environments that are below their lower critical temperature much better than they can in those above their upper critical temperature.

THE VERTEBRATE THERMOSTAT

The thermoregulatory mechanisms and adaptations we have already discussed are the controlled systems for the regulation of body temperature. These controlled systems must receive commands from a regulatory system that integrates information relevant to the regulation of body temperature. A convenient name for the regulatory system in this case is thermostat. All animals that thermoregulate, both vertebrate and invertebrate, must have regulatory systems, but here we will focus on the vertebrate thermostat.

Where is the vertebrate thermostat? The major integrative center is at the bottom of the brain in a structure called the **hypothalamus**. If you slide your tongue back as far as possible along the roof of your mouth, it will be just a few centimeters below your hypothalamus. The hypothalamus is a part of many regulatory systems, so we will refer to it many times in the chapters to come. If the hypothalamus of a mammal's brain is damaged, the animal loses its ability to regulate its body temperature, which then rises in warm environments and falls in cold ones.

Set Points and Feedback

What information does the vertebrate thermostat use? In many species the temperature of the hypothalamus itself is a major source of feedback to the thermostat. Cooling the hypothalamus causes fishes and reptiles to seek a warmer environment, and heating the hypothalamus causes them to seek a cooler environment. In mammals, cooling the hypothalamus can stimulate constriction of blood vessels to the skin and increase metabolic heat production. Because of the activation of these thermoregulatory responses, body temperature rises when the hypothalamus is cooled. Conversely, warming of the hypothalamus stimulates dilation of blood vessels to the skin and sweating or panting, and the overall body temperature falls when the hypothalamus is warmed (Figure 35.21). The hypothalamus appears to generate a set point like a thermostat setting. When the temperature of the hypothalamus exceeds or drops below that set point, thermoregulatory responses (the controlled system) are activated to reverse the direction of temperature change. Hence hypothalamic temperature is a negative feedback signal.

An animal has separate set points for activating different thermoregulatory responses. If the hypothalamus of a mammal is heated and cooled, the vessels supplying blood to the skin constrict at a specific hypothalamic temperature. A slightly lower hypothalamic temperature initiates shivering, and a hypothalamic temperature two or three degrees higher initiates panting. We can describe the characteristics of hypothalamic control of each response. For example, if we measure metabolic heat production while heating and cooling the hypothalamus (see Figure 35.21), we can describe the results graphically (Figure 35.22). Within a certain range of hypothalamic temperatures, metabolic heat production remains low and constant, but cooling the hypothalamus below a certain level—the set point—stimulates increased metabolic heat production. The increase in heat production is proportional to how much the hypothalamus is cooled below the set point. This regulatory system is much more sophisticated than a simple on–off thermostat like the one in a house.

The vertebrate thermostat integrates other sources of information in addition to hypothalamic temperature. It integrates information about the temperature of the environment as registered by temperature sensors in the skin. Changes in skin temperature shift the hypothalamic set points for responses. As Figure 35.22 shows, in a warm environment you might have to cool the hypothalamus of a mammal to stimulate it to shiver, but in a cold environment you would

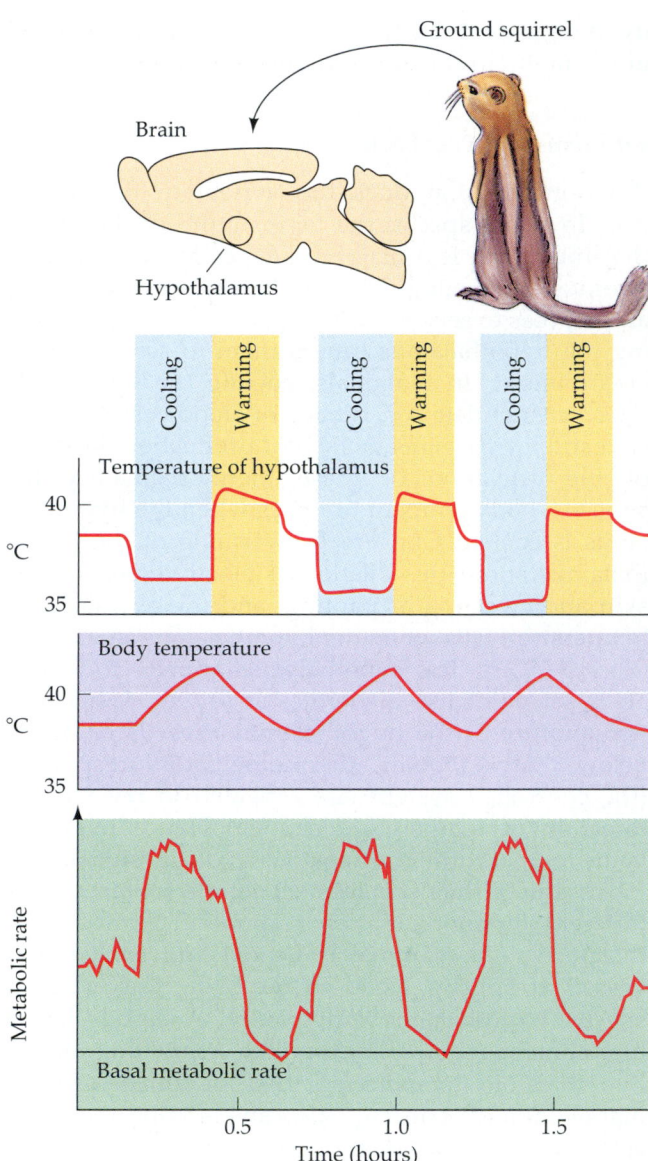

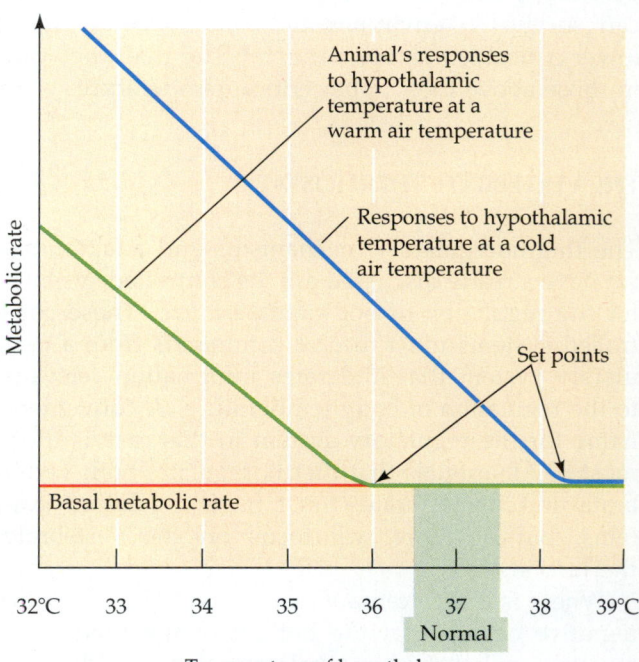

35.21 The Hypothalamus Acts as a Thermostat

In an experiment a ground squirrel was maintained at low environmental temperatures so that its initial metabolic heat production was high. Cooling the hypothalamus increased metabolic heat production even further and the animal's body temperature rose. Heating the hypothalamus reduced metabolic rate, and the animal's body temperature fell.

the heat production response. If you are the unlucky person developing a fever, you may feel unbearably cold (the chills) even though your body temperature is normal. You shiver, put on more clothes, and turn up your room temperature or electric blanket. As a result your body temperature rises until it matches the new set point. At the higher body temperature you no longer feel cold, and you may not feel as if you are hot, but someone touching your forehead will say that you are "burning up." If you take an aspirin, it lowers your set point to normal. Now you feel hot, take off clothes, and even sweat until your elevated body temperature returns to normal. Extreme fevers can be dangerous and must be reduced, but there is evidence that moderate fevers help the body fight an infection (Box 35.A). Perhaps we should not be too hasty in using medication to counter shifts in our hypothalamic set points.

35.22 Adjustable Set Points

Vertebrates have different a set point for the metabolic heat production response to hypothalamic temperature at different environmental temperatures. Other factors, such as being asleep or awake, the time of day, or the presence of a fever can also affect the set point.

have to warm the hypothalamus of the same animal to stop its shivering. The set point for the metabolic heat production response is higher when the skin is cold and lower when the skin is warm. Information from the skin can be considered feedforward that adjusts the hypothalamic set point. Many other factors also shift hypothalamic set points for responses. Set points are higher during wakefulness than during sleep, and they are higher during the active part of the daily cycle than during the inactive part, even if the animal is awake at both times.

Fever also causes shifts in set points. Fevers are rises in body temperature in response to substances called **pyrogens** derived from bacteria or viruses that invade the body. Injections of the killed bacteria or even the purified cell walls of the killed bacteria can also cause fever. The presence of the pyrogen in the body causes a rise in the hypothalamic set point for

Fevers and "Feeling Crummy"

You respond to many infectious illnesses by getting a fever and feeling crummy. You lose your appetite, you have no energy, your joints and muscles ache, you get the chills, and you feel like just putting on flannel pajamas and getting into bed. Are these well-known symptoms simply unfortunate side effects? To the contrary, scientists are beginning to think that getting a fever and feeling crummy are adaptive responses that help us fight diseases.

The immediate causes of these responses are chemical messages from the immune system. When an infectious virus or bacterium invades the body, scavenger cells called macrophages grab it. One of the things macrophages do is release chemicals called interleukins that sound the

alarm to other cells of the immune system throughout the body. Interleukins cause many other responses as well. They make neurons transmitting pain more sensitive, and so we ache. They make us sleepy. They stimulate the hypothalamus to release corticotropin-releasing hormone, which initiates the stress responses of the body. Interleukins also cause a rise in the hypothalamic set point for thermoregulatory responses. Intracellular messengers activated by the interleukins include prostaglandins. A potent inhibitor of prostaglandin synthesis is aspirin, thus explaining how this drug can reduce fever and make us feel better. But is it a good idea to reduce fever and feel less crummy if these are adaptive responses to infection?

Some convincing evidence that fever is an adaptive response to infection came from experiments on lizards by Matthew Kluger at the University of Michigan. Lizards having access to a heat lamp maintain their body temperature at about 38°C

by shuttling in and out of the vicinity of light. When injected with pathogenic bacteria, the lizards spend more time under the light and raise their body temperatures to between 40 and 42°C—they develop fevers. Does a fever help the lizard fight infection? To answer this question, groups of lizards receiving equal inoculations of bacteria were kept in incubators at 34, 36, 38, 40, and 42°C. All of the lizards at 34 and 36°C died, about 25 percent at 38°C survived, and about 75 percent at 40 and 42°C survived. Fevers do help.

Even though fever clearly seems to be an adaptive response to infection, high fevers can be dangerous and even lethal. Even modest fevers can be dangerous to people with weakened hearts and people who are chronically ill. A fetus can be endangered when a pregnant woman has a high fever. Drugs that reduce fever may be important in such cases, but perhaps they should not be taken by most people at the first sign of aches or chills.

TURNING DOWN THE THERMOSTAT

Having learned that fever results from turning up the hypothalamic thermostat, we can ask, Are there cases of turning down the thermostat so that body temperature is regulated at a lower level? The answer is yes, but not all decreases in body temperature are regulated. Hypothermia is the condition in which body temperature is below normal. It can result from a natural turning down of the thermostat, or from traumatic events such as starvation (lack of fuel), exposure, serious illness, or anesthesia. Because of Q_{10} effects, hypothermia slows metabolism, slows the heart, weakens muscle contractions (including those of the heart), decreases nerve conduction, and causes unconsciousness. This is not a happy state of affairs for most endotherms and can lead to death, but it can also be somewhat protective.

There are cases in which drowning victims have been under water for 10 or 15 minutes or more and have shown no pulse when pulled out of the water. Nevertheless, some—mostly small children drowned

in cold water—were revived by paramedics and recovered to a remarkable extent. The rapid fall in their body temperature slowed metabolism and thereby slowed the progress of cell damage caused by lack of oxygen. In this way, the hypothermia prevented irreversible brain damage even though it was a pathological condition induced by drowning. We will see next that some animals can anticipate unfavorable circumstances and induce the hypothermia of torpor or hibernation as an adaptive, protective mechanism by turning down their thermostats.

Shallow Torpor

Hypothermia conserves metabolic energy. Many species of birds and mammals use regulated hypothermia as a means of surviving periods of cold and food scarcity. Because of their extremely high surface-to-volume ratios, very small endotherms such as hummingbirds and pocket mice may exhaust their metabolic reserves just to get through one day without food if they are at normal body temperature. Animals of such species can extend the period over which they can survive without food by dropping body temperature. This adaptive hypothermia is called **shallow torpor** or daily torpor because it usually occurs on a daily basis, with body temperature falling at the time of day the animal normally becomes inactive. Body temperature can drop 10 to 15°C during shallow torpor, resulting in an enormous saving of metabolic energy.

A small bird, the willow tit, studied by R. Reinertsen in Norway, provides an example of shallow torpor that shows how well regulated this process can be. Willow tits live through the winter above the Arctic Circle. In spite of their good thermal insulation, these tiny birds must become hypothermic to survive the long, cold Arctic nights. Each evening the bird lowers its metabolic rate to a level that it maintains all night, and its body temperature falls as a result of the decreased heat production. How low the bird's metabolic rate drops is different on different nights (Figure 35.23). The decrease in its metabolism depends on air temperature, on season (and hence on length of night), and on the bird's fat reserves at roosting time. Every morning the bird has depleted its fat reserves and must immediately feed on seeds it has stored nearby. This is living on the razor's edge! The brain of this small animal integrates all the relevant information, resulting in just the right resetting of its thermostat to get it through the night.

Hibernation

Regulated hypothermia can last for days or even weeks, with drops to very low temperatures; this phenomenon is called **hibernation**. Many diverse species of mammals hibernate, but only one species of bird, the poorwill, has been shown definitely to hibernate. For the deep sleep of hibernation, the body's thermostat is turned down to an extremely low level to maximize energy conservation. Body temperature falls during hibernation because the hypothalamic set point drops, and arousal from hibernation is due to a return of the hypothalamic set point to a normal mammalian level. Many hibernators maintain body temperatures around 2 to 4°C during hibernation. The metabolic rate needed to sustain an animal after this incredible drop in body temperature is only $\frac{1}{30}$ to $\frac{1}{50}$ of basal metabolic rate, an enormous saving of metabolic energy.

Animals hibernate when temperatures are low and food is scarce. Individual bouts of hibernation may last from less than a day to over a week (Figure 35.24). A bout terminates spontaneously when the hibernator's body temperature returns to normal. The animal may remain at its normal temperature for a few hours to a day, during which time it may eat. (Some hibernators store food in their well-insulated nests.) Then the animal enters another bout of hibernation.

The hibernation season is controlled by an internal biological clock (or calendar), which continues to run with a periodicity of about a year even when animals are kept under constant conditions in the laboratory. This is called a **circannual rhythm** (from the Latin *circa* = about and *annus* = year). A typical circannual cycle for a hibernator such as a ground squirrel includes an active season, during which it cannot hibernate even if exposed to cold temperatures and deprived of food. During the active season, usually spring through fall, the animals breed, raise their

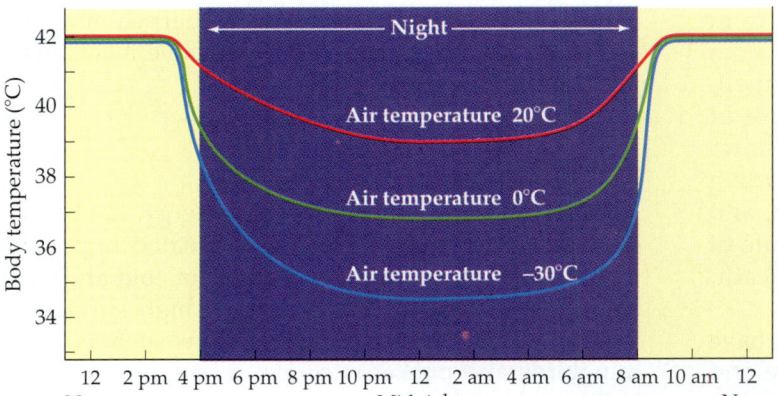

Willow tit

35.23 Hypothermia Deepens with Cold and Dark

The curves show how a willow tit's body temperature changes during long nights at different environmental temperatures. The colder the air, the deeper is the bird's hypothermia. Notice that the depth of hypothermia is set early in the night, and must therefore be a result of information available at that time—not simply a consequence of running out of fuel reserves faster during colder nights. If nights are made shorter (which is possible in the laboratory), the birds maintain higher body temperatures at these same air temperatures.

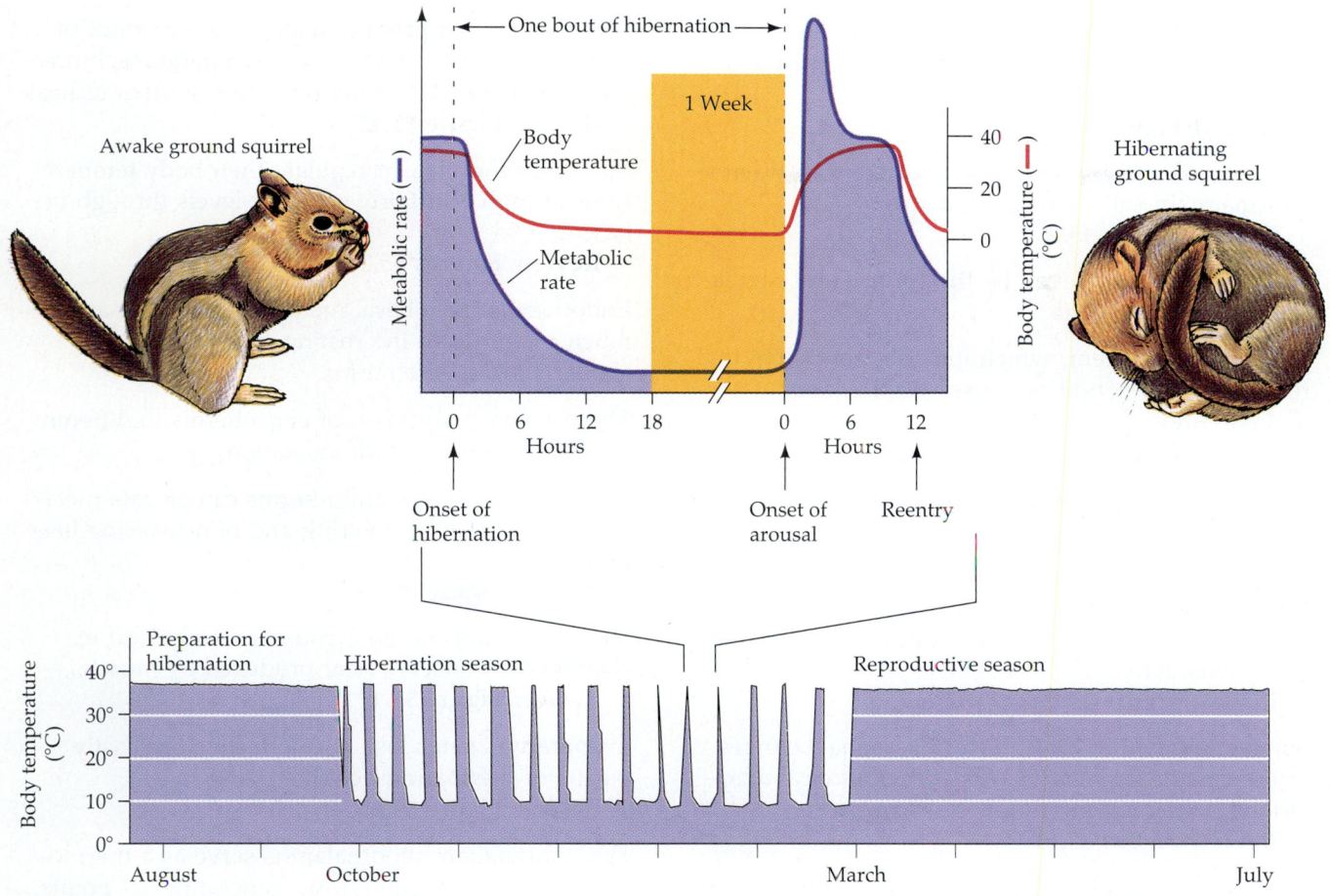

35.24 A Ground Squirrel Hibernates in Bouts
There are approximately three bouts per month during the hibernation season. A bout ends when the animal's body temperature returns to normal mammalian levels; it may remain there for a few hours or even a day or more before another bout begins.

young, prepare their nests for winter, fatten their bodies, and store food. During the hibernation season, as Figure 35.24 shows, animals hibernate in recurrent bouts. They progressively lose body weight even if food is available. Toward the end of the hi-

bernation season the reproductive organs grow and become functional. The ability of hibernators to reduce the set point so dramatically probably evolved as an extension of the set point decrease that accompanies sleep even in nonhibernating species of mammals and birds.

SUMMARY of Main Ideas about Physiology, Homeostasis, and Temperature Regulation

The organs of an animal maintain a constancy, or homeostasis, of the animal's internal environment.

Organs are made of four types of tissues: epithelial, connective, muscle, and nervous.
   Review Figure 35.1

Organs can be grouped into systems with common or complementary functions.

The nervous and endocrine systems process information and control physiological functions and behavior.
   Review Figure 35.2

Organs that function in protection, support, and movement include the skin, the skeletal system, and the muscular system.
   Review Figure 35.3

The reproductive system produces gametes, achieves fertilization, and supports and protects the developing fetus.
   Review Figure 35.4

The organs of nutrition include the lungs and the digestive system.
   Review Figure 35.5

The circulatory system provides internal transport of nutrients, wastes, heat, hormones, and elements of the immune system.
    Review Figure 35.6

The excretory organs eliminate nitrogenous wastes and maintain salt and water balance.
    Review Figure 35.7

Homeostasis is achieved by the control and regulation of organ systems.

A regulatory system, which includes set points that reflect optimal conditions, uses feedback about actual conditions.
    Review Figures 35.8 and 35.9

Life can exist only within the range of temperatures between the freezing point of water and the temperature that denatures proteins.

Most biological processes and reactions are temperature-sensitive, and the Q_{10} is a measure of temperature sensitivity.
    Review Figure 35.10

Some animals that cannot avoid seasonal changes in their body temperatures have biochemical adaptations that compensate for the changes.
    Review Figure 35.11

Animals have evolved behavioral and physiological adaptations for controlling their body temperatures.

Homeotherms maintain a fairly constant body temperature most of the time; poikilotherms do not.

Endotherms can produce a significant amount of metabolic heat to elevate body temperature, but ectotherms depend on environmental sources of heat.
    Review Figure 35.12

Many ectotherms can regulate their body temperatures at high and fairly constant levels through behavior.
    Review Figures 35.13 and 35.15

Endotherms have basal metabolic rates that are much higher than the resting metabolic rates of similarly sized ectotherms.

The primary adaptation of endotherms to different climates is their level of insulation.

In response to cold, endotherms can elevate metabolic rate through shivering and nonshivering heat production.
    Review Figure 35.19

Some fish have vascular countercurrent heat exchangers to conserve heat produced by muscle.
    Review Figure 35.17

Evaporative water loss is an effective but costly means of dissipating heat.
    Review Figure 35.18

The mammalian hypothalamus serve as a thermostat by sensing temperature, generating set points, and sending commands to effector organs.
    Review Figures 35.21 and 35.22

The mammalian thermostat is turned down during shallow torpor and deep hibernation, thereby conserving energy.
    Review Figure 33.24

SELF-QUIZ

1. If the Q_{10} of the metabolic rate of an animal is 2, then
   a. the animal is better acclimatized to a cold environment than if its Q_{10} were 3.
   b. the animal is an ectotherm.
   c. the animal consumes half as much oxygen per hour at 20°C as it does at 30°C.
   d. the animal's metabolic rate is not at basal levels.
   e. the animal produces twice as much heat at 20°C than at 30°C.

2. Which of the following statements is *true* of brown fat?
   a. It produces heat without producing ATP.
   b. It insulates animals acclimatized to cold.

   c. It is a major source of heat production for birds.
   d. It is found only in hibernators.
   e. It provides fuel for muscle cells responsible for shivering.

3. What is the most important and most general difference between mammals and birds adapted to cold climates in comparison to species adapted to warm climates?
   a. Higher basal metabolic rates
   b. Higher Q_{10}'s
   c. Brown fat
   d. Greater insulation
   e. Ability to hibernate

4. Which of the following would cause a *decrease* in the hypothalamic temperature set point for metabolic heat production?

   a. Entering a cold environment
   b. Taking an aspirin when you have a fever
   c. Arousing from hibernation
   d. Getting an infection that causes a fever
   e. Cooling the hypothalamus

5. Mammalian hibernation
   a. occurs when animals run out of metabolic fuel.
   b. is a regulated decrease in body temperature.
   c. is less common than hibernation in birds.
   d. can occur at any time of year.
   e. lasts for a period of several months, during which body temperature remains close to environmental temperature.

6. Which of the following is an important difference between an ectotherm and an endotherm of similar body size?
   a. Ectotherms have higher Q_{10}'s.
   b. Only ectotherms use behavioral thermoregulation.
   c. Only endotherms can constrict and dilate the blood vessels to the skin to alter heat flow.
   d. Only endotherms can get fevers.
   e. At body temperatures of 37°C, the ectotherm has a lower metabolic rate than the endotherm.

7. The function of the countercurrent heat exchanger in "hot" fish is:
   a. to trap heat in the muscles.
   b. to produce heat.
   c. to heat the blood returning to the heart.
   d. to dissipate excess heat generated by powerful swimming muscles.
   e. to cool the skin.

8. What is the difference between a winter- and a summer-acclimatized fish that is termed "metabolic compensation"?
   a. The winter-acclimatized fish has a higher Q_{10}.
   b. The winter-acclimatized fish develops greater insulation.
   c. The winter-acclimatized fish hibernates.
   d. The summer-acclimatized fish has a countercurrent heat exchanger.
   e. The summer-acclimatized fish has a lower metabolic rate at any given water temperature than does the winter fish.

9. Which of the following is an important characteristic of epithelial cells?
   a. They generate electric signals.
   b. They contract.
   c. They have an extensive extracellular matrix.
   d. They have secretory functions.
   e. They are found only on the surface of the body.

10. Negative feedback
    a. works in opposition to positive feedback to achieve homeostasis of a physiological variable.
    b. always turns off a process.
    c. reduces an error signal in a regulatory system.
    d. is responsible for metabolic compensation.
    e. is a feature of the thermoregulatory systems of endotherms but not of ectotherms.

FOR STUDY

1. Make a table that lists all of the properties of the internal environment that you think are critical to keep the cells of the body healthy. Next to each property list the organs or organ system responsible for maintaining it.

2. What are the major differences between ectotherms and endotherms? Compare and contrast their major thermoregulatory adaptations.

3. Why is an environment above the upper critical temperature of an endotherm more dangerous for that animal than is an environment below its lower critical temperature?

4. Why is it difficult for a fish to be endothermic? How do "hot" fish overcome these difficulties?

5. If the temperature of the hypothalamus of a mammal is the feedback information for its thermostat, why does the hypothalamic temperature scarcely change when that animal moves between environments hot enough and cold enough to stimulate the animal to pant and to shiver, respectively?

READINGS

Crawshaw, L. I., B. P. Moffitt, D. E. Lemons and J. A. Downey. 1981. "The Evolutionary Development of Vertebrate Thermoregulation." *American Scientist*, vol. 69, pages 543–550. All vertebrates thermoregulate, and the nervous system mechanisms involved appear to have a common origin even though the effector mechanisms may differ.

French, A. R. 1986. "The Patterns of Mammalian Hibernation." *American Scientist*. vol. 76, pages 569–575. Body size has important consequences for energy metabolism; therefore, a variety of patterns of hibernation have evolved, as this article discusses.

Heinrich, B. 1981. "The Regulation of Temperature in the Honeybee Swarm." *Scientific American*, June. When honeybees leave their hive in a swarm, they thermoregulate.

Heller, H. C., L. I. Crawshaw and H. T. Hammel. 1978. "The Thermostat of Vertebrate Animals." *Scientific American*, August. This article describes research on and properties of the brain mechanisms responsible for thermoregulation in vertebrates and the adaptations in those mechanisms that make hibernation possible.

Schmidt-Nielsen, K. 1981. "Countercurrent Systems in Animals." *Scientific American*, May. Countercurrent exchanges are the basis for a variety of physiological adaptations, some of which are presented in this article. Developed in special detail is the case of water conservation in the camel's nose.

Schmidt-Nielsen, K. 1990. *Animal Physiology: Adaptation and Environment*, 4th Edition. Cambridge University Press, New York. An excellent advanced textbook on comparative animal physiology. Chapter 8, "Temperature Regulation," expands on many of the topics presented in this chapter.

A Wimpy and a Macho Male Cichlid

36

Animal Hormones

A species of cichlid fish that lives in Lake Tanganyika in east central Africa gives biological meaning to "macho" and "wimpy." In shallow pools around the edge of the lake, big, brightly colored males stake out and vigorously defend territories against neighboring males. These "macho" males constantly patrol their territories and display their sexual adornments for the benefit of females who assemble in groups at the edge of the colony. The females are hard to see because they are inactive and protectively colored. When a female is ready to spawn and is impressed by a male's territory and display, she enters his territory and lays her eggs in a spawning pit the male has prepared. The male then fertilizes the eggs. At any one time, only about 10 percent of the males in the population are displaying and holding territories. All the other males are small and nondescript like the females, nonaggressive, and incapable of fertilizing eggs—that is, "wimpy." If, however, a macho male is removed by a predator, a group of wimpy males fight over the vacated territory. The winner rapidly turns into a macho male, brightly colored, big, aggressive, and able to attract females and fertilize eggs.

What accounts for this dramatic change in lifestyle of cichlid males? Soon after the wimpy male's victory, certain cells in its brain enlarge and secrete a chemical message that triggers a cascade of other cells to secrete chemical messages. These molecules circulating around the body help bring about a variety of changes in cells and tissues that convert the once wimp into a macho male. This is one example of how chemical messages, or hormones, released in this case by a behavioral stimulus, can produce and coordinate major developmental, physiological, and behavioral changes in an animal.

CHEMICAL MESSAGES

In this chapter we will examine the science of **endocrinology**—the study of hormones and their actions. A **hormone** is a substance that serves as a chemical message between cells of a multicellular organism. A chemical communication system that uses a hormone is made up of at least two cells: One cell produces and releases the hormone—the message—and a second cell with appropriate **receptors** receives the message. The receiving cell is called the **target cell**. The receipt of the message activates mechanisms within the target cell that interpret the message and respond to it. The response may be developmental, physiological, or behavioral.

A simple way to classify chemical communication systems is according to the distance over which the messages operate: Are their effects local or are they distributed throughout the body? (The effects of some chemical messages are even exerted on other organisms in the environment. We will learn about those substances—pheromones—in Chapter 45.)

Local Hormones

Some chemical messages, released from secreting cells into surrounding extracellular fluids, exert their effects locally (Figure 36.1a). These hormones are inactivated so rapidly by enzymes in the extracellular environment or taken up so completely by cells in the immediate vicinity that they usually do not exert effects on distant cells.

An example of a local hormone is **histamine**, one of the mediators of inflammation, a tissue response that can help protect the body from invasion by foreign organisms or materials. Histamine is released in damaged tissues by specialized cells called mast cells. When the skin is cut by a dirty object, the area around the cut becomes inflamed—red, hot, and swollen. Histamine causes this response by dilating the local blood vessels and making them more permeable, or leaky, allowing blood plasma, including protective blood proteins and white blood cells, to move into the damaged tissue.

Local responses to histamine are protective, but histamine responses spread over large areas of the body can cause problems, such as the symptoms of hay fever. When a person sensitized to a type of pollen inhales that pollen, it causes cells in the respiratory passages to release histamine. The histamine causes the tissues of the passages to swell and to increase their secretions of mucus, leading to congestion, coughing, sneezing, and a runny nose. Such allergic reactions are unpleasant but rarely dangerous. In a person who is extremely allergic, however, or in a person who has a blood-borne infection, histamine and other mediators of inflammation may be released in such large amounts that they enter the blood and circulate around the body. The resulting expansion of blood vessels and leakage of fluid from the circulatory system can cause blood pressure to drop severely. Fluid may leak into the lungs and the airways may become severely congested. The resulting failure of the circulatory and respiratory systems, termed **anaphylactic shock**, can be lethal. Some highly sensitive people can go into anaphylactic shock from a single bee sting, an injection of an antibiotic to which they are allergic, or ingestion of a food to which they are allergic.

Circulating Hormones

Traditionally we have thought of a hormone as a chemical message secreted by cells and distributed throughout the body by the circulatory system (Figure 36.1b). Wherever such a hormone encounters a

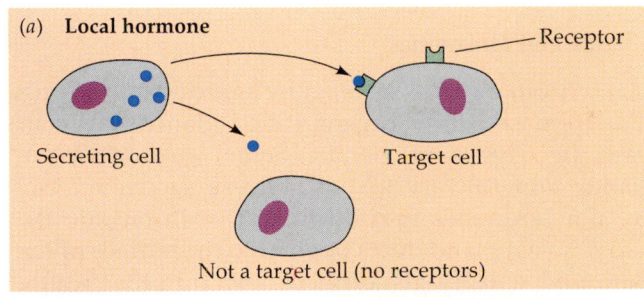

(a) **Local hormone**

Receptor

Secreting cell

Target cell

Not a target cell (no receptors)

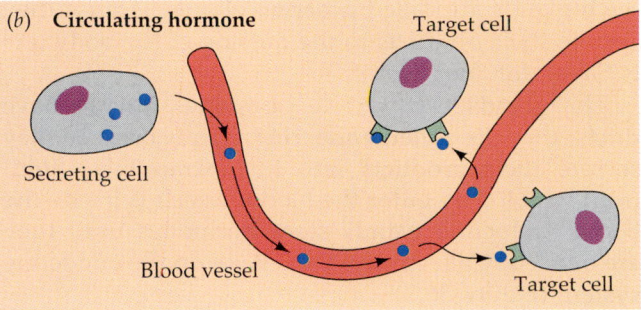

(b) **Circulating hormone**

Target cell

Secreting cell

Blood vessel

Target cell

36.1 Chemical Signaling Systems
(a) Many cells in the body secrete hormones that influence only nearby target cells. (b) Some cells secrete hormones into the bloodstream, which carries them to target cells elsewhere in the body.

cell with a receptor to which it can bind, it triggers a response. The nature of the response depends on the responding cell. The same hormone can cause different responses in different types of cells. For example, consider the hormone epinephrine (adrenaline). If a lion creeps up behind you and roars, your brain sends signals through your nervous system to epinephrine-containing cells, which immediately release epinephrine. The hormone diffuses into the blood and rapidly circulates around your body. What does it do for you?

Epinephrine activates receptors in the heart to make the heart beat faster and pump more blood. Epinephrine activates receptors in the vessels supplying blood to your digestive tract, causing those vessels to constrict—you can digest lunch later. Your heart is pumping more blood, and a greater percentage of that blood is going to the muscles needed for your escape. In the liver, epinephrine stimulates the breakdown of glycogen into glucose for a quick energy supply. In fatty tissue, epinephrine stimulates the breakdown of fats as another source of energy. These are some of the many actions triggered by this one hormone. They all contribute to increasing your chances of escaping the lion. Whether a cell in your body responds to the surge of epinephrine depends on whether it has epinephrine receptors. The specific response of each cell with epinephrine receptors depends on the type of cell it is.

Glands and Hormones

Many hormones are secreted by aggregations of cells that form secretory organs called **glands**. Animals have two types of glands. Some, such as sweat glands and salivary glands, release secretions that are not hormones into ducts that open outside the body. Sweat gland ducts open onto the surface of the skin, and salivary gland ducts open into the mouth. Such glands are called **exocrine glands** because they secrete their products to the outside of the body (*exo* is Greek for "outside of").

The glands that secrete hormones do not have ducts; they are called **endocrine glands** because they secrete their products into the extracellular fluid, from which they enter the blood, which is inside the body. Endocrine glands store hormones until they are needed. Collectively, they make up the endocrine system (Figure 36.2).

Not all hormones are secreted by endocrine glands. Some are secreted by cells that are dispersed among other cells rather than being clumped together to form glands. For example, many hormones of the digestive tract are produced and secreted by cells in the lining of the tract. These hormones enter the blood and, like epinephrine, circulate throughout the body and activate cells that have the appropriate receptors. Many hormones are secreted by cells in the nervous system. These hormones, called **neurohormones**, can act locally, but some are taken up by the circulatory system and act on cells at some distance from their site of release.

HORMONAL CONTROL IN INVERTEBRATES

The hormones of invertebrate animals have been well studied, and these studies have made major contributions to our knowledge of how endocrine systems function. Many invertebrate hormones have molecular structures similar or identical to hormones of vertebrates, but their functions may differ. In this chapter we cannot begin to do justice to the diversity of hormones in invertebrates, but we'll discuss two important aspects of the lives of many invertebrates that are controlled by hormonal mechanisms: molt and metamorphosis.

Molting in Insects

The British physiologist Sir Vincent Wigglesworth was a pioneer in the study of hormonal control of growth and development in insects. He conducted experiments on the blood-sucking bug *Rhodnius*. Upon hatching from the egg, *Rhodnius* is nearly a miniature version of the adult, but it lacks some adult features. Because insects have rigid exoskeletons,

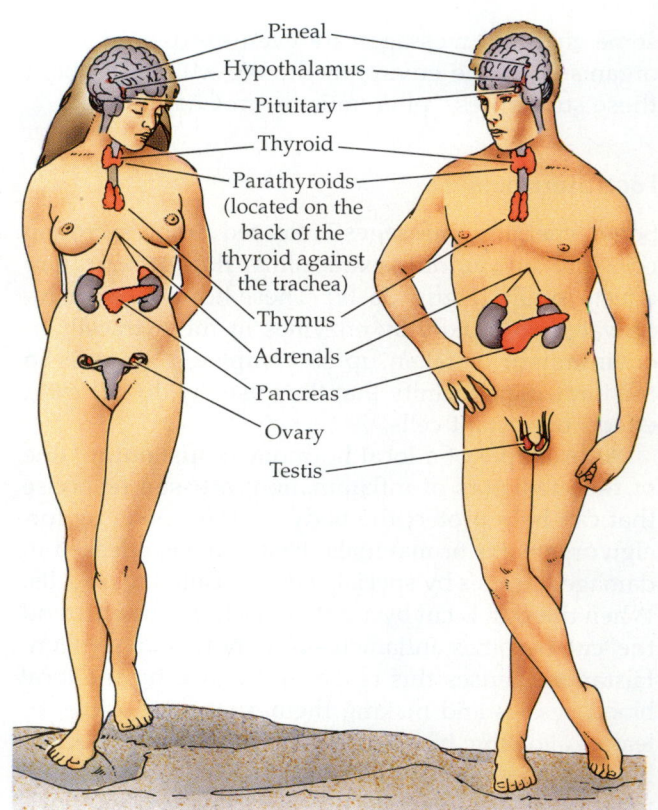

36.2 The Endocrine Glands of Mammals

their growth is episodic, punctuated with molts (shedding) of the exoskeleton. Each growth stage between two molts is called an instar. A blood meal triggers each episode of molting and growth in *Rhodnius*, which molts five times before developing into a complete adult. *Rhodnius* is a hardy experimental animal; it can live a long time even after it is decapitated. If decapitated soon after it has a blood meal, a *Rhodnius* may live up to a year, but it does not molt. If decapitated a week after its blood meal, it will molt (Figure 36.3a). These observations led to the hypothesis that something diffusing slowly from the region of the head controls the molt.

The proof that one or more diffusing substances cause the molt came from a clever experiment in which Wigglesworth decapitated two *Rhodnius*: one that had just had its blood meal and another that had had its blood meal one week earlier. The two decapitated bodies were connected with a short piece of glass tubing—and they both molted (Figure 36.3b). Thus one or more substances from the bug fed earlier crossed through the glass tube and stimulated the molting process in the other bug.

We now know that two hormones regulate molting. Cells in the brain produce a substance, simply called **brain hormone**, that is transported to a pair of neuroendocrine structures attached to the brain called the **corpora cardiaca** (singular: corpus cardiacum). After appropriate stimulation—which for

36.3 A Diffusible Substance Controls Molt

(a) Whether a decapitated *Rhodnius* will molt depends on the interval between a blood meal and the decapitation. *(b)* Apparently a diffusible substance from the head region is necessary for molt; this is demonstrated by connecting two bugs—decapitated at different times—with a glass tube.

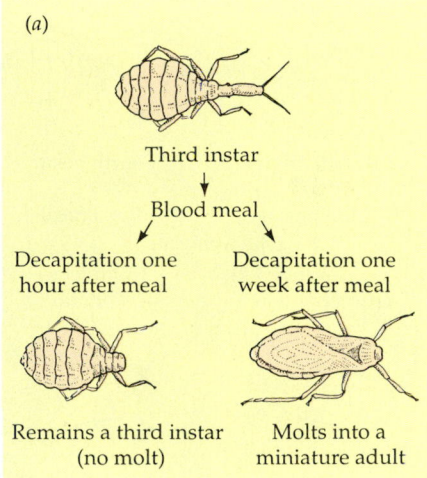

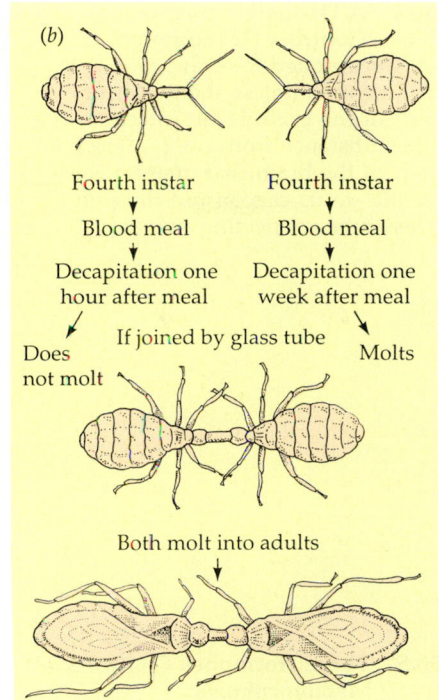

Rhodnius is a blood meal—the corpora cardiaca release brain hormone, which diffuses to an endocrine gland, the **prothoracic gland**. The prothoracic gland then produces and releases a hormone called **ecdysone**, which directly stimulates the molt.

The control of molting by brain hormone and ecdysone is a general mechanism in insects. The nervous system receives various types of information that may be relevant in determining the optimal timing for growth and development. It makes sense, therefore, that the nervous system should control the endocrine glands that produce the hormones that orchestrate all the physiological processes involved in development and molting. Later in this chapter we will see similar links between the nervous system and certain endocrine glands in vertebrates.

Development in Insects

The *Rhodnius* decapitation experiments just described yielded a curious result. Regardless of the instar used, the decapitated bug always molted directly into an adult form. Additional experiments by Wigglesworth demonstrated that a hormone other than those responsible for molting, and located in the rear of the head, must determine whether, under natural conditions, a bug molts into another juvenile instar or into an adult. The head of *Rhodnius* is long, so it was possible to remove just the front part of the head, which contains the neuroendocrine cells that secrete and release brain hormone, while leaving intact the rear part, which contains two other endocrine structures called the **corpora allata** (singular: corpus allatum). When fourth-instar bugs were partially decapitated so that the corpora allata remained intact, they molted into fifth instars and not into adults (Figure 36.4a).

This experiment indicating a role of the posterior part of the head in maintaining juvenile status was followed up by more glass tube experiments. When an unfed, completely decapitated, fifth-instar bug was connected to a fourth-instar bug that had been fed but had had only the front part of its head removed, both bugs molted into juvenile forms (Figure 36.4b). Some substance coming from the rear part of the head of the fourth-instar bug prevented the expected result that both bugs would molt into adult forms. We now know that the substance is **juvenile hormone** and that it comes from the corpora allata. As long as juvenile hormone is present, *Rhodnius* molts into another juvenile instar. The corpora allata normally stop producing juvenile hormone during the fifth instar. If juvenile hormone is absent, the bug molts into the adult form.

The control of development by juvenile hormone is more complex in insects that undergo complete metamorphosis. Complete metamorphosis includes a pupal stage that metamorphoses into the adult (see Chapter 17). An excellent example is the silkworm moth, *Hyalophora cecropia* (Figure 36.5). As long as juvenile hormone is present in high concentrations, larvae molt into larvae. When levels of juvenile hormone fall, larvae molt into pupae. Because no juvenile hormone is produced in the pupae, they molt into adults. In our perpetual war against insects, juvenile hormone is a new weapon. Synthetic forms of juvenile hormone can be distributed in the environment to prevent the development of adult insects capable of reproduction.

The existence and function of insect hormones was

36.4 A Diffusible Substance Controls Development

(a) Decapitated *Rhodnius* molt into adult forms unless the posterior part of the brain is left intact. *(b)* The substance from the posterior part of the brain that maintains juvenile status can diffuse through a glass tube connecting two bugs.

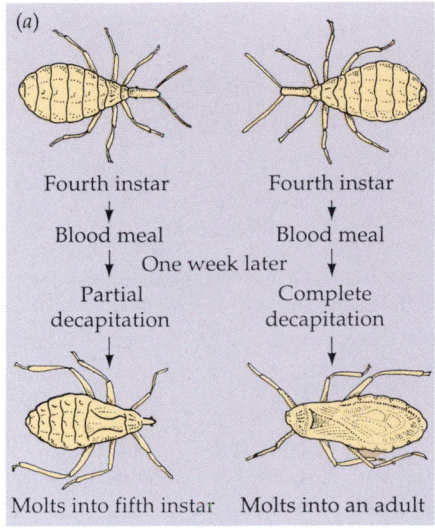

(a)

Fourth instar → Blood meal → One week later → Partial decapitation → Molts into fifth instar

Fourth instar → Blood meal → Complete decapitation → Molts into an adult

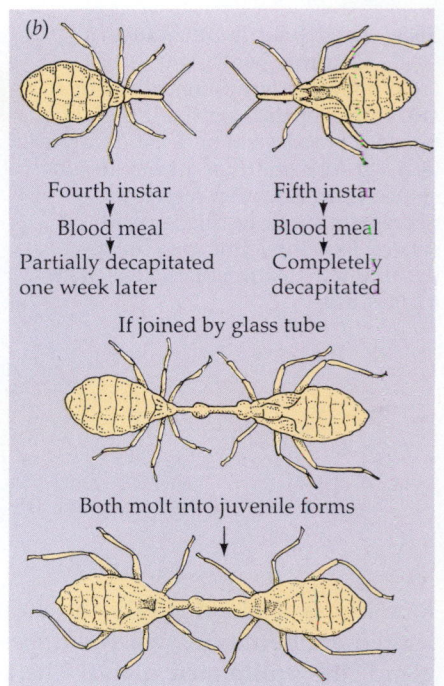

(b)

Fourth instar → Blood meal → Partially decapitated one week later

Fifth instar → Blood meal → Completely decapitated

If joined by glass tube → Both molt into juvenile forms

36.5 Three Hormones Control Molt and Metamorphosis

Neurosecretory cells in the brain of the silkworm moth produce brain hormone that is stored in and released from the corpora cardiaca. Brain hormone stimulates the prothoracic gland to produce ecdysone. The corpora allata produce juvenile hormone. As long as juvenile hormone is abundant, the larva molts into a larger larva in response to ecdysone. As juvenile hormone wanes, the larva molts into a pupa. The pupa does not produce juvenile hormone, so it metamorphoses into an adult. The release of ecdysone is episodic, and each release stimulates a molt.

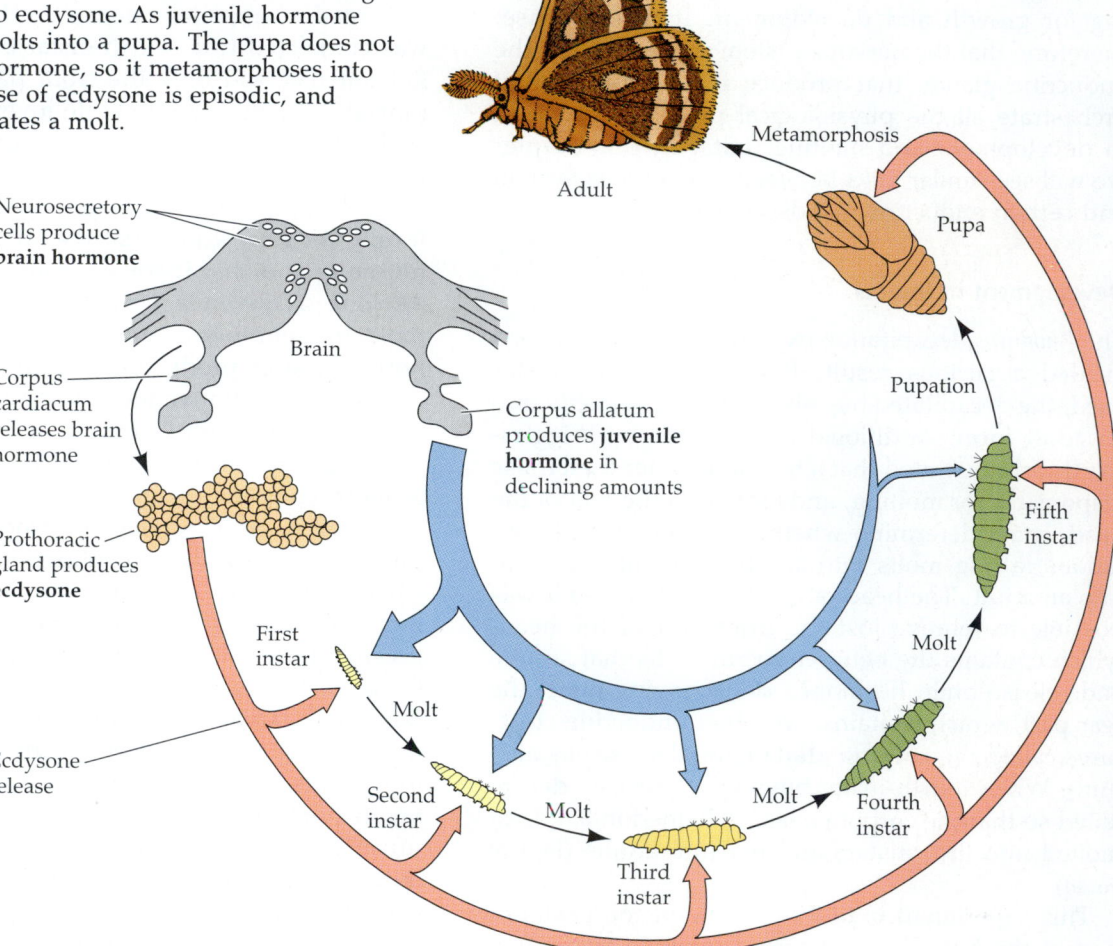

Neurosecretory cells produce **brain hormone**

Corpus cardiacum releases brain hormone

Prothoracic gland produces **ecdysone**

Brain

Corpus allatum produces **juvenile hormone** in declining amounts

Adult

Metamorphosis

Pupa

Pupation

Fifth instar

Molt

Ecdysone release

First instar

Molt

Second instar

Molt

Third instar

Molt

Fourth instar

experimentally demonstrated many years before the hormones were identified chemically. That is not surprising when you consider the tiny amounts of certain hormones that exist in an organism. In one of the earliest studies of ecdysone, biochemists produced only 250 mg of pure ecdysone (about one-fourth the weight of an apple seed) from 4 tons of silkworms!

VERTEBRATE HORMONES

As endocrine systems have evolved, the same chemical messages have become coupled to new physiological responses. In many cases the same chemical substance is a hormone in widely divergent species but has completely different actions. Many vertebrate hormones have molecular structures similar or identical to those of invertebrate hormones, but their functions are different. The hormone thyroxine, for example, is found in animal species ranging from tunicates (sea squirts) to humans, but thyroxine's function differs greatly among these species. In mammals it elevates cellular metabolic rate; in frogs it is essential for metamorphosis from tadpole to adult. Another example is the hormone prolactin, which stimulates milk production in female mammals after they give birth. In pigeons and doves prolactin stimulates the production of crop milk for nourishment of the young. Crop milk is really not a milk secretion at all, but a sloughing off of cells lining the upper digestive tract. In amphibians prolactin causes the animals to prepare for reproduction by seeking water, and in fishes, such as salmon, that migrate between salt and fresh water, prolactin regulates the mechanisms that maintain osmotic balance with the changing environment.

The endocrine systems of vertebrates are varied and complex. We recognize at least nine endocrine glands (see Figure 36.2), most of which produce and release more than one hormone. In addition, cells in many other organs produce and release hormones. The list of chemical messages in the bodies of vertebrates is long and growing longer. To make the subject more manageable, we will focus mostly on the hormones of mammals—how they function and how they are controlled. Table 36.1 presents an overview of the hormones of humans. Notice that the column listing the target tissues of these hormones includes every organ system of the body.

We begin our survey by examining the "master gland," the **pituitary**, which produces many hormones, some of which target other hormone-secreting cells elsewhere in the body. The pituitary gland sits in a depression at the bottom of the skull just over the back of the roof of the mouth (Figure 36.6). It is attached to the part of the brain called the hy-

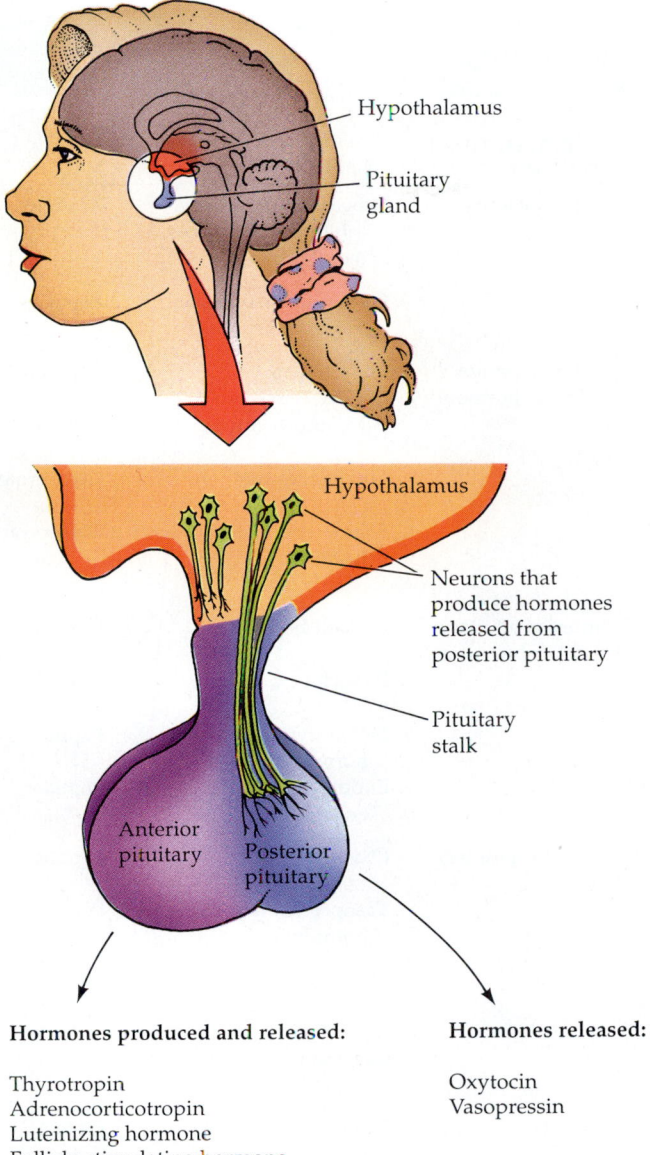

Hormones produced and released:

Thyrotropin
Adrenocorticotropin
Luteinizing hormone
Follicle-stimulating hormone

Growth hormone
Prolactin
Melanocyte-stimulating hormone
Endorphins
Enkephalins

Hormones released:

Oxytocin
Vasopressin

36.6 A Powerful Gland
The human pituitary gland is only the size of a blueberry, yet it secretes many hormones. The posterior pituitary secretes two hormones produced in the hypothalamus. Among the hormones produced and secreted by the anterior pituitary are four (listed first in the figure) that stimulate their target cells to secrete other hormones.

pothalamus, which we discovered in Chapter 35 is the location of the body's thermostat. The hypothalamus is involved in many homeostatic regulatory systems, including endocrine systems.

The pituitary has distinct anterior and posterior divisions that have separate origins during develop-

TABLE 36.1
Principal Hormones of Humans

SECRETING TISSUE OR GLAND	HORMONE	CHEMICAL NATURE	TARGETS	IMPORTANT PROPERTIES OR ACTIONS
Hypothalamus	Releasing and release-inhibiting hormones (see Table 36.2)	Peptides	Anterior pituitary	Control secretion of hormones of anterior pituitary
	Oxytocin, vasopressin	Peptides	(See Posterior pituitary)	Stored and released by posterior pituitary
Anterior pituitary: Tropic hormones	Thyrotropin	Glycoprotein	Thyroid gland	Stimulates synthesis and secretion of thyroxine
	Adrenocorticotropin	Polypeptide	Adrenal cortex	Stimulates release of hormones from adrenal cortex
	Luteinizing hormone	Glycoprotein	Gonads	Stimulates secretion of sex hormones from ovaries and testes
	Follicle-stimulating hormone	Glycoprotein	Gonads	Stimulates growth and maturation of eggs in females; stimulates sperm production in males
Anterior pituitary: Other hormones	Growth hormone	Protein	Bones, liver, muscles	Stimulates protein synthesis and growth
	Prolactin	Protein	Mammary glands	Stimulates milk production
	Melanocyte-stimulating hormone	Peptide	Melanocytes	Controls skin pigmentation
	Endorphins and enkephalins	Peptides	Spinal cord neurons	Decreases painful sensations
Posterior pituitary	Oxytocin	Peptide	Uterus, breasts	Induces birth by stimulating labor contractions; causes milk flow
	Vasopressin (antidiuretic hormone)	Peptide	Kidneys	Stimulates water reabsorption
Thyroid	Thyroxine	Iodinated amino acid derivative	Many tissues	Stimulates and maintains metabolism necessary for normal development and growth
	Calcitonin	Peptide	Bones	Stimulates bone formation; lowers blood calcium
Parathyroids	Parathormone	Protein	Bones	Absorbs bone; raises blood calcium
Thymus	Thymosins	Peptides	Immune system	Activate immune responses of T cells in the lymphatic system
Pancreas	Insulin	Protein	Muscles, liver, fat, other tissues	Stimulates uptake and metabolism of glucose; increases conversion of glucose to glycogen and fat
	Glucagon	Protein	Liver	Stimulates breakdown of glycogen and raises blood sugar

ment. The **anterior pituitary** originates as an outpocketing of the mouth region of the embryonic digestive tract, and the **posterior pituitary** originates as an outpocketing of the developing brain in the region that becomes the hypothalamus. Thus the posterior pituitary derives from nervous-system tissue, and the hormones it stores and releases are neurohormones produced in the hypothalamus. The anterior pituitary consists of epithelial cells that develop endocrine functions.

Posterior Pituitary Hormones

The posterior pituitary releases two neurohormones, **vasopressin** (also called antidiuretic hormone, or ADH) and **oxytocin**. These are small peptides synthesized in nerve cells in the hypothalamus. Vasopressin and oxytocin move down long extensions of these nerve cells that stretch down the pituitary stalk into the posterior pituitary, where the hormones are stored in the nerve endings (see Figure 36.6).

TABLE 36.1
Principal Hormones of Humans

SECRETING TISSUE OR GLAND	HORMONE	CHEMICAL NATURE	TARGETS	IMPORTANT PROPERTIES OR ACTIONS
Pancreas (continued)	Somatostatin	Peptide	Digestive tract; other cells of the pancreas	Inhibits insulin and glucagon release; decreases secretion, motility, and absorption in the digestive tract
Adrenal medulla	Adrenaline, noradrenaline	Modified amino acids	Heart, blood vessels, liver, fat cells	Stimulate "fight-or-flight" reactions: increase heart rate, redistribute blood to muscles, raise blood sugar
Adrenal cortex	Glucocorticoids (cortisol)	Steroids	Muscles immune system, other tissues	Mediate response to stress; reduce metabolism of glucose, increase metabolism of proteins and fats; reduce inflammation and immune responses
	Mineralocorticoids (aldosterone)	Steroids	Kidneys	Stimulates excretion of potassium ions and reabsorption of sodium ions
Stomach lining	Gastrin	Peptide	Stomach	Promotes digestion of food by stimulating release of digestive juices; stimulates stomach movements that mix food and digestive juices
Lining of small intestine	Secretin	Peptide	Pancreas	Stimulates secretion of bicarbonate solution by ducts of pancreas
	Cholecystokinin	Peptide	Pancreas, liver, gall bladder	Stimulates secretion of digestive enzymes by pancreas and other digestive juices from liver; stimulates contractions of gall bladder and ducts
	Enterogastrone	Polypeptide	Stomach	Inhibits digestive activities in the stomach
Pineal	Melatonin	Modified amino acid	Hypothalamus	Involved in biological rhythms
Ovaries	Estrogens	Steroids	Breasts, uterus, other tissues	Stimulate development and maintenance of female characteristics and sexual behavior
	Progesterone	Steroid	Uterus	Sustains pregnancy; helps to maintain secondary female sexual characteristics
Testes	Androgens	Steroids	Various tissues	Stimulate development and maintenance of male sexual behavior and secondary male sexual characteristics; stimulate sperm production
Most cells	Prostaglandins	Modified fatty acids	Various tissues	Have many diverse actions
Heart	Atrial natriuretic hormone	Peptide	Kidneys	Increases sodium ion excretion

The posterior pituitary increases its release of vasopressin whenever blood pressure falls or the blood becomes too salty. The main action of vasopressin is to increase the amount of water conserved by the kidneys. When vasopressin secretion is high, the kidneys reabsorb more water and produce only a small volume of highly concentrated urine. When vasopressin secretion is low, the kidneys produce a large volume of dilute urine. We will discuss the mechanism of vasopressin action in Chapter 44.

When a woman is about to give birth, her posterior pituitary releases oxytocin, which stimulates the contractions of the muscles that push the baby out of her body. Oxytocin also brings about the flow of milk from her breasts. The baby's suckling stimulates nerve cells in the mother, causing secretion of oxytocin. Even the sight and sounds of her baby can cause a nursing mother to secrete oxytocin and release milk from her breasts.

Anterior Pituitary Hormones

The anterior pituitary produces and secretes many peptide hormones, each of which is produced by a different type of pituitary cell. Four of these hormones that control the activities of other endocrine glands are **tropic hormones**. The four tropic hormones are thyrotropin, adrenocorticotropin, luteinizing hormone, and follicle-stimulating hormone. We will say more about these tropic hormones when we describe their target glands (thyroid, adrenal cortex, testes, and ovaries) later in this chapter and in the next. The other hormones produced by the anterior pituitary influence tissues that are not endocrine glands. These hormones are growth hormone, prolactin, melanocyte-stimulating hormone, and endorphins and enkephalins.

Growth hormone consists of about 200 amino acids and acts on a wide variety of tissues to promote growth directly and indirectly. One of its important direct effects is to stimulate cells to take up amino acids. Growth hormone promotes growth indirectly by stimulating the liver to produce growth-regulating chemical messages called **somatomedins**, which circulate in the blood and stimulate the growth of bone and cartilage. Overproduction of growth hormone in children causes gigantism, and underproduction causes dwarfism (Figure 36.7). High levels of growth hormone in adults cannot cause increased height because the shafts and the growth plates of the long bones have fused. Rather, abnormally high levels of growth hormone in adults cause thickening of the hands, feet, jaw, nose, and ears—a condition known as acromegaly.

36.7 Effects of Abnormal Amounts of Growth Hormone
(a) Gigantism results from the overproduction of growth hormone in childhood. In this news photo from 1939, a young man and his father visit New York on business; the son is over 8 feet tall, whereas his father is of average height: 5 feet, 11 inches. (b) When the anterior pituitary does not produce enough growth hormone during childhood, pituitary dwarfism results. The man on the left is P. T. Barnum, circus entrepreneur. With him is Charles Stratton, a dwarf who appeared in Barnum's circus under the name General Tom Thumb.

(a)

(b)

Beginning in the late 1950s, children diagnosed as having a serious deficiency of growth hormone, and therefore destined to become dwarfs, were treated with human growth hormone extracted from human pituitaries from cadavers. The treatment was successful in stimulating substantial growth, but it was extremely costly; a year's supply of human growth hormone for one individual required up to 50 pituitaries. In the mid-1980s scientists using genetic engineering technology isolated the gene for human growth hormone and introduced it into bacteria, which produced enough of the hormone to make it commercially available.

Preventing pituitary dwarfism is now feasible and affordable, but the availability of growth hormone raises new questions. Should every child at the lower end of the height charts be treated? Should a normal child whose parents want him or her to play basketball be given growth hormone? These types of questions are impossible to answer with scientific data alone. The controversy around growth hormone has become even more complex because of a recent study suggesting that when growth hormone is administered to older persons, it reverses some of the effects of aging. In comparison to a control group not receiving growth hormone, a group of elderly persons receiving growth hormone decreased their body fat, increased their muscle mass, and reported feeling more energetic. It is not known, however, if these changes will last beyond the period of treatment, or if the treatment has side effects.

Earlier in the chapter we described the evolutionary diversity of the functions of **prolactin**, another hormone produced by the anterior pituitary. In human females the major function of prolactin is to stimulate the production and secretion of milk. In some mammals prolactin also functions as an important hormone during pregnancy. In human males prolactin plays a role along with other pituitary hormones in controlling the endocrine function of the testes.

Melanocyte-stimulating hormone is produced in very low amounts by the human anterior pituitary, and its functions in humans are not well understood. Melanocytes are cells that contain melanin, a black pigment. In fishes, amphibians, and reptiles that can change their color, melanocyte-stimulating hormone changes the way melanin is distributed in the melanocytes, thereby darkening or lightening the tissue containing them.

Endorphins and **enkephalins**, the remaining hormones of the anterior pituitary, are referred to as the body's "natural opiates." These molecules help control pain. Interestingly, the production of endorphins and enkephalins in the pituitary is encoded by the same gene as are two other pituitary hormones. The gene codes for a large parent molecule called pro-opiomelanocortin. This large molecule is cleaved to produce several peptides, some of which have hormonal functions. Adrenocorticotropin, melanocyte-stimulating hormone, endorphins, and enkephalins all result from the cleavage of pro-opiomelanocortin.

Hypothalamic Neurohormones

The idea of the anterior pituitary as the "master gland" received quite a blow with the discovery that it is really a "middleman" controlled by the hypothalamus. The hypothalamus receives information about conditions in the body and in the external environment through the nervous system. If the connection between the hypothalamus and the pituitary is cut, pituitary hormones are no longer released in response to changes in the environment or in the body. If pituitary cells are maintained in culture, extracts of hypothalamic tissue stimulate some of those cells to release their hormones into the culture medium. Therefore, scientists hypothesized that secretions of the hypothalamic cells control the activities of the cells of the anterior pituitary. Although hypothalamic neurons do not extend into the anterior pituitary as they do into the posterior pituitary, a possible route by which such chemical messages could reach the anterior pituitary was known: a special set of blood vessels called **portal blood vessels** that run between the hypothalamus and the anterior pituitary (Figure 36.8). It was thus proposed that secretions from nerve endings in the hypothalamus are absorbed into the blood and are conducted down the portal vessels to the anterior pituitary, where they cause the release of anterior pituitary hormones.

In the 1960s two large teams of scientists, led by Roger Guillemin and Andrew Schally, initiated the search for the hypothalamic releasing neurohormones. Because the amounts of such hormones in any individual mammal would be tiny, "bucket biochemistry" was called for. The scientists set up teams in slaughterhouses to collect massive numbers of hypothalami from pigs and sheep. The resulting *tons* of tissue were shipped to laboratories in refrigerated trucks. One effort began with the hypothalami from 270,000 sheep and yielded only 1 mg of purified thyrotropin-releasing hormone. Biochemical analysis of this pure sample revealed that thyrotropin-releasing hormone contains only three amino acids; it is a tripeptide. Soon after discovering thyrotropin-releasing hormone, the scientists identified gonadotropin-releasing hormone, which controls the release of follicle-stimulating hormone and luteinizing hormone from the anterior pituitary. For these discoveries Guillemin and Schally received the 1972 Nobel prize in medicine. Because isolation techniques have been improved enormously, we now need only a few milligrams of tissue to isolate and characterize a peptide.

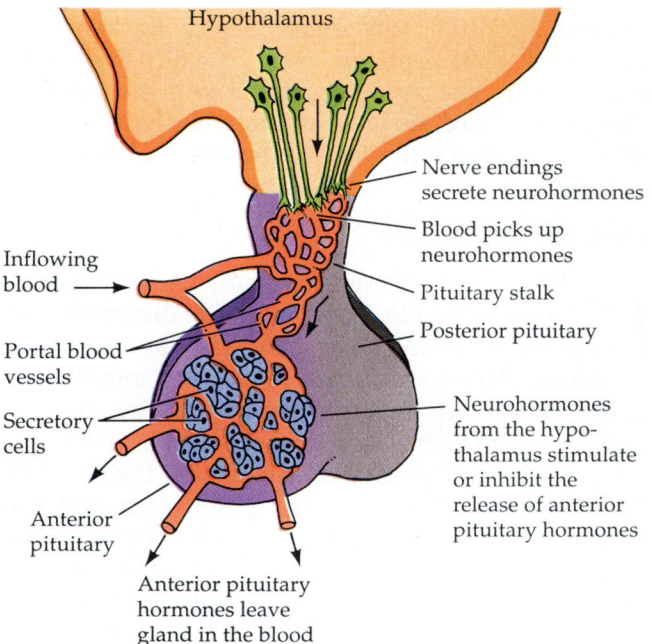

**36.8   The Hypothalamus Controls the
Anterior Pituitary**

A system of blood vessels runs between the hypothalamus and the anterior pituitary. Tropic neurohormones produced by hypothalamic cells enter these blood vessels and are transported to the anterior pituitary, where they control the activity of the pituitary cells that synthesize and release hormones.

Many more hypothalamic neurohormones are now known, and they include **releasing hormones** and **release-inhibiting hormones** (Table 36.2).

Thyroid Hormones

The **thyroid gland** consists of two lobes, one on either side of the trachea (windpipe), connected by a strip of thyroid tissue that wraps around the front of the

trachea. If you gently place your thumb and forefinger on either side of your trachea just below your Adam's apple and swallow, you will feel your thyroid gland move up and down under your fingertips.

The thyroid gland produces the hormones thyroxine and calcitonin. The **thyroxine** molecule consists of two molecules of the amino acid tyrosine and four atoms of iodine. (Another form of thyroid hormone has only three iodine atoms, but for convenience we will refer to both as thyroxine.) Thyroxine in mammals plays many roles in regulating cell metabolism. It elevates the metabolic rates of most cells and tissues and promotes the use of carbohydrates rather than fats for fuel. Exposure to cold for several days leads to an increased release of thyroxine and an increase in basal metabolic rate. Thyroxine is especially crucial during development and growth. It promotes amino acid uptake and protein synthesis by cells. Insufficient thyroxine in a human fetus or growing child greatly retards physical and mental growth, resulting in a condition known as cretinism.

Malfunction of the thyroid gland causes goiter, a condition in which the thyroid gland becomes very large (Figure 36.9). Goiter can be associated with either **hyperthyroidism** (high levels of thyroxine) or **hypothyroidism** (low levels of thyroxine). To understand how these opposite conditions can lead to the same symptom, we must consider the control of thyroid activity by the pituitary and the hypothalamus.

The tropic hormone **thyrotropin**, which is secreted into the blood by the anterior pituitary, determines the activity of the thyroid gland. Thyrotropin activates the thyroid gland cells that produce thyroxine. Thyrotropin-releasing hormone produced in the hypothalamus and transported to the pituitary through the portal blood vessels activates the thyrotropin-producing pituitary cells. The brain uses environmental information such as temperature or day length to determine whether to increase or decrease

TABLE 36.2
Releasing and Release-Inhibiting Neurohormones of the Hypothalamus

NEUROHORMONE	ACTION
Thyrotropin-releasing hormone (TRH)	Stimulates thyrotropin release
Gonadotropin-releasing hormone	Stimulates release of follicle-stimulating hormone and luteinizing hormone
Prolactin release-inhibiting hormone	Inhibits prolactin release
Prolactin-releasing hormone	Stimulates prolactin release
Somatostatin (growth hormone release-inhibiting hormone)	Inhibits growth hormone release; interferes with thyrotropin release
Growth hormone-releasing hormone	Stimulates growth hormone release
Adrenocorticotropin-releasing hormone	Stimulates adrenocorticotropin release
Melanocyte-stimulating hormone release–inhibiting hormone	Inhibits melanocyte-stimulating hormone release

Hypothalamus

Nerve endings secrete neurohormones

Blood picks up neurohormones

Pituitary stalk

Posterior pituitary

Inflowing blood

Portal blood vessels

Secretory cells

Neurohormones from the hypothalamus stimulate or inhibit the release of anterior pituitary hormones

Anterior pituitary

Anterior pituitary hormones leave gland in the blood

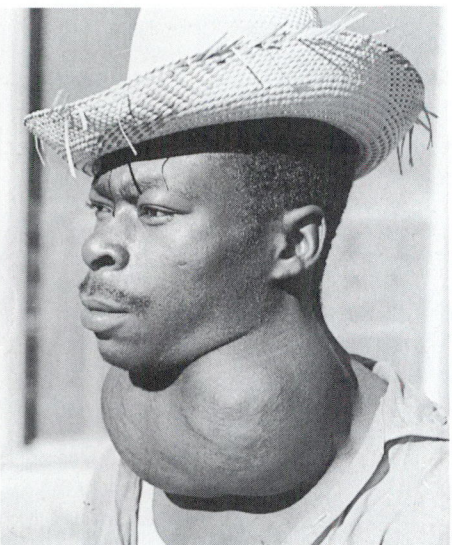

36.9 Goiter
A goiter is a greatly enlarged thyroid gland. Worldwide, goiter affects about 5 percent of the population. The addition of iodine to table salt has greatly reduced the incidence of the condition in industrialized nations, but goiter is still common in the less developed countries of the world.

the secretion of thyrotropin-releasing hormone. There is a very important negative feedback loop in this sequence of steps: Circulating thyroxine inhibits the response of the pituitary cells to thyrotropin-releasing hormone. Therefore, when thyroxine levels are high less thyrotropin is released, and when thyroxine levels are low more thyrotropin is released (Figure 36.10).

Hyperthyroid goiter results when the pituitary cells are not turned off by thyroxine. Thyrotropin levels remain high and the thyroid gland is activated so much that it grows bigger. Production of thyroxine by the thyroid stays abnormally high. Hyperthyroid patients have high metabolic rates, are jumpy and nervous, usually feel hot, and may have a buildup of fat behind the eyeballs, causing their eyes to bulge.

Hypothyroid goiter results when there is not enough circulating thyroxine to turn off thyrotropin production. Its most common cause is a deficiency of dietary iodide, without which the thyroid gland cannot make *functional* thyroxine. Without thyroxine, thyrotropin levels remain high, so the thyroid continues to produce large amounts of nonfunctional thyroxine and gets very large. The symptoms of hypothyroidism are low metabolism, intolerance of cold, and general physical and mental sluggishness. Hypothyroid goiter used to be extremely common in mountainous areas and regions far from the oceans, where there is little iodide in the soil or water. The addition of iodide to table salt has greatly reduced the incidence of the disease.

Another hormone of the mammalian thyroid gland is **calcitonin**. It is not produced by the same cells that produce thyroxine. Calcitonin helps regulate the levels of calcium circulating in the blood (Figure 36.11). Bone is a huge repository of calcium in the body and is continually being remodeled. Cells called **osteoclasts** break down bone and release calcium; **osteoblasts**, on the other hand, use circulating calcium to deposit new bone. Calcitonin decreases the activity of osteoclasts and stimulates the activity of osteoblasts, thus shifting the balance from adding calcium ions to the blood to removing calcium ions from the blood. The regulation of blood calcium levels is influenced more strongly by parathormone, which we consider in the next section, than it is by calcitonin, but calcitonin plays an important role in preventing bone loss in women during pregnancy.

Parathormone

The **parathyroid glands** are four tiny structures embedded on the surface of the thyroid gland. Their single hormone product is parathyroid hormone, or **parathormone**, a critical control element in the regulation of blood calcium levels. Growth and remodeling of bone require calcium; so do many cellular processes, such as nerve and muscle functions, which are sensitive to changes in calcium concentration. Muscle contraction and nerve function are severely impaired if the blood calcium level rises or falls by as little as 30 percent of normal values. A fall in blood calcium triggers the release of parathormone, which in turn stimulates actions that add calcium to the blood. Parathormone stimulates osteo-

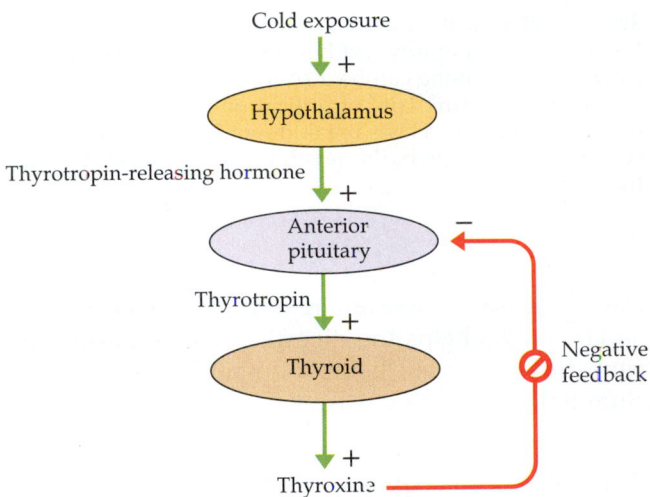

36.10 Regulation of Thyroid Function
Environmental cues such as exposure to cold stimulate the hypothalamus to produce thyrotropin-releasing hormone, which initiates a cascade of events that stimulate the thyroid to release thyroxine. High thyroxine levels act as negative feedback.

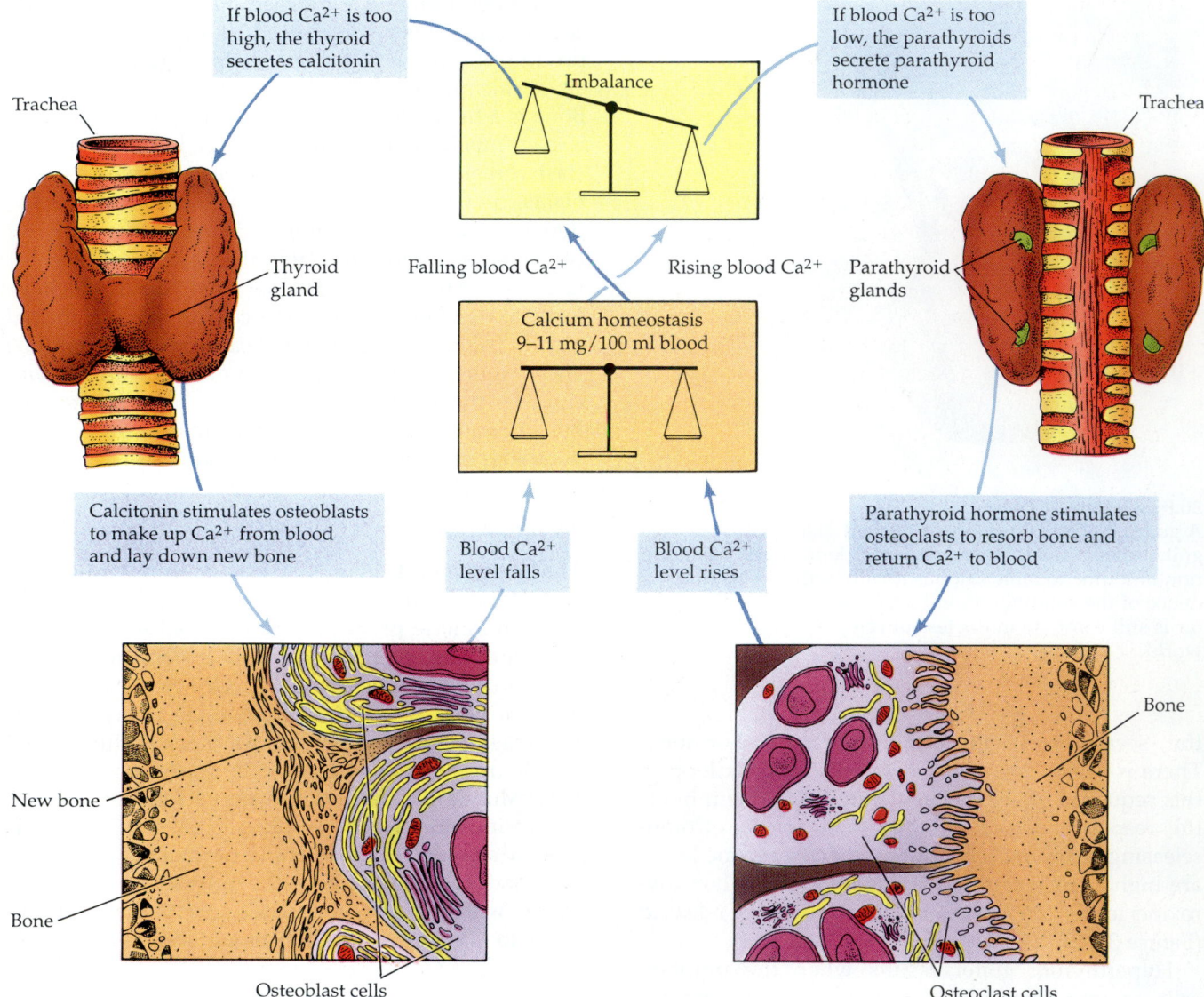

If blood Ca²⁺ is too high, the thyroid secretes calcitonin

If blood Ca²⁺ is too low, the parathyroids secrete parathyroid hormone

Trachea

Trachea

Imbalance

Thyroid gland

Falling blood Ca²⁺

Rising blood Ca²⁺

Parathyroid glands

Calcium homeostasis 9–11 mg/100 ml blood

Calcitonin stimulates osteoblasts to make up Ca²⁺ from blood and lay down new bone

Blood Ca²⁺ level falls

Blood Ca²⁺ level rises

Parathyroid hormone stimulates osteoclasts to resorb bone and return Ca²⁺ to blood

Bone

New bone

Bone

Osteoblast cells

Osteoclast cells

36.11   Calcium Balance
Calcitonin and parathyroid hormone help regulate blood calcium levels. Bone can be a source (site of production) of calcium or a sink (site of utilization or storage) for excess calcium. Osteoclasts break down bone and release calcium; osteoblasts build new bone using calcium from the blood.

clasts to dissolve bone and release calcium (see Figure 36.11). It also helps the digestive tract to absorb calcium from food and helps the kidneys reabsorb calcium before excreting wastes.

Pancreatic Hormones

Before the 1920s, diabetes mellitus was a fatal disease, characterized by weakness, lethargy, and body wasting. The disease was known to be connected with a gland located just below the stomach, the **pancreas**, and with abnormal glucose metabolism,

but the link was not clear. Today we know that diabetes mellitus is caused by a lack of the hormone **insulin**. Insulin replacement therapy makes it possible for more than 1.5 million diabetics in the United States to lead almost normal lives.

Most cells in an untreated diabetic's body cannot use glucose in the blood for metabolic fuel. As a result, glucose accumulates in the blood until it is lost in the urine. High blood glucose causes water to move from cells into the blood by osmosis, and the kidneys increase urine output to excrete the excess fluid volume from the blood. The name *diabetes* refers to copious production of urine, and *mellitus* (from the Greek for "honey") reflects the fact that the urine of the untreated diabetic is sweet. Since the cells of the body cannot use blood glucose for fuel, they must burn fat and protein. As a result, the body of the untreated diabetic wastes away, and critical tissues and organs are damaged.

The change in the outlook for diabetics came almost overnight in 1921, when medical doctor Frederick Banting and medical student Charles Best of the University of Toronto discovered that they could reduce the symptoms of diabetes with an extract they prepared from pancreatic tissue. The work of Banting and Best led to enormous relief of human suffering. The active component of the extract Banting and Best prepared was a small protein hormone, insulin, consisting of 51 amino acids. Insulin is produced in clusters of cells in the pancreas, called **islets of Langerhans** after a German medical student who discovered them. Other cells in the islets produce two other hormones, glucagon and somatostatin. The rest of the pancreas produces enzymes and secretions that travel through ducts to the intestine, where they play roles in digestion. Thus the pancreas is both an endocrine and an exocrine gland.

Following a meal, the concentration of glucose in the blood rises as glucose is absorbed from the gut. This rise in glucose concentration stimulates the pancreas to release insulin. Insulin stimulates cells to use glucose as fuel and to convert it into storage products such as glycogen and fat. When there is no longer food in the gut, the blood's glucose concentration falls, and the pancreas stops releasing insulin. As a result, most of the cells of the body shift to using glycogen and fat rather than glucose for fuel. If the concentration of glucose in the blood falls below normal, cells in the islets release the hormone **glucagon**, which stimulates the liver to convert glycogen back to glucose to resupply the blood. These effects and conversions will be discussed in greater detail in Chapter 43.

Somatostatin, a hormone released from the pancreas in response to rapid rises of glucose and amino acids in the blood, inhibits the release of both insulin and glucagon and slows the digestive activities of the gut. Pancreatic somatostatin extends the period of time over which nutrients are absorbed from the gut and used by the cells of the body. Somatostatin is also secreted by cells in a different part of the body, and these secretions serve a different hormonal function. First discovered as a hypothalamic neurohormone that *inhibits* the release of growth hormone by the pituitary, it was called growth hormone release-inhibiting hormone, but somatostatin is a more convenient name.

Adrenal Hormones

An adrenal gland sits above each kidney. Functionally and anatomically an adrenal gland is a gland within a gland (Figure 36.12). The core, called the **adrenal medulla**, produces the hormone **epinephrine** (also known as adrenaline) and, to a lesser degree, **norepinephrine** (or noradrenaline). Surrounding the

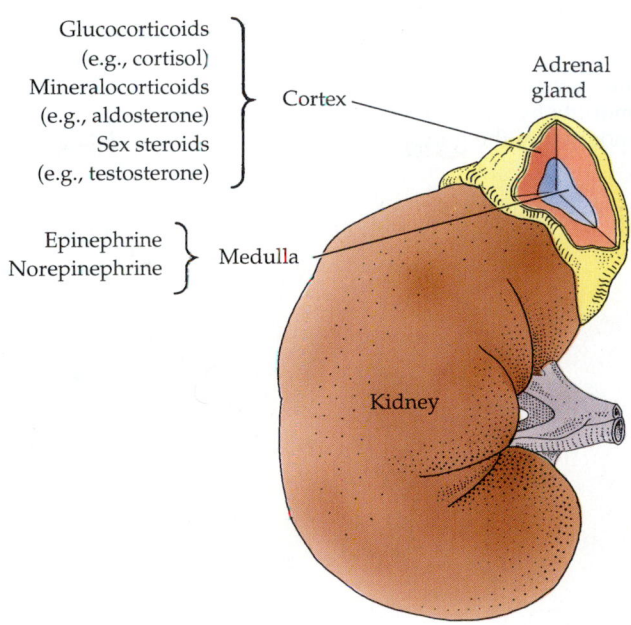

Hormones released into blood:

Glucocorticoids (e.g., cortisol)
Mineralocorticoids (e.g., aldosterone)
Sex steroids (e.g., testosterone) } Cortex

Epinephrine
Norepinephrine } Medulla

Adrenal gland

Kidney

36.12 The Adrenals: Two Glands in One
The adrenal medulla and the adrenal cortex produce different hormones. Together they form the adrenal gland.

medulla (as an apricot surrounds its pit) is the **adrenal cortex**, which produces other hormones. The medulla develops from nervous system tissue and is under the control of the nervous system; the cortex is under hormonal control, largely by adrenocorticotropin from the anterior pituitary.

By producing epinephrine, which arouses the body to action, the adrenal medulla is involved in "fight-or-flight" reactions. As we saw earlier in the chapter, in stressful situations epinephrine increases heart rate, breathing rate, and blood pressure, and it diverts blood flow to active skeletal muscles and away from the gut.

All hormones produced by the adrenal cortex are steroids synthesized from cholesterol (Figure 36.13). In general they are called the **corticosteroids**, and they are divided into three functional classes. The **glucocorticoids** influence blood glucose concentrations as well as other aspects of fat, protein, and carbohydrate metabolism. The **mineralocorticoids** influence the ionic balance of the extracellular fluids. The **sex steroids** stimulate sexual development and reproductive activity. Sex steroids are secreted in only small amounts by the adrenal cortex and will be discussed further in the section on the gonads. Of the 30 or so different steroids produced by the adrenal cortex, the only two of great importance in human physiological functions under normal conditions are the mineralocorticoid aldosterone (which helps regulate salt concentration in the blood; see Chapter 44) and the glucocorticoid cortisol.

36.13   Steroid Hormones Begin as Cholesterol

Different side groups on the sterol backbone (gray) confer different properties on steroid hormones. This simplified outline of steroid hormone biosynthesis leaves out many intermediate steps. Sex steroids are produced in small amounts by the adrenal cortex and in much greater amounts by the gonads.

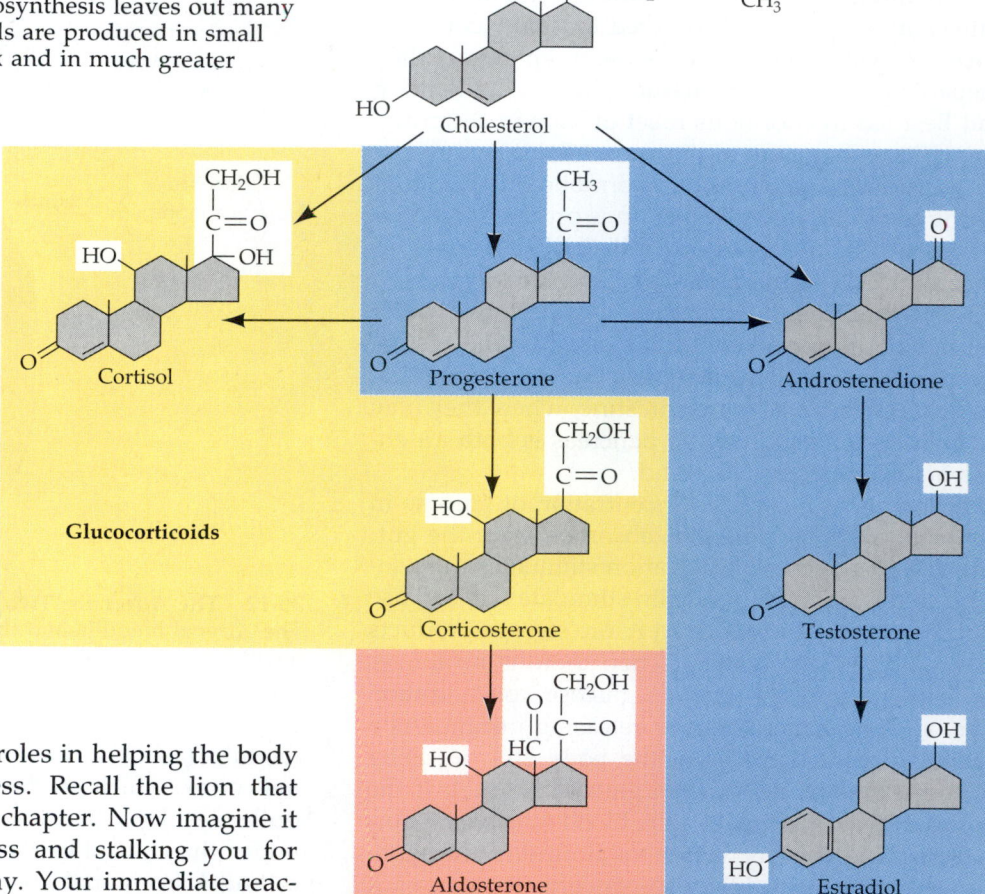

Cholesterol

Cortisol

Glucocorticoids

Progesterone

Androstenedione

Corticosterone

Testosterone

Aldosterone
Mineralocorticoid

Estradiol
Sex steroids

Cortisol plays important roles in helping the body respond to short-term stress. Recall the lion that roared at you earlier in the chapter. Now imagine it chasing you into high grass and stalking you for hours as you try to get away. Your immediate reaction is stimulated by your nervous system and by the release of epinephrine. Your heart is beating faster, you are breathing faster, and your running muscles are getting maximal supplies of oxygen and glucose. This is not a sustainable situation, so within about 5 minutes cortisol levels rise and help you sustain your escape. Because you need a high level of blood glucose for your brain to function, cortisol stimulates the other cells of your body to decrease their use of glucose and start to metabolize fats and proteins for energy. This is not a time to feel sick, have allergic reactions, or heal wounds, so cortisol blocks immune system reactions. This is why cortisol is useful for reducing inflammations and allergies.

The effects of epinephrine and cortisol in reactions to short-term stress are beneficial, but the prolongation of these effects by the chronic, long-term stresses of modern life can be very damaging to the body. High blood pressure, poor gastrointestinal function, inhibition of protein synthesis, fat mobilization, and inhibition of the immune system are not healthy responses over long periods of time. Details on the interactions of stress, aging, and the cortisol response are discussed in Box 36.A.

Cortisol release is controlled by the pituitary hormone **adrenocorticotropin**, which in turn is controlled by the hypothalamic adrenocorticotropin-re-

leasing hormone. Because the cortisol response to stress has this chain of steps, each involving secretion, diffusion, circulation, and cell activation, it is much slower than the epinephrine response to stress.

The Sex Hormones

The testes of the male and the ovaries of the female (that is, the gonads) produce hormones as well as gametes. Most of the gonadal hormones are steroids synthesized from cholesterol (see Figure 36.13). The male steroids are collectively called **androgens**, and the dominant one is **testosterone**. The female steroids are **estrogens** and **progesterone**. The dominant estrogen is **estradiol**. The sex steroids have important developmental effects: They determine whether a fetus develops into a female or a male. (A fetus is the latter stage of an embryo; a human embryo is called a fetus from the eighth week of pregnancy to the moment of birth.) After birth the sex steroids control the maturation of the reproductive organs and the development and maintenance of secondary sexual characteristics, such as breasts and facial hair.

The sex steroids begin to exert developmental ef-

BOX 36.A

The Rat Race

We usually assume that excessive stress leads to premature aging. Stress reactions *increase* blood pressure and fat metabolism while they *inhibit* digestion and immune system function. When prolonged, the effects of stress can contribute to cardiovascular disease, strokes, ulcers, and susceptibility to diseases. Therefore, assuming that stress accelerates aging is not unreasonable. The exact physiological interactions between stress and aging have only recently been elucidated, however, and they are quite interesting. Research by Robert Sapolsky at Stanford University has shown that old rats can initiate a stress response just as effectively as young rats, but they cannot turn it off as rapidly. All the harmful effects of stress persist longer in older rats than in younger ones. Why?

The answer involves the way negative feedback controls stress responses. A region of the brain called the hippocampus has cells with receptors for cortisol. When a rat experiences stress, cortisol levels rise in the blood. The cortisol activates these hippocampal cells, which inhibit the secretion of adrenocorticotropin-releasing hormone by the hypothalamus. The drop in adrenocorticotropin-releasing hormone causes a decline in adrenocorticotropin release from the pituitary, which induces the adrenal cortex to stop producing cortisol.

As rats age, they lose the hippocampal cells that function in this negative feedback loop. Apparently cortisol contributes to the demise of these cells. The more stress experienced, the more of these crucial hippocampal cells are lost. The result is the loss of the negative feedback mechanism that protects the body from the harmful effects of sustained stress responses. In rats the increased incidence of stress responses leads to many disorders usually associated with aging—strokes, cardiovascular disease, digestive system malfunction, and impaired immune system function that increases susceptibility to cancers and other diseases. These relationships between stress and aging are also believed to pertain to humans.

---

fects at about the seventh week of gestation in the human embryo. Until that time, the embryo can develop into either sex. The ultimate instructions for sex determination reside in the genes. Individuals receiving two X chromosomes normally become females, and individuals receiving an X and a Y chromosome normally become males. These genetic instructions, however, are carried out through the production and action of the sex steroids, and the potential for error exists.

The presence of the Y chromosome normally causes the embryonic, undifferentiated gonads to begin producing androgens in the seventh week. In response to the androgens, the reproductive system develops into that of a male (Figure 36.14). If the androgens are not produced at that time, or the androgen receptors do not function, the female reproductive structures develop even if the fetus is a male genetically. In humans, female development is the default, or neutral, course; a fetus develops female characteristics unless androgens are present to trigger male development. The opposite situation exists in some other vertebrates—male development is the default condition, which is switched to female development if estrogens are present.

Occasionally the hormonal control of sexual development does not work perfectly and intersex individuals are produced. The most extreme (but rare) case is a true **hermaphrodite**, who has both testes and ovaries. **Pseudohermaphrodites** have the gonads of one sex and the external sex organs of the other. For example, an XY fetus will develop testes, but if his tissues are insensitive to the androgens produced by those testes, the external sex organs and the secondary sexual characteristics of a female develop (Figure 36.15).

Sex steroids have dramatic effects at the time of puberty. Throughout childhood the production of sex steroids by the gonads is extremely low. As puberty approaches, the hypothalamus begins to produce and secrete gonadotropin-releasing hormone, which causes the pituitary to produce two gonadotropins—**luteinizing hormone** and **follicle-stimulating hormone**. In the preadolescent male, the increased level of luteinizing hormone stimulates groups of cells in the testes to synthesize androgens, which in turn initiate the profound physiological, anatomical, and psychological changes associated with adolescence. The voice deepens, hair begins to grow on the face and body, the testes and the penis grow, and skeletal muscles enlarge. Even an active program of weight lifting will not lead to massive muscle development in preadolescent boys or in women because such increase requires a level of androgens not normally found in individuals other than males past puberty. Taking synthetic androgens to enhance muscle development is a dangerous practice and a serious type of substance abuse (Box 36.B).

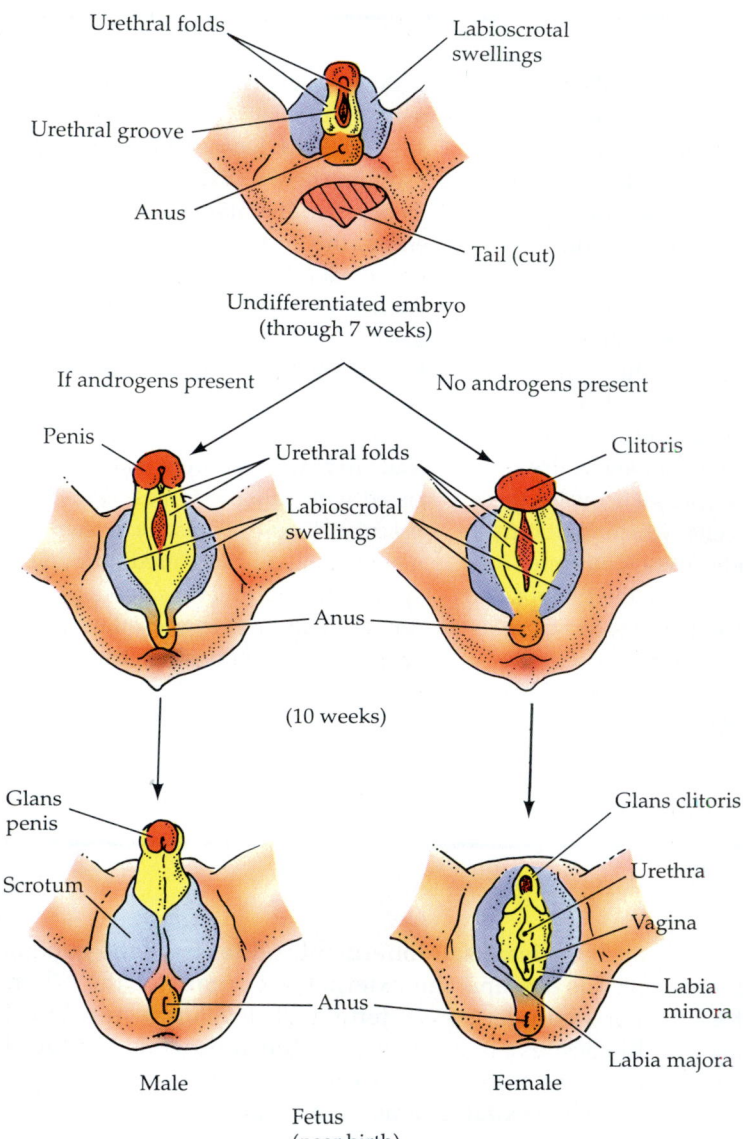

36.14   Sex Development
Hormones direct sex development along one of two pathways. The sex organs of early human embryos are similar (top). The testes of genetic males begin to secrete androgens about seven weeks after fertilization. Under the influence of androgens, a penis and scrotum form. Without the influence of androgens, female external organs develop.

Puberty in the female is also ushered in by an increased release of gonadotropin-releasing hormone by the hypothalamus. The levels of luteinizing hormone and follicle-stimulating hormone rise, stimulating the ovaries to begin producing the female sex hormones. The increased circulating levels of these sex steroids initiate the development of the traits characteristic of a sexually mature woman: enlarged breasts, vagina, and uterus; a broad pelvis; increased subcutaneous fat; pubic hair; and the initiation of the menstrual cycle.

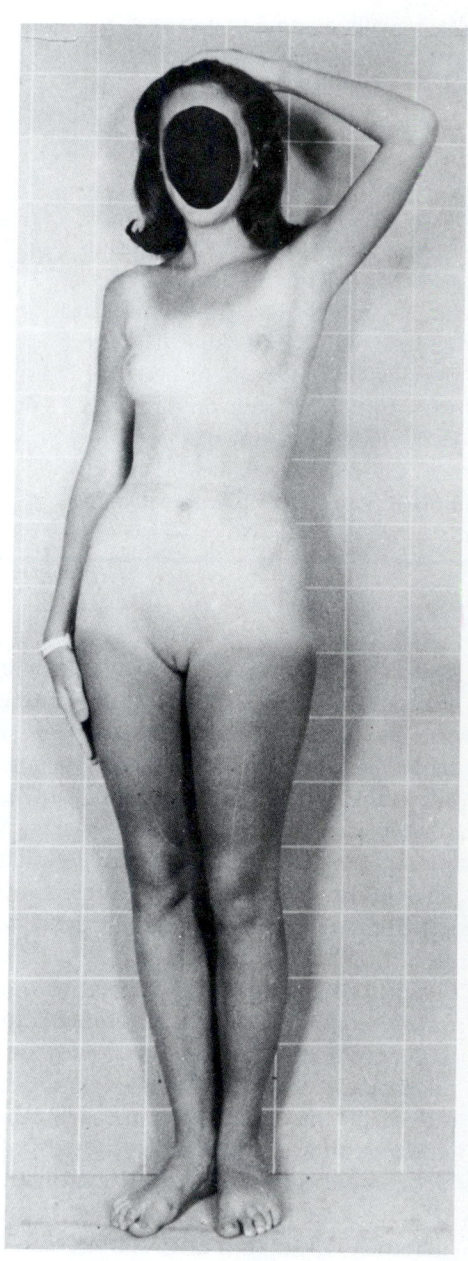

36.15   A Pseudohermaphrodite
This person, who is genetically a male with an XY genotype, carries a mutation that leaves body cells unresponsive to male sex hormones. During early fetal life, the developing testes produced normal amounts of testosterone, but because of the mutation, the cells forming the sex organs could not respond. Development thus followed the female pattern. Fully functional testes developed within the body, but external female organs developed. The person does not have a uterus or ovaries, however, and the vagina ends blindly.

BOX 36.B

Muscle-Building Anabolic Steroids

The androgen testosterone helps skeletal muscles grow, especially when they are exercised regularly. The bulging biceps, triceps, pectorals, and deltoids of body builders are extreme examples of the skeletal muscle growth that occurs in every male past puberty. Natural muscular development can be exaggerated by both men and women who want to increase their maximum strength in athletic competition if they take synthetic androgens—**anabolic steroids.** However, anabolic steroids have serious negative side effects. In women,

the use of artificial androgens causes the breasts and uterus to shrink, the clitoris to enlarge, menstruation to become irregular, facial and body hair to grow, and the voice to deepen. In men, the testes shrink, hair loss increases, the breasts enlarge, and sterility can result. Other effects are even more serious. For example, taking anabolic steroids greatly increases the risk of heart disease, certain cancers, kidney damage, and personality disorders such as depression, mania, psychoses, and extreme aggression. Most official athletic organizations, including the International Olympic Committee, ban anabolic steroid use. Competing athletes are tested frequently for the presence of steroids in their blood. In the 1988 Olympics, Canadian track competitor Ben Johnson was forced to forfeit his gold medal and the rec-

ords he set because tests indicated that he had used anabolic steroids.

Other Hormones

We have discussed all of the major endocrine glands and "classical" hormones in this chapter, but there are many hormones we have not mentioned. Examples include the numerous hormones produced in the digestive tract that help organize the way the gut processes food. Even the heart has endocrine functions. When blood pressure rises and causes the walls of the heart to stretch, certain cells in the walls of the heart release **atrial natriuretic hormone**. This hormone increases the excretion of sodium ions and water by the kidneys, thereby lowering blood volume and blood pressure. As we discuss the physiology of the organ systems of the body in the chapters that follow, we will frequently mention hormones that those organs produce.

RECEIVING AND RESPONDING TO HORMONES

In discussing the functions of many hormones in this chapter, we have not yet addressed the question of how these chemical messages are read by the cells that receive them. How do hormones induce their actions at the cellular and molecular levels? We start to answer this question by dividing hormones into two groups according to where they bind their re-

ceptors. **Water-soluble hormones**, such as peptide and protein hormones, do not cross cell membranes readily, and their receptors are membrane proteins with binding regions that project from the cell surface. **Lipid-soluble hormones**, such as the steroids and thyroxine, can easily pass through cell membranes, and their receptors are in the cytoplasm or the nucleus of the cell.

Receptors for Water-Soluble Hormones and Second Messengers

Water-soluble hormones bind with receptors on the surface of the target cells. The receptors are glycoproteins that have a binding domain that projects beyond the outside of the cell membrane, a transmembrane domain, and a catalytic domain that extends into the cytoplasm of the cell. Directly or indirectly, the catalytic domain of most receptors of water-soluble hormones initiates cell responses by activating a protein kinase. There are many different kinds of protein kinases, but they all phosphorylate proteins; they catalyze the transfer of phosphate groups from ATP to specific proteins. In some cases the added phosphate activates the protein, and in other cases phosphorylation inactivates the protein. Thus the binding of the hormone to its receptor can result in activation or inhibition of a cellular process.

Some receptors for water-soluble hormones have their own protein kinase sites on their catalytic domains. When hormones bind to these receptors, the protein kinase sites are activated to catalyze the phosphorylation of cytoplasmic proteins. In this case the catalytic domain acts directly as a protein kinase.

Most receptors for water-soluble hormones act indirectly by way of **second messengers**. The hormone itself is the first messenger. When it binds to its receptor, the hormone stimulates a chain of reactions, producing small, diffusible molecules that serve as second messengers within the cell. The second messengers activate protein kinases. Cyclic AMP, or **cAMP** (cyclic adenosine monophosphate) is a well-studied second messenger that activates a wide range of protein kinases in many different kinds of cells.

The second messenger role of cAMP was discovered by E. W. Sutherland of Washington University, who began this work in the 1950s. He was investigating how epinephrine stimulates liver cells to break down carbohydrate storage molecules and liberate glucose. It became evident that there were a number of steps between the binding of the hormone by the receptor and the liberation of glucose. Sutherland was able to show that one of these steps was the control of enzyme activity through phosphorylation. This was the first demonstration of this common mechanism for regulating enzyme function. Sutherland then discovered that epinephrine could stimulate disrupted liver cells to release glucose as long as fragments of their plasma membranes were present. Showing hormone action in a cell-free system was also a major landmark in biochemistry. The third major discovery in this research program was that the hormone interacted with the membrane fragments to produce a small molecule that could stimulate the phosphorylation of enzymes in a liver cell mixture free of membranes. This molecule was identified as cAMP. For this work, E. W. Sutherland received the Nobel prize in 1971.

In the years after Sutherland's work, the list of systems known to be activated by cAMP grew rapidly. Many hormones in vertebrate tissues act via this second messenger. These systems include epinephrine's stimulation of the breakdown of stored carbohydrates and fats, adrenocorticotropin's stimulation of the production of glucocorticoids in the adrenals, luteinizing hormone's stimulation of androgen synthesis, and many more.

How can the same second messenger induce completely different responses in different cells? Different target cells have different protein kinases that are activated by the second messenger. The specificity of hormone action depends not only on the receptors that determine which cells respond to a given hormone, but also on what responding mechanisms a cell has.

The Role of G-Proteins in the Production of Second Messengers

A complex set of reactions takes place between binding of the hormone and production of the second messenger. When a receptor binds a hormone molecule, the receptor's shape (the tertiary structure; see Chapter 3) changes. In its new form, the receptor can interact with a second membrane protein, enabling that protein to bind a molecule of guanosine triphosphate (GTP). Proteins that bind GTP are called **G-proteins**.

Next, the subunit of the G-protein that is bound to the GTP separates and moves to another membrane protein, the enzyme **adenylate cyclase**. The complex consisting of G-protein and GTP activates the adenylate cyclase. The active adenylate cyclase catalyzes the conversion of ATP to cAMP within the cell. Thus, the G-protein is the link between the receptor and the production of the second messenger (Figure 36.16). As we will see in the next section, by activating a protein kinase that is otherwise inactive, the second messenger cAMP takes the next step leading to the target cell's response to the hormone.

There are many kinds of G-proteins. Some are even inhibitory, inactivating adenylate cyclase. The G-protein subunit that binds the GTP eventually hydrolyzes the GTP to GDP, thus inactivating itself and helping terminate hormone-induced responses.

G-proteins are important control elements in many cells and are the targets of some pathogenic organisms. Examples are the bacteria that cause the diseases cholera, whooping cough, and some forms of "traveler's diarrhea." Cholera bacteria produce a toxin that prevents the G-protein from hydrolyzing the GTP and inactivating itself. Thus this toxin causes cells to continue to produce high levels of cAMP. In the lining of the intestine where the cholera bacteria attack, cAMP stimulates the active transport of sodium ions into the gut. The toxin therefore causes massive sodium loss, with water following, resulting in severe, rapid dehydration and ionic imbalance that can lead to death in a day or two.

cAMP Targets and Response Cascades

The action of epinephrine on liver cells is an example of how cAMP-dependent protein kinases work (Figure 36.17). Epinephrine binds to its receptor on the plasma membrane, adenylate cyclase is activated, and cAMP is formed. cAMP activates a specific protein kinase that acts on two other enzymes, adding a phosphate group from ATP to each one. One of the newly phosphorylated proteins is the enzyme glycogen synthase. This enzyme catalyzes the joining of glucose molecules to synthesize the energy-storing molecule glycogen, but it is inactivated by the addi-

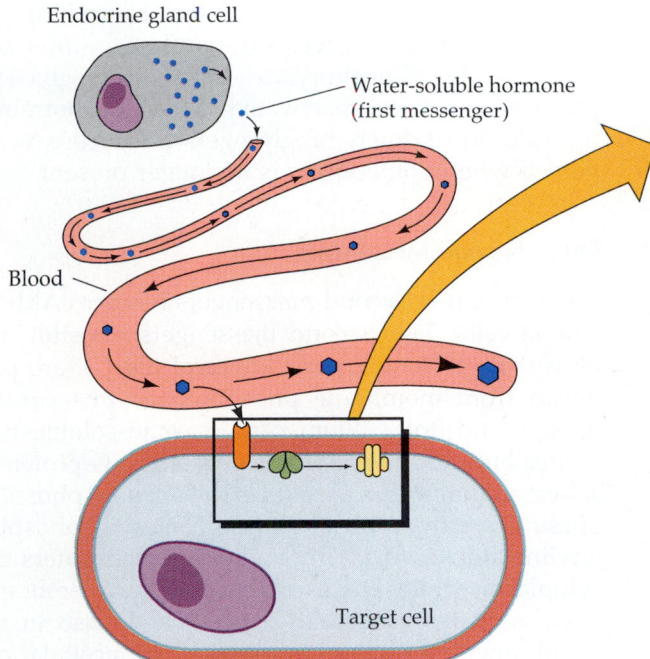

36.16 Second Messengers
Water-soluble hormones bind to receptors on the surface of their target cells. For receptors that act indirectly to phosphorylate protein kinases, this binding begins a chain of reactions that involves G-proteins and the production of a second messenger, such as cAMP.

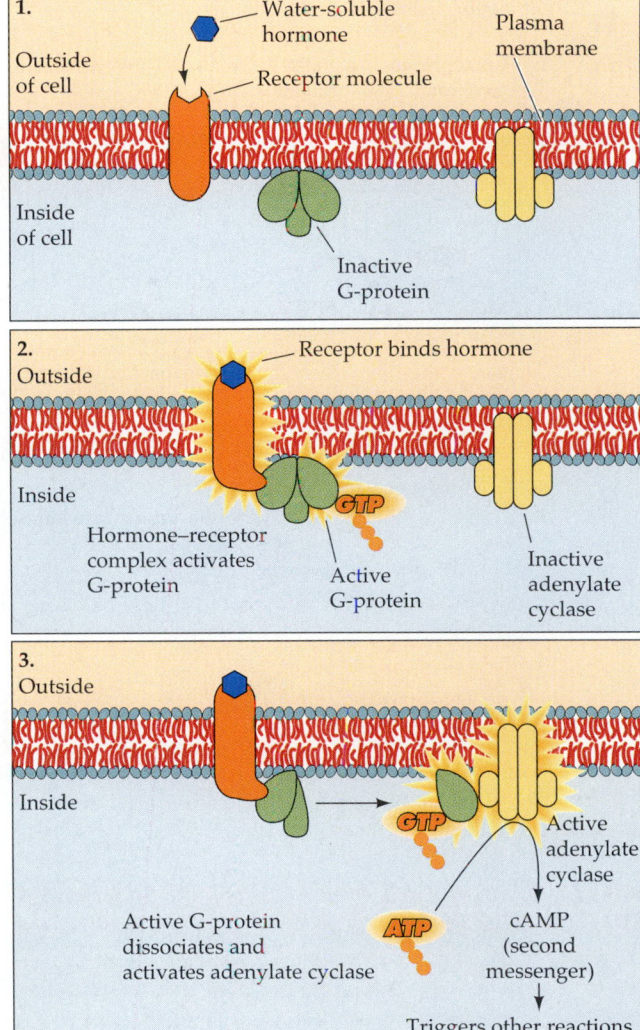

tion of the phosphate group. The other enzyme phosphorylated by the cAMP-activated protein kinase is phosphorylase kinase, which is activated by the addition of the phosphate group. Phosphorylase kinase, itself a protein kinase, catalyzes the phosphorylation of the enzyme glycogen phosphorylase. Glycogen phosphorylase participates in the breakdown of glycogen to glucose. Thus cAMP, through its effects on two protein kinases, inhibits the storage of glucose as glycogen and promotes the release of glucose through glycogen breakdown. Both of these effects increase glucose levels in liver cells and hence in the blood as well.

This cascade of regulatory steps amplifies the effect of a single hormone molecule. A molecule of epinephrine binds to a single receptor molecule, but the activated receptor activates many molecules (let's say ten) of the G-protein. Each activated G-protein activates one molecule of adenylate cyclase, but adenylate cyclase is an enzyme and can catalyze the production of perhaps 100 molecules of cAMP. Each molecule of cAMP activates only one protein kinase molecule, but each protein kinase may activate 100 phosphorylase kinase molecules, each phosphorylase kinase may activate 100 glycogen phosphorylase molecules, and each glycogen phosphorylase may catalyze the production of 100 molecules of glucose from glycogen. Thus an amplification of $10 \times 100 \times 100 \times$

100×100 is achieved; that is, each molecule of epinephrine can cause the production of about 1 billion molecules of glucose.

Unless there is a continuing supply of the hormone, it diffuses away from the receptor or is enzymatically degraded, allowing the receptor to revert to its inactive tertiary structure. In turn, the concentration of the complex of G-protein and GTP decreases as the enzyme inactivates itself by breaking down the GTP, and cAMP is no longer formed. The cAMP still present is quickly removed by the action of specific phosphodiesterases, enzymes that catalyze the conversion of cAMP to an inactive product.

In one other way the decrease in the second messenger, cAMP, causes the actions originally induced by the hormone to cease suddenly. Besides activating protein kinases, cAMP inhibits phosphoprotein phosphatase, whose role is to remove phosphate groups from proteins phosphorylated by protein kinases. Thus when cAMP levels fall, phosphoprotein phosphatase is no longer inhibited and is free to

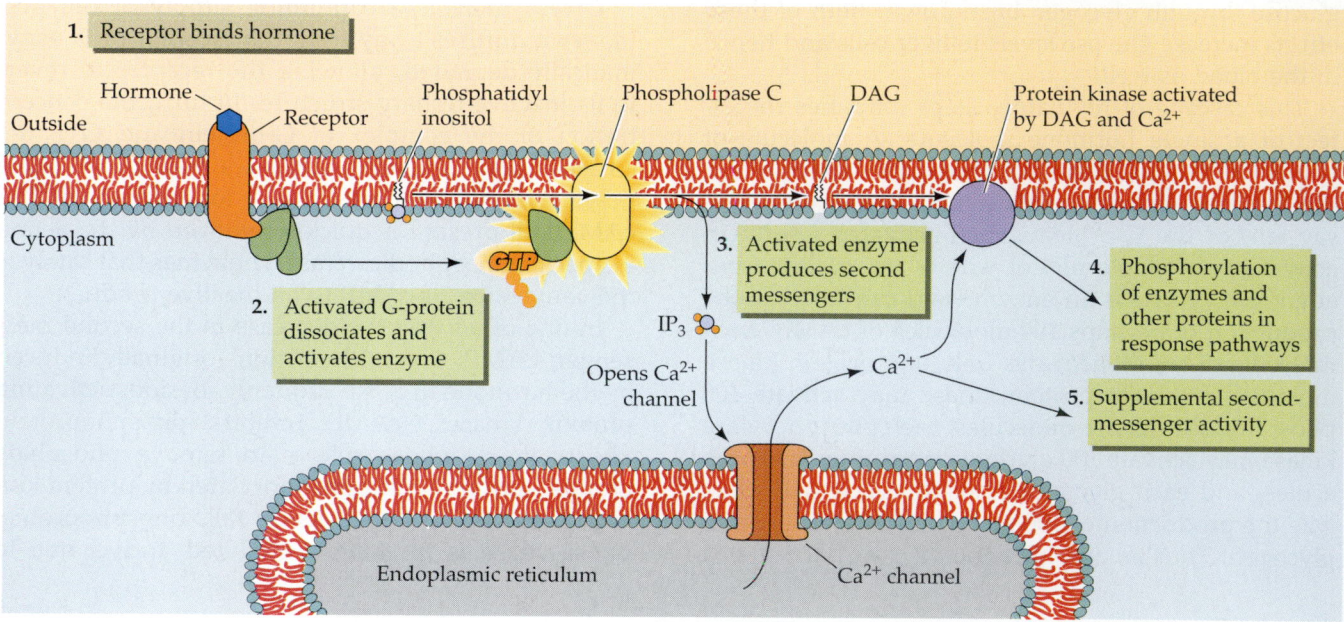

36.17   Epinephrine and cAMP
Epinephrine stimulates production of cAMP in liver cells. In liver cells cAMP then triggers a cascade of events that stimulates breakdown of glycogen to glucose and inhibits glycogen synthesis.

dephosphorylate the three enzymes controlling glycogen metabolism. Glycogen synthase is thus activated while phosphorylase kinase and glycogen phosphorylase are inactivated. Thus the hormone-induced breakdown of glycogen to glucose is reversed when epinephrine is no longer present.

Other Second Messengers

There are other second messengers besides cAMP in animal cells. Two second messengers, **inositol trisphosphate** (IP3) and **diacylglycerol** (DAG), are produced from membrane phospholipids that contain phosphoinositols. When certain water-soluble hormones bind to their receptor proteins, a G-protein is activated that in turn activates the enzyme phospholipase C. Active phospholipase C cleaves phosphatidylinositol to form IP3 and DAG. IP3 enters the cytoplasm while DAG remains in the membrane. DAG activates a protein kinase that is also in the membrane. IP3 causes the release of intracellular calcium ions that can stimulate the protein kinase as well as regulate other cellular functions that depend on calcium ions (Figure 36.18). Some hormones that act through the IP3 and DAG second-messenger system are norepinephrine and vasopressin.

Another second messenger, active in some target cells, is **cGMP** (cyclic guanosine monophosphate), a close chemical analog of cAMP. In many cases, cGMP acts in opposition to cAMP by activating a phosphodiesterase that breaks down cAMP. The effects of

36.18   The IP3 and DAG Second-Messenger System
A G-protein activates the enzyme phospholipase C, which catalyzes the hydrolysis of a membrane phospholipid to form inositol trisphosphate (IP3) and diacylglycerol (DAG). DAG stays in the membrane and IP3 enters the cytoplasm. Both serve as second messengers.

insulin on some target cells are mediated by cGMP; that is, insulin is the first messenger and cGMP the second messenger.

Calcium Ions

Calcium ions (Ca^{2+}) mediate many responses of different kinds of cells. As we shall see in Chapters 38 and 40, Ca^{2+} ions play crucial roles in the functions of both nerve cells and muscle cells. In some cases, such as activating protein kinase C (see Figure 36.18) and controlling some membrane ion channels, Ca^{2+} acts directly. In other situations, however, Ca^{2+} must combine with a calcium-binding protein, of which the most widely distributed is **calmodulin**. Calmodulin is activated by binding with Ca^{2+}, and the active complex can then trigger cell responses including activation of more protein kinases, smooth muscle contraction, microtubule assembly, protein synthesis, and various secretory events. Another calcium-binding protein, troponin, regulates a key reaction in the contraction of the skeletal muscles of vertebrates.

Lipid-Soluble Hormones

Steroid hormones—such as estrogens, progesterone, and the hormones of the adrenal cortex—as well as thyroxine, generally do not react with receptors on the target-cell surface (although it is now known that there are some steroid receptors bound to plasma membranes). These hormones are all lipid-soluble, which, as you may recall from Chapter 5, means that they pass readily through the lipid-rich plasma membrane. They act by stimulating the synthesis of new kinds of proteins through gene activation rather than by altering the activity of proteins already present in the target cells.

Once inside a cell, a lipid-soluble hormone binds to a receptor protein in the cytoplasm (Figure 36.19). The presence of a receptor protein is what distinguishes a responsive cell from a nonresponsive cell. A receptor protein is specific for a particular hormone, and it changes shape when it binds its hormone. The hormone–receptor complex associates with acidic chromosomal proteins, and thus with the DNA of the chromosomes. The receptor protein itself cannot bind the acidic chromosomal proteins unless it has already bound a hormone molecule and undergone the necessary change in structure. Once associated with the chromosomal proteins, the hormone activates the transcription of certain genes into messenger RNAs, which are exported to the cytoplasm and translated into specific proteins.

The actions of lipid-soluble hormones are slower and last longer than the actions of water-soluble hormones. When water-soluble hormones bind to their receptors on cell surfaces, some induce changes in membrane permeabilities to ions and some activate

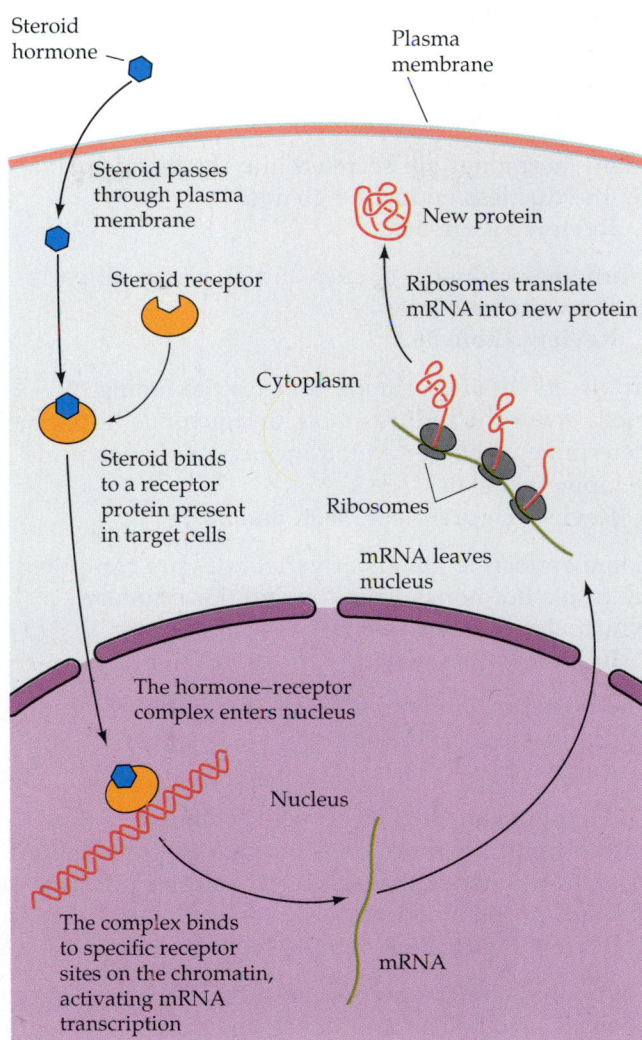

36.19   Action of Lipid-Soluble Hormones
The receptors of lipid-soluble hormones, such as steroids, are inside of cells. These hormones activate gene transcription.

or inactivate enzymes within the cell. In general, these are much more rapid and fleeting responses than the changes in gene expression induced by lipid-soluble hormones.

Hormones in Control and Regulation

In the last chapter we discussed the importance of control and regulation in animal physiology. Control and regulation require information, and that information is coded as electrical signals in the nervous system (see Chapter 38) or as chemical signals in endocrine systems. Throughout the rest of this section, we will see how these two types of information are used in physiological systems. In the next chapter, however, we will build on what we have learned about hormones by studying reproduction, since reproductive systems depend very heavily on hormonal mechanisms for control and regulation.

SUMMARY of Main Ideas about Animal Hormones

Hormones are chemical messages from one cell to another.

Many hormones are secreted into the bloodstream by the ductless endocrine glands.
Review Figure 36.2

Hormones influence tissues and organs in all parts of the vertebrate body.
Review Table 36.1

In insects, brain hormone controls the timing of molt, ecdysone induces molt, and juvenile hormone determines whether a molt includes a change in developmental state.
Review Figures 36.3, 36.4, and 36.5

Some endocrine glands of vertebrates are controlled by tropic hormones from the anterior pituitary, which also secretes several other hormones.
Review Figure 36.6

The posterior pituitary secretes two hormones produced in the hypothalamus.
Review Figure 36.6

The hypothalamus controls the secretion of some anterior pituitary hormones through the production of neurohormones that reach the anterior pituitary through portal blood vessels.
Review Figure 36.8 and Table 36.2

Thyroxine, a hormone produced by the thyroid, influences metabolism and development.
Review Figure 36.10

Parathormone, a hormone from the parathyroid glands, and calcitonin, from the thyroid, control calcium metabolism.
Review Figure 36.11

Insulin and glucagon, hormones from the pancreas, regulate blood sugar.

Hormones of the adrenal medulla stimulate "fight-or-flight" responses.

Hormones of the adrenal cortex control stress responses and salt balance.
Review Figures 36.12 and 36.13

Hormones of the gonads control sexual development, reproduction, and sexual behavior.
Review Figure 36.14

A hormone is either water-soluble or lipid-soluble.

Water-soluble hormones do not cross cell membranes readily, and their receptors project from the surface of the target cell.
Review Figure 36.16

After a water-soluble hormone binds to its receptor, the receptor relays the message to a second messenger within the cell.
Review Figures 36.16 and 36.18

Lipid-soluble hormones pass easily through cell membranes, and their receptors are in the cytoplasm or nucleus.
Review Figure 36.19

cAMP is an important second messenger that can trigger a cascade of intracellular events that amplify the response of the cell to a hormone molecule.
Review Figure 36.17

SELF-QUIZ

1. Which of the following statements is true for all hormones?
   a. They are secreted by glands.
   b. They have receptors on cell surfaces.
   c. They may stimulate different responses in different cells.
   d. They target cells distant from their site of release.
   e. When the same hormone occurs in different species, it has the same action.

2. The hormone ecdysone
   a. is released from the posterior pituitary.
   b. stimulates molt and metamorphosis in insects.
   c. maintains an insect in larval stages unless brain hormone is present.
   d. stimulates secretion of juvenile hormone from the prothoracic glands.
   e. keeps the insect exoskeleton flexible to permit growth.

3. The posterior pituitary
   a. produces oxytocin.
   b. is under the control of hypothalamic releasing neurohormones.
   c. secretes tropic hormones.
   d. secretes neurohormones.
   e. is under feedback control by thyroxine.

4. Growth hormone
   a. can cause adults to grow taller.
   b. stimulates protein synthesis.
   c. is released by the hypothalamus.
   d. can be obtained only from cadavers.
   e. is a steroid.

5. Both epinephrine and cortisol are secreted in response to stress. Which of the following statements is also *true for both* of these hormones?
   a. They act to increase blood glucose.
   b. The receptors are on the surfaces of target cells.
   c. They are secreted by the adrenal cortex.
   d. Their secretion is stimulated by adrenocorticotropin.
   e. They are secreted into the blood within seconds of the onset of stress.

6. Prior to puberty
   a. the pituitary is secreting luteinizing hormone and follicle-stimulating hormone, but the gonads are unresponsive.
   b. the hypothalamus does not secrete much gonadotropin-releasing hormone.
   c. males can stimulate massive muscle development through a vigorous training program.

d. testosterone plays no role in development of the male sex organs.
   e. genetic females will develop male genitals unless estrogen is present.

7. Which of the following is *not* true of cyclic AMP?
   a. It is broken down by adenylate cyclase.
   b. It is involved in the chain of events whereby epinephrine stimulates liver cells to break down glycogen.
   c. It is a second messenger mediating intracellular responses to many hormones.
   d. Many of its effects are mediated by protein kinases.
   e. A molecule of cAMP activates only a single protein kinase molecule.

8. Steroid hormones
   a. are all produced by the adrenal cortex.

b. have only cell surface receptors.
   c. are lipophobic.
   d. act through altering the activity of proteins in the target cell.
   e. act through stimulating the production of new proteins in the target cell.

9. Which is a likely cause of goiter?
   a. The thyroid gland is producing too much parathormone.
   b. Circulating levels of thyrotropin are too low.
   c. An inadequate supply of functional thyroxine.
   d. An oversupply of functional thyroxine.
   e. Too much iodine in the diet.

10. Parathormone
    a. stimulates osteoblasts to lay down new bone.
    b. reduces blood calcium levels.
    c. stimulates calcitonin release.
    d. is produced by the thyroid gland.
    e. is released when blood calcium levels fall.

FOR STUDY

1. Compare the mechanisms of action of peptide and steroid hormones.

2. Explain how both hyperthyroidism and hypothyroidism can cause goiter. Include the roles of the hypothalamus and the pituitary in your answers.

3. Explain the developmental abnormalities that can produce a genetic male with female secondary sexual characteristics. Describe the gonads of such an individual.

4. How did Sutherland's experiments demonstrate that the result of epinephrine combining with a membrane-bounded receptor is the production of a second messenger?

5. How can cAMP working through a protein kinase activate one enzyme while inactivating another enzyme in the same cell?

READINGS

Atkinson, M. A. and N. K. MacLaren. 1990. "What Causes Diabetes?" *Scientific American*, July. This paper reveals how malfunctions of the immune system cause insulin-dependent diabetes.

Berridge, M. J. 1985. "The Molecular Basis of Communication within the Cell." *Scientific American*, October. An authoritative account of second messengers and their roles in biological phenomena.

Bloom, F. E. 1981. "Neuropeptides." *Scientific American*, October. A description of the discovery, synthesis, distribution, and actions of peptides that serve as chemical messengers in the nervous system and as hormones in the body. Focuses on vasopressin, oxytocin, endorphins, enkephalins, and a few others.

Cantin, M. and J. Genest. 1986. "The Heart as an Endocrine Gland." *Scientific American*, February. Interesting account of the discovery and characterization of a hormone half a century after its existence was predicted.

Carafoli, E. and J. T. Penniston. 1985. "The Calcium Signal." *Scientific American*, November. Calcium as a second messenger; the roles of calcium-binding proteins.

Eckert, R., D. Randall and G. Augustine. 1988. *Animal Physiology*, 3rd Edition. W. H. Freeman, San Francisco. An excellent textbook; particularly useful with respect to second messengers and regulatory physiology.

Fernald, R. D. 1993. "Cichlids in Love." *The Sciences*, July/August. A fascinating study of "wimpy" and "macho" behavior among cichlid fish.

Snyder, S. H. 1985. "The Molecular Basis of Communication between Cells." *Scientific American*, October. An overview of the relationships between the nervous and endocrine systems; the focus is on chemical messengers and their molecular biology.

Vander, A. J., J. H. Sherman and D. S. Luciano. 1994. *Human Physiology: The Mechanisms of Body Function*, 6th Edition. McGraw-Hill, New York. Chapter 10 deals specifically with hormonal regulation.

Reproduction: An Essential Feature of Animal Life
Like humans, birds reproduce sexually. The cycle of birth, sexual maturation, and reproduction provides a continuous line of genetic information from generation to generation.

37

Animal Reproduction

We are sexually reproducing animals. Usually we think of eggs and sperm as the means for reproducing ourselves, but let's consider another viewpoint. Very early in the life of a new human embryo, during the first cell divisions, arises a population of cells that wanders through the body until the sex organs form, at which time these nomadic cells migrate to the sex organs. Eventually these cells give rise to the eggs and sperm that will produce the next generation. It is as if we the organisms are devices created by our sex cells to reproduce themselves. The sex cells themselves are part of a lineage of cells called the germ line, which is punctuated each generation by meiosis and recombination. This view of the relation between the organism and the sex cells reveals the centrality of reproductive processes in the lives of animals. Because these processes depend extensively on hormonal mechanisms of control and integration, we take up the topic of animal reproduction immediately following our discussion of animal hormones and before we learn about the nervous system.

ASEXUAL REPRODUCTION

Sexual reproduction is a nearly universal trait of animals, but many species can reproduce asexually as well. Offspring produced asexually are genetically identical to one another. Asexual reproduction is highly efficient because there is no mating, which requires energy and involves risks. In addition, all individuals of the population can convert resources into offspring, allowing the population to grow as rapidly as resources permit. However, asexual reproduction does not generate genotypic diversity. An asexually reproducing population does not have a wide variety of genotypes on which natural selection can act as the environment changes. Nevertheless, some animals reproduce asexually.

Budding

A common mode of asexual reproduction in simple multicellular animals is for a new individual to arise as an outgrowth of an older one—a process called **budding.** Some sponges form buds of undifferentiated cells on the outsides of their bodies. These buds grow by mitotic cell division, and the cells undergo differentiation before the buds break away from the parents and become independent sponges. Many freshwater sponges produce internal buds, or **gemmules.** A gemmule consists of several undifferentiated cells. Eventually the gemmules escape from the parent and become free-living individuals, genetically identical to the parent. Budding is part of the life cycle of some cnidarians, such as *Hydra* (Fig-

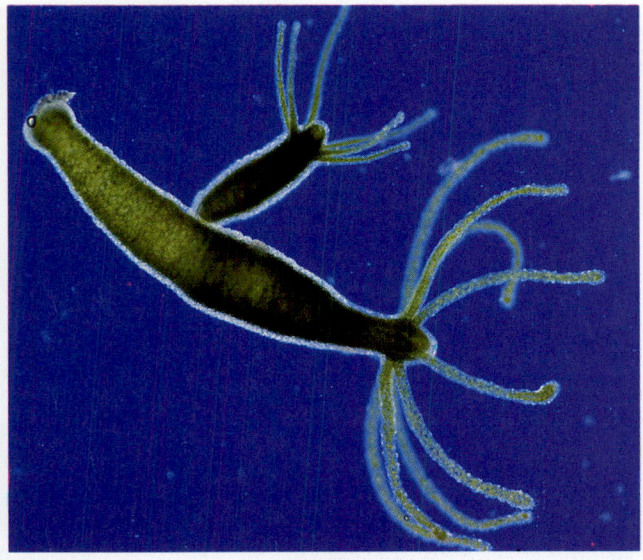

(b)

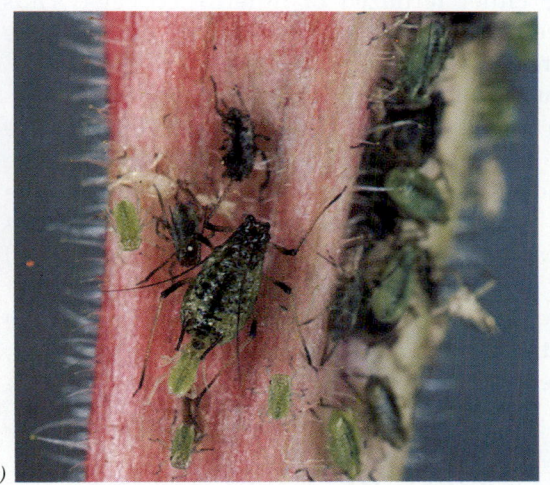

(a)

(c)

37.1 Asexual Reproduction in Animals
(a) Budding: A new individual forms as an outgrowth from an adult *Hydra*. (b) Regeneration: A single amputated arm from the sea star *Asterias rubens* develops into a new animal. (c) Parthenogensis: Aphids can hatch from unfertilized eggs.

ure 37.1a). The bud resembles the parent and may grow as large as the parent before it becomes an independent organism.

Regeneration

The cells in sponges and certain cnidarians that initiate budding are **totipotent;** that is, they have the ability to give rise to new, complete organisms. Totipotency is not a characteristic of most cells of most animals (see the discussion on determination and differentiation in Chapter 17). There are notable exceptions, however, in which pieces of animals can develop into whole animals. A dramatic example of such **regeneration** was unwittingly produced by a group of public officials who tried to protect oyster beds by instituting a search-and-destroy mission aimed at sea stars that were preying on the oysters. They "killed" these echinoderm predators by cutting them into pieces and dumping the pieces back into the sea. Sea stars, however, have remarkable abilities of regeneration. If they lose arms they regenerate new ones, and if a severed arm includes a portion of the central disc of the animal's body, it can regenerate into a complete sea star (Figure 37.1b). In their attempt to eliminate the sea stars, the public officials created large numbers of pieces capable of regeneration, and the predator population increased enormously.

Regeneration usually follows an animal's being broken by an outside force, but in some species the breakage is a normal event initiated by the animal itself. Certain species of segmented worms (annelids) develop segments with rudimentary heads bearing sensory organs, then they break apart. Each fragmented segment forms a new worm.

Parthenogenesis

Parthenogenesis is a type of asexual reproduction in which the offspring develop from unfertilized eggs (Figure 37.1c). Many animals, especially arthropods, reproduce parthenogenetically, as do some species of fish, amphibians, and reptiles. Most species that reproduce parthenogenetically also engage in sexual reproduction or sexual behavior. Aphids, for example, are parthenogenetic in the spring and summer, multiplying rapidly while conditions are favorable. Some of the unfertilized eggs laid in spring and summer develop into male aphids, others into females. As conditions become less favorable, the aphids mate and the females lay fertilized eggs. These eggs do not hatch until the following spring, and they yield only females. Species capable of parthenogenesis frequently switch from asexual to sexual reproduction when environmental conditions change. Parthenogenesis is used when conditions are stable and favorable; sexual reproduction introduces genotypic variability when conditions are changing, stressful, or unpredictable.

In some species parthenogenesis is part of the mechanism that determines sex. For example, in ants and in most species of bees and wasps, females develop from unfertilized eggs and are haploid and males develop from fertilized eggs and are diploid. Most females are sterile workers, but a select few become fertile queens. After a queen mates, she has a supply of sperm that she controls, enabling her to produce either fertilized or unfertilized eggs. Thus the queen determines when and how much of the colony resources are expended on males.

Parthenogenetic reproduction in some species requires a sex act even though this act does not fertilize the egg. The eggs of parthenogenetically reproducing ticks and mites, for example, develop only after the animals have mated, but the eggs remain unfertilized. Some species of beetles have no males at all and can reproduce only parthenogenetically, yet their eggs require sperm to trigger development. These beetles therefore mate with males of closely related, but different, species.

SEXUAL REPRODUCTIVE SYSTEMS OF ANIMALS

The enormous genotypic diversity in most sexually reproducing species derives from the independent assortment of chromosomes and the recombination of alleles on those chromosomes. As you know from Chapter 10, an animal has two alleles for each gene, and many pairs of alleles are heterozygous. A sexually reproducing animal packages single alleles from each pair into reproductive cells called **gametes.** Because each gamete receives one or the other allele of a gene, each gamete contains a complete set of genes, with a unique assortment of alleles. Sexual reproduction requires **fertilization**—the fusion of two gametes (almost always from different individuals) to form a **zygote.** The zygote receives half of its alleles from each parent and therefore has a new, unique genotype. Natural selection acts on the genotypic diversity produced by this process. Individuals that have genotypes best suited to environmental conditions are the most likely to survive and produce the largest number of offspring.

Both sexes, female (♀) and male (♂), produce haploid gametes from germ cells. The tiny gametes of males are called **sperm;** they move by beating their flagella. The much larger female gametes are called **eggs,** or **ova** (singular: ovum) and are nonmotile (Figure 37.2). Sperm and eggs are produced in the primary sex organs, the **gonads.** Male gonads are **testes** (singular: testis), and female gonads are **ovaries.** In addition to primary sex organs, most animals (except sponges and cnidarians) have accessory sex organs, including ducts, glands, and structures that deliver

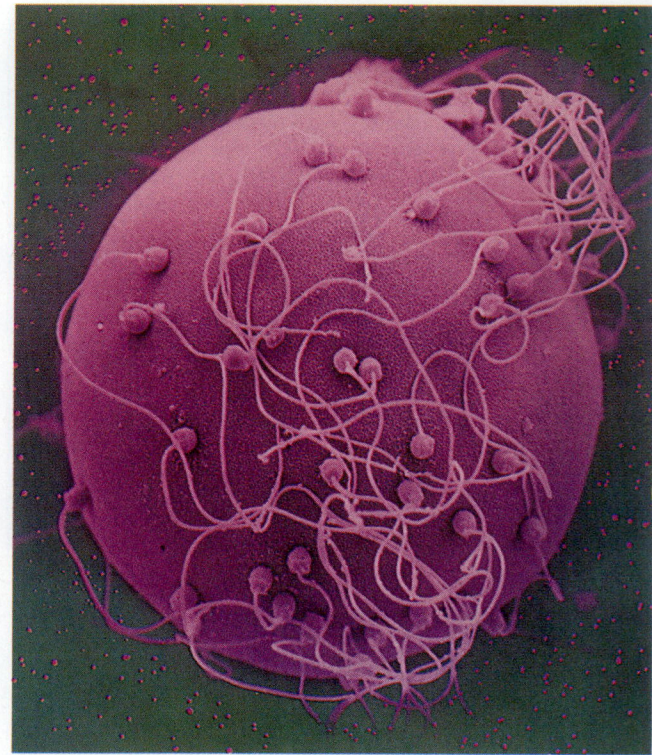

37.2   Gametes Differ in Size
Male gametes—the sperm—are small and motile, propelled by long flagella. The female gamete—the egg—is large and provisioned to nourish the early stages of the embryo's development. In this micrograph of mammalian fertilization, many sperm have attached to a single egg, but only one sperm cell will enter and fertilize it.

and receive gametes. The primary and accessory sex organs of an animal constitute its reproductive system.

Gametogenesis

Gametogenesis is the formation of gametes, and in all animal species except sponges, it takes place in the gonads. As we described in the introduction to this chapter, the gametes derive from a special lineage of cells called the **germ line.** Those cells are not produced by the gonads; they come to reside in the gonads only after the gonads have formed in the embryo. The germ cells are diploid, and they proliferate by mitosis. The cells resulting from the mitotic proliferation of germ cells in the gonads of females are called **oogonia** (singular: oogonium), and those in the gonads of males are called **spermatogonia** (singular: spermatogonium). Meiosis, the next step in gametogenesis, reduces the chromosomes to the haploid number, and the haploid cells mature into sperm and ova. Meiosis is central to the formation of both sperm and ova; you might review the discussion of meiosis in Chapter 9 before reading further.

Spermatogenesis is the process by which sperm form from germ cells. In mammals, sperm are produced in the tubules within the testes. The process begins when the diploid spermatogonia near the wall of a seminiferous tubule increase in size and divide by mitosis to become **primary spermatocytes.** The primary spermatocytes undergo the first meiotic division to form **secondary spermatocytes,** which are haploid. (Recall that the first meiotic division halves the number of chromosomes.) These cells remain connected by bridges of cytoplasm after each division. The second meiotic division produces four hap-

loid **spermatids** for each primary spermatocyte that entered meiosis (Figure 37.3a).

Spermatids differ from one another genetically because the random orientation of chromosomes at the first meiotic metaphase shuffles the parental genomes. A given spermatid contains some maternal chromosomes and some paternal chromosomes; the particular combination is a matter of chance. Crossing over during the first meiotic division also contributes to the genetic differences among spermatids.

As the spermatocytes develop into spermatids and the spermatids develop into sperm, they move pro-

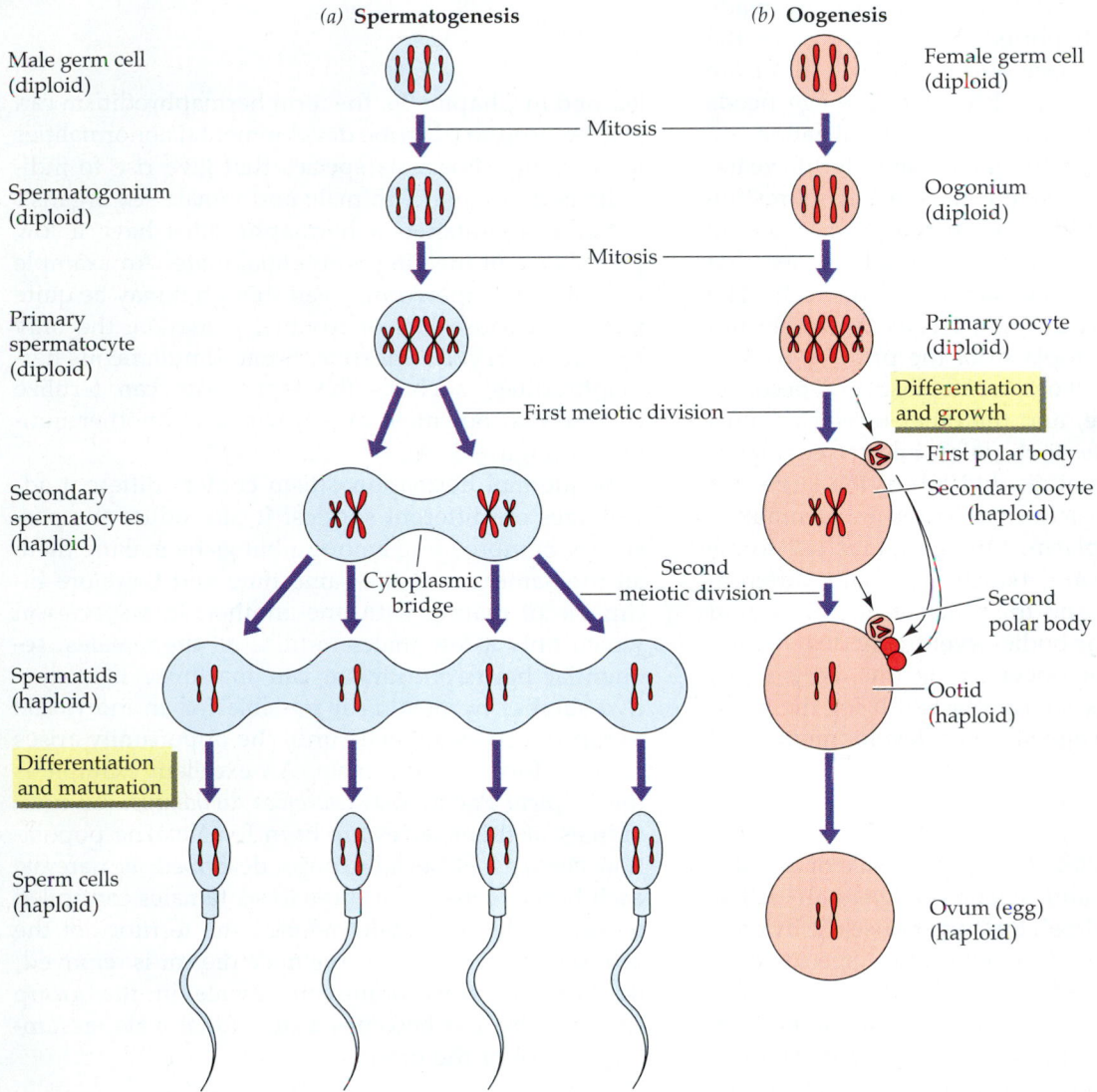

(a) **Spermatogenesis**   *(b)* **Oogenesis**

Male germ cell (diploid) — Mitosis — Female germ cell (diploid)

Spermatogonium (diploid) — Mitosis — Oogonium (diploid)

Primary spermatocyte (diploid) — First meiotic division — Primary oocyte (diploid)

Differentiation and growth

Secondary spermatocytes (haploid) — Second meiotic division — First polar body / Secondary oocyte (haploid)

Cytoplasmic bridge — Second polar body

Spermatids (haploid) — Ootid (haploid)

Differentiation and maturation

Sperm cells (haploid) — Ovum (egg) (haploid)

37.3   Gametogenesis
(a) The formation of haploid spermatids from diploid spermatogonia. Spermatids, all of which are different genetically, will differentiate into sperm. A mature mammalian sperm cell is seen in Figure 37.2 and diagrammed in Figure 37.6. *(b)* Diploid oogonia develop into larger primary oocytes that grow and accumulate materials and energy. The first meiotic division produces a haploid secondary oocyte and a small, adjacent, nucleus-containing polar body. The second meiotic division produces another polar body (the first polar body may also divide at this time) and the haploid egg.

gressively from the outermost region of the seminiferous tubule toward the center. The fully differentiated sperm are finally released from the Sertoli cells. The entire process takes about ten weeks. Each day a human male produces about 30 million sperm.

Oogenesis is the process of meiosis and development of the oogonia into eggs. Some oogonia develop into **primary oocytes,** which enter the first meiotic division but arrest in prophase I (Figure 37.3b). During this arrest, the primary oocytes enlarge, gaining yolk, ribosomes, cytoplasmic organelles, and energy stores. They accumulate rRNA, mRNA, and tRNA, as well as materials from the blood, and they form follicle cells, which surround them. Many of the lipids and proteins stored by vertebrate ova are made in the liver and transported to the ovaries in the bloodstream. The primary oocyte acquires all of the energy, raw materials, and RNA that the egg needs to survive its first cell divisions after fertilization.

Each month during a human female's fertile years, at least one primary oocyte comes out of its resting stage and matures into an egg. As this primary oocyte resumes meiosis, the nucleus completes its first meiotic division near the surface of the cell. The daughter cells of this division receive a grossly unequal share of the cytoplasm of the primary oocyte. One receives almost all of the cytoplasm and becomes the **secondary oocyte,** and the other receives almost none and forms the **first polar body** (see Figure 37.3b). The second meiotic division of the large secondary oocyte is also accompanied by an asymmetric division of the cytoplasm. One daughter cell forms the large, haploid **ootid,** which eventually differentiates into an ovum, and the other forms the **second polar body.** The polar bodies eventually degenerate, so the end result of oogenesis is one very large, haploid ovum that is well provisioned for the rapid divisions of the cleavage stage of development.

Sex Types

Most animals are a distinct sex type—male or female. Species having male and female members are called **dioecious** (from the Greek for "two houses"). By contrast, a single individual of some other species may possess both female and male reproductive systems. Such species are called **monoecious** ("one house") or **hermaphroditic** (from the name of a male, Hermaphroditus, whose body, according to Greek myth, was joined with that of a nymph). Almost all invertebrate groups have hermaphroditic species. An earthworm is an example of a **simultaneous hermaphrodite,** meaning that it is both male and female at the same time. When two earthworms mate, both are fertilized and produce offspring. Some animals are **sequential hermaphrodites,** being male and female at different times in their life cycles. As we

37.4 Hermaphroditic Mating
Although some hermaphroditic species can fertilize themselves, most must mate with another individual; the sea slug *Aplysia* often mates in groups. Here sea slugs form a mating chain in which each animal is functioning as a female for the animal behind and as a male for the animal in front.

learned in Chapter 36, the term hermaphroditism can also be used to describe developmental abnormalities in normally dioecious species that give rise to individuals that have both male and female sex organs.

Some simultaneous hermaphrodites have a low probability of meeting a potential mate. An example is a parasitic tapeworm. Even though it may be quite large and cause lots of trouble, it may be the only tapeworm in your intestine. Some simultaneous hermaphrodites, such as the tapeworm, can fertilize themselves, but most must mate with another individual (Figure 37.4).

Sequential hermaphroditism confers different advantages on different species. It can reduce the possibility of inbreeding among siblings by making them all the same sex at the same time and therefore incapable of mating with one another. In a species in which only a few males fertilize all the females, sequential hermaphroditism can maximize reproductive success by making it possible for an individual to reproduce as a female until the opportunity arises for it to function as a male. An excellent example is the tropical Pacific fish *Labroides dimidiatus*. All individuals of this species are born female. The population consists of social groups described as harems; each harem consists of three to six females controlled by one male. The male defends the territory of the group from intruders. If the male dies or is removed, the largest, most dominant female in the group changes sex and becomes a functional male, assuming control of the group.

Getting Eggs and Sperm Together

Fertilization can be external or internal. Sexually reproducing animals may release their gametes into the environment, where the meeting of gametes results in fertilization, or the male gametes may be inserted into the female's reproductive tract, where fertilization occurs. Animals that fertilize externally reproduce in aquatic habitats where gametes are not in danger of drying out. External fertilization is the more common pattern among simpler animals, especially those that are sessile.

External fertilization favors the evolution of traits that increase the probability that male and female gametes will meet. One simple adaptation is the production of huge numbers of gametes. A female oyster, for example, may produce 100 million eggs per year, and the number of sperm produced by a male oyster is astronomical. Numbers alone do not guarantee that gametes will meet, however. Also crucial are mechanisms of timing that synchronize the reproductive activities of the males and the females of a population. Seasonal breeders may use photoperiod cues, changes in temperature, or changes in the weather to time their production and release of gametes.

Sexual behavior plays an important role in bringing gametes together. Many species travel great distances to congregate with potential mates and release their gametes at the same time in a suitable environment. An excellent example is the remarkable migration of salmon. These fish hatch and go through juvenile stages in fresh water. They then migrate to the ocean, where they live and grow for three to five years. When finally ready to breed (spawn), they migrate back to the stream in which they were hatched, where they spawn and die.

Because gametes released into a dry environment die quickly, internal fertilization is a major adaptation for terrestrial life. Many aquatic species also practice internal fertilization. A great advantage of internal fertilization is the protection it provides for the early developmental stages of the organism. Animals have evolved an incredible diversity of sexual behaviors and accessory sex organs that facilitate internal fertilization. In general, a tubular structure, the **penis,** enables the male to deposit sperm in the female's accessory sex organ, the **vagina,** or in some species, the **cloaca** (plural: cloacae; a cavity common to the digestive, urinary, and reproductive systems).

Copulation is an act that permits sperm to move directly from the male's reproductive system into the female's reproductive system. The transfer of sperm can also be indirect. The males of some species of mites and scorpions (among the arthropods) and salamanders (among the vertebrates) deposit **spermatophores**—containers filled with their sperm—in the environment. When a female mite finds a spermatophore, she straddles it and opens a pair of plates in her abdomen so that the tip of the spermatophore enters her reproductive tract and allows the sperm to enter. Some female salamanders use the lips of their cloacae to scoop up the portion of the gelatinous spermatophore containing the sperm.

Male squids and spiders play a more active role in spermatophore transfer. The male spider secretes a drop containing sperm into a bit of web; then, with a special structure on a foreleg, he picks up the sperm-containing web and inserts it through the female's genital opening. The male squid uses one special tentacle to pick up a spermatophore and insert it into the female's genital opening. In the process, the tip of his tentacle may break off and remain in the female's body along with the sperm.

Most male insects copulate and transfer spermatophores to the female's vagina through a tubular penis. The genitalia (external parts of the sex organs) of insects often have species-specific shapes that match in a lock-and-key fashion (Figure 37.5). The incredible morphological diversity in genitalia of some groups of species has led to the hypothesis that the lock-and-key fit is a reproductive isolating mechanism (see Chapter 19). This idea is controversial, but at a minimum, the fit between male and female genitalia assures a tight, secure fit between the mating pair during the prolonged period of sperm transfer. The males of some insect species use elaborate structures on their penises to scoop out the female's reproductive tract, removing sperm deposited there by other males. Following this cleaning, a male transfers his own sperm into the tract.

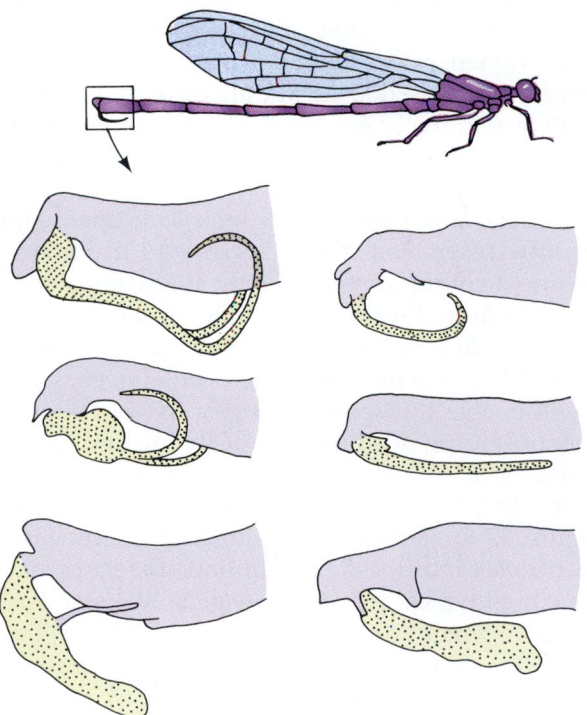

37.5 Species-Specific Penis Shapes
The penises of six species of *Argia*, a genus of damselflies, are remarkably different. Each fits into the corresponding female organ as a key in a lock to facilitate a tight union during a prolonged copulation. The structures shown in color are used to clear the female reproductive tract of sperm deposited by other males.

REPRODUCTIVE SYSTEMS IN HUMANS AND OTHER MAMMALS

So far we have seen only a small sampling of the fascinating diversity of animal reproductive systems. We will now look at the mammalian reproductive system in greater depth, using the human as our model. Many of the details are the same for other vertebrates.

The Male

The paired testes of mammals, except those of bats, elephants, and aquatic mammals, are lodged outside the body cavity in a pouch of skin, the **scrotum** (Figure 37.6a). Spermatogenesis takes place in most mammals only at a temperature slightly lower than normal body temperature. The scrotum keeps the testes at a temperature optimal for spermatogenesis. Muscles in the scrotum contract in a cold environment, bringing the testes closer to the warmth of the body; in a hot environment they relax, and the testes are suspended farther from the body.

A testis consists of tightly coiled **seminiferous tubules** within which spermatogenesis takes place. The tubule walls are lined with spermatogonia. In going from the tubule wall toward the center, you find germ cells in successive stages of spermatogenesis (see Figure 37.3a). Fully differentiated spermatids are shed into the lumen of the tubule. These germ cells are intimately associated with **Sertoli cells**, which nurture them (Figure 37.6b). Between the seminiferous tubules are clusters of cells that produce the male sex hormones.

Just after being produced by meiosis, a spermatid bears little resemblance to a sperm. As it differentiates into a sperm, its nucleus becomes compact, its motile flagellum develops into a tail, and most of its cytoplasm is lost. As the head of the sperm forms, it is capped by an **acrosome,** which contains enzymes that will enable the sperm to digest its way into an egg. Between the head and tail of the mature sperm is a midpiece containing two centrioles and mitochondria to provide energy for locomotion (bottom of Figure 37.6). The microtubules that extend from the centrioles into the flagellum have the same pattern as in all typical eukaryotic flagella, the standard 9 + 2 arrangement described in Chapter 4.

From the seminiferous tubules, sperm move into a storage structure called the **epididymis,** where they mature and become motile. The epididymis connects to the **urethra** by a tube called the **vas deferens.** The urethra comes from the bladder, runs through the penis, and opens to the outside of the body at the tip of the **penis.** The urethra is the common duct for the urinary and reproductive systems (see Figure 37.6a).

The shaft of the penis is covered with normal skin, but the tip, or **glans penis,** is covered with thinner, more sensitive skin that is especially responsive to sexual stimulation. A fold of skin called the foreskin covers the glans of the human penis. The practice of circumcision removes a portion of the foreskin. There is no strong rationale or justification for circumcision based on health, yet it remains a cultural or religious tradition for many people.

The penis becomes hard and erect during sexual arousal because blood fills shafts of spongy tissue that run the length of the penis (see Figure 37.6a). The presence of this blood creates pressure and closes off the vessels that normally drain the penis. Thus, the penis becomes engorged with blood, facilitating insertion into the vagina. Some species of mammals, but not humans, have a bone in the penis; even those species, however, depend on erectile tissue for copulation.

The culmination of the male sex act propels sperm through the vas deferens and the urethra. This process of sperm movement has two steps, **emission** and **ejaculation.** During emission, sperm and the secretions of several accessory glands move into the urethra at the base of the penis. Together, the sperm and these secretions constitute **semen,** the fluid that is ejaculated into the female's vagina. About 60 percent of the volume of the semen is seminal fluid, which comes from the **seminal vesicles.** Seminal fluid is thick because it contains mucus and protein. It also contains fructose, which serves as an energy reserve for the sperm, and modified fatty acids called prostaglandins that stimulate contractions in the female reproductive tract. Another source of secretions is the **prostate gland,** which produces a thin, milky, alkaline fluid. The prostate fluid helps to neutralize the acidity of the urethra and the female reproductive tract to create a favorable environment for the sperm. Prostate fluid also contains a clotting enzyme that works on the protein in the seminal fluid to convert the semen into a gelatinous mass.

Ejaculation, which follows emission, is caused by wavelike contractions of muscles at the base of the penis surrounding the urethra. The rigidity of the erect penis allows these contractions to force the semen through the urethra and out of the body.

The Female

Eggs (ova) are produced and released by the female gonads, the ovaries. The ovaries are paired structures in the lower part of the body cavity. Ovulation releases an egg from the ovary directly into the body cavity. Before it can float away, the released egg is swept into the fringed end of one of the paired tubes called **oviducts** (also known as fallopian tubes). Cilia lining the oviduct propel the egg slowly toward the

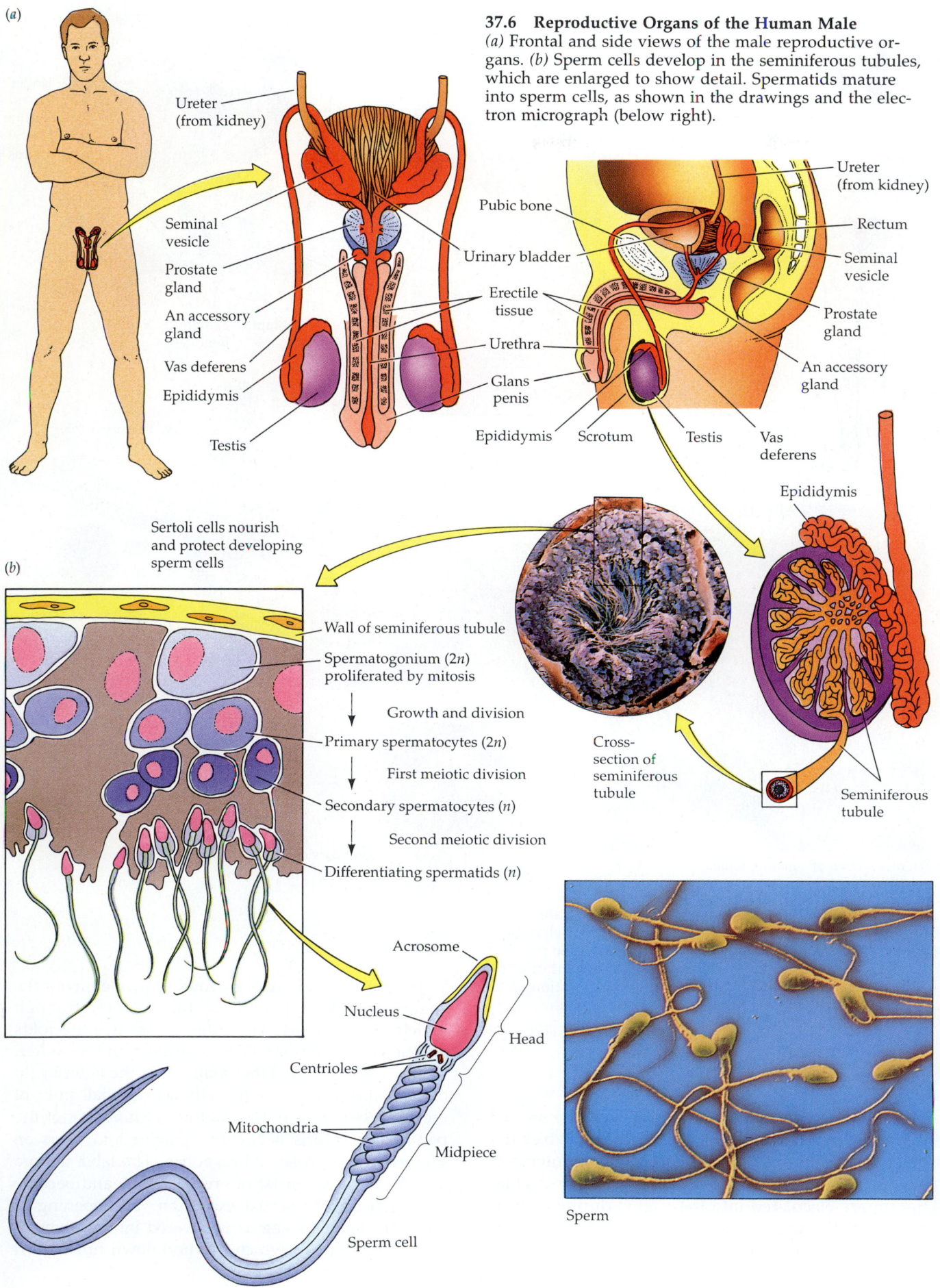

(a)

37.6 Reproductive Organs of the Human Male
(a) Frontal and side views of the male reproductive organs. (b) Sperm cells develop in the seminiferous tubules, which are enlarged to show detail. Spermatids mature into sperm cells, as shown in the drawings and the electron micrograph (below right).

Ureter (from kidney)

Seminal vesicle

Prostate gland

An accessory gland

Vas deferens

Epididymis

Testis

Pubic bone

Urinary bladder

Erectile tissue

Urethra

Glans penis

Epididymis   Scrotum   Testis   Vas deferens

Ureter (from kidney)

Rectum

Seminal vesicle

Prostate gland

An accessory gland

(b)

Sertoli cells nourish and protect developing sperm cells

Epididymis

Wall of seminiferous tubule

Spermatogonium (2n) proliferated by mitosis

Growth and division

Primary spermatocytes (2n)

First meiotic division

Secondary spermatocytes (n)

Second meiotic division

Differentiating spermatids (n)

Cross-section of seminiferous tubule

Seminiferous tubule

Acrosome

Nucleus

Centrioles

Mitochondria

Head

Midpiece

Sperm cell

Sperm

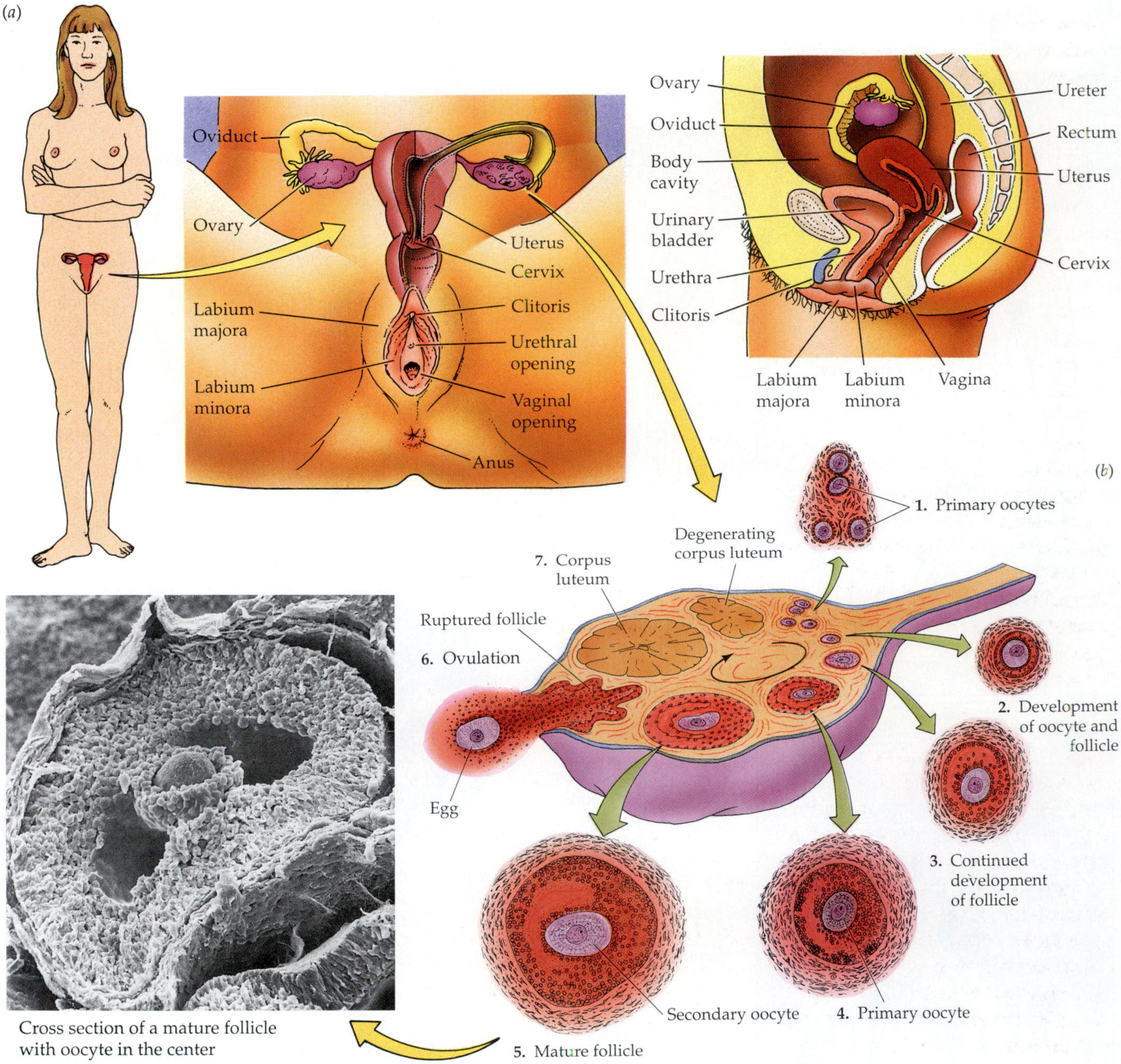

(a)

Oviduct

Ovary

Labium
majora

Labium
minora

Uterus
Cervix
Clitoris
Urethral
opening
Vaginal
opening

Anus

Ovary
Oviduct
Body
cavity
Urinary
bladder
Urethra
Clitoris

Ureter
Rectum
Uterus

Cervix

Labium   Labium   Vagina
majora   minora

(b)

1. Primary oocytes

7. Corpus
luteum

Ruptured follicle

6. Ovulation

Egg

Degenerating
corpus luteum

2. Development
of oocyte and
follicle

3. Continued
development
of follicle

Secondary oocyte   4. Primary oocyte

5. Mature follicle

Cross section of a mature follicle
with oocyte in the center

37.7 Reproductive Organs of the Human Female
(a) Frontal and side views of the female reproductive organs. *(b)* The developing follicle in the ovary. The progression of stages in the ovary is (1) development of a follicle, (2) ovulation, and (3) growth and degeneration of the corpus luteum. The micrograph shows a mature mammalian follicle; the oocyte is in the center.

uterus, or womb, which is a muscular, thick-walled cavity shaped like an upside-down pear. Babies develop in the uterus. At the bottom of the uterus is an opening, the **cervix,** that leads into the **vagina.** Sperm are ejaculated into the vagina during copulation, and the baby passes through it during birth. Figure 37.7*a* shows the female reproductive organs.

Two sets of skin folds surround the opening of the vagina and the opening of the urethra, through which urine passes. The inner, more delicate folds are the **labia** (singular: labium) **minora** and the outer, thicker folds are the **labia majora.** At the anterior tip of the labia minora is the **clitoris,** a small bulb of erectile tissue that is the anatomical homolog of the penis. The clitoris is highly sensitive and plays an important role in sexual response. The labia minora and the clitoris consist of erectile tissue and become engorged during sexual excitation. The opening of an infant female's vagina is covered by a thin membrane, the hymen, which has no known function. It is eventually ruptured by vigorous physical activity or first intercourse, but it can make first intercourse difficult or painful for the female.

To fertilize an egg, sperm swim from the vagina up through the cervix, the uterus, and most of the oviduct. The egg is fertilized in the upper region of the oviduct. The resulting zygote undergoes its first cell divisions, becoming a **blastocyst,** as it continues to move down the oviduct. When the blastocyst reaches the uterus it implants itself in the uterine lining, the **endometrium.** Some of the tissues of the blastocyst interact with the endometrium to form a structure called the **placenta,** which exchanges nutrients and waste products between the mother's blood and the baby's blood.

The Ovarian Cycle

At birth, a female has about a million primary oocytes in each ovary. By the time she reaches puberty (sexual maturity), she has only about 200,000 primary oocytes in each ovary; the rest have degenerated. During a woman's fertile years, only about 450 of these oocytes will mature completely into eggs and be released. When she is about 50 years old, she reaches **menopause,** the end of fertility. Only a few oocytes are then left in each ovary. Throughout a woman's life, oocytes are degenerating, and no new ones are produced.

A layer of cells surrounds each egg in the ovary. These cells, together with the eggs, constitute the functional unit of the ovary, the **follicle** (Figure 37.7b). Between puberty and menopause, 6 to 12 follicles mature within the ovaries of a human female each month. In each of these follicles, the egg enlarges and the surrounding cells proliferate. After about a week one of these follicles is larger than the rest and continues to grow, while the others cease to develop and shrink. In the remaining follicle, the follicular cells nurture the growing egg, supplying it with nutrients and even with macromolecules that it will use in early stages of development if it is fertilized. After two weeks of growth, the follicle ruptures and releases an egg. Following **ovulation,** as this release is called, the follicular cells continue to proliferate and form a mass of endocrine tissue about the size of a marble. This structure, which remains in the ovary, is the **corpus luteum.** It functions as an endocrine gland, producing estrogen and progesterone for about two weeks. It then degenerates unless the egg meets a sperm and is fertilized. We will return to the corpus luteum later in this chapter.

The Menstrual Cycle

Ovulation is part of the regular reproductive cycle in female animals. The human reproductive cycle is called the menstrual cycle because it ends conspicuously with **menstruation,** the sloughing off of the endometrium, the uterine lining. This sloughed-off tissue and blood from the uterine wall are lost through the vagina. The menstrual cycle consists of two coordinated cycles, one in the ovary, which results in the release of an egg each month, and one in the uterus, which prepares the endometrium to receive a blastocyst. The human reproductive cycle has a period of about 28 days or one month (a synonym for menstruation is menses, the Latin word for "months.") Some mammals have shorter ovarian cycles and others have longer ones. Rats and mice have ovarian cycles of about four days; many other mammalian species have only one cycle per year.

Most mammals do not end their cycles with menstruation; instead, the uterine lining is reabsorbed rather than being sloughed off. In these species the reproductive cycle is called the **estrous cycle** because its most striking event is the sexual receptivity of the female at the time of ovulation, called estrus, or "heat." When the female comes into estrus, she actively solicits male attention and may be aggressive to other females. She attracts males by releasing chemical signals as well as through behavior. The human female is unusual among mammals in that she is potentially sexually receptive throughout her reproductive cycle and at all seasons of the year.

Hormonal Control of the Menstrual Cycle

The ovarian and uterine cycles of human females are coordinated and timed by hormones. Gonadotropins secreted by the anterior pituitary are the central elements of this control. Prior to puberty, the secretion of gonadotropins is low, and the ovaries are inactive. At puberty, the hypothalamus increases its release of gonadotropin-releasing hormone, thus stimulating the anterior pituitary to secrete follicle-stimulating hormone and luteinizing hormone (Figure 37.8a). In response to these two gonadotropins, ovarian tissue grows and produces estrogen, and the follicles go through early stages of maturation. The rise in estrogen causes the development of secondary sexual characteristics, including the maturation of the uterus. Between puberty and menopause (at which time menstrual cycles cease), the interactions of gonadotropin-releasing hormone, the gonadotropins, and the sex steroids control the reproductive cycle.

Menstruation (menses) marks the beginning of the uterine and ovarian cycles (Figure 37.8b–d). A few days before menstruation begins, the anterior pituitary begins to increase its secretion of follicle-stimulating hormone and luteinizing hormone. In response to these gonadotropins, follicles mature in the ovaries and estrogen levels rise slowly. After about a week of growth, all but one of these follicles wither away.

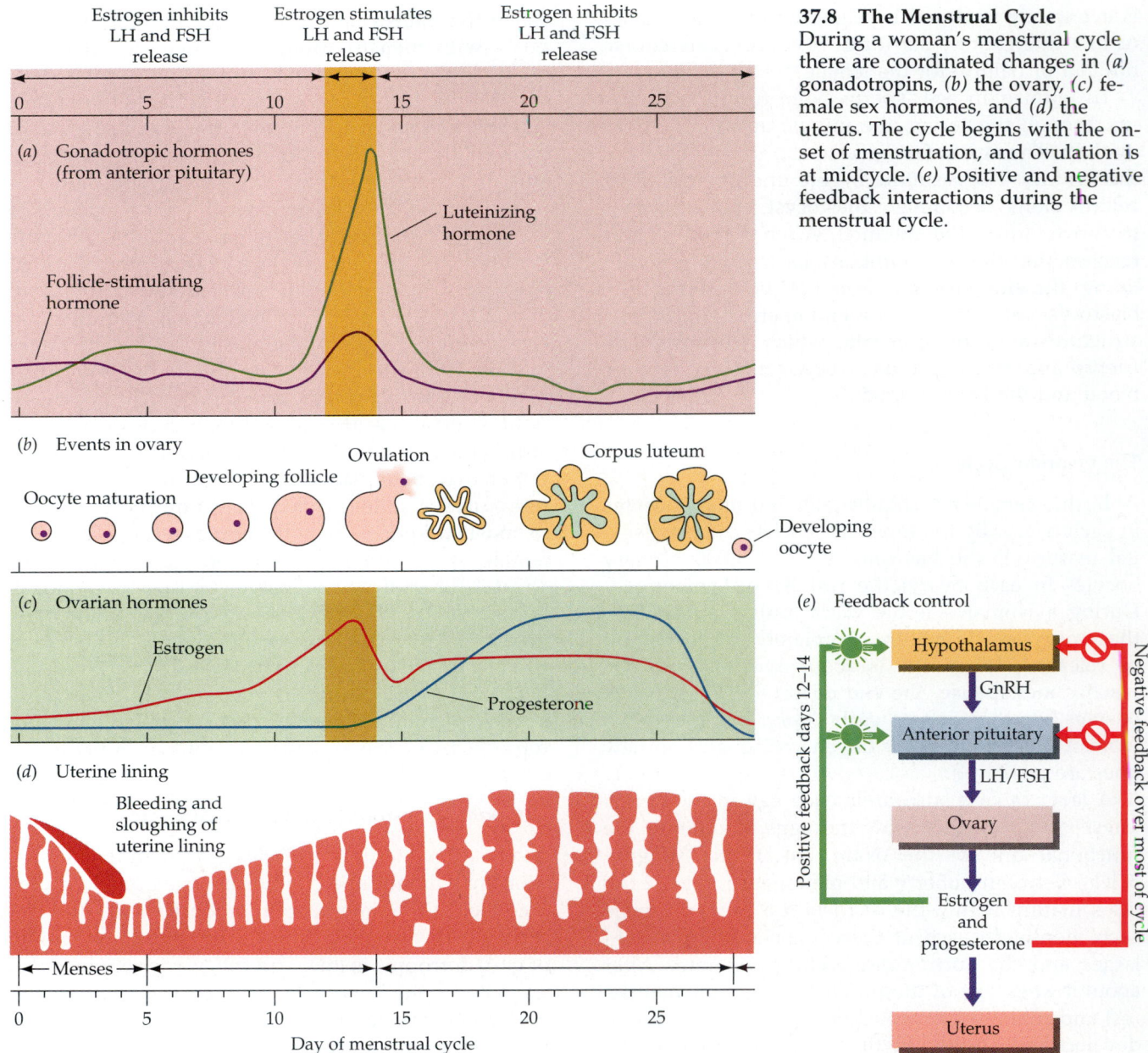

37.8   The Menstrual Cycle
During a woman's menstrual cycle there are coordinated changes in (a) gonadotropins, (b) the ovary, (c) female sex hormones, and (d) the uterus. The cycle begins with the onset of menstruation, and ovulation is at midcycle. (e) Positive and negative feedback interactions during the menstrual cycle.

The one follicle that is still growing secretes increasing amounts of estrogen, stimulating the endometrium to grow. Estrogen exerts a negative feedback on gonadotropin release by the pituitary during the first 12 days of the cycle. Then, on about day 12, estrogen exerts a positive rather a negative feedback on the pituitary (Figure 37.8e). As a result, there is a great surge of luteinizing hormone and, to a lesser extent, follicle-stimulating hormone. The luteinizing hormone surge triggers the mature follicle to rupture and release the egg, and it stimulates the follicle cells to develop into the corpus luteum and to secrete estrogen and progesterone.

Estrogen and especially progesterone secreted by the corpus luteum following ovulation are crucial to the continued development and maintenance of the endometrium. In addition, these sex steroids have negative feedback on the pituitary, inhibiting gonadotropin release and thus preventing new follicles from beginning to mature. If the egg is not fertilized, the corpus luteum degenerates on or about day 26 of the cycle. Without the production of steroids by the corpus luteum, the endometrium sloughs off. The decrease in circulating steroids also relieves the negative feedback on the hypothalamus and pituitary so that gonadotropin-releasing hormone, follicle-stimulating hormone, and luteinizing hormone all increase. The increase in these hormones induces the next round of follicles to develop, and the cycle begins again.

If the egg is fertilized, a zygote is created. The zygote undergoes numerous cell divisions, becoming a blastocyst as it travels down the oviduct. When the blastocyst arrives in the uterus and implants in the endometrium, a new hormone comes into play. A layer of cells covering the blastocyst secretes **human chorionic gonadotropin,** which keeps the corpus luteum functional. These same tissues also produce estrogen and progesterone. Eventually these tissues derived from the blastocyst take over for the corpus luteum. Continued high levels of estrogen and progesterone prevent the pituitary from secreting gonadotropins, and the ovarian cycle ceases for the duration of the pregnancy. This same mechanism is exploited by birth control pills, which contain synthetic hormones resembling estrogen and progesterone that prevent the ovarian cycle through negative feedback to the hypothalamus and pituitary.

Human Sexual Responses

The sexual responses of both women and men consist of four phases: excitement, plateau, orgasm, and resolution. As sexual excitement begins in a woman, her heart rate and blood pressure rise, muscular tension increases, her breasts swell, and her nipples become erect. Her external genitals, including the sensitive clitoris, swell as they become filled with blood, and the walls of the vagina secrete lubricating fluid that facilitates copulation.

As a woman's sexual excitement increases, she enters the plateau phase. Her blood pressure and heart rate rise further, her breathing becomes rapid, and the glans and shaft of the clitoris begin to retract—the greater the excitement, the greater the retraction. The sensitivity that once focused in the clitoris spreads over the external genitals, and the clitoris itself becomes even more sensitive.

Orgasm begins with a contraction of the outer third of the vagina lasting two to four seconds, followed by shorter contractions approximately one second apart. Orgasm may last as long as a few minutes, and, unlike men, some women can experience several orgasms in rapid succession. During the resolution phase, blood drains from the genitals and body physiology returns to close to normal. The resolution phase lasts approximately five to ten minutes after orgasm; if she does not experience orgasm, a woman's resolution phase may take 30 minutes or longer.

The cycle of a man's sexual responses is very similar to that of the woman. The excitement phase is marked by an increase in blood pressure, heart rate, and muscle tension. The penis fills with blood and becomes hard and erect. In the plateau phase, breathing becomes rapid, the diameter of the glans increases, and a clear lubricating fluid oozes from the penis. The testes also swell and the scrotum tightens. Pressure and friction against the nerve endings in the glans and in the skin along the shaft of the penis eventually trigger orgasm. Massive spasms of the muscles in the genital area and contractions in the accessory reproductive organs result in ejaculation. Within a few minutes after ejaculation, the penis shrinks to its normal size, and body physiology returns to resting conditions.

Unlike the sexual response of a female, the male sexual response includes a refractory period immediately after orgasm. During this period, which may last 20 minutes or more, a man cannot achieve a full erection or another orgasm, regardless of the intensity of sexual stimulation. Figure 37.9 shows the male and female response cycles.

37.9 Human Sexual Responses
The dashed lines show that both males and females may have repeated orgasms, but in the male they are separated by refractory periods during which sexual excitement cannot be maintained. Females have a greater diversity of response cycles, as shown by the three sets of lines. The cycle most similar to that of the male is shown in white. Alternatively, a female may experience sustained multiple orgasms (black curve) or may omit the plateau phase in a surge toward a very intense orgasm (red curve). Females do not have refractory periods.

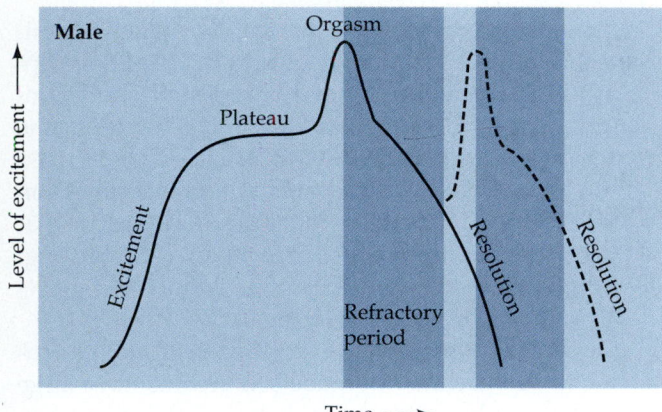

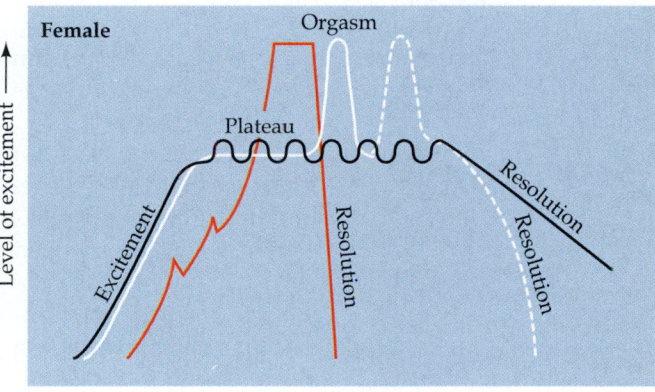

FERTILIZATION

As you learned in Chapter 17, the union of sperm and egg, or fertilization of the egg, results in a diploid zygote and initiates the development of the embryo. Fertilization is not a single event, but a complex series of processes. It begins with the juxtaposition of sperm and egg, accomplished in most species by sexual behavior. The final distance between sperm and egg must be bridged by the motility of sperm because eggs are universally unable to move. When egg and sperm finally meet, several events take place in sequence: the sperm is activated, the sperm gains access to the plasma membrane of the egg, sperm and egg membranes fuse, and the egg is activated. Egg activation sets up blocks to entry by additional sperm, stimulates the final meiotic division of the egg nucleus, and initiates the first stages of development. The last event of fertilization is fusion of the egg and sperm nuclei to create the diploid nucleus of the zygote. We will now look at each of these steps.

Activating the Sperm

Mammalian sperm face a formidable task after they are ejaculated into the female's reproductive tract. They must swim up from the vagina, through the uterus, and into the oviducts, where they might find an egg. They are aided in their journey by waves of muscular contractions of the vagina that are part of the female response to sexual stimulation and by stimulation from prostaglandins in the semen. Sperm can reach the upper ends of the oviducts within ten minutes of ejaculation. The mammalian egg, like any other cell, is bounded by a plasma membrane. Immediately surrounding the plasma membrane is a glycoprotein envelope called the **zona pellucida.** Surrounding all of that is a layer called the **cumulus,** consisting of follicle cells in a jelly matrix (Figure 37.10). When sperm are first deposited in the vagina, they are not capable of penetrating all these barriers to fertilize the egg. In the uterine environment, the sperm undergo **capacitation;** that is, they become capable of interacting with the egg and its barriers. Because the response of a capacitated sperm to an egg is mediated by the acrosome of the sperm, it is called the **acrosomal reaction.**

The acrosomal reaction is initiated in different places and at different times depending on the species. In all cases, however, the first step is the breakdown of the membranes bounding the sperm head and the acrosome, which releases the enzymes contained in the acrosome. One such enzyme, **hyaluronidase,** helps disperse the cumulus cells surrounding the egg by digesting the hyaluronic acid in the extracellular matrix that binds the cumulus cells together.

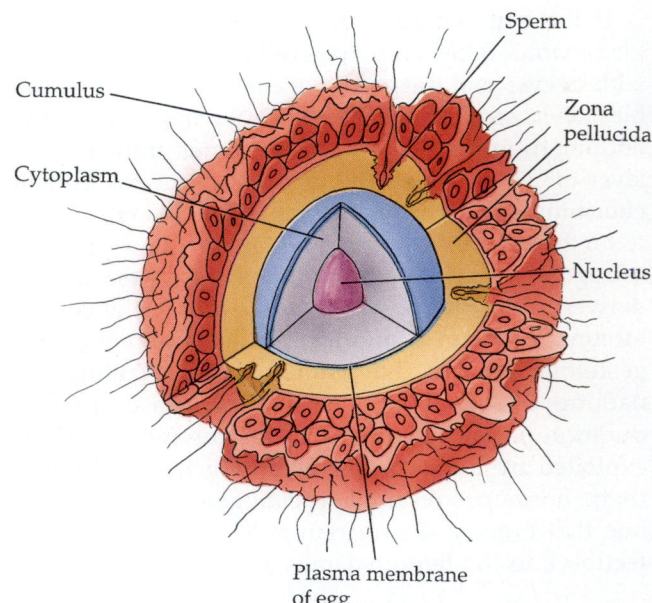

37.10   Barriers to a Sperm Cell
The human egg has several layers that the sperm cell must penetrate to reach the nucleus and fertilize the egg. Both the cumulus and the zona pellucida must be penetrated before the sperm can come into contact and eventually fuse with the egg's plasma membrane.

Other enzymes released from the acrosome also help disrupt the cumulus. Even though only one sperm fuses with the egg, acrosomal enzymes released from many sperm make the plasma membrane of the egg more accessible.

After the sperm penetrates the cumulus, the sperm head reacts enzymatically with the zona pellucida. The surface of the sperm head contains enzyme molecules and the zona pellucida contains substrate molecules. The enzyme binds to the substrate, linking the sperm to the egg. Acrosomal enzymes then digest a path through the zona pellucida so that the sperm can come into contact and eventually fuse with the plasma membrane of the egg.

Activating the Egg and Blocking Polyspermy

The unfertilized egg is metabolically sluggish, conserving its resources for the early stages of development. The binding of the sperm to the plasma membrane of the egg and the entry of the sperm into the egg activates the egg and initiates a programmed sequence of events. The first responses to fertilization are **blocks to polyspermy,** that is, mechanisms that prevent more than one sperm from entering the egg. If more than one sperm enters the egg, the resulting embryo will probably not survive.

Blocks to polyspermy have been studied intensively in sea urchins. Because sea urchins have large eggs that can be fertilized in dishes of seawater, they are excellent experimental subjects for studying fertilization. Within a tenth of a second after the first sperm enters a sea urchin egg, the egg takes in sodium ions, which changes the electric potential across the egg's plasma membrane. This change prevents the entry of additional sperm and is called the fast block to polyspermy.

There is also a slow block to polyspermy that takes 20 to 30 seconds (Figure 37.11). A sea urchin egg has a membranous structure called a **vitelline envelope,** rather than a zona pellucida, surrounding its plasma membrane. The vitelline envelope is bonded to the plasma membrane and has sperm-binding receptors on its surface. Just under the plasma membrane are cortical vesicles filled with enzymes. The sea urchin egg, like all other animal eggs, contains calcium stored in organelles within the cell. When a sperm enters, the egg releases calcium into its own cytoplasm. The increase in calcium causes the cortical vesicles to fuse with the plasma membrane and release their enzymes, which break the bonds between the vitelline envelope and the plasma membrane. Water then flows by osmosis into the space between the vitelline envelope and the plasma membrane, raising the vitelline envelope away from the plasma membrane to form the **fertilization membrane.** The enzymes from the cortical granules remove the sperm-binding receptors from the surface of the fertilization membrane and cause it to harden, preventing the passage of additional sperm through it.

The release of calcium ions within the egg following fertilization activates the egg metabolically. The pH of the cytoplasm increases, oxygen consumption rises, and protein synthesis increases. The fusion of sperm and egg nuclei does not take place until some time after the sperm enters the egg's cytoplasm— about 1 hour in sea urchins and about 12 hours in mammals. The egg nucleus must complete its second meiotic division before egg and sperm nuclei unite.

Most methods of birth control are focused on events surrounding fertilization. Physical barriers and behavioral changes are used to prevent the meeting of sperm and egg. Hormonal manipulations are employed to disrupt the ovarian cycle and prevent ovulation. Most recently, a chemical means of preventing implantation of the fertilized egg in the endometrium has been developed. Birth control methods and their relative effectiveness are discussed in Box 37.A.

CARE AND NURTURE OF THE EMBRYO

After development begins, the embryo requires access to oxygen, removal of carbon dioxide, a continuous source of nutrients, and a suitable physical environment. Two general patterns of care and nurture

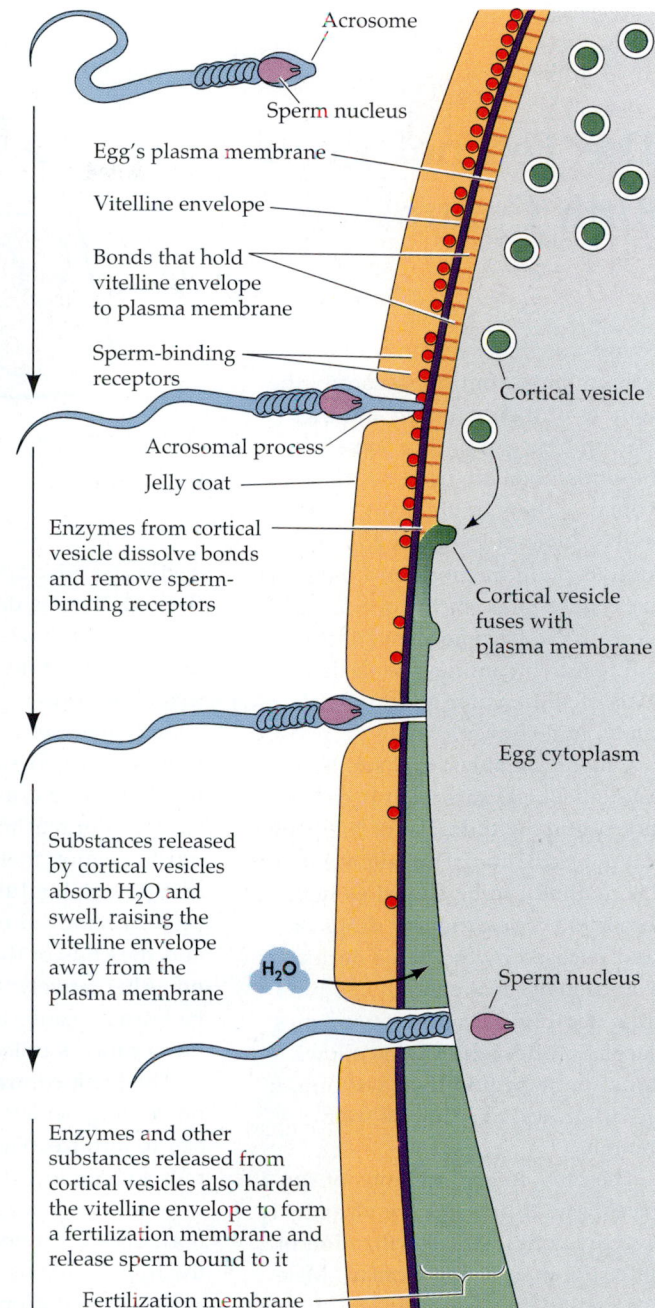

37.11 Slow Block to Polyspermy
As a sea urchin sperm approaches an egg to fertilize it, the contents of cortical vesicles remove the sperm-binding receptors and raise the vitelline envelope, which becomes the fertilization membrane. The fertilization membrane prevents the entry of any other sperm.

of the embryo have evolved in animals: oviparity and viviparity.

Oviparity

Oviparous animals lay eggs in the environment; their offspring go through the embryonic stages outside

BOX 37.A

The Technology of Birth Control

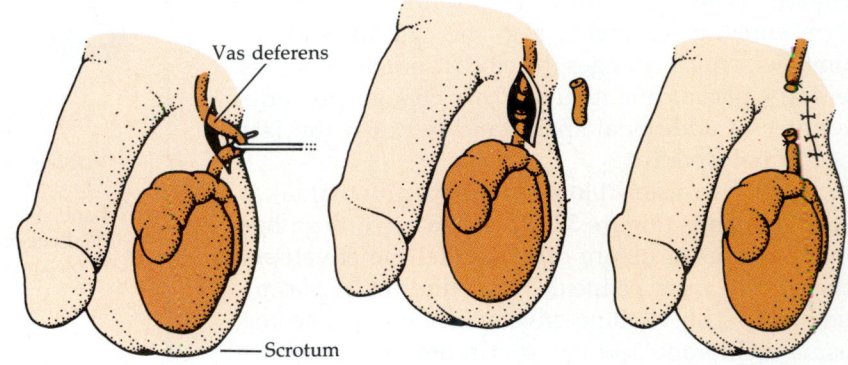

A vasectomy is a minor operation in which each vas deferens is cut and the cut ends are tied closed. The pathway from the testes to the penis is thus interrupted, and the man's ejaculate will not contain sperm.

People use many methods of contraception (birth control) to control the number of their children and the spacing between births. Some of these methods are used by the woman, others by the man. Here we review some of the most common contraception methods and their relative failure rates. Birth control techniques have become very effective, and research continues with the goals of still-greater effectiveness, safety, and convenience.

One of the oldest and simplest methods of contraception is **coitus interruptus,** withdrawal of the penis before ejaculation. The failure rate of this method can be almost as high as 40 percent. Even the few drops of fluid released by the penis during arousal may carry enough sperm to bring about fertilization. Sometimes ejaculation near the vagina allows some sperm to find their way into it, or withdrawal may not be soon enough.

The only certain methods of contraception (at present) are virtually irreversible ones—the **sterilization** of either the man or the woman. Male sterilization by vasectomy is a simple operation performed under a local anesthetic in a doctor's office. As the figure shows, each vas deferens is cut and the cut ends are then tied off. After this minor surgery, the man's ejaculate no longer contains

sperm, because sperm cannot pass through the vas deferens after leaving the epididymis, but the operation does not affect his hormone levels or sexual responses.

In female sterilization, the aim is to make it impossible for the egg to travel to the uterus and to block sperm from reaching the egg. The most common method is tubal ligation ("tying the tubes"). A small piece is removed from each oviduct, and the ends of the oviduct are tied off. Alternatively, the oviducts may be burned (cauterized) to seal them off, a process called endoscopy.

The **birth control pill** works by preventing ovulation, so there is no egg to fertilize. The most common pills contain low doses of synthetic estrogens and progesterones. These hormones exert negative feedback on the hypothalamus and the pituitary so that gonadotropin release is not sufficient to permit development of the ovum and the follicle. The ovarian cycle is suspended. On about day 5 of her menstrual cycle, the woman starts taking a pill each day, thus raising her hormone levels. After 20

or 21 days, the pill is discontinued, the lining of the uterus disintegrates, and slight menstrual bleeding occurs. There has been much discussion of negative side effects of oral contraceptives. These side effects include increased risk of blood clot formation, heart attack, and stroke, but they were mostly associated with pills containing higher hormone concentrations than are used in current pills. For pills in use today, there is a very low risk of these side effects, except for women over 35 years who smoke, for whom the risk is significantly greater. Risk of death from using the pill is less than that associated with a full-term pregnancy, and the pill is the most effective method of contraception other than sterilization.

Another highly effective method of contraception (with a failure rate varying from 1 percent to about 7 percent) is the intrauterine device, or **IUD.** The IUD is a small piece of plastic or copper that is inserted in the uterus. The IUD probably works by preventing implantation of the fertilized egg. Complications that can

the mother's body. Eggs are always much larger than sperm because they contain stored nutrients, or yolk, on which the entire course of development depends. Oviparous terrestrial animals, such as reptiles, birds, and insects, coat their eggs with tough, waterproof

membranes or shells to keep them from drying out and to protect against predators. The protective coverings of terrestrial eggs must, however, be permeable to oxygen and carbon dioxide.

Oviparous animals may engage in various forms

Common Methods of Birth Control

METHOD	MODE OF ACTION	FAILURE RATE (PREGNANCIES PER 100 WOMEN PER YEAR)
Coitus interruptus	Withdrawal of penis before ejaculation	10–40
Vasectomy	Prevents release of sperm	0.0–0.15
Tubal ligation	Prevents egg from entering uterus	0.0–0.05
"The pill"	Prevents ovulation	0–3
RU486	Prevents development of fertilized egg	0–15
Intrauterine device (IUD)	Prevents implantation of fertilized egg	0.5–6
Condom	Prevents sperm from entering vagina	3–20
Diaphragm/jelly; sponge	Prevents sperm from entering uterus; kills sperm	3–25
Vaginal jelly or foam	Kills sperm; blocks sperm movement	3–30
Rhythm method	Abstinence near time of ovulation	15–35
Douche	Supposedly flushes sperm from vagina	80
(Unprotected)	(No form of birth control)	(85)

arise from its use, including uterine infections and unintended sterility, have led many women to consider other options. Lawsuits against IUD manufacturers and subsequent insurance considerations have resulted in a decline in its use and manufacture in the United States, although it is still widely used in many countries.

Two primary mechanical methods of contraception have been in use for over a century. The **condom** ("rubber," or "prophylactic") is a sheath made of latex or of lamb intestinal material that can be fitted over the erect penis. A condom traps the ejaculate so that sperm do not enter the vagina. Condoms also help prevent the spread of sexually transmitted diseases such as AIDS, syphilis, and gonorrhea. In theory, the use of a condom can be highly effective, with a failure rate near zero; in practice, the failure rate is about 15 percent, because of faulty technique.

The **diaphragm** is a dome-shaped piece of rubber with a firm rim that fits over the woman's cervix and thus blocks sperm from entering the uterus. Smaller than the diaphragm is the **cervical cap,** which fits snugly just over the tip of the cervix. Both are treated first with contraceptive jelly or cream and then inserted through the vagina before intercourse. Failure rates are about the same as for condoms. A device simpler than the diaphragm is the contraceptive vaginal sponge. It is a circular, highly absorbent, polyurethane sponge permeated with a spermicide. Placed in the upper region of the vagina, it blocks, absorbs, and kills sperm. The sponge stays effective for about a day. It is easier to use than the diaphragm but has about the same failure rate.

Used alone, spermicidal foams, jellies, and creams have a failure rate of 25 percent or more. About all that

can be said for them is that they are more effective than nothing. Douching (flushing the vagina with liquid after intercourse) is, in spite of popular belief, essentially useless as a method of birth control. Remember that sperm can reach the upper regions of the oviducts ten minutes after ejaculation.

Some people attempt to avoid pregnancy by the **rhythm method.** The couple avoids sex from day 10 to day 20 of the menstrual cycle, when the woman is fertile. The use of a calendar to track the cycle may be supplemented by the basal body temperature method, which is based on the observation that a woman's body temperature drops on the day of ovulation and rises sharply on the day afterward. Other methods of predicting the time of ovulation are under development; significant improvements must be made if the rhythm method's failure rate (between 15 and 35 percent) is to be reduced. Added to the uncertainty of the timing of ovulation is the fact that the ovum remains viable for two to three days and sperm remain viable for up to six days in the female reproductive tract.

A recent addition to birth control technology is a drug, **Ru486,** developed in France. Ru486 is not a contraceptive pill, but a *contragestational* pill. It opposes the actions of progesterone produced by the corpus luteum. Progesterone is essential for maintenance of the uterine lining. If Ru486 is administered (usually with prostaglandins) at the time of the first missed menses after fertilization, it causes the uterine lining to be sloughed off along with the embryo, which is in very early stages of development and implantation.

of protective parental behavior focused on their eggs—nest construction and incubation are good examples—but until the eggs hatch, the embryos are entirely dependent on the nutrients stored in the egg at the time of fertilization. After leaving the protective coverings of the egg, the offspring may receive continuing parental care as it completes its development into a mature organism. Among mammals, only the monotremes—the spiny anteater and the duck-billed platypus—are oviparous (Figure 37.12a).

37.12 Animals Can be Classified by Where Development Takes Place
(a) Oviparous animals such as the duck-billed platypus lay eggs. (b) Marsupials are viviparous even though their infants are born in an extremely premature state. The infants develop in their mother's marsupium, or pouch, as this baby kangaroo (known as a "joey") has done.

(a)

(b)

Viviparity

Viviparous animals retain the embryo within the mother's body for part of its development. During this time, the embryo depends on nutrients supplied by the mother, not on nutrients stored in the egg. Viviparous animals are said to give birth to "live offspring"—a curious choice of words, because the offspring of oviparous animals are certainly not dead. Most viviparous animals are mammals, and most mammals are viviparous. Viviparous mammals have an enlarged and thickened portion of the female reproductive tract that holds the developing embryo; as you know, this structure is called the uterus.

In marsupials, the order of mammals that includes kangaroos and opossums, the uterus simply holds the embryo and does not have special adaptations to supply it with nutrients. Marsupials are very immature when born. They crawl into a pouch called a marsupium on the mother's belly, attach firmly to the nipple of a mammary gland, and complete their development outside of the mother's body (Figure 37.12b).

Mammals other than monotremes and marsupials are called **eutherian mammals.** A distinguishing feature of eutherian mammals is the intimate association of blood supplies of mother and embryo in the placenta. Nutrients pass from mother to embryo and wastes pass from embryo to mother through the placenta. We will discuss the structure of the placenta in the next section.

The eggs of some fishes, amphibians, and reptiles are fertilized internally and then retained within the body of the female until they hatch. The young then leave the mother's body. In such cases, the developing embryos receive all their nutrition from the yolk stored in the eggs, which makes this very different from viviparity. This reproductive pattern is called **ovoviviparity.**

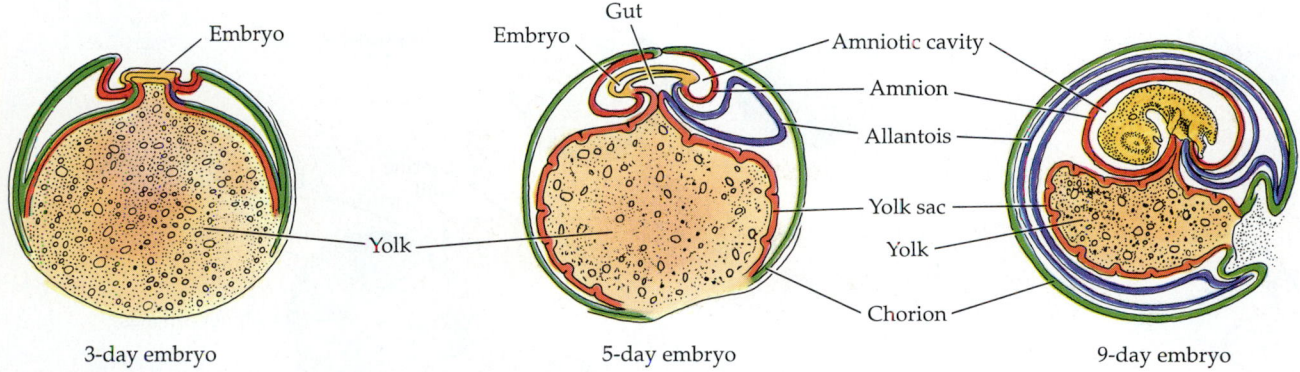

37.13 The Amnion, Allantois, and Chorion: Extraembryonic Membranes
These specialized membranes enclose the embryo inside the egg of a chicken.
Reptiles and mammals also have these three extraembryonic membranes.

The Extraembryonic Membranes and the Beginning of Development

The embryos of reptiles, birds, and mammals are surrounded by a series of membranes. Anything that reaches the embryo from the environment must pass through these membranes. Figure 37.13 uses the chicken egg to demonstrate how these extraembryonic membranes—amnion, allantois, and chorion—form. The bird embryo starts out as a disc of cells sitting on top of an enormous body of yolk. Early cell division and movements produce the three basic tissue layers of the embryo: the ectoderm, the mesoderm, and the endoderm. Besides making up the body of the embryo, these tissue layers grow out from the embryo to form cavities. The endoderm just over the yolk, along with its associated mesoderm, form a **yolk sac,** which absorbs nutrients from the yolk. Another outgrowth of endoderm and mesoderm becomes the **allantois,** which forms a cavity to receive wastes from the embryo. A growth of ectoderm with associated mesoderm becomes the **amnion,** forming a cavity immediately surrounding the embryo. A more extensive outgrowth of ectoderm and mesoderm becomes the **chorion,** which lines the inside surface of the egg shell.

These extraembryonic membranes are the basic features of the **amniotic egg,** which was a major step in the evolution of reptiles from amphibian ancestors about 300 million years ago. The amniotic egg was the adaptation that freed terrestrial vertebrates from dependence on an aquatic environment for reproduction. Fish or amphibian eggs rapidly dry out if they are exposed to air, but the amniotic egg provides an aqueous environment within which the embryo can develop.

The same extraembryonic membranes found inside birds' eggs also form in mammals. The mammalian blastocyst is a hollow ball of cells that has a central fluid-filled cavity called the blastocoel (Figure 37.14). The first membrane to appear is the chorion; it is apparent by the fifth cell division after fertilization and completely surrounds the blastocyst. It takes more than three days for the human blastocyst to travel down the oviducts to the uterus, where it lives free for the next two to three days. About the sixth day after fertilization, the blastocyst attaches to the lining of the uterus. The chorion plays an important role in implantation by inducing responses in the endometrium. As the blastocyst invades it, the endometrium proliferates and develops more blood vessels. The interaction of the chorion with the wall of the uterus is the beginning of the placenta, which will grow and become the site of exchange of nutrients and wastes between mother and embryo.

A compact inner mass of cells within the blastocyst forms the embryo. As we saw in the bird egg (see Figure 37.13), the amnion surrounds the embryo, creating a fluid-filled cavity within which the developing embryo floats. The allantois forms a stalk or cord that connects the embryo with the chorion at the location where the placenta will form. This allantoic stalk becomes the **umbilical cord.** Blood vessels from the embryo grow down the umbilical cord to carry nutrients from and wastes to the placenta. It's easy to see why astronauts taking space walks refer to the cables and the air hoses attaching them to the spacecraft as their umbilical cords.

Cells slough off the embryo and float in the amniotic fluid that bathes the embryo. Later in development, a small sample of the amniotic fluid may be withdrawn with a needle as the first step of a process called **amniocentesis** (Figure 37.15). Some of these cells can be cultured and used for biochemical and genetic analyses that can reveal the sex of the fetus as well as genetic markers for diseases such as cystic fibrosis, Tay–Sachs disease, and Down syndrome. Amniocentesis usually is not performed until after the fourteenth week of pregnancy, and the tests re-

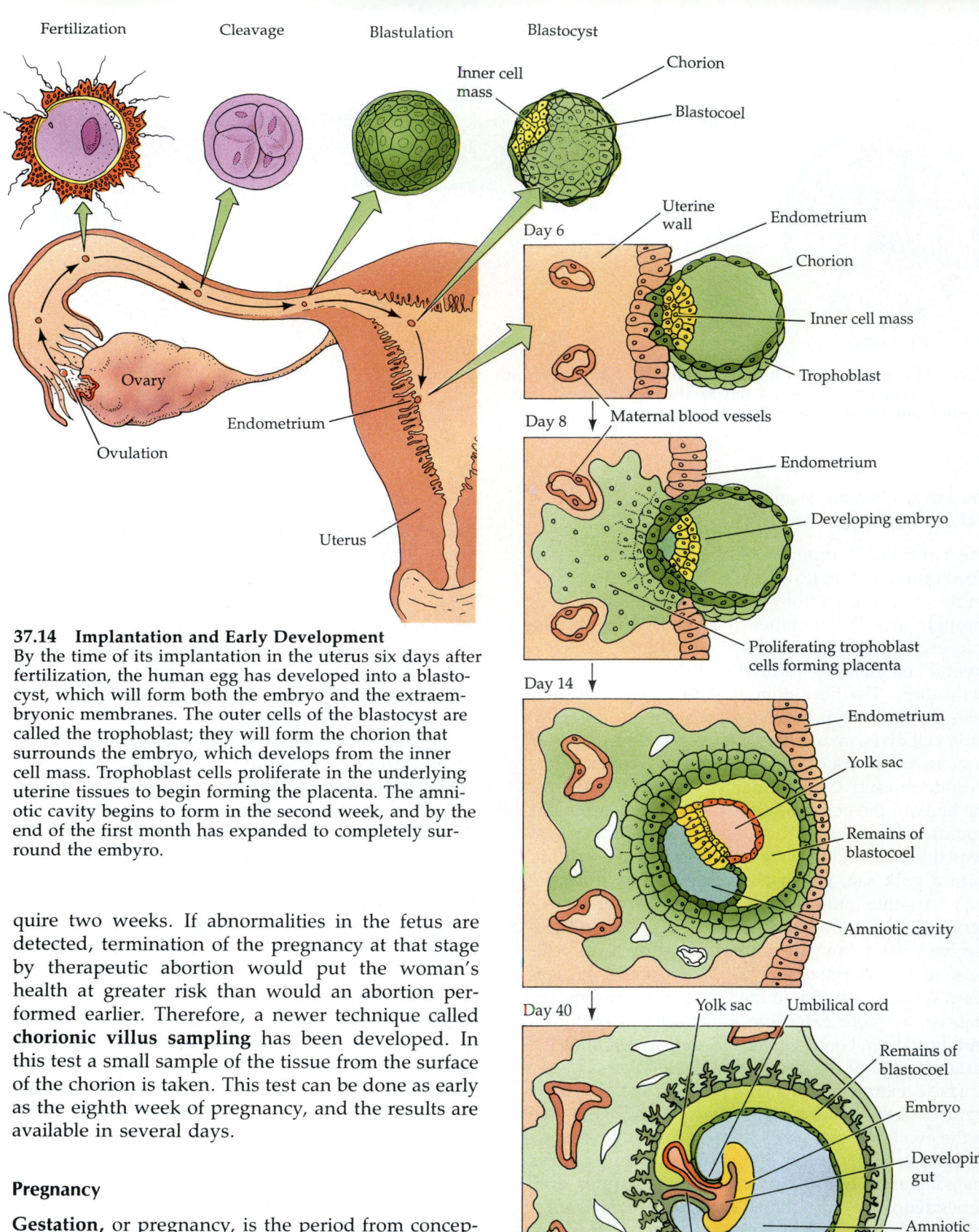

Fertilization Cleavage Blastulation Blastocyst

Inner cell mass — Chorion
Blastocoel

Ovary

Ovulation

Endometrium

Uterus

Day 6 — Uterine wall — Endometrium — Chorion — Inner cell mass — Trophoblast

Day 8 — Maternal blood vessels — Endometrium — Developing embryo

Proliferating trophoblast cells forming placenta

Day 14 — Endometrium — Yolk sac — Remains of blastocoel — Amniotic cavity

Day 40 — Yolk sac — Umbilical cord — Remains of blastocoel — Embryo — Developing gut — Amniotic cavity — Amnion — Chorion — Endometrium

Placenta Allantois

37.14 Implantation and Early Development
By the time of its implantation in the uterus six days after fertilization, the human egg has developed into a blastocyst, which will form both the embryo and the extraembryonic membranes. The outer cells of the blastocyst are called the trophoblast; they will form the chorion that surrounds the embryo, which develops from the inner cell mass. Trophoblast cells proliferate in the underlying uterine tissues to begin forming the placenta. The amniotic cavity begins to form in the second week, and by the end of the first month has expanded to completely surround the embyro.

quire two weeks. If abnormalities in the fetus are detected, termination of the pregnancy at that stage by therapeutic abortion would put the woman's health at greater risk than would an abortion performed earlier. Therefore, a newer technique called **chorionic villus sampling** has been developed. In this test a small sample of the tissue from the surface of the chorion is taken. This test can be done as early as the eighth week of pregnancy, and the results are available in several days.

Pregnancy

Gestation, or pregnancy, is the period from conception (fertilization of egg by sperm) to birth. During gestation the embryo develops in the uterus. In general, the duration of pregnancy in mammals correlates positively with body size; in mice it is about 21 days, in cats and dogs about 60 days, in humans about 266 days, in horses about 330 days, and in elephants about 600 days. In discussing the events

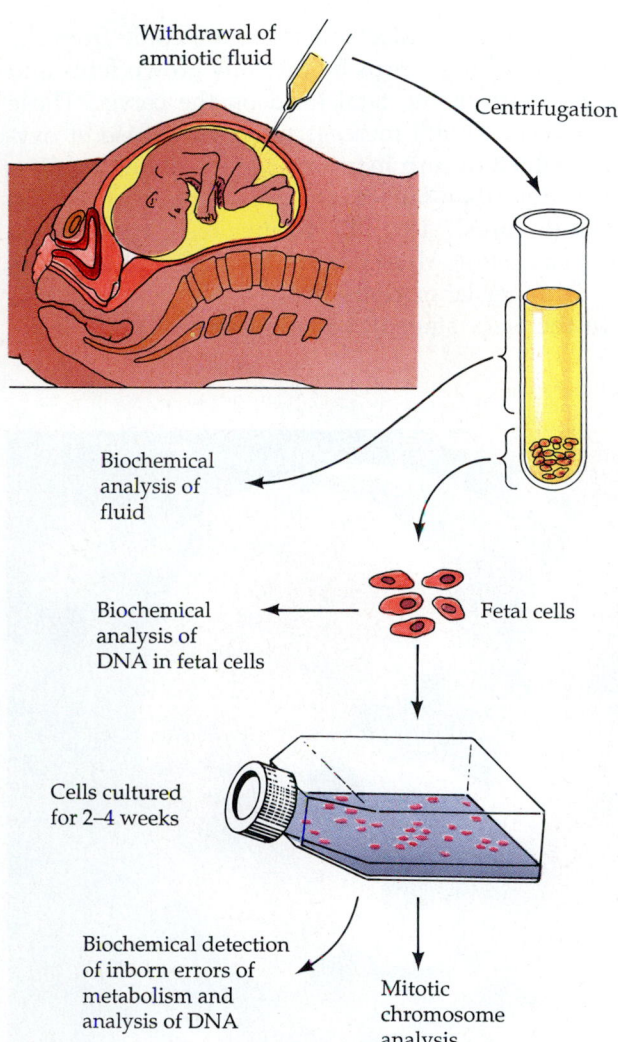

Withdrawal of amniotic fluid

Centrifugation

Biochemical analysis of fluid

Biochemical analysis of DNA in fetal cells

Fetal cells

Cells cultured for 2–4 weeks

Biochemical detection of inborn errors of metabolism and analysis of DNA

Mitotic chromosome analysis

37.15 Amniocentesis Enables Genetic Analysis of the Fetus

Genetic information, including the sex of the fetus, can be gained by amniocentesis—the withdrawal and analysis of a small amount of amniotic fluid. The procedure is usually performed late in the third or early in the fourth month of pregnancy.

though the fetus at the end of that time is still so small that it would fit into a teaspoon. Soon after the blastocyst implants, it begins to secrete human chorionic gonadotropin (HCG), the hormone that stimulates the corpus luteum to continue producing estrogen and progesterone. It is HCG that is detected in pregnancy tests. The high levels of estrogen and progesterone prevent menstruation, which would abort the embryo, and exert negative feedback on the hypothalamus and the pituitary, inhibiting the release of follicle-stimulating hormone and luteinizing hormone and preventing a new round of ovulation. Side effects of these hormonal shifts are the well-known symptoms of pregnancy: morning sickness, mood swings, changes in the senses of taste and smell, and swelling of the breasts.

During the second trimester the fetus grows rapidly to about 600 grams, and the mother's abdomen enlarges considerably. The limbs of the fetus elongate, and the fingers, toes, and facial features become well formed (Figure 37.16b). Fetal movements are first felt by the mother early in the second trimester, and they become progressively stronger and more coordinated. By the end of the second trimester, the fetus may suck its thumb.

The production of estrogen and progesterone by the placenta increases during the second trimester. The placental tissues do not produce these steroids directly from cholesterol. The tissues receive androgens through the circulation and convert them to estrogen and progesterone. These androgens come from two sources: the adrenal cortex of the mother and the adrenal cortex of the fetus itself. As placental production of these hormones increases, the level of human chorionic gonadotropin and the activity of the corpus luteum decrease. The corpus luteum degenerates by the second trimester, but ovulation and menstruation are still inhibited by the high levels of steroids secreted by the placenta. Along with these hormonal changes, the unpleasant symptoms of early pregnancy usually disappear.

The fetus and the mother continue to grow rapidly during the third trimester. As the fetus approaches its full size, pressure on the mother's internal organs can cause indigestion, constipation, frequent urination, shortness of breath, and swelling of the legs and ankles. Throughout pregnancy the circulatory system of the fetus has been functioning, and as the

of human pregnancy, we divide it into three trimesters of about three months each.

The first trimester begins with fertilization. The blastocyst, as we know, goes through a series of rapid cell divisions, known as cleavage, before implantation. After implantation, the differentiation of tissues and organs we discussed in Chapter 17 begins; the first trimester is the main period of **organogenesis.** The heart begins to beat in week 4, and limbs form by week 8. Figure 37.16a shows an embryo midway between these two developmental events. Most organs are present in at least primitive form by the end of the first trimester. Because the first trimester is a time of rapid cell division and differentiation, it is the period during which the embryo is most sensitive to radiation, drugs, and chemicals that can cause birth defects. An embryo can be damaged before the mother even knows she is pregnant. By the end of the first trimester, the embryo appears to be a miniature version of the adult and is called a **fetus.**

Hormonal changes cause major and noticeable responses in the mother during the first trimester, even third trimester approaches its end, other internal organs mature. The digestive system begins to function, the liver stores glycogen, the kidneys produce urine, and the brain undergoes cycles of sleep and waking.

BIRTH

Throughout pregnancy the uterus periodically undergoes slow, weak, rhythmic contractions called Braxton-Hicks contractions. These contractions become gradually stronger during the third trimester and are sometimes called false labor contractions. True labor contractions usually mark the beginning of childbirth, or **parturition.** In some women, however, the first signs of labor are the discharge of the mucous plug that blocks the uterus during pregnancy ("a bloody show") or the rupture of the amnion and the loss of the amniotic fluid ("waters breaking").

Labor

Many factors contribute to the onset of labor. Hormonal and mechanical stimuli increase the contractility of the uterus. Progesterone inhibits and estrogen stimulates contractions of uterine muscle. Toward the end of the third trimester the estrogen–progesterone ratio shifts in favor of estrogen. Oxytocin stimulates uterine contraction; its secretion by the pituitaries of both mother and fetus increases at the time of labor. Mechanical stimuli come from the stretching of the uterus by the fully grown fetus and the pressure of the fetal head on the cervix. These mechanical stimuli increase pituitary release of oxytocin, which in turn increases the activity of the uterine muscle that causes even more pressure on the cervix (Figure 37.17). This positive feedback converts the weak, slow, rhythmic contractions of the uterus into stronger labor contractions.

In the early stage of labor, the contractions of the

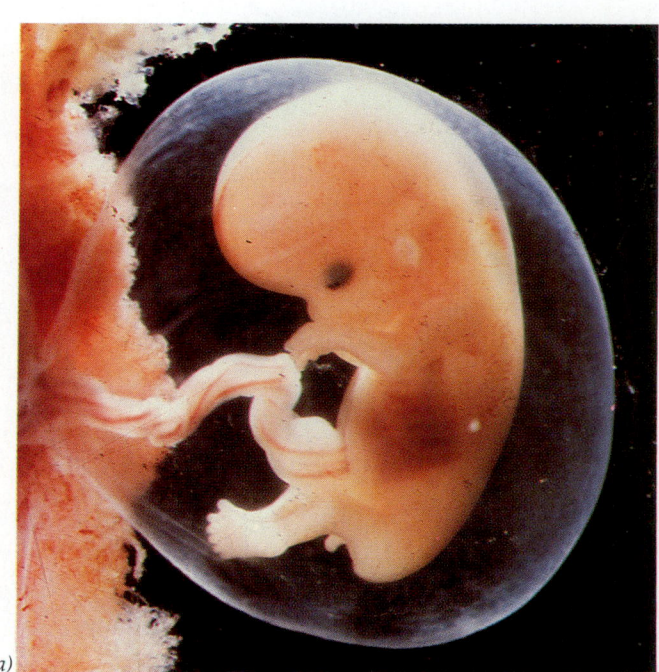

(a)

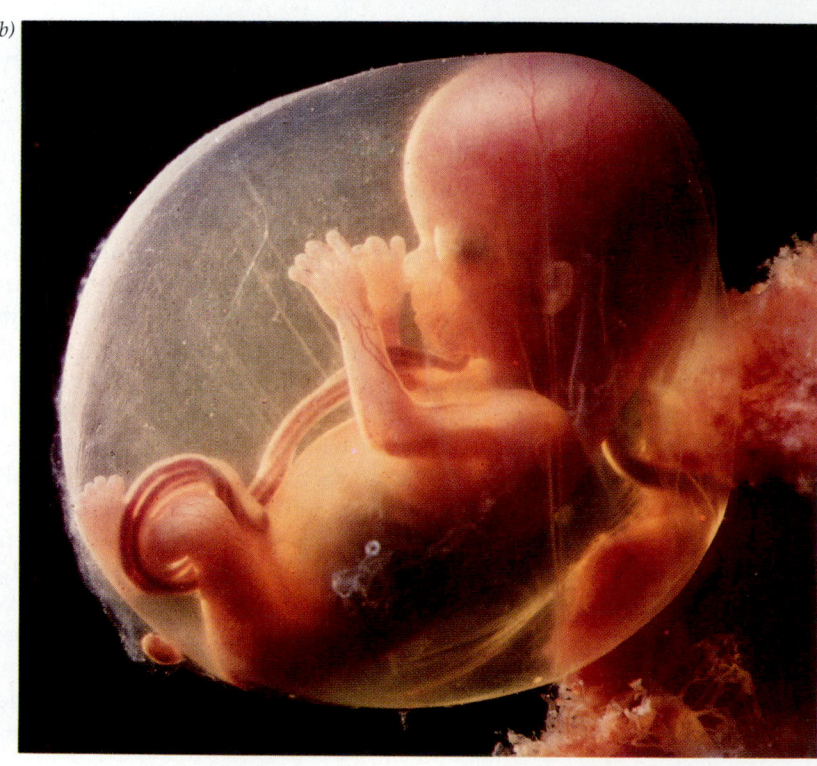

(b)

37.16 A Human Embryo
(a) The first trimester of pregnancy is a period of rapid cell division and differentiation; the organs and body structures of this six-week-old embryo are forming rapidly. (b) At four months the fetus moves freely within its protective amniotic membrane. The fingers and toes are fully formed.

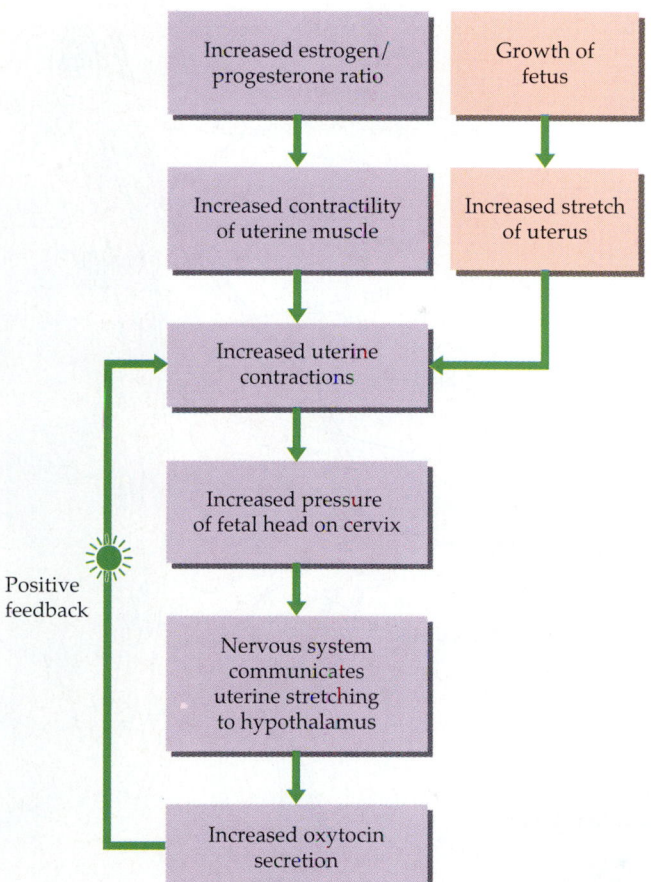

37.17 Increasing Oxytocin Production and Birth
Release of oxytocin by the pituitary gland increases uterine contractions during labor and birth. Note the positive feedback loop.

uterus are 15 to 20 minutes apart, and each lasts 45 to 60 seconds. During this time the contractions pull the cervix open until it is large enough to allow the baby to pass through. This stage of labor lasts an average of 12 to 15 hours in a first pregnancy and 8 hours or less in subsequent ones. Gradually the contractions become more frequent and more intense.

Delivery

In the second stage of labor, which begins when the cervix is fully dilated, the baby's head moves into the vagina and becomes visible from the outside. The usual head-down position of the baby at the time of delivery comes about when the fetus shifts its orientation during the seventh month. If the fetus fails to reorient head down, the birth is more difficult. Passage of the fetus through the vagina is assisted by the woman's bearing down with her abdominal and other muscles to help push the baby along. Once the head and shoulders of the baby clear the cervix, the rest of its body eases out rapidly, but it is still connected to the placenta in the mother by the um-

bilical cord. This second stage of labor may take as little as a minute, or up to half an hour or more in a first pregnancy.

As soon as the baby clears the birth canal, it can start breathing and become independent of its mother's circulation. The umbilical cord may then be clamped and cut. The segment still attached to the baby dries up and sloughs off in a few days, leaving behind its distinctive signature, the belly button—more properly called the umbilicus. The detachment and expulsion of the placenta and fetal membranes takes from a few minutes to an hour, and may be accompanied by uterine contractions.

A **caesarian section** is the surgical extraction of the baby from the uterus (the term comes from Julius Caesar, who was supposedly born this way). It may be necessary for a number of reasons: if the fetus is large and the mother's pelvis small, if the first stage of labor lasts too long, if the cervix fails to dilate sufficiently, or if there is a sudden threat to the health of the baby or the mother.

Lactation

Throughout pregnancy, the high circulating levels of estrogen and progesterone cause the mammary glands of the breasts to develop in preparation for lactation (the secretion of milk). Prolactin secretion from the anterior pituitary also increases progressively during pregnancy, but its effect is countered by estrogen and progesterone, which inhibit the production of milk. Just before birth, the breasts may secrete a few milliliters of fluid each day. This fluid, called **colostrum,** contains little fat, and its rate of production is very low. With expulsion of the placenta, the estrogen and progesterone levels in the mother's bloodstream fall rapidly, and within a few days the well-developed mammary glands are producing milk. This milk does not flow readily into the ducts of the breasts, however. If it did, it would dribble out continuously, rather than just flowing out when the baby suckled. Oxytocin plays an important role in controlling lactation (Figure 37.18a). When the baby suckles, stimulation of the breast causes the release of oxytocin from the posterior pituitary. In about 30 seconds this oxytocin reaches the breasts and stimulates contraction of the muscle cells that surround the milk-secreting cells. The milk is thereby "let down," or ejected, into the ducts of the breasts (Figure 37.18b).

Oxytocin also has a role just after the delivery of the baby. If the newborn baby is placed at the mother's breast, even though the breast cannot deliver more than colostrum at this time, the suckling of the infant causes oxytocin release, which stimulates continued uterine contractions that help to expel the placenta and inhibit bleeding from the uterine wall.

(a)

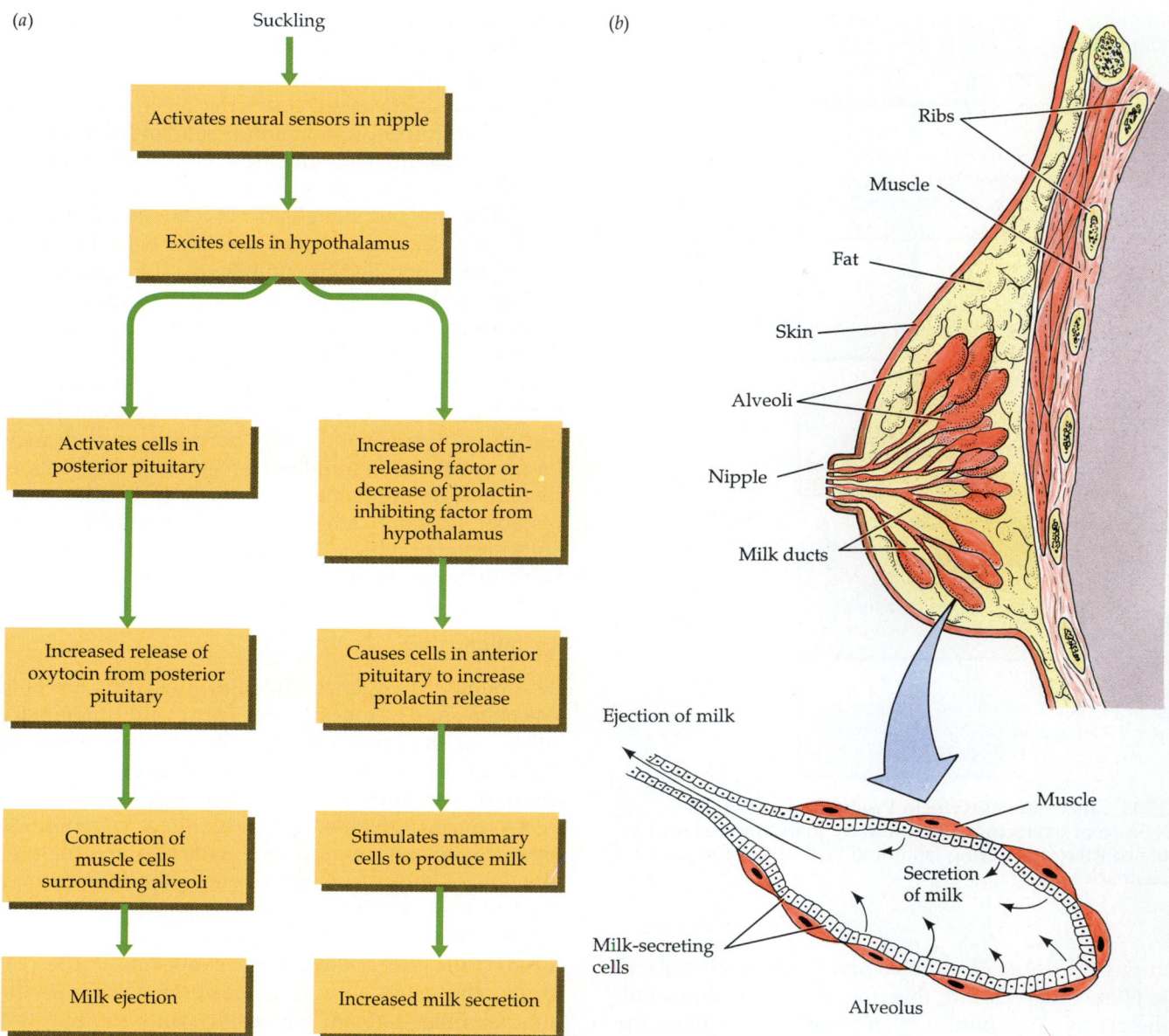

Suckling

Activates neural sensors in nipple

Excites cells in hypothalamus

| Activates cells in posterior pituitary | Increase of prolactin-releasing factor or decrease of prolactin-inhibiting factor from hypothalamus |

| Increased release of oxytocin from posterior pituitary | Causes cells in anterior pituitary to increase prolactin release |

| Contraction of muscle cells surrounding alveoli | Stimulates mammary cells to produce milk |

| Milk ejection | Increased milk secretion |

(b)

Ribs

Muscle

Fat

Skin

Alveoli

Nipple

Milk ducts

Ejection of milk

Muscle

Secretion of milk

Milk-secreting cells

Alveolus

37.18   Lactation is Controlled by Hormones
(a) The role of hormones in lactation. *(b)* The anatomy of the mammary gland. The alveoli are reservoirs for milk.

SUMMARY of Main Ideas about Animal Reproduction

Some simple animals reproduce asexually by budding, regeneration, or parthenogenesis, but most animals reproduce sexually.

Sex produces genotypic diversity through recombination.

The male testes and the female ovaries are the internal reproductive organs in which gametes form.

Both spermatogenesis and oogenesis require mitotic proliferation of primary germ cells, meiosis, and maturation of gametes.
   Review Figure 37.3

The reproductive systems of humans include the gonads where gametogenesis takes place, ducts that deliver mature gametes to the site of fertilization, structures for copulation, and in the female, a structure, the uterus, that provides an environment for the developing fetus.
   Review Figures 37.6 and 37.7

Hormones control the ovarian and menstrual cycles of the human female so that an ovarian follicle matures each month and releases an ovum at a time when the uterus is prepared to receive it if it becomes fertilized.
   Review Figure 37.8

Sexual behavior brings eggs and sperm together, and it involves strong physiological responses and conscious sensations.
Review Figure 37.9

As sperm and egg interact in fertilization, the sperm penetrates the egg's outer layers and binds to the egg plasma membrane. The egg reacts to prevent multiple sperm entry and becomes activated to complete meiosis and begin development.
Review Figures 37.10 and 37.11

Within the human uterus, the placenta, which includes tissues from the mother and from the em-

bryo, supplies the embryo with oxygen and nutrients and removes its waste products.
Review Figure 37.14

Human childbirth is assisted by a positive feedback hormonal mechanism in which stretching of the uterine wall stimulates oxytocin release, which in turn causes uterine muscles to contract.
Review Figure 37.17

Oxytocin and prolactin help control lactation following childbirth.
Review Figure 37.18

SELF-QUIZ

1. Match each of the following modes of asexual reproduction with the statement or description that characterizes it. (Each letter may be used more than once, and more than one letter may apply to each statement.)
   a. Budding
   b. Regeneration
   c. Parthenogenesis
   (i) A form of asexual reproduction that usually follows an animal being broken by an external force, but it can also be initiated by the animal itself.
   (ii) Many freshwater sponges produce clusters of undifferentiated cells which eventually "escape" the parent and become free-living organisms genetically identical to the parent.
   (iii) Offspring develop from unfertilized eggs.
   (iv) The process requires totipotent cells.
   (v) Species that reproduce this way may also engage in sexual reproduction.

2. A species in which the individual possesses both male and female reproductive systems is termed (choose all that apply)
   a. dioecious.
   b. parthenogenetic.
   c. hermaphroditic.
   d. diploid.
   e. monoecious.

3. The major advantage of internal fertilization is that

   a. it ensures paternity.
   b. it permits the fertilization of many gametes.
   c. it reduces the incidence of destructive competitive interactions between the members of a group.
   d. it results in the formation of a stable pair-bond between mates.
   e. it allows the developing organism to enjoy a greater degree of protection during the early phases of development.

4. Which one of the following statements about oocytes is true?
   a. At birth, the human female has produced all the oocytes she will ever produce.
   b. At the onset of puberty, ovarian follicles produce new ones in response to hormonal stimulation.
   c. At the onset of menopause, the human female stops producing them.
   d. They are produced by the human female throughout adolescence.
   e. Those produced by the female are stored in the seminiferous tubules.

5. Spermatogenesis and oogenesis differ in that
   a. spermatogenesis produces gametes with greater energy stores than those produced by oogenesis.
   b. spermatogenesis produces four equally functional diploid cells per meiotic event and oogenesis does not.

   c. oogenesis produces four equally functional haploid cells per meiotic event and spermatogenesis does not.
   d. spermatogenesis produces many gametes with meager energy reserves, whereas oogenesis produces relatively few, well-provisioned gametes.
   e. in humans, spermatogenesis begins before birth, whereas oogenesis does not start until the onset of puberty.

6. The acrosome of the sperm
   a. carries genetic information.
   b. provides energy for movement.
   c. carries the enzymes that facilitate fertilization.
   d. induces ovulation.
   e. prevents polyspermy.

7. During oogenesis in mammals, the second meiotic division occurs
   a. after capacitation.
   b. after implantation.
   c. before ovulation.
   d. before the acrosomal reaction.
   e. after a sperm enters the egg.

8. One of the major differences between the sexual response cycles in human males and females is
   a. the increase in blood pressure in males.
   b. the increase in heart rate in females.
   c. the presence of a refractory period in females after orgasm.
   d. the presence of a refractory period in males after orgasm.
   e. the increase in muscle tension in males.

9. Which of the following membranes is part of the embryonic contribution to placenta formation?
   a. Amnion
   b. Chorion
   c. Uterine membrane
   d. Fertilization membrane
   e. Zona pellucida

10. Contractions of muscles in the uterine wall and in the breasts are stimulated by
   a. progesterone.
   b. estrogen.
   c. prolactin.
   d. oxytocin.
   e. human chorionic gonadotropin.

FOR STUDY

1. Compare and contrast spermatogenesis and oogenesis in terms of the products of and the timetable for each process.

2. Describe how the events during sperm activation and egg activation lead to successful fertilization.

3. Describe the mechanisms controlling lactation.

4. Ovarian and uterine events in the month following ovulation differ depending on whether or not fertilization occurs. Describe the differences and explain their hormonal controls.

5. Explain how positive feedback plays a role in birth.

READINGS

Beaconsfield, P., G. Budwood and R. Beaconsfield. 1980. "The Placenta." *Scientific American*, August. Describes implantation and development of the organ that is the intermediary between fetus and mother. The many functions of the placenta are described.

Epel, D. 1977. "The Program of Fertilization." *Scientific American*, November. The initial events in the sperm–ovum interaction.

Gilbert, S. F. 1994. *Developmental Biology*, 4th Edition. Sinauer Associates, Sunderland, MA. This excellent text on animal development includes chapters on fertilization, as well as on the germ line and gametogenesis.

Johnson, M. H. and B. J. Everitt. 1988. *Essential Reproduction*, 3rd Edition. Blackwell Scientific, Oxford. A concise and comprehensive technical account of the biology of gametogenesis, fertilization, and pregnancy.

Katchadourian, H. A. 1989. *Fundamentals of Human Sexuality*, 5th Edition. Saunders, Philadelphia. An introductory text that covers the anatomy and physiology of sex and reproduction in the first two chapters and then discusses developmental, behavioral, and social aspects of sex.

Short, R. V. 1984. "Breast Feeding." *Scientific American*, April. Breast feeding has hormonal consequences that have contraceptive effects. Trends toward bottle feeding in many developing nations may be causing rises in their rates of population growth.

Wassarman, P. M. 1988. "Fertilization in Mammals." *Scientific American*, December. Examines the molecular and cellular events that surround the fusion of sperm and egg.

The human brain weighs about 1.5 kilograms, is mostly water, and has the consistency and color of vanilla custard. The complexity of this small mass of tissue, however, exceeds that of any other known matter. The work of the brain is to process and store information and to control the physiology and behavior of the body. The brain is constantly receiving information from all the senses, integrating and interpreting that information, and generating commands to the muscles and organs of the body. The brain senses the need to act, decides on the appropriate action, orchestrates it, initiates and coordinates it, monitors it, and remembers it. The essence of individuality and personality resides in the brain. You could imagine remaining yourself after replacing any organ of your body except your brain.

Some of the actions controlled by the brain are conscious, or voluntary; others, such as the functions of the heart, lungs, and gut are involuntary. What is remarkable is that so many voluntary and involuntary actions are all going on simultaneously. Every second of its life, the brain processes thousands of bits of information. The brain does not function alone, however; it is part of an essential system—the nervous system—of a human or other animal. Like the hormones we considered in Chapter 36, the nervous system provides communication among cells. In this chapter we discuss nervous systems, particularly the structure and function of the human nervous system, as well as how nerve cells communicate with each other.

COMMUNICATION AND COMPLEXITY

A nervous system uses **sensors** (see Chapter 39), such as the eyes and ears, to transduce (convert) stimuli into messages that it can process. To cause behavior or physiological responses, the nervous system must communicate messages to **effectors**, such as muscles and glands (see Chapter 40). Together the brain and spinal cord make up the **central nervous system**, and nerve cells that carry information to and from the central nervous system make up the **peripheral nervous system**.

The information that flows through the nervous system consists of electrical and chemical messages. The electrical messages are nerve impulses. A nerve impulse is a rapid change in electrical charge across a small portion of the plasma membrane of the nerve cell. The nerve impulse travels along the membrane of a nerve cell, and that can be a long distance in the body because nerve cells have long extensions. When the nerve impulse reaches a point where the nerve cell makes contact with another nerve cell, or a muscle or gland cell, a chemical message is released that

The Human Brain
This horizontal section has been colored by computer to show the different brain regions. The front of the brain is at the top.

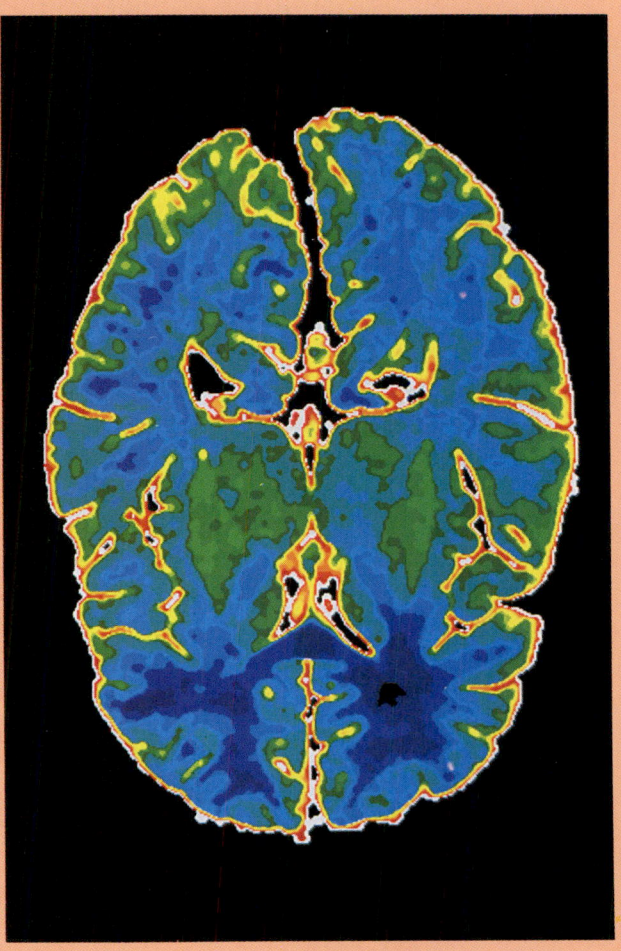

38

Neurons and the Nervous System

849

communicates across the gap between the two cells. In most cases the nerve impulse cannot spread from cell to cell without the intervention of a chemical message.

Throughout the animal kingdom, nervous systems vary in complexity, ranging from the complex nervous system of humans to the simple nerve nets of cnidarians, which seem to do little more than detect food or danger and cause tentacles to retract and the body to constrict (Figure 38.1). In all cases, however, nervous systems control behavior and the functions of the body. In general, the more complex the behavior and physiological capabilities of a species, the larger is its nervous system. Sometimes size belies capacity, however. Consider, for example, the nervous systems of small spiders that have programmed within them the thousands of precise movements necessary to construct a beautiful web without any prior experience. One of the greatest challenges of biology is to understand how the human brain functions, but it would be a major breakthrough even to understand a much simpler nervous system. Much progress in neurobiology (the science that studies nervous systems) has come from research on the simpler nervous systems of invertebrates. In this chapter we will frequently use the human nervous system as our model, but our knowledge comes from research on a long list of animal species. The way nerve cells function is almost identical in animals as different as squids and humans.

38.1 Nervous Systems Vary in Complexity
As we compare animals that have increasingly complex sensory and behavioral abilities, we find (moving clockwise from the sea anemone to the earthworm) information processing increasingly centralized in ganglia (collections of nerve cells) or in a brain. The brain and spinal cord in the human constitute the central nervous system, which communicates with the body through nerves that make up the peripheral nervous system.

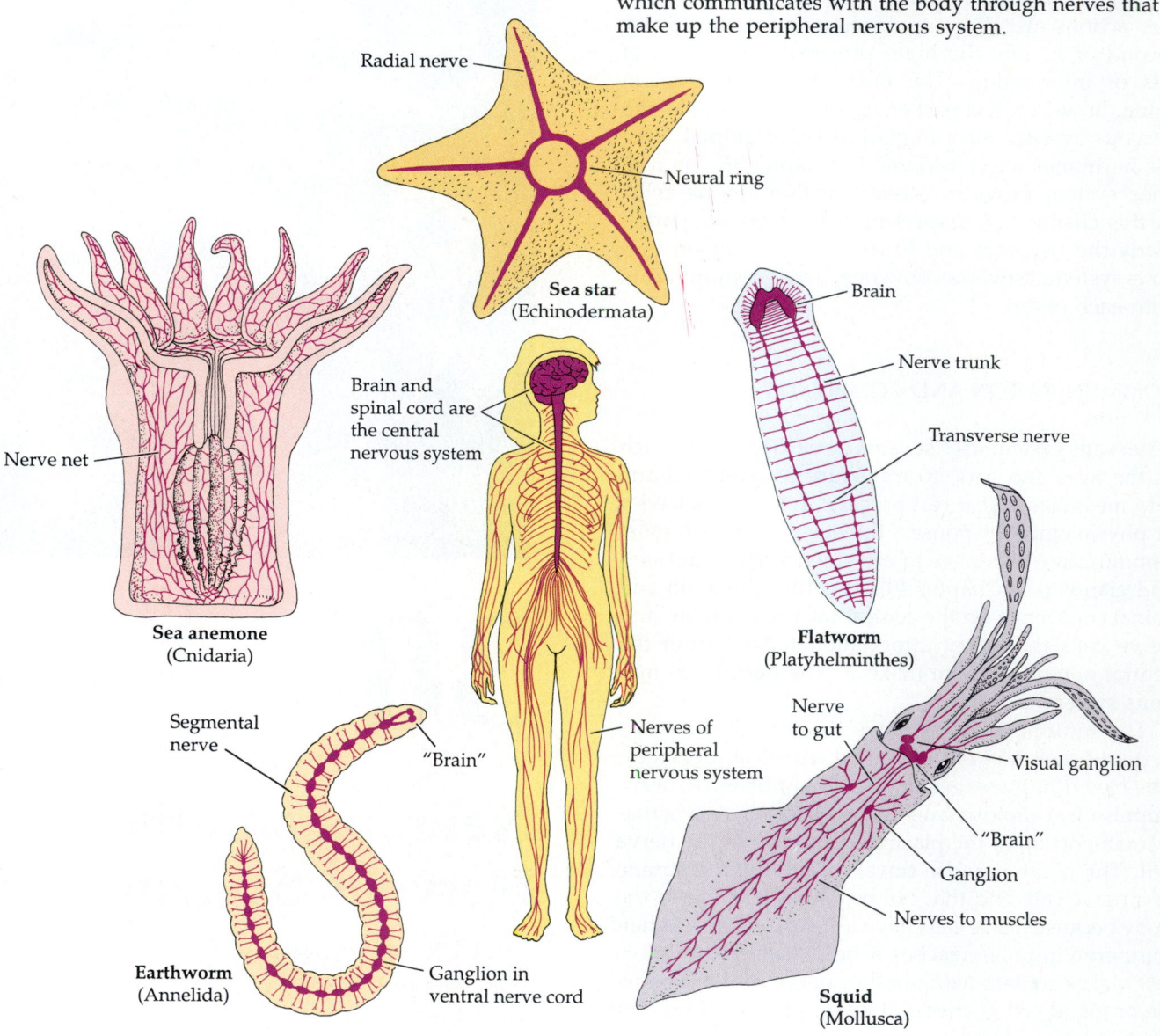

Radial nerve

Neural ring

Sea star
(Echinodermata)

Brain

Nerve trunk

Transverse nerve

Flatworm
(Platyhelminthes)

Nerve net

Brain and spinal cord are the central nervous system

Sea anemone
(Cnidaria)

Segmental nerve

"Brain"

Nerves of peripheral nervous system

Nerve to gut

Visual ganglion

"Brain"

Ganglion

Nerves to muscles

Earthworm
(Annelida)

Ganglion in ventral nerve cord

Squid
(Mollusca)

CELLS OF THE NERVOUS SYSTEM

The functions of the brain depend on the properties of its cells. **Neurons** are the cells that make it possible for the brain and the rest of the nervous system to transmit and integrate information. The important property of neurons is that their plasma membranes can generate electrical signals called nerve impulses, or **action potentials**. Their plasma membranes can also conduct these electrical signals rapidly from one location on a cell to the most distant reaches of that cell—a distance that can be more than a meter for some neurons. Where a neuron contacts another neuron or a muscle or gland cell, special structures called synapses transmit the message carried by the action potential in one neuron to the next cell. Much

of neurobiology focuses on the structure and function of neurons, but they are not the only type of cell in the nervous system. In fact, there are more **glial cells** than neurons in the brain. As we will see shortly, glial cells do not generate or conduct action potentials; they have supporting roles for neurons.

Neurons

Most neurons have four regions: a cell body, dendrites, an axon, and axon terminals (Figure 38.2), but

38.2 Neurons
(a) A generalized diagram of a neuron includes a cell body, dendrites that collect input, an axon that conducts action potentials, and axon terminals that make synapses with the target cells (red cell). (b) Neurons that are specific to various parts of the body.

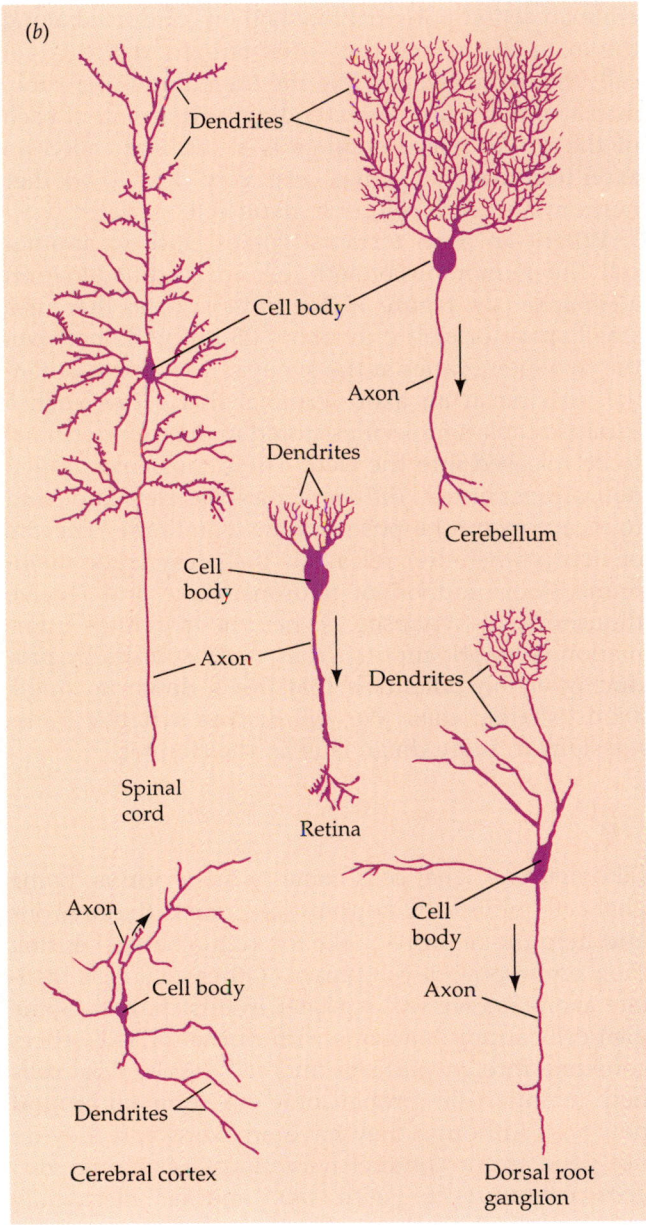

the variation in different types of neurons is considerable. The **cell body** contains the nucleus and most of the cell's organelles. Many projections may sprout from the cell body. Most nerve-cell projections are bushlike **dendrites** (from the Greek *dendron* = "tree"), which bring information from other neurons or sensory cells to the neuron's cell body. In most neurons one projection is much longer than the others and is called the **axon**. Axons usually carry information away from the cell body. The length of the axon varies greatly in different types of neurons—as does the degree of branching of the dendrites. The axons of some neurons are remarkably long. The cell body of the neuron that causes your little toe to flex is located in the spinal cord in the middle of your back, and its axon goes all the way down your leg to the muscles in the little toe. Axons are the "telephone lines" of the nervous system. Information received by the dendrites can influence the cell body to generate an action potential that is then conducted along the axon to the cell that is its target. At the target cell, the axon divides, like the frayed end of a rope, into a spray of fine nerve endings. At the tip of each of these tiny nerve endings is a swelling called an **axon terminal** that comes very, very close to another neuron, a muscle cell, or a gland cell.

Where an axon terminal comes close to another cell, the membranes of both cells are modified to form a **synapse** (see Figure 38.2*a*). In most cases there is a small space or cleft only about 25 nm wide between the two membranes at the synapse. An action potential arriving at an axon terminal that forms such a synapse causes molecules stored in the axon terminal to be released into the cleft. These molecules, called neurotransmitters, diffuse across the cleft and bind to receptors on the postsynaptic membrane. The site of neurotransmitter release is the presynaptic membrane. Most individual neurons make and receive thousands of synapses. A neuron integrates information (synaptic inputs) from many sources by producing action potentials that travel down its single axon to target cells. We will discuss synaptic transmissions in more detail later in the chapter.

Glia

Like neurons, glial cells come in many forms. Some glial cells physically support and orient the neurons and help the neurons make the right contacts during their embryonic development. Other glial cells insulate axons, as we will see later in this chapter. Some glial cells supply neurons with nutrients; still others consume foreign particles and cell debris. Glial cells help maintain the proper ionic environment around neurons. Although they have no axons and they do not generate or conduct nerve impulses, some glial cells communicate with one another electrically through a special type of contact called a **gap junction**, a connection that enables ions to flow between cells (see Chapter 5). Gap junctions can also exist between neurons and between muscle cells. In these cases action potentials can cross between the cells without being converted into chemical messages.

Glial cells called **astrocytes** (because they look like stars) create the **blood–brain barrier**. Blood vessels throughout the body are very permeable to many chemicals, some of which are toxic. The brain is better protected from toxic substances than most tissues of the body because of the blood–brain barrier. Astrocytes form this barrier by surrounding the smallest, most permeable blood vessels in the brain. Protection of the brain is crucial because, unlike other tissues of the body, the brain cannot recover from damage by generating new cells. Shortly after birth, neurons in the brain cease cell division, and at that time we have the greatest number of neurons we will ever have in our lives. Throughout the rest of life neurons are progressively lost as they die. Without the blood–brain barrier, the rate of neuron loss could be much greater. The barrier is not perfect, however. Because it consists mostly of cell membranes, it is most permeable to fat-soluble substances. Anesthetics are fat-soluble chemicals and so is alcohol; both have well-known effects on the brain.

Cells in Circuits

The human brain contains about 100 billion neurons, and each neuron may make synapses with 1,000 or more other neurons. Thus there may be as many as a million billion synapses in the human brain. Therein lies the incredible ability of the brain to process information. The thousands of circuits in the nervous system serve many different functions. Specific regions of the nervous system contain the circuits for particular functions.

FROM STIMULUS TO RESPONSE: INFORMATION FLOW IN THE NERVOUS SYSTEM

In vertebrates the brain and spinal cord are the central nervous system. The peripheral nervous system, made up of cranial nerves and spinal nerves, conducts information between the central nervous system and the other parts of the body. Each **nerve** is a bundle of axons (Figure 38.3). A nerve carries information about many things simultaneously. Some axons in a nerve may be carrying information to the central nervous system while other axons in the same nerve are carrying information from the central nervous system to the organs of the body. The peripheral nervous system extends to every tissue of the body.

38.3 Many Axons Make Up a Nerve

Some of the axons in a nerve conduct information from the body's cells, organs and tissues to the central nervous system, while others conduct information from the central nervous system to body parts. The scanning electron micrograph shows a cross section of part of a nerve.

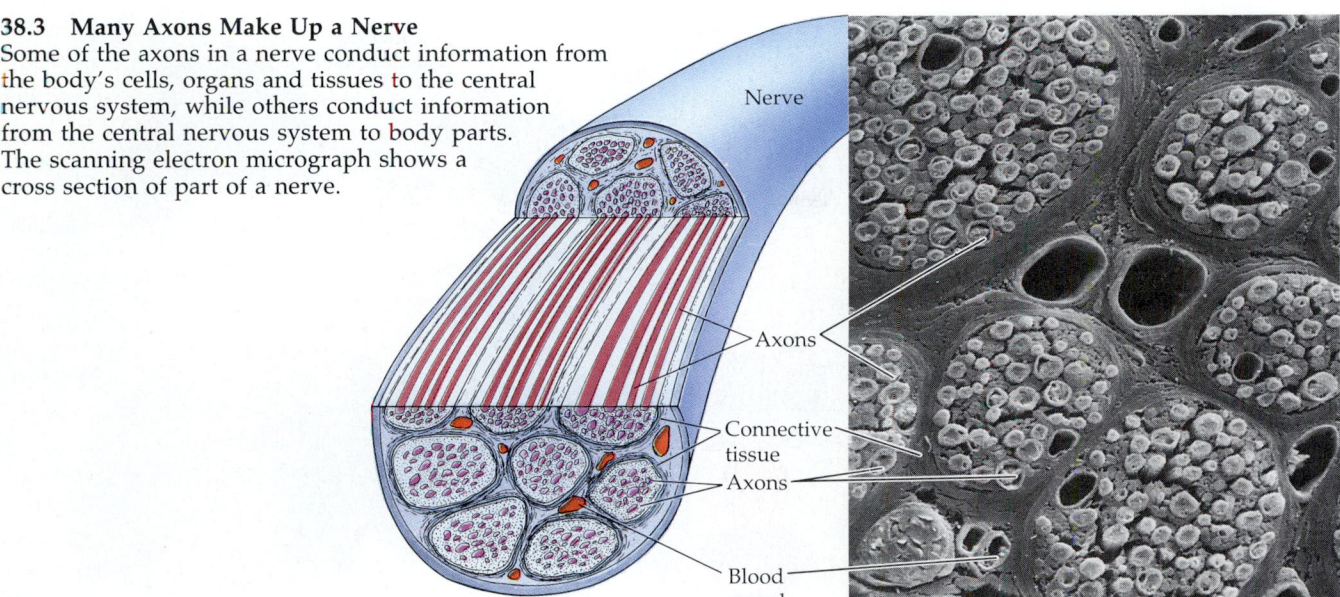

Nerve

Axons

Connective tissue

Axons

Blood vessel

Conceptually, the peripheral nervous system has two sides: an **afferent** side that brings information to the central nervous system, and an **efferent** side that carries information away from the central nervous system (Figure 38.4). The afferent side has a division that carries information from our conscious senses and a division that carries information of which we are unaware. We are consciously aware of vision, hearing, touch, taste, pain, balance, and the position of the limbs of our body. We are not consciously aware of most physiological conditions, such as our blood pressure, body temperature, blood sugar level, and oxygen supply. The efferent side of the peripheral nervous system also has two divisions: a voluntary division that executes our conscious movements and an involuntary, or autonomic, division that controls involuntary physiological functions. You must also keep in mind that hormones provide information to the central nervous system, and that neurohormones are important output messages. To translate this conceptual blueprint into an understanding of the anatomical structures of the nervous system, we will first consider their development.

Development of the Vertebrate Nervous System

The nervous system of a vertebrate begins as a hollow tube of neural tissue. The tube runs the length of the embryo on its dorsal side. At the head end of the embryo, this neural tube forms three swellings that become the basic divisions of the brain: the hindbrain, the midbrain, and the forebrain. The rest of the neural tube forms the spinal cord. The cranial and spinal nerves, which are the peripheral nervous system, sprout from the neural tube and grow throughout the embryo.

Each of the three regions in the embryonic brain develops into several structures in the adult brain. From the hindbrain come the **medulla**, the **pons**, and the **cerebellum**. The medulla is continuous with the spinal cord, the pons is in front of the medulla, and the cerebellum is an outgrowth of the pons. The medulla and pons control some physiological functions, such as breathing and circulation. The cerebellum orchestrates and refines behavior patterns.

From the embryonic midbrain come structures that

38.4 Organization of the Nervous System

The peripheral nervous system (colored backgrounds) carries information both to and from the central nervous system.

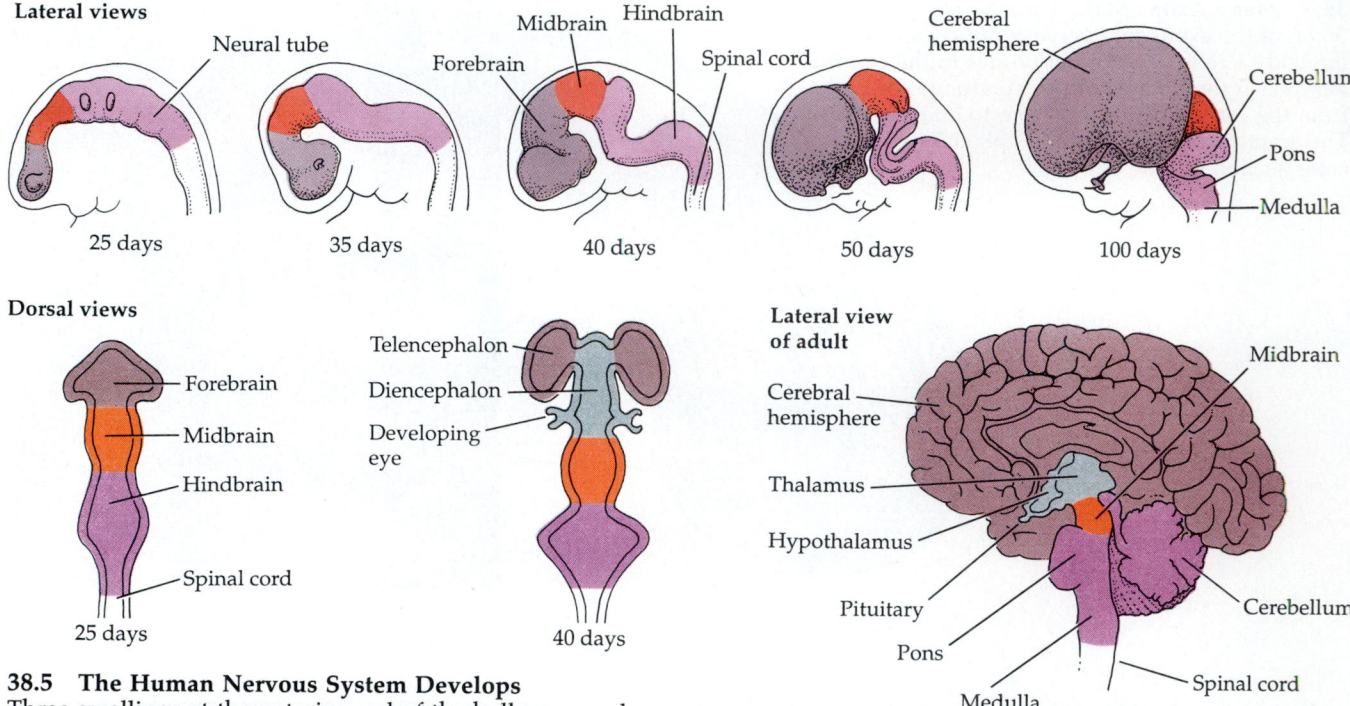

Lateral views

25 days   35 days   40 days   50 days   100 days

Dorsal views

Forebrain
Midbrain
Hindbrain
Spinal cord
25 days

Telencephalon
Diencephalon
Developing eye
40 days

Lateral view of adult

Cerebral hemisphere
Thalamus
Hypothalamus
Pituitary
Pons
Medulla
Midbrain
Cerebellum
Spinal cord

38.5 The Human Nervous System Develops
Three swellings at the anterior end of the hollow neural tube in the early embryo develop into the parts of the adult brain. The final view is an adult brain section (cut down the brain midline).

process aspects of visual and auditory (hearing) information. The embryonic forebrain develops into the **diencephalon** and the surrounding **telencephalon**, which consists of two **cerebral hemispheres**, also referred to collectively as the **cerebrum**. In mammals the telencephalon plays major roles in sensory perception, learning, memory, and conscious behavior. The diencephalon forms the **thalamus**—the final relay station for sensory information going to the telencephalon—and the **hypothalamus**, which regulates many physiological functions and biological drives (see Chapters 35 and 36).

Figure 38.5 shows how the hollow neural tube of the human embryo develops into hindbrain, midbrain, and forebrain. Note the somewhat linear arrangement of the structures. Information moves up and down this linear neural axis. A communication from the spinal cord to the telencephalon travels through the medulla, pons, midbrain, and diencephalon. These four structures are referred to collectively as the **brain stem**. In general, more-primitive and autonomic (involuntary) functions are located farther down the neural axis, and more-complex and evolutionarily advanced functions are higher on the neural axis.

As we go up the vertebrate phylogenetic scale from fish to mammals, the telencephalon increases in size, complexity, and importance. The forebrain dominates the nervous systems of mammals; when it is damaged, severe impairment or even coma results. A shark, by contrast, can swim almost normally with its telencephalon removed.

The Spinal Cord

The spinal cord conducts information between the brain and the organs of the body, and it processes and integrates information. A cross section of the spinal cord reveals a central area of gray matter in the shape of a butterfly surrounded by an area of white matter (Figure 38.6). The gray matter contains the cell bodies of the spinal neurons, and the white matter contains the axons that conduct information up and down the spinal cord. Spinal nerves leave the spinal cord at regular intervals on each side. Each spinal nerve has two roots, one connecting with the **dorsal horn** of the gray area, the other connecting with the **ventral horn** of the gray area. Each spinal nerve carries both afferent information from the sense organs and efferent information to the muscles and glands of the body. Afferent information enters the spinal cord through the dorsal roots of the nerves, and efferent information leaves the spinal cord through the ventral roots.

Information entering the dorsal horn can be transmitted to neurons that will carry it to the brain, to neurons that send commands directly out to muscles, or to cells called **interneurons** that reside entirely in

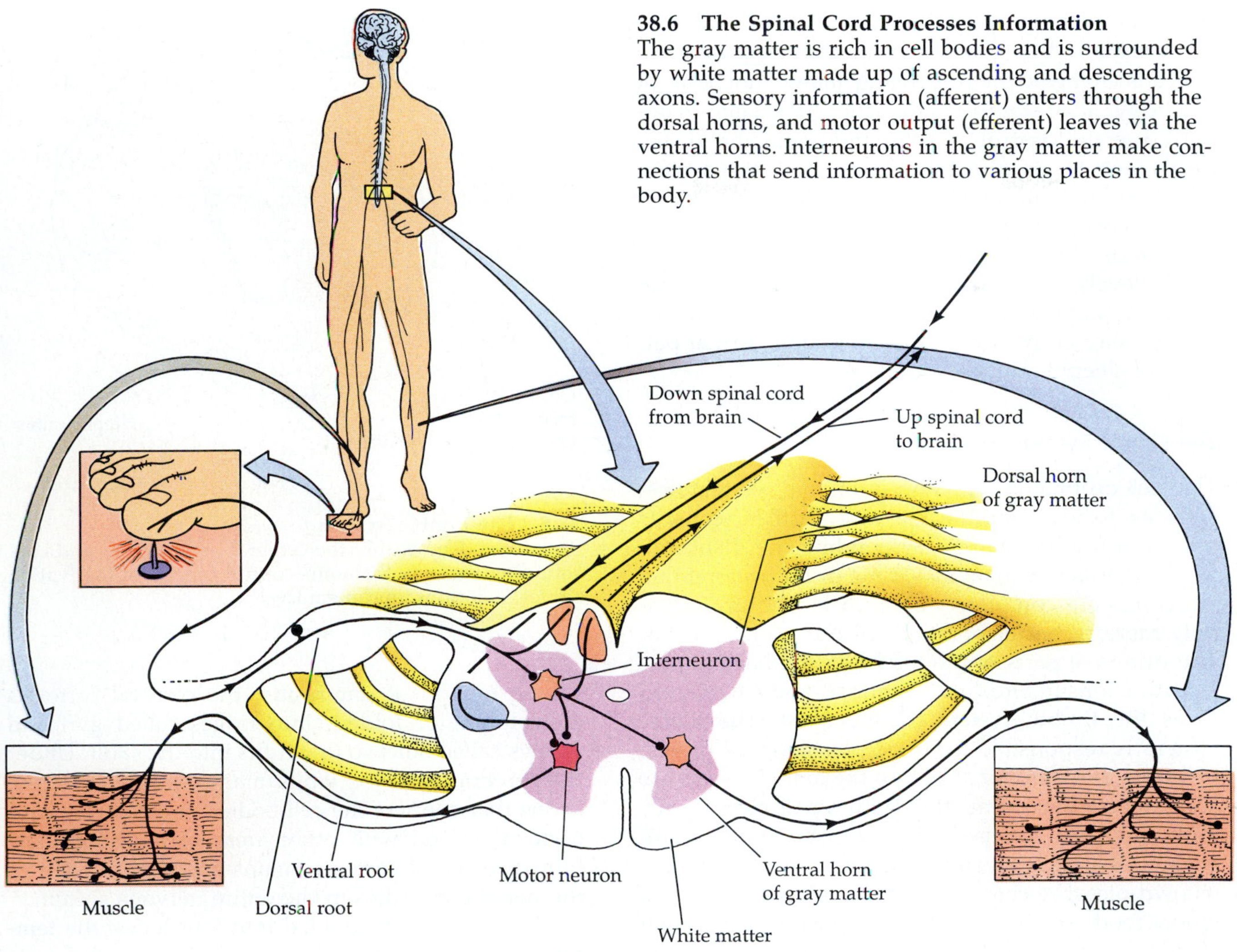

38.6 The Spinal Cord Processes Information
The gray matter is rich in cell bodies and is surrounded by white matter made up of ascending and descending axons. Sensory information (afferent) enters through the dorsal horns, and motor output (efferent) leaves via the ventral horns. Interneurons in the gray matter make connections that send information to various places in the body.

Down spinal cord from brain

Up spinal cord to brain

Dorsal horn of gray matter

Interneuron

Muscle

Ventral root

Dorsal root

Motor neuron

Ventral horn of gray matter

White matter

Muscle

the gray matter of the spinal cord. Interneurons connect with efferent neurons in the ventral horns and communicate with other spinal neurons up and down the spinal axis to amplify the response and to generate more-complex motor patterns. The spinal cord processes and integrates a lot of information. For example, when you step on a tack, spinal circuits coordinate the rapid pulling back that is carried out by many muscles on both sides of your body. Spinal circuits can also generate repetitive motor patterns such as those of walking.

The Reticular System

The **reticular system** extends throughout the core of the medulla, pons, and midbrain (Figure 38.7). In-

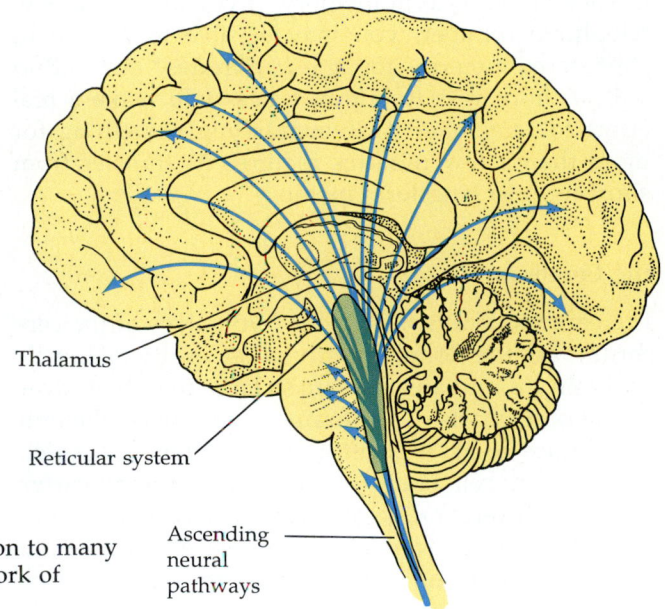

Thalamus

Reticular system

Ascending neural pathways

38.7 The Reticular System
Neurons in the core of the brain stem relay afferent information to many brain centers, and neuronal activity within this reticular network of neurons (blue) controls levels of alertness and sleep.

formation coming up the neural axis passes through the reticular system, where connections are made to neurons involved in controlling many functions of the body. For example, the reticular system is involved in the control of sleep and wakefulness. Because high levels of activity in the reticular system maintain the brain in a waking condition, this structure is sometimes called the reticular activating system. If the brain stem is damaged at midbrain or higher levels, the person is likely to remain in a coma, but if the damage is below the reticular system, the person may be paralyzed but will have normal patterns of sleeping and waking.

The Limbic System

The telencephalon of more-primitive vertebrates, such as fishes, amphibians, and reptiles, consists only of a few structures surrounding the diencephalon. In birds and mammals these primitive forebrain structures are completely covered by the evolutionarily more recent elaborations of the telencephalon. The primitive parts of the forebrain still have important functions in birds and mammals and are referred to as the **limbic system** (Figure 38.8). The limbic system is responsible for basic physiological drives, instincts, and emotions. Within the limbic system are areas that when stimulated with small electric currents can cause intense sensations of pleasure, pain, or rage. If a rat is given the opportunity to stimulate its own pleasure centers by pressing a switch, it will ignore food, water, and even sex, pushing the switch until it is exhausted. Pleasure and pain centers in the limbic system are believed to play roles in learning and in physiological drives.

One part of the limbic system, the **hippocampus**, is necessary in humans for the transfer of short-term memory to long-term memory. If you are told a new telephone number, you may be able to hold it in short-term memory for a few minutes, but within half an hour it is forgotten unless you make a real effort to remember it. Remembering something for more than a few minutes requires its transfer from short-term to long-term memory.

The Cerebrum

The cerebral hemispheres, the two halves of the cerebrum, are the dominant structures in the mammalian brain. In humans they are so large that they cover all the other parts of the brain except the cerebellum (see Figure 38.5). A sheet of gray matter (tissue rich in neuronal cell bodies) called the **cerebral cortex** covers each cerebral hemisphere. The cortex is about 4 mm thick and covers a total surface area over both hemispheres of one square meter. Since it would be rather inconvenient to have flat structures a meter

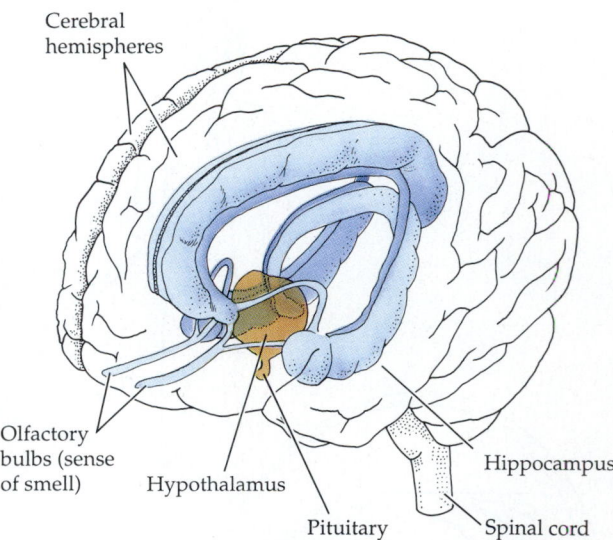

38.8   The Limbic System
Structures deep within the cerebral hemispheres and surrounding the hypothalamus control aspects of motivation, drives, emotions, and memory.

square on top of our heads, the cerebral cortex is convoluted, or folded, into ridges called gyri and valleys called sulci so that it fits into the skull. Under the cerebral cortex is white matter, made up of the axons that connect the cell bodies in the cortex with each other and with other areas of the brain. The human cerebral cortex contains about 80 percent of the nerve cell bodies in the entire nervous system.

The cerebrum is divided into four lobes: the **temporal, frontal, parietal**, and **occipital** lobes (Figure 38.9a). Specific regions of each lobe govern certain functions (Figure 38.9b). The temporal lobe processes auditory (hearing) information and is responsible for language. An area of cortex on the underside of the temporal lobe is specialized to recognize faces. If this part of your temporal lobe were damaged, you would remember people's names but would not be able to match the names with the correct faces. The occipital lobe processes visual information.

The frontal and the parietal lobes are separated by a deep valley called the **central sulcus**. The strip of parietal lobe cortex just behind the central sulcus is the **primary somatosensory cortex**. This area receives information through the thalamus about touch and pressure sensations. The whole body surface is represented in this strip of cortex, with the head at the bottom and the legs at the top (Figure 38.10). Areas of the body that have many sensory neurons and are capable of making fine distinctions in touch (such as the lips and the fingers) have disproportionately large representation. If a very small area of the somatosensory cortex is stimulated electrically, the subject reports feeling specific sensations, such as touch, from a very specific part of the body.

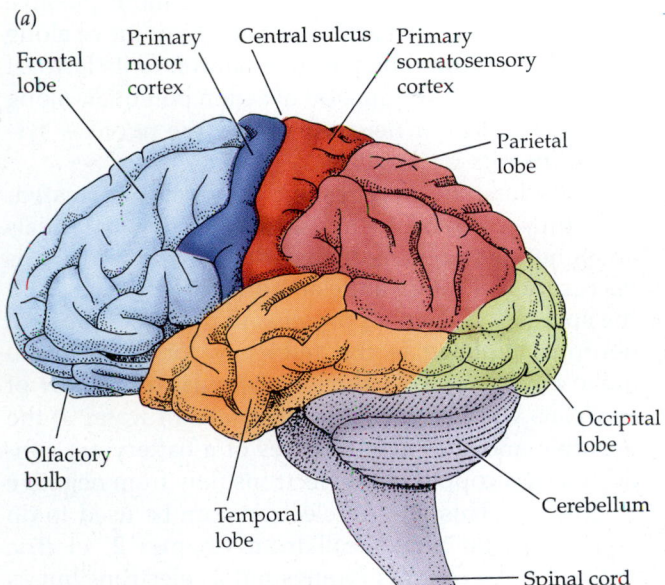

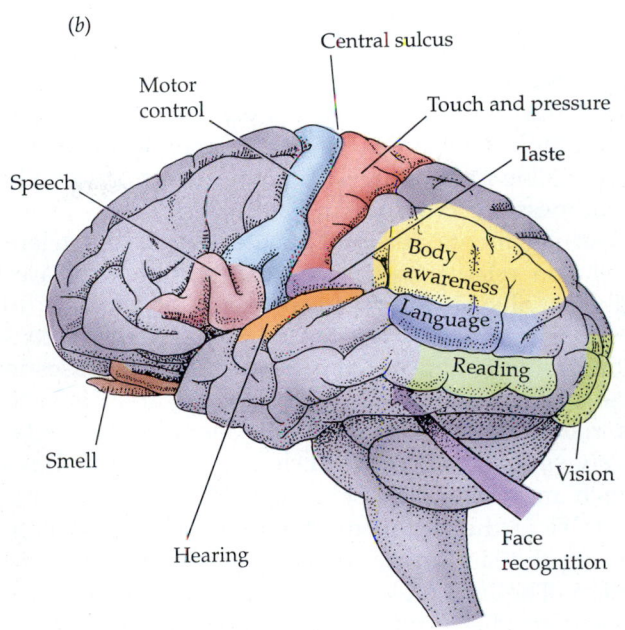

38.9   The Human Cerebrum
The highly convoluted halves of the cerebrum, viewed here from the left cerebral hemisphere, cover most of the other structures of the brain. (a) Each cerebral hemisphere is divided into four lobes. (b) Different functions are located in particular areas of these lobes.

A strip of the frontal lobe cortex just in front of the central sulcus is the **primary motor cortex**. The cells in this region have axons that extend to muscles in specific parts of the body. Once again, parts of the body are represented in the primary motor cortex, with the head region on the lower side and the lower part of the body in the top areas; fine motor control has the greatest representation. If a very small region of the frontal lobe is electrically stimulated, the response is the twitch of a muscle, but not a coordinated, complex behavior.

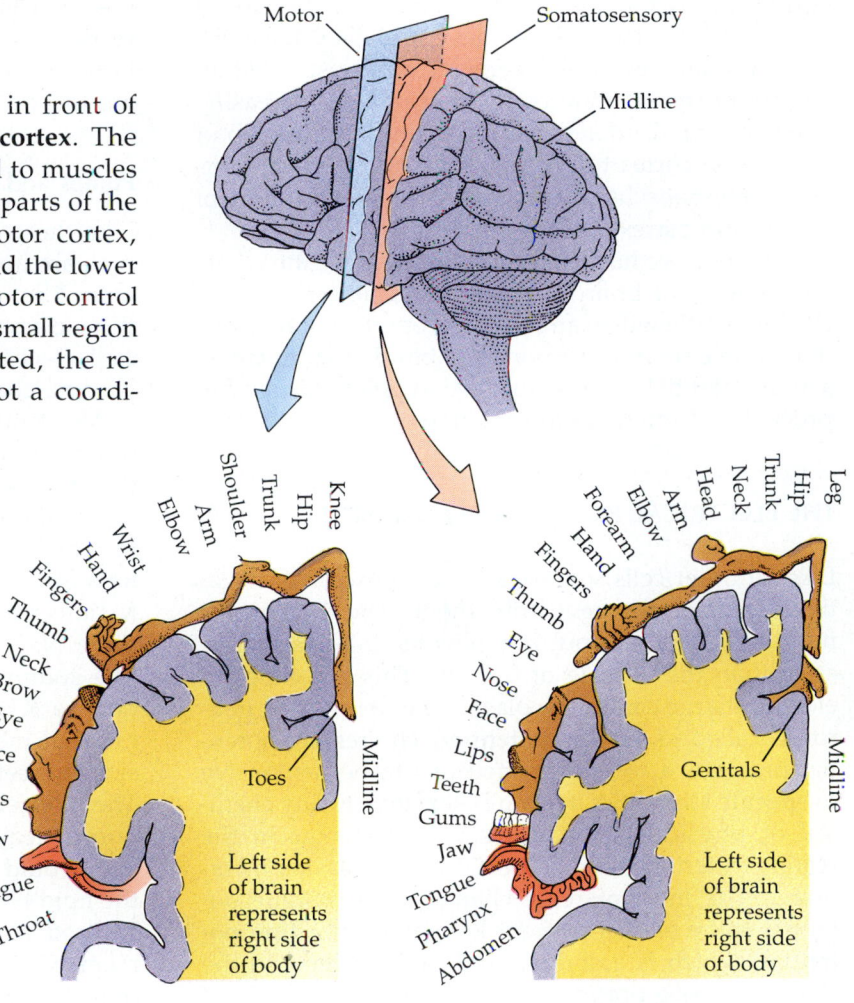

38.10   The Body Is Represented in the Primary Motor and Somatosensory Cortexes
Cross sections through the primary somatosensory cortex and primary motor cortex can be represented as maps of the human body. Body parts are shown in relation to the brain area devoted to them. The left side of the body is represented in the right cerebral cortex, and the right side of the body is represented in the left cerebral cortex.

Large areas of the cerebral cortex do not receive direct sensory input and are designated as **association cortex**. Cells in the association cortex receive input from multiple sensory areas and send output to multiple motor areas. Areas of association cortex responsible for reading, language, speech, and body awareness are indicated in Figure 38.9*b*.

As we mentioned earlier, the size of the telencephalon relative to the rest of the brain increases substantially as we go from fish to amphibian, to reptile, to birds and mammals. Even when we consider only mammals, the cerebral cortex increases in size and complexity when going from animals such as rodents, whose behavioral repertoires are relatively simple, to animals such as primates that have much more complex behavior. The most dramatic increase in the size of the cerebral cortex took place during the last several million years of human evolution. The incredible intellectual capacities of *Homo sapiens* are the result of the enlargement of the cerebral cortex. Humans do not have the largest brains in the animal kingdom; elephants, whales, and porpoises have larger brains in terms of mass. If we compare brain size to body size, however, humans and dolphins top the list. Humans have the largest ratio of brain size to body size, and they have the most highly developed cerebral cortexes. Another feature of the cerebral cortex that reflects increasing behavioral and intellectual capabilities is the ratio of association cortex to primary sensory and motor cortexes. Humans have the largest relative amount of association cortex.

Now that we have some appreciation for the whats and wheres of brain functions, it is time to turn to the hows. We will examine the properties of neurons that enable them to respond to stimulation, to create action potentials, to conduct action potentials, and to process and integrate information.

THE ELECTRICAL PROPERTIES OF NEURONS

Like all other cells, neurons have an excess of negative electrical charge within them. The inside of a neuron is usually about 70 millivolts (mV) more negative than the outside of the cell. This difference in electric charge across the plasma membrane of a neuron is called its **resting potential**. The resting potential provides a means for neurons to be responsive to specific stimuli. A neuron is sensitive to any chemical or physical factor that causes a change in the resting potential across a portion of its plasma membrane. The most extreme change in membrane potential is the electrical event known as an **action potential**, which is a sudden and rapid reverse in the charge across a portion of the membrane. For a brief moment, only one or two milliseconds, the inside of

that part of the membrane becomes more positive than the outside. An action potential can move along a membrane from one part of a neuron to its farthest extensions. This conduction of action potentials along the membranes of neurons is how the nervous system transmits information.

To understand how resting potentials are created, how they are perturbed, and how action potentials are generated and conducted along membranes, it is necessary to know a little about electricity, ions, and the special ion-channel proteins in the membranes of neurons. Voltage is the tendency for electrons to move between two points. Voltage is to the flow of electrons what pressure is to the flow of water. If the negative and the positive poles of a battery are connected by a copper wire, electrons flow from negative to positive. This flow of electrons can be used to do work. As you may recall from Chapter 2, electric charges cross cell membranes not as electrons but as charged ions. The major ions that carry electric charges across the membranes of neurons are sodium (Na^+), chloride (Cl^-), potassium (K^+), and calcium (Ca^{2+}). It is also important to remember that ions with opposite charges attract each other. With these basics of bioelectricity in mind, we can ask how the resting potential of the membrane is maintained, and how the flows of ions through membrane channels are turned on and off to generate action potentials.

Pumps and Channels

Like those of all other cells, the membranes of neurons are lipid bilayers that are rather impermeable to ions. The membranes of neurons, however, contain three classes of proteins that give neurons their special electrical properties. These proteins act as pumps, channels, and receptors.

Membrane pumps use energy to move ions or other molecules against their concentration gradients. The major pump in neuronal membranes is the sodium–potassium pump that we learned about in Chapter 5. The action of this pump expels Na^+ ions from the inside of the cell, exchanging them for K^+ ions from outside the cell. The pump expels about three Na^+ ions for every two K^+ ions it brings in. The sodium–potassium pump keeps the concentration of K^+ inside the cell greater than that of the external medium, and the concentration of Na^+ inside the cell less than that of the external medium. The concentration differences established by the pump mean that K^+ would diffuse out of the cell and Na^+ would diffuse into the cell if the ions could cross the lipid bilayer. By itself, the unequal distribution of K^+ and Na^+ ions on the two sides of the plasma membrane does not create the resting potential of the cell. To explain the resting potential, we must introduce ion channels.

Ion channels are pores formed by proteins in the lipid bilayer. These water-filled pores allow ions to pass through. They are selective—that is, they may allow only one type of ion to pass through; thus there are potassium channels, sodium channels, chloride channels, and calcium channels. Ions can move in either direction through a channel. Most ion channels of neurons behave as if they contain a "gate" that opens to allow ions to pass under some conditions, but closes under other conditions. **Voltage-gated** channels open or close in response to the voltage across the membrane, and **chemically gated** channels open or close depending on the presence or absence of a specific chemical that binds to the channel protein or to a separate receptor that in turn alters the channel protein. Both voltage-gated and chemically gated channels play important roles in neuronal functions, as we will see, but first we will see how nongated channels, channels that are always open, are responsible for maintaining the resting potential in neurons and other cells.

The Resting Potential

Potassium channels are the most common type of open channel in the plasma membranes of neurons. These channels make neurons much more permeable to K^+ than to any other ions. As Figure 38.11 shows, this characteristic is what explains the resting potential. Because the neuron's plasma membrane is permeable to K^+, and because the sodium–potassium pump keeps the concentration of K^+ inside the cell much higher than that outside the cell, K^+ tends to diffuse out of the cell. If a K^+ ion leaves the cell, it leaves behind an unmatched negative charge. The tendency of the K^+ ions to diffuse out of the cell through the open channels thus causes the inside of the cell to become negatively charged in comparison to the external medium. Negative charges attract positive charges, such as those carried by K^+ ions. Eventually, a balance is established between the tendency for K^+ ions to diffuse down their concentration gradient to the outside of the cell and the attraction of the unmatched negative charges to pull them back inside the cell. The resting potential is the voltage difference that creates that balance. If you know the K^+ concentrations inside and outside the cell, you can calculate the resting potential of the membrane (Box 38.A).

Changes in Membrane Potentials

Because K^+ ions tend to leave the cell, the plasma membrane is polarized—that is, regions of unequal electric charge are separated, resulting in a resting potential. If a stimulus perturbs this resting potential, and the inside of the cell becomes *more* negative, the

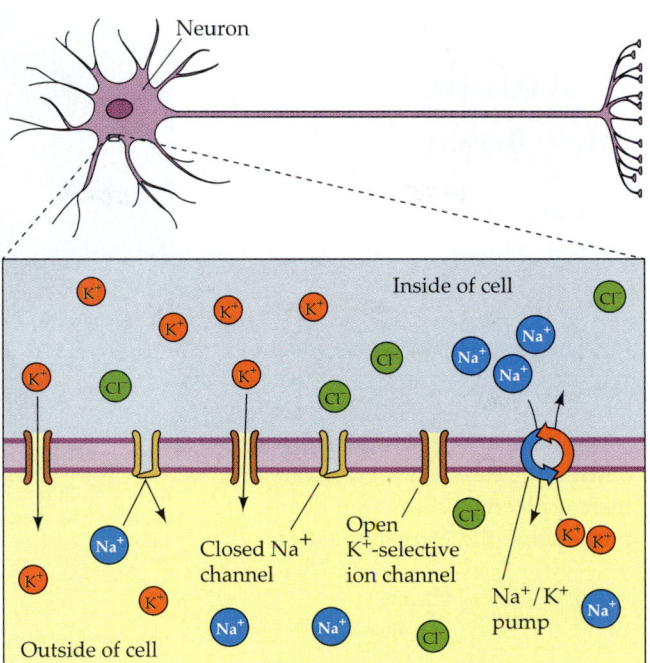

38.11 Ion Pumps and Ion Channels Produce the Resting Potential
The sodium–potassium pumps of a neuron establish ion concentration gradients across its plasma membrane. Under resting conditions more K^+ channels are open in the membrane than are Na^+ channels. Because of the diffusion of K^+ ions out of the cell, the membrane has an excess of positive charges on the outside and negative charges on the inside. This difference between the two charges is the resting potential.

membrane is **hyperpolarized**. If the inside of the cell becomes *less* negative, the membrane is **depolarized**. Changes in the gated channels can cause a membrane to hyperpolarize or depolarize. Think, for example, what would happen if some sodium channels in the cell membrane opened. Na^+ ions would diffuse into the cell because of their higher concentration on the outside of the membrane, and they would also be attracted into the cell by the excess negative charge. As a result of the entry of Na^+ ions, the membrane would become depolarized in comparison to its resting condition.

There are also gated Cl^- channels in neuronal membranes, and the concentration of Cl^- ions is greater in the extracellular fluid than in the intracellular fluid. What would happen to membrane polarity if some of these Cl^- channels opened? The opening and closing of ion channels resulting in changes in the polarity of the membrane is the basic mechanism by which neurons respond to electrical, chemical, or other stimuli.

What good does it do for a neuron to undergo a change in its resting membrane potential at a particular location? Can that information be passed on to

BOX 38.A

Calculating Membrane Potentials

This stellate ganglion from a squid is almost 6 millimeters at its widest point. Giant axons extend from the huge neurons.

If we assume that the membrane of a cell is permeable to only one type of ion, such as K^+, and if we know the concentration of that ion inside and outside the cell, we can calculate the resting potential across the membrane using the **Nernst equation**:

$$E_K = \frac{RT}{zF} \ln \frac{[K^+]_{out}}{[K^+]_{in}}$$

where E_K is the **potassium equilibrium potential** in millivolts (mV). This is the potential difference across the membrane that will prevent a net diffusion of K^+ ions down their concentration gradient. R is the universal gas constant, T is the absolute temperature, z is the number of charges carried by each ion, F is the Faraday constant, and $\ln [K^+]_{out}/[K^+]_{in}$ is the natural logarithm of the K^+ concentration difference.

This equation may seem complex, but think of it this way: When the cell is in equilibrium, the amount of work necessary to move a K^+ ion into the cell against its concentration gradient equals the amount of work necessary to move a positive electric charge out of the cell against the voltage difference. Rather than consider just the tiny amounts of work associated with moving single ions or charges, we calculate in terms of moles of ions or charges. The work

to move one mole of K^+ ions is equal to $(RT) (\ln [K^+]_{out}/[K^+]_{in})$, and the work to move one mole of positive charges is equal to $E_K zF$.

Using the Nernst equation is easier if we assume a temperature of 20°C (293 K), combine all constants, change to base 10 logarithms, and convert the equation to the following form:

$$E_K = (58 \text{ mV}) \left(\ln \frac{[K^+]_{out}}{[K^+]_{in}} \right)$$

For example, consider the huge neuron from a squid, shown in the photograph. The concentration of K^+ outside the neuron is 20 mM, whereas the concentration of K^+ inside the neuron is 400 mM. The base 10 logarithm of 20/400 is -1.3. Plugging this number into the simplified Nernst equation gives us a potassium equilibrium potential of -75 mV.

This calculated value is not too far from the measured resting potential in this cell, -60 mV. The Nernst equation can be used to calculate the equilibrium potential for any ion to which the membrane is permeable.

Why is the calculated potassium equilibrium potential using the Nernst equation only close to, and not the same as, the resting potential recorded from this cell using an electrode? In reality, the membrane is somewhat permeable to other ions in addition to K^+. Those other ions also influence the membrane potential, but less so because the membrane is much less permeable to these ions than it is to K^+ ions. Another equation, the Goldman equation, includes all of the ions to which the membrane is permeable, along with the relative permeabilities of the membrane to those ions. When the Goldman equation is used, the calculated membrane potential matches the measured potential extremely well. The Goldman equation for the squid axon would be:

$$V_m = \frac{RT}{F} \ln \frac{P_K[K^-]_{out} + P_{Na}[Na^+]_{out} + P_{Cl}[Cl^-]_{in}}{P_K[K^-]_{in} + P_{Na}[Na^+]_{in} + P_{Cl}[Cl^-]_{out}}$$

The P's in the equation stand for relative membrane permeabilities; for the resting squid neuron they are $P_K : P_{Na} : P_{Cl} = 1 : 0.04 : 0.45$. When these values are used to calculate V_m, the resting potential for the squid neuron, the result is -60 mV, the same as the measured resting potential.

Nernst Equilibrium Potentials of Ions in and around Squid Neurons

ION	CONCENTRATION IN CYTOPLASM (mM)	CONCENTRATION IN EXTRACELLULAR FLUID (mM)	NERNST POTENTIAL (mV)
K^+	400	20	-75
Na^+	50	440	$+55$
Cl^-	52	560	-60

other parts of the cell? A local perturbation of membrane potential causes electric currents to flow, and the flow spreads the change in membrane potential. The nerve cell is a poor conductor of electricity, however, and the change in membrane polarity diminishes and disappears before it gets very far from the site of stimulation. Communication of a stimulus by the flow of electric current is useful only over very short distances. As we will see, however, electric current is an important part of the mechanisms by which synapses and sensory stimuli generate action potentials, and it is also involved in the propagation of an action potential along an axon.

Action Potentials

Action potentials enable neurons to convey information over long distances with no loss of the signal. An action potential is a sudden and major change in membrane potential that lasts for only one or two milliseconds. It is conducted along the axon of a neuron at speeds up to 100 m/s, which is equivalent to running the length of a football field in one second. Using an electrode—a fine, electrically insulated wire or a very thin glass pipette containing a solution that conducts electric charges—we can record very tiny, local, electrical events that occur across cell membranes. If we place the tips of a pair of electrodes on the two sides of the membrane of a resting axon and measure the voltage difference, it is about -70 mV (Figure 38.12). If these electrodes are exposed to an action potential traveling down the axon, they register a rapid change in membrane potential, from -70 mV to about $+40$ mV. The membrane potential rapidly returns to its resting level of -70 mV as the action potential passes by. At any location along this axon we could insert another pair of electrodes and record the same action potential. The height of the action potential does not change as it travels along the axon. The action potential is an all-or-nothing, self-regenerating event.

Voltage-gated sodium channels are primarily responsible for the action potential. At a normal membrane resting potential these channels are mostly closed. If a stimulus or a synaptic input makes the membrane less negative—that is, depolarizes it—these Na^+ channels have a higher probability of flipping open briefly—for less than a millisecond. Because of the action of the sodium–potassium pump, Na^+ concentration is much higher outside the axon than inside, so whenever the sodium channels open in a part of the membrane, Na^+ ions from the outside enter the cell at that location. The entering Na^+ ions make the inside of the membrane more positive. Eventually a membrane potential is reached at which so many Na^+ channels open that the membrane potential suddenly rises to a very positive value.

The opening of the Na^+ channels causes the rise of the action potential—what neurobiologists call the spike. What causes the return of the membrane to resting potential? The main reason for the drop is that, after opening briefly, the sodium channels close and remain inactive for a few milliseconds. This is long enough for the membrane to return to resting potential. Some axons also have voltage-gated potassium channels. Because these channels open more slowly than the sodium channels and stay open longer, they help return the voltage across the membrane to its resting level by allowing K^+ ions to carry excess positive charges out of the cell.

The behavior of the voltage-gated sodium channels can be explained by assuming that they have two voltage-sensitive gates, an activation gate and an inactivation gate. Under resting conditions the activation gate is closed and the inactivation gate is open. Depolarization of the membrane to threshold causes both gates to change state, but the activation gate responds faster. As a result, the channel is open for the passage of Na^+ ions for a brief period of time between the opening of the activation gate and the closing of the inactivation gate. The inactivation gates remain closed for a few milliseconds before they spontaneously open again, thus explaining why the membrane has a **refractory period** (a period during which it cannot act) before it can fire another action potential. When the inactivation gates finally open, the activation gates are closed and the membrane is poised to respond once again to a depolarizing stimulus by firing another action potential.

The difference in concentration of Na^+ ions across the plasma membrane of neurons is the "battery" that drives the action potential. How rapidly does the battery run down? It might seem that a substantial number of Na^+ and K^+ ions would have to cross the membrane for the membrane potential to go from -70 mV to $+40$ mV, and back to -70 mV again. In fact, only about one Na^+ (or K^+) ion in 10 million actually moves through the channels during the passage of an action potential. Thus the effect of a single action potential on the concentration ratios of Na^+ or K^+ is very small. Even hundreds of action potentials barely change the concentration differences of Na^+ and K^+ on the two sides of the membrane. Thus it is not difficult for the sodium–potassium pump to keep the "battery" charged, even when the cell is generating many action potentials every second.

Propagation of the Action Potential

How does an action potential move over long distances? When one part of an axon fires an action potential, the adjacent regions of membrane also become depolarized because of the spread of local electric current. Such movements of ions depolarize the

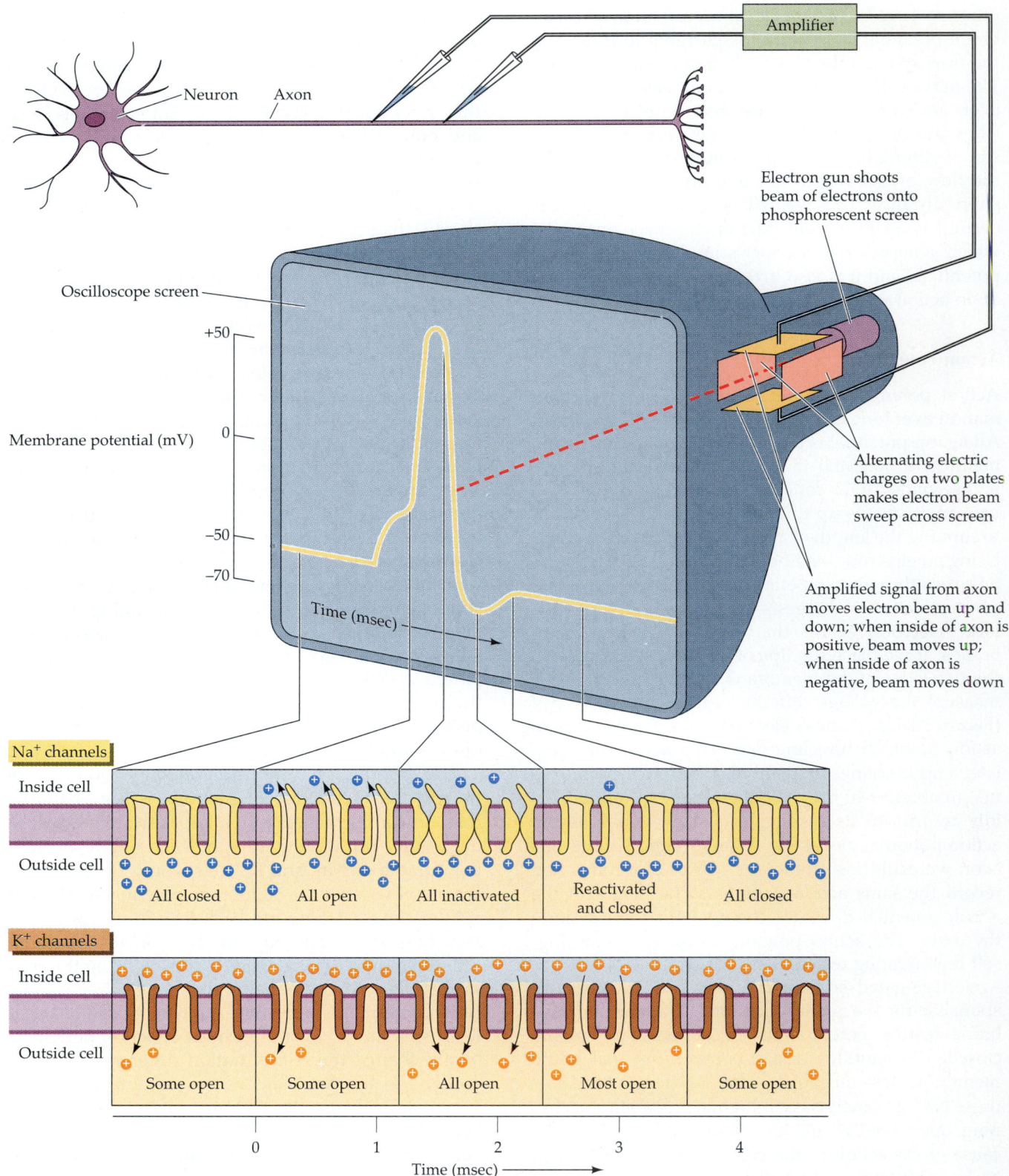

Amplifier

Neuron   Axon

Electron gun shoots beam of electrons onto phosphorescent screen

Oscilloscope screen

Membrane potential (mV)

+50

0

−50
−70

Time (msec)

Alternating electric charges on two plates makes electron beam sweep across screen

Amplified signal from axon moves electron beam up and down; when inside of axon is positive, beam moves up; when inside of axon is negative, beam moves down

Na⁺ channels

Inside cell

Outside cell

All closed   All open   All inactivated   Reactivated and closed   All closed

K⁺ channels

Inside cell

Outside cell

Some open   Some open   All open   Most open   Some open

0   1   2   3   4

Time (msec)

38.12   The Action Potential Can Be Visualized on an Oscilloscope

A pair of electrodes detects an action potential as a voltage change across the membrane of an axon. The signal from the electrodes is amplified and fed into an oscilloscope. A beam of electrons sweeps across the screen in a set period of time. That beam is deflected up if the signal from the electrodes is positive and down if the signal is negative. Thus the action potential is seen on the screen as a change in membrane potential through time. The action potential is created in the axon by voltage-gated Na⁺ and K⁺ channels opening and closing, as depicted at the bottom of the figure. The membrane potential at any given time depends on which and how many channels are open.

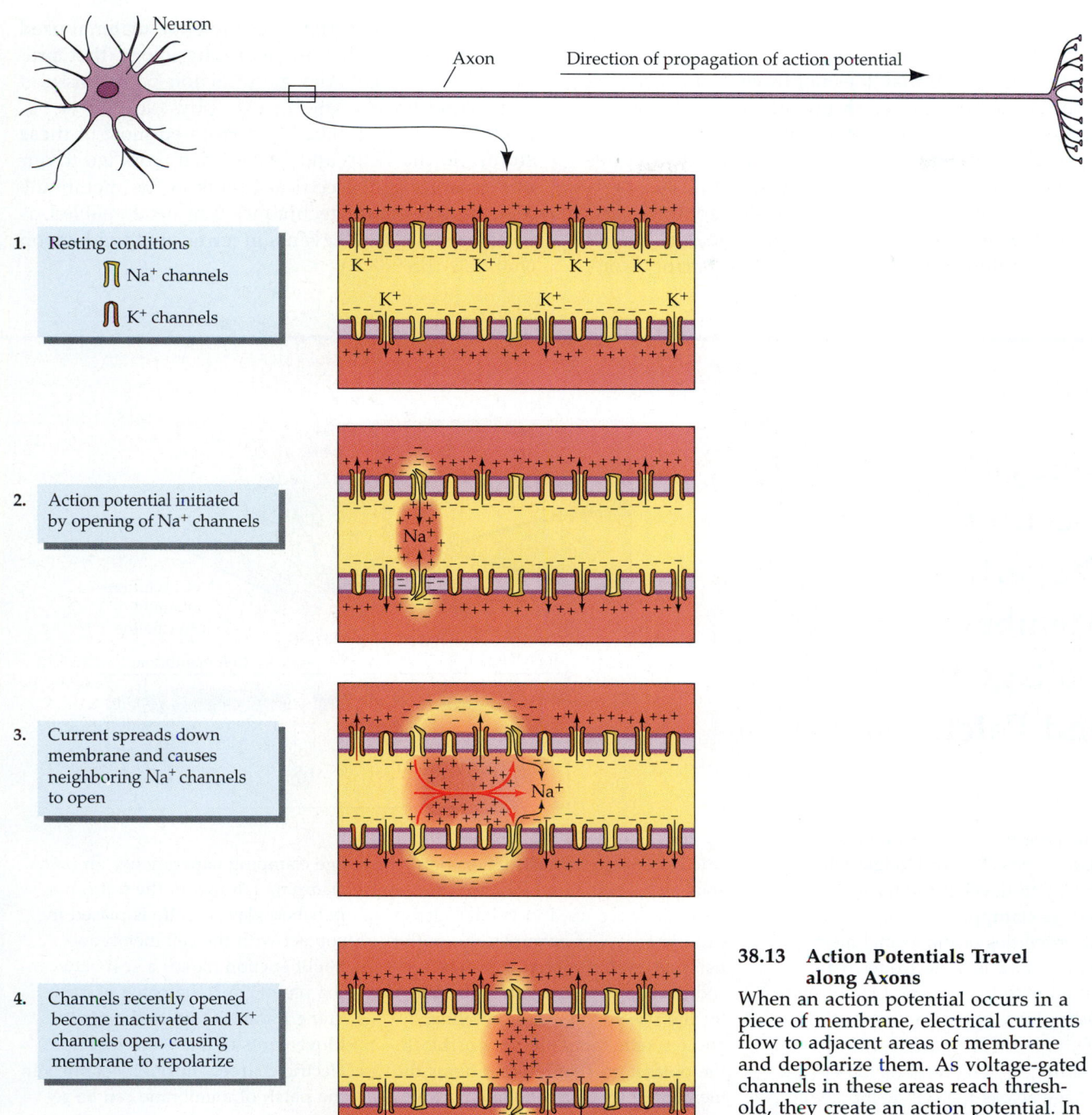

Neuron

Axon   Direction of propagation of action potential

1. Resting conditions

Π Na⁺ channels

Π K⁺ channels

2. Action potential initiated by opening of Na⁺ channels

Na⁺

3. Current spreads down membrane and causes neighboring Na⁺ channels to open

Na⁺

4. Channels recently opened become inactivated and K⁺ channels open, causing membrane to repolarize

5. Action potential continues to be regenerated along axon as steps **3** and **4** are repeated

38.13 Action Potentials Travel along Axons
When an action potential occurs in a piece of membrane, electrical currents flow to adjacent areas of membrane and depolarize them. As voltage-gated channels in these areas reach threshold, they create an action potential. In this way, an action potential continuously regenerates itself along the axon.

region of the membrane adjacent to that experiencing the action potential (Figure 38.13). Outside the axon, positive ions flow rapidly toward the depolarized region, attracted by the negative charge balance there. Inside the cell, positive charges move *away* from the depolarized region, where they are more abundant, and *toward* the adjacent, more negative regions. The net result for the membrane is a ten-

dency to repolarize at the point of the existing action potential and to depolarize at the adjacent regions. If the depolarization of an adjacent region of membrane brings it to the threshold level that causes massive opening of sodium channels, an action potential is generated. Because an action potential always brings the area of membrane adjacent to it to threshold, it is self-regenerating and propagates itself along the axon. The action potential propagates itself in only one direction; it cannot reverse itself because the part of the membrane it came from is undergoing its refractory period.

The action potential does not travel along all axons at the same speed. Action potentials travel faster in large-diameter axons than in small-diameter axons. Among invertebrates, the axon diameter determines the rate of conduction, and axons that transmit messages involved in escape behavior are very thick. The axons that enable squid to escape predators are almost a millimeter in diameter. These giant axons were the most important experimental material used in the classic studies in neurophysiology that produced basic discoveries about action potentials and their conduction. The British physiologists A. L. Hodgkin and A. F. Huxley performed most of these studies in the 1940s and 1950s. Their work led to our understanding of electrical events in the membrane and to more-recent techniques that have enabled us to understand these events in terms of individual ion channels (Box 38.B).

BOX 38.B

Studying the Electrical Properties of Membranes with Voltage Clamps and Patch Clamps

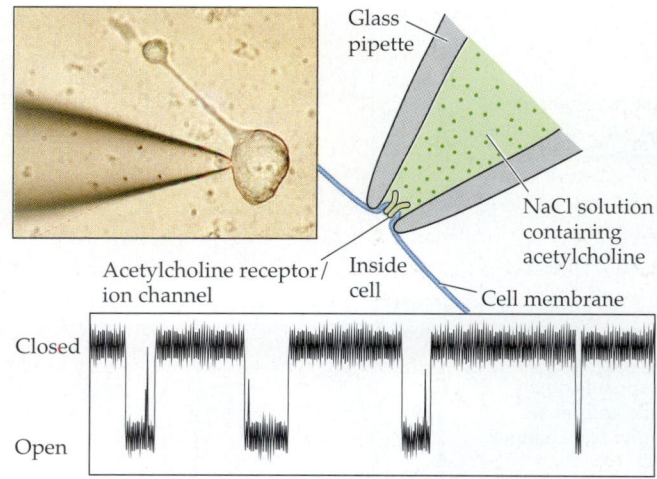

The large size of the squid axon made it possible for Hodgkin and Huxley to develop the technique of **voltage clamping** to study the electrical properties of the axonal membranes. Two fine electrodes were placed within the axon. One of these electrodes was used to measure voltage, the other to pass electric current into the cell to control (or clamp) the voltage across the cell membrane. Monitoring the change in the current necessary to maintain the voltage clamp at a particular level gave a direct measure of the ionic currents across the membrane. Thus, when the membrane was depolarized by an action potential, it was possible to measure the changing electric currents due to Na^+ ions and K^+ ions flowing across the membrane. Hodgkin and Huxley were able to determine the separate contributions of the Na^+ and K^+ ions by changing the concentrations of these ions in-

side or outside the cell. Later, poisons and drugs that block specific channels were used in voltage clamp experiments. For example, the puffer fish poison, tetrodotoxin, blocks the sodium channel that is responsible for the rise in the action potential, and tetraethylammonium ions block the potassium channel that helps the membrane repolarize following the action potential. Voltage clamp studies led to the first explanations of action potentials in terms of the properties of ion channels, but those ion channels were only postulated because there were no means of observing them directly.

A technique called **patch clamping**, developed in the 1980s by Bert Sakmann and Erwin Neher, made it possible to study single ion channels in membranes and therefore the molecular basis for the electrical properties of membranes revealed by volt-

age clamping experiments. In patch clamping (shown in the figure), a polished glass pipette is placed in contact with the cell membrane. Slight suction makes a seal between the pipette and the patch of membrane under the tip of the pipette. Movements of ions, and therefore electric charges, through channels in the patch of membrane can be recorded through the pipette. The solution filling the pipette determines the ion concentrations on the outside of the patch and may also contain chemicals to bind to receptors. If the patch is torn loose from the cell, the ion concentrations on the interior side of the membrane can also be changed. A patch may contain only one or a few ion channels; thus the electrical recording from that patch can show individual channels opening and closing. Neher and Sakmann received the Nobel prize in 1991.

In nervous systems more complex than those of invertebrates, increasing the speed of action potentials by increasing the diameter of axons would result in enormous nerves. The optic nerve from each of our eyes contains about one million axons. If we used plain, simple axons to build optic nerves that conduct information from the eye to the brain as fast as ours do, each optic nerve would have to be about the diameter of the eyeball itself. Other groups of axons in the brain, the spinal cord, and the peripheral nervous system would be equally unwieldy. Evolution

has increased propagation velocity in vertebrate axons in a different way. The axons of vertebrates are insulated with membranous wrappings produced by specialized glial cells called **Schwann cells**. The membranous wrapping is called **myelin**. It is the myelin that gives the light, shiny appearance to "white matter," which is any nervous system tissue containing mostly axons. The myelination of an axon is not continuous. At regular intervals called **nodes of Ranvier** the bare axon is exposed (Figure 38.14a).

How do the nodes of Ranvier and the myelin insulation of the axon increase propagation velocity? At the nodes of Ranvier an axon can fire action potentials, but in the adjacent regions of the axon that are insulated with myelin, electric charges cannot accumulate or cross the membrane. Therefore, the local depolarization caused by the action potential at a node causes electric current to flow to the next place in the membrane where charges can accumulate and cross—the next node of Ranvier. Electric current flows very fast compared to how long ion channels take to open and close. Thus the depolarization caused by an action potential at one node spreads

(a) Myelination

38.14 Saltatory Action Potentials
(a) Some axons are myelinated by the wrappings of Schwann cell membranes. (b) Action potentials can occur only at gaps in the myelin wrap (nodes of Ranvier), but electric currents created by an action potential can flow to adjacent nodes and depolarize them. As a result, the action potential is conducted down the axon by jumping from node to node.

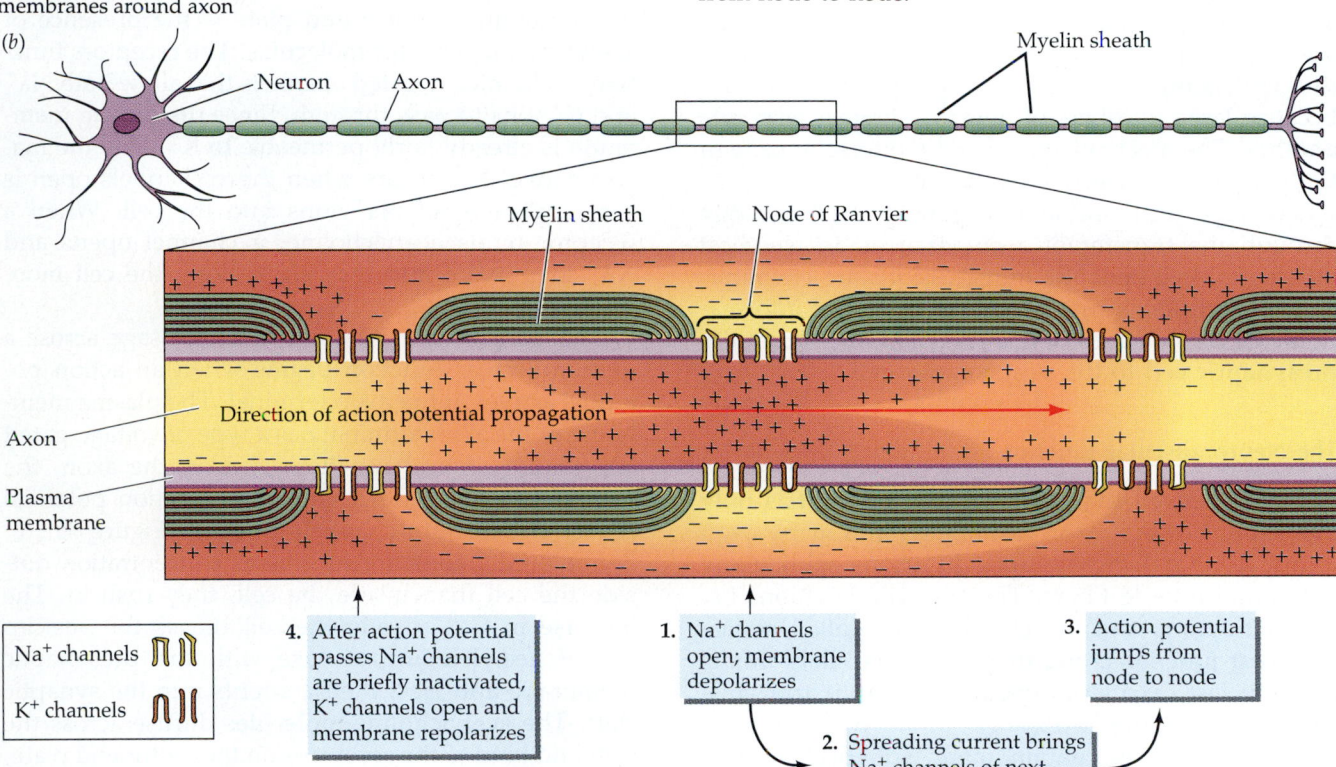

very rapidly to the next node. When that node reaches threshold, it fires an action potential, and so forth, with the action potentials jumping from node to node down the axon (Figure 38.14*b*). This form of impulse propagation is called **saltatory** (jumping) **conduction** and is much quicker than continuous impulse conduction down an unmyelinated axon.

You have probably experienced the difference in the velocity at which action potentials travel down myelinated and unmyelinated axons. If you touch a very hot or very cold object, you experience a sharp pain before you sense whether the object is hot or cold. With the sensing of the temperature also comes a burning pain different from the first sharp pain. Sensory axons carrying sharp-pain sensation are myelinated, but most axons carrying information about temperature, as well as axons carrying the sensation of burning, aching pain, are unmyelinated. As a result, you know that something is wrong before you know what it is or how bad it is. The unmyelinated temperature and burning-pain axons conduct impulses at only 1 to 2 meters per second, whereas the myelinated sharp-pain axons conduct impulses at velocities up to 5 or 6 m/s. The largest myelinated axons in the human nervous system conduct impulses at velocities up to 120 meters per second.

SYNAPTIC TRANSMISSION

The most remarkable abilities of nervous systems stem from the interactions of neurons. These interactions process and integrate information. Our nervous systems can orchestrate complex behaviors, deal with complex concepts, and learn and remember because large numbers of neurons interact with one another. The mechanisms for these interactions lie in the synapses between cells. **Synapses** are junctions where one cell influences another cell directly through the transfer of a chemical or an electrical message. Most synaptic transmissions are chemical, and we will focus first on those. Chemical information crosses synapses in one direction only, from the **presynaptic cell** to the **postsynaptic cell**.

The Neuromuscular Junction

A motor neuron that innervates a muscle has only one axon, but that axon can branch into many axon terminals that form synaptic junctions with many individual muscle fibers. The synaptic junctions between neurons and muscle fibers are called neuromuscular junctions, and they are an excellent model for how fast, excitatory chemical synaptic transmissions work (Figure 38.15).

Axon terminals contain many spherical vesicles filled with chemical messenger molecules called **neurotransmitters**. The neurotransmitter of all motor

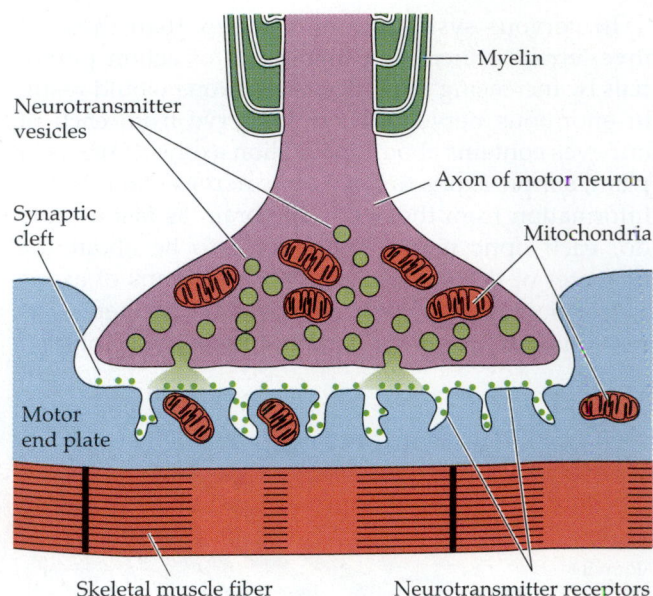

38.15   The Neuromuscular Junction Is a Chemical Synapse
A motor neuron communicates chemically with muscle cells at the neuromuscular junction when transmitter (green) crosses the synaptic cleft.

neurons that innervate vertebrate skeletal muscles is acetylcholine. The postsynaptic membrane is part of the muscle cell's plasma membrane, but it is slightly modified in the area of the synapse and is called a **motor end plate**. The modification that makes a patch of membrane a motor end plate is the presence of acetylcholine receptor molecules. The receptors function as chemically gated channels that allow both Na^+ and K^+ ions to pass through. Since the resting membrane is already fairly permeable to K^+ ions, the major change that occurs when these channels open is the movement of Na^+ ions into the cell. When a receptor binds acetylcholine, a channel opens and Na^+ ions move into the cell, making the cell more positive inside.

The transmission of a chemical message across a neuromuscular junction begins when an action potential arrives at the axon terminal. The plasma membrane of the axon terminal has a type of voltage-gated ion channel found nowhere else on the axon: the voltage-gated calcium channel. The action potential causes the calcium channels to open (Figure 38.16). Because Ca^{2+} ions are in greater concentration outside the cell than inside the cell, they rush in. The increase in Ca^{2+} inside the cell causes the vesicles full of acetylcholine to fuse with the presynaptic membrane and eject their contents into the synaptic cleft. The acetylcholine molecules diffuse across the cleft and bind to the receptors on the motor end plate, causing the sodium channels to open briefly and depolarize the postsynaptic cell membrane.

1. Action potential arrives at axon terminal (Na⁺ channels open, depolarizing terminal membrane)

2. Depolarization of terminal membrane causes voltage-gated Ca²⁺ channels to open. Ca²⁺ enters cell and triggers fusion of transmitter vesicles with presynaptic membrane

3. Neurotransmitter molecules diffuse across synaptic cleft and bind to receptors on post-synaptic membrane. Activated receptors open chemically gated Na⁺ channels and depolarize postsynaptic membrane

4. Neurotransmitter is broken down and taken back up to be reused in acetylcholine synthesis

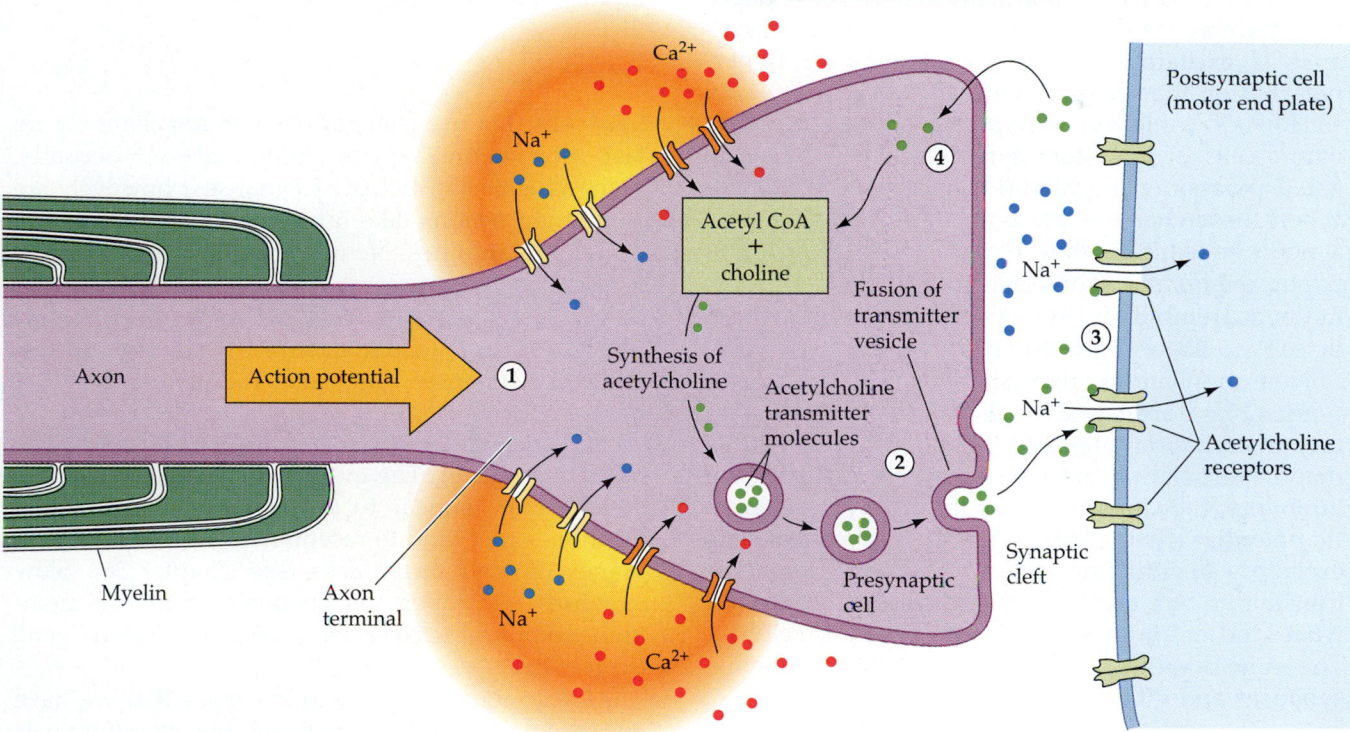

38.16 Synaptic Transmission Begins with the Arrival of an Action Potential

Events in the Postsynaptic Membrane

The postsynaptic membranes differ from the presynaptic membranes in an important way: Because motor end plates have very few voltage-gated sodium channels, they do not fire action potentials. This is true not only of motor end plates, but also of dendrites and of most regions of nerve cell bodies. The binding of neurotransmitter to receptors and the resultant opening of chemically gated ion channels perturbs the resting potential of the postsynaptic membrane. This local change in membrane potential spreads to neighboring regions of the plasma membrane of the postsynaptic cell. Eventually, the spreading depolarization may reach an area of membrane that does contain voltage-gated channels. The entire plasma membrane of a skeletal muscle fiber, except for the motor end plates, has voltage-gated sodium channels. If the axon terminal of a presynaptic cell releases sufficient amounts of neurotransmitter to depolarize a motor end plate enough to bring the surrounding membrane to threshold, action potentials are fired in those areas of membrane. These action potentials are then conducted throughout the muscle

fiber's system of membranes, causing the fiber to contract. (We'll learn about the coupling of muscle membrane action potentials and contraction of muscle fibers in Chapter 40.)

How much neurotransmitter is enough? Neither a single acetylcholine molecule nor the contents of an entire vesicle (about 10,000 acetylcholine molecules) is enough to bring the plasma membrane of a muscle cell to threshold. A single action potential in an axon terminal, however, releases about 100 vesicles, which is enough to fire an action potential in the muscle fiber and cause it to twitch.

Excitatory and Inhibitory Synapses

In vertebrates, the synapses between motor neurons and skeletal muscle are always excitatory; that is, motor end plates always respond to acetylcholine by depolarizing. There are several different kinds of synapses between neurons, however. Recall that a given neuron may have many dendrites. Axon terminals from many other neurons may make synapses with those dendrites and with the cell body. The axon terminals of different presynaptic neurons may store and release different neurotransmitters, and membranes of the dendrites and cell body of a postsynaptic neuron may have receptors to a variety of neurotransmitters. Thus a given postsynaptic neuron can receive various chemical messages. If the postsynaptic neuron's response to a neurotransmitter is depolarization, as at the neuromuscular junction, the synapse is excitatory, but if the response is hyperpolarization the synapse is inhibitory.

How do inhibitory synapses work? The postsynaptic cells in inhibitory synapses have chemically gated potassium or chloride channels as receptors. When these channels are activated by binding with a neurotransmitter, they hyperpolarize the postsynaptic membrane. Thus the release of neurotransmitter at an inhibitory synapse makes the postsynaptic cell *less* likely to fire an action potential.

Neurotransmitters that depolarize the postsynaptic membrane are excitatory and bring about an **excitatory postsynaptic potential** (EPSP). Neurotransmitters that hyperpolarize the postsynaptic membrane are inhibitory; they bring about an **inhibitory postsynaptic potential** (IPSP). However, whether a synapse is excitatory or inhibitory depends not on the neurotransmitter but on the postsynaptic receptors—on what kind of ion channels the postsynaptic cell has. The same neurotransmitter can be excitatory at some synapses and inhibitory at others.

Summation

Individual neurons "decide" whether or not to fire action potentials by summing excitatory and inhibitory postsynaptic potentials. This summation ability of neurons is the major mechanism by which the nervous system integrates information. Each neuron may receive 10,000 or more synaptic inputs, yet it has only one output—an action potential in a single axon. All the information contained in the thousands of inputs a neuron receives is reduced to the rate at which that neuron generates action potentials in its axon. For most neurons the critical area for "decision making" is the **axon hillock**, the region of the cell body at the base of the axon. The plasma membrane of the axon hillock is not insulated by glia and has many voltage-gated channels. Excitatory and inhibitory postsynaptic potentials from anywhere on the dendrites or the cell body spread to the axon hillock. If the resulting combined potential depolarizes this area of membrane to threshold, the axon fires an action potential. Because postsynaptic potentials decrease as they spread from the site of the synapse, all postsynaptic potentials do not have equal influences on the axon hillock. A synapse at the end of a dendrite has less influence than a synapse on the cell body near the axon hillock.

Excitatory and inhibitory postsynaptic potentials can be summed over space or over time. Spatial summation adds up the simultaneous influences of synapses from different sites on the postsynaptic cell (Figure 38.17a). Temporal summation adds up postsynaptic potentials generated at the same site in a rapid sequence (Figure 38.17b).

Other Synapses

Synapses that use chemically gated ion channels are fast; their actions happen within a few milliseconds. Some chemically mediated synapses, however, are slow; their actions take hundreds of milliseconds or even many minutes. Neurotransmitters at these slow synapses activate second-messenger systems, rather than directly controlling ion channels in the postsynaptic cell. Presynaptic events are the same in fast and slow synapses, but when the neurotransmitter of a slow synapse binds to a receptor, it activates a second messenger such as cAMP (cyclic adenosine monophosphate). The mechanisms of slow synapses are therefore similar to the mechanisms of certain hormones that bind to receptors in the plasma membranes of their target cells (see Chapter 36). Slow synapses may open ion channels, influence membrane pumps, activate enzymes, and induce gene expression.

All the neuron-to-neuron synapses that we have discussed up to now are between the axon terminals of a presynaptic cell and the cell body or dendrites of a postsynaptic cell. Synapses can also form between the axon terminals of one cell and the axon of another cell. Such a synapse can modulate how much neurotransmitter the second cell releases in response to action potentials traveling down its axon. We refer to this mechanism of regulating synaptic strength as **presynaptic excitation** or **presynaptic inhibition**.

Electrical synapses, or gap junctions, are completely different from chemical synapses because they directly couple neurons electrically. At gap junctions, the presynaptic and postsynaptic cell membranes are separated by a space of 2 to 3 nm, but the membrane proteins of the two neurons form connexons—molecular tunnels that bridge the two cells—through which ions and small molecules can readily pass (see Figure 5.8). Electrical transmission across gap junctions is very fast and can proceed in either direction; that is, stimulation of either neuron can result in an action potential in the other. Gap junctions are less common in the complex nervous systems of vertebrates than they are in the simple nervous systems of invertebrates, for two very important reasons. First, electrical continuity between neurons does not allow summation of synaptic inputs, which is what enables complex nervous systems to integrate information. Second, an effective electrical synapse requires a large area of contact between the presynaptic and postsynaptic cells. This condition rules out the possibility of thousands of synaptic inputs to a single neuron—which is the norm in complex nervous systems.

Neurotransmitters and Receptors

At present more than 25 neurotransmitters are recognized, and more will surely be discovered. No others are as thoroughly understood as **acetylcholine**, the neurotransmitter at all synapses between motor neurons and skeletal muscles (in the voluntary division of the peripheral nervous system). Acetylcholine and norepinephrine are the neurotransmitters of the other efferent division of the peripheral nervous system, the autonomic, which controls involuntary physiological functions. These two neurotransmitters also play roles in the central nervous system, but they constitute only a small percentage of the neurotransmitter content of the brain.

The workhorse neurotransmitters of the brain are simple amino acids. Glutamic acid and aspartic acid are excitatory, whereas glycine and gamma-aminobutyric acid (GABA) are inhibitory. Another important group of neurotransmitters is the monoamines, which are derivatives of amino acids. They include dopamine and norepinephrine (derivatives of tyrosine) and serotonin (a derivative of tryptophan). A number of peptides also function as neurotransmitters. A very exciting recent discovery is that two gases, carbon monoxide and nitric oxide, are used by neurons as intercellular messengers.

A neurotransmitter may have several different types of receptors in different tissues and may induce different actions. For example, acetylcholine has two well-known receptor types, called **muscarinic** and **nicotinic** because of other compounds that also bind to them. Nicotine, the active ingredient in tobacco, binds to acetylcholine receptors in the skeletal mus-

38.17   The Postsynaptic Membrane Integrates Information
Postsynaptic potentials can be summed over space or time. When the sum exceeds a threshold, action potentials are generated. (a) Spatial summation occurs when several postsynaptic potentials arrive at the axon hillock simultaneously. (b) Temporal summation means that postsynaptic potentials created at the same site in rapid succession can also be summed.

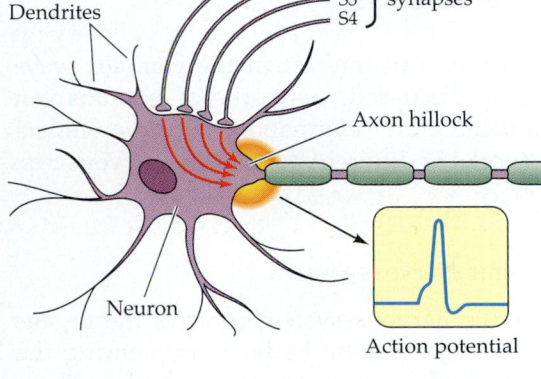

(a)   **Spatial summation**
Several EPSPs arriving simultaneously raise the action potential above threshold

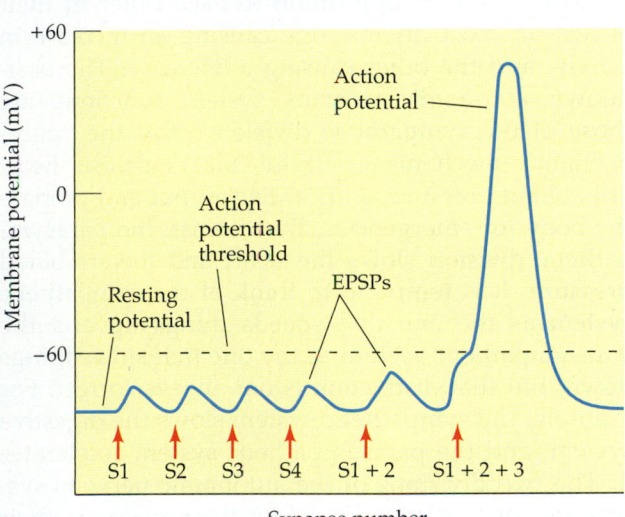

Synapse number

(b)   **Temporal summation**
Two or more EPSPs in rapid succession raise the action potential above threshold

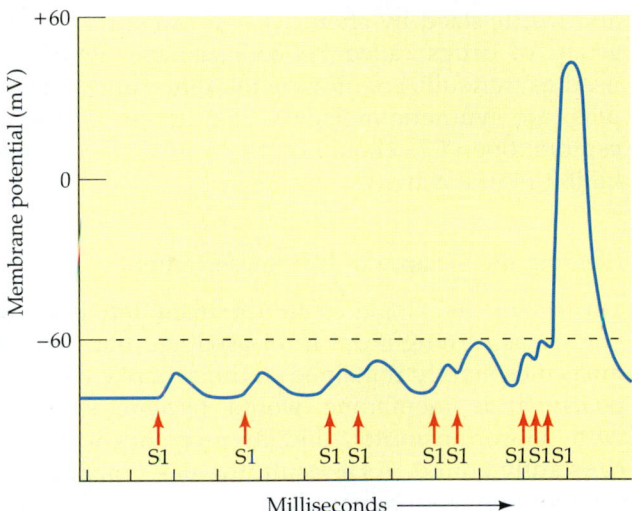

Milliseconds ⟶

cles, but not to those in heart muscle or in the autonomic nervous system. Muscarine, a compound found in the deadly poisonous mushroom *Amanita muscaria*, binds to the acetylcholine receptors in heart muscle and in the autonomic nervous system, but not to those in skeletal muscle. Both types of acetylcholine receptors are found in the central nervous system, where nicotinic receptors tend to be excitatory and muscarinic tend to be inhibitory. These receptors are the reason that smoking tobacco has behavioral and physiological effects and is addictive and why a number of cultures around the world have used *Amanita* mushrooms as hallucinogenic drugs.

The drug **curare**, extracted from the bark of a South American plant and used by native peoples to make poisoned darts and arrows, binds to nicotinic receptors but does not activate them. Therefore, skeletal muscles in an animal poisoned by curare cannot respond to motor neuron activation. The animal goes into flaccid (relaxed) paralysis and dies because it stops breathing. Curare is used medically to treat severe muscle spasms and to prevent muscle contractions that would interfere with surgery. Another compound, **atropine**, which is extracted from the plant *Atropa belladonna*, binds to muscarinic receptors and prevents acetylcholine from activating them. Atropine is used medically to increase heart rate, decrease secretions of digestive juices, and decrease spasms of the gut. Most people have encountered atropine; it is what the eye doctor uses to dilate the pupils for eye examinations. In the past atropine was used cosmetically to make the eyes look big and dark—hence the plant's species name, belladonna, meaning "beautiful lady." Of course, these beautiful ladies could not see very well.

The ability of compounds extracted from plants and animals to bind to certain neurotransmitter receptors is the basis for neuropharmacology, the study and development of drugs that influence the nervous system. Natural products are still an important source of drugs, but today many drugs are designed and synthesized by chemists. For example, a major group of drugs called benzodiazepines, which are used as tranquilizers, muscle relaxants, and sleeping pills, are synthetic molecules that act on GABA receptors, open Cl⁻ channels, hyperpolarize cells, and inhibit neural activity.

Clearing the Synapse of Neurotransmitter

Turning off the action of neurotransmitters is as important as turning it on. If released neurotransmitter molecules simply remained in the synaptic cleft, the postsynaptic membrane would become saturated with neurotransmitter, and its receptors would be constantly bound. As a result, the postsynaptic neuron would remain hyperpolarized or depolarized and would be unresponsive to short-term changes in the

presynaptic neuron. Thus neurotransmitter must be cleared from the synaptic cleft shortly after it is released by the axon terminal.

Neurotransmitter action is terminated in one of several ways. First, enzymes may destroy the neurotransmitter. For example, acetylcholine is rapidly destroyed by the enzyme **acetylcholinesterase**, which is present in the synaptic cleft in close association with the acetylcholine receptors on the postsynaptic membrane. Some of the most deadly nerve gases that were developed for chemical warfare work by inhibiting acetylcholinesterase. As a result, acetylcholine lingers in the synaptic clefts, causing the victim to die of spastic muscle paralysis. Some agricultural insecticides, such as malathion, also inhibit acetylcholinesterase and can poison farm workers if used without safety precautions. Second, neurotransmitter may simply diffuse away from the cleft. Third, neurotransmitter may be taken up via active transport by nearby cell membranes. Each of these mechanisms—enzymatic destruction, diffusion, and active transport—can clear the synaptic cleft so that a new, discrete signal can pass through the synapse.

NEURONS IN CIRCUITS

Because neurons can interact in the complex ways we have just discussed, networks of neurons can process and integrate information. Next we will examine networks in different parts of the nervous system.

The Autonomic Nervous System

The autonomic nervous system controls the organs and organ systems of the body by influencing the activities of glands and involuntary muscles. There are two divisions of the autonomic nervous system, the **sympathetic** and the **parasympathetic**. These two divisions work in opposition to each other in their effects on most organs, one causing an increase in activity and the other causing a decrease. The best-known autonomic nervous system functions are those of the sympathetic division called the "fight-or-flight" mechanisms—those that increase heart rate, blood pressure, and cardiac output and prepare the body for emergencies. By contrast, the parasympathetic division slows the heart and lowers blood pressure. It is tempting to think of the sympathetic system as the one that speeds things up and the parasympathetic system as the one that slows things down, but that distinction is not always correct. For example, the sympathetic system slows the digestive system, and the parasympathetic system accelerates it. The two divisions of the autonomic nervous system are easily distinguished by their anatomy, their neurotransmitters, and their actions (Figure 38.18).

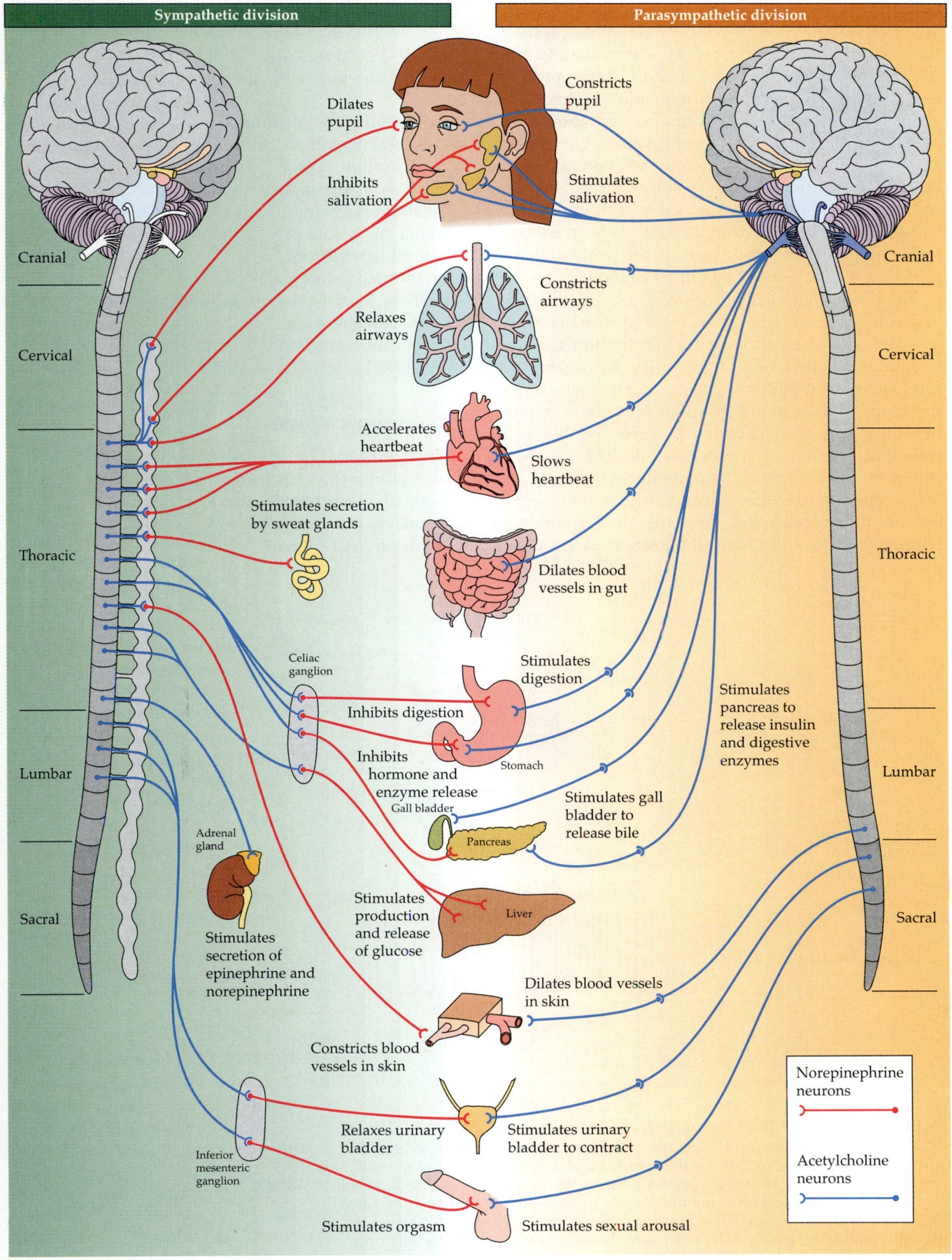

Sympathetic division

Parasympathetic division

Cranial

Cervical

Thoracic

Lumbar

Sacral

Dilates
pupil

Inhibits
salivation

Relaxes
airways

Accelerates
heartbeat

Stimulates secretion
by sweat glands

Celiac
ganglion

Inhibits digestion

Inhibits
hormone and
enzyme release

Adrenal
gland

Stimulates
secretion of
epinephrine and
norepinephrine

Stimulates
production
and release
of glucose

Constricts blood
vessels in skin

Relaxes urinary
bladder

Inferior
mesenteric
ganglion

Stimulates orgasm

Constricts
pupil

Stimulates
salivation

Constricts
airways

Slows
heartbeat

Dilates blood
vessels in gut

Stimulates
digestion

Stomach

Gall bladder

Pancreas

Liver

Stimulates
pancreas to
release insulin
and digestive
enzymes

Stimulates gall
bladder to
release bile

Dilates blood vessels
in skin

Stimulates urinary
bladder to contract

Stimulates sexual arousal

Cranial

Cervical

Thoracic

Lumbar

Sacral

Norepinephrine
neurons

Acetylcholine
neurons

38.18   Organization of the Autonomic Nervous System

Both divisions of the autonomic nervous system are efferent pathways of the central nervous system. Each autonomic efferent begins with a neuron that has its cell body in the brain stem or spinal cord and uses acetylcholine as its neurotransmitter. These cells are called **preganglionic neurons** because the second neuron in each autonomic output pathway resides in a **ganglion** (a collection of neuron cell bodies) that is outside the central nervous system. The second neuron in an autonomic nervous system output pathway is a **postganglionic neuron** because its axon extends from the ganglion. The axons of the postganglionic cells end on the cells of the target organs.

The postganglionic neurons of the sympathetic division use norepinephrine as their neurotransmitter, and those of the parasympathetic division use acetylcholine as their neurotransmitter. In organs that receive both sympathetic and parasympathetic input, the target cells respond in opposite ways to norepinephrine and acetylcholine. A region of the heart called the pacemaker, which generates the heartbeat, provides an example. Norepinephrine increases the firing rate of pacemaker cells and causes the heart to beat faster. Acetylcholine decreases the firing rate of pacemaker cells and causes the heart to beat slower.

In another example, norepinephrine causes muscle cells in the digestive tract to hyperpolarize, which slows digestion. Acetylcholine depolarizes muscle cells in the gut, which accelerates digestion (Figure 38.19).

Anatomy also distinguishes the sympathetic from the parasympathetic division of the autonomic nervous system (see Figure 38.18). The preganglionic neurons of the parasympathetic division come from the brain stem and the lowest (sacral) segment of the spinal cord. The preganglionic neurons of the sympathetic division come from the upper regions of the spinal cord below the neck—the thoracic and lumbar regions. Most of the ganglia of the sympathetic nervous system are lined up in two chains, one on either side of the spinal cord. The parasympathetic ganglia are close to, sometimes sitting on, the target organs.

Monosynaptic Reflexes

Much information is processed through neural circuits within the spinal cord. The simplest example of a spinal neural circuit that controls behavior is the **monosynaptic reflex loop**. This type of reflex depends on neural circuits made up of a sensory neuron

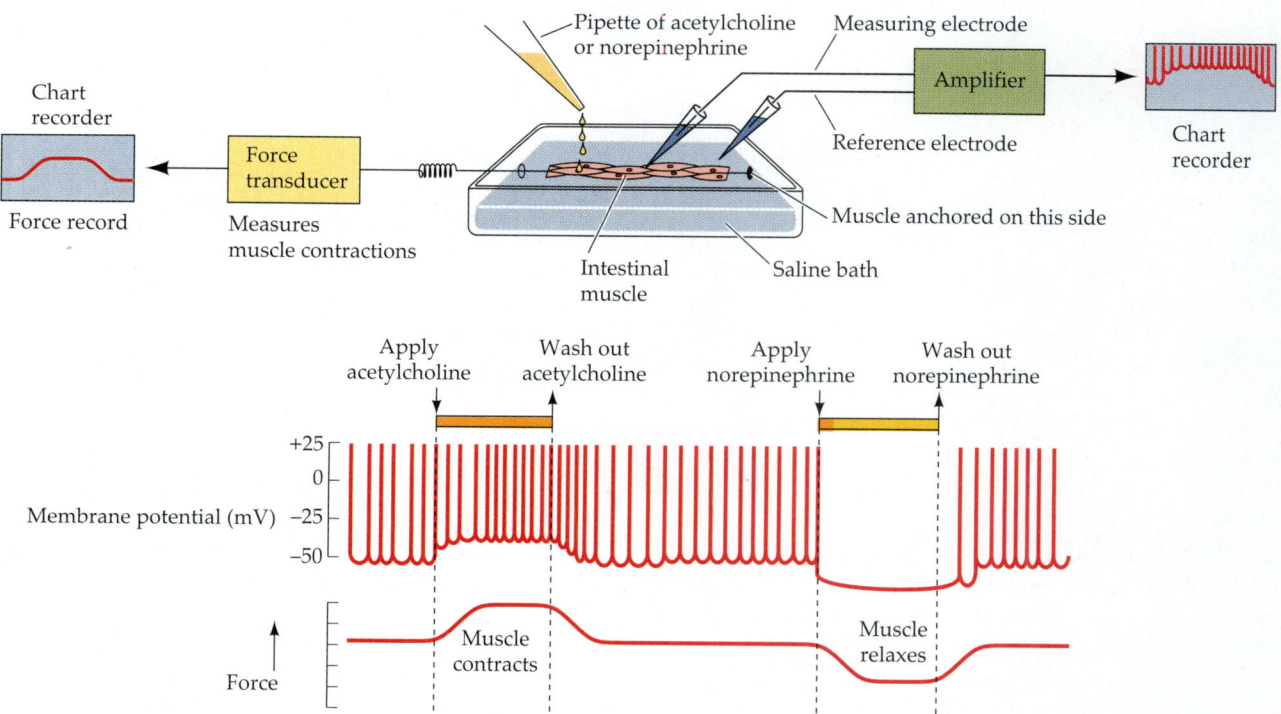

38.19 Responses to Postganglionic Transmitter
In this experiment, a strip of intestinal muscle is mounted in a saline bath that allows electric current to flow so that the force of contractions of the muscle can be measured. An electrode records action potentials in a muscle cell. When acetylcholine is dripped onto the muscle, the cells depolarize, fire action potentials more rapidly, and increase their force of contraction. Norepinephrine, on the other hand, causes the cells to hyperpolarize, decrease their rate of firing, and decrease their force of contraction.

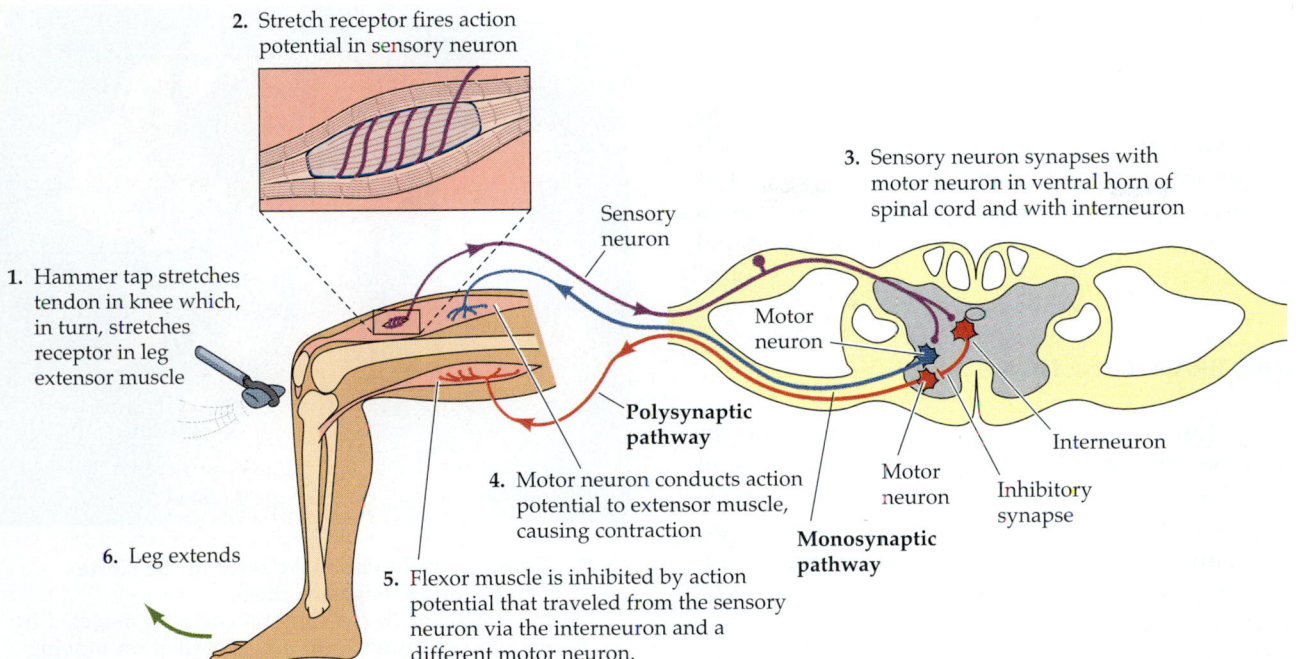

2. Stretch receptor fires action potential in sensory neuron

3. Sensory neuron synapses with motor neuron in ventral horn of spinal cord and with interneuron

Sensory neuron

1. Hammer tap stretches tendon in knee which, in turn, stretches receptor in leg extensor muscle

Motor neuron

Polysynaptic pathway

Interneuron

4. Motor neuron conducts action potential to extensor muscle, causing contraction

Motor neuron

Inhibitory synapse

Monosynaptic pathway

6. Leg extends

5. Flexor muscle is inhibited by action potential that traveled from the sensory neuron via the interneuron and a different motor neuron.

38.20 Monosynaptic and Polysynaptic Spinal Reflexes
The knee-jerk reflex is an example of a *monosynaptic* reflex loop in action. Because muscles work in antagonistic pairs, however, one must relax while the other contracts. Thus the knee-jerk reflex is accompanied by a parallel reflex that involves spinal interneurons and therefore more than one synapse—that is, it is *polysynaptic*.

and a motor neuron with just one synapse (hence the term *mono*synaptic) between them. It is common in a visit to a physician to have your reflexes tested by being struck below the kneecap with a rubber mallet. The response is a rapid, involuntary extension of the leg—the knee-jerk reflex. Similar reflexes can be elicited by sharp blows to tendons of the wrist, elbow, ankle, and other joints. Figure 38.20 shows how the knee-jerk reflex works.

The mallet blow on the tendon causes a quick stretch of the muscle attached to that tendon. Within the muscle are modified muscle fibers wrapped in connective tissue. These **muscle spindles** are stretch sensors that activate sensory neurons when they are stretched. The number of nerve impulses per second carried by the sensory neuron signals the degree of stretch. The cell body of the sensory neuron from the muscle spindle is in a ganglion on the dorsal root of the spinal cord, and the axon of this neuron extends all the way to the gray matter of the ventral horn of the spinal cord. There the sensory fiber branches and forms synapses with motor neurons for the same muscle from which the sensory neuron originated. Each motor neuron sends impulses along its axon, which leaves the spinal cord through the ventral root. The axon of the motor neuron synapses on the stretched muscle, causing it to contract. The function

of this reflex loop is to adjust the contraction in the muscle to changing loads. An increased load on the limb stretches the muscle, and the stretch reflex returns it to the desired position by increasing the strength of contraction of the muscle.

Even though the knee-jerk reflex is involuntary, you are aware of being struck on the knee. This means the information also travels to your brain. Branches of the sensory neuron form synapses with interneurons in the dorsal horn of the spinal cord. These interneurons send axons up the dorsal white matter tracts to the thalamus and on to the cerebral cortex. Motor commands from the cerebral cortex descend the spinal cord in other white matter tracts to form synapses with the same motor neurons involved in the reflex loop. Thus the same muscle can be controlled both by involuntary reflexes and by conscious commands.

Polysynaptic Reflexes

Most involuntary reflex circuits include more than one synapse. A number of interneurons in the central nervous system are necessary for more-complex responses. For example, a limb can move in opposite directions because muscles work in pairs. One muscle of a pair is an extensor and the other is a flexor. At the same time a flexor contracts, the extensor must relax, or the limb cannot move. A polysynaptic reflex is added to the monosynaptic circuit controlling the stretch reflex to cause relaxation of the opposing muscle (see Figure 38.20).

Much more complex polysynaptic reflexes are responsible for coordinated escape movements. If you step on a tack, many muscles in your foot and leg work together to produce a coordinated withdrawal response of that limb, while muscles on the opposite side of the body cause extension of your other leg to support your body weight and maintain your balance. The large number of muscle contractions and relaxations required for this sequence of movements are initially orchestrated by interneurons in the spinal cord.

HIGHER BRAIN FUNCTIONS

Very few functions of the nervous system have been worked out to the point of identifying the neural circuits that underlie them. Brain processes responsible for phenomena such as thought, perception, memory, and learning are extremely complex. Nevertheless, neurobiologists using a wide range of techniques are making considerable progress in understanding some of the neural mechanisms involved in these higher functions of the nervous system. For example, there have been rapid advances in understanding how the nervous system processes visual information, which we will discuss in the next chapter. The remainder of this chapter focuses on several complex aspects of brain and behavior that present challenges. Neurobiologists want to learn how individual neurons function and how neurons interact in circuits to produce these behaviors.

Sleeping and Dreaming

A dominant feature of our behavior is the daily cycle of being asleep and awake. All birds and mammals sleep, and probably all other vertebrates sleep as well. We spend one-third of our lives sleeping, yet we do not know why or how. We do know, however, that we need to sleep. Loss of sleep impairs alertness and performance. Most people in our society—certainly most college students—are chronically sleep-deprived. Many of the accidents and serious mistakes that endanger lives can be attributed to impaired alertness due to sleep loss. Yet insomnia (difficulty in falling asleep) is one of the most common medical complaints. Thus it is important to learn more about the neural control of sleep.

A common tool of sleep researchers is the **electroencephalogram** (EEG), a record of the electrical activity occurring primarily in the cerebral cortex (Figure 38.21). Sleep researchers also record the electrical activity of one or more skeletal muscles as an **electromyogram** (EMG) on the same moving chart. EEG and EMG patterns reveal the transition from being awake to being asleep; they also reveal that there are different states of sleep. Mammals other than humans have two major sleep states: **slow-wave sleep** and

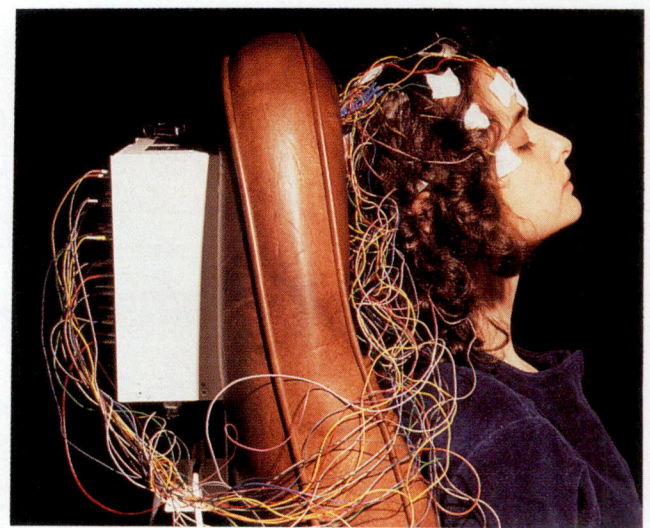

(a)

38.21 Patterns of Electrical Activity in the Cortex Characterize Stages of Sleep
(a) Electrical activity of the cerebral cortex is detected by electrodes placed on the scalp and recorded on moving chart paper by a polygraph. The resulting record is the electroencephalogram (EEG). *(b)* The pattern of the EEG is different in wakefulness and in different stages of sleep. *(c)* During a night we cycle through the different stages of sleep. We experience our deepest slow-wave sleep (stage 4) during the first half of the night. We dream during REM sleep, which usually occurs in four or five episodes during the night. We often awaken briefly after REM sleep.

rapid eye-movement sleep, commonly referred to as **REM sleep**. In humans we characterize sleep as REM sleep or non-REM sleep. Human non-REM sleep can be further divided into four stages, and only the two deepest stages are considered true slow-wave sleep. When you fall asleep at night you enter non-REM sleep and progress from stage 1 to stage 4, with stages 3 and 4 being deep, restorative, slow-wave-sleep. After this first episode of non-REM sleep, you enter an episode of REM sleep. Throughout the night, you have four or five cycles of non-REM/REM sleep (see Figure 38.21). About 80 percent of human sleep is non-REM sleep and 20 percent is REM sleep.

Vivid dreams and nightmares occur during REM sleep, which gets its name from the jerky movements the eyeballs make during this state. The most remarkable feature of REM sleep is that commands from the brain almost completely paralyze the skeletal muscles. Occasional muscle twitches break through the paralysis, as in a dog that appears to be trying to run in its sleep. If you look closely at a sleeping dog when its legs and paws are twitching, you will be able to see the rapid eye movements. The function of muscle paralysis during REM sleep is probably to prevent the acting out of dreams. Sleepwalking occurs during non-REM sleep.

(b)

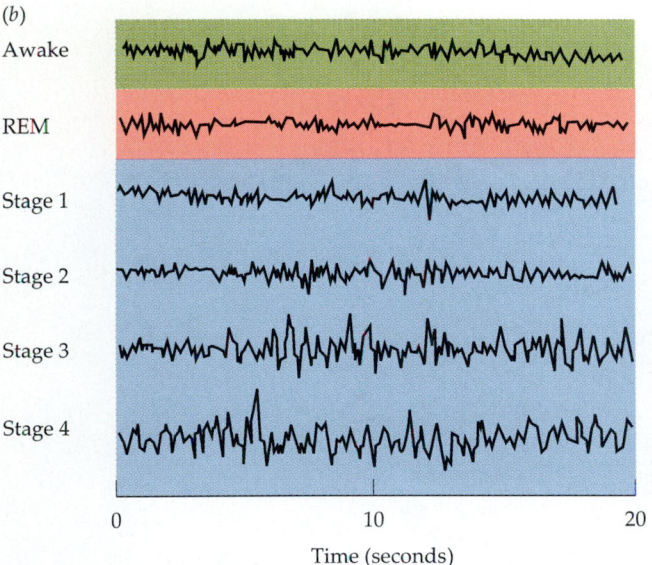

(c)

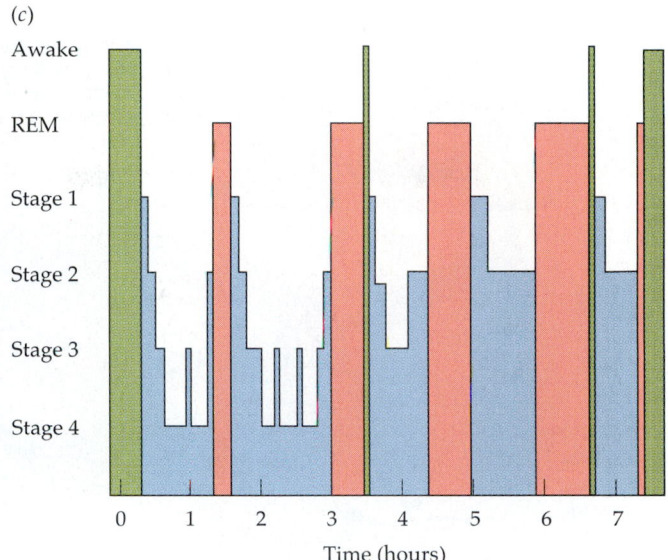

Learning and Memory

Learning is the modification of behavior by experience. **Memory** is the ability of the nervous system to retain what is learned and what is experienced. Even very simple animals can learn and remember, but these two abilities are developed the most in humans. Language, culture, artistic creativity, and scientific progress are made possible by these abilities. Consider the amount of information associated with learning a language. The capacity of human memory and the rate at which items can be retrieved are remarkable features of the nervous system. Is it possible to understand these phenomena in terms of the cells and molecules that make up the brain?

Habituation and sensitization are simple forms of learning that can be studied in all nervous systems, from the nerve nets of cnidarians to the human central nervous system. **Habituation** is learning to ignore a repeated stimulus that conveys little information. **Sensitization** is learning to be especially aware of a stimulus that conveys important information. For example, humans can habituate to noisy, busy, crowded environments, ignoring most of the barrage of sensory information they receive from those environments. Hearing your name spoken in a crowded, noisy room, however, immediately gets your attention.

Habituation and sensitization can be understood as processes that rely on individual synapses. The synaptic basis of habituation and sensitization has been studied extensively by Eric Kandel and his colleagues at Columbia University. The animal they use in their experiments is a marine mollusk called a sea slug or a sea hare (*Aplysia californica*). Because the sea slug does not have a shell, it is very vulnerable to predators. When the sea slug is undisturbed, its siphon is extended to take in water to ventilate its gill

membranes. If the siphon is touched, the animal withdraws it (Figure 38.22a). If the siphon is touched repeatedly, the animal habituates to the stimulus and no longer withdraws it. The researchers found that the siphon-withdrawal reflex depends on sensory neurons and motor neurons with only one synapse between them. By studying the characteristics of that synapse, they learned that habituation was due to a decrease in the amount of neurotransmitter released by the axon terminals of the presynaptic cells. This reduced release of neurotransmitter could last a considerable time and was not due simply to fatigue or depletion of neurotransmitter.

The researchers also studied sensitization. To sensitize the sea slug, they applied mild electric stimulation to its tail just before gently touching its siphon. Now the animal responded to the touch of its siphon with a much more vigorous withdrawal. Study of the synapse in the withdrawal reflex pathway showed that sensitization was due to the action of a third neuron that formed synapses with the axon terminals of the sensory cell. This sensitizing neuron caused the axon terminal of the sensory neuron to release more transmitter in response to each action potential (Figure 38.22b).

Another form of learning that is widespread among animal species is **associative learning**, in which two unrelated stimuli are linked to the same response. The simplest example of associative learning is the **conditioned reflex**, discovered by the Russian physiologist Ivan Pavlov. Pavlov was studying control of digestive functions in dogs and observed that a dog salivates at the sight or smell of food—a simple autonomic reflex. He discovered that if he rang a bell just before the food was presented to the dog, after a few trials the dog would salivate at the sound of the bell, even if no food followed. The

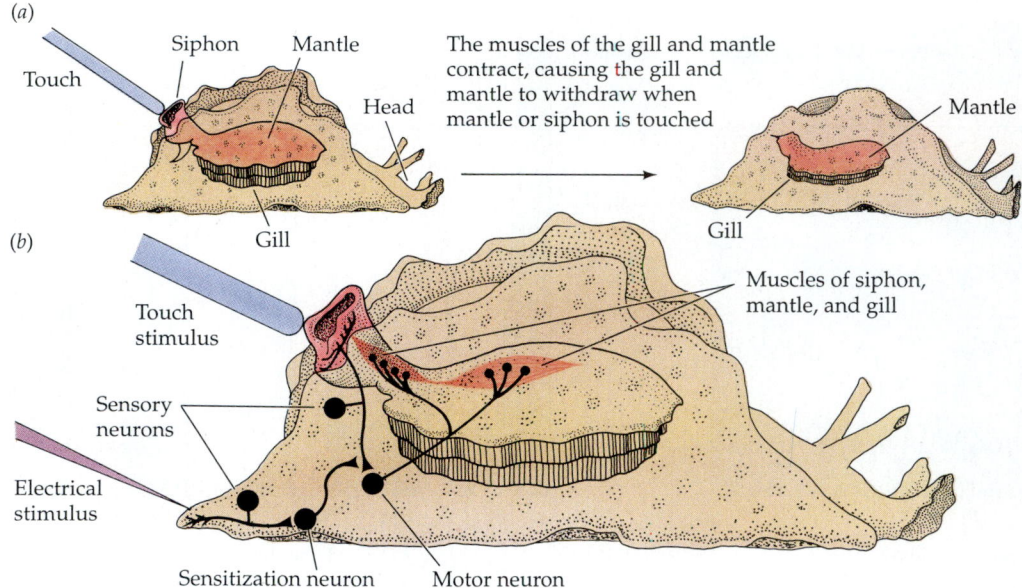

38.22 Habituation and Sensitization in a Slug
The siphon-withdrawal reflex of the sea slug *Aplysia*, shown in *(a)*, can be overcome by habituation. *(b)* The opposite effect, sensitization, can also be achieved. Applying a mild electrical shock to the animalUs tail just prior to touching the siphon intensifies the siphon-withdrawal reflex. Habituation and sensitization can be understood in terms of the neurons and synapses diagrammed here.

salivation reflex was conditioned to be associated with the sound of a bell, which normally is unrelated to feeding and digestion. This simple form of learning has been studied extensively in efforts to understand its underlying neural mechanisms.

Attempts to treat human diseases have sometimes led to increases in scientific understanding. Epilepsy is a disorder characterized by uncontrollable increases in neural activity in specific parts of the brain. The resulting epileptic fits can endanger the afflicted individual. At one time, serious cases of epilepsy were sometimes treated by destroying the part of the brain from which the surge of activity originated. To find the right area, the surgery was done under local anesthesia and different regions of the brain were electrically stimulated with fine electrodes while the patient described the resulting sensations. When some regions of association cortex were stimulated, patients sometimes reported vivid memories, as if reliving events from the past. Such observations were the first evidence that memories have anatomical locations in the brain and exist as properties of neurons and networks of neurons. Because the destruction of a small area of the brain does not completely erase a memory, however, it is postulated that memory is a function distributed over many brain regions and that a memory may be stimulated via many different routes.

You should be able to recognize several forms of memory from your own experience. There is **immediate memory** for events that are happening now.

Immediate memory is almost perfectly photographic, but it lasts only seconds. **Short-term memory** contains less information, but it lasts longer—on the order of 10 to 15 minutes. If you are introduced to a group of new people, you may remember most of their names for 5 or 10 minutes, but will have forgotten them in an hour or so if you have not repeated them, written them down, or used them in a conversation lasting longer than the round of introductions. Repetition, use, or reinforcement by something that gets your attention (such as the title President or Queen) facilitates the transfer of short-term memory to **long-term memory**, which can last for days, months, or years.

Knowledge about neural mechanisms for the transfer of short-term memory to long-term memory has come from observations of patients who have lost parts of the limbic system, notably the hippocampus. A famous case is that of H. M., who had his hippocampus removed on both the left and right sides of his brain in an effort to control severe epilepsy. Following his surgery, H. M. could not transfer any information to long-term memory. If someone were introduced to him, had a conversation with him, and then left the room for an hour, when that person returned, he or she was unknown to H. M., and it was as if the previous conversation had never taken place. H. M. had normal memory for events that happened before his surgery, but he could remember post-surgery events for only 10 or 15 minutes.

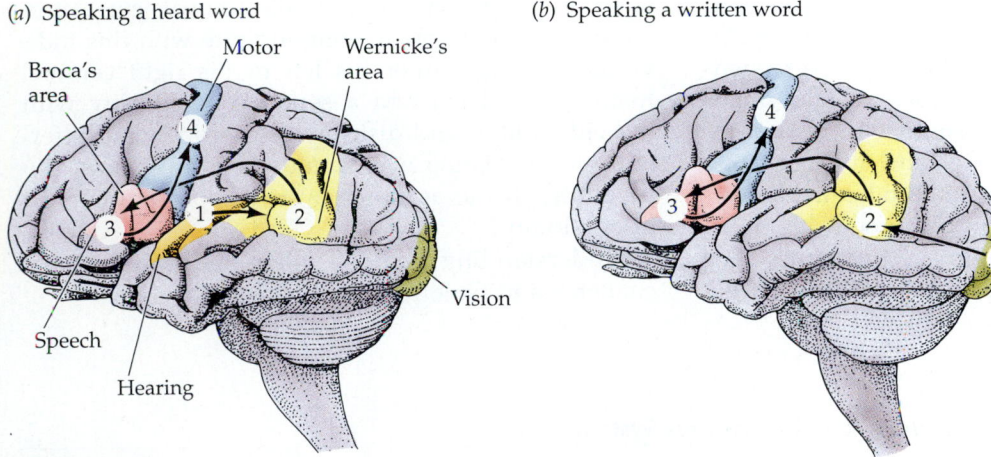

(a) Speaking a heard word
(b) Speaking a written word

Broca's area
Motor
Wernicke's area
Speech
Hearing
Vision

38.23 Language Areas
Different regions of the left cerebral cortex participate in the use of (a) spoken and (b) written language.

Language, Lateralization, and Human Intellect

No aspect of brain function is as fundamentally related to consciousness and intellect as is language. Therefore, studies of the brain mechanisms underlying the acquisition and use of language are extremely interesting to neuroscientists. A curious fact of language abilities is that they are located mostly in one cerebral hemisphere—the left hemisphere in 97 percent of all people. This phenomenon is referred to as the **lateralization** of language functions. Some of the most fascinating research on this subject has been done by Roger Sperry and his colleagues at the California Institute of Technology; Sperry received the 1981 Nobel prize in medicine for this work. The two cerebral hemispheres are connected by a white-matter tract called the **corpus callosum**. In one severe form of epilepsy, bursts of action potentials travel from hemisphere to hemisphere across the corpus callosum. Cutting the tract eliminates the problem, and patients function quite normally following the surgery.

Experiments have revealed interesting deficits in the language abilities of these "split-brain" patients, however. With the connections between the two hemispheres cut, the knowledge or experience of the right hemisphere can no longer be expressed in language, nor can language be used to communicate with the right hemisphere. Sensory input from the right hand goes to the left cerebral hemisphere, and sensory input from the left hand goes to the right cerebral hemisphere. If a split-brain patient is blindfolded and a familiar tool is placed in his or her right hand, the patient can identify the tool and describe its use. If the tool is placed in the left hand, however, the patient can use the tool correctly, but cannot name it or describe its use. In split-brain individuals, the right hemisphere has lost access to the language

abilities that reside predominantly in the left hemisphere. Language is said to be *lateralized* to the left hemisphere.

The brain mechanisms of language in the left hemisphere have been the focus of much research. The experimental subjects are persons who have suffered damage to some region of the left hemisphere and are left with one of many forms of **aphasias**, deficits in the abilities to use or understand words. The known language areas of the left hemisphere are shown in Figure 38.23a. Broca's area, in the frontal lobe just in front of the motor cortex, is essential for speech. Damage to Broca's area results in halting, slow, unclear speech or even complete loss of speech, but the patient can still read and understand language. In the temporal lobe, close to its border with the occipital lobe, is Wernicke's area, which is more involved with sensory than with motor aspects of language. Damage to Wernicke's area can cause a person to lose the ability to speak sensibly while retaining the abilities to form the sounds of normal speech and to imitate its rhythm. Moreover, such a patient cannot understand spoken or written language. Near Wernicke's area is the angular gyrus, which is believed to be essential for integration of spoken and written language.

Normal language ability depends on the flow of information between various areas of the left cerebral cortex (Figure 38.23). Input from spoken language travels from the primary auditory (hearing) cortex to Wernicke's area. Input from reading language travels from the primary visual cortex to the angular gyrus to Wernicke's area. Commands to speak are formulated in Wernicke's area and travel to Broca's area and from there to the primary motor cortex. Damage to any one of these areas or the pathways between them can result in aphasia.

Young children can recover remarkably from even severe damage to the left cerebral hemisphere because lateralization of language abilities to the left hemisphere has not fully developed, and the right hemisphere can take over all of the language-related functions. One such patient in Sperry's split-brain studies produced provocative results with respect to the relationship between language and intellect. This patient had language functions in both hemispheres. Because of left-brain damage in childhood, his right hemisphere had developed language functions, and over time his left hemisphere had recovered. Then, to treat epilepsy, his corpus callosum was cut. Afterward it was possible to communicate with this individual through either his left or his right cerebral hemisphere. Each had a separate personality with individual likes and dislikes. Each responded differently to evaluating events and projecting plans into the future. It was as if there were two persons housed in one brain.

Understanding the brain will be one of the greatest challenges in biology for many years to come.

SUMMARY of Main Ideas about Neurons and the Nervous System

Nervous systems process and integrate information received from sensors and communicate commands to effectors.

Nervous systems are composed of cells called neurons, which have cell bodies, dendrites, and axons.
Review Figure 38.2

Neurons communicate with other cells at synaptic junctions.
Review Figures 38.15 and 38.16

A nerve is a bundle of many axons carrying information to and from the central nervous system.
Review Figure 38.3

The brain and spinal cord are the central nervous system, and the cranial and spinal nerves are the peripheral nervous system.
Review Figure 38.4

From a hollow tube the vertebrate nervous system develops a hindbrain, a midbrain, and a forebrain.
Review Figure 38.5

Different regions of the central nervous system have different functions.

The spinal cord receives and sends information.
Review Figure 38.6

The reticular system controls sleeping and waking.
Review Figure 38.7

The limbic system functions in emotion, drive, instinct, and memory.
Review Figure 38.8

The convoluted cerebral hemispheres are the dominant structures of the human brain, and regions within them are responsible for motor functions and particular kinds of sensory information.
Review Figures 38.9, 38.10, and 38.23

The resting potential of a neuron is the difference prevailing most of the time between the electric charge on the inside of the plasma membrane and that on the outside.
Review Figure 38.11

Action potentials are rapid reverses in charge across portions of the plasma membrane.
Review Figure 38.12

Action potentials are self-regenerating, all-or-nothing events that transmit information by traveling down axons.
Review Figure 38.13

Myelinated axons propagate action potentials faster than other axons do.
Review Figure 38.14

Synaptic inputs added together generate action potentials when the sum exceeds a threshold.
Review Figure 38.17

Neurons arranged in circuits control the activities of most organs of the body.
Review Figures 38.18 and 38.20

The two states of human sleep are REM sleep and non-REM sleep, and non-REM sleep can be divided into four stages based on electrical activity in the brain.
Review Figure 38.21

Habituation and sensitization are simple forms of learning and memory that can be studied in the less complex nervous systems of invertebrates.
Review Figure 38.22

Human use of language involves communication between different areas of the cortex.
Review Figure 38.23

SELF-QUIZ

1. In the nervous system, the *most* abundant cell type is the
   a. motor neuron.
   b. sensory neuron.
   c. preganglionic parasympathetic neuron.
   d. glial cell.
   e. preganglionic sympathetic neuron.

2. Within the nerve cell, information moves from
   a. dendrite to cell body to axon.
   b. axon to cell body to dendrite.
   c. cell body to axon to dendrite.
   d. axon to dendrite to cell body.
   e. dendrite to axon to cell body.

3. Which of the following statements is *not* true?
   a. Sensory afferents carry information of which we are consciously aware.
   b. Visceral afferents carry information about physiological functions of which we are not consciously aware.
   c. The voluntary motor division of the efferent side of the peripheral nervous system executes conscious movements.
   d. The cranial nerves and spinal nerves are parts of the peripheral nervous system.
   e. Afferent and efferent axons never travel in the same nerve.

4. Which of the following statements is *not* true?
   a. In the spinal cord, the white matter contains the axons that conduct information.
   b. The limbic system is involved in basic physiological drives, instincts, and emotions.
   c. The limbic system consists of primitive forebrain structures.
   d. Most nerve cell bodies in the human nervous system are contained within the limbic system.
   e. In humans, a part of the limbic system is necessary for the transfer of short-term memory to long-term memory.

5. Which of the following statements accurately describes an action potential?
   a. Its magnitude *increases* along the axon.
   b. Its magnitude *decreases* along the axon.
   c. All action potentials in a single neuron are of the same magnitude.
   d. During an action potential the transmembrane potential of a neuron remains constant.
   e. It permanently shifts a neuron's transmembrane potential away from its resting value.

6. A neuron that has just fired an action potential cannot be immediately restimulated to fire a second action potential. The short interval of time during which restimulation is not possible is called
   a. hyperpolarization.
   b. the resting potential.
   c. depolarization.
   d. repolarization.
   e. the refractory period.

7. The rate of propagation of an action potential depends on
   a. whether or not the axon is myelinated.
   b. the axon's diameter.
   c. whether or not the axon is insulated by glial cells.
   d. the cross-sectional area of the axon.
   e. all of the above

8. The binding of neurotransmitter to the postsynaptic receptors in an inhibitory synapse results in
   a. depolarization of the transmembrane potential.
   b. generation of an action potential.
   c. hyperpolarization of the transmembrane potential.
   d. increased permeability of the membrane to sodium ions.
   e. increased permeability of the membrane to calcium ions.

9. Whether a synapse is excitatory or inhibitory depends on
   a. the type of neurotransmitter.
   b. the presynaptic terminal.
   c. the size of the synapse.
   d. the nature of the postsynaptic neurotransmitter receptors.
   e. the concentration of neurotransmitter in the synaptic space.

10. The part of the brain that differs the most in complexity between mammals and amphibians is
    a. the midbrain.
    b. the forebrain.
    c. the cerebellum.
    d. the limbic system.
    e. the hippocampus.

FOR STUDY

1. Compare and contrast the two divisions of the autonomic nervous system. Emphasize distinctions with respect to their anatomical organization, neurotransmitters used, and general effects on the functions of specific organ systems.

2. Outline the development of the vertebrate nervous system. Where on the neural axis are the more evolutionarily primitive and advanced functions located?

3. Describe the electrochemical and structural elements involved in the establishment and maintenance of the neuron's transmembrane resting potential.

4. Describe the processes and structures involved in (a) the initiation and propagation of an action potential, and (b) synaptic transmission.

5. Define and describe the synaptic basis for habituation and sensitization in *Aplysia*.

READINGS

Camhi, J. M. 1984. *Neuroethology: Nerve Cells and the Natural Behavior of Animals.* Sinauer Associates, Sunderland, MA. This advanced text covers the properties and functions of neurons but emphasizes aspects relevant to sensory abilities of animals and the neural control of behavior. Examples are taken from a wide variety of vertebrates and invertebrates.

Kandel, E. R., J. H. Schwartz and T. M. Jessell. 1991. *Principles of Neural Science,* 3rd Edition. Elsevier, New York. A very thorough, advanced text in neurobiology.

Llinas, R. R. 1988. *The Biology of the Brain: From Neurons to Networks,* and Llinas, R. R. 1990. *The Workings of the Brain: Development, Memory, and Perception.* W. H. Freeman, New York. These two volumes are a rich collection of articles on aspects of neuroscience that were published in *Scientific American* since 1976. They provide a broad yet selectively in-depth survey of modern neurobiology.

Nicholls, J. G., A. R. Martin and B. G. Wallace. 1992. *From Neuron to Brain,* 3rd Edition. Sinauer Associates, Sunderland, MA. An advanced text on cellular neurobiology.

Shepherd, G. M. 1994. *Neurobiology,* 3rd Edition. Oxford University Press, New York. A comprehensive advanced text that covers the full range of neurobiology from molecular mechanisms to human behavior.

Thompson, R. F. 1985. *The Brain: An Introduction to Neuroscience.* W. H. Freeman, New York. A well-written and easy-to-understand introductory text on neuroscience, covering topics from membrane events to the neural basis of behavior.

"Angela, you *look* great. I *hear* that you and Carl are going to the Khyber Pass for dinner."

"Yes, you raved so much about the wonderful *aromas* and complex, spicy *flavors* that we couldn't resist. After all, you have good *taste*, my friend."

"You'll also like the *feel* of the place. The tables are low and you sit on *soft* cushions. The walls are covered with *brightly colored* hangings. Stay in *touch*, I'd like to know how you liked it."

"I just hope the food isn't too *hot*, or you'll *hear* from me sooner than you think."

We cannot discuss anything for very long without using words that refer to our senses. Our senses are the window through which we view the world, and the world is what our senses tell us it is. Because different species look through different sensory windows, their views of the world differ. Dogs do not see color, but they have keener senses of hearing and smell than humans do. As you gaze at a beautiful sunset, your dog may be sniffing around at your feet and pricking up its ears as it hears small animals moving in the underbrush. Bees can see patterns on flowers that reflect ultraviolet light; we cannot. In environments that are totally dark to us, some snakes can "see" the infrared radiation emitted by bodies warmer than the environment. Bats can use reflected sound to avoid obstacles and catch small insects. The sounds bats emit are extremely intense, but they are beyond our range of hearing. In murky waters, the duck-billed platypus uses its sensitive bill to feel and taste food items, and electric fishes detect other fishes by the electric fields they create. How the environment "looks" to any animal depends on what information that animal receives from its sensors.

WHAT IS A SENSOR?

Sensors are cells of the nervous system that transduce (convert) physical or chemical stimuli into signals that are transmitted to other parts of the nervous system for processing and interpretation. Most sensors are modified neurons, but some are other types of cells closely associated with neurons. Sensors are specialized for specific types of stimuli. In this chapter we will examine chemosensors, which respond to specific molecular structures; mechanosensors, which respond to mechanical forces; and photosensors, which respond to light. In general, the sensor possesses a membrane protein that detects the stimulus and responds by altering the flow of ions across the cell membrane (Figure 39.1). The resulting change in membrane potential causes the sensor either to fire action potentials itself or to secrete neurotransmitter

A Keen Sense of Smell Put to Work
A search dog trained to sniff out drugs, alcohol, and guns patrols school lockers.

39

Sensory Systems

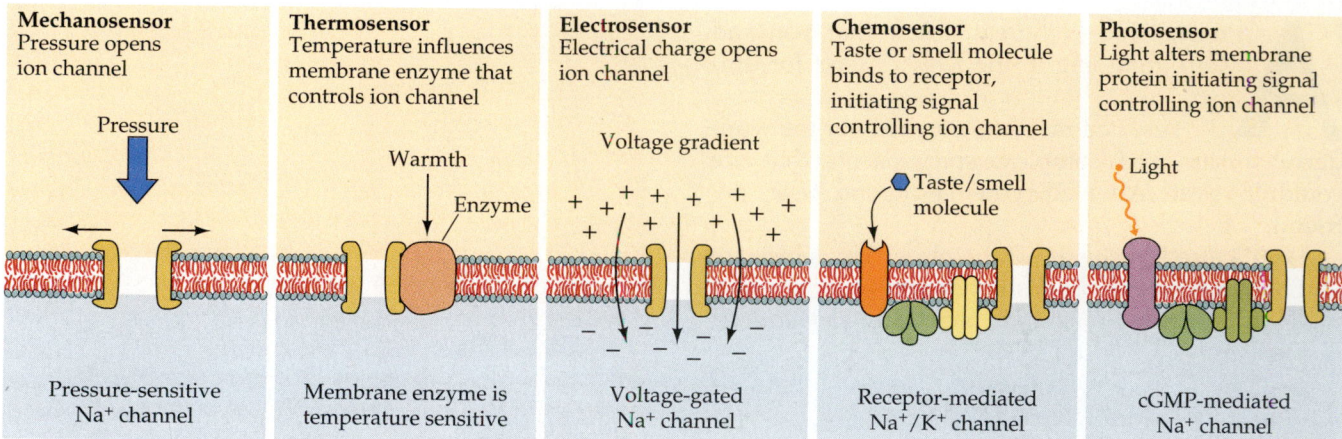

39.1 Different Proteins in Sensors Alter Ion Channels
Receptor proteins are embedded in the plasma membranes of sensors. Sensory stimuli such as pressure or warmth modify the receptor proteins, which in turn modify ion channels. The receptor proteins in the plasma membranes of mechanosensors, thermosensors, and electrosensors are themselves the ion channels. The receptor proteins in the plasma membranes of chemosensors and photosensors initiate biochemical cascades that eventually open or close ion channels.

onto an associated cell that fires action potentials. Ultimately the stimulus is transduced into the universal message of the nervous system—action potentials (see Chapter 38).

Sensation

If the messages derived from all sensors are the same, how can we perceive different sensations? Sensations such as temperature, itch, pressure, pain, light, smell, and sound differ because the messages from sensors arrive at different places in the central nervous system. Action potentials arriving in the visual cortex are interpreted as light, in the auditory cortex as sound, in the olfactory bulb as smell, and so forth. A small patch of skin on your arm contains sensors that increase their firing rates when the skin is warmed and others that increase activity when the skin is cooled. Other types of sensors in the same patch of skin respond to touch, movement of hairs, irritants such as mosquito bites, and pain from cuts or burns. As we learned in Chapter 38, these sensors in your arm transmit their messages through axons that enter the central nervous system through the dorsal horn of the spinal cord. The synapses made by those axons in the central nervous system and the subsequent pathways of transmission determine whether the stimulation of the patch of skin on your arm is perceived as warmth, cold, pain, touch, itch, or tickle.

The specificity of sensory circuits is dramatically illustrated in persons who have had a limb or part of a limb amputated. Although the sensors from that region are gone, the axons that came from those sensors to the spinal cord may remain. If those axons are stimulated, the person feels specific sensations as if they were coming from the limb that is no longer there—a phenomenon known as a phantom limb.

The messages from some sensors communicate information about internal conditions in the body, but we may not be consciously aware of that information. The brain receives continuous information about such things as body temperature, blood sugar, blood carbon dioxide and oxygen, arterial pressure, muscle tensions, and positions of limbs. All this information is important for maintaining homeostasis, but thankfully we don't have to think about it—if we did, we would have no time to think about anything else. Sensors produce information that the nervous system can use, but that information does not always result in conscious sensation.

Sensory Organs

Some sensors are assembled with other types of cells into sensory organs such as eyes, ears, and noses that enhance the ability of the sensors to collect, filter, and amplify stimuli. For example, a photosensor detects electromagnetic radiation of only a particular range of wavelengths and therefore filters out radiation of other wavelengths. This filtering is the basis for color vision, and the specificity of photosensors explains why some insects can see ultraviolet light and some snakes can see infrared radiation. In some simple organisms photosensors sense only the presence of light, but in more-complex animals, photosensors are combined with other cell types into eyes. We'll learn how eyes collect light and focus it onto sheets of photosensors so that patterns of light can be detected. The basis of vision is the ability of the eye to filter available light information for patterns and colors.

Similarly, we'll see that the sense of hearing depends on mechanosensors, but the accessory structures that constitute the ear make it possible to amplify low levels of sound and filter it so that it also conveys directional information. Some sensory organs can reduce the level of stimulus energy that reaches sensors. For example, the pupillary reflex of vertebrate eyes varies the amount of light falling on the photosensors, and tiny muscles in ears can dampen the energy from loud sounds before it reaches the sensitive mechanosensors.

Sensory Transduction

In this chapter we will examine several sensor types and the sensory organs with which they are associated. A general question in each case is, how does the sensor transduce stimulus energy into action potentials? Although different for different sensors, the details of sensory transduction all fit into a general pattern. Figure 39.1 illustrated the first steps of sensory transduction: A receptor protein is activated by a specific stimulus. The activated receptor protein opens or closes specific ion channels in the plasma membrane of the sensor by one of several mechanisms. The receptor protein may be part of the ion channel and by changing its conformation may open or close the channel directly. Alternatively, the activated receptor protein may set in motion intracellular events that eventually affect the ion channel. Figure 39.2 reviews these first steps of sensory transduction and outlines the subsequent steps.

The opening or closing of ion channels in response to a stimulus changes the sensor's membrane potential, which is called the **receptor potential**. Such changes in membrane potential can spread electrotonically over short distances, but to travel long distances in the nervous system receptor potentials must be converted into action potentials. Interestingly, the intracellular events involved in converting the original, stimulus-induced alteration of the ion channels into the generation of action potentials can amplify the signal. In other words, the energy in the output of the sensor can be much greater than the energy in the stimulus.

Receptor potentials produce action potentials in two ways: by generating action potentials within the sensors or by causing the release of neurotransmitter that induces an associated neuron to generate action potentials. In the first case, the sensor has a region of plasma membrane with voltage-gated sodium channels. A receptor potential in such a cell is called a **generator potential** because it generates action potentials by causing the voltage-gated Na$^+$ channels to open.

A good example of generator potentials is found in stretch sensors of crayfish (Figure 39.3). By placing

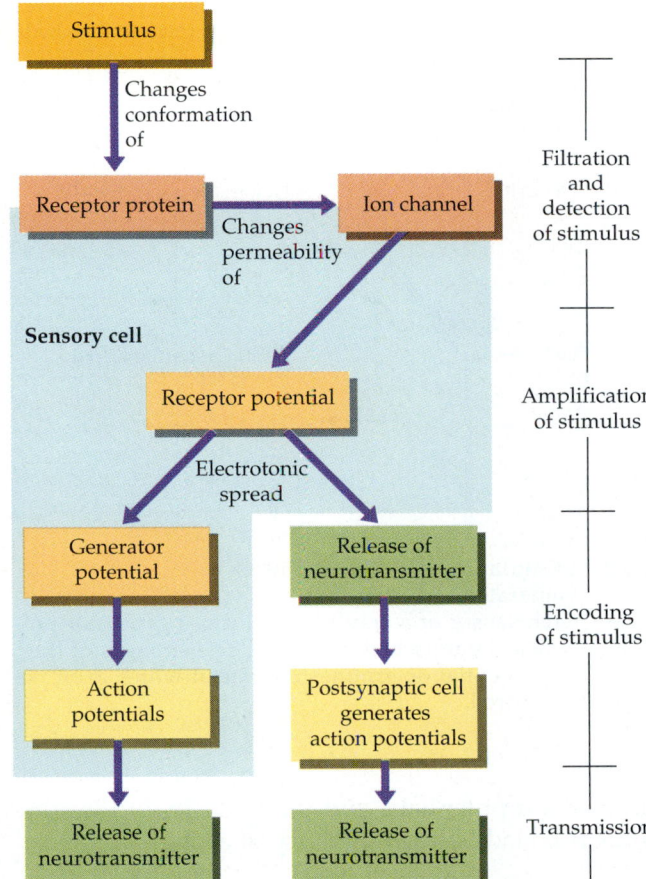

39.2 Sensory Transduction Involves Several Steps
The detection of a stimulus and subsequent change in ion channels was described in Figure **39.1**. Sensors process and amplify the stimulus, either producing action potentials or releasing neurotransmitters that induce associated neurons to produce action potentials.

an electrode in the cell body of the crayfish stretch sensor, we can record the changes in the receptor potential that result from stretching the muscle to which the dendrites of the cell are attached. These changes in receptor potential become a generator potential at the base of the sensor's axon, where there are voltage-gated Na$^+$ channels. Action potentials generated here travel down the axon to the central nervous system. The rate at which the axon fires action potentials depends on the magnitude of the generator potential, which in turn depends on how much the muscle is stretched.

In sensors that do not fire action potentials, the spreading receptor potential reaches a presynaptic patch of plasma membrane and induces the release of neurotransmitter. The neurotransmitter can then activate ligand-gated ion channels on a postsynaptic membrane and cause the postsynaptic cell to fire action potentials. An excellent example is the photosensor that we will study in some detail later in this chapter. In either case, the stimulus is transduced

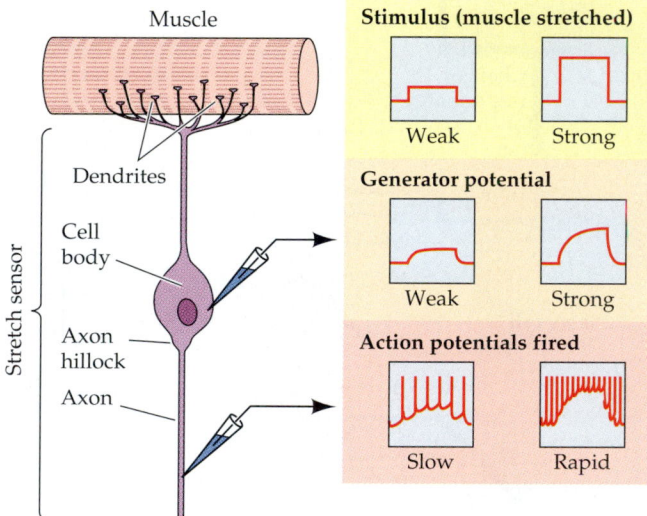

39.3 Stimulating a Sensor Produces a Generator Potential
The stretch sensor of a crayfish produces a generator potential when the muscle is stretched. The strength of the generator potential determines the rate at which axon potentials are fired.

into action potentials, and the intensity of the stimulus is coded by the frequency of action potentials.

Changing Sensitivity

An important characteristic of many sensors is that they can stop being excited by a stimulus that initially caused them to be active. In other words, they adapt to the stimulus. **Adaptation** enables an animal to ignore background or unchanging conditions while remaining sensitive to changes or to new information. (Note that this use of "adaptation" is different from its application in an evolutionary context.) When you dress, you feel each item of clothing touch your skin, but the sensation of clothes touching your skin is not constantly on your mind throughout the day. You are immediately aware, however, when a seam rips, your shoe comes untied, or someone touches your back ever so lightly. The ability of animals to discriminate between important and unimportant stimuli is partly due to the fact that some sensors adapt; it is also due to information processing by the central nervous system. Some sensors adapt very little or do so slowly; examples are pain sensors and sensors for balance.

In the rest of this chapter we will learn how sensory systems gather and filter stimuli, transduce stimuli into action potentials, and transmit action potentials to the central nervous sytem, as well as how the central nervous sytem processes that information to yield perceptions about the internal and external worlds. Sensors are also important components of

autonomic regulatory mechanisms, as we will see in subsequent chapters.

CHEMOSENSORS

Animals receive information about chemical stimuli through **chemosensors**, which respond to specific molecules in the environment. Chemosensors are responsible for smell, taste, and the monitoring of aspects of the internal environment such as the level of carbon dioxide in the bloodstream. Chemosensitivity is universal among animals. A colony of corals responds to a small amount of meat extract in the seawater around it by extending bodies and tentacles and searching for food. A single amino acid can stimulate this response. If, however, an extract from an injured individual of the colony is released into the water, the colony members retract their tentacles and bodies to avoid danger. Humans have similar reactions to chemical stimuli. Upon smelling freshly baked bread we salivate and feel hungry, but we gag and retch when we smell rotting meat. Information from chemosensors can cause powerful behavioral and autonomic responses.

Chemosensation in Arthropods

Arthropods use chemical signals to attract mates. These signals, called **pheromones**, demonstrate the sensitivity of chemosensory systems. The female silkworm moth releases a pheromone called bombykol from a gland at the tip of her abdomen. The male silkworm moth has sensors for this molecule on his antennae (Figure 39.4). Each feathery antenna carries about 10,000 bombykol-sensitive hairs, and each hair has a dendrite of a sensor cell at its core. A single molecule of bombykol is sufficient to activate a dendrite and generate action potentials in the antennal nerve that transmits the signal to the central nervous system. When approximately 200 hairs per second are activated, the male flies upwind in search of the female. Because of the male's high degree of sensitivity, the sexual message of a female moth is likely to reach any male within a downwind area stretching over several kilometers. Because the rate of firing in the male's sensory nerves is proportional to bombykol concentrations in the air, the male can follow a concentration gradient and home in on the emitting female.

Many arthropods have chemosensory hairs, with each hair containing one or more specific types of sensor. For example, crabs and flies have chemosensory hairs on their feet; they taste potential food by stepping in it. These hairs have sensors for sugars, amino acids, salts, and other molecules (Figure 39.5). After a fly steps in a drop of sugar water and tastes

(a)　(b)

39.4 A Scent That Travels Several Kilometers
Mating in silkworms of the genus *Bombyx* is coordinated by a chemical attractant. (a) The female moth releases the attractant from a gland at the tip of her abdomen. (b) From as far away as several kilometers, a male moth detects this chemical attractant in the air passing over his antennae, which are covered with chemosensitive hairs.

it, its proboscis (a tubular feeding structure) extends and it feeds. Potential food items stimulate extension of the proboscis; other substances do not.

Olfaction

The sense of smell, called **olfaction**, is another form of chemosensation. In vertebrates, the smell sensors are neurons embedded in a layer of epithelial cells at the top of the nasal cavity (Figure 39.6). The axons of smell sensors project to the olfactory bulb of the brain, and their dendrites end in olfactory hairs that project to the surface of the nasal epithelium. A protective layer of mucus covers the epithelium. Molecules must diffuse through this mucus to get to the receptors on the olfactory hairs. When you have a

cold or an attack of hay fever, the amount of mucus increases and the epithelium swells. With this in mind, study Figure 39.6 and you will easily understand why you lose your sense of smell at those times. A dog has up to 40 million nerve endings per square centimeter of nasal epithelium, many more than humans do. Although humans have a sensitive olfactory system, we are unusual among mammals in that we depend more on vision than on olfaction (we tend to join bird-watching societies more often than mammal-smelling societies). Whales and porpoises have no olfactory sensors and hence no sense of smell.

How does an olfactory sensor transduce the structure of a molecule into action potentials? A molecule that triggers an olfactory sensor is called an odorant molecule. Odorant molecules bind to receptors on

39.5 Tasting with the Feet
Using sensory hairs on their feet, flies such as this fruit fly (*Drosophila melanogaster*) can identify a potential food source by stepping in it.

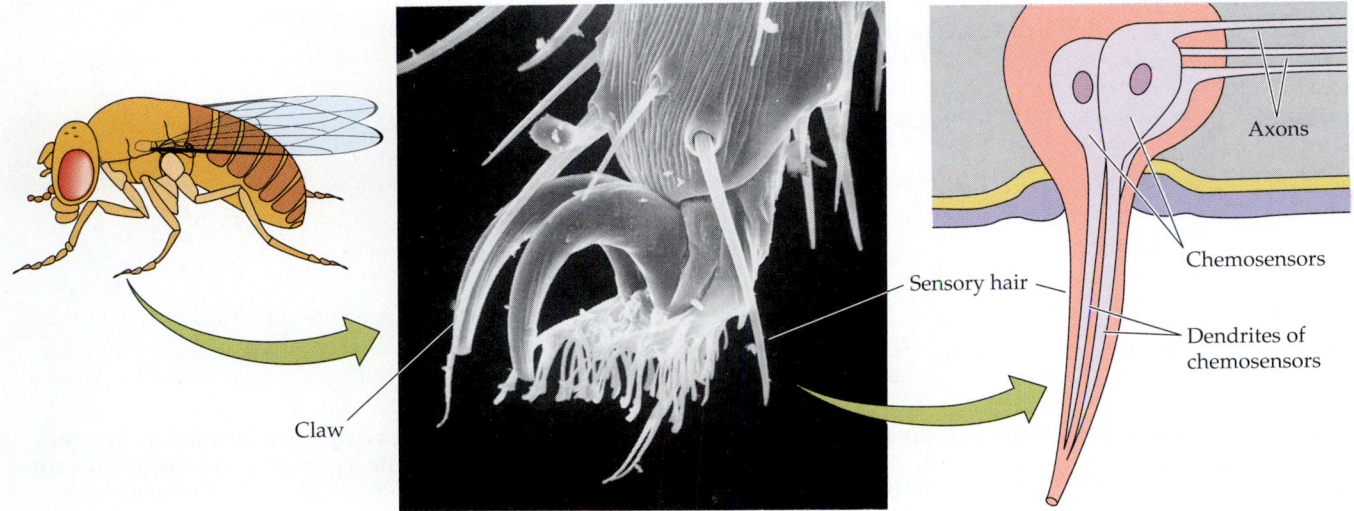

Claw

Sensory hair

Axons

Chemosensors

Dendrites of chemosensors

Olfactory bulb

Brain

Olfactory bulb

Olfactory bulb

Axon

Bone

Nasal cavity

Basal cell

Olfactory sensor

Supporting cell

Air

Mucus film

Olfactory hairs

39.6   Olfactory Sensors Communicate Directly with the Brain

The sensors of the human olfactory system are embedded in the tissue lining the nasal cavity. The sensors' receptors are on the olfactory hairs. The axons of the sensors project to the olfactory bulb of the brain.

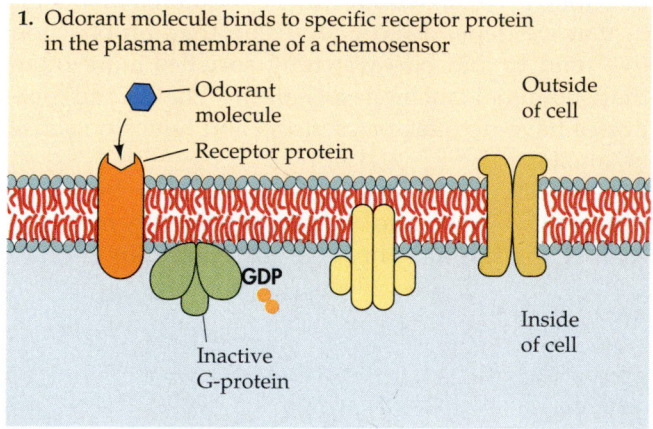

1. Odorant molecule binds to specific receptor protein in the plasma membrane of a chemosensor

Odorant molecule

Receptor protein

Outside of cell

GDP

Inside of cell

Inactive G-protein

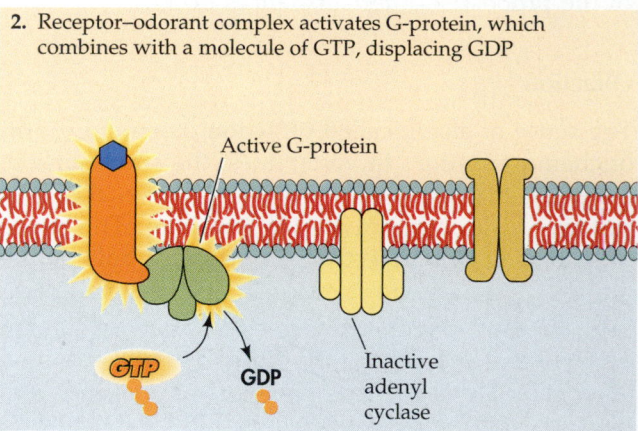

2. Receptor–odorant complex activates G-protein, which combines with a molecule of GTP, displacing GDP

Active G-protein

GTP

GDP

Inactive adenyl cyclase

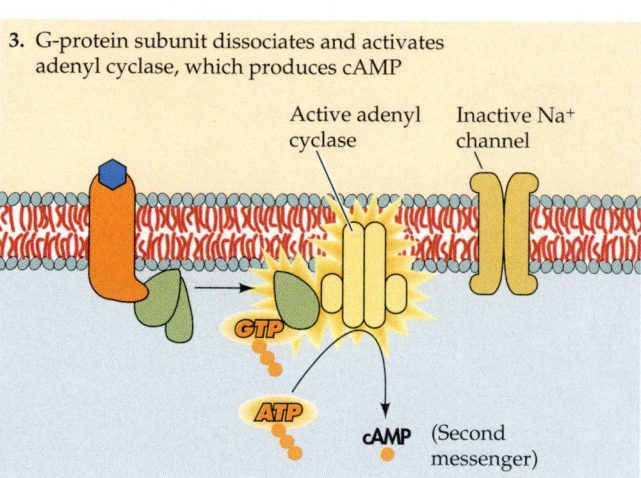

3. G-protein subunit dissociates and activates adenyl cyclase, which produces cAMP

Active adenyl cyclase

Inactive Na$^+$ channel

GTP

ATP

cAMP (Second messenger)

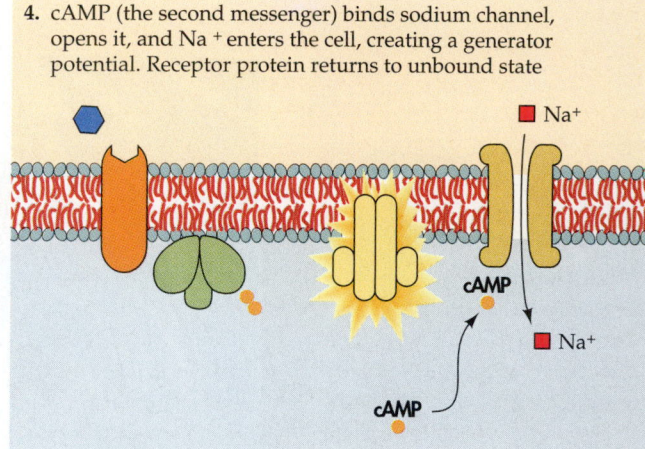

4. cAMP (the second messenger) binds sodium channel, opens it, and Na$^+$ enters the cell, creating a generator potential. Receptor protein returns to unbound state

Na$^+$

cAMP

Na$^+$

cAMP

39.7   Olfactory Receptor Proteins Activate the cAMP Cascade

In order for us to smell something, odorant molecules must bind to receptors on the olfactory hairs of sensors and initiate the cAMP cascade by activating a G-protein. The cascade amplifies the signal so that a single odorant molecule can cause the generation of action potentials that are transmitted to the brain.

the olfactory hairs of the sensors. Olfactory receptors are specific for particular odorant molecules, and they work like a lock-and-key mechanism does. If a "key" (an odorant molecule) fits the "lock" (the receptor), a G-protein is activated, which in turn activates an enzyme (adenyl cyclase, for example) that causes an increase of a second messenger in the cytoplasm of the sensor. The second messenger binds with sodium channel proteins in the sensor's plasma membrane and opens the channels, causing an influx of Na$^+$. The sensor thus depolarizes to threshold and fires action potentials (Figure 39.7).

The olfactory world has an enormous number of "keys"—molecules that produce distinct smells. Are there a correspondingly large number of "locks"—receptor proteins? Indeed there are. Researchers have recently discovered an enormous family of genes that code for olfactory receptor proteins.

How does the sensor signal the intensity of a smell? It responds in a graded fashion to the concentration of odorant molecules: The more odorant molecules that bind to receptors, the more action potentials that are generated and the greater the intensity of the perceived smell.

Gustation

The sense of taste, called **gustation**, in humans and other vertebrates depends on clusters of sensors called **taste buds**. The taste buds of terrestrial vertebrates are in the mouth, but some fish have taste buds in the skin that enhance their ability to taste their environment. Some fish living in murky water are sensitive to small amounts of amino acids in the water around them and can find food without the use of vision. The duck-billed platypus, a monotreme mammal, has similar abilities because it has taste buds on the sensitive skin of its bill. What is a taste bud and how does it work?

A taste bud is a cluster of many taste sensors. A human tongue has approximately 10,000 or so taste buds. The taste buds are embedded in the epithelium of the tongue or are found on the raised papillae of the tongue. Look at your tongue in a mirror; the papillae make it look fuzzy. Each papilla has many taste buds. A taste bud's outer surface has a pore that exposes the tips of the taste sensors (Figure 39.8). Microvilli (tiny hairlike projections) increase the surface areas of the sensors where their tips converge at the taste pore. Taste sensors, unlike olfactory sensors, are not neurons. At their bases, taste sensors form synapses with dendrites of sensory neurons.

Gustation begins at receptors in the membranes of the microvilli. As with olfactory transduction, receptors on the sensors bind molecules, and the binding causes changes in the membrane polarity of the sensors. Because the taste sensors are not neurons, how-

Taste bud with taste pore

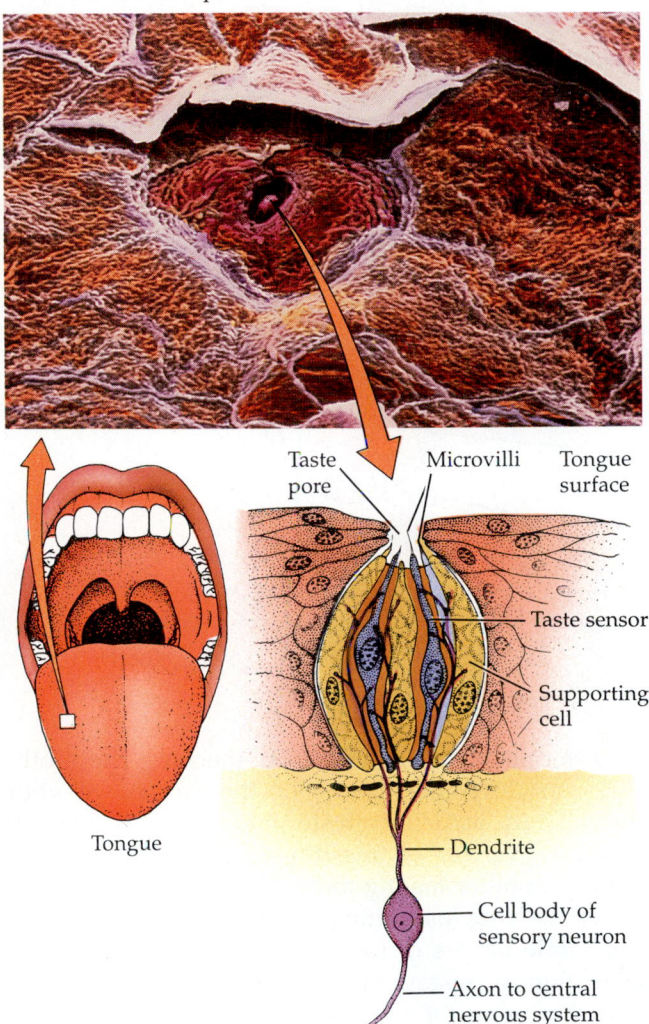

39.8 Taste Buds Are Clusters of Taste Sensors
The scanning electron micrograph of a papilla on the tongue shows the taste pore, where microvilli of the sensors that make up the taste bud come into contact with stimuli. The drawing to the right is a cross section of a taste bud. Note that the taste sensors are not neurons.

ever, they do not fire action potentials. Instead, they release neurotransmitter onto the dendrites of the sensory neurons. The sensory neurons respond to the neurotransmitter by firing action potentials that are conducted to the central nervous system. The tongue does a lot of hard work, so its epithelium is shed and replaced at a rapid rate. Taste buds last only a few days before they are replaced, but the sensory neurons associated with them live on, always forming new synapses as new taste buds form.

You may have heard it said that humans can perceive only four tastes: sweet, salty, sour, and bitter. Particular regions of the tongue have taste buds responsible for these general categories of taste, but the regions overlap to a large extent (Figure 39.9). You can map your own tongue by dipping toothpicks

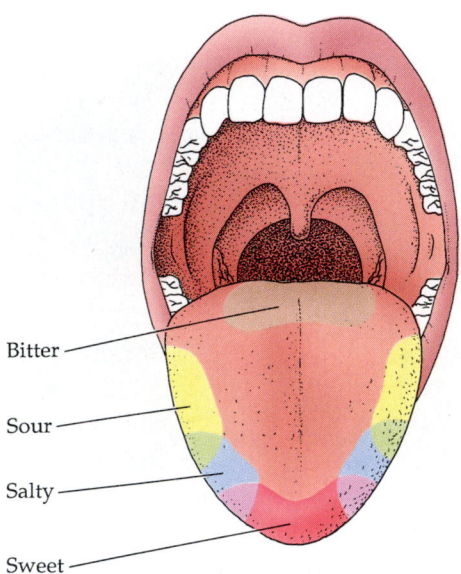

39.9   A Taste Map
Sensors for sweet, salty, sour, and bitter in the human tongue reside in specific regions that overlap in some areas.

in different solutions and then touching the toothpicks to different regions of the surface of your tongue. In actuality, taste buds can distinguish among a variety of sweet-tasting molecules and a variety of bitter-tasting molecules. The full complexity of the chemosensitivity that enables us to enjoy the subtle flavors of food comes from the combined activation of gustatory and olfactory sensors; hence you lose some of your sense of "taste" when you have a cold.

Why does a snake continually sample the air with its forked tongue darting in and out? If the snake, like us, tasted only sweet, salty, sour, and bitter with its tongue, it would not get much useful information by tasting the air. The forks of the snake's tongue fit into cavities in the roof of its mouth that are richly endowed with olfactory sensors. The tongue samples the air and presents the sample directly to these sensors. Thus the snake is really using its tongue to smell its environment, not to taste it. Why doesn't the snake use the flow of air to and from its lungs as we do to smell the environment? Air flow to and from the lungs is slow and intermittent in reptiles, but the tongue can dart in and out many times in a second. It is a quick source of olfactory information.

MECHANOSENSORS

Mechanosensors are specialized cells sensitive to mechanical forces that distort their membranes. A variety of mechanosensors in the skin are responsible for the perception of touch, pressure, and tickle. Stretch

sensors in muscles, tendons, and joints give information about the position of the parts of the body in space and the forces acting on them. Stretch sensors in the walls of blood vessels signal blood pressure. "Hair" cells with extensions that are sensitive to being bent are incorporated into mechanisms for hearing and mechanisms for signaling the body's position with respect to gravity. Physical distortion of a mechanosensor's plasma membrane causes ion channels to open and alters the resting potential of the cell; this change leads to the generation of action potentials (see Figure 39.3). The rates of action potentials in the sensory nerves tell the central nervous system the strengths of the stimuli that are exciting the mechanosensor.

Touch and Pressure

Objects touching our skin generate varied sensations because our skin is packed with diverse mechanosensors (Figure 39.10). The outer layers of skin, especially hairless skin such as lips and fingertips, contain many whorls of nerve endings enclosed in connective tissue capsules. These very sensitive mechanosensors are called **Meissner's corpuscles**, and they respond to objects that touch the skin even lightly. Meissner's corpuscles adapt very rapidly, however. That is one reason why you roll a small object between your fingers, rather than holding it still, to discern its shape and texture. As you roll it,

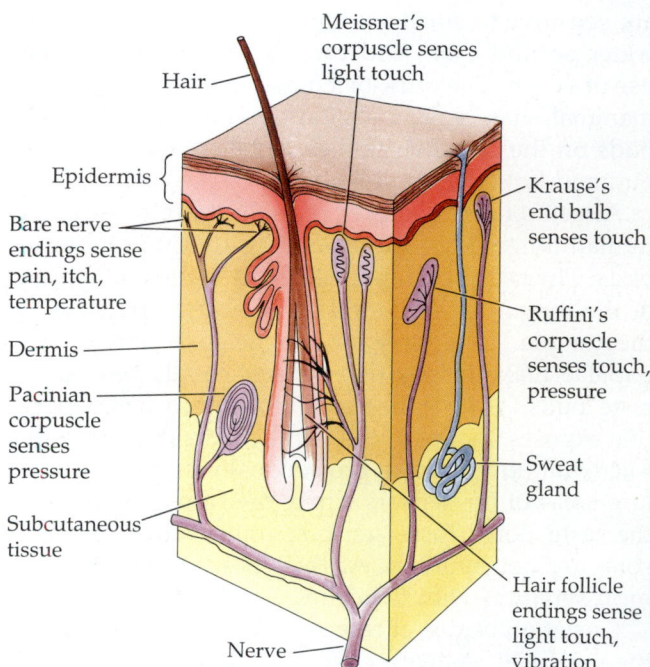

39.10   The Skin Feels Many Sensations
Even a very small patch of skin contains a diversity of sensors that send information to the brain.

you continue to stimulate sensors anew. When you hold an object still, the sensors originally activated adapt. Try it. Also in the outer regions of the skin are **expanded-tip tactile sensors** of various kinds. They differ from Meissner's corpuscles in that they adapt only partially and slowly. They are useful for providing steady-state information about objects that continue to touch the skin.

The density of the tactile sensors varies across the surface of the body. A two-point discrimination test demonstrates this fact. If you lightly touch someone's back with two toothpicks, you can determine how far apart the two stimuli have to be before the person can distinguish whether he or she was touched with one or two points. The same test applied to the person's lips or fingertips reveals a finer spatial discrimination; that is, the person can identify as separate two stimuli that are closer together than those on the back.

Deep in the skin, extensions of neurons wrap around hair follicles. When the hairs are displaced, those neurons are stimulated. Also deep within the skin is another type of mechanosensor, the **Pacinian corpuscles**. These sensors look like onions because they are made up of concentric layers of connective tissue cells encapsulating an extension of a sensory neuron. Pacinian corpuscles respond especially well to vibrations applied to the skin, but they adapt rapidly to steady pressure. The connective tissue capsule is important in the adaptation of these sensors. An initial pressure distorts the corpuscle and the membrane of the neuron at its core, but the layers of the capsule rapidly rearrange to redistribute the force, thus eliminating the distortion of the membrane of the neuron.

Stretch Sensors

An animal receives information from **stretch sensors** about the position of its limbs and the stresses on its muscles and joints. These mechanosensors are activated by being stretched. They continuously feed information to the central nervous system, and that information is essential for the coordination of movements. We encountered one important type of stretch sensor, the muscle spindle, when we discussed the monosynaptic reflex loop in Chapter 38. Muscle spindles are embedded in connective tissue within skeletal muscles and consist of modified muscle fibers that are innervated in the center with extensions of sensory neurons. Whenever the muscle is stretched, the spindle cells are also stretched, and the neurons transmit action potentials to the central nervous system. Earlier in this chapter (see Figure 39.3), we learned how crayfish stretch sensors transduce physical force into action potentials.

Another stretch sensor is found in tendons and ligaments. It is called the **Golgi tendon organ**. Its role is to provide information about the force generated by a contracting muscle. When a contraction becomes too forceful, the information from the Golgi tendon organ feeds into the spinal cord, inhibits the motor neuron, and causes the contracting muscle to relax, thus protecting the muscle from tearing.

Hair Cells

Hair cells are mechanosensors that are not neurons. From one surface they have projections called **stereocilia** that look like a set of organ pipes. When these stereocilia are bent, they alter receptor proteins in the hair cell's plasma membrane. When the stereocilia of some hair cells are bent in one direction, the receptor potential becomes more negative, and when they are bent in the opposite direction, it becomes more positive. When the receptor potential becomes more positive, the hair cells release neurotransmitter to the sensory neurons associated with them, and the sensory neurons send action potentials to the brain.

Hair cells are found in the lateral line sensory system of fishes. The lateral line consists of a canal just under the surface of the skin that runs down each side of the fish. The canal has numerous openings to the external environment. Many structures called cupulae project into the lateral line canal. Each cupula contains hair cells whose stereocilia are embedded in gelatinous (jellylike) material. Movements of water in the lateral line canal move the cupulae and stimulate the hair cells (Figure 39.11). Thus the lateral line provides information about movements of the fish through the water as well as information about the moving objects, such as predators or prey, that cause pressure waves in the surrounding water.

Many invertebrates have equilibrium organs called **statocysts** that use sensory hairs to signal the position of the animal with respect to gravity. In the case of the lobster, the statocyst is a chamber lined with hollow, nonliving hairs made of chitin. Each hair receives the dendrite of a sensory neuron. In the center of the statocyst is a dense **statolith** consisting of grains of sand (Figure 39.12). Due to gravity, the statolith stimulates the sensory hairs that are lowest, as determined by the position of the animal. When a scientist replaced the statoliths of lobsters with iron filings and held a magnet over the animals, they swam upside down. When he held the magnet to their sides, they swam on their sides. The behavior of the lobsters proved the role of the statoliths.

Vertebrates also have equilibrium organs. The mammalian inner ear, for example has two organs of equilibrium that also use hair cells to detect the position of the body with respect to gravity. In the next section we will examine the structure of the ear. For the moment it is enough to know that the inner ear

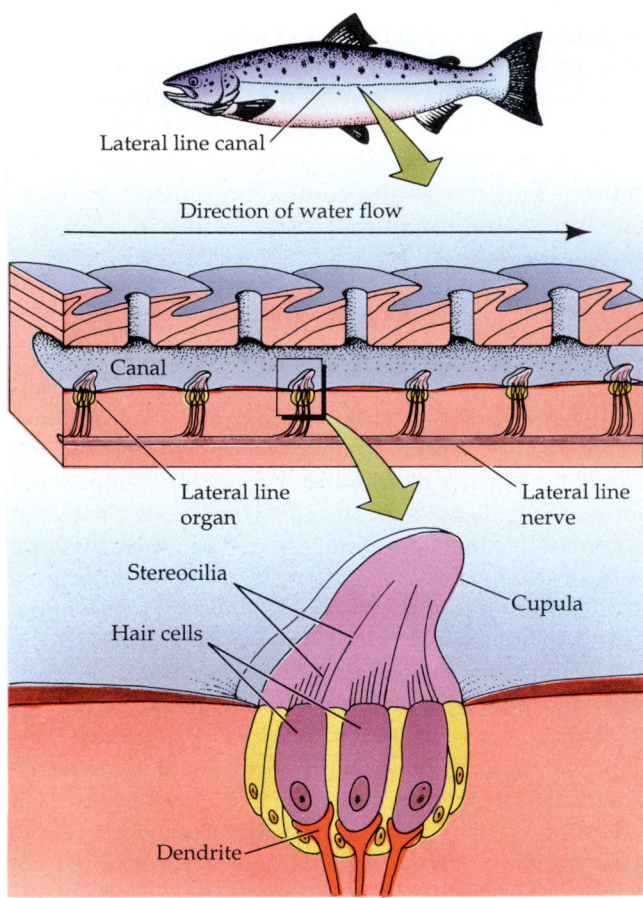

39.11 The Lateral Line System Contains Mechanosensors
Hair cells in the lateral line organs of a fish detect movement of the water around the animal, giving the fish information about its own movements and the movements of objects nearby.

contains three **semicircular canals** at right angles to one another (Figure 39.13). Each semicircular canal has a swelling called an **ampulla**, which contains a group of hair cells with their stereocilia embedded in a gelatinous cupula. The canals are filled with fluid. As an animal's head changes position, the fluid in its semicircular canals moves, puts pressure on the cupulae, and bends the stereocilia of the hair cells. The second equilibrium organ is found in the vestibule. This organ, the **vestibular apparatus**, has two chambers whose function is like that of the statocysts of invertebrates. Hair cells line the floors of the chambers; their stereocilia are embedded in a layer of gelatinous material. On top of this layer are **otoliths**

39.12 How a Lobster Knows Which Way Is Up
The statocyst is a sense organ found in many invertebrates. The force of gravity moves statoliths within the sensory hair-lined statocyst, giving the animal information about its position.

(literally, "ear stones"), which are granules of calcium carbonate. As the head moves, gravity pulls on the dense otoliths, which bend the stereocilia of the hair cells.

Auditory Systems

The stimuli that animals perceive as sound are pressure waves. Auditory systems use mechanosensors to transduce pressure waves into action potentials. These systems include special structures to gather sound, direct it to the sensors, and amplify it. Human hearing provides good examples of these aspects of auditory systems. The organs of hearing are the ears. The two prominent structures on the sides of our heads usually thought of as ears are the **ear pinnae**. The pinna of an ear collects sound waves and directs them into the auditory canal leading to the actual hearing apparatus in the middle ear and the inner ear. If you have seen a rabbit or a horse change the orientation of its ear pinnae to focus on a particular sound, then you have witnessed the role of ear pinnae in hearing.

The human ear is diagrammed at progressively higher levels of magnification in Figure 39.14. The eardrum, or **tympanic membrane**, covers the end of the auditory canal. The tympanic membrane vibrates in response to pressure waves traveling down the auditory canal. The chamber of the middle ear, an air-filled cavity, lies on the other side of the tympanic membrane. The middle ear is open to the throat at

Statocyst is located at the base of the antennule

Dense object (statolith)
Sensory hairs
Sensory neurons

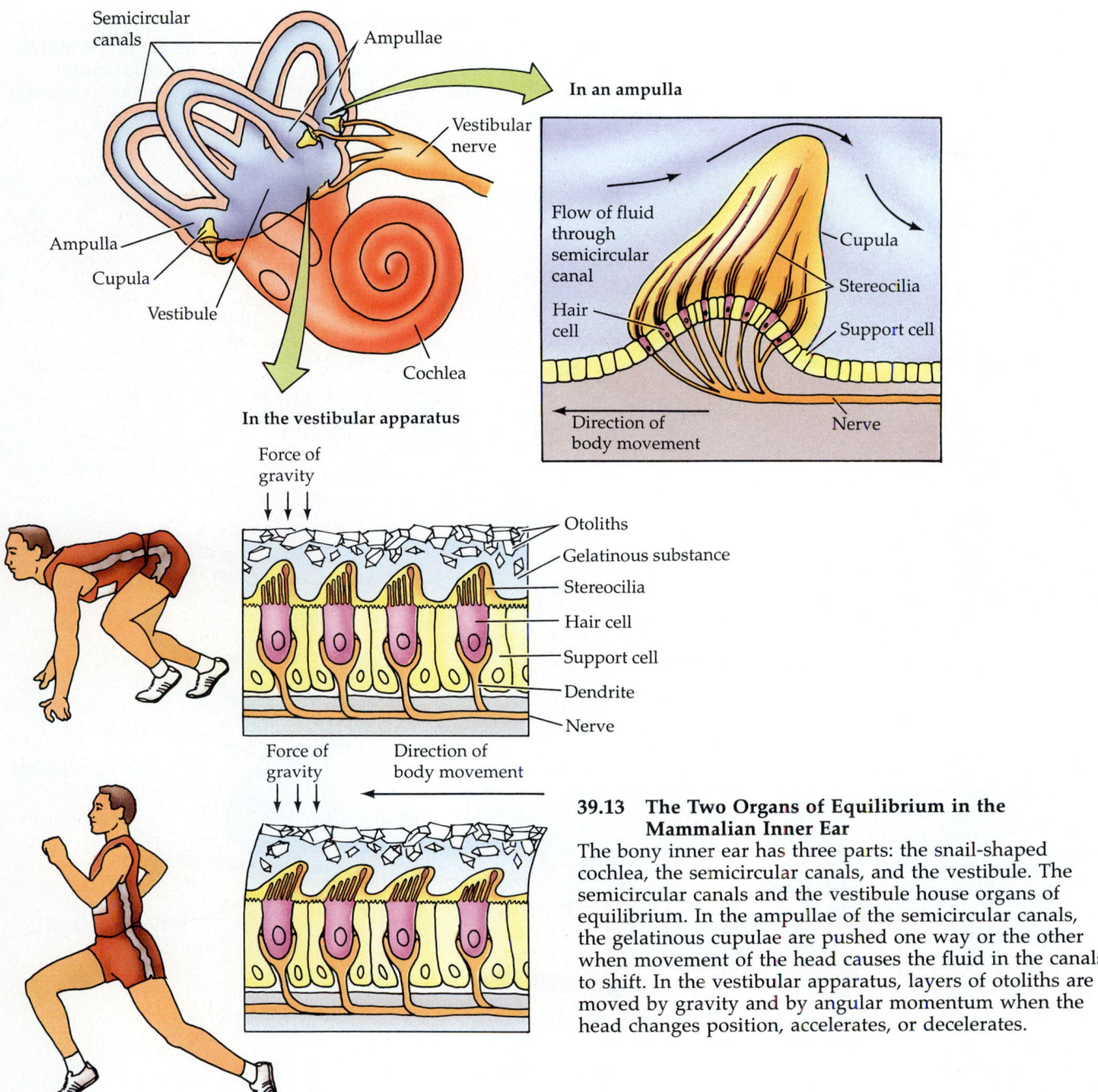

39.13 The Two Organs of Equilibrium in the Mammalian Inner Ear
The bony inner ear has three parts: the snail-shaped cochlea, the semicircular canals, and the vestibule. The semicircular canals and the vestibule house organs of equilibrium. In the ampullae of the semicircular canals, the gelatinous cupulae are pushed one way or the other when movement of the head causes the fluid in the canals to shift. In the vestibular apparatus, layers of otoliths are moved by gravity and by angular momentum when the head changes position, accelerates, or decelerates.

the back of the mouth through the eustachian tube. Because air flows through the **eustachian tube**, pressure equilibrates between the middle ear and the outside world. When you have a cold or hay fever, the tube becomes blocked by mucus or by tissue swelling, and you have difficulty "clearing your ears," or equilibrating the pressure in the middle ear with the outside air pressure. Then the flexible tympanic membrane bulges in or out, dampening your hearing and sometimes causing earaches.

The middle ear contains three delicate bones called

the **ear ossicles**, individually named the **malleus** (hammer), **incus** (anvil), and **stapes** (stirrup). The ossicles transmit the vibrations of the tympanic membrane to the fluid-filled inner ear, where they will be transduced into action potentials. The leverlike action of the ossicles amplifies the vibrations about twentyfold. The malleus is attached to the center of the tympanic membrane and at the other end of the chain of ossicles, the stapes is attached to a smaller membrane called the **oval window**, which covers an opening into the inner ear. The incus serves as a pivot point. When the tympanic membrane moves in, the lever action of the ossicles pushes the stapes, and the

39.14 Major Structures of the Human Ear
Pressure waves travel through the auditory canal of the outer ear to the middle ear. The ossicles of the middle ear (the malleus, incus, and stapes) transmit these vibrations to the cochlea, where they are transduced into action potentials by the hair cells of the organ of Corti.

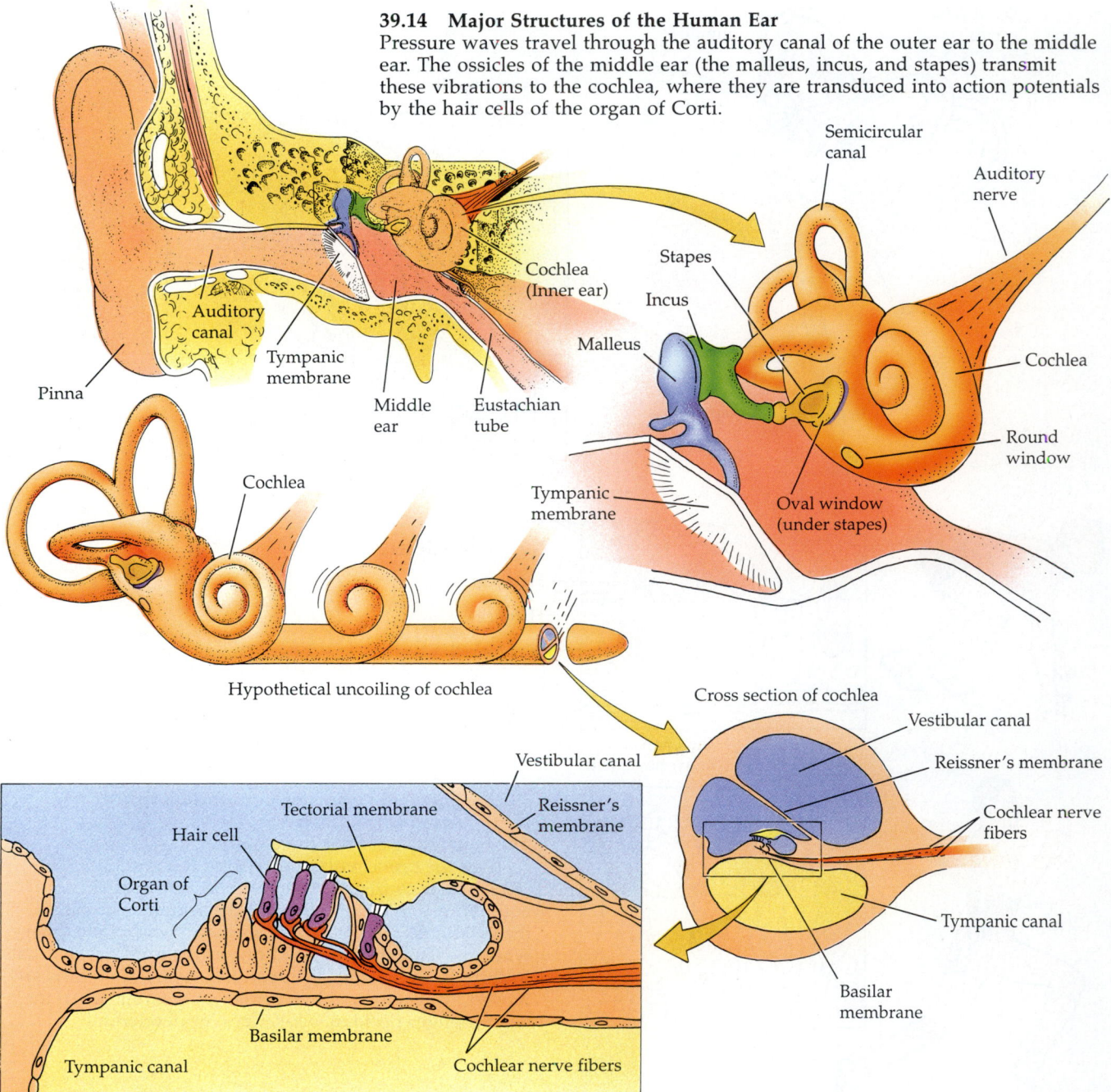

oval window bulges into the inner ear. When the tympanic membrane moves out, the stapes and the oval window are also pulled out. In this way, pressure waves in the auditory canal are converted into pressure waves in the fluid-filled inner ear.

Pressure waves are transduced into action potentials in the inner ear. The inner ear is a long, narrow, coiled chamber called the **cochlea** (from the Latin and Greek words for "snail" or "shell"). The cross section of the cochlea in Figure 39.14 reveals that it is composed of three parallel canals separated by two membranes: Reissner's membrane and the basilar membrane. Sitting on the basilar membrane is the **organ**

of Corti, the apparatus that transduces pressure waves into action potentials in the auditory nerve—the nerve that conveys information from the ear to the brain. The organ of Corti contains hair cells whose stereocilia are in contact with an overhanging, rigid shelf called the tectorial membrane. Whenever the basilar membrane flexes, the tectorial membrane bends the hair cell stereocilia. As a consequence, the hair cells depolarize or hyperpolarize, altering the rate of action potentials transmitted to the brain by their associated sensory neurons.

What causes the basilar membrane to flex, and how does this mechanism distinguish sounds of different frequencies? In Figure 39.15 the cochlea is shown uncoiled to make it easier to understand its structure and function. To simplify matters we have left out Reissner's membrane, thus combining the upper and the middle canals into one upper canal. The purpose of Reissner's membrane is to contain a specific aqueous environment for the organ of Corti separate from the aqueous environment in the rest of the cochlea. This role is important for the nutrition of the sensitive organ of Corti, but it has nothing to do with the transduction of sound waves. The simplified diagram of the cochlea in Figure 39.15 reveals two additional features that are important for its function. First, the upper and lower chambers separated by the basilar membrane are joined at the end of the cochlea farthest from the oval window, making one continuous canal that folds back on itself. Second, just as the oval window is a flexible membrane at the beginning of the cochlea, the **round window** (see also Figure 39.14) is a flexible membrane at the end of the long cochlear canal.

Air is highly compressible, but fluids are not. Therefore, a sound pressure wave can travel through air without much displacement of the air, but a sound pressure wave in fluid causes displacement of the fluid. Imagine holding a screen-door spring slightly stretched between your two hands. Someone could grab the spring in the center and move it back and forth without moving its ends—it is compressible. Now imagine holding a broomstick in the same way. If someone grabs its middle and moves it back and forth, obviously the ends will move too—the broomstick is incompressible.

How does this comparison of springs and broomsticks relate to the inner ear? When the stapes pushes the oval window in, the fluid in the upper canal of the cochlea is displaced. Think about what happens if the oval window moves in very slowly. The cochlear fluid displacement travels down the upper canal, round the bend, and back through the lower chamber. At the end of the lower canal the displacement is absorbed by the round window membrane's bulging outward. Now what happens if the oval window vibrates in and out rapidly? The waves of fluid displacement do not have enough time to travel all the way to the end of the upper canal and back through the lower canal. Instead, they take a shortcut by crossing the basilar membrane, causing it to flex. The more rapid the vibration, the closer to the oval and round windows the wave of displacement will flex the basilar membrane. Thus different pitches of sound will flex the basilar membrane at different locations and activate different sets of hair cells (see Figure 39.15). This ability of the basilar membrane to respond to vibrations of different frequencies is enhanced by its structure. Near the oval and round windows (the proximal end) it is narrow and stiff,

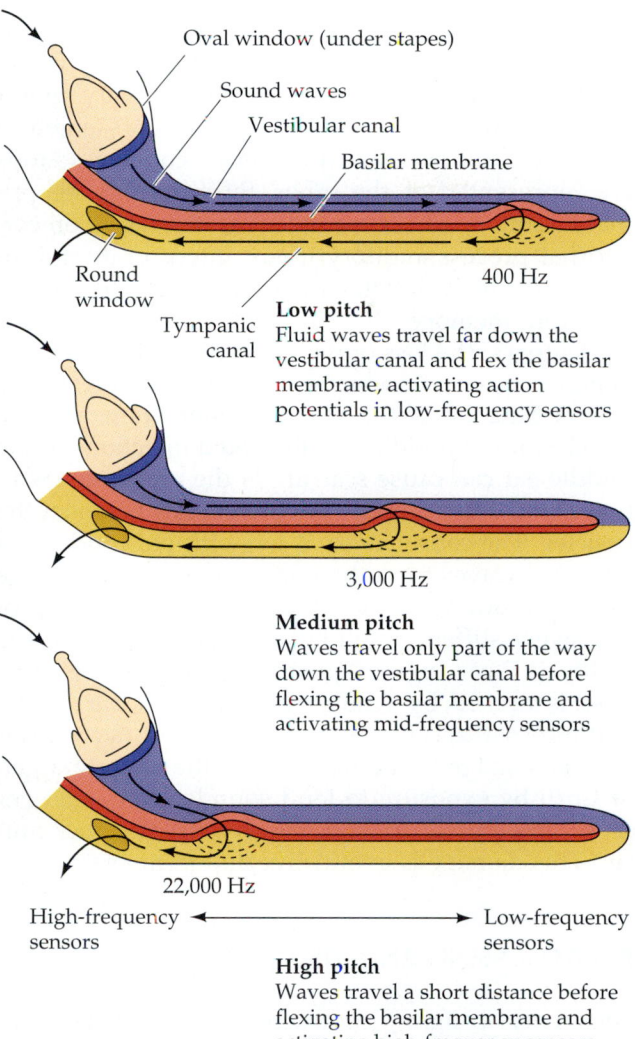

39.15 Sound Pressure Waves in the Inner Ear
For simplicity, this diagram illustrates the cochlea as uncoiled. Pressure waves of different frequencies flex the basilar membrane at different locations. Information about sound frequency is specified by which hair cells—the sensors—are activated.

but it gradually gets wider and more flexible toward the opposite (distal) end. So it is easier for the proximal basilar membrane to resonate with high frequencies and for the distal basilar membrane to resonate with lower frequencies. A complex sound made up of many frequencies will distort the basilar membrane at many places simultaneously and activate a unique subset of hair cells.

Action potentials generated by the mechanosensors at different places along the organ of Corti travel to the brain stem along the auditory nerve. The auditory pathways make several synapses in the brain stem and send off collateral fibers to the reticular activating system. That is why sudden or loud noises wake us up or get our attention. Eventually the auditory pathways reach the temporal lobes of the cerebral cortex. The primary auditory cortex contains tone maps analogous to the maps of the body found in the primary somatosensory and primary motor cortices. High-frequency sounds are represented at one end of this patch of cortex, low-frequency sounds are represented at the other. Surrounding the primary auditory cortex are the areas of association cortex that process auditory input, interpret it, and integrate the information with inputs from other senses and from memory.

Deafness, the loss of the sense of hearing, has two general causes. **Conduction deafness** is due to the loss of function of the tympanic membrane and the ossicles of the middle ear. Repeated infections of the middle ear can cause scarring of the tympanic membrane and stiffening of the connections between the ossicles. The consequence is less-efficient conduction of sound waves from the tympanic membrane to the oval window. With increasing age, the ossicles progressively stiffen, resulting in a gradual loss of the ability to hear high-frequency sounds. **Nerve deafness** is caused by damage to the inner ear or the auditory pathways. A common cause of nerve deafness is damage to the hair cells of the delicate organ of Corti by exposure to loud sounds such as jet engines, pneumatic drills, or highly amplified rock music. This damage is cumulative and permanent.

PHOTOSENSORS AND VISUAL SYSTEMS

Sensitivity to light—photosensitivity—confers upon the simplest animals the ability to orient to the sun and sky and gives more-complex animals instantaneous detailed information about objects in the environment. It is not surprising that simple and complex animals can sense and respond to light. What is remarkable is that across the entire range of animal species evolution has conserved the same basis for photosensitivity: the molecule rhodopsin. In this section we will learn how rhodopsin responds when stimulated by light energy, how that response is transduced into neural signals, and how the brains of vertebrates process those signals to result in vision. We will also examine the structures of eyes, the organs that gather and focus light energy onto photosensitive cells. Finally, we will learn how the brain uses action potentials from the retina to create our mental image of the visual world.

Rhodopsin

Photosensitivity depends on the ability of a molecule to absorb photons of light and to respond by changing its conformation. The molecule that does this in the eyes of all animals is **rhodopsin** (Figure 39.16). Rhodopsin consists of a protein, **opsin**, which by itself does not absorb light, and a light-absorbing group, **11-*cis* retinal**. The light-absorbing group is cradled in the center of the opsin, and the entire rhodopsin molecule sits in the plasma membrane of a photosensor cell. When the 11-*cis* retinal absorbs a photon of light energy, its shape changes into a different isomer of retinal—all-*trans* retinal. This conformational change puts a strain on the bonds between retinal and opsin, and the two components

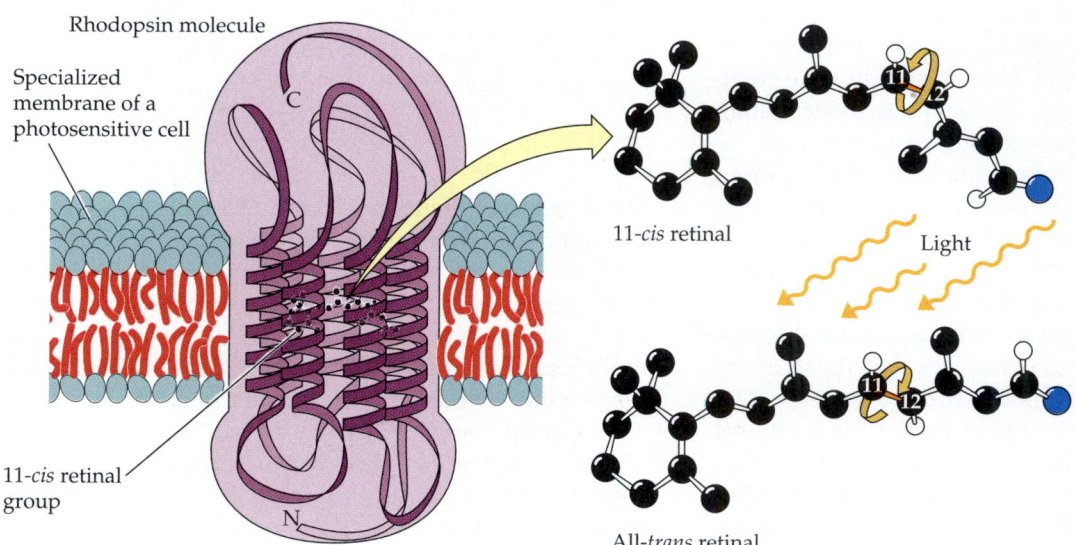

39.16 A Photosensitive Molecule
Rhodopsin is a transmembrane protein (opsin) that contains a light-responsive group (11-*cis* retinal). When 11-*cis* retinal absorbs a photon of light energy, it changes shape, becoming all-*trans* retinal, which is not responsive to light. The molecule returns spontaneously to the 11-*cis* conformation.

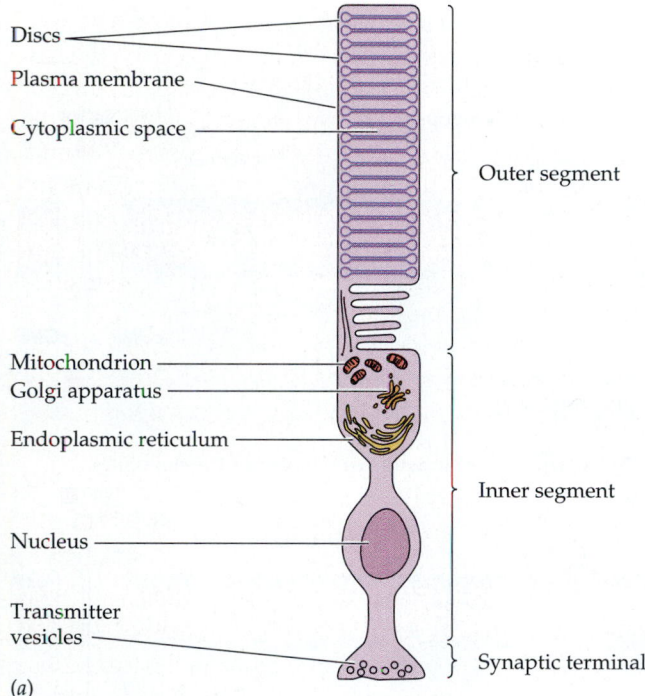

Discs
Plasma membrane
Cytoplasmic space

Outer segment

Mitochondrion
Golgi apparatus
Endoplasmic reticulum

Inner segment

Nucleus

Transmitter vesicles

Synaptic terminal

(a)

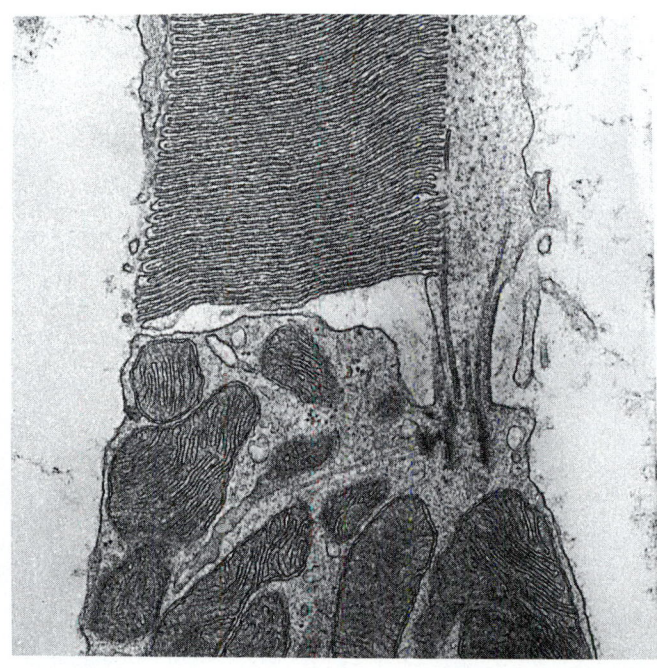

(b)

39.17 A Vertebrate Photosensor
(a) The rod cell of the vertebrate retina is a neuron modified for photosensitivity. The membranes of a rod cell's discs are densely packed with rhodopsin. (b) A transmission electron micrograph of a section through a photosensor.

break apart. As a result of this disassociation of retinal and opsin, referred to as bleaching, the molecule loses its photosensitivity. The retinal spontaneously returns to its 11-*cis* isomer and recombines with opsin to become, once again, photosensitive rhodopsin.

How does the light-induced change in rhodopsin's conformation transduce light into a cellular response? As retinal goes from the 11-*cis* to the all-*trans* forms, its interactions with opsin pass through several unstable intermediate stages. One stage is known as photoexcited rhodopsin because it triggers a cascade of reactions that changes ion flows, producing the alteration of membrane potential that is the photosensor's response to light. Let's explore these events of transduction in more detail.

The rhodopsin molecule sits in the membrane of a photosensor. How does this molecule communicate to the cell that it has absorbed a photon? And how does the sensor then communicate to the nervous system that its rhodopsin molecules are receiving light? To answer these questions, we must see how photosensors respond to light. A good example of a photosensor cell is a vertebrate **rod cell**, which is a modified neuron (Figure 39.17). A dense layer of photosensor cells at the back of the eye forms the **retina**, which, as we will see, is the structure that transduces the visual world into the language of the

nervous system. Each rod cell in the retina has an inner segment and an outer segment. The inner segment contains the usual organelles of a cell and has a synaptic terminal at its base where the cell communicates with other neurons. The outer segment is highly specialized and contains a stack of discs. These discs form by the invagination (folding inward) and pinching off of the plasma membrane. The membranous discs are densely packed with rhodopsin; their function is to capture photons of light passing through the rod cell.

To see how the rod cell responds to light, we can penetrate a single rod cell with an electrode. Through this electrode we can record the receptor potential of the rod cell in the dark and in the light (Figure 39.18). From what we have learned about other types of sensors, we might expect stimulation of the photosensor with light would make its receptor potential less negative, but the opposite is true. When the rod cell is kept in the dark, it already has a very high resting potential in comparison with other neurons. In fact, the plasma membrane of the rod cell is fairly permeable to Na^+ ions, so these positive charges are continually entering the cell. When a light is flashed on the dark-adapted rod cell, its receptor potential becomes more negative—it hyperpolarizes. The rod cell itself *does not* generate action potentials. The rod cell does change its rate of neurotransmitter release, however, as its membrane polarity changes. Later we will learn how other cells in the visual pathway respond to neurotransmitter released from the photosensors so that information is communicated to the brain in the form of action potentials.

39.18 A Rod Cell Responds to Light
The receptor potential of a rod cell hyperpolarizes (becomes more negative) in response to a flash of light.

How does the absorption of light by rhodopsin hyperpolarize the rod cell? When rhodopsin is excited by light, it initiates a cascade of events. The photoexcited rhodopsin combines with and activates another protein, a G-protein called transducin. Activated transducin in turn activates a phosphodiesterase. Active phosphodiesterase converts cyclic GMP (cGMP) to 5'-GMP. What was that cGMP doing before it was converted to 5'-GMP? It was holding open the sodium channels and keeping the cell depolarized. As cGMP is destroyed, the sodium channels close, and the cell hyperpolarizes. This may seem like a roundabout way of doing business, but its significance is its enormous amplification ability. Each molecule of photoexcited rhodopsin can activate about 500 transducin molecules, thus activating about 500 phosphodiesterase molecules. The catalytic prowess of a molecule of phosphodiesterase is great; it can hydrolyze over 4,000 molecules of cGMP per second. The bottom line is that a single photon of light can cause the closing of over a million sodium channels and thereby change the rod cell's receptor potential (Figure 39.19). Now let's see how photosensors work in animals.

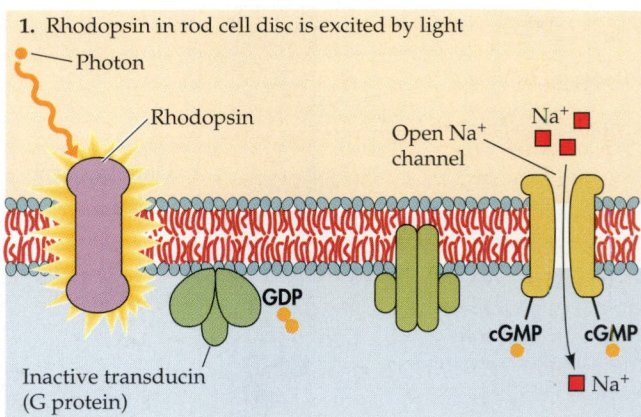

1. Rhodopsin in rod cell disc is excited by light

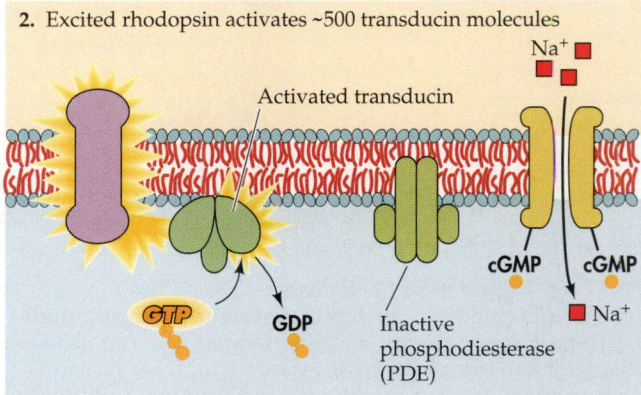

2. Excited rhodopsin activates ~500 transducin molecules

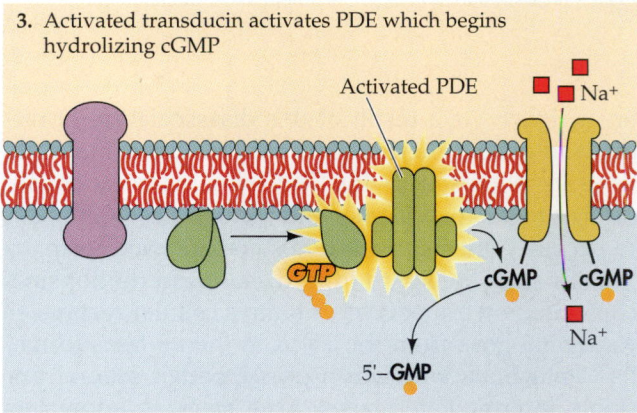

3. Activated transducin activates PDE which begins hydrolizing cGMP

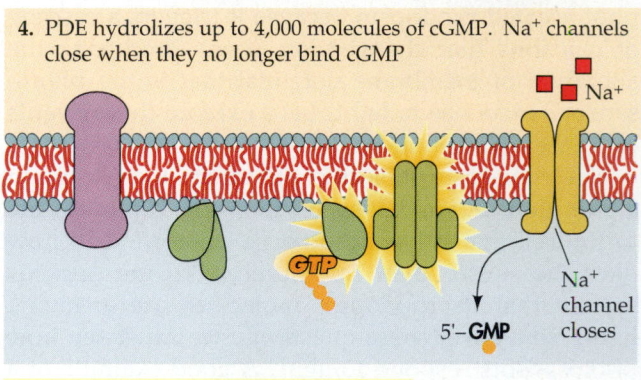

4. PDE hydrolizes up to 4,000 molecules of cGMP. Na⁺ channels close when they no longer bind cGMP

One photon of light can result in closing over **one million** Na⁺ channels!

◀ 39.19 Light Absorption Closes Na⁺ Channels

A key player in these events is a G-protein called transducin. Inactive transducin has three subunits, one of which binds GDP. Activated rhodopsin causes the GDP to be replaced by GTP, and the transducin molecule splits. The subunit with GTP can activate a phosphodiesterase (PDE) molecule. Active phosphodiesterase degrades cGMP bound to the Na⁺ channel and causes Na⁺ channels to close.

Visual Systems of Invertebrates

Flatworms, simple multicellular animals, obtain directional information about light from photosensitive cells organized into eye cups (Figure 39.20). The eye cups are bilateral structures, and each is partially shielded from light by a layer of pigmented cells lining the cup. Because the openings of the eye cups face in opposite directions, the photosensors on the two sides of the animal are unequally stimulated unless the animal is facing directly toward or away from a light source. Using directional information about light sources, the flatworm moves away from light.

Arthropods (such as crustaceans, spiders, and insects) have evolved **compound eyes** that provide them with information about patterns or images in the environment. Each compound eye consists of many optical units called **ommatidia** (singular: ommatidium). The number of ommatidia in a compound eye varies from only a few in some ants, to 800 in fruit flies, to 10,000 in some dragonflies. Each ommatidium has a lens structure that directs light onto photosensors called retinula cells. Flies have seven elongated retinula cells in each ommatidium. The inner borders of the retinula cells are covered with microvilli that contain rhodopsin and constitute a light trap. Since the microvilli of the different retinula cells overlap, they appear to form a central rod, called

a rhabdom, down the center of the ommatidium. Axons from the retinula cells communicate with the nervous system (Figure 39.21). Because each ommatidium of a compound eye is directed at a slightly different part of the visual world, only a crude, or perhaps broken, image of the visual field can be communicated from the compound eye to the central nervous system.

Vertebrate and Cephalopod Eyes

Vertebrates and cephalopod mollusks have evolved eyes with exceptional abilities to form images of the visual world. These eyes operate like cameras, and considering that they evolved independently of each other, their high degree of similarity is remarkable (Figure 39.22). The vertebrate eye is a spherical, fluid-filled structure bounded by a tough connective tissue layer called the sclera. At the front of the eyeball, the sclera forms the transparent **cornea** through which light enters the eye. Just inside the cornea is the pigmented **iris**, which gives the eye its color. The important function of the iris is to control the amount of light that reaches the photosensors at the back of the eyeball, just as the diaphragm of the camera controls the amount of light reaching the film. The central opening of the iris is the **pupil**. The iris is under control of the autonomic nervous system. In bright light the iris constricts and the pupil is very small, but as light levels fall, the iris relaxes and the pupil enlarges.

Behind the iris is the crystalline lens, which helps focus images on the photosensitive layer—the ret-

39.20 A Simple Photosensory System Gives Directional Information

Although flatworms do not "see" as we understand it, the eye cups of this flatworm enable it to move away from a light source to an area where it may be less visible to predators.

Flatworm responds to light by moving directly away from the source toward darkness

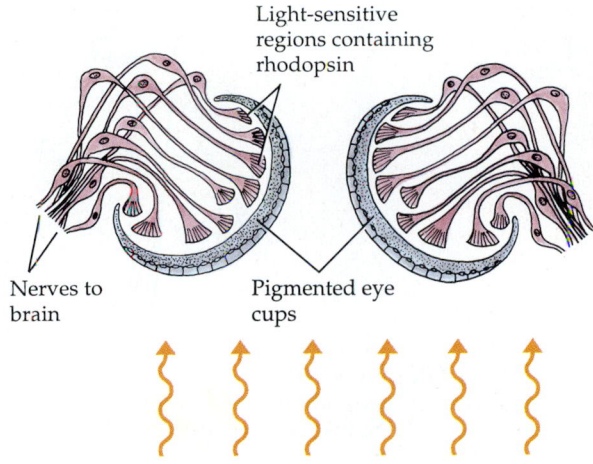

Light-sensitive regions containing rhodopsin

Nerves to brain

Pigmented eye cups

39.21 Eyes of an Insect

(a) The compound eyes of a horsefly; each eye contains hundreds of ommatidia. (b) Each ommatidium focuses light on a rhabdom consisting of overlapping, light-sensitive plasma membranes of a few photosensors.

(a)

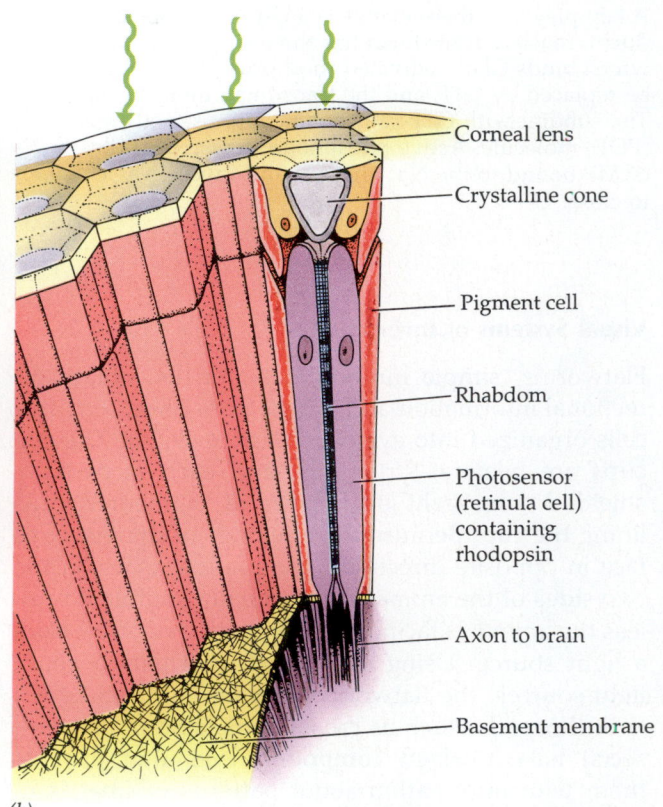

Corneal lens

Crystalline cone

Pigment cell

Rhabdom

Photosensor (retinula cell) containing rhodopsin

Axon to brain

Basement membrane

(b)

ina—at the back of the eye. The cornea and the fluids of the eye chambers also help focus light on the retina, but the lens is responsible for the ability to accommodate—to focus on objects at various locations in the near visual field. To focus a camera on objects close at hand, you must adjust the distance between the lens and the film. Fishes, amphibians,

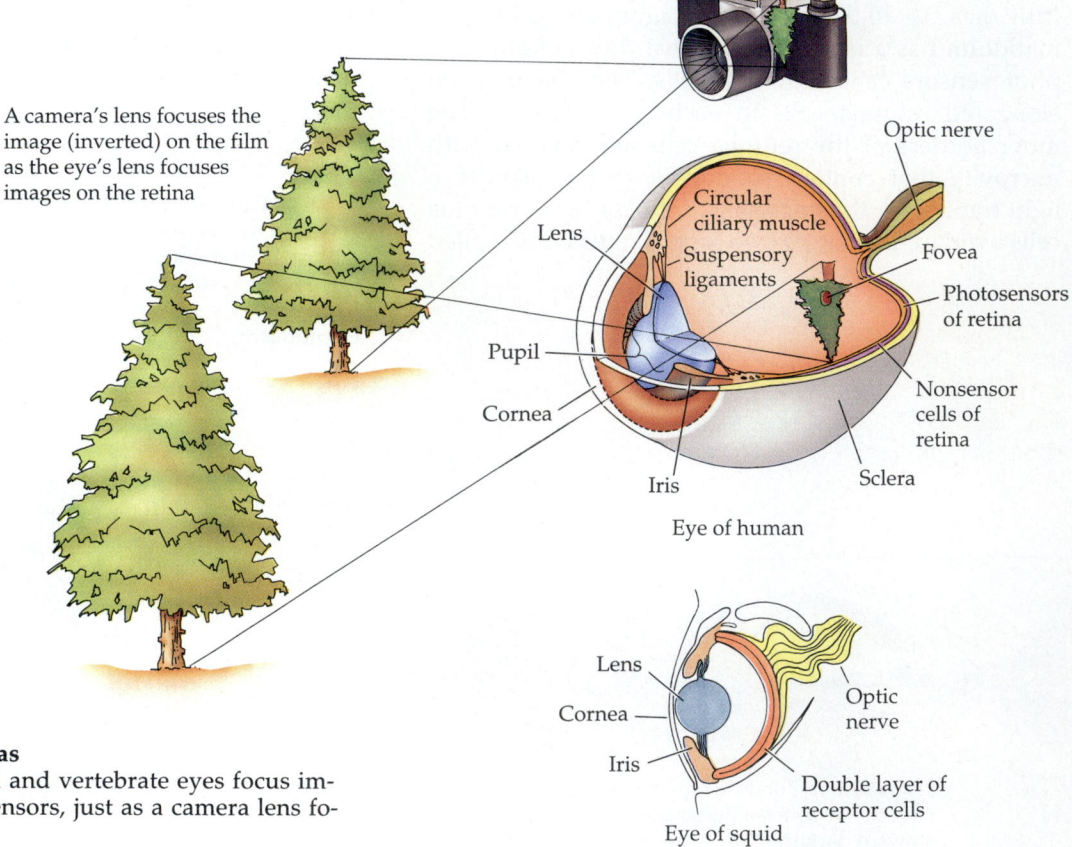

A camera's lens focuses the image (inverted) on the film as the eye's lens focuses images on the retina

Optic nerve

Lens

Circular ciliary muscle

Suspensory ligaments

Fovea

Photosensors of retina

Pupil

Nonsensor cells of retina

Cornea

Iris

Sclera

Eye of human

Lens

Cornea

Optic nerve

Iris

Double layer of receptor cells

Eye of squid

39.22 Eyes Like Cameras

The lenses of cephalopod and vertebrate eyes focus images on layers of photosensors, just as a camera lens focuses images on film.

and reptiles accommodate in a similar manner, moving the lenses of their eyes closer to or farther from their retinas. Mammals and birds use a different method; they alter the shape of the lens.

The lens is contained in a connective tissue sheath that tends to keep it in a spherical shape, but it is also suspended by suspensory ligaments that pull it into a flatter shape. Circular muscles called the ciliary muscles counteract the pull of the suspensory ligaments and permit the lens to round up. With the ciliary muscles at rest, the flatter lens has the correct optical properties to focus distant images on the retina, but not close images. Contracting the ciliary muscles rounds up the lens, changing its light-bending properties to bring close images into focus (Figure 39.23). As we get older, our lenses become less elastic, and we lose the ability to focus on objects close at hand without the help of corrective lenses. Prolonged concentration on small, close objects (such as the type on these pages) tires and strains the eyes by overworking the ciliary muscles.

The Vertebrate Retina

The retina is an extension of the brain. During development, neural tissue grows out from the brain to form the retina. In addition to a layer of photosensors, the retina includes layers of cells that process the visual information from the photosensors and transmit it to the brain in the form of action potentials in the optic nerves. A curious feature of the anatomy of the retina is that the outer segments of the photosensors are all the way at the back of the retina so light must pass through all the layers of retinal cells before reaching the place where photons are captured by rhodopsin. Later we will examine in detail how the cells of the retina process information, but first let's describe some general features of retinal organization.

The density of photosensors is not the same across the entire retina. Light coming from the center of the field of vision falls on an area of the retina called the **fovea**, where the density of sensors is the highest. The human fovea has about 160,000 sensors per square millimeter. A hawk has about 1 million sensors per square millimeter of fovea, making its vision

about eight times sharper than ours. In addition, the hawk has two foveas in each eye. One fovea receives light from straight ahead; the other receives light from below. Thus while the hawk is flying, it sees both its projected flight path and the ground below, where it might detect a mouse scurrying in the grass.

The fovea of a horse is a long, vertical patch of retina. The horse's lens is not good at accommodation, but it focuses distant objects that are straight ahead on one part of this long fovea and close objects that are below the head on another part of the fovea. When horses are startled by an object close at hand, they pull their heads back and rear up to bring the object into focus on the close-vision fovea.

Where blood vessels and the bundle of axons going to the brain pass through the back of the eye, there is a blind spot on the retina. You are normally not aware of your blind spot, but you can find it. Stare straight ahead, holding a pencil in your outstretched hand so that the eraser is in the center of your field of vision. While continuing to stare straight ahead, slowly move the pencil to the side until the eraser disappears. When this happens, the light from the eraser is focused directly on your blind spot.

Until now we have referred to only one type of photosensor, the rod cell. There are, however, two major types of photoreceptors, both named for their shapes: rod cells and **cone cells** (Figure 39.24). A human retina has about 3 million cones and about 100 million rods. Rod cells are more sensitive to light

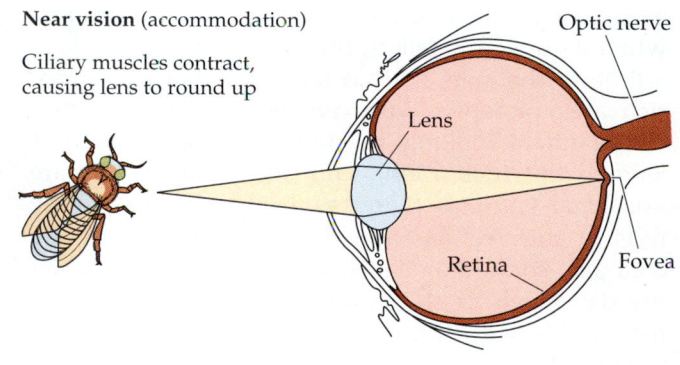

Near vision (accommodation)

Ciliary muscles contract, causing lens to round up

Optic nerve

Lens

Retina   Fovea

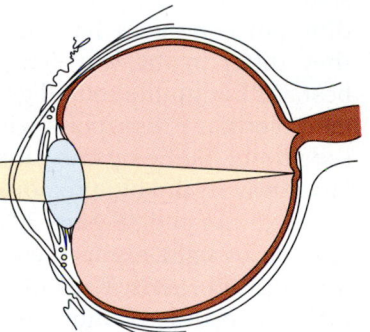

Distant vision

Ciliary muscles relax and suspensory ligaments pull lens to a flatter shape

39.23 Coming into Focus
Mammals and birds focus their eyes on close objects by changing the shape of the lens.

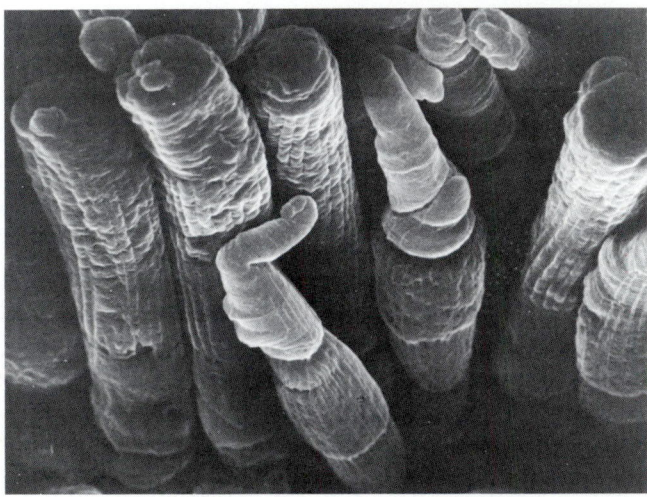

39.24 Rods and Cones
This scanning electron micrograph of photosensors in the retina of a mud puppy, an amphibian, shows cylindrical rods and tapered cones.

but do not contribute to color vision. Cones are responsible for color vision but are less sensitive to light. Cones are also responsible for our sharpest vision because, even though there are many more rods than cones in human retinas, our foveas contain mostly cones.

Because cones have low sensitivity to light, they are of no use at night. At night our vision is not very sharp, and we see only in black and white. You may have trouble seeing a small object such as a keyhole at night when you are looking straight at it—that is, when its image is falling on your fovea. If you look a little to the side, so that the image falls on a rod-rich area of retina, you can see the object better. Astronomers looking for faint objects in the sky learned this trick a long time ago. Animals that are nocturnal (such as flying squirrels) may have only rods in their retinas and have no color vision. By contrast, some animals that are active only during the day (such as chipmunks and ground squirrels) have only cones in their retinas.

How do cone cells enable us to see color? There are at least three kinds of cone cells, each possessing slightly different types of opsin molecules. Because different cone cells have different opsin molecules, they differ in the wavelengths of light they absorb best. Although the retinal group is the light absorber, its molecular interactions with opsin tune its spectral sensitivity. Some opsins cause retinal to absorb most efficiently in the blue region, some in the green, and some in the red (Figure 39.25). Intermediate wavelengths of light excite these classes of cones in different proportions. Recently the genes coding for the different opsins of humans were identified: one for blue-sensitive opsin, one for red-sensitive opsin, and several for green-sensitive opsin.

The human retina is organized into five layers of neurons that receive visual information and process it before sending it to the brain (Figure 39.26). Strangely enough, as mentioned earlier, the layer of photosensors is all the way at the back of the retina, so light has to traverse all the other layers before reaching the rods and cones. The disc-containing ends of the rods and cones are partly buried in a layer of pigmented epithelium that absorbs photons not captured by rhodopsin and prevents any backward scattering of light that might decrease visual sharpness. Nocturnal animals, such as cats, have a highly reflective layer, called the tapetum, behind the photosensors. Photons not captured on their first pass through the photosensors are reflected back, thus increasing visual sensitivity (but not sharpness) in low-light conditions. The reflective tapetum is what makes cats' eyes appear to glow in the dark.

A first step in investigating how the human retina processes visual information is to study how its five layers of neurons are interconnected and how they influence one another. The neurons at the back of the retina are the photosensors. As we know, the photosensors hyperpolarize in response to light, but they do not generate action potentials. The cells at the front of the retina are ganglion cells. Ganglion cells fire action potentials, and their axons form the optic nerves that travel to the brain. The photosensors and ganglion cells are connected by bipolar cells. The changes in membrane potential of rods and cones in response to light alter the rate at which the rods and cones release neurotransmitter at their synapses with the bipolar cells. Like rods and cones, bipolar cells do not fire action potentials. In response to neurotransmitter from the photosensors, the membrane potentials of bipolar cells change, altering the rate at which they release neurotransmitter onto ganglion cells. The ganglion cells generate action poten-

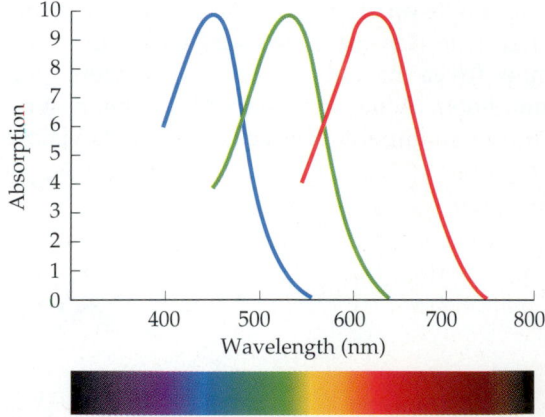

39.25 Absorption Spectra of Cone Cells
Human color vision is based on three kinds of cone cells. Each absorbs a different band of wavelengths most effectively.

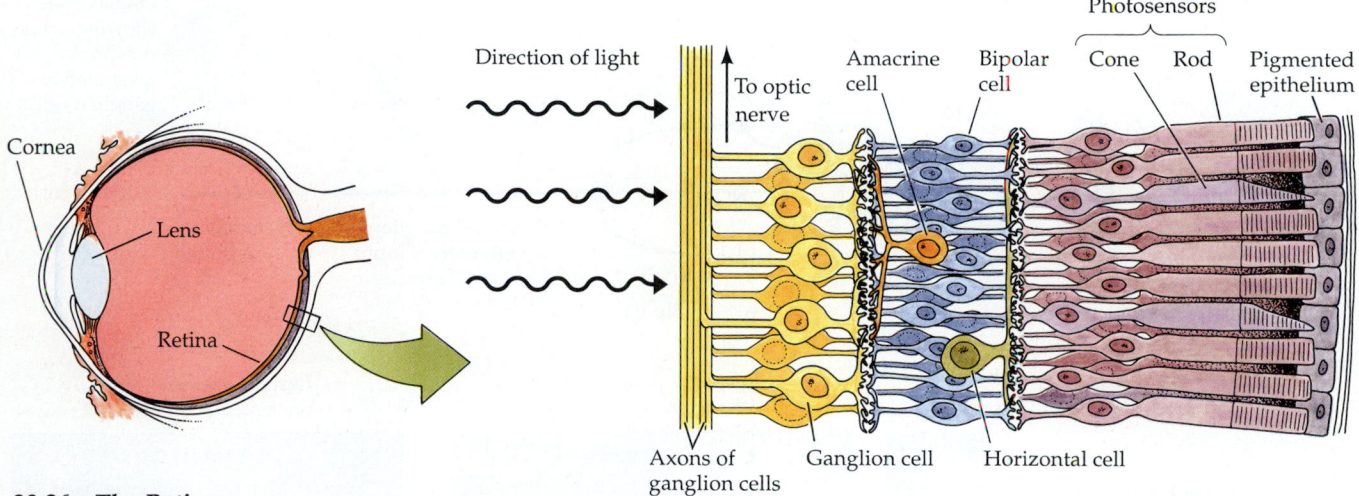

39.26 The Retina
Light travels through layers of transparent neurons—ganglion, amacrine, bipolar, and horizontal cells—and is absorbed by discs in the rods and cones (the photosensory layer) at the back of the retina. The visual information is then processed forward through several layers of neurons, with final convergence on ganglion cells that send their axons to the brain.

tials, and the rate of neurotransmitter release from the bipolar cells determines the rate at which ganglion cells fire action potentials. Thus the direct flow of information in the retina is from photosensor to bipolar cell to ganglion cell. Ganglion cells send the information to the brain.

What do the other two layers (the horizontal cells and the amacrine cells; see Figure 39.26) do? They communicate laterally across the retina. Horizontal cells connect neighboring pairs of photosensors and bipolar cells. Thus the communication between a photosensor and its bipolar cell can be influenced by the amount of light absorbed by neighboring photosensors. This lateral flow of information sharpens the perception of contrast between light and dark patterns falling on the retina. Amacrine cells connect neighboring pairs of bipolar cells and ganglion cells. One role of amacrine cells is to adjust the sensitivity of the eyes according to the overall level of light falling on the retina.

Knowing the paths of information flow through the retina still doesn't tell us how that information is processed. What does the eye tell the brain in response to a pattern of light falling on the retina? One aspect of information processing in the retina is information reduction. There are over 100 million photosensors in each retina, but only about 1 million ganglion cells sending messages to the brain. How is the information from all those photosensors reduced to the messages sent to the brain by the ganglion cells? This question was addressed in some elegant, classic experiments in which scientists used electrodes to record the activity of single ganglion cells in living animals while their retinas were stimulated with spots of light. They found that each ganglion cell has a well-defined **receptive field** that consists of

a specific group of photosensors. Stimulating these photosensors with light activates the ganglion cell (Figure 39.27a). Information from many photosensors is reduced in this way to a single message.

The receptive fields of ganglion cells are all circular, but the way a spot of light influences the activity of the ganglion cell depends on where in the receptive field it falls. Each ganglion cell's receptive field is divided into two concentric areas, called the center and the surround. There are two kinds of receptive fields, on-center and off-center. Stimulating the center of an on-center receptive field excites the ganglion cell; stimulating the surround inhibits it. Stimulating the center of an off-center receptive field inhibits the ganglion cell, and stimulating the surround excites it (Figure 39.27b). Center effects are always stronger than surround effects. The response of a ganglion cell to stimulation of the center of its receptive field depends on how much of the surround area is also stimulated. A small dot of light directly on the center has the maximal effect, a bar of light hitting the center and pieces of the surround has less of an effect, and a large, uniform patch of light falling equally on center and surround has no effect. Ganglion cells communicate information about light-dark contrasts falling on their receptive fields to the brain.

How are receptive fields related to the connections between the neurons of the retina? The photosensors in the center of a ganglion cell's receptive field are connected to that ganglion cell by bipolar cells. The photosensors in the surround send information to the center photosensors and thus to the ganglion cell through the lateral connections of horizontal cells. Thus the receptive field of a ganglion cell is due to a pattern of synapses between photosensors, horizontal cells, bipolar cells, and ganglion cells. The recep-

(a) **Experimental design**

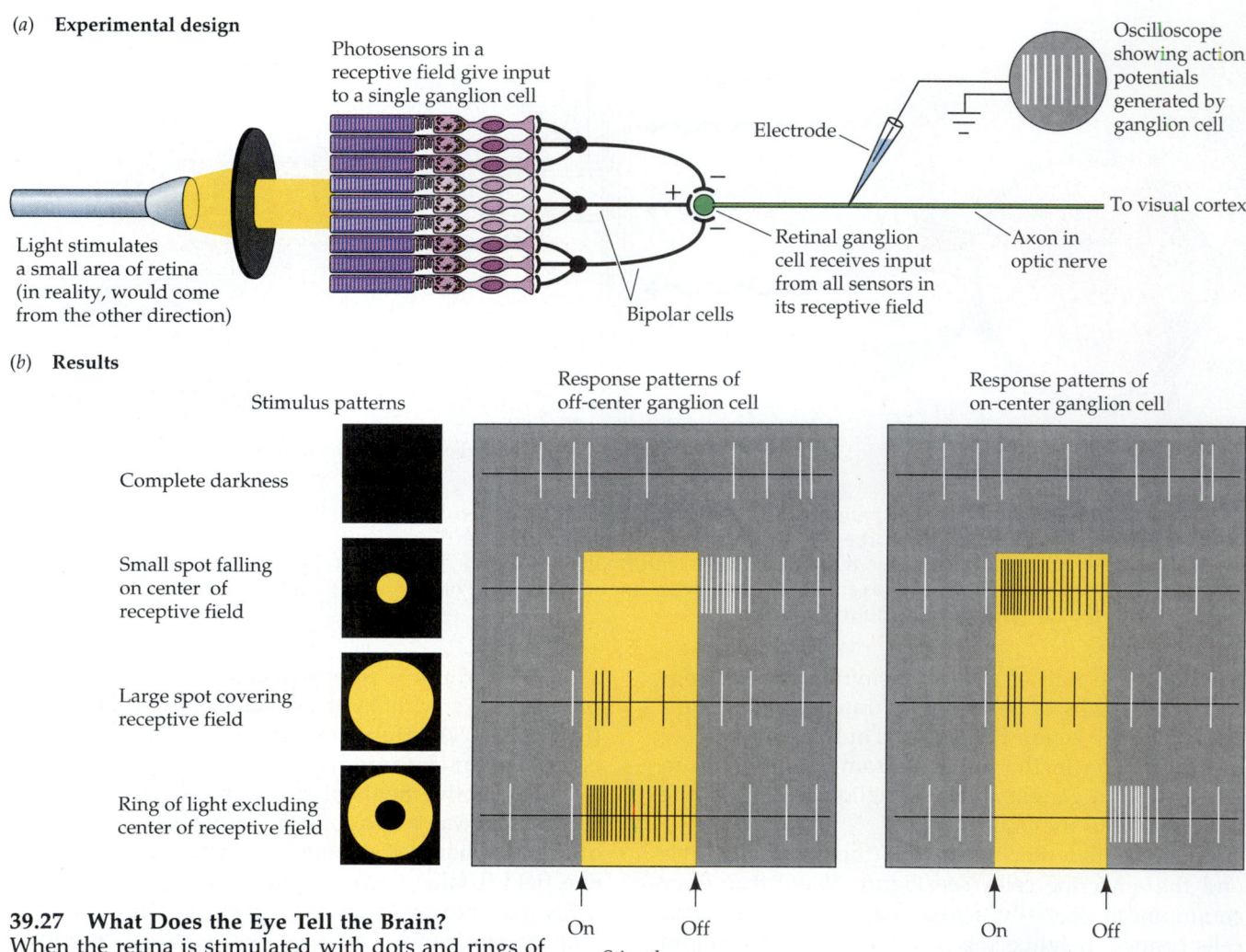

(b) Results

39.27 What Does the Eye Tell the Brain?
When the retina is stimulated with dots and rings of light, individual ganglion cells show different response properties. Each ganglion cell is spontaneously active and responds to light shining on a small circular area of retina in its receptive field. Some ganglion cells are stimulated and some are inhibited by a spot of light falling on the receptive field centers. An "on-center" ganglion cell is inhibited by a ring of light falling on the peripheral area of its receptive field. The opposite is true of "off-center" ganglion cells. Cells may show a brief response when the stimulus is turned off.

tive fields of neighboring ganglion cells can overlap greatly; a given photosensor can be connected to several ganglion cells.

The eye sends the brain simple messages about the pattern of light intensities falling on small, circular patches of retina. How does the brain create a mental image of the visual world from this information? David Hubel and Torsten Wiesel of Harvard University tackled this question. Information from the retina is transmitted through the optic nerves to a relay station in the thalamus and then to the brain's visual processing area, in the occipital cortex at the back of the cerebral hemispheres. Hubel and Wiesel

recorded the activities of single cells in the visual processing areas of the brains of living animals while they stimulated the animals' retinas with spots and bars of light. They found that, like ganglion cells, cells in the visual cortex have receptive fields—specific areas of the retina that when stimulated by light influence the rate at which the cortical cells fire action potentials.

Hubel and Wiesel learned that cells in the visual cortex have different types of receptive fields. One type of cell, called the **simple cell**, is maximally stimulated by a bar of light with a certain orientation falling on one place on the retina. Simple cells probably receive input from a number of ganglion cells that have their circular receptive fields lined up in a row. Another type of cortical cell, the **complex cell**, is also maximally stimulated by a bar of light with a particular orientation, but the bar may fall anywhere on a large area of retina described as that cell's receptive field. Complex cells seem to receive input from a number of simple cells that share a certain stimulus orientation but that have receptive fields in

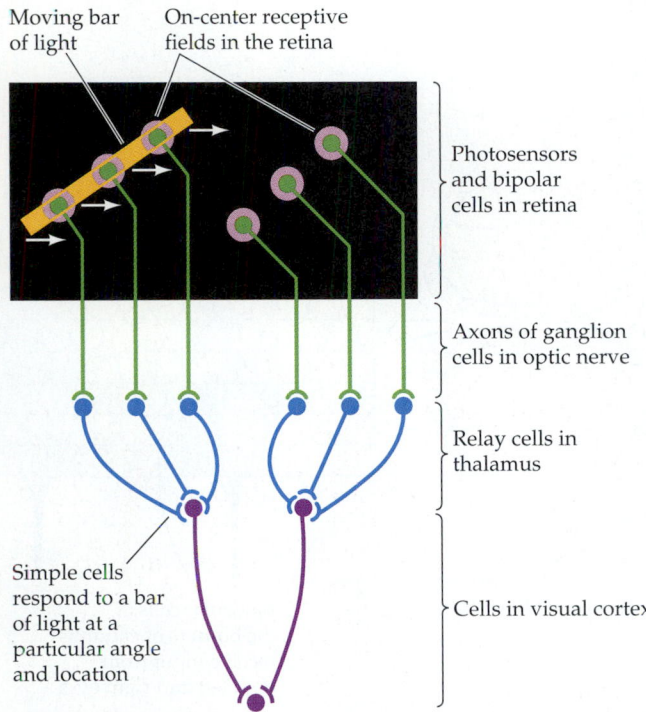

Moving bar of light

On-center receptive fields in the retina

Photosensors and bipolar cells in retina

Axons of ganglion cells in optic nerve

Relay cells in thalamus

Cells in visual cortex

Simple cells respond to a bar of light at a particular angle and location

Complex cell responds to light at a particular angle at any location on the retina, or to a bar of light moving across the retina

39.28 Receptive Fields of Cells in the Visual Cortex
Cells in the visual cortex respond to specific patterns of light falling on the retina. Ganglion cells that project information about circular receptive fields converge on simple cells in the cortex in such a way that simple cells have linear receptive fields. Simple cells project to complex cells in such a way that complex cells can respond to linear stimuli falling on different areas of the retina.

different places on the retina (Figure 39.28). Some complex cells respond best when the bar of light moves in a particular direction, perhaps depending on the combination of on-center and off-center receptive fields.

The concept that emerges from these experiments is that the brain assembles a mental image of the visual world by analyzing edges of light patterns falling on the retina. This analysis is conducted in a massively parallel fashion. Each retina sends 1 million axons to the brain, but there are at least 200 million neurons in the visual cortex. A bit of information from a retinal ganglion cell is received by hundreds of cortical cells, each responsive to a different combination of orientation, position, and even movement of contrasting lines in the pattern of light falling on the retina.

Binocular Vision

How do we see objects in three dimensions? The quick answer is that our two eyes see overlapping, yet slightly different, fields. Turn a typical conical flowerpot upside down and look down at it so that the bottom of the pot is exactly in the center of your overall field of vision. You see the bottom of the pot, and you see equal amounts of the sides and rim of the pot as concentric circles around the bottom. Now if you close your left eye, you see more of the right

side and right rim of the pot. With your right eye closed you see more of the left side and left rim of the pot. The discrepancies in the information coming from your two eyes are interpreted by the brain to provide information about the depth and the three-dimensional shape of the flowerpot. If you are blind in one eye, you have great difficulty discriminating distances. Animals whose eyes are on the sides of their heads have nonoverlapping fields of vision and, as a result, poor depth vision, but they can see predators creeping up from behind!

The story of how the brain integrates information from two eyes begins with the paths of the optic nerves. If you look at the underside of the brain, the optic nerves from the two eyes appear to join together just under the hypothalamus and then separate again. The place where they join is called the **optic chiasm**. Axons from the half of each retina closest to your nose cross in the optic chiasm and go to opposite sides of your brain. The axons from the other half of each retina go to the same side of the brain. The result of this division of axons in the optic chiasm is that all visual information from your left visual field (everything left of straight ahead) goes to the right side of your brain, as shown in red in Figure 39.29. All visual information from your right visual field goes to the left side of your brain, as indicated in green. Both eyes transmit information about a specific spot in your right visual field, for example, to the same place in the left visual cortex. How are the two sources of information integrated?

Cells in the visual cortex are organized in columns. These columns alternate: left eye, right eye, left eye, right eye, and so on. Cells closest to the border between two columns receive input from both eyes and are therefore **binocular cells**. Binocular cells interpret distance by measuring the disparity between where the same stimulus falls on the two retinas. What is disparity? Hold your finger out in front of you and look at it, closing one eye and then the other. Your finger appears to jump back and forth because its image falls on a different position on each retina. Repeat the exercise with an object at a distance. It doesn't appear to jump back and forth as much because there is less disparity in the positions of the image on the two retinas. Certain binocular cells respond optimally to a stimulus falling on both retinas

Human brain (viewed from underneath)

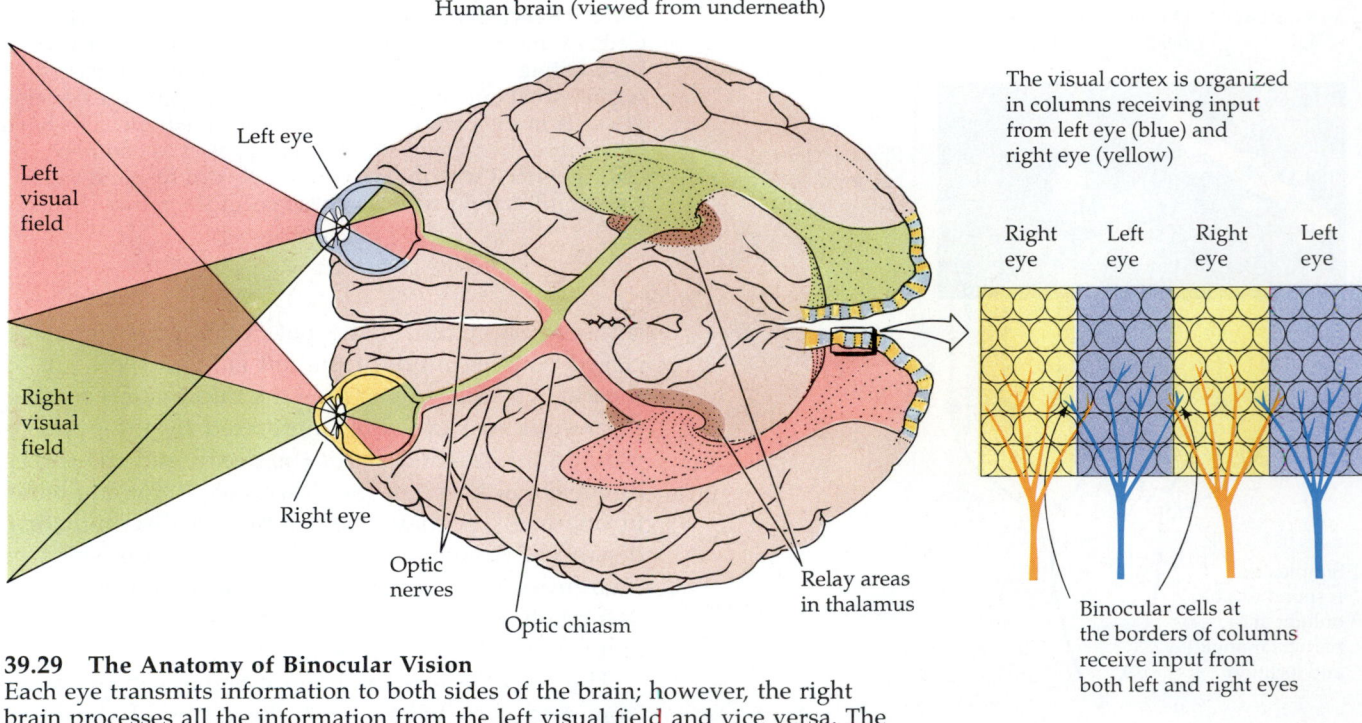

Left eye

Left visual field

Right visual field

Right eye

Optic nerves

Optic chiasm

Relay areas in thalamus

The visual cortex is organized in columns receiving input from left eye (blue) and right eye (yellow)

Right eye   Left eye   Right eye   Left eye

Binocular cells at the borders of columns receive input from both left and right eyes

39.29   The Anatomy of Binocular Vision
Each eye transmits information to both sides of the brain; however, the right brain processes all the information from the left visual field and vice versa. The visual cortex sorts visual field information according to whether it comes from the right or left eye. Binocular cells in the visual cortex receive the same visual information from both eyes.

with a particular disparity. Which set of binocular cells is stimulated depends on how far away the stimulus is.

When we look at something, we can detect shape, color, depth, and movement. Where does all this information come together? Is there a single cell that fires only when a red sports car drives by? Probably not. A specific visual experience comes from simultaneous activity in many cells. To add to the complexity, a visual experience is not strictly visual, but is enhanced by information from the other senses and from memory.

OTHER SENSORY WORLDS

After emphasizing the incredible neural complexity of sensory integration, we must recognize that humans make use of only a subset of the information available to us in the environment. Other animals have sensory systems that enable them to use different subsets and different types of information.

Infrared and Ultraviolet Detection

When discussing vision we use the term "visible spectrum," but what we really mean is light visible to us. The human visible spectrum is a very narrow region of the entire, continuous range of electromag-

netic radiation in the environment (see Figure 8.6). For example, we cannot see ultraviolet radiation, but many insects can. One of the seven photosensors in each ommatidium of a fruit fly is sensitive to ultraviolet light. The visual sensitivity of many pollinating insects includes the ultraviolet part of the spectrum. Some flowers have patterns that are invisible to us but that show up if we photograph them with film that is sensitive to ultraviolet light (see Figure 48.23). Those patterns provide information to prospective pollinators, but humans are not equipped to receive that information.

At the other end of the spectrum is infrared radiation, which we sense as heat. Other animals extract much more information from the infrared radiation—especially infrared radiation emitted by potential prey. A group of snakes known as pit vipers have pit organs, one just in front of each eye, that can sense infrared radiation (Figure 39.30). In total darkness these snakes can locate a prey item such as a mouse, orient to it, and strike it with great accuracy based on the directional information that comes to them from the warmth of the mouse's body.

Echolocation

Some species emit intense sounds and create images of their environments from the echoes of those sounds. Bats, porpoises, dolphins, and, to a lesser

39.30 Pit Organs "See" Heat
The eyelash viper of the Costa Rican rainforest is a pit viper. The "hole" just below and in front of its eye is a pit organ that senses infrared radiation. Such snakes can locate prey in total darkness based on directional information they receive through their pit organs.

39.31 Flying in the Dark
Using echolocation, this bat avoids an obstacle course of thin wires while flying in complete darkness.

extent, whales, are accomplished echolocators. Some species of bats have elaborate modifications of their noses to direct the sounds they emit, as well as impressive ear pinnae to collect the returning echoes. The sounds they emit as pulses (about 20 to 80 per second) are above our range of hearing, but they are extremely loud in contrast to their faint echoes bouncing off small insects. An echolocating bat is similar to a construction worker who is trying to overhear a whispered conversation while using a pneumatic drill. To avoid deafening themselves, bats use muscles in their middle ears to dampen their sensitivity while they emit sounds, then relax them quickly enough to hear the echoes. The ability of bats to use echolocation to "see" their environment is so good that in a totally dark room strung with fine wires, bats can capture tiny flying insects while navigating around the wires (Figure 39.31).

Detection of Electric Fields

We already discussed the mechanosensors in the lateral lines of fishes (see Figure 39.11). The lateral lines of some species, especially ones that live in murky waters (catfish, for example) also contain *electro*sensors. These sensors enable the fish to detect weak electric fields, which can help them locate prey. The use of electrosensors is even more sophisticated in species called electric fishes. These animals have evolved electric organs in their tails that generate a continuous series of electric pulses, creating a weak electric field around their bodies (Figure 39.32). Any objects in the environment, such as rocks, logs, plants, or other fish, disrupt the electric fish's electric field, and the electrosensors of the lateral line detect those disruptions. In some species of electric fish,

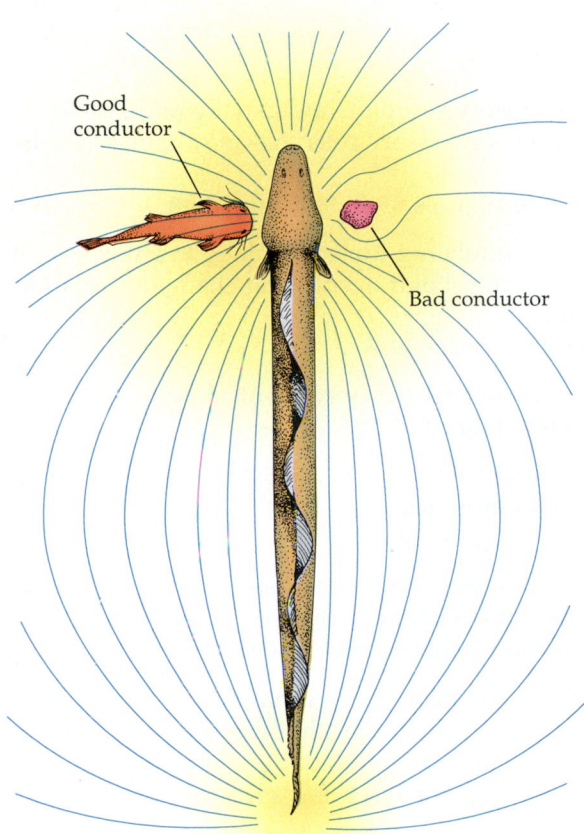

Good conductor

Bad conductor

39.32 Sensing with an Electric Field
Fish such as the electric eel generate personal electric fields and sense perturbations within their fields. Objects that are poor conductors of electricity perturb the field greatly, whereas good conductors perturb it only slightly or not at all.

each individual in a group emits electric pulses at a different frequency. If a new fish is added to the group, they all adjust their frequencies.

Magnetic Sense

As astonishing as it may seem, the magnetic lines of force of the planet provide directional information for some animals. From experiments with homing pigeons, biologists have accumulated evidence of this phenomenon. Normally, homing pigeons released from locations distant from their homes can orient and find their way home, even if they are fitted with translucent contact lenses that eliminate all directional information from the sun. However, pigeons that are released on cloudy days with tiny magnets glued to their heads become disoriented and cannot navigate. The magnetic sense is poorly understood at present, but single magnetosensory neurons have been discovered.

SUMMARY of Main Ideas about Sensory Systems

Sensory systems gather information about an animal's external and internal environments and transmit it to the central nervous system.

Sensors are cells with membrane proteins that alter ion channels in response to specific stimuli, creating receptor potentials and eventually action potentials.
Review Figures 39.1, 39.2, and 39.3

Sensory systems can adapt to constant levels of stimulation.

Chemosensors for smell, taste, and sensing of pheromones have receptor molecules that bind to specific molecules.

Chemosensory receptors are usually on hairlike projections of sensory cells that may or may not be neurons.
Review Figures 39.5, 39.6, and 39.8

When a stimulating molecule binds to a receptor, it initiates intracellular events that change membrane potential.
Review Figure 39.7

The skin contains many types of sensors.
Review Figure 39.10

Mechanosensors respond to distortions of their plasma membranes.
Review Figure 39.3

A stretch sensor is a mechanosensor that provides information about the position and movement of muscles and joints or about blood pressure.

Hair cells (mechanosensors that are not neurons) have stereocilia that alter the receptor potential of the plasma membrane when they are bent.
Review Figures 39.11 and 39.12

The organs of equilibrium in the mammalian inner ear use hair cells to sense position and movement.
Review Figure 39.13

Hair cells in the human ear convert pressure waves into action potentials.
Review Figures 39.14 and 39.15

Photosensitivity in animals depends on the capture of photons of light by rhodopsin.
Review Figure 39.16

Rhodopsin is contained in the membranes of specialized photosensitive cells.
Review Figures 39.17, 39.20, and 39.21

Excitement of rhodopsin by light initiates a cascade of intracellular events that alters receptor potential in the sensor.
Review Figures 39.18 and 39.19

Vertebrate photosensors are rod cells, responsible for dim light and black-and-white vision; and cone cells, responsible for color vision.
Review Figures 39.24 and 39.25

Animals see with eyes that focus patterns of light onto layers of photosensors.
Review Figures 39.22 and 39.23

Layers of neurons in the retina process information from photosensors and send resulting action potentials to the brain.
Review Figure 39.26

The vertebrate retina has receptive fields that communicate responses to spots of light to the visual cortex.
Review Figure 39.27

Cells in the visual cortex integrate information from the retina; they have receptive fields that respond to patterns of light falling on the retina.
Review Figure 39.28

Binocular vision is due to cells in the visual cortex that receive input from receptive fields in both retinas.
Review Figure 39.29

Using sensory abilities that we do not share, bats echolocate, insects see ultraviolet radiation, pit vipers "see" infrared radiation, some fish sense electric fields, and homing pigeons orient to Earth's magnetic fields.

SELF-QUIZ

1. Which of the following statements is *not* true?
   a. Sensory transduction in vertebrate sensory systems involves the conversion (direct or indirect) of a physical or chemical stimulus into nerve impulses.
   b. In general, a stimulus causes a change in the flow of ions across the plasma membrane of a sensor.
   c. The term adaptation is given to the process by which a sensory system becomes insensitive to a continuing source of stimulation.
   d. The more intense a stimulus, the greater the magnitude of each action potential fired by a receptor.
   e. Sensory adaptation plays a role in the ability of organisms to discriminate between important and unimportant information.

2. The female silkworm moth releases a chemical called bombykol from a gland at the tip of her abdomen. Bombykol is
   a. a sex hormone.
   b. detected by the male only when present in large quantities.
   c. not species-specific.
   d. detected by hairs on the antennae of male silkworm moths.
   e. a chemical that is basic to the taste process in arthropods.

3. Which of the following statements is *not* true?
   a. Dogs are unusual among mammals in that they depend more on olfaction than on vision as their dominant sensory modality.
   b. Olfactory stimuli are recognized by the interaction between the stimulus and a specific macromolecule on olfactory hairs.
   c. The more odorant molecules that bind to receptors, the more

action potentials that are generated.
   d. The greater the number of action potentials generated by an olfactory receptor, the greater the intensity of the perceived smell.
   e. The perception of different smells results from the activation of different combinations of olfactory receptors.

4. The touch receptors that are located very close to the skin surface are
   a. relatively insensitive to light touch.
   b. very quick to adapt to stimuli.
   c. uniformly distributed throughout the surface of the body.
   d. called Pacinian corpuscles.
   e. slow and only partially able to adapt to stimuli.

5. The membrane that gives us the ability to discriminate different pitches of sound is the
   a. round window.
   b. oval window.
   c. tympanic membrane.
   d. tectorial membrane.
   e. basilar membrane.

6. Which of the following statements is *not* true?
   a. A rod cell's transmembrane potential becomes more negative when the rod cell is exposed to light after a period of darkness.
   b. A photosensor releases the most neurotransmitter (per unit time) when in total darkness.
   c. Whereas in vision the intensity of a stimulus is encoded by the degree of hyperpolarization of photosensors, in hearing the intensity of a stimulus is encoded by changes in firing rates of sensory cells.
   d. Stiffening of the ossicles in the middle ear can lead to deafness.

e. The interactions between hammer (malleus), anvil (incus), and stirrup (stapes) conducts sound waves across the fluid-filled middle ear.

7. In primates the region of the retina where the central part of the visual field falls is called the
   a. central ganglion cell.
   b. fovea.
   c. optic nerve.
   d. cornea.
   e. pupil.

8. The region of the vertebrate eye where the optic nerve passes out of the retina is called the
   a. fovea.
   b. iris.
   c. blind spot.
   d. pupil.
   e. optic chiasm.

9. Which of the following statements about the cones in a human eye is *not* true?
   a. They are responsible for sharp vision.
   b. They encode for color vision.
   c. They are more sensitive to light than rods are.
   d. They are fewer in number than rods are.
   e. They exist in high numbers at the fovea.

10. The color in vision results from the
   a. ability of each cone to absorb all wavelengths of light equally.
   b. lens of the eye acting like a prism and separating the different wavelengths by light.
   c. different absorption of wavelengths of light by different classes of cones.
   d. three different isomers of retinal in different classes of cone cells.
   e. absorption of different wavelengths of light by amacrine and horizontal cells.

FOR STUDY

1. Drawing on your knowledge of the structure and function of the human auditory system, describe how the brain is able to distinguish sounds of different frequencies.

2. Describe the molecular mechanisms whereby a single photon is able to cause hyperpolarization of a rod cell.

3. Compare and contrast the functioning of olfactory sensors and photosensors. What is the basis whereby each system discriminates between an apple and an orange?

4. Describe and contrast two sensory systems that enable animals to "see" in the dark. What problems or limitations are inherent in these systems in comparison with vision?

5. Describe what is meant by a receptor potential and how it functions to encode intensity of stimulus. Use a specific sensor as your example.

READINGS

Camhi, J. M. 1984. *Neuroethology: Nerve Cells and the Natural Behavior of Animals.* Sinauer Associates, Sunderland, MA. This text is particularly good for putting basic neurophysiology in the context of whole animals and their behaviors. Part II deals with the sensory worlds of animals.

Hubel, D. H. 1988. *Eye, Brain, and Vision.* Scientific American Library Series No. 22. W. H. Freeman, New York. A comprehensive and beautifully illustrated book about the neurophysiology and neuroanatomy of vision. It is very readable for the nonexpert, yet it presents the depth and breadth of knowledge and experience of someone who has been a major contributor to this area of research.

Hudspeth, A. J. 1983. "The Hair Cells of the Inner Ear." *Scientific American,* January. The inner ear transduces mechanical forces of pressure waves into action potentials transmitted to the brain. This paper describes the cells that accomplish the transduction and explain how they do it.

Knudsen, E. I. 1981. "The Hearing of the Barn Owl." *Scientific American,* June. The barn owl can use its remarkably precise and sensitive auditory system to locate prey in complete darkness. This paper explains the neurophysiological basis for its extreme accuracy.

Konishi, M. 1993. "Listening with Two Ears." *Scientific American,* April. Continuing the research reported in the article by Knudsen in 1981, this paper shows how the brain of the barn owl integrates signals from its two ears to create a map of its auditory environment.

Koretz, J. F. and G. H. Handelman, 1988. "How the Human Eye Focuses." *Scientific American,* July. In addition to describing in detail how the eye accommodates for distance, this article explains the changes that occur with aging.

Newman, E. A. and P. H. Hartline. 1982. "The Infrared 'Vision' of Snakes." *Scientific American,* March. Some snakes are able to use infrared radiation emitted by objects in their environment to construct a sensory world.

Stryer, L. 1987. "The Molecules of Visual Excitation." *Scientific American,* July. This paper describes the molecular chain of events that transduce photons of light falling on the retina into action potentials that are transmitted to the brain.

Suga, N. 1990. "Biosonar and Neural Computation in Bats." *Scientific American,* June. The use of echolocation by bats to construct a view of their environment requires complex neural processing of information by their auditory systems.

Information from sensors is not of much value to an animal unless it can respond. An obvious response is movement, and a fascinating array of adaptations have evolved that enable animals to move. Consider jumping. When bending down to smell a flower you see a spider. Startled by the spider, neural signals from your brain routed through spinal circuits activate muscles in your legs to contract, causing your legs to extend, and you jump. At the same time, the spider jumps in the opposite direction. Did its nervous system also cause muscles to contract and extend its jumping legs? No; the spider doesn't have such muscles. Instead, extracellular fluids were squeezed into its hollow jumping legs, and the increased pressure in the legs caused them to extend and the spider to jump.

While this drama was playing out, a flea sensed your body heat and jumped 20 centimeters onto your leg. The flea's jump involves still a different mechanism, one that works in the same way as a slingshot. The flea is so small and its initial acceleration so great that no muscle can contract fast enough to cause such a movement. At the base of its jumping legs is an elastic material that is compressed by slow muscles while the flea is resting. When a trigger mechanism is released, the elastic material recoils like a slingshot and propels the flea a distance perhaps 200 times its body length.

Another remarkable jumper that combines the mechanisms used by humans with those used by the flea is a kangaroo. As you run faster and faster, the number of strides you take per minute and the energy you expend per minute increases rapidly. Neither effect is true for the kangaroo. When running at speeds from about 5 to 25 kilometers per hour, the kangaroo has the same number of strides per minute and no increase in its metabolic rate. How can this be so? As we will learn in this chapter, muscles are attached to bones by tendons. Like the material at the base of the flea's legs, tendons can be elastic. The kangaroo's tendons stretch when it lands, and their recoil helps power the next jump. To move faster, the kangaroo increases its stride length, thereby increasing the stretch on its tendons each time it lands and the magnitude of the recoil at the initiation of each jump.

HOW ANIMALS RESPOND TO INFORMATION FROM SENSORS

Jumping is one way an animal can respond to information received by its sensors. In the examples of jumping behavior just discussed, muscles generate mechanical forces, and structures such as bones, elastic tissues, exoskeletons, and fluid-filled cavities apply mechanical force to cause behavior. Adaptations

A Spider's Jump Seems Effortless

40

Effectors

that animals use to respond to information that is sensed, integrated, and transmitted by its neural and endocrine systems are called **effectors**. This broad definition includes the internal organs and organ systems that the animal uses to control its internal environment; these effectors are the subjects of subsequent chapters. In this chapter we will focus on the mechanisms of creating mechanical forces and using those forces to change shape and to move—the basis for most animal behavior. A fish swims, an earthworm crawls, a mosquito flies, and a kangaroo jumps because cells can move.

We will focus first on the molecular basis for movement, then on how movements of molecules move cilia, flagella, and muscle filaments. We will then examine different muscle types and how neurons enable muscles to contract. Since muscles with no rigid supports to pull against would be nothing but quivering masses of tissue, we will also study the skeletal systems that enable animals to use the contractile forces of muscles for specific tasks.

CILIA, FLAGELLA, AND MICROTUBULES

Two cellular structures, microtubules and microfilaments (see Figure 4.25) create cell movement. Both of these structures cause movement by the sliding of long protein molecules past one another. Microtubules generate the small movements of cilia and flagella. Microfilaments, as we'll see later, reach their highest level of organization in muscle cells, which generate large-scale movements.

Ciliated Cells

Certain animal cells have tiny, hairlike appendages called cilia. Each cilium is tiny, about 0.25 μm in diameter, but cilia occur in dense patches. Animals of most species use ciliated cells to move liquids and particles over cell surfaces. Many invertebrates use ciliated cells to obtain food and oxygen. Cilia circulate a current of water across the gill surfaces of some mollusks, for example. Oxygen diffuses across the gill membranes, and food, consisting of tiny organisms and detritus, is filtered from the water by the netlike gills and ingested (Figure 40.1). Cilia around the mouths of rotifers sweep microorganisms and detritus directly into the gut.

The airways of many animals are lined with and cleaned by ciliated cells (Figure 40.2). In humans, the cilia continuously sweep a layer of mucus from deep down in the lungs, up through the windpipe, and into the throat. The mucus carries particles of dirt and dead cells. We can then either swallow or spit out the mucus, and with it the trapped detritus. Ciliated cells lining the female reproductive tract create currents that sweep eggs from the ovaries into the oviducts and all the way down to the uterus (see Chapter 37).

A cilium pushes with the same basic motion as a swimmer's arms during the breaststroke (Figure 40.3a). During the **power stroke**, the cilium projects stiffly outward and moves backward, propelling the cell forward (or the medium backward). During the **recovery stroke**, the cilium folds as it returns to its

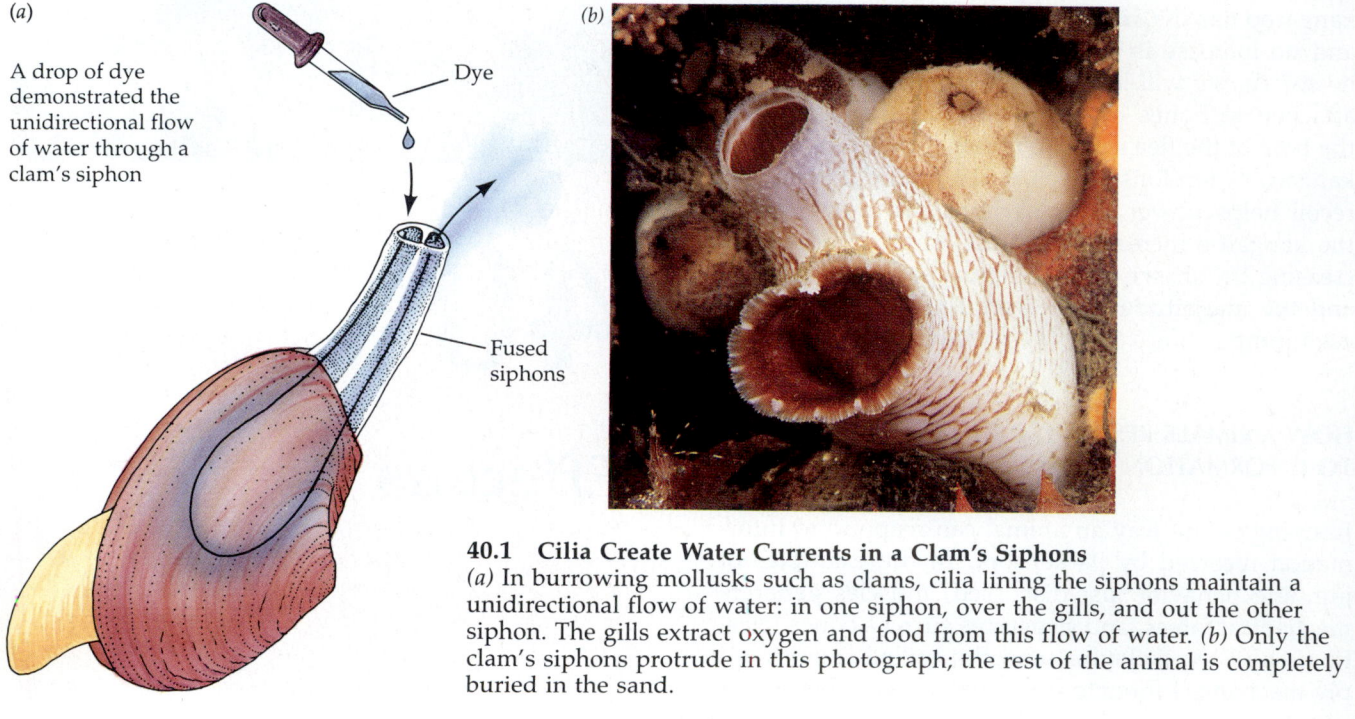

(a)

A drop of dye demonstrated the unidirectional flow of water through a clam's siphon

Dye

Fused siphons

40.1 Cilia Create Water Currents in a Clam's Siphons
(a) In burrowing mollusks such as clams, cilia lining the siphons maintain a unidirectional flow of water: in one siphon, over the gills, and out the other siphon. The gills extract oxygen and food from this flow of water. *(b)* Only the clam's siphons protrude in this photograph; the rest of the animal is completely buried in the sand.

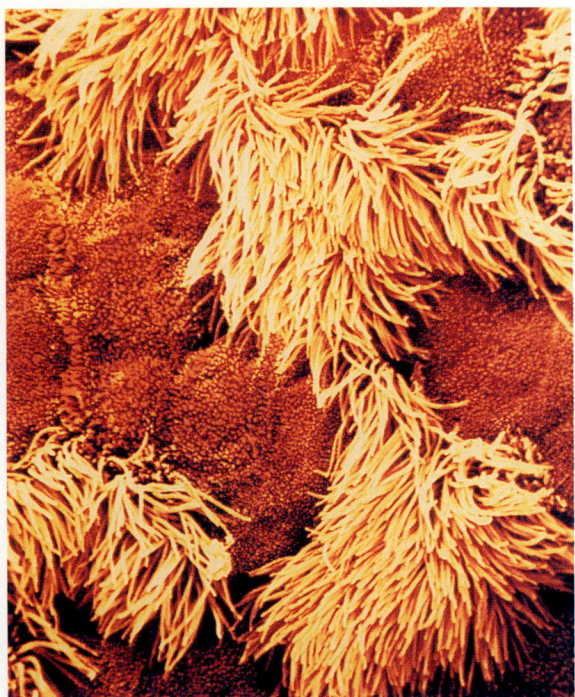

40.2 Cilia Line Respiratory Passages
A scanning electron micrograph of a rabbit's airway.

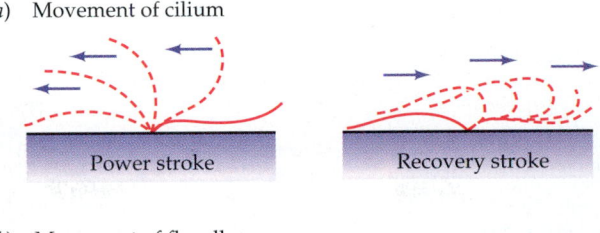

(a) Movement of cilium

Power stroke

Recovery stroke

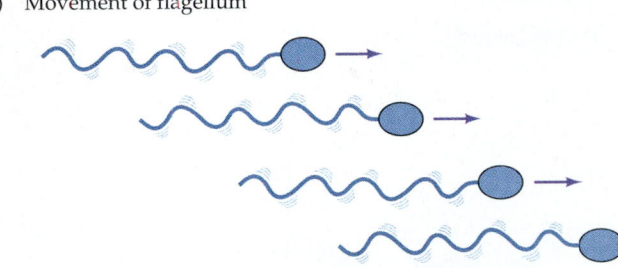

(b) Movement of flagellum

40.3 Cilia and Flagella Move Differently
(a) A cilium moves in a pattern similar to an arm of a swimmer doing the breaststroke. (b) Flagella are much longer than cilia. A flagellum moves in an undulating, whiplike pattern.

original position. The power stroke is fast and the recovery stroke is slow. As you know from moving your arm or leg in water, there is less resistance the slower you move. The resistance of the medium to the recovery stroke thus is slight compared with its resistance to the power stroke. Fluids exposed to the beating of cilia are propelled in the direction of the power stroke. Cilia typically beat in coordinated waves. At any particular moment, some cilia of a cell are moving through the power stroke and others are recovering.

Flagellated Cells

The **flagella** of eukaryotes are identical to cilia except that they are longer and occur singly or in groups of only a few on any one cell. Flagellated cells maintain a flow of water through the bodies of sponges, bringing in food and oxygen and removing carbon dioxide and wastes. Flagella power the movement of the sperm of most species. Because of their greater length, flagella have a whiplike stroke pattern rather than the "swimming" stroke pattern of cilia (Figure 40.3b).

How Cilia and Flagella Move

The central structure of a cilium or a flagellum, called the **axoneme**, contains a ring of nine pairs of microtubules. In the center of the ring may be an additional pair of microtubules, a single microtubule, or no microtubule (see Figure 4.26). Microtubules are hollow tubes formed from polymerization of the globular polypeptide **tubulin**. Other proteins in the axoneme form spokes, side arms, and cross-links (Figure 40.4a). Side arms composed of the protein **dynein** generate force. Dynein is a mechanoenzyme that catalyzes the hydrolysis of ATP and uses the released energy to change its orientation, thereby generating mechanical force. When the dynein arms on one microtubule pair contact a neighboring microtubule pair and bind to it, ATP is broken down, and the resulting conformational changes in the dynein molecules cause the arms to point toward the base of the axoneme. This action pushes the microtubule pair ahead in relation to its neighbor. The dynein arms then detach from the neighboring pair and reorient to their starting horizontal position. As the cycle is repeated, adjacent microtubule pairs try to "row" past each other, with the dynein side arms acting as "oars" (Figure 40.4b). Because the microtubules are anchored at the bottom, the axoneme bends, instead of elongating, as the microtubule pairs slide past one another (Figure 40.4c). In ways not fully understood, the central microtubule and the other proteins that bind the axoneme together control the dynein action so that not all the microtubule pairs are "rowing" at the same time.

Scientists have investigated the motile mechanisms of the axoneme by selectively removing its proteins. Axonemes severed from cells continue to flex in a normal pattern if exposed to calcium ions (Ca^{2+}) and ATP, demonstrating that the motile mechanism is part of the axoneme itself. Gently treating

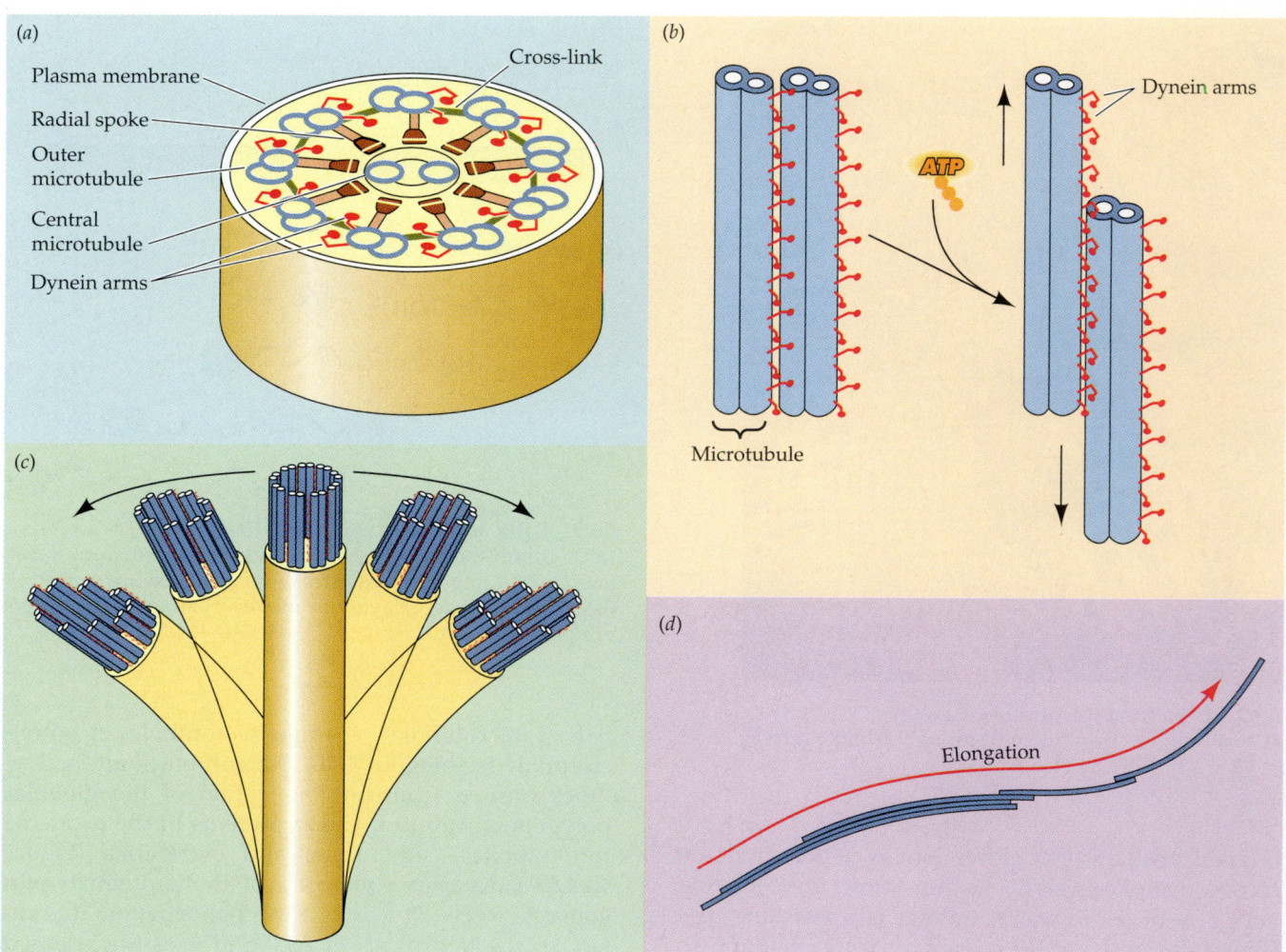

(a)
Plasma membrane
Radial spoke
Outer microtubule
Central microtubule
Dynein arms
Cross-link

(b)
Dynein arms
ATP
Microtubule

(c)

(d)
Elongation

40.4 Microtubules Create Motion by Pushing against Each Other

Cilia and flagella move because of the actions of proteins in the axoneme. (a) A cross section of an axoneme. The microtubules occur in doublets that run the length of the axoneme. (b) Dynein arms of the doublets generate force by making and breaking cross-links with the neighboring microtubule pair. (c) When microtubule pairs try to slide past each other, the axoneme bends because the microtubules are anchored together at the base. (d) If all links between microtubules except the dynein arms are eliminated, ATP causes the microtubules to slide past each other, and the axoneme elongates.

the axoneme with enzymes that hydrolyze proteins, however, disrupts the spokes and cross-links, leaving only the microtubules and dynein arms intact. If the isolated microtubules are then exposed to Ca^{2+} and ATP, they row past one another and the whole structure elongates manyfold (Figure 40.4d), demonstrating that the forces moving microtubule pairs along one another are the basis for the bending of the intact axoneme. When dynein (and only dynein) is removed from isolated axonemes, the microtubules lose their ATPase activity and their motility. Restoring purified dynein to the axonemes restores ATPase activity and motility.

Microtubules as Intracellular Effectors

Microtubules play important roles in cell movements. As components of the cytoskeleton, microtubules also contribute to the cell's shape. Cells can change shape and move by polymerizing and depolymerizing the tubulin in their microtubules. During mitosis, the spindle that moves chromosomes to the mitotic poles at anaphase forms by the polymerization of tubulin. Another example of microtubule involvement in cell movement is the growth of the axons of neurons in the developing nervous system. Neurons find and make their appropriate connections by sending out long extensions that search for the correct contact cells. If polymerization of tubulin is chemically inhibited, the neurons do not extend. Microtubules are important intracellular effectors for changing cell shape, moving organelles, and enabling cells to respond to the environment.

MICROFILAMENTS AND CELL MOVEMENT

Like microtubules, protein **microfilaments** change cell shape and cause cell movements. The dominant

40.5 Cell Projections Supported by Microfilaments
(a) The cells lining the gut have numerous fingerlike projections called microvilli. (b) Some mechanosensors, such as these hair cells in the organ of Corti, have stereocilia. Both microvilli and hair cells are stiffened by microfilaments.

(a)

(b)

microfilament in cells is the protein **actin**, and bundles of cross-linked actin strands form important structural components of cells. For example, the microvilli (tiny projections) that increase the absorptive surface area of the cells lining the gut are stiffened by actin microfilaments (Figure 40.5a), as are the stereocilia of the sensory hair cells mentioned in Chapter 39 (Figure 40.5b). Like microtubules, actin microfilaments can change the shape of a cell simply by polymerizing and depolymerizing. The projections sent out by phagocytic cells (see Figure 16.2) are an example of this process.

Together with the protein myosin, actin microfilaments generate the contractile forces responsible for many aspects of cell locomotion and changes in cell shape. For example, the contractile ring that divides a cell undergoing mitosis into two daughter cells is composed of actin microfilaments in association with myosin. The mechanisms that many cells employ to engulf materials (phagocytosis and pinocytosis; see Chapter 4) also rely on actin microfilaments and myosin. Nets of actin and myosin beneath the cell membrane change a cell's shape during phagocytosis.

The movement of certain cells in multicellular animals is due to the activity of actin microfilaments and myosin. During development many cells migrate by such **amoeboid movement**, and throughout an animal's life phagocytic cells (see Chapter 16) circulate in the blood, squeeze through the walls of the blood vessels, and wander through the tissues by amoeboid movement. The mechanisms of amoeboid movement have been studied extensively in the protist for which this type of movement was named—

the amoeba, which lives in freshwater streams and ponds.

The amoeba moves by extending lobe-shaped projections called pseudopods and then seemingly squeezing itself into them (Figure 40.6). The cytoplasm in the core of the amoeba is relatively liquid and is called **plasmasol**, but just beneath the plasma membrane the cytoplasm is much thicker and is called **plasmagel**. To form a pseudopod, the thick plasmagel in a certain area of the cell thins, allowing a bulge to form. Just under the cell surface, in the plasmagel, is a network of actin microfilaments that interacts with myosin to squeeze plasmasol into the bulge, thus forming a pseudopod. As the network continues to contract, cytoplasm streams in the direction of the pseudopod. Eventually the cytoplasm at the leading edge of the pseudopod converts to gel

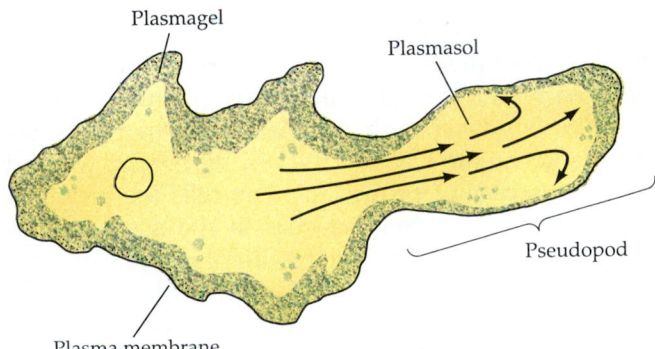

Plasmagel

Plasmasol

Pseudopod

Plasma membrane

40.6 Squeezing into a Pseudopod
Many animal cells engage in amoeboid movement but an amoeba (a protist; see page 508) does it best.

and the pseudopod stops forming. Thus the basis for amoeboid motion is the ability of the cytoplasm to cycle through sol and gel states and the ability of the microfilament network under the cell membrane to contract and cause the cytoplasmic streaming that pushes out a pseudopod.

MUSCLES

Muscle contraction is the most important effector mechanism that animals have for responding to their environments. All behavioral and most physiological responses depend on muscle cells. Muscle cells are specialized for contraction and have high densities of actin microfilaments and myosin. Such cells are found throughout the animal kingdom. They account for the thrashing movements of nematodes, the expansion–contraction movements of earthworms, the pulsating movements of jellyfish, and the limb movements of arthropods and vertebrates. Muscle cells are found in the walls of blood vessels, guts, bladders, and hearts. Wherever contraction of whole tissues takes place in animals, muscle cells are responsible. In all cases, the molecular mechanism of contraction is the same, but there are many specializations of muscle cells fitting them to the wide variety of functions they serve. We begin our study of muscle by looking at the three types of muscle cells found in vertebrates: smooth muscle, skeletal muscle, and cardiac (heart) muscle.

Smooth Muscle

Smooth muscle provides the contractile forces for most of our internal organs, which are under the control of the autonomic nervous system. Smooth muscle moves food through the digestive tract, controls the flow of blood through blood vessels, and empties the urinary bladder. Structurally, smooth muscle cells are the simplest muscle cells. They are usually long and spindle-shaped, and each cell has a single nucleus. Because the filaments of actin and myosin in smooth muscle are not as regularly arranged as those in the other muscle types, the contractile machinery is not obvious when the cells are viewed under the light microscope (Figure 40.7a).

If we study smooth muscle from a particular organ, such as the walls of the digestive tract, we find it has interesting properties. The cells are arranged in sheets, and individual cells in the sheets are in electrical contact with one another through gap junctions. As a result, an action potential generated in the membrane of one smooth muscle cell can spread to all the cells in the sheet of tissue. Another interesting property of a smooth muscle cell is that the resting potential of its membrane is sensitive to being

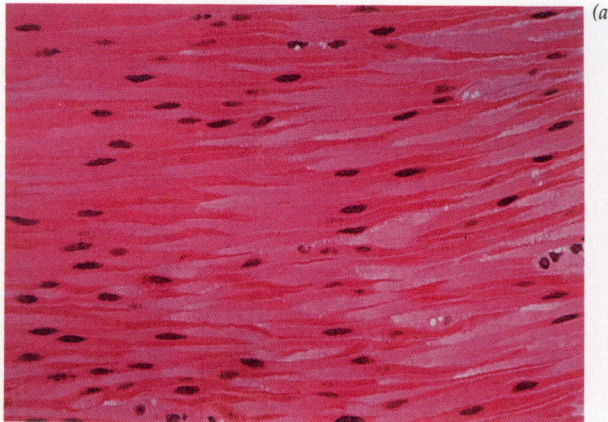
(a)

(b)

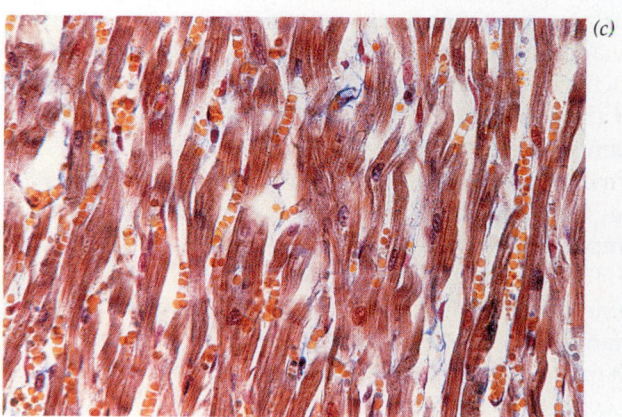
(c)

40.7 The Three Types of Vertebrate Muscle Tissue
(a) Smooth muscle cells are usually arranged in sheets such as those that make up the walls of the stomach and the intestine. The dark structures in this sheet of muscle cells are nuclei. (b) Skeletal, or striated, muscle appears to be striped, or banded, because of its highly regular arrangement of contractile filaments. (c) Cardiac muscle is also striated, but cardiac muscle fibers branch and create a meshwork that resists tearing or breaking.

stretched. If the wall of the digestive tract is stretched in one location (such as by receiving a mouthful of food), the membranes of the stretched cells depolarize, reach threshold, and fire action potentials that cause the cells to contract. Thus smooth muscle contracts after being stretched, and the harder it is stretched, the stronger the contraction. Later in this chapter we will discuss how membrane depolarization triggers contraction.

Skeletal Muscle

Skeletal muscle carries out all voluntary movements, such as running or playing a piano, and generates the movements of breathing. Skeletal muscle is called **striated muscle** because the highly regular arrangement of its actin microfilaments and myosin gives it a striped appearance (Figure 40.7b). Skeletal muscle cells, or **muscle fibers**, are large, and they have many nuclei because they develop through the fusion of many individual cells. A muscle such as your biceps (which bends your arm) is composed of many muscle fibers bundled together by connective tissue.

What is the relation between a muscle fiber and the actin and myosin filaments responsible for contraction? Each muscle fiber is composed of **myofibrils**: bundles of contractile filaments made up of actin and myosin (Figure 40.8). Within each myofibril are thin filaments, which are actin microfilaments, and thick filaments, which are composed of myosin. If we cut across the myofibril at certain locations, we see only thick filaments, in other locations only thin filaments, but in most regions of the myofibril, each thick myosin filament is surrounded by six thin actin filaments.

A longitudinal view of a myofibril reveals the striated appearance of skeletal muscle. The band pattern of the myofibril is due to repeating units called **sarcomeres**, which are the units of contraction (Figure 40.8a). Each sarcomere is made of overlapping filaments of actin and myosin. As the muscle contracts, the sarcomeres shorten, and the appearance of the band pattern changes.

The observation that the widths of the bands in the sarcomeres change when a muscle contracts led two British biologists, Hugh Huxley and Andrew Huxley, to propose a molecular mechanism of muscle contraction. Let's look at the myofibril's band pattern in detail (Figure 40.8b). Each sarcomere is bounded by Z lines that anchor the thin actin filaments. Centered in the sarcomere is the A band, which contains all the myosin filaments. The H zone and the I band, which appear light, are regions where actin and myosin filaments do not overlap in the relaxed muscle. The dark stripe within the H zone is called the M band; it contains proteins that help hold the myosin filaments in their regular hexagonal arrangement. When the muscle contracts, the sarcomere shortens. The H zone and I band become much narrower, and the Z lines move toward the A band as if the actin filaments were sliding into the region occupied by the myosin filaments. This observation led

Huxley and Huxley to propose the **sliding-filament theory** of muscle contraction: Actin and myosin filaments slide past each other as the muscle contracts.

To understand what makes the filaments slide, we must examine the structure of actin and of myosin. The myosin molecule consists of two long polypeptide chains coiled together, each ending in a globular head (Figure 40.8c). The myosin filament is made up of many myosin molecules arranged in parallel with their heads projecting laterally from one or the other end of the filament (Figure 40.8d). The actin filament consists of a helical arrangement of two chains of monomers like two strands of pearls twisted together (Figure 40.8e). Twisting around the actin chains are two strands of another protein, **tropomyosin**. The myosin heads have sites that can bind to actin and thereby form bridges between the myosin and the actin filaments. The myosin heads also have ATPase activity; they bind and hydrolyze ATP. The energy released changes the orientation of the myosin head.

These details explain the cycle of events that cause actin and myosin filaments to slide past one another and shorten the sarcomere (Figure 40.8f). A myosin head binds to an actin filament. Upon binding, the head changes its orientation with respect to the myosin filament, exerting a force that causes the actin and myosin filaments to slide about 5 to 10 nanometers. Next, the myosin head binds a molecule of ATP, which causes it to release the actin. When the ATP is hydrolyzed, the energy released causes the myosin head to return to its original conformation, in which it can bind again to actin. The hydrolysis of the ATP is like cocking the hammer of a pistol; binding with actin pulls the trigger.

We have been discussing the cycle of contraction in terms of a single myosin head. Don't forget that each myosin filament has many myosin heads at both ends and is surrounded by six actin filaments; thus the contraction of the sarcomere involves a great many cycles of interaction between actin and myosin molecules. That is why when a single myosin head breaks its contact with actin, the actin filaments do not slip backwards.

An interesting aspect of this contractile mechanism is that ATP is needed to break the actin–myosin bonds but not to form them. Thus muscles require ATP to stop contracting. This fact explains why muscles stiffen soon after animals die, a condition known as rigor mortis. Death stops the replenishment of the ATP stores of muscle cells, so the myosin–actin bridges cannot be broken, and the muscles stiffen. Eventually the proteins begin to lose their integrity and the muscles soften. Because these events have regular time courses that differ somewhat for different regions of the body, an examination of the stiffness of the muscles of a corpse sometimes helps a coroner estimate the time of death.

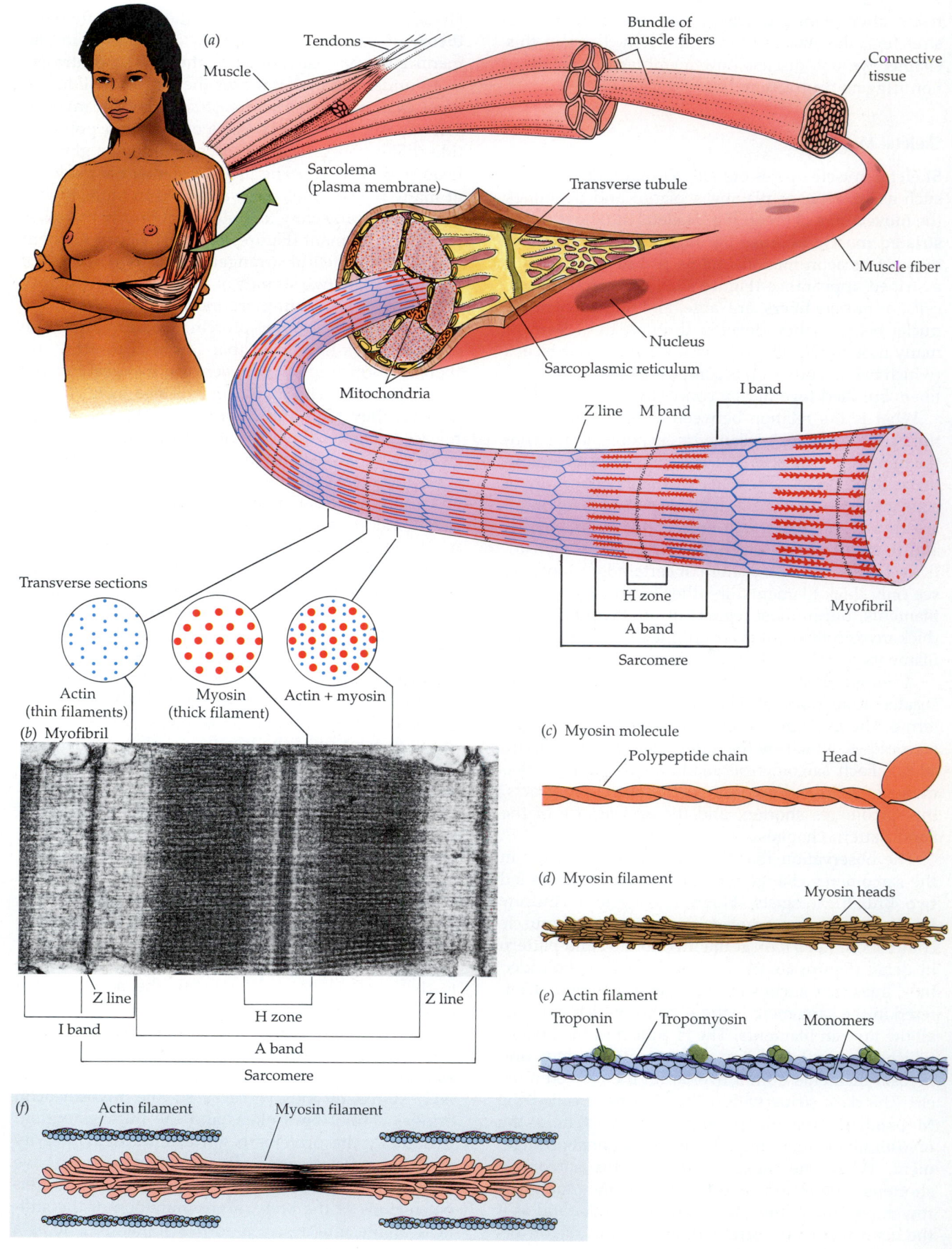

(a)

Tendons

Muscle

Bundle of muscle fibers

Connective tissue

Sarcolema (plasma membrane)

Transverse tubule

Muscle fiber

Nucleus

Sarcoplasmic reticulum

Mitochondria

Z line   M band

I band

H zone

A band

Sarcomere

Myofibril

Transverse sections

Actin (thin filaments)

Myosin (thick filament)

Actin + myosin

(b) Myofibril

Z line

H zone

Z line

I band

A band

Sarcomere

(c) Myosin molecule

Polypeptide chain

Head

(d) Myosin filament

Myosin heads

(e) Actin filament

Troponin   Tropomyosin   Monomers

(f)

Actin filament   Myosin filament

40.8 The Structure of Skeletal Muscle from Tissue Down to Molecules

(a) A skeletal muscle is made up of bundles of muscle fibers. Each muscle fiber is a multinucleate cell containing numerous myofibrils, which are highly ordered assemblages of thick myosin and actin filaments. Within each muscle fiber the nuclei, the mitochondria, and the sarcoplasmic reticulum surround the myofibrils. (b) The structure of the myofibrils gives muscle fibers their characteristic striated appearance. Where there are only actin filaments the myofibril appears light; where there are both actin and myosin filaments the myofibril appears dark. (c,d) Myosin filaments consist of bundles of molecules with long polypeptide tails and globular heads. (e) Actin filaments consist of two chains of monomers twisting around each other. Two polypeptide chains of tropomyosin twist around the actin chains. (f) Actin and myosin filaments overlap in the sarcomere.

Controlling the Actin–Myosin Interaction

Muscle contractions are initiated by nerve action potentials arriving at the neuromuscular junction. Motor neurons are generally highly branched and can innervate up to 100 muscle fibers each. All the fibers innervated by a single motor neuron are a **motor unit** and contract simultaneously in response to the action potentials fired by that motor neuron. To understand the fine control the nervous system has over the sliding of actin and myosin filaments, we must examine the membrane system of the muscle fiber and some additional protein components of the actin filaments.

Like neurons, vertebrate skeletal muscle fibers are excitable cells: When they are depolarized to a threshold that opens their voltage-gated sodium channels, their plasma membranes generate action potentials, just as the membranes of axons do. The initial depolarization that spreads across the muscle fiber plasma membrane is generated at the neuromuscular junction—the synapse between the motor neuron and the muscle fiber. As we saw in Chapter 38, neurotransmitter from the motor neuron binds to receptors in the postsynaptic membrane, causing ion channels to open. The depolarization of the postsynaptic membrane spreads to the surrounding plasma membrane of the muscle fiber which contains voltage-gated ion channels. When threshold is reached, the plasma membrane fires an action potential that is rapidly conducted to all points on the surface of the muscle fiber.

The plasma membrane of the muscle fiber is continuous with a system of tubules that descends into and branches throughout the cytoplasm (also called the sarcoplasm) of the muscle fiber. These are the transverse tubules, or **T-tubules**, and they communicate with a network of membranes, called the **sarcoplasmic reticulum**, that surrounds every myofibril (see Figure 40.8a). The wave of depolarization that spreads over the plasma membrane of the muscle fiber also spreads through the T-tubule system. Calcium pumps in the membranes of the sarcoplasmic reticulum cause this membrane-bounded compartment of the fiber to take up and sequester Ca^{2+}. As a result, there is a high concentration of Ca^{2+} in the sarcoplasmic reticulum and low concentration of Ca^{2+} in the sarcoplasm surrounding the filaments. When a wave of depolarization spreads through the T-tubule system, it induces Ca^{2+} channels in the sarcoplasmic reticulum to open, resulting in the diffusion of Ca^{2+} ions out of the sarcoplasmic reticulum and into the sarcoplasm surrounding the microfilaments. The Ca^{2+} stimulates the interaction of actin and myosin and the sliding of the filaments. How does it do so?

Remember that an actin filament is a helical arrangement of two strands of actin monomers. Lying in the grooves between the two actin strands is the two-stranded protein tropomyosin (Figure 40.9a). At regular intervals the filament also includes another globular protein, **troponin**. The troponin molecule has three subunits; one binds actin, one binds tropomyosin, and one binds Ca^{2+}. When Ca^{2+} is sequestered in the sarcoplasmic reticulum, the tropomyosin strands block the sites where myosin heads can bind to the actin. When the T-tubule system depolarizes, Ca^{2+} is released into the sarcoplasm, where it binds to the troponin, changing the shape of the troponin molecule. Because the troponin is also bound to the tropomyosin, this conformational change of the troponin twists the tropomyosin enough to expose the actin–myosin binding sites. This initiates the cycle of making and breaking actin–myosin bridges; the filaments are pulled past one another, and the muscle fiber contracts. When the T-tubule system repolarizes, the calcium pumps remove the Ca^{2+} ions from the sarcoplasm, causing the tropomyosin to return to the position in which it blocks the binding of the myosin heads to the actin strands, and the muscle fiber returns to its resting condition. Figure 40.9b reviews the cycle.

Twitches, Graded Contractions, and Tonus

In vertebrate skeletal muscle, the arrival of an action potential at the neuromuscular junction causes an action potential in the muscle fiber. The spread of the action potential through the membrane system of the muscle fiber causes a minimum unit of contraction, called a **twitch**. A twitch can be measured in terms of the tension, or force, it generates (Figure 40.10). If action potentials in the muscle fiber are adequately separated in time, each twitch is a discrete, all-or-none phenomenon. If action potentials are fired more rapidly, however, new twitches are triggered before the filaments have had a chance to return to their

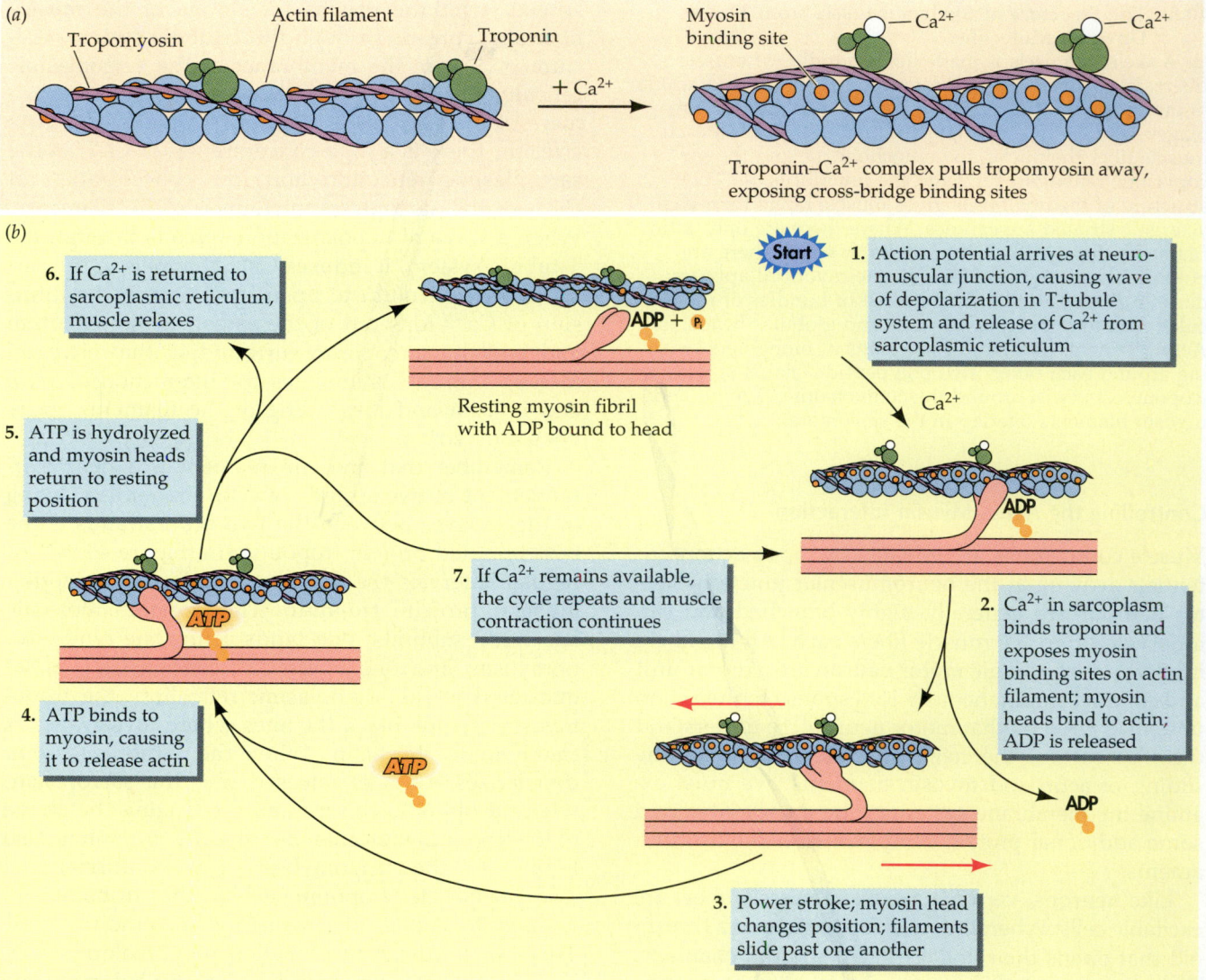

(a) Tropomyosin — Actin filament — Troponin — Myosin binding site — Ca²⁺ — Ca²⁺

+ Ca²⁺ →

Troponin–Ca²⁺ complex pulls tropomyosin away, exposing cross-bridge binding sites

(b)

Start

1. Action potential arrives at neuromuscular junction, causing wave of depolarization in T-tubule system and release of Ca²⁺ from sarcoplasmic reticulum

6. If Ca²⁺ is returned to sarcoplasmic reticulum, muscle relaxes

Resting myosin fibril with ADP bound to head

ADP + P$_i$

Ca²⁺

5. ATP is hydrolyzed and myosin heads return to resting position

ADP

7. If Ca²⁺ remains available, the cycle repeats and muscle contraction continues

2. Ca²⁺ in sarcoplasm binds troponin and exposes myosin binding sites on actin filament; myosin heads bind to actin; ADP is released

ATP

4. ATP binds to myosin, causing it to release actin

ATP

ADP

3. Power stroke; myosin head changes position; filaments slide past one another

40.9 Tropomyosin and Troponin

Tropomyosin and troponin molecules on actin filaments control the formation of crossbridges. (a) When Ca²⁺ binds to troponin it exposes crossbridge binding sites. (b) As long as crossbridge sites and ATP are available, the cycle of actin and myosin interactions continues and the filaments slide.

resting condition. As a result, the twitches sum and the tension generated by the fiber increases and becomes more continuous. Thus the individual muscle fiber can show a graded response to increased levels of stimulation by its motor neuron. At high levels of stimulation, the calcium pumps in the sarcoplasmic reticulum can no longer remove Ca²⁺ ions from the sarcoplasm between action potentials, and the contractile machinery generates maximum tension—a condition known as **tetanus**. (Do not confuse this condition with the disease tetanus, which is caused by a bacterial toxin and characterized by spastic contractions of skeletal muscles.)

How long a muscle fiber can maintain a tetanic

contraction depends on its supply of ATP. Eventually the fiber will fatigue. It may seem paradoxical that the lack of ATP causes fatigue, since the action of ATP is to break the actin–myosin bonds. But remember that the energy released from the hydrolysis of ATP "recocks" the myosin heads, allowing them to cycle through another power stroke. The situation is like rowing a boat upstream: You cannot maintain your position relative to the stream bank by just holding the oars out against the current; you have to keep rowing.

The ability of a whole muscle to generate different levels of tension depends also on how many muscle fibers in that muscle are activated. Whether a muscle contraction is strong or weak depends both on how many motor neurons to that muscle are firing and on the rate at which those neurons are firing. These two factors can be thought of as spatial and temporal summation, respectively. Both types of summation increase the strength of contraction of the muscle as

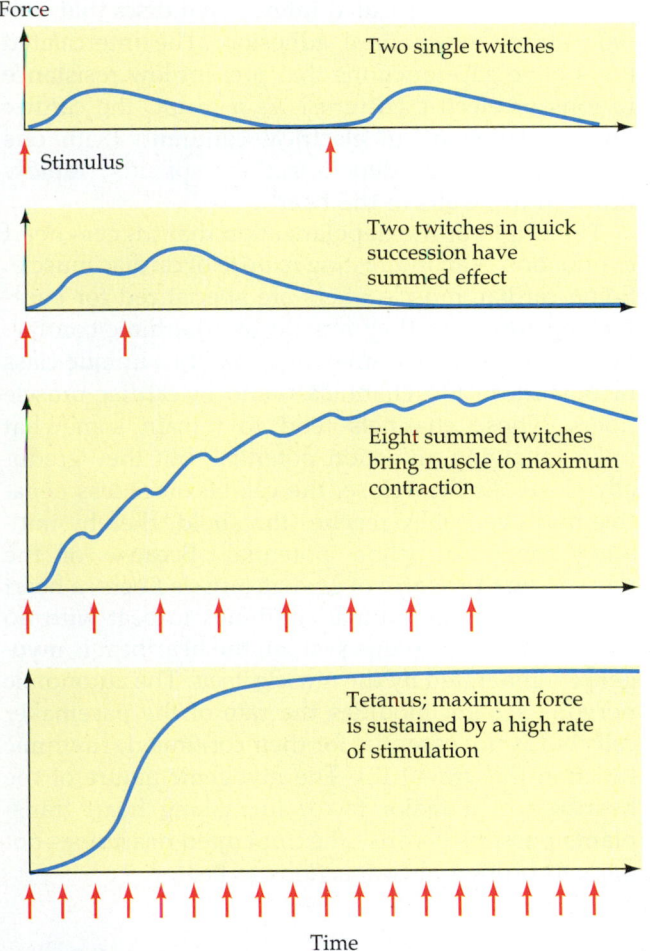

Force

Two single twitches

Stimulus

Two twitches in quick succession have summed effect

Eight summed twitches bring muscle to maximum contraction

Tetanus; maximum force is sustained by a high rate of stimulation

Time

40.10 Twitches and Tetanus
Each red arrow represents a single, brief electric pulse to the nerve that is innervating the muscle. This stimulus elicits a twitch, the minimum unit of contraction of a muscle. Twitches that occur in rapid succession have a summed effect. Tetanus is the maximum state of tension a muscle can achieve.

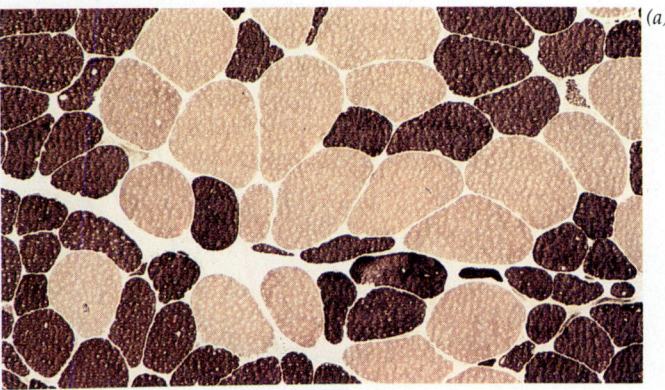
(a)

40.11 Two Types of Muscle Fibers are Specialized For Fast or For Sustained Contraction
(a) Skeletal muscle consists of fast- and slow-twitch fibers. In this stained micrograph, which is a cross section of a skeletal muscle, slow-twitch fibers are dark and fast-twitch fibers are light. (b) World-class athletes in different sports have different distributions of fiber types. Slow-

a whole. Faster twitching of individual fibers causes temporal summation (Figure 40.10), and an increase in the number of motor units involved in the contraction causes spatial summation. (Remember that a motor unit is all the muscle fibers innervated by a single neuron, and that a single muscle consists of many motor units.)

Many muscles of the body maintain a low level of tension called **tonus** even when the body is at rest. For example, the muscles of our neck, trunk, and limbs that maintain our posture against the pull of gravity are always working, even when we are standing or sitting still. Muscle tonus comes from the activity of a small but changing number of motor units in a muscle; at any one time some of the muscle's fibers are contracting and others are relaxed. Tonus is constantly being readjusted by the nervous system.

Fast- and Slow-Twitch Fibers

Not all skeletal muscle fibers are alike in how they twitch, and a single muscle may contain more than one type of fiber. The two major types of skeletal muscle fibers are **slow-twitch fibers** and **fast-twitch fibers** (Figure 40.11a). Slow-twitch fibers are also

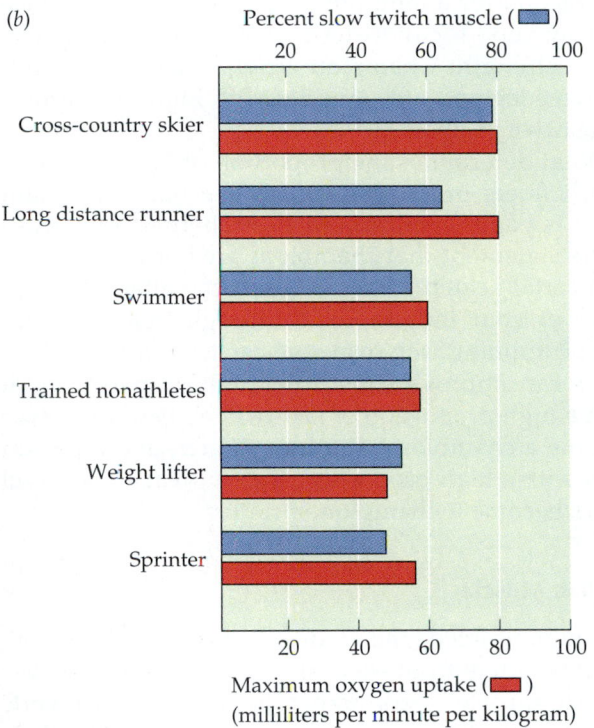
(b)

Percent slow twitch muscle (■)

Cross-country skier

Long distance runner

Swimmer

Trained nonathletes

Weight lifter

Sprinter

Maximum oxygen uptake (■)
(milliliters per minute per kilogram)

twitch fibers are better adapted for sustained aerobic activity. Fast-twitch fibers can generate maximum tension quickly, but they also fatigue quickly. For comparison, we include here average values for amateur athletes in good condition. The maximum oxygen uptake is a measure of the maximum level of aerobic metabolism the person can sustain for a period of time greater than a few minutes.

called red muscle because they have lots of the oxygen-binding molecule myoglobin, they have lots of mitochondria, and they are well supplied with blood vessels. A single twitch of a slow-twitch fiber produces low tension. The maximum tension a slow-twitch fiber can produce is low and develops slowly, but these fibers are highly resistant to fatigue. Because slow-twitch fibers have substantial reserves of glycogen and fat, their abundant mitochondria can maintain a steady, prolonged production of ATP if oxygen is available. Muscles with high proportions of slow-twitch fibers are good for long-term, aerobic work (work that requires lots of oxygen). Champion long-distance runners, cross-country skiers, swimmers, and bicyclists have leg and arm muscles consisting mostly of slow twitch fibers (Figure 40.11b).

Fast-twitch skeletal muscle fibers are also called white muscle because they have fewer mitochondria, little or no myoglobin, and fewer blood vessels than slow-twitch fibers do. The white meat of domestic chickens is composed of fast-twitch fibers. Fast-twitch fibers can develop maximum tension more rapidly than slow-twitch fibers can, and that maximum is greater, but they fatigue rapidly. The myosin of fast-twitch fibers has high ATPase activity, so they can put the energy of ATP to work very rapidly, but they cannot replenish it rapidly enough to sustain contraction for a long time. Fast-twitch fibers are especially good for short-term work that requires maximum strength. Champion weight lifters and sprinters have leg and arm muscles with high proportions of fast-twitch fibers.

What determines the proportion of fast- and slow-twitch fibers in your muscles? The most important factor is your genetic heritage, so there is some truth to the statement that champions are born, not made. To a certain extent, however, you can alter the properties of your muscle fibers through training. With aerobic training, the oxidative capacity of fast-twitch fibers can improve substantially. But a person born with a high proportion of fast-twitch fibers will never become a champion marathon runner, and a person born with a high proportion of slow-twitch fibers will never become a champion sprinter.

Cardiac Muscle

Although striated, the cardiac (heart) muscle of vertebrates differs from skeletal muscle in several ways. Cardiac muscle fibers branch to create a meshwork of contractile elements (see Figure 40.7c). Unlike skeletal muscles, which can easily be torn along the length of the muscle fibers, the meshwork of the cardiac muscle cannot be separated in this way; thus the heart walls can withstand high pressures without danger of tearing or forming leaks. Unlike skeletal muscle fibers, each cardiac muscle fiber is a uninucleate cell. Each cell is joined to other cardiac muscle fibers by structures called **intercalated discs** that provide strong mechanical adhesion. The intercalated discs have gap junctions that present low resistance to ions or electric currents. As a result, the cardiac muscle fibers are in electrical continuity with one another, and a depolarization spreads rapidly through the walls of the heart.

The origin of the depolarization that triggers heart contraction is an interesting feature of cardiac muscle. Some cardiac muscle fibers are specialized for pacemaking function; they initiate the rhythmic contraction of the heart. Pacemaking is due to a unique class of potassium ion channels found in cardiac muscle fibers. These channels tend to remain somewhat open following an action potential, but they gradually close. As they close, the cell becomes less negative and eventually reaches threshold, thereby initiating the next action potential. Because of the pacemaking property of cardiac muscle fibers, a heart removed from an animal continues to beat with no input from the nervous system; the heartbeat is **myogenic**—generated by the muscle itself. The autonomic nervous system modifies the rate of the pacemaker cells but is not essential for their continued, rhythmic function (Figure 40.12). The myogenic nature of the heartbeat is a major factor in making heart transplants possible because the implanted heart does not depend on neural connections to beat.

SKELETAL SYSTEMS

Muscles can only contract and relax. Without something rigid to pull against, muscles can do little more

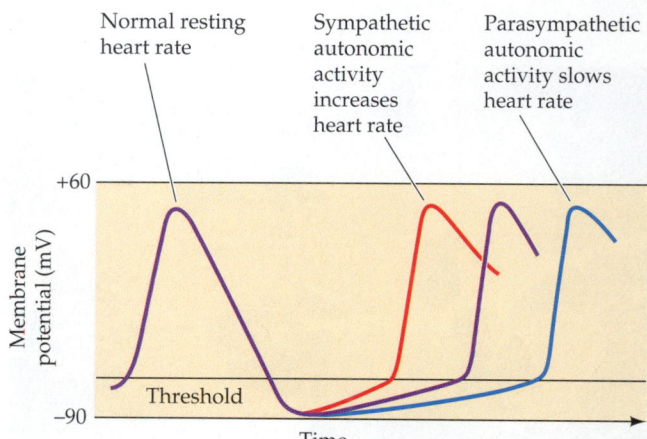

40.12 Pacemaker Cells Generate Action Potentials Spontaneously
The resting potentials of the plasma membranes of pacemaker cells spontaneously depolarize to threshold and fire action potentials. The rate of spontaneous depolarization determines heart rate. Sympathetic stimulation increases the rate of depolarization, whereas parasympathetic stimulation decreases the rate of depolarization.

than lie in a formless mass that twitches and changes shape. **Skeletal systems** provide rigid supports against which muscles can pull, thereby creating directed movements. The three types of skeletal systems found in animals are hydrostatic skeletons, exoskeletons, and endoskeletons.

Hydrostatic Skeletons

The simplest type of skeleton is the **hydrostatic skeleton** of cnidarians, annelids, and many other soft-bodied invertebrates. It consists of a volume of incompressible fluid (water) enclosed in a body cavity surrounded by muscle. When muscles oriented in a certain direction contract, the fluid-filled body cavity bulges out in the opposite direction. The sea anemone has a hydrostatic skeleton (see Figure 26.14). Its body cavity is filled with seawater. To extend its body and its tentacles, the anemone closes its mouth and constricts muscle fibers that are arranged in circles around its body. Contraction of these circular muscles puts pressure on the liquid in the body cavity, and that pressure forces the body and tentacles to extend. If alarmed, the anemone retracts its tentacles and body by contracting muscle fibers that are arranged longitudinally in the body wall and in the long dimension of the tentacles.

The hydrostatic skeletons of some animals have become adapted for locomotion. An annelid such as the earthworm uses its hydrostatic skeleton to crawl (see Figure 26.24). The earthworm's body cavity is divided into many separate segments. The body wall has a muscle layer in which the muscle fibers are arranged in circles around the body cavity, and another muscle layer in which the muscle fibers run lengthwise. A closed compartment in each segment of the worm is filled with fluid. If the circular muscles in a segment contract, the compartment in that segment narrows and elongates. If the lengthwise (longitudinal) muscles of a segment contract, the compartment shortens and bulges outward. Alternating contractions of the the circular and longitudinal muscles create waves of narrowing and widening, lengthening and shortening, that travel down the body of the earthworm. The bulging, short segments serve as anchors as the narrowing, expanding segments project forward, and longitudinal contractions pull other segments forward. Bristles help the widest parts of the body to hold firm against the substrate. The alternating waves of contraction and extension along its body allow the earthworm to make fairly rapid progress through or over the soil (Figure 40.13).

Another type of locomotion made possible by adaptation of the hydrostatic skeleton is the jet propulsion used by the squid and the octopus. Muscles surrounding a water-filled cavity in these cephalopods contract, putting the water under pressure and expelling it from the animal's body. As the water shoots out under pressure, it propels the animal in the opposite direction.

Exoskeletons

An **exoskeleton** is a hardened outer surface to which internal muscles can be attached. Contractions of those muscles cause jointed segments of the exoskeleton to move relative to each other. The simplest

40.13 A Hydrostatic Skeleton and Locomotion
(a) An earthworm's hydrostatic skeleton consists of fluid-filled compartments separated by septa. (b) Contractions of circular muscles cause a compartment (and its corresponding segment) to elongate, and contractions of longitudinal muscles cause a compartment (and its segment) to shorten. Alternating waves of elongation and contraction move the earthworm through the soil. Bristles prevent parts of the worm from moving backward as waves of contractions pass by.

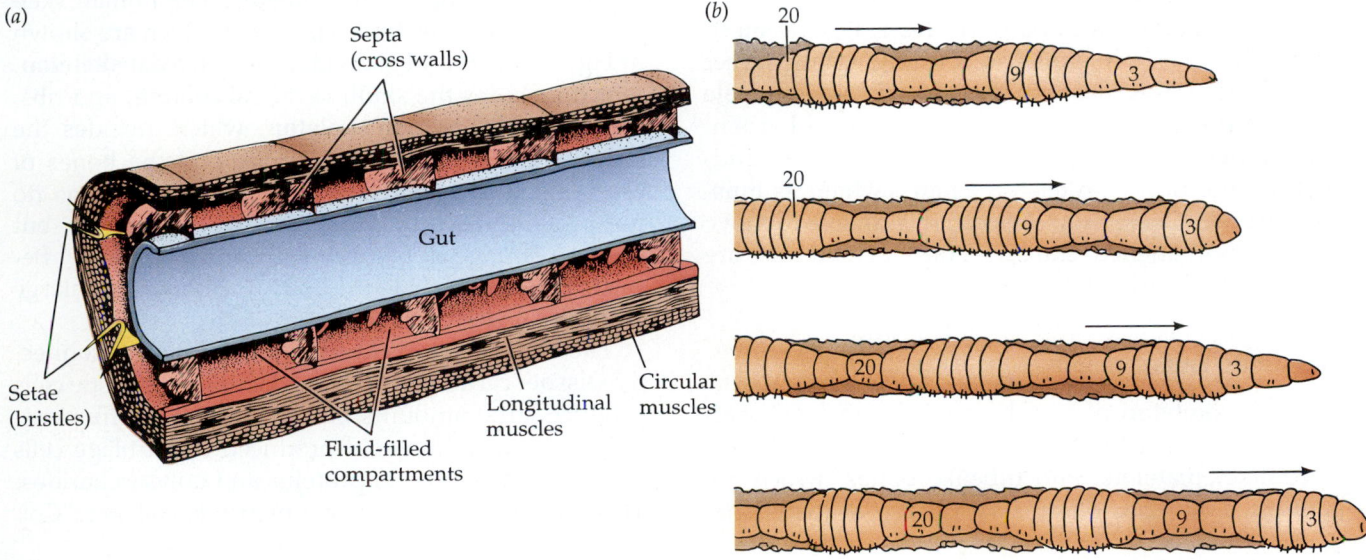

(a)

Septa (cross walls)

Gut

Setae (bristles)

Fluid-filled compartments

Longitudinal muscles

Circular muscles

(b)

example of an exoskeleton is the shell of a mollusk, which generally consists of just one or two pieces. Some marine bivalves and snails have shells composed of protein strengthened by crystals of calcium carbonate (a rock-hard material). These shells can be massive, affording significant protection against predators. The shells of land snails generally lack the hard mineral component and are much lighter. Molluscan shells can grow as the animal grows, and growth rings are usually apparent on the shells. The soft parts of the molluscan body have a hydrostatic skeleton as well. The hydrostatic skeleton is used in locomotion; the exoskeleton mainly provides protection. (Some scallops, however, swim by opening their shells and snapping them shut—another version of jet propulsion.)

The most complex exoskeletons are found among the arthropods. Plates of exoskeleton, or **cuticle**, cover all the outer surfaces of the arthropod's body and all its appendages. The plates are secreted by a layer of cells just below the exoskeleton. A continuous, layered waxy coating covers the entire body. The skeleton contains stiffening materials everywhere except at the joints, where flexibility must be retained. The layers of cuticle include an outer, thin, waxy epicuticle that protects the bodies of terrestrial arthropods from drying out, and a thicker, inner endocuticle that forms most of the structure. The endocuticle is a tough, pliable material found only in arthropods. It consists of a complex of protein and **chitin**, a nitrogen-containing polysaccharide. In marine crustaceans the endocuticle is further toughened by insoluble calcium salts. The thickness of the exoskeleton varies, forming a very efficient armor. Muscles attached to the inner surfaces of the arthropod exoskeleton move its parts around the joints (Figure 40.14).

An exoskeleton protects all the soft tissues of the animal, but it is itself subject to damage such as abrasion and crushing. The greatest drawback of the arthropod exoskeleton is that it cannot grow. Therefore, if the animal is to become larger, it must **molt**, shedding its exoskeleton and forming a new, larger one. During this process the animal is vulnerable because the new exoskeleton takes time to harden. The animal's body is temporarily unprotected and, without the firm exoskeleton against which its muscles can exert maximum tension, it is unable to move rapidly. Soft-shelled crabs, a gourmet delicacy, are crabs caught when they are molting.

Vertebrate Endoskeletons

The **endoskeleton** of vertebrates is an internal scaffolding to which the muscles attach. It is composed of rodlike, platelike, and tubelike bones, which are connected to each other at a variety of joints that

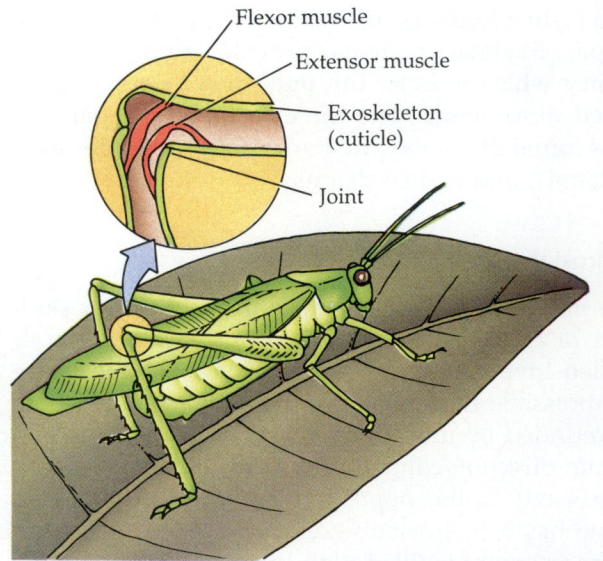

40.14   An Insect's Exoskeleton
Muscles attached to the exoskeleton move parts around flexible joints. The insect in the photograph is a katydid.

allow a wide range of movements. The human skeleton consists of 206 bones (some of which are shown in Figure 40.15) and is divided into an **axial skeleton**, which includes the skull, vertebral column, and ribs, and an **appendicular skeleton**, which includes the pectoral girdle, the pelvic girdle, and the bones of the arms, legs, hands, and feet. Endoskeletons do not provide the protection that exoskeletons do, but their advantage is that bones continue to grow. Because bones are inside the body, the body can enlarge without shedding its skeleton.

The endoskeleton consists of two kinds of connective tissue: cartilage and bone. Connective tissue cells produce large amounts of extracellular matrix material. The matrix material produced by cartilage cells is a rubbery mixture of proteins and polysaccharides. The principal protein in the matrix is collagen. Col-

40.15 The Human Endoskeleton

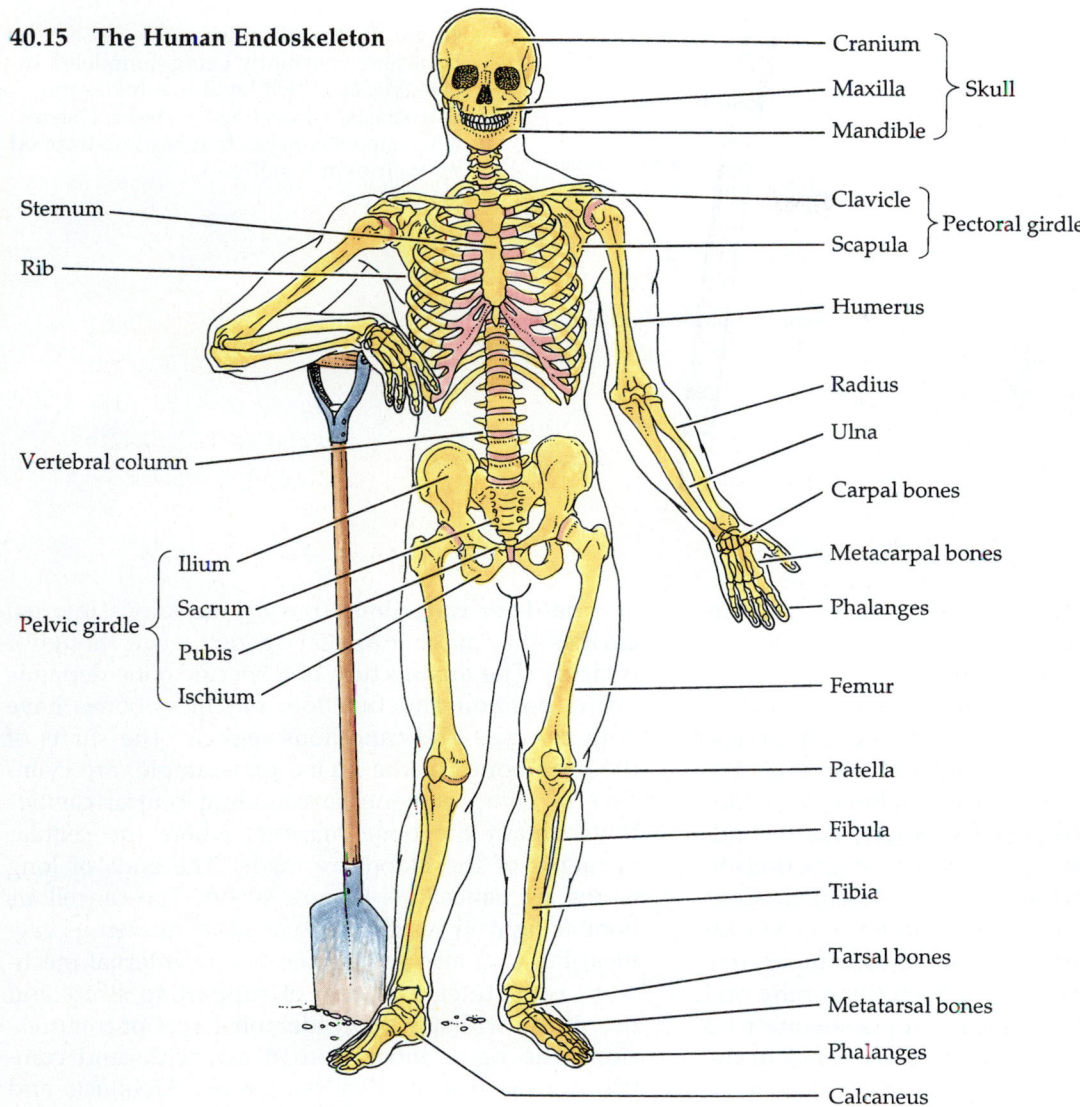

Cranium ⎫
Maxilla ⎬ Skull
Mandible ⎭

Clavicle ⎫ Pectoral girdle
Scapula ⎭

Humerus

Radius

Ulna

Carpal bones

Metacarpal bones

Phalanges

Femur

Patella

Fibula

Tibia

Tarsal bones

Metatarsal bones

Phalanges

Calcaneus

Sternum

Rib

Vertebral column

Pelvic girdle ⎰ Ilium
Sacrum
Pubis
Ischium

lagen fibers run in all directions through the gel-like matrix and give it the well-known strength and resiliency of "gristle." Cartilage is found in parts of the endoskeleton where both stiffness and resiliency are required, such as on the surfaces of joints, where bones move against each other. Cartilage is also the supportive tissue in stiff but flexible structures such as the larynx (voice box), the nose, and the ears. Sharks and rays are called cartilaginous fishes (see Figure 27.14) because their skeletons are composed entirely of cartilage. In all other vertebrates, cartilage is the principal component of the embryonic skeleton, but over the course of development it is gradually replaced by bone.

Bone consists mostly of extracellular matrix material that contains collagen fibers as well as crystals of insoluble calcium phosphate, which give bone its rigidity and hardness. The skeleton serves as a reservoir of calcium for the rest of the body and is in dynamic equilibrium with soluble calcium in the extracellular fluids of the body. This equilibrium is under hor-

monal control by calcitonin and parathyroid hormone (see Figure 36.11). If too much calcium is taken from the skeleton, the bones are seriously weakened.

The living cells of bone—called osteoblasts, osteocytes, and osteoclasts—are responsible for the dynamic remodeling of bone structure that is constantly under way. **Osteoblasts** lay down new matrix on bone surfaces. These cells gradually become surrounded by matrix and eventually become enclosed within the bone, at which point they cease laying down matrix but continue to exist within small lacunae (cavities) in the bone. In this state they are called **osteocytes**. In spite of the vast amounts of matrix between them, osteocytes remain in contact with one another through long cellular extensions that run through tiny channels in the bone. Communication between osteocytes is believed to be important in controlling the activities of the cells that are laying down new bone or eroding it away.

The cells that erode or reabsorb bone are the **osteoclasts**. They are derived from the same cell lineage

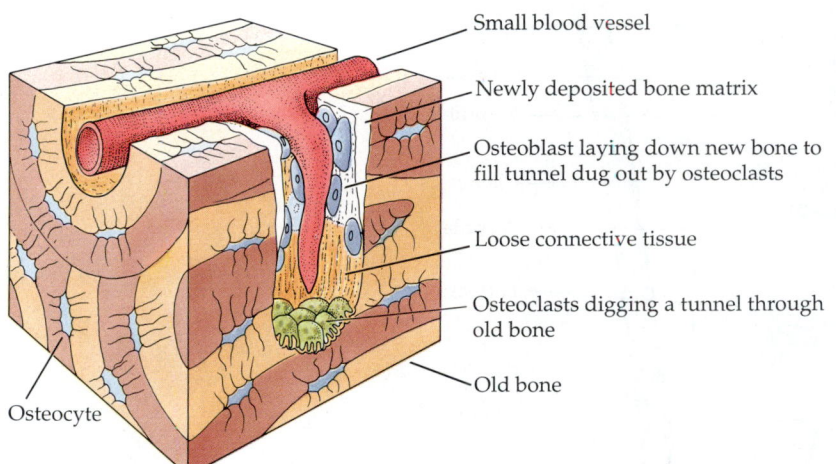

Small blood vessel

Newly deposited bone matrix

Osteoblast laying down new bone to fill tunnel dug out by osteoclasts

Loose connective tissue

Osteoclasts digging a tunnel through old bone

Old bone

Osteocyte

40.16 Renovating Bone
Bones are constantly being remodeled by osteoblasts, which lay down bone, and osteoclasts, which dissolve bone. Osteocytes are osteoblasts that become trapped by their own handiwork.

that produces the white blood cells. Osteoclasts burrow into bone, forming cavities and tunnels. Osteoblasts follow osteoclasts, depositing new bone (Figure 40.16). Thus the interplay of osteoblasts and osteoclasts constantly replaces and remodels the bones. How the activities of these cells are coordinated is not understood, but stress placed on bones provides information used in the process. A remarkable finding in studies of astronauts spending long periods in zero gravity was that their bones decalcified. Conversely, certain bones of athletes can become considerably thicker than they were prior to training or than the same bones in nonathletes. Both thickening and thinning of bones are experienced by someone who has a leg in a cast for a long time. The bones of the uninjured leg carry the person's weight and thicken, while the bones of the inactive leg in the cast thin. The jawbones of people who lose their teeth experience less compressional force during chewing and become considerably remodeled (Figure 40.17).

Types of Bone

Bones are divided into two types, based on how they develop. Membranous bone forms on a scaffolding of connective tissue membrane; cartilage bone forms first as cartilaginous structures and is gradually hardened (ossified) to become bone. The outer bones of the skull are membranous bones; the bones of the limbs are cartilage bones. Cartilage bones can grow throughout ossification. In the long bones of the legs and arms, for example, ossification occurs first at the centers and later at each end (Figure 40.18). Growth can continue until these areas of ossification join. The membranous bones forming the skull cap grow until their edges meet. The fontanel, or "soft spot," on the top of a baby's head is where the skull bones have not yet joined.

The composition of bone may be **compact** (solid

and hard) or **cancellous** (having numerous internal cavities that make it appear spongy, even though it is rigid). The architecture of a specific bone depends on its position and function, but most bones have both compact and cancellous regions. The shafts of the long bones of the limbs, for example, are cylinders of compact bone surrounding central cavities that contain the bone marrow, where the cellular elements of the blood are made. The ends of long bones are cancellous (Figure 40.19). The cancellous bone is light in weight because of its numerous cavities, but it is also strong because its internal meshwork constitutes a system of supporting struts and ties. It can withstand considerable forces of compression. The rigid, tubelike shaft can withstand compression as well as bending forces. Architects and nature alike use hollow tubes as lightweight structural elements. In a solid rod that is subjected to a bending force, one side of the rod is compressed while the other side is stretched, and both help to resist the force. Because the center of a solid rod contributes very little to its ability to resist bending, hollowing out a rod reduces its weight but not its strength.

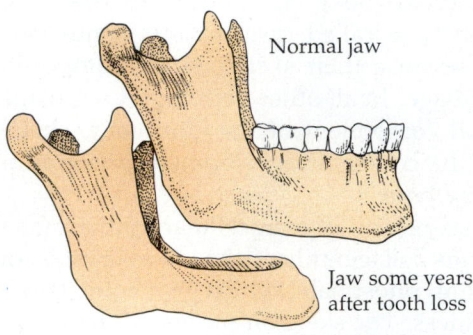

Normal jaw

Jaw some years after tooth loss

40.17 Tooth Loss and Jawbone Structure
Human jawbone is reabsorbed after tooth loss because of lack of compressional forces.

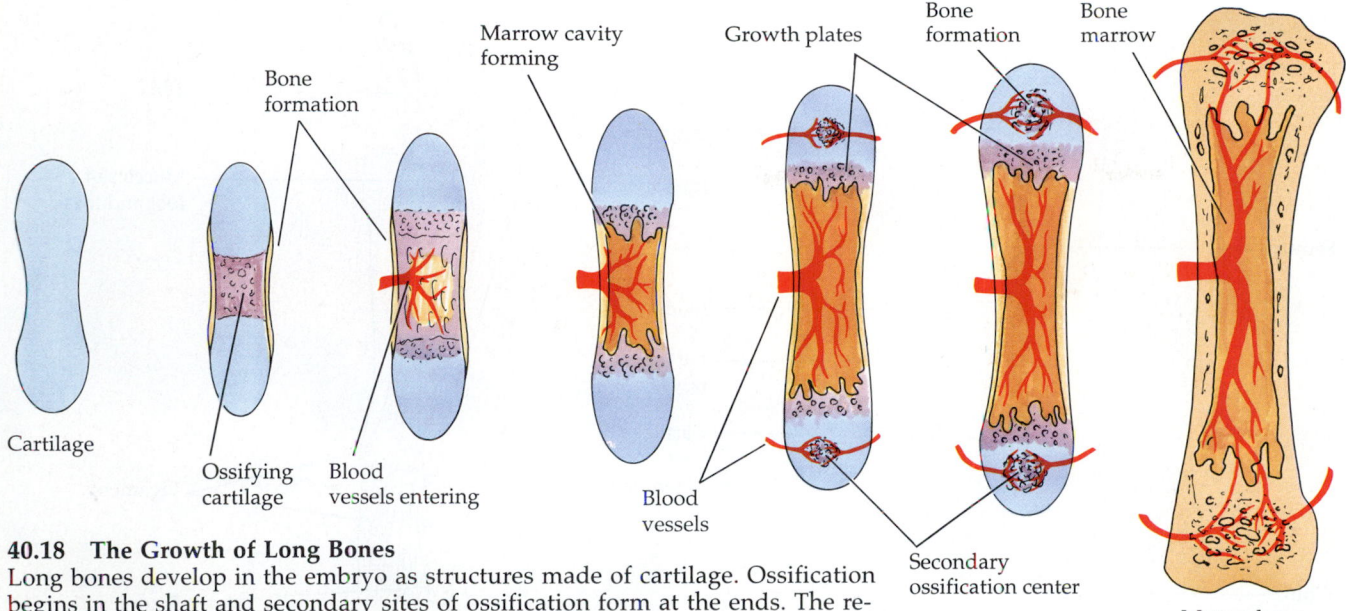

40.18 The Growth of Long Bones
Long bones develop in the embryo as structures made of cartilage. Ossification begins in the shaft and secondary sites of ossification form at the ends. The regions between the ossified parts can grow. Eventually the areas of ossification fuse and elongation of the bone ceases.

Most of the compact bone in mammals is called Haversian bone because it is composed of structural units called **Haversian systems** (Figure 40.20). Each system is a set of thin, concentric bony cylinders, between which are the osteocytes in their lacunae. Through the center of each Haversian system runs a narrow canal containing blood vessels (see Figure 40.16). The osteocytes in one Haversian system connect only with osteocytes in the same system; no channels cross the boundaries (called glue lines) between systems. An important feature of Haversian bone is its resistance to fracturing. If a crack forms in one Haversian system, it tends to stop at the nearest glue line.

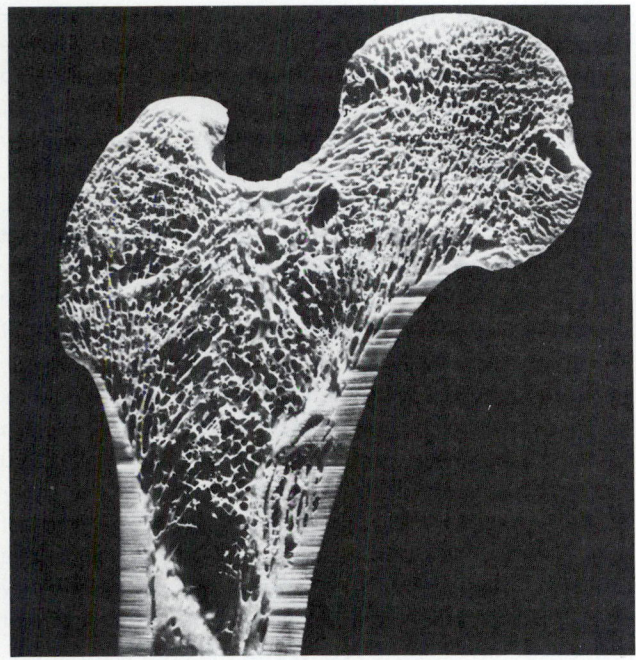

40.19 Internal Architecture of Bone
Bone may have cancellous (spongy) and compact regions. The ends of long bones are cancellous and the shafts are tubes of compact bone.

40.20 Haversian Systems in Bone
Osteoblasts lay down bone in layers. In long bones these layers form concentric tubes parallel to the long axis of the bone. At the center of the tube is a canal containing blood vessels and nerves. This micrograph is colored because the bone cross section was illuminated with polarized light.

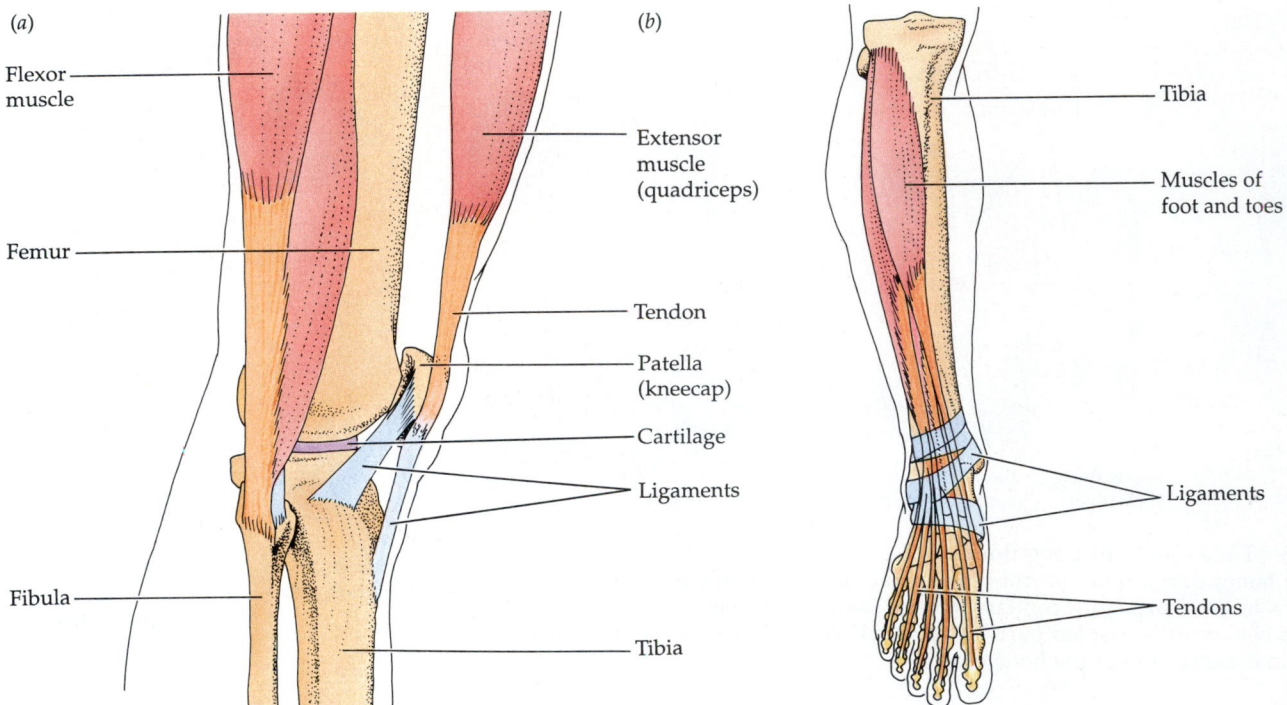

(a)

Flexor muscle

Femur

Fibula

(b)

Extensor muscle (quadriceps)

Tendon

Patella (kneecap)

Cartilage

Ligaments

Tibia

Tibia

Muscles of foot and toes

Ligaments

Tendons

40.21 Joints, Ligaments, and Tendons
(a) This side view of the human knee shows tendons that attach muscle to bone and ligaments that attach bone to bone. Flexor and extensor muscles work antagonistically to operate the joint. *(b)* Tendons connect muscles in the front of the leg to foot and toe bones. These tendons, which pass under strapping ligaments at the front of the ankle, elevate the foot and toes.

Joints and Levers

Muscles and bones work together around **joints**, where two or more bones come together. Since muscles can only contract and relax, they create movement around joints by working in antagonistic pairs: When one contracts, the other relaxes. With respect to a particular joint, such as the knee, we can refer to the muscle that flexes the joint as the **flexor** and the muscle that extends the joint as the **extensor** (Figure 40.21*a*). The bones that meet at the joint are held in place by **ligaments**, which are flexible bands of connective tissue. Other straps of connective tissue, **tendons**, attach the muscles to the bones. In many kinds of joints, only the tendon spans the joint, sometimes moving over the surfaces of the bone like a rope over a pulley. It is the tendon of the quadriceps muscle traveling over the knee joint that is tapped to elicit the knee-jerk reflex.

Ligaments can also hold tendons in place and change the direction of the force they exert. For example, many of the muscles that extend your toes are actually in your lower leg. The tendons from these muscles travel over the front of your ankle, across the upper surfaces of the foot bones, and attach to the bones of your toes. Straps of ligaments over the ankle hold these tendons in place and allow them to bend at a right angle at the ankle (Figure 40.21*b*).

The human skeleton has a wide variety of joints with different ranges of movement. The knee joint is a simple hinge that has almost no rotational movement and can flex in one direction only. At the shoulders and hips are ball-and-socket joints that allow movement in almost any direction. A pivotal joint between the two bones of the forearm where they meet at the elbow allows the smaller bone, the radius, to rotate when the wrist is twisted from side to side. Several kinds of joints permit some rotation, but not in all directions as do the ball-and-socket joints. Examples of these joints are found in the bones of the hands, and they give the hands a wide range of possible movements (Figure 40.22).

Bones around joints and the muscles that work with these bones can be thought of as levers. A lever has a power arm and a load arm that work around a fulcrum (pivot). The ratio of these two arms determines whether a particular lever can exert a lot of force over a short distance or is better at translating force into big or fast movements. Compare the jaw joint and the knee joint (Figure 40.23). The power arm of the jaw is long relative to the load arm, allowing the jaw to apply great pressures over a small distance, such as when you crack a nut with your teeth. The power arm of the lower leg, on the other hand, is short relative to the load arm, so you can run fast, jump high, and deliver swift kicks, but you

40.22 Types of Joints
The designs of joints are similar to mechanical counterparts and enable a variety of movements.

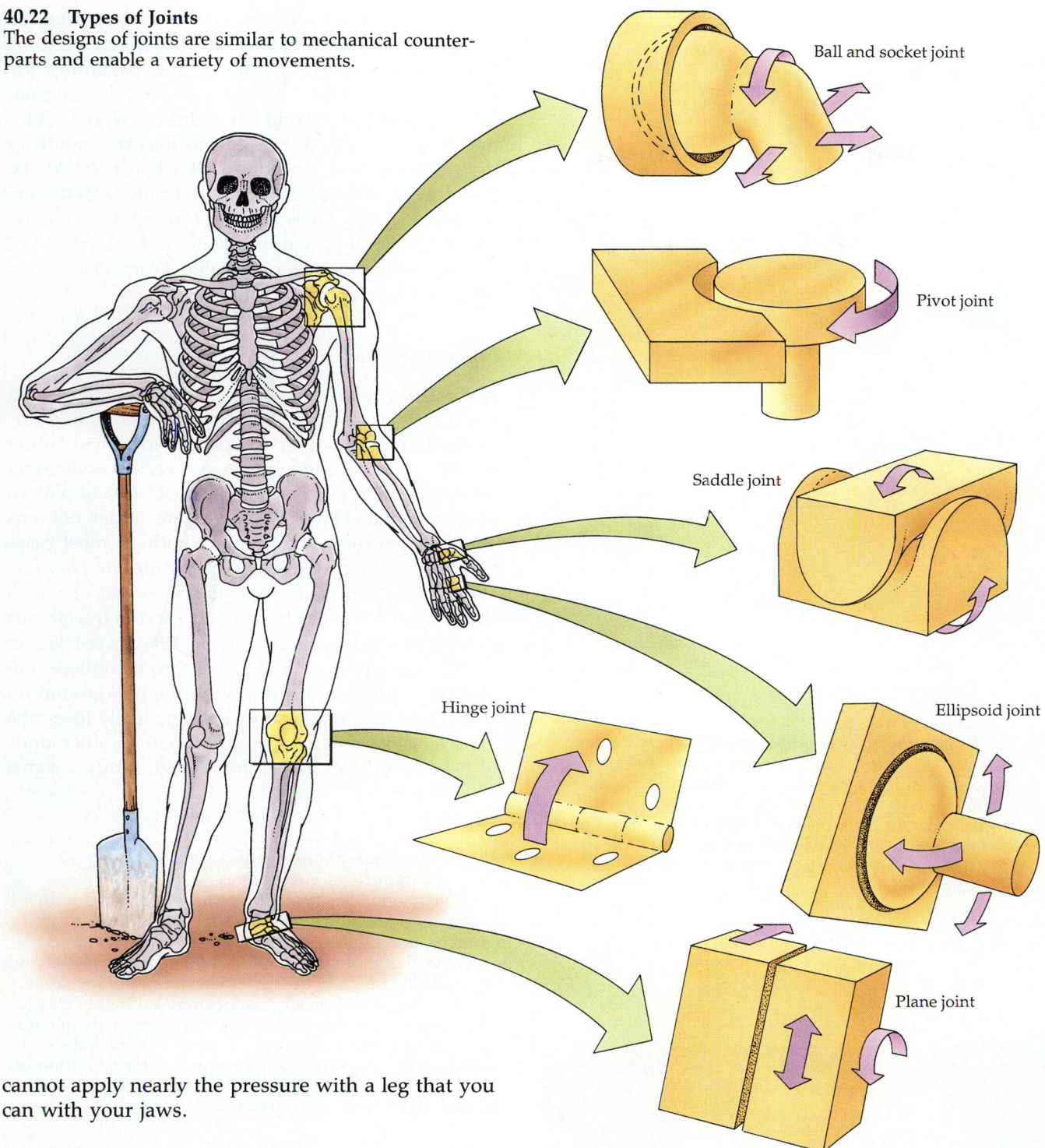

Ball and socket joint

Pivot joint

Saddle joint

Ellipsoid joint

Hinge joint

Plane joint

cannot apply nearly the pressure with a leg that you can with your jaws.

OTHER EFFECTORS

Muscles in animals serve almost universally as effectors. Other effectors are more specialized and are not shared by many animal species. Some specialized effectors are used for defense, some for communication, some for capture of prey or avoidance of predators. A discussion of all the effectors animals use would take an entire book, but we can briefly mention a few here to give a sampling of their evolutionary diversity.

Some animals possess highly specialized organs that are fired like miniature missiles to capture prey and repel enemies. **Nematocysts** are elaborate cellular structures produced only by hydras, jellyfish, and other cnidarians. They are concentrated in huge numbers on the outer surface of the tentacles of the animal. Each nematocyst is made up of a slender thread coiled tightly within a capsule, which is armed with a spinelike trigger projecting to the outside (see Figure 26.10). When a potential prey organism

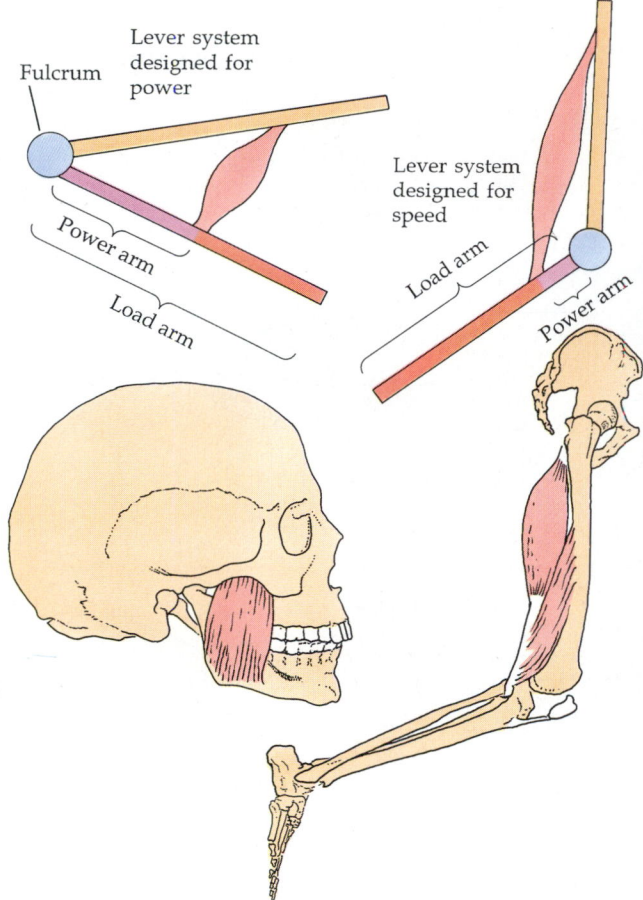

40.23   Bones and Joints as Systems of Levers
A lever works around a fulcrum and has a power arm and a load arm. If the ratio of the load arm to the power arm is low, the lever can generate much force over a small distance. An example is the human jaw. If the ratio of the load arm to the power arm is high, the lever can rapidly move small weights over a long distance. An example is the human leg.

brushes the trigger, the nematocyst fires, turning the thread inside out and exposing little spines along its base. The thread either entangles or penetrates the body of the victim, and a poison may be simultaneously released around the point of contact. Once the prey is subdued, it is pulled into the mouth of the cnidarian and swallowed. A jellyfish called the Portuguese man-of-war has tentacles that can be several meters long. These animals can capture, subdue, and devour full-grown mackerel, and the poison of their nematocysts is so potent that it can kill a human who gets tangled in the tentacles.

Chromatophores

A change in body color is an effector response that some animals use to camouflage themselves in a particular environment or to communicate with other animals. **Chromatophores** are pigment-containing cells in the skin that can change the color and pattern of the animal. Chromatophores are under nervous system or hormonal control, or both; in most cases they can effect a change within minutes or even seconds. In squids, soles, and flounders, all of which spend much time on the sea floor, and in the famous chameleons (a group of African lizards; see Figure 27.20b) and a few other animals, chromatophores enable the animal to blend in with the background on which it is resting and thus be more likely to escape discovery by predators (Figure 40.24a). In other kinds of fishes and lizards, a color change sends a signal

40.24   Chromatophores Help Animals Camouflage Themselves
(a) This octopus has adapted its chromatophores so that it is almost indistinguishable from its background of coral, sponges, and other animals. The octopus is the largest element in the photograph; its eyes are partially closed and its head fills much of the lower left quarter of the scene. (b) The chromatophores of the octopus are highly elastic. Radiating muscle fibers change them from small spheres of pigment to large sheets of pigment. Different chromatophores have different pigments, so the coloration of the animal depends on which chromatophores are extended and which are contracted.

(a)

(b)

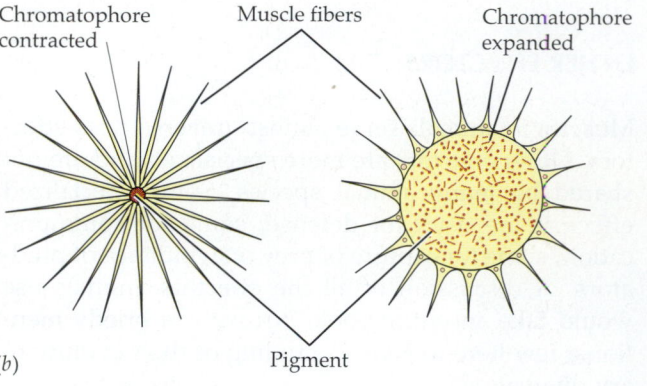

to potential mates and territorial rivals of the same species.

There are three principal types of chromatophore cells. The most common type has fixed cell boundaries, within which pigmented granules may be moved about by microfilaments. When the pigment is concentrated in the center of each chromatophore, the animal is pale; the animal turns darker when the pigment is dispersed throughout the cell. Some other chromatophores are capable of amoeboid movement. They can mold themselves into shapes with a minimal surface area, leaving the tissue relatively pale, or they can flatten out to make the tissue appear darker. Cephalopods have chromatophores that can undergo shape changes as a result of the action of muscle fibers radiating outward from the cell. When the muscles are relaxed, the chromatophores are small and compact and the animal is pale. To darken the animal, the muscles contract and spread the chromatophores over more of the surface (Figure 40.24b). Chromatophores with different pigments enable animals to assume different hues or to become mottled to match the background more precisely.

Glands

Glands are effector organs that produce and release chemicals. Some glands produce chemicals that are responsible for communication among animal cells (see Chapter 36). Other glands produce chemicals that are used defensively or to capture prey. Certain snakes, frogs, salamanders, spiders, mollusks, and fish have poison glands. Many of these poisons have proven to be of practical use to humans. The poison dendrotoxin, which certain tribes of the Amazonian jungles use on the tips of their arrows, comes from the skin of a frog (see Figure 27.17b) and blocks certain potassium channels. The snake venom bungarotoxin inactivates the neuromuscular acetylcholine receptors. The pufferfish poison tetrodotoxin blocks voltage-gated sodium channels. A poison from a mollusk, conotoxin, blocks calcium channels. There are many such examples, and as you can imagine, they are useful research tools for neurobiologists.

Not all defensive secretions are poisonous. A well-known example is the odoriferous chemical mercaptan, which is sprayed by skunks. Human olfaction is more sensitive to mercaptan than to any other compound. The bombardier beetle (see Figure 48.8a) is another spectacular example of an animal that releases an irritating defensive secretion.

The glands that produce and release pheromones, chemicals used in communication among animals—which we mentioned when we discussed chemical sensors in Chapter 39 and which we will discuss further in Chapter 45—are another category of effectors. Still other effector glands produce secretions necessary to facilitate physiological functions. Examples are salivary gland secretions that aid digestion and sweat gland secretions that are an effective means of heat loss.

Sound and Light Producers

The ability to produce sound is an extremely important effector for humans, since speech is one of the most distinctive features of our species. Mammals, birds, and amphibians have evolved a variety of organs that create sound by passing streams of air over structures that vibrate. Insects such as cicadas produce sounds by rubbing together rough surfaces on their appendages. Sound production is by no means a universal effector, however. Most species of animals do not produce sound.

Production of light is even more rare. The classic example of a light-producing animal is the firefly. In an organ at the tip of the firefly's abdomen, an enzyme, luciferase, catalyzes the reaction of a protein, luciferin, with ATP, releasing light energy. The primary function of the light is to attract mates. Bioluminescence is used by a number of plant and animal species (see Box 7.A), and in many cases its function is not known.

Electric Organs

A number of fishes can generate electricity, including the electric eel, the knife fish, the torpedo (a type of ray), and the electric catfish. The electric fields they generate are used for sensing the environment (see Figure 39.32), for communication, and for stunning potential predators or prey. The electric organs of these animals evolved from muscle, and they produce electric potentials in the same general way as nerves and muscles do. Electric organs consist of very large, disc-shaped cells arranged in long rows like stacks of coins. When the cells discharge simultaneously, the electric organ can generate far more current than can nerve or muscle. Electric eels, for example, can produce up to 600 volts with an output of approximately 100 watts—enough to light a row of light bulbs or to temporarily stun a person.

SUMMARY of Main Ideas about Effectors

Effectors enable animals to respond to stimulation from their internal and external environments.

Most effector mechanisms generate mechanical forces that move or change the shapes of cells or whole animals.

Cellular movement comes from two structures: microtubules and microfilaments.

Microtubules and microfilaments depend on long protein molecules that can slide past each other.
Review Figure 40.4

Microtubules move cilia and flagella.
Review Figure 40.3

Microfilaments are responsible for amoeboid movement and for muscle cell contractions.
Review Figure 40.6

The three types of vertebrate muscle are smooth, skeletal, and cardiac.
Review Figure 40.7

Skeletal muscle contracts when actin and myosin filaments in muscle cells slide past each other.
Review Figure 40.8

In a skeletal muscle cell, the microfilaments are organized into myofibrils; myofibrils are made up of units called sarcomeres that contain overlapping myosin and actin filaments.

Myosin has globular heads that bind with actin, change their orientation, and release, causing the filaments to slide past each other and contract the muscle fiber

Action potentials cause the sarcoplasmic reticulum to release calcium ions (Ca^{2+}) into the cytoplasm of the muscle fiber.

Ca^{2+} binds to troponin molecules on the actin filaments and thereby exposes the sites to which myosin can bind.
Review Figure 40.9

In skeletal muscle, a single action potential causes a minimum unit of contraction called a twitch.
Review Figure 40.10

Fast-twitch muscle fibers can generate maximum tension quickly but also fatigue rapidly; slow-twitch fibers generate less tension and do so more slowly but are resistant to fatigue.
Review Figure 40.11

Some cardiac muscle depolarizes spontaneously giving it the ability to serve as a pacemaker.
Review Figure 40.12

Skeletal systems provide rigid supports against which muscles can pull.

Hydrostatic skeletons have fluid-filled cavities that can be squeezed by muscles.
Review Figure 40.13

Exoskeletons are hardened outer surfaces to which internal muscles are attached.
Review Figure 40.14

Endoskeletons are internal, jointed systems of rodlike, platelike, and tubelike rigid supports consisting of bone and cartilage to which muscles attach.
Review Figure 40.15

Bone is continually being remodeled by cells that lay down new bone (osteoblasts) and cells that absorb bone (osteoclasts).
Review Figure 40.16

Bones, which ossify (harden), can grow until centers of ossification meet.
Review Figure 40.18

Muscles and bones work together around joints as systems of levers.
Review Figures 40.22 and 40.23

Tendons connect muscles to bones; ligaments connect bones to each other and help direct the forces generated by muscles by holding tendons in place.
Review Figure 40.21

Other effector organs include nematocysts, chromatophores, glands, and structures that produce sound, light, and electric pulses.

SELF-QUIZ

1. The movement of cilia and flagella is due to
   a. polymerization and depolymerization of tubulin.
   b. making and breaking of crossbridges between actin and myosin.
   c. contractions of microtubules.
   d. changes in conformations of dynein molecules.
   e. the spokes of the axoneme using energy of ATP to contract.

2. Smooth muscle differs from both cardiac and skeletal muscle in that
   a. it can act as a pacemaker for rhythmic contractions.
   b. contractions of smooth muscle are not due to interactions between neighboring microfilaments.
   c. neighboring cells can be in electrical continuity through gap junctions.
   d. neighboring cells are tightly coupled by intercalated discs.
   e. the membranes of smooth muscle cells are depolarized by stretching.

3. Fast-twitch fibers differ from slow-twitch fibers in that
   a. they are more common in the leg muscles of champion sprinters than in the same muscles of marathon runners.
   b. they have more mitochondria.
   c. they fatigue less rapidly.
   d. their abundance is more a product of genetics than of training.
   e. they are more common in the leg muscles of champion cross-country skiers than in the same muscles of weight lifters.

4. The role of Ca^{2+} in the control of muscle contraction is
   a. to cause depolarization of the T-tubule system.
   b. to change the conformation of troponin, thus exposing myosin binding sites.
   c. to change the conformation of myosin heads, thus causing microfilaments to slide past each other.
   d. to bind to tropomyosin and break actin–myosin cross-bridges.
   e. to block the ATP binding site on myosin heads, enabling muscle to relax.

5. Which of the following statements about muscle contractions is *not* true?
   a. A single action potential at the neuromuscular junction is sufficient to cause a muscle to twitch.
   b. Once maximum muscle tension is achieved, no ATP is required to maintain that level of tension.
   c. An action potential in the muscle cell activates contraction by releasing Ca^{2+} into the sarcoplasm.
   d. Summation of twitches leads to a graded increase in the tension that can be generated by a single muscle fiber.
   e. The tension generated by a muscle can be varied by controlling how many of its motor units are active.

6. Which of the following statements about the structure of skeletal muscle is true?
   a. The bright bands of the sarcomere are the regions where actin and myosin filaments overlap.
   b. When a muscle contracts, the A bands of the sarcomere (dark regions) lengthen.
   c. The myosin filaments are anchored in the Z lines.
   d. When a muscle contracts, the H bands of the sarcomere (light regions) shorten.
   e. The sarcoplasm of the muscle cell is contained within the sarcoplasmic reticulum.

7. The long bones of our arms and legs are strong and can resist both compressional and bending forces because
   a. they are solid rods of compact bone.
   b. their extracellular matrix contains crystals of calcium carbonate.
   c. their extracellular matrix consists mostly of collagen and polysaccharides.
   d. they have a very high density of osteoclasts.
   e. they consist of lightweight cancellous bone with an internal meshwork of supporting elements.

8. If we compare the jaw joint with the knee joint as lever systems,
   a. the jaw joint can apply greater compressional forces.
   b. their ratios of power arm to load arm are about the same.
   c. the knee joint has greater rotational abilities.
   d. the knee joint has a greater ratio of power arm to load arm.
   e. only the jaw is a hinged joint.

9. Which of the following statements about skeletons is true?
   a. They can consist only of cartilage.
   b. Hydrostatic skeletons can be used only for amoeboid locomotion.
   c. An advantage of exoskeletons is that they can continue to grow throughout the life of the animal.
   d. External skeletons must remain flexible, so they never include calcium carbonate crystals as do bones.
   e. Internal skeletons consist of four different types of bones: compact, cancellous, dermal, and cartilage.

10. Chemicals used by neurophysiologists to block voltage-gated sodium channels have come from
    a. chromatophores.
    b. nematocytes.
    c. electric eels.
    d. luciferase.
    e. poison glands of fish.

FOR STUDY

1. Describe in outline form all the events that occur between the arrival of an action potential at a motor nerve terminal and the contraction of a muscle fiber.

2. How do we know that the basis for the movement of cilia resides in the dynein components of the axoneme?

3. Wombats are powerful digging animals and kangaroos are powerful jumping animals. How do you think the structure of their legs would compare in terms of their designs as lever systems?

4. Maria and Margaret are identical twin sisters. Their mother was an Olympic marathon runner and their father was on the varsity rowing team in college. Maria has become a serious cross-country skier and Margaret has joined the track team as a sprinter. Which one do you think will have the greatest chance of becoming a champion in her sport, and why?

5. If an adolescent breaks a leg bone close to the ankle joint, after the break heals that leg may not grow as long as the other one. Explain why, including an explanation of why the leg grows at all.

READINGS

Alexander, R. M. 1991. "How Dinosaurs Ran." *Scientific American*, April. An example of how physics and engineering approaches can be used to study the effectors of extinct animals.

Cameron, J. N. 1985. "Molting in the Blue Crab." *Scientific American*, May. How an arthropod deals with its exoskeleton in order to grow.

Caplan, A. I. 1984. "Cartilage." *Scientific American*, October. A component of the vertebrate skeletal system, cartilage plays a surprisingly diverse group of roles in the developing and mature animal.

Carafoli, E. and J. T. Penniston. 1985. "The Calcium Signal." *Scientific American*, November. Calcium as a second messenger; calcium in muscle contraction.

Cohen, C. 1975. "The Protein Switch of Muscle Contraction." *Scientific American*, November. Interaction of muscle proteins and calcium ions.

Eckert, R., D. Randall and G. Augustine. 1988. *Animal Physiology*, 3rd Edition. W. H. Freeman, New York. An excellent advanced textbook. Chapter 10 deals with muscle and Chapter 11 with cell motility.

Gans, C. 1974. *Biomechanics: An Approach to Vertebrate Biology*. University of Michigan Press, Ann Arbor. A small classic on the architecture of animals and how their structure is adapted to their environment and lifestyle.

Hadley, N. F. 1986. "The Arthropod Cuticle." *Scientific American*, July. Properties of the exoskeleton of the most abundant and diversified phylum of animals.

Lazarides, E. and J. P. Revel. 1979. "The Molecular Basis of Cell Movement." *Scientific American*, May. Microtubules and microfilaments in action.

Schmidt-Nielsen, K. 1990. *Animal Physiology: Adaptation and Environment*, 4th Edition. Cambridge University Press, New York. Chapter 11 gives a marvelous treatment of skeletons, muscles, and other effectors.

Vander, A. J., J. H. Sherman and D. S. Luciano. 1994. *Human Physiology: The Mechanisms of Body Function*, 6th Edition. McGraw-Hill, New York. Chapter 11 deals with the structure and function of human muscle.

The wail that heralds the birth of an infant and brings joy to its parents also initiates the infant's breathing, a process that must continue every minute of its life. Whether awake or asleep, exercising or resting, talking or eating, we must breathe. For brief moments, such as when we swim under water, we can hold our breath, but the urge to breathe mounts rapidly and soon becomes overwhelming. By ignoring a petulant child who threatens to hold its breath until its demands are met, we do not endanger the child's life. Events such as drowning or choking, however, that prevent breathing but are not under a person's own control can lead quickly to death.

BREATHING AND LIFE

Breathing is absolutely coupled with life. We explained the reason for this connection in Chapter 7. Cells need a constant supply of energy to carry out their functions. This energy comes in the form of ATP produced through the oxidation of nutrient molecules. An animal's production of ATP can be sustained only where oxygen gas (O_2) is present. Some cells, such as muscle cells, can survive short periods without O_2 by deriving ATP from glycolysis only and incurring an O_2 debt that has to be paid back later. Brain cells, however, have little capacity to function in the absence of O_2. In the short run, lack of O_2 causes loss of consciousness. As the length of time without O_2 increases, cell functions grind to a halt, cells die, and tissues and organs suffer irreversible damage. Breathing provides the body with the O_2 required to support the energy metabolism of all its cells. Breathing also eliminates one of the waste products of cell metabolism, carbon dioxide (CO_2).

For humans and other large animals, breathing makes possible the exchange of CO_2 and O_2 between the body and the environment. Gas molecules diffuse across membranes that separate the internal environment of the body from the external environment. Some animals, especially small and inactive ones, accomplish gas exchange without breathing because the contact between their gas-exchange membranes and the environment is adequate to support diffusion of O_2 in and diffusion of CO_2 out without their expending energy to move the environment over those membranes, which is what breathing does. Gas exchange is commonly referred to as respiration, as in artificial respiration, but whole-animal gas exchange should not be confused with cellular respiration, even though the two processes are tightly linked.

Animals differ greatly in the rates of gas exchange necessary to support their energy metabolism. At room temperature, a frog consumes about 0.01 liters of O_2 per hour, and a resting human consumes about

Portable Life Support
In space there is no oxygen to breathe. Because humans cannot live without oxygen, astronauts who leave their spaceship take their own atmosphere to support respiration.

41

Gas Exchange in Animals

15 liters of O₂ per hour. A marathon runner consumes about 150 liters of O₂ per hour during a race. The runner, however, can be outdistanced and outlasted by fish breathing water and by birds flying at very high altitudes, where there is little O₂. We'll learn why in this chapter.

LIMITS TO GAS EXCHANGE

Diffusion is the only means by which respiratory gases are exchanged. There are no active transport mechanisms for respiratory gases. This fact is true for all gas-exchange systems, despite their diverse structures and functions. To help you understand the adaptations of gas-exchange systems that we will cover in this chapter, review the discussion of diffusion in Chapter 5. Because diffusion is strictly a physical phenomenon, it is limited by physical factors such as whether animals breathe air or water.

Breathing Air or Water

Animals that breathe water must have a more efficient gas-exchange system than animals that breathe air. O₂ can be exchanged more easily in air than in water for several reasons. First, the oxygen content of air is much higher than the oxygen content of an equal volume of water. The maximum O₂ content of a rapidly flowing stream splashing over rocks and tumbling over waterfalls is less than 10 ml of O₂ per liter of water. The O₂ content of fresh air is about 200 ml of O₂ per liter of air. Second, O₂ diffuses much more slowly in water than in air. In a still pond, the O₂ content of the water may be zero only a few millimeters below the surface if the water has not been disturbed. Finally, when an animal breathes, it performs work to move air or water over its gas-exchange surfaces. More energy is required to move water than to move air because water is denser and thicker than air.

The slow diffusion of O₂ molecules in water imposes a gas-exchange constraint on air-breathing animals as well as on water-breathing animals. Eukaryotic cells respire in their mitochondria, which are in the cytoplasm—an aqueous medium. Cells are bathed with extracellular fluid—also an aqueous medium. The slowness of O₂ diffusion in water limits the efficiency of O₂ distribution from gas-exchange membranes to the sites of cellular respiration in both

41.1 Keeping in Touch with the Medium
(a) No cell in the leaflike body of this marine flatworm is more than a millimeter away from seawater. (b) The same is true of sponges; they have body walls perforated by many channels lined with flagellated cells. These channels communicate with the outside world and with a central cavity. The flagella maintain currents of water through the channels, through the central cavity, and out of the animal. Every cell in the sponge is very close to the respiratory medium. (c) The gills of this newt project like a feathery fringe and provide a large surface area for gas exchange. Blood circulating through the gills comes into close contact with the respiratory medium.

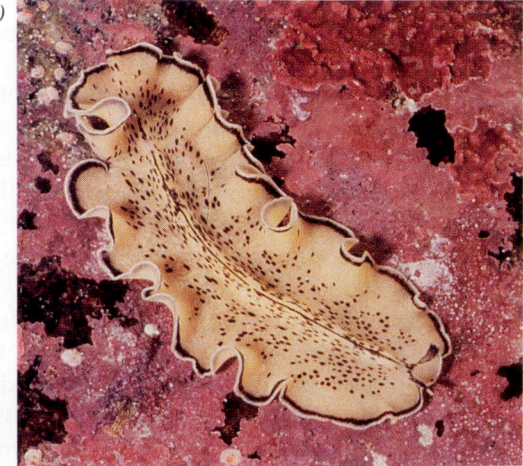
(a)

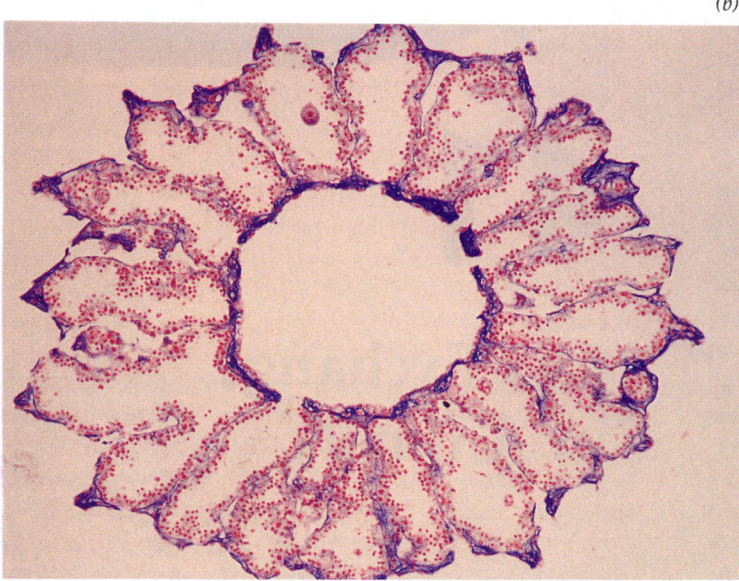
(b)

(c)

air-breathing and water-breathing animals. In water, a supply of O_2 that is adequate to support the metabolism of a typical animal cell can be obtained only if the diffusion path is shorter than about 1 millimeter. Therefore, in an animal that lacks an internal system for transporting gases, no cell may be more than about 1 millimeter from the outside world. This is a severe size limit, but one way to accommodate it and still grow bigger is to have a flat, leaflike body plan, which is common among simple invertebrates (Figure 41.1a). Another way is to have a very thin body built around a central cavity through which water circulates (Figure 41.1b). Otherwise, specialized structures are required to provide an increased surface area for diffusion, and an internal circulatory system is needed to carry gases to and from these exchange structures (Figure 41.1c).

Temperature

Temperature is a crucial factor influencing gas exchange in animals that breathe water. Almost all water breathers are ectotherms. The body temperatures of aquatic ectotherms are closely tied to the temperature of the water around them. As the temperature of the water rises, so does their body temperature, and because of Q_{10} effects (see Chapter 35), energy expenditure and oxygen demand rise exponentially. But warm water holds less gas than cold water does. (Just think of what happens when you open a warm bottle of beer or soda.) Thus aquatic ectotherms are in a double bind: As the temperature of their environment goes up, so does their demand for O_2; but the availability of O_2 in their environment goes down (Figure 41.2). If the animal performs work to move water across its gas-exchange surfaces (as fish do, for example), the energy the animal must expend increases as water temperature rises. Thus as water temperature goes up, the water breather must extract more O_2 from the environment, or it must decrease its energy expenditures for activities other than breathing.

Altitude

Just as a rise in temperature reduces the supply of O_2 available for aquatic animals, an increase in altitude reduces the O_2 supply for air breathers. The amount of O_2 in the atmosphere decreases with increasing altitude. One way of expressing the concentration of gases in air and in water is by their **partial pressures**. At sea level, the pressure exerted by the atmosphere is the equivalent to that produced by a column of mercury 760 mm high. We therefore say that the **barometric pressure** (atmospheric pressure) is 760 mm of mercury (Hg). Because dry air is 20.9 percent O_2, the partial pressure of oxygen (P_{O_2}) at

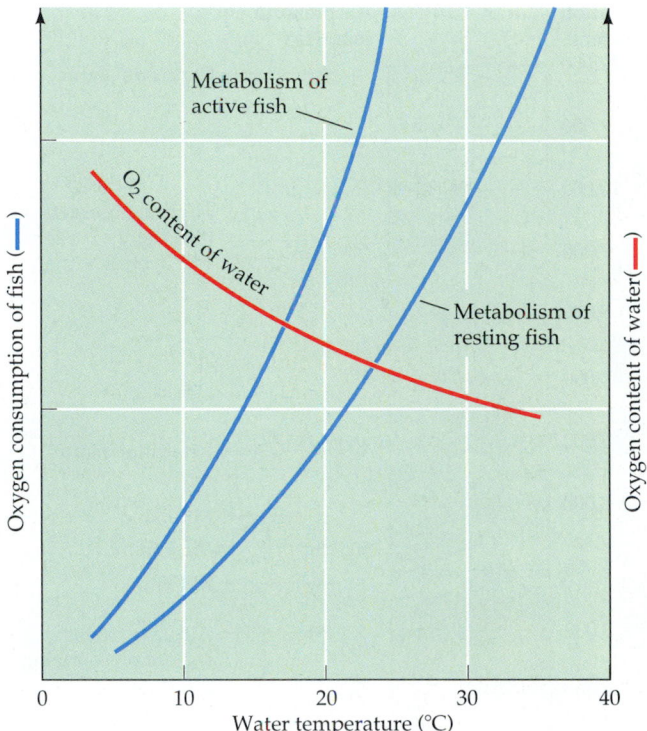

41.2 The Double Bind of Water Breathers
As water temperature increases, so do the body temperatures of water-breathing ectotherms, and therefore their oxygen needs increase. However, warm water carries less oxygen in solution than does cold water.

sea level is 20.9 percent of 760 mm Hg, or about 159 mm Hg. As you go higher in elevation, there is less and less air above you, so barometric pressure declines. At an altitude of 5,300 meters, barometric pressure is only half as much as it is at sea level, so the P_{O_2} at that altitude is only 80 mm Hg. At the summit of Mount Everest (8,848 meters) the P_{O_2} is only about 50 mm Hg or roughly one-third what it is at sea level. Remember that diffusion of O_2 into the body is dependent on O_2 concentration differences between the air and the body fluids, so the drastically reduced O_2 concentration in the air at a high altitude constrains O_2 uptake. This low O_2 concentration is why mountain climbers who venture to the heights of Mount Everest breathe O_2 from pressurized bottles they carry with them (Figure 41.3).

Carbon Dioxide Exchange with the Environment

Respiratory gas exchange is a two-way street. CO_2 diffuses out of the body as O_2 diffuses in. Given the same concentration gradient, CO_2 and O_2 molecules diffuse at about the same rate whether in air or in water. However, the concentration gradients for diffusion of O_2 and CO_2 across gas-exchange membranes are generally not the same. The concentration of CO_2 in the atmosphere is so low, and its solubility

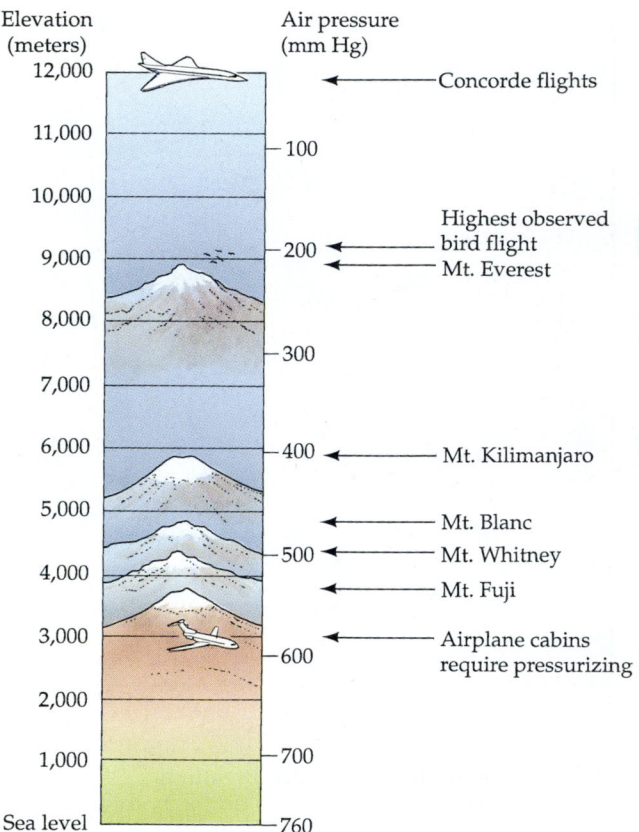

Elevation (meters) / Air pressure (mm Hg)

- 12,000 — Concorde flights
- 11,000 — 100
- 10,000
- 9,000 — 200 — Highest observed bird flight / Mt. Everest
- 8,000 — 300
- 7,000
- 6,000 — 400 — Mt. Kilimanjaro
- 5,000 — Mt. Blanc
- 4,000 — 500 — Mt. Whitney / Mt. Fuji
- 3,000 — 600 — Airplane cabins require pressurizing
- 2,000
- 1,000 — 700
- Sea level — 760

41.3 Scaling Heights
The oxygen content of the atmosphere decreases with altitude. Therefore, airplane cabins must be pressurized and mountain climbers must carry pressurized containers of oxygen. Birds, however, have been observed flying over even the highest peaks.

in the aquatic environment is so high, that diffusion of CO_2 from an animal is usually not a problem. Transporting CO_2 from where it is produced in the body to where it diffuses into the environment, however, can be a limiting factor in gas exchange and hence metabolism.

RESPIRATORY ADAPTATIONS

Animals have evolved a great diversity of adaptations to maximize their rates of gas exchange. All of these adaptations, however, work through influencing a few physical parameters that are described by a simple equation called **Fick's law of diffusion**:

$$Q = DA \frac{C_1 - C_2}{L}$$

Fick's law describes the rate, Q, at which a substance diffuses between two locations. D is the diffusion coefficient, which is a characteristic of a particular substance diffusing in a particular medium at a particular temperature. For example, perfume has a

higher D than does motor oil, and substances diffuse faster in air than in water. A is the cross-sectional area over which the substance is diffusing. C_1 and C_2 are the concentrations of the substance at two locations, and L is the distance between those locations. Therefore, $(C_1 - C_2)/L$ is a concentration gradient. Animals can maximize D for the respiratory gases by using air rather than water for the gas-exchange medium whenever possible. All other adaptations for maximizing respiratory gas exchange must influence the surface area for exchange or the concentration gradient across that surface area.

Surface Area

There are many anatomical adaptations that maximize specialized body surface areas over which gases can diffuse (Figure 41.4). **External gills** are highly branched and folded elaborations of the body surface that provide a large surface area for gas exchange. They consist of thin, delicate membranes that minimize the path length traversed by diffusing molecules of O_2 and CO_2 (see Figure 41.1c). Because external gills are vulnerable to damage and are tempting morsels for carnivorous organisms, it is not surprising that protective body cavities for gills have evolved. Many mollusks, arthropods, and fishes have **internal gills** in such cavities.

Like water-breathers, air-breathing animals have adapted by increasing their surface area for gas exchange, but their structures are quite different from gills. First, gas-exchange surfaces in air breathers

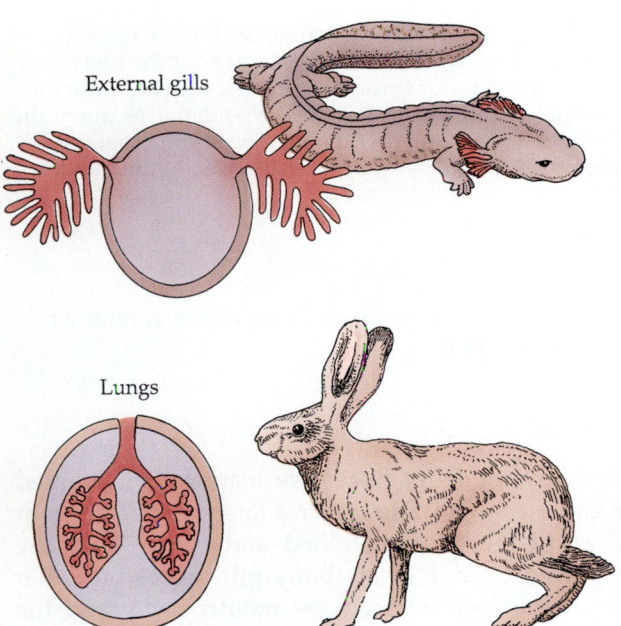

External gills

Lungs

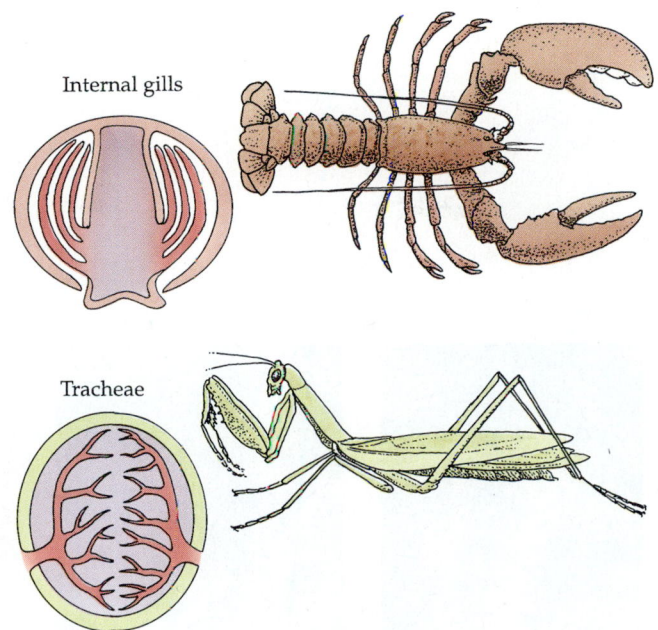

Internal gills

Tracheae

41.4 Gas-Exchange Organs
Increased surface area for the diffusion of respiratory gases is a common feature. Gills are adaptations for breathing water; lungs and tracheae are adaptations for breathing air.

must be in moist internal cavities to prevent drying out. Second, surface elaborations such as gills work only in water because without water for support, they collapse and stick together like the pages of a wet magazine and thus lose effective surface area. That is why a fish suffocates in air in spite of the much higher O_2 concentration. The gas-exchange structures, or **lungs**, of most air-breathing vertebrates are highly divided, elastic air sacs, which we describe in greater detail later in the chapter. The gas-exchange structures of insects are highly branched systems of air-filled tubes that branch through all the tissues of the insect's body. In both cases the surface areas for gas exchange are greatly enhanced by their division into many small units.

Ventilation and Perfusion

Fick's law of diffusion points to another way besides increasing surface area for increasing respiratory gas exchange. Animals can maximize the concentration gradients for the respiratory gases across the gas-exchange membranes in several ways. First, gill and lung membranes can be very thin so that the path length for diffusion (L) is small. Second, the environmental side of the exchange surfaces can be exposed to fresh respiratory medium (air or water) with the highest possible O_2 concentration and the lowest possible CO_2 concentration. Third, the opposite conditions—the lowest possible O_2 concentration and

highest possible CO_2 concentration—can be maintained on the internal sides of the exchange surfaces. Mechanisms that move substances over the gas-exchange surfaces are important for maximizing concentration gradients.

External gills are exposed to fresh respiratory medium as they wave around in the environment. Gas-exchange surfaces that are enclosed in body cavities, however, must be **ventilated**; that is, the animal must move fresh respiratory medium over internal gills or lungs. Breathing consists of the movements that ventilate gills or lungs. **Perfusion** is the movement of blood across the internal side of the gas-exchange membranes. Blood carries O_2 away as it diffuses across from the environmental side, and it brings CO_2 to the exchange surfaces so that it can diffuse in the opposite direction.

An animal's **gas-exchange system** is made up of its gas-exchange surfaces and the mechanisms it uses to ventilate and perfuse those surfaces. The following sections describe four gas-exchange systems. First we will look at the unique gas-exchange system of insects. Then we will describe two remarkably efficient systems: fish gills and bird lungs. Finally, we will discuss mammalian lungs, which in comparison to fish gills and bird lungs are a relatively inefficient gas-exchange system.

Insect Respiration

Respiratory gases diffuse through air most of the way to and from every cell of an insect's body. This diffusion is achieved through a system of air tubes, or **tracheae**, that open to the outside environment through holes called **spiracles** in the sides of the

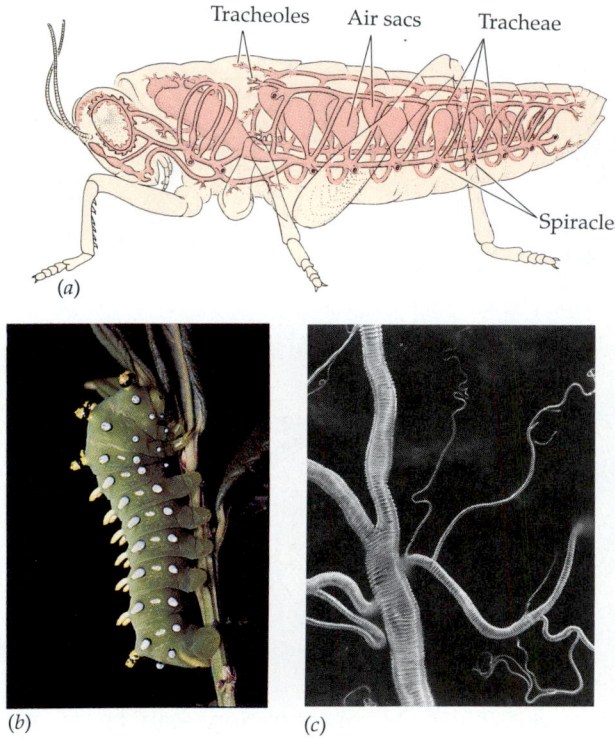

(a)

(b)   (c)

41.5   The Tracheal Gas-Exchange System of Insects
(a) The tracheal system extends throughout the body and opens to the exterior through spiracles. (b) The spiracles of the larva of a silk moth look like golden eyes down the side of the animal. (c) A scanning electron micrograph of part of the tracheal system shows tracheoles and air capillaries.

is enough to meet the O_2 demand of the animal while it is under water.

Fish Gills

The internal gills of fishes are marvelously adapted for gas exchange. They offer a large surface area for gas exchange between blood and water. They are supported by five or six bony **gill arches** on either side of the fish between the mouth cavity and the protective **opercular flaps** (Figure 41.6a). Water flows into the fish's mouth and out from under the opercular flaps. Each gill arch is lined with hundreds of leaf-shaped gill filaments arranged in two columns. These columns of gill filaments point toward the opercular opening, which is the direction of water flow (Figure 41.6b). The tips of the gill filaments of adjacent arches interlock. The upper and lower flat surfaces of each gill filament have rows of evenly spaced folds, or **lamellae**, which greatly increase the gill surface area. The surface area of the lamellae is the site of gas exchange. The interlocking network of gill filaments and lamellae directs the flow so that practically all water that passes across the gills comes into close contact with the gas-exchange surfaces.

The flow of blood perfusing the inner surfaces of the lamellae is unidirectional because of the arrangement of the **afferent blood vessels**, which bring blood to the gills, and the **efferent blood vessels** which take blood away from the gills (Figure 41.6c). Blood flow through the lamellae is in the opposite direction to the water flow over the lamellae. Such **countercurrent flow** makes gas exchange much more efficient than parallel flow (Box 41.A). Countercurrent exchange is an important principle in a number of different physiological systems.

The very delicate structure of the lamellae minimizes the path length for diffusion of gases between blood and water. Blood travels in blood vessels through the gill arches and the gill filaments, but in the lamellae the blood flows between the two surfaces of the lamellae as a sheet not much more than one red blood cell thick. The surfaces of the lamellae consist of highly flattened epithelial cells with almost no cytoplasm, so the water and the red blood cells are separated by little more than 1 or 2 μm.

Besides a large surface area and a short diffusion path length, what more can be done to maximize the

abdomen. The tracheae branch into even finer tubes, or tracheoles, until they end in tiny **air capillaries** (Figure 41.5). In the insect's flight muscle and other highly active tissue, no mitochondrion is more than a few micrometers away from an air capillary. Because the diffusion rate of oxygen is about 300,000 times higher in air than in water, air capillaries enable insects to supply oxygen to their cells at high rates. Many insects metabolize at extremely high rates, but this relatively simple gas-exchange system is well able to provide them with the oxygen they need. The rate of diffusion in insect tracheae and air capillaries is limited, however, by the small diameter and by the length of these dead-end airways, so insects are relatively small animals.

Some species of bugs that dive and stay under water for long periods make use of an interesting variation on diffusion. These bugs carry with them a bubble of air. A small bubble may not seem like a very large reservoir of oxygen, yet these bugs can stay under water almost indefinitely with their small air tanks. The secret has to do with the partial pressure of O_2 in the bubble. When the bug dives, the air bubble contains about 80 percent nitrogen and 20 percent O_2. As the insect consumes the O_2 in its bubble, the bubble shrinks, but its nitrogen concentration increases, and its O_2 concentration decreases. When the partial pressure of O_2 in the bubble falls below the partial pressure of O_2 in the surrounding water, O_2 diffuses into the bubble. For many of these small bugs, the rate of O_2 diffusion into the bubble

41.6 Internal Gills in Fish Enable Countercurrent Exchange

(a) Opercular flaps protect and help ventilate the gills. (b) Each gill arch supports two rows of gill filaments; each filament is folded into many thin, flat lamellae that are the gas-exchange surfaces. O_2 diffuses from water into the blood. (c) Blood flow through the lamellae is countercurrent to the flow of water over the lamellae.

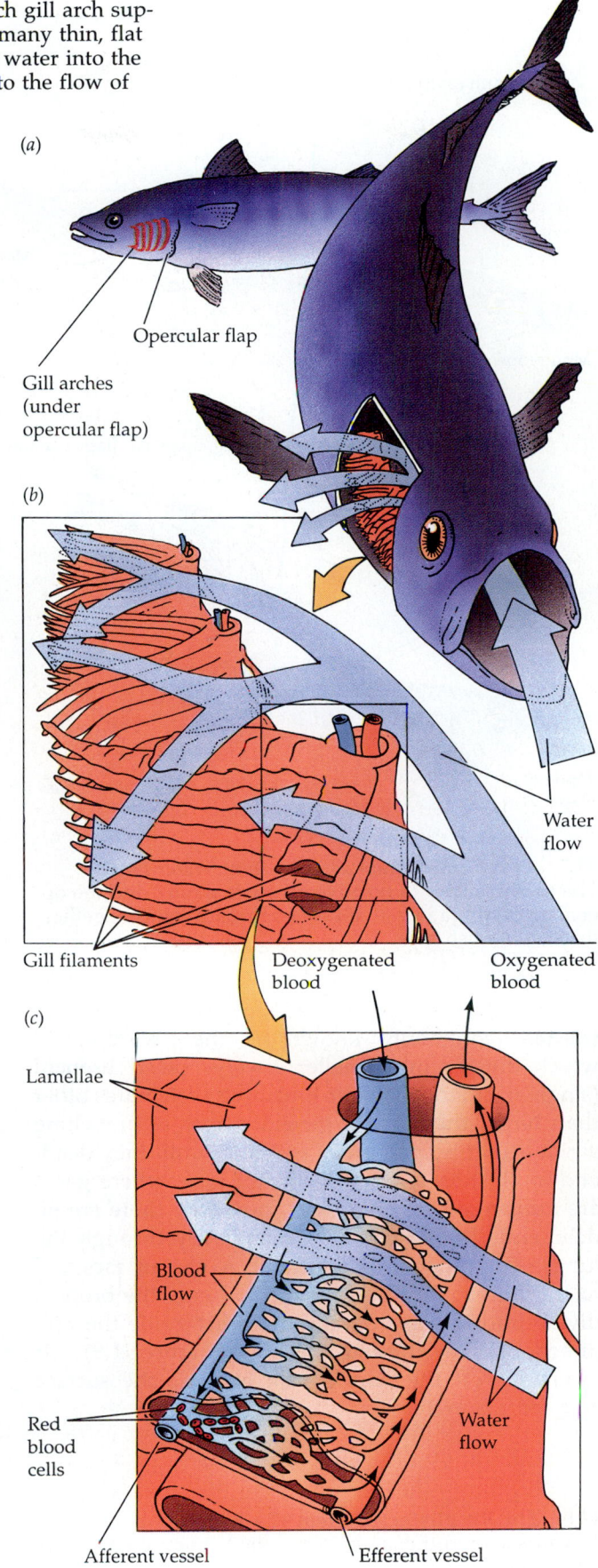

(a)

Opercular flap

Gill arches (under opercular flap)

(b)

Water flow

Gill filaments

Deoxygenated blood

Oxygenated blood

(c)

Lamellae

Blood flow

Red blood cells

Water flow

Afferent vessel

Efferent vessel

rate of diffusion? The concentration difference of O_2 between water and blood can be maximized. Fish accomplish this task by ventilating the external surface and perfusing the internal surface of the lamellae. A constant flow of water moving over the gills maximizes the O_2 concentration on the external surfaces. On the internal side, the circulation of blood minimizes the concentration of O_2 by sweeping the O_2 away as rapidly as it diffuses across.

Most fishes ventilate the external surfaces of their gills by means of a two-pump mechanism that maintains a unidirectional and constant flow of water over the gills. The closing and contracting of the mouth cavity acts as a **positive-pressure pump**, pushing water over the gills. The opening and closing of the opercular flaps acts as a **negative-pressure pump**, or suction pump, pulling water over the gills. Because these pumps are slightly out of phase, they maintain an almost continuous flow of water across the gills (Figure 41.7).

In summary, fish can extract an adequate supply of O_2 from meager environmental sources by maximizing the surface area for diffusion, minimizing the path length for diffusion, and maximizing oxygen extraction efficiency by constant, unidirectional, countercurrent flow of blood and water over the opposite sides of the gas-exchange surfaces.

Bird Lungs

Birds can sustain extremely high levels of activity for much longer periods than mammals can, and they can do so at very high altitudes, where mammals cannot even survive because the oxygen content of the air is so low. Yet the lungs of a bird are smaller than those of a mammal of a similar size. A unique feature of birds is that in addition to lungs they have **air sacs** at several locations in their bodies. The air sacs connect with one another, with the lungs, and with air spaces in some of the bones of the bird. Even though the air sacs as well as the lungs receive inhaled air, the air sacs are not gas-exchange surfaces. If a sample of air or pure oxygen is tied off in an air sac, its composition does not change rapidly, as it would if the O_2 were diffusing into the blood and CO_2 were diffusing into the air sac.

The anatomy of the bird lung is unique among air-breathing vertebrates. As in other air-breathing vertebrates, air enters and leaves the system through a

(a) Mouth open, mouth cavity expanding, opercular flaps closed, opercular cavity expanding

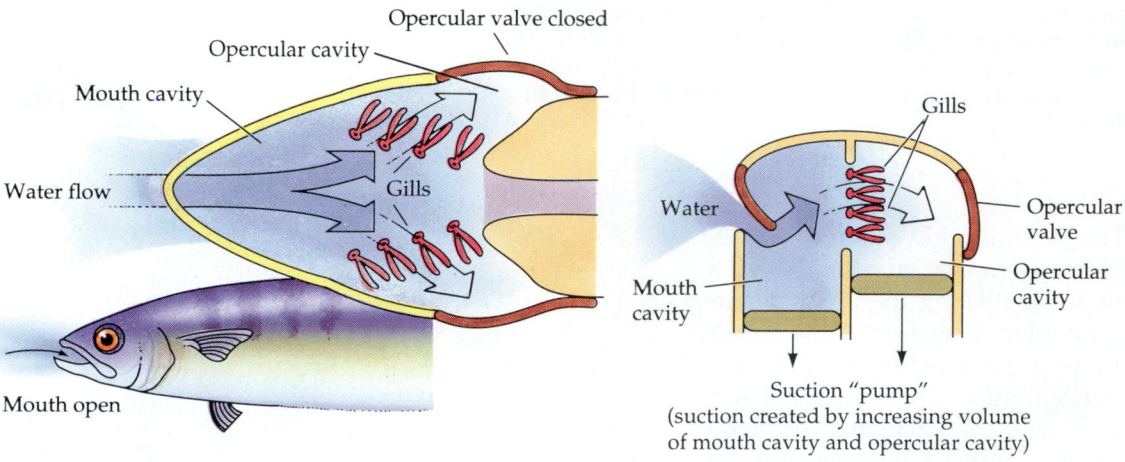
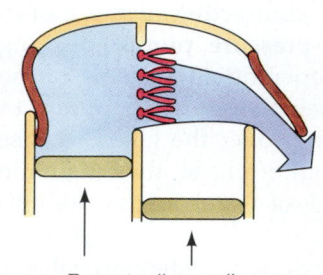

Suction "pump"
(suction created by increasing volume
of mouth cavity and opercular cavity)

(b) Mouth closed, mouth cavity contracting, opercular flaps opening, opercular cavity contracting

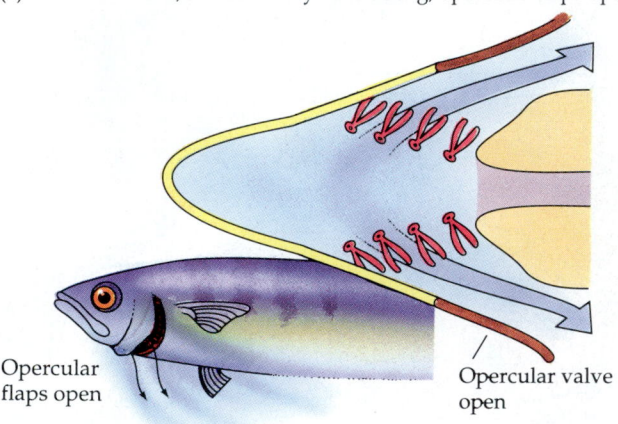

Pressure "pump"
(pressure created by decreasing volume
of mouth cavity and opercular cavity)

41.7 Two Pumps Maintain Constant Water Flow
The mouth cavity (a positive-pressure pump) and the opercular cavity (a negative-pressure pump) work together to ventilate fish gills.

trachea (commonly known as the "windpipe"), which divides into smaller airways called **bronchi** (singular, *bronchus*). In air-breathing vertebrates other than birds, the bronchi generate trees of branching airways that become finer and finer until they dead-end in clusters of microscopic air sacs, where gases are exchanged. In bird lungs, however, there are no dead-ends, so air can flow completely through the lungs. The bronchi distribute air to the air sacs and to the lungs (Figure 41.8). In the lungs the bronchi divide into tubelike **parabronchi** that guide the unidirectional flow through the lungs (Figure 41.9). Air capillaries off the parabronchi increase the surface area for gas exchange.

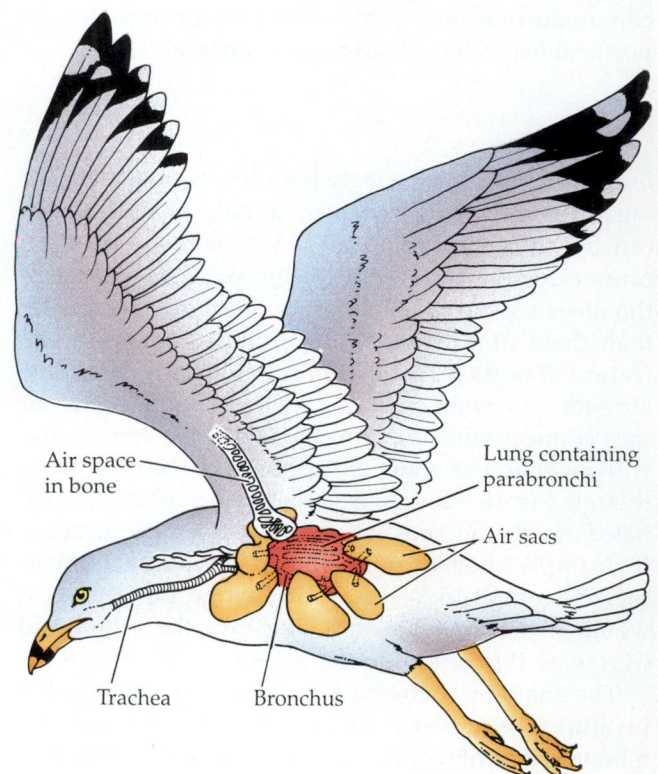

41.8 The Respiratory System of a Bird
The air sacs and the air spaces in the bones are unique to bird anatomy.

BOX 41.A

Countercurrent Exchangers

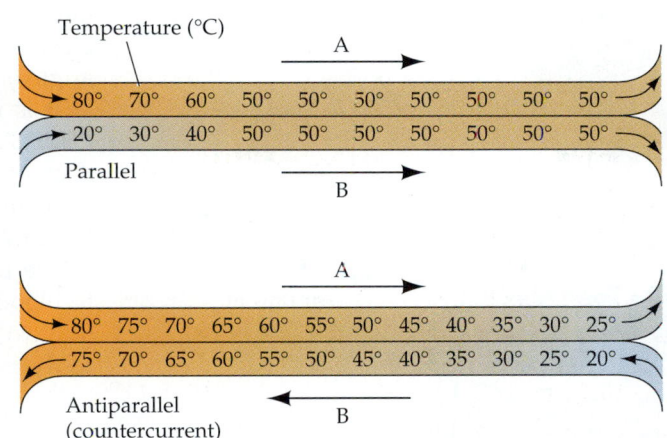

Temperature (°C)

Consider two pipes, side by side, in which parallel streams of water flow in the same direction. In pipe A, the water that enters has a temperature of 80°C; in B, the water that enters has a temperature of 20°C. If we assume that heat may be exchanged between the two pipes but may not be lost to the environment, then what happens is clear: Heat is transferred from A to B until the temperatures in the two pipes are identical—approximately 50°C in this example. The second law of thermodynamics (see Chapter 6) tells us that heat cannot flow spontaneously from a cooler to a warmer system; therefore, once both pipes reach 50°C, no further heat can be transferred from A to B.

Suppose now that we make a minor change; we have the water flow in opposite directions (antiparallel) in pipes A and B. As before, heat flows from A to B, but in this case the transfer of heat is much more complete—instead of about half of the

heat being transferred, almost all of it passes from A to B. As water flows through pipe B, it gets hotter and hotter, but it is always cooler than the water in pipe A. As water approaches the end of pipe A, it is relatively cool, but it is still warmer than that in pipe B, so it continues to give up heat. Thus heat transfer occurs along the entire region of overlap of the two pipes. The only difference between these two examples is that in the first case the flow is parallel and in the second case antiparallel. The antiparallel system is usually called a **countercurrent exchanger**.

Countercurrent systems are common in the animal kingdom. In fish gills the efficiency of the oxygenation of blood is maximized by having

blood flow in the direction opposite to that of the oxygen-bearing water (see Figure 41.6c). When the water leaves the gill, it has lost much of its oxygen, and the blood has become maximally oxygenated. If blood flowed in the same direction as the water, oxygen exchange would be much less complete. In Chapter 44 we will see the operation of a countercurrent exchange system in the vertebrate kidney—in a structure called the loop of Henle—allowing the formation of a steep salt concentration gradient within the kidney. In Chapter 35 we encountered heat exchangers in "hot fish" (see Figure 35.17b) and in the extremities of desert animals (see Figure 35.20a).

Another unusual feature of bird lungs is that in comparison to mammalian lungs they expand and contract relatively little during a breathing cycle. To make things even more puzzling, the bird lungs contract during inhalation and expand during exhalation!

The puzzle of how birds breathe was solved when researchers used very small gas sensors placed at different locations in the air sacs and airways to follow the path of air flow. They discovered that the air sacs can be divided into an anterior group and a posterior group, and that these two groups act as bellows to maintain a continuous, unidirectional flow of air through the lungs (Figure 41.10). When the bird inhales, the fresh air coming in through the trachea goes primarily to the posterior air sacs and into the portion of the lungs closest to these air sacs.

Simultaneously, air that was in the lungs flows into the anterior air sacs. When the bird exhales, the air in the posterior air sacs flows through the lungs to the anterior air sacs and continues out through the trachea. Thus the flow of air through the bird respiratory system during a breathing cycle is in and out through the trachea but unidirectional through the posterior air sacs, through the parabronchi of the lungs, and through the anterior air sacs. The advantages of this unique gas-exchange system are similar to those of fish gills. Because air from the outside flows unidirectionally and practically continuously over the gas-exchange surfaces, the concentration of O_2 on the environmental side of those surfaces is maximized. Furthermore, the unidirectional flow of air through the system makes possible a pattern of blood flow to minimize the O_2 concentration on the internal side of the exchange surfaces. However, in birds, the flow appears to be crosscurrent (at right angles) rather than countercurrent to the airflow.

It is now clear how birds can fly over Mount Everest. A bird supplies its gas-exchange surfaces with a continuous flow of fresh air that has an oxygen concentration close to that of the ambient air. Even though the P_{O_2} of the ambient air is only slightly above the P_{O_2} of the blood, diffusion of O_2 from air to blood can take place.

Next we will see why mammals would not be able to fly over Mount Everest—even if they could fly!

Breathing in Mammals

Vertebrate lungs have their origins in outpocketings of the digestive tract (Figure 41.11). At the beginning of their evolution, lungs were dead-end sacs, and

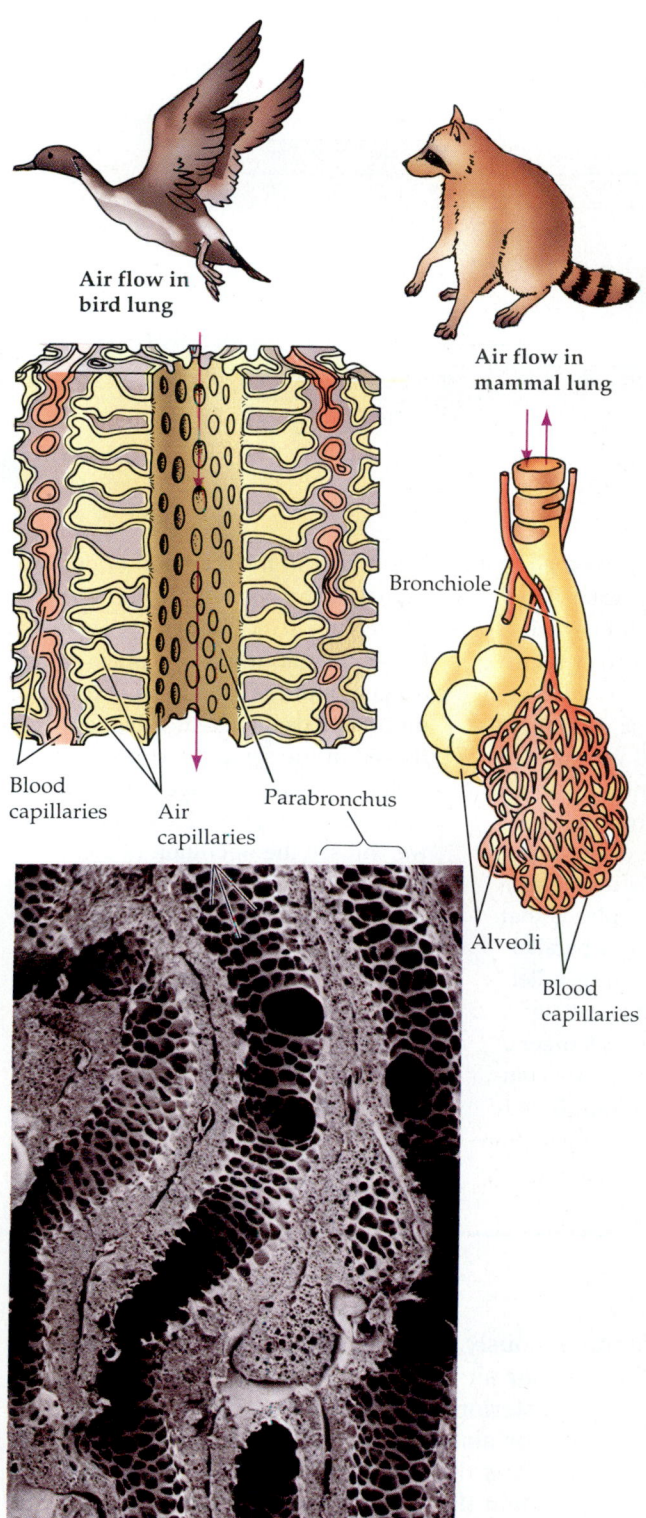

41.9   Air Flow Through Bird Lungs Is Constant and Unidirectional
The gas-exchange surfaces of mammals are alveoli, which are blind sacs, so air flow must be tidal. The gas-exchange surfaces of birds are air capillaries branching off the parabronchi that run through the lungs; these structures are shown in the scanning electron micrograph.

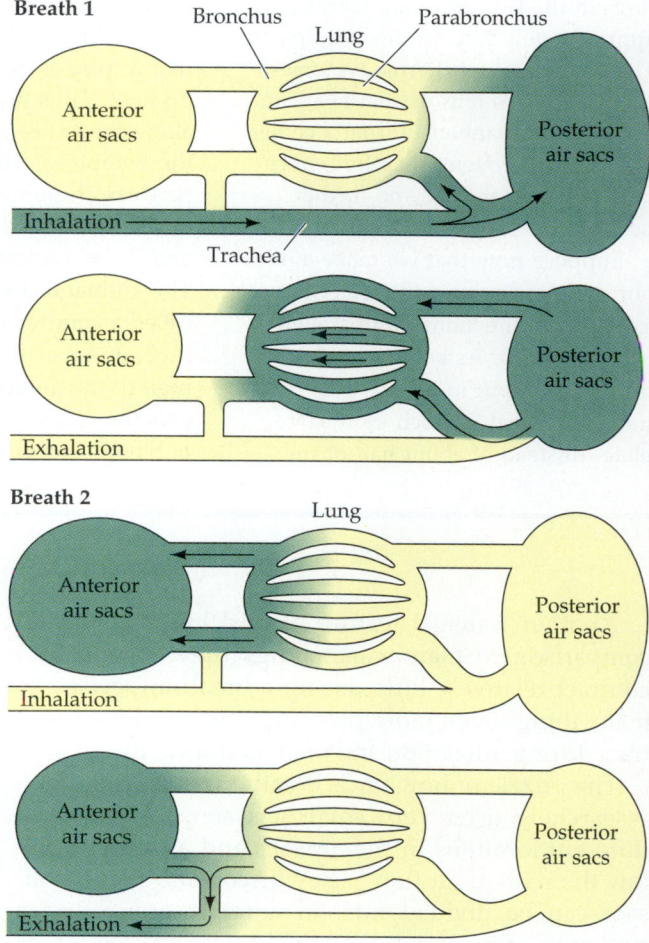

41.10   The Path of Air Flow through Bird Lungs
The fresh air a bird takes in with one breath (green) will travel through the lungs in one direction, from the posterior to the anterior air sacs. Two cycles of inhalation and exhalation are required for the air to travel the full length of the bird's respiratory tract.

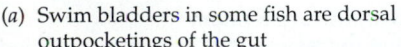

(*a*) Swim bladders in some fish are dorsal outpocketings of the gut

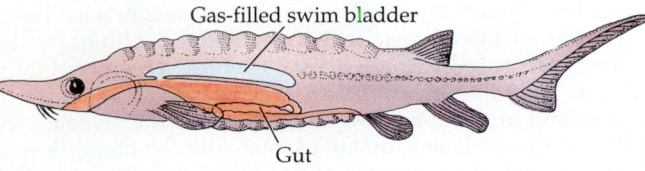

Gas-filled swim bladder

Gut

(*b*) Lungfish lungs are a ventral outpocketing of the gut

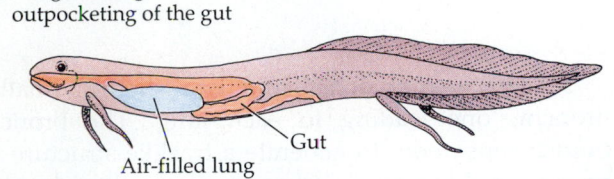

Gut

Air-filled lung

(*c*) Amphibian lungs are ventral outpocketings of the gut, though they lie dorsal to it

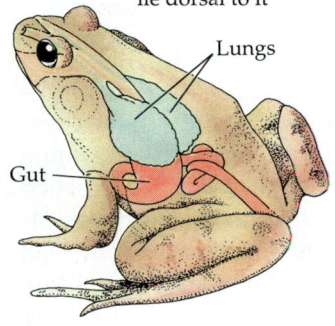

Lungs

Gut

41.11   Lung Evolution

(*a*) Outpocketings of the digestive tract evolved into swim bladders in some fish. In the lungfish (*b*) and terrestrial vertebrates such as amphibians (*c*), gut outpockets evolved into lungs.

they remain so today in all air-breathing vertebrates except birds. Because lungs are dead-end sacs, ventilation cannot be constant and unidirectional, but must be tidal: Air comes in and then flows out by the same route. A spyrometer shows how we use our lung capacity in breathing (Figure 41.12). When we are at rest, the amount of air that our normal breathing cycle moves per breath is called the **tidal volume** (about 500 ml for an average human adult). We can breathe much more deeply and inhale more air than our resting tidal volume, and the additional volume of air we can take in above normal tidal volume is our **inspiratory reserve volume**. Conversely, we can forcefully exhale more air than we normally do during a resting exhalation. This additional amount of air that can be forced out of the lungs is the **expiratory reserve volume**. Even after the most extreme exhalation possible, however, some

41.12   Measuring Lung Ventilation with a Spyrometer

Breathing from a closed reservoir of air and measuring the changes in the volume of that reservoir demonstrate the characteristics of mammalian breathing.

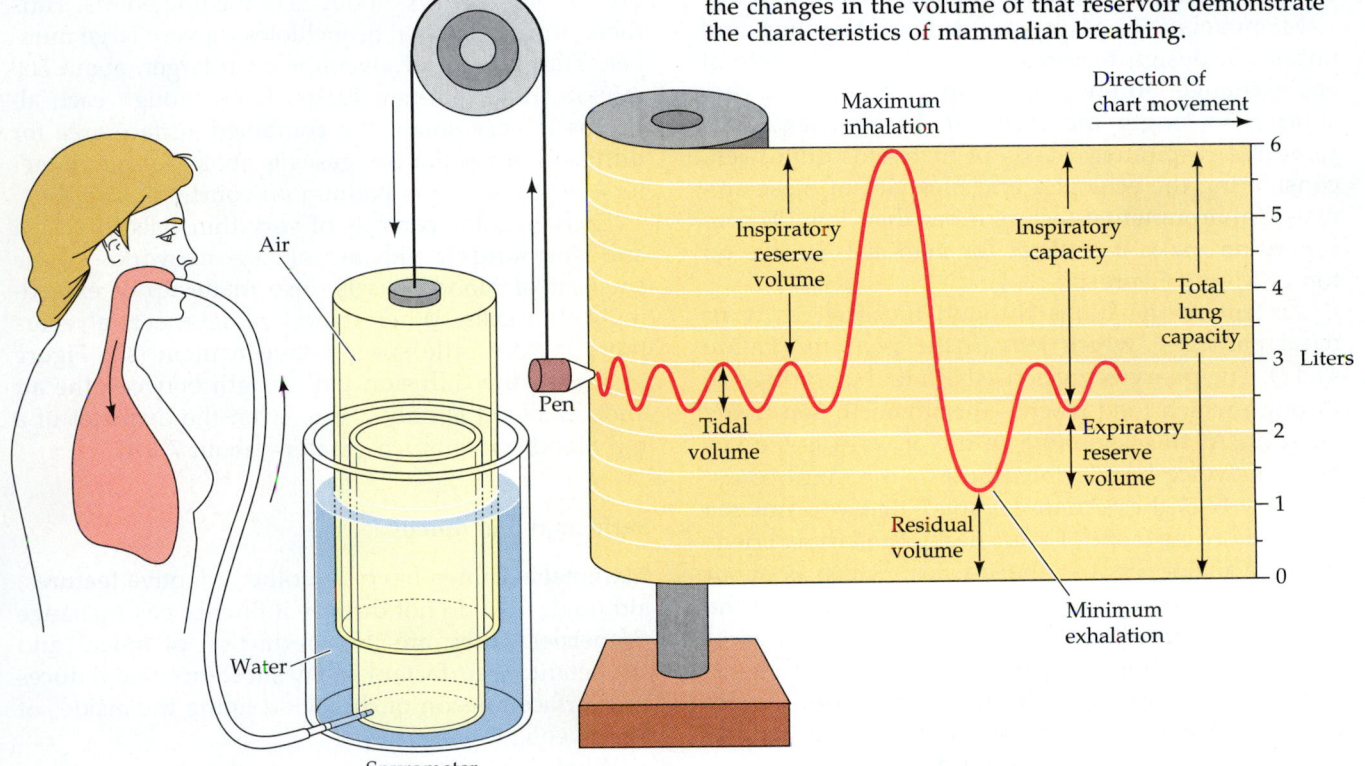

Air

Pen

Water

Spyrometer

Maximum inhalation

Direction of chart movement

Inspiratory reserve volume

Inspiratory capacity

Total lung capacity

Tidal volume

Expiratory reserve volume

Residual volume

Minimum exhalation

6
5
4
3   Liters
2
1
0

air remains in the lungs. The lungs and airways cannot be collapsed completely; they always contain a **residual volume**. The **total lung capacity** is the sum of the residual volume, expiratory reserve volume, tidal volume, and inspiratory reserve volume.

Tidal breathing severely limits the concentration difference driving the diffusion of O_2 from air into the blood. Fresh air is not moving into the lungs during half of the respiratory cycle; therefore the average O_2 concentration of air in the lungs is less than it is in the air outside the lungs. The incoming air also mixes with the stale air that was not expelled by the previous exhalation. The lung volume that is not ventilated with fresh air is **dead space**. This dead space consists of the residual volume and, depending on the depth of breathing, some or all of the expiratory reserve volume. The scale in Figure 41.12 tells us that a tidal volume of 500 ml of fresh air mixes with up to 2,000 ml of stale moist air before reaching the gas-exchange surfaces. When the P_{O_2} in the ambient air is 150 mm Hg, the P_{O_2} of the air that reaches the gas-exchange surfaces is only about 100 mm Hg. By contrast, the P_{O_2} in the water bathing the lamellae of the fish gills or in the air flowing through the air capillaries of the bird lung is the same as the P_{O_2} in the outside water or air.

As well as reducing the concentration difference, tidal breathing reduces the efficiency of gas exchange in another way. It does not allow countercurrent gas exchange between air and blood. Because the air enters and leaves the gas-exchange structures by the same route, there is no anatomical way that blood can flow parallel to and countercurrent to the air flow.

Mammalian lungs possess some interesting and important design features that maximize the rate of gas exchange: an enormous surface area and a very short path length for diffusion. Mammalian lungs serve the respiratory needs of mammals quite well, considering the ecologies and lifestyles of these animals. Environmental factors other than low O_2 concentration make it difficult for mammals to live on top of Mount Everest!

Air enters the lungs through the oral cavity or nasal passages, which join in the pharynx (Figure 41.13). The pharynx gives rise both to the esophagus, through which food reaches the stomach, and to the airways. At the beginning of the airways is the **larynx**, or "voice box," which houses the vocal cords. The larynx is the "Adam's apple" that you can see and feel on the front of your neck. The larynx opens into the major airway, the trachea, which is about the diameter of a garden hose. The thin walls of the trachea are prevented from collapsing by rings of cartilage that support them as air pressure changes during the breathing cycle. If you run your fingers down the front of your neck just below your larynx, you can feel a couple of these rings of cartilage.

41.13 The Human Respiratory System
The lungs lie within the thoracic cavity, which is bounded by the ribs and the diaphragm. Pleural membranes line the part of the thoracic cavity containing the lungs, so the lungs are actually in the pleural cavities. Air enters the lungs from the oral cavity or nasal passages via the trachea and bronchi, and eventually reaches the alveoli. There, the air is in intimate contact with the blood flowing through the networks of fine blood vessels surrounding the alveoli.

The trachea branches into two slightly smaller bronchi, one leading to each lung. The bronchi branch repeatedly to generate a treelike structure of progressively smaller airways going to all regions of the lungs. Structurally, each of these bronchi is a smaller version of the trachea; they all have supporting cartilage rings. As the branching of the bronchial tree continues to produce still smaller airways, the cartilage supports eventually disappear, marking the transition to **bronchioles**. The branching continues until the bronchioles are less than the diameter of a pencil lead, at which point the tiny, thin-walled air sacs called **alveoli** begin. Alveoli resemble clusters of grapes on a system of stems (see Figure 41.13). The "stems" are the bronchioles, which have about six more branch points. Finally, terminal bronchioles end in alveoli. The alveoli are the sites of gas exchange. Because the airways only conduct the air to and from the alveoli, their volume is physiological dead space. If you trace an airway from the primary bronchus leaving the trachea down to the very last terminal bronchiole, you pass about 23 branching points. Thus there are 2^{23} terminal bronchioles—a very large number. The number of alveoli is even larger, about 300 million in the human lungs. Even though each alveolus is very small, the combined surface area for diffusion of respiratory gases is about 70 square meters, or the size of a badminton court.

Each alveolus consists of very thin cells. Between and surrounding the alveoli are networks of the smallest of blood vessels, also made up of exceedingly thin cells. Where blood vessel meets alveolus there is very little space between them (see Figure 41.13), so the diffusion path length between the air and the blood is only 2 μm. Even the diameter of a red blood cell is much greater—about 7 μm.

Surfactant and Mucus

Mammalian lungs have two other adaptive features, although they do not directly influence gas-exchange properties. They are the production of mucus and surfactant. A surfactant is any substance that reduces the surface tension of the liquid lining the insides of the alveoli.

What is surface tension and why do we need to

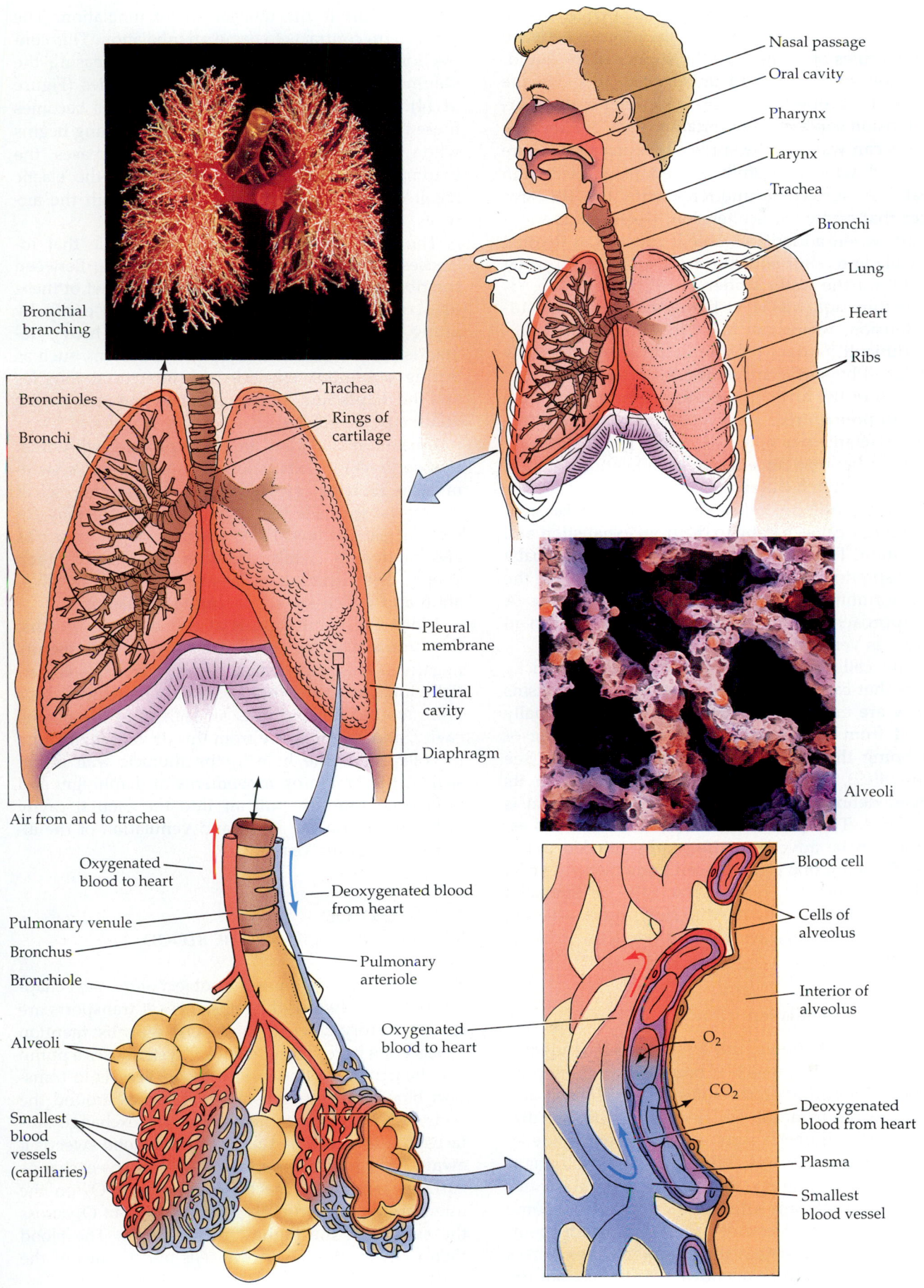

Bronchial branching

Nasal passage

Oral cavity

Pharynx

Larynx

Trachea

Bronchi

Lung

Heart

Ribs

Bronchioles

Bronchi

Trachea

Rings of cartilage

Pleural membrane

Pleural cavity

Diaphragm

Alveoli

Air from and to trachea

Oxygenated blood to heart

Pulmonary venule

Bronchus

Bronchiole

Alveoli

Smallest blood vessels (capillaries)

Deoxygenated blood from heart

Pulmonary arteriole

Oxygenated blood to heart

Blood cell

Cells of alveolus

Interior of alveolus

O_2

CO_2

Deoxygenated blood from heart

Plasma

Smallest blood vessel

consider it as we study lung function? Surface tension arises from the attractive (cohesive) forces between the molecules of a liquid. At the surface of the liquid, these cohesive forces are unbalanced and give the surface the properties of an elastic membrane. Surface tension explains why certain insects called water striders can walk on the surface of water (see Figure 2.18) and why a carefully placed razor blade can "float." A surfactant interferes with the cohesive forces that create surface tension. Detergent is a surfactant; when added to water it causes the floating razor blade to sink and can make walking on water difficult for the water strider.

The thin, aqueous layer lining the alveoli has surface tension, which can make inflation of the lungs very difficult. Surface tension in the alveoli normally is reduced by surfactant molecules produced by certain cells in the alveoli. If a baby is born more than a month prematurely, its alveoli may not be producing surfactant. Such a baby has great difficulty breathing because an enormous inhalation effort is required to stretch the alveoli against the surface tension. This condition, called **respiratory distress syndrome**, may cause a baby to die from exhaustion and suffocation. The common treatment is to put the baby on a respirator to assist its breathing and to give the baby hormones to speed its lung development. A new approach, applying surfactant to the lungs in an aerosol, is very promising.

Many cells lining the airways produce a sticky mucus that captures bits of dirt and microorganisms as they are inhaled. The mucus must be continually cleared from the airways, however; the beating of cilia lining the airways accomplishes this task (see Figure 40.2). The cilia move the mucus with its trapped debris up toward the pharynx, where it is swallowed. This phenomenon, called the mucus escalator, can be adversely affected by inhaled pollutants. Smoking one cigarette can immobilize the cilia of the airways for hours. Hacking, or smoker's cough, results from the need to clear the obstructing mucus from the airways when the mucus escalator is out of order.

Mechanics of Ventilation

As Figure 41.13 shows, the lungs are suspended in the **thoracic cavity**, which is bounded on the top by the shoulder girdle, on the sides by the rib cage, and on the bottom by a domed sheet of muscle, the **diaphragm**. The thoracic cavity is lined on the inside by the **pleural membranes**, which divide it into right and left **pleural cavities**. Because the pleural cavities are closed spaces, any effort to increase their volume creates negative pressure—suction—inside them. Negative pressure within the pleural cavities causes the lungs to expand as air flows into them from the

outside. This is the mechanism of inhalation. The diaphragm contracts to begin an inhalation. This contraction pulls the diaphragm down, increasing the volume of the thoracic and pleural cavities (Figure 41.14). As pressure in the pleural cavities becomes more negative, air enters the lungs. Exhaling begins when the contraction of the diaphragm ceases, the diaphragm relaxes and moves up, and the elastic recoil of the lungs pushes air out through the airways.

The diaphragm is not the only muscle that increases the volume of the thoracic cavity. Between the ribs are **intercostal muscles**, and one set of these intercostal muscles expands the thoracic cavity by lifting the ribs up and outward. When heavy demands are placed on the respiratory system, such as during strenuous exercise, the intercostal muscles and the diaphragm contract together and increase the volume of air inhaled.

Inhalation is always an active process, with muscles contracting; exhalation is usually passive, with muscles relaxing. However, there is another set of intercostal muscles that can be called into play for forceful exhalations. Place your hands on your ribs and abdomen while breathing shallowly, then deeply. Feel which muscles are active during inhalation and which are active during exhalation.

When the diaphragm is at rest between breaths, the pressure in the pleural cavities is still slightly negative. This slight suction keeps the alveoli partially inflated. If the thoracic wall is punctured, by a knife wound for example, air leaks into the pleural cavity, and the pressure from this air causes the lung to collapse. If the hole in the thoracic wall is not sealed, the breathing movements of diaphragm and intercostal muscles pull air into the pleural cavity rather than into the lung, and ventilation of the alveoli in that lung ceases.

TRANSPORT OF RESPIRATORY GASES BY THE BLOOD

The circulatory system is the subject of the next chapter, but since two of the substances it transports are the respiratory gases (O_2 and CO_2), we must mention aspects of it here. The circulatory system uses a pump (the heart) and a network of blood vessels to transport blood and the substances it carries around the body. As O_2 diffuses across the gas-exchange surfaces into the vessels, the circulating blood sweeps it away. This internal perfusion of the gas-exchange surfaces minimizes the concentration of O_2 on the internal side and promotes the diffusion of O_2 across the surface at the highest possible rate. The blood then delivers this O_2 to the cells and tissues of the body.

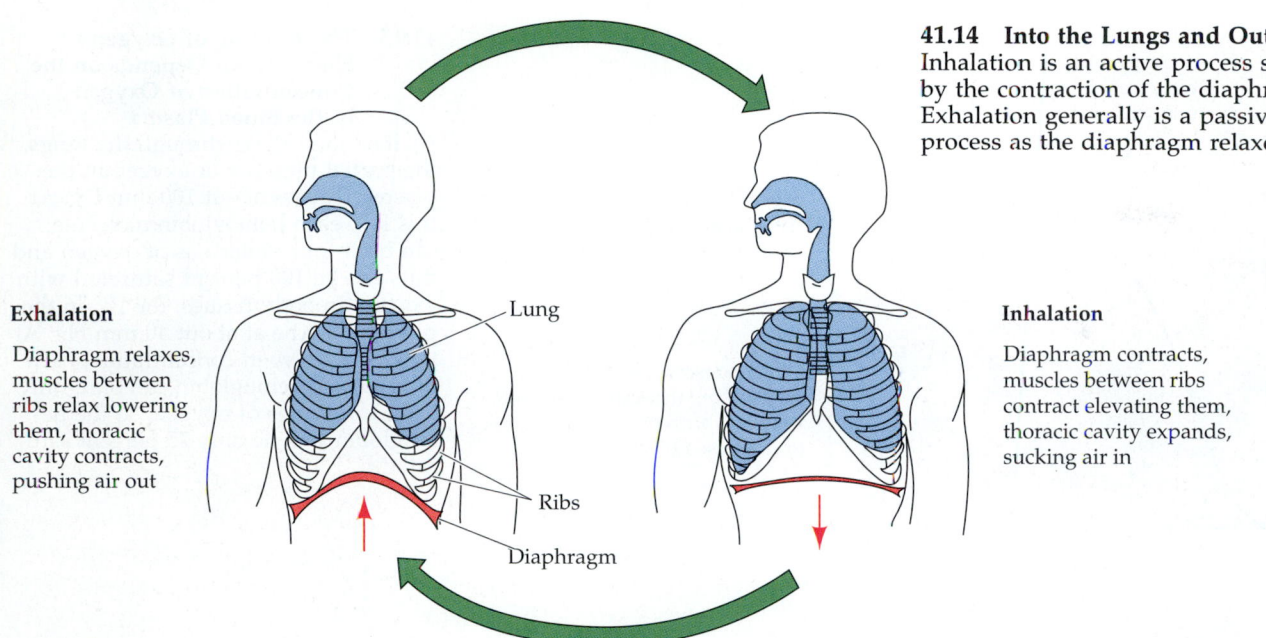

41.14 Into the Lungs and Out Again
Inhalation is an active process spurred by the contraction of the diaphragm. Exhalation generally is a passive process as the diaphragm relaxes.

Exhalation

Diaphragm relaxes, muscles between ribs relax lowering them, thoracic cavity contracts, pushing air out

Inhalation

Diaphragm contracts, muscles between ribs contract elevating them, thoracic cavity expands, sucking air in

Lung

Ribs

Diaphragm

The liquid part of the blood, the **blood plasma**, carries some O_2 in solution, but the ability of the blood to pick up and transport O_2 would be quite limited if plasma were the only means available. Blood plasma carries about 0.3 ml of oxygen per 100 ml. To support the O_2 needs of a person at *rest*, the heart would have to pump about 5,000 liters of blood plasma *per hour* (enough to fill the gas tanks of about 100 cars). Fortunately, the blood also contains **red blood cells** (see page 293), which are red because they are loaded with the oxygen-binding pigment hemoglobin. Hemoglobin increases the capacity of the blood to transport oxygen by about 60-fold.

Hemoglobin

Red blood cells contain enormous numbers of hemoglobin molecules. Hemoglobin is a protein consisting of four polypeptide subunits (see Figure 3.20). Each of these polypeptides surrounds a heme group—an iron-containing ring structure that can reversibly bind a molecule of O_2. As O_2 diffuses into the red blood cells, it binds to hemoglobin. Once O_2 is bound, it cannot diffuse back across the cell membrane. By mopping up O_2 molecules as they enter the red blood cells, hemoglobin maximizes the concentration difference driving the diffusion of O_2 into the red blood cells. In addition, hemoglobin enables the red blood cells to carry a large amount of O_2 for use by the tissues of the body.

The ability of hemoglobin to pick up or release O_2 depends on the concentration or partial pressure of O_2 in its environment. When the P_{O_2} of the blood plasma is high, as it usually is in the lung capillaries, each molecule of hemoglobin can carry its maximum load of four molecules of O_2. As the blood circulates through capillary beds elsewhere in the body, it encounters lower P_{O_2}'s. At these lower P_{O_2}'s the hemoglobin releases some of the O_2 it is carrying. The lower the P_{O_2} of the environment, the more O_2 is released from hemoglobin to diffuse out of the red blood cells and into the tissues. The relation between P_{O_2} and the amount of O_2 bound to hemoglobin is not linear, however. This relationship is described by a sigmoid (S-shaped) curve (Figure 41.15), and it is an important property of hemoglobin.

Remember that the hemoglobin molecule consists of four subunits, each of which can bind one molecule of O_2. At very low P_{O_2}'s, one subunit will bind an O_2 molecule. As a result, the shape of this subunit changes, causing an alteration in the quarternary structure of the whole hemoglobin molecule (see Chapter 3). This structural change makes it easier for the other subunits to bind a molecule of O_2; that is, their O_2 affinity is increased. Only small increases in the P_{O_2} cause most hemoglobin molecules to pick up a second and then a third molecule of O_2. The influence of the binding of O_2 by one subunit on the binding affinity of the other subunits is called **positive cooperativity**, because binding of the first molecule makes binding of the second easier, and so forth. After the binding of the third molecule of O_2 a large increase in P_{O_2} is required for the fourth subunits of all the hemoglobin molecules to be loaded because there are fewer and fewer available binding sites as more and more of the hemoglobin molecules become fully saturated.

The significance of the interactions of the hemoglobin subunits that result in the sigmoid shape of the hemoglobin–oxygen binding curve in Figure

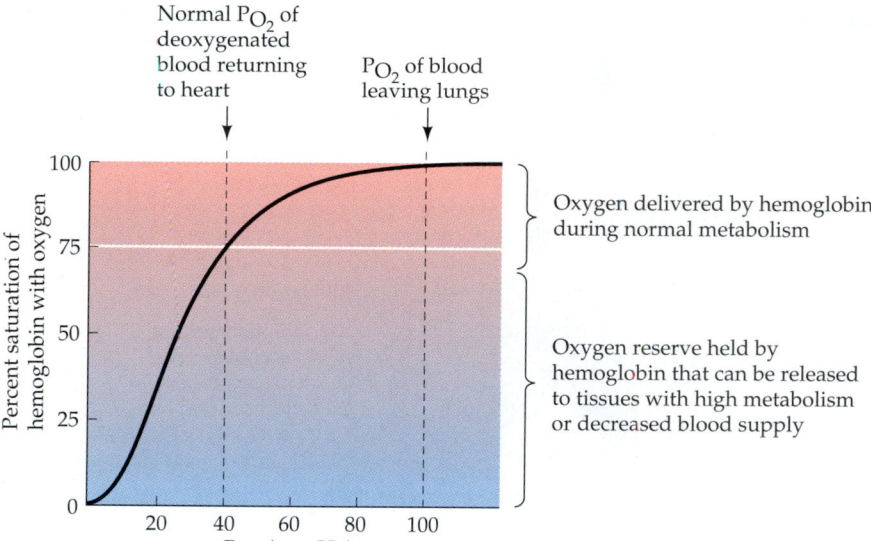

41.15   The Binding of Oxygen to Hemoglobin Depends on the Concentration of Oxygen in the Blood Plasma
As the blood flows through the lungs, the partial pressure of oxygen in the plasma reaches about 100 mm Hg. At this P_{O_2}, each hemoglobin molecule can bind four molecules of oxygen and thus can be 100 percent saturated with oxygen. In body tissues, the P_{O_2} in the plasma may be at about 40 mm Hg. At this lower oxygen concentration, each molecule of hemoglobin can bind only three molecules of oxygen, so the hemoglobin will be only 75 percent saturated.

41.15 is best appreciated by considering the dynamics of unloading the O_2 in the tissues. The P_{O_2} that normally exists in the alveoli of the lungs is about 100 mm Hg, and at this P_{O_2} the hemoglobin is 100 percent saturated (each molecule of hemoglobin carrying four molecules of O_2). The P_{O_2} in mixed venous blood is usually about 40 mm Hg. Thus the hemoglobin returning to the heart from the body is still about 75 percent saturated. That means that most hemoglobin molecules drop only one of their four O_2 molecules as they circulate through the body.

This system may seem very inefficient for delivery of oxygen to the tissues, but it is actually extremely adaptive. When a tissue becomes oxygen-starved and its local P_{O_2} falls below 40 mm Hg, the hemoglobin flowing through that tissue will drop much more of its oxygen load with only small additional decreases in P_{O_2}. The steep portion of the sigmoid hemoglobin–oxygen binding curve comes into play when tissue P_{O_2} falls below the normal 40 mm Hg. Thus the cooperative oxygen-binding property of hemoglobin is very effective in making O_2 available to the tissues precisely when and where it is most needed.

Myoglobin

Muscle cells have their own oxygen-binding molecule, **myoglobin**. Myoglobin consists of just one polypeptide chain associated with an iron-containing ring structure that can bind one molecule of oxygen (see Figure 3.19). Myoglobin has a higher affinity for O_2 than hemoglobin does (Figure 41.16), so it picks up and holds oxygen at P_{O_2}'s at which hemoglobin is releasing its bound O_2. Myoglobin provides a reserve of oxygen for the muscle cells for times when metabolic demands are high and blood flow is interrupted as contracting muscles constrict blood vessels.

When hemoglobin has no more O_2 to give up, and tissue P_{O_2} falls even lower, myoglobin releases its bound O_2. Diving mammals such as seals that can remain active under water for many minutes have high concentrations of myoglobin in their muscles. Muscles called upon for extended periods of work frequently have more myoglobin than muscles that are used for short, intermittent periods. This is one of the reasons for the difference in appearance of the "white" and "dark" meat of chickens and turkeys. These birds are not long-distance fliers, and their flight muscles (the white meat) have little myoglobin. Ducks and geese, however, come from distinguished lineages of long-distance fliers. Their flight muscles have much myoglobin, as well as more mitochondria and more blood vessels, and thus appear dark.

Regulation of Hemoglobin Function

The various factors that influence the oxygen-binding properties of hemoglobin also influence oxygen delivery to tissues. For example, there are variations in the chemical composition of the polypeptide chains that form the hemoglobin molecule. The normal hemoglobin of adult humans has two each of two kinds of polypeptide chains—two α chains and two β chains—and has the oxygen-binding characteristics shown in Figure 41.16. Before birth, the fetus has a different form of hemoglobin consisting of two α chains and two γ chains. The chemical composition of fetal hemoglobin enables fetal blood to pick up O_2 from maternal blood when both are at the same P_{O_2}. Fetal hemoglobin thus has an oxygen-binding curve that plots to the left of the adult curve. This difference between maternal and fetal hemoglobin facilitates the transfer of O_2 from the mother's blood to the blood of the fetus in the placenta.

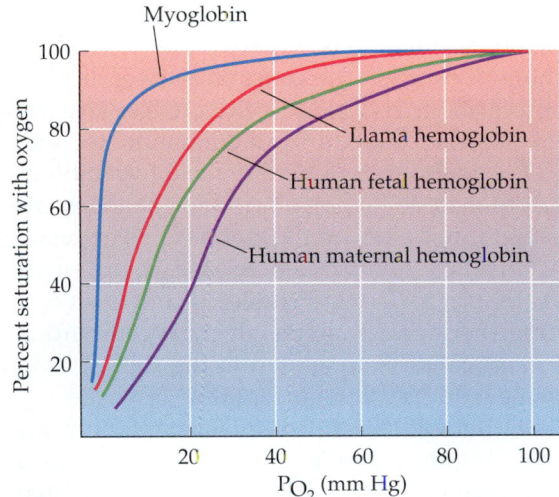

41.16 Oxygen-Binding Adaptations
Llamas are used as pack animals in the high Andes
Mountains because they are so well adapted to high altitudes, where the partial pressure of oxygen is low. Evolution has adapted the oxygen-binding properties of different hemoglobins and of myoglobin. Llama hemoglobin
has such a high affinity for oxygen that it is 100 percent
saturated even at the low P_{O_2}'s found in the Andes. The
higher oxygen affinity of fetal hemoglobin in comparison
to maternal hemoglobin enables fetal blood to pick up
oxygen from maternal blood when both are at the same
P_{O_2}. Myoglobin can serve as an oxygen reservoir because
it remains 100 percent saturated until the P_{O_2} falls so low
that hemoglobin has given up most of its oxygen.

Llamas and vicuñas are mammals native to high
altitudes in the Andes mountains of South America
(see Figure 41.16). The hemoglobins of these animals,
like those of the human fetus, must pick up O_2 in an
environment with a low P_{O_2}. In the animal's natural
habitat, over 5,000 meters above sea level, the P_{O_2} is
below 85 mm Hg, and the P_{O_2} in their lungs is about
50 mm Hg. The hemoglobins of llamas and vicuñas
have oxygen-binding curves much to the left of the
curves of hemoglobins of most other mammals—in
other words, they can become saturated with O_2 at
lower P_{O_2}'s than those of other animals can.

The oxygen-binding properties of normal adult hemoglobin are influenced by physiological conditions.
The influence of pH on the function of hemoglobin
has been well studied and is known as the **Bohr
effect**. As the pH of the blood plasma falls, the oxygen-binding curve shifts to the right (Figure 41.17).
This shift means that the hemoglobin will release
more O_2 to the tissues. Where does hemoglobin encounter a decreased pH as it circulates through the
body? In tissues with very high metabolic rates the
pH is reduced by the release of acidic metabolites
such as lactic acid, fatty acids, and CO_2, which combines with water to form carbonic acid. Because of
the Bohr effect, hemoglobin releases more of its

bound oxygen in these tissues—another way that O_2
is supplied where and when it is most needed.

Diphosphoglyceric acid is a normal intermediate
metabolite that plays an important role in regulating
hemoglobin function. The mature mammalian red

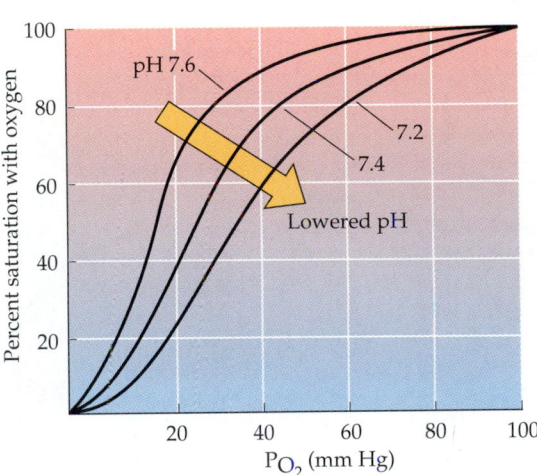

**41.17 The Oxygen-Binding Properties of Hemoglobin
Can Change**
Changes in pH affect the oxygen-binding capacity of hemoglobin. Lowering pH shifts the binding curve to the
right; more oxygen is then being released to the tissues.

blood cell is a simple cell. It is little more than a sac of hemoglobin, but it has a very high content of diphosphoglyceric acid. The concentration of diphosphoglyceric acid in red blood cells increases in response to exercise and during acclimation to high altitude. Diphosphoglyceric acid reversibly combines with deoxygenated hemoglobin and changes its shape so that it has a lower affinity for O_2. The result is that at any P_{O_2}, hemoglobin releases more of its bound O_2 than it otherwise would. In other words, diphosphoglyceric acid shifts the oxygen-bindng curve of mammalian hemoglobin to the right.

The llama and the human employ opposite adjustments of hemoglobin function as adaptations for life at high altitudes. The llama's hemoglobin has a left-shifted oxygen-binding curve, which means that it can become 100 percent saturated with O_2 at the low P_{O_2}'s at high altitude. As a consequence, the llama's tissues must operate at a lower P_{O_2}. By contrast, human hemoglobin acquires, through acclimation, a right-shifted oxygen-binding curve. Human hemoglobin never becomes fully saturated with O_2 at high altitude, but more of the O_2 carried by that hemoglobin is released to the tissues.

Transport of Carbon Dioxide from the Tissues

Delivering O_2 to the tissues is only half of the respiratory function of the blood. The blood also must take metabolic waste products away from the tissues. Because we are concerned with respiratory gases in this chapter, the metabolic waste product we will consider is carbon dioxide. CO_2 is highly soluble and readily diffuses through cell membranes, moving from its site of production in a cell into the blood,

where its concentration is lower. Very little CO_2 is transported by the blood in this dissolved form, however. Most CO_2 produced by tissues is transported to the lungs in the form of the **bicarbonate ion**, HCO_3^-. How and where CO_2 becomes HCO_3^-, is transported, and then is converted back to CO_2 is an interesting story.

When CO_2 dissolves in water, some of it slowly reacts with the water molecules to form carbonic acid (H_2CO_3), some of which then dissociates into a proton (H^+) and a bicarbonate ion (HCO_3^-). This sequence of events is expressed as follows:

$$CO_2 + H_2O \rightleftharpoons H_2CO_3 \rightleftharpoons H^+ + HCO_3^-$$

In the blood plasma, the reaction between CO_2 and H_2O goes too slowly to have much of an effect. It is different, however, in the red blood cells, where the enzyme **carbonic anhydrase** speeds up the conversion of CO_2 to H_2CO_3. The newly formed carbonic acid dissociates, and the resulting bicarbonate ion diffuses back out into the plasma (Figure 41.18). This action of carbonic anhydrase in the red blood cells creates a sink for CO_2, thus facilitating the diffusion of CO_2 from tissue cells to plasma to red blood cells. Most CO_2 is transported by the blood as bicarbonate ions in the plasma. Some CO_2 is also carried in chemical combination with deoxygenated hemoglobin as **carboxyhemoglobin**.

41.18   Carbon Dioxide Is Transported as Bicarbonate Ions
In tissues, CO_2 diffuses from cells into plasma and into the red blood cells. In the red blood cells, CO_2 is rapidly converted to bicarbonate ions because carbonic anhydrase is present. Bicarbonate ions leave red blood cells in exchange for chloride ions. In the lungs, these processes are reversed. Some CO_2 combines with hemoglobin (Hb).

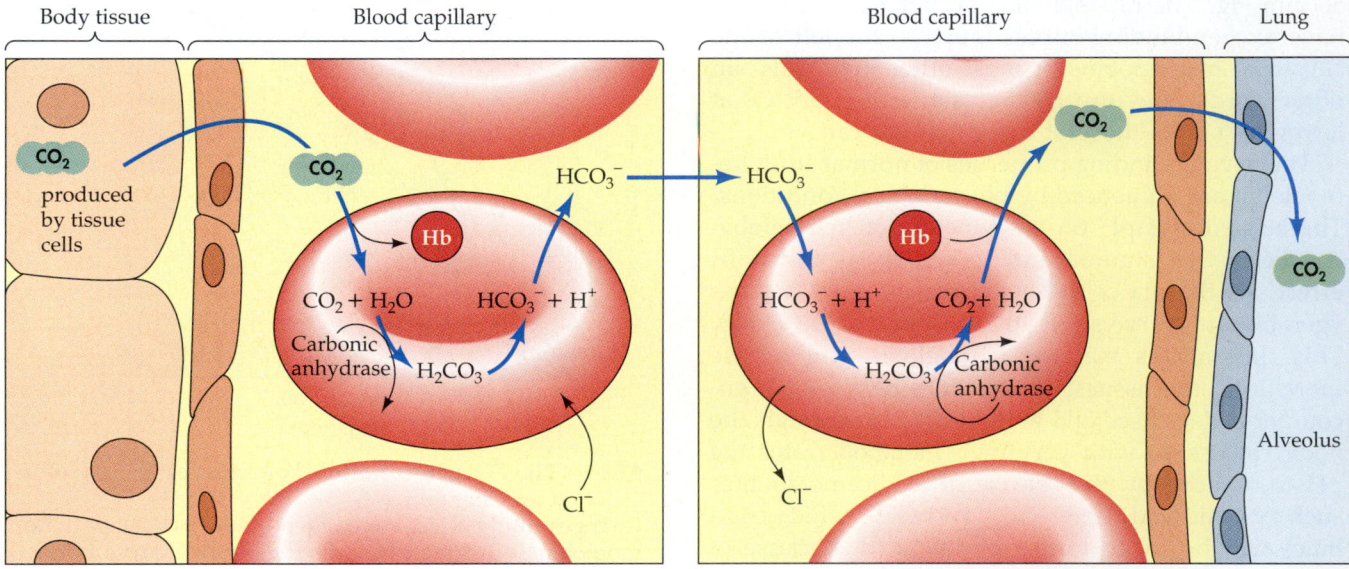

Diffusion of CO_2 into blood and conversion to HCO_3^-

Diffusion of CO_2 out of blood into lungs

In the lungs, the reactions involving CO_2 and bicarbonate are reversed. CO_2 diffuses from the blood plasma into the air in the alveoli and is exhaled. Breathing keeps CO_2 concentrations in the alveoli low, so CO_2 diffuses from red blood cells to the plasma. The loss of CO_2 from the red blood cells shifts the equilibrium between CO_2 and bicarbonate. As the HCO_3^- in the red blood cells is converted back to CO_2, more HCO_3^- moves into the red blood cells from the plasma. Remember that an enzyme like carbonic anhydrase only speeds up a reversible reaction; it does not determine its direction. Direction is determined by concentrations of reactants and products (see Chapter 6).

REGULATION OF VENTILATION

We must breathe every minute of our lives. We don't worry about our need to breathe or even think about it very often because breathing is an autonomic function of the nervous system. The breathing pattern easily adjusts itself around other activities such as speech and eating. Most impressively, our breathing rates change to match the metabolic demands of our bodies. Now we will learn how the regular respiratory cycle is generated and how it is controlled so that we get the oxygen we need and eliminate the carbon dioxide we produce as our levels of activity change.

The Ventilatory Rhythm

Breathing is an involuntary function. The complex, coordinated movements of the diaphragm and other muscles do not require conscious thought. Automatically, the central nervous system maintains a ventilatory rhythm and modifies its depth and frequency to meet the demands of the body for O_2 supply and CO_2 elimination. The ventilatory rhythm ceases if the spinal cord is severed in the neck region, showing that the rhythm is generated in the brain. If the brain stem is cut just above the medulla, the segment of the brain stem just above the spinal cord, a crude ventilatory rhythm remains (Figure 41.19).

Groups of neurons within the medulla increase their firing rates just before an inhalation begins. As more and more of these neurons fire, and fire faster and faster, the inhalation muscles contract. Suddenly the neurons stop firing, the inhalation muscles relax, and exhalation begins. Exhalation is usually a passive process that depends on the elastic recoil of the lung tissues. When respiratory demand is high, however, as during strenuous exercise, exhalation neurons in the medulla increase their firing rates and accelerate the ventilatory rhythm by adding an active component to the exhalation phase of the cycle. Brain areas

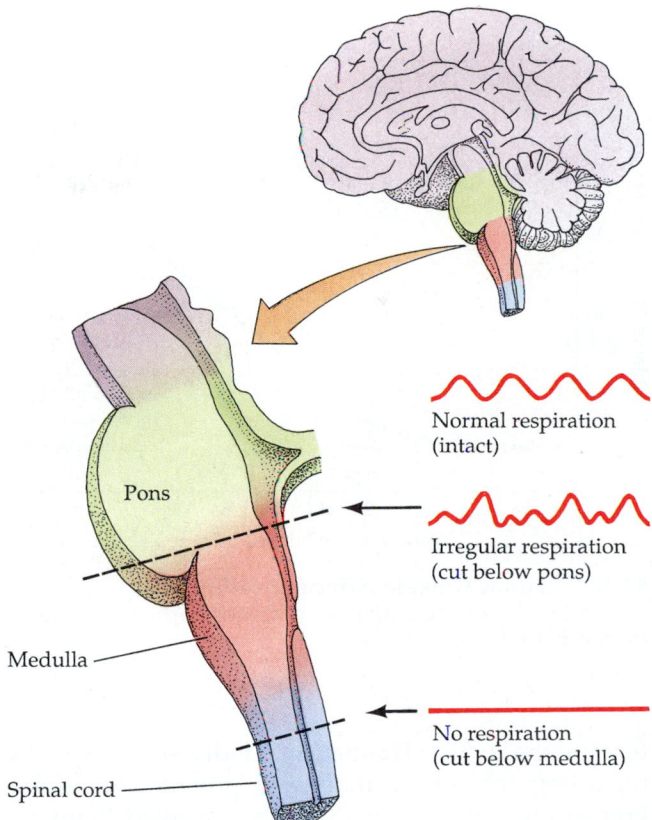

41.19 The Brain Stem Generates and Controls Breathing Rhythm
Severing the brain stem at different levels reveals that the basic breathing rhythm is generated in the medulla and modified by neurons in or above the pons.

above the medulla can also modify the ventilatory rhythm to accommodate speech, ingestion of food, coughing, and emotional states.

As respiratory demands increase, the activities of the inhalation neurons also increase, thus contributing to greater depth of inhalation. An override reflex, however, prevents the ventilatory muscles from overdistending and damaging the lung tissue. This reflex is named the Hering–Breuer reflex (after the two physiologists who discovered it). It begins with stretch sensors in the lung tissue. When stretched, these sensors send impulses via the vagus nerve that inhibit the inhalation neurons.

Matching Ventilation to Metabolic Needs

When the partial pressure of O_2 and the partial pressure of CO_2 in the blood change, the respiratory rhythm changes to return these values to normal levels. An early experimental approach to understanding gas exchange in humans addressed the reasonable expectation that the blood P_{O_2}, or P_{CO_2}, or both, should provide feedback to the respiratory cen-

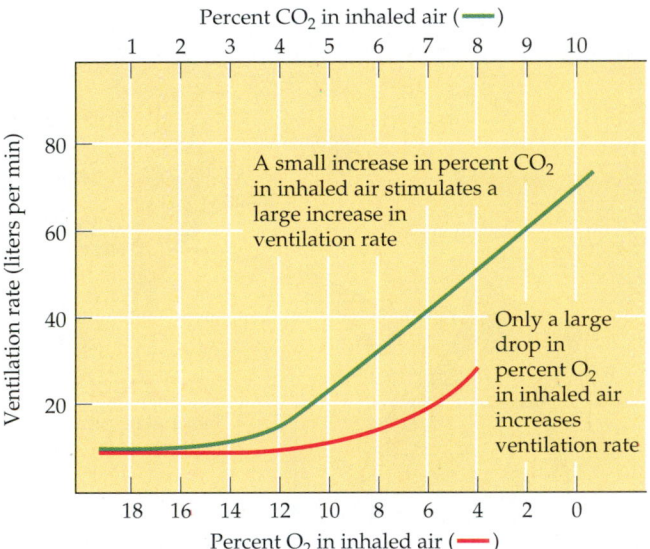

Percent CO_2 in inhaled air (—)

A small increase in percent CO_2 in inhaled air stimulates a large increase in ventilation rate

Only a large drop in percent O_2 in inhaled air increases ventilation rate

Ventilation rate (liters per min)

Percent O_2 in inhaled air (—)

41.20 Carbon Dioxide Affects Breathing
Breathing is more sensitive to increased carbon dioxide than it is to decreased oxygen.

P_{O_2}'s and metabolic rates. Normal fluctuations in metabolism and ventilation have very little effect on the maximum amount of O_2 carried by the blood. By contrast, small changes in metabolism and alveolar P_{CO_2} do influence the concentration of CO_2 in the blood. Changes in blood P_{CO_2} are a much finer index of energy demands and respiratory performance than is the O_2 content of the blood.

ters in the brain. Dramatic and disastrous insight regarding this expectation was provided by three French physiologists in 1875. They wanted to investigate the physiological effects of breathing low concentrations of O_2. Sophisticated gas pumps and pressure chambers did not exist in 1875, so the three decided to go up in a balloon to very high altitudes and observe the effects of the rarefied atmosphere on one another. They noted no ill effects and continued to throw out ballast, going higher than 8,000 meters. Then all three became unconscious. The balloon finally descended on its own, and one of the physiologists regained consciousness to find his two colleagues dead. This infamous flight of the balloon *Zenith* is tragic proof that the human body is not very good at sensing its own need for O_2.

Humans and other mammals are very sensitive, however, to increases in the P_{CO_2} of the blood, whether caused by energy demands or by the composition of the air breathed. If you re-breathe a small volume of air, thereby increasing the P_{CO_2} of that reservoir, your breathing becomes deeper and more rapid, and you become anxious and agitated. You react the same way even if pure O_2 is continually released into the reservoir to keep its P_{O_2} constant. Typical ventilatory responses to changes in blood P_{O_2} and P_{CO_2} are shown in Figure 41.20.

It makes sense that CO_2 rather than O_2 is the dominant feedback stimulus for ventilation. As we have seen, animals have evolved respiratory systems and hemoglobin properties that work to keep the blood that is leaving the gas-exchange surfaces fully saturated with O_2 over a broad range of alveolar

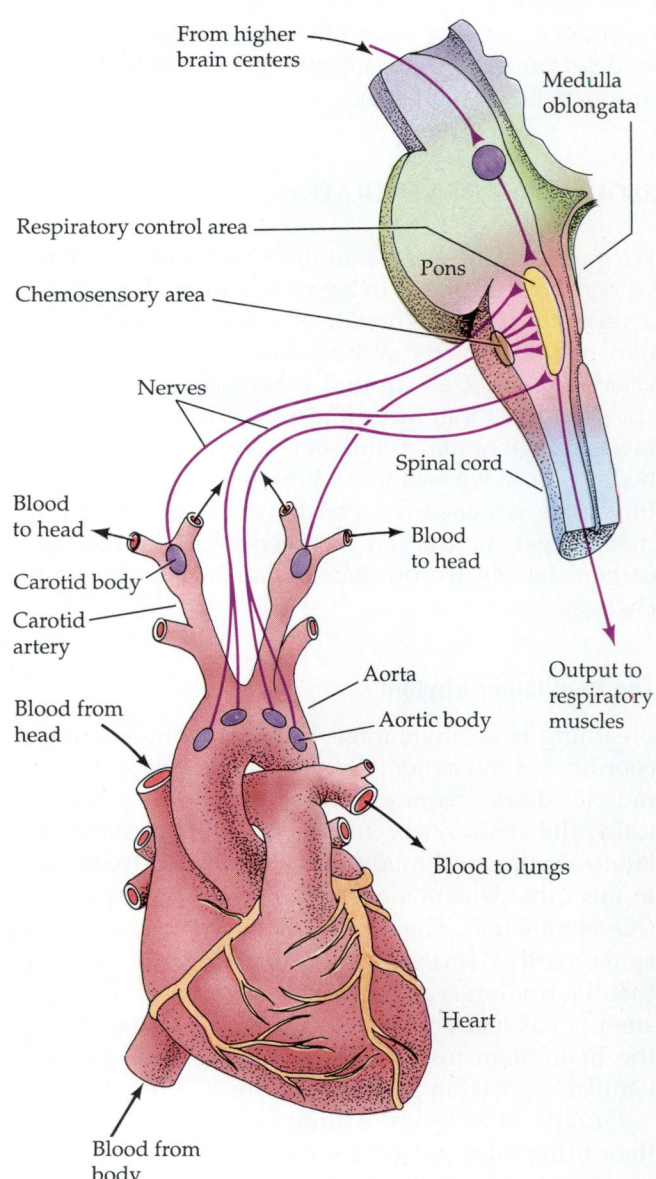

From higher brain centers

Medulla oblongata

Respiratory control area

Chemosensory area

Pons

Nerves

Spinal cord

Blood to head

Blood to head

Carotid body

Carotid artery

Aorta

Output to respiratory muscles

Blood from head

Aortic body

Blood to lungs

Heart

Blood from body

41.21 Chemosensors Sense Changes in Gas Concentrations
Chemosensors on large blood vessels leaving the heart are sensitive to the partial pressure of oxygen in the blood; other chemosensors on the surface of the medulla are sensitive to the partial pressure of carbon dioxide in the blood. The body uses information from these chemosensors to match breathing rate to metabolic demand.

Where are gas concentrations in the blood sensed? The major site of CO_2 sensitivity is an area on the ventral surface of the medulla, not far from the groups of neurons that generate the ventilatory rhythm. Sensitivity to the O_2 concentration of the blood resides in nodes of tissue on the large blood vessels leaving the heart, the aorta and the carotid arteries (Figure 41.21). These carotid and aortic bodies receive enormous supplies of blood, and they

contain chemosensory nerve endings. If the blood supply to these structures decreases, or if the P_{O_2} of the blood falls dramatically, the chemosensors are activated and send impulses to the respiratory centers. Although we are not very sensitive to changes in blood P_{O_2}, the carotid and aortic bodies can stimulate increases in ventilation during exposure to very high altitude or when blood volume or blood pressure are very low.

SUMMARY of Main Ideas about Gas Exchange in Animals

Respiratory gas-exchange systems of animals promote diffusion of O_2 from the environment into the body and diffusion of CO_2 from the body to the environment.

Maximizing the surface area for diffusion of respiratory gases is an adaptation that takes many forms in different species, including flat body shapes, external gills, internal gills, lungs, and both air-filled and water-filled channels that branch throughout the body.
 Review Figures 41.1 and 41.4

In comparison to air-breathing animals, water breathers are under the constraints of lower O_2 content in their environment, decreasing O_2 availability but increasing O_2 need as temperature rises, and the fact that more work is required to move water than to move air over gas-exchange surfaces.
 Review Figure 41.2

The O_2 content of air decreases as barometric pressure decreases.
 Review Figure 41.3

The concentration gradient for diffusion of respiratory gases is maximized by ventilating the environmental side of exchange membranes with air or water and perfusing the internal side with blood.

The tracheal gas-exchange system of insects supports the high metabolic rate of insect tissue by distributing O_2 to cells through air capillaries.
 Review Figure 41.5

Respiratory systems of fish and birds are highly efficient because of adaptations that enable continuous, unidirectional flow over their gas-exchange surfaces.
 Review Figures 41.6, 41.7, 41.8, 41.9 and 41.10

Countercurrent flow of water and blood on opposite sides of respiratory gas-exchange surfaces augments the efficiency of respiratory gas exchange in fish.
 Review Figure 41.6 and Box 41.A

The lungs of amphibians, reptiles, and mammals require tidal ventilation, which limits their gas-exchange effiency by limiting the concentration gradients for respiratory gas diffusion.
 Review Figures 41.12 and 41.14

The alveoli of lungs present large surface areas for gas exchange.
 Review Figure 41.13

Hemoglobin binds O_2, carries it in the bloodstream, and releases it where it is needed.
 Review Figures 41.15, 41.16, and 41.17

CO_2 produced in tissues leaves the body via the blood and the lungs.
 Review Figure 41.18

The breathing rhythm is generated in the medulla and modified by inputs from other brain regions.
 Review Figure 41.19

CO_2 concentration in the blood is a major feedback signal controlling the rate of breathing; O_2 concentration has a lesser effect.
 Review Figure 41.20

Chemosensors on the suface of the medulla detect CO_2 and chemosensors on blood vessels leaving the heart detect O_2 concentration.
 Review Figure 41.21

SELF-QUIZ

1. Which of the following statements is *not* true?
   a. Respiratory gases are exchanged by diffusion only.
   b. Oxygen has a lower rate of diffusion in water than in air.
   c. The oxygen content of water falls as the temperature of water rises, all other things being equal.
   d. The amount of oxygen in the atmosphere decreases with increasing altitude.
   e. Birds have evolved active transport mechanisms to augment their respiratory gas exchange.

2. Which of the following statements about the respiratory system of birds is *not* true?
   a. Respiratory gas exchange does not occur in the air sacs.
   b. The respiratory system of birds can achieve more complete exchange of O_2 from air to blood than that of humans.
   c. Air passes through birds' lungs in only one direction.
   d. The gas-exchange surfaces in bird lungs are the alveoli.
   e. A breath of air remains in the bird respiratory system for two breathing cycles.

3. In a countercurrent exchange system
   a. the fluids in two tubes flow in opposite directions.
   b. gas exchange between two streams is less complete than if they flowed in parallel.
   c. a gas and a liquid flow in parallel tubes.
   d. the materials in solution move against the current.
   e. the fluids in two tubes are funneled into a final common vessel.

4. In the human respiratory system
   a. the lungs and airways are completely collapsed after a forceful exhalation.
   b. the average O_2 concentration of air inside the lungs is always less than it is in the air outside the lungs.
   c. the P_{O_2} of the blood leaving the lungs is greater than the P_{O_2} of the exhaled air.
   d. the amount of air that is moved per breath during normal, at-rest breathing is termed the total lung capacity.
   e. oxygen and carbon dioxide are actively transported across the alveolar–capillary membranes.

5. Which of the following statements about the human respiratory system is *not* true?
   a. During inhalation there is a negative pressure in the space between the lung and the thoracic wall.
   b. Smoking one cigarette can immobilize the cilia lining the airways for hours.
   c. The respiratory control center in the medulla responds more strongly to changes in arterial O_2 concentration than to changes in arterial CO_2 concentrations.
   d. Without surfactant, the work of breathing is greatly increased.
   e. The diaphragm contracts during inhalation and relaxes during exhalation.

6. The hemoglobin of a human fetus
   a. is the same as that of an adult.
   b. has a higher affinity for O_2 than that of an adult.
   c. has only two protein subunits instead of four.
   d. is supplied by the mother's red blood cells.
   e. has a lower affinity for O_2 than that of an adult.

7. As blood flows through an active muscle, it
   a. becomes saturated with oxygen.
   b. takes up only a small amount of oxygen.
   c. unloads more of its oxygen than it would in a resting muscle.
   d. tends to decrease the partial pressure of oxygen in the muscle tissues.
   e. is denatured.

8. Most carbon dioxide is carried in the blood
   a. in red blood cell cytoplasm.
   b. dissolved in the plasma.
   c. in the plasma as bicarbonate ions.
   d. bound to plasma proteins.
   e. in red blood cells bound to hemoglobin.

9. Myoglobin
   a. binds O_2 at P_{O_2}'s at which hemoglobin is releasing its bound O_2.
   b. has a lower affinity for O_2 than hemoglobin does.
   c. consists of four polypeptide chains, just as hemoglobin does.
   d. provides an immediate source of O_2 for muscle cells at the onset of activity.
   e. can bind four O_2 molecules at once.

10. When the level of CO_2 in the blood becomes greater than the set operating range,
    a. the rate of respiration decreases.
    b. the pH of the blood rises.
    c. the respiratory centers become dormant.
    d. the rate of respiration increases.
    e. the blood becomes more alkaline.

FOR STUDY

1. Compare and contrast the respiratory systems of birds, fish, and humans. Why can birds and fish outperform mammals in environments where the concentration of O_2 is low?

2. What does the following chemical equation represent and how does it relate to gas exchange in human lungs and in active tissues?

$$CO_2 + H_2O \rightleftharpoons H_2CO_3 \rightleftharpoons H^+ + HCO_3^-$$

3. Describe and contrast the adaptations of llamas and humans for gas exchange at high altitude.

4. Describe, in terms of the structure of the human respiratory system, how air is brought into and then expelled from the lungs.

5. Workers A and B must inspect two large gas-storage tanks. Tank 1 contains 100 percent N_2 (nitrogen gas); tank 2 contains 100 percent CO_2. The tanks are *not* flushed out before inspection. Worker A goes into tank 1, becomes unconscious, and dies. Worker B goes into tank 2, feels strangely short of breath, and leaves the tank while feeling somewhat dizzy. Explain why worker A died whereas worker B lived.

READINGS

Eckert, R., D. Randall and G. Augustine. 1988. *Animal Physiology: Mechanisms and Adaptations*, 3rd Edition. W. H. Freeman, New York. An outstanding textbook of animal physiology. Chapter 14 covers gas exchange.

Feder, M. E. and W. W. Burggren. 1985. "Skin Breathing in Vertebrates." *Scientific American*, November. Not all breathing involves lungs or gills. This article discusses adaptations possessed by many vertebrates for gas exchange through the skin.

Perutz, M. F. 1978. "Hemoglobin Structure and Respiration." *Scientific American*, December. An authoritative article on hemoglobin structure and function by the man who received the Nobel prize for his work on this subject.

Schmidt-Nielsen, K. 1971. "How Birds Breathe." *Scientific American*, December. Describes the complex adaptations of the avian respiratory system.

Schmidt-Nielsen, K. 1990. *Animal Physiology: Adaptation and Environment*, 4th Edition. Cambridge University Press, New York. An outstanding textbook that emphasizes the comparative approach.

Vander, A. J., J. H. Sherman and D. S. Luciano. 1994. *Human Physiology: The Mechanisms of Body Function*, 6th Edition. McGraw-Hill, New York. Chapter 15 deals with human respiration.

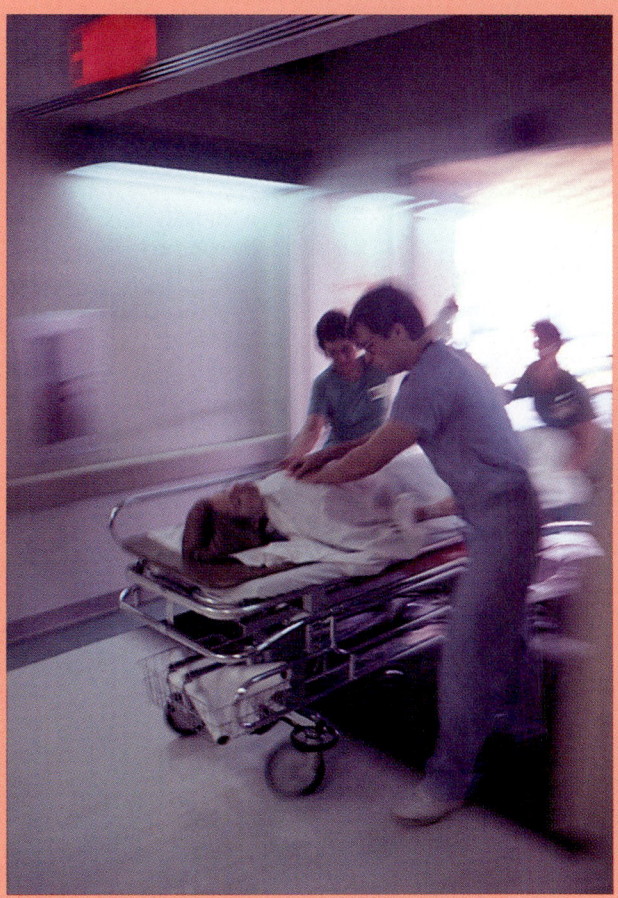

Heart Attack!
Hospital emergency rooms deal with the aftermath of sudden heart attacks. Here an emergency team attempts to stimulate a failed heart to resume beating.

42

Internal Transport and Circulatory Systems

Sweating, severe chest pain, fainting, 911, paramedics, flashing lights, emergency room, intensive care—heart attack. We all will experience this sequence of events directly or indirectly. More than one-third of the deaths in the United States are due to heart failure, and many such deaths are those of a sudden nature that we call heart attacks. Why is it so serious when the heart fails? The answer is that all organs of the body depend on the heart.

The needs of all the cells in the body of an animal are served by the internal fluid environment (see Chapter 35). Cells take up nutrients from the fluid that bathes them, and they release their waste products into that fluid. The activities of the cells change the internal fluid environment in ways that are not healthy for the cells themselves. The organs return different aspects of the internal environment to optimal levels. The lungs take in oxygen and eliminate carbon dioxide, the digestive tract takes in nutrients, the liver controls nutrient levels and eliminates toxic compounds, and the kidneys control salt concentrations and eliminate toxic wastes. The cells of each organ depend on the activities of all the other organs, and only through a system of transport, or circulation, can the activities of each organ influence the internal environment of the entire body. So when the heart stops, transport stops, the internal environment deteriorates, and cells get sick and die.

TRANSPORT SYSTEMS

A heart is not all there is to a transport system; it is just the pump. In addition, there must be a vehicle (blood) to transport materials through the system and a series of conduits (blood vessels) through which the materials can be pumped around the body. Heart, blood, and vessels together constitute a circulatory system, also known as a cardiovascular system (from the Greek *kardia* = "heart" and the Latin *vasculum* = "small vessel"). In this chapter we will explore adaptations of animals that directly serve the needs of their cells by bringing them nutrients and removing their wastes. Although this chapter focuses on cardiovascular systems, the simplest transport systems are not vascular; they do not use vessels. Some do not even use a pump to move things around. But all carry out the essential task of transport systems: transporting substances to and away from every cell of the bodies of animals.

Gastrovascular Cavities

A circulatory system is unnecessary if all the cells of an organism are close enough to the external environment that nutrients, respiratory gases, and wastes

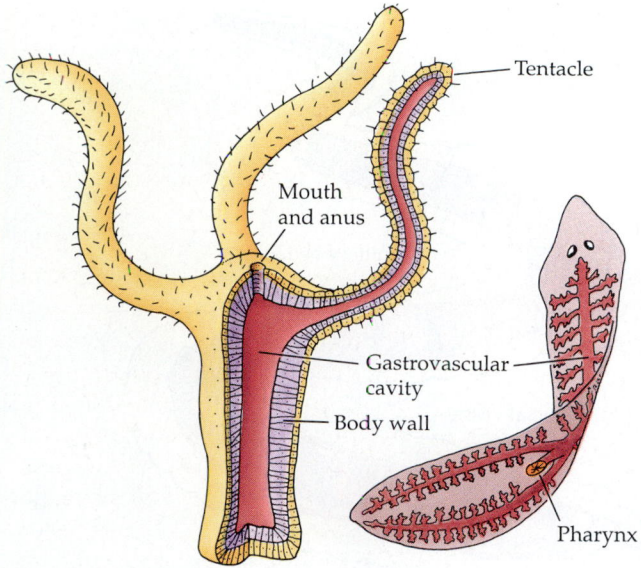

(a) Cnidarian (*Hydra*)   (b) Flatworm (*Planaria*)

42.1 Gastrovascular Cavities
Gastrovascular cavities in animals without circulatory systems serve the metabolic needs of the innermost cells of the body. (a) The gastrovascular cavity of a cnidarian, *Hydra*, extends into the tentacles. No cell of the hydra is more than one cell away from either the gastrovascular cavity or the external medium. (b) The gastrovascular cavity of a flatworm, *Planaria*, extends into all regions of the animal's flattened body.

can diffuse directly between the cells and the outside environment. Small aquatic invertebrates have various structures and shapes that permit direct diffusional exchange. The hydra, a cnidarian, is a good example (see Figure 26.12). This aquatic animal is cylindrical and only two cell layers thick. Each of the hydra's cells contacts water that is either surrounding the animal or circulating through its **gastrovascular cavity**, a dead-end sac that serves both for digestion ("gastro") and transport ("vascular") (Figure 42.1a). The cells of some other invertebrates are served by diffusion from highly branched gastrovascular systems. Flattened body shapes minimize the diffusion path length—the distance that molecules have to diffuse between cells and the external environment (Figure 42.1b). A central gastrovascular system cannot, however, serve the needs of larger animals with many layers of cells. Transport in such animals requires a circulatory system. All terrestrial animals require circulatory systems because none of their cells are served by an external medium; all of their cells must be served by the interstitial fluids.

Open Circulatory Systems

In the simplest circulatory systems the interstitial fluid is simply squeezed through intercellular spaces as the animal moves. In these **open circulatory sys-**

tems there is no distinction between interstitial fluid and blood. Usually a muscular pump, or heart, assists the distribution of the fluid throughout the tissues. The contractions of the heart propel the interstitial fluid through vessels leading to different regions of the body, but the fluid leaves those vessels to trickle through the tissues and eventually to return to the heart. In the arthropod shown in Figure 42.2a the fluid returns to the heart through valved holes called **ostia** when the heart relaxes. In the mollusk in Figure 42.2b, open vessels aid in the return of interstitial fluid to the heart.

Closed Circulatory Systems

In a **closed circulatory system** some components of the blood never leave the vessels. Blood circulates through the vascular system, pumped by one or more muscular hearts. The system keeps the circulating blood separate from the interstitial fluid.

A simple example of a closed circulatory system is

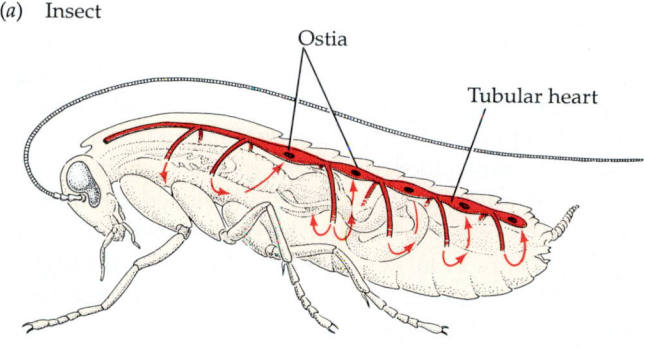

42.2 Open Circulatory Systems
The open circulatory systems of an arthropod (a) and a mollusk (b). In both, blood is pumped by a tubular heart and directed to regions of the body through vessels that open into interstitial spaces. In arthropods the blood, or hemolymph, reenters the heart through the ostia, and in the mollusk a system of vessels drains the interstitial spaces and returns the blood to the heart.

that of the common earthworm, an annelid (see Figure 26.24). One large blood vessel on the ventral side of the earthworm carries blood from its anterior end to its posterior end. In each segment of the worm, smaller vessels branch off and transport the blood to even smaller vessels in the tissues of that segment. Here, respiratory gases, nutrients, and metabolic wastes diffuse between the blood and the interstitial fluids. The blood then flows into larger vessels that lead into a single large vessel on the dorsal side of the worm. The dorsal vessel carries the blood from the posterior to the anterior end. Five pairs of vessels connect the large dorsal and ventral vessels in the anterior end, thus completing the circuit (Figure 42.3). The dorsal vessel and the five pairs of connecting vessels serve as hearts for the earthworm; their contractions keep the blood circulating. The direction of circulation is determined by one-way valves in the dorsal vessel and in the five pairs of connecting vessels.

Closed circulatory systems have several advantages over open ones. First, closed systems can deliver oxygen and nutrients to the tissues and carry away metabolic wastes more rapidly than open systems can. Second, closed systems can direct blood to specific tissues. Third, cellular elements and large molecules that function within the vascular system can be kept within it; examples are red blood cells and large molecules that help in the distribution of hormones and nutrients. Overall, closed circulatory systems can support higher levels of metabolic activity, especially in larger animals. How then do highly active insect species achieve high levels of metabolic output with their open circulatory systems? The key to answering this question is to remember something you learned in the previous chapter: Insects do not depend on their circulatory systems for respiratory gas exchange (see Figure 41.5); oxygen diffuses directly to their muscles through tracheae and air capillaries.

All vertebrates have closed circulatory systems and chambered hearts. Chambered hearts have valves that prevent the backflow of blood when the heart contracts. From fishes to amphibians to reptiles to birds and mammals, the complexity and the number of chambers of the heart increase. An important consequence of this increased complexity is the gradual separation of the circulation into two circuits, one to the lungs and one to the rest of the body. In fishes, blood is pumped from the heart to the gills and then to the tissues of the body and back to the heart. In birds and mammals, blood is pumped from the heart to the lungs and back to the heart in the **pulmonary circuit**, and from the heart to the rest of the body and back to the heart in the **systemic circuit**. We will see how the separation of the circulation into two

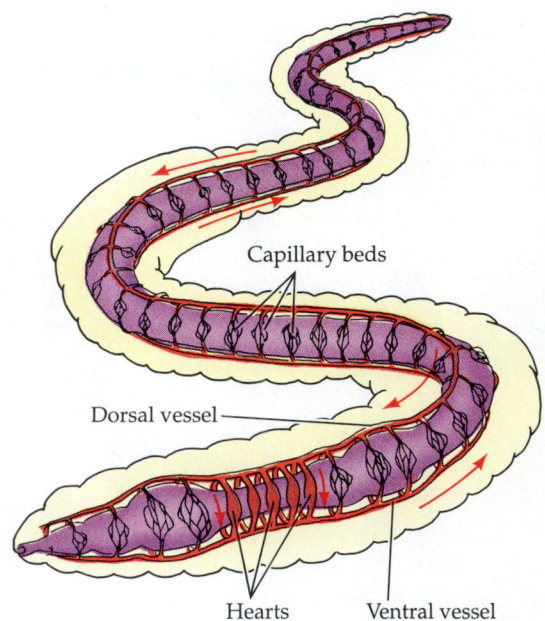

42.3 A Closed Circulatory System
In a closed circulatory system, blood is confined to the blood vessels, kept separate from the interstitial fluid, and is pumped by one or more muscular hearts. The earthworm, with large dorsal and ventral blood vessels and a branching network of smaller vessels, exemplifies this system.

circuits improves the efficiency and capacity of the circulatory system.

The vascular system includes **arteries** that carry blood away from the heart, and **veins** that carry blood back to the heart. **Arterioles** are small arteries, and **venules** are small veins. **Capillaries** are very small, thin-walled vessels that connect arterioles and venules. Exchanges between the blood and the interstitial fluid take place only across capillary walls.

Circulatory Systems of Fishes

Fish hearts have two chambers. A less muscular chamber called the **atrium** receives blood from the body and pumps it into a more muscular chamber, the **ventricle**, which then pumps the blood to the gills, where gas exchange takes place. Blood leaving the gills collects in a large dorsal artery, the **aorta**, which distributes it to smaller arteries and arterioles leading to all the organs and tissues of the body. In the tissues the blood flows through capillary beds, then collects in venules and veins, and is eventually returned to the heart (Figure 42.4a). Most of the pressure imparted to the blood by the contraction of the ventricle is dissipated by the high resistance of the many tiny, narrow spaces the blood flows through in the gills. As a result, the blood entering the aorta

42.4 Vertebrate Circulatory Systems

All vertebrates have closed circulatory systems. *(a)* Fishes have a heart with two chambers: a single atrium and a single ventricle. *(b)* The lungfish heart has two atria, one receiving oxygenated blood from the lung and one deoxygenated blood from the body. *(c)* The pulmonary and systemic circuits are partially separated in adult amphibians. The heart is three-chambered, with two atria and one ventricle. *(d)* The ventricle of the reptilian heart is partially divided by a septum to direct the flow of oxygenated blood to the body and deoxygenated blood to the lungs. An advantage of this incomplete division of the ventricle is that most of the blood can be directed to the systemic circuit when the animal is not breathing. *(e)* Birds and mammals have four-chambered hearts. Their pulmonary and systemic circuits are totally separate.

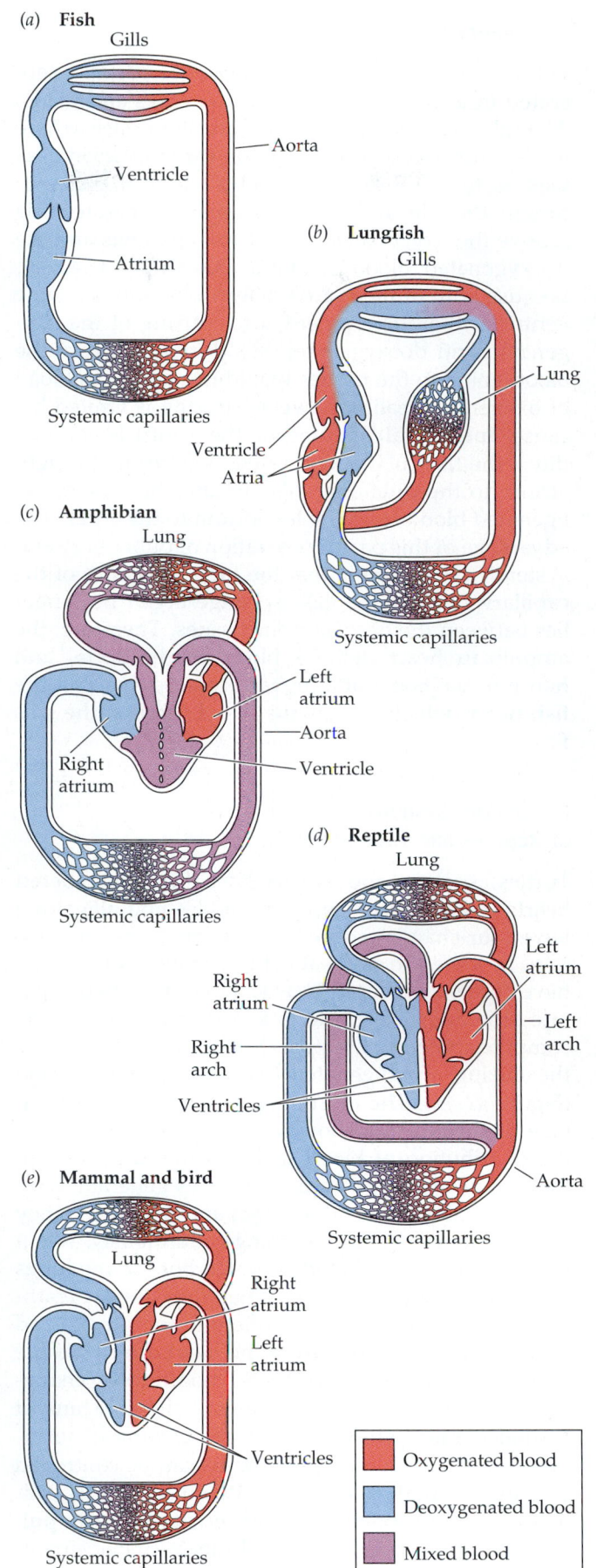

of the fish is under low pressure, limiting the ability of the fish circulatory system to supply the tissues with oxygen and nutrients. This limitation on arterial blood pressure does not seem to hamper the performance of many rapidly swimming species, such as tuna and marlin.

An important evolutionary step, however, is reflected in the circulatory systems of African lungfish. These fish are periodically exposed to water with low oxygen content or to situations in which their aquatic environment dries up. Their adaptation for dealing with these conditions is an outpocketing of the gut that serves as a lung (see Figure 41.12*b*). This lung contains many thin-walled blood vessels, so deoxygenated blood flowing through those vessels can pick up oxygen from air gulped into the lung. What blood vessels serve this new organ? The last pair of gill arteries is modified to carry blood to the lung, and a new vessel carries oxygenated blood from the lung back to the heart. In addition, two other gill arches have lost their gill filaments, and their blood vessels deliver blood from the heart directly to the dorsal aorta (Figure 42.4*b*). Because a few of the gill arches retain gill filaments, African lungfish can breathe either air or water.

In the evolution of vertebrate circulatory systems, the lungfish reveals the transition step leading to separate pulmonary and systemic circuits. Adaptations of the lungfish heart are also evolutionarily important. Unlike the hearts of other fish, the lungfish heart has a partially divided atrium with the left side receiving oxygenated blood from the lungs and the right side receiving deoxygenated blood from the other tissues. The two bloodstreams stay mostly separate as they flow through the ventricle and the large vessel leading to the gill arches, so the oxygenated blood goes to the gill arteries leading to the dorsal aorta, and the deoxygenated blood goes to the arches with functional gill filaments and to the lung.

Circulatory Systems of Amphibians

Pulmonary and systemic circulation are partially separated in adult amphibians such as frogs and toads. A single ventricle pumps blood to the lungs, where it picks up oxygen and dumps carbon dioxide, as well as to the rest of the body, where it picks up carbon dioxide and dumps oxygen. Separate atria receive the oxygenated blood from the lungs and the deoxygenated blood from the body (Figure 42.4c). Because both of these atria deliver blood to the same ventricle, there is a potential for mixing of the oxygenated and deoxygenated blood in which case the blood going to the tissues would not carry a full load of oxygen. In reality, however, mixing is limited because anatomical features of the ventricle tend to direct the flow of deoxygenated blood from the right atrium to the pulmonary circuit and the flow of oxygenated blood from the left atrium to the aorta. The advantage of this partial separation of pulmonary and systemic circulation is that the high resistance of the capillary beds of the gas-exchange organ no longer lies between the heart and the tissues. Therefore, the amphibian heart delivers blood to the aorta, and hence to the body, at a higher pressure than can the fish heart, which pumps the blood through the gills first.

Circulatory Systems of Reptiles and Crocodilians

Turtles, snakes, and lizards have three-chambered hearts, while crocodilians (crocodiles and alligators) have four-chambered hearts. It is an oversimplification, however, to say that turtles, snakes, and lizards have three-chambered hearts. They have two separate atria and a ventricle that is partially divided by complex anatomical features that probably influence the mixing of oxygenated and deoxygenated blood (Figure 42.4d). The most important and unusual feature of their hearts, however, is the ability to alter the distribution of blood going to the lungs and to the rest of the body. Because reptiles generally have lower metabolic rates than birds and mammals, they can get along without breathing continuously. When they are breathing, blood is routed both to the lungs and to the rest of the body. When they aren't breathing, they can decrease blood flow to their lungs and send most of the blood from the heart directly to the body. They apparently redirect blood flow by changing the resistance in the pulmonary circuit. Think of a single pump with several hoses coming off of it. The flow through individual hoses can be controlled by valves that alter resistance. When a snake, lizard, or turtle is not breathing, the resistance in its pulmonary circuit is higher than in its systemic circuit,

and most of the blood pumped out of the ventricle goes to the body.

Have crocodilians, with their four-chambered hearts, lost this adaptation to control distribution of blood to lungs and body? Not really. All fish, amphibians, and reptiles have two aortas leaving their ventricle. In crocodilians, one aorta comes from the right ventricle and the other from the left ventricle, but they are connected to each other soon after they leave the heart. Because the ventricles are completely separate, they can generate different pressures when the heart contracts. When the animal is breathing, the pressure in the left ventricle and its aorta is higher than the pressure in the right ventricle. This higher pressure is communicated to the aorta leaving the right ventricle, where it creates a back pressure on the valve between the right ventricle and its aorta. Thus when the animal is breathing, all of the blood pumped out of its right ventricle goes to the lungs. When the animal is not breathing, the pressures in the ventricles equalize, and blood from both ventricles flows into the aortas, increasing the proportion of blood going to the body.

Crocodilians have sophisticated adaptations that enable them (like mammals and birds) to maintain a systemic blood pressure that is higher than the pulmonary blood pressure when they are breathing, but enables them (like reptiles) to redirect pulmonary blood to the systemic circuit when they are not breathing. From the circulatory perspective, it is no wonder that reptiles and crocodilians have been such a successful group of animals for such a long time.

Circulatory Systems of Birds and Mammals

Birds and mammals have four-chambered hearts and fully separated pulmonary and systemic circuits (Figure 42.4e). Several advantages arise from this design. First, since oxygenated and deoxygenated bloodstreams cannot mix, the systemic circuit is always receiving arterial blood with the highest oxygen content. Second, respiratory gas exchange is maximized because the blood with the lowest oxygen content and highest CO_2 content is sent to the lungs. Third, because the systemic and the pulmonary circuits are completely separate, they can operate at different pressures. Why is this important? Mammalian and bird tissues have high nutrient demands and thus a very high density of the smallest vessels, the capillaries. Many small vessels present lots of resistance to the flow of blood. Therefore, higher pressure is required to maintain adequate blood flow in the systemic circuit. The pulmonary circuit does not have such a large number of capillaries and such a high resistance, so it doesn't require such high pressure.

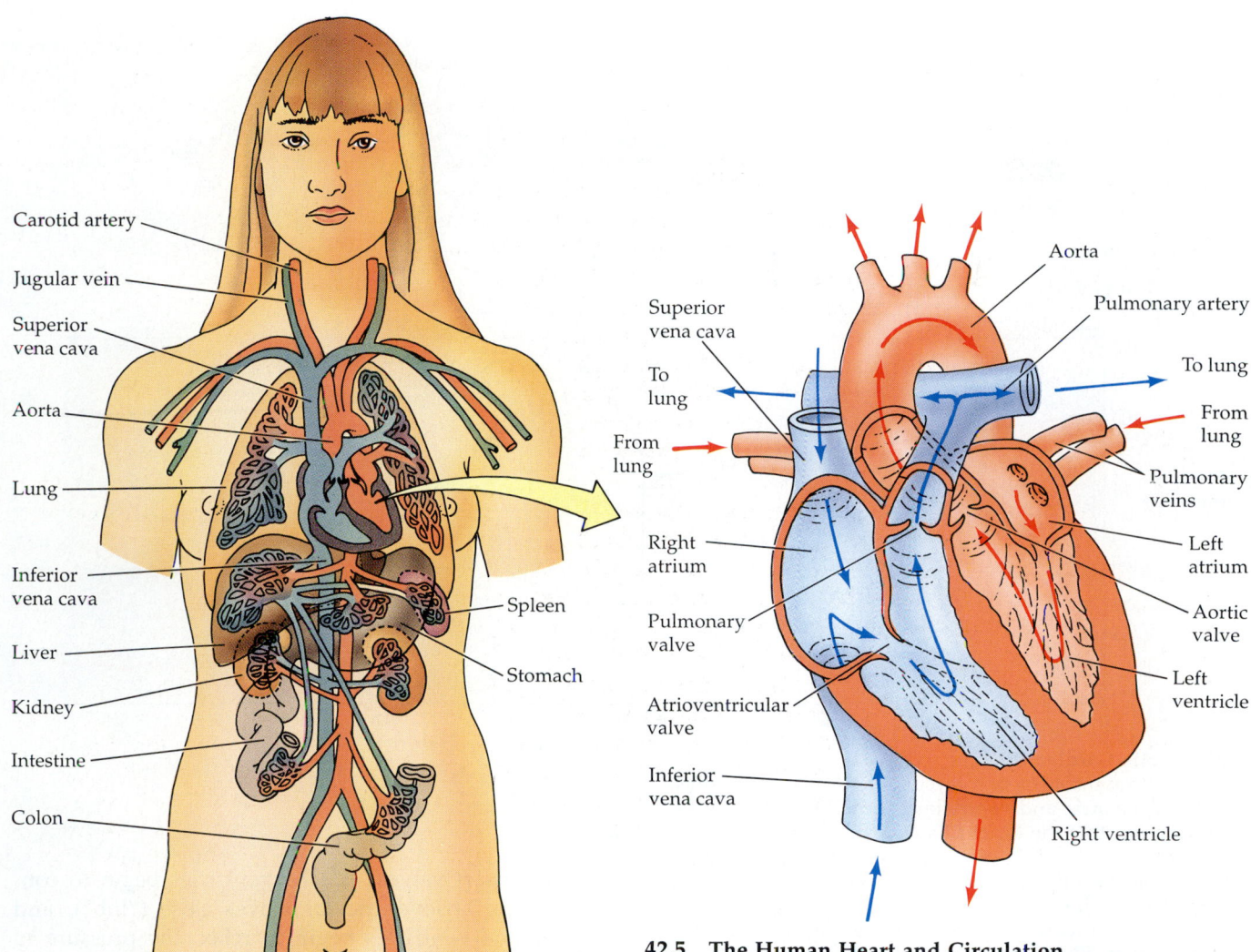

Carotid artery

Jugular vein

Superior
vena cava

Aorta

Lung

Inferior
vena cava

Liver

Kidney

Intestine

Colon

Spleen

Stomach

Superior
vena cava

To
lung

From
lung

Right
atrium

Pulmonary
valve

Atrioventricular
valve

Inferior
vena cava

Aorta

Pulmonary artery

To lung

From
lung

Pulmonary
veins

Left
atrium

Aortic
valve

Left
ventricle

Right ventricle

42.5   The Human Heart and Circulation
Deoxygenated blood from the tissues of the body enters the right atrium and flows through an atrioventricular valve into the right ventricle. The right ventricle pumps the blood into the pulmonary circuit, from which it returns to the left atrium and flows into the left ventricle through an atrioventricular valve. The left ventricle pumps blood into the systemic circuit. The atrioventricular valves prevent blood from flowing back into the atria when the ventricles contract. Pulmonary and aortic valves prevent blood from flowing back into ventricles from the arteries when the ventricles relax.

THE HUMAN HEART

Structure and Function

Like all mammalian hearts, the human heart has four chambers, two atria and two ventricles (Figure 42.5). The atrium and ventricle on the right side of your body are called the right atrium and right ventricle. The atrium and ventricle on the left side of your body are called the left atrium and left ventricle. Each atrium pumps blood into its respective ventricle, and the ventricles pump blood into arteries. The right ventricle pumps blood through the pulmonary circuit, and the left ventricle pumps blood through the systemic circuit. Valves between the atria and ventricles, the **atrioventricular valves**, prevent backflow of blood into the atria when the ventricles contract. The **pulmonary valves** and the **aortic valves** positioned between the ventricles and the arteries prevent the backflow of blood into the ventricles.

Let's follow the circulation of the blood through the heart. The right atrium receives blood from the **superior vena cava** and the **inferior vena cava**, large veins that collect blood from the upper and lower body, respectively. The veins of the heart itself also drain into the right atrium. The right ventricle pumps blood into the **pulmonary artery**, which transports it to the lungs. The **pulmonary veins** return the oxygenated blood from the lungs to the left atrium, from which it enters the left ventricle. The walls of the left ventricle are powerful muscles that contract around the blood with a wringing motion starting from the

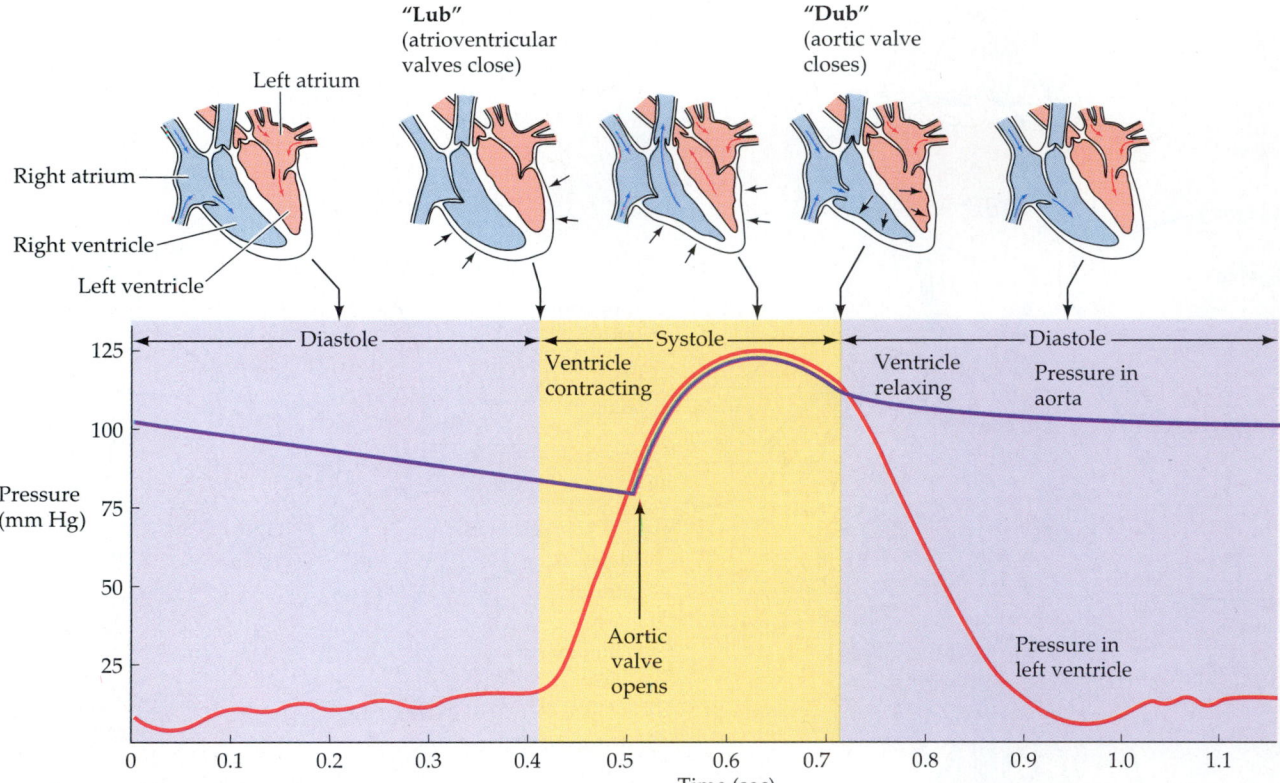

42.6   The Cardiac Cycle

The rhythmic contraction (systole) and relaxation (diastole) of the atria and ventricles is called the cardiac cycle. During diastole the heart fills with blood. Just before systole, the atria contract and maximize the filling of the ventricles. When ventricles contract, the atrioventricular valves shut (first heart sound, "lub") and pressure in the ventricles builds up until the aortic and pulmonary valves open. Blood then leaves the ventricles. At the end of systole the ventricles relax, pressure in the ventricles falls because pressure is now greater in the aorta, and the aortic and pulmonary valves slam shut (second heart sound, "dub"). Pressure in the ventricles continues to fall until the atrioventricular valves open and the heart refills.

bottom. When pressure in the left ventricle is high enough to push open the aortic valve, the blood rushes into the aorta to begin its circulation throughout the body and eventually back to the right atrium. Note in Figure 42.5 how much more massive the left ventricle is than the right ventricle. Because there are many more arterioles and capillaries in the systemic circuit than in the pulmonary circuit, the resistance is higher in the systemic circuit and the left ventricle must squeeze with greater force than the right, even though both are pumping the same volume of blood.

The pumping of the heart—contraction of the two atria followed by contraction of the two ventricles—is the **cardiac cycle**. Contraction of the ventricles is called **systole** and relaxation of the ventricles is called **diastole** (Figure 42.6). The sounds of the cardiac cycle, the "lub-dub" heard through a stethoscope placed on the chest, are created by the slamming shut

of the heart valves. As the ventricles begin to contract, the atrioventricular valves close ("lub"), and when the ventricles begin to relax, the pressure in the aorta and pulmonary artery causes the aortic and pulmonary valves to bang shut ("dub"). Defective valves produce the sounds of heart murmurs. For example, if an atrioventricular valve is defective, blood will flow back into the atria with a "whoosh" sound following the "lub."

The cardiac cycle can be felt in the pulsation of arteries such as the one that supplies blood to your hand. You can feel your pulse by placing two fingers from one hand lightly over the wrist of the other hand just below the thumb. During systole, blood surges through the arteries of your arm and hand and you can feel that as a pulsing of the artery in your wrist.

You can measure blood-pressure changes associated with the cardiac cycle in the large artery in your arm by using an inflatable pressure cuff called a sphygmomanometer and a stethoscope (Figure 42.7). When the inflation pressure of the cuff exceeds maximum (systolic) blood pressure in the artery, blood flow in the artery stops. As the pressure in the cuff is gradually released, a point is reached when blood pressure at the peak of systole is greater than the pressure in the cuff. At this point, a little blood squirts through the closed artery and the artery slams

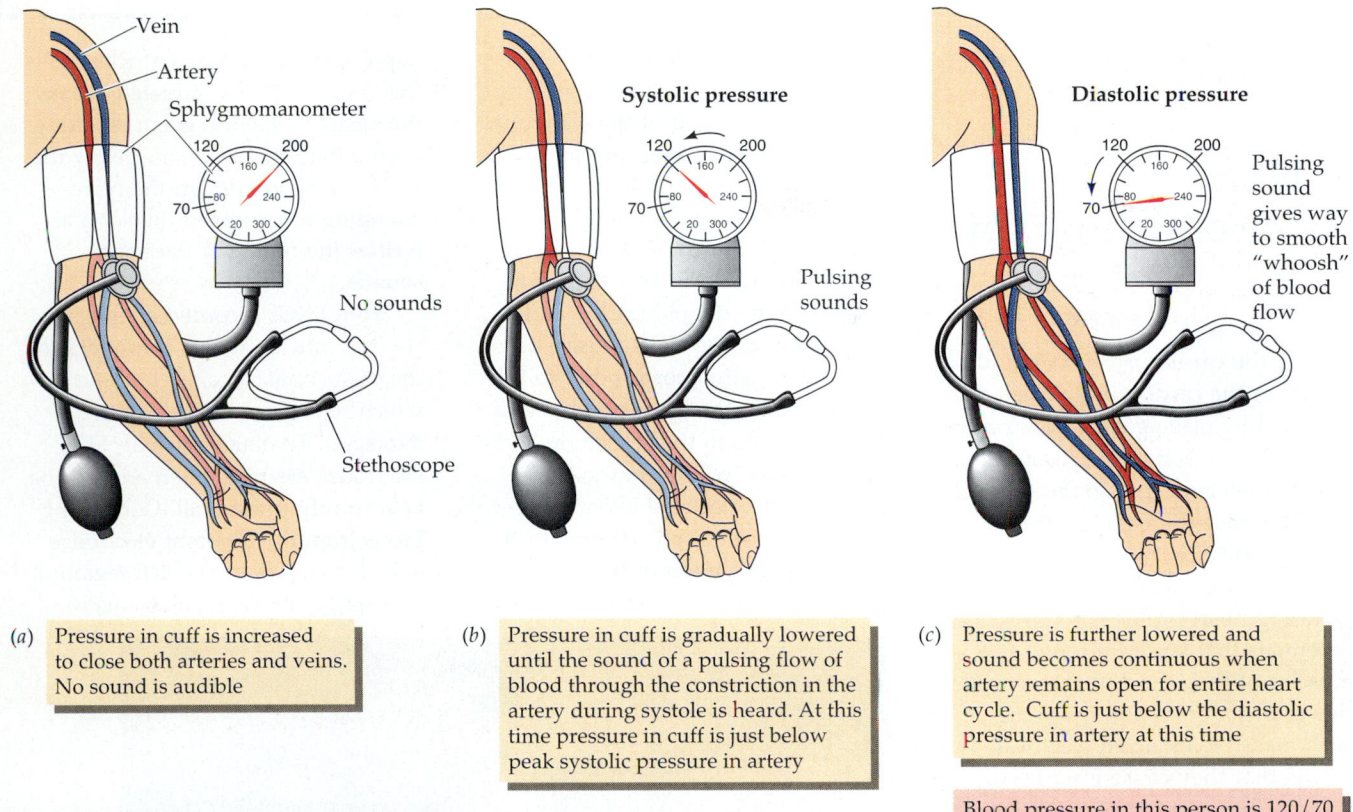

Vein
Artery
Sphygmomanometer
No sounds
Stethoscope

Systolic pressure
Pulsing sounds

Diastolic pressure
Pulsing sound gives way to smooth "whoosh" of blood flow

(a) Pressure in cuff is increased to close both arteries and veins. No sound is audible

(b) Pressure in cuff is gradually lowered until the sound of a pulsing flow of blood through the constriction in the artery during systole is heard. At this time pressure in cuff is just below peak systolic pressure in artery

(c) Pressure is further lowered and sound becomes continuous when artery remains open for entire heart cycle. Cuff is just below the diastolic pressure in artery at this time

Blood pressure in this person is 120/70

42.7 Blood Pressure Measured in an Artery of the Arm
Blood pressure in a major artery can be measured with an inflatable pressure cuff called a sphygmomanometer.

shut, producing a sound that can be heard through a stethoscope applied to the arm. The pressure at which these slamming sounds are first heard is the systolic blood pressure. As the cuff pressure is reduced even more, the slamming sounds gradually disappear to be replaced by a "whoosh" sound as the blood flow in the artery becomes more continuous. The pressure at which slamming sounds are no longer heard is the diastolic blood pressure. In a conventional blood-pressure reading, the systolic value is placed over the diastolic value. Normal values for a young adult might be 120 mm of mercury (Hg) during systole and 80 mm Hg during diastole, or 120/80.

Cardiac Muscle and the Heartbeat

As we saw in Chapter 40, some unique properties help cardiac muscle function as an effective pump. One such property is that the cardiac muscle cells are in electrical continuity with one another. Special junctions called gap junctions enable action potentials to spread rapidly from cell to cell. (Recall what we learned about gap junctions in Chapters 5 and 38.) Because a spreading action potential stimulates contraction, large groups of cardiac muscle cells contract in unison; this coordinated contraction is important for pumping blood. The massive electrical events associated with the heartbeat can be measured on the body surface as an electrocardiogram or EKG (Box 42.A).

Another important property of cardiac muscle cells is that some of them have the ability to initiate action potentials and therefore can contract spontaneously without stimulation from the nervous system. These cells stimulate neighboring cells to contract, thereby acting as pacemakers. The important characteristic of a pacemaker cell is that its resting membrane potential gradually becomes less negative (depolarizes) until it reaches the threshold voltage for initiating an action potential (see Figure 40.12). The nervous system controls the heartbeat (speeds it up or slows it down) by influencing the rate at which pacemaker cells undergo their gradual depolarization between action potentials.

Under normal circumstances the pacemaker activity of the heart originates from modified cardiac muscle cells located at the junction of the superior vena cava and right atrium, in the **sinoatrial node** (Figure 42.8). An action potential spreads from the sinoatrial node across the atrial walls, causing the two atria to contract in unison. There are no gap junctions, how-

BOX 42.A

The Electrocardiogram

During the cardiac cycle, electrical events in the cardiac muscle can be recorded by placing electrodes on the surface of the body. The recording is called an **electrocardiogram**, or EKG (EKG because the Greek word for heart is *kardia*, but ECG is also used). The EKG is an important tool for diagnosing heart problems. The action potentials that sweep through the muscles of the atria and the ventricles prior to their contraction are such massive, localized electrical events that they cause electric currents to flow outward from the heart to all parts of the body. Electrodes placed on the surface of the body at different locations—usually on the wrists and ankles—detect those electric currents at different times because the heart is positioned asym-

metrically in the chest cavity. The appearance of the EKG depends on the exact placement of the electrodes used for the recording. Placing them on the right wrist and left ankle produced the EKG shown here.

The waves of the EKG are designated as P, Q, R, S, and T. P corresponds to the depolarization and contraction of the atrial muscles; Q, R, and S together correspond to the depolarization of the ventricles; and T corresponds to the relaxation and repolarization of the ventricles. Where is the electrical event corresponding to the repolarization of the atria? Repolarization of the atria occurs at the same time as the massive

depolarization of the ventricles, so the QRS complex completely masks the smaller electrical event resulting from atrial repolarization. Below the EKG tracing are drawn the corresponding pressures in the aorta as well as the timing of the heart sounds.

From EKGs recorded after a person has a heart attack, cardiologists (heart specialists) can determine which regions of the heart were damaged. To obtain such an EKG, electrodes are positioned around the heart on the chest wall. Comparing EKGs from the different electrodes tells the cardiologist which region of the heart is behaving abnormally.

The electrocardiogram is used to monitor heart function during an exercise tolerance test.

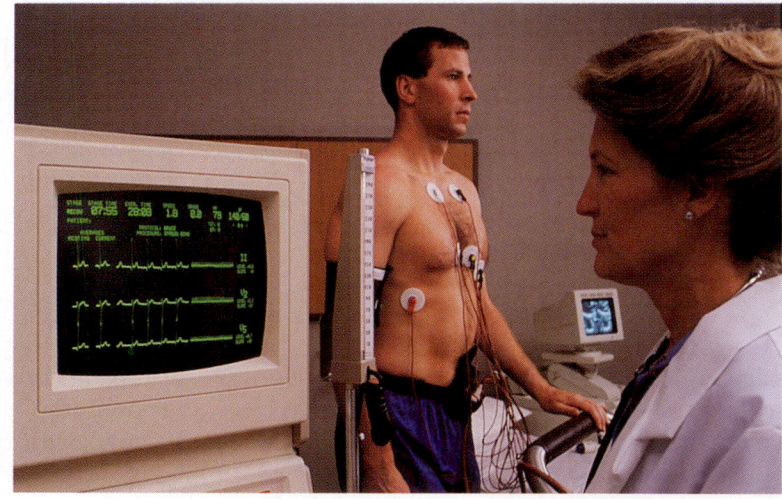

A normal electrocardiogram (EKG)

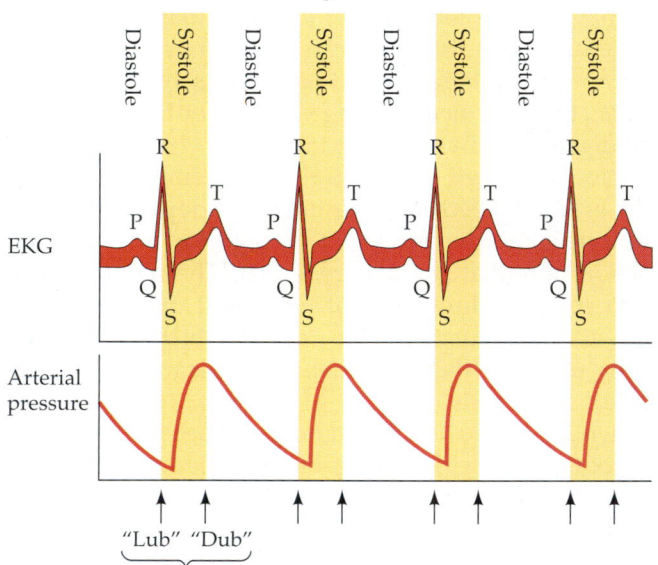

"Lub" "Dub"
The sounds heard through a stethoscope occur at the beginning and end of systole

Some abnormal EKGs

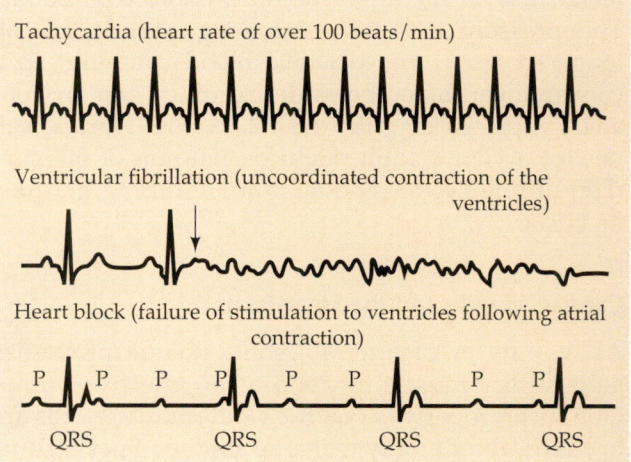

Tachycardia (heart rate of over 100 beats/min)

Ventricular fibrillation (uncoordinated contraction of the ventricles)

Heart block (failure of stimulation to ventricles following atrial contraction)

Besides detecting rhythmic irregularities in heartbeat (arrhythmias), EKGs can detect damage to the heart muscle (infarctions) or decreased blood supply to the heart muscle (ischemias) by changes in the size and shape of the EKG curves

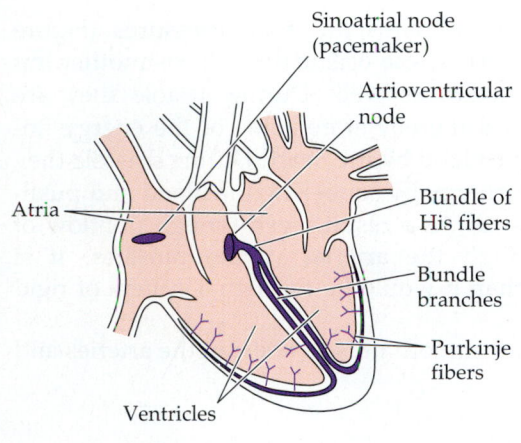

Sinoatrial node (pacemaker)

Atrioventricular node

Atria

Bundle of His fibers

Bundle branches

Purkinje fibers

Ventricles

Heart at rest

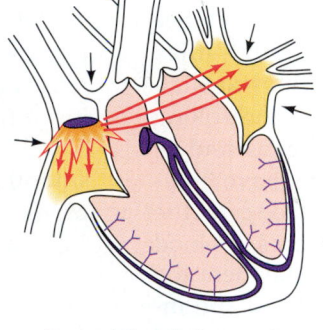

Sinoatrial node fires, action potentials spread through atria which contract

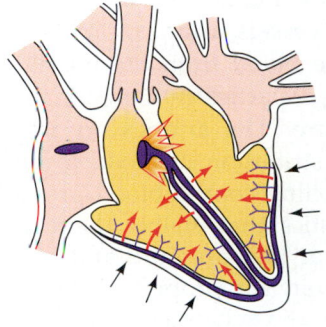

Atrioventricular node fires, sending impulses along conducting fibers; ventricles contract

42.8 The Heartbeat

Pacemaker cells in the sinoatrial node initiate action potentials that spread through the walls of the atria, causing them to contract. Because the walls of the ventricles are not in electrical continuity with the atrial muscle tissue, the action potentials initiated in the pacemaker must pass through the atrioventricular node. When cells of the atrioventricular node fire action potentials, they spread rapidly through the bundle of His and Purkinje fibers to all regions of the ventricular muscle, causing it to contract.

ever, between the atria and the ventricles. The action potential initiated in the atria passes to the ventricles through another node of modified cardiac muscle cells, the **atrioventricular node**. The atrioventricular node passes the action potential on to the ventricles via modified muscle fibers called the **bundle of His**. The bundle of His divides into right and left bundle branches, which connect with **Purkinje fibers** that branch throughout the ventricular muscle.

The timing of the spread of the action potential from atria to ventricles is important. The atrioventricular node imposes a short delay in the spread of the action potential from atria to ventricles. Then the action potential spreads very rapidly throughout the ventricles, causing them to contract. Thus the atria contract before the ventricles do, so the blood passes progressively from the atria to the ventricles to the arteries.

Control of the Heartbeat

The activity of pacemaker cells, and therefore the heart rate, is altered by acetylcholine and norepinephrine released by the autonomic nervous system (Figure 42.9; see also Figure 40.12). Parasympathetic nerves can release acetylcholine onto the sinoatrial and atrioventricular nodes. Acetylcholine slows the pace of action potential generation, thereby slowing the heartbeat. Overactivity of the parasympathetic system can even lead to fainting, which is referred

to as a **vagal reaction** because parasympathetic fibers reach the heart from the brain via a nerve called the vagus nerve. A vagal reaction can be stimulated by deeply felt grief or by having blood withdrawn from a vein.

The opposite effect—speeding up of the heartbeat—is stimulated when sympathetic nerves release norepinephrine onto the cells of the sinoatrial and atrioventricular nodes. An increase in sympathetic nervous system activity also elevates the level of epinephrine (adrenaline) in the blood, which contributes to the excitatory effect on the heart. Norepinephrine and epinephrine also strengthen the contractions of the cardiac muscle cells so that more blood is ejected per beat.

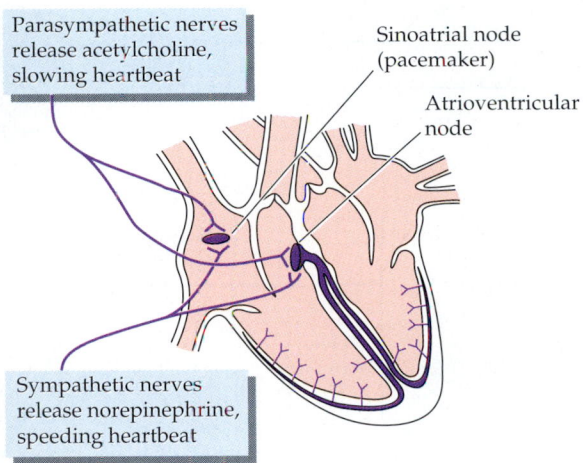

Parasympathetic nerves release acetylcholine, slowing heartbeat

Sinoatrial node (pacemaker)

Atrioventricular node

Sympathetic nerves release norepinephrine, speeding heartbeat

42.9 Changing the Heartbeat through the Autonomic Nervous System

Parasympathetic nerves slow the pacemaker; sympathetic nerves speed up the pacemaker.

THE VASCULAR SYSTEM

The blood circulates around the body in a system of blood vessels: arteries, capillaries, and veins. Arteries receive blood from the heart, and accordingly they have properties that enable them to withstand high pressure. The arteries are important in controlling the distribution of blood to different organs and in controlling central blood pressure. Veins have characteristics that enable them to return blood to the heart at low pressure and also to serve as a blood reservoir. The properties of capillaries make them the site of all exchanges between the blood and the internal environment. It is important to understand how the structure of the different vessels supports their functions.

Arteries and Arterioles

Blood pressure is highest in the vessels that carry blood away from the heart—the arteries and arterioles—and their structure reflects this fact. The walls of the large arteries have many elastic fibers that

enable them to withstand high pressures (Figure 42.10, top left). These elastic fibers have another important function as well. During systole they are stretched and thereby store some of the energy imparted to the blood by the heart. During diastole they return this energy by squeezing the blood and pushing it forward. As a result, even though the flow of blood through the arterial system pulsates, it is smoother than it would be through a system of rigid pipes.

Abundant smooth muscle fibers in the arteries and

42.10 Anatomy of Blood Vessels
The anatomical characteristics of blood vessels match their functions. Arteries have lots of elastic fibers and muscle fibers (top left). They must withstand high pressures. Arterioles have muscle fibers that control blood flow to different capillary beds. The total cross sectional area of capillaries is larger than for any other class of vessels and they are more permeable, thus suiting them for their function of exchange of nutrients and wastes with the extracellular fluids. Because veins (top right) operate under low pressure, they have valves to prevent backflow of blood. As veins get larger they have more elastic fibers, enabling them to accommodate changing volumes of blood.

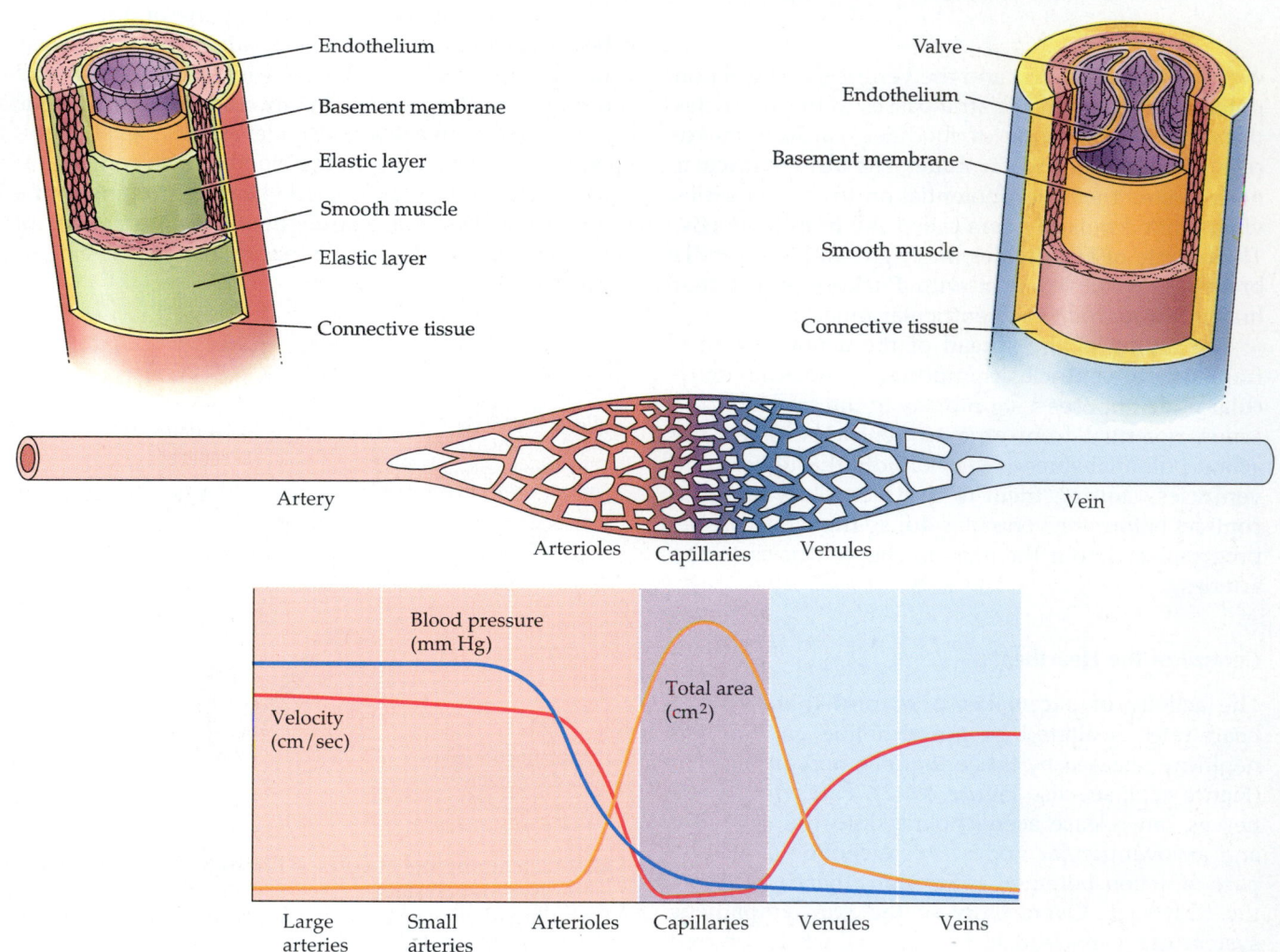

BOX 42.B

Vascular Disease

Vascular disease is by far the largest single killer in the developed Western world; it is responsible for about half the deaths each year. The immediate cause of most of these deaths is heart attack or stroke, but those events are the end result of a disease called **atherosclerosis** (hardening of the arteries) that begins many years before symptoms are detected. Hence atherosclerosis is called the silent killer. What is atherosclerosis, and how can it be prevented?

Healthy arteries have a smooth internal lining of endothelial cells (see Figure 42.10). This lining can be damaged by chronic high blood pressure, smoking, a high-fat diet, and other causes. Fatty deposits called **plaque** begin to form at sites of endothelial damage. First the endothelial cells at the damaged site swell and proliferate; then they are joined by smooth muscle cells migrating from below. Lipids, especially cholesterol, are deposited in these cells so that the plaque becomes fatty. Fibrous connective tissue invades the plaque, and along with deposits of calcium makes the artery wall less elastic; this process is what gives us the terms artherosclerosis and hardening of the arteries.

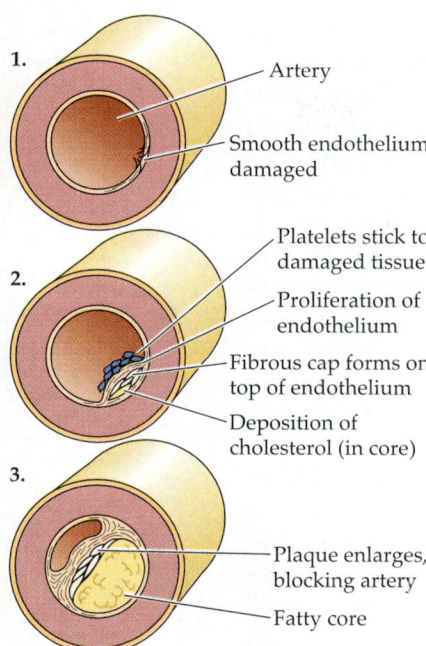

The development of atherosclerotic plaque in an artery.

1.
— Artery
— Smooth endothelium damaged

2.
— Platelets stick to damaged tissue
— Proliferation of endothelium
— Fibrous cap forms on top of endothelium
— Deposition of cholesterol (in core)

3.
— Plaque enlarges, blocking artery
— Fatty core

The growing plaque narrows the artery and causes turbulence in the blood flowing over it. Blood platelets, which are discussed later in this chapter, stick to the plaque and initiate the formation of a blood clot, called a **thrombus**, that further blocks the artery. The blood supply to the heart itself flows through the **coronary arteries**. These arteries are highly susceptible to atherosclerosis; as they narrow, blood flow to the heart muscles decreases. Chest pains and shortness of breath during mild exertion are symptoms of this condition. A person with atherosclerosis is at high risk of forming a thrombus in a coronary artery. Such a **coronary**

thrombosis can totally block the vessel, causing a heart attack, or **coronary infarction**. A piece of a thrombus breaking loose, called an **embolus**, is likely to travel to and become lodged in a vessel of smaller diameter, blocking its flow (an **embolism**). Arteries already narrowed by plaque formation are likely places for an embolus to lodge. If an embolism occurs in an artery in the brain, the cells fed by that artery die. This is called a stroke. The specific damage resulting from a stroke, such as memory loss, speech impairment, or paralysis, depends on the location of the blocked artery.

The most important solution to vascular disease is prevention, not treatment. The risk factors for developing atherosclerosis are: high-fat and high-cholesterol diet, smoking, a sedentary lifestyle, **hypertension** (high blood pressure), obesity, certain medical conditions such as diabetes, and genetic predisposition. There is not much you can do about the genes you inherit or about some forms of diabetes, but the other risk factors can be avoided, thus decreasing the significance of genetic predisposition. It is never too early to take steps to prevent atherosclerosis. Many American children are overweight, and up to 25 percent may have cholesterol levels that are too high. Many teenagers already have well-developed plaques in their arteries. Changes in diet and behavior can prevent and reverse these trends and help to fend off the silent killer.

---

arterioles contract and expand the diameter of those vessels. These changes in diameter alter the resistance of the vessels, which controls the flow of blood. By increasing and decreasing the resistance of these vessels, neural and hormonal mechanisms control the distribution of blood to different tissues of the body and control central blood pressure. The arteries and arterioles are called the **resistance vessels** because their resistance varies. Diseases of the arteries

cause about half of the human deaths each year in developed countries (Box 42.B).

Capillaries

Beds of capillaries connect arterioles to venules. No cell of the body is more than a couple of cell diameters away from a capillary. The needs of cells are served by the exchange of materials between blood and interstitial fluid. This exchange takes place across the capillary walls. It is possible because capillaries have thin, permeable walls and because blood flows through them slowly under very low pressure (see Figure 42.10). To anyone who has played with a garden hose, it may seem strange that big arteries have high pressure and fast flow and that when the blood flows into the small capillaries the pressure and flow decrease. When you restrict the diameter of the garden hose by placing your thumb over the opening, the pressure in the hose increases, which in turn increases the velocity of the water spraying out of the hose. This puzzle is resolved by one more piece of information. Arterioles branch into so many capillaries that the total cross-sectional area of capillaries is much greater than that of any other class of vessels. Even though each capillary is so small that the red blood cells pass through in single file (Figure 42.11), each arteriole gives rise to such a large number of capillaries that together they have a much greater capacity for blood than do the arterioles. An analogy is a fast-flowing river dividing up into many small rivulets flowing across a flat, broad delta. Each rivulet may be small and its flow sluggish, yet all together they accommodate all of the water poured into the delta by the river.

Exchange in Capillary Beds

The walls of capillaries are permeable to water and small molecules, but not to large molecules such as proteins. Blood pressure therefore tends to squeeze water and small molecules out of the capillaries and into the surrounding interstitial spaces. This process is filtration. The large molecules that cannot cross the capillary wall create an osmotic potential (also called osmotic pressure) that tends to draw water back into the capillary.

Blood pressure is highest on the arterial side of a

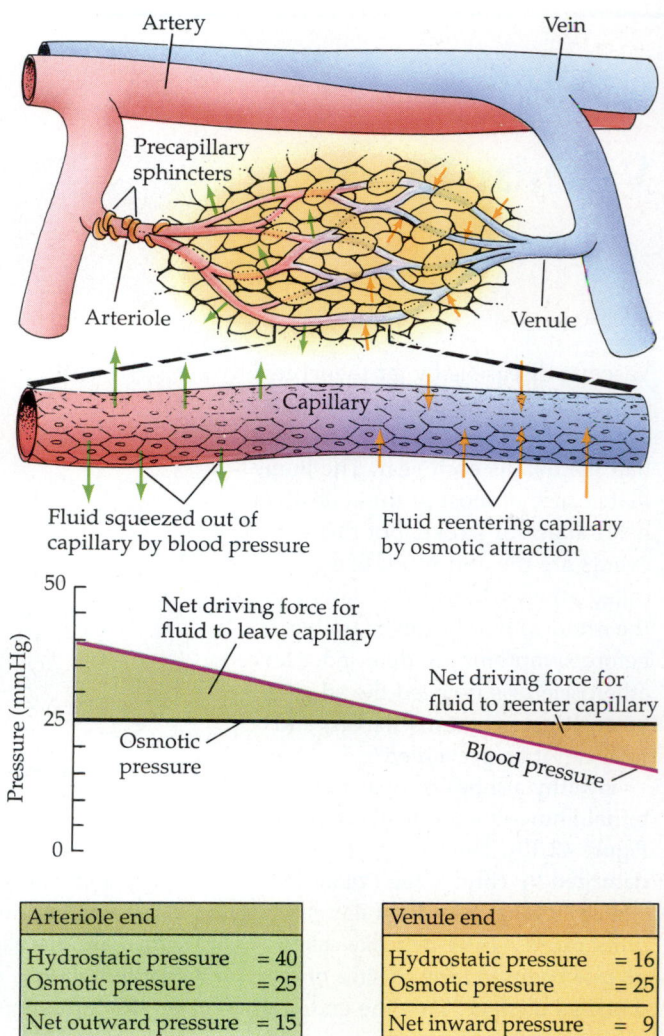

Fluid squeezed out of capillary by blood pressure

Fluid reentering capillary by osmotic attraction

Arteriole end		Venule end	
Hydrostatic pressure	= 40	Hydrostatic pressure	= 16
Osmotic pressure	= 25	Osmotic pressure	= 25
Net outward pressure	= 15	Net inward pressure	= 9

42.12 A Balance of Forces Controls the Exchange of Fluids between Blood Vessels and Interstitial Space
Fluids are squeezed out of capillaries by blood pressure and pulled back in by osmotic pressure created by large molecules that cannot leave the capillaries.

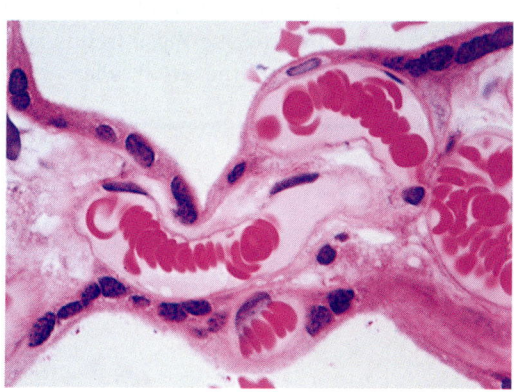

42.11 A Narrow Lane
Red blood cells—the disclike structures—pass through capillaries slowly and in single file.

capillary bed and steadily decreases as the blood flows to the venous side. Therefore, more water is squeezed out of the capillaries on the arterial side of the bed. The osmotic potential pulling water back into the capillary gradually becomes the dominant force as the blood flows toward the venous side of the bed. The interactions of the two opposing forces—the blood pressure versus the osmotic potential—determines the net flow of water (Figure 42.12).

The balance between blood pressure and osmotic potential changes if the blood pressure in the arterioles and the permeability of the capillary walls change. An example of such a change is associated with the inflammation that accompanies injuries to the skin or allergic reactions. The inflamed area becomes hot and red because blood flow to the area increases. The inflamed tissue also swells. The major cause of these events is a chemical called **histamine** that is released mainly by certain white blood cells flowing through the damaged tissue. Histamine makes blood vessels expand, thus increasing blood flow to the area and increasing pressure in the capillaries. Because histamine also increases the permeability of the capillaries and venules, more water leaves the capillaries and venules, and the tissue swells due to the accumulation of interstitial fluids, a condition known as **edema**. The use of drugs called **antihistamines** can alleviate inflammation and allergic reactions.

The loss of water from the capillaries increases if the osmotic potential of the blood decreases, as is seen in the disease kwashiorkor. This disease is caused by severe protein starvation. When the body has no amino acids available for the synthesis of essential proteins, it begins to break down its own blood proteins. Thus fewer molecules are available in the blood to maintain the osmotic potential that pulls water back into the capillaries. The result is that interstitial fluids build up, swelling the abdomen and the extremities (Figure 42.13).

Whether specific small molecules cross a capillary wall depends on the architecture of the capillary, the type of substance, and the concentration difference between the blood and the interstitial fluid. Capillary walls are membranous, and lipid-soluble substances pass freely across a membrane from the area of higher concentration to that of lower concentration (see Chapter 5). Consider what happens in the capillary beds of skeletal muscle tissue. Because the concentration of oxygen is high in the blood coming from the arteriole but very low in active skeletal muscle tissue, oxygen readily moves from the blood into the muscle. At the same time, carbon dioxide rapidly moves into the blood because its concentration is high in the working muscle but low in the blood. The concentrations of these gases in the blood thus change rapidly as the blood travels through the capillary beds.

Small molecules in the blood generally can pass through the capillary walls, but the capillaries in different tissues are differentially selective to the sizes of molecules that can pass through them. In all capillaries, O_2, CO_2, glucose, lactate, and small ions such as Na^+ and Cl^- can cross. In the capillaries of the brain, not much else can cross unless it is a lipid-soluble substance such as alcohol; this high selectivity of brain capillaries is known as the blood–brain barrier (see Chapter 38). In other tissues the capillaries are much less selective and even have pores to permit the passage of large molecules. Such capillaries are found in the digestive tract, where nutrients are absorbed, and in the kidneys, where wastes are filtered. Some capillaries have large gaps that permit the

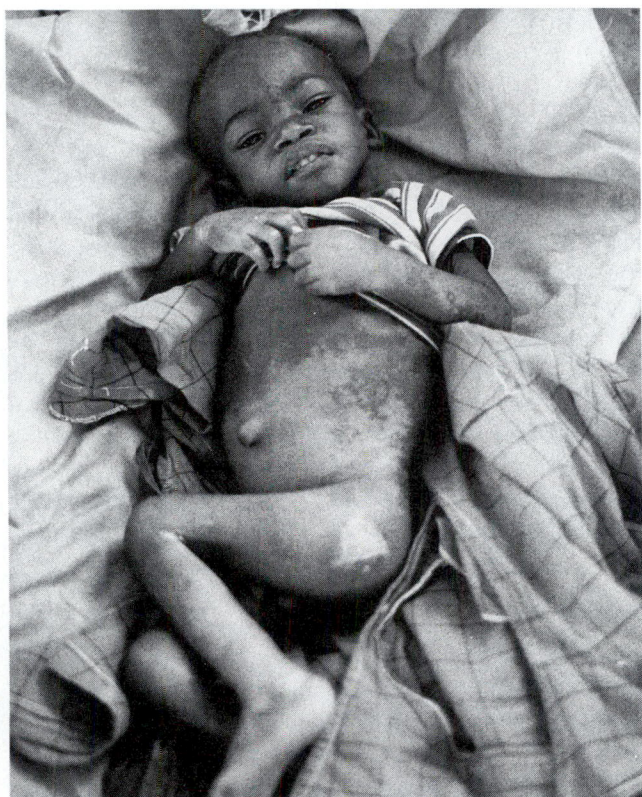

42.13 Kwashiorkor, "The Rejected One"
A child may show the symptoms of protein malnutrition soon after breast-feeding is terminated if the new diet lacks protein. Swollen abdomen, face, hands, and feet due to edema and spindly limbs are hallmarks of serious protein starvation. The limbs are spindly because the body breaks down muscle tissue, as well as blood proteins, to obtain needed amino acids.

movement of even larger substances. These capillaries are found in the bone marrow, spleen, and liver. Substances can move across many capillary walls by endocytosis (see Chapter 4).

The Lymphatic System

The interstitial fluid that accumulates outside the capillaries contains water and small molecules but no red blood cells and less protein than is in blood. A separate system of vessels—the **lymphatic system**—returns the interstitial fluid to the blood. After entering the lymphatic vessels, the interstitial fluid is called **lymph**. Fine lymphatic capillaries merge progressively into larger and larger vessels and end in a major vessel—the **thoracic duct**—which empties into the superior vena cava returning blood to the heart (see Figure 16.3). Lymphatic vessels have one-way valves that keep the lymph flowing toward the thoracic duct. The propelling force moving the lymph is pressure on the lymphatic vessels from the contractions of nearby skeletal muscles.

Mammals and birds have lymph nodes along the major lymphatic vessels. Lymph nodes are an important component of the defensive machinery of the body. They are a major site of lymphocyte production and of the phagocytic action that removes microorganisms and other foreign materials from the circulation. The lymph nodes also act as mechanical filters. Particles become trapped there and are digested by the phagocytes that are abundant in the nodes. Lymph nodes swell during infection. Some of them, particularly those on the side of the neck or in the armpit, become noticeable at such times. The nodes also trap metastasized cancer cells, that is, those that have broken free of the original tumor. Because such cells may start additional tumors, surgeons often remove the neighboring lymph nodes when they excise a malignant tumor.

Venous Return

Blood flows back to the heart through the veins, but what propels it? The pressure of the blood flowing from capillaries to venules is extremely low, so it cannot be the beating of the heart that propels blood through the veins. If the veins are above the level of the heart, gravity helps; below the level of the heart, blood must be moved against the pull of gravity. In actuality, blood tends to accumulate in veins, and the walls of veins are more expandable than the walls of arteries. As much as 80 percent of the total blood volume may be in the veins at any one time. Veins are called **capacitance vessels** because of their high capacity to store blood.

Blood must be returned from the veins so that circulation can continue. If too much blood remains in the veins, then too little blood returns to the heart, and thus too little blood is pumped to the brain; a person may faint as a result. Fainting is self-correcting because a fainting person falls, thereby changing from the position in which gravity caused blood to accumulate in the lower body. There are means other than fainting, however, by which blood is moved from the tissues back to the heart.

The most important of the forces that propel venous and lymphatic return from the regions of the body below the heart is the milking action caused by skeletal muscle contraction around the vessels. As muscles contract, the vessels are squeezed and the blood moves through them. Blood flow might be temporarily obstructed during a prolonged muscle contraction, but with relaxation of the muscles the blood is free to move again. Within the veins are valves that prevent the backflow of blood. Thus whenever a vein is squeezed, blood is propelled forward because the valves prevent it from flowing backward. In this way blood is gradually pushed toward the heart (Figure 42.14). As we already noted, the lymphatic vessels have similar valves.

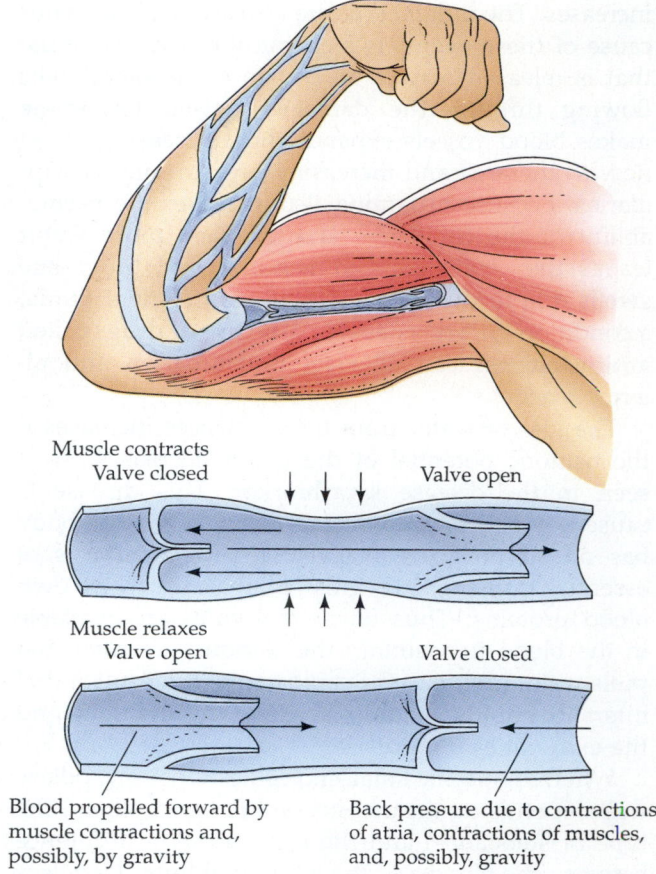

Muscle contracts
Valve closed   Valve open

Muscle relaxes
Valve open   Valve closed

Blood propelled forward by muscle contractions and, possibly, by gravity   Back pressure due to contractions of atria, contractions of muscles, and, possibly, gravity

42.14 One-Way Flow
Contractions of skeletal muscles squeeze the veins. This squeezing moves the blood in the veins toward the heart because of one-way valves that prevent backflow.

People who must stand still for prolonged periods, thereby accumulating blood in the veins of the lower body, appreciate the role of muscle contraction in venous return. The guards at Buckingham Palace in England, for instance, must shift their weight and contract their leg muscles periodically to prevent the extreme lowering of blood flow to the heart that causes fainting. Gravity causes edema as well as blood accumulation in veins. The back pressure that builds up in the capillaries when blood accumulates in the veins shifts the balance between blood pressure and osmotic potential so that there is a net movement of fluid into the interstitial spaces. This is why you have trouble putting your shoes back on after you sit for a long time with your shoes off, such as on an airline flight. In persons with very expandable veins, the veins may become so stretched that the valves can no longer prevent backflow. This condition produces varicose (swollen) veins. Draining these veins is highly desirable and can be aided by wearing support hose and periodically elevating the legs above the level of the heart.

During exercise, the milking action of muscles speeds blood toward the heart to be pumped to the lungs and then to the respiring tissues. As an animal runs, its legs act as auxiliary vascular pumps, returning blood to the heart from the veins of the lower body. As a greater volume of blood returns to the heart it contracts more forcefully, and its pumping action becomes more effective. This strengthening of the heartbeat is due to a property of cardiac muscle fibers referred to as the **Frank–Starling law**: If the fibers are stretched, as they are when the volume of returning blood increases, they contract more forcefully. This principle holds (within a certain range) whenever venous return increases, by any mechanism.

The actions of breathing also help return venous blood to the heart. The ventilatory muscles create suction that pulls air into the lungs (see Chapter 41), and this suction also pulls blood and lymph toward the chest, increasing venous return to the right atrium.

Some smooth muscle in the walls of the veins moves venous blood back to the heart by constricting the veins and moving the blood forward. These muscles are rare in most of the veins and are totally absent from lymphatic vessels in humans. They do not play a major role in venous return. However, in the largest veins closest to the heart, smooth muscle contraction at the onset of exercise can suddenly increase venous return and stimulate the heart in accord with the Frank–Starling law, thus increasing cardiac output.

THE BLOOD

We have considered the circulation of the blood in detail without looking closely at the blood itself. Blood is a tissue; it has cellular elements suspended in an aqueous medium of specific, yet complex, composition. The cells of the blood can be separated from the aqueous medium, called **plasma**, by centrifugation (Figure 42.15). If we take a 100-milliliter sample of blood and spin it in a centrifuge, all the cells move to the bottom of the tube, leaving the straw-colored, clear plasma on top. The **packed cell volume** or **hematocrit**, is the percentage of the blood volume made up by cells. Normal hematocrit is about 38 percent for women and 42 percent for men, but the values can vary considerably. They are usually higher, for example, in people living and doing heavy work at high altitude. We will consider next the three classes of cellular elements in blood: the red blood cells (erythrocytes); the white blood cells (leukocytes); and the platelets, which are pinched-off fragments of cells.

Red Blood Cells

Most of the cells in the blood are **erythrocytes**, or red blood cells. The function of red blood cells is to transport the respiratory gases (see Chapter 41). There are about 5 million red blood cells per milliliter of blood but only 5,000 to 10,000 white blood cells in the same volume. Red blood cells form from special cells called

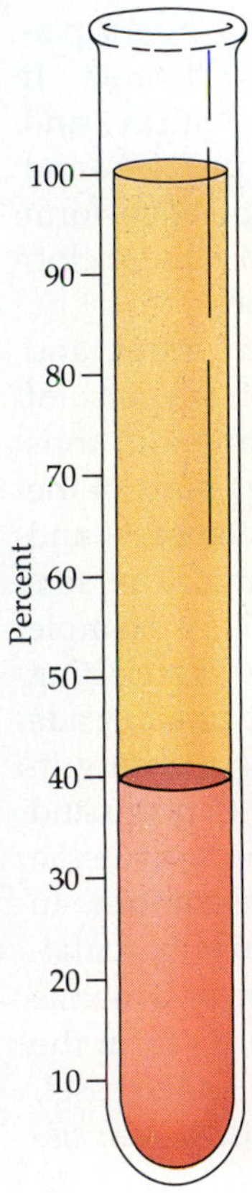

42.15 The Composition of Blood
Blood consists of a complex aqueous solution, numerous cell types, and cell fragments.

Withdraw blood from arm, place in test tube, and centrifuge

Plasma portion

Components	Water	Salts Sodium, potassium, calcium, magnesium, chloride, bicarbonate	Plasma proteins Albumin Fibrinogen Immunoglobulins	Transported by blood Nutrients (e.g. glucose, vitamins) Waste products of metabolism Respiratory gases (O_2 and CO_2) Hormones Heat
Functions	Solvent	Osmotic balance, pH buffering, regulation of membrane potentials	Osmotic balance, pH buffering, clotting, immune responses	

Cellular portion (hematocrit)

Components	Erythrocytes (red blood cells)	Leukocytes (white blood cells) Basophil, Eosinophil, Neutrophil, Lymphocyte, Monocyte	Platelets
Number per mm³ of blood	5–6 million	5,000–10,000	250,000–400,000
Functions	Transport oxygen and carbon dioxide	Destroy foreign cells, produce antibodies; roles in allergic responses	Blood clotting

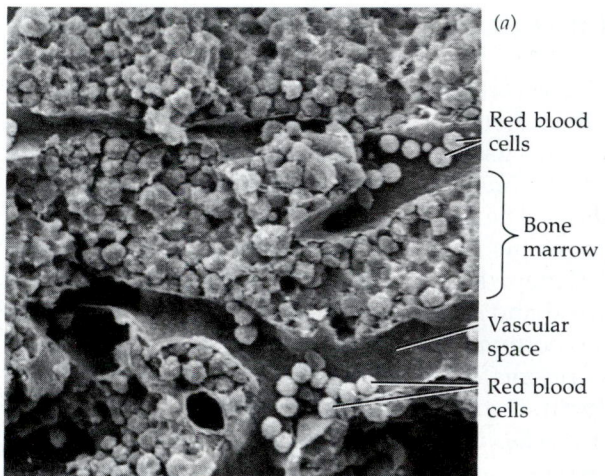

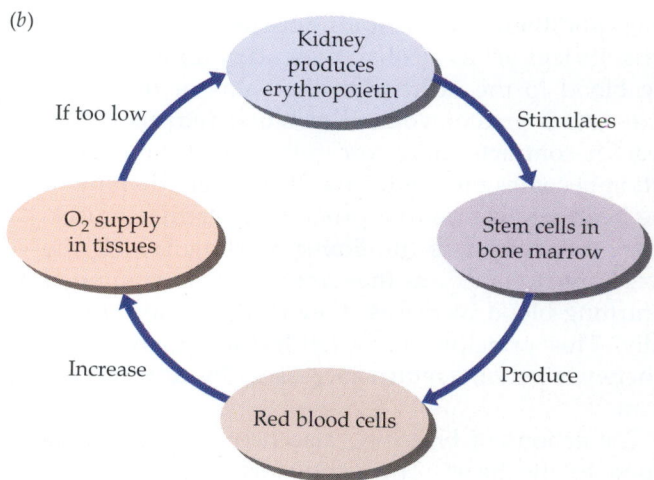

42.16 Red Blood Cells Form in the Bone Marrow
(a) In this scanning electron micrograph of a section of bone, the marrow is surrounded by fine blood vessels. As new red blood cells mature, they squeeze through the endothelium lining the vessels and enter the blood. *(b)* Erythropoietin stimulates stem cells in the bone marrow to produce red blood cells.

stem cells. Stem cells are found in the bone marrow, particularly in the ribs, breastbone, pelvis, and vertebrae (Figure 42.16*a*). Red blood cell production is controlled by a hormone, **erythropoietin**, which is released by cells in the kidney in response to insufficient oxygen (Figure 42.16*b*).

Erythropoietin stimulates stem cells to produce red blood cells. Under normal conditions your bone marrow produces about 2 million red blood cells every second. The developing red blood cells divide many times while still in the bone marrow, and during this time they are producing hemoglobin. When the hemoglobin content of a red blood cell approaches about 30 percent, its nucleus, endoplasmic reticulum, Golgi apparatus, and mitochondria begin to break down. This process is almost complete when the new red blood cell squeezes through pores in the endothelial walls of blood vessels and enters the circulation. Each red blood cell circulates for about 120 days and is then broken down. The iron from its hemoglobin molecules is recycled to the bone marrow. Mature red blood cells are biconcave flexible discs packed with hemoglobin. Their shape gives them a large surface area for gas exchange, and their flexibility enables them to squeeze through the capillaries (see Figures 41.16 and 42.11).

White Blood Cells

Leukocytes, or white blood cells, defend the body against infection (see Chapter 16). Some leukocytes search for and destroy foreign cells; some are phagocytes that consume bacteria, debris, and even dead or damaged cells from our own bodies; and some manufacture antibodies. Leukocytes squeeze through capillary walls and spend a great deal of time outside the vascular system. They move about by amoeboid motion and travel to sites of infection and cell damage by following cues from chemicals released by dead or sick cells.

Platelets and Blood Clotting

Platelets bud off from large cells in the bone marrow. A platelet is just a tiny fragment of a cell with no nucleus, but it is packed with enzymes and chemicals necessary for its function of sealing leaks in the blood vessels—that is, clotting the blood. When a vessel is damaged, collagen fibers are exposed. When a platelet encounters collagen fibers, it is activated. It swells, becomes irregularly shaped and sticky, and releases chemicals that activate other platelets and initiate the clotting of blood. The sticky platelets form a plug at the damaged site, and the subsequent clotting forms a stronger patch on the vessel.

The clotting of blood requires many steps and many **clotting factors**. The absence of any one of these factors can cause excessive bleeding and thus can be lethal. Because the liver produces most of the clotting factors, liver diseases such as hepatitis and cirrhosis can result in excessive bleeding. The sex-linked trait hemophilia (see Chapter 10) is an example of a genetic inability to produce one of the clotting factors. Blood clotting factors participate in a cascade of steps that activate other substances circulating in the blood. The cascade begins with cell damage and platelet activation and continues with the conversion of an inactive circulating enzyme, **prothrombin**, to its active form, **thrombin**. Thrombin causes circulating protein molecules called **fibrinogen** to polymerize and form **fibrin** threads. The fibrin threads form the meshwork that clots the blood cells, seals the vessel, and provides a scaffold for the formation of scar tissue (Figure 42.17).

Injury to the lining of a blood vessel exposes collagen fibers; platelets adhere and get sticky

Platelets release substances that cause the vessel to contract. Sticky platelets form a plug and initiate formation of a fibrin clot

The fibrin clot seals the wound until the vessel wall heals

Red blood cell

Collagen fibers

Platelet

Platelet plug

Fibrin meshwork

(a)

Clotting factors
1. Released from platelets and injured tissue
2. Plasma proteins synthesized in liver and circulating in inactive form

Prothrombin circulating in plasma

Thrombin

Fibrinogen circulating in plasma

Fibrin

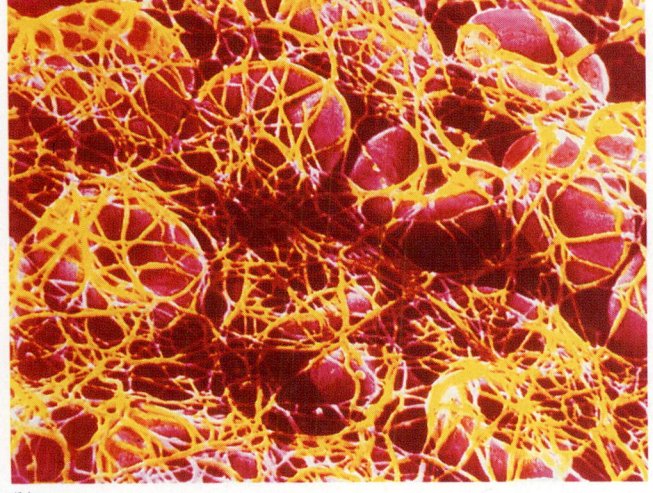

(b)

42.17 Blood Clotting
(a) Damage to a blood vessel initiates a cascade of events that produces a fibrin meshwork. (b) As the fibrin forms, red blood cells become enmeshed and form a clot, as this scanning electron micrograph shows.

Plasma

Plasma is a complex solution of gases, ions, nutrient molecules, and proteins. Most of the ions are Na^+ and Cl^- (hence the salty taste of blood), but many other ions are also present. Nutrient molecules in plasma include glucose, amino acids, lipids, cholesterol, and lactic acid. The circulating proteins have many functions. We have just noted proteins that function in blood clotting; others of interest include albumin, which is largely responsible for the osmotic potential in capillaries that prevents a massive loss of water from plasma to interstitial spaces; antibodies (the immunoglobulins); hormones; and various carrier molecules, such as **transferrin**, which carries iron

from the gut to where it is stored or used. Plasma is very similar to interstitial fluid in composition, and most of its components move readily between these two fluid compartments of the body. The main difference between the two fluids is the higher concentration of proteins in the plasma.

CONTROL AND REGULATION OF CIRCULATION

The circulatory system is controlled and regulated at many levels. Every tissue requires an adequate supply of blood that is saturated with O_2, that carries essential nutrients, and that is relatively free of waste products. The nervous system cannot monitor and control every capillary bed in the body. Instead, each bed regulates its own blood flow through **autoregulatory mechanisms** that cause the arterioles supplying the bed to constrict or dilate.

The autoregulatory actions of every capillary bed in every tissue influence the pressure and composition of the arterial blood leaving the heart. For example, if many arterioles suddenly dilate, allowing blood to flow through many more capillary beds, arterial blood pressure falls. If all the newly filled capillary beds contribute metabolic waste products to the blood, the concentration of wastes in the blood returning to the heart increases. Thus events in all capillary beds throughout the body produce combined effects on arterial blood pressure and composition. The nervous and endocrine systems respond to changes in arterial blood pressure and composition by changing breathing, heart rate, and blood distribution to match the metabolic needs of the body.

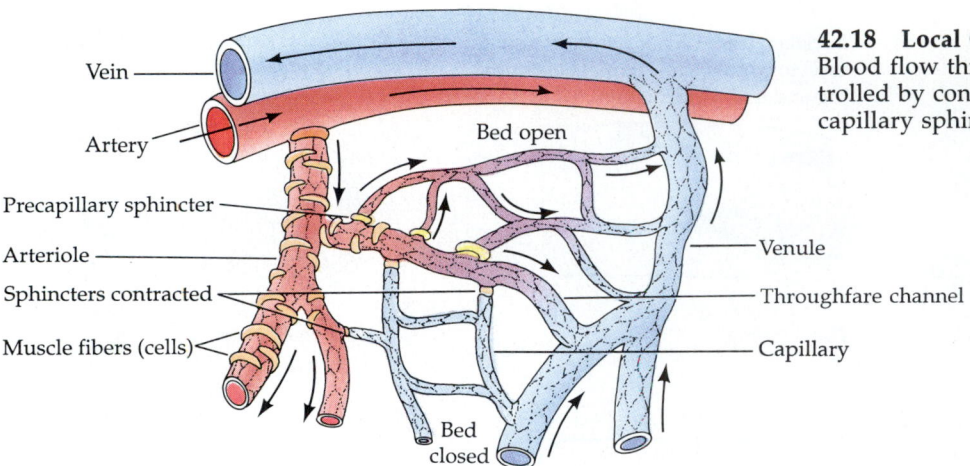

42.18   Local Control of Blood Flow
Blood flow through a capillary bed is controlled by constriction of arterioles and precapillary sphincters.

Autoregulation

The autoregulatory mechanisms that adjust the flow of blood to a tissue are part of the tissue itself, but they can be influenced by the nervous system and certain hormones. The amount of blood that flows through a capillary bed is controlled by the degree of contraction of the smooth muscles of the arteries and arterioles feeding that bed: As the muscles contract, they constrict the vessels, thereby decreasing the flow. The flow of blood in a typical capillary bed is diagrammed in Figure 42.18. Blood flows into the bed from an arteriole. Smooth-muscle "cuffs," or precapillary sphincters, on the arteriole can completely shut off the supply of blood to the capillary bed. When the precapillary sphincters are relaxed and the arteriole is open, the arterial blood pressure pushes blood into the capillaries.

Autoregulation depends on the sensitivity of the smooth muscle to the composition of its chemical environment. Low O_2 concentrations and high CO_2 concentrations cause the smooth muscle to relax, thus increasing the supply of blood, which brings in more O_2 and carries away CO_2. Increases in the concentration of products of metabolism other than CO_2, such as lactate, hydrogen ions, potassium, and adenosine, also promote increased blood flow by this mechanism. Hence activities that increase the metabolism of a tissue also increase the blood flow to that tissue.

Control and Regulation by the Endocrine and Nervous Systems

The same smooth muscles of arteries and arterioles that respond to autoregulatory stimuli also respond to signals from the endocrine and central nervous systems. Most arteries and arterioles are innervated by the autonomic nervous system, particularly the sympathetic division. Most sympathetic neurons release norepinephrine, which causes the smooth muscle fibers to contract, thus constricting the vessels and increasing their resistance to blood flow. An exception is found in skeletal muscle, where specialized sympathetic neurons release acetylcholine and cause the smooth muscles of the arterioles to relax and the vessels to dilate, causing more blood to flow to the muscle.

Hormones also can cause arterioles to constrict. Epinephrine has actions similar to those of norepinephrine; it is released from the adrenal medulla during massive sympathetic activation—the fight-or-flight response. Angiotensin, produced when blood pressure to the kidneys falls, causes arterioles to constrict. Vasopressin, released by the posterior pituitary, has similar effects (Figure 42.19). These hormones influence arterioles located for the most part in peripheral tissues (extremities) or in tissues whose functions need not be maintained continuously (such as the gut). By reducing blood flow in those arterioles, the hormones increase the central blood pressure and blood flow to essential organs such as the heart, brain, and kidneys.

The autonomic nervous system activity that controls heart rate and constriction of blood vessels originates in cardiovascular centers in the medulla of the brain stem. Many inputs converge on this central integrative network and influence the commands it issues via parasympathetic and sympathetic fibers (Figure 42.20). Of special importance is information about changes in blood pressure from stretch sensors in the walls of the great arteries—the aorta and the carotid arteries. Increased activity in the stretch sensors indicates rising blood pressure and inhibits sympathetic nervous system output. As a result, the heart slows and arterioles in peripheral tissues dilate. If pressure in the great arteries falls, the activity of the stretch sensors decreases, stimulating sympathetic output. Increased sympathetic output causes the heart to beat faster and the arterioles in peripheral tissues to constrict. When arterial pressure falls, the change in stretch-sensor activity also causes the hypothalamus to release vasopressin, which helps to

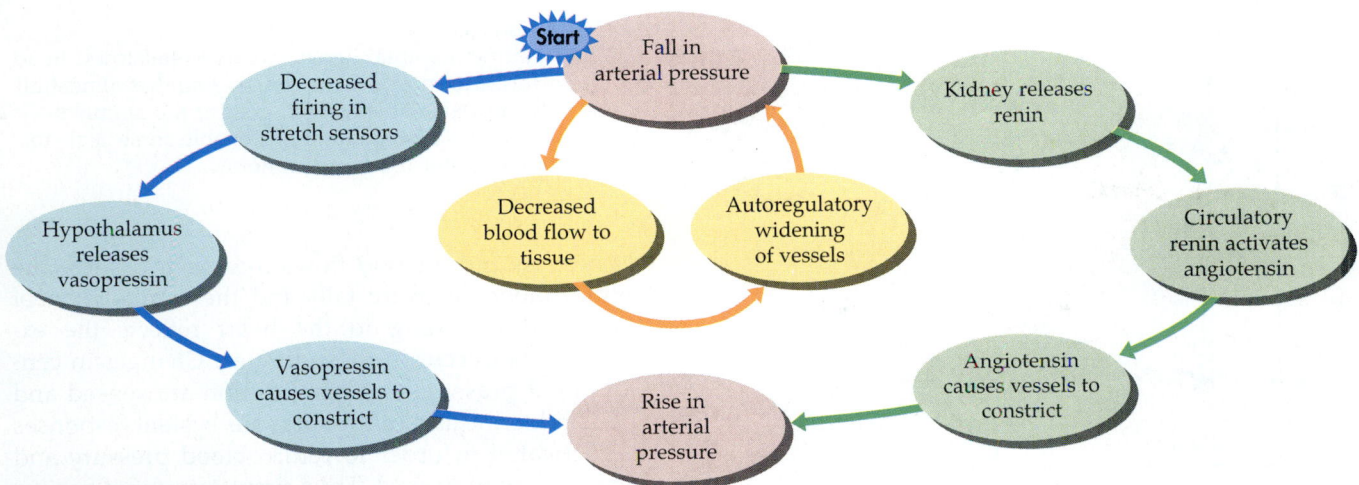

42.19 Controlling Blood Pressure through Vascular Resistance
A fall in arterial pressure reduces blood flow to tissues, resulting in local accumulation of metabolic wastes. This change in the extracellular environment stimulates autoregulatory opening of the arteries that would lead to a further fall in central blood pressure if this were not prevented by the negative feedback mechanisms shown in this diagram, which work through promoting constriction of arteries in less essential tissues.

increase blood pressure by stimulating peripheral arterioles to constrict.

Other information that causes the medullary regulatory system to increase heart rate and blood pressure comes from the carotid and aortic bodies (see Figure 41.23). These nodules of modified smooth muscle tissue are chemosensors that respond to inadequate O_2 supply. If arterial blood flow slows or the O_2 content of the arterial blood falls drastically, these sensors are activated and send signals to the regulatory center.

The regulatory center also receives input from other brain areas. Emotions and the anticipation of intense activity, such as at the start of a race, can cause the center to increase heart rate and blood pressure. A reflex that slows the heart is the so-called diving reflex, which is highly developed in marine mammals (Figure 42.21). Humans also have a diving reflex, which causes the heart to beat more slowly when the face is immersed in water.

A question we can ask about any physiological system is, "What is being regulated and how?" In the respiratory system (see Chapter 41), it is primarily the CO_2 concentration of the blood, and to a lesser extent the O_2 concentration, that is regulated by changes in the depth and frequency of breathing. Regulation in the circulatory system is more complex. The blood flow to individual tissues is regulated by local, autoregulatory mechanisms that cause dilation of local arterioles and precapillary sphincters when the tissue needs more oxygen or has accumulated

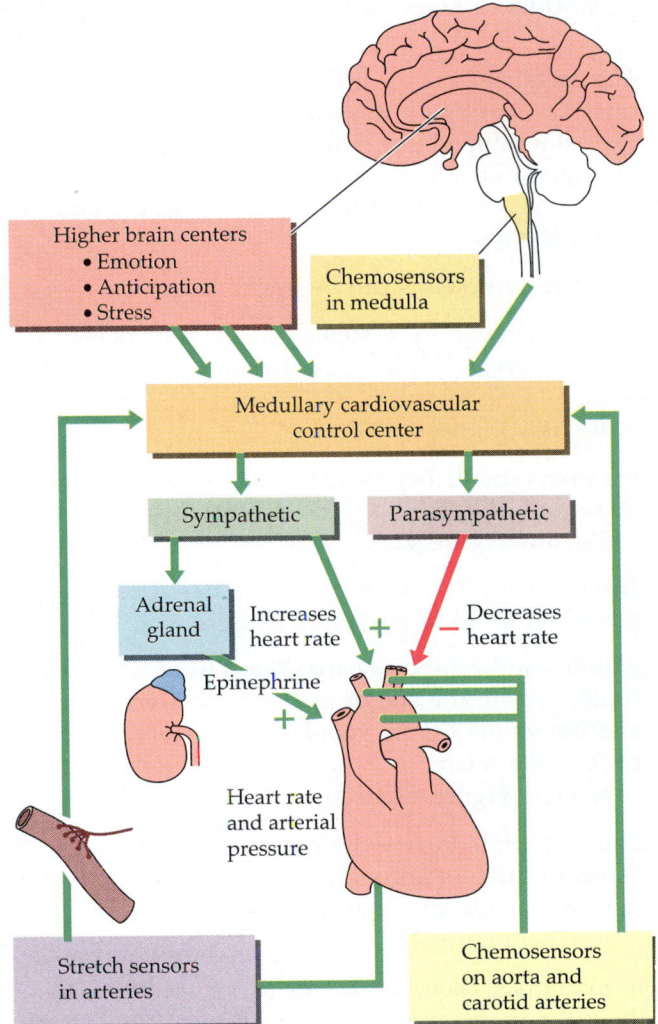

42.20 Regulating Blood Pressure
The autonomic nervous system controls heart rate in response to information about blood pressure and blood composition that is integrated by regulatory centers in the medulla.

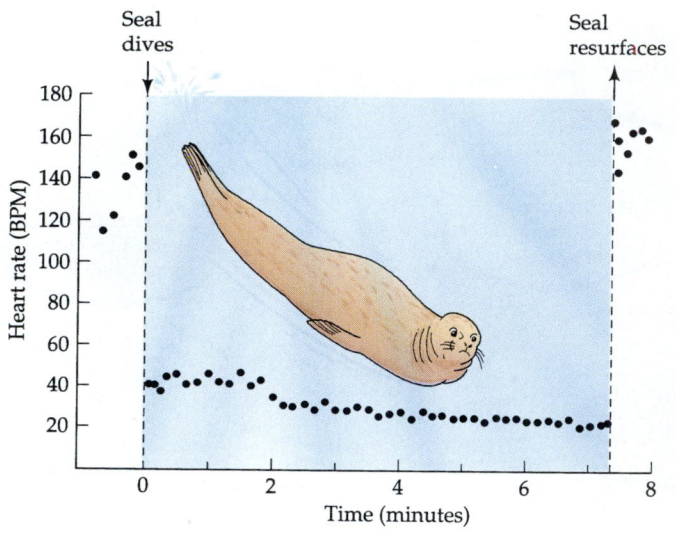

Seal dives | Seal resurfaces

Heart rate (BPM)

Time (minutes)

42.21 Master Divers
When a marine mammal dives, its heart rate slows. In addition, arteries to most organs constrict so that almost all blood flow and available oxygen goes to the animal's heart and brain. These adaptations enable some seals to remain under water for up to an hour.

wastes. As more blood flows into such tissues, the central blood pressure falls and the composition of the blood returning to the heart reflects the exchanges that occur in those tissues. Changes in central blood pressure and composition are sensed and both endocrine and central nervous system responses are activated in order to return blood pressure and composition to normal. Thus circulatory functions are matched to the regional and overall needs of the body.

SUMMARY of Main Ideas about Internal Transport and Circulatory Systems

The metabolic needs of the cells of very small animals are met by direct exchange of materials with the external medium. The metabolic needs of cells of larger animals are met by a circulatory system that transports nutrients, respiratory gases, and metabolic wastes.

Circulatory systems can be open or closed.

In open circulatory systems the blood or hemolymph leaves vessels and percolates through tissues.
   Review Figure 42.2

In closed circulatory systems the blood is contained in a system of vessels.
   Review Figure 42.3

The circulatory systems of vertebrates consist of a heart and a closed system of vessels.

Arteries and arterioles carry blood from the heart; capillaries are the site of exchange between blood and interstitial fluids; venules and veins carry blood back to the heart.
   Review Figure 42.10

The vertebrate heart evolved from two chambers in fishes to three in amphibians and reptiles, and to four in crocodilians, mammals, and birds.
   Review Figure 42.4

In mammals, blood circulates through two circuits: the pulmonary circuit and the systemic circuit.
   Review Figure 42.5

The cardiac cycle has two components: systole, in which the ventricles contract and the "lub" sound is heard; and diastole, in which the ventricles relax and the "dub" sound is heard
   Review Figure 42.6

A pacemaker (the sinoatrial node) control the cardiac cycle by initiating a wave of depolarization in the atria, which is conducted to the ventricles through the atrioventricular node.
   Review Figure 42.8

The autonomic nervous system controls the pacemaker.
   Review Figure 42.9

The measurement of blood pressure using a sphygmomanometer and a stethoscope is based on the cardiac cycle.
   Review Figure 42.7

Capillary beds are the site of fluid exchange between the blood and the interstitial fluids.

The exchange of fluids between the blood and interstitial fluids is determined by the balance between blood pressure and osmotic potential in the capillaries.
   Review Figure 42.12

Blood can be divided into a plasma portion (water, salts, and proteins) and a cellular portion (red blood cells, white blood cells, and platelets).
   Review Figure 42.15

Blood cells form in the bone marrow.

Red blood cells transport oxygen.
Review Figure 42.16

White blood cells defend the body from foreign substances.

Platelets (cell fragments), along with circulating proteins, are involved in clotting responses.
Review Figure 42.17

Blood flow through capillary beds is controlled by local conditions, hormones, and the autonomic nervous system.
Review Figures 42.18 and 42.19

Heart rate is controlled by the autonomic nervous system, which responds to information about blood pressure and blood composition that is integrated by regulatory centers in the brain stem.
Review Figure 42.20

SELF-QUIZ

1. An open circulatory system is characterized by
   a. the absence of a heart.
   b. the absence of blood vessels.
   c. blood with a composition different from that of interstitial fluid.
   d. the absence of capillaries.
   e. a higher pressure circuit through gills than to other organs.

2. Which of the following statements about vertebrate circulatory systems is *not* true?
   a. In fish, oxygenated blood from the gills returns to the heart through the left atrium.
   b. In mammals, deoxygenated blood leaves the heart through the pulmonary artery.
   c. In amphibians, deoxygenated blood enters the heart through the right atrium.
   d. In reptiles, the blood in the pulmonary artery has a lower oxygen content than the blood in the aorta.
   e. In birds, the pressure in the aorta is higher than the pressure in the pulmonary artery.

3. Which of the following statements about the human heart is true?
   a. The walls of the right ventricle are thicker than the walls of the left ventricle.
   b. Blood flowing through atrioventricular valves is always deoxygenated blood.
   c. The second heart sound is due to the closing of the aortic valve.
   d. Blood returns to the heart from the lungs in the vena cava.
   e. During systole the aortic valve is open and the pulmonary valve is closed.

4. Pacemaker actions of cardiac muscle
   a. are due to opposing actions of norepinephrine and acetylcholine.
   b. are localized in the bundle of His.
   c. depend on the gap junctions between cells that make up the atria and those that make up the ventricles.
   d. are due to spontaneous depolarization of the plasma membranes of some cardiac muscle cells.
   e. result from hyperpolarization of cells in the sinoatrial node.

5. Blood flow through capillaries is slow because
   a. lots of blood volume is lost from the capillaries.
   b. the pressure in venules is high.
   c. the total cross-sectional area of capillaries is larger than that of arterioles.
   d. the osmotic pressure in capillaries is very high.
   e. red blood cells are bigger than capillaries and must squeeze through.

6. How are lymphatic vessels like veins?
   a. Both have nodes where they join together into larger common vessels.
   b. Both carry blood under low pressure.
   c. Both are capacitance vessels.
   d. Both have valves.
   e. Both carry fluids rich in plasma proteins.

7. The production of red blood cells
   a. ceases if the hematocrit falls below normal.
   b. is stimulated by erythropoietin.
   c. is about equal to the production of white blood cells.
   d. is inhibited by prothrombin.
   e. occurs in bone marrow before birth and in lymph nodes after birth.

8. Which of the following does *not* increase blood flow through a capillary bed?
   a. High concentrations of CO_2
   b. High concentrations of lactate and hydrogen ions
   c. Histamine
   d. Vasopressin
   e. Increase in arterial pressure

9. The clotting of the blood
   a. is impaired in hemophiliacs because they don't produce platelets.
   b. is initiated when platelets release fibrinogen.
   c. involves a cascade of factors produced in the liver.
   d. is initiated by leukocytes forming a meshwork.
   e. requires conversion of angiotensinogen to angiotensin.

10. Autoregulation of blood flow to a tissue is due to
    a. sympathetic innervation.
    b. the release of vasopressin by the hypothalamus.
    c. increased activity of baroreceptors.
    d. chemosensors in carotid and aortic bodies.
    e. the effect of local environment on arterioles.

FOR STUDY

1. How is cardiac output increased at the beginning of a race? Include the Frank–Starling law in your answer.

2. The final stages of alcoholism involve loss of liver function and accumulation of fluids in extremities and the abdominal cavity. Explain how these two consequences of alcoholism are related.

3. A sudden and massive loss of blood results in a fall in blood pressure. Describe several mechanisms that help return blood pressure to normal.

4. You can describe the cycle of events in a ventricle of the heart by a graph that plots the pressure in the ventricle on the y axis and the volume of blood in the ventricle on the x axis. What would such a graph look like? Where would the heart sounds occur on this graph? How would the graph differ for the left and the right ventricles?

5. Why doesn't diastolic blood pressure fall to zero between heartbeats? Why does systolic blood pressure increase with (a) sympathetic activity, (b) increased venous return, and (c) age?

READINGS

Eckert, R., D. Randall and G. Augustine. 1988. *Animal Physiology: Mechanisms and Adaptations*, 3rd Edition. W. H. Freeman, New York. An outstanding textbook of animal physiology, with excellent coverage of the circulatory system.

Golde, D. W. and J. C. Gasson. 1988. "Hormones That Stimulate the Growth of Blood Cells." *Scientific American*, July. Now made by recombinant DNA methods, hemopoietins promise to transform the practice of medicine.

Robinson, T. F., S. M. Factor and E. H. Sonnenblick. 1986. "The Heart as a Suction Pump." *Scientific American*, July. A new proposal concerning the filling of the heart, along with interesting general information on cardiac muscle and the connective tissues of the heart.

Scholander, P. F. 1963. "The Master Switch of Life." *Scientific American*, December. Delightful, classic description of the discovery of the diving adaptations of marine mammals.

Vander, A. J., J. H. Sherman and D. S. Luciano. 1994. *Human Physiology: The Mechanisms of Body Function*, 6th Edition. McGraw-Hill, New York. Chapter 14 deals with circulation.

Zapol, W. M. 1987. "Diving Adaptations of the Weddell Seal." *Scientific American*, June. These breath-holding master divers provide an opportunity to study the diving reflex.

Zucker, M. 1980. "The Functioning of Blood Platelets." *Scientific American*, June. The role of platelets in blood clotting.

"Shark!" That single word can send chills down the spines of ocean bathers and surfers around the world. Why? Because these predators can take us as prey. The image of the great white shark is that of a veritable eating machine. It can be more than 6 meters long and can take a single bite over half a meter in diameter. It attacks at considerable speed and in the final moments of its approach opens its enormous mouth, turns its head upward, and embeds rows of triangular teeth into its prey. The shark then shakes violently to rip away a huge chunk of flesh by the sawing actions of its teeth. The attack can be so violent that some of its teeth are left in the carcass. The loss of teeth is no problem, however, because new teeth continually form and advance forward as if on conveyer belts to take the place of those that are lost.

Now imagine a shark three times bigger than the great white—18 meters long, with a mouth 2 meters wide. Could such a terrifying animal exist only in horror films? No. These are the dimensions of whale sharks, which are not terrifying at all. These enormous creatures are gentle, slow swimmers that divers have been known to hitch rides on by holding onto a fin or a tail. Their huge mouths take in thousands of tons of water per hour, which they filter through their gills to extract the millions of small organisms that are their food. When feeding, the contents of a whale shark's stomach may weigh half a ton or more. What animals eat and how they eat it are some of their most distinguishing characteristics.

DEFINING ORGANISMS BY WHAT THEY EAT

Animals must eat to stay alive, and in a sense they are what they eat. Animals are **heterotrophs**: They derive both their energy and their structural molecules from their food. **Autotrophs** (most plants, some bacteria, and some protists), on the other hand, can trap solar energy through photosynthesis and use it to synthesize all of their structures from inorganic materials. Almost all life runs on solar energy, but heterotrophs receive theirs indirectly, via the autotrophs. Photosynthetic autotrophs provide the nutritional foundation for all other known ecosystems, and heterotrophs have evolved an enormous diversity of adaptations for exploiting that continual source of life (Figure 43.1).

Heterotrophic nutritional lifestyles span a great range. **Saprophytes**, such as fungi, simply absorb organic molecules from dead organisms. Rather than feeding passively, as saprophytes do, some animals actively feed on the remains of dead organisms. **Detritivores**, such as earthworms, process environmental deposits containing organic matter for their nutri-

Shark!
To this great white shark, we are food.

43

Animal Nutrition

(a)

(b)

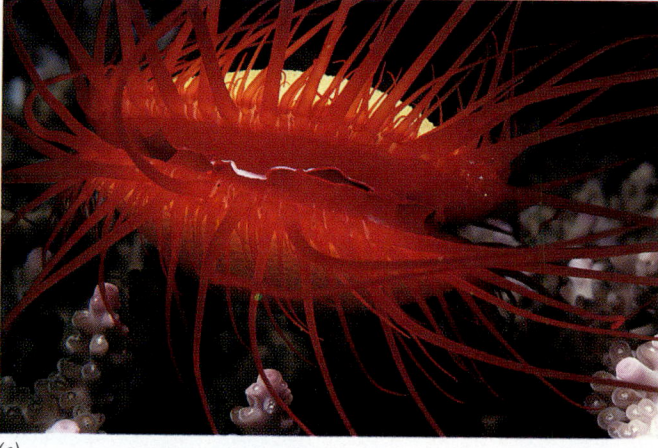

(c)

(d)

tional needs. All animals that feed on other living organisms can be considered **predators**. **Herbivores** are predators that prey on plants; **carnivores** prey on other animals. **Omnivores**, including humans, prey on both plants and animals. **Filter feeders**, such as clams and blue whales, prey on small organisms by filtering them out of the environmental medium. We are only too familiar with **fluid feeders**, which include mosquitoes, aphids, and leeches, as well as birds that feed on plant nectar. In this chapter we will examine some of the anatomical adaptations for this diversity of nutritional lifestyles.

The statement "We are what we eat" is true only in a limited sense. Although some parasitic animals can absorb all the basic nutrients they need from their environment, most animals must break down the complex molecules in their food into simple units

43.1 A Focus on the Consumers
Heterotrophs have evolved an amazing range of adaptations for exploiting sources of energy. (a) The manatee is a herbivore whose source of food is aquatic vegetation in tropical and subtropical rivers and lagoons. (b) The long bill of this Australian spiny-cheeked honeyeater, a fluid feeder, enables it to harvest the tiny amounts of nectar in individual flowers. (c) The red file clam of the South Pacific islands obtains food from the ocean water it constantly filters through its system. (d) The carnivorous polar bear is a fearsome predator. A strong swimmer, it feeds on fish and seals from the frigid Arctic Ocean; its feet, with hairy soles, are adapted to moving swiftly and surely across the ice packs, so it is equally able to bring down prey, including birds and mammals, on land.

such as amino acids, fatty acids, and sugars. These simplest nutrient units can then be used for synthesis of new molecules or metabolized as an energy source. The breakdown of food molecules is **digestion**. The cells of some animals, such as sponges, engulf particles of food and digest them intracellularly, but most animals process their food through **extracellular digestion**, in a digestive cavity called a gut. The gut of simple animals, such as flatworms and jellyfish, is a saclike structure with only one opening to the environment. More-complex animals have a tubular gut with a separate entrance and exit for food. Such animals ingest (take in) food through a mouth, where it may be broken up. The food then passes through the tubular digestive tract, where digestive enzymes break it down. Throughout this process the food is really outside the body because it has not crossed any cell membranes. Digestion occurs *outside* the cells. The products of digestion are then absorbed into the body and taken up by the cells.

In this chapter we will first consider the nutrients that organisms require and why. Then we will look at how animals procure nutrients, and once ingested, how nutrients are processed, digested, and absorbed. Finally we will learn how the body regulates its traffic in molecules used for metabolic fuel.

NUTRIENT REQUIREMENTS

Animals need food to supply the energy for metabolism and the carbon skeletons that they cannot synthesize but must have in order to build larger organic molecules. Mineral nutrients, such as iron and calcium, that animals require to build functional and structural molecules also come from food. The diet also provides complex organic molecules called vitamins that are needed in small quantities as cofactors for enzymes and for other purposes.

Nutrients as Fuel

In Chapters 6 and 7 we learned that energy in the chemical bonds of food molecules is transferred to the high-energy phosphate bonds of ATP. ATP then provides energy for active transport, biosynthesis of molecules, degradation of molecules, muscle contraction, and other work. Because they are never 100 percent efficient, these energy conversions produce heat as a by-product. Even the useful energy conversions eventually are reduced to heat, as molecules that were synthesized are broken down and energy of movement is dissipated by friction. In time, all the energy that is transferred to ATP from the chemical bonds of food molecules is released to the environment as heat. It is convenient, therefore, to talk about the energy requirements of animals and the energy

content of food in terms of a measure of heat energy: the **calorie**. A calorie is the amount of heat necessary to raise the temperature of one gram of water one degree centigrade. Since this value is a tiny amount of energy in comparison to the energy requirements of many animals, physiologists commonly use the **kilocalorie** (kcal) as a unit (1,000 calories = 1 kcal). Nutritionists also use the kilocalorie as a standard unit of energy, but they traditionally refer to it as the **Calorie**, which is always capitalized to distinguish it from the single calorie. A person on a diet of 1,000 Calories per may consume up to 1,000 kcal/day. Such confusion of terms is unfortunate, but we live with it. (It should be noted that physiologists are gradually abandoning the calorie as an energy unit as they switch to the International System of Units. In this system the basic unit of energy is the joule; 1 calorie = 4.184 joules.)

The metabolic rate of an animal (see Chapter 35) is a measure of the overall energy needs that must be met by an animal's ingestion and digestion of food. The components of food that provide energy are fats, carbohydrates, and proteins. Fat yields 9.5 kcal per gram when it is metabolically oxidized, carbohydrate yields 4.2 kcal per gram, and protein yields about 4.1 kcal per gram. The basal metabolic rate of a human is about 1,300 to 1,500 kcal a day for an adult female and 1,600 to 1,800 kcal a day for an adult male. Physical activity adds to this basal energy requirement. Some equivalences of food, energy, and exercise are shown in Figure 43.2.

Although the cells of the body use energy continuously, most animals do not eat continuously. Humans generally eat several meals a day, a lion may eat once in several days, a boa constrictor may eat once a month, and hibernating animals may go five to six months without eating. Therefore, animals must store fuel molecules that can be released as needed between meals. Carbohydrate is stored in liver and muscle cells as glycogen (see Chapter 3), but the total glycogen stores are usually not more than the equivalent of a day's energy requirements. Fat is the most important form of stored energy in the bodies of animals. Not only does fat have the highest energy content per gram, but it can be stored with little associated water, making it more compact. If migrating birds had to store energy as glycogen rather than fat to fuel long flights, they would be too heavy to fly! Protein is not used to store energy, although body protein can be metabolized as an energy source as a last resort.

If an animal takes in too little food to meet its needs for metabolic energy, it is **undernourished** and must make up the shortfall by metabolizing some of the molecules of its own body. Consumption of self for fuel begins with the storage compounds glycogen and fat. Protein loss is minimized for as long as pos-

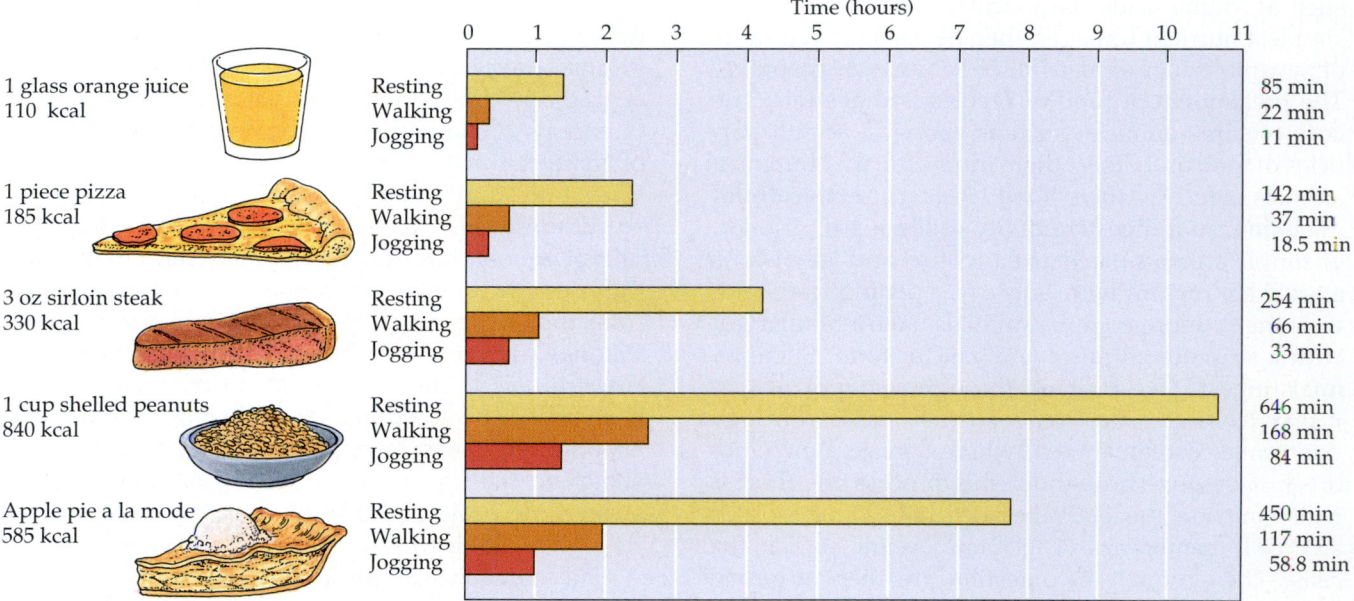

43.2   Food Energy and How Fast We Burn It
The energy in kilocalories for several common food items is shown on the left. The graph indicates about how long it would take a person with a basal metabolic rate of about 1,800 kilocalories a day to utilize the equivalent amount of energy while involved in various activities.

sible, but eventually the body has to use its own proteins for fuel. The breakdown of body proteins impairs body functions and eventually leads to death. Blood proteins are among the first to go, resulting in loss of fluid to the interstitial spaces (edema; see Chapter 42). Muscles atrophy (waste away) and eventually even brain protein is lost. Figure 43.3 shows the course of starvation. Undernourishment is rampant among people in underdeveloped and war-torn nations, and a billion people—one-fifth of the world's population—are undernourished. (Ironically, one cause of life-threatening undernourishment in Western developed nations is a self-imposed starvation called **anorexia nervosa** that results from a psychological aversion to body fat.)

When an animal consistently takes in more food than it needs to meet its energy demands, it is **overnourished**. The excess nutrients are stored as increased body mass. First, glycogen reserves build up; then additional dietary carbohydrate, fat, and protein are converted to body fat. In some species, such as hibernators, seasonal overnutrition is an important adaptation for surviving periods when food is unavailable. In humans, however, overnutrition can be a serious health hazard, increasing the risk of high blood pressure, heart attack, diabetes, and other disorders. A common clay building brick weighs about 5 pounds, so a person who is 50 pounds overweight is constantly carrying around the equivalent of ten bricks. That alone is quite a strain on the heart, but in addition, each extra pound of body tissue includes

miles of additional blood vessels through which the heart must pump blood. Obesity is a health hazard, but so are poorly planned fad or crash diets that can lead to malnutrition (discussed in the next section). People spend billions of dollars every year on schemes to lose weight, even though all they need to do is follow a simple rule: Take in fewer calories than your body burns, but maintain a balanced diet.

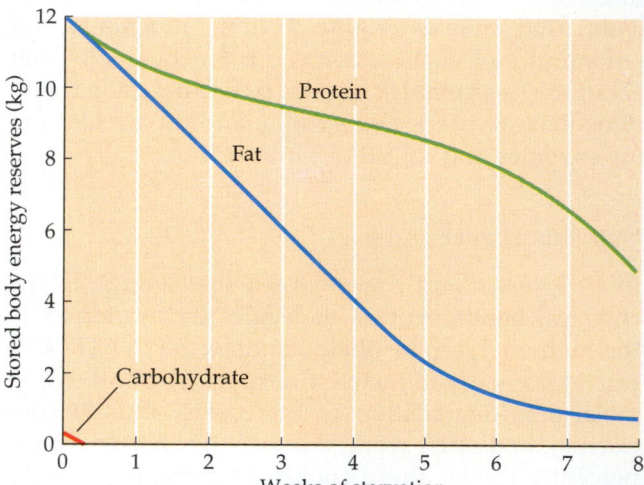

43.3   Depletion of Body Energy Reserves during Starvation
The carbohydrate reserves of our bodies are meager and are depleted by only a single day without food intake. Our major energy reserve is fat; even a person of average body weight has enough fat to survive four or five weeks without food. When most body fat has been exhausted, the only remaining fuel is protein, which is lost at an accelerating rate, with serious consequences and often death.

Nutrients as Building Blocks

Every animal requires certain basic organic molecules (carbon skeletons) that it cannot synthesize for itself but must have in order to build the complex organic molecules needed for life. An example of a required carbon skeleton is the acetyl group (Figure 43.4). Animals cannot make acetyl groups from carbon, oxygen, and hydrogen molecules; they obtain acetyl groups by metabolizing carbohydrates, fats, or proteins. From these acquired acetyl groups, animals create a wealth of other necessary compounds, including fatty acids, steroid hormones, electron carriers for cellular respiration, certain amino acids, and, indirectly, legions of other compounds. The three major classes of nutrients—carbohydrates, fats, and proteins—provide both the energy and the carbon skeletons for biosynthesis.

Because the acetyl group can be derived from the metabolism of virtually any food, it is unlikely ever to be in short supply for an animal with an adequate food supply. Other carbon skeletons, however, are derived from more-limited sources, and an animal can suffer a deficiency of these materials even if its caloric intake is adequate. This state of deficiency is called **malnutrition**. Amino acids, the building blocks of protein, are a good example of such substances. Humans obtain amino acids by digesting protein from food and absorbing the resulting amino acids. The body then synthesizes its own protein molecules, as specified by its DNA, from these dietary amino acids. Another source of amino acids is the breakdown of existing body proteins, which are in constant turnover as the tissues of the body undergo normal remodeling and renewal.

Animals can synthesize some of their own amino acids by taking carbon skeletons synthesized from acetyl or other groups and transferring to them amino groups ($-NH_2$) derived from other amino acids. Most animals, however, cannot synthesize all the amino acids they need. Each species has certain **essential amino acids** that must be obtained from food. Different species have different essential amino acids and, in general, herbivores have fewer essential amino acids than do carnivores. If an animal does not take in one of its essential amino acids, its protein synthesis is impaired. Think of protein synthesis as typing a story. If the typewriter is missing a key, the story either comes to a stop or has an error in it wherever the letter represented by that key is needed. In protein synthesis, the story usually comes to a stop and a functional protein is not produced.

Humans require eight essential amino acids in their diet: isoleucine, leucine, lysine, methionine, phenylalanine, threonine, tryptophan, and valine (see Table 3.1). All eight are available in milk, eggs, or meat; however, no plant food contains all eight. A strict vegetarian thus runs a risk of protein malnutrition. An appropriate dietary *mixture* of plant foods, however, supplies all eight essential amino acids (Figure 43.5). Wheat, corn, rice, and other grains are deficient in lysine and isoleucine but are well stocked with most of the others. Beans, lentils, and other legumes have lots of lysine but are low in methionine and tryptophan. Eating only grains or only beans would lead to a serious deficiency of one or more amino acids. If grains and beans are eaten together, however, the diet includes all the essential amino acids.

In general, grains are complemented by legumes or by milk products; legumes are complemented by grains and by seeds and nuts. Long before the chemical basis for this complementarity was understood, societies with little access to meat learned appropriate

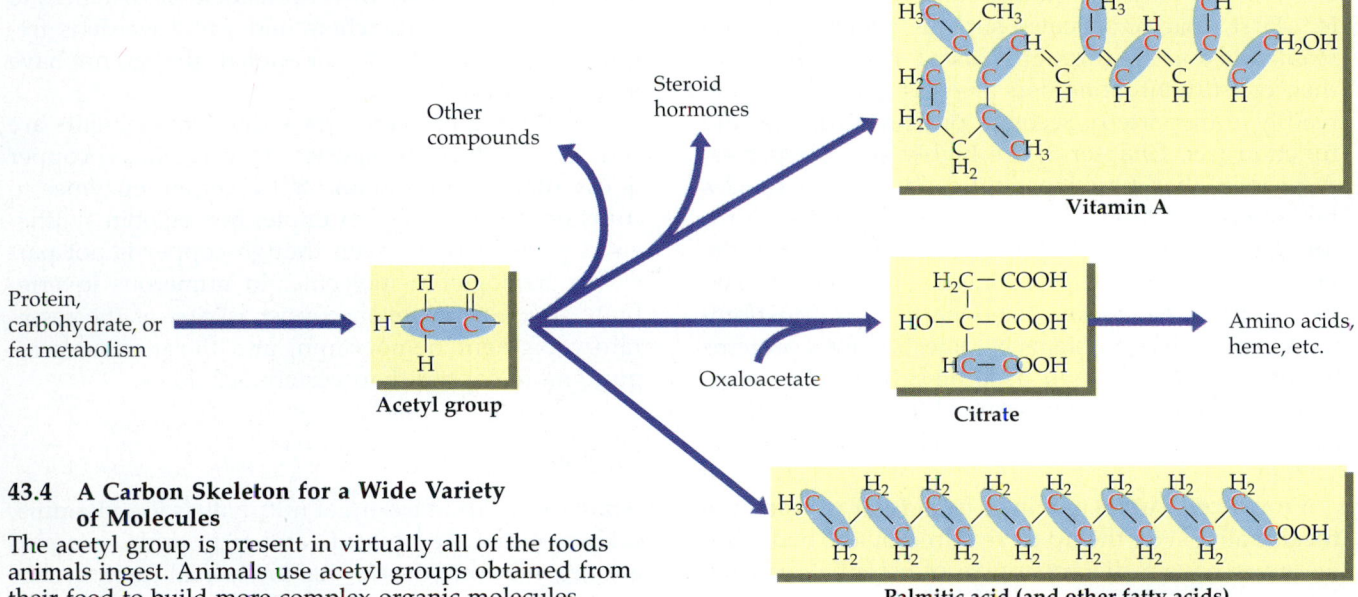

43.4 A Carbon Skeleton for a Wide Variety of Molecules

The acetyl group is present in virtually all of the foods animals ingest. Animals use acetyl groups obtained from their food to build more-complex organic molecules.

43.5 A Strategy for Vegetarians
By combining cereal grains and legumes, a vegetarian can obtain all eight essential amino acids.

dietary practices through trial and error. Many Central and South American peoples traditionally ate beans with corn, and the native peoples of North America complemented their beans with squash. Remember that we do not retain great stores of free amino acids in our bodies, yet we synthesize proteins continuously. It makes little nutritional sense to eat grains one day and beans the next; they must be eaten together for proper amino acid balance. Excess amino acids are burned for fuel, converted to fat, or excreted as waste.

Why are dietary proteins completely digested to their constituent amino acids before being used by the body? Wouldn't it be more energy-efficient to reuse some dietary proteins directly? There are several reasons why ingested proteins are not used "as is." First, macromolecules such as proteins are not readily taken up through plasma membranes, but their constituent monomers (such as amino acids) are readily transported. Second, protein structure and function (see Chapter 3) are highly species-specific. A protein that functions optimally in one species might not function well in another species. Third, foreign proteins entering the body directly from the gut would be recognized as invaders and be attacked by the immune system (see Chapter 16). Most animals avoid these problems by digesting food proteins extracellularly and then absorbing the amino acids into the body. The new proteins formed from these amino acids are recognized as "self" by the immune system.

From acetyl units obtained from food we can synthesize almost all the lipids required by the body, but we must have a dietary source of essential fatty ac-

ids—notably linoleic acid. Essential fatty acids are necessary components of membrane phospholipids, and a deficiency can lead to problems such as infertility and impaired lactation.

Mineral Nutrients

Table 43.1 lists the principal mineral elements required by animals. Certain species require additional elements. Elements required in large amounts are known as **macronutrients**; elements required in only tiny amounts are called **micronutrients**. Some essential elements are required in such minute amounts that deficiencies are never observed, but these elements are nevertheless essential.

Animals need calcium and phosphorus in great quantity. Calcium phosphate is the principal structural material in bones and teeth. Muscle contraction, nerve function, and many other intracellular functions in animals require calcium. Phosphorus is an integral component of nucleic acids. We learned in Chapter 7 about the role of phosphate groups in biological energy transfers. Sulfur is part of the structure of two amino acids and is therefore found in almost all proteins. Other essential compounds also contain sulfur. Iron is the oxygen-binding atom in both hemoglobin and myoglobin, the oxygen-carrying proteins in vertebrate blood and muscle. In addition, iron undergoes redox reactions in some of the electron-carrying proteins of cellular respiration. A number of mineral nutrients—among them magnesium, manganese, zinc, and cobalt—act as cofactors for enzymes. Potassium, sodium, and chloride ions are particularly important in the osmotic balance of tissues and in the electrical properties of membranes, including resting potentials and action potentials.

Animals require large amounts of both sodium and chloride ions. Because plants contain few of these ions, herbivores may travel considerable distances to natural salt licks. Ranchers and game wardens frequently supply salt licks for animals that do not have access to natural sources.

Specific requirements for individual elements are different in different species. In vertebrates, copper is essential in trace amounts for certain enzymes to function properly. For example, hemoglobin synthesis requires copper, even though copper is not part of the hemoglobin molecule. In numerous invertebrate species, however, copper is part of the respiratory pigment hemocyanin, and those animals require more copper than vertebrates do.

Vitamins

Another group of essential nutrients is the **vitamins**. Like essential amino acids and fatty acids, vitamins are organic compounds that an animal cannot make

TABLE 43.1
Mineral Elements Required by Animals

ELEMENT	SOURCE IN HUMAN DIET	MAJOR FUNCTIONS
Macronutrients		
Calcium (Ca)	Dairy foods, eggs, green leafy vegetables, whole grains, legumes, nuts	Found in bones and teeth; blood clotting; nerve and muscle action; enzyme activation
Chlorine (Cl)	Table salt (NaCl), meat, eggs, vegetables, dairy foods	Water balance; digestion (as HCl); principal negative ion in fluid around cells
Magnesium (Mg)	Green vegetables, meat, whole grains, nuts, milk, legumes	Required by many enzymes; found in bones and teeth
Phosphorus (P)	Dairy foods, eggs, meat, whole grains, legumes, nuts	Found in nucleic acids, ATP, and phospholipids; bone formation; buffers; metabolism of sugars
Potassium (K)	Meat, whole grains, fruits, vegetables	Nerve and muscle action; protein synthesis; principal positive ion in cells
Sodium (Na)	Table salt, dairy foods, meat, eggs, vegetables	Nerve and muscle action; water balance; principal positive ion in fluid around cells
Sulfur (S)	Meat, eggs, dairy foods, nuts, legumes	Found in proteins and coenzymes; detoxification of harmful substances
Micronutrients		
Chromium (Cr)	Meat, dairy foods, whole grains, dried beans, peanuts, brewer's yeast	Glucose metabolism
Cobalt (Co)	Meat, tap water	Found in vitamin B_{12}; formation of red blood cells
Copper (Cu)	Liver, meat, fish, shellfish, legumes, whole grains, nuts	Found in active site of many redox enzymes and electron carriers; production of hemoglobin; bone formation
Fluorine (F)	Most water supplies	Resistance to tooth decay
Iodine (I)	Fish, shellfish, iodized salt	Found in thyroid hormones
Iron (Fe)	Liver, meat, green vegetables, eggs, whole grains, legumes, nuts	Found in active sites of many redox enzymes and electron carriers, hemoglobin, and myoglobin
Manganese (Mn)	Organ meats, whole grains, legumes, nuts, tea, coffee	Activates many enzymes
Molybdenum (Mo)	Organ meats, dairy foods, whole grains, green vegetables, legumes	Found in some enzymes
Selenium (Se)	Meat, seafood, whole grains, eggs, chicken, milk, garlic	Fat metabolism
Zinc (Zn)	Liver, fish, shellfish, and many other foods	Found in some enzymes and some transcription factors; insulin physiology

for itself but that are required for its normal growth and metabolism (Box 43.A). Most vitamins function as coenzymes or parts of coenzymes and are required in very small amounts compared with essential amino acids and fatty acids that have structural roles. The list of required vitamins varies from species to species. For example, ascorbic acid (vitamin C) is not a vitamin for most mammals because they can make it themselves. Primates (including humans), however, do not have this ability, so ascorbic acid is a vitamin. If we do not get it in our diet, we develop the disease known as scurvy (Box 43.B). There are 13 such compounds that humans cannot synthesize in sufficient quantities (Table 43.2). They are divided into two groups: Water-soluble vitamins and fat-soluble vitamins.

Water-soluble vitamins (the B complex and vitamin C) play roles in both vertebrates and invertebrates. The B vitamins are coenzymes or parts of coenzymes. The B vitamin niacin, for example, we have encountered already under another name, nicotinamide (see Chapter 7). It is the portion of NAD (nicotinamide adenine dinucleotide) and NADP that undergoes oxidation and reduction in the respiratory chain and in other key redox systems in all living things. Riboflavin (vitamin B_2) similarly is the site of oxidation and reduction in the respiratory chain intermediates FAD (flavin adenine dinucleotide) and FMN (flavin mononucleotide). Vitamin C (ascorbic acid) has a number of functions, among them an essential role in the formation of the structural protein collagen. Collagen is a fibrous protein that is a

TABLE 43.2
Vitamins in the Human Diet

VITAMIN	SOURCE	FUNCTION	DEFICIENCY SYMPTOMS
Water-Soluble			
B₁, thiamin	Liver, legumes, whole grains, yeast	Coenzyme in cellular respiration	Beriberi, loss of appetite, fatigue
B₂, riboflavin	Dairy foods, organ meats, eggs, green leafy vegetables	Coenzyme in cellular respiration (in FAD and FMN)	Lesions in corners of mouth, eye irritation, skin disorders
Niacin (nicotinamide, nicotinic acid)	Meat, fowl, liver, yeast	Coenzyme in cellular metabolism (in NAD and NADP)	Pellagra, skin disorders, diarrhea, mental disorders
B₆, pyridoxine	Liver, whole grains, dairy foods	Coenzyme in amino acid metabolism	Anemia, slow growth, skin problems, convulsions
Pantothenic acid	Liver, eggs, yeast	In acetyl CoA	Adrenal problems, reproductive problems
Biotin	Liver, yeast, bacteria in gut	In coenzymes	Skin problems, loss of hair
B₁₂ cobalamin	Liver, meat, dairy foods, eggs	Coenzyme in formation of nucleic acids and proteins, and in red blood cell formation	Pernicious anemia
Folic acid	Vegetables, eggs, liver, whole grains	Coenzyme in formation of heme and nucleotides	Anemia
C, ascorbic acid	Citrus fruits, tomatoes, potatoes	Aids formation of connective tissues; prevents oxidation of cellular constituents	Scurvy, slow healing, poor bone growth
Fat-Soluble			
A, retinol	Fruits, vegetables, liver, dairy foods	In visual pigments	Night blindness, damage to mucous membranes
D, calciferol	Fortified milk, fish oils, sunshine	Absorption of calcium and phosphorus	Rickets
E, tocopherol	Meat, dairy foods, whole grains	Muscle maintenance, prevents oxidation of cellular components	Anemia
K, menadione	Intestinal bacteria, liver	Blood clotting	Blood-clotting problems (in newborns)

major constituent of bone, cartilage, tendons, ligaments, and skin. The water-soluble compounds that are vitamins for humans are essential to all animals. Some species, however, can make some of those compounds in sufficient quantity that they do not require the compounds in their diet.

Fat-soluble vitamins—vitamins A, D, E, and K—have diverse functions. Vitamin A (retinol) is a precursor of retinal, the visual pigment in our eyes. Vitamin D (calciferol) regulates the absorption and metabolism of calcium. Although vitamin D may be obtained in the diet, it can also be produced in human skin by the action of ultraviolet wavelengths of sunlight on certain lipids already present in the body. Thus vitamin D is only a vitamin for individuals with inadequate exposure to the sun, such as people living in cold climates where clothing usually covers most of the body and where the sun may not shine for long periods of time.

The need for vitamin D may have been an important factor in the evolution of skin color. Human races adapted to equatorial and low latitudes have dark skin pigmentation as a protection against the damaging effects of ultraviolet radiation. These peoples generally have extensive skin areas exposed to the sun on a regular basis, so adequate synthesis of vitamin D occurs in their skins. In general, races that became adapted to higher latitudes lost dark skin pigmentation. Presumably, lighter skin facilitates vitamin D production in the relatively small areas of skin exposed to sunlight during the short days of winter. An exception to the correlation between latitude and skin pigmentation is the Inuit peoples of the Arctic. These dark-skinned people obtain plenty of vitamin D from the large amounts of fish oils in their diets; for them, exposure to sunlight is not a factor in obtaining this vitamin.

Vitamin E is poorly understood. Its principal function may be to protect unsaturated fatty acids in cellular membranes from oxidation. Vitamin K functions

BOX 43.A

Beriberi and the Vitamin Concept

Beriberi is Sinhalese (a language of Sri Lanka) for "extreme weakness." This disease of humans is found wherever unbalanced diets are common. It became particularly prevalent in Asia in the nineteenth century, when it became standard practice to mill rice to a high, white polish and discard the hulls that are present in brown rice. There are several forms of beriberi, with different symptoms, but the heart is generally adversely affected.

In 1897 Christiaan Eijkman, working in what is now Indonesia, discovered that chickens developed beriberi-like symptoms when fed a diet of polished rice. During the next decade, other investigators found that people in Malaysia on a polished rice diet developed beriberi, whereas those who ate brown rice did not. Finally, in 1912, Polish-born Casimir Funk showed that pigeons with beriberi could be cured of their symptoms by feeding them a concentrate of rice polishings—the hulls that were discarded to make the rice more "appealing." He went on to suggest that beriberi and some other diseases are dietary in origin, and that they result from deficiencies in specific substances, for which he coined the term *vitamines* because he mistakenly thought that all these substances were amines vital for life. In 1926 thiamin (vitamin B_1)—the substance lost in the rice milling process—was the first vitamin to be isolated in pure form; in 1936 its structure was determined and it was synthesized for the first time.

Eijkman's and Funk's investigations with birds were the first in a series of experiments on animals that established that diseases can result from dietary deficiencies. Before these experiments, all diseases were thought to be caused by microorganisms.

---

in blood clotting following an injury and hence plays a crucial role in the protection of the body. The fat-soluble vitamins generally are required by vertebrates but not by invertebrates.

When water-soluble vitamins are ingested in excess of bodily needs, they are simply eliminated in the urine. (This is the fate of much of the vitamin C that people take in excessive doses.) The fat-soluble vitamins, however, accumulate in body fat and may build up to life-threatening levels if taken in excess. (Inuit peoples generally do not eat polar bear liver, which is unusually high in fat-soluble vitamin A.)

BOX 43.B

Scurvy and Vitamin C

In 1498 Vasco da Gama sailed around the Cape of Good Hope. In the process, he lost 100 of his 160 crew members to a disease that was becoming all too familiar on ocean voyages lasting several months. The symptoms of **scurvy** include general debility, hemorrhaging and decay of skin and flesh, bleeding gums and loss of teeth, and finally, death.

More than a century later, the British physician James Lind found that shipboard scurvy could be prevented by having sailors eat citrus fruit or sauerkraut. The effects were dramatic. In 1760 one British naval hospital treated 1,754 cases of scurvy. Beginning in 1795, the Royal Navy required all sailors to take a daily ration of lemon juice—and in 1806 the same hospital treated exactly one case of scurvy. The navy switched from lemons to limes in 1865, and since that time British sailors have been referred to as "limeys."

In 1907 researchers induced scurvy in guinea pigs by giving them a diet of dried hay and oats, with no fresh plant material. By supplementing that diet with various foods, the researchers were able to determine which foods prevented scurvy. At last, in 1932, the anti-scurvy factor was isolated from lemon juice. The anti-scurvy factor is ascorbic acid, a compound identified in plant extracts six years earlier by Albert Szent-Györgyi and now commonly called vitamin C.

Vitamin deficiency diseases are rare among primitive societies living according to long-established tradition, but they frequently occur when new habits are thrust upon people by civilization or technology. Scurvy and other deficiency diseases may also arise when people are cut off from their normal diets by such things as ocean voyages or imprisonment, or when armies are on campaign or cities are under siege.

Nutritional Deficiency Diseases in Humans

In humans, chronic shortage of a nutrient produces a characteristic deficiency disease. If the deficiency is not remedied, death may follow. An example is kwashiorkor. Kwashiorkor results from protein deficiency, which causes swelling of the extremities, distension of the abdomen (see Figure 42.13), immune system breakdown, degeneration of the liver, mental retardation, and other problems.

Shortage of any of the vitamins results in specific deficiency symptoms (see Table 43.2). Two deficiency diseases, beriberi and scurvy, were discussed in Boxes 43.A and 43.B. **Pellagra**, which results from a deficiency of the B vitamin niacin, is a common and severe disease in many poor areas. It also occurs frequently in conjunction with chronic alcoholism. Its symptoms include diarrhea, itching skin, abdominal pain, and other problems. Vitamin D deficiency decreases the absorption and use of calcium, leading to softening of the bones and a distortion of the skeleton. This deficiency disease is known as **rickets**. Vitamin B_{12} (cobalamin) is produced by microorganisms that live in our intestines and use the cobalt in our diet. Cobalamin is present in all foods of animal origin. Plants neither use nor produce vitamin B_{12}, and a strictly vegetarian diet (not supplemented by vitamin pills) can lead to **pernicious anemia**, the B_{12} deficiency disease.

Inadequate mineral nutrition can also lead to deficiency diseases. Iodine, for example, is a constituent of the hormone thyroxin, which is produced in the thyroid gland. If insufficient iodine is obtained in the diet, the thyroid gland grows larger in an attempt to compensate for the inadequate production of thyroxin. The swelling that results is called a **goiter** (see Figure 36.9). Goiters are common in mountain areas such as the Andes of South America because of low iodine levels in the soil and hence in the crops grown there. Goiters were once common in Switzerland and in the Great Lakes area of the United States, but the problem was solved by adding small amounts of iodine to table salt or to drinking water.

ADAPTATIONS FOR FEEDING

The ways an animal acquires its nutrients and its adaptations for doing so are frequently its most distinguishing characteristics. The role that a species plays in nature is described as its ecological niche, and its feeding specializations and adaptations are major dimensions of that ecological niche (see Chapter 48). The crucial adaptations that enable a species to exploit a particular source of nutrition are frequently physiological and biochemical. For example, the Australian koala eats nothing but leaves of eucalyptus trees. Eucalyptus leaves are tough, low in nutrient content, and loaded with pungent, toxic compounds that evolved to protect the trees from predators. Yet the koala's gut can digest and detoxify the leaves and absorb all of the nutrients the animal needs from this formidably specialized diet. The feeding adaptations that are most obvious to us, however, are the anatomical and behavioral features that animals use to acquire and ingest their food.

Food Acquisition by Carnivores

The predatory behaviors of many carnivores are legendary. One need only call to mind the hunting skills of hawks, wolves, or any member of the cat family. Carnivores have evolved stealth, speed, power, large jaws, sharp teeth, and strong gripping appendages. A cheetah, for instance, first stalks its prey stealthily from downwind, aided by its natural camouflage. When close enough, it dashes after the prey at speeds as fast as 110 kilometers per hour. It then brings the prey down with its sharp, powerful claws and teeth. Carnivores also have evolved remarkable means of detecting prey. Bats use echolocation, pit vipers sense infrared radiation from the warm bodies of their prey, and certain fishes detect electric fields created in the water by their prey (see Chapter 39).

Adaptations for killing and ingesting prey are diverse and specialized. These adaptations can be especially important when the prey species are capable of inflicting damage on the predators. Many species of snakes take relatively large prey that are well equipped with sharp teeth and claws. A snake may strike with poisonous fangs and immobilize its prey before ingesting it. A boa or python kills its prey by squeezing it with coils of its powerful body. To swallow large prey, a snake's lower jaw disengages from its joint with the skull. The tentacles of jellyfish, corals, squid, and octopus, the long, sticky tongues of frogs and chameleons, and the webs of spiders are other examples of fascinating adaptations for capturing and immobilizing prey.

Because some prey items are impossible for a predator to ingest, digestion is sometimes accomplished externally. Sea stars evert their stomachs (turn them inside out) and digest their molluscan prey while they are still in their shells (Figure 43.6). Spiders usually prey on insects with indigestible exoskeletons. The spider can inject its prey with digestive enzymes and then suck out the liquefied contents, leaving behind the empty exoskeletons frequently seen in old spider webs.

Food Acquisition by Herbivores

Herbivores obtain food less dramatically than predators do. Cows or sheep graze in grassy meadows

43.6 Inside-Out Digestion
This sea star is eating two mollusks. It is holding them with its arms as the tissue from its everted stomach digests them.

while caterpillars munch steadily on leaves. Some herbivores have striking adaptations for feeding, such as the trunk (a flexible, gripping nose) of the elephant, the long neck of the giraffe, or the wing design that enables the hovering flight hummingbirds use to gather nectar swiftly from large numbers of flowers.

Behavior that can almost be described as agricultural is an adaptation of some herbivores. There are species of termites and tropical leaf-cutter ants that prepare and tend subterranean fungus gardens (Figure 43.7). Individuals of these species forage for plant

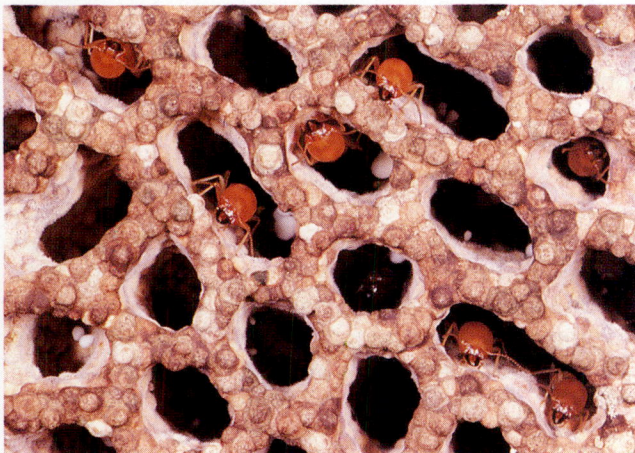

43.7 Fungus Gardens
These termites harvest plant matter, which they cannot digest, and spread it on the walls of their underground chambers. The fungus that grows in these carefully prepared beds helps break down the woody matter for the termites and also serves as food.

material, which they bring into their nests and process into a spongy comb structure of undigested cellulose. By depositing their feces into the comb they innoculate it with fungal spores that send fungal mycelia throughout the comb. The mycelia produce enzymes that break down the comb's cellulose, providing nutrients for the fungus which then forms fruiting bodies. The insects eat the fruiting bodies, as well as the older comb riddled with mycelia. Through this symbiotic relationship, the fungus helps the insects derive nutrition from the forage they collect. There are also termite species with symbiotic protists in their guts that provide a similar service of producing enzymes that break down cellulose.

Food Acquisition by Filter Feeders

Filter feeders strain out particles or small organisms suspended in water. Most sessile (stationary) aquatic animals, such as sponges, corals, barnacles, and bivalve mollusks, feed in this way. There are also mobile filter feeders, including the baleen whales, tadpoles, mosquito larvae, hundreds of fishes (such as herring, sardines, menhaden, and as we learned at the beginning of the chapter, whale sharks), and even some birds. All filter feeders have some way of passing great volumes of water through a filterlike device at the front end of the gut. Flamingos sift through water and mud with their grooved bills, capturing insect larvae, worms, seeds, bacteria, and other matter. Many stationary filter feeders employ mucus to extract particles from water. Oysters, clams, and other bivalves draw water over their gills, where sticky mucus traps food particles. Their gills are densely covered with cilia that convey the mucus and its trapped particles toward the mouth, where coarse matter is rejected and the rest allowed to enter.

The most impressive filter feeders are the baleen whales, the largest of which (in fact, the largest animal ever to live on this planet) is the blue whale (*Balaenoptera musculus*). A blue whale may grow to a length of 30 meters—the equivalent of several school buses placed end to end. Its tongue is the size of an elephant and its heart is the size of a small car, yet the blue whale eats tiny crustaceans called krill, which it filters from seawater. If you run your tongue along the roof of your mouth you will feel ridges. These ridges are greatly enlarged in baleen whales, forming **baleen plates** with fringed edges (Figure 43.8). The whale approaches a swarm of krill and gulps it in, along with thousands of liters of water. As it closes its mouth, its tongue forces the seawater out through the fringe of the baleen plates, leaving the krill behind. Krill are very abundant in cold, nutrient-rich waters and make it possible for the whale to support its metabolic rate of about a million kilocalories per day.

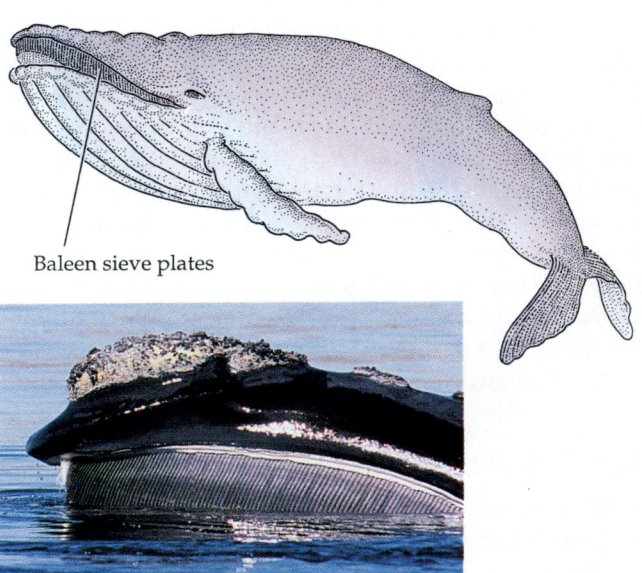

Baleen sieve plates

43.8   The Largest Eating the Smallest
Visible in this feeding whale are the baleen plates that
hang from the roof of its mouth and filter small animals
from the huge volumes of water that pass through them.
The front of its snout is encrusted with barnacles, which
are also filter feeders.

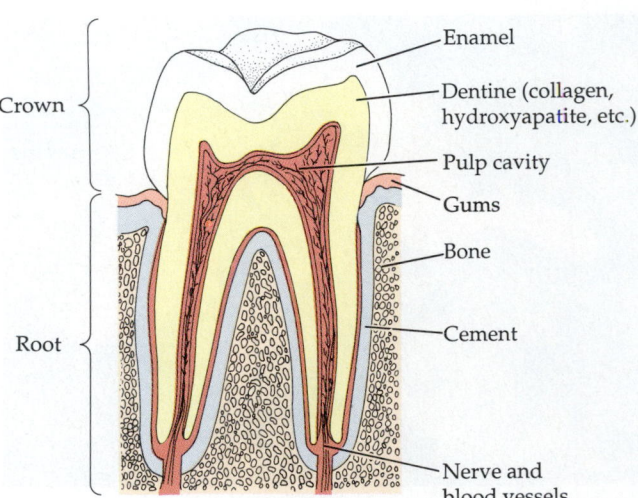

43.9   A Mammalian Tooth Has Three Layers
A section through a mammalian tooth shows that the
crown above the gums is covered with hard enamel. Below the enamel is a thick dentine layer that extends into
the skull to form the root, and inside the dentine is the
pulp cavity containing the tooth's supply of blood vessels
and nerves. The tooth is held in its bony socket in the
skull by a fairly soft "cement." The teeth of other vertebrates, such as lizards, are not entrenched in bony sockets
and consequently must be replaced frequently.

Vertebrate Teeth

Many vertebrate species have distinctive teeth. Teeth
are adapted for the acquisition and initial processing
of specific types of foods, and because they are one
of the hardest structures of the body, they remain in
the environment long after the animals die. Paleontologists use teeth to identify animals that lived in
the distant past and to understand their behavior.

All mammalian teeth have a general structure consisting of three layers (Figure 43.9). An extremely
hard material called **enamel**, composed principally of
calcium phosphate, covers the crown of the tooth.
Both the crown and the root contain a layer of a bony
material called **dentine**, within which is a pulp cavity
containing blood vessels, nerves, and the cells that
produce the dentine. The shapes and organization of
mammalian teeth, however, can be very different,
since they are adapted to specific diets (Figure 43.10).
In general, incisors are teeth used for cutting, chopping, or gnawing; canines are teeth used for stabbing,
ripping, and shredding; and molars and premolars
(the cheek teeth) are used for shearing, crushing, and
grinding.

The highly varied diet of humans is reflected by
our multipurpose set of teeth, as is common among
omnivores. Children first develop a set of 20 "milk
teeth," which are also called deciduous teeth because
they are lost and replaced by permanent teeth. The
permanent teeth of adults include four upper and
four lower incisors for biting, and two upper and two
lower canines for tearing. Behind the canines, on
each side of the jaw, are four upper and four lower
premolars and six upper and six lower molars for
crushing and grinding. This is a total of 32 teeth
before the dentist starts extracting them. The last set
of molars, or the rearmost tooth on each side of the
upper and lower jaw, usually does not erupt through
the gums until the person reaches 18 years of age or
older. These "wisdom teeth" frequently present
problems because the jaws may be too small to accommodate them.

DIGESTION

Most animals digest food extracellularly. Animals
take food into a body cavity that is continuous with
the outside environment and then secrete digestive
enzymes into that cavity. The enzymes act on the
food, reducing it to nutrient molecules that can be
absorbed by the cells lining the cavity. Only after
they are absorbed by the cells are the nutrients within
the body of the animal. The simplest digestive system
is a gastrovascular cavity that connects to the outside
world through a single opening. After a cnidarian
captures a prey with its stinging nematocysts (see
Figure 26.10), its tentacles cram the prey into the
gastrovascular cavity (see Figure 42.1*a*) where en-

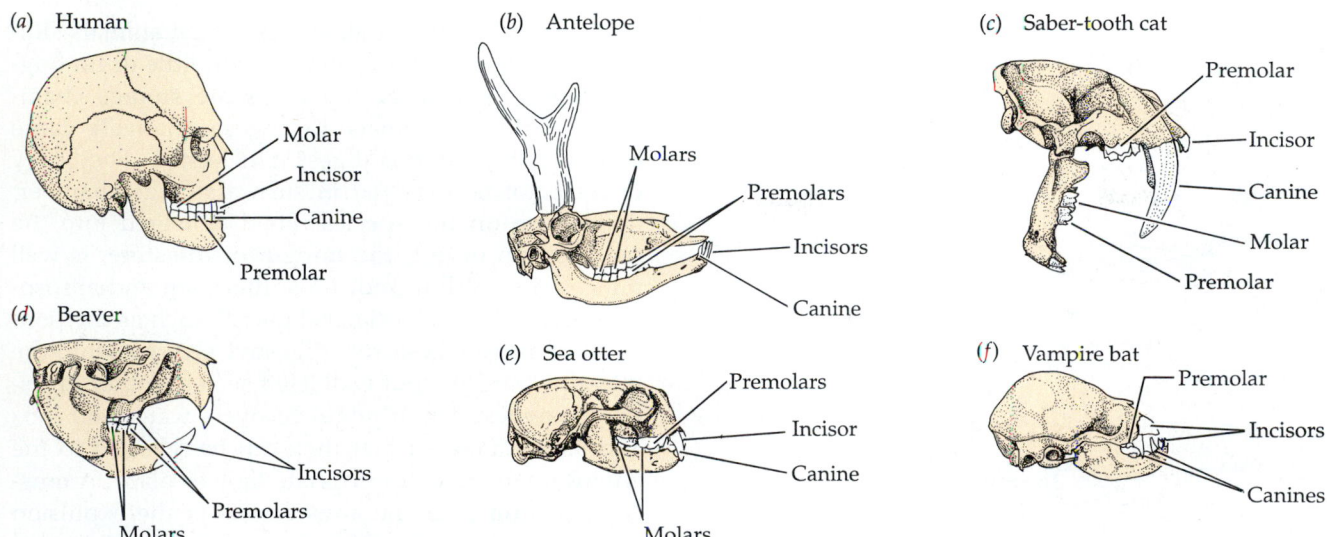

(a) Human

Molar
Incisor
Canine
Premolar

(b) Antelope

Molars
Premolars
Incisors
Canine

(c) Saber-tooth cat

Premolar
Incisor
Canine
Molar
Premolar

(d) Beaver

Incisors
Premolars
Molars

(e) Sea otter

Premolars
Incisor
Canine
Molars

(f) Vampire bat

Premolar
Incisors
Canines

43.10 Mammalian Teeth Are Specialized for Different Diets
(a) A human is an omnivore and has a generalized set of teeth. (b) The pronghorn antelope grazes on tough grasses, herbs, and shrubs. Its upper jaw has no incisors or canines, and in the lower jaw they are far forward and are used to tear the leaves off plants. The antelope's cheek teeth are highly adapted for grinding coarse vegetation. (c) Carnivores such as the extinct saber-toothed cat have greatly enlarged canine teeth for gripping, killing, and tearing their prey. The incisors are used for scraping muscle off of bone, and the cheek teeth are used for shearing flesh and crushing bones. (d) Rodents such as the beaver have enlarged incisors that they use for gnawing. The beaver cuts down and strips the bark from trees with its incisors. The incisor teeth of rodents grow continuously to compensate for wear and tear. The beaver has no canines, but its cheek teeth are well adapted for grinding fibrous plant material. (e) The sea otter has well-developed cheek teeth for crushing the shells of marine invertebrates. (f) The upper incisors and canines of the vampire bat are triangular and sharp. They are used to make incisions in the skin of large mammals so the bat can drink their blood.

zymes partially digest it. Extracellular digestion in cnidarians is supplemented by intracellular digestion: Cells lining the gut take in some small food particles by endocytosis.

The gastrovascular cavity of flatworms is more complex than that of cnidarians, but it also has only one opening through which food enters and waste products exit. Where the gastrovascular cavity meets the mouth, it narrows to form a tubular **pharynx**. This muscular structure can be pushed out through the mouth during feeding. In addition, extensive branching of the gastrovascular cavity increases the efficiency of absorption of nutrient molecules; the branches increase the surface area through which absorption can occur (Figure 43.11).

Some multicellular animals have no digestive systems. Many of these are internal parasites such as

tapeworms. They live in an environment so rich in already digested nutrient molecules that they just absorb them directly into their cells.

Tubular Guts

The guts of all animal groups other than sponges, cnidarians, and flatworms are tubular, with an opening at either end. A mouth takes in food; solid digestive wastes, or feces, are excreted through an anus. Different regions in the tubular gut are specialized for particular functions (Figure 43.12). Remember as we discuss these regions that all locations within the tubular gut are really outside the body of the animal.

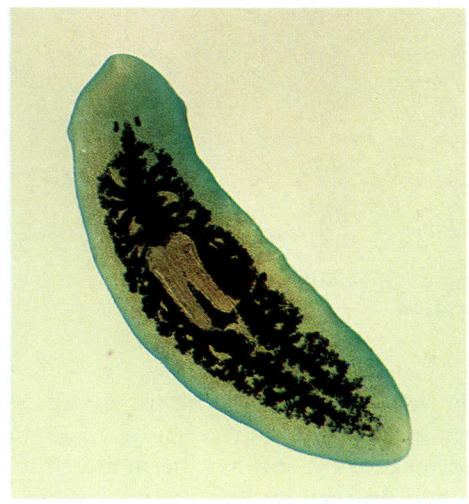

43.11 Digestive Tract of a Flatworm
The gastrovascular system of this flatworm has been stained to reveal all of its branches. This system communicates to the outside through the muscular pharynx visible in the center of the animal.

Nematode

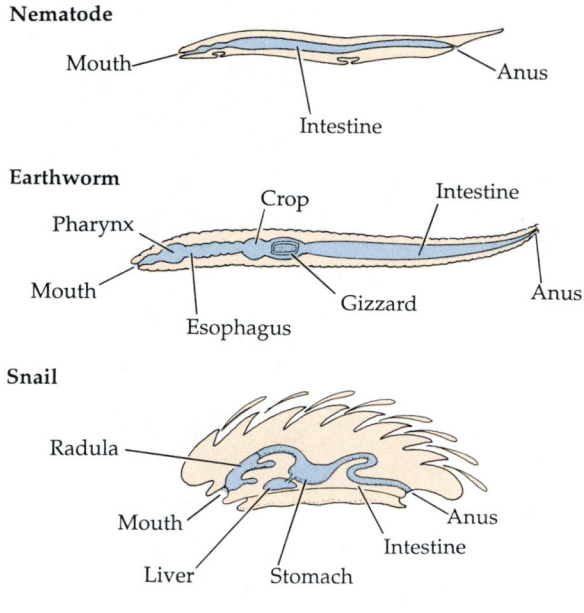

Earthworm

Snail

Cockroach

Rabbit

43.12 Compartments Specialized for Digestion and Absorption

Only by crossing the membranes lining the gut does a nutrient molecule enter the body.

At the anterior end of the gut are the mouth (the opening itself) and **buccal cavity** (mouth cavity). Food may be broken up by teeth (in some vertebrates), by the radula in snails, or by mandibles (in insects), or somewhat further along the gut by structures such as the gizzards of birds and earthworms, where muscular contractions of the gut grind the

food together with small stones. Some animals simply ingest large chunks of food with little or no fragmentation. **Stomachs** and **crops** are storage chambers, enabling animals to ingest relatively large amounts of food and digest it at leisure. Food may or may not be digested in such a storage chamber, depending on the species. Food delivered into the next section of gut, the **midgut** or **intestine**, is well minced and well mixed. Most digestion and absorption occurs here. Specialized glands such as the pancreas in mammals secrete digestive enzymes into the intestine, and the gut wall itself secretes other digestive enzymes. The **hindgut** recovers water and ions and stores feces so that they can be released to the environment at an appropriate time or place. A muscular **rectum** near the anus assists in the expulsion of undigested wastes, the process of **defecation**.

Within the hindguts of many species are colonies of bacteria that live in cooperation (symbiosis) with their hosts. The bacteria obtain their own nutrition from the food passing through the host's gut while contributing to the digestive processes of the host. Members of the leech genus *Hirudo* produce no enzymes that can digest the proteins in the blood they suck from vertebrates; however, a colony of gut bacteria produces the enzymes necessary to break down those proteins into amino acids, which are subsequently used by both the leeches and the bacteria. Many animals obtain vitamins from the bacteria in their hindguts. Herbivores such as rabbits, cattle, termites, and cockroaches depend on microorganisms in their guts for the digestion of cellulose. In some, specialized regions of their guts may even serve as microbial fermenters. An example is the caecum of the rabbit (see Figure 43.12).

In many animals, the parts of the gut that absorb nutrients have evolved extensive surface areas for absorption. The earthworm has a long, dorsal infolding of its intestine, called the typhlosole (Figure 43.13*a*), that provides extra absorptive surface area. The shark's intestine has a spiral valve, forcing food to take a longer path and thus encounter more absorptive surface (Figure 43.13*b*). In many vertebrates the wall of the gut is richly folded, with the individual folds bearing legions of tiny fingerlike projections called **villi**. The villi in turn have microscopic projections called microvilli, on the cells that line their surfaces (see Figure 43.14). Microvilli present an enormous surface area for the absorption of nutrients.

Digestive Enzymes

Protein, carbohydrate, and fat macromolecules are broken down into their simplest monomeric units by digestive enzymes. All of these enzymes cleave the chemical bonds of macromolecules through a reaction that adds a water molecule at the site of cleavage;

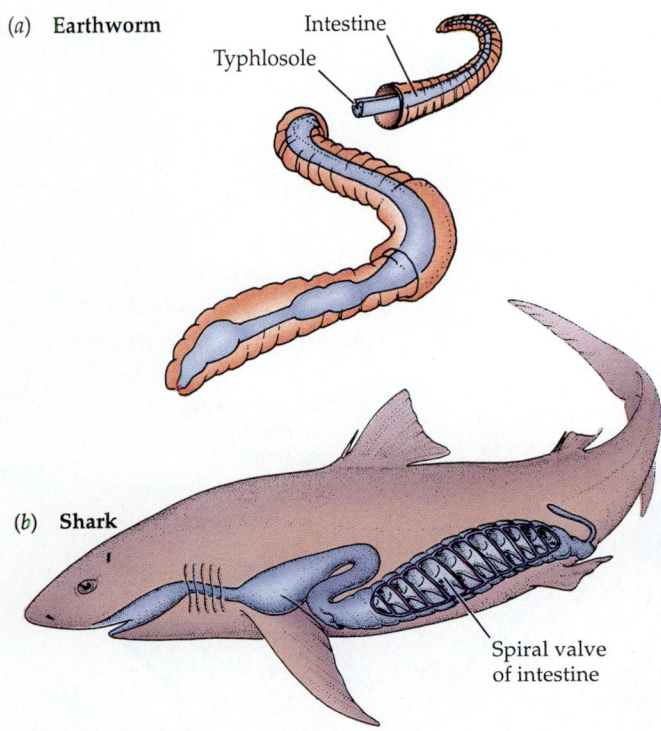

(a) Earthworm   Intestine
Typhlosole

(b) Shark

Spiral valve
of intestine

**43.13   Greater Intestinal Surface Area Means
More Absorption**
(a) In earthworms: the adaptation is a typhlosole: a simple longitudinal infolding of the intestinal wall. (b) Sharks have evolved a spiral valve that increases the surface area of the intestine.

hence they are generally called **hydrolytic enzymes**. Examples of hydrolytic cleavage are the breaking of the bonds between adjacent amino acids of a protein or peptide (see Figure 6.14) and the breaking of the bonds between adjacent glucose units in starch (see Figure 6.7).

Digestive enzymes are classified according to the substances they hydrolyze: carbohydrases hydrolyze carbohydrates; proteases, proteins; peptidases, peptides; lipases, fats; and nucleases, the nucleic acids. The prefixes *exo-* (outside) and *endo-* (within) indicate where the enzyme cleaves the molecule. Thus an endoprotease hydrolyzes a protein at an internal site along the polypeptide chain, and an exoprotease snips away amino acids at the ends of the molecule (see Figures 6.16 and 6.17).

How can an organism produce enzymes to digest biological macromolecules without digesting itself? The answer is that the digesting, as you know, is usually done *outside* the animal. The gut is simply a tunnel through the animal; food in the gut is outside the body and hence can receive treatment (such as high acidity or potent enzymes) that would be intolerable within a cell or a tissue. (In the cases in which digestion is intracellular, such as in cnidarians, the hydrolytic reactions are localized within food vacu-

oles.) Most digestive enzymes are produced in an inactive form, known as a **zymogen**. Thus they do not act on the cells that produce them. When secreted into the gut, the zymogens are activated, sometimes by exposure to a different pH but more often by the action of another enzyme.

The gut itself is not digested by activated enzymes because it is protected by a covering of mucus, a slimy material secreted by special cells in the lining of the gut. The mucus also lubricates the gut and protects it from abrasion. If mucus production is inadequate, digestive enzymes or stomach acid can act upon the gut, producing sores called **ulcers**. Insects rely on a different trick to prevent digestion of the gut lining. Within the gut they secrete a thin tube of chitin, a modified polysaccharide (see Figure 3.13) that is also found in the insect's protective exoskeleton. The chitin tube encloses the food and enzymes and protects the gut from abrasion and self-digestion.

STRUCTURE AND FUNCTION
OF THE VERTEBRATE GUT

The separate compartments that have specific functions in the digestive tracts of vertebrates are all part of a continuous tube that runs from mouth to anus. The specific functions must be coordinated so that they occur in proper sequence and at appropriate rates. Let's take a tour of the vertebrate gut to see how structure and function work together to move food through the gut and bring about its sequential digestion and the absorption of nutrients.

The Tissue Layers

The cellular architecture of the tube that forms the vertebrate gut follows a common plan throughout. Four major layers of different cell types form the wall of the tube (Figure 43.14). These layers differ somewhat from compartment to compartment, but they are always present. Starting in the cavity, or **lumen**, of the gut, the first tissue layer is the **mucosa**. Nutrients are absorbed across the membranes of the mucosal cells; in some regions of the gut, those membranes have many folds that increase their surface area. Mucosal cells also have secretory functions. Some secrete mucus that lubricates the food and protects the walls of the gut. Others secrete digestive enzymes, and still others in the stomach secrete hydrochloric acid (HCl). At the base of the mucosa are some smooth muscle cells and just outside the mucosa is the second layer of cells, the **submucosa**. Here we find the blood and lymph vessels that carry absorbed nutrients to the rest of the body. The submucosa also contains a network of nerves; these neurons are both sensory (responsible for stomach

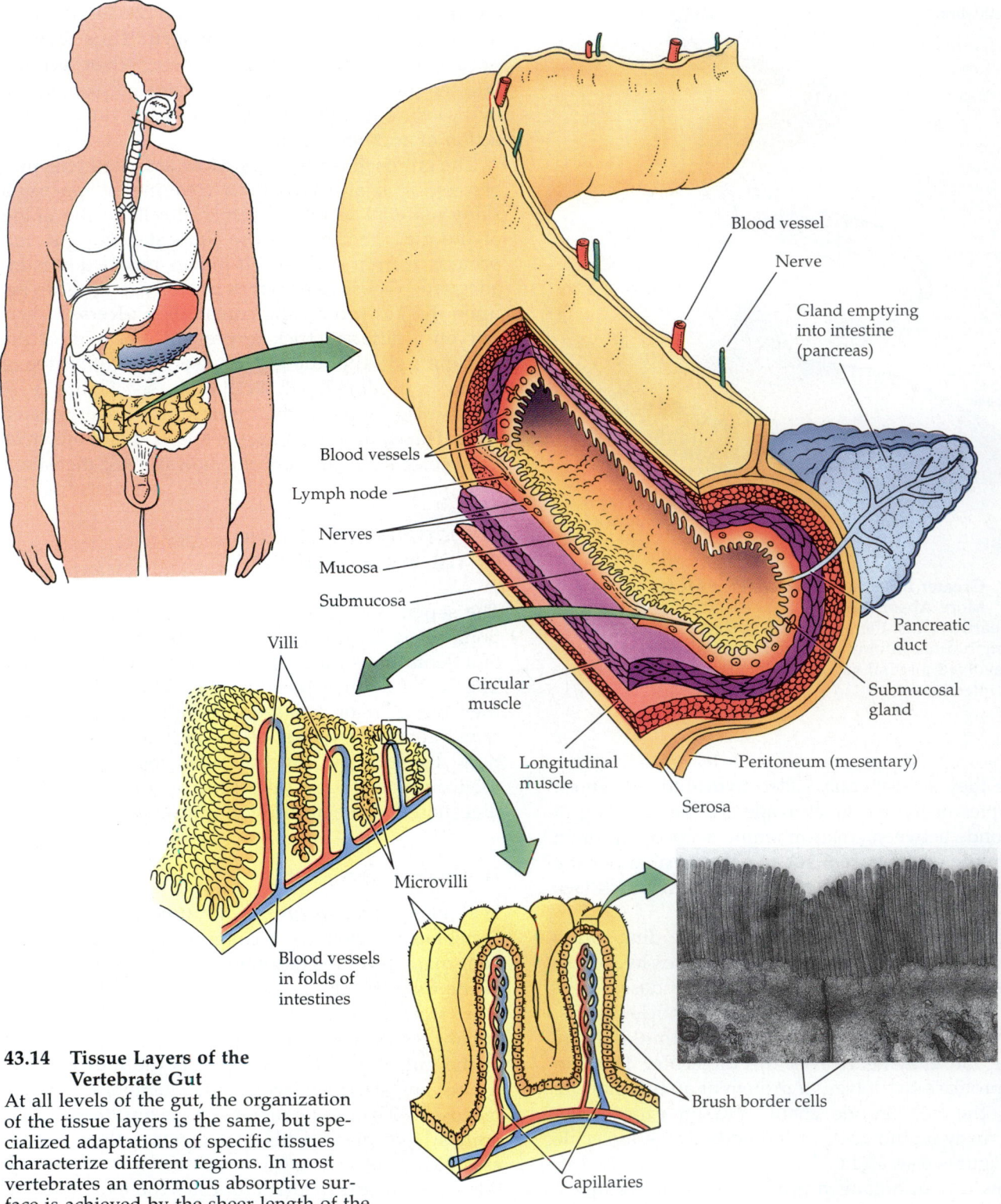

Blood vessel

Nerve

Gland emptying
into intestine
(pancreas)

Blood vessels

Lymph node

Nerves

Mucosa

Submucosa

Pancreatic
duct

Submucosal
gland

Villi

Circular
muscle

Longitudinal
muscle

Peritoneum (mesentary)

Serosa

Microvilli

Blood vessels
in folds of
intestines

Brush border cells

Capillaries

**43.14 Tissue Layers of the
Vertebrate Gut**
At all levels of the gut, the organization
of the tissue layers is the same, but specialized adaptations of specific tissues
characterize different regions. In most
vertebrates an enormous absorptive surface is achieved by the sheer length of the
tubular small intestine and the folding of
its linning. Finally, the thin, fingerlike
microvilli that cover the villi increase
absorptive surface area enormously.

aches!) and regulatory (controlling the various secretory functions of the gut).

External to the submucosa are two layers of
smooth muscle cells responsible for the movements
of the gut. Because the cells of the innermost layer
are oriented *around* the gut, this layer is called **circular
muscle**. It constricts the lumen. The cells of the outermost layer of the tube's wall, the **longitudinal mus-**

cle, are arranged along the length of the gut. When this layer contracts, the gut shortens. Between these two layers of muscle is a network of nerves that controls the movements of the gut, coordinating the different regions with one another.

Surrounding the gut is a fibrous coat called the **serosa**. Like other abdominal organs, the gut is also covered and supported by a tissue, the **peritoneum**.

Movement of Food in the Gut

Food entering the mouths of most vertebrates is chewed and mixed with the secretions of salivary glands. The muscular tongue then pushes the mouthful, or bolus, of food toward the back of the mouth cavity. By making contact with the soft tissue at the back of the mouth, the food initiates a complex series of neural reflex actions known as swallowing. Stand in front of a mirror and gently touch this tissue at the back of your mouth with the eraser of your pencil or with a cotton swab. You may gag slightly, but you will also experience an uncontrollable urge to swallow. Swallowing involves many muscles doing a variety of jobs that propel the food through the **pharynx** and into the **esophagus** without allowing any of it to enter the windpipe (trachea) or nasal passages (Figure 43.15).

Once the food is in the esophagus, peristalsis takes over and pushes the food toward the stomach. **Peristalsis** is a wave of smooth muscle contraction that moves progressively down the gut from the pharynx toward the anus. The smooth muscle of the gut contracts in response to being stretched. Swallowing a bolus of food stretches the upper end of the esophagus, and this stretching initiates a wave of contraction that slowly pushes the contents of the gut toward the anus. Peristalsis can occasionally run in the opposite direction, however. When your stomach is very full, pressure on your abdomen or a sudden movement can push some stomach contents into the lower end of your esophagus. This can initiate peristaltic movements that bring the acidic, partially digested food into your mouth. When you vomit, contractions of the abdominal muscles explosively force stomach contents out through the esophagus. Prior to vomiting, waves of reverse peristalsis can even bring the contents of the upper regions of the intestine back into the stomach to be expelled.

The movement of food from the stomach into the esophagus is normally prevented by a thick ring of circular smooth muscle at the junction of the esophagus and the stomach. This ring of muscle, called a **sphincter**, is normally constricted. Waves of peristalsis cause it to relax enough to let food pass through. Sphincter muscles are found elsewhere in the digestive tract as well. The pyloric sphincter governs the passage of stomach contents into the intestine. Another important sphincter surrounds the anus.

Digestion in the Mouth and Stomach

In addition to physically disrupting food, the mouth initiates the digestion of carbohydrates through the

43.15 Swallowing and Peristalsis
Food pushed to the back of the mouth triggers the swallowing reflexes. Once food enters the esophagus, peristalsis propels it to the stomach.

Swallowing

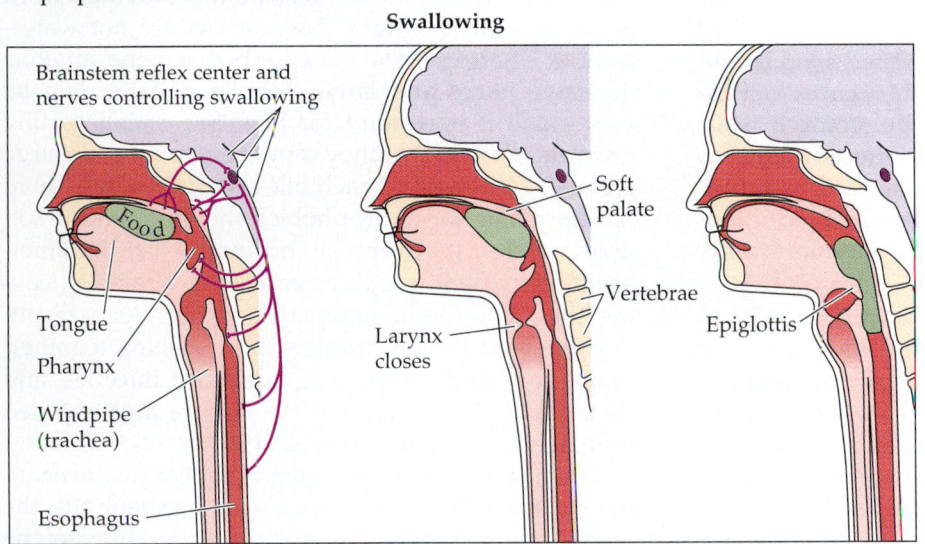

Peristalsis

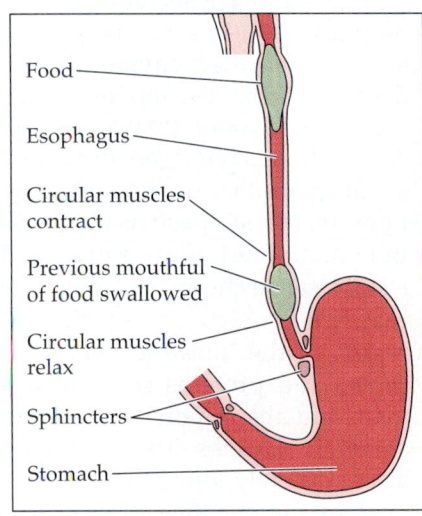

1. Food is chewed and tongue pushes bolus of food to back of mouth. Sensory nerves initiate the swallowing reflex

2. Soft palate is pulled up, vocal cords close larynx

3. Larynx pulled up and forward and covered by epiglottis; bolus of food enters esophagus

4. Peristaltic contractions propel food to stomach

action of the enzyme amylase, which is secreted with the saliva and mixed with the food as it is chewed. Amylase is a carbohydrase; it hydrolyzes the bonds between the 6-carbon sugar units that make up the long-chain starch molecules. The action of amylase is what makes a piece of bread or cracker taste sweet if you chew it long enough.

Most vertebrates can rapidly consume a large volume of food, but digesting that food is a long, slow process. The stomach stores the food devoured in the course of a meal and continues breaking it down physically. The secretions of the stomach kill microorganisms taken in with the food and begin the digestion of proteins. The major enzyme produced by the stomach is an endopeptidase called **pepsin**. Pepsin is secreted as the zymogen **pepsinogen** by cells in the **gastric pits**—deep folds in the stomach lining. Other cells in the gastric pits produce hydrochloric acid, and still others near the openings of the gastric pits and throughout the stomach mucosa secrete mucus.

Hydrochloric acid (HCl) maintains the stomach fluid (the gastric juice) at a pH between 1 and 3. This low pH activates the conversion of pepsinogen to pepsin. This conversion is amplified as the newly formed pepsin activates other pepsinogen molecules, a process called **autocatalysis**. Hydrochloric acid also provides the right pH for the enzymatic action of pepsin. In addition, the low pH helps dissolve the intercellular substances holding the ingested tissues together. Breakdown of the ingested tissues exposes more food surface area to the action of digestive enzymes. Mucus secreted by the stomach mucosa coats and protects the walls of the stomach from being eroded and digested by the HCl and pepsin.

Contractions of the muscles in the walls of the stomach churn its contents, thoroughly mixing them with the stomach secretions. The acidic, fluid mixture of digestive juices and partially digested food in the stomach is called **chyme**. A few substances can be absorbed from the chyme across the stomach wall, including alcohol (hence its rapid effects), aspirin, and caffeine, but even these are absorbed in rather small quantities in the stomach. Peristaltic contractions of the stomach walls push the chyme toward the bottom end of the stomach. These waves of peristalsis cause the pyloric sphincter to relax briefly so that little squirts of the chyme can enter the first region of the intestine, where digestion of carbohydrates and proteins continues, digestion of fats begins, and absorption of nutrients begins. The human stomach empties itself gradually over a period of approximately four hours. This slow passage of food enables the intestine to work on a little material at a time and prolongs the digestive and absorptive processes throughout much of the time between meals.

The Small Intestine, Gall Bladder, and Pancreas

Although the **small intestine** takes its name from its diameter, it is really a very large organ and is the site of the major events of digestion and absorption. The small intestine of an adult human is more than 6 meters long; its coils fill much of the lower abdominal cavity (Figure 43.16). Because of its length, and because of the folds, villi, and microvilli of its lining, its inner surface area is enormous: about 550 square meters, or roughly the size of a tennis court. Across this surface the small intestine absorbs the nutrient molecules from food. The small intestine has three sections: The initial section—the **duodenum**—is the site of most digestion; the **jejunum** and the **ileum** carry out 90 percent of the absorption of nutrients.

Digestion requires many specialized enzymes as well as several other secretions. Two accessory organs that are not part of the digestive tract—the liver and the pancreas—provide many of these enzymes and secretions. The liver synthesizes a substance called **bile** from cholesterol. Bile emulsifies fats just as soap emulsifies grease on your clothes or hands. Bile secreted from the liver flows through the **hepatic duct**. A side branch of the hepatic duct delivers bile to the gallbladder, where it is stored until it is needed to assist in fat digestion (Figure 43.17). Below the branching point to the gallbladder, the hepatic duct is called the **common bile duct**. When undigested fats enter the duodenum, the gallbladder releases bile, which flows down the common bile duct and enters the duodenum.

To understand the role of bile in fat digestion, think of the oil in salad dressing; it is not soluble in water (it is hydrophobic) and tends to aggregate together in large globules. The enzymes that digest fat, the **lipases**, are water-soluble and must do their work in an aqueous medium, but the fats are not water-soluble. Therefore, the interface between the aqueous digestive juices and large globules of fat would be very small if it weren't for bile. Bile stabilizes tiny droplets of fat so that they cannot aggregate into large globules. One end of each bile molecule is soluble in fat (lipophilic, or hydrophobic); the other end is soluble in water (hydrophilic, or lipophobic). Bile molecules bury their lipophilic ends in fat droplets, leaving their lipophobic ends sticking out. As a result, they prevent the fat droplets from sticking together. These very small fat particles are called **micelles**, and their small size maximizes the surface area exposed to lipase action (see Figure 43.19).

The **pancreas** is a large gland that lies just beneath the stomach. It has both endocrine (secreting to the inside of the body) and exocrine (secreting to the outside of the body) functions. Here we will consider the exocrine products, which it delivers to the gut

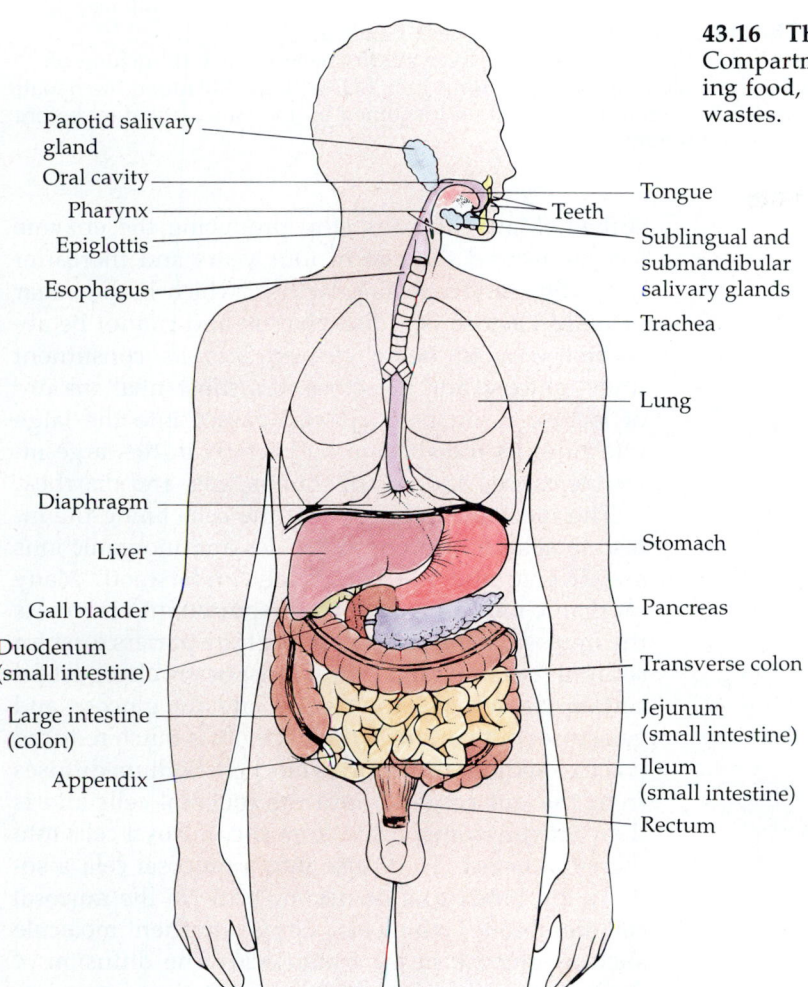

43.16 The Human Gut
Compartments within this long tube specialize in digesting food, absorbing nutrients, and storing and expelling wastes.

Parotid salivary gland
Oral cavity
Pharynx
Epiglottis
Esophagus
Diaphragm
Liver
Gall bladder
Duodenum (small intestine)
Large intestine (colon)
Appendix

Tongue
Teeth
Sublingual and submandibular salivary glands
Trachea
Lung
Stomach
Pancreas
Transverse colon
Jejunum (small intestine)
Ileum (small intestine)
Rectum

through the pancreatic duct. The pancreatic duct joins the common bile duct before entering the duodenum (see Figure 43.17). The pancreas produces a host of digestive enzymes (Table 43.3). As in the stomach, these enzymes are released as zymogens; otherwise they would digest the pancreas and its ducts before they ever reached the duodenum. Once in the duodenum, one of these inactive enzymes, **trypsinogen**, is activated by **enterokinase**, which is produced by cells lining the duodenum (Figure 43.18). Active **trypsin** can cleave other trypsinogen molecules to release even more active trypsin (another example of autocatalysis). Similarly, trypsin acts on the other zymogens secreted by the pancreas and releases their active enzymes. The mixture of zymogens produced by the pancreas can be very dangerous if the pancreatic duct is blocked or if the pancreas is injured by an infection or a severe blow to the abdomen. A few trypsinogen molecules spontaneously converting to trypsin can initiate a chain reaction of enzyme activation that digests the pancreas in a very short period of time, destroying both its endocrine and exocrine functions.

The pancreas produces, in addition to digestive enzymes, a secretion rich in bicarbonate ions (HCO_3^-). Bicarbonate ions neutralize the pH of the chyme that enters the duodenum from the stomach. This neutralization is essential because intestinal enzymes function best at a neutral or slightly alkaline pH.

Absorption in the Small Intestine

Only the smallest products of digestion pass through the mucosa of the small intestine and into the blood and lymphatic vessels that lie in the submucosa. The final digestion of proteins and carbohydrates that produces these absorbable products takes place among the microvilli. The mucosal cells with microvilli produce dipeptidase, which cleaves larger peptides into tripeptides, dipeptides, and individual amino acids that the cells can absorb. These cells also produce the enzymes maltase, lactase, and sucrase, which cleave the common disaccharides into their constituent, absorbable monosaccharides—glucose, galactose, and fructose. Disaccharides are not ab-

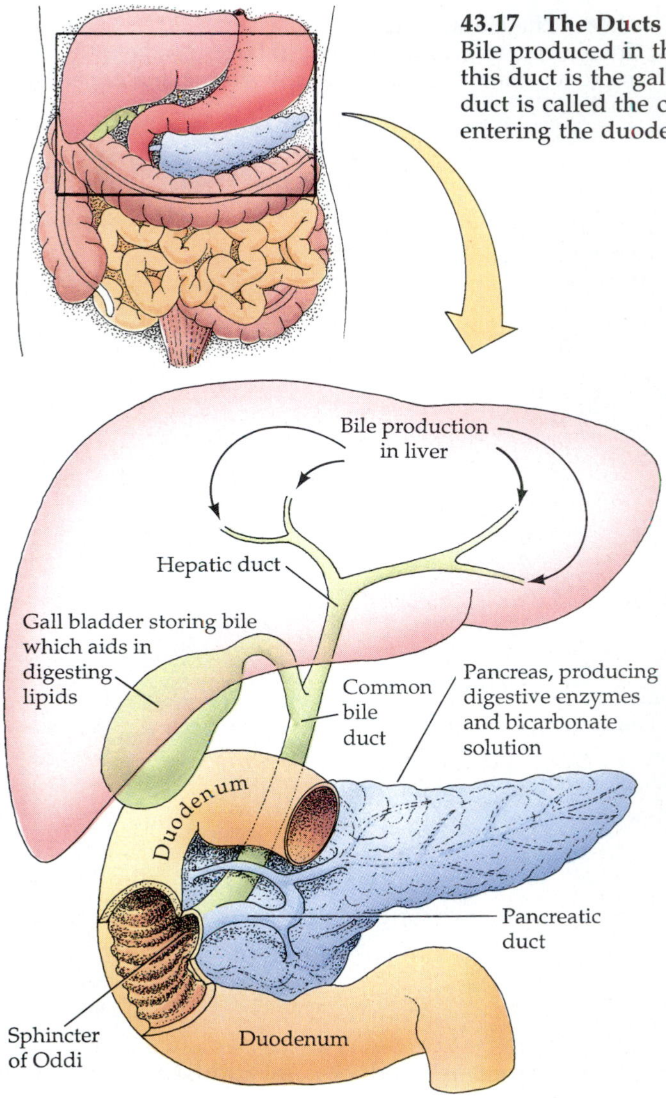

43.17 The Ducts of the Gallbladder and Pancreas
Bile produced in the liver leaves the liver via the hepatic duct. Branching off this duct is the gallbladder, which stores bile. Below the gallbladder, the hepatic duct is called the common bile duct and is joined by the pancreatic duct before entering the duodenum.

sorbed. Many humans stop producing the enzyme lactase around the age of four years and thereafter have difficulty digesting lactose, which is the sugar in milk. Lactose is a disaccharide and cannot be absorbed without being cleaved into its constituent units, glucose and galactose. If a substantial amount of lactose is unabsorbed and passes into the large intestine, its metabolism by bacteria in the large intestine causes abdominal cramps, gas, and diarrhea.

The mechanisms by which the cells lining the intestine absorb nutrient molecules and inorganic ions are diverse and not completely understood. Many inorganic ions are actively transported into or across the mucosa. For example, active transporters exist for sodium, calcium, and iron. Transporters also exist for certain classes of amino acids and for glucose and galactose, but curiously their activity is much reduced if active sodium transport is blocked. Sodium diffuses from the gut contents into the mucosal cells and is then actively transported from the mucosal cells into the submucosa. To diffuse into a mucosal cell, a sodium ion binds to a carrier molecule in the mucosal cell membrane, which also binds a nutrient molecule such as glucose or an amino acid. The diffusion of the sodium ion, driven by a concentration difference, therefore drives the absorption of the nutrient molecule. This mechanism is called **sodium cotransport** (see Figure 5.15).

TABLE 43.3
Sources and Functions of the Major Digestive Enzymes of Humans

ENZYME	SOURCE	ACTION	SITE OF ACTION
Salivary amylase	Salivary glands	Starch → Maltose	Mouth
Pepsin	Stomach	Proteins → Peptides; autocatalysis	Stomach
Pancreatic amylase	Pancreas	Starch → Maltose	Small intestine
Lipase	Pancreas	Fats → Fatty acids and glycerol	Small intestine
Nuclease	Pancreas	Nucleic acids → Nucleotides	Small intestine
Trypsin	Pancreas	Proteins → Peptides; activation of zymogens	Small intestine
Chymotrypsin	Pancreas	Proteins → Peptides	Small intestine
Carboxypeptidase	Pancreas	Peptides → Peptides and amino acids	Small intestine
Aminopeptidase	Small intestine	Peptides → Peptides and amino acids	Small intestine
Dipeptidase	Small intestine	Dipeptides → Amino acids	Small intestine
Enterokinase	Small intestine	Trypsinogen → Trypsin	Small intestine
Nuclease	Small intestine	Nucleic acids → Nucleotides	Small intestine
Maltase	Small intestine	Maltose → Glucose	Small intestine
Lactase	Small intestine	Lactose → Galactose and glucose	Small intestine
Sucrase	Small intestine	Sucrose → Fructose and glucose	Small intestine

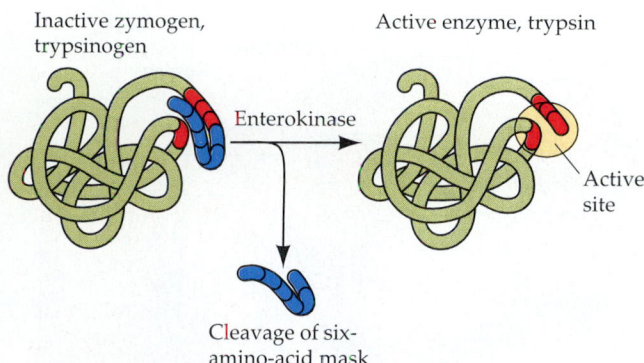

Inactive zymogen, trypsinogen

Active enzyme, trypsin

Enterokinase

Active site

Cleavage of six-amino-acid mask

43.18 Zymogen Activation
Powerful digestive enzymes often exist as inactive zymogens until their catalytic activity is required. The zymogen trypsinogen, for example, is secreted from the pancreas; when it reaches the small intestine, the enzyme enterokinase cleaves a chain of six amino acids that masked the active site, transforming trypsinogen into the active digestive enzyme trypsin.

The absorption of the products of fat digestion is much simpler (Figure 43.19). Lipases break down fats into fatty acids and monoglycerides, which are lipid-soluble and are thus able to dissolve in the membranes of the microvilli and diffuse into the mucosal cells. Once in the cells, they are resynthesized into triglycerides, combined with cholesterol and phospholipids, and coated with protein to form water-soluble **chylomicrons**, which are really little particles of fat. The chylomicrons pass into the lymphatic vessels in the submucosa and into the bloodstream through the thoracic duct. Following a meal rich in fats, the chylomicrons can be so abundant in the blood that they give it a milky appearance.

The bile that emulsifies the fats is not absorbed along with the monoglycerides and the fatty acids, but is recycled back and forth between the gut contents and the microvilli. Finally, in the ileum, bile is actively reabsorbed and returned to the liver via the bloodstream. Bile is synthesized in the liver from cholesterol. Cholesterol is also synthesized in the liver, and additionally comes from our diets.

Remember that high cholesterol levels contribute to arterial plaque formation and therefore to cardiovascular disease (see Box 42.B). The body has no way of breaking down excess cholesterol, so high dietary intake or high levels of synthesis create problems. One major way cholesterol leaves the body is through the elimination of unreabsorbed bile in the feces. The rationale for including certain kinds of fiber in our diet is that the fiber binds the bile, decreases its reabsorption in the ileum, and thus helps to lower body cholesterol.

The Large Intestine

Peristalsis gradually pushes the contents of the small intestine into the large intestine, or **colon**. The rate of peristalsis is controlled so that food passes through the small intestine slowly enough for digestion and absorption to be complete, but quickly enough to ensure an adequate supply of nutrients for the body. The material that enters the colon has had most of its nutrients removed, but it contains a lot of water and inorganic ions. The colon reabsorbs water and ions, producing semisolid feces from the slurry of indigestible materials it receives from the small intestine. If too much water is reabsorbed, constipation

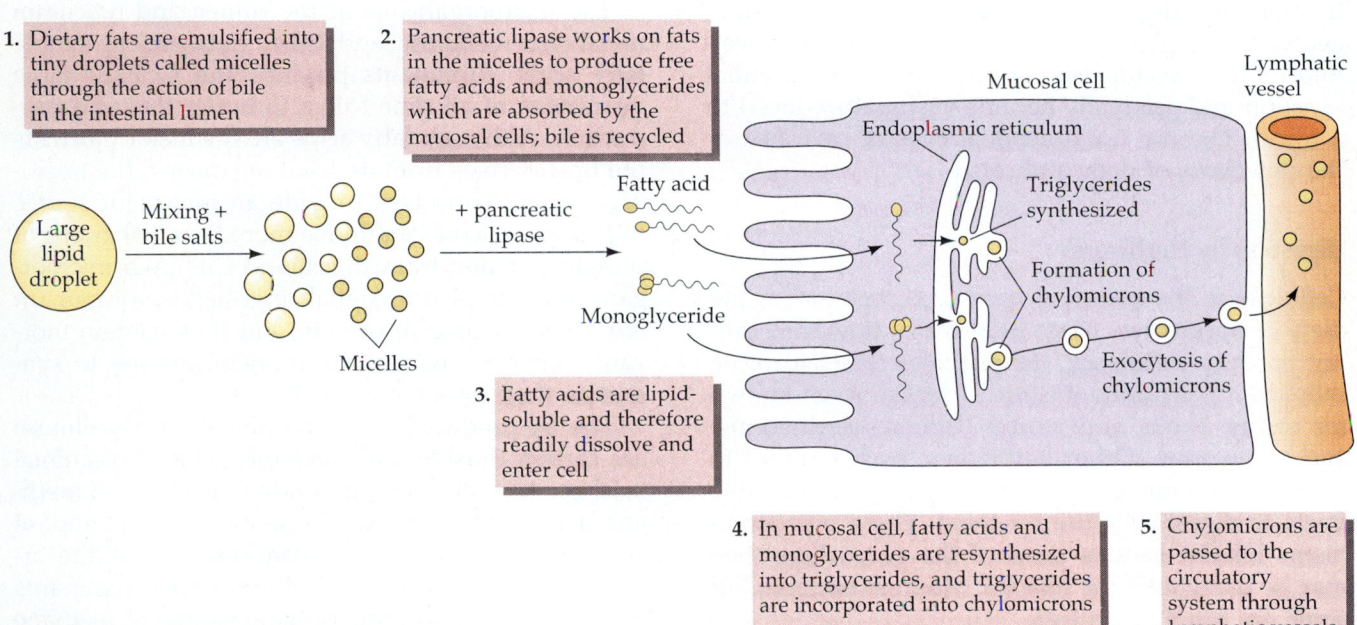

1. Dietary fats are emulsified into tiny droplets called micelles through the action of bile in the intestinal lumen

2. Pancreatic lipase works on fats in the micelles to produce free fatty acids and monoglycerides which are absorbed by the mucosal cells; bile is recycled

Mucosal cell

Lymphatic vessel

Endoplasmic reticulum

Fatty acid

Triglycerides synthesized

Large lipid droplet

Mixing + bile salts

+ pancreatic lipase

Formation of chylomicrons

Monoglyceride

Micelles

Exocytosis of chylomicrons

3. Fatty acids are lipid-soluble and therefore readily dissolve and enter cell

4. In mucosal cell, fatty acids and monoglycerides are resynthesized into triglycerides, and triglycerides are incorporated into chylomicrons

5. Chylomicrons are passed to the circulatory system through lymphatic vessels

43.19 Digestion and Absorption of Fats

can result. The opposite condition, diarrhea, results if too little water is reabsorbed or if water is secreted into the colon. (Both constipation and diarrhea can be induced by toxins from microorganisms.) Feces are stored in the last segment of the colon and periodically excreted.

Immense populations of bacteria live within the colon. One of the resident species is *Escherichia coli*, the bacterium so popular with researchers in biochemistry, genetics, and molecular biology (see Box 4.B). This inhabitant of the colon lives on matter indigestible to humans and produces some products useful to the host. For example, vitamin K and biotin are synthesized by *E. coli* and absorbed across the wall of the colon. Many species of mammals maximize the nutritional benefits from such bacterial activity by reingesting their own feces, a behavior called **coprophagy**. Excessive or prolonged intake of antibiotics can lead to vitamin deficiency because the antibiotics kill the normal intestinal bacteria at the same time they are killing the disease-causing organisms for which they are intended. The intestinal bacteria produce gases such as methane and hydrogen sulfide as by-products of their largely anaerobic metabolism. Humans expel gas after eating beans because the beans are rich in carbohydrates that bacteria—but not humans—can break down.

The large intestine of humans has a small, finger-like pouch called the **appendix**, which is best known for the trouble it causes when it becomes infected. The human appendix plays no essential role in digestion, but it does contribute to immune system function. It can be surgically removed without serious consequences. The part of the gut that forms the appendix in humans forms the much larger caecum in herbivores (see Figure 43.12), where it functions in cellulose digestion. As our primate ancestors evolved to exploit diets less rich in indigestible cellulose, the caecum no longer served an essential function and gradually became **vestigial** (reduced to a trace), like the nonfunctional eyes of cave fish or the dewclaws of dogs and cats.

Digestion by Herbivores

Cellulose is the principal organic compound in the diets of herbivores. Most herbivores, however, cannot produce **cellulases**, the enzymes that hydrolyze cellulose. Exceptions include silverfish (well known for eating books and stored papers), earthworms, and shipworms. Other herbivores, from termites to cattle, rely on microorganisms living in their digestive tracts to digest cellulose for them. These microorganisms inhabit various parts of the gut, where they may be present by the billions. Most are bacteria, but some are fungi or protists.

The digestive tracts of **ruminants** such as cattle,

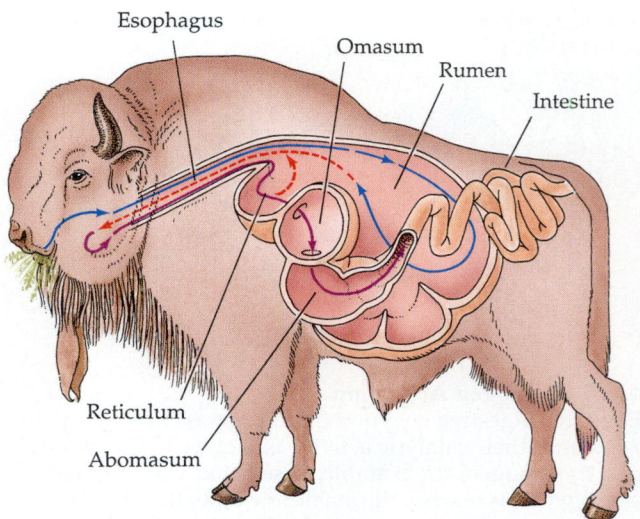

43.20 The Ruminant Stomach
Specialized stomach compartments—the rumen, reticulum, omasum, and abomasum—enable ruminants to digest and subsist on protein-poor plant material.

goats, and sheep are specialized to maximize benefits from microorganisms. In place of the usual mammalian stomach, ruminants have a large, four-chambered organ (Figure 43.20). The first and largest of these chambers is the **rumen**, the second is the **reticulum**. Both are packed with anaerobic microorganisms that break down cellulose. These two chambers serve as fermentation vats for the digestion of cellulose. The ruminant periodically regurgitates the contents of the rumen (the cud) into the mouth for rechewing. When the more thoroughly ground-up vegetable fibers are swallowed again, they present more surface area to the microorganisms for their digestive actions.

The microorganisms in the rumen and reticulum metabolize cellulose and other nutrients to simple fatty acids. Ruminants produce and swallow large quantitites of alkaline saliva to buffer this acid production. Although fatty acids are the major nutrients the host derives from its microorganisms, the microorganisms themselves provide an important source of protein. A cow can derive more than 100 grams of protein per day from digestion of its own microorganisms. The plant materials ingested by a ruminant are a poor source of protein, but they contain inorganic nitrogen that the microorganisms use to synthesize their own amino acids.

The by-products of the fermentation of cellulose are carbon dioxide and methane, which the animal belches. A single cow can produce 400 liters of methane a day. Methane is the second most abundant "greenhouse gas," whose concentration in the atmosphere is increasing, and domesticated ruminants are second only to industry as a source of methane emitted into the atmosphere.

The food leaving the rumen carries with it enormous numbers of the cellulose-fermenting microorganisms. This mixture passes through the **omasum**, where it is concentrated by water reabsorption. It then enters the true stomach, the **abomasum**, which secretes hydrochloric acid and proteases. The microorganisms are killed by the acid, digested by the proteases, and passed on to the small intestine for further digestion and absorption. The rate of multiplication of microorganisms in the rumen is great enough to offset their loss, so a well-balanced, mutually beneficial relationship is maintained.

As we mentioned earlier, mammalian herbivores other than ruminants have microbial farms and cellulose fermentation vats in a branch off the large intestine called the caecum. Rabbits and hares are good examples (see Figure 43.12). Since the caecum empties into the large intestine, the absorption of the nutrients produced by the microorganisms is inefficient and incomplete. Therefore, these animals practice coprophagy. They frequently produce two kinds of feces, one of pure waste (which they discard), and one consisting mostly of caecal material, which they ingest directly from the anus. As this caecal material passes through the stomach and small intestine, the nutrients it contains are digested and absorbed.

CONTROL AND REGULATION OF DIGESTION

The vertebrate gut could be described as an assembly line in reverse. As with a standard assembly line, control and coordination of sequential processes is critical. Both neural and hormonal controls govern gut functions.

Neural Reflexes

Everyone has experienced salivation stimulated by the sight or smell of food. That response is a neural reflex, as is the act of swallowing following tactile stimulation at the back of the mouth. Many neural reflexes coordinate activities in different regions of the digestive tract so that it works in a properly timed, assembly line manner. For example, loading the stomach with food stimulates increased activity in the colon, which can lead to a bowel movement. This phenomenon is the **gastrocolic reflex**. The digestive tract is unusual in that it has an intrinsic (that is, its own) nervous system. So in addition to neural reflexes involving central nervous system, such as salivation and swallowing, neural messages can travel within the digestive tract without being processed by the central nervous system.

About one hundred years ago, the Russian physiologist Ivan Pavlov, in an effort to explain all regulation and control of the digestive tract in terms of neural reflexes, discovered a very basic form of learning, the **conditioned reflex**. When a dog is presented with food, it salivates—an unconditioned reflex. If a bell is rung whenever the dog is presented with food, after a number of trials the dog will salivate whenever the bell is rung, even if no food is present—a conditioned reflex. Pavlov tried to explain the control of the secretory activity of the pancreas in terms of a neural reflex. He showed that the presence in the duodenum of acidic chyme from the stomach stimulated the pancreas to secrete digestive juices. Even hydrochloric acid alone applied to the duodenum stimulated the pancreas to secrete. When Pavlov tried to discover the path of the neural reflex controlling this response, however, he ran into trouble. The response could not be eliminated by destroying neural connections between the gut, the central nervous system, and the pancreas. Other researchers even removed a section of duodenum from the digestive tract of an animal and sutured it into a closed loop with no neural connections. They placed the loop back in the abdominal cavity, and when acid was injected into it, the pancreas was stimulated to secrete.

You have probably guessed the explanation of Pavlov's dilemma, but the answer was first demonstrated by two British physiologists, William Bayliss and Ernest Starling, who were working at the same time as Pavlov. They removed a section of small intestine from an animal and scraped off its mucosal lining. They ground the mucosa with sand, filtered it, and injected the extract into the bloodstream of an animal. The recipient animal secreted pancreatic juices. This was the first demonstration of a chemical message traveling through the circulatory system and having an effect on a specific tissue. They named their newly discovered chemical message **secretin** and also coined the general term "hormone." The moral of the story is that control of the functions of the digestive tract is not by neural reflexes alone; it is also by hormonal mechanisms.

Hormonal Controls

Several hormones control the activities of the digestive tract and its accessory organs (Figure 43.21). Bayliss and Starling's mucosal cell extract probably contained multiple hormones, but the dominant one, secretin, is responsible primarily for stimulating the pancreas to secrete a solution rich in bicarbonate ions. In response to fats and proteins in the chyme, the mucosa of the small intestine also secretes **cholecystokinin**, a hormone that stimulates the gallbladder to release bile and the pancreas to release digestive enzymes. Cholecystokinin and secretin also slow down the movements of the stomach, which slows the delivery of chyme into the small intestine.

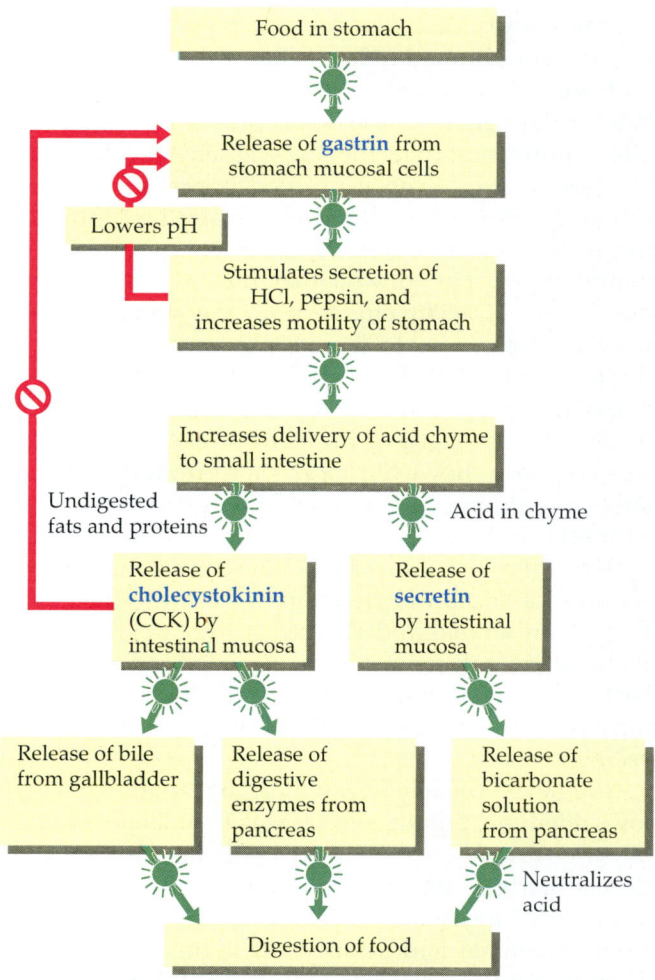

43.21   Hormones Control Digestion
Several hormones (blue type) are involved in feedback loops that control the sequential processing of food in the digestive tract.

The stomach secretes a hormone called **gastrin** into the blood. Cells in the lower region of the stomach release gastrin when they are stimulated by the presence of food. Gastrin circulates in the blood until it reaches cells in the upper areas of the stomach wall, where it stimulates the secretions and movements of the stomach. Gastrin release is inhibited when the stomach contents become too acidic—another example of negative feedback.

CONTROL AND REGULATION OF FUEL METABOLISM

Most animals do not eat continuously. There are times when food is in the gut and nutrients are being absorbed and are readily available to supply energy and molecular building blocks. When nutrients are not being absorbed, however, the continuous processes of energy metabolism and biosynthesis must

run off of internal reserves. For this reason nutrient traffic must be controlled so that reserves accumulate during absorption and those reserves are used appropriately when the gut is empty.

The Role of the Liver

The liver directs the traffic of nutrient molecules used in energy metabolism (fuel molecules). When nutrients are abundant in the circulatory system, the liver can store them in the forms of glycogen (animal starch) and fat. The liver also synthesizes plasma proteins from circulating amino acids. When the availability of fuel molecules in the bloodstream declines, the liver delivers glucose and fats back to the blood. The liver has an enormous capacity to interconvert fuel molecules. Liver cells can convert monosaccharides into either glycogen or fat, and vice versa. Certain amino acids and some other molecules, such as pyruvate and lactate, can be converted into glucose—a process called **gluconeogenesis**. The liver is also the major controller of fat metabolism through its production of lipoproteins (Box 43.C).

Hormonal Control of Fuel Metabolism

The **absorptive period** refers to the time that food is in the gut and nutrients are being absorbed and circulated in the blood. During this time the liver converts glucose to glycogen and fat, the body fat tissues convert glucose and fatty acids to stored fat, and the cells of the body preferentially use glucose for metabolic fuel. When food is no longer in the gut—the **postabsorptive period**—these processes reverse. The liver breaks down glycogen to supply glucose to the blood; the liver and the fat tissues supply fatty acids to the blood; and most of the cells of the body preferentially use fatty acids for metabolic fuel. The major exception to this rule is the cells of the nervous system, which require a constant supply of glucose for their energetic needs. Although the nervous system can use other fuels to a limited extent, its overall dependence on glucose is the reason it is so important for other cells of the body to shift to fat metabolism during the postabsorptive period. This shift preserves the available glucose and glycogen stores for the nervous system for as long as possible.

What directs the traffic in fuel molecules? Two hormones produced and released by the pancreas, **insulin** and **glucagon**, are largely responsible for controlling the metabolic directions fuel molecules take (Figure 43.22). The most important of these hormones is insulin, which is produced in response to high blood glucose levels. The pancreas releases insulin into the circulatory system when blood glucose rises above the normal postabsorptive level of about 90 milligrams of glucose per 100 milliliters of plasma.

BOX 43.C

Lipoproteins: The Good, the Bad, and the Ugly

In the intestine, bile solves the problem of processing hydrophobic fats in an aqueous medium. The *transportation* of fats in the circulatory system presents the same problem, but the solution is lipoproteins. A **lipoprotein** is a particle made up of a core of fat and cholesterol and a covering of protein that makes it hydrophilic. The largest lipoprotein particles are the chylomicrons produced by the cells lining the intestine to transport dietary fat and cholesterol into the circulation (see Figure 43.19). As lipoproteins circulate through the liver and the fat tissues around the body, receptors on the capillary walls recognize the protein coat, and lipases begin to hydrolyze the fats, which are then absorbed into fat or liver cells. Thus the protein coat of the lipoprotein makes it water-soluble and serves as an "address" that targets the lipoprotein to a specific tissue.

Other lipoproteins originate in the liver and are classified according to their density. Because fat has a low density (it floats in water), the more fat a lipoprotein contains, the lower its density. Very-low-density lipoprotein (VLDL) produced by the liver contains mostly triglyceride fats that are being transported to fat cells in tissues around the body. Low-density lipoproteins (LDL) contain mostly cholesterol, which they transport to tissues around the body for use in biosynthesis and to be deposited. High-density lipoproteins (HDL) serve as acceptors of cholesterol and are believed to remove cholesterol from tissues and return it to the liver, where it can be used to synthesize bile. Because of their differing functions in cholesterol regulation, LDL is sometimes called "bad cholesterol" and HDL "good cholesterol," but those designations are somewhat controversial at present. We do know, however, that a high ratio of LDL to HDL in a person's blood is a risk factor for atherosclerotic heart disease. Cigarette smoking lowers HDL levels, and regular exercise increases HDL levels.

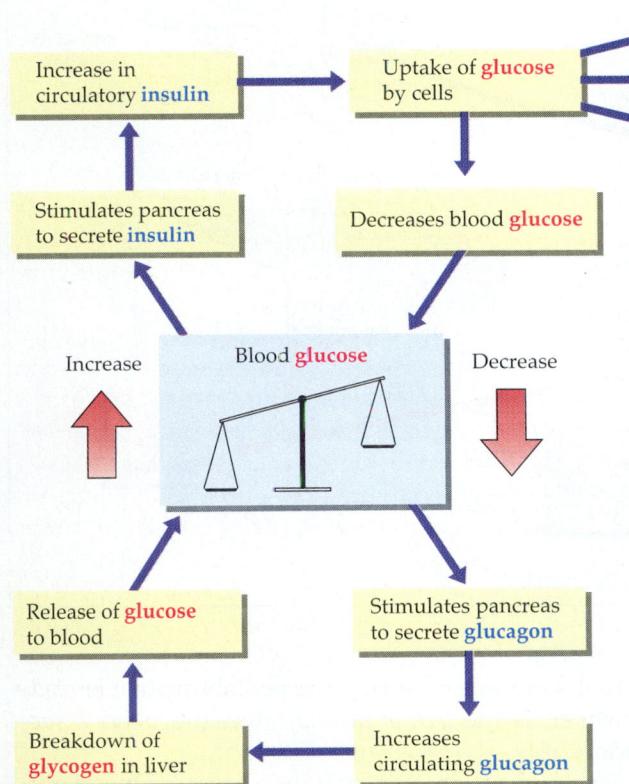

43.22 Regulation of Blood Glucose
Insulin and glucagon maintain the homeostasis of blood glucose.

Insulin facilitates the entry of glucose into most cells of the body. Thus when insulin is present, most cells burn glucose as their metabolic fuel, fat cells use glucose to make fat, and liver cells convert glucose to glycogen and fat. As soon as blood glucose falls back to postabsorptive levels, insulin release diminishes rapidly, and the entry of glucose into cells other than those of the nervous system is inhibited. Without a supply of glucose, cells switch to using glycogen and fat as their metabolic fuels. In the absence of insulin, the liver and fat cells stop synthesizing glycogen and fat and begin breaking them down. As a result, the liver supplies glucose to the blood rather than taking it from the blood, and both the liver and the fat tissues supply fatty acids to the blood.

The pancreas releases glucagon when the blood glucose concentration falls below the normal postabsorptive level. Glucagon has the opposite effect of insulin; it stimulates liver cells to break down glycogen and carry out gluconeogenesis, thus releasing glucose into the blood. The major hormonal control

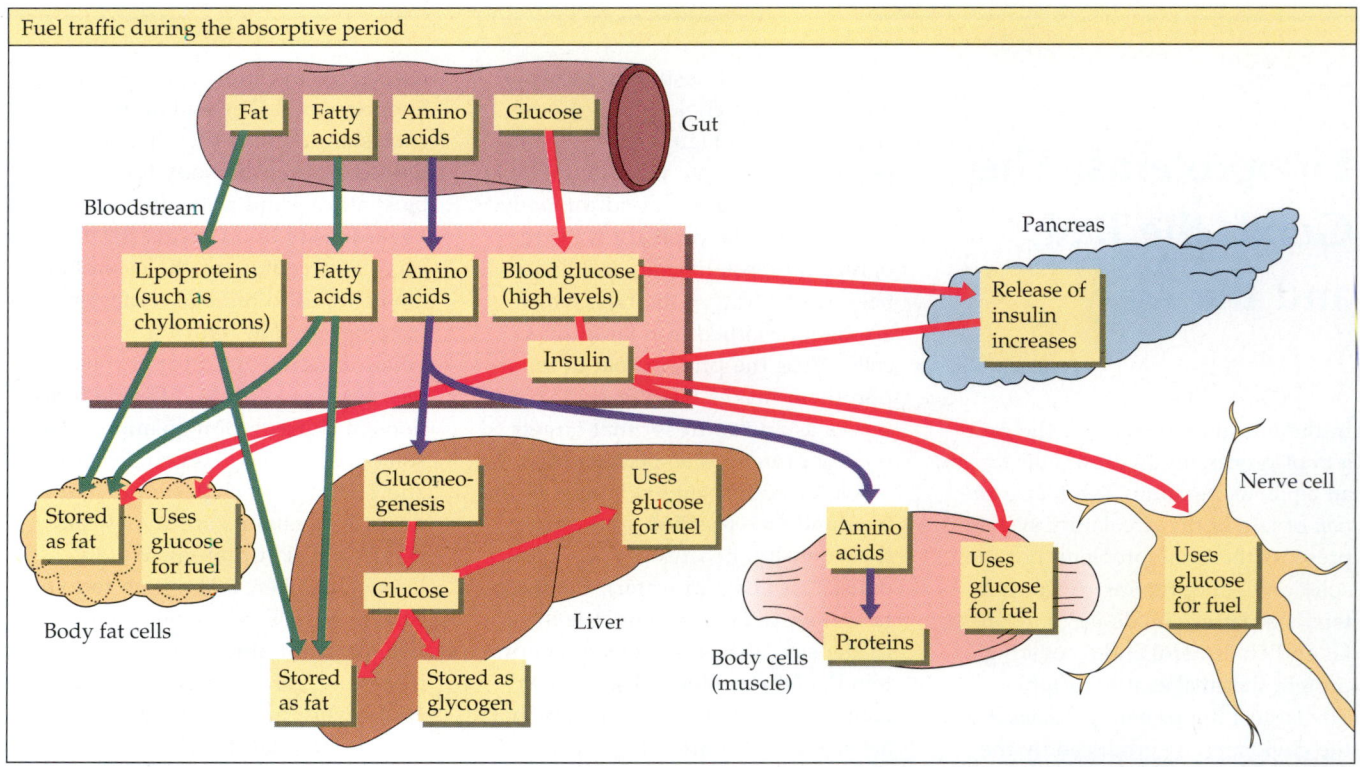

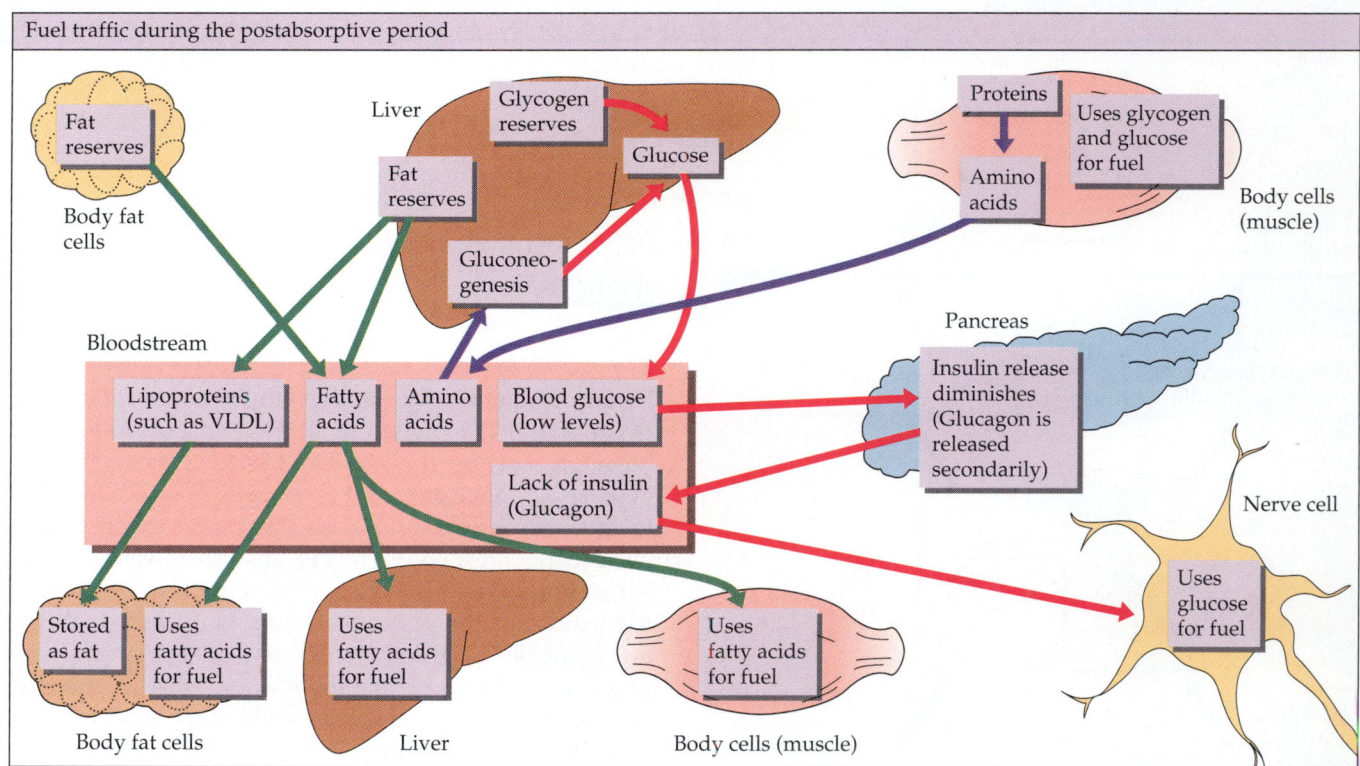

43.23 Fuel Molecule Traffic during the Absorptive and Postabsorptive Periods

Insulin promotes glucose uptake by liver, muscle, and fat cells during the absorption period. During the postabsorptive period the lack of insulin blocks glucose uptake by these same tissues and promotes fat and glycogen breakdown to supply metabolic fuel. Red arrows = carbohydrate traffic; blue arrows = protein/amino acid traffic; green arrows = fat/fatty acid traffic.

of fuel metabolism during the postabsorptive period, however, is the *lack of insulin*; glucagon plays a secondary role.

Two other hormones that help preserve blood glucose levels during the postabsorptive period are epinephrine and the glucocorticoid cortisol. Low blood glucose is a stress that triggers glucose-sensitive cells in the hypothalamus to signal the adrenal medulla, through the sympathetic nervous system, to secrete epinephrine. Epinephrine has an effect on liver cells similar to that of glucagon. Through the cAMP second-messenger cascade (see Chapter 36), it increases the breakdown of glycogen and increases gluconeo-genesis. The stress of low blood glucose also stimulates the adrenal cortex to secrete cortisol. Cortisol inhibits glucose metabolism in many cells while promoting metabolism of fats and proteins.

The traffic of fuel molecules during the absorptive and postabsorptive periods is summarized in Figure 43.23. The steps controlled by insulin and glucagon are indicated. Note that during the absorptive period, the direction of traffic in all fuel molecules is toward storage, and glucose is the preferred energy source for all cells. During the postabsorptive period, most cells switch to metabolizing fat so that blood glucose reserves are saved for the nervous sytem. The level of circulating glucose is maintained through glycogen breakdown and gluconeogenesis.

SUMMARY of Main Ideas about Animal Nutrition

Carbohydrates, fats, and proteins in food supply animals with metabolic energy and carbon skeletons for biosynthesis.
Review Figures 43.2 and 43.4

An animal with insufficient caloric intake is under-nourished and must metabolize its own carbohydrate, fat, and finally its own protein for energy.
Review Figure 43.3

Proper nutrition requires specific molecules such as essential amino acids, essential fatty acids, and vitamins.
Review Figure 43.5 and Table 43.2

Certain mineral elements are required nutrients.
Review Table 43.1

Inadequate supplies of required nutrients result in malnutrition and may lead to deficiency diseases.

Animals have evolved a great variety of adaptations for exploiting sources of food.

Carnivores eat other animals, herbivores eat plants, omnivores eat other animals and plants, and detritivores eat the decomposition products of dead organisms.

Some animals are deposit feeders, some are fluid feeders, and others are filter feeders.

Teeth are a general but highly variable feeding adaptation of vertebrates.
Review Figures 43.9 and 43.10

Guts are cavities or tubes where digestion and absorption occur.

Guts are usually divided into separate compartments with specialized functions but with the same basic tissue structure throughout.
Review Figures 43.12 and 43.14

Some compartments of the gut may have large populations of microorganisms that aid in digesting molecules which otherwise would be indigestible to the host.
Review Figure 43.20

In vertebrate guts, food passes from the mouth through the esophagus to the stomach, and then to the small intestine, followed by the large intestine.
Review Figure 43.16

The stomach stores food and initiates some aspects of digestion.

Digestive enzymes reduce carbohydrate, fat, and protein molecules to the simplest units that can be absorbed by cells of the small intestine and circulated to the body.
Review Figure 43.19 and Table 43.3

Bile from the liver, bicarbonate ions from the pancreas, and enzymes from the pancreas (in the form of inactive zymogens) are secreted into the small intestine through the common bile duct.
Review 43.17 and 43.18

Most nutrients are absorbed in the small intestine.

In the large intestine, or colon, water and ions are reabsorbed into the body, and feces are formed and stored.

The activities of the various compartments of the gut are controlled and coordinated by hormones and by neural reflexes involving the central nervous system and the intrinsic nervous system of the gut.
Review Figure 43.21

Fuel metabolism is regulated mainly by the pancreatic hormones insulin and glucagon.
Review Figure 43.22

The liver manages the traffic in fuel molecules.
Review Figure 43.23

During the absorptive period the liver absorbs the excess fuel molecules, interconverting the different types, and stores them as glycogen and fat.

During the postabsorptive period the liver breaks down its glycogen and fat to supply the blood with glucose and fatty acids.

SELF-QUIZ

1. Most of the metabolic energy a bird requires for a long-distance migratory flight is stored as
   a. glycogen.
   b. fat.
   c. protein.
   d. carbohydrate.
   e. ATP.

2. Which of the following statements about essential amino acids is true?
   a. They are not found in vegetarian diets.
   b. They are stored by the body for when they are needed.
   c. Without them one is undernourished.
   d. All animals require the same ones.
   e. Humans can acquire all of theirs by eating milk, eggs, and meat.

3. Which of the following statements about vitamins is true?
   a. They are essential inorganic nutrients.
   b. They are required in larger amounts than are essential amino acids.
   c. Many serve as coenzymes.
   d. Vitamin D can be acquired only by eating meat or dairy products.
   e. When vitamin C is eaten in large quantities, the excess is stored in fat for later use.

4. The digestive enzymes of the small intestine
   a. do not function best at a low pH.
   b. are produced and released in response to circulating secretin.
   c. are produced and released under neural control.
   d. are all secreted by the pancreas.
   e. are all activated by an acidic environment.

5. Which of the following statements about nutrient absorption across the gut epithelium is true?
   a. Carbohydrates are absorbed as disaccharides.
   b. Fats are absorbed as fatty acids and monoglycerides.
   c. Amino acids move across only by diffusion.
   d. Bile salts transport fats across.
   e. Most nutrients are absorbed in the duodenum.

6. Chylomicrons are like the tiny particles of dietary fat in the lumen of the small intestine in that
   a. both are coated with bile salts.
   b. both are lipid-soluble.
   c. both travel in lacteals.
   d. both contain triglyceride.
   e. both are coated with lipoproteins.

7. Microbial fermentation in the guts of cattle
   a. produces fatty acids as a major nutrient for the cattle.
   b. occurs in specialized regions of the small intestine.
   c. occurs in the caecum, from which food is regurgitated to be chewed again and swallowed into the true stomach.
   d. produces methane as a major nutrient.
   e. is possible because the stomach wall does not secrete hydrochloric acid.

8. Which of the following is stimulated by cholecystokinin?
   a. Stomach motility
   b. Release of bile
   c. Secretion of hydrochloric acid
   d. Secretion of bicarbonate ions
   e. Secretion of mucus

9. During the absorptive period
   a. breakdown of glycogen supplies glucose to blood.
   b. glucagon secretion is high.
   c. the number of circulating lipoproteins is low.
   d. glucose is the major metabolic fuel.
   e. synthesis of fats and glycogen in muscle is inhibited.

10. During the postabsorptive period
    a. glucose is the major metabolic fuel.
    b. glucagon stimulates the liver to produce glycogen.
    c. insulin facilitates the uptake of glucose by brain cells.
    d. the major metabolic fuel is fatty acids.
    e. liver functions slow down because of low insulin levels.

FOR STUDY

1. From what you have learned about nutrition in this chapter, discuss some of the problems with "crash" or "fad" diets. What should one take into account when considering or planning a diet aimed at weight reduction?

2. The digestive tract must move food slowly enough to enable digestion and absorption but fast enough to supply the animal's energetic needs. Describe controls that speed up and slow down the activities of the digestive tract.

3. Describe the role of the liver in the homeostasis of blood glucose. What are the controlling factors?

4. Why is obstruction of the common bile duct so serious? Consider in your answer the multiple functions of the pancreas and the way in which digestive enzymes are processed.

5. Trace the history of a fatty acid molecule from being on a piece of buttered toast to being in a plaque on a coronary artery. What possible forms and structures could it have passed through in the body? Describe a direct and an indirect route it could have taken.

READINGS

Atkinson, M. A. and N. K. MacLaren. 1990. "What Causes Diabetes?" *Scientific American*, July. The body's own immune system can cause this serious disease that destroys the ability to regulate fuel metabolism.

Brown, M. S. and J. L. Goldstein. 1984. "How LDL Receptors Influence Cholesterol and Atherosclerosis." *Scientific American*, November. A detailed discussion of what does and does not happen to the cholesterol and fatty acids we consume.

Davenport, H. 1972. "Why the Stomach Does Not Digest Itself." *Scientific American*, January. The stomach contains hydrochloric acid but usually does not digest its own lining.

Degabriele, R. 1980. "The Physiology of the Koala." *Scientific American*, July. An amazing adaptation to an unusual and limited diet.

Lienhard, G. E., J. W. Slot, D. E. James and M. M. Mueckler. 1992. "How Cells Absorb Glucose." *Scientific American*, January. Research on the glucose transporter and how it is regulated by insulin.

Moog, F. 1981. "The Lining of the Small Intestine." *Scientific American*, November. How nutrients are absorbed.

Orci, L., J.-D. Vassalli and A. Perrelet. 1988. "The Insulin Factory." *Scientific American*, September. The molecular biology of the synthesis and release of insulin by beta cells of the pancreas.

Sanderson, S. L. and R. Wassersug. 1990. "Suspension-Feeding Vertebrates." *Scientific American*, March. By filtering small particles out of the water, some animals exploit an abundant food resource.

Schmidt-Nielsen, K. 1990. *Animal Physiology: Adaptation and Environment*, 4th Edition. Cambridge University Press, New York. An outstanding textbook, emphasizing the comparative approach.

Scrimshaw, N. S. 1991. "Iron Deficiency." *Scientific American*, October. A discussion of one of the most common dietary deficiencies in humans and its consequences.

Uvnas-Moberg, K. 1989. "The Gastrointestinal Tract in Growth and Reproduction." *Scientific American*, July. A concise treatment of hormonal controls of digestive tract function and how they change to accommodate special needs of pregnancy and lactation.

Vander, A. J., J. H. Sherman and D. S. Luciano. 1994. *Human Physiology: The Mechanisms of Body Function*, 6th Edition. McGraw-Hill, New York. Chapters 17 and 18 provide a readily understandable treatment of digestion and absorption of food.

Water Is Essential for Life
During the dry season on the African savanna, animals must congregate around the scarce sources of water.

44

Salt and Water Balance and Nitrogen Excretion

Blood, sweat, and tears all taste salty, like seawater. Life evolved in the sea, and if a complex animal is to exist elsewhere, it must carry its own internal sea to bathe the cells of its body. This statement sometimes is taken to mean that the mineral ion, or salt, composition of our blood is the same as that of the ancient seas. That conclusion is wrong. Although all animals on Earth share the same distant origins, there is great diversity in the composition of their body fluids, and no animal has an internal environment just like the environment in which life evolved. However, all animals do require both water and salts.

The availability of salts and water can be the critical characteristic of an environment that determines which organisms can live there, in what numbers, and when. Some animals with extreme adaptations can live in the hottest, driest deserts on Earth, rarely, if ever, experiencing liquid water in their environment, but their populations are very sparse. Many species living in habitats that are seasonally dry migrate long distances to find water. During the dry season on the African plains, predators and prey alike congregate at scarce water holes. The availability of salts and water in the external environment is only part of the story, however. Most animals must regulate the salt and water composition of their internal environments within narrow limits. How they do so is the focus of this chapter.

INTERNAL ENVIRONMENT

The extracellular fluids (including interstitial fluid and the blood plasma) that we carry with us service all our cells (see Chapter 35). In addition to supplying cells with oxygen and nutrients and carrying away waste products, these extracellular fluids determine the water balance of the cells. To understand what is meant by water balance, recall that cell membranes are permeable to water and that the movement of water across membranes depends on differences in osmotic potential. (You may find it useful to review the discussion of osmosis in Chapter 5.) If the osmotic potential of the extracellular fluid is less negative (that is, the fluid contains fewer solutes) than that of the intracellular fluids, water moves into the cells, causing them to swell and possibly burst. If the osmotic potential of the extracellular fluid is more negative (the fluid contains more solutes) than that of the intracellular fluids, the cells lose water and shrink. The osmotic potential of the extracellular environment determines both the volume and osmotic potential of the intracellular environment.

Excretory systems consist of the organs that help regulate the osmotic potential and the volume of the extracellular fluids. In addition, excretory systems

SALT AND WATER BALANCE AND NITROGEN EXCRETION **1009**

regulate the composition of the extracellular fluids by excreting molecules that are in excess (such as NaCl when we eat lots of salty popcorn) and conserving those that are valuable or in short supply (such as glucose and amino acids). In terrestrial organisms, excretory systems also eliminate the toxic waste products of nitrogen metabolism.

Exactly what the excretory system of a particular species must do depends on the environment in which that species lives. In this chapter we will examine excretory systems that maintain salt and water balance and eliminate nitrogen in marine, freshwater, and terrestrial habitats. In spite of the evolutionary diversity of the anatomical and physiological details, all these systems obey a common rule and employ common mechanisms. The common rule is that there is no active transport of water; water must be moved either by pressure or by a difference in osmotic potential. The common mechanisms derive from the fact that most excretory organs consist of systems of tubules that receive extracellular fluid and alter its composition to produce **urine**—the fluid waste product that is excreted. The extracellular fluid enters the excretory tubules by **filtration**, and its composition is changed by processes of active **secretion** and **reabsorption** of specific solute molecules by the cells of the tubules. These same three mechanisms—filtration, secretion, and reabsorption—are used both in systems that excrete water and conserve salts and in systems that do the opposite, conserving water and excreting salts. The intestine (see Chapter 43) is an important excretory organ. It absorbs water and controls the loss of water with the solid wastes of digestion. Intestinal cells influence the solute composition of the internal environment through the selective transport of ions.

WATER, SALTS, AND THE ENVIRONMENT

We think of marine and freshwater environments as being distinctly different, one salty and the other not. In reality, aqueous environments grade continuously from fresh to extremely salty. Consider a place where a river enters the sea through a bay or a marsh. Aqueous environments within that bay or marsh range in salinity (salt content) from the fresh water of the river to the open sea. Evaporating tide pools can become much saltier than the sea. Animals live in all these environments.

Most marine invertebrates can adjust to a wide range of environmental salinities by allowing their body fluids to have the same osmotic potential as the environment; they thus avoid the risk of being burst or shrunk by osmotic movement of water. Such animals are called **osmoconformers**. There are limits to osmoconformity, however. No animal could have the

same osmotic potential as fresh water and survive; nor could animals survive with internal salt concentrations as high as those that may be reached in an evaporating tide pool. Such concentrations cause proteins to denature. Animals that maintain an osmotic potential of their internal fluids different from that of their environment are called **osmoregulators**. Even animals that osmoconform over a wide range of osmotic potentials must osmoregulate at the extremes of environmental salinity.

To osmoregulate in fresh water, animals must continuously excrete the water that invades their bodies by osmosis, but while doing so they must conserve solutes; hence they produce large amounts of dilute urine. To osmoregulate in salt water, animals must conserve water and excrete salts, thus tending to produce small amounts of urine.

The brine shrimp *Artemia* is adapted to live in environments of almost any salinity. *Artemia* are found in huge numbers in the most saline environments possible, such as the Great Salt Lake in Utah or coastal evaporation ponds where salt is obtained for commercial purposes (Figure 44.1). The salinity of such water reaches 300 grams per liter (normal seawater contains about (35 g/l). *Artemia* are harvested from these environments and sold for fish food. *Artemia* cannot survive for long in fresh water, but they can live in very dilute seawater, in which they maintain the osmotic potential of their body fluids above the osmotic potential of the environment. Under these conditions, *Artemia* is a **hypertonic osmoregulator**, meaning that it regulates the concentration (the osmolarity) of its body fluids so that they are more concentrated than (hypertonic to) the environment (Figure 44.2). At high environmental salinities, *Artemia* is an exceptionally effective hypotonic osmoregulator, keeping the osmotic potential of its body fluids well below that of the water in which it is living. Very few organisms can survive in the crystallizing brine in which *Artemia* thrives. The main mechanism this small crustacean uses for osmoregulation is the active transport of NaCl across its gill membranes.

Osmoconformers can be **ionic conformers**; they can control ionic composition, as well as the osmolarity, of their body fluids to match that of the environment. Most osmoconformers, however, are **ionic regulators** to some degree: They employ active transport mechanisms to maintain specific ions in their body fluids at concentrations different from those in the environment.

The terrestrial environment presents entirely different sets of problems for salt and water balance. Because the terrestrial environment is extremely desiccating (drying), most terrestrial animals must conserve water. Exceptions are animals such as muskrats and beavers that practically live in water. Terrestrial

 (a)

 (b)

44.1 Evaporating Salt Ponds Vary in Salinity
(a) Impoundments of water around San Francisco Bay yield salt when they evaporate. Because different species of bacteria and algae grow best at different salinities, the colors of the salt ponds change as they evaporate. (b) One animal that can live in these ponds at all salinities is a tiny crustacean, the brine shrimp (*Artemia*).

animals obtain their salts from their food. Plants generally have low concentrations of sodium, so most herbivores must be very effective in conserving sodium ions. As we mentioned in Chapter 43, some terrestrial herbivores travel long distances to naturally occurring salt licks to supplement their dietary intake of sodium. By contrast, birds that feed on marine animals must excrete the large excess of sodium they ingest with their food. Their **nasal salt glands** excrete a concentrated solution of sodium chloride via a duct that empties into the nasal cavity. Birds such as penguins and sea gulls that have nasal salt glands can be seen frequently sneezing or shaking their heads to get rid of the very salty droplets that form (Figure 44.3).

THE EXCRETION OF NITROGENOUS WASTES

The end products of the metabolism of carbohydrates and fats are water (H_2O) and carbon dioxide (CO_2); they present no problems for excretion. Proteins and nucleic acids, however, contain nitrogen in addition to carbon, hydrogen, and oxygen. The metabolism of proteins and nucleic acids produces nitrogenous waste in addition to H_2O and CO_2. Most of that waste is ammonia (NH_3). Ammonia is highly toxic and must be excreted continuously to prevent accumulation, or

detoxified by conversion into other molecules for excretion. Those molecules are principally **urea** and **uric acid** (Figure 44.4).

Ammonia excretion is relatively simple for aquatic animals. Ammonia diffuses and is highly soluble in water. Animals that breathe water continuously lose ammonia from their blood to the environment by diffusion across their gill membranes. Animals that excrete nitrogen as ammonia are called **ammonotelic;**

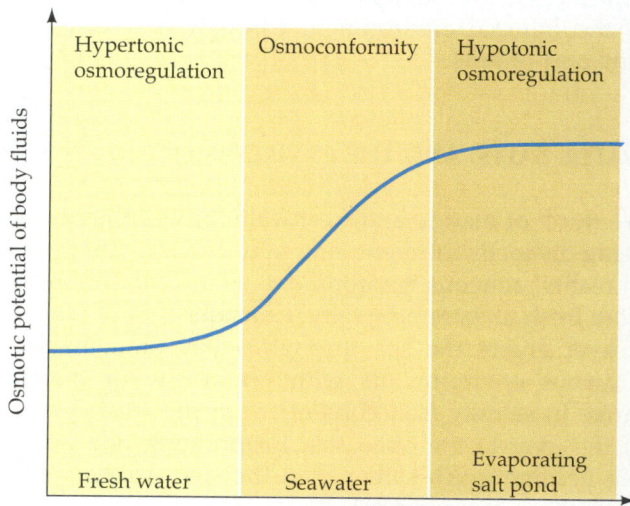

44.2 Osmoregulation and Osmoconformity
Over a broad range of environmental salinities, many marine invertebrates are osmoconformers. Animals that live at the extremes of environmental salinities, however, can display osmoregulatory abilities. They become hypertonic osmoregulators in very dilute water, or hypotonic osmoregulators in very saline water.

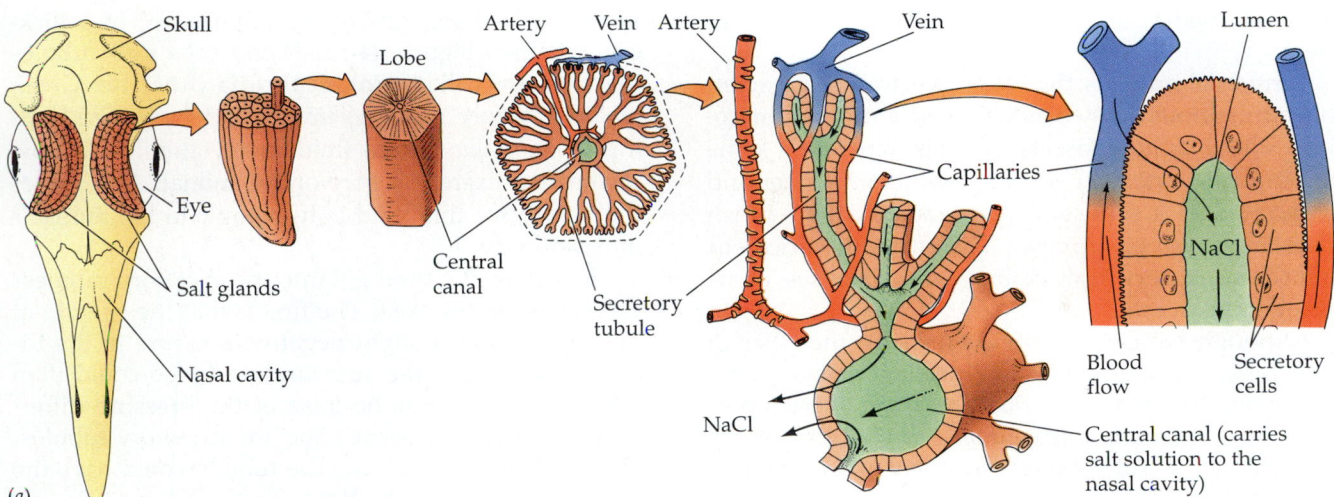

(a)

44.3   Nasal Glands Excrete Excess Salt

(a) Marine birds such as sea gulls and penguins have nasal salt glands adapted to excrete the excess salt from the seawater they consume with their food. Salt glands lie in grooves in the skull above the eyes. (b) This Adélie penguin has returned from a feeding trip at sea and is excreting salt through its nasal salt gland. A patch of evaporated salt lies on the rock below the bird's beak.

(b)

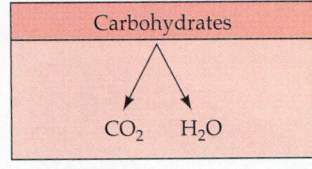

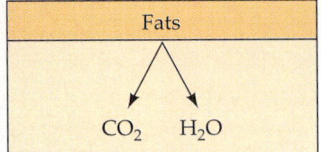

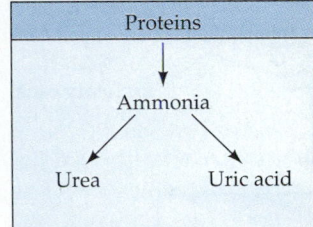

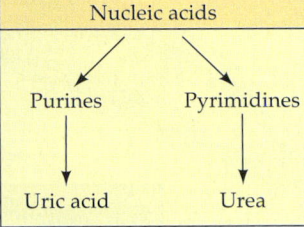

44.4   Waste Products of Metabolism

Whereas the metabolism of carbohydrates and fats yields only water and carbon dioxide, the metabolism of proteins and nucleic acids produces the nitrogenous wastes ammonia, uric acid, and urea. The metabolism of nucleic acids begins with their breakdown into their constituent bases—pyrimidines and purines (see Chapter 3).

they include aquatic invertebrates and bony fishes. Crocodiles and amphibian tadpoles are also ammonotelic.

Ammonia is a more dangerous metabolite for terrestrial animals that have limited access to water. In mammals, ammonia is lethal when it reaches only 5 milligrams per 100 milliliters of blood. Therefore, terrestrial (and some aquatic) animals convert ammonia into either urea or uric acid. **Ureotelic** animals, such as mammals, amphibians, and cartilaginous fishes (sharks and rays), excrete urea as their principal nitrogenous waste product. Urea is quite soluble in water, but excretion of urea solutions at low concentrations could result in a large loss of water that many terrestrial animals could ill afford. Later in the chapter we will see that mammals have evolved excretory systems that produce urine, which contains urea and is hypertonic to their body, thereby conserving water while excreting the urea. The cartilaginous fishes are another story. These marine species maintain their body fluids hypertonic to the marine environment by retaining high concentrations of urea. Because water moves into their bodies by osmosis and must be excreted, water conservation is not a problem for them.

Terrestrial animals that conserve water by excreting nitrogenous wastes as uric acid are called **uricotelic**. These include insects, reptiles, birds, and some amphibians. Uric acid is very insoluble in water and is excreted as a semisolid (for example, the whitish material in bird droppings). Therefore, the uricotelic animal loses very little water as it disposes of its nitrogenous waste.

Although we can classify animals on the basis of their major nitrogenous waste product as being ammonotelic, ureotelic, or uricotelic, most species produce more than one nitrogenous waste. Humans are ureotelic, yet we also excrete uric acid and ammonia, as anyone who has changed diapers knows. (Actually, most of the ammonia in diapers is produced by the bacterial breakdown of urea. The bacterium that performs this reaction was first isolated by a microbiologist from a diaper of his child.) The uric acid in human urine comes largely from the metabolism of nucleic acids and of caffeine. In the classical disease gout, uric acid levels in the body fluids increase, and uric acid precipitates in the joints and elsewhere, causing swelling and pain.

In some species, different developmental forms live in quite different habitats and have different forms of nitrogen excretion. Tadpoles of frogs and toads, for example, excrete ammonia across their gill membranes, but when they develop into adult frogs or toads they generally excrete urea. Some adult frogs and toads that live in arid habitats excrete uric acid. These examples show the considerable evolutionary flexibility in how nitrogenous wastes are excreted.

INVERTEBRATE EXCRETORY SYSTEMS

Most marine invertebrates are osmoconformers, so they have few adaptations for salt and water balance other than the active transport of specific ions. For nitrogen excretion they can passively lose ammonia by diffusion to the seawater. Freshwater and terrestrial invertebrates, however, display a variety of fascinating adaptations for maintaining salt and water balance and excreting nitrogen. Although diverse, all these adaptations are based on the same basic principles: filtration of body fluids and active secretion and reabsorption of specific ions.

Protonephridia

Many flatworms, such as **Planaria** (see Figure 43.11), live in fresh water and excrete water through an elaborate network of tubules running throughout their bodies. The tubules end in **flame cells**, so called because each flame cell has a tuft of cilia beating

inside the tubule, giving the appearance of a flickering flame (Figure 44.5). Flame cells and tubules together are called **protonephridia** (from the Greek *proto* = "before" and *nephros* = "kidney"). The beating of the cilia causes fluid in the tubules to flow toward the excretory pore of the animal. As it leaves the flatworm, this fluid is hypotonic to the animal's internal body fluids.

How does the fluid get into the excretory tubules? Two possibilities exist. The first is that the beating of the cilia creates a slight negative pressure in the tubule. Water from the surrounding tissue could then filter into the tubule because of the pressure difference between the tissues and the excretory tubules. The fluid that filtered into the tubule would have the same osmolarity as the tissue fluids, but as it flowed down the tubules the cells of the tubules could actively reabsorb solutes from this tubular fluid or urine. The second possibility is that ions are actively transported into the excretory tubules at their upper ends. Water would follow these ions because of the increased osmotic potential of the tubular fluid. Then, as the fluid moved down the tubules, ions could be actively reabsorbed. In either case, the urine leaving the protonephridia would be hypotonic to the internal fluids of the flatworm.

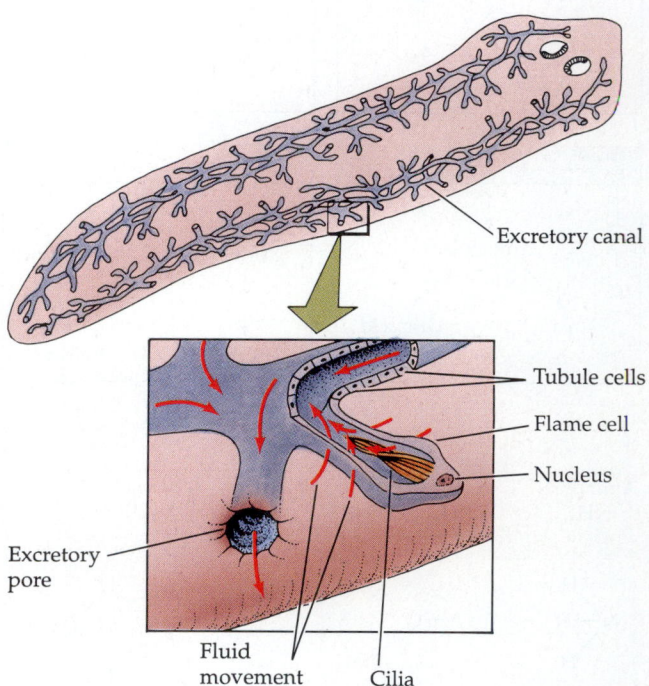

44.5   Protonephridia in Flatworms
The protonephridia of the flatworm *Planaria* consist of tubules ending in flame cells. Body fluids enter the space enclosed by the flame cell and are driven down the tubules toward the excretory pore by the beating of the cilia in the flame cells. The tubule cells modify the composition of this fluid.

Metanephridia

Filtration of body fluids and tubular processing of urine are highly developed in annelid worms, such as the earthworm. Recall that annelids have fluid-filled body cavities called coeloms (see Figure 26.24) and closed circulatory systems through which blood is pumped under pressure (see Figure 42.3). The pressure causes the blood to be filtered across the thin, permeable capillary walls into the coelom. This process is called filtration because the cells and large protein molecules of the blood stay behind in the capillaries while water and small molecules leave the capillaries and enter the coelom. Where does this coelomic fluid go?

Each segment of the earthworm contains a pair of **metanephridia** (*meta* is Greek for "akin to"). Each metanephridium begins in one segment as a ciliated, funnel-like opening in the coelom called a **nephrostome**, which leads into a tubule in the next segment. The tubule ends in a pore called the nephridiopore, which opens to the outside of the animal (Figure 44.6). Coelomic fluid enters the metanephridia through the nephrostomes. As the coelomic fluid passes through the tubules, the cells of the tubules actively reabsorb certain molecules from it and actively secrete other molecules into it. What leaves the animal through the nephridiopores is a hypotonic urine containing nitrogenous wastes, among other solutes.

Why should earthworms, which are terrestrial, excrete a hypotonic urine? Earthworms live in moist soil, an environment with 100 percent relative humidity, and they are in constant contact with a film of water covering the soil particles. Thus, like fresh-water aquatic animals, an earthworm must excrete the excess water that continually enters its body by osmosis.

In the metanephridium we see all the basic processes used in the excretory systems of vertebrates that will be discussed later in this chapter: filtration of the body fluids, and tubular processing of the filtrate by active secretion and reabsorption of ions and molecules.

Malpighian Tubules

Insects have remarkable systems for excreting nitrogenous wastes with very little loss of water. These animals can live in the driest habitats on Earth. The insect excretory system consists of blind tubules (from 2 to more than 100) attached to the gut between the midgut and hindgut and projecting into the fluid-filled coelom (Figure 44.7). The cells of these **Malpighian tubules** actively transport uric acid, potassium ions, and sodium ions from the coelomic fluid into the tubules. As solutes are secreted into the tubules, water follows because of the difference in osmotic potential. The walls of the Malpighian tubules have muscle fibers whose contractions help to move the contents of the tubules toward the hindgut.

In the hindgut the tubular fluid continues to change in composition. The hindgut contents are more acidic than the tubular fluids, and as a result the uric acid becomes less soluble and precipitates out of solution as it approaches and enters the rectum. The cells of the walls of the hindgut and rectum actively transport sodium and potassium ions from the gut contents back into the coelom. Because the uric acid molecules have precipitated out of solution, water is free to follow the reabsorbed salts back into the coelom through osmosis. Remaining in the rectum are crystals of uric acid mixed with undigested food, and this dry matter is all that the insect eliminates. The Malpighian tubule system is a highly effective mechanism for excreting nitrogenous wastes and some salts without giving up any significant fraction of the animal's precious water supply.

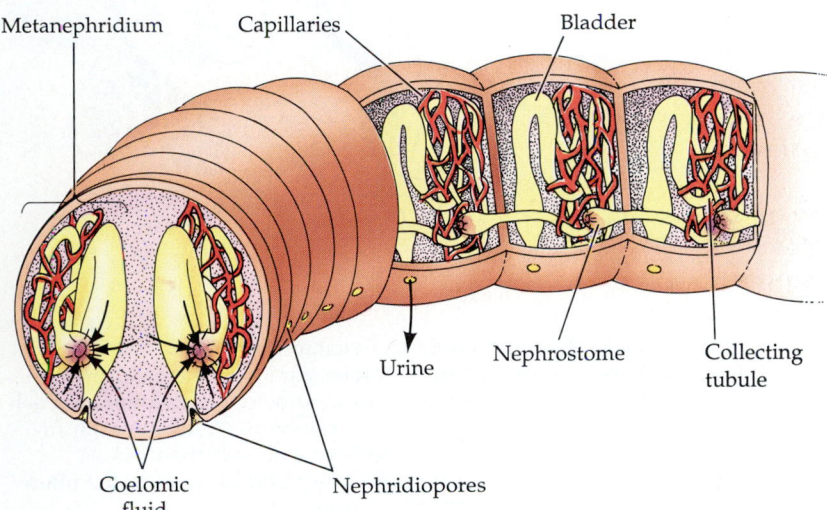

Metanephridium · Capillaries · Bladder · Urine · Nephrostome · Collecting tubule · Coelomic fluid · Nephridiopores

44.6 Metanephridia in Earthworms
The metanephridia of annelids are arranged segmentally. The cross section (left) shows a pair of metanephridia. Longitudinal sections (right) show only one metanephridium of the two in each segment. Coelomic fluid enters the metanephridium through a nephrostome. The tubule cells of the metanephridium alter the composition of this fluid, producing a dilute urine that is excreted through the nephridiopore.

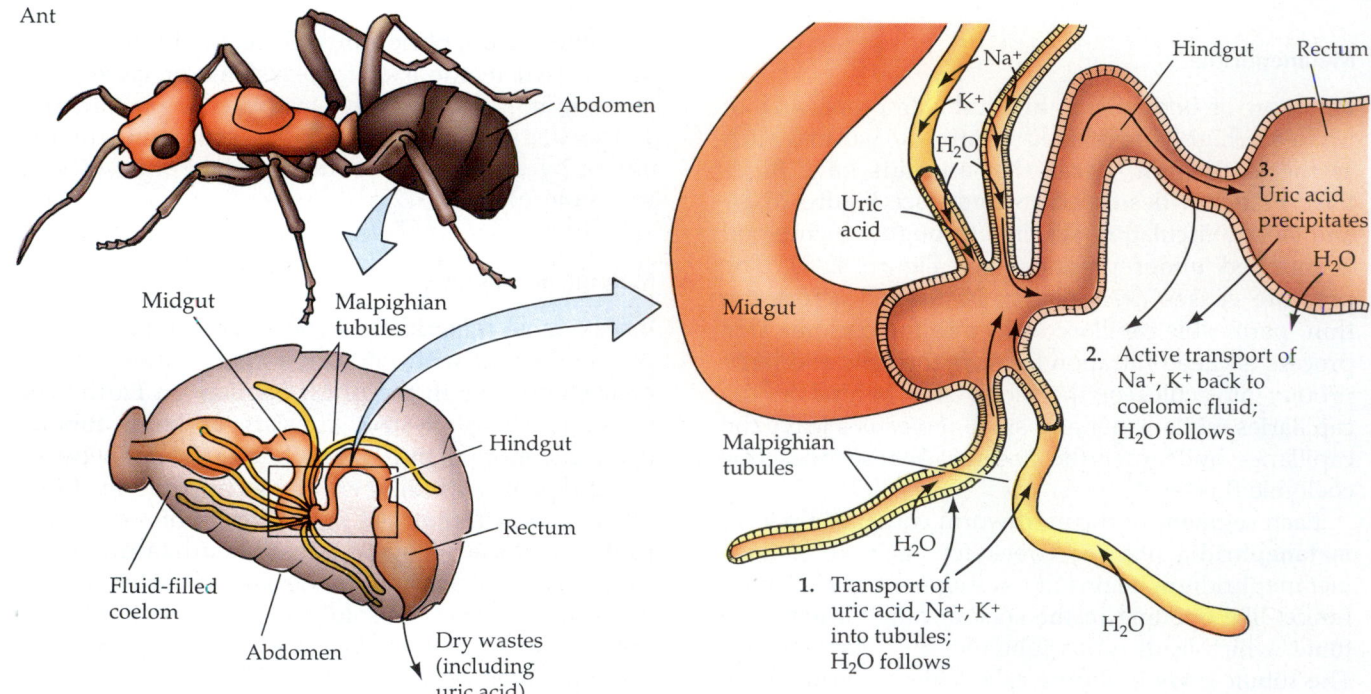

Ant

44.7 Malpighian Tubules in Insects
Blind, thin-walled Malpighian tubules are attached to the junction of the insect's mid- and hindgut.

Green Glands

Malpighian tubules are found in some arthropods, such as insects and spiders, but not in others, such as the crustaceans (crabs, lobsters, crayfish, and their relatives). The crustacean excretory system consists of an **end sac** and **labyrinth** connected by a **nephridial canal** to a bladder that empties to the outside by way of an excretory pore (Figure 44.8). This entire assembly in a crayfish or lobster is called a **green gland** and is regarded by many people as a gourmet treat. A key feature of this excretory system is that extracellular fluid (called hemolymph in these animals with open circulatory systems) is filtered into the end sac by the high pressure within the coelom. The volume and composition of the filtrate may be altered in the labyrinth, but it remains at the same osmolarity as the coelomic fluid. From the labyrinth, the fluid passes through a nephridial canal into a bladder, where it is stored until it is excreted through the excretory pore.

Freshwater crayfish produce large volumes of hypotonic urine; marine crayfish do not. Moreover, when marine crayfish are placed in dilute seawater, they cannot produce hypotonic urine. The difference between the freshwater and marine species is the length of their nephridial canals. All of this information taken together points to the nephridial canals as being responsible for producing hypotonic urine by actively reabsorbing salts from filtered coelomic fluids.

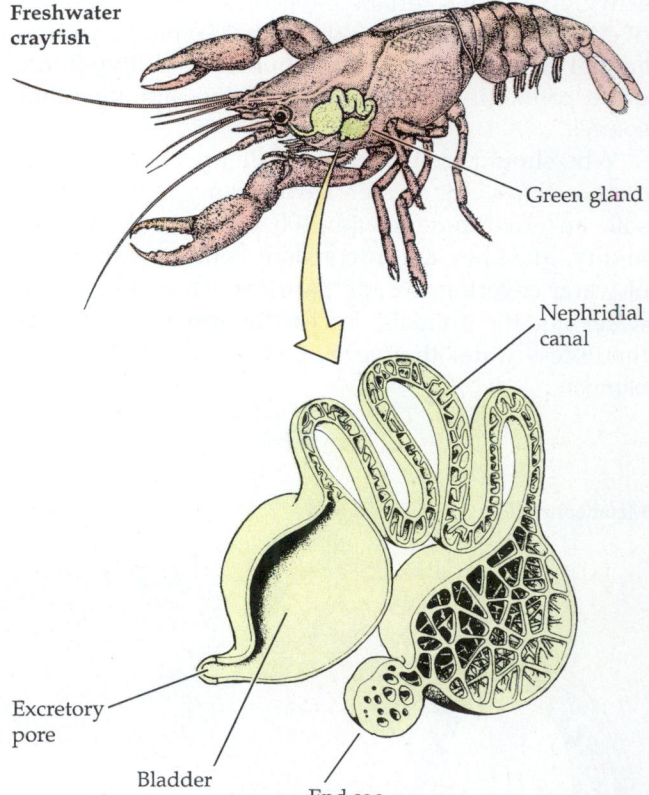

Freshwater crayfish

44.8 Green Glands in Crustaceans
Wastes empty from the green glands to the exterior through excretory pores located below each antenna. Each green gland is bathed in hemolymph. A hemolymph filtrate is forced into the bulbous end sac, then passes through the labyrinth and a nephridial canal to the bladder and out the excretory pore.

VERTEBRATE EXCRETORY SYSTEMS

The major vertebrate organ for salt and water balance and nitrogen excretion is the **kidney**. The functional unit of the kidney is the **nephron**. Each human kidney has about 1 million nephrons. To understand how the kidney works, you must understand the structure and function of the nephron. A remarkable fact about the kidneys of vertebrates is that in different species the same basic organ serves different needs. The kidneys of freshwater fish excrete water, whereas the kidneys of most mammals conserve water. To understand how the kidney can fulfill opposite functions in different animals, we need to look at the nephron.

The Structure and Functions of the Nephron

Each nephron has a vascular and a tubular component (Figure 44.9). The vascular component is unusual in consisting of two capillary beds between the arteriole that supplies it and the venule that drains it. The first capillary bed is a dense knot of very permeable vessels called a **glomerulus** (Figure 44.10a). Blood enters the glomerulus through what is called an **afferent arteriole** and exits through an **efferent arteriole**. The efferent arteriole gives rise to the second set of capillaries, the **peritubular capillaries**, which surround the tubular component of the nephron.

The tubular component of the nephron begins with **Bowman's capsule**, which encloses the glomerulus. The glomerulus appears to be pushed into Bowman's capsule much like a fist pushed into an inflated balloon. Together, the glomerulus and its surrounding Bowman's capsule are called the **renal corpuscle**. The cells of the capsule that come into direct contact with the glomerular capillaries are called **podocytes**. These highly specialized cells have numerous armlike extensions, each with hundreds of fine, fingerlike projections. The podocytes wrap around the capillaries so that their fingerlike projections cover the capillaries completely (Figure 44.10b,c).

The glomerulus filters the blood to produce a tubular fluid without cells and large molecules. The cells of the capillaries and the podocytes of Bowman's capsule participate in filtration. The walls of the capillaries have pores that allow water and small molecules to leave the capillary but that are too small to permit red blood cells and very large protein molecules to pass. Even smaller than the pores in the capillaries are the narrow slits between the fingerlike projections of the podocytes. The result is that water and small molecules pass from the capillary blood and enter the tubule of the nephron (Figure 44.10d), but red blood cells and proteins remain in the capillaries.

The force that drives filtration in the glomerulus is the pressure of the arterial blood. As in every

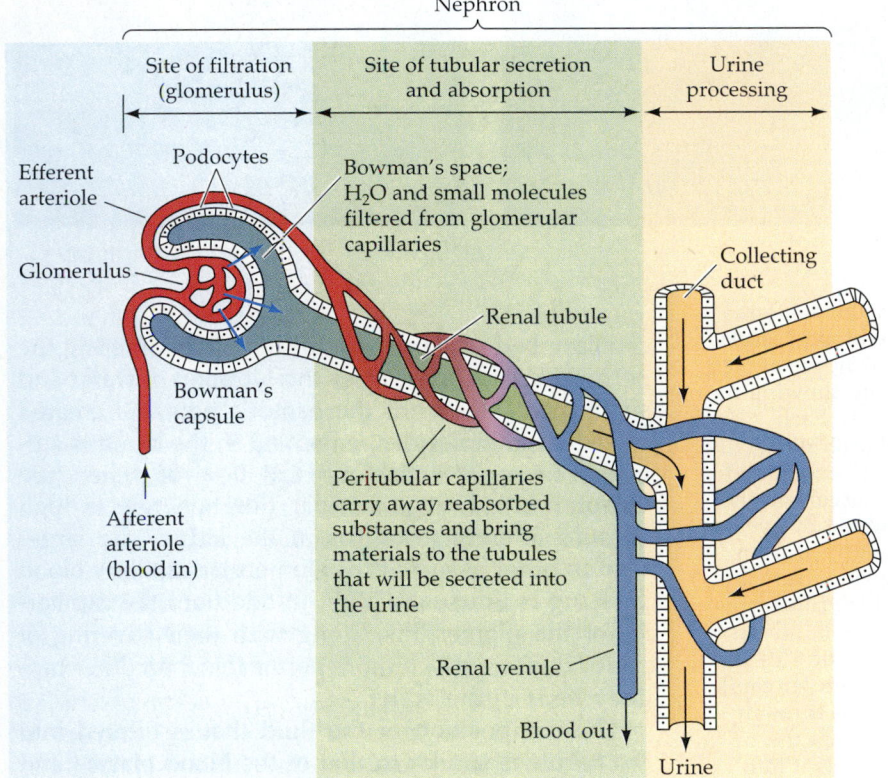

44.9 The Vertebrate Nephron The nephron is a system of tubules closely associated with a system of blood vessels. An afferent arteriole supplies blood to the glomerulus, a knot of capillaries. The glomerulus is the site of blood filtration. It is drained by an efferent arteriole, which gives rise to the peritubular capillaries, which surround the tubules of the nephron. A venule drains the peritubular capillaries. The end of the renal tubule system envelops the glomerulus so that the filtrate from the glomerular capillaries enters the tubules. The processed filtrate (urine) of the individual nephrons enters collecting ducts and is delivered to a common duct leaving the kidney.

(a)

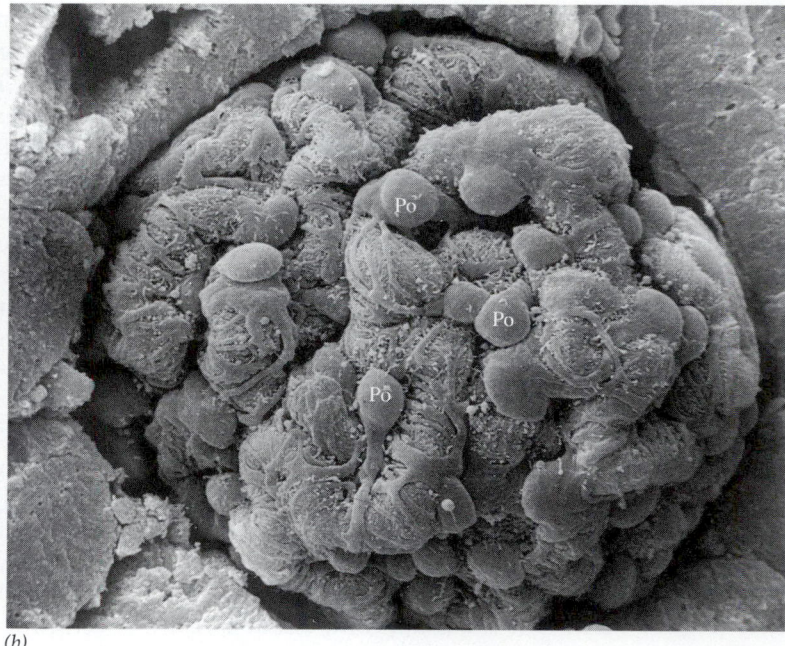

(b)

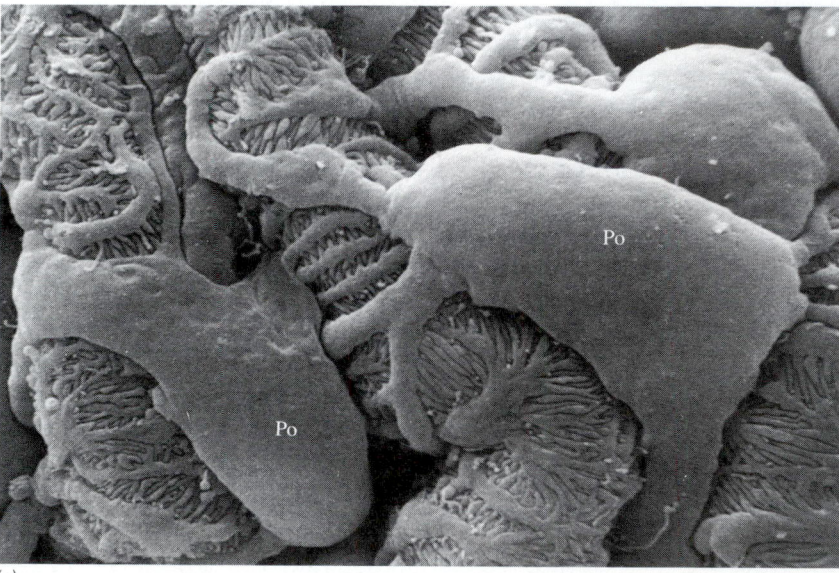

(c)

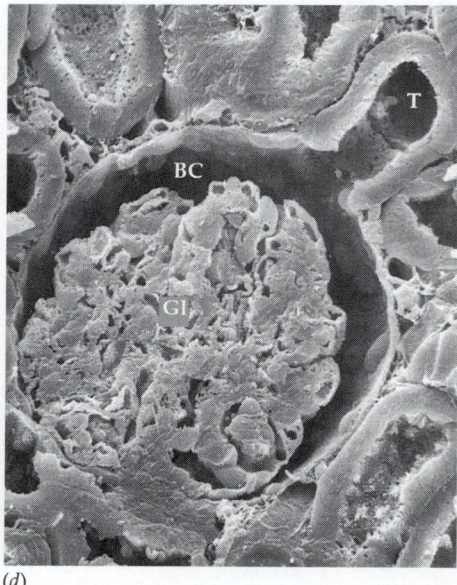

(d)

44.10   An SEM Tour of the Nephron

These scanning electron micrographs show the anatomical bases for kidney function. (*a*) When the blood vessels are filled with latex and all tissue etched away, we are left with a cast of the blood vessels in the kidney showing the knots of capillaries that are the glomeruli (Gl). Each glomerulus has an afferent and an efferent arteriole (Ar). Peritubular capillaries (Pt) are looser networks surrounding the tubules of the nephron. (*b*) In a live organism the capillaries of the glomeruli are tightly wrapped by specialized tubule cells called podocytes (Po) derived from the cells of the inner wall of Bowman's capsule. Here and in parts (*c*) and (*d*) we are looking at the glomerulus from inside Bowman's space. (*c*) Each podocyte has hundreds of tiny, fingerlike projections that create filtration slits between them. Anything passing from the glomerular capillaries into the tubule of the nephron must pass through these slits. (d) Bowman's capsule (BC) surrounds the glomerulus (Gl), collects the filtrate, and funnels it into the tubule (T) of the nephron.

capillary bed, the pressure of the blood entering the permeable capillary causes the filtration of water and small molecules until the osmotic potential created by the large molecules remaining in the blood is sufficient to counter the outward flow of water (see Chapter 42). The glomerular filtration rate is high because afferent arterioles in the kidney are larger than in other tissues; thus glomerular capillary blood pressure is unusually high. In addition, the capillaries of the glomerulus, along with their covering of podocytes, are much more permeable than other capillary beds in the body.

The composition of the fluid that is filtered into the tubule is similar to that of the blood plasma and different from that of the urine. This filtrate contains glucose, amino acids, ions, and nitrogenous wastes in the same concentrations as in the blood plasma, but it lacks the plasma proteins. As this fluid passes down the tubule, its composition changes as the cells of the tubule actively reabsorb certain molecules from the tubular fluid and secrete other molecules into it. When the tubular fluid leaves the kidney as urine, its composition is very different from that of the original filtrate. The function of the **renal tubules**—the tubules of the nephrons (see Figure 44.9)—is to control the composition of the urine through active secretion and reabsorption of specific molecules. The peritubular capillaries serve the needs of the renal tubules by bringing to them the molecules to be secreted into the tubules and carrying away the molecules that are reabsorbed from the tubules back into the blood.

The Evolution of the Nephron

It is believed that the earliest vertebrates lived in fresh water and that the nephron evolved as a structure to excrete the excess water constantly entering these animals by osmosis. Does a study of the evolution of the vertebrate nephron support this view? Since those earliest vertebrates are long extinct and the soft tissues of kidneys are not preserved in the fossil record, we must employ indirect means to reveal the evolution of the nephron. We can study the kidneys of present-day members of the oldest of the vertebrate groups with the expectation that they may retain some primitive features of structure and organization. Another approach is to study the embryonic stages of development in such species, because those developmental stages frequently reveal primitive structure and organization not evident in the adult organism. The embryonic development of the kidneys of fish and amphibians offers suggestions about the sequence of evolutionary stages of the vertebrate nephron.

The earliest nephrons to appear in embryonic fish

and amphibians open directly into the coelomic cavity through a nephrostome (a ciliated opening; Figure 44.11a). This arrangement is very similar to the organization of the annelid metanephridium (see Figure 44.6). Near the nephrostome is a knot of capillaries that protrudes up under the membrane lining the coelomic cavity, where the blood filtrate can pass into the coelom. The resulting coelomic fluid flows through the nephrostome into tubules that convert the fluid into urine.

In a slightly more advanced stage of kidney development, the knot of capillaries (the glomerulus) does not protrude up under the coleomic lining; instead it is encapsulated by an elaboration of the renal tubule (Bowman's capsule) as it leaves the nephrostome (Figure 44.11b). In the most advanced stage of nephron development, the nephrostome is lost and all of the tubular fluid is derived directly from filtration in the glomerulus (Figure 44.11c).

The function of the earliest vertebrate nephron apparently was to eliminate coelomic fluid while conserving important molecules. Subsequent stages in the evolution of the nephron enhanced its ability to handle a large volume of filtrate derived directly from the blood. From this evidence we can conclude that the original function of the nephron was most likely to bail excess water out of animals while conserving valuable molecules. This conclusion supports the notion that the earliest vertebrates lived in fresh water

44.11 The Evolution of the Nephron
This model of the evolution of the vertebrate nephron is based on studies of the kidneys (especially embryological stages of those kidneys) of present-day descendants of the oldest vertebrate groups. (a) A schematic cross section of a vertebrate body shows that the most primitive nephron had a tubule with the nephrostome opening into the coelom; blood filtrate entered the coelom from knots of capillaries along its borders. (b) The next evolutionary stage probably involved a specialization of the tubule to enable filtration directly into the tubule, but the nephrostome remained. (c) The final stage eliminates the nephrostome entirely.

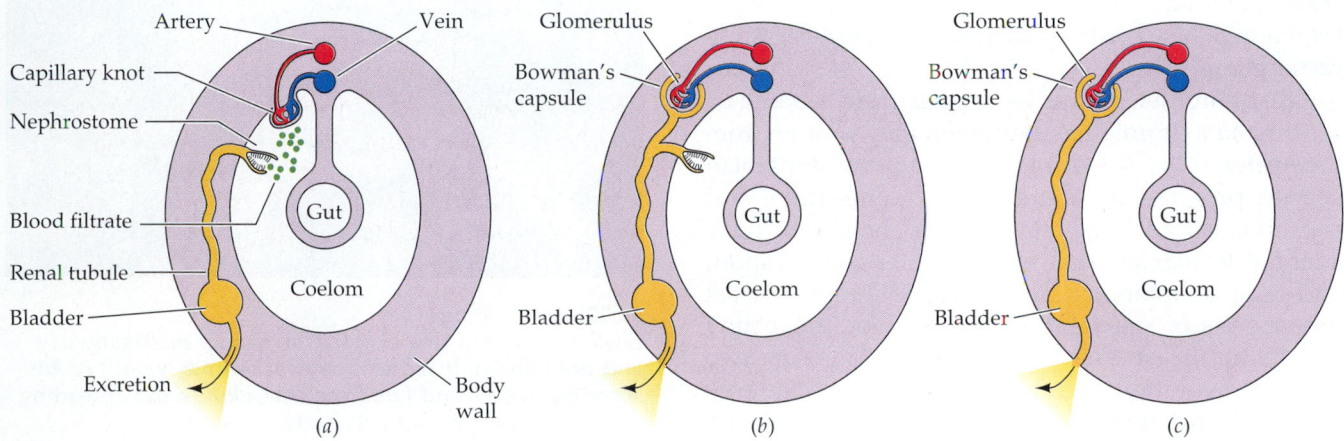

and that some of their descendents had to evolve secondary adaptations to enable them to live in habitats where it was necessary to conserve water.

Water Conservation in Vertebrates

If the vertebrate nephron evolved as a structure to excrete water while conserving salts and essential small molecules, how then have vertebrates adapted to environments where water must be conserved and salts excreted? The answer to this question differs for each vertebrate group. Even among marine fish, the bony fish have a different set of adaptations than do the cartilaginous fishes.

Marine bony fish cannot produce urine more concentrated than their body fluids, but they osmoregulate their body fluids to only one-fourth to one-third the osmotic potential of seawater. They prevent excessive loss of water by producing very little urine. Their urine production is low because their kidneys have fewer glomeruli than do the kidneys of freshwater fish. In some species of marine bony fish, the kidneys have no glomeruli at all! Even though the glomeruli are reduced or absent, renal tubules with closed ends are retained for active excretion of ions and certain molecules. Marine bony fish meet their water needs by drinking seawater, but this practice results in a large salt load. The fish handle salt loads by actively excreting ions from gill membranes and from the renal tubules. Nitrogenous wastes are lost as ammonia from the gill membranes.

Cartilaginous fish are osmoconformers but not ion conformers. Unlike marine bony fish, cartilaginous fish convert nitrogenous waste to urea and retain large amounts of that urea in their body fluids so that their body fluids have the same osmotic concentration as seawater. In some cases these fluids are even slightly hypertonic to seawater, causing the water to move into the body of the fish by osmosis. These species have adapted to a concentration of urea in the body fluids that would be fatal to other vertebrates. Sharks and rays have the problem of excreting the large amounts of salts that they take in with their food. They have several sites of active secretion of NaCl, but the major one is a salt-secreting **rectal gland**.

Most amphibians live in or near fresh water and are limited to humid habitats when they venture from the water. Like freshwater fish, typical amphibian species produce large amounts of dilute urine and conserve salts. Some amphibians, however, have adapted to habitats that require water conservation, and their adaptations are diverse. There is at least one species of saltwater amphibian, the crab-eating frog of Southeast Asia. For marine fish there are two different evolutionary solutions to the osmotic problems of living in salt water. Which one is employed by the crab-eating frog? Like the cartilaginous fish, the crab-eating frog retains urea in its body fluids to the extent that it is slightly hypertonic to the seawater in the mangrove swamps where it lives. Only adult frogs have this adaptation, however, so the species requires fresh water in order to reproduce.

Amphibians from very dry terrestrial environments have been studied by Vaughan Shoemaker at the University of California. An important adaptation in these species is a reduction in the water permeability of their skins. Some secrete a waxy substance that they spread over the skin to waterproof it (Figure 44.12). Several species of frogs that live in arid regions of Australia have remarkable adaptations. These animals burrow deep in the ground and estivate during long dry periods. **Estivation** is a state of very low metabolic activity. When it rains, these frogs come out of estivation, feed, and reproduce. Their most interesting adaptation, however, is that they have enormous urinary bladders. Prior to entering estivation, they fill their bladders with very dilute urine which may make up one-third of their body weight. This dilute urine serves as a water reservoir that they use gradually during the long periods of estivation. Australian aboriginal peoples use these estivating frogs as an emergency source of drinking water.

Reptiles occupy habitats ranging from aquatic to extremely hot and dry. Three major adaptations have freed the reptiles from maintaining the close association with water that is necessary for amphibians. First, reptiles do not need fresh water to reproduce because they employ internal fertilization and lay eggs with shells that retard evaporative water loss.

44.12 Waxy Frogs
Phyllomedusa is a tree frog that lives in a seasonally dry and hot habitat. It reduces its evaporative water loss by secreting waxes and fats from skin glands and spreading these secretions all over its body.

Second, they have scaly, dry skins that are much less permeable to water than is the skin of most amphibians. Third, they excrete nitrogenous wastes as uric acid solids and lose little water in the process.

Birds have the same adaptations for water conservation as reptiles have: internal fertilization, shelled eggs, skin that retards water loss, and uric acid as the nitrogenous waste product. In addition, some birds can produce a urine that is hypertonic to their body fluids. This last ability is much more highly developed in mammals.

STRUCTURE AND FUNCTION OF THE MAMMALIAN KIDNEY

The ability of mammals and birds to produce urine that is hypertonic to their body fluids represents a major step in kidney evolution. In these species we see for the first time the kidney playing the major role in water conservation. A structure that originally evolved to excrete water has been converted to a structure to do the opposite, conserve water. To understand how this evolutionary switch occurred, we must examine the structure and function of the whole kidney.

Anatomy

We will focus on humans as an example of the mammalian excretory system. Humans have two kidneys at the rear of the abdominal cavity at the level of the midback (Figure 44.13a). Each kidney releases the urine it produces into a tube (the **ureter**) that leads to the **urinary bladder**, where the urine is stored until it is excreted through the urethra. The **urethra** is a short tube opening to the outside at the end of the penis in males or just anterior to the vagina in females. Two sphincter muscles surrounding the base of the urethra control the timing of urination. One of these sphincters is a smooth muscle and is controlled by the parasympathetic division of the autonomic nervous system. When the bladder is full, it activates stretch sensors in its wall and a spinal reflex relaxes this sphincter. This reflex is the only control of urination in infants, but the reflex gradually comes under the influence of higher centers in the nervous system as a child grows older. The other sphincter is a skeletal muscle and is controlled by the voluntary, or conscious, nervous system. When the bladder is *very* full, only serious concentration prevents urination.

The kidney is shaped like a kidney bean; when cut down its long axis and split open as a bean splits open, its important anatomical features are revealed (Figure 44.13b). The ureter and the **renal artery** and **renal vein** enter the kidney on its concave (punched-

in) side. The ureter divides into several branches, the ends of which envelop projections of kidney tissue called **renal pyramids**. The renal pyramids make up the internal core, or **medulla**, of the kidney. The medulla is capped by a distinctly different tissue called the **cortex**. The renal artery and vein give rise to many arterioles and venules in the region between the cortex and the medulla.

The secret of the ability of the mammalian kidney to produce concentrated urine is in the relationship between the structures of the nephron (the functional unit of the kidney) and the anatomy of the kidney. Each human kidney contains about 1 million nephrons, and their organization within the kidney is very regular (see Figure 44.13b). All of the glomeruli are located in the cortex. The initial segment of the tubule of a nephron is called the **proximal convoluted tubule**—"proximal" because it is closest to the glomerulus and "convoluted" because it is twisted (Figure 44.13c) All the proximal convoluted tubules are also located in the cortex. At a certain point the proximal tubule takes a dive directly down into the medulla, giving rise to the portion of the tubule called the **loop of Henle**, which runs straight down into the medulla, makes a hairpin turn, and comes straight back to the cortex. This ascending limb of the loop of Henle becomes the **distal convoluted tubule** in the cortex. The distal convoluted tubules of many nephrons join a common **collecting duct** in the cortex. The collecting ducts then run parallel with the loops of Henle down through the medulla, and empty into the ureter at the tips of the renal pyramids.

The organization of the blood vessels of the kidney closely parallels the organization of the nephrons (Figure 44.13c). Arterioles branch from the renal arteries and radiate into the cortex. An *afferent* arteriole carries blood to each glomerulus. Draining each glomerulus is an *efferent* arteriole that gives rise to the peritubular capillaries, which surround mostly the cortical portions of the tubules. A few peritubular capillaries run into the medulla in parallel with the loops of Henle and the collecting ducts. These capillaries are the **vasa recta**. All the peritubular capillaries from a nephron join back together into a venule that joins with venules from other nephrons and eventually leads to the renal vein, which takes blood from the kidney. Remember that anything coming into the kidney comes through the renal artery, and everything that comes into the kidney must leave either through the renal vein or the ureter (there is also some drainage of lymph from the kidney, but it is minor). Only a small, selective percentage of everything that is filtered leaves the kidney in the urine. To understand kidney function, you must understand how most of the substances and water filtered from the blood in the glomerulus return to the venous blood draining the kidney.

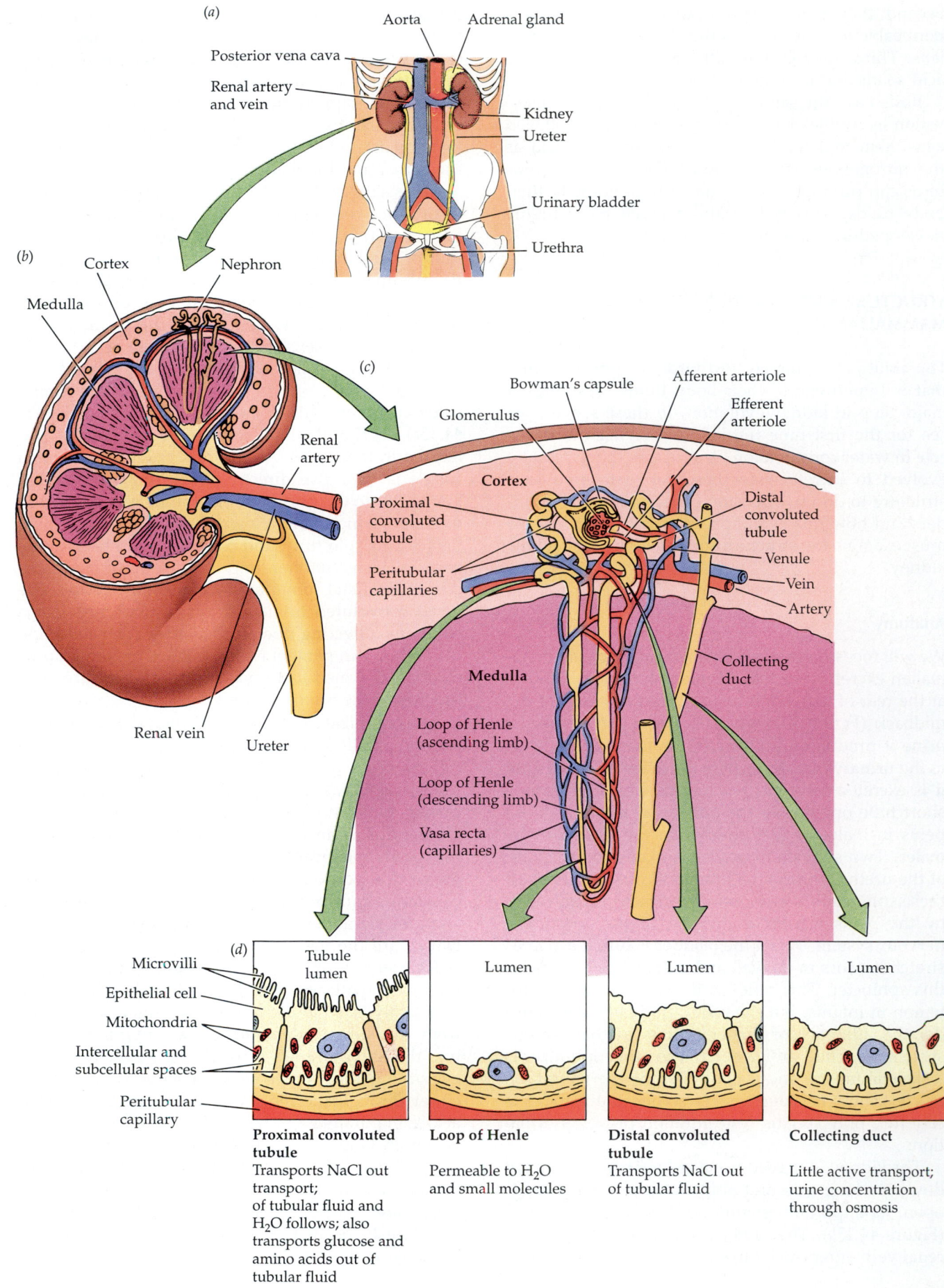

(a)

Aorta
Adrenal gland
Posterior vena cava
Renal artery and vein
Kidney
Ureter
Urinary bladder
Urethra

(b)

Cortex
Nephron
Medulla
Renal artery
Renal vein
Ureter

(c)

Bowman's capsule
Afferent arteriole
Glomerulus
Efferent arteriole
Cortex
Proximal convoluted tubule
Distal convoluted tubule
Peritubular capillaries
Venule
Vein
Artery
Medulla
Collecting duct
Loop of Henle (ascending limb)
Loop of Henle (descending limb)
Vasa recta (capillaries)

(d)

Microvilli
Epithelial cell
Mitochondria
Intercellular and subcellular spaces
Peritubular capillary

Tubule lumen
Lumen
Lumen
Lumen

Proximal convoluted tubule
Transports NaCl out transport; of tubular fluid and H_2O follows; also transports glucose and amino acids out of tubular fluid

Loop of Henle
Permeable to H_2O and small molecules

Distal convoluted tubule
Transports NaCl out of tubular fluid

Collecting duct
Little active transport; urine concentration through osmosis

44.13 The Human Excretory System

(a) The human excretory system consists of two kidneys positioned in the upper rear of the abdominal cavity. The urine they produce is conducted to the urinary bladder through the ureters. The urethra drains the bladder. (b) A longitudinal section of the kidney reveals an internal structure that includes a cortex and, beneath it, a medulla. Urine leaves the kidney from the inner surface of the medulla and is collected in branches of the ureter. (c) A closer look at the cortex and medulla reveals that the glomeruli, the proximal convoluted tubules, and the distal convoluted tubules are in the cortex. The loops of Henle and the vasa recta are in the medulla. Collecting ducts run from the cortex to the tips of the medulla. (d) The structures of the cells of the renal tubules reflect the functions of the different tubule segments. The cells of the proximal convoluted tubule have many mitochondria; a well-developed border of microvilli lines the inside of the tubule to increase the surface area available for the absorption of substances from the filtrate. Intercellular spaces and indentations at the basal end of the cells increase the area of cell contact with interstitial fluids. Distal tubule cells also have many mitochondria and extensively folded basal surfaces. Collecting-duct cells are less adapted for active secretion and reabsorption, and cells in the thin regions of the loop of Henle are flat, with few mitochondria or surface indentations.

Glomerular Filtration Rate

Most of the water and solutes filtered in the glomerulus are reabsorbed and do not appear in the urine. We reach this conclusion by comparing the huge daily filtration volume with the volume of urine produced each day. The kidneys receive about 20 percent of the blood pumped into arteries by the heart. The cardiac output of a human at rest is about 5 liters per minute, so the kidney receives over 1,400 liters of blood per day—an enormous volume. How much of this huge volume is filtered? The answer is about 12 percent. This is still a large number—180 liters per day! Since we normally urinate 2 to 3 liters per day, 98 to 99 percent of the fluid volume that is filtered in the glomerulus is reabsorbed into the blood. Where and how is this enormous fluid volume reabsorbed from the renal tubules back into the blood?

Tubular Reabsorption

Most of the water and solutes in the glomerular filtrate are reabsorbed in the proximal convoluted tubule. The cells of this section of the renal tubule are cuboidal, and their surfaces facing into the tubule have thousands of **microvilli**, which increase their surface area for reabsorption (Figure 44.13d). These cells have lots of mitochondria—an indication that they are biochemically active. They transport NaCl and other solutes, such as glucose and amino acids, out of the tubular fluid. Virtually all glucose molecules and amino acid molecules that are filtered from the blood are actively reabsorbed across the cells of

the proximal convoluted tubules and into the interstitial fluids. This movement of solutes into the interstitial fluid makes it hypertonic to the tubular fluid, and water flows from the tubular fluid in response to this difference. The water and solutes that are moved across the tubular cells by this process are taken up by the peritubular capillaries and returned to the venous blood leaving the kidney.

In spite of the large volume of reabsorption of water and solutes by the proximal convoluted tubule, the overall concentration, or osmotic potential, of the fluid that enters the loop of Henle is not different from that of the blood plasma, even though their compositions are quite different. Next we have to consider how the kidney produces urine that is hypertonic to the blood plasma.

The Countercurrent Multiplier

Humans can produce urine that is four times more concentrated than their plasma. Some mammals that live in very dry deserts, such as kangaroo rats, are able to conserve water so well that their urine may be 12 to 15 times more concentrated than their blood plasma. How does the structure of the mammalian kidney enable it to produce urine that is more concentrated than the blood plasma? This remarkable ability is due to the loops of Henle, which function as a **countercurrent multiplier system**. "Countercurrent" refers to the direction of urine flow in the descending versus the ascending limbs of the loop (see Box 41.A) and "multiplier" refers to the ability of this system to create a concentration gradient in the renal medulla. The loops of Henle do not concentrate the tubular fluid, but they increase the osmotic concentration of the surrounding extracellular environment in the renal medulla. Let's see how they do it.

The cells of the descending limb of the loop of Henle, and the initial cells of the ascending limb, are unspecialized. These cells are flat, with no microvilli and few mitochondria. The part of the tubule made up of these cells is permeable to water and small molecules. Partway up the ascending limb, however, the cells become specialized for transport again. They are cuboidal, with lots of mitochondria, and have some microvilli on their surfaces facing into the tubule (see Figure 44.13d). The portion of the ascending limb made up of these cells is impermeable to water, but the cells actively transport NaCl out of the tubular fluid. As a result, the urine becomes more dilute as it flows toward the distal convoluted tubule (Figure 44.14). Where does the NaCl that is transported out of the ascending limb go? It enters the interstitial fluid in the renal medulla, from which it can diffuse back into the descending limb of the loop of Henle. The NaCl that enters the descending limb flows around the loop and back up the ascending limb, where it is

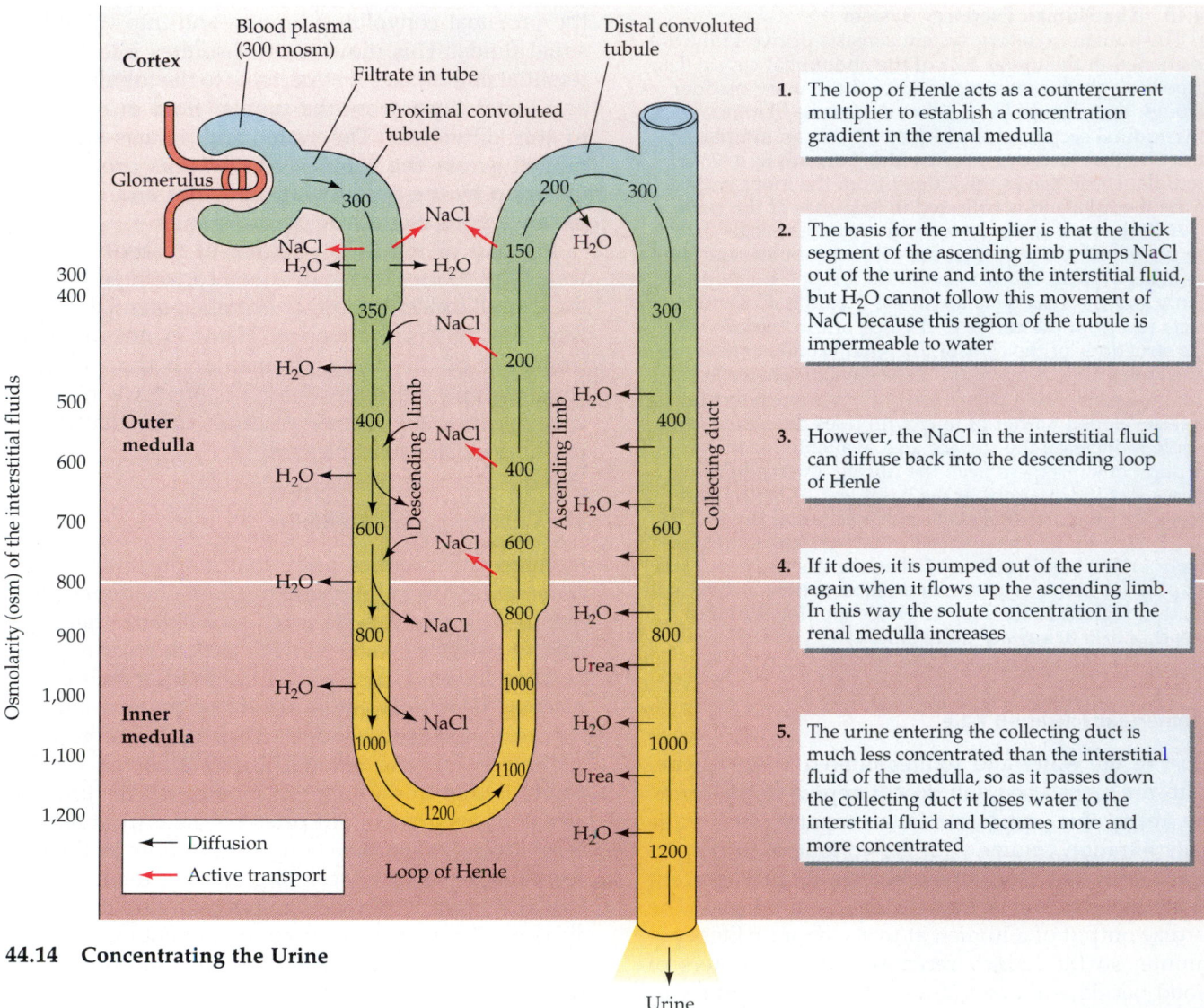

44.14 Concentrating the Urine

transported out of the urine once again. As a result, NaCl accumulates in the interstitial fluid of the renal medulla, with the highest concentration near the tips of the renal pyramids.

How does this action of the loop of Henle concentrate the urine? In Figure 44.13, we can see that the urine is less concentrated when it leaves the loop than when it entered. The active secretion of substances to be excreted and reabsorption of substances to be conserved continues in the distal convoluted tubules. The urine becomes concentrated in the collecting duct. The collecting duct runs from the cortex, where it receives filtrate from the distal tubules, down through the medulla, to the tip of the renal pyramids, where it discharges into the ureter. Over this distance it is surrounded by increasingly concentrated interstitial fluid. The collecting duct is permeable to water, but not to ions, so the osmotic potential of the interstitial environment draws water from the fluid in the collecting duct and leaves behind an increasingly concentrated solution. The urine that

leaves the collecting duct at the tip of a renal pyramid can be almost as concentrated as the highest interstitial concentration established by the countercurrent multiplier system. The water reabsorbed from the collecting duct exits the renal medulla via the vasa recta, which are highly permeable to salts and water.

In summary, the mammalian kidney works in the following manner: The glomeruli filter large volumes of blood plasma. The proximal convoluted tubules reabsorb most of this volume along with valuable molecules such as glucose and amino acids, actively transporting NaCl and other solutes from the tubular fluid. Water follows because of the local difference in osmotic potential created by the transport of the solutes. The loops of Henle create a concentration gradient in the medulla of the kidney. As the urine flows in the collecting ducts through this concentration gradient, water is reabsorbed, thus creating a urine hypertonic to the blood plasma.

REGULATION OF KIDNEY FUNCTIONS

The function of kidneys is to regulate the volume, the osmolarity, and the chemical composition of extracellular fluids. If the fluids do not have the right composition to meet the needs of the body cells, the cells cannot survive. Thus, if the kidneys fail, death will ensue—unless the victim has access to an artificial kidney machine, which can cleanse the blood through **dialysis** (Box 44.A). Multiple systems control and regulate kidney functions. Although we will discuss them separately, they are always working together in an integrated fashion to match kidney function to the needs of the body.

BOX 44.A

Artificial Kidneys

Sudden and complete loss of kidney function is called acute renal failure. It results in the retention of salts and water (leading to high blood pressure), as well as in the retention of urea and metabolic acids. A patient with acute renal failure will die in one to two weeks if not treated. It is now possible to compensate for renal failure and even surgical removal of the kidneys by using artificial kidneys. In an artificial kidney, or dialysis unit, the blood of the patient and a dialyzing fluid come into very close contact, separated only by a semipermeable membrane. This membrane allows small molecules to diffuse from the patient's blood into the dialysis fluid. Because molecules and ions diffuse from an area of high concentration to an area of lower concentration, the composition of the dialysis fluid is crucial. The concentrations of the molecules or ions we want to conserve, such as glucose or sodium, must be the same in the dialysis fluid as in the plasma. The concentrations of molecules and ions we want to clear from the plasma, such as urea and sulfate, must be zero in the dialysis fluid. The total must equal the osmotic potential of the plasma.

The figure shows a schematic drawing of a dialysis machine. Arterial blood flows between semipermeable membranes, which are surrounded with dialysis fluid at body temperature. The "cleansed" blood is returned to the body through a vein and the used dialysis fluid is discarded. At any one time only about 500 milliliters of blood are in the dialysis unit, and it processes several hundred milliliters of blood per minute. A patient with no kidney function must be on the dialysis machine for 4 to 6 hours three times a week.

This man is monitoring the flow of his blood through a dialysis unit that eliminates metabolic waste products normally removed from the blood by the kidneys. The mechanisms of the dialysis unit are illustrated in the diagram.

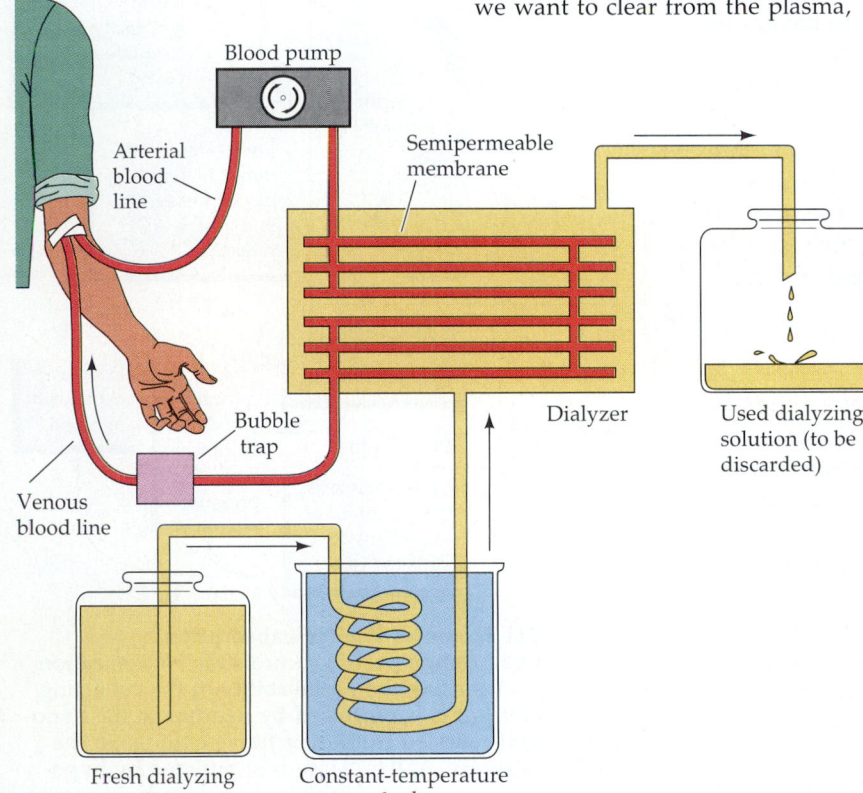

Blood pump

Arterial blood line

Semipermeable membrane

Dialyzer

Used dialyzing solution (to be discarded)

Bubble trap

Venous blood line

Fresh dialyzing solution

Constant-temperature bath

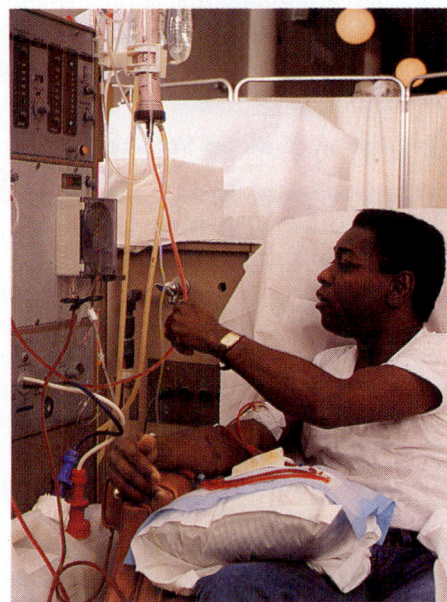

Autoregulation of the Glomerular Filtration Rate

If the kidneys stop filtering the blood, they cannot accomplish any of their functions. Therefore, there are mechanisms that keep the blood filtering through the glomeruli at a constant high rate, regardless of what is happening elsewhere in the body. Because these adaptations of the kidney are to maintain its own functions, these mechanisms are called autoregulatory. The glomerular filtration rate (GFR) depends on an adequate blood supply to the kidneys at an adequate blood pressure. The autoregulatory mechanisms compensate for decreases in cardiac output or decreases in blood pressure so that the GFR remains high (Figure 44.15).

One autoregulatory mechanism is the dilation (expansion) of the afferent renal arterioles when blood pressure falls. This dilation decreases the resistance in the arterioles and helps maintain blood pressure in the glomerular capillaries. If that response does not keep the GFR from falling, then the kidney releases an enzyme, **renin**, into the blood. Renin acts on a circulating protein to begin converting this protein into an active hormone called **angiotensin**. Angiotensin has several effects that help restore the GFR to normal. First, angiotensin causes the efferent renal arteriole to constrict, which elevates blood pressure in the glomerular capillaries. Second, angiotensin causes peripheral blood vessels all over the body to constrict—an action that elevates central blood pressure. Third, angiotensin stimulates the adrenal cortex to release the hormone **aldosterone**. Aldosterone stimulates sodium reabsorption by the kidney, thereby making the reabsorption of water more effective. Enhanced water reabsorption helps maintain blood volume and therefore central blood pressure. Finally, angiotensin acts on structures in the brain to stimulate thirst. Increased water intake in response to thirst increases blood volume and blood pressure.

Regulation of Blood Volume

When you lose blood, your blood pressure tends to fall. Besides activating the kidney autoregulatory mechanisms described in the previous section, a drop in blood pressure decreases the activity of stretch sensors in the walls of the large arteries such as the aorta and the carotids. These stretch sensors provide information to cells in the hypothalamus that produce **antidiuretic hormone** (also called vasopressin) and send it down their axons to the posterior pituitary gland (see Chapter 36). As stretch sensor activity falls, the production and release of this hormone increases (Figure 44.16).

Antidiuretic hormone (ADH) acts on the collecting ducts of the kidney by increasing their permeability

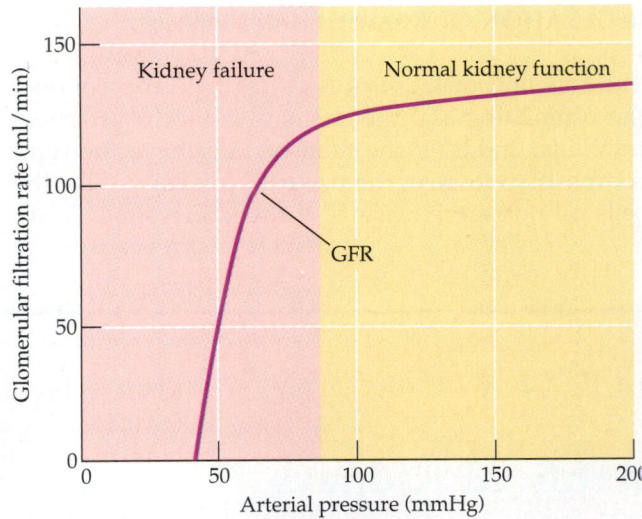

44.15 The Glomerular Filtration Rate Remains Constant
Glomerular filtration is driven by arterial pressure, yet because of autoregulatory mechanisms in the kidney the glomerular filtration rate (GFR) is independent of arterial pressure over a wide range. When arterial pressure falls too low, however, the kidney fails to produce urine.

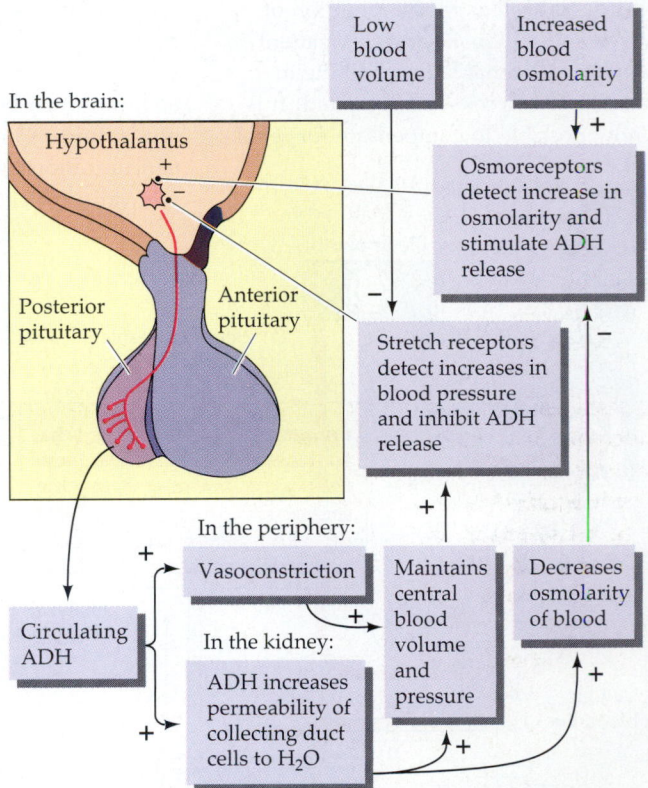

44.16 ADH Promotes Water Reabsorption
Antidiuretic hormone (ADH) controls the concentration of urine by increasing the permeability of the collecting duct to water. ADH is produced by neurons in the hypothalamus and released from their nerve endings in the posterior pituitary. ADH release is stimulated by hypothalamic osmosensors and inhibited by stretch sensors in the great arteries.

to water. When there is a high circulating level of ADH, the collecting ducts are very permeable to water, more water is reabsorbed from the urine, and only small quantities of concentrated urine are produced, thus conserving blood volume and blood pressure. Without ADH, water cannot be reabsorbed from the collecting ducts, and lots of very dilute urine is produced. Diabetes insipidus is a disease that results from a lack of ADH ("insipidus" derives from the dilute, or "tasteless" character of the urine). Diabetes mellitus, caused by an inability of cells to take up glucose from the blood, also causes copious urine

production, but the urine tastes sweet ("mellitus" means honeylike). Caffeine and alcohol inhibit the actions of ADH and increase the production of urine. The resulting dehydrating effect of alcohol creates "hangover" headaches by decreasing the amount of cerebrospinal fluid that is providing a cushion (waterbed) for the brain.

Regulation of Blood Osmolarity

Sensors in the hypothalamus monitor the osmotic potential of the blood. If blood osmolarity increases,

---

BOX 44.B

Water Balance in the Vampire Bat

The Australian vampire bat (*Desmodus*) is a small mammal that feeds at night on the blood of sleeping large mammals, such as cattle. Blood is a liquid, high-protein diet. To process this diet, the renal system of the vampire bat must shift from drought conditions to flood conditions and back to drought conditions in minutes. At sunset, when the bat has

not had a meal for many hours, it is producing a highly concentrated urine at a low rate to conserve its precious body water. If it is successful in finding prey, the bat must process as much blood in as short a time as possible, before the victim wakes up. To maximize its nutrient intake, the bat concentrates its blood meal by rapidly excreting its water content. Accordingly, within minutes the bat produces copious amounts of very dilute urine. The warm fluid running down the victim's neck and waking it is not blood!

As soon as the meal is ended—usually abruptly—the bat begins to digest the concentrated blood in its

gut. Because the concentrated blood is mostly protein, a large amount of nitrogenous waste is produced and must be excreted as urea in solution. But now water is in short supply. The bat must limit its water loss because a long time may pass before the next meal. Consequently, the bat's kidneys produce small amounts of extremely concentrated urine. This urine can be more than 20 times the concentration of the bat's plasma. Humans, in comparison, can produce a urine only about 4 times as concentrated as their plasma. In this way the remarkable regulatory abilities of the vampire bat kidney enable the animal to process its unusual diet.

The graph shows the changes in a vampire bat's urine concentration and urine flow rate before and after its meal of blood. In the photograph, two vampire bats roost on the ceiling of their cave during the day.

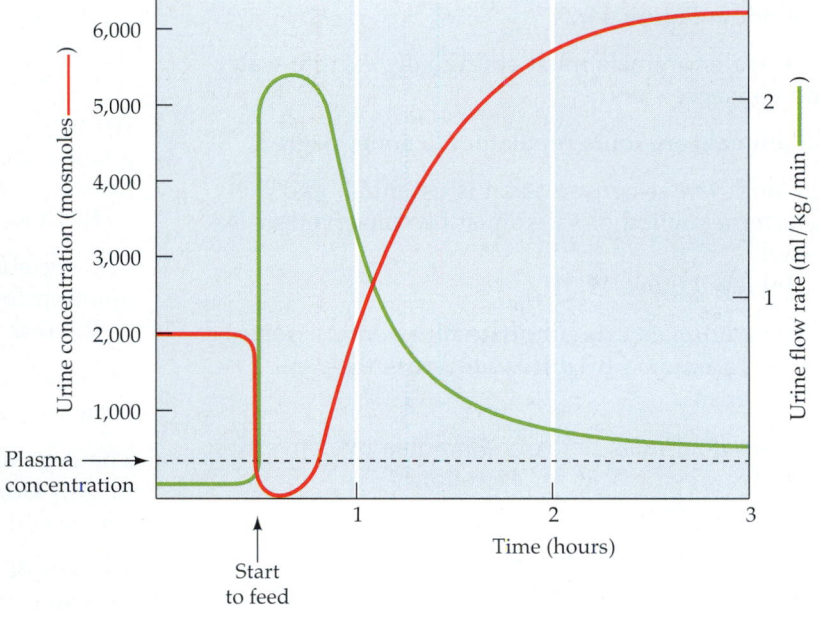

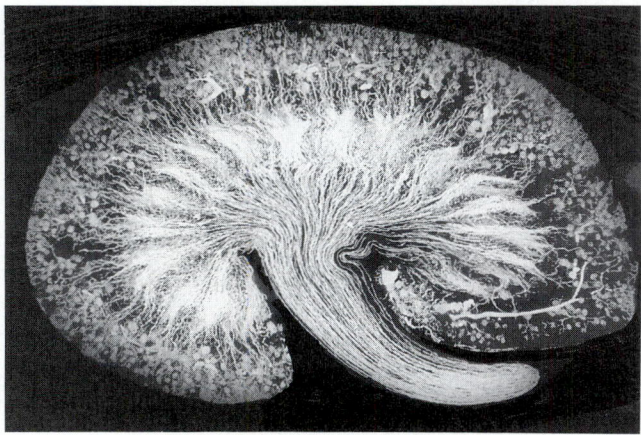

44.17 The Ability to Concentrate

The ability of the mammalian kidney to concentrate urine depends on the lengths of its loops of Henle relative to the overall size of the kidney. Some desert rodents have single renal pyramids that are so long they protrude out of the kidney and into the ureter, as shown here.

these **osmosensors** stimulate increased release of antidiuretic hormone to enhance water reabsorption from the kidney. The osmosensors also stimulate thirst. The resulting increased water intake dilutes the blood as it expands blood volume.

Regulatory Flexibility

The ability of the mammalian kidney to produce a concentrated urine has made it possible for mammals to inhabit some of the most arid habitats on Earth. Some of these animals, such as the desert gerbil, have such extremely long loops of Henle that their renal pyramid (each of their kidneys has only one in contrast to ours) extends far out of the concave surface of the kidney (Figure 44.17). These animals are so effective in conserving water that they can survive on the water released by the metabolism of their dry food; they do not need to drink! The concentrating ability of the mammalian kidney, coupled with the remarkable flexibility of its regulatory systems, enables it to adapt to rapidly changing conditions. This regulatory flexibility is quite pronounced in the vampire bat, which can display the full range of extremes of mammalian salt and water balance mechanisms in a matter of minutes (Box 44.B).

SUMMARY of Main Ideas about Salt and Water Balance and Nitrogen Excretion

The problems of salt and water balance and nitrogen excretion that animals face depend on their environments.

Marine animals can be osmoconformers or osmoregulators.
   Review Figure 44.2

Freshwater animals must continually excrete water and conserve salts.

All animals are ionic regulators to some degree.

On land, water conservation is essential, and diet determines whether salts must be conserved or excreted.
   Review Figure 44.3

Aquatic animals can eliminate nitrogenous wastes such as ammonia by diffusion across their gill membranes.

Terrestrial animals detoxify ammonia by converting it to urea or uric acid for excretion.
   Review Figure 44.4

In excretory systems pressure drives filtration of body fluids into a system of tubules, whose cells alter the composition of the filtrate.

Protonephridia of flatworms consist of flame cells and excretory tubules.
   Review Figure 44.5

Metanephridia of segmented worms take in coelomic fluid and alter its composition through active secretion and active reabsorption of solutes by tubule cells.
   Review Figure 44.6

Arthropod excretory organs include the green glands of crustaceans and the Malpighian tubules of insects.
   Review Figures 44.7 and 44.8

The vertebrate nephron originally evolved as an adaptation for excreting water and conserving solutes.
   Review Figures 44.9 and 44.11

Mammals and birds produce urine that is hypertonic to their body fluids.

The organization of the renal tubules in the mammalian kidney is the basis for its ability to produce concentrated urine.

Glomeruli in the cortex of the kidney filter water and solutes from the blood.

Proximal convoluted tubules reabsorb most of the kidney-filtered water and many of the solutes.
Review Figure 44.13

The loop of Henle creates a concentration gradient in the tissues of the renal medulla, and the collecting duct concentrates the urine by allowing the osmotic loss of water to the surrounding interstitial fluid of the medulla.
Review Figure 44.14

Kidney function in mammals is regulated by autoregulatory mechanisms for maintaining a constant high glomerular filtration rate even if blood pressure varies.
Review Figure 44.15

An important autoregulatory mechanism is the release of renin by the kidney when blood pressure falls; renin activates angiotensin, which causes peripheral vasoconstriction, causes release of aldosterone, and stimulates thirst.

Changes in blood pressure and blood osmolarity influence the release of antidiuretic hormone, which controls the permeability of the collecting duct to water.
Review Figure 44.16

SELF-QUIZ

1. Which of the following statements about osmoregulators is true?
   a. Most marine invertebrates are osmoregulators.
   b. All freshwater invertebrates are hypertonic osmoregulators.
   c. Cartilaginous fish are hypotonic osmoregulators.
   d. Bony marine fish are hypertonic osmoregulators.
   e. Mammals are hypotonic osmoregulators.

2. The excretion of nitrogenous wastes
   a. by humans can be in the form of urea and uric acid.
   b. by mammals is never in the form of uric acid.
   c. by marine fish is in the form of urea.
   d. does not contribute to the osmotic potential of the urine.
   e. requires more water if the waste product is the rather insoluble uric acid.

3. How are earthworm metanephridia like mammalian nephrons?
   a. Both process coelomic fluid.
   b. Both take in fluid through a ciliated opening.
   c. Both produce hypertonic urine.
   d. Both employ tubular secretion and reabsorption to control urine composition.
   e. Both deliver urine to a urinary bladder.

4. What is the role of renal podocytes?
   a. They control the glomerular filtration rate by changing resistances of renal arterioles.
   b. They reabsorb most of the glucose that is filtered from the plasma.
   c. They prevent red blood cells and large molecules from entering the renal tubules.
   d. They provide a large surface area for tubular secretion and reabsorption.
   e. They release renin when the glomerular filtration rate falls.

5. Which of the following are *not* found in a renal pyramid?
   a. Collecting ducts
   b. Vasa recta
   c. Peritubular capillaries
   d. Convoluted tubules
   e. Loops of Henle

6. Which part of the nephron is responsible for most of the difference in mammals between the glomerular filtration rate and the urine production rate?
   a. The glomerulus
   b. The proximal convoluted tubule
   c. The loop of Henle
   d. The distal convoluted tubule
   e. The collecting duct

7. For mammals of the same size, what feature of their excretory systems would give them the greatest ability to produce a hypertonic urine?
   a. Higher glomerular filtration rate
   b. Longer convoluted tubules
   c. Increased number of nephrons
   d. More-permeable collecting ducts
   e. Longer loops of Henle

8. Which of the following would *not* be a response stimulated by a large drop in blood pressure?
   a. Constriction of afferent renal arterioles
   b. Increased release of renin
   c. Increased release of antidiuretic hormone
   d. Increased thirst
   e. Constriction of efferent renal arterioles

9. Which of the following statements about angiotensin is true?
   a. It is secreted by the kidney when the glomerular filtration rate falls.
   b. It is released by the posterior pituitary when blood pressure falls.
   c. It stimulates thirst.
   d. It increases permeability of the collecting ducts to water.
   e. It decreases glomerular filtration rate when blood pressure rises.

10. Birds that feed on marine animals ingest a lot of salt, but excrete most of it by means of
    a. Malpighian tubules.
    b. rectal salt glands.
    c. green glands.
    d. hypertonic urine.
    e. nasal salt glands.

FOR STUDY

1. What do marine fish, reptiles, mammals, and insects have in common with respect to water balance? Compare their physiological adaptations for dealing with their common problem.

2. What are the relative advantages and disadvantages of ammonia, urea, and uric acid as nitrogenous waste products of animals?

3. Explain how the kidney is able to maintain a constant glomerular filtration rate over a wide range of arterial blood pressures. Referring back to what you learned about regulation of circulatory function in Chapter 42, how can a decrease in glomerular filtration rate cause an increase in cardiac output?

4. Inulin is a molecule that is filtered in the glomerulus, but it is not secreted or reabsorbed by the renal tubules. If you injected inulin into a subject and after a brief time measured the concentration of inulin in the blood and in the urine of the subject, how could you determine the subject's glomerular filtration rate? Assume that the rate of urine production is 1 milliliter per minute.

5. Explain the roles of the loop of Henle and the collecting duct in producing a hypertonic urine in mammals. How is this mechanism controlled in response to changes in osmolarity of the blood and in blood pressure?

READINGS

Cantin, M. and J. Genest. 1986. "The Heart as an Endocrine Gland." *Scientific American*, February. Heart tissue secretes a hormone that helps control salt and water balance.

Eckert, R., D. Randall and G. Augustine. 1988. *Animal Physiology: Mechanisms and Adaptations*, 3rd Edition. W. H. Freeman, New York. An outstanding textbook of animal physiology. Chapter 12 covers water and salt balance and excretion.

Heatwole, H. 1978. "Adaptations of Marine Snakes." *American Scientist*, vol. 66, pages 594–604. A variety of adaptations, including means of maintaining salt and water balance, allows several groups of snakes to exploit the marine environment.

McClanahan, L. L., R. Ruibal and V. H. Shoemaker. 1994. "Frogs and Toads in Deserts." *Scientific American*, March. Recent research on amphibians adapted to arid environments.

Schmidt-Nielsen, K. 1990. *Animal Physiology: Adaptation and Environment*, 4th Edition. Cambridge University Press, New York. An excellent textbook, emphasizing the comparative approach.

Smith, H. W. 1961. *From Fish to Philosopher*. Doubleday, Garden City, NJ. A classic, using salt balance and excretory physiology as the organizing principle for a survey of vertebrate evolution.

Stricker, E. M. and J. G. Verbalis. 1988. "Hormones and Behavior: The Biology of Thirst and Sodium Appetite." *American Scientist*, vol. 76, page 261. The control of water and salt intake is an important part of osmoregulation.

Vander, A. J., J. H. Sherman and D. S. Luciano. 1994. *Human Physiology: The Mechanisms of Body Function*, 6th Edition. McGraw-Hill, New York. Chapter 16 deals with the regulation of water and salt balance.

Spider webs are objects of beauty and marvels of engineering. The construction of a classic web used to capture prey requires complex behavior. For example, a garden spider, as immortalized in the children's story *Charlotte's Web* by E. B. White, spins a new web every day in the early morning hours before dawn. From an initial attachment point, she strings a horizontal thread. From the middle of that thread she drops a vertical thread to a lower attachment point. Pulling it taut creates a Y, the center of which will be the hub of the finished web. The spider adds a few more radial supports and a few surrounding "framing" threads. Then she fills in all the radial spokes according to a set of rules. Finally, she lays down a spiral of sticky threads with regular spacing and attachment points to the radial spokes. This remarkable feat of construction takes only half an hour, but it requires thousands of specific movements performed in just the right sequence. Where is the blueprint for Charlotte's web? How does she acquire the construction skills needed to build it?

The blueprint is coded in the genes and built into the spider's nervous system as a motor "program," or score. Learning plays no role in the expression of that complex blueprint. Newly hatched spiders disperse to new locations and usually spin their first webs without ever having experienced a web built by an adult of their species. Nevertheless, they build perfect webs the first time; each of the thousands of movements happens in just the right sequence. It is remarkable that the genetic code and the simple nervous system of a spider can contain and express behavior as complex as spinning an orb web.

Web spinning by spiders is an animal behavior—an act or set of acts performed with respect to another animal or the environment. Behavior falls into three general classes: acts to acquire food, acts to avoid environmental threats, and acts to reproduce. In studying any behavior we can ask *what*, *how*, and *why*. *What* questions focus on the details of behavior, including the circumstances that influence when an animal acts in a certain way. Some *how* questions refer to the underlying neural, hormonal, and anatomical mechanisms that we have been studying in Part Six. Other *how* questions refer to the means by which an animal acquires a behavior. Some behaviors, such as web spinning, are genetically determined; others are learned. Many behaviors involve complex interactions of inheritance and learning.

Animal behavior is a large field of study. For this chapter we have selected some interesting examples of approaches to answering *what* and *how* questions. *Why* questions have to do with the evolution of behavior and will be the major focus of the next chapter. After a discussion of genetically determined behavior and behavior that results from a combination of genetics and learning, such as bird song, we will turn

An Orb-Weaving Garden Spider and Its Web

45

Animal Behavior

to animal communication. Studies of animal communication reveal the constraints that environment places on behavior. We will continue with a look at the timing of behaviors, or biological rhythms. Finally, we will discuss how animals find their way through unfamiliar territory. Throughout the chapter use what you read to raise your own questions about human behavior.

GENETICALLY DETERMINED BEHAVIOR

A behavior that is genetically determined rather than learned is also called a **fixed action pattern**. Such behavior is highly stereotypic, that is, performed the same way every time. It is also species-specific; there is very little variation in the way different individuals of the same species perform the behavior. The behavior is expressed differently, however, in even closely related species. For example, different species of spiders spin webs of different designs.

Fixed action patterns require no learning or prior experience for their expression, and they are generally not modified by learning. Another spider example illustrates this point. Spiders spin other structures in addition to webs for capturing prey. Most spiders lay their eggs in a cocoon that they form by spinning a base plate, building up the walls (inside of which they lay the eggs), and spinning a lid to close the cocoon. Although this behavior requires thousands of individual movements, it is performed exactly the same way every time and is not modified by experience. If the spider is moved to a new location after she finishes the base plate, she will continue to spin the sides of the cocoon, lay her eggs (which fall out the bottom), and spin the lid. If she is placed on her previously completed base plate the next time she is ready to begin a cocoon, she will spin a new base plate over the old one as if it were not there. If she is nutritionally deprived and runs out of silk in the middle of spinning a cocoon, she will complete all the thousands of movements in a pantomime of cocoon building. Once started, the cocoon-building motor score runs from beginning to end, and it can be started only at the beginning.

Fixed action patterns are good material for studying the mechanisms of animal behavior. We can study their genetics and the sequence of events whereby gene expression eventually results in a behavior; the influence of hormones on the development and expression of the behavior; and the detailed neurophysiology that underlies the behavior. First, however, we must demonstrate that a given behavior *is* genetically determined. One powerful way of proving genetic determination is to deprive the animal of any opportunity to learn the behavior in question, then see if that behavior is expressed.

Deprivation Experiments

In a **deprivation experiment**, an animal is reared so that it is deprived of all experience relevant to the behavior under study. For example, a tree squirrel was reared in isolation, on a liquid diet, and in a cage without soil or other particulate matter. When the young squirrel was given a nut, it put the nut in its mouth and ran around the cage. Eventually it oriented toward a corner of the cage and made stereotypic digging movements, placed the nut in the corner, went through the motions of refilling the imaginary hole, and ended by tamping the nonexistent soil with its nose. The squirrel had never handled a food object and had never experienced soil, yet the fixed action patterns involved in burying its nut were fully expressed.

Deprivation experiments occur naturally. Many species, especially insects living in seasonal environments, have life cycles of one year and the generations do not overlap: The adults lay eggs and die before the eggs hatch or the young mature into adults. Learning from adults of the parental generation is impossible in such species, so the complex behavior necessary for survival and reproductive success must be genetically programmed. Web spinning by spiders is an example of complex behavior in species that may have no opportunity to learn from other members of their species.

The courtship behavior of the triangular web spider is a similar example. The male spider must approach a female in her web. If he simply blundered into the web, he would give the same signals as a prey item caught in the web and the female would probably kill him and eat him. To avoid having his reproductive effort cut short, the male is genetically programmed to approach an anchor strand of the web and pluck it in just the right way to send a courtship message to the female. If the message is correct and the female is receptive, he can enter the web and mate with the female rather than serve merely as her dinner. In some species, the female eats the male anyway after they mate, but at least he has achieved reproductive success before becoming nutriment for his own offspring, and he is supplying energy for eggs fertilized only by him.

Triggering Fixed Action Patterns

A behavior that is not expressed during a deprivation experiment may nonetheless be genetically programmed. The right conditions may not have been available to stimulate the behavior during the experiment. Thus, the squirrel had to be given a nut to trigger its digging and burying behaviors. Specific stimuli are usually required to elicit the expression of fixed action patterns. These stimuli are called sign

45.1 Triggering Aggressive Behavior
A mounted immature male robin (right) with no red feathers does not stimulate aggression from a territorial adult male. The formless clump of red feathers on the left, however, does trigger aggressive territoriality.

stimuli or **releasers**. Two **ethologists** (scientists who study animal behavior), Konrad Lorenz and Niko Tinbergen, did classic studies of releasers and fixed action patterns. Their work provided insights into the properties of releasers.

Releasers are usually a simple subset of all the sensory information available to an animal. Male European robins have red feathers on their breasts. During the breeding season, the sight of a male robin stimulates another male robin to sing, make aggressive displays, and attack the intruder if he does not heed the warnings. An immature male robin, whose feathers are all brown, does not elicit aggressive behavior. A tuft of red feathers on a stick, however, will elicit an attack (Figure 45.1). A patch of red in certain locations is a sufficient releaser for male aggressive behavior in robins.

Just as the motor score of the fixed action pattern is genetically programmed, so is the information that enables the animal to recognize the releaser for that fixed action pattern. Evolving a genetic mechanism to respond to a simple stimulus is more feasible than evolving a mechanism to recognize a complex set of stimuli. The simplicity of most releasers has resulted in some curious discoveries.

Tinbergen and A. C. Perdeck carefully examined the releasers and fixed action patterns involved in the interactions between herring gulls and their chicks during feeding. The adult gull has a red dot

at the end of its bill (Figure 45.2). When it returns to its nest, the chicks peck at the red dot, and the adult regurgitates food for the chicks to eat. Tinbergen and Perdeck asked what were the essential characteristics of the parent gull that released food solicitation behavior in the chicks. They made paper cutout models of gull heads and bills but varied the colors and the shapes. Then they rated each model according to how many pecks it received from chicks. The surprising results were that the shape or color of the head made no difference. In fact, a head was not even necessary; the chicks responded just as well to models of bills alone. The color of the bill and the dot also were not critical as long as there was a contrast between the two. Surprisingly, the most effective releaser for chick pecking was a long, thin object with a dark tip that had no resemblance to an adult herring gull (Figure 45.3).

The simplicity of the properties of releasers makes possible the existence of a **supernormal releaser**, one that is more effective than the natural condition in eliciting a fixed action pattern. In the case of a bird called the oystercatcher, the sight of its clutch of eggs releases incubation behavior. If an oystercatcher is given the choice between its own clutch of two eggs and a clutch of three artificial eggs, it will sit on the larger clutch of artificial eggs. When given the choice of its own clutch of two eggs or one very large artificial egg, it will try to incubate the large egg, even if it can hardly straddle it. The abnormally large clutch and egg are supernormal releasers. These choices would not occur in nature, so counterselection has not prevented the evolution of such maladaptive behavior.

Supernormal releasers, which are created by scientists, are a curiosity, but natural selection has produced some dramatic results by favoring the exag-

45.2 The Dot Marks the Spot
This gull is incubating eggs. When the young hatch they will peck at the parent's red bill spot, stimulating the parent to regurgitate food.

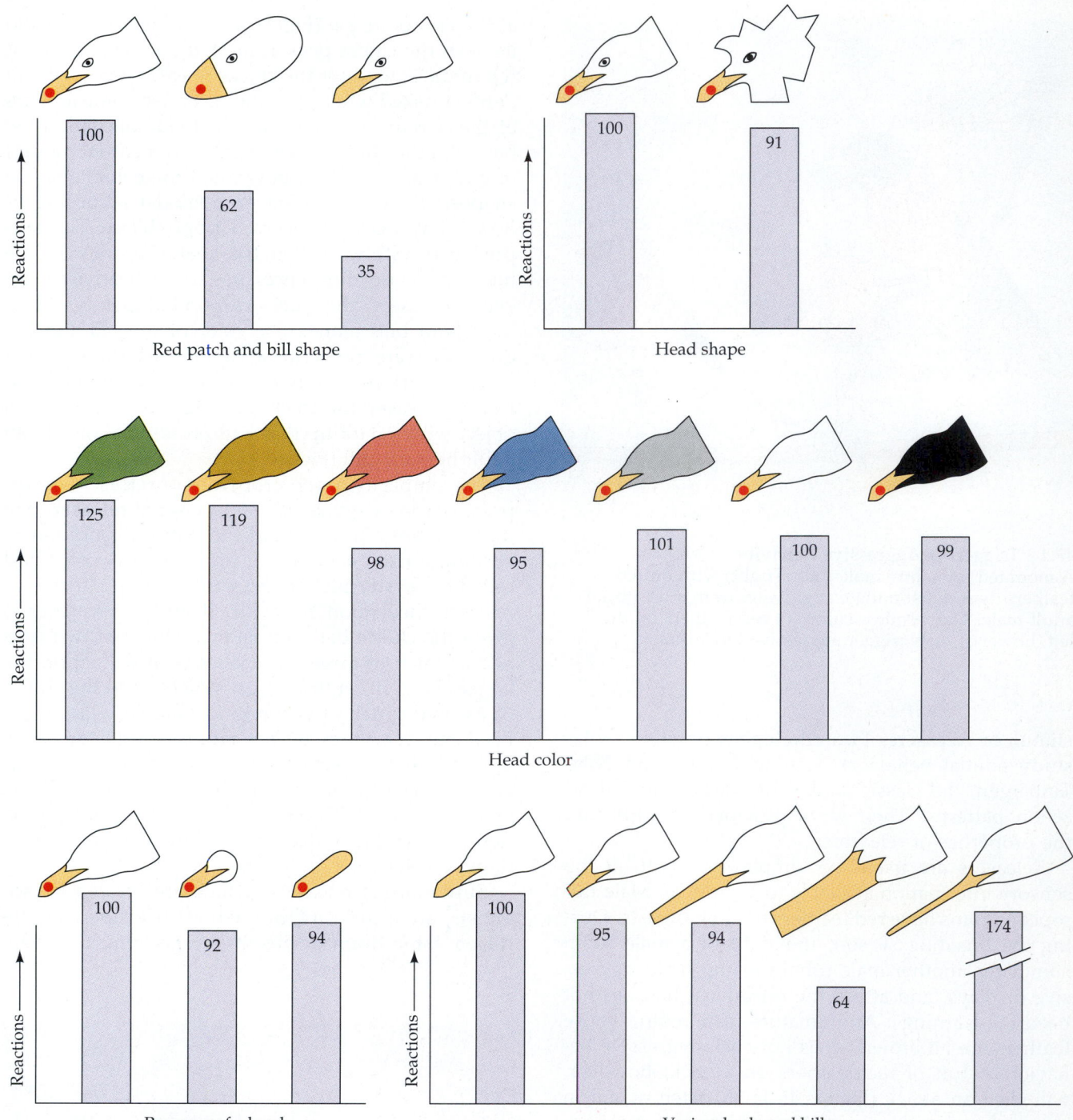

45.3 Releasing the Food Solicitation Response
A series of experiments rated the pecking response of herring gull chicks. The presence of the red dot and the shape of the bill seem to be releasers; the shape and color of the adult's head have little or no influence on the response.

geration of releasers. Many of the elaborate behavior patterns and physical attributes used by species in courtship displays have arisen through natural selection favoring more-effective releasers. Bowerbird males even use colorful objects collected from the environment to enhance their courtship displays (see Figure 46.13).

Motivation

Another reason that a fixed action pattern may not be expressed in a deprivation experiment is that the animal is not in the appropriate developmental or physiological state. Juvenile animals do not show courtship behavior even if the appropriate releasers are present. An adult animal may not engage in aggressive display or courtship display when it is not in reproductive condition. The same animals that may be highly aggressive to one another during the reproductive season may ignore one another at other times of the year.

The internal conditions of an animal determine its motivational state, and the motivational state determines which fixed action pattern is expressed at any particular time. The total behavior of an animal is not simply a random sequence of fixed action patterns depending on what releasers it happens to encounter. Depending on its motivational state, an animal may search for the appropriate releaser and ignore many others. This search behavior may depend heavily on previous experience; hence, it is evidence of learning. For example, a hungry predator is likely to return to a site where it encountered prey in the past.

Fixed Action Patterns versus Learned Behavior

The ability to learn and to modify behavior as a result of experience is often highly adaptive. Most of human behavior is the result of learning. Why then are so many behavior patterns in so many species genetically determined and not modifiable? We've already considered one answer to that question. If role models and opportunities to learn are not available, there is no alternative to programmed behavior. Fixed action patterns also can be adaptive when mistakes are costly or dangerous. Mating with a member of the wrong species is a costly mistake; the function of much of courtship behavior is to guarantee correct species recognition. In an environment in which incorrect as well as correct models exist, learning the wrong pattern of courtship behavior is possible. Fixed action patterns governing mating behavior can prevent such mistakes.

In behavior patterns such as predator avoidance or capture of dangerous prey, there is no room for mistakes. If the behavior is not performed promptly and accurately the first time, there may not be a second chance. Rattlesnakes prey on kangaroo rats. A kangaroo rat that has never encountered a rattlesnake can avoid the snake in total darkness because the *sound* of the snake moving through the air to strike releases the powerful escape jump of the kangaroo rat (Figure 45.4). A kangaroo rat does not have the luxury of learning what a rattlesnake sounds like when it is striking. Let's look at another example. Although king snakes eat rattlesnakes, they are not immune to rattlesnake venom. When a king snake first encounters a rattlesnake, it strikes in such a way that its jaws clamp shut the mouth of the rattlesnake; then it begins the long process of swallowing. A king snake that grabs a rattlesnake at any other place on its body will not have the opportunity to learn by

45.4 Some Things Can't Be Learned by Trial and Error
The sound of a striking rattlesnake triggers an automatic escape jump in a kangaroo rat; the rat does not have to learn this behavior.

trial and error. Thus, *genetically programmed behavior is highly adaptive for species that have little opportunity to learn, for species that might learn the wrong behavior, and in situations where mistakes are costly or dangerous.*

Many behavior patterns are intricate interactions of genetically programmed elements and elements modified by experience. One example that has been the subject of elegant experiments is bird song. Adult male passerine birds use a species-specific song in territorial display and courtship. A few species, such as song sparrows, express their species-specific song even during deprivation experiments but most do not. For most species, such as the white-crowned sparrow, learning is an essential step in the acquisition of song. If the eggs of white-crowned sparrows are hatched in an incubator and the young male birds are reared in isolation from the song of their species, their adult songs will be an unusual assemblage of sounds (Figure 45.5a,b). This species cannot express its species-specific song without hearing that song as a nestling.

Although the white-crowned sparrow must hear the song of its own species during its nestling period, it does not sing as a juvenile. It apparently uses auditory input as a nestling to form a template in its nervous system. It then matches the template through trial and error when it reaches sexual maturity the following spring. If a bird that has heard its correct song as a juvenile is deafened before it begins to express its song, it will never develop its species-specific song (Figure 45.5c). The bird must be able to hear itself to match the template stored in its nervous system. If it is deafened *after* it expresses its correct song, it will continue to sing like a normal bird. Two periods of learning are essential: the first in the nestling stage, and the second at the onset of sexual maturity.

Studies of what birds can learn and when they can

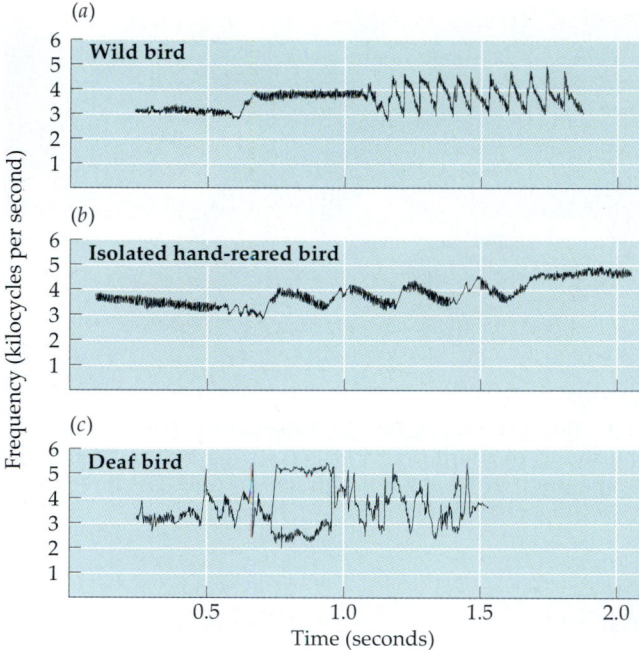

45.5   Song Learning in White-Crowned Sparrows
These sonograms visually record sound frequencies and plot them over time. (a) The species-specific song of a male white-crowned sparrow in its natural state. (b) In a deprivation experiment, a bird hatched in an incubator and reared in isolation from other birds does not learn the species-specific song. (c) In a variation on the experiment, a young bird that heard the correct song but was deafened before it reached maturity could not reproduce the song.

learn, however, reveal strong genetic limits to the modifiability of their behavior through experience. In the case of a white-crowned sparrow, it must hear its species' song within a narrow **critical period** during its development. Once this critical period has passed, the bird cannot learn to sing its species-specific song, regardless of how many role models it experiences. What a bird can learn during its critical period is also severely limited, as revealed by experiments on hand-reared chaffinches that were played various tape recordings of bird song during their critical periods. If exposed to the songs of other species, the chaffinches did not learn them. They also did not learn a chaffinch song played backward or with the elements scrambled. Even if they heard a chaffinch song played in pure tones, they did not form a template. If they heard a normal chaffinch song along with all these other sounds, however, they developed templates and learned to sing the proper song the following spring. Thus the chaffinch is genetically programmed to recognize the appropriate song to learn and when to learn it.

What advantage are genetic limits on what a bird can learn and when? A bird's acoustic environment can be quite complex. Many species of songbirds may be singing in the same area. The critical period limits

learning to the period of most intimate contact between the young bird and its parents to ensure that the father's song is the one experienced most intensively. Further limits on nestling song sensitivity help guarantee that the template it forms is not contaminated with other sounds it hears.

The learning of a song template by a nestling bird is an example of **imprinting**. We mentioned earlier that releasers generally are simple subsets of available information because there are limits to what can be programmed genetically. Imprinting makes it possible to encode complex information in the nervous system rather than in the genes. Offspring can imprint on their parents and parents on their offspring to ensure individual recognition, even in a crowded situation such as a colony or a herd. If a mother goat does not nuzzle and lick her newborn within 5 to 10 minutes after birth, she will not recognize it as her own later. In this case imprinting depends on olfactory cues, and the critical period is determined by the high levels of circulating oxytocin at the time of birth.

HORMONES AND BEHAVIOR

All behavior depends on the nervous system for initiation, coordination, and execution. In addition, as we have discussed, fixed action patterns are built into the nervous system as a motor score. Yet that motor score is expressed only under certain conditions. The endocrine system, through its controlling influences on development and physiological state of the animal, has a large role in determining when a particular motor pattern can be and is performed. In the previous section we saw that learning can play a role in the acquisiton of a species-specific behavior, but there may be narrow developmental windows or critical periods during which certain defined learning can occur. Hormones control the complex interaction between genetically determined behavior and learning during specific stages of development. We have already seen one example of hormones controlling behavior: high levels of oxytocin in the female goat's blood at the time she gives birth determine a window of time during which she can imprint on her infant. Next we will study two more-complex cases in which hormones control the development, learning, and expression of species-specific behavior.

Sexual Behavior in Rats

Differences in the behavior of males and females of a species are clear examples of genetically determined behavior. Such differences in sexual behavior have been shown to be due to the actions of the sex steroids on the developing brain and on the mature brain. Like most other animals, rats behave sexually in accord with fixed action patterns. When a female rat is in estrus (receptive to males), she responds to a tactile stimulus of her hindquarters with a stereo-typic posture called **lordosis**. She lowers her front legs, extends her hind legs, arches her back, and deflects her tail to one side. When a male encounters a female in estrus, he copulates with her with the following sequence of behaviors: He mounts her from the rear, clasps her hindquarters, inserts his penis into her vagina, and thrusts. The roles of genotype and sex hormones in the development of the fixed action patterns of lordosis and copulation have been investigated through manipulating the exposure of the developing rat brain to sex steroids.

If a female rat has her ovaries removed (that is, is spayed), *either as a newborn or as an adult,* she will not show lordosis unless she is injected with female sex

steroids. The hormones are necessary for the expression of the female sexual behavior. If this same adult female is injected with testosterone, she does not show male sexual behavior (first two panels of Figure 45.6). There is a surprising variation on this experiment. If a *newborn* female rat has her ovaries removed and is injected with testosterone, she will not show lordosis when treated with female sex hormones as

45.6 Hormonal Control of Sexual Behavior
Sex steroids control both the development and expression of sexual behavior in rats. The presence of testosterone in newborn rats of both sexes whose reproductive organs (ovaries or testes) have been removed establishes male behavior patterns, and its absence establishes female patterns. Injections of sex steroids in gonadectomized adult rats stimulate expression of the sexual behavior pattern that developed in response to genotype and early steroid exposure.

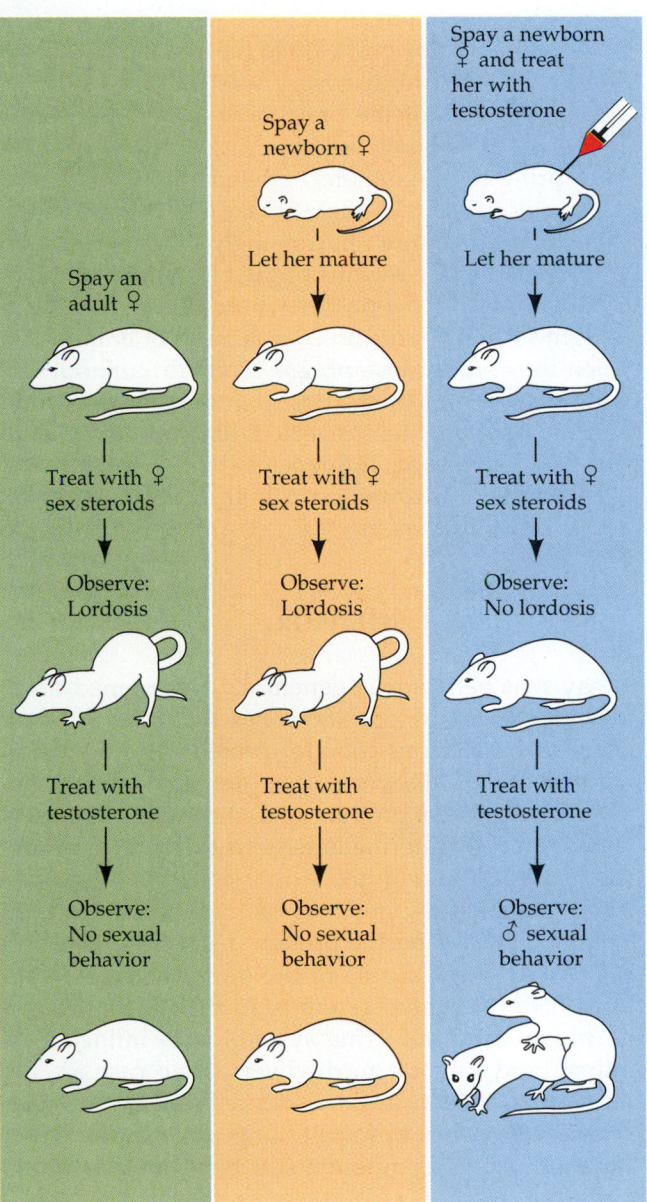
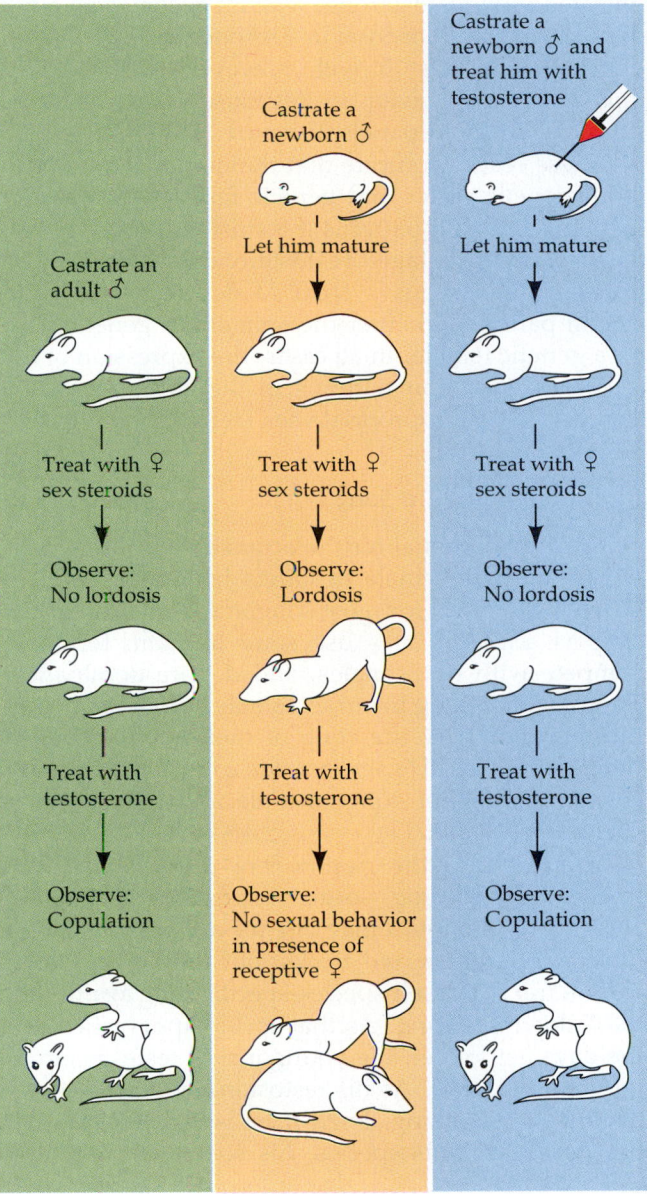

an adult. But if this *genetically female* adult rat is treated with testosterone, she will mount other females in estrus and show the male fixed action patterns associated with copulation. The presence of testosterone in the newborn ovariectomized female masculinizes her developing nervous system. When she reaches adulthood, her nervous system responds to male rather than to female steroids and generates male fixed action patterns (third panel of Figure 45.6).

Similar experiments on genetic males do not yield entirely reciprocal results. Castration (removal of the testes) of an adult male does not alter its response to treatment with sex steroids. Such a castrated male does not show lordosis when treated with female hormones, and it does show male sexual behavior when injected with testosterone. If a *newborn* male is castrated, it *will* show lordosis when injected with female hormones as an adult, but it will *not* show male sexual behavior when injected with testosterone as an adult. If the newborn male is castrated *and* injected with testosterone, when it becomes an adult it will not show lordosis in response to injections of female hormones. It will, however, show normal male sexual behavior in response to testosterone (three right-hand panels of Figure 45.6).

These results indicate that the nervous systems of both genetic males and genetic females develop female fixed action patterns if testosterone is not present at an early stage. Testosterone in the newborn causes the nervous system to develop male fixed action patterns whether the animal is a genetic male or a genetic female. In all cases, the expression of the sexual fixed action patterns in the adult requires a certain level of appropriate sex steroids.

Bird Brains and Bird Song

Learning is essential for the acquisition of bird song. Both male and female nestlings hear their species-specific song, but for most songbirds, only the males sing as adults. Males use song to claim territory, compete with other males, and declare dominance. They also use song to attract females, suggesting that the females know the song of their species even if they do not sing. Do sex steroids control the learning and expression of song in male and female songbirds?

After leaving the nest where they experienced their father's singing, young songbirds from temperate and arctic habitats may migrate and associate with other species in mixed flocks, but they do not sing and do not hear their species-specific song again until the following spring. As that spring approaches and the days get longer, the young male's testes begin to grow and mature. As his testosterone level rises, he begins to try to sing. Even if he is isolated from all other males of his species, his song will gradually

improve until it is a proper rendition of his species-specific song. At that point the song is **crystallized**— the bird expresses it in similar form every spring thereafter. The juvenile bird's brain learns the pattern of the song by hearing the father. During the subsequent spring, under the influence of testosterone, the bird learns to express that song—a behavior that then becomes rigidly fixed in its nervous system.

Why don't the females sing? Can't they learn the patterns of their species-specific song? Don't they have the muscular or nervous system capabilities necessary to sing? Or do they simply lack the hormonal stimulus for developing the behavior? To answer these questions, investigators injected female songbirds with testosterone in the spring. In response to these injections, the females developed their species-specific song and sang just as the males do. Apparently females learn the song pattern of their species when they are nestlings and have the capability to express it, but they normally lack the hormonal stimulation.

What does testosterone do to the brain of the songbird? A remarkable discovery was that testosterone causes the parts of the brain necessary for learning and expressing song to grow larger (Figure 45.7). Each spring certain regions of the males' brains grow. The individual cells increase in size, they grow longer extensions and—most surprisingly—the *numbers* of brain cells in those regions of the bird brain increase. Prior to these discoveries, newborn vertebrates were thought to have their full complement of brain cells, which they would lose progressively throughout life without replacing them. Research on the neurobiology of bird song has revealed that hormones can control behavior by influencing brain structure as well as brain functioning on both a developmental and a seasonal basis.

THE GENETICS OF BEHAVIOR

To say that behavior is genetically determined does not mean there are specific genes that code for specific behavior. Genes code for protein, and there are many complex steps between the expression of a gene as a protein product and the expression of a behavior. Many intermediate steps exist between any gene and its phenotypic expression, but the complexity is especially great when the phenotypic trait is a behavior. A specific protein may affect behavior by playing a critical role in the development of patterns in the nervous system or in the functioning of the nervous or endocrine system; such influence is indirect and difficult to discover. In no case are all the steps between a gene and a behavior known. Nevertheless, the approaches of genetics clearly substantiate genetic components of behavior and bring

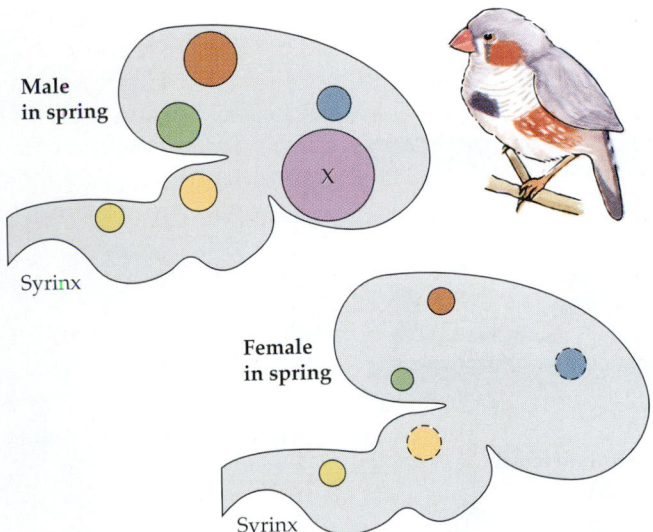

45.7 Effects of Testosterone on Bird Brains
Testosterone induces growth in the regions (colored circles) of a songbird's brain that are responsible for song. During the nonbreeding season, the brains of male and female zebra finches are similar, but in spring rising testosterone levels in the male cause the song regions of its brain to develop. The sizes of the circles are proportional to the volume of the brain occupied by that region; dashed circles indicate estimated volumes. The area labeled "X" is not found in the brains of female finches.

us closer to understanding the underlying mechanisms of inheritance of fixed action patterns. The genetic approaches we will discuss are hybridization, artificial selection, crossing of selected strains, and molecular analysis of genes and gene products.

Hybridization Experiments

The material for genetic analysis is variability, and variability is most pronounced between species. Closely related species frequently show large differences in fixed action patterns, and if such species can be hybridized, the offspring reveal interesting dis-

ruptions of their behavior. A classic case is nest building in lovebirds of the genus *Agapornis*. One species, *A. roseicollis*, carries nesting material tucked under its tail feathers. Another species, *A. fischeri*, carries nesting material in its beak (Figure 45.8). Are these simple behavior patterns learned or are they genetically programmed? When the two species are crossed, the hybrid offspring display a maladaptive combination of the two carrying methods. The hybrid picks up nesting material and tucks it into its tail feathers, but does not release the object immediately. As a result, the hybrid inevitably pulls the nesting material out of its tail feathers and drops it. With years of experience, hybrids learn to carry material in their beaks, but they always make an intention movement toward their tail feathers when they pick up nesting material. This hybridization study indicates that the ways in which birds of the two species carry nesting materials are genetically determined.

Konrad Lorenz conducted hybridization experiments on ducks to investigate the genetic determinants of their elaborate courtship. Dabbling ducks such as mallards, teals, pintails, and gadwalls are closely related and can interbreed, but they rarely interbreed in nature because of the specificity of their courtship displays. Each male duck performs a carefully choreographed water ballet (Figure 45.9), and the female probably will not accept his advances unless the entire display is successfully completed. The displays of dabbling duck species consist of about 20 components altogether. The display of each species includes a subset of these components put together in a certain sequence. When Lorenz crossbred the species, he found that the hybrids expressed some components of the display of each parent in new combinations. Most interesting, the hybrids sometimes showed display components that were not in the repertoire of either parent, but were characteristic of the displays of other species. These hybridization studies demonstrated that the motor patterns of the courtship displays were genetically programmed. Fe-

(a)   *(b)*

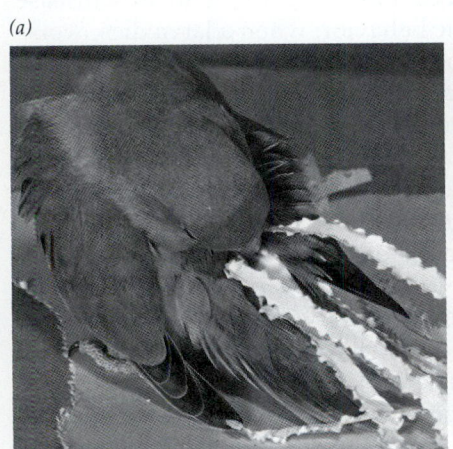 

45.8 Nest Building Behavior Is in the Genes
(a) Peach-faced lovebirds (*Agapornis roseicollis*) carry nest-building materials tucked in their back feathers. *(b)* Fischer's lovebirds (*A. fischeri*) carry the objects in their bills. Hybrid offspring of the two species display a confused combination of the two behaviors, indicating that the behaviors are genetically programmed.

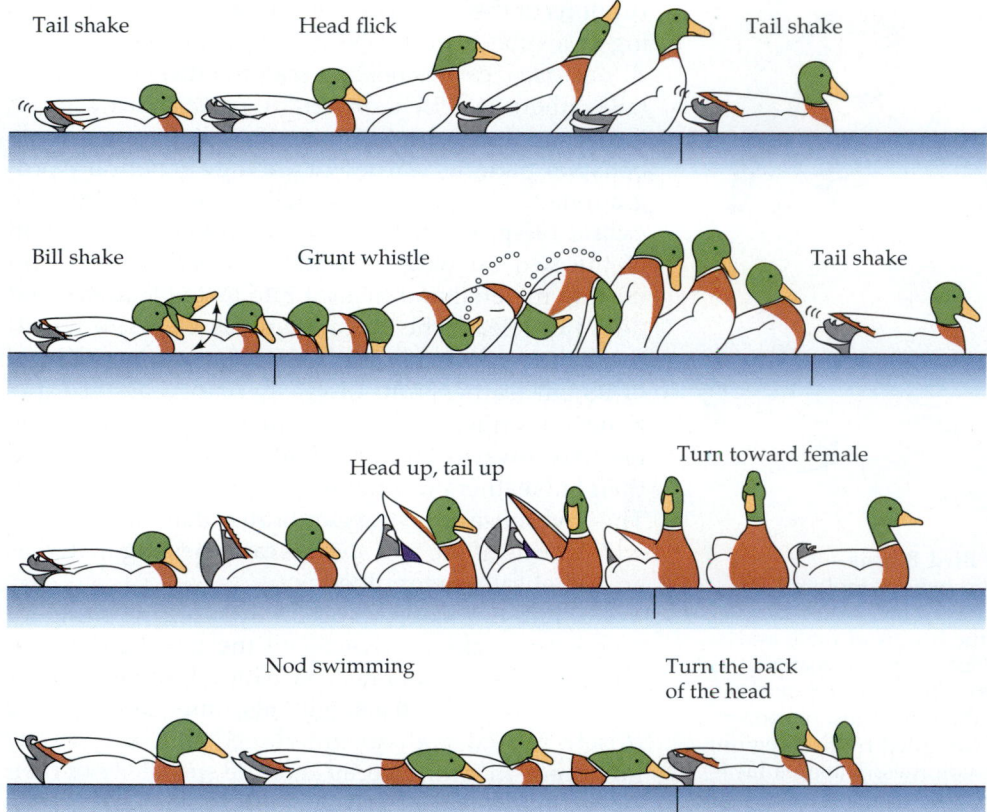

Tail shake   Head flick   Tail shake

Bill shake   Grunt whistle   Tail shake

Head up, tail up   Turn toward female

Nod swimming   Turn the back
of the head

45.9 Courtship Ballet of the Mallard
The courtship display of the male mallard duck contains about ten elements. Closely related duck species may display some of the same ten elements, but they will have other elements not displayed by mallards. The elements of the courtship display and their sequence are species-specific and act to prevent hybridization.

males were not interested in males showing the hybrid displays, thus demonstrating the adaptive significance of the species-specific fixed action patterns.

Selection and Crossing of Selected Strains

Domesticated animals provide abundant evidence that artificial selection of mating pairs on the basis of their behavior can result in strains with distinct behavioral as well as anatomical characteristics. Among dogs, consider retrievers, pointers, and shepherds. Each has a particular behavioral tendency that can be honed to a fine degree by training, whereas other strains cannot be trained in this way. Dogs and other large animals, however, are not the best subjects for genetic studies. Most controlled selection experiments in behavioral genetics have been done on more convenient laboratory animals with short life cycle times and high numbers of offspring. A favorite subject for such studies has been the fruit fly, genus *Drosophila*. Artificial selection has been successful in

shaping a variety of behavior patterns in fruit flies, especially aspects of their courtship and mating behavior. Crossing of selected strains reveals that behavioral differences produced by artificial selection are usually due to multiple genes that probably influence the behavior indirectly by altering general properties of the nervous system.

Few behavioral genetic studies reveal simple Mendelian segregation of behavioral traits. One case that does is nest cleaning behavior in honeybees. Nest cleaning counteracts a bacterium that infects and kills the larvae of honeybees. Hives or colonies of one strain of honeybee that is resistant to this disease practices hygienic behavior; when a larva dies, workers uncap its brood cell and remove the carcass from the hive. Another strain of honeybee does not show hygienic behavior and is, therefore, more susceptible to the spread of the disease. When these two strains were crossed, the results indicated that the hygienic behavior was controlled by two recessive genes. Colonies of the F_1 generation are all nonhygienic, indicating that the behavior is controlled by recessive genes. Backcrossing the F_1 with the pure hygienic strain produced the typical 3:1 ratio expected for a two-gene trait. The behavior of the nonhygienic colonies is very interesting. One-third of them show no hygienic behavior at all; one-third uncap the cells of dead larvae but do not remove them; and one-third do not uncap cells but will remove carcasses if the cells are open (Figure 45.10).

Even though these results appear to indicate a gene for uncapping and a gene for removal, these behavior patterns are complex. They involve sensory mechanisms, orientation movements, and motor patterns, each of which depends on multiple properties of many cells. The genetic deficits of nonhygienic bees could influence very small, specific, yet critical properties of some cells. Lacking one critical property, such as a critical synapse or a particular sensory receptor, the whole behavior would not be expressed. The responsible gene, then, is not a specific gene that codes for the entire behavior.

Molecular Genetics of Behavior

The powerful techniques of molecular genetics enable investigation of the functions of specific genes that influence behavior. For example, in the marine mollusk *Aplysia*, egg laying involves a sequence of fixed action patterns. The eggs are extruded from the animal in long strings by contractions of the muscles of the reproductive duct. The animal stops whatever it is doing (usually eating or crawling) and takes the

egg string in its mouth. With a series of stereotypic head movements, it pulls the egg string from the duct and coils it into a mass glued together by secretions from its mouth. Finally, with a strong head movement, it affixes the entire mass of eggs to a solid substrate. *Aplysia* has a very simple nervous system, and it was discovered that specific cells in its nervous system produce a peptide that can elicit certain aspects of egg laying. This peptide is called egg-laying hormone. The amino acid sequence of egg-laying hormone was determined, and then molecular genetic techniques were used to find the gene that coded for it. The surprising discovery was that the gene codes for a precursor molecule that has almost 300 amino acids, whereas egg-laying hormone has only 36 amino acids. The precursor molecule also contains other peptides that function as neural signals controlling aspects of egg-laying behavior. One gene could thus code for a set of neural signals necessary to elicit the coordinated motor scores involved in egg-laying behavior. This example is about as close as we can get to making connections between a specific gene and a specific behavior, but the connections depend on the existence of a highly organized nervous system, which of course is a product of many genes.

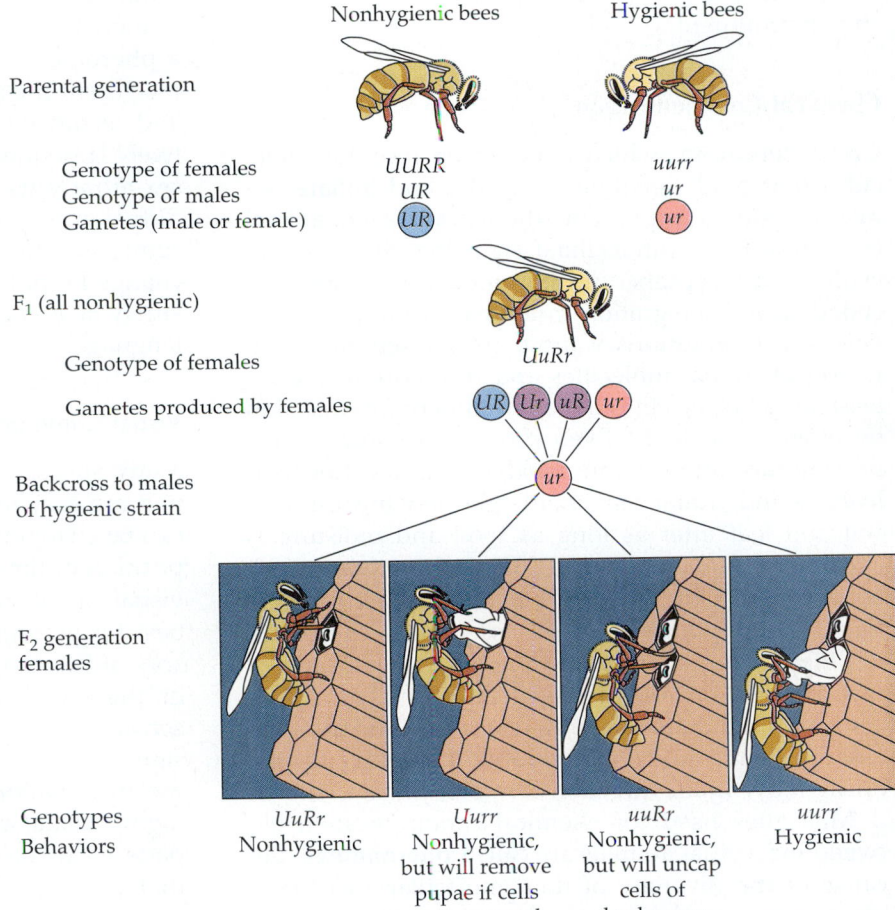

45.10 Genes and Hygienic Behavior in Honeybees
Some honeybee strains make a practice of removing the carcasses of dead larvae from their nests. This behavior seems to have two components—uncapping the larval cell (*u*), and removing the carcass (*r*)—each of which is under the control of a recessive gene. Honeybee females are diploid and males are haploid. The workers in a colony are all females.

COMMUNICATION

Having discussed methods of investigating animal behavior, let's focus now on specific types of behavior. As investigations into animal communication illustrate, a diversity of issues can arise in studies of even a specific type of behavior.

Communication is behavior that influences the actions of other individuals. It consists of **displays** or **signals** that can be perceived by other individuals. Displays and signals are behaviors, anatomical features, or physiological responses that convey information to another individual. The information they convey may be secondary or even incidental to their original function. If the transfer of information benefits the animal generating the signal or display, selection can shape it to enhance its information content. The displays or signals that an animal can generate depend of course on its physiology and anatomy. Its ability to perceive displays or signals depends on its sensory physiology and on the environment through which the display or signal must be transmitted. We learned in Chapter 39 that sensory physiology includes chemosensation, tactile sensation, audition, vision, and electrosensation. These are the forms of animal communication. Studies of communication can be complex because they must take into account the sender, the receiver, and the environment.

Chemical Communication

Chemosensation, which is based on receptor molecules that bind signaling molecules and initiate cellular responses, is probably the oldest form of animal communication. Intracellular signaling involves molecules and receptors, and hormonal integration preceded neural integration in animal evolution. It is a very short evolutionary jump from internal coordination by signal molecules and receptors to the release of molecules into the environment for detection by other individuals. Even protists use such means of communication. Slime molds spend most of their lives as individual amoeboid cells moving through soil and leaf litter as long as food and moisture is adequate (see Chapter 23). When the environment becomes less favorable, the individuals aggregate, form a fruiting body, and release spores. The chemical signal that coordinates aggregation is cAMP, a well-known intracellular signaling molecule (see Chapter 36). Stressed cells release cAMP into the environment, and other individuals sensing the cAMP move in the direction of higher concentrations.

Molecules used for chemical communication between individual animals are called **pheromones**. Because of the diversity of their molecular structures, pheromones can communicate very specific messages that contain a great deal of information. For example, when a female gypsy moth is ready to be inseminated, she releases a pheromone called gyplure. Male gypsy moths downwind by as much as thousands of meters are informed by these molecules that a female of their species is sexually receptive. By orienting to the wind direction and the concentration gradient, they can find her (Figure 45.11a). Territory marking is another example in which detailed information is conveyed by chemical communication (Figure 45.11b). The receiver of a message from a male cheetah detects information about the animal leaving the message: species, individual identity, reproductive status, height of message (indicating size of the animal that left it), and strength of scent (indicating time elapsed since the message was left).

An important feature of pheromones is that once they are released, they remain in the environment for a long time. A pheromonal message can act as a territorial marker long after the animal claiming the territory has left. By contrast, the signals of a vocal or visual territorial display of a song bird disappear as soon as the bird stops singing and displaying. The durability of pheromonal signals enables them to be used to mark trails, as ants do, or to indicate directionality, as in the case of the gypsy moth sex attractant. The chemical nature and the size of the pheromonal molecule determine its diffusion coefficient. The greater its diffusion coefficient the more rapidly a pheromone diffuses, the farther the message will reach, but the sooner it will disappear. Trail-marking and territory-marking pheromones tend to be relatively large molecules with low diffusion coefficients; sex attractants tend to be small molecules with high diffusion coefficients. A disadvantage of pheromonal communication is that the message cannot be changed rapidly. A discussion based on smells would surely be less effective than one using speech or sign language.

Visual Communication

Many species use visual communication. Visual signals are easy to produce, come in an endless variety, can be changed very rapidly, and clearly indicate the position of the signaler. The extreme directionality of visual signals means, however, that they are not the best for getting the attention of a receiver. The sensors of the receiver must be focused on the signaler or the message will be missed. Most animals are sensitive to light and can therefore receive visual signals, but sharpness of vision limits the detail of the information that can be transmitted. Birds are highly visual and have evolved a vast diversity of patterns of colored feathers and body appendages that can be incorporated into complex displays used in communication (Figure 45.12).

(a)

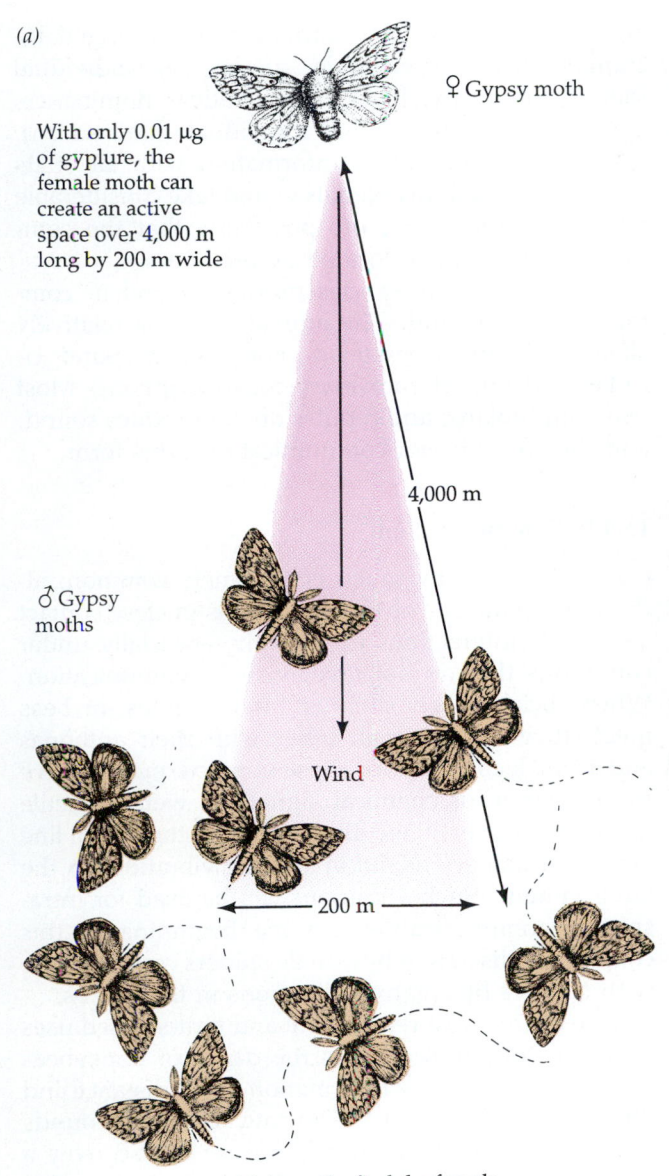

♀ Gypsy moth

With only 0.01 μg of gyplure, the female moth can create an active space over 4,000 m long by 200 m wide

4,000 m

♂ Gypsy moths

Wind

200 m

Male moths find the female by flying upwind against the gyplure gradient

(b)

45.11 Many Animals Communicate with Pheromones
(a) A female gypsy moth secretes the pheromone gyplure which can attract males thousands of meters downwind when it binds to sensors on their antennas. (b) To mark his territory, this male cheetah is spraying pheromonal secretions from a scent gland in his hindquarters onto a tree. Other cheetahs passing the spot will know that the area is "claimed," and they will know something about who claimed it.

Because visual communication requires light, it is not useful for many species at night or in environments that lack light, such as caves and the ocean depths. Some species have surmounted this constraint on visual communication by evolving their own light-emitting mechanisms. Fireflies use a luciferin/luciferase mechanism to create flashes of light. By emitting flashes in species-specific patterns, fireflies can advertise for mates at night by sending visual signals.

Firefly communication raises another interesting issue. Any system of communication is vulnerable to

45.12 Selection for Effective Display Has Greatly Modified the Anatomy of Some Animals
Male peafowl—peacocks—have evolved brilliant tail plumage, which they display to the females (peahens) during courtship. These elaborate tail feathers are folded when the bird is not displaying.

exploitation by illicit senders or receivers. Predators are commonly illicit receivers. When an animal emits a message in any form, it tends to signal its own position, and the information can be used by predators. Predators of fireflies can locate them in the dark by the flashes of light they emit. Some species of fireflies are themselves predators of other species of fireflies and have evolved the "deceitful" behavior of emitting signal patterns that mimic the mating messages of other species. When a prospective suitor homes in on the signal, it is eaten.

Auditory Communication

Humans are very familiar with communicating by sound. In Chapter 39 we discussed the physical properties of sound and the sensory structures that transduce sound pressure waves into neural signals. Compared with visual communication, auditory communication has several obvious advantages and disadvantages. Sound can be used at night and in dark environments. Sound can go around objects that would interfere with visual signals. Sound is better than visual signals in getting the attention of a receiver because the sensors do not have to be focused on the signaler for the message to be received. Like visual signals, sound can provide directional information, as long as the receiver has at least two sensors spaced somewhat apart. Differences in the sound intensity, in the time of arrival, and in the phase of the pressure wave reaching the two sensors can provide information about the direction of the sound source. By maximizing or minimizing these features of the sounds they emit, animals can make their location easier or more difficult to determine. Pure tones with sudden onsets and offsets are much easier for the receiver to localize than complex sounds with gradual onsets and offsets.

Sound is good for communicating over long distances. Even though the intensity of sound decreases with distance from the source, loud sounds can be used to communicate over distances much greater than those possible with visual signals. An extreme example is the communication of whales. Some whales, such as the humpback, have very complex songs. When these sounds are produced at a certain depth (around 1000 meters), the sound waves are channeled between the thermocline (a sudden temperature change in the water column) and much deeper waters so they can be heard hundreds of kilometers away. In this way, humpback whales use sound communication to locate each other over vast areas of ocean.

Auditory signals cannot convey complex information as rapidly as visual signals can. A well-known expression states, "A picture is worth a thousand words." When individuals are in visual contact, an enormous amount of information is exchanged instantaneously (for example, species, sex, individual identity, maturity, level of motivation, dominance, vigor, alliances with other individuals, and so on). Coding that amount of information with all of its subtleties as auditory signals would take considerable time, thus increasing the possibility that the communicators could be located by predators.

Remarkably few species produce sound or communicate by sound. The animal world is relatively silent. Most invertebrates do not produce sound; cicadas and crickets are marvelous exceptions. Most fish, amphibians, and reptiles do not produce sound, and therefore do not communicate via this form.

Tactile Communication

Communication by touch is extremely common, although not always obvious. Animals in close contact use tactile interactions extensively, especially under conditions that do not favor visual communication. When social insects such as ants, termites, or bees meet, they contact each other with their antennas and front legs. Some of these contacts may involve the exchange of chemical signals as well as tactile ones. In Chapter 39 we discussed how the lateral line organs of fish are useful in sensing vibrations in the environment. Such vibrations can be used for intraspecific communication. At the beginning of this chapter we discussed how male spiders communicate with females by creating vibrations in their webs.

One of the most remarkable and best-studied uses of tactile communication is the dance of honeybees that is used to convey information about distance and direction to a food source. Dancing bees make sounds and carry odors on their bodies, but they convey a great deal of information by dance movement. The dances are monitored by other bees, who follow and touch the dancer. When a foraging bee finds food, she returns to the hive and communicates her discovery by dancing in the dark on the vertical surface of the honeycomb. If the food is less than 80 to 100 meters from the hive, she performs a **round dance**, running rapidly in a circle and reversing her direction after each circumference. The odor on her body indicates the flower to be looked for, but the dance contains no information about the direction in which to go. If the food source is farther than 80 meters, she performs a waggle dance, which conveys information about both the direction and the distance of the food source. In the waggle dance, a bee repeatedly traces out a figure-eight pattern as she runs on the vertical surface. She alternates half circles to the left and right, with vigorous wagging of her abdomen in the short, straight crossover between turns. The angle of the straight line indicates the direction of the food sources relative to the direction of the sun (Fig-

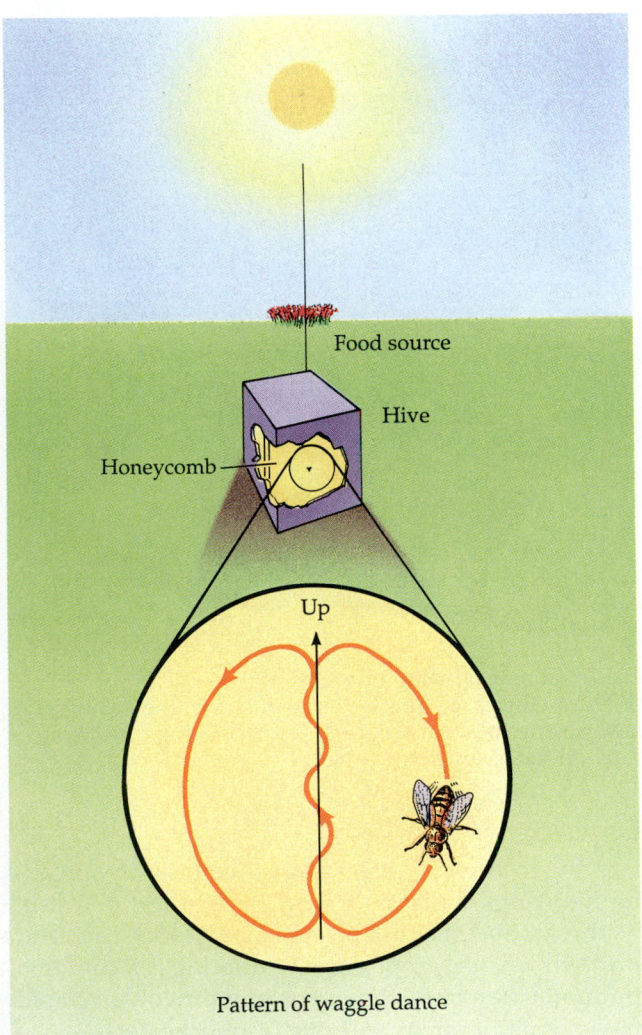

45.13   The Waggle Dance of the Honeybee

By running straight up on the surface of the honeycomb in a dark hive, a honeybee tells her hive mates that there is a food source in the direction of the sun and at least 80 meters from the hive. The rate of circling indicates distance to the food source, while the intensity of the wag- gling indicates the richness of the food source. If the food source were in the opposite direction from the sun, she would orient her waggle runs straight down. When her waggle dance is run at an angle from the vertical, the other bees know that the same angle separates the direction of the food source from the direction of the sun.

ure 45.13). The speed of the dancing indicates the distance to the food source: the farther away it is, the slower the waggle run. The dances of honeybees are unusual because they are based on an arbitrary convention: straight down could just as well indicate the direction of the sun as straight up. Arbitrary, symbolic conventions like this have been developed to an extreme degree in human language.

Electrocommunication

In Chapter 39 we learned about electrosensors of fish species living in murky water and about electric fish that generate electric fields by emitting series of electrical pulses. We discussed how these trains of electrical pulses can be used for sensing objects in the immediate surroundings of the fish, but they can also be used for communication. An electrode connected to an amplifier and a speaker can be used to "listen" to the signal generated by each fish in a tank holding numerous electric fish. The amplifier reveals that each fish emits a pulse at a different frequency, and the frequency each fish uses relates to its status in the population. Glass knife fish (*Eigenmannia*) males emit lower frequencies than females. The most dominant male has the lowest frequency, and the most dominant female has the highest frequency. When a new individual is introduced into the population, the other individuals adjust their frequencies so that they do not overlap, and the signal of the new individual indicates its position in the population. In their natural environment—the murky waters of tropical rainforests—these fish can tell the identity, sex, and social position of another member of the population by its electric signals. When two individuals interact directly, they interrupt their constant frequency emissions and modulate them to produce more-complex, "chirplike" signals that perhaps communicate even more information.

Origins of Communication Signals

Included among the constraints on the evolution of communication signals are the anatomical and physiological characteristics of a species that are available to be shaped by natural selection for the purpose of conveying information to other individuals. Charles Darwin was the first person to give serious attention to this problem of the origin and evolution of communication signals. In 1872 he published the results of his detailed studies in a book entitled *The Expression of the Emotions in Man and Animals*. In this perceptive book Darwin identified several important means by which communication signals originate.

One source of raw material for the evolution of signals is **intention movements**: movements that precede a particular behavior. For example, a bird ready to take off flexes its legs, sleeks its feathers, and raises its wings slightly at the shoulders. A bird in a threatening situation, such as facing a neighbor challenging its territorial boundary, will experience a conflict between motivation to flee and motivation to attack. Thus its intention movements are exaggerated and mixed. Raising its wings at the shoulders might make it appear bigger to its adversary, thus augmenting its attack intention movements. The combination of behavior patterns will convey information to the adversary about the degree of motivation of the defender. Selection can work to enhance this threat display that originated in intention movements. In red-winged blackbirds, for example, the red shoulder patches are most effectively displayed when the wings are slightly raised in the threat posture (see Figure 46.15*b*). Darwin pointed out the threat posture of the domestic cat, which appears to result from conflicting motivations. The forelegs seem to push back while the hind legs push forward, resulting in the hunched back that makes the cat look bigger and more threatening (Figure 45.14).

Autonomic responses (see Chapter 38) are another possible source of material upon which selection can operate to produce displays. Urination and defecation are autonomic responses that have been extensively used as signals. The erection of fur or feathers under sympathetic nervous system control is another example. A particularly picturesque example is the mating display of the male frigate bird. Frigate birds are large and black and nest on oceanic islands near the equator. The male builds a nest and then sits on

45.14 Threat Display of a Cat
This drawing from Charles Darwin shows the arched back posture of a cat when it is experiencing conflicting motivations to attack or to flee.

it, trying to attract females flying overhead by spreading his wings, inflating a huge, bright red pouch in his throat, and shaking his head and pouch back and forth while vocalizing. Throat fluttering is a common thermoregulatory response of birds involving rapid movements of air in and out of pouches of skin on the front of the neck. The bizarre and dramatic courtship display of the male frigate bird probably originated through exaggeration of autonomic thermoregulatory behaviors (Figure 45.15).

Displacement behavior is a third class of behavior that can be shaped by natural selection into communication displays. In a tense situation that involves highly conflicting motivations such as attack and escape, an animal sometimes does something completely irrelevant. It might groom itself, feed, or attack some object in the vicinity. If such behavior enhances the display of the animal, it can be favored by selection and incorporated into the display. For example, male three-spined stickleback fish defend small territories when they build a nest to attract potential mates. At the boundary of his territory, the male stickleback can experience equally strong motivations to attack and to flee from a neighboring male. During boundary disputes, the males engage in head-down threat displays that resemble nest-building postures, but no nest building takes place. This threat display probably evolved from nest-building displacement activities resulting from the approach–avoidance conflicts of boundary disputes.

This discussion of animal communication has fo-

45.15 Frigate Bird Display
A male frigate bird displays his red throat pouch for the female next to him. This courtship display may have evolved from the throat-fluttering behavior birds use in thermoregulation.

cused on examples of how natural selection shapes the behavior of a species. In Chapter 46 we will study in more detail the evolution and functions of displays in the context of social behavior.

THE TIMING OF BEHAVIOR: BIOLOGICAL RHYTHMS

An important aspect of behavior is its temporal organization. The neurophysiology of most behavior is poorly understood, but the study of biological rhythms has led to major discoveries about brain mechanisms.

Circadian Rhythms

Our planet turns on its axis once every 24 hours, creating a cycle of environmental conditions that has existed throughout the evolution of life. Many organisms thus evolved rhythmicity. In Chapter 34 we encountered rhythmicity in plants. Indeed, daily cycles are a characteristic of almost all organisms. What is surprising, however, is that daily rhythmicity does not depend on the 24-hour cycle of light and dark. If animals are kept under absolutely constant environmental conditions, such as constant dark and constant temperature, with food and water available all the time, they still demonstrate daily cycles of activity: sleeping, eating, drinking, and just about anything else that can be measured. This persistence of the cycle in the absence of light/dark cycles suggests that the animal has an internal (endogenous) clock. Without time cues from the environment,

however, these daily cycles are not exactly 24 hours. They are therefore called **circadian rhythms** (*circa* = "about," *dies* = "day").

To discuss biological rhythms, we must introduce some terminology. A rhythm can be thought of as a series of cycles, and the length of one of those cycles is the **period** of the rhythm. Any point on the cycle is a **phase** of that cycle. Hence when two rhythms completely match, they are in phase, and if a rhythm is shifted (as in resetting a clock), it is phase-advanced or phase-delayed. Since the period of a circadian rhythm is not exactly 24 hours, it must be phase-advanced or phase-delayed every day to remain in phase with the daily cycle of the environment. This process of the resetting of the rhythm by environmental cues is called **entrainment**. An animal kept in constant conditions will not be entrained to the 24-hour cycle of the environment, and its circadian clock will free-run with its natural periodicity. If its period is less than 24 hours, the animal will begin its activity a little earlier each day (Figure 45.16).

Animals that have free-running circadian rhythms can be used in experiments to investigate stimuli that phase-shift or entrain the clock. Under natural conditions, environmental cues such as the onset of light or dark entrain the free-running rhythm to the 24-hour cycle of the real world. In the laboratory it is possible to entrain circadian rhythms in animals held under constant conditions with short pulses of light or dark administered every 24 hours. Researchers can also entrain animals to light or dark pulses given at intervals not equal to 24 hours, as long as those intervals are not too short or too long. The range of entrainment of the endogenous clock is limited.

When you fly across several time zones, your circadian clock is out of phase with the real world at your destination and jet lag results. Gradually your endogenous rhythm synchronizes itself with the real world as it is reentrained every day by environmental cues. Since your internal rhythm cannot be shifted by more than 30 to 60 minutes each day, it takes a number of days to reentrain your endogenous clock to real time in your new location. This period of reentrainment is when you have the symptoms of jet lag, because your endogenous rhythm is waking you up, making you sleepy, initiating activities in your digestive tract, and stimulating many other physiological functions at inappropriate times of the day.

Where is the circadian clock? In mammals the master circadian clock is located in two tiny groups of cells just above the optic chiasm, the place where the two optic nerves cross (see Figure 39.29). Hence these structures are called the **suprachiasmatic nuclei**. If these two little groups of cells are destroyed, the animal loses circadian rhythmicity. Under constant conditions the animal is equally likely to be active or

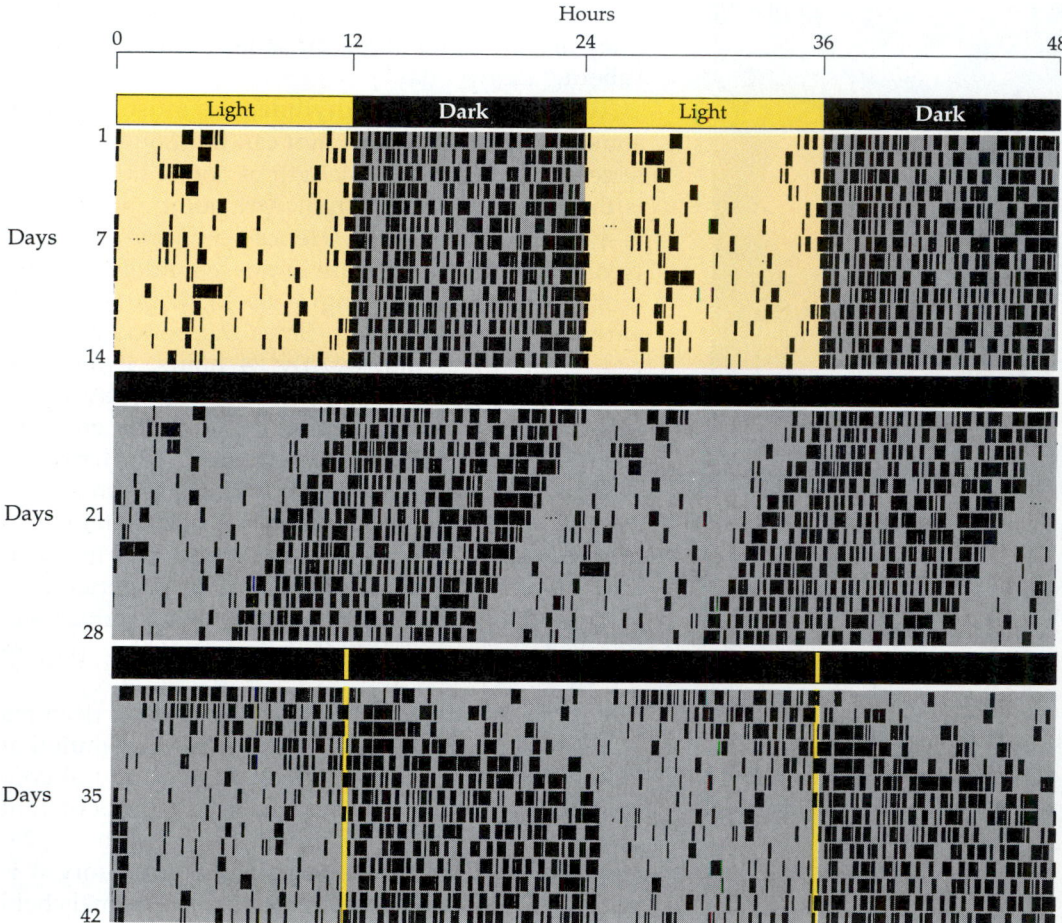

45.16   Circadian Rhythms
The black marks indicate that a mouse is running on an activity wheel. Two days are recorded on each line, so the data for each day are plotted twice, once to the right side of the 24-hour mark and again at the start of the next line down. This double plotting makes patterns easier to see. Changes in the schedule of light and dark are indicated by shading. First the mouse sees 12 hours of light and 12 hours of dark every day, then it is placed in total darkness, and finally it is given a 10-minute exposure to light each day. In constant darkness the circadian rhythm is free-running, but a 10-minute flash of light can entrain it. This figure is idealized but represents results from real experiments.

asleep, eat or drink, at any time of day. Its activities are randomly distributed (Figure 45.17). Experiments first done by Patricia deCoursey at the University of South Carolina show that circadian rhythmicity can be restored in an animal whose suprachiasmatic nuclei have been destroyed by transplanting those nuclei from another animal. In no other known case can a brain tissue transplant restore a complex behavior! Since the restored rhythm has the period of the animal donating the tissue, the transplant clearly controls the recipient's behavior and does not just provide a permissive factor.

Because circadian rhythms are found in virtually every animal group, as well as in protists, plants, and fungi, the molecular mechanisms for generating circadian rhythms must be very general properties of

cells. A diversity of master clocks using these mechanisms have been produced by natural selection. Invertebrates, for example, do not have suprachiasmatic nuclei, but they certainly have circadian rhythmicity. In the mollusk *Aplysia*, cells in the eyes are the circadian clock. Circadian control systems are diverse even among vertebrates. The circadian clock of birds resides in the pineal gland, a mass of tissue between the cerebral hemispheres that produces the hormone melatonin. If the pineal gland of a bird is removed, the bird loses circadian rhythmicity. In mammals, light entrains the circadian clock via photosensors in the eyes, but in birds, the pineal gland itself is sensitive to light and is sometimes called the third eye. If a small amount of black ink is injected under the skin on the top of a bird's head to blacken

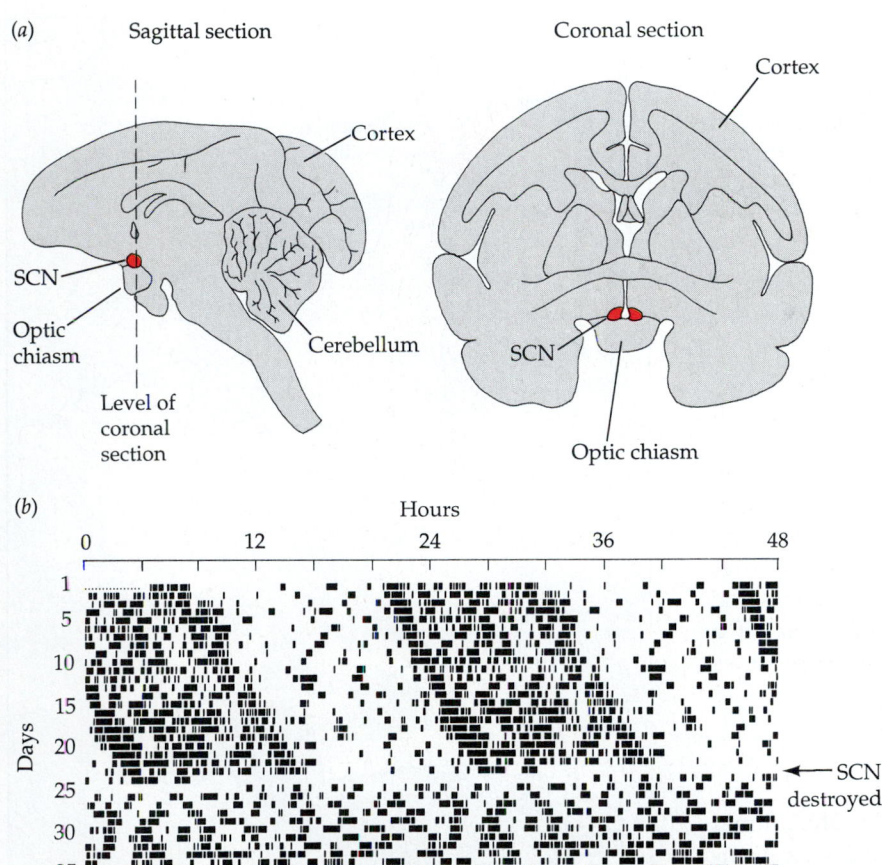

45.17 Where the Clock Is
(a) The circadian clock of mammals is in the suprachiasmatic nuclei (SCN) of the brain. These nuclei are located just above the site on the bottom of the brain where the optic nerves join. (b) If its suprachiasmatic nuclei are destroyed, a mammal loses its circadian rhythm.

the skull over the pineal gland, the bird will not entrain to the cycle of light and dark, but will have a free-running circadian rhythm.

Circannual Rhythms

In addition to turning on its axis every 24 hours, our planet also revolves around the sun once every 365 days. Earth is tilted on its axis, so its revolution around the sun results in seasonal changes in day length at all locations except at the equator. These changes in day length secondarily create seasonal changes in temperature, humidity, weather, and other variables. Because the behavior of animals must adapt to these seasonal changes, animals must be able to anticipate seasons and adjust their behavior accordingly. Most animals should not come into reproductive condition and mate just prior to winter, because their offspring would be born during a time of little food and harsh weather conditions. For many species, the change in day length is an excellent and absolutely reliable indicator of seasonal changes to come. If photoperiod has a direct effect on the physiology and behavior of a species, that species is said to be photoperiodic. For example, if male deer are held in captivity and subjected to two cycles of day-

length change in one year, they will grow and drop their antlers twice during that year. Many species of birds are also photoperiodic.

For some animals, changing day length is not a reliable cue. Hibernators, for example, spend long months in dark burrows underground but have to be physiologically prepared to breed almost as soon as they emerge in the spring. The timing of their breeding is important because their young must have time to grow and fatten before the next winter. Other examples of animals that receive little or ambiguous information from changes in day length are resident tropical species, migratory species that overwinter near the equator, or migratory species (mostly birds) that migrate across the equator. For a bird overwintering in the tropics, there is no change in photoperiod that it can use to time its migration north to the breeding grounds. A bird that crosses the equator must fly south as day length decreases at one time of year but fly north when day length decreases at another time of year. If change in day length is not a reliable cue for seasonal behavior, what else could be used as a calendar?

Hibernators and equatorial migrants have endogenous annual rhythms, called **circannual rhythms**. Their nervous systems have built-in calendars. Just as circadian rhythms are not exactly 24 hours, circannual rhythms are not exactly 365 days. The circannual rhythm of an animal under constant conditions may be 360 days, or 345 days. Rarely is it longer than 365 days, however, because being late for an annual event such as breeding would be a very costly mistake.

HOW DO THEY FIND THEIR WAY?

Within a local environment, finding your way is not a problem. You remember landmarks and orient yourself with respect to those reference points. **Orientation** is a very common animal behavior. It simply means that the animal organizes its activity spatially with respect to reference points such as objects in the environment, a predator or prey, a mate or offspring, a nest or food source, or even a signal such as the call or display of another individual. But what if a destination is at a considerable distance? How does the animal orient to it and find its way?

Piloting

In most cases the answer is quite simple: The animal knows and remembers the structure of its environment. It uses landmarks to find its nest, a safe hiding place, or a food source. Orienting by means of landmarks is called **piloting**. Even long-distance migrations of animals can be achieved by piloting that does not depend on specific landmarks. For example, the gray whales that spend the summer feeding in the Gulf of Alaska and the Bering Sea and migrate south in the winter to breed in lagoons on the Pacific coast of Baja California can find their way by following two simple rules (Figure 45.18): Keep the land to the left in the fall and to the right in the spring. By following the west coast of North America, they can travel from summer to winter areas and back again by piloting. Coastlines, mountain chains, rivers, water currents, and wind patterns serve as piloting cues for many species. Yet there are remarkable cases of long-distance orientation and movement that cannot be explained on the basis of piloting by landmarks.

Homing

The ability of an animal to return to a nest site, burrow, or any other specific location is **homing**. In most cases homing is merely piloting in a known environment, but some animals are capable of much more sophisticated feats of navigation. People who breed and race homing pigeons take the pigeons from their home loft and release them at a remote site where they have never been before. The first pigeon home wins. Data on departure directions, known

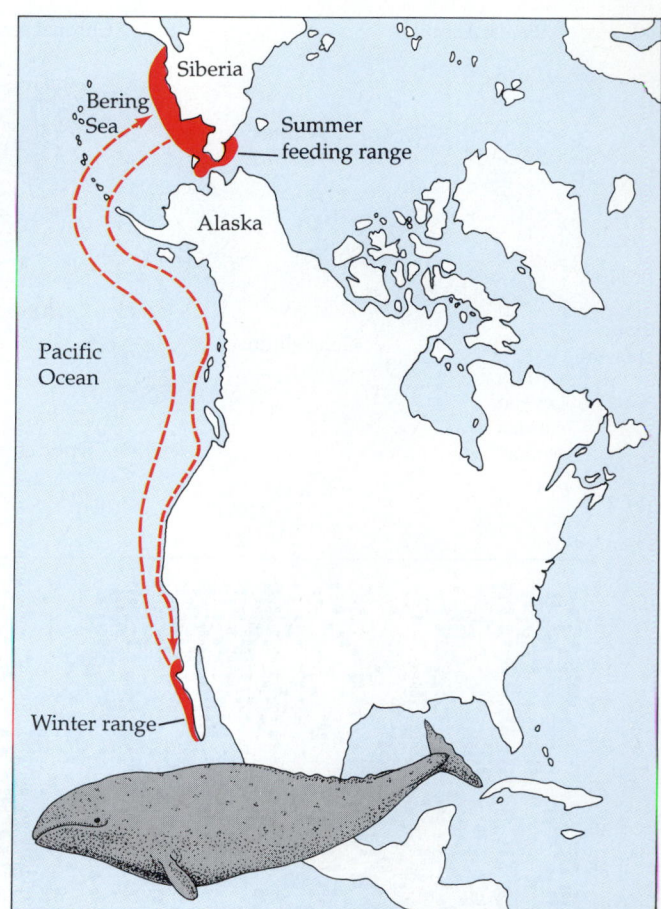

45.18 Piloting
Gray whales migrate south in winter, from the Bering Sea to the coast of Baja California. They follow a landmark: the west coast of North America. Such navigation is called piloting.

flying speed, and distance traveled show that the pigeons fly fairly directly from the point of release to home. They do not randomly search until they encounter familiar territory. Scientists have used homing pigeons to investigate the mechanisms of animal navigation. In one series of experiments the pigeons were fitted with frosted contact lenses. They could see no details other than degree of light and dark. These pigeons still homed and fluttered down to the ground in the vicinity of their loft. They were able to navigate without visual images of the landscape.

Marine birds provide many dramatic cases of homing over great distances in an environment where landmarks are rare. In daily feeding trips, many marine birds fly over hundreds of miles of featureless ocean and then return directly to a nest site on a tiny island. Remarkable feats of homing are demonstrated by albatrosses. When a young albatross leaves its nest on an oceanic island, it flies widely over the southern oceans for eight or nine years before it reaches reproductive maturity. At that time it flies back to the

45.19   Coming Home
A pair of black-browed albatrosses engage in courtship display over their partially completed mud nest. Many albatrosses return to the site of their own birth to find a mate, and will return to that site year after year.

island where it was raised to select a mate and build a nest (Figure 45.19). After the first mating season the pair separates, and each bird resumes its solitary wanderings over the oceans. The next year they return to the same nest site at the same time, reestablish their pair bond, and breed. Thereafter they return to the nest to breed every other year, spending many months in between at sea. These long-distance, synchronous homing trips are amazing feats of navigation and timing.

Migration

Ever since humans inhabited temperate and subpolar latitudes, they must have been aware of the fact that whole populations of animals, especially birds, disappear and reappear seasonally. It was not until the early nineteenth century, however, that patterns of migration were established by marking individual birds with identification bands around their legs. Being able to identify individual birds in a population made it possible to demonstrate that the same birds and their offspring returned to the same breeding grounds year after year, and that these same birds were found during the nonbreeding season at distant locations hundreds or even thousands of kilometers from the breeding grounds.

How do migrants find their way over such great distances? A reasonable hypothesis is that young birds on their first migration follow experienced birds and learn the landmarks by which they pilot in subsequent years. However, adult birds of many species leave the breeding grounds before the young have finished fattening and are ready to begin their first migration. These naive birds must be able to navigate accurately on their own and with little room for mistakes. Some species of small songbirds breed in the

high latitudes of North America, fly to the coast for fattening, and then fly over the North Atlantic Ocean on a direct route to South America (Figure 45.20). They cannot land on water and their fuel reserves are limited by their small size. Considering distance, flight speed, and metabolic rate, they must be extremely efficient and accurate in navigating to their landfall on the coast of South America.

Navigation

Homing and migrating animals find their way by several mechanisms of navigation. Although piloting is a type of navigation, it cannot explain the abilities of many species to take direct routes to their destinations through areas they have never experienced.

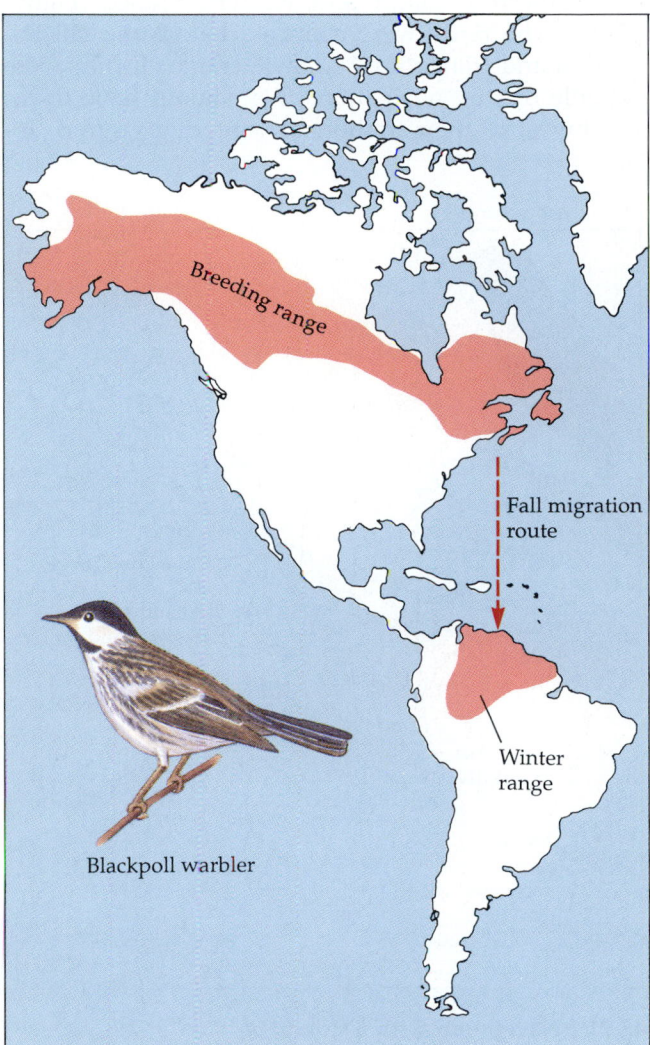

45.20   Songbirds Migrate over the Atlantic
The blackpoll warbler is one of many species that breeds over the northern United States and Canada and winters in South America. Its fall migration is first to the northeast coast of North America, where it feeds in preparation for the nonstop overwater flight to South America.

Humans use two systems of navigation that differ in complexity: distance and direction navigation and bicoordinate navigation. Distance and direction navigation involves knowing the direction to reach the destination and knowing how far away that destination is. With a compass to determine direction and a means of measuring distance, humans can navigate. Bicoordinate navigation, also known as true navigation, involves knowing the latitude and longitude (the map coordinates) of position and destination. From that information a route can be plotted to the destination. Do animals have these sophisticated abilities to navigate?

Researchers conducted an experiment with European starlings to determine their method of navigation. These short-distance migrants travel between breeding grounds in the Netherlands and northern Germany and wintering grounds to the southwest, in southern England and western France (Figure 45.21). The birds were captured before the fall migration and transported to Switzerland. If they were capable of true navigation, they should have flown northwest to their traditional wintering ground. In-

45.22   Raring to Go
A captive bird ready to migrate shows migratory restlessness in a circular cage. The cage is lined with a paper funnel, and on the floor is an ink pad. The bird's feet mark the orientation of its activity.

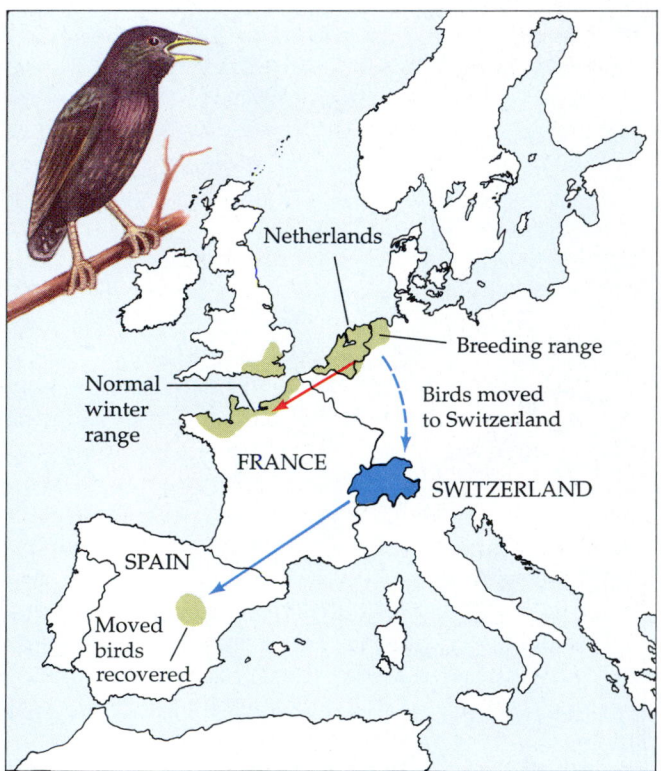

45.21   Navigation with a Compass
European starlings normally make a short winter migration in a southwesterly direction, from the Netherlands to coastal France and southern England (red arrow). Experimental populations of starlings moved to a site in Switzerland did not fly northwest to their traditional grounds, but followed the same southwesterly route (blue arrow), which took them to Spain.

stead they flew in their normal southwest direction and landed in Spain. The researchers concluded that the starlings used distance and direction for navigation.

How do animals determine distance and direction? In many instances, distance is not a problem as long as the animal recognizes its destination. Homing animals recognize landmarks and can pilot once they reach familiar areas. Some evidence suggests that biological rhythms play a role in determining migration distances for some species. Birds kept in captivity display increased and oriented activity at the time of year when they would normally migrate (Figure 45.22). Such **migratory restlessness** has a definite duration, which corresponds to the usual duration of migration of the species. Since distance is determined by how long an animal moves in a given direction, the programming of the duration of migratory restlessness can set the distance for its migration.

Two obvious candidates for determining direction are the sun and the stars. During the day the sun is an excellent compass, as long as time is known. In the northern hemisphere the sun rises in the east, sets in the west, and points south at noon. Animals can tell the time of day by means of their circadian clocks. Furthermore, clock-shifting experiments demonstrate that animals use circadian clocks to determine direction from the position of the sun. Researchers placed birds in a circular cage that enabled them to see the sun and sky but no other visual cues. Food bins were arranged around the sides of the cage, and the birds were trained to expect food in the bin in one particular direction, for example south.

After training, no matter when they were fed, and even with the cage rotated between feedings, they always went to the bin at the southern end of the cage for food (Figure 45.23a). Next the birds were placed in a room with a controlled light cycle and their circadian rhythms were phase-shifted. For example, in the controlled light room the lights were turned on at midnight. After a couple of weeks the circadian clocks of the birds were phase-advanced by six hours. The birds were returned to the circular cage under natural light conditions with sunrise at 6 A.M. Because of the shift in their circadian rhythms, their endogenous clocks were indicating noon at the time the sun came up. If food was always in the south, and it was sun-up, they should have oriented 90° to the right of the direction of the sun. But since their circadian clocks were telling them it was noon, they looked for food in the direction of the sun—the east bin (Figure 45.23b). The six-hour phase shift in their circadian clocks resulted in a 90° error in their orientation. These types of experiments on many species have shown that animals can orient by means of a time-compensated solar compass.

Many animals are normally active at night; in addition, many day-active species of birds migrate at

night and cannot use the sun to determine direction. Two sources of information about direction are available from the stars. The positions of constellations change because the Earth is rotating. With a star map and a clock, direction can be determined from any constellation. One point in the sky, however, does not change position during the night: the point directly over the axis on which the Earth turns. In the northern hemisphere, a star called Polaris or the North Star lies in that position and always indicates north. Stephen Emlen at Cornell University investigated whether birds use these sources of directional information from the stars. Emlen raised young birds in a planetarium, where star patterns are projected on the ceiling of a large, domed room. The star patterns in the planetarium could be slowly rotated to simulate the rotation of Earth. When the star patterns were not rotated, birds caught in the wild could still orient perfectly well in the planetarium, but birds raised in the planetarium under a nonmoving sky could not. If the star patterns in the planetarium were rotated each night as the young birds matured, they were able to orient in the planetarium, showing that birds can learn to use star patterns for orientation if the sky rotates. No evidence was found, however, that the birds used their circadian clocks to derive directional information from the star patterns. Experienced birds were not confused by a still sky or a sky that rotated faster than normal. The birds were orienting to the fixed point in the sky, the North Star. Young birds raised under a sky that rotated around a different star imprinted on that star and oriented to it as if it were the North Star (Figure 45.24). These studies showed that birds raised in the northern hemisphere learn a star map that they can use for orientation at night by imprinting on the fixed point in the sky.

Animals cannot use sun and star compasses when the sky is overcast, yet they still home and migrate under such conditions. Are there other sources of information they can use for orientation? There appears to be considerable redundancy in animals' abilities to sense direction. Pigeons home perfectly well on overcast days, but this ability is severely impaired if small magnets are attached to their heads. These experiments and subsequent ones with more-sophisticated ways of disrupting the magnetic field around the bird have demonstrated a magnetic sense. Cells have been found that contain small particles of the magnetic mineral magnetite, but the neurophysiology of the magnetic sense is largely unknown. Another cue is the plane of polarization of light, which can give directional information even under heavy cloud cover. Very low frequencies of sound can give information about coastlines and mountain chains. Weather patterns can also provide considerable directional information.

(a) Pigeon placed in a circular cage from which it can see the sky (but not the horizon) can be trained to seek food in one direction, even when cage is rotated between trials

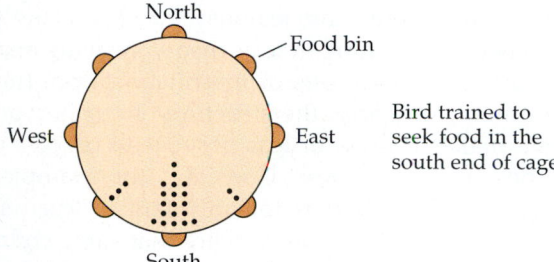

(b) Pigeon placed on altered light–dark cycle and its circadian rhythm phase-advanced by 6 hours. Bird is then returned to training cage under natural sky

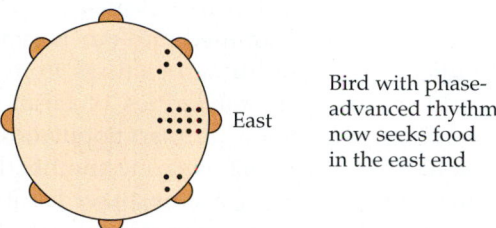

45.23 The Time-Compensated Solar Compass
(a) In the circular cage experiments, pigeons were trained to search for food in the south by filling only the southernmost food bin. Each dot represents a peck in search of food. (b) Birds whose circadian rhythms were phase-shifted forward by 6 hours oriented as though the dawn sun was at its noon position, searching for the south food bin in the east.

45.24 Star Patterns Can Be Altered in a Planetarium
This scientist has placed birds in orientation cages (see Figure 45.22) in a planetarium. By changing the positions or movements of the stars projected on the planetarium ceiling, he can investigate what information the birds use to orient their migratory restlessness.

Much less is known about the mechanisms of bicoordinate navigation than about distance and direction navigation, but some animals have the ability both to sense their geographical positions and to know where they should go. In other words, they have a map sense. Distance and direction capabilities do little good without a map. Information about longitude and latitude is available from natural cues, but the evidence that animals can or do use those sources of information is meager. Longitude can be determined by position of the sun and time of day: If the sun comes up earlier than expected, the animal must be east of home, and if the sun comes up later than expected, then it is west of home. Time and sun position also give information about latitude. At a given time of day in the northern hemisphere, a sun position higher in the sky than expected indicates that an animal is south of home, and if it is lower in the sky than expected the animal is north of home. Other information about longitude and latitude can come from sensing Earth's magnetic lines of force and from the positions of the stars. To pinpoint home using these sources of information, an animal would have to have extremely precise and accurate sensory capabilities that have yet to be demonstrated. Perhaps, however, pinpointing home is not required of an animal's bicoordinate navigational abilities. If an animal gets anywhere near home, piloting can take over, and piloting can use a variety of long-distance cues—such as coastlines, smells, and low-frequency sounds—that have greater ranges than specific visual landmarks have.

HUMAN BEHAVIOR

The behavior of an animal is a mixture of components that are genetically programmed and components that can be molded by learning. Even some aspects of learned behavior patterns, however, may have genetic determinants in terms of what can be learned and when it can be learned. Thus natural selection shapes not only the physiology and morphology of a species, but also its behavior. In some situations natural selection favors fixed action patterns; in others learned behavior is favored. In many cases a mixture of fixed and learned behavioral components is the optimal adaptation. Given these considerations, how would we characterize human behavior?

An important characteristic of human behavior is the extent to which it can be modified by experience. Transmission of learned behavior from generation to generation is culture, the hallmark of humans. Nevertheless, the structure and many functions of our brain are coded in our genome, including drives, limits to and propensities for learning, and even some motor patterns. Biological drives such as hunger, thirst, sexual desire, and sleepiness are inherent to our nervous systems. Is it reasonable, therefore, to expect that emotions such as anger, aggression, fear, love, hate, and jealousy are solely the consequences of learning? Our sensory systems enable us to use certain subsets of information from the environment; similarly, the structure of our nervous system makes it more or less possible to process certain types of information. Consider, for example, how basic and simple it is for an infant to learn spoken language, yet how many years that same child must struggle to master reading and writing. Verbal communication is deeply rooted in our evolutionary past, whereas reading and writing are relatively recent products of human culture.

Finally, the evidence indicates that some motor patterns are programmed into our nervous systems. Studies of diverse human cultures from around the world reveal basic similarities of facial expressions and body language in human populations that have had little or no contact with one another. Infants born blind smile, frown, and show other facial expressions at appropriate times, even though they have never observed such expressions in others. Acknowledging that our behavior has been shaped through evolution in no way detracts from the value we place on the learning abilities of humans. The genetic determination of human behavior is in terms of its broad outline rather than its fine detail.

SUMMARY of Main Ideas about Animal Behavior

Numerous behavior patterns of many species are genetically determined and expressed without prior experience; they are called fixed action patterns.

Deprivation experiments demonstrate whether a given behavior is a fixed action pattern.

An animal must be in the appropriate stage of development or motivation, and required stimuli or releasers must be present for a fixed action pattern to be expressed.

A releaser is a stimulus that elicits a fixed action pattern; since the effective features of a releaser are simple, supernormal releasers are possible.
Review Figure 45.3

Fixed action patterns are adaptive in situations where there are no opportunities to learn, where it is possible to learn the wrong behavior, and where mistakes are costly and dangerous.

Learned behavior, such as bird song, may be shaped by natural selection in terms of what can be learned and when it can be learned.
Review Figure 45.5

Hormones play a part in controlling behavior.

Exposure to sex steroids during development can determine whether male or female fixed action patterns develop for sexual behavior, and sex steroid levels in the adult control the expression of those fixed action patterns.
Review Figure 45.6

Genetic experiments demonstrate the heritability of behaviors, the fact that artificial selection can change behaviors, and possible molecular mechanisms connecting genes to behavior.
Review Figures 45.9 and 45.10

Communicative behaviors convey information to other individuals, who then alter their behavior.

All sensory systems have been exploited by selection for purposes of communication.
Review Figures 45.11 and 45.13

Endogenous circadian rhythms drive daily patterns of behavior.

Although the free-running period of a circadian rhythm is not precisely 24 hours, the rhythm can be entrained to a 24-hour cycle.
Review Figure 45.16

Circadian clock mechanisms have been localized to specific brain structures.
Review Figure 45.17

Long-distance movements of animals require abilities to navigate. Mechanisms of navigation used by animals include piloting, distance and direction navigation, and bicoordinate navigation.
Review Figure 45.21

Time-compensated solar compasses and star maps are used for directional information.
Review Figures 45.23 and 45.24

Most human behavior is due to or influenced strongly by learning, but behavioral drives, emotions, some motor patterns, and propensities and abilities to learn certain types of information may be genetically determined.

SELF-QUIZ

1. The building of a web by a spider is an example of
   a. a fixed action pattern.
   b. a releaser.
   c. displacement behavior.
   d. imprinting.
   e. a learned behavior.

2. If you do not see courtship behavior in a deprivation experiment, you can conclude that
   a. the animal is not sexually mature.
   b. the animal has low sexual drive.
   c. it is the wrong time of year.
   d. the appropriate releaser is not present.
   e. None of the above

3. Which of the following statements about releasers is true?
   a. The appropriate releaser always triggers a fixed action pattern.
   b. A releaser is a simple subset of sensory cues available to the animal.
   c. Releasers are learned through imprinting.
   d. A releaser triggers a learned behavior pattern.
   e. An animal responds to a releaser only when it is sexually mature.

4. Which of the following statements about the genetics of behavior is true?
   a. About 20 genes control the courtship displays of male dabbling ducks.
   b. One gene can code for several chemical signals involved in controlling a behavior.
   c. Genes for retrieving, pointing, and herding have been described in dogs.
   d. A single gene causes lovebirds to carry nesting material tucked in their tail feathers.
   e. Hygienic behavior in bees has been shown to be due to two dominant genes.

5. A display or signal is a behavior that
   a. has evolved to influence the behavior of other individuals.
   b. stimulates one or more types of sensors.
   c. stimulates the endocrine or reproductive systems of other individuals.
   d. began as an intention movement.
   e. began as a displacement behavior.

6. If the sun were to come up earlier than expected on the basis of a circadian rhythm
   a. it could cause symptoms of jet lag.
   b. it could phase-advance the circadian rhythm.
   c. the animal could be east of home.
   d. it could entrain the circadian rhythm.
   e. All of the above

7. To have the ability to pilot, an animal must
   a. have a time-compensated solar compass.
   b. orient to a fixed point in the night sky.
   c. be able to know the distance between two points.
   d. know landmarks.
   e. know its longitude and latitude.

8. Birds that migrate at night
   a. inherit a star map.
   b. determine direction by knowing the time and the position in the sky of a star constellation.
   c. orient to the fixed point in the sky.
   d. imprint on one or more key constellations.
   e. determine distance, but not direction, from the stars.

9. The most likely explanation for the observation that humans from entirely different societies smile when they greet a friend is that
   a. they share a common culture.
   b. they have imprinted on smiling faces when they were infants.
   c. they have learned that smiling does not stimulate aggression.
   d. smiling is a fixed action pattern.
   e. smiling is a behavior that has spread around the world.

10. If (1) a bird is trained to seek food on the western side of a cage open to the sky, (2) the bird's circadian rhythm is then phase-delayed by 6 hours, and (3) after phase-shifting the bird is returned to the open cage at noon real time, it seek food in the
    a. north.
    b. south.
    c. east.
    d. west.

FOR STUDY

1. Critique this statement: Hygienic behavior of bees is controlled by two genes as demonstrated by hybridization and backcrossing experiments.

2. Photoperiod (day length) can provide information about season (time of year), so why do some birds have circannual rhythms?

3. If you raised a songbird in a deprivation experiment and it did not sing the song of its species the following fall, what possible hypotheses could you formulate about this result and how could you test them?

4. Male dogs lift a hind leg when they urinate; female dogs squat. If a male puppy receives an injection of estrogen when it is a newborn, it will never lift its leg to urinate for the rest of its life; it will squat. How might this result be explained?

5. Pick an animal (other than a human) that you think would have mostly learned behavior and another animal that you think would have mostly genetically determined behavior. What differences in their biological characteristics could account for the differences in their behavioral repertoires?

READINGS

Alcock, J. 1993. *Animal Behavior*, 5th Edition. Sinauer Associates, Sunderland, MA. A balanced textbook, recommended to readers searching for a good next step into the subject.

Emlen, S. 1975. "The Stellar-Orientation System of a Migratory Bird." *Scientific American*, August. Experiments on stellar-orientation mechanisms of birds done in a planetarium.

Gould, J. L. and P. Marler. 1987. "Learning by Instinct." *Scientific American*, January. The interactions of learning and instinct, focusing on bees and on bird song.

Gwinner, P. 1986. "Internal Rhythms in Bird Migration." *Scientific American*, April. Circannual rhythms play critical roles in long-distance migration.

Kirchner, W. H. and W. F. Towne. 1994. "The Sensory Basis of the Honeybee's Dance Language." *Scientific American*, June. New experiments test what components of the honeybee's dance communicate information.

Lorenz, K. 1958. "The Evolution of Behavior." *Scientific American*, December. An essay on the evolution of releasers.

Scheller, R. H. and R. Axel. 1984. "How Genes Control an Innate Behavior." *Scientific American*, March. One gene codes for a number of neural signals.

Tinbergen, N. 1960. *The Herring Gull's World*. Doubleday, Garden City, NJ. A delightful account of the behavior of one species from the pen of one of the founders of modern ethology.

Tinbergen, N. 1952. "The Curious Behavior of the Stickleback." *Scientific American*, December. A classic study of releasers and fixed action patterns.

PART SEVEN

Ecology and Biogeography

Adult male elephant seals are huge—several times larger than adult females. Males have large canine teeth with which they fight and thick skin that serves as a shield. Their odd elephantlike snouts are resonating chambers that help exaggerate the roars that accompany their threat displays. A male elephant seal defends a small area of beach by threatening other males and fighting with challengers. Because larger males generally defeat smaller ones, natural selection has favored large size in males. Females congregate on a small number of good-quality beaches, where they give birth to their pups and again become pregnant. Males that control good pupping beaches achieve most of the matings. Less than 10 percent of male elephant seals ever mate with a female. Of those that do, some mate with more than 100 females during their lifetime. By contrast, nearly all females that survive to adulthood breed, but a female rarely weans as many as ten pups during her lifetime.

The life of an elephant seal can be viewed as a sequence of responses to the physical environment, to members of other species, and to members of its own species. Many things can influence these responses. A male elephant seal can reproduce only if he survives to reproductive age, finds a pupping beach, and defends a piece of it from other males. If he cannot defend an area on the beach, he may patrol the water off the beach, attempting to intercept and mate with females coming ashore. A female also must survive to reproductive age and migrate to a pupping beach. She accepts as her mating partner the male that controls the place on the beach where she gives birth to her pup. What makes male and female elephant seals behave the way they do?

The primary goal of behavioral ecology, the subject of this chapter, is to investigate the ultimate causes of behavior in terms of the selective pressures that shape its evolution. Behavioral ecology is part of the broader science of ecology, the subject of Part Seven. **Ecology** is the branch of biology that investigates the interactions of organisms with one another and with their environments.

The term *environment*, as used by ecologists, includes physical factors, such as water, nutrients, light, temperature, and wind. It also includes many biological factors, such as all the organisms that influence the lives of individuals. Because different species are adapted for life in different environments, their interactions with the physical and biological environments also differ. An environmental factor that exerts a strong influence on individuals of one species may have no influence on those of another species.

Interactions between organisms and their environments are two-way processes: Organisms both influence and are influenced by their environments. Man-

A Male Elephant Seal Displays His Strength

46
Behavioral Ecology

aging the environmental changes caused by our own species is one of the major problems of the modern world. For this reason, ecologists are often asked to help analyze causes of environmental problems and to assist in finding solutions for them. But don't confuse the science of ecology with the term "ecology" as it is often used in popular writing to refer to the functioning of nature. Pollution does not destroy the ecology of an area, but it may well modify, damage, or destroy important ecological processes.

Ecologists study patterns of distribution and abundance of organisms to determine how those patterns are established and maintained, and how they change over short and long time periods. From its roots in descriptive natural history, ecology has developed into a complex field of inquiry dealing with levels of organization ranging from relations of individual organisms with their physical and biological environments to the structure of communities and ecosystems. This complex subject can be approached in many different ways. We touched on some aspects of interactions between individual animals and their physical and biological environments in Part Six. In this chapter we discuss how animals make the decisions that influence their survival and reproductive success. The use of the word *decision* does not imply that the choices animals make between alternative responses are conscious, but that these choices influence the survival and reproductive success of the animals and thus are molded by natural selection. In the following chapter we turn to the structure and dynamics of populations of a single species. In later chapters we add levels of complexity and discuss the functioning of all the species living in a region and their interactions with the physical environment.

COSTS AND BENEFITS OF BEHAVIOR

Fitness is a central concept in the study of evolution (see Chapter 19). The fitness of a genotype or phenotype is its reproductive contribution to subsequent generations *relative* to the contribution of other genotypes or phenotypes. How an individual behaves exerts a major influence on its survival and reproductive success; for example, a male elephant seal that cannot defeat his rivals will never mate. Which behavior patterns evolve depends on how they help an animal compared with other types of behavior in the same circumstances.

Ecologists interested in the evolution of behavior analyze their observations in terms of costs and benefits. Such analyses are based on the principle that an animal has only a limited amount of time and resources to devote to different kinds of acts. A certain behavior may be costly to the animal that performs it, but it may also benefit the animal. The **energetic cost** of a behavior is the difference between the energy the animal would have expended had it rested and the energy expended in performing the behavior. A male elephant seal that rises on his forelimbs, roars, and fights with rivals expends more energy than a resting male. He therefore exhausts his fat reserves faster than if he had not attempted to defend a parcel of beach.

The **risk cost** of a behavior is the increased chance of being injured or killed as a result of performing it, compared with resting. A displaying male elephant seal is more likely to be injured by a rival than a male that avoids fights (Figure 46.1). The **opportunity cost** of a behavior is the sum of the benefits the animal forfeits by not being able to perform other behaviors during the same time interval. A male elephant seal cannot search for food while he defends a section of beach, and the longer he stays the more time he needs to regain his energy reserves once he returns to the sea.

An animal generally does not perform a behavior for which the total costs are greater than the sum of the **benefits**: the improvements in survival and reproductive success that the animal achieves by performing the behavior. Measuring these costs and benefits directly is difficult, but ecologists can find out which costs and benefits are most important for different species by observing how their behavior changes when environmental conditions change. Ecologists believe not that animals consciously calculate costs and benefits, but that through many generations, natural selection molds behavior in accordance with costs and benefits.

To see how scientists can infer the costs and benefits of a behavior, consider how small birds react to foreign eggs in their nests. Normally the eggs in a

46.1 Fighting Is Risky
The male elephant seal on the right has been defeated and wounded; he will probably never sire offspring.

bird's nest were laid by the attending female, but females of brood parasites—species that do not build their own nests or raise their own offspring—lay their eggs in another bird's nest. In North America, female brown-headed cowbirds lay their eggs in the nests of many species of small birds. Individuals of species that are too small to grasp and remove the cowbird egg have only two choices. They can either abandon the nest when a parasitic egg appears in it or continue to incubate. The cost of abandonment includes building a new nest and laying a new clutch of eggs (which may also be parasitized). Accepting the egg also entails a cost: In nests with young cowbirds, fewer of the host offspring survive than in unparasitized nests because the young cowbird takes food that would otherwise be eaten by the host nestlings. A female prothonotary warbler is more likely to continue to incubate her eggs after being parasitized by a cowbird if there are few other good nesting cavities on her territory than if there are many alternative nesting sites (Figure 46.2). From this fact we can infer that the cost of abandonment usually is lower than the cost of continuing to incubate only if other good nesting sites are available.

Changes in social behavior may also reveal components of costs and benefits. Small birds protect themselves from predators by forming flocks, but individuals in flocks interfere with one another's foraging. Individuals of some species change their flocking behavior when the weather or the risk of predation changes. For example, yellow-eyed juncos, small seed-eating finches, fight with one another over food. Dominant individuals usually defend an area from which they attempt to exclude all other juncos. Juncos also spend time watching for predators, principally by looking up and scanning. When they are fighting or scanning, they cannot look for food. When they are feeding or fighting, they are vulnerable to surprise attack by a hawk. The best way to avoid predators is to watch all the time, but that approach would lead to starvation. The risk of surprise attack decreases as the number of juncos in a flock increases because the chance that at least one bird is scanning at any moment increases, even if each individual spends only a small amount of time scanning. The larger the flock, the less each bird looks up, and the more time each bird spends feeding. The larger the flock, however, the more time each bird also spends fighting.

Juncos need to eat more on cold days than they do on warm days, so they have less time to defend space and scan for predators. Attesting to this fact is the observation that on cold days, junco flocks wintering in Arizona contained an average of 7 birds, whereas on warm days they contained an average of only 2 birds. When a behavioral ecologist flew a trained hawk over the canyon in which juncos were

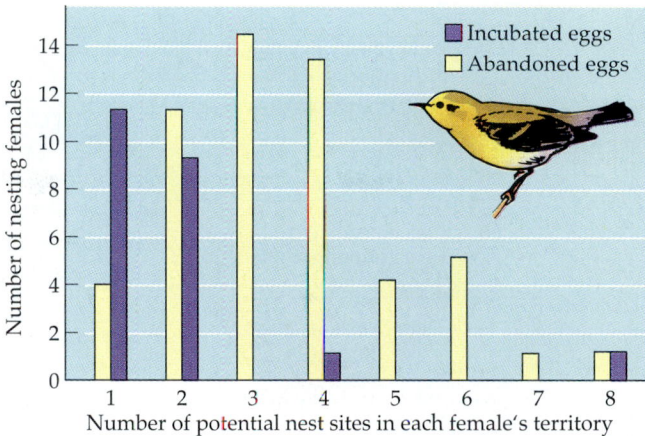

46.2 To Abandon or Not To Abandon
Female prothonotary warblers are likely to abandon a parasitized nest only if other good nest sites are available.

feeding, mean flock size increased from 3.9 birds before the hawk flew by to 7.3 birds afterward. The birds tolerated more interference with foraging when the need for protection from predation was higher.

A cost–benefit approach provides a framework within which behavioral ecologists design experiments and make observations that enable them to understand why behavior patterns evolve in such rich and varied ways. This framework is what stimulated ecologists to measure the number of alternative nest sites in territories of prothonotary warblers and to alter risks of predation to foraging juncos by flying a hawk over the flocks. Even though costs and benefits were not calculated in those studies, observations revealed which ones were important.

DEALING WITH THE NONSOCIAL ENVIRONMENT

The environment in which an organism normally lives is called its **habitat**. Everything that the organism needs must be provided by its habitat. Suitable habitat may be broken into patches, some of which are large enough to support one or more individuals for their lifetimes, or at least for a complete breeding season. In most areas patches of suitable habitat are separated by patches of unsuitable habitat. Selecting a place to live can be viewed as a sequence of decisions, each of which limits the options for successive choices (Figure 46.3). Once a habitat is chosen, foraging options are restricted to patches within the habitat.

The cues organisms use to select suitable habitats are as varied as the organisms themselves. A young red abalone begins its life as an egg that is fertilized in the open ocean. About 14 hours after fertilization the egg hatches, but the larva emerges with enough

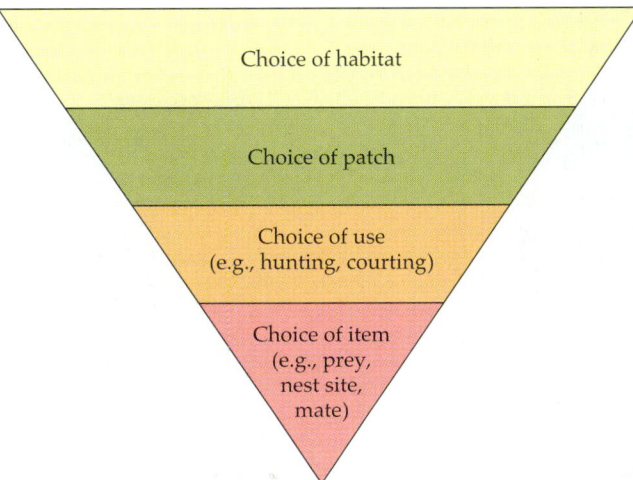

46.3 Hierarchical Choices
Choices of where to live and what behaviors to engage in
there are hierarchical.

yolk to continue developing for another seven days
without eating. At the end of seven days the larva
stops developing, settles to the bottom, and metamorphoses into a young juvenile, after which it can
no longer swim back up into the water. Red abalone
larvae settle only on a substrate made up of coralline
algae, on which they feed (Figure 46.4). They recognize coralline algae by a small, water-soluble peptide containing about ten amino acids that all coral-

line algae produce. In the laboratory, abalone larvae
will settle on any substrate that has this molecule—
but in nature *only* coralline algae have it. This simple
cue ensures that the larvae always settle on a substrate that is suitable for their future development.

Individuals of other species use different cues.
When a young female poplar aphid hatches from an
egg in a crevice in the bark of a cottonwood tree, she
walks out to small branches and selects a leaf. By
injecting a chemical into the leaf she induces formation of a gall, a hollow ball of tissue, in which she
lives and produces offspring from unfertilized eggs.
Females on large leaves produce more and heavier
offspring than do females on small leaves (Figure
46.5). Females prefer to settle on large leaves, which
are soon fully occupied. Females that have already
settled attempt to prevent other females from settling
near them. Late arrivals must settle either on smaller
leaves or farther out on the central rib of an already
occupied leaf. A female can do about as well being a
second female to arrive on a large leaf as she can as
the first female on a smaller leaf (Table 46.1). As we
would expect, aphid females double up only on large
leaves, although such leaves are a small fraction of
the leaves on a cottonwood tree.

Although settling individuals use a variety of cues,
all habitat selection cues have a common feature:
They are good predictors of general conditions suitable for future survival and reproduction. Young red
abalones grow faster on coralline algae than on other
kinds of algae. Poplar aphids use leaf size because it
is a good predictor of reproductive success.

After choosing a habitat, individuals use the resources of the area, such as shelter from the physical
environment, nest sites, and food. Because food is

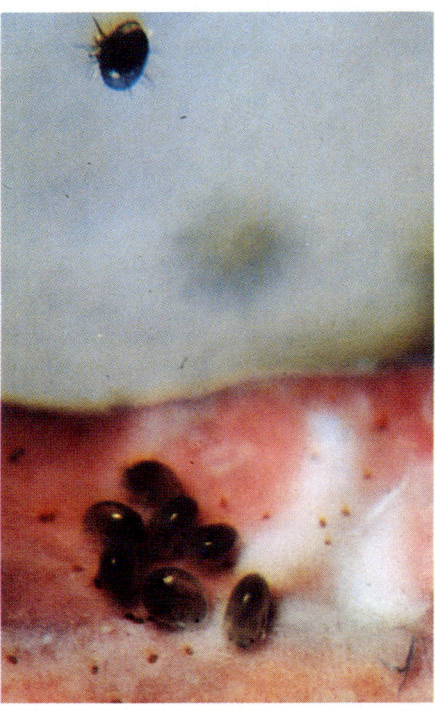

46.4 Algal Chemistry Provides the Cue
These red abalone larvae (dark ovals) will settle only
when they recognize the chemical composition of a
coralline algal substrate.

46.5 Big Leaves Are Better
A female poplar aphid forms a hollow gall at the base of
a large leaf. Within the gall she will produce several hundred offspring.

TABLE 46.1
Effects of Leaf Size and Position of Gall on Reproductive Success of Female Poplar Aphids

NUMBERS OF GALLS/LEAF	MEAN LEAF SIZE (cm)	MEAN NUMBER OF OFFSPRING PRODUCED BY		
		FIRST FEMALE	SECOND FEMALE	THIRD FEMALE
1	10.2	80	—	—
2	12.3	95	74	—
3	14.6	138	75	29

so important, we consider it here in some detail. When an animal is looking for food, how much time should it spend searching before moving to another site? When many different types of prey are available, which ones should a predator take and which ones should it ignore? Foraging theory was developed to help answer this type of question. Consider the second question. When it encounters a potential prey item, a predator either pursues the prey or ignores it. If the prey is pursued and captured, it must be subdued and eaten. Finally, if the predator must pause to digest its prey (a good example is a snake swallowing a large prey), it may not be able to resume foraging immediately. The time and energy required for each of these different activities must be accounted for if we wish to understand why predators forage as they do.

In one study ecologists tried to determine why predators take certain prey while ignoring others that are suitable as food and could be captured. To develop a model of foraging behavior, the ecologists made some assumptions about how natural selection might operate. The ecologists reasoned that if one predator in a population managed to obtain food more efficiently than other members of the population did, it would have more time for other activities, such as reproduction, than the others would. The superior predator might therefore produce more offspring. If there were a genetic basis for the superior foraging ability, the ability could, over evolutionary time, be selected for and become the characteristic foraging behavior of the species.

The ecologists also assumed that a predator chooses prey in a way that maximizes its energy intake rate. They characterized each type of prey by the amount of time it took the predator to pursue, capture, and consume an individual, and by the amount of energy an individual prey contains. The ecologists then ranked the prey according to the energy the predator obtains relative to the amount of time the predator spends capturing the prey. The top-ranked prey type is the one that yields the most energy per time invested.

With this information the ecologists calculated the rate at which the predator would take in energy if it selected prey in different ways. They found that if the top-ranked prey is sufficiently abundant, a predator gains the most energy per unit time spent foraging by taking only that prey type and ignoring all others. As the abundance of the top-ranked prey type decreases, an energy-maximizing predator adds lower-ranked prey items to its diet in order of the energy per unit time that those prey yield. The surprising result is that whether or not a prey item is included in an energy-maximizing predator's diet depends not on the abundance of that prey type, but only on the abundances of other prey types that yield more energy per unit time spent capturing and eating them.

The ecologists tested this foraging model using bluegill sunfish. They performed laboratory experiments with bluegills to measure energy contents of different prey, handling time for prey, energy spent searching for and handling prey, and actual encounter rates with prey under different prey densities. Using these values, the investigators predicted the proportions of large, medium, and small water fleas that bluegills would capture in environments stocked with different densities and proportions of water fleas of those sizes. For example, they predicted that in an environment stocked with a low density and equal proportions of the three sizes of prey, the fish would take equal proportions of all three sizes. To test their predictions, the investigators put the bluegills in three different environments and observed the proportions of the water fleas of different sizes that the fish actually captured (Figure 46.6a).

The proportions of large, medium, and small water fleas taken by the fish were very close to those predicted by the model. The major difference between predicted and observed results was that the fish ate some small prey that the model predicted they should have ignored. The most likely reason is that the fish could not estimate accurately the true sizes of some prey. Determining sizes of objects is particularly difficult for a predator in open water, where few clues about the distance to a prey are available. A smaller prey close to the predator and a

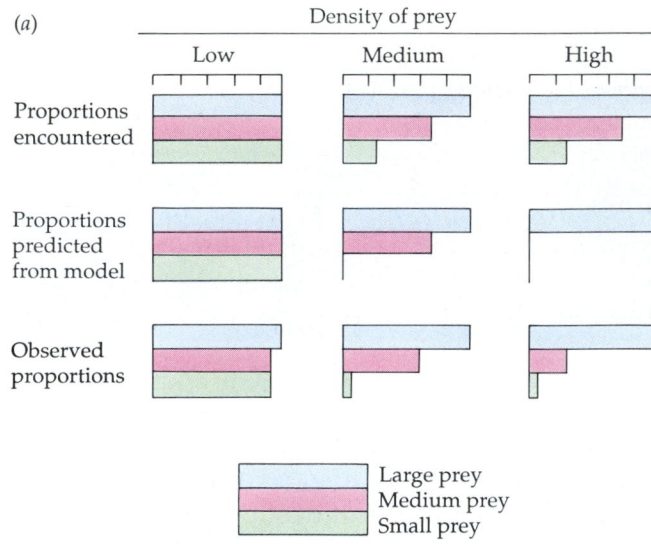

(a)

Density of prey

	Low	Medium	High
Proportions encountered			
Proportions predicted from model			
Observed proportions			

Large prey
Medium prey
Small prey

(b)

Smaller, closer prey
appear to be as
large as larger,
farther prey

46.6 Bluegills Are Energy Maximizers
(a) The widths of the bars showing observed proportions in the diet are very similar to those showing predicted proportions, supporting the hypothesis that bluegills select prey to maximize energy intake. (b) How an energy maximizer can make a mistake.

46.7 Group Hunting Improves Foraging Efficiency
(a) Wild dogs in Africa hunt cooperatively to capture prey that is much larger than any single dog could bring down. Here they attack a wildebeest. (b) White pelicans forage as a group, forming a line to drive prey into a vul- nerable area. (c) Having killed their prey, a band of *Homo habilis* drives rival predators—spotted hyenas and saber- tooth cats—from a fallen dinothere (an extinct relative of modern elephants).

(a)

(b)

(c)

larger prey farther away make images of the same size on the predator's retina (Figure 46.6b). Fish may thus attack and eat small prey that are very close to them when first seen. This is what the bluegills did.

DEALING WITH INDIVIDUALS OF THE SAME SPECIES

Individuals of sexually reproducing species must mate in order to produce offspring. Associations for reproduction may consist of little more than a coming together of eggs and sperm, but individuals of many species associate for longer times to provide care for offspring. Associating with individuals of one's own species may also improve survival for reasons unrelated to reproduction.

Social behavior evolves when the cooperation of individuals of the same species results in a higher rate of survival and more offspring than are possible for solitary individuals. Determining the effects of group living on survival and reproductive success is not easy, however. A given behavior pattern may be advantageous for individuals of one species but disadvantageous for those of another species, or even for the same individuals at a different time or place.

Benefits of Social Life

Social life may improve hunting success or expand the range of prey that can be captured. For example, by hunting together, animals of some species, such as African hunting dogs (Figure 46.7a), are able to capture prey too large for any one of them to subdue alone. White pelicans cooperate to maneuver prey into places where they are easier to catch (Figure 46.7b). Cooperative hunting was a key component of the evolution of human sociality. By hunting in groups, our ancestors were able to kill large mammals they could not have subdued as individual hunters (Figure 46.7c). These social humans could also defend their prey and themselves from other carnivores.

Individuals of many species are better protected from predators if they live in groups. Predators may be able to find a group of animals more easily than they can a solitary animal, but a group may defend itself better when it is found. When attacked by wolves, musk oxen form a circle, with the young animals inside (Figure 46.8). The wolves have great difficulty penetrating the formidable barrier of the large heads and massive horns of the adult animals. Musk ox group size is larger in areas where wolves are abundant than where wolves are scarce. The cost of grouping is that feeding efficiency is poorer in larger groups; the benefit of grouping is that predator defense is better.

Many small birds form tight flocks when they are attacked by a hawk. Clumping deters a hawk, because the predator risks injury if it hits one of the prey with its wing when dashing into a compact group. Single individuals are also difficult to follow in a rapidly moving group. To investigate the importance of flocking as an antipredator adaptation, scientists in England released a trained goshawk near flocks of wood pigeons and measured the percentage of attacks that were successful. They found that the hawk was most successful when it attacked solitary pigeons; success in capturing a pigeon decreased as the number of pigeons in the flock increased (Figure 46.9). As the figure shows, the main reason was that larger flocks reacted to the hawk's approach while it was farther away.

46.8 Defensive Postures of Musk Oxen
This compact circle of formidable adults encloses young oxen.

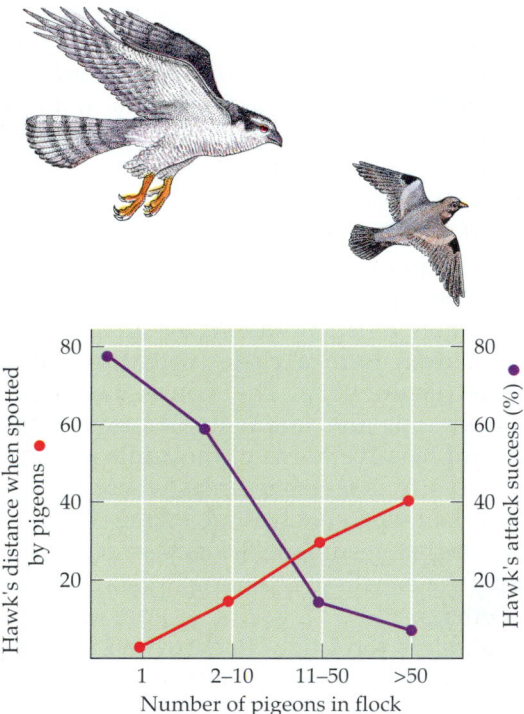

46.9 Flocking Gives Protection from Predators
Goshawks are less successful when they attack wood pigeons in flocks than when they attack solitary pigeons because large flocks react sooner to the hawk's approach.

Costs of Group Living

An almost universal cost associated with group living is higher exposure to disease and parasites. Long before the causes of diseases were known, people sensed that association with sick persons increased the chances of contracting the illness. Quarantine has been employed as a means of combating the spread of illness for as long as we have had written records. The diseases of wild animals are not well known, but most of those that have been studied are also spread by close contact.

Like the benefits, the costs of group living depend on circumstances. Individuals in groups may compete for food, interfere with one another's foraging, injure one another's offspring, or inhibit one another's reproduction. The effects of group living on survival and reproductive success of an individual also depend on its age, sex, size, and physical condition. Individuals may be larger or smaller than the average for their age and sex. Variation in skills, competitive abilities, and attractiveness to potential mates is often associated with these size differences.

The largest males of *Centris pallida*, a solitary bee of the American southwest, are three times the size of the smallest males (Figure 46.10). These size differences, although environmentally determined, are fixed for an individual's life. Large males search for females about to emerge from their buried pupae.

When they detect a female, they dig her up and copulate with her, but the digging takes several minutes and often attracts other males. These new arrivals fight with the original male, and the largest male usually wins. If a small male searched and dug for buried females, he would probably just serve as a female finder for larger males. Instead, smaller males patrol potential pupation sites and wait for females that emerge without being discovered by large males. Intermediate-sized males sometimes dig and sometimes patrol.

A large, dominant male that controls access to all females greatly increases his reproductive success by living in a group; by contrast a small, subordinate male that achieves no copulations may survive better by living in the group, but he has no reproductive success. An infant that would quickly die if it were separated from its mother benefits differently from group living than a subadult capable of surviving on its own. *Social groups are organized collections of individuals whose survival and reproductive success are influenced differently by group living.*

TYPES OF SOCIAL ACTS

Individuals living in social groups perform many different types of acts. These acts can be grouped into four categories according to their effects on individuals (Table 46.2). An **altruistic act** benefits another individual at a cost to the performer. A **selfish act** benefits the performer and inflicts a cost on another individual. A **cooperative act** benefits both the performer and the recipient. A **spiteful act** inflicts costs on both. These terms are purely descriptive; they do not imply conscious motivation or awareness on the part of the animal.

46.10 Size Variation in Male Bees
The large male on the right is digging a female out of the soil while the small male on the left waits for a female to emerge.

TABLE 46.2
Types of Social Acts

	Act benefits the recipient +	Act costs the recipient −
Act benefits the performer +	Cooperative	Selfish
Act costs the performer −	Altruistic	Spiteful

Many current studies of the social behavior of animals attempt to measure the relative costs and benefits of social acts, how the effects of the acts are distributed among the individuals of the group, and how individuals are related genetically. If a genetic basis for a cooperative or selfish act exists, and if performing it increases the fitness of the performer, then the genes governing the act will increase in frequency in the lineage. In other words, cooperative or selfish behavior will evolve.

Altruism has long been the subject of a lively debate among biologists interested in animal behavior, and understanding how altruistic behavior patterns could have evolved is not easy. Charles Darwin was puzzled by reproductive altruism in social insects, discussed later in this chapter. How can an act that *lowers* the performer's chances for survival or for passing on its own genes evolve into a behavior pattern? The key lies in genetic relatedness: Altruistic behaviors may evolve when performers and recipients are genetically related. As we learned in Chapter 19, an individual may influence its fitness in two different ways. First, it may produce its own offspring, contributing to its **individual fitness**. Second, it may help the survival of relatives that bear the same alleles because they are descended from a common ancestor. This process is called **kin selection**. Together, individual fitness and kin selection determine the **inclusive fitness** of the individual. Altruistic acts eventually may evolve into altruistic behavior patterns. When the benefits of increasing the reproductive success of related individuals exceed the costs of decreasing the altruist's own reproductive success, then the altruist's inclusive fitness is enhanced. Box

46.A shows a way to calculate the relatedness of individuals.

Many social groups consist of some individuals that are close relatives and others that are unrelated or are distantly related. Individuals of some of these species recognize their relatives and adjust their behavior accordingly. White-fronted bee-eaters are colonial, cooperatively breeding African birds in which most pairs are assisted by nonbreeding adults who help incubate eggs and feed nestlings (Figure 46.11). Both males and females help, and individuals whose nests fail may help at other nests later in the same breeding season. However, about half the nonbreeding individuals who could help do not do so. Nearly all helper individuals assist close relatives. When helpers have a choice of two nests at which to help, about 95 percent of the time they choose the nest with the young most closely related to them. Helping among white-throated bee-eaters is altruistic because individuals that help are not more successful when they become breeders than those who do not help, nor do they appear to gain any other advantage from helping. Extending help to nonrelatives does not improve fitness; observations show that white-fronted bee-eaters rarely help nonrelatives.

46.11 White-Fronted Bee-Eaters Help Close Relatives
These bee-eaters are perched at and near the entrances to their nests in a river bank. The colored tags on the birds enable investigators to identify individuals.

Calculating the Coefficient of Relatedness

The **coefficient of relatedness, r,** is the probability that an allele in one individual is an identical copy, by descent, of an allele in another individ-ual. To calculate r we construct a diagram showing the individuals and their common ancestors, linked across generations by arrows. Because meiosis takes place at each generation link, the probability that a copy of an allele gets passed on is 0.5. For k generation links, the probability is $(0.5)^k$. To calculate r we sum this value for all possible pathways between the two individuals; that is, $r = \Sigma(0.5)^k$.

Some examples are diagrammed here. Values of r are calculated between two individuals, represented by solid circles. Other relatives are indicated by open circles. The generation links in the calculations are represented by solid lines. The dashed lines represent other links in the pedigree. Full siblings and cousins can inherit identical genes by two pathways, from mother (blue) or father (red).

Parent and offspring

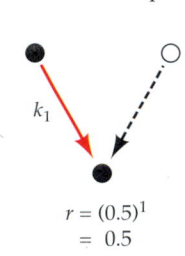

$r = (0.5)^1$
$= 0.5$

Grandparent and grandchild

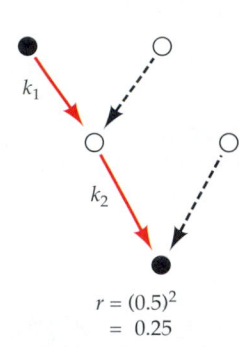

$r = (0.5)^2$
$= 0.25$

Sibling

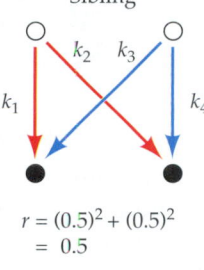

$r = (0.5)^2 + (0.5)^2$
$= 0.5$

Full cousins

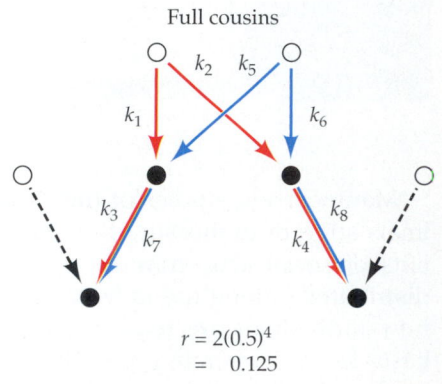

$r = 2(0.5)^4$
$= 0.125$

Cooperative behavior can evolve relatively easily among related individuals. More difficult to understand are danger warnings, food sharing, and grooming behavior between unrelated individuals of the same species or between members of different species. How can we explain the evolution of such cooperative behavior? The proposed answer is a model called **reciprocal altruism**. According to the model, acts of reciprocal altruism evolve if the performer is in turn the recipient of beneficial acts from the individuals it has helped. If there is a genetic basis for the acts, natural selection may increase the frequency of alleles governing the cooperative behavior.

CHOOSING MATING PARTNERS

Although social behavior is complex, it involves only a small set of choices. Individuals choose their associates, how to interact with them, and when to leave them. Variations in how those decisions are made can result in an amazing variety of social systems. The most widespread choice of associates, made by most sexually reproducing species, is choosing a mate. The most basic mating decision is choosing a partner of the correct species. Once the correct species has been determined, additional decisions can be based on the qualities of a potential mate, on the resources it controls, or on a combination of the two. Among species in which individuals do not control resources, traits of the partner are the only criteria for mate selection.

The reproductive behavior of males and females is often very different. Males usually initiate courtship and often fight for possession of females. Females seldom fight over males and often reject courting males. Why do males and females approach courtship and copulation so differently? The answer lies in the large difference in size between sperm and eggs (see Figure 37.2). Because sperm are so small, one male produces enough to sire a very large number of offspring—usually many more than the number of eggs a female can produce or the number of young she can nourish. Therefore, males of most species can increase their reproductive success by mating with many females. Eggs, on the other hand, are typically very large relative to sperm. Many bird eggs weigh 15 to 20 percent of the female's body weight. It would take a huge number of sperm to account for an equal percentage of a male's body weight. Consequently, a female is unlikely to increase her reproductive output by increasing the number of males with which she mates. The reproductive success of a female depends primarily upon the resources her mate controls, the amount of assistance he provides in the care of her offspring, and the quality of the genes she receives from him.

Male Mating Tactics

Males employ a variety of tactics to induce females to copulate with them. Generally, courtship behavior signals in some way that the male is in good health, that he is a good provider of parental care, or that he has a good genotype. Males of some species of hanging flies defend dead insects upon which receptive females feed. A female hanging fly will mate with a male only if he provides her with a morsel of food. The bigger the food item, the longer she copulates with him and the more of her eggs he fertilizes (Figure 46.12). A female gains from this behavior because she obtains a better supply of energy for egg production. Also, because males fight for possession of dead insects, a male with a large insect is likely to be a good fighter and in good health.

Males of the satin bowerbird of Australia build and decorate elaborate structures called bowers that are used only for courting and mating with females (Figure 46.13). Males do not associate with females after mating, and they provide no parental care. Adjacent bowers are often close enough for males to steal materials from one another's bowers. The ability of a male to steal materials from his neighbors is posi-

46.13 Satin Bowerbird at His Bower
The male satin bowerbird in this photo prefers blue objects, some of which he has stolen from neighbors.

tively correlated with his dominance at feeding sites. Therefore, females can use the number of objects a male has accumulated at his bower as an indication of his competitive ability.

Male satin bowerbirds prefer blue objects. Because blue flowers are rare in Australian rainforests, the number of blue objects in a male's bower is a better indicator of his raiding ability than is the number of nonblue objects. The more objects in his bower, the more matings a male obtains. Females that mate with males who have elaborately decorated bowers produce sons who will, in turn, build and control attractive bowers. Female satin bowerbirds are also able to assess the number of parasites a male carries and avoid mating with more heavily parasitized males.

Whether a male fertilizes the eggs of a female with whom he has copulated depends on when they copulate and whether she copulates with other males. Males have evolved behavior patterns that increase the probability that it is *their* sperm that fertilize the eggs. The simplest method is to remain with the female and prevent other males from copulating, but this method has high opportunity costs because a male cannot do anything else while he is guarding a female.

Males of many species have evolved mechanisms that are more elaborate but take less time. A male black-winged damselfly who grabs a receptive female uses his penis to scrub out sperm deposited by other males in the female's sperm storage chamber. Inves-

46.12 A Male Wins His Mate by Demonstrating Foraging Prowess
The male hanging fly on the left has just presented a moth to his mate, thus demonstrating his foraging skills. She feeds on the moth while they copulate.

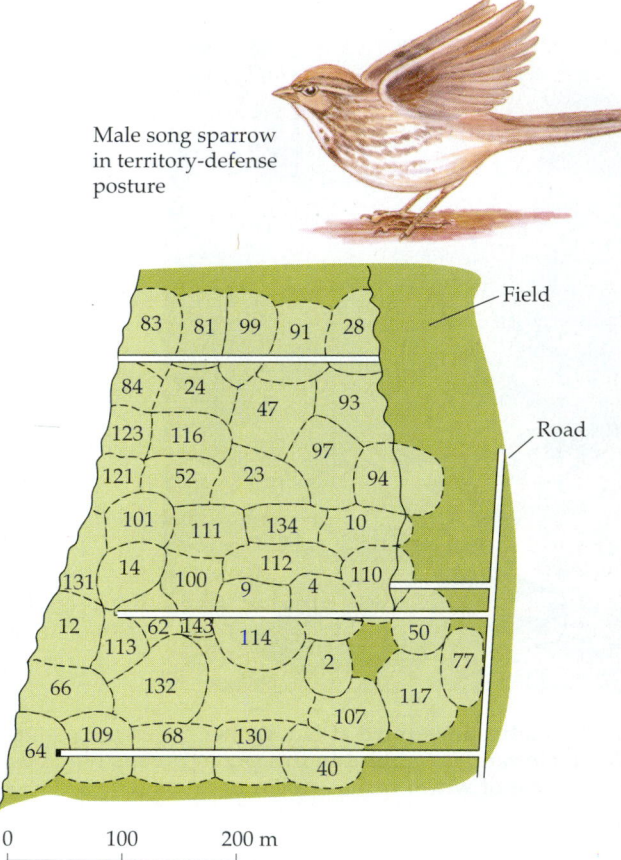

Male song sparrow in territory-defense posture

46.14   Type A Territories Provide Everything
Even the smallest of the territories staked out by male song sparrows in an Ohio study provides all the resources a pair needs to rear its offspring. Each number identifies an individual male bird.

ing young. Many aquatic birds, such as gulls, penguins, and cormorants, breed in dense colonies within which individuals defend small areas around their nests (Figure 46.16). The colonies are usually located on islands where the nests are safe from terrestrial predators from the mainland, particularly mammals.

(a)

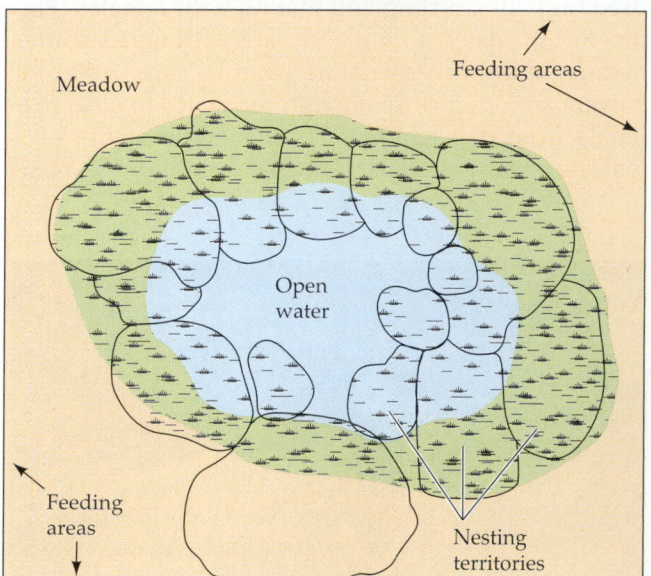

(b)

46.15   Type B Territories Provide Nesting Sites and Some Food
(a) A male red-winged blackbird defends his territory. This display, which is accompanied by a song, warns other males to stay away from the territory but attracts females searching for breeding sites. (b) Each enclosed area is a territory of a male red-winged blackbird. Because territories on the edge of the marsh are less contested than those toward the center, they are larger. Feeding areas extend outside the territory boundaries.

tigators have found that a copulating male removes between 90 and 100 percent of competing sperm before he inserts his own sperm into the chamber (see Figure 37.6). Males of many insects deposit a plug that effectively seals the opening to the female's genital chamber and prevents other sperm from entering.

Like the elephant seals we saw at the beginning of this chapter, males of some species defend **territories** that contain food, nesting sites, or other resources. Behavioral ecologists recognize four types of territories, labeled types A, B, C, and D. Type A territories are all-purpose; they provide mating sites, nesting sites, and the food necessary to rear offspring (Figure 46.14). Type B territories include a large breeding and nesting area but do not supply most of the food necessary to rear young. Male red-winged blackbirds defend type B territories in emergent vegetation in marshes (Figure 46.15). The territories provide nesting sites over water that are protected from some terrestrial predators, but both males and females get most of their food from upland areas near the marshes.

Type C territories are strictly for nesting and rear-

46.16   Type C Territories Can Be Small
The spacing of individual breeding territories of cape gannets is determined by how far an incubating bird can reach to peck its neighbors without getting off its eggs.

Type D territories are small parcels of land used only for courtship display and mating. These display grounds, called **leks**, may be used for many years. Within a lek each male defends a small courting space. Females come to the display grounds to

46.17   Type D Territories Are Display Grounds
This male Uganda kob (right) has attracted several potential mates.

choose a mate. During the breeding season male Uganda kobs gather on small traditional display grounds on the African plains. Males battle intensely for possession of central territories on these display grounds. Females come to the display grounds and they usually mate with males holding central territories (Figure 46.17). The females then leave the area and raise their young with no help from the males.

If a male controls sufficient resources, he may attract more than one female, a phenomenon known as **resource defense polygyny**. The number of females attracted to a red-winged blackbird territory depends very little on the traits of the male holding the area. Rather, females respond to the quality of nest sites and the quantity of food in and close to his territory. The probability that a female returns to a particular territory for the next breeding season is just as high when the male has changed as when her previous mate is still present. A female continues her nesting activities even if the territorial male changes within a breeding season.

Female Mating Tactics

Females can improve their reproductive success if they can assess the genetic quality and health of potential mates, the quantity of parental care they may provide, and the quality of the resources they control. But how can females make such assessments when all males attempt to signal that they are good in all three of these traits? The answer is that by paying particular attention to the traits for which males cannot cheat, females have favored the evolution of "reliable" signals. Possession of a large dead insect signals good fighting ability in a male hanging fly. A male satin bowerbird in possession of a good bower with many blue objects is almost certainly a dominant male.

Females should avoid mating with a sick male because he may transmit parasites directly to his mate and offspring; in addition, a healthy male may have genetic resistance to diseases that can be inherited by his offspring. The role of parasites in mate choice has been studied with barn swallows, a monogamous bird species in which males assist their mates in feeding the offspring. Barn swallow nests are often infested with blood-sucking mites that reduce the weights of nestlings. The lower the weight of a swallow when it leaves its nest, the less likely it is to survive to reproductive age.

Investigators inoculated some barn swallow nests with mites and killed mites in other nests by spraying them with a mite killer. They found that males raised in mite-free nests grew longer tails than those raised in infested nests. In spring, males with longer tails have fewer mites than males with shorter tails, and offspring of heavily parasitized males are more likely

46.18 Longer Tails Mean Healthier Males
Female barn swallows prefer long-tailed males, which have lower parasite loads than males with short tails.

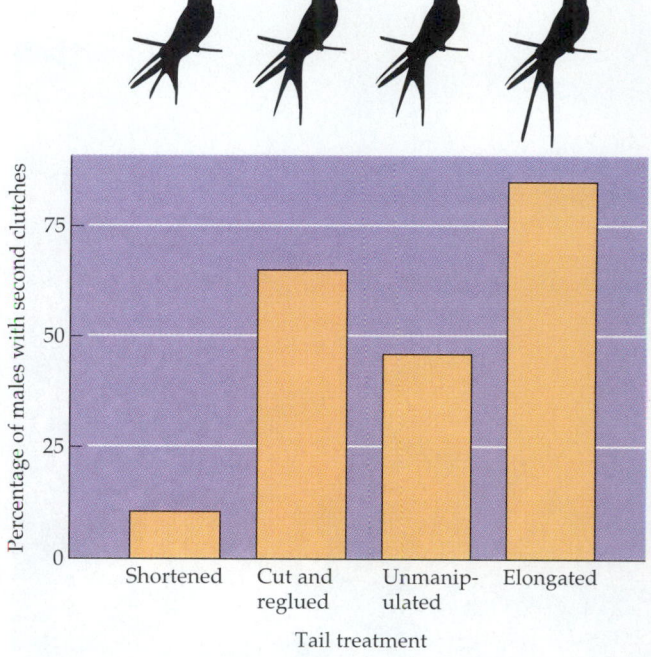

to become infested, even if they are raised in foster nests away from their fathers, suggesting that susceptibility to mites is partly genetically determined. Male barn swallows prominently display their tails to females during courtship flights prior to pair formation. Females prefer males with longer tails (Figure 46.18) and profit from this preference by producing healthier offspring.

A female's choice of a sperm donor does not necessarily end at copulation; she may be able to choose from among the sperm she has received. A female of the moth *Utethesia ornatrix*, for example, usually mates with four or five males. From the spermatophores of these males the female receives nutrition and defensive chemicals derived from the food plant upon which the males fed when they were larvae. Regardless of the order in which a female mates with males, how long she copulates with each one, or the interval between copulations, most of her eggs are fertilized by the largest male with which she copulated. The largest male sired 100 percent of the offspring of 36 of 53 females studied in Florida. This selection appears to be entirely under the control of the female, who is able to block movement of sperm to the storage chamber. Males can do nothing to influence their success because they do not leave plugs after copulating and they are unable to remove rival males' sperm.

Mate Choice by Hermaphrodites

Individuals in some species of annelids, slugs, and fishes are simultaneous hermaphrodites; that is, each one produces both eggs and sperm (see Figure 37.5). During courtship each hermaphrodite produces its own eggs and fertilizes the eggs of its partner. Such a system, however, is vulnerable to individuals that fertilize the eggs of their partners but lay no eggs of their own. Because these individuals can use the energy they save to find and fertilize other hermaphrodites, they are likely to have higher fitness than individuals that lay all their eggs at the end of one courtship bout. How can such "cheating" be prevented?

In one method, called "egg trading," the pair alternates sex roles over the course of mating. Each time an individual acts as a female, it lays only some of its eggs. It does not lay more eggs until its partner has also laid some (Figure 46.19). Egg trading forces each individual to demonstrate its commitment to egg production; neither one can fertilize a large batch of eggs and then leave.

ROLES OF THE SEXES

After copulating, individuals may invest time and energy in caring for offspring. **Parental investment** is an act of parental care that increases the chances of an offspring's survival but reduces the ability of the parent to produce additional offspring. Parental investment may lower the chances of survival of the parent itself because the parent could use this time

46.19 Hermaphrodites Take Turns Laying Eggs
Individual hamlet fish (*Hypoplectrus*) may alternate sexual roles as many as four times during a single mating. In this photo, the individual acting as the male curves its body around the relatively motionless female of the moment.

(a)   (b)

46.20 Some Female Birds Are Brighter Than Males
Female red phalaropes (*Phalaropus fulicarius*) are larger and more brightly colored than males. (*a*) A dull-plumaged male takes a break from incubation in order to feed himself. (*b*) After laying her eggs, the more brightly colored female abandons the male.

to engage in other activities that would improve its own status. Parental investment evolves when its benefits exceed the costs it inevitably imposes.

An individual is expected to stop investing in one group of offspring when the costs, measured as lower survival of the abandoned individuals, are less than the benefits achieved by starting to invest in the next group of offspring. Among most species of birds, it is difficult for one parent to rear a brood of nestlings successfully. An unassisted parent, deserted by its mate, usually can raise few or no offspring. Therefore, opportunities for additional matings must be very good for a deserting parent to improve its fitness. Scientists believe that opportunities seldom are good enough because most species of birds breed monogamously; males and females remain together and share in caring for their offspring.

Under special circumstances an individual bird can increase its fitness by deserting its mate while the offspring are still receiving care. The conditions for the evolution of desertion by females are especially stringent, but they have been met in a few species of sandpipers with reversed sexual roles. The females of these species, unlike those of most bird species, are more brightly colored than the males (Figure 46.20). These females compete with one another for males, lay three or four large eggs, and then leave the incubation of the eggs and the care of the young to the males. The eggs hatch into young that are able to walk and feed themselves immediately upon birth. One parent can guard these few offspring and lead them to good foraging sites. Females can improve their reproductive success by terminating parental care early and gathering more energy. They can then lay additional clutches of eggs for the same male if the first clutch is lost to predators, or they can lay additional clutches for other males.

Males and females often differ strikingly in the kinds and amounts of parental investment they make. Differences begin with gamete production. Males produce tiny sperm whereas females produce much larger eggs. Males and females may also differ in the type of care they are able to give fertilized eggs and offspring. For example, only female mammals have functional mammary glands. Male mammals

cannot produce milk, and males of most species contribute nothing to offspring nutrition. Birds do not produce milk. Among birds, all aspects of reproduction except for laying the eggs can be performed equally well by males and females.

Sex roles among fishes differ from those of birds and mammals because most species of fishes do not feed their young. Parental care consists primarily of guarding eggs and young from predators (Figure 46.21). In many fish species, males are the primary guards. A male can guard a clutch of eggs while attracting additional females to lay eggs in his nest. A female, on the other hand, can produce another clutch of eggs sooner if she resumes foraging immediately after mating rather than spending time guarding eggs.

SEXUAL SELECTION

Traits may evolve among individuals of one sex as a result of **sexual selection**, the spread of traits that confer advantages to their bearers during courtship

46.21 Damselfish Guard Their Young
These two damselfish are guarding their large brood of tiny offspring in Lake Tanganyika, Tanzania. The young are vigorously defended by both parents for as long as 48 days.

or when they compete for mates or resources. The successful competitors have exclusive access to mates that are attracted to those resources. Traits that improve success in courtship evolve as a result of mating preferences by individuals of the opposite sex.

Sexual selection is responsible for the evolution of the remarkable tails of the African long-tailed widow bird, which are longer than their heads and bodies combined. A behavioral ecologist shortened the tails of some males by cutting them and lengthened the tails of other males by gluing on additional feathers. Both short-tailed and long-tailed males successfully defended their display sites, indicating that the long tail does not confer an advantage in male–male competition. However, males with artificially elongated tails attracted about four times more females than males with shortened tails did.

Mate selection and competition among males for access to females have favored large sizes among males of many vertebrate species. A comparison of many species reveals correlations between size differences between the sexes and mating system. Among American blackbirds, size differences between males and females are greatest among polygynous species (species in which each male mates with more than one female; Table 46.3). This pattern suggests that competition among males for breeding opportunities is a key factor leading to larger males throughout this group of nearly 100 species of birds. This interpretation of the general pattern is reinforced by detailed studies of individual species. For example, in the Montezuma oropendola, a species in which males are much larger than females, heavier males are better fighters than lighter males, and they achieve most of the copulations.

THE EVOLUTION OF ANIMAL SOCIETIES

The decisions animals make about where to settle, with whom to mate, what to invest in a reproductive effort, and when to terminate investment all help determine the type of social system they have. Today's social systems are the result of long periods of evolution, but there are few records of past social systems. Behavior leaves few traces in the fossil record; possible routes of the evolution of social systems must be inferred from current patterns of social organization. Fortunately, many stages of complexity exist among species, and the simpler systems provide clues about the stages through which the more complex ones may have passed.

You have already seen that animals may benefit in a number of ways from joining and remaining in groups—protection from predators, expanded foraging opportunities, and improved reproductive success. The most widespread form of social system is the family, an association of one or more adults and their dependent offspring. If parental care is extended or if the breeding season is longer than the time it takes for the young to mature, the adults may still be caring for younger offspring when older offspring reach the age at which they could help their parents. Many communal breeding systems probably evolved by this route.

Florida scrub jays, for example, live all year in territories, each of which has a breeding pair and up to six helpers (Figure 46.22). About three-fourths of the helpers are offspring from the previous breeding season that remain with their parents. If both adults have survived, these individuals are full siblings of the young they are helping. If one parent has died

46.22 Cooperation among Scrub Jays
Florida scrub jay helpers, most of whom are offspring from the previous breeding season, help feed nestlings and defend the nest against predators such as the approaching snake.

TABLE 46.3
Sexual Size Dimorphism and Mating Systems among American Blackbirds

Group	Mating system	Spacing pattern	% Size difference between sexes (mean and range)	♂	♀
Oropendolas	Polygynous	Colonial	25 (15–35)		
Caciques	Polygynous	Colonial	22 (21–23)		
Caciques	Monogamous	Territorial	12 (10–15)		
Orioles	Monogamous	Territorial	6 (0–14)		
Marsh nesting species	Polygynous	Territorial	16 (12–20)		
Marsh nesting species	Monogamous	Territorial	7 (6–14)		
Grackles	Monogamous	Territorial	13 (11–14)		
Grackles	Polygynous	Territorial	21 (20–22)		

and the living partner has remated, they are half siblings. Most helpers spend 1 to 3 years with their parents and then leave to establish their own territories. Young males remain longer with their parents and help more than young females do. They also gain more from helping because they may take over the parental territory if their father dies or they may help to enlarge the territory and claim a portion for themselves. Helpers assist their parents by defending the nest, by giving alarm calls, and by feeding the

nestlings. As a result, more young survive in nests with helpers than in nests without helpers. The young could do better by breeding on their own if territories were available to them. They remain with their parents because all suitable areas are occupied and they must wait for a vacancy before they can become breeders.

Most mammals evolved sociality via the extended family route. In simple mammalian social systems, solitary females or male–female pairs care for their

46.23 Jackals and Their Helpers
A male jackal has been helping his parents by bringing
food to his younger siblings; his chances of breeding independently are probably poor at this time. Here he is
groomed by the mated pair.

young. As the period of parental care increases, earlier young are still present when the next generation
is born, and they often help rear their younger siblings (Figure 46.23). In most mammal species, female
offspring remain in their **natal group**—the group in
which they were born—but males tend to leave, or
are driven out, and must seek other social groups.
Therefore, among mammals most helpers are females.

EUSOCIAL BEHAVIOR

All adult individuals in the social systems we have
discussed are capable of reproducing, even though

they may refrain from doing so for periods of time.
In contrast, colonies of termites, ants, some bees and
wasps, and one species of mammal, contain large
numbers of sterile individuals. These workers defend
the group against predators and bring food to the
colony, but they do not reproduce. A worker may be
killed while defending the colony; a bee, for example,
dies when her stinger remains embedded in her victim's skin (Figure 46.24*a*). Some species have specialized soldiers with large defensive weapons (Figure 46.24*b*). Species with social systems having sterile
classes are said to be **eusocial**.

How could such extreme altruism evolve? The first
clue is that sterile workers are prevalent among species of ants, bees, and wasps. These members of the
order Hymenoptera have an unusual sex-determination system in which males are haploid, with only
one set of chromosomes, but females are diploid. A
fertilized egg hatches into a female; an unfertilized
egg hatches into a male. If a female mates with only
one male, all the sperm she receives are identical
because males have only one set of chromosomes, all
of which are transmitted to each sperm cell. Therefore, a female's daughters share all of these genes.
They also share, on average, half of the genes they
receive from their mother. As a result, sisters share
75 percent of their alleles rather than 50 percent as
they would if both parents were diploid. Sisters are
genetically more similar to one another ($r = 0.75$; see
Box 46.A) than the mother is to her daughters and
sons ($r = 0.50$).

The British evolutionist W. D. Hamilton first suggested that eusociality evolved because worker female hymenopterans benefit by helping to raise their
sisters to which they are more similar genetically ($r = 0.75$) than they would be to any offspring they produced on their own ($r = 0.50$). One way to test this
hypothesis is to measure sex ratios in colonies of

46.24 Sterile Classes in Eusocial Insects
Eusocial insect species have classes of individuals who are sterile. These sterile
individuals defend and provide for the colony but do not reproduce. (*a*) This
honeybee has left its stinger in the skin of its enemy. (*b*) The bodies of these sol-
dier army ants from Panama have evolved to contain powerful weaponry.

eusocial insects. If Hamilton's hypothesis is correct, there should be a conflict between workers and their queen mother. The queen, who is equally related ($r = 0.50$) to both her sons and daughters should maximize her fitness by producing equal numbers of sons and daughters. Her daughters, however, are more closely related to their sisters ($r = 0.75$) than they are to their brothers ($r = 0.25$). The daughters, therefore, should maximize their fitness with an investment ratio of 75:25 (3:1) in favor of sisters. The queen can control the sex of the eggs she lays, but the workers who care for and feed the larvae could skew the sex ratio of surviving offspring by giving more food to their sisters than to their brothers. In fact, among species in which the queen normally mates only once, there is a strong skewing of investment in favor of females. If the founding queen is removed, one of the daughters typically becomes the new queen. She produces offspring who are nieces and nephews of the workers produced by the first queen ($r = 0.375$). As predicted by Hamilton's hypothesis, these manipulated colonies produce more males than do colonies with the original queen.

In spite of some strong support for Hamilton's hypothesis, available data do not support other predictions from the hypothesis. First, the hypothesis can explain eusociality only if the founding queen mates with only one male; but queens of many eusocial species, including the honeybee, mate repeatedly during their nuptial flights. If so, workers may not be more closely related to their sisters than to their own offspring. The hypothesis also requires colonies to be founded by single queens, but multiple queens are common among eusocial species.

Finally, Hamilton's hypothesis predicts eusociality only in species with a haplo-diploid sex-determination system. Termite colonies may have many thousands of sterile workers, but both male and female termites are diploid. And one species of mammal—the naked mole rat—is also eusocial and diploid. These unusual mammals live in underground colonies of 70 to 80 individuals for which tunnel systems are maintained by sterile workers. Breeding is restricted to a single large queen and several kings that live in a nest chamber in the center of the colony. Other females and males are sterile. Naked mole rats and termites tell us that haplo-diploidy is not necessary for eusociality to evolve.

The inability of Hamilton's haplo-diploid hypothesis to explain many aspects of eusocial behavior stimulated a search for environmental factors that might favor helping. One clue is that nearly all eusocial animals construct elaborate nests or burrow systems within which their offspring are reared (Figure 46.25). Forming a new colony and constructing these nests and tunnels is difficult, and most founding events fail. If opportunities for independent reproduction are poor, remaining at home and helping may enhance fitness. Poor chances of breeding independently characterize many of the species of birds and mammals in which offspring stay with their parents to help rear future siblings. Helpers are especially common among species of carnivorous mammals that depend on dens that are often in limited supply and that feed on prey that are difficult to capture. These species are not eusocial, but helping behavior may be a necessary prerequisite for the evolution of eusociality.

Although much progress has been made in understanding eusocial behavior, biologists still do not understand why eusociality evolved in some species but not in others. For example, all six other species

46.25 Termite Mounds are Large and Complex
These immense Australian termite mounds require elaborate construction and maintenance by the entire large colony. Elaborate nests or burrows are a common characteristic of nearly all eusocial animals.

of African mole rats live in underground burrows, but only one of those species has a social system similar to that of the naked mole rat. Why the other five have not evolved eusocial behavior is a mystery.

ECOLOGY AND SOCIAL ORGANIZATION

The type of social organization a species evolves is strongly related to the environment in which it lives. Among the African weaverbirds, species that live in forests eat insects, feed alone, and build well-hidden nests. Most of these species are monogamous, and males and females have identical plumage. In marked contrast, weaverbirds that live in the tree-studded grasslands called savannas eat primarily seeds, feed in large flocks, and nest in colonies, usually in isolated *Acacia* trees, where their nests are large and conspicuous (Figure 46.26). Most of these species are polygynous, with brightly colored males and dull females.

These striking differences probably evolved because nest sites and insects in forests are widely dispersed. Solitary pairs can use these resources more efficiently than animals in groups can. In savannas, the *Acacia* trees are widely scattered. As a result, good nesting sites are scarce and are highly clumped. Nests are difficult to hide in the isolated trees, and the insulation provided by the bulky nests built by many species of weaverbirds protects the young from the hot sun to which the nests are exposed most of the day. Males compete for these limited nest sites; the males that hold the better sites attract more females. Males spend their time attempting to attract additional mates rather than helping to rear the offspring they already have, which explains the evolution of brighter plumage among males.

All species of African hooved animals eat plants; their social organization and feeding ecology are correlated with the size of the animal (Table 46.4). Smaller animals have higher metabolic demands per

46.26 Weaverbirds Nest Communally
Many African weaverbirds nest in colonies in isolated trees. Although these nests are highly conspicuous, it is difficult for most avian, mammalian, and reptilian predators to get to them.

unit of body weight than do larger animals (see Chapter 35). Smaller hooved animals are therefore very selective in what they eat. They feed preferentially on high-protein foods such as buds, young leaves, and fruits. These foods are widely scattered in forested environments where it is possible to hide from predators. Hiding is an effective tactic for solitary animals. The largest hooved mammal species are able to eat lower-quality food, but they must process great quantities of it each day. They feed in grasslands with high standing crops of herbaceous vegetation, follow the rains to areas where grass growth is best, and live in large herds (Figure 46.27). Living

---

TABLE 46.4
Social Organization and Feeding Ecology in African Hooved Mammals

SPECIES	BODY WEIGHT (kg)	FEEDING ECOLOGY[a]	GROUP SIZE	SOCIAL ORGANIZATION
Dik-dik, duiker	3–60	Selective browsing and grazing	1 or 2	Pair
Reedbuck, gerenuk	20–60	Selective browsing and grazing	2–12	Male with small harem
Impala, gazelles	20–250	Grazing and browsing	2–100	Territory-defending male with harem
Wildebeest, hartebeest	90–270	Grazing	Up to thousands	Herd in which males defend females
Eland, buffalo	300–600	Unselective grazing	Up to thousands	Herd with male dominance hierarchy

[a] "Browsing" means eating the leaves and other portions of woody plants; "grazing" means eating grasses and other herbaceous plants.

46.27 Living in Large Herds
East African wildebeests live in large herds that move from place to place to obtain the fresh, green grass they eat. Their major predators—lions—live in permanent territories and often have little to eat when the wildebeest herds are far away.

interact cooperatively, by grooming and defending one another.

We have described only a few animal social systems, but the sample demonstrates some important concepts. First, social systems are best studied not by asking why they benefit the species, but by asking how the individuals that join together to form a social system benefit by the association. Second, social systems are dynamic; individuals constantly communicate with one another and adjust their relationships. Third, how individuals relate to one another depends in part upon the degree of their genetic relatedness. Certain types of helping behavior evolve much more readily among close relatives than among more distantly related individuals. Finally, social systems evolve in relation to the animals' size, diet, and the environments in which they live. The great variety of social organization among animals and the highly dynamic nature of these systems demand that students of social systems be good natural historians. They must know a great deal about the lives of the species they study and must be good field observers.

in herds makes it possible for males to compete among themselves for the control of females and for dominant males to defend medium to large harems. In these open environments hiding is impossible, but it is also difficult for predators to approach the herds undetected. Species of intermediate sizes have feeding ecologies and social systems intermediate between those of the smaller and larger species.

Differences in primate social systems are also correlated with environmental differences. Nocturnal forest-dwelling insectivores, such as lorises, some lemurs, and the owl monkey, live in pairs and are usually solitary foragers. Many species that are active during the day take insects and other animal food when they are available, but most of them eat fruits, seeds, and leaves. In Africa and Asia, troop sizes are smallest among arboreal forest-dwelling species, whatever their diets, and largest among the ground-dwelling savanna species, such as baboons (Figure 46.28).

Like other female mammals, female primates usually remain with their natal troop, whereas young males are driven away by older males before they become reproductively active. In troops with more than one male, strong dominance hierarchies exist among the males, and one or two of the males do most of the copulating. Females may have dominance relationships, and young females may assume the status of their mothers when they mature. Among some species, such as vervets, females frequently

46.28 Baboons in Groups
Baboons forage in open savannas and travel in large groups. Typically, adult males are on the outer edges of the group while females and infants are toward the center. If a predator approaches, the formidable males cooperate in defending the group.

SUMMARY of Main Ideas about Behavioral Ecology

During their lives, animals decide where to settle, what resources to use, with whom to mate, how much to invest in reproduction, and when to terminate investments. Many of these choices are hierarchical
   Review Figure 46.3

Foraging theory helps us understand why animals eat what they do.
   Review Figure 46.6

Animals form groups because they survive and reproduce better in them than they do in a solitary lifestyle.

One advantage of group living is enhanced foraging success,
   Review Figure 46.7

Groups offer better protection against predators.
   Review Figures 46.8 and 46.9

Depending on its effect on performer and recipient, a social act is categorized as altruistic, selfish, cooperative, or spiteful.
   Review Table 46.2

Because they usually increase their fitness by maximizing the number of copulations they achieve, males typically compete for females by advertising their qualities or the qualities of the resources they control.
   Review Figures 46.12 and 46.13

Males often compete for breeding opportunities by defending space.
   Review Figures 46.14, 46.15, 46.16, and 46.17

Many of the traits that distinguish males from females evolved because they improved male reproductive success.
   Review Figure 46.18

Altruistic behavior can evolve via kin selection or via reciprocal altruism.

Eusocial behavior is especially prevalent among insects having a haplo-diploid sex-determination mechanism, but it has also evolved in termites and in one mammal.

Social systems of animals evolve in relation to diet and the type of habitat in which they live.
   Review Table 46.4

SELF-QUIZ

1. Which of the following is *not* a component of the cost of performing a behavioral act?
   a. Its energy cost
   b. The risk of being injured
   c. Its opportunity cost
   d. The risk of being attacked by a predator
   e. Its information cost

2. An almost universal cost associated with group living is
   a. increased risk of predation.
   b. interference with foraging.
   c. higher exposure to disease and parasites.
   d. poorer access to mates.
   e. poorer access to sleeping sites.

3. An act is said to be altruistic if it
   a. benefits the performer by inflicting a cost on another individual.
   b. benefits both the performer and another individual.
   c. imposes a cost on the performer and on another individual.
   d. benefits another individual at a cost to the performer.
   e. imposes a cost on a performer without benefiting any other individual.

4. Which of the following statements about male and female roles in social systems is *not* correct?
   a. Females invest more in gamete production but they may invest more or less than males in the care of offspring.
   b. Care by both parents is prevalent among birds.
   c. Males of most mammal species help feed offspring.
   d. Males with a high probability of being the parent invest more in parental care than males that are less certainly related to the offspring of their mates.
   e. Among fishes, if parental care by individuals of the two sexes is unequal, the male nearly always does more than the female.

5. Male and female mating tactics usually differ because
   a. males are typically larger than females.
   b. males do not contribute as many genes to their offspring as females do.

   c. males but not females usually can increase their fitness by mating with more than one female.
   d. males can control copulations to their advantage.
   e. males and females occupy different positions when they copulate.

6. Choice of mating partner may be based on
   a. the inherent qualities of a potential mate.
   b. the resources held by a potential mate.
   c. both the inherent qualities of a potential mate and the resources it holds.
   d. the success of individuals of the opposite sex in courtship.
   e. all of the above.

7. When a male holds a territory in which more than one female breeds,
   a. the male is an individual with a high coefficient of relatedness.
   b. the male is an individual with a high inclusive fitness.

c. the behavior is known as resource defense polygyny.

d. the behavior is known as sexual selection

e. the behavior is known as resource defense polyandry

8. Among social birds, there are usually more male than female helpers in species with helpers because

a. males are better helpers than females.

b. males typically receive greater benefits from helping.

c. males have lower mortality rates than females.

d. a higher percentage of young males help.

e. males often must wait longer than females to obtain a breeding opportunity.

9. Among social mammals most helpers are females because

a. only females have functional mammary glands.

b. males mature too slowly to help.

c. females form closer ties with their mothers than males do.

d. young males leave their social group or are driven out by their fathers.

e. young males are too aggressive to help.

10. Smaller African hooved mammals are usually solitary because

a. they feed on scattered, high-quality foods in forested environments.

b. the low quality of their food does not permit them to assemble in groups.

c. they are too small to defend themselves against predators.

d. they are too small to follow the rains to areas where grass growth is best.

e. they are usually driven from their natal groups.

FOR STUDY

1. Most hawks are solitary hunters. Swallows often hunt in groups. What are some plausible explanations for this difference? How could you test your ideas?

2. Because the costs and benefits of a behavior can seldom be measured directly, behavioral ecologists often use indirect measures. What are the strengths and weaknesses of some of these measures?

3. Among birds, males of promiscuous species that display communally are usually much larger and more brightly colored than females, whereas among monogamous species, males are usually similar in size to females, whether or not they are more brightly colored. What hypotheses may explain this difference?

4. Polyandry is a mating system in which one female has a "harem" of several males. Why is polyandry much rarer among both birds and mammals than polygyny, in which one male has several females?

5. Many animals defend space, but the sizes of the territories they defend and the resources the areas provide vary enormously. Why don't all animals defend the same type and size of territory?

6. Among frogs, a male clasps a gravid female (one that is full of eggs) behind her front legs and stays with her until she lays the eggs, at which time he fertilizes them. In most species of frogs, the male remains clasped to the female for a short time, usually no longer than a few hours. In some species,

however, pairs may remain together for up to several weeks. Given that a male cannot court or mate with any other female while clasping one female and that a female lays only a single clutch of eggs, why is it advantageous for males to behave in this way? What can you guess about the breeding ecology of frogs that remain clasped for long periods? Why should females permit males to clasp them for so long? (Females do not struggle!)

7. Among vertebrates, helpers are individuals capable of reproducing, and most of them later breed on their own. Among eusocial insects, sterile classes have evolved repeatedly. What differences between vertebrates and insects might explain the failure of sterile classes to evolve in the former?

READINGS

Alcock, J. 1993. *Animal Behavior*, 5th Edition. Sinauer Associates, Sunderland, MA. An excellent account of animal behavior from an evolutionary perspective. Contains additional material on all topics discussed in this chapter.

Krebs, J. R. and N. B. Davies. 1993. *An Introduction to Behavioural Ecology*, 3rd Edition. Blackwell Scientific Publications, Oxford. A succinct summary of the methods and results of modern studies of behavioral ecology.

Rubenstein, D. I. and R. W. Wrangham (Eds.). 1986. *Ecological Aspects of Social Evolution*. Princeton University Press, Princeton, NJ. The results of long-term field studies of 18 species of birds and mammals are treated in an evolutionary ecological context. Rich in natural history details.

Trivers, R. 1985. *Social Evolution*, Benjamin/Cummings, Menlo Park, CA. An excellent treatment of social behavior, with numerous references to human social behavior.

Wilson, E. O. 1971. *The Insect Societies*. Harvard University Press, Cambridge, MA. A classic detailed account of the social insects and other arthropods.

Wilson, E. O. 1975. *Sociobiology: The New Synthesis*. Harvard University Press, Cambridge, MA. An impressive work covering all aspects of social evolution and animal societies.

An Immature Red-Spotted Newt
The toxic red eft represents one stage of the salamander's complex life cycle.

47

Structure and Dynamics of Populations

The red-spotted newt is a colorful salamander that breeds in small ponds throughout eastern North America. In most parts of its range, the red-spotted newt has a complex life cycle. Adults, which are olive-colored and have red spots, feed and lay their eggs in ponds. The eggs hatch into larvae that grow and develop in the ponds for several months. Eventually the larvae lose their gills and change into efts, immature newts that leave the water to live terrestrial lives for four to nine years. The red efts are highly toxic to their natural predators. Most efts return to the ponds in which they were born to reproduce, but some travel long distances and colonize new ponds.

Newly created farm ponds in rural areas and wildlife management ponds in national forests are populated quickly by red-spotted newts, showing that they are good colonizers of new habitats. The survival of eggs and larvae to the eft stage depends on the intensity of disease and leech attacks on them. In most ponds very few larvae survive to become efts. Only ponds where disease-causing organisms and leeches are not well established can produce many efts. Long-term studies show that the sizes of populations of red-spotted newts in ponds fluctuate widely across years.

Population numbers of many other species also fluctuate considerably, but some species seem to be just about as abundant in one year as in another. Some species also are consistently much more common than others. These observations stimulate a variety of questions: What causes a species to be common or rare? Why does the size of some populations fluctuate yearly and seasonally? Why is a species common in some parts of its geographic range, rare in others, and absent outside its range?

Answering such questions is the work of population ecology. At the beginning of a study, the first things a population ecologist might want to find out are whether the population is increasing or decreasing in number and how its members are distributed in space. The second step would be to investigate causes of the observed population growth or decline and distribution. To do this, in a study of red-spotted newts or any other species, the population ecologist would collect detailed information about members of the species throughout their life cycles.

The complete life cycle consists of birth, growth to maturity, reproduction, and death. During its life an individual organism ingests nutrients or food, grows, interacts with other individuals of the same and other species, reproduces, and usually moves or is moved so that it does not die exactly where it was born. How fast the members of a population grow, which individuals they interact with, when they reproduce, and when they die vary. The sum of these varied activities of the individuals in a population determines whether the population as a whole is increasing, decreasing, or maintaining a steady size.

LIFE HISTORIES

The stages an individual typically goes through during its life constitute the **life history** of a species. Although life histories are very diverse, they are all based on a small set of traits: size at birth; how fast individuals grow; how long they live; how many times they reproduce; how dramatically and when they change form; the ages at which they die; how much they move around; and the number, size, and sex composition of their offspring.

Growth

For at least part of their lives, all organisms grow by gathering and assimilating energy and nutrients. Some organisms gather energy and nutrients throughout their lives, but in many species, energy gathering is confined to a particular stage. For example, most moths feed only when they are larvae. The adults, which lack mouth parts and digestive tracts, live on energy gathered by the larvae and survive only long enough to disperse, mate, and lay eggs.

Change of Form

Many organisms have resting or reorganization stages in their life cycles. Resting stages, such as spores, seeds, and many eggs, have low metabolic rates and usually are highly resistant to changes in the physical environment. The most striking reorganization stages are those of insects such as beetles, flies, moths, butterflies, and bees, which undergo radical metamorphoses from their larval to their adult forms (see Chapter 26). These metamorphoses take place during the pupal stage: Larval tissues and organs are broken down, and adult tissues and organs are constructed from the larval material.

Dispersal

At some times in their lives all organisms disperse. Some, such as plants and sessile animals, disperse as spores or seeds before much growth has taken place. Others, such as insects and birds, disperse primarily as adults. Still others may disperse during several different stages. Individuals of some species can change their locations many times during their lives in response to environmental changes. Others must remain in the first place they settle.

Life history activities often overlap. For example, an organism may be growing, reproducing, and moving at the same time (Figure 47.1). Nonetheless, understanding how different types of organisms express these activities and how performing one activity constrains an individual's ability to perform others provides a useful perspective for thinking about the lives of organisms and the dynamics of their populations. The life history patterns in Figure 47.1 illustrate the great diversity in nature. The patterns include ones in which reproduction terminates the life of the individual (annual plant, moth, salmon), one in which growth and reproduction overlap broadly (tree), one in which growth is limited to a short period but reproduction continues intermittently for a long time after the individual has reached full size (marmot), and one in which both growth and reproduction are extended but do not overlap during the life of an individual (red-spotted newt).

Life History Trade-Offs

All life history traits are molded by natural selection under the constraint that changes that improve fitness via one life history trait often reduce fitness via another trait. The most important of these trade-offs are the subject of this section.

Every newborn individual begins to grow with energy and nutrients from its maternal parent, but how much energy and nutrients individuals receive from their parents varies greatly from species to species. Orchid seeds receive very little, grass seeds slightly more. Coconuts, birds, and newborn placental mammals receive large amounts of maternal energy. The larger the amount of energy provided to each offspring, the larger it can grow before it must gather its own energy but the fewer offspring a parent can produce for a given amount of energy—a major trade-off.

During their growth periods, individuals of many species are completely independent of their parents, obtaining all their own energy and providing their own protection. In some animal species, however, the parents provide additional care and protection that may extend, as it does in many birds and mammals, until the individual has reached adult size. The more parental care the parents provide, the fewer offspring they can produce for the same investment in reproduction. Therefore, there is also trade-off between number of offspring produced and the amount of care parents provide to the offspring after they hatch.

Different species produce their offspring at different times and produce different numbers of offspring in a given batch (known as a clutch or litter in animals or a seed crop in vascular plants). Some organisms reproduce only once and then die. A bacterium that divides to form two daughter cells may be considered

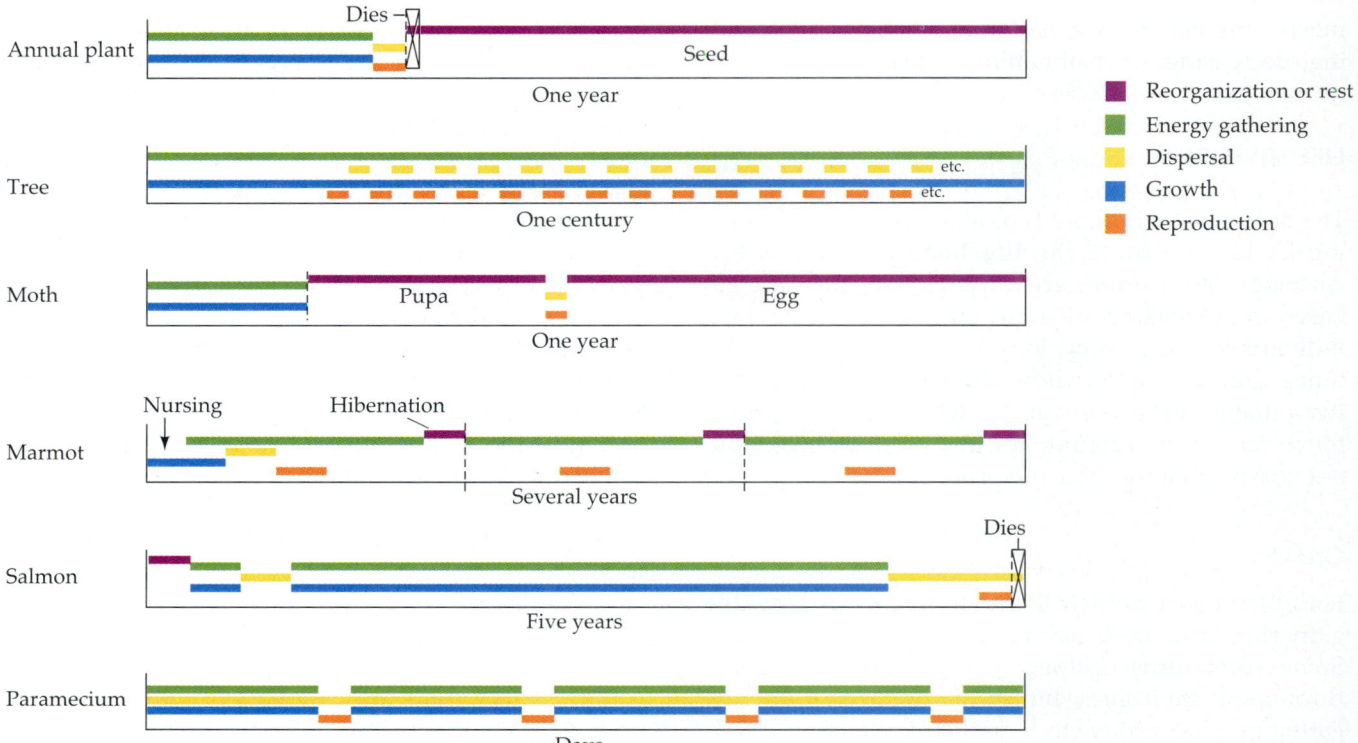

Annual plant — One year

Tree — One century

Moth — One year

Marmot — Several years

Salmon — Five years

Paramecium — Days

Legend:
- Reorganization or rest
- Energy gathering
- Dispersal
- Growth
- Reproduction

47.1 Life Histories Differ

All life histories have stages of growth, dispersal, reproduction, energy gathering, and reorganization or resting; but these stages, how long they last, and how much they overlap varies. The six patterns shown here are some of the common ones. To make comparisons among them easier, all six are started at the beginning of a growth period.

to have died when it split. Annual plants invest so much energy in seed production that they do not survive long after reproducing. Most insects and spiders live less than one year and die soon after reproducing.

Some longer-lived organisms also reproduce only once in their lifetimes and die very soon afterward. Pacific salmon (genus *Oncorhynchus*) hatch in fresh water, spend a number of years at sea, return to fresh water, spawn, and die. Most agaves (century plants) of the American Southwest store up energy for many years before producing a large flowering stalk, forming many seeds, and dying (Figure 47.2). Yucca plants, which grow in the same environments, appear similar, but they invest less in each reproduction and live to reproduce many times. If two individuals, one of which reproduces only once, the other several times, have the same amount of energy to invest in reproduction, the former can produce more offspring in a single clutch than the latter because it reserves no energy for its own future survival.

Even among organisms that live to reproduce more than once, there is a trade-off between reproduction and survival. Studies of many species, in-

47.2 Agaves Reproduce Once and Die

This century plant has mobilized the energy stored during its long life to produce a large flowering stalk with hundreds of flowers, literally reproducing itself to death.

cluding plants, shrimp, snails, fishes, birds, and mammals, show that engaging in reproduction reduces adult survival rates. Female red deer that produce fawns and nurse them die at a younger age than females that do not produce fawns. The more an adult invests in reproduction, the shorter its lifetime. In the extreme case, death immediately follows reproduction.

Members of many species do not begin to reproduce until they have reached full size, but others, such as most plants, start while still relatively small and continue to reproduce as they grow. Among continuously growing animals, such as many mollusks, fishes, and reptiles, growth and reproduction overlap extensively during the lives of individuals. Reproduction usually reduces growth because these two processes compete for the limited amount of energy an individual has at its disposal. For example, beech trees in Germany grow more slowly during years when they produce large seed crops than they do in years when their seed crops are small (Figure 47.3). Reproduction at the expense of growth is a third major life history trade-off.

Timing of Reproduction

The timing of reproduction influences its contribution to future generations. We can compare the production of offspring to earning interest on money deposited in a bank. It pays to deposit money into the bank as soon as possible so that it can begin earning interest. Offspring produced early in an adult's life likewise "yield interest" quickly because they too will begin to reproduce at a young age. If juvenile survival is very poor, however, and reproduction greatly reduces the life span of the parent, natural selection may favor the delay of reproduction until the parents are older. For example, most birds begin to reproduce when they are one year old, but gulls, penguins, and albatrosses do not breed until they are three to nine years old. Although individuals in these three groups reach full size within a year, at one year old they are not efficient foragers and cannot gather enough energy to reproduce without jeopardizing their survival and future reproduction.

Reproductive value is a measure of the contribution an individual of a particular age is likely to make to the growth rate of the population. Many newborn individuals die before they have a chance to reproduce. The probability that an individual will survive long enough to reproduce increases with its age. The reproductive value of an individual thus steadily increases until it begins to reproduce. After maturity, reproductive value usually declines and reaches zero when the individual has finished reproducing.

Because reproductive value decreases with age in a mature individual, the power of natural selection acting on alleles that produce their phenotypic effects only at older ages is increasingly weaker. After the last offspring have been born and raised, natural selection cannot influence phenotypic traits, even those that are highly detrimental to survival. Alleles that delay the phenotypic expression of harmful alleles, however, *are* favored by natural selection. The result is that the expression of many harmful alleles is delayed. These alleles increasingly express themselves phenotypically as individuals age, causing increased mortality, especially after reproduction has ceased.

The fact that phenotypic effects of undesirable alleles are often expressed late in life poses serious social problems for people in modern industrial societies. As a result of improved hygiene and nutrition, most people in such societies are now spared the serious childhood infections that cause the high death rates of people in nonindustrial societies. Most people live to the time when the so-called genetic diseases of old age begin to afflict them. Cancer and heart disease, the main killers in industrialized societies, are much more difficult to deal with than the infectious diseases that used to cause death. The social costs of extending life by a few more years for persons six decades old and older are great. For this reason, despite the expenditure of enormous resources to extend life, the average age at death in the United States has changed very little during the past 30 years. As one killer is eliminated, another takes its place (Figure 47.4). In the context of life history theory and natural selection, there appears to be little hope of escaping from this situation.

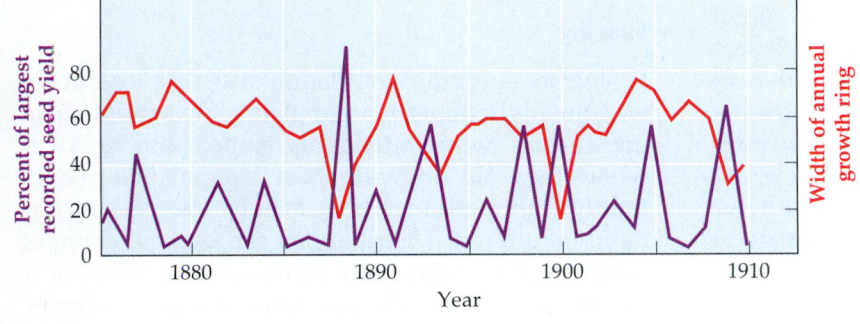

47.3 Reproduction Slows Growth Rates
The width of their growth rings (plotted in red) reveals that beech trees grow more slowly when they produce large crops of beech nuts than they do during years of low reproduction.

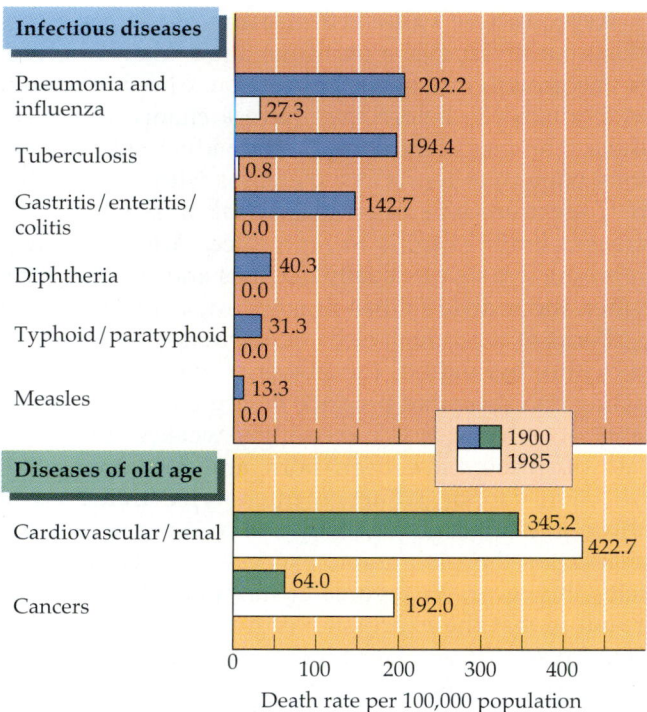

Infectious diseases

Pneumonia and influenza	202.2 / 27.3
Tuberculosis	194.4 / 0.8
Gastritis/enteritis/colitis	142.7 / 0.0
Diphtheria	40.3 / 0.0
Typhoid/paratyphoid	31.3 / 0.0
Measles	13.3 / 0.0

■ 1900
□ 1985

Diseases of old age

Cardiovascular/renal	345.2 / 422.7
Cancers	64.0 / 192.0

0   100   200   300   400

Death rate per 100,000 population

47.4 Causes of Human Death
During the twentieth century the major causes of death in the United States have shifted from contagious diseases to the diseases of old age.

Phylogenetic Constraints on Life History Evolution

Life histories are constrained not only by trade-offs between offspring size and number, between amount of parental care and number of offspring, and between growth and reproduction, but also by phylogeny. For example, all birds lay eggs and all mammals nurse their offspring with milk produced by mammary glands. There is probably no appropriate genetic variation among birds to allow natural selection to produce birds with mammary glands. Therefore, ecologists try to determine why some birds lay more eggs than others, but they do not expect birds to evolve mammary glands.

POPULATION STRUCTURE

At any given moment an individual organism occupies only one spot and is of one particular age, but the members of a population are distributed over space and differ in age and size. These features determine **population structure.** To determine the population structure of a species, an ecologist asks, about its members, How many are there? Where are they? What are their ages? Ecologists study population structure at scales ranging from local subpopulations to the entire species. Geneticists and evolutionists also study population structure, but they are interested primarily in distributions of genotypes and their degree of isolation from one another because that component of structure influences how populations evolve (see Chapter 19).

Population Density

The number of individuals of a species per unit of area (or volume) is its **population density.** One reason ecologists are interested in population density is that dense populations may often exert strong influences on populations of other species. Others—people who work in agriculture, conservation, or medicine, for example—often wish to manage species to raise their densities (crop plants, aesthetically attractive species, threatened or endangered species) or reduce their densities (agricultural pests, disease organisms). To manipulate densities we must know what factors make populations grow and shrink and how these factors work.

Because organisms and their environments differ, densities are measured in more than one way. Ecologists usually measure the densities of organisms in terrestrial environments as the number of individuals per unit of area, but number per unit of volume is generally more useful for organisms living in water. For species whose members differ markedly in size, as with most plants and some animals (such as mollusks, fish, and reptiles), the total mass of individuals—their **biomass**—may be the most useful measure.

Sometimes individuals can be counted directly without missing any of them or counting any of them twice, but this process is usually impossible or too laborious. We thus commonly estimate densities by sampling the population in representative areas and extending our findings to the whole area, or by marking and recapturing individuals. For example, if we capture and mark 100 individuals in a population, we can estimate the total size of the population by determining what percentage of individuals captured in a later sample are already marked. If 10 percent of the individuals are already marked, we would conclude that the population contains about 1,000 individuals.

Spacing

Ecologists studying population structure look at the way the individuals in a population are spaced. Spacing reveals why individuals settled and survived where they did. Individuals of a population may be tightly clumped together, evenly spaced, or randomly scattered. Distributions can become clumped when young individuals settle close to their birth places, when suitable environments are "islands"

Age Distribution

Populations are composed of individuals ranging from newborns to postreproductive adults. The proportions of individuals in each age group in a population make up its **age distribution.** Whereas the density and spacing of a population are spatial attributes, age distribution is a temporal (time-oriented) attribute. The timing and rates of births and deaths determine age distributions. If birth rates and death rates are both high, a population is dominated by young individuals, as illustrated by the human population of Mexico in 1984 (Figure 47.6*a*). If birth rates and death rates are low, there is a relatively even distribution of individuals of different ages, as illustrated by the human population of the United States (Figure 47.6*b*). The age distribution of a population can reveal much about the recent history of births and deaths in the population.

The timing of births and deaths may influence age distributions for many years in populations of long-lived species. The population of the United States is a good example. Between 1947 and 1964, the U.S. experienced what is called the post-World War II baby boom. During these years average family size grew from 2.5 to 3.8 children; an unprecedented 4.3 million babies were born in 1957. Birth rates declined during the 1960s, but Americans born during the baby boom will constitute the dominant age class into the twenty-first century (Figure 47.7*a*). "Baby boomers" became parents in the 1980s, producing another bulge in the age distribution—a baby boom echo—but they had, on average, fewer children than their parents had, so the bulge is not as large. Fish pop-

47.5 Even Spacing
Each of these king cormorants placed its nest far enough from the nests of its neighbors to be out of pecking range of the neighbors, at least while sitting on the nest to incubate their eggs.

separated by unsuitable areas, or by chance. The even spacing of some plants is a result of competition for light, water, and soil nutrients or because they rub against one another when moved by wind or water currents. Among animals, defense of space is the most common cause of even distribution (Figure 47.5). Random distributions may result when many factors interact to influence where individuals settle and survive.

47.6 Age Distributions Summarize Birth and Death Patterns
(*a*) The human population of Mexico in 1984. (*b*) The human population of the United States in 1985. The length of the bars indicates the percentage of the population in each age class.

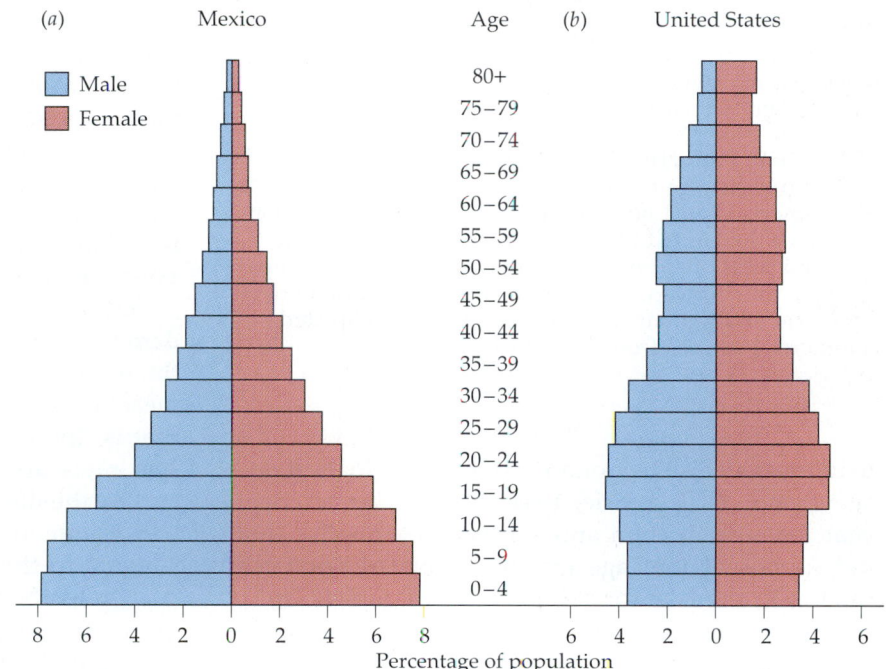

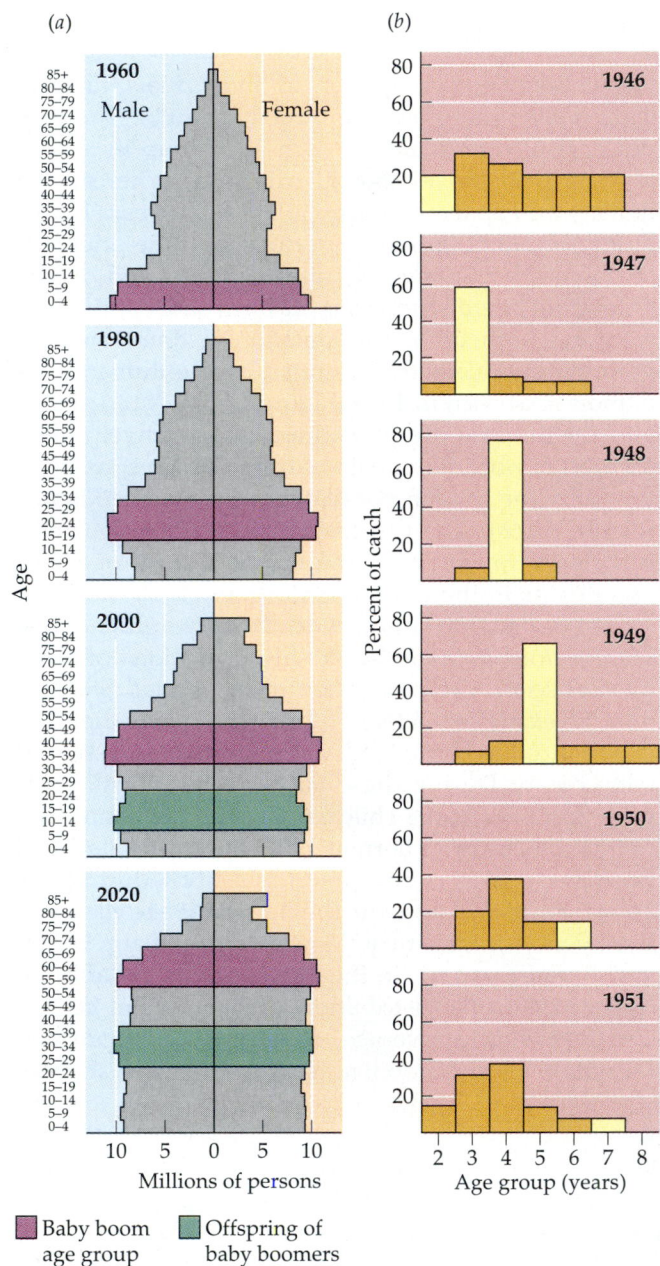

Baby boom age group (pink) **Offspring of baby boomers** (green)

47.7 Some Populations Are Dominated by Specific Age Groups
(a) "Baby boomers" now dominate age classes in the population of the United States, and projections show that they and their offspring will be the dominant age groups in the decades ahead. (b) In the whitefish population of Lake Erie, individuals that hatched in 1944 dominated commercial catches from 1947 through 1949.

ulations may also be dominated by individuals of one age group. In Lake Erie, 1944 was such an excellent year for reproduction and survival of whitefish that individuals of that age group dominated whitefish catches in the lake for several years (Figure 47.7b).

The age distributions of populations reveal much about their history and their potential for future

growth. The echo effect shows how the high birth rates in the United States of the 1950s and 1960s created a momentum for future growth. Age distributions also reveal why there is a long delay between the time a nation achieves a two-child family average and the time at which births equal deaths and that nation's population growth stops.

POPULATION DYNAMICS

At any given moment a population has a particular structure determined by the spatial distribution of its members and their ages. As we have just seen, however, population structure is not static. Changes in population structure influence whether the population will increase or decrease; that is, they determine the dynamics of the population. Let's examine how ecologists use information on the structure of populations, combined with life history traits, to determine population dynamics and their causes.

Births and Deaths

The simple knowledge of when individuals are born and when they die can provide a surprising amount of information about a population. Because of its great practical importance death was the first aspect of population dynamics to be studied. When a person dies, some expenses, such as the cost of burial and support for surviving relatives, usually are borne by society at large. Not surprisingly, then, societies have been interested in the probability that individuals of different ages will die because that determines when and how much money must be spent. The cost of modern life insurance—really death insurance, but we all prefer the euphemism—is based on elaborate and accurate estimates of death schedules, but the practice of insuring for death has ancient roots. Roman societies provided funeral insurance for their members, and they developed and used the first tables showing ages of death of their citizens. These tables—death schedules—were inaccurate, but the fact that they were constructed indicates their importance 2,000 years ago.

Births, deaths, immigration, and emigration are **demographic events**—that is, events that determine the number of individuals in a population. Ecologists measure the rates at which these events take place, that is, the number of such events per unit of time. The rates are influenced by environmental factors and by the life history traits of the species. The number of individuals in a population at any given time is equal to the number present at some time in the past, plus the number born between then and now, minus the number that died, plus the number that immigrated, minus the number that emigrated. That

TABLE 47.1
A Life Table for a Cohort of 843 Individuals of *Poa annua*, a Short-Lived Grass

AGE INTERVAL[a]	NUMBER ALIVE AT BEGINNING	PROPORTION ALIVE AT BEGINNING	NUMBER DYING	DEATH RATE	NUMBER OF SEEDS PRODUCED
0–3	843	1.000	121	0.144	0
3–6	722	0.857	195	0.270	300
6–9	527	0.625	211	0.400	620
9–12	316	0.375	172	0.544	430
12–15	144	0.171	95	0.625	210
15–18	54	0.064	39	0.722	60
18–21	15	0.018	12	0.800	30
21–24	3	0.004	3	1.000	10
24	0	—	—	—	—

[a] Period between two exact ages (in months).

is, the number of individuals at a given time, N_1, is given by the equation

$$N_1 = N_0 + B - D + I - E$$

where N_1 = number of individuals at time 1, N_0 = number of individuals at time zero, B = number born, D = number that died, I = number that immigrated, and E = number that emigrated, all counted between time zero and time 1.

Life Tables

Life tables can help us visualize patterns of births and deaths in a population. We construct a life table by determining for a group of individuals born at the same time—known as a **cohort**—the number still alive at specific times and the number of offspring they produced during each time interval. An example is shown in Table 47.1. As you can see from the table, members of this cohort of the grass *Poa annua* began producing seeds some time after they were three months old and continued to produce seeds the rest of their lives. By the end of two years, all members of the cohort were dead. Note that life tables include numbers observed as well as rates calculated from those numbers. We can calculate the probability of dying during the age interval 6–9 months by dividing the number that died by the number alive at the beginning: $^{211}/_{527} = 0.4$. The data in Table 47.1 show that the number of seeds produced per individual peaks at 6 months of age and that individuals produce very few seeds after they are a year old.

Ecologists often use graphs to highlight the most important changes in birth and death rates in populations. Graphs of survivorship—the mirror image of death rate—in relation to age show when individuals survive well and when they do not. To interpret survivorship data, ecologists have found it useful to compare real data with several hypothetical curves that illustrate a range of possibilities. A useful type of graph plots the proportion of individuals of a cohort that are still alive at different times during their total potential life span (Figure 47.8*a*). At one extreme, nearly all individuals survive for their entire potential life span and die almost simultaneously (hypothetical curve I). At the other extreme, the survivorship of young individuals is very low, but survivorship is high for the remainder of the life span (hypothetical curve III). An intermediate possibility is that survivorship is the same throughout the life span (hypothetical curve II).

Survivorship curves from real populations often resemble one of these hypothetical curves. For example, survivorship of *Poa annua* seedlings is very high for the first six months, but then, as in hypothetical curve I, it declines significantly in older individuals (Figure 47.8*b*). Many wild birds have survivorship curves similar to hypothetical curve II; the probability of their surviving is about the same over most of the life span once individuals are a few months old (Figure 47.8*c*). A more common type of survivorship pattern, especially among organisms that produce many offspring, each of which receives little energy and no subsequent parental care, is one with low survivorship of young individuals, followed by high survivorship during the middle part of the life span, and then low survivorship again toward the end of the life span. Dall sheep, although they do not have a high birth rate, have such a survivorship curve (Figure 47.8*d*), one that combines the first part of the type III curve with the middle and late parts of the type I curve. Survivorship curves help us understand how birth and death rates change through time, but all such curves are incomplete.

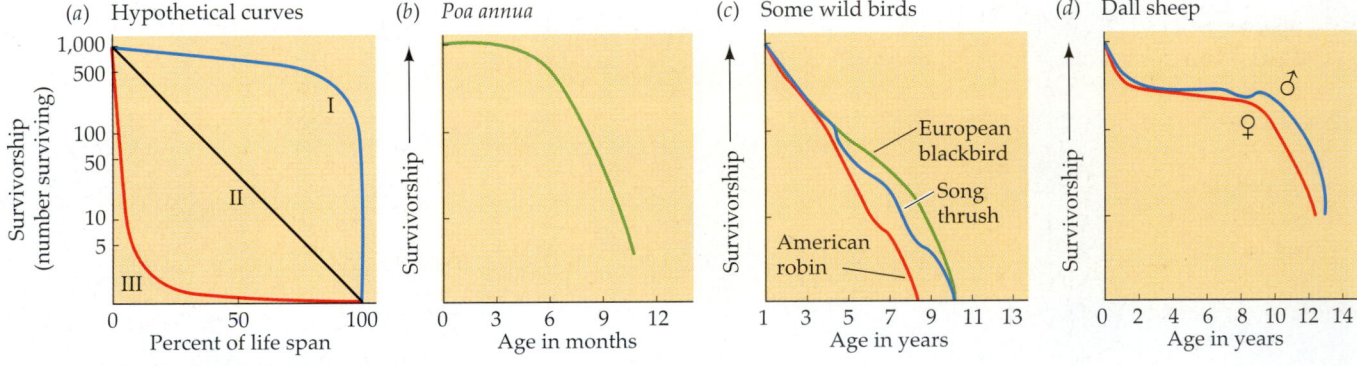

47.8 Survivorship Curves
Survivorship curves show the number of individuals of the cohort still surviving at different times in life. (a) The range of possible patterns. (b) *Poa annua*. (c) Three species of thrushes. (d) Male and female dall sheep.

None of them includes deaths of zygotes prior to birth or of seeds prior to germination.

Males and females of a single species may have different survivorship curves. For example, survivorship of adult female red deer is nearly constant once they reach reproductive age (two to three years old), although females that reproduce do not live as long as nonreproductive females. The death rates of males, however, rise sharply when they reach their breeding age of seven to eight years (Table 47.2) because the males engage in heavy combat with each other during the breeding season.

Modular Organisms

Like the preceding discussion, analyses of population dynamics usually focus on the number of individuals. For many kinds of organisms, called **unitary organisms,** individuals are easy to distinguish and most adult members of the population are similar in size and shape. All populations of mollusks, echinoderms, insects, and vertebrates, for example, consist of unitary organisms.

Individuals of **modular organisms,** organisms whose bodies consist of repeated units, are not easy to distinguish, and the members of a population have highly varied sizes and shapes. Most plants are modular, and there are many important groups of modular protists, fungi, and animals (for example, sponges, corals, moss animals, colonial tunicates). The fertilized egg of a modular organism develops into a unit of construction, a module, which then produces additional modules much like itself (Figure 47.9). Usually these modules remain attached to one another, and the aggregate is usually immobile.

The limits of modular individuals are often difficult to determine. Even more important for the study of ecology, the impact of modular organisms on their environment often depends less on the number of genetically distinct individuals than on the number and size of modules. A large tree with many branches is an organism very different from a small sapling with just a few branches. The modules of a single organism may also differ markedly in size and age. Therefore, students of modular organisms are often concerned primarily with the number, size, and shape of modules rather than with the number of genetically distinct individuals.

Because modular organisms grow by adding modules, they can actually shift their positions even when the individuals are attached to the substrate. Figure 47.10 shows the annual change in distribution of modules of an understory shrub, *Piper pseudobumbratum*, in a Costa Rican rainforest. This shrub produces new shoots (modules) that sprout from its roots. Shoots that encounter bright light grow and survive better than those that encounter dim light. A tree

TABLE 47.2
Death Rates of Male and Female Red Deer on Rhum Island, Scotland[a]

| | DEATH RATE | |
AGE	FEMALES	MALES
0–6 months	12.4	12.4
6–12 months	15.0	20.8
Yearlings	7.4	13.0
2 years	1.1	1.8
3–4 years	3.6	1.7
5–6 years	3.8	2.2
7–8 years	2.3	6.1
9–10 years	2.8	16.3
11–12 years	8.7	37.0

[a] Percentage of those in each age class alive at the beginning of a particular year that died during that year.

47.9 A Single Modular Organism May Look Like a Population
Each clump of these quaking aspens consists of a single genetic individual that has spread by underground roots and has sent up many stems, each of which appears to be a separate tree.

that fell in 1988 on the left side of the area shown in the figure created a light gap. Modules survived well in this brighter area but poorly in the right-hand side of the area, which remained dark. Thus the population "migrated" about 6 meters in five years.

POPULATION GROWTH WHEN RESOURCES ARE ABUNDANT

If a single bacterium selected at random from the surface of this book, and all its descendants, were able to grow and reproduce in an unlimited environment, an explosive population growth would result. In a month this bacterial colony would weigh more than the visible universe and would be expanding outward at the speed of light. Similarly, a single pair of Atlantic cod and their descendants reproducing without hindrance would in six years fill the Atlantic Ocean with their bodies. Humans mature and breed very slowly, but if the existing human population could achieve the impossible feat of continuing to increase at its present rate (which is less than its maximum potential), in 2,000 years the human population would weigh as much as the entire Earth.

And 4,000 years later, it would weigh as much as the visible universe and be expanding at the speed of light!

All populations have this potential for explosive growth because as the number of individuals in the population increases, the number of new individuals added per unit time accelerates, even though the rate of growth per individual, called the per capita growth rate, remains constant. This form of increase is called **exponential growth** (Figure 47.11*a*). If for the moment we ignore immigration and emigration, such a growth pattern is expressed mathematically in the following way:

$$
\begin{pmatrix} \text{Rate of} \\ \text{increase in} \\ \text{number of} \\ \text{individuals} \end{pmatrix} = \begin{pmatrix} \text{Average} \\ \text{per capita} \\ \text{birth} \\ \text{rate} \end{pmatrix} - \begin{pmatrix} \text{Average} \\ \text{per capita} \\ \text{death} \\ \text{rate} \end{pmatrix} \times \begin{pmatrix} \text{Number} \\ \text{of} \\ \text{individuals} \end{pmatrix}
$$

or, more concisely:

$$
\frac{dN}{dt} = (b - d)N
$$

where dN/dt is the rate of change in size of the population (dN = change in number of individuals; dt = change in time). The difference between the average per capita birth rate (b) and the average per capita death rate (d) is called r. When conditions are optimal for the population, r has its highest value, called r_{max},

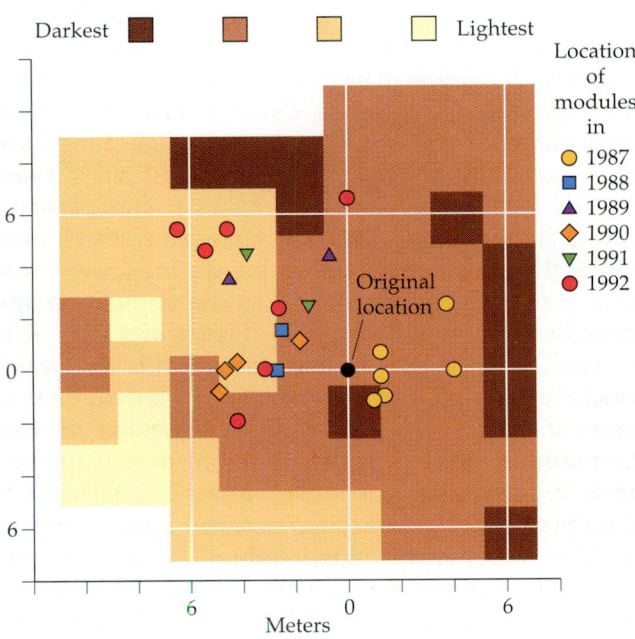

47.10 Modular Organisms May Shift Positions
When a treefall at the left side of the region shown opened up a light gap, a modular population of the shrub *Piper pseudobumbratum* shifted some six meters toward the area of greater light. The symbols indicate locations of modules in different years.

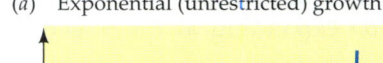

(a) Exponential (unrestricted) growth

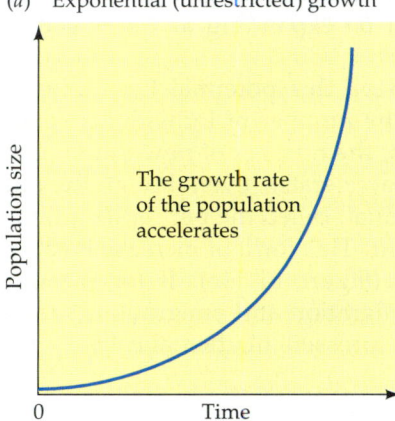

The growth rate of the population accelerates

(b) Logistic (restricted) growth

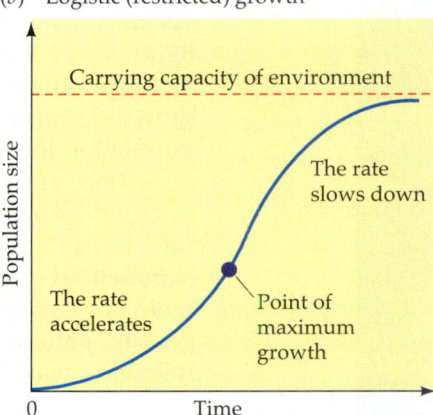

Carrying capacity of environment

The rate slows down

The rate accelerates

Point of maximum growth

47.11 Population Growth Curves
(a) Theoretically, a population in an imaginary environment with unlimited resources could grow like this. (b) Typically, a population in an environment with limited resources grows like this.

the **intrinsic rate of increase;** r_{max} has a characteristic value for each species. Therefore, the rate of growth of a population under optimal conditions is $dN/dt = r_{max}N$.

POPULATION GROWTH WHEN RESOURCES ARE SCARCE

Populations may grow rapidly for short times, but no population can maintain exponential growth for very long because environmental limitations cause birth rates to drop and death rates to rise. In fact, over long time periods the size of most populations fluctuates around a relatively constant number. The simplest way to picture the limits imposed by the environment is to recognize that an environment can support no more than a certain number of individuals of any particular species. This number, called the environmental **carrying capacity,** is determined by the availability of resources—food, nest sites, shelter—as well as by disease, predators, and perhaps social interactions. These limitations mean that, rather than being exponential, population growth follows a curve that flattens out as the population approaches the carrying capacity (Figure 47.11b).

The S-shaped growth pattern, which is characteristic of many populations growing in environments with limited resources, can be represented mathematically by adding to the equation for exponential growth a term that slows the population's growth as it approaches the carrying capacity. The simplest such equation is for **logistic growth,** in which each individual depresses population growth equally:

$$\frac{dN}{dt} = r\left(\frac{K - N}{K}\right)N$$

where K is the carrying capacity and the other symbols are the same as in the equation for exponential growth. The biological assumption in this equation is that each additional individual makes things

slightly worse for the others because it competes with them for available resources or for other reasons. Population growth stops when $N = K$ because then $(K - N) = 0$, so $(K - N)/K = 0$, and thus $dN/dt = 0$.

The logistic growth equation contains some important simplifications that are not characteristic of most populations. The most critical assumptions are that each individual exerts its effects immediately at birth and that all individuals produce equal effects. In nature, however, organisms grow during their lives, and their effects on others normally increase with age, so there may be a delay between the birth of an individual and the time at which it begins to affect the other members of the population. A seedling tree, for example, exerts a much smaller effect on its neighbors than a large adult tree does.

Logistic growth can be demonstrated readily in the laboratory by introducing a few individuals into an environment with abundant (but not unlimited) resources. Under such circumstances populations often grow rapidly, then slow down and eventually begin to maintain a size that fluctuates around K (the carrying capacity), whose value can be varied by altering the amount or availability of resources.

DYNAMICS OF SPECIES RANGES

Environmental conditions, births, deaths, and dispersal determine the density and distribution of a species. Within the geographic range of most species, population densities are higher toward the center of the range and decline toward the periphery, becoming zero beyond the edge of the range of the species. The scissor-tailed flycatcher, a bird that winters in Central America and breeds in the southern Great Plains of the United States, illustrates this pattern of population density (Figure 47.12). Species that are abundant—that is, species that attain high local population densities—usually have larger ranges than species that are not common anywhere.

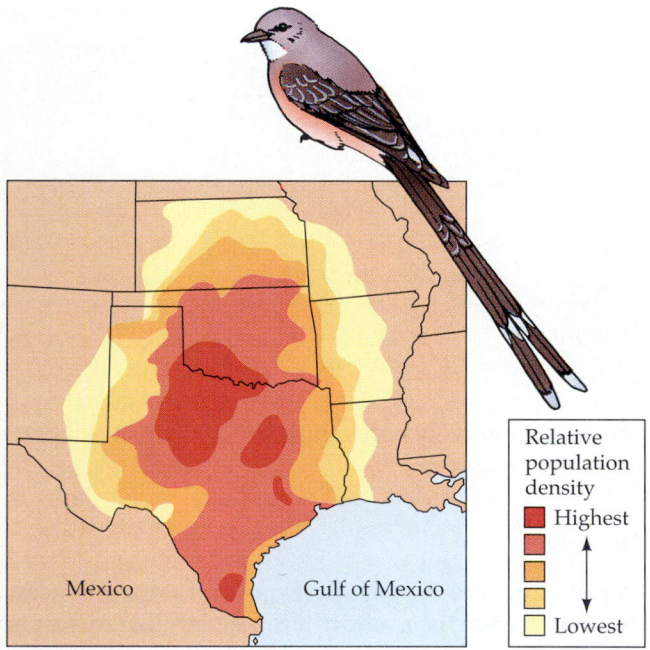

47.12 Population Densities Are Greater toward the Center of a Range
Local population densities of the scissor-tailed flycatcher are highest toward the center of its breeding range and decline gradually toward the periphery. Tan areas are outside the breeding range.

Relative population density

Highest

↕

Lowest

Mexico

Gulf of Mexico

Ecologists believe that the reason for the general pattern of declining abundance toward the periphery of a species' range is that suitable conditions are found more frequently toward the center of a species' range than at the periphery. Similarly, the ranges of rare species are smaller than those of common species because rare species have more-stringent environmental requirements, which are seldom met.

Some species, however, achieve their highest population densities at the margins of their ranges. This situation often arises when range boundaries are at the edges of land masses where there is an abrupt transition from suitable terrestrial habitats to unsuitable marine habitats. Competition between ecologically similar species can also result in abrupt changes in population densities at the edges of ranges, as we will discuss in Chapter 48.

POPULATION REGULATION

As we have seen, some species are more common than others, and population densities are typically higher in the center of a species' range and decline toward its periphery. Populations of some species fluctuate substantially in abundance, but many populations fail to attract our attention precisely because their numbers do not fluctuate very much. This con-

stancy suggests that in nature population size is regulated in some way. The activities of individuals that we have already discussed—births, deaths, and dispersal—as they are influenced by the density of individuals and by disturbances that affect their environments, are what regulate population size.

Density Dependence

A population may be more likely to decrease in density when its members are common and to increase in density when they are rare. Regulation of a population by changes in per capita birth or death rates in response to density is said to be **density-dependent**. Death or birth rates may be density-dependent for several reasons. First, as a species increases in abundance, it may deplete its food supply, reducing the amount that each individual gets. This decreased nutrition may increase death rates and lower birth rates. Second, predators may be attracted to regions where densities of their prey have increased. If the predators are then able to capture a larger proportion of the prey than when the prey were scarce, the per capita death rate of the prey rises. Third, diseases, which may increase death rates, spread more easily in dense populations than in sparse populations.

If the per capita birth and death rates in a population are unrelated to the population's density, population regulation is said to be **density-independent**. For example, a very cold spell in winter may kill a large proportion of individuals of a species regardless of the population's density. Even seemingly density-*independent* environmental factors, however, may result indirectly in density-*dependent* population regulation. The cold weather may not kill organisms directly but may increase the amount of food individuals need to eat each day. Individuals pushed by population density into poor foraging areas may then be more likely to die than those in better foraging areas. Or the death rate may be related to the quality of sleeping places. If the population density is high, a larger fraction of individuals may be forced to sleep in places that expose them more to the cold.

Density dependence is an important concept for understanding population regulation. Various combinations of density-dependent and density-independent events can regulate a population. The graphs in Figure 47.13 show how birth and death rates can change in relation to population density. If birth or death rates or both are density-dependent, the population responds to increases or decreases in its density by returning to an equilibrium density. If neither rate is density-dependent, there is no equilibrium, but, as we have just discussed, some of the agents influencing birth and death rates are likely to act in a density-dependent manner. The abundance of a species is determined by the combined effects of all

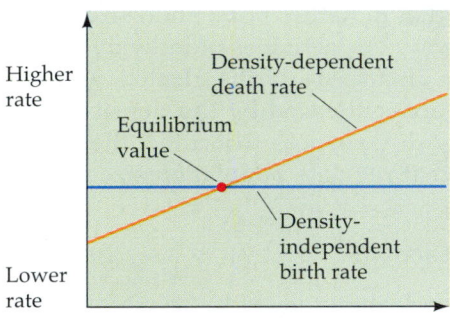
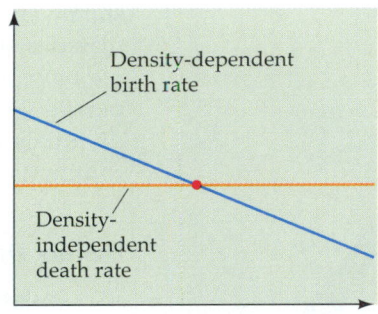
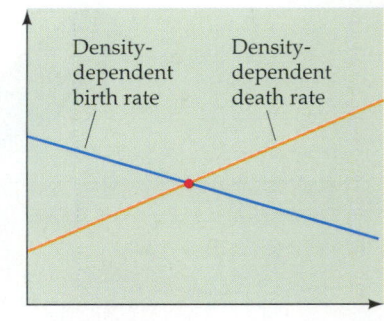

Population density

47.13 Density Dependence Regulates Population Size
If birth rate or death rate or both are density-dependent, a population's size tends to fluctuate around an equilibrium value.

the factors and processes, density-dependent and density-independent, impinging upon its subpopulations.

An experiment to test whether a population was controlled by density-dependent death rates was performed on a gall-forming fly (*Eurosta solidaginis*) that infests goldenrod (*Solidago altissima*), a common herbaceous plant in fields in eastern North America. *Eurosta* is widespread, but it exists at moderately low and relatively constant densities. Adult flies emerge in early June from galls (swellings of the plant stem) in which they overwintered. A female fly lays her eggs in a bud at the tip of a growing goldenrod stem. After it hatches, the larva bores into the stem and develops inside a gall, which forms rapidly, reaching full size in early July (Figure 47.14). In September the fully developed larva, which has fed on the gall tissue, pupates within the gall after excavating an exit tunnel through which it will emerge. There is only one larva per gall, and most stems have only one gall.

Eurosta larvae and pupae are attacked by a suite of predators that includes a parasitoid wasp, a burrowing beetle, and birds. To determine whether predation rates act on *Eurosta* in a density-dependent manner, an ecologist collected all the galls from a number of plots, kept them in an unheated room during the winter, and released the flies and their predatory wasps into the fields at the appropriate time the following spring. On some plots more flies were released than had been present the previous year; on other plots fewer flies were released. That is, population densities were augmented or reduced. At the end of the following winter all galls were collected, and survival rates of *Eurosta* were determined by dissecting the galls. Death rates during the year were strongly density-dependent; that is, they were highest in the experimental plots where the greatest density of flies was released. Thus populations of *Eurosta* in the study area appear to be main-

tained at relatively constant levels by density-dependent predation.

Disturbance

Populations are repeatedly exposed to disturbances. A **disturbance** is a short-term event that disrupts populations, communities, or ecosystems by changing their environment. Common physical disturbances are fire, hurricanes, ice storms, floods, landslides, and lava flows. Biological disturbances include

47.14 Density-Dependent Predation Regulates an Insect Population
This goldenrod was stimulated to form multiple galls when a female *Eurosta* fly laid her eggs on the stem. A larval fly will hatch from each gall. The death rate from predation on *Eurosta* larvae and pupae is greatest when the fly's population density is greatest, and lower when density is low; thus the *Eurosta* population is maintained at a relatively constant level.

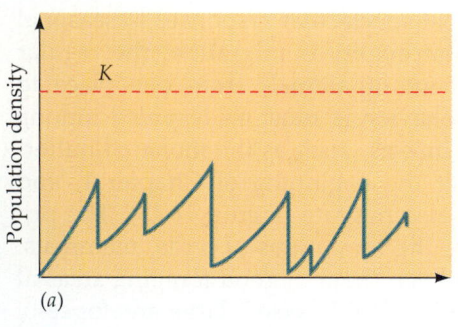
(a)

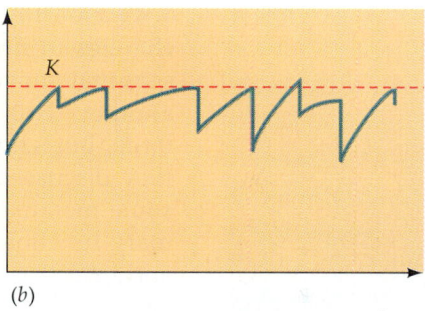
(b)

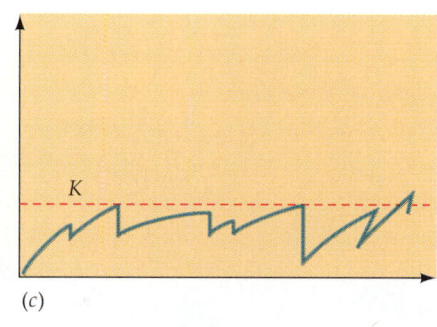
(c)

Time ⟶

47.15 Disturbances Regulate Population Size
(a) Dynamics of a population dominated by phases of population growth after repeated disasters. The population density is well below *K* (the carrying capacity) most of the time. (b) Dynamics of a population dominated by limitations on environmental carrying capacity. The population density is close to *K* most of the time. (c) Same as (b) but with a much lower carrying capacity.

tree falls, diseases, and the burrowing and trampling activities of animals. Disturbances differ in spatial distribution, frequency, predictability, and severity.

In general, small organisms are affected more by environmental disturbances than large ones are. For example, most insect populations in the temperate zone are constantly recovering from disturbances. Many bird species, on the other hand, appear to be close to the environmental carrying capacity much of the time. Insect populations often behave according the pattern in Figure 47.15a, whereas birds generally follow a pattern that is similar to those in Figures 47.15b and c.

Responses of organisms depend upon the frequency and severity of disturbances. Organisms respond behaviorally and physiologically to disturbances that occur regularly and repeatedly during their lifetimes. Animals seek shelter in storms and go into hibernation in winter. Trees drop their leaves in winter and change physiologically so that they can tolerate frosts. If a disturbance is unusually severe, however, the tolerances of individuals may be exceeded and many may die. An organism is more likely to have adaptations for tolerating a particular type of disturbance if the disturbance is frequent relative to the organism's life span.

Adaptations to disturbance are widespread among organisms. Many plants resprout after their aboveground parts are destroyed by fire. Some plants of fire-prone environments mature their seeds in tough capsules or cones that remain closed until heated to very high temperatures. The heat of a passing fire causes the capsules to open and release the seeds within a few days. The seeds then fall and germinate in the bed of ashes (Figure 47.16).

Although generally we do not think of organisms as affecting disturbances, organisms can influence

the frequency of some disturbances. Immediately after a fire there is not enough combustible organic matter to carry another fire. As vegetation grows back, however, dead wood, branches, and leaves accumulate, gradually increasing the supply of fuel to support another fire. Thus the frequency of fires may be proportional to the rate at which vegetation accumulates as fuel. As many trees age, their roots become weakened by fungal infections. Old, large trees are thus susceptible to being toppled by high winds. Therefore, the likelihood of a major blowdown increases with forest age. The waves of blowdowns that move across fir forests in northern New England generating a complex pattern of forests at different ages of regrowth are a striking example (Figure 47.17).

Coping with Habitat Changes

A common response within animal populations to disturbances or other changes in environments is the

47.16 Opened by Fire
These capsules of *Eucalyptus pyriformis*, a tree native to Australia, were opened by the heat of a brush fire. They scattered their seeds onto the bed of ashes, where some of the seeds germinated.

47.17 Waves of Mortality among Fir Trees
The green strips are young trees growing where old trees
were killed by being blown over some time in the past.
The gray strips are the remains of trees blown over more
recently.

dispersal of some members. If habitat quality declines
greatly, individuals may be able to improve their survival and reproduction by going elsewhere. If regularly repeated seasons cause temporal changes in a
habitat, organisms may adjust their life cycles in
equally regular ways, appearing to anticipate the
changes.

One of the most spectacular responses to seasonal
changes in habitat quality is **migration,** the regular
seasonal movement of animals from one place to
another. This behavior is most widespread among
birds, but some insects, such as the monarch butterflies discussed at the beginning of Chapter 1, and
some mammals also migrate (Figure 47.18). The primary function of the migrations of birds, mammals,
and insects is to keep them in good foraging areas at
all times of the year. Wildebeests, large antelopes of
East Africa, follow the rains to places where there is
fresh growth of grasses. Most insectivorous birds migrate in winter from high latitudes to more favorable
wintering grounds at low latitudes.

Movements of animals that are less regular than
annual migrations may be caused by **irruptions,**
buildups of large populations when food supplies are
favorable. Irruptions are often followed by mass dispersal when supplies diminish. The most famous
case is that of "migratory" locusts in Africa that irrupt
in certain centers and spread out over areas where
they do not normally live, devouring nearly everything in their path (Figure 47.19).

POPULATION REGULATION UNDER HUMAN MANAGEMENT

For many centuries, humans have tried to decrease
populations of species they consider undesirable and
to increase populations of desirable species. Desire
for such control has motivated many studies of natural populations. The desirability of species changes
as new uses for their products are discovered or old
products lose their markets. For example, guayule
(*Parthenium argentatum*) is a shrub of the deserts of
the southwestern United States and Mexico that produces a milky sap with many of the properties of
rubber (Figure 47.20). Once considered to have no
commercial value, guayule has rapidly become a
plant of great interest. It is now being cultivated, and
the rubberlike properties of its sap are the subject of
intensive research.

Although other examples of a shift in our attitude
regarding the value of a particular organism exist, in
general our perceptions of the desirability or unde-
sirability of species have remained stable. Animals
that eat crop plants, for instance, have long been

47.18 Animals Migrate to Remain in Suitable Environments
These caribou in the American Arctic are migrating from
the open tundra to winter feeding grounds at the edge of
the boreal forest where food, some of it in the form of
lichens on the branches of trees, is more readily available
when the ground is covered with snow.

high birth and growth rates. Hunting seasons for birds and mammals are determined with this objective in mind.

Certain populations of organisms can sustain their yields despite a high rate of harvest. Such populations (many species of fish, for example) reproduce at high rates, each female laying thousands or millions of eggs. Another characteristic of these high-yielding populations is that the growth of offspring is density-dependent and lasts a long time. If prereproductive individuals are harvested at a high rate, the growth rates of the remaining individuals normally increase. Most fish populations can be harvested heavily on a sustained basis because only a small number of females must survive to reproductive age to produce the eggs needed to maintain the population.

Fish can, of course, be overharvested. Many populations have been greatly reduced because so many individuals were harvested that too few reproductive adults survived to maintain the population. The Georges Bank off the coast of New England—a source of cod, halibut, and other prime food fishes—is a good example. As fish populations were reduced, more effective equipment had to be used for their harvest, more effort had to be expended to get the same yield, income to fishermen dropped, and the price of fish to the consumer increased.

The whaling industry has also engaged in excessive harvest. The blue whale, Earth's largest animal, was the first whale species to be hunted nearly to extinction. The industry then turned successively to smaller and smaller species of whales, the only ones still numerous enough to support commercially viable whaling operations (Figure 47.21). Management of whale populations is difficult for two reasons. First, whales reproduce at very low rates because they have long prereproductive periods, produce only one offspring at a time, and have long intervals between births. Thus many whales are needed to produce even a small number of offspring. Second, because whales are distributed widely throughout Earth's oceans, they are an international resource whose conservation and wise management depends on cooperative action by *all* whaling nations. Recovery of whale populations will require observance of the moratorium on fishing for whales currently in effect.

The same principles apply if we wish to reduce the size of populations of undesirable species and

47.19 Irruptions Stimulate Locusts to Migrate
A swarm of migratory locusts like this one over a cotton field in Ethiopia can totally defoliate a field within minutes. These insects are captured and eaten by people; they are considered a delicacy in some areas.

considered undesirable, whereas many kinds of fish, game birds, and mammals are consistently thought to be desirable.

We base our policies for controlling and managing populations of organisms on our understanding of how populations grow and are regulated. A general principle from population studies is that the total number of births and the growth rates of individuals tend to be highest when a population is well below its carrying capacity. Therefore, if we wish to maximize the number of individuals that can be harvested, we should manage the populations so that they are far enough below carrying capacity to have

47.20 Guayule Is Now a Valuable Crop
A desert shrub with no commercial value until recently, guayule is now cultivated because it produces a sap that can be refined to yield materials that are rubber substitutes.

Blue whales

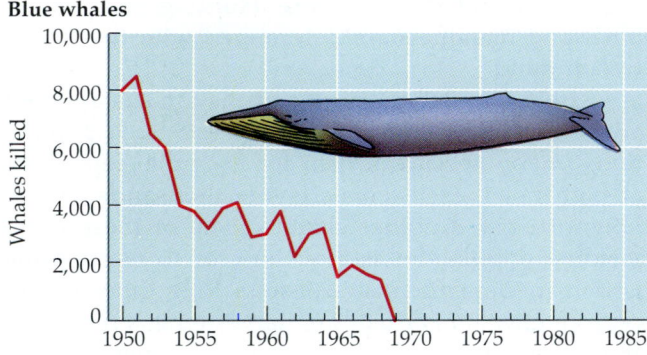

Fin whales

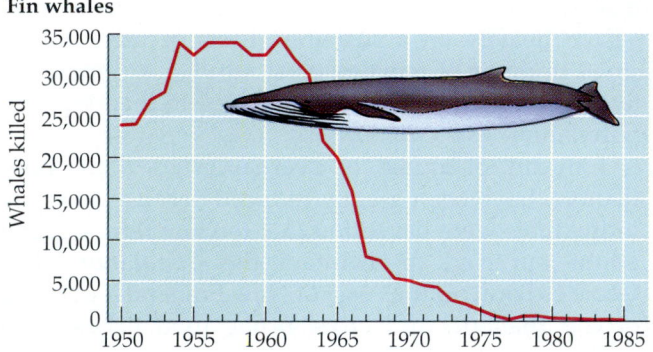

Sei whales

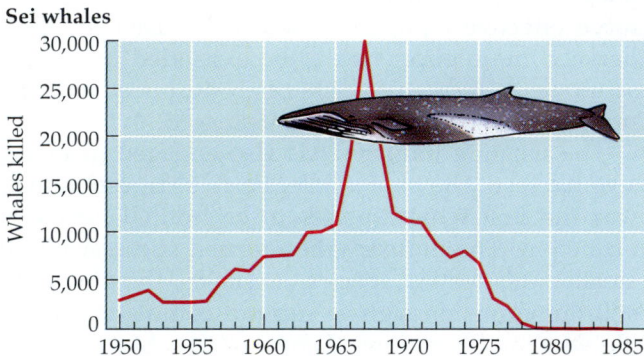

Sperm whales

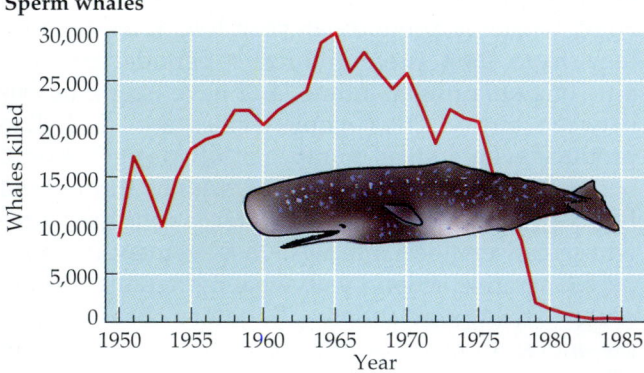

47.21   Overexploitation of Whales
These graphs show the number of whales of four species killed each year from 1969 to 1985. The graph for the largest species is on the top; the one for the smallest species is at the bottom. Many whale species were hunted almost to extinction in the nineteenth and early twentieth centuries. The alarming decline in whale populations led to a moratorium on commercial whaling in 1985, which most of the world's nations abide by.

keep them at low densities. At densities well below carrying capacity, populations have high birth rates and can therefore withstand higher death rates than they could close to carrying capacity. Killing part of the population is only a temporary measure that the organisms, by their resulting increased reproduction, usually counteract swiftly. A far more effective approach to reducing the population of a species is to remove its resources, thereby lowering the carrying capacity of its environment. We can rid our dumps and cities of rats more easily by making garbage unavailable (reducing the carrying capacity of the rats' environment) rather than by poisoning the rats.

Similarly, if we wish to preserve a rare species, the most important step is to provide it with suitable habitat. If habitat is available, the species will usually reproduce at rates sufficient to maintain the population. If the habitat is insufficient, preserving the species usually requires expensive and continuing intervention, such as providing extra food.

Managing the Human Population

Managing our own population has become an urgent ecological issue because human population growth is responsible for most environmental problems, from pollution to extinctions of other species. For thousands of years, Earth's carrying capacity for human populations was determined by food supply, water supply, and disease—that is, by the technology available to garner resources and to combat diseases. The carrying capacity was then at a low level. Domestication of plants and animals and cultivation of land enabled our ancestors to increase dramatically the resources at their disposal. These developments stimulated rapid population growth to the next carrying capacity limit, which was determined by the agricultural productivity possible with only human- and animal-powered tools. Agricultural machines and artificial fertilizers, made possible by the tapping of fossil fuels, greatly increased agricultural productivity, raising Earth's carrying capacity for people. The development of modern medicine reduced the effectiveness of diseases as limiting factors on human populations, raising the global carrying capacity still further (Figure 47.22). In combination with hygiene, medicine has allowed people to live in large numbers in areas where diseases formerly kept numbers very low.

What is Earth's present carrying capacity for people? We have arrived at the stage where carrying capacity is determined by Earth's ability to absorb the by-products of our enormous consumption of fossil fuel energy and by whether we are willing to cause the extinction of millions of species to accommodate our increasing use of environmental resources. We must choose how to limit the total size of our population by considering the kinds of lifestyles possible at different densities and the fates of other organisms on Earth. We will explore some of the consequences of high human population densities for the survival of other species in Chapter 51.

47.22 Human Population Growth

For thousands of years Earth's human population was relatively stable. Recently the human population has grown nearly exponentially as Earth's carrying capacity for people increased. Current rates of growth, however, have taxed even this increased carrying capacity.

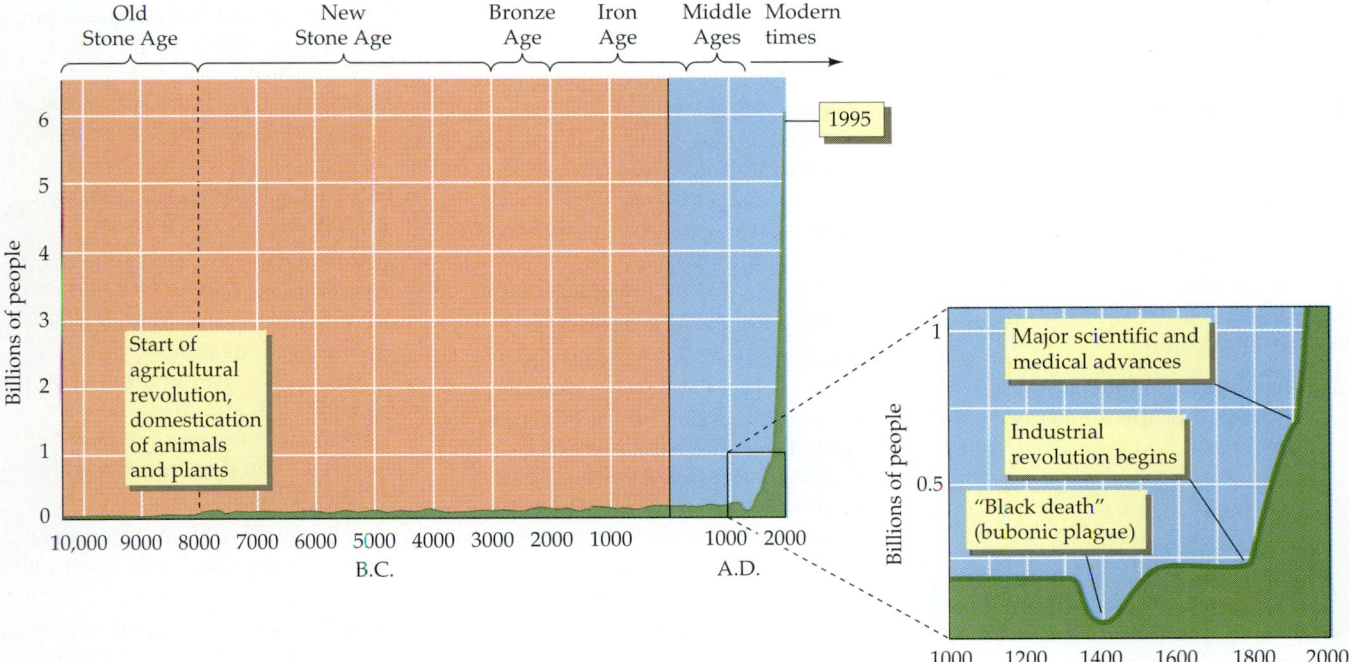

SUMMARY of Main Ideas about the Structure and Dynamics of Populations

All life histories have stages of growth, dispersal, reproduction, energy and nutrient gathering, and resting or reorganization. How long these stages last and how much they overlap vary from species to species.
    Review Figure 47.1

Life history evolution is limited by trade-offs that exist because increasing performance via one trait often reduces performance via others.
    Review Figure 47.3

A population's structure is determined by the density, spacing, and age distribution of its members.
    Review Figure 47.6

Life tables record the numbers of individuals born at a given time that are still alive at subsequent time intervals and the number of offspring born to them during each of these intervals.
    Review Table 47.1

A population with unlimited resources could grow exponentially.
    Review Figure 47.11*a*

A population with limited resources may grow logistically.
    Review Figure 47.11*b*

Population density is often greater toward the center of a species' range than toward its periphery.
    Review Figure 47.12

Many populations are regulated by factors that act in a density-dependent manner.
    Review Figure 47.13

The dynamics of many populations are dominated by repeated disturbances that cause high death rates.
    Review Figures 47.15*a* **and 47.17**

People often manage populations to increase the sizes of desirable populations and decrease the sizes of undesirable ones.

Managing the human population has become the most urgent environmental and social problem of the present time.
    Review Figure 47.22

SELF-QUIZ

1. The number of individuals of a species per unit of area is known as its
   a. population size.
   b. population density.
   c. population structure.
   d. subpopulation.
   e. biomass.

2. The age distribution of a population is determined by
   a. the timing of births.
   b. the timing of deaths.
   c. the timing of births and deaths.
   d. the rate at which the population is growing.
   e. the rate at which the population is decreasing.

3. Which of the following is *not* a component of the life history of all organisms?
   a. Growth
   b. Dispersal
   c. Reproduction
   d. Reorganization
   e. Energy gathering

4. Which of the following is *not* a demographic event?
   a. Growth
   b. Birth
   c. Death
   d. Immigration
   e. Emigration

5. A group of individuals born at the same time is known as a
   a. deme.
   b. subpopulation.
   c. Mendelian population.
   d. cohort.
   e. taxon.

6. A population grows at its intrinsic rate of increase when
   a. its birth rate is the highest.
   b. its death rate is the lowest.
   c. conditions are optimal.
   d. it is close to the environmental carrying capacity.
   e. it is well below the environmental carrying capacity.

7. Some organisms reproduce only once in their lifetime because they
   a. invest so much in reproduction that they have insufficient reserves for survival.
   b. produce so many offspring at one time that they do not need to survive longer.
   c. don't have enough eggs to reproduce again.
   d. don't have enough sperm to reproduce again.
   e. have stopped growing.

8. Which of the following is *not* true of reproductive value?
   a. It is a measure of the contribution an individual of age *x* can make to the growth rate of the population.

   b. It increases until an individual begins to reproduce.
   c. It reaches its maximum when an individual completes reproduction.
   d. It usually reaches its maximum when an individual begins to reproduce.
   e. It usually declines during the reproductive life of an individual.

9. Density-dependent population regulation results when
   a. only birth rates change in response to density.
   b. only death rates change in response to density.
   c. diseases spread in a population.
   d. both birth and death rates change in response to density.
   e. population densities fluctuate very little.

10. The best way to reduce the population of an undesirable species is to
    a. reduce the carrying capacity of the environment for the species.
    b. selectively kill reproducing adults.
    c. selectively kill prereproductive individuals.
    d. attempt to kill individuals of all ages.
    e. sterilize individuals.

FOR STUDY

1. Huntington's disease is a severe disorder of the human nervous system that generally results in death. It is caused by a dominant allele that does not usually express itself phenotypically until its bearer is 35 to 40 years old. How quickly is the gene that causes Huntington's disease likely to be eliminated from the human population? How would your answer change if the gene expressed itself when its bearer was 20 years old? 10 years old?

2. Many people have improperly formed wisdom teeth and must spend a lot of money to have them removed. Assuming, as is probably the case, that the presence or absence of wisdom teeth and their mode of development are partly under genetic control, will we gradually lose our wisdom teeth by evolutionary processes?

3. Some organisms, such as oysters and elm trees, produce vast quantities of offspring, nearly all of which die before they reach adulthood. What fraction of such deaths are likely to be selective, that is, dependent on the genotypes of the individuals that die? If most such deaths are nonselective, what does that imply for the rates of evolution of oysters and elms?

4. Most organisms whose population densities we wish to increase are long-lived and have low reproductive rates, whereas most organisms whose populations we attempt to reduce are short-lived but have high reproductive rates. What is the significance of this difference for management strategies and effectiveness of management practices?

5. In the mid-nineteenth century, the human population of Ireland depended largely on a single food crop, the potato. When a disease caused the potato crop to fail, the Irish population declined drastically for three reasons: (1) A large

percentage of the population emigrated to the United States and to other countries, (2) the average age of a woman at marriage increased from about 20 to about 30 years, and (3) many families starved to death. None of these

social changes was planned at the national level, yet all contributed to adjusting population size to the new carrying capacity. Discuss the ecological strategies involved, using examples from other species.

6. From a purely ecological standpoint, can the problem of world hunger ever be overcome by improved agriculture alone? What other components must a hunger-control policy include?

READINGS

Begon, M. and M. Mortimer. 1986. *Population Ecology: A Unified Study of Animals and Plants*, 2nd Edition. Blackwell Scientific Publications, Oxford. An excellent introduction to population dynamics that stresses the differences between plants and animals.

Begon, M., J. L. Harper and C. R. Townsend. 1990. *Ecology: Individuals, Populations, and Communities*, 2nd Edition. Blackwell Scientific Publications, Cambridge, MA. An excellent basic text for all aspects of contemporary ecology.

Charnov, E. L. 1993. *Life History Invariants: Some Explorations of Symmetry in Evolutionary Ecology*. Oxford University Press, Oxford. An excellent but technical treatment of phylogenetic constraints and trade-offs in life histories.

Harper, J. L. 1977. *The Population Biology of Plants*. Academic Press, New York. The most complete review of the literature on plant populations; an excellent advanced reference for most topics in plant population biology.

Hutchinson, G. E. 1978. *An Introduction to Population Ecology*. Yale University Press, New Haven. An excellent advanced text with a particularly good historical account of the development of modern ideas.

Ricklefs, R. E. 1990. *Ecology*, 3rd Edition. W. H. Freeman, New York. An excellent text that covers both dynamic and evolutionary aspects of ecology.

Stearns, S. C. 1992. *The Evolution of Life Histories*. Oxford University Press, Oxford. A thorough coverage of all aspects of life history evolution including a discussion of the major analytical tools used in life history analyses.

Wilson, E. O. and W. H. Bossert. 1971. *A Primer of Population Biology*. Sinauer Associates, Sunderland, MA. A self-teaching book that covers the basic principles of population ecology.

A female parasitic wasp bores through the syconium of a fig tree to lay her eggs on the fig wasp larvae inside. Wasp and fig life cycles are closely tied to one another.

48

Interactions within Ecological Communities

Most species of fig trees can reproduce only with the help of certain wasps. Wasps of most species visit only one fig species and most fig species are visited by only one wasp species. A fig tree begins reproduction by producing a large number of closed inflorescences called syconia. Inside each syconium many female flowers form. Female fig wasps pollinate these flowers. A female fig wasp bearing both her own fertilized eggs and fig pollen enters a syconium through a small hole that soon seals. She pollinates receptive female flowers, lays her eggs in the ovaries of some of the flowers, and dies. Each wasp larva develops within—and eats—one seed. Not all of these larvae develop, however; some are parasitized by another wasp species. The parasitic female punctures the fig with her long ovipositor and lays her eggs on the fig wasp larvae.

As the syconium ripens, pollen-bearing male flowers mature. Just before the fruit ripens, the surviving male fig wasps chew their way out of the seeds in which they developed and crawl around inside the syconium searching for seeds housing female wasps. The males chew open these seeds and mate with the females. Females emerge from their seeds and collect pollen from male flowers. Females of most species have specialized structures on their thoraxes to hold the pollen. The females leave the syconium through a hole cut in its wall by males, fly to another fig tree, and begin the reproductive cycle again. The syconium finishes ripening, becoming a sweet fig that may eventually be eaten by a bird or mammal.

Fig wasps depend completely on figs to complete their life cycles. To produce offspring, a female fig wasp must carry pollen to a receptive fruit and pollinate the flowers. Otherwise the ovaries of the flowers will not develop into seeds that can be eaten by wasp larvae. In some fig species, only one female wasp normally enters a syconium. If she is not carrying pollen, none of her offspring survive.

A fig pays a price to get its flowers pollinated: Wasp larvae consume many potential fig seeds. Seeds eaten by fig wasps are part of the investment a fig makes in order to receive and donate pollen. If a syconium is not visited by a fig wasp bearing pollen, no seeds mature and the fruit is aborted.

Organisms that live together in a small area and interact with one another form an **ecological community.** The figs and fig wasps, for example, are part of a community that includes other plants and animals, fungi, protists, bacteria, and archaebacteria. Each species interacts in unique ways with other species in its community. The study of such interactions, and how they determine which and how many species live in a place, is the focus of community ecology and the subject of this chapter.

TYPES OF ECOLOGICAL INTERACTIONS

The relationship between figs and fig wasps shows that populations of organisms interact with one another in complex and surprising ways. These interactions can be grouped into a few major types according to their consequences. One organism, by its activities, may benefit itself while harming another. Such a relationship is typical when one species uses another as its source of resources. The eater is called a predator or parasite, and the eaten is its prey or host. Their interaction is known as a **predator–prey** or **host–parasite** interaction. Alternatively, two organisms may mutually harm one another. This type of interaction is common if the two organisms use the same resources and those resources are insufficient to supply their combined needs. Such organisms are called competitors, and their interactions constitute **competition.** If both participants benefit from an interaction, we call them mutualists, and their interaction is a **mutualism.** If one participant benefits but the other is unaffected, the interaction is a **commensalism.** If one participant is harmed but the other is unaffected, the interaction is an **amensalism.** Table 48.1 summarizes these major types of interactions.

As the interaction between the fig and the fig wasp illustrates, these types of interactions are not clearcut, both because the strengths of interactions vary and because many cases do not fit the categories neatly. Both figs and fig wasps are completely dependent on their mutualism, but the fig wasp is also a predator on fig seeds. And other organisms also prey on these seeds. Some of the birds and mammals that eat figs kill fig seeds, whereas others pass seeds unharmed through their digestive tracts, thus dispersing them to new locations. Although we can list exceptions, many interactions fit well within the categories in the simple scheme shown in Table 48.1, so we will use these categories as a guide for exploring interactions among species in this chapter, beginning with predation and competition.

RESOURCES AND CONSUMERS

Many interactions between organisms within communities center on resources and consumers. A **resource** is any substance directly consumed by an organism that can potentially lead to the growth of its population. The two key properties of a resource are that its amount or availability is reduced by being consumed and that it is used by a consumer for maintenance and growth. We usually think first of resources that can be consumed by being eaten, but organisms also consume space, hiding places, and nest sites by occupying them. Factors such as temperature, humidity, salinity, and pH, even though they may strongly affect population growth, are not resources because they are not consumed. Some resources, such as space, are not altered by being used. Occupied space may be unavailable to other organisms, but it immediately becomes available for occupancy when the user leaves. Other resources must regenerate after being reduced in quantity by consumers if they are to be available in the community in the future.

Ecologists say that each species has an **ecological niche,** an abstraction that sums up the ecological position of the species within the community and the range of physical conditions and resources required by the species. A species' **fundamental niche** is the range of conditions under which it could survive if there were no competitors, predators, or other negative influences. If other species are added, however, the new species might reduce the resources used by the first species, or the new species might prey upon the first species so successfully in certain places that the first species cannot survive there. In the presence of other species, then, the fundamental niche of a species is reduced to its **realized niche.**

The interaction between two species of barnacles, *Balanus balanoides* and *Chthamalus stellatus*, illustrates the differences between fundamental and realized niches. These two species live in the intertidal zone of rocky North Atlantic shores. Adult *Chthamalus*

TABLE 48.1
Types of Ecological Interactions

| | | EFFECTS ON ORGANISM 2 | | |
		BENEFIT	HARM	NO EFFECT
EFFECTS ON ORGANISM 1	BENEFIT	Mutualism	Predation or parasitism	Commensalism
	HARM	Predation or parasitism	Competition	Amensalism
	NO EFFECT	Commensalism	Amensalism	—

generally live higher in the intertidal zone than do adult *Balanus*, but young *Chthamalus* settle in large numbers in the *Balanus* zone. In the absence of *Balanus*, young *Chthamalus* survive and grow well in the *Balanus* zone, but if *Balanus* are present the *Chthamalus* in the *Balanus* zone become eliminated by being smothered, crushed, or undercut by the larger, rapidly growing species. Young *Balanus* may settle in the higher part of the intertidal zone, but they grow poorly because they lose water rapidly when exposed to air; therefore *Chthamalus* compete successfully with them there. The result is intertidal zonation, with *Chthamalus* growing above *Balanus* (Figure 48.1).

The realized niche of *Chthamalus* is more restricted than its fundamental niche because of competition with *Balanus* in the lower region, and the realized niche of *Balanus* is less than its fundamental niche because of the effects of desiccation coupled with competition with *Chthamalus* in the higher region. Experiments have shown that the vertical ranges of adults of both barnacles are greater if the other species is removed. Figure 48.1 shows the differences between fundamental and realized niches of the barnacles for only one dimension: height in the intertidal zone. Other important niche dimensions for barnacles include the type of substrate, the amount of wave action, and the water temperature.

Ecologists studying niches direct their attention toward the differences in environmental requirements among coexisting species. Even though a species needs a resource, that resource may not be significant for understanding the species' functioning in the community. For example, most terrestrial animals have a strict but similar requirement for a certain minimum level of oxygen. Studying the use of oxygen, however, reveals very little about the structure of terrestrial communities because the supply of oxygen is nearly always above that minimum level; animals seldom deplete the supply. The primary resources that influence distributions and abundances of terrestrial species are those that are depletable and renew slowly, such as food. In aquatic environments, however, where oxygen supply is highly variable, organisms regularly deplete dissolved oxygen. Aquatic ecologists, unlike terrestrial ecologists, pay careful attention to oxygen as a key niche variable.

Limiting resources—resources whose supply is less than the demand made upon them by organisms—often influence the boundaries of species' niches. As the previous discussion shows, which resources are limiting may differ among environments, but some resources, such as food supplies, are usually limiting for many organisms in most environments. Therefore, we will examine interactions that relate to food next.

48.1 Fundamental and Realized Niches of Barnacles The width of the red and gold bars is proportional to the density of the populations. The width of the green and purple wedges indicates the importance of the factor in limiting population density. The lower end of the distribution of adult *Chthamalus* is strongly curtailed by interspecific competition with *Balanus*; the upper end of the distribution of adult *Balanus* contracts slightly because of competition with *Chthamalus* and desiccation.

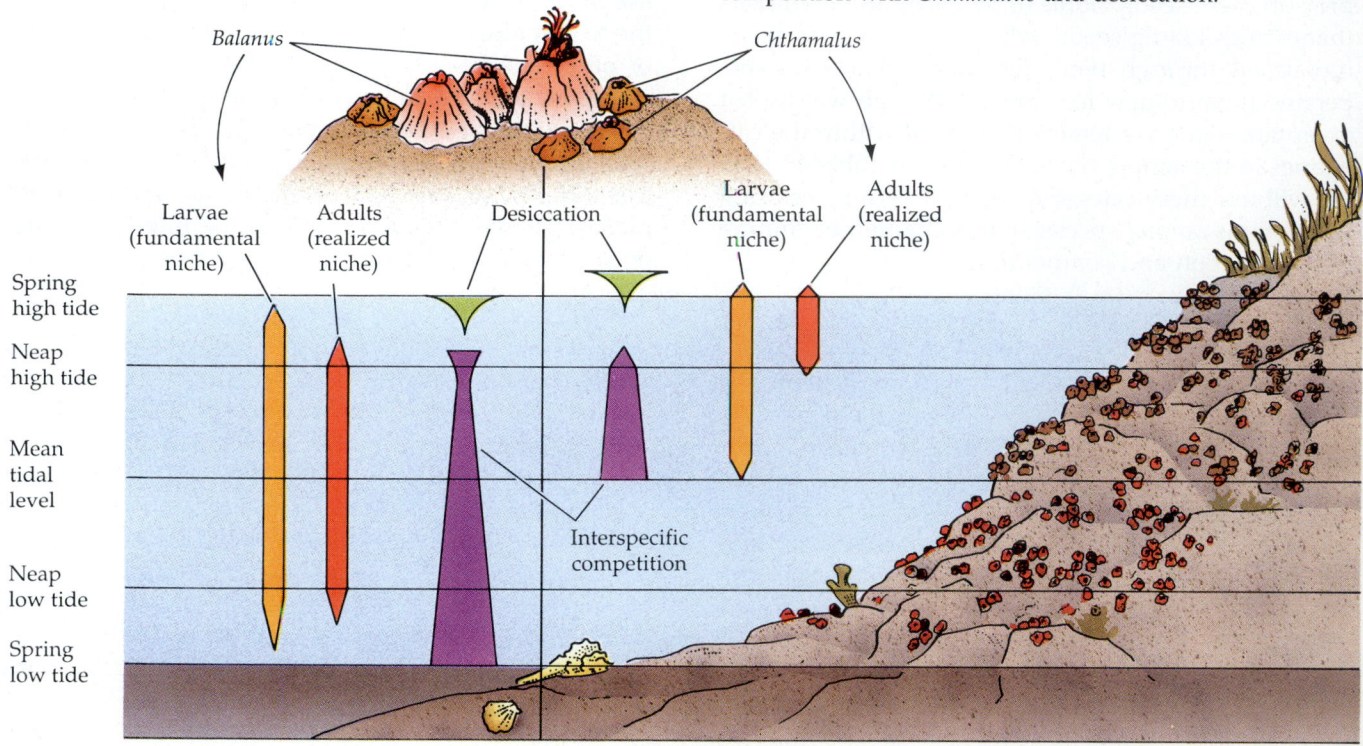

48.2   Parasites and Parasitoids
(a) This Caribbean soldierfish is host to the parasitic isopod attached to its head between its eyes. The fish has no way to remove the isopod, which feeds on its body tissues. (b) A parasitoid wasp of the genus *Ephialtes* lays an egg on a beetle larva deep within a log. Her long ovipositor has extremely sensitive chemosensors that help her locate the larva and determine whether it is already parasitized. The larva that hatches from her egg will consume the beetle larva.

(a)

(b)

PREDATOR–PREY INTERACTIONS

The study of interactions between food resources and their consumers—predator–prey interactions—is a major part of ecological research. In some environments, populations of prey are replaced immediately after being reduced by predators. On rocky marine shorelines, each wave brings a new and undiminished supply of planktonic food to suspension-feeding animals such as barnacles. The numbers of many prey, however, increase only as fast as they can reproduce. If their populations are reduced, their numbers remain low until the next breeding season. How consistently prey are available determines the kinds of behavior predators can employ. Sessile predators, such as web-spinning spiders, mussels, and barnacles, live primarily where their prey are available for long time periods.

The relative sizes of predators and prey strongly influence their interactions because they determine how the predator captures and handles the prey. If the predator is much larger than its prey, the prey are handled in bulk. The world's largest predators, baleen whales, feed on very small prey, which they filter from the water. Predators that are only moderately larger than their prey usually pursue and capture their prey one at a time. Many predators are as small as or even smaller than their prey and feed upon them internally or after attaching to them externally. Such predators are called **parasites** (Figure 48.2*a*). Insects whose larvae are parasites that eventually kill their hosts are known as **parasitoids** (Figure 48.2*b*). The larvae of parasitoid wasps and flies prey on many other species of insects.

Many parasitoids and other parasites are small enough relative to their prey to obtain all the food they need for a complete life cycle from a single prey individual. The parasites choose foods only once or a few times during a life cycle. Usually the egg-laying adult females make the choice. The eggs are laid in food-rich environments, but hosts have defenses against parasites and may be able to destroy the eggs. The ability of a host to defend itself depends upon its condition and on the number and type of parasites attacking it. If a host is already weakened by stresses imposed by the physical environment or shortage of food, parasites are more likely to gain a foothold.

Bark beetles attack pine trees by tunneling into the trunks of trees. They lay their eggs in those tunnels and their larvae excavate more tunnels as they consume the nutritive layers just below the outer bark. A tree defends itself by exuding the sticky pitch for which pines are famous. Weakened trees exude less pitch than healthy trees do, and the more beetles that attack, the greater the probability that they will overcome the tree's defenses. Each colonizing beetle releases a powerful pheromone that attracts other individuals to the same tree. If enough beetles are attracted, their tunneling larvae may eventually kill the tree (Figure 48.3).

Types of Predators

Herbivores eat the tissues of plants. Plant tissues are abundant, but many of them offer poor nutrition for other organisms. Wood is primarily cellulose, itself very difficult to break down, impregnated with lignins, which are even more difficult to digest. Because wood is hard to digest, few organisms attack branches and trunks unless the plant is already weak or dead. Plants also produce roots, leaves, flowers, nectar, pollen, fruits, and seeds, which differ in chemistry, size, structure, and pattern of production. The organisms specialized for eating different plant tissues are correspondingly diverse.

Typical predators, many of which are **carnivores**—

48.3 Pine Attacked by Bark Beetles
Masses of egg-laying bark beetles have attacked this tree. The eggs hatched, and the galleries under the bark were created by the developing larvae burrowing through the tissues, eating as they went.

organisms that eat animals—are the animals we usually think of when predation comes to mind. They are generally moderately larger than their prey, and they pursue and capture their prey one by one. Predators eat many prey individuals during their lives. A small bird in a temperate forest during winter, for example, must eat an average-sized insect or seed every few seconds during the day to maintain itself.

Suspension feeders, as we saw in Chapters 26 and 27, are found in many animal phyla (Brachiopoda, Mollusca, Annelida, the arthropod phyla, Phoronida, Ectoprocta, Echinodermata, Chordata). They feed upon prey much smaller than themselves that are suspended in the water or air. Every suspension feeder has an apparatus with which it extracts prey from the medium. The structure of this apparatus determines the upper and lower size limits of prey that can be captured. Prey that are too small pass through the mesh of the structure; prey that are too large bounce off it. Most suspension feeders depend on movement of the surrounding medium through their filtering apparatus, which can be accomplished by movement of the medium or of the animal. An important feature of suspension feeding is that most of the work of feeding is finished *before* the prey contact the predator. The predator cannot save much energy by rejecting prey; in fact, it must usually expend energy to remove unwanted prey. For this reason most suspension feeders eat most prey that are retained by their collecting devices.

Short-Term Predator–Prey Dynamics

When a typical predator captures and eats a prey individual, it reduces the size of the prey population by one. To understand how numbers of predators and prey fluctuate, consider a simple model consisting of a homogeneous environment in which one predator species eats only one prey species. The predators can find enough to eat if the rate at which they encounter prey is above a certain threshold value. Below that threshold they will lose weight and eventually starve. Nevertheless, the predators may continue to eat prey even when they are scarce, reducing the prey population to an even lower level. Eventually the numbers of predators may be reduced by starvation or emigration, which may allow the prey population to increase its numbers. This increase of prey may, in turn, permit the predators to increase. Because of this pattern, the recovery of a predator population often lags behind the recovery of its prey population. Thus predator–prey interactions often change the population densities of both species, producing oscillations—decreases and increases that appear wavelike when graphed (Figure 48.4a).

Population density oscillations of small mammals living at high latitudes and their predators are the best-known examples of predator–prey oscillations. Populations of Arctic lemmings and their chief predators—snowy owls, jaegers, and Arctic foxes—oscillate with a three- to four-year periodicity. Populations of Canadian lynx and their principal prey—snowshoe hares—oscillate on a 9- to 11-year cycle (Figure 48.4b). Oscillations are most striking and regular in systems in which a few predators are highly dependent upon one prey. Oscillations of populations of predators and their prey are typically reduced if predators have alternate prey.

A detailed field study of oscillations in the densities of a predator population and that of its principal prey was carried out on Isle Royale in Lake Superior. During the hard winter of 1949 a few wolves crossed the ice from Ontario to find a dense population of moose, which had been living in a predator-free environment from the time when they colonized the island early in the century. Populations of both moose and wolves, which have been monitored since 1958, have oscillated in the expected manner (Figure 48.5), but this pattern may be coming to an end. The wolf population, which numbered 50 animals in 1980, has been declining rapidly, and the moose population has been increasing. Only four wolf pups have been born in the past two years, all to the same female in one pack. The other two packs on the island have only two wolves each; the total wolf population is now only 13 animals. Restriction enzyme analysis of the wolves' mitochondrial DNA shows that the

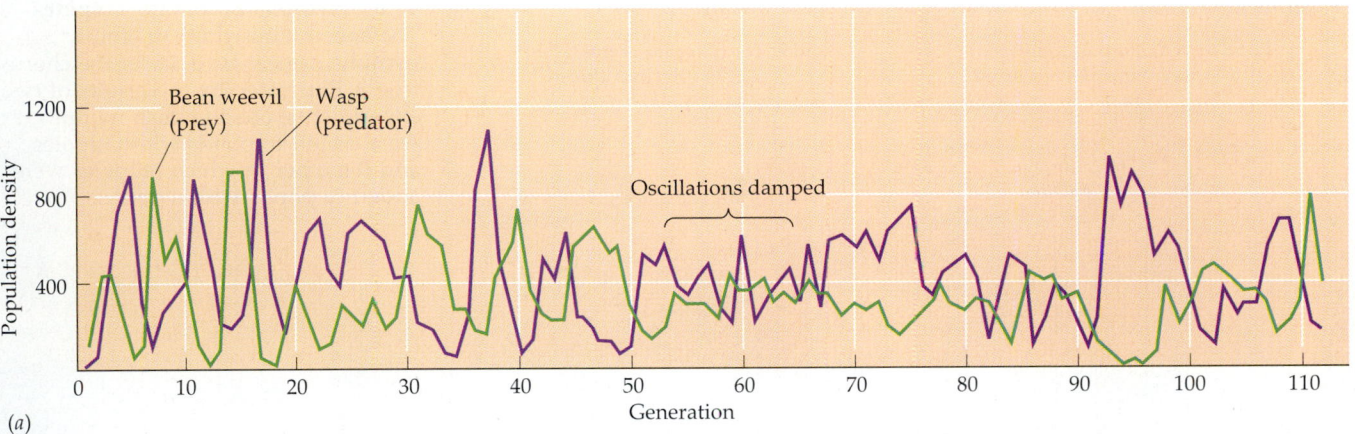

(a)

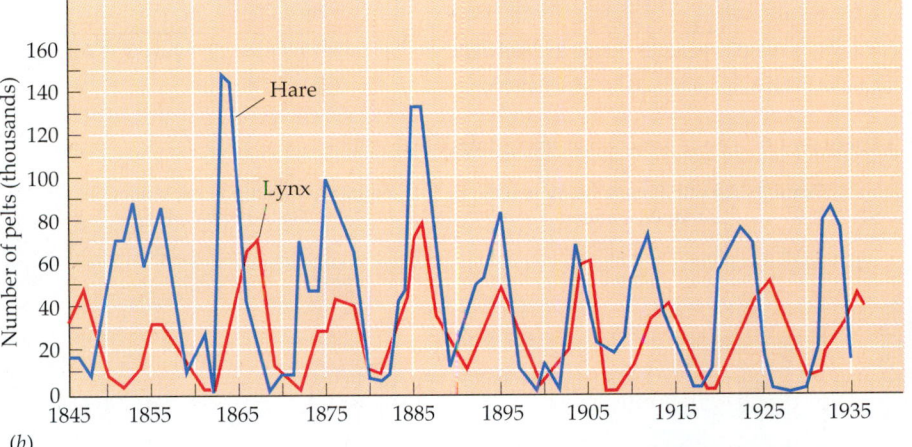

48.4 Predator–Prey Oscillations
(a) Oscillations in sizes of laboratory populations of the azuki bean weevil and its predator, a parasitoid wasp. (b) The 9- to 11-year population cycles of the snowshoe hare and its predator, the lynx, in Canada were revealed by the number of pelts taken by fur trappers and sold to the Hudson Bay Company.

(b)

wolves are all descended from a single female; they have only half the genetic variation of mainland wolves. This lack of genetic variation may make them especially susceptible to a parvovirus, which is responsible for the population decline.

In heterogeneous environments, prey may be eliminated in places where they are more vulnerable to predators but survive where they are better protected. In ponds on islands in Lake Superior, predation restricts the chorus frog (*Pseudacris triseriata*) to

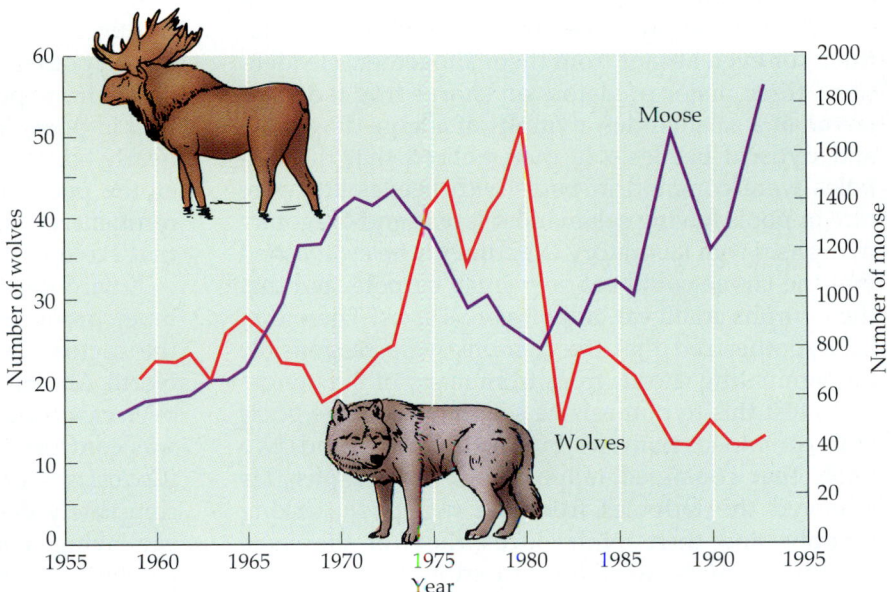

48.5 Wolf Numbers Follow Moose Numbers
The changes in the sizes of moose and wolf populations on Isle Royale between 1965 and 1990 illustrate the complexities of predator–prey dynamics.

Experiment A: Ponds 1 and 2 had no dragonfly nymphs and many tadpoles

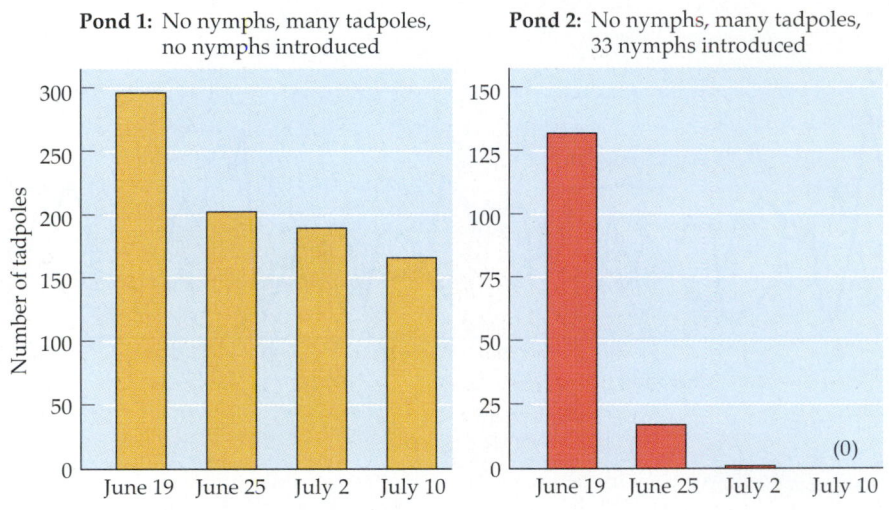

Pond 1: No nymphs, many tadpoles, no nymphs introduced

Pond 2: No nymphs, many tadpoles, 33 nymphs introduced

Number of tadpoles

48.6   Nymphs Eliminate Tadpoles
The speed with which dragonfly nymphs can eat tadpoles of the chorus frog is illustrated by the results of two experiments, one in which nymphs were added to ponds with tadpoles and the other in which tadpoles were added to ponds with nymphs.

Experiment B: Ponds 3 and 4 had many dragonfly nymphs and no tadpoles

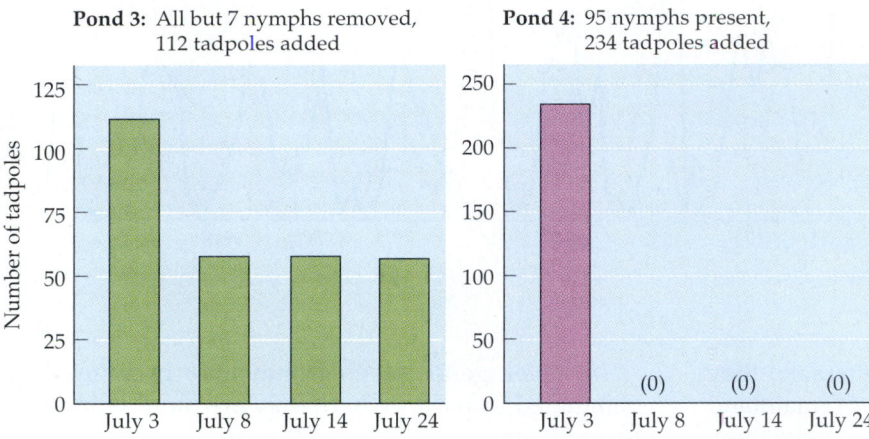

Pond 3: All but 7 nymphs removed, 112 tadpoles added

Pond 4: 95 nymphs present, 234 tadpoles added

Number of tadpoles

a subset of the habitats that would otherwise be suitable for it. An ecologist observed that chorus frogs were rare in many pools that seemed suitable for them, or even absent from them altogether. He identified three major predators on chorus frog tadpoles: larvae of a salamander, nymphs of a large dragonfly, and dytiscid beetles. He then noticed that the tadpoles were common in pools with beetles but were rare in pools having salamander larvae and dragonfly nymphs. From laboratory experiments he established that the larvae could eat only small tadpoles but that the nymphs could eat tadpoles of all sizes. Therefore, he hypothesized that the nymphs were responsible for eliminating chorus frogs from many of the ponds.

To test this hypothesis he selected two large pools that contained nymphs but no tadpoles and two pools that contained tadpoles but no nymphs. He removed the tadpoles from the two pools lacking nymphs and then reintroduced them at the same density. Nymphs were introduced into one of these

pools at typical densities. He also removed nearly all nymphs from the pools that had them and then introduced tadpoles. The dramatic results supported the hypothesis (Figure 48.6). Tadpoles were eliminated from pools with nymphs, but they survived well in pools from which the nymphs were absent or nearly so. Why dragonfly nymphs are not found in all the ponds is unknown. In this example the environment is heterogeneous over space, a condition that ecologists refer to as spatial patchiness.

Environments may also be heterogeneous over time, producing temporal patchiness. For example, the cactus *Opuntia*, introduced into Australia from North America, spread rapidly and became common over vast expanses of valuable sheep-grazing land. It was controlled by the introduction of a moth species (*Cactoblastis cactorum*) whose larvae hatch into and completely destroy patches of *Opuntia* found by the egg-laying females (Figure 48.7). However, new patches of cactus arise in other places from seeds

48.7 A Predator Eliminates Its Prey Locally
Cactoblastis caterpillars consume an *Opuntia* cactus. Voracious *Cactoblastis* larvae will consume this plant and all the other *Opuntia* individuals in the local patch.

dispersed by birds. These new patches flourish until they are found and destroyed by *Cactoblastis*. Today, over a large region, the numbers of both *Opuntia* and *Cactoblastis* are fairly constant and low, but in the local areas that make up the whole, there are vigorous oscillations resulting from extermination of first the prey and then the predator.

Long-Term Predator–Prey Interactions

So far we have considered short-term dynamic behavior of predator–prey systems. As might be expected, predators are agents of evolution as well as agents of mortality. Prey have evolved a rich variety of responses to predation that make them more difficult to capture, subdue, and eat. Among these responses are toxic hairs and bristles, tough spines, noxious chemicals and the means for ejecting them (Figure 48.8), and mimicry of inedible objects or of larger or dangerous organisms. Plants have also evolved defenses, as we will see in a later section. By evolving, the prey become agents of selection too. Predators, in turn, evolve to be more effective in overcoming prey defenses.

MIMICRY. Among the best studied of evolved responses to predation is **mimicry**: taking on the appearance of an inedible or unpalatable item. In **Batesian mimicry** a palatable species mimics a noxious or harmful model. Examples are the mimicry of ants by spiders, of bees and wasps by many different insects (Figure 48.9), of poisonous coral snakes by some spe-

cies of harmless snakes, and of toxic salamanders by palatable ones. Batesian mimicry works because a predator that captures an individual of the unpalatable model species learns to avoid any prey that appears similar. If a predator captures a palatable mimic, however, it is rewarded, and it learns to associate palatability with prey of that appearance. As a result, models are attacked more often than they would be if there were no mimics. Because models that differ from their mimics more than the average are less likely to be attacked by predators that have eaten a mimic, directional selection causes models to evolve away from mimics. Batesian mimicry systems are stable only if a mimic evolves toward a model faster than the model evolves away from it, which

(a)

(b)

48.8 Defenses of Animal Prey
(a) A bombardier beetle (*Brachinus*) ejects a noxious spray at the temperature of boiling water in the direction of an antagonist. The spray is ejected in high-speed pulses more than 20 times in succession. *(b)* The Indo-Pacific lionfish *Pterois volitans* is among the most toxic of all reef fishes. Glands at the base of its spines can inject poison into an attacker; its bright markings are thought to warn potential predators of this capability.

48.9 A Batesian Mimic
This mantispid fly is an effective mimic of a wasp.

all species, including the predator, benefit when inexperienced predators eat individuals of any of the species, because the predators learn rapidly that all similar species are unpalatable. Some of the most spectacular tropical butterflies are members of mimicry systems (Figure 48.10), as are many kinds of bees and wasps.

PARASITE VIRULENCE. The ability of parasites to infect new hosts depends on the ability of the parasite to overcome the defenses of the host, its rate of population growth within a host, and the length of time an infected host survives. A virulent parasite has a more severe effect on its host and may kill it faster than a less virulent parasite. If parasites kill their hosts quickly, they can be transferred to new hosts for only a short time, unless they can continue to be transferred after the host dies. A parasite that multiplies slowly within its host, however, may be outcompeted by faster-multiplying individuals of its own or other species. The level of virulence evolved by a parasite depends on the relative importance of these effects.

Parasites transmitted to new hosts directly or by other organisms benefit greatly if their hosts live long

usually requires the mimic to be less common than the model.

The convergence over evolutionary time in the appearance of two or more unpalatable species is called **Müllerian mimicry.** In a Müllerian mimicry system,

- ■ Highly unpalatable (blue)
- ■ Moderately unpalatable (green)
- ■ Highly palatable (red)
- ■ Palatability not yet tested with birds (gray)

48.10 Müllerian and Batesian Mimics
By converging in appearance, the unpalatable Müllerian mimics among these Costa Rican butterflies and moths reinforce each other in deterring predators; the palatable Batesian mimics get a free ride.

because they have more time during which they can be transferred between hosts. Parasite species with such transmission methods may therefore evolve in ways that make them less virulent to their hosts. By contrast, parasites that are transmitted by physical agents, such as water, may not be affected adversely if their host dies quickly because they can continue to be transmitted after the host has died. Such species seldom evolve in ways that reduce their virulence because parasite genotypes that multiply fast and quickly dominate populations within hosts are readily transmitted after their hosts die. It is thus not surprising that the pathogens responsible for the most severe human diseases, such as dysentery, cholera, and hepatitis, are borne by water.

Parasites that are transmitted to vertebrates by arthropods grow and reproduce only in vertebrates; the arthropods are simply dispersal agents. Vertebrates offer large amounts of tissues that can be converted to parasite growth and reproduction without killing the host. By contrast, the small arthropod vectors would be destroyed quickly if the parasites grew within their bodies. Thus parasites are usually more benign in arthropod vectors than in their vertebrate hosts. The malaria-causing protist *Plasmodium* (see Figure 23.10) has little effect on the mosquitoes that transmit it.

Fig wasps and their nematode parasites illustrate how the ease with which parasites can move to new hosts influences the evolution of parasite virulence. When we introduced the biology of fig wasps at the beginning of the chapter, we did not mention that fig wasps are parasitized by nematodes. Each of 11 species of fig wasps studied by an ecologist in Panama is infected by a single species of nematode of the genus *Parasitodiplogaster*. One of these nematode species parasitizes two wasp species; the other nine parasitize only one each. Immature nematodes crawl onto newly emerged female fig wasps and are carried by them to the next fig. After the fig wasp has laid her eggs, the nematodes enter her body and begin to consume it. Later, about six or seven adult nematodes emerge from the body of the dead wasp, mate, and lay eggs within the same fig in which the host wasp laid her eggs. The nematodes' eggs hatch at the same time the next generation of fig wasps emerges.

If only a single female fig wasp has laid eggs in a fig, her offspring provide the only opportunity for survival of the nematodes' offspring. If broods are founded by more than one female fig wasp, however, a nematode's larvae can survive by parasitizing offspring of other females. The more founding female fig wasps (foundresses), the greater the number of individuals through which the parasites can be transmitted, and the less important are the offspring of the female wasp on which the nematodes arrived.

Based on the theory we have described, the ecologist predicted that nematodes parasitizing fig wasp species in which usually only one female enters a fig should be less virulent than nematodes parasitizing species in which several females usually enter a single fig. Because the bodies of the foundresses are still present in a fig when it ripens, by cutting open figs the ecologist could determine the number of foundresses, and by allowing the brood to complete development he could also count the proportion of fig wasps parasitized by nematodes. As predicted, the nematodes that infested species with multiple foundresses infested a higher proportion of their hosts than those that infested species typically having single foundresses.

PLANT DEFENSES. Leaves are the prey of many predators. Herbivores that eat leaves of herbaceous (nonwoody) plants are called grazers. Herbivores that eat leaves of woody plants are called browsers. Many leaves have physical defenses: They are tough or have hairs or spines. Most leaves also contain chemicals called **secondary compounds** that have negative effects on predators. Many of these substances fit into one or the other of two groups.

One group of defensive secondary compounds, the acute toxins, interferes with herbivore metabolic functions. Some of these toxins, such as nicotine, interfere with transmission of nerve impulses to muscles. Others imitate insect hormones, blocking insect metamorphosis. Still other toxins are unusual amino acids that become incorporated into herbivore proteins and interfere with their functioning (see Box 31.A and Figure 31.13).

Defensive chemicals of the second group make leaves difficult to digest, thus reducing their suitability as food for herbivores. The most common of these substances are tannins, which are present in the leaves of some herbaceous and most woody species. As most leaves age, their concentrations of tannins increase, and the leaves become tougher. Tannins may be present in such large quantities that waters draining from areas dominated by tanniferous plants are tea-colored. The most famous of such "blackwater rivers" is the Rio Negro in Brazil (Figure 48.11).

Some plants respond to being grazed or browsed by increasing the concentrations of defensive chemicals in their leaves. Mountain birches in northern Finland are eaten by caterpillars of the moth *Oporinia autumnata*. When caterpillars attack a birch, the tree responds by increasing the concentrations of defensive chemicals in its leaves, sometimes within a few days of the initial attack. The larvae of *Oporinia*, of another moth, and of two sawflies—all of which feed on birches—develop more slowly when they feed on leaves growing close to leaves heavily damaged by

48.11   The Rio Negro
The dark waters of the Rio Negro (right) contrast strikingly with those of the Solimoes River where they join near Manaus, Brazil.

48.12   Shrubs Dominate Sonoran Desert Communities
Competition for soil moisture keeps the density of shrubs in this community low.

their feeding during the *same* growing season. Caterpillars to which ecologists fed leaves from birch trees that had been severely damaged the previous year grew more slowly than caterpillars fed leaves from birches only lightly damaged the previous year.

COMPETITION

The Sonoran Desert of the southwestern United States and adjacent parts of Mexico is a region with mild winters and very hot summers. Throughout the area, trees grow only along rivers, where their roots can reach subsurface water. Evergreen shrubs, well separated from each other because their roots compete for water, dominate the vegetation (Figure 48.12). Following the first heavy rains of a wet period, shrubs produce new leaves, and seeds of annual plants germinate. The annuals grow rapidly, flower, and produce large quantities of seeds, which fall to the soil at the end of the rainy season. Many animals in the Sonoran Desert eat vegetative tissues of the perennial and annual plants. Some species visit flowers for nectar and pollen; others eat the flowers themselves. Ants, rodents, and birds harvest seeds after they are shed. Birds primarily exploit local dense patches of seeds and take a relatively small proportion of the total seeds produced. Ants and rodents, on the other hand, consume large quantities of seeds.

Ecologists interested in how species interact studied ants and rodents in the Sonoran Desert. They found, as might be expected, that ants on the average eat somewhat smaller seeds than rodents do but that there is much overlap in the sizes of seeds taken by the two kinds of animals (Figure 48.13). The ecologists removed ants from some sites, rodents from other sites, and both ants and rodents from a third set of sites. If either ants or rodents were removed, population densities of the other group increased (Table 48.2). These experimental results demonstrate

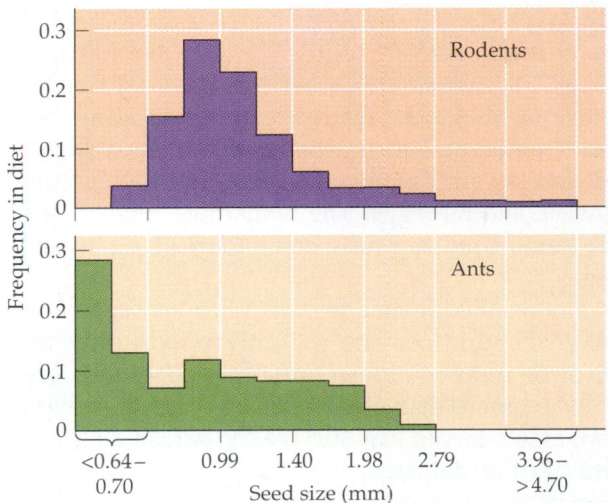

48.13   Sizes of Seeds Harvested by Ants and Rodents
Frequency histograms show the proportions of seeds of each size that make up the diets of ants and rodents.

TABLE 48.2
Experiments Show How Ants and Rodents Interact with Their Food Supply

	RODENTS REMOVED	ANTS REMOVED	RODENTS AND ANTS REMOVED	CONTROL
Number of ant colonies	543	0	0	318
Number of rodents	0	144	0	122
Density of seeds relative to control	1.0	1.0	5.5	1.0

that competition for food links the ants and the rodents, and that ants and rodents greatly reduce seed densities.

If resources are reduced to a great degree by the users—as seeds in the Sonoran Desert are by ants and rodents—the interaction is called **exploitation competition.** When an individual, by its behavior, directly prevents other individuals from gaining access to a resource, the interaction is called **interference competition.** Behaviorally complex animals often interfere with one another; plants and simpler animals usually compete by reducing the supply of resources.

Individuals of the same species or of different species may compete for resources. **Intraspecific competition,** competition among individuals of the same species, may result in reduced growth and reproductive rates of some individuals, may exclude some individuals from better habitats, and may cause the death of others, as we noted in Chapter 47. Few shrubs in Figure 48.12 are very close to others because intraspecific competition for water has thinned the population. **Interspecific competition,** competition among different species, affects individuals in the same way, but in addition, an entire species may be excluded from habitats where it cannot compete successfully. In extreme cases, a competitor may cause the extinction of another species.

As Charles Darwin first pointed out, competition is often most intense among individuals of the same species because they are so similar to one another in size, shape, and requirements. Nonetheless, organisms of different species often overlap greatly in the resources they use, so interspecific competition may also be intense. Interspecific competition is especially strong among plants because most species require the same mineral nutrients—although they need them in different proportions—and all are powered by sunlight.

The first experiments to test interspecific competition were performed by the Russian ecologist G. F. Gause and reported in an influential book, *The Struggle for Existence,* published in 1934. Gause began by conducting experiments with microorganisms in test

tubes within which the environment was homogeneous. In every experiment, the winner completely excluded the loser, an outcome called **competitive exclusion.** Which of two species was the competitive winner depended on conditions in the test tube. The results of one of Gause's experiments with two species of *Paramecium,* in which *Paramecium aurelia* was the winner, are shown in Figure 48.14. In simple laboratory environments such as these, competitive exclusion is common; it is not unusual for a species

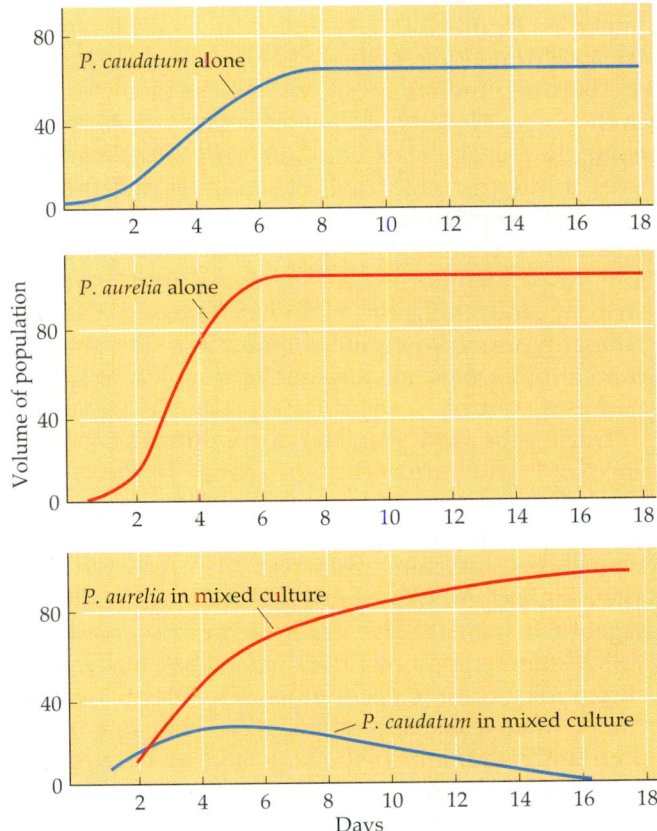

48.14 Competition between Two Protists
Populations of *Paramecium caudatum* and *Paramecium aurelia* each flourish alone. When they are grown together in the same container, however, one species soon eliminates the other.

to be a loser in an environment in which it survives well when alone.

In later experiments Gause was able to prevent competitive exclusion by providing a heterogeneous medium containing some places where one species did better and other places where its competitor did better. For example, *Paramecium bursaria* and *Paramecium aurelia* exclude one another in homogeneous environments, but they form stable mixtures if there is deoxygenated water at the bottom of the containers because *P. bursaria*, which has mutualistic algae within its cell, can feed in deoxygenated water, whereas *P. aurelia* cannot. Although competitive exclusion is a common outcome of laboratory experiments in homogeneous environments, many species with similar ecological requirements live together in most natural communities. A species may persist in an area even if its population densities are reduced by competition.

One way to detect interspecific competition and determine its effects is to remove individuals of one or more potentially competing species and observe the responses of the remaining species. This method is how ecologists identified exploitation competition between ants and rodents in the Sonoran Desert (see Table 48.2) and between barnacles on rocky shorelines (see Figure 48.1). A similar removal experiment was performed to test the hypothesis that the boundary between the ranges of two species of chipmunks in the Sierra Nevada of California was maintained by competition. The least chipmunk (*Eutamias minimus*) lives in the dry sagebrush community, and the yellow-pine chipmunk (*Eutamias amoenus*) lives in the adjacent community of pine and mountain mahogany. Least chipmunks were removed from part of a strip of land along the boundary between the two habitat types; yellow-pine chipmunks were removed from another part; all chipmunks were left in a third part as a control.

During the next year, the distributions of the two species of chipmunks did not change in the control part of the strip, but least chipmunks invaded the pine and mountain mahogany habitat from which the yellow-pine chipmunks had been removed. Yellow-pine chipmunks, however, did not invade the sagebrush habitat. The experimenter observed that yellow-pine chipmunks dominated the smaller least chipmunks during encounters. Therefore, he concluded that least chipmunks were kept out of the pine and mountain mahogany habitat because the yellow-pine chipmunks drove them out, a clear-cut case of interspecific competitive exclusion. Because the absence of the yellow-pine chipmunks from the sagebrush habitat seemed unrelated to interspecific competition, the experimenter hypothesized that the sagebrush habitat was too dry for yellow-pine chipmunks.

Investigators studying plant competition can readily manipulate the densities of the plants and availability of light, water, and nutrients, and can measure growth and reproduction. Ecologists studied competition among two species of weedy wild oats (*Avena fatua* and *Avena ludoviciana*) and domestic oats (*Avena sativa*). They sowed a set of plots, each containing the crop species (*A. sativa*) at the same density. In each plot they also sowed one of the weeds at either high or low density. They also sowed plots with only the two weeds. Some of the plants were harvested when young, others when mature. The plants were dried and weighed, and the investigators compared the weights from the different plots. They found *A. fatua* to be the most vigorous interspecific competitor of the three species under the experimental conditions. *A. fatua* depressed the weight of stems, leaves, and flower heads of the other species more than the other species depressed it. Ranked by competitive success, the three species are *A. fatua* > *A. sativa* > *A. ludoviciana*.

We have discussed predation and competition as if they were very different processes, but they are intimately related. Competitors are actually rival predators that are trying to capture the same "prey," that is, the same resources. Interactions among competitors may also affect their relationships with their own predators. A competitor may force its rival to forage in more dangerous places, thereby exposing it to predators. Or a predator, by preying upon a dominant competitor, may create openings for species that would otherwise be competitively excluded. The sea star *Pisaster ochraceous* is an abundant predator in rocky intertidal communities on the Pacific coast of North America. Its preferred prey are mussels (*Mytilus californianus*), which, in the absence of sea star predation, dominate a broad belt of the intertidal zone and push out other competitors. By feeding heavily on mussels, *Pisaster* creates bare spaces that are taken over by a variety of other species (Figure 48.15).

The role of *Pisaster* was demonstrated experimentally by removing it from selected parts of the intertidal zone repeatedly over a five-year period. The removals resulted in two major changes. First, the lower edge of the mussel bed extended farther down into the intertidal zone where sea stars were removed, showing that sea stars were able to eliminate the mussels completely where they were covered with water most of the time. Second, and more dramatically, 28 species of animals and algae disappeared in the removal zone, until only *Mytilus*, the dominant competitor, occupied the entire substrate. Competition and predation, in combination with physical factors, such as desiccation and wave action, determine which species live in these rocky intertidal communities.

48.15 Its Prey Was a Dominant Competitor
This sea star, *Pisaster ochraceous*, is resting on rocks from
which it has harvested all the mussels. Other organisms,
including the algae visible in the photograph, will colonize the site.

OTHER INTERSPECIFIC INTERACTIONS

During predator–prey and competitive interactions,
one or both partners in the interaction are harmed.
Amensalism causes harm to one of the partners without affecting the other (see Table 48.1). In the other
two types of interspecific interactions—commensalism and mutualism—neither partner is harmed, but
one or both may benefit.

Amensalism and Commensalism

An individual may harm another organism without
benefiting itself by doing so. By trampling, for example, mammals create bare spaces around water
holes. They benefit by drinking water but not by
trampling the plants they kill. Leaves and branches
falling from trees damage smaller plants beneath
them (Figure 48.16). The trees benefit by dropping
old structures whether or not they damage other
plants. Such interactions—amensalisms—are widespread and important. Herbs, shrubs, and small trees
in forests often are damaged more by falling objects
than they are by herbivores.

Commensalism benefits one partner but has no
effect on the other. An example is the relationship
between cattle egrets and grazing mammals. Cattle
egrets are found throughout the tropics and subtropics. They are most often found foraging on the
ground around cattle or other large mammals (Figure
48.17). They concentrate their attention near the
heads of animals, catching insects flushed by the

animals' feet and mouths. Cattle egrets foraging close
to grazing mammals capture more food for less effort
than egrets foraging away from grazing mammals.
The benefit to the egrets is clear; the mammals neither gain nor lose.

Mutualism

Mutualisms—interactions that benefit both participants—are important among virtually all plant and
animal groups. Mutualistic interactions exist between
plants and microorganisms, protists and fungi, plants
and insects, and among plants. Animals also have
mutualistic interactions with protists and with one
another. The evolution of eukaryotic organisms is
believed to be the result of mutualistic interactions
between previously free-living mitochondria and
chloroplasts and the cells they originally infected (see
Chapters 4 and 18).

Some of the most complex and ecologically impor-

48.16 Damage Comes from Above
Shrubs and herbaceous plants are often destroyed by
dead branches falling from tall trees.

48.17 Commensalism
Cattle egrets, such as these individuals foraging around Cape buffalo in East Africa, catch more insects with less work than do egrets foraging away from the larger beasts.

tant mutualisms are between members of different kingdoms. Nitrogen-fixing bacteria of the genus *Rhizobium* receive protection and nutrients from their host plant and provide the plant with nitrogen (see Figure 32.10). Nitrogen for plant growth is often in short supply in terrestrial environments. Many plants have mutualistic associations with fungi attached to their roots (see Box 24.A).

Lichens are compound organisms consisting of highly modified fungi that harbor either cyanobacteria or green algae among their hyphae (see Chapter 24). The fungi absorb water and nutrients and provide a supporting structure; the microorganisms conduct photosynthesis. This mutualistic combination is especially successful at occupying inhospitable habitats such as rock surfaces, tree bark, and bare, hard ground.

Animals have important mutualistic interactions with protists. Corals and some tunicates gain most of their energy from photosynthetic protists that live within their tissues. In exchange, they provide the protists with nutrients from small animals they capture. Termites have nitrogen-fixing protists in their guts that help them digest cellulose in the wood they eat. Young termites must acquire their protists by eating the feces of other termites; if prevented from doing so, they soon die. The protists are given a suitable living environment and an abundant supply of wood.

ANIMAL–ANIMAL MUTUALISMS. Many species of ants have mutualistic relationships with aphids. Ants "milk" these small, plant-sucking insects by stroking them with their forelegs and antennae. The aphids respond by secreting droplets of partly digested plant sap that has passed through their guts. The ants protect the aphids from predatory wasps, beetles, and other natural enemies. The aphids lose nothing,

because plant sap has an excess of sugar relative to the amino acids aphids need.

Some coral reef fishes and shrimps obtain their energy by eating parasites from the scales and gills of larger fish (Figure 48.18). In Africa a species of wading bird removes parasites from among crocodiles' teeth. These mutualisms are particularly interesting because the cleaners are potentially suitable prey that enter the mouths of dangerous predators. The predators refrain from attacking the cleaners, but such restraint could not have been present when the interactions first began to evolve. Biologists suspect that such cleaning began with the cleaners removing parasites from less-dangerous locations.

48.18 An Animal–Animal Mutualism
This prawn (*Lysmata grabhami*) is cleaning a coral reef fish known as sweetlips.

(a)

(b)

(c)

48.19   A Plant–Animal Mutualism
Acacia trees have large, swollen, hollow thorns (a) that house ants. The ants patrol the trees, eating the eggs and larvae of herbivorous insects and cutting
away tips of vines and branches of neighboring plants that would otherwise
smother the acacia. In an experiment, small acacia trees were cut down and
ants were allowed to recolonize some trees but not others. Those with ant
colonies (b) grew back quickly, but those without ants (c) were heavily attacked
by other insects and regained their leaves very slowly.

PLANT–ANIMAL MUTUALISMS.   Terrestrial plants have
many mutualistic interactions with animals. A complex mutualism between trees and ants that live in
Central America illustrates some benefits from such
interaction. Trees of the species *Acacia cornigera* have
large, hollow thorns in which ants of the genus *Pseu-
domyrmex* construct their nests and raise their young
(Figure 48.19). These ants live only on acacias. The
trees have special nutritive bodies on their leaves
upon which the ants feed, as they do on nectar produced at the bases of the leaf petioles. The ants attack
and drive off leaf-eating insects; they even bite and
sting browsing mammals. They also cut down other
plants, particularly vines that grow over their host

tree. In this mutualism the animals get room and
board and the plants get protection against predators
and competitors.
    Many vascular plants depend upon animals to
move their pollen and seeds (Table 48.3). In pollination, the plants benefit by having their pollen carried
to other plants and by receiving pollen to fertilize
their ovules. The animals benefit by obtaining food
in the form of nectar and pollen. Animals are not the
only means by which pollen is transferred among
plants, but such interactions have been favored by
natural selection because they are efficient. Plants
provide animals with attractive rewards; movement
to another flower of the same species is encouraged

TABLE 48.3
Mutualistic Relationships of Plants with Animals

BENEFIT TO PLANTS	BENEFIT TO ANIMALS	SOME EXAMPLES
Animals disperse pollen	Animals feed on pollen or nectar	Most plants with brightly colored flowers
Animals disperse seeds	Animals feed on fleshy rewards surrounding or attached to seeds	Conifers such as junipers and yews
Animals disperse both pollen and seeds	Animals feed on both floral and fruit rewards	Most tropical trees, shrubs at all latitudes

(a)

(b)

48.20 Fruits Attract Different Frugivores
(a) Bright red fruits are attractive to many birds, such as this Bohemian waxwing. (b) A *Formica* ant removes a ripe seed from a pod of a golden snake plant in the Colorado Rockies. The ant-attracting elaiosome is the white tissue wrapped around the black seed.

by the existence of similar rewards there. As a result, the animals transfer pollen efficiently to the stigmas of plants belonging to the same species. But there is a price: The energy and materials the plant spends to produce rewards for animals cannot be used for growth or seed production.

Animals are induced to move seeds by the presence of nutritive rewards attached to or surrounding them. Many seeds are surrounded by fleshy fruits that are eaten by animals, which regurgitate or defecate the seeds some time later away from the plant (Figure 48.20a). A nutritive body called an elaiosome is attached to many seeds that are dispersed by bats or ants. The ants carry the elaiosome and the seed back to their nests for later consumption (Figure 48.20b). Although ants carry seeds only short distances, they often bury them in good germination sites where they are protected from fire and other predators.

Interactions between plants and their pollinators and seed dispersers are mutualistic, but they are not without conflict. The fig wasps discussed at the beginning of this chapter are both pollinators and seed predators. Many seed dispersers are also seed predators that destroy some of the seeds they remove from plants. Many animals visit flowers without transferring any pollen, sometimes cutting holes in them to get to nectar-producing regions at the base of the corolla. Some plants, in turn, attract insects without providing any rewards. The flowers of certain orchids, for example, mimic female insects, enticing the male insects to copulate with the flowers (Figure 48.21). The male insects sire no offspring in this way, nor do they obtain any reward.

INTERACTION AND COEVOLUTION

The richness and abundance of interactions among organisms that we have been discussing in this chapter demonstrate that traits of all species have been influenced by interactions with other coexisting species. That is, species have **coevolved** with one another. The traits of predators influence those of their prey. Parasites coevolve with their hosts, and mutualists coevolve with one another. What is much less

48.21 Some Orchids Mimic Female Insects
Flowers of some orchids so closely resemble female wasps that males are fooled into attempting to copulate with the flowers, as this male wasp is trying to do.

(a)

(b)

48.22   Yuccas and Yucca Moths Coevolved
(a) The Joshua tree (*Yucca brevifolia*) is pollinated only by (b) the yucca moth (*Tegeticula yuccasella*), shown here on a yucca flower.

clear, however, is the extent to which particular traits of species are the result of interactions with only one other species.

Such species-specific coevolution has been clearly demonstrated in only a few cases. One such example is the coevolution between figs and fig wasps. Another is the relationship between species of yucca and the moths of the genus *Tegeticula* that pollinate them. A female moth enters a yucca flower and lays one to five eggs on the ovary. When the eggs hatch, the larvae burrow into the ovary and feed upon the developing seeds. This is typical predation. After she has laid her eggs, however, the female moth scrapes pollen from the anthers in the flower, rolls it into a small ball, flies to another yucca plant, and places the pollen ball on the stigma of the flower before laying another batch of eggs. *Yucca* has no other pollinators, *Tegeticula* larvae eat no other food, and each yucca species has a specific moth species associated with it (Figure 48.22). The relationship shows other signs of close coevolution. The moth refrains from laying more than a few eggs on any one ovary, and yucca pollen is unusually sticky and readily formed into a ball.

Although species-specific coevolution is relatively rare, **diffuse coevolution,** in which traits of a species are influenced by interactions with a wide variety of predators, parasites, prey, and mutualists, is wide-spread. Most flowers are pollinated by a number of pollinators, and most pollinators visit many species of flowers. Most flowers adapted for bird pollination are red, a color that attracts most birds, not just a few species. Many flowers adapted for insect pollination have honey guides, contrasting colors that lead to the entrances to the flowers. These lines, which are conspicuous when viewed under ultraviolet light (Figure 48.23), are visible to bees and butterflies but not to birds. Bat-pollinated flowers open at night and have wide openings into which the head of a bat can enter, making the floral rewards accessible to many species of bats.

Diffuse coevolution also accounts for the traits of the fleshy fruits that surround many seeds. Most bird-dispersed fruits are red or some combination of red and another color. Mammal-dispersed fruits are typically purple. Very few fruits, however, are adapted for dispersal by only a few species of birds or mammals.

The traits of flowers and fruits are the result of diffuse coevolution between plants and animals because flower visitors and fruit dispersers must use many different plant species to survive throughout the year. Most plant species produce flowers and fruits for only a few weeks or months. The animals must travel to where flowers and fruits are available and switch to feeding on whichever plant species are flowering or bearing fruit.

(a)

(b)

48.23   Bees See Flowers by Reflected Ultraviolet Light
(a) Under normal sunlight these black-eyed susans appear familiar to us. (b) Bees, however, see the same flowers more as they appear in this ultraviolet photograph, in which part of the outer ray petals blends with the central flowers in the heads to make a larger visual target.

HOW SPECIES AFFECT ECOLOGICAL COMMUNITIES

Species influence the communities in which they live in many different ways. They may alter microclimate, soil structure, and water movement. Such changes in the physical environment change its suitability as habitat for other organisms. As we have just seen, species also change the amount and distribution of resources, and they consume one another.

The Role of Plants

In terrestrial communities, plants are the major modifiers of physical environments. They also provide the pathway through which energy and nutrients enter communities. Because of their variety and complexity, and the sizes of some of them, plants also form most of the structural environment for other organisms. Anyone who has walked into the shade of a tree on a hot, sunny day knows that the climate near the ground is strongly influenced by vascular plants. Except where temperatures are very cold or moisture is scarce, trees grow large enough to shade smaller plants beneath them and the surface of the soil. Temperatures fluctuate less between day and night underneath trees than they do in the open, and light levels are much lower there. The leaves of trees intercept and evaporate much of the rain that falls on them, so less reaches the ground than in open areas. The rain that does reach the ground, however, evaporates more slowly inside a forest than in the open because in forests temperatures are lower, humidities are higher, and it is less windy.

In addition to modifying climates, vascular plants provide most of the structure of terrestrial environments. A typical forest is a mixture of trees, shrubs, and herbaceous plants (Figure 48.24). The crowns of the trees form the canopy of the forest. Exposed to sun, wind, and precipitation, the canopy is the major modifier of the local climate. Beneath the canopy is the understory, a mixture of smaller trees, shrubs, and herbs. Some of these plants are young trees that may eventually become members of the canopy, but many of them belong to species that never grow so large. These plants live in the climate modified by the trees above them and are often damaged by leaves, branches, and fruits falling from the taller trees.

In addition to size, there are many other characteristics that differ among plants of a forest: bark, number and angles of branches, sizes and shapes of leaves, kinds of flowers and fruits, and how much they die back during unfavorable seasons. It is in such structurally complex plant communities that members of the other kingdoms live. The types of food available to these other organisms, when food is available, and the ways in which it can be found and captured depend upon the structure of the plant community. This crucial role of plants is the reason that ecologists spend much time measuring the structure of plant communities and observing how animals move and find their food in them.

Succession

A complex forest like the one in Figure 48.24 does not develop quickly. Some of the large trees may be several hundred years old. Other individuals of the same and different species may have occupied the site before today's plants grew there. Long ago people noticed that following the destruction of most of the plants on a site by a disturbance such as a fire or landslide, the first invading plants differed from

48.24 Forests Have Complex Structure
This tropical evergreen forest in Guatemala has trees of different ages and sizes, vines, and many shrubs.

those that colonized the site later. The gradual process by which the species composition of a community changes is called **ecological succession.** Patterns and causes of succession differ according to climate and soils, but during all successions the soil of the site and the conditions under which individuals of later-arriving species must grow are modified by the early colonists.

Succession may begin at sites that support no organisms. Consider, for example, the succession of plants in Glacier Bay, Alaska, following rapid retreat

48.25 Soil Properties Change During Vegetation Succession
As the plant community occupying a moraine changed from pioneering plants to a spruce forest, nitrogen accumulated in both the forest floor and in the mineral soil. Alder trees fixed nitrogen in the soil, paving the way for the coniferous forest.

of the glaciers there during the last 200 years. The pattern of succession of plants and changes in soil nitrogen content at Glacier Bay are shown in Figure 48.25. The glacier retreated about 100 kilometers during this period, leaving a series of moraines—gravel deposits formed where the glacial front was stationary for a number of years.

Moraines were first invaded by small pioneer organisms such as bacteria, fungi and algae, which were followed by lichens, mosses, and a few species of shallow-rooted herbs that can grow on the nutrient-poor moraine (see Figure 32.8). The pioneering plants were followed by shrubby willows, then by alders, and eventually conifers. The succession was determined in part by changes in the soil caused by the plants themselves. Alder trees have nitrogen-fixing fungi in nodules on their roots. Because nitrogen is virtually absent from glacial moraines, it required a century for alders to fix enough nitrogen in the soil to allow good growth of conifers. The conifers then outcompeted and displaced the alders.

Succession may be initiated after disturbances, such as fires and logging, that remove only parts of the community. The first invaders of disturbed sites are usually fast-growing plants that photosynthesize well in full sunlight. However, these plants create shade in which their own seedlings grow more poorly than do seedlings of shade-adapted plants. Thus plants that grow well under full sun are replaced by plants that germinate and grow better in deeper shade.

Another type of succession—degradative succession—takes place when all or part of the dead body of a plant or animal is decomposed. Degradative succession is driven by activities of early species that remove some nutrients and change others, making available new resources for later species. Degradative

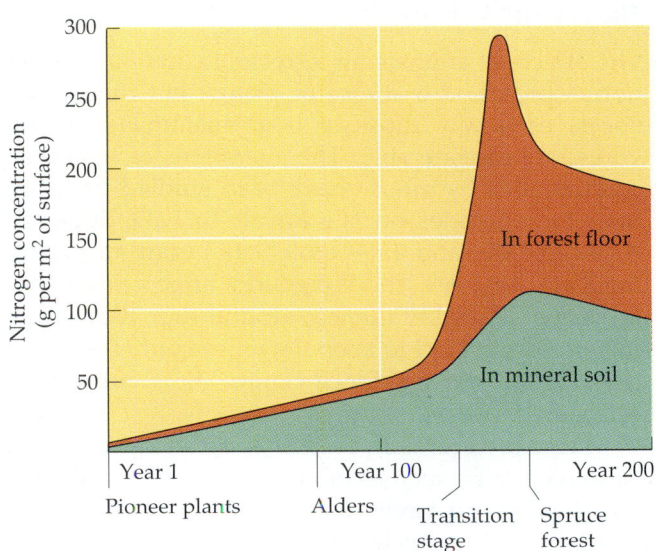

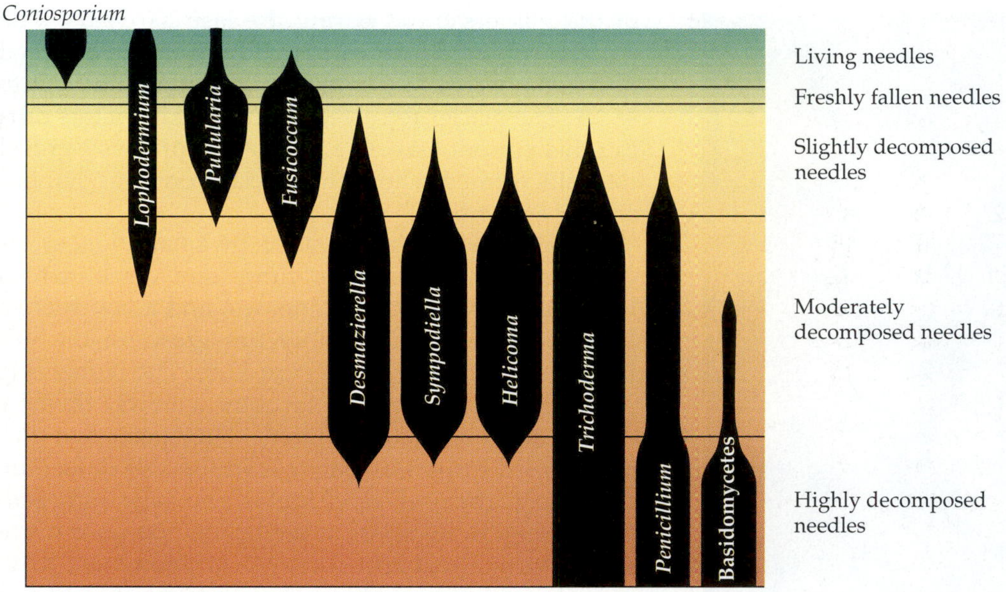

Coniosporium

48.26 Degradative Succession on Pine Needles
As indicated by the widths of the black bars, the abundances of ten types of
fungi in pine litter change with time and according to the layer.

succession terminates when the resource is completely consumed.

The succession of fungal species decomposing the pine needles beneath Scots pine (*Pinus sylvestris*) is shown in Figure 48.26. New litter is continuously deposited under pines, so the surface layer is young and deeper layers of litter are progressively older. Degradative succession begins when the first group of organisms starts consuming the needles as they fall and continues for about seven years, after which the last group of organisms—basidiomycetes—have decomposed the remaining cellulose and lignin. By then, the remains are no longer recognizable as pine needles.

The Role of Animals

The effects of animals on terrestrial communities are not as profound as those of plants, but their influences, especially those of large mammalian herbivores, are considerable. The alterations of ecological communities by moose have been studied for more than four decades on Isle Royale in Lake Superior. Between 1948 and 1950 ecologists began an experiment to evaluate the long-term effects of moose browsing. They built fences around four 100-square-meter plots of land to keep the moose out. They then laid out but did not fence control plots outside the exclosures.

Moose preferentially feed on early successional deciduous plants, such as mountain ash, mountain maple, aspen, and birch. They rarely eat white spruce and balsam fir, species that replace deciduous trees

during succession, because the foliage of these conifers has high concentrations of indigestible resins and low concentrations of nitrogen. In the control plots, deciduous species were so heavily eaten by moose that spruce and fir were the only plants that grew above the height at which moose feed. Inside the exclosures, on the other hand, deciduous trees remained abundant. Thus, by reducing the abundances of deciduous trees, the moose accelerate the successional change from deciduous trees to conifers.

Beavers strongly influence ecological communities by cutting trees and building dams. On the Kabetogama Peninsula in northern Minnesota, logging and fires in the late 1930s caused large increases in aspen trees, which grow well in cleared, disturbed areas. Because aspen trees are the favorite food of beavers, the number of beaver dams in the area increased from 71 in 1940 to 835 in 1986. Less than 1 percent of the peninsula was covered by beaver ponds in 1940, but this number increased to 13 percent in 1986. As a result, the amount of open water, marsh, bog, standing dead trees, and seasonally flooded lands, all of which are aquatic ecological communities created by beaver dams, increased greatly (Figure 48.27). Beavers also modify terrestrial communities by cutting trees in the vicinity of their dams. A beaver family typically cuts about a metric ton of wood within 100 meters of its pond each year. In northern Minnesota, beavers often virtually clear-cut stands of aspen, favoring the succession to alders and hazels and, eventually, to white spruce and balsam fir.

Species whose influences on the structure and functioning of ecological communities are greater

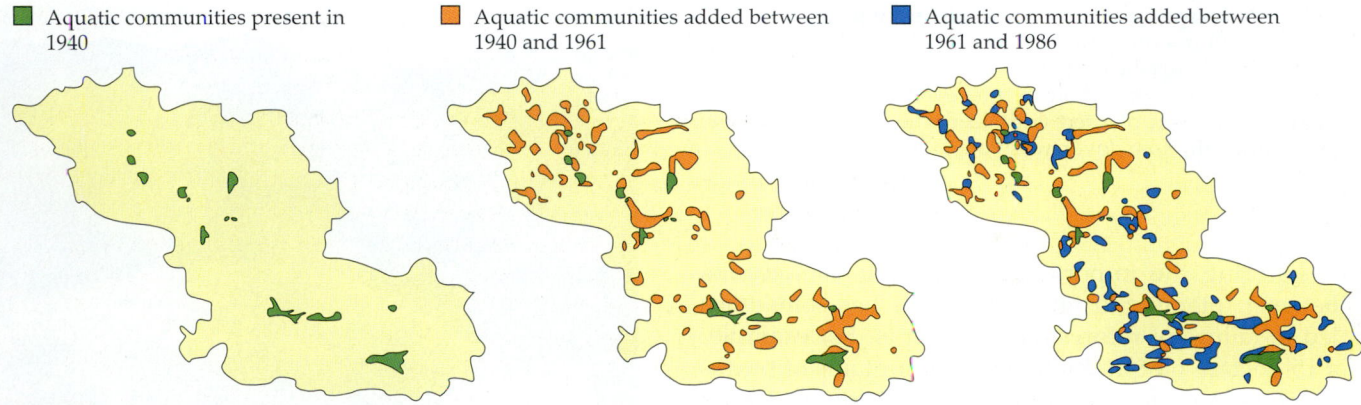

■ Aquatic communities present in 1940

■ Aquatic communities added between 1940 and 1961

■ Aquatic communities added between 1961 and 1986

48.27 Beavers Create New Communities
These maps of a 45-square-kilometer watershed in Minnesota show the increase of aquatic ecological communities between 1940 and 1986 due to dam building by beavers.

than would be expected on the basis of their abundance are called **keystone species.** Moose on Isle Royale and beavers in northern Minnesota are keystone species that exert their effects by changing vegetation structure. Some keystone species influence ecological communities in other ways. For example, in the Rocky Mountains of Colorado, red-naped sapsuckers (*Sphyrapicus nuchalis*) drill holes into spruce, aspen, and especially willow trees. They feed on the sap that wells up in the holes and the insects that become entangled in it. More than 40 other species—including hummingbirds, warblers, chipmunks, squirrels, mice, butterflies, moths, and wasps—also feed on the sugary sap (Figure 48.28). Sapsuckers excavate nesting holes in aspen trees, and they use a hole for only one breeding season. Two species of swallows that are unable to excavate their own holes use the abandoned holes of sapsuckers as nest sites. Aspen groves that are far from willows, and hence lack sapsuckers, have no nesting swallows. The importance of sapsuckers in the Rocky Mountains is not immediately evident, but when we look closely, we see that these birds exert major influences on many other species.

The Role of Microorganisms

Because they are tiny, microorganisms are not obvious, but their role in ecological communities is enormous. Wood is broken down almost solely by microorganisms, especially fungi. Nitrogen fixation, the only source of biological nitrogen, is carried out only by prokaryotic microorganisms. Plants and the microorganisms with which they share the soils have an intricate relationship. Plants allocate a high proportion of photosynthate to roots, much of which is diverted to mycorrhizal fungi living there.

48.28 The Red-Naped Sapsucker Is a Keystone Species
(a) The red-naped sapsucker, *Sphyrapicus nuchalis*, drills holes in trees, releasing the sticky sap. Once the sapsucker has drilled a hole, many different species come to feed on the sap. (b) A broad-tailed hummingbird drinks sap from a hole in a willow stem. (c) A butterfly and a common housefly obtain sap from another willow tree.

(a)

(b)

(c)

48.29   Nonregenerating Clear-Cuts in Oregon
Because soil microorganisms were eliminated by burning
and herbicide application, no conifers are growing in
these 15- to 20-year-old clear-cuts, even though each
clearing has been planted with seedlings three to five
times since the tree cover was cut.

Ignoring the importance of the interconnections
between plants and microorganisms has sometimes
led to failure of reforestation attempts. For example,
a 15-hectare plot in the Klamath Mountains of southern Oregon, clear-cut in 1968, has been replanted
four times. All the plantings were failures, even
though forests in this area regenerate readily after
wildfires. The reason for the failures is that the site
was both burned and treated with herbicides to open
it up for better growth of conifer seedlings. The early
successional deciduous trees and shrubs that were
killed by this treatment, however, support soil organisms and improve temperatures and moisture,
thereby maintaining the populations of soil microorganisms. When those plants were eliminated, most
soil microorganisms died, and conifers then were
unable to grow. Many such nonregenerating clear-
cuts dot high mountains in this region (Figure 48.29).

PATTERNS OF SPECIES RICHNESS

All the interactions among all the species living together in an ecological community could theoretically
be described in detail. Because there are many thousands of species in most communities, however, such
a description would be as complex as nature itself.
Even if all the necessary information were available,
such descriptions would not help us identify the most
important interactions. To build comprehensible pictures of complex ecological communities we need
terms that describe aggregate features of groups of
organisms. When we choose our aggregate terms
wisely, they capture important information about
ecological communities.

An important aggregate feature of any community
is **species richness**—most simply defined as the number of species living in the community. Often, however, we want to know not only the number of species but also their relative abundance or rarity. Other
times we want to know about the relative sizes of
members of the different species. For such purposes,
we need a weighted measure of species richness,
sometimes called **species diversity.** Species are usually weighted by their abundance, total weight, or
energy production and consumption because these
traits influence community dynamics. For example,
consider two imaginary forest stands, each of which
has 100 trees belonging to ten species. In one stand,

there are ten individuals of each species; in the other
stand there are 91 individuals of one species and one
of each of the others. Although the number of species
is the same (ten) in the two stands, the stand in which
most individuals are members of one species is less
diverse.

Ecological communities differ dramatically in the
number of species that live in them. Within a region,
forests typically have more species than grasslands
have, and tropical communities have more species
than temperate communities have. The number of
species in a community depends on factors that act
on different time and space scales. Competition,
predation, disease, and mutualisms often produce
their effects very quickly. They operate primarily at
local levels. Immigration, emigration, and habitat
fragmentation usually produce their effects more
slowly and at regional scales. Evolutionary changes
are even slower. In this chapter we concentrate on
local and regional factors affecting patterns of species
richness. The role of long-term historical factors will
be explored in Chapter 50.

Local Species Richness

What determines how many species can fit into a
community? Part of the answer to this question depends on the variety of resources available to organisms in a community, how many of those resources
are available throughout the year, the range of resources used by each species (that is, how specialized
the species are), and the amount of overlap between
species in resource use. If dependent on resource
availability, the number of species in a community
will be positively correlated with the variety of resources. More species may fit if each one uses a

relatively small part of the resources, if they overlap extensively in their use of resources, or if the range of resources is fully used (rather than if parts of it are unexploited). The number of species that can live together may also depend on predators, disease, and the kinds and frequencies of disturbances. Let's look at some examples of how these factors influence local species richness.

The structural complexity of a community affects its species richness. Communities dominated by structurally complex plants support more species of freshwater fishes, birds, and insects than do communities dominated by structurally simple plants. Ecologists studied the richness of fish species in 18 freshwater lakes in Wisconsin and found it to be positively correlated with the structural complexity of submerged vegetation. More-complex vegetation offers fish more hiding places from predators and more places to seek food. Similarly, terrestrial communities with structurally more-complex plants offer birds a greater range of resources and more ways they can be exploited. Trees have trunks, branches of various sizes, and leaves from which prey can be captured, and there are many ways to move through trees. Grasses provide only a few kinds of food, and there are only a few ways for animals to move through grasses.

Structurally complex terrestrial communities have more species of insects because they provide both a greater range of food resources and a greater diversity of hiding places. Insects are highly vulnerable to predators if they sit conspicuously on leaves or branches. Many insects mimic leaves or twigs, hide in crevices in bark, or resemble their backgrounds in ways that make them harder to see. Structurally complex plants provide more different kinds of hiding places than simple plants do. Therefore, more kinds of insects can and do hide on them.

If not too severe, disturbance often increases species richness in ecological communities. An example is grazing. When rabbits were removed from the Ro-thamsted Experimental Station in England, the number of plant species declined slowly but steadily because the more vigorous competitors, no longer subjected to heavy grazing, eliminated less-competitive species. Plants may create disturbances that allow more species to persist. Among the most important are tree falls, which create gaps in the canopy within which many tree species germinate and grow. Many forest trees cannot reach adult size unless they grow in such gaps, where there is more light and nutrients than in the closed forest.

Influence of Regional Species Richness on Local Species Richness

Much has been learned about factors influencing species richness from studies of small areas of a hectare or less. If species richness is determined only by local events, however, then habitats that have similar structures and rates of photosynthesis, and are subjected to similar disturbances, should support biological communities with similar species richness. Numerous studies have demonstrated that this is not the case. The nature of the differences shows that regional processes also affect local species richness.

Local species richness is often correlated with the number of species found regionally. For example, ecologists have found that the number of gall wasps on a particular species of oak in the state of California is positively correlated with the total number of gall wasp species known to feed on that oak throughout its range (Figure 48.30a). Thus the number of gall-wasp species on an oak within a small area is determined, in part, by the number of gall-wasp species living on that oak species in a larger region. A similar pattern has been revealed by studies of songbirds on Caribbean islands. The more species on an island, the greater the number of species in a single habitat on the island (Figure 48.30b).

The more widely distributed and extensive a habitat, the more likely organisms are to evolve adapta-

48.30 Local Richness Is Related to Regional Richness
The number of species in a local environment correlates positively with the number of species in the larger region. (a) Gall-wasp species on a species of California oak. (b) The number of songbird species in patches of secondary forest on some Caribbean islands and on the Panamanian mainland.

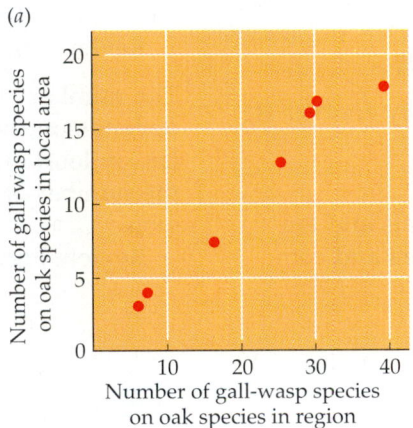

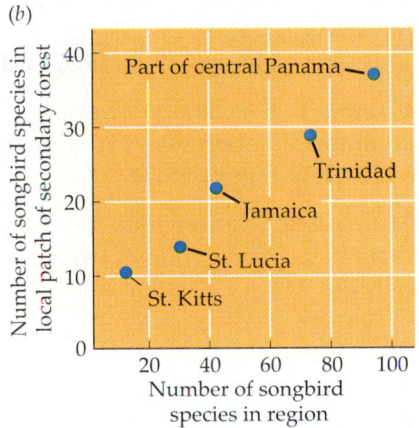

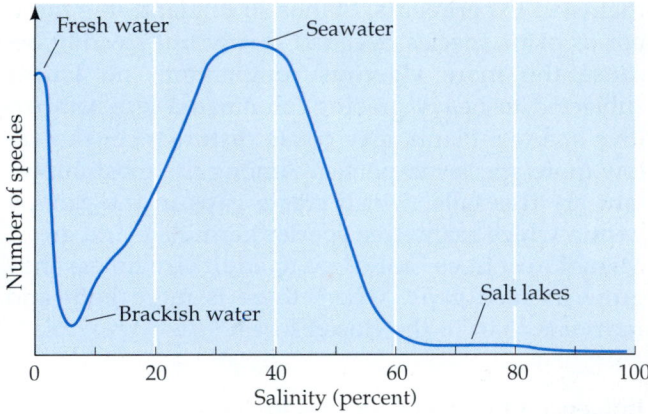

48.31 Larger Habitats, More Species
Many species of invertebrates live in fresh water, and many live in seawater. Brackish water and salt lakes—rare habitats compared to fresh water and seawater—are species-poor.

tions to that habitat. The salinity of seawater differs greatly from that of fresh water, but worldwide the salinities of each are relatively uniform. Waters of intermediate salinity and waters that are more saline than the oceans are relatively rare. The species richness of aquatic invertebrates throughout the world reflects this pattern (Figure 48.31).

A similar pattern is found among plants growing on outcrops of serpentine rocks. Soils derived from serpentine are highly deficient in several essential plant nutrients and are often regarded as highly unfavorable for plants. In western North America, where serpentine outcrops are small and scattered among other rock types, serpentine areas do indeed support relatively few plant species. In parts of South Africa, however, where serpentine is the most common rock type, plant species richness is higher on serpentine than on soil derived from other rock types. To understand many aspects of species richness, then, we need to know the geographic extent of different habitats.

SUMMARY of Main Ideas about Interactions within Ecological Communities

Organisms interact with one another in varied and complex ways.
Review Table 48.1

The resources a species uses and its interactions with its predators and competitors define its niche.

Most species could exist in a wider range of environments than they do if competitors and predators were not present.
Review Figure 48.1

Predator–prey interactions often lead to wide oscillations in abundances of both predators and their prey.
Review Figures 48.4 and 48.5

Predators influence the evolution of their prey, leading to mimicry and possession of chemical and mechanical defenses.
Review Figures 48.8, 48.9, and 48.10

Competitive exclusion of one species by another is common in laboratory experiments.
Review Figure 48.14

Mutualistic relationships between plants and animals help protect plants from predators and competitors.

Plants have mutualistic interactions with animals that transport their gametes and seeds.
Review Figures 48.21 and 48.22

Species change the communities in which they live by altering the physical environment, by providing structure, by modifying resource abundances and distributions, and by consuming one another.

Plants are major modifiers of climate and providers of community structure, and they are the pathway through which energy enters ecosystems. The structure of plant communities develops slowly by succession, during which species replace one another over time.
Review Figures 48.25, 48.26, and 48.27

Animals, especially large mammals, modify the structure and functioning of biological communities.
Review Figure 48.28

The number of species in an ecological community depends on the extent and complexity of the habitat, on local events (especially disturbances), and on the number of species in the larger surrounding region.
Review Figures 48.30 and 48.3

SELF-QUIZ

1. Two organisms that use the same resources when those resources are in short supply are said to be
   a. predators.
   b. competitors.
   c. mutualists.
   d. commensalists.
   e. amensalists.

2. Which of the following is *not* a resource?
   a. Food
   b. Space
   c. Hiding places
   d. Nest sites
   e. Temperature

3. A fundamental niche is the range of conditions under which a species could survive if
   a. there were no predators or competitors.
   b. there were no predators or other negative influences.
   c. there were no competitors, predators, or other negative influences.
   d. environmental conditions were ideal.
   e. the environment were fundamentally different.

4. An animal that is about the same size as its prey and that attacks it from the outside is called a
   a. predator.
   b. parasite.
   c. commensalist.
   d. competitor.
   e. parasitoid.

5. Which of the following factors tends to stabilize populations of predators and their prey?
   a. A high birth rate of the prey
   b. A high birth rate of the predator
   c. The ability of predators to reduce prey further when they are scarce
   d. The ability of predators to search widely for prey
   e. Environmental heterogeneity

6. The convergence over evolutionary time in the appearance of two or more unpalatable species is called
   a. cladism.
   b. mutual adaptation.
   c. Müllerian mimicry.
   d. Batesian mimicry.
   e. convergent mimicry.

7. When an individual, by its behavior, directly prevents other individuals from using a resource, the interaction is called
   a. interference competition.
   b. exploitation competition.
   c. behavioral competition.
   d. intraspecific competition.
   e. competitive exclusion.

8. Damage caused to shrubs by branches falling from overhead trees is an example of
   a. interference competition.
   b. partial predation.
   c. amensalism.
   d. commensalism.
   e. diffuse coevolution.

9. Ecological succession is
   a. the changes in species over time.
   b. the gradual process by which the species composition of a community changes.
   c. the changes in a forest as the trees grow larger.
   d. the process by which a species becomes abundant.
   e. the buildup of soil nutrients.

10. More species may be fit into a community if
    a. each species uses a narrow range of resources.
    b. species overlap extensively in their use of resources.
    c. all resources in the community are fully used.
    d. a greater range of resources is available in the community.
    e. all of the above.

11. Relatively few species of animals live in very salty lakes because
    a. it is difficult to adapt to high salt concentrations.
    b. there is little to eat in very salty lakes.
    c. organisms are less likely to adapt to rare habitats.
    d. organisms evolved in the ocean.
    e. organisms evolved in fresh water.

FOR STUDY

1. A general rule commonly accepted by ecologists states that a "jack of all trades is master of none." Yet most ecological communities are mixtures of jacks and masters, that is, generalists and specialists. Under what conditions would you expect jacks to be more successful? Masters? Why?

2. What features of predator–prey interactions tend to generate instabilities that lead to fluctuations in densities of both species? Given that instabilities are expected, what keeps populations of either predator or prey from fluctuating to extinction?

3. Pests and pathogens usually have generation times much shorter than those of their hosts. Consequently, they should be able to evolve faster. What prevents them from evolving so fast that they completely overcome the resistances of their hosts and exterminate them?

4. The wind disperses pollen randomly across the countryside rather than directing it toward stigmas of plants belonging to the same species. Given this inefficiency, why are there so many wind-pollinated plants? If seeds that land close to the parent plant are less likely to survive than those that are carried farther away, why do so many plants produce seeds lacking dispersal devices?

5. On the eastern side of the Sierra Nevada in California, four species of chipmunks occupy adjacent habitats from which they exclude each other by direct aggressive interference. In the San Jacinto Mountains of southern California, three other chipmunk species similarly occupy adjacent habitats, but no interspecific aggression is observed. Each species simply remains in its own habitat. Which of these two assemblages do you think is the older one? Why?

6. Some direct interactions between two species benefit only one of those species. Give examples of such "one-way" benefits in each of the following cases.
   a. Between two species of plants (give one example of energetic support and one example of physical support)
   b. Between a plant and a browser
   c. Between a predator and its prey

7. Wood is an abundant food source that has been available for millions of years. Why have so few animals evolved the ability to eat wood?

8. In the text we discussed several cases showing that large mammals modify the communities in which they live. Give some examples in which smaller animals modify their communities.

READINGS

Barbour, M. G., H. J. Burk and W. D. Pitts. 1980. *Terrestrial Plant Ecology.* Benjamin/Cummings, Menlo Park, CA. A general text on the interactions between plants and their surroundings, including several good chapters on communities and vegetation types.

Begon, M., J.L. Harper and C. R. Townsend. 1990. *Ecology: Individuals, Populations, and Communities,* 2nd Edition. Blackwell Scientific Publications, Cambridge, MA. A comprehensive treatment of all aspects of ecology.

Futuyma, D. J. and M. Slatkin (Eds.). 1983. *Coevolution.* Sinauer Associates, Sunderland, MA. An excellent collection of essays that summarizes current knowledge of coevolutionary relationships among all types of organisms.

Handel, S. N., and A. J. Beattie. 1990. "Seed Dispersal by Ants." *Scientific American,* August. Thousands of plant species rely on ants to disperse their seeds. With special food lures and other adaptations, a plant can induce the insects to carry away its seeds without harming them.

Harborne, J. B. 1982. *Introduction to Ecological Biochemistry,* 2nd Edition. Academic Press, New York. An excellent presentation of the types of chemicals produced by living organisms and their roles in ecological interactions.

Rennie, J. 1992. "Living Together." *Scientific American,* January. Parasites and their hosts have devised many odd strategies—perhaps even sex—in their endless game of adaptive one-upsmanship. Yet sometimes they seem to cooperate.

Ricklefs, R. E. 1990. *Ecology,* 3rd Edition. W. H. Freeman, New York. An excellent text covering all aspects of ecology.

Thompson, J. N. 1982. *Interaction and Coevolution.* John Wiley & Sons, New York. A review of patterns in and conditions favoring the evolution of close interactions among species.

Wickler, W. 1968. *Mimicry in Plants and Animals.* Wiedenfeld and Nicholson, London. A general and straightforward presentation of the evolution of close resemblances among organisms that are not closely related to one another.

As we saw in Chapter 18, organisms have been influencing the composition of the atmosphere, and hence Earth's climate, for billions of years, but never before has a single species caused such large changes in Earth's functioning as humans are causing today. Scientists have been aware of and have anticipated some of the environmental changes humans are causing, but they were unprepared for the discovery, in 1985, that the high-altitude ozone layer, which shields organisms from harmful ultraviolet radiation, had, during part of the year, thinned greatly over Antarctica. Ozone is produced by sunlight in tropical regions and transported to high latitudes, where it is destroyed. Many processes are involved in the destruction of ozone, but chlorine compounds, produced mainly by human activity, appear to be primarily responsible for the current unusually high rates of elimination of atmospheric ozone. The depletion of ozone is now almost complete at the latitudes where conditions favor its destruction by chlorofluorocarbons (CFCs), compounds released to the atmosphere by human activity. How can the production of what appear to be small amounts of a new chemical result in large changes in Earth's atmosphere? To understand why the ozone hole exists, and why other products of human activity are also altering global-scale processes, we must study ecological processes on a global scale.

ENERGY FLOW THROUGH ECOSYSTEMS

The organisms living in a particular area, together with the physical environment with which they interact, constitute an **ecosystem**. Ecosystems can be recognized and studied at many different spatial scales, ranging from local units, such as lakes, to the entire globe. At the global scale, Earth is a single ecosystem.

Although many different species interact in all ecosystems, we can understand how ecosystems function and how humans can exert powerful influences on them if we know how the energy and materials captured by organisms are transformed and transferred to other organisms when they eat one another. Organisms depend on the input of energy (in the form of sunlight or high-energy molecules), water, and minerals for metabolism and growth. Almost all energy utilized by organisms comes (or once came) from the sun. Even the fossil fuels—coal, oil, and natural gas—upon which the economy of modern civilization is based are reserves of captured solar energy locked up in the remains of organisms that lived millions of years ago.

Only a small portion (about 5 percent) of the solar energy that arrives on Earth is captured by photo-

The Ozone Layer Is Thinning
Satellite photographs of the Northern Hemisphere in March 1993 indicate broad areas (dark blue through purple to black) in which the ozone layer has become thin.

Percent Difference

-25.0   -15.0   -5.0   0.0   5.0

49

Ecosystems

synthesis. The remaining energy is either radiated back into the atmosphere as heat (especially in places where Earth's surface is bare because there is too little water to support plant growth) or, where the surface is covered in green leaves, consumed by the evaporation of water from plants. The energy that *is* captured powers the "metabolism" of ecosystems.

Gross Primary Production

Energy flow in most ecosystems originates with photosynthesis. The rate of photosynthesis depends on the amount of solar radiation, the availability of water, the abundance of mineral nutrients, and the temperature, among other factors. The total amount of energy that plants assimilate by photosynthesis is called **gross primary production.**

The rate of gross primary production depends on the amount of water available because water must move through plants when the stomata are open as they photosynthesize. The rate of photosynthesis depends on temperature because most chemical reactions proceed faster with increasing temperature. Temperature and moisture interact to determine the rate of gross primary production because the evaporative power of air, which affects transpiration and thus the flow of water within plants, is less at lower temperatures than at higher temperatures. In many areas on Earth, the annual amount of gross primary production is determined by available soil moisture and by soil fertility. Shortage of water limits primary production during much of the year in most arid regions. Production in aquatic systems is limited by light, which decreases rapidly with depth; by nutrients, which sink and must be replaced by upwelling of water; and by temperature.

Net Primary Production

Plants use most of the energy they capture to maintain themselves—to fuel their respiration—but some of the energy produces new tissues that can be eaten by herbivores, or, after the plant dies, by organisms that eat their remains. Because much of the captured energy goes to power their own metabolism, plants always contain much less energy than the total amount they assimilated. The amount that remains after we subtract the energy the plants use for maintenance and for building tissues is known as **net primary production** (Figure 49.1). How is net primary production harvested by animals?

Trophic Levels

When we study the flow of energy through ecosystems, it is useful to group organisms according to their source of energy. The organisms that obtain their energy from a common source constitute a **trophic level.** Photosynthetic plants get their energy directly from sunlight. Collectively, they constitute the trophic level called photosynthesizers. Organisms that eat plants constitute the trophic level called herbivores. Organisms that eat herbivores are called primary carnivores. Those that eat primary carnivores are called secondary carnivores, and so on. Organisms that eat the dead bodies of organisms or their remains are called detritivores. Many organisms, such as ourselves, obtain their food from more than one trophic level below them. Such organisms are called omnivores. Table 49.1 summarizes the major trophic levels.

49.1 Limits to Net Primary Production
Ecologists harvest, dry, and weigh a plant's parts to estimate how much energy the plant has stored in its tissues. (a) Net primary production increases rapidly with precipitation but reaches a plateau because a forest can transpire only so much water, no matter how much falls. (b) Net primary production increases steadily with increasing mean annual temperature.

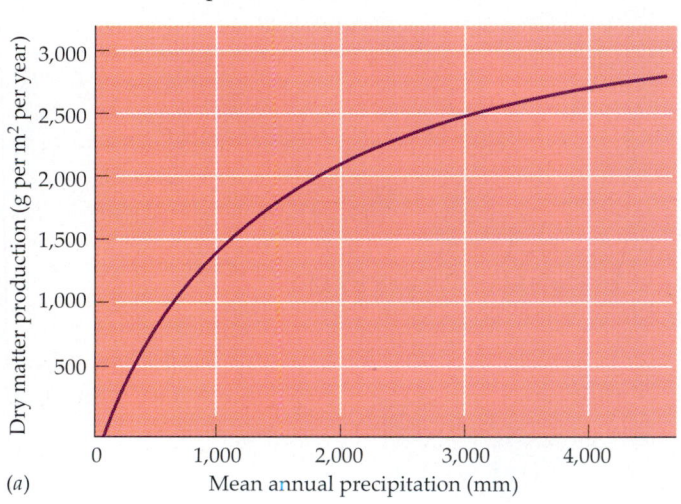

(a)

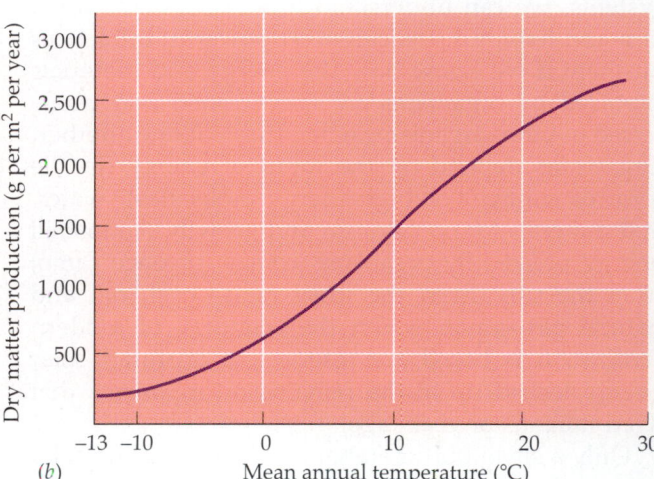

(b)

**TABLE 49.1
The Major Trophic Levels**

TROPHIC LEVEL	EXAMPLES	SOURCE OF ENERGY
Photosynthesizers	Green plants, photosynthetic bacteria and protists	Solar energy
Herbivores	Termites, grasshoppers, water fleas, anchovies, deer, geese	Tissues of photosynthesizers
Primary carnivores	Spiders, warblers, wolves, copepods	Herbivores
Secondary carnivores	Tuna, falcons, killer whales	Primary carnivores
Omnivores	Humans, oppossums, crabs, robins	Several trophic levels
Detritivores	Fungi, many bacteria, worms, millipedes, vultures	Dead bodies and waste products of other organisms

Organisms in a particular trophic level occupy a position in an ecological community that is determined by the number of steps through which energy passes to reach them. Photosynthesizers are often called **primary producers** because they produce the energy-rich organic molecules upon which all other organisms feed. All other organisms are often called **consumers** because they consume, either directly or indirectly, the energy-rich organic molecules produced by photosynthetic organisms.

A set of linkages in which a plant is eaten by an herbivore, which is in turn eaten by a primary carnivore, and so on, is called a **food chain.** Food chains are usually interconnected to make a **food web.** The arrows in representations of food webs show who eats whom. A simplified food web, not including detritivores, for Gatun Lake, Panama, is shown in Figure 49.2. Part of the food web in an Australian sandy desert is shown in Figure 49.3. This food web differs from that of Gatun Lake in that a single species, the pygmy monitor lizard, is the major predator on all of the species in the trophic level immediately below it. Food webs are a useful summary of predator–prey interactions within a community. A complete food web showing the position of every species in an ecosystem would be confusingly complex because most biological communities have so many species. Therefore, similar species, especially those at lower trophic levels, are usually lumped together, as they are in the diagrams of both of these food webs.

The Maintenance of Organisms and Energy Flow

The energy that organisms use to maintain themselves is dissipated as heat, a form of energy that cannot be used by other organisms. The energy content of an organism's net production—its growth plus reproduction—is the amount of energy available to organisms at the next trophic level (Figure 49.4). The efficiency of energy transfer through food webs de-

pends on the fraction of net production at one trophic level that is eaten by organisms at the next level, and how those organisms divide the ingested energy between production and maintenance (respiration). We can calculate the efficiency of energy transfer of a species or group of animals as

$$E = P/(P + R)$$

where E is efficiency, P is net production, and R is respiration. The E values for different animal taxa reveal two patterns (Table 49.2). First, birds and mammals have very low efficiencies because they expend so much energy maintaining constant high body temperatures. Second, herbivores are less efficient than carnivores because plant tissues generally take more energy to digest than animal tissues do.

Even when efficiencies of energy transfer and con-

**TABLE 49.2
Average Production Efficiencies for Various Groups of Animals**

GROUP	PRODUCTION EFFICIENCY, $P/(P + R)$ (PERCENT)
Insectivores (mammals)	0.9
Birds	1.3
Small mammals	1.5
Large mammals	3.1
Fishes and social insects	10.0
Invertebrates other than insects	
Herbivores	21
Carnivores	28
Detritivores	36
Nonsocial insects	
Herbivores	39
Carnivores	56
Detritivores	47

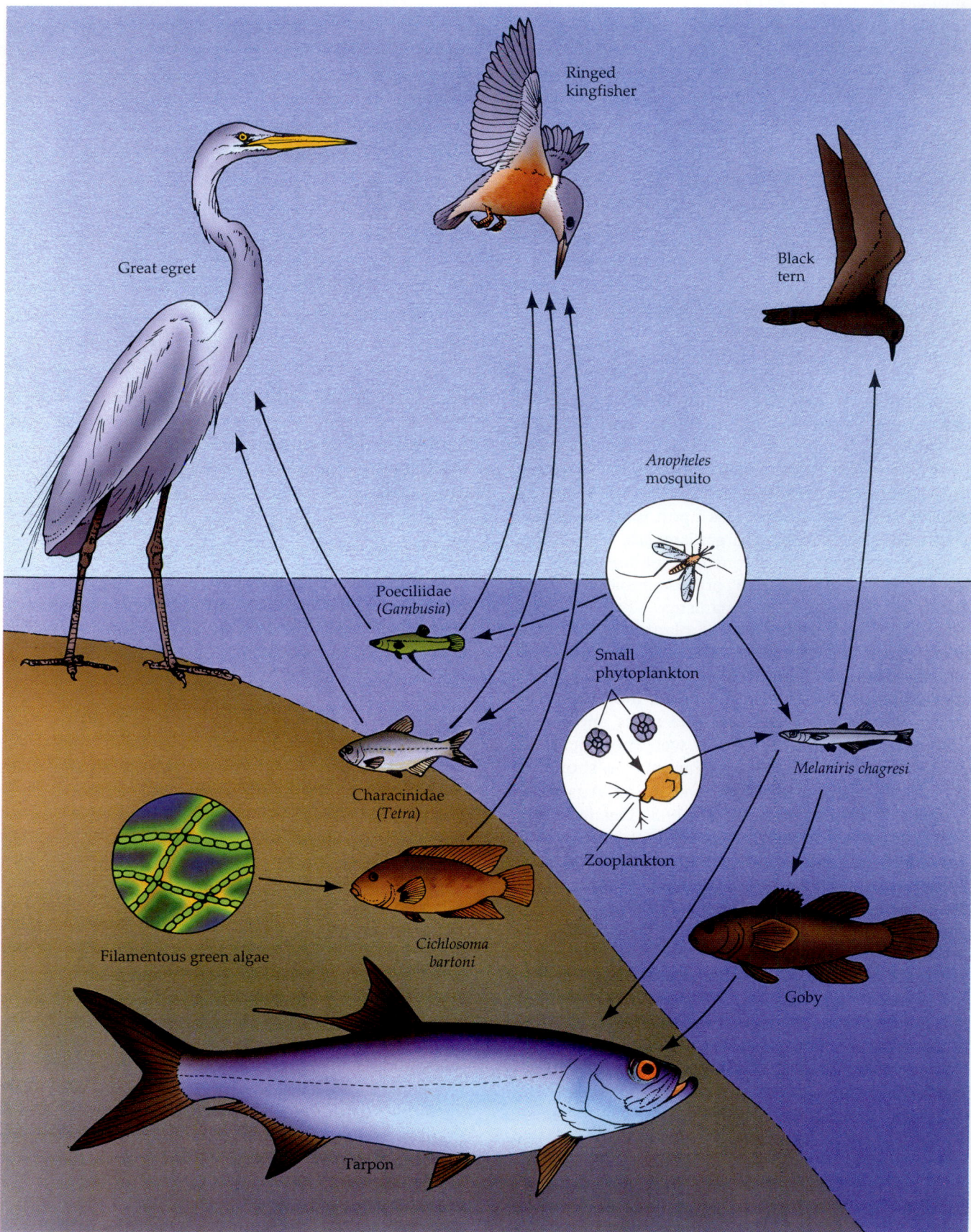

49.2   A Food Web in Gatun Lake, Panama
The lowest trophic levels contain many species of phytoplankton and zooplankton. Notice that three species of fish each eat the same prey species and that most of the secondary carnivores consume more than one prey species.

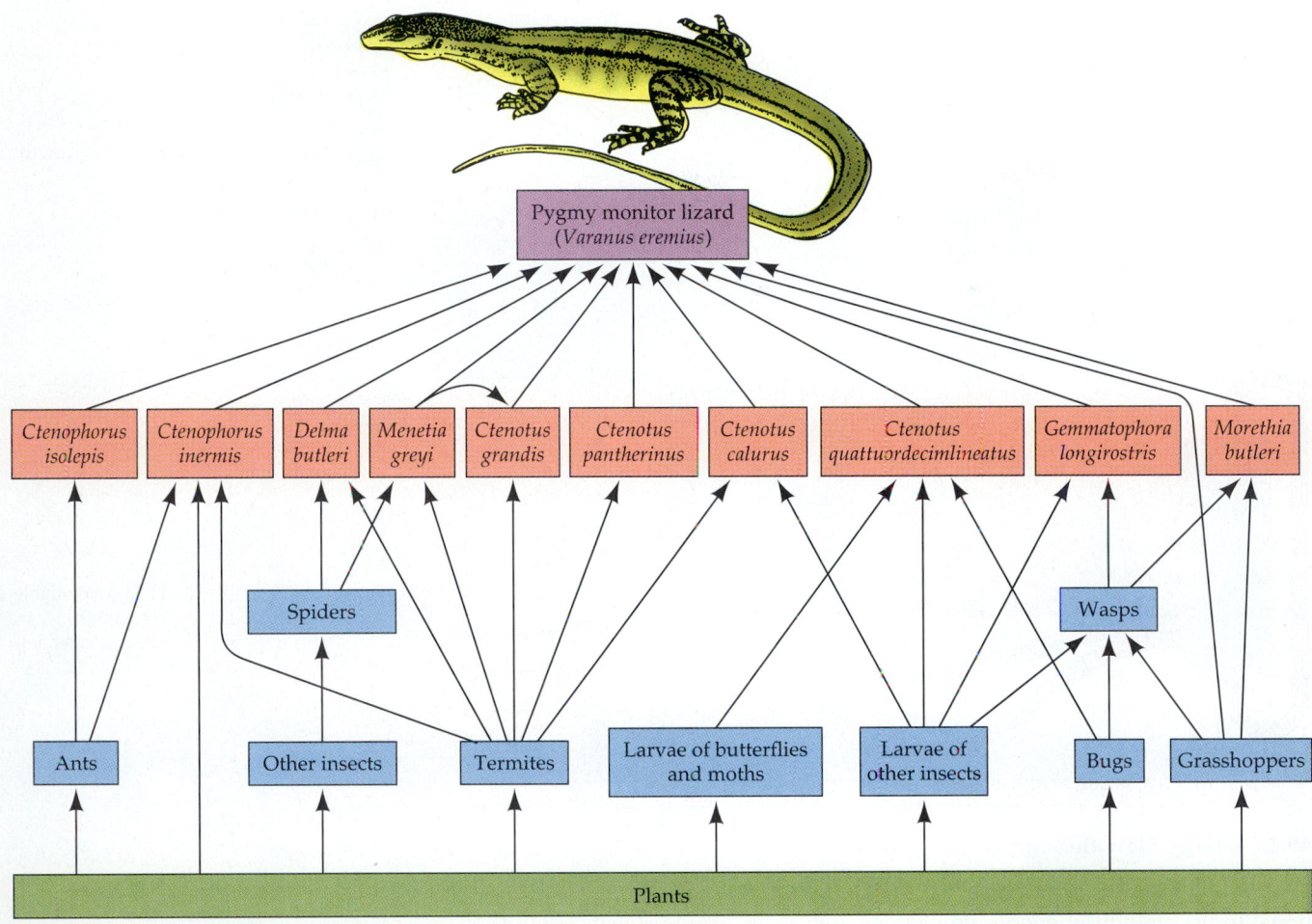

49.3 Who Eats Whom in an Australian Desert
The top predator—a pygmy monitor lizard—feeds mainly on other lizards (ten
species of them, shown in the red boxes) and on grasshoppers.

sumption rates are both high, seldom is as much as 20 percent of the energy assimilated by a trophic level converted to production at the next trophic level. The amount of energy reaching an upper trophic level is determined by the net primary production and the efficiencies with which food energy is converted to biomass (the total weight of organisms) in the trophic levels below it. To show how energy decreases in moving from lower to higher trophic levels, ecologists construct diagrams that are **pyramids of energy.** A **pyramid of biomass,** which shows the standing biomass of organisms at different trophic levels, illustrates the amount of biomass that is available for organisms of the next trophic level.

Pyramids of energy and biomass for an ecosystem usually have similar shapes, but sometimes they do not. The patterns depend on the structures of organisms and how they allocate their energies. On land, dominant photosynthetic plants store energy for moderately long periods, and they dominate both energy flow and standing biomass (Figure 49.5a.) In aquatic communities, on the other hand, the dominant photosynthesizers are bacteria and protists. They have such high rates of cell division that a small standing biomass can feed a much larger biomass of herbivores that grow and reproduce much more slowly. This can produce an inverted pyramid of biomass, even though the pyramid of energy for the same ecosystem has the typical shape (Figure 49.5b).

ENERGY FLOW IN GRASSLANDS. Although most terrestrial ecosystems are dominated by large plants, they differ strikingly in pattern of energy flow, as is evident when we compare grasslands and forests. Grassland plants produce few woody tissues and can support many herbivores. Mammals—wild or domestic—may consume 30 to 40 percent of the aboveground biomass, that is, 30 to 40 percent of the annual aboveground net primary production. Insects may consume an additional 5 to 15 percent. Soil animals, primarily nematodes, may consume 6 to 40 percent of the belowground biomass.

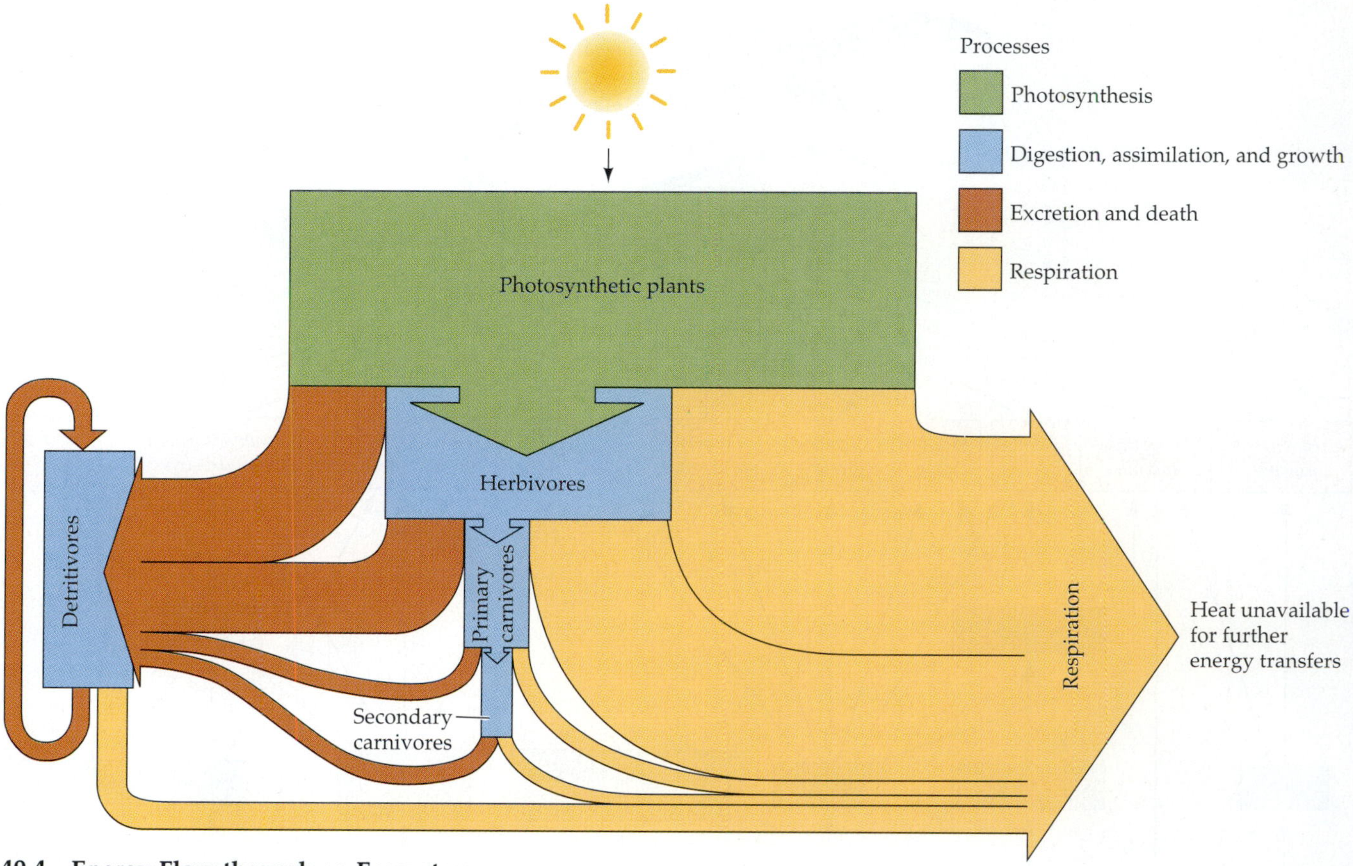

Processes

Photosynthesis

Digestion, assimilation, and growth

Excretion and death

Respiration

Photosynthetic plants

Herbivores

Detritivores

Primary carnivores

Secondary carnivores

Respiration

Heat unavailable for further energy transfers

49.4   Energy Flow through an Ecosystem
The quantities of energy flowing through an ecosystem can be visualized using a diagram like this one, in which the width of the channels is roughly proportional to the amount of energy flowing through them. Trophic levels are indicated by blocks of green or blue; energy channels are in tan or brown, and the direction of energy flow is shown by arrows.

Consumption rates in local areas may be even higher. For example, in certain prairies, grazers harvest 60 to 80 percent of the biomass. Within prairie dog colonies, the prairie dogs take most of the harvest, but bison, elk, and pronghorn, which graze there part of the year, take some as well (Figure 49.6). Typically, the adjacent areas are only lightly grazed. As a result, aboveground plant biomass within colonies is much less than it is outside prairie dog colonies.

This intense grazing has little effect on net primary

49.5   Pyramids of Energy Flow Rates and Biomass
Ecosystems can be compared in terms of the amount of material present in organisms at different trophic levels (left), and in terms of energy flow (right). (a) A grassland is typical of most terrestrial ecosystems. Most of the biomass is found in the green plants, and most of the energy flows through them. (b) A marine community produces an inverted pyramid of biomass. The producers here are unicellular algae, which divide so rapidly that a small biomass can support a much larger biomass of herbivores.

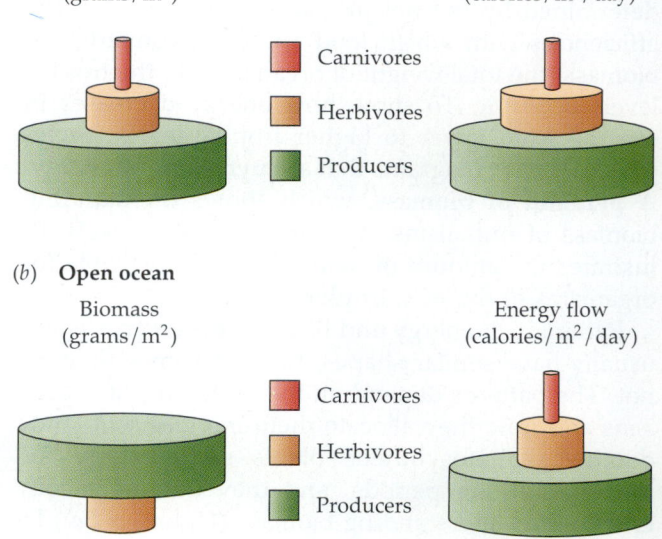

(a) **Grassland**

Biomass (grams / m²)

Energy flow (calories / m² / day)

Carnivores

Herbivores

Producers

(b) **Open ocean**

Biomass (grams / m²)

Energy flow (calories / m² / day)

Carnivores

Herbivores

Producers

49.6 Bison Graze in Prairie Dog Colonies
Bison prefer to graze within this prairie dog colony because the plants there have higher protein levels than plants outside the colony. Notice that bison have cropped the grass short within the colony.

production within the colonies, which remains very similar to that of adjacent areas, but the grazing by herbivores changes the patterns of nutrient cycling and competition between soil microorganisms and plants. Plants in prairie dog colonies allocate most of their energy to regrowth of aboveground tissues and relatively little to roots. In addition, because soil is more exposed within colonies, surface soil temperatures are several degrees higher in summer than in adjacent areas, so conversion of organic nitrogen-containing compounds to inorganic compounds by soil microorganisms proceeds faster. As a result, plant tissues in prairie dog colonies have higher concentrations of proteins than elsewhere—a desirable outcome from the perspective of the prairie dogs.

ENERGY FLOW IN FORESTS. In contrast to those in grasslands, the dominant plants in forests allocate substantial energy to forming wood, which accumulates at high rates in growing forests. As we have seen, wood is heavily defended by the plants and is rarely eaten unless a plant is diseased or otherwise weakened. And, in most forests leaves fall to the ground relatively undamaged at the end of the growing season. Although there are outbreaks of defoliating insects in forests, browsing rates are generally so low that forest ecologists often ignore losses to herbivores when calculating forest production.

DETRITIVORES. As shown in diagrams of energy flow in ecosystems, much of the energy ingested by organisms is converted to biomass that is eventually consumed by detritivores (see Figure 49.4). Detritivores transform the remains of organisms (detritus) into carbon dioxide, water, and mineral nutrients that can be taken up by plants again. If there were no detritivores, most nutrients would eventually be tied up in dead bodies, where they would be unavailable to plants. Therefore, continued ecosystem productivity depends on rapid decomposition of detritus. Under the warm, wet conditions found in tropical forests, detritus is decomposed within a few weeks or months, and no litter accumulates on the soil surface. Rates of decomposition are slower under colder and drier conditions. On cold mountains at high latitudes, decomposition of leaf litter may take decades; decomposition of trunks may take more than a century.

CHEMICAL CYCLING IN ECOSYSTEMS

As we have just seen, energy is not fully recycled in ecosystems because at each transformation some of it is dissipated as heat, a form that cannot be used by organisms. Chemical elements, on the other hand, are not lost when they are incorporated into new trophic levels; they cycle through organisms and the physical environment. This movement depends in large part upon the activities of organisms. Thus the quantities of elements available to organisms are strongly influenced by how organisms get them, how long they hold onto them, and what they do with them while they have them. Carbon, nitrogen, phosphorus, calcium, sodium, sulfur, hydrogen, and oxygen, together with smaller amounts of other elements, are the prime materials of which organisms are constructed. They are vital to the functioning of life today, as they were during its evolutionary past and will be in its future.

To understand the cycling of elements, it is convenient to divide the global ecosystem into four compartments—oceans, fresh waters, atmosphere, and land—because the physical environments in each compartment and the types of organisms living there are different. The amounts of elements found in the different compartments, what happens to those elements, and the rates at which they enter and leave the compartments differ strikingly. After we have described these compartments, we will consider them together to illustrate how the elements cycle through the global ecosystem.

Oceans

The oceans are deep in many places, but they exchange materials with the atmosphere only at their surface. Oceans receive materials from land as runoff from rivers. On time scales of hundreds to thousands

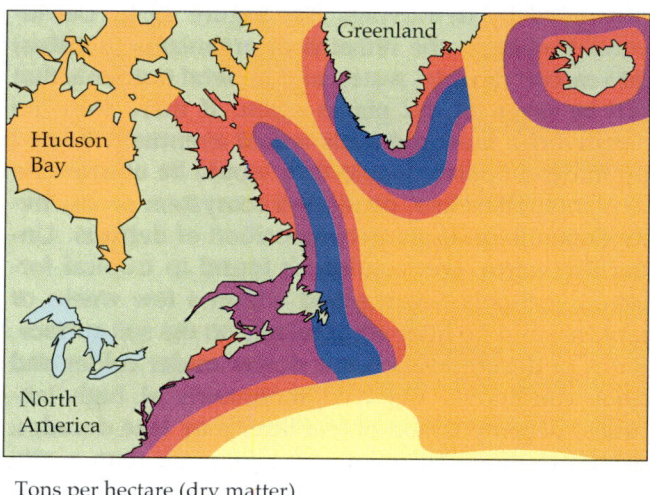

Tons per hectare (dry matter)

| ■ 0.0–2.2 | ■ 2.3–3.7 | ■ 3.8–5.5 | ■ 5.6–7.5 | ■ >7.5 |

49.7 Primary Production Is High in Upwelling Zones
Primary production in the North Atlantic is highest adjacent to continents in areas where surface waters, driven by the prevailing winds, are moving offshore and being replaced by nutrient-rich water from below.

of years, oceans are also the ultimate repository of most materials produced by human activity, even though the immediate receivers are often other compartments of the global ecosystem. Because of their huge size and because they exchange materials with the atmosphere only at their surface, oceans respond very slowly to outside disturbances. Except on continental shelves, ocean waters mix very slowly and are strongly stratified.

Elements that enter the oceans from other compartments gradually sink to the seafloor unless they are brought back to the surface by cool water that rises, as it does near the coasts of continents (Figure 49.7). Such zones of upwelling are rich in nutrients, and many of the world's great fisheries are concentrated there. Over most of the oceans, nutrient concentrations are very low. Oxygen is usually present at all depths, because even slow mixing suffices to replenish the oxygen consumed by the respiration and decomposition of the few organisms that can live in the nutrient-poor water. Most of the elements that enter the oceans settle to the bottom and remain there until the bottom sediments are elevated above sea level by movements of Earth's crust. Many millions of years may elapse in the meantime.

Fresh Waters

Lakes and rivers contain much less water than oceans do, but because these bodies of water are relatively small, most elements entering them are not buried in bottom sediments for long periods of time. Some elements enter fresh waters in rainfall, but most are released by weathering of rocks and are carried to lakes and rivers via groundwater (the water that resides in the soil and in rocks) or by surface flow.

After entering rivers, elements are usually carried rapidly to the oceans. In lakes, however, they are taken up by organisms and incorporated into their cells. These organisms eventually die and sink to the bottom, where decomposition of their tissues uses up the oxygen. Surface waters of lakes thus quickly become depleted of nutrients, and deeper waters become depleted of oxygen. The decrease in photosynthesis rates in lakes that normally would result from nutrient depletion, however, is countered by vertical movements of water that bring nutrients to the surface and oxygen to deeper water. In shallow lakes wind is an important mixing agent, but in deeper lakes it usually mixes only surface waters.

In temperate regions a very important mixing process in lakes is overturning, which occurs twice every year (Figure 49.8). This process is assisted by wind, but it depends on the fact that water is most dense at 4°C; above and below that temperature it expands. In spring in the temperate zone, the sun warms the surface layer of a lake. Initially, the warm layer is very shallow, but as spring and summer progress, its depth gradually increases. However, there is still a well-defined zone—the thermocline—where the temperature drops abruptly to about 4°C. Only if a lake is shallow enough to warm to the bottom does the temperature of the deepest water rise above 4°C. As the surface of a lake cools in autumn, the cooler surface water, which is denser than the warmer water below it, sinks, and is replaced by warmer water from below. This process continues until the entire water column has reached 4°C. At this point, the density of the water is uniform throughout the lake, and even modest winds readily mix the entire water column. Colder weather then cools the surface water below 4°C. This water becomes less dense than the 4°C water below it and floats at the top. Another turnover occurs in spring, when surface layers above the thermocline warm to 4°C, and the water column, again being of uniform density throughout, is easily mixed by wind.

Deep tropical and subtropical lakes may become permanently stratified because they never become cool enough to have uniformly dense water. Their bottom waters lack oxygen because decomposition quickly depletes any oxygen reaching them. Many tropical lakes, however, are overturned at least periodically by strong winds so that their deeper waters are occasionally oxygenated.

Atmosphere

The atmosphere is a thin sphere of gases surrounding Earth. About 80 percent of the mass of the atmo-

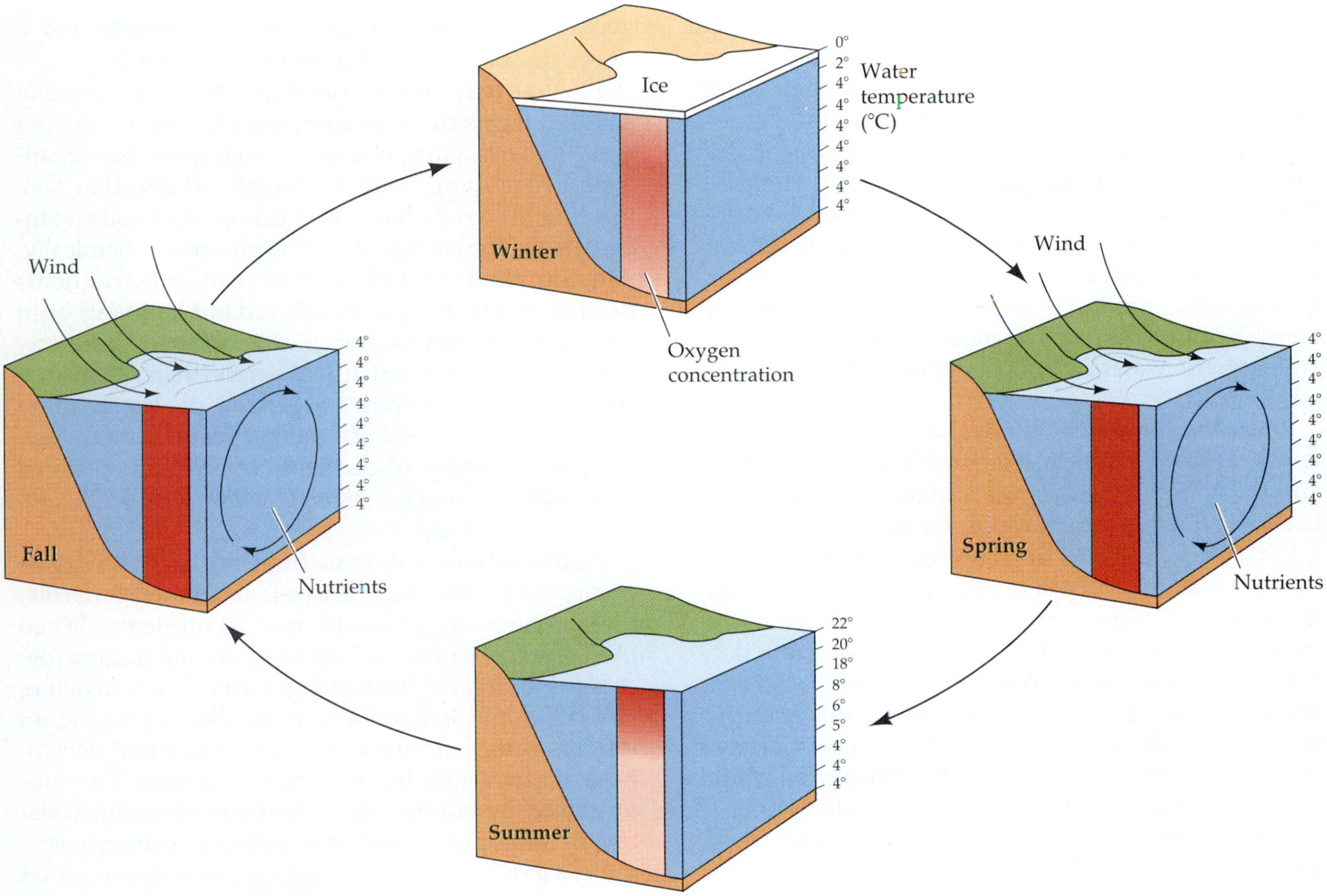

49.8 Annual Temperature and Oxygen Cycles in a Temperate Lake

These vertical temperature profiles are typical of temperate zone lakes that freeze in winter. Turnovers that occur in spring and fall allow nutrients and oxygen to become evenly distributed in the water column. How much surface waters warm during summer varies with the size and depth of the lake and with the local climate. Oxygen concentrations are shown by the intensity of the red in the vertical bands.

sphere lies in its lowest layer, the troposphere, which extends upward from Earth's surface about 17 kilometers in the tropics and subtropics, but only about 10 kilometers at higher latitudes (Figure 49.9). Most global air circulation takes place within the troposphere, and virtually all atmospheric water vapor is located there. Above the troposphere, the stratosphere, which extends upward to about 50 kilometers above Earth's surface, contains about 99 percent of the remaining atmospheric mass, but it is extremely dry. Materials enter the stratosphere from the troposphere near the equator, where the air rises to high altitudes, and they tend then to remain there for a relatively long time because stratospheric air circulation is horizontal. The ozone (O_3) in the stratosphere absorbs most shorter wavelengths of biologically damaging ultraviolet radiation, which is why

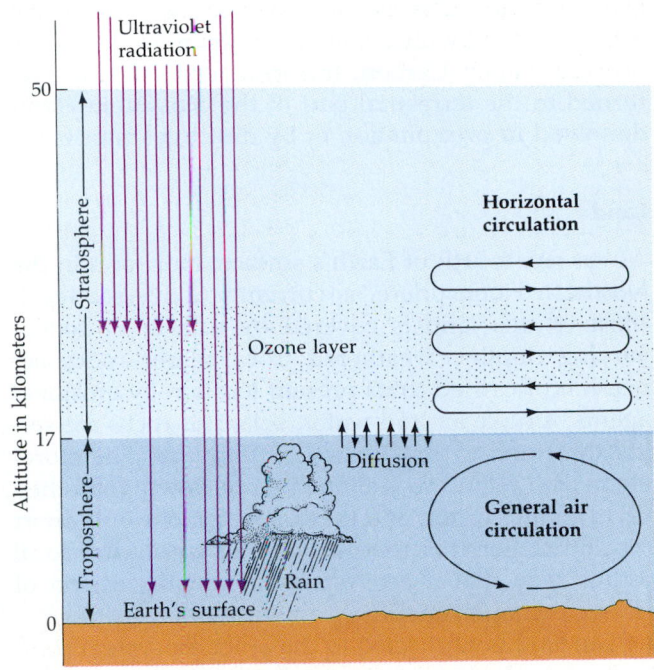

49.9 Earth's Atmosphere

The two lowest atmospheric layers, the troposphere and the stratosphere, differ from each other in their circulation, the amount of moisture they contain, and the amount of ultraviolet radiation they receive.

the thinning of the ozone layer, which we discussed at the beginning of this chapter, is of great concern.

The atmosphere is 78.08 percent nitrogen, 20.95 percent oxygen, 0.93 percent argon, and 0.03 percent carbon dioxide and contains traces of hydrogen gas, neon, helium, krypton, xenon, ozone, and methane. The atmosphere contains Earth's biggest pool of nitrogen and large supplies of oxygen. Although carbon dioxide constitutes a very small fraction of the atmosphere, it is the source of carbon used by terrestrial photosynthetic organisms. Concentrations of atmospheric water vapor are highly variable in space and time.

The atmosphere plays a decisive role in regulating temperatures at and close to Earth's surface. Without an atmosphere, the average surface temperature on Earth would be about $-18°C$ rather than its actual $+17°C$. Earth remains at this warm temperature because the atmosphere is relatively transparent to visible light but traps a large part of the outgoing infrared radiation (heat), the main radiation emitted by a cool body like Earth. Water vapor, carbon dioxide, and ozone are especially important in this trapping of infrared radiation. This is why, as we will discuss later, increased concentrations of atmospheric carbon dioxide may lead to important climatic changes.

The atmosphere is a transport medium for many gases, as well as for airborne particles containing carbon, nitrogen, sulfur, phosphorus, and other nutrient elements. Many of these compounds are oxidized in the atmosphere by photochemical reactions involving the —OH radical, which forms when ozone is bombarded by ultraviolet radiation in the presence of water vapor. Carbon, nitrogen, and sulfur are returned to the terrestrial part of the ecosystem either dissolved in precipitation or by direct gas transfer.

Land

About one-fourth of Earth's surface, most of it in the Northern Hemisphere, is currently above sea level. Most of this land is covered by a layer of soil of varying depths, weathered from parent rocks beneath it or carried to its present location by erosional agents. Unlike air and water, soils and rocks are solids that tend to remain where they are. Therefore, elements on land move much more slowly than they do in air and water, and they usually move only short distances. For this reason, we will emphasize local rather than global ecosystems in our discussion of the land compartment.

The land is connected to the other ecosystem compartments because terrestrial organisms take elements from and release elements to the air. In addition, elements in soils are carried in solution into the groundwater and eventually into rivers and oceans, where they are lost to organisms until an episode of

geological uplifting raises marine sediments and a new cycle of erosion and weathering begins.

As you may recall from Chapter 32, the type of soil that forms in an area depends on the underlying rock, as well as on climate, topography, the organisms living there, and the length of time that soil-forming processes have been acting. As a soil weathers, its clay particles slowly decompose chemically. After hundreds of thousands of years, most nutrients needed by plants have weathered out and have been carried by groundwater to the oceans. Therefore, very old soils are much less fertile than are young soils. Figure 32.6 shows a general picture of these changes over a period of 1 million years. Even though the global supply of nutrients is constant, regional and local deficiencies strongly affect ecosystem processes on the land.

Plants extract nutrients from the soil and incorporate them into their tissues. If the soil is fertile, these tissues are generally rich in nutrients. When the tissues then fall and decompose, the **humus** (decomposed organic matter) that forms is rich, alkaline, and dark, and is known as mull. Plants growing on infertile soils form nutrient-poor tissues that decompose to an acidic humus known as mor. The mor produced by conifers is particularly resistant to decay. Because needle decomposition takes many years, there is a permanent thick layer of partly decomposed needles on the soil surface under conifers.

BIOGEOCHEMICAL CYCLES

Earth is essentially a closed system with respect to carbon, hydrogen, oxygen, nitrogen, phosphorus, and sulfur, the elements organisms need in large quantities. These elements cycle through organisms to the environment and back again through organisms, without input from extraterrestrial sources. The carbon and nitrogen atoms of which life is composed today are the same atoms that were in dinosaurs, insects, and trees in the Mesozoic era. Some of these elements circulate continually, but large quantities of the molecules of other elements are temporarily lost from circulation through deposition in deep-sea sediments. The pattern of movement of an element through organisms and reservoirs in the physical environment is called its **biogeochemical cycle**.

Each element has a distinctive biogeochemical cycle whose properties depend on the physical and chemical nature of the element and how it is used by organisms. All elements cycle quickly through organisms because no individual, even of the longest-lived species, lives very long in geological terms. Elements, such as carbon and nitrogen, that exist in the atmosphere as a gas cycle faster than elements that are not gaseous. We will illustrate the properties of bio-

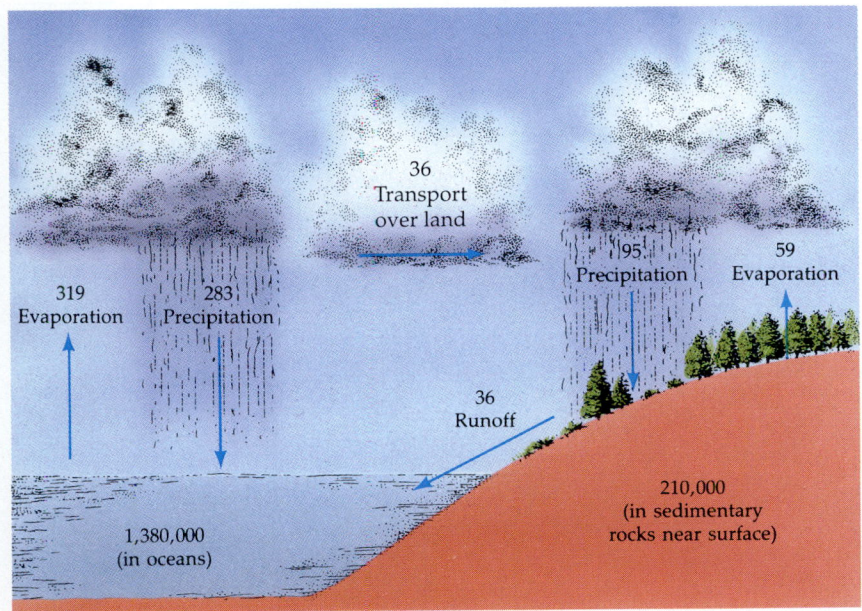

49.10 The Global Hydrological Cycle
The numbers give the relative amounts of water (expressed as units of 10^{18} g) held or exchanged. Notice that the greatest exchanges take place at the ocean surface. Although rocks contain large quantities of water, this "locked-in" water plays very little role in the hydrological cycle.

geochemical cycles by discussing the elements that are needed by organisms in large amounts.

The Hydrological Cycle

Although water is a compound, not an element, because of its importance to life we will discuss its cycle here together with those of the important elements. The movement of water from land to ocean is obvious. Its return to land from the oceans is not as evident, even though the amount moving in each direction is the same. The cycling of water through the oceans, atmosphere, and land, known as the **hydrological cycle,** begins with the evaporation of water, most of it from ocean surfaces. Some water returns to the ocean as precipitation, but much less falls back on the ocean than is evaporated from it. The remaining evaporated water is carried by winds over the land, where it falls as rain or snow. Water also evaporates from the soil, from freshwater lakes and rivers, and from the leaves of plants, but the total amount evaporated is less than the amount that falls as precipitation. The excess water eventually returns to the oceans via rivers, coastal runoff, and subterranean discharge (Figure 49.10).

The Carbon Cycle

Organisms are triumphs of carbon chemistry and must, to survive, have access to carbon atoms. As we have seen, nearly all carbon in organisms comes from carbon dioxide in the atmosphere or from carbonate ions in the water. In the cells of photosynthetic bacteria and algae and in leaves of plants it is incorporated into organic molecules by photosynthesis. All organisms in other kingdoms get their carbon by eating other organisms or their remains.

Although atmospheric carbon dioxide is the immediate source of carbon for terrestrial organisms, only a small part of Earth's carbon is in the atmosphere. Most of it is nongaseous, dissolved carbon (bicarbonate and carbonate ions) in oceans and carbonate minerals in rocks. Sedimentary rocks hold most of Earth's carbon that is in rocks. Movement of carbon between rocks and other reservoirs of carbon is very slow. The quantities of carbon in each ecosystem compartment and the yearly carbon fluxes between compartments are shown in Figure 49.11. Notice that although marine organisms have very little carbon, they have a profound influence on the distribution of carbon in the seas. They convert soluble carbonate ions from seawater to insoluble ocean sediments by depositing carbon in their shells and skeletons, which eventually sink to the bottom. Biological processes redistribute carbon between atmospheric and terrestrial compartments, removing it from the atmosphere during photosynthesis and returning it to the atmosphere during respiration. Growing plants at middle and high latitudes in the Northern Hemisphere incorporate so much carbon into their bodies during the summer that they reduce the concentration of atmospheric carbon dioxide from about 350 parts per million in winter to 340 parts per million during midsummer. This carbon is released back into the atmosphere by decomposition in the autumn.

At times in the remote past, large quantities of carbon were removed from the global carbon cycle when organisms died in large numbers in environments without oxygen. In such environments decay

49.11 The Global Carbon Cycle

The two large reservoirs of carbon are dissolved carbon in the oceans and carbonate minerals in rocks (not shown). The numbers show the quantities of carbon (expressed as units of 10^{15} g) in organisms and various carbon reservoirs and the amounts that move between the various compartments. The widths of the arrows are proportional to the size of the flux.

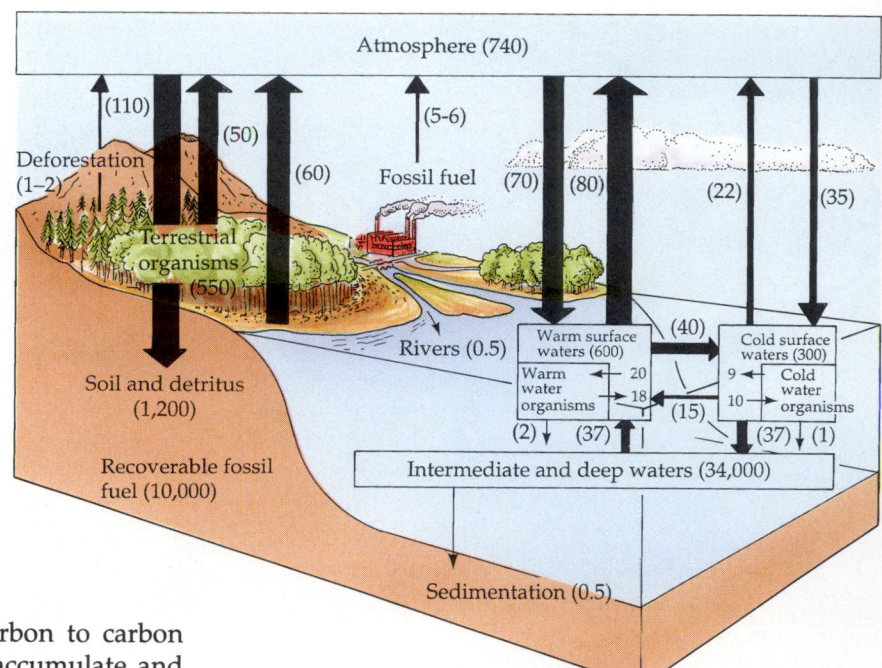

organisms do not reduce organic carbon to carbon dioxide. Instead, organic molecules accumulate and eventually become oil, natural gas, coal, or peat. Humans have discovered and used these deposits, known as fossil fuels, at ever-increasing rates during the past 150 years. As a result, carbon dioxide, the final product of burning those fuels, is being released into the atmosphere faster than it is being removed by living organisms and transferred to the oceans and into terrestrial biomass (see Figure 49.16). This addition to the carbon cycle is also shown in Figure 49.11.

The buildup of atmospheric carbon dioxide will warm the global climate during the next century because carbon dioxide is transparent to sunlight but opaque to radiated heat. Carbon dioxide thus permits sunlight to strike and warm Earth, but it traps some of the heat that Earth radiates back toward space. Enough carbon is released by burning fossil fuels to alter the heat balance of Earth, even though the absolute quantity is small relative to other components of the carbon cycle.

The Nitrogen Cycle

Nitrogen is an essential component of many organic molecules, such as nucleic acids and proteins. Nitrogen is an abundant, chemically inert gas that makes up 78 percent of the atmosphere, but unlike carbon dioxide, which photosynthesizers use directly, nitrogen cannot be used by most organisms in its gaseous form. Nitrogen can be converted into biologically useful forms by only a few species of bacteria and cyanobacteria (see Chapter 22). Therefore, despite its abundance, usable nitrogen is often in short supply in ecosystems. This is why nearly all commercial fertilizers contain biologically useful compounds of nitrogen.

Just as organisms other than nitrogen fixers cannot take up nitrogen gas directly from the atmosphere, they also do not respire nitrogen back to the atmosphere. Instead, organic molecules containing nitrogen are converted to inorganic molecules in several stages by different organisms. Most of these nitrogen-containing compounds, such as nitrates or ammonia, are again taken up by plants. This movement of nitrogen among organisms accounts for about 95 percent of all nitrogen fluxes on Earth (Figure 49.12).

The quantity of nitric oxides and nitrate that humans produce by burning fossil fuels and by manufacturing and using fertilizers is about half the amount the rest of the living world produces. The use of fertilizers has raised agricultural productivity, but nitrogen fertilization also contributes to the problem of acid rain and to the declines in biodiversity associated with it.

The Phosphorus Cycle

The phosphorus cycle differs from those of carbon and nitrogen—as well as those of sulfur, oxygen, and hydrogen—in that it lacks a gaseous phase. Some phosphorus is transported on dust particles, but in general the atmosphere plays a very minor role in the phosphorus cycle. Phosphorus exists primarily as phosphate (PO_4^{3-}) or similar compounds. Most phosphate deposits are of marine origin. Biological materials that are decomposing on the seafloor release phosphate, which enters sediment pores and is eventually incorporated into a type of rock known as apatite. Such deposits are forming today on conti-

49.12 The Global Nitrogen Cycle
Unicellular organisms in the soil and
oceans create the fluxes, which are expressed as 10^9 kg of nitrogen per year.
The widths of the arrows are proportional to the size of the flux. The several stages of inorganic nitrogen are nitrate (NO_3), nitrite (NO_2), and
ammonium (NH_4).

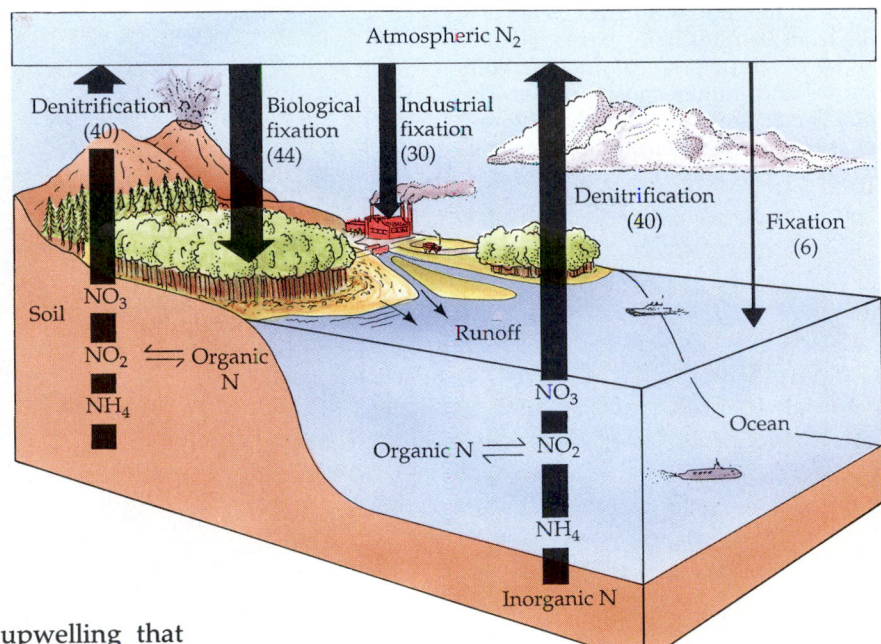

nental margins beneath regions of upwelling that
support large concentrations of organisms (see Figure
49.7).

In our discussions of carbon and nitrogen we have
seen that the reservoirs of elements and fluxes between reservoirs tell us a lot about how the elements
cycle. Understanding how phosphorus cycles does
not require an illustration; simply study the numbers
in Table 49.3. Notice how much more rapidly phosphorus is cycled through marine organisms (ocean
biota) than through terrestrial organisms (land biota).
Phosphorus also moves readily between the surface
and the bottom of the oceans. On average, a phosphorus atom is cycled about 50 times between deep
waters and surface waters before being removed to
the sediments. Each time a phosphorus atom reaches
surface waters, it is cycled between the oceanic biota
and the dissolved phosphates in the water about 25
times before it returns to deep water. As a result, the

average phosphorus atom is incorporated into marine
organisms about 1,250 times during its stay in the
ocean! The overall phosphorus cycle is fairly constant, but phosphorus is often a limiting nutrient in
soils and lakes. This is why phosphate is a component of most fertilizers and why adding phosphate
to lakes causes marked increases in their biological
productivity.

The Sulfur Cycle

Emissions of sulfur dioxide and hydrogen sulfide
from volcanoes and fumaroles (vents for hot gases)
are the only significant natural nonbiological flux of

TABLE 49.3
**Reservoirs, Fluxes, and Residence Times of Phosphorus in the Global
Ecosystem**

RESERVOIR	AMOUNT IN RESERVOIR (10^6 METRIC TONS)	FLUX (10^6 METRIC TONS/YEAR)	RESIDENCE TIME (YEARS)
Atmosphere	0.0028	4.5	0.0006 (53 hrs)
Land biota	3,000	63.5	47.2
Land	2,000,000	88–100	2,000
Shallow ocean	2,710	1,058	2.56
Ocean biota	138	1,040	0.1327 (48 days)
Deep ocean	87,100	60	1,452
Sediments	4×10^9	214	1.87×10^8
Total ocean ecosystem	89,810	1.9	47,270

49.13   The Global Sulfur Cycle

The transfer rates here, expressed as 10^9 kg of sulfur per year, include both natural and human-caused fluxes. The total fluxes are now more than twice what they were a century ago, primarily because of fossil fuel combustion.

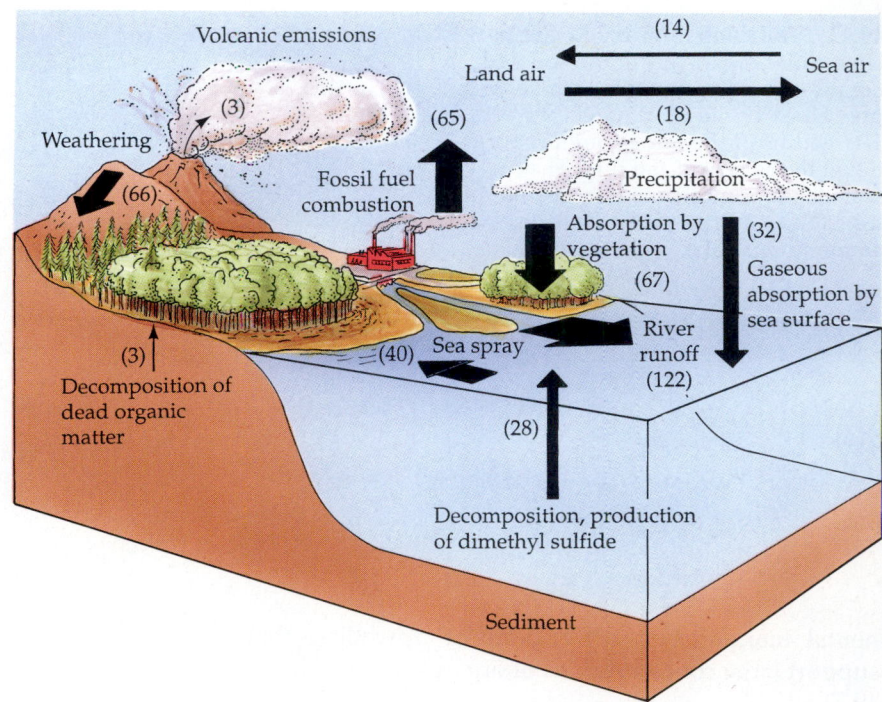

sulfur. These emissions release, on average, between 10 and 20 percent of the total natural flux of gaseous sulfur to the atmosphere, but they vary greatly in time and space. Large eruptions spread great quantities of sulfur over broad areas, but they are rare events. Volatile sulfur compounds are also emitted by both terrestrial and marine organisms. Certain marine algae produce large amounts of dimethyl sulfide. Why they do so is not certain, but because coastal algae produce the largest amounts, researchers believe that dimethyl sulfide protects the algae from desiccation and excess salinity. Dimethyl sulfide released by algae accounts for about half of the sulfur in the biotic part of the sulfur cycle; the other half is produced by terrestrial organisms. On land, the breakdown of organic sulfur compounds during fermentation is the most important mechanism of sulfur release (Figure 49.13).

Sulfur is apparently always abundant enough to meet the needs of living organisms. It also plays an important role in global climate. Even if air is moist, clouds do not form readily unless there are nuclei around which water can condense. Dimethyl sulfide is the major source of such nuclei. Therefore, increases or decreases in sulfur emissions can change cloud cover and hence climate. Sulfur emitted by human activities now equals natural sulfur fluxes, and the increased sulfur cycle is already influencing global climates. The increase in atmospheric concentrations of carbon dioxide has not warmed Earth as much during the past century as expected because cloud cover has increased along with increases in atmospheric concentrations of dimethyl sulfide, thus allowing less sunlight to reach Earth.

HUMAN ALTERATIONS OF BIOGEOCHEMICAL CYCLES

The elements circulating in the biogeochemical cycles we have just discussed are essential for the metabolism of organisms. The interaction between organisms and elemental cycles is reciprocal. Organisms influence the rates at which elements cycle, and recycling rates, in turn, influence the rates of metabolic processes.

Human activity has greatly modified the quantities of elements being cycled and where they enter and exit ecosystems. These changes can increase metabolic rates if they increase the availability of nutrients, or they can decrease metabolic rates if levels of elements become high enough to be toxic to organisms or if high levels cause other detrimental environmental changes. We will now consider several examples of consequences resulting from human modifications of biogeochemical cycles. These consequences range from local to global, and they include both increases and decreases of metabolic rates.

Lake Eutrophication: A Local Effect

The most striking and best-studied example of local effects of altered biogeochemical cycles is **eutrophication**—the addition of nutrients, especially phosphorus, to fresh water. Humans, which tend to be concentrated around water, dump large quantities of nutrients directly and indirectly into lakes and rivers. Most of these nutrients come from domestic and industrial sewage, but many come from leaching of fertilizers and pesticides from agricultural lands draining into rivers and lakes. Some nutrients arrive in precipitation.

In fresh water, photosynthesis is most often limited by supplies of phosphorus. In eutrophic (enriched) lakes, the extra phosphorus provided by fertilizers and detergents allows algae and bacteria to multiply, forming blooms that turn water green. Usually spherical green algae are replaced by filamentous cyanobacteria as eutrophication proceeds (see Figure 22.15c). The decomposition of dead cells produced by the increased biological activity uses up all oxygen in the lake. Not until the water column overturns again (see Figure 49.8) does the water contain oxygen. Anaerobic organisms thus come to dominate the sediments. These organisms are unable to break down organic compounds all the way to carbon dioxide, and many of the end products of their activities have unpleasant odors. The decrease in oxygen in the water also harms some aquatic animals.

Lake Erie is a eutrophic lake. Two hundred years ago Lake Erie had moderate levels of photosynthesis and clear, oxygenated water. Today more than 15 million people live in the Lake Erie basin. Nearby cities pour more than 250 billion liters of domestic and industrial wastes into the lake annually. The entire basin is intensely farmed and heavily fertilized. In the early part of this century, nutrients in the lake increased greatly and algae proliferated. At the water filtration plant at Cleveland, for example, algae increased from 81 per milliliter in 1929 to 2,423 per milliliter in 1962. Algal blooms and populations of bacteria also increased. The numbers of *Escherichia coli* increased enough to cause the closing of many of the lake's beaches because of health hazards.

As the oxygen level dropped in deeper lake waters, many native species that thrive only in oxygenated water declined. They were replaced by species of clams, snails, and midges that can survive in relatively anaerobic environments. For example, nymphs of the mayfly genus *Hexagenia* were replaced by worms as dominant organisms of the lake bottom (Figure 49.14). The fish population also changed. Prior to the turn of the century, dominant fishes in Lake Erie were lake herring, blue pike, carp, yellow perch, sauger, whitefish, and walleye. Lake trout were common in deeper waters. By 1925, herring had become too scarce to support the herring industry. After 1945, blue pike, sauger, and whitefish became very scarce and lake trout disappeared. Currently the lake's fishing industry depends upon yellow perch, smelt, sheepshead, white bass, carp, catfish, and walleye, most of which are less valuable commercially than the species that declined.

Since 1972 the United States and Canada have invested more than 7.5 billion dollars to improve

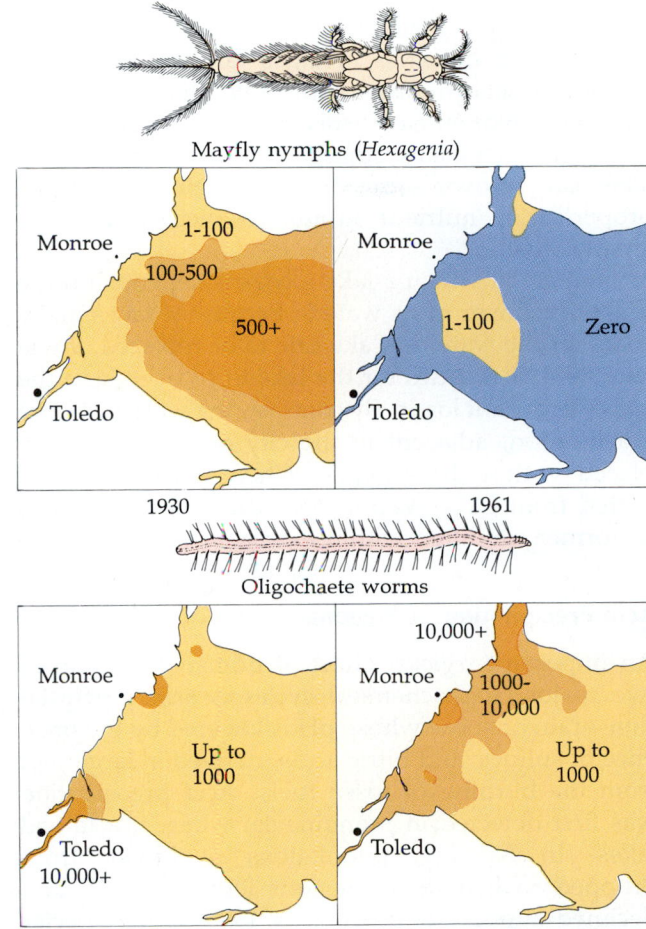

49.14 Eutrophication Changed Lake Erie
As Lake Erie became more polluted during the middle part of the present century, the original mayfly population decreased (upper maps) but oligochaete worms increased (lower maps). The numbers represent the number of individuals per square meter.

municipal waste facilities and reduce discharges of phosphorus into Lake Erie. As a result, the amount of phosphates added to Lake Erie decreased more than 80 percent from the maximum level, and phosphorus concentrations in the lake have declined substantially. Since 1985, estimated inputs of phosphorus to Lake Erie have been reduced to about the target goal of 11,000 metric tons per year. Deeper waters of Lake Erie still become poor in oxygen during summer months, but the rate of oxygen depletion is declining. Algal blooms have decreased, as have populations of small fishes, such as alewife, spottail shiner, and emerald shiner, which feed on the algae.

Lake Erie cannot be restored to its historical assemblage of organisms because some species, such as the blue pike, became extinct. Others, such as Pacific salmon and zebra mussels, have been introduced, the former deliberately, the latter accidentally. Zebra mussels, which now exist in densities as high as

70,000 per square meter, are responsible for greatly reducing the densities of algae and cyanobacteria. Zebra mussels probably cannot be eliminated from Lake Erie. Nonetheless, this example, together with those of other lakes adjacent to cities, shows that lakes can recover some of their original functional properties if nutrient inputs to their waters are greatly reduced.

The rate at which a lake recovers depends on the rate of turnover in its waters. Because it takes many years for the water of Lake Erie to be replaced, it will take several decades for the lake to recover from the heavy pollutant loads. By contrast, the waters of Lake Washington, adjacent to the city of Seattle, are replaced within three years. When sewage was diverted from Lake Washington, the lake returned to its former condition within a decade.

Acid Precipitation: A Regional Effect

An important *regional* effect of human alteration of two major biogeochemical cycles is **acid precipitation**—rain or snow whose pH is lowered by the presence of sulfuric and nitric acids, derived in large part from the burning of fossil fuels. Acid precipitation was first detected in Scandinavia, where it acidified lakes. In Norway, populations of brown trout dropped to half their previous levels by 1978 and declined another 40 percent by 1983. Other species of fish were similarly affected. Acid precipitation is now a phenomenon of all major industrial countries and is particularly widespread in eastern North America (Figure 49.15). The normal pH of precipitation in this region is about 5.6, but precipitation in New England now averages about pH 4.1, and there are occasional storms with a precipitation pH as low as 3.0. Precipitation with a pH of about 3.5 or lower causes direct damage to the leaves of plants. In central Europe, acid precipitation has contributed to the moderate to severe damage suffered by 15 to 20 percent of the growing stock of harvestable forest.

Ecologists in Canada studied the effects of acid precipitation by adding enough sulfuric acid to reduce the pH of two lakes from about 6.6 to 5.2. In both lakes, nitrifying bacteria failed to adapt to these moderately acidic conditions, with the result that the nitrogen cycle was blocked and ammonium accumulated in the water. When the ecologists stopped adding acid to one lake, the pH increased to 5.4 and nitrification resumed after a lag of about one year. These experiments show that lakes are very sensitive to acidification but that they can recover quickly when pH returns to normal values.

Oxides of nitrogen and sulfur entering the atmosphere from the burning of fossil fuels may travel hundreds of kilometers before they settle to Earth in precipitation or as dry particles. The source of acid precipitation in New England is primarily the Ohio Valley; that of Scandinavia is primarily England and Germany. When pollution originates in one area but causes problems in another, solving the problem may be politically difficult. Acid precipitation is caused by generation of the energy upon which modern societies depend. Oxides of nitrogen and sulfur can be removed from smokestack gases of large installations, but costs rise sharply as the percentage removed rises above 90 percent. The number of sources emitting oxides is now so great that almost complete removal will be necessary to correct the problem, even if no new sources are added.

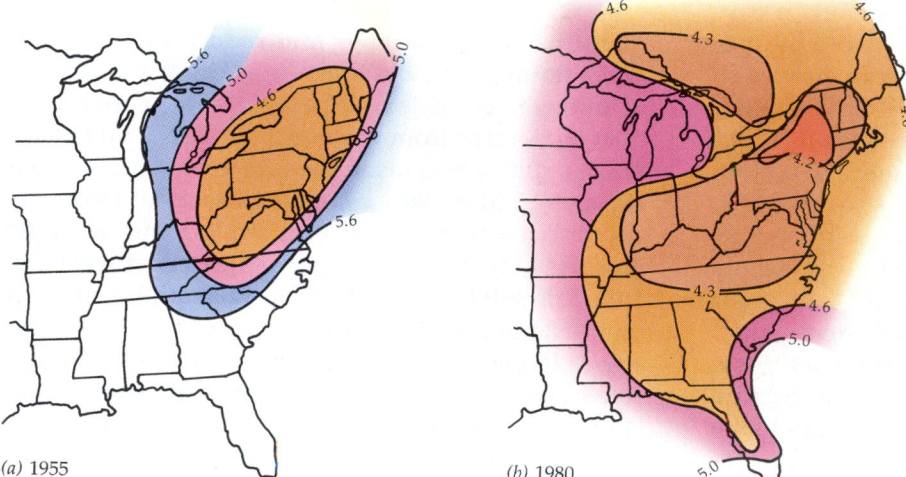

(a) 1955

(b) 1980

49.15   Increases in Acid Precipitation in Eastern North America
These maps show the annual average pH of precipitation in eastern North America. The oxides of nitrogen and sulfur—the principal contributors to acid precipitation—travel far enough from their sources that the effects of many sources blend together to produce the pattern shown here.

Alterations of the Carbon Cycle: A Global Effect

The carbon cycle is the biogeochemical cycle most seriously disturbed globally by human activity. Climatologists have made measurements of atmospheric concentrations of carbon dioxide on top of Mauna Loa in Hawaii since 1958. These measurements reveal a slow but steady increase in atmospheric carbon dioxide concentrations (Figure 49.16). From a variety of calculations, atmospheric scientists believe that 150 years ago, before the Industrial Revolution, the concentration of atmospheric carbon dioxide was probably about 265 parts per million. Today it is 350 parts per million. This increase has been caused primarily by combustion of fossil fuels and secondarily by the burning of forests. If current trends in both these activities continue, atmospheric carbon dioxide is expected to double its 1900 value of 290 parts per million by the middle of the twenty-first century. This carbon dioxide will eventually be transferred to the oceans and deposited in sediments as calcium carbonate ($CaCO_3$), but the rate of this process is much slower than the rate at which humans are introducing carbon dioxide into the atmosphere.

Climatologists have analyzed air trapped in the Antarctic and Greenland ice caps and found that concentrations of carbon dioxide in the atmosphere have varied considerably in the past. During the last ice age—between 15,000 and 30,000 years ago—the concentration was as low as 200 parts per million; during a warm interval 5,000 years ago it may have been slightly higher than it is today. Global climate models developed by climatologists can predict the likely consequences of further increases in concentrations of atmospheric carbon dioxide. Predictions from these models are imprecise because nature is much more complicated than the models. Nonetheless, if the concentration of atmospheric carbon dioxide doubles its current level, the mean temperature of Earth is expected to increase 3 to 5°C, with larger increases at higher latitudes. Thus today's climates will shift to higher latitudes. A carbon dioxide doubling would probably cause droughts in the central regions of continents and would increase precipitation in coastal areas. Global warming may result in melting of the Greenland and Antarctic ice caps and will warm the oceans, which will thus expand, raising sea levels and flooding coastal cities and agricultural lands.

Because carbon dioxide is carried by air movements to places thousands of kilometers away from where it enters the atmosphere, the problem is global. Although it is very difficult for societies to decide how much they should invest today to avert potential future climatic problems, many nations have committed themselves to efforts at reducing their emissions of carbon dioxide. Such reductions will not be easy because carbon dioxide is the inevi-

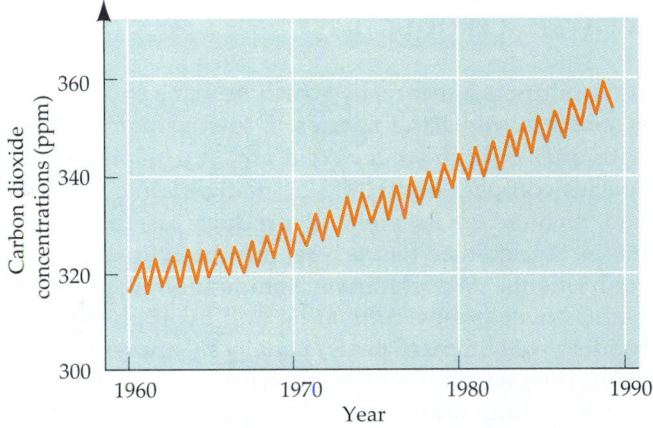

49.16 Atmospheric Carbon Dioxide Concentrations Are Increasing
These concentrations, expressed as parts per million by volume of dry air, were recorded on top of Mauna Loa, Hawaii. Each year CO_2 concentrations rise during the winter, when respiration exceeds photosynthesis, and fall during the summer, when photosynthesis exceeds respiration. However, the steady rise in overall concentrations is very apparent.

table end product of fossil fuel combustion and modern societies are powered by fossil fuels. The amount of carbon dioxide emitted by a power plant burning fossil fuel cannot be reduced by cleansing the gases leaving the smokestack. Because so much energy is currently wasted, however, many steps can be taken to increase energy-use efficiency so that we get more valuable services for the same amount of fuel burned. In addition, we can substitute energy sources that do not contain carbon (solar, geothermal, nuclear) for fossil fuels.

Alteration of the Chlorine Cycle

Organisms require chlorine in only small amounts. For this reason, at first thought it might seem that alteration of the natural chlorine cycle is a minor matter of no global significance. However, modern technological societies use and release to the environment large quantities of chlorine-containing compounds. The most important of these compounds are the long-lived chlorofluorocarbons (CFCs), which are now found throughout the globe and which have caused a significant depletion of the stratospheric ozone layer, as we saw in the beginning of this chapter. Scientists did not expect CFCs to have a major environmental effect because they are so inert. In fact, it was precisely because CFCs are so inactive chemically that their use was readily adopted. The widening ozone hole is ample testimony, however, to the broad-reaching effects of these seemingly harmless compounds.

AGRICULTURE AND ECOSYSTEM PRODUCTIVITY

Agriculture is a means by which humans exploit ecosystems by replacing species of low economic value with ones of high value. We do this by helping some species compete better and by manipulating the system to increase its yield of products useful to humans. Agriculture has several intricately intertwined components: We eliminate competition with other plants—weeds—by cultivating and by applying herbicides; we eliminate pests, usually by applying toxic chemicals; we augment photosynthesis by fertilizing and irrigating; and we develop special high-yielding strains of plants that respond to additional fertilizer by increasing their growth rates. All these components must work together because "miracle" strains do not actually yield more than other strains unless they are provided with fertilizers and protected from competitors and herbivores. Agriculture also depends on energy from outside the system for cultivating and harvesting. Traditional and modern agriculture provide this energy in different ways (Figure 49.17).

Although human manipulations of agricultural systems have spectacularly increased food production per hectare, they have also created problems. Insecticides and herbicides pollute lakes, rivers, and groundwaters in most industrialized countries. Many agricultural pests have evolved resistances to pesticides. In response, we increase the use of pesticides, creating more severe pollution problems. Agriculturalists are combatting these problems by developing

new methods of pest control in agriculture. Generally known as **integrated pest management**, these methods combine chemical approaches with cultural practices—such as crop rotation, mixed plantings of crop plants, and mechanical tillage of the soil—and biological methods—such as development of pest-resistant strains of crops, use of predators and parasites rather than pesticides, and use of chemical attractants. Under integrated pest management, farmers use chemicals sparingly enough to avoid most pollution problems. The reduced use of chemicals also serves the goals of integrated pest management in a different way: It reduces the chance that pests will evolve resistance to pesticides.

Organisms and materials move into and out of all natural ecosystems. Some ecosystems, such as rivers and the deep sea, are powered primarily by organic compounds and materials imported from elsewhere. In most ecosystems, however, losses of materials through one component are made up by gains through another. Agricultural ecosystems differ from natural ecosystems because humans *extract* as many energy-rich compounds as possible from the systems and consume them elsewhere. Preindustrial agricultural people lived near their fields and returned most wastes, including their own excretory products, to the fields. The flow of nutrients changed dramatically when most people moved to cities. Today, most humans eat their food far from where it was produced, and we release large quantities of our wastes into small areas. Modern agricultural systems require heavy applications of fertilizers to replace those elements removed when crops are harvested and transported elsewhere.

Agriculture developed in temperate regions, but the methods designed for temperate regions often work poorly in the tropics. Lush forests grow even on ancient, impoverished tropical soils. Their annual

49.17 Agriculture Requires Energy
(a) In traditional agriculture, people supply most of the energy, as in this Bengali rice paddy. (b) Modern agriculture is based on high rates of consumption of fossil fuels. Often it involves toxification of the environment as well, as in the spraying of pesticides shown here.

(a)

(b)

photosynthesis per unit of ground area is often not much less than that of forests on much better soils. This high rate of growth on poor soil is possible because nutrients in tropical forests are very tightly cycled. Minerals in tropical forests are concentrated within the plants themselves. Trees produce long-lived leaves and shed them a few at a time, over long periods rather than all at once. Plants remove many minerals from their leaves before they drop them, and microorganisms and tree roots quickly break down litter, capturing mineral nutrients before they can leach down through the soil. Removing the natural vegetation disrupts this tight cycling. When a tropical forest is cleared and temperate-zone agriculture takes its place, the tropical soils may become exhausted within only a few years. By contrast, large concentrations of minerals reside in temperate forest soils, which is why these soils can provide good agricultural yields for many years. Unless replaced, however, the nutrients removed by agriculture, even in temperate soils, will eventually be exhausted.

CLIMATES ON EARTH

As well as being the source of energy for photosynthesis and ecosystem energetics, the sun drives the global circulation patterns of air and ocean waters. The warming and cooling of moving masses of air and water, in turn, explain much of Earth's climatic patterns. Climates vary greatly from place to place on Earth primarily because different places receive different amounts of solar energy and because the monthly amount of incident solar energy is nearly constant at the equator but varies dramatically at high latitudes.

Solar Energy Inputs

Every place on Earth receives the same total number of hours of sunlight each year—an average of 12 hours per day—but not the same amount of *heat*. The rate at which heat arrives per unit of ground area depends primarily on the angle of sunlight. If the sun is low in the sky, a given amount of solar energy is spread over a larger area (and is thus less intense) than if the sun is directly overhead. In addition, when the sun is low in the sky, the sunlight must pass through more atmosphere. If sunlight must pass through more atmosphere, more of its energy is absorbed and reflected before it reaches the ground. At higher latitudes, there is more variation in both day length and the angle of arriving solar energy in the course of a year than at latitudes closer to the equator.

On average, the mean annual air temperature of Earth decreases about 0.4°C for every degree of latitude (about 110 kilometers) at sea level. Air temperature also decreases with elevation. The effect of elevation on temperature is due to the properties of gases. As a parcel of air rises, it expands, its pressure drops, and energy is expended in pushing molecules apart. With that loss of energy, the temperature of the air drops. When the parcel of air descends, it is compressed, its pressure rises, the same amount of energy is recovered, and its temperature increases.

When wind patterns bring air into contact with a mountain range, the air must rise to pass over mountains, and it cools as it does so. Because cool air cannot hold as much moisture as warm air can, unless the air is very dry, clouds form and moisture is released as rain or snow. For this reason, the windward side of a mountain range generally receives more rainfall than the leeward side does. On the leeward side, the air descends and warms, and the dry air picks up rather than releases moisture. Places on the leeward side of a mountain range that receive little rainfall for this reason are said to be in a **rain shadow** (Figure 49.18).

Global Atmospheric Circulation

Earth's climates are strongly influenced by global air circulation patterns that result from the differences in input of solar energy and the properties of air that we have already discussed. Air rises not only when it crosses mountains, but also when it is heated by the sun. The height to which the air rises, which is proportional to the amount it is heated by the sun, is greatest in the tropics. When air rises at the equator, air flows toward the equator from the north and south to take its place. That air, in turn, is replaced by air from aloft that descends after having traveled away from the equator at great heights. At roughly 30° north and south latitudes, air that cooled and lost its moisture while rising at the equator descends and warms. Many of Earth's deserts, such as the Sahara and the Australian deserts, are located at those latitudes. At about 60° north and south latitudes air rises again, for reasons that are not well understood. Cold, dense air descends at the poles, where there is little input of solar energy. The black arrows around the edge of Figure 49.19 show these vertical patterns, which are one component of Earth's winds.

The spinning of Earth on its axis influences surface winds because Earth's velocity is very rapid at the equator but very slow close to the poles. An air mass at a particular latitude has the same velocity as Earth has at that latitude. As an air mass moves toward the equator, it confronts a faster and faster spin, and, because its velocity increases slowly, it slows down relative to Earth beneath it. As an air mass moves poleward, it confronts a slower and slower spin, and it speeds up relative to Earth beneath it. Therefore, air masses moving latitudinally are deflected to the

49.18 A Rain Shadow
Average annual rainfall is lower in the lee of a mountain range oriented at right angles to the prevailing winds.

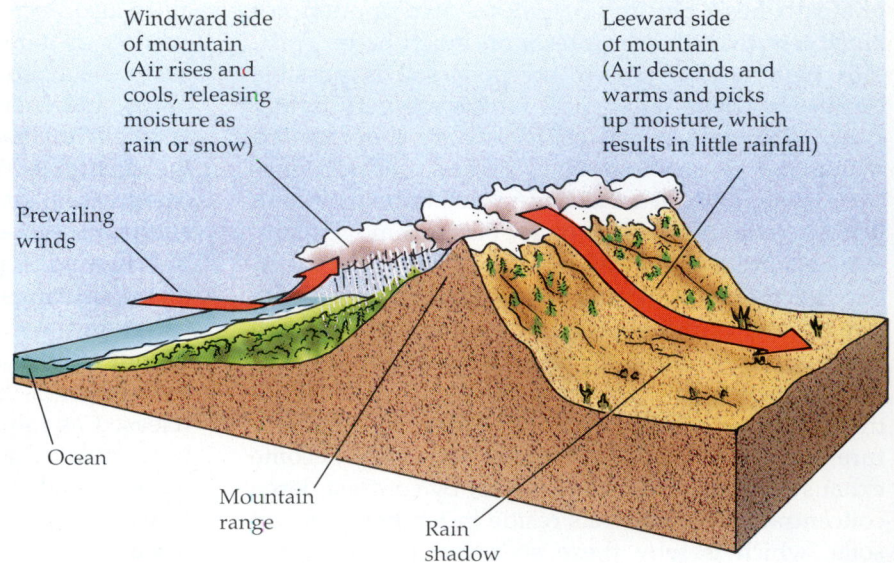

Windward side of mountain (Air rises and cools, releasing moisture as rain or snow)

Leeward side of mountain (Air descends and warms and picks up moisture, which results in little rainfall)

Prevailing winds

Ocean

Mountain range

Rain shadow

right in the Northern Hemisphere and to the left in the Southern Hemisphere. Winds blowing toward the equator from the north and south veer to become the northeast and southeast trade winds, respectively. Winds blowing away from the equator also veer and become the westerlies that prevail at midlatitudes. These surface winds are shown by the blue arrows in Figure 49.19.

Because Earth's axis is tilted, the amount of solar energy that reaches a given region, which is at its maximum at the time of year when the sun is directly overhead at noon, varies seasonally as Earth orbits the sun. The **intertropical convergence zone**—the location of greatest solar energy input and the site where trade winds converge and air rises most strongly—thus shifts with the season. This zone shifts to the north during the northern summer (southern winter) and to the south during the southern summer (northern winter), as far as the Tropic of Cancer and the Tropic of Capricorn, respectively. However, the intertropical convergence zone lags behind the overhead passage of the sun by a bit more than a month because it takes that long to heat the surface mass of Earth. Seasonal changes in climate in the tropics and subtropics are associated with movement of the intertropical convergence zone because whenever an area is in the zone, air rises and heavy rains fall (see Figure 50.20). When the convergence zone is to the north or south of a tropical region, the prevailing winds are trade winds, which seldom yield rain unless forced to rise over mountains.

49.19 Circulation of Earth's Atmosphere
If we could stand outside Earth and observe the movement of the air, we would see vertical movements like those indicated by the black arrows and surface winds like those shown by the blue arrows.

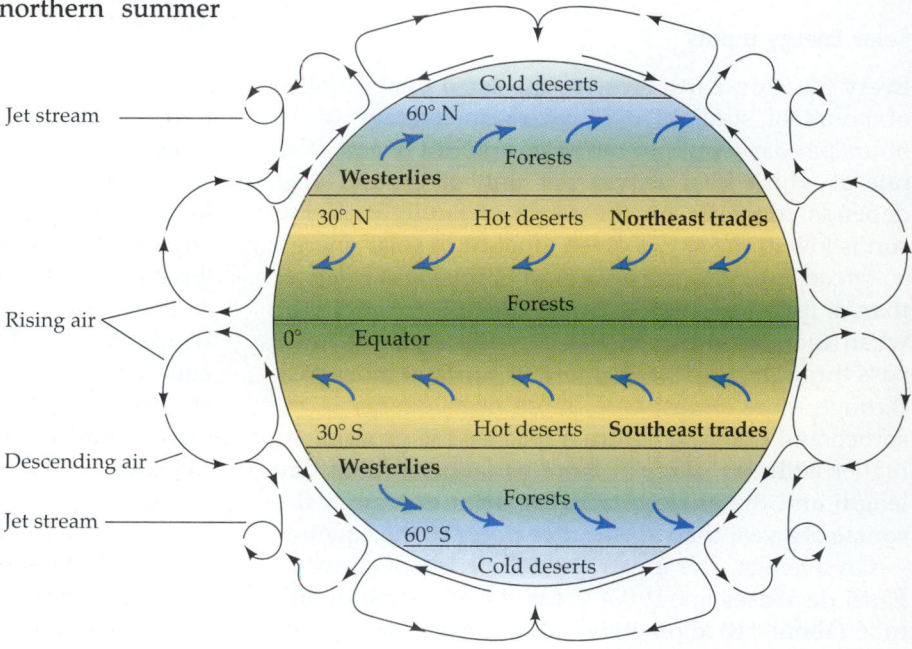

Jet stream

Rising air

Descending air

Jet stream

Cold deserts
60° N
Forests
Westerlies
30° N   Hot deserts   **Northeast trades**
Forests
0°   Equator
30° S   Hot deserts   **Southeast trades**
Westerlies
Forests
60° S
Cold deserts

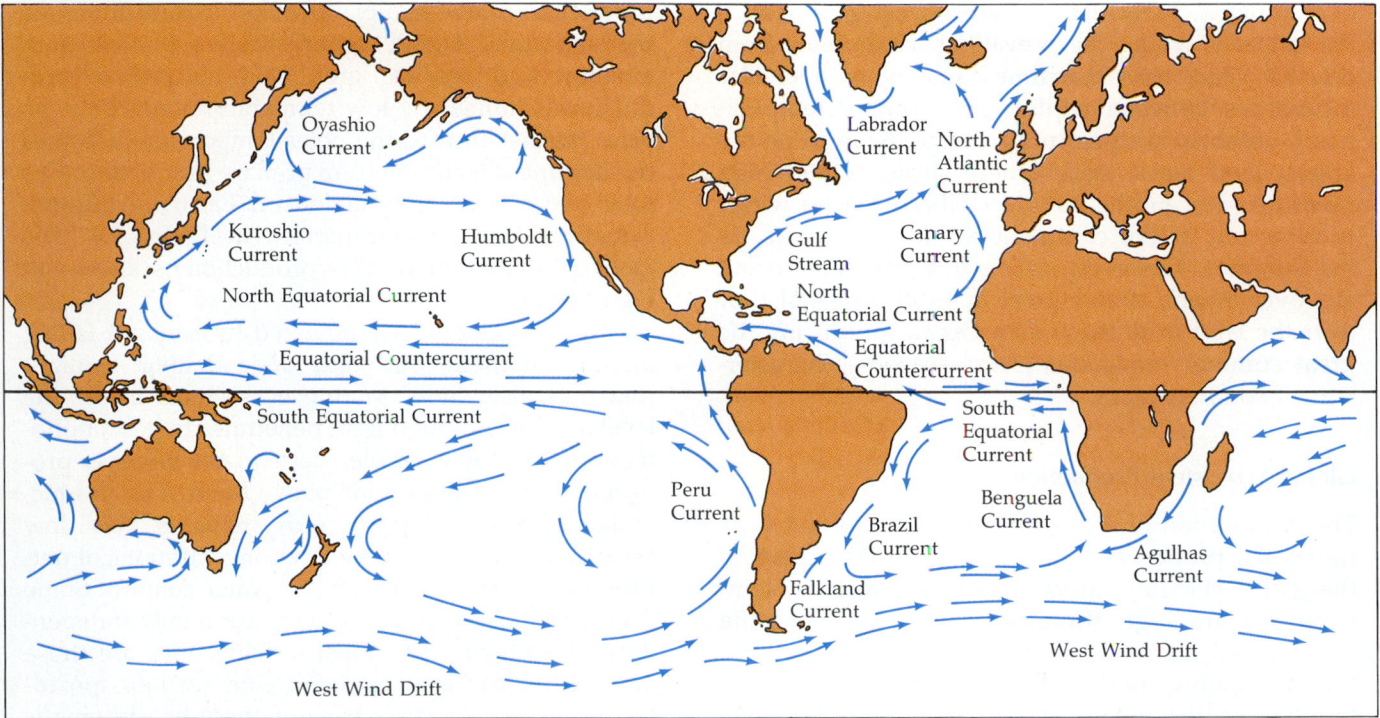

49.20 Global Oceanic Circulation
To see that ocean currents are driven primarily by the wind, compare the surface currents shown here with the prevailing surface winds shown in Figure 49.19. Also no- tice how the continents deflect ocean currents. Currents in the deep oceans differ strikingly from the surface ones shown here.

Global Oceanic Circulation

The oceans play an important role in world climates because their waters move long distances and because water has a high specific heat. The **specific heat** of a substance is the amount of energy required to raise the temperature of 1 gram of the substance 1°C. For water, this value is 1 cal/g at 15°C. Similarly, 1 gram of water cooling 1°C gives off 1 cal/g. Air and land surfaces have much lower specific heats. Consequently, in comparison with continents, oceans warm up more slowly in summer because it takes more heat to raise their temperature and cool off more slowly in winter because more heat must be released to cool them.

The global pattern of wind circulation, determined by the solar energy input to a rotating Earth (as we saw in the previous section) drives the circulation of ocean waters. Ocean water generally moves in the direction of the prevailing winds (Figure 49.20). Winds blowing toward the equator from the northeast and southeast cause water to converge at the equator and move westward until it is blocked by a continental land mass. At that point the water splits, some of it moving north and some of it moving south along continental shores. This poleward movement of ocean water is a major mechanism of heat transfer to high latitudes. As it moves toward the poles, the

water veers right in the Northern Hemisphere and left in the Southern Hemisphere. Thus water turns eastward until it is blocked by another continent and is deflected laterally along its shores. In both hemispheres, water flows toward the equator along the west sides of continents, continuing to veer right and left, respectively, until it meets at the equator and flows westward again.

The typical large circulation pattern of ocean water, clockwise in the Northern Hemisphere and counterclockwise in the Southern Hemisphere, is modified by the positions of continents. In the North Pacific, for example, movement of water into the Arctic Ocean is mostly blocked by land and very shallow continental shelves. Most water turns south along the west coast of North America. In the North Atlantic, however, there is a wide gap between Iceland and the British Isles through which large amounts of warm water flow poleward. At latitude 60° south, where no continental land masses impede water flow, there is a powerful eastward-flowing ocean current driven by the strong westerly winds that blow most of the time in that region.

At high latitudes, the temperature of the interior of large continents fluctuates greatly with the seasons, becoming very cold in winter and hot in summer, a pattern called a **continental climate.** The coasts of continents, particularly those on west sides at middle latitudes where the prevailing winds blow from ocean to land, have **maritime climates,** with smaller differences between winter and summer temperatures. Seasonal temperatures change the most on the largest land mass, Asia, where strong winter high pressure (descending air) over Siberia causes winds to blow out of the continent toward the coasts. In the summer, however, strong low pressure (rising air) over Siberia draws great quantities of moist air over the land from the Indian Ocean, producing the great summer monsoons (rainy seasons) characteristic of southern Asia.

Global Ecosystem Production

The distribution of ecological communities and their functional properties is determined in large part by the global climate pattern. In tropical regions, temperatures are high throughout the year, and the water supply is adequate most of the time. In these climates highly productive forests thrive. At lower middle latitudes, where dry air descends and warms, plant growth is limited by lack of moisture, primary production is low, and plants of low stature dominate the landscape. At still higher latitudes, there is more moisture and trees can grow well, but primary production is limited by low temperatures much of the year. Further toward the poles temperature becomes the dominant factor limiting primary production. The total net primary production on Earth in different climatic regions is summarized in Table 49.4. The global distribution of this production is shown in Figure 49.21.

Most ecosystems are powered by sunlight falling directly on them, but some depend upon sunlight that falls elsewhere. Marine ecosystems below the level at which enough light penetrates to permit photosynthesis, for example, depend on biomass produced in the well-lit zone above them. The productivity of most deep-sea ecosystems is very low because it depends on the low concentrations of detritus that descend through the water column. Some deep-sea ecosystems, however, are totally independent of sunlight. The most striking ones are those around hot springs associated with seafloor spreading zones. The energy base of these ecosystems is chemoautotrophy by sulfur-oxidizing bacteria that obtain energy by oxidizing hydrogen sulfide emitted

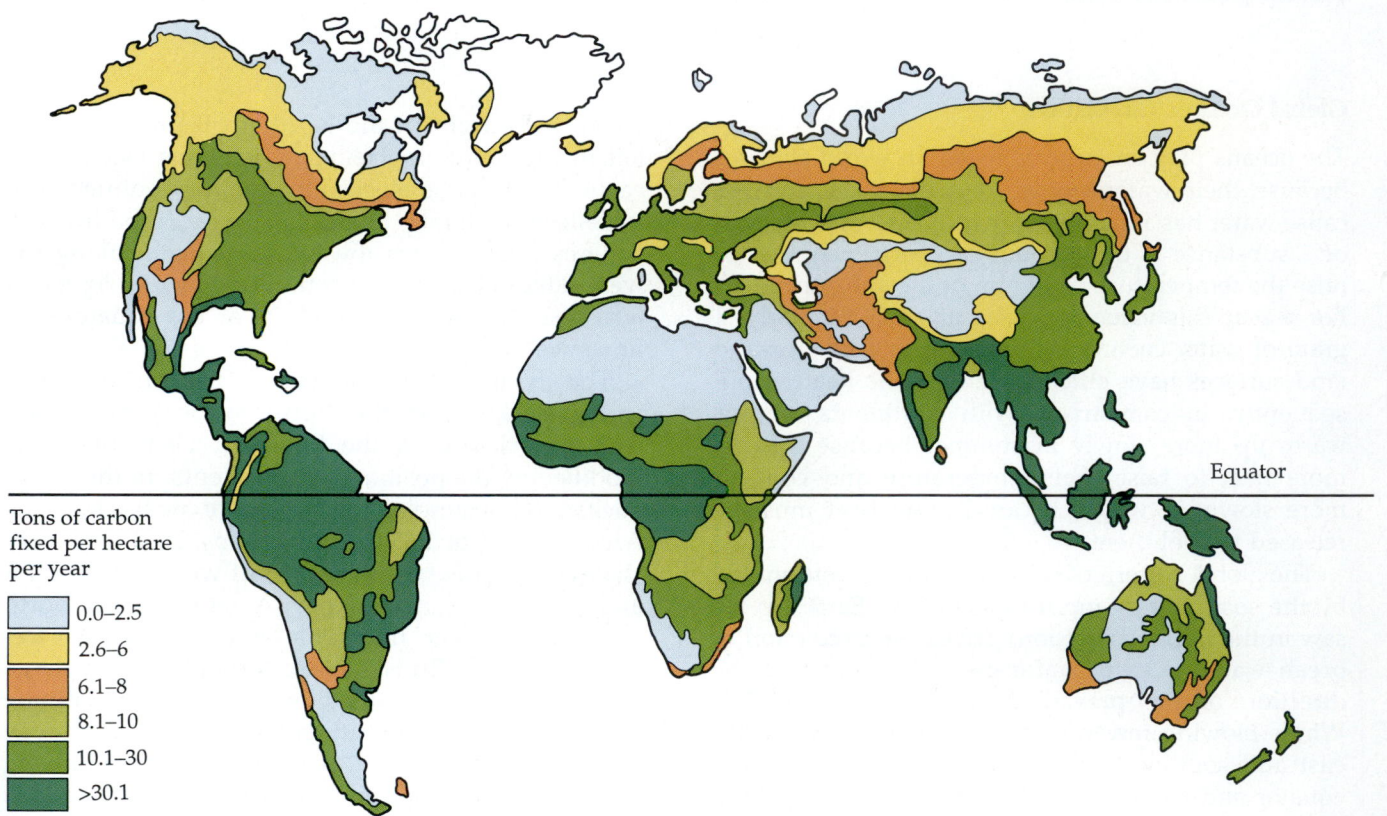

Tons of carbon
fixed per hectare
per year

- 0.0–2.5
- 2.6–6
- 6.1–8
- 8.1–10
- 10.1–30
- >30.1

Equator

49.21 Biological Production of Terrestrial Ecosystems
Areas of high annual production are in wet tropical and subtropical regions and the wetter parts of temperate latitudes. Low production characterizes the hot subtropical deserts (where moisture is limiting) and high latitudes (where cool temperatures lower photosynthetic rates).

TABLE 49.4
Areas, Biomass of Plants, and Net Primary Production of Earth's Major Vegetation Zones

VEGETATION ZONE	AREA 10^6 KM^2	PERCENT	MASS OF PLANTS 10^9 TONS	PERCENT	NET PRIMARY PRODUCTION 10^9 TONS	PERCENT
Polar	8.05	1.6	13.77	0.6	1.33	0.6
Conifer forest	23.20	4.5	439.06	18.3	15.17	6.5
Temperate	22.53	4.5	278.67	11.5	17.97	7.7
Subtropical	24.26	4.8	323.90	13.5	34.55	14.8
Tropical	55.85	10.8	1,347.10	56.1	102.53	44.2
Total land	133.89	26.2	2,402.5	100	171.55	73.8
Glaciers	13.9	2.7	0	0	0	0
Lakes and rivers	2.0	0.4	0.04	<0.01	1.0	0.4
All continents	149.79	29.3	2402.54	100	172.55	74.2
Ocean	361.0	70.7	0.17	<0.001	60.0	25.8
Earth total	510.79	100	2,402.71	100	232.55	100

from the vents. Most of the other organisms of these ecosystems live directly or indirectly on the sulfur-oxidizing bacteria (see Figure 22.2).

This overview of the global pattern of biological production on Earth is sufficient to identify which processes limit primary production and nutrient cycling in different climatic zones and how they operate, but it does not give you a picture of what these ecosystems look like and how they function. Describing ecosystems is one of the goals of the next chapter.

SUMMARY of Main Ideas about Ecosystems

Ecosystems are powered by solar energy that first enters living organisms via photosynthesis at rates controlled by temperature and precipitation.
Review Figure 49.1

Food webs summarize who eats whom in ecological communities.
Review Figures 49.2 and 49.3

Because much of the energy taken in by an organism is used for maintenance and is eventually dissipated as heat, the efficiency of energy transfer to higher trophic levels is usually very low.
Review Figures 49.4 and 49.5

The main elements of living organisms—carbon, nitrogen, phosphorus, sulfur, hydrogen, and oxygen—cycle between organisms and other compartments of the global ecosystem.
Review Figures 49.10, 49.11, 49.12, 49.13, and Table 49.3

Human activity greatly modifies cycles of basic minerals on local, regional, and global scales.
Review Figures 49.14, 49.15, and 49.16

Earth's climate is determined primarily by the pattern of solar energy input at different latitudes and by Earth's rotation on its axis.

The directions of prevailing winds differ over the surface of Earth.
Review Figure 49.19

Surface winds drive global oceanic circulation.
Review Figure 49.20

The distribution of primary production on Earth is determined primarily by Earth's climate.
Review Figure 49.21

SELF-QUIZ

1. Which of the following is true about the amount of sunlight and heat arriving on Earth?
   a. Every place on Earth receives the same annual number of hours of sunlight and the same amount of heat.
   b. Every place on Earth receives the same annual number of hours of sunlight but not the same amount of heat.
   c. Every place on Earth receives the same annual amount of heat but not the same number of hours of sunlight.
   d. Both the annual amount of sunlight and the amount of heat received vary over the surface of Earth.
   e. None of the above.

2. When an area is within the intertropical convergence zone,
   a. the northeast trade winds blow steadily.
   b. the southeast trade winds blow steadily.
   c. air is descending and it seldom rains.
   d. air is rising and heavy rains fall frequently.
   e. westerly winds blow steadily.

3. Zones of marine upwelling are important because
   a. they help scientists measure the chemistry of deep ocean water.
   b. they bring to the surface organisms that are difficult to observe elsewhere.
   c. ships can sail faster in these zones.
   d. they increase marine productivity by bringing nutrients back to surface ocean waters.
   e. they bring oxygenated water to the surface.

4. Which of the following is *not* true of the troposphere?
   a. It contains nearly all atmospheric water vapor.
   b. Materials enter it primarily at the intertropical convergence zone.
   c. It is about 17 km deep in the tropics.
   d. Most global atmospheric circulation takes place there.
   e. It contains about 80 percent of the mass of the atmosphere.

5. The humus formed from decomposition of nutrient-rich leaves is called
   a. the thermocline.
   b. mor.
   c. mull.
   d. eutrophication.
   e. dirt.

6. Carbon dioxide is called a greenhouse gas because
   a. it is used in greenhouses to increase plant growth rates.
   b. it is transparent to heat radiation but opaque to sunlight.
   c. it is transparent to sunlight but opaque to heat radiation.
   d. it is transparent to both sunlight and to heat radiation.
   e. it is opaque to both sunlight and heat radiation.

7. The phosphorus cycle differs from those of carbon and nitrogen in that
   a. it lacks a gaseous phase.
   b. it lacks a liquid phase.
   c. only phosphorus is cycled through marine organisms.
   d. living organisms do not need phosphorus.
   e. The phosphorus cycle does not differ importantly from the carbon and nitrogen cycles.

8. Acid precipitation results from human modifications of
   a. the carbon and nitrogen cycles.
   b. the carbon and sulfur cycles.
   c. the carbon and phosphorus cycles.
   d. the nitrogen and sulfur cycles.
   e. the nitrogen and phosphorus cycles.

9. The total amount of energy that plants assimilate by photosynthesis is called
   a. gross primary production.
   b. net primary production.
   c. biomass.
   d. a pyramid of energy.
   e. eutrophication.

10. The amount of energy reaching an upper trophic level is determined by
    a. net primary production.
    b. net primary production and the efficiencies with which food energy is converted to biomass.
    c. gross primary production.
    d. gross primary production and the efficiencies with which food energy is converted to biomass.
    e. gross primary production and net primary production.

11. Which of the following is *not* a component of integrated pest management?
    a. Use of cultural strategies such as crop rotation and mixed plantings
    b. Use of pest-resistant strains of crops
    c. Use of predators and parasites of crop pests
    d. Use of chemical attractants
    e. Use of chemical pesticides whenever pests are discovered

FOR STUDY

1. How would you expect temperature and oxygen profiles to appear in a broad, shallow tropical lake? In a very deep tropical lake? Why?

2. The waters of Lake Washington, adjacent to the city of Seattle, rapidly returned to their preindustrial levels of purity when sewage was diverted from the lake to Puget Sound, an arm of the Pacific Ocean. Would all lakes being polluted with sewage clean up as rapidly as Lake Washington if pollutant input were stopped? What characteristics of a lake are most important to its rate of recovery following removal of pollutant inputs? What is the diverted sewage likely to do to Puget Sound?

3. Tropical forests currently are being cut at a very rapid rate. Does this necessarily mean that deforestation is a major source of input of carbon dioxide to the atmosphere? If not, why not?

4. The two drawings below represent biomass pyramids for (*a*) an old field in Georgia and (*b*) the English Channel. Explain the significance of the inversion of the second pyramid compared with the first.

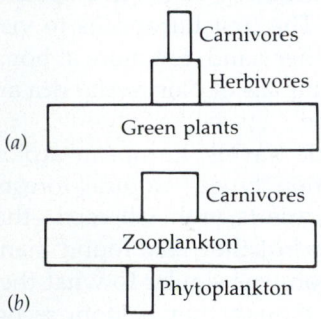

(*a*)
- Carnivores
- Herbivores
- Green plants

(*b*)
- Carnivores
- Zooplankton
- Phytoplankton

5. The amount of energy flowing through a food chain declines more or less rapidly depending upon the nature of the organisms in the chain. Which of the following simplified food chains is likely to be more efficient? Why? What criteria of efficiency are you using?
- *a.* phytoplankton → zooplankton → herring
- *b.* shrubs → deer → wolf

6. A government official authorizes the construction of a large power plant in a former wilderness area. Its smokestacks discharge into the air great quantities of waste resulting from the combustion of coal. List and describe all *likely* ecological results, at local, regional, and global levels. Suppose the wastes were thoroughly scrubbed from the stack gases. Which of the ecological results you have just outlined would still happen?

READINGS

Bazzaz, F. A. and E. D. Fajer. 1992. "Plant Life in a CO$_2$-Rich World." *Scientific American*, January. Even without considerations of global warming, increasing atmospheric levels of carbon dioxide may greatly alter the structure and function of ecosystems. These changes will not necessarily benefit plants.

Berner, R. A. and A. C. Lasaga. 1989. "Modeling the Geochemical Carbon Cycle." *Scientific American*, March. Natural geochemical processes that result in the slow buildup of atmospheric carbon dioxide may have caused past geologic intervals of global warming through the greenhouse effect.

Gates, D. M. 1993. *Climate Change and Its Biological Consequences*. Sinauer Associates, Sunderland, MA. An introduction to Earth's climates, past and present, and the effect climate change has on organisms and ecosystems. An accessible explanation of the various computer models that predict human activity will cause massive climate warming in the next century, the text discusses what the effects of such warming might be.

Jordan, C. F. 1985. *Nutrient Cycling in Tropical Forest Ecosystems*. John Wiley & Sons, New York. A useful summary of nutrient cycling patterns in tropical forests and how they differ from those of temperate forests.

Kusler, J. A., W. J. Mitsch and J. S. Larson. 1994. "Wetlands." *Scientific American*, January. Wetlands are vital ecosystems: They purify our water, protect us from floods, and are incubators for aquatic life. This article assesses what happens when the protection of an essential natural resource must be weighed against society's demand for real estate, fossil fuels, and agricultural land.

Makarewicz, J. C. and P. Bertram. 1991. "Evidence for the Restoration of the Lake Erie Ecosystem." *BioScience* vol. 41, pages 216–223. Details of changes in water quality, oxygen levels, and population dynamics of microorganisms, plants, and animals in Lake Erie.

Mohnen, V. A. 1988. "The Challenge of Acid Rain." *Scientific American*, August. Acid rain's effects in soil and water leave no doubt about the need to control its causes. Now advances in technology have yielded environmentally and economically attractive solutions.

Rambler, M. B., L. Margulis, and R. Fester (Eds.). 1989. *Global Ecology: Towards a Science of the Biosphere*. Academic Press, Boston. A collection of essays covering a wide variety of the interactions between organisms, global biogeochemical cycles, and global climate.

Raven, P. H., L. R. Berg and G. B. Johnson. 1993. *Environment*. Saunders College Publishing, Philadelphia. A thorough textbook of environmental studies, illustrated in color and with case studies and "sidelights" of interest.

Schneider, S. H. 1989. "The Changing Climate." *Scientific American*, September. Global warming should be unmistakable within a decade or two. Prompt emission cuts could slow the buildup of heat-trapping gases and limit this risky planetwide experiment.

A Unique Australian Mammal
Koalas were a strange sight to Australian settlers coming from western Europe, where no similar animal exists.

50

Biogeography

When the first Europeans arrived in Australia, they saw plants and animals that differed in perplexing ways from the ones they had known at home: flowers pollinated by brush-tongued parrots, and small, bearlike animals living exclusively in the treetops and feeding on toxic leaves. The first Europeans to visit North America, on the other hand, felt more at home because the plants and animals of North America are similar to those of Europe.

During their worldwide travels, European explorers found many vegetation types—tropical forests, mangrove forests, and deserts with tall cacti—that were unfamiliar to them, but they also found many areas where the vegetation was similar to what they knew back home, even though they seldom recognized any familiar species. The study of the diversity of organisms over space and time began when those eighteenth-century travelers, who first noted intercontinental differences in distributions of organisms, attempted to understand those differences.

THE GOALS OF BIOGEOGRAPHY

Biogeography is the study of distributions of organisms, both past and present. Superficially, explaining species' distributions seems to be a simple matter because the question of why a species is or is not found in a certain location has only a few possible answers. If a species occupies an area, either it evolved there or it evolved elsewhere and dispersed to the area. If a species is *not* found in a particular area, either it evolved elsewhere and never dispersed to the area, or it was once present in the area but no longer lives there.

The key problem for a biogeographer is to determine which distribution patterns can be explained by where the taxa evolved and which ones resulted from dispersal. Although the problem sounds simple, finding the answers requires understanding the evolutionary histories of the species in question. Such information comes from fossils and from phylogenies based on the traits of the species (see Chapter 21). Knowledge of how Earth changed during the time the organisms were evolving is also important because it tells us where the continents were when lineages of taxa split. Biogeographers must draw upon and interpret a broad array of knowledge to explain the distributions of organisms.

We have described the information a modern biogeographer uses, but early biogeographers had no phylogenies, and they believed that Earth was no more than a few thousand years old. The notion that the continents might have moved was not proposed until 1912, when it was introduced by the German meteorologist Alfred Wegener. Wegener based his

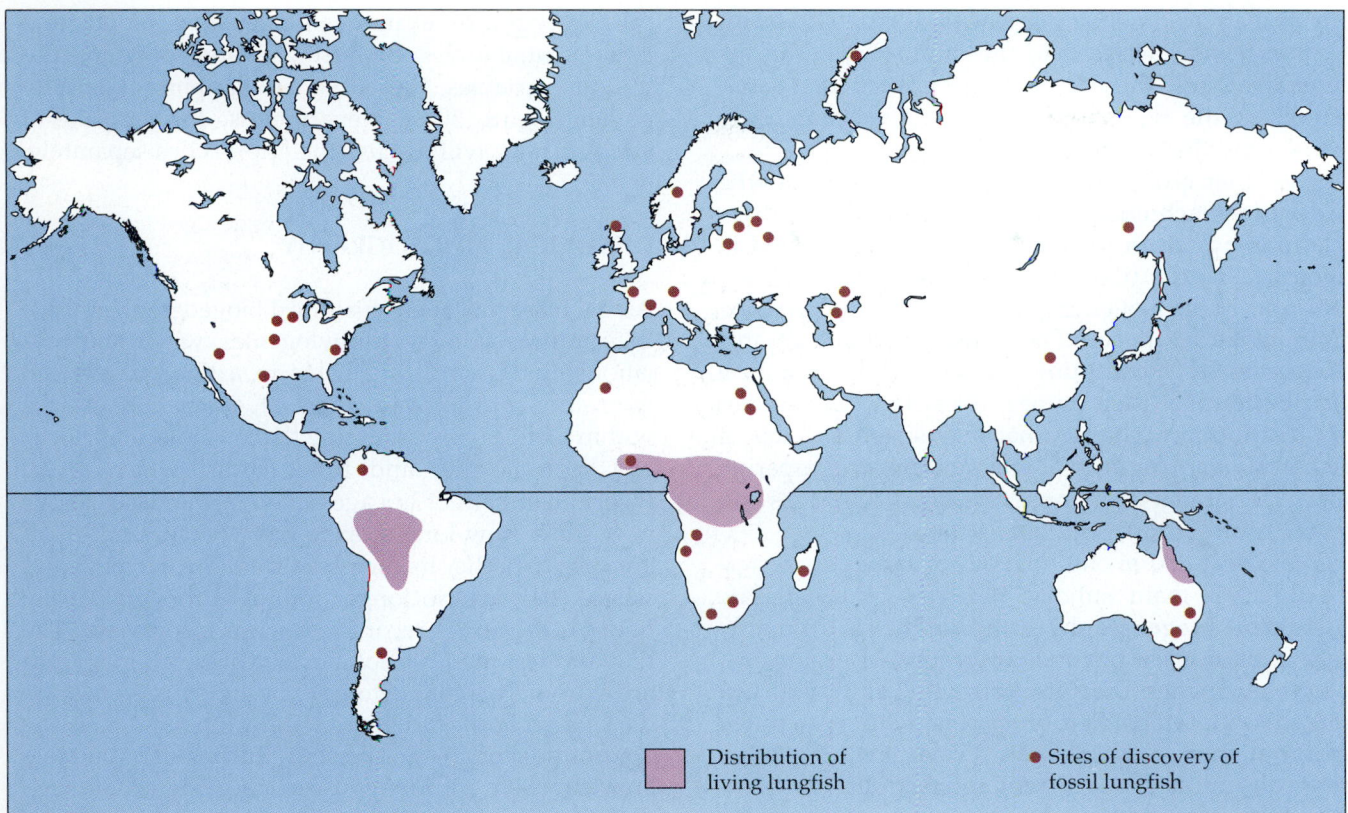

Distribution of living lungfish

● Sites of discovery of fossil lungfish

50.1 Putting the Past and Present Together
Lungfish live today only in South America, Africa, and Australia, but plotting the sites where their fossils have been found suggests that they evolved before the breakup of Pangaea.

ideas on the shapes of continents, whose outlines seem to fit together like pieces of a puzzle, and on biogeography—distributions of plants and animals that were hard to explain if one assumed that the continents had always been where they are now. When Wegener proposed his ideas, few scientists took them seriously, primarily because there were no known mechanisms by which continents could have moved and because no convincing geological evidence of such movements existed. As we learned in Chapter 28, geological evidence and plausible mechanisms are both available today, and the broad pattern of continental movement is now clear.

Although some continents, such as Africa and South America, began to drift apart many millions of years ago, the effects of this movement on species' distributions are considerable because by the early Mesozoic era (about 245 million years ago), when the continents were still very close to one another, many groups of nonmarine organisms, including insects, freshwater fishes, and frogs, had already evolved. Some organisms that are found on most continents today may have been present on those land masses when they were all part of Pangaea (Figure 50.1). By the close of the Mesozoic era (about 65 million years

ago), Pangaea was breaking apart and the continents were drifting away from each other (see Figure 28.2). At first, the water gaps between continents increased, but eventually drifting, which continues today, brought India in contact with southern Asia, Australia closer to Southeast Asia, and South America in contact with North America. Continental drift thus has influenced biogeographic patterns throughout the history of life on Earth.

The early biogeographers not only believed in a relatively constant Earth, but they also thought that Earth was very young, and they did not understand how new species were formed. Much valuable information was assembled by the early biogeographers, but their interpretations were constrained by their belief that organisms had been created in their current forms in their current ranges.

More recently, when the great age of Earth and the fact of evolution became understood, two groups of investigators developed new methods for generating testable hypotheses that could lead to generalizations about geographical distributions. One group consisted of ecologists who studied how current distributions were influenced by interactions among species and by interactions between species and their physical environments. Because local habitats always contain fewer species than the number living in the general region, these ecologists believed that interactions of the types discussed in Chapter 48 could explain the distributions of organisms.

The other group consisted of biogeographers who added new techniques, especially improved cladistic methods by which phylogenies were constructed in rigorous, standardized ways (see Chapter 21), to their tool box. These biogeographers recognized that a taxonomic cladogram could be transformed into an area cladogram by substituting the species' geographical distributions for their names. Area cladograms identify distribution patterns that may suggest routes of dispersal or appearance of major barriers. Comparing area cladograms of many evolutionary lineages can reveal common distributional patterns.

Ecological and evolutionary approaches characterize the two main subdisciplines of biogeography. **Historical biogeography** concerns itself primarily with evolutionary histories of groups of organisms. Where and when did they originate? How did they spread? What does their present-day distribution tell us about their past histories? **Ecological biogeography** concentrates on current interactions of organisms with the physical environment and with one another. Ecological biogeographers seek to understand how ecological relationships influence where species and higher taxa are found today. The integration of these subdisciplines of biogeography is essential to a full understanding of the geographic distributions of organisms.

The names of these two subdisciplines are somewhat misleading. All biogeography is historical. Ecological biogeography concentrates on recent history, current interactions, and changes within the past few thousand years. It also concentrates on patterns of distributions within local areas and regions. Histori-cal biogeography examines longer time periods and larger spatial scales. Because time and space are continuous variables, these two subdisciplines blend together (Figure 50.2). Nevertheless, for the sake of simplicity we will discuss the approaches separately.

HISTORICAL BIOGEOGRAPHY

As we have just seen, historical biogeographers base their interpretations on phylogenies, which show relationships among organisms in a lineage. We can infer the approximate times of separation of taxa within a lineage by assuming that a "molecular clock" has been ticking, and from fossils, which reveal where members of lineages lived in the past. Fossils may show how long a taxon has been present in an area and whether its members formerly lived in areas where they are no longer found. Although usually helpful, the fossil record is always incomplete. The first and last members of a taxon that lived in an area are extremely unlikely to have become fossils that are discovered and described. A third type of data is the distribution of living species. Much more complete and extensive information can be gathered on such distributions than will ever be available from fossils. Much can be learned by examining distribution patterns of *many different groups* of living organisms, looking for similarities that provide clues about past events.

Parsimony

Consider the distribution of the New Zealand flightless weevil *Lyperobius huttoni,* a species that is found

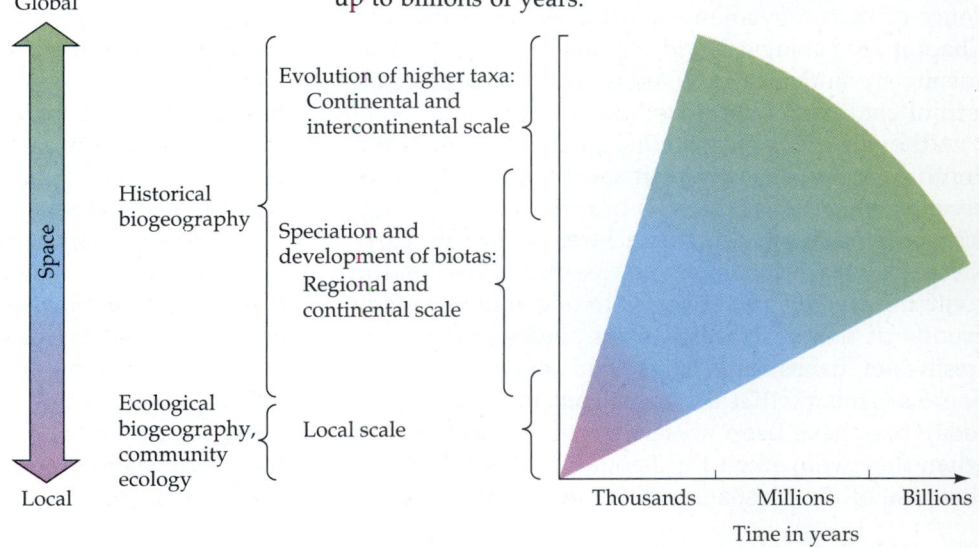

50.2 Time and Space in Biogeography
Biogeographical processes are studied at local, regional, and global spatial scales and at temporal scales ranging up to billions of years.

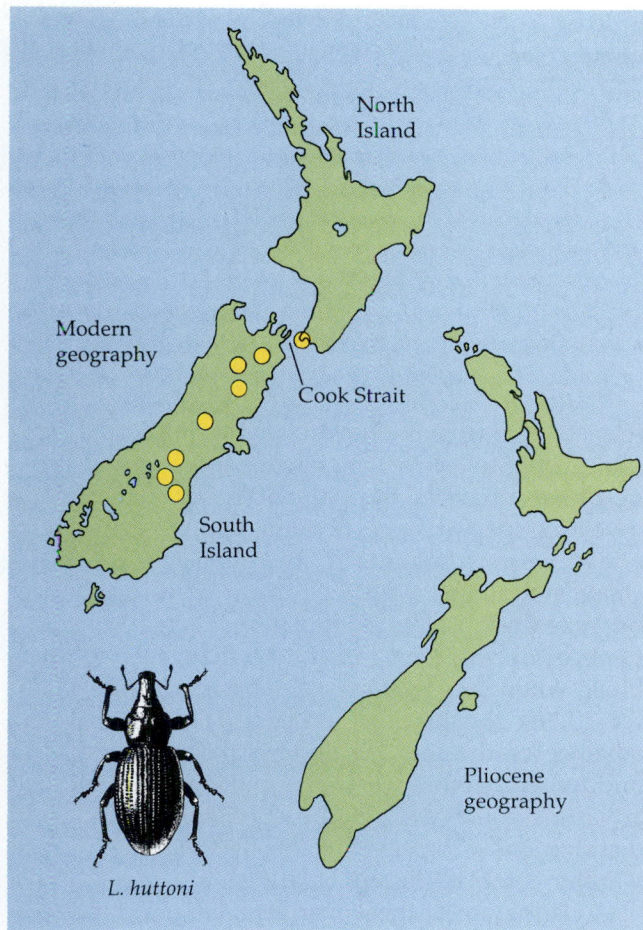

50.3 Distributed across Two Islands
Yellow circles indicate the current distribution of the weevil *Lyperobius huttoni*. Compare the present New Zealand geography with that of the Pliocene.

in the mountains of South Island and on sea cliffs at the extreme southwest corner of North Island (Figure 50.3). If you knew only its current distribution and the current positions of the two islands, you might guess that *L. huttoni* had somehow managed to cross Cook Strait, which separates the two islands, even though this weevil cannot fly. More than 60 other animal and plant species, however, including other species of flightless insects, share this disjunct distribution. The assumption that all of these species made the same ocean crossing is unlikely. In fact, that assumption is unnecessary because recent geological evidence, summarized by the Pliocene map in Figure 50.3, indicates that the present-day southwest tip of North Island was formerly united with South Island and was separated by a water gap from a then smaller North Island. Therefore, none of those 60 species need have made a water crossing.

This example illustrates an important method commonly used in biogeographic analysis. Although organisms *do* cross major oceanic and terrestrial barriers, biogeographers often apply the rule of

parsimony when interpreting distribution patterns, just as evolutionists do when reconstructing phylogenies (see Chapter 21). A parsimonious interpretation of a distribution pattern requires us to assume the smallest number of unobserved events to account for it. You do not need to postulate a dispersal event for *L. huttoni* if you assume that it and the other species with similar disjunct distributions already lived on what has become the southwest tip of North Island when it was part of South Island. One geological event that separated many formerly continuous distributions can account for today's distributions. Of course, some of the species having disjunct distributions between South and North Island, such as birds and flying insects, may indeed have dispersed across Cook Strait, but in the absence of evidence indicating that they did, most biogeographers favor the more parsimonious interpretation. The principle of parsimony is a useful operating rule, but you should remember that evolutionary history has not necessarily always been parsimonious.

Vicariance and Dispersal

We have noted that a species can be found in an area either because it evolved there or because it dispersed to the area. If a species lives in two distinct areas now separated by a barrier but it lived in the entire area before the barrier was imposed, it is said to have a **vicariant distribution.** A species that exists in both areas because it crossed the barrier is said to have a **dispersal distribution.** In principle, vicariant and dispersal distributions are distinct; in actuality, however, they blur because barriers vary greatly in their width and in the difficulty of crossing them. Cook Strait is a modest barrier compared to the distance between Australia and New Zealand or between New Zealand and South America. Because all organisms disperse from their birth places to where they eventually live, the question, How far must an organism move before its distribution should be attributed to dispersal?, has no simple answer.

Species, genera, and families found in only one region are said to be **endemic** to that location. As far as we know, all species are endemic to Earth. Some species are endemic to one continent. Others are restricted to very small areas, such as tiny islands or single mountain tops. A species may disperse widely and then die out where it originated, so biogeographers cannot assume that a species endemic to a region formed there. Endemic taxa can be very old ones that are becoming extinct or very young taxa that have recently evolved in a restricted area.

The longer an area has been isolated from other areas by a vicariant event, such as continental drift, the more endemic taxa it is likely to have, because there has been more time for evolutionary divergence

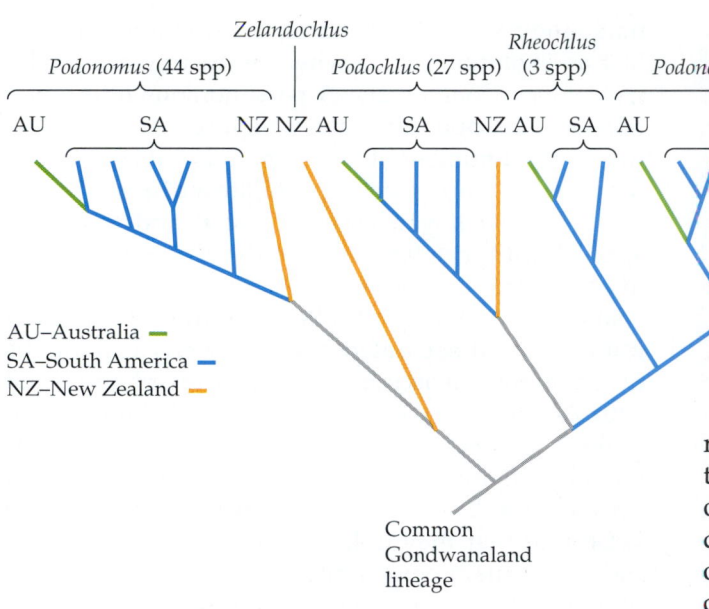

AU–Australia
SA–South America
NZ–New Zealand

Common
Gondwanaland
lineage

50.4 Phylogeny of Some Midges
By comparing the cladogram of some midges with their current geographical distributions, we can infer that New Zealand became isolated from Australia while midges were still dispersing between the two continents.

to take place. Australia, which has been separated the longest from other continents—about 65 million years—has the most distinct biota. South America has the next most distinct biota, having been isolated from other continents for nearly 60 million years. North America and Eurasia, which were joined together for much of Earth's history, have very similar biotas. That is why the early European travelers felt more at home in North America than in Australia.

Reconstructing Biogeographic Histories

Two examples will illustrate how biogeographers use modern methods to reconstruct the biogeographic history of taxa. First we consider the distribution of some midges in a family of flies (Chironomidae) with aquatic larvae. Many species of midges live in the cool, temperate areas of the Southern Hemisphere. Figure 50.4 shows a cladogram of five of the genera, together with their distributions in Australia, South America, and New Zealand. One genus is restricted to New Zealand; the lineages giving rise to modern New Zealand representatives of two other genera separated from the lineages leading to the South American and Australian species before the latter two lineages separated from one another. These distributions of midges suggest that New Zealand became isolated from Australia when Australia and South America were still exchanging midges. The closest relatives of many New Zealand species other than midges are somewhere other than Australia, supporting the notion that New Zealand separated from the other southern continents early during the breakup of Gondwanaland (see Figure 28.2).

In the second example, biogeographers have investigated the possibility that recent advances and

retreats of glaciers in North America initiated speciation events in many songbird genera. During times of glacial advance, the populations of many forest-dwelling species were divided into two segments, one in western mountains, the other in the Appalachian mountains, both well to the south of the ice. Because these isolated populations then evolved differences during their separation, they failed to interbreed when they again became sympatric (see Chapter 20) as the glaciers retreated. During the next advance of the ice, the eastern species could have budded off another western population. Several such advances could explain the speciation and current distributions of species in the black-throated green warbler complex (Figure 50.5).

A cladogram of these warblers, based on their mitochondrial DNA, shows that the black-throated gray warbler represents the earliest branch from the black-throated green lineage. This split could have been caused by a glacial advance, as could the separation of the black-throated green warbler from the ancestor of the hermit and Townsend's warblers. Hermit and Townsend's warblers, however, are sister species (Figure 50.6). The separation of these two sister species cannot be explained by a separation of eastern and western populations by glacial advance. They probably became isolated on different western mountain ranges, perhaps because of climate changes associated with glacial advances.

Major Terrestrial Biogeographic Regions

All continents have been isolated from one another long enough to have evolved distinct biotas. The differences among continental biotas, first recognized more than a century ago, formed the basis for dividing Earth into the major biogeographic regions shown in Figure 50.7. Notice that, with the exception of the Australian region, these regions today are not completely separated from each other by water, although they were in the past (see Figure 28.2). The biological distinctness of these biogeographic regions is maintained today in part by mountain and desert barriers to dispersal.

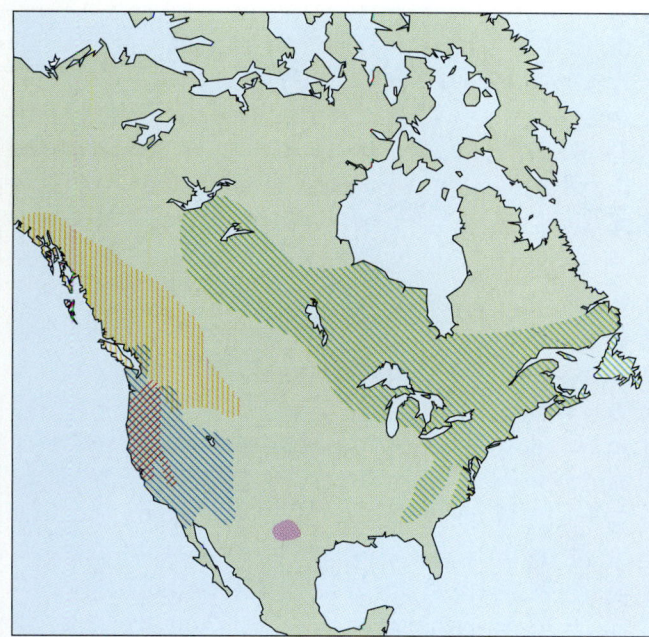

Hermit warbler

Black-throated green warbler

Black-throated grey warbler

Townsend's warbler

Golden-cheeked warbler

50.5 Why Are There Five Species?
The present-day breeding ranges of five closely related species of warbler support the hypothesis that glaciers repeatedly isolated western populations of a widespread ancestral species, opening possibilities for the evolution of differences, including reproductive isolation.

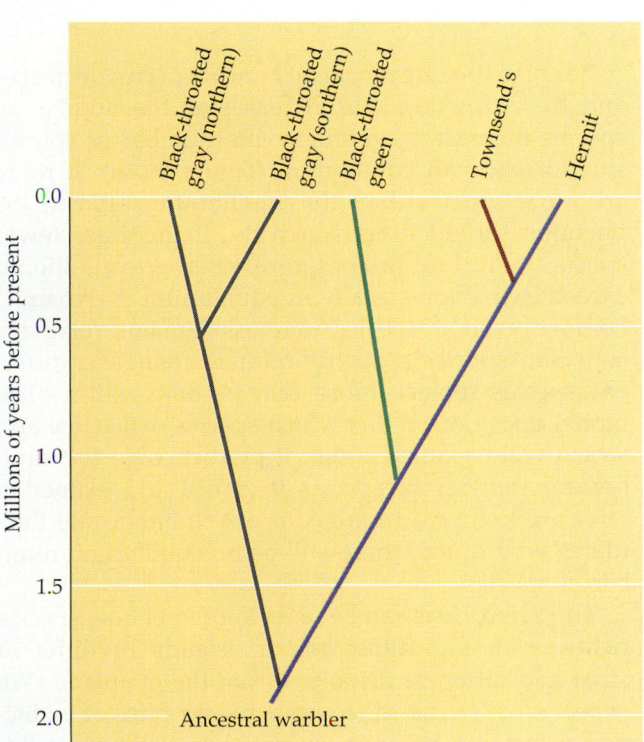

50.6 Cladistic Analysis Aids Biogeographic Reconstruction
This cladogram, constructed from mitochondrial DNA data, adds information used to understand the biogeographic ranges of warblers shown in Figure 50.5.

ECOLOGICAL BIOGEOGRAPHY

Ecological biogeographers use the wealth of information on current distributions of organisms to test theories that explain the number of species in different communities, how species disperse, and how effective different types of barriers are. They can also use experiments to test hypotheses, something that historical biogeographers cannot do. As an example, we will discuss a model that attempts to account for the species richness of an area, and we will look at an experiment that was conducted to test this model.

An Equilibrium Model of Species Richness

What determines the species richness of an area? It is obvious that new species immigrating into an area increase richness and that species becoming extinct decrease it. How might biogeographers consider the effects of these two processes to account for species richness? It is easiest to visualize the effects if we consider, as did Robert MacArthur of Princeton University and Edward O. Wilson of Harvard University, oceanic islands that initially have no species. Consider a newly formed oceanic island that receives colonists from a mainland area. Let's call the list of species on the mainland that might possibly invade

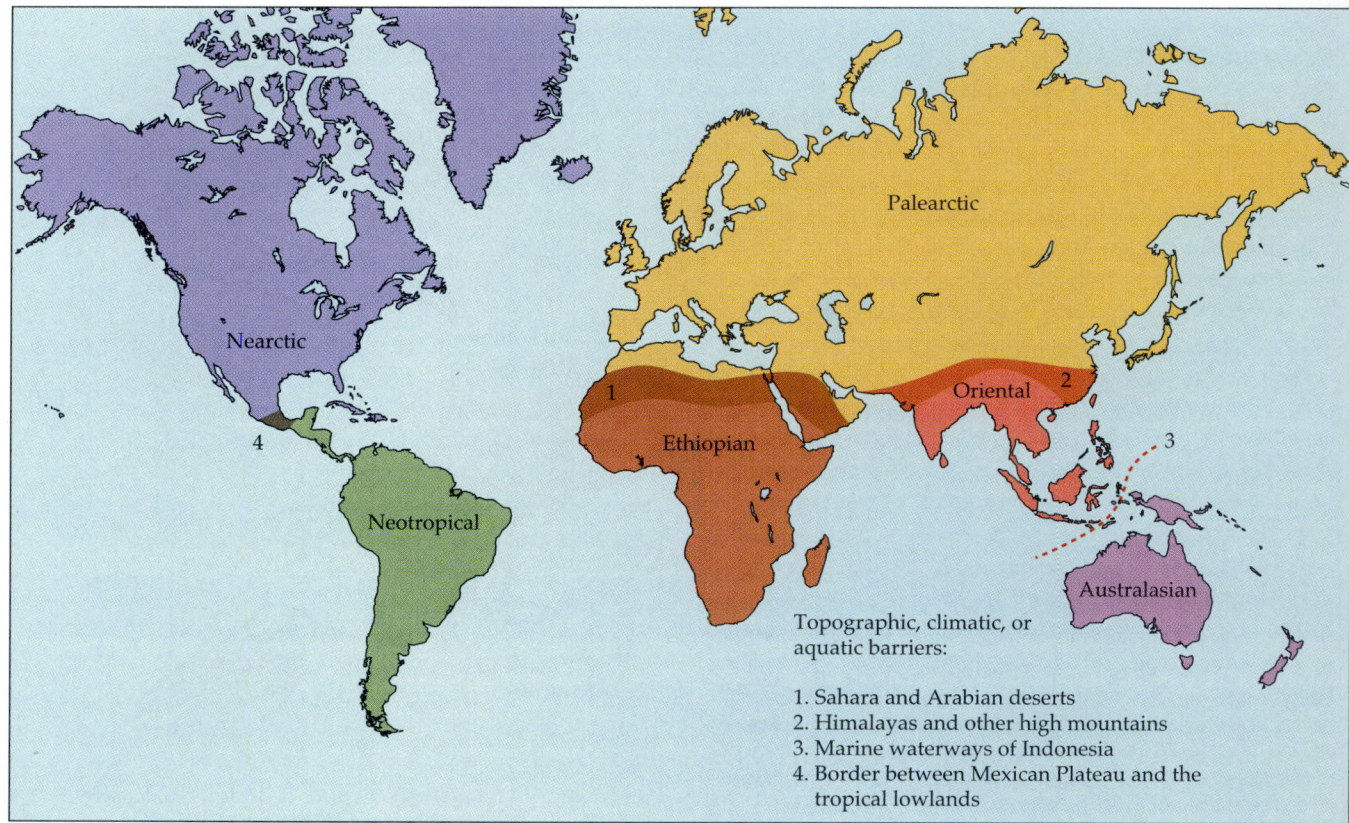

50.7 Major Biogeographic Regions
Barriers separating the biogeographic regions are shown as broad bands rather than lines because biotas change gradually rather than abruptly from one region to another.

the island the species pool. The first colonists to arrive on the island are all "new" species because there are no species there already. As the number of species on the island increases, a larger fraction of immigrants are members of species already present, so even if the same number of species arrives as before, the rate of arrival of *new* species decreases, until it reaches zero when the island has all the species in the pool.

Now think about extinction rates. First there will be only a few species on the island, and their populations may grow large. As more species arrive and their numbers increase, the resources of the island will have to be divided among more species. We therefore expect the average population size to become smaller as the number of species becomes larger. The smaller a population, the more likely it is to become extinct. In addition, the number of species that can become extinct increases as species accumulate on an island. New arrivals to an island may include pathogens and predators that increase the probability of extinction of other species, further increasing the number of species becoming extinct per unit of time.

Because the rate of arrival of new species decreases and the extinction rate increases as the number of species increases, eventually the number of species should reach an equilibrium (Figure 50.8a). If there are more species than the equilibrium number, extinctions should exceed arrivals. If there are fewer species than the equilibrium number, arrivals should exceed extinctions. Such an equilibrium is dynamic, because even if species richness remains relatively constant, species composition may change as different species replace those that become extinct. The model does not predict which species will arrive and which will become extinct. It predicts only the equilibrium number of species if arrival and extinction rates are known and are constant. If either rate fluctuates very much, there will be no equilibrium number of species.

The same ideas can be used to predict how species richness should differ among islands of different sizes and different distances from the mainland. We expect smaller islands to have higher rates of extinction of species because species' populations would, on average, be smaller there. Similarly, we expect fewer colonizers to reach islands more distant from the mainland. Figure 50.8b gives relative species richness equilibria for islands of different size and distance from the mainland.

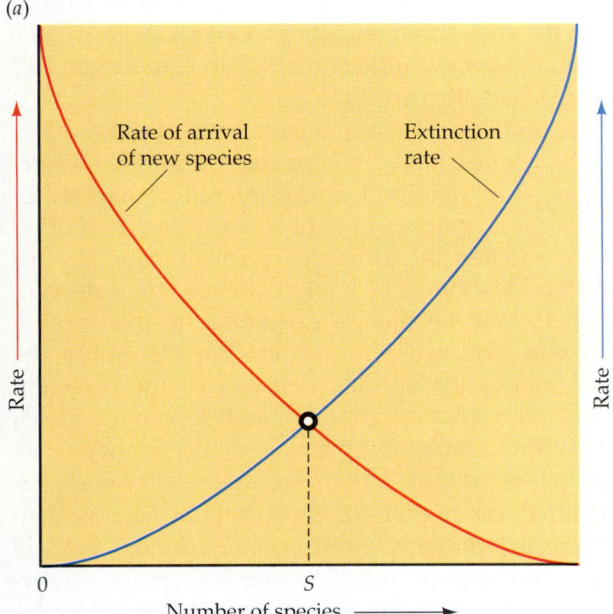

(a)

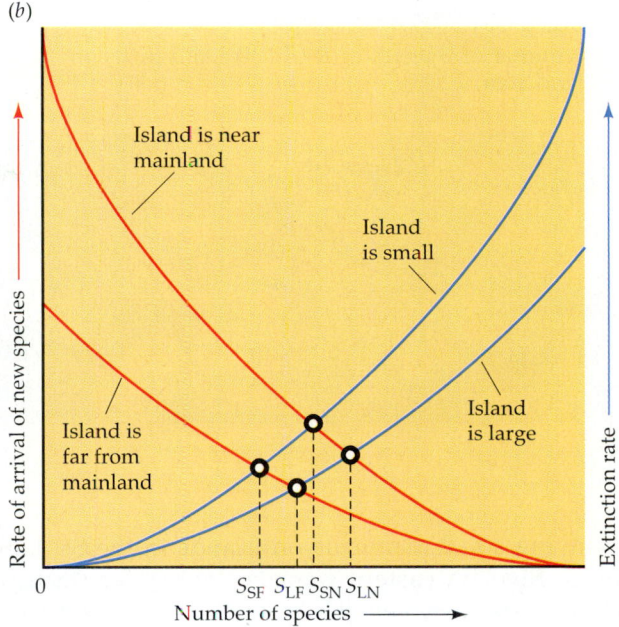

(b)

50.8 A Model of Species Richness Equilibrium
Rates of arrival of new species and extinction of species already present determine the number of species. (a) The number of species reaches an equilibrium (S) when the number of new species arriving equals the number becoming extinct. (b) At equilibrium, small islands far from the mainland (S_{SF}) have the fewest species; large islands near the mainland (S_{LN}) have the largest number of species.

Tests of the Species Richness Equilibrium Model

The equilibrium model predicts that the number of species should be positively correlated with island size and negatively correlated with distance from the mainland. Bird species on islands of the Pacific Ocean exhibit this pattern (Figure 50.9). New Guinea is the mainland for most of these islands. Similar patterns are known for plants, insects, lizards, and mammals.

Natural disturbances sometimes permit immigration and extinction rates to be estimated directly. The eruption of Krakatau in August 1883, which we described in Chapter 28, destroyed all life on the island's surface. After the lava cooled, Krakatau was colonized rapidly by plants and animals from nearby islands. By 1933 Krakatau was again covered with a tropical evergreen forest, and 271 species of plants and 27 species of resident land birds were found. The 1920s, a period when a forest canopy was developing, was apparently a time of high immigration rates of both birds (Table 50.1) and plants to Krakatau. Birds probably brought seeds of many of those plants because, between 1908 and 1934, both the percentage of plants with bird-dispersed seeds increased (from 20 to 25), as did the absolute number of species dispersed by birds (from 21 to 54). The numbers of species of plants and birds are not now increasing as

fast as they did during the 1920s, but equilibrium has not yet been reached for these, or any other, taxa on Krakatau. Future biological censuses of Krakatau will continue to measure directly the arrival and extinction rates of species and will show if and when different taxa reach the equilibrium number of species.

Colonization rarely can be observed directly. Usually it must be estimated from repeated censuses of an area, but such censuses do not detect species that arrive but fail to establish colonies that persist until the next census. Under special circumstances, however, dispersers leave traces of their movements. For example, the tracks of small mammals in fresh snow form a record of their movements. A biogeographer

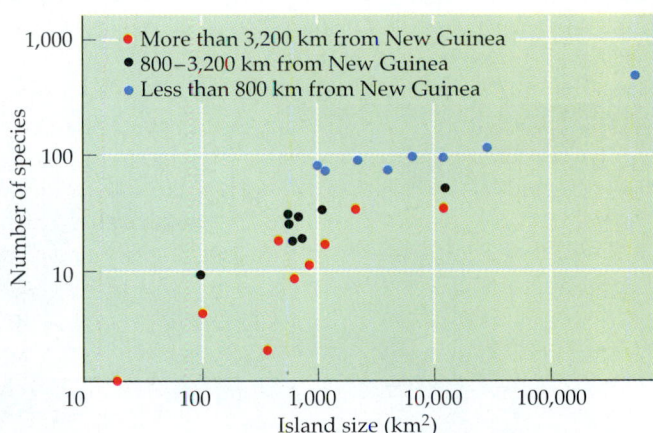

50.9 Small, Distant Islands Have Fewer Bird Species
The dots give numbers of land and freshwater bird species on islands of different sizes in the Moluccas, Melanesia, Micronesia, and Polynesia. These islands have been divided into three groups according to their distance from the relevant mainland, New Guinea.

TABLE 50.1
Number of Species of Resident Land Birds on Krakatau

PERIOD	NUMBER OF SPECIES	EXTINC-TIONS	COLONIZA-TIONS
1908	13		
1908–1919		2	17
1919–1921	28		
1921–1933		4	3
1933–1934	27		
1934–1951		1	9
1951	35		

studying small mammals on islands in the St. Lawrence River in eastern Canada, looked for tracks in fresh snow between the mainland and the islands during the winter when the river was frozen. He found that many more mice and shrews started to cross the ice than arrived at the islands. The fraction that made successful crossings was higher among those species that were commonly found on the islands than among those species rarely found on the islands.

An experiment testing the equilibrium model of island biogeography was carried out off the southern tip of Florida, a region dotted by thousands of small islands consisting entirely of red mangrove trees rooted in shallow water. Islands that are about 12 meters in diameter each contain 20 to 45 species of arthropods. Smaller islands have fewer species, larger ones more. Six islands were fumigated with methyl bromide, which destroyed all arthropods on them. Methyl bromide decomposes rapidly and does not inhibit recolonization. The rates of recolonization

of the islands by arthropods were very high. Within a year the fumigated islands had about their original number of species, indicating that an equilibrium had been reached (Figure 50.10).

The models and examples we have discussed pertain to oceanic islands, where water provides barriers to dispersal. Mainland areas are full of **habitat islands,** that is, patches of habitat separated from other similar patches by different types of habitats. For many species living in habitat islands, the intervening areas may be just as unsuitable to live in as if they were covered with water. We can apply the island model for species richness to these habitat islands, recognizing that some intervening areas, though unsuitable for permanent occupancy, may nonetheless permit a brief stopover. Therefore, arrival rates are higher for most habitat islands than they are for similarly sized oceanic islands. We will discuss the importance of habitat islands for conservation of species richness in the next chapter.

TERRESTRIAL BIOMES

Ecological communities dominated by evergreen shrubs with tough leaves—chaparral—are found in five regions of the world (the Mediterranean basin, California, central Chile, extreme southern Africa, and southwestern Australia) that have mild, wet winters and hot, dry summers. These similarities among plants living in so-called Mediterranean climates are due to convergent evolution (the evolution of similarities among species that were originally very different from one another). These species converged because soil moisture levels are most favorable in winter when temperatures are lowest, and least fa-

50.10 Experimental Island Biogeography
Scaffolding is erected by scientists to enclose a small mangrove island in the Florida Keys. Methyl bromide introduced into the enclosure killed all arthropods inside it. When the enclosure was removed, arthropods quickly recolonized the island.

vorable when temperatures are higher. Under this regime, natural selection favors the ability to photosynthesize in winter, when leaves may be exposed to occasional freezing weather, and the ability to extract water from relatively dry soils. The tough, drought-resistant leaves of chaparral shrubs function well when it is cool and moist and when it is hot and dry.

Chaparral vegetation is just one example of the convergence in appearance and ecological characteristics of plants living in different parts of Earth that have similar climates. Ecologists apply the name **biomes** to these major types of ecosystems that are characterized by their dominant vegetation. The vegetation of a biome appears similar wherever it is found, but the plants in these communities, despite their similarities, may not be closely related phylogenetically. Although biomes are identified and named by their characteristic vegetation, sometimes supplemented by their location or climate, each biome has many species of organisms in other kingdoms adapted to the physical environment and the physical structure provided by the plants.

Because climate plays a key role in determining which types of plants are in a given environment, the distribution of biomes on Earth is strongly influenced by annual temperature ranges and rainfall amounts (Figure 50.11). In certain biomes, such as tropical rainforest, temperatures are nearly constant but rainfall varies seasonally. In others, such as tem-

perate deciduous forest, precipitation is relatively constant throughout the year, but temperature varies strikingly between summer and winter. In still others, both temperature and precipitation change seasonally. The general distribution of terrestrial biomes in relation to mean annual temperature and mean annual precipitation is shown in Figure 50.12. The geographical distribution of the biomes is shown in Figure 50.13. Referring to these figures as we discuss the biomes will help you see how each one fits into global climatic types.

Biomes are identified on the basis of their dominant ecological communities, but many other community types are found within each biome. For example, the deciduous forest biome contains, among other things, streams, lakes, marshes, salt flats, dry slopes with shallow soils, moist valleys with deep soils, farmlands, pastures, and cities. Human activity has greatly altered the dominant vegetation of many of Earth's biomes. Biomes usually blend into and intermingle with one another; sharp boundaries are rare. Nonetheless, by recognizing major biomes we draw attention to the ecosystems that would predominate if natural processes had not been disturbed.

Keep in mind the relationships between vegetation and climate as we survey Earth's major terrestrial biomes. We will begin at high latitudes and work toward the equator, describing the climate, dominant vegetation, seasonal changes in ecological processes, and some characteristic animals of each biome.

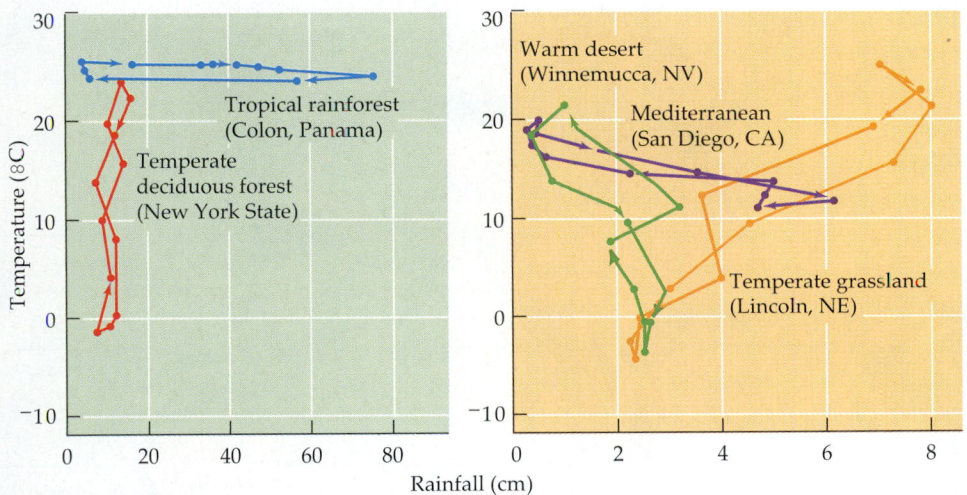

50.11 Climates of Some Biomes
These graphs indicate seasonal changes in temperature and precipitation. From the graph on the left, we see that little fluctuation in temperature and heavy rainfall during most months gives the tropical rainforest biome a very different climate from that of the temperate deciduous forest biome, which has relatively constant rainfall but marked temperature variations. On the right we see that the temperate grassland biome has great temperature fluctuations and rainfall that peaks in summer; the warm desert biome has sharp seasonal temperature differences and low rainfall; and the Mediterranean biome has a narrow temperature range and a winter rainfall peak.

Arctic, Alpine
Tundra
Boreal forest
Cold temperate
Cold desert
Temperate deciduous forest
Temperate rainforest
Woodland
Shrubland
Grassland
Tropical montane forest
Warm temperate
Savanna
Tropical
Thorn forest
Hot desert
Tropical dry forest
Tropical evergreen forest
Mean annual temperature (8C)
Mean annual precipitation (cm)
–15
–10
–5
0
5
10
15
20
25
30
50
100
150
200
250
300
350
400
450

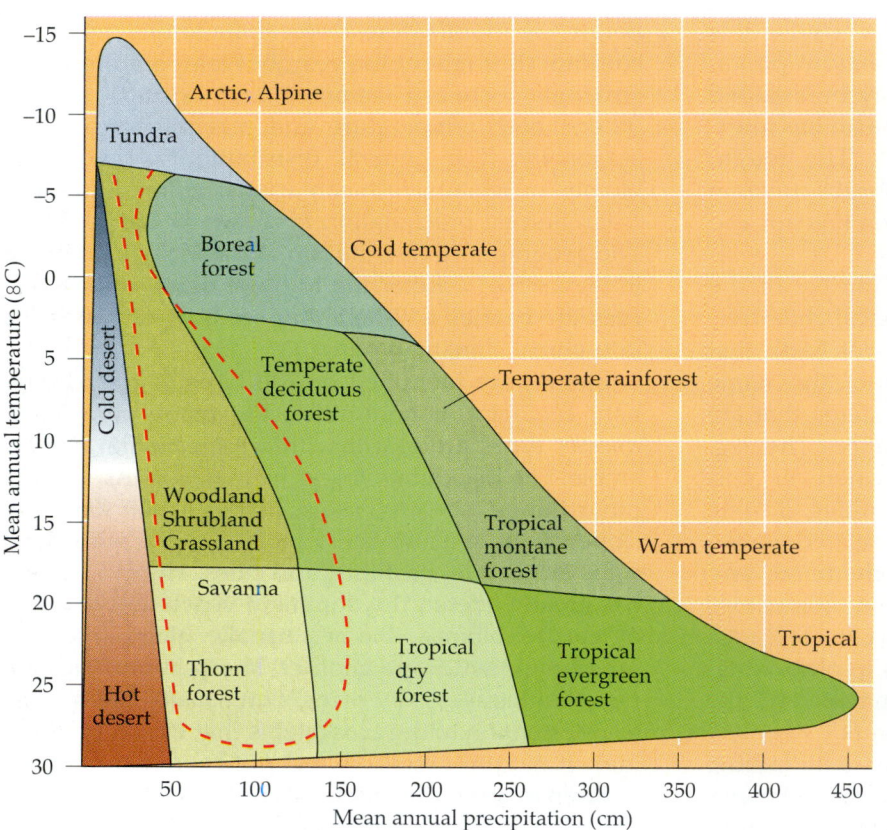

50.12 Most Biomes Have a Distinct Temperature and Precipitation Range
In regions having the ranges indicated within the red dashed lines, other factors—such as seasonality of drought, fire, and grazing—strongly affect which biome is present. The chaparral biome is not included because its distribution depends on the seasonal distribution of rainfall, a climate feature not shown here.

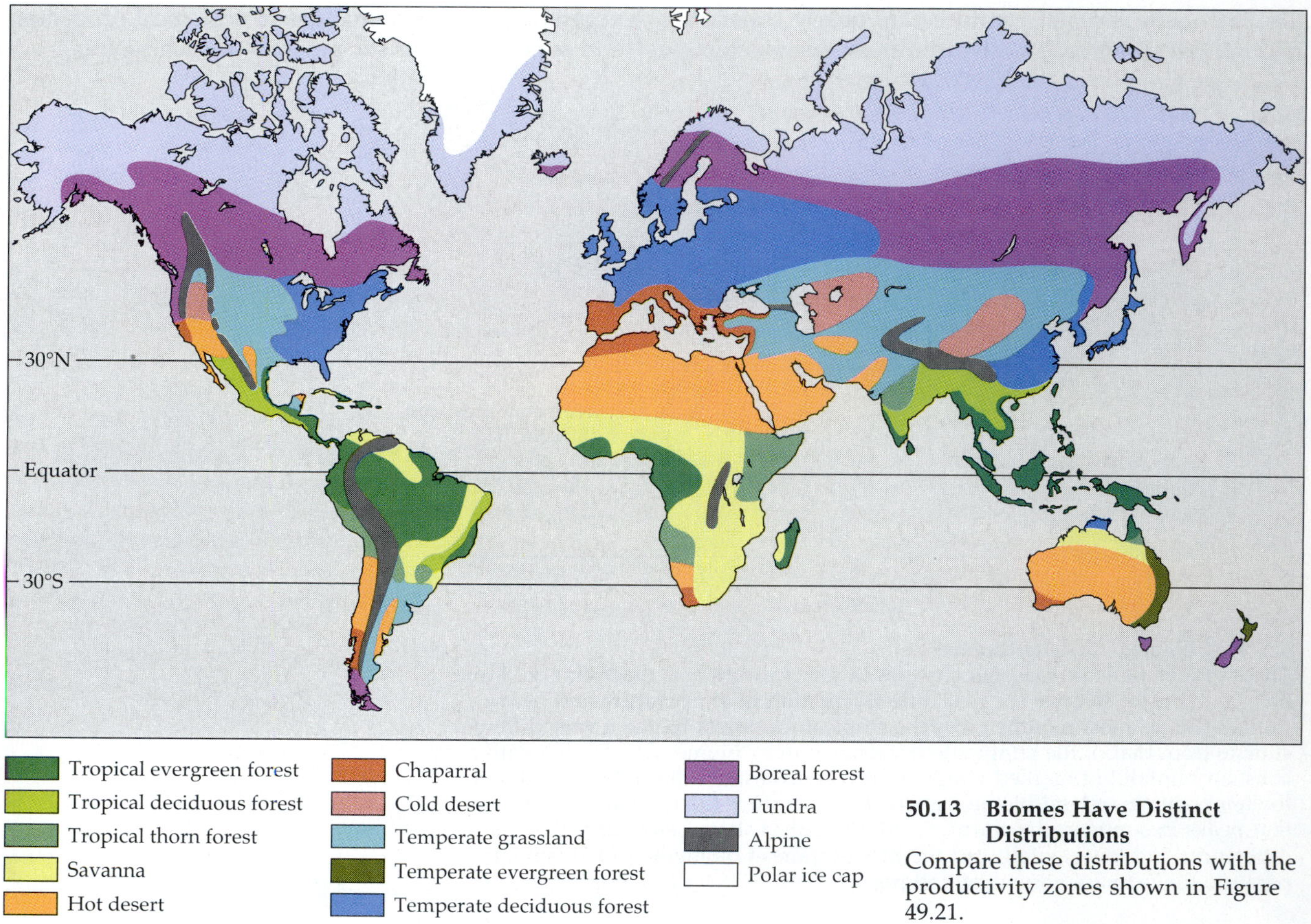

30°N
Equator
30°S
Tropical evergreen forest
Tropical deciduous forest
Tropical thorn forest
Savanna
Hot desert
Chaparral
Cold desert
Temperate grassland
Temperate evergreen forest
Temperate deciduous forest
Boreal forest
Tundra
Alpine
Polar ice cap

50.13 Biomes Have Distinct Distributions
Compare these distributions with the productivity zones shown in Figure 49.21.

Tundra

The tundra biome is found in the Arctic and high in mountains at all latitudes. Because the climate in these areas is too cold for trees, tundra vegetation is dominated by low-growing perennial plants (Figure 50.14). Permanently frozen soil—**permafrost**—underlies Arctic tundra. The top few centimeters of soil thaw during the short but often warm summers, during which the sun shines 24 hours every day. Plants grow actively in the shallow, water-logged soils for a few months each year. Most lowland Arctic tundra is very wet even though precipitation is low because water cannot drain down through the frozen soil. Montane (upland) tundra in the Arctic is better drained than lowland tundra, and south-facing slopes may not be underlain by permafrost.

Tropical alpine tundra has 12-hour days and 12-hour nights all year. At these high altitudes the temperature is never very high and drops to freezing on most clear nights. In tropical alpine communities plants photosynthesize slowly all year, and most animals are year-round residents. Most Arctic tundra animals either migrate into the area for the summer and go elsewhere in the winter, or they are dormant most of the year.

Boreal Forest

Moving from the Arctic tundra toward the equator and to a lower elevation on temperate-zone mountains, we find the boreal forest biome, which is dominated by coniferous evergreen trees (Figure 50.15a). Over most of the boreal forest biome, winters are long and very cold and summers are short but often warm. The shortness of the summer favors trees with evergreen leaves that live for several years. These trees are ready to photosynthesize as soon as temperatures become favorable in the spring. Evergreen coniferous forests also grow along the west coasts of continents at middle to high latitudes where winters are mild but very wet and summers are cool and relatively dry. The dominant trees photosynthesize actively during mild weather in the winter but often must close their stomata during summer droughts. These forests have Earth's tallest trees, and they support the highest standing biomasses of wood of all ecological communities.

Boreal forests in the Southern Hemisphere are dominated by southern beech trees (*Nothofagus*), angiosperms with small leaves; they appear similar to conifers at a distance (Figure 50.15b). Some species of *Nothofagus* are evergreen; others are deciduous.

Boreal forests have only a few tree species, nearly all of which are wind-pollinated and have wind-dispersed seeds. The dominant animals—such as in-

(a)

(b)

50.14 Arctic and Alpine Tundra
(a) This picture was taken in Denali National Park, Alaska. In winter the tundra is covered with snow, which remains for eight months or more. (b) This alpine area in Glacier National Park, Montana is dominated by wildflowers in the spring.

50.15   Boreal Forest
(a) In mountainous areas, boreal forests grow up to timberline, the upper limit of tree growth. This northern-hemisphere forest is in Jasper National Park, Alberta, Canada. (b) This southern boreal forest along the banks of the Arrayanes River in the Andes Mountains of Patagonia, Argentina, is dominated by southern beeches (*Nothofagus*).

sects, moose, and hares—eat leaves. Because there are so few species, many prey have a single major predator species, and population densities of both predators and prey may fluctuate markedly, as we saw in Chapter 48. The seeds in the cones of the conifers are eaten both by species that store seeds, such as squirrels, jays, and nutcrackers, and those, such as crossbills, that eat the seeds while the cones still hang on the trees.

Temperate Deciduous Forest

The temperate deciduous forest biome is found in eastern North America and eastern Asia and in parts of western Europe. These are regions where temperatures fluctuate dramatically between summer and winter and ample precipitation falls throughout the year. Deciduous trees, which lose their leaves during the cold winter but produce leaves that photosyn-

thesize rapidly during the warm, moist summers, dominate these environments (Figure 50.16). There are many more tree species in deciduous forests than in boreal forests, and many deciduous trees have animal-dispersed pollen and fruits. The temperate forests richest in species are in the southern Appalachian Mountains of the United States and in eastern China and Japan, areas that were little disturbed by Pleistocene glaciation.

Deciduous forests have striking seasonal cycles of activity, the most conspicuous of which are the changes in the leaves. Many birds migrate into this biome in summer when insects are abundant. Understory plants often flower and fruit in early spring before the canopy above them has leafed out. Most trees and shrubs produce their fruits in autumn, at the end of the growing season. Annual primary production in these forests may equal that in many tropical forests.

(a)

(b)

50.16 Temperate Deciduous Forest
(a) The trees in these Rhode Island woods are mostly oaks and maples. This photo catches them in July, with a full complement of leaves. *(b)* The same scene six months later, in early January. The trees have shed their leaves and are in a condition of winter dormancy.

Grassland

The grassland biome is found in many climates, but all of them are relatively dry much of the year. In some grasslands most precipitation falls in the winter; in others, most of it falls in the summer. Tropical grasslands are hot all year, but winters are cold in most temperate grasslands. Grasslands are found in so many different climates because grasses are able to survive disturbances. Grasses store much of their energy underground and quickly resprout after they are heavily grazed or burned. As we saw in Chapter 49, grasslands typically support large populations of grazing mammals, and fires are common. Grasslands are structurally simple, but they are rich in species of grasses, sedges, and **forbs**—broad-leaved herbaceous species that live together with grasses and sedges (Figure 50.17). Many forbs have showy flowers, and grasslands are often riots of color when forbs are in full bloom.

50.17 Temperate Grassland
This grassland in Kansas has many species of grasses and forbs. In this late-summer scene, the flowering stalks of forbs project above the shorter grasses.

50.18   Cold Desert
This cold desert in California is on the eastern slopes of the Sierra Nevada. A few species of low-growing shrubs dominate, with small grasses and forbs growing among them.

Cold Desert

The cold desert biome is found in dry regions at middle to high latitudes, especially in the interiors of large continents. Many cold deserts are in the rain shadows of mountain ranges, where seasonal changes in temperature are great and where most of the little rain that falls does so in the winter. Cold deserts are dominated by a few species of low-growing shrubs (Figure 50.18). The surface soil layers are recharged with moisture in the winter, and plant growth is concentrated in the spring. By early summer these deserts are often barren; so little rain falls that plants cannot photosynthesize much during the hot summers. Cold deserts are relatively poor in species in most taxonomic groups, but because the plants produce many seeds, seed-eating birds, ants, and rodents are common.

Hot Desert

The hot desert biome is found in two belts centered at 30° north and 30° south latitudes, respectively, where air from aloft descends, warms, and picks up moisture (see Figure 49.19). Hot deserts receive most of their rainfall in summer, when the intertropical convergence zone (see Chapter 49) moves poleward, but they also receive winter rains from storms that form over mid-latitude oceans. The driest regions, where neither summer nor winter rains penetrate regularly, are in the center of Australia and the middle of the Sahara Desert. Except in the driest areas, hot deserts have a richer and structurally more diverse vegetation than do cold deserts. Succulent plants that store large quantities of water in their expandable stems and that photosynthesize primarily with their stems rather than with leaves are conspic-

uous in many hot deserts (Figure 50.19). Annual plants are abundant during rainy periods. Animal pollination and animal dispersal of fruits are common, and great quantities of seeds are produced. The richest bee fauna on Earth is found in the hot deserts

50.19   Hot Desert
In this scene in the Sonoran Desert of Arizona, large saguaro and cholla cactus share the stage with brilliant yellow brittlebrush. Note the structural richness of this desert.

of southwestern United States and adjacent Mexico. Population densities of rodents and ants are often remarkably high, and lizards and snakes are typically very common.

Chaparral

The chaparral, or Mediterranean, biome is found on the west sides of continents at moderate latitudes where winters are cool and wet and summers are hot and dry. As we have seen, such climates are found in the Mediterranean region, California, central Chile, extreme southern Africa, and southwestern Australia. The chaparral biome is dominated by low-growing shrubs with tough evergreen leaves. The shrubs carry out most of their growth and photosynthesis in early spring, which is when insects are active and birds breed.

Many shrubs of Northern Hemisphere chaparral produce bird-dispersed fruits that ripen in late fall or winter, when large numbers of migrant birds arrive from the north. One such fruit—the olive—has played a very important role in human history, providing a rich food source for people at a period of otherwise low food availability. Annual plants grow abundantly in chaparral climates, providing seeds that store well during the hot, dry summers. This biome thus also supports large populations of small rodents, most of which store seeds in underground burrows.

Thorn Forest and Tropical Savanna

The remaining biomes are found at low latitudes where seasonal temperature fluctuations are small and annual cycles are dominated by wet and dry seasons. In general, the length of time that a region is close to the intertropical convergence zone, and hence receives rainfall, increases toward the equator. Because the intertropical convergence zone shifts latitudinally in a seasonably predictable way, there is a characteristic latitudinal pattern of distribution of rainy and dry seasons in tropical and subtropical regions (Figure 50.20).

Thorn forests, which are found on the equatorial sides of hot deserts, contain many plants similar to those in deserts (Figure 50.21a). Little or no rain falls for eight or nine months during the winter, but rainfall may be heavy during the summer wet season.

Dry, tropical and subtropical regions of Africa, South America, and Australia have extensive areas of savannas, dominated by grasses and grasslike plants but with scattered trees. By their grazing and browsing, large populations of mammals maintain the savanna vegetation in Africa (Figure 50.21b). If these areas are not grazed, browsed, or burned, they revert to dense thorn forest.

Tropical Deciduous Forest

As the length of the rainy season increases toward the equator, thorn forests are replaced by tropical deciduous forests. Tropical deciduous forests are taller, have fewer succulent plants, and are much richer in species than are thorn forests. Most of the trees, except those growing along rivers, lose their leaves during the dry season (Figure 50.22); many of them flower while they are leafless. The community is very rich in species of both plants and animals. The long dry season is very hot and often windy.

Because tropical deciduous forest soils are less leached of nutrients than are soils in wetter areas, they are some of the best tropical soils for agriculture. As a result, most tropical deciduous forests have been cleared for grazing cattle and growing crops. Where water is available for irrigation during the dry season, farmers can grow both a dry-season and a rainy-season crop.

Tropical Evergreen Forest

The tropical evergreen forest biome is found in equatorial regions where total rainfall exceeds 250 cm of

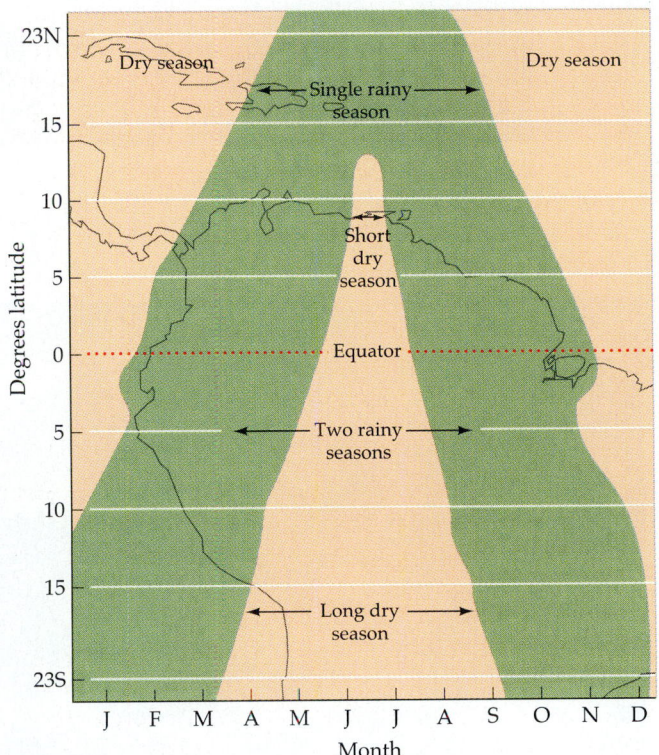

50.20 Rainy Seasons Change with Latitude
In the tropics and subtropics, which months are rainy and which are dry is highly predictable based on the region's latitude.

50.21 Thorn Forest and Savanna
(a) Thorn forests, such as this one in Sonora, Mexico, are dominated by large cacti and short trees, many of which are leafless during the long dry season. (b) African savannas support large herds of grazing and browsing mammals, such as these impalas.

(a)

(b)

50.22 Tropical Deciduous Forest
Only a few trees, growing in better-watered sites, still have leaves during the late hot, dry season (March) in this Costa Rican forest. As is typical for that time of year in northwestern Costa Rica, no significant rain had fallen since December.

50.23 Tropical Evergreen Forest
This photo, taken from a low-flying plane in Borneo, reveals an example of the most productive biome on Earth. You can tell that this forest is rich in tree species by looking at the variety of canopy shapes.

rain annually and the dry season lasts no more than a few months. This is the richest of all biomes in species of both plants and animals, and it has the highest overall productivity of all ecological communities (Figure 50.23). As we saw in Chapter 49, however, this productivity is based on very tight cycling of mineral nutrients, most of which are tied up in vegetation. Except where they are very young, the soils are deeply weathered and usually cannot support agriculture without massive applications of fertilizers.

These wet tropical forests may have up to 500 species of trees per square kilometer. Most of the species are rare, and nearly all of them rely on animals to transport their pollen and disperse their fruits. Food webs are extremely complex. Most species of invertebrates living in these forests have not been described or named by scientists. Human activity currently is destroying this biome at a very high rate.

Tropical Montane Forest

Temperature decreases about 6°C for each 1,000-meter increase in elevation on the slopes of tropical mountains. Trees in tropical montane (upland) forests are shorter than are lowland tropical trees. Leaves are also smaller, and epiphytes—plants that derive their nutrients and moisture from air and from water and that usually grow on other plants—are more luxuriant, especially at elevations where clouds form regularly, bathing the forest in moisture on most days. Photosynthesis in tropical montane forests is depressed because of low temperatures and because the leaves are wet most of the time.

AQUATIC ECOSYSTEMS

Aquatic communities are built upon the same predator–prey, competitive, and mutualistic interactions among their component species as are terrestrial communities, but the differences between water and air have caused different outcomes from those interactions. First, water, compared with air, is a very dense medium, and it provides much more support for organisms living in it. At the same time, moving water creates much greater forces than moving air does. Submerged plants depend upon the water to support them and are flexible so that they move with water currents rather than resisting them. Second, movement through water is more difficult than movement through air. Third, because sunlight penetrates only short distances through water, communities at moderate depths in lakes and oceans receive little light. Fourth, whereas most terrestrial environments have abundant oxygen, oxygen is often in short supply in aquatic environments.

Other features of aquatic communities are related to these basic properties of water. Except in shallow shoreline waters, most aquatic photosynthesizers are tiny; all their tissues are eaten readily by animals. These small photosynthesizers, most of which are bacteria and algae, provide little physical structure to aquatic communities. Water currents carry large amounts of food to organisms on the margins of oceans, leading to communities in which animals compete directly with one another and with algae and plants for space. In terrestrial communities animals rarely become abundant enough to fill up space. In many aquatic communities, animals, rather than plants, are the largest and longest-lived organisms.

(a)

(b)

50.24 Freshwater Biomes: Still and Moving
(a) Several species of photosynthetic plants grow in placid water near the shore of a New Jersey lake. Insects eat the plants and are in turn fed upon by sunfish. (b) Plants cannot obtain a foothold in the fast-moving waters of the Adams River in British Columbia, and the bottom remains rocky. These sockeye salmon matured at sea, and will spawn and die without feeding in the river.

because of currents, only plants that are firmly attached to the bottom can retain a fixed position. As a result, plant densities are low and there is little photosynthesis in most streams and rivers (Figure 50.24b. The organisms living in them depend on food that falls into the water from terrestrial communities. Attached plants in shallow, slow-moving water support grazing animals, but most stream animals eat food drifting downstream.

Marine Biomes

Oceans cover about 71 percent of Earth's surface, and there are no absolute barriers to movement of marine organisms among all the ocean basins. Ocean water moves in great circular patterns—clockwise in the Northern Hemisphere and counterclockwise in the Southern Hemisphere (see Figure 49.20)—carrying with it organisms that have limited swimming abilities. Despite the lack of barriers, recognizable biogeographic regions exist in the sea, especially along coasts. Boundaries between them are determined by the temperature at the surface of the sea during the coldest months of the year.

The distances to which the eggs and larvae of marine organisms can move are determined in large part by the length of time they live as part of the plankton community, during which time they are carried passively by ocean currents. The period between egg laying and metamorphosis—the larval dispersal period—is inversely proportional to egg size. Therefore, the larvae from small eggs can disperse over greater distances than larvae from large eggs can. For the most part, species sedentary as adults that have managed to colonize the intertidal and subtidal zones of isolated islands in the Pacific Ocean are species with small eggs.

Attached multicellular plants are limited to a very thin strip along the edges of oceans. Most of the photosynthesis of the oceans is carried out by single-celled organisms, particularly bacteria, diatoms, and dinoflagellates (see Chapter 23). These organisms are most abundant in zones of coastal upwelling, where they are maintained in surface waters by upward-moving currents and where nutrient and light levels are high. The most important grazers of phytoplankton in the sea are species of crustaceans that differ from those that dominate fresh waters.

Freshwater Biomes

Green algae and filamentous cyanobacteria conduct most of the photosynthesis in lakes. Vascular plants grow only in shallow waters near shores (Figure 50.24a). Rotifers, crustaceans, insects, and fishes live in lakes. Insects in particular occupy a wide variety of ecological niches. There are also freshwater clams, cnidarians, various worms, crustaceans, and sponges, but, except for crustaceans, they do not assume the prominence in lakes that their relatives achieve in the sea. Many groups of insects, such as dragonflies, damselflies, mayflies, caddisflies, and true flies, have aquatic larvae that become terrestrial adults. Their emergence and metamorphosis link aquatic and terrestrial communities.

Rivers and streams are dynamic bodies of water whose properties change dramatically with seasons as well as between their sources in highlands and their coastal floodplains. Rivers are small and usually clear near their sources and become larger, warmer, and murkier toward their mouths. The large quantities of solids carried in suspension in many rivers reduce the depth to which sunlight penetrates, which limits possibilities for photosynthesis. In addition,

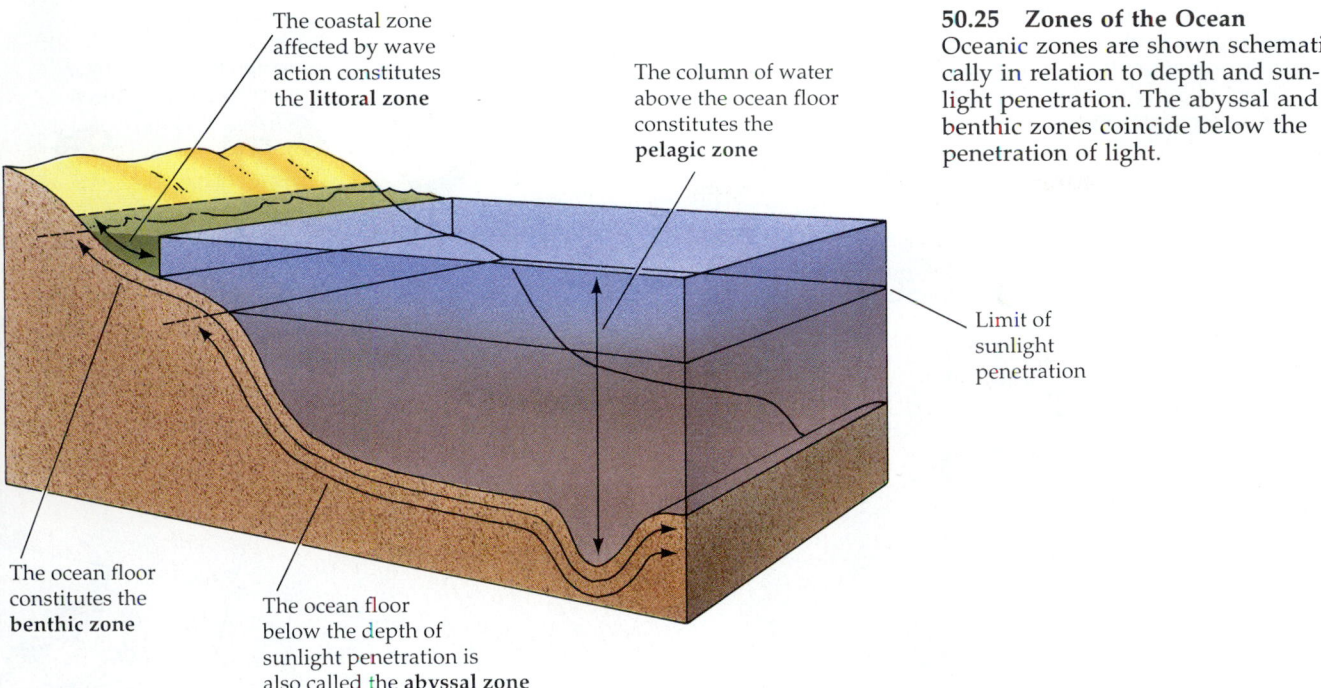

The coastal zone affected by wave action constitutes the **littoral zone**

The column of water above the ocean floor constitutes the **pelagic zone**

Limit of sunlight penetration

The ocean floor constitutes the **benthic zone**

The ocean floor below the depth of sunlight penetration is also called the **abyssal zone**

50.25 Zones of the Ocean
Oceanic zones are shown schematically in relation to depth and sunlight penetration. The abyssal and benthic zones coincide below the penetration of light.

For convenience we have named different parts of the oceans (Figure 50.25). The bottom of the ocean is the **benthic zone;** the open water is the **pelagic zone.** The **abyssal zone** is the part of the benthic zone that receives no sunlight. Animals living in the abyssal zone (except for those living around thermal vents) depend, directly or indirectly, on the remains of organisms drifting down from sunlit parts of the ocean. Food is distributed very sparsely on the bottom of deep oceans, and organisms living there have remarkable adaptations enabling them to devour prey items that are very large compared with themselves (Figure 50.26a).

(a)

50.26 Abyssal and Pelagic Fishes
(a) In the sunless depths of the abyssal ocean, a viperfish lures opossum shrimp using bioluminescence. The very large mouth and small eyes of the viperfish are typical of many deep-sea fishes. (b) In the open ocean, where there are no hiding places, fish must be able to swim rapidly to escape predators; the streamlined body shape of these pelagic horse-eye jacks is an adaptation for swift movement. (b)

50.27 A Littoral Community

In this depiction of bottom-dwelling species in an Atlantic littoral zone, the yellow area is always above water; the green area is the intertidal; the blue area is permanently under water but is subject to turbulence from wave action.

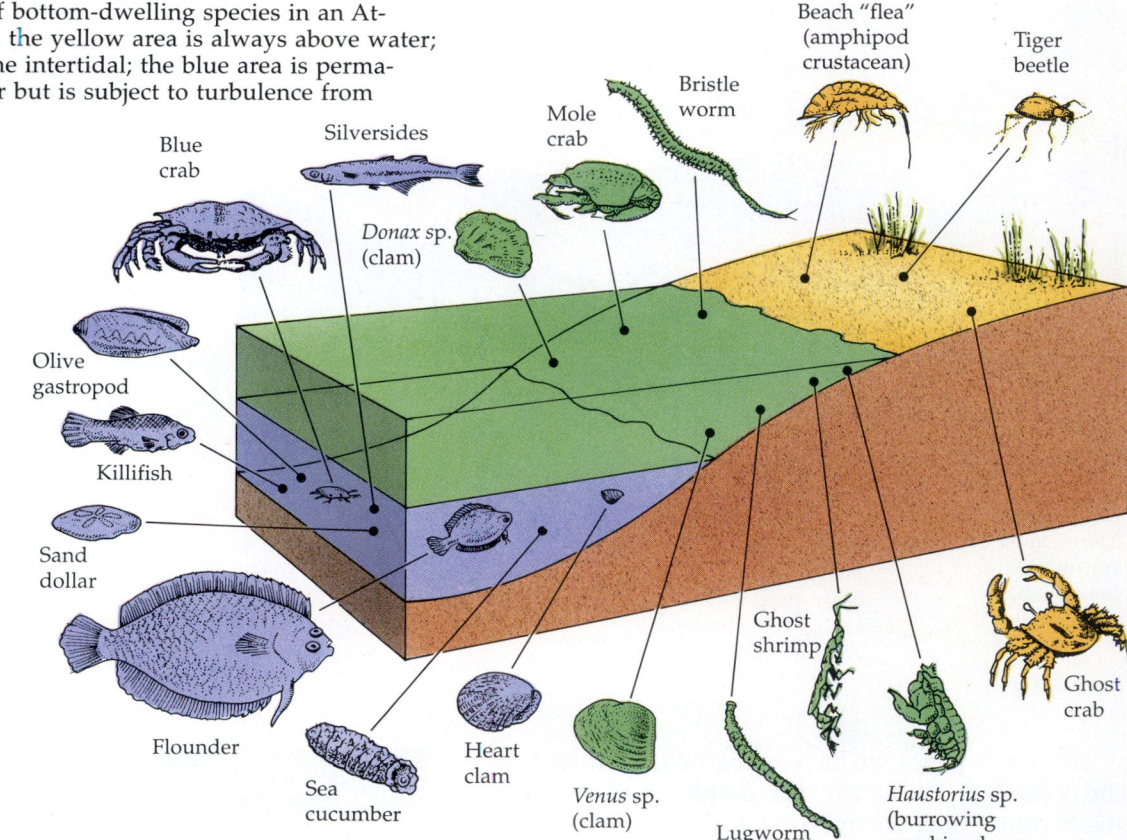

Pelagic communities are relatively simple, consisting of zooplankton and phytoplankton, supplemented by planktonic larvae of benthic organisms during brief periods of the year. These small organisms are eaten by planktivores ranging in size from small fish to giant whales. Most pelagic fish live in large schools and are streamlined for rapid movement through the open water (Figure 50.26b).

Communities of the **littoral zone**—the coastal zone from the uppermost limits of tidal action down to the depth where the water is thoroughly stirred by wave action—are richer in species and more complex in structure than pelagic communities are. Communities on and in mud and sand consist mostly of burrowing animals. Some feed by extracting prey from the water column, others construct nets on the mud surface, and still others burrow through the substrate, extracting very small prey or seeking out larger, suspension-feeding organisms (Figure 50.27).

Communities on rocky substrates in the littoral zone are sites of intense competition for space because waves carry such a plentiful supply of food to the animals living there. Among competitors for space in a single small area may be algae from several different phyla and animals from diverse phyla: sponges, cnidarians, mollusks, annelids, moss animals, echinoderms, and chordates (Figure 50.28).

50.28 A Crowded Rocky Intertidal Community

This community on the coast of British Columbia is dominated by mussels (dark) and barnacles (light), among which clumps of algae are growing. Many of the barnacles are growing directly on the shells of the mussels. Many smaller organisms not visible in this photograph live in the crevices among the mussels.

50.29 A Coral Reef Community
The many coral species in this Polynesian lagoon provide a haven for fish, such as these butterfly fish, as well as many other species. A clam gapes at the right; the black spines of a sea urchin peek out at the center of the scene.

Muddy intertidal shores in tropical regions are dominated by low-stature forests composed of a number of different trees collectively called mangroves (see Figure 50.10). Mangroves are among the few trees able to grow in salt water, and by trapping sediments among their roots and branches, they contribute to the building up of tropical shorelines. Many animal species live permanently among mangroves, and many pelagic species come there to spawn.

The richest and structurally most complex of all marine communities are coral reefs, which fringe coasts of tropical continents and islands (Figure 50.29). Like the trees of terrestrial communities, corals provide structures within which many different ways of life are possible. The richness of coral reef life, which may include over 100 species of corals alone, can only be hinted at even in the best of illustrations. Fishes of a wider variety of shapes and colors than those of the pelagic zone feed on a diversity of prey and hide from predators in the wide variety of structures formed by corals.

Continental shores are important as breeding grounds for many pelagic species. All pelagic species that lay large eggs move to coastal waters, particularly sheltered bays and estuaries, to breed. The vitality of many populations of marine organisms, particularly fish and arthropods, depends upon the existence of these spawning grounds. Unfortunately, these are often excellent areas for commercial and residential developments, and they also receive pollutants from the rivers that flow into them. If estuaries are not managed wisely, many marine species may become extinct.

SUMMARY of Main Ideas about Biogeography

Biogeographers study processes at different temporal and spatial scales to explain the past and present distributions of organisms on Earth.
   Review Figure 50.2

Data from current distributions and reconstructed phylogenies are integrated to explain distributions of organisms.
   Review Figures 50.4, 50.5, and 50.6

Biogeographers recognize six major biogeographic regions.
   Review Figure 50.7

The number of species in an area is the result of a dynamic balance between extinction and arrival of new species.
   Review Figure 50.8

The distributions of biomes are determined primarily by climate.
   Review Figures 50.11, 50.12, and 50.13

Which months are rainy and which are dry is correlated with latitude in tropical and subtropical regions.
   Review Figure 50.20

SELF-QUIZ

1. Biogeography as a science began when
   a. eighteenth-century travelers first noted intercontinental differences in the distributions of organisms.
   b. Europeans went to the Middle East during the Crusades.
   c. cladistic methods were developed.
   d. the fact of continental drift was accepted.
   e. Charles Darwin proposed the theory of natural selection.

2. Historical and ecological biogeography differ in that
   a. only historical biogeography is concerned with history.
   b. historical biogeography is concerned with longer time periods and larger space scales.
   c. both are concerned with the same time scales, but historical biogeography deals with larger space scales.
   d. both are concerned with the same space scales, but historical biogeography deals with longer time scales.
   e. historical biogeography is not concerned with the current distributions of organisms.

3. Marine biogeographic regions are less distinct than terrestrial ones because
   a. the ocean biota is more poorly known than the terrestrial biota.
   b. there are currently fewer barriers to dispersal of marine organisms.
   c. most marine families and higher taxa evolved before the oceans were separated by continental drift.

   d. we know less about distributions of marine organisms.
   e. oceanic circulation is faster than atmospheric circulation.

4. A parsimonious interpretation of a distribution pattern is one that
   a. requires the smallest number of vicariant events.
   b. requires the smallest number of dispersal events.
   c. requires the smallest total number of vicariant plus dispersal events.
   d. accords with the cladogram of a group.
   e. accounts for centers of endemism.

5. The only major biogeographic region that today is isolated by water from other regions is
   a. Greenland.
   b. Africa.
   c. South America.
   d. Australasia.
   e. None of the major regions is isolated by water today.

6. Equilibrium species richness is reached in the MacArthur–Wilson model when
   a. immigration rates of new species and extinction rates of species are equal.
   b. immigration rates of all species and extinction rates of species are equal.
   c. the rate of vicariant events equals the rates of dispersal.
   d. the rate of island formation equals the rate of island loss.
   e. No equilibrium number of species exists in that model.

7. Chaparral vegetation is dominated by
   a. deciduous trees.
   b. evergreen trees.
   c. deciduous shrubs.
   d. evergreen shrubs.
   e. grasses.

8. Which of the following is *not* true of tropical evergreen forests?
   a. They have large numbers of species of trees.
   b. Most plant species are animal-pollinated.
   c. Most plant species have animal-dispersed fruits.
   d. Biological energy flow is very high.
   e. Productivity depends on a rich supply of soil nutrients.

9. Which of the following is *not* true of river ecosystems?
   a. Only plants that are attached to the bottom can remain in a single location.
   b. Sunlight does not penetrate very far into river water.
   c. There is little photosynthesis, and animals depend on imported food.
   d. Animal communities are dominated by mollusks and cnidarians.
   e. Stream properties change dramatically between the headwaters and lowlands.

10. At all depths, the bottom of the ocean is known as the
    a. benthic zone.
    b. abyssal zone.
    c. pelagic zone.
    d. interoceanic convergence zone.
    e. subtidal zone.

FOR STUDY

1. Horses evolved in North America but subsequently became extinct there. They survived to modern times only in Africa and Asia. In the absence of a fossil record we would probably infer that horses originated in the Old World. Today, the Hawaiian Islands have by far the greatest richness of species of fruit flies (*Drosophila*). Would you conclude that the genus *Drosophila* originally evolved in Hawaii and spread to other regions? Under what circumstances do you think it is safe to conclude that a group of organisms evolved close to where the greatest number of species live today?

2. For nearly every ecological community, the number of species present is much fewer than the number potentially available to colonize it. Why is this phenomenon evidence for species equilibrium but not for species saturation? Is it really very good evidence for equilibrium? What do you consider the strongest evidence for species equilibrium?

3. A well-known legend states that Saint Patrick drove the snakes out of Ireland. Give some alternative explanations, based on sound biogeographic principles, for the absence of indigenous snakes in that country.

4. What are some significant present-day human concerns whose solutions involve biogeographic considerations? What kinds of biogeographic knowledge are most important for each one?

5. Most of the world's flightless birds are either nocturnal and secretive (for example, the kagu of New Caledonia) or large, swift, and well-armed (for example, the ostrich of Africa). The exceptions are found primarily on islands, and many of these island species have become extinct with the arrival of humans and their domestic animals. What special biogeographic conditions on islands might permit the survival of flightless birds? Why has human colonization so often resulted in the extinction of such birds? The power of flight has been lost secondarily in representatives of many groups of birds; what are some possible evolutionary advantages of flightlessness that might offset its obvious disadvantages?

READINGS

Brown, J. H. and A. C. Gibson. 1983. *Biogeography.* Mosby, St. Louis. A comprehensive treatment of both ecological and historical biogeography.

Humphries, C. J. and L. R. Parenti. 1986. *Cladistic Biogeography.* Clarendon Press, Oxford. A concise treatment of the ways in which cladistic methods are used to determine the causes of current distributions of organisms.

MacArthur, R. H. 1972. *Geographical Ecology: Patterns in the Distribution of Species.* Harper and Row, New York. The best introduction to quantitative theories of ecological biogeography.

MacArthur, R. H. and E. O. Wilson. 1967. *The Theory of Island Biogeography.* Princeton University Press, Princeton, NJ. The classic book that launched modern investigations of ecological biogeography.

Myers, A. A. and P. S. Giller. 1988. *Analytical Biogeography: An Integrated Approach to the Study of Animal and Plant Distributions.* Chapman & Hall, London. Contains chapters by different authors on many aspects of ecological and historical biogeography, including discussions of modern methods and their significance.

Terborgh, J. 1993. *Diversity and the Tropical Rain Forest.* W. H. Freeman, San Francisco. An exciting account of the richness of life in tropical evergreen forests.

Source of a Life-Saving Drug
A drug for combating leukemia was derived from the Madagascar rosy periwinkle.

51

Conservation Biology

Through much of human history people have cut forests, but rates of forest destruction have never before been as high as they are today. Forests in northwestern North America and in many areas of the tropics have been logged at especially high rates during recent decades. The loss of forests and other natural habitats is of great concern because, despite our ability to alter and restructure our surroundings, human lives depend on other species in many ways. We depend on other organisms for essential ecosystem services, such as the purification of the air we breathe and the water we drink, the maintenance of the soil that grows our food, and the safe disposal of much of our waste. More than half the medical prescriptions written in the United States contain a natural plant or animal product. Because the search for and exploitation of such products from the living world has barely begun, many other drugs from wild species probably will be discovered in the future. Unfortunately, however, many species may be eliminated by forest destruction before we find out if they might be sources of useful products. As fossil fuels—from which we manufacture most artificial substitutes for natural products—become more expensive, natural plant and animal materials may once again increase in importance. In addition, we derive enormous aesthetic pleasure from interacting with other organisms. Many people would see a world with far fewer species as a less desirable one in which to live.

Despite the value of other organisms to the quality of human life, our activities are causing the extinction of many species and the loss of many ecosystems. We do not know how many species will become extinct during the next 50 years, but some scientists estimate that more than 25 percent of the world's biota may be lost within the next century if corrective measures are not taken. The number of extinctions will depend both on what we do and on unexpected events. To develop sound strategies for preserving biological diversity, we need to understand current trends and the causes of current extinctions.

Conservation biology is a relatively new discipline that has developed in response to the **biodiversity crisis**—the accelerating rate at which species are now being lost because of human activities. It studies the causes of species richness and the means by which genes, species, communities, and ecosystems can be conserved. The science of conservation biology draws heavily on concepts and knowledge from population genetics, evolution, ecology, biogeography, and wildlife management. In turn, the needs of conservation are stimulating new research in those fields. As a result, the basic sciences of ecology, biogeography, and evolutionary biology are being enriched while they help to solve important conservation problems.

CAUSES OF EXTINCTIONS

Human activities have caused extinctions of species for thousands of years, but today we have more powerful tools than our ancestors had for decimating species, and billions more of us are using them. Important causes of extinctions are overexploitation, the introduction of predators and diseases, the loss of mutualists, and habitat destruction. In the future, climate modification from global warming may be added to this list.

Overexploitation

Until recently, humans caused extinctions primarily by overhunting or by introducing predators that, in turn, overexploited native species. Great numbers of species were lost from islands, where animals had evolved in relatively predator-free conditions and thus had few defenses against human hunters or any other predators. For example, when Polynesian people settled in Hawaii about 2,000 years ago, they quickly exterminated, probably by overhunting, at least 39 species of endemic land birds, including 7 species of geese, 2 species of flightless ibises, a sea eagle, a small hawk, 7 flightless rails, 3 species of owls, 2 large crows, a honeyeater, and at least 15 finches (Figure 51.1). No humans were numbered among the organisms inhabiting New Zealand until about 1,000 years ago, when the Polynesian ancestors of the Maori colonized the island. Hunting by Maori caused the extinction of many species of large flightless birds, including all of the many species of moas, some of which were larger than ostriches. Birds were easily hunted to extinction on many islands because they had repeatedly evolved flightless forms on predator-free islands.

On the continents, large mammals have been the greatest sufferers from human activities. When humans arrived in North America over the Bering Land Bridge, about 20,000 years ago, they encountered a rich fauna of large mammals. Most of those species were exterminated within a few thousand years. A similar extermination of large animals followed the human colonization of Australia, about 30,000 years ago. At that time Australia had 13 genera of marsupials larger than 50 kg, a genus of gigantic lizards, and a genus of heavy flightless birds. All the species in 13 of those 15 genera had become extinct by 18,000 years ago.

Other North American species did not survive the European immigration of the nineteenth century. The passenger pigeon, the most numerous bird in North America in the early 1800s, became extinct by 1914, largely due to overhunting. Russian whalers exterminated the unusual Steller's sea cow of the North Pacific in the late 1800s. The American bison was on

51.1 Extinct Flightless Hawaiian Birds
The goose (top), flightless ibis (lower left), and flightless rail (lower right) pictured here were among the many species of Hawaiian birds exterminated by the Polynesian settlers of the islands.

the brink of extinction at the beginning of this century and might well be extinct today if its hunting had not been outlawed. Loss of species through overhunting continues today. Elephants and rhinoceroses are threatened in Africa because poachers kill them for their valuable tusks and horns. At present rates of killing, these two species will disappear from most of their already reduced ranges in 20 years.

Introduced Pests, Predators, and Competitors

Deliberately and accidentally, people move many species of organisms from one continent to another. Pheasants and partridges were introduced into North America for hunting. European settlers took their crops and domesticated animals to Australia. They introduced other species, such as rabbits and foxes, for sport. Weed seeds were carried around the world in the soil used as ballast in sailing ships and as contaminants in sacks of crop seeds. Despite quarantines, disease organisms spread rapidly, carried by infected plants, animals, and people.

A species that has evolved over time in a community with certain predators and competitors may be vulnerable to a newly introduced predator or competitor. Introduced species have caused the extinctions of thousands of native species worldwide. Nearly half of the small marsupials and rodents of Australia have been wiped out during the last 100 years by a combination of competition with introduced rabbits for forage and predation by introduced cats and foxes. Black rats carried to remote oceanic islands on ships are especially destructive predators. Native rice rats survive in the Galapagos archipelago only on islands not invaded by black rats. On some Galapagos islands, introduced pigs and rats regularly excavate all nests of the giant Galapagos tortoises and devour the eggs. Populations of some tortoises are maintained today only by conservationists who remove eggs and rear the young tortoises in captivity until they are large enough to defend themselves against pigs and rats.

Outbreaks of pests in new environments have quickly followed the intercontinental introductions of the pests by humans. Forest trees in eastern North America, for example, have been attacked by several European diseases. The chestnut blight, caused by a fungus originally from Europe, virtually eliminated the American chestnut, once a dominant tree in forests of the Appalachian Mountains. Some individuals still resprout, but sprouts are soon found by the blight and killed. Nearly all American elms over large areas of the East and Midwest have been killed by Dutch elm disease, caused by the fungus *Ceratocystis ulmi*. The disease is thought to have originated in Asia, and it was first recorded in western Europe about 1920; the first infestation in North America was reported in 1930. Ecologists suspect that intercontinental movement of disease organisms caused extinctions in the past, but evidence of disease outbreaks is not usually preserved in the fossil record.

Loss of Mutualists

Many plants have mutualistic relationships with pollinators, but usually the mutualisms are not highly species-specific (see Chapter 48). On islands, however, where ecological communities have relatively few species, plant–pollinator interactions are often highly specific. *Lobelia* is a genus of showy-flowered plants that has hundreds of species. A single lobelia species colonized the Hawaiian Islands and eventually gave rise there to 110 of the world's 350 species.

51.2 Coevolved Mutualists
As the iiwi, a Hawaiian honeycreeper, inserts its bill into the corolla tube to extract nectar from the lobelia flower, it deposits pollen from flowers it visited previously. Declining populations of the honeycreeper also threaten the plant, which is left without a pollinator.

Also in the Hawaiian Islands a single colonizing species of songbird gave rise to at least 47 species of honeycreepers, some of which have long, slender, curved bills. These nectar-feeding birds were the only pollinators of many species of lobelias, which evolved long, curved tubular flowers that match the shapes of their pollinators' bills (Figure 51.2).

Today, half of the nectar-feeding birds of Hawaii are extinct, leaving many lobelias without pollinators. Because some perennial plants live much longer than birds, many lobelias still survive, but populations of some species have been reduced to only a few individuals. A few species reproduce now only with human assistance: Biologists transfer pollen among individuals to pollinate them.

Habitat Destruction

The 5.5 billion humans that exist on Earth today are fed, clothed, and housed by agricultural and forestry industries that convert natural ecological communities containing many species into highly modified communities dominated by one or a few species of plants. Within these communities, humans discourage the presence of other species by applying chemicals that kill competing plants, bacteria, fungi, nematodes, insects and other arthropods, and vertebrates. Although agricultural ecosystems have reduced species richness for thousands of years, early agriculture cultivated many economically valuable species in mixed plantings. Instead of planting enormous tracts of land in single crops, traditional farmers planted different crops together, maintaining some of the diversity that is key to the success of natural communities. The traditional systems also

BOX 51.A

Forest Analogs

As long as people continue to use the land, a forest that has been logged cannot be restored to its original state. It is feasible, however, to create "forest analogs" that retain ecological functions such as nutrient recycling, water cycling, and erosion control that were performed by the original forest, while commercially valuable species are being harvested in a sustainable manner. These analogs can have enough variety of plant life forms—shrubs, climbing vines, canopy trees—to support a rich com-

Strip-cutting, Colombia

Cacao plantation, Costa Rica

munity of animals. The small cacao farms of Latin America are a good example. On these farms, a diverse understory of maize, bananas, and other crops grows under the cacao trees, which are the source of chocolate. The cacao trees protect the soil from sunlight and torrential rains, and there is enough vegetational structure to support many more species of animals than would be found in a monoculture—the cultivation of a single crop species.

Careful planning of harvests can also greatly reduce the damage logging does to a forest. In tropical forests a typical logging operation extracts only about 10 percent of the trees—those with high market value—but destroys about 50 percent of the canopy in the process. Such

simple changes as identifying the valuable trees in advance, designing the least-obstructed paths for extracting the logs, and cutting canopy vines before felling a tree can reduce damage to neighboring plants by 20 to 40 percent. These preliminary steps also reduce logging costs. Another promising approach that reduces damage to other plants is strip-cutting, in which loggers cut 30- to 40-meter-long strips in the forest and remove the commercially valuable trees. Local people then use the less-valuable trees for making poles, firewood, or charcoal. The swaths that are cut are narrow enough that the forest will regenerate rapidly from seeds carried to the strip from the surrounding forest by wind or by animals.

supported many other species incidentally (Box 51.A). When ecosystems, such as agricultural lands and plantation forests, are managed so as to divert most of their primary production to certain species intended for human use, we say that their production is **co-opted.** Agriculture and forestry today are so extensive that more than 30 percent of all terrestrial production is co-opted for human use (Table 51.1). All the other species on Earth have only two-thirds of the total global terrestrial production available for

their use, and the fraction is steadily decreasing.

Because of increasing habitat modification and the co-option of primary production, habitat loss is certain to be the most important cause of species extinction during the next century. The habitats required by some species are being completely destroyed. Other habitats, particularly old-growth forests, natural grasslands, and estuaries, are being reduced to small, widely separated patches that may be thought of as habitat islands.

TABLE 51.1.
Co-option of Net Primary Production by Human Manipulations of Ecosystems

HABITAT CATEGORY	NET PRIMARY PRODUCTION CO-OPTED[a]
Cultivated land	15.0
Grazing land	11.6
Forest land	13.6
Human-occupied areas	0.4
Total	40.6
Total net primary production	132.1
Percent co-opted	30.7

[a] Values are in petagrams (one petagram = 10^{15} grams).

Populations of native predators and parasites may increase as a result of habitat modifications, thereby endangering species that they did not previously threaten. As we mentioned in Chapter 46, brown-headed cowbirds, which are native to North America, lay their eggs in other birds' nests; a female cowbird usually removes one host egg when she lays her own. The hosts incubate the cowbird eggs and feed the cowbird nestlings. In the nest of many host species that are smaller than cowbirds, only the cowbird nestling survives. Cowbird populations have increased dramatically in many parts of North America as a result of land clearing and associated high densities of domestic livestock. Losses to nest parasitism are added to heavy predation by crows, jays, native

mammals, and domestic cats and dogs, especially in suburban areas. Many species of songbirds are declining in eastern North America because they cannot produce enough offspring to replace adult losses (Figure 51.3).

STUDIES OF INDIVIDUAL SPECIES

Studies undertaken by conservation biologists range from those focused on preserving a single species to those with the goal of preserving entire communities and ecosystems. Often single species of special economic, aesthetic, or ecological value serve as surrogates for entire communities. The protection of species that require large areas of suitable habitat, such as elephants and eagles, may ensure the survival of most or all other species living in that ecosystem.

The Probability of Survival

Local populations can be reduced to very small numbers by the loss of habitat, because the remaining patches can accommodate only a few individuals,

51.3 Declining Songbird Populations in the Eastern U.S.
Graphs show the results of long-term observations carried out by researchers and volunteers in specific forest patches in the Middle Atlantic states. High predation rates on songbird nests are making it difficult for some populations to replace themselves. Migrant birds (as opposed to birds who are resident in the forest all year) are hardest hit, because many of them are faced with the destruction of the tropical and subtropical habitats in which they normally spend the winters.

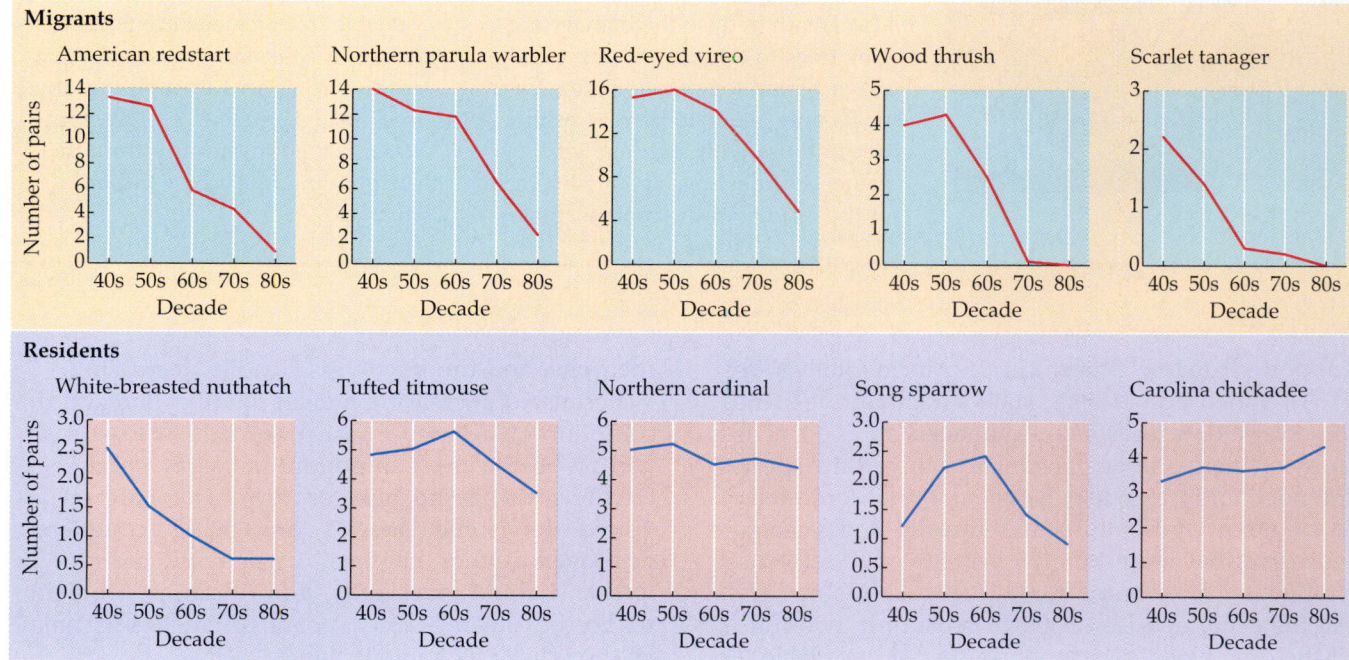

and by the deterioration of habitat quality, because the remaining habitat cannot support as high a density of individuals. Several factors contribute to the increased probability that a population will become extinct when its numbers are reduced to low levels. Populations with only a few individuals are highly susceptible to harmful effects of local disturbances such as fires, unusual weather, disease, and predators.

Estimating the risks faced by small populations is an important component of preservation analyses. The development of the concept of a **minimum viable population** (MVP), now widely used to estimate a population's risk of extinction, was stimulated by the National Forest Management Act of 1976, which requires the U.S. Forest Service to maintain viable populations of all native vertebrate species in each national forest. A minimum viable population is the estimated density or number of individuals necessary for the species to maintain or increase its numbers in a region. There is no sharp threshold above which populations are viable and below which they are not, but an MVP analysis can estimate a population's risk of extinction over decades and centuries, time frames that are appropriate for management plans.

A **population vulnerability analysis** (PVA) is carried out to estimate how the size of a population influences its risk of becoming extinct in a specified time period—for example, 100 years. A PVA is based on knowledge of the interactions between the genetic variation, morphology, physiology, and behavior of a population, and its environment, both physical and biological. One component of a PVA is estimation of the extent and significance of **demographic stochasticity,** that is, the amount of random variation in birth and death rates. In a small population, extinction is likely when high death rates coincide with low birth rates. Estimates of the sizes of local populations at high risk of immediate extinction due to demographic stochasticity range from 10 individuals for microorganisms reproducing by fission to about 50 for sexually reproducing animals with lengthy prereproductive periods. Larger populations also may be at high risk because the same environmental conditions that cause low birth rates are likely to cause high death rates.

Another component of a PVA is analysis of **genetic stochasticity,** fluctuations in a population's level of genetic variation. In small populations inbreeding is increased, which puts the fitness of inbred offspring at risk. The threat of decreased fitness due to inbreeding may influence management strategies. A genetic survey of more than 200 structural loci among cheetahs of southern Africa revealed almost no genetic variation (Figure 51.4). In fact, all the cheetahs in the southern African population are so similar that, unlike most vertebrates, they can accept skin grafts from

51.4 Cheetahs of Southern Africa
Mother and offspring look alike and are almost identical genetically.

one another. This tissue compatibility indicates that there may be no genetic variation at the major histocompatibility complex (MHC), a normally highly polymorphic genetic region associated with disease resistance as well as with graft rejection (see Chapter 16). For these reasons, conservation biologists are especially concerned about the vulnerability of wild cheetah populations and are planning to crossbreed cheetahs from eastern Africa with southern African cheetahs to increase the genetic variation in the southern populations.

Knowledge of population genetics is important for planning the reintroduction of individuals from captivity or the translocation of individuals from other geographical regions to reestablish extinct populations. For example, when goats called ibexes (*Capra ibex ibex*) were wiped out in Czechoslovakia by overhunting, the species was successfully reestablished using ibexes of the same subspecies from nearby Austria. Some years later, ibexes from Turkey (*Capra ibex aegagrus*) and from the Sinai (*Capra ibex nubiana*) were also introduced to the herd. Animals from these populations can interbreed, but resulting hybrid females came into heat in early fall instead of winter and gave birth in mid-winter, when the offspring could not survive. The resulting reproductive failures were so severe that the entire population died out.

An attempt to avoid this type of genetically caused disaster is being made in the Ozark Mountains of southern Missouri. Prairie plant communities from

(a)   (b)

51.5   Both the Glades and the Lizards are Endangered
(a) Open glades exist only as small patches on south-
facing slopes in the Ozark Mountains. (b) The collared
lizards of the Ozarks, which live only in these open
glades, are genetically different from those elsewhere in
the range of the species.

the southwestern United States became established
in these mountains about 8,000 years ago during an
unusually hot, dry period. When the climate became
cooler and moister again about 4,000 years ago, the
prairie vegetation survived only in isolated, fire-
prone, open glades on south-facing slopes with shallow soils. Animals, such as the collared lizard, that
depend on these plant communities were reduced to
local, isolated populations, and their habitat has been
further reduced by agriculture and fire prevention
(Figure 51.5).

Collared lizards had been wiped out from all the
glades in the area where the Missouri Conservation
Commission is performing reintroduction experiments, so individuals had to be imported from other
places. The Ozark populations, which have been isolated for about 2,000 lizard generations, are genetically distinct from those elsewhere in the range of
the species. Therefore, a decision was made to use
lizards from other Missouri glades. Genetic analyses
revealed that the lizards from a single glade are all
genetically identical, the result of 2,000 generations
of genetic drift in populations whose sizes are nearly
always below 50 individuals. Founding lizards could
have been taken from a single glade, but a minimum
of ten mature lizards was considered necessary to
achieve a high probability of success, and donor populations would have been threatened by removing
such a large fraction of their members at one time.
Therefore, lizards from at least five different glades
were released together. Donor populations were selected to have distinct, maternally inherited mitochondrial DNA markers so that, by sampling off-

spring over the years, investigators will be able to
measure reproductive success of the released lizards
and their descendants.

Captive Propagation

Species being threatened by overexploitation, loss of
habitat, or environmental degradation through pollution can be preserved in captivity while the external
threats to their existence are reduced or removed.
Success at this venture requires research on nutrition
and the preparation of suitable diets, on the use of
vaccinations and antibiotics, and on the control and
enhancement of reproduction by both behavioral and
technical means (artificial insemination, embryo
transfers).

Captive propagation is only a temporary measure
that buys time. Existing zoos and botanical gardens
simply do not have enough space to maintain adequate populations of more than a small fraction of
rare and endangered species. In addition, a species
maintained in captivity can no longer evolve together
with the other species in its ecological community.
Nonetheless, captive propagation plays an important
role in maintaining species during critical periods and
in providing a source of individuals to be reintroduced into the wild. Captive propagation projects in
zoos have been very effective in raising public awareness of the biodiversity crisis.

A successful example of captive propagation is the
reintroduction of the peregrine falcon (Figure 51.6)
to parts of its range from which it had been wiped
out, primarily because of the widespread use of or-

51.6 Release to the Wild
A young peregrine falcon is released to forage on its own and, eventually, to breed.

ganochlorine pesticides, such as DDT and dieldrin. These persistent pesticides gradually accumulate in animals at the top of the food chain such as the falcon. High concentrations of the pesticide interfered with the deposition of calcium in eggshells, causing the falcons' eggs to break easily and resulting in a high rate of reproductive failure. Much of the habitat from which peregrines had been eliminated became suitable for these birds again when some nations greatly restricted the use of organochlorine pesticides.

In 1942, about 350 pairs of peregrines bred in the United States east of the Mississippi River. This breeding population disappeared entirely by 1960, and no peregrines are known to have reproduced in this region during the next 20 years. Captive breeding of peregrines began at Cornell University in 1970, and by the end of 1986 more than 850 peregrines reared in captivity had been released in 13 eastern states, with spectacular success (Figure 51.7).

Captive propagation is also playing a vital role in attempts to save the California condor from extinction. With its 9.5-foot wing span the California condor is North America's largest bird. Two hundred years ago, condors ranged from southern British Columbia to northern Mexico, but by the 1940s they were confined to a small region in the mountains and foothills north of Los Angeles. By 1978, only 25 to 30 birds remained, and the wild population was plunging toward extinction. Six of the remaining 15 wild birds disappeared in 1985 alone!

In the hope of saving the condor from extinction, a captive propagation program was initiated in 1983. To maximize genetic variation in the captive population, all wild birds were captured, the last one in April 1987. The first chick that was conceived in captivity hatched in 1988, and by 1993, nine captive pairs were producing chicks and the captive population

had increased to more than 60 birds. The captive population was large enough to release six captive-bred birds in the mountains north of Los Angeles in 1992 (Figure 51.8). These birds are being provided with contaminant-free food in remote areas, and three of them were still alive in the spring of 1994. The released birds are using the same roosting sites, bathing pools, and mountain ridges as their predecessors did. Introductions of captive-reared birds into Montana and New Mexico are planned for 1994. It is still too early to pronounce the program a success, but without captive propagation, the California condor would probably be extinct today.

Captive propagation efforts are expensive. The condor rehabilitation program costs about 1 million dollars a year. The Peregrine Fund at Cornell has

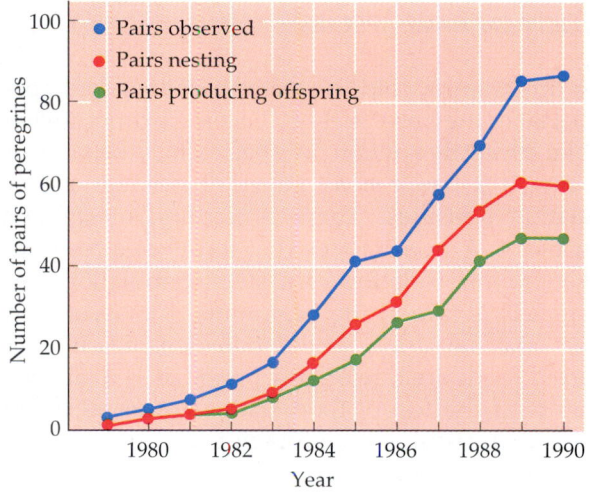

51.7 Peregrine Falcon Populations Have Been Reestablished
Throughout the eastern United States, many pairs now attempt to reproduce, most of them successfully.

51.8 Back in the Wild
Captive propagation of the California condor is providing the individuals that are being reintroduced into the species' former range.

spent nearly 3 million dollars during the past 25 years, and the expenses of other cooperating agencies add at least another half million to the total. These amounts seem small, however, when compared with costs of other human activities. It is estimated that the work needed to restore all of the world's threatened birds of prey could be accomplished with 5 million dollars per year, the approximate cost of one armored tank.

BIOLOGY OF RARE SPECIES

Much attention is paid to rare and endangered species because they are likely to become extinct unless we take protective action. A species may be considered rare if it has a small geographic range or low population densities. There is nothing mysterious about rarity. A new species that forms as a result of the geographic isolation of a small population (see Chapter 20) is rare at the beginning of its existence. Species may have low population densities if their preferred foods are rare or only periodically available. Predators at the top of food chains are usually rare because they require large foraging areas.

Interestingly, rarity is an inevitable consequence of high species richness. Whereas a typical tropical forest contains several hundred species of trees, a temperate forest may have only a few dozen species. The number of individual trees per hectare in tropical forests, however, is about the same as that in temperate forests. Therefore, in tropical forests the same number of individuals is divided among a much larger number of species, so many of the species must be rare. Similar arguments apply for most taxa of organisms. Thus the regions of the world with the greatest species richness inevitably have the largest number of rare species. For the same reason, species-rich genera tend to have more rare species than do species-poor genera.

Do rare species differ from common species other than having small ranges and low population densities? We might expect so because natural selection may favor different traits in a plant with a restricted range or a low population density. For example, if most of a rare plant's neighbors belong to other species, as is likely, its success in interspecific competition will probably affect its survival more than its success in intraspecific competition will. Herbivores and pollinators are less likely to evolve specialized relationships with a rare plant than with a common one. Therefore, rare species are more likely to defend themselves against generalized herbivores and to have adaptations for placing pollen precisely on the bodies of generalized pollinators (Table 51.2). Orchids have extreme adaptations for precision of pollen placement: Their unusual flower shapes restrict pollinator access to a single route, ensuring that the pollinator comes in contact with the flower's pollen. The pollen of orchids is packaged in compact sacs (called pollinia) that are placed precisely on the bodies of the pollinators so that the pollinia are less likely to contact the stigmas or other surfaces when the pollinators visit flowers of other species (Figure 51.9). It is not unusual for individuals of some species of orchid bees, the major pollinators of many tropical

51.9 Orchid Bee Carries a Pollinium
The pollen of this Ecuadoran orchid is "packaged" into a neat little sac—the yellow structure—that an orchid bee (*Eulaema*) now carries on its back.

TABLE 51.2.
Probable Results of Natural Selection Acting on Rare and Common Plants

| TRAIT | TRAITS OF PLANT POPULATION | | |
	SMALL RANGE, LOCALLY DENSE	LARGE RANGE, LOCALLY RARE	LARGE RANGE, LOCALLY COMMON
Competitive abilities	Intraspecific > interspecific	Interspecific > intraspecific	Intraspecific > interspecific
Target of chemical defenses	Generalist herbivores	Generalist herbivores	Specialists and generalists
Flower longevity	Short	Long	Short
Reward to pollinators per flower	Moderate	Large or none	Moderate
Number and placement of stamens	Many, scattered	Few, precisely placed	Number and location variable

orchids, to carry pollinia of as many as half a dozen species of orchids.

Differences between rare and common species have important implications for conservation practices. Species that have been rare for a long time may evolve traits that enable them to reproduce even though their members are widely separated. Species that have been common for a long time, however, may become vulnerable to extinction if their population densities or ranges are suddenly reduced. Therefore, newly rare species may require special management if they are to persist at densities much lower than those under which they evolved.

CONSERVATION AND CLIMATE CHANGE

As a result of global warming, average temperatures in North America will probably increase 2 to 5°C by the end of the twenty-first century. Conservation biologists are attempting to predict the effects of this warming trend on North American deciduous forest trees. Each rise of 1°C translates into a range shift northward of about 150 km; that is, as the climate warms by 1°C, the average temperature formerly found at a certain location will instead be found 150 km north of that location. An organism that survives best at that average temperature will need to shift its range 150 km north to remain in the same climate. Therefore, trees might need to shift their ranges as much as 500 to 800 km in a single century.

Most deciduous forest trees grow for long periods before they begin to reproduce, and their seeds move only very short distances. The American beech appears to be especially ill-equipped for shifting its range. If temperatures were to increase, adult beeches would begin to produce fewer seeds. Within a few decades, seedlings in the forest understory would begin dying. Adult trees would survive much longer, giving the appearance of a healthy population long after reproduction had ceased. The beech,

whose seeds are dispersed primarily by jays, advanced at the frontiers of its range only about 20 km per century during past climatic warming. Beeches would thus have to migrate 40 times faster than they did in the past to keep up with the anticipated rate of climate change. Even though there might be areas of suitable climate, beeches probably could not reach them without human assistance (Figure 51.10). To maintain beech forests we may need to intervene by moving seeds and by assisting seedling establishment.

Forest models project difficulties for the Kirtland's warbler as well, an endangered species that nests only in young stands of jack pine on sandy soils in Michigan (Figure 51.11). The current population of Kirtland's warblers is less than 1,000 individuals. The stands of jack pines depend on periodic forest fires for their persistence because the cones of jack pines remain closed on the branches until they are heated by a hot fire. Only when heated do they open and release their seeds, which germinate in the ash on the floor of the burned forest. Kirtland's warblers nest only in jack pine forests that are 8 to 18 years old. To ensure that the birds will have new stands of jack pine for nesting habitat in the years immediately ahead, conservation biologists are managing controlled burns in jack pine forests. They are also removing brown-headed cowbirds, which are heavily parasitizing the warblers.

In the current climate jack pines grow rapidly between fires, but growth rates would decrease if the climate warmed. Central Michigan is at the southern boundary of the range of jack pines; at this boundary growth rates are currently depressed during warm summers. During a period of steady climatic warming, jack pines would be replaced by white pines, red pines, and sugar maples, trees able to tolerate warmer climates. Northward movement of the jack pine forests would be difficult because a broad intervening region lacks the coarse, sandy soils that both jack pines and Kirtland's warblers need. The loss of

(a)

51.10 Threatened by Climate Warming
(a) Seedlings and saplings abound in this healthy beech forest. (b) If the climate of eastern North America warms, about half the potential future range of beech trees will be beyond the northernmost extent of the current range.

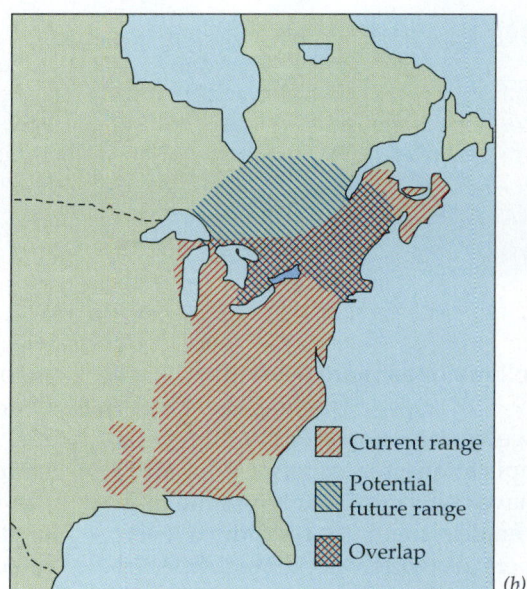

Current range

Potential future range

Overlap

(b)

its required habitat could easily wipe out the Kirtland's warbler.

In the past, global climate changed at a much slower rate than that predicted for the coming century. Most organisms were able to shift their ranges rapidly enough to keep up with climate changes. In addition, because habitats were more continuous, migration routes were not blocked by extensive areas of unsuitable habitats, as they are today. Organisms may therefore have much more difficulty dealing

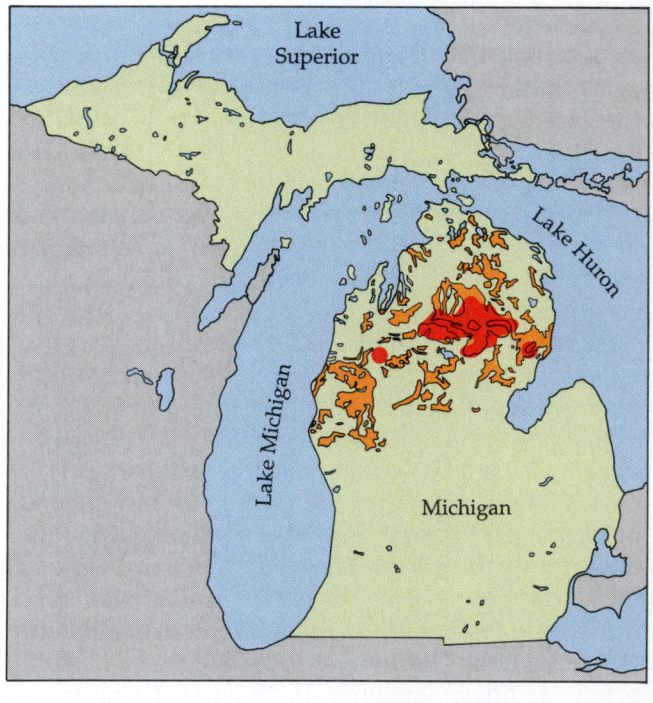

51.11   A Male Kirtland's Warbler in a Young Jack Pine
The map to the right shows the warbler's breeding range.

The breeding range of Kirtland's warbler

Distribution of sandy soils

with climate changes during the next century than they did during glacial periods.

If the globe warms, climatic zones will not simply shift northward; new climates will develop and some existing climates will disappear. New climates are certain to develop in tropical lowlands. All models predict that the climate will warm less in tropical regions than at high latitudes, but a warming of even 2°C would result in lowland climates hotter than those found anywhere in the humid tropics today. Adaptation to those climates may prove difficult for many tropical organisms.

COMMUNITY-LEVEL CONSERVATION

Because many species can survive only in the ecological communities in which they evolved, conservation biologists are as much concerned about preserving complete ecological communities as they are about maintaining individual species. Although the types of research conducted for these two purposes overlap, community-level conservation research has its own distinct themes.

Endemism

Because tropical ecosystems are generally richer in species than are ecosystems at higher latitudes (see Chapter 50), loss of tropical habitats threatens more species than does loss of comparable areas of temperate habitats. The number of species that become extinct as a result of habitat destruction also depends on how many local species are **endemic**—that is, are found nowhere else. For example, nearly all the mammals and birds of Madagascar are found only on that island. Therefore, if the small fragments of tropical forests remaining on Madagascar are destroyed, the species dependent on them are certain to be wiped out in the wild (Figure 51.12).

(a)

(b)

(c)

51.12 Madagascar Abounds with Endemic Species Among the many species found only on the island of Madagascar are: *(a)* Certain *Euphorbia* (with leaves) and *Alluaudia* (leafless) species of the dry forest; *(b)* an entire primate group, the lemurs, exemplified here by the black lemur, and *(c)* the Madagascar day gecko *Phelsuma madagascariensis*, a reptile.

Endemism is especially marked on islands, but some mainland regions also have high degrees of endemism. For example, the Rift Valley lakes of Africa harbor more than 1,000 species of fishes, most of which live in only one lake. The Atlantic coastal forests of southeastern Brazil are another center of endemism. Because only about 1 percent of the original extent of those forests remains, many species there have become extinct or are in danger of immediate extinction. Mountainous regions have many endemic species because temperature and rainfall change rapidly with elevation, creating many distinct habitats within a small area.

Centers of endemism are not the same for all groups of organisms. The Cape region at the southern tip of Africa, for example, has a flora of 8,500 species, 80 percent of which are endemic, but only 4 of the 187 species of birds found there are endemic. The reason is that the Cape region, an area of only 90,000 square kilometers, is too small for geographic speciation among birds, but plants readily speciate sympatrically in small areas by means of polyploidy (see Chapter 20).

Keystone Species

Keystone species exert strong influences on the structure and functioning of the ecological communities in which they live. The sea star *Pisaster ochraceous*, discussed in Chapter 48, is a keystone species because it suppresses populations of mussels, the dominant competitors in the rocky intertidal zone. Certain mutualistic relationships may be central to a community, and these relationships may ultimately depend on keystone species. For example, in Peruvian forests, only a dozen species of figs and palms support an entire community of large frugivorous (fruit-eating) birds and mammals during the part of the year when fruits are least available. Loss of those few tree species would probably eliminate most of the frugivores, even if hundreds of other tree species remained. In turn, loss of the frugivores might seri-

ously impair the dispersal of the seeds of many other species of trees. Thus many mutualistic relationships are probably maintained by a few keystone tree species that constitute only a small fraction of the 2,000 species of trees in the forest.

Other keystone species include herbivores that alter habitat structure and change vegetational succession (for example, termites and large mammals, such as moose and elephants), species that maintain particular landscape features (such as beavers), and parasites and pathogenic microorganisms. Because the extinction of keystone species could result in the extinction of many species in their communities, conservation biologists need to identify keystone species quickly and take action to preserve them.

Habitat Fragmentation

Human population growth and the co-option of a large fraction of global terrestrial production has resulted in extreme fragmentation of natural habitats. Small habitat patches differ from larger patches in ways that affect the survival of species. Small habitat patches are influenced by **size effects,** that is, they are qualitatively different from larger patches of the same habitat. Small patches cannot support populations of species that require large areas, and they can harbor only small populations of many of the species that can survive there. In addition, the fraction of a patch that is influenced by **edge effects**—phenomena that occur where one type of habitat meets another—increases rapidly as patch size decreases (Figure 51.13). What are some examples of edge effects? Close to the edges of forest patches winds are stronger, temperatures are higher, humidities are lower, and light levels are higher than they are farther inside the forest. Species from surrounding habitats can invade the edges of patches to compete with or prey upon the species living there.

Patch size and the size of animal home ranges interact to influence the likelihood that animal species can persist in a habitat fragment of a certain size. At one extreme, if an animal's home range—the area it must use to find enough food—is larger than the size of an available patch of suitable habitat, the animal either disappears from that patch or greatly expands its home range to include more than one patch. In such a situation, animals must repeatedly cross areas

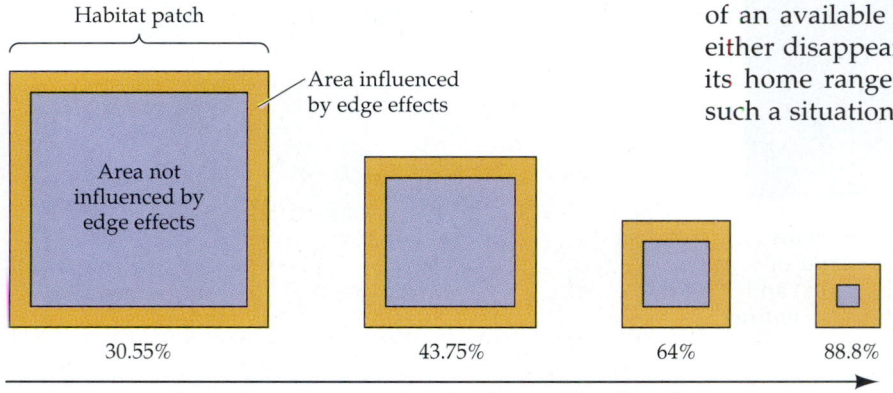

Habitat patch

Area influenced by edge effects

Area not influenced by edge effects

30.55%   43.75%   64%   88.8%

Increasing percentage of patch influenced by edge effects

51.13 Edge Effects
These diagrams show the proportions of square areas influenced by edge effects if such effects penetrate equal distances in patches of all sizes. The smaller the patch, the greater the edge effects.

22222222222

(a)　　(b)

51.14 Old-Growth Forests and Owls
(a) Old-growth coniferous forests of the Pacific Northwest have trees of all ages and there are many large logs on the forest floor. (b) The northern spotted owl depends on these forests for successful reproduction.

of unsuitable habitat, where they may be considered pests or where they may be at great risk of predation or other dangers. For example, in Florida the current number of black bears killed by automobiles while they are moving between small patches of suitable habitat is higher than the number of bears born, so the population is declining.

In the northwestern United States more than 70 percent of the old-growth coniferous forests of the region have been cut. Because current forestry practices are based on clear-cutting stands every 50 to 80 years, second-growth forests do not become old enough to acquire key characteristics of old growth, such as having trees of all ages (some of these conifers live over 500 years), having many dead trees and snags, and having large logs on the forest floor (Figure 51.14a). Among the species that require old growth to maintain viable populations are salamanders, which live inside rotting logs, the only microhabitat that retains moisture during the dry summers. Another is the spotted owl, a species that hunts for rodents that live in mature forests (Figure 51.14b). As the fraction of the area remaining in old growth is reduced, the home range of the owls increase; hunting success, and hence reproductive success, decreases, and juvenile mortality during dispersal increases. Thus the proportion of remaining suitable habitat occupied by spotted owls decreases. The species faces a high risk of being wiped out in most of Washington and Oregon within the next 50 years if clear-cutting of old-growth forests continues.

The endangerment of species dependent on old growth is the result of logging the coniferous rainforests of the Pacific Northwest at unsustainable rates for the past 50 years. If these species and the forest industry are to survive, major changes in forestry practices will be required. Some of these changes are now being initiated.

Information on the distribution and abundance of organisms in a place before its habitats become fragmented is usually not available. Near Manaus, Brazil, however, a major research project was launched before logging took place. Land owners agreed to preserve forest patches of certain sizes and locations, and censuses of those patches were conducted before logging began, while the areas were still part of continuous forest. Within a few years of clearing the surrounding areas, species were already disappearing from the isolated patches (Figure 51.15). The first species to be eliminated were monkeys with large home ranges, such as the black spider monkey, the tufted capuchin, and the bearded saki. Birds that follow swarms of army ants to capture insects flushed by the ants also disappeared quickly from small patches (Figure 51.16). A particular colony of army ants is a useful resource for the birds only when it is raiding, about 27 days of the 35-day period between colony moves. Therefore, the birds must have access to a number of colonies to be guaranteed that always at least one is in the raiding stage. Smaller patches have so few ant colonies that there are periods when none are raiding.

51.15 Brazilian Forest Fragments Studied for Species Loss
Isolated patches lose species much more quickly than patches connected to the main forest do, even if the isolated patches, such as the one in the foreground, are larger than the connected ones.

HABITAT AND ECOSYSTEM MANAGEMENT

An important component of conservation is the establishment of parks, sanctuaries, and reserves whose primary function is to maintain species and ecosystems relatively free of human disturbance. The National Park system of the United States plays a vital role, not only in species preservation, but also in providing opportunities for people to enjoy natural environments and develop an appreciation for them. More parks are being added to systems in many countries, but it is unlikely that their size and number will ever be equal to the task of ecosystem and species preservation. There will never be enough parks, and

51.16 Extinction in Patches
The white-plumed antbird, a species common in Brazilian forests, has become extinct in isolated forest plots.

they are, and will be, too small to maintain all species or to permit evolutionary adaptations to continue within their boundaries. Moreover, most parks have been established around areas of monumental geological features, not areas of high species richness. National parks were not designed to function as conservation units, and only recently have they begun to assume an important conservation role.

The United States model of national parks must be modified before it is exported to most other countries. Parks in the United States were established primarily in areas where Europeans had not yet settled and where the indigenous people had already been exterminated. In most countries, parks must be established in already settled areas. The people living there cannot be evicted, nor is it possible, in most cases, to prevent hungry people from settling in or hunting in the parks. The high rates of human population growth in most tropical countries guarantee that pressures on parks from agricultural settlers will increase rather than decrease.

For these reasons, lands exploited for food, medicines, and fiber must play an important role in management for species preservation. These lands are far more extensive than parks and reserves, and they include climates and ecosystems not represented in the parks. Fortunately, many patterns of economic exploitation of lands are compatible with the preservation of most species of organisms. Only a few species, such as predators on people and on domestic animals, or large, destructive herbivores, are incompatible with human uses of land.

Megareserves

A key development in conservation practice is the concept of **megareserves.** A megareserve is a large area of land that includes, at its core, an undisturbed natural area. Surrounding that core are buffer areas in which economic activity that does not destroy the ecosystem is permitted. Such activity may include sustainable harvesting of animal populations and plant products, such as rubber, fruits, nuts, and wood. On the edges of the megareserve is a zone in which more intensive land use, such as agriculture or plantation forestry, is permitted. Costa Rica has pioneered the development of megareserves. It has consolidated its parks and reserves into eight megareserves that should maintain about 80 percent of the country's biodiversity (Figure 51.17). Each megareserve includes natural areas and areas managed for economically valuable products. Some of the reserves remain the homes of indigenous people who will continue to use the environment in their traditional ways.

The largest of the Costa Rican megareserves is La Amistad Biosphere Reserve, an area of more than

51.17 The Megareserves of Costa Rica
The eight areas outlined in red are being managed for
both biodiversity and economic activities. The green areas
are the current reserves.

51.18 The Reason for a Reserve
Creating a sanctuary for the rare black howler monkey
was the impetus for turning an area of Belize into a reserve.

500,000 hectares that includes three national parks, a
biological reserve, five Indian reservations, and two
forest reserves. Altitudes within the reserve range
from 100 to 3,819 meters. It contains the largest tract
of highland vegetation in Central America, and it has
considerable hydroelectric generating potential. All
the native large predators still survive there. However, the reserve faces threats from surrounding agriculturists, overhunting, and logging.

A strategy for the conservation and development
of La Amistad has been completed and is being implemented. If managed properly, the reserve can provide drinking water, electricity, forest products, and
nature tourism, protect indigenous cultures, and preserve species. In combination, these benefits out-
weigh what could be gained by logging the steep
slopes and converting them to low-productivity agricultural systems.

An unusual reserve, created with the help of local
farmers and landowners, is the Community Baboon
Sanctuary in Belize. The original purpose of the reserve was to protect the black howler monkey,
known locally as "baboon," a species with a restricted
range in Mexico, Belize, and Guatemala (Figure
51.18). The sanctuary was planned in consultation
with people living in the area, and private landowners have agreed to use their land in accordance with
standards established by the reserve. The formal plan
grew out of extensive discussions among local people. Talks were given at local schools, and a booklet,
Baboons of Belize, was distributed. Publicity for the
program locally, nationally, and internationally increased local pride and helped generate tourism, an
important goal. The sanctuary, which has been

slowly expanding, now occupies 47 square kilometers
of land, including seven villages along 32 km of riv-
erine forest and an 0.8-km strip on either side of the
river. In the future the reserve will link up with two
other sanctuaries to form a megareserve. A reserve
manager has been hired and a system of administra-
tion has been established.

Forest Reserves

Forest reserves in which economically valuable products are harvested can combine species preservation
and economic development. Such a plan was proposed at the first annual meeting of rubber tappers
in Brasilia, Brazil, in 1985. The plan is being implemented in Acre, the Brazilian state where the rubber
tappers are strongest and best organized. In Acre,
the value of wild rubber, Brazil nuts, and several
other forest products harvested in 1980 was estimated
at over 26 million dollars. The estimated current value
of the sustainable harvest of such products is less
than the short-term value of products of cattle ranching and agriculture, but grazing and agriculture are
not sustainable in those regions, whereas forests can
yield harvested products indefinitely.

Whether such forest reserves can survive politically is uncertain. Officially recognized title to the
land is difficult to obtain and maintain, and the tappers are caught between their debts to middlemen
and increasing threats to their livelihood from forest
clearing for timber production and ranching. The lat-

ter activities are currently highly subsidized by the Brazilian government to the benefit of wealthy absentee landlords. Political organization of the tappers and other extractive users, cooperation from the Brazilian federal government, and foreign assistance are vital ingredients for the success of forest reserves. If successful, Amazonian forest reserves may serve as models for other regions. Forest reserves could play a major role in conservation efforts worldwide.

Restoration Ecology

Many areas that could be incorporated into megareserves have been highly altered by human activities. Some of these areas must be restored to their original state if they are to play their intended roles in biodiversity conservation. In addition, many small reserves need to be restored and managed so that species that formerly lived there and have already disappeared can return. For these reasons, a subdiscipline of conservation biology, known as **restoration ecology,** is growing rapidly. Research on methods of restoring species and ecosystems is needed because many ecological communities will not recover, or will do so only very slowly, if there is no creative intervention in the recovery process.

The world's largest restoration project is under way in Guanacaste National Park in northwestern Costa Rica (Figure 51.19). The goal is to restore a large area of tropical deciduous forest, the most threatened ecosystem in Central America, from small fragments that remain in an area converted primarily to cattle grazing. One method of restoration would be to exclude fire and domestic livestock from the park and let nature take its course. However, although grass patches less than 120 hectares would be clothed by woody vegetation within 20 years, large expanses of pasture would require 50 to 200 years to become completely covered by forest because of the slow rates of movement of tree seeds into large areas. We can speed up reforestation by manipulating the habitat, which is what Daniel Janzen, architect of the restoration project, has chosen to do. To design his project, Janzen is using basic ecological information about the abilities of different plant species to germinate and grow in the degraded pastures that surround most of the forest fragments, information about the ability of existing seed dispersers to move seeds from the forest into the abandoned pastures, and information about the populations of pastureland species that are serious predators on seeds and seedlings of forest trees.

The single most important threat to Guanacaste National Park during the coming decades is fire. Occasionally fires in the region are due to natural causes, but most fires are started by people. Moreover, the introduced grasses, if not grazed, produce highly flammable dense stands that can produce hot fires that penetrate far into surrounding forests. Domestic livestock keep these grasses under control by grazing, and they are also important dispersers of the seeds of some native trees that are good at invading pastures and in whose shade many other species germinate. Therefore, the restoration program will encourage grazing by domestic livestock in the park until woody succession has progressed to the point where grass no longer poses serious competition to the woody species and is no longer sufficiently dense to carry hot fires.

(a)

(b)

51.19 Restoring a Tropical Deciduous Forest
(a) Most of the tropical deciduous forests in Central America have been cut, burned, and converted to cattle pasture. (b) In Guanacaste National Park, the cattle grazing under the trees are being used to disperse seeds of trees and to reduce grass cover during the forest restoration process.

Restoration of damaged and degraded habitats is an important activity, but unwarranted faith in the success of restoration efforts can lead to a loss of natural ecosystems. In the United States, belief that existing ecosystems can be replaced or relocated has made it easy to get permits for development. Because the number of mitigation projects to create replacement wetlands has soared recently, many inexperienced people are undertaking the task of constructing functional wetlands. Even the most experienced wetland ecologists, however, are having great difficulties in this endeavor. A restored wetland was conceived as part of the compensation agreement that allowed the California Department of Transportation to widen Interstate Highway 5—a project that damaged a marsh and jeopardized two endangered birds, the light-footed clapper rail and the least tern, and an endangered plant, the salt marsh bird's beak. Despite stringent, court-imposed standards and the involvement of wetland experts, the endangered birds are still not breeding in the "restored" marsh nine years after it was created. The restoration ecologists have not given up, but advice given by a recent National Research Council committee on wetland restoration needs to be heeded: "Wetland restoration should not be used to mitigate avoidable destruction of other wetlands until it can be scientifically demonstrated that the replacement ecosystems are of equal or better functioning."

ECOSYSTEM SERVICES

Natural ecosystems provide many essential services for humans without which we could not survive, except at great cost, through artificial means. Ecosystems absorb carbon dioxide and other gases, emit oxygen, cleanse water, regulate stream flows, and provide recreational and aesthetic services. The economic value of ecosystem services is hard to quantify, but some estimates have been made. The cost of building a system that would duplicate the treatment of wastewater and the fish production now provided by a hectare of Louisiana wetland is estimated to be about $200,000. This figure does not include the value provided by the wetland for sulfate reduction, carbon dioxide fixation, oxygen release, or waterfowl production. The storage and purification of water, binding of soil, and fertilization of woodland provided by

a hectare of streamside vegetation in Georgia was calculated to be worth $2,000 per year.

Ecologist F. H. Bormann has aptly summarized the cost of substituting technologies for lost ecosystem services:

We must find replacements for wood products, build erosion control works, enlarge reservoirs, upgrade air pollution control technology, install flood control works, improve water purification plants, increase air conditioning, and provide new recreational facilities. These substitutes represent an enormous tax burden, a drain on the world's supply of natural resources, and increased stress on the natural system that remains. Clearly, the diminution of solar-powered natural systems and the expansion of fossil-powered human systems are currently locked in a positive feedback cycle. Increased consumption of fossil energy means increased stress on natural systems, which in turn means still more consumption of fossil energy to replace lost natural functions if the quality of life is to be maintained.

Some ecosystem services, such as aesthetic benefits, cannot be replaced with technological inventions. One of the largest sources of foreign income in Kenya is nature tourism. The loss of a single species probably would not reduce the flow of tourists to Kenya, but if elephants, rhinoceroses, lions, leopards, and buffalo were all to disappear, fewer people would be likely to pay the high price of a Kenyan vacation. Populations of these species can be maintained only if large tracts of the ecosystems in which they live are preserved.

The preservation of biological diversity and ecosystem services is one of the greatest challenges facing humankind. Many of the scientific tools needed for the task are already available, but effective implementation of these tools requires important changes in people's attitudes toward other species. If species are valued only because they are economically useful to us, increased losses of species are inevitable because other uses of natural habitats are likely to be seen as more profitable, at least in the short run. Even though a wetland has great value, some people usually can make enough money in the short run by destroying the wetland to be motivated to do so. Only when we highly value biological diversity and ecosystem functioning as the heritage of all humankind, a heritage to be passed along to our descendants as rich and full as possible, will we begin to reduce the current alarming rates of ecosystem destruction and extinctions.

SUMMARY of Main Ideas about Conservation Biology

Humans have caused extinctions of species by overhunting, by introducing predators and competitors, and by destroying habitat.

Human-caused global warning may threaten species and habitats in the twenty-first century.
Review Figure 51.10

Conservation biologists try to estimate the extinction risk faced by small populations.

Rare species may be particularly vulnerable to extinction and require special management.

Fragmentation increases the portion of habitats influenced by edge effects.

Some species are being propagated in captivity in

the hope that reintroductions may increase natural populations in the future.

We need to preserve habitats and complete ecological communities.

Megareserves can protect natural habitats while accommodating many human activities.
Review Figure 51.17

SELF-QUIZ

1. Which of the following is *not* currently a major cause of species extinctions?
    a. Habitat destruction
    b. Climate change
    c. Overexploitation
    d. Introduction of predators
    e. Introduction of diseases

2. When ecosystems are managed to favor strongly those species intended for human use, we say that their production is:
    a. modified.
    b. diverted.
    c. co-opted.
    d. channeled.
    e. managed.

3. A minimum viable population is
    a. the estimated number of individuals necessary for the species to maintain genetic diversity.
    b. the estimated number of individuals necessary for the species to persist in all U.S. national forests.
    c. the estimated number of individuals necessary for the species to survive for several decades.
    d. the estimated number of individuals necessary for a species to maintain or increase its numbers in a region.
    e. the minimum density required for individuals to find mating partners.

4. Which of the following is *not* a component of a population vulnerability analysis?
    a. Spatial structure of a population
    b. Sex ratio within a population
    c. Amount of variation in birth and death rates
    d. Amount of heterozygosity and genetic variance
    e. Captive propagation of individuals

5. Orchids cross-pollinate successfully even though they usually are very rare because
    a. their unusual shapes restrict pollinator access to one route.
    b. the pollen of orchids is packaged in sacs called pollinia.
    c. pollen is placed very precisely on the bodies of visitors.
    d. pollinia seldom brush off on the stigmas of other species.
    e. all of the above

6. Conservation biologists are concerned about global warming because
    a. the rate of change in climate is projected to be faster than the rate at which ranges of many species can move.
    b. it is already too hot in the tropics.
    c. climates have been so stable for thousands of years that many species lack abilities to tolerate variable temperatures.
    d. climate change will be especially harmful to rare species.
    e. none of the above

7. A species that is found only in a particular region is said to be
    a. an indicator species for that region.
    b. a restricted species.
    c. a vulnerable species.
    d. endemic to that region.
    e. demographically constrained.

8. A keystone species is one that
    a. preys heavily on a particular species.
    b. is especially vulnerable to extinction.
    c. is restricted to a small geographic area.
    d. experiences considerable demographic stochasticity.
    e. strongly influences the structure and functioning of its ecological community.

9. As a habitat patch gets smaller it
    a. cannot support populations of species that require large areas.
    b. supports only small populations of many species.
    c. is influenced to an increasing degree by edge effects.
    d. is invaded by species from surrounding habitats.
    e. all of the above.

10. Restoration ecology is an important discipline because
    a. many areas being incorporated into megareserves have been highly degraded.
    b. many areas being incorporated into megareserves are vulnerable to global climate change.
    c. many species suffer from demographic stochasticity.
    d. many species are genetically impoverished.
    e. fire is a threat to many reserves.

11. Which of the following is *not* an ecosystem service?
    a. Production of carbon dioxide
    b. Flood control
    c. Water purification
    d. Air purification
    e. Preservation of biological diversity

FOR STUDY

1. Most species driven to extinction by people in the past were large vertebrates. Do you expect this pattern to persist into the future? If not, why not?

2. Species endangered as a result of global climatic warming might be preserved if we move individuals from areas that are becoming unsuitable to those likely to be better for them in the future. What are the major difficulties associated with such interventions? For what types of species would they work well? Poorly?

3. Conservation biologists have debated extensively which is better: many small reserves or a few large ones. What biological processes should be evaluated in making judgments about size and location of reserves? To what extent should we be concerned with preserving the largest number of species rather than those species judged to be of unusual importance?

4. During World War I, French doctors adopted a "triage" system of dealing with wounded soldiers. The wounded were divided into three categories: those almost certain to die no matter what was done to help them, those likely to recover even if not assisted, and those whose probability of survival was greatly increased if they were given medical attention. The limited resources available to the doctors were directed primarily at the third category. What would be the implications of adopting a similar attitude toward species preservation?

5. Utilitarian arguments dominate discussions about the importance of preserving the biological richness of the planet. In your opinion, what role should moral arguments play?

READINGS

Defenders of Wildlife. 1989. *Preserving Communities & Corridors*. Washington, D.C. A short collection of essays exploring the roles of corridors in preserving wildlife and how we can best implement important conservation legislation.

Gradwohl, J. and R. Greenberg. 1988. *Saving the Tropical Forests*. Island Press, Washington, D.C. A good account of the causes of tropical forest destruction and of successful projects throughout the world where local communities have averted forest destruction while reaping social and financial benefits.

Meffe, G. K. and C. R. Carroll (Eds.). 1994. *Principles of Conservation Biology*. Sinauer Associates, Sunderland, MA. A multiauthored text that provides an extensive overview of conservation biology.

National Research Council. 1986. *Ecological Knowledge and Environmental Problem-Solving*. National Academy Press, Washington, D.C. A thorough review of ecological knowledge and how it has been useful in helping to solve environmental problems. Includes descriptions and analyses of 13 successful projects.

Primack, R. B. 1993. *Essentials of Conservation Biology*. Sinauer Associates, Sunderland, MA. An introductory text that combines theory with basic and applied research to explain the connections between conservation biology and other disciplines.

Repetto, R. 1990. "Deforestation in the Tropics." *Scientific American*, April. Government policies that encourage exploitation (in particular excessive logging and clearing for ranches and farms) are largely to blame for the accelerating destruction of tropical forests.

Tattersall, I. 1993. "Madagascar's Lemurs." *Scientific American*, January. The proper study of humans is the lemur. Of all living creatures, none more closely resembles the ancestor from which humans and great apes branched 50 million years ago. But the lemurs' diverse Madagascan habitats are disappearing fast, and so are they.

Terborgh, J. 1992. "Why American Songbirds Are Vanishing." *Scientific American*, May. An avian chorus still heralds the beginning of spring in North America, but the number of singers has declined sharply of late. The trend will be difficult to reverse.

Western, D. and M. Pearl (Eds.). 1989. *Conservation for the Twenty-First Century*. Oxford University Press, New York. A rich set of essays by conservationists, governmental decision makers, and wildlife managers that identify gaps in knowledge and propose agendas for conservation action worldwide.

Wilson, E. O. 1989. "Threats To Biodiversity." *Scientific American*, September. Habitat destruction, mostly in the tropics, is driving thousands of species each year to extinction. The consequences will be dire—unless the trend is reversed.

Wilson, E. O. 1992. *The Diversity of Life*. Belknap Press of Harvard University Press, Cambridge, MA. A readable book that outlines the processes that created the diversity of life, explains the threats to that diversity, and shows what we must do to preserve biodiversity.

Abdomen (ab´ duh mun) [L.: belly]   In arthropods, the posterior portion of the body; in mammals, the part of the body containing the intestines and most other internal organs, posterior to the thorax.

Abomasum (ab´ oh may´ sum)   The true stomach of ruminants (animals such as cattle, sheep, and goats).

Abscisic acid (ab sighs´ ik) [L. *abscissio*: breaking off]   A plant growth substance having growth-inhibiting action. Causes stomata to close.

Abscission (ab sizh´ un) [L. *abscissio*: breaking off]   The process by which leaves, petals, and fruits separate from a plant.

Absolute temperature scale   A temperature scale in which the degree is the same size as in the Celsius (centigrade) scale, and zero is the state of no molecular motion. Absolute zero is –273° on the Celsius scale.

Absorption   (1) Of light: complete retention, without reflection or transmission. (2) Of liquids: soaking up (taking in through pores or cracks).

Absorption spectrum   A graph of light absorption versus wavelength of light; shows how much light is absorbed at each wavelength.

Abyssal zone (uh biss´ ul) [Gr. *abyssos*: bottomless]   That portion of the deep ocean where no light penetrates.

Abzyme   An immunoglobulin (antibody) with catalytic activity.

Accessory fruit   A fruit derived from parts in addition to the ovary and seeds. (Contrast with simple fruit, aggregate fruit, multiple fruit.)

Accessory pigments   Pigments that absorb light and transfer energy to chlorophylls for photosynthesis.

Acclimatization   Changes in an organism that improve its ability to tolerate seasonal changes in its environment.

Acellular   Not composed of cells.

Acetylcholine   A neurotransmitter substance that carries information across vertebrate neuromuscular junctions and some other synapses. **Acetylcholinesterase** is an enzyme that breaks down acetylcholine.

Acetyl CoA (acetyl coenzyme A)   Compound that reacts with oxaloacetate to produce citrate at the beginning of the citric acid cycle; a key metabolic intermediate in the formation of many compounds.

Acid [L. *acidus*: sharp, sour]   A substance that can release a proton. (Contrast with base.)

Acid precipitation   Precipitation that has a lower pH than normal as a result of acid-forming precursors introduced into the atmosphere by human activities.

Acidic   Having a pH of less than 7.0 (a hydrogen ion concentration greater than 10^{-7} molar).

Acoelomate   Lacking a coelom.

Acquired Immune Deficiency Syndrome   See AIDS.

Acrosome (a´ krow soam) [Gr. *akros*: highest or outermost + *soma*: body]   The structure at the forward tip of an animal sperm which is the first to fuse with the egg membrane and enter the egg cell.

ACTH (adrenocorticotropin)   A pituitary hormone that stimulates the adrenal cortex.

Actin [Gr. *aktis*: a ray]   One of the two major proteins of muscle; it makes up the thin filaments. Forms the microfilaments found in most eukaryotic cells.

Action potential   An impulse in a neuron taking the form of a wave of depolarization or hyperpolarization imposed on a polarized cell surface.

Action spectrum   A graph of biological activity versus wavelength of light. It compares the effectiveness of light of different wavelengths.

Activation energy   The energy barrier that blocks the tendency for a set of chemical substances to react. A reaction is speeded up if this energy barrier is surmounted by adding heat energy, or if the barrier is lowered by providing a different reaction pathway with the aid of a catalyst. Designated by the symbol E_a.

Active site   The region on the surface of an enzyme where the substrate binds, and where catalysis occurs.

Active transport   The transport of a substance across a biological membrane against a concentration gradient—that is, from a region of low concentration (of that substance) to a region of high concentration. Active transport requires the expenditure of energy and is a saturable process. (Contrast with facilitated diffusion, free diffusion; see primary active transport, secondary active transport.)

Adaptation (a dap tay´ shun)   In evolutionary biology, a particular structure, physiological process, or behavior that makes an organism better able to survive and reproduce. Also, the evolutionary process that leads to the development or persistance of such a trait.

Adenosine triphosphate   See ATP.

Adenylate cyclase   Enzyme catalyzing the formation of cyclic AMP from ATP.

Adhesion molecules   See cell adhesion molecules.

Adrenal (a dree´ nal) [L. *ad-*: toward + *renes*: kidneys]   An endocrine gland located near the kidneys of vertebrates, consisting of two glandular parts, the cortex and medulla.

Adrenaline   See epinephrine.

Adrenocorticotropin   See ACTH.

Adsorption   Binding of a gas or a solute to the surface of a solid.

Aerenchyma (air eng´ kyma) [Gr. *aer*: air + *enchyma*: infusion]   Modified parenchyma tissue, with many air spaces, found in shoots of some aquatic plants. (See parenchyma.)

Aerobic (air oh´ bic) [Gr. *aer*: air + *bios*: life] In the presence of oxygen, or requiring oxygen.

Afferent (af´ ur unt) [L. *ad*: to + *ferre*: to bear] To or toward, as in a neuron that carries impulses to the central nervous system, or a blood vessel that carries blood to a structure. (Contrast with efferents.)

Age distribution The proportion of individuals in a population belonging to each of the age categories into which the population has been divided. The number of divisions is arbitrary.

Aggregate fruit A fruit developing from several carpels of a single flower. (Contrast with simple fruit, accessory fruit, multiple fruit.)

AIDS (Acquired immune deficiency syndrome) Condition in which the body's helper T lymphocytes are destroyed, leaving the victim subject to opportunistic diseases. Caused by the HIV-I virus.

Air sacs Structures in the avian respiratory system that facilitate unidirectional flow of air through the lungs.

Alcohol An organic compound with one or more hydroxyl (–OH) groups.

Aldehyde (al´ duh hide) A compound with a –CHO functional group. Many sugars are aldehydes. (Contrast with ketone.)

Aldosterone (al dahs´ ter own) A steroid hormone produced in the adrenal cortex of mammals. Promotes secretion of potassium and reabsorption of sodium in the kidney.

Aleurone layer (al´ yur own) [Gr. *aleuron*: wheat flour] In grass seeds, a specialized cell layer just between the seed coat and the endosperm, synthesizing hydrolytic enzymes under the influence of gibberellin, and thus helping mobilize reserves for the developing embryo.

Alga (al´ gah) (plural: algae) [L.: seaweed] Any one of a wide diversity of protists belonging to the phyla Pyrrophyta, Chrysophyta, Phaeophyta, Rhodophyta, and Chlorophyta (and, formerly, Cyanophyta— "blue-green algae"). Most live in the water, where they are the dominant autotrophs; most are unicellular, but a minority are multicellular ("seaweeds" and similar protists).

Allele (a leel´) [Gr. *allos*: other] The alternate forms of a genetic character found at a given locus on a chromosome.

Allele frequency The relative proportion of a particular allele in a specific population.

Allergy [Ger. *allergie*: altered reaction] An overreaction to an antigen in amounts that do not affect most people; often involves IgE antibodies.

Allometric growth A pattern of growth in which some parts of the body of an organism grow faster than others, resulting in a change in body proportions as the organism grows.

Allopatric (al´ lo pat´ rick) [Gr. *allos*: other + *patria*: fatherland] Pertaining to populations that occur in different places.

Allopatric speciation See geographical speciation.

Allostery (al´ lo steer´ y) [Gr. *allos*: other + *stereos*: structure] Regulation of the activity of an enzyme by binding, at a site other than the catalytic active site, of an effector molecule that does not have the same structure as any of the enzyme's substrates.

Alpha helix Type of protein secondary structure; a right-handed spiral.

Alternation of generations The succession of haploid and diploid phases in a sexually reproducing organism. In most animals (male wasps and honey bees are notable exceptions), the haploid phase consists only of the gametes. In fungi, algae, and plants, however, the haploid phase may be the more prominent phase (as in fungi and mosses) or may be as prominent as the diploid phase (see the life cycle of *Ulva*, for example). In vascular plants, the diploid phase is more prominent.

Altruistic act A behavior whose performance harms the actor but benefits other individuals.

Alveolus (al ve´ o lus) (plural: alveoli) [L. *alveus*: cavity] A small, baglike cavity, especially the blind sacs of the lung.

Amensalism (a men´ sul ism) Interaction in which one animal is harmed and the other is unaffected. (Contrast with commensalism, mutualism.)

Amine An organic compound with an amino group (see Amino acid).

Amino acid An organic compound of the general formula H_2N–CHR–COOH, where R can be one of 20 or more different side groups. An amino acid is so named because it has both a basic amine group, $–NH_2$, and an acidic carboxyl group, –COOH. Proteins are polymers of amino acids.

Ammonotelic (am moan´ o teel´ ic) [Gr. *telos*: end] Describes an organism in which the final product of breakdown of nitrogen-containing compounds (primarily proteins) is ammonia. (Contrast with ureotelic, uricotelic.)

Amniocentesis A medical procedure in which cells from the fetus are obtained from the amniotic fluid. The genetic material of the cells is then examined. (Contrast with chorionic villus sampling.)

Amniotic egg The eggs of birds and reptiles, which can be incubated in air because the embryo is enclosed by a fluid-filled sac.

Amoeba (a mee´ bah) [Gr. *amoibe*: change] Any one of a large number of different kinds of unicellular protists belonging to the phylum Rhizopoda, characterized among other features by its ability to change shape frequently through the protrusion and retraction of cytoplasmic extensions called pseudopods.

Amoeboid (a mee´ boid) Like an amoeba; constantly changing shape by the protrusion and retraction of pseudopodia.

Amphi- [Gr.: both] Prefix used to denote a character or kind of organism that occupies two or more states. For example, amphibian (an animal that lives both on the land and in the water).

Amphibian (am fib´ ee an) A member of the vertebrate class Amphibia, such as a frog, toad, or salamander.

Amphipathic (am´ fi path´ ic) [Gr. *amphi*: both + *pathos*: emotion] Of a molecule, having both hydrophilic and hydrophobic regions.

amu (atomic mass unit, or dalton) The basic unit of mass on an atomic scale, defined as one-twelfth the mass of a carbon-12 atom. There are 6.023×10^{23} amu in one gram. This number is known as Avogadro's number.

Amylase (am´ ill ase) Any of a group of enzymes that digest starch.

Anabolism (an ab´ uh liz´ em) [Gr. *ana*: up, throughout + *ballein*: to throw] Synthetic reactions of metabolism, in which complex molecules are formed from simpler ones. (Contrast with catabolism.)

Anaerobic (an ur row´ bic) [Gr. *an*: not + *aer*: air + *bios*: life] Occurring without the use of molecular oxygen, O_2.

Anagenesis See vertical evolution.

Analogy (a nal´ o jee) [Gr. *analogia*: resembling] A resemblance in function, and often appearance as well, between two structures which is due to convergence in evolution rather than to common ancestry. (Contrast with homology.)

Anaphase (an´ a phase) [Gr. *ana*: indicating upward progress] The stage in nuclear division at which the first separation of sister chromatids (or, in the first meiotic division, of paired homologues) occurs. Anaphase lasts from the moment of first separation to the time at which the moving chromosomes converge at the poles of the spindle.

Anaphylactic shock A precipitous drop in blood pressure caused by loss of fluid from capillaries because of an increase in their permeability stimulated by an allergic reaction.

Ancestral trait Trait shared by a group of organisms as a result of descent from a common ancestor.

Androgens (an´ dro jens) The male sex steroids.

Aneuploid (an´ you ploy dee) A condition in which one or more chromosomes or pieces of chromosomes are either lacking or present in excess.

Angiosperm (an´ jee oh spurm) [Gr. *angion*: vessel + *sperma*: seed] One of the flowering plants; literally, one whose seed is carried in a "vessel," which is the fruit. (See fruit.)

Angiotensin (an´ jee oh ten´ sin) A peptide hormone that raises blood pressure by causing peripheral vessels to constrict; maintains glomerular filtration by constricting efferent glomerular vessels; stimulates thirst; and stimulates the release of aldosterone.

Animal [L. *animus*: breath, soul]   A member of the kingdom Animalia. In general, a multicellular eukaryote that obtains its food by ingestion.

Animal pole   In some eggs, zygotes, and embryos, the pole away from the bulk of the yolk (contrast with vegetal pole).

Anion (an´ eye one)   An ion with one or more negative charges. (Contrast with cation.)

Anisogamy (an´ eye sog´ a mee) [Gr. *aniso*: unequal + *gamos*: marriage]   The existence of two dissimilar gametes (egg and sperm).

Annelid (an´ el id)   A member of the phylum Annelida; one of the segmented worms, such as an earthworm or leech.

Annual   Referring to a plant whose life cycle is completed in one growing season. (Contrast with biennial, perennial.)

Anorexia nervosa (an or ex´ ee ah) [Gr. *an*: not + *orexis*: appetite]   Severe malnutrition and body wasting brought on by a psychological aversion to food.

Anterior   Toward the front.

Anterior pituitary   The portion of the vertebrate pituitary gland that derives from gut epithelium and produces tropic hormones.

Anther (an´ thur) [Gr. *anthos*: flower]   A pollen-bearing portion of the stamen of a flower.

Antheridium (an´ thur id´ ee um) (plural: antheridia) [Gr. *antheros*: blooming]   The multicellular structure that produces the sperm in bryophytes and ferns.

Antibody   One of millions of blood proteins, produced by the immune system, that specifically recognizes a foreign substance and initiates its removal from the body.

Anticodon   A "triplet" of three nucleotides in transfer RNA that is able to pair with a complementary triplet (a codon) in messenger RNA, thus aligning the transfer RNA on the proper place on the messenger. The codon (and, reciprocally, the anticodon) codes for a specific amino acid.

Antidiuretic hormone   A hormone that controls water reabsorption in the mammalian kidney. Also called vasopressin.

Antigen (an´ ti jun)   Any substance that stimulates the production of an antibody or antibodies upon introduction into the body of a vertebrate.

Antigenic determinant   A specific region of an antigen, which is recognized by and binds to a specific antibody.

Antiparallel   Parallel but running in opposite directions. The two strands of DNA are antiparallel.

Antipodals (an tip´ o dulls) [Gr. *anti*: against + *podus*: foot]   Cells (usually three) of the mature embryo sac of a flowering plant, located at the end opposite the egg (and micropyle).

Antiport   A membrane transport protein that carries one substance in one direction

and another in the opposite direction. (Contrast with symport.)

Antisense nucleic acid   A single-stranded RNA or DNA complementary to and thus targeted against the mRNA transcribed from a harmful gene such as an oncogene.

Anus (a´ nus)   Opening through which digestive wastes are expelled, located at the posterior end of the gut.

Aorta (a or´ tuh) [Gr. *aorte*: aorta]   The main trunk of the arteries leading to the systemic (as opposed to the pulmonary) circulation.

Apex (a´ pecks)   The tip or highest point of a structure, as the apex of a growing stem or root.

Apical (a´ pi kul)   Pertaining to the apex, as the apical meristem, which is the actively growing tissue at the tip of a stem or root.

Apomixis (ap oh mix´ is) [Gr. *apo*: away from + *mixis*: sexual intercourse]   The asexual production of seeds.

Apoplast (ap´ oh plast)   in plants, the continuous meshwork of cell walls and extracellular spaces through which material can pass without crossing a plasma membrane. (Contrast with symplast.)

Appendix   A vestigial portion of the human gut at the junction of the ileum with the colon.

Apterous   Lacking wings. (Contrast with alate: having wings.)

Aquatic [L. *aqua*: water]   Living in or on water, or taking place in or on water.

Aqueous [L. *aqua*: water]   Containing water, or dissolved in water.

Archaebacteria (ark´ ee bacteria) [Gr. *archaios*: ancient]   One of the two kingdoms of prokaryotes; the archaebacteria possess distinctive lipids and lack peptidoglycan. Most live in extreme environments. (Contrast with eubacteria.)

Archegonium (ar´ ke go´ nee um) [Gr. *archegonos*: first of a kind]   The multicellular structure that produces eggs in bryophytes, ferns, and gymnosperms.

Archenteron (ark en´ ter on) [Gr. *archos*: beginning + *enteron*: bowel]   The earliest primordial animal digestive tract.

Arteriole   One of the branches of an artery.

Arteriosclerosis   See atherosclerosis.

Artery   A muscular blood vessel carrying oxygenated blood away from the heart to other parts of the body. (Contrast with vein.)

Artifact [L. *ars, artis*: art + *facere*: to make]   Something made by human effort or intervention. In biology, something that was not present in the living cell or organism, but was unintentionally produced by an experimental procedure.

Ascospore (ass´ ko spor)   A fungus spore produced within an ascus.

Ascus (ass´ cuss) [Gr. *askos*: bladder]   In fungi belonging to the class Ascomycetes (sac fungi), the club-shaped sporangium within which spores are produced by meiosis.

Asexual   Without sex.

Associative learning   "Pavlovian" learning, in which an animal comes to associate a previously neutral stimulus (such as the ringing of a bell) with a particular reward or punishment.

Assortative mating   A breeding system under which mates are selected on the basis of a particular trait or group of traits. Results in more pairs of individuals sharing traits than would be the case if mating were random.

Assortment (genetic)   The random separation during meiosis of nonhomologous chromosomes and of genes carried on nonhomologous chromosomes. For example, if genes *A* and *B* are borne on nonhomologous chromosomes, meiosis of diploid cells of genotype *AaBb* will produce haploid cells of the following types in equal numbers: *AB*, *Ab*, *aB*, and *ab*.

Asymmetric   The state of lacking any plane of symmetry.

Asymmetric carbon atom   In a molecule, a carbon atom to which four different atoms or groups are bound.

Atherosclerosis (ath´ er oh sklair oh´ sis)   A disease of the lining of the arteries characterized by fatty, cholesterol-rich deposits in the walls of the arteries. When fibroblasts infiltrate these deposits and calcium precipitates in them, the disease become arteriosclerosis, or "hardening of the arteries."

Atmosphere   The gaseous mass surrounding our planet. Also: a unit of pressure, equal to the normal pressure of air at sea level.

Atom [Gr. *atomos*: indivisible]   The smallest unit of a chemical element. Consists of a nucleus and one or more electrons.

Atomic mass unit   See amu.

Atomic number   The number of protons in the nucleus of an atom, also equal to the number of electrons around the neutral atom. Determines the chemical properties of the atom.

Atomic weight   The average weight of an atom of an element on the amu scale. (The average depends upon the relative amounts of different isotopes of an element on Earth.)

ATP (adenosine triphosphate)   A compound containing adenine, ribose, and three phosphate groups. When it is formed, useful energy is stored; when it is broken down (to ADP or AMP), energy is released to drive endergonic reactions. ATP is a universal energy storage compound.

Atrium (a´ tree um)   A body cavity, as in the hearts of vertebrates. The thin-walled chamber(s) entered by blood on its way to the ventricle(s). Also, the outer ear.

Autocatalysis   An enzymatic reaction in which the inactive form of an enzyme is converted into its active form by the enzyme itself.

Autoimmune disease   A disorder in which the immune system attacks the animal's own body.

Autonomic nervous system The system (which in vertebrates comprises sympathetic and parasympathetic subsystems) that controls such involuntary functions as those of guts and glands.

Autoradiography The detection of a radioactive substance in a cell or organism by putting it in contact with a photographic emulsion and allowing the material to "take its own picture." The emulsion is developed, and the location of the radioactivity in the cell is seen by the presence of silver grains in the emulsion.

Autoregulatory mechanism A feedback mechanism that enables a structure to regulate its own function.

Autosome Any chromosome (in a eukaryote) other than a sex chromosome.

Autotroph (au´ tow trow´ fik) [Gr. *autos*: self + *trophe*: food] An organism that is capable of living exclusively on inorganic materials, water, and some energy source such as sunlight or chemically reduced matter. (Contrast with heterotroph.)

Auxin (awk´ sin) [Gr. *auxein*: increase] In plants, a substance (indoleacetic acid) that regulates growth and various aspects of development.

Auxotroph (awks´ o trofe) [Gr. *auxanein*: to grow + *trophe*: food] A mutant form of an organism that requires a nutrient or nutrients not required by the wild-type, or reference, form of the organism. (Contrast with prototroph.)

Avogadro's number The conversion factor between atomic mass units and grams. More usefully, the number of atoms in that quantity of an element which, expressed in grams, is numerically equal to the atomic weight in amu; 6.023×10^{23} atoms. (See mole.)

Axon [Gr.: axle] Fiber of a neuron which can carry action potentials. Carries impulses away from the cell body of the neuron; releases a neurotransmitter substance.

Axon hillock The junction between an axon and its cell body; where action potentials are generated.

Axon terminals The endings of an axon; they form synapses and release neurotransmitter.

Axoneme (ax´ oh neem) The complex of microtubules and their crossbridges that forms the motile apparatus of a cilium.

Bacillus (buh sil´ us) [L.: little rod] Any of various rod-shaped bacteria.

Bacteriophage (bak teer´ ee o fayj) [Gr. *bakterion*: little rod + *phagein*: to eat] One of a group of viruses that infect bacteria and ultimately cause their disintegration.

Bacterium (bak teer´ ee um) (plural: bacteria) [Gr. *bakterion*: little rod] A prokaryote. An organism with chromosomes not contained in nuclear envelopes.

Balanced polymorphism [Gr. *polymorphos*: having many forms] The maintenance of more than one form, or the maintenance at a given locus of more than one allele, at frequencies of greater than one percent in a population. Often results when heterozygotes are superior to both homozygotes.

Baroreceptor [Gr. *baros*: weight] A pressure-sensing cell or organ.

Barr body In mammals, an inactivated X chromosome.

Basal body Centriole found at the base of a eukaryotic flagellum or cilium.

Basal metabolic rate The minimum rate of energy turnover in an awake (but resting) bird or mammal that is not expending energy for thermoregulation.

Base A substance which can accept a proton (H^+). (Contrast with acid.) In nucleic acids, a nitrogen-containing base (purine or pyrimidine) is attached to each sugar in the backbone.

Base pairing See complementary base pairing.

Basic having a pH greater than 7.0 (having a hydrogen ion concentration lower than 10^{-7} molar).

Basidium (bass id´ ee yum) In fungi of the class Basidiomycetes, the characteristic sporangium in which four spores are formed by meiosis and then borne externally before being shed.

Batesian mimicry Mimicry by a relatively harmless kind of organism of a more dangerous one, by which the mimic enjoys protection from predators that mistake it for the dangerous model. (Contrast with Müllerian mimicry.)

B cell A type of lymphocyte involved in the humoral immune response of vertebrates. Upon recognizing an antigenic determinant, a B cell develops into a plasma cell, which secretes an antibody. (Contrast with a T cell.)

Benefit An improvement in survival and reproductive success resulting from a behavior. (Contrast with cost.)

Benthic zone [Gr. *benthos*: bottom of the sea] The bottom of the ocean. (Contrast with pelagic zone.)

Beta-pleated sheet Type of protein secondary structure; results from hydrogen bonding between polypeptide regions running antiparallel to each other.

Biennial Referring to a plant whose life cycle includes vegetative growth in the first year and flowering and senescence in the second year. (Contrast with annual, perennial.)

Bilateral symmetry The condition in which only the right and left sides of an organism, divided exactly down the back, are mirror images of each other. (Contrast with radial symmetry.)

Bile A secretion of the liver delivered to the small intestine via the common bile duct. In the intestine, bile emulsifies fats.

Binocular cells Neurons in the visual cortex that respond to input from both retinas; involved in depth perception.

Binomial (bye nome´ ee al) Consisting of two names; for example, the binomial nomenclature of biology which gives the name of the genus followed by the name of the species.

Biodiversity crisis The current high rate of loss of species, caused primarily by human activities.

Biogenesis [Gr. *bios*: life + *genesis*: source] The origin of living things from other living things.

Biogeochemical cycles Movement of elements through living organisms and the physical environment.

Biogeography The scientific study of the geographic distribution of organisms. Ecological biogeography is concerned with the habitats in which organisms live, historical biogeography with the complete geographic ranges of organisms and the historical circumstances that determine the ranges.

Biological species concept The view that a species is most usefully defined as a population or series of populations within which there is a significant amount of gene flow under natural conditions, but which is genetically isolated from other populations.

Biology [Gr. *bios*: life + *logos*: discourse] The scientific study of life in all its forms.

Bioluminescence The production of light by biochemical processes in an organism.

Biomass The total weight of all the living organisms, or some designated group of living organisms, in a given area.

Biome (bye´ ome) A major division of the ecological communities of Earth; characterized by distinctive vegetation.

Biota (bye oh´ tah) All of the organisms, including animals, plants, fungi, and microorganisms, found in a given area.

Biotic (bye ah´ tik) Pertaining to any aspect of life, especially to characteristics of entire populations or ecosystems.

Bipedal locomotion (by ped´ ul) [L. *bipes*: two-footed] Walking on two feet.

Biradial symmetry Radial symmetry modified so that only two planes can divide the animal into similar halves.

Blastocoel (blass´ toe seal) [Br. *blastos*: sprout + *koilos*: hollow] The central, hollow cavity of a blastula.

Blastodisc (blass´ toe disk) A disk of cells forming on the surface of a large yolk mass, comparable to a blastula, but occurring in forms in which the massive yolk restricts cleavage to one side of the egg only.

Blastomere A cell produced by the division of a fertilized egg.

Blastopore The opening from the archenteron to the exterior of a gastrula.

Blastula (blass´ chu luh) [Gr. *blastos*: sprout] An early stage in animal embryology; in many species, a hollow sphere of cells surrounding a central cavity.

Blood–brain barrier A property of the blood vessels of the brain that prevents most chemicals from diffusing from the blood into the brain.

Bloom A sudden increase in the density of phytoplankton, especially in a freshwater lake.

Body plan An entire animal, its organ systems, and the integrated functioning of its parts.

Bohr effect (boar) The reduction in affinity of hemoglobin for oxygen caused by acidic conditions, usually as a result of increased CO_2.

Bolting In rosetted angiosperms, a dramatic elongation of the stem, usually followed by flowering.

Bottleneck A combination of environmental conditions that causes a serious reduction in the size of the population.

Bowman's capsule An elaboration of kidney tubule cells that surrounds a know of capillaries (the glomerulus). Blood is filtered across the walls of these capillaries and the filtrate is collected into Bowman's capsule.

Brain A structure of nervous systems that provides the highest level of integration, control, and regulation.

Brain stem The portion of the vertebrate brain between the spinal cord and the forebrain.

Bronchus (plural: bronchi) The major airway(s) branching off the trachea into the vertebrate lung.

Browser An animal that feeds on the tissues of woody plants.

Bryophyte (bri´ uh fite´) [Gr. *bruon*: moss + *phyton*: plant] Any nonvascular plant, including mosses, liverworts, and hornworts.

Bud primordium [L. *primordium*: the beginning] In plants, a small mass of potentially meristematic tissue found in the angle between the leaf stalk and the shoot apex. Will give rise to a lateral branch under appropriate conditions.

Budding Asexual reproduction in which a more or less complete new organism simply grows from the body of the parent organism and eventually detaches itself.

Buffering A process by which a system resists change—particularly in pH, in which case added acid or base is partially converted to another form.

Bulb In plants, an underground storage organ composed principally of enlarged and fleshy leaf bases.

Bundle sheath In C_4 plants, a layer of photosynthetic cells between the mesophyll and a vascular bundle of a leaf.

C_3 photosynthesis The form of photosynthesis in which 3-phosphoglycerate is the first stable product, and ribulose bisphosphate is the CO_2 receptor.

C_4 photosynthesis The form of photosynthesis in which oxaloacetate is the first stable product, and phosphoenolpyruvate is the CO_2 acceptor. C_4 plants also perform the reactions of C_3 photosynthesis.

Caecum (see´ cum) [L. *caecus*: blind] A blind branch off the large intestine. In many nonruminant mammals, the caecum contains a colony of microorganisms that contribute to the digestion of food.

Calcitonin A hormone produced by the thyroid gland; it lowers blood calcium and promotes bone formation. (Contrast with parathormone.)

Callus [L. *calleo*: thick-skinned] In plants, wound tissue, of relatively undifferentiated proliferating cell mass, frequently maintained in cell culture.

Calmodulin (cal mod´ joo lin) A calcium-binding protein found in all animal and plant cells; mediates many calcium-regulated processes.

Calorie [L. *calor*: heat] The amount of heat required to raise the temperature of one gram of water by one degree Celsius (1°C) from 14.5°C to 15.5°C. In nutrition studies, "Calorie" (spelled with a capital C) refers to the kilocalorie (1 kcal = 1,000 cal), the amount of heat required to raise the temperature of one kilogram of water by 1°C.

Calvin–Benson cycle The stage of photosynthesis in which CO_2 reacts with RuBP to form 3PG, 3PG is reduced to a sugar, and RuBP is regenerated, while other products are released to the rest of the plant.

Calyptra (kuh lip´ tra) [Gr. *kalyptra*: covering for the head] A hood or cap found partially covering the apex of the sporophyte capsule in many moss species, formed from the expanded wall and neck of the archegonium.

Calyx (kay´ licks) [Gr. *kalyx*: cup] All of the sepals of a flower, collectively.

CAM See crassulacean acid metabolism.

Cambium (kam´ bee um) [L. *cambiare*: to exchange] A meristem that gives rise to radial rows of cells in stem and root, increasing them in girth; commonly applied to the vascular cambium which produces wood and phloem, and the cork cambium, which produces bark.

cAMP (cyclic AMP) A compound, formed from ATP, that mediates the effects of numerous animal hormones. Also needed for the transcription of catabolite-repressible operons in bacteria. Used for communication by cellular slime molds.

Canopy The leaf-bearing part of a tree. Collectively the aggregate of the leaves and branches of the larger woody plants of an ecological community.

Capacitance vessels Refers to veins because of their variable capacity to hold blood.

Capillaries [L. *capillaris*: hair] Very small tubes, especially the smallest blood-carrying vessels of animals between the termination of the arteries and the beginnings of the veins.

Capping In eukaryote RNA processing, the addition of a modified G at the 5´ end of the molecule.

Capsid The protein coat of a virus.

Capsule In bryophytes, the spore case. In some bacteria, a gelatinous layer exterior to the cell wall.

Carbohydrates Organic compounds with the general formula $C_nH_{2m}O_m$. Common examples are sugars, starch, and cellulose.

Carbon budget The amount of atmospheric carbon (from carbon dioxide) incorporated into organic molecules by a plant.

Carboxylic acid (kar box sill´ ik) An organic acid containing the carboxyl group, –COOH, which dissociates to the carboxylate ion, –COO⁻.

Carcinogen (car sin´ oh jen) A substance that causes cancer.

Cardiac (kar´ dee ak) [Gr. *kardia*: heart] Pertaining to the heart and its functions.

Carnivore [L. *carn*: flesh + *vovare*: to devour] An organism that feeds on animal tissue. (Contrast with detritivore, herbivore, omnivore.)

Carotenoid (ka rah´ tuh noid) [L. *carota*: carrot] A yellow, orange, or red lipid pigment commonly found as an accessory pigment in photosynthesis; also found in fungi.

Carpel (kar´ pel) [Gr. *karpos*: fruit] The organ of the flower that contains one or more ovules.

Carrier In facilitated diffusion, a membrane protein that binds a specific molecule and transports it through the membrane. In genetics, a person heterozygous for a recessive trait. In respiratory and photosynthetic electron transport, a participating substance such as NAD that exists in both oxidized and reduced forms.

Carrying capacity In ecology, the largest number of organisms of a particular species that can be maintained indefinitely in a given part of the environment.

Cartilage In vertebrates, a tough connective tissue found in joints, the outer ear, and elsewhere. Forms the entire skeleton in some animal groups.

Casparian strip A band of cell wall containing suberin and lignin, found in the endodermis. Restricts the movement of water across the endodermis.

Catabolism [Ge. *kata*: down + *ballein*: to throw] Degradational reactions of metabolism, in which complex molecules are broken down. (Contrast with anabolism.)

Catabolite repression The decreased synthesis of many enzymes that tend to provide glucose for a cell; caused by the presence of excellent carbon sources, particularly glucose.

Catalyst (cat´ a list) [Gr. *kata-*, implying the breaking down of a compound] A chemical substance that accelerates a reaction without itself being consumed in the overall course of the reaction. Catalysts lower the activation energy of a reaction. Enzymes are biological catalysts.

Cation (cat´ eye on)   An ion with one or more positive charges. (Contrast with anion.)

Caudal [L. *cauda*: tail]   Pertaining to the tail, or to the posterior part of the body.

cDNA   See complementary DNA.

Cell adhesion molecules   Molecules on animal cell surfaces that affect the selective association of cells during development of the embryo.

Cell cycle   The stages through which a cell passes between one division and the next. Includes all stages of interphase and mitosis.

Cell theory   The theory, well established, that organisms consist of cells, and that all cells come from preexisting cells.

Cell wall   A relatively rigid structure that encloses cells of plants, fungi, many protists, and most bacteria. The cell wall gives these cells their shape and limits their expansion in hypotonic media.

Cellular immune system   That part of the immune system that is based on the activities of T cells. Directed against parasites, fungi, intracellular viruses, and foreign tissues (grafts). (Contrast with humoral immune system.)

Cellular respiration   See respiration.

Cellulose (sell´ you lowss)   A straight-chain polymer of glucose molecules, used by plants as a structural supporting material. **Cellulase** is an enzyme that hydrolyzes cellulose.

Central dogma of molecular biology   The statement that information flows from DNA to RNA to polypeptide (in retroviruses, there is also information flow from RNA to cDNA).

Central nervous system   That part of the nervous system which is condensed and centrally located, e.g., the brain and spinal cord of vertebrates; the chain of cerebral, thoracic and abdominal ganglia of arthropods.

Centrifuge [L. *fugere*: to flee]   A device in which a sample can be spun around a central axis at high speed, creating a centrifugal force that mimics a very strong gravitational force. Used to separate mixtures of suspended materials.

Centriole (sen´ tree ole)   A paired organelle that helps organize the microtubules in animal and protist cells during nuclear division.

Centromere (sen´ tro meer) [Gr. *centron*: center + *meros*: part]   The region where sister chromatids join.

Cephalization (sef´ uh luh zay´ shun) [Gr. *kephale*: head]   The evolutionary trend toward increasing concentration of brain and sensory organs at the anterior end of the animal.

Cephalopod (sef´ a low pod)   A member of the mollusk class Cephalopoda, such as a squid or an octopus.

Cerebellum (sair´ uh bell´ um) [L.: diminutive of *cerebrum*: brain]   The brain region that controls muscular coordination; located at the anterior end of the hindbrain.

Cerebral cortex   The thin layer of gray matter (neuronal cell bodies) that overlays the cerebrum.

Cerebrum (su ree´ brum) [L.: brain]   The dorsal anterior portion of the forebrain, making up the largest part of the brain of mammals. In mammals, the chief coordination center of the nervous system; consists of two **cerebral hemispheres**.

Cervix (sir´ vix) [L.: neck]   The opening of the uterus into the vagina.

cGMP (cyclic guanosine monophosphate)   An intracellular messenger that is part of signal transmission pathways involving G-proteins. (See G-protein.)

Channel   A membrane protein that forms an aqueous passageway though which specific solutes may pass by simple diffusion; some channels are gated: they open and close in response to binding of specific molecules.

Character   In taxonomy, any trait of an organism used in creating a classification system.

Chemical bond   An attractive force stably linking two atoms.

Chemiosmotic mechanism   According to this model, ATP formation in mitochondria and chloroplasts results from a pumping of protons across a membrane (against a gradient of electrical charge and of pH), followed by the return of the protons through a protein channel with ATPase activity.

Chemoautotroph   An organism that uses carbon dioxide as a carbon source and obtains energy by oxidizing inorganic substances from its environment. (Contrast with chemoheterotroph, photoautotroph, photoheterotroph.)

Chemoheterotroph   An organism that must obtain both carbon and energy from organic substances. (Contrast with chemoautotroph, photoautotroph, photoheterotroph.)

Chemosensor   A cell or tissue that senses specific substances in its environment.

Chemosynthesis   Synthesis of food substances, using the oxidation of reduced materials from the environment as a source of energy.

Chiasma (kie az´ muh) (plural: chiasmata) [Gr.: cross]   An "x"-shaped connection between paired homologous chromosomes in prophase I of meiosis. A chiasma is the visible manifestation of crossing-over between homologous chromosomes.

Chitin (kye´ tin) [Gr. *chiton*: tunic]   The characteristic tough but flexible organic component of the exoskeleton of arthropods, consisting of a complex, nitrogen-containing polysaccharide. Also found in cell walls of fungi.

Chlorophyll (klor´ o fill) [Gr. *chloros*: green + *phyllon*: leaf]   Any of a few green pigments associated with chloroplasts or with certain bacterial membranes; responsible for trapping light energy for photosynthesis.

Chloroplast [Gr. *chloros*: green + *plast*: a particle]   An organelle bounded by a double membrane containing the enzymes and pigments that perform photosynthesis. Chloroplasts occur only in eukaryotes.

Choanocyte (cho´ an oh cite)   The collared, flagellated feeding cells of sponges.

Cholecystokinin (ko´ lee sis to kai nin)   A hormone produced and released by the lining of the duodenum when it is stimulated by undigested fats and proteins. It stimulates the gallbladder to release bile and slows stomach activity.

Chorion (kor´ ee on) [Gr. *khorion*: afterbirth]   The outermost of the membranes protecting mammal, bird, and reptile embryos; in mammals it forms part of the placenta.

Chorionic villus sampling   A medical procedure that extracts a portion of the chorion from a pregnant woman to enable genetic and biochemical analysis of the embryo. (Contrast with amniocentesis.)

Chromatid (kro´ ma tid)   Each of a pair of new sister chromosomes from the time at which the molecular duplication occurs until the time at which the centromeres separate at the anaphase of nuclear division.

Chromatin   The nucleic acid–protein complex found in eukaryotic chromosomes.

Chromatography   Any one of several techniques for the separation of chemical substances, based on differing relative tendencies of the substances to associate with a mobile phase or a stationary phase.

Chromatophore (krow mat´ o for) [Gr. *chroma*: color + *phoreus*: carrier]   A pigment-bearing cell that expands or contracts to change the color of the organism.

Chromosomal aberration   Any large change in the structure of a chromosome, including duplication or loss of chromosomes or parts thereof, usually gross enough to be detected with the light microscope.

Chromosome (krome´ o sowm) [Gr. *chroma*: color = *soma*: body]   In bacteria and viruses, the DNA molecule that contains most or all of the genetic information of the cell or virus. In eukaryotes, a structure composed of DNA and proteins that bears part of the genetic information of the cell.

Chromosome walking   A technique based on recognition of overlapping fragments; used as a step in DNA sequencing.

Chylomicron (ky low my´ cron)   Particles of lipid coated with protein, produced in the gut from dietary fats and secreted into the extracellular fluids.

Chyme (kime) [Gr. *chymus*, juice]   Created in the stomach; a mixture of ingested food with the digestive juices secreted by the salivary glands and the stomach lining.

Ciliate (sil´ ee ate)   A member of the protist phylum Ciliophora, unicellular organisms that propel themselves by means of cilia.

Cilium (sil´ ee um) (plural: cilia) [L. *cilium*: eyelash]   Hairlike organelle used for locomotion by many unicellular organisms and for moving water and mucus by many multicellular organisms. Generally shorter than a flagellum.

Circadian rhythm (sir kade´ ee an) [L. *circa*: approximately + *dies*: day] A rhythm in behavior, growth, or some other activity that recurs about every 24 hours under constant conditions.

Circannual rhythm (sir can´ you al) [L. *circa*: approximately + *annus*: year) A rhythm of behavior, growth, or some other activity that recurs on a yearly basis.

Citric acid cycle A set of chemical reactions in cellular respiration, in which acetyl CoA reacts with oxaloacetate to form citric acid, and oxaloacetate is regenerated. Acetyl CoA is oxidized to carbon dioxide, and hydrogen atoms are stored as NADH and $FADH_2$.

Clade (clayd) [Gr. *klados*: branch] All of the organisms, both living and fossil, descended from a particular common ancestor.

Cladistic classification A classification based entirely on the phylogenetic relationships among organisms.

Cladogenesis (clay doh jen´ e sis) [Gr. *klados*: branch + *genesis*: source] The formation of a new species by the splitting of an evolutionary lineage.

Cladogram Graphic representation of a cladistic relationship.

Class In taxonomy, the category below the phylum and above the order; a group of related, similar orders.

Clathrin A fibrous protein on the inner surfaces of animal cell membranes that strengthens coated vesicles and thus participates in receptor-mediated endocytosis.

Clay A soil constituent comprising particles smaller than 2 micrometers in diameter.

Cleavages First divisions of the fertilized egg of an animal.

Climax In ecology, a community that terminates a succession and which tends to replace itself unless it is further disturbed or the physical environment changes.

Climograph (clime´ o graf) Graph relating temperature and precipitation with time of year.

Cline A gradual change in the traits of a species over a geographical gradient.

Clitoris (klit´ er us, kilte´ er us) A structure in the human female reproductive system that is homologous with the male penis and is involved in sexual stimulation.

Cloaca (klo ay´ kuh) [L. *cloaca*: sewer] In some invertebrates, the posterior part of the gut; in many vertebrates, a cavity receiving material from the digestive, reproductive, and excretory systems.

Clonal deletion In immunology, the inactivation or destruction of lymphocyte clones that would produce immune reactions against the animal's own body.

Clonal selection The mechanism by which exposure to antigen results in the activation of selected T-cell or B-cell clones, resulting in an immune response.

Clone [Gr. *klon*: twig, shoot] Genetically identical cells or organisms produced from a common ancestor by asexual means.

Clutch The number of offspring produced in a given batch.

Coacervate (ko as´ er vate) [L. *coacervare*: to heap up] An aggregate of colloidal particles in suspension.

Coacervate drop Drops formed when a mixture of large proteins and polysaccharides is shaken in water. The interiors of these drops, which are often very stable, contain most of the proteins and polysaccharides.

Coated vesicle Vesicle, sometimes formed from a coated pit, with characteristic "bristly" surface; its membrane contains distinctive proteins, including clathrin.

Coccus (kock´ us) [Gr. *kokkos*: berry, pit] Any of various spherical or spheroidal bacteria.

Cochlea (kock´ lee uh) [Gr. *kokhlos*: a land snail] A spiral tube in the inner ear of vertebrates; it contains the sensory cells involved in hearing.

Codominance A condition in which two alleles at a locus produce different phenotypic effects and both effects appear in heterozygotes.

Codon A "triplet" of three nucleotides in messenger RNA that directs the placement of a particular amino acid into a polypeptide chain. (Contrast with anticodon.)

Coefficient of relatedness The probability that an allele in one individual is an identical copy, by descent, of an allele in another individual.

Coelom (see´ lum) [Gr. *koiloma*: cavity] The body cavity of certain animals, which is lined with cells of mesodermal origin.

Coelomate Having a coelom.

Coenocyte (seen´ a sight) [Gr.: common cell] A "cell" bounded by a single plasma membrane, but containing many nuclei.

Coenzyme A nonprotein molecule that plays a role in catalysis by an enzyme. The coenzyme may be part of the enzyme molecule or free in solution. Some coenzymes are oxidizing or reducing agents, others play different roles.

Coevolution Concurrent evolution of two or more species that are mutually affecting each other's evolution.

Cohort (co´ hort) [L. *cohors*: company of soldiers] A group of similar-age organisms, considered as it passes through time.

Coitus (koe´ i tus) [L. *coitus*: a coming together] The act of sexual intercourse.

Coleoptile (koe´ lee op´ til) [Gr. *koleos*: sheath + *ptilon*: feather] A pointed sheath covering the shoot of grass seedlings.

Collagen [Gr. *kolla*: glue] A fibrous protein found extensively in bone and connective tissue.

Collecting duct In vertebrates, a tubule that receives urine produced in the nephrons of the kidney and delivers that fluid to the ureter for excretion.

Collenchyma (cull eng´ kyma) [Gr. *kolla*: glue + *enchyma*: infusion] A type of plant cell, living at functional maturity, which lends flexible support by virtue of primary cell walls thickened at the corners. (Contrast with parenchyma, sclerenchyma.)

Colon [Gr. *kolon*: large intestine] The large intestine.

Colostrum (koh los´ trum) Substance secreted by the mammary glands around the time of an infant's birth. It contains protein and lactose but little fat, and its rate of production is less than the rate of milk production two or three days after birth.

Commensalism The form of symbiosis in which one species benefits from the association, while the other is neither harmed nor benefited.

Common bile duct A single duct that delivers bile from the gallbladder and secretions from the pancreas into the small intestine.

Communication Action on the part of one organism (or cell) that alters the pattern of behavior in another organism (or cell) in an adaptive fashion.

Community Any ecologically integrated group of species of microorganisms, plants, and animals inhabiting a given area.

Companion cell Specialized cell found adjacent to a sieve tube element in some flowering plants.

Comparative analysis An approach to studying evolution in which hypotheses are tested by measuring the distribution of states among a large number of species.

Compensation point The light intensity at which the rates of photosynthesis and of cellular respiration are equal.

Competitive inhibitor A substance, similar in structure to an enzyme's substrate, that binds the active site and thus inhibits a reaction.

Competition In ecology, use of the same resource by two or more species, when the resource is present in insufficient supply for the combined needs of the species.

Competitive exclusion A result of competition between species for a limiting resource in which one species completely eliminates the other.

Competitive inhibitor A substance, similar in structure to an enzyme's substrate, that binds the active site and inhibits a reaction.

Complement system A group of eleven proteins that play a role in some reactions of the immune system. The complement proteins are not immunoglobulins.

Complementary base pairing The A–T (or A–U), T–A (or U–A), C–G and G–C pairing of bases in double-stranded DNA, in transcription, and between tRNA and mRNA.

Complementary DNA (cDNA) DNA formed by reverse transcriptase acting with an RNA template; essential intermediate in the reproduction of retroviruses; used as a tool in recombinant DNA technology; lacks introns.

Complete metamorphosis A change of state during the life cycle of an organism in which the body is almost completely rebuilt to produce an individual with a completely different body form. Characteristic of insects such as butterflies, moths, beetles, ants, wasps, and flies.

Compound (1) A substance made up of atoms of more than one element. (2) Made up of many units, as the compound eyes of arthropods (as opposed to the simple eyes of the same group of organisms).

Compression wood See reaction wood.

Condensation reaction A reaction in which two molecules become connected by a covalent bond, and a molecule of water is released. ($AH + BOH \rightarrow AB + H_2O$.)

Cones (1) In the vertebrate retina: photoreceptors responsible for color vision. (2) In gymnosperms: reproductive structures consisting of many sporophylls packed relatively tightly.

Conidium (ko nid´ ee um) [Gr. *konis*: dust] An asexual fungus spore borne singly or in chains either apically or laterally on a hypha.

Conifer (kahn´ e fer) [Gr. *konos*: cone + *phero*: carry] One of the cone-bearing gymnosperms, mostly trees, such as pines and firs.

Conjugation (kahn´ jew gay´ shun) [L. *conjugare*: yoke together] The close approximation of two cells during which they exchange genetic material, as in *Paramecium* and other ciliates, or during which DNA passes from one to the other through a tube, as in bacteria.

Connective tissue An animal tissue that connects or surrounds other tissues; its cells are embedded in a collagen-containing matrix.

Connexon In a gap junction, a protein channel linking adjacent animal cells.

Consensus sequences Short stretches of DNA that appear, with little variation, in many different genes.

Constitutive enzyme An enzyme that is present in approximately constant amounts in a system, whether its substrates are present or absent. (Contrast with inducible enzyme.)

Consumer An organism that eats the tissues of some other organism.

Continental climate A pattern, typical of the interiors of large continents at high latitudes, in which bitterly cold winters alternate with hot summers. (Contrast with maritime climate.)

Continental drift The gradual drifting apart of the world's continents that has occurred over a period of billions of years.

Contractile vacuole An organelle, often found in protists, which pumps excess water out of the cell and keeps it from being "flooded" in hypotonic environments.

Cooperative act Behavior in which two or more individuals interact to their mutual benefit. No conscious awareness by the actors of the effects of their behavior is implied.

Cooption The act of capturing something for a particular use. In ecology refers to the diversion of ecological production for human use. Such production is said to be coopted.

Copulation Reproductive behavior that results in a male depositing sperm in the reproductive tract of a female.

Corepressor A low molecular weight compound that unites with a protein (the repressor) to prevent transcription in a repressible operon.

Cork A waterproofing tissue in plants, with suberin-containing cell walls. Produced by a cork cambium.

Corm A conical, underground stem that gives rise to a new plant. (Contrast with bulb.)

Corolla (ko role´ lah) [L.: diminutive of *corona*: wreath, crown] All of the petals of a flower, collectively.

Coronary (kor´ oh nair ee) Referring to the blood vessels of the heart.

Corpus luteum (kor´ pus loo´ tee um) [L. *corpus*: body + *luteum*: yellow] A structure formed from a follicle after ovulation; it produces hormones important to the maintenance of pregnancy.

Cortex [L.: bark or rind] (1) In plants: the tissue between the epidermis and the vascular tissue of a stem or root. (2) In animals: the outer tissue of certain organs, such as the adrenal cortex and cerebral cortex.

Corticosteroids Steroid hormones produced and released by the cortex of the adrenal gland.

Cost See energetic cost, opportunity cost, risk cost.

Cotyledon (kot´ ul lee´ dun) [Gr. *kotyledon*: a hollow space] A "seed leaf." An embryonic organ which stores and digests reserve materials; may expand when seed germinates.

Covalent bond A chemical bond that arises from the sharing of electrons between two atoms. Usually a strong bond.

Crassulacean acid metabolism (CAM) A metabolic pathway enabling the plants that possess it to store carbon dioxide at night and then perform photosynthesis during the day with stomata closed.

Crista (plural: cristae) A small, shelflike projection of the inner membrane of a mitochondrion; the site of oxidative phosphorylation.

Critical night length In the photoperiodic flowering response of short-day plants, the length of night above which flowering occurs and below which the plant remains vegetative. (The reverse applies in the case of long-day plants.)

Critical period The age during which some particular type of learning must take place or during which it occurs much more easily than at other times. Typical of song learning among birds.

Cross-pollination The pollination of one plant by pollen from another plant. (Contrast with self-pollination.)

Cross (transverse) section A section taken perpendicular to the longest axis of a structure.

Crossing over The mechanism by which linked markers undergo recombination. In general, the term refers to the reciprocal exchange of corresponding segments between two homologous chromatids. However, the reciprocity of crossing-over is problematical in prokaryotes and viruses; and even in eukaryotes, very closely linked markers often recombine by a nonreciprocal mechanism.

CRP The cAMP receptor protein that interacts with the promoter to enhance transcription; a lowered cAMP concentration results in catabolite repression.

Crustacean (crus tay´ see an) A member of the phylum Crustacea, such as a crab, shrimp, or sowbug.

Cryptic appearance The resemblance of an animal to some part of its environment, which helps it to escape detection by predators.

Culture A laboratory association of organisms under controlled conditions. Also the collection of knowledge, tools, values, and rules that characterize a human society.

Cuticle A waxy layer on the outer surface of a plant or an insect, tending to retard water loss.

Cutin (cue´ tin) [L. *cutis*: skin] A mixture of long, straight-chain hydrocarbons and waxes secreted by the plant epidermis, providing a water-impermeable coating on aerial plant parts.

Cyanobacteria (sigh an´ o bacteria) [Gr. *kuanos*: the color blue] A division of photosynthetic bacteria, formerly referred to as blue-green algae; they lack sexual reproduction, and they use chlorophyll *a* in their photosynthesis.

Cyclic AMP See cAMP.

Cyclins Proteins that activate maturation-promoting factor, bringing about transitions in the cell cycle.

Cyst (sist) [Gr. *kystis*: pouch] (1) A resistant, thick-walled cell formed by some protists and other organisms. (2) An abnormal sac, containing a liquid or semisolid substance, produced in response to injury or illness.

Cytochromes (sy´ toe chromes) [Gr. *kytos*: container + *chroma*: color] Iron-containing red proteins, components of the electron-transfer chains in photophosphorylation and respiration.

Cytokinesis (sy´ toe kine ee´ sis) [Gr. *kytos*: container + *kinein*: to move] The division of the cytoplasm of a dividing cell. (Contrast with mitosis.)

Cytokinin (sy´ toe kine´ in) [Gr. *kytos*: container + *kinein*: to move] A member of a class of plant growth substances playing roles in senescence, cell division, and other phenomena.

Cytoplasm The contents of the cell, excluding the nucleus.

Cytoplasmic determinants In animal development, gene products whose spatial distribution may determine such things as embryonic axes.

Cytoskeleton The network of microtubules and microfilaments that gives a eukaryotic cell its shape and its capacity to arrange its organelles and to move.

Cytosol The fluid portion of the cytoplasm, excluding organelles and other solids.

Cytotoxic T cells Cells of the cellular immune system that recognize and directly eliminate virus-infected cells. (Contrast with helper T cells, suppressor T cells.)

Dalton See amu.

Deciduous (de sid´ you us) [L. *decidere*: fall off] Referring to a plant that sheds its leaves at certain seasons. (Contrast with evergreen.)

Degeneracy The situation in which a single amino acid may be represented by any of two or more different codons in messenger RNA. Most of the amino acids can be represented by more than one codon.

Degradative succession Ecological succession occuring on the dead remains of the bodies of plants and animals, as when leaves or animal bodies rot.

Dehydration See condensation reaction.

Deletion (genetic) A mutation resulting from the loss of a continuous segment of a gene or chromosome. Such mutations never revert to wild-type. (Contrast with duplication, point mutation.)

Deme (deem) [Gr. *demos*: common people] Any local population of individuals belonging to the same species and among which mating is random.

Demographic processes The events—such as births, deaths, immigration, and emigration—that determine the number of individuals in a population.

Demographic stochasticity Random variations in the factors influencing the size, density, and distribution of a population.

Demography The study of dynamical changes in the sizes, densities, and distributions of populations.

Denaturation Loss of activity of an enzyme or nucleic acid molecule as a result of structural changes induced by heat or other means.

Dendrite [Gr. *dendron*: a tree] A fiber of a neuron which often cannot carry action potentials. Usually much branched and relatively short compared with the axon, and commonly carries information to the cell body of the neuron.

Denitrification Metabolic activity by which inorganic nitrogen-containing ions are reduced to form nitrogen gas and other products; carried on by certain soil bacteria.

Density dependence Change in the severity of action of agents affecting birth and death rates within populations. Such changes may be directly or inversely related to population density.

Density independence The state where the severity of action of agents affecting birth and death rates within a population does not change with the density of the population.

Deoxyribonucleic acid See DNA.

Depolarization A change in the electric potential across a membrane from a condition in which the inside of the cell is more negative than the outside to a condition in which the inside is less negative, or even positive, with reference to the outside of the cell. (Contrast with hyperpolarization.)

Desmosome (dez´ mo sowm) [Gr. *desmos*: bond + *soma*: body] An adhering junction between animal cells.

Derived trait A trait found among members of a lineage that was not present in the ancestors of that lineage.

Dermal tissue system The outer covering of a plant, consisting of epidermis in the young plant and periderm in a plant with extensive secondary growth. (Contrast with ground tissue system and vascular tissue system.)

Determinate cleavage A pattern of early embryological development in which the potential of cells is determined very early such that separated cells develop only into partial embryos. (Contrast with indeterminate cleavage.)

Determination Process whereby an embryonic cell or group of cells becomes fixed into a predictable developmental pathway.

Detritivore (di try´ ti vore) [L. *detritus*: worn away + *vorare*: to devour] An organism that eats the dead remains of other organisms.

Deuterium An isotope of hydrogen possessing one neutron in its nucleus. Deuterium oxide is called "heavy water."

Deuterostome One of two major lines of evolution in animals, characterized by radial cleavage, enterocoelous development, and other traits.

Deuterium An isotope of hydrogen, possessing one neutron in its nucleus; deuterium oxide is called "heavy water."

Development Progressive change, as in structure or metabolism; in most kinds of organisms, development continues throughout the life of the organism.

Dialysis (dye ahl´ uh sis) [Gr. *dialyein*: separation] The removal of ions or small molecules from a solution by their diffusion across a semipermeable membrane to a solvent where their concentration is lower.

Diaphragm (dye´ uh fram) [Gr. *diaphrassein*, to barricade] (1) A sheet of muscle that separates the thoracic and abdominal cavities in mammals; responsible for the action of breathing. (2) A method of birth control in which a sheet of rubber is fitted over the woman's cervix, blocking the entry of sperm.

Diastole (dye ahs´ toll ee) [Gr.: dilation] The portion of the cardiac cycle when the heart muscle relaxes. (Contrast with systole.)

Dicot (short for dicotyledon) [Gr. *dis*: two + *kotyledon*: a cup-shaped hollow] Any member of the angiosperm class Dicotyledones, flowering plants in which the embryo produces two cotyledons prior to germination. Leaves of most dicots have major veins arranged in a branched or reticulate pattern.

Differentiation Process whereby originally similar cells follow different developmental pathways. The actual expression of determination.

Diffuse coevolution The situation in which the evolution of a lineage is influenced by its interactions with a number of species, most of which exert only a small influence on the evolution of the focal lineage.

Diffusion Random movement of molecules or other particles, resulting in even distribution of the particles when no barriers are present.

Digestion Enzyme-catalyzed process by which large, usually insoluble, molecules (foods) are hydrolyzed to form smaller molecules of soluble substances.

Dihybrid cross A mating in which the parents differ with respect to the alleles of two loci of interest.

Dikaryon (di care´ ee ahn) [Gr. *dis*: two + *karyon*: kernel] A cell or organism carrying two genetically distinguishable nuclei. Common in fungi.

Dioecious (die eesh´ us) [Gr.: two houses] Organisms in which the two sexes are "housed" in two different individuals, so that eggs and sperm are not produced in the same individuals. Examples: humans, fruit flies, oak trees, date palms. (Contrast with monoecious.)

Diploblastic Having two cell layers. (Contrast with triploblastic.)

Diploid (dip´ loid) [Gr. *diploos*: double] Having a chromosome complement consisting of two copies (homologues) of each chromosome. A diploid individual (or cell) usually arises as a result of the fusion of two gametes, each with just one copy of each chromosome. Thus, the two homologues in each chromosome pair in a diploid cell are of separate origin, one derived from the female parent and one from the male parent.

Diplontic life cycle A life cycle in which every cell except the gametes is diploid.

Directional selection Selection in which phenotypes at one extreme of the population distribution are favored. (Contrast with disruptive selection; stabilizing selection.)

Disaccharide A carbohydrate made up of two monosaccharides (simple sugars).

Dispersal stage Stage in its life history at which an organism moves from its birthplace to where it will live as an adult.

Displacement activity Apparently irrelevant behavior performed by an animal under conflict situations, especially when tendencies to attack and escape are closely balanced.

Display A behavior that has evolved to influence the actions of other individuals.

Disruptive selection Selection in which phenotypes at both extremes of the population distribution are favored. (Contrast with directional selection; stabilizing selection.)

Distal Away from the point of attachment or other reference point. (Contrast with proximal.)

Disturbance A short-term event that disrupts populations, communities, or ecosystems by changing the environment.

Diverticulum (di ver tic´ u lum) [L. *divertere*: turn away] A small cavity or tube that connects to a major cavity or tube.

Division A term used by some microbiologists and formerly by botanists, corresponding to the term phylum.

DNA (deoxyribonucleic acid) The fundamental hereditary material of all living organisms. In eukaryotes, stored primarily in the cell nucleus. A nucleic acid using deoxyribose rather than ribose.

DNA hybridization A process by which DNAs from two species are mixed and heated so that interspecific double helixes are formed.

DNA ligase Enzyme that unites Okazaki fragments of the lagging strand during DNA replication; also mends breaks in DNA strands. It connects pieces of a DNA strand and is used in recombinant DNA technology.

DNA methylation Addition of methyl groups to DNA; plays role in regulation of gene expression; protects a bacterium's DNA against its restriction endonucleases.

DNA polymerase Any of a group of enzymes that catalyze the formation of DNA strands from a DNA template.

Dominance In genetic terminology, the ability of one allelic form of a gene to determine the phenotype of a heterozygous individual, in which the homologous chromosome carries both it and a different allele. For example, if *A* and *a* are two allelic forms of a gene, *A* is said to be dominant to *a* if *AA* diploids and *Aa* diploids are phenotypically identical and are distinguishable from *aa* diploids. The *a* allele is said to be recessive.

Dominance hierarchy The set of relationships within a group of animals, usually established and maintained by aggression, in which one individual has precedence over all others in eating, mating, and other activities; a second individual has precedence over all but the highest-ranking individual, and so on down the line.

Dormancy A condition in which normal activity is suspended, as in some seeds and buds.

Dorsal [L. *dorsum*: back] Pertaining to the back or upper surface. (Contrast with ventral.)

Double fertilization Process virtually unique to angiosperms in which one sperm nucleus combines with the egg to produce a zygote, and the other sperm nucleus combines with the two polar nuclei to produce the first cell of the triploid endosperm.

Double helix Of DNA: molecular structure in which two complementary polynucleotide strands, antiparallel to each other, form a right-handed spiral.

Duodenum (doo´ uh dee´ num) The beginning portion of the vertebrate small intestine. (Contrast with ileum, jejunum.)

Duplication (genetic) A mutation resulting from the introduction into the genome of an extra copy of a segment of a gene or chromosome. (Contrast with deletion, point mutation.)

Dynein [Gr. *dunamis*: power] A protein that undergoes conformational changes and thus plays a part in the movement of eukaryotic flagella and cilia.

Ear pinnae (pin´ ee) [L. *wings*] External ear structures that surround the auditory canals.

Ecdysone (eck die´ sone) [Gr. *ek*: out of + *dyo*: to clothe] In insects, a hormone that induces molting.

Echinoderm (e kine´ oh durm) A member of the phylum Echinodermata, such as a seastar or sea urchin.

Ecological biogeography The study of the distributions of organisms from an ecological perspective, usually concentrating on migration, dispersal, and species interactions.

Ecological community The species living together at a particular site.

Ecological niche (nitch) [L. *nidus*: nest] The functioning of a species in relation to other species and its physical environment.

Ecology [Gr. *oikos*: house + *logos*: discourse, study] The scientific study of the interaction of organisms with their environment, including both the physical environment and the other organisms that live in it.

Ecosystem (eek´ oh sis tum) The organisms of a particular habitat, such as a pond or forest, together with the physical environment in which they live.

Ecto- (eck´ toh) [Gr.: outer, outside] A prefix used to designate a structure on the outer surface of the body. For example, ectoderm. (Contrast with endo- and meso-.)

Ectoderm [Gr. *ektos*: outside + *derma*: skin] The outermost of the three embryonic tissue layers first delineated during gastrulation. Gives rise to the skin, sense organs, nervous system, etc.

Ectotherm [Gr. *ektos*: outside + *thermos*: heat] An animal unable to control its body temperature. (Contrast with endotherm.)

Edema (i dee´ mah) [Gr. *oidema*: swelling] Tissue swelling caused by the accumulation of fluid.

Edge effect The changes in ecological processes in a community caused by physical and biological factors originating in an adjacent community.

Effector Any organ, cell, or organelle that moves the organism through the environment or else alters the environment to the organism's advantage. Examples include muscle, bone, and a wide variety of exocrine glands.

Efferent [L. *ex*: out + *ferre*: to bear] Away from, as in neurons that conduct action potentials out from the central nervous system, or arterioles that conduct blood away from a structure. (Contrast with afferent.)

Egg In all sexually reproducing organisms, the female gamete; in birds, reptiles, and some other vertebrates, a structure within which early embryonic development occurs.

Elasticity The property of returning quickly to a former state after a disturbance.

Electrocardiogram (EKG) A graphic recording of electrical potentials from the heart.

Electroencephalogram (EEG) A graphic recording of electrical potentials from the brain.

Electromyogram (EMG) A graphic recording of electrical potentials from muscle.

Electron (e lek´ tron) [L. *electrum*: amber (associated with static electricity), from Gr. *slektor*: bright sun (color of amber)] One of the three most important fundamental particles of matter, with mass approximately 0.00055 amu and charge −1.

Electron microscope An instrument that uses an electron beam to form images of minute structures; the transmission electron microscope is useful for thinly-sliced material, and the scanning electron microscope gives surface views of cells and organisms.

Electrophoresis (e lek´ tro fo ree´ sis) [L. *electrum*: amber + Gr. *phorein*: to bear] A separation technique in which substances are separated from one another on the basis of their electric charges and molecular weights.

Electrotonic potential In neurons, a hyperpolarization or small depolarization of the membrane potential induced by the application of a small electric current. (Contrast with action potential, resting potential.)

Elemental substance A substance composed of only one type of atom.

Embolus (em´ buh lus) [Gr. *embolos*: inserted object; stopper] A circulating blood clot. Blockage of a blood vessel by an embolus or by a bubble of gas is referred to as an **embolism**. (Contrast with thrombus.)

Embryo [Gr. *en-*: in + *bryein*: to grow] A young animal, or young plant sporophyte, while it is still contained within a protective structure such as a seed, egg, or uterus.

Embryo sac In angiosperms, the female gametophyte. Found within the ovule, it consists of eight or fewer cells, membrane bounded, but without cellulose walls between them.

Emergent property A property of a complex system that is not exhibited by its individual component parts.

Emigration The deliberate and usually oriented departure of an organism from the habitat in which it has been living.

Endemic (en dem´ ik) [Gr. *endemos*: dwelling in a place] Confined to a particular region, thus often having a comparatively restricted distribution.

Endergonic reaction One for which energy must be supplied. (Contrast with exergonic reaction.)

Endo- [Gr.: within, inside] A prefix used to designate an innermost structure. For example, endoderm, endocrine. (Contrast with ecto-, meso-.)

Endocrine gland (en´ doh krin) [Gr. *endon*: inside + *krinein*: to separate] Any gland, such as the adrenal or pituitary gland of vertebrates, that secretes certain substances, especially hormones, into the body through the blood.

Endocrinology The study of hormones and their actions.

Endocytosis A process by which liquids or solid particles are taken up by a cell through invagination of the plasma membrane. (Contrast with exocytosis.)

Endoderm [Gr. *endon*: within + *derma*: skin] The innermost of the three embryonic tissue layers first delineated during gastrulation. Gives rise to the digestive and respiratory tracts and structures associated with them.

Endodermis [Gr. *endon*: within + *derma*: skin] In plants, a specialized cell layer marking the inside of the cortex in roots and some stems. Frequently a barrier to free diffusion of solutes.

Endomembrane system Endoplasmic reticulum plus Golgi apparatus plus, when present, lysosomes; thus, a system of membranes that exchange material with one another.

Endometrium (en do mee´ tree um) [Gr. *endon*: within + *metrios*: womb] The epithelial cells lining the uterus of mammals.

Endoplasmic reticulum [Gr. *endon*: within + L. *plasma*: form; L. *reticulum*: little net] A system of membrane-bounded tubes and flattened sacs, often continuous with the nuclear envelope, found in the cytoplasm of eukaryotes. Exists as rough ER, studded with ribosomes, and smooth ER, lacking ribosomes.

Endorphins Naturally occurring, opiatelike substances in the mammalian brain.

Endoskeleton A skeleton covered by other, soft body tissues. (Contrast with exoskeleton.)

Endosperm [Gr. *endon*: within + *sperma*: seed] A specialized triploid seed tissue found only in angiosperms; contains stored food for the developing embryo.

Endosymbiosis [Gr. *endon*: within + *syn*: together + *bios*: life] The living together of two species, with one living inside the body (or even the cells) of the other.

Endosymbiotic theory Theory that the eukaryotic cell evolved from a prokaryote that contained other, endosymbiotic prokaryotes.

Endotherm [Gr. *endon*: within + *thermos*: hot] An animal that can control its body temperature by the expenditure of its own metabolic energy. (Contrast with ectotherm.)

Energetic cost The difference between the energy an animal would have expended had it rested, and that expended in performing a behavior.

Energy The capacity to do work.

Enhancer In eukaryotes, a DNA sequence, lying on either side of the gene it regulates, that stimulates a specific promoter.

Enkephalins [Gr. *en-*: in + *kephale*: head] Two of the endorphins. (See endorphin.)

Enterocoelous development A pattern of development in which the coelum is formed by an outpocketing of the embryonic gut (enteron).

Enterokinase (ent uh row kine´ ase) An enzyme secreted by the mucosa of the duodenum. It activates the zymogen trypsinogen to create the active digestive enzyme trypsin.

Entrainment With respect to circadian rhythms, the process whereby the period is adjusted to match the 24-hour environmental cycle.

Entropy (en´ tro pee) [Gr. *en*: in + *tropein*: to change] A measure of the degree of disorder in any system. A perfectly ordered system has zero entropy; increasing disorder is measured by positive entropy. Spontaneous reactions in a closed system are always accompanied by an increase in disorder and entropy.

Environment An organism's surroundings, both living and nonliving; includes temperature, light intensity, and all other species that influence the focal organism.

Enzyme (en´ zime) [Gr. *en*: in + *zyme*: yeast] A protein, on the surface of which are chemical groups so arranged as to make the enzyme a catalyst for a chemical reaction.

Eon The largest division of geological time.

Epi- [Gr.: upon, over] A prefix used to designate a structure located on top of another; for example: epidermis, epiphyte.

Epicotyl (epp´ i kot´ il) [Gr. *epi*: upon + *kotyle*: something hollow] That part of a plant embryo or seedling that is above the cotyledons.

Epidermis [Gr. *epi*: upon + *derma*: skin] In plants and animals, the outermost cell layers. (Only one cell layer thick in plants.)

Epididymis (epuh did´ uh mus) [Gr. *epi*: upon + *didymos*: testicle] Coiled tubules in the testes that store sperm and conduct sperm from the seiminiferous tubules to the vas deferens.

Epinephrine (ep i nef´ rin) [Gr. *epi*: upon + *nephros*: a kidney] The "fight or flight" hormone. Produced by the medulla of the adrenal gland, it also functions as a neurotransmitter. Also known as adrenaline.

Epiphyte (ep´ e fyte) [Gr. *epi*: upon + *phyton*: plant] A specialized plant that grows on the surface of other plants but does not parasitize them.

Episome A plasmid that may exist either free or integrated into a chromosome. (See plasmid.)

Epistasis An interaction between genes, in which the presence of a particular allele of one gene determines whether another gene will be expressed.

Epithelium In animals, a layer of cells covering or lining an external surface or a cavity.

Equilibrium (1) In biochemistry, a state in which forward and reverse reactions are proceeding at counterbalancing rates, so there is no observable change in the concentrations of reactants and products. (2) In evolutionary genetics, a condition in which allele and genotype frequencies in a population are constant from generation to generation.

Era The second largest division of geological time.

Error signal In physiology, the difference between a set-point and a feedback signal that results in a corrective response.

Erythrocyte (ur rith´ row sight) [Gr. *erythros*: red + *kytos*: hollow vessel] A red blood cell.

Esophagus (i soff´ i gus) [Gr. *oisophagos*: gullet] That part of the gut between the pharynx and the stomach.

Essential amino acid An amino acid an animal cannot synthesize for itself and must obtain from its diet.

Essential element An irreplaceable mineral element without which normal growth and reproduction cannot proceed.

Estivation (ess tuh vay´ shun) [L. *aestivalis*: summer] A state of dormancy and hypometabolism that occurs during the summer; usually a means of surviving drought and/or intense heat. Contrast with hibernation.

Estrogen Any of several steroid sex hormones, produced chiefly by the ovaries in mammals.

Estrous cycle The cyclical changes in reproductive physiology and behavior in female mammals (other than some primates), culminating in estrus.

Estrus (es´ truss) [L. *oestrus*: frenzy] The period of heat, or maximum sexual receptivity, in some female mammals. Ordinarily, the estrus is also the time of release of eggs in the female.

Ethology (ee thol´ o jee) [Gr. *ethos*: habit, custom + *logos*: discourse] The study of whole patterns of animal behavior in natural environments, stressing the analysis of adaptation and evolution of the patterns.

Ethylene One of the plant hormones, the gas $H_2C{=\!=}CH_2$.

Etiolation Plant growth in the absence of light.

Eubacteria (yew bacteria) Kingdom including the great majority of bacteria, such as the gram negative bacteria, gram positive bacteria, mycoplasmas, etc. (Contrast with Archaebacteria.)

Euchromatin Chromatin that is diffuse and non-staining during interphase; may be transcribed. (Contrast with heterochromatin.)

Eukaryotes (yew car´ ry otes) [Gr. *eu*: true + *karyon*: kernel or nucleus] Organisms whose cells contain their genetic material inside a nucleus. Includes all life other than the viruses, Archaebacteria, and Eubacteria.

Eusocial Term applied to insects, such as termites, ants, and many bees and wasps, in which individuals cooperate in the care of offspring, there are sterile castes, and generations overlap.

Eutrophication (yoo trofe´ ik ay´ shun) [Gr. *eu-*: well + *trephein*: to flourish] The addition of nutrient materials to water. Especially in lakes, the subsequent flourishing of algae and microorganisms can result in oxygen depletion and the eventual stifling of life in the water.

Evergreen A plant that retains its leaves through all seasons. (Contrast with deciduous.)

Evolution Any gradual change. Organic evolution, often referred to as evolution, is any genetic and resulting phenotypic change in organisms from generation to generation.

Evolutionary agent Any factor that influences the direction and rate of evolutionary changes.

Evolutionary biology The collective branches of biology that study evolutionary process and their products—the diversity and history of living things.

Evolutionary conservative Traits of organisms that evolve very slowly.

Evolutionary radiation The proliferation of species within a single evolutionary lineage.

Excitatory postsynaptic potential (EPSP) A change in the resting potential of a postsynaptic membrane in a positive (depolarizing) direction. (Contrast with inhibitory postsynaptic potential.)

Excretion Release of metabolic wastes by an organism.

Exergonic reaction A reaction in which free energy is released. (Contrast with endergonic reaction.)

Exo- (eks´ oh) Same as ecto-.

Exocrine gland (eks´ oh krin) [Gr. *exo*: outside + *krinein*: to separate] Any gland, such as a salivary gland, that secretes to the outside of the body or into the gut.

Exocytosis A process by which a vesicle within a cell fuses with the plasma membrane and releases its contents to the outside. (Contrast with endocytosis.)

Exon A portion of a DNA molecule, in eukaryotes, that codes for part of a polypeptide. (Contrast with intron.)

Exoskeleton (eks´ oh skel´ e ton) A hard covering on the outside of the body; the exoskeleton of insects and other arthropods has many of the same functions as the bony internal skeleton of vertebrates. (Contrast with endoskeleton.)

Experiment A scientific method in which particular factors are manipulated while other factors are held constant so that the potential influences of the manipulated factors can be determined.

Exploitation competition Competition that occurs because resources are depleted. (Contrast with interference competition.)

Exponential growth Growth, especially in the number of organisms in a population, which is a simple function of the size of the growing entity: the larger the entity, the faster it grows. (Contrast with logistic growth.)

Expressivity The degree to which a genotype is expressed in the phenotype— may be affected by the environment.

Extensor A muscle the extends an appendage.

Extinction The termination of a lineage of organisms.

Extrinsic protein A membrane protein found only on the surface of the membrane. (Contrast with intrinsic protein.)

F_1 generation The immediate progeny of a mating; the first filial generation.

F_2 generation The immediate progeny of a mating between members of the F_1 generation.

F-duction Transfer of genes from one bacterium to another, using the F-factor as a vehicle.

F-factor In some bacteria, the fertility factor; a plasmid conferring "maleness" on the cell that contains it.

Facilitated diffusion Passive movement through a membrane involving a specific carrier protein; does not proceed against a concentration gradient. (Contrast with active transport, free diffusion.)

Facultative Capable of occurring or not occurring, as in facultative aerobes. (Contrast with obligate.)

Family In taxonomy, the category below the order and above the genus; a group of related, similar genera.

Fat A triglyceride that is solid at room temperature. (Contrast with oil.)

Fatty acid A molecule with a long hydrocarbon tail and a carboxyl group at the other end. Found in many lipids.

Fauna (faw´ nah) All of the animals found in a given area. (Contrast with flora.)

Feces [L. *faeces*: dregs] Waste excreted from the digestive system.

Feedback control Control of a particular step of a multistep process, induced by the presence or absence of a product of one of the later steps. A thermostat regulating the flow of heating oil to a furnace in a home is a negative feedback control device.

Fermentation (fur men tay´ shun) [L. *fermentum*: yeast] The degradation of a substance such as glucose to smaller molecules with the extraction of energy, without the use of oxygen (i.e., anaerobically). Involves the glycolytic pathway.

Fertilization Union of gametes. Also known as syngamy.

Fertilization membrane A membrane surrounding an animal egg which becomes rapidly raised above the egg surface within seconds after fertilization, serving to prevent entry of a second sperm.

Fetus The latter stages of an embryo that is still contained in an egg or uterus; in humans, the unborn young from the eighth week of pregnancy to the moment of birth.

Fiber An elongated and tapering cell of vascular plants, usually with a thick cell wall. Serves a support function.

Fibrin A protein that polymerizes to form long threads that provide structure to a blood clot.

Filter feeder An organism that feeds upon much smaller organisms, that are suspended in water or air, by means of a straining device.

Filtration In the excretory physiology of some animals, the process by which the initial urine is formed; water and most solutes are transferred into the excretory tract, while proteins are retained in the blood or hemolymph.

First law of thermodynamics Energy can be neither created nor destroyed.

Fission Reproduction of a prokaryote by division of a cell into two comparable progeny cells.

Fitness The contribution of a genotype or phenotype to the composition of subsequent generations, relative to the contribution of other genotypes or phenotypes. (See inclusive fitness.)

Fixed action pattern A behavior that is genetically programmed.

Flagellate (flaj´ el late) A member of the phylum Mastigophora, unicellular eukaryotes that propel themselves by flagella.

Flagellin (fla jell´ in) The protein from which prokaryotic (but not eukaryotic) flagella are constructed.

Flagellum (fla jell´ um) (plural: flagella) [L. *flagellum*: whip] Long, whiplike appendage that propels cells. Prokaryotic flagella differ sharply from those found in eukaryotes.

Flexor A muscle that flexes an appendage.

Flora (flore´ ah) All of the plants found in a given area. (Contrast with fauna.)

Florigen A plant hormone (not yet isolated) involved in the conversion of a vegetative shoot apex to a flower.

Flower The total reproductive structure of an angiosperm; its basic parts include the calyx, corolla, stamens, and carpels.

Fluorescence The emission of a photon of visible light by an excited atom or molecule.

Follicle [L. *folliculus*: little bag] In female mammals, an immature egg surrounded by nutritive cells.

Follicle-stimulating hormone A gonadotropic hormone produced by the anterior pituitary.

Food chain A portion of a food web, most commonly a simple sequence of prey species and the predators that consume them.

Food web The complete set of food links between species in a community; a diagram indicating which ones are the eaters and which are consumed.

Forb Any broad-leaved (dicotyledonous), herbaceous plant. Especially applied to such plants growing in grasslands.

Fossil Any recognizable structure originating from an organism, or any impression from such a structure, that has been preserved over geological time.

Founder effect Random changes in allele frequencies resulting from establishment of a population by a very small number of individuals.

Fovea [L. *fovea*; a small pit] The area, in the vertebrate retina, of most distinct vision.

Frame-shift mutation A mutation resulting from the addition or deletion of a single base pair in the DNA sequence of a gene. As a result of this, mRNA transcribed from such a gene is translated normally until the ribosome reaches the point at which the mutation has occurred. From that point on, codons are read out of proper register and the amino acid sequence bears no resemblance to the normal sequence. (Contrast with missense mutation, nonsense mutation.)

Free diffusion Diffusion directly across a membrane without the involvement of carrier molecules. Free diffusion is not saturable and cannot cause the net transport from a region of low concentration to a region of higher concentration. (Contrast with facilitated diffusion and active transport.)

Free energy That energy which is available for doing useful work, after allowance has been made for the increase or decrease of disorder. Designated by the symbol G (for Gibbs free energy), and defined by: $G = H - TS$, where H = heat, S = entropy, and T = absolute (Kelvin) temperature.

Frequency-dependent selection Selection that changes in intensity when the proportion of individuals under selection increases or decreases.

Fruit In angiosperms, a ripened and mature ovary (or group of ovaries) containing the seeds. Sometimes applied to reproductive structures of other groups of plants, and includes any adjacent parts which may be fused with the reproductive structures.

Fruiting body A structure that bears spores.

Fundamental niche The range of condition under which an organism could survive if it were the only one in the environment. (Contrast with realized niche.)

Fungus (fung´ gus) A member of the kingdom Fungi, a (usually) multicellular eukaryote with absorptive nutrition.

G_1 phase In the cell cycle, the gap between the end of mitosis and the onset of the S phase.

G_2 phase In the cell cycle, the gap between the S (synthesis) phase and the onset of mitosis.

G-protein A membrane protein involved in signal transduction; characterized by binding guanyl nucleotides. The activation of certain receptors activates the G-protein, which in turn activates adenylate cyclase. G-protein activation involves binding a GTP molecule in place of a GDP molecule.

Gametangium (gam i tan´ gee um) [Gr. *gamos*: marriage + *angeion*: vessel or reservoir] Any plant or fungal structure within which a gamete is formed.

Gamete (gam´ eet) [Gr. *gamete*: wife, *gametes*: husband] The mature sexual reproductive cell: the egg or the sperm.

Gametocyte (ga meet´ oh site) [Gr. *gamete*: wife, *gametes*: husband + *kytos*: cell] The cell that gives rise to sex cells, either the eggs or the sperm. (See oocyte and spermatocyte.)

Gametogenesis (ga meet´ oh jen´ e sis) [Gr. *gamete*: wife, *gametes*: husband + *genesis*: source] The specialized series of cellular divisions that leads to the production of sex cells (gametes). (Contrast with oogenesis and spermatogenesis.)

Gametophyte (ga meet´ oh fyte) In plants with alternation of generations, the haploid phase that produces the gametes. (Contrast with sporophyte.)

Ganglion (gang´ glee un) [Gr.: tumor] A group or concentration of neuron cell bodies.

Gap junction A 2.7-nanometer gap between plasma membranes of two animal cells, spanned by protein channels. Gap junctions allow chemical substances or electrical signals to pass from cell to cell.

Gas exchange In animals, the process of taking up oxygen from the environment and releasing carbon dioxide to the environment.

Gastrovascular cavity Serving for both digestion (gastro) and circulation (vascular); in particular, the central cavity of the body of jellyfish and other cnidarians.

Gastrula (gas´ true luh) [Gr. *gaster*: stomach] An embryo forming the characteristic three cell layers (ectoderm, endoderm, and mesoderm) which will give rise to all of the major tissue systems of the adult animal.

Gastrulation Development of a blastula into a gastrula.

Gated channel A channel (membrane protein) that opens and closes in response to binding of specific molecules or to changes in membrane potential.

Gene [Gr. *gen*: to produce] A unit of heredity. Used here as the unit of genetic function which carries the information for a single polypeptide.

Gene amplification Creation of multiple copies of a particular gene, allowing the production of large amounts of the RNA transcript (as in rRNA synthesis in oocytes).

Gene cloning Formation of a clone of bacteria or yeast cells containing a particular foreign gene.

Gene family A set of identical, or once-identical, genes, derived from a single parent gene; need not be on the same chromosomes; classic example is the globin family in vertebrates.

Gene flow The exchange of genes between different species (an extreme case referred to as hybridization) or between different populations of the same species caused by migration following breeding.

Gene pool All of the genes in a population.

Gene therapy Treatment of a genetic disease by providing patients with cells containing wild-type alleles for the genes that are nonfunctional in their bodies.

Generative nucleus In a pollen tube, a haploid nucleus that undergoes mitosis to produce the two sperm nuclei that participate in double fertilization. (Contrast with tube nucleus.)

Generator potential A stimulus-induced change in membrane resting potential in the direction of threshold for generating action potentials.

Genet The genetic individual of a plant that is composed of a number of nearly identical but repeated units.

Genetic drift Changes in gene frequencies from generation to generation in a small population as a result of random processes.

Genetic stochasticity Variation in the frequencies of alleles and genotypes in a population over time.

Genetics The study of heredity.

Genetic structure The frequencies of alleles and genotypes in a population.

Genome (jee´ nome) The genes in a complete haploid set of chromosomes.

Genome project An effort to map and sequence the entire genome of a species.

Genotype (jean´ oh type) [Gr. *gen*: to produce + *typos*: impression] An exact description of the genetic constitution of an individual, either with respect to a single trait or with respect to a larger set of traits. (Contrast with phenotype.)

Genus (jean´ us) (plural: genera) [Gr. *genos*: stock, kind] A group of related, similar species.

Geographical (allopatric) speciation
Formation of two species from one by the interposition of (or crossing of) a physical barrier. (Contrast with parapatric, sympatric speciation.)

Geotropism See gravitropism.

Germ cell A reproductive cell or gamete of a multicellular organism.

Germination The sprouting of a seed or spore.

Gestation (jes tay´ shun) [L. *gestare*: to bear] The period during which the embryo of a mammal develops within the uterus. Also known as **pregnancy**.

Gibberellin (jib er el´ lin) [L. *gibberella*: hunchback (refers to shape of a reproductive structure of a fungus that produces gibberellins)] One of a class of plant growth substances playing roles in stem elongation, seed germination, flowering of certain plants, etc. Named for the fungus *Gibberella*.

Gill An organ for gas exchange in aquatic organisms.

Gill arch A skeletal structure that supports gill filaments and the blood vessels that supply them.

Gizzard (giz´ erd) [L. *gigeria*: cooked chicken parts] A very muscular port of the stomach of birds that grinds up food, sometimes with the aid of fragments of stone.

Gland An organ or group of cells that produces and secretes one or more substances.

Glans penis Sexually sensitive tissue at the tip of the penis.

Glia (glee´ uh) [Gr.: glue] Cells, found only in the nervous system, which do not conduct action potentials.

Glomerulus (glo mare´ yew lus) [L. *glomus*: ball] Sites in the kidney where blood filtration takes place. Each glomerulus consists of a knot of capillaries served by afferent and efferent arterioles.

Glucocorticoids Steroid hormones produced by the adrenal medulla. Secreted in response to ACTH, they inhibit glucose uptake by many tissues in addition to mediating other stress responses.

Glucagon A hormone produced and released by cells in the islets of Langerhans of the pancreas. It stimulates the breakdown of glycogen in liver cells.

Gluconeogenesis The biochemical synthesis of glucose from other substances, such as amino acids, lactate, and glycerol.

Glucose (glue´ kose) [Gr. *gleukos*: sweet wine mash for fermentation] The most common sugar, one of several monosaccharides with the formula $C_6H_{12}O_6$.

Glycerol (gliss´ er ole) A three-carbon alcohol with three hydroxyl groups, the linking component of phospholipids and triglycerides.

Glycogen (gly´ ko jen) A branched-chain polymer of glucose, similar to starch (which is less branched and may be of lower molecular weight). Exists mostly in liver and muscle;

the principal storage carbohydrate of most animals and fungi.

Glycolysis (gly kol´ li sis) [from glucose + Gr. *lysis*: loosening] The enzymatic breakdown of glucose to pyruvic acid. One of the oldest energy-yielding machanisms in living organisms.

Glycosidic linkage The connection in an oligosaccharide or polysaccharide chain, formed by removal of water during the linking of monosaccharides.by root pressure.

Glyoxysome (gly ox´ ee soam) A type of microbody, found in plants, in which stored lipids are converted to carbohydrates.

Golgi apparatus (goal´ jee) A system of concentrically folded membranes found in the cytoplasm of eukaryotic cells. Plays a role in the production and release of secretory materials such as the digestive enzymes manufactured in the pancreas. First described by Camillo Golgi (1844–1926).

Gonad (go´ nad) [Gr. *gone*: seed, that which produces seed] An organ that produces sex cells in animals: either an ovary (female gonad) or testis (male gonad).

Gonadotropin A hormone that stimulates the gonads.

Grade The level of complexity found in an animal's body plan.

Gram stain A differential stain useful in characterizing bacteria.

Granum Within a chloroplast, a stack of thylakoids.

Gravitropism A directed plant growth response to gravity.

Grazer An animal that eats the vegetative tissues of herbaceous plants.

Green gland An excretory organ of crustaceans.

Gross morphology The sizes and shapes of the major body parts of a plant or animal.

Gross primary production The total energy captured by plants growing in a particular area.

Ground meristem That part of an apical meristem that gives rise to the ground tissue system of the primary plant body.

Ground tissue system Those parts of the plant body not included in the dermal or vascular tissue systems. Ground tissues function in storage, photosynthesis, and support.

Groundwater Water present deep in soils and rocks; may be stationary or flow slowly eventually to discharge into lakes, rivers, or oceans.

Group transfer The exchange of atoms between molecules.

Growth Irreversible increase in volume (probably the most accurate definition, but at best a dangerous oversimplification).

Growth factors A group of proteins that circulate in the blood and trigger the normal growth of cells. Each growth factor acts only on certain target cells.

Growth stage That stage in the life history of an organism in which it grows to its adult size.

Guard cells In plants, paired epidermal cells which surround and control the opening of a stoma (pore).

Gut An animal's digestive tract.

Guttation The extrusion of liquid water through openings in leaves, caused by root pressure.

Gymnosperm (jim´ no sperm) [Gr. *gymnos*: naked + *sperma*: seed] A plant, such as a pine or other conifer, whose seeds do not develop within an ovary (hence, the seeds are "naked").

Habit The form or pattern of growth characteristic of an organism.

Habitat The environment in which an organism lives.

Habituation (ha bich´ oo ay shun) The simplest form of learning, in which an animal presented with a stimulus without reward or punishment eventually ceases to respond.

Hair cell A type of mechanosensor in animals.

Half-life The time required for half of a sample of a radioactive isotope to decay to its stable, nonradioactive form.

Halophyte (hal´ oh fyte) [Gr. *halos*: salt + *phyton*: plant] A plant that grows in a saline (salty) environment.

Haploid (hap´ loid) [Gr. *haploeides*: single] Having a chromosome complement consisting of just one copy of each chromosome. This is the normal "ploidy" of gametes or of asexual spores produced by meiosis or of organisms (such as the gametophyte generation of plants) that grow from such spores without fertilization.

Haplontic life cycle A life cycle in which the zygote is the only diploid cell.

Hardy–Weinberg rule The rule that the basic processes of Mendelian heredity (meiosis and recombination) do not alter either the frequencies of genes or their diploid combinations. The Law also states how the percentages of diploid combinations can be predicted from a knowledge of the proportions of alleles in the population.

Haustorium (haw stor´ ee um) [L. *haustus*: draw up] A specialized hypha or other structure by which fungi and some parasitic plants draw food from a host plant.

Haversian systems Units of organization in compact bone that reflect the action of intercommunicating osteoblasts.

Helper T cells T cells that participate in the activation of B cells and of other T cells; targets of the HIV-I virus, the agent of AIDS. (Contrast with cytotoxic T cells, suppressor T cells.)

Hematocrit (heme at o krit) [Gr. *haima*: blood + *krites*: judge] The proportion of 100 cc of blood that consists of red blood cells.

Hemizygous(hem´ ee zie´ gus) [Gr. *hemi*: half + *zygotos*: joined]   In a diploid organism, having only one allele for a given trait, typically the case for X-linked genes in male mammals and Z-linked genes in female birds. (Contrast with homozygous, heterozygous.)

Hemoglobin (hee´ mo glow´ bin) [Gr. *haima*: blood + L. *globus*: globe]   The colored protein of vertebrate blood (and blood of some invertebrates) which transports oxygen.

Hepatic (heh pat´ ik) [Gr. *hepar*: liver]   Pertaining to the liver.

Hepatic duct   The duct that conveys bile from the liver to the gallbladder.

Herbicide (ur´ bis ide)   A chemical substance that kills plants.

Herbivore [L. *herba*: plant + *vorare*: to devour]   An animal which eats the tissues of plants. (Contrast with carnivore, detritivore, omnivore.)

Heritable   Able to be inherited; in biology usually refers to genetically determined traits.

Hermaphroditism (her maf´ row dite´ ism) [Gr. *hermaphroditos*: a person with both male and female traits]   The coexistence of both female and male sex organs in the same organism.

Hertz (abbreviated as Hz)   Cycles per second.

Hetero- [Gr.: other, different]   A prefix used in biology to mean that two or more different conditions are involved; for example, heterotroph, heterozygous.

Heterochromatin   Chromatin that retains its coiling during interphase; generally not transcribed. (Contrast with euchromatin.)

Heterocyst   A large, thick-walled cell in the filaments of certain cyanobacteria; performs nitrogen fixation.

Heterogeneous nuclear RNA (hnRNA)   The product of transcription of a eukaryotic gene, including transcripts of introns.

Heterokaryon (het´ er oh care´ ee ahn) [Gr. *heteros*: different + *karyon*: kernel]   A cell or organism carrying a mixture of genetically distinguishable nuclei. A heterokaryon is usually the result of the fusion of two cells without fusion of their nuclei.

Heteromorphic (het´ er oh more´ fik) [Gr. *heteros*: different + *morphe*: form]   having a different form or appearance, as two heteromorphic life stages of a plant. (Contrast with isomorphic.)

Heterosporous (het´ er os´ por us)   Producing two types of spores, one of which gives rise to a female megaspore and the other to a male microspore. Heterosporous plants produce distinct female and male gametophytes. (Contrast with homosporous.)

Heterotherm   An animal that regulates its body temperature at a constant level at some times but not others, such as a hibernator.

Heterotroph (het´ er oh trof) [Gr. *heteros*: different + *trophe*: food]   An organism that requires preformed organic molecules as food. (Contrast with autotroph.)

Heterozygous (het´ er oh zie´ gus) [Gr. *heteros*: different + *zygotos*: joined]   Of a diploid organism having different alleles of a given gene on the pair of homologues carrying that gene. (Contrast with homozygous.)

Hexose   A six-carbon sugar, such as glucose or fructose.

Hfr (for "high frequency of recombination")   Donor bacterium in which the F-factor has been integrated into the chromosome. This produces a bacterium that transfers its chromosomal markers at a very high frequency to recipient (F$^-$) cells.

Hibernation [L. *hibernus*: winter]   The state of inactivity of some animals during winter; marked by a drop in body temperature and metabolic rate.

Hippocampus   A part of the forebrain that takes part in long-term memory formation.

Histamine (hiss; tah meen)   A substance released within a damaged tissue by a type of white blood cell. Histamines are responsible for aspects of allergice reactions, including the increased vascular permeability that leads to edema (swelling).

Histology   The study of tissues.

Histone   Any one of a group of basic proteins forming the core of a nucleosome, the structural unit of a eukaryotic chromosome. (See nucleosome.)

Historical biogeography   The study of the distributions of organisms from a long-term, historical perspective.

hnRNA   See heterogeneous nuclear RNA.

Holdfast   In many large attached algae, specialized tissue attaching the plant to its substratum.

Homeobox   A segment of DNA, found in a few genes, perhaps regulating the expression of other genes and thus controlling large-scale developmental processes.

Homeostasis (home´ ee o sta´ sis) [Gr. *homos*: same + *stasis*: position]   The maintenance of a steady state, such as a constant temperature or a stable social structure, by means of physiological or behavioral feedback responses.

Homeotherm (home´ ee o therm) [Gr. *homos*: same + *therme*: heat]   An animal that maintains a constant body temperature by virtue of its own heating and cooling mechanisms. (Contrast with heterotherm, poikilotherm.)

Homeotic genes (home´ ee ott´ ic)   Genes that determine what entire segments of an animal become.

Homeotic mutation   A drastic mutation causing the transformation of body parts in *Drosophila* metamorphosis. Examples include the *Antennapedia* and *ophthalmoptera* mutants.

Homolog (home´ o log´) [Gr. *homos*: same + *logos*: word]   One of a pair, or larger set, of chromosomes having the same overall genetic composition and sequence. In diploid organisms, each chromosome inherited from one parent is matched by an identical (except for mutational changes) chromosome—its homolog—from the other parent.

Homology (ho mol´ o jee) [Gr. *homologi(a)*: agreement]   A similarity between two structures that is due to inheritance from a common ancestor. The structures are said to be homologous. (Contrast with analogy.)

Homoplasy (home´ uh play zee) [Gr. *homos*: same + *plastikos*: to mold]   The presence in several species of a trait not present in their most common ancestor. Can result from convergent evolution, reverse evolution, or parallel evolution.

Homosporous   Producing a single type of spore that gives rise to a single type of gametophyte, bearing both female and male reproductive organs. (Contrast with heterosporous.)

Homozygous (home´ o zie´ gus) [Gr. *homos*: same + *zygotos*: joined]   Of a diploid organism having identical alleles of a given gene on both homologous chromosomes. An organism may be a "homozygote" with respect to one gene and, at the same time, a "heterozygote" with respect to another. (Contrast with heterozygous.)

Hormone (hore´ mone) [Gr. *hormon*: excite, stimulate]   A substance produced in one part of a multicellular organism and transported to another part where it exerts its specific effect on the physiology or biochemistry of the target cells.

Host   An organism that harbors a parasite and provides it with nourishment.

Host–parasite interaction   The dynamic interaction between populations of a host and the parasites that attack it.

Humoral immune system   The part of the immune system mediated by B cells; it is mediated by circulating antibodies and is active against extracellular bacterial and viral infections.

Humus (hew´ muss)   The partly decomposed remains of plants and animals on the surface of a soil. Its characteristics depend primarily upon climate and the species of plants growing on the site.

Hyaluronidase (hill yew ron´ uh dase)   An enzyme that digests proteoglycans. Found in sperm cells, it helps digest the coatings surrounding an egg so the sperm can penetrate the egg cell membrane.

Hybrid (high´ brid) [L. *hybrida*: mongrel]   The offspring of genetically dissimilar parents.

Hybridoma   A cell produced by the fusion of an antibody-producing cell with a myeloma cell; it produces monoclonal antibodies.

Hydrocarbon   A compound containing only carbon and hydrogen atoms.

Hydrogen bond   A chemical bond which arises from the attraction between the slight positive charge on a hydrogen atom and a slight negative charge on a nearby fluorine, oxygen, or nitrogen atom. Weak bonds, but found in great quantities in proteins, nucleic acids, and other biological macromolecules.

Hydrological cycle The sum total of movement of water from the oceans to the atmosphere, to the soil, and back to the oceans. Some water is cycled many times within compartments of the system before completing one full circuit.

Hydrolyze (hi´ dro lize) [Gr. *hydro*: water + *lysis*: cleavage] To break a chemical bond, as in a peptide linkage, with the insertion of the components of water, –H and –OH, at the cleaved ends of a chain. The digestion of proteins is a hydrolysis.

Hydrophilic [Gr. *hydro*: water + *philia*: love] Having an affinity for water. (Contrast with hydrophobic.)

Hydrophobic [Gr. *hydro*: water + *phobia*: fear] Molecules and amino acid side chains, which are mainly hydrocarbons (compounds of C and H with no charged groups or polar groups), have a lower energy when they are clustered together than when they are distributed through an aqueous solution. Because of their attraction for one another and their reluctance to mix with water they are called "hydrophobic." Oil is a hydrophobic substance; phenylalanine is a hydrophobic animo acid in a protein. (Contrast with hydrophilic.)

Hydrophobic interaction A weak attraction between highly nonpolar molecules or parts of molecules suspended in water.

Hydrostatic skeleton The incompressible internal liquids of some animals that transfer forces from one part of the body to another when acted upon by the surrounding muscles.

Hydroxyl group The –OH group, characteristic of alcohols.

Hymenopteran (high´ man op´ ter an) A member of the insect order Hymenoptera, such as a wasp, bee, or ant.

Hyperpolarization A change in the resting potential of a membrane so the inside of a cell becomes more electronegative. (Contrast with depolarization.)

Hypertension High blood pressure.

Hypertonic [Gr.: higher tension] Having a more negative osmotic potential, as a result of having a higher concentration of osmotically active particles. Said of one solution as compared with another. (Contrast with hypotonic, isotonic.)

Hypha (high´ fuh) (plural: hyphae) [Gr. *hyphe*: web] In the fungi, any single filament. May be multinucleate (zygomycetes, ascomycetes) or multicellular (basidiomycetes).

Hypocotyl That part of the embryonic or seedling plant shoot that is below the cotyledons.

Hypothalamus The part of the brain lying below the thalamus; it coordinates water balance, reproduction, temperature regulation, and metabolism.

Hypothetico-deductive method A method of science in which hypotheses are erected, predictions are made from them, and experiments and observations are performed to test the predictions. The process may be repeated many times in the course of answering a question.

Hypotonic [Gr.: lower tension] Having a less negative osmotic potential, as a result of having a lower concentration of osmotically active particles. Said of one solution as compared with another. (Contrast with hypertonic, isotonic.)

Imaginal disc In insect larvae, groups of cells that develop into specific adult organs.

Imbibition [L. *imbibo*: to drink] The binding of a solvent to another molecule. Dry starch and protein will imbibe water.

Immunoglobulins A class of proteins, with a characteristic structure, active as receptors and effectors in the immune system.

Immunological tolerance A mechanism by which an animal does not mount an immune response to the antigenic determinants of its own macromolecules.

Imprinting A rapid form of learning, in which an animal comes to make a particular response, which is maintained for life, to some object or other organism.

Inclusive fitness The sum of an individual's own fitness (the effect of producing its own offspring: the individual selection component) plus its influence on fitness in relatives other than direct descendants (the kin selection component).

Incomplete dominance Condition in which the heterozygous phenotype is intermediate between the two homozygous phenotypes.

Incomplete metamorphosis Insect development in which changes between instars are gradual.

Incus (in´ kus) [L. *incus*: anvil] The middle of the three bones that conduct movements of the eardrum to the oval window of the inner ear. (See malleus, stapes.)

Indeterminate cleavage A pattern of development in which individual cells retain the potential to develop into complete organisms if separated from one another well into development.

Individual fitness That component of inclusive fitness that results from an organism producing its own offspring. (Contrast with kin selection component.)

Indoleacetic acid See auxin.

Induced fit A change in the tertiary structures of some enzymes, caused by binding of substrate to the active site.

Inducer In enzyme systems, a small molecule which, when added to a growth medium, causes a large increase in the level of some enzyme. Generally it acts by binding to repressor and changing its conformation so that the repressor does not bind to the operator. In embryology, a substance that causes a group of target cells to differentiate in a particular way.

Inducible enzyme An enzyme that is present in much larger amounts when a particular compound (the inducer) has been added to the system. (Contrast with constitutive enzyme.)

Inflammation A nonspecific defense against pathogens; characterized by redness, swelling, pain, and increased temperature.

Inflorescence A structure composed of several flowers.

Ingestion Taking in of food by swallowing.

Inhibitor A substance which binds to the surface of an enzyme and interferes with its action on its substrates.

Inhibitory postsynaptic potential A change in the resting potential of a postsynaptic membrane in the hyperpolarizing (negative) direction.

Initiation complex Combination of a ribosomal light subunit, an mRNA molecule, and the tRNA charged with the first amino acid coded for by the mRNA; formed at the onset of translation.

Inositol triphosphate (IP3) An intracellular second messenger derived from membrane phospholipids.

Insertion sequence A large piece of DNA that can give rise to copies at other loci; a type of transposable genetic element.

Instar (in´ star) [L.: image, form] An immature stage of an insect between molts.

Instinct Behavior that is relatively highly sterotyped and self-differentiating, that develops in individuals unable to observe other individuals performing the behavior or to practice the behavior in the presence of the objects toward which it is usually directed.

Insulin (in´ su lin) [L. *insula*: island] A hormone, synthesized in islet cells of the pancreas, that promotes the conversion of glucose to the storage material, glycogen.

Integrase An enzyme that integrates retroviral cDNA into the genome of the host cell.

Integrated pest management A method of control of pests in which natural predators and parasites are used in conjunction with sparing use of chemical methods to achieve control of a pest without causing serious adverse environmental side effects.

Integument [L. *integumentum*: covering] A protective surface structure. In gymnosperms and angiosperms, a layer of tissue around the ovule which will become the seed coat. Gymnosperm ovules have one integument, angiosperm ovules two.

Intention movement The preparatory motions that animals go through prior to a complete behavior response; for example, the crouch before flying, the snarl before biting, etc.

Intercalary meristem A meristematic region in plants which occurs not apically, but between two regions of mature tissue. Intercalary meristems occur in the nodes of grass stems, for example.

Intercostal muscles   Muscles between the ribs that can augment breathing movements by elevating and suppressing the rib cage.

Interference competition   Competition resulting from direct behavioral interactions between organisms. (Contrast with exploitation competition.)

Interferon   A glycoprotein produced by virus-infected animal cells; increases the resistance of neighboring cells to the virus.

Interkinesis   The phase between the first and second meiotic divisions.

Interleukins   Regulatory proteins, produced by macrophages and lymphocytes, that act upon other lymphocytes and direct their development.

Intermediate filaments   Fibrous proteins that stabilize cell structure and resist tension.

Internode   Section between two nodes of a plant stem.

Interphase   The period between successive nuclear divisions during which the chromosomes are diffuse and the nuclear envelope is intact. It is during this period that the cell is most active in transcribing and translating genetic information.

Interspecific competition   Competition between members of two or more species.

Interstitial fluid   In vertebrates, the fluid filling the spaces between cells.

Intertropical convergence zone   The tropical region where the air rises most strongly; moves north and south with the passage of the sun overhead.

Intraspecific competition   Competition among members of a single species.

Intrinsic protein   A membrane protein that is embedded in the phospholipid bilayer of the membrane. (Contrast with extrinsic protein.)

Intrinsic rate of increase   The rate at which a population can grow when its density is low and environmental conditions are highly favorable.

Intron   A portion of a DNA molecule that, because of RNA splicing, is not involved in coding for part of a polypeptide molecule. (Contrast with exon.)

Invagination   An infolding.

Invasiveness   Ability of a bacterium to multiply within the body of a host.

Inversion (genetic)   A rare mutational event that leads to the reversal of the order of genes within a segment of a chromosome, as if that segment had been removed from the chromosome, turned 180°, and then reattached.

Invertebrate   Any animal that is not a vertebrate, that is, whose nerve cord is not enclosed in a backbone of bony segments.

In vitro [L.: in glass]   In a test tube, rather than in a living organism. (Contrast with in vivo.)

In vivo [L.: in the living state]   In a living organism. Many processes that occur in vivo can be reproduced in vitro with the right selection of cellular components. (Contrast with in vitro.)

Ion (eye´ on) [Gr.: wanderer]   An atom or group of atoms with electrons added or removed, giving it a negative or positive electrical charge.

Ionic channel   A membrane protein that can let ions pass across the membrane. The channel can be ion-selective, and it can be voltage-gated or ligand-gated.

Ionic bond   A chemical bond which arises from the electrostatic attraction between positively and negatively charged ions. Usually a strong bond.

Iris (eye´ ris) [Gr. iris: rainbow]   The round, pigmented membrane that surrounds the pupil of the eye and adjusts its aperture to regulate the amount of light entering the eye.

Irruption   A rapid increase in the density of a population. Often followed by massive emigration.

Islets of Langerhans   Clusters of hormone-producing cells in the pancreas.

Isogamy (eye sog´ ah mee) [Gr. isos: equal + gamos: marriage]   A kind of sexual reproduction in which the gametes (or gametangia) are not distinguishable on the basis of size or morphology.

Isolating mechanism   Geographical, physiological, ecological, or behavioral mechanisms that lead to a reduction in the frequency of hybrid matings.

Isomers   Molecules consisting of the same numbers and kinds of atoms, but differing in the way in which the atoms are combined.

Isomorphic (eye´ so more´ fik) [Gr. isos: equal + morphe: form]   having the same form or appearance, as two isomorphic life stages. (Contrast with heteromorphic.)

Isotonic [Gr.: same tension]   Having the same osmotic potential. Said of two solutions. (Contrast with hypertonic, hypotonic.)

Isotope (eye´ so tope) [Gr. isos: equal + topos: place]   Two isotopes of the same chemical element have the same number of protons in their nuclei, but differ in the number of neutrons.

Isozymes   Chemically different enzymes that catalyze the same reaction.

Jejunum (jih jew´ num)   The middle division of the small intestine, where most absorption of nutrients occurs. (See duodenum, ileum.)

Joule (jool, or jowl)   A unit of energy, equal to 0.24 calories.

Juvenile hormone   In insects, a hormone maintaining larval growth and preventing maturation or pupation.

Karyogamy (care´ ee og´ uh me) [Gr. karyon: kernel, nut + gamos: marriage]   Fusion of gamete nuclei.

Karyotype   The number, forms, and types of chromosomes in a cell.

Kelvin temperature scale   See absolute temperature scale.

Keratin (ker´ a tin) [Gr. keras: horn]   A protein which contains sulfur and is part of such hard tissues as horn, nail, and the outermost cells of the skin.

Ketone (key´ tone)   A compound with a C=O group attached to two other groups, neither of which is an H atom. Many sugars are ketones. (Contrast with aldehyde.)

Keystone species   A species that exerts a major influence on the composition and dynamics of the community in which it lives.

Kidneys   A pair of excretory organs in vertebrates.

Kin selection component   The component of inclusive fitness resulting from helping the survival of relatives containing the same alleles by descent from a common ancestor.

Kinase (kye´ nase)   An enzyme that transfers a phosphate group from ATP to another molecule. Protein kinases transfer phosphate from ATP to specific proteins, playing important roles in cell regulation.

Kinesis (ki nee´ sis) [Gr.: movement]   Orientation behavior in which the organism does not move in a particular direction with reference to a stimulus but instead simply moves at an increasing or decreasing rate until it ends up farther from the object or closer to it. (Contrast with taxis.)

Kinetochore (kin net´ oh core) [Gr. kinetos: moving + khorein: to move]   Specialized structure on a centromere to which microtubules attach.

Kingdom   The highest taxonomic category in the Linnaean system.

Knockout mouse   A genetically engineered mouse in which one or more functioning alleles have been replaced by defective alleles.

Lactic acid   The end product of fermentation in vertebrate muscle and some microorganisms.

Lagging strand   In DNA replication, the daughter strand that is synthesized discontinuously.

Lamella   Layer.

Larynx (lar´ inks)   A structure between the pharynx and the trachea that includes the vocal cords.

Larva (plural: larvae) [L.: ghost, early stage]   An immature stage of any invertebrate animal that differs dramatically in appearance from the adult.

Lateral   Pertaining to the side.

Lateral inhibition   In visual information processing in the arthropod eye, the mutual inhibition of optic nerve cells; results in enhanced detection of edges.

Laterization (lat´ ur iz ay shun)   The formation of a nutrient-poor soil that is rich in insoluble iron and aluminum compounds.

Law of independent assortment Alleles of different, unlinked genes assort independently of one another during gamete formation, Mendel's second law.

Law of segregation Alleles segregate from one another during gamete formation, Mendel's first law.

Leader sequence A sequence of amino acids at the N-terminal end of a newly synthesized protein, determining where the protein will be placed in the cell.

Leading strand In DNA replication, the daughter strand that is synthesized continuously.

Leaf axil The upper angle between a leaf and the stem, site of lateral buds which under appropriate circumstances become activated to form lateral branches.

Leaf primordium [L.: the beginning] A small mound of cells on the flank of a shoot apical meristem that will give rise to a leaf.

Lek A traditional courtship display ground, where males display to females.

Lenticel Spongy region in a plant's periderm, allowing gas exchange.

Leucoplast A colorless plastid that stores starch or fat.

Leukocyte (loo´ ko sight) [Gr. *leukos*: clear + *kutos*: hollow vessel] A white blood cell.

Leuteinizing hormone A peptide hormone produced by pituitary cells that stimulates follicle maturation in females.

Lichen (lie´ kun) [Gr. *leikhen*: licker] An organism resulting from the symbiotic association of a true fungus and either a cyanobacterium or a unicellular alga.

Life cycle The entire span of the life of an organism from the moment of fertilization (or asexual generation) to the time it reproduces in turn.

Life history The stages an individual goes through during its life.

Life table A table showing, for a group of equal-aged individuals, the proportion still alive at different times in the future and the number of offspring they produce during each time interval.

Ligament A band of connective tissue linking two bones in a joint.

Ligand (lig´ and) A molecule that binds to a receptor site of another molecule.

Light compass reaction A reaction of many invertebrates in which the angle between the direction of movement and the direction of the sun is kept constant.

Lignin The principal noncarbohydrate component of wood, a polymer that binds together cellulose fibrils in some plant cell walls.

Limiting resource The required resource whose supply most strongly influences the size of a population.

Linkage In genetics, association between markers on the same chromosome such that they do not show random assortment.

Linked markers recombine with one another at frequencies less than 0.5; the closer the markers on the chromosome, the lower the frequency of recombination.

Lipase (lip´ ase; lye´ pase) An enzyme that digests fats.

Lipids (lip´ ids) [Gr. *lipos*: fat] Substances in a cell which are easily extracted by organic solvents; fats, oils, waxes, steroids, and other large organic molecules, including those which, with proteins, make up the cell membranes. (See phospholipids.)

Litter The partly decomposed remains of plants on the surface and in the upper layers of the soil.

Littoral zone The coastal zone from the upper limits of tidal action down to the depths where the water is thoroughly stirred by wave action.

Liver A large digestive gland. In vertebrates, it secretes bile and is involved in the formation of blood.

Lobes Regions of the human cerebral hemispheres; includes the temporal, frontal, parietal, and occipital lobes.

Locus In genetics, a specific location on a chromosome. May be considered to be synonymous with "gene."

Logistic growth Growth, especially in the size of an organism or in the number of organisms that constitute a population, which slows steadily as the entity approaches its maximum size. (Contrast with exponential growth.)

Loop of Henle (hen´ lee) Long, hairpin loop of the mammalian renal tubule that runs from the cortex down into the medulla, and back to the cortex. Creates a concentration gradient in the interstitial fluids in the medulla.

Lophophore A U-shaped fold of the body wall with hollow, ciliated tentacles that encircles the mouth of animals in several different phyla. Used for filtering prey from the surrounding water.

Lordosis (lor doe´ sis) [Gk. *lordosis*: curving forward] A posture assumed by females of some mammalian species (especially rodents) to signal sexual receptivity.

Lumen (loo´ men) [L.: light] The cavity inside any tubular part of an organ, such as a piece of gut or a kidney tubule.

Lungs A pair of saclike chambers within the bodies of some animals, functioning in gas exchange.

Luteinizing hormone A gonadotropin produced by the anterior pituitary. It stimulates the gonads to produce sex hormones.

Lymph [L. *lympha*: water] A clear, watery fluid that is formed as a filtrate of blood; it contains white blood cells; it collects in a series of special vessels and is returned to the bloodstream.

Lymphocyte A major class of white blood cells. Includes T cells, B cells, and other cell types important in the immune response.

Lysis (lie´ sis) [Gr.: a loosening] Bursting of a cell.

Lysogenic The condition of a bacterium that carries the genome of a virus in a relatively stable form. (Contrast with lytic.)

Lysosome (lie´ so soam) [Gr. *lysis*: a loosening + *soma*: body] A membrane-bounded inclusion found in eukaryotic cells (other than plants). Lysosomes contain a mixture of enzymes that can digest most of the macromolecules found in the rest of the cell.

Lysozyme (lie´ so zyme) An enzyme in saliva, tears, and nasal secretions that attacks bacterial cell walls, as one of the body's nonspecific defense mechanisms.

Lytic Condition in which a bacterium lyses shortly after infection by a virus; the viral genome does not become stabilized within the bacterial cell. (Contrast with lysogenic.)

Macro- (mack´ roh) [Gr. *makros*: large, long] A prefix commonly used to denote something large. (Contrast with micro-.)

Macroevolution Evolutionary changes occurring over long time spans and usually involving changes in many traits. (Contrast with microevolution.)

Macroevolutionary time The time required for macroveolutionary changes in a lineage.

Macromolecule A giant polymeric molecule. The macromolecules are proteins, polysaccharides, and nucleic acids.

Macronutrient A mineral element required by plant tissues in concentrations of at least 1 milligram per gram of their dry matter.

Major histocompatibility complex (MHC) A complex of linked genes, with multiple alleles, that control a number of immunological phenomena; it is important in graft rejection.

Malleus (mal´ ee us) [L. *malleus*: hammer] The first of the three bones that conduct movements of the eardrum to the oval window of the inner ear. (See incus, stapes.)

Malpighian tubule (mal pee´ gy un) A type of protonephridium found in insects.

Mammal [L. *mamma*: breast, teat] Any animal of the class Mammalia, characterized by the production of milk by the female mammary glands and the possession of hair for body covering.

Mantle A sheet of specialized tissues that covers most of the viscera of mollusks; provides protection to internal organs and secretes the shell.

Mapping In genetics, determining the order of genes on a chromosome and the distances between them.

Marine [L. *mare*: sea, ocean] Pertaining to or living in the ocean. (Contrast with aquatic, terrestrial.)

Maritime climate Weather pattern typical of coasts of continents, particularly those on the western sides at mid latitudes, in which the difference between summer and winter is relatively small. (Contrast with continental climate.)

Marsupial (mar soo´ pee al)   A mammal belonging to the subclass Metatheria, such as opossums and kangaroos. Most have a pouch (marsupium) that contains the milk glands and serves as a receptacle for the young.

Mass extinctions   Geological periods during which rates of extinction were much higher than during intervening times.

Mass number   The sum of the number of protons and neutrons in an atom's nucleus.

Maternal inheritance (cytoplasmic inheritance)   Inheritance in which the phenotype of the offspring depends on factors, such as mitochondria or chloroplasts, that are inherited from the female parent through the cytoplasm of the female gamete.

Mating type   In some bacteria, fungi, and protists, sexual reproduction can occur only between partners of different mating type. "Mating type" is not the same as "sex," since some species have as many as 8 mating types; mating may also be between hermaphroditic partners of opposite mating type, with both partners acting as both "male" and "female" in terms of donating and receiving genetic information.

Maturation   The automatic development of a pattern of behavior, which becomes increasingly complex or precise as the animal matures. Unlike learning, the development does not require experience to occur.

Mechanosensor   A cell that is sensitive to physical movement and generates action potentials in response.

Medulla (meh dull´ luh) [L.: narrow]   (1) The inner, core region of an organ, as in the adrenal medulla (adrenal gland) or the renal medulla (kidneys). (2) The portion of the brain stem that connects to the spinal cord.

Medusa (meh doo´ suh)   The tentacle-bearing, jellyfish-like, free-swimming sexual stage in the life cycle of a cnidarian.

Mega- [Gr. *megas*: large, great]   A prefix often used to denote something large. (Contrast with micro-.)

Megareserve   A large park or reserve; usually has associated buffer areas in which human use of the environment is restricted to activities that do not destroy the functioning of the ecosystem.

Megasporangium   The special structure (sporangium) that produces the megaspores.

Megaspore [Gr. *megas*: large + *spora*:seed]   In plants, a haploid spore that produces a female gametophyte. In many cases the megaspore is larger than the male-producing microspore.

Meiosis (my oh´ sis) [Gr.: diminution]   Division of a diploid nucleus to produce four haploid daughter cells. The process consists of two successive nuclear divisions with only one cycle of chromosome replication.

Membrane potential   The difference in electrical charge between the inside and the outside of a cell, caused by a difference in the distribution of ions.

Mendelian population   A local population of individuals belonging to the same species and exchanging genes with one another.

Menopause   The time in a human female's life when the ovarian and menstrual cycles cease.

Menstrual cycle   The monthly sloughing off of the uterine lining if fertilization does not occur in the female. Occurs between puberty and menopause.

Meristem [Gr. *meristos*: divided]   Plant tissue made up of actively dividing cells.

Mesenchyme (mez´ en kyme) [Gr. *mesos*: middle + *enchyma*: infusion]   Embryonic or unspecialized cells derived from the mesoderm.

Meso- (mez´ oh) [Gr.: middle]   A prefix often used to designate a structure located in the middle, or a stage that appears at some intermediate time. For example, mesoderm, Mesozoic.

Mesoderm [Gr. *mesos*: middle + *derma*: skin]   The middle of the three embryonic tissue layers first delineated during gastrulation. Gives rise to skeleton, circulatory system, muscles, excretory system, and most of the reproductive system.

Mesoglea   The jelly-like middle layer that constitutes the bulk of the bodies of the medusae of many cnidarians; not a true cell layer.

Mesophyll (mez´ a fill) [Gr. *mesos*: middle + *phyllon*: leaf]   Chloroplast-containing, photosynthetic cells in the interior of leaves.

Mesosome (mez´ o soam´) [Gr. *mesos*: middle + *soma*: body]   A localized infolding of the plasma membrane of a bacterium.

Messenger RNA (mRNA)   A transcript of one of the strands of DNA, it carries information (as a sequence of codons) for the synthesis of one or more proteins.

Meta- [Gr.: between, along with, beyond]   A prefix used in biology to denote a change or a shift to a new form or level; for example, as used in metamorphosis.

Metabolic compensation   Changes in biochemical properties of an organism that render it less sensitive to temperature changes.

Metabolic pathway   A series of enzyme-catalyzed reactions so arranged that the product of one reaction is the substrate of the next.

Metabolism (meh tab´ a lizm) [Gr. *metabole*: to change]   The sum total of the chemical reactions that occur in an organism, or some subset of that total (as in "respiratory metabolism").

Metamorphosis (met´ a mor´ fo sis) [Gr. *meta*: between + *morphe*: form, shape]   A radical change occurring between one developmental stage and another, as for example from a tadpole to a frog or an insect larva to the adult.

Metaphase (met´ a phase) [Gr. *meta*: between]   The stage in nuclear division at which the centromeres of the highly supercoiled chromosomes are all lying on a plane (the metaphase plane or plate) perpendicular to a line connecting the division poles.

Metastasis (meh tass´ tuh sis)   The spread of cancer cells from their original site to other parts of the body.

Methanogen   Any member of a group of Archaebacteria that release methane as a metabolic product. This group is considered to be an extremely ancient one.

MHC   See major histocompatibility complex.

Micelles (my sells´) [L. *mica*: grain, crumb]   The small particles of fat in the small intestine, resulting from the emulsification of dietary fat by bile.

Micro- (mike´ roh) [Gr. *mikros*: small]   A prefix often used to denote something small. (Contrast with macro-, mega-.)

Microbiology [Gr. *mikros*: small + *bios*: life + *logos*: discourse]   The scientific study of microscopic organisms, particularly bacteria, unicellular algae, protists, and viruses.

Microbody   A small organelle, bounded by a single membrane and possessing a granular interior. Peroxisomes and glyoxysomes are types of microbodies.

Microevolution   The small evolutionary changes typically occurring over short time spans; generally involving a small number of traits and minor genetic changes. (Contrast with macroevolution.)

Microevolutionary time   The time required for microevolutionary changes within a lineage of organisms.

Microfilament   Minute fibrous structure generally composed of actin found in the cytoplasm of eukaryotic cells. They play a role in the motion of cells.

Micromorphology   The structure of the macromolecules of an organism.

Micronutrient   A mineral element required by plant tissues in concentrations of less than 100 micrograms per gram of their dry matter.

Microorganism   Any microscopic organism, such as a bacterium or one-celled alga.

Micropyle (mike´ roh pile) [Gr. *mikros*: small + *pyle*: gate]   Opening in the integument(s) of a seed plant ovule through which pollen grows to reach the female gametophyte within.

Microsporangium   The special structure (sporangium) that produces the microspores.

Microspores [Gr. *mikros*: small + *spora*: seed]   In plants, a haploid spore that produces a male gametophyte. In many cases the microspore is smaller than the female-producing megaspore.

Microtubules   Minute tubular structures found in centrioles, spindle apparatus, cilia, flagella, and other places in the cytoplasm of eukaryotic cells. These tubules play roles in the motion and maintenance of shape of eukaryotic cells.

Microvilli (singular: microvillus) The projections of epithelial cells, such as the cells lining the small intestine, that increase their surface area.

Middle lamella A layer of derivative polysaccharides that separates plant cells; a common middle lamella lies outside the primary walls of the two cells.

Migration The regular, seasonal movements of animals between breeding and nonbreeding ranges.

Mimicry (mim´ ik ree) The resemblance of one kind of organism to another, or to some inanimate object; serves the function of making the organism difficult to find, of discouraging potential enemies or of attracting potential prey. (See Batesian mimicry and Müllerian mimicry.)

Mineral An inorganic substance other than water.

Mineralocorticoid A hormone produced by the adrenal cortex that influences mineral ion balance; aldosterone.

Minimal medium A medium for the growth of bacteria, fungi, or tissue cultures, containing only those nutrients absolutely required for the growth of wild-type cells.

Minimum viable population. The smallest number of individuals required for a population to persist in a region.

Missense mutation A mutation that changes a codon for one amino acid to a codon for a different amino acid. (Contrast with frame-shift mutation, nonsense mutation.)

Mitochondrial matrix The fluid interior of the mitochondrion, enclosed by the inner mitochondrial membrane.

Mitochondrion (my´ toe kon´ dree un) (plural: mitochondria) [Gr. mitos: thread + chondros: cartilage, or grain] An organelle that occurs in eukaryotic cells and contains the enzymes of the ctric acid cycle, the respiratory chain, and oxidative phosphorylation. A mitochondrion is bounded by a double membrane.

Mitosis (my toe´ sis) [Gr. mitos: thread] Nuclear division in eukaryotes leading to the formation of two daughter nuclei each with a chromosome complement identical to that of the original nucleus.

Mitotic center Cellular region that organizes the microtubules for mitosis. In animals a centrosome serves as the mitotic center.

Mobbing Gathering of calling animals around a predator; their calls and the confusion they create reduce the probability that the predator can hunt successfully in the area.

Modular organism An organism which grows by producing additional units of body construction that are very similar to the units of which it is already composed.

Mole A quantity of a compound whose weight in grams is numerically equal to its molecular weight expressed in atomic mass units. Avogadro's number of molecules: 6.023×10^{23} molecules.

Molecular clock See radiometric clock.

Molecular formula A representation that shows how many atoms of each element are present in a molecule.

Molecular weight The sum of the atomic weights of the atoms in a molecule.

Molecule A particle made up of two or more atoms joined by covalent bonds or ionic attractions.

Mollusk (mol´ lusk) A member of the phylum Mollusca, such as a snail, clam, or octopus.

Molting The process of shedding part or all of an outer covering, as the shedding of feathers by birds or of the entire exoskeleton by arthropods.

Monecious (mo nee´ shus) [Gr.: one house] Organisms in which both sexes are "housed" in a single individual, which produces both eggs and sperm. (In some plants, these are found in different flowers within the same plant.) Examples: corn, peas, earthworms, hydras. (Contrast with dioecious, perfect flower.)

Moneran (moh neer´ un) A bacterium. This term was coined when both archaebacteria and eubacteria were considered to be members of a single kingdom, Monera.

Mono- [Gr. monos: one] Prefix denoting a single entity. (Contrast with poly.)

Monoclonal antibody Antibody produced in the laboratory from a clone of hybridoma cells, each of which produces the same specific antibody.

Monocot (short for monocotyledon) [Gr. monos: one + kotyledon: a cup-shaped hollow] Any member of the angiosperm class Monocotyledones, plants in which the embryo produces but a single cotyledon (seed leaf). Leaves of most monocots have their major veins arranged parallel to each other.

Monohybrid cross A mating in which the parents differ with respect to the alleles of only one locus of interest.

Monomer A small molecule, two or more of which can be combined to form oligomers (consisting of a few monomers) or polymers (consisting of many monomers).

Monophyletic (mon´ oh fih leht´ ik) [Gk. monos: single + phylon: tribe] Being descended from a single ancestral stock.

Monosaccharide A simple sugar. Oligosaccharides and polysaccharides are made up of monosaccharides.

Monosynaptic reflex A neural reflex that begins in a sensory neuron and makes a single synapse before activating a motor neuron.

Morphogens Diffusible substances whose concentration gradients determine patterns of development in animals and plants.

Morphogenesis (more´ fo jen´ e sis) [Gr. morphe: form + genesis: origin] The development of form. Morphogenesis is the overall consequence of determination, differentiation, and growth.

Morphology (more fol´ o jee) [Gr. morphe: form + logos: discourse] The scientific study of organic form, including both its development and function.

Mosaic development Pattern of animal embryonic development in which each blastomere contributes a specific part of the adult body. (Contrast with regulative development.)

Motor end plate The modified area on a muscle cell membrane where a synapse is formed with a motor neuron.

Motor neuron A neuron carrying information from the central nervous system to an effector such as a muscle fiber.

Motor unit A motor neuron and the set of muscle fibers it controls.

mRNA (See messenger RNA.)

Mucosa (mew koh´ sah) An epithelial membrane containing cells that secrete mucus. The inner cell layers of the digestive and respiratory tracts.

Müllerian mimicry The resemblance of two or more unpleasant or dangerous kinds of organisms to each other; the mimicry gives each added protection because potential enemies that learn to avoid members of one group tend to avoid members of the others even though they lack prior experience with them.

Multicellular [L. multus: much + cella: chamber] Consisting of more than one cell, as for example a multicellular organism. (Contrast with unicellular.)

Multiple fruit A fruit formed from an inflorescence. (Contrast with accessory fruit, aggregate fruit, simple fruit.)

Muscle fiber A single muscle cell. In the case of striated muscle, a syncitial, multinucleate cell.

Muscle spindle Modified muscle fibers encased in a connective sheat and functioning as stretch sensors.

Muscle tissue Contractile tissue containing actin and myosin organized into polymeric chains called microfilaments. In vertebrates, the tissues are either cardiac muscle, smooth muscle, or striated (skeletal) muscle.

Mutagen (mute´ ah jen) [L. mutare: change + Gr. genesis: source] An agent, especially a chemical, that increases the mutation rate.

Mutation In the broad sense, any discontinuous change in the genetic constitution of an organism. In the narrow sense, the word usually refers to a "point mutation," a change along a very narrow portion of the nucleic acid sequence.

Mutation pressure Evolution (change in gene proportions) by different mutation rates alone.

Mutualism The type of symbiosis, such as that exhibited by fungi and algae or cyanobacteria in forming lichens, in which both species profit from the association.

Mycelium (my seel´ ee yum) [Gr. *mykes*: fungus]   In the fungi, a mass of hyphae.

Mycorrhiza (my´ ka rye´ za) [Gr. *mykes*: fungus + *rhiza*: root]   An association of the root of a plant with the mycelium of a fungus.

Myelin (my´ a lin)   A material forming a sheath around some axons. It is formed by Schwann cells that wrap themselves about the axon. It serves to insulate the axon electrically and to increase the rate of transmission of a nervous impulse.

Myofibril (my´ oh fy´ bril) [Gr. *mys*: muscle + L. *fibrilla*: small fiber]   A polymeric unit of actin or myosin in a muscle.

Myogenic (my oh jen´ ik) [Gr. *mys*: muscle + *genesis*: source]   Originating in muscle.

Myoglobin (my´ oh globe´ in) [Gr. *mys*: muscle + L. *globus*: sphere]   An oxygen-binding molecule found in muscle. Consists of a heme unit and a single globiin chain, and carrys less oxygen than hemoglobin.

Myosin [Gr. *mys*: muscle]   One of the two major proteins of muscle, it makes up the thick filaments. (See actin.)

NAD (nicotinamide adenine dinucleotide)   A compound found in all living cells, existing in two interconvertible forms: the oxidizing agent NAD⁺ and the reducing agent NADH.

NADP (nicotinamide adenine dinucleotide phosphate)   Like NAD, but possessing another phosphate group; plays similar roles but is used by different enzymes.

Natal group   The group into which an individual was born.

Natural killer cell   A small leukocyte that nonspecifically kills certain tumor cells and virus-infected cells in tissue cultures.

Natural selection   The differential contribution of offspring to the next generation by various genetic types belonging to the same population. The mechanism of evolution proposed by Charles Darwin.

Nauplius (no´ plee us) [Gk. *nauplios*: shellfish]   The typical larva of crustaceans. Has three pairs of appendages and a median compound eye.

Negative control   The situation in which a regulatory macromolecule (generally a repressor) functions to turn off transcription. In the absence of a regulatory macromolecule, the structural genes are turned on.

Negative feedback   A pattern of regulation in which a change in a sensed variable results in a correction that opposes the change.

Nekton [Gr. *nekhein*: to swim]   Animals, such as fish, that can swim against currents of water. (Contrast with plankton.)

Nematocyst (ne mat´ o sist) [Gr. *nema*: thread + *kystis*: cell]   An elaborate, threadlike structure produced by cells of jellyfish and other cnidarians, used chiefly to paralyze and capture prey.

Nephridium (nef rid´ ee um) [Gr. *nephros*: kidney]   An organ which is involved in excretion, and often in water balance, involving a tube that opens to the exterior at one end.

Nephron (nef´ ron) [Gr. *nephros*: kidney]   The basic component of the kidney, which is made up of numerous nephrons. Its form varies in detail, but it always has at one end a device for receiving a filtrate of blood, and then a tubule that absorbs selected parts of the filtrate back into the bloodstream.

Nephrostome (nef´ ro stome) [Gr. *nephros*: kidney + *stoma*: opening]   An opening in a nephridium through which body fluids can enter.

Nerve   A structure consisting of many neuronal axons and connective tissue.

Net primary production   Total photosynthesis minus respiration by plants.

Neural plate   A thickened strip of ectoderm along the dorsal side of the early vertebrate embryo; gives rise to the central nervous system.

Neurohormone   A hormone produced and secreted by neurons.

Neuron (noor´ on) [Gr. *neuron*: nerve, sinew]   A cell derived from embryonic ectoderm and characterized by a membrane potential that can change in response to stimuli, generating action potentials. Action potentials are generated along an extension of the cell (the axon), which makes junctions (synapses) with other neurons, muscle cells, or gland cells.

Neurotransmitter   A substance, produced in and released by one neuron, that diffuses across a synapse and excites or inhibits the postsynaptic neuron.

Neurula (nure´ you la) [Gr. *neuron*: nerve]   Embryonic stage during formation of the dorsal nerve cord by two ectodermal ridges.

Neutral alleles   Alleles that differ so slightly that the proteins for which they code function identically.

Neutron (new´ tron) [E.: neutral]   One of the three most fundamental particles of matter, with mass approximately 1 amu and no electrical charge.

Nicotinamide adenine dinucleotide   (See NAD.)

Nicotinamide adenine dinucleotide phosphate   (See NADP.)

Nitrification   The oxidation of ammonia to nitrite and nitrate ions, performed by certain soil bacteria.

Nitrogenase   In nitrogen-fixing organisms, an enzyme complex that mediates the stepwise reduction of atmospheric N_2 to ammonia.

Nitrogen fixation   Conversion of nitrogen gas to ammonia, which makes nitrogen available to living things. Carried out by certain prokaryotes, some of them free-living and others living within plant roots.

Node [L. *nodus*: knob, knot]   In plants, a (sometimes enlarged) point on a stem where a leaf or bud is or was attached.

Node of Ranvier   A gap in the myelin sheath covering an axon, where the axonal membrane can fire action potentials.

Noncompetitive inhibitor   An inhibitor that binds the enzyme at a site other than the active site. (Contrast with competitive inhibitor.)

Nondisjunction   Failure of sister chromatids to separate in meiosis II or mitosis, or failure of homologous chromosomes to separate in meiosis I. Results in aneuploidy.

Nonpolar molecule   A molecule whose electric charge is evenly balanced from one end of the molecule to the other.

Nonsense (chain-terminating) mutation   Mutations that change a codon for an amino acid to one of the codons (UAG, UAA, or UGA) that signal termination of translation. The resulting gene product is a shortened polypeptide that begins normally at the amino-terminal end and ends at the position of the altered codon. (Contrast with frameshift mutation, missense mutation.)

Nonvascular plants   Those plants lacking well-developed vascular tissue; the liverworts, hornworts, and mosses. (Contrast with vascular plants.)

Normal flora   The bacteria and fungi that live on animal body surfaces without causing disease.

Norepinephrine   A neurotransmitter found in the central nervous system and also at the postganglionic nerve endings of the sympathetic nervous system. Also called noradrenaline.

Notochord (no´ tow kord) [Gr. *notos*: back + *chorde*: string]   A flexible rod of gelatinous material serving as a support in the embryos of all chordates and in the adults of tunicates and lancelets.

Nuclear envelope   The surface, consisting of two layers of membrane, that encloses the nucleus of eukaryotic cells.

Nucleic acid (new klay´ ik) [E.: nucleus of a cell]   A long-chain alternating polymer of deoxyribose or ribose and phosphate groups, with nitrogenous bases—adenine, thymine, uracil, guanine, or cytosine (A, T, U, G, or C)—as side chains. DNA and RNA are nucleic acids.

Nucleoid (new´ klee oid)   The region that harbors the chromosomes of a prokaryotic cell. Unlike the eukaryotic nucleus, it is not bounded by a membrane.

Nucleolar organizer (new klee´ o lar)   A region on a chromosome that is associated with the formation of a new nucleolus following nuclear division. The site of the genes that code for ribosomal RNA.

Nucleolus (new klee´ oh lus) [from L. diminutive of *nux*: little kernel or little nut]   A small, generally spherical body found within the nucleus of eukaryotic cells. The site of synthesis of ribosomal RNA.

Nucleoplasm (new´ klee o plazm)   The fluid material within the nuclear envelope of a cell, as opposed to the chromosomes, nucleoli, and other particulate constituents.

Nucleosome   A portion of a eukaryotic chromosome, consisting of part of the DNA molecule wrapped around a group of histone molecules, and held together by another type of histone molecule. The chromosome is made up of many nucleosomes.

Nucleotide   The basic chemical unit (monomer) in a nucleic acid. A nucleotide in RNA consists of one of four nitrogenous bases linked to ribose, which in turn is linked to phosphate. In DNA, deoxyribose is present instead of ribose.

Nucleus (new´ klee us) [from L. diminutive of *nux*: kernel or nut]   (1) In chemistry, the dense central portion of an atom, made up of protons and neutrons, with a positive charge. Surrounded by a cloud of negatively charged electrons. (2) In cells, the centrally located chamber of eukaryotic cells that is bounded by a double membrane and contains the chromosomes. The information center of the cell.

Nutrient   A food substance; or, in the case of mineral nutrients, an inorganic element required for completion of the life cycle of an organism.

Obligate (ob´ li gut)   Necessary, as in obligate anaerobe. (Contrast with facultative.)

Obligate anaerobe   An animal that can live only in oxygenated environments.

Observational analysis   A scientific method in which data are gathered in unmanipulated situations to test hypotheses. Often employed in the field where experimental manipulations are difficult or impossible.

Oceanic zone   The deeper ocean basins.

Oil   A triglyceride that is liquid at room temperature. (Contrast with fat.)

Okazaki fragments   Newly formed DNA strands making up the lagging strand in DNA replication. DNA ligase links the Okazaki fragments to give a continuous strand.

Olfactory   Having to do with the sense of smell.

Oligomer   A compound molecule of intermediate size, made up of two to a few monomers. (Contrast with monomer, polymer.)

Omasum (oh may´ sum)   The third division of the ruminant stomach. Its function is mostly the absorption of wastes. (Contrast with abomasum, rumen.)

Ommatidium [Gr. *omma*: an eye]   One of the units which, collected into groups of up to 20,000, make up the compound eye of arthropods.

Omnivore [L. *omnis*: all, everything + *vorare*: to devour]   An organism that eats both animal and plant material. (Contrast with carnivore, detritivore, herbivore.)

Oncogenic (ong´ co jen´ ik) [Gr. *onkos*: mass, tumor + *genes*: born]   Causing cancer.

Ontogeny (on toj´ e nee) [Gr. *onto*: from "to be" + *genesis*: source]   The development of a single organism in the course of its life history.

Oocyte (oh´ eh site) [Gr. *oon*: egg + *kytos*: cell]   The cell that gives rise to eggs in animals.

Oogenesis (oh´ eh jen e sis) [Gr. *oon*: egg + *genesis*: source]   Female gametogenesis, leading to production of the egg.

Oogonium (oh´ eh go´ nee um)   In some algae and fungi, a cell in which an egg is produced.

Operator   The region of an operon that acts as the binding site for the repressor.

Operon   A genetic unit of transcription, typically consisting of several structural genes that are transcribed together; the operon contains at least two control regions: the promoter and the operator.

Opportunity cost   The sum of the benefits an animal forfeits by not being able to perform some other behavior during the time when it is performing a given behavior.

Opsin (op´ sin) [Gr. *opsis*: sight]   The protein protion of the visual pigment rhodopsin. (See rhodopsin.)

Optic chiasm   Stucture on the lower surface of the vertebrate brain where the two optic nerves come together.

Optical isomers   Isomers that differ in the configuration of the four different groups attached to a single carbon atom; so named because solutions of the two isomers rotate the plane of polarized light in opposite directions. The two isomers are mirror images of one another.

Order   In taxonomy, the category below the class and above the family; a group of related, similar families.

Organ   A body part, such as the heart, liver, brain, root, or leaf, composed of different tissues integrated to perform a distinct function for the body as a whole.

Organ of Corti   Structure in the inner ear that transforms mechanical forces produced from pressure waves ("sound waves") into action potentials that are sensed as sound.

Organelles (or´ gan els´) [L.: little organ]   Organized structures that are found in or on cells. Examples: ribosomes, nuclei, mitochrondria, chloroplasts, cilia, and contractile vacuoles.

Organic   Pertaining to any aspect of living matter, e.g., to its evolution, structure, or chemistry. The term is also applied to any chemical compound that contains carbon.

Organism   Any living creature.

Organizer, embryonic   A region of an embryo which directs the development of nearby regions. In amphibian early gastrulas, the dorsal lip of the blastopore.

Osmoregulation   Regulation of the chemical composition of the body fluids of an organism.

Osmosensor   A neuron that converts changes in the osmotic potential of interstial fluids into action potentials.

Osmosis (oz mo´ sis) [Gr. *osmos*: to push]   The movement of water through a differentially permeable membrane from one region to another where the water potential is more negative. This is often a region in which the concentration of dissolved molecules or ions is higher, although the effect of dissolved substances may be offset by hydrostatic pressure in cells with semi-rigid walls.

Osmotic potential   A property of any solution, resulting from its solute content; it may be zero or have a negative value. A negative osmotic potential tends to cause water to move into the solution; it may be offset by a positive pressure potential in the solution or by a more negative water potential in a neighboring solution. (Contrast with pressure potential.)

Ossicle (ah´ sick ul) [L. *os*: bone]   The calcified construction unit of echinoderm skeletons.

Osteoblasts   Cells that lay down the protein matrix of bone. (Contrast with osteoclasts.)

Osteoclasts   Cells that dissolve bone. (Contrast with osteoblasts.)

Otolith (oh´ tuh lith) [Gk.*otikos*: ear + *lithos*: stone[Structures in the vertebrate vestibular apparatus that mechanically stimulate hair cells when the head moves or changes position.

Outgroup   A taxon that separated from another taxon, whose lineage is to be inferred, before the latter underwent evolutionary radiation.

Oval window   The flexible membrane which, when moved by the bones of the middle ear, produces pressure waves in the inner ear

Ovary (oh´ var ee)   Any female organ, in plants or animals, that produces an egg.

Oviduct [L. *ovum*: egg + *ducere*: to lead]   In mammals, the tube serving to transport eggs to the uterus or to outside of the body.

Oviparous (oh vip´ uh rus)   Reproduction in which eggs are released by the female and development is external to the mother's body. (Contrast with viviparous.)

Ovulation   The release of an egg from an ovary.

Ovule (oh´ vule) [L. *ovulum*: little egg]   In plants, an organ that contains a gametophyte and, within the gametophyte, an egg; when it matures, an ovule becomes a seed.

Ovum (oh´ vum) [L.: egg]   The egg, the female sex cell.

Oxidation (ox i day´ shun)   Relative loss of electrons in a chemical reaction; either outright removal to form an ion, or the sharing of electrons with substances having a greater affinity for them, such as oxygen. Most oxidation, including biological ones, are associated with the liberation of energy. (Contrast with reduction.)

Oxidative phosphorylation   ATP formation in the mitochondrion, associated with flow of electrons through the respiratory chain. (Contrast with substrate-level phosphorylation.)

Oxidizing agent   A substance that can accept electrons from another. The oxidizing agent becomes reduced; its partner becomes oxidized.

P generation   The individuals that mate in a genetic cross. Their immediate offspring are the F_1 generation.

Pacemaker   That part of the heart which undergoes most rapid spontaneous contraction, thus setting the pace for the beat of the entire heart. In mammals, the sinoatrial (SA) node. Also, an artificial device, implanted in the heart, that initiates rhythmic contraction of the organ.

Pacinian corpuscle   A sensory neuron surrounded by sheaths of connective tissue. Found in the deep layers of the skin, where it senses touch and vibration.

Paleobiology   The study of fossil evidence, the comparative biochemistry of living organisms, and conditions on the early Earth to determine the stages in the evolution of life.

Paleobotany   The scientific study of fossil plants and all aspects of extinct plant life.

Paleontology (pale´ ee on tol´ oh jee) [Gr. *palaios*: ancient, old + *logos*: discourse]   The scientific study of fossils and all aspects of extinct life.

Palisade parenchyma   In leaves, one or several layers of tightly packed, columnar photosynthetic cells, frequently found just below the upper epidermis.

Pancreas (pan´ cree us)   A gland, located near the stomach of vertebrates, that secretes digestive enzymes into the small intestine and releases insulin into the bloodstream.

Pangaea (pan jee´ uh) [Gk. *pan*: all, every]   The single land mass formed when all the continents came together in the Permian period.

Parabronchi   Passages in the lungs of birds through which air flows.

Paradigm   A general framework within which some scientific discipline (or even the whole Earth) is viewed and within which questions are asked and hypotheses are developed. Scientific revolutions usually involve major paradigm changes.

Parapatric speciation   Development of reproductive isolation among members of a continuous population in the absence of a geographical barrier. (Contrast with geographic, sympatric speciation.)

Paraphyletic taxon   A taxon that includes some, but not all, of the descendants of a single ancestor.

Parasite   An organism that attacks and consumes parts of an organism much larger than itself. Parasites sometimes, but not always, kill the host.

Parasitoid   A parasite that is so large relative to its host that only one individual or at most a few individuals can live within a single host.

Parasympathetic nervous system   A portion of the autonomic (involuntary) nervous system. Activity in the parasympathetic nervous system produces effects such as decreased blood pressure and decelerated heart beat. The neurotransmitter for this system is acetylcholine. (Contrast with sympathetic nervous system.)

Parathormone   Hormone secreted by the parathyroid glands. Stimulates osteoclast activity and raises blood calcium levels.

Parathyroids   Four glands on the posterior surface of the thyroid that produce and release parathormone.

Parenchyma (pair eng´ kyma) [Gr. *para*: beside + *enchyma*: infusion]   A plant tissue composed of relatively unspecialized cells without secondary walls.

Parental investment   Investment in one offspring or group of offspring that reduces the ability of the parent to assist other offspring.

Parsimony   The principle of preferring the simplest among a set of plausible explanations of a phenomenon. Commonly employed in evolutionary and biogeographic studies.

Parthenocarpy   Formation of fruit from a flower without fertilization.

Parthenogenesis (par´ then oh jen´ e sis) [Gr. *parthenos*: virgin + *genesis*: source]   The production of an organism from an unfertilized egg.

Partial pressure   The portion of the barometric pressure of a mixture of gases that is due to one component of that mixture. For example, the partial pressure of oxygen at sea level is 20.9% of barometric pressure.

Parturition (part uh rish un) [L. *parturire*, to give birth]   Childbirth.

Pasteur effect   The sharp decrease in rate of glucose utilization when

Pastoralism   A nomadic form of human culture based on the tending of herds of domestic animals.conditions become aerobic.

Patch clamping   A technique for isolating a tiny patch of membrane to allow the study of ion movement through a particular channel.

Pathogen (path´ o jen) [Gr. *pathos*: suffering + *gignomai*: causing]   An organism that causes disease.

Pattern formation   In animal embryonic development, the organization of differentiated tissues into specific structures such as wings.

Pedigree   The pattern of transmission of a genetic trait in a family.

Pelagic zone (puh ladj´ ik) [Gr. *pelagos*: the sea]   The open waters of the ocean.

Pellicle (pell´ ik el) [L. *pellis*: skin]   A thin, filmy covering.

Penetrance   Of a genotype, the proportion of individuals with that genotype who show the expected phenotype.

Penis (pee´ nis) [L.: tail, penis]   The male organ inserted into the female during coitus (sexual intercourse).

Pentose (pen´ tose) [Gk. *penta*: five]   A five-carbon sugar, such as ribose or deoxyribose.

PEP carboxylase   The enzyme that combines carbon dioxide with PEP to form a 4-carbon dicarboxylic acid at the start of C_4 photosynthesis or of Crassulacean acid metabolism (CAM).

Pepsin [Gr. *pepsis*: digestion]   An enzyme, in gastric juice, that digests protein.

Peptide linkage   The connecting group in a protein chain, –CO–NH–, formed by removal of water during the linking of amino acids, –COOH to –NH$_2$. Also called an amide linkage.

Peptidoglycan   The cell wall material of many prokaryotes, consisting of a single enormous molecule that surrounds the entire cell.

Perennial (per ren´ ee al) [L. *per*: through + *annus*: a year]   Referring to a plant that lives from year to year. (Contrast with annual, biennial.)

Perfect flower   A flower with both stamens and carpels, therefore hermaphroditic.

Pericycle [Gr. *peri*: around + *kyklos*: ring or circle]   In plant roots, tissue just within the endodermis, but outside of the root vascular tissue. Meristematic activity of pericycle cells produces lateral root primordia.

Periderm   The outer tissue of the secondary plant body, consisting primarily of cork.

Period   (1) A minor category in the geological time scale. (2) The duration of a cyclical event, such as a circadian rhythm.

Peripheral nervous system   Neurons that transmit information to and from the central nervous system and whose cell bodies reside outside the brain or spinal cord.

Peristalsis (pair´ i stall´ sis) [Gr. *peri*: around + *stellein*: place]   Wavelike muscular contractions proceeding along a tubular organ, propelling the contents along the tube.

Peritoneum   The mesodermal lining of the coelom among coelomate animals.

Permafrost   Soil that remains frozen for many years.

Permease   A protein in membranes that specifically transports a compound or family of compounds across the membrane.

Peroxisome   A microbody that houses reactions in which toxic peroxides are formed. The peroxisome isolates these peroxides from the rest of the cell.

Petal   In an angiosperm flower, a sterile modified leaf, nonphotosynthetic, frequently brightly colored, and often serving to attract pollinating insects.

Petiole (pet´ ee ole) [L. *petiolus*: small foot]   The stalk of a leaf.

pH   The negative logarithm of the hydrogen ion concentration; a measure of the acidity of a solution. A solution with pH = 7 is said to be neutral; pH values higher than 7 characterize basic solutions, while acidic solutions have pH values less than 7.

Phage (fayj)   Short for bacteriophage.

Phagocyte   A white blood cell that ingests microorganisms by endocytosis.

Phagocytosis [Gr.: *phagein* to eat; cell-eating]   A form of endocytosis, the uptake of a solid particle by forming a pocket of plasma membrane around the particle and pinching off the pocket to form an intracellular particle bounded by membrane. (Contrast with pinocytosis.)

Pharyngeal slits   Slits in the pharynx that originally functioned in gas exchange but became modified for other purposes among vertebrates.

Pharynx [Gr.: throat]   The part of the gut between the mouth and the esophagus.

Phenogram   Graphic representation of phenetic similarities.

Phenotype (fee´ no type) [Gr. *phanein*: to show + *typos*: impression]   The observable properties of an individual as they have developed under the combined influences of the genetic constitution of the individual and the effects of environmental factors. (Contrast with genotype.)

Pheromone (feer´ o mone) [Gr. *phero*: carry + *hormon*: excite, arouse]   A chemical substance used in communication between organisms of the same species.

Phloem (flo´ um) [Gr. *phloos*: bark]   In vascular plants, the food-conducting tissue. It consists of sieve cells or sieve tubes, fibers, and other specialized cells.

Phosphate group   The functional group $-OPO_3H_2$; the transfer of energy from one compound to another is often accomplished by the transfer of a phosphate group.

Phosphodiester linkage   The connection in a nucleic acid strand, formed by linking two nucleotides.

3-Phosphoglycerate   The first product of photosynthesis, produced by the reaction of ribulose bisphosphate with carbon dioxide.

Phospholipids   Cellular materials that contain phosphorus and are soluble in organic solvents. An example is lecithin (phosphatidyl choline). Phospholipids are important constituents of cellular membranes. (See lipids.)

Phosphorylation   The addition of a phosphate group.

Photoautotroph   An organism that obtains energy from light and carbon from carbon dioxide. (Contrast with chemoautotroph, chemoheterotroph, photoheterotroph.)

Photoheterotroph   An organism that obtains energy from light but must obtain its carbon from organic compounds. (Contrast with chemoautotroph, chemoheterotroph, photoautotroph.)

Photon (foe´ tohn) [Gr. *photos*: light]   A quantum of visible radiation; a "packet" of light energy.

Photoperiod (foe´ tow peer´ ee ud)   The duration of a period of light, such as the length of time in a 24-hour cycle in which daylight is present. The regulation of processes such as flowering by the changing length of day (or of night) is known as **photoperiodism**.

Photophosphorylation   Photosynthetic reactions in which light energy trapped by chlorophyll is used to produce ATP and, in noncyclic photophosphorylation, is used to reduce $NADP^+$ to NADPH.

Photorespiration   Light-driven uptake of oxygen and release of carbon dioxide, the carbon being derived from the early reactions of photosynthesis.

Photosensor   A cell that senses and responds to light energy.

Photosynthesis (foe tow sin´ the sis) [literally, "synthesis out of light"]   Metabolic processes, carried out by green plants, by which visible light is trapped and the energy used to synthesize compounds such as ATP and glucose.

Phototropism [Gr. *photos*: light + *trope*: a turning]   A directed plant growth response to light.

Phylogenetic tree   Graphic representation of lines of descent among organisms.

Phylogeny (fy loj´ e nee) [Gr. *phylon*: tribe, race + *genesis*: source]   The evolutionary history of a particular group of organisms; also, the diagram of the "family tree" that shows genetic linkages between ancestors and descendants.

Phylum [Gr. *phylon*: tribe, stock]   In taxonomy, a high-level category just beneath kingdom and above the class; a group of related, similar classes.

Physiological time   The time required for significant changes in the physiological processes or states within an organism.

Physiology (fiz´ ee ol´ o jee) [Gr. *physis*: natural form + *logos*: discourse, study]   The scientific study of the functions of living organisms and the individual organs, tissues, and cells of which they are composed.

Phytoalexins   Substances toxic to fungi, produced by plants in response to fungal infection.

Phytochrome (fy´ tow krome) [Gr. *phyton*: plant + *chroma*: color]   A plant pigment regulating a large number of developmental and other phenomena in plants; can exist in two different forms, one of which is active and the other is not. Different wavelengths of light can drive it from one form to the other.

Phytoplankton (fy´ tow plangk´ ton) [Gr. *phyton*: plant + *planktos*: wandering]   The autotrophic portion of the plankton, consisting mostly of algae.

Pigment   A substance that absorbs visible light.

Piloting   Finding one's way by means of landmarks.

Pilus (pill´ us) [Lat. *pilus*: hair]   A surface appendage by which some bacteria adhere to one another during conjugation.

Pinocytosis [Gr.: drinking cell]   A form of endocytosis; the uptake of liquids by engulfing a sample of the external medium into a pocket of the plasma membrane followed by pinching off the pocket to form an intracellular vesicle. (Contrast with phagocytosis and endocytosis.)

Pistil [L. *pistillum*: pestle]   The female structure of an angiosperm flower, within which the ovules are borne. May consist of a single carpel, or of several carpels fused into a single structure. Usually differentiated into ovary, style, and stigma.

Pith   In plants, relatively unspecialized tissue found within a cylinder of vascular tissue.

Pituitary   A small gland attached to the base of the brain in vertebrates. Its hormones control the activities of other glands. Also known as the hypophysis.

Placenta (pla sen´ ta) [Gr. *plax*: flat surface]   The organ, found in most mammals, that provides for the nourishment of the fetus and elimination of the fetal waste products. It is formed by the union of membranes of the mother's uterine lining with the membranes from the fetus.

Placental (pla sen´ tal)   Pertaining to mammals of the subclass Eutheria, a group that is characterized by the presence of a placenta and that contains the majority of living species of mammals.

Plankton [Gr. *planktos*: wandering]   The free-floating organisms of the sea and fresh water that for the most part move passively with the water currents. Consisting mostly of microorganisms and small plants and animals. (Contrast with nekton.)

Plant   A member of the kingdom Plantae. Multicellular, gaining its nutrition by photosynthesis.

Planula (plan´ yew la) [L. *planum*: something flat]   The free-swimming, ciliated larva of the cnidarians.

Plaque (plack) [Fr.: a metal plate or coin]   (1) A circular clearing in a turbid layer (lawn) of bacteria growing on the surface of a nutrient agar gel. Produced by successive rounds of infection initiated by a single bacteriophage. (2) An accumulation of prokaryotic organisms on tooth enamel. Acids produced by the metabolism of these microorganisms can cause tooth decay.

Plasma (plaz´ muh) [Gr. *plassein*: to mold]   The liquid portion of blood, in which blood cells and other particulates are suspended.

Plasma cell   An antibody-secreting cell that developed from a B cell. The effector cell of the humoral immune system.

Plasma membrane The membrane that surrounds the cell, regulating the entry and exit of molecules and ions. Every cell has a plasma membrane.

Plasmid A DNA molecule distinct from the chromosome(s); that is, an extrachromosomal element. May replicate independently of the chromosome.

Plasmodesma (plural: plasmodesmata) [Gr. *plasma*: formed or molded + *desmos*: band] A cytoplasmic strand connecting two adjacent plant cells.

Plasmodium In the noncellular slime molds, a multinucleate mass of protoplasm surrounded by a membrane; characteristic of the vegetative feeding stage.

Plasmolysis (plaz mol´ i sis) Shrinking of the cytoplasm and plasma membrane away from the cell wall, resulting from the osmotic outflow of water. Occurs only in cells with rigid cell walls.

Plastid Organelle in plants that serves for food manufacture (by photosynthesis) or food storage; bounded by a double membrane.

Platelet A membrane-bounded body without a nucleus, arising as a fragment of a cell in the bone marrow of mammals. Important to blood-clotting action.

Pleiotropy (plee´ a tro pee) [Gr. *pleion*: more] The determination of more than one character by a single gene.

Pleural membrane [Gk. *pleuras*: rib, side] The membrane lining the outside of the lungs and the walls of the thoracic cavity. Inflammation of these membranes is a condition known as **pleurisy.**

Podocytes Cells of Bowman's capsule of the nephron that cover the capillaries of the glomerulus, forming filtration slits.

Poikilotherm (poy´ kill o therm) [Gr. *poikilos*: varied + *therme*: heat] An animal whose body temperature tends to vary with the surrounding environment. (Contrast with homeotherm, heterotherm.)

Point mutation A mutation that results from a small, localized alteration in the chemical structure of a gene. Such mutations can give rise to wild-type revertants as a result of reverse mutation. In genetic crosses, a point mutation behaves as if it resided at a single point on the genetic map. (Contrast with deletion.)

Polar body A nonfunctional nucleus produced by meiosis, accompanied by very little cytoplasm. The meiosis which produces the mammalian egg produces in addition three polar bodies.

Polar molecule A molecule in which the electric charge is not distributed evenly in the covalent bonds.

Polar nucleus One of two nuclei derived from each end of the angiosperm embryo sac, both of which become centrally located. They fuse with a male nucleus to form the primary triploid nucleus that will prduce the endosperm tissue of the angiosperm seed.

Pollen [L.: fine powder, dust] The fertilizing element of seed plants, containing the male gametophyte and the gamete, at the stage in which it is shed.

Pollination Process of transferring pollen from the anther to the receptive surface (stigma) of the ovary in plants.

Poly- [Gr. *poly*: many] A prefix denoting multiple entities.

Polygamy [Gr. *poly*: many + *gamos*: marriage] A breeding system in which an individual acquires more than one mate. In polyandry, a female mates with more than one male, in polygyny, a male mates with more than one female.

Polygenes Multiple loci whose alleles increase or decrease a continuously variable phenotypic trait.

Polymer A large molecule made up of similar or identical subunits called monomers. (Contrast with monomer, oligomer.)

Polymerase chain reaction (PCR) A technique for the rapid production of millions of copies of a particular stretch of DNA.

Polymerization reactions Chemical reactions that generate polymers by means of condensation reactions.

Polymorphism (pol´ lee mor´ fiz um) [Gr. *poly*: many + *morphe*: form, shape] (1) In genetics, the coexistence in the same population of two distinct hereditary types based on different alleles. (2) In social organisms such as colonial cnidarians and social insects, the coexistence of two or more functionally different castes within the same colony.

Polyp The sessile, asexual stage in the life cycle of most cnidarians.

Polypeptide A large molecule made up of many amino acids joined by peptide linkages. Large polypeptides are called proteins.

Polyploid (pol´ lee ploid) A cell or an organism in which the number of complete sets of chromosomes is greater than two.

Polysaccharide A macromolecule composed of many monosaccharides (simple sugars). Common examples are cellulose and starch.

Polysome A complex consisting of a threadlike molecule of messenger RNA and several (or many) ribosomes. The ribosomes move along the mRNA, synthesizing polypeptide chains as they proceed.

Polytene (pol´ lee teen) [Gr. *poly*: many + *taenia*: ribbon] An adjective describing giant interphase chromosomes, such as those found in the salivary glands of fly larvae. The characteristic, reproducible pattern of bands and bulges seen on these chromosomes has provided a method for preparing detailed chromosome maps of several organisms.

Pons [L. *pons*: bridge] Region of the brain stem anterior to the medulla.

Population Any group of organisms coexisting at the same time and in the same place and capable of interbreeding with one another.

Population density The number of individuals (or modules) of a population in a unit of area or volume.

Population dynamics The sum of the activities of the members of a population.

Population structure The proportions of individuals in a population belonging to different age classes (age structure). Also, the distribution of the population in space and the amount of migration between subpopulations.

Population vulnerability analysis A determination of the risk of extinction of a population given its current size and distribution.

Portal vein A vein connecting two capillary beds, as in the hepatic portal system.

Positive control The situation in which a regulatory macromolecule is needed to turn transcription of structural genes on. In its absence, transcription will not occur.

Positive cooperativity Occurs when a molecule can bind several ligands and each one that binds alters the conformation of the molecule so that it can bind the next ligand more easily. The binding of four molecules of O_2 by hemoglobin is an example of positive cooperativity.

Positive feedback A regulatory system in which an error signal stimulates responses that increase the error.

Postabsorptive period When there is no food in the gut and no nutrients are being absorbed.

Posterior Toward or pertaining to the rear.

Postsynaptic cell The cell whose membranes receive the neurotransmitter released at a synapse.

Postzygotic isolating mechanism Any factor that reduces the viability of zygotes resulting from matings between individuals of different species.

Predator An organism that kills and eats other organisms. Predation is usually thought of as involving the consumption of animals by animals, but in the broad usage of ecology it can also mean the eating of plants.

Pressure potential The actual physical (hydrostatic) pressure within a cell. (Contrast with osmotic potential, water potential.)

Presynaptic excitation/inhibition Occurs when a neuron modifies activity at a synapse by releasing a neurotransmitter onto the presynaptic nerve terminal.

Prey [L. *praeda*: booty] An organism hunted or caught as an energy source.

Prezygotic isolating mechanism A mechanism that reduces the probability that individuals of different species will mate.

Primary active transport Form of active transport in which ATP is hydrolyzed, yielding the energy required to transport ions against their concentration gradients. (Contrast with secondary active transport.)

Primary growth In plants, growth produced by the apical meristems. (Contrast with secondary growth.)

Primary producer   A photosynthetic or chemosynthetic organism that synthesizes complex organic molecules from simple inorganic ones.

Primary succession   Succession that begins in an areas initially devoid of life, such as on recently exposed glacial till or lava flows.

Primary structure   The specific sequence of amino acids in a protein.

Primary wall   Cellulose-rich cell wall layers laid down by a growing plant cell.

Primate (pry´ mate)   A member of the order Primates, such as a lemur, monkey, ape, or human.

Primer   A short, single-stranded segment of DNA serving as the necessary starting material for the synthesis of a new DNA strand, which is synthesized from the 3´ end of the primer.

Primitive streak   A line running axially along the blastodisc, the site of inward cell migration during formation of the three-layered embryo. Formed in the embryos of birds and fish.

Primordium [L. *primordium*: origin]   The most rudimentary stage of an organ or other part.

Principle of superposition   The generalization that younger rocks lie on top of older rocks unless Earth movements have altered their positions.

Pro- [L.: first, before, favoring]   A prefix often used in biology to denote a developmental stage that comes first or an evolutionary form that appeared earlier than another. For example, prokaryote, prophase.

Probe   A segment of single stranded nucleic acid used to identify DNA molecules containing the complementary sequence.

Procambium   Primary meristem that produces the vascular tissue.

Progesterone [L. *pro*: favoring + *gestare*: to bear]   A vertebrate female sex hormone that maintains pregnancy.

Prokaryotes (pro kar´ ry otes) [L. *pro*: before + Gk. *karyon*: kernel, nucleus]   Organisms whose genetic material is not contained within a nucleus. The bacteria. Considered an earlier stage in the evolution of life than the eukaryotes.

Prometaphase   The phase of nuclear division that begins with the disintegration of the nuclear envelope.

Promoter   The region of an operon that acts as the initial binding site for RNA polymerase.

Prophage (pro´ fayj)   The noninfectious units that are linked with the chromosomes of the host bacteria and multiply with them but do not cause dissolution of the cell. Prophage can later enter into the lytic phase to complete the virus life cycle.

Prophase (pro´ phase)   The first stage of nuclear division, during which chromosomes condense from diffuse, threadlike material to discrete, compact bodies.

Proplastid [Gr. *pro*: before + *plastos*: molded]   A plant cell organelle which under appropriate conditions will develop into a plastid, usually the photosynthetic chloroplast. If plants are kept in the dark, proplastids may become quite large and complex.

Prostaglandin   Any one of a group of specialized lipids with hormone-like functions. It is not clear that they act at any considerable distance from the site of their production.

Prosthetic group   Any nonprotein portion of an enzyme.

Protease (pro´ tee ase)   See proteolytic enzyme.

Protein (pro´ teen) [Gr. *protos*: first]   One of the most fundamental building substances of living organisms. A long-chain polymer of amino acids with twenty different common side chains. Occurs with its polymer chain extended in fibrous proteins, or coiled into a compact macromolecule in enzymes and other globular proteins.

Proteolytic enzyme   An enzyme whose main catalytic function is the digestion of a protein or polypeptide chain. The digestive enzymes trypsin, pepsin, and carboxypeptidase are all proteolytic enzymes (proteases).

Protist   A member of the kingdom Protista, which consists of those eukaryotes not included in the kingdoms Animalia, Fungi, or Plantae. Many protists are unicellular. The kingdom Protista includes protozoa, algae, and fungus-like protists.

Protoderm   Primary meristem that gives rise to epidermis.

Proton (pro´ ton) [Gr. *protos*: first]   One of the three most fundamental particles of matter, with mass approximately 1 amu and an electrical charge of +1.

Proton motive force   The proton gradient and electric charge difference produced by chemiosmotic proton pumping. It drives protons back across the membrane, with the concomitant formation of ATP.

Protonema (pro´ tow nee´ mah) [Gr. *protos*: first + *nema*: thread]   The hairlike growth form that constitutes an early stage in the development of a moss gametophyte.

Proto-oncogenes   The normal alleles of genes possessing oncogenes (cancer-causing genes) as mutant alleles. Proto-oncogenes encode growth factors and receptor proteins.

Protoplast   A cell that would normally have a cell wall, from which the wall has been removed by enzymatic digestion or by special growth conditions.

Protostome   One of two major lines of animal evolution, characterized by spiral, determinate cleavage of the egg, and by schizocoelous development. (Contrast with deuterostome.)

Prototroph (pro´ tow trofe´) [Gr. *protos*: first + *trophein*: to nourish]   The nutritional wild-type, or reference form, of an organism. Any deviant form that requires growth nutrients not required by the prototrophic form is said to be a nutritional mutant, or auxotroph.

Protozoa   A group of single-celled organisms classified by some biologists as a single phylum; includes the flagellates, amoebas, and ciliates. This textbook follows most modern classifications in elevating the protozoans to a distinct kingdom (Protista) and each of their major subgroups to the rank of phylum.

Provincialized   A biogeographic term referring to the separation, by environmental barriers, of the biota into units with distinct species compositions.

Provirus   See prophage.

Proximal   Near the point of attachment or other reference point. (Contrast with distal.)

Pseudocoelom   A body cavity not surrounded by a peritoneum. Characteristic of nematodes and rotifers.

Pseudogene   A DNA segment that is homologous to a functional gene but contains a nucleotide change that prevents its expression.

Pseudoplasmodium [Gr. *pseudes*: false + *plasma*: mold or form]   In the cellular slime molds such as *Dictyostelium*, an aggregation of single amoeboid cells. Occurs prior to formation of a fruiting structure.

Pseudopod (soo´ do pod) [Gr. *pseudes*: false + *podos*: foot]   A temporary, soft extension of the cell body that is used in location, attachment to surfaces, or engulfing particles.

Pulmonary   Pertaining to the lungs.

Pupa (pew´ pa) [L.: doll, puppet]   In certain insects (the Holometabola), the encased developmental stage that intervenes between the larva and the adult.

Pupil   The opening in teh vertebrate eye through which light passes.

Purine (pure´ een)   A type of nitrogenous base. The purines adenine and guanine are found in nucleic acids.

Purkinje fibers   Specialized heart muscle cells that conduct excitation throughout the ventricular muscle.

Pyramid of biomass   Graphical representation of the total masses at different trophic levels in an ecosystem.

Pyramid of energy   Graphical representation of the total energy contents at different trophic levels in an ecosystem.

Pyrimidine (peer im´ a deen)   A type of nitrogenous base. The pyrimidines cytosine, thymine, and uracil are found in nucleic acids.

Pyrogen   A substance that causes fever.

Pyruvate   A three-carbon acid; the end product of glycolysis and the raw material for the citric acid cycle.

Q$_{10}$ A value that compares the rate of a biochemical process or reaction over a 10°C range of temperature. A process that is not temperature-sensitive has a Q$_{10}$ of 1. Values of 2 or 3 mean the reaction speeds up as temperature increases.

Quantum (kwon´ tum) [L. *quantus*: how great]   An indivisible unit of energy.

Quaternary structure   Of aggregating proteins, the arrangement of polypeptide subunits.

R factor (resistance factor)   A plasmid that contains one or more genes that encode resistance to antibiotics.

Radial symmetry   The condition in which two halves of a body are mirror images of each other regardless of the angle of the cut, providing the cut is made along the center line. Thus, a cylinder cut lengthwise down its center displays this form of symmetry. (Contrast with bilateral symmetry.)

Radioisotope   A radioactive isotope of an element. Examples are carbon-14 (^{14}C) and hydrogen-3, or tritium (^{3}H).

Radiometric clock   The use of the regular, known rates of decay of radioisotopes of elements to determine dates of events in the distant past.

Radiotherapy   Treatment, as of cancer, with X- or gamma rays.

Radula   The toothed feeding organ of many mollusks. Used to scrape prey from hard substrates.

Rain shadow   A region of low precipitation on the leeward side of a mountain range.

Ramet   The repeated morphological units of sessile, modular organisms. (Contrast with genet.)

Random drift   Evolution (change in gene proportions) by chance processes alone.

Rate constant   Of a particular chemical reaction, a constant which, when multiplied by the concentration(s) of reactant(s), gives the rate of the reaction.

Reactant   A chemical substance that enters into a chemical reaction with another substance.

Reaction, chemical   A process in which atoms combine or change bonding partners.

Reaction wood   Modified wood produced in branches in response to gravitational stimulation. Gymnosperms produce compression wood that tends to push the branch up; angiosperms produce tension wood that tends to pull the branch up.

Realized niche   The actual niche occupied by an organism; it differs from the fundamental niche because of the presence of other species.

Receptacle [L. *receptaculum*: reservoir]   In an angiosperm flower, the end of the stem to which all of the various flower parts are attached.

Receptive field   Of a neuron, the area on the retina from which the activity of that neuron can be influenced.

Receptor-mediated endocytosis   A form of endocytosis in which macromolecules in the environment bind specific receptor proteins in the plasma membrane and are brought into the cell interior in coated vesicles.

Receptor potential   The change in the resting potential of a sensory cell when it is stimulated.

Recessive   See dominance.

Reciprocal altruism   The exchange of altruistic acts between two or more individuals. The acts may be separated considerably in time.

Reciprocal crosses   A pair of crosses, in one of which a female of genotype A mates with a male of genotype B and in the other of which a female of genotype B mates with a male of genotype A.

Recombinant   An individual, meiotic product, or single chromosome in which genetic materials originally present in two individuals end up in the same haploid complement of genes. The reshuffling of genes can be either by independent segragation, or by crossing over between homologous chromosomes. For example, a human may pass on genes from both parents in a single haploid gamete.

Recombinant DNA technology   The application of genetic tools (restriction endonucleases, plasmids, and transformation) to the production of specific proteins by biological "factories" such as bacteria.

Rectum   The terminal portion of the gut, ending at the anus.

Redirected activity   The direction of some behavior, such as aggression, away from the primary target and toward another, less appropriate object.

Redox reaction   A chemical reaction in which one reactant becomes oxidized and the other becomes reduced.

Reducing agent   A substance that can donate electrons to another substance. The reducing agent becomes oxidized, and its partner becomes reduced.

Reduction (re duk´ shun)   Gain of electrons; the reverse of oxidation. Most reductions lead to the storage of chemical energy, which can be released later by an oxidation reaction. Energy storage compounds such as sugars and fats are highly reduced compounds. (Contrast with oxidation.)

Reflex   An automatic action, involving only a few neurons (in vertebrates, often in the spinal cord), in which a motor response swiftly follows a sensory stimulus.

Refractory period   Of a neuron, the time interval after an action potential, during which another action potential cannot be elicited.

Region   In biogeography, a major division of the world distinguished by its peculiar animals or plants. For example, Africa south of the Sahara is recognized as constituting the Ethiopian region.

Regulative development   A pattern of animal embryonic development in which the fates of the first blastomeres are not absolutely fixed. (Contrast with mosaic development.)

Regulatory gene   A gene that contains the information for making a regulatory macromolecule, often a repressor protein.

Releaser   A sensory stimulus that triggers a fixed action pattern.

Releasing hormone   One of several hypothalamic hormones that stimulates the secretion of anterior pituitary hormone.

REM sleep   A sleep state characterized by dreaming, skeletal muscle relaxation, and rapid eye movements.

Renal [L. *renes*: kidneys]   Relating to the kidneys.

Replica plating   A technique used in the selection of colonies of cells with a desired genotype.

Replication fork   A point at which a DNA molecule is replicating. The fork forms by the unwinding of the parent molecule.

Repressible enzyme   An enzyme whose synthesis can be decreased or prevented by the presence of a particular compound.

Repressor   A protein coded by the regulatory gene. The repressor can bind to a specific operator and prevent transcription of the operon.

Reproductive isolating mechanism   Any trait that prevents individuals from two different populations from producing fertile hybrids.

Reproductive isolation   The condition in which a population is not exchanging genes with other populations of the same species.

Reproductive value   The expected contribution of an individual of a particular age to the future growth of the population to which it belongs.

Resolving power   Of an optical device such as a microscope, the smallest distance between two lines that allows the lines to be seen as separate from one another.

Resource   Something in the environment required by an organism for its maintenance and growth that is consumed in the process of being used.

Resource defense polygamy   A breeding system in which individuals of one sex (usually males) defend resources that are attractive to individuals of the other sex (usually females); individuals holding better resources attract more mates.

Respiration (res pi ra´ shun) [L. *spirare*: to breathe]   (1) Cellular respiration; the oxidation of the end products of glycolysis with the storage of much energy in ATP. The oxidant in the respiration of eukaryotes is oxygen gas. Some bacteria can use nitrate or sulfate instead of O_2. (2) Breathing.

Respiratory chain   The terminal reactions of cellular respiration, in which electrons are passed from NAD or FAD, through a series of intermediate carriers, to molecular oxygen, with the concomitant production of ATP.

Respiratory uncoupler A substance that allows protons to cross the inner mitochondrial membrane without the concomitant formation of ATP, thus uncoupling respiration from phosphorylation.

Resting potential The membrane potential of a living cell at rest. In cells at rest, the interior is negative to the exterior. (Contrast with action potential, electrotonic potential.)

Restoration ecology The science and practice of restoring damaged or degraded ecosystems.

Restriction endonuclease Any one of several enzymes, produced by bacteria, that break foreign DNA molecules at very specific sites. Some produce "sticky ends." Extensively used in recombinant DNA technology.

Restriction map A partial genetic map of a DNA molecule, showing the points at which particular restriction endonuclease recognition sites reside.

Retina (rett´ in uh) [L. *rete*: net] The light-sensitive layer of cells in the vertebrate or cephalopod eye.

Retinal The light-absorbing portion of visual pigment molecules. Derived from β-carotene.

Retrovirus An RNA virus that contains reverse transcriptase. Its RNA serves as a template for cDNA production, and the cDNA is integrated into a chromosome of the mammalian host cell.

Reverse transcriptase An enzyme that catalyzes the production of DNA (cDNA), using RNA as a template; essential to the reproduction of retroviruses.

Reversion (genetic) A mutational event that restores wild-type phenotype to a mutant.

RFLP (Restriction fragment length polymorphism) Coexistence of two or more patterns of restriction fragments (patterns produced by restriction enzymes), as revealed by a probe. The polymorphism reflects a difference in DNA sequence on homologous chromosomes.

Rhizoids (rye´ zoids) [Gr. *rhiza*: root] Hairlike extensions of cells in mosses, liverworts, and a few vascular plants that serve the same function as roots and root hairs in vascular plants. The term is also applied to branched, rootlike extensions of some fungi and algae.

Rhizome (rye´ zome) [Gr. *rhizoma*: mass of roots] A special underground stem (as opposed to root) that runs horizontally beneath the ground.

Rhodopsin A photopigment used in the visual process of transducing photons of light into changes in the membrane potential of photosensory cells.

Ribonucleic acid See RNA.

Ribose (rye´ bose) A sugar of chemical formula $C_5H_{10}O_5$, one of the building blocks of ribonucleic acids.

Ribosomal RNA (rRNA) Several species of RNA that are incorporated into the ribosome.

Ribosome A small organelle that is the site of protein synthesis.

Ribozyme An RNA molecule with catalytic activity.

Ribulose 1,5-bisphosphate (RuBP) The compound in chloroplasts which reacts with carbon dioxide in the first reaction of the Calvin-Benson cycle.

Risk cost The increased chance of being injured or killed as a result of performing a behavior, compared to resting.

RNA (ribonucleic acid) A nucleic acid using ribose. Various classes of RNA are involved in the transcription and translation of genetic information. RNA serves as the genetic storage material in some viruses.

RNA polymerase An enzyme that catalyzes the formation of RNA from a DNA template.

RNA splicing The last stage of RNA processing in eukaryotes, in which the transcripts of introns are excised through the action of small nuclear ribonucleoprotein particles (snRNP).

Rods Light-sensitive cells (photosensors) in the retina. (Contrast with cones.)

Root cap A thimble-shaped mass of cells, produced by the root apical meristem, that protects the meristem and that is the organ that perceives the gravitational stimulus in root gravitropism.

Root hair A specialized epidermal cell with a long, thin process that absorbs water and minerals from the soil solution.

Round dance The dance performed on the vertical surface of a honeycomb by a returning honeybee forager when she has discovered a food source less than 100 meters from the hive.

Round window A flexible membrane between the middle and inner ear that distributes pressure waves in the fluid of the inner ear.

rRNA See ribosomal RNA.

Rubisco (RuBP carboxylase) Enzyme that combines carbon dioxide with ribulose bisphosphate to produce 3-phosphoglycerate, the first product of C_3 photosynthesis. The most abundant protein on Earth.

Rumen (rew´ mun) The first division of the ruminant stomach. It stores and initiates bacterial fermentation of food. Food is regurgitated from the rumen for further chewing. (Contrast with abomasum, omasum.)

Ruminant An herbivorous, cud-chewing mammal such as a cow, sheep, or deer, having a stomach consisting of four compartments.

S phase In the cell cycle, the stage of interphase during which DNA is replicated. (Contrast with G_1 phase, G_2 phase.)

Sap An aqueous solution of nutrients, minerals, and other substances that passes through the xylem of plants.

Saprobe [Gr. *sapros*: rotten + *bios*: life] An organism (usually a bacterium or fungus) that obtains its carbon and energy directly from dead organic matter.

Sarcomere (sark´ o meer) [Gr. *sark*: flesh + *meros*: a part] The contractile unit of a skeletal muscle.

Saturated hydrocarbon A compound consisting only of carbon and hydrogen, with the hydrogen atoms connected by single bonds.

Schizocoelous development Formation of a coelom during embryological development by a splitting of mesodermal masses.

Schwann cell A glial cell that wraps around part of the axon of a peripheral neuron, creating a myelin sheath.

Sclereid A type of sclerenchyma cell, commonly found in nutshells, that is not elongated.

Sclerenchyma (skler eng´ kyma) A plant tissue composed of cells with heavily thickened cell walls, dead at functional maturity. The principal types of sclerenchyma cells are fibers and sclereids.

Scrapie-associated fibril A type of protein fibril found in nervous tissues of mammals infected with certain diseases, notably scrapie, kuru, and Creutzfeld-Jacob disease. Little is known about these fibrils, including whether they are the causal agents of the diseases.

Scrotum (skrote´ um) A sac of skin that contains the testicles in most species of mammals.

Secondary active transport Form of active transport in which ions or molecules are transported against their concentration gradient using energy obtained by relaxation of a gradient of sodium ion concentration rather than directly from ATP. (Contrast with primary active transport.)

Secondary compound A compound synthesized by a plant that is not needed for basic cellular metabolism. Typically has an antiherbivore or antiparasite function.

Secondary growth In plants, growth produced by vascular and cork cambia, contributing to an increase in girth. (Contrast with primary growth.)

Secondary structure Of a protein, localized regularities of structure, such as the α-helix and the β-pleated sheet.

Secondary wall Wall layers laid down by a plant cell that has ceased growing; often impregnated with lignin or suberin.

Second law of thermodynamics States that in any real (irreversible) process, there is a decrease in free energy and an increase in entropy.

Second messenger A compound, such as cyclic AMP, that is released within a target cell after a hormone or other "first messenger" has bound to a surface receptor on a cell; the second messenger triggers further reactions within the cell.

Secretin (si kreet´ in)   A peptide hormone secreted by the upper region of the small intestine when acidic chyme is present. Stimulates the pancreatic duct to secrete bicarbonate ions.

Section   A thin slice, usually for microscopy, as a tangential section or a transverse section.

Seed   A fertilized, ripened ovule of a gymnosperm or angiosperm. Consists of the embryo, nutritive tissue, and a seed coat.

Seed crop   The number of seeds produced by a plant during a particular bout of reproduction.

Seedling   A young plant that has grown from a seed (rather than by grafting or by other means.)

Segmentation genes   In insect larvae, genes that determine the number and polarity of larval segments.

Segregation (genetic)   The separation of alleles, or of homologous chromosomes, from one another during meiosis so that each of the haploid daughter nuclei produced by meiosis contains one or the other member of the pair found in the diploid mother cell, but never both.

Selective permeability   A characteristic of a membrane, allowing certain substances to pass through while other substances are excluded.

Self-differentiating   Behavior that develops without experience with the normal objects toward which it is usually directed and without any practice. (See also instinct.)

Selfish act   A behavioral act that benefits its performer but harms the recipients.

Self-pollination   The fertilization of a plant by its own pollen. (Contrast with cross-pollination.)

Semelparous organism   An organism that reproduces only once in its lifetime. (Contrast with iteroparous.)

Semen (see´ men) [L.: seed]   The thick, whitish liquid produced by the male reproductive organ in mammals, containing sperm.

Semicircular canals   Part of the vestibular system of mammals.

Semiconservative replication   The common way in which DNA is synthesized. Each of the two partner strands in a double helix acts as a template for a new partner strand. Hence, after replication, each double helix consists of one old and one new strand.

Seminiferous tubules   The tubules within the testes within which sperm production occurs.

Senescence [L. senescere: to grow old]   Aging; deteriorative changes with aging.

Sensor   A sensory cell; a cell transduces a physical or chemical stimulus into a membrane potential change.

Sensory neuron   A neuron leading from a sensory cell to the central nervous system. (Contrast with motor neuron.)

Sepal (see´ pul)   One of the outermost structures of the flower, usually protective in function and enclosing the rest of the flower in the bud stage.

Septum [L.: partition]   A membrane or wall between two cavities.

Sertoli cells   Cells in the seminiferous tubules that nurture the developing sperm.

Serum   That part of the blood plasma that remains after clots have formed and been removed.

Sessile (sess´ ul) [L. sedere: to sit]   Permanently attached; not moving.

Sertoli cells   Cells in the seminiferous tubules that nurture the developing sperm.

Set point   In a regulatory system, the threshold sensitivity to the feedback stimulus.

Sex chromosome   In organisms with a chromosomal mechanism of sex determination, one of the chromosomes involved in sex determination. One sex chromosome, the X chromosome, is present in two copies in one sex and only one copy in the other sex. The autosomes, as opposed to the sex chromosomes, are present in two copies in both sexes. In many organisms, there is a second sex chromosome, the Y chromosome, that is found in only one sex—the sex having only one copy of the X.

Sexduction   See F-duction.

Sex linkage   The pattern of inheritance characteristic of genes located on the sex chromosomes of organisms having a chromosomal mechanism for sex determination. The sex that is diploid with respect to sex chromosomes can assume three genotypes: homozygous wild-type, homozygous mutant, or heterozygous carrier. The other sex, haploid for sex chromosomes, is either hemizygous wild-type or hemizygous mutant.

Sexuality   The ability, by any of a multitude of mechanisms, to bring together in one individual genes that were originally carried by two different individuals. The capacity for genetic recombination.

Sexual selection   Selection by one sex of characteristics in individuals of the opposite sex. Also, the favoring of characteristics in one sex as a result of competition among individuals of that sex for mates.

Shoot   The aerial part of a vascular plant, consisting of the leaves, stem(s), and flowers.

Sibling   A brother or sister.

Sieve plate   In sieve tubes, the highly specialized end walls in which are concentrated the clusters of pores through which the protoplasts of adjacent sieve tube elements are interconnected.

Sieve tube   A column of specialized cells found in the phloem, specialized to conduct organic matter from sources (such as photosynthesizing leaves) to sinks (such as roots). Found principally in flowering plants.

Sieve tube element   A single cell of a sieve tube, containing cytoplasm but relatively few organelles, with highly specialized perforated end walls leading to elements above and below.

Sign stimulus   The single stimulus, or one out of a very few stimuli, by which an animal distinguishes key objects, such as an enemy, or a mate, or a place to nest, etc.

Signal   A component of a behavior that transmits information to another individual that influences the future behavior of the receiver.

Signal sequence   N-terminal sequence of a protein that directs the protein through a particular cellular membrane.

Simple development   In insects and other arthropods, development in which eggs hatch into juveniles that are similar in form to adults.

Simple fruit   A fruit that develops from a single ovary. (Contrast with accessory fruit, aggregate fruit, multiple fruit.)

Sinoatrial node (sigh´ no ay´ tree al)   The pacemaker of the mammalian heart.

Sinus (sigh´ nus) [L. sinus: a bend, hollow]   A cavity in a bone, a tissue space, or an enlargement in a blood vessel.

Skeletal muscle   See striated muscle.

Sliding filament theory   A proposed mechanism of muscle contraction based on formation and breaking of crossbridges between actin and myosin filaments, causing them to slide together.

Small intestine   The portion of the gut between the stomach and the colon, consisting of the duodenum, the jejunum, and the ileum.

Small nuclear ribonucleoprotein particle (snRNP)   A complex of an enzyme and a small nuclear RNA molecule, functioning in RNA splicing.

Smooth muscle   One of three types of muscle tissue. Usually consists of sheets of mononucleated cells innervated by the autonomic nervous system.

Social insect   One of the kinds of insect that form colonies with reproductive castes and worker castes; in particular, the termites, ants, social bees, and social wasps.

Society   A group of individuals belonging to the same species and organized in a cooperative manner; in the broadest sense, includes parents and their offspring.

Sodium cotransport   Carrier-mediated transport of molecules across membranes driven by sodium ions binding to the same carrier and moving down their concentration gradient.

Sodium–potassium pump   The complex protein in plasma membranes that is responsible for primary active transport; it pumps sodium ions out of the cell and potassium ions into the cell, both against their concentration gradients.

Solute A substance that is dissolved in a liquid (solvent).

Solution A liquid (solvent) and its dissolved solutes.

Solvent A liquid that has dissolved or can dissolve one or more solutes.

Somatic [Gr. *soma*: body] Pertaining to the body, or body cells (rather than to germ cells).

Somite (so´ might) One of the segments into which an embryo becomes divided longitudinally, leading to the eventual segmentation of the animal as illustrated by the spinal column, ribs, and associated muscles.

Southern blotting Transfer of DNA fragments from an electrophoretic gel to a sheet of paper or other absorbent material for analysis with a probe.

Spatial summation In the production or inhibition of action potentials in a postsynaptic neuron, the interaction of depolarizations and hyperpolarizations produced by several terminal boutons.

Spawning The direct release of sex cells into the water.

Speciation (spee´ shee ay´ shun) The process of splitting one population into two populations that are reproductively isolated from one another.

Species (spee´ shees) [L.: kind] The basic lower unit of classification, consisting of a population or series of populations of closely related and similar organisms. The more narrowly defined "biological species" consists of individuals capable of interbreeding freely with each other but not with members of other species.

Species diversity A weighted representation of the species of organisms living in a region; large and common species are given greater weight than are small and rare ones. (Contrast with species richness.)

Species pool All the species potentially available to colonize a particular habitat.

Species richness The number of species of organisms living in a region. (Contrast with species diversity.)

Specific heat The amount of energy that must be absorbed by a gram of a substance to raise its temperature by one degree centigrade. By convention, water is assigned a specific heat of one.

Sperm [Gr. *sperma*: seed] A male reproductive cell.

Spermatocyte (spur mat´ oh site) [Gr. *sperma*: seed + *kytos*: cell] The cell that gives rise to the sperm in animals.

Spermatogenesis (spur mat´ oh jen´ e sis) [Gr. *sperma*: seed + *genesis*: source] Male gametogenesis, leading to the production of sperm.

Spermatogonia Undifferentiated germ cells that give rise to primary spermatocytes and hence to sperm.

Spermatophore A package of sperm deposited in the environment by an invertebrate male, and then either inserted by him

into the reproductive tract of the female or taken up by the female herself.

Sphincter (sfingk´ ter) [Gr. *sphinkter*: that which binds tight] A ring of muscle that can close an orifice, for example at the anus.

Spindle apparatus An array of microtubules stretching from pole to pole of a dividing nucleus and playing a role in the movement of chromosomes at nuclear division. Named for its shape.

Spiracle (spy´ rih kel) [L. *spirare*: to breathe] An opening of the treacheal respiratory system of terrestrial arthorpods.

Spiteful act A behavioral act that harms both the actor and the recipient of the act.

Spongy parenchyma In leaves, a layer of loosely packed photosynthetic cells with extensive intercellular spaces for gas diffusion. Frequently found between the palisade parenchyma and the lower epidermis.

Spontaneous generation The idea that life is generated continually from nonliving matter. Usually distinguished from the current idea that life evolved from nonliving matter under primordial conditions at an early stage in the history of earth.

Spontaneous reaction A chemical reaction which will proceed on its own, without any outside influence. A spontaneous reaction need not be rapid.

Sporangiophore [Gr. *phore*: to bear] Any branch bearing one or more sporangia.

Sporangium (spor an´ gee um) [Gr. *spora*: seed + *angeion*: vessel or reservoir] In plants and fungi, any specialized stucture within which one or more spores are formed.

Spore [Gr. *spora*: seed] Any asexual reproductive cell capable of developing into an adult plant without gametic fusion. Haploid spores develop into gametophytes, diploid spores into sporophytes. In prokaryotes, a resistant cell capable of surviving unfavorable periods.

Sporophyll (spor´ o fill) [Gr. *spora*: seed + *phyllon*: leaf] Any leaf or leaflike structure that bears sporangia; refers to carpels and stamens of angiosperms and to sporangium-bearing leaves on ferns, for example.

Sporophyte (spor´ o fyte) [Gr. *spora*: seed + *phyton*: plant] In plants with alternation of generations, the diploid phase that produces the spores. (Contrast with gametophyte.)

Stabilizing selection Selection against the extreme phenotypes in a population, so that the intermediate types are favored. (Contrast with disruptive selection.)

Stamen (stay´ men) [L.: thread] A male (pollen-producing) unit of a flower, usually composed of an anther, which bears the pollen, and a filament, which is a stalk supporting the anther.

Starch [O.E. *stearc*: stiff] An α-linked polymer of glucose; used by plants as a means of storing energy and carbon atoms.

Stasis Period during which little or no evolutionary change takes place within a lineage or groups of lineages.

Statocyst (stat´ oh sist) [Gk. *statos*: stationary + *kystos*: pouch] An organ of equilibrium in some invertebrates.

Statolith (stat´ oh lith) [Gk. *statos*: stationary + *lithos*: stone] A solid object that responds to gravity or movement and stimulates the mechanosensors of a statocyst.

Stele (steel) [Gr. *stele*: pillar] The central cylinder of vascular tissue in a plant stem.

Stem cell A cell capable of extensive proliferation, generating more stem cells and a large clone of differentiated progeny cells, as in the formation of red blood cells.

Step cline A sudden change in one or more traits of a species along a geographical gradient.

Steroid Any of numerous lipids based on a 17-carbon atom ring system.

Sticky ends On a piece of two-stranded DNA, short, complementary, one-stranded regions produced by the action of a restriction endonuclease. Sticky ends allow the joining of segments of DNA from different sources.

Stigma [L.: mark, brand] The part of the pistil at the apex of the style, which is receptive to pollen, and on which pollen germinates.

Stimulus Something causing a response; something in the environment detected by a receptor.

Stolon A horizontal stem that forms roots at intervals.

Stoma (plural: stomata) [Gr. *stoma*: mouth, opening] Small opening in the plant epidermis that permits gas exchange; bounded by a pair of guard cells whose osmotic status regulates the size of the opening.

Stratosphere The part of the atmosphere above the troposphere; extends upward to approximately 50 kilometers above the surface of the earth; contains very little water.

Stratum (plural strata) A layer or sedimentary rock laid down at a particular time in a past.

Striated muscle Contractile tissue characterized by multinucleated cells containing highly ordered arrangements of actin and myosin microfilaments. Also known as **skeletal muscle**.

Strobilus (strobe´ a lus) [Gr. *strobilos*: a cone] The cone, or characteristic multiple fruit, of the pine and other gymnosperms. Also, a cone-shaped mass of sprophylls found in club mosses.

Stroma The fluid contents of an organelle, such as a chloroplast.

Stromatolite A composite, flat-to-domed structure composed of successive mineral layers. Some are known to be produced by the action of bacteria in salt or fresh water, and some ancient ones are considered to be evidence for early life on the earth.

Structural formula A representation of the positions of atoms and bonds in a molecule.

Structural gene A gene that encodes the primary structure of a protein.

Style [Gr. *stylos*: pillar or column] In flowering plants, a column of tissue extending from the tip of the ovary, and bearing the stigma or receptive surface for pollen at its apex.

Sub- [L.: under] A prefix often used to designate a structure that lies beneath another or is less than another. For example, subcutaneous, subspecies.

Suberin A waxy material serving as a waterproofing agent in cork and in the Casparian strips of the endodermis in plants.

Submucosa (sub mew koe´ sah) The tissue layer just under the epithelial lining of the lumen of the digestive tract. (Contrast with mucosa.)

Substrate (sub´ strayte) The molecule or molecules on which an enzyme exerts catalytic action.

Substrate level phosphorylation ATP formation resulting from direct transfer of a phosphate group to ADP from an intermediate in glycolysis. (Contrast with oxidative phosphorylation.)

Succession In ecology, the gradual, sequential series of changes in species composition of a community following a disturbance.

Sulfhydryl group The –SH group.

Summation The ability of a neuron to fire action potentials in response to numerous subthreshold postsynaptic potentials arriving simultaneously at differentiated places on the cell, or arriving at the same site in rapid succession.

Supercoiling Coiling on coiling, as in DNA during prophase.

Supernormal stimulus Any stimulus, or any intensity of a variable stimulus, that is preferred by animals over the natural sign stimulus.

Suppressor T cells T cells that inhibit the res-ponses of B cells and other T cells to antigens. (Contrast with cytotoxic T cells, helper T cells.)

Surface tension A measure of the cohesiveness of the surface of a liquid. As a result of hydrogen bonding, water has a very high surface tension, allowing some insects to walk on the water surface.

Surface-to-volume ratio For any cell, organism, or geometrical solid, the ratio of surface area to volume; this is an important factor in setting an upper limit on the size a cell or organism can attain.

Surfactant A substance that decreases the surface tension of a liquid. Lung surfactant, secreted by cells of the alveoli, is mostly phospholipid and decreases the amount of work necessary to inflate the lungs.

Survivorship curve A plot of the logarithm of the fraction of individuals still alive, as a function of time.

Suspensor In plants, a cell or group of cells derived from the zygote, but not actually part of the embryo proper, which in some seed plants pushes the young embryo deeper into nutritive gametophyte tissue or endosperm by its growth.

Swim bladder An internal gas-filled organ that helps fishes maintain their position in the water column; later evolved into an organ for gas exchange in some lineages.

Symbiosis (sim´ bee oh´ sis) [Gr.: to live together] The living together of two or more species in a prolonged and intimate ecological relationship. (See parasitism, commensalism, mutualism.)

Symmetry In biology, the property that two halves of an object are mirror images of each other. (See bilateral symmetry and radial symmetry.)

Sympathetic nervous system A division of the autonomic (involuntary) nervous system. Its activities include increasing blood pressure and acceleration of the heartbeat. The neurotransmitter at the sympathetic terminals is epinephrine or norepinephrine. (Contrast with parasympathetic nervous system.)

Sympatric (sim pat´ rik) [Gr. *syn*: together + *patria*: homeland] Referring to populations whose geographic regions overlap at least in part.

Sympatric speciation Formation of new species even though members of the daughter species overlap in their distribution during the speciation process. (Contrast with geographic, parapatric speciation.)

Symplast The continuous meshwork of the interiors of living cells in the plant body, resulting from the presence of plasmodesmata. (Contrast with apoplast.)

Symport A membrane transport protein that carries two substances in the same direction across the membrane. (Contrast with antiport.)

Synapse (sin´ aps) [Gr. *syn*: together + *haptein*: to fasten] The narrow gap between the terminal bouton of one neutron and the dendrite or cell body of another.

Synapsis (sin ap´ sis) The highly specific parallel alignment (pairing) of homologous chromosomes during the first division of meiosis.

Synaptic vesicle A membrane-bounded vesicle, containing neurotransmitter, which is produced in and discharged by the presynaptic neuron.

Synergids (sin nur´ jids) Two cells found close to the egg cell in the angiosperm embryo sac; they disappear shortly after fertilization.

Syngamy (sing´ guh mee) [Gr. *sun*-: together + *gamos*: marriage] Union of gametes. Also known as fertilization.

Syrinx (sear´ inks) [Gr.: pipe, cavity] A specialized structure at the junction of the trachea and the primary bronchi leading to the lungs. The vocal organ of birds.

Systematics The scientific study of the diversity of organisms.

Systemic circulation The part of the circulatory system serving those parts of the body other than the lungs or gills.

Systole (sis´ tuh lee) [Gr.: contraction] Contraction of a chamber of the heart, driving blood forward in the circulatory system.

T cell A type of lymphocyte, involved in the cellular immune response. The final stages of its development occur in the thymus gland. (Contrast with B cell; see also cytotoxic T cell, helper T cell, suppressor T cell.)

T cell receptor A protein on the surface of a T cell that recognizes the antigenic determinant for which the cell is specific.

Target cell A cell which has the appropriate receptors to bind and respond to a particular hormone or other chemical mediator.

Taste bud A structure in the epithelium of the tongue that includes a cluster of chemosensors innervated by sensory neurons.

TATA box An eight-base-pair sequence, found about 25 base pairs before the starting point for transcription in many eukaryotic promoters, that binds a transcription factor and thus helps initiate transcription.

Taxis (tak´ sis) [Gr. *taxis*: arrange, put in order] The movement of an organism in a particular direction with reference to a stimulus. A taxis usually involves the employment of one sense and a movement directly toward or away from the stimulus, or else the maintenance of a constant angle to it. Thus a positive phototaxis is movement toward a light source, negative geotaxis is movement upward (away from gravity), and so on.

Taxon A unit in a taxonomic system.

Taxonomy (taks on´ oh me) [Gr. *taxis*: arrange, classify] The science of classification of organisms.

Telophase (tee´ lo phase) [Gr. *telos*: end] The final phase of mitosis or meiosis during which chromosomes became diffuse, nuclear envelopes reform, and nucleoli begin to reappear in the daughter nuclei.

Template In biochemistry, a molecule or surface upon which another molecule is synthesized in complementary fashion, as in the replication of DNA. In the brain, a pattern that responds to a normal input but not to incorrect inputs.

Template strand In a stretch of double-stranded DNA, the strand that is transcribed.

Temporal summation In the production or inhibition of action potentials in a postsynaptic neuron, the interaction of depolarizations or hyperpolarizations produced by rapidly repeated stimulation of a single point.

Tendon A collagen-containing band of tissue that connects a muscle with a bone.

Tension wood See reaction wood.

Tepal In an angiosperm flower, a sterile modified leaf. This term is used to refer to such flower parts when one is unable to distinguish between petals and sepals.

Terrestrial (ter res´ tree al) [L. *terra*: earth] Pertaining to the land. (Contrast with aquatic, marine.)

Territory A fixed area from which an animal or group of animals excludes other members of the same species by aggressive behavior or display.

Tertiary structure In reference to a protein, the relative locations in three-dimensional space of all the atoms in the molecule. The overall shape of a protein. (Contrast with primary, secondary, and quaternary structures.)

Test cross A cross of a dominant-phenotype individual (which may be either heterozygous or homozygous) with a homozygous-recessive individual.

Testis (tes´ tis) (plural: testes) [L.: witness] The male gonad; that is, the organ that produces the male sex cells.

Testosterone (tes toss´ tuhr own) A male sex steroid hormone.

Tetanus [Gr. *tetanos*: stretched] (1) In physiology, a state of sustained, maximal muscular contraction caused by rapidly repeated stimulation. (2) In medicine, an often-fatal disease ("lockjaw") caused by the bacterium *Clostridium tetani*.

Thalamus A region of the vertebrate forebrain; involved in integration of sensory input.

Thallus (thal´ us) [Gr.: sprout] Any algal body which is not differentiated into root, stem, and leaf.

Thermocline In a body of water, the zone where the temperatures change abruptly to about 4°C.

Thermoneutral zone The range of temperatures over that an endotherm does not have to expend extra energy to thermoregulate.

Thermosensor A cell or structure that responds to changes in temperature.

Thoracic cavity The portion of the mammalian body cavity bounded by the ribs, shoulders, and diaphragm. Contains the heart and the lungs.

Thorax In an insect, the middle region of the body, between the head and abdomen. In mammals, the part of the body between the neck and the diaphragm.

Thrombin An enzyme that converts fibrinogen to fibrin, thus triggering the formation of blood clots.

Thrombus (throm´ bus) [Gk. *thrombos*: clot] A blood clot that forms within a blood vessel and remains attached to the wall of the vessel. (Contrast with embolus.)

Thylakoid A flattened sac within a chloroplast. The membranes of the numerous thylakoids contain all of the chlorophyll in a plant, in addition to the electron carriers of photophosphorylation. Thylakoids stack to form grana.

Thymus A ductless, glandular portion of the lymphoid system, involved in development of the immune system of vertebrates.

Thyroid [Gr. *thyreos*: door-shaped] A two-lobed gland in vertebrates. Produces the hormone thyroxin.

Thyrotropic hormone A hormone that is produced in the pituitary gland of amphibia such as frogs and transported in the bloodstream to the thyroid gland, inducing the thyroid gland to produce the thyroid hormone that regulates metamorphosis from tadpole to adult frog.

Tight junction A junction between epithelial cells, in which there is no gap whatever between the adjacent cells. Materials may get through a tight junction only by entering the epithelial cells themselves.

Tissue A group of similar cells organized into a functional unit and usually integrated with other tissues to form part of an organ such as a heart or leaf.

Tonus A low level of muscular tension that is maintained even when the body is at rest.

Tornaria (tor nare´ e ah) [L. *tornus*: lathe] The free-swimming ciliated larva of certain echinoderms and hemichordates; its existence indicates the evolutionary relationship of these two groups.

Totipotency In a cell, the condition of possessing all the genetic information and other capacities necessary to form an entire individual.

Toxigenicity The ability of a bacterium to produce chemical substances injurious to the tissues of the host organism.

Trachea (tray´ kee ah) [Gr. *trakhoia*: a small rough artery] A tube that carries air to the bronchi of the lungs of vertebrates, or to the cells of arthropods.

Tracheid (tray´ kee id) A distinctive conducting and supporting cell found in the xylem of nearly all vascular plants, characterized by tapering ends and walls that are pitted but not perforated.

Trade winds The winds that blow toward the intertropical convergence zone from the northeast and southeast.

Transcription The synthesis of RNA, using one strand of DNA as the template.

Transcription factors Proteins that assemble on a eukaryotic chromosome, allowing RNA polymerase II to perform transcription.

Transduction (1) Transfer of genes from one bacterium to another, with a bacterial virus acting as the carrier of the genes. (2) In sensory cells, the transformation of a stimulus (e.g., light energy, sound pressure waves, chemical or electrical stimulants) into action potentials.

Transfection Uptake, incorporation, and expression of recombinant DNA.

Transfer cells A modified parenchyma cell that transports mineral ions from its cytoplasm into its cell wall, thus moving the ions from the symplast into the apoplast.

Transfer RNA (tRNA) A category of relatively small RNA molecules (about 75 nucleotides). Each kind of transfer RNA is able to accept a particular activated amino acid from its specific activating enzyme, after which the amino acid is added to a growing polypeptide chain.

Transformation Mechanism for transfer of genetic information in bacteria in which pure DNA extracted from bacteria of one genotype is taken in through the cell surface of bacteria of a different genotype and incorporated into the chromosome of the recipient cell. By extension, the term has come to be applied to phenomena in other organisms in which specific genetic alterations have been produced by treatment with purified DNA from genetically marked donors.

Transgenic Containing recombinant DNA incorporated into its genetic material.

Translation The synthesis of a protein (polypeptide). This occurs on ribosomes, using the information encoded in messenger RNA.

Translocation (1) In genetics, a rare mutational event that moves a portion of a chromosome to a new location, generally on a nonhomologous chromosome. (2) In vascular plants, movement of solutes in the phloem.

Transpiration [L. *spirare*: to breathe] The evaporation of water from plant leaves and stem, driven by heat from the sun, and providing the motive force to raise water (plus ions) from the roots.

Transposable element A segment of DNA that can move to, or give rise to copies at, another locus on the same or a different chromosome. May be a single insertion sequence or a more complex structure (transposon) consisting of two insertion sequences and one or more intervening genes.

Trichocyst (trick´ o sist) [Gr. *trichos*: hair + *kystis*: cell] A threadlike organelle ejected from the surface of ciliates, used both as a weapon and as an anchoring device.

Triglyceride A simple lipid in which three fatty acids are combined with one molecule of glycerol.

Triplet See codon.

Triplet repeat Occurrence of repeated triplet of bases in a gene, often leading to genetic disease, as does excessive repetition of CGG in the gene responsible for fragile-X syndrome.

Triploblastic Having three cell layers. (Contrast with diploblastic.)

Trisomic Containing three, rather than two members of a chromosome pair.

tRNA See transfer RNA.

Trochophore (troke´ o fore) [Gr. *trochos*: wheel + *phoreus*: bearer] The free-swimming larva of some annelids and mollusks, distinguished by a wheel-like band of cilia around the middle, and indicating an evolutionary relationship between these two groups.

Trophic level A group of organisms united by obtaining their energy from the same part of the food web of a biological community.

Tropic hormones Hormones of the anterior pituitary that control the secretion of hormones by other endocrine glands.

Tropism [Gr. *tropos*: to turn]   In plants, growth toward or away from a stimulus such as light (phototropism) or gravity (gravitropism).

Tropomyosin (troe poe my´ oh sin)   A protein that, along with actin, constitutes the thin filaments of myofibrils. It controls the interactions of actin and myosin necessary for muscle contraction.

Troposphere   The atmospheric zone reaching upward approximately 17 km in the tropics and subtropics but only to about 10 km at higher latitudes. The zone in which virtually all the water vapor in the atmosphere is located.

Trypsin   A protein-digesting enzyme. Secreted by the pancreas in its inactive form (trypsinogen), it becomes active in the duodenum of the small intestine.

T-tubules   A set of transverse tubes that penetrates skeletal muscle fibers and terminates in the sarcoplasmic reticulum. The T-system transmits impulses to the sacs, which then release CA^{2+} to initiate muscle contraction.

Tube foot   In echinoderms, a part of the water vascular system. It grasps the substratum, prey, or other solid objects.

Tube nucleus   In a pollen tube, the haploid nucleus that does not participate in double fertilization. (Contrast with generative nucleus.)

Tuber [L.: swelling]   A short, fleshy underground stem, usually much enlarged, and serving a storage function, as in the case of the potato.

Tubulin   A protein that polymerizes to form microtubules.

Tumor   A disorganized mass of cells, often growing out of control. Malignant tumors spread to other parts of the body.

Turgor   See pressure potential.

Twitch   A single unit of muscle contraction.

Tympanic membrane [Gr. *tympanum*: drum] The eardrum.

Umbilical cord   Tissue made up of embryonic membranes and blood vessels that connects the embryo to the placenta in eutherian mammals.

Uncoupler   See respiratory uncoupler.

Understory   The aggregate of smaller plants growing beneath the canopy of dominant plants in a forest.

Unicellular (yoon´ e sell´ yer ler) [L. *unus*: one + *cella*: chamber]   Consisting of a single cell; as for example a unicellular organism. (Contrast with multicellular.)

Uniport   A membrane transport protein that carries a single substance. (Contrast with antiport, symport.)

Unitary organism   An organism that consists of only one module.

Unsaturated hydrocarbon   A compound containing only carbon and hydrogen atoms. One or more pairs of carbon atoms are connected by double bonds.

Upwelling   The upward movement of nutrient-rich, cooler water from deeper layers of the ocean.

Urea   A compound serving as the main excreted form of nitrogen by many animals, including mammals.

Ureotelic   Describes an organism in which the final product of the breakdown of nitrogen-containing compounds (primarily proteins) is urea. (Contrast with ammonotelic, uricotelic.)

Ureter (your´ uh tur) [Gr. *ouron*: urine]   A long duct leading from the vertebrate kidney to the urinary bladder or the cloaca.

Urethra (you ree´ thra) [Gr. *ouron*: urine]   In most mammals, the canal through which urine is discharged from the bladder and which serves as the genital duct in males.

Uric acid   A compound that serves as the main excreted form of nitrogen in some animals, particularly those which must conserve water, such as birds, insects, and reptiles.

Uricotelic   Describes an organism in which the final product of the breakdown of nitrogen-containing compounds (primarily proteins) is uric acid. (Contrast with ammonotelic, ureotelic.)

Urinary bladder   A structure structure that receives urine from the kidneys via the ureter, stores it, and expels it periodically through the urethra.

Urine (you´ rin) [Gk. *ouron*: urine]   In vertebrates, the fluid waste product containing the toxic nitrogenous by-products of protein and amino acid metabolism.

Uterus (yoo´ ter us) [L.: womb]   The uterus or womb is a specialized portion of the female reproductive tract in certain mammals. It receives the fertilized egg and nurtures the embryo in its early development.

Vaccination   Injection of virus or bacteria or their proteins into the body, to induce immunization. The injected material is usually attenuated (weakened) before injection.

Vacuole (vac´ yew ole) [Fr.: small vacuum] A liquid-filled cavity in a cell, enclosed within a single membrane. Vacuoles play a wide variety of roles in cellular metabolism, some being digestive chambers, some storage chambers, some waste bins, and so forth.

Vagina (vuh jine´ uh) [L.: sheath]   In female mammals, the passage leading from the external genital orifice to the uterus; receives the copulatory organ of the male in mating.

Van der Waals interaction   A weak attraction between atoms resulting from the interaction of the electrons of one atom with the nucleus of the other atom. This attraction is about one-fourth as strong as a hydrogen bond.

Vascular (vas´ kew lar)   Pertaining to organs and tissues that conduct fluid, such as blood vessels in animals and phloem and xylem in plants.

Vascular bundle   In vascular plants, a strand of vascular tissue, including conducting cells of xylem and phloem as well as thick-walled fibers.

Vascular plants Those plants with xylem and phloem, including psilophytes, club mosses, horsetails, ferns, gymnosperms, and angiosperms. (Contrast with nonvascular plants.)

Vascular ray   In vascular plants, radially oriented sheets of cells produced by the vascular cambium, carrying materials laterally between the wood and the phloem.

Vascular tissue system   The conductive system of the plant, consisting primarily of xylem and phloem. (Contrast with dermal tissue system, ground tissue system.)

Vasopressin   See antidiuretic hormone.

Vector   (1) An agent, such as an insect, that carries a pathogen affecting another species. (2) A plasmid or virus that carries an inserted piece of DNA into a bacterium for cloning purposes in recombinant DNA technology.

Vegetal pole   In some eggs, zygotes, and embryos, the pole near the bulk of the yolk. (Contrast with animal pole.)

Vegetative   Nonreproductive, or nonflowering, or asexual.

Vein [L. *vena*: channel]   A blood vessel that returns blood to the heart. (Contrast with artery.)

Vena cava [L.: hollow vein]   One of a pair of large veins that carry blood from the systemic circulatory system into the heart.

Ventral [L. *venter*: belly, womb]   Toward or pertaining to the belly or lower side. (Contrast with dorsal.)

Ventricle   A muscular heart chamber that pumps blood through the body.

Vernalization [L. *vernalis*: belonging to spring]   Events occurring during a required chilling period, leading eventually to flowering. Vernalization may require many weeks of below-freezing temperatures.

Vertebral column   The jointed, dorsal column that is the primary support structure of vertebrates.

Vertebrate   An animal whose nerve cord is enclosed in a backbone of bony segments, called vertebrae. The principal groups of vertebrate animals are the fishes, amphibians, reptiles, birds, and mammals.

Vertical evolution   Evolutionary change in a single lineage over time. Also called anagenesis.

Vessel [L. *vasculum*: a small vessel]   In botany, a tube-shaped portion of the xylem consisting of hollow cells (vessel elements) placed end to end and connected by perforations. Together with tracheids, vessel elements conduct water and minerals in the plant.

Vestibular apparatus (ves tib´ yew lar) [L. *vestibulum*: an enclosed passage]   Structures associated with the vertebrate ear; these structures sense changes in position or momentum of the head, affecting balance and motor skills.

Vestigial (ves tij´ ee al) [L. *vestigium*: footprint, track]   The remains of body structures that are no longer of adaptive value to the organism and therefore are not maintained by selection.

Vicariance (vye care´ ee unce) [L. *vicus*: change]   The splitting of the range of a taxon by the imposition of some barrier to dispersal of its members. May lead to cladogenesis.

Vicariant distribution   A distribution resulting from the disruption of a formerly continuous range by a vicariant event.

Villus (vil´ lus) (plural: villi) [L.: shaggy hair]   A hairlike projection from a membrane; for example, from many gut walls.

Virion (veer´ e on)   The virus particle, the minimum unit capable of infecting a cell.

Viroid (vye´ roid)   An infectious agent consisting of a single-stranded RNA molecule with no protein coat; produces diseases in plants.

Virus [L.: poison, slimy liquid]   Any of a group of ultramicroscopic infectious particles constructed of nucleic acid and protein (and, sometimes, lipid) that can reproduce only in living cells.

Visceral mass   The major internal organs of a mollusk.

Vitamin [L. *vita*: life]   Any one of several structurally unrelated organic compounds that an organism cannot synthesize itself, but nevertheless requires in small quantity for normal growth and metabolism.

Viviparous (vye vip´ uh rus) [L. *vivus*: alive]   Reproduction in which fertilization of the egg and development of the embryo occur inside the mother's body. (Contrast with oviparous.)

Waggle dance   The running movement of a working honey bee on the hive, during which the worker traces out a repeated figure eight. The dance contains elements that transmit to other bees the location of the food.

Water potential   In osmosis, the tendency for a system (a cell or solution) to take up water from pure water, through a differentially permeable membrane. Water flows toward the system with a more negative water potential. (Contrast with osmotic potential, pressure potential.)

Water vascular system   The array of canals and tubelike appendages that serves as the circulatory system, locomotory system, and food capturing system of many echinoderms; is in direct connection with the surrounding sea water.

Wavelength   The distance between successive peaks of a wave train, such as electromagnetic radiation.

Wild-type   Geneticists' term for standard or reference type. Deviants from this standard, even if the deviants are found in the wild, are said to be mutant.

Xanthophyll (zan´ tho fill) [Gr. *xanthos*: yellowish-brown + *phyllon*: leaf]   A yellow or orange pigment commonly found as an accessory pigment in photosynthesis, but found elsewhere as well. An oxygen-containing carotenoid.

X chromosome   See sex chromosome.

Xerophyte (zee´ row fyte) [Gr. *xerox*: dry + *phyton*: plant]   A plant adapted to an environment with a limited water supply.

Xylem (zy´ lum) [Gr. *xylon*: wood]   In vascular plants, the woody tissue that conducts water and minerals; xylem consists, in various plants, of tracheids, vessel elements, fibers, and other highly specialized cells.

Y chromosome   See sex chromosome.

Yolk   The stored food material in animal eggs, usually rich in protein and lipid.

Z-DNA   A form of DNA in which the molecule spirals to the left rather than to the right.

Zooplankton (zoe´ o plang ton) [Gr. *zoon*: animal + *planktos*: wandering]   The animal portion of the plankton.

Zoospore (zoe´ o spore) [Gr. *zoon*: animal + *spora*: seed]   In algae and fungi, any swimming spore. May be diploid or haploid.

Zygospore   A highly resistant type of fungal spore produced by the zygomycetes (conjugating fungi).

Zygote (zye´ gote) [Gr. *zygotos*: yoked]   The cell created by the union of two gametes, in which the gamete nuclei are also fused. The earliest stage of the diploid generation.

Zymogen   An inactive precursor of a digestive enzyme secreted into the lumen of the gut, where a protease cleaves it to form the active enzyme. Zymogens make it unnecessary for some digestive enzymes to be formed inside cells, which active ezymes might digest.

ANSWERS TO SELF-QUIZZES

Chapter 2

1. b	6. a
2. e	7. c
3. c	8. b
4. c	9. e
5. d	10. d

Chapter 3

1. e	6. a
2. d	7. c
3. c	8. e
4. d	9. a
5. b	10. d

Chapter 4

1. a	6. e
2. e	7. a
3. c	8. d
4. e	9. b
5. c	10. d

Chapter 5

1. e	6. b
2. d	7. c
3. a	8. b
4. d	9. e
5. c	10. c

Chapter 6

1. c	6. a
2. e	7. e
3. b	8. b
4. c	9. d
5. c	10. e

Chapter 7

1. a	6. d
2. d	7. a
3. c	8. e
4. e	9. c
5. c	10. e

Chapter 8

1. c	6. d
2. b	7. c
3. d	8. d
4. b	9. b
5. e	10. b

Chapter 9

1. e	6. a
2. c	7. e
3. b	8. d
4. d	9. b
5. c	10. a

Chapter 10*

1. d	6. d
2. a	7. b
3. e	8. a
4. d	9. b
5. d	10. c

Chapter 11

1. c	6. d
2. a	7. b
3. c	8. d
4. b	9. d
5. e	10. a

Chapter 12

1. c	6. d
2. d	7. c
3. b	8. a
4. b	9. b
5. e	10. d

Chapter 13

1. d	6. b
2. c	7. c
3. a	8. d
4. b	9. e
5. e	10. c

Chapter 14

1. c	6. c
2. b	7. a
3. e	8. b
4. d	9. c
5. a	10. d

Chapter 15

1. d	6. c
2. a	7. e
3. d	8. d
4. e	9. b
5. c	10. c

Chapter 16

1. d	6. a
2. b	7. d
3. e	8. d
4. e	9. a
5. c	10. e

Chapter 17

1. d	6. c
2. c	7. d
3. e	8. b
4. a	9. a
5. b	10. b

Chapter 18

1. e	6. c
2. d	7. e
3. d	8. c
4. e	9. a
5. e	10. a

Chapter 19

1. d	6. e
2. c	7. b
3. d	8. e
4. b	9. d
5. d	10. d

Chapter 20

1. c	6. a
2. a	7. b
3. e	8. a
4. d	9. c
5. c	10. a

Chapter 21

1. e	6. d
2. c	7. a
3. a	8. d
4. c	9. b
5. a	10. d

Chapter 22

1. e	6. a
2. e	7. c
3. b	8. c
4. d	9. b
5. b	10. d

Chapter 23

1. a	6. d
2. e	7. c
3. c	8. b
4. d	9. b
5. a	10. d

Chapter 24

1. b	6. a
2. d	7. e
3. e	8. a
4. c	9. c
5. d	10. c

Chapter 25

1. d	6. b
2. c	7. b
3. e	8. c
4. b	9. d
5. c	10. c

Chapter 26

1. c	7. d
2. d	8. c
3. b	9. d
4. e	10. e
5. e	11. a
6. b	

Chapter 27

1. b	7. d
2. c	8. e
3. a	9. a
4. c	10. e
5. c	11. c
6. b	12. c

*Answers to the Genetics "For Study" questions in Chapter 10 appear at the end of this section

Chapter 28

1. d	6. b
2. e	7. c
3. d	8. e
4. a	9. e
5. a	10. b

Chapter 29

1. d	6. b
2. b	7. b
3. e	8. c
4. e	9. a
5. a	10. d

Chapter 30

1. c	6. d
2. d	7. e
3. b	8. a
4. e	9. d
5. b	10. d

Chapter 31

1. d	6. a
2. a	7. c
3. b	8. c
4. c	9. e
5. d	10. b

Chapter 32

1. d	6. e
2. c	7. a
3. a	8. b
4. e	9. d
5. c	10. e

Chapter 33

1. a	6. c
2. e	7. e
3. c	8. c
4. d	9. a
5. b	10. b

Chapter 34

1. c	6. b
2. d	7. d
3. d	8. e
4. b	9. a
5. e	10. b

Chapter 35

1. c	6. e
2. a	7. a
3. d	8. e
4. b	9. d
5. b	10. c

Chapter 36

1. c	6. b
2. b	7. a
3. d	8. e
4. b	9. c
5. a	10. e

Chapter 37

1. (i) b	4. a
(ii) a	5. d
(iii) c	6. c
(iv) a,b,c	7. e
(v) a,b,c	8. d
2. c,e	9. b
3. e	10. d

Chapter 38

1. d	6. e
2. a	7. e
3. e	8. c
4. d	9. d
5. c	10. b

Chapter 39

1. d	6. e
2. d	7. b
3. a	8. c
4. b	9. c
5. e	10. c

Chapter 40

1. d	6. d
2. e	7. b
3. a	8. a
4. b	9. a
5. b	10. e

Chapter 41

1. e	6. b
2. d	7. c
3. a	8. c
4. b	9. a
5. c	10. d

Chapter 42

1. d	6. d
2. a	7. b
3. c	8. d
4. d	9. c
5. c	10. e

Chapter 43

1. b	6. d
2. e	7. a
3. c	8. b
4. a	9. d
5. b	10. d

Chapter 44

1. b	6. b
2. a	7. e
3. d	8. a
4. c	9. c
5. d	10. e

Chapter 45

1. a	6. c
2. e	7. d
3. b	8. c
4. b	9. d
5. a	10. a

Chapter 46

1. e	6. e
2. c	7. c
3. d	8. e
4. c	9. d
5. c	10. a

Chapter 47

1. b	6. c
2. c	7. a
3. d	8. c
4. a	9. d
5. d	10. a

Chapter 48

1. b	7. a
2. e	8. c
3. c	9. b
4. e	10. e
5. e	11. c
6. c	

Chapter 49

1. b	7. a
2. d	8. d
3. d	9. a
4. b	10. b
5. c	11. e
6. c	

Chapter 50

1. a	6. a
2. b	7. d
3. b	8. e
4. c	9. d
5. d	10. a

Chapter 51

1. b	7. d
2. c	8. e
3. d	9. e
4. e	10. a
5. e	11. a
6. a	

Answers to Genetics "For Study" Questions, Chapter 10

1. Each of the eight boxes should contain *Tt*.

2. See Figure 10.3.

3. The trait is autosomal. Mother *dp dp*, father *Dp dp*. If the trait were sex-linked, all daughters would be wild-type and sons would be *dumpy*.

4. All females wild-type; all males spotted.

5. F$_1$ all wild-type, *PpSwsw*; F$_2$ 9:3:3:1 in phenotypes. See Figure 10.7 for analogous genotypes.

6a. Ratio of phenotypes in F$_2$ is 3:1 (double dominant to double recessive). See Figure 10.13.
6b. The F$_1$ are *Pby pBy*; they produce just two kinds of gametes (*Pby* and *pBy*). Combine them carefully and see the 1:2:1 phenotypic ratio fall out in the F$_2$.
6c. Pink-blistery.
6d. See Figures 9.16 and 9.17. Crossing over took place in the F$_1$ generation.

7. The genotypes are *PpSwsw*, *Ppswsw*, and *ppswsw* in a ratio of 1:1:1:1.

8a. 1 black:2 blue:1 splashed white.
8b. Always cross black with splashed white.

9a. $w^+ > w^e > w$
9b. Parents $w^e w$ and w^+Y. Progeny w^+w^e, w^+w, w^eY, and wY.

10. All will have normal vision because they inherit Dad's wild-type X chromosome, but half of them will be carriers.

11. Agouti parent: *AaBb*. Albino offspring *aaBb* and *aabb*; black offspring *Aabb*; agouti offspring *AaBb*.

12. The purple parent must be *AaB–*. If it were *AAB–*, there would be no white progeny—all the progeny contain *B* from the white parent. To be purple, that parent must contain at least one *B*; from the data given we cannot tell whether it is *AaBB* or *AaBb*.

13. Yellow parent = s^Ys^b; offspring 3 yellow (s^Y–): 1 black (s^bs^b). Black parent = s^bs^b; offspring all black (s^bs^b). Orange parent = s^Os^b; offspring 3 orange (s^O–): 1 black (s^bs^b). Both s^O and s^Y are dominant to s^b.

ILLUSTRATION CREDITS

Texas. 4.10: R. Rodewald, Univ. of Virginia/ BPS. 4.11: Jim Solliday/BPS. 4.14: Hilton Mollenhauer, U.S.D.A. Research Unit, College Station, TX. 4.15: Runk/ Schoenberger from Grant Heilman Photography, Inc. 4.16: E. H. Newcomb & W. P. Wergin, Univ. of Wisconsin/BPS. 4.17: D. J. Wrobel, Monterey Bay Aquarium/BPS. 4.19: B. F. King, Univ. of California, Davis, School of Medicine/BPS. 4.20a: G. T. Cole, Univ. of Texas, Austin/BPS. 4.21c: H. S. Pankratz, Michigan State Univ./BPS. 4.22a: R. Rodewald, Univ. of Virginia/BPS. 4.23: E. H. Newcomb & S. E. Frederick, Univ. of Wisconsin/BPS. 4.24: M. C. Ledbetter, Brookhaven National Laboratory. 4.25 left: Gopal Murti, Science Photo Library/Photo Researchers, Inc. 4.25 center: R. Alexley/Peter Arnold, Inc. 4.25 right: Gopal Murti, Science Photo Library/Photo Researchers, Inc. 4.26a,b: W. L. Dentler, Univ. of Kansas/BPS. 4.27a: B. F. King, Univ. of California, Davis, School of Medicine/BPS. 4.28a: J. R. Waaland, Univ. of Washington/BPS. 4.28b: E. H. Newcomb, Univ. of Wisconsin/BPS. Box 4.A: After N. Campbell, 1990, Biology, 2nd Ed., Benjamin Cummings Publishing Co.

Chapter 5 Opener: J. David Robertson, Duke Univ. Medical Center. 5.2: After L. Stryer, 1981, Biochemistry, 2nd Ed., W. H. Freeman. 5.4a: L. A. Staehelin, Univ. of Colorado/BPS. 5.4b: J. D. Robertson, Duke Univ. 5.7c: G. T. Cole, Univ. of Texas, Austin/ BPS. 5.8 top: D. S. Friend, Univ. of California, San Francisco. 5.8 center: Darcy E. Kelly, Univ. of Washington. 5.8 bottom: Courtesy of C. Peracchia. 5.19a–d: M. M. Perry, J. Cell Sci. 39, p. 266, 1979.

Chapter 6 Opener: Jim Merli. 6.1: Nuridsany et Perennou/Photo Researchers, Inc. 6.9, 6.16b, 6.17b: Richard Alexander, Univ. of Pennsylvania

Chapter 7 Opener: Runk/Schoenberger from Grant Heilman Photography, Inc. 7.10a: Hilton Mollenhauer, U.S.D.A. Research Unit, College Station, Texas. 7.10b: E. Racker, Cornell Univ. Box 7.Aa: Ken Lucas/BPS. Box 7.Ab: Michael Fogden DRK PHOTO. Box 7.Ac: G. M. Thomas & G. Poinar, Univ. of California, Berkeley. Box 7.Ad: K. V. Wood, courtesy of M. DeLuca, Univ. of California, San Diego.

Chapter 8 Opener: Art Wolfe. 8.1, bottom: J. H. Troughton and L. A. Donaldson. 8.1, top: Runk/Schoenberger from Grant Heilman Photography, Inc. 8.19: J. R. Waaland, Univ. of Washington/BPS. 8.20a: J. A. Bassham, Lawrence Berkeley Lab., Univ. of California. 8.20b, 8.27b: E. H. Newcomb & S. E. Frederick, Univ. of Wisconsin/BPS.

Chapter 9 Opener: Scott Spiker/Adventure Photo. 9.1a: Phil Gates, Univ. of Durham/

BPS. 9.1b: R. Rodewald, Univ. of Virginia/BPS. 9.2a: G.F. Bahr, Armed Forces Inst. of Pathology. 9.3b: A. L. Olins, Univ. of Tennessee, Oak Ridge Grad. School of Biomedical Science. 9.5 insert: David Ward, Yale Univ. School of Medicine. 9.8: Andrew S. Bajer, Univ. of Oregon. 9.9: J. B. Rattner & S. G. Phillips, J. Cell Biol. 57, p. 359, 1973. 9.10b, c: C. L. Rieder, New York State Dept. of Health/BPS. 9.11a: T. E. Schroeder, Univ. of Washington/BPS. 9.11b: B. A. Palevitz & E. H. Newcomb, Univ. of Wisconsin/BPS. 9.12: G. T. Cole, Univ. of Texas, Austin/BPS. 9.14a,b: David Ward, Yale Univ. School of Medicine. 9.15, 9.17: C. A. Hasenkampf, Univ. of Toronto/BPS. 9.19: B. Schuh, Monmouth Medical Center. 9.20: © Ruth Kavenoff, Designergenes Ltd., P.O. Box 100, Del Mar, CA 90214. 9.21: J.J. Cardamone, Jr., Univ. of Pittsburgh/BPS.

Chapter 10 Opener: Runk/Schoenberger from Grant Heilman Photography, Inc. 10.12: Carl W. May/BPS. 10.17: Namboori B. Raju, Stanford Univ., Eur. J. Cell Biol. 23, p. 208, 1980. 10.20: Peter J.Bryant/BPS. 10.23: After N. Campbell, 1990, Biology, 2nd Ed., Benjamin Cummings Publishing Co. 10.24: © Walter Chandoha, 1991.

Chapter 11 Opener: Dan Richardson. 11.3 right, 11.5: Dan Richardson. 11.15: G. W. Willis, Ochsner Medical Instution/BPS. 11.19a: Dan Richardson. 11.22b: Courtesy of J. E. Edstrom and EMBO J.

Chapter 12 Opener: A. B. Dowsett, Science Photo Library/Photo Researchers, Inc. 12.1: Richard Humbert/BPS. 12.4: L. Caro & R. Curtiss. 12.17: Brian Matthews, Univ. of Oregon.

Chapter 13 Opener: Ken Edward, Science Source/Photo Researchers Inc. 13.7: Karen Dyer, Vivigen. 13.8: Joseph Gall, Carnegie Institution of Washington. 13.9: O. L. Miller, Jr., & B. R. Beatty. 13.12: After W. T. Keeton and J. L. Gould, Biological Science, 5th Edition, W. W. Norton & Co.

Chapter 14 Opener: Hank Morgan/Photo Researchers, Inc. 14.5a: N.Y. State Agricultural Experiment Station, Cornell Univ. 14.5b: J. S. Yun & T.E. Wagner, Ohio Univ. 14.11b: Mike Tincher, courtesy of Agracetus, Inc. (a subsidiary of W.R. Grace & Co.) 14.13: M. L. Pardue & J. G. Gall, Chromosomes Today 3, p. 47, 1972. 14.17a: Phil Gates, Univ. of Durham/BPS. 14.18: Paul F. Umbeck, courtesy of Agracetus, Inc. (a subsidiary of W.R. Grace & Co.) 14.19: N.Y. State Agricultural Experiment Station, Cornell Univ. 14.20: Advanced Genetic Sciences. 14.21: Larry Lefever from Grant Heilman Photography, Inc.

Chapter 15 Opener: Chip Mitchell. 15.1: From C. Harrison et al., J. Med. Genet. 20, p. 280, 1983. 15.9: Elaine Rebman/Photo Researchers, Inc. 15.10: P. P. H. DeBruyn, Univ. of Chicago.

Chapter 16 Opener: Painting by Keith Haring. 16.1: Z. Skobe, Forsyth Dental Center/BPS. 16.2: Courtesy of Lennart Nilsson. © Boehringer Ingelheim GmbH. 16.4: G. W. Willis, Ochsner Medical Institution/BPS. 16.9: R. Rodewald, Univ. of Virginia/BPS. 16.10b: Arthur J. Olson, Scripps Research Institute. 16.14: L. Winograd, Stanford Univ. 16.20: A. Liepins, Sloan-Kettering Research Inst. 16.25: A. Calin, Stanford Univ. School of Medicine. 16.26: After R. C. Gallo, The AIDS Virus, © 1987: by Scientific American, Inc.

Chapter 17 Opener: Norbert Wu. 17.1 sea urchin embryo: George Watchmaker. 17.1 sea urchin: D. J. Wrobel, Monterey Bay Aquarium/BPS. 17.1 tadpole, frog, chick embryo: © E. R. Degginger. 17.1 rooster: John Colwell from Grant Heilman Photography, Inc. 17.3a: After J. E. Sulston and H. R. Horvitz, Dev. Biol. 56, p. 110, 1977. 17.10: George M. Malacinski and A. W. Neff. 17.12: Peter J. Bryant/BPS. 17.19: Susan Strome. 17.28: C. Rushlow and M. Levine. 17.23a: From B. Alberts et al., 1983, Molecular Biology of the Cell, Garland Publishing Co. 17.24: F. R. Turner, Indiana Univ. 17.26: E. B. Lewis.

Chapter 18 Opener: Larry Ulrich/DRK PHOTO. 18.3: Stanley M. Awramik, U. of California/BPS.

Chapter 19 Opener: Frans Lanting/Minden Pictures. 19.6: Frank S. Balthis, Nature's Design. 19.7: Harold W. Pratt/BPS. 19.11a,b: Gary J.James/BPS. 19.12: After D. Futuyma, Evolutionary Biology, 2nd Ed., Sinauer Associates, Inc., 1987. 19.16a,b: Richard Alexander, Univ. of Pennsylvania.

Chapter 20 Opener: Raymond A. Mendez. 20.1: Des and Jen Bartlett, Bruce Coleman, Inc. 20.2a: Edward Ely/BPS. 20.2b: © John Shaw/NHPA. 20.4: Anthony D. Bradshaw, Univ. of Liverpool. 20.6: © E. R. Degginger. 20.7a–c: R.W. VanDevender. 20.8: Paul A. Johnsgard, Univ. of Nebraska. 20.9a: Peter J. Bryant/BPS. 20.9b: Kenneth Y. Kaneshiro, Univ. of Hawaii at Manoa. 20.9c: Peter J. Bryant/BPS. 20.11b: Heather Angel, BIOFO-TOS. 20.11a: Virginia P. Weinland/Photo Researchers, Inc. 20.12a: Gary J. James/BPS. 20.12b,c: © Jim Denny.

Chapter 21 Opener: Joe McDonald. 21.1a: Helen E. Carr/BPS. 21.1b: Barbara J. Miller/BPS. 21.1c: © L. Campbell/NHPA. 21.5a: Jon Stewart/BPS. 21.5b: Barbara J. Miller/BPS. 21.9a,b: Peter J. Bryant/BPS. 21.10: Illustration by Marianne Collins.

21.12*a,b*: Paul A. Johnsgard, Univ. of Nebraska. 21.14*a,b*: Art Wolfe.

Chapter 22 *Opener*: Tony Brain, Science Photo Library/Photo Researchers, Inc. 22.1: Alfred Pasieka/Science Photo Library Photo Researchers, Inc. 22.2: H. W. Jannasch, Woods Hole Oceanographic Institution. 22.3, 22.4, 22.5: T. J. Beveridge, Univ. of Guelph/BPS. 22.6*a*: Leonard Lessin , Peter Arnold Inc. 22.7*b*: C. Forsberg & T. J. Beveridge, Univ. of Guelph/BPS. 22.7*a*: Paul W. Johnson/BPS. 22.7*b*: K. Stephens, Stanford Univ./BPS. 22.8*a*: J.A. Breznak & H. S. Pankratz, Michigan State Univ./BPS. 22.8*b*: G. W. Willis, Ochsner Medical Institution/BPS. 22.9: T. J. Beveridge, Univ. of Guelph/BPS. 22.10: G. W. Willis, Ochsner Medical Institution/BPS. 22.11: D. A. Glawe, Univ. of Illinois/BPS. 22.12: G. W. Willis, Ochsner Medical Institution/BPS. 22.13*a*: W. Burgdorfer, Rocky Mountain Lab. 22.13*b*: Nat. Animal Disease Center, Ames, IA. 22.14: S. C. Holt, Univ. of Texas Health Science Center, San Antonio/BPS. 22.15*a*: Paul W. Johnson/BPS. 22.15*b*: H. S. Pankratz, Michigan State Univ./BPS. 22.15*c*: © E. R. Degginger. 22.16*a,b*: Leon J. Le Beau/BPS. 22.17*a*: G. W. Willis, Ochsner Medical Institution/BPS. 22.17*b*: G. W. Willis, Ochsner Medical Institution/BPS. 22.18: Centers for Disease Control, Atlanta. 22.19: M. G. Gabridge, cytoGraphics, Inc./BPS. 22.20: Arthur J. Olson, Scripps Research Institute. 22.21: D. T. Brown et al., *J. Virol.* 10, p. 524, 1972. 22.22*a*: D. L. D. Caspar, Brandeis Univ. 22.22*b*: D. S. Goodsell & A. J. Olson, Scripps Research Institute. 22.22*c*: S. C. Holt, Univ. of Texas Health Science Center, San Antonio/BPS. 22.22*d*: F. A. Murphy, Centers for Disease Control, Atlanta. Box 22.A1 *upper left*: Centers for Disease Control, Atlanta. Box 22.A *upper center*: S. C. Holt, Univ. of Texas Health Science Center, San Antonio/BPS. Box 22.A *lower left*: Leon J. Le Beau/BPS. Box 22.A *lower center*: A J. J. Cardamone, Jr., Univ. of Pittsburgh/BPS.

Chapter 23 *Opener*: Jan Hinsch, Science Photo Library/Photo Researchers, Inc. 23.3*a*: Animals Animals/Oxford Scientific Films. 23.3*b*: © E. R. Degginger. 23.3*c*: James Solliday/BPS. 23.4: Eric V. Gravé, Science Source/Photo Researchers, Inc. 23.6: G. W. Willis, M.D./BPS. 23.26*a*: Dennis D. Kunkel/BPS. 23.7: Paul W.Johnson/BPS. 23.9*a*: Jim Solliday/BPS.23.9*b*: Eric V. Gravé/Photo Researchers, Inc. 23.11*a*: Jim Solliday/BPS. 23.11*b–d*: Paul W. Johnson/BPS. 23.12*b*: M. A. Jakus, NIH.23.14: Eric V. Gravé. 23.16*a*: Barbara J. Miller/BPS. 23.16*b*: Henry Aldrich, Institute of Food and Agricultural Sciences, Univ. of Florida. 23.17*a*: D. W. Francis, Univ. of Delaware. 23.17*b*: © David Scharf. 23.18: J. R. Waaland, Univ. of Washington/BPS. 23.19: Dwight R. Kuhn. 23.20*a*: Paul W. Johnson/BPS. 23.20*b*: Gary J.James/BPS. 23.21*a*: Charles Gellis/Photo Researchers, Inc.

23.21*b*: V. Cassie. 23.24*a*: J.N. A. Lott, McMaster Univ./BPS. 23.24*b*: J. R. Waaland, Univ. of Washington/BPS. 23.25*a*: Maria Schefter/BPS. 23.25*b*: J. N. A. Lott, McMaster Univ./BPS. 23.26*a*: Dennis D. Kunkel/BPS. 23.26*b*: Harold W. Pratt/BPS. 23.26*c*: J. R. Waaland, Univ. of Washington/BPS. Box 23.A*a*: Gerald Corsi, Tom Stack and Associates. Box 23.A*b*: J. N. A. Lott, McMaster Univ./BPS.

Chapter 24 *Opener*: Stephen J. Kraseman/DRK PHOTO. 24.1*a*: D. A. Glawe, Univ. of Illinois/BPS. 24.1*b*: L. E. Gilbert, Univ. of Texas, Austin/BPS. 24.1*c*: G. L. Barron, Univ. of Guelph/BPS. 24.2*b*: G. T.Cole, Univ. of Texas, Austin/BPS. 24.3: D. A. Glawe, Univ. of Illinois/BPS. 24.4: G. L. Barron, Univ. of Guelph/BPS. 24.5: Barbara J. Miller/BPS. 24.7 *upper & lower left*: W. F.Schadel, Small World Enterprises/BPS. 24.8: J. R. Waaland, Univ. of Washington/BPS. 24.9*b*: D. A. Glawe, Univ. of Illinois/BPS. 24.10: W. F. Schadel, Small World Enterprises/BPS. 24.11*a*: Jim Solliday/BPS. 24.11*b*: D. A. Glawe, Univ. of Illinois/BPS. 24.11*c*: Richard Humbert/BPS. 24.12: Centers for Disease Control, Atlanta. 24.13*a*: Michael Fogden DRK PHOTO. 24.13*b*: Photography by Rannels, Grant Heilman Photography, Inc. 24.13*c*: M. Graybill and J. Hodder/BPS. 24.14 *inset*: © Biophoto Associates. 24.15*a*: E. I. Friedmann, Florida State Univ. 24.15*b*: Barbara J. O'Donnell/BPS. 24.16*a*: Grant Heilman, Grant Heilman Photography, Inc. 24.16*b*: J. N. A. Lott, McMaster Univ./BPS. 24.16*c*: Barbara J. Miller/BPS. 24.17*a*: J. N. A. Lott, McMaster Univ./BPS. Box 24.A: R. L. Peterson, Univ. of Guelph/BPS.

Chapter 25 *Opener*: Art Wolfe. 25.1: J. Robert Stottlemeyer/BPS. 25.3: Gary J. James/BPS. 25.4*a,b*: J. R. Waaland/BPS. 25.5*a*: © E. R. Degginger. 25.5*b*: Runk/Schoenberger from Grant Heilman Photography, Inc. 25.5*c*: J. N. A Lott, McMaster Univ./BPS. 25.7: J. H. Troughton. 25.8: © E. R. Degginger. 25.9: Fig. information provided by Prof. Hermann Pfefferkorn, Dept. of Geology, Univ. of Pennsylvania. Original oil painting by John Woolsey. 25.14*a*: Runk/Schoenberger from Grant Heilman Photography, Inc. 25.14*b*: J. N. A. Lott, McMaster Univ./BPS. 25.15*a*: Carl W. May/BPS. 25.15*b*, 25.16: J. N. A. Lott, McMaster Univ./BPS. 25.17*a*: Barbara J. Miller/BPS. 25.17*b*: J. N.A. Lott, McMaster Univ./BPS. 25.17*c*: Art Wolfe. 25.18: J. N. A. Lott, McMaster Univ./BPS. 25.21: Phil Gates, Univ. of Durham/BPS. 25.22*a*: John Cancalosi/DRK PHOTO. 25.22*b*: Ken Lucas/BPS. 25.22*c*: Gary J. James/BPS. 25.22*d*: Joel Simon. 25.25*a*: B. Miller/BPS. 25.25*b*: Roger de la Harpe/BPS. 25.25*c*: Grant Heilman Photography 25.26*a*: Barbara J. Miller/BPS. 25.26*b*: Barbara J. Miller/BPS. 25.28*a*: J. N.A. Lott, McMaster Univ./BPS. 25.28*b*: J. N.A. Lott, McMaster Univ./BPS. 25.28*c*: Catherine M. Pringle/BPS. 25.28*d*:

C.S. Lobban/BPS. 25.30*a*: Jon Mark Stewart/BPS. 25.30*b*: Lara Hartley, TERRAPHOTOGRAPHICS/BPS. 25.30*c*: © E. R. Degginger. 25.31*a*: Jon Mark Stewart/BPS. 25.31*b*: Lefever/Grushow from Grant Heilman Photography, Inc. 25.31*c*: Jon Mark Stewart/BPS. 25.32: Barbara J. O'Donnell/BPS.

Chapter 26 *Opener*: Norbert Wu. 26.8*a*: Ken Lucas/BPS. 26.8*b*: Robert Brons/BPS. 26.8*c*: Joel Simon. 26.10, 26.11, 26.12, 26.13: Adapted from F. M. Bayerand, H. B. Owre, 1968, *The Free-Living Lower Invertebrates*, Macmillan Pubishing Co. 26.14: After G. and R. Brusca, 1990, *Invertebrates*, Sinauer Associates, Inc. 25.14*a*: Andrew J. Martinez/Photo Researchers,Inc. 25.14*b*: Douglas Faulkner/Photo Researchers, Inc. 26.16*a*: Robert Brons/BPS. 26.16*b,c*: After G. and R. Brusca, 1990, *Invertebrates*, Sinauer Associates, Inc. 26.17*b*, 26.18*a*: D. J. Wrobel, Monterey Bay Aquarium/BPS. 26.18*c*, 26.21*b*: Robert Brons/BPS. 26.21*c*: Jim Solliday/BPS. 26.22*a*: After G. and R. Brusca, 1990, *Invertebrates*, Sinauer Associates, Inc. 26.22*b*: Jim Solliday/BPS. 26.23: R. R. Hessler, Scripps Institute of Oceanography. 26.25*a*: Andrew J. Martinez/Photo Researchers, Inc. 26.25*c*: Roger K. Burnard/BPS. 26.25*d*: © Robert & Linda Mitchell. 26.27*a*: M. P. L. Fogden/Bruce Coleman, Inc. 26.27*b*: From Kristensen and Hallas, 1980. 26.29*a*: Joel Simon. 26.28: J. N. A. Lott, McMaster Univ./BPS. 26.29*b*: Barbara J. Miller/BPS. 26.30*a*: Ken Lucas/BPS. 26.30*b*: Peter J. Bryant/BPS. 26.30*c*: L. E. Gilbert, Univ. of Texas, Austin/BPS. 26.30*d*: Robert Brons/BPS. 26.32*a*: Gregory Ochocki/Photo Researchers, Inc. 26.32*b*: Peter J.Bryant/BPS. 26.32*c*: D. J. Wrobel, Monterey Bay Aquarium/BPS. 26.32*d*: C. R. Wyttenbach, Univ. of Kansas/BPS. 26.33*a*: Ken Lucas/BPS. 26.33*b*: Roger K. Burnard/BPS. 26.34*a*: Richard Humbert/BPS. 26.34*b*: Peter J. Bryant/BPS. 26.34*c- g*: Peter J. Bryant/BPS. 26.34*h*: © E. R. Degginger. 26.36*a*: Ken Lucas/BPS. 26.36*b*: Harold W. Pratt/BPS. 26.36*c*: D. J. Wrobel, Monterey Bay Aquarium/BPS. 26.36*d,e,g*: Ken Lucas/BPS. 26.36*f*: J. W. Porter, Univ. of Georgia/BPS. 26.37: D. J. Wrobel, Monterey Bay Aquarium/BPS.

Chapter 27 *Opener*: Heather Angel, BIOFOTOS. 27.2*a*: D. J. Wrobel, Monterey Bay Aquarium/BPS. 27.2*b*: Robert Brons/BPS. 27.4: D. J. Wrobel, Monterey Bay Aquarium/BPS. 27.5*b*: C.R. Wyttenbach, Univ. of Kansas/BPS. 27.8*a*: Doug Perrine/DRK PHOTO. 27.8*b*: Joel Simon 27.8*c–e*: D. J. Wrobel, Monterey Bay Aquarium/BPS. 27.9: Robert Brons/BPS. 27.10*b*: © Heather Angel, BIOFOTOS. 27.12: Tom Stack/Tom Stack and Associates. 27.14*a*: Tom McHugh/Photo Researchers, Inc. 27.14*b*: D. J. Wrobel, Monterey Bay Aquarium/BPS. 27.15*a*: Ken Lucas/BPS. 27.16*a*: Peter Scoones, Planet Earth Pictures. 27.15*b–d*: Ken Lucas/BPS.

27.15e: Animals Animals/Breck P. Kent. 27.16a: Peter Scoones/Planet Earth Pictures. 27.17a: Ken Lucas/BPS. 27.17b: Art Wolfe. 27.17c: E.D. Brodie, Jr., Univ. of Texas, Arlington/BPS. 27.20a: Doug Perrine DRK PHOTO. 27.20b: Carl Gans, Univ. of Michigan/BPS. 27.20c: Joe McDonald, Bruce Coleman Inc. 27.20d: Michael P. Fogden, Bruce Coleman Inc. 27.21a: © E. R. Degginger. 27.21b: Wayne Lankinen/DRK PHOTO. 27.22: Courtesy of Carnegie Museum of Natural History, Pittsburgh. 27.24a: Johnny Johnson/DRK PHOTO. 27.24b: D. Cavagnaro/DRK PHOTO. 27.24c,d: Stephen J. Kraseman/DRK PHOTO. 27.26a: John Cancalosi/Tom Stack and Associates. 27.26b: Joel Simon. 27.26c: M. P. L. Fogden, Bruce Coleman, Inc. 27.27a: J. N. A. Lott, McMaster Univ./BPS. 27.27b: Merlin D. Tuttle, Bat Conservation International. 27.27c: Stephen J. Kraseman/DRK PHOTO. 27.27d: Robert Stottlemyer/BPS. 27.29a: Art Wolfe. 27.29b: Stanley Breeden/DRK PHOTO. 27.29c: Frans Lanting/Minden Pictures. 27.30a,b: Steve Kaufman/DRK PHOTO. 27.31a: © E. R. Degginger. 27.31b: © Peter Drowne, E. R. Degginger. 27.31c: Kennan Ward/DRK PHOTO. 27.31d: © Peter Drowne, E. R. Degginger. 27.34a: Edward S. Ross. 27.34b,c: Robert E. Ford, TERRAPHO-TOGRAPHICS/BPS. Box 27A: Anne Marie Weber/Adventure Photo.

Chapter 28 *Opener*: Frans Lanting/Minden Pictures. 28.3: Peter Ward, Univ. of Washington 28.4: Mark R. Meyer. 28.7a: Ken Lucas/BPS. 28.7b: S. M. Awramik, Univ. of California/BPS. 28.8a: S. Conway Morris. 23.8b: From S. J.Gould, 1989, Wonderful Life, © W. W. Norton. 28.9: Courtesy of the Natural History Museum of London. 28.10: Courtesy of the Smithsonian Institution. 28.11: Ken Lucas/BPS. 28.12: Painting by Rudolph Zallinger; courtesy of the Peabody Museum of Natural History, Yale Univ. 28.14: Painting by Chip Clark; courtesy of the Smithsonian Institution.

Chapter 29 *Opener*: Ed Reschke, Peter Arnold, Inc. 29.1: Joel Simon. 29.2: Michael P. Gadomski/Bruce Coleman, Inc. 29.3: Gary J. James/BPS. 29.4: Thomas Hovland from Grant Heilman Photography, Inc. 29.7a: Runk/Schoenberger from Grant Heilman Photography, Inc. 29.7b: © E. R. Degginger. 29.8: Grant Heilman Photography. 29.9a: J. R. Waaland, Univ. of Washington/BPS. 29.9b: © E. R. Degginger. 29.9c: Carl W. May/BPS. 29.13a,b: Phil Gates, Univ. of Durham/BPS. 29.13c: Runk/Schhoenberger from Grant Heilman Photography, Inc. 29.13d: Phil Gates, Univ. of Durham/BPS. 29.13e: © E. R. Degginger. 29.13f, 29.15b, 29.17 top & bottom: J. R. Waaland, Univ. of Washington/BPS. 29.18a: L. Elkin, Hayward State Univ./BPS. 29.18b,c: Jim Solliday/BPS. 29.18d: Phil Gates, Univ. of Durham/BPS. 29.20 top: Dwight R. Kuhn. 29.20 bottom: © E. R. Degginger. 29.20: J. R. Waaland, Univ. of

Washington/BPS. 29.21a,b: Phil Gates, Univ. of Durham/BPS. 29.23: J. N.A. Lott, McMaster Univ./BPS. 29.24: Jim Solliday/BPS. 29.25: J. N. A. Lott, McMaster Univ., BPS. 29.26a,b: Phil Gates, Univ. of Durham/BPS. 29.27b: W. F. Schadel, Small World Enterprises/BPS. 29.27c: E. J. Cable/Tom Stack and Associates.

Chapter 30 *Opener*: Gary Gray/DRK PHOTO. 30.2a,b: Runk/Schoenberger from Grant Heilman Photography, Inc. 30.7: Phil Gates, Univ. of Durham/BPS. 30.10: J. H. Troughton & L. A. Donaldson. 30.12: Jon Stewart/BPS. 30.13: Thomas Eisner, Cornell Univ. 30.17: Larry Lefever from Grant Heilman Photography, Inc.

Chapter 31 *Opener*: Thomas Eisner, Cornell Univ. 31.1: Jon Mark Stewart/BPS. 31.2: J. N. A. Lott, McMaster Univ./BPS. 31.3a: J. N. A. Lott, McMaster Univ./BPS. 31.3b: J. N. A. Lott, McMaster Univ./BPS. 31.4: Carl W. May/BPS. 31.5: Gary J. James/BPS. 31.6: Joel Simon. 31.7: J. N. A. Lott, McMaster Univ./BPS. 31.8: © Robert & Linda Mitchell. 31.9: J. Antonovics, Duke Univ. 31.10: Jane Grushow from Grant Heilman Photography, Inc. 31.13: Richard Alexander, Univ. of Pennsylvania.

Chapter 32 *Opener*: Runk/Schoenberger from Grant Heilman Photography, Inc. 32.1: Lou Jacobs, Jr. from Grant Heilman Photography, Inc. 32.4: © William E. Ferguson. 32.7: Runk/Schoenberger from Grant Heilman Photography, Inc. 32.8: Barbara J. O'Donnell/BPS. 32.10: E. H. Newcomb & S. R. Tandon, Univ. of Wisconsin/BPS. 32.12, 32.13: Runk/Schoenberger from Grant Heilman Photography, Inc. Box 32.A: Aladar A. Szalay, Univ. of Alberta.

Chapter 33 *Opener*: Andrew Taylor Photography. 33.4: Art Wolfe. 33.5: Barbara J. Miller/BPS. 33.6: Barry L. Runk from Grant Heilman Photography, Inc. 33.10: Runk/Schoenberger from Grant Heilman Photography, Inc. 33.18 inset: Biophoto Associates/Photo Researchers, Inc. 33.20: Grant Heilman Photography. 33.22: J. N. A. Lott, McMaster Univ./BPS. 33.21: Grant Heilman from Grant Heilman Photography, Inc. 33.23: J. N. A. Lott, McMaster Univ./BPS. 33.28: R. Last, Cornell Univ. Courtesy of the Society for Plant Physiology.

Chapter 34 *Opener*: Art Wolfe. 34.1 bottom: J. R. Waaland, Univ. of Washington/BPS. 34.1 top: Jim Solliday/BPS. 34.2: Dennis D. Kunkel/BPS. 34.3: J. N. A. Lott, McMaster Univ./BPS. 34.7a: Gary J. James/BPS. 34.7b: John Colwell from Grant Heilman Photography, Inc. 34.16a: J. N. A. Lott, McMaster Univ./BPS. 34.16b: Phil Gates, Univ. of Durham/BPS. 34.18: Plant Genetics, Inc.

Chapter 35 *Opener*: Animals Animals/Gerald L. Kooyman. 35.14a: © Cherry Alexander/NHPA. 35.14b: Belinda Wright/DRK PHOTO. 35.19b,c: G. W. Willis, Ochsner Medical Institution/BPS. 35.19d Fran Thomas, Stanford Univ. 35.20a: Stephen J. Kraseman/DRK PHOTO. 35.20b: Art Wolfe.

Chapter 36 *Opener*: R. D. Fernald, Stanford Univ. 36.7a: AP Wide World Photos. 36.7b: The Bettmann Archive Inc. 36.9: S. H. Ingbar, Harvard Medical School. Box 36.B: James Sugar, Black Star.

Chapter 37 *Opener*: Belinda Wright/DRK Photo. 37.1a: © M. Walker/NHPA. 37.1b: J. Greenfield, Planet Earth Pictures. 37.1c: Geoff du Feu, Planet Earth Pictures. 37.2: David M. Phillips/Photo Researchers, Inc. 37.6, center insert: P. Motta, Univ. La Sapienza, Rome, Science Photo Library/Photo Researchers, Inc. 37.6, bottom insert: David M. Phillips/Photo Researchers, Inc. 37.7: P. Bagavandoss. 37.12a: Animals Animals/Fritz Prenzel. 37.12b (embryo): John Cancalosi/DRK PHOTO. 37.12b (adult): Animals Animals/Mickey Gibson. 37.16a,b: From A Child Is Born. Photos © Lennart Nilsson, Bonnier Fakta.

Chapter 38 *Opener*: Scott Camazine/Photo Researchers, Inc. 38.3: From R.G. Kessel and R. H. Kardon, Tissues and Organs. © 1979, W. H. Freeman and Co. 38.21: Dan McCoy, Rainbow. Boxes 38.A and 38.B: William F. Gilly, Hopkins Marine Station.

Chapter 39 *Opener*: Shelly Katz/Time Magazine. 39.4a: R. A. Steinbrecht. 39.4b: Animals Animals/G. I. Bernard, Oxford Scientific Films. 39.5 center: Peter J. Bryant/BPS. 39.8 top: P. Motta, Univ. La Sapienza, Rome, Science Photo Library/Photo Researchers, Inc. 39.17 right: S. Fisher, Univ. of California, Santa Barbara. 39.21a: Animals Animals/G. I. Bernard, Oxford Scientific Films. 39.24: © E. R. Lewis, Y. Y. Zeevi & F. S. Werblin, Univ. of California, Berkeley/BPS. 39.30: Michael Fogden/DRK PHOTO. 39.31: © Stephen Dalton/NHPA.

Chapter 40 *Opener*: Mik Dakin/Bruce Coleman, Inc. 40.1: D. J. Wrobel, Monterey Bay Aquarium/BPS. 40.2: CNRI Science Photo Library Photo Researchers. 40.5a: Secchi-Lecaque-Roussel, UCLAF/CNRI, Science Photo Library/Photo Researchers, Inc. 40.5b: P. Motta, Univ. La Sapienza, Rome, Science Photo Library/Photo Researchers, Inc. 40.7a: M. I. Walker, Science Source/Photo Researchers, Inc. 40.7b: Michael Abbey, Science Source/Photo Researchers, Inc. 40.7c: CNRI, Science Photo Library/Photo Researchers, Inc. 40.8b: F. A. Pepe, Univ. of Pennsylvania School of Medicine/BPS. 40.11: G.W. Willis, Ochsner Medical Institution/BPS. 40.14: John Dudak/Phototake. 40.19: G. Mili. 40.20: Robert Brons/BPS. 40.24a: David J. Wrobel/BPS.

Chapter 41 *Opener*: Courtesy of NASA. 41.1*a*: © Sea Studios, Inc. 41.1*b*: Robert Brons/BPS. 41.1*c*: © Robert & Linda Mitchell. 41.3: © 1991, Eric Reynolds/Adventure Photo. 41.5*b*: Peter J. Bryant/BPS. 41.5*c*: Thomas Eisner, Cornell Univ. 41.9: Walt Tyler, Univ. of California, Davis. 41.13 *top insert*: Science Photo Library, Science Source/Photo Researchers, Inc. 41.13 *bottom insert*: P. Motta, Univ. La Sapienza, Rome, Science Photo Library/Photo Researchers, Inc. 41.16: George Holton/Photo Researchers, Inc.

Chapter 42 *Opener*: Larry Mulvehill, Science Source/Photo Researchers, Inc. 42.11: © Ed Reschke. 42.13: UNICEF, Maggie Murray-Lee. 42.15: After N. Campbell, 1990, Biology, 2nd Ed., Benjamin Cummings Publishing Co. 42.16*a*: From R. G. Kessel & R. H. Kardon, *Tissues and Organs*, © 1979, W. H. Freeman and Co. 42.17*b*: Secchi-Lecaque-Roussel, UCLAF/CNRI, Science Photo Library/Photo Researchers, Inc. Box 42.A: Jon Feingersh/Tom Stack and Associates.

Chapter 43 *Opener*: Animals Animals/Carl Roessler. 43.1*a*: Timothy O'Keefe/Tom Stack and Associates. 43.1*b*: Animals Animals/A. G.(Bert) Wells, Oxford Scientific Films. 43.1*c*: Animals Animals/Bruce Watkins. 43.1*d*: Jack Stein Grove/Tom Stack and Associates. 43.6: D. J. Wrobel, Monterey Bay Aquarium/BPS. 43.7: © Robert & Linda Mitchell. 43.8: © James D. Watt. 43.11: © Ed Reschke. 43.14 *insert*: E. S. Strauss.

Chapter 44 *Opener*: Art Wolfe. 44.1*a*: Helen E. Carr/BPS. 44.1*b*: © E. R. Degginger. 44.3*b*: Marc Chappell, Univ. of California, Riverside. 44.10 *a–d*: From R. G. Kessel & R. H. Kardon, *Tissues and Organs*, © 1979, W.H. Freeman and Co. 44.12: Gregory G. Dimijian, M.D./Photo Researchers, Inc. 44.17: Lise Bankir, I.N.S.E.R.M. Unit, Hopital Necker, Paris. Box 44.A: © Custom Medical Stock Photos. Box 44.B: © Robert & Linda Mitchell.

Chapter 45 *Opener*: John Gerlach/DRK PHOTO. 45.2: Marc Chappell, Univ. of California, Riverside. 45.8: W. C. Dilger. 45.11*b*: Anup & Manos Shah/Planet Earth Pictures. 45.12: Anthony Mercieca/Photo Researchers, Inc. 45.15: H. Craig Heller. 46.17: Animals Animals/Michael Dick. 45.19:

H. Craig Heller. 45.22, 45.24: © Jonathan Blair, Woodfin Camp & Associates.

Chapter 46 *Opener*: © E. R. Degginger. 46.1: Kennan Ward/DRK PHOTO. 46.4: Aileen N. C. Morse, Marine Science Institute, Univ of California at Santa Barbara. 46.5: T. G. Whitham, Northern Arizona Univ. 46.7*a*: Art Wolfe. 46.7*b*: Robert and Jean Pollock/BPS. 46.7*c*: Drawing by Sally Landry. 46.8: Erwin and Peggy Bauer. 46.10: John Alcock, Arizona State Univ. 46.12: John Alcock, Arizona State Univ. 46.11: Natalie J. Demong. 46.13: Michael Fogden/DRK PHOTO. 46.15*a*: Elizabeth N. Orians. 46.16: Clem Haagner/Bruce Coleman, Inc. 46.19: S. G. Hoffman. 46.20: J. Erckmann. 46.21: Georgette Douwma/Planet Earth Pictures. 46.23: Patricia Moehlman. 46.24*a*: Jeremy Burgess, Science Photo Library/Photo Researchers, Inc. 46.24*b*: D. Houston/Bruce Coleman, Inc. 46.25: Stanley Breeden/DRK PHOTO. 46.26: Art Wolfe. 46.27: Jonathan Scott, Planet Earth Pictures. 46.28: K. and K. Ammann/Bruce Coleman, Inc.

Chapter 47 *Opener*: © E. R. Degginger. 47.2: Frans Lanting/Minden Pictures. 47.5: © Brian Hawkes/NHPA. 47.9: Dennis Johns. 47.14: Richard Root/Cornell Univ. 47.16: Elizabeth N. Orians. 47.17: Douglas Sprugel, College of Forest Resources, Univ. of Washington. 47.18: Michio Hoshino/Minden Pictures. 47.19: G.Tortoli, Food and Agriculture Organization of the United Nations. 47.20: Animals Animals Earth Scenes/George H. H. Huey.

Chapter 48 *Opener*: Michael Fogden/DRK PHOTO. 48.2*a*: S. K. Webster, Monterey Bay Aquarium/BPS. 48.2*b*: E. S. Ross. 48.3: Mark Mattock/Planet Earth Pictures. 48.7: John R. Hosking, NSW Agriculture, Australia. 48.8*a*: Thomas Eisner, Cornell Univ. 48.8*b*: Ken Lucas/BPS. 48.9: © James Carmichael/NHPA. 48.10: Thomas Eisner, Cornell Univ. 48.11: M.Freeman/Bruce Coleman, Inc. 48.12: Kim Heacox Photography/DRK PHOTO. 48.15: Charlie Ott/Photo Researchers, Inc. 48.16: G. T. Bernard, Oxford Scientific Films/Animals Animals Earth Scenes 48.17: Jonathan Scott/Planet Earth Pictures. 48.18: A. Kerstich/Planet Earth Pictures. 48.19*a*: Larry E. Gilbert, Univ. of Texas, Austin. 48.19*b,c*: Daniel Janzen, Univ. of

Pennsylvania. 48.20*a*: S. Nielsen/DRK PHOTO. 48.20*b*: © Raymond A. Mendez. 48.21: John Alcock, Arizona State Univ. 48.22*a*: © Larry Ulrich. 48.22*b*: E. S. Ross. 48.23*a*: Thomas Eisner, Cornell Univ. 48.24: Joel Simon. 48.26: After M. Begon, J. Harper, and C. Townsend, 1986, *Ecology*, Blackwell Scientific Publications. 48.28*a–c*: Robert and Jean Pollock/BPS.

Chapter 49 *Opener*: Courtesy of NASA. 49.6: Animals Animals/Len Rue, Jr. 49.17*a*: Brian A. Whitton. 49.17*b*: J. N. A. Lott, McMaster Univ./BPS.

Chapter 50 *Opener*: Norman Owen Tomalin/Bruce Coleman, Inc. 50.10: E. O. Wilson, Harvard Univ. 50.14*a,b*: Art Wolfe. 50.15*a*: J. N. A. Lott, McMaster Univ./BPS. 50.15*b*: M. Sutton/Tom Stack and Associates. 50.16*a,b*: Paul W. Johnson/BPS. 50.17: D. W. Kaufman, Kansas State Univ. 50.18: Edward Ely/BPS. 50.19: Kim Heacox Photography/ DRK PHOTO. 50.21*a*: Elizabeth N. Orians. 50.21*b*: Frans Lanting/Minden Pictures. 50.22: Elizabeth N. Orians. 50.23: Art Wolfe. 50.24*a*: © E. R. Degginger. 50.24*b*: Jett Britnell/DRK PHOTO. 50.26*a*: Norbert Wu. 50.26*b*: M. C. Chamberlain/DRK PHOTO. 50.28: Gary Gray/DRK PHOTO. 50.29: Art Wolfe.

Chapter 51 *Opener*: Kevin Schafer, Tom Stack and Associates. 51.1: Paintings by H. Douglas Pratt. Courtesy of the Bernice P. Bishop Museum, Honolulu, Hawaii. 51.2: David S. Boynton. 51.4: Joe McDonald/Tom Stack and Associates. 51.5: Danny R. Billings, Missouri Dept. of Conservation. 51.6: Animals Animals/Fred Whitehead. 51.8: Kenneth W. Fink/Bruce Coleman, Inc. 51.9: Edward S. Ross. 51.10*a*: John Gerlach/DRK PHOTO. 51.11: Richard P. Smith, Tom Stack and Associates. 51.12*a*: © Walt Anderson. 51.12*b*: Frans Lanting/Minden Pictures. 51.12*c*: Animals Animals/Paul Freed. 51.14*a,b*: Art Wolfe. 51.15, 51. 16: Courtesy of the Smithsonian Institution, Office of Environmental Awareness/Richard Bierregaard, photographer. 51.18: Animals Animals/Peter Weimann. 51.19*a,b*: © Bill Gabriel, BIOGRAPHICS. Box 51.A *top*: Steven L. Hilty/Bruce Coleman, Inc. Box 51.A *bottom*: N. H. Cheatham/DRK PHOTO.

INDEX

and volume, *10*
Weinberg, W., 430
Welwitschia, 561, *562*
Went, Frits W., *742–743*
Wernicke's area, *877–878*
West coasts, 1163, 1167
Westerly winds, *1146*
Western blotting, 321
Wetlands
  adaptations by plants to, 706–707, *708*, 709
  economic value of, 1193
  restoration of, 1193
Whale shark, 621, 979
Whales
  communication, 1042
  filter feeding, 641, 989–*990*, 1103
  navigation, *1048*
  overhunting, 1095–*1096*
"What" questions, 1029
Wheat, vernalization of, 769–770
Wheat rust, 534–536, *535*, 539
Whelks, 604
Whip grafting, *771*
Whisk ferns, 557–558
White blood cells, 971, *972*
  genes, 367–370, 371
  in immune responses, 342, 355, 356–357
  See also Lymphocytes; Natural killer (NK) cells; Phagocytes
White clover, 224
White-crowned sparrows, 1033–*1034*
White-fronted bee-eater, *1065*
White-handed gibbon, *635*
White-headed duck, *474*
White light, 167
White matter, 854–*855*, 856, 865
White muscle, *919–920*, 948
White pelican, *1062–*1063
White-plumed antbird, 1189–*1190*
Whitefish, 1086
Whitham, Thomas, 711–712
Whorls, 764

"Why" questions, 4, 6, 1029
Widowbirds, 1072
Wiesel, Torsten, 902
Wigglesworth, Sir Vincent, 802–804
Wild dogs, *1062–1063*
Wild-type, defined, 224
Wildebeests, *1076–1077*
  migration, 1094
Wilfarth, 724
Wilkesia, 458
Wilkins, Maurice, 243
Willow tit, *796*
Willows, *714*
Wilson, E. O., 1157
Wilting, 684, 690–*691*
Wind
  global patterns, 1145–*1146*, 1147
  and lake turnovers, 1134–*1135*
  and pollination, 561, 563, 758, *760*
Windpipe. *See* Trachea
Wine
  grapes, 772
  and yeast, 538
Wingless insects, 600, *601*, 602
Wings
  of birds, *3*, 400–*401*, 627–628
  of insects, 600, *601*, 602
Winter dormancy, 1093
  in animals. *See* Hibernation
  in plants, 735, *750*, *751*, 766
Winter survival, 1091, 1093
Winter wheat, 769–770
Wisdom teeth, 990
Woese, Carl, 485, 489
Wolf spiders, *598*
Wollman, Elie, 279, 285
Wolpert, Lewis, 400
Wolves, 792, *1104–1105*
Womb. *See* Uterus
Wood
  biomass, 1163
  decomposition, 1121
  defenses, 356, 714, 1103, 1133

growth of, 673, 681–*683*
  as supporting tissue, 684–*686*
Wood pigeons, *1063–1064*
Wood rose, *570*
Wood ticks, *598*
World Health Organization, 359
Worms, 591. *See also* Annelids; Earthworms; Flatworms; Nematodes

X chromosomes, *227–230*
  inactivation, 300–*302*
  interaction with hormones, 815
X-linked traits, 335–336, 338
X-ray crystallography, 124, 243
X rays, 165, *166*
  as mutagens, 255–256, 276, 345, 349
ψX174 virus, 265
Xanthophylls, in green algae, 523
Xenon, atmospheric, 1136
Xeroderma pigmentosum, 254
Xerophytes, *705–707*
Xgal, 317–318
Xiphosura, *597–598*
Xylem, 548
  cell types, 674–676, *677*
  growth of, 681–683
  of roots, 679–680
  transport in, 693–696, *701*

Y chromosomes, *227–230*, 301
  interaction with hormones, 815
Yeasts, *62*, 538
  fermentation by, 156, 538
  genetic material, *324*
  mating types, *308*
  in recombinant DNA technology, 311, 314, 331
  regulatory mechanisms, 157
Yellow-eyed junco, *1059*
Yellow-pine chipmunk, 1112
Yersinia, 492

Yolk, *383–385*
  in amniotic eggs, *625*, *838–841*
  and polarity, 394
Yolk plug, 387, *388–390*
Yolk sac, *841*
Yucca, 1082, *1117*

Z-DNA, 246, *248*
z gene, 316–318
Zea mays, in genetics studies, 226–227, 230
Zeatin, *739*, 749
Zebra finches, *1037*
Zebra mussels, 1141–1142
Zebroids, *447*
Zeevaart, Jan A. D., 769
Zenith tragedy, 952
Zinc
  as animal nutrient, 984–*985*
  in enzymes, *126*
  as plant nutrient, 719, 720
Zinc fingers, *305*
Zona pellucida, *836*
Zone of polarizing activity (ZPA), 400–*401*
Zones of upwelling. *See* Upwelling
Zooflagellates, *507–508*
Zoomastigophora, *507–508*
Zooplankton, 580
Zoos, 1182
Zoospores
  in funguslike protists, 516
  in green algae, 523–525
ZPA, 400–*401*
Zygomycetes, *529*, 536–*537*
Zygospores, 536–537
Zygotes, 202–*203*, 826, 835, 836
  in alternating generations, 506, 760–762
  cleavage, 382–385
  polarity, 394–396
  See also Embryos; Gametes
Zymogens, 993–*999*